FOURTH EDITION

HANDBOOK OF
BIOCHEMISTRY AND
MOLECULAR
BIOLOGY

FOURTH EDITION

HANDBOOK OF
BIOCHEMISTRY AND
MOLECULAR
BIOLOGY

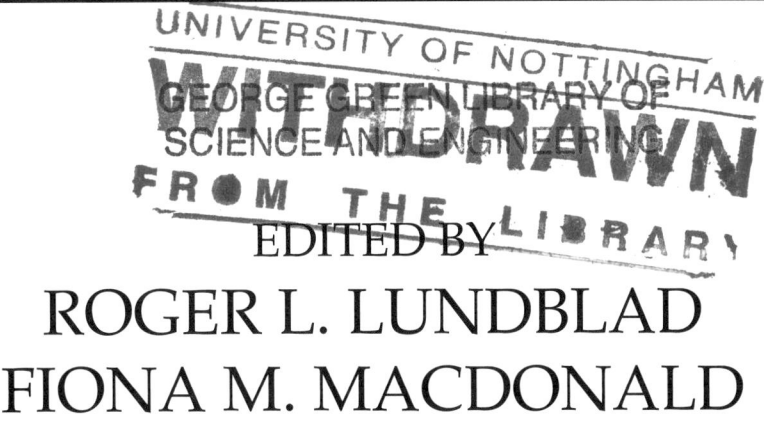

EDITED BY

ROGER L. LUNDBLAD
FIONA M. MACDONALD

 **CRC Press**
Taylor & Francis Group
Boca Raton   London   New York

CRC Press is an imprint of the
Taylor & Francis Group, an **informa** business

CRC Press
Taylor & Francis Group
6000 Broken Sound Parkway NW, Suite 300
Boca Raton, FL 33487-2742

Printed in the United States of America on acid-free paper
10 9 8 7 6 5 4 3 2 1

International Standard Book Number: 978-0-8493-9168-2 (Hardback)

Library of Congress Cataloging-in-Publication Data

Handbook of biochemistry and molecular biology / edited by Roger L. Lundblad and Fiona M. Macdonald.
　　p. ; cm.
　Includes bibliographical references and index.
　ISBN-13: 978-0-8493-9168-2 (alk. paper)
　ISBN-10: 0-8493-9168-7 (alk. paper)
　1. Biochemistry--Handbooks, manuals, etc. 2. Molecular biology--Handbooks, manuals, etc. I. Lundblad, Roger L. II. Macdonald, Fiona.
　[DNLM: 1. Biochemical Phenomena--Tables. QU 16 H2363 2010]

QH345.H347 2010
572--dc22
　　　2009042039

Visit the Taylor & Francis Web site at
http://www.taylorandfrancis.com

and the CRC Press Web site at
http://www.crcpress.com

This work is dedicated to the many scientists of the "The Greatest Generation" who contributed to the base of our knowledge of biochemistry and molecular biology.

Roger L. Lundblad, Ph.D.

To my parents, Pat and Paul Macdonald, whose unwavering love and support has been my guiding light.

Fiona M. Macdonald, Ph.D., F.R.S.C.

TABLE OF CONTENTS

Table of Contents   ix

FOREWORD

For almost a century, CRC Press has been a leader in providing concise compilations of scientific data for researchers, teachers, and students. The *CRC Handbook of Chemistry and Physics,* which first appeared in 1913 and is now in its 90th edition, is the basic source of physical science data that most chemists, physicists, and engineers turn to. Other widely used handbooks from CRC Press cover materials science, engineering, and mathematics. Many specialized handbooks have also appeared under the CRC imprint, ranging from semiconductors to lipids.

The 1968 publication of the *CRC Handbook of Biochemistry,* edited by Herbert A. Sober, marked a milestone in bioscience data. Appearing just 15 years after Watson and Crick elucidated the structure of DNA, the subtitle of this work, *Selected Data for Molecular Biology,* was a recognition that molecular biology was the new frontier of the biosciences. This was followed by the multivolume *Handbook of Biochemistry and Molecular Biology,* edited by Gerald D. Fasman, and its single volume abridged version *Practical Handbook of Biochemistry and Molecular Biology,* which appeared in 1989. The intervening 20 years has seen an explosion of data in this field and an exponential growth in the translation of recent discoveries into new technology. This new *Handbook of Biochemistry,* edited by Roger Lundblad and Fiona M. Macdonald, is thus a welcome addition to the list of handbooks available through CRC Press. I am sure it will find heavy use in both basic research and biotechnology.

David R. Lide
Editor-in-Chief, CRC Handbook of Chemistry and Physics

PREFACE

This is the fourth edition of the *Handbook of Biochemistry and Molecular Biology*. The first edition was published as a single volume in 1968 under the guidance of Herbert Sober. The second edition appeared in 1970 and the third, with Gerald Fasman as editor, appeared in eight volumes published in 1975–6. This increase in size reflected the rapid advances in knowledge in the then relatively new field of molecular biology.

It is intended that current *Handbook of Biochemistry and Molecular Biology* be a companion volume to the *CRC Handbook of Chemistry and Physics*—a single volume ready-reference work that will find a home on the bookshelves of biochemists and molecular biologists everywhere.

This fourth edition contains materials from the previous editions as well as extensive new material. Staying within the confines of a single volume has meant difficult decisions on which tables to include, and the editors welcome feedback from readers. The advent of electronic media allows for more frequent updating and it is hoped that any infelicities in our selection may be readily rectified. Additionally, suggestions on new topics for this Handbook and notification of errors are always appreciated. Address all comments to Editor, *Handbook of Biochemistry and Molecular Biology*, Taylor & Francis Group, 6000 Broken Sound Parkway NW, Suite 300, Boca Raton, FL 33487.

Much of the current content is derived from the research of the giants of biochemistry and molecular biology in the three decades following World War II. While it seems the vogue to develop new names and descriptions for old, established concepts, biochemistry and molecular biology continue to be the mainstays of current biomedical research.

Roger L. Lundblad
lundbladr@bellsouth.net

Fiona M. Macdonald
fiona.macdonald@taylorandfrancis.com
March 2010

ACKNOWLEDGMENTS

This work would not have been possible without the outstanding support of Jill Jurgensen and Glen Butler of Taylor & Francis. The support of Professor Edward A. Dennis is acknowledged. The help of various research librarians at the University of North Carolina at Chapel Hill is also acknowledged. Professors Charles Craik of the University of California at San Francisco and Bryce Plapp of the University of Iowa provided advice on selection of materials to be included. However, the editors take all responsibility for the selection of content.

EDITORS

Roger L. Lundblad, PhD

Roger L. Lundblad is a native of San Francisco, California. He received his undergraduate education at Pacific Lutheran University and his PhD degree in biochemistry at the University of Washington. After postdoctoral work in the laboratories of Stanford Moore and William Stein at the Rockefeller University, he joined the faculty of the University of North Carolina at Chapel Hill. He joined the Hyland Division of Baxter Healthcare in 1990. Currently Dr. Lundblad is an independent consultant and writer in biotechnology in Chapel Hill, North Carolina. He is an adjunct professor of Pathology at the University of North Carolina at Chapel Hill and editor-in-chief of the Internet Journal of Genomics and Proteomics.

Fiona M. Macdonald, PhD, F.R.S.C.

Fiona M. Macdonald received her BSc in chemistry from Durham University, UK. She obtained her PhD in inorganic biochemistry at Birkbeck College, University of London, studying under Peter Sadler. Having spent most of her career in scientific publishing, she is now at Taylor & Francis and is involved in developing chemical information products.

Section I
Amino Acids, Peptides, and Proteins

PROPERTIES OF AMINO ACIDS

This table gives selected properties of some important amino acids and closely related compounds. The first part of the table lists the 20 "standard" amino acids that are the basic constituents of proteins. The second part includes other amino acids and related compounds of biochemical importance. Within each part of the table the compounds are listed by name in alphabetical order. Structures are given in the following table.

Symbol : Three-letter symbol for the standard amino acids
M_r: Molecular weight
t_m: Melting point
pK_a, pK_b, pK_c, pK_d : Negative of the logarithm of the acid dissociation constants for the COOH and NH$_2$ groups (and, in some cases, other groups) in the molecule (at 25°C)
pI: pH at the isoelectric point
S: Solubility in water in units of grams of compound per kilogram of water; a temperature of 25°C is understood unless otherwise stated in a superscript. When quantitative data are not available, the notations sl.s. (for slightly soluble), s. (for soluble), and v.s. (for very soluble) are used.
V_2^0: Partial molar volume in aqueous solution at infinite dilution (at 25°C)

Data on the enthalpy of formation of many of these compounds are included in the table "Heat of Combustion, Enthalpy and Free Energy of Formation of Amino Acids and Related Compounds" on p. 69 of this Handbook. Absorption spectra and optical rotation data can be found in Reference 3. Partial molar volume is taken from Reference 5; other thermodynamic properties, including solubility as a function of temperature, are given in References 3 and 5. Most of the pK values come from References 1, 6, and 7.

References

1. Dawson, R. M. C., Elliott, D. C., Elliott, W. H., and Jones, K. M., *Data for Biochemical Research*, Third Edition, Clarendon Press, Oxford, 1986.
2. O'Neil, Maryadele J., Ed., *The Merck Index, Fourteenth Edition*, Merck & Co., Rahway, NJ, 2006.
3. Sober, H. A., Ed., *CRC Handbook of Biochemistry. Selected Data for Molecular Biology*, CRC Press, Boca Raton, FL, 1968.
4. Voet, D., and Voet, J. G., *Biochemistry, Second Edition*, John Wiley & Sons, New York, 1995.
5. Hinz, H. J., Ed., *Thermodynamic Data for Biochemistry and Biotechnology*, Springer–Verlag, Heidelberg, 1986.
6. Fasman, G. D., Ed. *Practical Handbook of Biochemistry and Molecular Biology*, CRC Press, Boca Raton, FL, 1989.
7. Smith, R. M., and Martell, A. E., *NIST Standard Reference Database 46: Critically Selected Stability Constants of Metal Complexes Database*, Version 3.0, National Institute of Standards and Technology, Gaithersburg, MD, 1997.
8. Ramasami, P., *J. Chem. Eng. Data*, 47, 1164, 2002.

Symbol	Name	Mol. Form	M_r	t_m/°C	pK_a	pK_b	pK_c	pK_d	pI	S/g kg^{-1}	V_2^0/cm^3 mol^{-1}
Ala	L-Alanine	C$_3$H$_7$NO$_2$	89.09	297	2.33	9.71			6.00	166.9	60.54
Arg	L-Arginine	C$_6$H$_{14}$N$_4$O$_2$	174.20	244	2.03	9.00	12.10		10.76	182.6	127.42
Asn	L-Asparagine	C$_4$H$_8$N$_2$O$_3$	132.12	235	2.16	8.73			5.41	25.1	78.0
Asp	L-Aspartic acid	C$_4$H$_7$NO$_4$	133.10	270	1.95	9.66	3.71		2.77	5.04	74.8
Cys	L-Cysteine	C$_3$H$_7$NO$_2$S	121.16	240	1.91	10.28	8.14		5.07	v.s.	73.45
Gln	L-Glutamine	C$_5$H$_{10}$N$_2$O$_3$	146.14	185	2.18	9.00			5.65	42.5	
Glu	L-Glutamic acid	C$_5$H$_9$NO$_4$	147.13	160	2.16	9.58	4.15		3.22	8.6	89.85
Gly	Glycine	C$_2$H$_5$NO$_2$	75.07	290	2.34	9.58			5.97	250.2	43.26
His	L-Histidine	C$_6$H$_9$N$_3$O$_2$	155.15	287	1.70	9.09	6.04		7.59	43.5	98.3
Ile	L-Isoleucine	C$_6$H$_{13}$NO$_2$	131.17	284	2.26	9.60			6.02	34.2	105.80
Leu	L-Leucine	C$_6$H$_{13}$NO$_2$	131.17	293	2.32	9.58			5.98	22.0	107.77
Lys	L-Lysine	C$_6$H$_{14}$N$_2$O$_2$	146.19	224	2.15	9.16	10.67		9.74	5.8	108.5
Met	L-Methionine	C$_5$H$_{11}$NO$_2$S	149.21	281	2.16	9.08			5.74	56	105.57
Phe	L-Phenylalanine	C$_9$H$_{11}$NO$_2$	165.19	283	2.18	9.09			5.48	27.9	121.5
Pro	L-Proline	C$_5$H$_9$NO$_2$	115.13	221	1.95	10.47			6.30	1622	82.76
Ser	L-Serine	C$_3$H$_7$NO$_3$	105.09	228	2.13	9.05			5.68	250	60.62
Thr	L-Threonine	C$_4$H$_9$NO$_3$	119.12	256	2.20	8.96			5.60	98.1	76.90
Trp	L-Tryptophan	C$_{11}$H$_{12}$N$_2$O$_2$	204.23	289	2.38	9.34			5.89	13.2	143.8
Tyr	L-Tyrosine	C$_9$H$_{11}$NO$_3$	181.19	343	2.24	9.04	10.10		5.66	0.46	
Val	L-Valine	C$_5$H$_{11}$NO$_2$	117.15	315	2.27	9.52			5.96	88	90.75
	N-Acetylglutamic acid	C$_7$H$_{11}$NO$_5$	189.17	199						s.	
	N^6-Acetyl-L-lysine	C$_8$H$_{16}$N$_2$O$_3$	188.22	265	2.12	9.51					
	β-Alanine	C$_3$H$_7$NO$_2$	89.09	200	3.51	10.08				723.6	58.28

PROPERTIES OF AMINO ACIDS (Continued)

Name	Mol. Form	M_r	$t_m/°C$	pK_a	pK_b	pK_c	pK_d	pI	S/g kg^{-1}	V_2^0/cm^3 mol^{-1}
2-Aminoadipic acid	$C_6H_{11}NO_4$	161.16	207	2.14	4.21	9.77		3.18	2.2[40]	
DL-2-Aminobutanoic acid	$C_4H_9NO_2$	103.12	304	2.30	9.63			6.06	210	75.6
DL-3-Aminobutanoic acid	$C_4H_9NO_2$	103.12	194.3	3.43	10.05			7.30	1250	76.3
4-Aminobutanoic acid	$C_4H_9NO_2$	103.12	203	4.02	10.35				971	73.2
10-Aminodecanoic acid	$C_{10}H_{21}NO_2$	187.28	188.5							167.3
7-Aminoheptanoic acid	$C_7H_{15}NO_2$	145.20	195						v.s.	120.0
6-Aminohexanoic acid	$C_6H_{13}NO_2$	131.17	205					7.29	863	104.2
L-3-Amino-2-methylpropanoic acid	$C_4H_9NO_2$	103.12	185						s.	
2-Amino-2-methylpropanoic acid	$C_4H_9NO_2$	103.12	335	2.36	10.21			5.72	137	77.55
9-Aminononanoic acid	$C_9H_{19}NO_2$	173.26	191							151.3
8-Aminooctanoic acid	$C_8H_{17}NO_2$	159.23	192							136.1
5-Amino-4-oxopentanoic acid	$C_5H_9NO_3$	131.13	118	4.05	8.90					
5-Aminopentanoic acid	$C_5H_{11}NO_2$	117.15	157 dec						s.	87.6
o-Anthranilic acid	$C_7H_7NO_2$	137.14	146	2.05	4.95				3.5[14]	
Azaserine	$C_5H_7N_3O_4$	173.13	150		8.55				v.s.	
Canavanine	$C_5H_{12}N_4O_3$	176.17	172	2.50	6.60	9.25		7.93	v.s.	
L-γ-Carboxyglutamic acid	$C_6H_9NO_6$	191.14	167	1.70	9.90	4.75	3.20			
Carnosine	$C_9H_{14}N_4O_3$	226.23	260	2.51	9.35	6.76			322	
Citrulline	$C_6H_{13}N_3O_3$	175.19	222	2.32	9.30			5.92	s.	
Creatine	$C_4H_9N_3O_2$	131.13	303	2.63	14.30				16	
L-Cysteic acid	$C_3H_7NO_5S$	169.16	260	1.89	8.70	1.30			v.s.	
L-Cystine	$C_6H_{12}N_2O_4S_2$	240.30	260	1.50	8.80	2.05	8.03		0.11	
2,4-Diaminobutanoic acid	$C_4H_{10}N_2O_2$	118.13	118.1	1.85	8.24	10.44		9.27	s.	
3,5-Dibromo-L-tyrosine	$C_9H_9Br_2NO_3$	338.98	245						2.72	
3,5-Dichloro-L-tyrosine	$C_9H_9Cl_2NO_3$	250.08	247						1.97	
3,5-Diiodo-L-tyrosine	$C_9H_9I_2NO_3$	432.98	213	2.12	9.10	6.16			0.62	
Dopamine	$C_8H_{11}NO_2$	153.18			10.36	8.88			s.	
L-Ethionine	$C_6H_{13}NO_2S$	163.24	273	2.18	9.05	13.10				
N-Glycylglycine	$C_4H_8N_2O_3$	132.12	263	3.13	8.10				225	
Guanidinoacetic acid	$C_3H_7N_3O_2$	117.11	282	2.82					5	
Histamine	$C_5H_9N_3$	111.15	83		9.83	6.11			v.s.	
L-Homocysteine	$C_4H_9NO_2S$	135.19	232	2.15	8.57	10.38		5.55	s.	
Homocystine	$C_8H_{16}N_2O_4S_2$	268.35	264	1.59	9.44	2.54	8.52		0.2	
L-Homoserine	$C_4H_9NO_3$	119.12	203	2.27	9.28			6.17	1100	
3-Hydroxy-DL-glutamic acid	$C_5H_9NO_5$	163.13	209					3.28		
5-Hydroxylysine	$C_6H_{14}N_2O_3$	162.19		2.13	8.85	9.83		9.15		
trans-4-Hydroxy-L-proline	$C_5H_9NO_3$	131.13	274	1.82	9.47			5.74	361	84.49
L-3-Iodotyrosine	$C_9H_{10}INO_3$	307.08	205	2.20	9.10	8.70			sl.s.	
L-Kynurenine	$C_{10}H_{12}N_2O_3$	208.21	194						sl.s.	
L-Lanthionine	$C_6H_{12}N_2O_4S$	208.24	294						1.5	
Levodopa	$C_9H_{11}NO_4$	197.19	277	2.32	8.72	9.96	11.79		5[20]	
L-1-Methylhistidine	$C_7H_{11}N_3O_2$	169.18	249	1.69	8.85	6.48			200	
L-Norleucine	$C_6H_{13}NO_2$	131.17	301	2.31	9.68			6.09	15	107.7
L-Norvaline	$C_5H_{11}NO_2$	117.15	307	2.31	9.65				107	91.8
L-Ornithine	$C_5H_{12}N_2O_2$	132.16	140	1.94	8.78	10.52		9.73	v.s.	
O-Phosphoserine	$C_3H_8NO_6P$	185.07	166	2.14	9.80	5.70				
L-Pyroglutamic acid	$C_5H_7NO_3$	129.12	162	3.32						
Sarcosine	$C_3H_7NO_2$	89.09	212	2.18	9.97				428	
Taurine	$C_2H_7NO_3S$	125.15	328	-0.3	9.06				105	
L-Thyroxine	$C_{15}H_{11}I_4NO_4$	776.87	235	2.20	10.01	6.45			sl.s.	

STRUCTURES OF COMMON AMINO ACIDS

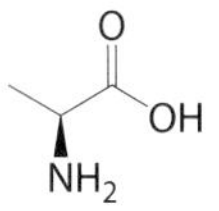

L-Alanine (Ala)

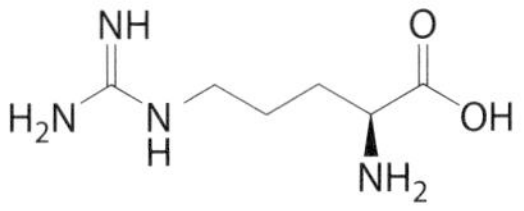

L-Arginine (Arg)

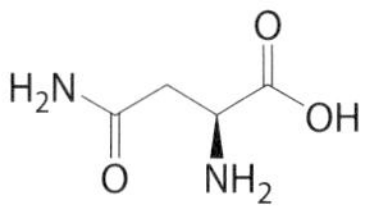

L-Asparagine (Asn)

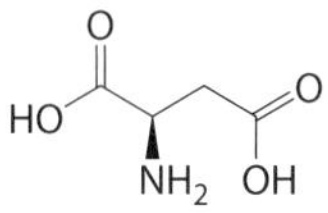

L-Aspartic acid (Asp)

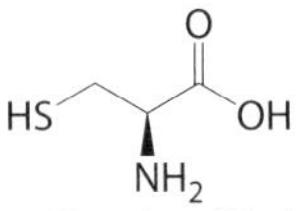

L-Cysteine (Cys)

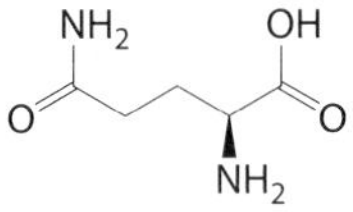

L-Glutamine (Gln)

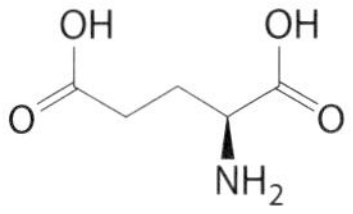

L-Glutamic acid (Glu)

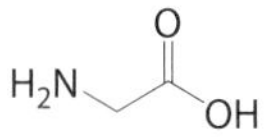

Glycine (Gly)

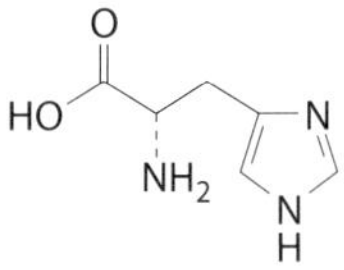

L-Histidine (His)

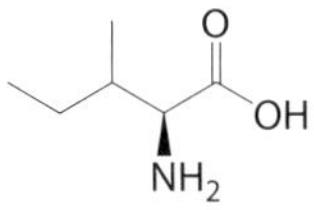

L-Isoleucine (Ile)

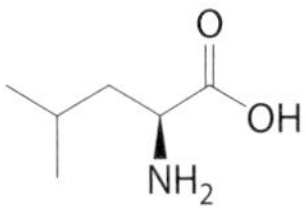

L-Leucine (Leu)

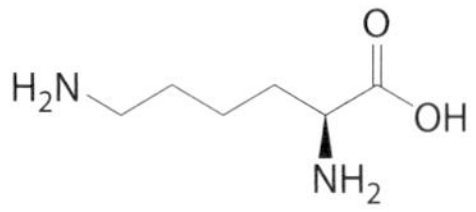

L-Lysine (Lys)

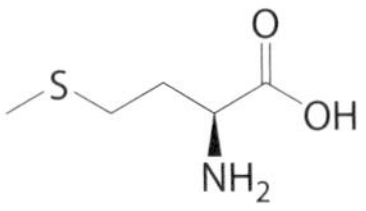

L-Methionine (Met)

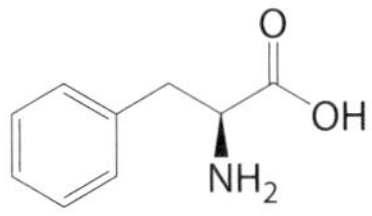

L-Phenylalanine (Phe)

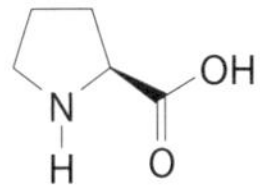

L-Proline (Pro)

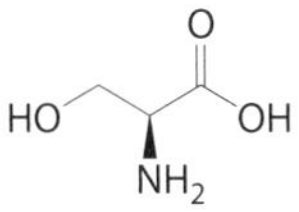

L-Serine (Ser)

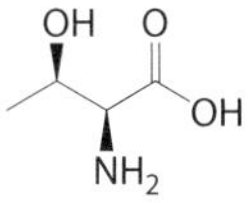

L-Threonine (Thr)

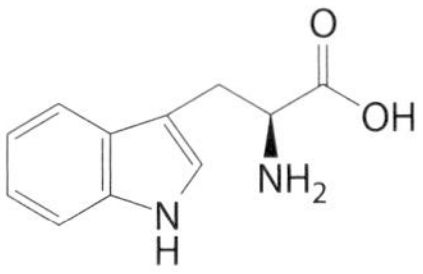

L-Tryptophan (Trp)

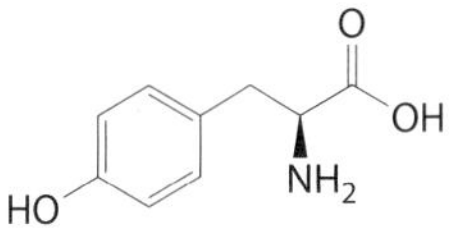

L-Tyrosine (Tyr)

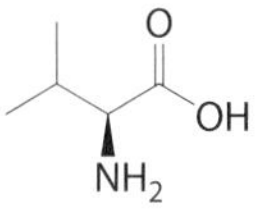

L-Valine (Val)

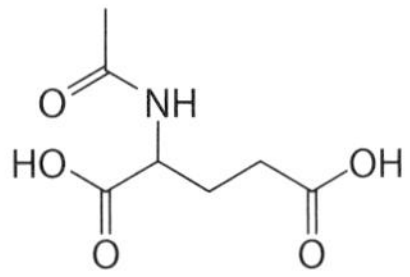

N-Acetylglutamic acid

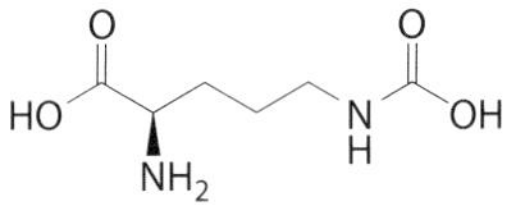

N^6-Acetyl-L-lysine

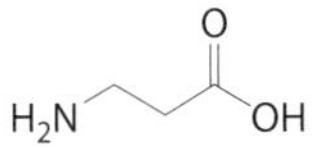

β-Alanine

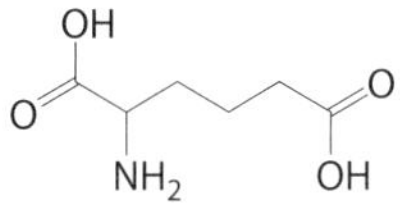

2-Aminoadipic acid

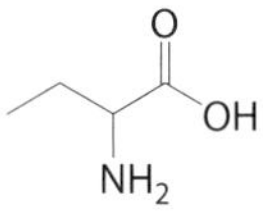

DL-2-Aminobutanoic acid

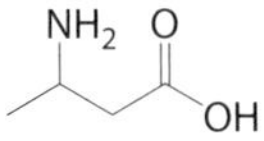

DL-3-Aminobutanoic acid

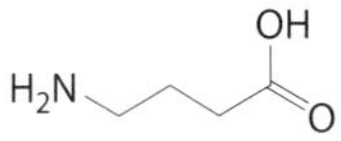

4-Aminobutanoic acid

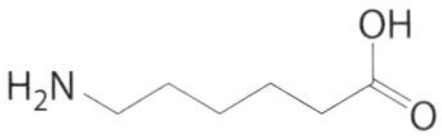

6-Aminohexanoic acid

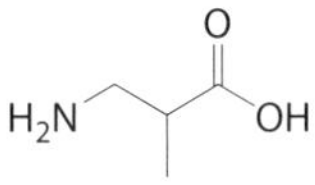

L-3-Amino-2-methylpropanoic acid

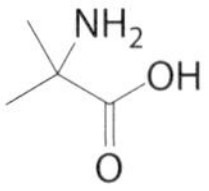

2-Amino-2-methylpropanoic acid

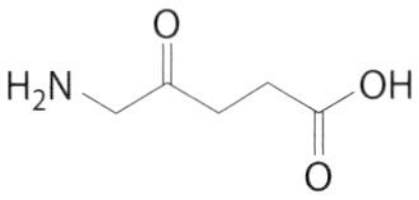

5-Amino-4-oxopentanoic acid

5-Aminopentanoic acid

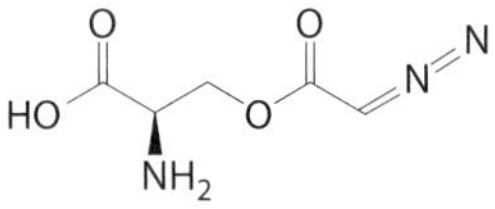

Azaserine

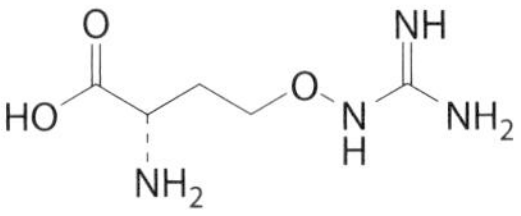

Canavanine

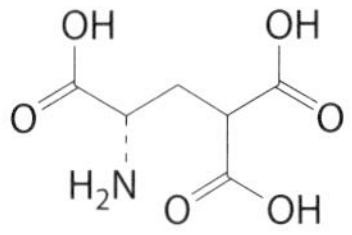

L-γ-Carboxyglutamic acid

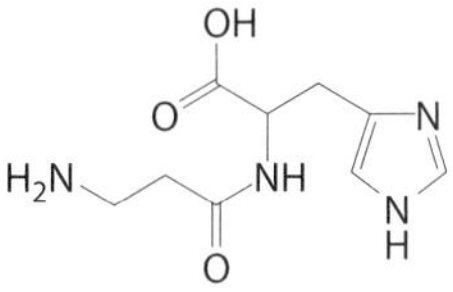

Carnosine

STRUCTURES OF COMMON AMINO ACIDS (Continued)

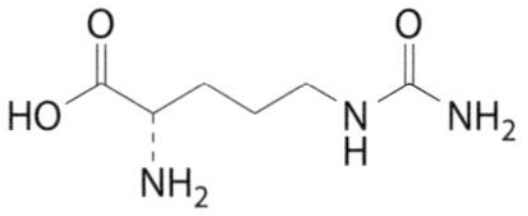

Citrulline

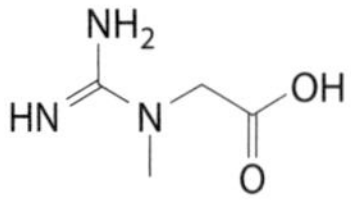

Creatine

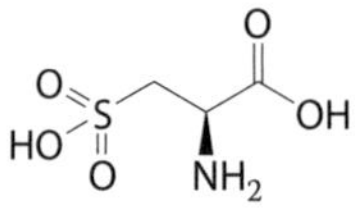

L-Cysteic acid

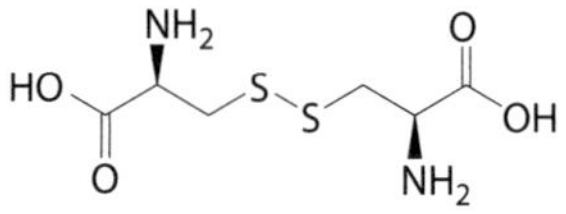

L-Cystine

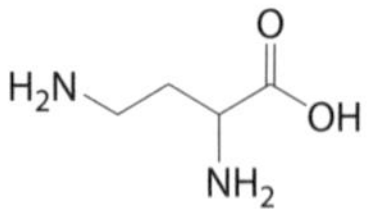

2,4-Diaminobutanoic acid

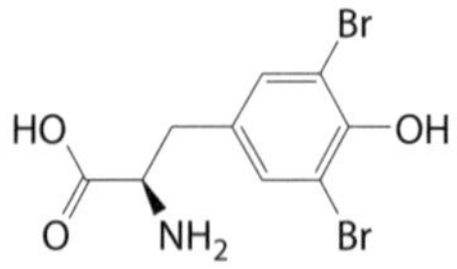

3,5-Dibromo-L-tyrosine

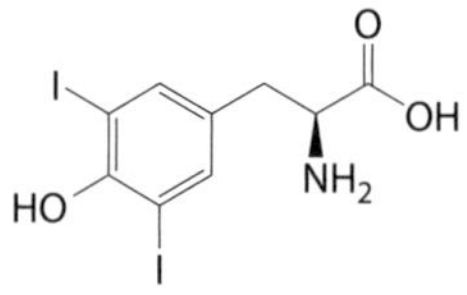

3,5-Diiodo-L-tyrosine

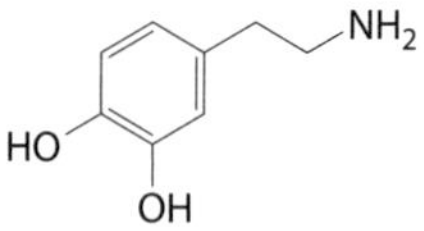

Dopamine

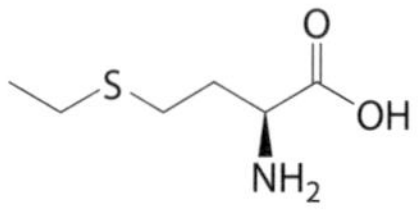

L-Ethionine

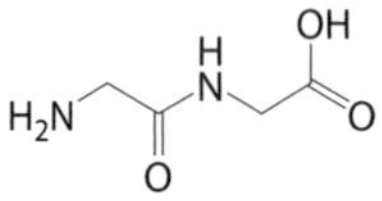

N-Glycylglycine

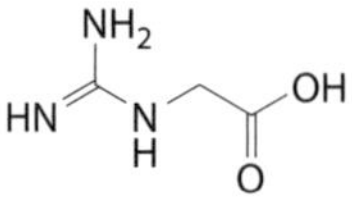

Guanidinoacetic acid

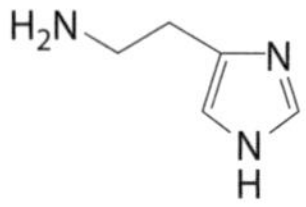

Histamine

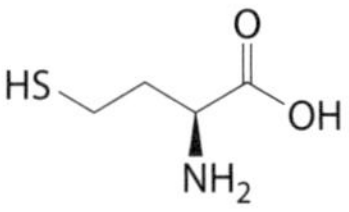

L-Homocysteine

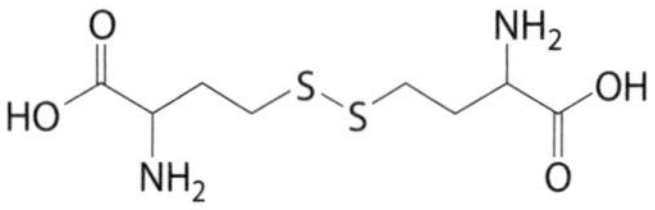

Homocystine

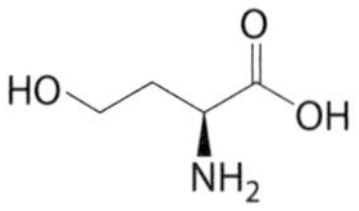

L-Homoserine

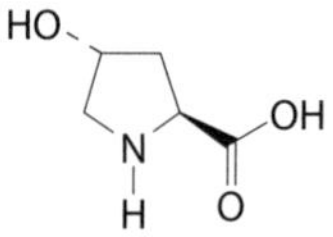

trans-4-Hydroxy-L-proline

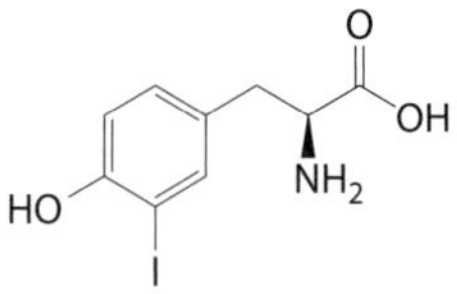

L-3-Iodotyrosine

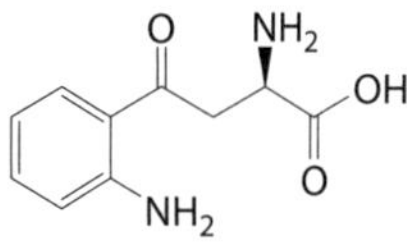

L-Kynurenine

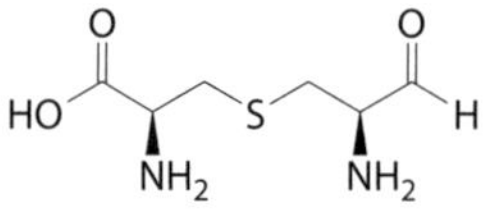

L-Lanthionine

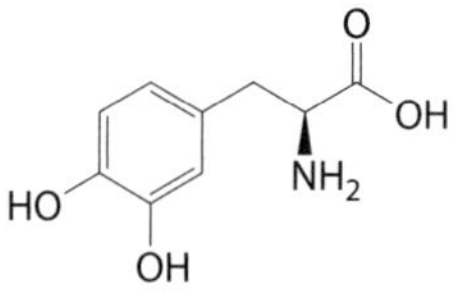

Levodopa

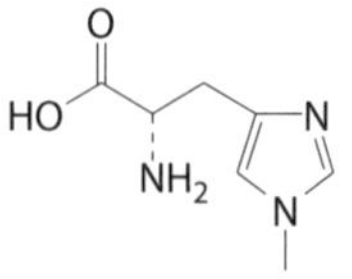

L-1-Methylhistidine

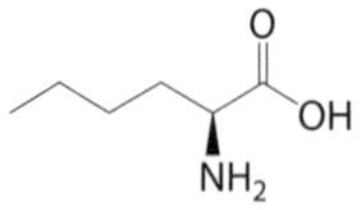

L-Norleucine

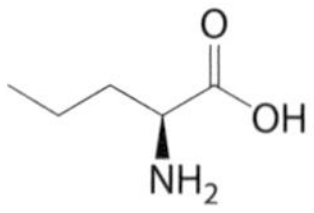

L-Norvaline

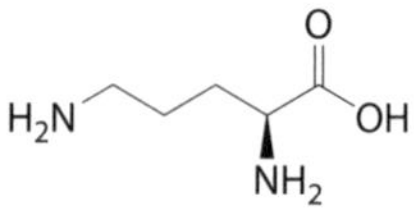

L-Ornithine

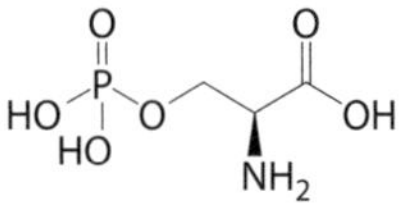

O-Phosphoserine

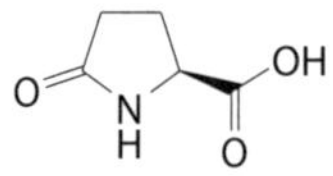

L-Pyroglutamic acid

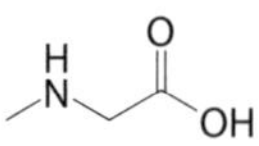

Sarcosine

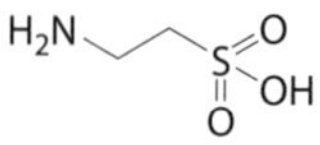

Taurine

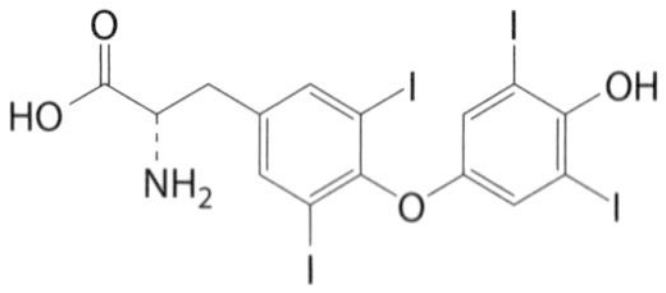

L-Thyroxine

DATA ON THE NATURALLY OCCURRING AMINO ACIDS

Elizabeth Dodd Mooz

The amino acids included in these tables are those for which reliable evidence exists for their occurrence in nature. These tables are intended as a guide to the primary literature in which the isolation and characterization of the amino acids are reported. Originally, it was planned to include more factual data on the chemical and physical properties of these compounds; however, the many different conditions employed by various authors in measuring these properties (i.e., chromatography and spectral data) made them difficult to arrange into useful tables. The rotation values are as given in the references cited; unfortunately, in some cases there is no information given on temperature, solvent, or concentration.

The investigator employing the data in these tables is urged to refer to the original articles in order to evaluate for himself the reliability of the information reported. These references are intended to be informative to the reader rather than to give credit to individual scientists who published the original reports. Thus not all published material is cited.

The compounds listed in Sections A to N are known to be of the L configuration. Section O contains some of the D amino acids which occur naturally. This last section is not intended to be complete since most properties of the D amino acids correspond to those of their L enantiomorphs. Therefore, emphasis was placed on including those D amino acids whose L isomers have not been found in nature. The reader will find additional information on the D amino acids in the review by Corrigan[263] and in the book by Meister.[1]

Compilation of data for these tables was completed in December 1974. Appreciation is expressed to Doctors L. Fowden, John F. Thompson, Peter Müller, and M. Bodanszky who were helpful in supplying recent references and to Dr. David Pruess who made review material available to me prior to its publication. A special word of thanks to Dr. Alton Meister who made available reprints of journal articles which I was not able to obtain.

DATA ON THE NATURALLY OCCURRING AMINO ACIDS (Continued)

No.	Amino Acid (Synonym)	Source	Formula (Mol Wt)	Melting Point °C[a]	$[\alpha]_D$[b]	pK_a	References: Isolation and Purification	References: Chromatography	References: Chemistry	References: Spectral Data
	A. L-MONOAMINO, MONOCARBOXYLIC ACIDS									
1	Alanine (α-aminopropionic acid)	Silk fibroin	$C_3H_7NO_2$ (89.09)	297°	$+1.8^{25}$ (*c* 2, H_2O) (1) $+14.6^{25}$ (*c* 2, 5 *N* HCl) (1)	2.34 9.69	2	3	4	4
2	β-Alanine (β-aminopropionic acid)	*Iris tingitana*	$C_3H_7NO_2$ (89.09)	196° (dec)	–	3.55 10.24	5	5	5	–
3	α-Aminobutyric acid	Yeast protein	$C_4H_9NO_2$ (103.12)	292° (dec)	$+20.5^{25}$ (*c* 1–2, 5 *N* HCl) (290) $+9.3^{25}$ (*c* 1–2, H_2O) (290) $+42^{25}$ (*c* 1–2, gl acetic) (290)	2.29 9.83	6	7	6	–
4	γ-Aminobutyric acid (piperidinic acid)	Bacteria	$C_4H_9NO_2$ (103.12)	203° (dec)	–	4.03 10.56 (290)	8–10	9, 10	11	–
5	1-Aminocyclopropane-1-carboxylic acid	Pears and apples	$C_4H_7NO_2$ (101.11)	–	–	–	11	11	12	12
6	2-Amino-3-formyl-3-pentenoic acid	*Bankera fulgineoalba* (a mushroom)	$C_6H_9NO_3$ (143.15)	–	–	–	13	13	13	13
7	α-Aminoheptanoic acid	*Claviceps purpures*	$C_7H_{15}NO_2$ (145.21)	–	–	–	14	14	–	–
7a	2-Amino-4,5-hexadienoic acid	*Amanita solitaria*	$C_6H_9NO_2$ (127.16)	200° (dec) (14a)	–	–	14a	–	–	14a
8	2-Amino-4-hexenoic acid	Ilamycin	$C_6H_{11}NO_2$ (129.17)	–	–	–	15	15	–	–
8a	2-Amino-4-hydroxy-hept-6-ynoic acid	*Euphoria longan*	$C_7H_{11}NO_3$ (157.19)	–	-27^{20} (*c* 2, H_2O) -8^{20} (*c* 1, 5 *N* HCl) (15a)	–	15a	15a	–	15a
8b	2-Amino-6-hydroxy-4-methyl-4-hexenoic acid	*Aesculus California* seeds	$C_7H_{13}NO_3$ (159.21)	–	-31^{20} (*c* 2.2, H_2O) $+2^{20}$ (*c* 1.1, 5 *N* HCl) (23b)	–	23b	23b	–	15b
8c	2-Amino-4-hydroxy-5-methyl hexenoic acid	*Euphoria longan*	$C_7H_{11}NO_3$ (157.19)	–	-27^{20} (*c* 2, H_2O) -13^{20} (*c* 1, 5 *N* HCl) (15a)	–	15a	–	15a	15a
8d	2-Amino-3-hydroxy-methyl-3-pentenoic acid	*Bankera fuligineoalba*	$C_6H_{11}NO_3$ (145.18)	160–161° (dec) (13)	$+182^{25}$ (*c* 0.8, H_2O) $+201^{25}$ (*c* 0 8, 0 3 *N* HCl) (13)	–	13	13	13	13

DATA ON THE NATURALLY OCCURRING AMINO ACIDS (Continued)

No.	Amino Acid (Synonym)	Source	Formula (Mol Wt)	Melting Point °C[a]	$[\alpha]_D$[b]	pK_a	References			
							Isolation and Purification	Chromatography	Chemistry	Spectral Data
9	α-Aminoisobutyric acid	Iris tingitana, muscle protein	C₄H₉NO₂ (103.12)	200° (dec)	–	2.36 10.21 (290)	16	16	–	–
10	β-Aminoisobutyric acid	Iris tingitana	C₄H₉NO₂ (103.12)	179° (17)	-21^{26} (c 0.43, H₂O) (17)	–	17	17	17	17
10a	2-Amino-4-methoxy-trans-3-butenoic acid	Pseudomonas aeruginosa	C₆H₉NO₃ (131.15)	–	–	–	17a	17b	–	17a, 17b
11	γ-Amino-α-methylene butyric acid	Arachis hypogaea (groundnut plants)	C₄H₉NO₂ (115.13)	152° (18)		–	18	18	18	–
12	2-Amino-4-methyl-hexanoic acid (homoisoleucine)	Aesculus californica seeds	C₇H₁₅NO₂ (145.21)	–	-2^{20} (c 1, H₂O) (19) $+24^{20}$ (c 0.87, 5 N HCl) (19)	–	19	19	19	19
13	2-Amino-4-methyl-4-hexenoic acid	Aesculus californica seeds	C₇H₁₃NO₂ (143.19)	–	-61^{20} (c 2.4, H₂O) (19) -36 (c 1.2, 6 N HCl) (19)	–	19	19	19	19
13a	2-Amino-4-methyl-5-hexenoic acid	Streptomyces species	C₇H₁₃NO₂ (143.21)	260° (dec) (19a)	$-9.6°$ (c 1.78, H₂O) $+5.7°$ (c 0.7, 1 N HCl) (99a)	–	19a	19a	–	19a
14	2-Amino-5-methyl-4-hexenoic acid	Leucocortinarius bulbiger	C₇H₁₃NO₂ (143.19)	260–270° (dec) (22a)	-45.9^{23} (c 0.47, H₂O) -7^{23} (c 0.4, 1 N HCl) (22a)	–	22, 22a	22a	–	22a
14a	2-Amino-4-methyl-5-hexenoic acid	Euphoria longan	C₇H₁₁NO₂ (141.19)	–	-33^{20} (c 2, H₂O) -27^{20} (c 1, 5 N HCl) (15a)	–	15a	–	15a	15a
15	α-Amino-octanoic acid	Aspergillus atypigue	C₈H₁₇NO₂ (159.23)	–	–	–	23	23	23	23
15a	2-Amino-4-pentynoic acid	Streptomyces sp. #8–4	C₅H₇NO₂ (113.13)	241–242° (dec) (23a)	-31.1^{25} (c 1, H₂O) -5.5^{25} (c 1, 5 N HCl) (23a)	–	23a	23a	23a	23a
15a′	cis-α-(Carboxy-cyclopropyl)glycine	Aesculus parviflora	C₆H₉NO₄ (159.16)	–	$+25^{20}$ (c 1, H₂O) $+58$ (c 0.5, 5 N HCl) (23a′)	–	23a′	23a′	–	23a′
15b	trans-α-(Carboxy-cyclopropyl)glycine	Blighia sapida	C₆H₉NO₄ (159.16)	–	$+107^{20}$ (c 2, H₂O) $+146^{20}$ (c 1, 5 N HCl) (23a′)	–	23a′	23a′	–	23a′

DATA ON THE NATURALLY OCCURRING AMINO ACIDS (Continued)

No.	Amino Acid (Synonym)	Source	Formula (Mol Wt)	Melting Point °C[a]	$[\alpha]_D$[b]	pK_a	Isolation and Purification	Chromatography	Chemistry	Spectral Data
15b'	trans-α-(2-Carboxy-methylcyclopropyl)glycine	Blighia unijugata	$C_7H_{11}NO_4$ (173.19)	—	$+12^{20}$ (c 1, H_2O) $+45^{20}$ (c 0.5, 5 N HCl) (99a)	—	99a	99a	—	99a
15c	γ-Glutamyl-2-amino-4-methylhex-4-enoic acid	Aesculus californica seeds	$C_{12}H_{20}N_2O_5$ (272.34)	—	$+17^{20}$ (c 3, H_2O) (23b)	—	23b	23b	—	—
B. L-MONOAMINO, DICARBOXYLIC ACIDS										
16	Glycine (α-aminoacetic acid)	Gelatin hydrolyzate	$C_2H_6NO_2$ (75.07)	290° (dec) (1)	—	2.35 9.78 (290)	24	3	25	25
17	Hypoglycin A [α-amino-β-(2-methylene cyclopropyl)propionic acid]	Blighia sapida	$C_7H_{11}NO_2$ (141.18)	280–284° (26)	$+9.2$ (c 1, H_2O) (26)	—	26	26	27	27
18	Isoleucine (α-amino-β-methylvaleric acid)	Sugar beet molasses	$C_6H_{13}NO_2$ (131.17)	284° (1)	$+39.5^{25}$ (c 1, 5 N HCl) (290) $+12.4^{25}$ (c 1, H_2O) (290)	2.36 9.68	3	9	29	1
19	Leucine (α-aminoisocaproic acid)	Muscle fiber, wool	$C_6H_{13}NO_2$ (131.17)	337° (1)	-11^{25} (c 2, H_2O) $+16^{25}$ c 2, 5 N HCl) (1)	2.36 960 (1)	30	3	31	31
19a	N-Methyl-γ-methyl-alloisoleucine	Etamycin	$C_8H_{17}NO_2$ (159.26)	—	—	—	31a	31a	—	—
19b	β-Methyl-β-(methylene-cyclopropyl)alanine	Aesculus californica seeds	$C_8H_{13}NO_2$ (155.22)	—	1.5^{20} (c 2, H_2O) $+45^{20}$ (c 1, 5 N HCl) (23b)	— 960 (1)	23b	23b	—	15b
20	α-(Methylene cyclopropyl)glycine	Litchi chinensis	$C_6H_9NO_2$ (127.15)	—	$+43^{22.5}$ (c 0.5, 5 N HCl) (32)	—	32	32	32	32
21	β-(Methylene-cyclopropyl)-β-methylalanine	Aesculus californica	$C_8H_{13}NO_2$ (155.19)	—	1.5^{20} (c 2, H_2O) $+45^{20}$ (c 1, 5 N HCl)	—	19	19	19	19, 21
21a	β-Methylenenorleucine	Amanita vaginata	$C_7H_{13}NO_2$ (143.21)	171° (35a)	$+158^{20}$ (c 0.51, 1 N HCl) $+149^{20}$ (c 0.56, H_2O) (35a)	—	—	35a	35a	—

DATA ON THE NATURALLY OCCURRING AMINO ACIDS (Continued)

No.	Amino Acid (Synonym)	Source	Formula (Mol Wt)	Melting Point °C[a]	$[\alpha]_D$[b]	pK_a	References Isolation and Purification	Chromatography	Chemistry	Spectral Data
22	Valine (α-amino isovaleric acid)	Casein	$C_5H_{11}NO_2$ (117.15)	292–295° (1)	$+28.3^{25}$ (c 1, 2, 5 N HCl) $+5.63^{25}$ (c 1–2, H₂O) (290)	2.32 9.62 (1)	35	3	35, 36	36
23	α-Aminoadipic acid	Pisum sativum	$C_6H_{11}NO_4$ (161.18)	195° (37)	$+3.2^{25}$ (c 2, H₂O (290) $+23^{22}$ (c 2, 6 N HCl) (37)	2.14 4.21 9.77 (290)	37	37	37	—
24	3-Aminoglutaric acid	Chondria armata	$C_5H_9NO_4$ (162.13)	280–282° (38)	$\pm0^c$ (c 2, 5 N HCl) (38)	—	38	38	38	38
25	α-Aminopimelic acid	Asplenium septentrionale	$C_7H_{13}NO_4$ (175.19)	204° (39)	—	—	39	39	39	—
26	Asparagine (α-aminosuccinamic acid)	Asparagus	$C_4H_8N_2O_3$ (132.12)	236° (1)	$+5.06^{25}$ (c 2, H₂O) (290) $+33.2$ (3 N HCl) (1)	2.02 8.80 (1)	40	3	41	42
27	Aspartic acid (α-aminosuccinic acid)	Conglutin, legumin	$C_4H_7NO_4$ (133.10)	270° (1)	$+5.05^{25}$ (c 2, H₂O) $+25.4^{25}$ (c 2, 5 N HCl) (1)	1.88 3.65 9.60 (1)	43	3	41	41
28	Ethylasparagine	Ecballium elaterium, Bryonia dioica	$C_6H_{12}N_2O_3$ (160.19)	—	—	—	45	45	45	45
29	γ-Ethylideneglutamic acid	Mimosa	$C_7H_{11}NO_4$ (173.18)	—	$+21^{20}$ (c 2.8, H₂O) (47) $+38.3^{20}$ (c 1.4, 6 N HCl) (47)	—	46	46	—	46
30	N-Fumarylalanine	Penicillium recticulosum	$C_7H_9NO_5$ (187.16)	229° (48)	—	—	48	—	48	—
31	Glutamic acid (α-aminoglutaric acid)	Gluten-fibrin hydrolyzates	$C_5H_9NO_4$ (147.13)	249° (1)	$+12^{25}$ (c 2, H₂O) $+31.8^{28}$ (c 2, 5 N HCl) (1)	2.19 4.25 9.67 (1)	49	3	50	50
32	Glutamine (α-aminoglutaramic acid)	Beet juice	$C_5H_{10}N_2O_3$ (146.15)	185° (1)	$+6.3^{25}$ (c 2, H₂O) $+31.8^{25}$ (c 2, 1 N HCl) (1)	2.17 9.13 (1)	51	3	50	50

DATA ON THE NATURALLY OCCURRING AMINO ACIDS (Continued)

No.	Amino Acid (Synonym)	Source	Formula (Mol Wt)	Melting Point °C[a]	$[\alpha]_D$[b]	pKa	Isolation and Purification	Chromatography	Chemistry	Spectral Data
33	N^5-Isopropylglutamine	Lunaria annua	$C_8H_{16}N_2O_3$ (188.23)	–	+7.1²² (c 1.7, H₂O) (53)	–	53	53	53	–
34	N^4-Methylasparagine	–	$C_5H_{10}N_2O_3$ (146.15)	241–244° (54)	–4.2 (c 5.5, H₂O) (54)	–	–	54	–	–
35	β-Methylaspartic acid	Clostridium tetanomorphum	$C_5H_9NO_4$ (147.13)	–	–10 (c 0.4, H₂O) +12.4 (c 3, 1 N HCl) +13.3 (c 3, 5 N HCl) (55)	3.5 9.9 (290)	55	55	55	–
36	γ-Methylglutamic acid	Phyllitis scolopendrium	$C_6H_{11}NO_4$ (141.17)	–	–	–	56	56	–	–
37	γ-Methyleneglutamic ac.d	Arachis hypogaea	$C_6H_9NO_4$ (159.15)	196° (57)	–	–	57	57	57	–
38	γ-Methyleneglutamine	Arachis hypogaea	$C_6H_{10}N_2O_3$ (158.17)	173–182° (57)	–	–	57	57	57	–
39	Theanine (α-amino-γ-N-ethylglutaramic acid)	Xerocomus badius	$C_7H_{14}N_2O_3$ (174)	–	–	–	59	59	59	59
39a	β-N-Acetyl-α,β-diaminopropionic acid (β-acetamido-L-alanine)	Acacia armata seeds	$C_5H_{10}N_2O_3$ (146.17)	–	–87²⁰ (c 8, H₂O) –35²⁰ (c 4, 6 N HCl) (59a)	–	59a	59a	59a	59a
				C. L-DIAMINO, MONOCARBOXYLIC ACIDS						
40	N-Acetylornithine	Asplenium species	$C_7H_{14}N_2O_3$ (174.11)	200° (dec) (60)	–	–	60	60	60	–
41	α-Amino-γ-N-acetylaminobutyric acid	Latex of Euphorbia pulcherrima	$C_6H_{12}N_2O_3$ (160.18)	220–222° (dec) (61)	–	4.45 (33)	61	61	–	–
42	N-ε-(2-Amino-2-carboxyethyl)lysine	Alkali-treated protein	$C_9H_{19}N_3O_4$ (233.28)	–	–	2.2 6.5 8.8 9.9 (62)	62	62	62	–
43	N-δ-(2-Amino-2-carboxyethyl)ornithine	Alkali-treated wool	$C_8H_{17}N_3O_4$ (273.72)	–	–	–	63	63	63	–
44	2-Amino-3-di-methylaminopropionic acid	Streptomyces neocaliberis	$C_5H_{12}N_2O_2$ (117.15)	–	–17.8²⁵ (c 1, H₂O) +18.1 (c 1, HCl pH 3) (64)	–	64	64	64	64
45	α-Amino-β-methyl-aminopropionic acid	Cycas circinalis	$C_4H_{11}N_2O_2$ (105.15)	165–167° (65)	–	–	65	65	65	65

DATA ON THE NATURALLY OCCURRING AMINO ACIDS (Continued)

No.	Amino Acid (Synonym)	Source	Formula (Mol Wt)	Melting Point °C[a]	$[\alpha]_D$[b]	pKa	References			
							Isolation and Purification	Chroma-tography	Chemistry	Spectral Data
45a	α-Amino-β-oxalyl-aminopropionic acid	*Crotalaria*	$C_5H_8N_2O_5$ (176.15)	—	—	—	—	65a	—	—
46	Canaline	*Canavalia ensiformis*	$C_4N_{10}N_2O_3$ (134.14)	—	—	2.40 3.70 9.20 (20)	66	—	67	—
46a	*threo*-α,β-Diaminobutyric acid	Amphomycin hydrolyzate	$C_4H_{10}N_2O_2$ (118.16)	213–214° (dec) (67a)	$+27.1^{25}$ (c 2, 5 N HCl) (67a)	—	67b	67b	67a	67b
47	α,γ-Diaminobutyric acid (γ-aminobutyrine)	Glumamycin	$C_4H_{10}N_2O_2$ (134.14)	—	$+7.2^{25}$ (c 2, H₂O) $+14.6^{18}$ (c 3.67, H₂O)[g] (290) $+14.6^{10}$ (c 3.67, H₂O)[g] (290)	1.85 10.50 (20) 8.28	68	68	68	—
48	3,5-Diaminohexanoic acid	*Clostridium stricklandii*	$C_6H_{14}N_2O_2$ (146.19)	204–208° (69)	—	—	69	69	69	69
48a	2,6-Diamino-7-hydroxyazelaic acid	*Bacillus brevis* (edeine A and B)	$C_9H_{18}N_2O_{15}$ (234.29)	—	—	—	69a	69a	69a	—
49	α,β-Diaminopropionic acid (β-aminoalanine)	Mimosa	$C_3H_8N_2O_2$ (104.11)	—	—	1.23 6.73 9.56 (20)	70	70	70	—
49a	$N^\varepsilon,N^\varepsilon$-Dimethyllysine	Human urine	$C_8H_{18}N_2O_2$ (174.28)	214–216° (dec) (70a)	—	—	70a	70a	—	70a
49b	N^5-Iminoethylornithine	Streptomyces broth	$C_7H_{15}N_3O_2$ (173.25)	226–229° (70b)	$+20.6^{25}$ (c 1, 5 N HCl) (70b)	1.97 8.86 11.83 (70b)	70b	70b	—	70b
50	Lathyrus factor (β-N-(γ-glutamyl)-aminopropionitrile)	*Lathyrus pusillus*	$C_8H_{13}N_3O_3$ (199.22)	193–194° (72)	$+28^{18}$ (c 1, 6 N HCl) (72)	2.2 9.14	71	72	72	72
51	Lysine (α,ε-diaminocaproic acid)	Casein	$C_6H_{14}N_2O_2$ (146.19)	224–225° (dec) (73)	$+14.6^{20}$ (H₂O) (73)	2.16 9.18 10.79 (290)	74	3	75	—
52	β-Lysine (isolysine; β,ε-diaminocaproic acid)	Viomycin	$C_6H_{14}N_2O_2$ (146.19)	240–241° (76)	—	—	76	76	76	76

DATA ON THE NATURALLY OCCURRING AMINO ACIDS (Continued)

No.	Amino Acid (Synonym)	Source	Formula (Mol Wt)	Melting Point °C[a]	$[\alpha]_D$[b]	pK_a	References			
							Isolation and Purification	Chromatography	Chemistry	Spectral Data
53	Lysopine N^2-(D-1-carboxyethyl)-lysine	Calf thymus histone	$C_9H_{18}N_2O_4$ (204.25)	157–160° (77)	+18 (c 1.4, H_2O) (77)	–	78	78	77	77
54	ε-N-Methyllysine	Calf thymus histone	$C_7H_{16}N_2O_2$ (160.23)	–	–	–	79	79	–	79
55	Ornithine (α,δ-diaminovaleric acid)	Asplenium nidus	$C_5H_{12}H_2O_2$ (132.16)	–	+12.1 (c 2, H_2O) +28.4 (c 2, 5 N HCl) (1)	1.71 8.69 10.76 (290)	60	60	81	81
55a	4-Oxalysine	Streptomyces	$C_5N_{12}N_2O_3$ (148.19)	–	–	–	81a	81a	–	81a
56	β-N'-Oxalyl-α,β-diaminopropionic acid	Lathyrus sativus	$C_5H_8N_2O_3$ (176.13)	206° (dec) (82)	-36.9^{27} (c 0.66, 4 N HCl) (82)	1.95 2.95 9.25 (82)	82	82	82	82
56a	β-Putreanine [N-(4-aminobutyl)-3-aminopropionic acid]	Bovine brain	$C_7H_{16}N_2O_2$ (160.25)	250–251° (dec) (82a)	–	3.2 9.4 11.2 (82a)	82a	82a	–	82a

D. L-DIAMINO, DICARBOXYLIC ACIDS

No.	Amino Acid (Synonym)	Source	Formula (Mol Wt)	Melting Point °C[a]	$[\alpha]_D$[b]	pK_a	Isolation and Purification	Chromatography	Chemistry	Spectral Data
57	Acetylenic dicarboxylic acid diamide	Streptomyces chibaensis	$C_4H_4N_2O_2$ (112.09)	216–218° (dec)	–	–	83	83	83	83
58	α,ε-Diaminopimelic acid	Pine pollen	$C_7H_{14}N_2O_4$ (190.20)	–	$+8.1^{25}$ (c 5, H_2O) $+45^{26}$ (c 1, 1 N HCl) $+45.1^{24}$ (c 2.6, 5 N HCl) (290)	1.8 2.2 9.9 8.8 (290)	84	84	85	–
59	2,3-Diaminosuccinic acid	Streptomyces rimosus	$C_4H_8N_2O_4$ (148.10)	240–290° (dec)	–	–	86	86	86	86

E. L-KETO, HYDROXY, AND HYDROXY SUBSTITUTED AMINO ACIDS

No.	Amino Acid (Synonym)	Source	Formula (Mol Wt)	Melting Point °C[a]	$[\alpha]_D$[b]	pK_a	Isolation and Purification	Chromatography	Chemistry	Spectral Data
60	O-Acetylhomoserine	Pisum	$C_6H_{11}NO_4$ (161.17)	–	–	–	87	87	87	–
60a	Threo-α-amino-β,γ-dihydroxybutyric acid	Streptomyces	$C_4H_9NO_4$ (135.14)	210°(dec)	-13.3^{25} (c 1, H_2O) -1.1^{25} (c 1, 2.2 N HCl) (87a)	–	87a	–	–	–
61	2-Amino-4,5-dihydroxy pentanoic acid	Lunaria annua	$C_5H_{11}NO_4$ (149.15)	–	–	–	88	88	88	–
61a	2-Amino-3-formyl-3-pentenoic acid	Bankera fuligineoalba	$C_6H_9NO_3$ (143.16)	–	–	–	88a	88a	88a	88a
62	α-Amino-γ-hydroxy-adipic acid	Vibrio comma	$C_6H_{11}NO_5$ (177.17)	–	–	–	89	–	–	–

DATA ON THE NATURALLY OCCURRING AMINO ACIDS (Continued)

No.	Amino Acid (Synonym)	Source	Formula (Mol Wt)	Melting Point °C[a]	$[\alpha]_D$[b]	pKa	Isolation and Purification	Chromatography	Chemistry	Spectral Data
								References		
63	2-Amino-6-hydroxy-aminohexanoic acid	*Mycobacterium phlei*	$C_6H_{14}N_2O_3$ (162.19)	—	$+6.3^{20}$ (c 5, H_2O) $+23.9^{10}$ (c 5.1, 1 N HCl) (90)	—	90	90	90	—
64	α-Amino-γ-hydroxy-butyric acid	*Escherichia coli* mutants	$C_4H_9NO_3$ (119.12)	199° (91)	—	—	91	91	91	—
65	γ-Amino-β-hydroxy-butyric acid	*Escherichia coli* mutants	$C_4H_9NO_3$ (119.12)	—	—	—	—	92	—	—
66	2-Amino-6-hydroxy-4-methyl-4-hexenoic acid	*Aesculus californica*	$C_7H_{13}NO_3$ (159.19)	—	-30^{20} (c 2.2, H_2O) $+2^{20}$ (c 1.1, 5 N HCl (19)	—	19	19	19	19, 21
67	2-Amino-3-hydroxy-methyl-3-pentenoic acid	*Bankera fuligineoalba*	$C_6H_{11}NO_3$ (145.17)	160–161° (13)	$+182^{25}$ (c 0.8, H_2O) $+201^{25}$ (c 0.8, 0.3 N HCl (13)	—	13	13	13	13
68	α-Amino-γ-hydroxy-pimelic acid	*Asplenium septentrionale*	$C_7H_{13}NO_5$ (191.19)	—	—	—	96	96	—	—
69	α-Amino-δ-hydroxy-valeric acid	*Canavalia ensiformis*	$C_5H_{11}NO_3$ (133.15)	216° (dec) (97)	$+6^{25}$ (c 2.65, H_2O) $+2.4^{25}_{365}$ (c 2.65, H_2O) (97)	—	97	97	97	97
70	α-Amino-β-ketobutyric acid	Mikramycin A	$C_4H_7NO_3$ (117.11)	—	—	—	98	—	—	—
71	α-Amino-β-methyl-γ,δ-dihydroxyisocaproic acid	Phalloidin	$C_7H_{15}NO_4$ (177.21)	208–210° (99)	—	—	99	99	99	99
71a	2-Amino-5-methyl-6-hydroxyhex-4-enoic acid	*Blighia unijugata*	$C_7H_{13}NO_3$ (159.21)	—	—	—	99a	99a	—	99a
72	O-Butylhomoserine	Soil bacterium	$C_8H_{17}NO_3$ (175.23)	267° (100)	—	—	100	100	100	100
72a	Dihydrorhizobitoxine[O-(2-amino-3-hydroxypropyl)-homoserine]	*Rhizobium japonicum*	$C_7H_{16}N_2O_4$ (192.25)	—	—	7.2 8.6 (100b)	100a	—	—	100
73	β,γ-Dihydroxyglutamic acid	*Rheum rhaponticum*	$C_5H_9NO_6$ (179.13)	—	—	—	101	101	101	—
74	β,γ-Dihydroxyisoleucine	Thiostrepton	$C_6H_{15}N0_4$ (165.20)	—	—	—	102	—	—	—
75	γ,δ-Dihydroxyleucine	Phalloin	$C_6H_{13}NO_4$ (163.18)	—	—	—	103	—	103	103
75a	δ,ε-Dihydroxy-norleucine	Bovine tendon	$C_6H_{13}NO_4$ (163.20)	—	—	—	103a	—	103a	—
76	O-Ethylhomoserine	Soil bacterium	$C_6H_{13}NO_3$ (147.18)	262° (100)	-14^{30} (c 2.5, H_2O) (100)	—	100	100	100	100

DATA ON THE NATURALLY OCCURRING AMINO ACIDS (Continued)

No.	Amino Acid (Synonym)	Source	Formula (Mol Wt)	Melting Point °C[a]	$[\alpha]_D^b$	pKa	References			
							Isolation and Purification	Chroma-tography	Chemistry	Spectral Data
76a	β-Guanido-γ-hydroxyvaline	Viomycin	$C_6H_{14}N_4O_3$ (190.24)	182° (dec) (104a)	–	–	–	–	–	104a
77	Homoserine (α-amino-γ-hydroxybutyric acid)	Pisum sativum	$C_4H_9NO_3$ (119.12)	–	-8.8^{25} (c 1–2, H_2O) $+18.3^{26}$ (c 2, 2 N HCl) (290)	2.71 9.62 (290)	105	105	–	–
78	α-Hydroxyalanine	Peptides of ergot	$C_3H_7NO_3$ (105.10)	–	–	–	106	–	–	–
79	α-Hydroxy-γ-amino-butyric acid	E. coli mutants	$C_4H_9NO_3$ (119.12)	–	–	–	91	91	–	–
80	β-Hydroxy-γ-amino-butyric acid	Mammalian brain	$C_4H_9NO_3$ (119.12)	–	–	–	–	108	–	–
81	α-Hydroxy-ε-amino-caproic acid	Neurospora crassa	$C_6H_{13}NO_3$ (147.18)	–	–	–	109	109	–	–
82	β-Hydroxyasparagine	Human urine	$C_4H_8O_4N_2$ (148.12)	238–240° (dec) (110)	–	2.09 8.29 (20)	110	110	110	110
83	β-Hydroxyaspartic acid	Azotobacter	$C_4H_7NO_5$ (149.10)	–	$+41.4$ (c 2.42, H_2O) $+53.0$ (c 2.46, 1 N HCl) (290)[d]	1.91 3.51 9.11 (20)	111	111	–	–
83a	N-(2-Hydroxyethyl)-alanine	Rumen protozoa	$C_5H_{11}NO_3$ (133.17)	–	–	–	111a	–	–	111a
84	N^4-(2-Hydroxyethyl)-asparagine	Bryonia dioica	$C_6H_{12}N_2O_4$ (176.20)	199–200° (112)	-2.9^{20} (c 5, H_2O) (112)	–	112	112	112	112
85	N^5-(2-Hydroxyethyl)-glutamine	Lunaria annua	$C_7H_{14}N_2O_4$ (190.23)	–	$+5.8^{19}$ (c 1.8, H_2O) (88)	–	88	88	88	–
86	β-Hydroxyglutamic acid	Mycobacterium tuberculosis	$C_5H_9NO_5$ (163.13)	187° (dec) (290)	$+8.69$ (H_2O) $+30.8^{20}$ (c 2, 20% HCl) (290)[d]	–	114	114	–	–
87	γ-Hydroxyglutamic acid	Linaria vulgaris	$C_5H_9NO_5$ (163.13)	–	–	–	115	115	115	–
88	γ-Hydroxyglutamine	Phlox decussata	$C_5H_{10}N_2O_4$ (162.15)	163–164° (dec) (116)	–	–	116	116	116	–
89	ε-Hydroxylamino-norleucine (α-amino-ε-hydroxyaminohexanoic acid)	Mycobacterium phlei	$C_5H_{12}N_2O_3$ (148.17)	223–225° (dec) (117)	$+6.3^{20}$ (c 5, H_2O) $+23.9^{18}$ (c 5.1, N HCl) (117)	–	117	117	117	–
89a	4-Hydroxyisoleucine	Trigonella foenumgraecum	$C_6H_{13}NO_3$ (147.20)	–	$+31^{20}$ (c 1, H_2O) (117a)	–	117a	117a	–	117a

DATA ON THE NATURALLY OCCURRING AMINO ACIDS (Continued)

No.	Amino Acid (Synonym)	Source	Formula (Mol Wt)	Melting Point °C[a]	$[\alpha]_D$[b]	pK_a	References: Isolation and Purification	References: Chromatography	References: Chemistry	References: Spectral Data
90	δ-Hydroxy-γ-ketonorvaline	*Streptomyces akiyoshiensis* novo	$C_5H_9NO_4$ (147.13)	–	-8.2^{17} (*c* 3.4, H_2O) (118)	2.0 9.1	118	118	118	118
91	δ-Hydroxyleucenine (δ-ketoleucine)	Phalloidin	$C_6H_{11}NO_3$ (145.17)	–	–	–	119	119	119	119
92	β-Hydroxyleucine	Antibiotic from *Paecilomyces* strain	$C_6H_{13}NO_3$ (147.17)	–	–	–	16	16	16	–
93	δ-Hydroxyleucine	*Paecilomyces*	$C_6H_{13}NO_3$ (147.17)	–	–	–	121	121	121	121
94	*threo*-β-Hydroxyleucine	*Deutzia gracilis*	$C_6H_{13}NO_3$ (147.17)	–	–	–	122	122	122	122
95	α-Hydroxylysine (α,ε-diamino-α-hydroxycaproic acid)	*Silvia officinalis*	$C_6H_{15}N_2O_3$ (162.19)	–	–	2.13 8.62 9.67 (20)	123	123	–	–
96	δ-Hydroxylysine (α,ε-diamino-δ-hydroxycaproic acid)	Fish gelatin	$C_6H_{15}N_2O_3$ (162.19)	–	–	–	124	111	–	–
96a	β-Hydroxynorvaline	*Streptomyces* species	$C_5H_{11}NO_3$ (133.17)	244° (dec) (124a)	–	–	124a	124a	–	124a
97	γ-Hydroxynorvaline	*Lathyrus odoratus*	$C_5H_{11}NO_3$ (133.15)	–	$+22^{20}$ (*c* 5, H_2O) +32 (*c* 2.5, gl. acetic) (126)	–	126	126	126	–
98	γ-Hydroxyornithine	*Vicia sativa*	$C_5H_{12}N_2O_3$ (148.16)	–	–	–	127	127	–	–
99	α-Hydroxyvaline	Ergot	$C_5H_{11}NO_3$ (133.15)	–	–	2.55 9.77 (20)	106	–	–	–
100	γ-Hydroxyvaline	*Kalanchoe daigremontiana*	$C_5H_{11}NO_3$ (133.15)	228° (dec) (129)	+1020 (H_2O) (129)	–	129	129	–	–
100a	Hypusine	Bovine brain	$C_{10}H_{23}N_3O_3$ (233.27)	234–238° (dec) (129a)	–	–	129a	–	129a	129a
100b	Isoserine	*Bacillus brevis* (edeine A and B)	$C_3H_7NO_3$ (105.11)	–	–	–	129b	129b	129b	–
101	4-Ketonorleucine (2-amino-4-keto-hexanoic acid)	*Citrobacter freundii*	$C_6H_{11}NO_3$ (145.17)	142–143° (130)	–	–	131	131	130	130
102	γ-Methyl-γ-hydroxy glutamic acid	*Phyllitis scolopendrium*	$C_6H_{11}NO_5$ (157.17)	–	–	–	56	56	–	–

DATA ON THE NATURALLY OCCURRING AMINO ACIDS (Continued)

No.	Amino Acid (Synonym)	Source	Formula (Mol Wt)	Melting Point °Cᵃ	$[\alpha]_D^b$	pKₐ	References Isolation and Purification	References Chromatography	References Chemistry	References Spectral Data
103	Pantonine (α-amino-β,β-dimethyl-γ-hydroxybutyric acid)	Escherichia coli	$C_6H_{13}NO_3$ (147.18)	–	–	–	132	132	132	–
103a	threo-β-Phenylserine	Canthium eurysides	$C_9H_{11}NO_3$ (181.21)	–	–	–	–	–	–	252a
103b	Pinnatanine [N^5-(2-hydroxy-methylbutadienyl)-allo-γ-hydroxyglutamine]	Staphylea pinnata	$C_{10}H_{16}N_2O_5$ (244.28)	165° (dec) (132a)	+3.2²⁷ (c 0.5, H_2O) (132a)	9.1 (132a)	132a	132a	–	132a
104	O-Propylhomoserine	Soil bacterium	$C_7H_{15}NO_3$ (161.21)	265° (100)	–11³⁰ (c 2, H_2O) (100)	–	100	100	100	100
104a	Rhizobitoxine [2-amino-4-(2-amino-3-hydroxypropoxy)but-3-enoic acid]	Rhizobium japonicum	$C_7H_{15}N_2O_4$ (191.24)	–	–	–	132b	–	–	132c
105	Serine (α-amino-β-hydroxypropionic acid)	Silk fibroin	$C_3H_7NO_3$ (105.09)	228° (dec) (1)	–7.5²⁵ (c2, H_2O) +15.1²⁵ (c2.5 N HCl) (290)	2.19 9.21 (290)	134	3	134	134
106	O-Succinylhomoserine	Escherichia coli	$C_8H_{13}NO_6$ (219.20)	180–181° (135)	–	4.4 9.5 (135)	135	135	135	–
107	Tabtoxinine (α,ε-diamino-β-hydroxypimelic acid)	Pseudomonas tabaci	$C_7H_{14}N_2O_5$ (186.20)	–	–	–	136	137	137	–
108	Threonine (α-amino-β-hydroxybutyric acid)	Fibrin hydrolyzate	$C_4H_9NO_3$ (119.12)	253° (1)	–28²⁵ (c 1–2, H_2O) –15²⁵ (c 1–2, 5 N HCl) (290)	2.09 9.10 (290)	138	3	138, 139	139

F. L-AROMATIC AMINO ACIDS

No.	Amino Acid (Synonym)	Source	Formula (Mol Wt)	Melting Point °Cᵃ	$[\alpha]_D^b$	pKₐ	Isolation and Purification	Chromatography	Chemistry	Spectral Data
109	α-Amino-β-phenyl-butyric acid	Streptomyces bottropensis	$C_{10}H_{13}NO_2$ (187)	176–177° (140)	–	–	140	140	140	140
109a	β-Amino-β-phenyl-propionic acid	Roccella canariensis hydrolyzate	$C_9H_{11}NO_2$ (165.21)	–	–	–	–	–	–	140a
110	3-Carboxy-4-hydroxy-phenylalanine (m-carboxytyrosine)	Reseda odorata	$C_{10}H_{11}NO_5$ (165.15)	–	–7.7²⁵ (c 0.9, 1 N NaOH) –29.9²⁴ (c 0.6, 0.2 M PO_4, pH 7) (297)	2 3.4 9.3 12–13 (297)	141	141	141	141

DATA ON THE NATURALLY OCCURRING AMINO ACIDS (Continued)

No.	Amino Acid (Synonym)	Source	Formula (Mol Wt)	Melting Point °C[a]	$[\alpha]_D$[b]	pK$_a$	Isolation and Purification	Chromatography	Chemistry	Spectral Data
							References			
111	m-Carboxyphenyl-alanine	Iris bulbs	C$_{10}$H$_{11}$NO$_4$ (191.26)	—	—	1.5 3.9 (142)	142	142	142	142
111a	2,3-Dihydroxy-N-benzoylserine	Escherichia coli	C$_{10}$H$_{11}$NO$_6$ (241.22)	193–194° (142a)	—	—	142a	142a	—	142a
112	2,4-Dihydroxy-6-methylphenylalanine	Agrostemma githago	C$_{10}$H$_{13}$NO$_4$ (211.23)	252° (144)	+19.7^{18} (1 N HCl) (144)	—	144	144	144	144
113	3,4-Dihydroxy-phenylalanine (DOPA)	Vicia faba	C$_9$H$_{11}$NO$_4$ (209.21)	—	−14.3 (1 N HCl) (146)	2.32e 8.68e 9.88e	145	1	146	—
114	3,5-Dihydroxy-phenylglycine	Euphorbia helioscopia	C$_8$H$_9$NO$_4$ (183.16)	230–232° (147)	—	—	147	147	147	147
114a	γ-Glutaminyl-4-hydroxybenzene	Agaricus bisporus	C$_{11}$H$_{14}$N$_2$O$_4$ (238.27)	225–226° (147a)	+42.5^{25} (c 0.67, 0.1 N NaOH) (147a)	—	147a	—	—	147a
114b	3-Hydroxymethyl-phenylalanine	Caesalpinia tinctoria	C$_{10}$H$_{13}$NO$_3$ (195.24)	—	—	—	147b	147b	—	147b
114c	4-Hydroxy-3-hydroxymethyl-phenylalanine	Caesalpinia tinctoria	C$_{10}$H$_{13}$NO$_4$ (211.24)	—	−36^{20} (c 1, H$_2$O) −4^{20} (c 0.67,1 N NaOH) (147b)	—	147b	147b	—	147b
115	3-Hydroxykynurenine	Human urine	C$_{10}$H$_{12}$N$_2$O$_4$ (224.21)	—	—	—	148	148	—	—
115a	p-Hydroxymethyl-phenylalanine	Escherichia coli	C$_{10}$H$_{13}$NO$_3$ (195.24)	231–233° (dec) (148a)	−32.5^{20} (148a)	—	—	148b	148b	148a
116	m-Hydroxyphenyl-glycine	Euphorbia helioscopia	C$_8$H$_9$NO$_3$ (177.16)	212–214° (147)	—	—	147	147	147	147
117	Kynurenine (β-anthraniloyl-α-aminopropionic acid)	Rabbit urine	C$_{10}$H$_{12}$N$_2$O$_3$ (208.21)	191° (dec) (290)	−30.5^{25} (c 1, H$_2$O) (290)	—	148	—	148	148
118	O-Methyltyrosine (β-(p-methoxyphenyl)-alanine)	Puromycin	C$_{10}$H$_{13}$NO$_3$ (195.22)	191° (150)	−5 – 9$_{546}$ (HCl) (150) −3.2$_{546}$ (1 N NaOH) (150)	—	149	—	150	—
119	Phenylalanine	Lupinus luteus	C$_9$H$_{11}$NO$_2$ (165.19)	284° (1)	−34.5^{25} (c 1–2, H$_2$O) (290) −4.5^{25} (c 1–2, 5 N HCl) (290)	2.16 9.18 (290)	151	3	152	152

DATA ON THE NATURALLY OCCURRING AMINO ACIDS (Continued)

No.	Amino Acid (Synonym)	Source	Formula (Mol Wt)	Melting Point °C[a]	$[\alpha]_D^b$	pKa	References			
							Isolation and Purification	Chromatography	Chemistry	Spectral Data
120	Tyrosine (α-amino-β-hydroxyphenyl propionic acid)	Casein, alkaline hydrolyzate	$C_9H_{11}NO_3$ (181.19)	344° (1)	-10^{25} (c 2, 5 N HCl)(1)	2.20, 9.11 (1), 10.13 (290)	153	3	154	154
120a	β-Tyrosine	*Bacillus brevis* (edeine A and B)	$C_9H_{11}NO_3$ (181.21)	–	–	–	129b	129b	129b	–
121	m-Tyrosine	*Euphorbia myrsinites* L	$C_9H_{11}NO_3$ (181.19)	272–274° (155)	-14.5^{22} (70% EtOH) $+8.9$ (70% EtOH, 2 N HCl) (155)	–	155	155	155	155
						G. L-UREIDO AND GUANIDO AMINO ACIDS				
121a	N-Acetylarginine	Cattle brain	$C_7H_{16}N_4O_3$ (204.27)	270° (155a)	–	–	155a	155a	155a	155a
122	Albizziine (2-amino-3-ureidopropionic acid)	*Mimosaceae*	$C_4H_9NO_3$ (119.12)	–	-66^{26} (c 2, H$_2$O) (157)	–	156	157	157	157
123	Arginine (amino-δ-guanidinovaleric acid)	*Lupinus luteus*	$C_6H_{14}N_4O_2$	238° (1)	$+12.5^{25}$ (c 2, H$_2$O) $+27.6^{25}$ (c 2, 5 N HCl) (1)	1.82, 8.99, 12.48 (290)	158	3	159	159
124	Canavanine (α-amino-(O-guanidyl)-γ-hydroxybutyric acid)	*Canavalia ensiformis*	$C_5H_{12}N_4O_3$ (176.19)	172° (160)	$+18.6^{18.5}$ (c 7.8, H$_2$O) (160)	2.50, 6.60, 9.25 (33)	161	160	160	162
125	Canavanosuccinic acid	*Canavalia ensiformis*	$C_9H_{16}N_4O_7$ (292.27)	–	–	–	163	163	–	–
126	Citrulline	Watermelon	$C_6H_{13}N_3O_3$	202° (164)	$+4^{25}$ (c 2, H$_2$O) $+24.2^{25}$ (c 2, N HCl) $+10.8$ (c 1.1, 0.1 N NaOH) (290)	2.43, 9.41 (290)	164	165	164	–
127	Desaminocanavanine	Canavanine	$C_5H_9N_3O_3$	256–257° (166)	$+26.61^{21}$ (H$_2$O) (166)	–	166	–	166	–
127a	N^G,N^G-Dimethylarginine	Bovine brain	$C_8H_{18}N_4O_2$ (202.30)	198–201° (70a)	–	–	166a	166a	–	166a
127b	N^G,N'^G-Dimethylarginine	Bovine brain	$C_8H_{18}N_4O_2$ (202.30)	237–239° (dec) (70a)	–	–	166a	166a	70a	166a

DATA ON THE NATURALLY OCCURRING AMINO ACIDS (Continued)

No.	Amino Acid (Synonym)	Source	Formula (Mol Wt)	Melting Point °C[a]	$[\alpha]_D^b$	pK$_a$	Isolation and Purification	Chromatography	Chemistry	Spectral Data
128	Gigartinine (α-amino-δ-(guanylureido)valeric acid)	Gymnogongrus flabelliformis	C$_7$H$_{15}$N$_5$O$_3$ (217.25)	–	–	–	167	167	167	167
129	Homoarginine	Lathyrus species	C$_7$H$_{16}$N$_4$O$_2$ (188.25)	–	+42.4° (c 0.452, 1.02 N HCl)f (168)	–	168	168	168	168
130	Homocitrulline	Human urine	C$_7$H$_{14}$N$_3$O$_3$ (189.22)	–	–	–	169	169	–	–
131	γ-Hydroxyarginine	Vicia sativa	C$_6$H$_{14}$N$_4$O$_3$ (190.20)	–	–	–	170	170	170	170
131a	N^5-Hydroxyarginine	Bacillus species	C$_6$H$_{14}$N$_4$O$_3$ (190.24)	206–212° (dec) (170a)	+21^{25} (c 1, 5 N HCl) (170a)	–	170a	170a	170a	170a
132	γ-Hydroxyhomoarginine (α-amino-ε-guanidino-γ-hydroxyhexanoic acid)	Lathyrus tingitanus	C$_7$H$_{16}$N$_4$O$_3$ (204.23)	–	–	–	171	171	171	171
133	Indospicine (α-amino-ε-amidino caproic acid)	Endigofera spicata	C$_7$H$_{15}$N$_3$O$_2$ (173.23)	131–134° (173)	+18^{22} (c 1.1, 5 N HCl) (173)	–	173	173	173	–
133a	ω-N-Methylarginine (guanidinomethylarginine)	Bovine brain	C$_7$H$_{16}$N$_4$O$_2$ (188.27)	–	–	–	166a	166a	–	166a
				H. L-AMINO ACIDS CONTAINING OTHER NITROGENOUS GROUPS						
134	Alanosine [2-amino-3-(N-nitrosohydroxyamino) propionic acid]	Streptomyces alanosinicus	C$_3$H$_7$N$_3$O$_4$ (149.11)	–	+8 (1 N HCl) –46 (0.1 N NaOH) (174)	(174)	174	–	174	174
135	Azaserine (O-diazoacetylserine)	Streptomyces	C$_5$H$_7$N$_3$O$_4$ (173.14)	146–162° (175)	–0.5$^{27.5}$ (c 8.46, H$_2$O) (175)	8.55 (34)	175	–	175	175
136	β-Cyanoalanine	Vicia sativa	C$_4$H$_6$N$_2$O$_2$ (114.11)	214.5° (176)	–2.9^{26} (c 1.4, 1 N acetic acid) (177)	1.7.7.4 (177)	176	176	177	177
136a	γ-Cyano-α-aminobutyric acid	Chromobacterium violaceum	C$_5$H$_8$N$_2$O$_2$ (128.15)	221–223° (177a)	+32.1^{21} (c 0.38, 1 N HOAc) (177a)	–	177a	177a	177a	177a
137	ε-Diazo-δ-ketonorleucine	Streptomyces	C$_6$H$_9$N$_3$O$_3$ (171.17)	145–155° (178)	21^{26} (c 5.4, H$_2$O) (178)	–	178	178	178	178
138	Hadacidin (N-formyl-N-hydroxyaminoacetic acid)	Penicillium frequentans	C$_3$H$_5$NO$_4$ (119.08)	205–210° (179)	–	–	179	–	179	179
				I. L-HETEROCYCLIC AMINO ACIDS						
138a	2-Alanyl-3-isoxazolin-5-one	Pea seedlings	C$_6$H$_8$N$_2$O$_4$ (172.16)	203–205° (dec) (180b)	–	–	180a	180b	180b	180b
139	Allohydroxyproline	Santalum album	C$_5$H$_9$NO$_3$ (131.13)	248° (180)	–57^{25} (c 0.65, H$_2$O) (180)	–	180	180	180	–
140	Allokainic acid (3-carboxymethyl-4-isopropenyl proline)	Digenea simplex	C$_{10}$H$_{15}$NO$_4$ (213.24)	–	+8^{26} (H$_2$O) (184, 185)	–	184, 185	–	184, 185	184, 185

DATA ON THE NATURALLY OCCURRING AMINO ACIDS (Continued)

No.	Amino Acid (Synonym)	Source	Formula (Mol Wt)	Melting Point °C[a]	$[\alpha]_D$[b]	pK_a	References Isolation and Purification	Chromatography	Chemistry	Spectral Data
140a	1-Amino-2-nitrocyclopentane carboxylic acid	Aspergillus wentii	$C_6H_{10}N_2O_4$ (174.18)	150° (dec) (185a)	–	–	185a	185a	185a	185a
141	4-Aminopipecolic acid	Strophanthus scandens	$C_6H_{11}N_2O_2$ (143.18)	–	–	–	186	186	–	–
141a	cis-3-Aminoproline	Morchella esculenta	$C_5H_{10}N_2O_2$ (130.17)	215° (dec) (186a)	$+5.8^{20}$ (c 2, H2O) $+23.0^{20}$ (c 2, 5 N HCl) (186a)	–	186a	186a	–	186a
141b	Anticapsin	Streptomyces griseoplanus	$C_9H_{13}NO_4$ (199.23)	240° (dec) (186b)	$+125^{25}$ (c 1, H2O) (186b)	4.3 10.1 (186b)	186b	186b	–	186b
142	Ascorbigen	Cabbage	$C_{15}H_{15}NO_6$ (305.30)	–	–	–	187	188	189	–
143	Azetidine-2-carboxylic acid	Convallaria majalis	$C_4H_7NO_2$ (101.11)	–	–	–	190	190	190	190
143a	Azirinomycin (3-methyl-2-hydroazirine carboxylic acid)	Streptomyces aureus	$C_4H_5NO_2$ (99.10)	–	–	–	189a	189a	–	189a
144	Baikiain (1,2,3,6-tetrahydropyridine-α-carboxylic acid)	Baikiaea plurijuga	$C_6H_9NO_2$ (127.15)	273–274° (183)	–	–	181	182	183	183
144a	N-Carbamoyl-2-(p-hydroxyphenyl)glycine	Vicia faba	$C_9H_{10}N_2O_4$ (210.21)	194–195° (dec) (190a)	–	–	190a	–	–	190a
144b	3-Carboxy-6,7-dihydroxy-1,2,3,4-tetrahydroisoquinoline	Mucuna mutisiana	$C_{10}H_{11}NO_4$ (209.22)	286–288° (190b)	-114.9^{25} (c 1.65, 20% HCl) (190b)	–	190b	–	190b	190b
144c	Clavicipitic acid	Claviceps (ergot fungus)	$C_{16}H_{18}N_2O_2$ (270.36)	262° (dec) (190c)	–	–	190c	–	–	190c
145	Cucurbitine (3-amino-3-carboxy-pyrrolidine)	Cucurbita moschata	$C_5H_{10}NO_2$ (116.14)	–	-19.76^{27} (c 9.3, H2O) (191)	–	191	191	191	191
145a	N-Dihydro jasmonoylisoleucine	Gibberella fujikuroi	$C_{18}H_{31}NO_4$ (325.50)	140–141° (191a)	–	–	–	191a	191a	191a
145b	2,5-Dihydrophenylalanine (1,4-cyclohexadiene-1-alanine)	Streptomycete X-13, 185	$C_9H_{13}NO_2$ (167.23)	206–208° (191b)	-33.7^{25} (c 1, 5 N HCl) (191b)	–	191b	–	–	191b
145c	2-N,6-N-Di-(2,3-dihydroxybenzoyl)lysine	Azobacter vinelandii	$C_{20}H_{22}N_2O_8$ (418.44)	–	–	4.8 9 (191c)	191c	191c	–	191c
145d	cis-3,4-trans-3,4-Dihydroxyproline	Diatom cell walls	$C_5H_9NO_4$ (147.15)	262° (dec) (191d)	-61.2^{20} (c 0.5, H2O) (191d)	–	191d	191d	–	191e
145e	β-(2,6-Dihydroxypyrimidin-1-yl) alanine	Pea seedlings	$C_7H_9N_3O_4$ (199.19)	230° (dec) (191f)	–	–	191f	–	191f	191f

DATA ON THE NATURALLY OCCURRING AMINO ACIDS (Continued)

No.	Amino Acid (Synonym)	Source	Formula (Mol Wt)	Melting Point °C[a]	$[\alpha]_D$[b]	pKa	Isolation and Purification	Chromatography	Chemistry	Spectral Data
								References		
145f	4,6-Dihydroxyquinoline-2-carboxylic acid	Tobacco leaves	$C_{10}H_7NO_4$ (205.18)	287° (dec) (191g)	–	–	191g	191g	191g	191g
146	Dihydrozanthurenic acid (8-hydroxy-1,2,3,4-tetrahydro-4-ketoquinaldic acid)	Lepidoptera	$C_{10}H_9NO_4$ (207.19)	185–190° (192)	-45^{20} (c 0.9, MeOH) $+18^{20}$ (c 0.9, MeOH-HCl) (192)	–	192	192	–	192
147	Domoic acid (2-carboxy-3-carboxymethyl-4-1-methyl-2-carboxy-1,3-hexadienyl-pyrrolidine)	*Chondria armata*	$C_{15}H_{21}NO_6$ (311.35)	217° (193)	-109.6^{12} (c 1.314, H₂O) (193)	2.20 3.72 4.93 9.82 (193)	193	193	193	193
148	Echinine (2-*tert*-pentenyl-5,7-diisopentenyltryptophan)	*Aspergillus glaucus*	$C_{26}H_{36}N_2O_2$ (408.50)	169–172° (194)	–	–	194	–	194	194
148a	Enduracidine [α-amino-β-(2-iminoimidazolidinyl)-propionic acid]	Enduracidin hydrolyzate	$C_6H_{12}N_4O_2$ (172.22)	–	$+63.3^{22}$ (1 M HCl) $+57.6^{22}$ (1 M NaOH) (194a)	2.5 8.3 12 (94a)	–	194a	194a	194a
148a'	Furanomycin (α-amino-(2,5-dihydro-5-methyl)furan-2-acetic acid)	*Streptomyces* L-803	$C_7H_{11}NO_3$ (157.19)	220–223° (dec) (194a')	$+136.1^{27}$ (c 1, H₂O) (194a')	2.4 9.1 (194a')	194a'	194a'	194a'	194a'
148a"	Furosine [ε-N-(2-furoylmethyl)lysine]	Heated milk	$C_{12}H_{18}N_2O_4$ (254.32)	–	–	–	194a"	194b	–	194a"
148b	γ-Glutaminyl-3,4-benzoquinone	*Agaricus bisporus*	$C_{11}H_{12}N_2O_5$ (252.25)	–	–	–	194b	194b	–	194b
149	Guvacine	*Areca cathecu*	$C_6H_9NO_2$ (127.15)	–	–	–	195, 196	–	–	–
150	Histidine	Protamine from sturgeon sperm	$C_6H_9N_3O_2$ (155.16)	277° (1)	-38.5^{25} (H₂O) $+11.8$ (5 N HCl(1))	1.82 6.00 9.17 (1)	197	3	198	198
150a	β-Hydroxyhistidine	Bleomycin A₂ (antibiotic)	$C_6H_9N_3O_3$ (171.18)	205° (dec) (198a)	$+40^{28}$ (c 1, H₂O) (198a)	<2.0 198a 5.5 8.8 (198a)	198a	–	198a	–
151	4-Hydroxy-4 methylproline	Apples	$C_6H_{11}NO_2$ (145.16)	–	–	–	199	199	–	200
152	Hydroxyminaline	*Penicillium aspergillus*	$C_5H_5NO_3$ (127.10)	–	–	–	201	–	201	201

DATA ON THE NATURALLY OCCURRING AMINO ACIDS (Continued)

No.	Amino Acid (Synonym)	Source	Formula (Mol Wt)	Melting Point °C[a]	$[\alpha]_D$[b]	pK_a	References Isolation and Purification	Chromatography	Chemistry	Spectral Data
153	4-Hydroxypipecolic acid	*Acacia pentadenia*	$C_6H_{11}NO_3$ (145.16)	250–270° (202)	−12.5[21] (H_2O) +0.34[21] (1 N HCl) −18.5[21] (1 N NaOH) (202)	–	202	202	202	203
154	5-Hydroxypipecolic acid	*Rhapis excelsa*	$C_6H_{11}NO_3$ (145.16)	–	–	–	203	203	–	203
154a	5-Hydroxypiperidazine-3-carboxylic acid	Monamycin hydrolyzate	$C_5H_{10}N_2O_3$ (146.17)	201–202° (DNP deriv.) (204a)	+21.4[24] (c 0.39, H_2O) (204b)	–	204a	204a	204a	204, 204a
155	3-Hydroxyproline (3-hydroxypyrrolidine-2-carboxylic acid)	Telomycin	$C_5H_9NO_3$ (131.13)	225–235° (206)	−17.4[20] (c 1, H_2O) +13.3[20] (c 0.5, 1 N HCl) (206a)	–	206	206	206	206a
156	4-Hydroxyproline (4-hydroxypyrrolidine-2-carboxylic acid)	Gelatin hydrolyzate	$C_5H_9NO_3$ (131.13)	273–274° (290)	−76.0[25] (c 2, H_2O)	1.82 9.66 (290)	207	3	208	208
157	2-Hydroxytryptophan	Phalloidin	$C_{11}H_{12}N_2O_3$ (220.22)	257° (210)	−50.5[25] (c 2, 5 N HCl)(290) +40.8 (c 4.2, 1 N NaOH) (210)	–	209	–	210	–
158	5-Hydroxytryptophan	*Chromobacterium violaceum*	$C_{11}H_{12}N_2O_3$ (220.22)	273° (dec) (290)	−32.5[22] (c 1, H_2O) +16[22] (c 1, 4 N HCl) (290)	–	211	211	–	–
159	Ibotenic acid (α-amino-3-keto-4-isoxazoline-5-acetic acid)	*Amanita strobiliformis*	$C_5H_6N_2O_4$ (114.10)	177–178° (212)	±0° (212)	5.1 8.2 (212)	212	–	212	212
160	4-Imidazoleacetic acid	*Polyporus sulfureus*	$C_5H_7N_2O_2$ (125.11)	–	–	2.96 7.35 (20)	213	–	–	–
161	N-(Indole-3-acetyl)aspartic acid	Magnolia	$C_{14}H_{14}N_2O_5$ (290.29)	–	–	–	214	214	–	–
162	Indole-3-acetyl-ε-lysine	*Pseudomonas savastanoi*	$C_{16}H_{21}N_3O_3$ (303.36)	259–261° (215)	+22[23] (2 N HCl) (215)	–	215	215	215	215
162a	N-Jasmonoylisoleucine	*Gibberella fujikuroi*	$C_{18}H_{29}NO_4$ (323.48)	147–149° (191a)	–	–	–	191a	191a	191a

DATA ON THE NATURALLY OCCURRING AMINO ACIDS (Continued)

No.	Amino Acid (Synonym)	Source	Formula (Mol Wt)	Melting Point °C[a]	$[\alpha]_D$[b]	pKa	Isolation and Purification	Chromatography	Chemistry	Spectral Data
163	Kainic acid (3-carboxymethyl-4-isopropenyl proline)	*Digenea simplex*	$C_{10}H_{15}NO_4$ (213.25)	251° (216)	-15^{17} (H_2O) (185, 216)	–	216	216	–	–
164	4-Keto-5-methyl-proline	Actinomycin	$C_6H_9NO_3$ (143.18)	215° (217)	–	–	217	217	217	217
165	4-Ketopipecolic acid	Staphylomycin	$C_6H_9NO_3$ (143.15)	–	–	–	218	–	–	–
166	4-Ketoproline	Actinomycin	$C_5H_9N_4O_2$ (181.00)	–	–	–	219	–	219	–
167	Lathytine (tingitanine)	*Lathyrus tingitanus*	$C_7H_{10}N_4O_2$ (182.00)	215° (220)	-55.9^{21} (H_2O) (220)	2.4 / 4.1 / 9.0 (220)	220	–	–	220
167a	exo-3,4-Methanoproline	*Aesculus parviflora*	$C_6H_9NO_2$ (127.16)	–	-132^{20} (c 2, H_2O) / -104^{20} (c 1, 5 N HCl) (23a)	–	23a	23a	–	23a
168	4-Methyleneproline	*Eriobotrya japonice*	$C_6H_9NO_2$ (127.15)	225° (221)	$\pm O^c$	–	221	221	222	–
168a	1-Methyl-6-hydroxy-1,2,3,4-tetrahydroisoquinoline-3-carboxylic acid	*Euphorbia myrisinites*	$C_{11}H_{13}NO_3$ (207.25)	–	–	–	222a	222a	–	222a
169	4-Methylproline	Apples	$C_6H_{11}NO_2$ (129.16)	232–234° (224)	-52 (c 0.3, H_2O) (224)	–	223	223	224	224
170	N-Methylproline (hygric acid)	Apples	$C_6H_{11}NO_2$	–	–	–	223	223	223	–
170a	N-Methylstreptolidine	Streptothricin	$C_7H_{14}N_4O_3$ (202.25)	–	–	–	224a	224a	–	–
171	β-Methyltryptophan	Telomycin	$C_{12}H_{14}N_2O_2$ (218.25)	–	–	–	226	–	226	226
172	Mimosine	*Mimosa pudica*	$C_8H_{10}N_2O_4$ (198.18)	228–229° / 227	-21^{22} (H_2O) / $+10^{22}$ (1% HCl) (227)	–	227a	–	227	227
173	Minaline (pyrrole-2-carboxylic acid)	Diastase	$C_5H_5NO_2$ (111.10)	180° (228)	–	–	228	–	228	228
174	Muscazone	*Amanita muscaria*	$C_5H_6N_2O_5$ (174.12)	190° (229)	–	–	229	–	230	231
174a	Nicotianamine [1-(3'-(γ-amino-α-γ-dicarboxypropylamino)-propylazetidine-2-carboxylic acid]	Tobacco leaves	$C_{12}H_{21}N_3O_6$ (303.36)	240° (dec) (231a)	-60.5^{23} (c 2.7, H_2O) (231a)	–	231a	–	231a	231a
175	β-3-Oxindolylalanine	Phalloidin	$C_{11}H_{12}N_2O_3$ (220.24)	249–253° (209)	$+39.2^{20}$ (1 N NaOH) (209)	–	209	–	209	–

DATA ON THE NATURALLY OCCURRING AMINO ACIDS (Continued)

No.	Amino Acid (Synonym)	Source	Formula (Mol Wt)	Melting Point °C[a]	$[\alpha]_D$[b]	pK_a	Isolation and Purification	Chromatography	Chemistry	Spectral Data
176	Pipecolic acid	Apples	$C_6H_{11}NO_2$ (129.17)	260° (dec) (290)	-25.4^{18} (c 5, H₂O) -13.3^{25} (c2, 5 N HCl)(290)	–	233	233	m	–
176a	Piperidazine-3-carboxylic acid	Monamycin hydrolyzate	$C_5H_{10}N_2O_2$ (130.17)	153–155° (DNP deriv.) (204b)	$+307^{25}$ (c 0.18, CH₃OH) (204b)	–	–	–	–	204a
177	Proline (pyrrolidine-2-carboxylic acid)	Casein hydrolyzate	$C_5H_9NO_2$ (115.13)	222° (1)	-86.2^{25} (c 1–2, H₂O) -60.4 (c 1–2, 5 N HCl) (1)	1.95 10.64 (290)	234	3	235	235
178	β-Pyrazol-1-ylalanine	Citrullus vulgaris	$C_6H_9N_3O_2$ (155.17)	236–238° (236)	-7.3^{20} (c 3.4, H₂O) (236)	2.2 (236)	236	236	236	236
178a	Pyridosine [ε-(1,4-dihydro-6-methyl-3-hydroxy-4-oxo-1-pyridyl)-lysine]	Heated milk	$C_{12}H_{19}N_3O_4$ (269.34)	–	–	–	236a	–	–	236a
178b	Pyrrolidone carboxylic acid (5-oxoproline)	Human plasma	$C_5H_7NO_3$ (129.13)	–	–	–	236b	236b	–	–
179	Roseanine	Roseothricin	$C_6H_{12}N_4O_3$ (188.21)	–	$+56.8^{22}$ (c 2.35, H₂O)[c] (238)	–	237	238	238	–
179a	Stendomycidine	Stendomycin (from Streptomyces)	$C_8H_{16}N_4O_2$ (200.28)	–	–	–	238a	238a	238a	238a, 238b
180	Stizolobic acid [β-(3-carboxypyran-5-yl)alanine]	Stizolobium hassjoo	$C_9H_9NO_6$ (227.18)	231–233° (239)	–	–	239	239	239	239
181	Stizolobinic acid [β-(6-carboxy-α-pyran-3-yl)alanine]	Stizolobium hassjoo	$C_9H_9NO_6$ (207.18)	–	–	–	239	239	240	240
181a	Streptolidine	Streptothricin	$C_6H_{12}N_4O_3$ (188.22)	215° (dec) (224a)	$+55.3^{25}$ (c 1.01, H₂O) (224a)	–	224a	224a	224a	224a
182	Tricholomic acid (α-amino-3-keto-5-isoxazolidine acetic acid)	Tricholoma muscarium	$C_5H_8N_4O_4$ (160.13)	207° (242)	m	6.0 8.6 (242)	242	m	242	242
183	Tryptophan (α-amino-β-3-indolepropionic acid)	Casein	$C_{11}H_{12}N_2O_2$ (204.24)	282° (1)	-33.7^{25} (c 1–2, H₂O) $+2.8^{25}$ (c 1–2, 1N HCl) (1)	2.43 9.44 (290)	243	3	244	244
183a	Tuberactidine (2-imino-4-hydroxyhexahydro-6-pyrimidinylglycine)	Streptomyces griseoverticillattus	$C_6H_{12}N_4O_3$ (188.22)	182° (dec) (244a)	-25.8^{15} (c 0.5, H₂O) (244a)	–	244a	–	–	244a

DATA ON THE NATURALLY OCCURRING AMINO ACIDS (Continued)

No.	Amino Acid (Synonym)	Source	Formula (Mol Wt)	Melting Point °C[a]	$[\alpha]_D$[b]	pK$_a$	Isolation and Purification	Chromatography	Chemistry	Spectral Data
184	Viomycidine (guanidine-1-pyrroline-5-carboxylic acid)	Viomycin	$C_6H_{10}N_4O_2$ (170.19)	181–182° (dec) (246a)	-151^{322} (c 1.25, H$_2$O) $-38^{22.2}$ (c 0.8, HCl) (246a)	1.3 5.5 12.6 (246)	76, 246a	246, 246a	246, 246a	246a
185	Willardine [3-(1-uracyl)alanine]	Mimosa	$C_7H_9N_3O_4$ (199.18)	–	-12.1^{22} (c 1.2, 1 N HCl) (248)	–	247	247	247, 248	248
				1. L-N-SUBSTITUTED AMINO ACIDS						
186	Abrine (N-methyltryptophan)	Abrus precatorius	$C_{12}H_{14}N_2O_3$ (218.41)	297° (dec) (249)	–	–	250	–	249	–
186a	N-Acetylalanine	Human brain	$C_5H_9NO_3$ (131.15)	–	–	–	250a	250a	250a	–
186a′	N-Acetylglutamic acid	Mammalian liver	$C_7H_{11}NO_5$ (189.19)	–	–	–	250a′	250a′	–	–
186b	N-Acetylaspartic acid	Extract of cat brain	$C_6H_9NO_5$ (175.16)	123–125° (250b)	-11.8^{27} (2% methyl cellosolve) (250b)	–	250b	250b	–	250b
187	2,3-Dihydro-3,3-dimethyl-1 H-pyrrolo[1,2-α]- indole-9-alanine		$C_{16}H_{20}N_2O_2$ (272.34)	170–175° (dec) (251)	–	–	251	251	251	251
188	β,N-Dimethylleucine	Etamycin	$C_8H_{17}NO_2$ (159.23)	315–316° (dec) (252)	$+33.15$ (c 2, H$_2$O) $+39.2^{29}$ (c 2.2, 5 N HCl)(252)	–	252	252	252	–
188a	N,N-Dimethylphenylalanine	Canthium eurysides	$C_{11}H_{15}NO_2$ (193.27)	–	–	–	–	–	–	252a
189	γ-Formyl-N-methylnorvaline	Ilamycin	$C_7H_{13}NO_3$ (159.19)	–	–	1.8 10.2 (253)	253	–	253	253
190	Fusarinine [δ-N-(cis-5-hydroxy-3-methylpent-2-enoyl)-δ-N-hydroxyornithine]	Fusarium	$C_{11}H_{20}N_2O_4$ (244.30)	–	–	–	254	254	254	254
191	Homarine (N-methylpicolinic acid)	Arenicola marina	$C_7H_7NO_2$ (137.15)	–	–	–	255	255	–	–
192	4-Hydroxy-N-methylproline	Aformosia elata heartwood	$C_6H_{11}NO_3$ (145.16)	238–240° (dec) (256)	-86.6 (c 1.5, H$_2$O) (256)	–	256	–	256	256
193	Merodesmosine	Elastin	$C_{18}H_{34}N_4O_6$ (420)	–	–	–	257	257	257	257
194	N-Methylalanine	Dichapetalum cymosum (Gifblaar)	$C_4H_9NO_2$ (103.12)	–	–	–	258	258	258	258
195	1-Methylhistidine	Cat urine	$C_7H_{11}N_3O_2$ (169.18)	245–247° (259)	-25.8^{18} (c 3.9, H$_2$O) (290)	1.69 6.48 (imidazole) 8.85 (290)	259	259	–	–

DATA ON THE NATURALLY OCCURRING AMINO ACIDS (Continued)

No.	Amino Acid (Synonym)	Source	Formula (Mol Wt)	Melting Point °C[a]	$[\alpha]_D^b$	pK_a	References: Isolation and Purification	Chromatography	Chemistry	Spectral Data
196	3-Methylhistidine	Human urine	$C_7H_{11}N_3O_2$ (169.18)	—	-26.5^{26} (c 2.1, H₂O) (290) $+13.5^{27}$ (c 1.9, 1 N HCl)(260)	—	260	260	260	260
197	N-Methylisoleucine	Enniatin A	$C_7H_{15}NO_2$ (145.21)	—	$+28.6^{22}$ (c 1.034, H₂O) $+44.8^{22}$ (c 1.162, 5 N HCl) (261)	—	261	261	261	—
198	N-Methylleucine	Enniatin A	$C_7H_{15}NO_2$ (145.21)	—	$+21.4^{15}$ (c 0.77, H₂O) $+31.3^{15}$ (c 0.86, 5 N HCl)(262)	—	262	262	262	—
199	N-Methyl-β-methylleucine	Etamycin	$C_8H_{17}NO_2$ (159.23)	—	$+26^{30}$ (c 1.8, H₂O) $+33.2^{30}$ (c 1.9, 5 N HCl)(252)	—	252	252	252	—
200	N-Methyl-O-methylserine	Myobacterium butyricum	$C_5H_{11}NO_3$ (133.14)	203–205° (dec) (264)	—	—	264	264	264	264
201	N-Methylphenylglycine (α-phenylsarcosine)	Etamycin	$C_9H_{12}NO_2$ (166.21)	—	$+118^{31}$ (c 4.8, 1 N, HCl) (252)	—	252	252	252	—
201a	N-Methylthreonine	Stendomycin hydrolyzale	$C_5H_{11}NO_3$ (133.17)	—	-17^{25} (c 2, 5 N HCl) (264a)	—	—	—	—	264a
202	N-Methyltyrosine (Surinamine)	Andira	$C_{10}H_{13}NO_3$ (195.22)	280–300° (267)	-18.6 (267)	—	266	—	267	—
203	N-Methylvaline	Actinomycin	$C_6H_{13}NO_2$ (131.18)	—	+17.5 (H₂O) +30.9 (5N HCl) (269)	—	268	268	269	—
204	Saccharopine (N⁶-(2-glutaryl)lysine)	Saccharomyces	$C_{11}H_{20}N_2O_6$ (276.30)	—	$+33.6^{23}$ (c 1, 0.5 N HCl) (271)	2.6 4.1 9.2 10.3 (270)	270	270	271	271
205	Sarcosine (N-methylglycine)	Cladonia silvatica	$C_3H_7NO_2$ (89.10)	210° (dec) (290)	—	2.21 10.20 (270)	272	272	—	—
	K. L-SULFUR AND SELENIUM CONTAINING AMINO ACIDS									
206	N-Acetyldjenkolic acid	Acacia farnesiana	$C_9H_{16}N_2O_5S$ (264.32)	170° (273)	-49.0^{25} (c 2, 1% HCl) -60.2^{25} (c 1, 1 N HCl)(273)	—	273	273	273	273
207	Alliin	Allium sativum	$C_6H_{11}NO_3S$ (177.24)	163–165° (274)	$+62.8^{21}$ (H₂O) (274)	—	274	—	356	—
208	S-Allylcysteine	Allium sativum	$C_6H_{11}NO_2S$ (161.24)	218° (275)	-8.7^{20} (275)	—	275	275	275	—

DATA ON THE NATURALLY OCCURRING AMINO ACIDS (Continued)

No.	Amino Acid (Synonym)	Source	Formula (Mol Wt)	Melting Point °C[a]	$[\alpha]_D$[b]	pK$_a$	Isolation and Purification	Chroma-tography	Chemistry	Spectral Data
								References		
209	S-Allylmercaptocysteine	Allium sativum	$C_6H_{11}NO_2S_2$ (193.31)	188° (276)	$-95.3\pm10^{23.5}$ (c 0.19, 6 N HCl) (276)	–	276	276	276	276
210	α-Amino-β-(2-amino-2-carboxyethylmercaptobutyric acid)	Subtilin	$C_7H_{14}N_2O_4S$ (222.28)	–	-34.7^{24} (c 5.4, 1 N HCl) (277)	–	277	277	277	277
211	Cystathionine S-(2-amino-2-carboxyethyl-homocysteine)	Human brain	$C_7H_{14}N_2O_4S$ (222.28)	301° (279) 312° (290)	$+(26.4^{25}$ (c 0.8, 1 N HCl) (279)	–	279	279	279	279
212	3-Amino-(3-carboxy-propyldimethylsulfonium)	Cabbage	$C_6H_{14}NO_2S$ (164.26)	–		–	280	280	280	–
212a	α-Amino-γ-[2-(4-carboxy)-thiazolylbutyric acid	Xeromus subtomentosus (mushroom)	$C_8H_{10}N_2O_4S$ (230.26)	237–238° (280a)	–	3.7 (280a)	280a	280a	280a	280a
213	Carbamyltaurine	Cat urine	$C_3H_8N_2O_3S$ (152.17)		–	–	281	–	–	–
214	2-[S-(β-Carboxy-β-aminoethyltryptophan)]	Amanita phalloides	$C_{14}H_{17}N_3O_4S$ (323.40)		–	–	–	119	–	–
215	S-(β-Carboxyethyl)cysteine	Albizia julibrissin	$C_6H_{11}NO_4S$ (193.17)	218° (282)	-9.33^{20} (c 3, 1 N HCl) (283)	–	283	283	282	283
216	N-(1-Carboxyethyl)taurine	Red algae	$C_5H_{11}NO_5S$ (197.22)	258° (283)	-1.15^{13} (c 5, 1 N NaOH) (284)	–	284	284	284	–
216a	S-(Carboxymethyl)-homocysteine	Human urine	$C_6H_{11}NO_4S$ (193.24)	223–225° (dec) (284a)	–	–	284a	–	–	284a
217	S-(2-Carboxyisopropyl)-cysteine	Acacia	$C_7H_{13}NO_4S$ (207.26)	202° (284)	$+6.6^{22}$ (c 2, H$_2$O) $+31^{25}$ (c 1.94, 1 N NaOH) (285)	–	285	285	285	285
218	S-(2-Carboxypropyl)cysteine	Onions	$C_7H_{13}NO_4S$ (207.26)	191–194° (285) (286)	-50.1^{21} (H$_2$O) (286)	–	286, 287	286, 287	286, 287	286, 287
219	Chondrine (1,4-thiazane-5-carboxylic acid 1-oxide)	Chondria crassicaulis	$C_5H_9NO_3S$ (163.20)	255–257° (287)	$+20.9^{16}$ (c 2, H$_2$O) (288) $+30.2^{16}$ (c 2, 6 n HCl) (288)	–	288	288	288	–
220	Cycloallin (3-methyl-1,4-thiazane-5-carboxylic acid oxide)	Onions	$C_6H_{11}NO_3S$ (177.24)	–	-17.4^{20} (H$_2$O) (289)	–	289	289	289	–

DATA ON THE NATURALLY OCCURRING AMINO ACIDS (Continued)

No.	Amino Acid (Synonym)	Source	Formula (Mol Wt)	Melting Point °C[a]	$[\alpha]_D$[b]	pK_a	References: Isolation and Purification	References: Chromatography	References: Chemistry	References: Spectral Data
221	Cysteic acid (β-sulfo-α-aminopropionic acid)	Sheep's fleece	$C_3H_7NO_5S$ (169.17)	289° (dec) (290)	+8.66 (*c* 7.4, H_2O (290)	1.3 (SO_3H) 1.9 8.7 (290)	291	292	–	–
222	Cysteine (α-amino-β-mercaptopropionic acid	Cystine	$C_3H_7NO_2S$ (121.15)	178° (1)	-16.5^{25} (*c* 2, H_2O) $+6.5^{25}$ (*c* 2, 5 *N*	1.92 8.35 10.46 (SH)	293	3	294	294
223	Cysteine sulfinic acid (β-sulfinyl-α-aminopropionic acid)	Rat brain	$C_3H_7NO_4S$ (153.17)	–	+11 (H_2O) +24 (*c* 1, 1 *N* HCl)	ca 2.1 (290)	295	295	–	–
224	Cystine [β,β′-dithiodi-(α-aminopropionic acid))	Urinary calculi	$C_6H_{12}N_4O_4S_2$ (240.29)	261° (1)	-232^{25} (*c* 1, 5 *N* HCl) (1)	<1 2.1 8.02 8.71 (290)	296	3	294	294
225	Cystine disulfoxide	Rat tissue	$C_6H_{12}N_2O_6S_2$ (272.33)	–	–	–	–	298	298	298
226	*S*-(1,2-Dicarboxyethyl)-cysteine	Bovine lens	$C_7H_{11}NO_6S$ (237.25)	–	–	–	299	299	299	
227	Dichrostachinic acid [*S*-(β-hydroxy-β-carboxyethanesulfonyl methyl)cysteine]	*Mimosa*	$C_7H_{13}NO_7S$ (255.27)	201° (300)	$+9.2^{24}$ (*c* 2.2, 1 *N* HCl) (300)	–	300	300	300	300
228	Dihydroallin (*S*-propylcysteine sulfoxide)	*Allium cepa*	$C_6H_{13}NO_3S$ (179.26)	–	–	–	301	301	301	
229	Djenkolic acid	Djenkol beans	$C_7H_{14}N_2O_4S_2$ (254.32)	300–350° (303)	$-65^{20.5}$ (*c* 1, 1 *N* HCl) (303)	–	302	–	303	–
230	Ethionine	*Escherichia coli*	$C_6H_{13}NO_2S$ (163.23)	–	–	–	304	304	–	–
231	Felinine	Cat urine	$C_8H_{17}NO_3S$ (207.30)	177° (305)	$+23^{20}$ (*c* 2.2, H_2O) (305)	–	305	305	305	–
232	*N*-Formyl methionine	*Escherichia coli*	$C_6H_{11}NO_3$ (177.22)	–	–	–	306	306	–	–
233	Glucobrassicin [*S*-β-1-(glucopyranosyl)-3-indolylacetothiohydroximy-*O*-sulfate]	*Brassica* species	$C_{16}H_{20}N_2O_9S_2$ (447.47)	–	-13.3^{23} (*c* 3, H_2O) (189)	–	189	189	189	189
234	Guanidotaurine	*Arenicola marina*	$C_3H_9N_3O_3S$ (167.20)	228–230° (308)	–	–	308	308	308	–
235	Homocysteine	*Neurospora*	$C_4H_9NO_2S$ (135.18)	–	–	2.22 8.87 10.86 (290)	278	310	–	–

DATA ON THE NATURALLY OCCURRING AMINO ACIDS (Continued)

No.	Amino Acid (Synonym)	Source	Formula (Mol Wt)	Melting Point °C[a]	$[\alpha]_D{}^b$	pK_a	References			
							Isolation and Purification	Chromatography	Chemistry	Spectral Data
236	Homocystine	Human urine	$C_8H_{16}N_2O_4S_2$ (268.36)	282–3° (dec) (290)	-16^{21} (c 0.06, H$_2$O) $+78^{25}$ (c 1–2, 5N HCl) (290)	1.59 2.54 8.52 9.44 (290)	311	311	–	–
237	Homocysteinecysteine disulfide	Human urine	$C_7H_{14}N_2O_4S_2$ (254.35)	–	-52.2^{25} (1 N HCl) (312)	–	312	–	312	312
238	Homolanthionine	Escherichia coli	$C_8H_{16}N_2O_4S$ (236.29)	–	$+37.3^{24}$ (C 1, 1 N HCl) (363)	–	313	313	313	–
239	Homomethionine (5-methylthionorvaline)	Cabbage	$C_6H_{13}NO_2S$ (163.25)	223–225° (314)	$+(21^{25.5}$ (c 0.3, 6 N HCl)(314)	–	314	314	314	314
239a	β-6-(4-Hydroxy-benzothiazolyl)alanine	Chicken feather pigment	$C_{10}H_9N_2O_3S$ (237.27)	–	–	–	314a	314a	–	314a
239a′	S-(2-Hydroxy-2-carboxyethyl)-homocysteine	Human urine	$C_7H_{13}NO_5S$ (223.27)	–	–	–	284a	–	–	–
239a″	S-(2-Hydroxy-2-carboxyethylthio)-homocysteine	Human urine	$C_7H_{13}NO_5S_2$(223.27)	–	–	–	377	377	377	–
239b	S-(3-Hydroxy-3-carboxy-n-propylthio)cysteine	Human urine	$C_7H_{13}NO_5S_2$ (255.33)	176–177° (dec) (377)	-96.3^{26} (c 2.3, 1 N HCl) (377)	–	377	377	377	377
239c	S-(3-Hydroxy-3-carboxy-n-propyl)cysteine	Human urine	$C_7H_{13}NO_5S$ (223.27)	–	–	–	284a	–	–	–
239c′	S-(3-Hydroxy-3-carboxy-n-propylthio)homocysteine	Human urine	$C_8H_{15}NO_5S_2$ (269.36)	187–188° (dec) (377)	$+8.1^{25}$ (c 15.2, 1 N HCl) (377)	–	377	377	377	377
239d	α-Hydroxycysteinecysteine disulfide	Human urine	$C_6H_{12}N_2O_5S_2$ (256.32)	181° (dec) (378)	-263^{23} (c 2.2, H$_2$O) (378)	–	–	378	378	–
240	Hypotaurine (2-aminoethane-sulfinic acid)	Rat brain	$C_2H_7NO_2S$ (109.14)	–	–	–	295	295	–	–
241	Isovalthine (isopropylcarboxy-methylcysteine)	Human urine	$C_8H_{15}NO_4S$ (221.28)	–	–	–	316	317	317	317
242	Lanthionine [β,β′-thiodi-(α-aminopropionic acid)]	Wool	$C_6H_{12}N_2O_4S$ (208.23)	270–304° (318)	–	–	318	–	319	–
242a	β-Mercaptolactatecysteine disulfide	Human urine	$C_6H_{11}NO_5S_2$ (241.30)	–	–	–	–	319a	319a	–
243	S-Methylcysteine	Phaseolus vulgaris	$C_4H_9NO_2S$ (135.19)	220° (320)	-26^{25} (c 2.5) (320)	8.75 (20)	320	320	320	320
243a	S-Methylcysteine sulfoxide	Cabbage	$C_4H_9NO_3S$ (151.20)	173° (320a)	–	–	–	–	–	–
244	Methionine (α-amino-γ-methylthiobutyric acid)	Casein hydrolyzate	$C_5H_{11}NO_2S$ (149.21)	283° (1)	-10^{25} (H$_2$O) $+23.2^{25}$ (5 N HCl)(1)	2.28 9.21 (1)	321	3	322	322

DATA ON THE NATURALLY OCCURRING AMINO ACIDS (Continued)

No.	Amino Acid (Synonym)	Source	Formula (Mol Wt)	Melting Point °C [a]	$[\alpha]_D$ [b]	pKa	References			
							Isolation and Purification	Chromatography	Chemistry	Spectral Data
245	3,3'-(2-Methylethylene-1,2-dithio)dialanine	*Allium shoenoprasum*	$C_9H_{17}N_2O_4S_2$ (282.40)	–	–	–	323	323	323	323
246	β-Methyllanthionine	Yeast	$C_7H_{14}N_2O_4S$ (222.28)	–	–	–	277	–	277	–
247	S-Methylmethionine (α-aminodimethyl-γ-butyrothetin)	Asparagus	$C_6H_{13}NO_2S$ (164.24)	–	–	–	325	325	280	325
248	β-Methylselenoalanine (β-methylselenocysteine)	*Stanleya pinnata*	$C_4H_9NO_2Se$ (182.08)	–	–	–	327	327	–	–
249	Neoglucobrassicin	*Brassica napus*	$C_{17}H_{21}N_2O_{10}S_2$ (476.48)	175° (328)	–	–	328	328	328	328
250	S-(Prop-1-enyl)cysteine	*Allium sativum*	$C_6H_{11}NO_2S_2$ (161.24)	–	–	–	329	329	–	–
251	S-(Prop-1-enyl)cysteine sulfoxide	Onions	$C_6H_{11}NO_3S$ (177.22)	146–148° (330)	–	–	330	330	330	330
252	S-n-Propylcysteine	*Allium sativum*	$C_6H_{13}NO_2S$ (163.26)	–	–	–	331	331	–	–
253	S-n-Propylcysteine sulfoxide	Onions	$C_6H_{13}NO_3S$ (179.26)	–	–	–	301	301	–	–
254	Selenocystathionine	*Stanleya pinnata*	$C_7H_{14}N_2O_4Se$ (269.07)	–	–	–	327	327	327	–
255	Selenocystine	*Astragalus pectinatus*	$C_6H_{12}N_2O_4Se_2$ (334.11)	263–265° (334)	–	–	334	–	334	–
255a	Selenomethionine	*Escherichia coli* (hydrolyzate)	$C_5H_{11}NO_2Se$ (196.13)	–	–	–	334a	334a	–	–
256	Selenomethylselenocysteine	*Astragalus bisulcatus*	$C_3H_6NO_2Se$ (167.03)	–	–	–	335	335	335	–
257	Taurine (2-aminoethane-sulfonic acid)	Plant and animal tissue	$C_2H_7NO_3S$ (125.15)	320° (dec) (290)	–	−0.3, 9.06 (290)	–	292	–	–
258	β-(2-Thiazole)-β-alanine	*Bottromycin*	$C_6H_8N_2O_2S$ (172.22)	197.5–201.5° (337)	–	–	337	337	337	337
259	Thiolhistidine	Erythrocytes and microorganisms	$C_6H_9N_2O_2S$ (173.23)	–	-10^{25} (c 2, 1 N HCl) (339)	1.84, 8.47, 11.4 (290)	338	–	339	–
260	Thiostreptine	*Thiostrepton*	$C_9H_{14}N_2O_4S$ (246.32)	–	-4^{25} (c 1, 1 N acetic acid) (102)	–	102	102	102	102
261	Tyrosine-O-sulfate	Human urine	$C_9H_{11}NO_6S$ (261.26)	–	–	–	341	342	–	–

DATA ON THE NATURALLY OCCURRING AMINO ACIDS (Continued)

No.	Amino Acid (Synonym)	Source	Formula (Mol Wt)	Melting Point °C[a]	$[\alpha]_D^b$	pK$_a$	Isolation and Purification	Chromatography	Chemistry	Spectral Data
							References			
L. L-HALOGEN-CONTAINING AMINO ACIDS										
262	2-Amino-4,4-dichlorobutyric acid	*Streptomyces armentosus*	$C_4H_7Cl_2NO_2$ (172.02)	—	+6.7^{25} (c 0.74, H$_2$O) +26.2^{25} (c 0.74, 1 N HCl)(64)	—	64	64	64	64
262a	5-Chloropiperidazine-3-carboxylic acid	Monamycin hydrolyzate	$C_5H_9ClN_2O_2$ (164.61)	83–85° (DNP deriv)(204a)	+157^{33} (c 0.18, CHCl$_3$) (204b)	—	204b	204b	—	204a, 204b
263	3,5-Dibromotyrosine	*Gorgona* species	$C_9H_9NO_3Br_2$ (339.01)	245° (344)	—	2.17 6.45 (OH) 7.60	344	—	344	—
263a	2,4-Diiodohistidine	Human urine	$C_6H_7I_2N_3O_2$ (406.96)	—	—	—	344a	344a	—	—
264	3,3′-Diiodothyronine	Bovine thyroid gland	$C_{15}H_{13}I_2NO_4$ (525.11)	233–234° (dec) (345)	—	—	345	345	345	—
265	3,5-Diiodotyrosine (iodogorgoic acid)	Coral protein	$C_9H_9I_2NO_3$ (433.01)	194° (dec) (290)	+2.9 (1.1 N HCl)(1)	2.12 6.48 (OH) 7.82 (290)	346	347	348	—
266	3-Monobromotyrosine	Sea fans and sponges	$C_9H_{10}NO_3Br$ (260.10)	—	—	—	349	349	—	—
266a	3-Monobromo-5-monochlorotyrosine	*Buccinum undatum*	$C_9H_9ClNO_3Br$ (294.55)	—	—	—	349a	349a	—	349a
266b	5-Monochlorotyrosine	*Buccinum undatum*	$C_9H_{10}ClNO_3$ (215.65)	—	—	—	349b	349b	—	349b
267	2-Monoiodohistidine	Rat thyroid gland	$C_6H_8IN_3O_2$ (281.02)	—	—	—	350	350	—	—
268	Monoiodotyrosine	*Nereocystis luetkeana* (an alga)	$C_9H_{10}INO_3$ (307.11)	—	—	—	351	351	—	—
269	Thyroxine	Thyroid gland	$C_{15}H_{11}I_4NO_4$ (776.88)	236° (dec) (290) 95% EtOH (290)	+15 (c 5, 1 N HCl, (OH) 10.1 (290)	2.2 6.45	352	353	348	348
270	3,5,3′-Triiodothyronine	*Phaseolus vulgaris*	$C_{15}H_{12}I_3NO_4$ (650.98)	233–234° (dec) (290)	+23.6^{24} (c 5, 1 N HCl-EtOH) (290)	2.2 8.40 (OH) 10.1 (290)	355	347	—	—
M. L-PHOSPHORUS-CONTAINING AMINO ACIDS										
271	α-Amino-β-phosphonopropionic acid	*Tetrahymena pyriformis*; *Zoanthus sociatus*	$C_2H_8NO_5P$ (157.07)	228° (dec) (357)	—	2.2 4.5 8.8 11.0 (357)	358	358	357	—
272	Ciliatine (2-aminoethyl phosphonic acid)	Sea anemone	$C_2H_8NO_3P$ (125.07)	280–281° (dec) (359)	—	6.4 (359)	359	359	359	359
273	2-Dimethylamino-ethylphosphonic acid	Sea anemone	$C_4H_{12}NO_3P$ (153.13)	249.5° (360)	—	—	360	360	360	360

DATA ON THE NATURALLY OCCURRING AMINO ACIDS (Continued)

No.	Amino Acid (Synonym)	Source	Formula (Mol Wt)	Melting Point °C[a]	$[\alpha]_D^b$	pK$_a$	Isolation and Purification	Chromatography	Chemistry	Spectral Data
273a	1-Hydroxy-2-amino-ethylphosphonic acid	*Acanthamoeba castellanii* (plasma membrane)	$C_2H_8NO_4P$ (141.08)	–	–	–	360a	360a	–	360a
274	Lombricine (2-amino-2-carboxyethyl-2-guanidine-ethyl hydrogen phosphate)	Earthworm	$C_6H_{15}N_4O_6P$ (270.21)	223–224° (361)	$+14.5^{23.5}$ (c 0.93, H$_2$O) (362)	8.9 (20)	361	361	362	362
275	2-Methylamino-ethylphosphonic acid	Sea anemone	$C_3H_{10}NO_3P$ (139.10)	291° (360)	–	–	360	360	360	360
276	O-Phosphohomoserine	*Lactobacillus*	$C_4H_{10}NO_6P$ (199.11)	178° (dec) (364)	$+6.25^{22.5}$ (c 2.4, H$_2$O) (364)	–	364	364	364	–
277	O-Phosphoserine	Casein	$C_3H_8NO_6P$ (185.08)	–	+7.2 (H$_2$O) (366)	–	365	366	366	–
278	2-Trimethylaminoethyl-betaine phosphonic acid	Sea anemone	$C_5H_{14}NO_3P$ (167.16)	250–252° (360)	–	–	360	360	360	360

N. L-BETAINES

No.	Amino Acid (Synonym)	Source	Formula (Mol Wt)	Melting Point °C[a]	$[\alpha]_D^b$	pK$_a$	Isolation and Purification	Chromatography	Chemistry	Spectral Data
279	N-(3-Amino-3-carboxypropyl)-β-carboxypyridinium betaine	Tobacco leaves	$C_{10}H_{14}N_2O_5$ (242.24)	241–243° (dec) (368)	$+24^{24}$ (c 2, H$_2$O) (368)	–	368	368	368	368
280	Betonicine (4-hydroxyproline betaine)	*Stachys (Betonica) officinalis*	$C_7H_{13}NO_3$ (159.19)	243–244° (dec) (369)	-36.6^{15} (369)	–	369	–	369	–
281	γ-Butyrobetaine (γ-aminobutyric acid betaine)	Rat brain	$C_7H_{15}NO_2$ (144.20)	180–184° (370)	–	–	370	370	–	–
282	Carnitine (γ-amino-β-hydroxybutyric acid betaine)	Vertebrate muscle	$C_7H_{15}NO_3$ (161.21)	195–197° (371)	$+23.5^{22}$ (c 5, H$_2$O) (374)	–	372	373	367, 371	–
283	Desmosine	Bovine elastin	$C_{24}H_{39}N_5O_8$ (879.43)	–	–	1.70 2.40 8.80 9.85 (375)	375	–	257, 375	375
284	Ergothioneine (betaine of thiol histidine)	Ergot	$C_9H_{15}N_3O_2$ (229.29)	290° (340)	$+115^{27.5}$ (c 1, H$_2$O) (290)	–	336	340	340	340
285	Hercynin (histidine betaine)	Mushrooms	$C_9H_{15}N_3O_2$ (203.22)	–	–	–	332	–	333	–
286	Homobetaine (β-alanine betaine)		$C_6H_{13}NO_2$ (131.18)	–	–	–	324	–	–	–
287	Homostachydrine (pipecolic acid betaine)	Alfalfa	$C_8H_{14}NO_2$ (156.21)	–	–	–	315	309	–	–
288	3-Hydroxystachydrine (3-hydroxyproline betaine)	*Courbonia virgata*	$C_7H_{13}NO_3$ (159.20)	210–212° (307)	$+10^{30}$ (c 2.9, H$_2$O) (307)	–	307	–	307	–

DATA ON THE NATURALLY OCCURRING AMINO ACIDS (Continued)

No.	Amino Acid (Synonym)	Source	Formula (Mol Wt)	Melting Point °C[a]	$[\alpha]_D$[b]	pK_a	References Isolation and Purification	Chromatography	Chemistry	Spectral Data
289	Hypaphorin (tryptophan betaine)	*Erythrina subumbrans*	$C_{14}H_{18}N_2O_2$ (246.29)	–	–	–	58	–	282	–
290	Isodesmosine	Bovine elastin	$C_{24}H_{39}N_5O_8$ (879.43)	–	–	–	375	–	257	–
291	Luminine (α-amino-ε-trimethylamino adipic acid)	*Laminaria angustata*	$C_9H_{20}N_2O_4$ (220.29)	–	–	–	245	–	245	245
292	Lycin (glycine betaine)	*Lycium barbarum*	$C_5H_{11}NO_2$ (118.16)	–			234	–	–	–
293	Miokinine (ornithine betaine)	Human skeletal muscle	$C_{11}H_{25}N_2O_2$ (217.33)	–			232	–	–	–
294	Nicotianine (N-3-amino-3-carboxypropyl-β-carboxypyridinium betaine)	Tobacco leaves	$C_{10}H_{14}N_2O_5$	241–243° (dec) (156)	+28.4²⁴ (c 5, H₂O) (156)	–	156	156	156	156
295	Stachydrine (proline betaine)	*Stachys tuberifera*	$C_7H_{13}NO_2$ (157.22)	–	–20.7²⁵ (H₂O) (125)	–	128	–	128	–
296	Trigonelline (coffearin)	*Foenum graceum*	$C_7H_7NO_2$ (137.15)	–	–	–	120	107	–	–
296a	ε-N-Trimethyl-δ-hydroxylysine betaine	Diatom cell walls	$C_9H_{20}N_2O_3$ (204.31)	243° (dec) (381)	+15.0²⁴ (c 0.84, H₂O) +23.1²⁴ (c 0.84, 2 N HCl) (381)	–	381	381	–	–
297	ε-N'-Trimethyllysine betaine	Histone of murine ascites cells	$C_9H_{20}N_2O_2$ (188.28)	225–226° (dec) (70a)	+10.8¹⁸ (c 5, H₂O) (379)	–	44, 379	44, 380	380	379

O. D-AMINO ACIDS

No.	Amino Acid (Synonym)	Source	Formula (Mol Wt)	Melting Point °C[a]	$[\alpha]_D$[b]	pK_a	Isolation and Purification	Chromatography	Chemistry	Spectral Data
297a	D-Alloenduracididine [α-amino-β-(2-iminoimidazolidinyl)-propionic acid]	Enduracidin hydrolyzate	$C_6H_{12}N_4O_2$ (172.22)	–	+8.7²³ (1M HCl) +13.3²³ (1M NaOH) (194)	2.5 8.3 12 (194a)	–	194a	–	194a
297b	D-Alloisoleucine	Stendomycin hydrolyzate	$C_6H_{13}NO_2$ (131.20)	–	–	–	–	264a	–	–
298	D-Allothreonine	*Actinomycetales* species	$C_4H_9NO_3$ (119.12)	–	–	2.11 9.10 (290)	52	52	–	–
299	1-Amino-D-proline	Flax seed	$C_5H_{10}N_2O_2$ (130.15)	155° (dec) (80)	+113²⁵ (c 2, 0.5 N HCl)(80)	–	80	80	80	80
299a	O-Carbamyl D-serine	*Streptomyces* strain	$C_4H_8N_2O_4$ (148.14)	238° (dec) (384)	–19.6 (c 2, 1 N HCl) +2 (c 2, H₂O) (384)	–	384	384	384	–

DATA ON THE NATURALLY OCCURRING AMINO ACIDS (Continued)

No.	Amino Acid (Synonym)	Source	Formula (Mol Wt)	Melting Point °C[a]	$[\alpha]_D$[b]	pK_a	References Isolation and Purification	Chromatography	Chemistry	Spectral Data
300	D-(3-Carboxy-4-hydroxyphenyl) glycine	Reseda luteola	$C_9H_9NO_5$ (211.18)	<250° (dec) (93)	-121^{22} (c 0.75, 1 N HCl) (93)	–	93	93	93	93
301	D-Cycloserine (oxamycin, D-4-amino-3-isoxazolinone)	Streptomyces orchidaceus	$C_3H_5N_2O_2$ (101.09)	156° (dec) (94)	-89^{25} (c 0.6, H_2O)(94)	(94)	94	–	94	94
302	m-Carboxyphenylglycine	Iris bulbs	$C_9H_9NO_4$ (195.18)	215° (143)	$+112^{25}$ (c 5, 2 N HCl)(326)	4.4 7.3 (326)	143	143	143	143
302a	N-α-Malonyl-D-alanine	Pea seedlings	$C_6H_9NO_5$ (175.16)	138–140° (dec) (382)	$+33^{28}$ (c 0.38, H_2O) (382)	–	382	382	382	382
303	N-α-Malonyl-D-methionine	Tobacco	$C_8H_{13}NO_5S$ (235.30)	–	–	–	95	95	95	–
304	N-α-Malonyl-D-tryptophan	Spinach	$C_{14}H_{13}N_2O_5$ (289.27)	–	–	–	104	–	104	–
304a	N-Methyl-D-leucine	Griselimycin hydrolyzate	$C_7H_{15}NO_2$ (145.23)	–	-19.6^{20} (c 0.9, H_2O) (376)	–	–	–	–	376
305	D-α-Methylserine	Streptomyces	$C_4H_9NO_3$ (119.12)	–	–	2.3 9.4 (113)	113	–	113	113
306	D-Octopine [N-α-(1-Carboxyethyl-arginine]	Octopus muscle	$C_9H_{18}N_4O_4$ (246.28)	–	$+20.6^{24}$ (c 1, H_2O) 20^{25} (c 2, 5 N HCl) (290)	1.36 2.40 8.76 11.3 (290)	133	–	205	–
307	D-Penicillamine	Penicillin	$C_5H_{11}NO_2S$ (149.22)	–	(225)	1.8 7.9 10.5	225	–	225	–
308	Turcine (D-allohydroxyproline betaine)	Stachys (Betonica) officinalis	$C_7H_{13}NO_3$ (159.19)	249° (241)	$+36.26^{15}$ (241)	–	241	–	241	–

a   Melting point or decomposition point in degrees, C.
b   c, grams/100 mL of solution at 20 to 25°C unless specified; wavelength as subscript in millimicrons; temperature as superscript. References are 1, 3, and 290 unless indicated.
c   The isolated amino acid appears to be a racemic DL mixture.
d   The naturally occurring isomer is the erythro-L-form 290.
e   Thermodynamic values.
f   As hydrochloride of amino acid.
g   For cycloalliin sulfoxide.

References

1. Meister, *Biochemistry of the Amino Acids*, 2nd ed., Academic Press, New York, 1965, 28.
2. Schutzenberger and Bourgeois, *C.R. Hebd. Seances Acad. Sci. (Paris)*, 81, 1191 (1875).
3. Greenstein and Winitz, *Chemistry of the Amino Acids*, Vol.2, John Wiley & Sons, New York, 1961, 1382.
4. Greenstein and Winitz, *Chemistry of the Amino Acids*, Vol.2, John Wiley & Sons, New York, 1961, 1819.
5. Morris and Thompson, *Nature*, 190, 718 (1961).
6. Abderhalden and Bahn, *Hoppe Seyler's Z. Physiol. Chem.*, 245, 246 (1937).
7. Virtanen and Miettinen, *Biochim. Biophys. Acta*, 12, 181 (1953).
8. Ackerman, *Hoppe Seyler's Z. Physiol. Chem.*, 69, 273 (1910).
9. Work, *Bull. Soc. Chim. Biol*, 31, 138 (1949).
10. Steward, *Science*, 110, 439 (1949).
11. Gabriel, *Chem. Ber.*, 23, 1767 (1890).
12. Vahatalo and Virtanen, *Acta Chem. Sc.*, 11, 741 (1957).
13. Doyle and Levenberg, *Biochemistry*, 7, 2457 (1968).
14. Steiner and Hartmann, *Biochem. Z*, 340, 436 (1964).
14a. Chilton, Tsou, Kirk, and Benedict, *Tetrahedron Lett.*, 6283 (1968).
15. Takita, *J. Antibiot. (Tokyo)*, 17, 264 (1964).
15a. Sung, Fowden, Millington, and Sheppard, *Phytochemistry*, 8, 1227 (1969).
15b. Millington and Sheppard, *Phytochemistry*, 7, 1027 (1968).
16. Kenner and Sheppard, *Nature*, 181, 48 (1958).
17. Asen, *J. Biol. Chem.*, 234, 343 (1959).
17a. Scannell, Pruess, Demny, Sello, Williams, and Stempel, *J. Antibiot. (Tokyo)*, 25, 122 (1972).
17b. Sahm, Knobloch, and Wagner, *J. Antibiot. (Tokyo)*, 26, 389 (1973).
18. Fowden and Done, *Biochem. J.*, 55, 548 (1953).
19. Fowden and Smith, *Phytochemistry*, 7, 809 (1968).
19a. Kelly, Martin, and Hanka, *Can. J. Chem.*, 47, 2504 (1969).
20. Perrin, *Dissociation Constants of Organic Bases in Aqueous Solution.* Butterworths, London, 1965.
21. Millington and Sheppard, *Phytochemistry*, 7, 1027 (1968).
22. Dardenne, Casimar, and Jadot, *Phytochemistry*, 7, 1401 (1968).
22a. Dardenne, Casimar, and Jadot, *Phytochemistry*, 7, 1401 (1968).
23. Staron, Allard, and Xuong, *C.R. Hebd. Seances Acad. Sci. (Paris)*, 260, 3502 (1965).
23a. Scannell, Pruess, Demny, Weiss, Williams, and Stempel, *J. Antibiot. (Tokyo)*, 24, 239 (1971).
23a'. Fowden, Smith, Millington, and Sheppard, *Phytochemistry*, 8, 437 (1969).
23b. Fowden and Smith, *Phytochemistry*, 7, 809 (1968).
24. Braconnot, *Ann. Chim. Phys.*, 13, 113 (1820).
25. Shorey, *J. Am. Chem. Soc.*, 19, 881 (1897).
26. Hassall and Keyle, *Biochem. J.*, 60, 334 (1955).
27. Carbon, *J. Am. Chem. Soc.*, 80, 1002 (1958).
28. Ehrlich, *Chem. Ber.*, 37, 1809 (1904).
29. Greenstein and Winitz, *Chemistry of the Amino Acids*, Vol.3, John Wiley & Sons, New York, 1961, 2043.
30. Proust, *Ann. Chim. Phys.*, 10, 29 (1819).
31. Greenstein and Winitz, *Chemistry of the Amino Acids*, Vol.3, John Wiley & Sons, New York, 1961, 2075.
31a. Walker, Bodanszky, and Perlman, *J. Antibiot. (Tokyo)*, 23, 255 (1970).
32. Gray and Fowden, *Biochem. J.*, 82, 385 (1961).
33. Kortrum, Vogel, and Andrussow, *Dissociation Constants of Organic Acids in Aqueous Solutions, Butterworths, London*, 1961.
34. Yukowa, Handbook of Organic Structural Analysis. Benjamin, New York, 1965, 584.
35. Fischer, *Hoppe Seyler's Z. Physiol. Chem.*, 33, 151 (1901).
35a. Vervier and Casimir, *Phytochemistry*, 9, 2059 (1970).
35b. Levenberg, *J. Biol. Chem.*, 243, 6009 (1968).
36. Greenstein and Winitz, *Chemistry of the Amino Acids*, Vol.3, John Wiley & Sons, New York, 1961, 2368.
37. Hatanaka and Virtanen, *Acta Chem. Scand.*, 16, 514 (1962).
38. Takemoto and Sai, *J. Pharm. Soc. Jap.*, 85, 33 (1965).
39. Virtanen and Berg, *Acta Chem. Scand.*, 8, 1085 (1954).
40. Vauqelin and Robiquet, *Ann. Chim.*, 57, 88 (1806).

41. Greenstein and Winitz, *Chemistry of the Amino Acids*, Vol.3, John Wiley & Sons, New York, 1961, 1856.
42. Davies and Evans, *J. Chem. Soc.*, p.480 (1953).
43. Ritthausen, *J. Prakt. Chem.*, 103, 233 (1868).
44. Hempel, Lange, and Berkofer, *Naturwissenschaften*, 55, 37 (1968).
45. Gray and Fowden, *Nature*, 189, 401 (1961).
46. Gmelin and Larsen, *Biochim. Biophys. Acta*, 136, 572 (1967).
46a. Nulu and Bell, *Phytochemistry*, 11, 2573 (1972).
47. Fowden, *Biochem. J.*, 98, 57 (1966).
48. Birkinshaw, Raistick, and Smith, *Biochem. J.*, 36, 829 (1942).
49. Ritthausen, *J. Prakt. Chem.*, 99, 454 (1866).
50. Greenstein and Winitz, *Chemistry of the Amino Acids*, Vol.3, John Wiley & Sons, New York, 1961, 1929.
51. Schulze and Bosshard, *Landwirtsch. Vers. Stn.*, 29, 295 (1883).
52. Ikawa, Snell, and Lederer, *Nature*, 188, 558 (1961).
53. Larsen, *Acta Chem. Scand*, 19, 1071 (1965).
54. Fowden and Gray, *Amino Acid Pools*, Holden, Ed., Elsevier, Amsterdam, 1962, 46.
55. Barker, Smyth, Wawszkiewicz, Lee, and Wilson, *Arch. Biochem. Biophys.*, 78, 468 (1958).
56. Virtanen and Berg, *Acta Chem. Scand.*, 9, 553 (1955).
56a. Przybylska and Strong, *Phytochemistry*, 7, 471 (1968).
57. Done and Fowden, *Biochem. J.*, 51, 451 (1952).
58. Greshoff, *Chem. Ber.*, 23, 3537 (1890).
59. Casimir, Jadot, and Renard, *Biochim. Biophys. Acta*, 39, 462 (1960).
59a. Seneviratne and Fowden, *Phytochemistry*, 7, 1030 (1968).
60. Virtanena and Linko, *Acta Chem. Scand.*, 531 (1955).
61. Liss, *Phytochemistry*, 1, 87 (1962).
62. Bohak, *J. Biol. Chem.*, 239, 2878 (1964).
63. Ziegler, Melchert, and Lurken, *Nature*, 214, 404 (1967).
64. Argoudelis, Herr, Mason, Pyke, and Zieserl, *Biochemistry*, 6, 165 (1967).
65. Vega and Bell, *Phytochemistry*, 6, 759 (1967).
65a. Bell, *Nature*, 218, 197 (1968).
66. Damodaran and Narayanan, *Biochem. J.*, 34, 1449 (1940).
67. Kitagawa and Monobe, *J. Biochem. (Tokyo)*, 18, 333 (1933).
67a. Bodanszky and Bodanszky, *J. Antibiot. (Tokyo)*, 23, 149 (1970).
67b. Bodanszky, Chaturvedi, Scozzie, Griffith, and Bodanszky, *Antimicrob. Agents Chemother.*, p.135 (1969).
68. Fujino, Inoue, Ueyanagi, and Miyake, *Bull. Chem. Soc. Jap.*, 34, 740 (1961).
69. Tsai and Stadtman, *Arch. Biochem. Biophys.*, 125, 210 (1968).
69a. Hettinger and Craig, *Biochemistry*, 9, 1224 (1970).
70. Gmelin, Strauss, and Hasenmaiet, *Hoppe Seyler's Z. Physiol. Chem.*, 314, 28 (1959).
70a. Kakimoto and Akazawa, *J. Biol. Chem.*, 245, 5751 (1970).
70b. Scannell, Ax, Pruess, Williams, Demm, and Stempel, *J. Antibiot. (Tokyo)*, 25, 179 (1972).
71. Dupuy and Lee, *J. Am. Pharm. Assoc. (Sci. Ed.)*, 43, 61 (1954).
72. Schilling and Strong, *J. Am. Chem. Soc.*, 77, 2843 (1955).
73. Vickery and Leavenworth, *J. Biol. Chem.*, 76, 437 (1928).
74. Drechsel, *Z. Prakt. Chem.*, 39, 425 (1889).
75. Greenstein and Winitz, *Chemistry of the Amino Acids*, Vol.3, John Wiley & Sons, New York, 1961, 2477.
76. Haskell, Fusari, Frohardt, and Bartz, *J. Am. Chem. Soc.*, 74, 599 (1952).
77. Biemann, Lioret, Asselineau, Lederer, and Polonsky, *Bull. Soc. Chim. Biol.*, 42, 979; *Biochim. Biophys. Acta*, 40, 369 (1960).
78. Lioret, *C.R. Hebd Seances Acad Sci. (Paris)*, 244, 2171 (1957).
79. Murray, *Biochemistry*, 3, 10 (1964).
80. Klosterman, Lamoureux, and Parsons, *Biochemistry*, 6, 170 (1967).
81. Greenstein and Winitz, *Chemistry of the Amino Acids*, Vol 3, John Wiley & Sons, New York, 1961, 2477.
81a. Stapley, Miller, Mata, and Hendlin, *Antimicrob. Agents Chemother.*, p.401 (1967).
82. Rao, Adiga, and Sarma, *Biochemistry*, 3, 432 (1964).
82a. Shiba, Kubota, and Kaneto, *Tetrahedron*, 26, 4307 (1970).
83. Suzuki, Nakamura, Okuma, and Tomiyama, *J. Antibiot. (Tokyo)*, 11A(81), 84 (1958).
84. Cummings and Hudgins, *Am. J. Med. Sci.*, 236, 311 (1958).
85. Sorensen and Andersen, *Hoppe Seyler's Z. Physiol. Chem.*, 56, 250 (1908).

86. Hochstein, *J. Org. Chem.*, 24, 679 (1959).
87. Grobbelaar and Steward, *Nature*, 182, 1358 (1958).
87a. Westley, Pruess, Volpe, Demny, and Stempel, *J. Antibiot. (Tokyo)*, 24, 330 (1971).
88. Larsen, *Acta Chem. Scand.*, 21, 1592 (1967).
88a. Doyle and Levenberg, *Biochemistry*, 7, 2457 (1968).
89. Blass and Macheboeuf, *Helv. Chim. Acta*, 29, 1315 (1946).
90. Snow, *J. Chem. Soc.*, 2588, 4080 (1954).
91. Virtanen and Hietala, *Acta Chem. Scand.*, 9, 549 (1955).
92. Umbreit and Heneage, *J. Biol. Chem.*, 201, 15 (1953).
93. Kjaer and Larsen, *Acta Chem. Scand.*, 17, 2397 (1963).
94. Hidy, Hodge, Yound, Harned, Brewer, Phillips, Runge, Staveley, Pohland, Boaz, and Sullivan, *J. Am. Chem. Soc.*, 77, 2345 (1955).
95. Keglevic, Ladesic, and Pokorney, *Arch. Biochem. Biophys.*, 124, 443 (1968).
96. Virtanen, Uksila, and Matikkala, *Acte Chem. Scand.*, 1091 (1954).
97. Thompson, Morris, and Hunt, *J. Biol. Chem.*, 239, 1122 (1964).
98. Okabe, *J. Antibiot. Ser. A*, 13, 412 (1961).
99. Wieland and Haufer, *Justus Liebigs Ann. Chem.*, 619, 35 (1968).
99a. Fowden, MacGibbon, Mellon, and Sheppard, *Phytochemistry*, 11, 1105 (1972).
99a'. Rudzats, Gellert, and Halpern, *Biochem. Biophys. Res. Commun.*, 47, 290 (1972).
100. Murooka and Harada, *Agric. Biol. Chem.*, 31, 1035 (1967).
100a. Giovanelli, Owens, and Mudd, *Biochim. Biophys. Acta*, 227, 671 (1971).
100b. Owens, Thompson, and Fennessey, *Chem. Commun.*, p. 715 (1972).
101. Virtanen and Ettala, *Acta Chem. Scand.*, 11, 182 (1957).
102. Bodanszky, Alicino, Birkhimer, and Williams, *J. Am. Chem. Soc.*, 84, 2004 (1962).
103. Wieland and Schopf, *Justus Liebigs Ann. Chem.*, 626, 174 (1959).
103a. Mechanic and Tanzer, *Biochem. Biophys. Res. Commun.*, 41, 1597 (1970).
104. Good and Andreae, *Plant Physiol.*, 32, 561 (1957).
104a. Takita and Maeda, *J. Antibiot. (Tokyo)*, 22, 39 (1969).
105. Saarivirta and Virtanen, *Acta Chem. Scand.*, 19, 1008 (1965).
106. Craig, *J. Biol. Chem.*, 125, 289 (1938).
107. Joshi and Handler, *J. Biol. Chem.*, 235, 2981 (1961).
108. Setsuseo, *J. Osaka Univ.*, 833 (1957).
109. Schweet, Holden, and Lowy, *J. Biol Chem.*, 211, 517 (1954).
110. Tominaga, Hiwaki, Maekawa, and Yoshida, *J. Biochem. (Tokyo)*, 53, 227 (1963).
111. Wilding and Stahlmann, *Phytochemistry*, 1, 241 (1962).
111a. Kemp and Dawson, *Biochim. Biophys. Acta*, 176, 678 (1969).
112. Fowden, *Biochem. J.*, 81, 155 (1961).
113. Flynn, Hinman, Caron, and Woolf, *J. Am. Chem. Soc.*, 75, 5867 (1953).
114. Nagao, *Bull. Fac. Fish Hokkaido Univ.*, 2, 128 (1951).
115. Hatanaka, *Acta Chem. Scand.*, 16, 513 (1962).
116. Brandner and Virtanen, *Acta Chem. Scand.*, 17, 2563 (1963).
117. Snow, *J. Chem. Soc.*, 2589 (1954).
117a. Fowden, Pratt, and Smith, *Phytochemistry*, 12, 1707 (1973).
118. Miyake, *Chem. Pharm. Bull. (Tokyo)*, 1071 (1960).
119. Wieland and Schön, *Justus Liebigs Ann. Chem.*, 593, 157 (1955).
120. Jahns, *Chem. Ber.*, 18, 2518 (1885); 20, 2840 (1887).
121. Jadot and Casimir, *Biochim. Biophys. Acta*, 48, 400 (1961).
122. Jadot, Casimir, and Alderweireldt, *Biochim. Biophys. Acta*, 78, 500 (1963).
123. Brieskorn and Glasz, *Naturwissenschaften*, 51, 216 (1964).
124. Schryver, *Proc. R. Soc.*, B98, 58 (1925).
124a. Godtfredsen, Vangedal, and Thomas, *Tetrahedron*, 26, 4931 (1970).
125. Steenbock, *J. Biol. Chem.*, 35, 1 (1918).
126. Fowden, *Nature*, 209, 807 (1966).
127. Bell and Tirimanna, *Biochem. J.*, 91, 356 (1964).
128. Planta and Schulze, *Chem. Ber.*, 26, 939 (1893).
129. Pollard, Sondheimer, and Steward, *Nature*, 182, 1356 (1958).
129a. Shiba, Mizote, Kaneko, Nakajima, and Kakimoto, *Biochim. Biophys. Acta*, 244, 523 (1971).
129b. Hettinger and Craig, *Biochemistry*, 9, 1224 (1970).
130. Barry and Roark, *J. Biol. Chem.*, 239, 1541 (1964).
131. Barry, Chen, and Roark, *J. Gen. Microbiol.*, 33, 95 (1963).
132. Ackermann and Kirby, *J. Biol. Chem.*, 175, 483 (1948).
132a. Grove, Daxenbichler, Weisleder, and van Etlen, *Tetrahedron Lett.*, 4477 (1971).
132b. Owens, Guggenheim, and Hilton, *Biochim. Biophys. Acta*, 158, 219 (1968).
132c. Owens, Thompson, Pitcher, and Williams, *Chem. Commun.*, p. 714 (1972).
133. Morizawa, *Acta Sch. Med. Univer. Kioto*, 9, 285 (1927).
134. Cramer, *J. Prakt. Chem.*, 96, 76 (1865).
135. Flavin, Delavier-Klutchko, and Slaughter, *Science*, 143, 50 (1964).
136. Wooley, *J. Biol. Chem.*, 197, 409 (1952).
137. Wooley, *J. Biol. Chem.*, 198, 807 (1953).
138. Rose, McCoy, Meyer, and Carter, *J. Biol. Chem.*, 112, 283 (1935).
139. Greenstein and Winitz, *Chemistry of the Amino Acids*, Vol. 3, John Wiley & Sons, New York, 1961, 2238.
140. Wisvisz, Van der Hoever, and Nijenhuis, *J. Am. Chem. Soc.*, 79, 4522 (1957).
140a. Bohman, *Tetrahedron Lett.*, p. 3065 (1970).
141. Olesen-Larsen, *Biochim. Biophys. Acta*, 93, 200 (1964).
142. Thompson, Morris, Asen, and Irreverre, *J. Biol. Chem.*, 236, 1183 (1961).
142a. O'Brien, Cox, and Gibson, *Biochim. Biophys. Acta*, 177, 321 (1969).
143. Morris, Thompson, Asen, and Irreverre, *J. Am. Chem. Soc.*, 81, 6069 (1959).
144. Schneider, *Biochem. Z.*, 330, 428 (1958).
145. Guggenheim, *Z. Physiol. Chem.*, 88, 276 (1913).
146. Greenstein and Winitz, *Chemistry of the Amino Acids*, Vol. 3, John Wiley & Sons, New York, 1961, 2713.
147. Muller and Schulte, *Z. Naturforsch.*, 23b, 659 (1958).
147a. Weaver, Rajagopalan, Handler, Rosenthal, and Jeffs, *J. Biol Chem.*, 246, 2010 (1971).
147b. Watson and Fowden, *Phytochemistry*, 12, 617 (1973).
148a. Smith and Sloane, *Biochim. Biophys. Acta*, 148, 414 (1967).
148b. Sloane and Smith, *Biochim. Biophys. Acta*, 158, 394 (1968).
148. Makino, Satoh, Fujik, and Kawaguchi, *Nature*, 170, 977 (1952).
149. Waller, Fryth, Hutchings, and Williams, *J. Am. Chem. Soc.*, 75, 2025 (1953).
150. Behr and Clark, *J. Am. Chem. Soc.*, 54, 1630 (1932).
151. Schultze and Barbieri, *Chem. Ber.*, 14, 1785 (1881).
152. Greenstein and Winitz, *Chemistry of the Amino Acids*, Vol. 3, John Wiley & Sons, New York, 1961, 2156.
153. Liebig, *Justus Liebigs Ann. Chem.*, 57, 127 (1846).
154. Greenstein and Winitz, *Chemistry of the Amino Acids*, Vol. 3, John Wiley & Sons, New York, 1961, 2348.
155. Mothes, Schütte, Müller, Ardenne, and Tümmler, *Z. Naturforsch.*, 196, 1161 (1964).
155a. Ohkusu and Mori, *J. Neurochem.*, 16, 1485 (1969).
156. Noguchi, Sakuma, and Tamaki, *Phytochemistry*, 7, 1861 (1968).
157. Kjaer and Larsen, *Acta Chem. Scand.*, 13, 1565 (1959).
158. Schulze and Steiger, *Chem. Ber.*, 19, 1177 (1886); *Z. Physiol. Chem.*, 11, 43 (1886).
159. Greenstein and Winitz, *Chemistry of the Amino Acids*, Vol. 3, John Wiley & Sons, New York, 1961, 1841.
160. Fearon and Bell, *Biochem. J.*, 59, 221 (1955).
161. Kitagawa and Tomiyama, *J. Biochem. (Tokyo)*, 11, 265 (1929).
162. Bell, *Biochem. J.*, 75, 618 (1960).
163. Walker, *J. Biol. Chem.*, 204, 139 (1954).
164. Wada, *Biochem. Z.*, 224, 420 (1930).
165. Rogers, *Biochim. Biophys. Acta*, 29, 33 (1958).
166. Kitagawa and Tsukamoto, *J. Biochem. (Tokyo)*, 26, 373 (1937).
166a. Nakajima, Matsuoka, and Kakimoto, *Biochim. Biophys. Acta*, 230, 212 (1971).
167. Ito and Hashimoto, *Nature*, 211, 417 (1966).
168. Rao, Ramachandran, and Adiga, *Biochemistry*, 2, 298 (1962).
169. Gerritsen, Vaughn, and Waisman, *Arch. Biochem. Biophys.*, 100, 298 (1963).
170. Bell and Tirimanna, *Nature*, 197, 901 (1963).
170a. Maehr, Blount, Pruess, Yarmchuk, and Kellett, *J. Antibiot. (Tokyo)*, 26, 284 (1973).
171. Bell, *Biochem. J.*, 21, 358 (1964).
173. Hegarty and Pound, *Nature*, 217, 354 (1968).

174. Tamoni and Gallo, *Farmaco Ed. Sci.*, 21, 269 (1966).

175. Fusari, Bartz, and Elder, *Nature*, 173, 72; *J. Am. Chem. Soc.*, 76, 2878 (1954).

176. Ressler, *J. Biol. Chem.*, 237, 733 (1962).

177. Ressler and Ratzkin, *J. Org. Chem.*, 26, 3356 (1961).

177a. Brysk and Ressler, *J. Biol. Chem.*, 245, 1156 (1970).

178. Dion, Fusari, Jakubowski, Zora, and Bartz, *J. Am. Chem. Soc.*, 78, 3075 (1956).

179. Kaczka, Gitterman, Dulaney, and Folkers, *Biochemistry*, 1, 340 (1962).

180. Radhakrishnan and Giri, *Biochem. J.*, 58, 57 (1954).

180a. Lambein and Van Parijs, *Biochem. Biophys. Res. Commun.*, 32, 474 (1968).

180b. Lambein, Schamp, Vandendriessche, and Van Parijs, *Biochem. Biophys. Res. Commun.*, 37, 375 (1969).

181. King, *J. Chem. Soc.*, p. 3590 (1950).

182. Grobbelaar, *Nature*, 175, 703 (1955).

183. Dobson and Raphael, *J. Chem. Soc.*, p. 3642 (1958).

184. Tanaka, Miyamoto, Honjo, Morimoto, Sugawa, Uchibayshi, Sanno, and Tatsuoka, *Proc. Jap. Acad.*, 33, 47 (1957).

185. Tanaka, Miyamoto, Honjo, Morimoto, Sugawa, Uchibayshi, Sanno, and Tatsuoka, *Chem. Abstr.*, 51, 1788 (1957).

185a. Burrows and Turner, *J. Chem. Soc.*, 255 (1966).

186. Schenk and Schütte, *Naturwissenschaften*, 48, 223 (1961).

186a. Hatanaka, *Phytochemistry*, 8, 1305 (1969).

186b. Shah, Neuss, Gorman, and Boeck, *J. Antibiot. (Tokyo)*, 23, 613 (1970).

187. Prochazka, *Czech. Chem. Commun.*, 22, 333–654 (1957).

188. Pironen and Virtanen, *Acta Chem. Scand.*, 16, 1286 (1962).

189. Gmelin and Virtanen, *Ann. Acad. Sci. Fenn. (Med.)*, 107, 3 (1961).

189a. Miller, Tristram, and Wolf, *J. Antibiot. (Tokyo)*, 24, 48 (1971).

190. Fowden, *Nature*, 176, 347 (1955).

190a. Eagles, Laird, Matai, Self, and Synge, *Biochem. J.*, 121, 425 (1971).

190b. Bell, Nulu, and Cone, *Phytochemistry*, 10, 2191 (1971).

190c. Robbers and Floss, *Tetrahedron Lett.*, p. 1857 (1969).

191. Fang, Li, Nin, and Tseng, *Sci. Sinica*, 10, 845 (1961).

191a. Cross and Webster, *J. Chem. Soc. (Org.)*, 1839 (1970).

191b. Scannell, Pruess, Demny, Williams, and Stempel, *J. Antibiot. (Tokyo)*, 23, 618 (1970).

191c. Corbin and Bulen, *Biochemistry*, 8, 757 (1969).

191d. Nakajima and Volcani, *Science*, 164, 1400 (1969).

191e. Karle, Daly, and Witkop, *Science*, 164, 1401 (1969).

191f. Brown and Mangat, *Biochim. Biophys. Acta*, 177, 427 (1969).

191g. Macnicol, *Biochem. J.*, 107, 473 (1968).

192. Brown, *J. Am. Chem. Soc*, 87, 4202 (1965).

193. Daigo, *J. Pharm. Soc. Jap.*, 79, 353 (1959).

194. Casnati, Quilico, and Ricca, *Gazz. Chim. Ital.*, 93, 349 (1963).

194a. Horii and Kameda, *J. Antibiot. (Tokyo)*, 21, 665 (1968).

194a′. Katagiri, Tori, Kimura, Yoshida, Nagasaki, and Minato, *J. Med. Chem.*, 10, 1149 (1967).

194a″. Finot, Bricout, Viani, and Mauron, *Experimentia*, 24, 1097 (1968).

194b. Weaver, Rajagopalan, and Handler, *J. Biol. Chem.*, 246, 2015 (1971).

195. Jahns, *Chem. Ber.*, 24, 2615 (1891).

196. Freidenberg, *Chem. Ber.*, 51, 976 (1918).

197. Kossel, *Hoppe Seyler's Z. Physiol. Chem.*, 22, 176 (1896).

198. Greenstein and Winitz, *Chemistry of the Amino Acids*, Vol. 3, John Wiley & Sons, New York, 1961, 1971.

198a. Takita, Yoshioka, Muraoka, Maeda, and Umezawa, *J. Antibiot. (Tokyo)*, 24, 795 (1971).

199. Hulme, *Nature*, 174, 1055 (1954).

200. Biemann and Deffner, *Nature*, 191, 380 (1961).

201. Minagawa, *Proc. Imp. Acad. (Tokyo)*, 21, 33, 37 (1945).

202. Virtanen and Gmeim, *Acta Chem. Scand.*, 13, 1244 (1959).

203. Schoolery and Virtanen, *Acta Chem. Scand.*, 16, 2457 (1962).

204. Hassall, Morton, Ogihara, and Thomas, *Chem. Commun.*, p. 1079 (1969).

204a. Bevan, Davies, Hassall, Morton, and Phillips, *J. Chem. Soc. (Org.)*, p. 514 (1971).

205. Irvine and Wilson, *J. Biol. Chem.*, 127, 555 (1939).

206. Sheehan and Whitney, *J. Am. Chem. Soc.*, 84, 3980 (1962).

206a. Sung and Fowden, *Phytochemistry*, 7, 2061 (1968).

207. Fischer, *Chem. Ber.*, 35, 2660 (1902).

208. Greenstein and Winitz, *Chemistry of the Amino Acids*, Vol. 3, John Wiley & Sons, New York, 1961, 2018.

209. Wieland and Witkop, *Justus Liebigs Ann. Chem.*, 543, 171 (1940).

210. Kotake, Sakan, and Miwa, *Chem. Ber.*, 85, 690 (1952).

211. Mitoma, Weissbach, and Udenfriend, *Nature*, 175, 994 (1955).

212. Takemoto, Nakajima, and Yokobe, *J. Pharm. Soc. Jap.*, 84, 1232 (1964).

213. List, *Planta Med.*, 6, 424 (1958).

214. von Klämbt, *Naturwissenschaften*, 47, 398 (1960).

215. Hutzinger and Kosuge, *Biochemistry*, 7, 601 (1968).

216. Murakami, Takemoto, and Shimzu, *J. Pharm. Soc. Jap.*, 73, 1026 (1953).

217. Beockman and Staehler, *Naturwissenschaften*, 52, 391 (1965).

218. Vanderhaeghe and Parmetier, *17th Congr. Pure Appl. Chem.*, Butterworths, London, 1959, 56.

219. Brockmann, *Ann. N. Y. Acad. Sci.*, 89, 323 (1960).

220. Bell, *Biochim. Biophys. Acta*, 47, 602 (1961).

221. Gray and Fowden, *Nature*, 193, 1285 (1962).

222. Bethell, Kenner, and Shepperd, *Nature*, 194, 864 (1962).

222a. Müller and Schütte, *Z. Naturforsch.*, 236, 491 (1968).

223. Hulme and Arthington, *Nature*, 173, 588 (1954).

224. Burroughs, Dalby, Kenner, and Sheppard, *Nature*, 189, 394 (1961).

224a. Borders, Sax, Lancaster, Hausmann, Mitscher, Wetzel, and Patterson, *Tetrahedron*, 26, 3123 (1970).

225. Chain, *Ann. Rev. Biochem.*, 17, 657 (1948).

226. Sheehan, Drummond, Gardner, Maeda, Mania, Nakamura, Sen, and Stock, *J. Am. Chem. Soc.*, 85, 2867 (1963).

227. Adams and Johnson, *J. Am. Chem. Soc.*, 71, 705 (1949).

227a. Murakoshi, Kuramoto, Ohmiya, and Haginiwa, *Chem. Pharm. Bull. (Japan)*, 20, 855 (1972).

228. Minagawa, *Proc. Jap. Acad.*, 22, 130 (1946).

229. Eugster, Muller, and Good, *Tetrahedron Lett.*, 23, 1813 (1965).

230. Reiner and Eugster, *Helv. Chim. Acta*, 50, 128 (1967).

231. Fritz, Gagnent, Zbinden, Geigy, and Eugster, *Tetrahedron Lett.*, 25, 2075 (1965).

231a. Noma, Noguchi, and Tamaki, *Tetrahedron Lett.*, p. 2017 (1971).

232. Engeland and Biehler, *Hoppe Seyler's Z. Physiol. Chem.*, 123, 290 (1922).

233. Hulme and Arthington, *Nature*, 170, 659 (1952).

234. Husemann and Marme, *Justus Liebigs Ann. Chem.*, Suppl. 2, 382 (1863).

235. Greenstein and Winitz, *Chemistry of the Amino Acids*, Vol. 3, John Wiley & Sons, New York, 1961, 2316.

236. Noe and Fowden, *Biochem. J.*, 77, 543 (1960).

236a. Finot, Viani, Bricout, and Mauron, *Experimentia*, 25, 134 (1969).

236b. Wolfersberger and Tabachnik, *Experimenta*, 29, 346 (1973).

237. Nakanishi, Ito, and Hirata, *J. Am. Chem. Soc.*, 76, 2845 (1954).

238. Carter, Sweeley, Daniels, McNary, Schaffner, West, Tamelen, Dyer, and Whaley, *J. Am. Chem Soc.*, 83, 4296 (1961).

238a. Bodanszky, Marconi, and Bodanszky, *J. Antiobiot., (Tokyo)*, 22, 40 (1969).

238b. Marconi and Bodanszky, *J. Antibiot. (Tokyo)*, 23, 120 (1970).

239. Hattori and Komamine, *Nature*, 183, 1116 (1959).

240. Senoh, Imamato, Maeno, Tokyama, Sakan, Komamine, and Hattori, *Tetrahedron Lett.*, 46, 3431 (1964).

241. Schulze and Trier, *Hoppe Seyler's Z. Physiol. Chem.*, 76, 258; 79, 235 (1912).

242. Takemoto and Nakajima, *J. Pharm. Soc. Jap.*, 84, 1230 (1964).

243. Hopkins and Cole, *J. Physiol. (Lond.)*, 27, 418 (1901).

244. Greenstein and Winitz, *Chemistry of the Amino Acids*, Vol. 3, John Wiley & Sons, New York, 1961, 2316.

244a. Nakamiya, Shiba, Kaneko, Sakakibara, Take, and Abe, *Tetrahedron Lett.*, p. 3497 (1970).

245. Takemoto, Diago, and Takagi, *J. Pharm. Soc. Jap.*, 84, 1176 (1964).

246. Dyer, Hayes, Miller, and Nassar, *J. Am. Chem. Soc.*, 86, 5363 (1964).

246a. Büchi and Raleigh, *J. Org. Chem.*, 36, 873 (1971).

247. Gmelin, *Hoppe Seyler's Z. Physiol. Chem.*, 316, 164 (1959).

248. Kjaer, Knudsen, and Larsen, *Acta Chem. Scand.*, 15, 1193 (1961).

249. Gordon and Jackson, *J. Biol. Chem.*, 110, 151 (1935).

250. Ghatak and Kaul, *Chem. Zentralbl.*, 3730 (1932).
250a. Auditore and Wade, *J. Neurochem.*, 18, 2389 (1971).
250a′.Hall, Metzenberg, and Cohen, *J. Biol. Chem.*, 230, 1013 (1958).
250b. Tallan, Moore, and Stein, *J. Biol. Chem.*, 219, 257 (1956).
251. Takita, Naganawa, Maeda, and Umezawa, *J. Antibiot. (Tokyo)*, 17, 90 (9164).
252. Sheehan, Zachau, and Lawson, *J. Am. Chem. Soc.*, 80, 3349 (1958).
252a. Bouloin, Ottinger, Pais, and Chiurdoglu, *Bull. Soc. Chim. Belges*, 78, 583 (1969).
253. Takita, *J. Antibiot. (Tokyo)*, 16, 175 (1963).
254. Emery, *Biochemistry*, 4, 1410 (1965).
255. Ackerman, *Hoppe Seyler's Z. Physiol. Chem.*, 302, 80 (1955).
256. Morgan, *Chem. Ind. (Lond.)*, p. 542 (1964).
257. Starcher, Partridge, and Elsden, *Biochemistry*, 6, 2425 (1967).
258. Eloff and Grobbelaar, *J. S. Afr. Chem. Inst.*, 20, 190 (1967).
259. Searle and Westall, *Biochem. J.*, 48, 1 (P) (1951).
260. Tallan, Stein, and Moore, *J. Biol. Chem.*, 206, 825 (1954).
261. Plattner and Nager, *Helv. Chim. Acta*, 31, 665 (1948).
262. Plattner and Nager, *Helv. Chim. Acta*, 31, 2192 (1948).
263. Corrigan, *Science*, 164, 142 (1969).
264. Vilkas, Rojas, and Lederer, *C. R. Hebd. Seances Acad. Sci. (Paris)*, 261, 4258 (1965).
264a. Bodanszky, Muramatsu, Bodanszky, Lukin, and Doubler, *J. Antibiot. (Tokyo)*, 21, 77 (1968).
265. Thompson, Morris, and Smith, *Ann. Rev. Biochem.*, 38, 137 (1969).
266. Hiller-Bombien, *Arch. Pharm.*, 230, 513 (1892).
267. Winterstein, *Hoppe Seyler's Z. Physiol. Chem.*, 105, 20 (1919).
268. Brockmann and Grobhofer, *Naturwissenschaften*, 36, 376 (1949).
269. Plattner and Nager, *Helv. Chim. Acta*, 31, 2203 (1948).
270. Darling and Larsen, *Acta Chem. Scand.*, 15, 743 (1961).
271. Kjaer and Larsen, *Acta Chem. Scand.*, 15, 750 (1961).
272. Linko, Alfthan, Miettinen, and Virtanen, *Acta Chem. Scand.*, 7, 1310 (1953).
273. Gmelin, Kjaer, and Larsen, *Phytochemistry*, 1, 233 (1962).
274. Stoll and Seebeck, *Helv. Chim. Acta*, 31, 189 (1948).
275. Suzuki, *Chem. Pharm. Bull. (Tokyo)*, 9, 251 (1961).
276. Sugii, *Chem. Pharm. Bull. (Tokyo)*, 12, 1114 (1964).
277. Alderton, *J. Am. Chem. Soc.*, 75, 2391 (1953).
278. Horowitz, *J. Biol. Chem.*, 171, 255 (1947).
279. Tallan, Moore, and Stein, *J. Biol. Chem.*, 230, 707 (1958).
280. McRorie, Sutherland, Lewis, Barton, Glazener, and Shive, *J. Am. Chem. Soc.*, 76, 115 (1954).
280a. Jadot, Casimir, and Warin, *Bull. Soc. Chim. Belges*, 78, 299 (1969).
281. Salkowski, *Chem. Ber.*, 6, 744 (1873).
282. Romburgh and Barger, *J. Chem. Soc. (Lond.)*, 99, 2068 (1911).
283. Gmelin, Strauss, and Hasenmaler, *Z. Naturforsch.*, 13b, 252 (1958).
284. Kuriyama, *Nature*, 192, 969 (1961).
284a. Kodama, Yao, Kobayashi, Hirayama, Fujii, Mizuhara, Haraguchi, and Hirosawa, *Physiol. Chem. Phys.*, 1, 72 (1969).
285. Gmelin and Hietala, *Hoppe Seyler's Z. Physiol. Chem.*, 322, 278 (1960).
286. Virtanen and Matikkala, *Hoppe Seyler's Z. Physiol. Chem.*, 322, 8 (1960).
287. Mizuhara and Ohmori, *Arch. Biochem. Biophys.*, 92, 53 (1961).
288. Kuriyama, Takagi, and Murata, *Bull. Fac. Fish Hokkaido Univ.*, 11, 58 (1960).
289. Virtanen and Matikkala, *Acta Chem. Scand.*, 13, 623 (1959).
290. Dawson, Elliott, Elliott, and Jones, *Data for Biochemical Research*, Oxford University Press, Oxford, 1969, 2.
291. Martin and Synge, *Adv. Protein Chem.*, 2, 7 (1945).
292. Dent, *Biochem. J.*, 43, 169 (1948).
293. Baumann, *Hoppe Seyler's Z. Physiol. Chem.*, 8, 299 (1884).
294. Greenstein and Winitz, *Chemistry of the Amino Acids*, Vol. 3, John Wiley & Sons, New York, 1961, 1879.
295. Bergeret and Chatagner, *Biochim. Biophys. Acta*, 14, 297 (1954).
296. Wollaston, *Ann. Chim. Phys.*, 76, 21 (1810).
297. Larsen and Kjaer, *Acta Chem. Scand.*, 16, 142 (1962).
298. Sweetman, *Nature*, 183, 744 (1959).
299. Calam and Waley, *Biochem. J.*, 86, 226 (1963).
300. Gmelin, *Hoppe Seyler's Z. Physiol. Chem.*, 327, 186 (1962).
301. Virtanen and Matikkala, *Acta Chem. Scand.*, 13, 1898 (1959).
302. van Veen and Hijman, *Rec. Trav. Chem. Pays-Bas*, 54, 493 (1935).

303. Armstrong and du Vigneaud, *J. Biol. Chem.*, 168, 373 (1947).
304. Fisher and Mallette, *J. Gen. Physiol.*, 45, 1 (1961).
305. Westall, *Biochem. J.*, 55, 244 (1953).
306. Adams and Capecchi, *Proc. Natl. Acad. Sci. USA*, 55, 147 (1966).
307. Cornforth and Henry, *J. Chem. Soc.*, 597 (1952).
308. Thoai and Robin, *Biochim. Biophys. Acta*, 13, 533 (1954).
309. Robertson and Marion, *Can. J. Chem.*, 37, 1043 (1959).
310. Stock, Friedel, and Hambsch, *Hoppe Seyler's Z. Physiol. Chem.*, 305, 166 (1956).
311. Gerritson, Vaughn, and Waisman, *Biochem. Biophys. Res. Commun.*, 9, 493 (1962).
312. Frimpter, *J. Biol. Chem.*, 236, PC51 (1961).
313. Huang, *Biochemistry*, 2, 296 (1963).
314. Sugii, Suketa, and Suzuchi, *Chem. Pharm. Bull. (Tokyo)*, 12, 1115 (1964).
314a. Minale, Fattorusso, Cimino, DeStefano, and Nicolaus, *Gazzetta*, 97, 1636 (1967).
315. Wiehler and Marion, *Can. J. Chem.*, 36, 339 (1958).
316. Ohmori and Mizuhara, *Arch. Biochem. Biophys.*, 96, 179 (1962).
317. Ohmori, *Arch. Biochem. Biophys.*, 104, 509 (1964).
318. Horn, Jones, and Ringel, *J. Biol. Chem.*, 138, 141 (1941).
319. du Vigneaud and Brown, *J. Biol. Chem.*, 138, 151 (1941).
319a. Ampola, Bixby, Crawhall, Efron, Parker, Sneddon, and Young, *Biochem. J.*, 107, 16P (1968).
320. Thompson, *Nature*, 178, 593 (1956).
320a. Fujiwara, Itokawa, Uchino, and Inoue, *Experimenta*, 28, 254 (1972).
321. Mueller, *Proc. Soc. Exp. Biol. Med.*, 19, 161 (1922).
322. Greenstein and Winitz, *Chemistry of the Amino Acids*, Vol. 3, John Wiley & Sons, New York, 1961, 2125.
323. Matikkala and Virtanen, *Acta Chem. Scand.*, 17, 1799 (1963).
324. Guggenheim, *Die Biogenen. Amine*, S. Karger, Basel, (1951)
325. Challenger and Hayward, *Biochem. J.*, 58, 10 (1954).
326. Friis and Kjaer, *Acta Chem. Scand.*, 17, 2391 (1963).
327. Shrift and Virupaksha, *Biochim. Biophys. Acta*, 100, 65 (1965).
328. Gmelin and Virtanen, *Suomen Kemistilehti*, B35, 34 (1962); *Acta Chem. Scand.*, 16, 1378 (1962).
329. Matikkala and Virtanen, *Acta Chem. Scand.*, 16, 2461 (1962).
330. Spare and Virtanen, *Acta Chem. Scand.*, 17, 641 (1963).
331. Virtanen, Hatanaka, and Berlin, *Suomen Kemistilehti*, B35, 52 (1962).
332. Kutscher, *Zentralbl. Physiol.*, 24, 775 (1910).
333. Barger and Ewins, *Biochem.*, 7, 204 (1913).
334. Horn and Jones, *J. Biol. Chem.*, 139, 649 (1941).
334a. Tuve and Williams, *J. Biol. Chem.*, 236, 597 (1961).
335. Trelease, DiSomma, and Jacobs, *Science*, 132, 618 (1960).
336. Tanret, *C. R. Hebd. Seances Acad. Sci.(Paris)*, 149, 222 (1909).
337. Waisvisz, van der Hoeven, and Rijenhuis, *J. Am. Chem. Soc.*, 79, 4524 (1957).
338. Behre and Benedict, *J. Biol. Chem.*, 82, 11 (1929).
339. Greenstein and Winitz, *Chemistry of the Amino Acids*, Vol. 3, John Wiley & Sons, New York, 1961, 2671.
340. Heath, *Nature*, 166, 106 (1950).
341. Tallan, Bella, Stein, and Moore, *J. Biol. Chem.*, 217, 703 (1955).
342. Bettelheim, *J. Am. Chem. Soc.*, 76, 2838 (1954).
343. Partridge, Elsden, and Thomas, *Nature*, 197, 1297 (1963).
344. Morner, *Hoppe Seyler's Z. Physiol. Chem.*, 88, 138 (1913).
344a. Savoie, Massin, and Savoie, *J. Clin. Invest.*, 52, 116 (1973).
345. Gross and Pitt-Rivers, *Biochem. J.*, 53, 645 (1950).
346. Drechsel, *Z. Biol.*, 33, 96 (1896).
347. Greenstein and Winitz, *Chemistry of the Amino Acids*, Vol. 3, John Wiley & Sons, New York, 1961, 1426.
348. Greenstein and Winitz, *Chemistry of the Amino Acids*, Vol. 3, John Wiley & Sons, New York, 1961, 2259.
349. Low, *J. Mar. Res.*, 10, 239 (1951).
349a. Hunt and Breuer, *Biochim. Biophys. Acta*, 252, 401 (1971).
349b. Hunt, *FEBS Lett.*, 24, 109 (1972).
350. Roche, Lissitzky, and Michel, *C. R. Hebd. Seances Acad. Sci. (Paris)*, 232, 2047 (1951).
351. Roche and Yagi, *C. R. Soc. Biol.*, 146, 642 (1952).
352. Kendall, *J. Biol. Chem.*, 39, 125 (1919).
353. Coulson, *J. Sci. Food Agric. Abstr.*, 6, 674 (1955).

354. Butenandt, Weidel, Weicher, and Von Derjugen, *Hoppe Seyler's Z. Physiol. Chem.*, 279, 27 (1937).

355. Fowden, *Physiol. Plant*, 12, 657 (1959).

356. Stoll and Seebeck, *Helv. Chim. Acta*, 34, 481 (1951).

357. Chambers and Isbell, *J. Org. Chem.*, 29, 832 (1964).

358. Kittredge and Hughes, *Biochemistry*, 3, 991 (1964).

359. Kittredge, Roberts, and Simonsen, *Biochemistry,* 1, 624 (1962).

360. Kittredge, Isbell, and Hughes, *Biochemistry*, 6, 289 (1967).

360a. Korn, Dearborn, Fales, and Sokoloski, *J. Biol. Chem.*, 248, 2257 (1973).

361. Thoai and Robin, *Biochim. Biophys. Acta*, 14, 76 (1954).

362. Beatty, Magrath, and Ennor, *J. Am. Chem. Soc.*, 82, 4983 (1960); *J. Biol. Chem.*, 236, 1028 (1961).

363. Weiss and Stekol, *J. Am. Chem. Soc.*, 73, 2497 (1951).

364. Agren, *Acta Chem. Scand.*, 16, 1607 (1962).

365. Lipmann, *Biochem. Z.*, 262, 3 (1933).

366. Greenstein and Winitz, *Chemistry of the Amino Acids*, Vol. 3, John Wiley & Sons, New York, 1961, 2208.

367. Tomita and Sendju, *Hoppe Seyler's Z. Physiol. Chem.*, 169, 263 (1927).

368. Noguchi, Sakuma, and Tamaki, *Arch. Biochem. Biophys.*, 125, 1017 (1968).

369. Kung and Trier, *Hoppe Seyler's Z. Physiol. Chem.*, 85, 209 (1913).

370. Hosein and Proulx, *Nature*, 187, 321 (1960).

371. Carter and Bhattacharyya, *J. Am. Chem. Soc.*, 75, 2503 (1953).

372. Gulewitsch and Krimberg, *Hoppe Seyler's Z. Physiol. Chem.*, 45, 326 (1905).

373. Friedman, McFarlane, Bhattacharyya, and Fraenkel, *Arch. Biochem. Biophys.*, 59, 484 (1955).

374. Carter, Bhattacharyya, Weidman, and Fraenkel, *Arch. Biochem. Biophys.*, 38, 405 (1952).

375. Thomas, Elsden, and Partridge, *Nature*, 200, 651 (1963).

376. Terlain and Thomas, *Bull. Soc. Chim. Fr.*, p. 2349 (1971).

377. Kodama, Ohmori, Suzuki, Mizuhara, Oura, Isshiki, and Uemura, *Physiol. Chem. Phys.*, 3, 81 (1971).

378. Wälti and Hope, *J. Chem. Soc. (Org.)*, p. 2326 (1971).

379. Larsen, *Acta Chem. Scand.*, 22, 1369 (1968).

380. Delange, Glazer, and Smith, *J. Biol. Chem.*, 244, 1385 (1969).

381. Nakajima and Volcani, *Biochem. Biophys. Res. Commun.*, 39, 28 (1970).

382. Ogawa, Fukuda, and Sasaoka, *Biochim. Biophys. Acta*, 297, 60 (1973).

383. Zygmunt and Martin, *J. Med. Chem.*, 12, 953 (1969).

384. Hagemann, Pénasse, and Teillon, *Biochim. Biophys. Acta*, 17, 240 (1955).

STRUCTURES AND SYMBOLS FOR SYNTHETIC AMINO ACIDS INCORPORATED INTO SYNTHETIC POLYPEPTIDES

M. C. Khosla and W. E. Cohn

The amino acids included in this list are those that have been incorporated into biologically active peptides, e.g., angiotensin II,[a] to study structure-activity relationships. Most of these amino acids are synthetic and are either available commercially or have been synthesized by various investigators as structural variants of naturally occurring amino acids. However, a few of these are also naturally occurring.[b] The selection here is of those most widely used and whose representation by symbols in peptide sequences has caused problems for authors and editors. The symbols listed are those considered most in keeping with the system originated by the IUPAC-IUB Commission on Biochemical Nomenclature[c,d]

and have been chosen with an eye to internal consistency, ability to evoke the proper name, and suitability for use in sequences. Only one new one has been invented: ▲ for -yn- (triple bond), by analogy with Δ for -en- (double bond).

The list may also be useful in selecting suitable isosteres of natural amino acids. The bibliography may be helpful in synthesis, resolution, or studies of the effects of these substances on the biological activity of various peptides.

The following trivial names are listed under other names (given by the number of the entry): N-amidinoglycine (67), 6-aminocaproic acid (18), 2-aminoethanesulfonic acid (112), β,β-bis(trifluoromethyl)alanine (70), carbamoylglycine (73), 2-(2-carboxyhydrazino)propane (83), cycloleucine (15), diethylalanine (17), dihydrophenylalanine (46), dopa (57), glycocyamine (67), isolysine (52), β-lysine (52), mercaptovaline (60), α-methylalanine (21), penicillamine (60), 5-pyrrolidone-2-carboxylic acid (109), surinamine (94), tetrahydrophenylalanine (47), trimethylammoniocaproic acid (116).

---

[a] For a review of structure-activity relationships and a listing of various analogs in which a number of these amino acids have been incorporated in angiotensin molecule, see Khosla, Smeby, and Bumpus.[26]

[b] IUPAC Commission on the Nomenclature of α-Amino Acids (Recommendations 1974), *Biochemistry*, 14, 449 (1975).

[c] IUPAC-IUB Commission on Biochemical Nomenclature, Symbols for Amino-Acid Derivatives and Peptides (Recommendations 1971), *Biochemistry*, 11, 1726 (1972). See Nomenclature section.

[d] IUPAC Commission on Biochemical Nomenclature, Abbreviations and symbols for Nucleic Acids, Polynucleotides, and Their Constituents (Recommendations 1970), *J. Biol. Chem.*, 245, 5171 (1970). See Nomenclature Section.

STRUCTURE AND SYMBOLS FOR SYNTHETIC AMINO ACIDS INCORPORATED INTO SYNTHETIC POLYPEPTIDES (Continued)

No.	Structure	Name/Reference	Symbol
1	$H_2 C=C(NHCOCH_3)COOH$	2-Acetamidoacrylic acid	AcAacr
2	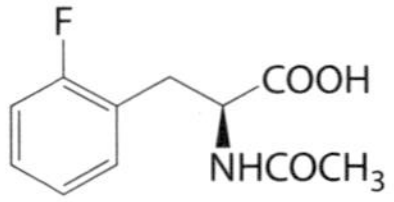 (structure: fluorophenyl with COOH, NHCOCH₃)	N^α-Acetyl-2-fluorophenylalanine	AcPhe(2F)
3	$H_2 N(CH_2)_4 CH(NHCOCH_3)CONHCH_3$	N^α-Acetyllysine-N-methylamide	Ac-Lys-NHMe
4	$H_2 NCH_2 CH_2 COOH$	β-Alanine[b]	βAla[c]
5	$CH_3 CH_2 CH(CH_3)CH(NH_2)COOH$	Alloisoleucine[b]	alle[c]
6	$HOOC(CH_3)CH(NH_2)COOH$	2-Aminoadipic acid	Aad[c]
7	$HOOCCH_2 CH_2 CH(NH_2)CH_2 COOH$	3-Aminoadipic acid	βAad[c]
8	$H_2 NCH_2 CH(CH_2 C_6 H_5)COOH$	3-Amino-2-benzylpropionic acid	βApr(αBzl)
9	$H_2 NCH(CH_2 C_6 H_5)CH_2 COOH$	3-Amino-3-benzylpropionic acid	βApr(βBzl)
10	$CH_3 CH_2 CH(NH_2)COOH$	2-Aminobutyric acid	Abu[c]
11	$CH_3 CH(NH_2)CH_2 COOH$	3-Aminobutyric acid	βAbu[c]
12	$H_2 N(CH_2)_3 COOH$	4-Aminobutyric acid	γAbu
13	$CH_3 CH=C(NH_2)COOH$	2-Aminocrotonic acid	ACrt
14	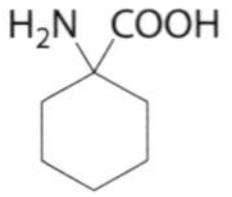 $H_2 N$ COOH (cyclohexane)	1-Aminocyclohexane-1-carboxylic acid (cyclonorleucine)	cHxA(αCx); cNle[b]
15	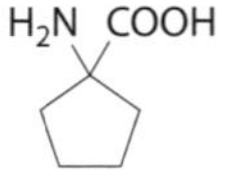 $H_2 N$ COOH (cyclopentane)	1-Aminocyclopentane-1-carboxylic acid (cycloleucine)	cPeA(αCx); cLeu[c]
16	$(CH_3)_2 NCH_2 C≡CCH_2 CH(NH_2)COOH$	2-Amino-6-dimethylamino-4-hexynoic acid (1)	$\alpha\varepsilon$ A₂ hx(▲γ, N^εMe₂)
17	$CH_3 CH_2 CH(CH_2 CH_3)CH(NH_2)COOH$	2-Amino-3-ethylvaleric acid (diethylalanine)	Ala(βEt₂)
18	$H_2 N(CH_2)_5 COOH$	6-Aminohexanoic acid (6-aminocaproic acid)	εAhx[c]
19	$(CH_3)_2 CHCH_2 CHCH_2 CH(NH_2)CH(OH)CH_2 COOH$	4-Amino-3-hydroxy-6-methylheptanoic acid (2,3)	γAhp(βOH, εMe)
20	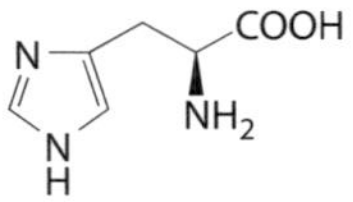 (imidazolyl structure with COOH, NH₂)	2-Amino-3-(2-imidazolyl)propionic acid	Apr(βIm-2)
21	$(CH_3)_2 C(NH_2)COOH$	2-Aminoisobutyric acid (α-methylalanine)	Ala(αMe)
22	$H_2 NCH_2 SO_3 H$	Aminomethanesulfonic acid	Ams
23	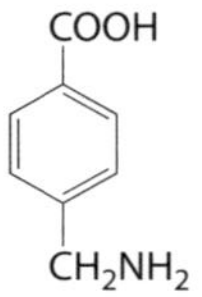 (benzene with COOH, CH₂NH₂)	4-Aminomethylbenzoic acid	Bz(4Ame); Bz(4CH₂ NH₂)

STRUCTURE AND SYMBOLS FOR SYNTHETIC AMINO ACIDS INCORPORATED INTO SYNTHETIC POLYPEPTIDES (Continued)

No.	Structure	Name/Reference	Symbol
24	$CH_3\ CH_2\ CH(CH_3)\ CH_2\ CH_2\ CH(NH_2)COOH$	2-Amino-4-methyl-hexanoic acid (4)	Ahx(γMe)
25	$CH_3\ CH{=}C(CH_3)CH_2\ CH_2\ CH(NH_2)COOH$	2-Amino-4-methyl-4-hexenoic acid (4)	Ahx($\Delta\gamma$, γMe)
26	$CH_2{=}C(CH_3)CH_2\ CH_2\ CH_2\ CH(NH_2)COOH$	2-Amino-5-methyl-5-hexenoic acid (4)	Ahx(Δ^δ, δMe)
27	$H_2\,N(CH_2)_7\ COOH$	8-Aminooctanoic acid	ωAoc
28	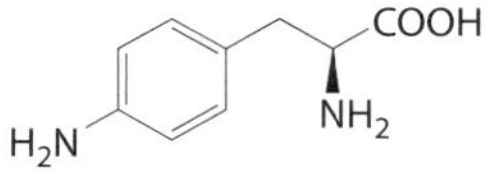	(4-Amino)phenylalanine (5)	Phe(4NH$_2$)
29	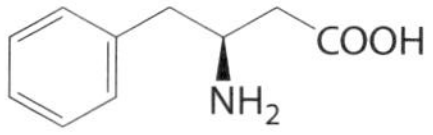	3-Amino-4-phenylbutyric acid	βAbu(γPh)
30	$HOOCCH(NH_2)(CH_2)_4\ COOH$	2-Aminopimelic acid	αApm
31	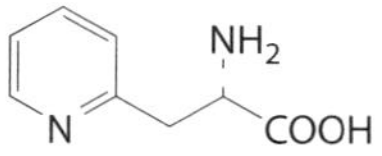	2-Amino-3-(2-pyridyl)propionic acid	Apr(βPrd-2)
32	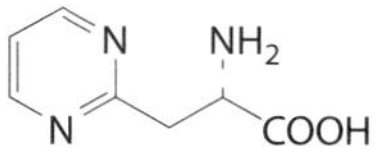	2-Amino-3-(2-pyrimidyl)propionic acid	Apr(βPyr-2)[c,d]
33	$(CH_3)_3\overset{+}{N}(CH_2)_4\ CH(NH_2)COOH$	2-Amino-6-(trimethylammonio)hexanoic acid	α,ε,A$_2$hx(N$^\varepsilon$Me$_3$)
34	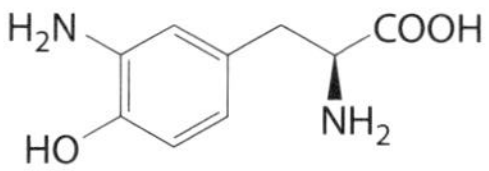	3-Aminotyrosine	Tyr(3NH2)
35	$HOOCCH_2\ CH(NH_2)CONHCH_3$	Aspartic α-methylamide	Asp-NHMe
36	$CH_3\ NHCOCH_2\ CH(NH_2)\ COOH$	Aspartic β-methylamide	Asn(Me); Asp(NHMe)
37	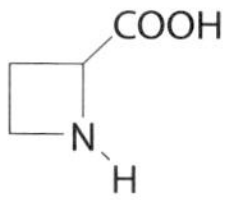	Azetidine-2-carboxylic acid	Azt
38	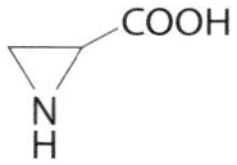	Aziridinecarboxylic acid	Azr
39	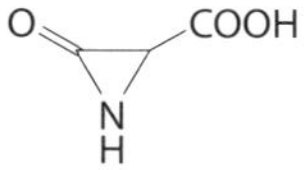	Aziridinonecarboxylic acid (6)	Azro
40	$(PhCH_2)_2C(NH_2)COOH$	(α-Benzyl)phenylalanine (7)	Phe(αBzl)
41	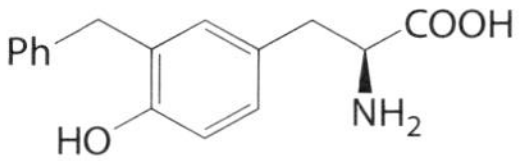	3-Benzyltyrosine (8)	Tyr(3Bzl)

STRUCTURE AND SYMBOLS FOR SYNTHETIC AMINO ACIDS INCORPORATED INTO SYNTHETIC POLYPEPTIDES (Continued)

No.	Structure	Name/Reference	Symbol
42	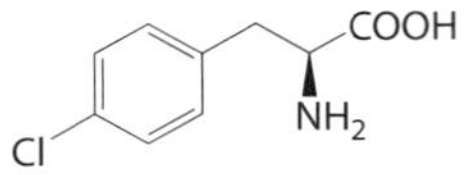	(4-Chloro)phenylalanine (5)	Phe(4Cl)
43	$H_2 NCONH(CH_2)_3 CH(NH_2)COOH$	Citrulline[b]	Ctr
44	$NCCH_2 CH(NH_2)COOH$	β-Cyanoalanine	Ala(βCN)
45	$NCCH_2 CH_2 NHCH_2 COOH$	N-(2-Cyanoethyl)glycine	(CNEt)Gly; CNEt-Gly
46	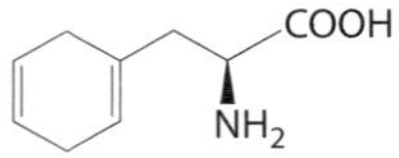	β-(1,4-Cyclohexadienyl)alanine (9) (dihydrophenylalanine)	Ala($βcHxΔ^1Δ^5$); Phe(H_2)
47	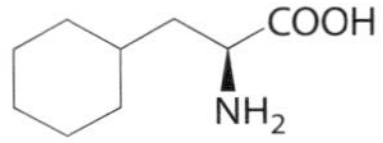	β-(Cyclohexyl)alanine (10, 20) (hexahydrophenylalanine)	Ala(βcHx); Phe(H_6)
48	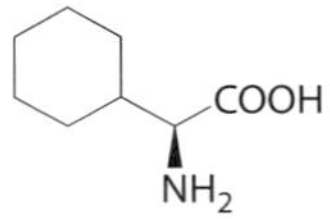	α-(Cyclohexyl)glycine	Gly(cHx)
49	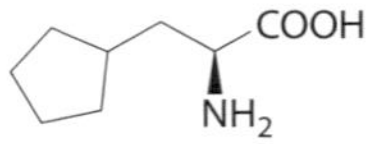	β-(Cyclopentyl)alanine	Ala(βcPe)
50	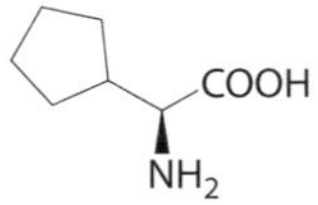	α-(Cyclopentyl)glycine	Gly(cPe)
51	$H_2 NCH_2 CH_2 CH(NH_2)COOH$	2,4-Diaminobutyric acid	A_2 bu [c]
52	$H_2 N(CH_2)_3 CH(NH_2)CH_2 COOH$	3,6-Diaminohexanoic acid (isolysine;[b] β-lysine[b])	βε A_2 hx; βLys
53	$H_2 NCH_2 C≡CCH_2 CH(NH_2)COOH$	2,6-Diamino-4-hexynoic acid (11)	αεA_2 hx($▲^γ$)
54	$HOOCCH(NH_2)(CH_2)_3 CH(NH_2)COOH$	2,2′-Diaminopimelic acid	A_2 pm [c]
55	$H_2 NCH_2 CH(NH_2)COOH$	2,3-Diaminopropionic acid	A_2 pr [c]
56	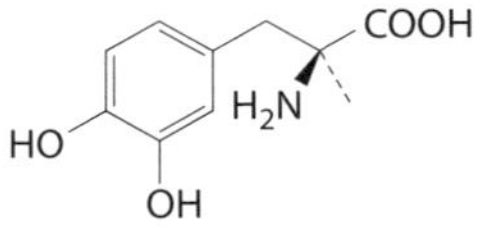	3,4-Dihydroxy-(α-methyl)phenylalanine [β-(3,4-Dihydroxyphenyl)-α-methylalanine]	Dopa(αMe)
57	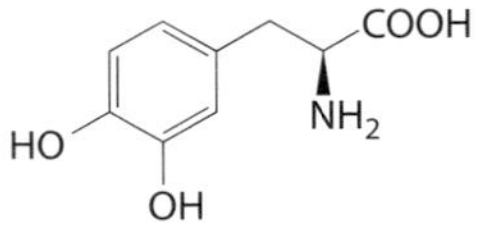	3,4-Dihydroxyphenylalanine[b]	Dopa [b]

STRUCTURE AND SYMBOLS FOR SYNTHETIC AMINO ACIDS INCORPORATED INTO SYNTHETIC POLYPEPTIDES (Continued)

No.	Structure	Name/Reference	Symbol
58	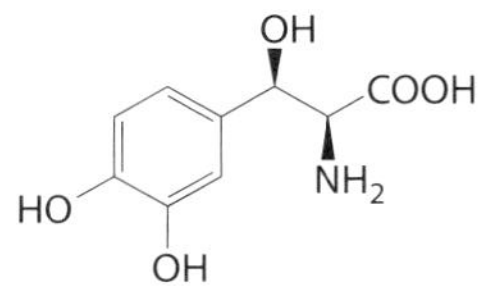	(3,4-Dihydroxyphenyl)serine	Dopa(βHO)
59	$HOOCCH_2\ CH(Me_2\ N{\rightarrow}O)COOH$	N,N-Dimethylaspartic N-oxide (12)	(O,Me$_2$)Asp; Me$_2$(O)Asp
60	$(CH_3)_2\ C(SH)CH(NH_2)COOH$	β,β-Dimethylcysteine (β-mercaptovaline; penicillamine[b])	Val(βSH); Cys(βMe$_2$)
61	$(CH_3)_2\ CHCH(CH_3)CH(NHCH_3)COOH\ CH_3$	*threo*-N, β-Dimethylleucine (13)	MeLeu(βMe)
62	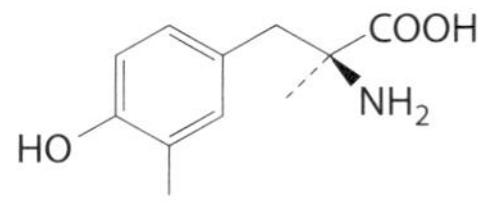	α,3-Dimethyltyrosine	Tyr(α,3-Me$_2$)
63	$(C_6H_5)_2\ C(NH_2)COOH$	α,α-Diphenylglycine	Gly(Ph$_2$)
64	$CH_3\ CH_2\ SCH_2\ CH_2\ CH(NH_2)COOH$	Ethionine[b]	Eth
65	$CH_3\ CH_2\ NHCH_2\ COOH$	N-Ethylglycine	EtGly
66	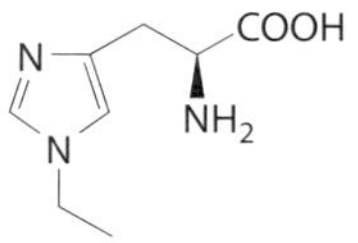	*tele*Ethylhistidine;[b,c] "1-Ethylhistidine" (14) (cf. 88, 89)	His(τEt)[b,c]
67	$H_2\ NC(=NH)NHCH_2\ CO{-}$	Guanidinoacetyl (N-amidinoglycyl; glycocyamine)	GdnAc-; AmdGly–
68	$H_2\ NC(=NH)NHCH_2\ CH(NH_2)COOH$	β-Guanidinoalanine	Ala(βGdn)
69	$H_2\ NC(=NH)NH(CH_2)_4\ COOH$	5-Guanidinovaleric acid	Vlr(δGdn)
70	$(CF_3)_2\ CHCH(NH_2)COOH$	χ-Hexafluorovaline [β,β-bis(trifluoromethyl)alanine]	Val(γF$_6$)
71	$H_2\ NC(=HN)NH(CH_2)_4\ CH(NH_2)COOH$	Homoarginine[b]	Har[c]
72	$H_2\ N(CH_2)_5\ CH(NH_2)COOH$	Homolysine[b] (15)	Hly[c]
73	$H_2\ NCONHCH_2\ COOH$	Hydantoic acid; (carbamoylglycine)	CbmGly
74	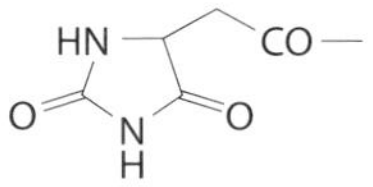	5-Hydantoinacetyl	HydAc-
75	$CH_3\ CH(OH)CH_2\ CH_2\ CH(NH_2)COOH$	ε-Hydroxynorleucine (16)	Nle(εOH)
76	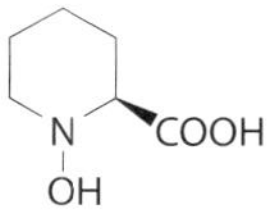	1-Hydroxypipecolic acid (17)	Pip(1HO)

STRUCTURE AND SYMBOLS FOR SYNTHETIC AMINO ACIDS INCORPORATED INTO SYNTHETIC POLYPEPTIDES (Continued)

No.	Structure	Name/Reference	Symbol
77	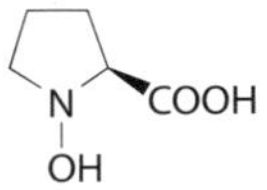	1-Hydroxyproline (17)	Pro(1HO)[c]
78	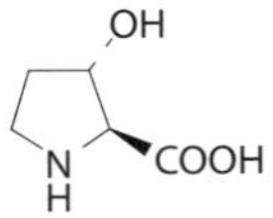	3-Hydroxyproline	Pro(3HO)[c]
79	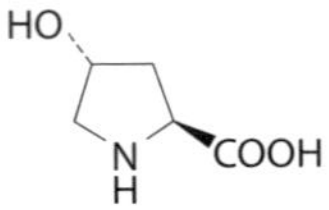	4-Hydroxyproline	Pro(4HO)[c]
80	$CH_3 N(OH)CH_2 COOH$	N-Hydroxysarcosine (17)	Sar(N-HO)
81	$HOOCCH_2 CH(NH_2)CONH_2$	Isoasparagine[b]	Asp-NH$_2$[c]
82	$HOOCCH_2 CH_2 CH(NH_2)CONH_2$	Isoglutamine[b]	Glu-NH$_2$[c]
83	$H_2 NN(CHMe_2)COOH$	2-Isopropylcarbazic acid [2-(1-carboxyhydrazino)propane]	Hdz(iPr)
84	$(CH_3)_2 CHNH(CH_2)_3 CH(NH_2)COOH$	N$^\delta$-Isopropylornithine	Orn(δPr)
85	$CH_3 C(SH)(NH_2)COOH$	α-Mercaptoalanine (18)	Ala(αSH)
86	$CH_3 CH(NHCH_3)COOH$	N-Methylalanine (19)	MeAla
87	$CH_3 CH_2 CH(CH_3)CH(NHCH_3)COOH$	N-Methylalloisoleucine (20)	(Me)alle
88	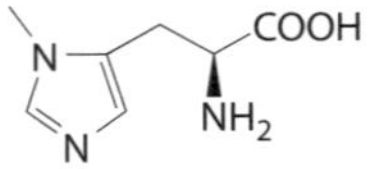	*tele*Methylhistidine;[b,c] "1-Methylhistidine" (14, 21) (cf. 89)	His(τMe)[b,c]
89	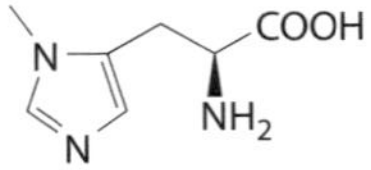	*pros*Methylhistidine;[b,c] "3-Methylhistidine" (21) (cf. 88)	His(πMe)[b,c]
90	$HOOCCH_2 CH(NHCH_3)CONH_2$	N-Methylisoasparagine (22)	MeAsp-NH$_2$
91	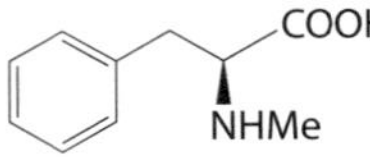	(N-Methyl)phenylalanine (23)	MePhe
92	$CH_3 OCH_2 CH(NH_2)COOH$	O-Methylserine	Ser(Me)[c]
93	$CH_3 CH(OCH_3)CH(NH_2)COOH$	O-Methylthreonine	Thr(Me)[c]
94	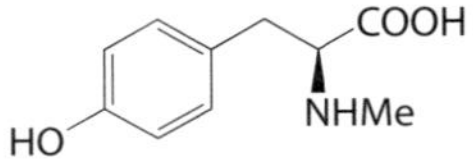	N-Methyltyrosine (22) (surinamine)[b]	MeTyr

STRUCTURE AND SYMBOLS FOR SYNTHETIC AMINO ACIDS INCORPORATED INTO SYNTHETIC POLYPEPTIDES (Continued)

No.	Structure	Name/Reference	Symbol
95	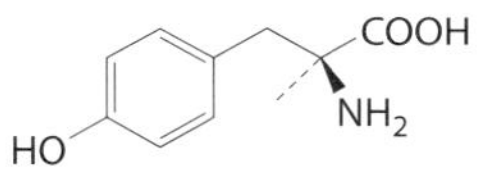	α-Methyltyrosine	Tyr(αMe)
96	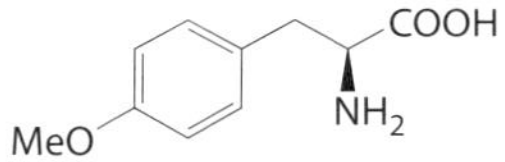	O-Methyltyrosine	Tyr(O^4Me); Phe(4-OMe)
97	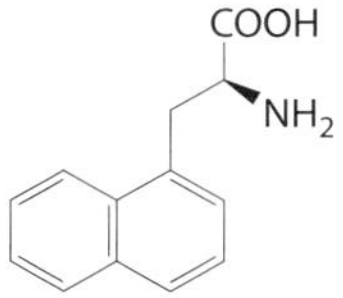	β-(1-Naphthyl)alanine	Ala(βNap-1)
98	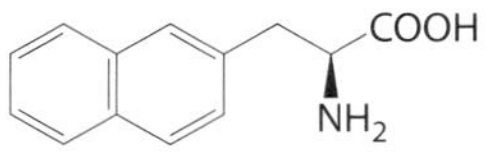	β-(2-Naphthyl)alanine	Ala(βNap-2)
99	O_2 NHNC(=NH)NHCH$_2$ CO—	Nitroguanidinoacetyl	NGdnAc-
100	CH$_3$ (CH$_2$)$_3$ CH(NH$_2$)COOH	Norleucine[b] (2-aminohexanoic acid)	Nle[c]
101	CH$_3$ CH$_2$ CH$_2$ CH(NH$_2$)COOH	Norvaline (2-aminovaleric acid)	Nva[c]
102	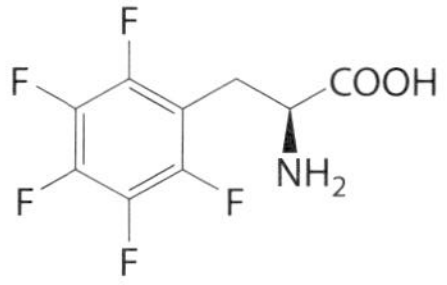	(Pentafluorophenyl)alanine	Ala(βPhF$_5$)
103	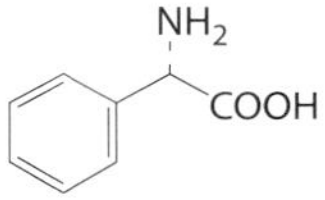	Phenylglycine	Gly(Ph)
104	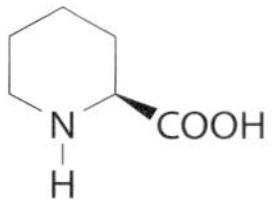	Pipecolic acid (piperidine-2-carboxylic acid)	Pip
105	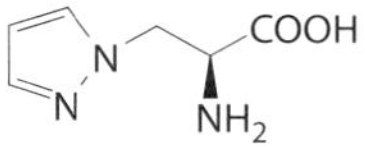	β-(1-Pyrazolyl)alanine	Ala(βPz1)
106	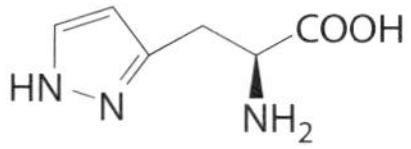	β-(3-Pyrazolyl)alanine (24, 25)	Ala (βPz3)

STRUCTURE AND SYMBOLS FOR SYNTHETIC AMINO ACIDS INCORPORATED INTO SYNTHETIC POLYPEPTIDES (Continued)

No.	Structure	Name/Reference	Symbol
107	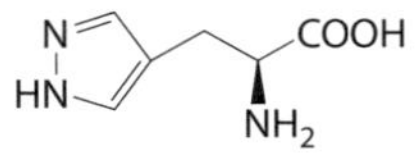	β-(4-Pyrazolyl)alanine (25)	Ala(βPz4)
108	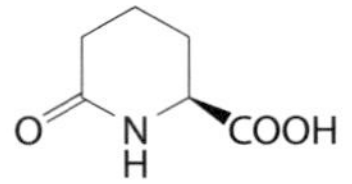	Pyro-2-aminoadipic acid	pAad; < Aad
109	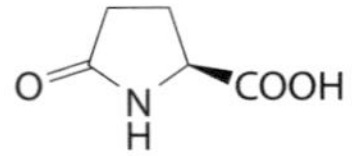	Pyroglutamic acid[b] 5-pyrrolidone-2-carboxylic acid	pGlu; <Glu[c]
110	CH_3 HNCH$_2$ COOH	Sarcosine[b]; (N-methylglycine)	Sar[c]; MeGly
111	HOOCCH$_2$ CH$_2$ CONH$_2$	Succinamic acid	Suc-NH$_2$
112	H$_2$ NCH$_2$ CH$_2$ SO$_3$ H	Taurine (2-aminoethanesulfonic acid)	Tau
113	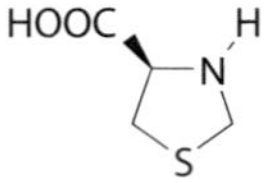	Thiazolidine-4-carboxylic acid	Tzl
114	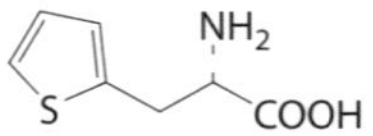	β-(2-Thienyl)alanine	Ala(βThi2)
115	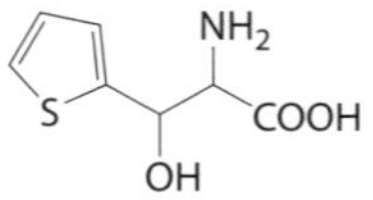	β-(2-Thienyl)serine	Ser(βThi2)
116	$(CH_3)_3 \overset{+}{N}(CH_2)_5$ COOH	ε-(Trimethylammonio)hexanoic acid [(ε-trimethylammonio)caproic acid]	εAhx(N$^\varepsilon$Me$_3$)
117	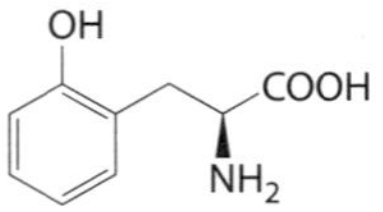	o-Tyrosine	Phe(2HO)
118	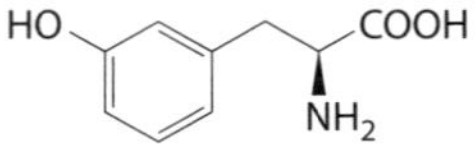	m-Tyrosine[b]	Phe(3HO)

Compiled by M. C. Khosla and W. E. Cohn.

References

1. Jansen, Weustink, Kerling, and Havinga, *Rec. Trav. Chim. Pays-Bas,* 88, 819 (1969).

2. Umezawa, Aoyagi, Morishima, Matsuzaki, Hamada, and Takeuchi, *J. Antiobiot. (Tokyo),* 23, 259 (1970).

3. Morishima, Takita, and Umezawa, *J. Antibiot.* (Tokyo), 26, 115 (1973).

4. Edelson, Skinner, Ravel, and Shive, *J. Am. Chem. Soc.,* 81, 5150 (1959).

5. Houghten and Rapoport, *J. Med. Chem.,* 17, 556 (1974).

6. Miyoshi, *Bull. Chem. Soc. Jap.,* 46, 1489 (1973).

7. Rydon, *J. Chem. Soc. Perkin Trans. 1,* 2634 (1972).

8. Erickson and Merrifield, *J. Am. Chem. Soc.,* 95, 3750 (1973).

9. Nagarajan, Diamond, and Ressler, *J. Org. Chem.,* 38, 621 (1973).

10. Khosla, Leese, Maloy, Ferreira, Smeby, an Bumpus, *J. Med. Chem.,* 15, 792 (1972).

11. Jansen, Kerling, and Havinga, *Rec. Trav. Chim. Pays-Bas,* 89, 861 (1970).

12. Ikutani, *Bull. Chem. Soc. Jap.,* 43, 3602 (1970).

13. Sheehan and Ledis, *J. Am. Chem. Soc.,* 95, 875 (1973).

14. Beyerman, Maat, and Van Zon, *Rec. Trav. Chim. Pays-Bas,* 91, 246 (1972).

15. Bodanszky and Lindeberg, *J. Med. Chem.,* 14, 1197 (1971).

16. Dreyfuss, *J. Med. Chem.,* 17, 252 (1974).

17. Nagasawa, Kohlhoff, Fraser, and Mikhail, *J. Med. Chem.,* 15, 483 (1972).

18. Patel, Currie, Jr., and Olsen, *J. Org. Chem.,* 38, 126 (1973).

19. Khosla, Hall, Smeby, and Bumpus, *J. Med. Chem.,* 17, 431 (1974).

20. Khosla, Hall, Smeby, and Bumpus, *J. Med. Chem.,* 17, 1156 (1974).

21. Needleman, Marshall, and Rivier, *J. Med. Chem.,* 16, 968 (1973).

22. Khosla, Smeby, and Bumpus, Abstr. 169th Natl. Meet. Am. Chem. Soc. Philadelphia, April 1975, MEDI 57.

23. Khosla, Smeby, and Bumpus, *J. Am. Chem. Soc.,* 94, 4721 (1972).

24. Hofmann and Bowers, *J. Med. Chem.,* 13, 1099 (1970).

25. Seeman, McGandy, and Rosenstein, *J. am. Chem. Soc.,* 94, 1717 (1972).

26. Khosla, Smeby, and Bumpus, in *Handbook of Experimental Pharmacology,* Vol. 37, Page and Bumpus, Eds., Springer-Verlag, Heidelberg, 1974, 126.

UNNATURAL AMINO ACIDS FOR INCORPORATION INTO PROTEINS

Amino acids can be divided into two groups. Proteinogenic amino acids are defined as amino acids which can be incorporated into proteins.[1–4] Proteogenic amino acids are differentiated from amino acids which are subject to post translational modification[5,6] such as the γ-carboxylation of glutamic acid. Non proteinogenic (non protein) amino acids[7] such as cyclopentenylglycine[8] and L-p-hydroxyphenylglycine[9] are involved in metabolism and in peptide antibiotics. There has been considerable interest in the development of unnatural amino acids. While the broad definition of an unnatural amino acid is any synthetic organic carboxylic acid which contains an amino or imino function, the definition for this section is that an unnatural amino acid must be capable of incorporation into a protein in a specific manner using a biological system.[10–17]

References

1. Hardy, P.M., The protein amino acids, in *Chemistry and Biochemistry of the Amino Acids,* ed. G.C. Barrett, Chapman & Hall, London, UK, Chapter 2, pps. 6-24, 1985
2. Szyperski, T., Biosynthetically directed fractional ^{13}C-labeling of proteinogenic amino acid. No efficient analytical tool to investigate intermediary metabolism, *Eur.J.Biochem.* 232, 433-448, 1995
3. Seebach, D., Beck, A.M., and Bierbaum, D.J., The world of β- and γ-peptides comprised of homologated proteinogenic amino acids and other components, *Chem.Biodivers.* 1, 1111-1239, 2004
4. Chan, W.C., Higton, A., and Davis, J.J., Amino acids, in *Amino Acids, Peptides, and Proteins,* Vol. 35, ed. J.S. Davies, RSC Publishing, Cambridge, UK, pps. 1-73, 2006
5. Walsh, C.T., Garneau-Tsodikova, S., and Gatto, G.J., Jr., Protein post-translational modification: the chemistry of proteome diversification, *Angew.Chem.Int.Ed.Engl.* 44, 7342-7373, 2005
6. Walsh, G. and Jeffries, R., Post-translational modifications in the context of therapeutic proteins, *Nat.Biotechnol.* 24, 1241-1252, 2006.
7. Hunt, S., The non-protein amino acids, in *Chemistry and Biochemistry of the Amino Acids,* ed. G.C. Barrett, Chapman & Hall, London, UK, Chapter 5, pps. 55-138, 1985
8. Cramer, U. and Spener, F. Biosynthesis of cyclopentenyl fatty acids. Cyclopentenylglycine, a non-proteinogenic amino acids as precursor of cyclic fatty acids in *Flacourtiaceae, Eur.J.Biochem.* 94, 495-500, 1977
9. Hubbard, B.K., Thomas, M.G., and Walsh, C.T., Biosynthesis of L-p-hydroxyphenylglycine, a non-proteinogenic amino acid constituent of peptide antibiotics, *Chem.Biol.* 7, 931-942, 2000
10. Brookes, P., Studies on the incorporation of an unnatural amino acid, p-di-(2-hydroxy[^{14}C$_2$]ethyl)amino-L-phenylalanine into proteins, *Brit. J.Cancer* 13, 313-317, 1959
11. Anthony-Cahill, S.J., Griffith, M.C., Noren, C.J., *et al.,* Site-specific mutagenesis with unnatural amino acids, *Trends Biochem.Sci.* 14, 400-403, 1989
12. Bain, J.D., Switzer, C., Chamberlin, A.R., and Benner, S.A., Ribosome-mediated incorporation of a non-standard amino acid into a peptide through expansion of the genetic code, *Nature* 356, 537-539, 1992
13. Noren, C.J., Anthony-Cahill, S.J., Suich, D.J., *et al., In vitro* expression of an amber mutation by a chemically aminoacylated transfer RNA prepared by runoff transcription, *Nucleic Acids Res.* 18, 83-88, 1990
14. Ibba, M. and Hennecke, H., Relaxing the substrate specificity of an aminoacyl-tRNA synthetase allows *in vitro* and *in vivo* synthesis of proteins containing unnatural amino acids, *FEBS Lett.* 364, 272-275, 1995
15. Bacher, J.M. and Ellington, A.D., The directed evolution of organismal chemistry: Unnatural amino acids and incorporation, in *Translation Mechanisms,* ed. J.Lapointe and L. Brackier-Gingras, Landes Bioscienc/Kluwer Academic, New York, NY, USA, Chapter 5, pps. 80-94, 2003
16. Magliery, T.J., Pastrnak, M., Anderson, J.C., In vitro tools and in vivo engineering: Incorporation of unnatural amino acids into proteins, in *Translation Mechanisms,* ed. J.Lapointe and L. Brackier-Gingras, Landes Bioscienc/Kluwer Academic, New York, NY, USA, Chapter 6, pps. 95-114, 2003
17. Liu, W., Brock, A., Chen, S., *et al.,* Genetic incorporation of unnatural amino acids into proteins in mammalian cells, *Nat.Methods* 4, 239-244, 2007

	Unnatural Amino Acid	Structure	References
1	*p*-Propargyloxyphenylalanine	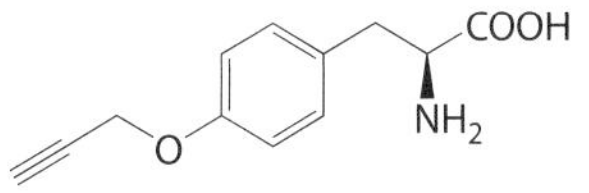	1–3
2	*p*-Azidophenylalanine	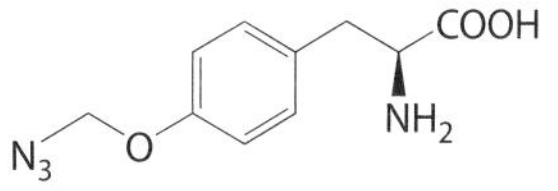	1,4,5
3	4-Fluorotryptophan	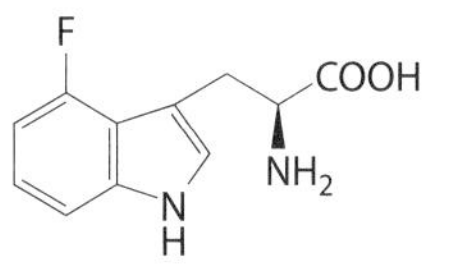	6–11
4	Azaleucine	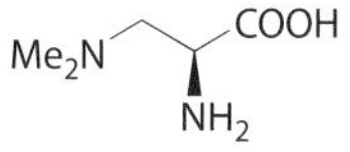	12
5	*p*-Fluorophenylalanine	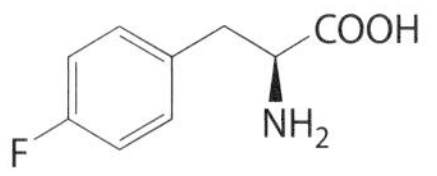	13–21
6	*O*-Methyltyrosine	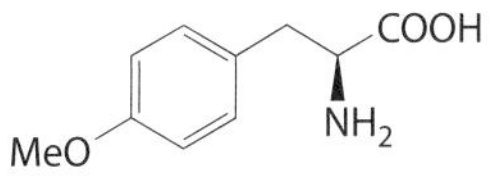	22–24

UNNATURAL AMINO ACIDS FOR INCORPORATION INTO PROTEINS (Continued)

	Unnatural Amino Acid	Structure	References
7	*p*-Acetylphenylalanine or *m*-acetylphenylalanine		25–28
8	*p*-Iodophenylalanine		29
9	*p*-Benzoylphenylalanine		30–31

References for table

1. Dieters, A., Cropp, A., Mukherji, M., *et al.*, Adding amino acids with novel reactivity to the genetic code of *Saccharomyces cerevisiae*, *J.Amer.Chem.Soc.* 125, 11782-11783, 2003

2. Dieters, A. and Schultz, P.G., In vivo incorporation of an alkyne into proteins in *Escherichia coli*, *Bioorg.Med.Chem.Lett.* 15, 1521-1524, 2005

3. Iida, S., Asakura, N., Tabata, K. *et al.*, Incorporation of unnatural amino acids into cytochrome c3 and specific viologen binding to the unnatural amino acid, *Chembiochem* 7, 1853-1855, 2006

4. Dieters, A., Cropps, T.A., Summerer, D., *et al.*, Site-specific PEGylation of proteins containing unnatural amino acids, *Bioorg.Med.Chem.Lett.* 14, 5743-5745, 2004

5. Carrico, I.S., Mackarinec, S.A., Heilshorn, S.C., Lithographic patterning of photoreactive cell-adhesive proteins, *J.Am.Chem.Soc.*, in press, 2007

6. Pratt, E.A. and Ho, C., Incorporation of fluorotryptophans into proteins of *Escherichia coli*, *Biochemistry* 14, 3035-3040, 1975

7. Browne, D.T. and Otvos, J.D., 4-Fluorotryptophan alkaline phosphatase from *E.coli*: preparation, properties, and ¹⁹F NMR spectrum, *Biochem.Biophys.Res.Commun.* 68, 907-913, 1976

8. Wong, C.Y. and Eftink, M.R., Incorporation of tryptophan analogues into staphylococcal nuclease: stability toward thermal and guanidine-HCl induced unfolding, *Biochemistry* 37, 8947-8953, 1998

9. Zhang, Q.S., Shen, L., Wang, E.D., and Wang, Y.L., Biosynthesis and characterization of 4-fluorotryptophan-labeled *Escherichia coli* arginyl tRNA synthetase, *J.Protein Chem.* 18, 187-192, 1999

10. Mohammed, F., Prentice, G.A., and Merrill, A.R., Protein-protein interaction using tryptophan analogues: novel spectroscopic probes for toxin-elongation factor-2 interactions, *Biochemistry* 40, 10273-10283, 2001

11 Bacher, J.D. and Ellington, A.D., Selection and characterization of *Escherichia coli* variants capable of growth on an otherwise toxic tryptophan analogue, *J.Bacteriol.* 183, 5414-5425, 2001

12. Lemeigan, B. Sonigo, P. and Marliere, P. Phenotypic suppression by incorporation of an alien amino acid, *J.Mol.Biol.* 231, 161-166, 1993

13. Dunn, T.F. and Leach, F.R., Incorporation of *p*-fluorophenylalanine into protein by a cell-free system, *J.Biol.Chem.* 242, 2693-2699, 1967

13. Rels, P.J. and Gillespie, J.M., Effects of phenylalanine and analogues of methionine and phenylalanine on the composition of wool and mouse hair, *Aust.J.Biol.Sci.* 38, 151-163, 1985

15. Kast, P. and Hennecke, H., Amino acid substrate specificity of *Escherichia coli* phenylalanyl-tRNA synthetase altered by distinct mutations, *J.Mol.Biol.* 222, 99-124, 1991

16. Lian, C., Le, H., Montez, B., *et al.*, Fluorine-19 nuclear magnetic resonance spectroscopic study of fluorophenylalanine and fluorotryptophan-labeled avian egg white lysozymes, *Biochemistry* 33, 5238-5245, 1994

17. Danielson, M.A., Biemann, M.P., Koshland, D.E.,Jr., and Falke, J.J., Attractant- and disulfide-induced conformational changes in the ligand binding domain of the chemotaxis aspartate receptor: a ¹⁹F NMR study, *Biochemistry* 33, 6100-6109, 1994

18 Furter, R., Expansion of the genetic code: site-directed *p*-fluoro-phenylalanine incorporation in *Escherichia coli*, *Protein Sci.* 7, 419-426, 1998

19. Minka, C., Nuber, R., Moroder, L., and Budisa, N., Noninvasive tracing of recombinant proteins with "fluorophenylalanine-fingers", *Anal. Biochem.* 284, 29-34, 2000

20. Bann, J.G. and Frieden, C., Folding and domain-domain interactions of the chaperone PapD measured by ¹⁹F NMR, *Biochemistry* 43, 13775-13786, 2004

21. Jackson, J.C., Hammill, J.T. , and Mehl, R.A., Site-specific incorporation of a ¹⁹F-amino acid into proteins as an NMR probe for characterizing protein structure and reactivity, *J.Am.Chem.Soc.* 129, 1160-1166, 2007

22. Wang, L., Brock, A., and Schultz, P.G., Adding L-3-(2-naphthalyl)alanine to the genetic code of *E.coli*, *J.Amer.Chem.Soc.* 124, 1836-1837, 2002

23 Zhang, D., Valdehi, N., Goddard, W.A., Jr., *et al.*, Structure-based design of mutant *Methanococcus jannaschii* tyrosyl-tRNA synthetase for incorporation of O-methyl-L-tyrosine, *Proc.Natl.Acad.Sci.USA* 99, 6579-6584, 2002

24. Zhang, Y, Wang, L., Schultz, P.G., and Wilson, I.A., Crystal structures of apo wild-type *M. jannashchii* tyrosyl-tRNA synthetase (TyrRS) and an engineered TyrRS specific for O-methyl-L-tyrosine, *Protein Sci.* 14, 1340-1349, 2005

25 Wang, L., Zhang, Z., Brock, A., and Schultz, P.G., Addition of the keto functional group to the genetic code of *Escherichia coli*, *Proc.Natl. Acad.Sci.USA* 100, 56-61, 2003

26. Zhang, Z., Smith, B.A., Wang, L., *et al.*, A new strategy for the site-specific modifications of proteins *in vivo*, *Biochemistry* 42, 6735-6746, 2003

27 Wu, N., Dieters, A., Cropp, T.A., *et al.*, A genetically encoded photo-caged amino acid, *J.Am.Chem.Soc.* 126, 14306-14307, 2004

28. Taki, M. and Sisido, M., Leucyl/phenylalanyl (L/F)-tRNA-protein transferase-mediated aminoacyl transfer of a nonnatural amino acid to the N-terminal of peptides and proteins and subsequent functionalization by bioorthogonal reactions, *Biopolymers* 88, 263-271, 2007

29 Black, K.M., Clark-Lewis, I., and Wallace, C.J., Conserved tryptophan in cytochrome c: importance of the unique side-chain features of the indole moiety, *Biochem.J.* 359, 715-720, 2001

30 Chin, J.W., Cropp, T.A., Anderson, J.C., *et al.*, An expanded eukaryotic genetic code, *Science* 301, 964-967, 2003

31 Farrell, I.S., Tornoney, R., Hazen, J.L., *et al.*, Photo-crosslinking interacting proteins with a genetically encoded benzophenone, *Nat. Methods* 2, 377-384, 2005

PROPERTIES OF THE *α*-KETO ACID ANALOGS OF AMINO ACIDS

| α-Keto Acid | α-Amino Acid Analog | 2,4-Dinitrophenylhydrazone | | Amino Acids After Hydrogenation (14) | Reduction by Lactic Dehydrogenase[b,g] | Decarboxylation by Yeast Decarboxylase[c] |
| | | Crystallization | | | | |
		M.p. (°C)	Solvent[a]			
Pyruvic	Alanine	216	h	Alanine	26,800	1,200
α-Ketoadipamic	α-Aminoadipamic acid (homoglutamine)					
α-Ketoadipic	α-Aminoadipic acid	208	h	α-Aminoadipic acid		
α-Ketobutyric	α-Aminobutyric acid	198	h	α-Aminobutyric acid	21,000	
α-Ketoheptylic	α-Aminoheptylic acid	130	e, l	α-Aminoheptylic acid	483	
α-Keto-ε-hydroxycaproic	α-Amino-ε-hydroxy-caproic acid	183	h	α-Amino-ε-hydroxy-caproic acid	181	25
Mesoxalic	α-Aminomalonic acid	205	hc	α-Aminomalonic acid, glycine		
α-Ketophenylacetic	α-Aminophenylacetic acid	193	h	α-Aminophenylacetic acid, cyclohexylglycine	2.6	0
DL-Oxalosuccinic	α-Aminotricarballylic acid					
α-Keto-δ-guanidinovaleric	Arginine	216, 267 (1)		Arginine	9.0	0
α-Ketosuccinamic	Asparagine	183		Asparagine, aspartic acid	8,930	
Oxalacetic	Aspartic acid	218	h	Aspartic acid, alanine, β-alanine	12.8	
α-Keto-δ-carbamidovaleric	Citrulline	190	h	Citrulline	4.3	0
β-Cyclohexylpyruvic	β-Cyclohexylalanine	189	h	β-Cyclohexylalanine	14.6	0
β-Sulfopyruvic	Cysteic acid	210	a	Cysteic acid, alanine	89.6	
β-Mercaptopyruvic	Cysteine	195–200 (2) 161–162 (3)		Alanine	27,000	750
α-Keto-γ-ethiolbutyric	Ethionine	131	h	Ethionine	1,650	121
α-Ketoglutaric	Glutamic acid	220	h	Glutamic acid	9.2	0
α-Ketoglutaric γ-ethyl ester	Glutamic acid γ-ethyl ester				49.0	721
α-Ketoglutaramic	Glutamine			Glutamine, glutamic acid		
Glyoxylic	Glycine	203	h	Glycine	21,100	0
β-Imidazolylpyruvic	Histidine	190–192, 240 (1)	hc, e, l	Histidine		
α-Keto-γ-hydroxybutyric	Homoserine					
DL-α-Keto-β-methylvaleric	DL-Isoleucine (or DL-alloisoleucine)	169	h	Isoleucine		
L-α-Keto-β-methylvaleric	L-Isoleucine (or D-alloisoleucine)	176	h	Isoleucine	5.0	1,000
D-α-Keto-β-methylvaleric	D-Isoleucine (or L-alloisoleucine)	176	h	Isoleucine	1.9	280
α-Ketoisocaproic	Leucine	162	h	Leucine	3.2	306
Trimethylpyruvic	tert-Leucine	180	h	tert-Leucine		
α-Keto-ε-aminocaproic	Lysine	212	h	Lysine, pipecolic acid		
α-Keto-γ-methiolbutyric	Methionine	150	h	Methionine	1,550	125
α-Keto-γ-methylsulfonylbutyric	Methionine sulfone	175	h	Methione sulfone		
α-Keto-δ-nitroguanidinovaleric	Nitroarginine	225	ac	Nitroarginine, arginine	42.6	0
α-Ketocaproic	Norleucine	153	h	Norleucine	560	
α-Ketovaleric	Norvaline	167	h	Norvaline	1,470	
α-Keto-δ-aminovaleric	Ornithine, proline	232–242 (4) 219 (5) 211–212 (6)		Ornithine, proline, pentahomoserine[d]		
Phenylpyruvic	Phenylalanine	162–164. 192–194 (7)		Phenylalanine	755	0

55

PROPERTIES OF THE α-KETO ACID ANALOGS OF AMINO ACIDS (Continued)

α-Keto Acid	α-Amino Acid Analog	2,4-Dinitrophenylhydrazone		Amino Acids After Hydrogenation (14)	Reduction by Lactic Dehydrogenase[b,g]	Decarboxylation by Yeast Decarboxylase[c]
		Crystallization				
		M.p. (°C)	Solvent[a]			
S-Benzyl-β-mercaptopyruvic	S-Benzylcysteine	150	a			
β-Hydroxypyruvic	Serine	162	e	Serine, alanine	26,000	0[h]
N-Succinyl-α-amino-ε-ketopimelic (15)	N-Succinyl-α,ε-diaminopimelic acid	137–143	h	N-Succinyl-α,ε-diaminopimelic acid		
DL-α-Keto-β-hydroxybutyric	DL-Threonine (or DL-allothreonine)	157–158 (8)		Threonine, α-aminobutyric acid	20,000	
β-[3,5-Diiodo-4-(3′,5′-diiodo 4′-hydroxyphenoxy) phenyl]	Thyroxine					
β-Indolylpyruvic	Tryptophan	169 (1)		Tryptophan	670	0
p-Hydroxyphenylpyruvic	Tyrosine	178	h	Tyrosine	345	0
α-Ketoisovaleric	Valine	196	h	Valine	103	922

[a] h = water; e = ethyl acetate; l = ligroin; hc = hydrochloric acid; ac = glacial acetic acid; a = ethanol.

[b] Mole × 10^{-8} of DPNH oxidized per mg of enzyme per minute at 26° (9, 10).

[c] μL. CO_2 per hour (10).

[d] α-Amino-δ-hydroxy-n-valeric acid.

[e] Originally designated d (11). Originally designated l (11).

[g] Additional data have been published on the reduction of α-keto acids by lactic dehydrogenase (12).

[h] This keto acid has been reported to be decarboxylated by yeast preparations; the reaction is much more rapid at pH 6.3 than at 5 (13).

From Meister, *Biochemistry of the Amino Acids*, 2nd ed., Academic Press, New York, 1965, 162–164. With permission.

References

1. Stumpf and Green, *J. Biol. Chem.*, 153, 387 (1944).
2. Schneider and Reinefeld, *Biochem. Z.*, 318, 507 (1948).
3. Meister, Fraser, and Tice, *J. Biol. Chem.*, 206, 561 (1954).
4. Krebs, *Enzymologia*, 7, 53 (1939).
5. Blanchard, Green, Nocito, and Ratner, *J. Biol Chem.*, 155, 421 (1944).
6. Meister, *J. Biol. Chem.*, 206, 579 (1954).
7. Fones, *J. Org. Chem.*, 17, 1534 (1952).
8. Sprinson and Chargaff, *J. Biol. Chem.*, 164, 417 (1947).
9. Meister, *J. Biol. Chem.*, 197, 309(1953).
10. Meister, *J. Biol. Chem.*, 184, 117(1950).
11. Meister, *J. Biol. Chem.*, 190,269 (1951).
12. Czok and Büchler, *Advanc. Protein. Chem.*, 15, 315 (1960).
13. Dickens and Williamson, *Nature*, 178, 1349 (1956).
14. Meister and Abendschein, *Anal. Chem.*, 28, 171 (1956).
15. Gilvarg, *J. Biol. Chem.*, 236, 1429(1961).

α,β-UNSATURATED AMINO ACIDS

α,β-Unsaturated amino acids with free α-amino groups are not stable; N-acylated α,β-unsaturated amino acids are stable compounds. The α,β-unsaturated amino acids listed in this table are present in natural products in which they are stabilized by peptide bond formation. The addition of mercaptans to α,β-unsaturated amino acids and the reversible conversion to keto acids and amides are of biological significance.[1,2,11]

No.	Amino Acid/Synonym	Source	Structure	Formula (Mol Wt)	References Chemistry	References Spectral Data
1	**Dehydroalanine**	Nisin Subtilin	$H_2C=C(NH_2)COOH$	$C_3H_5NO_2$ 87.08	1, 2 2	1, 2 2
2	β-**Methyldehydroalanine**	Nisin, Subtilin Stendomycin	$H_3CCH=C(NH_2)COOH$	$C_4H_7NO_2$ 101.10	1, 2 3	1, 2 3
3	**Dehydroserine**	Viomycin Capreomycin	$HOHC=C(NH_2)COOH$	$C_3H_5NO_3$ 103.08	4 4	4 4
4	**Dehydroleucine**	Albonoursin	$(H_3C)_2CHCH=C(NH_2)COOH$	$C_6H_{11}NO_2$ 129.16	5	
5	**Dehydrophenylalanine**	Albonoursin	$PhCH=C(NH_2)COOH$	$C_9H_9NO_2$ 163.17	5	
6	**Dehydrotryptophan**	Telomycin		$C_{11}H_{10}N_2O_2$ 202.21	6	6
7	**Dehydroarginine**	Viomycin Capreomycin	$H_2NC(=NH)$ $NHCH_2CH_2CH=C(NH_2)$ $COOH$	$C_6H_{12}N_4O_2$ 172.19	4 4	4 4
8	**Dehydroproline**	Ostreogrycin A		$C_5H_7NO_2$ 113.11	7	7
9	**Dehydrovaline**	Penicillin	$(H_3C)_2C=C(NH_2)COOH$	$C_5H_9NO_2$ 115.13	8	
10	**Dehydrocysteine**	Micrococcin Thiostrepton	$HSCH=C(NH_2)COOH$	$C_3H_5NO_2S$	9, 10	10

Compiled by Erhard Gross.

References

1. Gross and Morell, *J. Am. Chem. Soc.*, 89, 2791 (1967).
2. Gross and Morell, *Fed. Eur. Biochem. Soc. Lett.*, 2, 61 (1968).
3. Bodanszky, Izdebski, and Muramatsu, *J. Am. Chem. Soc.*, 91, 2351 (1969).
4. Bycroft, Cameron, Croft, Hassanali-Walji, Johnson, and Webb, *Tetrahedron Lett.*, p. 5901 (1968).
5. Khokhlov and Lokshin, *Tetrahedron Lett.*, p.1881 (1963).
6. Sheehan, Mani, Nakamura, Stock, and Maeda, *J. Am. Chem. Soc.*, 90, 462 (1968).
7. Delpierre, Eastwood, Gream, Kingston, Sarin, Todd, and Williams, *J. Chem. Soc.*, p. 1653 (1966).
8. Abraham and Newton, in *Antibiotics*, Gottlieb and Shawn, Eds., Springer-Verlag, New York, 1967.
9. Brookes, Clarke, Majhofer, Mijovic, and Walker, *J. Chem. Soc.*, p. 925 (1960).
10. Bodanszky, Sheehan, Fried, Williams, and Birkhimer, *J. Am. Chem. Soc.*, 82, 4747 (1960).
11. Gross, Morell, and Craig, *Proc. Natl. Acad. Sci. U.S.A.*, 62, 953 (1969).

This table originally appeared in Sober, Ed., *Handbook of Biochemistry and Selected Data for Molecular Biology*, 2nd ed., Chemical Rubber Co., Cleveland, 1970.

AMINO ACID ANTAGONISTS

Amino Acid	Analog	System	Reference
α-Alanine	*α*-Aminoethanesulfonic acid	Bacteria	1
		Mouse tumor	2
	Glycine	Bacteria	3
	α-Aminoisobutyric acid	Bacteria	4
	Serine	Bacteria	3
D-Alanine	D-Cycloserine	Bacterial cell wall	5–10
	O-Carbamyl-D-serine	Bacterial cell wall	11
	D-*α*-Aminobutyric acid	Bacterial cell wall	12
β-Alanine	*β*-Aminobutyric acid	Yeast	13
	Propionic acid	Bacteria	14
	Asparagine	Yeast	15
	D-Serine	Bacteria	16
Arginine	Canavanine	Yeast, *Neurospora*, bacteria	17, 18-22
		Carcinosarcoma	23
		Animals	24, 25
		Plants	26, 27
		Tissue culture	28, 29
	Lysine	Arginase	30
	Ornithine	Arginase	31
	Homoarginine	Bacteria	22, 32
Aspartic acid	Cysteic acid	Bacteria	1, 33, 34
		Bacteria	35
	β-Hydroxyaspartic acid	Bacteria	34, 36, 37
	Diaminosuccinic acid	Bacteria	36
	Aspartophenone	Bacteria, yeast	4
	α-Aminolevulinic acid	Bacteria, yeast	4
	α-Methylaspartic acid	Bacteria	38
	β-Aspartic acid hydrazide	Bacteria	38
	S-Methylcysteine sulfoxide	Bacteria	39
	β-Methylaspartic acid	Bacteria	40
	Hadacidin	Purine biosynthesis	41
Asparagine	2-Amino-2-carboxyethanesulfonamide	*Neurospora*	42
Cysteine	Allylglycine	Bacteria, yeast	43
α,*ε*-Diaminopimelic acid	*α*,*α*-Diaminosuberic acid	Bacteria	44
	α,*α*-Diaminosebacic acid	Bacteria	44
	β-Hydroxy-*α*,*ε*-diaminopimelic acid	*Escherichia coli*	45
	γ-Methyl-*α*,*ε*-diaminopimelic acid	*E. coli*	46
	Cystine	*E. coli*	47
Glutamic acid	Methionine sulfoxide	Bacteria	48, 49
		Glutamine synthesis	50
	γ-Glutamylethylamide	Bacteria	51
	β-Hydroxyglutamic acid	Bacteria	52, 53
	Methionine sulfoximine	Bacteria	54, 55
	α-Methylglutamic acid	Enzymes	56, 57
	γ-Phosphonoglutamic acid	Glutamine synthesis	58
	P-Ethyl-*γ*-phosphonoglutamic acid	Glutamine synthesis	58
	γ-Fluoroglutamic acid[a]	Glutamine synthesis	59
Glutamine	*S*-Carbamylcysteine	Bacteria	60
		Ascites cells	61
	O-Carbamylserine	Bacteria	62
	O-Carbazylserine	Bacteria	63
	3-Amino-3-carboxypropanesulfonamide	*E. coli*	64
	N-Benzylglutamine	*Streptococcus lactis*	65
	Azaserine	Enzymes	66
	6-Diazo-5-oxonorleucine	Enzymes	66
	γ-Glutamylhydrazide	*S. faecalis*	67
Glycine	*α*-Aminomethanesulfonic acid	Bacteriophage	68
		Vaccinia virus	69
		Bacteria	1
		E. coli	4

AMINO ACID ANTAGONISTS (Continued)

Amino Acid	Analog	System	Reference
Histidine	D-Histidine	Histidase	70
	Imidazole		70
	2-Thiazolealanine	*E. coli*	71
	1,2,4-Triazolealanine	*E. coli*	71, 72
		Salmonella	73
Isoleucine	Leucine	Bacteria	74
		Rats	75
	Methallylglycine	Bacteria, yeast	4, 43
	ω-Dehydroisoleucine	Bacteria	76
	3-Cyclopentene-1-glycine	Bacteria	77
	Cyclopentene glycine	Bacteria	78
	2-Cyclopentene-1-glycine	Bacteria	79
	O-Methylthreonine	Tumor cells	80
	β-Hydroxyleucine	Bacteria	37
Leucine	D-Leucine	Bacteria	81
	α-Aminoisoamylsulfonic acid	Bacteria	1, 33
		Mouse tumor	2
	Norvaline	Bacteria	4
	Norleucine	Bacteria	1, 4, 82
	Methallylglycine	Yeast, bacteria	4
	α-Amino-β-chlorobutyric acid	Yeast, bacteria	4
	Valine	Bacteria	83
	δ-Chloroleucine	*Neurospora*	84
	Isoleucine	Bacteria	85
	β-Hydroxynorleucine	Bacteria	86
	β-Hydroxyleucine	Bacteria	86
	Cyclopentene alanine	Bacteria	87
	3-Cyclopentene-1-alanine	Bacteria	88
	2-Amino-4-methylhexenoic acid	Bacteria	89
	5′,5′,5′-Trifluoroleucine	*E. coli*	90
	4-Azaleucine	*E. coli*	91
Lysine	α-Amino-ε-hydroxycaproic acid	Rat	92
	Arginine	*Neurospora*	93
	2,6-Diaminoheptanoic acid	Bacteria	94
	Oxalysine	Bacteria	95
	3-Aminomethylcyclohexane glycine	Bacteria	96
	3-Aminocyclohexane alanine	Bacteria	97
	trans-4-Dehydrolysine	Bacteria	98
	S-(β-Aminoethyl)cysteine	Bacteria	99
	4-Azalysine	Bacteria	
Methionine	2-Amino-5-heptenoic acid (crotylalanine)	*E. coli*	100
	2-Amino-4-hexenoic acid (crotylglycine)	*E. coli*	101
	Methoxinine	Bacteria	102
		Vaccinia virus	69
		Rats	103
	Norleucine	Bacteria	82, 104, 105
		Animal tissues	106
		Casein	107
	Ethionine	Bacteria, animals	28, 102, 105, 108–120
		Amylase	121, 122
		Yeast	123
		Tumors	124
		Pancreatic proteins	125
	Methionine sulfoximine	Bacteria	126
	Threonine	*Neurospora*	127
	Selenomethionine	*Chlorella*	128
		E. coli, yeast	129–132
Ornithine	α-Amino-δ-hydroxyvaleric acid	Bacteria	21
	Canaline	Bacteria	21
Phenylalanine	α-Amino-β-phenylethanesulfonic acid	Mouse tumor	2
	Tyrosine	Bacteria	133
	β-Phenylserine	Bacteria	134–136
	Cyclohexylalanine	Rats	137
	o-Aminophenylalanine	*E. coli*	138
	p-Aminophenylalanine	Bacteria	139, 140

AMINO ACID ANTAGONISTS (Continued)

Amino Acid	Analog	System	Reference
	Fluorophenylalanines	Fungi, bacteria	140–148
		Lysozyme, albumin	149
		Muscle enzymes	150
		Amylase	151
		Hemoglobin	152
		Rats	153
Phenylalanine	Chlorophenylalanines	Fungi	147
	Bromophenylalanines	Fungi	147
	β-2-Thienylalanine	Rat, bacteria, yeast	136, 154–164
		β-Galactosidase	165
	β-3-Thienylalanine	Bacteria, yeast	166
	β-2-Furylalanine	Bacteria, yeast	4
	β-3-Furylalanine	Bacteria, yeast	4
	β-2-Pyrrolealanine	Bacteria, yeast	167
	1-Cyclopentene-1-alanine	Bacteria	87
	1-Cyclohexene-1-alanine	Bacteria	87
	2-Amino-4-methyl-4-hexenoic acid	Bacteria	89
	S-(1,2-Dichlorovinyl)cysteine	E. coli	168
	β-4-Pyridylalanine	Bacteria	169
	Tryptophan	Bacteria	170
	β-2-Pyridylalanine	Bacteria	171
	β-4-Pyrazolealanine	Bacteria	171
	β-4-Thiazolealanine	Bacteria	171
	p-Nitrophenylalanine	Bacteria	140
Proline	Hydroxyproline	Fungi	172
	3,4-Dehydroproline	Bacteria, beans	173, 174
	Azetidine-2-carboxylic acid	Bacteria, beans	175, 176
		Actinomycin	177, 178
Serine	α-Methylserine	Bacteria	4
	Homoserine	Bacteria	4
	Threonine	Bacteria	179, 180
	Isoserine	Enzymes	181
Threonine	Serine	Bacteria	179, 180, 182
	β-Hydroxynorvaline	Bacteria	86, 183
	β-Hydroxynorleucine	Bacteria	86, 183
Thyroxine	Ethers of 3,5-diiodotyrosine	Tadpoles	184
Tryptophan	Methyltryptophans	Bacteria	141, 185–188
		Bacteriophage	189
	Naphthylalanines	Bacteria	190, 191
		Rat	4
	Indoleacrylic acid	Bacteria	192
	Naphthylacrylic acid	Bacteria	193
	β-(2-Benzothienyl)alanine	Bacteria	194
	Styrylacetic acid	Bacteria	193
	Indole	Bacteriophage	195
	α-Amino-β-3(indazole)propionic acid (Tryptazan)	Yeast	196
		Enzyme	197
		E. coli	198
	5-Fluorotryptophan	Enzyme	199, 200
	6-Fluorotryptophan	Enzyme	201
	7-Azatryptophan	E. coli	202–205
Tyrosine	Fluorotyrosines	Fungi	147, 206
		Rat	206
	p-Aminophenylalanine	Fungi	139
	m-Nitrotyrosine	Bacteria	140
	β-(5-Hydroxy-2-pyridyl)alanine	Bacteria	207
Valine	α-Aminoisobutanesulfonic acid	Bacteria	1, 2, 33
		Vaccinia	69
	α-Aminobutyric acid	Bacteria	4, 182
	Norvaline	Bacteria	4
	Leucine, isoleucine	Bacteria	83, 182
	Methallylglycine	Bacteria, yeast	4
	β-Hydroxyvaline	Bacteria	86, 183
	ω-Dehydroalloisoleucine	Bacteria	76

[a] Some of the observed inhibition may be due to fluoride ion present in the amino acid preparation or formed during incubation.[208]

Courtesy of Herbert M. Kagan, Boston University School of Medicine (from Meister, A., Ed., *Biochemistry of the Amino Acids,* 2nd ed., Vol. 1, 1965, 233–238).

References

1. McIlwain, *Brit. J. Exp. Pathol.* 22, 148 (1941).
2. Greenberg and Schulman, *Science* 106, 271 (1947).
3. Snell and Guirard, *Proc. Nat. Acad. Sci. (US)* 29, 66 (1943).
4. Dittmer, *Ann. N.Y. Acad. Sci.* 52, 1274 (1950).
5. Bondi, Kornblum and Forte, *Proc. Soc. Exp. Biol. Med.* 96, 270 (1957).
6. Zygmunt, *J. Bacteriol.* 84, 154 (1962); 85, 1217 (1963).
7. Strominger, Threnn and Scott, *J. Amer. Chem. Soc.* 81, 3803 (1959).
8. Strominger, Ito and Threnn, *J. Amer. Chem. Soc.* 82, 998 (1960).
9. Neuhaus and Lynch, *Biochem. Biophys. Res. Commun.* 8, 377 (1962).
10. Moulder, Novosel and Officer, *J. Bactierol.* 85, 707 (1963).
11. Tanaka, *Biochem. Biophys. Res. Commun.* 12, 68 (1963).
12. Snell, Radin and Ikawa, *J. Biol. Chem.* 217, 803 (1955).
13. Nielsen, *Naturwissenschaften* 31, 146 (1943).
14. Wright and Skeggs, *Arch. Biochem.* 10, 383 (1946).
15. Sarett and Cheldelin, *J. Bacteriol.* 49, 31 (1945).
16. Durham and Milligan, *Biochem. Biophys. Res. Commun.* 7, 342 (1962).
17. Richmond, *Biochem. J.* 73, 261 (1959).
18. Horowitz and Srb, *J. Biol. Chem.* 174, 371 (1948).
19. Teas, *J. Biol. Chem.* 190, 369 (1951).
20. Miller and Harrison, *Nature* 166, 1035 (1950).
21. Volcani and Snell, *J. Biol. Chem.* 174, 893 (1948).
22. Walker, *J. Biol. Chem.* 212, 207. 617 (1955).
23. Kruse, White, Carter and McCoy, *Cancer Res.* 19, 122 (1959).
24. Owaga, *J. Agr. Chem. Soc. Jap.* 10, 225 (1934); *Chem. Abstr.* 28, 4458 (1934).
25. Owaga, *J. Agr. Chem. Soc. Jap.* 11, 559 (1935); *Chem. Abstr.* 29, 740, 3379 (1935).
26. Steward, Pollard, Patchett and Witkop, *Biochim. Biophys. Acta.* 28, 308 (1958).
27. Bonner, *Amer. J. Bot.* 36, 323, 429 (1949).
28. Gros and Gros-Doulcet, *Exp. Cell. Res.* 14, 104 (1958).
29. Morgan, Morton and Pasieka, *J. Biol. Chem.* 233, 664 (1958).
30. Hunter and Downs, *J. Biol. Chem.* 157, 427 (1945).
31. Gross Z, *Physiol. Chem.* 112, 236 (1921).
32. Walker, *J. Biol. Chem.* 212, 617 (1955).
33. McIlwain, *J. Chem. Soc.* p 75 (1941).
34. Ifland and Shive, *J. Biol. Chem.* 223, 949 (1956).
35. Shive, Ackerman and Ravel, *Chem. Soc.* 69, 2567 (1947).
36. Shive and Macow, *J. Biol. Chem.* 162, 452 (1946).
37. Otani, *Arch. Biochem. Biophys.* 101, 131 (1963).
38. Roberts and Hunter, *Proc. Soc. Exp. Biol. Med.* 83, 720 (1953).
39. Arnold, Morris and Thompson, *Nature* 186, 1051 (1960).
40. Woolley *J. Biol. Chem.* 235, 3238 (1960).
41. Shigeura and Gordon, *J. Biol. Chem.* 237, 1932, 1937 (1962).
42. Heymann, Ginsberg, Gulick, Konopka and Mayer, *J. Amer. Chem. Soc.* 81, 5125 (1959).
43. Dittmer, Goering, Goodman and Cristol, *J. Amer. Chem. Soc.* 70, 2499 (1948).
44. Simmonds, *Biochem. J.* 58, 520 (1954).
45. Rhuland, *J. Bacteriol.* 73, 778 (1957).
46. Rhuland and Hamilton, *Biochim. Biophys. Acta.* 51, 525 (1961).
47. Meadow, Hoare and Work, *Biochem. J.* 66, 270 (1957).
48. Borek, Miller, Sheiness and Waelsch, *J. Biol. Chem.* 163, 347 (1946).
49. Borek and Waelsch, *Arch. Biochem.* 14, 143 (1947).
50. Elliot and Gale, *Nature* 161, 129 (1948).
51. Lichtenstein and Grossowicz, *J. Biol. Chem.* 171, 387 (1947).
52. Borek and Waelsch, *J. Biol. Chem.* 177, 135 (1949).
53. Ayengar and Roberts, *Proc. Soc. Exp. Biol. Med.* 79, 476 (1952).
54. Pace and McDermott, *Nature* 169, 415 (1952).
55. Heathcote and Pace, *Nature* 166, 353 (1950).
56. Roberts, *J. Biol. Chem.* 202, 359 (1953).
57. Lichtenstein, Ross and Cohen, *Nature* 171, 45 (1953); *J Biol. Chem.* 201, 117 (1953).
58. Mastalerz, *Arch. Immunol. Terap. Dos.* 7, 201 (1959).
59. Provided by Pattison [Buchanan, Dean and Pattison *Can. J. Chem.* 40, 1571 (1962)], and studied in the author's laboratory.
60. Ravel, McCord, Skinner and Shive, *J. Biol. Chem.* 232, 159 (1958).
61. Rabinovitz and Fisher, *J. Nat. Cancer Inst.* 28, 1165 (1962).
62. Skinner, McCord, Ravel and Shive, *J. Amer. Chem. Soc.* 78, 2412 (1956).
63. McCord, Ravel, Skinner and Shive, *J. Amer. Chem. Soc.* 80, 3762 (1958).
64. Reisner *J. Amer. Chem. Soc.* 78, 5102 (1956).
65. Edelson, Skinner and Shive, *J. Med. Pharm. Chem.* 1, 165 (1959).
66. Meister (1962) in *The Enzymes,* Boyer, Lardy and Myrback, Eds. Academic, New York, VI, p 247.
67. McIlwain, Roper and Hughes, *Biochem. J.* 42, 492 (1948).
68. Spizizen, *J. Infect. Dis.* 73, 212 (1943).
69. Thompson, *J. Immunol.* 55, 345 (1947).
70. Edlbacher, Baur and Becker, *Z. Physiol. Chem.* 265, 61 (1940).
71. Moyed *J. Biol. Chem.* 236, 2261 (1961).
72. Jones and Ainsworth, *J. Amer. Chem. Soc.* 77, 1538 (1955).
73. Levin and Hartman, *J. Bacteriol.* 86, 820 (1963).
74. Doudoroff, *Proc. Soc. Exp. Biol. Med.* 53, 73 (1943).
75. Harper, Benton, Winje and Elvehjem, *Arch. Biochem. Biophys.* 51, 523 (1954).
76. Parker, Skinner and Shive, *J. Biol. Chem.* 236, 3267 (1961).
77. Edelson, Fissekis, Skinner and Shive, *J. Amer. Chem. Soc.* 80, 2698 (1958).
78. Harding and Shive, *J. Biol. Chem.* 206, 401 (1954).
79. Dennis, Plant, Skinner, Sutherland and Shive, *J. Amer. Chem. Soc.* 77, 2362 (1955).
80. Rabinovitz, Olson and Greenberg, *J. Amer. Chem. Soc.* 77, 3109 (1955).
81. Fox, Fling and Bollenback, *J. Biol. Chem.* 155, 465 (1944).
82. Harding and Shive *J. Biol. Chem.* 174, 743 (1948).
83. Brickson, Henderson, Solhjell and Elvehjem, *J. Biol. Chem.* 176, 517 (1948).
84. Ryan, *Arch. Biochem. Biophys.* 36, 487 (1952).
85. Hirsh and Cohen, *Biochem. J.* 53, 25 (1953).
86. Buston and Bishop, *J. Biol. Chem.* 215, 217 (1955).
87. Pal, Skinner, Dennis and Shive, *J. Amer. Chem. Soc.* 78, 5116 (1956).
88. Edelson, Skinner, Ravel and Shive, *Arch. Biochem. Biophys.* 80, 416 (1959).
89. Edelson, Skinner, Ravel and Shive, *J. Amer. Chem. Soc.* 81, 5150 (1959).
90. Rennert and Anker, *Biochemistry* 2, 471 (1963).
91. Smith, Bayliss and McCord, *J. Arch. Biochem. Biophys.* 102, 313 (1963).
92. Page, Gaudry and Gringras, *J. Biol. Chem.* 171, 831 (1948).
93. Daermann, *Arch. Biochem.* 5, 373 (1944).
94. McLaren and Knight, *J. Amer. Chem. Soc.* 73, 4478 (1951).
95. McCord, Ravel, Skinner and Shive, *J. Amer. Chem. Soc.* 79, 5693 (1957).
96. Davis, Skinner and Shive, *Arch. Biochem. Biophys.* 87, 88 (1960).
97. Davis, Ravel, Skinner and Shive, *Arch. Biochem. Biophys.* 76, 139 (1958).
98. Davis, Skinner and Shive, *J. Amer. Chem. Soc.* 83, 2279 (1961).
99. Shiota, Folk and Tietze, *Arch. Biochem. Biophys.* 77, 372 (1958).
100. Goering, Cristol and Dittmer, *J. Amer. Chem. Soc.* 70, 3314 (1948).
101. Skinner, Edelson and Shive, *J. Amer. Chem. Soc.* 83, 2281 (1961).
102. Roblin, Lampen, English, Cole and Vaughan, *J. Amer. Chem. Soc.* 67, 290 (1945).
103. Shaffer and Critchfield, *J. Biol. Chem.* 174, 489 (1948).
104. Reisner, *J. Amer. Chem. Soc.* 78, 2132 (1956).
105. Porter and Meyers, *Arch. Biochem.* 8, 169 (1945).
106. Rabinovitz, Olson and Greenberg, *J. Biol. Chem.* 210, 837 (1954).
107. Black and Kleiber, *J. Amer. Chem. Soc.* 77, 6082 (1955).
108. Dyer *J. Biol. Chem.* 124, 519 (1938).
109. Harris and Kohn, *J, Pharmacol. Exp. Therap.* 73, 383 (1941).
110. Halvorson and Spiegelman, *J. Bacteriol.* 64, 207 (1952).
111. Simpson, Farber and Tarver, *J. Biol. Chem.* 182, 81 (1950).
112. Simmonds, Keller, Chandler and du Vigneaud, *J. Biol. Chem.* 183, 191 (1950).
113. Stekol and Weiss, *J. Biol. Chem.* 179, 1049 (1949).
114. Stekol and Weiss, *J. Biol. Chem.* 185, 577, 585 (1950).
115. Farber, Simpson and Tarver, *J. Biol. Chem.* 182, 91 (1950) .
116. Keston and Wortis, *Proc. Soc. Exp. Biol. Med.* 61, 439 (1946).
117. Levine and Tarver, *J. Biol. Chem.* 192, 835 (1951).
118. Tarver (1954) in *The Proteins,* Neurath and Bailey, Eds., Academic, New York IIB, p 1199.
119. Swendseid, Swanson and Bethell, *J. Biol. Chem.* 201, 803 (1953).
120. Levy, Montanez, Murphy and Dunn, *Cancer Res.* 13, 507 (1953).
121. Yoshida, *Biochim. Biophys. Acta.* 29, 213 (1958).
122. Yoshida and Yamasaki, *Biochim. Biophys. Acta.* 34, 158 (1959).
123. Parks, *J. Biol. Chem.* 232, 169 (1958).

124. Rabinovitz, Olson and Greenberg, *J. Biol. Chem.* 227, 217 (1957).
125. Hansson and Garzo, *Biochim. Biophys. Acta.* 61, 121 (1962).
126. Heathcote, *Lancet* 257, 1130 (1949).
127. Doudney and Wagner, *Proc. Nat. Acad. Sci. U.S.* 38, 196 (1952); 39, 1043 (1953).
128. Shrift, *Amer. J. Bot.* 41, 345 (l954).
129. Cowie and Cohen, *Biochim. Biophys. Acta.* 26, 252 (1957).
130. Cohen and Cowie, *Compt. Rend. Acad. Sci.* 244, 680 (1957).
131. Blau, *Biochim. Biophys. Acta.* 49, 389 (1961).
132. Tuve and Williams, *J. Amer. Chem. Soc.* 79, 5830 (1957).
133. Beerstecher and Shive, *J. Biol. Chem.* 167, 527 (1947).
134. Beerstecher and Shive, *J. Biol. Chem.* 164, 53 (1946).
135. Fox and Warner, *J. Biol. Chem.* 210, 119 (1954).
136. Miller and Simmonds, *Science* 126, 445 (1957).
137. Baltes, Elliott, Doisy and Doisy, *J. Biol. Chem.* 194, 627 (1952).
138. Davis, Lloyd, Fletcher, Bayliss and McCord, *Arch. Biochem. Biophys.* 102, 48 (1963).
139. Burckhalter and Stephens, *J. Amer. Chem. Soc.* 73. 56 (1951).
140. Bergmann, Sicher and Volcani, *Biochem. J.* 54, 1 (1953).
141. Munier and Cohen, *Biochim. Biophys. Acta.* 21, 592 (1956).
142. Munier and Cohen, *Biochim. Biophys. Acta.* 31, 378 (1959).
143. Cohen and Munier *Biochim. Biophys. Acta.* 31, 347 (1959).
144. Cowie, Cohen, Bolton and De Robichon-Szulmajster, *Biochim. Biophys. Acta.* 34, 39 (1959).
145. Cohen, Halvorson and Spiegelman (1958) *Microsomal Particles Protein Syn. Papers. Symp. Biophys. Soc. 1st,* Cambridge, Mass, Pergamon, New York, p 100.
146. Baker, Johnson and Fox, *Biochim. Biophys. Acta.* 28, 318 (1958).
147. Mitchell and Niemann, *J. Amer. Chem. Soc.* 69, 1232 (1947).
148. Atkinson, Melvin and Fox, *Arch. Biochem. Biophys.* 31, 205 (1951).
149. Vaughan and Steinberg, *Biochim. Biophys. Acta.* 40, 230 (1960).
150. Boyer and Westhead, *Abstr. Meeting. Amer. Chem. Soc.* Washington, DC. p 2C (1958).
151. Yoshida, *Biochim. Biophys. Acta.* 41, 98 (1960).
152. Kruh and Rose, *Biochim. Biophys. Acta.* 34, 561 (1959).
153. Armstrong and Lewis, *J. Biol. Chem.* 188, 91 (1951).
154. du Vigneaud. McKennis, Simmonds, Dittmer and Brown, *J. Biol. Chem.* 159, 385 (1945).
155. Dittmer, Ellis, McKennis and du Vigneaud, *J. Biol. Chem.* 164, 761 (1946).
156. Dittmer, Hertz and Chambers, *J. Biol. Chem.* 166, 541 (1946).
157. Garst, Campaigne and Day, *J. Biol. Chem.* 180, 1013 (1949).
158. Ferger and du Vigneaud, *J. Biol. Chem.* 179, 61 (1949).
159. Drea, *J. Bacteriol.* 56, 257 (1948).
160. Dunn and Dittmer, *J. Biol. Chem.* 188, 263 (1951).
161. Kihara and Snell, *J. Biol. Chem.* 212, 83 (1955).
162. Ferger and du Vigneaud, *J. Biol. Chem.* 174, 241 (1948).
163. Dunn, *J. Biol. Chem.* 233, 411 (1958).
164. Dunn, Ravel and Shive, *J. Biol. Chem.* 219, 809 (1956).
165. Janeček and Rickenberg, *Biochim. Biophys. Acta.* 81, 108 (1964).
166. Dittmer, *J. Amer. Chem. Soc.* 71, 1205 (1949).
167. Herz, Dittmer and Cristol, *J. Amer. Chem. Soc.* 70, 504 (1948).
168. Dickie and Schultz, *Arch. Biochem. Biophys.* 100, 279, 285 (1963).
169. Elliott, Fuller and Harington *J. Chem. Soc.* p. 85 (1948).
170. Beerstecher and Shive *J. Amer. Chem. Soc.* 69, 461 (1947).
171. Lansford and Shive, *Arch. Biochem. Biophys.* 38, 347 (1952) .
172. Robbins and McVeigh, *Amer. J. Bot.* 33, 638 (1946).
173. Fowden, Neale and Tristram, *Nature* 199, 35 (1963).
174. Smith, Ravel, Skinner and Shive, *Arch. Biochem. Biophys.* 99, 60 (1962).
175. Fowden and Richmond, *Biochim. Biophys. Acta.* 71, 459 (1963).
176. Fowden, *J. Exp. Bot.* 14, 387 (1963).
177. Katz and Goss, *Biochem. J.* 73, 458 (1959).
178. Katz, *Ann. N.Y. Acad. Sci.* 89, 304 (1960).
179. Meinke and Holland, *J. Biol. Chem.* 173, 535 (1948).
180. Holland and Meinke, *J. Biol. Chem.* 178, 7 (1949).
181. Leibman and Fellner, *J. Biol. Chem.* 237, 2213 (1962).
182. Gladstone, *Brit. J. Exp. Pathol.* 20, 189 (1939).
183. Buston, Churchman and Bishop, *J. Biol. Chem.* 204, 665 (1953) .
184. Woolley, *J. Biol. Chem.* 164, 11 (1946).
185. Trudinger and Cohen, *Biochem. J.* 62, 488 (1956).
186. Anderson, *Science* 101, 565 (1945).
187. Fildes and Rydon, *Brit. J. Exp. Pathol.* 28, 211 (1947).
188. Marshall and Woods, *Biochem. J.* 51, ii (1952).
189. Cohen and Anderson *J. Exp. Med.* 84, 525 (1946).
190. Erlenmeyer and Grubemann, *Helv. Chim. Acta.* 30, 297 (1947).
191. Dittmer, Herz and Cristol, *J. Biol. Chem.* 173, 323 (1948).
192. Fildes, *Brit. J. Exp. Pathol.* 22, 293 (1941).
193. Block and Erlenmeyer, *Helo. Chim. Acta.* 25, 694, 1063 (1942).
194. Avakian, Mars and Martin, *J. Amer. Chem. Soc.* 70, 3075 (1948).
195. Delbrück *J. Bacteriol.* 56, 1 (1948).
196. Halvorson, Spiegelman and Hinman, *Arch. Biochem. Biophys.* 55, 512 (1956).
197. Durham and Martin, *Biochim. Biophys. Acta.* 71, 481 (1963).
198. Brawerman and Yǔeas, *Arch. Biochem. Biophys.* 68, 112 (1957).
199. Moyed, *J. Biol. Chem.* 235, 1098 (1960).
200. Bergmann, Eschinazi, Sicher and Volcani, *Bull. Res. Soc. Israel* 2, 308 (1952).
201. Moyed and Friedmann, *Science* 129, 968 (1959).
202. Robison and Robison, *J. Amer. Chem. Soc.* 77, 457 (1955).
203. Pardee, Shore and Prestidge, *Biochim. Biophys. Acta.* 21, 406 (1956).
204. Pardee and Prestidge, *Biochim. Biophys. Acta.* 27, 330 (1958) .
205. Kidder and Dewey *Biochim. Biophys. Acta.* 17, 288 (1955).
206. Niemann and Rapport, *J. Amer. Chem. Soc.* 68, 1671 (1946).
207. Norton, Skinner and Shive, *J. Org. Chem.* 26, 1495 (1961).
208. Kagan, H., unpublished studies.

COEFFICIENTS OF SOLUBILITY EQUATIONS OF CERTAIN AMINO ACIDS IN WATER[a]

Amino Acid	a_1	$b_1 \times 10^2$	$c_1 \times 10^5$	a_2	a_3	$b_3 \times 10^2$	$c_3 \times 10^5$	a_4	$b_4 \times 10^2$	$c_4 \times 10^5$
L-Alanine	2.1048	0.4669	—	0.1551	-2.5792	1.075	—	-6.5150	1.037	—
D,L-Alanine	2.0830	0.5608	—	0.1333	-3.2199	1.291	—	-7.1317	1.245	—
L-Asparagine H$_2$O	0.9289	2.311	-4.981	-1.2475	-25.9584	1.059	-11.47	-30.2463	11.79	-11.84
L-Aspartic acid	0.3194	1.519	—	-1.8047	-13.7113	3.499	—	-17.7370	3.502	—
D,L-Aspartic acid	0.4181	2.016	-4.999	-1.7060	-25.1918	10.93	-11.51	-29.2797	10.98	-11.61
L-Cystine	-1.299	1.357	—	-3.680	-18.643	3.125	—	-21.023	3.125	—
D,L-Cystine[c]	-1.7959	0.8013	27.89	-4.1766	19.4912	-23.66	47.61	15.4747	-23.66	47.61
D,L-Cystine[c]	-2.1087	3.367	-22.56	-4.4894	-36.9568	14.48	-16.80	-40.9733	14.48	-16.80
Meso-cystine[c]	-1.7190	0.4514	27.39	-4.0997	34.7268	-33.38	63.01	30.7013	-33.38	63.01
Meso-cystine[c]	-2.6034	5.890	-49.41	-4.9841	-133.4125	75.4125	-113.8	-137.429	-75.73	-113.8
L-Dibromotyrosine (hydrated)	0.0839	1.627	—	-2.445	-15.881	3.753	—	-19.894	3.752	—
L-Dibromotyrosine (anhydrous)	0.188	0.9884	—	-2.343	-11.610	2.276	—	-15.537	2.247	—
L-Dichlorotyrosine	0.0065	1.038	4.648	-2.392	-4.058	-3.450	1.069	-8.426	-3.215	1.030
L-Diiodotyrosine	-0.690	1.92	—	-3.326	-19.745	4.42	—	-23.761	4.43	—
D,L-Diiodotyrosine	-0.827	1.43	—	-3.464	-16.989	3.30	—	-21.006	3.30	—
D-Glutamic acid	0.5331	1.613	—	-1.6345	-13.9054	3.714	—	-17.9095	3.709	—
D,L-Glutamic acid	0.9317	1.523	—	-1.2359	-12.4244	3.507	—	-16.4071	3.495	—
Glycine	2.1516	1.087	-4.114	0.2762	-13.2619	7.676	-9.473	-17.8976	8.171	-10.50
L-Hydroxyproline	2.4603	0.3891	—	0.3428	-1.6575	0.8959	—	-5.5906	0.8514	—
L-Isoleucine	1.5787	0.07682	2.594	-0.5389	2.7190	-3.081	5.972	-1.3913	-3.020	5.866
D,L-Isoleucine	1.2616	0.2512	3.794	-0.8560	2.9651	-4.193	8.736	-1.1373	-4.134	8.632
L-Leucine[d]	1.3561	0.02233	3.727	-0.7615	4.5073	-4.683	8.582	-0.7252	-3.814	7.198
D,L-Leucine	0.9013	0.2635	4.591	-1.2163	3.4260	-5.167	10.57	-0.6258	-5.143	10.53
D,L-Methionine	1.2597	1.108	-1.221	-0.9140	-11.1682	4.086	-2.811	-15.2099	4.111	-2.871
D,L-Norleucine	0.9258	0.4524	3.402	-1.1918	0.2523	-3.236	7.833	-4.2067	-2.941	7.340
L-Phenylalanine	1.2974	0.6982	—	-0.9204	-6.510	1.608	—	-10.5103	1.601	—
D,L-Phenylalanine	0.9986	0.5252	3.140	-1.2192	-0.7184	-2.739	7.229	-5.0876	-2.495	6.808
L-Proline	3.1050	0.4206	—	1.0441	-0.2407	0.9686	—	-3.8586	0.7586	—
D,L-Serine	1.3432	1.520	-3.548	-0.6782	-17.2153	7.963	-8.169	-21.4529	8.134	-8.504
Taurine	1.5945	1.916	-8.500	-0.5029	-27.8015	15.10	-19.57	-32.1283	15.35	-20.07
L-Tryptophan	0.9156	0.4834	2.988	-1.3942	-11.3824	4.872	6.881	-15.3928	4.869	-6.879
L-Tyrosine	-0.708	1.46	—	-2.966	-10.799	3.36	—	-20.062	3.37	—
D,L-Tyrosine	-0.833	1.51	—	-3.091	-16.562	3.46	—	-20.577	3.46	—
D,L-Valine[b]	1.9211	8.1515	8.589	-0.1456	0.6274	-8.927	1.978	-3.4570	-8.528	1.9058
L-Valine[b]	1.6675	97.75	80.22	-0.4011	-20.8468	123.4	18.47	-24.2293	119.4	17.85
L-Valine[b]	1.8847	11.42	4.799	-0.1839	-0.3175	-3.406	1.105	-3.6570	-8.125	1.910
L-Valine[b]	1.9227	—	—	-0.1459	-0.3359	—	—	-4.3653	—	—
L-Valine[b]	1.8836	—	—	-0.1850	-0.4260	—	—	-4.4542	—	—
D,L-Valine	1.7749	0.2389	2.607	-0.2966	-2.2921	-2.729	6.003	-1.7417	-2.705	5.928

[a] Solubility equations:

a. $\log S = a_1 + b_1 t + c_1 t^2$ (grams per 1,000 gms water)

b. $\log m = a_2 + b_1 t + c_1 t^2$ (moles per 1,000 gms water)

c. $\ln m = a_3 + b_3 T + c_3 T^2$

d. $\ln N_2 = a_4 + b_4 T + c_4 T^2$

[b] The five sets of values which are given under L-valine refer to the various crystal forms.

[c] The first set of values refer to 273.1 K to 303.1 K and the second set to 298.1 K to 323.5 K absolute.

[d] Values are not strictly accurate due to contamination of the leucine with a small amount of methionine.

Reprinted with slight modification from *The Chemistry of the Amino Acids and Proteins*, C. L. A. Schmidt, Ed. Charles C Thomas, Publisher, Springfield, Ill., 1944, 845. Courtesy of the publisher.

HEAT CAPACITIES, ABSOLUTE ENTROPIES, AND ENTROPIES OF FORMATION OF AMINO ACIDS AND RELATED COMPOUNDS

John O. Hutchens

Heat capacity (C_p°), absolute entropy (S°), and the entropy change for formation from the elements (ΔSf°) are given for a temperature of 298.15 K. The units are cal deg^{-1} mole^{-1} for all entries except for proteins where they are cal deg^{-1} g^{-1}. 1 cal = 4.1840 absolute joules and 0°C—273.15 K. International Atomic Weights of 1959 are employed. In calculating ΔSf° the entropies of the elements used are (in cal deg^{-1} mole^{-1}) C (graphite), 1.3609; H_2 (gas), 31.211; O_2 (gas), 49.003; N_2 (gas), 45.767; Cl_2 (gas), 53.31; and S (rhombic), 7.62.[1] S° for D_2 (gas) is 34.620 cal deg^{-1} mole^{-1}.[2]

The references cited give heat capacities at other temperatures, and, in most cases, the additional thermodynamic functions $(H^\circ - H_0^\circ)$, $(H^\circ - H_0^\circ)/T$ and $-(F^\circ - H_0^\circ)/T$. None of the stated values is more accurate than ± 0.2%. Values obtained by extrapolating below 90 K are uncertain by ± 1—2 cal deg^{-1} mole^{-1}. Entries are listed in alphabetical order under (1) amino acids, (2) peptides, (3) proteins, and (4) miscellaneous related substances.

Compound	C_p°	S°	$-\Delta Sf^\circ$	Reference	Remarks
		AMINO ACIDS			
L-Alanine	29.22	30.88	154.33	3, 4	
D-Arginine	55.8[a]	59.9	307.3[a]	5	Extrapolated below 90 K
L-Arginine · HCl	26.37	68.43	341.01	6	
L-Asparagine	38.3[a]	41.7	207.9[a]	7	
L-Asparagine · H_2O	49.69	50.10	255.17	8	
L-Aspartic Acid	37.09	40.66	194.91	8	
DL-Citrulline	55.2[a]	60.8	292.4[a]	9	Extrapolated below 90 K
L-Cysteine	38.8[a]	40.6	152.2[a]	10	Extrapolated below 90 K
L-Cystine	62.60	67.06	287.38	11	
L-Glutamic Acid	41.84	44.98	223.16	8	
D-Glutamic Acid · HCl	50.0[a]	59.3	251.1[a]	12	Extrapolated below 90 K
L-Glutamine	44.02	46.62	235.51	8	
Glycine	23.71	24.74	127.90	3, 4	
L-Histidine · HCl	59.64	65.99	242.54	6	
L-Hydroxyproline	36.79	41.19	202.45	13	Transitions at 21.9 and 31.5 K
L-Hydroxyproline (deuterated)	39.40	43.04	205.72	14	Transitions at 25.9 and 28.9 K. From 99.6% D_2O. Carboxyl, hydroxyl, and imido hydrogens presumably replaced
L-Isoleucine	45.00	49.71	233.21	4	—
L-Leucine	48.03	50.62	232.30	4	—
L-Lysine · HCl	57.10	63.21	300.46	6	—
L-Methionine	69.32	55.32	202.65	11	Transition at 305.5 K
DL-Ornithine	45.6[a]	46.2	242.6[a]	9	Extrapolated below 90 K
L-Phenylalanine	48.52	51.06	204.74	15	—
L-Proline	36.13	39.21	179.93	15	—
L-Serine	32.40	35.65	174.06	16	—
L-Threonine	35.2	36.5	205.8	13	Several samples tested. Difficult to dry. Anomalous behavior 250—300 K
L-Tryptophan	56.92	60.00	237.01	15	—
L-Tyrosine	51.73	51.15	229.15	15	—
L-Valine	40.35	42.75	207.60	4	—
		PEPTIDES			
DL-Alanylglycine	43.6[a]	51.0	231.1[a]	18	Extrapolated below 90 K. MW not stated in references. C_p° given per g in references
Glycylglycine	39.19	43.09	206.47	13	
DL-Leucylglycine	61.3[a]	67.2	312.6[a]	17	Extrapolated below 90 K. MW not stated in reference. C_p° given per g in reference
		PROTEINS			
Bovine Serum Albumin	—	—	—	18	
Native Hydrated	0.3161	0.3249	—	—	2.14% H_2O. No heat of fusion noted
Native Anhydrous	0.3049	0.3175	—	—	
Denatured Anhydrous	0.3096	0.3205	—	—	

Compound	C_p°	S°	$-\Delta Sf^\circ$	Reference	Remarks
Bovine Zinc Insulin	—	—	—	13	
Native Hydrated	0.3155	0.3252	—	—	4.0% H_2O. No heat of fusion noted
Native Anhydrous	0.2996	0.3144	—	—	—
Collagen	—	—	—	14	Bovine serosal collagen
Native Hydrated	0.3834	0.3589	—	—	13.53% H_2O. No heat of fusion noted
Native Anhydrous	0.2921	0.3081	—	—	
α-Chymotrypsinogen	—	—	—	18	Bovine source
Native Hydrated	0.3834	0.3635	—	—	10.7% H_2O. No heat of fusion noted
Native Anhydrous	0.3090	0.3227	—	—	—
MISCELLANEOUS RELATED SUBSTANCES					
Creatine	41.1[a]	45.3	218.2[a]	19	—
Creatinine	33.2[a]	40.0	167.8[a]	19	—
Urea	22.26	25.00	109.06	20	—

[a] Calculated by compiler, not in reference cited.

References

1. National Bureau of Standards Circular 500 (1961), Selected Values of Chemical Thermodynamic Properties, U.S. Gov. Print. Off., Washington, D.C.
2. National Bureau of Standards Report No. 8504 (1964), Preliminary Report on the Thermodynamic Properties of Selected Light-Element and Some Related Compounds, U.S. Gov. Print. Off., Washington, D.C.
3. Hutchens, Cole, and Stout, *J. Am. Chem. Soc.*, 82, 4813 (1960).
4. Hutchens, Cole, and Stout, *J. Phys. Chem.*, 67, 1128 (1963).
5. Huffman and Ellis, *J. Am. Chem. Soc.*, 59, 2150 (1937).
6. Cole, Hutchens, and Stout, *J. Phys. Chem.*, 67, 2245 (1963).
7. Huffman and Borsook, *J. Am. Chem. Soc.*, 54, 4297 (1932).
8. Hutchens, Cole, Robie, and Stout, *J. Biol. Chem.*, 238, 2407 (1963).
9. Huffman and Fox, *J. Am. Chem. Soc.*, 63, 3464 (1940).
10. Huffman and Ellis, *J. Am. Chem. Soc.*, 57, 46 (1935).
11. Hutchens, Cole, and Stout, *J. Biol. Chem.*, 239, 591 (1964).
12. Huffman, Ellis, and Borsook, *J. Am. Chem. Soc.*, 62, 297 (1940).
13. Hutchens, J. O., Cole, A. G., and Stout, J. W., unpublished data.
14. Hutchens, J. O., Kim, S., and Stout, J. W., unpublished data.
15. Cole, Hutchens, and Stout, *J. Phys. Chem.*, 67, 1852 (1963).
16. Hutchens, Cole, and Stout, *J. Phys. Chem.*, 239, 4194 (1964).
17. Huffman, *J. Am. Chem. Soc.*, 63, 688 (1941).
18. Hutchens, J. O., unpublished data.
19. Huffman and Borsook, *J. Am. Chem. Soc.*, 54, 4297 (1932).
20. Ruerhwein and Huffman, *J. Am. Chem. Soc.*, 68, 1759 (1946).

This table originally appeared in Sober, Ed., *Handbook of Biochemistry and Selected Data for Molecular Biology*, 2nd ed., Chemical Rubber Co., Cleveland, 1970.

HEAT OF COMBUSTION, ENTHALPY AND FREE ENERGY OF FORMATION OF AMINO ACIDS AND RELATED COMPOUNDS

John O. Hutchens

Heat of combustion (ΔH_c°) is given as the enthalpy change for the reaction of burning the compound at constant pressure to produce CO_2 (gas), H_2O (liquid), N_2 (gas), and S (rhombic). The enthalpy of formation (ΔHf°) is given for formation of the compound from C (graphite), H_2 (gas), O_2 (gas), N_2 (gas), and S (rhombic). In calculating ΔHf°, the enthalpies of formation of CO_2 (gas, $p = 0$) = 94.0518 Kcal mole^{-1} and of H_2O (liquid) = 68.3174 Kcal mole^{-1} were employed.[1] Most of the heats of combustion were originally reported as the enthalpy change for the combustion reaction with all gases at $p = 1$ atmosphere. No correction has been made for further expansion of the gases. However, all heats of combustion have been corrected so that International Atomic Weights of 1959 apply. The listings are for a temperature of 298.15°K. 0°C = 273.15°K and 1 cal = 4.1840 absolute joules. The units for all entries are kcal mole^{-1}.

In calculating ΔGf° the free energy change for formation of the compound from C (graphite), H_2 (gas, f = 1); O_2 (gas, f = 1), N_2 (gas, f = 1), and S (rhombic) the entropies of formation (ΔSf°) of amino acids and peptides listed in table on page 65 were employed. $\Delta Gf^\circ = \Delta Hf^\circ - T\Delta Sf^\circ$. Where available ΔSf° for the L-isomer of an amino acid was used in calculating ΔGf° for D- and DL-form on the assumption that this entropy value was more reliable. Available entropy data on D- and DL-forms depend on extrapolation of heat capacity data below 90°K while those for the L-amino acids in most cases do not.

The table is intended to be comprehensive rather than selective. All heats of combustion which have come to the compiler's attention have been included regardless of degree of reliability. Where more than one heat of combustion has been reported for a compound, ΔHf° and ΔGf° have been calculated only from the apparently more reliable data. Where no reasonable choice could be made between apparently equally reliable recent heats of combustion (since 1930) ΔHf° and ΔGf° have been calculated from both.

Compounds are listed alphabetically under (1) Amino Acids, (2) Peptides, and (3) Miscellaneous related compounds.

Compound	$-\Delta H_c^\circ$	Reference	$-\Delta Hf^\circ$	$-\Delta Gf^\circ$	Remarks
		AMINO ACIDS			
D-Alanine	387.1	2	134.2	88.2	—
	387.1	3			—
L-Alanine	386.8[a]	4, 5	134.5	88.4	—
DL-Alanine	386.6	6	134.7	88.7	—
	387.6	3			—
	387.8	7			—
	389.5	8			—
D-Arginine	893.5	6	149.0	57.4	—
L-Asparagine	460.8	2	188.7	126.7	—
	463.5	9			—
L-Asparagine · H_2O	458.1	2	259.7	183.6	—
	459.8	10			—
L-Aspartic Acid	382.6	2	232.6	174.5	—
	382.2	10			
	385.0	11			
	385.0	12			
	385.7	8			
L-Cysteine	394.6	13	126.7	81.3	—
L-Cystine	724.6	13, 14	249.6	163.9	—
D-Glutamic Acid	537.5	2	240.2	173.7	—
L-Glutamic Acid	536.4[a]	15, 16	241.3	174.8	—
	536.9	12	—	—	—
	542.6	8	—	—	—
L-Glutamine	614.3[a]	15, 16	197.5	127.3	—
Glycine	230.5[a]	4, 5	128.4	90.3	—
	232.6	6	126.3	88.2	—
	233.4	3	—	—	—
	234.5	9	—	—	—
L-Isoleucine	855.8[a]	4, 5	152.5	83.0	—
D-Leucine	856.0	6	152.4	83.1	—
L-Leucine	856.0	6	152.4	83.1	—
	853.7[a]	15	154.6	85.4	—
DL-Leucine	855.2	6	153.2	83.9	—
	856.0	9	—	—	—

Compound	$-\Delta H_c^{\circ}$	Reference	$-\Delta Hf^{\circ}$	$-\Delta Gf^{\circ}$	Remarks
L-Methionine	664.8[a]	4, 14	181.2	120.9	—
L-Phenylalanine	1110.6	4	111.6	50.6	—
DL-Phenylalanine	1111.9	8	110.9	49.9	—
	1112.3	17	—	—	
L-Serine	347.7	18	173.6	121.6	Calculated from structural factor -OH for -H on L-Alanine
Isoserine	343.8	3	177.5	125.5	Secondary alcohol. $-\Delta H_c^{\circ}$ lower than Serine expected
L-Threonine	490.7[a]	4	192.9	131.5	—
L-Tryptophan	1,345.2[a]	4	99.2	28.5	—
L-Tyrosine	1061.7	12	160.5	92.2	Agree with structural change -OH for -H on L-Phenylalanine
	1058.3	6	—	—	—
	1070.8	10	—	—	—
L-Valine	698.3[a]	15, 5	147.7	85.8	—
DL-Valine	701.1	3	144.9	83.0	—

PEPTIDES

Compound	$-\Delta H_c^{\circ}$	Reference	$-\Delta Hf^{\circ}$	$-\Delta Gf^{\circ}$	Remarks
DL-Alanylglycine	625.9	19	186.0	117.1	—
DL-Alanyl-DL-Phenylalanine	1505.4	20	169.8	77.4[b]	—
DL-Alanyl-DL-Phenylalanyl-glycine	1742.7	21	223.0	107.2[b]	—
Glycycl-DL-Alanyl-DL-phenylalanine	1744.2	20	221.5	105.7	—
Glycylglycine	471.4	19	178.1	116.6	—
	470.8	8	—	—	—
di-Glycylglycine	710.0	3	230.1	145.1[b]	—
tri-Glycylglycine	946.9	3	283.7	175.3[b]	—
Glycyl-DL-Phenylalanine	1349.2	20	163.6	79.1[b]	—
Glycyl-DL-Tryptophan	1586.2	21	148.8	54.7[b]	—
Glycyl-DL-Valine	937.0	20	199.6	115.3[b]	—
DL-Leucylglycine	1093.4	19	205.6	112.4	—
	1095.8	3	—	—	—
DL-Leucylglycylglycine	1333.7	8	258.8	139.7[b]	
DL-Serylserine	692.7	20	281.5	192.4[b]	
DL-Valyl-DL-Phenylalanine	1816.8	21	183.1	74.9[b]	

MISCELLANEOUS

Compound	$-\Delta H_c^{\circ}$	Reference	$-\Delta Hf^{\circ}$	$-\Delta Gf^{\circ}$	Remarks
Creatine	555.1	2	128.4	63.3	
Creatinine	558.1	2	57.0	7.0	
Urea	151.0	22	79.6	47.1	

[a] Authors give only ΔE for bomb process. ΔH_c° calculated by compiler.

[b] Calculated assuming $\Delta_{298.15} = 10.3$ cal deg^{-1} mole^{-1} for the reaction: Amino Acid (solid) + Amino Acid (solid) $\rightarrow$ Dipeptide (solid) + H_2O (liquid).

References

1. National Bureau of Standards Circular 500 (1961), *Selected Values of Chemical Thermodynamic Properties*, U.S. Gov. Print. Off., Washington, D.C.
2. Huffman, Ellis, and Fox, *J. Am. Chem. Soc.*, 58, 1728 (1936)
3. Wrede, *Z. Phys. Chem.*, 75, 81 (1910)
4. Tsuzuki, Harper, and Hunt, *J. Phys. Chem.*, 62, 1594 (1958).
5. Hutchens, Cole, and Stout, *J. Phys. Chem.*, 67, 1128 (1963).
6. Huffman, Fox, and Ellis, *J. Am. Chem. Soc.*, 59, 2144 (1937).
7. Landrieu, *Compt. Rend.*, 142, 580 (1906).
8. Fischer and Wrede, *Sitz. Kgl. Preuss. Akad. Wiss.*, p 687 (1904).
9. Stohmann and Langbein, *J. Prakt. Chem.*, 44, 336 (1891).
10. Emery and Benedict, *Am. J. Physiol.*, 28, 301 (1911).
11. Stohmann, *Z. Phys. Chem.*, 10, 410 (1892).
12. Oka, *Nippon Seirigaku Zasshi*, 9, 365 (1944).
13. Huffman and Ellis, *J. Am. Chem. Soc.*, 57, 41 (1935).
14. Hutchens, Cole, and Stout, *J. Biol. Chem.*, 239, 591 (1964).
15. Tsuzuki and Hunt, *J. Phys. Chem.*, 61, 1668 (1957).
16. Hutchens, Cole, Robie, and Stout, *J. Biol. Chem.*, 238, 2407 (1963).
17. Breitenbach, Derkosch, and Wessely, *Nature*, 169, 922 (1952).
18. Hutchens, Cole, and Stout, *J. Phys. Chem.*, 67, 1852 (1963).
19. Huffman, *J. Phys. Chem.*, 46, 885 (1952).
20. Ponomarev, Alekseeva, and Akimova, *Russ. J. Phys. Chem.* (Engl. Transl.), 36, 457 (1963).
21. Alekseeva and Ponomarev, *Russ. J. Phys. Chem.* (Engl. Transl.), 38, 731 (1964).
22. Huffman, *J. Am. Chem. Soc.*, 62, 1009 (1940).

This table originally appeared in Sober, Ed., *Handbook of Biochemistry and Selected Data for Molecular Biology*, 2nd ed., Chemical Rubber Co., Cleveland, 1970.

SOLUBILITIES OF AMINO ACIDS IN WATER AT VARIOUS TEMPERATURES

John O. Hutchens

The table gives the solubilities of the amino acids in grams per kilogram of water at various temperatures. None of the experimental observations were made at temperatures above 70°C. In most cases the experimental values were obtained both from a colder originally unsaturated solution and a warmer originally saturated solution. Even where four significant figures are given, many of the values probably are not reliable to better than 1 to 2% as judged by agreement between investigators. Lack of a value at a given temperature indicates that, in the opinion of the compiler, extrapolation of the curve from lower temperatures could not be justified.

SOLUBILITIES OF AMINO ACIDS IN WATER AT VARIOUS TEMPERATURES 0°–100° C (g/kg)

Substance	0°	10°	20°	30°	40°	50°	60°	70°	80°	90°	100°	References
L-Alanine	127.3	141.7	157.8	175.7	195.7	217.9	242.6	270.2	300.8	335.0	373.0	1, 2
DL-Alanine	121.1	137.8	156.7	178.3	202.9	230.9	262.7	299.0	340.1	387.0	440.4	1–3
L-Arginine · HCL	400.0	553.0	718.0	931.0	1240.0	—	—	—	—	—	—	2
L-Asparagine · H₂O	8.49	14.29	23.5	37.79	59.37	91.18	136.8	200.6	287.7	403.0	551.7	4
L-Aspartic Acid	1.72	2.82	4.18	5.94	8.38	11.99	17.01	24.14	34.25	48.59	68.93	1, 3
DL-Aspartic Acid	2.62	4.12	6.33	9.50	14.0	20.0	28.0	38.4	51.4	67.3	85.9	1, 3
L-Cystine	0.0502	0.0686	0.0938	0.1281	0.1751	0.2394	0.3272	0.4472	0.612	0.836	1.142	4
L-Diiodotyrosine	0.204	0.318	0.494	0.769	1.197	1.862	2.897	4.508	7.015	10.91	16.98	1
L-Glutamic Acid	3.29	4.98	7.20	10.19	14.70	—	—	—	—	—	—	2
DL-Glutamic Acid	8.55	12.13	17.22	24.47	34.75	49.34	70.06	99.50	141.3	200.8	284.9	1, 3
Glycine	141.8	180.4	225.2	275.9	331.6	391.0	452.6	513.9	572.7	626.2	671.7	1, 3
L-Hydroxyproline	288.6	315.6	345.2	377.6	413.0	451.7	494.1	540.4	591.0	646.5	706.9	5
L-Isoleucine	—ᵃ	32.0	33.6	35.4	37.5	—	—	—	—	—	—	2
DL-Isoleucine	18.26	19.52	21.23	23.50	26.47	30.3	35.39	42.01	50.75	62.37	78.02	1, 3
L-Leucine	22.70	23.01	23.74	24.90	26.58	28.87	31.89	35.84	40.98	47.65	56.38	1
DL-Leucine	7.97	8.56	9.39	10.51	12.03	14.06	16.78	20.46	25.46	32.38	42.06	1, 3
L-Lysine · HCL	462.0	556.0	666.0	799.0	965.0	—	—	—	—	—	—	2
L-Methionine	36.8	43.7	51.4	60.4	70.9	—	—	—	—	—	—	2
DL-Methionine	18.19	23.41	29.95	38.12	48.23	60.70	75.95	94.52	116.9	143.9	176.0	4
DL-Norleucine	8.43	9.43	10.71	12.36	14.49	17.27	20.88	25.7	32.02	40.60	52.29	1, 3
L-Phenylalanine	19.83	23.29	27.35	32.13	37.73	44.31	52.04	61.11	71.78	84.29	99.00	4
DL-Phenylalanine	9.97	11.33	13.07	15.29	18.15	21.87	26.71	33.12	41.66	53.16	68.8	1, 3
L-Proline	1274.0	1403.0	1545.0	1703.0	1876.0	2066.0	2277.0	2508.0	2764.0	3045.0	3355.0	5
L-Serine	133.0	247.0	362.0	476.0	592.0	—	—	—	—	—	—	2
DL-Serine	22.04	31.03	42.95	58.52	78.42	103.4	134.1	171.1	214.8	265.4	322.4	4
Taurine	39.31	59.92	87.84	123.8	167.8	218.8	274.2	330.5	383.1	427.0	457.6	4
L-Tryptophan	8.23	9.27	10.57	12.23	14.35	17.06	20.57	25.14	31.16	39.14	49.87	4
L-Tyrosine	0.196	0.274	0.384	0.537	0.752	1.052	1.473	2.1	2.884	4.036	5.7	1, 3
DL-Tyrosine	0.147	0.208	0.294	0.417	0.743	0.836	1.2	1.7	2.4	3.4	4.8	6
L-Valine	—ᵇ	53.7	56.5	59.7	63.6	—	—	—	—	—	—	2, 7
DL-Valine	59.6	63.3	68.1	74.2	81.7	91.1	102.8	117.4	135.8	158.9	188.1	1, 3

ᵃ Experimental value of 33 g/kg H₂O at 1°C.[2]
ᵇ Experimental value of 53.4 g/kg H₂O at 1°C.[2]

References

1. Dalton and Schmidt, *J. Biol. Chem.*, 103, 549 (1933).
2. Hade, Thesis, University of Chicago, December 1962.
3. Dunn, Ross, and Read, *J. Biol. Chem.*, 103, 579 (1933).
4. Dalton and Schmidt, *J. Biol. Chem.*, 109, 241 (1935).
5. Tomiyama and Schmidt, *J. Gen. Physiol.*, 19, 379 (1935).
6. Winnick and Schmidt, *J. Gen. Physiol.*, 18, 889 (1935).
7. Dalton and Schmidt, *J. Gen. Physiol.*, 19, 767 (1936).

This table originally appeared in Sober, Ed., *Handbook of Biochemistry and Selected Data for Molecular Biology*, 2nd ed., Chemical Rubber Co., Cleveland, 1970.

HEATS OF SOLUTION OF AMINO ACIDS IN AQUEOUS SOLUTION AT 25° C

John O. Hutchens

Enthalpy changes are given for: (1) Solution of the solid crystalline amino acid into the infinitely dilute solution $(\bar{H}_2^\circ - H_2^\circ)$, (2) solution of the solid crystalline substance into saturated solution $(\bar{H}_2(\text{sat}) - H_2^\circ)$, and (3) dilution of the saturated solution to infinite dilution $(\bar{H}_2^\circ - H_2(\text{sat}))$. All entries are Kcal mole^{-1} and are for 25° C. Columns headed "cal" indicate direct calorimetric measurement. Columns headed "soly" indicate that the value was calculated from solubility and activity data. Generally speaking the two methods agree within the limits of error of either method.

The data suffer from multiple defects. The purity of many of the amino acids used in older work[1] is subject to question. Calculations from solubility data are handicapped by poor knowledge of activity coefficients, particularly for sparingly soluble amino acids. Some amino acids form hydrates, and it is not always clear that the solubility data are for the crystalline anhydrous form or that equilibrium has been reached as regards either crystalline form or solubility.

Compound	$(\bar{H}_2^\circ - H_2^\circ)$ cal	$(\bar{H}_2^\circ - H_2^\circ)$ soly	$(\bar{H}_2^\circ(\text{sat}) - H_2^\circ)$ cal	$(\bar{H}_2^\circ(\text{sat}) - H_2^\circ)$ soly	$(\bar{H}_2^\circ - H_2(\text{sat}))$	Reference	Remarks
D-Alanine	—	1.83	—	1.83	0	1	Agrees with (2). Assumed $\partial \ln \gamma_m / \partial \ln m = 0$
L-Alanine	—	1.83	—	2.0	−0.2	2	Assumes $\partial \ln \gamma_N / \partial \ln N = 0.09$
DL-Alanine	2.0	2.2	2.2	2.2	−0.2	1	
D-Arginine	1.5	—	1.1	—	0.4	1	—
L-Arginine·HCl	—	7.9	—	7.0	0.9	2	Assumes $\partial \ln \gamma_N / \partial \ln N = -0.245$
L-Asparagine	5.8	—	5.8	–	0	1	—
L-Asparagine·H₂O	8.0	8.4	8.0	—	0	1	—
	—	—	—	7.7	—	2, 3	—
L-Aspartic Acid	6.0	6.2	5.8	5.6	0.2?	1	Questionable assumption that $\partial \ln \gamma_N / \partial \ln m = 0.549$
	—	6.3	—	6.3	0	2, 3	Assumes $\partial \ln \gamma_N / \partial \ln N = 0$
DL-Aspartic Acid	7.1	7.2	6.9	6.5	0.2?	1	Questionable assumption that $\partial \ln \gamma_m / \partial \ln m = 0.549$
L-Cystine	—	5.5	—	—	—	1	Too sparsely soluble for reliable activity measurements
L-Diiodotyrosine	—	7.8	—	7.8	0	1	—
D-Glutamic Acid	—	6.5	6.3	6.0	—	1	Questionable assumption that $\partial \ln \gamma_m / \partial \ln m = 0.539$
DL-Glutamic Acid	6.5	6.2	6.3	5.7	0.2?	1	Questionable assumption that $\partial \ln \gamma_m / \partial \ln m = 0.539$
Glycine	3.8	3.4	3.4	3.4	0.4	1	$\partial \ln \gamma_m / \partial \ln m = 0.06$
L-Histidine	3.3	—	3.2	—	0.1	1	$\bar{H}_2^\circ - \bar{H}_2$ (sat) from dilution of 0.5 M solution
L-Histidine·HCl	—	10.2	—	—	—	2	Activity data needed $\gamma_\pm \neq 1$
L-Hydroxyproline	1.4	1.4	1.4	—	<0.1	1	$\bar{H}_2^\circ - \bar{H}_2$ (sat) from dilution of 2 M solution
D-Isoleucine	—	0.8	—	—	—	1	—
L-Isoleucine	—	0.9	—	0.9	<0.1	2	—
DL-Isoleucine	—	1.8	—	1.8	—	1	
L-Leucine	—	0.8	—	0.8	—	1	Sample contained methionine
	—	1.0	—	1.0	<0.1	2, 3	—
DL-Leucine	—	2.0	—	2.1	−0.1	1	Questionable assumption that $\partial \ln \gamma_m / \partial \ln m = 0.38$
D-Lysine	−4.0	—	−3.5	–	−0.5	1	$\bar{H}_2^\circ - \bar{H}_2$ (sat) from dilution of 1 M solution
L-Lysine·HCl	—	5.0	—	7.0	−2.0	2	Assumes $\partial \ln \gamma_N \partial \ln N = 0.44$
L-Methionine	—	2.8	—	2.8	<0.1	2	
DL-Norleucine	—	2.5	—	2.5	—	1	—
L-Phenylalanine	—	2.8	—	–	—	1	—
	—	2.7	—	2.7	—	2, 3	Forms hydrate
DL-Phenylalanine	—	2.8	—	2.8	—	1	—
L-Proline	−0.8	1.3	>0.3	—	−1.0	1	$\bar{H}_2^\circ - \bar{H}_2$ (sat) from dilution of 8 M solution
L-Serine·H₂O	—	4.6	—	3.7	0.9	2	—
L-Serine	2.8	—	—	—	—	4	$\bar{H}_2^\circ - \bar{H}_2$ includes heat of hydration
DL-Serine	5.2	5.4	—	5.0	0.1	1	Must form hydrate in solution
L-Tryptophan	—	1.4	—	—	—	1	—
	—	—	—	2.5	—	3	Possibly forms hydrate
L-Tyrosine	—	6.0	—	6.0	—	1	—
D-Valine	—	0.5	—	—	—	1	—
L-Valine	—	0.9	—	1.0	−0.1	2, 3	—
DL-Valine	1.4	1.5	1.7	1.6	−0.3	1	Assumes $\partial \ln \gamma_m / \partial \ln m = -0.549$

References

1. Huffman and Borsook, in *Chemistry of the Amino Acids and Proteins*, Schmidt, Ed., C C Thomas, Springfield, Ill., 1938. A compilation, original references are cited.
2. Hade, Thesis, University of Chicago, December 1962.
3. Hade, E. P. K., personal communication.
4. Hutchens, J. O. and Hade, E. P. K., unpublished data.

This table originally appeared in Sober, Ed., *Handbook of Biochemistry and Selected Data for Molecular Biology*, 2nd ed., Chemical Rubber Co., Cleveland, 1970.

FREE ENERGIES OF SOLUTION AND STANDARD FREE ENERGY OF FORMATION OF AMINO ACIDS IN AQUEOUS SOLUTION AT 25° C

The table lists the molality (m, moles per kg of H_2O) of the saturated solution at 25°C (298.15 K); the appropriate molal activity coefficient (γ_m or $\gamma_m^{\pm}$); the free energy change for transporting one mole of the solute from the saturated solution to a hypothetical aqueous solution at an activity of 1 molal (ΔG soln); and the free energy change for formation of the substance in hypothetical 1 molal solution from the elements ($\Delta Gf°$). The units of ΔG soln and $\Delta Gf°$ are kcal mole^{-1}.

Substance	Saturated Solution m mole/kg	Reference	γ_m	Reference	ΔG soln kcal mole^{-1}	$-\Delta Gf°$ kcal mole^{-1}	Remarks
L-Alanine	1.862	1	1.045	2	−0.368	88.8	—
L-Arginine · HCl	4.061	1	0.587^a	3	−1.03	—	Assumes 1:1 electrolyte ΔG solution=2RT ln $(m_{\pm}\gamma_m^{\pm})$
L-Asparagine · H_2O	0.190	1	1.0	4	0.983	182.6	—
L-Aspartic Acid	0.0375	1	1.0	4	2.06	172.4	—
L-Cystine	4.57×10^{-4}	5	1.0	—	4.53	159.4	Activity coefficient assumed
L-Glutamic Acid	0.0586	1	1.0	4	1.77	173.0	—
L-Glutamine	0.291	6	1.0	4	0.731	126.6	—
Glycine	3.33	5	0.729	7	−0.525	{ 90.8 88.7	Heats of combustion disagree
L-Hydroxyproline	2.75	5	1.05	8	−0.629	—	—
L-Isoleucine	0.263	1	1.0	4	0.791	82.2	—
L -Leucine	0.165	1	1.0	4	1.07	{ 82.0 84.3	Heats of combustion disagree
L -Methionine	0.377	1	0.875	4	0.656	120.2	—
L -Phenylalanine	0.167	1	1.0	4	1.06	49.5	—
L -Proline	14.1	5	3.13	8	−2.24	—	Activity coefficient from extrapolation above m=7.3
L -Serine	4.02	1	0.602	3	−0.524	122.1	—
L -Tryptophan	0.0666	1	1.0	4	1.60	26.9	—
L -Tyrosine	2.51×10^{-3}	5	1.0	—	3.55	88.6	Activity coefficient assumed
L -Valine	0.496	1	0.923	4	0.461	85.3	—

^{a} Activity coefficient is $\gamma_m^{\pm}$.

References

1. Hade, Thesis, University of Chicago, December 1962.
2. Smith and Smith, *J. Biol. Chem.*, 121, 607 (1937).
3. Hutchens, Figlio, and Granite, *J. Biol. Chem.*, 238, 1419 (1963).
4. Hutchens, J. O. and Nancy Norton, unpublished data.
5. Borsook and Huffman, in *The Chemistry of the Amino Acids and Proteins*, Schmidt, Ed., C C Thomas, Springfield, Ill., 1938.
6. Weast, Ed., *Handbook of Chemistry and Physics*, Chemical Rubber Co., Cleveland, Ohio, 1956.
7. Smith and Smith, *J. Biol. Chem.*, 117, 209 (1937).
8. Smith and Smith, *J. Biol. Chem.*, 132, 57 (1940).

This table originally appeared in Sober, Ed., *Handbook of Biochemistry and Selected Data for Molecular Biology*, 2nd ed., Chemical Rubber Co., Cleveland, 1970.

FAR ULTRAVIOLET ABSORPTION SPECTRA OF AMINO ACIDS

The absorption of a chromophoric amino acid in this spectral region is due to the combined absorptions of the side chain chromophore and of the carboxylate group. Because the carboxylate group is consumed in polymerizing amino acids to polypeptides, an amino acid residue absorbs less intensely than a free amino acid. The magnitude of this difference can be estimated from the spectra of the nonchromophoric amino acids – leucine, proline, alanine, serine, and threonine – whose total absorption is due only to the carboxylate group. The variations between the absorptions of these amino acids reflect the variability of carboxylate absorption in slightly different environments.

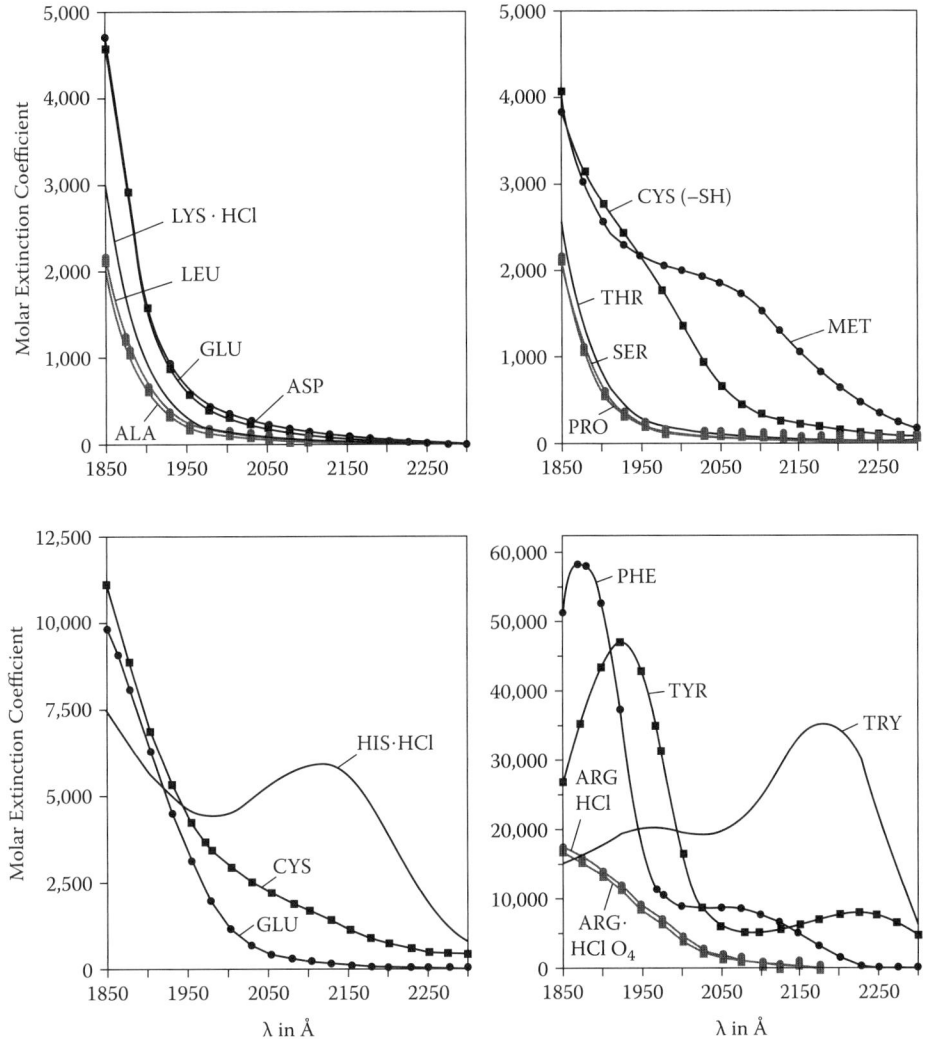

Far ultraviolet spectra of amino acids. All amino acids were in aqueous solution at pH 5, except cystine (pH 3). The dibasic amino acids were measured as hydrochlorides and the absorbance corrected by subtracting the absorbance contribution of chloride ion. Taken from Wetlaufer, *Adv. Protein Chem.*, 17, 320 (1962). With the permission of Academic Press and R. Sussman-McDiarmid and W. Gratzer.

FAR ULTRAVIOLET ABSORPTION SPECTRA OF AMINO ACIDS (Continued)

Amino Acid	$\lambda_{190.0}$	$\lambda_{197.0}$	$\lambda_{205.0}$	Maxima		Minima		Shoulder	
				λ	ε	λ	ε	λ	ε
IN NEUTRAL WATER									
Tryptophan	17.60	20.50	19.60	196.7	20.60	203.3	19.40	–	–
	–	–	–	218.6	46.70	–	–	–	–
Tyrosine	42.80	35.50	5.60	192.5	47.50	208.0	4.88	–	–
	–	–	–	223.2	8.26	–	–	–	–
Phenylalanine	54.50	12.30	9.36	187.7	59.60	202.5	8.96	–	–
	–	–	–	206.0	9.34	–	–	–	–
Histidine[b]	5.57	4.35	5.17	211.3	5.86	198.4	4.22	–	–
Cysteine[c]	2.82	1.94	0.730	–	–	–	–	195.2	2.18
1/2 Cystine	3.25	1.76	1.05	–	–	–	–	207.0	0.96
Methionine	2.69	2.11	1.86	–	–	–	–	204.7	1.89
Arginine[c]	13.1	6.61	1.36	–	–	–	–	–	–
Acids[d]	1.61	0.460	0.230	–	–	–	–	–	–
Amides[d]	6.38	2.06	0.400	–	–	–	–	–	–
Lysine[b]	0.890	0.200	0.110	–	–	–	–	–	–
Leueine[d]	0.670	0.190	0.100	–	–	–	–	–	–
Alanine[d]	0.570	0.150	0.070	–	–	–	–	–	–
Proline[d]	0.540	0.150	0.070	–	–	–	–	–	–
Serine[d]	0.610	0.160	0.080	–	–	–	–	–	–
Threonine[d]	0.750	0.180	0.100	–	–	–	–	–	–
IN 0.1 M SODIUM DODECYL SULFATE[d]									
Tryptophan	16.70	19.70	19.00	197.3	19.80	204.1	18.90	–	–
	–	–	–	220.0	46.60	–	–	–	–
Tyrosine	39.10	36.60	5.87	193.2	45.10	208.5	4.88	–	–
	–	–	–	223.7	7.86	–	–	–	–
Phenylalanine	54.10	11.30	8.45	188.3	57.0	201.5	7.79	–	–
	–	–	–	207.2	8.66	–	–	–	–
Histidine[b]	5.88	4.48	5.03	212.0	5.88	199.0	4.25	–	–
Cysteine[c]	2.66	1.79	0.650	–	–	–	–	194.5	2.15
1/2 Cystine	2.62	1.61	0.92	–	–	–	–	–	–
Methionine	2.67	2.10	1.84	–	–	–	–	204.1	1.89
Arginine[c]	12.50	5.70	0.94	–	–	–	–	–	

Compiled by Ruth McDiarmid.

[a] Molar extinctions, $\varepsilon \times 10^{-3}$. Wavelength, λ, in millimicrons.

[b] The absorptions of lysine and histidine were determined for the hydrochlorides and corrected for the absorption of the chloride ion. $\varepsilon_{Cl^-} = 0.740$ ($\lambda=190.0$), 0.050 ($\lambda=197.0$) and 0($\lambda=205.0$).

[c] The absorptions of cysteine and arginine were determined for the HClO$_4$ salt.

[d] The spectra of the carboxylic acids, the amides, the aliphatic, and the hydroxy amino acids and lysine are unchanged from those obtained in neutral water.

This figure and table originally appeared in Sober, Ed., *Handbook of Biochemistry and Selected Data for Molecular Biology*, 2nd ed., Chemical Rubber Co., Cleveland, 1970.

UV ABSORPTION CHARACTERISTICS OF *N*-ACETYL METHYL ESTERS OF THE AROMATIC AMINO ACIDS, CYSTINE, AND OF *N*-ACETYLCYSTEINE

	Water[a]		Ethanol[a]	
	λ	ε	λ	ε
Phenylalanine				
Inflection	(208)	10.20	(208)	10.40
Inflection	(217)	5.00	(217)	5.30
Minimum	240	0.080	244	0.088
Maximum	241.2	0.086	242.0	0.093
Maximum	246.5	0.115	247.3	0.114
Maximum	251.5	0.157	252.3	0.158
Maximum	**257.4**	**0.197**	**258.3**	**0.195**
Inflection	(260.7)	–	(261.2)	–
Maximum	263.4	0.151	264.2	0.155
Maximum	267.1	0.091	267.8	0.096
Tyrosine				
Maximum	193	51.70	–	–
Minimum	212	7.00	212	6.20
Maximum	224	8.80	227	10.20
Minimum	247	0.176	246	0.174
Maximum	**274.6**	**1.420**	**278.4**	**1.790**
Inflection	281.9	–	285.7	–
Tryptophan				
Minimum	205	21.40	206	21.30
Maximum	219	35.00	221	37.20
Minimum	245	1.900	245	1.560
Maximum	**279.8**	**5.600**	**282.0**	**6.170**
Maximum	288.5	4.750	290.6	5.330
Cystine				
Inflection	(250)	0.360	(253)	0.372
Inflection	260	0.280	260	0.320
Inflection	280	0.110	280	0.135
Inflection	300	0.025	300	0.035
Inflection	320	0.006	320	0.007
***N*-Acetylcysteine**				
	250	0.015	250	0.020
	280	0.005	280	0.005
	320	(nil)	320	(nil)

Compiled by W. B. Gtatzer.

[a] λ, wavelength in millimicrons; $\varepsilon \times 10^{-3}$, molar extinctions. Inflection denotes unresolved inflection.

From J. E. Bailey, Ph.D. Thesis, London University (1966).

NUMERICAL VALUES OF THE ABSORBANCES OF THE AROMATIC AMINO ACIDS IN ACID, NEUTRAL AND ALKALINE SOLUTIONS

Molecular Absorbances of Tyrosine

nm[c]	Neutral[a]	Alkaline[b]	nm[c]	Neutral[a]	Alkaline[b]
230	4980	7752±108	276	1367±0	1206±4
232	3449	8667±38	278	1260±2	1344±5
234	1833±14	9634±19	280	1197±0	1507±5
236	1014±43	10440±20	282	1112±2	1675±5
238	571±36	11000±10	284	845±8	1850±6
240	349±34	11300±20	286	506±7	2024±4
240.5 ↑	–	11340±30	288	248±8	2179±5
242	252+20	11230±40	290	113±0	2300±7
244	209±18	10760±50	292	50±1	2367±5
245.3 ↓	202±20	–	293.2 ↑	–	2381±6
246	205±17	9918±78	294	23±1	2377±8
248	218±15	8734±72	296	13±0	2317±10
250	246±14	7382±56	298	8±1	2195±16
252	287±13	5844±77	300	6±0	2006±23
254	341±14	4471±55	302	5±1	1747±29
256	401±12	3360±46	304	3±0	1445±27
258	485±10	2476±20	306	2±1	1107±35
260	582±9	1883±17	308	1±0	800±27
262	693±13	1467±7	310	1±0	547±21
264	821±13	1204±14	312	–	346±15
266	960±14	1054±16	314	–	206±12
268	1083±13	985±13	316	–	118±9
269.3 ↓	–	974±8	318	–	67±5
270	1197±9	979±9	320	–	32±3
272	1310±9	1019±8	322	–	15±3
274	1394±6	1094±8	324	–	6±2
274.8 ↑	1405±7	–	326	–	1±1

Compiled by Elmer Mihalyi.

[a] 0.1 M phosphate buffer, pH 7.1.
[b] 0.1 N KOH.
[c] Maxima, minima, and inflection points are indicated by ↑, ↓, and ~.

Reprinted with permission from *J. Chem. Eng. Data*, 13, 179 (1969). Copyright 1969 American Chemical Society.

Molecular Absorbances of Tryptophan

nm[c]	Neutral[a]	Alkaline[b]	nm[c]	Neutral[a]	Alkaline[b]
230	6818	13200	279.0 ↑	5579±14	–
232	4037±60	7470	280	5559±12	5377±43
234	2772±71	4354±81	280.4 ↑	–	5385±34
236	2184±64	2951±50	282	5323±10	5302±34
238	1904±55	2282±29	284	4762±11	4962±42
240	1764±52	1959±30	285.8 ↓	4471±6	–
242.0 ↓	1737±49	1813±25	286	4482±11	4596±22
244	1772±48	1773±29	286.8 ↓	–	4565±27
244.4 ↓	–	1763±29	287.8 ↑	4650±12	–
246	1869±40	1792±27	288	4646±16	4634±19
248	2018±35	1877±23	288.3 ↑	–	4639±28
250	2217±32	2013±25	290	3935±5	4393±32
252	2462±19	2187±37	292	2732±5	3551±46
254	2760±27	2410±38	294	1824±5	2666±27
256	3087±20	2664±25	296	1211±10	1990±24
258	3422±18	2953±39	298	797±4	1472±19
260	3787±17	3261±34	300	510±1	1064±19
262	4142±14	3586±46	302	314±3	755±16
264	4472±10	3895±32	304	184±2	517±10
266	4777±14	4212±48	306	112±4	333±6
268	5020±15	4481±46	308	55±9	217±4
270	5220±8	4742±37	310	27±11	129±5
272	5331±5	4933±45	312	11±8	84±8
272.1 ↑	5344±5	–	314	3±2	53±7
273.6 ↓	5329±10	–	316	–	31±7
274	5341±8	5025±34	318	–	17±4
274.5 ~	–	5062±38	320	–	8±2
276	5431±8	5108±39	322	–	3±4
278	5554±12	5275±46			

Compiled by Elmer Mihalyi.

[a] 0.1 *M* phosphate buffer, pH 7.1.
[b] 0.1 *N* KOH.
[c] Maxima, minima, and inflection points are indicated by ↑, ↓, and ~.

Reprinted with permission from *J. Chem. Eng. Data*, 13, 179 (1969). Copyright 1969 American Chemical Society.

Molecular Absorbances of Phenylalanine

nm[c]	Neutral[a]	nm[c]	Alkaline[b]	nm[c]	Neutral[a]	nm[c]	Alkaline[b]
230	32.8±1.5	230	161.9±1.9	257.6 ↑	195.1±1.5	257	188.4±2.8
232	32.1±1.6	232	99.2±1.9	258	193.4±1.3	258	209.1±0.3
234	35.6±2.1	234	70.7±2.4	259	171.9±1.0	258.2 ↑	209.6±0.2
236	42.8±2.1	236	63.3±2.7	260	147.0±0.6	260	184.2±1.0
238	48.5±2.3	238	62.3±2.6	261.9 ↓	127.7±1.5	260.7 ~	178.6±0.3
240	59.4±2.0	240	68.9±3.2	262	128.1±1.4	262	157.8±0.9
242 ~	72.2±2.3	242	83.0±2.8	263.7 ↑	151.5±0.6	262.7 ↓	105.5±1.3
		243 ~	85.4±2.9	264	148.7±0.4	263.9 ↑	161.2±1.0
244	80.1±2.1	244	89.0±3.0	265	119.8±1.3	264	160.0±2.1
246	102.0±0.6	246	108.9±2.8	266	91.8±1.4	266	114.3±1.6
247.4 ↑	110.7±2.2	247	120.9±1.5	266.8 ~	85.6±1.5	266.5 ↓	109.7±1.8
248	109.8±1.9	248.0 ↑	126.1±1.4			267.7 ↑	117.7±1.8
248.3 ↓	109.5±2.0	248.7 ↓	125.1±1.7	268	74.7±1.0	268	115.0±1.0
250	123.5±2.6	250	132.7±1.8	270	30.0±1.8	270	50.2±2.0
251	143.0±2.8	251	149.3±1.9	272	14.3±1.0	272	18.7±1.1
252	153.9±1.0	252	167.0±1.1	274	5.4±0.3	274	7.4±0.3
252.2 ↑	154.1±1.0	252.9 ↑	171.5±1.3	276	2.2±0.4	276	2.6±0.4
254	139.6±1.0	254	166.3±0.8	278	1.1±0.5	278	0.7±0.3
254.5 ↓	138.5±1.4	254.9 ↓	162.8±1.7	280	0.7±0.3	280	0.4±0.2
256	156.5±2.2	256	168.4±1.9				

Compiled by Elmer Mihalyi.

[a] 0.1 M phosphate buffer, pH 7.1.
[b] 0.1 N KOH.
[c] Maxima, minima, and inflection points are indicated by ↑, ↓, and ~.

Reprinted with permission from *J. Chem. Eng. Data*, 13, 179 (1969). Copyright 1969 American Chemical Society.

Alkaline[b] vs. Neutral[a] Difference Spectra of Tyrosine, Tryptophan, and Phenylalanine

nm	Tyrosine	Tryptophan	Phenylalanine	nm	Tyrosine	Tryptophan	Phenylalanine
230	3041	4135	123.9	280	315±1	−191±18	–
232	5440	3213	66.0	282	558±1	−24±4	–
234	7608	1621	35.4	284	994±10	194±3	–
236	9415	732±35	20.7	286	1513±11	110±10	–
238	10490	345±23	13.9	288	1936±1	11±3	–
240	11060	149±21	9.9	290	2196±14	467±8	–
242	11090	45±15	11.1	292	2331±15	802±3	–
244	10660	−40±15	8.8	294	2357±7	830±8	–
246	9844	−104±11	9.0	296	2307±6	755±10	–
248	8567	−172±10	16.4	298	2194±7	652±13	–
250	7205	−233±9	8.3	300	2002±3	527±5	–
252	5671	−298±16	15.3	302	1754±2	413±11	–
254	4344	−371±10	25.8	304	1437±2	300±14	–
256	3127	−435±14	11.3	306	1097±9	205±8	–
258	2142	−490±19	19.1	308	792±14	137±5	–
260	1368±37	−535±13	37.8	310	526±13	88±3	–
262	820±36	−564±8	26.9	312	334±9	55±8	–
264	420±25	−580±10	16.1	314	221±19	22±14	–
266	125±20	−573±12	22.4	316	101±7	16±10	–
268	−78±18	−539±14	42.8	318	62±2	5±2	–
270	−225	−486±12	17.4	320	28±4	0	–
272	−296	−394±12	4.0	322	12±5	–	–
274	−299	−308±7	1.7	324	3±4	–	–
276	−158	−312±16	0.3	326	1±1	–	–
278	89±5	−278±13	0	328	0	–	–

Compiled by Elmer Mihalyi.

[a] 0.1 M phosphate buffer, pH 7.1.
[b] 0.1 N KOH.

Reprinted with permission from *J. Chem. Eng. Data*, 13, 179 (1969). Copyright 1969 American Chemical Society.

Acid[b] vs. Neutral[a] Difference Spectra of Tyrosine, Tryptophan, and Phenylalanine

nm	Tyrosine	Tryptophan	Phenylalanine	nm	Tyrosine	Tryptophan	Phenylalanine
230	—	—	46.7	276	−40	128 ± 16	0
232	576	421 ± 49	34.1	278	−45	110 ± 10	—
234	441	610 ± 37	23.3	280	−34	71 ± 11	—
236	346	590 ± 31	16.0	282	−46	-23 ± 7	—
238	218	512 ± 29	10.8	284	−73	-92 ± 9	—
240	108	432 ± 23	6.9	286	−71	-3 ± 9	—
242	40	358 ± 19	3.9	288	−49	-26 ± 4	—
244	4	305 ± 20	3.3	290	−31	-250 ± 5	—
246	−13	263 ± 21	1.2	292	−20	-317 ± 9	—
248	−18	240 ± 16	−0.3	294	−16	-276 ± 9	—
250	−20	223 ± 14	2.5	296	−14	-227 ± 5	—
252	−16	216 ± 17	−1.8	298	−12	-177 ± 7	—
254	−12	219 ± 15	−1.9	300	−10	-131 ± 9	—
256	−7	222 ± 11	4.0	302	−9	-88 ± 6	—
258	−5	223 ± 14	−3.3	304	−7	-59 ± 8	—
260	−3	225 ± 14	−4.3	306	−6	-40 ± 10	—
262	0	232 ± 13	2.5	308	−4	-24 ± 10	—
264	0	232 ± 18	−3.1	310	−3	-13 ± 10	—
266	−4	224 ± 15	−3.4	312	−2	-7 ± 7	—
268	−8	214 ± 7	−4.4	314	−1	-4 ± 4	—
270	−11	190 ± 9	−2.0	316	−1	0	—
272	−13	159 ± 12	−0.7	318	−0	—	—
274	−20	127 ± 15	−0.3				

Compiled by Elmer Mihalyi.

[a] 0.1 M phosphate buffer, pH 7.1.
[b] 0.1 N HCl.

From *J. Chem. Eng. Data*, 13, 179 (1969). With permission of the author and copyright owner.

LUMINESCENCE OF THE AROMATIC AMINO ACIDS

Amino Acid	Solvent	pH	Temp.	λ_{ex}^a	$\lambda_{f,m}^b$	Q^c	τ^d (ns)	Ref.
Phe	H_2O	7	~25	254	282	0.04		1
	H_2O	6	23	260		0.024		2
	H_2O + 0.55% glucose	0	27		285	0.02		3
	H_2O + 0.55% glucose	6	27		285	0.03		3
	H_2O	7	20	248	282	0.025	6.8	4
	H_2O	7	25				6.4	5
Tyr	H_2O	7	~25	254	303	0.21		1
	H_2O	7	~25				3.2	5
	H_2O	6	23	275		0.14		2
	H_2O	7	~25				3.6	6
	H_2O	7	23				2.6	7
	H_2O	7	~25			0.09		8
Trp	H_2O	7	~25	254	348	0.20		1
	H_2O	Alkaline	~25	280		0.51		9
	H_2O	Acid	~25	280		0.085		9
	H_2O	7	~25				3.0	5
	H_2O	7	23				2.6	7
	H_2O	7	~25				2.5	10
	H_2O	1	~25	286	345	0.091		11
	H_2O	7	~25	287	352	0.149		11
	H_2O	11	~25	289	359	0.289		11
	H_2O	8.9	25		342	0.51		12
	H_2O	10.2	25				6.1	10
	H_2O	7	27			0.14		13
	H_2O	7	25			0.12		8
	H_2O	7	25				2.8	13
	H_2O	6	23	280		0.13		2
	H_2O + 0.55% glucose	0	27		350	0.05		3
	H_2O + 0.55% glucose	6	27		350	0.20		3
	H_2O + 0.55% glucose	12	27		355	0.17		3
	H_2O-ethylene glycol (1:1)	4.7	25				2.9	14

Compiled by R. F. Steiner.

[a] The excitation wavelength in nanometers. When not listed it is either not cited in the original reference or else encompasses a range of wavelengths.
[b] The wavelength, in nanometers, of maximum fluorescence intensity.
[c] The absolute quantum yield.
[d] The excited lifetime, in nanoseconds.

References

1. Teale and Weber, *Biochem. J.*, 65, 476 (1957).
2. Chen, *Anal. Lett.*, 1, 35 (1967)
3. Bishai, Kuntz, and Augenstein, *Biochem. Biophys. Acta*, 140, 381 (1967).
4. Leroy, Lami, and Laustriat, *Photochem. Photobiol.*, 13, 411 (1971).
5. Gladchenko, Kostko, Pikulik, and Sevchenko, *Dokl. Akad. Nauk Belorrus. SSR*, 9, 647 (1965).
6. Blumberg, Eisinger, and Navon, *Biophys. J.*, 8, A-106 (1968)
7. Chen, Vurek, and Alexander, *Science*, 156, 949 (1967).
8. Borresen, *Acta Chem. Scand.*, 21, 920 (1967).
9. Cowgill, *Arch. Biochem. Biophys.*, 100, 36 (1963).
10. Badley and Teale, *J. Mol.*, 44, 71 (1969).
11. Bridges and Williams, *Biochem. J.*, 107, 225 (1968).
12. Longworth, *Biopolymers*, 4, 1131 (1966).
13. Eisinger and Navon, *J. Chem. Phys.*, 50, 2069 (1969).
14. Weinryb and Steiner, *Biochemistry*, 7, 2488 (1968).

LUMINESCENCE OF DERIVATIVES OF THE AROMATIC AMINO ACIDS

Compound	Solvent	pH	Temp.	λ_{ex}^a	$\lambda_{f,m}^b$	Q_{rel}	τ^c(ns)	Ref.
				TRYPTOPHAN[d]				
L-Trp	H_2O	7	23	290		1.00	3.0	10
L-Trp	H_2O-Ethylene glycol (1:1)	4.7	25	290	357	1.00	2.9	1
Acetyl-DL-Trp	H_2O-Ethylene glycol (1:1)	7.5	25	290	361	1.76	5.2	1
Acetyl-Trp	H_2O	5	25	290		1.6	4.8	4
Acetyl-L-Trp	H_2O	7	~25	280	355	1.40		2
Acetyl-L-Trp-amide	H_2O-Ethylene glycol (1:1)	7.5	25	290	356	0.87	3.8	1
Acetyl-Trp-amide	H_2O	7	23	290			3.0	10
Acetyl-Trp-amide	H_2O	5	25	290		1.10	2.6	4
L-Trp-amide	H_2O-Ethylene glycol (1:1)	4.7	25	290	351	0.59	1.4	1
DL-Trp-amide	H_2O	7	25	280	355	1.00		2
Trp-amide	H_2O	7	25	280		0.70		5
Acetyl-Trp-methyl ester	H_2O	7	25	280	350		0.55	5
Trp-methyl ester	H_2O	7	25	280		0.25		5
L-Trp-ethyl ester	H_2O	7	25	280	355	0.16		2
L-Trp-ethyl ester	H_2O-Ethylene glycol (1:1)	4.7	25	290	349	0.17	0.50	1
L-Trp-Gly	H_2O	7	25	280		0.70		2
L-Trp-Gly	H_2O-Ethylene glycol (1:1)	4.7	25	290	350	0.65	1.60	1
L-Trp-Gly	H_2O	7	23	290			2.2	10
L-Trp-Gly-Gly	H_2O-Ethylene glycol (1:1)	4.7	25	290	348	0.48	1.8	1
Gly-L-Trp	H_2O-Ethylene glycol (1:1)	4.7	25	290	356	0.49	1.4	1
Gly-L-Trp	H_2O	7	25	280	355	0.29		2
Gly-Trp	H_2O	7	25	280		0.25		5
Gly-L-Trp	H_2O	7	23	290			1.5	10
Gly-Trp	H_2O	9.8	25	280		0.70		5
Gly-Gly-L-Trp	H_2O-Ethylene glycol (1:1)	4.7	25	290	362	0.65	2.0	1
Gly-Gly-Trp	H_2O	7	25	280		0.40		5
Gly-Gly-Trp	H_2O	9.8	25	280		0.50		5
Gly-Gly-Gly-L-Trp	H_2O-Ethylene glycol (1:1)	4.7	25	290	360	0.76	2.4	1
Trp-Gly-Trp-Gly	H_2O-Ethylene glycol (1:1)	4.7	25	290	352	0.69	2.0	1
L-Trp-L-Phe	H_2O-Ethylene glycol (1:1)	4.7	25	290	350	0.63	2.0	1
L-Trp-L-Trp	H_2O	7	~25	280		0.45		2
L-Trp-L-Trp	H_2O	7	23	290			1.6	10
L-Trp-L-Trp	H_2O-Ethylene glycol (1:1)	4.7	25	290	361	0.50	1.5	1
L-Trp-L-Tyr	H_2O-Ethylene glycol (1:1)	4.7	25	290	351	0.75	1.9	1
L-Trp-L-Tyr	H_2O	7	~25	280		0.60		2
Pro-Trp	H_2O	7	25	280		0.25		5
Pro-Trp	H_2O	10.2	25	280		0.95		5
Tryptamine	H_2O	5	25	290		2.10	6.0	4
				TYROSINE[e]				
DL-Tyr	H_2O	7	25	280	320	1.00		2
Tyramine	H_2O	7	25	280	320	0.88		2
L-Tyr-ethyl ester	H_2O	5.5	25	275		0.12		7
Tyr-amide	H_2O	5.5	25	275		0.25		7
N-Methyl-Tyr	H_2O	5.5	25	275		0.38		7
Acetyl-L-Tyr	H_2O	5.5	25	275		0.88		7
Acetyl-Tyr-amide	H_2O	5.5	25	275		0.45		7
L-Tyr-Gly	H_2O	7	25	280	320	0.35		2
L-Tyr-Gly	H_2O	5.5	25	275		0.33		7
L-Tyr-Gly-Gly	H_2O	5.5	25	275		0.22		7
L-Cystinyl-bis-L-Tyr	H_2O	8.5	25	270		0.08		11
Tyr-Cys-S-S-Cys	H_2O	8.5	25	270		0.08		11
Tyr-Ala	H_2O	6	25	270		0.43		3
Tyr-Phe	H_2O	6	25	270		0.38		3
Tyr-Tyr	H_2O	6	25	270		0.38		3
Gly-L-Tyr	H_2O	7	25	280	320	0.33		2

LUMINESCENCE OF DERIVATIVES OF THE AROMATIC AMINO ACIDS (Continued)

Compound	Solvent	pH	Temp.	λ_{ex}^a	$\lambda_{f,m}^b$	Q_{rel}	τ^c(ns)	Ref.
			TYROSINE[e]*(Continued)*					
Gly-Tyr	H_2O	5.5	25	280		0.38		8
Gly-Gly-L-Tyr	H_2O	5.5	25	275		0.54		7
Gly-Gly-Gly-L-Tyr	H_2O	5.5	25	275		0.58		7
Gly-L-Tyr-Gly	H_2O	5.5	25	275		0.22		7
Gly-L-Tyr-Gly-amide	H_2O	7	25	280	320	0.17		2
Gly-Tyr-Gly-amide	H_2O	6	25	270		0.18		3
Leu-L-Tyr	H_2O	7	25	280	320	0.50		2
Leu-Tyr	H_2O	5.5	25	280		0.48		12
Glu-Tyr	H_2O	5.5	25	280		0.45		8
Met-Tyr	H_2O	5.5	25	280		0.45		8
Arg-Tyr	H_2O	5.5	25	280		0.50		8
His-Tyr	H_2O	5.5	25	270		0.50		8
HS-Cys-Tyr	H_2O	8.5	25	270		0.28		11
			PHENYLALANINE[f]					
L-Phe	H_2O-Ethylene glycol (1:1)	4.7	25	250	287	1.00		1
Gly-DL-Phe	H_2O-Ethylene glycol (1:1)	4.7	25	250	287	1.15		1
DL-Ala-DL-Phe	H_2O-Ethylene glycol (1:1)	4.7	25	250	285	1.10		1
L-Lys-L-Phe	H_2O-Ethylene glycol (1:1)	4.7	25	250	287	0.98		1
L-His-L-Phe	H_2O-Ethylene glycol (1:1)	4.7	25	250	288	0.82		1
L-Arg-L-Phe	H_2-Ethylene glycol (1:1)	4.7	25	250	286	0.77		1
L-Met-L-Phe	H_2O-Ethylene glycol (1:1)	4.7	25	250	287	0.44		1

Compiled by R. F. Steiner.

[a] The excitation wavelength in nanometers. When not listed it is either not cited in the original reference, or else encompasses a range of wavelengths.

[b] The wavelength, in nanometers, of maximum fluorescence intensity.

[c] The excited lifetime, in nanoseconds.

[d] Q_{rel} = quantum yield, relative to that of tryptophan.

[e] Q_{rel} = quantum yield, relative to that of tyrosine.

[f] Q_{rel} = quantum yield, relative to that of phenylalanine.

References

1. Weinryb and Steiner, *Biochemistry*, 7, 2488 (1968).
2. Cowgill, *Arch. Biochem. Biophys.*, 100, 36 (1963).
3. Cowgill, *Biochem. Biophys. Acta*, 75, 272 (1963).
4. Kirby and Steiner, *J. Phys. Chem.*, 74, 4480 (1970).
5. Cowgill, *Biochim. Biophys. Acta*, 133, 6 (1967).
6. Edelhoch, Brand, and Wilchek, *Biochemistry*, 6, 547 (1967).
7. Edelhoch, Perlman, and Wilchek, *Biochemistry*, 7, 3893 (1968).
8. Russell and Cowgill, *Biochim. Biophys. Acta*, 154, 231 (1968).
9. Cowgill, *Biochim. Biophys. Acta*, 100, 37 (1967).
10. Chem, *Arch. Biochem. Biophys*, 158, 605 (1973).
11. Cowgill, *Biochim. Biophys. Acta*, 140, 37 (1967).
12. Cowgill, *Arch. Biochem. Biophys.*, 104, 84 (1964).

LUMINESCENCE OF PROTEINS LACKING TRYPTOPHAN

Protein	Solvent	pH	Temp.	λ_{ex}^a	$\lambda_{f,m}^b$	Q_{rel}^c	τ^d (ns)	Ref.
Angiotensin amide	H_2O	7	25			0.36 (0.075)		2
Insulin (bovine)	H_2O	7	25	277	304	0.19	1.4	1
Insulin B chain	H_2O	7	25			0.26 (0.055)		2
Oxytocin	H_2O	7.8	25			0.26		2, 3
Ribonuclease	H_2O	7	25	277	304	0.10	1.9	1

Compiled by R.F. Steiner.

[a] The excitation wavelength in nanometers.
[b] The wavelength of maximum fluorescence.
[c] The quantum yield relative to that of tyrosine (absolute yield in parentheses).
[d] The excited lifetime, in nanoseconds.

References

1. Longworth, in *Excited States of Proteins and Nucleic Acids*, Steiner and Weinryb, Eds., Plenum, New York, 1971.
2. Cowgill, *Biochim. Biophys. Acta,* 133, 6 (1967).
3. Cowgill, *Arch. Biochem. Biophys.,* 104, 84 (1964).

LUMINESCENCE OF PROTEINS CONTAINING TRYPTOPHAN

Protein	Solvent	pH	Temp.	λ_{ex}^a	$\lambda_{f,m}^b$	Q^c	τ^d (ns)	Ref.
F-Actin	H_2O	7	~25	297	332	0.24		1
Actomyosin	H_2O	7	~25	297	337	0.20		1
Albumin (bovine serum)	H_2O	5.5–7	~25	297	343	0.39		1
	H_2O	7	23		270		4.6	2
	H_2O		~25	280	342	0.15		3
	H_2O	7	25	295	343	0.26		4
	H_2O	7	25		270	0.51		5
	H_2O	7	25	280	342	0.21	4.6	5
Albumin (human serum)	H_2O	5.5–7	~25	297	342	0.31		1
	H_2O		~25	280	339	0.07		3
	H_2O	7	25	295	341	0.155		4
	H_2O					0.24	4.3	5
	H_2O					0.22	4.1	6
	H_2O	5.5	20	295	343	0.21	4.8	7
	H_2O	7	23				4.5	2
	H_2O	7	25	280	339	0.11	3.3	5
Aldolase	H_2O	7	25	280		0.10		4
	H_2O	7	25	280	328	0.10		5
	H_2O	7	25	295	328	0.12		4
β-Amylase	H_2O	7	~25	297	335	0.32		1
Arginase (bovine liver)	H_2O	7	~25	297	337	0.18		1
Avidin	H_2O	7	25	280	338			5
Azurin	H_2O	7	25	280	308	0.10		5
Carbonic anhydrase	H_2O	7	25	280	336	0.17	2.6	5
Carboxypeptidase A	H_2O	7	25	280	327	0.12		4
α-Chymotrypsin	H_2O	7	~25	297	336	0.14		1
	H_2O	7		280	334	0.095		3
	H_2O	7	25	295	332	0.144		4
	H_2O					0.13	3.0	5
	H_2O						3.0	6
	H_2O	7	23	270			3.4	2
	H_2O	7	25	280		0.13		4
Chymotrypsinogen A	H_2O	7	~25	297	333	0.127		1
	H_2O	7	~25	280	331	0.07		3
	H_2O	7	25	295	331	0.124		4
	H_2O					0.10	1.6	5
	H_2O						1.6	6
	H_2O						1.9	8
	H_2O	7	23	290			2.9	2
	H_2O	7	25	280		0.11		4
Corticotropin	H_2O	7	25	280	350	0.08		5
Elastase	H_2O	7	25	280	335			5
Endonuclease	H_2O	7	25	280	334			5
γI-Globulin	H_2O	6	25	297	335	0.08		1
(human)	H_2O	7	23				3.2	2
Glutamate dehydrogenase	H_2O	7	25	280	332	0.300	4.6	5
Glucagon	H_2O	7	25	280	345	0.14		5
Glyceraldehyde-3-phosphate dehydrogenase	H_2O	7	25	297	335.5	0.135		1
Growth hormone	H_2O	7	25	280	325	0.15		5
Hemoglobin	H_2O	7	25	280	335	0.001		5
Hyaluronidase (bovine testicle)	H_2O	7	~25	297	336	0.14		1
α-Lactalbumin	H_2O	7	25	295	328	0.06		4
	H_2O	7	25	280		0.05		4

LUMINESCENCE OF PROTEINS CONTAINING TRYPTOPHAN (Continued)

Protein	Solvent	pH	Temp.	λ_{ex}^{a}	$\lambda_{f,m}^{b}$	Q^{c}	τ^{d} (ns)	Ref.
Lactate dehydrogenase	H_2O	7	25	280	345	0.38		5
β-Lactoglobulin A	H_2O	7	25	280	330	0.08		5
β-Lactoglobulin AB	H_2O	6–8	~25	297	332	0.12		1
(bovine)	H_2O					0.15		5
	H_2O	7	25	295	333	0.082		4
	H_2O	7	25	280		0.08		4
	H_2O	7	25	280		0.08		4
Lysozyme	H_2O	7.5	25	295	337			9
(egg white)	H_2O	7	25	295	338	0.079		4
	H_2O		~25	280	341	0.06		3
	H_2O						2.6	5
	H_2O	7	23	270			2.0	2
	H_2O	7	25	280		0.07		4
Apomyoglobin	H_2O	8.5	~25	297	335.5	0.16		1
(sperm whale)	H_2O					0.15	2.9	10
	H_2O	7	15	288	328	0.12	2.8	11
	H_2O					0.16		5
	H_2O	7	23	290			3.0	2
Myosin (rabbit)	H_2O	7	~25	297	338	0.22		1
Ovalbumin	H_2O	7	~25	280	332	0.19	4.5	5
	H_2O	7	25	295	334	0.25		4
Papain	H_2O	9.5–10	~25	297	350	0.16		1
	H_2O	7.5–7.8	~25	297	347	0.16		1
	H_2O	7	25	295	342	0.16		4
	H_2O			280		0.15		12
	H_2O			280		0.13	4.6	13
	H_2O	7	25	280		0.14		4
	H_2O	4.5–5.3	~25	297	340	0.082		1
	H_2O					0.11		5
	H_2O					0.079		4
	H_2O	4.5	23	288		0.08		14
	H_2O			280		0.10	3.0	12
	H_2O			280		0.08	3.4	13
Pepsin	H_2O	5.2–5.5	~25	297	343	0.26		1
	H_2O		~25	280	342	0.13		3
	H_2O					0.22	4.6	15
	H_2O	7	25	295	339	0.185		4
	H_2O					0.31		5
	H_2O	4.5	23	288		0.22		14
	H_2O						4.5	8
Acid phosphatase (wheat bran)	H_2O	7	~25	297	337	0.16		1
Phosphocreatine kinase	H_2O	7	~25	297	333.5	0.11		1
Pseudoacetyl cholinesterase (human)	H_2O	7	~25	297	334			1
Pyruvate kinase	H_2O	7	~25	297	339	0.44		1
(rabbit)	H_2O					0.20		5
Staphylococcus aureus nuclease	H_2O	5	~25	270		0.46		18
Subtilisin carlsberg	H_2O	7	~25	280	305			5
Tobacco mosaic virus protein (depolymerized)	H_2O	7.4	~25	297	332	0.40		1
Tobacco mosaic virus protein (polymerized)	H_2O	6.4	~25	297	331.5	0.37		1

LUMINESCENCE OF PROTEINS CONTAINING TRYPTOPHAN (Continued)

Protein	Solvent	pH	Temp.	λ_{ex}^{a}	$\lambda_{f,m}^{b}$	Q^{c}	τ^{d} (ns)	Ref.
Trypsin	H_2O	7	~25	297	334.5	0.11		1
(bovine)	H_2O		~25	280	332	0.08		3
	H_2O	7	25	295	335	0.13		4
	H_2O					0.08		5
	H_2O					0.15		1.6
	H_2O						2.0	6
	H_2O						2.4	8
	H_2O	7	25	280		0.13		4
Trypsinogen	H_2O	7	25	295	332	0.14		4
	H_2O	7	25	280		0.12		4
Tryptophanyl tRNA synthetase	H_2O	5	~25	270		0.46		18

Compiled by R.F. Steiner.

[a] The excitation wavelength in nanometers.
[b] The wavelength of maximum fluorescence.
[c] The absolute quantum yield.
[d] The excited lifetime, in nanoseconds.

References

1. Burstein, Vedenkina, and Ivkova, *Photochem. Photobiol*, 18, 263 (1973).
2. Chen, Vurek, and Alexander, *Science*, 156, 949 (1967).
3. Teale, *Biochem. J.*, 76, 381 (1960).
4. Kronman and Holmes, *Photochem. Photobiol.*, 14, 113 (1971).
5. Longworth, in *Excited States of Proteins and Nucleic Acids*, Steiner and Weinryb, Eds., Plenum, New York (1971).
6. Konev, Kostko, Pikulik, and Chernitski, *Biofizika*, 11, 965 (1966).
7. DeLauder and Wahl, *Biochem. Biophsy. Res. Commun.*, 42, 398 (1971).
8. Konev, Kostvo, Pikulik, and Volotovski, *Dokl. Akad. Nauk Beloruss. SSR*, 10, 500 (1966).
9. Lehrer, *Biochemistry*, 10, 3254 (1971).
10. Kirby and Steiner, *J. Biol. Chem.*, 245, 6300 (1970).
11. Anderson, Brunori, and Weber, *Biochemistry*, 9, 4723 (1970).
12. Steiner, *Biochemistry*, 10, 771 (1971).
13. Weinryb and Steiner, *Biochemistry*, 9, 135 (1970).
14. Shinitzki and Goldman, *Eur. J. Biochem.*, 3, 139 (1967).
15. Badley and Teale, *J. Biol.*, 44, 71 (1969).
16. Barenboim, Sokolenko, and Turoverov, *Cytologiya*, 10, 636 (1968).
17. Edelhoch, Condliffe, Lippoldt, and Burger, *J. Biol. Chem.*, 241, 5205 (1966).
18. Edelhoch, Perlman, and Wilchek, *Ann. N. Y. Acad. Sci.*, 158, 391 (1969).

HYDROPHOBICITIES OF AMINO ACIDS AND PROTEINS

C. C. Bigelow and M. Channon

TABLE 1: Hydrophobicities of Amino Acids

The hydrophobicity of an amino acid side chain is derived from the measurement of the solubility of the amino acid in water and an organic solvent.[1,2] The data are converted into the molar free energy of transfer for the amino acid from an aqueous solution to a solution in the organic solvent at the same mole fraction at the limit of infinite dilution:

$$\Delta F_t = RT \ln \frac{N_w \gamma_w}{N_{org} \gamma_{org}}$$

The hydrophobicity of the side chain, $H\Phi$, is then determined by subtracting ΔF_t for glycine. The thermodynamic interpretation of the hydrophobicity has been discussed by Noaki and Tanford.[2]

Amino Acid	HΦ (cal/mol)	Amino Acid	HΦ (cal/mol)
Trp	3,400	Cys/2	1,000
Ile	2,950	Arg	750
Pro	2,600	His	500
Phe	2,500	Ala	500
Tyr	2,300	Thr	400
Leu	1,800	Gly	(0)
Val	1,500	Asp	0[a]
Lys	1,500	Glu	0[a]
Met	1,300	Ser	−300

Compiled by C. C. Bigelow and M. Channon.

[a] Assumed values.

References

1. Tanford, *J. Am. Chem. Soc.*, 84, 4240 (1962),
2. Nozaki and Tanford, *J. Biol. Chem.*, 246, 2211 (1971),

TABLE 2: Average Hydrophobicities of Selected Proteins

The values in the table have been calculated for proteins from amino acid compositions in the literature. Proteins have been included only if the content of all the amino acids is known. The selection has been guided by a desire to include proteins of different properties (globular and fibrous, aggregating and monomeric) as well as members of some families of related proteins.

$$H\Phi_{ave} = \frac{\sum_i n_i H\Phi_i}{\sum_i n_i}$$

where

n_i = the number of residues of the ith amino acid;
$H\Phi_i$ = its hydrophobicity from Table 1.[475, 476]

	Protein	Average Hydrophobicity	Reference
1	Acetoacetic acid decarboxylase (*Clostridium acetobutylicum*)	1,140	1
2	Acetylcholine receptor (*Electrophorus electricus*)	1,060	2
3	Acetylcholinesterase (*Electrophorus electricus*)	990	3
4	*N*-Acetyl-β-D-glucosaminidase-A (beef spleen)	930	5
6	*N*-Acetyl-β-D-glucosaminidase-B (beef spleen)	930	5
7	*N*-Acetyl-β-D-glucosaminidase-A (porcine kidney)	1,040	6
8	*N*-Acetyl-β-D-glucosaminidase-B (porcine kidney)	1,030	6
9	*N*-Acetyl-β-D-glucosaminidase (human plasma)	960	7
10	Acetylornithine γ-transaminase (*Escherichia coli*)	930	8
11	O-Acetylserine sulfhydrylase A (*Salmonella typhimurium*)	1,020	9

TABLE 2: Average Hydrophobicities of Selected Proteins (Continued)

	Protein	Average Hydrophobicity	Reference
12	α_1-Acid glycoprotein (human plasma)	1,040	10
13	Actin (*Acanthamoeba castellani*)	1,030	11
14	Actin (beef carotid)	1,000	12
15	Actin (beef heart muscle)	980	13
16	Actin (beef skeletal muscle)	980	13
17	Actin (brown trout)	1,000	14
18	Actin (chicken)	1,010	13
19	Actin (fish)	980	13
20	Actin (frog)	980	13
21	Actin (human platelet)	950	15
22	Actin (human uterus)	980	16
23	Actin (lamb)	980	13
24	Actin (mollusc)	910	13
25	Actin (pig)	990	13
26	Actin (rabbit muscle)	1,010	17
27	Actin (rabbit muscle)	1,000	18
28	Actin (sheep uterus)	990	16
29	Acylcarrier protein (*Escherichia coli*)	830	19
30	Acylphosphatase (bovine brain)	980	20
31	Acylphosphatase (horse muscle)	900	21
32	Adenosine and adenosine monophosphate deaminase (*Aspergillus oryzae*)	890	22
33	Adenosine deaminase (calf intestinal muscle)	1,010	23
34	Adenosine deaminase (calf spleen)	990	24
35	Adenosine monophosphate deaminase (rabbit muscle)	1,060	22
36	Adenosine monophosphate nucleosidase (*Azotobacter vinelandii*)	770	25
37	Adenosine triphosphatase, ($Na^+ + K^+$), large chain (canine renal medulla)	1,040	26
38	Adenosine triphosphatase, ($Na^+ + K^+$), small chain (canine renal medulla)	1,130	26
39	Adenosine triphosphate-creatine transphosphorylase (rabbit)	980	27
40	Adenylate cyclase (*Brevibacterium liquefaciens*)	880	28
41	Adenylosuccinate – AMP-lyase (*Neurospora* sp.)	940	29
42	Adrenodoxin (bovine)	850	30
43	Adrenodoxin reductase (bovine adrenal gland)	940	31
44	Aequorin (*Aequorea*)	990	32
45	Agglutinin (wheat germ)	680	33
46	Alanine amino transferase (rat liver)	990	34
47	Albumin (bovine plasma)	1,000	35
48	Albumin (dog plasma)	990	36
49	Albumin (human serum)	960	37
50	Albumin (rat serum)	980	38
51	Alcohol dehydrogenase (horse liver)	1,030	39
52	Aldehyde dehydrogenase, protein A (Baker's yeast)	1,020	40
53	Aldolase B (rabbit liver)	940	41
54	Aldolase C (rabbit brain)	980	41
55	Aldolase, fructose-1,6 -diphosphate (*Boa constrictor constrictors*)	970	42
56	Aldolase, fructose-1,6-diphosphate (*Discostichus mawsonii*)	980	43
57	Aldolase, fructose-1,6-dophosphate (*Gallus domesticus*, brain)	920	44
58	Aldolase, fructose-1,6-diphosphate (*Gallus domesticus*, breast muscle)	960	45
59	Aldolase, fructose-1,6-diphosphate (*Gallus domestricus*, liver)	970	46
60	Aldolase, fructose diphosphate (*Micrococcus aerogenes*)	940	47
61	Aldolase, fructose-1,6-diphosphate (ox)	970	48
62	Aldolase, fructose-1,6-diphosphate (porcine)	960	48
63	Aldolase, fructose-1,6-diphosphate (rabbit liver)	960	49
64	Aldolase, fructose-1,6-diphosphate (rabbit, muscle)	960	48
65	Aldolase, fructose-1,6-diphosphate (sturgeon)	990	48
66	Aldolase, fructose-1,6-diphosphate (*Trematomus borchgrevinki*)	980	43
67	Aldolase, fructose-1,6-diphosphate (yeast)	990	50
68	Aldolase, 2-keto-3-deoxy-6-phosphogluconate (*Pseudomonas putida*)	1,090	51
69	Alkaline phosphatase (*Bacillus licheniformis* MC 14)	940	52
70	Alkaline phosphatase (*Escherichia coli*)	810	53

TABLE 2: Average Hydrophobicities of Selected Proteins (Continued)

	Protein	Average Hydrophobicity	Reference
71	Alkaline proteinase-B (*Streptomyces rectus proteolyticus*)	820	54
72	ω-Amidase (rat liver)	980	55
73	Amino acid decarboxylase, aromatic (porcine kidney)	1,050	56
74	D-Amino acid oxidase (porcine kidney)	1,050	57
75	Aminopeptidase (*Aeromonas proteolytica*)	860	58
76	Aminotripeptidase, TP-2 (porcine kidney)	1,090	59
77	Amylase (*Bacillus macerans*)	910	60
78	Amylase (rat pancreas)	970	61
79	Amylase PI (rabbit pancreas)	940	62
80	Amylase PII (rabbit pancreas)	950	62
81	Amylase PIII (rabbit pancreas)	930	62
82	Amylase (rabbit parotid)	940	62
83	α-Amylase (*Aspergillus aryzae*)	990	63
84	α-Amylase (*Bacillus stearothermophilus*)	970	64
85	α-Amylase (*Bacillus subtilis*)	970	65
86	α-Amylase (porcine pancreas)	1,040	66
87	α-Amylase (human saliva)	940	67
88	α-Amylase-I (porcine pancreas)	960	68
89	α-Amylase-II (porcine pancreas)	970	68
90	Anthranylate synthetase, component I (*Pseudomonas putida*)	1,020	69
91	Anthranylate synthetase, component II (*Pseudomonas putida*)	950	69
92	Apo-high density lipoprotein (bovine plasma)	960	70
93	Apolipoprotein – alanine-I (human plasma)	830	71
94	Apolipoprotein – alanine-II (human plasma)	830	71
95	Apolipoprotein – glutamine-I (human plasma)	870	72
96	Apolipoprotein – glutamine-I (porcine plasma)	890	72
97	Apolipoprotein – glutamine-II (human plasma)	970	73
98	Apolipoprotein-valine (human plasma)	920	74
99	Apovitellinin-I (*Dromaeus novaehollandiae*)	1,210	75
100	α-L-Arabinofuranosidase (*Aspergillus niger* K1)	900	76
101	L-Arabinose-binding protein (*Escherichia coli* B/r)	1,020	77
102	L-Arabinose isomerase (*Escherichia coli*)	990	78
103	Arginine decarboxylase (*Escherichia coli*)	980	79
104	Arginine kinase (*Callinectus sapidus*)	940	80
105	Arginine kinase (*Homarus americanus*)	980	81
106	Arginine kinase, negative (*Limulus polyphemus*)	950	80
107	Arginine kinase, neutral (*Limulus polyphemus*)	970	80
108	Arginine kinase (*Pagurus bernhardus*)	970	80
109	Asparaginase (*Escherichia coli* B)	870	82
110	Asparaginase (guinea pig)	970	83
111	L-Asparaginase (*Proteus vulgaris*)	950	84
112	Aspartate aminotransferase (ox heart)	980	85
113	Aspartate aminotransferase (rat brain, cytoplasm)	1,050	86
114	Aspartate aminotransferase (rat brain, mitochondria, fraction II)	990	86
115	Aspartate aminotransferase (rat brain, mitochondria, fraction III)	990	86
116	Aspartate-β-decarboxylase (*Alcaligenes faecalis*)	1,010	87
117	Aspartate-β-decarboxylase (*Pseudomonas dacunhae*)	1,020	88
118	Aspartate transcarbamylase, catalytic chain (*Escherichia coli*)	990	89
119	Aspartate transcarbamylase, regulatory chain (*Escherichia coli*)	980	89
120	Aspartokinase, lysine sensitive (*Escherichia coli* K-12, HfrH)	940	90
121	Aspergilliopeptidase B (*Aspergillus oryzae*)	810	91
122	Avidin (chicken egg)	930	92
123	Azurin (*Pseudomonas fluorescens*)	880	93
124	Azurin (*Pseudomonas polymyxa*)	970	94
125	Biotin carboyxyl carrier protein (*Escherichia coli*)	1,300	95
126	Blastokinin (rabbit)	1,030	96
127	Bradykininogen (bovine)	980	97
128	α-Bungarotoxin (*Bungarus multicintus*)	1,040	98
129	C_1-inactivator (human plasma)	1,020	99

TABLE 2: Average Hydrophobicities of Selected Proteins (Continued)

	Protein	Average Hydrophobicity	Reference
130	Calcitonin (bovine)	1,080	100
131	Calcitonin (human)	1,100	101
132	Calcitonin (porcine)	1,100	102
133	Calcitonin (salmon)	860	103
134	Calcium-binding protein (bovine adrenal medulla)	780	104
135	Calcium-binding protein (chick duodenal mucosa)	930	105
136	Calcium-binding protein-B (parvalbumin) (*Cyprinus carpio* muscle)	890	106
137	Calcium-binding protein (porcine brain)	740	107
138	Carbamyl phosphate synthetase (*Escherichia coli* B)	960	108
139	Carbonic anhydrase (bovine)	1,000	109
140	Carbonic anhydrase (*Gallus domesticus*)	1,000	110
141	Carbonic anhydrase (parsley)	1,100	111
142	Carbonic anhydrase (sheep)	990	112
143	Carbonic anhydrase B (bovine)	950	113
144	Carbonic anhydrase B (equine)	940	114
145	Carbonic anhydrase B (*Macaca mulata*)	960	115
146	Carbonic anhydrase B (porcine)	1,000	116
147	Carbonic anhydrase C (equine)	960	114
148	Carbonic anhydrase C (porcine)	1,010	117
149	β-Carboxy-cis,cis-muconate lactonizing enzyme (*Pseudomonas putida*)	940	118
150	γ-Carboxymuconolactone decarboxylase (*Pseudomonas putida*)	1,170	119
151	Carboxypeptidase A (bovine pancreas)	1,010	120
152	Carboxypeptidase A (*Penaeus setiferus*)	910	121
153	Carboxypeptidase B (*Penaeus setiferus*)	940	121
154	Carboxypeptidase B (*Protopterus aethiopicus*)	1,010	122
155	Procarboxypeptidase B (*Protopterus aethiopicus*)	990	122
156	Carboxypeptidase B (dogfish pancreas)	1,070	123
157	Carboxypeptidase Y (yeast)	1,050	124
158	Carboxytransphosphorylase (propionic acid bacteria)	950	125
159	Catalase (bovine liver)	1,110	126
160	Colchicine-binding protein (sea urchin)	930	127
161	Colchicine-binding protein (porcine brain)	930	128
162	Chorionic gonadotropin-α (human)	960	126
163	Chorionic gonadotropin-β (human)	1,120	130
164	Chorismate mutase-prephenate dehydratase (*Escherichia coli* K-12)	990	131
165	Chymoelastase, lysine free (*Streptomyces griseus* K-1)	720	132
166	Chymoelastase, guanidine stable (*Streptomyces griseus* K-1)	760	132
167	Chymotrypsin anionic (fin whale)	1,060	133
168	Chymotrypsin II (human)	940	134
169	Chymotrypsinogen A (dogfish)	1,010	135
170	Chymotrypsinogen B (bovine pancreas)	950	136
171	Chymotrypsinogen B (porcine pancreas)	1,010	137
172	Chymotrypsinogen C (porcine pancreas)	970	137
173	Cocoonase (*Antheraea pernyi*)	840	138
174	Cocoonase (*Antheraea polyphemus*)	860	138
175	Procecoonase (*Antheraea polyphemus*)	840	138
176	Colicin I$_a$-CA53 (*Escherichia coli*)	880	139
177	Colicin I$_b$-P9 (*Escherichia coli*)	940	139
178	Collagen (chicken tendon)	880	140
179	Collagen, α-fraction (calf skin)	880	141
180	Collagen, spongin B (sponge)	760	142
181	Collagen (sturgeon swim bladder)	770	143
182	Collagenase (*Uca pugilator*)	920	144
183	Conalbumin (chicken)	980	145
184	Corrinoid protein (*Clostridium thermoaceticum*)	890	146
185	Creatine kinase (*Cyprinus carpio* L.)	960	147
186	Creatine kinase (human)	970	148
187	Crotonase (*Clostridium acetobutylicum*)	970	149

TABLE 2: Average Hydrophobicities of Selected Proteins (Continued)

	Protein	Average Hydrophobicity	Reference
188	Crotoxin (*Crotalus terrificus terrificus*)	970	150
189	Cystathionase (rat liver)	980	151
190	Cytochrome b$_s$ (bovine liver)	1,000	152
191	Cytochrome b$_s$ (equine liver)	970	152
192	Cytochrome b$_s$ (porcine liver)	900	152
193	Cytochrome c (horse heart)	1,050	153
194	Cytochrome c (chicken)	1,050	154
195	Cytochrome c (cow, pig, sheep)	1,020	154
196	Cytochrome c (dog)	1,070	154
197	Cytochrome c (kangaroo)	1,130	154
198	Cytochrome c (king penguin)	1,060	154
199	Cytochrome c (*Macaca mulata*)	1,060	154
200	Cytochrome c (moth)	1,060	154
201	Cytochrome c (pigeon)	1,070	154
202	Cytochrome c (rabbit)	1,070	154
203	Cytochrome c (rattlesnake)	1,050	154
204	Cytochrome c (*Saccharomyces*)	1,040	154
205	Cytochrome c (tuna)	1,050	154
206	Cytochrome c (turkey)	1,050	154
207	Cytochrome CA (*Humicola lanuginosa*)	910	155
208	Cytochrome C$_3$ (*Desulfovibrio desulfuricans*)	920	156
209	Cytochrome C$_3$ (*Desulfovibrio gigas*)	870	157
210	Cytochrome C$_3$ (*Desulfovibrio salexigens*)	840	157
211	Cytochrome C$_3$ (*Desulfovibrio vulgaris*)	840	158
212	Cytochrome 553 (*Monochysis lutheri*)	730	159
213	Daunorubicin reductase (rat liver)	1,100	160
214	3-Deoxy-D-arabinoheptulosonate 7-phosphate synthetase-chorismate mutase (*Bacillus subtillis*)	1,000	161
215	Deoxycytidylate deaminase (T$_2$ r$^+$ bacteriophage)	970	162
216	Deoxyribonuclease (bovine pancreas)	930	163
217	Deoxyribonuclease A (bovine pancreas)	970	164
218	Deoxyribonuclease B (bovine pancreas)	980	164
219	Deoxyribonuclease C (bovine pancreas)	980	164
220	Deoxyribonuclease II (porcine spleen)	1,050	165
221	Deoxyribonucleic acid ligase (*Escherichia coli*)	970	166
222	Deoxyribonucleotidase inhibitor II (calf spleen)	1,010	167
223	α-Dialkylamino transamidase (*Pseudomonas cepacia*)	930	168
224	Dihydrofolate reductase (T$_4$ bacteriophage)	1,030	169
225	Dihydrofolate reductase (*Escherichia coli* MB 1428)	1,030	170
226	Dihydrofolate reductase (*Lactobacillus casei*)	1,010	171
227	Dihydrofolate reductase (*Streptococcus faecium*)	1,060	172
228	3,4-Dihydroxyphenylacetate-2,3-oxygenase (*Pseudomonas ovalis*)	980	173
229	Dopamine-β-hydroxylase (bovine adrenal)	950	174
230	Elastin (bone aorta)	940	175
231	Elastin (bovine ear cartilage)	990	175
232	Elastin (bovine ligamentum nuchae)	980	175
233	Encephalitogenic A-1 protein (bovine brain)	850	176
234	Encephalitogenic A-1 protein (human brain)	820	176
235	Endopolygalacturonase (*Verticillium albo-atrum*)	850	177
236	Enolase (monkey muscle)	970	178
237	Enolase (*Onkorhynchus keta*)	980	179
238	Enolase (*Onkorhynchus kisutch*)		
239	Enolase (rabbit muscle)	990	180
240	Enolase (*Thermus X-1*)	900	181
241	Enolase (*Thermus aquaticus* YT-1)	770	182
242	Enolase (yeast)	890	183
243	Enterotoxin C (*Staphylococcus* 137)	990	184
244	Enterotoxin E (*Staphylococcus aureus* FRI-326)	940	185
245	Enterotoxin A (*Staphylococcus aureus* 13N-2909)	980	186

TABLE 2: Average Hydrophobicities of Selected Proteins (Continued)

	Protein	Average Hydrophobicity	Reference
246	Erythrocuprein (human)	740	187
247	Esterase (pig liver)	1,100	188
248	Estradiol dehydrogenase (human placenta)	940	189
249	Factor III lac (*Staphylococcus aureus*)	850	190
250	Factor VIII (bovine plasma)	1,030	191
251	Factor VIII (human plasma)	990	192
252	Ferredoxin (*Chlorobium thiosulfatophilum*)	910	193
253	Ferredoxin (*Clostridium pasteurianium*)	920	194
254	Ferredoxin (*Chromatium*)	870	195
255	Ferredoxin (*Cyperus rotundus*)	900	196
256	Ferredoxin (*Desulfoyibrio gigas*)	930	196
257	Ferredoxin (*Spinacea*)	850	197
258	Ferredoxin (*Scenedesmus*)	800	198
259	Ferredoxin I (*Azotobacter vinelandii*)	1,050	199
260	Ferredoxin I (*Bacillus polymxa*)	920	200
261	Ferredoxin II (*Bacillus polymxa*)	920	200
262	Ferritin (horse spleen)	900	201
263	Ferritin (tadpole red blood cell)	850	202
264	Fibrinogen B (lobster)	990	203
265	Fibroin (*Bombyx mori*)	330	204
266	Fibroin (*Tussah*)	430	204
267	Ficin II (*Ficus glabrata*)	930	205
268	Ficin III (*Ficus glabrata*)	960	205
269	Flagellin (*Proteus vulgaris*)	730	206
270	Flagellin (*Salmonella* SJ 25)	680	207
271	Flagellin (*Salmonella typhimurium*)	700	208
272	Flavodoxin (*Clostridium* MP)	1,010	209
273	Flavodoxin (*Clostridium pasteurianium*)	820	210
274	Flavodoxin (*Desulfovibrio gigas*)	890	211
275	Flavodoxin (*Desulfovibrio vulgaris*)	870	212
276	Flavodoxin (*Peptostreptococcus elsdenii*)	870	213
277	Follicle stimulating hormone (ovine)	900	214
278	*N*-formimino-L-glutamate iminohydrolase (*Pseudomonas* ATCC 11299b)	930	215
279	Fructose-1,6-diphosphatase (porcine kidney)	1,010	216
280	β-Galactosidase (*Escherichia coli*)	970	217
281	Gastricin (human)	920	218
282	Gastricin (porcine)	1,100	219
283	Gliadin SP 2-2 (Cappelle 1966 flour)	1,300	220
284	Gliadin SP 2-1 (wheat, Wichita 1963)	1,250	220
285	Gliadin SP 2-2 (wheat, Wichita 1963)	1,180	220
286	Gliadin SP 2-3 (wheat, Wichita 1963)	1,180	220
287	Globulin, 0.6 S γ_2 (human plasma)	1,090	221
288	Glucagon (bovine)	810	222
289	Glutamate decarboxylase (*Escherichia coli*)	1,080	223
290	Glutamate dehydrogenase (bovine liver)	1,000	224
291	Glutamate dehydrogenase (*Neurospora crassa*)	880	225
292	Glutaminase-asparaginase (*Achromobacteraceae*)	960	226
293	Glutamine synthetase (rate liver)	1,010	227
294	γ-Glutamyl cyclotransferase (porcine liver)	960	228
295	Glyceraldehyde-3-phosphate dehydrogenase (*Bacillus stearothermophilus*)	940	229
296	Glyceraldehyde-3-phosphate dehydrogenase (bovine liver)	980	230
297	Glycerol-3-phosphate dehydrogenase (rabbit muscle)	1,040	231
298	Glycocyamine kinase (*Nepthys coeca*)	990	232
299	Glycogen phosphorylase (Baker's yeast)	1,050	233
300	Glycogen phosphorylase (*Carcharhinus falciformis*)	1,080	234
301	Glycogen phosphorylase (human)	1,060	235
302	Glycogen phosphorylase (rabbit liver)	1,060	236
303	Glycogen phosphorylase (rat)	1,070	237
304	Glycogen synthetase (porcine kidney)	870	238

TABLE 2: Average Hydrophobicities of Selected Proteins (Continued)

	Protein	Average Hydrophobicity	Reference
305	γ-Glycoprotein (human plasma)	1,070	239
306	Growth hormone (human)	950	240
307	Haptoglobin, α_2-chain (human plasma)	960	241
308	Haptoglobin, α_{1S}-chain (human plasma)	990	241
309	Haptoglobin, β-chain (human plasma)	1,020	242
310	Hemagglutinin (*Lens esculenta Muench*)	1,000	243
311	Hemagglutinin I (*Pisum sativum L. var. pyram*)	990	244
312	Hemagglutinin II (*Pisum sativum L. var. pyram*)	930	244
313	Hemagglutinin (*Robina pseudoaccacia*)	990	245
314	Hemagglutinin II (snail)	1,070	246
315	Hemagglutinin (soy)	990	247
316	Hemagglutinin (*Ulex europeus*)	950	248
317	Hemerythrin (*Golfingia gouldii*)	1,170	249
318	Hemocyanin (*Cancer magister*)	1,000	250
319	Hemocyanin (*Crustaceae C. sapidus*)	980	251
320	Hemocyanin (*Crustaceae E. spinifrons*)	980	251
321	Hemocyanin (*Crustaceae H. vulgaris*)	980	251
322	Hemocyanin (*Crustaceae P. vulgaris*)	980	251
323	Hemocyanin (*Mollusca E. moschata*)	1,050	251
324	Hemocyanin (*Mollusca M. brandaris*)	1,000	251
325	Hemocyanin (*Mollusca M. tranculus*)	1,010	251
326	Hemocyanin (*Mollusca O. macropus*)	1,050	251
327	Hemocyanin (*Mollusca O. vulgaris*)	1,080	251
328	Hemocyanin (*Xiphosura L. polyphemus*)	980	251
329	Hemoglobin (*Ascaris* body wall)	980	252
330	Hemoglobin (*Entosphenus japonicus*)	1,010	253
331	Hemoglobin (*Glycera dibranchiata*)	870	254
332	Hemoglobin, α-chain (*Catostomus clarkii*)	1,100	255
333	Hemoglobin, α-chain (human)	960	256
334	Hemoglobin, γ-chain (bovine)	960	256
335	Hemoglobin, γ-chain (human)	960	256
336	Hemoglobin, F-1 (*Eptatretus burgeri*)	1,170	258
337	Hemoglobin, F-2 (*Eptatretus burgeri*)	1,160	258
338	Hemoglobin, F-3 (*Eptatretus burgeri*)	1,140	258
339	Hemoglobin, F-4 (*Eptatretus burgeri*)	1,040	258
340	Hemopexin (human)	1,020	259
341	Hemopexin (rabbit)	1,010	259
342	High potential iron sulfur protein (*Chromatium vinosum* D)	880	260
343	High potential iron sulfur protein (*Thiocapsa pfennigii*)	930	261
344	Histidine: ammonia lyase (*Pseudomonas* ATCC 11299b)	890	262
345	Histidine-binding protein (*Salmonella typhimurium*)	970	263
346	Histidine decarboxylase (*Lactobacillus* 30a)	980	264
347	L-Histidinol phosphate aminotransferase (*Salmonella typhimurium*)	970	265
348	Histidyl-tRNA synthetase (*Salmonella typhimurium*)	940	266
349	Histone III (Calf thymus)	990	267
350	Histone III (*Letiobus bubalus*)	980	268
351	Histone III (pea)	980	269
352	Histone III (rainbow trout)	950	270
353	Hyaluronidase (bovine testicular)	1,030	271
354	Hydrogenase (*Clostridium pasteurianium* W5)	1,160	272
355	3-Hydroxyacyl coenzyme A dehydrogenase (porcine heart muscle)	960	273
356	β-Hydroxybutyrate dehydrogenase (bovine heart mitochondria)	930	274
357	α^l-Hydroxysteroid dehydrogenase (*Pseudomonas testosteroni*)	900	275
358	17-β-Hydroxysteroid dehydrogenase (human placenta)	910	276
359	Immunoglobulin, Eu heavy chain (human plasma)	980	277
360	Immunoglobulin, Eu light chain (human plasma)	860	277
361	Immunoglobulin, New λ-chain (human plasma)	870	278
365	Inorganic pyrophosphatase (yeast)	1,130	279
363	Insulin (bovine)	1,020	280

TABLE 2: Average Hydrophobicities of Selected Proteins (Continued)

	Protein	Average Hydrophobicity	Reference
364	Insulin (cod)	1,110	281
365	Inverfase (*Saccharomyces* FH4C)	990	282
366	Isoamylase I (porcine pancreas)	960	283
367	Isoamylase II (porcine pancrease)	970	283
368	Isocitrate dehydrogenase (*Azotobacter vinelandii* ATCC 9104)	980	284
369	Isocitrate dehydrogenase (porcine liver)	990	285
370	Isomerase I (rabbit)	1,080	286
371	Isomerase II (rabbit)	1,100	286
372	Keratinase (*Trichophyton mentagrophytes*)	880	287
373	Δ^5-3-Ketosteroid isomerase (*Pseudomonas testosteronii*)	910	288
374	α-Lactalbumin (bovine milk)	1,050	289
375	Lactoferrin (bovine)	920	290
376	β-Lactoglobulin (caprid)	1,040	291
377	β-Lactoglobulin A (bovine)	1,070	291
378	β-Lactoglobulin A (ovine)	1,050	291
379	β-Lactoglobulin A (bovine)	1,060	291
380	β-Lactoglobulin A (ovine)	1,050	291
381	β-Lactoglobulin C (bovine)	1,070	292
382	Lactoperoxidase (bovine milk)	1,080	292
383	Lectin, α-D-galactosyl-binding (*Bandeiraea simplicifola*)	980	293
384	Leghemoglobin I (soyaben root nodule)	1,010	294
385	Leghemoglobin II (soyaben root nodule)	1,010	294
386	Leucine aminopeptidase (bovine lens)	1,010	295
387	Lipase (rat pancreas)	980	61
388	β-Lipolytic hormone (porcine pituitary)	950	296
389	γ-Lipolytic hormone (sheep)	780	297
390	α-Lipovitellin (chicken egg yolk)	980	298
391	Luciferase (*Renilla reniformis*)	940	299
392	Luciferase-α (MAV)	880	300
393	Luciferase-β (MAV)	960	300
394	Luciferase-α (*Photobacterium fischeri*)	960	300
395	Luciferase-β (*Photobacterium fischeri*)	990	300
396	Luteinizing hormone-α chain (equine)	1,070	301
397	Luteinizing hormone-β chain (equine)	1,120	301
398	Lysine decarboxylase (*Escherichia coli*)	1,080	302
399	L-Lysine monooxygenase (*Pseudomonas fluorescens*)	1,040	303
400	Lysostaphin (*Staphylococcus aureus*)	950	304
401	Lysozyme (*Chalaropsis*)	960	305
402	Lysozyme, chick type (black swan)	970	306
403	Lysozyme, goose type (black swan)	930	306
404	Lysozyme (chicken)	890	307
405	Lysozyme (papaya)	1,080	308
406	Lysozyme (turkey)	910	309
407	Lysyl: tRNA ligase (*Escherichia coli*)	990	310
408	Lysyl: tRNA ligase (yeast)	1,040	310
409	α-Lytic protease (*Sorangium* sp.)	780	311
410	β-Lytic protease (*Sorangium* sp.)	820	311
411	Melate dehydrogenase (*Bacillus subtillis*)	970	312
412	Malate-lactate transhydrogenase (*Micrococcus lactilyticus*)	960	313
413	Malate-vitamin K reductase (*Mycobacterium phleii*)	890	314
414	L-Malic enzyme (*Escherichia coli*)	1,020	315
415	β-Melanocyte stimulating hormone (bovine)	1,180	316
416	Melilotate hydroxylase (*Pseudomonas* sp.)	970	317
417	Methionyl-tRNA synthetase (*Escherichia coli*)	1,040	318
418	β-Methyl aspartase (*Clostridium tetanomorphum*)	980	319
419	β-Methylcrotonylcoenzyme A carboxylase (*Achromobacter*)	890	320
420	Methylmalonatesemialdehydedehydrogenase (*Pseudomonas aeruginosa*)	1,030	321
421	Molydbenum-iron protein (soyaben root nodule bacteroid)	1,010	322
422	Monellin (*Dioscoreophyllum cumminsii*)	1,140	323

TABLE 2: Average Hydrophobicities of Selected Proteins (Continued)

	Protein	Average Hydrophobicity	Reference
423	Motilin (porcine intestinal mucosa)	1,020	324
424	cis,cis-Muconate lactonizing enzyme (*Pseudomonas putida*)	990	325
425	Muconolactone isomerase (*Pseudomonas putida*)	1,070	325
426	Mutarotase (bovine kidney cortex)	850	326
427	Myoglobin (human)	1,000	327
428	Myoglobin (whale)	1,040	328
429	Myoglobin (*Zalophus californianus*)	1,030	329
430	Myohemery thrin (*Dendrostamum pyroides*)	1,090	330
431	Myosin (bovine heart)	880	331
432	Myosin (cod)	860	332
433	myosin (rabbit)	890	333
434	myosin (tuna)	880	333
435	Neocarzinostatin (*Streptomyces carzinostaticus* F-41)	750	334
436	Neurophysin I (bovine pituitary)	850	335
437	Neurophysin II (bovine pituitary)	830	335
438	Neurophysin III (bovine pituitary)	820	336
439	Neurotoxin I (*Androctonus australis*)	1,070	337
440	Neurotoxin II (*Androctonus australis*)	990	337
441	Neurotoxin III (*Androctonus australis*)	1,120	337
442	Neurotoxin I (*Leiurus quinquestriatus quinquestriatus*)	1,040	337
443	Neurotoxin II (*Leiurus quinquestriatus quinquestriatus*)	970	337
444	Neurotoxin III (*Leiurus quinquestriatus quinquestriatus*)	1,030	337
445	Neurotoxin IV (*Leiurus quinquestriatus quinquestriatus*)	1,050	337
446	Neurotoxin V (*Leiurus quinquestriatus quinquestriatus*)	1,020	337
447	Nuclease (Micrococcus sodonecis ATCC 11880)	880	338
448	Nuclease (*Staphylococcus*)	980	339
449	Nucleoside diphosphokinase (Brewer's yeast)	1,030	340
450	Nucleoside phosphotransferase-A-chain (carrot)	920	341
451	Nucleoside phosphotransferase-B-chain (carrot)	860	341
452	5'-Nucleotidase (*Escherichia coli*)	930	342
453	Ornithine transcarbamylase (bovine liver)	1,040	343
454	Ornithine transcarbamylase (*Streptococcus faecalis*)	990	343
455	Ovalbumin (chicken egg)	980	145
456	Ovomacroglobulin (chiken egg)	1,040	344
457	Ovomucin (chicken egg)	990	345
458	Ovomucoid (chicken egg)	830	145
459	Ovotransferrin (chicken egg)	960	346
460	2-Oxoglutarate dehydrogenase (porcine heart)	1,010	347
461	Oxytocin (mammalian)	1,290	348
462	Papain (papaya)	1,030	349
463	Parathyroid hormone (bovine)	900	350
464	Parvalbumin (rabbit skeletal muscle)	920	351
465	Penicillocarboxypeptidase-S (*Penicillium janthinellum*)	980	352
466	Pepsin (bovine)	940	353
467	Pepsin (human)	930	218
468	Pepsin I (*Rhizopus chinensis*)	910	354
469	Pepsin II (*Rhizopus chinensis*)	910	354
470	Pepsin A (bovine)	960	355
471	Pepsinogen A (*Mustelus canis*)	920	356
472	Pepsinogen A (porcine)	970	357
473	Pepsinogen C (porcine)	1,020	358
474	Phenylalanine ammonia lyase (*Solanum tuberosum*)	1,030	359
475	Phenylalanine ammonia lyase (*Zea mays* L.)	940	359
476	Phenylalanine hydroxylase (rat liver)	930	360
477	Phosphatidyl serine decarboxylase (*Escherichia coli*)	1,050	361
478	Phosphoenol pyruvate carboxykinase (yeast)	1,030	362
479	Phosphofructokinase (chicken liver)	940	363
480	Phosphoglucomutase (rabbit muscle)	1,020	364
481	Phosphoglucose isomerase (human)	1,020	365

TABLE 2: Average Hydrophobicities of Selected Proteins (Continued)

	Protein	Average Hydrophobicity	Reference
482	Phosphoglucose isomerase (rabbit muscle)	1,010	366
483	Phosphoglycerate kinase (human erythrocyte)	980	367
484	Phospholipase A (*Bitis gabonica* venom)	910	368
485	Phospholipase A$_1$ (*Crotalus adamanteus*)	1,030	369
486	Phospholipase A$_2$ (*Crotalus atrox*)	920	370
487	Phospholipase A (*Laticuda semifasciata*)	950	371
488	Phosphorylase (frog skeletal muscle)	1,070	372
489	Phycocyanin (*P. calothricoides*)	910	93
490	Phycocyanin (*S. lividicus*)	1,040	93
491	Phycoerythrin (*P. tenera*)	830	373
492	Phytochrome (oat)	1,010	374
493	Plastocyanin (parsley)	880	375
494	Prealbumin (human serum)	990	376
495	Prealbumin, Pt 1-1 (monkey)	990	377
496	Prealbumin, Pt 2-2 (monkey)	1,010	377
497	Procarboxypeptidase A (bovine)	1,010	378
498	Procarboxypeptidase A (*Squalus acanthias*)	1,010	379
499	Progesterone-binding protein (guinea pig plasma)	1,040	380
500	Prohistidine decarboxylase (*Lactobacillus* 30a)	990	381
501	Prolactin (ovine pituitary)	950	382
502	Protease (*Bacillus thermoprotolyticus*)	940	383
503	Protease (*Mucor miehei* CBS 370.65)	940	384
504	Protease (*Staphylococcus aureus* V8)	900	385
505	Protease – A$_1$ (*Streptomyces griseus*)	700	386
506	Protease-acid (*Rhizopus chinensis*)	950	387
507	Protein kinase modulator (lobster tail muscle)	720	388
508	Prothrombin (bovine)	960	389
509	Protyrosinase (Rana pipiens pipiens	1,100	390
510	Putidaredoxin (*Pseudomonas putida*)	910	391
511	Putrescine oxidase (*Micrococcus rubens*)	860	392
512	Pyruvate kinase (bovine skeletal muscle)	1,010	393
513	Quinonoid dihydropterin reductase (sheep liver)	890	394
514	Retinol-binding protein (human)	960	395
515	Retinol-binding protein (monkey)	960	396
516	Rhodopsin (bovine)	1,120	397
517	Riboflavin-binding protein (chicken)	910	398
518	Ribonuclease (*Aspergillus fumigatus*)	830	399
519	Ribonuclease (*Bacillus subtilis*)	1,000	400
520	Ribonuclease (bovine pancreas)	780	401
521	Rubonuclease (ovine pancreas)	750	402
522	Ribonuclease Ch. (*Chalaropsis sp.*)	940	403
523	Ribonuclease N (*Neurospora crassa*)	890	404
524	Ribonuclease R$_1$ (*Rhizopus oligosporus*)	970	405
525	Ribonuclease R$_2$ (*Rhizopus oligosporus*)	960	405
526	Ribonuclease T$_1$ (*Aspergillus oryzae*)	740	406
527	Ribonuclease U$_1$ (*Ustilago sphaerogena*)	760	407
528	Ribonucleotide reductase (*Lactobacillus leichmanii*)	960	408
529	Ribulose diphosphate carboxylase (*Hydrogenomonas eutropha*)	980	409
530	Ribulose diphosphate carboxylase (*Hydrogenomonas facilis*)	980	409
531	Ribulose diphosphate carboxylase (*Rhodospirillum rubrum*)	940	410
532	Rubredoxin (*Clostridium pasteurianum*)	990	411
533	Rubredoxin (*Micrococcus aerogenes*)	1,120	412
534	Rubredoxin (*Pseudomonas oleovorans*)	960	413
535	D-Serine dehydratase (*Escherichia coli*)	920	414
536	Streptokinase (human plasma)	980	415
537	Subtilisin BPN (*Bacillus subtilis*)	810	416
538	Succinyl coenzyme A synthetase (*Escherichia coli*)	970	417
539	Sucrose synthetase (*Phaseolus aureus*)	1,040	418
540	Superoxide dismutase (*Neurospora crassa*)	760	419

TABLE 2: Average Hydrophobicities of Selected Proteins (Continued)

	Protein	Average Hydrophobicity	Reference
541	Thermolysin (*Bacillus thermoproteolyticus*)	890	420
542	Thioredoxin (T-4 bacteriophage)	1,150	421
543	Thioredoxin II (yeast)	960	422
544	Thrombin (bovine)	1,040	423
545	Thymidylate synthetase (T-2 bacteriophage)	1,140	424
546	Thyroglobulin (calf)	950	425
547	Thyroglobulin (human)	950	425
548	Thyroglobulin (porcine)	950	425
549	Thyroglobulin (rabbit)	920	425
550	Thyroglobulin (sheep)	950	425
551	Thyroid stimulating hormone – α-chain (human)	1,010	426
552	Thyroid stimulating hormone – β-chain (human)	1,170	426
553	Toxin FVII (*Dendroaspis angusticeps*)	890	427
554	Toxin α (*Dendroaspis polylepsis*)	940	428
555	Toxin γ (*Dendroaspis polylepsis*)	1,000	428
556	Toxin 4 (*Enhydrina schistosa*)	760	429
557	Toxin 5 (*Enhydrina schistosa*)	710	429
558	Toxin (*Laticauda colubrina*)	790	430
559	Toxin (*Laticauda laticauda*)	790	430
560	Toxin a (*Laticauda semifasciata*)	880	431
561	Toxin b (*Laticauda semifasciata*)	930	431
562	Toxin (*Naja haje haje*)	830	432
563	Toxin b (*Naja melanoleuca*)	1,090	433
564	Toxin d (*Naja melanoleuca*)	830	433
565	Toxin (*Naja naja atra*)	680	434
566	Toxin (*Naja nigricollis*)	930	435
567	α-Toxin A (*Staphylococcus aureus* Woods 46)	890	436
568	α-Toxin B (*Staphylococcus aureus* Woods 46)	890	436
569	Transaminase B (*Salmonella typhimurium*)	980	437
570	Transcortin (guinea pig)	960	438
571	Transcortin (human)	970	438
572	Transcortin (rabbit)	960	440
573	Transcortin (rat)	1,030	440
574	Transferrin (bovine)	930	441
575	Transferrin (equine)	970	441
576	Transferrin (human)	930	442
577	Transferrin (porcine serum)	970	441
578	Transferrin (rabbit serum)	970	441
579	Triose phosphate dehydrogenase (*Bombus nevadensis*)	1,040	443
580	Triose phosphate dehydrogenase (honey bee)	1,020	443
581	Triose phosphate dehydrogenase (lobster)	980	443
582	Triose phosphate dehydrogenase (porcine)	990	443
583	Triose phosphate isomerase-I (human erythrocyte)	920	444
584	Triose phosphate isomerase-III (human erythrocyte)	940	444
585	Triose phosphate isomerase (rabbit muscle)	930	445
586	Triose phosphate isomerase (yeast)	970	446
587	Troponin-C (rabbit muscle)	780	447
588	Troponin-T (rabbit muscle)	1,040	447
589	Trypsin (*Evasterias trochelli*)	1,000	448
590	Trypsin (lungfish)	920	449
591	Trypsin (human)	870	450
592	Trypsin, anionic (human)	930	451
593	Trypsin (porcine)	970	452
594	Trypsin (sheep)	850	453
595	Trypsin (shrimp)	840	454
596	Trypsin (*Streptomyces griseus*)	820	455
597	Trypsin inhibitor (*Ascaris lumbricoides suis*)	970	456
598	Trypsin inhibitor (bovine pancreas)	1,150	457
599	Trypsin inhibitor (porcine pancreas)	900	458

TABLE 2: Average Hydrophobicities of Selected Proteins (Continued)

	Protein	Average Hydrophobicity	Reference
600	Trypsin inhibitor (bovine pancreas)	1,070	459
601	Trypsin inhibitor I (*Phaseolus vulgaris*)	800	460
602	Trypsin inhibitor II (*Phaseolus vulgaris*)	780	460
603	Trypsinogen (bovine)	910	461
604	Trypsinogen (dogfish)	950	462
605	Trypsinogen (lungfish)	930	449
606	Trypsinogen, anionic (porcine)	980	463
607	Trypsinogen (sheep)	900	464
608	Tryptophan synthetase A (*Escherichia coli*)	1,060	465
609	Tryptophan synthetase – α-chain (*Aerobacter aerogenes*)	1,060	466
610	Tryptophan synthetase – α-chain (*Bacillus subtilis*)	1,020	467
611	Tryptophan synthetase – α-chain (*Salmonella typhimurium*)	980	466
612	Tryptophan synthetase – β-chain (*Bacillus subtilis*)	940	468
613	Tryptophanase (*Bacillus alvei*)	980	469
614	Tryptophanase (*Escherichia coli* B/1t7-A)	1,030	470
615	Tyrosyl-tRNA synthetase (*Bacillus stearo thermophilus*)	1,060	471
616	Urease (jackbean meal)	990	472
617	Urokinase S-1 (human urine)	1,030	473
618	Urokinase S-2 (human urine)	990	473
619	Valine:tRNA ligase (yeast)	1,080	310
620	Vasopressin, lysine (mammalian)	1,210	474

Complied by C. C. Bigelow and M. Channon.

References

1. Lederer, Courts, Laursen, and Westheimer, *Biochemistry*, 5, 823 (1966).
2. Klett, Fulpius, Cooper, Smith, Reich, and Possani, *J. Biol. Chem.*, 248, 6841 (1973).
3. Rosenberry, Chang, and Chen, *J. Biol. Chem.*, 247, 1555 (1972).
4. Mega, Ikenaka, and Matsushima, *J. Biochem.* (Tokyo), 68, 109 (1970).
5. Verpoorte, *J. Biol. Chem.*, 247, 4787 (1972).
6. Wetmore and Verpoorte, *Can. J. Biochem.*, 50, 563 (1972).
7. Verpoorte, *Biochemistry*, 13, 793 (1974).
8. Forsyth, Theil, and Jones, *J. Biol. Chem.*, 245, 5354 (1970).
9. Becker, Kredich, and Tomkins, *J. Biol. Chem.*, 244, 2418 (1969).
10. Ikenaka, Ishiguro, Emura, Kaufman, Isemura, Bauer, and Schmid, *Biochemistry*, 11, 3817 (1972).
11. Weihing and Korn, *Biochemistry*, 10, 590 (1971).
12. Gosselin-Rey, Gerady, Gaspar-Godfroid, and Carsten, *Biochim. Biophys. Acta*, 175, 165 (1969).
13. Carsten and Katz, *Biochim. Biophys. Acta*, 90, 534 (1964).
14. Bridgen, *Biochem. J.*, 123, 591 (1971).
15. Booyse, Hoveke, and Rafelson, Jr., *J. Biol. Chem.*, 248, 4083 (1973).
16. Carsten, *Biochemistry*, 4, 1049 (1965).
17. Adelstein and Kuehl, *Biochemistry*, 9, 1355 (1970).
18. Elzinga, *Biochemistry*, 9 1365 (1970).
19. Vanaman, Wakil, and Hill, *J. Biol. Chem.*, 243, 6420 (1968).
20. Diederich and Grisolia, *J. Biol. Chem.*, 244, 2412 (1969).
21. Ramponi, Guerritore, Treves, Nassi, and Baccari, *Arch. Biochem. Biophys.*, 130, 362 (1969).
22. Wolfenden, Tomozawa, and Bamman, *Biochemistry*, 7, 3965 (1968).
23. Phelan, McEvoy, Rooney, and Brady, *Biochim. Biophys. Acta*, 200, 370 (1970).
24. Pfrogner, *Arch. Biochem. Biophys.*, 119, 147 (1967).
25. Schramm and Hochstein, *Biochemistry*, 11, 2777 (1972).
26. Kyte, *J. Biol. Chem.*, 247, 7642 (1972).
27. Noltmann, Mahowald, and Kuby, *J. Biol. Chem.*, 237, 1146 (1962).
28. Takai, Kurashina, Suzuki-Hori, Okamoto, and Hayashi, *J. Biol. Chem.*, 249, 1965 (1974).
29. Woodward and Braymer, *J. Biol. Chem.*, 241, 580 (1966).
30. Tanaka, Haniu, and Yasunobu, *J. Biol. Chem.*, 248, 1141 (1973).
31. Chu and Kimura, *J. Biol. Chem.*, 248, 2089 (1973).
32. Shimomura and Johnson, *Biochemistry*, 8, 3991 (1969).
33. Nagata and Burger, *J. Biol. Chem.*, 249, 3116 (1974).
34. Matsuzawa and Segal, *J. Biol. Chem.*, 243, 5929 (1968).
35. Pederson and Foster, *Biochemistry*, 8, 2357 (1969).
36. Allerton, Elwyn, Edsall, and Spahr, *J. Biol. Chem.*, 237, 85 (1962).
37. McMenamy, Dintzis, and Watson, *J. Biol. Chem.*, 246, 4744 (1971).
38. Peters, Jr., *J. Biol. Chem.*, 237, 2182 (1962).
39. Jornvall, *Eur. J. Biochem.*, 16, 25 (1970).
40. Steinman and Jakoby, *J. Biol. Chem.*, 243, 730 (1968).
41. Penhoet, Kochman, and Rutter, *Biochemistry*, 8, 4396 (1969).
42. Schwartz and Horecker, *Arch. Biochem. Biophys.*, 115, 407 (1966).
43. Komatsu and Feeney, *Biochim. Biophys. Acta*, 206, 305 (1970).
44. Marquardt, *Can. J. Biochem.*, 48, 322 (1970).
45. Marquardt, *Can. J. Biochem.*, 47, 527 (1969).
46. Marquardt, *Can. J. Biochem.*, 49, 658 (1971).
47. Lebherg, Bradshaw, and Rutter, *J. Biol. Chem.*, 248, 1660 (1973).
48. Anderson, Gibbons, and Perham, *Eur. J. Biochem.*, 11, 503 (1969).
49. Rutter, Woodfin, and Blostein, *Acta Chem. Scand.*, 17, S226 (1963).
50. Harris, Kobes, Teller, and Rutter, *Biochemistry*, 8, 2442 (1969).
51. Robertson, Hammerstedt, and Wood, *J. Biol. Chem.*, 246, 2075 (1971).
52. Hulett-Cowling and Campbell, *Biochemistry*, 10, 1364 (1971).
53. Christen, Vallee, and Simpson, *Biochemistry*, 10, 1377 (1971).
54. Mizusawa and Yoshida, *J. Biol. Chem.*, 247, 6978 (1972).
55. Hersh, *Biochemistry*, 10, 2884 (1971).
56. Christenson, Dairman, and Udenfriend, *Arch. Biochem. Biophys.*, 141, 356 (1970).
57. Tu and McCormick, *J. Biol. Chem.*, 248, 6339 (1973).
58. Prescott, Wilkes, Wagner, and Wilson, *J. Biol. Chem.*, 246, 1756 (1971).
59. Chenoweth, Brown, Valenzuela, and Smith, *J. Biol. Chem.*, 248, 1684 (1972).
60. DcPinto and Campbell, *Biochemistry*, 7, 114 (1968).
61. Vandermeers and Christophe, *Biochim. Biophys. Acta*, 154, 110 (1968).
62. Malacinski and Rutter, *Biochemistry*, 8, 4382 (1969).
63. Stein, Junge, and Fischer, *J. Biol. Chem.*, 235, 371 (1960).

64. Pfueller and Elliott, *J. Biol. Chem.*, 244, 48 (1967).
65. Junge, Stein, Neurath, and Fischer, *J. Biol. Chem.*, 234, 556 (1959).
66. Caldwell, Dickey, Hanrahan, Kung, Kung, and Misko, *J. Am. Chem. Soc.*, 76, 143 (1954).
67. Muus, *J. Am. Chem. Soc.*, 76, 5163 (1954).
68. Cozzone, Paséro, Beaupoil, and Marchis-Mouren, *Biochim. Biophys. Acta*, 207, 490 (1970).
69. Queener, Queener, Meeks, and Gunsalus, *J. Biol. Chem.*, 248, 151 (1973).
70. Jonas, *J. Biol. Chem.*, 247, 7767 (1972).
71. Morrisett, David, Pownall, and Gotto, Jr., *Biochemistry*, 12, 1290 (1973).
72. Jackson, Baker, Taunton, Smith, Garner, and Gotto, Jr., *J. Biol. Chem.*, 248, 2639 (1973).
73. Lux, John, and Brewer, Jr., *J. Biol. Chem.*, 247, 7510 (1972).
74. Brown, Levy, and Fredrickson, *J. Biol. Chem.*, 245, 6588 (1970).
75. Burley, *Biochemistry*, 12, 1464 (1973).
76. Kaji and Tagawa, *Biochim. Biophys. Acta*, 207, 456 (1970).
77. Parsons and Hogg, *J. Biol. Chem.*, 249, 3602 (1974).
78. Patrick and Lee, *J. Biol. Chem.*, 244, 4277 (1969).
79. Boeker, Fischer, and Snell, *J. Biol. Chem.*, 244, 5239 (1969).
80. Blethen and Kaplan, *Biochemistry*, 7, 2123 (1968).
81. Blethen and Kaplan, *Biochemistry*, 6, 1413 (1967).
82. Ho, Milikin, Bobbitt, Grinnon, Burch, Frank, Boeck, and Squires, *J. Biol. Chem.*, 245, 3708 (1970).
83. Yellin and Wriston, Jr., *Biochemistry*, 5, 1605 (1966).
84. Tosa, Sano, Yamamoto, Nakamura, and Chibata, *Biochemistry*, 12, 1075 (1973).
85. Marino, Scardi, and Zito, *Biochem. J.*, 99, 595 (1966).
86. Magee and Phillips, *Biochemistry*, 10, 3397 (1971).
87. Tate and Meister, *Biochemistry*, 7, 3240 (1968).
88. Tate and Meister, *Biochemistry*, 9, 2626 (1970).
89. Weber, *J. Biol. Chem.*, 243, 543 (1968).
90. Lafuma, Gros, and Patte, *Eur. J. Biochem.*, 15, 111 (1970).
91. Subramanian and Kalnitsky, *Biochemistry*, 3, 1868 (1964).
92. Green and Toms, *Biochem. J.*, 118, 67 (1970).
93. Berns, Scott, and O'Reilly, *Science*, 145, 1054 (1964).
94. Ambler and Brown, *J. Mol. Biol.*, 9, 825 (1964).
95. Fall and Vagelos, *J. Biol. Chem.*, 247 8005 (1972).
96. Krishnan and Daniel Jr., *Biochim. Biophys. Acta*, 168, 579 (1966).
97. Nagawa, Mizushima, Sato, Iwanaga, and Suziki, *J. Biochem.* (Tokyo), 60, 643 (1966).
98. Clark, Macmurchie, Elliott, Wolcott, Landed, and Raftery, *Biochemistry*, 11, 1663 (1972).
99. Haupt, Heimburger, Kranz, and Schwick, *Eur. J. Biochem.*, 17, 254 (1970).
100. Brewer, Jr., Schlueter, and Aldred, *J. Biol. Chem.*, 245, 4232 (1970).
101. Neher, Riniker, Maier, Byfield, Gudmundsson, and MacIntyre, *Nature* (Lond.), 220, 984 (1968).
102. Brewer, Jr., Keutman, Potts, Jr., Reisfeld, Schlueter, and Munson, *J. Biol. Chem.*, 243, 5739 (1968).
103. O'Dor, Parkes, and Copp, *Can. J. Biochem.*, 47, 823 (1969).
104. Brooks and Siegel, *J. Biol. Chem.*, 248, 4189 (1973).
105. Bredderman and Wasserman, *Biochemistry*, 13, 1687 (1974).
106. Coffee and Bradshaw, *J. Biol. Chem.*, 248, 3305 (1973).
107. Wolff and Siegel, *J. Biol. Chem.*, 247, 4180 (1972).
108. Foley, Poon, and Anderson, *Biochemistry*, 10, 4562 (1971).
109. Nyman and Lindskog, *Biochim. Biophys. Acta*, 85, 141 (1964).
110. Bernstein and Schraer, *J. Biol. Chem.*, 247, 1306 (1972).
111. Tobin, *J. Biol. Chem.*, 245, 2656 (1970).
112. Tanis and Tashian, *Biochemistry*, 10, 4852 (1971).
113. Wong and Tanford, *J. Biol. Chem.*, 248, 8518 (1973).
114. Furth, *J. Biol. Chem.*, 243, 4832 (1968).
115. Duff and Coleman, *Biochemistry*, 5, 2009 (1966).
116. Ashworth, Spencer, and Brewer, *Arch. Biochem. Biophys.*, 142, 122 (1971).
117. Tanis, Tashian, and Yu, *J. Biol. Chem.*, 245, 6003 (1970).
118. Patel, Meagher, and Ornston, *Biochemistry*, 12, 3531 (1973).
119. Parke, Meagher, and Ornston, *Biochemistry*, 12, 3537 (1973).
120. Smith and Stockell, *J. Biol. Chem.*, 207, 501 (1954).
121. Gates and Travis, *Biochemistry*, 12, 1867 (1973).
122. Reeck and Neurath, *Biochemistry*, 11, 3947 (1972).
123. Prahl and Neurath, *Biochemistry*, 5, 4137 (1966).
124. Haberland, Willard, and Wood, *Biochemistry*, 11, 712 (1972).
125. Hayashi, Moore, and Stein, *J. Biol. Chem.*, 248, 2296 (1973).
126. Schnuchel, *Z. Physiol. Chem.*, 303, 91 (1956).
127. Shelanski and Taylor, *J. Cell Biol.*, 38, 304 (1968).
128. Weisenbug, Borisy, and Taylor, *Biochemistry*, 7, 4466 (1968).
129. Bellisario, Carlsen, and Bahl, *J. Biol. Chem.*, 248, 6796 (1973).
130. Carlsen, Bahl, and Swaminathan, *J. Biol. Chem.*, 248, 6810 (1973).
131. Davidson, Blackburn, and Dopheide, *J. Biol. Chem.*, 247, 4441 (1972).
132. Siegel and Award, Jr., *J. Biol. Chem.*, 248, 3233 (1973).
133. Koide and Matsuoka, *J. Biochem.* (Tokyo), 68, 1 (1970).
134. Coan, Roberts, and Travis, *Biochemistry*, 10, 2711 (1971).
135. Prahl and Neurath, *Biochemistry*, 5, 2131 (1966).
136. Smillie, Enenkel, and Kay, *J. Biol. Chem.*, 241, 2097 (1966).
137. Gratecos, Guy, Rovery, and Desnuelle, *Biochim. Biophys. Acta*, 175, 82 (1969).
138. Kramer, Felsted, and Law, *J. Biol. Chem.*, 248, 3021 (1973).
139. Konisky, *J. Biol. Chem.*, 247, 3750 (1972).
140. Leach, *Biochem. J.*, 67, 83 (1957).
141. Piez, Weiss, and Lewis, *J. Biol, Chem.*, 235, 1987 (1960).
142. Piez and Gross, *Biochim. Biophys. Acta*, 34, 24 (1959).
143. Eastore, *Biochem. J.*, 65, 363 (1957).
144. Eisen, Henderson, Jeffrey, and Bradshaw, *Biochemistry*, 12, 1814 (1973).
145. Lewis, Snell, Hirschman, and Fraenkel-Conrat, *J. Biol. Chem.*, 186, 23 (1950).
146. Ljungdahl, LeGall, and Lee, *Biochemistry*, 12, 1802 (1973).
147. Gosselin-Rey and Gerday, *Biochim. Biophys. Acta*, 221, 241 (1970).
148. Kumdavalli, Moreland, and Watts, *Biochem. J.*, 117, 513 (1970).
149. Waterson, Castellino, Hass, and Hill, *J. Biol. Chem.*, 247, 5266 (1972).
150. Fischer and Dorfel, *Z. Physiol. Chem.*, 297, 278 (1954).
151. Loiselet and Chatagner, *Biochim. Biophys. Acta*, 130, 180 (1966).
152. Ozols, *Biochemistry*, 13, 426 (1974).
153. Stellwagen and Rysary, *J. Biol. Chem.*, 247, 8074 (1972).
154. Margoliash and Schejter, *Adv. Protein Chem.*, 21, 113 (1966).
155. Morgan, Hensley, Jr., and Riehm, *J. Biol. Chem.*, 247, 6555 (1972).
156. Drucker, Trousil, Campbell, Barlow, and Margoliash, *Biochemistry*, 9, 1515 (1970).
157. Drucker, Trousil, and Campbell, *Biochemistry*, 9, 3395 (1970).
158. Trousil, and Campbell, *J. Biol. Chem.*, 249, 386 (1974).
159. Laycock and Cragie, *Can. J. Biochem.*, 49, 641 (1971).
160. Felsted, Gee, and Bachur, *J. Biol. Chem.*, 249, 3672 (1974).
161. Huang, Nakatsukasa, and Nester, *J. Biol. Chem.*, 249, 4467 (1974).
162. Maley, Guarino, and Maley, *J. Biol. Chem.*, 247, 931 (1972).
163. Gehrmann, and Okada, *Biochim. Biophys. Acta*, 23, 621 (1957).
164. Salnikow, Moore, and Stein, *J. Biol. Chem.*, 245, 5685 (1970).
165. Oshima and Price, *J. Biol. Chem.*, 248, 7522 (1973).
166. Modrich, Anraku, and Lehaman, *J. Biol. Chem.*, 248, 7495 (1973).
167. Lindberg, *Biochemistry*, 6, 323 (1967).
168. Lamartiniere, Itoh, and Dempsey, *Biochemistry*, 10, 4783 (1971).
169. Erickson and Mathews, *Biochemistry*, 12, 372 (1973).
170. Poe, Greenfield, Hirshfield, Williams, and Hoogsteen, *Biochemistry*, 11, 1023 (1972).
171. Gundersen, Dunlap, Harding, Freisheim, Otting, and Huennekens, *Biochemistry*, 11, 1018 (1972).
172. D'Souza, Warwick, and Freisheim, *Biochemistry*, 11, 1528 (1972).
173. Senoh, Kita, and Kamimoto, in *Biological and Chemical Aspects of Oxygenases*, Bloch and Hayaishi, Eds. Maruzen, Tokyo, 1966, 378.
174. Craine, Daniels, and Kaufman, *J. Biol. Chem.*, 248, 7838 (1973).
175. Gotte, Stern, Elsden, and Partridge, *Biochem. J.*, 87, 344 (1963).
176. Oshiro and Eylar, *Arch. Biochem. Biophys.*, 138, 606 (1970).
177. Wang and Keen, *Arch. Biochem. Biophys.*, 141, 749 (1970).
178. Winstead, *Biochemistry*, 11, 1046 (1972).
179. Ruth, Soja, and Wold, *Arch. Biochem. Biophys.*, 140, 1 (1970).
180. Holt and Wold, *J. Biol. Chem.*, 236, 3227 (1961).
181. Barnes and Stellwagen, *Biochemistry*, 12, 1559 (1973).
182. Stellwagen, Cronlund, and Barnes, *Biochemistry*, 12, 1552 (1973).
183. Malmström, Kimmel, and Smith, *J. Biol. Chem.*, 234, 1108 (1959).

184. Huang, Shih, Borja, Avena, and Bergdoll, *Biochemistry*, 6, 1480 (1967).
185. Borja, Fanning, Huang, and Bergdoll, *J. Biol. Chem.*, 247, 2456 (1972).
186. Schantz, Roessler, Woodburn, Lynch, Jacoby, Silverman, Gorman, and Spero, *Biochemistry*, 11, 360 (1972).
187. Kimmel, Markowitz, and Brown, *J. Biol. Chem.*, 234, 46 (1959).
188. Barker and Jencks, *Biochemistry*, 8, 3879 (1969).
189. Burns, Engle, and Bethune, *Biochemistry*, 11, 2699 (1972).
190. Hays, Simoni, and Roseman, *J. Biol. Chem.*, 248, 941 (1973).
191. Schmer, Kirby, Teller, and Davie, *J. Biol. Chem.*, 247, 2512 (1972).
192. Legaz, Schmer, Counts, and Davie, *J. Biol. Chem.*, 248, 3946 (1973).
193. Buchanan, Matsubara, and Evans, *Biochim. Biophys. Acta*, 189, 46 (1969).
194. Tanaka, Nakashima, Benson, Mower, and Yasunobu, *Biochemistry*, 5, 1666 (1966).
195. Sasaki and Matsubara, *Biochem. Biophys. Res. Commun.*, 28, 467 (1967).
196. Lee, Travis, and Black, *Arch. Biochem. Biophys.*, 141, 676 (1970).
197. Matsubara, Sasaki, and Chain, *Proc. Natl. Acad. Sci. USA*, 57, 439 (1967).
198. Matsubara, *J. Biol. Chem.*, 243, 370 (1968).
199. Yoch and Arnon, *J. Biol. Chem.*, 247, 4514 (1972).
200. Stombaugh, Burris, and Orme-Johnson, *J. Biol. Chem.*, 248, 7951 (1973).
201. Harrison, Hofmann, and Mainwaring, *J. Mol. Biol.*, 4, 251 (1962).
202. Theil, *J. Biol. Chem.*, 248, 622 (1973).
203. Fuller and Doolittle, *Biochemistry*, 10, 1305 (1971).
204. Lucas, Shaw, and Smith, *J. Mol. Biol.*, 2, 339 (1960).
205. Kortt, Hamilton, Webb, and Zener, *Biochemistry*, 13, 2023 (1974).
206. Kobayashi, Rinker, and Koffler, *Arch, Biochem. Biophys.*, 84, 342 (1959).
207. Hotani, Ooi, Kagawa, Asakura, and Yamaguchi, *Biochim. Biophys. Acta*, 214, 206 (1970).
208. Joys and Rankis, *J. Biol. Chem.*, 247, 5180 (1972).
209. Tanaka, Haniu, and Yasunobu, *J. Biol. Chem.*, 249, 4393 (1974).
210. Knight, Jr. and Hardy, *J. Biol. Chem.*, 242, 1370 (1967).
211. Dubourdieu and LeGall, *Biochem. Biophys. Res. Commun.*, 38, 965 (1970).
212. Tanaka, Haniu, Matsueda, Yasunobu, Mayhew, and Massey, *Biochemistry*, 10, 3041 (1971).
213. Mayhew and Massey, *J. Biol. Chem.*, 244, 794 (1969).
214. Cahill, Shetlar, Payne, Endecott, and Li, *Biochim. Biophys. Acta*, 154, 40 (1968).
215. Wickner and Tabor, *J. Biol. Chem.*, 247, 1605 (1972).
216. Mendicino, Kratowich, and Oliver, *J. Biol. Chem.*, 247, 6643 (1972).
217. Fowler and Zabin, *J. Biol. Chem.*, 245, 5032 (1970).
218. Mills and Tang, *J. Biol. Chem.*, 242, 3093 (1967).
219. Tauber and Madison, *J. Biol. Chem.*, 240, 645 (1965).
220. Booth and Ewart, *Biochim, Biophys. Acta*, 181, 226 (1969).
221. Nimberg and Schmid, *J. Biol. Chem.*, 247, 5056 (1972).
222. Behrens and Bromer, *Vitam. Horm.* (New York), 16, 263 (1958).
223. Strausbauch and Fischer, *Biochemistry*, 9, 226 (1970).
224. Moon and Smith, *J. Biol. Chem.*, 248, 3082 (1973).
225. Jacobson, Strickland, and Barratt, *Biochim. Biophys. Acta*, 188, 283 (1969).
226. Roberts, Holcenberg, and Dolowy, *J. Biol. Chem.*, 247, 84 (1972).
227. Tate, Leu, and Meister, *J. Biol. Chem.*, 247, 5312 (1972).
228. Adamson, Sewczuk, and Connell, *Can. J. Biochem.*, 49, 218 (1971).
229. Singleton, Kimmel, and Amelunxen, *J. Biol. Chem.*, 224, 1623 (1969).
230. Heinz and Kulbe, *Z. Physiol. Chem.*, 351, 249 (1970).
231. Fondy, Ross, and Sollohub, *J. Biol. Chem.*, 244, 1631 (1969).
232. Pradel, Kassab, Conlay, and Thoai, *Biochim. Biophys. Acta*, 154, 305 (1968).
233. Fosset, Muir, Nielsen, and Fischer, *Biochemistry*, 10, 4105 (1971).
234. Assaf and Yunis, *Biochemistry*, 12, 1423 (1973).
235. Appelman, Yunis, Krebs, and Fischer, *J. Biol. Chem.*, 238, 1358 (1963).
236. Wolf, Fischer, and Krebs, *Biochemistry*, 9, 1923 (1970).
237. Sevilla and Fischer, *Biochemistry*, 8, 2161 (1969).
238. Issa and Mendicino, *J. Biol. Chem.*, 248, 685 (1973).
239. Boenisch and Alper, *Biochim. Biophys. Acta*, 214, 135 (1970).
240. Dixon and Li, *J. Gen. Physiol.*, suppl. 45, 176 (1962).
241. Black and Dixon, *Nature* (Lond.), 218, 736 (1968).
242. Barnett, Lee, and Bowman, *Biochemistry*, 11, 1189 (1972).
243. Ticha, Entlicher, Kostir, and Kocourek, *Biochim. Biophys. Acta*, 221, 282 (1970).
244. Entlicher, Kostir, and Kocourek, *Biochim. Biophys. Acta*, 221, 272 (1970).
245. Bourillon and Fort, *Biochim. Biophys. Acta*, 154, 28 (1968).
246. Hammarström and Kabat, *Biochemistry*, 8, 2696 (1969).
247. Wada, Pallansch, and Liener, *J. Biol. Chem.*, 233, 395 (1958).
248. Matsumoto and Osawa, *Biochim. Biophys. Acta*, 194, 180 (1969).
249. Groskopf, Holleman, Margoliash, and Klotz, *Biochemistry*, 5, 3779 (1966).
250. Carpenter and van Holde, *Biochemistry*, 12, 2231 (1973).
251. Ghiretti-Magaldi, Nuzzolo, and Ghiretti, *Biochemistry*, 5, 1943 (1966).
252. Okazaki, Wittenberg, Briehl, and Wittenberg, *Biochim. Biophys. Acta*, 140, 258 (1967).
253. Dohi, Sugita, and Yoneyama, *J. Biol. Chem.*, 248, 2354 (1973).
254. Imamura, Baldwin, and Riggs, *J. Biol. Chem.*, 247, 2785 (1972).
255. Powers and Edmundson, *J. Biol. Chem.*, 247, 6694 (1972).
256. Schroeder, Shelton, Shelton, Cormick, and Jones, *Biochemistry*, 2, 992 (1963).
257. Babin, Schroeder, Shelton, Shelton, and Robberson, *Biochemistry*, 5, 1297 (1966).
258. Bannai, Sugita, and Yoneyama, *J. Biol. Chem.*, 247, 505 (1972).
259. Hrkal and Muller-Eberhard, *Biochemistry*, 10, 1746 (1971).
260. Dus, Tedro, and Bartsch, *J. Biol. Chem.*, 248, 7318 (1973).
261. Tedro, Meyer, and Kamen, *J. Biol. Chem.*, 249, 1182 (1974).
262. Klee and Gladner, *J. Biol. Chem.*, 247, 8051 (1972).
263. Lever, *J. Biol. Chem.*, 247, 4317 (1972).
264. Chang and Snell, *Biochemistry*, 7, 2012 (1968).
265. Henderson and Snell, *J. Biol. Chem.*, 248, 1906 (1973).
266. DeLorenzo, Di Natale, and Schechter, *J. Biol. Chem.*, 249, 908 (1974).
267. Delange, Hooper, and Smith, *J. Biol. Chem.*, 248, 3261 (1973).
268. Hooper, Smith, Sommer, and Chalkley, *J. Biol. Chem.*, 248, 3275 (1973).
269. Patthy, Smith, and Johnson, *J. Biol. Chem.*, 248, 6834 (1973).
270. Bailey and Dixon, *J. Biol. Chem.*, 248, 5463 (1973).
271. Brunish and Hogberg, *C.R. Trav. Lab. Carlsberg*, 32, 35 (1960).
272. Nakos and Mortenson, *Biochemistry*, 10, 2442 (1971).
273. Noyes and Bradshaw, *J. Biol. Chem.*, 248, 3052 (1973).
274. Menzel and Hammes, *J. Biol. Chem.*, 248, 4885 (1973).
275. Squire, Delin, and Porath, *Biochim. Biophys. Acta*, 89, 409 (1964).
276. Jarabak and Street, *Biochemistry*, 10, 3831 (1971).
277. Edelman, *Biochemistry*, 9, 3197 (1970).
278. Chen and Poljak, *Biochemistry*, 13, 1295 (1974).
279. Heinrikson, Sterner, Noyes, Cooperman, and Bruchman, *J. Biol. Chem.*, 248, 4235 (1973).
280. Corfield and Robson, *Biochem. J.*, 84, 146 (1962).
281. Grant and Reid, *Biochem. J.*, 106, 531 (1968).
282. Neumann and Lampen, *Biochemistry*, 6, 468 (1967).
283. Cozzone, Paséro, and Marchis-Mouren, *Biochim. Biophys. Acta*, 200, 590 (1970).
284. Chung and Franzen, *Biochemistry*, 8, 3175 (1969).
285. Illingworth and Tipton, *Biochem. J.*, 118, 253 (1970).
286. Yoshida and Carter, *Biochim. Biophys. Acta*, 194, 151 (1969).
287. Yu, Harmon, Wachter, and Blank, *Arch. Biochem. Biophys.*, 135, 363 (1969).
288. Kawahara, Wang, and Talalay, *J. Biol. Chem.*, 237, 1500 (1962).
289. Gordon and Ziegler, *Arch. Biochem. Biophys.*, 57, 80 (1955).
290. Castellino, Fish, and Mann, *J. Biol. Chem.*, 245, 4269 (1970).
291. Bell, McKenzie, and Shaw, *Biochim. Biophys. Acta*, 154, 284 (1968).
292. Rombauts, Schroeder, and Morrison, *Biochemistry*, 6, 2965 (1967).
293. Hayes and Goldstein, *J. Biol. Chem.*, 249, 1094 (1974).
294. Elfolk, *Acta Chem. Scand.*, 15, 545 (1961).
295. Carpenter and Vahl, *J. Biol. Chem.*, 248, 294 (1972).
296. Gilardeau and Chretien, *Can. J. Biochem.*, 48, 1017 (1970).
297. Gráf, Cseh, and Medzihradszky-Schweiger, *Biochim. Biophys. Acta*, 175, 444 (1969).
298. Bernardi and Cook, *Biochim. Biophys. Acta*, 44, 96 (1960).

299. Karkhanis and Cormier, *Biochemistry*, 10, 317 (1971).
300. Hastings, Weber, Friedland, Eberbard, Mitchell, and Gunsalus, *Biochemistry*, 8, 4681 (1969).
301. Landefeld and McShan, *Biochemistry*, 13, 1389 (1974).
302. Sabo and Fischer, *Biochemistry*, 13, 670 (1974).
303. Flashner and Massey, *J. Biol. Chem.*, 248, 2579 (1974).
304. Trayer and Buckley, *J. Biol. Chem.*, 245, 4842 (1970).
305. Shih and Hash, *J. Biol. Chem.*, 246, 996 (1971).
306. Arnheim, Hindenburg, Begg, and Morgan, *J. Biol. Chem.*, 248, 8036 (1973).
307. Canfield and Anfinsen, *J. Biol. Chem.*, 238, 2684 (1963).
308. Smith, Kimmel, Brown, and Thompson, *J. Biol. Chem.*, 215, 67 (1955).
309. Larue and Speck, Jr., *J. Biol. Chem.*, 245, 1985 (1970).
310. Rymo, Lundvik, and Lagerkvist, *J. Biol. Chem.*, 247, 3888 (1972).
311. Jurášek and Whitaker, *Can. J. Biochem.*, 45, 917 (1967).
312. Yoshida, *J. Biol. Chem.*, 240, 1113 (1965).
313. Allen and Patil, *J. Biol. Chem.*, 247, 909 (1972).
314. Imai and Brodie, *J. Biol. Chem.*, 248, 7487 (1973).
315. Spina, Jr., Bright, and Rosenbloom, *Biochemistry*, 9, 3794 (1970).
316. Li, *Adv. Protein Chem.*, 12, 270 (1957).
317. Strickland and Massey, *J. Biol. Chem.*, 248, 2944 (1973).
318. Lawrence, *Eur. J. Biochem.*, 15, 436 (1970).
319. Hsiang, Myrtle, and Bright, *J. Biol. Chem.*, 242, 3079 (1967).
320. Apitz-Castro, Rehn, and Lynen, *Eur. J. Biochem.*, 16, 71 (1970).
321. Bannerjee, Sanders, and Sokatch, *J. Biol. Chem.*, 245, 1828 (1970).
322. Israel, Howard, Evans, and Russel, *J. Biol. Chem.*, 249, 500 (1974).
323. Morris, Mortenson, Deibler, and Cagan, *J. Biol. Chem.*, 248, 534 (1973).
324. Brown, Cook, and Dryburgh, *Can. J. Biochem.*, 51, 533 (1973).
325. Meagher and Ornston, *Biochemistry*, 12, 3523 (1973).
326. Fishman, Pentchev, and Bailey, *Biochemistry*, 12, 2490 (1973).
327. Perkoff, Hill, Brown, and Tyler, *J. Biol. Chem.*, 237, 2820 (1962).
328. Edmundson, *Nature*, (Lond.), 205, 883 (1965).
329. Vigna, Gurd, and Gurd, *J. Biol. Chem.*, 249, 4144 (1974).
330. Klippenstein, Riper, and Oosterom, *J. Biol. Chem.*, 247, 5956 (1972).
331. Tada, Bailin, Bárány, and Bárány, *Biochemistry*, 8, 4842 (1969).
332. Connell and Howgate, *Biochem. J.*, 71, 83 (1959).
333. Chung, Richards, and Olcott, *Biochemistry*, 6, 3154 (1967).
334. Samy, Atreyi, Maeda, and Meienhofer, *Biochemistry*, 13, 1007 (1974).
335. Hollenberg and Hope, *Biochem. J.*, 106, 557 (1968).
336. Furth and Hope, *Biochem. J.*, 116, 545 (1970).
337. Miranda, Kupeyan, Rochat, Rochat, and Lissitzky, *Eur. J. Biochem.*, 16, 514 (1970).
338. Berry, Johnson, and Campbell, *Biochim. Biophys. Acta*, 220, 269 (1970).
339. Omenn, Ontjes, and Anfinsen, *Biochemistry*, 9, 304 (1970).
340. Palmieri, Yue, Jacobs, Maland, Yu, and Kuby, *J. Biol. Chem.*, 248, 4486 (1973).
341. Rodgers and Chargaff, *J. Biol. Chem.*, 247, 5448 (1972).
342. Neu, *J. Biol. Chem.*, 242, 3896 (1967).
343. Marshall and Cohen, *J. Biol. Chem.*, 247, 1641 (1972).
344. Donovan, Mapes, Davis, and Hamburg, *Biochemistry*, 8, 4190 (1969).
345. Donovan, Davis, and White, *Biochim. Biophys. Acta*, 207, 190 (1970).
346. Phillips and Azari, *Biochemistry*, 10, 1160 (1971).
347. Koike, Hamada, Tanaka, Otsuka, Ogasahara, and Koike, *J. Biol. Chem.*, 249, 3836 (1974).
348. DuVigneaud, Ressler, Swan, Roberts, and Katsoyannis, *J. Am. Chem. Soc.*, 76, 3115 (1954).
349. Smith, Stockell, and Kimmel, *J. Biol. Chem.*, 207, 551 (1954).
350. Rasmussen and Craig, *J. Biol. Chem.*, 236, 759 (1961).
351. Lehky, Blum, Stein and Fischer, *J. Biol. Chem.*, 249, 4332 (1974).
352. Jones and Hofmann, *Can. J. Biochem.*, 50, 1297 (1972).
353. Lang and Kassell, *Biochemistry*, 10, 2296 (1971).
354. Graham, Sodek, and Hofmann, *Can. J. Biochem.*, 51, 789 (1973).
355. Chow and Kassell, *J. Biol. Chem.*, 243, 1718 (1968).
356. Merrett, Bar-Eli, and Van Vunakis, *Biochemistry*, 8, 3696 (1969).
357. Rajagopalan, Moore, and Stein, *J. Biol. Chem.*, 241, 4940 (1966).
358. Ryle and Hamilton, *Biochem. J.*, 101, 176 (1966).
359. Havir and Hanson, *Biochemistry*, 12, 1583 (1973).
360. Fisher, Kirkwood, and Kaufman, *J. Biol. Chem.*, 247, 5161 (1972).
361. Dowhan, Wickner, and Kennedy, *J. Biol. Chem.*, 249, 3079 (1974).
362. Cannata, *J. Biol. Chem.*, 245, 792 (1970).
363. Kono, Uyeda, and Oliver, *J. Biol. Chem.*, 248, 8592 (1973).
364. Harshman and Six, *Biochemistry*, 8, 3423 (1969).
365. Tilley and Gracy, *J. Biol. Chem.*, 249, 4571 (1974).
366. Pon, Schnackerz, Blackburn, Chatterjee, and Noltmann, *Biochemistry*, 9, 1506 (1970).
367. Yoshida and Watanabe, *J. Biol. Chem.*, 247, 440 (1972).
368. Botes and Viljoen, *J. Biol. Chem.*, 249, 3827 (1974).
369. Wells and Hanahan, *Biochemistry*, 8, 414 (1969).
370. Hachimori, Wells, and Hanahan, *Biochemistry*, 10, 4084 (1971).
371. Tu, Passey, and Toom, *Arch. Biochem. Biophys.*, 140, 96 (1970).
372. Metzger, Glaser, and Helmreich, *Biochemistry*, 7, 2021 (1968).
373. Kimmel and Smith, *Bull. Soc. Chem. Biol.*, 40, 2049 (1958).
374. Mumford and Jenner, *Biochemistry*, 5, 3657 (1966).
375. Graziani, Agrò, Rotilio, Barra, and Mondovi, *Biochemistry*, 13, 804 (1974).
376. Peterson, *J. Biol. Chem.*, 246, 34 (1971).
377. von Jaarsveld, Branch, Robbins, Morgan, Kanda, and Canfield, *J. Biol. Chem.*, 248, 7898 (1973).
378. Freisheim, Walsh, and Neurath, *Biochemistry*, 6, 3010 (1967).
379. Lacko and Neurath, *Biochemistry*, 9, 4680 (1970).
380. Milgrom, Allouch, Atger, and Baulieu, *J. Biol. Chem.*, 248, 1106 (1973).
381. Recsei and snell, *Biochemistry*, 12, 365 (1973).
382. Ma, Brovetto-Cruz, and Li, *Biochemistry*, 9, 2302 (1970).
383. Ohta, *J. Biol. Chem.*, 242, 509 (1967).
384. Rickert and Elliot, *Can. J. Biochem.*, 51, 1638 (1973).
385. Drapeau, Boily, and Houmard, *J. Biol. Chem.*, 247, 6720 (1972).
386. Johnson and Smillie, *Can. J. Biochem.*, 50, 589 (1972).
387. Tsuru, Hattori, Tsuji, and Fukumoto, *J. Biochem.* (Tokyo), 67, 415 (1970).
388. Donnelly, Jr., Kuo, Reyes, Liu, and Greengard, *J. Biol. Chem.*, 248, 190 1973).
389. Heldebrant, Butkowski, Bajaj, and Mann, *J. Biol. Chem.*, 248, 7149 (1973).
390. Barisas and McGuire, *J. Biol. Chem.*, 249, 3151 (1974).
391. Tanaka, Haniu, Yasunobu, Dus, and Gunsalus, *J. Biol. Chem.*, 249, 3689 (1974).
392. DeSa, *J. Biol. Chem.*, 247, 5527 (1972).
393. Cardenas, Dyson, and Strandholm, *J. Biol. Chem.*, 248, 6931 (1973).
394. Cheema, Soldin, Knapp, Hofmann, and Scrimgeour, *Can. J. Biochem.*, 51, 1229 (1973).
395. Rask, Vahlquist, and Peterson, *J. Biol. Chem.*, 246, 6638 (1971).
396. Vahlquist and Peterson, *Biochemistry*, 11, 4526 (1972).
397. Shields, Dinovo, Henriksen, Kimbel, Jr., and Millar, *Biochim. Biophys. Acta*, 147, 238 (1967).
398. Farrell, Jr., Mallette, Buss, and Clagett, *Biochim. Biophys. Acta*, 194, 433 (1969).
399. Glitz, Angel, and Eichler, *Biochemistry*, 11, 1746 (1972).
400. Lees and Hartley, Jr., *Biochemistry*, 5, 3951 (1966).
401. Yankeelov, Jr., *Biochemistry*, 9, 2433 (1970).
402. Becker, Halbrook, and Hirs, *J. Biol. Chem.*, 248, 7826 (1973).
403. Fletcher, Jr. and Hash, *Biochemistry*, 11, 4281 (1972).
404. Uchida and Egami, in *The Enzymes*, 3rd ed., Boyer, Ed., Academic Press, New York, 1971, 205.
405. Woodroof and Glitz, *Biochemistry*, 10, 1532 (1971).
406. Takahashi, *J. Biol. Chem.*, 240, PC 4117 (1965).
407. Kenney and Dekker, *Biochemistry*, 10, 4962 (1971).
408. Panagou, Orr, Dunstone, and Blakley, *Biochemistry*, 11, 2378 (1972).
409. Kuehn and McFadden, *Biochemistry*, 8, 2403 (1969).
410. Tabita and McFadden, *J. Biol. Chem.*, 249, 3459 (1974).
411. Lovenburg and Williams, *Biochemistry*, 8, 141 (1969).
412. Bachmayer, Benson, Yasunobu, Garrard, and Whiteley, *Biochemistry*, 7, 986 (1968).
413. Lode and Coon, *J. Biol. Chem.*, 246, 791 (1971).
414. Dowhan, Jr. and Snell, *J. Biol. Chem.*, 245, 4618 (1970).
415. Brockway and Castellino, *Biochemistry*, 13, 2063 (1974).
416. Matsubara, Kaspar, Brown, and Smith, *J. Biol. Chem.*, 240, 1125 (1965).
417. Leitzmann, Wu, and Boyer, *Biochemistry*, 8, 2338 (1970).

418. Delmer, *J. Biol. Chem.*, 247, 3822 (1972).
419. Misra and Fridovich, *J. Biol. Chem.*, 247, 3410 (1972).
420. Titani, Hermodson, Ericsson, Walsh, and Neurath, *Biochemistry*, 11, 2427 (1972).
421. Sjöberg, *J. Biol. Chem.*, 247, 8058 (1972).
422. Gonzalez, Baldesten, and Reichard, *J. Biol. Chem.*, 245, 2363 (1970).
423. Batt, Mihula, Mann, Guarracino, Altiere, Graham, Quigley, Wolf, and Zafonte, *J. Biol. Chem.*, 245, 4857 (1970).
424. Galivan, Maley, and Maley, *Biochemistry*, 13, 2282 (1974).
425. Spiro, *J. Biol. Chem.*, 245, 5820 (1970).
426. Cornell and Pierce, *J. Biol. Chem.*, 248, 4327 (1973).
427. Viljoen and Botes, *J. Biol. Chem.*, 248, 4915 (1973).
428. Strydom, *J. Biol. Chem.*, 247, 4029 (1972).
429. Karlsson, Eaker, Fryklund, and Kadin, *Biochemistry*, 11, 4628 (1972).
430. Sato, Abe, and Tamiya, *Biochem. J.*, 115, 85 (1969).
431. Tu, Hong, and Solie, *Biochemistry*, 10, 1295 (1971).
432. Botes and Strydom, *J. Biol. Chem.*, 244, 4147 (1969).
433. Botes, *J. Biol. Chem.*, 247, 2866 (1972).
434. Chang and Hayashi, *Biochem. Biophys. Res. Commun.*, 37, 841 (1969).
435. Karlsson, Eaker, and Porath, *Biochim. Biophys. Acta*, 127, 505 (1966).
436. Six and Harshman, *Biochemistry*, 12, 2672 (1973).
437. Lipscomb, Horton, and Armstrong, *Biochemistry*, 13, 2070 (1974).
438. Schneider and Slaunwhite, Jr., *Biochemistry*, 10, 2086 (1971).
439. Chader and Westphal, *J. Biol. Chem.*, 243, 928 (1968).
440. Chader and Westphal, *Biochemistry*, 7, 4272 (1968).
441. Hudson, Ohno, Brockway, and Castellino, *Biochemistry*, 12, 1047 (1973).
442. Mann, Fish, Cox, and Tanford, *Biochemistry*, 9, 1348 (1970).
443. Carlson and Brosemer, *Biochemistry*, 10, 2113 (1971).
444. Sawyer, Tilley, and Gracy, *J. Biol. Chem.*, 247, 6499 (1972).
445. Norton, Pfuderer, Stringer, and Hartman, *Biochemistry*, 9, 4952 (1970).
446. Krietsch, Pentchev, Klingenburg, Hofstatter, and Bucher, *Eur. J. Biochem.*, 14, 289 (1970).
447. Greaser and Gergely, *J. Biol. Chem.*, 248, 2125 (1973).
448. Winter and Neurath, *Biochemistry*, 9, 4673 (1970).
449. Reeck and Neurath, *Biochemistry*, 11, 503 (1972).
450. Travis and Roberts, *Biochemistry*, 8, 2884 (1969).
451. Mallery and Travis, *Biochemistry*, 12, 2847 (1973).
452. Travis and Liener, *J. Biol. Chem.*, 240, 1967 (1965).
453. Travis, *Biochem. Biophys. Res. Commun.*, 30, 730 (1968).
454. Gates and Travis, *Biochemistry*, 8, 4483 (1969).
455. Jurášek and Smillie, *Can. J. Biochem.*, 51, 1077 (1973).
456. Kucich and Peanasky, *Biochim. Biophys. Acta*, 200, 47 (1970).
457. Sherman and Kassell, *Biochemistry*, 7, 3634 (1968).
458. Tschesche and Wachter, *Eur. J. Biochem.*, 16, 187 (1970).
459. Kassell, Radicevic, Berlow, Peanasky, and Laskowski, Sr., *J. Biol. Chem.*, 238, 3274 (1966).
460. Wilson and Laskowski, Sr., *J. Biol. Chem.*, 248, 756 (1973).
461. Walsh and Neurath, *Proc. Natl. Acad. Sci. USA*, 52, 884 (1964).
462. Bradshaw, Neurath, Tye, Walsh, and Winter, *Nature* (Lond.), 226, 237 (1970).
463. Voytek and Gjessing, *J. Biol. Chem.*, 246, 508 (1971).
464. Schyns, Bricteaux-Grégoire, and Florkin, *Biochim. Biophys. Acta*, 175, 97 (1969).
465. Henning, Helinski, Chao, and Yanofsky, *J. Biol. Chem.*, 237, 1523 (1962).
466. Li and Yanofsky, *J. Biol. Chem.*, 248, 1830 (1973).
467. Hoch, *J. Biol. Chem.*, 248, 2999 (1973).
468. Hoch, *J. Biol. Chem.*, 248 2992 (1973).
469. Hoch and DeMoss, *J. Biol. Chem.*, 247, 1750 (1972).
470. Kamamiyama, Wada, Matsubara, and Snell, *J. Biol. Chem.*, 247, 1571 (1972).
471. Koch, *Biochemistry*, 13, 2307 (1974).
472. Milton and Taylor, *Biochem. J.*, 113, 678 (1969).
473. White, Barlow, and Mozen, *Biochemistry*, 5, 2160 (1966).
474. DuVigneaud, Bartlett, and Johl, *J. Am. Chem. Soc.*, 79, 5572 (1957).
475. Tanford, *J. Am. Chem. Soc.*, 84, 4240 (1962).
476. Bigelow, *J. Theor. Biol.*, 16, 187 (1967).

CHEMICAL SPECIFICITY OF REAGENTS FOR PROTEIN MODIFICATION[a]

Amino Acid	Reagent	Other Residues Modified
α-amino groups (N-terminal amino acids)[b]	Glyoxal	Slower reaction at lysine
α-amino groups (N-terminal amino acids)	Cyanate	Slower reaction at lysine
Arginine	Diones and glyoxals including 2,3-butanedione and phenylglyoxal; products usually not stable in the absence of borate buffers	Possible reaction with lysine
Asparagine	Mild acid with deamidation and potential cyclization with formation of β-aspartic acid	Glutamine, minor oxidation at cysteine, cystine; possible peptide bond cleavage which is sequence-dependent
Aspartic Acid	Carbodiimides	Minor reaction at hydroxyl functions; modification of C-terminal carboxyl group
Aspartic Acid	Woodward's reagent K	Minor reaction at hydroxyl functions; modification of C-terminal carboxyl group
Cysteine	Haloalkyl reagents such as bromoacetamide, bromoethylamine	Histidine, amino groups, carboxyl groups. Unless residue at active site, reaction is very slow compared to cysteine. Reaction does occur with thiourea
Cysteine	Alkylmethanesulfonates (modification reversible)	No other reactions demonstrated
Cysteine	Cyanylation (cleavage can be achieved after cyanylation)	Lysine although specificity can obtained with reagents other than cyanate such as 1-cyano-4-dimethylamino-pyridinium tetrafluoroborate
Cysteine	Maleimides such as N-ethylmaleimide	No other reactions demonstrated
Cysteine	Sodium tetrathionate (reversible)	No other reactions demonstrated
Cystine	Reducing agents such mercaptoethanol, dithiothreitol, phosphines	No other reactions demonstrated
Cystine	Performic acid oxidation for cleavage to cysteic acid	Methionine, histidine, tyrosine, tryptophan
Cystine	Sodium sulfite (oxidative sulfitolysis)	Cysteine
Glutamine	Mild acid	Deamidation; rate slower than asparagine
Glutamic Acid	Carbodiimides	Possible reaction at hydroxyl groups; modification of C-terminal
Histidine	Diethylpyrocarbonate	Tyrosine, possible amino groups
Histidine	Haloalkyl reactions such as iodoacetic acid	Cysteine will predominate unless histidine is active site residue
Histidine	Photooxidation	Cysteine, methionine, tyrosine, tryptophan
Lysine	Methyl acetamidate; methyl picolinimidate	Amino-terminal
Lysine	Cyanate	Cysteine, amino-terminal
Lysine	Trinitrobenzene sulfonic acid	Cysteine
Lysine	N-Hydroxysuccinimide	Amino-terminal
Lysine	Reducing sugars, glyoxal, methyl glyoxal, aldehyde	Arginine, histidine
Methionine	Oxidation with performic acid	Cysteine, cystine, histidine, tryptophan, tyrosine
Methionine	Alkylation with α-haloketoalkyl acids/amide such as iodoacetate at pH 3.0	No other reactions demonstrated
Methionine	Cyanogen bromide with peptide bond cleavage. The S-methyl group is lost in the reaction as methyl thiocyanate	Oxidation at cysteine possible
Tryptophan	N-Bromosuccinimide	Cysteine, methionine
Tryptophan	BPNS-skatole	Possible
Tryptophan	2-Hydroxy-5-nitrobenzyl bromide	Cysteine possible
Tryptophan	N-Bromosuccinimide	Cysteine and cystine possible
Tryptophan	Nitrophenyl sulfenyl chlorides	Cysteine and cystine possible
Tyrosine	N-Acetylimidazole; reversible at alkaline pH or in the presence of hydroxylamine	Lysine, amino groups which not reversible
Tyrosine	Tetranitromethane (converted by amine by reduction)	Sulfydryl, also cross-linking at tyrosine residues

[a] Specificity of reaction is rarely absolute

[b] The pKa of amino-terminal amino groups is lower than that of most lysine residues and specificity of modification can be enhanced by performing modification reactions pH below 7.

REAGENTS FOR THE CHEMICAL MODIFICATION OF PROTEINS

Chemical structures for these reagents may be found on p. 126.

	Reagent	Specificity/Conditions[a]	M.W.	References
1	Acetic anhydride	Lysine, α-amino groups, tyrosine hydroxyl; preferred reaction is at lysine; pH 8 or greater; reaction can be "driven" α-amino groups at pH less than 6.5. Avoid nucleophilic buffers such as Tris; hydrolysis of the reagent is an issue above pH 9.5. Acetic anhydride has been used for trace labeling in the study of protein conformation and more recently the deuterated derivative has been used in proteomics for differential isotope tagging	102.1	1-5
2	N-Acetylimidazole (1-acetylimidazole)	Tyrosine hydroxyl groups, lysine ε-amino groups, transient reaction at histidine; neutral pH	110.1	6-10
3	BNPS-skatole [3-bromo-3-methyl-2-(2-nitro-phenylmercapto)-3H-indole; 2-(2′-nitrophenylsulfenyl)-3-methyl-3 bromoindolenine	Tryptophan; 50-70% acetic acid; associated with peptide bond cleavage	363.3	11-15
4	Bromoacetamide (2-bromoacetamide)	Cysteine; reaction with active site histidine residues; also reaction with lysine, methionine and possibly carboxylic acids. Reaction at pH 5-9 but reaction with methionine at pH 3.0. Reaction rate below pH 7.5 is usually slow as the modification of cysteine requires thiolate anion (pKa[2] for cysteine is 8.7. Reaction is slower than iodoacetamide. A neutral reagent[b]	138	16-20
5	Bromoacetic acid (2-bromoacetic acid)	Reaction parameters similar to bromoacetamide except bromoacetic acid is a charged reagent at pH greater than 4 (pKa is 2.7 at 25°C). Amide and acid derivatives can show different reaction patterns[c]	139	21-25
6	Bromoethylamine	Modification of sulfhydryl groups; conversion of cysteine to lysine analogue (S-2-aminoethylcysteine); reaction with cysteine at alkaline pH (see bromoacetamide). Reaction is reasonably specific for cysteine with possible modification at the amino-terminal α-amino group and histidine	204.9 as HBr salt	26-30
7	N-Bromosuccinimide	Modification of tryptophan with some oxidative side reactions; pH 4-6	178	31-35
8	2,3-Butanedione (diacetyl)	Modification of arginine residues; reversible reaction with the product stabilized by the presence of borate; reaction at alkaline pH	86.1	36-40
9	Citraconic anhydride	Reversible modification of lysine residues; modification at pH above 8.0 and reversed below pH 6.0	112.1	41-45
10	Cyanate[d]	Carbamylation of α-amino groups in proteins at alkaline pH with some preference toward the modification of N-terminal α-amino groups. Reaction also occurs at cysteine residues using 2-nitro-5-thiocyanatobenzoic acid[d]	N/A	46-50
11	1,2-Cyclohexanedione	Modification of arginine in the presence of borate. At pH 7-9, the reaction is reversible; above pH 9, the reaction is irreversible with the formation of several products	112.1	51-55
12	DCC (1,3-dicyclohexylcarbodiimide)	Modification of carboxyl groups in proteins (activated carboxyl group which is then modified with a nucleophile such as glycine methyl ester; solubility issues have made EDC a more attractive reagent); also used for synthesis of phosphate ester bonds and peptide bonds; pH less than 5, modification requires protonated carboxyl group. The reagent has been used at more alkaline pH values. The modification of tyrosine and cysteine has been reported. DCC is an inhibitor of the proton-translocating ATPase in mitochondria and has been extensively used to characterize that activity	206.3	56-60
14	Diethylpyrocarbonate	Modification of histidine residues in proteins with transient modification of tyrosine; possible reaction at amino groups. Disubstitution of histidine results in ring-opening	162.1	61-65

REAGENTS FOR THE CHEMICAL MODIFICATION OF PROTEINS (Continued)

	Reagent	Specificity/Conditions[a]	M.W.	References
15	EDC [1-ethyl-3(3-dimethylaminopropyl)-carbodiimide]; N-(3-dimethylaminopropyl)-N′-ethylcarbodiimide	Modification of carboxyl groups in proteins frequently with N-hydroxysuccinimide. Used for zero-length cross-linking in proteins and for the coupling of proteins to matrices and for the preparation of protein conjugates	155.2	66-70
16	Ellman's reagent (5,5′-dithiobisnitrobenzoic acid	Modification and measurement of cysteine (sulfhydryl) groups in proteins	396.4	71-75
17	N-Ethylmaleimide	Modification of sulfhydryl groups via Michael addition to the maleimide ring. There are some proteins which are distinguished by their modification with N-ethylmaleimide including N-ethylmaleimide-sensitive fusion protein (NSF) and soluble N-ethylmaleimide-sensitive factor attachment protein receptors (SNARES)	125.1	76-80
18	Ethyleneimine (aziridine)	Early as modification reagent for cysteine but largely supplanted by bromoethylamine. Aziridine (ethyleneimine) serves a functional base for reagents used for the modification of cysteine in proteins	43.1	81-85
19	Formaldehyde[e,f] (reductive methylation)	Formation of a Schiff base with a primary amine (e.g., ε-amino group of lysine) which is reduced with sodium borohydride or more often with sodium cyanoborohydride which results in the formation, the in the case of lysine, of ε-methyllysine and ε-dimethyllysine. The modification is performed at alkaline pH; in a related reaction, the formylation of tryptophan occurs with formic acid under acidic conditions (HCOOH/HCl) which is reduced by base[g]	30.0	86-90
20	Gold	Reaction with cysteine. This process is used to bind proteins to solid matrices. The reaction of gold with proteins provides some of the basis for the use of gold and gold compounds in the treatment of arthritis	N/A	91-95
21	2-Hydroxy-5-nitrobenzyl bromide (Koshland's reagent)	Modification of tryptophan by alkylation of the indole ring. Reaction at acid pH (pH 2-6). Disubstitution can occur. This was one of the first "reporter groups"[h]	232.0	96-100
22	N-Hydroxysuccinimide	Functional group for modification of amino groups in proteins, frequently with carbodiimide for carboxyl modification and coupling in proteins	115.1	101-105
23	2-Iminothiolane (Traut's reagent)	Placement of a sulfhydryl groups by the modification of amino groups in proteins and other amino-containing materials	137.6 as the HCl	105-110
24	Iodoacetamide	Reaction is faster than with bromo or chloro derivatives. Reaction characteristics similar to bromoacetamide; as bromoacetamide, iodacetamide is neutral	185	111-115
25	Iodoacetic acid	See bromoacetic acid for reaction conditions. As with bromoacetic acid, reagent is charged at pH greater than 5 (pKa is 3.12 at 25°C)	186	116-120
26	2-Mercaptoethanol (β-mercaptoethanol; 2-thioethanol	Used to maintain proteins in reduced states, previously used for the reduction of disulfide bonds but largely replaced by TCEP (see below)	78.1	121-125
27	Mercuric chloride (HgCl$_2$)	Reaction with organic sulfhydryl groups such as cysteine. The chemistry of this reaction is poorly understood[i]. Mercuric chloride is used for the inhibition of membrane enzymes and is described as a specific inhibitor of aquaporin[j]	271.5	126-130
28	Methyl acetimidate	Modification of amino groups. Imido esters are the functional groups for a number of cross-linking agents such as dimethylsuberimidate[k]. One of the more interesting imido esters is methyl picolinimidate[l]. Reaction at pH 8-10	109.6 as HCl	131-135
29	Methyl methane-thiosulfonate	Methyl methanethiosulfonate is one of a group of alkyl methanethiosulfonate derivatives which reversibly modify cysteine residues in proteins[m]. Reaction occurs at slightly alkaline pH (pH 7.8)	126.2	136-140
30	Ninhydrin (1H-indene-1,2,3-trione monohydrate; 2,2-dihydroxy-1,3-indanedione)	Determination of amino groups; provides the basis for detection in amino acid analysis; modification of arginine residues[n]. Ninhydrin is also used for the detection of cyanide	178.1	141-145

REAGENTS FOR THE CHEMICAL MODIFICATION OF PROTEINS (Continued)

	Reagent	Specificity/Conditions[a]	M.W.	References
31	2-Nitrophenylsulfenyl chloride (o-nitrophenyl-sulfenyl chloride)	Modification of tryptophan residues in proteins; reaction occurs at acid pH; modified tryptophan can be converted to the 2-thioltryptophan derivatives. This modification has been used to purify tryptophan peptides from protein hydrolyzates. An analogue, 2-(trifluoromethyl)-benzenesulfenyl chloride has been developed for use in mass spectrometry[o]	189.6	146-150
32	Performic acid (peroxyformic acid)	Cleavage of disulfide bonds; oxidation of cysteine and methionine in proteins with side reactions at tryptophan with other minor modifications[p]	62	151-155
33	Phenylglyoxal	Modification of arginine residues in proteins; reaction accelerated in the presence of bicarbonate buffers; reaction at alkaline pH. p-Hydroxyphenylglyoxal and p-nitrophenylglyoxal are useful derivates[q]	134.1 as hydrate	156-160
34	Sodium cyanoborohydride	Reducing agent for Schiff bases in proteins; reduces ketones, aldehydes, hydrazones, and enamines but does not reduce lactones, amides, or disulfide bonds. Used to "stabilize" Schiff base linkages between amine "probes" and aldehyde-based matrices	62.8	161-165
35	Sodium sulfite	Oxidative sulfitolysis to cleave disulfide bonds; conversion of cysteine to S-sulfocysteine	126	166-170
36	Sodium tetrathionate ($Na_2S_4O_6$)	Modification of cysteine to S-sulfoderivatives; a reversible modification. Also a mild oxidizing agent which can, via the S-sulfoderivative, drive the formation of disulfide bonds thus serving a protein cross-linking reagent[r]	306.3	171-175
37	TCEP (tris[2-(carboxyethyl] phosphine)	Reduction of disulfide bonds in proteins; reagent of choice. The reagent is readily soluble in water and reduces disulfide bonds at low pH (3.0)	286.7 at HCl	176-180
38	Tetranitromethane (TNM)	Nitration of tyrosine residues in proteins with nitration and possible cross-linking; also reacts with sulfhydryl groups; possible reaction with indole ring of tryptophan. Reaction at alkaline pH; does introduce a "reporter group" in proteins. The nitrotyrosine function can be reduced to aminotyrosine with sodium dithionite (sodium hydrosulfite). The modification with peroxynitrite is a similar reaction[s]	196	181-185
39	TNBS (trinitrobenzenesulfonic acid)	Modification of amino groups in proteins; used for the determination of amino groups. Reaction at alkaline pH; also reacts with sulfhydryl groups	293.2	186-190
40	Vinyl pyridine	Modification of sulfhydryl groups with majority of use in protein structure analysis; both the 2-vinylpyridine and the 4-vinylpyridine are used	105.1	191-195
41	Woodward's reagent K (N-ethyl-5-phenylisoxazolium-3-sulfonate)	Reagent used for the modification of carboxyl groups in proteins. Reaction at acidic pH	253.3	196-200

[a] Absolute specificity of modification cannot be guaranteed. Conditions are most common with the caveat that specific buffer effects are observed (see Buffers)

[b] Chaiken, I.M. and Smith, E.L., Reaction of chloroacetamide with the sulfhydryl group of papain, J.Biol.Chem. 244, 5087-5094, 1969; Chaiken, I.M. and Smith, E.L., Reaction of the sulfhydryl group of papain with chloroacetic acid, J.Biol.Chem. 244, 5095-5099, 1969

[c] Gerwin, B.I., Properties of the single sulfhydryl group of streptococcal proteinase. A comparison of the rates of alkylation by chloroacetic acid and chloroacetamide, J.Biol. Chem. 242, 451-456, 1967

[d] Early concern regarding the reaction of cyanate with protein was directed at urea. This was followed with work using potassium cyanate. More recent work has used 2-nitro-5-thiocyanatobenzoic acid for the modification of cysteine residues in proteins (Degani, Y. and Patchornik, A., Cyanylation of sulfhydryl groups by 2-nitro-5-thiocyanato-benzoic acid. High-yield modification and cleavage of peptides at cysteine residues, Biochemistry 13, 1-11, 1974; Price, N.C., Alternative products in the reaction of 2-nitro-5-thiocyanatobenzoic acid with thiol groups, Biochem.J. 159, 177-180, 1976; Wu, J. and Watson, J.T., Optimization of the cleavage reaction for cyanylated cysteinyl proteins for efficient and simplified mass mapping, Anal.Biochem. 258, 268-276, 1998)

[e] Other aldehydes such 4-hydroxy-2-nonenal, glyceraldehydes, reducing sugars such as glucose, also react in a similar manner. The reaction with reducing sugars and 4-hydroxy-2-nonenal is more complex; the reaction with glucose and reducing sugars is part of the Maillard reaction.

[f] Formaldehyde (paraformaldehyde) is used for tissue fixation and the chemistry is complex with the formation of protein cross-links. Antigen retrieval techniques have been developed for immunohistocytochemistry (Chu, W.S., Furusato, B., Wong, K., et al., Ultrasound-accelerated formalin fixation of tissue improves morphology, antigen and mRNA preservation, Mod.Pathol. 18, 850-863, 2005; Namimatsu, S., Ghazizadeh, M., and Sugisaki, Y., Reversing the effects of formalin fixation with citraconic anhydride and heat: a universal antigen retrieval method, J.Histochem.Cytochem. 53, 3-11, 2005; Shi, S.R., Liu, C., Balgley, B.M., et al., Protein extraction from formalin-fixed, paraffin-embedded tissue sections: quality evaluation by mass spectrometry, J.Histochem.Cytochem. 54, 739-743, 2006; Sompuram, S.R., Vani, K., Hafer, L.J., and Bogen, S.A., Antibodies immunoreactive with formalin-fixed tissue antigens recognize linear protein epitopes, Am.J.Clin.Pathol. 125, 82-90, 2006; Yamashita, S., Heat-induced antigen retrieval: mechanisms and application to histochemistry, Prog.Histochem.Cytochem. 41, 141-200, 2007)

[g] Previero, A., Coletti-Previero, M.A., and Cavadore, J.-C., A reversible chemical modification of the tryptophan residue, Biochim.Biophys.Acta 147, 453-461, 1967; Strosberg, A.D. and Kanarek, L., Immunochemical studies on hen's egg-white lysozyme. Effect of formylation of the tryptophan residues, FEBS Lett. 5, 324-326, 1969; Magous, R., Bali, J.P., Moroder, L., and Previero, A., Effect on Nin-formylation of the tryptophan residue on gastrin (HG-13) binding and on gastric acid secretion, Eur.J.Pharmacol. 77, 11-16, 1982

[h] Burr, M. and Koshland, D.E., Jr., Use of "reporter groups" in structure-function studies of proteins, Proc.Nat.Acad.Sci.USA 52, 1017-1024, 1964

REAGENTS FOR THE CHEMICAL MODIFICATION OF PROTEINS (Continued)

[i] Mercuric sulfide is the principle ore of mercury and exists in two forms; αHgS (red hexagonal) which is known as cinnabar and βHgS (black amorphous) which is known metacinnabar. The solubility product of mercuric sulfide is approximately 10^{-52}. Mercury salts can exist as mercurous [mercury(I)] and mercuric [mercury(II)]. Mercurous exists as a polyanion Hg_2^{2+} which can under disproportionation to form Hg and Hg^{2+}. Mercuric ion can form tight complexes with free sulfhydryl group and disulfide bonds. The chemistry and structure of these derivatives are still poorly understood. The possibility of a Hg crosslink between sulfhydryl groups is a possibility. The term mercaptan is derived from the phrase "mercury capture." See *The Chemistry of Mercury*, ed. C.A. McAuliffe, MacMillan Press, Ltd., London, UK, 1977; Roberts, H.L., Some general aspects of mercury chemistry, *Adv.Inorg.Chem.Radiochem.* 11, 309-339, 1968; Grant, G.J., Mercury: Inorganic and Coordination Chemistry, in *Encyclopedia of Inorganic Chemistry*, ed. R.B. King, John Wiley & Sons, Ltd., Chichester, UK, Volume 4, pps 2136-2145, 1994

[j] See Martinez-Ballesta, M.C., Diaz, R., Martinez, V., and Carvajal, M., Different blocking effects of $HgCl_2$ and NaCl on aquaporins of pepper plants, *J.Plant Physiol.* 160, 1487-1492, 2003; Liu, K., Nagase, H., Huang, C., Purification and functional characterization of aquaporin-8, *Biol.Cell* 98, 153-161, 2006; Yang, B., Kim, J.K., and Verkman, A.S., Comparative efficacy of $HgCl_2$ with candidate aquaporin-1 inhibitors DMSO, gold, TEA^+ and acetazolamide, *FEBS Lett.* 580, 6679-6684, 2006

[k] Coggins, J.R., Hooper, E.A., and Perham, R.N., Use of dimethyl suberimidate and novel periodate-cleavable bis(imido)esters to study the quaternary structure of the pyruvate dehydrogenase multienzyme complex of *Escherichia coli*, *Biochemistry* 15, 2527-2533, 1976

[l] McKinley-McKee, J.S. and Morris, D.L., The lysines in liver alcohol dehydrogenase. Chemical modification with pyridoxal-5′-phosphate and methyl picolinimidate, *Eur.J.Biochem.* 28, 1-11, 1972; Fries, R.W., Bohlken, D.P., Blakley, R.T., and Plapp, B.V., Activation of liver alcohol dehydrogenase by imidoesters generated in solution, *Biochemistry* 14, 5233-5238, 1975; Shaw, A. and Marienetti, G.V., The effect of imidoesters, fluorodinitrobenzene and trinitrobenzenesulfonate on ion transport in human erythrocytes, *Chem.Phys.Lipids* 27, 329-335, 1980

[m] The first derivative was methyl methanethiosulfonate (Smith, D.J., Maggio, E.T., and Kenyon, G.L., Simple alkanethiol groups for temporary blocking of sulfhydryl groups of enzymes, *Biochemistry* 14, 766-771, 1975). There are a number of alkyl derivatives in use including ionic derivatives such as ethylsulfonato methanethiosulfonate or 2-carboxyethyl methanethiosulfonate and ethyltrimethylammonium methanthiosulfonate.

[n] See Takahashi, K., Specific modification of arginine residues in proteins with ninhydrin, *J.Biochem.* 80, 1173-1176, 1976; Chaplin, M.F., The use of ninhydrin as a reagent for the reversible modification of arginine residues in proteins, *Biochem.J.* 155, 457-459, 1976

[o] Li, C., Gawandi, V., Protos, A., *et al.*, A matrix-assisted laser desorption/ionization compatible reagent for tagging tryptophan residues, *Eur.J.Mass Spectrom.* 12, 213-221, 2006

[p] Dai, J., Zhang, Y., Wang, J., *et al.*, Identification of degradation products formed during performic oxidation of peptides and proteins by high-performance liquid chromatography with matrix-assisted laser desorption/ionization and tandem mass spectrometry, *Rapid Commun.Mass Spectrom.* 19, 1130-1138, 2005

[q] See Yamasaki, R.B., Shimer, D.A., and Feeney, R.E., Colorimetric determination of arginine residues in proteins by *p*-nitrophenylglyoxal, *Anal.Biochem.* 111, 220-226, 1981

[r] See Parker, D.J. and Allison, W.S., The mechanism of inactivation of glyceraldehyde 3-phosphate dehydrogenase by tetrathionate, *o*-iodosobenzoate, and iodine monochloride, *J.Biol.Chem.* 244, 180-189, 1969; Chung, S.I., and Folk, J.E., Mechanism of the inactivation of guinea pig liver transglutaminase by tetrathionate, *J.Biol.Chem.* 245, 681-689, 1970; Silva, C.M., and Cidlowski, J.A., Direct evidence for intra- and intermolecular disulfide bond formation in the human glucocorticoid receptor. Inhibition of DNA binding and identification of a new receptor-associated protein, *J.Biol.Chem.* 264, 6638-6647, 1989; Kaufmann, S.H., Brunet, G. Talbot, B., *et al.*, Association of poly(ADP-ribose) polymerase with the nuclear matrix: The role intermolecular disulfide bond formation, RNA retention, and cell type, *Exp.Cell Res.* 192, 524-535, 1991; Desnoyers, S., Kirkland, J.B., and Poirier, G.G., Association of poly(ADP-ribose) polymerase with nuclear subfractions catalyzed with sodium tetrathionate and hydrogen peroxide crosslinks, *Mol.Cell.Biochem.* 159, 155-161, 1996; Tramontano, F., di Meglio, S., and Quesada, P., Co-localization of poly(ADPR)polymerase 1 (PARP-1), poly(ADPR)polymerase 2 (PARP-2) and related proteins in rat testis nuclear matrix defined by chemical cross-linking, *J.Cell.Biochem.* 94, 58-66, 2005

[s] See Haddad, I.Y., Zhu, S., Ischiropoulos, H., and Matalon, S., Nitration of surfactant protein A results in decreased ability to aggregate lipids, *Am.J.Physiol.* 270, L281-L288, 1996; Greis, K.D., Zhu, S., and Matalon, S., Identification of nitration sites on surfactant protein A by tandem electrospray mass spectrometry, *Arch.Biochem.Biophys.* 335, 396-402, 1996; Petersson, A.B., Steen, H., Kalume, D.E., *et al.*, Investigation of tyrosine nitration by mass spectrometry, *J.Mass Spectrom.* 36, 6160625, 2001; Lee, W.I. and Fung, H.L., Mechanism-based partial inactivation of glutathione *S*-transferase by nitroglycerin: tyrosine nitration vs. sulfhydryl oxidation, *Nitric Oxide* 8, 103-110, 2003; Batthyany, C., Souza, J.M., Duran, R, *et al.*, Time course and site(s) of cytochrome c tyrosine nitration by peroxynitrite, *Biochemistry* 44, 8038-8046, 2003

General references for chemical modification of proteins

Kellam, B., de Bank, P.A., and Shakesheff, K.M., Chemical modification of mammalian cell surfaces, *Chem.Soc.Rev.* 32, 327-337. 2003

Lundblad, R.L., *Chemical Reagents for Protein Modification*, CRC Press, Boca Raton, FL, USA, 2004

Leitner, A. and Lindner, W., Chemistry meets proteomics: the use of chemical tagging reactions for MS-based proteomics, *Proteomics* 6, 5418-5434, 2006

Antos, J.M. and Francis, M.B., Transition metal catalyzed methods for site-selective protein modification, *Curr.Opin.Chem.Biol.* 10, 253-262, 2006

References for table– reagents for the chemical modification of proteins

1. Giedroc, D.P., Puett, D., Sinha, S.K., and Brew, K., Calcium effects on calmodulin lysine reactivation, *Arch.Biochem.Biophys.* 252, 136-144, 1987

2. Illy, C., Thielens, N.M., and Arlaud, G.J., Chemical characterization and location of ionic interactions involved in the assembly of the C1 complex of human complement, *J.Protein Chem.* 12, 771-781, 1993

3. Che, F.Y. and Fricker, L.D., Quantitation of neuropeptides in Cpe(fat)/Cpe(fat) mice using differential isotopic tags and mass spectrometry, *Anal.Chem.* 74, 3190-3198, 2002

4. Turner, B.T., Jr., Sabo, T.M., Wilding, D., and Maurer, M.C., Mapping of factor XIII solvent accessibility as a function of activation state using chemical modification methods, *Biochemistry* 43, 9755-9765, 2004

5. Nam, H.W., Lee, G.Y., and Kim, Y.S., Mass spectrometric identification of K210 essential for rat malonyl-CoA decarboxylase catalysis, *J.Proteome Res.* 5, 1398 1406, 2006

6. Scherer, H.J., Karthein, R., Strieder, S., and Ruf, H.H., Chemical modification of prostaglandin endoperoxide synthase by *N*-acetylimidazole. Effect on enzyme activities and EPR spectroscopic properties, *Eur.J.Biochem.* 205, 751-757, 1992

7. Cymes, G.D., Igelesias, M.M., and Wolfenstein-Todel, C., Chemical modification of ovine prolactin with *N*-acetylimidazole, *Int.J.Pept. Protein Res.* 42, 33-28, 1993

8. Vazeux, G., Iturrioz, X., Corvol, P., and Llorens-Cortes, C., A tyrosine residue essential for catalytic activity in aminopeptidase A, *Biochem.J.* 327, 883-889, 1997

9. Pal, J.K., Bera, S.K., and Ghosh, S.K., Acetylation of α-crystallin with *N*-acetylimidazole and its influence upon the native aggregate and subunit reassembly, *Curr.Eye Res.* 19, 368-367, 1999

10. Zhang, F., Gao, J., Weng, J., *et al.*, Structural and functional differentiation of three groups of tyrosine residues by acetylation of *N*-acetylimidazole in manganese stabilizing protein, *Biochemistry* 44, 719-725, 2005

11. Zeitler, H.J. and Kulitz, M, Improved preparation and structural elucidation of the tryptophanyl cleavage reagent 2-(2′-nitrophenylsulfenyl)-3-methyl-3-bromoindolenine (BNPS-skatole), *J.Clin. Chem.Clin.Biochem.* 16, 669-674, 1978

12. Russell, J., Katzhendler, J., Kowalski, K., *et al.*, The single tryptophan residue of human placental lactogen. Effects of modification and cleavage on biologic activity and protein conformation, *J.Biol.Chem.* 256, 304-307, 1981

13. Xue, H., Xue, Y., Doublie, S., and Carter, C.W., Jr., Chemical modification of *Bacillus subtilis* tryptophanyl-tRNA synthetase, *Biochem. Cell Biol.* 75, 709-715, 1997

14. Rahali, V. and Gueguen, J., Chemical cleavage of bovine β-lactoglobulin by BNPS-skatole for preparative purposes: comparative study of hydrolytic procedures and peptide characterization, *J.Protein Chem.* 18, 1-12, 1999

15. Kibbey, M.M, Jameson, M.J., Eaton, E.M., and Rosenzweig, S.A. Insulin-like growth factor binding protein-2: contributions of the C-terminal domain to insulin-like growth factor-1 binding, *Mol. Pharmcol.* 69, 833-845, 2006

16. Horn, A., Vandenberg, C.A., and Lange, K., Statistical analysis of single sodium channels. Effects of *N*-bromoacetamide, *Biophys.J.* 45, 323-335, 1984

17. Pallotta, B.S., *N*-Bromosuccinimide removes a calcium-dependent component of channel opening from calcium-activated potassium channels in rat skeletal muscle, *J.Gen.Physiol.* 86, 601-611, 1985

18. Huang, R.C., Novel pharmacological properties of transient potassium currents in central neurons revealed by *N*-bromosuccinimide and other chemical modifiers, *Mol.Pharmacol.* 48, 451-458, 1995

19. Qi, X., Lee, S.H., and Kwon, J.Y., Aminobromination of unsaturated phosphonates, *J.Org.Chem.* 68, 9140-9143, 2003

20. Wang, H., Vath, G.M., Gleason, K.J., *et al.*, Probing the mechanism of hamster arylamine *N*-acetyltransferase 2 acetylation by active site modification, site-directed mutagenesis, and pre-steady state and steady sate kinetic studies, *Biochemistry* 43, 8234-8246, 2004

21. Glick, D.M., Goren, H.J., and Barnard, E.A., Concurrent bromoacetate reaction at histidine and methionine residues in ribonuclease, *Biochem.J.* 102, 7c-10c, 1967

22. Lennette, E.P. and Plapp, B.V., Kinetics of carboxymethylation of histidine hydantoin, *Biochemistry* 18, 3933-3988, 1979

23. Shapiro, R., Strydom, D.J., Weremowicz, S., and Vallee, B.L., Sites of modification of human angiogenin by bromoacetate at pH 5.5, *Biochem.Biophys.Res.Commun.* 156, 530-536, 1988

24. Schelte, P., Boeckler, C., Frisch, B., and Schuber, F., Differential reactivity of maleimide and bromoacetyl functions with thiols: application to the preparation of liposomal diepitope constructs, *Bioconjug. Chem.* 11, 118-123, 2000

25. Chatani, E., Tanimizu, N., Ueno, H., and Hayashi, R., Structural and functional changes in bovine pancreatic ribonuclease A by the replacement of Phe120 with other hydrophobic residues, *J.Biochem.* 129, 917-922, 2001

26. Okazaki, K., Yamada, H., and Imoto, T., A convenient *S*-2-aminoethylation of cysteinyl residues in reduced proteins, *Anal. Biochem.* 149, 516-520, 1985

27. Planas, A. and Kirsch, J.F., Sequential protection-modification method for selective sulfhydryl group derivatization in proteins having more than one cysteine, *Protein Eng.* 3, 625-628, 1990

28. Bochar, D.A., Tabernero, L., Stauffacher, C.V., and Rodwell, V.W., Aminoethylcysteine can replace the function of the essential active site lysine of *Pseudomonas mevalonii* 3-hydroxy-3-methylglutaryl coenzyme A reductase, *Biochemistry* 38, 8879-8883, 1999

29. Thevis, M., Ogorzalek Loo, R.R., and Loo, J.A., In-gel derivatization of proteins for cysteine-specific cleavages and their analysis by mass spectrometry, *J.Proteome Res.* 2, 163-172, 2003

30. Hopkins, C.E., Hernandez, G., Lee, J.P., and Tolan, D.R., Aminoethylation in model peptides reveals conditions for maximizing thiol specificity, *Arch.Biochem.Biophys.* 443, 1-10, 2005

31. McAllister, K.A., Marrone, L., and Clarke, A.J., The role of tryptophan residues in substrate binding to catalytic domains A and B of xylanase C from *Fibrobacter succinogenes* S85, *Biochim.Biophys.Acta* 1400, 342-352, 2000

32. Takita, T., Nakagoshi, M., Inouye, K., and Tonomura, B., Changes observed in the amino acid activation reaction, *J.Mol.Biol.* 325, 677-685, 2003

33. Sargisova, Y., Pierfederici, F.M., Scire, A., *et al.*, Computational, spectroscopic, and resonant mirror biosensor analysis of the interaction of adenodoxin with native and tryptophan-modified NADPH-adrenodoxin reductase, *Proteins* 57, 302-310, 2004

34. Faridmoayer, A. and Scaman, C.H., Binding residues and catalytic domain of soluble *Saccharomyces cerevisiae* processing α-glucosidase I, *Glycobiology* 15, 1341-1348, 2005

35. Kumar, A., Tyagi, N.K., and Kinne, R.K., Ligand-mediated and conformational changes and positioning of tyrptophans in reconstituted human sodium/*d*-glucose cotransporter (hSGLT1) probed by tryptophan fluorescence, *Biophys.Chem.* 127, 69077, 2007

36. Leitner, A. and Lindner, W., Functional probing of arginine residues in proteins using mass spectrometry and an arginine-specific covalent tagging concept, *Anal.Chem.* 77, 4481-4488, 2005

37. Foettinger, A., Leitner, A. and Lindner, W., Solid-phase capture and release of arginine peptides by selective tagging and boronate affinity chromatography, *J.Chromatog.A.* 1079, 187-196, 2005

38. Saraiva, M.A., Borges, C.M., and Florencio, M.H., Reactions of a modified lysine with aldehydic and diketonic dicarbonyl compounds: an electrospray mass spectrometry structure/activity study, *J. Mass Spectrom.* 41, 216-228, 2006

39. Holm, A., Rise, F., Sessler, N., *et al.*, Specific modification of peptide-bound citrulline residues, *Anal.Biochem.* 352, 68-76, 2006

40. Leitner, A., Amon, S., Rizzi, A., and Lindner, W., Use of the arginine-specific butanedione/phenylboronic acid tag for analysis of peptides and protein digests using matrix-assisted laser desorption/ionization mass spectrometry, *Rapid Commun.Mass Spectrom.* 21, 1321-1330, 2007

41. de Cuyper, M., Hodenius, M., Lacava, E.G., *et al.*, Attachment of water-soluble proteins to the surface of (magnetizable) phospholipid colloids via NeutraAvidin-derivatized phospholipids, *J.Colloid Interface Sci.* 245, 274-280, 2002

42. Hosseinkhani, S., Ranjbar, B., Haderi-Manesh, H., and Nemat-Gorgani, M., Chemical modification of glucose oxidase: possible formation of molten globule-like intermediate structure, *FEBS Lett.* 561, 213-216, 2004

43. Habibib, A.E., Khajeh, K., and Nemat-Gorgani, M., Chemical modification of lysine residues in *Bacillus licheniformis* α-amylase: conversion of an endo- to an exo-type enzyme, *J.Biochem.Mol.Biol.* 37, 642-647, 2004

44. Dai, W., Sato, S., Ishizaki, M., *et al.*, A new antigen retrieval method using citraconic anhydride for immunoelectron microscopy: localization of surfactant pro-protein C (proSP-C) in the type II alveolar epithelial cells, *J.Submicrosc.Cytol.Pathol.* 36, 219-224, 2004

45. Mossavarali, S., Hosseinkhani, S. Rnajbar, B., and Miroliaei, M., Stepwise modification of lysine residues of glucose oxidase with citraconic anhydride, *Int.J.Biol.Macromol.* 39, 192-196, 2006

46. Griffey, R.H., Scavini, M., and Eaton, R.P., Characterization of the carbamino adducts of insulin, *Biophys.J.* 54, 295-300, 1988

47. Kraus, L.M., Miyamura, S., Pecha, B.R., and Kraus, A.F., Jr., Carbamoylation of hemoglobin in uremic patients determined by antibody specific for homocitrulline (carbamoylated ε-*N*-lysine), *Mol.Immunol.* 28, 459-463, 1991

48. Reyes, A.M. Bravo, M., Ludwig, H., *et al.*, Modification of Cys-128 of pig kidney fructose 1,6-bisphosphatase with different thiol reagents: size dependent effect on the substrate and fructose-2,6-bisphosphate interaction, *J.Protein Chem.* 12, 159-168, 1993

49. Lapko, V.N., Smith, D.L., and Smith, J.B., Methylation and carbamylation of human gamma-crystallins, *Protein Sci.* 12, 1762-1774, 2003

50. Jaisson, S., Lorimier, S., Ricard-Blum, S., *et al.*, Impact of carbamylation of type I collagen conformational structure and its ability to activate human polymorphonuclear neutrophils, *Chem.Biol.* 13, 149-159, 2006

51. Chang, L.S., Wu. P.F., Liou, J.C., *et al.*, Chemical modification of arginine residues of *Notechis scutatus scutatus* notexin, *Toxicon* 44, 491-497, 2004

52. Masuda, T., Ide, N., and Kitabatake, N., Structure-sweetness relationship in egg white lysozme: role of lysine and arginine residues on the elication of lysozyme sweetness, *Chem.Senses* 30, 667-681, 2005

53. Herrman, A., Svangard, E., Claeson, P., *et al.*, Key role of glutamic acid for the cytotoxic activity of the cyclotide cycloviolacin O2, *Chem.Mol.Life Sci.* 63, 235-245, 2006

54. Schwartz, M.P., Barlow, D.E., Russell, J.N., Jr., *et al.*, Semiconductor surface-induced 1,3-hydrogen shift: the role of covalent vs. zwitterionic character, *J.Am.Chem.Soc.* 128, 11054-11061, 2006

55. Daniel, J., Oh, T.J., Lee, C.M., and Kolattukudy, P.E., AccD6, a member of the Fas II locus, is a functional carboxyltranferase subunit of the acyl-coenzyme A carboxylase in *Mycobacterium tuberculosis*, *J.Bacteriol.* 189, 911-917, 2007

56. Azzi, A., Casey, R.P., and Nalecz, M.J., The effect of *N,N'*-dicyclohexylcarbodiimide on enzymes of bioenergetic relevance, *Biochim.Biophys.Acta* 768, 209-226, 1984

57. Dimroth, P., Matthey, U., and Kaim, G., Critical evaluation of the one- versus the two-channel model for the operation of the ATP synthase's F(o) motor, *Biochim.Biophys.Acta* 14589, 506-513, 2000

58. Aresta, M., Dibenedetto, A., Fracchiolla, E., *et al.*, Mechanism of formation of organic carbonates from aliphatic alcohols and carbon dioxide under mild conditions promoted by carbodiimides. DFT calculation an experimental study, *J.Org.Chem.* 70, 6177-6186, 2005

59. Vgenopoulou, L., Gemperli, A.C., and Steuber, J., Specific modification of a Na$^+$ binding site in NADH:quinine oxidoreductase from *Klebsiella pneumonia* with dicyclohexylcarbodiimide, *J.Bacteriol.* 188, 3264-3272, 2006

60. Ogino, S., Sato, Y., Yamamoto, G., *et al.*, Relation of the number of cross-links and mechanical properties of multi-walled carbon nanotube films formed by a dehydration condensation reaction, *J.Phys.Chem.B Condens.Matter Mater.Surf.Interfaces Biophys.* 110, 23159-23163, 2006

61. Follmer, C. and Carlini, C.R., Effect of chemical modification of histidine on the copper-induced oligomerization of jack bean urease, *Arch.Biochem.Biophys.* 435, 15-20, 2005

62. Colleluori, D.M. Reczkowski, R.S., Emig, F.A., *et al.*, Probing the role of hyper-reactive histidine residue of arginase, *Arch.Biochem. Biophys.* 444,15-26, 2005

63. Runquist, J.A., and Miziorko, H., Functional contribution of a conserved mobile loop histidine of phosphoribulokinase, *Protein Sci.* 15, 837-842, 2006

64. Wang, X.Y., Sun, M.L., Zhao, D.M. and Wang, M., Kinetics of inactivation of phytase (phy A) during modification of histidine residue by IAA and DEP, *Protein Pept.Lett.* 13, 565-570, 2006

65. Nakanishi, N., Takeuchi, F., Okamoto, H., *et al.*, Characterization of heme-coordinating histidyl residues of cytochrome b5 based on the reactivity with Diethylpyrocarbonate: a mechanism for the opening of axial imidazole rings, *J.Biochem.* 140, 561-571, 2006

66. Ghosh, M.K., Kildsig, D.O., and Mitra, A.K., Preparation and characterization of methotrexate-immunoglobulin conjugates, *Drug.Des. Deliv.* 4, 13-25, 1989

67. Shen, X., Lagergard, T., Yang, Y., *et al.*, Preparation and preclinical evaluation of experimental group B streptococcus type III polysaccharide-cholera toxin B subunit conjugate vaccine for intranasal immunization, *Vaccine* 19, 850-861, 2000

68. Hafemann, B., Ghofrani, K., Gattner, H.G., *et al.*, Cross-linking by 1-ethyl-3-(3-dimethylaminopropyl)-carbodiimide (EDC) of a collagen/elastin membrane meant to be used as a dermal substitute: effects on physical, biochemical and biological features *in vitro*, *J.Mater.Sci.Mater.Med.* 12, 437-446, 2001

69. Zhang, R., Tang, M., Bowyer, A., *et al.*, A novel pH- and ionic-strength-sensitive carboxy methyl dextran hydrogel, *Biomaterials* 26, 4677-4683, 2005

70. Li, D., He, Q., Cui, Y., *et al.*, Immobilization of glucose oxidase onto gold nanoparticles with enhanced thermostability, *Biochem.Biophys. Res.Commun.* 355, 488-493, 2007

71. Owusu-Apenten, R., Colorimetric analysis of protein sulfhydryl groups in milk: applications and processing effects, *Crit.Rev.Food Sci.Nutr.* 45, 1-23, 2005

72. Laragione, T., Gianazza, E., Tonelli, R., *et al.*, Regulation of redox-sensitive exofacial protein thiols in CHO cells, *Biol.Chem.* 387, 1371-1376, 2006

73. Landino, L.M., Koumas, M.T., Mason, C.E., and Alston, J.A., Ascorbic acid reduction of microtubule protein disulfides and its

74. de Araujo, A.D., Palomo, J.M., Cramer, J., *et al.*, Diels-Alder ligation of peptides and proteins, *Chemistry* 12, 6095-6109, 2006

75. Cliff, M.J., Alizadeh, T., Jelinska, C., *et al.*, A thiol labelling competition experiment as a probe for sidechain pakcing in the kinetic folding intermediate of N-PGK, *J.Mol.Biol.* 364, 810-823, 2006

76. Mollinedo, F., Calafat, J., Janssen, H., *et al.*, Combinatorial SNARE complexes modulate the secretion of cytoplsstmic granules in human neutrophils, *J.Immunol.* 177, 2831-2841, 2006

77. Guan, L. and Kabade, H.R., Site-directed alkylation of cysteine to test solvent accessibility of membrane proteins, *Nat. Protoc.*, 2, 2012-2017, 2007

78. Cuddihy, S.L., Baty, J.W., Brown, K.K., *et al.*, Proteomic detection of oxidized and reduced thiol proteins in cultured cells, *Methods Mol. Biol.*, 519, 363-375, 2009

79. Togneri, J., Cheng, Y.S., Munson, M., *et al.*, Specific SNARE complex binding mode of the Sec1/Munc-18 protein, Seclp, *Proc.Nat.Acad. Sci.USA* 103, 17730-17735, 2006

80. Yan, L.J., Yang, S.H., Shu, H., *et al.*, Histochemical staining and quantification of dihydrolipoamide dehydrogenase diaphorase activity using blue native PAGE, *Electrophoresis*, in press, 2007

81. Diaper, C.M., Sutherland, A., Pillai, B., *et al.*, The stereoselective synthesis of aziridine analogues of diaminopimelic acid (DAP) and their interactions with dap epimerase, *Org.Biomol.Chem.* 3, 4402-4411, 2005

82. Ponte-Sucre, A., Vicik, R., Schultheis, M., *et al.*, Aziridine-2,3-dicarboxylates, peptidomimetic cysteine protease inhibitors with antileishmanial activity, *Antimicrob.Agents Chemother.* 50, 2439-2447, 2006

83. Vicik, R., Busemann, M., Gelaus, C., *et al.*, Aziridine-based inhibitors of cathepsin L: synthesis, inhibition activity, and docking studies, *ChemMedChem* 1, 1126-1141, 2006

84. Vicik, R., Buseman, M., Bauman, K., and Schirmeister, T., Inhibitors of cysteine proteases, *Curr.Top.Med.Chem.* 6, 331-353, 2006

85. Mladenovic, M., Schirmeister, T., Thiel, S., *et al.*, The importance of the active site histidine for the activity of epoxide- or aziridine-based inhibitors of cysteine proteases, *ChemMedChem* 2, 120-128, 2007

86. Brubaker, G., Peng, D.Q., Somerlot, B., *et al.*, Apolipoprotein A-1 lysine modification: effects on helical content, lipid binding and cholesterol acceptor activity, *Biochim.Biophys.Acta* 1761, 64-72, 2006

87. Fu, Q. and Li, L., Fragmentation of peptides with *N*-terminal dimethylation and imine/methylol adduction at the tryptophan side-chain, *J.Am.Soc.Mass Spectrom.* 17, 859-866, 2006

88. Xu, J. and Bowden, E.F., Determination of the orientation of adsorbed cytochrome C on carboxyalkanethiol self-assembled monolayers by *in situ* differential modification, *J.Am.Chem.Soc.* 128, 6813-6822, 2006

89. Hsu, J.L., Huang, S.Y., and Chen, S.H., Dimethyl multiplexed labeling combined with microcolumn separation and MS analysis for time course study in proteomics, *Electrophoresis* 27, 3652-3660, 2006

90. Walter, T.S., Meier, C., Assenberg, R., *et al.*, Lysine methylation as a routine rescue strategy for protein crystallization, *Structure* 14, 1617-1622, 2006

91. Torrance, L., Ziegler, A., Pittman, H., *et al.*, Oriented immobilization of engineered single-chain antibodies to develop biosensors for virus detection, *J.Virol.Methods* 134, 164-170, 2006

92. Li, Z.P., Duan, X.R., Liu, C.H., and Du, B.A., Selective determination of cysteine by resonance light scattering technique based on self-assembly of gold nanoparticles, *Anal.Biochem.* 351, 18-25, 2006

93. Talib, J., Beck, J.L., and Ralph, S.F., A mass spectrometric investigation of the binding of gold antiarthritic agents and the metabolite $[Au(CN)_2]^-$ to human serum albumin, *J.Biol.Inorg.Chem.* 11, 559-570, 2006

94. Gautier, C. and Burgi, T., Chiral *N*-isobutyryl-cysteine protected gold nanoparticles: preparation, size selection, and optimal activity in the US-Vis and infrared, *J.Am.Chem.Soc.* 128, 11079-11087, 2006

95. Urbina, R.D., Debaene, F., Jost, B., *et al.*, Self-assembled small-molecule microarrays for protease screening and profiling, *ChemBioChem* 7, 1790-1797, 2006

96. Strohalm, M., Kodicek, M., and Pechar, M., Tryptophan modification by 2-hydroxy-5-nitrobenzyl bromide studied by MALDI-TOF mass spectrometry, *Biochem.Biophys.Res. Commun.* 312, 811-816, 2003

97. Strohalm, M., Santrucek, J., Hynek, R., and Kodicek, M., Analysis of tryptophan surface accessibility in proteins by MALDI-TOF mass spectrometry, *Biochem.Biophys.Res.Commun.* 323, 1134-1138, 2004

98. Jung, J.W., Kuk, J.H. Kim, K.Y., *et al.*, Purification and characterization of exo-β-D-glucosaminidase from *Aspergillus fumigatus* S-26, *Protein Expr.Purif.* 45, 125-131, 2006

99. Tashima, I., Yoshida, T., Asada, Y., and Ohmachi, T., Purification and characterization of a novel L-2-amino-Δ_2-thiazoline-4-carboxylic acid hydroase from *Pseudomonas* sp. Strain ON-4a expressed in *E. coli, Appl.Microbiol.Biotechnol.* 72, 499-507, 2006

100. Ma, S.F., Nishikawa, M., Yabe, Y., *et al.*, Role of tyrosine and tryptophan in chemically modified serum albumin on its tissue distribution, *Biol.Pharm.Bull.* 29, 1926-1930, 2006

101. Smith, G.P., Kinetics of amine modification of proteins, *Bioconjug. Chem.* 17, 501-506, 2006

102. Adden, K., Gamble, L.J., Castner, D.G., *et al.*, Phosphonic acid monolayers for binding of bioactive molecules to titanium surfaces, *Langmuir* 22, 8197-8204, 2006

103. Noti, C., de Paz, J.L., Polito, L., and Seeberger, P.H., Preparation and use of microarrays containing synthetic heparin oligosaccharides for the rapid analysis of heparin-protein interactions, *Chemistry* 12, 8664-8686, 2006

104. Kenawy, el-R., el-Newehy, M., Abdel-Hay, F., and Ottenbrite, R.M., A new degradable hydroxamate linkage for pH-controlled drug delivery, *Biomacromolecules* 8, 196-201, 2007

105. Pandey, P., Singh, S.P., Arya, S.K., *et al.*, Application of thiolated gold nanoparticles for the enhancement of glucose oxidase activity, *Langmuir* 23, 3333-3337, 2007

106. Gauvreau, V., Chevalier, P., Vallieres, K., *et al.*, Engineering surfaces for bioconjugation: developing strategies and quantifying the extent of the reactions, *Bioconjug.Chem.* 15, 1146-1156, 2004

107. Balthasar, S., Michaelis, K., Dinauer, N, *et al.*, Preparation and characterization of antibody modified gelatin nanoparticles as drug carrier system for uptake in lymphocytes, *Biomaterials* 26, 2723-2732, 2005

108. Kommareddy, S. and Amiji, M., Preparation and evaluation of thiol-modified gelatin nanoparticles for intracellular DNA delivery in response to glutathione, *Bioconjug.Chem.* 16, 1423-1432, 2005

109. Langoth, N., Kalbacher, H., Schoffmann, G., *et al.*, Thiolated chitosans: design and *in vivo* evaluation of a mucoadhesive buccal peptide drug delivery system, *Pharm.Res.* 23, 573-579, 2006

110. Fowler, J.M., Stuart, M.C., and Wong, D.K., Self-assembled layer of thiolated protein G as an immunosensor scaffold, *Anal.Chem.* 79, 350-354, 2007

111. Jao, S.C., English Ospina, S.M., Berdis, A.J., *et al.*, Computational and mutational analysis of human glutaredoxin (thioltransferase): probing the molecular basis of the low pKa of cysteine 22 and its role in catalysis, *Biochemistry* 45, 4785-4796, 2006

112. Nelson, K.J., Parsonage, D., Hall, A., *et al.*, Cysteine pka values for bacterial peroxiredoxin AhpC, *Biochemistry*, 47, 12860-12868, 2008

113. Rogers, L.K., Leinweber, B.L., and Smith, C.V., Detection of reversible protein thiol modification in tissues, *Anal.Biochem.* 358, 171-184, 2006

114. Kurono, S., Kurono, T., Komori, N., *et al.*, Quantitative proteome analysis using D-labeled N-ethylmaleimide and ^{13}C-labeled iodoacetamide by matrix-assisted laser desorption/ionization time-of-flight mass spectrometer, *Bioorg.Med.Chem.* 14, 8197-8209, 2006

115. Yang, E. and Attygalle, A.B., LC/MS characterization of undesired products formed during iodoacetamide derivatization of sulfhydryl groups of peptides, *J.Mass Spectrom.* 42, 233-243, 2007

116. Meng, T.C., Hsu, S.F., and Tonka, N.K. Development of a modified in-gel assay to identify protein tyrosine phosphatases that are oxidized and inactivated *in vivo*, *Methods* 35, 28-36, 2005

117. Morty, R.E., Shih, A.Y., Fulop, V., and Andrews, N.W., Identification of the reactive cysteine residues in oligopeptidase B from *Trypanosoma brucei*, *FEBS Lett.* 579, 2191-2196, 2005

118. Hasegawa, G., Kikuchi, M., Kobayashi, Y., and Saito, Y., Synthesis and characterization of a novel reagent containing dansyl group, which specifically alkyates sulfhydryl group: an example of application for protein chemistry, *J.Biochem.Biophys.Methods* 63, 33-42, 2005

119. Atsriku, C, Scott, G.K., Benz, C.C., and Baldwin, M.A., Reactivity of zinc finger cysteines: chemical modification with labile zinc fingers in estrogen receptors, *J.Am.Soc.Mass Spectrom.* 16, 2017-2026, 2005

120. Chao, C.C., Chelius, D., Zhang, T., *et al.*, Insight into the virulence of *Richettsia prowazekii* by proteomic analysis and comparison with an a virulent strain, *Biochim.Biophys.Acta* 1774, 373-381, 2007

121. Grigorian, A.L., Bustamante, J.J., Hernandez, P., *et al.*, Extraordinary stable disulfide-linked homodimer of human growth hormone, *Protein Sci.* 14, 902-913, 2005

122. Hedberg, J.J., Bjerneld, E.J., Cetinkaya, S., *et al.*, A simplified 2-D electrophoresis protocol with the aid of an organic disulfide, *Proteomics* 5, 3088-3096, 2005

123. Wojcik, A., Naumov, S., Marciniak, B., and Brede, O., Repair reactions of pyrimidine-derived radicals by aliphatic thiols, *J.Phys.Chem. B.Matter.Mater.Surf.Interact.Biophys.* 110, 12738-12748, 2006

124. Okun, I., Malarchuk, S., Dubrovskaya, E., *et al.*, Screening for caspace-3 inhibitors: effect of a reducing agent on identified hit chemotypes, *J.Biomol.Screen.* 11, 694-703, 2006

125. Okado-Matsumoto, A., Guan, E., and Fridovich, I., Modification of cysteine 111 in human Cu.Zn-superoxide dismutase, *Free.Radic. Boil.Med.* 41, 1837-1846, 2006

126. Hisatome, I., Kurata, Y., Sasaki, N., *et al.*, Block of sodium channels by divalent mercury: role of specific cysteinyl residues in the P-loop region, *Biophys.J.* 79, 1336-1345, 2000

127. Kinne-Saffran, E., and Kinne, R.K., Inhibition by mercuric chloride of Na-K-2Cl cotransport activity in rectal gland plasma membrane vesicles isolated from *Squalus scanthias*, *Biochim.Biophys.Acta* 1510, 442-451, 2001

128. Taoka, S., Green, S.L., Loehr, T.M., and Banerjee, R., Mercuric chloride-induced spin or ligation state changes in ferric or ferrous human cystathionine β-synthase inhibit enzyme activity, *J.Inorg.Biochem.* 87, 253-259, 2001

129. Alencar, J.L., Lobysheva, I., Geffard, M., *et al.*, Role of S-nitrosylation of cysteine residues in long-lasting inhibitory effect of nitric oxide on arterial tone, *Mol.Pharmacol.* 63, 1148-1158, 2003

130. Durand, A., Giardina, T.,Villard, C., *et al.*, Rat kidney acylase I: further characterization and mutation studies on the involvement of Glu147 in the catalytic process, *Biochimie* 85, 953-962, 2003

131. Liu, X., Alexander, C., Serrano, J., *et al.*, Variable reactivity of an engineered cysteine at position 338 in cystic fibrosis transmembrane conductance regulator reflects different chemical states of the thiol, *J.Biol.Chem.* 281, 8275-8285, 2006

132. Audia, J.P., Roberts, R.A., and Winkler, H.H., Cysteine-scanning mutagenesis and thiol modification of the *Rickettsia prowazekii* ATP/ADP translocase: characterization of the TMs IV-VII and IX-XII and their accessibility to the aqueous translocation pathway, *Biochemistry* 45, 2648-2656, 2006

133. Tombolato, F., Ferrarini, A., and Freed, J.H., Modeling the effects of structure and dynamics of the nitroxide side chain on the ESR spectra of spin-labeled proteins, *J.Phys.Chem.BCondens.Matter Mater. Surf.Interfaces Biophys.* 110, 26260-26271, 2006

134. Karala, A.R., and Ruddock, L.W., Does S-methyl methanethiosulfonate trap the thiol-disulfide state of proteins? *Antioxid.Redox.Signal.* 9, 527-531, 2007

135. Thonon, D., Jacques, V., and Desreux, J.F., A gadolinium triacetic monoamide DOTA derivative with a methanesulfonate anchor group. Relaxivity properties and conjugation with albumin and thiolated particles, *Contrast Media Mol.Imaging* 2, 24-34, 2007

136. Ishikawa, Y., Yamamoto, Y., Otsubo, M., *et al.*, Chemical modification of amine groups on PS II protein(s) retards photoassembly of the photosynthetic water-oxidizing complex, *Biochemistry* 41, 1972-1980, 2002

137. Shortreed, M.R., Lamos, S.M., Frey, B.L., Ionizable isotopic labeling reagent for relative quantification of amine metabolites by mass spectrometry, *Anal.Chem.* 78, 6398-6403, 2006

138. Poon, S.F., Stock, N., Payne, N.K., *et al.*, Novel approach to pro-drugs of lactones: water soluble imidate and ortho-ester derivatives of a furanone-based COX-2 selective inhibitor, *Bioorg.Med.Chem.Lett.* 15, 2259-2263, 2005

139. Xu, J., Degraw, A.J., Duckworth, B.P., *et al.*, Synthesis and reactivity of 6,7-dihydrogeranylazides: reagents for primary amine incorporation into peptides and subsequent Staudinger ligation, *Chem.Biol. Drug Des.* 68, 85-96, 2006

140. Takaku, H., Sato, J., Ishida, H.K., *et al.*, A chemical synthesis of UDP-LacNAc and its regioisomer for finding "oligonucleotide transferases", *Glycoconj.J.* 23, 565-573, 2006

141. Leane, M.M., Nankervis, R., Smith, A., and Illum, L., Use of the ninhydrin assay to measure the release of chitosan from oral solid dosage forms, *Int.J.Pharm.* 271, 241-249, 2004

142. Drochioiu, G., Mangalagiu, I., Avram, E., *et al.*, Cyanide reaction with ninhydrin: elucidation of reaction and interference mechanisms, *Anal.Sci.* 20, 1443-1447, 2004

143. Hansen, D.B. and Joullie, M.M., The development of novel ninhydrin analogues, *Chem.Soc.Rev.* 34, 408-417, 2005

144. Wu, Y., Hussain, M., and Fassihi, R., Development of a simple analytical methodology for determination of glucosamine release from modified release matrix tablets, *J.Pharm.Biomed.Anal.* 38, 263-269, 2005

145. Lipscomb, I.P., Pinchin, H.E., Collin, R., The sensitivity of approved ninhydrin and biuret tests in the assessment of protein contamination on surgical steel as an aid to prevent iatrogenic prion transmission, *J.Hosp.Infect.* 64, 288-292, 2006

146. Marche, P., Girma, J.P., Morgat, J.L., and Fromageot, P., Specific tritiation of indole derivatives by catalytic desulfenylation. Application to the labelling of tryptophan-containing peptides, *Eur.J.Biochem.* 50, 375-382, 1975

147. Sasagawa, T., Titani, K., and Walsh, K.A., Selective isolation of tryptophan-containing peptides by hydrophobicity modulation, *Anal. Biochem.* 134, 224-229, 1983

148. Hassani, O., Mansuella, P., Cestele, S., *et al.*, Role of lysine and tryptophan residues in the biological activity of toxin VII (Ts γ) from the scorpion *Tityus serrulatus*, *Eur.J.Biochem.* 260, 76-86, 1999

149. Ou, K., Kesuma, D., Ganesan, K., *et al.*, Quantitative profiling of drug-associated proteomic alterations by combined 2-nitrophenyl-sulfenyl chloride (NBS) isotope labeling and 2DE/MS identification, *J.Proteome Res.* 5, 2194-2206, 2006

150. Matsunaga, H. and Haginaka, J., Investigation of chiral recognition mechanism on chicken α(1)-acid glycoprotein using separation system, *J.Chromatog. A.* 1106, 124-130, 2006

151. Matthiesen, R., Bauw, G., and Welinder, K.G., Use of performic acid oxidation to expand the mass distribution of tryptic peptides, *Anal. Chem.* 76, 6848-6852, 2004

152. Dai, J., Wang, J., Zhang, Y., *et al.*, Enrichment and identification of cysteine-containing peptides from tryptic digests of performic oxidized proteins by strong cation exchange LC an MALDI-TOF/TOF MS, *Anal.Chem.* 77, 7594-7604, 2005

153. Cao, J., Wijaya, R., and Leroy, F., Unzipping the cuticle of the human hair shaft to obtain micron/nano keratin filaments, *Biopolymers* 83, 614-618, 2006

154. Kulkarni, A.D., Rai, D., Bartolotti, L.F., and Pathak, R.K., Interaction of performic acid with water molecules: A first-principles study, *J.Phys.Chem.A. Mol.Spectros.Kinet.Environ.Gen. Theory* 110, 11855-11861, 2006

155. Bosch, L., Algeria, A., and Farre, R., Amino acid contents of infant foods, *Int.J.Food Sci.Nutr.* 57, 212-218, 2006

156. Johans, M., Milanesi, E., Franck, M., *et al.*, Modification of permeability transition pore arginine(s) by phenylglyoxal derivatives in isolated mitochondria and mammalian cells. Structure-function relationship of arginine ligands, *J.Biol.Chem.* 280, 12130-12136, 2005

157. Zhang, Q., Crosland, E., and Fabris, D., Nested Arg-specific bifunctional crosslinkers for MS- based structural analysis of proteins and protein assemblies, *Anal. Chim. Acta*, 627, 117-128, 2008

158. Takazaki, S., Abe, Y., Kang, D., *et al.*, The functional role of arginine 901 at the C-terminus of the human anion transporter band 3 protein, *J.Biochem.* 139, 903-912, 2006

159. Greig, N., Wyllie, S., Vickers, T.J., and Fairlamb, A.N. Trypanothione-dependent glyoxalase I in *Trypanosoma cruzi*, *Biochem.J.* 400, 217-223, 2006

160. Ye, M. and English, A.M., Binding of polyaminocarboxylate chelators to the active-site copper inhibits the GSNO-reductase activity but not the superoxide dismutase activity of Cu, Zn-superoxide dismutase, *Biochemistry* 45, 12723-12732, 2006

161. Peelen, D. and Smith, L.M., Immobilization of the amine-modified oligonucleotides on aldehyde-terminated alkanethiol monolayers on gold, *Langmuir* 21, 266-271, 2005

162. Kim, H.S. and Wainer, I.W., The covalent immobilization of microsomal uridine diphospho-glucuronosyltransferase (UDPGT): initial synthesis and characterization of an UDPGT immobilized enzyme reactor for the on-line study of glucuronidation, *J.Chromatog.B Analyt.Technol.Biomed.Life Sci.* 823, 158-166, 2005

163. Wildsmith, K.R., Albert, C.J., Hsu, F.F., *et al.*, Myeloperoxidase-derived 2-chlorohexadecanal forms Schiff bases with primary amines of ethanolamine glycerophospholipids and lysine, *Chem. Phys.Lipids* 139, 157-170, 2006

164. Mirzaei, H. and Regnier, F., Enrichment of carbonylated peptides using Girard P reagent and strong cation exchange chromatography, *Anal.Chem.* 78, 770-778, 2006

165. Boersema, P.J., Aye, T.T., van Veen, T.A., *et al.*, Triplex protein quantification based on stable isotope labeling by peptide dimethylation applied to cell and tissue lysates, *Proteomics*, 8, 4624-4632, 2008

166. Damodaran, S., Estimation of disulfide bonds using 2-nitro-5-thiosulfobenzoic acid: limitations, *Anal.Biochem.* 145, 200-204, 1985

167. Martin de Llano, J.J. and Gaviulanes, J.G., Increased electrophoretic mobility of sodium sulfite-treated jack bean urease, *Electrophoresis* 13, 300-304, 1992

168. Emerson, D. and Ghiorse, W.C., Role of disulfide bonds in maintaining the structural integrity of the sheath of the *Leptothrix discophora* SP-6, *J.Bacteriol.* 175, 7819-7827, 1993

169. Mukhopadhyay, A., Reversible protection of disulfide bonds followed by oxidativate folding render recombinant hCGβ highly immunogenic, *Vaccine* 18, 1802-1810, 2000

170. Raftery, M.J., Selective detection of thiosulfate-containing peptides using tandem mass spectrometry, *Rapid Commun.Mass Spectrom.* 19, 674-682, 2005

171. Sakoh, M., Okazaki, T., Nagaoka, Y., and Asami, K., N-terminal insertion of alamethicin in channel formation studied using its covalent dimer N-terminally linked by disulfide bond, *Biochim.Biophys. Acta.* 1612, 117-121, 2003

172. Tie, J.K., Jin, D.Y., Loiselle, D.R., *et al.*, Chemical modification of cysteine residues is a misleading indicator of their status as active site residues in the vitamin K-dependent γ-glutamyl carboxylation reaction, *J.Biol.Chem.* 279, 54079-54087, 2004

173. Voslar, M., Matejka, P., and Schreiber, I., Oscillatory reactions involving hydrogen peroxide and thiosulfate-kinetics of the oxidation of tetrathionate by hydrogen peroxide, *Inorg.Chem.* 45, 2824-2834, 2006

174. Hahn, S.K., Kim, J.S., and Shimoobouji, T., Injectable hyaluronic acid microhydrogels for controlled release formulation of erythropoietin, *J.Biomed.Mater.Res.A.* 80, 916-924, 2007

175. Hahn, S.K., Park, J.R., Tomimatsu, T., and Shimoboji, T., Synthesis and degradation test of hyaluronic acid hydrogels, *Int.J.Biol. Macromol.* 40, 374-380, 2007

176. Cline, D.J., Redding, S.E., Brohawn, S.G., *et al.* New water-soluble phosphines as reductants of peptide and protein disulfide bonds: reactivity and membrane permeability, *Biochemistry* 43, 15195-15203, 2004

177. Xu, G., Kiselar, J., He, Q., and Chance, M.R., Secondary reactions and strategies to improve quantitative protein footprinting, *Anal.Chem.* 77, 3029-3037, 2005

178. Willis, M.S., Hogan, J.K., Prabhakar, P., *et al.*, Investigation of protein refolding using a fractional factorial screen: a study of reagent effects and interactions, *Protein Sci.* 14, 1818-1826, 2005

179. Valcu, C.M. and Schlink, K., Reduction of proteins during sample preparation and two-dimensional gel electrophoresis of woody plant samples, *Proteomics* 6, 1599-1605, 2006

180. Boga, C., Fiume, L., Baglioni, M., *et al.*, Characteristics of the conjugate of the (6-maleimdiocaproyl)hydrazone derivative of doxorubicin with lactosaminated human albumin by ^{13}C NMR spectroscopy, *Eur. J. Pharm. Sci.*, 38, 262-269, 2009

181. Santrucek, J., Strohalm, M., Kadlcik, V., *et al.* Tyrosine residue modification studied by MALDI-TOF mass spectrometry, *Biochem. Biophys.Res.Commun.* 323, 1151-1156, 2004

182. Negrerie, M., Martin, J.L., and Njgiem, H.O., Functionality of nitrated acetylcholine receptor: the two-step formation of nitrotyrosines reveals their differential role in effector binding, *FEBS Lett.* 579, 2643-2647, 2005

183. Carven, G.J. and Stern, L.J., Probing the ligand-induced conformational change in HLA-DR1 by selective chemical modification and mass spectrometric mapping, *Biochemistry* 44, 13625-13637, 2005

184. Gruijthuijsen, Y.K., Grieshuber, I., Stocklinger, A., *et al.*, Nitration enhances the allergic potential of proteins, *Int.Arch.Allergy Immunol.* 141, 265-275, 2006

185. Ghesquiere, B., Goethals, M., van Damme, J., *et al.*, Improved tandem mass spectrometric characterization of 3-nitrotyrosine sites in peptides, *Rapid Commun.Mass Spectrom.* 20, 2885-2893, 2006

186. Korn, C., Scholz, S.R., Gimadutdinow, O., *et al.*, Involvement of conserved histidine, lysine, and tyrosine residues in the mechanism of DNA cleavage by the caspase-3 activated DNase CAD, *Nucleic Acids Res.* 30, 1325-1332, 2002

187. Metz, B., Jiskoot, W., Hennink, W.E., *et al.*, Physicochemical and immunochemical techniques predict the quality of diphtheria toxoid vaccines, *Vaccine* 22, 156-167, 2003

188. Lin, J.C., Chen, Q.X., Shi, Y., *et al.*, The chemical modification of the essential groups of β-*N*-acetyl-D-glucosaminidase from *Turbo corutus Solander*, *IUBMB Life* 55, 547-552, 2003

189. Jagtap, S. and Rao, M., Conformation and microenvironment of the active site of a low-molecular weight 1,4-β-D-glucan glucanohydrolase from an alkalothermophilic *Thermomonospora* sp.: involvement of lysine and cysteine residues, *Biochem.Biophys.Res.Commun.* 347, 428-432, 2006

190. Chang, L.S., Cheng, Y.C., and Chen, C.P., Modification of Lys-6 and Lys-65 affects the structural stability of Taiwan cobra phospholipase A₂, *Protein J.* 25, 127-134, 2006

191. Bingham, J.P., Broxton, N.M., Livett, B.G., *et al.*, Optimizing the connectivity in disulfide-rich peptides: α-conotoxin SII as a case study, *Anal.Biochem.* 338, 48-61, 2005

192. Maeda, K., Finnie, C., and Svensson, B., Identification of thioredoxin h-reducible disuphides in proteomes by differential labelling of cysteines: insight into recognition and regulation of proteins in barley seeds by thioredoxin h, *Proteomics* 5, 1634-1644, 2005

193. Winnik, W.M., Continuous pH/salt gradient and peptide score for strong cation exchange chromatography in 2D-nano-LC/MS/MS peptide identification for proteomics, *Anal.Chem.* 77, 4991-4998, 2005

194. Walter, J.K., Ruekert, C., Voss, M., *et al.*, The oligomerization of the coiled coil-domain of occludin is redox sensitive, *Ann. N.Y. Acad. Sci.*, 1165, 19-27, 2009

195. Chowdhury, S.M., Munske, G.R., Ronald, R.C., and Bruce, J.E., Evaluation of low energy CID and ECD fragmentation behavior of mono-oxidized thio-ether bonds in peptides, *J.Am.Soc.Mass Spectrom.* 18, 493-501, 2007

196. Salhany, J.M., Sloan, R.L., and Cordes, K.S., The carboxyl side chain of glutamate 681 interacts with a chloride binding modifier site that allosterically modulates the dimeric conformational state of band 3 (AE1). Implications for the mechanism of anion/proton cotransport, *Biochemistry* 42, 1589-1602, 2003

197. Kosters, H.A. and de Jongh, H.H., Spectrophotometric tool for the determination of the total carboxylate content in proteins: molar extinction coefficient of the enol ester from Woodward's reagent K reacted with protein carboxylates, *Anal.Chem.* 75, 22512-2516, 2003

198. Carvajal, N., Uribe, E., Lopez, V., and Salas, M., Inactivation of human liver arginase by Woodward's reagent K: evidence for reaction with His141, *Protein J.* 23, 179-183, 2004

199. Jennings, M.L., Evidence for a second binding/transport site for chloride in erythrocyte anion transporter AE1 modified at glutamate 681, *Biophys. J.* 88, 2681-2691, 2005

200. SinhaRoy, S., Banerjee, S., Ray, M., and Ray, S., Possible involvement of glutamic and/or aspartic residue(s) and requirement of mitochrondrial integrity for the protective effect of creatine against inhibition of cardiac mitochondrial respiration by methylglyoxal, *Mol. Cell.Biochem.* 271, 167-176, 2005

Acetic Anhydride

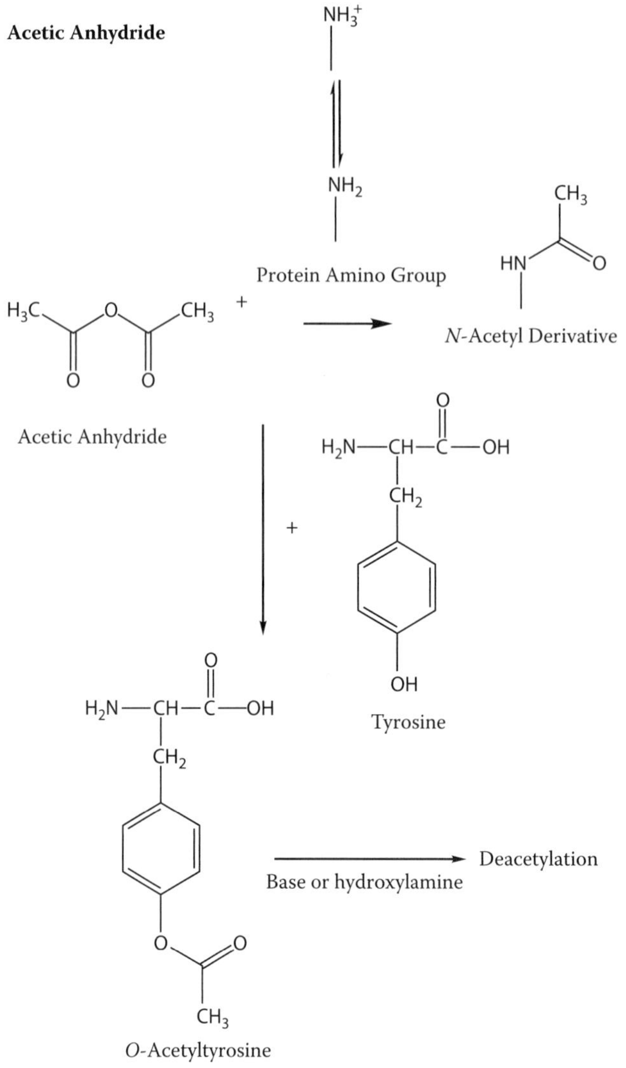

NH_3^+
NH_2
Protein Amino Group
CH_3
HN
O
N-Acetyl Derivative
H_3C
O
CH_3
+
Acetic Anhydride
$H_2N—CH—C—OH$
CH_2
+
OH
Tyrosine
Base or hydroxylamine
Deacetylation
O-Acetyltyrosine

***N*-Acetylimidazole**

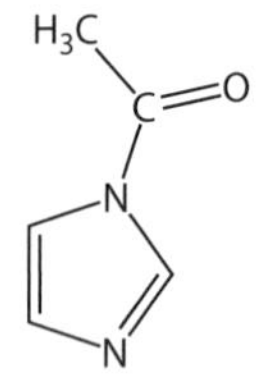

H_3C
C=O
N
N

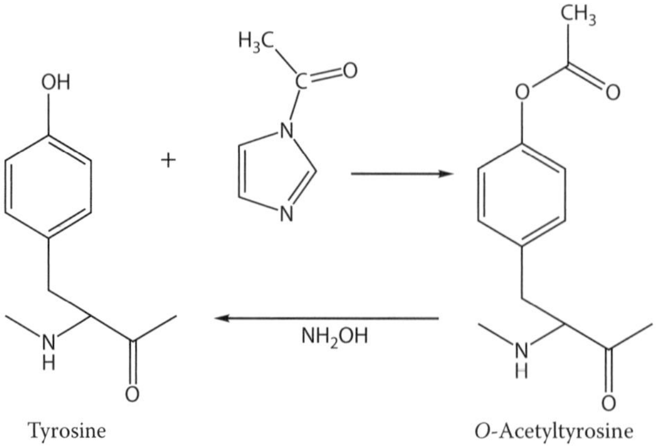

OH
+
H_3C
C=O
N
N
CH_3
O
N
H
NH_2OH
Tyrosine
O-Acetyltyrosine

BNPS-Skatole

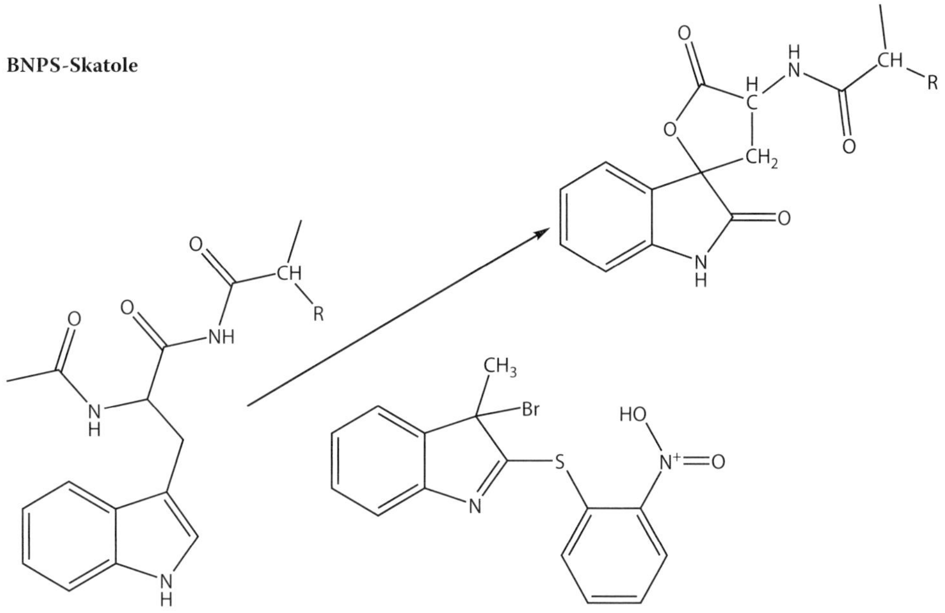

Tryptophan

BNPS-Skatole

Cyanate Cysteine

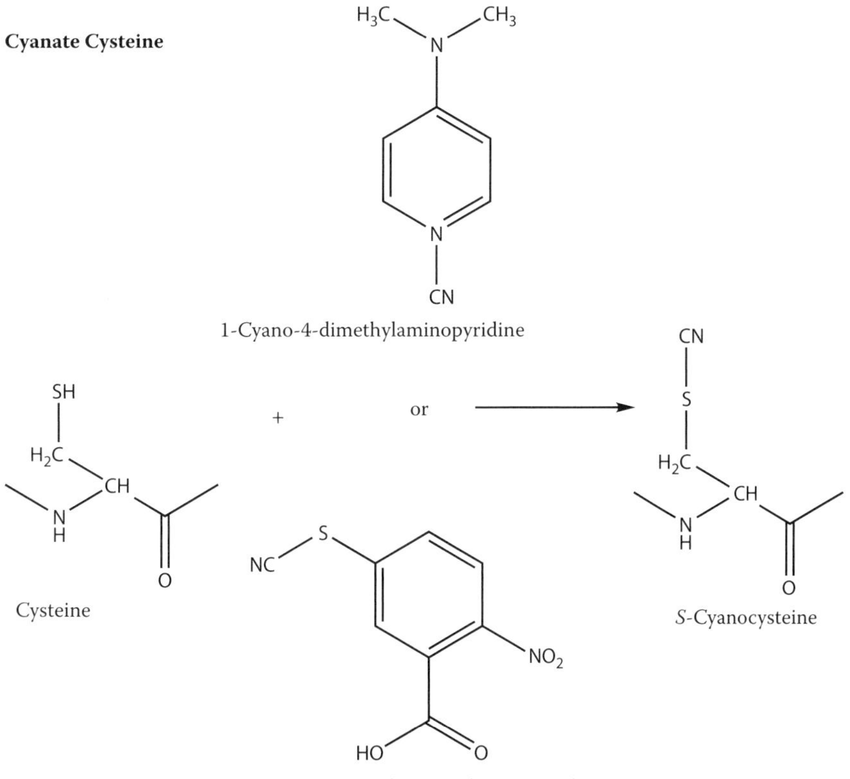

1-Cyano-4-dimethylaminopyridine

Cysteine

+   or

S-Cyanocysteine

2-Nitro-5-thiocyanobenzoic acid

Bromoethylamine

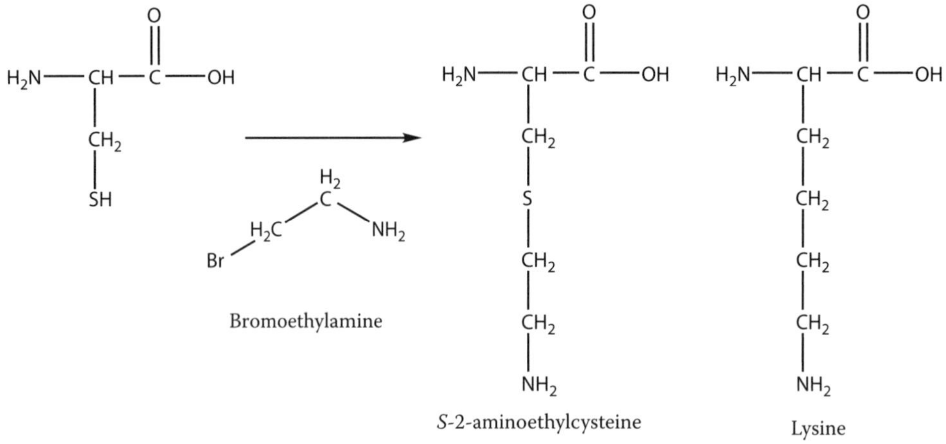

S-2-aminoethylcysteine   Lysine

N-Bromosuccinimide

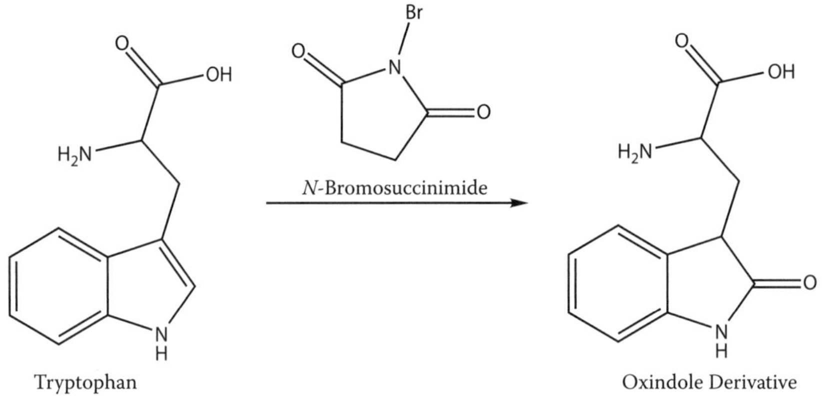

N-Bromosuccinimide

Tryptophan   Oxindole Derivative

2-3-Butanedione (diacetyl)

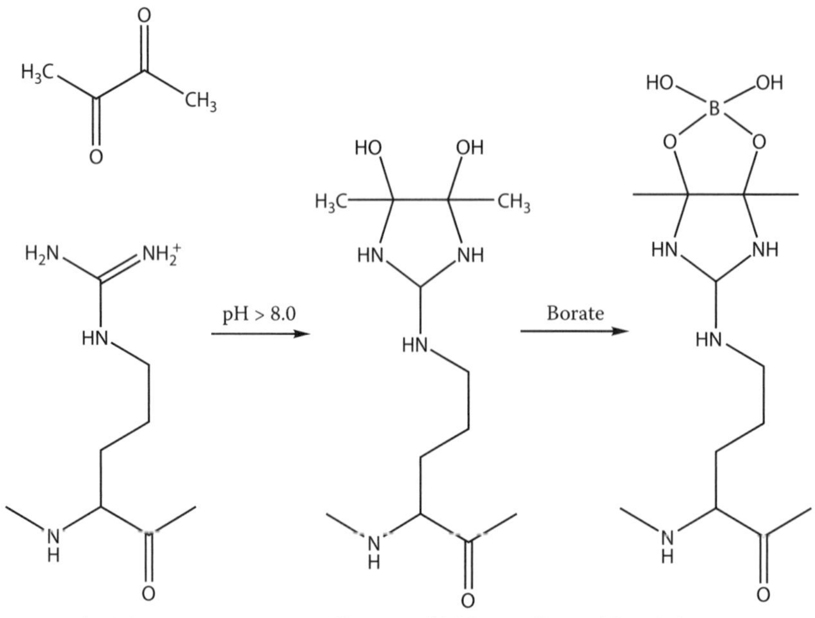

pH > 8.0   Borate

Arginine   Reaction of 2,3-butanedione with arginine

Citraconic Anhydride

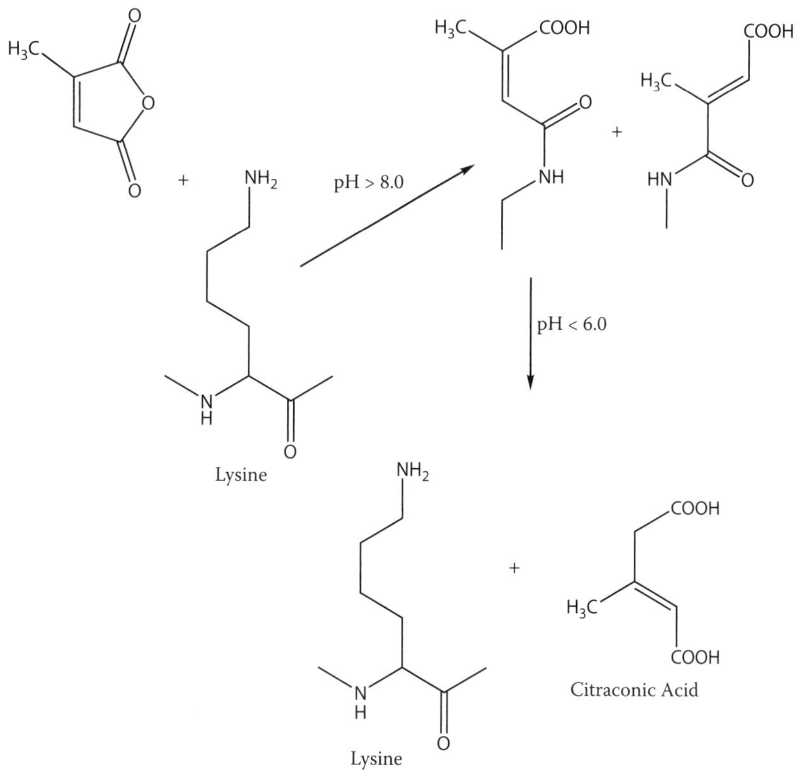

Lysine

pH > 8.0

+

pH < 6.0

Citraconic Acid

Lysine

Cyanate

NCO⁻

Lysine

Homocitrulline

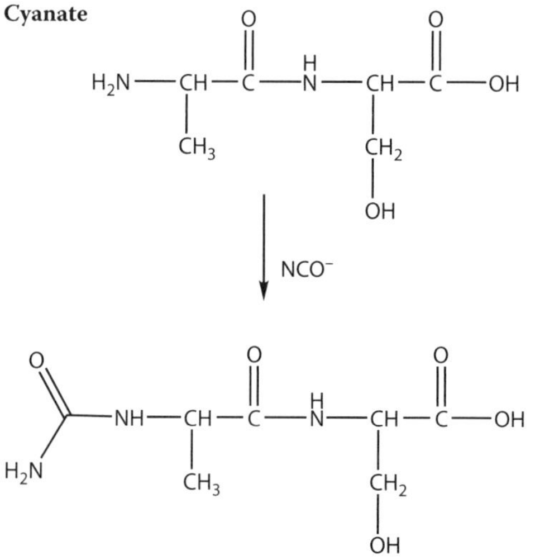

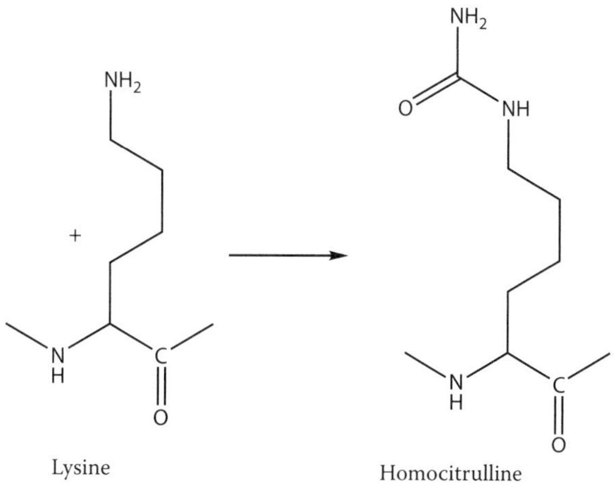

Cyclohexanedione

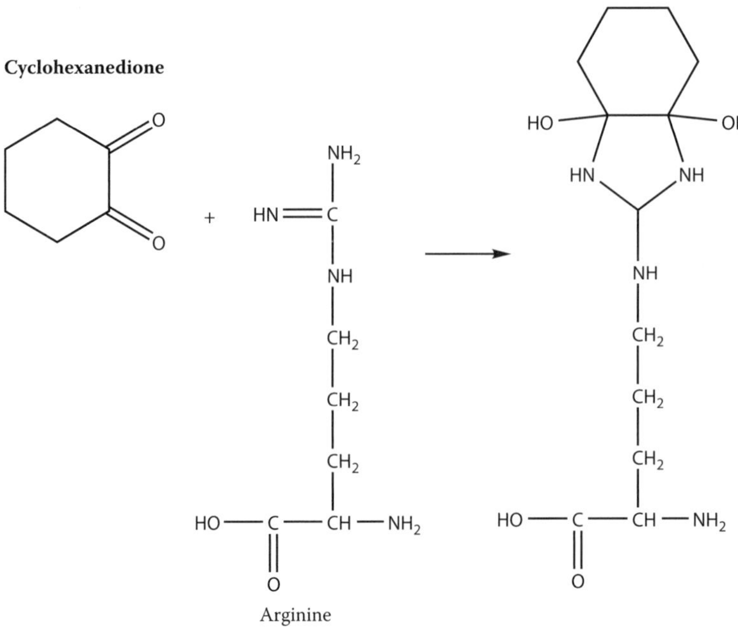

Arginine

Diethylpyrocarbonate (DEPC)
Ethoxyformic Anhydride

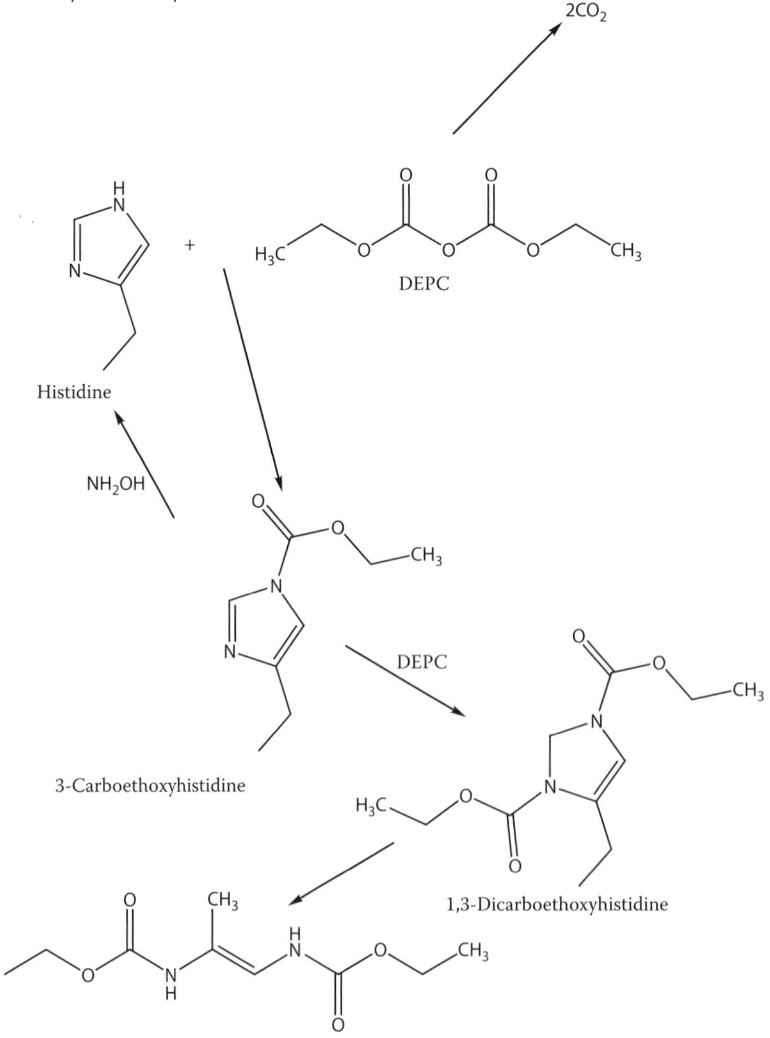

$2CH_3CH_2OH$
+
$2CO_2$

Histidine

NH_2OH

DEPC

3-Carboethoxyhistidine

1,3-Dicarboethoxyhistidine

Ellman's Reagent
5,5-dithio-*bis*-nitrobenzoic acid

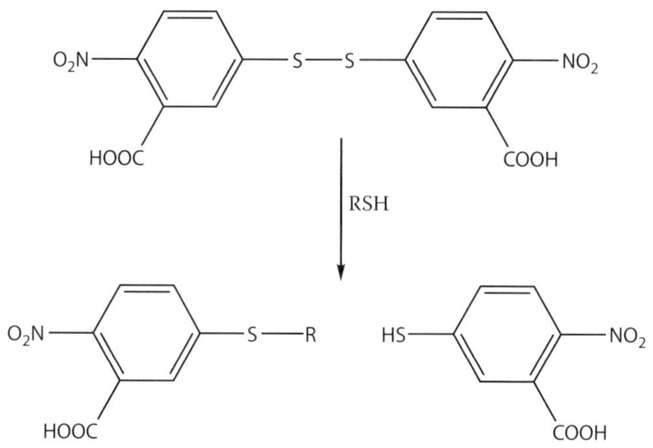

N-Ethylmaleimide

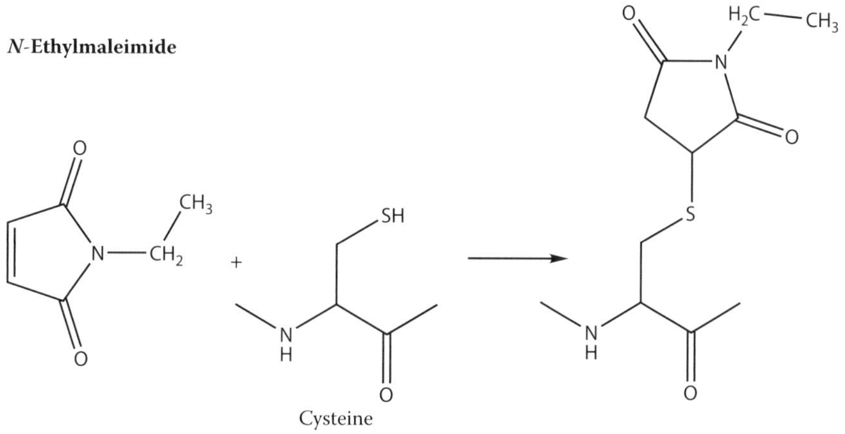

Cysteine

Gold

H_2 or H_2O

$1/2 H_2$

H_2 or H_2O

Ethyleneimine
 (Aziridine)

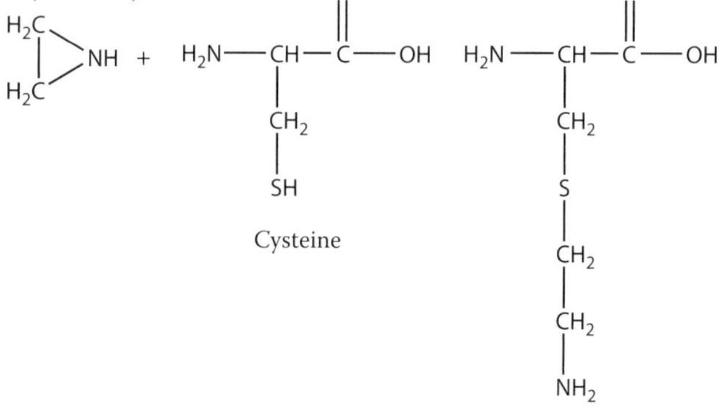

Cysteine

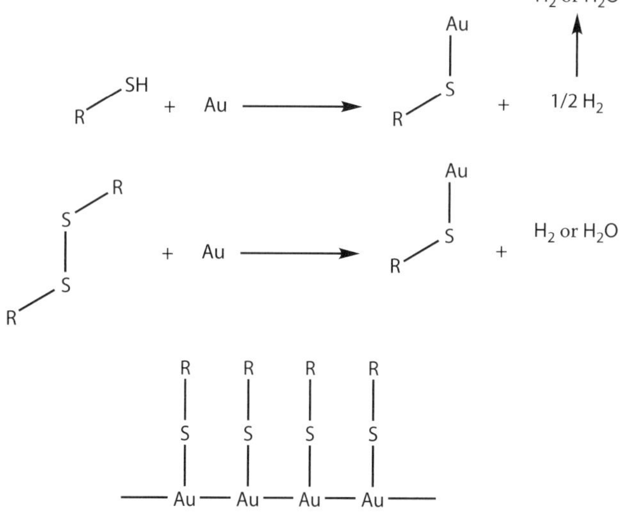

Self-assembled monolayer (SAM)

2-Hydroxy-5-nitrobenzyl bromide

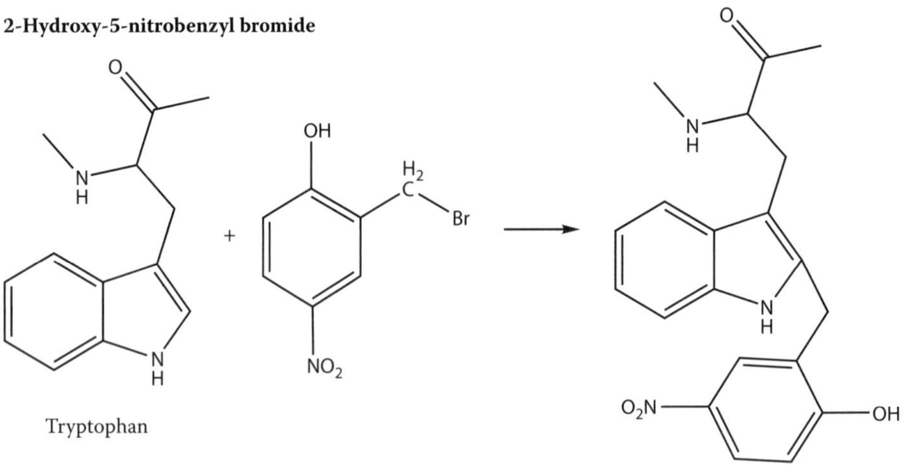

Tryptophan

N-Hydroxysuccinimide (NHS)

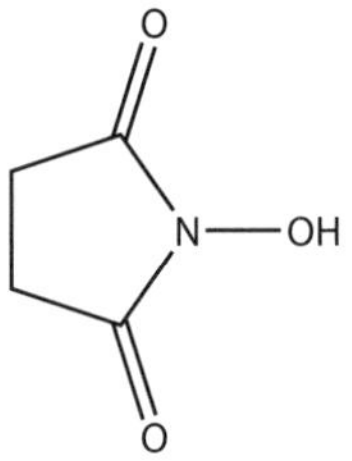

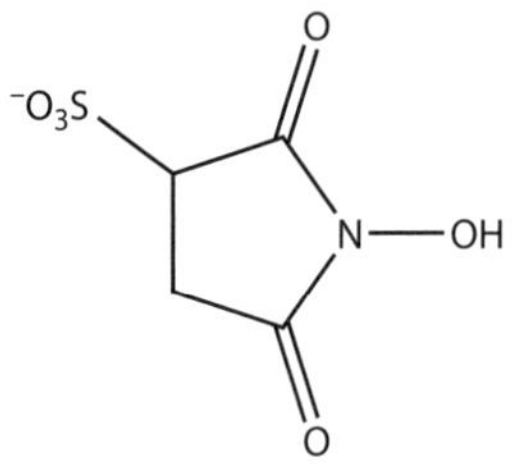

Sulfo-N-hydroxysuccinimide

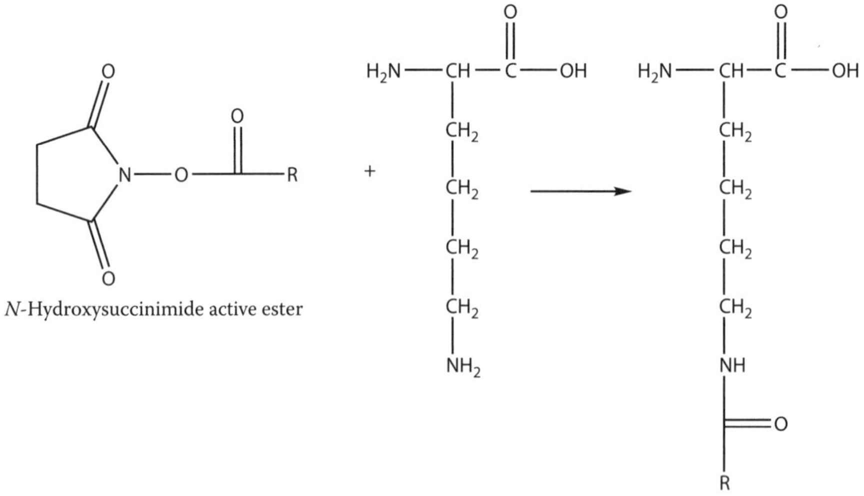

N-Hydroxysuccinimide active ester

1-Iminothiolane (Traut's Reagent)

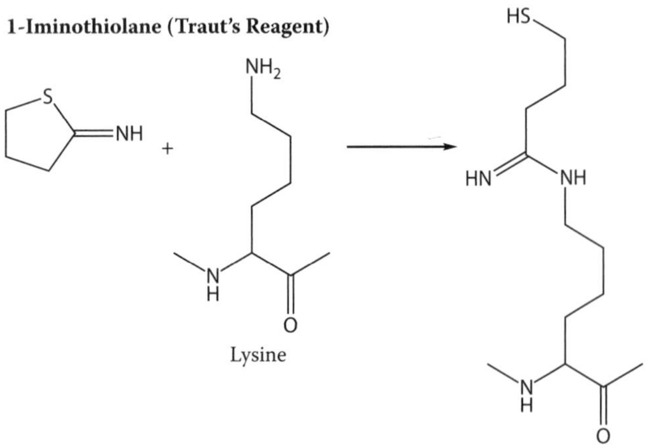

Lysine

Mercuric Chloride

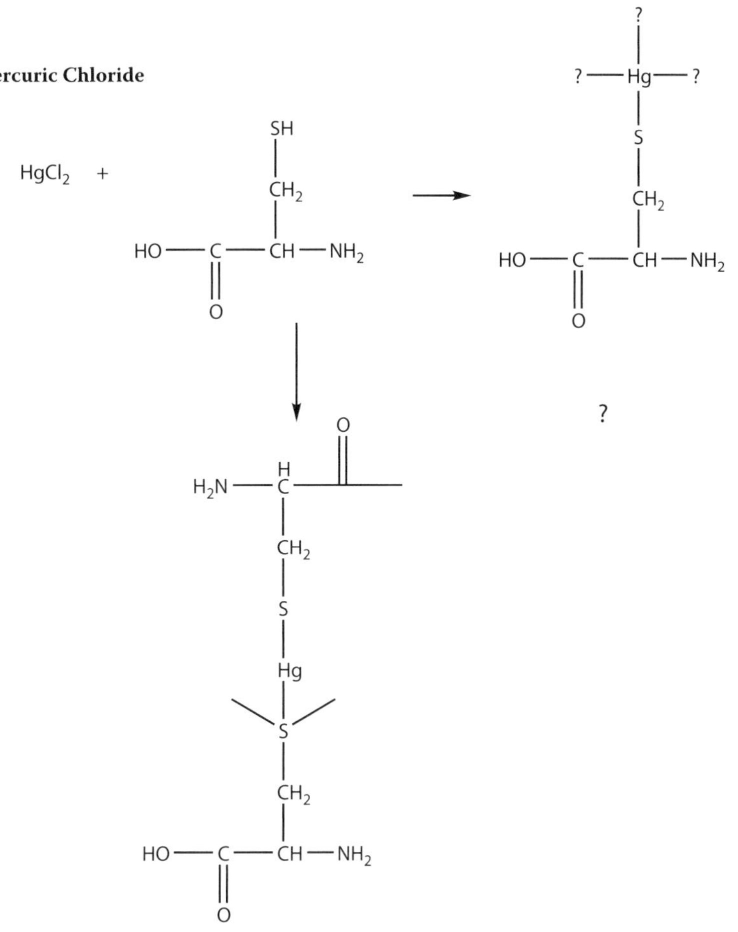

$HgCl_2$ +

SH
|
CH_2
|
HO—C—CH—NH_2
‖
O

?
|
?—Hg—?
|
S
|
CH_2
|
HO—C—CH—NH_2
‖
O

?

O
‖
H_2N—C—C——
| H
|
CH_2
|
S
|
Hg
|
S
|
CH_2
|
HO—C—CH—NH_2
‖
O

Ninhydrin

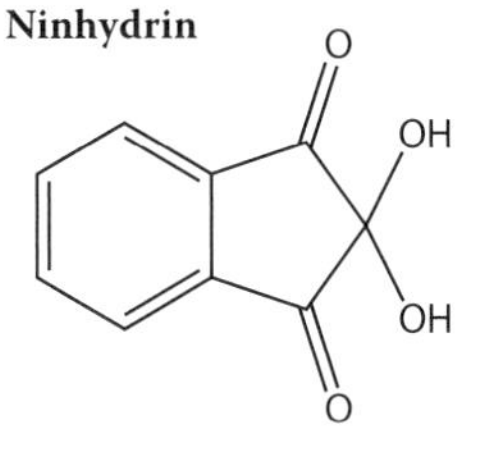

Phenylglyoxal

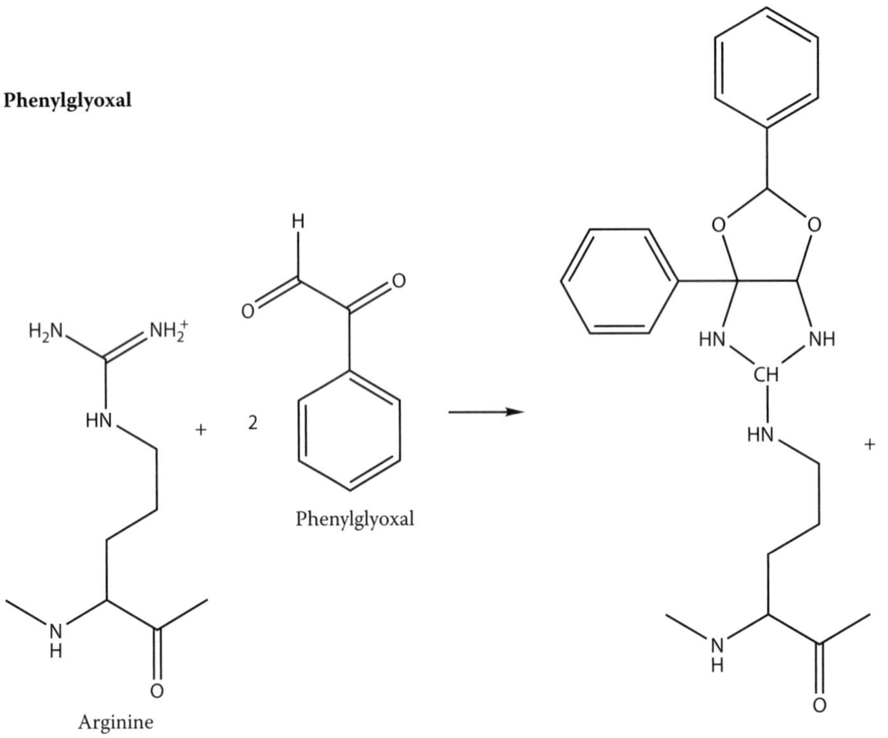

Phenylglyoxal

Arginine

A Scheme for the Reaction of Arginine with Phenylglyoxal

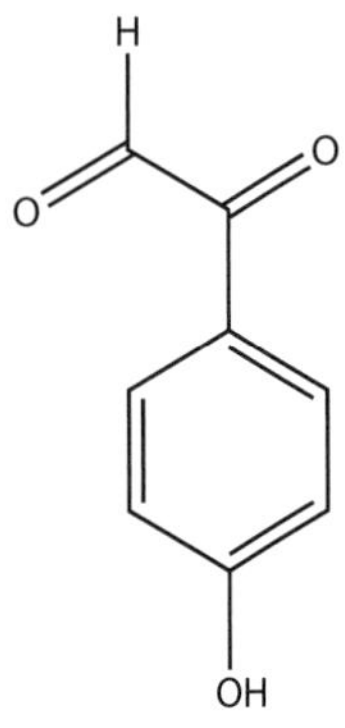

p-hydroxyphenylglyoxal

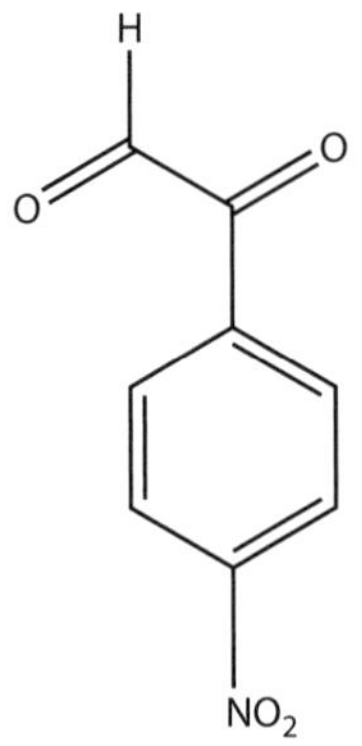

p-nitrophenylglyoxal

Sodium Sulfite

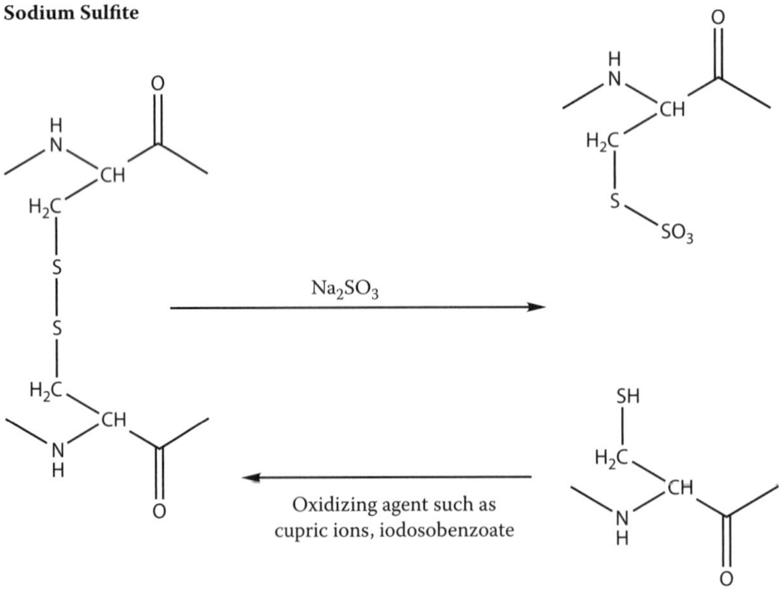

Na_2SO_3

Oxidizing agent such as
cupric ions, iodosobenzoate

Sodium Tetrathionate

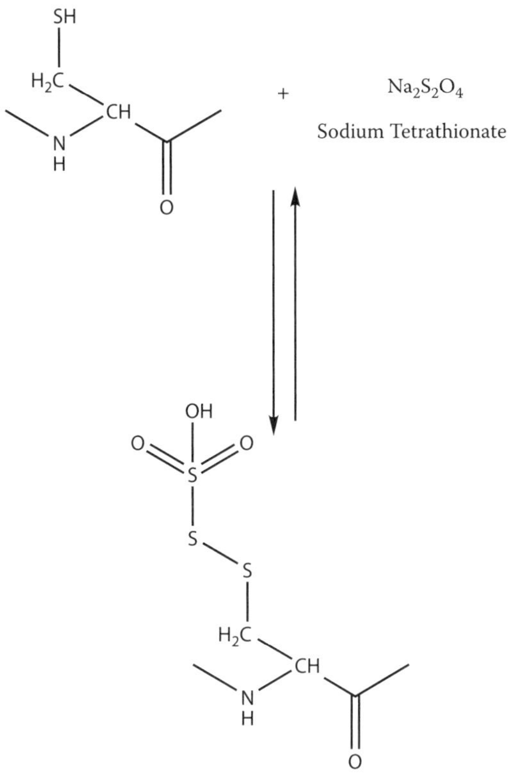

Vinylpyridine

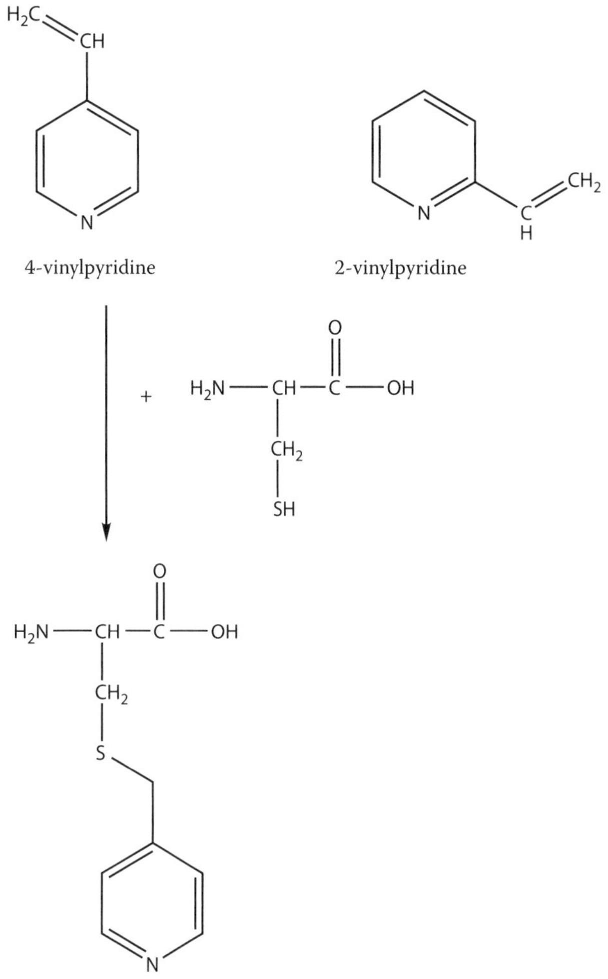

$Na_2S_2O_4$

Sodium Tetrathionate

4-vinylpyridine

2-vinylpyridine

Tetranitromethane

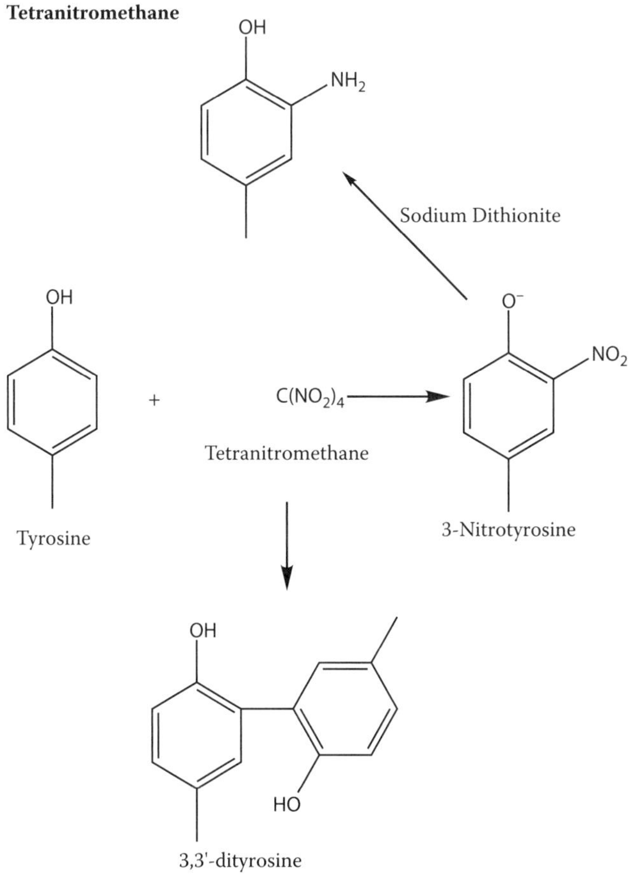

Sodium Dithionite

$C(NO_2)_4$

Tetranitromethane

Tyrosine

3-Nitrotyrosine

3,3'-dityrosine

Woodward's Reagent K

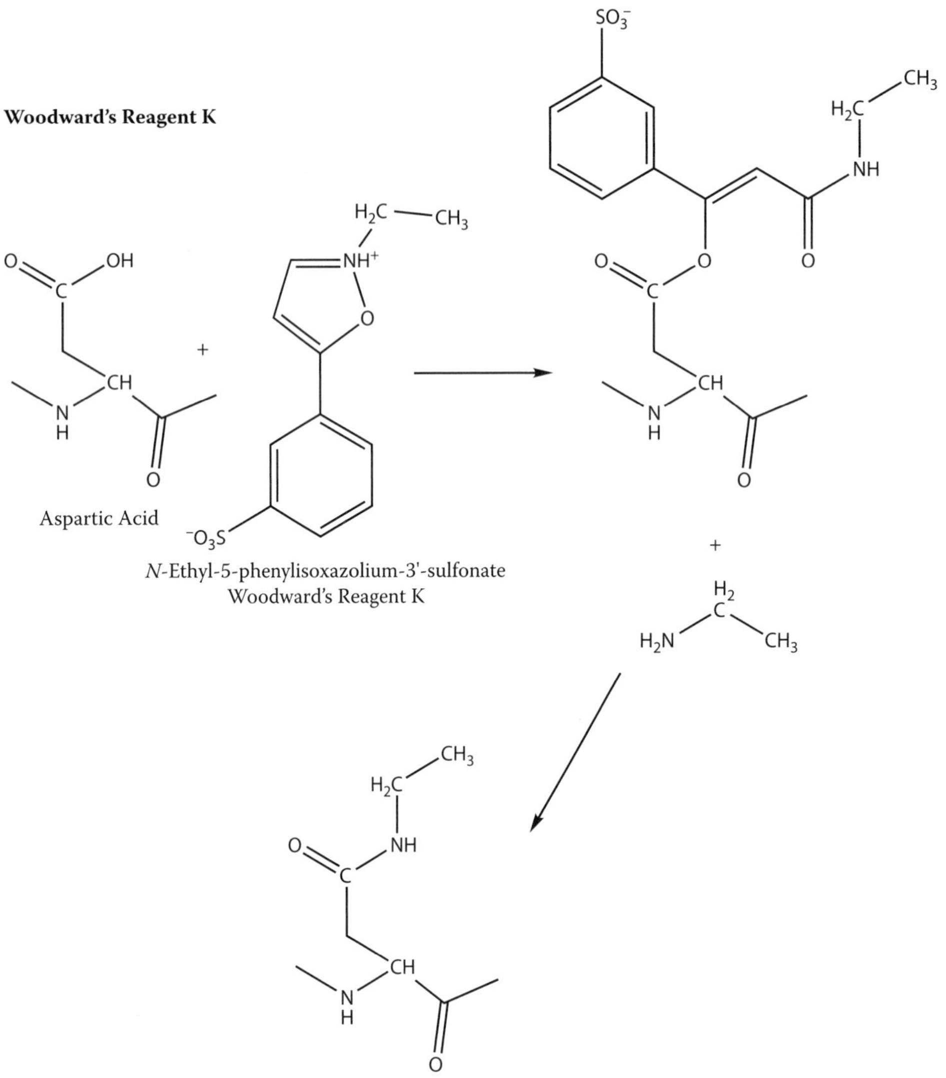

Aspartic Acid

N-Ethyl-5-phenylisoxazolium-3'-sulfonate
Woodward's Reagent K

PROTEIN pK VALUES

Lynne H. Botelho and Frank R. N. Gurd

The general techniques for determining individual pK values in proteins usually depend on NMR,[1,2] absorption,[3] or kinetic[4] procedures. Effects of neighboring groups may be evident in chemical shift influences[5–7] or in electrostatic influences on the hydrogen ion equilibria proper.[8]

1. Markley, Finkenstadt, Dugas, Leduc, and Drapeau, *Biochemistry*, 14, 998 (1975).
2. Markley, *Acc. Chem. Res.*, 8, 70 (1975).
3. Tanford, Hanenstein, and Rands, *J. Am. Chem. Soc.*, 77, 6409 (1955).
4. Garner, Bogardt, and Gurd, *J. Biol. Chem.*, in press .
5. Sachs, Schechter, and Cohen, *J. Biol. Chem.*, 246, 6576 (1971).
6. Shrager, Cohen, Heller, Sachs, and Schechter, *Biochemistry*, 11, 541 (1972).
7. Deslauriers, McGregor, Sarantakis, and Smith, *Biochemistry*, 13, 3443 (1974).
8. Roxby and Tanford, *Biochemistry*, 10, 3348 (1971).

TABLE 1A: Specific His pK Assignments in Ribonuclease A

Protein	pK Values	Reference
Bovine		
His 12	6.3	1
His 48	5.8	1
His 73	6.4	1
His 105	6.7	1
Rat		
His 12	6.6	1
His 48	6.2	1
His 73	7.6	1
His 105	6.3	1
His 119	6.1	1

References

1. Migchelsen and Beintema, *J. Mol Biol.*, 79, 25 (1973).

TABLE 1B: Nonspecific His pK Values for Ribonuclease A

Protein	pK Values				Reference
Coypu	5.8	6.3	6.3	8.0	1
Chinchilla	4.9	6.0	6.1	7.2	1
Bovine	6.01	6.17	6.72	6.9	2

References

1. Migchelsen and Beintema, *J. Mol. Biol.*, 79, 25 (1973).
2. Markley, *Acc. Chem. Res.*, 8, 70 (1975).

TABLE 1C: pK Values for Histidine Residues in Myoglobin

Species	pK Observed		Reference
Sperm whale	5.37	5.33	1, 2
	5.53	5.39	
	6.34	6.21	
	6.44	6.31	
	6.65	6.55	
	6.83	6.72	
	8.05	7.97	
Horse	5.7	5.5	1, 2
	6.0	5.8	
	6.6	6.5	
	6.9	6.8	
	7.0	6.9	
	7.6	7.6	
California grey whale	5.7		2
	6.2		
	6.6		
	6.8		
	7.8		
Inia geoffrensis	5.53		2
	5.95		
	6.17		
	6.31		
	6.45		
	6.66		
	8.05		
Tursiops truncatns	5.50		2
	5.95		
	6.24		
	6.26		
	6.42		
	6.60		
	7.82		
Balaenoptera acutorostrata	5.46		2
	5.65		
	6.10		
	6.23		
	6.41		
	6.59		
	7.86		

References

1. Cohen, Hagenmaier, Pollard, and Schechter, *J. Mol. Biol.*, 71, 513 (1972).
2. Botelho, Hanania, and Gurd, unpublished observations.

TABLE 1D: Histidine pK Values in Human Hemoglobin

Hemoglobin	pK Values	Reference
Human	6.8	1, 2
	7.0	
	7.0	
	7.1	
	7.2	
	7.13	
	7.7	
	8.1	
	8.1	

References

1. Donovan, *Methods Enzymol.*, 27, 497 (1973).
2. Mandel, *Proc. Natl. Acad. Sci. U.S.A.*, 52, 736 (1964).

TABLE 1E: pK Value for Human Hb His 146 β

Protein	His 146 β pK Value	Reference
Human hemoglobin		
Deoxy, His 146 β	8.0	1
	8.1	2
	7.4	3
Human hemoglobin		
Carboxy, His 146 β	7.1	1
	6.8	2

References

1. Kilmartin, Breen, Roberts, and Ho, *Proc. Natl. Acad. Sci. U.S.A.*, 70, 1246 (1973).
2. Greenfield and Williams, *Biochim. Biophys. Acta*, 257, 187 (1972).
3. Huestis and Raftery, *Biochemistry*, 11, 1648 (1972).

TABLE 1F: Specific Histidine pK Value for Cytochrome c

Species	pK Values	Reference
Horse		
His 33	6.41	1
Yeast		
His 33	6.74	2
His 39	6.56	2

References

1. Cohen, Fisher, and Schechter, *J. Biol. Chem.*, 249, 1113 (1974).
2. Cohen and Hayes, *J. Biol. Chem.*, 249, 5472 (1974).

TABLE 1G: pK Values for Histidine Residues in Carbonic Anhydrase

Species	pK Values			Reference
Human, B	5.91	5.88	6	1–3
	6.04	6.09	6.98	
	7.00	6.93	7.23	
	7.23	7.23	8.2	
		8.2	8.24	
Human, C	5.87	5.74		2, 3
	5.96	6.43		
	6.10	6.49		
	6.20	6.5		
	6.63	6.57		
	7.20	6.63		
	7.28	7.25		

References

1. King and Roberts, *Biochemistry*, 10, 558 (1971).
2. Cohen, Yem, Kandel, Gornall, Kandell, and Friedman, *Biochemistry*, 11, 327 (1972).
3. Pesando, *Biochemistry*, 14, 675 (1975).

TABLE 1H: Histidine PK Values from Nuclease

Protein	pK Values		Reference
Staphylococcus aureus			
Foggi	5.46	5.37	1, 2
	5.76	5.71	
	5.66, 5.74, 6.54[a]	5.74	
	6.57	6.50	
Staphylococcus aureus			
V8	5.55		1
	5.80, 6.10[a]		
	6.50		

[a] pK values of one histidine existing in multiple conformational forms of the enzyme which slowly interconvert.

References

1. Markley, *Acc. Chem. Res.*, 8, 70 (1975).
2. Cohen, Shrager, McNeel, and Schechter, *Nature*, 228, 642 (1970).

TABLE 1I: Histidine pK Values in Various Proteins

Protein	Number of His resolved	pK	Specific Assignment	Reference
Adenylate kinase (pig)	2 of 2	<5.5		1
		6.3		
Chymotrypsin A_δ (cow)	1 of 2	7.2	His 57	2, 3
Chymotrypsinogen A (cow)	1 of 2	7.2	His 57	2, 3
Lysozyme				
chicken		5.8		4, 5
human		7.1		4, 6
Neurophysin II (cow)	1 of 1	6.87		7
Ovomucoid (chicken)	4 of 4	5.94		8
		6.71		
		6.75		
		8.07		
Protease				
α-Lyter (Myxobacter 495)	1 of 1	<4		9, 10
(Staphylococcus aureus, V8)	3 of 3	6.69		11
		6.85		
		7.19		
Ribonuclease				
T_1 (Aspergillus oryzae)	2 of 3	7.9		12
		8.0		
Serine esterase		6.5–7.5		13
Trypsin (pig, β form)	4 of 4	5.0		14
		6.54		
		6.66		
		7.20		
Trypsin inhibitor (soybean, Kunitz)	2 of 2	5.27		15
		7.00		

References

1. Cohn, Leigh, and Reed, *Cold Spring Harbor Symp. Quant. Biol.*, 36, 533 (1972).
2. Robillard and Shulman, *J. Mol. Biol.*, 71, 507 (1972).
3. Robillard and Shulman, *Ann. N.Y. Acad. Sci.*, 69, 599 (1972).
4. Meadows, Markley, Cohen, and Jardetsky, *Proc. Natl. Acad. Sci. U.S.A.*, 58, 1307 (1967).
5. Cohen, Hagenmaier, Pollard, and Schechter, *J. Mol. Biol.*, 71, 513 (1972).
6. Cohen, *Nature* (Lond.), 223, 43 (1969).
7. Cohen, Griffen, Camier, Caizergues, Fromageot, and Cohen, *FEBS Lett.*, 25, 282 (1972).
8. Markley, *Ann. N.Y. Acad. Sci.*, 222, 347 (1973).
9. Hunkapiller, Smallcombe, Whitaker, and Richards, *J. Biol. Chem.*, 248, 8306 (1973).
10. Hunkapiller, Smallcombe, Whitaker, and Richards, *Biochemistry*, 12, 4732 (1973).
11. Markley, Finkenstadt, Dugas, Leduc, and Drapeau, *Biochemistry*, 14, 998 (1975).
12. Riterjans and Pongs, *Eur. J. Biochem.*, 18, 313 (1971).
13. Polgar and Bender, *Proc. Natl. Acad. Sci. U.S.A.*, 69, 599 (1972).
14. Markley, *Acc. Chem. Res.*, 8, 70 (1975).
15. Markley, *Biochemistry*, 12, 2245 (1973).

TABLE 2: pK Values for α-Amino Groups in Proteins

Protein	pK' Value	Reference
Human hemoglobin		
Carboxy		
α chain	6.72	1
	6.95	2
β chain	7.05	2
Cyano		
α chain	6.74	2
β chain	6.93	2
Deoxy		
α chain	7.79	2
β chain	6.84	2
Myoglobin		
Sperm whale	7.77	2
	7.96	3
California grey whale	7.74	3
Pilot whale	7.43	3
Dall porpoise	7.22	3
Harbor seal	7.66	3
Bovine pancreatic ribonuclease A	8.14	4
Horse hemoglobin		
Oxy		
α chain	7.3	5
Deoxy		
α chain	7.7	5

References

1. Hill and Davis, *J. Biol. Chem.*, 242, 2005 (1967).
2. Garner, Bogardt, Jr., and Gurd, *J. Biol. Chem*, in press.
3. Garner, Garner, and Gurd, *J. Biol. Chem.*, 248, 5451 (1973).
4. Carty and Hirs, *J. Biol. Chem.*, 243, 5254 (1968).
5. Kilmartin and Rossi-Bernardi, *Biochem. J.*, 124, 31 (1971).

TABLE 3: pK Values for ε-Amino Groups in Proteins

Protein	pK' Value	Method	Reference
Bovine pancreatic ribonuclease A			
Lys-41	9.11	Kinetic	1
Other Lys	10.1	Kinetic	1
All Lys	10.2	Titration	2
Sperm whale myoglobin			
All Lys except one	10.6	Titration (intrinsic pK)	3
Hen egg white Lysozyme			
Lys 97	10.1	NMR	4
Lys 116	10.2		
Lys 13	10.3		
Lys 33	10.4		
Lys 1	10.6		
Lys 96	10.7		

References

1. Carty and Hirs, *J. Biol. Chem.*, 243, 5254 (1968).
2. Tanford and Hanenstein, *J. Am. Chem. Soc.*, 78, 5287 (1956).
3. Shire, Hanania, and Gurd, *Biochemistry*, 13, 2967 (1974).
4. Bradbury and Brown, *Eur. J. Biochem.*, 40, 565 (1973).

TABLE 4A: Tyrosine pK Values

Protein	No of Groups	pK	Reference
Ribonuclease A	3 of 6	9.9	1
Insulin		9.7	2
Pepsin		9.5	3
Serum albumin		10.35	3
Lysozyme		10.8	4
Trypsin inhibitor (BPTI)			
Bovine, tyrosines		10.6	5
		10.8	
		11.1	
		11.6	

References

1. Tanford, Hauenstein, and Rands, *J. Am. Chem. Soc.*, 77, 6409 (1955).
2. Tanford and Epstein, *J. Am. Chem. Soc.*, 76, 2163 (1954).
3. Tanford and Roberts, Jr., *J. Am. Chem. Soc.*, 74, 2509 (1952).
4. Fromageot and Schnek, *Biochem. Biophys. Acta*, 6, 113 (1950): Tanford and Wagner, *J. Am. Chem. Soc.*, 76, 2331 (1954).
5. Karplus, Snyder, and Sykes, *Biochemistry*, 12, 1323 (1973).

TABLE 4B: Tyrosine pK Values in Hb

Species	No. of Residues	pK	Specific Residue pK	Reference
Horse	8 of 12	10.6		1
	4 of 12	>12		
Human A	8 of 12	10.6		1
	4 of 12	>12		
Human A carboxy	8 of 12	10.60	β0145 10.6	2
			β3130 10.6	
	4 of 12	>10.6	β335 >10.6	
Human a deoxy	6 of 12	10.77		2
Human F carboxy	6 of 10	10.45		2
Human F deoxy	4 of 10	10.65		2

References

1. Hermans, Jr., *Biochemistry*, 1, 193 (1962).
2. Nagel, Ranney, and Kucinskis, *Biochemistry*, 5, 1934 (1966).

TABLE 4C: Tyrosine pK Values in Mb

Species	pK	Reference
Sperm whale	10.3	1
	11.5	
	>12.8	
Horse	10.3	1
	11.5	
	>12.8	

Reference

1. Hermans, Jr., *Biochemistry*, 1., 193 (1962).

TABLE 5: pK Values for Human Hb Cys β 93 SH

Human hemoglobin	pK	Reference
Deoxy, cys β 93 SH	>11	1
	>10	2
	>9.5	3
Carboxy, cys β 93 SH	>11	1

References

1. Janssen, Willekens, De Bruin, and van Os, *Eur. J. Biochem.*, 45, 53 (1974).
2. Snow, *Biochem. J.*, 84, 360 (1962).
3. Guidotti, *J. Biol. Chem.*, 242, 3673 (1967).

TABLE 6: Carboxyl Side Chain pK Values Estimated in Lysozymes

Residue	Range of pK Values[1]
Gin 35	6–6.5
Asp 101	4.2–4.7
Asp 66	1.5–2
Asp 52	3–4.6

Reference

1. Imoto, Johnson, North, Phillips, and Rupley, in *The Enzymes*, Vol. VII, 3rd ed. Bayer, Ed., Academic Press, New York, 1972, 665.

PROTEASE INHIBITORS AND PROTEASE INHIBITOR COCKTAILS

While protease inhibitor cocktails have been in use for some time,[1] there are few rigorous studies examining their effect on proteolysis and very few concerned with proteolytic degradation during the processing of material for analysis or during purification.[2] It is usually assumed that proteolysis can be a problem and protease inhibitors or protease inhibitor cocktails are usually included as part of a protocol without the provision of justification. There are several excellent review articles in this area. Salveson and Nagase[3] discuss the inhibition of proteolytic enzymes in great detail including much practical information that should be considered in experimental design. The discussion of the relationship between inhibitor concentration, inhibitor/enzyme binding constants (association constants, binding constants, $t_{1/2}$, inhibition constants, etc.), and enzyme inhibition is of particular importance. For example, with a reversible enzyme inhibitor (such as benzamidine), if the K_i value is 100 nM, a 100 μM concentration of inhibitor would be required to decrease protease activity by 99.9%. Salveson and Nagase[3] also note the well-known differences in the reaction rates of inhibitors such as DFP and PMSF with the active site of serine proteases. DFP is much faster than PMSF with trypsin but equivalent rates are seen with chymotrypsin. PMSF is included in commercial protease inhibitor cocktails because of its lack of toxicity compared to DFP; 3,4-dichloroisocoumarin (3,4-DCI), as described by Powers and colleagues,[4] is faster than either DFP or PMSF. Also enzyme inhibition occurs in the presence of substrate (proteins), which will influence the effectiveness of both irreversible and reversible enzyme inhibitors. In addition, some protease inhibitor cocktails include both PMSF and benzamidine. Benzamidine is a competitive inhibitor of trypticlike serine proteases and slows the rate of inactivation of such enzymes by reagents such as PMSF.[5] The investigator is also advised to consider the modification of proteins and other biological compounds by protease inhibitors in reactions not associated with proteases such as the modification of tyrosine by DFP or PMSF.[6] In addition, some of the protease inhibitors such as DFP and PMSF are subject to hydrolysis under conditions (pH ≥ 7.0) used for modification. For those unfamiliar with the history of DFP, DFP is a potent neurotoxin (inhibitor of acetyl cholinesterase) and should be treated with considerable care; a prudent investigator has a DFP repair kit in close proximity (weak base and pralidoxime-2-chloride [2-PAM]). Given these various issues, it is critical to validate that, in fact, the sample is being protected against proteolysis.

References

1. Takei, Y., Marzi, I., Kauffman, F.C. et al., Increase in survival time of liver transplants by protease inhibitors and a calcium channel blocker, nisoldipine. *Transplantation* 50, 14–20, 1990.
2. Pyle, L.E., Barton, P., Fujiwara, Y., Mitchell, A., and Fidge, N., Secretion of biologically active human proapolipoprotein A-1 in a baculovirus-insect cell system: protection from degradation by protease inhibitors, *J. Lipid Res.* 36, 2355–2361, 1995.
3. Salveson, G. and Nagase, H., Inhibition of proteolytic enzymes, in *Proteolytic Enyzymes: Practical Approaches,* 2nd ed., R. Benyon and J.S. Bond, Eds., Oxford University Press, Oxford, UK, pp. 105–130, 2001.
4. Harper, J.W., Hemmi, K., and Powers, J.C., Reaction of serine proteases with substituted isocoumarins: discovery of 3,4-dichloroisocoumarin, a new general mechanism-based serine protease inhibitor, *Biochemistry* 24, 1831–1841, 1985.
5. Lundblad, R.L., A rapid method for the purification of bovine thrombin and the inhibition of the purified enzyme with phenylmethylsulfonyl fluoride, *Biochemistry* 10, 2501–2506, 1971.
6. Lundblad, R.L., *Chemical Reagents for Protein Modification,* CRC Press, Boca Raton, FL, 2004.

Characteristics of Selected Protease Inhibitors, Which Can be Used in Protease Inhibitor Cocktails[a]

Common Name	Other Nomenclature	M.W.	Primary Design
Amastatin 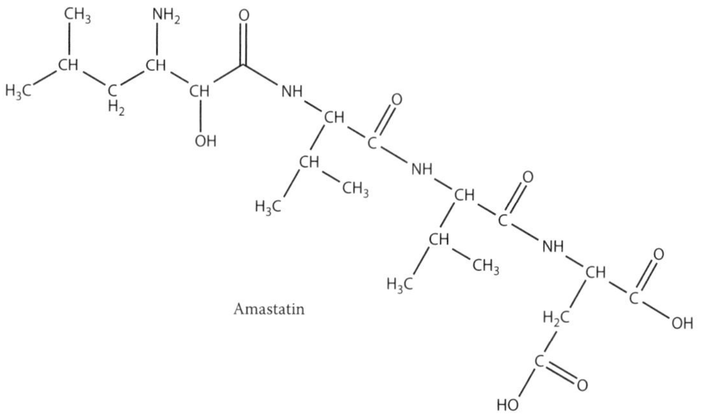	*N*-[(2*S*,3*R*)-3-amino-2-hydroxy-5-methyl hexanoyl]-L-valyl-L-valyl-L-aspartic acid	529.0	Inhibitor of some aminopeptidases.

Amastatin

Characteristics of Selected Protease Inhibitors, Which Can be Used in Protease Inhibitor Cocktails (Continued)

Common Name	Other Nomenclature	M.W.	Primary Design

Amastatin is a complex peptidelike inhibitor of aminopeptidases obtained from *Actinoycetes* culture. Amastatin is a competitive inhibitor of aminopeptidase A, aminopeptidase M, and other aminopeptidases. Amastatin has been used for the affinity purification of aminopeptidases. Amastatin has been shown to inhibit amino acid iosomerases. Amastatin is structurally related to bestatin and has been described as an immunomodulatory factor. See Aoyagi, T., Tobe, H., Kojima, F. et al., Amastatin, an inhibitor of aminopeptidase A, produced by actinomycetes, *J. Antibiot.* 31, 636–638, 1978; Tobe, H., Kojima, F., Aoyagi, T., and Umezawa, H., Purification by affinity chromatography using amastatin and properties of he aminopeptidase A from pig kidney, *Biochim. Biophys. Acta* 613, 459–468, 1980; Rich, D.H., Moon, B.J., and Harbeson, S., Inhibition of aminopeptidases by amastatin and bestatin derivatives. Effect of inhibitor structure on slow-binding processes, *J. Med. Chem.* 27, 417–422 , 1984; Meisenberg, G. and Simmons, W.H., Amastatin potentiates the behavioral effects of vasopressin and oxytocin in mice, *Peptides* 5, 535–539, 1984; Wilkes, S.H. and Prescott, J.M., The slow, tight binding of bestatin and amastatin to aminopeptidases, *J. Biol. Chem.* 260, 13154–13162, 1985; Matsuda, N., Katsuragi, Y., Saiga, Y. et al., Effects of aminopeptidase inhibitors actinonin and amastatin on chemotactic and phagocytic responses of human neutrophils, *Biochem. Int.* 16, 383–390, 1988; Orawski, A.T. and Simmons, W.H., Dipeptidase activities in rat brain synaptosomes can be distinguished on the basis of inhibition by bestatin and amastatin: identification of a kyotrophin (Tyr-Arg)-degrading enzyme, *Neurochem. Res.* 17, 817–820, 1992; Kim, H. and Lipscomb, W.N., X-ray crystallographic determination of the structure of bovine lens leucine aminopeptidase complexed with amastatin: formation of a catalytic mechanism, featuring a gem-diolate transition state, *Biochemistry* 32, 8365–8378, 1993; Bernkop-Schnurch, A., The use of inhibitory agents to overcome the enzymatic barrier to perorally administered therapeutic peptides and proteins, *J. Control. Release* 52, 1–16, 1998; Fortin, J.P., Gera, L., Bouthillier, J. et al., Endogenous aminopeptidase N decreases the potency of peptide agonists and antagonists of the kinin B1 receptors in the rabbit aorta, *J. Pharmacol. Exp. Ther.* 312, 1169–1176, 2005; Olivo Rdo, A., Teixeira Cde, R., and Silveira, P.F., Representative aminopeptidases and prolyl endopeptidase from murin macrophages; comparative activity levels in resident and elicited cells, *Biochem. Pharmacol.* 69, 1441–1450, 2005; Gera. L., Fortin, J.P., Adam, A. et al., Discovery of a dual-function peptide that combines aminopeptidase N inhibition and kinin B1 receptor antagonism, *J. Pharmacol. Exp. Ther.* 317, 300–308, 2006; Krsyanovic, M., Brgles, M., Halassy, B. et al., Purification and characterization of the *l*,(*l*/*d*)-aminopeptidase from guinea pig serum, *Prep. Biochem. Biotechnol.* 36, 175–195, 2006; Torres, A.M., Tsampazi, M., Tsampazi, C. et al., Mammalian *l* to *d*-amino-acid-residue isomerase from platypus venom, *FEBS Lett.* 580, 1587–1591, 2006.

Aprotinin   6512   Protein protease inhibitor.

Basic pancreatic trypsin inhibitor; Kunitz pancreatic trypsin inhibitor; Trasylol®. This protein inhibits some but not all trypticlike serine proteinases and is included in some protease inhibitor cocktails. See Hulsemann, A.R., Jongejan, R.C., Rolien Raatgeep, H. et al., Epithelium removal and peptidase inhibition enhance relaxation of human airways to vasoactive intestinal peptide, *Am. Rev. Respir. Dis.* 147, 1483–1486, 1993; Cornelius, R.M. and Brash, J.L., Adsorption from plasma and buffer of single- and two-chain high molecular weight kininogen to glass and sulfonated polyurethane surfaces, *Biomaterials* 20, 341–350, 1999; Lafleur, M.A., Handsley, M.M., Knauper, V. et al., Endothelial tubulogenesis with fibrin gels specifically requires the activity of membrane-type-matrix metalloproteinases (MT-MMPs), *J. Cell Sci.* 115, 3427–3438, 2002; Shah, R.B., Palamakula, A., and Khan, M.A., Cytotoxicity evaluation of enzyme inhibitors and absorption enhancers in Caco-2 cells for oral delivery of salmon calcitonin, *J. Pharm. Sci.* 93, 1070–1982; Spens, E. and Häggerström, L., Protease activity in protein-free (NS) myeloma cell cultures, In Vitro *Cell Dev. Biol.* 41, 330–336, 2005. As it is a potent inhibitor of plasmin, aprotinin is frequently included in fibrin gel-based cultures to preserve the fibrin gel structure. See Ye, Q., Zund, G., Benedikt, P. et al., Fibrin gel as a three-dimensional matrix in cardiovascular tissue engineering, *Eur. J. Cardiothorac. Surg.* 17, 587–591, 2000; Krasna, M., Planinsek, F., Knezevic, M. et al., Evaluation of a fibrin-based skin substitute prepared in a defined keratinocyte medium, *Int. J. Pharm.* 291, 31–37, 2005; Sun, X.T., Ding, Y.T., Yan, X.G. et al., Antiangiogenic synergistic effect of basic fibroblast growth factor and vascular endothelial growth factor in an *in vitro* quantitative microcarrier-based three-dimensional fibrin angiogenesis system, *World J. Gastroenterol.* 10, 2524–2528, 2004; Gille, J., Meisner, U., Ehlers, E.M. et al., Migration pattern, morphology and viability of cells suspended in or sealed with fibrin glue: a histomorphology study, *Tissue Cell* 37, 339–348, 2005; Yao, L., Swartz, D.D., Gugino, S.F. et al., Fibrin-based tissue-engineered blood vessels: differential effects of biomaterial and culture parameters on mechanical strength and vascular reactivity, *Tissue Eng.* 11, 991–1003, 2005. Aprotinin is used therapeutically in the inhibition of plasmin activity both as a freestanding product and as a component of fibrin sealant products.

Benzamidine HCI   156.61   Inhibitor of trypticlike serine
                                                                                                                              proteases.

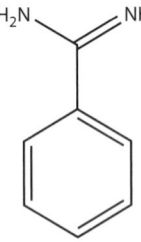

Benzamidine

An aromatic amidine derivative (Markwardt, F., Landmann, H., and Walsmann, P., Comparative studies on the inhibition of trypsin, plasmin, and thrombin by derivatives of benzylamine and benzamidine, *Eur. J. Biochem.* 6, 502–506, 1968; Guvench, O., Price, D.J., and Brooks, C.L., III, Receptor rigidity and ligand mobility in trypsin-ligand complexes, *Proteins* 58, 407–417, 2005), which is used as a competitive inhibitor of trypticlike serine proteases. It is not a particularly tight-binding inhibitor and is usually used at millimolar concentrations. Ensinck, J.W., Shepard, C., Dudl, R.J., and Williams, R.H., Use of benzamidine as a proteolytic inhibitor in the radioimmunoassay of glucagon in plasma, *J. Clin. Endocrinol. Metab.* 35, 463–467, 1972; Bode, W. and Schwager, P., The refined crystal structure of bovine beta-trypsin at 1.8 Å resolution. II. Crystallographic refinement, calcium-binding site, benzamidine-binding site, and active site at pH 7.0., *J. Mol. Biol.* 98, 693–717, 1975; Nastruzzi, C., Feriotto, G., Barbieri, R. et al., Differential effects of benzamidine derivatives on the expression of *c-myc* and HLA-DR alpha genes in a human B-lymphoid tumor cell line, *Cancer Lett.* 38, 297–305, 1988; Clement, B., Schmitt, S., and Zimmerman, M., Enzymatic reduction of benzamidoxime to benzamidine, *Arch. Pharm.* 321, 955–956, 1988; Clement, B., Immel, M., Schmitt, S., and Steinman, U., Biotransformation of benzamidine and benzamidoxime *in vivo*, *Arch. Pharm.* 326, 807–812, 1993; Renatus, M., Bode, W., Huber, R. et al., Structural and functional analysis of benzamidine-based inhibitors in complex with trypsin: implications for the inhibition of factor Xa, tPA, and urokinase, *J. Med. Chem.* 41, 5445–5456, 1998; Henriques, R.S., Fonseca, N., and

Characteristics of Selected Protease Inhibitors, Which Can be Used in Protease Inhibitor Cocktails (Continued)

Common Name	Other Nomenclature	M.W.	Primary Design

Ramos, M.J., On the modeling of snake venom serine proteinase interactions with benzamidine-based thrombin inhibitors, *Protein Sci.* 13, 2355–2369, 2004; Gustavsson, J., Farenmark, J., and Johansson, B.L., Quantitative determination of the ligand content in Benzamidine Sepharose® 4 Fast Flow media with ion-pair chromatography, *J. Chromatog. A* 1070, 103–109, 2005. Concentrated solutions of benzamidine will require pH adjustment prior to use.

Bestatin

Bestatin

N-[(2*S*,3*R*)-3-amino-2-hydroxy-ʟ-oxo-4-phenylbutyl]-ʟ-leucine

344.8

Aminopeptidase inhibitor; also described as a metalloproteinase inhibitor.

Bestatin is an inhibitor of some aminopeptidases and it was isolated from *Actinomycetes* culture. Bestatin was subsuently shown to have immunomodulatory activity and induces apoptosis in tumor cells. Bestatin is included in some proteaseinhibitor cocktails and has been demonstrated to inhibit intracellular protein degradation. See Umezawa, H., Aoyagi, T., Suda, H. et al., Bestatin, an inhibitor of aminopeptidase B, producted by actinomycetes, *J. Antibiot.* 29, 97–99, 1976; Suda, H., Takita, T., Aoyagi, T., and Umezawa, H., The structure of bestatin, *J. Antibiot.* 29, 100–101, 1976; Saito, M., Aoyagi, T., Umezawa, H., and Nagai, Y., Bestatin, a new specific inhibitor of aminopeptidases, enhances activation of small lymphocytes by concanavalin A, *Biochem. Biophys. Res. Commun.* 76, 526–533, 1976; Botbot, V. and Scornik, O.A., Degradation of abnormal proteins in intact mouse reticulocytes: accumulation of intermediates in the presence of bestatin, *Proc. Natl. Acad. Sci. USA* 76, 710–713, 1979; Botbol, V. and Scornik, O.A., Peptide intermediates in the degradation of cellular proteins. Bestatin permits their accumulation in mouse liver *in vivo*, *J. Biol. Chem.* 258, 1942–1949, 1983; Rich, D.H., Moon, B.J., and Harbeson, S., Inhibition of aminopeptidases by amastatin and bestatin derivatives. Effect of inhibitor structure on slow-binding processes, *J. Med. Chem.* 27, 417–422, 1984; Wilkes, S.H. and Prescott, J.M., The slow, tight binding of bestatin and amastatin to aminopeptidases, *J. Biol. Chem.* 260, 13154–13160, 1985; Patterson, E.K., Inhibition by bestatin of a mouse ascites tumor dipeptidase. Reversal by certain substrates, *J. Biol. Chem.* 264, 8004–8011, 1989; Botbol, V. and Scornik, O.A., Measurement of instant rates of protein degradation in the livers of intact mice by the accumulation of bestatin-induced peptides, *J. Biol. Chem.* 266, 2151–2157, 1991; Tieku, S. and Hooper, N.M., Inhibition of aminopeptidases N, A, and W. A re-evaluation of the actions of bestatin and inhibitors of angiotensin converting enzyme, *Biochem. Pharmacol.* 44, 1725–1730, 1992; Taylor, A., Peltier, C.Z., Torre, F.J., and Hakamian, N., Inhibition of bovine lens leucine aminopeptidase by bestatin: number of binding sites and slow binding of this inhibitor, *Biochemistry* 32, 784–790, 1993; Schaller, A., Bergey, D.R., and Ryan, C.A., Induction of wound response genes in tomato leaves by bestatin, an inhibitor of aminopeptidases, *Plant Cell* 7, 1893–1898, 1995; Nemoto, H., Ma, R., Suzuki, I.I., and Shibuya, M., A new one-pot method for the synthesis of alpha-siloxyamides from aldehydes or ketones and its application to the synthesis of (-)bestatin, *Org. Lett.* 2, 4245–4247, 2000; van Hensbergen, Y., Brfoxterman, H.J., Peters, E. et al., Aminopeptidase inhibitor bestatin stimulates microvascular endothelial cell invasion in a fibrin matrix, *Thromb. Haemost.* 90, 921–929, 2003; Stamper, C.C., Bienvenue, D.L., Bennett, B. et al., Spectroscopic and X-ray crystallographic characterization of bestatin bound to the aminopeptidase from *Aeromonas*(*Vibrio*)*proteolytica*, *Biochemistry* 43, 9620–9628, 2004; Zheng, W., Zhai, Q., Sun, J. et al., Bestatin, an inhibitor of aminopeptidases, provides a chemical genetics approach to dissect jasmonate signaling in *Aribidopsis*, *Plant Physiol.* 141, 1400–1413, 2006; Hui, M. and Hui, K.S., A novel aminopeptidase with highest preference for lysine, *Neurochem. Res.* 31, 95–102, 2006.

Cystatins

Protein Inhibitors of Cysteine Proteases

Inhibitors of cysteine proteinases.

Cystatin refers to a diverse family of protein cysteine protease inhibitors. There are three general types of cystatins: Type 1 (stefens), which are primarily found in the cytoplasm but can appear in extracellular fluids; Type 2, which are secreted and found in most extracellular fluids; and Type 3, which are multidomain protease inhibitors containing carbohydrates and that include the kininogens. Cystatin 3 is used to measure renal function in clinical chemistry. See Barrett, A.J., The cystatins: a diverse superfamily of cysteine peptidase inhibitors, *Biomed. Biochim. Acta* 45, 1363–1374, 1986; Katunuma, N., Mechanisms and regulation of lysosomal proteolysis, *Revis. Biol. Cellular* 20, 35–61, 1989; Gauthier, F., Lalmanach, G., Moeau, T. et al., Cystatin mimicry by synthetic peptides, *Biol. Chem. Hoppe Seyler* 373, 465–470, 1992; Bobek, L.A. and Levine, M.J., Cystatins — inhibitors of cysteine proteineases, *Crit. Rev. Oral Biol. Med.* 3, 307–332, 1992; Calkins, C.C., and Sloane, B.F., Mammalian cysteine protease inhibitors: biochemical properties and possible roles in tumor progression, *Biol. Chem. Hoppe Seyler* 376, 71–80, 1995; Turk, B., Turk, V., and Turk, D., Structural and functional aspects of papainlike cysteine proteinases and their protein inhibitors, *Biol. Chem.* 378, 141–150, 1997; Kos, J., Stabuc, B., Cimerman, N., and Brunner, N., Serum cystatin C, a new marker of glomerular filtration rate, is increased during malignant progression, *Clin. Chem.* 44, 2556–2557, 1998; Vray, B., Hartman, S., and Hoebeke, J., Immunomodulatory properties of cystatins, *Cell. Mol. Life Sci.* 59, 1503–1512, 2002; Arai, S., Matsumoto, I., Emori, Y., and Abe, K., Plant seed cystatins and their target enzymes of endogenous and exogenous origin, *J. Agric. Food Chem.* 50, 6612–6617, 2002; Abrahamson, M., Alvarez-Fernandez, M., and Nathanson, C.M., Cystatins, *Biochem. Soc. Symp.* 70, 179–199, 2003; Dubin, G., Proteinaceous cysteine protease inhibitors, *Cell. Mol. Life Sci.* 62, 653–669, 2005; Righetti, P.G., Castagna, A., Antonucci, F. et al., Proteome analysis in the clinical chemistry laboratory: myth or reality? *Clin. Chim. Acta* 357, 123–139, 2005; Overall, C.M. and Dean, R.A., Degradomics: systems biology of the protease web. Pleiotropic roles of MMPs in cancer, *Cancer Metastasis Rev.* 25, 69–75, 2006; Kotsylfakis, M., Sá-Nunes, A., Francischetti, I.M.B. et al., Anti-inflammatory and immunosuppressive activity of sialostatin L, a salivary cystatin from Tick *Ixodes scapularis*, *J. Biol. Chem.* 281, 26298–26307, 2006. DCI was developed by James C. Powers and coworkers at Georgia Institute of Technology (Harper, J.W., Hemmi, K., and Powers, J.C., Reaction of serine proteases with substituted isocoumarins: discovery of 3,4-dichloroisocoumarin, a new general mechanism-based serine protease inhibitor, *Biochemistry* 24, 1831–1841, 1985). This inhibitor is reasonably specific, although side reactions have been described. As with the sulfonyl fluorides and DFP, the modification is slowly reversible and enhanced by basic solvent conditions and/or nucleophiles. DCI has been used as a proteosome inhibitor. See Rusbridge, N.M. and Benyon, R.J., 3,4-dichloroisocoumarin, a serine protease inhibitor, inactivates glycogen phosphorylase b, *FEBS Lett.* 30, 133–136, 1990; Weaver,

Characteristics of Selected Protease Inhibitors, Which Can be Used in Protease Inhibitor Cocktails (Continued)

Common Name	Other Nomenclature	M.W.	Primary Design
3,4-Dichloroisocoumarin	DCI	215	Mechanism-based inhibitor of serine proteases.

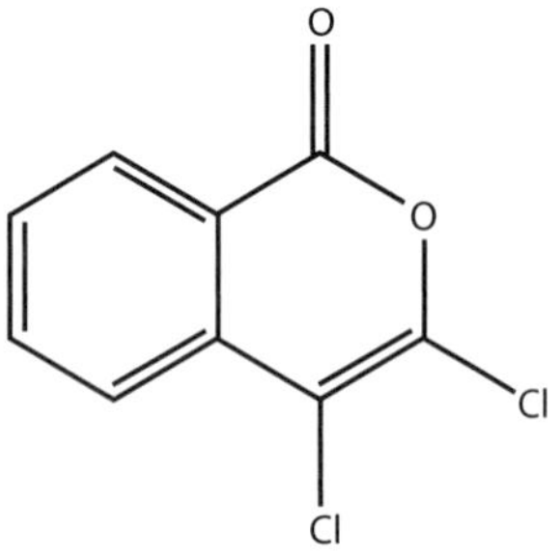

3,4-dichloroisocoumarin

V.M., Lach, B., Walker, P.R., and Sikorska, M., Role of proteolysis in apoptosis: involvement of serine proteases in internucleosomal DNA fragmentation in immature thymocytes, *Biochem. Cell Biol.* 71, 488–500, 1993; Garder, A.M., Aviel, S., and Argon, Y., Rapid degradation of an unassembled immunoglobulin light chain is mediated by a serine protease and occurs in a pre-Golgi compartment, *J. Biol. Chem.* 268, 25940–25947, 1993; Lu, Q. and Mellgren, R.L., Calpain inhibitors and serine protease inhibitors can produce apoptosis in HL-60 cells, *Arch. Biochem. Biophys.* 334, 175–181, 1996; Adams, J. and Stein, R., Novel inhibitors of the proteosome and their therapeutic use in inflammation, *Annu. Rep. Med. Chem.* 31, 279–288, 1996; Olson, S.T., Swanson, R., Patston, P.A., and Bjork, I., Apparent formation of sodium dodecyl sulfate-stable complexes between serpins and 3,4-dichloroisocoumarin-inactivated proteinases is due to regeneration of active proteinase from the inactivated enzyme, *J. Biol. Chem.* 272, 13338–13342, 1997; Mesner, P.W., Bible, K.C., Martins, L.M. et al., Characterization of caspase processing and activation in HL-60 cell cytosol under cell-free conditions — nucleotide requirement and inhibitor profile, *J. Biol. Chem.* 274, 22635–22645, 1999; Kam, C.M., Hudig, D., and Powers, J.C., Granzymes (lymphocyte serine proteases): characterization with natural and synthetic substrates and inhibitors, *Biochim. Biophys. Acta* 1477, 307–323, 2000; Rivett, A.J. and Gardner, R.C., Proteosome inhibitors: from *in vitro* uses to clinical trials, *J. Pep. Sci.* 6, 478–488, 2000; Bogyo, M. and Wang, E.W., Proteosome inhibitors: complex tools for a complex enzyme, *Curr. Top. Microbiol. Immunol.* 268, 185–208, 2002; Powers, J.C., Asgian, J.L., Ekici, O.D., and James, K.E., Irreversible inhibitors of serine, cysteine, and threonine proteases, *Chem. Rev.* 102, 4639–4740, 2002; Pochet, L., Frederick, R., and Masereei, B., Coumarin and isocoumarin as serine protease inhibitors, *Curr. Pharm. Des.* 10, 3781–3796, 2004.

Diisopropyl Phosphorofluoridate	DFP; Diisopropyl Fluorophosphate	184	Reaction at active site serine.

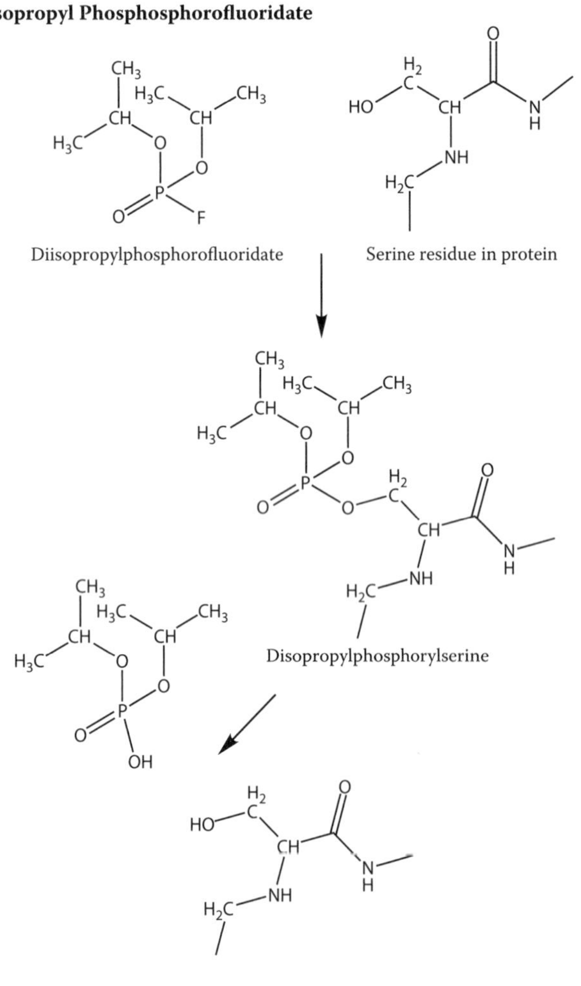

Diisopropylphosphorofluoridate   Serine residue in protein

Disopropylphosphorylserine

Characteristics of Selected Protease Inhibitors, Which Can be Used in Protease Inhibitor Cocktails (Continued)

Common Name	Other Nomenclature	M.W.	Primary Design

DFP was developed during World War II as a neurotoxin. DFP reacts with the active serine of serine proteases and was used to define the presence of this amino acid at the active sites of trypsin and chymotrypsin. DFP has been replaced by PMSF as a general reagent for inhibition of proteases although it is still used on occasion because of the ease of identification of the phosphoserine derivative. See Jansen, E.F., Jang, R., and Balls, A.K., The inhibition of purified, human plasma cholinesesterase with diisopropylfluorophosphate, *J. Biol. Chem.* 196, 247–253, 1952; Gladner, J.A. and Neurath, H.A., C-terminal groups in chymotrypsinogen and DFP-alpha-chymotrypsin in relation to the activation process, *Biochim. Biophys. Acta* 9, 335–336, 1952; Schaffer, N.K., May, S.C., Jr., and Summerson, W.H., Serine phosphoric acid from diisopropylphosphoryl chymotrypsin, *J. Biol. Chem.* 202, 67–76, 1953; Oosterbaan, R.A., Kunst, P., and Cohen, J.A., The nature of the reaction between diisopropylfluorophosphate and chymotrypsin, *Biochim. Biophys. Acta.* 16, 299–300, 1955; Wahlby, S., Studies on *Streptomyces griseus* protease. I. Separation of DFP-reacting enzymes and purification of one of the enzymes, *Biochim. Biophys. Acta* 151, 394–401, 1968; Hoskin, R.J. and Long, R.J., Purification of a DFP-hydrolyzing enzyme from squid head ganglion, *Arch. Biochem. Biophys.* 150, 548–555, 1972; Craik, C.S., Roczniak, S., Largman, C., and Rutter, W.J., The catalytic role of the active aspartic acid in serine proteases, *Science* 237, 909–913, 1987; D'Souza, C.A., Wood, D.D., She, Y.M., and Moscarello, M.A., Autocatalytic cleavage of myelin basic protein: an alternative to molecular mimicry, *Biochemistry* 44, 12905–12913, 2005. DFP is a potent neurotoxin and attention should be given to antidotes to organophosphates (Tuovinen, K., Kaliste-Korhonen, E., Raushel, F.M., and Hanninen, O., Phosphotriesterase, pralidoxime-2-chloride (2-PAM), and eptastigmine treatments and their combinations in DFP intoxication, *Toxicol. Appl. Pharmacol.* 141, 555–560, 1996; Auta, J., Costa, E., Davis, J., and Guidotti, A., Imidazenil: a potent and safe protective agent against diisopropyl fluorophosphate toxicity, *Neuropharmacology* 46, 397–403, 2004; Tuovinen, K., Organophosphate- induced convulsions and prevention of neuropathological damages, *Toxicology* 196, 31–39, 2004).

E-64	L-*trans*-epoxysuccinyl-leucylamide-(4-guanido)butane or *N*-[*N*-(L-*trans*-carboxyoxiran-2-carbonyl)-L-leucyl]-agmatine	357.4	Inhibitor of sulfhydryl proteases.

E-64 from *Aspergillus japonicus*

E-64 is a reasonably specific inhibitor of sulfhydryl proteases and it functions by forming a thioether linkage with the active site cysteine. E-64 is frequently referred to as an inhibitor of lysosomal proteases and antigen processing. See Hashida, S., Towatari, T., Kominami, E., and Katunuma, N., Inhibition by E-64 derivatives of rat liver cathepsins B and cathepsin L *in vitro* and *in vivo*, *J. Biochem.* 88, 1805–1811, 1980; Grinde, B., Selective inhibition of lysosomal protein degradation by the thiol proteinease inhibitors E-64, Ep-459, and Ep-457 in isolated rat hepatocytes, *Biochim. Biophys. Acta* 701, 328–333, 1982; Barrett, A.J., Kembhavi, A.A., Brown, A.A. et al., L-*trans*-epoxysuccinyl-leucylamiodo (4-guanidino) butane (E-64) and its analogues as inhibitors of cysteine proteinases including cathepsins B, H, and L, *Biochem. J.* 201, 189–198, 1982; Ko, Y.M., Yamanaka, T., Umeda, M., and Suzuki, Y., Effects of thiol protease inhibitors on intracellular degradation of exogenous β-galactosidase in cultured human skin fibroblasts, *Exp. Cell Res.* 148, 525–529, 1983; Tamai, M., Matsumoto, K., Omura, S. et al., *In vitro* and *in vivo* inhibition of cysteine proteinases by EST, a new analog of E-64, *J. Pharmacobiodyn.* 9, 672–677, 1986; Shaw, E., Cysteinyl proteinases and their selective inactivation, *Adv. Enzymol. Relat. Areas Mol. Biol.* 63, 271–347, 1990; Mehdi, S., Cell-penetrating inhibitors of calpain, *Trends Biochem. Sci.* 16, 150–153, 1991; Min, K.S., Nakatsubo, T., Fujita, Y. et al., Degradation of cadmium metallothionein *in vitro* by lysosomal proteases, *Toxicol. Appl. Pharmacol.* 113 299–305, 1992; Schirmeister, T. and Klackow, A., Cysteine protease inhibitors containing small rings, *Mini Rev. Med. Chem.* 3, 585–596, 2003.

EACA	ε-aminocaproic acid; 6-aminocaproic acid; 6-aminohexanoic acid; Amicar™	131.2	Analogue of lysine; inhibitor of trypsinlike enzymes such as plasmin.

Epsilon-aminocaproic acid
6-aminohexanoic acid

Lysine

EACA is an inhibitor of trypticlike serine proteases. It has been used as a hemostatic agent that functions by inhibiting fibrinolysis. It is included in some protease inhibitor cocktails. See Soter, N.A., Austen, K.F., and Gigli, I., Inhibition by epsilon-aminocaproic acid of the activation of the first component of the complement system, *J. Immunol.* 114, 928–932, 1975; Burden, A.C., Stacey, R., Wood, R.F., and Bell, P.R., Why do protease inhibitors enhance leukocyte migration inhibition to the antigen PPD? *Immunology* 35, 959–962, 1978; Nakagawa, H., Watanabe, K., and Sato, K., Inhibitory action of synthetic proteinase inhibitors and substrates on the chemotaxis of rat polymorphonuclear leukocytes *in vitro*, *J. Pharmacobiodyn.* 11, 674–678, 1988; Hill, G.E., Taylor, J.A., and Robbins, R.A., Differing effects of aprotinin and ε-aminocaproic acid on cytokine-induced inducible nitric oxide synthase expression, *Ann. Thorac. Surg.* 63, 74–77, 1997; Stonelake, P.S., Jones, C.E., Neoptolemos, J.P., and Baker, P.R., Proteinase inhibitors reduce basement membrane degradation by human breast cancer cell lines, *Br. J. Cancer* 75, 951–959, 1997; Sun, Z., Chen, Y.H., Wang, P. et al., The blockage of the high-affinity lysine-binding sites of plasminogen by EACA significantly inhibits prourokinase-induced plasminogen activation, *Biochim. Biophys. Acta* 1596, 182–192, 2002.

Characteristics of Selected Protease Inhibitors, Which Can be Used in Protease Inhibitor Cocktails (Continued)

Common Name	Other Nomenclature	M.W.	Primary Design
Ecotin			Broad-spectrum protease inhibitor derived from *Escherichia coli*.

Ecotin is a broad-spectrum inhibitor of serine proteases that can be engineered to enhance inhibition of specific enzymes. See McGrath, M.E., Hines, W.M., Sakanari, J.A. et al., The sequence and reactive site of ecotin. A general inhibitor of pancreatic serine proteases from *Escherichia coli*, *J. Biol. Chem.* 266, 6620–6625, 1991; Erpel, T., Hwang, P., Craik, C.S. et al., Physical map location of the new *Escherichia coli* gene eco, encoding the serin protease inhibitor ecotin, *J. Bacteriol.* 174, 1704, 1992; Wang, C.I., Yang, Q., and Craik, C.S., Isolation of a high affinity inhibitor of urokinase-type plasminogen activator by phage display of ecotin, *J. Biol. Chem.* 270, 12250–12256, 1995; Yang, S.Q., Wang, C.T., Gilmor, S.A. et al., Ecotin: a serine protease inhibitor with two distinct and interacting binding sites, *J. Mol. Biol.* 279, 945–957, 1998; Gilmor, S.A., Takeuchi, T., Yang, S.Q. et al., Compromise and accommodation in ecotin, a dimeric macromolecular inhibitor of serine proteases, *J. Mol. Biol.* 299, 993–1003, 2000; Eggers, C.T., Wang, S.X., Fletterick, R.J., and Craik, C.S., The role of ecotin dimerization in protease inhibition, *J. Mol. Biol.* 308, 975–991, 2001; Wang, B., Brown, K.C., Lodder, M. et al., Chemical-mediated site-specific proteolysis. Alteration of protein–protein interaction, *Biochemistry* 41, 2805–2813, 2002; Stoop, A.A. and Craik, C.S., Engineering of a macromolecular scaffold to develop specific protease inhibitors, *Nat. Biotechnol.* 21, 1063–1068, 2003; Eggers, C.T., Murray, I.A., Delmar, V.A. et al., The periplasmic serine protease inhibitor ecotin protects bacteria against neutrophil elastase, *Biochem. J.* 379, 107–118, 2004.

Ethylenediamine Tetraacetic Acid	EDTA	292.2	Metal ion chelator; inhibitor of metalloenzymes.

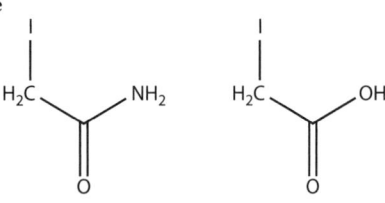

Edetic acid; EDTA; ethylenediaminetetraacetic acid;
N, N'-1, 2-ethanediaminediylbis[N-(carboxymethylglycine)]

(Ethylenedinitrilo)tetraacetic acid (ethylenediamine tetraacetic acid) chelates metal ions with a preference for divalent cations. EDTA functions as an inhibitor of metalloproteinases. See Manna, S.K., Bhattacharya, C., Gupta, S.K., and Samanta, A.K., Regulation of interleukin-8 receptor expression in human polymorphonuclear neutrophils, *Mol. Immunol.* 32, 883–893, 1995; Martin-Valmaseda, E.M., Sanchez-Yague, Y., Marcos, R., and Lianillo, M., Decrease in platelet, erythrocyte, and lymphocyte acetylcholinesterase activities due to the presence of protease inhibitors in the storage buffers, *Biochem. Mol. Biol. Int.* 41, 83–91, 1997; Oh-Ishi, M., Satoh, M., and Maeda, T., Preparative two-dimensional gel electrophoresis with agarose gels in the first dimension for high molecular mass proteins, *Electrophoresis* 21, 1653–1669, 2000; Shah, R.B., Palamakula, A., and Khan, M.A., Cytotoxicity evaluation of enzyme inhibitors and absorption enhancers in Caco-2 cells for oral delivery of salmon calcitonin, *J. Pharm. Sci.* 93, 1070–1082, 2004; Pagano, M.R., Paredi, M.E., and Crupkin, M., Cytoskeletal ultrastructure and lipid composition of I-Z-I fraction in muscle from pre- and post-spawned female hake (*Meriluccius hubbsi*), *Comp. Biochem. Physiol. B Biochem. Mol. Biol.* 141, 13–21, 2005; Wei, G.X. and Bobek, L.A., Human salivary mucin MUC7 12-mer-L and 12-mer-D peptides: antifungal activity in saliva, enhancement of activity with protease inhibitor cocktail or EDTA, and cytotoxicity to human cells, *Antimicrob. Agents Chemother.* 49, 2336–2342, 2005.

Iodoacetamide		185	Primary reaction with sulfhydryl groups and slower reaction with other protein nucleophiles.

Iodoacetic acid and iodoacetamide can both be used to modify nucleophiles in proteins. The chloro- and bromo-derivatives can be used as well but the rate of modification is slower. The haloacetyl function can also be used as the reactive function for more complex derivatives. Iodoacetamide is neutral compared to iodoacetic acid and is less influenced by the local environment of the reactive nucleophile. See Janatova, J., Lorenz, P.E., and Schechter, A.N., Third component of human complement: appearance of a sulfhydryl group following chemical or enzymatic inactivation, *Biochemistry* 19, 4471–4478, 1980; Haas, A.L., Murphey, K.E., and Bright, P.M., The inactivation of ubiquitin accounts for the inability to demonstrate ATP, ubiquitin-dependent proteolysis in liver extracts, *J. Biol. Chem.* 260, 4694–4703, 1985; Molla, A., Yamamoto, T., and Maeda, H., Characterization of 73 kDa thiol protease from *Serratia marcescens* and its effect on plasma proteins, *J. Biochem.* 104, 616–621, 1988; Wingfield, P., Graber, P., Turcatti, G. et al., Purification and characterization of a methionine-specific aminopeptidase from *Salmonella tyrphimurium*, *Eur. J. Biochem.* 180, 23–32, 1989; Kembhavi, A.A., Buttle, D.J., Rauber, P., and Barrett, A.J., Clostripain: characterization of the active site, *FEBS Lett.* 283, 277–280, 1991; Jagels, M.A., Travis, J., Potempa, J. et al., Proteolytic inactivation of the leukocyte C5a receptor by proteinases derived from *Porphyromas gingivalis*, *Infect. Immun.* 64, 1984–1991, 1996; Tanksale, A.M., Vernekar, J.V., Ghatge, M.S., and Deshpande, V.V., Evidence for tryptophan in proximity to histidine and cysteine as essential to the active site of an alkaline protease, *Biochem. Biophys. Res. Commun.* 270, 910–917, 2000; Karki, P., Lee, J., Shin, S.Y. et al., Kinetic comparison of procapase-3 and caspases-3, *Arch. Biochem. Biophys.* 442, 125–132, 2005. The haloalkyl derivatives do react with thiourea and are perhaps less reliable than maleimides.

Characteristics of Selected Protease Inhibitors, Which Can be Used in Protease Inhibitor Cocktails (Continued)

Common Name	Other Nomenclature	M.W.	Primary Design
LBTI	Lima Bean Trypsin Inhibitor	6500	Protein protease inhibitor.

Lima bean trypsin inhibitor is a protein/peptide with unusual stability. It is stable to heat (90°C for 15 minutes at pH 7 with no loss of activity) and acid (the original purification uses extraction with ethanol and dilute sulfuric acid). This is a reflection of the high content of cystine resulting in a "tight" structure. As a Bowman−Birk inhibitor, LBTI has seven disulfide bonds (Weder, J.K.P. and Hinkers, S.C., Complete amino acid sequence of the Lentil trypsin-chymotrypin inhibitor LCI-1.7 and a discussion of atypical binding sites of Bowman−Birk inhibitors, *J. Agric. Food Chem.* 52, 4219−4226, 2004). LBTI also inhibits both trypsin and chymotrypsin (Krahn, J. and Stevens, F.C., Lima bean trypsin inhibitor. Limited proteolysis by trypsin and chymotrypsin, *Biochemistry* 27, 1330−1335, 1970) as well as various other serine proteases. For additional information, see Fraenkel-Conrat, H., Bean, R.C., Ducay, E.D., and Olcott, H.S., Isolation and characterization of a trypsin inhibitor from lima beans, *Arch. Biochem. Biophys.* 37, 393−407, 1952; Stevens, F.C. and Doskoch, E., Lima bean protease inhibitor: reduction and reoxidation of the disulfide bonds and their reactivity in the trypsin-inhibitor complex, *Can. J. Biochem.* 51, 1021−1028, 1973; Nordlund, T.M., Liu, X.Y., and Sommer, J.H., Fluorescence polarization decay of tyrosine in lima bean trypsin inhibitor, *Proc. Natl. Acad. Sci. USA* 83, 8977−8981, 1986; Hanlon, M.H. and Liener, I.E., A kinetic analysis of the inhibition of rat and bovine trypsins by naturally occurring protease inhibitors, *Comp. Biochem. Physiol. B* 84, 53−57, 1986; Xiong, W., Chen, L.M., Woodley-Miller, C. et al., Identification, purification, and localization of tissue kallikrein in rat heart, *Biochem. J.* 267, 639−646, 1990; Briseid, K., Hoem, N.O., and Johannesen, S., Part of prekallikrein removed from human plasma together with IgG-immunoblot and functional tests, *Scand. J. Clin. Lab. Invest.* 59, 55−63, 1999; Yamasaki, Y., Satomi, S., Murai, N. et al., Inhibition of membrane-type serine protease 1/matriptase by natural and synthetic protease inhibitors, *J. Nutr. Sci. Vitaminol.* 49, 27−32, 2003.

| **Leupeptin** | (ac/pr-LeuLeuArginal) | | Transition-state inhibitor of proteinase. |

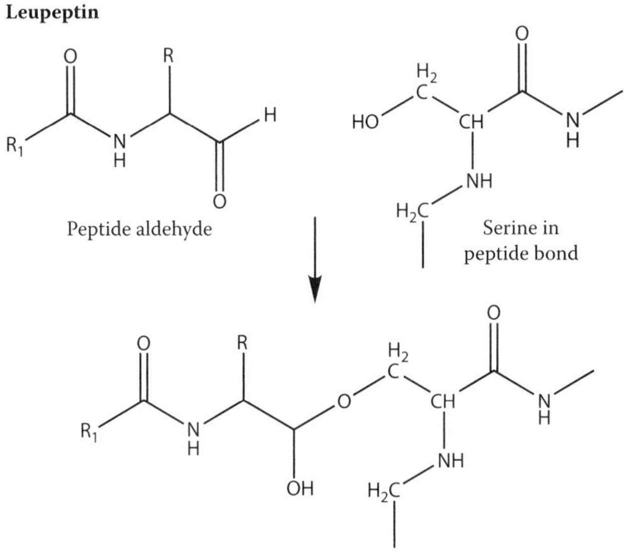

Peptide aldehyde

Serine in peptide bond

Stabilized tetrahedral aldol

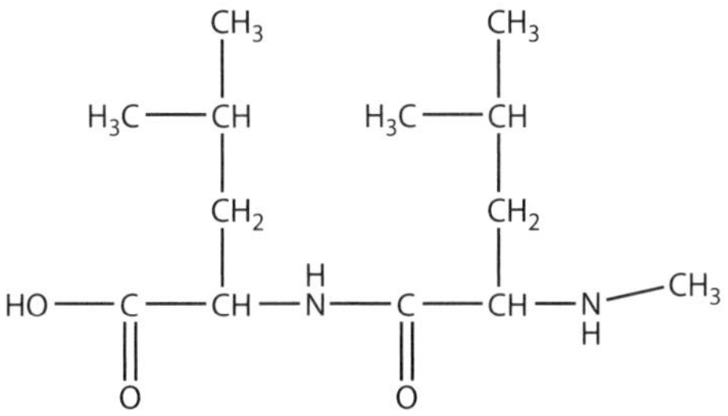

Leupeptide A

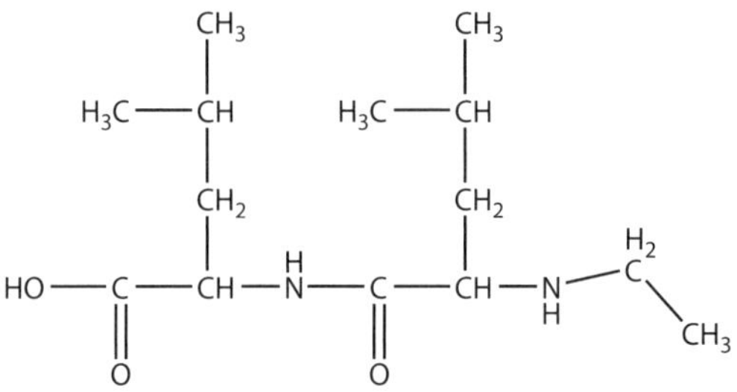

Leupeptin B

Characteristics of Selected Protease Inhibitors, Which Can be Used in Protease Inhibitor Cocktails (Continued)

Common Name	Other Nomenclature	M.W.	Primary Design

A tripeptide aldehyde (ac/pr-LeuLeuArginal) proteinase inhibitor isolated from *Actinomycetes*. It is a relatively common component of protease inhibitor cocktails used to preserve proteins during storage and purification. See Alpi, A. and Beevers, H., Proteinases and enzyme stability in crude extracts of castor bean endosperm, *Plant Physiol.* 67, 499–502, 1981; Ratjazak, T., Luc, T., Samec, A.M., and Hahnel, R., The influence of leupeptin, molybdate, and calcium ions on estrogen receptor stability, *FEBS Lett.* 136, 115–118, 1981; Takei, Y., Marzi, I., Kauffman, F.C. et al., Increase in survival time of liver transplants by protease inhibitors and a calcium channel blocker, nisoldipine, *Transplantation* 50, 14–20, 1990; Satoh, M., Hosoi, S., Miyaji, M. et al., Stable production of recombinant pro-urokinase by human lymphoblastoid Namalwa KJM-1 cells: host-cell dependency of the expressed-protein stability, *Cytotechnology* 13, 79–88, 1993; Hutchesson, A.C., Hughes, C.V., Bowden, S.J., and Ratcliffe, W.A., *In vitro* stability of endogenous parathyroid hormone-related protein in blood and plasma, *Ann. Clin. Biochem.* 31, 35–39, 1994; Agarwal, S. and Sohal, R.S., Aging and proteolysis of oxidized proteins, *Arch. Biochem. Biophys.* 309, 24–28, 1994; Yamada, T., Shinnoh, N., and Kobayashi, T., Proteinase inhibitors suppress the degradation of mutant adrenoleukodytrophy proteins but do not correct impairment of very long chain fatty acid metabolism in adrenoleukodystrophy fibroblasts, *Neurochem. Res.* 22, 233–237, 1997; Bi, M. and Singh, J., Effect of buffer pH, buffer concentration, and skin with or without enzyme inhibitors on the stability of [Arg(9)]-vasopressin, *Int. J. Pharm.* 197, 87–93, 2000; Bi, M. and Singh, J., Stability of luteinizing hormone-releasing hormone: effects of pH, temperature, pig skin, and enzyme inhibitors, *Pharm. Dev. Technol.* 5, 417–422, 2000; Ratnala, V.R., Swarts, H.G., VanOostrum, J. et al., Large-scale overproduction, functional purification, and ligand affinities of the His-tagged human histamine H1 receptor, *Eur. J. Biochem.* 271, 2636–2646, 2004.

(*p*-Amidinophenyl) Methanesulfonyl Fluoride	aPMSF	163	Reaction at active site serine.

(*p*-amidinophenyl) methanesulfonyl fluoride

(*p*-Amidinophenyl) methanesulfonyl fluoride was developed by Bing and coworkers (Laura, R., Robison, D.J., and Bing, D.H., [*p*-Amidinophenyl] methanesulfonyl fluoride, an irreversible inhibitor of serine proteases, *Biochemistry* 19, 4859–4864, 1980) to improve the specificity of PMSF for trypticlike enzymes. aPMSF readily reacts with trypsin but is only poorly reactive with chymotrypsin. See Katz, I.R., Thorbecke, G.J., Bell, M.K. et al., Protease-induced immunoregulatory activity of platelet factor 4, *Proc. Natl. Acad. Sci. USA* 83, 3491–3495, 1986; Unson, C.G. and Merrifield, R.B., Identification of an essential serine residue in glucagon: implications for an active site triad, *Proc. Natl. Acad. Sci. USA* 91, 454–458, 1994; Nikai, T., Komori, Y., Kato, S., and Sugihara, H., Bioloical properties of kinin-releasing enzyme from *Trimeresurus okinavensis(himehabu)* venom, *J. Nat. Toxins* 7, 23–35, 1998; Ishidoh, K., Takeda-Ezaki, M., Watanabe, S. et al., Analysis of where and which types of proteinases participate in lysosomal proteinase processing using balifomycin A1 and *Helicobacter pylori* Vac A toxin, *J. Biochem.* 125, 770–779, 1999; Komori, Y., Tatematsu, R., Tanida, S., and Nikai, T., Thrombin-like enzyme, flavovilase, with kinin-releasing activity from *Trimesurus flavoviridis(habu)* venom, *J. Nat. Toxins* 10, 239–248, 2001; Luo, L.Y., Shan, S.J., Elliott, M.B. et al., Purification and characterization of human kallikrein 11, a candidate prostate and ovarian cancer biomarker, from seminal plasma, *Clin. Cancer Res.* 12, 742–750, 2006. Reaction at a residue other than a serine has not been demonstrated although it is not unlikely that, as with DFP and PMSF, reaction could occur at a serine residue.

p-(Aminoethyl) Benzene Sulfonyl Fluoride	AEBSF; 4-(2-aminoethyl)-benzenesulfonyl fluoride (Pefabloc™ SC)	165	Reaction at active site serine.

4-(2-aminoethyl)benzenesulfonyl fluoride

This reagent was developed to improve the reactivity of PMSF. It was originally considered to be somewhat more effective than PMSF; however, AEBSF has been shown to be somewhat promiscuous in its reaction pattern and care is suggested in its use during sample preparation. See Su, B., Bochan, M.R., Hanna, W.L. et al., Human granzyme B is essential for DNA fragmentation of susceptible target cells, *Eur. J. Immunol.* 24, 2073–2080, 1994; Helser, A., Ulrichs, K., and Muller-Ruchholtz, W., Isolation of porcine pancreatic islets: low trypsin activity during the isolation procedure guarantees reproducible high islet yields, *J. Clin. Lab. Anal.* 8, 407–411, 1994; Dentan, C., Tselepis, A.D., Chapman, M.J., and Ninio, E., Pefabloc, 4-[2-aminoethyl'benzenesulfonyl fluoride, is a new potent nontoxic and irreversible inhibitor of PAF-degrading acetylhydrolase, *Biochim. Biophys. Acta* 1299, 353–357, 1996; Sweeney, B., Proudfoot, K., Parton, A.H. et al., Purification of the T-cell receptor zeta-chain: covalent modification by 4-(2-aminoethyl)-benzenesulfonyl fluoride, *Anal. Biochem.* 245, 107–109, 1997; Diatchuk, V., Lotan, O., Koshkin, V. et al., Inhibition of NADPH oxidase activation by 4-(2-aminoethyl)benzenesulfonyl fluoride and related compounds, *J. Biol. Chem.* 272, 13292–13301, 1997; Chu, T.M. and Kawinski, E., Plasmin, subtilisin-like endoproteases, tissue plasminogen activator, and urokinase plasminogen activator are involved in activation of latent TGF-beta 1 in human seminal plasma, *Biochem. Biophys. Res. Commun.* 253, 128–134, 1998; Guo, Z.J., Lamb, C., and Dixon, R.A., A serine protease from suspension-cultured soybean cells, *Phytochemistry* 47, 547–553, 1998; Wechuck, J.B., Goins, W.F., Glorioso, J.C., and Ataai, M.M., Effect of protease inhibitors on yield of HSV-1-based viral vectors, *Biotechnol. Prog.* 16, 493–496, 2000; Baszk, S., Stewart, N.A., Chrétien, M., and Basak, A., Aminoethyl benzenesulfonyl fluoride and its hexapeptide (AC-VFRSLK) conjugate are both *in vitro* inhibitors of subtilisin kexin isozyme-1, *FEBS Lett.* 573, 186–194, 2004; King, M.A., Halicka, H.D., and Dzrzynkiewicz, Z., Pro- and anti-apoptotic effects of an inhibitor of chymotrypsin-like serine proteases, *Cell Cycle* 3, 1566–1571, 2004; Odintsova, E.S., Buneva, V.N, and Nevinsky, G.A., Casein-hydrolyzing activity of sIGA antibodies from human milk, *J. Mol. Recog.* 18, 413–421, 2005; Solovyan, V.T. and Keski-Oja, J., Proteolytic activation of latent TGF-beta precedes caspase-3 activation and enhances apoptotic death of lung epithelial cells, *J. Cell Physiol.* 207, 445–453, 2006.

Characteristics of Selected Protease Inhibitors, Which Can be Used in Protease Inhibitor Cocktails (Continued)

Common Name	Other Nomenclature	M.W.	Primary Design
Pepstatin		685.9	Acid protease inhibitor.

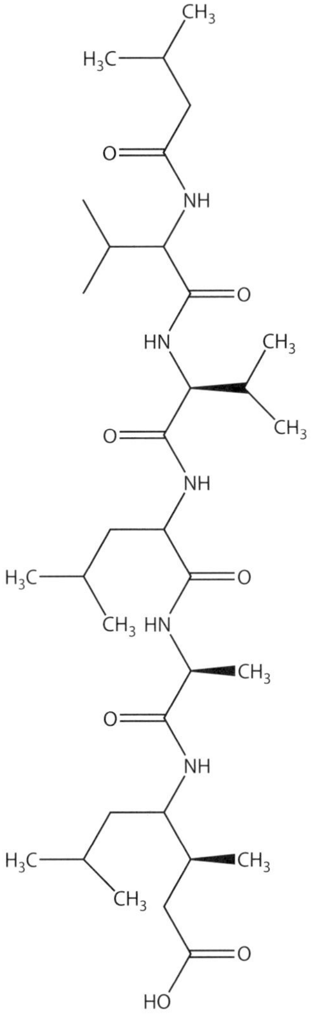

Pepstatin

A group of pentapeptide acid protease inhibitors isolated from *Streptomeyces* (Umezawa, H., Aoyagi, T., Morishima, H. et al., Pepstatin, a new pepsin inhibitor produced by *Actinomycetes*, *J. Antibiot.* 23, 259–262, 1970; Aoyagi, T., Kunimoto, S., Morichima, H. et al., Effect of pepstatin on acid proteases, *J. Antibiot.* 24, 687–694, 1971). Pepstatins are frequently included in protease inhibitor cocktails and used for the stabilization of proteins during extraction, storage, and purification. See Takei, Y., Marzi, I., Kaufmann, F.C. et al., Increase in survival time of liver transplants by protease inhibitors and a calcium channel blocker, nisoldipine, *Transplantation* 50, 14–20, 1990; Liang, M.N., Witt, S.N., and McConnell, H.M., Inhibition of class II MHC-peptide complex formation by protease inhibitors, *J. Immunol. Methods* 173, 127–131, 1994; Deng, J., Rudick, V., and Dory, L., Lysosomal degradation and sorting of apolipoprotein E in macrophages, *J. Lipid Res.* 36, 2129–2140, 1995; Wang, Y.K., Lin, H.H., and Tang, M.J., Collagen gel overlay induces two phases of apoptosis in MDCK cells, *Am. J. Physiol. Cell Physiol.* 280, C1440–C1448, 2001; Lafleur, M.A., Handsley, M.M., Knaupper, V. et al., Endothelial tubulogenesis within fibrin gels specifically requires the activity of membrane-type-matrix-metalloproteinases (MT-MMPs), *J. Cell Sci.* 155, 3427–3438, 2002.

Common Name	Other Nomenclature	M.W.	Primary Design
Phenanthroline Monohydrate	1,10-phenanthroline	198.2	Metal ion chelator; inhibitor of metalloenzymes; specificity for zinc-metalloenzymes.

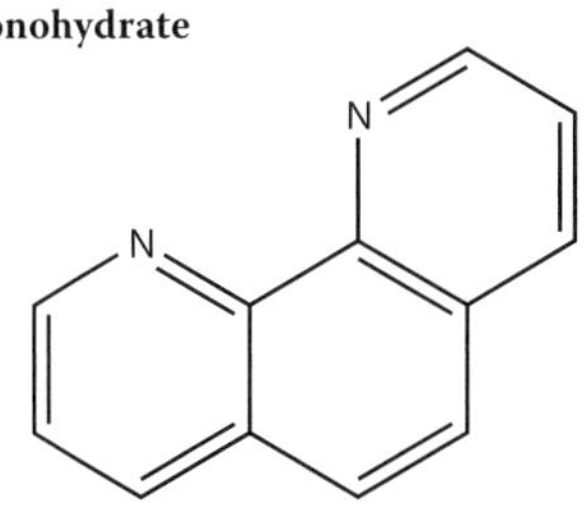

o-Phenanthroline
1,10-Phenanthroline

Characteristics of Selected Protease Inhibitors, Which Can be Used in Protease Inhibitor Cocktails (Continued)

Common Name	Other Nomenclature	M.W.	Primary Design

1,10-phenanthroline, *o*-phenanthroline: an inhibitor of metalloproteinases and a reagent for the detection of ferrous ions. See Felber, J.P., Cooobes, T.L., and Vallee, B.L., The mechanism of inhibition of carboxypeptidase A by 1,10-phenanthroline, *Biochemistry* 1, 231–238, 1962; Hakala, M.T. and Suolinna, E.M., Specific protection of folate reductase against chemical and proteolytic inactivation, *Mol. Pharmacol.* 2, 465–480, 1966; Latt, S.A., Holmquist, B., and Vallee, B.L., Thermolysin: a zinc metalloenzyme, *Biochem. Biophys. Res. Commun.* 37, 333–339, 1969; Berman, M.B. and Manabe, R., Corneal collagenases: evidence for zinc metalloenzymes, *Ann. Ophthalmol.* 5, 1993–1995, 1973; Seltzer, J.L., Jeffrey, J.J., and Eisen, A.Z., Evidence for mammalian collagenases as zinc ion metalloenzymes, *Biochim. Biophys. Acta* 485, 179–187, 1977; Krogdahl, A. and Holm, H., Inhibition of human and rat pancreatic proteinases by crude and purified soybean trypsin inhibitor, *J. Nutr.* 109, 551–558, 1979; St. John, A.C., Schroer, D.W., and Cannavacciuolo, L., Relative stability of intracellular proteins in bacterial cells, *Acta. Biol. Med. Ger.* 40, 1375–1384, 1981; Kitjaroentham, A., Suthiphongchai, T., and Wilairat, P., Effect of metalloprotease inhibitors on invasion of red blood cells by *Plasmodium falciparum*, *Acta Trop.* 97, 5–9, 2006; Thwaite, J.E., Hibbs, S., Tritall, R.W., and Atkins, T.P., Proteolytic degradation of human antimicrobioal peptide LL-37 by *Bacillus anthracis* may contribute to virulence, *Antimicrob. Agents Chemother.* 50, 2316–2322, 2006.

Phenylmethylsulfonyl Fluoride	PMSF	174	Reaction at active site serine.

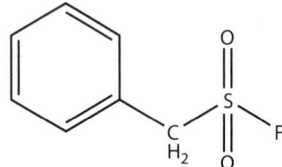

Phenylmethylsulfonyl fluoride (PMSF)

Phenylmethylsulfonyl fluoride was developed by David Fahrney and Allen Gold and inhibits serine proteases such as trypsin and chymotrypsin in a manner similar to DFP. The rate of modification of trypsin and chymotrypsin with PMSF is similar to that observed with DFP; however, the reaction with acetylcholinesterase with PMSF is much less than that of DFP (>6.1 × 10^{-2} M^{-1}min^{-1} vs. 1.3 × 10^4 M^{-1}min^{-1})(Fahrney, D.E. and Gold, A.M., Sulfonyl fluorides as inhibitors of esterases. I. Rates of reaction with acetylcholinesterase, α-chymotrypsin, and trypsin, *J. Amer. Chem. Soc.* 85, 997–1000, 1963). For other applications see Lundblad, R.L., A rapid method for the purification of bovine thrombin and the inhibition of the purified enzyme with phenylmethylsulfonyl fluoride, *Biochemistry* 10, 2501–2506, 1971; Pringle, J.R., Methods for avoiding proteolytic artefacts in studies of enzymes and other proteins from yeasts, *Methods Cell Biol.* 12, 149–184, 1975; Bendtzen, K., Human leukocyte migration inhibitory factor (LIF). I. Effect of synthetic and naturally occurring esterase and protease inhibitors, *Scand. J. Immunol.* 6, 125–131, 1977; Carter, D.B., Efird, P.H., and Chae, C.B., Chromatin-bound proteases and their inhibitors, *Methods Cell Biol.* 19, 175–190, 1978; Hubbard, J.R. and Kalimi, M., Influence of proteinase inhibitors on glucocorticoid receptor properties: recent progress and future perspectives, *Mol. Cell. Biochem.* 66, 101–109, 1985; Kato, T., Sakamoto, E., Kutsana, H. et al., Proteolytic conversion of STAT3alpha to STAT3gamma in human neutrophils: role of granule-derived serine proteases, *J. Biol. Chem.* 279, 31076–31080, 2004; Cho, I.H., Choi, E.S., Lim, H.G., and Lee, H.H., Purification and characterization of six fibrinolytic serine proteases from earthworm *Lumbricus rubellus*, *J. Biochem. Mol. Biol.* 37, 199–205, 2004; Khosravi, J., Diamandi, A., Bodani, U. et al., Pitfalls of immunoassay and sample for IGF-1: comparison of different assay methodologies using fresh and stored serum samples, *Clin. Biochem.* 38, 659–666, 2005; Shao, B., Belaaouaj, A., Velinde, C.L. et al., Methionine sulfoxide and proteolytic cleavage contribute to the inactivation of cathepsin G by hypochlorous acid: an oxidative mechanism for regulation of serine proteinases by myeloperoxidase, *J. Biol. Chem.* 260, 29311–29321, 2005; Pagano, M.R., Paredi, M.E., and Crupkin, M., Cytoskeletal ultrastructural and lipid composition of 1-Z-1 fraction in muscle from pre- and post-spawned female hake (*Merluccius hubbsi*), *Comp. Biochem. Physiol. B Biochem. Mol. Biol.* 141, 13–21, 2005. Although PMSF is reasonably specific for reaction with the serine residue at the active site of serine proteinases, as with DFP, reaction at tyrosine has been reported (De Venditis, E., Ursby, T., Rullo, R. et al., Phenylmethanesulfonyl fluoride inactivates an archeael superoxide dismutase by chemical modification of a specific tyrosine residue. Cloning, sequencing, and expression of the gene coding for *Sulfolobus solfataricus* dismutase, *Eur. J. Biochem.* 268, 1794–1801, 2001). PMSF does have solubility issues and usually ethanol or another suitable water-miscible organic solvent is used to introduce this reagent. On occasion, the volume of ethanol required influences the reaction (see Bramley, T.A., Menzies, G.S., and McPhie, C.A., Effects of alcohol on the human placental GnRH receptor system, *Mol. Hum. Reprod.* 5, 777–783, 1999).

SBTI	Soybean Trypsin Inhibitor	21,500	Protein protease inhibitor.

Soybean trypsin inhibitor (SBTI, STI) usually refers to the inhibitor first isolated by Kunitz (Kunitz, M., Crystalline soybean trypsin inhibitor, *J. Gen. Physiol.* 29, 149–154, 1946; Kunitz, M., Crystalline soybean trypsin inhibitor. II. General properties, *J. Gen. Physiol.* 30, 291–310, 1947). This material is described as the Kunitz inhibitor and is reasonably specific for trypticlike enzymes. There are other protease inhibitors derived from soybeans; the Bowman–Birk inhibitor (Birk, Y., The Bowman–Birk inhibitor. Trypsin and chymotrypsin-inhibitor from soybeans, *Int. J. Pept. Protein Res.* 25, 113–131, 1985; Birk, Y., Protein proteinase inhibitors in legume seeds — overview, *Arch. Latinoam. Nutr.* 44 (4 Suppl. 1), 26S–30S, 1996) is the best known and, unlike the Kunitz inhibitor, inhibits both trypsin and chymotrypsin; the Bowman–Birk inhibitor is also a double-headed inhibitor having two reactive sites (see Frattali, V. and Steiner, R.F., Soybean inhibitors. I. Separation and some properties of three inhibitors from commercial crude soybean trypsin inhibitor, *Biochemistry* 7, 521–530, 1968; Frattali, V. and Steiner, R.F., Interaction of trypsin and chymotrypsin with a soybean proteinase inhibitor, *Biochem. Biophys. Res. Commun.* 34, 480–487, 1969; Krogdahl, A. and Holm, H., Inhibition of human and rat pancreatic proteinases by crude and purified soybean trypsin inhibitor, *J. Nutr.* 109, 551–558, 1979). Soybean trypsin inhibitor (Kunitz) is used as a model protein (Liu, C.L., Kamei, D.T., King, J.A. et al., Separation of proteins and viruses using two-phase aqueous micellar systems, *J. Chromatog. B* 711, 127–138, 1998; Higgs, R.E., Knierman, M.D., Gelfanova, Y. et al., Comprehensive label-free method for the relative quantification of proteins from biological samples, *J. Proteome Res.* 4, 1442–1450, 2005). The broad specificity of the Kunitz inhibitor for trypticlike serine proteases provides the basis for its use in the demonstration of protease processing steps (Hansen, K.K., Sherman, P.M., Cellars, L. et al., A major role for proteolytic and proteinase-activated receptor-3 in the pathogenesis of infectious colitis, *Proc. Natl. Acad. Sci. USA* 102, 8363–8368, 2005).

Characteristics of Selected Protease Inhibitors, Which Can be Used in Protease Inhibitor Cocktails (Continued)

Common Name	Other Nomenclature	M.W.	Primary Design
Tosyl-lysine Chloromethyl Ketone	TLCK; 1-chloro-3-tosylamido-7-amino-2-heptanone	369.2 (HCl)	Reaction at active site histidine residues of trypsinlike serine proteases.

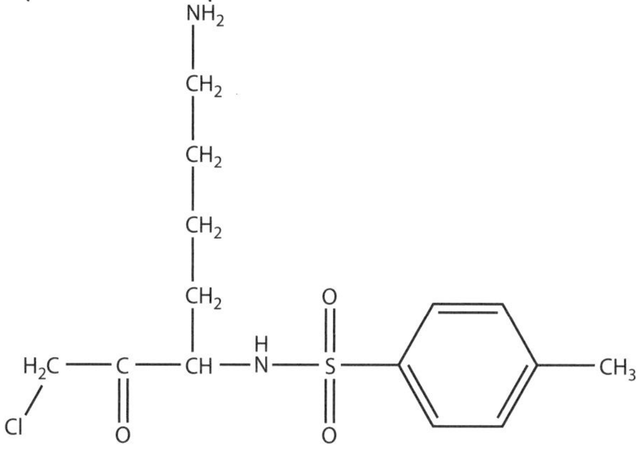

Tosyl-lysine chloromethyl ketone

Tosyl-lysine chloromethyl ketone (TLCK) was developed by Elliott Shaw and colleagues (Shaw, E., Mares-Guia, M., and Cohen, W., Evidence of an active center histidine in trypsin through use of a specific reagent, 1-chloro-3-tosylamido-7-amido-2-heptanone, the chloromethyl ketone derived from *N*-α tosyl-L-lysine, *Biochemistry* 4, 2219–2224, 1965). As with TPCK, reaction is not absolutely specific for trypticlike serine proteases (Earp, H.S., Austin, K.S., Gillespie, G.Y. et al., Characterization of distinct tyrosine-specific protein kinases in B and T lymphocytes, *J. Biol. Chem.* 260, 4351–4356, 1985; Needham, L. and Houslay, M.D., Tosyl-lysyl chloromethylketone detects conformational changes in the catalytic unit of adenylate cyclase induced by receptor and G-protein stimulation, *Biochem. Biophys. Res. Commun.* 156, 855–859, 1988). Reaction of this chloroalkyl compound with sulfydryl groups would be expected and it is possible that other protein nucleophilic centers would react, although this has not been unequivocally demonstrated. Attempts to synthesize the direct arginine analogue were unsuccessful; it was possible to make more complex arginine derivatives such as Ala-Phe-Arg-CMK, which was more effective with human plasma Kallikrein than the corresponding lysine derivatives (Ki = 0.078 μM vs. M vs. 4.9 μM) (Kettner, C. and Shaw, E., Synthesis of peptides of arginine chloromethyl ketone. Selective inactivation of human plasma kallikrein, *Biochemistry* 17, 4778–4784, 1978).

Tosyl-phenylalanine Chloromethyl Ketone	TPCK; L-1-tosylamido-2-phenylethyl chloromethyl ketone)	351.9	Reaction at active site histidine residues of chymotrypsinlike serine proteases.

Tosyl phenylalanine chloromethylketone

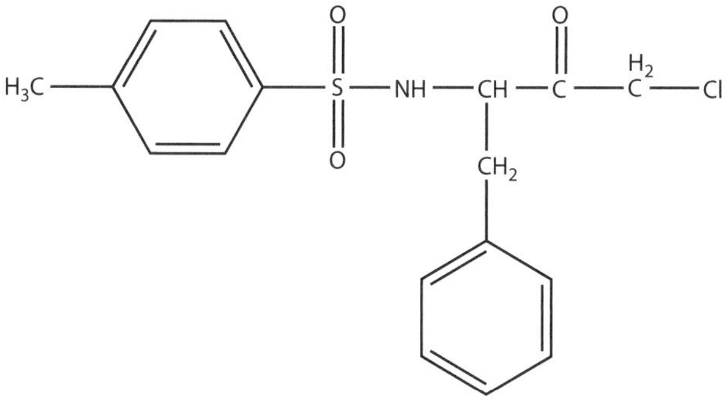

Tosyl-phenylalanine chloromethyl ketone (TPCK) was developed by Guenther Schoellmann and Elliott Shaw (Schoellmann, G. and Shaw, E., Direct evidence for the presence of histidine in the active center of chymotrypsin, *Biochemistry* 2, 252–255, 1963). TPCK was developed as an affinity label (Plapp, B.V., Application of affinity labeling for studying structure and function of enzymes, *Methods Enzymol.* 87, 469–499, 1982) where binding to chymotrypsin is driven by the phenyl function with subsequent alkylation of the active site histidine. The chloroalkyl function was selected to reduce reactivity with other protein nucleophiles such as cysteine. TPCK does undergo a slow rate of hydrolysis to form the corresponding alcohol. TPCK inactivates proteases with chymotrypsinlike specificity. The rate of inactivation is relatively slow but is irreversible; reaction rates can be enhanced by a more elaborate peptide chloromethyl ketone structure. In the case of cucumisin, a plant serine proteinase, TPCK did not result in inactivation while inactivation was achieved with Z-Ala-Ala-Pro-Phe-chloromethyl ketone (Yonezawa, H., Uchikoba, T., and Kaneda, M., Identification of the reactive histidine of cucumisin, a plant serine protease: modification with peptidyl chloromethyl ketone derivative of peptide substrate, *J. Biochem.* 118, 917–920, 1995). There is, however, significant reaction of TPCK with other proteins at residues other than histidine (see Rychlik, I., Jonak, J., and Sdelacek, J., Inhibition of the EF-Tu factor by L-1-tosylamido-2-phenylethyl chloromethyl ketone, *Acta Biol. Med. Ger.* 33, 867–876, 1974); TPCK has been described as an inhibitor of cysteine proteinases (Bennett, M.J., Van Leeuwen, E.M., and Kearse, K.P., Calnexin association is not sufficient to protect T cell receptor proteins from rapid degradation in CD4+CD8+ thymocytes, *J. Biol. Chem.* 273, 23674–23680, 1998). TPCK has been suggested to react with a lysine residue in aminoacylase (Frey, J., Kordel, W., and Schneider, F., The reaction of aminoacylase with chloromethylketone analogs of amino acids, *Z. Naturforsch.* 32, 769–776, 1966). Other reactions continue to be described (McCray, J.W. and Weil, R., Inactivation of interferons: halomethyl ketone derivatives of phenylalanine as affinity labels, *Proc. Natl. Acad. Sci. USA* 79, 4829–4833, 1982; Conseiller, E.C. and Lederer, F., Inhibition of NADPH oxidase by aminoacyl chloromethane protease inhibitors in phorbol-ester-stimulated human neutrophils: A reinvestigation. Are proteases really involved? *Eur. J. Biochem.* 183, 107–114, 1989; Borukhov, S.I. and Strongin, A.Y.,

Characteristics of Selected Protease Inhibitors, Which Can be Used in Protease Inhibitor Cocktails (Continued)

Common Name	Other Nomenclature	M.W.	Primary Design

Chemical modification of the recombinant human α-interferons and β-interferons, *Biochem. Biophys. Res. Commun.* 167, 74–80, 1990; Gillibert, M., Dehry, Z., Terrier, M. et al., Another biological effect of tosylphenylalanylchloromethane (TPCK): it prevents p47(phox) phosphorylation and translocation upon neutrophil stimulation, *Biochem. J.* 386, 549–556, 2005).

Peptide Halomethyl Ketones: While TPCK and TLCK represented a major advance in modifying active site residues in serine proteases, slow and relatively nonspecific reaction was a problem. The development of tripeptide halomethyl ketones provided a major advance in the value of such derivatives as presented in some specific examples below. However, even with these derivatives, reactions occur with "unexpected" enzymes. More general information can be obtained from the following references: Poulos, T.L., Alden, R.A., Freer, S.T. et al., Polypeptide halomethyl ketones bind to serine proteases as analogs of the tetrahedral intermediate. X-ray crystallographic comparison of lysine- and phenylalanine-polypeptide chloromethyl ketone-inhibited subtilisin, *J. Biol. Chem.* 251, 1097–1103, 1976; Powers, J.C., Reaction of serine proteases with halomethyl ketones, *Methods Enzymol.* 46, 197–208, 1977; Navarro, J., Abdel Ghany, M., and Racker, E., Inhibition of tyrosine protein kinases by halomethyl ketones, *Biochemistry* 21, 6138–6144, 1982; Conde, S., Perez, D.I., Martinez, A. et al., Thienyl and phenyl α-halomethyl ketones: new inhibitors of glycogen synthase kinase (GSK-3β) from a library of compound searching, *J. Med. Chem.* 46, 4631–4633, 2003.

Peptide Fluoromethyl Ketones: Fluoroalkyl derivatives of the peptide chloromethyl ketones have been prepared in an attempt to improve specificity by reducing nonspecific alkylation at cysteine residues (Rasnick, D., Synthesis of peptide fluoromethyl ketones and the inhibition of human cathepsin B, *Anal. Biochem.* 149, 461–465, 1985). Nonspecific reaction with sulfydryl groups such as those in glutathione was reduced; there was still reaction with active site cysteine although at a slower rate than with the chloroalkyl derivative (16,200 $M^{-1}s^{-1}$ vs. 45,300 $M^{-1}s^{-1}$; $t_{1/2}$ 21.9 min. vs. 5.1 min.). Reaction also occurred with serine proteases (Shaw, E., Angliker, H., Rauber, P. et al., Peptidyl fluoromethyl ketones as thiol protease inhibitors, *Biomed. Biochim. Acta* 45, 1397–1403, 1986) where the modification occurred at a histidine residue (Imperiali, B. and Abeles, R.H., Inhibition of serine proteases by peptide fluoromethyl ketones, *Biochemistry* 25, 3760–3767, 1986). The trifluoromethyl derivative was also an inhibitor but formed a hemiacetal derivative. The peptide fluoromethyl ketone, z-VAD-FMK, has proved to be a useful inhibitor of caspases

D-Phe-Pro-Arg-chloromethyl Ketone	PPACK		Reaction at active site histidine residues of trypsinlike serine proteases.

D-Phe-Pro-Arg-chloromethyl ketone was one of the first complex peptide halomethyl ketones synthesized. These derivatives have the advantage of increased reaction rate and specificity (see Williams, E.B. and Mann, K.G., Peptide chloromethyl ketones as labeling reagents, *Methods Enzymol.* 222, 503–513, 1993; Odake, S., Kam, C.M., and Powers, J.C., Inhibition of thrombin by arginine-containing peptide chloromethyl ketones and bis chloromethyl ketone-albumin conjugates, *J. Enzyme Inhib.* 9, 17–27, 1995; Lundblad, R.L., Bergstrom, J., De Vreker, R. et al., Measurement of active coagulation factors in Autoplex®-T with colorimetric active site-specific assay technology, *Thromb. Haemostas.* 80, 811–815, 1998). With chymotrypsin, CHO-PheCH$_2$Cl, $k_{obsv}/[I]$ = 0.55 $M^{-1}s^{-1}$ and Boc-Ala-Gly-Phe-CH$_2$Cl, $k_{obsv}/[I]$ = 3.34 $M^{-1}s^{-1}$ (Kurachi, K., Powers, J.C., and Wilcox, P.E., Kinetics of the reaction of chymotrypsin A α with peptide chloromethyl ketones in relation to subsite specificity, *Biochemistry* 12, 771–777, 1973. See also Ketter, C. and Shaw, E., The selective affinity labeling of factor Xa by peptides of arginine chloromethyl ketone, *Thromb. Res.* 22, 645–652, 1981; Shaw, E., Synthetic inactivators of kallikrein, *Adv. Exp. Med. Biol.* 156, 339–345, 1983; McMurray, J.S. and Dyckes, D.F., Evidence for hemiketals as intermediates in the inactivation of serine proteinases with halomethyl ketones, *Biochemistry* 25, 2298–2301, 1986). There is a similar peptide chloromethyl ketone, PPACK II (D-Phe-Phe-Arg-CMK), which has been used to stabilize B-type natriuretic peptide (BNP) in plasma samples (Belenky, A., Smith, A., Zhang, B. et al., The effect of class-specific protease inhibitors on the stabilization of B-type natriuretic peptide in human plasma, *Clin. Chim. Acta* 340, 163–172, 2004).

z-VAD-FMK	Benzyloxycarbonyl-Val-Ala-Asp(OMe) fluoromethyl ketone		Inhibitor of caspases.

z-VADFMK

Benzyloxycarbonyl-Val-Ala-Asp(OMe) fluoromethyl ketone (z-VAD-FMK) is a peptide halomethyl ketone used for the inhibition of caspases and related enzymes. Because z-VAD-FMK is neutral, it passes the cell membrane and can inhibit intracellular proteolysis and is useful in understanding the role of caspases and related enzymes in cellular function. See Zhu, H., Fearnhead, H.O., and Cohen, G.M., An ICE-like protease is a common mediator of apoptosis induced by diverse stimuli in human monocytes THP.1 cells, *FEBS Lett.* 374, 303–308, 1995; Mirzoeva, O.K., Yaqoob, P., Knox, K.A., and Calder, P.C., Inhibition of ICE-family cysteine proteases rescues murine lymphocytes from lipoxygenase inhibitor-induced apoptosis, *FEBS Lett.* 396, 266–270, 1996; Slee, E.A., Zhu, H., Chow, S.C. et al., Benzyloxycarbonyl-Val-Ala-Asp(OMe) fluoromethylketone (z-VAD. FMK) inhibits apoptosis by blocking the processing of CPP32, *Biochem. J.* 315, 21–24, 1996; Gottron, F.J., Ying, H.S., and Choi, D.W., Caspase inhibition selectively reduces the apoptotic component of oxygen-glucose deprivation-induced cortical neuronal cell death, *Mol. Cell. Neurosci.* 9, 159–169, 1997; Longthorne, V.L. and Williams, G.T., Caspase activity is required for commitment to Fas-mediated apoptosis, *EMBO J.* 16, 3805–3812, 1997; Hallan, E., Blomhoff, H.K., Smeland, E.B., and Long, J., Involvement of ICE (Caspase) family in gamma-radiation-induced apoptosis of normal B lymphocytes, *Scand. J. Immunol.* 46, 601–608, 1997; Polverino, A.J. and Patterson, S.D., Selective activation of caspases during apoptotic induction in HL-60 cells. Effects of a tetrapeptide inhibitor, *J. Biol. Chem.* 272, 7013–7021, 1997; Cohen, G.M., Caspases: the executioners of apoptosis, *Biochem. J.* 328, 1–16, 1997; Sarin, A., Haddad, E.K., and Henkart, P.A., Caspase dependence of target cell damage induced by cytotoxic lymphocytes, *J. Immunol.* 161, 2810–2816, 1998; Nicotera, P., Leist, M., Single, B., and Volbracht, C., Execution of apoptosis: converging or diverging pathway? *Biol. Chem.* 380, 1035–1040, 1999; Grfaczyk, P.P., Caspase inhibitors as anti-inflammatory and antiapoptotic

Characteristics of Selected Protease Inhibitors, Which Can be Used in Protease Inhibitor Cocktails (Continued)

Common Name	Other Nomenclature	M.W.	Primary Design

agents, *Prog. Med. Chem.* 39, 1–72, 2002; Blankenberg, F., Mari, C., and Strauss, H.W., Imaging cell death *in vivo, Q. J. Nucl. Med.* 47, 337–348, 2003; Srivastava, A., Henneke, P., Visintin, A. et al., The apoptotic response to pneumolysin in Toll-like receptor 4 dependent and protects against pneumococcal disease, *Infect. Immun.* 73, 6479–6489, 2005; Clements, K.M., Burton-Wurster, N., Nuttall, M.E., and Lust, G., Caspase-3/7 inhibition alters cell morphology in mitomycin-C treated chondrocytes, *J. Cell Physiol.* 205, 133–140, 2005; Coward, W.R., Marie, A., Yang, A. et al., Statin-induced proinflammatory response in mitrogen-activated peripheral blood mononuclear cells through the activation of caspases-1 and IL-18 secretion in monocytes, *J. Immunol.* 176, 5284–5292, 2006.

[a] The protease inhibitor cocktails referred to herein are not to be confused with the protease inhibitor cocktails that are used for therapy for patients who have Acquired Immune Deficiency Syndrome (AIDS).

General references for inhibitors of proteolytic enzymes

Albeck, A. and Kliper, S., Mechanism of cysteine protease inactivation by peptidyl epoxides, *Biochem. J.* 322, 879–884, 1997.

Banner, C.D. and Nixon, R.A., Eds., *Proteases and Protease Inhibitors in Alzheimer's Disease Pathogenesis*, New York Academy of Sciences, New York, 1992.

Barrett, A.J. and Salvesen, G., Eds., *Protease Inhibitors*, Elsevier, Amsterdam, NL, 1986.

Bernstein, N.K. and James, M.N., Novel ways to prevent proteolysis — prophytepsin and proplasmepsin II, *Curr. Opin. Struct. Biol.* 9, 684–689, 1999.

Birk, Y., Ed., *Plant Protease Inhibitors: Significance in Nutrition, Plant Protection, Cancer Prevention, and Genetic Engineering*, Springer, Berlin, 2003.

Cheronis, J.C.D. and Repine, J.E., *Proteases, Protease Inhibitors, and Protease-Derived Peptides: Importance in Human Pathophysiology and Therapeutics*, Birkhäuser Verlag, Basel, Switzerland, 1993.

Church, F.C., Ed., *Chemistry and Biology of Serpins*, Plenum Press, New York, 1997.

Frlan, R. and Gobec, S., Inhibitors of cathepsin B, *Curr. Med. Chem.* 13, 2309–2327, 2006.

Giglione, C., Boularot, A., and Meinnel, T., Protein *N*-terminal excision, *Cell. Mol. Life Sci.* 61, 1455–1474, 2004.

Johnson, S.L. and Pellechhia, M., Structure- and fragment-based approaches to protease inhibition, *Curr. Top. Med. Chem.* 6, 317–329, 2006.

Kim, D.H., Chemistry-based design of inhibitors for carboxypeptidase A, *Curr. Top. Med. Chem.* 4, 1217–1226, 2004.

Lowther, W.T., and Matthews, B.W., Structure and function of the methionine aminopeptidases, *Biochim. Biophys. Acta* 1477, 157–167, 2000.

Magnusson, S., Ed., *Regulatory Proteolytic Enzymes and Their Inhibitors*, Pergamon Press, Oxford, UK, 1986.

Powers, J.C. and Harper, J.W., Inhibition of serine proteinases, in *Proteinase Inhibitors*, Barrett, A.J. and Salvesen, G., Eds., Elsevier, Amsterdam, NL, chapter 3, pp. 55–152.

Saklatvala, J., and Nagase, H., Eds., *Proteases and the Regulation of Biological Processes*, Portland Press, London, UK, 2003.

Shaw, E., Cysteinyl proteinases and their selective inactivation, *Adv. Enzymol. Relat. Areas Mol. Biol.* 63, 271–347, 1990.

Stennicke, H.R. and Salvesen, G.S, Chemical ligation — an unusual paradigm in protease inhibition, *Mol. Cell.* 21, 727–728, 2006.

Tam, T.F., Leung-Toung, R., Li, W. et al., Medicinal chemistry and properties of 1,2,4-thiadiazoles, *Mini Rev. Med. Chem.* 5, 367–379, 2005.

Vogel, R., Trautschold, I., and Werle, E., *Natural Proteinase Inhibitors*, Academic Press, New York, 1968.

ASSAY OF SOLUTION PROTEIN CONCENTRATION

The determination of protein concentration is a somewhat overlooked procedure that is critical for the determination of the specific biological/therapeutic activity of most biopharmaceuticals, the "standardization" or normalization of samples for proteomic analysis and the comparison of cell homogenates. As such, it is unfortunate that most investigators do not recognize the limitations of the various procedures. The reader is recommended to some recent reviews of protein assay methods[1-3]. The purpose of this short section is to describe some commonly used techniques for the determination of protein concentration. Care must be taken with the use of these techniques several of the more frequently used techniques depend on protein quality as well as quantity. Thus the technique which is facile might not be accurate. It is noted that accuracy is an attribute in assay validation while facile is not. There are two issues which are common to any of the below assays. The first is the standard and the second is the solvent. The standard should be representative of the sample; albumin might not be the best choice. The concentration of the standard protein cannot be verified by preparation but must be verified by analysis. In other words, accurate dispensing of the standard protein and subsequent dissolution to a given volume does not ensure an accurate standard. The final concentration of a standard solution must be verified by analysis. For well-characterized proteins it is possible to employ ultraviolet spectroscopy using the known extinction coefficient for the standard protein. Thus the A280 of a 1 mg/mL of bovine serum albumin is 0.66 in a cuvette of 1 cm pathlength [126]. It is important to correct for any light scattering due to aggregated material, dust, etc., by recording the baseline over the range 400 to 310 nm where the protein does not absorb. While this procedure may seem somewhat tedious, it is necessary. It is possible to prepare a standard solution which can be used for a substantial period of time. The standard solution is best stored frozen in small aliquots, each of which is used once to calibrate the assay. The precise storage conditions used would require validation.

Solvent can have an effect on the analytical response and should be selected for (1) lack of an effect on the signal and (2) the ability to be used for both standard and samples. The reader is directed to the List of Buffers (p. 695) for a discussion of the effect of various buffers on protein analysis.

Biuret assay

The biuret assay measures the formation of a purple complex between copper salts and two or more peptide bonds under alkaline conditions. The assay was developed by Gornall and coworkers[4] and modified to a microplate format by Jenzano and coworkers[5]. The biuret assay is not available in kit form and the preparation of the reagents requires some skill. The biuret assay also lacks the sensitivity of many of the other assays. The biuret assay is accurate as it is insensitive to protein quality[2,3].

Selected references on the use of this method with various proteins are provided in Table 1.

Bicinchoninic acid (BCA) protein assay

This assay was developed by Smith and coworkers[6]. A modification for microplate use was developed by Jenzano and coworkers[5], This procedure is a modification of the Lowry et al.[7] reaction, but it is significantly easier and somewhat more sensitive[6]. The reaction is based on the formation of a complex of BCA with cuprous ion (Brenner, A.J. and Harris, E.D., A quantitative test for copper using

bicinchoninic acid, *Anal.Biochem.* 226, 80-84, 1995). The BCA reaction has the advantage of being able to measure protein bound to surfaces. This reaction is quite sensitive but it does reflect qualitative differences in proteins. As a reflection of the dependence on protein quality,[2] it is critical to select a standard that is qualitatively similar, if not identical, with the samples. This is obviously difficult when the assay is used with heterogeneous mixtures such as saliva or serum. Selected references to the use of this method are given in Table 2. Information on the use of the Lowry assay is presented in Table 3.

Dye-binding assay for protein using Coomassie Brilliant Blue G-250 (Bradford assay)

The dye-binding assay for proteins using Coomassie Brilliant Blue G-250 is likely the most sensitive and most extensively used protein assay at this time. It is also extremely easy to perform. The technique, as noted below, is extremely dependent on the quality of the protein. The procedure was developed by Bradford[8]. A modification for microplate technology is given by Jenzano and coworkers[5]. As noted above, this assay technique is likely the most sensitive and facile of the currently available procedures. Rigorous application of the dye-binding assay to the quantitative determination of a broad spectrum of proteins is difficult because of the marked influence of protein quality on the reaction. This is reflected by various studies attempting to modify the assay system to eliminate dependence on the quality of the protein.[9-14] Examples of the application of the Coomassie Blue dye-binding assay are presented in Table 4.

Kjeldahl assay

The Kjeldahl assay was developed in 1883[15] and is based on the determination of ammonia after hydrolysis of the sample in sulfuric acid. Most recent references to the use of this method relate to its use in food and environmental sciences[16-20]. It is our view that the Kjeldahl method remains a "gold" standard for protein assays[21] but we also appreciate the issues of technical complexity and lack of sensitivity which make routine use difficult for biopharmaceuticals. In addition, problems can arise in the analysis of proteins which contain impurities which themselves contain significant quantities of nitrogen, or in the analysis of proteins of unusual amino acid composition where the usual conversion factors may not apply. There are numerous commercial sources for support of the Kjeldahl assay[22-27]. Zellmer et al.,[28] have recently described an assay system which appears to have the accuracy of the Kjeldahl method with greatly improved sensitivity. Recent applications of the Kjeldahl assay are presented in Table 5.

Total amino acid analysis

Current technology for total amino acid analysis certainly has the various analytical attributes (sensitivity, accuracy, ruggedness) and is sufficiently rapid for use in the analysis of protein concentration[29-34] and has been suggested as a reference procedure for the determination of total protein concentration[35,36] With a characterized biopharmaceutical such as a growth factor, the concentration of the protein can be determined by measuring the amounts of specific stable and abundant amino acids such as alanine and lysine with reference to an added internal standard such as norleucine[37]. Application of amino acid analysis for total protein concentration are presented in Table 6.

Amido schwartz

The above approaches are certainly worth considering. However, it is somewhat remarkable that more attention has not been given to the amido black (Amidoschwarz 10B) assay developed by Schaeffner and Weissman[38]. This assay is based on the quantitative precipitation of protein from solution with trichloroacetic acid, the capture of the precipitated material by filtration followed by quantitative measurement of captured protein with amido schwarz dye (amido black is the preferred term). This study has been cited 2356 times (ISI) since its publication in 1973. The original assay used the addition of an equal amount of 60% trichloroacetic acid to a final concentration of 10%. The precipitate is capture by filtration and stained with amido black dye in methanol/glacial acetic acid/H_2O. The protein is visualized as a blue spot on an almost colorless background. The spot is excised from the filter, eluted with 25 mM NaOH-0.05 mM EDTA in 50% aqueous ethanol. The absorbance of the eluate at 630 nm is determined and concentration is determined by comparison with the results obtained with a standard protein. This assay has been used for the determination of protein concentration in grape juices and wines[39], low concentration of protein in phospholipids[40], and the *Escherichia coli* multidrug transporter EmrE[41] in the presence of detergents. Of direct relevance to proteomic analyses are the studies of Eliane[42] and coworkers on the determination of protein concentration of a *Medicago truncatula* root microsomal fraction with the amido black assay in a solution composed of 7.0 M urea-2.0 M thiourea-4% CHAPS (w/v)-0.1% (w/v) Triton X-100-2 mM tributylphosphine-2% ampholines. The reader is also commended to the study by Tate and coworkers[41] who validated the amido black assay with quantitative amino acid analysis which has been suggested as a method of choice for accurate determination of protein concentration[34,35]. A list of some other applications of amido black for protein assay is given in Table 7.

There has been limited application of fluorescent dyes for the determination of protein concentration (Table 8).

References

1. Dawnay, A. B. StJ., Hirst, A. D., Perry, D. E., and Chambers, R. E., A critical assessment of current analytical methods for the routine assay of serum total protein and recommendations for their improvement, *Ann. Clin. Biochem.*, 28, 556, 1991.
2. Sapan, C. V., Price, N. C., and Lundblad, R. L., Colorimetric protein assay techniques, *Biotechnol. Appl. Biochem.*, 29, 99-108, 1999.
3. Lundblad, R. L. and Price, N. C., Protein concentration determination. The Achilles' heel of cGMP, *Bioprocess International*, January, 2004, 1-7.
4. Gornall, A. G., Bardawill, C. J., and David, M. M., Determination of serum proteins by means of the biuret reaction, *J. Biol. Chem.*, 177, 751, 1949.
5. Jenzano, J. W., Hogan, S. L., Noyes, C. M., Featherstone, G. L., and Lundblad, R. L., Comparison of five techniques for the determination of protein content in mixed human saliva, *Anal. Biochem.*, 159, 370, 1986.
6. Smith, P. K., Krohn, R. I., Hermanson, G. T., Mallia, A. K., Gartner, F. H., Provenzano, M. D., Fujimoto, E. K., Goeke, N. M., Olson, B. J., and Klenk, D. C., Measurement of protein using bicinchoninic acid, *Anal. Biochem.*, 150, 76, 1985.
7. Lowry, O. H., Rosebrough, N. J., Farr, A. L., and Randall, R. J., Protein measurement with the folin phenol reagent, *J. Biol. Chem.*, 193, 265, 1951.
8. Bradford, M. M., A rapid and sensitive method for the determination of microgram quantities of protein utilizing the principle of protein-dye binding, *Anal. Biochem.*, 72, 248, 1976.
9. Wilmsatt, D. K. and Lott, J. A., Improved measurement of urinary total protein (including light-chain proteins) with a Coomassie Brilliant Blue G-250-sodium dodecyl sulfate reagent, *Clin. Chem.*, 33, 2100, 1987.
10. Tal, M., Silberstein, A., and Nusser, E., Why does Coomassie Brilliant Blue R interact differently with different proteins, *J. Biol. Chem.*, 260, 9976, 1985.
11. Pierce, J. and Suelter, C. H., An evaluation of the Coomassie Brilliant Blue G-250 dye-binding method for quantitative protein determination, *Anal. Biochem.*, 81, 478, 1977.
12. Read, S. M. and Northcote, D. H., Minization of variation in the response to different proteins of the coomassie blue G dye-binding assay for protein, *Anal. Biochem.*, 116, 53, 1981.
13. Sedmak, J. J. and Grossberg, S. F., A rapid, sensitive, and versatile assay for protein using coomassie brilliant blue G250, *Anal. Biochem.*, 79, 544, 1977.
14. Stoscheck, C. M., Increased uniformity in the response of the Coomassie blue G protein assay to different proteins, *Anal. Biochem.*, 184, 111, 1990.
15. Kjeldahl, J. Z., *Zeitschrift für Analytische Chemie*, 22, 366-382, 1883.
16. McPherson, T. N., Burian, S. J., Turin, H. J., Stenstrom, M. K. and Suffet, I. H., *Water Sci. Technol.* 45, 255-261, 2003.
17. Belloque, J. and Ramos, M., *J. Dairy Res.*, 69, 411-418, 2002, 2002
18. Shan, S. B., Bhumbla, D. K., Basden, T. J. and Lawrence, L. D., *J. Environ. Sci. Hlth.*, B 37, 493-505, 2002
19. Matttila, P., Salo-Vaananen, P., Konko, P., Aro, H. and Jalava, T. J., *Agricul. Food Chem.*, 50, 6419-6422, 2002.
20. Thompson, M., Owen, L., Wilkinson, K., Wood, R. and Damant, A., *Analyst* 127, 1666-1668, 2002.
21. Johnson, A. M., Rohlfs, E. M. and Silverman, L. M., (1999), Proteins, in *Teitz Textbook of Clinical Chemistry*, ed. C. A. Burhs and E. R. Ashwood, W. B. Saunders Co., Philadelphia, PA., Chapter 20, pp. 524-525.
22. http://www.calixo.net/braun/biochimie/kjeldahl.htm.
23. http://www.labconco.com/pdf/kjeldahl/index.shtml.
24. http://www.buchi-analytical.com/haupt.asp?nv=3759.
25. http://www.storesonline.com/site/251298.page/73181.
26. http://www.slrsystems.com/products.htm.
27. http://www.voigtglobal.com/kjeldahl_flasks.htm.
28. Zellmer, S., Kaltenborn, G., Rothe, U., Lehnich, H., Lasch, J., and Pauer, H.-D., *Anal. Biochem.*, 273, 163-167, 1999.
29. Anders, J. C., Parton, B. F., Petrie, G. E., Marlowe, R. L., and McEntire, J. E., *Biopharm International*, February, 30-37, 2003.
30. Weiss, M., Manneberg, M., Juranville, J.-F., Lahm, H.-W., and Fountaoulakis, M. Effect of the hydrolysis of method on the determination of the amino acid composition of proteins, *J. Chromatog. A.*, 795, 263-275, 1998.
31. Fountoulakis, M. and Lahm, H.-W., Hydrolysis and amino acid composition of proteins, *J. Chromatog. A.*, 826, 109-134, 1998.
32. Engelhart, W. G., Microwave hydrolysis of peptides and proteins for amino acid analysis, *Am. Biotechnol. Lab.*, 8, 30-34, 1990.
33. Chiou, S. H. and Wang, K. T., Peptide and protein hydrolysis by microwave irradiation, *J. Chromatog.*, 491, 424-431, 1989.
34. Bartolomeo, M. P. and Malsano, F., Validation of a reversed-phase method for quantitative amino acid analysis, *J. Biomol. Tech.*, 17, 131-137, 2006.
35. Sittampalam, G. S., Ellis, R. M., Miner, D. J., *et al.*, Evaluation of amino acid analysis as reference method to quantitate highly purified proteins, *J. Assoc. Off. Anal. Chem.*, 71, 833-838, 1988.
36. Henderson, L. O., Powell, M. R., Smith, S. J., *et al.*, Impact of protein measurements on standardization of assays of apolipoproteins A-I and B1, *Clin.Chem.*, 36, 1911-1917, 1990.
37. Price, N. C. (1996) The determination of protein concentration, in *Enzymology Labfax*, ed. P. C. Engel, Bios Scientific Publishers, Oxford, UK, pp. 34-41.

38. Schaeffner, W. and Weissman, C., A rapid, sensitive, and specific method for the determination of protein in dilute solutions, *Anal. Biochem.*, 56, 502-514, 1973.
39. Weiss, K. C. and Bisson, L. F., Optimisation of the Amido Black assay for the determination of protein content of grape juices and wines, *J. Science Food Agriculture*, 81, 583-589, 2001.
40. Bergo, H. O. and Christiansen, C., Determination of low levels of protein impurities in phospholipids samples, *Analyt. Biochem.*, 288, 225-227, 2001.
41. Butler, P. J. G., Ubarretxena-Belandia, I., Warne, T., and Tate, C. G., The *Escherichia coli* multidrug transporter EmrE is a dimer in the detergent solubilized state, *J. Mol. Biol.*, 340, 797-808, 2004.
42. Valot, B., Gianinazzi, S., and Elaine, D.-G., Sub-cellular proteomic analysis of a *Medicago truncatula* root microsomal fraction, *Phytochemistry*, 65, 1721-1732, 2004.

TABLE 1: Biuret Assay

Application	Reference
Dextran interferes with biuret assay of serum proteins	1
Interference of amino acids with the biuret reaction for urinary peptides. These investigators showed that, contrary to "conventional wisdom", the biuret assay showed cross-reaction with some amino acids and other compounds forming 5-membered and 6-membered complexes with copper.	2
Measurement of protein content of apple homogenates in an allergenicity study	3
Measurement of protein on surgical instruments	4
Measurement of protein concentration in serum and synovial fluid from human patients with arthropathies	5

TABLE 2: Bicinchoninic Acid Assay for Protein

Application	Reference
Protein concentration in synovial fluid	6
Measurement of protein concentration in liposomes	7
Measurement of cells attached to hyaluronic surfaces	8
Measurement of tear protein concentration	9
Proteins released by venom digestion	10

TABLE 3: Lowry Assay for Protein Concentration

Application	References
Determination of salivary protein concentration	11
Determination of protein concentration in aqueous humor	12
Determination of protein concentration in human milk	13
Determination of protein in human lens	14
Protein assay in cell-based toxicity studies	15

TABLE 4: Coomassie Blue Dye-Binding Assay (Bradford Assay)

Application	Reference
Measured soy protein extraction from various sources including soybean meal, soyprotein concentrate, and textured soy flake	16
Measured protein in aqueous phase from oil-water distribution. The amount of protein measured with Coomassie blue dye correlated ($r^2 = 0.91$) with protein concentration determined by tryptophan emission spectra (fourth derivative)	17
Measured glomalin extraction from soil	18
Automation of Coomassie dye-binding assay	19
Resonance light scattering with Bordeaux red correlates with Coomassie method for the determination of protein concentration in human serum, saliva, and urine.	20
Measured protein concentration in phenol extracts of plant tissues after precipitation with ammonium acetate in methanol	21
Measured total protein concentration in rat tissue (pancreas, parotid gland, submandibular gland, lacrimal gland) extracts	22
Measure protein release from alginate-dextran microspheres	23
Protein release from *Candida albicans* secondary to microwave irradiation	24

TABLE 5: Kjeldahl Assay for Protein Concentration

Application	Reference
Measurement of protein concentration in therapeutic protein concentrate. Kjeldahl used as the "gold standard." The biuret assay gave comparable values while dye-binding was lower. Specific activity differed with the protein concentration	25
Measurement of crude microbial protein derived from carbohydrate fermentation	26
Measurement of polylysine coating on alginate beads	27
Measurement of IgG concentration in the presence of nonionic surfactants and glycine	28
Manure protein concentration	29
Whey protein concentration	30

TABLE 6: Amino Acid Analysis for Protein Concentration

Application	Reference
Use of amino acid analysis as a primary method for the determination of the concentration of poly ADP-ribose polymerase 1 (PARP-1)	31
Determination of the concentration of NADH:ubiquinone oxidoreductase and establishment of coenzyme (FMN) and iron-sulfur cluster stoichiometry	32
Determination of the concentration of immobilized protein	33
Determination of IgG protein concentration	34
Protein concentration of troponin in standard reference preparations	35

TABLE 7: Amido Black (Schwarz) Method for Protein Assay

Application	Reference
Measurement of protein concentration by binding to a nitrocellulose membrane by filtration followed by staining with amido black. Protein concentration determined by densitometry	36
Measurement of protein in antibody-polysaccharide complexes	37
Measurement of anchorage-dependent cells	38
Measurement of cell viability in an immortalized keratinocyte cell line	39
Determination of protein concentration after transfer to nitrocellulose from SDS-PAGE gel	40

TABLE 8: Fluorescent Dyes for Determination of Protein Concentration

Dye	References
NanoOrange	41-45

References

1. Delanghe, J.R., Hamers, N., Taes, Y.E., and Libeer, J.C., Interference of dextran in biuret-type assays of serum proteins, *Clin.Chem.Lab. Med.* 43, 71-74, 2005

2. Hortin, G.L. and Meilinger, B., Cross-reactivity of amino acids and other compounds in the biuret reaction: interference with urinary peptide measurements, *Clin.Chem.* 51, 1411-1419, 2005

3. Carnes, J., Ferrer, A., and Fernandez-Caldas, E., Allergenicity of 10 different apple varieties, *Ann.Allergy Asthma Immunol.* 96, 564-570, 2006

4. Lipscomb, I.P., Pinchin, H.E., Collin, R., *et al.*, The sensitivity of approved Ninhydrin and Biuret tests in the assessment of protein contamination on surgical steel as an aid to prevent iatrogenic prion transmission, *J.Hosp.Infect.* 64, 288-292, 2006

5. Popko, J. Marciniak, J., Zalewska, A., *et al.*, Activity of *N*-acetyl-β-hexosaminidase and its isoenzymes in serum and synovial fluid from patients with different arthropathies, *Clin.Exp.Rheumatol.* 24, 690-693, 2006

6. Uehara, J., Kuboki, T, Fujisawa, T., *et al.*, Soluble tumour necrosis factor receptors in synovial fluids from tempromandibular joints with painful anterior disc displacement without reduction and osteoarthritis, *Arch.Oral.Biol.* 49, 133-142, 2004

7. Were, L.M., Bruce, B., Davidson, P.M., and Weiss, J., Encapsulation of nisin and lysozyme in liposomes enhances efficacy against *Listeria monocytogenes*, *J.Food Prot.* 67, 622-627, 2004

8. Cen, L., Neoh, K.G., Li, Y., and Kang, E.T., Assessment of *in vitro* bioactivity of hyaluronic acid and sulfated hyaluronic acid functionalized electroactive polymer, *Biomacromolecules* 5, 2238-2246, 2004

9. Yamada, M., Mochizuki, H., Kawai, M., *et al.*, Decreased tear lipocalin concentration in patients with meibomian gland dysfunction, *Br.J.Ophthalmol.* 89, 803-805, 2005

10. Nicholson, J., Mirtschin, P., Madaras, F., *et al.*, Digestive properties of the venom of the Australian Costal Taipan, *Oxyranus scutellatus* (Peters, 1867), *Toxicon* 48, 422-428, 2006

11. Yarat, A., Tunali, T., Pisiriciler, R., *et al.*, *Clin.Oral Investig.* 8, 36-39, 2004

12. Kawai, K., Sugiyama, K., and Kitazawa, Y., The effect of α_2-agonist on IOP rise following Nd-YAG laser iridotomy, *Tokai J.Exp.Clin.Med.* 29, 23-26, 2004

13. Milnewowicz, H. and Chmarek, M., Influence of smoking on metallothionein level and other proteins binding essential metals in human milk, *Acta Pediatr.* 94, 402-406, 2005

14. Raitelaitiene, R., Paunksnis, A., Ivanov, L., and Kurapkiene, S., Ultrasound and biochemical evaluation of human diabetic lens, *Medicina* (Kaunas) 41, 641-648, 2005

15. Dierickx, P., Prediction of human acute toxicity by the hep G2/24-hour/total protein assay, with protein measurement by the CBQCA method, *Altern.Lab.Anim.* 33, 207-213, 2005

16. Koppelman, S.J., Lakemond, C.M., Vlooswijk, R., and Hefle, S.L., Detection of soy protein in processed foods: literature overview and new experimental work, *J.AOAC Int.* 87, 1398-1407, 2004

17. Granger, C., Barey, P., Toutain, J., and Cansell, M., Direct quantification of protein partitioning in oil-to-water emulsion by front-face fluorescence: avoiding the need for centrifugation, *Colloids Surf. B Biointerfaces* 43, 158-162, 2005

18. Wright, S.F., Nichols, K.A., and Schmidt, W.F., Comparison of efficacy of three extractants to solubilize glomalin on hyphae and in soil, *Chemosphere* 64, 1219-1224, 2006

19. da Silva, M.A. and Arruda, M.A., Mechanization of the Bradford reaction for the spectrophotometric determination of total protein, *Anal.Biochem.* 351, 1551-157, 2006

20. Feng, S., Pan, Z., and Pan, J., Determination of proteins at nanogram levels with Bordeaux red based on the enhancement of resonance light scattering, *Sprectrochim.Acta A Mol.Biomol. Spectrosc.* 64, 574-579, 2006

21. Faurobert, M., Pelpoir, E., and Chaib, J., Phenol extraction of proteins for proteomic studies of recalcitrant plant tissues, *Methods Mol.Biol.* 359, 9-14, 2007

22. Changrani, N.R., Chonkar, A., Adeghate, E., and Singh, J., Effects of streptoozotocin-induced type 1 diabetes mellitus on total protein concentrations and cation contents in the isolated pancreas, parotid, submandibular, and lacrimal glands of rats, *Ann.N.Y.Acad.Sci.* 1084, 503-519, 2006

23. Reis, C.P. ,Ribeiro, A.J., Huong, S., *et al.*, Nanoparticulate delivery system for insulin: design, characterization and in vitro/in vivo bioactivity, *Eur.J.Pharm.Sci.* 30, 392-397, 2007

24. Campanha, N.H., Pavarina, A.C., Brunetti, I.L., *et al.*, *Candida albicans* inactivation and cell membrane integrity damage by microwave irradiation, *Mycoses* 50, 140-147, 2007

25. Lof, A.L., Gustafsson, G., Novak, V., *et al.*, Determination of total protein in highly purified factor IX concentrates, *Vox Sang.* 63, 172-177, 1992

26. Hall, M.B. and Herejk, C., Differences in yields of microbial crude protein from *in vitro* fermentation of carbohydrates, *J.Dairy Sci.* 84, 2486-2493, 2001

27. Simsek-Ege, F.A., Bond, G.M., and Stringer, J., Matrix molecular weight cut-off for encapsulation of carbonic anhydrase in polyelectrolyte beads, *J.Biomater.Sci.Polym.Ed.* 13, 1175-1187, 2002

28. Vidanovic, D., Milic Askrabic, J., Stankovic, M., and Poprzen, V., Effects of nonionic surfactants on the physical stability of immunoglobulin G in aqueous solution during mechanical agitation, *Pharmazie* 58, 399-404, 2003

29. Leek, A.B., Hayes, E.T., Curran, T.F., *et al.*, The influence of manure composition on emissions of odour and ammonia from finishing pigs fed different concentrations of dietary crude protein, *Bioresour. Technol.*, in press, 2006

30. Cheison S.C., Wang, Z., and Xu, S.Y., Preparation of whey protein hydrolysates using a single- and two-stage enzymatic membrane reactor and their immunological and antioxidant properties: Characterization by multivariate data analysis, *J.Agric.Food Chem.*, in press, 2007

31. Knight, M.I. and Chambers, P.J., Problems associated with determining protein concentration: a comparison of techniques for protein estimations, *Mol.Biotechnol.* 23, 19-28, 2003

32. Albracht, S.P., van der Linden, E., and Faber, B.W., Quantitative amino acid analysis of bovine NADH:ubiquinone oxidoreductase (Complex I) and related enzymes. Consequences for the number of prosthetic groups, *Biochim.Biophys.Acta* 1557, 41-49, 2003

33. Salchert, K., Pompe, T., Sperling, C., and Wenner, C., Quantitative analysis of immobilized proteins and protein mixtures by amino acid analysis, *J.Chromatog.A* 1005, 113-122, 2003

34. Schauer, U., Stemberg, F., Rieger, C.H., *et al.,* IgG subclass concentrations in certified reference material 470 and reference values for children and adults determined with the binding site reagents, *Clin. Chem.* 49, 1924-1929, 2003

35. Bunk, D.M. and Welch, M.J., Characterization of a new certified reference material for human cardiac troponin I, *Clin.Chem.* 52, 212-219, 2006

36. Nakamura, K., Tanaka, T., Kuwahara, A., and Takeo, K., Microassay for proteins on nitrocellulose filter using protein dye-staining procedure, *Anal.Biochem.* 148, 311-319, 1985

37. Cabrera, M.M. and Lund, F.A., Determination of protein in polysaccharide-antibody complexes, *Ann.Inst.Pasteur Immunol.* 137C, 51-55, 1986

38. Everitt, E. Wohlfart, C., Spectrophotometric quanitation of anchorage-dependent cell numbers using extraction of naphthol blue-black-stained cellular protein, *Anal.Biochem.* 162, 122-129, 1987

39. White, P.J., Fogarty, R.D., Werther, G.A., and Wraight, C.J., Antisense inhibition of IGF receptor expression in HaCaT keratinocytes: a model for antisense strategies in keratinocytes, *Antisense Nucleic Acid Drug Dev.* 10, 195-203, 2000

40. Himmelfarb, J. and McMonagle, E., Albumin is the major plasma protein target of oxidant stress in uremia, *Kidney Int.* 60, 358-363, 2001

41. Liu, T., Foote, R.S., Jacobson, S.C., *et al.,* Electrophoretic separation of proteins on a microchip with noncovalent, postcolumn labeling, *Anal.Chem.* 72, 4608-4613, 2000

42. Harvey, M.D., Bablekis, V., Banks, P.R., and Skinner, C.D., Utilization of the non-covalent fluorescent dye, NanoOrange, as a potential clinical diagnostic tool. Nanomolar human serum albumin quantitation, *J.Chromatog.B.Biomed.Sci.Appl.* 754, 345-356, 2001

43. Jones, L.J, Haugland, R.P., and Singer, V.L., Development and characterization of the NanoOrange protein quantitation assay: a fluorescence-based assay of proteins in solution, *BioTechniques* 34, 850-854, 2003

44. Stoyanov, A.V., Fan, Z.H., Das, C., *et al.,* On the possibility of applying noncovalent dyes for protein labeling in isoelectric focusing, *Anal.Biochem.* 350, 263-267, 2006

45. Williams, J.C, Jr., Zarse, C.A., Jackson, M.E., *et al.,* Variability of protein content in calcium oxalate monohydrate stones, *J.Endourol.* 20, 560-564, 2006

SPECTROPHOTOMETRIC DETERMINATION OF PROTEIN CONCENTRATION IN THE SHORT-WAVELENGTH ULTRAVIOLET

W. B. Gratzer

Whereas the extinction coefficients of proteins in the aromatic absorption band at 280 nm vary widely, the spectrum at shorter wavelengths is dominated by the absorption of the peptide bond and, therefore, has only a secondary dependence on amino acid composition and conformation. Measurements in this region can therefore serve for approximate concentration determinations of any protein. The following relations are available:

1. Scopes, *Anal. Biochem.*, 59, 277 (1974):
   a. E(1 mg/mL; 1 cm) at 205 nm = 31 with a stated error of 5%.
   b. This can be improved by applying a correction for the relatively strongly absorbing aromatic residues, by measuring the absorbance also at 280 nm. Two forms of this correction are

$$E(1 \text{ mg/ml}; 1 \text{ cm}) \text{ at } 205 \text{ nm} = 27.0 + 120 \times (A^{280}/A^{205})$$

or

$$E(1 \text{ mg/ml}; 1 \text{ cm}) \text{ at } 205 \text{ nm} = \frac{27.0}{1 - 3.85(A^{280}/A^{205})}$$

where the bracket term is the ratio of the absorbances measured at 280 and 205 nm; stated error, 2%.

2. Tombs, Soutar, and Maclagan, *Biochem. J.*, 73, 167 (1959):

$$E(1 \text{ mg/mL}; 1 \text{ cm}) \text{ at } 210 \text{ nm} = 20.5$$

3. Waddell, *J. Lab. Clin. Med.*, 48, 311 (1956): To avoid wavelength-setting error on steeply sloping curves, measurements are made at two wavelengths 10 nm apart and the absorbance difference is used to give the concentration:

$$C(\text{mg/ml}) = 0.144(A^{215} - A^{225})$$

where A^{215} and A^{225} are the absorbances read in 1 cm at 215 and 225 nm.

Note that the longer the wavelength, the lower the sensitivity of the spectrophotometric memethod method, but the hazard of interference from ultraviolet absorbing contaminants is less.

Section II
Lipids

A COMPREHENSIVE CLASSIFICATION SYSTEM FOR LIPIDS[1]

Eoin Fahy,* Shankar Subramaniam,[†] H. Alex Brown,[§] Christopher K. Glass,** Alfred H. Merrill, Jr.,[††] Robert C. Murphy,[§§] Christian R. H. Raetz,*** David W. Russell,[†††] Yousuke Seyama,[§§§] Walter Shaw,**** Takao Shimizu,[††††] Friedrich Spener,[§§§§] Gerrit van Meer,***** Michael S. VanNieuwenhze,[†††††] Stephen H. White,[§§§§§] Joseph L. Witztum,****** and Edward A. Dennis[2,††††††]

San Diego Supercomputer Center,* University of California, San Diego, 9500 Gilman Drive, La Jolla, CA 92093-0505; Department of Bioengineering,[†] University of California, San Diego, 9500 Gilman Drive, La Jolla, CA 92093-0412; Department of Pharmacology,[§] Vanderbilt University Medical Center, Nashville, TN 37232-6600; Department of Cellular and Molecular Medicine,** University of California, San Diego, 9500 Gilman Drive, La Jolla, CA 92093-0651; School of Biology,[††] Georgia Institute of Technology, Atlanta, GA 30332-0230; Department of Pharmacology,[§§] University of Colorado Health Sciences Center, Aurora, CO 80045-0508; Department of Biochemistry,*** Duke University Medical Center, Durham, NC 27710; Department of Molecular Genetics,[†††] University of Texas Southwestern Medical Center, Dallas, TX 75390-9046; Faculty of Human Life and Environmental Sciences,[§§§] Ochanomizu University, Tokyo 112-8610, Japan; Avanti Polar Lipids, Inc.,**** Alabaster, AL 35007; Department of Biochemistry and Molecular Biology,[††††] Faculty of Medicine, University of Tokyo, Tokyo 113-0033, Japan; Department of Molecular Biosciences,[§§§§] University of Graz, 8010 Graz, Austria; Department of Membrane Enzymology,***** Institute of Biomembranes, Utrecht University, 3584 CH Utrecht, The Netherlands; Department of Chemistry and Biochemistry,[†††††] University of California, San Diego, 9500 Gilman Drive, La Jolla, CA 92093-0358; Department of Physiology and Biophysics,[§§§§§] University of California at Irvine, Irvine, CA 92697-4560; Department of Medicine,****** University of California, San Diego, 9500 Gilman Drive, La Jolla, CA 92093-0682; and Department of Chemistry and Biochemistry and Department of Pharmacology,[††††††] University of California, San Diego, La Jolla, CA 92093-0601

Abstract Lipids are produced, transported, and recognized by the concerted actions of numerous enzymes, binding proteins, and receptors. A comprehensive analysis of lipid molecules, "lipidomics," in the context of genomics and proteomics is crucial to understanding cellular physiology and pathology; consequently, lipid biology has become a major research target of the postgenomic revolution and systems biology. To facilitate international communication about lipids, a comprehensive classification of lipids with a common platform that is compatible with informatics requirements has been developed to deal with the massive amounts of data that will be generated by our lipid community. As an initial step in this development, we divide lipids into eight categories (fatty acyls, glycerolipids, glycerophospholipids, sphingolipids, sterol lipids, prenol lipids, saccharolipids, and polyketides) containing distinct classes and subclasses of molecules, devise a common manner of representing the chemical structures of individual lipids and their derivatives, and provide a 12 digit identifier for each unique lipid molecule. The lipid classification scheme is chemically based and driven by the distinct hydrophobic and hydrophilic elements that compose the lipid.[jlr] This structured vocabulary will facilitate the systematization of lipid biology and enable the cataloging of lipids and their properties in a way that is compatible with other macromolecular databases.—Fahy, E., S. Subramaniam, H. A. Brown, C. K. Glass, A. H. Merrill, Jr., R. C. Murphy, C. R. H. Raetz, D. W. Russell, Y. Seyama, W. Shaw, T. Shimizu, F. Spener, G. van Meer, M. S. VanNieuwenhze, S. H. White, J. L. Witztum, and E. A. Dennis. **A comprehensive classification system for lipids.** *J. Lipid Res.* 2005. **46:** 839–861.

Supplementary key words lipidomics • informatics • nomenclature • chemical representation • fatty acyls • glycerolipids • glycerophospholipids • sphingolipids • sterol lipids • prenol lipids • saccharolipids • polyketides

The goal of collecting data on lipids using a "systems biology" approach to lipidomics requires the development of a comprehensive classification, nomenclature, and chemical representation system to accommodate the myriad lipids that exist in nature. Lipids have been loosely defined as biological substances that are generally hydrophobic in nature and in many cases soluble in organic solvents (1). These chemical properties cover a broad range of molecules, such as fatty acids, phospholipids, sterols, sphingolipids, terpenes, and others (2). The LIPID MAPS (LIPID Metabolites And Pathways Strategy; http://www.lipidmaps.org), Lipid Library (http://lipidlibrary.co.uk), Lipid Bank (http://lipidbank.jp), LIPIDAT (http://www.lipidat.chemistry.ohiostate.edu), and Cyberlipids (http://www.cyberlipid.org) websites provide useful online resources for an overview of these molecules and their structures. More accurate definitions are possible when lipids are considered from a structural and biosynthetic perspective, and many different classification schemes have been used over the years. However, for the purpose of comprehensive classification, we define lipids as hydrophobic or amphipathic small molecules that may originate entirely or in part by carbanion-based condensations of thioesters (fatty acids, polyketides, etc.) and/or by

Manuscript received 22 December 2004 and in revised form 4 February 2005.

Published, JLR Papers in Press, February 16, 2005. DOI 10.1194/jlr. E400004-JLR200

[1] The evaluation of this manuscript was handled by the former Editor-in-Chief Trudy Forte.

[2] To whom correspondence should be addressed.
E-mail: edennis@ucsd.edu

TABLE 1: Lipid Categories and Examples

Category	Abbreviation	Example
Fatty acyls	FA	dodecanoic acid
Glycerolipids	GL	1-hexadecanoyl-2-(9 Z-octadecenoyl)-sn-glycerol
Glycerophospholipids	GP	1-hexadecanoyl-2-(9 Z-octadecenoyl)-sn-glycero-3-phosphocholine
Sphingolipids	SP	N-(tetradecanoyl)-sphing-4-enine
Sterol lipids	ST	cholest-5-en-3β-ol
Prenol lipids	PR	2E,6E-farnesol
Saccharolipids	SL	UDP-3-O-(3R-hydroxy-tetradecanoyl)-αD-N-acetylglucosamine
Polyketides	PK	aflatoxin B_1

carbocation-based condensations of isoprene units (prenols, sterols, etc.). Additionally, lipids have been broadly subdivided into "simple" and "complex" groups, with simple lipids being those yielding at most two types of products on hydrolysis (e.g., fatty acids, sterols, and acylglycerols) and complex lipids (e.g., glycerophospholipids and glycosphingolipids) yielding three or more products on hydrolysis. The classification scheme presented here organizes lipids into well-defined categories that cover eukaryotic and prokaryotic sources and that is equally applicable to archaea and synthetic (man-made) lipids.

Lipids may be categorized based on their chemically functional backbone as polyketides, acylglycerols, sphingolipids, prenols, or saccharolipids. However, for historical and bioinformatics advantages, we chose to separate fatty acyls from other polyketides, the glycerophospholipids from the other glycerolipids, and sterol lipids from other prenols, resulting in a total of eight primary categories. An important aspect of this scheme is that it allows for subdivision of the main categories into classes and subclasses to handle the existing and emerging arrays of lipid structures. Although any classification scheme is in part subjective as a result of the structural and biosynthetic complexity of lipids, it is an essential prerequisite for the organization of lipid research and the development of systematic methods of data management. The classification scheme presented here is chemically based and driven by the distinct hydrophobic and hydrophilic elements that constitute the lipid. Biosynthetically related compounds that are not technically lipids because of their water solubility are included for completeness in this classification scheme.

The proposed lipid categories listed in **Table 1** have names that are, for the most part, well accepted in the literature. The fatty acyls (FA) are a diverse group of molecules synthesized by chain elongation of an acetyl-CoA primer with malonyl-CoA (or methylmalonyl-CoA) groups that may contain a cyclic functionality and/or are substituted with heteroatoms. Structures with a glycerol group are represented by two distinct categories: the glycerolipids (GL), which include acylglycerols but also encompass alkyl and 1 Z-alkenyl variants, and the glycerophospholipids (GP), which are defined by the presence of a phosphate (or phosphonate) group esterified to one of the glycerol hydroxyl groups. The sterol lipids (ST) and prenol lipids (PR) share a common biosynthetic pathway via the polymerization of dimethylallyl pyrophosphate/isopentenyl pyrophosphate but have obvious differences in terms of their eventual structure and function. Another well-defined category is the sphingolipids (SP), which contain a long-chain base as their core structure. This classification does not have a glycolipids category per se but rather places

glycosylated lipids in appropriate categories based on the identity of their core lipids. It also was necessary to define a category with the term "saccharolipids" (SL) to account for lipids in which fatty acyl groups are linked directly to a sugar backbone. This SL group is distinct from the term "glycolipid" that was defined by the International Union of Pure and Applied Chemists (IUPAC) as a lipid in which the fatty acyl portion of the molecule is present in a glycosidic linkage. The final category is the polyketides (PK), which are a diverse group of metabolites from plant and microbial sources. Protein modification by lipids (e.g., fatty acyl, prenyl, cholesterol) occurs in nature; however, these proteins are not included in this database but are listed in protein databases such as GenBank (http://www.ncbi.nlm.nih.gov) and SwissProt (http://www.ebi.ac.uk/swissprot/).

Lipid nomenclature

A naming scheme must unambiguously define a lipid structure in a manner that is amenable to chemists, biologists, and biomedical researchers. The issue of lipid nomenclature was last addressed in detail by the International Union of Pure and Applied Chemists and the International Union of Biochemistry and Molecular Biology (IUPAC-IUBMB) Commission on Biochemical Nomenclature in 1976, which subsequently published its recommendations (3). Since then, a number of additional documents relating to the naming of glycolipids (4), prenols (5), and steroids (6) have been released by this commission and placed on the IUPAC website (http://www.chem.qmul.ac.uk/iupac/). A large number of novel lipid classes have been discovered during the last three decades that have not yet been systematically named. The present classification includes these new lipids and incorporates a consistent nomenclature.

In conjunction with our proposed classification scheme, we provide examples of systematic (or semisystematic) names for the various classes and subclasses of lipids. The nomenclature proposal follows existing IUPAC-IUBMB rules closely and should not be viewed as a competing format. The main differences involve a) clarification of the use of core structures to simplify systematic naming of some of the more complex lipids, and b) provision of systematic names for recently discovered lipid classes.

Key features of our lipid nomenclature scheme are as follows:

a) The use of the stereospecific numbering (sn) method to describe glycerolipids and glycerophospholipids (3). The glycerol group is typically acylated or alkylated at the sn-1 and/or *sn*-2 position, with the exception of some lipids that contain more than one glycerol group and archaebacterial lipids in which *sn*-2 and/or *sn*-3 modification occurs.

b) Definition of sphinganine and sphing-4-enine as core structures for the sphingolipid category, where the D-*erythro* or 2*S*,3*R* configuration and 4*E* geometry (in the case of sphing-4-enine) are implied. In molecules containing stereochemistries other than the 2*S*,3*R* configuration, the full systematic names are to be used instead (e.g., 2*R*-amino-1,3*R*-octadecanediol).

c) The use of core names such as cholestane, androstane, and estrane for sterols.

d) Adherence to the names for fatty acids and acyl chains (formyl, acetyl, propionyl, butyryl, etc.) defined in Appendices A and B of the IUPAC-IUBMB recommendations (3).

e) The adoption of a condensed text nomenclature for the glycan portions of lipids, where sugar residues are represented by standard IUPAC abbreviations and where the anomeric carbon locants and stereochemistry are included but the parentheses are omitted. This system has also been proposed by the Consortium for Functional Glycomics (http://web.mit.edu/glycomics/consortium/main.shtml).

f) The use of *E/Z* designations (as opposed to *trans/cis*) to define double bond geometry.

g) The use of *R/S* designations (as opposed to α/β or D/L) to define stereochemistries. The exceptions are those describing substituents on glycerol (*sn*) and sterol core structures and anomeric carbons on sugar residues. In these latter special cases, the α/β format is firmly established.

h) The common term "lyso," denoting the position lacking a radyl group in glycerolipids and glycerophospholipids, will not be used in systematic names but will be included as a synonym.

i) The proposal for a single nomenclature scheme to cover the prostaglandins, isoprostanes, neuroprostanes, and related compounds, where the carbons participating in the cyclopentane ring closure are defined and where a consistent chain-numbering scheme is used.

j) The "d" and "t" designations used in shorthand notation of sphingolipids refer to 1,3-dihydroxy and 1,3,4-trihydroxy long-chain bases, respectively.

Lipid structure representation

In addition to having rules for lipid classification and nomenclature, it is important to establish clear guidelines for drawing lipid structures. Large and complex lipids are difficult to draw, which leads to the use of shorthand and unique formats that often generate more confusion than clarity among lipidologists. We propose a more consistent format for representing lipid structures in which, in the simplest case of the fatty acid derivatives, the acid group (or equivalent) is drawn on the right and the hydrophobic hydrocarbon chain is on the left (**Figure 1**). Notable exceptions are found in the eicosanoid class, in which the hydrocarbon chain wraps around in a counterclockwise direction to produce a more condensed structure. Similarly, with regard to the glycerolipids and glycerophospholipids, the radyl chains are drawn with the hydrocarbon chains to the left and the glycerol group depicted

horizontally with stereochemistry at the *sn* carbons defined (if known). The general term "radyl" is used to denote either acyl, alkyl, or 1-alkenyl substituents (http://www.chem.qmul.ac.uk/iupac/lipid/lip1n2.html), allowing for coverage of alkyl and 1 *Z*-alkenylglycerols. The sphingolipids, although they do not contain a glycerol group, have a similar structural relationship to the glycerophospholipids in many cases and may be drawn with the C1 hydroxyl group of the long-chain base to the right and the alkyl portion to the left. This methodology places the head groups of both sphingolipids and glycerophospholipids on the right side. Although the structures of sterols do not conform to these general rules of representation, the sterol esters may conveniently be drawn with the acyl group oriented according to these guidelines. In addition, the linear prenols or isoprenoids are drawn in a manner analogous to the fatty acids, with the terminal functional group on the right side. Inevitably, a number of structurally complex lipids, such as acylaminosugar glycans, polycyclic isoprenoids, and polyketides, do not lend themselves to these simplified drawing rules. Nevertheless, we believe that the adoption of the guidelines proposed here will unify chemical representation and make it more comprehensible.

Databasing lipids, annotation, and function

A number of repositories, such as GenBank, SwissProt, and ENSEMBL (http://www.ensembl.org), support nucleic acid and protein databases; however, there are only a few specialized databases [e.g., LIPIDAT (7) and Lipid Bank (8)] that provide a catalog, annotation, and functional classification of lipids. Given the importance of these molecules in cellular function and pathology, there is an imminent need for the creation of a well-organized database of lipids. The first step toward this goal is the establishment of an ontology of lipids that is extensible, flexible, and scalable. Before establishing an ontology, a structured vocabulary is needed, and the IUPAC nomenclature of the 1970s was an initial step in this direction.

The ontology of lipids must contain definitions, meanings, and interrelationships of all objects stored in the database. This ontology is then transformed into a well-defined schema that forms the foundation for a relational database of lipids. The LIPID MAPS project is building a robust database of lipids based on the proposed ontology.

Our database will provide structural and functional annotations and have links to relevant protein and gene data. In addition, a universal data format (XML) will be provided to facilitate exportation of the data into other repositories. This database will enable the storage of curated information on lipids in a web-accessible format and will provide a community standard for lipids.

An important database field will be the LIPID ID, a unique 12 character identifier based on the classification scheme described here. The format of the LIPID ID, outlined in **Table 2**, provides a systematic means of assigning unique IDs to lipid molecules and allows for the addition of large numbers of new categories, classes, and subclasses in the future, because a maximum of 100 classes/subclasses (00 to 99) may be specified. The last four characters of the ID constitute a unique identifier within a particular subclass and are randomly assigned. By initially using numeric characters, this allows 9,999 unique IDs per subclass, but with the additional use of 26 uppercase alphabetic characters, a total of 1.68 million possible combinations can be generated, providing ample scalability within each subclass. In cases in which lipid structures are obtained from

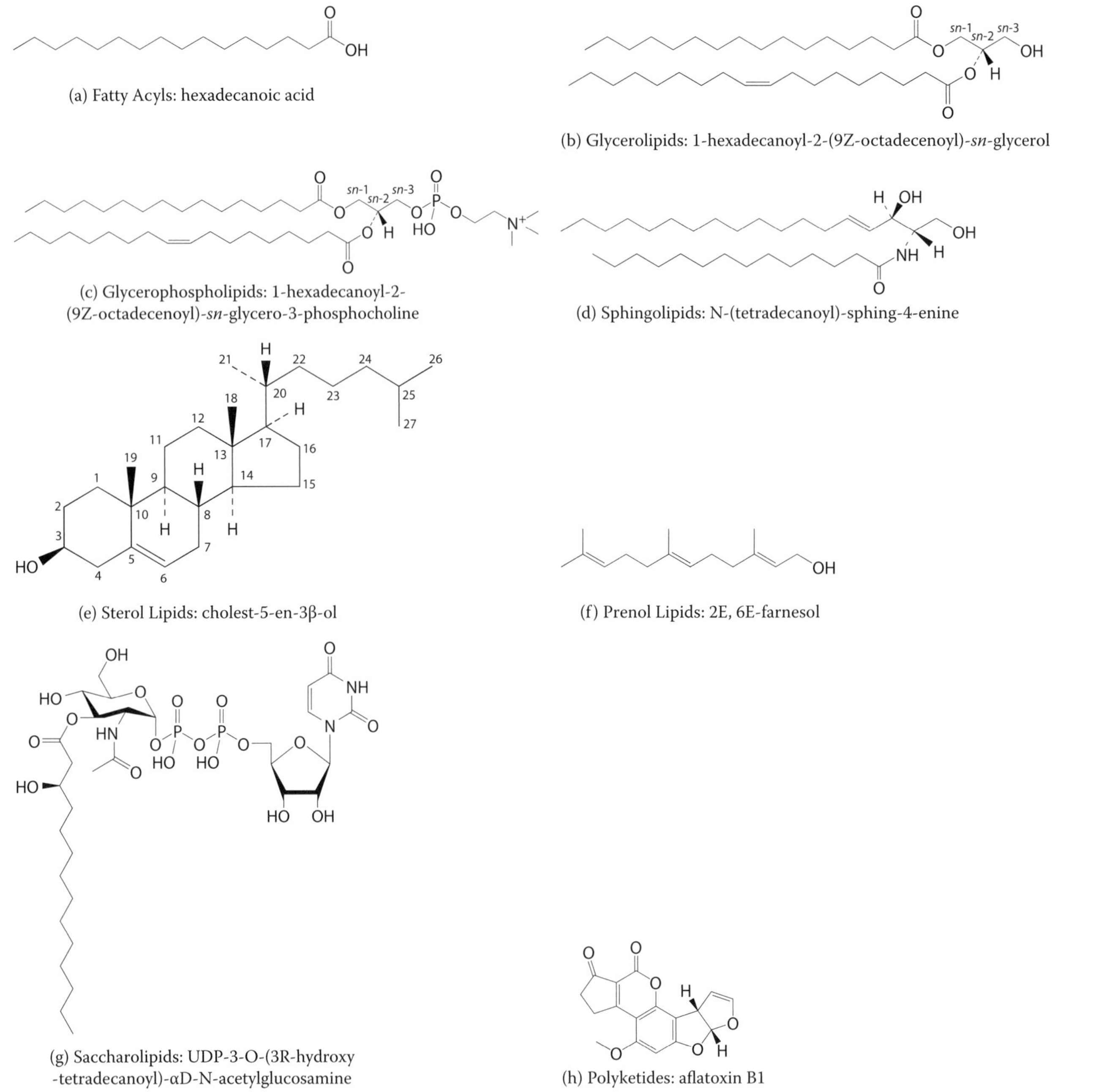

(a) Fatty Acyls: hexadecanoic acid

(b) Glycerolipids: 1-hexadecanoyl-2-(9Z-octadecenoyl)-*sn*-glycerol

(c) Glycerophospholipids: 1-hexadecanoyl-2-(9Z-octadecenoyl)-*sn*-glycero-3-phosphocholine

(d) Sphingolipids: N-(tetradecanoyl)-sphing-4-enine

(e) Sterol Lipids: cholest-5-en-3β-ol

(f) Prenol Lipids: 2E, 6E-farnesol

(g) Saccharolipids: UDP-3-O-(3R-hydroxy-tetradecanoyl)-αD-N-acetylglucosamine

(h) Polyketides: aflatoxin B1

FIGURE 1   Representative structures for each lipid category.

TABLE 2: Format of 12 Character LIPID ID

Characters	Description	Example
1–2	Fixed database designation	LM
3–4	Two letter category code	FA
5–6	Two digit class code	03
7–8	Two digit subclass code	02
9–12	Unique four character identifier within subclass	7312

other sources such as LipidBank or LIPIDAT, the corresponding IDs for those databases will be included to enable cross-referencing. The first two characters of the ID contain the database identifier (e.g., LM for LIPID MAPS), although other databases may choose to use their own two character identifier (at present, LB for Lipid Bank and LD for LIPIDAT) and assign the last four or more characters uniquely while retaining characters 3 to 8, which pertain to classification. The corresponding IDs of the other databases will always be included to enable cross-referencing. Further details regarding the numbering

TABLE 3: Shorthand Notation for Selected Lipid Categories

Category	Abbreviation	Class or Subclass	Example[a]
GP	GPCho	Glycerophosphocholines	GPCho (16:0/9Z,12Z-18:2)
GP	GPnCho	Glycerophosphonocholines	
GP	GPEtn	Glycerophosphoethanolamines	
GP	GPnEtn	Glycerophosphonoethanolamines	
GP	GPSer	Glycerophosphoserines	
GP	GPGro	Glycerophosphoglycerols	
GP	GPGroP	Glycerophosphoglycerophosphates	
GP	GPIns	Glycerophosphoinositols	
GP	GPInsP	Glycerophosphoinositol monophosphates	
GP	GPInsP$_2$	Glycerophosphoinositol bis-phosphates	
GP	GPInsP$_3$	Glycerophosphoinositol tris-phosphates	
GP	GPA	Glycerophosphates	
GP	GPP	Glyceropyrophosphates	
GP	CL	Glycerophosphoglycerophosphoglycerols	
GP	CDP-DG	CDP-glycerols	
GP	[glycan] GP	Glycerophosphoglucose lipids	
GP	[glycan] GPIns	Glycerophosphoinositolglycans	EtN-P-6Manα1–2Manα1–6 Manα1–4GlcNα1-6GPIns (14:0/14:0)
SP	Cer	Ceramides	Cer (d18:1/9E-16:1)
SP	SM	Phosphosphingolipids	SM (d18:1/24:0)
SP	[glycan]Cer	Glycosphingolipids	NeuAcα2–3Galβ1–4Glcβ-Cer (d18:1/16:0)
GL	MG	Monoradyl glycerols	MG (16:0/0:0/0:0)
GL	DG	Diradyl glycerols	DG (18:0/16:0/0:0)
GL	TG	Triradyl glycerols	TG (12:0/14:0/18:0)

[a] Shorthand notation for radyl substituents in categories GP and GL are presented in the order of *sn*-1 to *sn*-3. Shorthand notation for category SP is presented in the order of long-chain base and *N*-acyl substituent. Numbers separated by colons refer to carbon chain length and number of double bonds, respectively.

system will be decided by the International Lipids Classification and Nomenclature Committee (see below). In addition to the LIPID ID, each lipid in the database will be searchable by classification (category, class, subclass), systematic name, synonym(s), molecular formula, molecular weight, and many other parameters that are part of its ontology. An important feature will be the databasing of molecular structures, allowing the user to perform web-based substructure searches and structure retrieval across the database. This aim will be accomplished with a chemistry cartridge software component that will enable structures in formats such as MDL molfile and Chemdraw CDX to be imported directly into Oracle database tables.

Furthermore, many lipids, in particular the glycerolipids, glycerophospholipids, and sphingolipids, may be conveniently described in terms of a shorthand name in which abbreviations are used to define backbones, head groups, and sugar units and the radyl substituents are defined by a descriptor indicating carbon chain length and number of double bonds. These shorthand names lend themselves to fast, efficient text-based searches and are used widely in lipid research as compact alternatives to systematic names. The glycerophospholipids in the LIPIDAT database, for example, may be conveniently searched with a shorthand notation that has been extended to handle side chains with acyl, ether, branched-chain, and other functional groups (7). We propose the use of a shorthand notation for selected lipid categories (**Table 3**) that incorporates a condensed text nomenclature for glycan substituents. The abbreviations for the sugar units follow the current IUPAC-IUBMB recommendations (4).

Lipid classes and subclasses

Fatty acyls [FA]

The fatty acyl structure represents the major lipid building block of complex lipids and therefore is one of the most

fundamental categories of biological lipids. The fatty acyl group in the fatty acids and conjugates class (**Table 4**) is characterized by a repeating series of methylene groups that impart hydrophobic character to this category of lipids. The first subclass includes the straight-chain saturated fatty acids containing a terminal carboxylic acid. It could also be considered the most reduced end product of the polyketide pathway. Variants of this structure have one or more methyl substituents and encompass quite complex branched-chain fatty acids, such as the mycolic acids. The longest chain in branched-chain fatty acids defines the chain length of these compounds. A considerable number of variations on this basic structure occur in all kingdoms of life (9–12), including fatty acids with one or more double bonds and even acetylenic (triple) bonds. Heteroatoms of oxygen, halogen, nitrogen, and sulfur are also linked to the carbon chains in specific subclasses. Cyclic fatty acids containing three to six carbon atoms as well as heterocyclic rings containing oxygen or nitrogen are found in nature. The cyclopentenyl fatty acids are an example of this latter subclass. The thia fatty acid subclass contains sulfur atom(s) in the fatty acid structure and is exemplified by lipoic acid and biotin. Thiols and thioethers are in this class, but the thioesters are placed in the ester class because of the involvement of these and similar esters in fatty acid metabolism and synthesis.

Separate classes for more complex fatty acids with multiple functional groups (but nonbranched) are designated by the total number of carbon atoms found in the critical biosynthetic precursor. These include octadecanoids and lipids in the jasmonic acid pathway of plant hormone biosynthesis, even though jasmonic acids have lost some of their carbon atoms from the biochemical precursor, 12-oxophytodienoic acid (13). Eicosanoids derived from arachidonic acid include prostaglandins, leukotrienes, and other structural derivatives (14). The docosanoids contain 22 carbon atoms and derive from a common precursor, docosahexaenoic acid (15). Many members of these separate subclasses of more complex fatty acids have distinct biological activities.

TABLE 4: Fatty acyls [FA] Classes and Subclasses

Fatty acids and conjugates [FA01]
  Straight-chain fatty acids [FA0101]
  Methyl branched fatty acids [FA0102]
  Unsaturated fatty acids [FA0103]
  Hydroperoxy fatty acids [FA0104]
  Hydroxy fatty acids [FA0105]
  Oxo fatty acids [FA0106]
  Epoxy fatty acids [FA0107]
  Methoxy fatty acids [FA0108]
  Halogenated fatty acids [FA0109]
  Amino fatty acids [FA0110]
  Cyano fatty acids [FA0111]
  Nitro fatty acids [FA0112]
  Thia fatty acids [FA0113]
  Carbocyclic fatty acids [FA0114]
  Heterocyclic fatty acids [FA0115]
  Mycolic acids [FA0116]
  Dicarboxylic acids [FA0117]
Octadecanoids [FA02]
  12-Oxophytodienoic acid metabolites [FA0201]
  Jasmonic acids [FA0202]
Eicosanoids [FA03]
  Prostaglandins [FA0301]
  Leukotrienes [FA0302]
  Thromboxanes [FA0303]
  Lipoxins [FA0304]
  Hydroxyeicosatrienoic acids [FA0305]
  Hydroxyeicosatetraenoic acids [FA0306]
  Hydroxyeicosapentaenoic acids [FA0307]
  Epoxyeicosatrienoic acids [FA0308]
  Hepoxilins [FA0309]
  Levuglandins [FA0310]
  Isoprostanes [FA0311]
  Clavulones [FA0312]
Docosanoids [FA04]
Fatty alcohols [FA05]
Fatty aldehydes [FA06]
Fatty esters [FA07]
  Wax monoesters [FA0701]
  Wax diesters [FA0702]
  Cyano esters [FA0703]
  Lactones [FA0704]
  Fatty acyl-CoAs [FA0705]
  Fatty acyl-acyl carrier proteins (ACPs) [FA0706]
  Fatty acyl carnitines [FA0707]
  Fatty acyl adenylates [FA0708]
Fatty amides [FA08]
  Primary amides [FA0801]
  N-Acyl amides [FA0802]
  Fatty acyl homoserine lactones [FA0803]
  N-Acyl ethanolamides (endocannabinoids) [FA0804]
Fatty nitriles [FA09]
Fatty ethers [FA10]
Hydrocarbons [FA11]
Oxygenated hydrocarbons [FA12]
Other [FA00]

TABLE 5: Glycerolipids [GL] Classes and Subclasses

Monoradylglycerols [GL01]
  Monoacylglycerols [GL0101]
  Monoalkylglycerols [GL0102]
  Mono-(1 *Z*-alkenyl)-glycerols [GL0103]
  Monoacylglycerolglycosides [GL0104]
  Monoalkylglycerolglycosides [GL0105]
Diradylglycerols [GL02]
  Diacylglycerols [GL0201]
  Alkylacylglycerols [GL0202]
  Dialkylglycerols [GL0203]
  1 *Z*-Alkenylacylglycerols [GL0204]
  Diacylglycerolglycosides [GL0205]
  Alkylacylglycerolglycosides [GL0206]
  Dialkylglycerolglycosides [GL0207]
  Di-glycerol tetraethers [GL0208]
  Di-glycerol tetraether glycans [GL0209]
Triradylglycerols [GL03]
  Triacylglycerols [GL0301]
  Alkyldiacylglycerols [GL0302]
  Dialkylmonoacylglycerols [GL0303]
  1 *Z*-Alkenyldiacylglycerols [GL0304]
  Estolides [GL0305]
Other [GL00]

amides, and many simple amides have interesting biological activities in various organisms. Fatty acyl homoserine lactones are fatty amides involved in bacterial quorum sensing (16).

Hydrocarbons are included as a class of fatty acid derivatives because they correspond to six electron reduction products of fatty acids that may have been generated by loss of the carboxylic acid from a fatty acid or fatty acyl moiety during the process of diagenesis in geological samples. Long-chain ethers also have been observed in nature. Chemical structures of the fatty acyls are shown in **Figure 2.**

Glycerolipids [GL]

The glycerolipids essentially encompass all glycerol-containing lipids. We have purposely made glycerophospholipids a separate category because of their abundance and importance as membrane constituents, metabolic fuels, and signaling molecules. The glycerolipid category **(Table 5)** is dominated by the mono-, di- and tri-substituted glycerols, the most well-known being the fatty acid esters of glycerol (acylglycerols) (17, 18). Additional subclasses are represented by the glycerolglycans, which are characterized by the presence of one or more sugar residues attached to glycerol via a glycosidic linkage (19). Examples of structures in this category are shown in **Figure 3.** Macrocyclic ether lipids also occur as glycerolipids in the membranes of archaebacteria (20).

Glycerophospholipids [GP]

The glycerophospholipids are ubiquitous in nature and are key components of the lipid bilayer of cells. Phospholipids may be subdivided into distinct classes **(Table 6)** based on the nature of the polar "head group" at the *sn*-3 position of the glycerol backbone in eukaryotes and eu-bacteria or the *sn*-1 position in the case of archaebacteria (21). In the case of the glycerophosphoglycerols and glycerophosphoglycerophosphates, a second glycerol unit constitutes part of the head group, whereas for the glycerophosphoglycerophosphoglycerols (cardiolipins), a third glycerol unit is typically acylated at the *sn*-1' and *sn*-2' positions to create a pseudosymmetrical molecule. Each head group class is further differentiated on the basis of the *sn*-1 and *sn*-2 substituents on

Other major lipid classes in the fatty acyl category include fatty acid esters such as wax monoesters and diesters and the lactones. The fatty ester class also has subclasses that include important biochemical intermediates such as fatty acyl thioester-CoA derivatives, fatty acyl thioester-acyl carrier protein (ACP) derivatives, fatty acyl carnitines (esters of carnitine), and fatty adenylates, which are mixed anhydrides. The fatty alcohols and fatty aldehydes are typified by terminal hydroxy and oxo groups, respectively. The fatty amides are also *N*-fatty acylated amines and unsubstituted

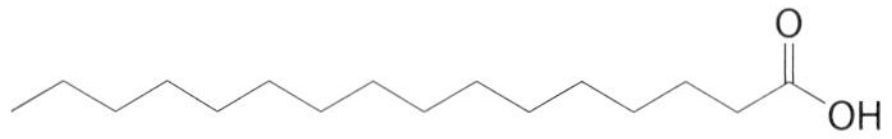

(a) Straight chain fatty acids: hexadecanoic acid

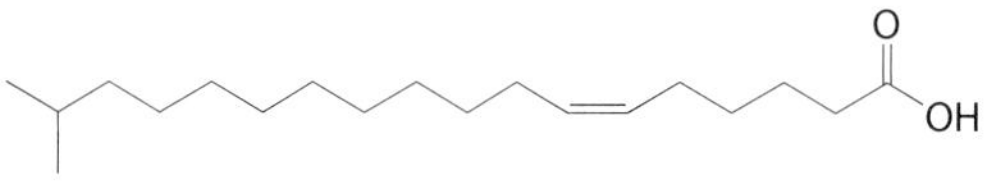

(b) Methyl branched fatty acids: 17-methyl-6Z-octadecenoic acid

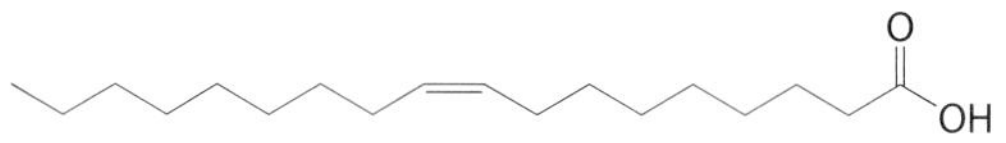

(c) Unsaturated fatty acids: 9Z-octadecenoic acid

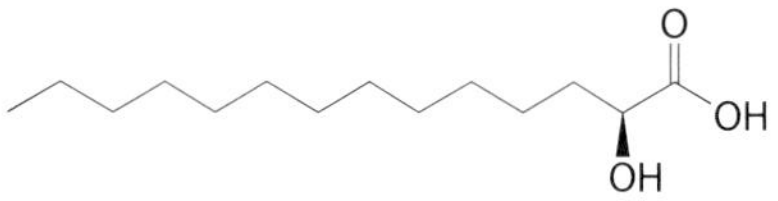

(d) Hydroxy fatty acids:
2S-hydroxytetradecanoic acid

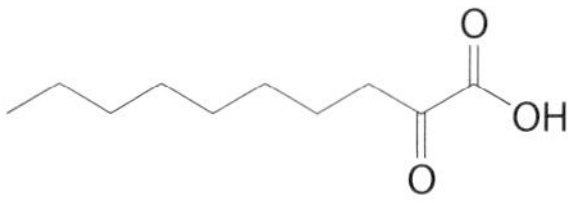

(e) Oxo fatty acids: 2-oxodecanoic acid

(f) Epoxy fatty acids: 6R,7S-epoxy-octadecanoic acid

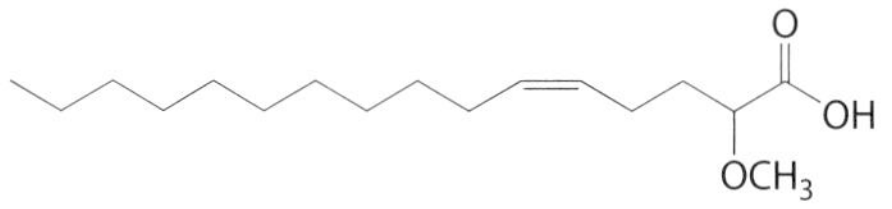

(g) Methoxy fatty acids:
2-methoxy-5Z-hexadecenoic acid

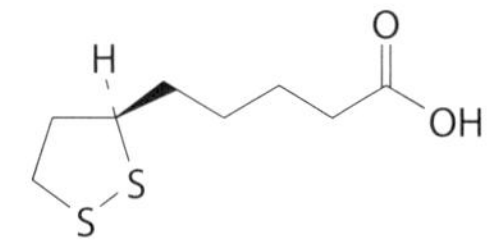

(h) Thia fatty acids: R-Lipoic acid;
1,2-dithiolane-3R-pentanoic acid

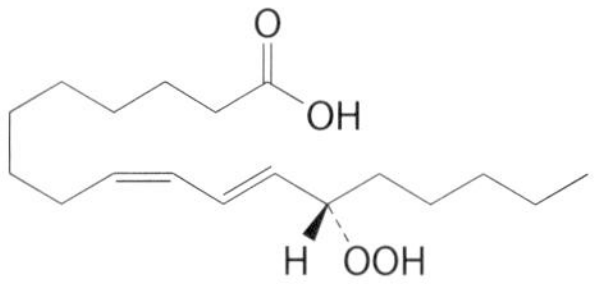

(i) Hydroperoxy fatty acids:
13S-hydroperoxy-9Z,
11E-octadecadienoic acid

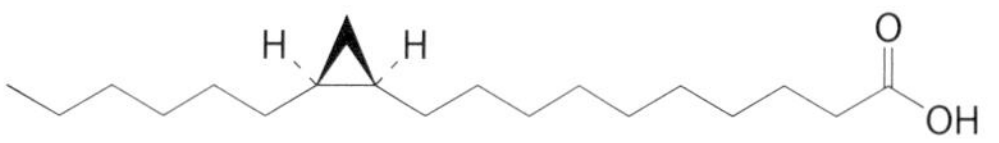

(j) Carbocyclic fatty acids: lactobacillic acid;
11R,12S-methyleneoctadecanoic acid

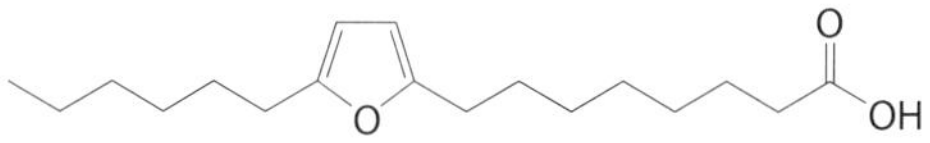

(k) Heterocyclic fatty acids:
8-(5-hexylfuran-2-yl)-octanoic acid

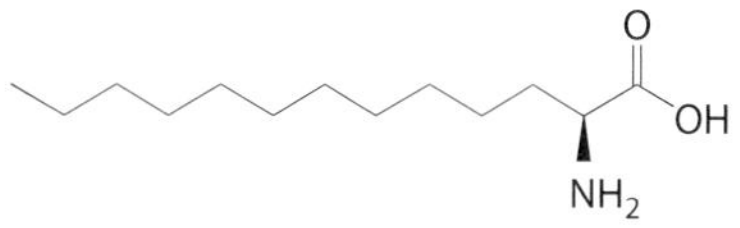

(l) Amino fatty acids: 2S-aminotridecanoic acid

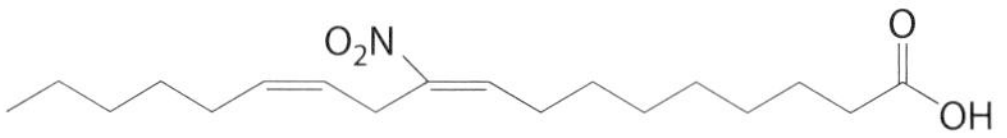

(m) Nitro fatty acids: 10-nitro, 9Z, 12Z-octadecadienoic acid

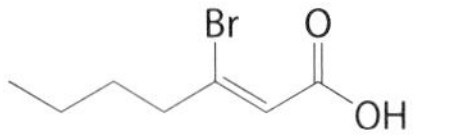

(n) Halogenated fatty acids:
3-bromo-2Z-heptenoic acid

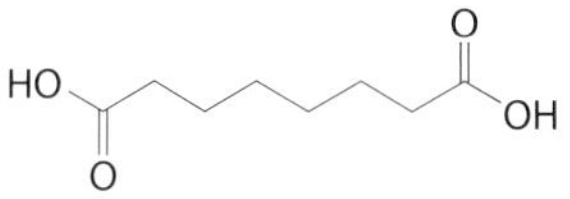

(o) Dicarboxylic acids:
1,8-octanedioic acid

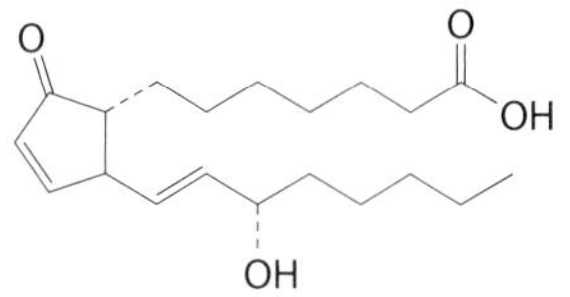

(p) Prostaglandins: Prostaglandin A1;
15S-hydroxy-9-oxo-10Z,13E-prostadienoic acid

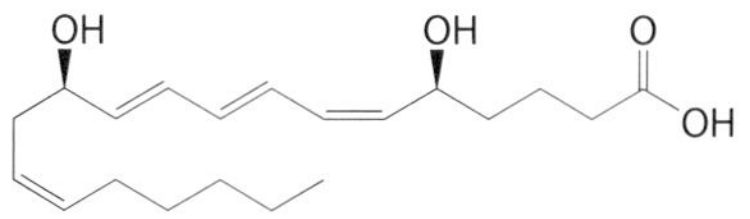

(q) Leukotrienes: Leukotriene B4;
5S,12R-dihydroxy-6Z,8E,10E,14Z-eicosatetraenoic acid

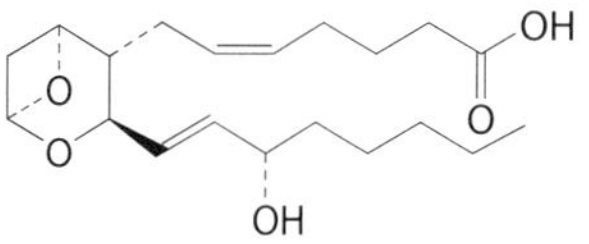

(r) Thromboxanes: Thromboxane A2;
9S,11S-epoxy,15S-hydroxythromboxa-5Z,13E-dien-1-oic acid

FIGURE 2 Representative structures for fatty acyls.

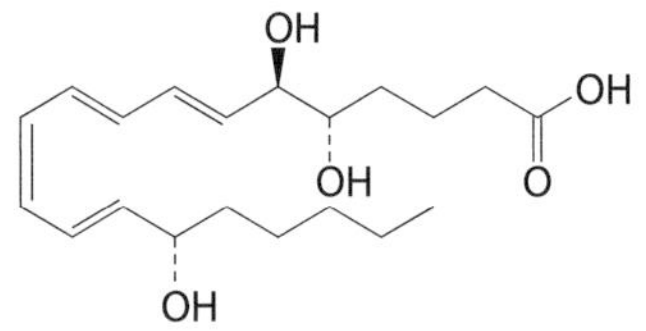

(s) Lipoxins: Lipoxin A4;
5S,6R,15S-trihydroxy-7E,9E,11Z,13E-eicosatetraenoic acid

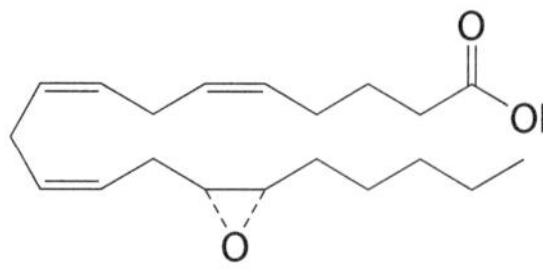

(t) Epoxyeicosatrienoic acids:
14R,15S-epoxy-5Z,8Z,11Z-eicosatrienoic acid

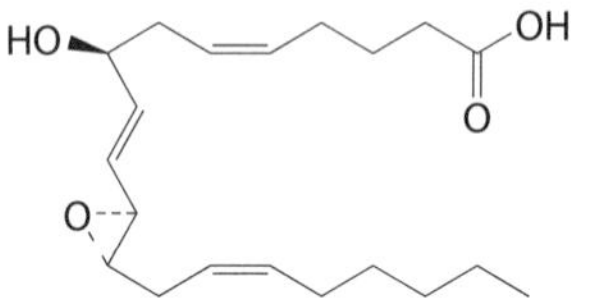

(u) Hepoxilins: Hepoxilin A3;
8R-hydroxy-11R,12S-epoxy-5Z,9E,14Z-eicosatrienoic acid

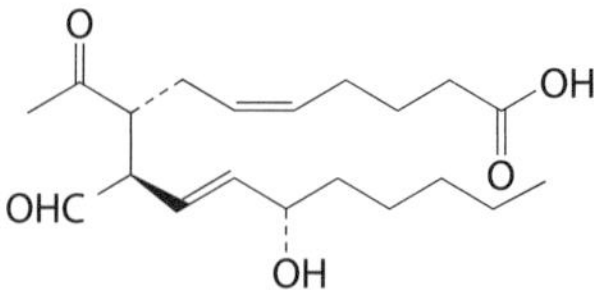

(v) Levuglandins: LGE2;
10,11-seco-9,11-dioxo-15S-hydroxy-5Z,13E-prostadienoic acid

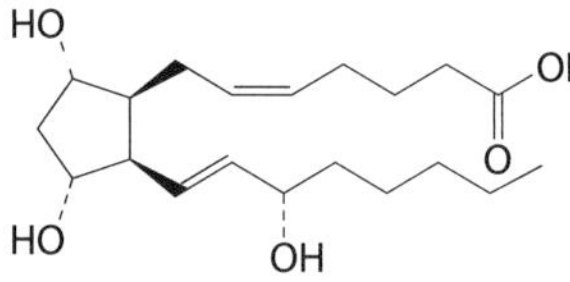

(w) Isoprostanes: 9S,11S,15S-trihydroxy-5Z,
13E-prostadienoic acid-cyclo[8S,12R]

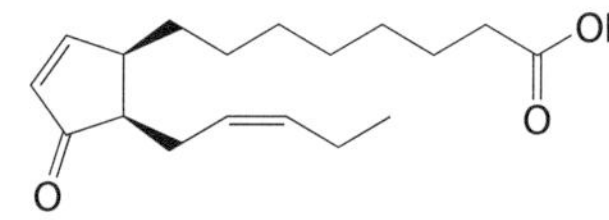

(x) Octadecanoids: 12-oxophytodienoic acid metabolites;
(9R,13R)-12-oxo-phyto-10Z,15Z-dienoic acid

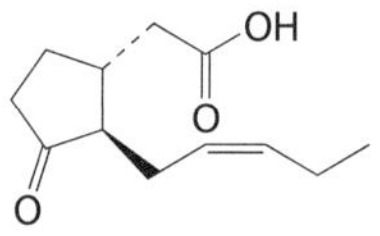

(y) Octadecanoids: Jasmonic acids: jasmonic acid;
(1R,2R)-3-oxo-2-(pent-2Z-enyl)-cyclopentaneacetic acid

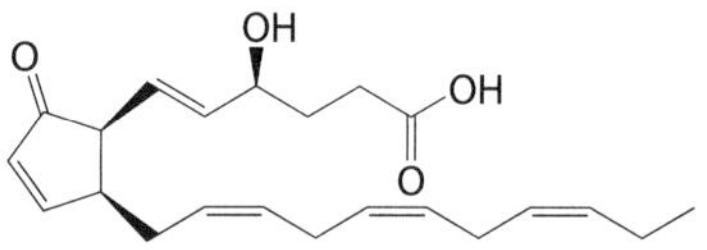

(z) Docosanoids: Neuroprostanes; 4S-hydroxy-8-oxo-
(5E,9Z,13Z,16Z,19Z)-neuroprostapentaenoic acid-cyclo[7S,11S]

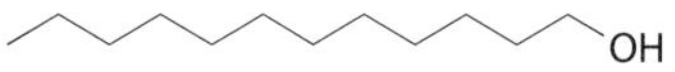

(aa) Fatty alcohols: dodecanol

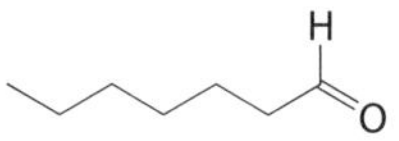

(ab) Fatty aldehydes: heptanal

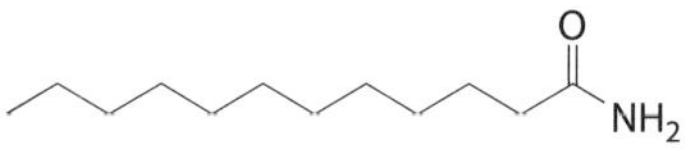

(ac) Fatty amides: N-acyl amides: dodecanamide

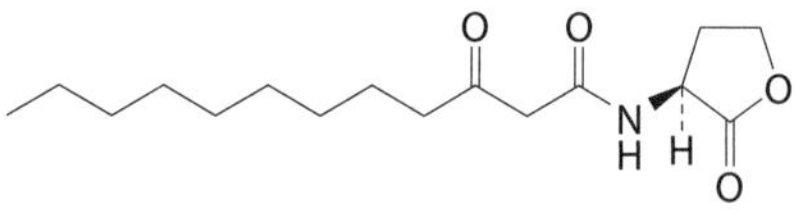

(ad) Fatty amides: Fatty acyl homoserine lactones:
N-(3-oxodecanoyl) homoserine lactone

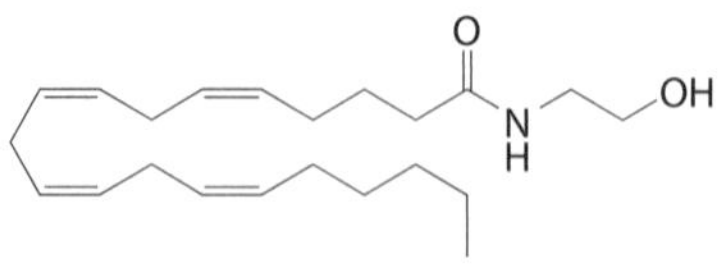

(ae) Fatty amides: N-acyl ethanolamides (endocannabinoids):
Anandamide; N-(5Z,8Z,11Z,14Z-eicosatetraenoyl)-ethanolamine

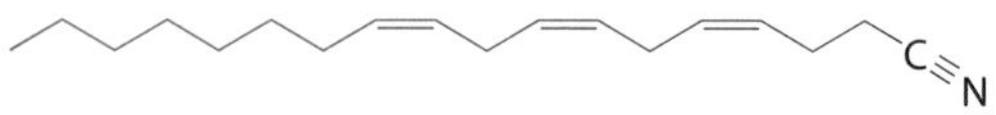

(af) Fatty nitriles: 4Z,7Z,10Z-octadecatrienenitrile

(ag) Hydrocarbons: tridecane

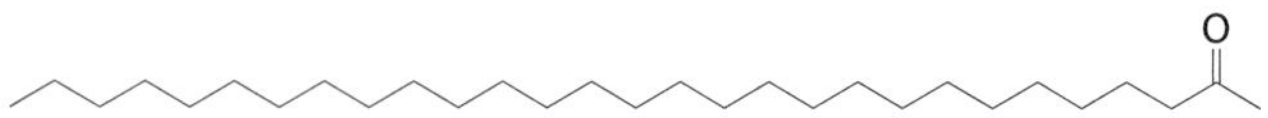

(ah) Oxygenated hydrocarbons: nonacosan-2-one

FIGURE 2   Representative structures for fatty acyls (Continued).

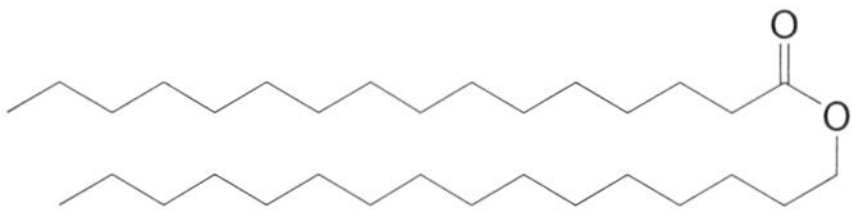

(ai) Wax monoesters:
1-hexadecyl hexadecanoate

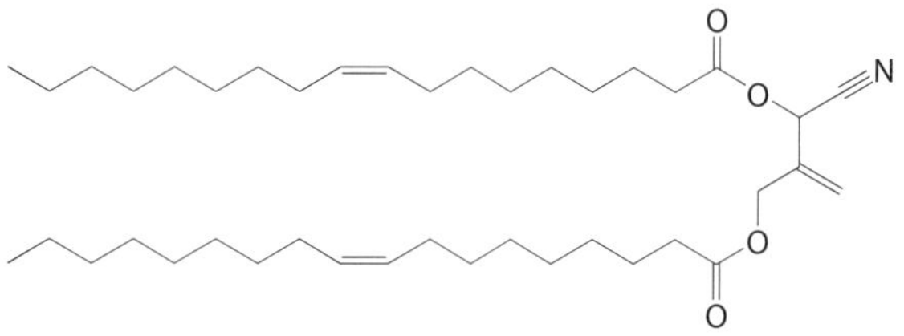

(aj) Cyano esters: 1,3-di-(octadec-9Z-enoyl)-1-cyano-2-methylene-propane-1,3-diol

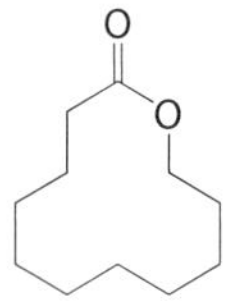

(ak) Lactones:
11-undecanolactone

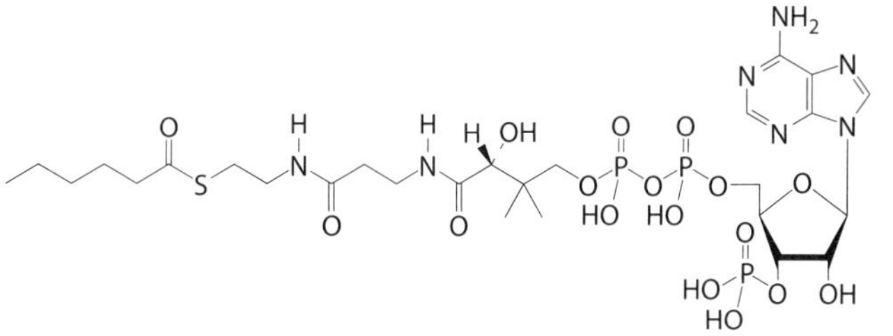

(al) Fatty acyl CoAs: R-hexanoyl CoA

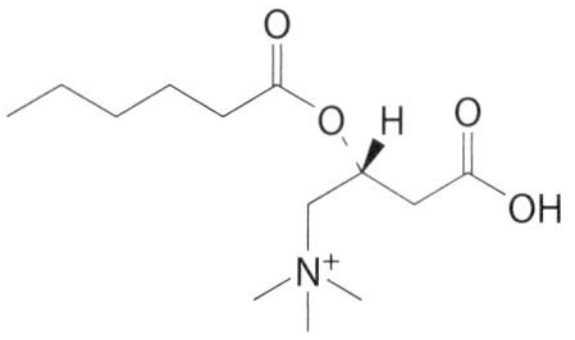

(am) Fatty acyl carnitines:
O-hexanoyl-R-carnitine

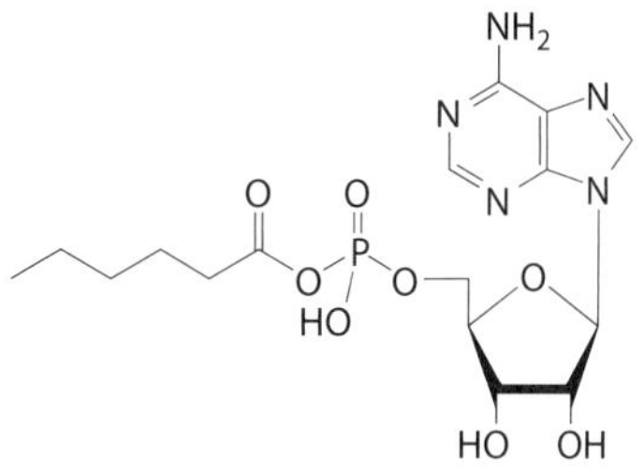

(an) Fatty acyl adenylates:
O-hexanoyladenosine monophosphate

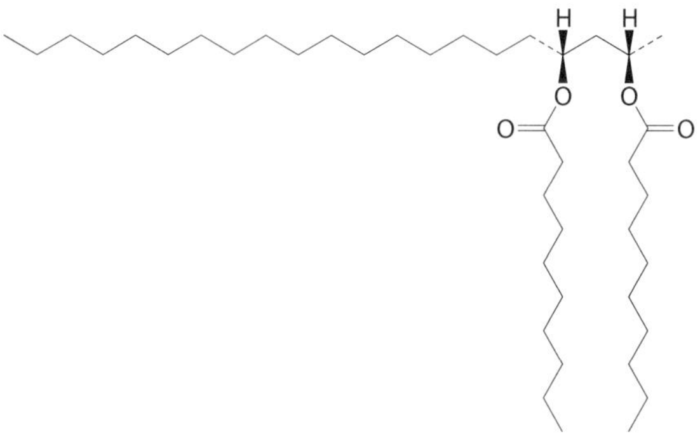

(ao) Wax diesters: 2S,3R-didecanoyl-docosane-2,3-diol

FIGURE 2. Representative structures for fatty acyls (Continued).

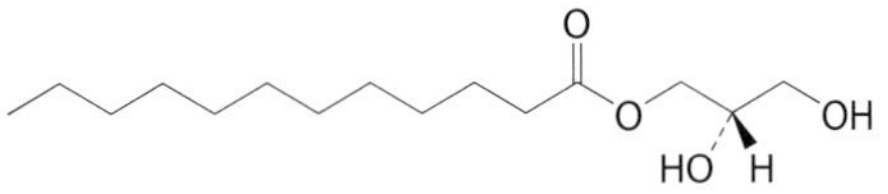

(a) Monoradylglycerols: Monoacylglycerols:
1-dodecanoyl-*sn*-glycerol

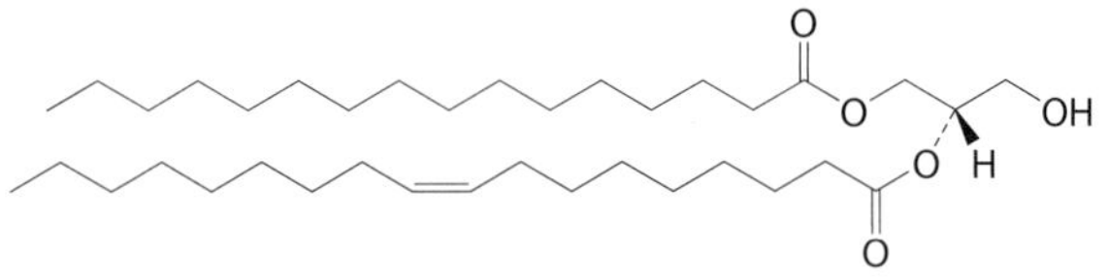

(b) Diradylglycerols: Diacylglycerols:
1-hexadecanoyl-2-(9Z-octadecenoyl)-*sn*-glycerol

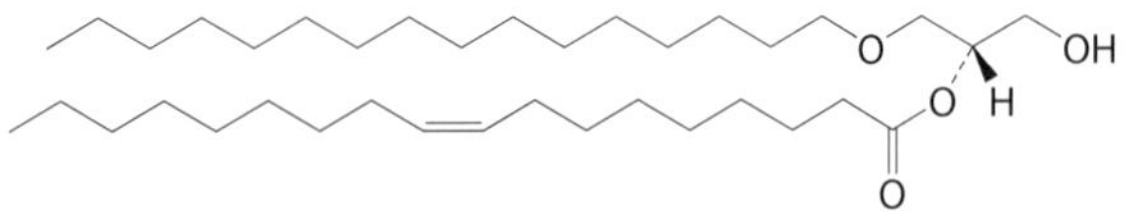

(c) Diradylglycerols: Alkylacylglycerols:
1-O-hexadecyl-2-(9Z-octadecenoyl)-*sn*-glycerol

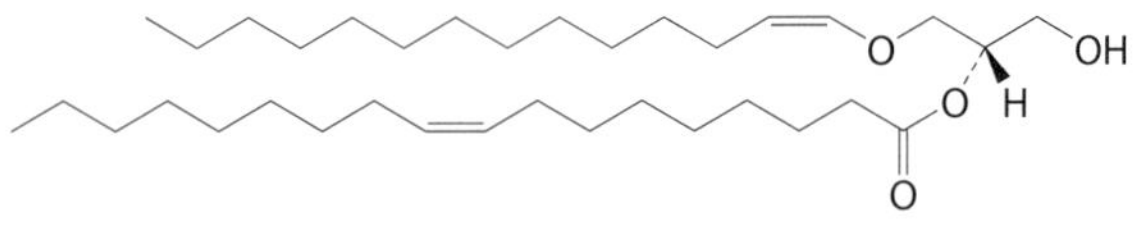

(d) Diradylglycerols: 1Z-alkenylacylglycerols:
1-O-(1Z-tetradecenyl)-2-(9Z-octadecenoyl)-*sn*-glycerol

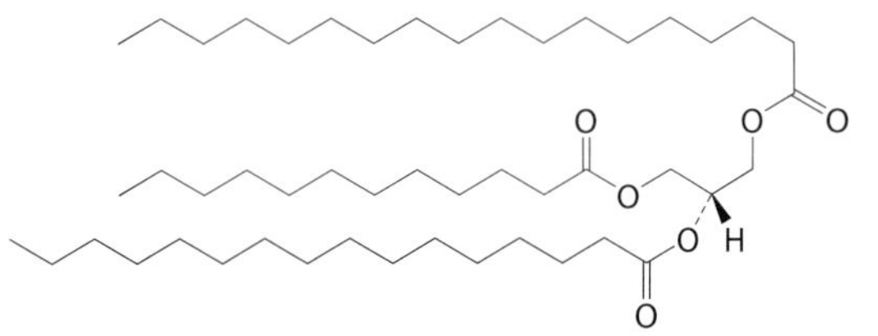

(e) Triradylglycerols: Triglycerols:
1-dodecanoyl-2-hexadecanoyl-3-octadecanoyl-*sn*-glycerol

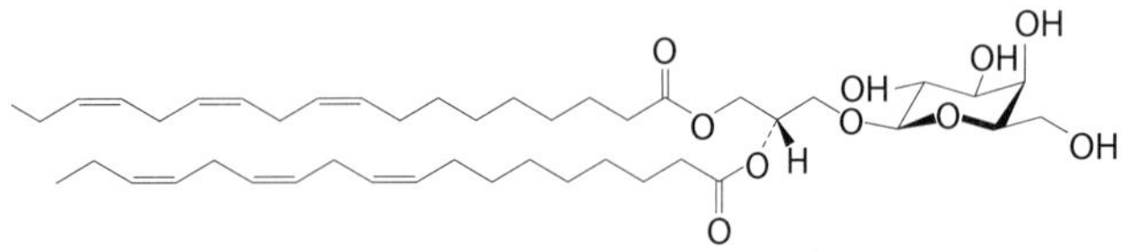

(f) Diradylglycerols: Diacylglycerol glycans:
1,2-di-(9Z,12Z,15Z-octadecatrienoyl)-3-O-β-D-galactosyl-*sn*-glycerol

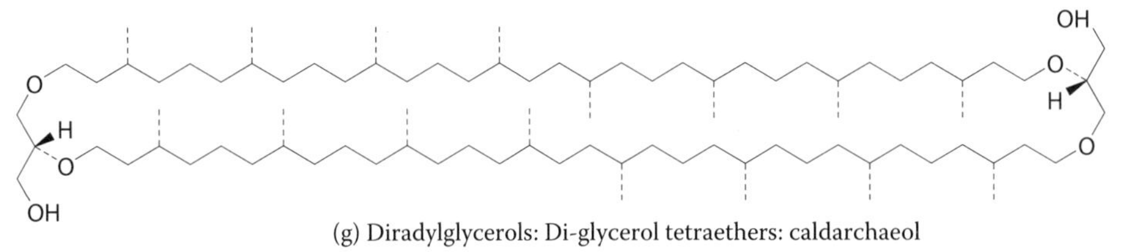

(g) Diradylglycerols: Di-glycerol tetraethers: caldarchaeol

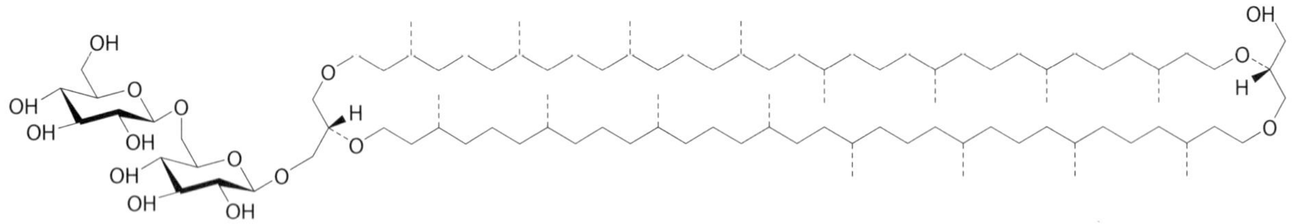

(h) Diradylglycerols: Di-glycerol tetraether glycans: gentiobiosylcaldarchaeol; Glcβ1-6Glcβ-caldarchaeol

FIGURE 3 Representative structures for glycerolipids.

TABLE 6: Glycerophospholipids [GP] Classes and Subclasses

Glycerophosphocholines [GP01]
  Diacylglycerophosphocholines [GP0101]
  1-Alkyl,2-acylglycerophosphocholines [GP0102]
  1 Z-Alkenyl,2-acylglycerophosphocholines [GP0103]
  Dialkylglycerophosphocholines [GP0104]
  Monoacylglycerophosphocholines [GP0105]
  1-Alkyl glycerophosphocholines [GP0106]
  1 Z-Alkenylglycerophosphocholines [GP0107]
Glycerophosphoethanolamines [GP02]
  Diacylglycerophosphoethanolamines [GP0201]
  1-Alkyl,2-acylglycerophosphoethanolamines [GP0202]
  1 Z-Alkenyl,2-acylglycerophosphoethanolamines [GP0203]
  Dialkylglycerophosphoethanolamines [GP0204]
  Monoacylglycerophosphoethanolamines [GP0205]
  1-Alkyl glycerophosphoethanolamines [GP0206]
  1 Z-Alkenylglycerophosphoethanolamines [GP0207]
Glycerophosphoserines [GP03]
  Diacylglycerophosphoserines [GP0301]
  1-Alkyl,2-acylglycerophosphoserines [GP0302]
  1 Z-Alkenyl,2-acylglycerophosphoserines [GP0303]
  Dialkylglycerophosphoserines [GP0304]
  Monoacylglycerophosphoserines [GP0305]
  1-Alkyl glycerophosphoserines [GP0306]
  1 Z-Alkenylglycerophosphoserines [GP0307]
Glycerophosphoglycerols [GP04]
  Diacylglycerophosphoglycerols [GP0401]
  1-Alkyl,2-acylglycerophosphoglycerols [GP0402]
  1 Z-Alkenyl,2-acylglycerophosphoglycerols [GP0403]
  Dialkylglycerophosphoglycerols [GP0404]
  Monoacylglycerophosphoglycerols [GP0405]
  1-Alkyl glycerophosphoglycerols [GP0406]
  1 Z-Alkenylglycerophosphoglycerols [GP0407]
  Diacylglycerophosphodiradylglycerols [GP0408]
  Diacylglycerophosphomonoradylglycerols [GP0409]
  Monoacylglycerophosphomonoradylglycerols [GP0410]
Glycerophosphoglycerophosphates [GP05]
  Diacylglycerophosphoglycerophosphates [GP0501]
  1-Alkyl,2-acylglycerophosphoglycerophosphates [GP0502]
  1 Z-Alkenyl,2-acylglycerophosphoglycerophosphates [GP0503]
  Dialkylglycerophosphoglycerophosphates [GP0504]
  Monoacylglycerophosphoglycerophosphates [GP0505]
  1-Alkyl glycerophosphoglycerophosphates [GP0506]
  1 Z-Alkenylglycerophosphoglycerophosphates [GP0507]
Glycerophosphoinositols [GP06]
  Diacylglycerophosphoinositols [GP0601]
  1-Alkyl,2-acylglycerophosphoinositols [GP0602]
  1 Z-Alkenyl,2-acylglycerophosphoinositols [GP0603]
  Dialkylglycerophosphoinositols [GP0604]
  Monoacylglycerophosphoinositols [GP0605]
  1-Alkyl glycerophosphoinositols [GP0606]
  1 Z-Alkenylglycerophosphoinositols [GP0607]
Glycerophosphoinositol monophosphates [GP07]
  Diacylglycerophosphoinositol monophosphates [GP0701]
  1-Alkyl,2-acylglycerophosphoinositol monophosphates [GP0702]
  1 Z-Alkenyl,2-acylglycerophosphoinositol monophosphates [GP0703]
  Dialkylglycerophosphoinositol monophosphates [GP0704]
  Monoacylglycerophosphoinositol monophosphates [GP0705]
  1-Alkyl glycerophosphoinositol monophosphates [GP0706]
  1 Z-Alkenylglycerophosphoinositol monophosphates [GP0707]
Glycerophosphoinositol bisphosphates [GP08]
  Diacylglycerophosphoinositol bisphosphates [GP0801]
  1-Alkyl,2-acylglycerophosphoinositol bisphosphates [GP0802]
  1 Z-Alkenyl,2-acylglycerophosphoinositol bisphosphates [GP0803]
  Monoacylglycerophosphoinositol bisphosphates [GP0804]
  1-Alkyl glycerophosphoinositol bisphosphates [GP0805]
  1 Z-Alkenylglycerophosphoinositol bisphosphates [GP0806]
Glycerophosphoinositol trisphosphates [GP09]
  Diacylglycerophosphoinositol trisphosphates [GP0901]
  1-Alkyl,2-acylglycerophosphoinositol trisphosphates [GP0902]
  1 Z-Alkenyl,2-acylglycerophosphoinositol trisphosphates [GP0903]
  Monoacylglycerophosphoinositol trisphosphates [GP0904]
  1-Alkyl glycerophosphoinositol trisphosphates [GP0905]
  1 Z-Alkenylglycerophosphoinositol trisphosphates [GP0906]

TABLE 6: Glycerophospholipids [GP] Classes and Subclasses (Continued)

Glycerophosphates [GP10]
  Diacylglycerophosphates [GP1001]
  1-Alkyl,2-acylglycerophosphates [GP 1002]
  1 Z-Alkenyl,2-acylglycerophosphates [GP1003]
  Dialkylglycerophosphates [GP 1004]
  Monoacylglycerophosphates [GP1005]
  1-Alkyl glycerophosphates [GP1006]
  1 Z-Alkenylglycerophosphates [GP1007]
Glyceropyrophosphates [GP11]
  Diacylglyceropyrophosphates [GP 1101]
  Monoacylglyceropyrophosphates [GP1102]
Glycerophosphoglycerophosphoglycerols (cardiolipins) [GP12]
  Diacylglycerophosphoglycerophosphodiradylglycerols [GP1201]
  Diacylglycerophosphoglycerophosphomonoradylglycerols [GP1202]
  1-Alkyl,2-acylglycerophosphoglycerophosphodiradylglycerols [GP1203]
  1-Alkyl,2-acylglycerophosphoglycerophosphomonoradylglycerols [GP1204]
  1 Z-Alkenyl,2-acylglycerophosphoglycerophosphodiradylglycerols [GP1205]
  1 Z-Alkenyl,2-acylglycerophosphoglycerophosphomonoradylglycerols [GP1206]
  Monoacylglycerophosphoglycerophosphomonoradylglycerols [GP1207]
  1-Alkyl glycerophosphoglycerophosphodiradylglycerols [GP1208]
  1-Alkyl glycerophosphoglycerophosphomonoradylglycerols [GP1209]
  1 Z-Alkenylglycerophosphoglycerophosphodiradylglycerols [GP1210]
  1 Z-Alkenylglycerophosphoglycerophosphomonoradylglycerols [GP1211]
CDP-glycerols [GP13]
  CDP-diacylglycerols [GP1301]
  CDP-1-alkyl,2-acylglycerols [GP1302]
  CDP-1 Z-alkenyl,2-acylglycerols [GP1303]
  CDP-dialkylglycerols [GP1304]
  CDP-monoacylglycerols [GP1305]
  CDP-1-alkyl glycerols [GP1306]
  CDP-1 Z-alkenylglycerols [GP1307]
Glycerophosphoglucose lipids [GP14]
  Diacylglycerophosphoglucose lipids [GP1401]
  1-Alkyl,2-acylglycerophosphoglucose lipids [GP1402]
  1 Z-Alkenyl,2-acylglycerophosphoglucose lipids [GP1403]
  Monoacylglycerophosphoglucose lipids [GP1404]
  1-Alkyl glycerophosphoglucose lipids [GP1405]
  1 Z-Alkenylglycerophosphoglucose lipids [GP1406]
Glycerophosphoinositolglycans [GP15]
  Diacylglycerophosphoinositolglycans [GP 1501]
  1-Alkyl,2-acylglycerophosphoinositolglycans [GP1502]
  1 Z-Alkenyl,2-acylglycerophosphoinositolglycans [GP1503]
  Monoacylglycerophosphoinositolglycans [GP1504]
  1-Alkyl glycerophosphoinositolglycans [GP1505]
  1 Z-Alkenylglycerophosphoinositolglycans [GP1506]
Glycerophosphonocholines [GP16]
  Diacylglycerophosphonocholines [GP 1601]
  1-Alkyl,2-acylglycerophosphonocholines [GP1602]
  1 Z-Alkenyl,2-acylglycerophosphonocholines [GP1603]
  Dialkylglycerophosphonocholines [GP1604]
  Monoacylglycerophosphonocholines [GP1605]
  1-Alkyl glycerophosphonocholines [GP1606]
  1 Z-Alkenylglycerophosphonocholines [GP1607]
Glycerophosphonoethanolamines [GP 17]
  Diacylglycerophosphonoethanolamines [GP1701]
  1-Alkyl,2-acylglycerophosphonoethanolamines [GP1702]
  1 Z-Alkenyl,2-acylglycerophosphonoethanolamines [GP1703]
  Dialkylglycerophosphonoethanolamines [GP 1704]
  Monoacylglycerophosphonoethanolamines [GP1705]
  1-Alkyl glycerophosphonoethanolamines [GP1706]
  1 Z-Alkenylglycerophosphonoethanolamines [GP1707]
Di-glycerol tetraether phospholipids (caldarchaeols) [GP18]
Glycerol-nonitol tetraether phospholipids [GP19]
Oxidized glycerophospholipids [GP20]
Other [GP00]

the glycerol backbone. Although the glycerol backbone is symmetrical, the second carbon becomes a chiral center when the sn-1 and sn-3 carbons have different substituents. A large number of trivial names are associated with phospholipids. In the systematic nomenclature, mono/di-radylglycerophospholipids with different acyl or alkyl substituents are designated by similar conventions for naming of classes (see below) and are grouped according to the common polar moieties (i.e., head groups).

Typically, one or both of these hydroxyl groups are acylated with long-chain fatty acids, but there are also alkyl-linked and 1 Z-alkenyl-linked (plasmalogen) glycerophospholipids, as well as dialkylether variants in prokaryotes. The main biosynthetic pathways for the formation of GPCho and GPEtn (see Table 3 for shorthand notation) were elucidated through the efforts of Kennedy and co-workers (22) in the 1950s and 1960s, and more detailed interconversion pathways to form additional classes of phospholipids were described more recently. In addition to serving as a primary component of cellular membranes and binding sites for intracellular and intercellular proteins, some glycerophospholipids in eukaryotic cells are either precursors of, or are themselves, membrane-derived second messengers. A separate class, called oxidized glycerophospholipids, is composed of molecules in which one or more of the side chains have been oxidized. Several overviews are available on the classification, nomenclature, metabolism, and profiling of glycerophospholipids (18, 23–26). Structures from this category are shown in **Figure 4.**

Sphingolipids [SP]

Sphingolipids are a complex family of compounds that share a common structural feature, a sphingoid base backbone that is synthesized de novo from serine and a long-chain fatty acyl-CoA, then converted into ceramides, phosphosphingolipids, glycosphingolipids, and other species, including protein adducts (27, 28). A number of organisms also produce sphingoid base analogs that have many of the same features as sphingolipids (such as long-chain alkyl and vicinal amino and hydroxyl groups) but differ in other features. These have been included in this category because some are known to function as inhibitors or antagonists of sphingolipids, and in some organisms, these types of compounds may serve as surrogates for sphingolipids.

Sphingolipids can be divided into several major classes (**Table 7**): the sphingoid bases and their simple derivatives (such as the 1-phosphate), the sphingoid bases with an amide-linked fatty acid (e.g., ceramides), and more complex sphingolipids with head groups that are attached via phosphodiester linkages (the phosphosphingolipids), via glycosidic bonds (the simple and complex glycosphingolipids such as cerebrosides and gangliosides), and other groups (such as phosphono- and arseno-sphingolipids). The IUPAC has recommended a systematic nomenclature for sphingolipids (3).

The major sphingoid base of mammals is commonly referred to as "sphingosine," because that name was affixed by the first scientist to isolate this compound (29). Sphingosine is (2S,3R,4E)-2-aminooctadec-4-ene-1,3-diol (it is also called D-erythro-sphingosine and sphing-4-enine). This is only one of many sphingoid bases found in nature, which vary in alkyl chain length and branching, the number and positions of double bonds, the presence of additional hydroxyl groups, and other features. The structural variation has functional significance; for example, sphingoid bases in the dermis have additional hydroxyls at position 4 (phytoceramides) and/or 6 that can interact with neighboring molecules, thereby strengthening the permeability barrier of skin.

Sphingoid bases are found in a variety of derivatives, including the 1-phosphates, lyso-sphingolipids (such as sphingosine 1-phosphocholine as well as sphingosine 1-glycosides), and N-methyl derivatives (N-methyl, N,N-dimethyl, and N,N,N-trimethyl). In addition, a large number of organisms, such as fungi and sponges, produce compounds with structural similarity to sphingoid bases, some of which (such as myriocin and the fumonisins) are potent inhibitors of enzymes of sphingolipid metabolism.

Ceramides (N-acyl-sphingoid bases) are a major subclass of sphingoid base derivatives with an amide-linked fatty acid. The fatty acids are typically saturated or monounsaturated with chain lengths from 14 to 26 carbon atoms; the presence of a hydroxyl group on carbon 2 is fairly common. Ceramides sometimes have specialized fatty acids, as illustrated by the skin ceramide in **Figure 5i**, which has a 30 carbon fatty acid with a hydroxyl group on the terminal (ω) carbon. Ceramides are generally precursors of more complex sphingolipids. The major phosphosphingolipids of mammals are sphingomyelins (ceramide phosphocholines), whereas insects contain mainly ceramide phosphoethanolamines and fungi have phytoceramidephosphoinositols and mannose-containing head groups.

Glycosphingolipids (4) are classified on the basis of carbohydrate composition: 1) neutral glycosphingolipids contain one or more uncharged sugars such as glucose (Glu), galactose (Gal), N-acetylglucosamine (GlcNAc), N-acetylgalactosamine (GalNAc), and fucose (Fuc), which are grouped into families based on the nature of the glyco-substituents as shown in the listing; 2) acidic glycosphingolipids contain ionized functional groups (phosphate or sulfate) attached to neutral sugars or charged sugar residues such as sialic acid (N-acetyl or N-glycoloyl neuraminic acid). The latter are called gangliosides, and the number of sialic acid residues is usually denoted with a subscript letter (i.e., mono-, di- or tri-) plus a number reflecting the subspecies within that category; 3) basic glycosphingolipids; 4) amphoteric glycosphingolipids. For a few glycosphingolipids, historically assigned names as antigens and blood group structures are still in common use (e.g., Lewis x and sialyl Lewis x). Some aquatic organisms contain sphingolipids in which the phosphate is replaced by a phosphono or arsenate group. The other category includes sphingolipids that are covalently attached to proteins; for example, ω-hydroxyceramides and ω-glucosylceramides are attached to surface proteins of skin, and inositol-phosphoceramides are used as membrane anchors for some fungal proteins in a manner analogous to the glycosylphosphatidylinositol anchors that are attached to proteins in other eukaryotes. Some examples of sphingolipid structures are shown in Figure 5.

Sterol lipids [ST]

The sterol category is subdivided primarily on the basis of biological function. The sterols, of which cholesterol and its derivatives are the most widely studied in mammalian systems, constitute an important component of membrane lipids, along with the glycerophospholipids and sphingomyelins (30). There are many examples of unique sterols from plant, fungal, and marine sources that are designated as distinct subclasses in this schema (**Table 8**). The steroids, which also contain the same fused four ring core structure, have different biological roles as hormones and signaling molecules (31). These are subdivided on the basis of the number of carbons in the core skeleton. The C_{18} steroids include the estrogen family, whereas the C_{19} steroids comprise the androgens such as testosterone and androsterone.

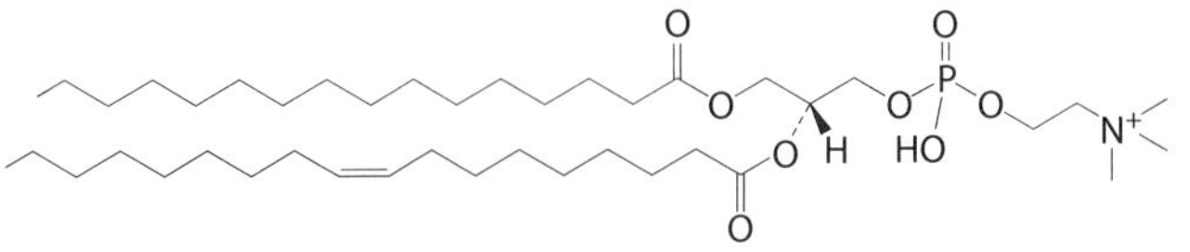

FIGURE 4 Representative structures for glycerophospholipids.

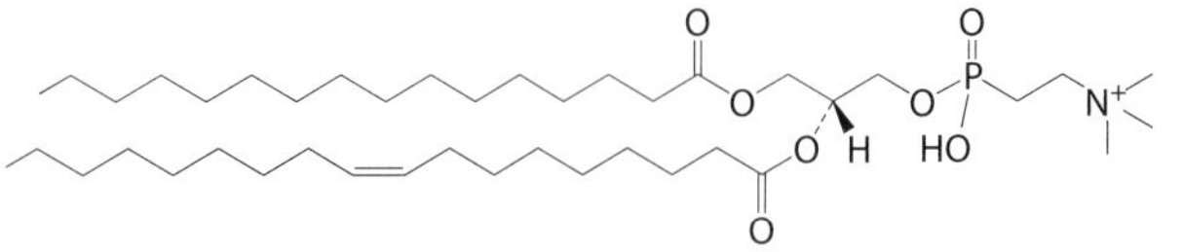

The C_{21} subclass, containing a two carbon side chain at the C_{17} position, includes the progestogens as well as the glucocorticoids and mineralocorticoids. The secosteroids, comprising various forms of vitamin D, are characterized by cleavage of the B ring of the core structure, hence the "seco" prefix (32). Additional classes within the sterols category are the bile acids (33), which in mammals are primarily derivatives of cholan-24-oic acid synthesized from cholesterol in the liver and their conjugates (sulfuric acid, taurine, glycine, glucuronic acid, and others). Sterol lipid structures are shown in **Figure 6.**

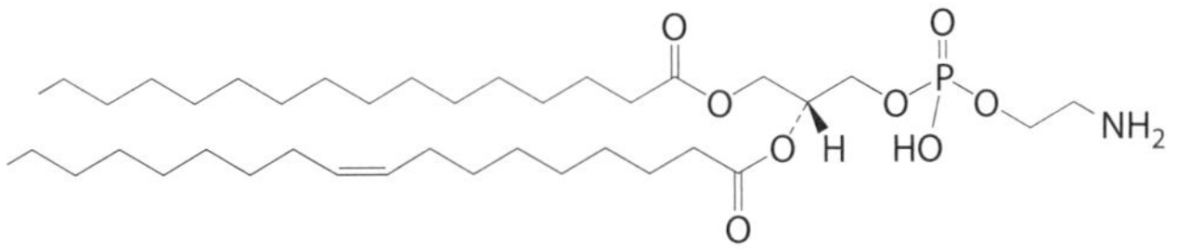

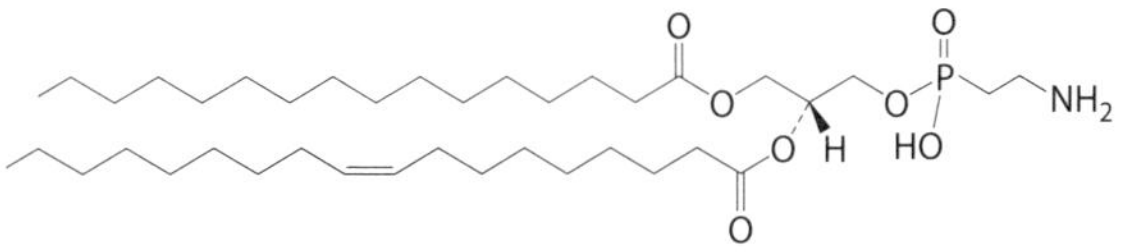

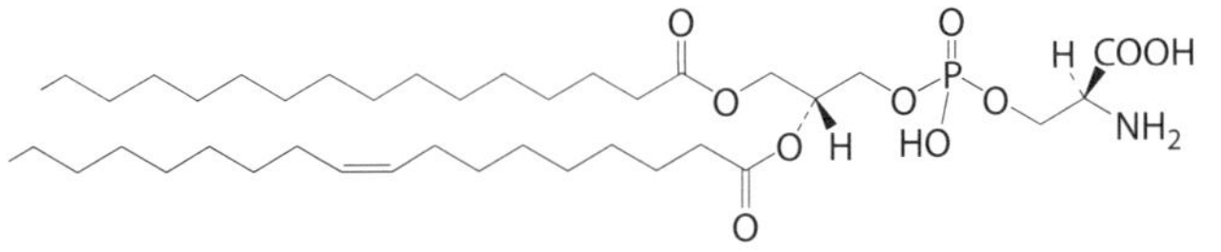

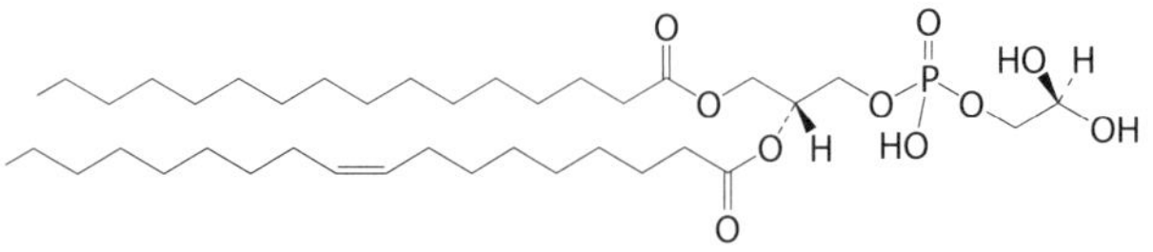

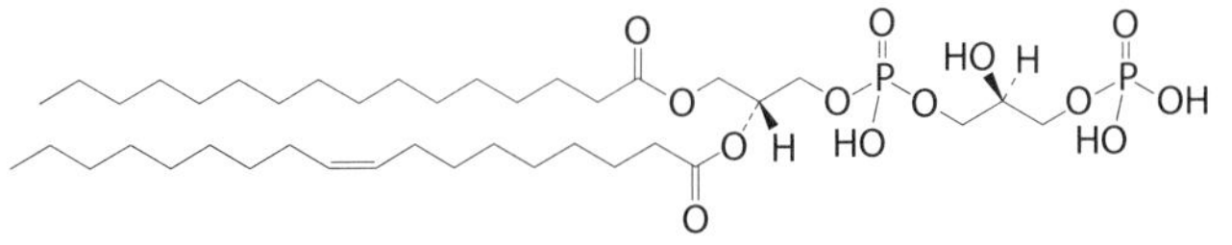

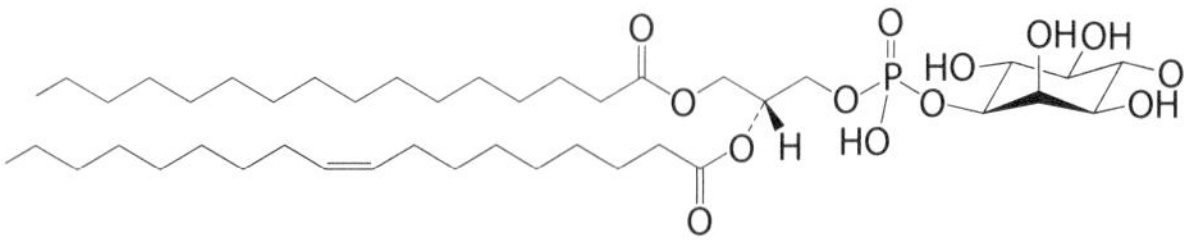

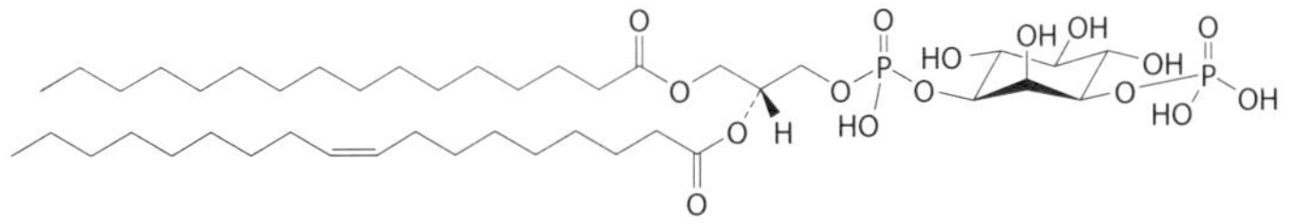

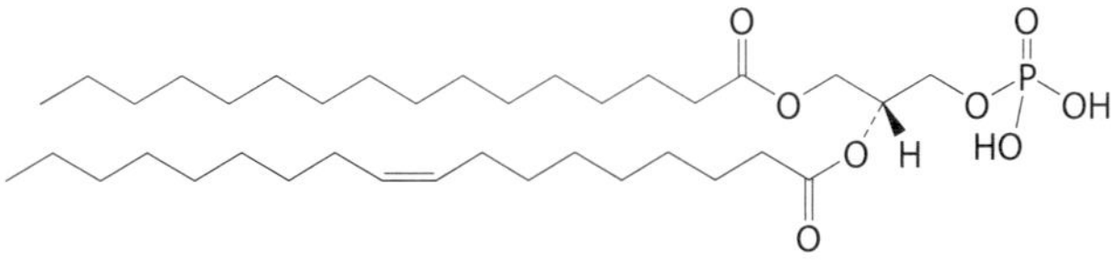

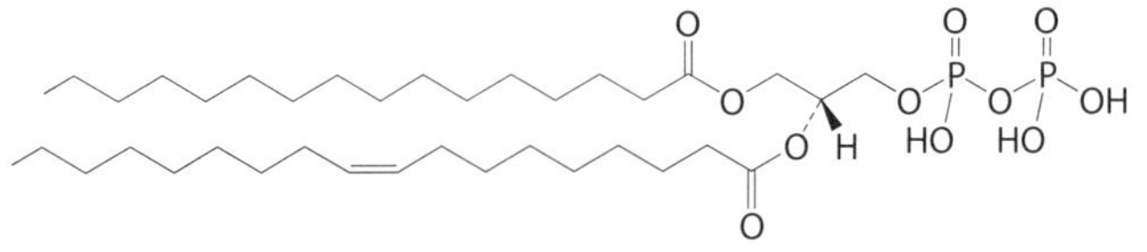

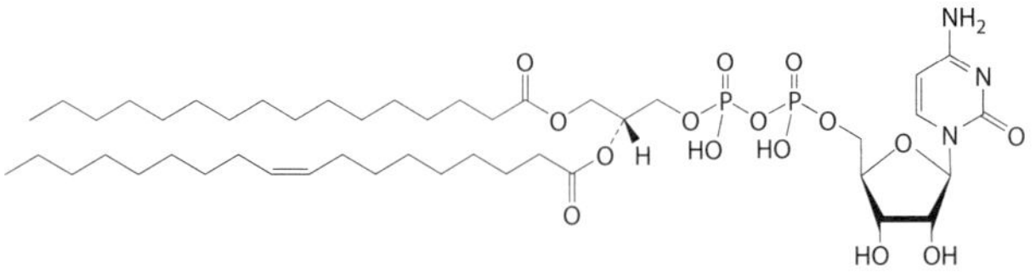

Prenol lipids [PR]

Prenols are synthesized from the five carbon precursors isopentenyl diphosphate and dimethylallyl diphosphate that are produced mainly via the mevalonic acid pathway (34). In some bacteria (e.g., *Escherichia coli*) and plants, isoprenoid precursors are made by the methylerythritol phosphate pathway (35). Because the simple isoprenoids (linear alcohols, diphosphates, etc.) are formed by the successive addition of C_5 units, it is convenient to classify them in this manner (**Table 9**), with a polyterpene subclass for those structures containing more than 40

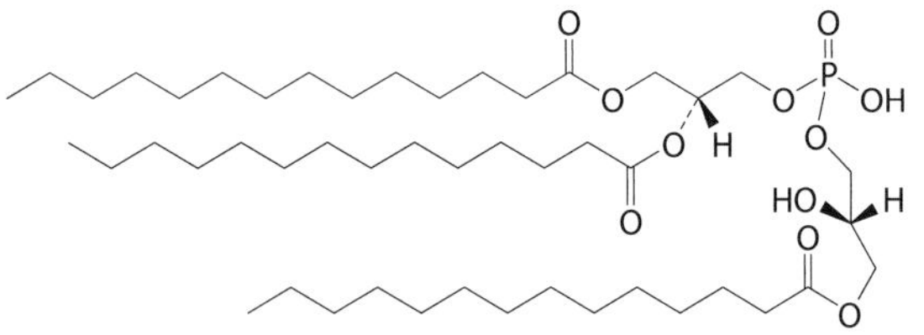

(m) Diacylglycerophosphomonoradylglycerols:
1,2-ditetradecanoyl-*sn*-glycero-3-phospho-
(3'-tetradecanoyl-1'-*sn*-glycerol)

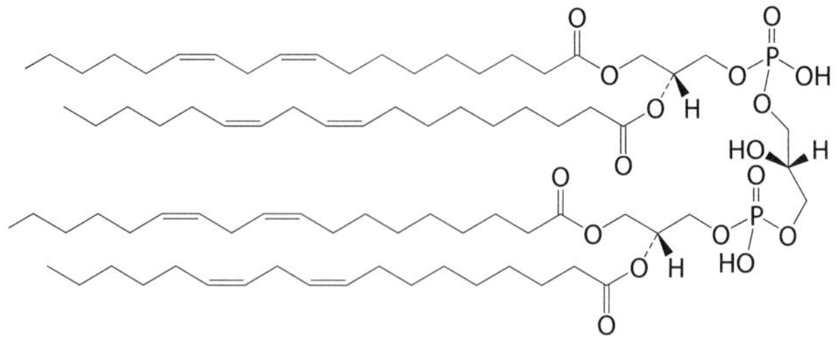

(n) Diacylglycerophosphoglycerophosphodiradylglycerols:
1',3'-Bis[1,2-Di-(9Z,12Z-octadecadienoyl)-
sn-glycero-3-phospho]-*sn*-glycerol

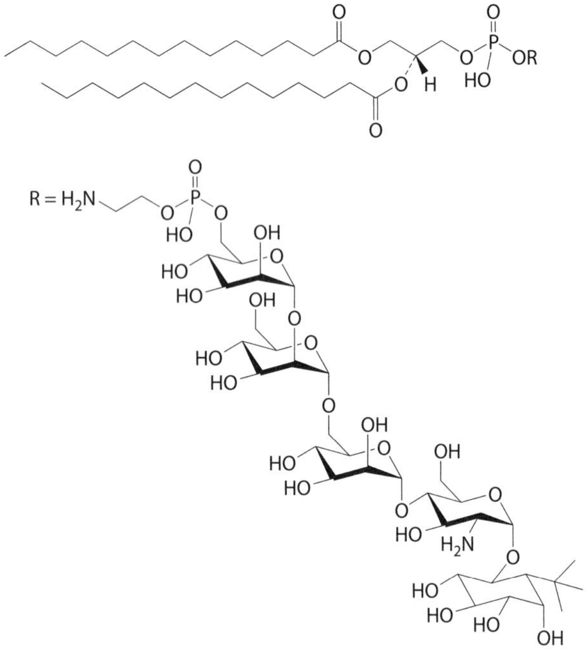

R = H₂N...

(o) Diacylglycerophosphoinositolglycans:
EtN-P-6Manα1-2Manα1-6Manα1-4GlcNα1-6GPIns(14:0/14:0)

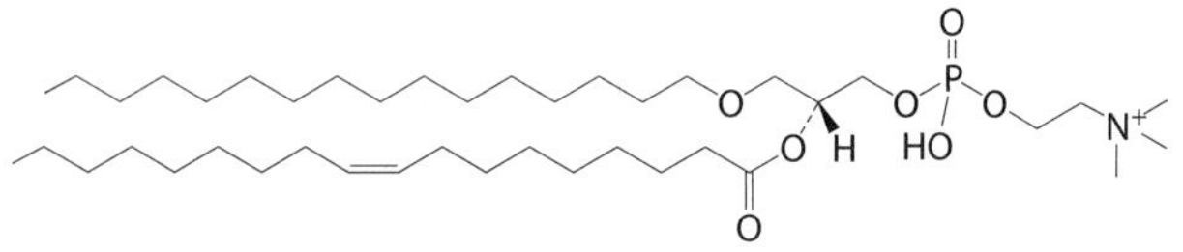

(p) 1-alkyl, 2-acylglycerophosphocholines:
1-O-hexadecyl-2-(9Z-octadecenoyl)-*sn*-glycero-3-phosphocholine

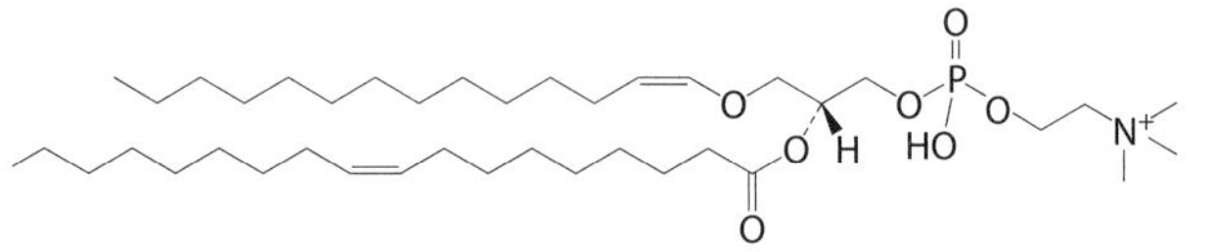

(q) 1Z-alkenyl, 2-acylglycerophosphocholines:
1-O-(1Z-tetradecenyl)-2-(9Z-octadecenoyl
-*sn*-glycero-3-phosphocholine

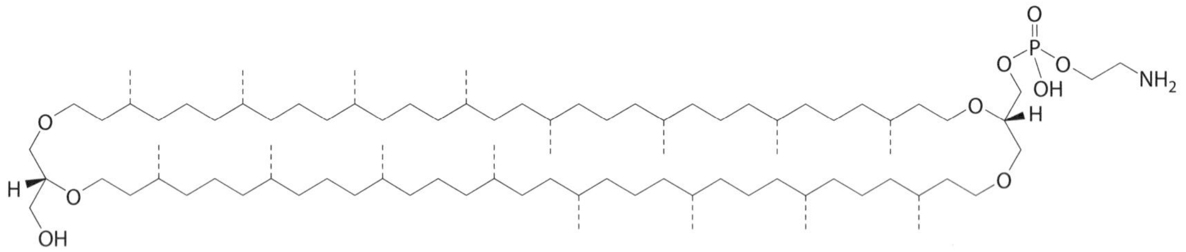

(r) Di-glycerol tetraether phospholipids (caldarchaeols): *sn*-caldarchaeo-1-phosphoethanolamine

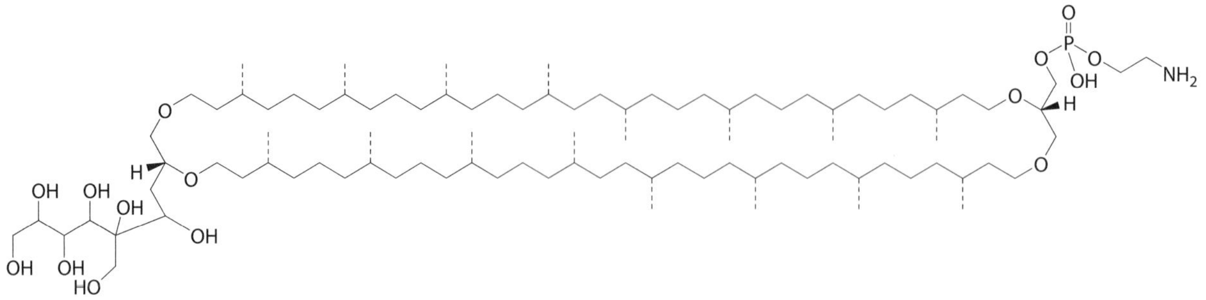

(s) Glycerol-nonitol tetraether phospholipids: *sn*-caldito-1-phosphoethanolamine

FIGURE 4 Representative structures for glycerophospholipids (Continued).

TABLE 7: Sphingolipids [SP] Classes and Subclasses

Sphingoid bases [SP01]
  Sphing-4-enines (sphingosines) [SP0101]
  Sphinganines [SP0102]
  4-Hydroxysphinganines (phytosphingosines) [SP0103]
  Sphingoid base homologs and variants [SP0104]
  Sphingoid base 1-phosphates [SP0105]
  Lysosphingomyelins and lysoglycosphingolipids [SP0106]
  N-Methylated sphingoid bases [SP0107]
  Sphingoid base analogs [SP0108]
Ceramides [SP02]
  N-Acylsphingosines (ceramides) [SP0201]
  N-Acylsphinganines (dihydroceramides) [SP0202]
  N-Acyl-4-hydroxysphinganines (phytoceramides) [SP0203]
  Acylceramides [SP0204]
  Ceramide 1-phosphates [SP0205]
Phosphosphingolipids [SP03]
  Ceramide phosphocholines (sphingomyelins) [SP0301]
  Ceramide phosphoethanolamines [SP0302]
  Ceramide phosphoinositols [SP0303]
Phosphonosphingolipids [SP04]
Neutral glycosphingolipids [SP05]
  Simple Glc series (GlcCer, LacCer, etc.) [SP0501]
  GalNAcβ1-3Galα1-4Galβ1-4Glc- (globo series) [SP0502]
  GalNAcβ1-4Galβ1-4Glc- (ganglio series) [SP0503]
  Galβ1-3GlcNAcβ1-3Galβ1-4Glc- (lacto series) [SP0504]
  Galβ1–4GlcNAcβ1-3Galβ1-4Glc- (neolacto series) [SP0505]
  GalNAcβ1-3Galα1-3Galβ1-4Glc- (isoglobo series) [SP0506]
  GlcNAcβ1-2Manα1-3Manβ1-4Glc- (mollu series) [SP0507]
  GalNAcβ1-4GlcNAcβ1-3Manβ1-4Glc- (arthro series) [SP0508]
  Gal- (gala series) [SP0509]
  Other [SP0510]
Acidic glycosphingolipids [SP06]
  Gangliosides [SP0601]
  Sulfoglycosphingolipids (sulfatides) [SP0602]
  Glucuronosphingolipids [SP0603]
  Phosphoglycosphingolipids [SP0604]
  Other [SP0600]
Basic glycosphingolipids [SP07]
Amphoteric glycosphingolipids [SP08]
Arsenosphingolipids [SP09]
Other [SP00]

carbons (i.e., >8 isoprenoid units) (36). Note that vitamin A and its derivatives and phytanic acid and its oxidation product pristanic acid are grouped under C_{20} isoprenoids. Carotenoids are important simple isoprenoids that function as antioxidants and as precursors of vitamin A (37). Another biologically important class of molecules is exemplified by the quinones and hydroquinones, which contain an isoprenoid tail attached to a quinonoid core of nonisoprenoid origin. Vitamins E and K (38, 39) as well as the ubiquinones (40) are examples of this class.

Polyprenols and their phosphorylated derivatives play important roles in the transport of oligosaccharides across membranes. Polyprenol phosphate sugars and polyprenol diphosphate sugars function in extracytoplasmic glycosylation reactions (41), in extracellular polysaccharide biosynthesis [for instance, peptidoglycan polymerization in bacteria (42)], and in eukaryotic protein *N*-glycosylation (43, 44). The biosynthesis and function of polyprenol phosphate sugars differ significantly from those of the polyprenol diphosphate sugars; therefore, we have placed them in separate subclasses. Bacteria synthesize polyprenols (called bactoprenols) in which the terminal isoprenoid unit attached to oxygen remains unsaturated, whereas in animal polyprenols (dolichols) the terminal isoprenoid is reduced. Bacterial polyprenols are typically 10 to 12 units long (40), whereas dolichols

usually consist of 18 to 22 isoprene units. In the phytoprenols of plants, the three distal units are reduced. Several examples of prenol lipid structures are shown in **Figure 7**.

Saccharolipids [SL]

We have avoided the term "glycolipid" in the classification scheme to maintain a focus on lipid structures. In fact, all eight lipid categories in the present scheme include important glycan derivatives, making the term glycolipid incompatible with the overall goal of lipid categorization. We have, in addition, coined the term "saccharolipids" to describe compounds in which fatty acids are linked directly to a sugar backbone, forming structures that are compatible with membrane bilayers. In the saccharolipids (**Table 10**), a sugar substitutes for the glycerol backbone that is present in glycerolipids and glycerophospholipids. Saccharolipids can occur as glycan or as phosphorylated derivatives. The most familiar saccharolipids are the acylated glucosamine precursors of the lipid A component of the lipopolysaccharides in Gram-negative bacteria (41). Typical lipid A molecules are disaccharides of glucosamine, which are derivatized with as many as seven fatty acyl chains (41, 45). Note that in naming these compounds, the total number of fatty acyl groups are counted regardless of the nature of the linkage (i.e., amide or ester). The minimal lipopolysaccharide required for growth in *E. coli* is a hexa-acylated lipid A that is glycosylated with two 3-deoxy-D-mannooctulosonic acid residues (see below). In some bacteria, the glucosamine backbone of lipid A is replaced by 2,3-diamino-2,3-dideoxyglucose (46); therefore, the class has been designated "Acylaminosugars." Included also in this class are the Nod factors of nitrogen-fixing bacteria (47), such as *Sinorhizobium meliloti*. The Nod factors are oligosaccharides of glucosamine that are usually derivatized with a single fatty acyl chain. Additional saccharolipids include fatty acylated derivatives of glucose, which are best exemplified by the acylated trehalose units of certain mycobacterial lipids (11). Acylated forms of glucose and sucrose also have been reported in plants (48). Some saccharolipid structures are shown in **Figure 8**.

Polyketides [PK]

Polyketides are synthesized by classic enzymes as well as iterative and multimodular enzymes with semiautonomous active sites that share mechanistic features with the fatty acid synthases, including the involvement of specialized acyl carrier proteins (49, 50); however, polyketide synthases generate a much greater diversity of natural product structures, many of which have the character of lipids. The class I polyketide synthases form constrained macrocyclic lactones, typically ranging in size from 14 to 40 atoms, whereas class II and III polyketide synthases generate complex aromatic ring systems (**Table 11**). Polyketide backbones are often further modified by glycosylation, methylation, hydroxylation, oxidation, and/or other processes. Some polyketides are linked with nonribosomally synthesized peptides to form hybrid scaffolds. Examples of the three polyketide classes are shown in **Figure 9**. Many commonly used antimicrobial, antiparasitic, and anticancer agents are polyketides or polyketide derivatives. Important examples of these drugs include erythromycins, tetracylines, nystatins, avermectins, and antitumor epothilones. Other polyketides are potent toxins. The possibility of recombining and reengineering the enzymatic modules that assemble polyketides has recently stimulated the search for novel "unnatural" natural products, especially in the antibiotic arena (51, 52).

We consider this minimal classification of polyketides as the first step in a more elaborate scheme. It will be important ultimately to include as many polyketide structures as possible in a

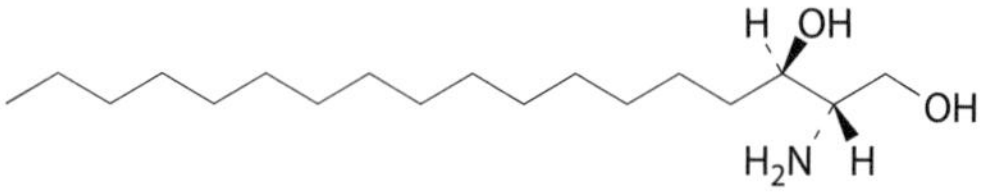

(a) Sphinganines: sphinganine

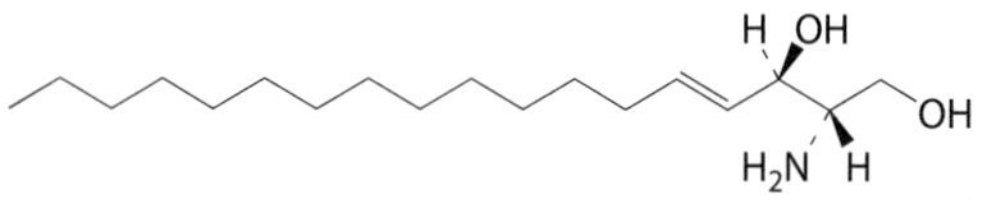

(b) Sphingosines: sphing-4-enine

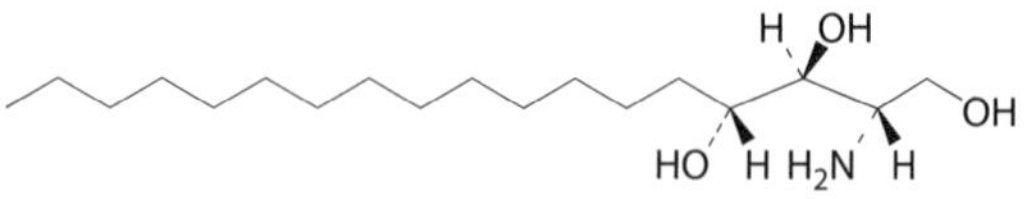

(c) Phytosphingosines: 4-hydroxysphinganine

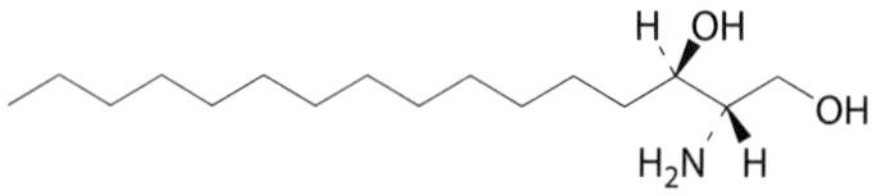

(d) Sphingoid base homologs and variants: hexadecasphinganine

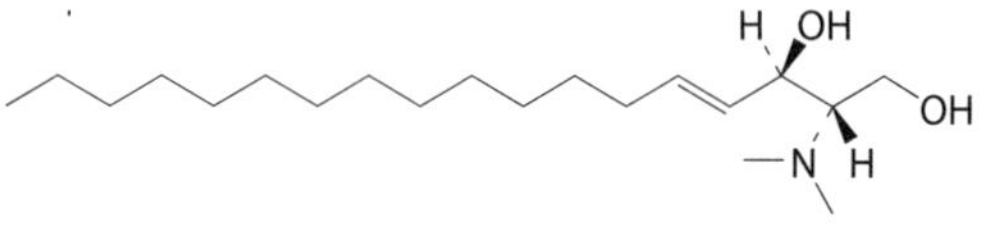

(e) N-methylated sphingoid bases: N,N-dimethylsphing-4-enine

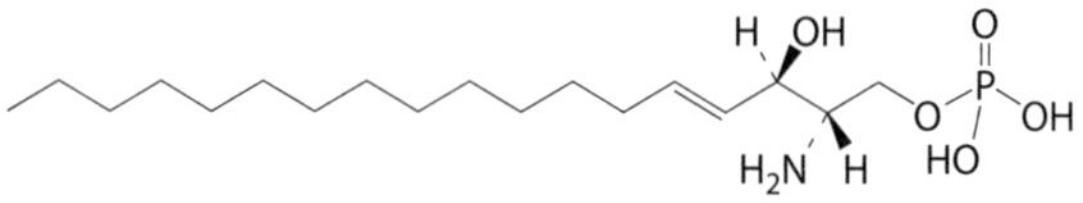

(f) Sphingoid base 1-phosphates: sphing-4-enine-1-phosphate

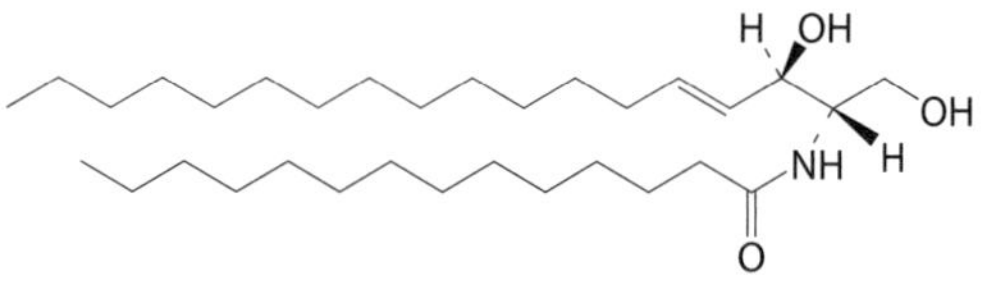

(g) N-acylsphingosines (ceramides): N-(tetradecanoyl)-sphing-4-enine

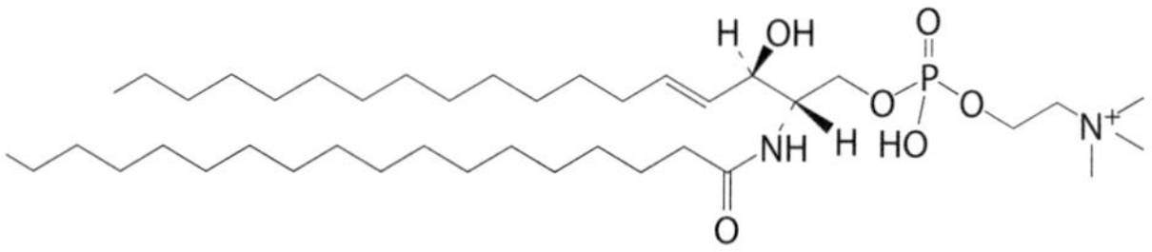

(h) Ceramide phosphocholines (sphingomyelins): N-(octadecanoyl)-sphing-4-enine-1-phosphocholine

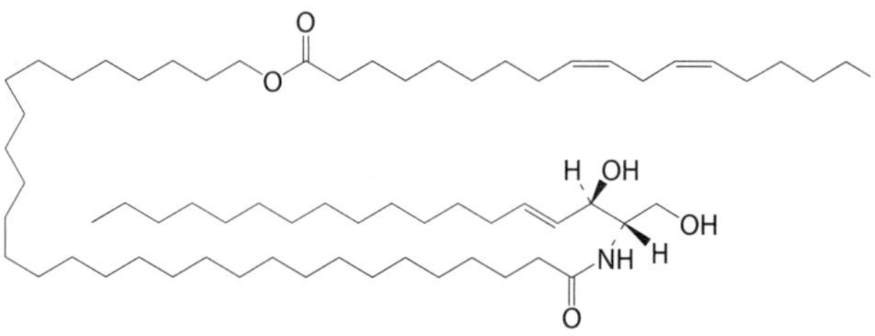

(i) Acylceramides: N-(30-(9Z,12Z-octadecadienoyloxy)-tricontanoyl)-sphing-4-enine

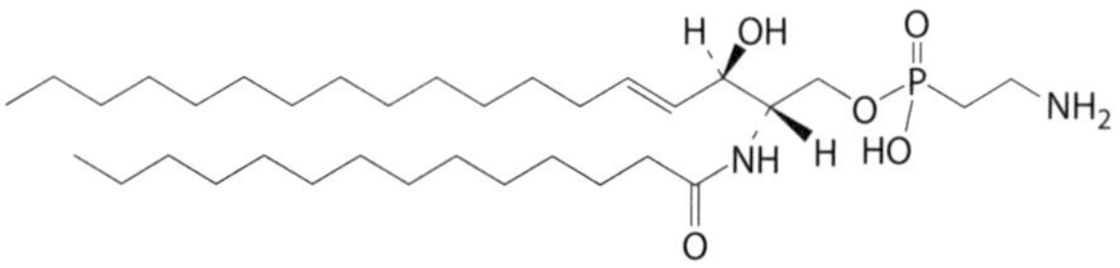

(j) Phosphonosphingolipids: N-(tetradecanoyl)-sphing-4-enine-1-(2-aminoethylphosphonate)

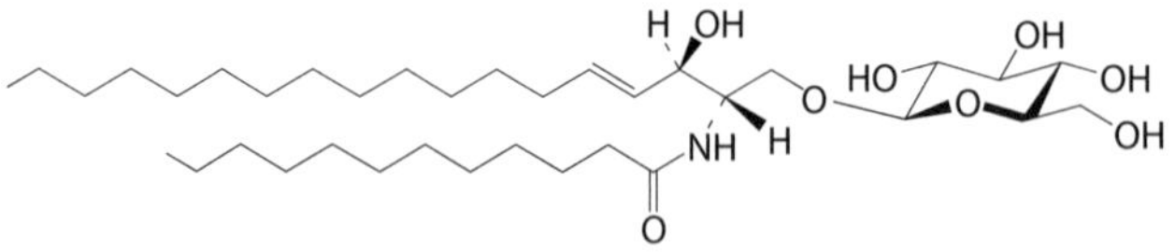

(k) Neutral Glycosphingolipids: Simple Glc series: Glcβ-Cer(d18:1/12:0)

FIGURE 5   Representative structures for sphingolipids.

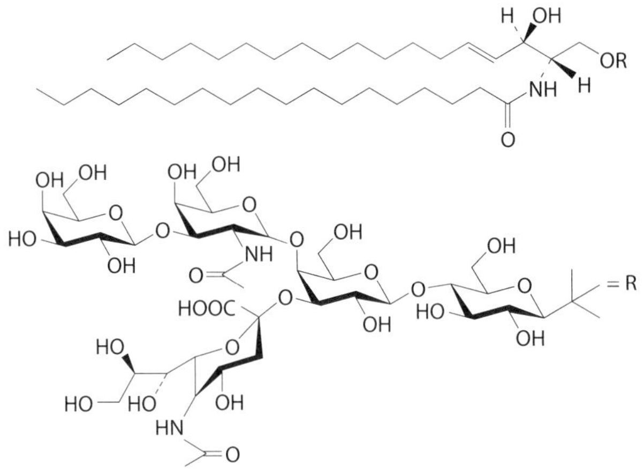

(l) Acidic Glycosphingolipids: Gangliosides:
Galβ1–3GalNAcβ1–4(NeuAcα2–3)Galβ1–4Glcβ-Cer(d18:1/18:0)

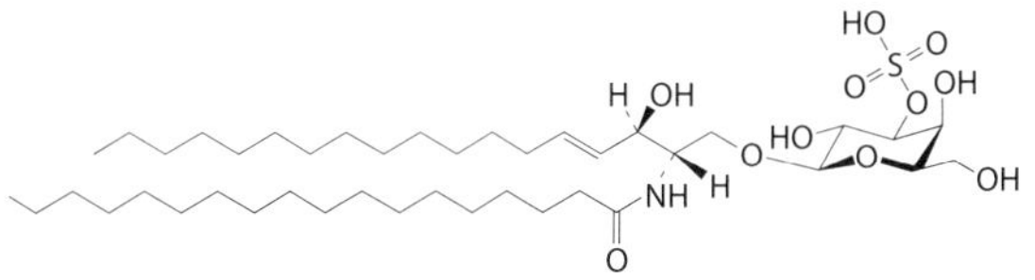

(m) Acidic Glycosphingolipids: Sulfosphingolipids:
(3'-sulfo)Galβ-Cer(d18:1/18:0)

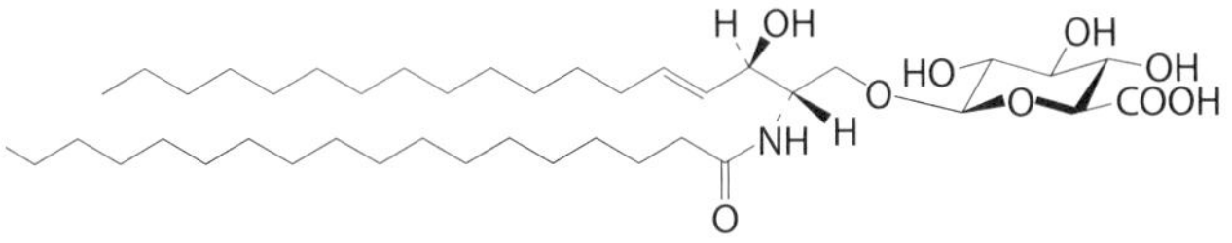

(n) Acidic Glycosphingolipids: Glucuronosphingolipids:
GlcUAβ-Cer(d18:1/18:0)

FIGURE 5 Representative structures for sphingolipids (Continued)

TABLE 8: Sterol Lipids [ST] Classes and Subclasses

Sterols [ST01]
Cholesterol and derivatives [ST0101]
  Cholesteryl esters [ST0102]
  Phytosterols and derivatives [ST0103]
  Marine sterols and derivatives [ST0104]
  Fungal sterols and derivatives [ST0105]
Steroids [ST02]
  C_{18} steroids (estrogens) and derivatives [ST0201]
  C_{19} steroids (androgens) and derivatives [ST0202]
  C_{21} steroids (gluco/mineralocorticoids, progestogins) and derivatives [ST0203]
Secosteroids [ST03]
  Vitamin D_2 and derivatives [ST0301]
  Vitamin D_3 and derivatives [ST0302]
Bile acids and derivatives [ST04]
  C_{24} bile acids, alcohols, and derivatives [ST0401]
  C_{26} bile acids, alcohols, and derivatives [ST0402]
  C_{27} bile acids, alcohols, and derivatives [ST0403]
  C_{28} bile acids, alcohols, and derivatives [ST0404]
Steroid conjugates [ST05]
  Glucuronides [ST0501]
  Sulfates [ST0502]
  Glycine conjugates [ST0503]
  Taurine conjugates [ST0504]
Hopanoids [ST06]
Other [ST00]

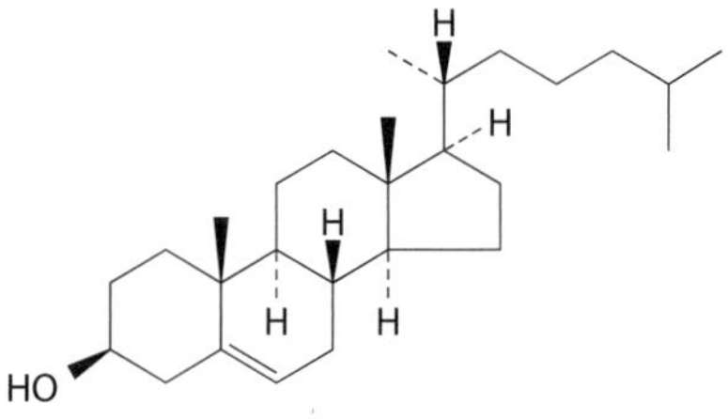

(a) Cholesterol and derivatives:
cholesterol; cholest-5-en-3β-ol

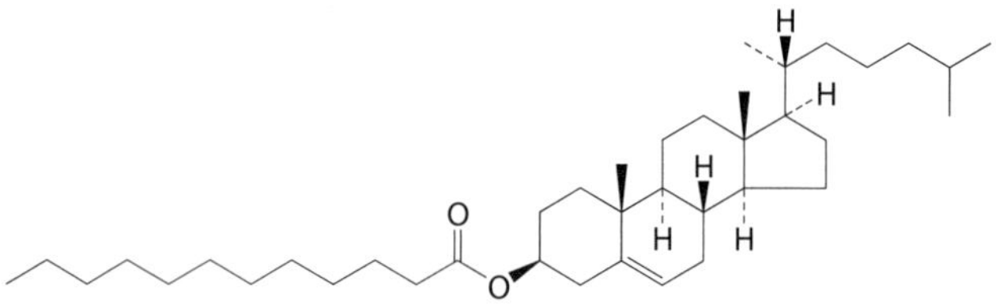

(b) Cholesteryl esters:
cholest-5-en-3β-yl dodecanoate

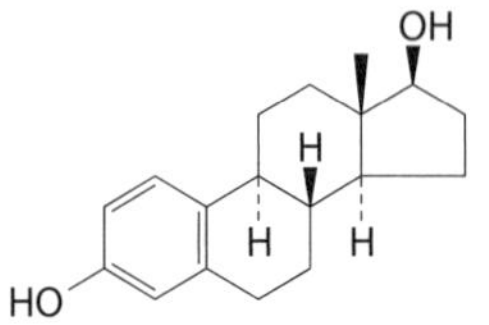

(c) C_{18} steroids (estrogens) and derivatives:
β-estradiol; 1,3,5[10]-estratriene-3,17β-diol

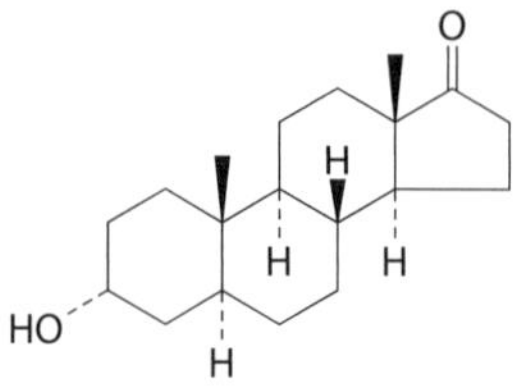

(d) C_{19} steroids (androgens) and derivatives:
androsterone; 3α-hydroxy-5α-androstan-17-one

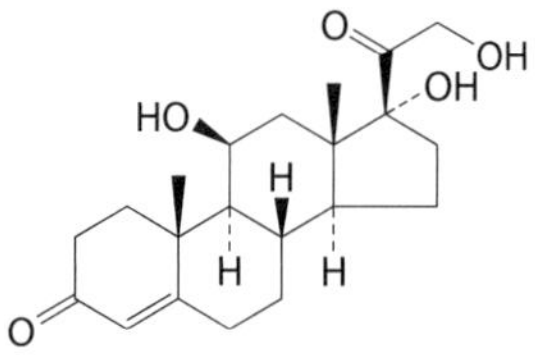

(e) C_{21} steroids and derivatives:
cortisol;11β,17α,21-trihydroxypregn-4-ene-3,20-dione

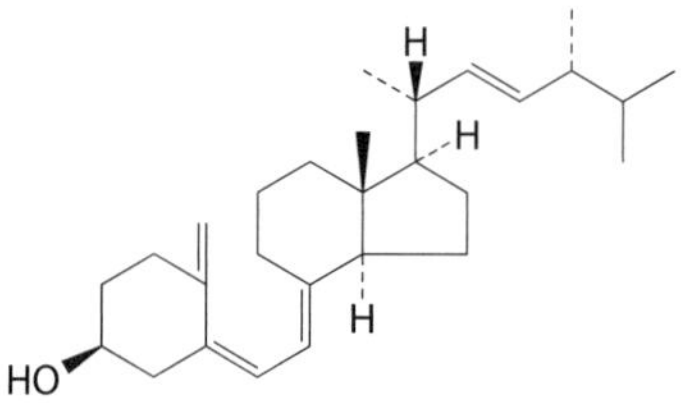

(f) Secosteroids: Vitamin D_2 and derivatives:
vitamin D_2; (5Z,7E,22E)-(3S)-9,10-seco-5,7,10(19),
22-ergostatetraen-3-ol

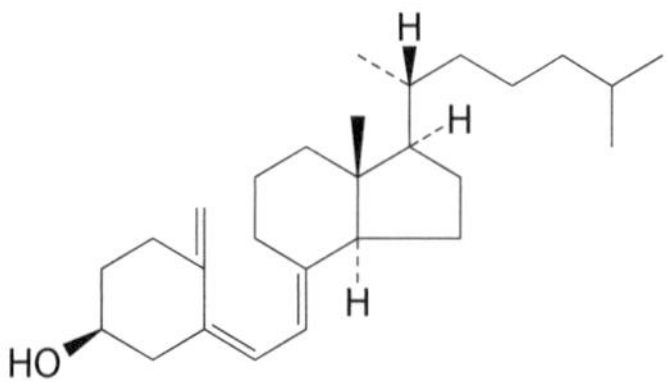

(g) Secosteroids: Vitamin D_3 and derivatives: vitamin D_3;
(5Z,7E)-(3S)-9,10-seco-5,7,10(19)-cholestatrien-3-ol

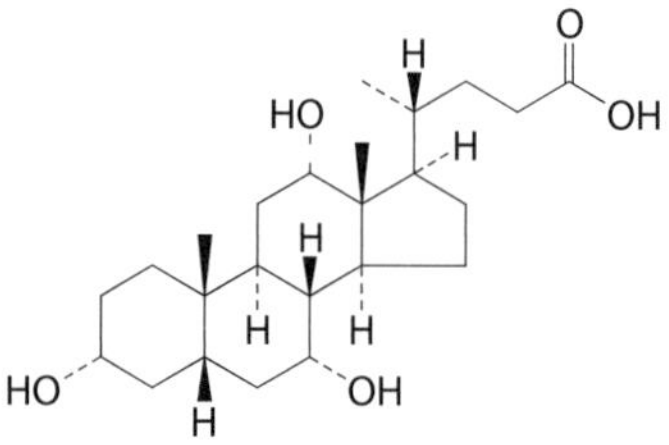

(h) C_{24} bile acids, alcohols, and derivatives:
cholic acid; 3α,7α,12α-trihydroxy-5β-cholan-24-oic acid

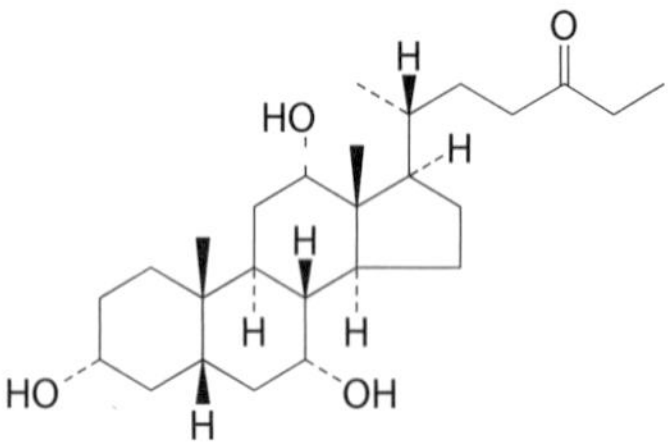

(i) C_{26} bile acids, alcohols, and derivatives:
3α,7α,12α-trihydroxy-27-nor-5β-cholestan-24-one

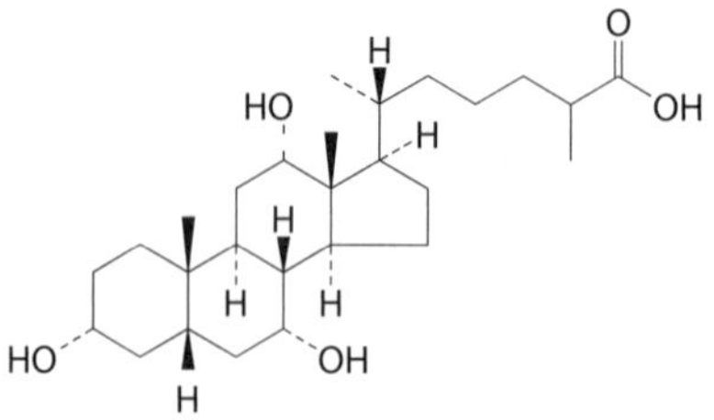

(j) C_{27} bile acids, alcohols, and derivatives:
3α,7α,12α-trihydroxy-5β-cholestan-26-oic acid

FIGURE 6   Representative structures for sterol lipids.

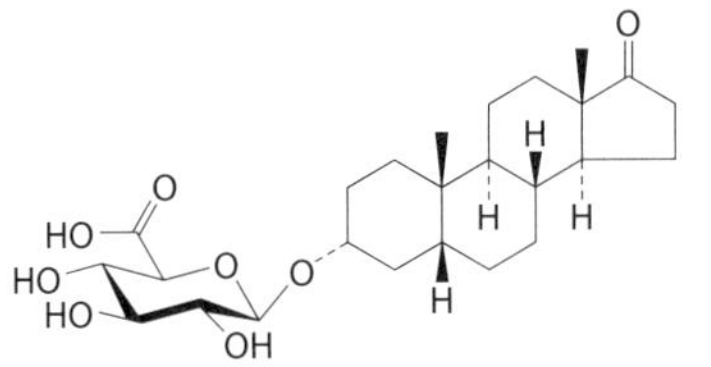

(k) Steroid conjugates: Glucuronides: 5α-androstane-3α-ol-17-one glucuronide

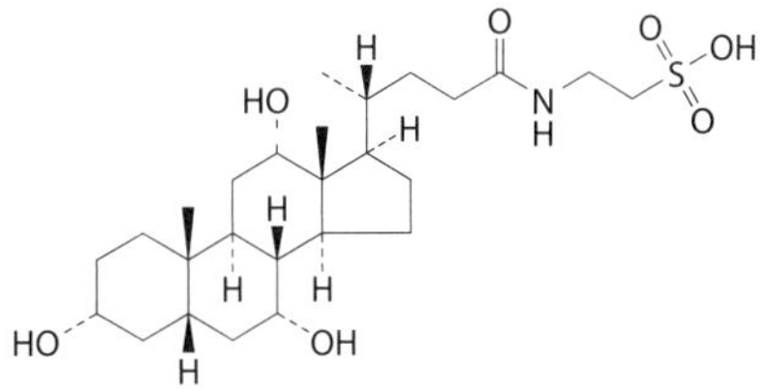

(l) Steroid conjugates: Taurine conjugates: taurocholic acid; N-(3α,7α,12α-trihydroxy-5β-cholan-24-oyl)-taurine

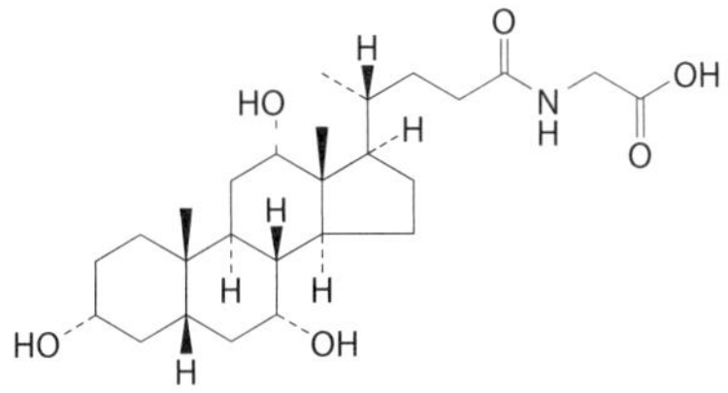

(m) Steroid conjugates: Glycine conjugates: glycocholic acid; N-(3α,7α,12α-trihydroxy-5β-cholan-24-oyl)-glycine

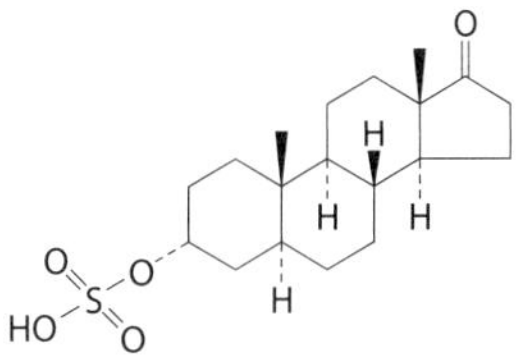

(n) Steroid conjugates: Sulfates: 5α-androstane-3α-ol-17-one sulfate

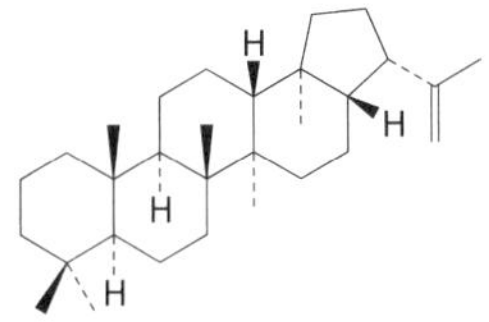

(o) Hopanoids: diploptene; hop-22(29)-ene

FIGURE 6 Representative structures for sterol lipids (Continued).

TABLE 9: Prenol Lipids [PR] Classes and Subclasses

Isoprenoids [PR01]
  C_5 isoprenoids [PR0101]
  C_{10} isoprenoids (monoterpenes) [PR0102]
  C_{15} isoprenoids (sesquiterpenes) [PR0103]
  C_{20} isoprenoids (diterpenes) [PR0104]
  C_{25} isoprenoids (sesterterpenes) [PR0105]
  C_{30} isoprenoids (triterpenes) [PR0106]
  C_{40} isoprenoids (tetraterpenes) [PR0107]
  Polyterpenes [PR0108]
Quinones and hydroquinones [PR02]
  Ubiquinones [PR0201]
  Vitamin E [PR0202]
  Vitamin K [PR0203]
Polyprenols [PR03]
  Bactoprenols [PR0301]
  Bactoprenol monophosphates [PR0302]
  Bactoprenol diphosphates [PR0303]
  Phytoprenols [PR0304]
  Phytoprenol monophosphates [PR0305]
  Phytoprenol diphosphates [PR0306]
  Dolichols [PR0307]
  Dolichol monophosphates [PR0308]
  Dolichol diphosphates [PR0309]
Other [PR00]

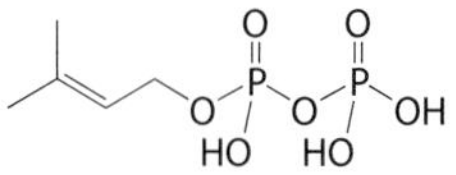

(a) C$_5$ isoprenoids: dimethylallyl pyrophosphate;
3-methylbut-2-enyl pyrophosphate

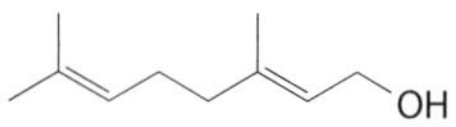

(b) C$_{10}$ isoprenoids; 2E-geraniol

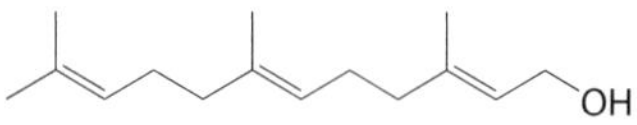

(c) C$_{15}$ isoprenoids; 2E,6E-farnesol

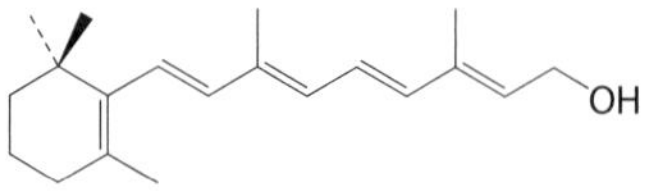

(d) C$_{20}$ isoprenoids; retinol: vitamin A

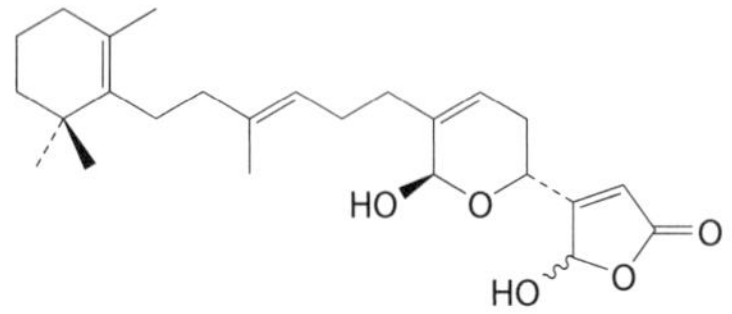

(e) C$_{25}$ isoprenoids: manoalide

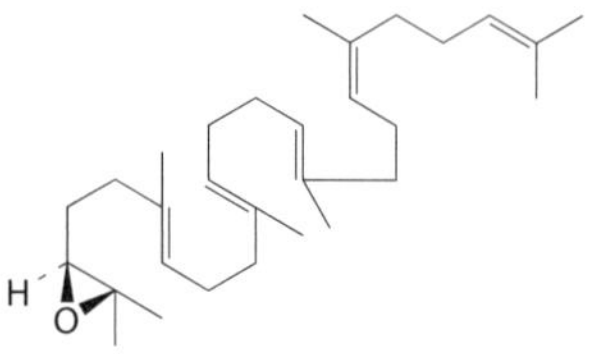

(f) C$_{30}$ isoprenoids: 3S-squalene-2,3-epoxide

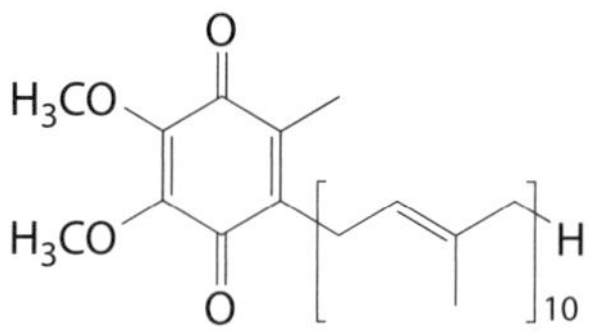

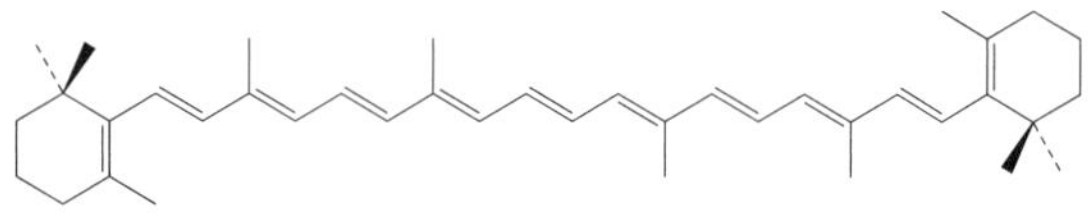

(g) C$_{40}$ isoprenoids: β-carotene

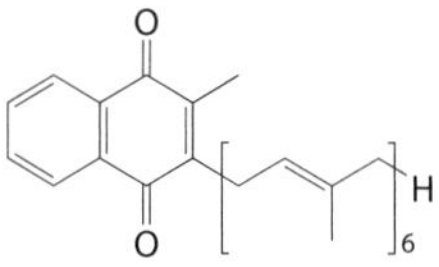

(h) Ubiquinones: ubiquinone-10 (Co-Q10);
2-methyl-3-decaprenyl-5,6-dimethoxy-1,4-benzoquinone

(i) vitamin K: vitamin K$_2$(30):
2-methyl, 3-hexaprenyl-1,4-naphthoquinone; menaquinone-6

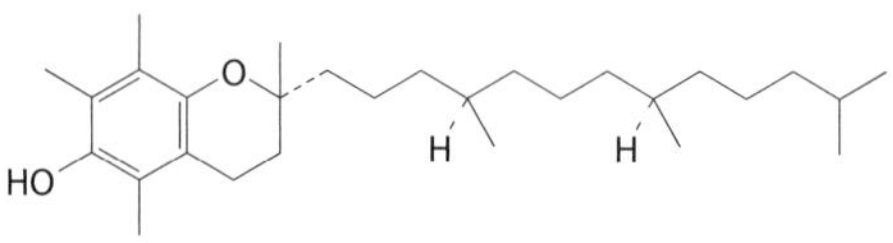

(j) vitamin E: (2R,4'R,8'R)-α-tocopherol

FIGURE 7   Representative structures for prenol lipids.

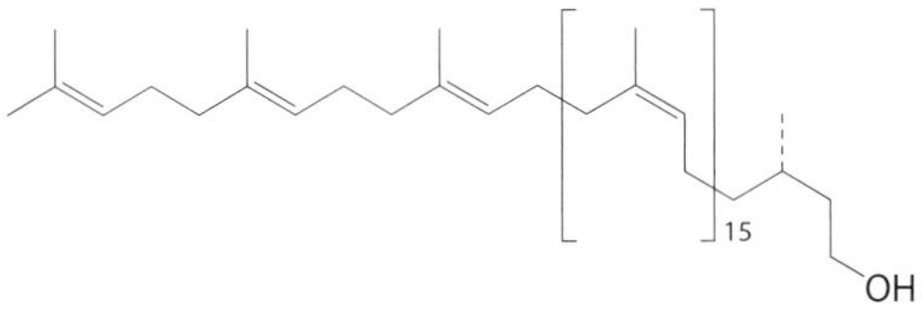

(k) Dolichols: Dol-19; α-dihydrononadecaprenol

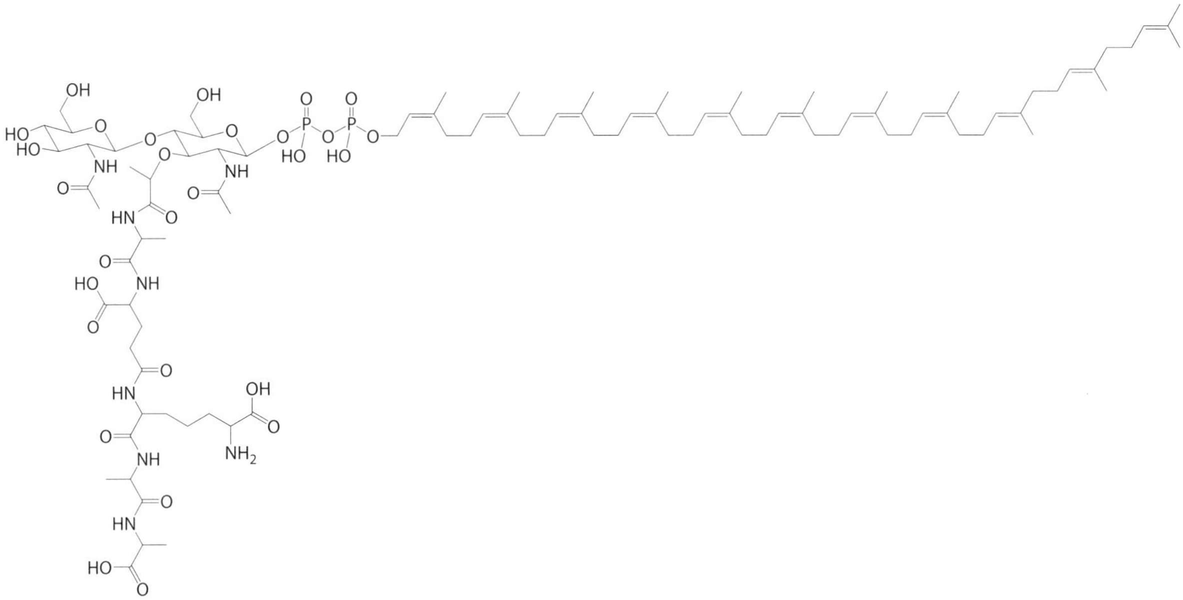

(l) Bactoprenol diphosphates: The Lipid II peptidoglycan precursor in *E. coli*; undecaprenyl diphosphate glycan

FIGURE 7 Representative structures for prenol lipids (Continued).

TABLE 10: Saccharolipids [SL] Classes and Subclasses

Acylaminosugars [SL01]
  Monoacylaminosugars [SL0101]
  Diacylaminosugars [SL0102]
  Triacylaminosugars [SL0103]
  Tetraacylaminosugars [SL0104]
  Pentaacylaminosugars [SL0105]
  Hexaacylaminosugars [SL0106]
  Heptaacylaminosugars [SL0107]
Acylaminosugar glycans [SL02]
Acyltrehaloses [SL03]
Acyltrehalose glycans [SL04]
Other [SL00]

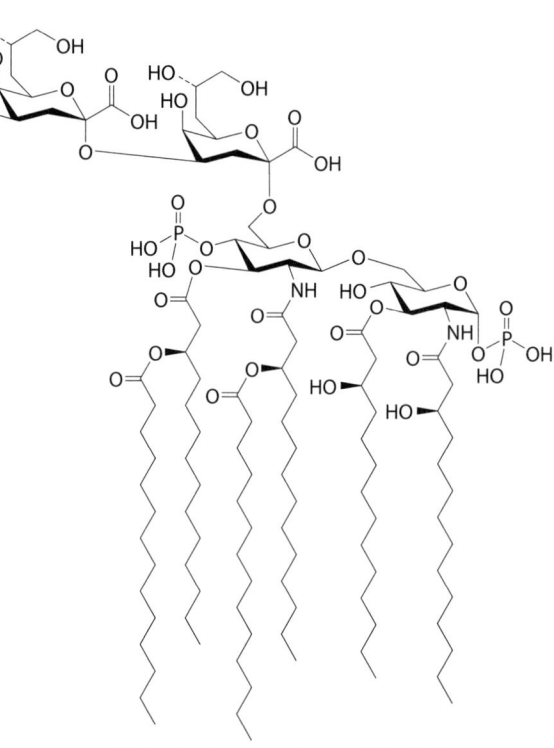

(a) Acylaminosugars: Monoacylaminosugars:
UDP-3-O-(3R-hydroxy-tetradecanoyl)-GlcN

(b) Acylaminosugars: Diacylaminosugars: lipid X

(c) Acylaminosugars: Tetraacylaminosugars: lipid IV$_A$

(d) Acylaminosugar glycans: Kdo$_2$ lipid A

FIGURE 8   Representative structures for saccharolipids.

TABLE 11: Polyketides [PK] Classes and Subclasses

Macrolide polyketides [PK01]
Aromatic polyketides [PK02]
Nonribosomal peptide/polyketide hybrids [PK03]
Other [PK00]

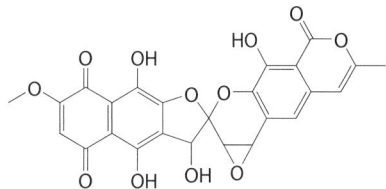

(a) Macrolide polyketides: 6-deoxyerythronolide B

(b) Aromatic polyketides: griseorhodin A

(c) Polyketide hybrids: epothilone D

FIGURE 9 Representative structures for polyketides.

lipid database that can be searched for substructure and chemical similarity.

Discussion

The goals of the LIPID MAPS initiative are to characterize known lipids and identify new ones, to quantitate temporal and spatial changes in lipids that occur with cellular metabolism, and to develop bioinformatics approaches that establish dynamic lipid networks; the goals of Lipid Bank (Japan) are to annotate and curate lipid structures and the literature associated with them; and the goals of the European Lipidomics Initiative are to coordinate and organize scientific interactions and workshops associated with lipid research. To coordinate the independent efforts from three continents and to facilitate collaborative work, a comprehensive classification of lipids with a common platform that is compatible with informatics requirements must be developed to deal with the massive amounts of data that will be generated by the lipid community. The proposed classification, nomenclature, and chemical representation system was initially designed to accommodate the massive data that will result from the LIPID MAPS effort, but it has been expanded to accommodate as many lipids as possible. We also have attempted to make the system compatible with existing lipid databases and the lipids currently annotated in them. It is designed to be expandable should new categories, classes, or subclasses be required in the future, and updates will be maintained on the LIPID MAPS website. The development of this system has been enriched by interaction with lipidologists across the world in the hopes that this system will be internationally accepted and used.[jlr]

The authors appreciate the agreement of the International Lipids Classification and Nomenclature Committee to advise on future issues involving the maintenance of these recommendations. This committee currently includes Edward A. Dennis (chair), Christian Raetz and Robert Murphy representing LIPID MAPS, Friedrich Spener representing the International Conference on the Biosciences of Lipids, Gerrit van Meer representing the European Lipidomics Initiative, and Yousuke Seyama and Takao Shimizu representing the LipidBank of the Japanese Conference on the Biochemistry of Lipids. The authors are most

appreciative of informative discussions and encouragement of this effort with Professor Richard Cammack, King's College, London, who is the Chairman of the Nomenclature Committee of IUBMB and the IUPAC/IUBMB Joint Commission on Biochemical Nomenclature. The authors thank the Consortium for Functional Glycomics (headed by Ram Sasisekharan at the Massachusetts Institute of Technology) for providing us with their text nomenclature for glycosylated structures. We are grateful to Dr. Jean Chin, Program Director at the National Institutes of General Medical Sciences, for her valuable input to this effort. This work was supported by the LIPID MAPS Large-Scale Collaborative Grant GM-069338 from the National Institutes of Health.

References

1. Smith, A. 2000. Oxford Dictionary of Biochemistry and Molecular Biology. 2nd edition. Oxford University Press, Oxford, UK.
2. Christie, W. W. 2003. Lipid Analysis. 3rd edition. Oily Press, Bridgewater, UK.
3. IUPAC-IUB Commission on Biochemical Nomenclature (CBN). The nomenclature of lipids (recommendations 1976). 1977. *Eur. J. Biochem.* **79:** 11–21; 1977. *Hoppe-Seylers Z. Physiol. Chem.* **358:** 617–631; 1977. *Lipids.* **12:** 455–468; 1977. *Mol. Cell. Biochem.* **17:** 157–171; 1978. *Chem. Phys. Lipids.* **21:** 159–173; 1978. *J. Lipid Res.* **19:** 114–128; 1978. *Biochem. J.* **171:** 21–35 (http://www.chem.qmul.ac.uk/iupac/lipid/).
4. IUPAC-IUB Joint Commission on Biochemical Nomenclature (JCBN). Nomenclature of glycolipids (recommendations 1997). 2000. *Adv. Carbohydr. Chem. Biochem.* **55:** 311–326; 1988. *Carbohydr. Res.* **312:** 167–175; 1998. *Eur. J. Biochem.* **257:** 293–298; 1999. *Glycoconjugate J.* **16:** 1–6; 1999. *J. Mol. Biol.* **286:** 963–970; 1997. *Pure Appl. Chem.* **69:** 2475–2487 (http://www.chem.qmul.ac.uk/iupac/misc/glylp.html).
5. IUPAC-IUB Joint Commission on Biochemical Nomenclature (JCBN). 1987. Nomenclature of prenols (recommendations 1987). *Eur. J. Biochem.* **167:** 181–184 (http://www.chem.qmul.ac.uk/iupac/misc/prenol.html).
6. IUPAC-IUB Joint Commission on Biochemical Nomenclature (JCBN). 1989. Nomenclature of steroids (recommendations 1989). *Eur. J. Biochem.* **186:** 429–458 (http://www.chem.qmul.ac.uk/iupac/steroid/).

7. Caffrey, M., and J. Hogan. 1992. LIPIDAT: a database of lipid phase transition temperatures and enthalpy changes. *Chem. Phys. Lipids.* **61**: 1–109 (http://www.lipidat.chemistry.ohio-state.edu).

8. Watanabe, K., E. Yasugi, and M. Ohshima. 2000. How to search the glycolipid data in "Lipidbank for web," the newly-developed lipid database in Japan. *Trends Gycosci. Glycotechnol.* **12**: 175–184.

9. Vance, D. E., and J. E. Vance, editors. 2002. Biochemistry of Lipids, Lipoproteins and Membranes. 4th edition. Elsevier Science, New York.

10. Small, D. M. 1986. The Physical Chemistry of Lipids. Handbook of Lipid Research. Vol. 4. D. J. Hanahan, editor. Plenum Press, New York.

11. Brennan, P. J., and H. Nikaido. 1995. The envelope of mycobacteria. *Annu. Rev. Biochem.* **64**: 29–63.

12. Ohlrogge, J. B. 1997. Regulation of fatty acid synthesis. *Annu. Rev. Plant Physiol. Plant Mol. Biol.* **48**: 109–136.

13. Agrawal, G. K., S. Tamogami, O. Han, H. Iwahasi, and R. Rakwal. 2004. Rice octadecanoid pathway. *Biochem. Biophys. Res. Commun.* **317**: 1–15.

14. Murphy, R. C., and W. L. Smith. 2002. The eicosanoids: cyclooxygenase, lipoxygenase, and epoxygenase pathways. *In* Biochemistry of Lipids, Lipoproteins and Membranes. 4th edition. D. E. Vance and J. E. Vance, editors. Elsevier Science, New York. 341–371.

15. Bazan, N. G. 1989. The metabolism of omega-3 polyunsaturated fatty acids in the eye: the possible role of docosahexaenoic acid and docosanoids in retinal physiology and ocular pathology. *Prog. Clin. Biol. Res.* **312**: 95–112.

16. Roche, D. M., J. T. Byers, D. S. Smith, F. G. Glansdorp, D. R. Spring, and M. Welch. 2004. Communications blackout? Do N-acylhomoserine-lactone-degrading enzymes have any role in quorum sensing? *Microbiology.* **150**: 2023–2028.

17. Stam, H., K. Schoonderwoerd, and W. C. Hulsmann. 1987. Synthesis, storage and degradation of myocardial triglycerides. *Basic Res. Cardiol.* **82 (Suppl. 1)**: 19–28.

18. Coleman, R. A., and D. P. Lee. 2004. Enzymes of triacylglycerol synthesis and their regulation. *Prog. Lipid Res.* **43**: 134–176.

19. Pahlsson, P., S. L. Spitalnik, P. F. Spitalnik, J. Fantini, O. Rakotonirainy, S. Ghardashkhani, J. Lindberg, P. Konradsson, and G. Larson. 1998. Characterization of galactosyl glycerolipids in the HT29 human colon carcinoma cell line. *Arch. Biochem. Biophys.* **396**: 187–198.

20. Koga, Y., M. Nishihara, H. Morii, and M. Akagawa-Matsushita. 1983. Ether polar lipids of methanogenic bacteria: structures, comparative aspects and biosyntheses. *Microbiol. Rev.* **57**: 164–182.

21. Pereto, J., P. Lopez-Garcia, and D. Moreira. 2004. Ancestral lipid biosynthesis and early membrane evolution. *Trends Biochem. Sci.* **29**: 469–477.

22. Kennedy, E. P. 1962. The metabolism and function of complex lipids. *Harvey Lecture Series.* **57**: 143–171.

23. G. Cevc, editor. 1993. Phospholipids Handbook. Marcel Dekker Inc., New York.

24. Forrester, J. S., S. B. Milne, P. T. Ivanova, and H. A. Brown. 2004. Computational lipidomics: a multiplexed analysis of dynamic changes in membrane lipid composition during signal transduction. *Mol. Pharmacol.* **65**: 813–821.

25. Ivanova, P. T., S. B. Milne, J. S. Forrester, and H. A. Brown. 2004. Lipid arrays: new tools in the understanding of membrane dynamics and lipid signaling. *Mol. Interventions.* **4**: 86–96.

26. Cronan, J. E. 2003. Bacterial membrane lipids: where do we stand? *Annu. Rev. Microbiol.* **57**: 203–224.

27. Merrill, A. H., Jr., and K. Sandhoff. 2002. Sphingolipids: metabolism and cell signaling. *In* New Comprehensive Biochemistry: Biochemistry of Lipids, Lipoproteins, and Membranes. D. E. Vance and J. E. Vance, editors. Elsevier Science, New York. 373–407.

28. Taniguchi, N., K. Honke, and M. Fukuda. 2002. Handbook of Glycosyltransferases and Related Genes. Springer-Verlag, Tokyo.

29. Thudichum, J. L. W. 1884. A Treatise on the Chemical Constitution of Brain. Bailliere, Tindall, and Cox, London.

30. Bach, D., and E. Wachtel. 2003. Phospholipid/cholesterol model membranes: formation of cholesterol crystallites. *Biochim. Biophys. Acta.* **1610**: 187–197.

31. Tsai, M. J., and B. W. O'Malley. 1994. Molecular mechanisms of action of steroid/thyroid receptor superfamily members. *Annu. Rev. Biochem.* **63**: 451–486.

32. Jones, G., S. A. Strugnell, and H. F. DeLuca. 1998. Current understanding of the molecular actions of vitamin D. *Physiol. Rev.* **78**: 1193–1231.

33. Russell, D. W. 2003. The enzymes, regulation, and genetics of bile acid synthesis. *Annu. Rev. Biochem.* **72**: 137–174.

34. Kuzuyama, T., and H. Seto. 2003. Diversity of the biosynthesis of the isoprene units. *Nat. Prod. Rep.* **20**: 171–183.

35. Rodriguez-Concepcion, M. 2004. The MEP pathway: a new target for the development of herbicides, antibiotics and antimalarial drugs. *Curr. Pharm. Res.* **10**: 2391–2400.

36. Porter, J. W., and S. L. Spurgeon. 1981. Biosynthesis of Isoprenoid Compounds. Vol. 1. John Wiley & Sons, New York.

37. Demming-Adams, B., and W. W. Adams. 2002. Antioxidants in photosynthesis and human nutrition. *Science.* **298**: 2149–2153.

38. Ricciarelli, R., J. M. Zingg, and A. AzziI. 2001. Vitamin E: protective role of a Janus molecule. *FASEBJ.* **15**: 2314–2325.

39. Meganathan, R. 2001. Biosynthesis of menaquinone (vitamin K2) and ubiquinone (coenzyme Q): a perspective on enzymatic mechanisms. *Vitam. Horm.* **61**: 173–218.

40. Meganathan, R. 2001. Ubiquinone biosynthesis in microorganisms. *FEMS Microbiol. Lett.* **203**: 131–139.

41. Raetz, C. R. H., and C. Whitfield. 2002. Lipopolysaccharide endotoxins. *Annu. Rev. Biochem.* **71**: 635–700.

42. Lazar, K., and S. Walker. 2002. Substrate analogues to study cell-wall biosynthesis and its inhibition. *Curr. Opin. Chem. Biol.* **6**: 786–793.

43. Schenk, B., F. Fernandez, and C. J. Waechter. 2001. The ins(ide) and out(side) of dolichyl phosphate biosynthesis and recycling in the endoplasmic reticulum. *Glycobiology.* **11**: 61R–71R.

44. Helenius, J., and M. Aebi. 2001. Intracellular functions of N-linked glycans. *Science.* **291**: 2364–2369.

45. Zähringer, U., B. Lindner, and E. T. Rietschel. 1999. Chemical structure of Lipid A: recent advances in structural analysis of biologically active molecules. *In* Endotoxin in Health and Disease. H. Brade, S. M. Opal, S. N. Vogel, and D. C. Morrison, editors. Marcel Dekker, New York. 93–114.

46. Sweet, C. R., A. A. Ribeiro, and C. R. Raetz. 2004. Oxidation and transamination of the 3'-position of UDP-N-acetylglucosamine by enzymes from *Acidithiobacillus ferrooxidans*. Role in the formation of lipid A molecules with four amide-linked acyl chains. *J. Biol. Chem.* **279**: 25400–25410.

47. Spaink, H. P. 2000. Root nodulation and infection factors produced by rhizobial bacteria. *Annu. Rev. Microbiol.* **54**: 257–288.

48. Ghangas, G. S., and J. C. Steffens. 1993. UDP glucose: fatty acid transglucosylation and transacylation in triacylglucose biosynthesis. *Proc. Natl. Acad. Sci. USA.* **90**: 9911–9915.

49. Walsh, C. T. 2004. Polyketide and nonribosomal peptide antibiotics: modularity and versatility. *Science.* **303**: 1805–1810.

50. Khosla, C., R. Gokhale, J. R. Jacobsen, and D. E. Cane. 1999. Tolerance and specificity of polyketide synthases. *Annu. Rev. Biochem.* **68**: 219–253.

51. Reeves, C. D. 2003. The enzymology of combinatorial biosynthesis. *Crit. Rev. Biotechnol.* **23**: 95–147.

52. Moore, B. S., and C. Hartweck. 2002. Biosynthesis and attachment of novel bacterial polyketide synthase starter units. *Nat. Prod. Rep.* **19**: 70–99.

PROPERTIES OF FATTY ACIDS AND THEIR METHYL ESTERS

This table gives the names and selected properties of some important fatty acids and their methyl esters. It includes most of the acids that are significant constituents of naturally occurring oils and fats. Compounds are listed first by number of carbon atoms and, secondly, by the degree of unsaturation. Both the systematic name and the common or trivial name are given, as well as the Chemical Abstracts Service Registry Number and the shorthand acid code that is frequently used. The first number in this code gives the number of carbon atoms; the number following the colon is the number of unsaturated centers (mainly double bonds). The location and orientation of the unsaturated centers follow. The symbols used are: c = *cis*; t = *trans*; a = acetylenic center; e = ethylenic center at end of chain; ep = *epoxy*. Thus 9c,11t indicates a double bond with *cis* orientation at the No. 9 carbon and another with *trans* orientation at the No. 11 carbon. More details on the codes can be found in Reference 1.

The table gives the molecular weight and melting point of the acid and the melting and boiling points of the methyl ester of the acid when available. A superscript on the boiling point indicates the pressure in mmHg (torr); if there is no superscript, the value refers to one atmosphere (760 mmHg). The references cover many other fatty acids beyond those listed here and give additional properties.

We are indebted to Frank D. Gunstone for advice on the content of the table.

References

1. Gunstone, F. D., Harwood, J. L., and Dijkstra, A. J., eds., *The Lipid Handbook, Third Edition,* CRC Press, Boca Raton, FL, 2006.
2. Gunstone, F. D., and Adlof, R. O., *Common (non-systematic) Names for Fatty Acids,* www.aocs.org/member/division/analytic/fanames. asp, 2003.
3. Firestone, D., *Physical and Chemical Characteristics of Oils, Fats, and Waxes,* Second Edition, AOCS Press, Urbana, IL, 2006.
4. Dawson, R. M. C., Elliott, D. C., Elliott, W. H., and Jones, K. M., *Data for Biochemical Research,* Third Edition, Clarendon Press, Oxford, 1986.
5. Altman, P. L., and Dittmer, D. S., eds., *Biology Data Book,* Second Edition, Vol. 1, Federation of American Societies for Experimental Biology, Bethesda, MD, 1972.
6. Fasman, G. D., Ed., *Practical Handbook of Biochemistry and Molecular Biology,* CRC Press, Boca Raton, FL, 1989.

Systematic Name	Common Name	Mol. Form.	Acid Code	CAS RN	Mol. Weight	mp/°C	Methyl Ester mp/°C	Methyl Ester bp/°C
Butanoic acid	Butyric acid	$C_4H_8O_2$	4:0	107-92-6	88.106	−5.1	−85.8	102.8
Pentanoic acid	Valeric acid	$C_5H_{10}O_2$	5:0	109-52-4	102.132	−33.6		127.4
3-Methylbutanoic acid	Isovaleric acid	$C_5H_{10}O_2$	4:0 3-Me	503-74-2	102.132	−29.3		116.5
Hexanoic acid	Caproic acid	$C_6H_{12}O_2$	6:0	142-62-1	116.158	−3	−71	149.5
Heptanoic acid	Enanthic acid	$C_7H_{14}O_2$	7:0	111-14-8	130.185	−7.2	−56	174
Octanoic acid	Caprylic acid	$C_8H_{16}O_2$	8:0	124-07-2	144.212	16.5	−40	192.9
Nonanoic acid	Pelargonic acid	$C_9H_{18}O_2$	9:0	112-05-0	158.238	12.4		213.5
Decanoic acid	Capric acid	$C_{10}H_{20}O_2$	10:0	334-48-5	172.265	31.4	−18	224
9-Decenoic acid	Caproleic acid	$C_{10}H_{18}O_2$	10:1 9e	14436-32-9	170.249	26.5		120[20]
Undecanoic acid		$C_{11}H_{22}O_2$	11:0	112-37-8	186.292	28.6		123[10]
Dodecanoic acid	Lauric acid	$C_{12}H_{24}O_2$	12:0	143-07-7	200.318	43.8	5.2	267
cis-9-Dodecenoic acid	Lauroleic acid	$C_{12}H_{22}O_2$	12:1 9c	2382-40-3	198.302			
Tridecanoic acid		$C_{13}H_{26}O_2$	13:0	638-53-9	214.344	41.5	6.5	92[1]
Tetradecanoic acid	Myristic acid	$C_{14}H_{28}O_2$	14:0	544-63-8	228.371	54.2	19	295
cis-9-Tetradecenoic acid	Myristoleic acid	$C_{14}H_{26}O_2$	14:1 9c	13147-06-3	226.355	−4		
Pentadecanoic acid		$C_{15}H_{30}O_2$	15:0	1002-84-2	242.398	52.3	18.5	153.5
Hexadecanoic acid	Palmitic acid	$C_{16}H_{32}O_2$	16:0	57-10-3	256.424	62.5	30	417
cis-9-Hexadecenoic acid	Palmitoleic acid	$C_{16}H_{30}O_2$	16:1 9c	373-49-9	254.408	0.5		140[5]
Heptadecanoic acid	Margaric acid	$C_{17}H_{34}O_2$	17:0	506-12-7	270.451	61.3	30	185[9]
Octadecanoic acid	Stearic acid	$C_{18}H_{36}O_2$	18:0	57-11-4	284.478	69.3	39.1	443
cis-6-Octadecenoic acid	Petroselinic acid	$C_{18}H_{34}O_2$	18:1 6c	593-39-5	282.462	29.8		
cis-9-Octadecenoic acid	Oleic acid	$C_{18}H_{34}O_2$	18:1 9c	112-80-1	282.462	13.4	−19.9	218.5[20]
trans-9-Octadecenoic acid	Elaidic acid	$C_{18}H_{34}O_2$	18:1 9t	112-79-8	282.462	45	13.5	218[24]
cis-11-Octadecenoic acid	*cis*-Vaccenic acid	$C_{18}H_{34}O_2$	18:1 11c	506-17-2	282.462	15		163[0.1]
trans-11-Octadecenoic acid	Vaccenic acid	$C_{18}H_{34}O_2$	18:1 11t	693-72-1	282.462	44		172[3]
cis-12,13-Epoxy-*cis*-9-octadecenoic acid	Vernolic acid	$C_{18}H_{32}O_3$	18:1 12,13-ep,9c	503-07-1	296.445	32.5		
12-Hydroxy-*cis*-9-octadecenoic acid	Ricinoleic acid	$C_{18}H_{34}O_3$	18:1 12-OH,9c	141-22-0	298.461	5.5		226[15]
cis,trans-9,11-Octadecadienoic acid	Rumenic (CLA)	$C_{18}H_{32}O_2$	18:2 9c,11t	1839-11-8	280.446	20		
cis,cis-9,12-Octadecadienoic acid	Linoleic acid	$C_{18}H_{32}O_2$	18:2 9c,12c	60-33-3	280.446	−7	−35	215[20]

PROPERTIES OF FATTY ACIDS AND THEIR METHYL ESTERS (Continued)

Systematic Name	Common Name	Mol. Form.	Acid Code	CAS RN	Mol. Weight	mp/°C	Methyl Ester mp/°C	Methyl Ester bp/°C
trans,cis-10,12-Octadecadienoic acid	(CLA)	$C_{18}H_{32}O_2$	18:2 10t,12c	22880-03-1	280.446	23	−12	
cis-9-Octadecen-12-ynoic acid	Crepenynic acid	$C_{18}H_{30}O_2$	18:2 9c,12a	2277-31-8	278.430			
cis,cis,cis-5,9,12-Octadecatrienoic acid	Pinolenic acid	$C_{18}H_{30}O_2$	18:3 5c,9c,12c	27213-43-0	278.430			
trans,cis,cis-5,9,12-Octadecatrienoic acid	Columbinic acid	$C_{18}H_{30}O_2$	18:3 5t,9c,12c	2441-53-4	278.430			
cis,cis,cis-6,9,12-Octadecatrienoic acid	γ-Linolenic acid	$C_{18}H_{30}O_2$	18:3 6c,9c,12c	506-26-3	278.430			$162^{0.5}$
trans,trans,cis-8,10,12-Octadecatrienoic acid	Calendic acid	$C_{18}H_{30}O_2$	18:3 8t,10t,12c	28872-28-8	278.430	40		
cis,trans,cis-9,11,13-Octadecatrienoic acid	Punicic acid	$C_{18}H_{30}O_2$	18:3 9c,11t,13c	544-72-9	278.430	45		
cis,trans,trans-9,11,13-Octadecatrienoic acid	α-Eleostearic acid	$C_{18}H_{30}O_2$	18:3 9c,11t,13t	506-23-0	278.430	49		148^1
trans,trans,cis-9,11,13-Octadecatrienoic acid	Catalpic acid	$C_{18}H_{30}O_2$	18:3 9t,11t,13c	4337-71-7	278.430	32		
trans,trans,trans-9,11,13-Octadecatrienoic acid	β-Eleostearic acid	$C_{18}H_{30}O_2$	18:3 9t,11t,13t	544-73-0	278.430	71.5	13	162^1
cis,cis,cis-9,12,15-Octadecatrienoic acid	α-Linolenic acid	$C_{18}H_{30}O_2$	18:3 9c,12c,15c	463-40-1	278.430	−11.3	−52	$109^{0.018}$
6,9,12,15-Octadecatetraenoic acid, all *cis*	Stearidonic acid	$C_{18}H_{28}O_2$	18:4 6c,9c,12c,15c	20290-75-9	276.414	−57		
cis,trans,trans,cis-9,11,13,15-Octadecatetraenoic acid	Parinaric acid	$C_{18}H_{28}O_2$	18:4 9c,11t,13t,15c	593-38-4	276.414	86		
Nonadecanoic acid		$C_{19}H_{38}O_2$	19:0	646-30-0	298.504	69.4	41.3	190^4
Eicosanoic acid	Arachidic acid	$C_{20}H_{40}O_2$	20:0	506-30-9	312.531	76.5	54.5	215^{10}
3,7,11,15-Tetramethylhexadecanoic acid	Phytanic acid	$C_{20}H_{40}O_2$	16:0 3,7,11,15-tetramethyl	14721-66-5	312.531	−65		
cis-5-Eicosenoic acid		$C_{20}H_{38}O_2$	20:1 5c	7050-07-9	310.515	27		
cis-9-Eicosenoic acid	Gadoleic acid	$C_{20}H_{38}O_2$	20:1 9c	29204-02-2	310.515	24.5		
cis-11-Eicosenoic acid	Gondoic acid	$C_{20}H_{38}O_2$	20:1 11c	2462-94-4	310.515	24		
cis,cis,cis-8,11,14-Eicosatrienoic acid	Dihomo-γ-linolenic acid	$C_{20}H_{34}O_2$	20:3 8c,11c,14c	1783-84-2				
5,8,11,14-Eicosatetraenoic acid, all *cis*	Arachidonic acid	$C_{20}H_{32}O_2$	20:4 5c,8c,11c,14c	506-32-1	304.467	−49.5		$195^{0.7}$
5,8,11,14,17-Eicosapentaenoic acid, all *cis*	Timnodonic acid, EPA	$C_{20}H_{30}O_2$	20:5 5c,8c,11c,14c,17c	10417-94-4	302.451	−54		
Heneicosanoic acid		$C_{21}H_{42}O_2$	21:0	2363-71-5	326.557	82	49	207^4
Docosanoic acid	Behenic acid	$C_{22}H_{44}O_2$	22:0	112-85-6	340.583	81.5	54	
cis-11-Docosenoic acid	Cetolic acid	$C_{22}H_{42}O_2$	22:1 11c	506-36-5	338.567	33		
cis-13-Docosenoic acid	Erucic acid	$C_{22}H_{42}O_2$	22:1 13c	112-86-7	338.567	34.7		221^5
trans-13-Docosenoic acid	Brassidic acid	$C_{22}H_{42}O_2$	22:1 13t	506-33-2	338.567	61.9	35	
cis,cis-5,13-Docosadienoic acid		$C_{22}H_{40}O_2$	22:2 5c,13c	676-39-1	336.552	−4		
7,10,13,16,19-Docosapentaenoic acid, all *cis*		$C_{22}H_{34}O_2$	22:5 7c,10c,13c,16c,19c					
4,7,10,13,16,19-Docosahexaenoic acid, all *cis*	Cervonic acid, DHA	$C_{22}H_{32}O_2$	22:6 4c,7c,10c,13c,16c,19c	2091-24-9		−45		
Tricosanoic acid		$C_{23}H_{46}O_2$	23:0	2433-96-7		79.6	53.4	
Tetracosanoic acid	Lignoceric acid	$C_{24}H_{48}O_2$	24:0	557-59-5	368.637	87.5	60	
cis-15-Tetracosenoic acid	Nervonic acid	$C_{24}H_{46}O_2$	24:1 15c	506-37-6	366.621	43	15	$165^{0.02}$
Pentacosanoic acid		$C_{25}H_{50}O_2$	25:0	506-38-7	382.664	77.5	62	
Hexacosanoic acid	Cerotic acid	$C_{26}H_{52}O_2$	26:0	506-46-7	396.690	88.5	63.8	286^{15}
Heptacosanoic acid		$C_{27}H_{54}O_2$	27:0	7138-40-1		87.6	64	
Octacosanoic acid	Montanic acid	$C_{28}H_{56}O_2$	28:0	506-48-9	424.744	90.9	67	
Nonacosanoic acid		$C_{29}H_{58}O_2$	29:0	4250-38-8	438.770	90.3	69	
Triacontanoic acid	Melissic acid	$C_{30}H_{60}O_2$	30:0	506-50-3	452.796	93.6	72	
Hentriacontanoic acid		$C_{31}H_{62}O_2$	31:0	38232-01-8	466.823	93.1		
Dotriacontanoic acid	Lacceric acid	$C_{32}H_{64}O_2$	32:0	3625-52-3	480.849	96.2		$192^{0.01}$

DENSITIES, SPECIFIC VOLUMES, AND TEMPERATURE COEFFICIENTS OF FATTY ACIDS FROM C_8 TO C_{12}

Acid	Temperature, °C	Density,[a] g/cc	Specific Volume, l/d	Temp. Coeff. per °C
Caprylic	10.0	1.0326	0.9685	0.00098
	15	1.0274	0.9733	—
	20	0.9109	1.0979	0.00046
	20.02	0.9101[b]	—	—
	25	0.9090	1.1002	—
	50.27	0.8862[b]	—	0.00099
Nonanoic	5.0	0.9952	1.0048	0.00074
	10	0.9916	1.0085	—
	15	0.9097	1.0993	0.00104
	15.00	0.9087[b]	—	—
	25	0.9011	1.1097	—
Capric	15.0	1.0266	0.9741	0.00085
	25	1.0176	0.9827	—
	35	0.8927	1.1202	0.00128
	35.05	0.8884[b]	—	—
	40	0.8876	1.1266	—
Hendecanoic	0.12	1.0431	0.9587	0.00054
	10.0	1.0373	0.9640	—
	20	0.9948	1.0052	0.00079
	25	0.9905	1.0096	—
	30	0.8907	1.1227	0.00093
	30.00	0.8889[b]	—	—
	35	0.8871	1.1273	—
	50.15	0.8741[b]	—	0.00095
Lauric	35.0	1.0099	0.9902	0.00087
	40	1.0055	0.9945	—
	45	0.8767	1.1406	0.00142
	45.10	0.8744[b]	—	0.00095
	50	0.8713	1.1477	—
	50.25	0.8707[b]	—	0.00095

[a] By air thermometer method unless specified otherwise.
[b] By pycnometer method.

From Markley, Klare S., *Fatty Acids*, 2nd ed., Part 1, Interscience Publishers, Inc., New York, 1960, 535. With permission of copyright owners.

COMPOSITION AND PROPERTIES OF COMMON OILS AND FATS

This table lists some of the most common naturally occurring oils and fats. The list is separated into those of plant origin, fish and other marine life origin, and land animal origin. The oils and fats consist mainly of esters of glycerol (i.e., triglycerides) with fatty acids of 10 to 22 carbon atoms. The four fatty acids with the highest concentration are given for each oil; concentrations are given in weight percent. Because there is often a wide variation in composition depending on the source of the oil sample, a range (or sometimes an average) is generally given. More complete data on composition, including minor fatty acids, sterols, and tocopherols, can be found in the references.

The acids are labeled by the codes described in the previous table, "Properties of Fatty Acids and Their Methyl Esters," which gives the systematic and common names of the acids. Thus 18:2 9c,12c indicates a C_{18} acid with two double bonds in the 9 and 12 positions, both with a *cis* configuration (*cis,cis*-9,12-octadecadienoic acid, or linoleic acid).

The density and refractive index of the oils are typical values; superscripts indicate the temperature in °C.

Notes:

- The composition figure given for oleic acid (18:1 9c) often includes low levels of other 18:1 isomers.

- In some oils where a concentration is given for 18:2 9c,12c (linoleic acid), other isomers of 18:2 may be included.
- Likewise, where a concentration is given for 18:3 9c,12c,15c (α-linolenic acid), other isomers of 18:3 may be included.
- The acid 20:5 6c,9c,12c,15c,17c, which is prevalent in many fish oils, is often abbreviated as 20:5 ω-3 or 20:5 n-3.

The assistance of Frank D. Gunstone in preparing this table is gratefully acknowledged.

References

1. Firestone, D., *Physical and Chemical Characteristics of Oils, Fats, and Waxes*, Second Edition, AOCS Press, Urbana, IL, 2006.
2. Gunstone, F. D., Harwood, J. L., and Dijkstra, A. J., eds., *The Lipid Handbook*, Third Edition, CRC Press, Boca Raton, FL, 2006.
3. Dawson, R. M. C., Elliott, D. C., Elliott, W. H., and Jones, K. M., *Data for Biochemical Research*, Third Edition, Clarendon Press, Oxford, 1986.
4. Altman, P. L., and Dittmer, D. S., eds., *Biology Data Book*, Second Edition, Vol. 1, Federation of American Societies for Experimental Biology, Bethesda, MD, 1972.

Type of Oil	Principal Fatty Acid Components in Weight %				mp/ °C	Density/ g cm^{-3}	Refractive Index	Iodine Value	Saponification Value
			PLANTS						
Almond kernel oil	18:1 9c	43–70%	18:2 9c,12c	24–30%		0.910^{25}	1.467^{26}	89–101	188–200
	16:0	4–13%	18:0	1–10%					
Apricot kernel oil	18:1 9c	58–66%	18:2 9c,12c	29–33%		0.910^{25}	1.469^{25}	97–110	185–199
	16:0	4.6–6%	18:0	1%					
Argan seed oil	18:1 9c	42–55%	18:2 9c,12c	30–34%		0.912^{20}	1.467^{20}	92–102	189–195
	16:0	12–16%	18:0	2–7%					
Avocado pulp oil	18:1 9c	56–74%	18:2 9c,12c	10–17%		0.912^{25}	1.466^{25}	85–90	177–198
	16:0	9–18%	16:1 9c	3–9%					
Babassu palm oil	12:0	40–55%	14:0	11–27%	24	0.914^{25}	1.450^{40}	10–18	245–256
	18:1 9c	9–20%	16:0	5.2–11%					
Blackcurrant oil	18:2 9c,12c	45–50%	18:3 6c,9c,12c	14–20%		0.923^{20}	1.480^{20}	173–182	185–195
	18:3 9c,12c,15c	12–15%	18:1 9c	9–13%					
Borage (star-flower) oil	18:2 9c,12c	36–40%	18:3 6c,9c,12c	17–25%				141–160	189–192
	18:1 9c	14–21%	16:0	9.4–12%					
Borneo tallow	18:0	39–43%	18:1 9c	34–37%	38	0.855^{100}	1.456^{40}	29–38	189–200
	16:0	18–21%	20:0	1.0%					
Cameline oil	18:3 9c,12c,15c	33–38%	18:2 9c,12c	15–16%		0.924^{15}	1.477^{20}	127–155	180–190
	20:1 total	14–16%	18:1 9c	12–24%					
Canola (rapeseed) oil (low linolenic)	18:1 9c	59–66%	18:2 9c,12c	24–29%	−10			91	
	16:0	4–5%	18:3 9c,12c,15c	2–3%					
Canola (rapeseed) oil (low erucic)	18:1 9c	52–67%	18:2 9c,12c	16–25%	−10	0.915^{20}	1.466^{40}	110–126	182–193
	18:3 9c,12c,15c	6–14%	16:0	3.3–6.0%					
Caraway seed oil	18:1 9c	40%	18:2 9c,12c	30%			1.471^{35}	128	178
	18:1 6c	26%	16:0	3%					
Cashew nut oil	18:1 9c	57–80%	18:2 9c,12c	16–22%		0.914^{15}	1.463^{40}	79–89	180–196
	16:0	4–17%	18:0	2–12%					

COMPOSITION AND PROPERTIES OF COMMON OILS AND FATS (Continued)

Type of Oil	Principal Fatty Acid Components in Weight %				mp/ °C	Density/ g cm^{-3}	Refractive Index	Iodine Value	Saponification Value
	PLANTS (Continued)								
Castor oil	18:1 12-OH,9c	88%	18:2 9c,12c	3–5%	−18	0.952^{25}	1.475^{25}	81–91	176–187
	18:1 9c	2.9–6%	22:0	2.1%					
Cherry kernel oil	18:2 9c,12c	42–45%	18:1 9c	35–49%		0.918^{25}	1.468^{40}	110–118	190–198
	16:0	4–9%	18:3 9c,11t,13t	3–10%					
Chinese vegetable tallow	16:0	58–72%	18:1 9c	20–35%	44	0.887^{25}	1.456^{40}	16–29	200–218
	18:0	1–8%	14:0	0.5–3.7%					
Cocoa butter	18:0	31–37%	18:1 9c	31–35%	34	0.974^{25}	1.457^{40}	32–40	192–200
	16:0	25–27%	18:2 9c,12c	2.8–4.0%					
Coconut oil	12:0	45–51%	14:0	17–21%	25	0.913^{40}	1.449^{40}	5–13	248–265
	16:0	7.7–10.2%	18:1 9c	5.4–9.9%					
Cohune nut oil	12:0	44–48%	14:0	16–17%		0.914^{25}	1.450^{40}	9–14	251–260
	18:1 9c	8–10%	16:0	7–10%					
Coriander seed oil	18:1 6c	53%	18:1 9c	32%		0.908^{25}	1.464^{25}	86–100	182–191
	18:2 9c,12c	7–14%	16:0	3–8%					
Corn oil	18:2 9c,12c	40–66%	18:1 9c	20–42%	−20	0.919^{20}	1.472^{25}	107–135	187–195
	16:0	9–16%	18:0	0–3%					
Cottonseed oil	18:2 9c,12c	47–58%	16:0	18–26%	−1	0.920^{20}	1.462^{40}	96–115	189–198
	18:1 9c	14–22%	18:0	2.1–3.3%					
Crambe oil	22:1 13c	55–60%	18:1 9c	12–15%		0.906^{25}	1.470^{25}	87–113	
	18:2 9c,12c	8–10%	18:3 9c,12c,15c	6–7%					
Cuphea seed oil (caprylic acid rich)	8:0	65–78%	10:0	19–24%					
	18:2 9c,12c	1–4%	16:0	0.6–3%					
Euphorbia lagascae seed oil	18:1 12,13-ep,9c	64%	18:1 other	19%		0.952^{25}	1.473^{25}	102	
	18:2 9c,12c	9%	16:0	4%					
Evening primrose oil	18:2 9c,12c	65–80%	18:3 6c,9c,12c	8–14%			1.479^{20}	147–155	193–198
	16:0	6–10%	18:1 9c	5–12%					
Grape seed oil	18:2 9c,12c	58–78%	18:1 9c	12–28%		0.923^{20}	1.475^{40}	130–138	188–194
	16:0	5.5–11%	18:0	3–6%					
Hazelnut oil (Chilean)	18:1 9c	39%	16:1 11c	22.7%					
	20:1 total	9.7%	22:1 total	9.5%					
Hazelnut oil (Filbert)	18:1 9c	72–84%	18:2 9c,12c	5.7–22%		0.909^{25}	1.473^{25}	83–90	188–197
	16:0	4.1–7.2%	18:0	1.5–2.4%					
Hempseed oil	18:2 9c,12c	45–60%	18:3 9c,12c,15c	15–30%		0.921^{25}	1.472^{40}	145–166	190–195
	18:1 9c	11–16%	16:0	6–12%					
Illipe (mowrah) butter	18:1 9c	34%	16:0	23%	27	0.862^{100}	1.460^{40}	53–70	188–207
	18:0	23%	18:2 9c,12c	14%					
Jojoba oil[a]	20:1 total	66–74%	22:1 undefined	9–19%					
	18:1 9c	5–12%	24:1 15c	1–5%					
Kapok seed oil[b]	18:1 9c	45–65%	16:0	10–28%	30	0.926^{15}	1.469^{25}	86–110	189–197
	18:2 9c,12c	7–35%	18:0	2–9%					
Kokum butter	18:0	49–56%	18:1 9c	39–49%	41		1.456^{40}	33–37	192
	16:0	2–5%	18:2 9c,12c	1–2%					
Kusum oil	18:1 9c	57–62%	20:0	20–25%			1.461^{40}	48–58	220–230
	16:0	5–8%	18:0	2–6%					
Linola oil	18:2 9c,12c	72%	18:1 9c	16%				142	
	16:0	5.6%	18:0	4.0%					
Linseed oil	18:3 9c,12c,15c	52–58%	18:1 9c	18–20%	−24	0.924^{25}	1.480^{25}	170–203	188–196
	18:2 9c,12c	17%	18:2 9c,12c	16%					
Macadamia nut oil	18:1 9c	56–59%	16:1 9c	21–22%					
	16:0	8–9%	18:0	2–4%					
Mango seed oil	18:1 9c	38–50%	18:0	31–49%		0.912^{15}	1.461^{25}	39–48	188–195
	18:2 9c,12c	3–6%	20:0	2–6%					

COMPOSITION AND PROPERTIES OF COMMON OILS AND FATS (Continued)

Type of Oil	Principal Fatty Acid Components in Weight %				mp/ °C	Density/ g cm^{-3}	Refractive Index	Iodine Value	Saponification Value
	PLANTS (Continued)								
Meadowfoam seed oil	20:1 5c	58–77%	22:1 total	8–24%			1.464^{40}	86–91	168
	22:2 5c,13c	7–15%	18:1 9c	1–3%					
Melon oil	18:2 9c,12c	67% (av.)	18:1 9c	12% (av.)					
	16:0	11% (av.)	18:0	9% (av.)					
Moringa peregrina seed oil	18:1 9c	70%	16:0	9%		0.903^{24}	1.460^{40}	70	185
	18:0	3.8%	22:0	2.4%					
Mustard seed oil	22:1 13c	43%	22:1 13c	22–50%		0.913^{20}	1.465^{40}	92–125	170–184
	18:3 9c,12c,15c	12%	18:2 9c,12c	10–24%					
Neem oil	18:1 9c	49–62%	18:0	14–24%	−3	0.912^{30}	1.462^{40}	68–71	195–205
	16:0	13–18%	18:2 9c,12c	7–15%					
Niger seed oil	18:2 9c,12c	52–78%	16:0	5–12%		0.924^{15}	1.468^{40}	126–135	188–193
	18:1 9c	4–10%	18:0	2–12%					
Nutmeg butter	14:0	76–83%	18:1 9c	5–11%	45		1.468^{40}	48–85	170–190
	16:0	4–10%	12:0	3–6%					
Oat oil	18:2 9c,12c	24–48%	18:1 9c	18–53%		0.917^{25}	1.467^{40}	105–116	190–199
	16:0	13–39%	18:0	0.5–4%					
Oiticica oil	18:3 9c,11t,13t, 4-oxo	70–80%	16:0	7%		0.972^{20}	1.514^{25}	140–150	188–193
	18:0	5%	18:1 9c	4–7%					
Olive oil	18:1 9c	55–83%	18:2 9c,12c	9%	−6	0.911^{20}	1.469^{20}	75–94	184–196
	16:0	7.5–20%	18:2 9c,12c	3.5–21%					
Palm kernel oil	12:0	40–55%	14:0	14–18%	24	0.922^{15}	1.450^{40}	14–21	230–250
	18:1 9c	12–21%	16:0	6.5–10%					
Palm oil	16:0	40–48%	18:1 9c	36–44%	35	0.914^{15}	1.455^{40}	49–55	190–209
	18:2 9c,12c	6.5–12%	18:0	3.5–6.5%					
Palm olein	18:1 9c	40–44%	16:0	38–43%		0.91^{40}	1.459^{40}	>56	194–202
	18:2 9c,12c	10–13%	18:0	3.7–4.8%					
Palm stearin	16:0	48–74%	18:1 9c	16–36%		0.884^{60}	1.449^{40}	<48	193–205
	18:0	3.9–5.6%	18:2 9c,12c	3.2–9.8%					
Parsley seed oil	18:1 6c	69–76%	18:1 9c	12–15%			1.4800^{40}	110–120	
	18:2 9c,12c	6–14%	16:0	2%					
Peanut oil	18:1 9c	36–67%	18:2 9c,12c	14–43%	3	0.914^{20}	1.463^{40}	86–107	187–196
	16:0	8.3–14%	22:0	2.1–4.4%					
Perilla oil	18:3 9c,12c,15c	59%	18:2 9c,12c	14–18%		0.924^{25}	1.477^{25}	192–208	188–197
	18:1 9c	11–13%	16:0	6–9%					
Phulwara butter	16:0	57–61%	18:1 9c	30–36%	43	0.862^{100}	1.458^{40}	40–51	188–200
	18:2 9c,12c	3–4%	18:0	3–4%					
Pine nut oil	18:2 9c,12c	47–51%	18:1 9c	36–39%		0.919^{15}		118–121	193–197
	16:0	6–8%	18:0	2–3%					
Poppy seed oil	18:2 9c,12c	62–73%	18:1 9c	16–30%	−15	0.916^{25}	1.469^{40}	132–146	188–196
	16:0	7–11%	18:0	1–4%					
Rice bran oil	18:1 9c	38–48%	18:2 9c,12c	16–36%		0.916^{25}	1.472^{25}	92–108	181–189
	16:0	16–28%	18:0	2–4%					
Safflower seed oil	18:2 9c,12c	68–83%	18:1 9c	8.4–30%		0.924^{15}	1.474^{25}	136–148	186–198
	16:0	5.3–8.0%	18:0	1.9–2.9%					
Safflower seed oil (high oleic)	18:1 9c	74–80%	18:2 9c,12c	13–18%		0.921^{20}	1.470^{25}	91–95	
	16:0	5–6%	18:0	1.5–2.0%					
Sal fat	18:0	33–57%	18:1 9c	31–52%	33		1.456^{40}	31–45	175–192
	16:0	6–23%	20:0	1–8%					
Sesame seed oil	18:2 9c,12c	40–51%	18:1 9c	33–44%	−6	0.917^{20}	1.467^{40}	104–120	187–195
	16:0	7.9–10.2%	18:0	4.4–6.7%					
Sheanut butter	18:1 9c	45–50%	18:0	36–41%	38	0.863^{100}	1.465^{40}	52–66	178–198
	16:0	4–8%	18:2 9c,12c	4–8%					

COMPOSITION AND PROPERTIES OF COMMON OILS AND FATS (Continued)

Type of Oil	Principal Fatty Acid Components in Weight %				mp/ °C	Density/ g cm^{-3}	Refractive Index	Iodine Value	Saponification Value
PLANTS (Continued)									
Soybean oil	18:2 9c,12c	50–57%	18:1 9c	18–28%	−16	0.920[20]	1.468[40]	118–139	189–195
	16:0	9–13%	18:3 9c,12c,15c	5.5–9.5%					
Stillingia seed kernel oil[c]	18:3 total	41–54%	18:2 9c,12c	24–30%		0.937[25]	1.483[25]	169–191	202–212
	18:1 9c	7–10%	16:0	6–9%					
Sunflower seed oil	18:2 9c,12c	48–74%	18:1 9c	13–40%	−17	0.919[20]	1.474[25]	118–145	188–194
	16:0	5–8%	18:0	2.5–7.0%					
Sunflower oil, high-oleic (HO)	18:1 9c	80%	18:2 9c,12c	10%					
	18:0	4.4%	16:0	3.5%		0.911[25]	1.468[25]	81	
Sunflower oil, mid-Oleic (NuSun oil)	18:1 9c	65%	18:2, 18:3	25%					
	16:0, 18:0	10%							
Tall oil	18:2 9c,12c	41–52%	18:1 9c	41–48%		0.969[25]	1.494[25]	140–180	154–180
	16:0	5–6%	18:0	2–3%					
Tung oil	18:3 9c,11t,13t	71–82%	18:2 9c,12c	8–15%	−2	0.912[25]	1.517[25]	160–175	189–195
	18:1 9c	4–10%	18:0	3%					
Ucuhuba butter oil	14:0	64–73%	12:0	13–15%		0.870[100]	1.451[50]	11–17	221–229
	18:1 9c	6–8%	16:0	3–9%					
Vernonia seed oil	18:1 12,13-ep,9c	62–72%	18:2 9c,12c	9–17%		0.901[30]	1.486[32]	55	176
	16:0	3–7%	18:0	2–6%					
Walnut oil	18:2 9c,12c	56–60%	18:1 9c	17–19%		0.921[25]	1.474[25]	138–162	189–197
	18:3 9c,12c,15c	13–14%	16:0	6–8%					
Wheatgerm oil	18:2 9c,12c	50–59%	18:1 9c	13–23%		0.926[25]	1.479[25]	100–128	179–217
	16:0	12–20%	18:3 9c,12c,15c	2–9%					
MARINE ANIMALS									
Anchovy oil	20:5 6c,9c,12c,15c,17c	22%	16:0	17%				163–169	191–194
	16:1 undefined	13%	18:1 undefined	10%					
Capelin oil[d]	20:1 undefined	17%	22:1 undefined	15%			1.463[50]	94–164	185–202
	18:1 undefined	14%	16:0	10%					
Cod liver oil	18:1 undefined	24%	20:1 undefined	13%		0.924[15]	1.482[25]	142–176	180–192
	22:6 4c,7c,10c,13c,16c,19c	11%	16:0	10%					
Herring oil	22:1 undefined	19%	16:0	17%		0.914[20]	1.474[25]	115–160	161–192
	20:1 undefined	15%	18:1 undefined	14%					
Mackerel oil	22:1 undefined	15%	16:0	14%		0.929[15]	1.481[20]	136–167	
	18:1 undefined	13%	20:1 undefined	12%					
Menhaden oil	16:0	19%	20:5 6c,9c,12c,15c,17c	14%		0.920[15]		150–200	192–199
	16:1 undefined	12%	18:1 undefined	11%					
Salmon oil	22:6 4c,7c,10c,13c,16c,19c	18%	20:5 6c,9c,12c,15c,17c	13%		0.924[15]	1.475[25]	130–160	183–186
	16:0	9.8%	16:1	4.8%					
Sardine oil	16:0	18%	20:5 6c,9c,12c,15c,17c	16%		0.915[25]	1.464[65]	159–192	188–199
	18:1 undefined	13%	16:1 undefined	10%					
Seal blubber oil, harp	18:1 9c	21%	20:1	12%					
	22:6 4c,7c,10c,13c,16c,19c	7.6%	20:5 6c,9c,12c,15c,17c	6.4%					
Shark liver oil	18:1 undefined	45%	16:0	21%		0.917[25]	1.476[25]	150–300	170–190
	20:1	12%	22:1	9%					

COMPOSITION AND PROPERTIES OF COMMON OILS AND FATS (Continued)

Type of Oil	Principal Fatty Acid Components in Weight %				mp/ °C	Density/ g cm⁻³	Refractive Index	Iodine Value	Saponification Value
MARINE ANIMALS (Continued)									
Tuna oil	22:6 4c,7c,10c,13c, 16c,19c	22%	16:0	22%					
	18:1 undefined	21%	20:5 6c,9c,12c,15c, 17c	6%					
Trout lipids	16:0	21–24%	18:1 undefined	18–31%					
	18:2	7–16%	16:1	4–10%					
Whale oil, minke	18:1 undefined	18%	20:1	17%					
	22:1	11%	16:1	9%					
LAND ANIMALS									
Beef tallow	18:1 undefined	31–50%	18:0	25–40%	47	0.902^{25}	1.454^{40}	33–47	190–200
	16:0	20–37%	14:0	1–6%					
Butterfat	16:0	28.1% (av.)	18:1 9c	20.8% (av.)	32	0.934^{15}	1.455^{40}	26–40	210–232
	14:0	10.8% (av.)	18:0	10.6% (av.)					
Chicken egg lipids, yolk	16:0	28%	18:1 9c	25%					
	18:0	17%	18:2 9c,12c	16%					
Chicken fat	18:1 undefined	37%	16:0	22%		0.918^{15}	1.456^{40}	76–80	
	18:2	20%	18:0	6%					
Milk fats, cow	16:0	28.2% (av.)	18:1 9c	21.4% (av.)					
	18:0	12.6% (av.)	14:0	10.6% (av.)					
Milk fats, human	18:1 9c	31.1% (av.)	16:0	21.6% (av.)					
	18:2 9c,12c	11.7% (av.)	14:0	6.6% (av.)					
Mutton tallow	18:1 undefined	30–42%	18:0	22–34%	48	0.946^{15}	1.455^{40}	35–46	
	16:0	20–27%	14:0	2–4%					
Pork lard	18:1 undefined	35–62%	16:0	20–32%	30	0.898^{20}			
	18:0	5–24%	18:2	3–16%					

[a] Jojoba oil consists primarily of wax esters of the acids listed here and long-chain alcohols.
[b] Kapok oil also contains up to 15% cyclopropene acids.
[c] Stillingia oil also contains 5–10% *trans*, *cis*-2,4-decadienoic acid (stillingic acid, 10:2 2t,4c).
[d] Capelin oil also contains about 10% 16:1.

Chemical Name, Synonym(s)	Structure	CAS Registry Number	Molecular Formula	Molecular Weight	UV	Melting Point °C	Optical Rotation
17β-Hydroxyestr-4-en-3-one, 19-Nortestosterone, Nandrolone	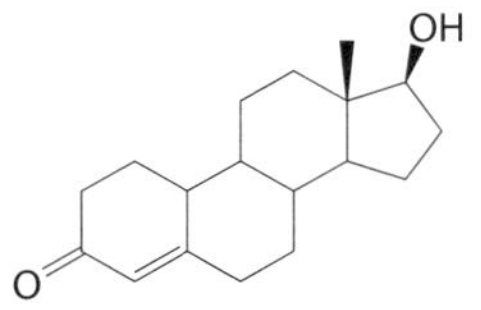	434-22-0	$C_{18}H_{26}O_2$	274.402	λ_{max} 241 (ε 17000) (EtOH)	Mp 111-112°. Mp 124° (double Mp)	$[\alpha]_D + 55$ (c, 0.93 in $CHCl_3$)
Androsta-1,4-diene-3,17-dione	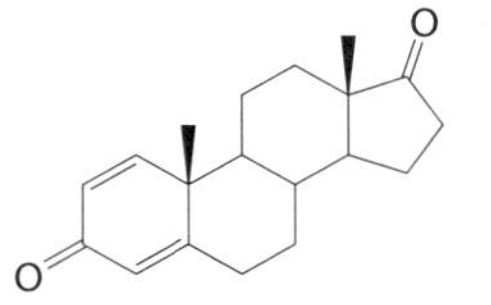	897-06-3	$C_{19}H_{24}O_2$	284.397	λ_{max} 240 (ε 7652) (MeOH)	Mp 141-142°	$[\alpha]_D + 115$ ($CHCl_3$)
5α-Androst-1-ene-3,17-dione	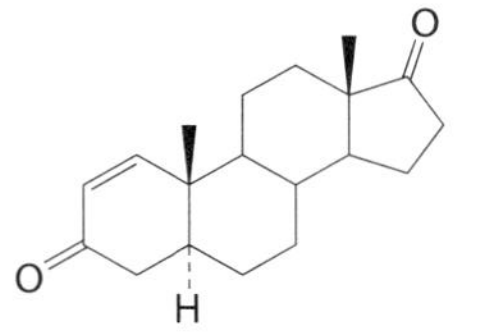	571-40-4	$C_{19}H_{26}O_2$	286.413	λ_{max} 230 (ε 10700) (MeOH)	Mp 141-143°	$[\alpha]^{20}_D + 148.5$ (EtOH)
3β-Hydroxyandrost-5-en-17-one, Dehydroepiandrosterone, Prasterone	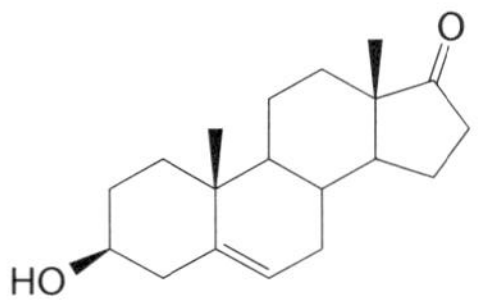	53-43-0	$C_{19}H_{28}O_2$	288.429		Mp 140-141° (needles). Mp 152-153° (leaflets)	$[\alpha]^{23}_D - 3$ ($CHCl_3$)
17β-Hydroxyandrost-4-en-3-one, Testosterone	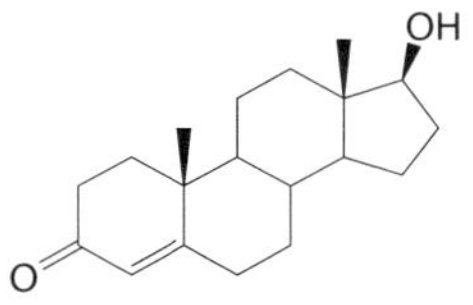	58-22-0	$C_{19}H_{28}O_2$	288.429	λ_{max} 240 (ε 16800) (MeOH)	Mp 155°	$[\alpha]_D + 109$ (c, 4 in EtOH)

ANDROGENS (Continued)

Chemical Name, Synonym(s)	Structure	CAS Registry Number	Molecular Formula	Molecular Weight	UV	Melting Point °C	Optical Rotation
Androst-4-ene-3β,17β-diol	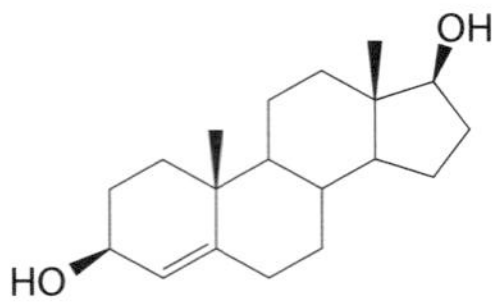	1156-92-9	$C_{19}H_{30}O_2$	290.445		Mp 168° (140-142°)	$[\alpha]^{20}_D$ +48 (c, 1.0 in CHCl$_3$)
3α-Hydroxy-5α-androstan-17-one, Androsterone	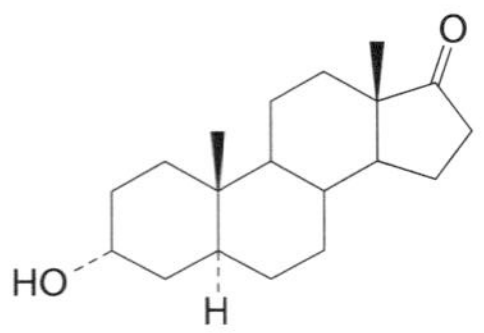	53-41-8	$C_{19}H_{30}O_2$	290.445		Mp 185-185.5°	$[\alpha]^{20}_D$ +94.6 (c, 0.7 in EtOH)
17β-Hydroxy-5α-androstan-3-one, Stanolone	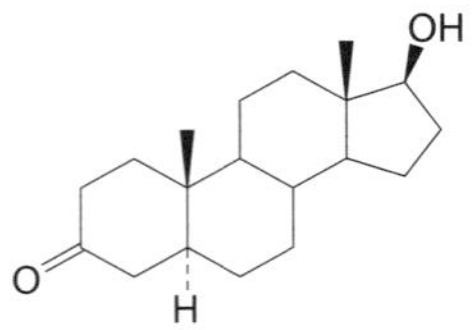	521-18-6	$C_{19}H_{30}O_2$	290.445		Mp 181°	$[\alpha]^{20}_D$ +32.4 (c, 1.0 in EtOH)
17β-Hydroxy-17α-methylandrost-4-en-3-one, 17α-Methyltestosterone	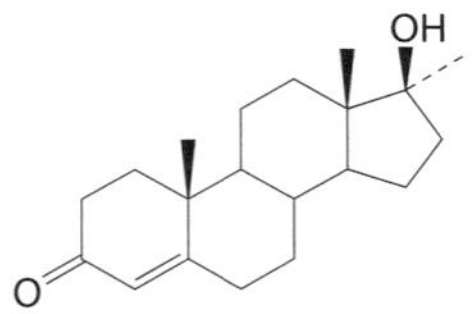	58-18-4	$C_{20}H_{30}O_2$	302.456	λ_{max} 241 (ε 16700) (MeOH)	Mp 164-166°	$[\alpha]_D$ +82 (CHCl$_3$)

Compiled by Fiona Macdonald. Data taken from *The Combined Chemical Dictionary on DVD*, Version 12:1, CRC Press, 2008, www.crcpress.com

BILE ACIDS

Chemical Name, Synonym(s)	Structure	CAS Registry Number	Molecular Formula	Molecular Weight	Melting Point °C	Optical Rotation
3α-Hydroxy-5β-cholan-24-oic acid, Lithocholic acid		434-13-9	$C_{24}H_{40}O_3$	376.578	Mp 186°	$[\alpha]_D^{20}$ +33.7 (c, 1.5 in EtOH)
3α,12α-Dihydroxy-5β-cholan-24-oic acid, Deoxycholic acid		83-44-3	$C_{24}H_{40}O_4$	392.578	Mp 187-189°	$[\alpha]_D$ +53 (CHCl$_3$)
3α,7α-Dihydroxy-5β-cholan-24-oic acid, Chenodeoxycholic acid, Chenodiol		474-25-9	$C_{24}H_{40}O_4$	392.578	Mp119°, 145-146°	$[\alpha]_D$ +11 (EtOH)
3α,6α-Dihydroxy-5β-cholan-24-oic acid, Hyodeoxycholic acid		83-49-8	$C_{24}H_{40}O_4$	392.578	Mp 197-198°	$[\alpha]_D$ +8 (MeOH)
3α,7β-Dihydroxy-5β-cholan-24-oic acid, Ursodeoxycholic acid		128-13-2	$C_{24}H_{40}O_4$	392.578	Mp 203° (198-200°)	$[\alpha]_D^{20}$ +58 (c, 2.0 in EtOH)
3α,7α,12α-Trihydroxy-5β-cholan-24-oic acid, Cholic acid		81-25-4	$C_{24}H_{40}O_5$	408.577	Mp 197°	$[\alpha]_D^{20}$ +37 (EtOH)

Compiled by Fiona Macdonald. Data taken from *The Combined Chemical Dictionary on DVD*, Version 12:1, CRC Press, 2008, www.crcpress.com

CORTICOIDS

Chemical Name, Synonym(s)	Structure	CAS Registry Number	Molecular Formula	Molecular Weight	Melting Point °C	UV	Optical Rotation
17α,21-Dihydroxypregna-1,4-diene-3,11,20-trione, Prednisone	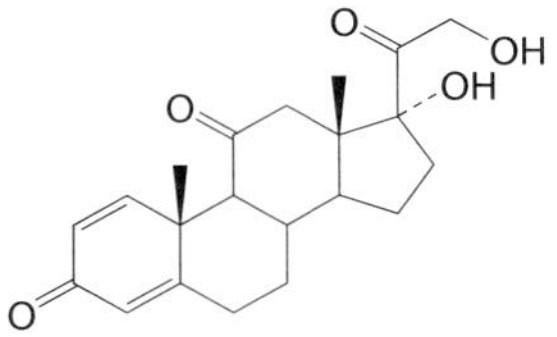	53-03-2	$C_{21}H_{26}O_5$	358.433	Mp 233-235° dec.		$[\alpha]^{25}_D$ +172 (dioxan)
21-Hydroxy-17β-pregn-4-ene-3,11,20-trione, 11-Dehydrocorticosterone	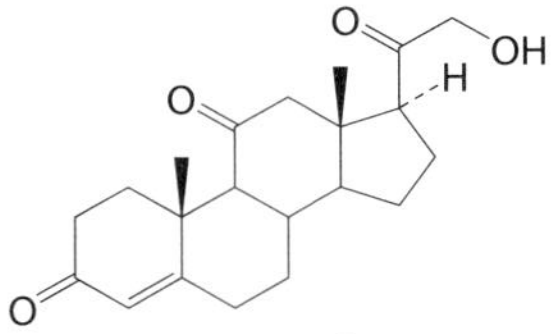	72-23-1	$C_{21}H_{28}O_4$	344.45	Mp 183-183.5°		$[\alpha]_D$ +239 (dioxan)
17α,21-Dihydroxypregn-4-ene-3,11,20-trione, Cortisone	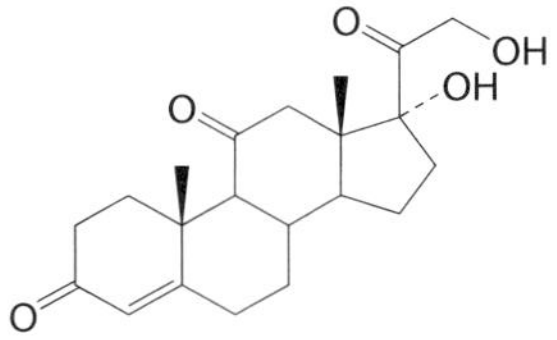	53-06-5	$C_{21}H_{28}O_5$	360.449	Mp 220-224°	λ_{max} 238 (ε 15400) (MeOH)	$[\alpha]_D$ +209 ($CHCl_3$)
11β,17α,21-Trihydroxypregna-1,4-diene-3,20-dione, Prednisolone	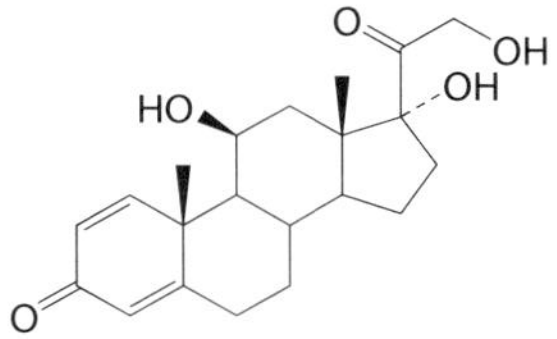	50-24-8	$C_{21}H_{28}O_5$	360.449	Mp 240-241° dec.	λ_{max} 242 (ε 15200) (MeOH)	$[\alpha]^{25}_D$ +102 (dioxan)
9α-Fluoro-11β,17α,21-trihydroxypregn-4-ene-3,20-dione, Fludrocortisone	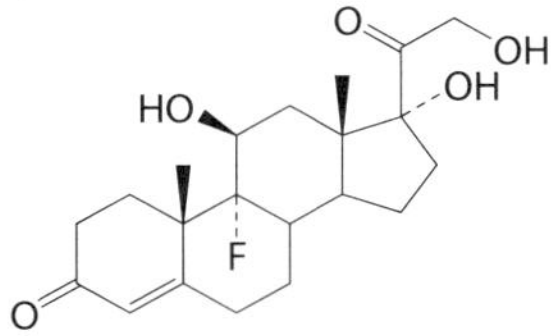	127-31-1	$C_{21}H_{29}FO_5$	380.455	Mp 260-262° dec.	λ_{max} 238 (ε 17100) (MeOH)	$[\alpha]^{23}_D$ +139 (c, 0.55 in EtOH)

CORTICOIDS (Continued)

Chemical Name, Synonym(s)	Structure	CAS Registry Number	Molecular Formula	Molecular Weight	Melting Point °C	UV	Optical Rotation
21-Hydroxy-17β-pregn-4-ene-3,20-dione, 11-Deoxycorticosterone	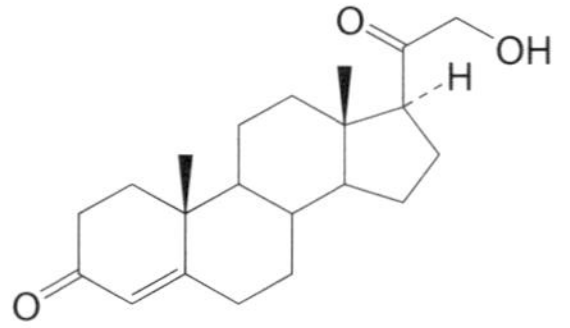	64-85-7	$C_{21}H_{30}O_3$	330.466	Mp 141-142°	λ_{max} 240 (ε 17200) (MeOH)	$[\alpha]^{23}_D$ +178 (EtOH)
11β,21-Dihydroxypregn-4-ene-3,20-dione, Corticosterone	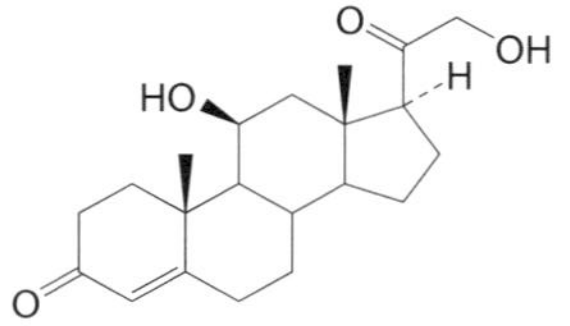	50-22-6	$C_{21}H_{30}O_4$	346.466	Mp 180-182°	λ_{max} 240 (ε 16700) (MeOH)	$[\alpha]^{15}_D$ +223 (c, 1 in EtOH)
11β,17α,21-Trihydroxypregn-4-ene-3,20-dione, Hydrocortisone, Cortisol	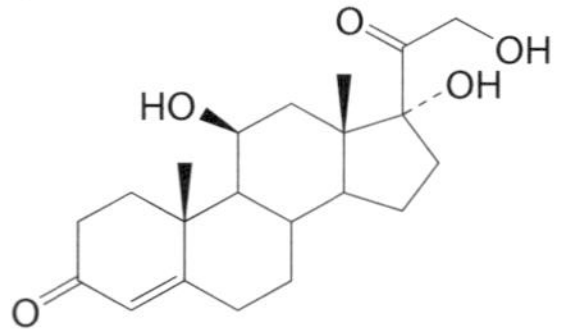	50-23-7	$C_{21}H_{30}O_5$	362.465	Mp 220° dec.	λ_{max} 240 (ε 16300) (MeOH)	$[\alpha]_D$ +167 (CHCl$_3$)
9α-Fluoro-11β,17α,21-trihydroxy-16α-methylpregna-1,4-diene-3,30-dione, Dexamethasone	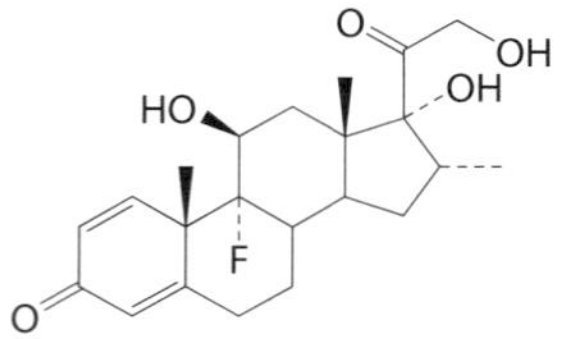	50-02-2	$C_{22}H_{29}FO_5$	392.466	Mp 262-264°	λ_{max} 238 (ε 15400) (MeOH)	$[\alpha]^{25}_D$ +77.5 (dioxan)
9α-Fluoro-11β,17α,21-trihydroxypregna-1,4-diene-3,20-dione 21-acetate, Isoflupredone acetate	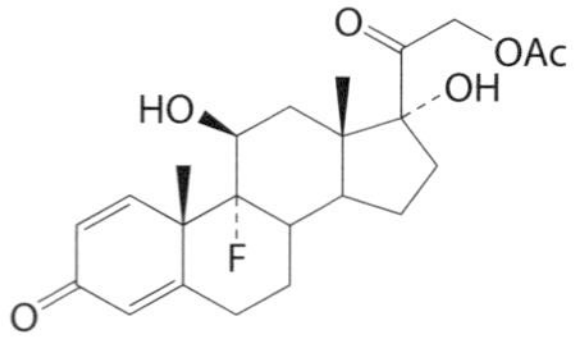		$C_{23}H_{29}FO_6$	420.477	Mp 242-247°		$[\alpha]_D$ +103.5 (Me$_2$CO)

Data taken from *The Combined Chemical Dictionary on DVD*, Version 12:1, CRC Press, 2008, www.crcpress.com

ESTROGENS

Chemical Name, Synonyms(s)	Structure	CAS Registry Number	Molecular Formula	Molecular Weight	Melting Point °C	UV	Optical Rotation
3-Hydroxyestra-1,3,5,7,9-pentaen-17-one, Equilenin		517-09-9	$C_{18}H_{18}O_2$	266.339	Mp 258-259° (250-251°)		$[\alpha]^{16}_D$ +87 (dioxan)
3-Hydroxyestra-1,3,5(10),7-tetraen-17-one, Equilin		474-86-2	$C_{18}H_{20}O_2$	268.355	Mp 238-240°		$[\alpha]^{25}_D$ +325 (c, 2 in EtOH)
3-Hydroxyestra-1,3,5(10)-trien-17-one, Estrone		53-16-7	$C_{18}H_{22}O_2$	270.371	Mp 254°, 256°, 259° (triple Mp)	λ_{max} 280 (ε 2080) (MeOH)	$[\alpha]_D$ +165 (CHCl$_3$)
Estra-1,3,5(10)-triene-3,17-diol; 17β-form, Estradiol		50-28-2	$C_{18}H_{24}O_2$	272.386	Mp 178°	λ_{max} 281 (ε 2120) (MeOH)	$[\alpha]^{18}_D$ +78 (EtOH)
Estra-1,3,5(10)-triene-3,16,17-triol; (16α,17βOH)-form, Estriol		50-27-1	$C_{18}H_{24}O_3$	288.386	Mp 214.6°	λ_{max} 280 (ε 2150) (MeOH)	
Ethinylestradiol		57-63-6	$C_{20}H_{24}O_2$	296.408	Mp 145-146°	λ_{max} 280 (ε 2130) (MeOH)	$[\alpha]_D$ +1 (dioxan)
Ethinylestradiol methyl ether			$C_{21}H_{26}O_2$	310.435		λ_{max} 278 (ε 1050) (MeOH)	$[\alpha]_D$ +3 (CHCl$_3$)

Data taken from *The Combined Chemical Dictionary on DVD, Version 12:1, CRC Press, 2008, www.crcpress.com*

205

PROGESTOGENS

Chemical Name, Synonym(s)	Structure	CAS Registry Number	Molecular Formula	Molecular Weight	Melting Point °C	UV	Optical Rotation
17β-Hydroxy-19-norpregn-4-en-20-yn-3-one, Norethisterone, Norethindrone	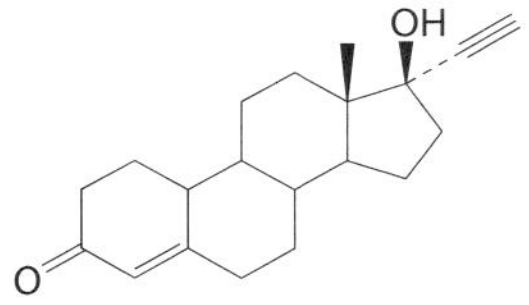	68-22-4	$C_{20}H_{26}O_2$	298.424	Mp 203-204°	λ_{max} 240 (ε 17600) (MeOH)	$[\alpha]^{20}_D$ − 31.7 (CHCl$_3$)
17β-Hydroxy-19-norpregn-5(10)-en-20-yn-3-one, Norethynodrel	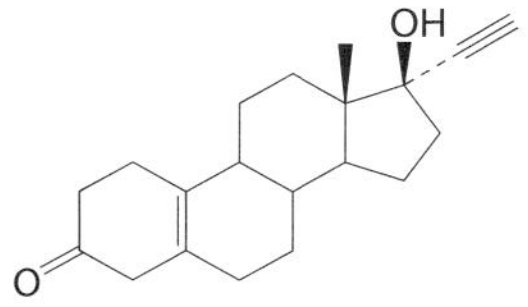	68-23-5	$C_{20}H_{26}O_2$	298.424	Mp 180-181.5° (169-170°)		$[\alpha]_D$ +123.2 (c, 1.00 in dioxan)
Chlormadinone	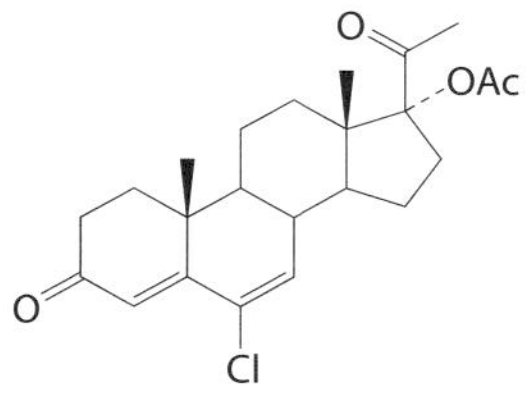	1961-77-9	$C_{21}H_{27}ClO_3$	362.895	Mp 212-214°	λ_{max} 284 (ε 22700) (MeOH)	$[\alpha]_D$ +68 (dioxan)
17β-Hydroxypregn-4-en-20-yn-3-one, Ethisterone	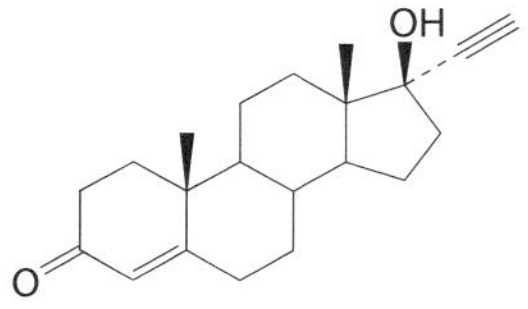	434-03-7	$C_{21}H_{28}O_2$	312.451	Mp 270-272°	λ_{max} 241 (ε 16900) (MeOH)	$[\alpha]^{20}_D$ +22.5 (dioxan)
13-Ethyl-17β-hydroxy-18,19-dinorpregn-4-en-20-yn-3-one, Levonorgestrel	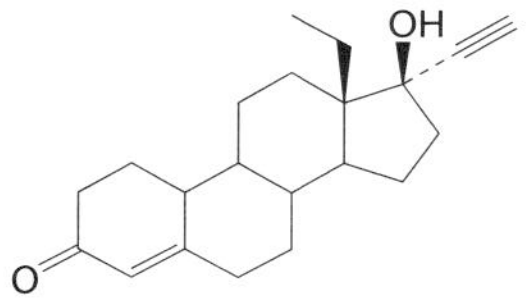	797-63-7	$C_{21}H_{28}O_2$	312.451	Mp 238-242°		$[\alpha]_D$ − 32.4 (c, 0.5 in CHCl$_3$)
Pregn-4-ene-3,20-dione, Progesterone	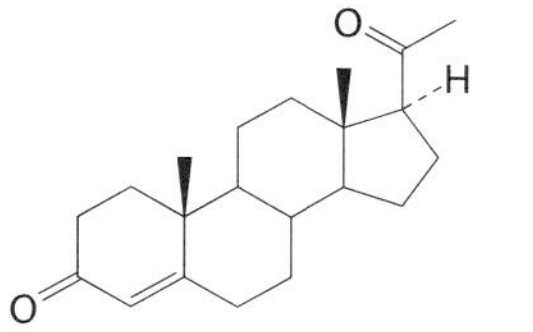	57-83-0	$C_{21}H_{30}O_2$	314.467	Mp 121-122° . Mp 127-131° (dimorph.)	λ_{max} 240 (ε 17000) (MeOH)	$[\alpha]^{20}_D$ +192 (c, 2 in dioxan)

PROGESTOGENS (Continued)

Chemical Name, Synonym(s)	Structure	CAS Registry Number	Molecular Formula	Molecular Weight	Melting Point °C	UV	Optical Rotation
17β-Acetoxy-19-norpregn-4-en-20-yn-3-one, Norethisterone acetate		51-98-9	$C_{22}H_{28}O_3$	340.461	Mp 161-163°	λ_{max} 239 (ε 17700) (MeOH)	$[\alpha]_D$ − 33 (CHCl$_3$)
17α-Acetoxy-6-methylpregna-4,6-diene-3,20-dione, Megestrol acetate		595-33-5	$C_{24}H_{32}O_4$	384.514	Mp 214-216°	λ_{max} 289 (ε 24000) (MeOH)	$[\alpha]_D$ +10 (c, 5 in CHCl$_3$)
17α-Acetoxy-6α-methylpregn-4-ene-3,20-dione, Medroxyprogesterone acetate		71-58-9	$C_{24}H_{34}O_4$	386.53	Mp 205-209°	λ_{max} 240 (ε 16000) (MeOH)	$[\alpha]_D$ +66 (CHCl$_3$)
17α-Acetoxy-6-methyl-16-methylenepregna-4,6-diene-3,20-dione, Melengestrol acetate		2919-66-6	$C_{25}H_{32}O_4$	396.525	Mp 224-226°	λ_{max} 287 (ε 22400) (MeOH)	$[\alpha]^{22}_D$ − 127 (c, 0.31 in CHCl$_3$)

Data taken from *The Combined Chemical Dictionary on DVD*, Version 12:1, CRC Press, 2008, www.crcpress.com

STEROLS

Chemical Name, Synonym(s)	Structure	CAS Registry Number	Molecular Formula	Molecular Weight	Melting Point °C	Optical Rotation
Cholesta-3,5-dien-7-one	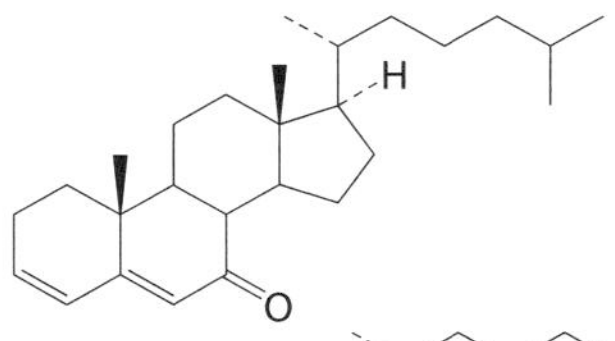	567-72-6	$C_{27}H_{42}O$	382.628	Mp 116°	$[\alpha]^{20}_D - 30$ (c, 1.1 in CHCl$_3$)
Cholesta-4,6-dien-3-one	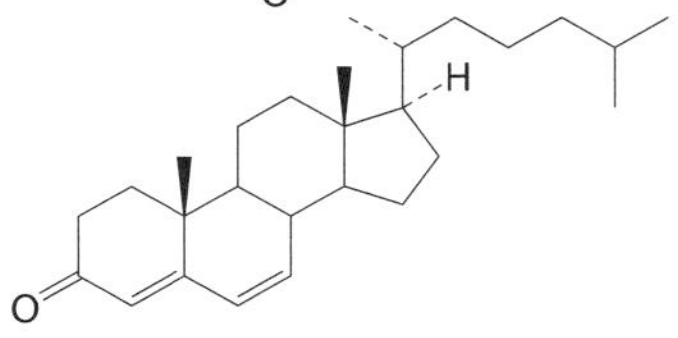	566-93-8	$C_{27}H_{42}O$	382.628	Mp 81-83°	$[\alpha]_D + 36$ (c, 1.3 in CHCl$_3$)
Cholesta-5,7,22-trien-3β-ol	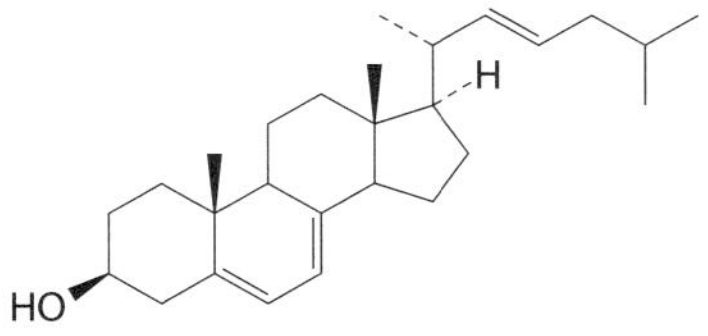		$C_{27}H_{42}O$	382.628		
Cholesta-5,7-dien-3β-ol, 7-Dehydrocholesterol	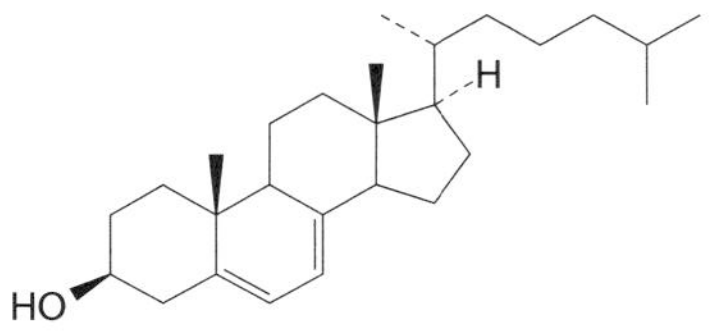	434-16-2	$C_{27}H_{44}O$	384.644	Mp 150-151° (anhyd.)	$[\alpha]^{20}_D - 113.6$ (CHCl$_3$)
Cholesta-5,24-dien-3β-ol, Desmosterol	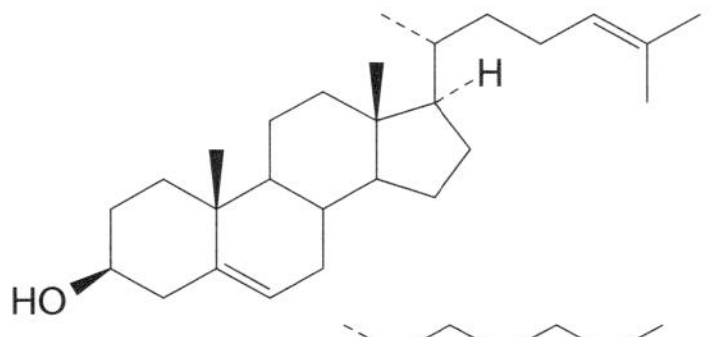	313-04-2	$C_{27}H_{44}O$	384.644	Mp 121.5-122.5°	$[\alpha]^{20}_D - 31.2$ (c, 1.1 in CHCl$_3$)
Cholest-4-en-3-one	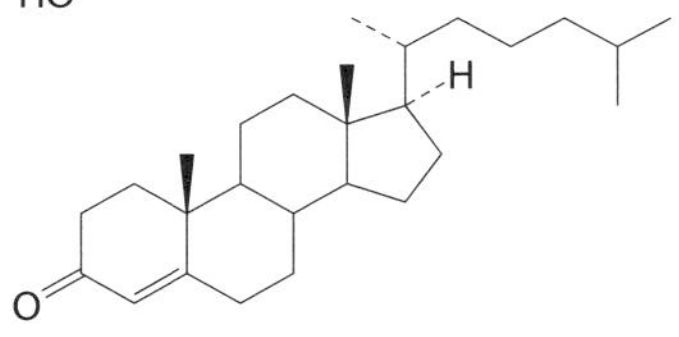	601-57-0	$C_{27}H_{44}O$	384.644	Mp 80-81.5°	$[\alpha]^{22}_D + 87$ (c, 1 in CHCl$_3$)

STEROLS (Continued)

Chemical Name, Synonym(s)	Structure	CAS Registry Number	Molecular Formula	Molecular Weight	Melting Point °C	Optical Rotation
Cholesta-5,22-dien-3β-ol	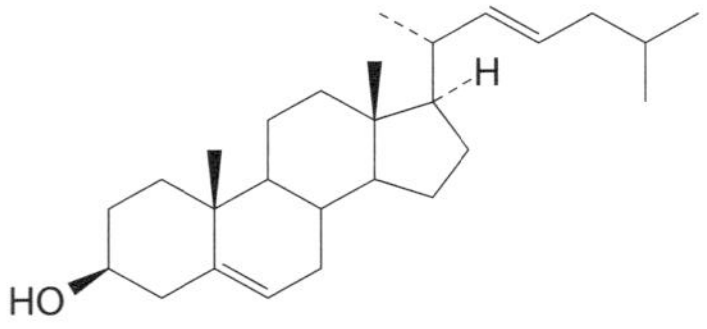	34347-28-9	$C_{27}H_{44}O$	384.644	Mp 133.5-134°	$[\alpha]^{23}_D - 60$ (c, 0.1 in $CHCl_3$)
5α-Cholesta-8,24-dien-3β-ol, Zymosterol	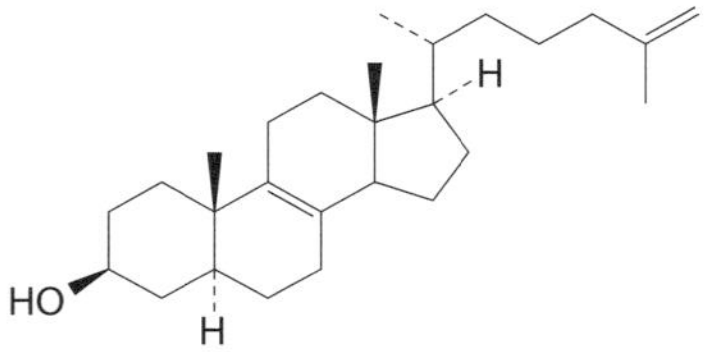	128-33-6	$C_{27}H_{44}O$	384.644	Mp 110°	$[\alpha]^{20}_D +49$ (c, 1 in $CHCl_3$)
5α-Cholestane-3,6-dione	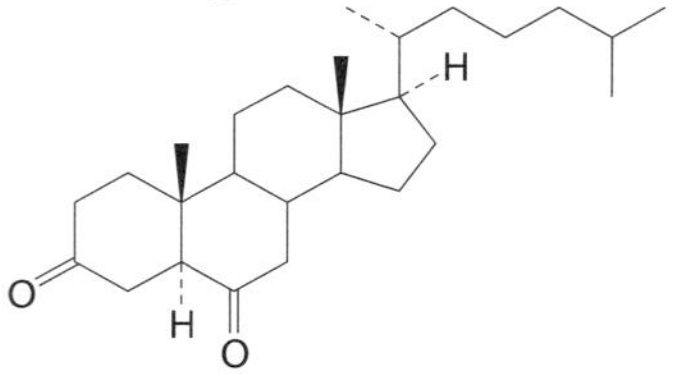	2243-09-6	$C_{27}H_{44}O_2$	400.643	Mp 172-173°	$[\alpha]^{25}_D +4$ (c, 1.6 in $CHCl_3$)
3β-Hyd-oxycholest-5-en-7-one, 7-Oxocholesterol	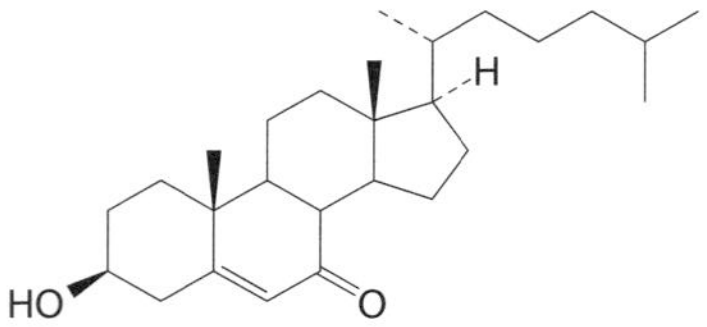	566-28-9	$C_{27}H_{44}O_2$	400.643	Mp 157° and 170° (double Mp)	$[\alpha]_D -108$ (c, 0.9 in $CHCl_3$)
Cholest-5-en-3β-ol, Cholesterol	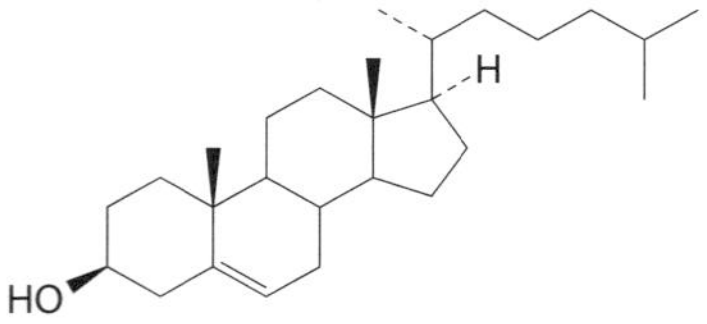	57-88-5	$C_{27}H_{46}O$	386.66	Mp 148.5° (anhyd.).	$[\alpha]_D - 38.2$ (Et_2O)

STEROLS (Continued)

Chemical Name, Synonym(s)	Structure	CAS Registry Number	Molecular Formula	Molecular Weight	Melting Point °C	Optical Rotation
5α-Cholest-7-en-3β-ol, Lathosterol		80-99-9	$C_{27}H_{46}O$	386.66	Mp 125–127°	$[\alpha]^{23}_D +4.8$ (c, 0.90 in $CHCl_3$)
5β-Cholestan-3-one, Coprostanone		601-53-6	$C_{27}H_{46}O$	386.66	Mp 63°	$[\alpha]_D +37$ ($CHCl_3$)
Cholest-4-ene-3β,6β-diol		570-88-7	$C_{27}H_{46}O_2$	402.659	Mp 257–258°	$[\alpha]^{20}_D +9$ (c, 0.6 in $CHCl_3$)
Cholest-5-ene-3β,7α-diol, 7α-Hydroxycholesterol		566-26-7	$C_{27}H_{46}O_2$	402.659	Mp 188–189°	$[\alpha]_D -91$ (c, 0.9 in $CHCl_3$)
Cholest-5-ene-3β,7β-diol, 7β-Hydroxycholesterol		566-27-8	$C_{27}H_{46}O_2$	402.659	Mp 172–176°. Mp 180–181° (double Mp)	$[\alpha]_D +3.3$ ($CHCl_3$)

STEROLS (Continued)

Chemical Name, Synonym(s)	Structure	CAS Registry Number	Molecular Formula	Molecular Weight	Melting Point °C	Optical Rotation
20S-Cholest-5-ene-3β,22R-diol, 22R-Hydroxycholesterol, Narthesterol	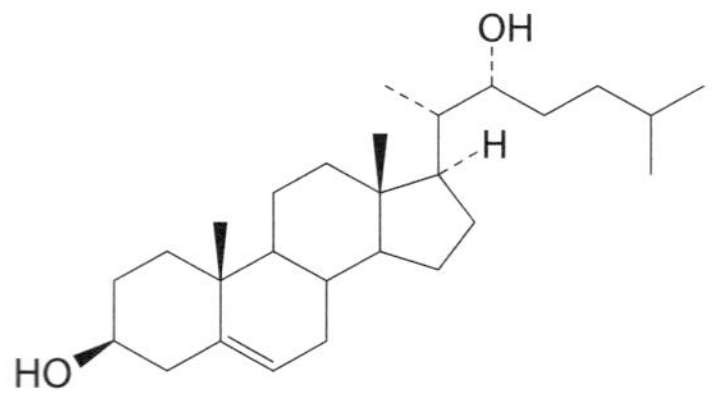	17954-98-2	$C_{27}H_{46}O_2$	402.659	Mp 184-185°	$[\alpha]_D - 38 \ (CHCl_3)$
3β-Hydroxy-5α-cholestan-6-one	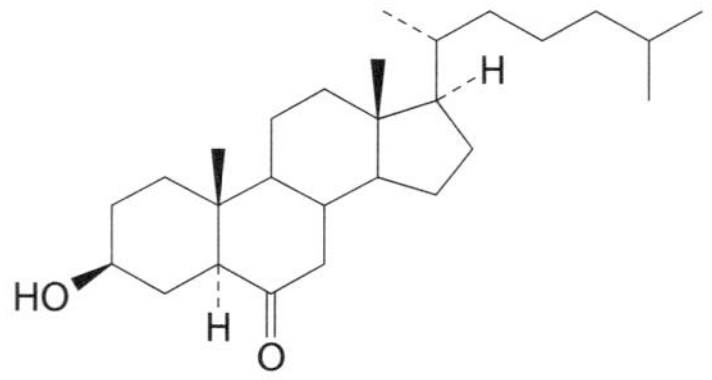	1175-06-0	$C_{27}H_{46}O_2$	402.659	Mp 142-143°	$[\alpha]^{24}_D - 5.1 \ (CHCl_3)$
3β,5α-Dihydroxycholestan-6-one	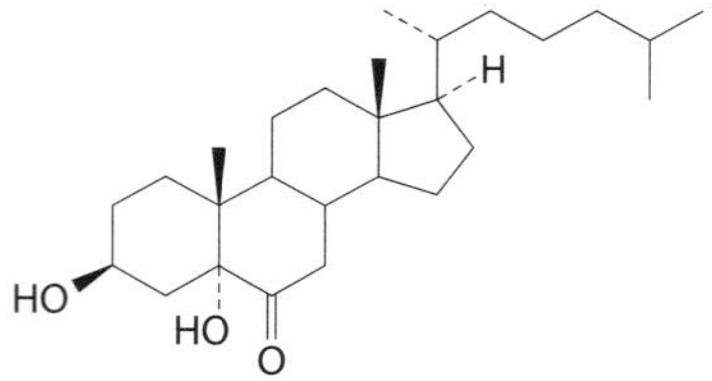	13027-33-3	$C_{27}H_{46}O_3$	418.659	Mp 231-232°	$[\alpha]_D - 32 \ (CHCl_3)$
5β-Cholestan-3β-ol, Coprostanol	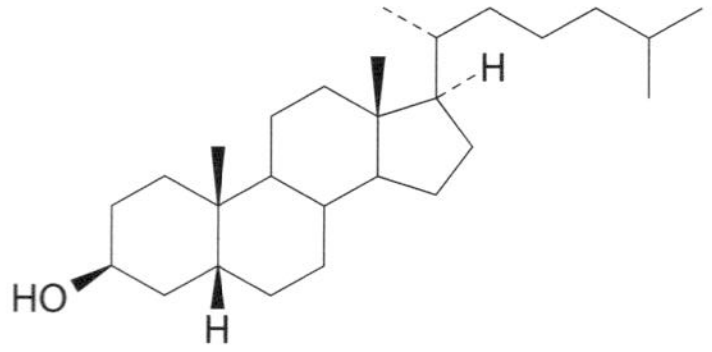	360-68-9	$C_{27}H_{48}O$	388.676	Mp 96-100°	$[\alpha]_D +26 \ (CHCl_3)$
5α-Cholestan-3β-ol, Cholestanol	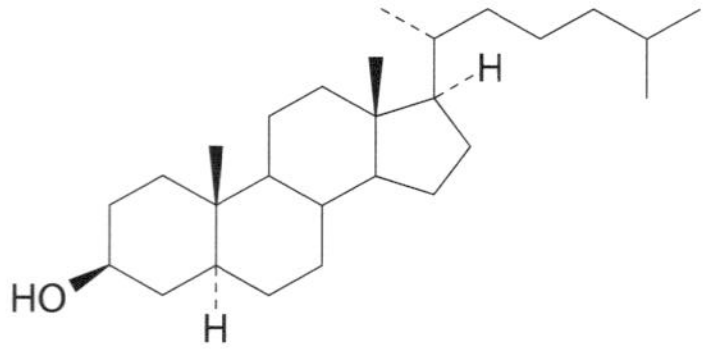	80-97-7	$C_{27}H_{48}O$	388.676	Mp 142-143°	$[\alpha]^{22}_D +27.4 \ (CHCl_3)$

STEROLS (Continued)

Chemical Name, Synonym(s)	Structure	CAS Registry Number	Molecular Formula	Molecular Weight	Melting Point °C	Optical Rotation
5β-Cholestan-3α-ol, Epicoprostanol	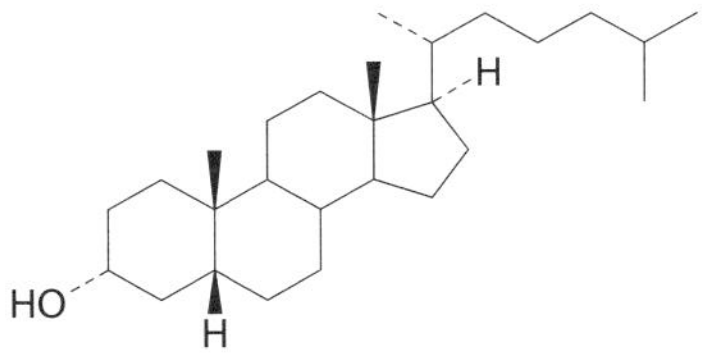	516-92-7	$C_{27}H_{48}O$	388.676	Mp 107-108°	$[\alpha]_D +31$ (CHCl$_3$)
Cholestane-3β,5β,6β-triol	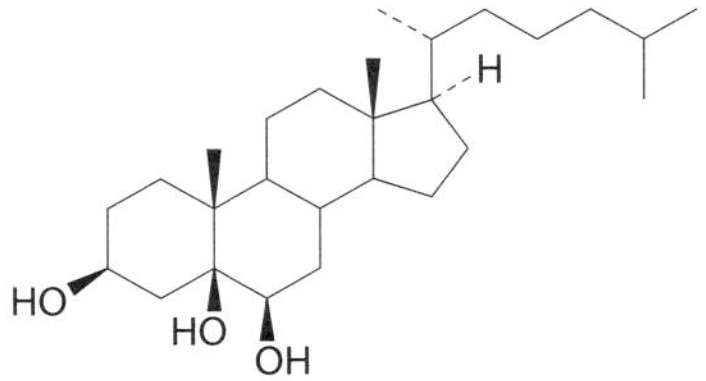	79254-30-1	$C_{27}H_{48}O_3$	420.674	Mp 127-128°	$[\alpha]_D^{20} +22$ (c, 1.0 in CHCl$_3$)
24R-Ergosta-5,7,9(11),22E-tetraen-3β-ol	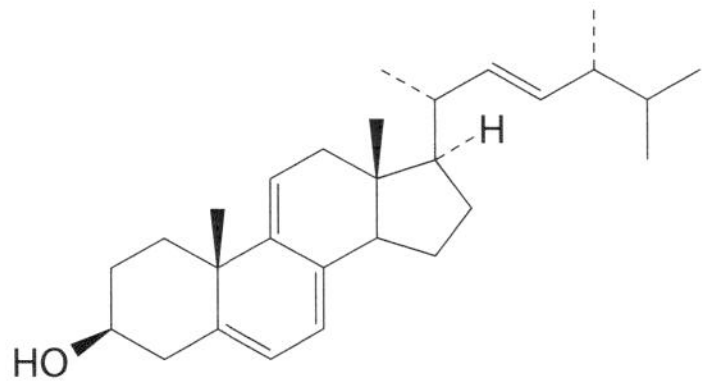	516-85-8	$C_{28}H_{42}O$	394.639	Mp 145-147°	$[\alpha]_D^{20} +147$ (CHCl$_3$)
24S-Ergosta-5,7,14,22E-tetraen-3β-ol, 14-Dehydroergosterol	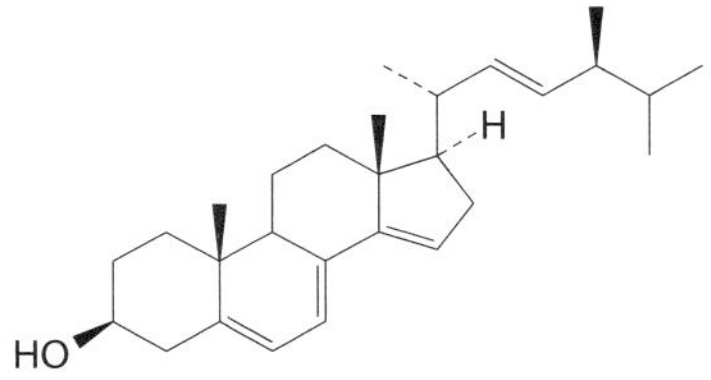		$C_{28}H_{42}O$	394.639	Mp 198-201° dec.	$[\alpha]_D -396$ (c, 0.21 in CCl$_4$)
Ergosta-5,7,22E,24(28)-tetraen-3β-ol	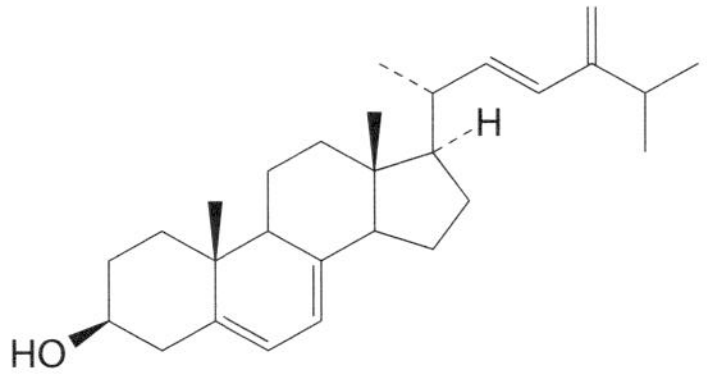	29560-24-5	$C_{28}H_{42}O$	394.639	Mp 118-119°	$[\alpha]_D^{25} -78$ (CHCl$_3$)

STEROLS (Continued)

Chemical Name, Synonym(s)	Structure	CAS Registry Number	Molecular Formula	Molecular Weight	Melting Point °C	Optical Rotation
24R-Ergosta-5,7,22E-trien-3β-ol, Ergosterol	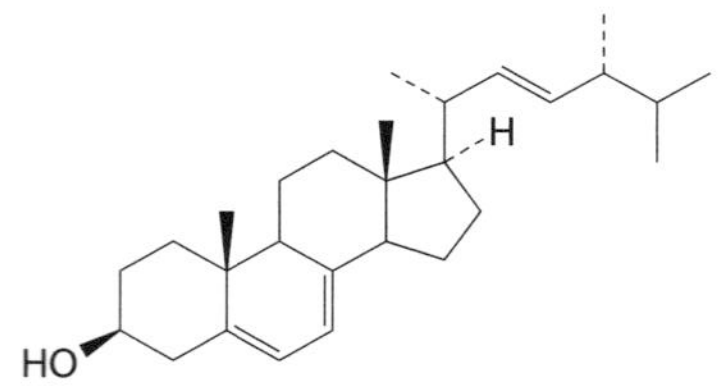	57-87-4	$C_{28}H_{44}O$	396.655	Mp 168°	$[\alpha]_D - 133$ (CHCl$_3$)
24S-Ergosta-5,7-dien-3β-ol, 22,23-Dihydroergosterol Provitamin D4	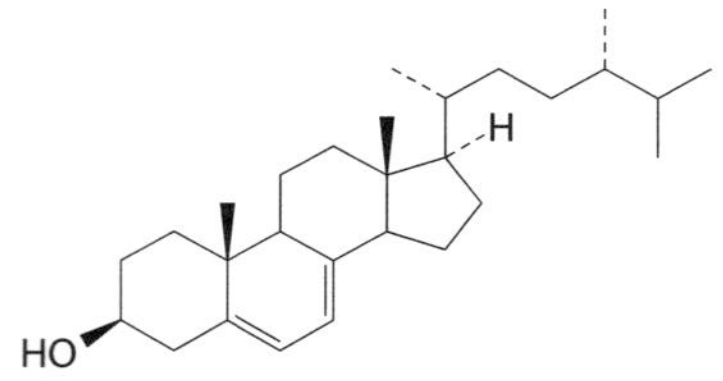	516-79-0	$C_{28}H_{46}O$	398.671	Mp 152-153°	$[\alpha]_D - 128.7$ (c, 0.4 in CHCl$_3$)
24R-Ergosta-5,22E-dien-3β-ol, Brassicasterol	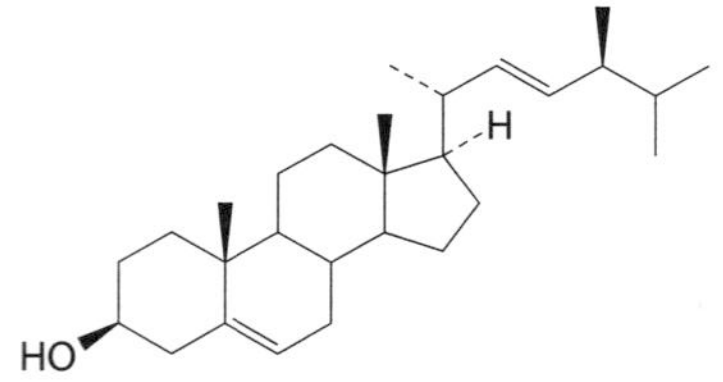	474-67-9	$C_{28}H_{46}O$	398.671	Mp 157-158° (148°)	$[\alpha]^{19}_D - 54$ (CHCl$_3$)
Ergosta-5,24(28)-dien-3β-ol, 24-Methylenecholesterol	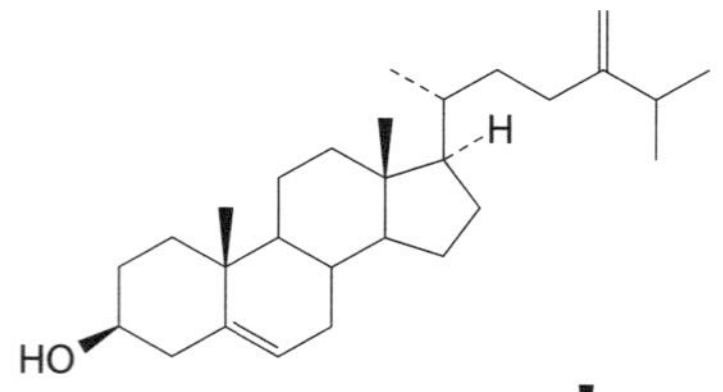	474-63-5	$C_{28}H_{46}O$	398.671	Mp 142-143°	$[\alpha]^{20}_D - 43.6$ (CHCl$_3$)
5α,24S-Ergosta-7,22E-dien-3β-ol, α-Dihydroergosterol	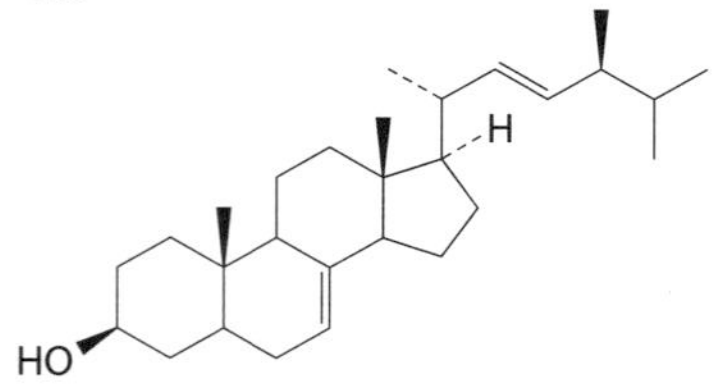	50364-22-2	$C_{28}H_{46}O$	398.671	Mp 173.5-174°	$[\alpha]_D - 23.4$

STEROLS (Continued)

Chemical Name, Synonym(s)	Structure	CAS Registry Number	Molecular Formula	Molecular Weight	Melting Point °C	Optical Rotation
5α-Ergosta-7,24(28)-dien-3β-ol, Episterol	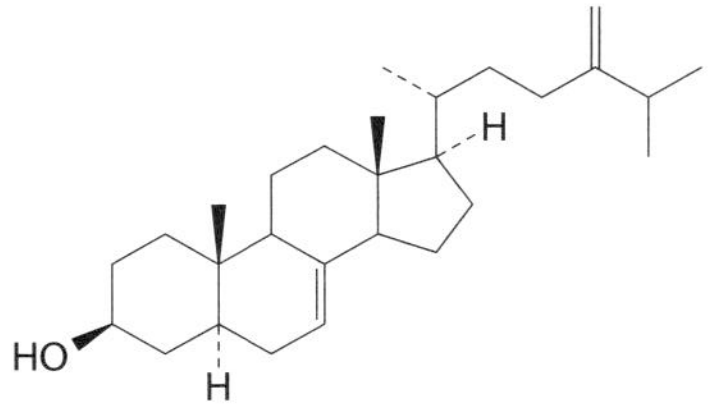	474-68-0	$C_{28}H_{46}O$	398.671	Mp 127.5-128.5° (150-151°)	$[\alpha]_D$ +4.4 (c, 2.3 in $CHCl_3$)
5α-Ergosta-8,23E-dien-3β-ol, Ascosterol	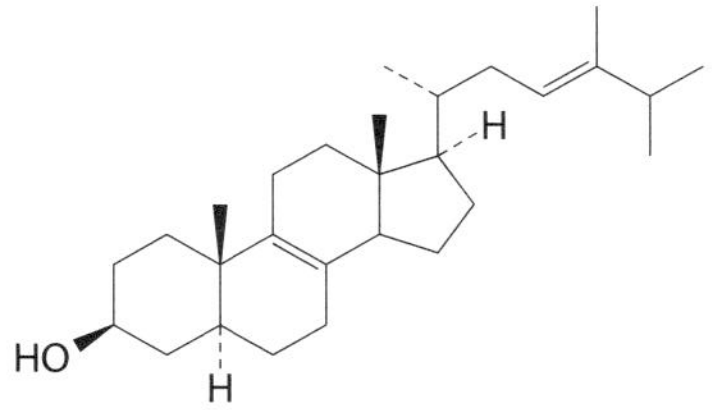	474-72-6	$C_{28}H_{46}O$	398.671	Mp 152-154°	$[\alpha]_D^{20}$ +37.5 ($CHCl_3$)
5α-Ergosta-8,24(28)-dien-3β-ol, Fecosterol	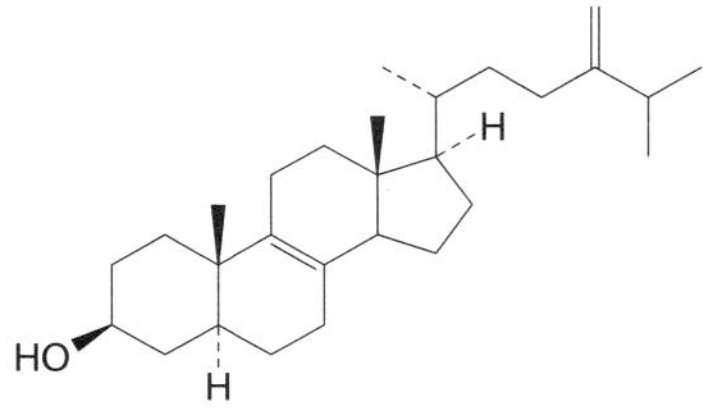	516-86-9	$C_{28}H_{46}O$	398.671	Mp 126-130°	$[\alpha]_D$ +44.8 ($CHCl_3$)
24R-Ergosta-7,22E-diene-3β,5α,6β-triol, Cerevisterol	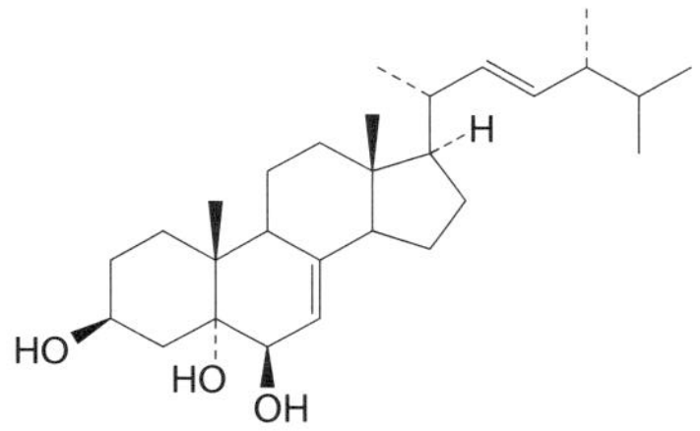	516-37-0	$C_{28}H_{46}O_3$	430.67	Mp 254-256°	$[\alpha]_D^{21}$ −79.9 (c, 1.35 in Py)
5α,24S-Ergost-7-en-3β-ol, Fungisterol, γ-Ergostenol	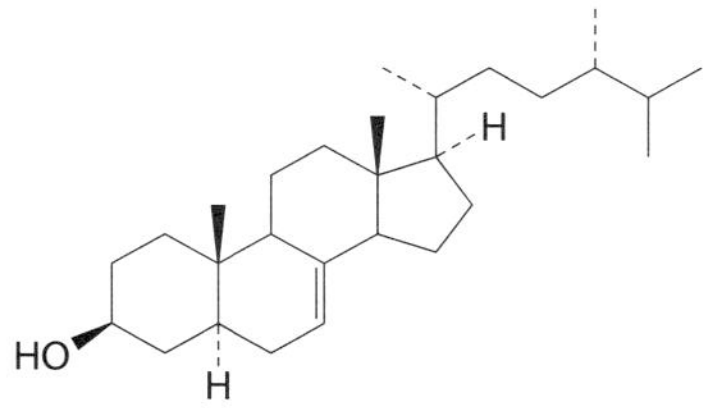	516-78-9	$C_{28}H_{48}O$	400.687	Mp 152°	$[\alpha]_D$ − 0.2 ($CHCl_3$)

STEROLS (Continued)

Chemical Name, Synonym(s)	Structure	CAS Registry Number	Molecular Formula	Molecular Weight	Melting Point °C	Optical Rotation
24R-Ergost-5-en-3β-ol, Campesterol	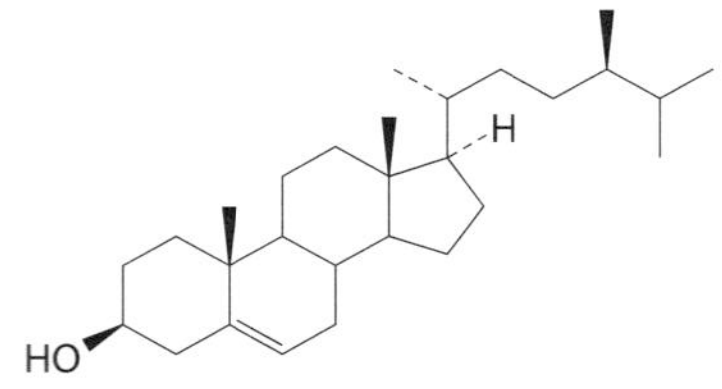	474-62-4	$C_{28}H_{48}O$	400.687	Mp 157-158°	$[\alpha]^{23}_{D} -33$ (CHCl$_3$)
5α,24R-Ergost-22E-en-3β-ol, Neospongosterol	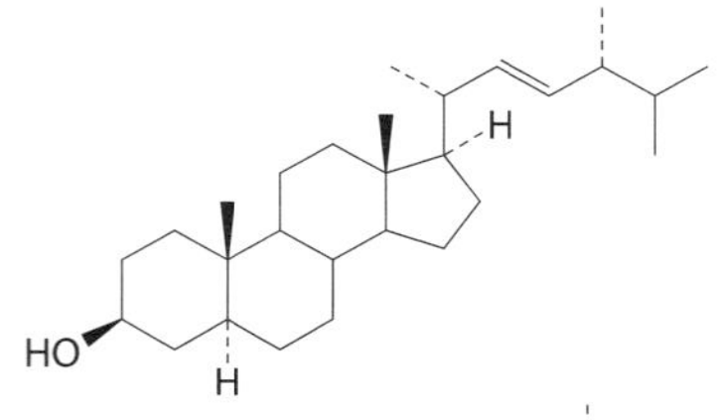	6538-02-9	$C_{28}H_{48}O$	400.687	Mp 153°	$[\alpha]^{24}_{D} +10$ (CHCl$_3$)
5α,24S-Ergostan-3β-ol, Ergostanol	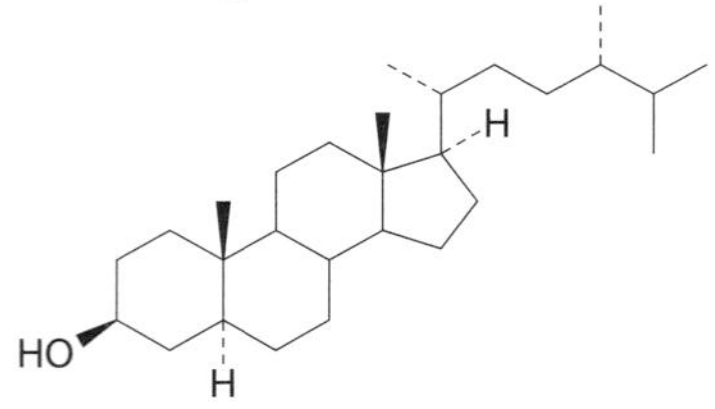		$C_{28}H_{50}O$	402.702	Mp 144-145°	$[\alpha]_{D} +15.94$ (CHCl$_3$)
25-Methylergosta-5,7,22E-trien-3β-ol, Haliclonasterol	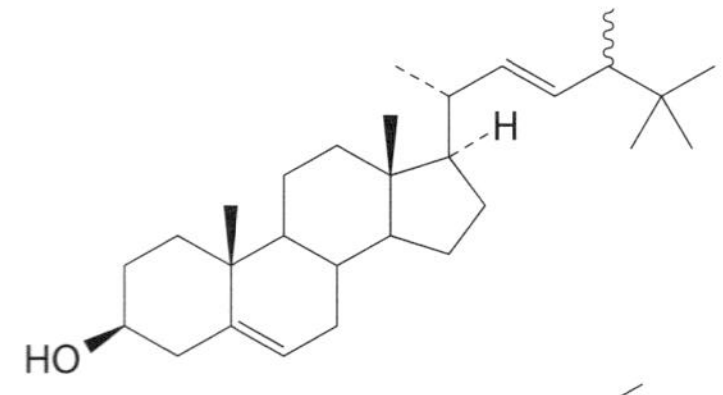	32352-65-1	$C_{29}H_{46}O$	410.682	Mp 140.5-141°	$[\alpha]^{27}_{D} -41.5$ (CHCl$_3$)
24R-Stigmasta-5,7,22-trien-3β-ol, Corbisterol	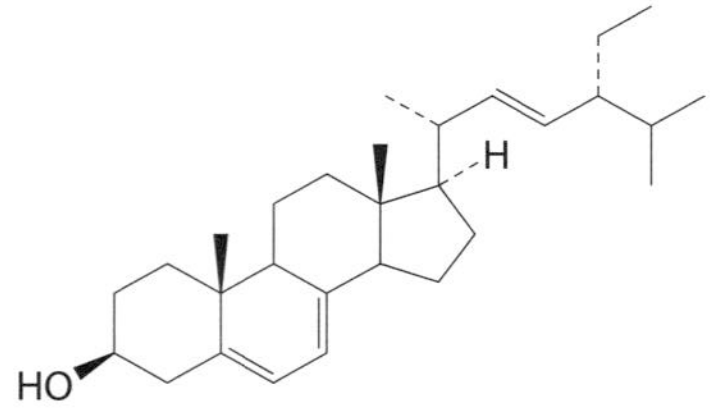	481-19-6	$C_{29}H_{46}O$	410.682	Mp 151-152°	$[\alpha]_{D} -105.5$ (EtOH)

STEROLS (Continued)

Chemical Name, Synonym(s)	Structure	CAS Registry Number	Molecular Formula	Molecular Weight	Melting Point °C	Optical Rotation
24S-Stigmasta-5,22E-dien-3β-ol, Stigmasterol	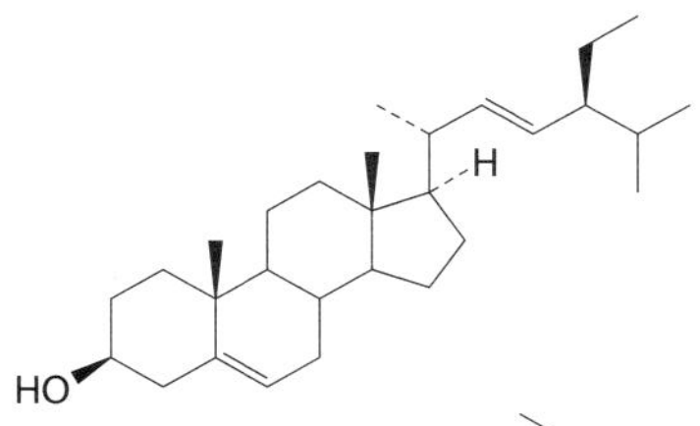	83-48-7	$C_{29}H_{48}O$	412.698	Mp 170°	$[\alpha]^{22}_{D} - 57$ (CHCl$_3$)
Stigmasta-5,24(28)E-dien-3β-ol, Fucosterol	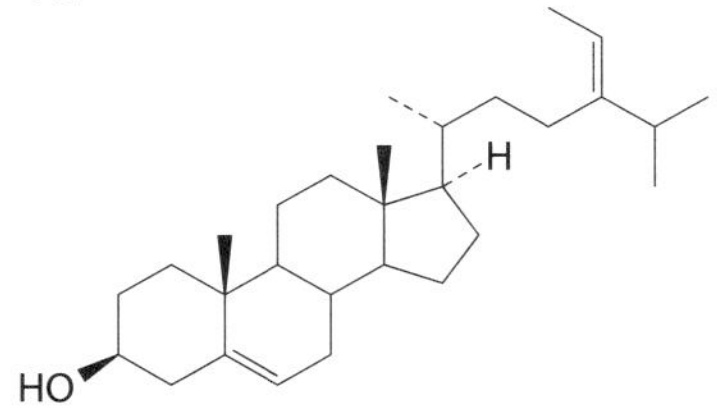	17605-67-3	$C_{29}H_{48}O$	412.698	Mp 124°	$[\alpha]^{20}_{D} - 38.4$ (CHCl$_3$)
5α,24R-Stigmasta-7,22E-dien-3β-ol, Chondrillasterol	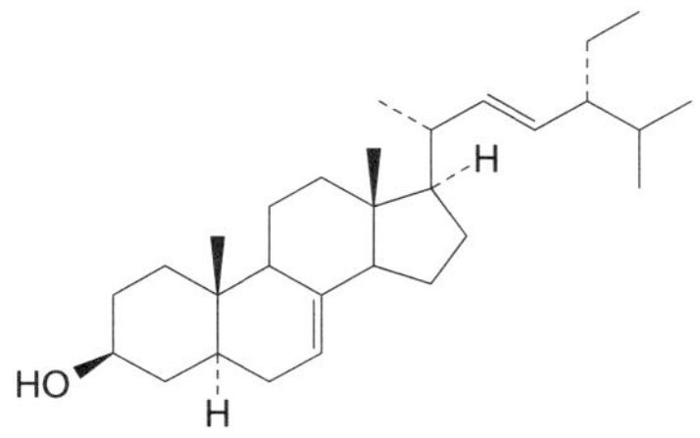	481-17-4	$C_{29}H_{48}O$	412.698	Mp 174.5-175.5°	$[\alpha]^{20}_{D} - 2$ (CHCl$_3$)
24R-Stigmasta-5,22E-dien-3-ol, Poriferasterol	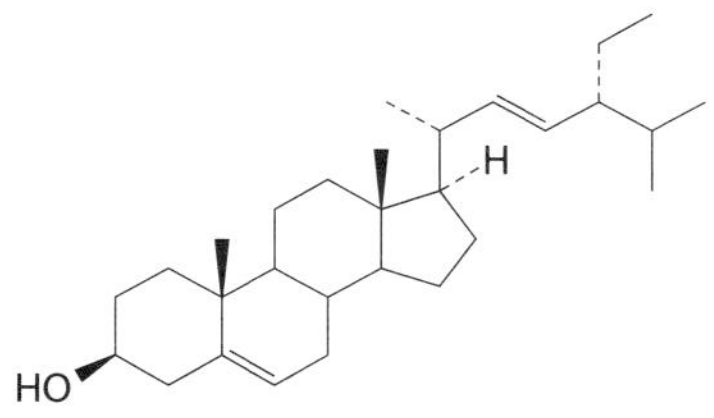	481-16-3	$C_{29}H_{48}O$	412.698	Mp 156°	$[\alpha]_{D} - 46$ (CHCl$_3$)
20S-Stigmasta-5,24(28)E-dien-3β-ol, Sargasterol	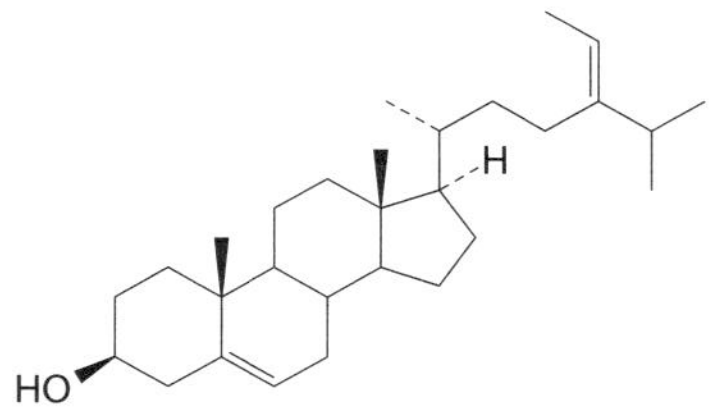	481-15-2	$C_{29}H_{48}O$	412.698		$[\alpha]_{D} - 47.5$ (CHCl$_3$)

STEROLS (Continued)

Chemical Name, Synonym(s)	Structure	CAS Registry Number	Molecular Formula	Molecular Weight	Melting Point °C	Optical Rotation
Stigmasta-5,24(28)-dien-3β-ol, Δ⁵-Avenasterol	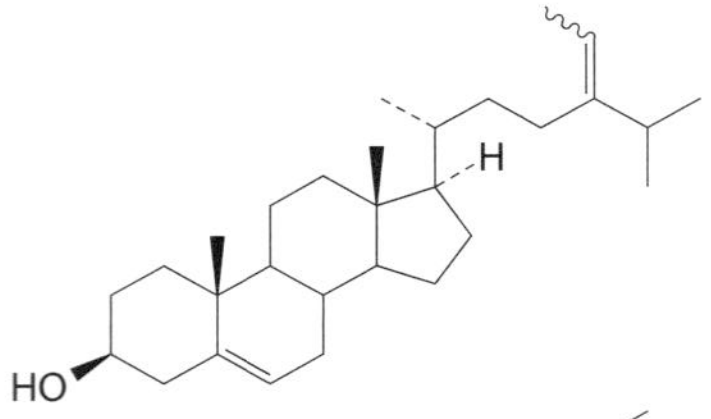	18472-36-1	$C_{29}H_{48}O$	412.698	Mp 137°	$[\alpha]^{26}_{D} - 37.6$ (CHCl₃)
5α-Stigmasta-7,24(28)Z-dien-3β-ol, Δ⁷-Avenasterol	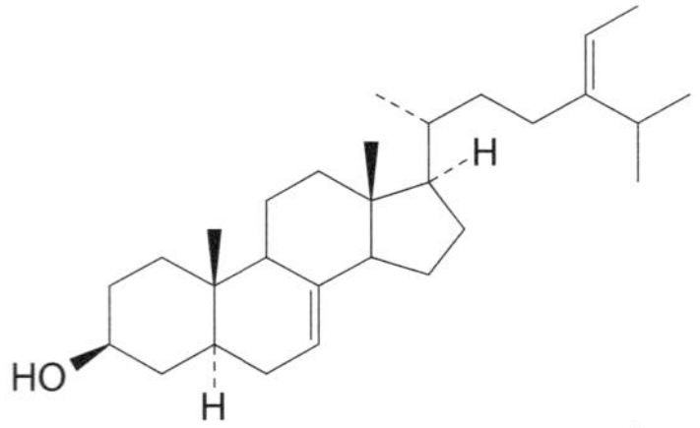	23290-26-8	$C_{29}H_{48}O$	412.698	Mp 148–151°	$[\alpha]^{22}_{D} +12.7$ (c, 1.28 in CHCl₃)
5α,24S-Stigmasta-7,22E-dien-3β-ol, α-Spinasterol	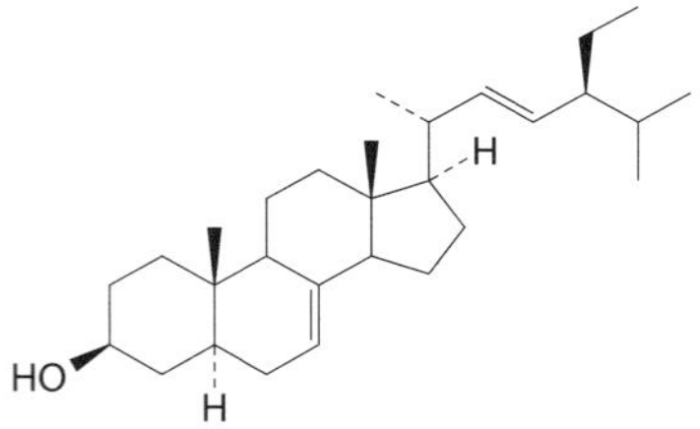	481-18-5	$C_{29}H_{48}O$	412.698	Mp 171–173°	$[\alpha]^{22}_{D} - 2.5$ (CHCl₃)
24R-Stigmast-5-en-3β-ol, β-Sitosterol	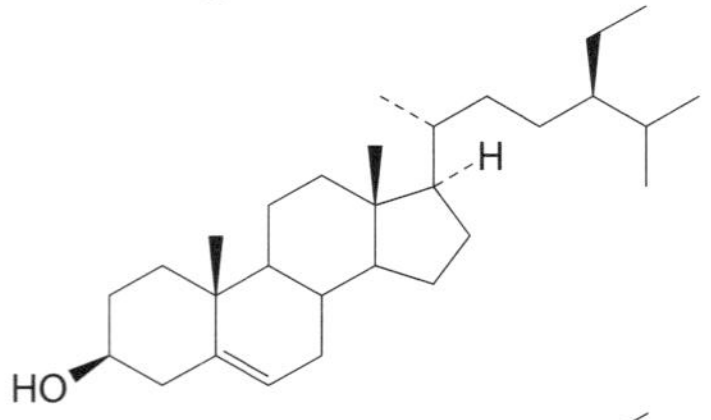	83-46-5	$C_{29}H_{50}O$	414.713	Mp 136–137°	$[\alpha]^{22}_{D} - 35$ (CHCl₃)
24S-Stigmast-5-en-3β-ol, γ-Sitosterol	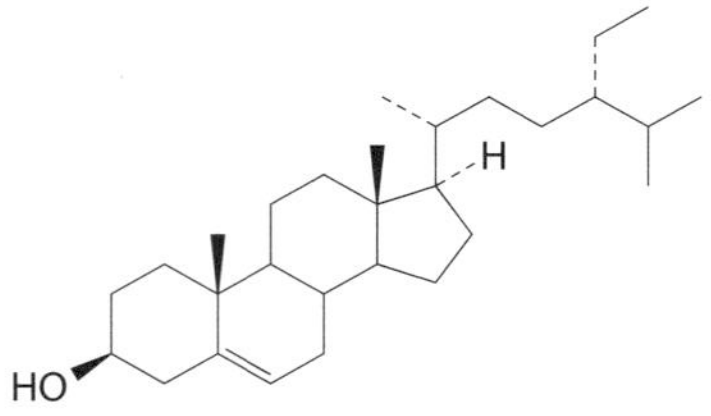	83-47-6	$C_{29}H_{50}O$	414.713	Mp 147–148°	$[\alpha]_{D} - 47.7$ (CHCl₃)

STEROLS (Continued)

Chemical Name, Synonym(s)	Structure	CAS Registry Number	Molecular Formula	Molecular Weight	Melting Point °C	Optical Rotation
$5\alpha,24R$-Stigmast-7-en-3β-ol	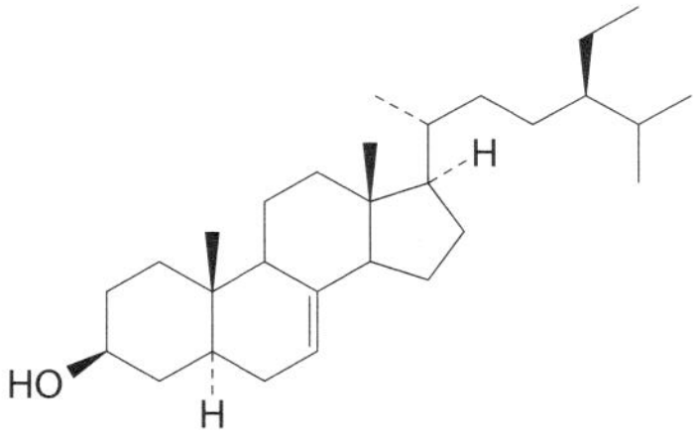	521-03-9	$C_{29}H_{50}O$	414.713	Mp 153-154° (137-140°)	$[\alpha]^{30}_D$ +9.1 (c, 0.95 in $CHCl_3$)
$5\alpha,24S$-Stigmast-22Z-en-3β-ol	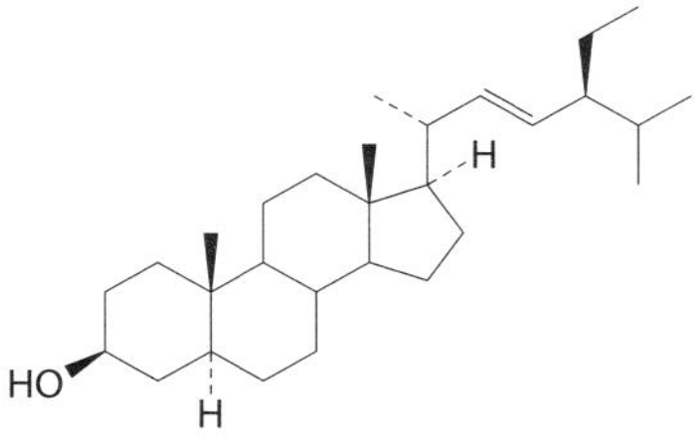	4736-56-5	$C_{29}H_{50}O$	414.713	Mp 158-159°	$[\alpha]_D$ +3.3 (c, 1.5 in $CHCl_3$)
$5\alpha,24R$-Stigmastan-3β-ol, Fucostanol	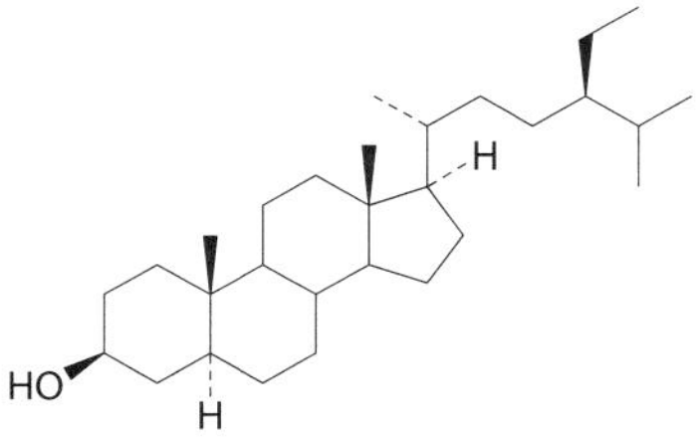	83-45-4	$C_{29}H_{52}O$	416.729	Mp 136-137°	$[\alpha]^{20}_D$ +24.8 ($CHCl_3$)

Data taken from *The Combined Chemical Dictionary on DVD*, **Version 12:1, CRC Press, 2008, www.crcpress.com**

PROSTAGLANDINS AND RELATED FATTY-ACID-DERIVED MATERIALS

Eicosanoid

The term eicosanoid is derived from "icos" referring to 20. The term eicosanoid refers to a diverse group of long-chain derivatives of eicosanoic acid which is more commonly referred to as arachidic acid. Arachidic acid is a saturated C_{20} fatty acid. Eicosanoid is used to include leukotriene, prostaglandins, lipoxins, and other derivatives of arachidonic acid. See Famaey, J.P., Phospholipases, eicosanoid production and inflammation, *Clin. Rheumatol.* 1, 84-94, 1982; Weber, P.C., Membrane phospholipid modification by dietary *n*-3 fatty acids: effects on eicosanoid formation and cell function, *Prog. Clin. Biol. Res.* 282, 263-294, 1988; Decker, K., Signal paths and regulation of superoxide, eicosanoid and cytokine formation in macrophages of rat liver, *Adv. Exp. Med. Biol.* 283, 507-520, 1991; Granström, E. and Kindahl, H., A critical approach to eicosanoid assay, *Adv. Prostaglandin Thromboxane Leukot. Res.* 21A, 295-302, 1991; Clissold, D. and Thickitt, C., Recent eicosanoid chemistry, *Nat. Prod. Rep.* 11, 621-637, 1994; Breyer, R.M., Kennedy, C.R., Zhang, Y., and Breyer, M.D., Structure-function analyses of eicosanoid receptors. Physiologic and therapeutic implications, *Ann. N. Y. Acad. Sci.* 905, 221-231, 2000; Leslie, C.C., Regulation of arachidonic acid availability for eicosanoid production, *Biochem. Cell. Biol.* 82, 1-17, 2004; and Serhan , C.N., Novel eicosanoid and docosanoid mediators: resolvins, docosatrienes, and neuroprotectins, *Curr. Opin. Clin. Nutr. Metab. Care* 8, 115-121, 2005.

Leukotriene

A group of biologically active lipids containing three conjugated double bonds derived from arachidonic acid. These materials were first obtained from leukocytes. Certain leukotrienes (C4, D4, and LTE4) have been designated as cysteinyl leukotrienes (Evans, J.F., The cysteinyl leukotriene receptors, *Prostaglandins Leukot. Essent. Fatty Acids* 69, 117-122, 2003). See also Samuelsson, B., Borgeat, P., Hammarström, S., and Murphy, R.C., Introduction of a nomenclature: leukotrienes, *Prostaglandins* 17, 785-787, 1979; Borgeat, P. and Samuelsson, B., Transformation of arachidonic acid by rabbit polymorphonuclear leukocytes,. Formation of a novel dihydroxyeicosatetraenoic acid, *J. Biol. Chem.* 254, 2643-2646, 1979; Lam, B.K. and Austen, K.F., Leukotriene C4 synthase: a pivotal enzyme in cellular biosynthesis of the cysteinyl leukotrienes, *Prostaglandins Other Lipid Mediat,* 68-69, 511-520, 2002; Capra, V., Molecular and functional aspects of human cyteinyl leukotriene receptors, *Pharmacol. Res.* 50, 1-11, 2004; Norel, X. and Brink, C., The quest for new cysteinyl-leukotriene and lipoxin receptors: recent clues, *Pharmacol. Ther.* 103, 81-94, 2004; Currie, G.P. and McLaughlin, K., The expanding role of leukotriene receptor antagonists in chronic asthma, *Ann. Allergy Asthma Immunol.* 97, 731-741, 2006; and Zaitsu, M., Imbalance between leukotriene synthesis and catabolism contributes to the pathogenesis of allergic diseases, *Med. Chem.* 3, 365-368, 2007

Lipoxin

A family of trihydroxyeicosatetraenoic acids derived from arachidonic acid which are mediators of inflammation and infection. The lipoxin receptor (ALX) is also considered a chemoattractant receptor. See Ramstedt, U., Ng, J., Wigzell, H., *et al.,* Action of novel eicosanoids lipoxin A and B on human natural killer cell cytotoxicity: effects on intracellular cAMP and target cell binding, *J. Immunol.* 135, 3434-3438, 1985; Hansson, A., Serhan, C.N., Haeggström, J., *et al.,* Activation of protein kinase C by lipoxin A and other eicosanoids. Intracellular action of oxygenation products of arachidonic acid, *Biochem. Biophys. Res. Commu.* 134, 1215-1222, 1986; Morris, J. and Wishka, D.G., Synthesis of Lipoxin B, *Adv. Prostaglandin Thromboxane Leukot. Res.* 16, 99-109, 1986; Palmblad, J., Gyllenhammar, H., Rignertz, B., *et al.,* The effects of lipoxin A and lipoxin B on functional responses of human granulocytes, *Biochem. Biophys. Res. Commun.* 145, 168-175, 1987; Fiore, S., Brezinski, M.E., Sheppard, K.A., and Serhan, C.N., The lipoxin biosynthetic circuit and their actions with human neutrophils, *Adv. Exp. Med. Biol.* 314, 109-132, 1991; Serhan, C.N., Lipoxin biosynthesis and its impact in inflammatory and vascular events, *Biochim. Biophys.Acta* 1212, 1-25, 1994; Serhan, C.N., Takano, T., and Maddox, J.F., Aspirin-triggered 15-*epi*-lipoxin A_4 are potent inhibitors of acute inflammation. Receptors and pathways, *Adv. Exp. Med. Biol.* 447, 133-149, 1999; Machado, F.S., and Aliberti, J., Impact of lipoxin-mediated regulation on immune response to infectious disease, *Immunol.Res.* 35, 209-218, 2006; Parkinson, J.F., Lipoxin and synthetic analogs: an overview of anti-inflammatory functions and new concepts in immunomodulation, *Inflamm. Allergy Drug Targets* 5, 91-106, 2006; and Chiang, N., Arita, M., and Serhan, C.N., *Prostaglandins Leukot. Essent. Fatty Acids* 73, 163-177, 2005.

Prostaglandins

A group of unsaturated, oxygenated long-chain fatty acids having diverse biological function. The material was first isolated by von Euler from the prostate gland (von Euler, U.S., On the specific vaso-dilating and plain muscle stimulating substance from accessory genital glands in man and certain animals (prostagland and vesioglandin), *J. Physiol.* 88, 213-234, 1936). See also Morrow, J.D., Chen, Y., Brame, C.J., *et al.,* The isoprostanes: unique prostaglandin-like products of free radical-initiated lipid peroxidation, *Drug Metab. Rev.* 31, 117-139, 1999; Hylton, C. and Robin, A.L., Update on prostaglandin analogs, *Curr. Opin. Ophthalmol.* 14, 65-69, 2003; Serhan, C.N. and Levy, B., Success of prostaglandin E2 in structure-function is a challenge for structure-based therapeutics, *Proc. Nat. Acad. Sci. USA* 100, 8609-8611, 2003; Simmons, D.L., Botting, R.M., and Hla, T., Cyclooxygenase isozymes: The biology of prostaglandin synthesis and inhibition, *Pharmacol. Rev.* 56, 387-437, 2004; Samuelsson, B., Morgenstern, R., and Jakobsson, P.J., Membrane prostaglandin E synthase-1: a novel therapeutic target, *Pharmacol. Rev.* 59, 207-224, 2007.

Section III
Vitamins and Coenzymes

PROPERTIES OF VITAMINS

Compound	Formula	Properties	Solubility (g/100 ml)	Stability
VITAMIN A				
Retinol Vitamin A alcohol Vitamin A$_1$ Axeropthol Anti-infective vitamin Antixerophthalmia factor Lard factor Biosterol Oleovitamin A 3,7-Dimethyl-9-(2,6,6-trimethyl-1-cyclohexen-1-yl)-2,4,6,8-nonatetraen-1-ol 2,6,6-Trimethyl-1-8′-hydroxy-3′,7′ dimethylnona-1′,3′,5′,7′-tetraenyl)cyclohex-1-ene	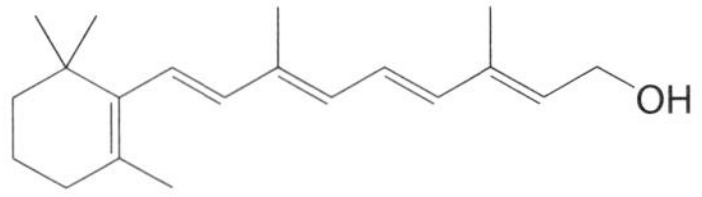 $C_{20}H_{30}O$ mol wt 286.44	Pale yellow prism crystals; mp 62 to 64°C (all *trans*); λ_{max} = 325 nm (ethanol); $E_{1\,cm}^{1\%}$ = 1,835; optically inactive; 1 mg = 3,333 IU or USP units	Sol in most organic solvents; sol in fats and oils; insol in water and glycerol	Unstable to oxygen, air, and UV light; esters relatively more stable
		—	—	—
		—	—	—
Dehydroretinol Retinol$_2$ Vitamin A$_2$ 3-Dehydroretinol	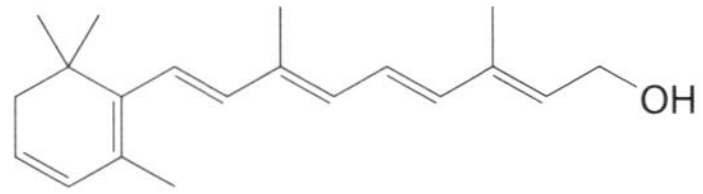 $C_{20}H_{28}O$ mol wt 284.42	Yellow crystals or oil; mp 17 to 19°C (all *trans*); λ_{max} = 288, 352 nm (ethanol); $E_{1\,cm}^{1\%}$ = 820, 1,450; optically inactive; 1 mg = 1,333 IU or USP units	Sol in most organic solvents; sol in fats and oils; insol in water and glycerol	Unstable to oxygen
Retinaldehyde[a,b] Retinal[a] Vitamin A aldehyde Retinene Axerophthal	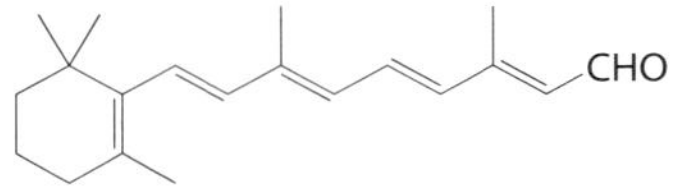 $C_{20}H_{28}O$ mol wt 284.42	Orange crystals; mp 61 to 64°C (all *trans*); λ_{max} = 373 nm (cyclohexane); $E_{1\,cm}^{1\%}$ = 1,548; all *trans* retinal has about 90% of the biopotency of all *trans* vitamin A acetate	Sol in most organic solvents; sol in fats and oils; practically insol in water	Unstable to oxygen
Retinoic acid Vitamin A acid Tretinoin 3,7-Dimethyl-9-(2,6,6-trimethyl-1-cyclohexen-1-yl)-2,4,6,8-nonatetraenoic acid	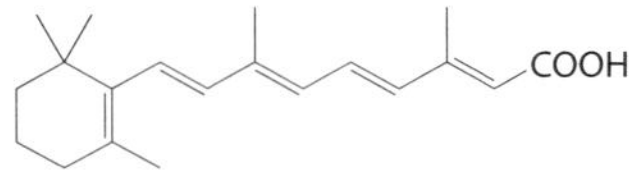 $C_{20}H_{28}O_2$ mol wt 300.44	Crystals; mp 180 to 182°C (all *trans*); λ_{max} = 350 nm (ethanol); $E_{1\,cm}^{1\%}$ = 1,510; fully active for growth but not for visual function or reproduction	Sol in most organic solvents and oils	Unstable and subject to oxidation but more stable than alcohol or esters

PROPERTIES OF VITAMINS (Continued)

Compound	Formula	Properties	Solubility (g/100 ml)	Stability
VITAMIN A (Continued)				
Retinyl acetate — Vitamin A acetate	OCOCH$_3$; $C_{22}H_{32}O_2$ mol wt 328.5	Yellow prismatic crystals; mp 57 to 60°C (all *trans*); $\lambda_{max} = 326$ nm (ethanol) $E_{1cm}^{1\%} = 1,550$; 1 mg = 2,907 IU or USP units	Like retinol; insol in water	Unstable but more stable than alcohol; can be stabilized
Retinyl palmitate — Vitamin A palmitate	OCO(CH$_2$)$_{14}$CH$_3$; $C_{36}H_{60}O_2$ mol wt 524.8	Yellow amorphous or crystalline; mp 27 to 29°C (all *trans*) $\lambda_{max} = 325$ to 328 nm (ethanol); $E_{1cm}^{1\%} = 940$ to 975; 1 mg = 1,820 IU or USP units	Like retinol; insol in water	Unstable but more stable than alcohol or acetate; can be stabilized
PROVITAMIN A CAROTENOIDS				
β-Carotene	$C_{40}H_{56}$ mol wt 536.89	Deep purple prisms or red leaflet crystals; mp 180 to 182°C (all *trans*); $\lambda_{max} = 273, 453, 481$ (petroleum ether); $E_{1cm}^{1\%} = 383, 2,592, 2,268$; optically inactive; 1 mg = 1,667 to 333 IU (depending on species fed)	Slightly sol in most organic solvents; very slightly sol in alcohol and oils; insol in glycerol and water	Unstable to oxygen; oxidizes to colorless products; can be used, however, as food color
β-Apo-8′-carotenal	CHO; $C_{30}H_{40}O$ mol wt 416.65	Dark violet crystals; mp 136 to 140°C (all *trans*); $\lambda_{max} = 461$ and 488 nm (cyclohexane); $E_{1cm}^{1\%} = 2,640$ at 461; 1 mg = 1,220 IU (rat)	Sol in most organic solvents; slightly sol in alcohol; insol in glycerol and water	Unstable to oxygen; can be used, however, as food color
Ergocalciferol — Vitamin D$_2$; Activated ergosterol; Calciferol; Oleovitam·n D$_2$; Viosterol; 9,10-Secoergosta-5,7,10(19),-22-tetraen-3-ol; Irradiated ergosta-5,7,22-trien-3β-ol	HO; $C_{28}H_{44}O$ mol wt 396.66	*VITAMIN D* White prism crystals; mp 115 to 118°C; $\lambda_{max} = 264$ nm (hexane); $E_{1cm}^{1\%} = 459$; $[\alpha]_D^{25} = +103$ to 106 (alcohol); 1 mg = 40,000 IU or USP units	Sol in most organic solvents; slightly sol in oils; insol in water	Crystals relatively unstable to air, accelerated by unsaturated fat and trace mineral contact; can be stabilized

PROPERTIES OF VITAMINS (Continued)

Compound	Formula	Properties	Solubility (g/100 ml)	Stability
		VITAMIN D (Continued)		
Cholecalciferol Vitamin D_3 Activated 7-dehydrocholesterol Oleovitamin D_3 22,23-Dihydro-24-demethylcalciferol Activated 5,7-cholestadien-3β-ol	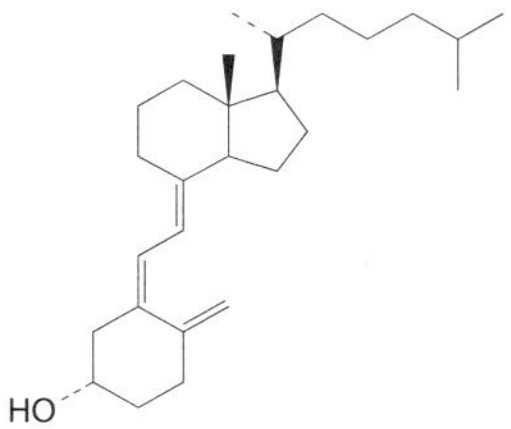 $C_{27}H_{44}O$ mol wt 384.65	White, fine needle crystals; mp 84 to 88°C; $\lambda_{max} = 264$ nm (hexane); $E_{1\,cm}^{1\%} = 450\,490$; $[\alpha]_D^{25} = +105$ to 112° (alcohol); 1 mg = 40,000 IU or USP units; $E_{mol} = 18,300$ (ethanol)	Like D_2	Similar to D_2
25-Hydroxycholecalciferol	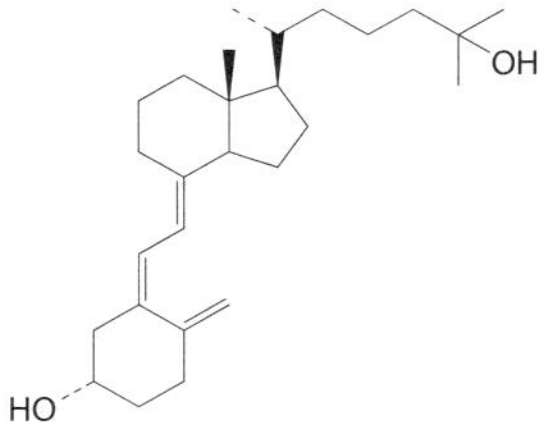 $C_{27}H_{44}O_2$ mol wt 416.65	Colorless, fine, needlelike crystals; mp 97 to 108°C; $[\alpha]_{589} = +78$ (dioxane); $\lambda_{max} = 264$ nm $E_{1\,mol} = 17,400$ (ethanol)	Similar to D_3	Similar to D_3
1α,25-Dihydroxycholecalciferol	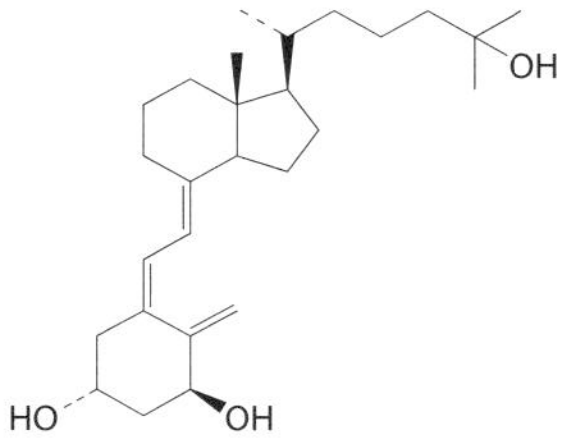 $C_{27}H_{44}O_2$ mol wt 400.65	Colorless, fine, needlelike crystals; mp 113 to 114°C; $[\alpha]_D^{25} = +48$ (ethanol); $\lambda_{max} = 264$ nm; $E_{1\,mol} = 14,270$ (ethanol)	Sol in esters, ethers, and alcohols	Similar to D_3
		VITAMIN E		
d-α-Tocopherol 2D,4'D,8'D-α-Tocopherol 2R,4'R,8'R-α-Tocopherol Antisterility factor 5,7,8-Trimethyltocol 2,5,7,8-Tetramethyl-2-(4',8',12'-trimethyltridecyl)-6-chromanol Abbreviation: α-T	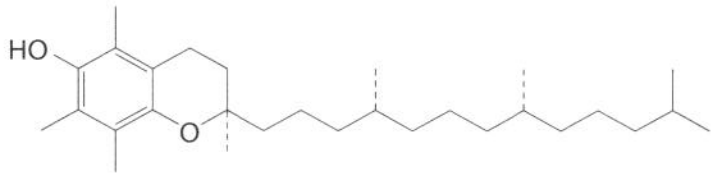 $C_{29}H_{50}O_2$ mol wt 430.69	Pale yellow viscous oil; mp 2.5 to 3.5°C; bp 200 to 220°C; (0.1 mm); $\lambda_{max} = 292$ (ethanol); $E_{1\,cm}^{1\%} = 74$ to 76; $[\alpha]_D^{25} = 0.32°$ (ethanol), $K_3Fe\,(CN)_6$ oxidation product: $[\alpha]_D^{25} = +26°$ (isooctane); 1 mg = 1.49 IU	Very sol in oils, fats, and many organic solvents; insol in water	Free tocopherol very unstable to air, iron salts, and bleaching agents; can be stabilized; esters quite stable

PROPERTIES OF VITAMINS (Continued)

VITAMIN E (Continued)

Compound	Formula	Properties	Solubility (g/100 ml)	Stability
dl-α-Tocopherol 2DL,4'DL,8'DL-α-Tocopherol 2RS,4'RS,8'RS-α-Tocopherol all-rac-α-Tocopherol		Pale yellow viscous oil; λ_{max} = 292 (ethanol); $E_{1\,cm}^{1\%}$ = 72 to 76; 1 mg = 1.10 IU	Very sol in oils, fats, and many organic solvents; insol in water	Free tocopherol very unstable to air, iron salts, and bleaching agents; can be stabilized; esters quite stable
d-α-Tocopheryl acetate d,α-Tocopherol acetate 2,5,7,8-Tetramethyl-2-(4'8',12'-trimethyltridecyl)-6-chromanol acetate 2D,4'D,8'D-α-tocopheryl acetate 2R,4'R,8'R-α-tocopheryl acetate	H_3COCO — (structure) $C_{31}H_{52}O_3$ mol wt 472.73	Pale yellow crystals; mp 28°C; λ_{max} = 284 nm (ethanol); $E_{1\,cm}^{1\%}$ = 43.6; $[\alpha]_D^{25}$ = + 3.2° (ethanol); 1 mg = 1.36 IU	Like α-tocopherol	Relatively stable
dl-α-Tocopheryl acetate 2DL,4'DL,8'DL-α-tocopheryl acetate 2RS,4'RS,8'RS-α-tocopheryl acetate all-rac-α-Tocopheryl acetate		Yellow viscous oil; bp 184°C (0.01 mm), 194°C (0.025 mm), 224°C (0.3 mm); λ_{max} = 284 nm (ethanol); $E_{1\,cm}^{1\%}$ = 43.6; 1 mg = 1.00 IU		
d-α-Tocopheryl acid succinate 2D,4'D,8'D-s,q 2D,4'D,8'D'-α-Tocopheryl acid succinate 2R,4'R,8'R-α-Tocopheryl acid succinate		White crystals; mp 73 to 78°C; λ_{max}=286 nm (ethanol); $E_{1\,cm}^{1\%}$ =38.5; 1 mg=1.21 IU	Less sol than acetate	Relatively stable
dl-α-Tocopheryl acid succinate 2DL,4'DL,8'DL-α-Tocopheryl acid succinate 2RS,4'RS,8'RS-α- Tocopheryl acid succinate all-rac-α-Tocopheryl acid succinate	$HOOCH_2CH_2COCO$ — (structure) $C_{33}H_{54}O_5$ mol wt 530.80	White crystals; λ_{max}=284 nm; $E_{1\,cm}^{1\%}$ =38.5; 1 mg=0.89 IU	Less sol than acetate	Relatively stable

PROPERTIES OF VITAMINS (Continued)

VITAMIN K

Compound	Formula	Properties	Solubility (g/100 ml)	Stability
Phytylmenaquinone Vitamin K_1 Antihemorrhagic vitamin Phylloquinone Phytylmenadione Phytonadione Coagulation vitamin Prothrombin factor 2-Methyl-3-phytyl-1,4-naphthoquinone Abbreviation: K^a or PMQ^b	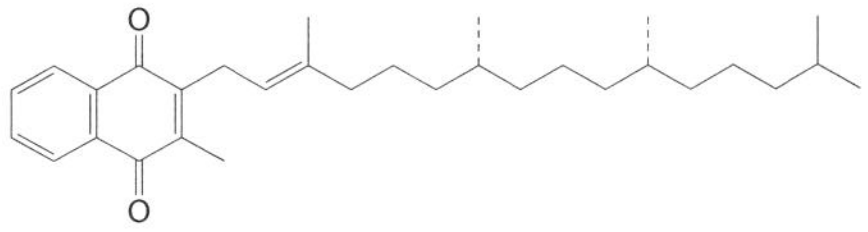 $C_{31}H_{46}O_2$ mol wt 450.71	Yellow viscous oil; mp −20°C; λ_{max}=243, 248, 261, 269, 325 nm (isooctane); $E_{1\,cm}^{1\%}$ =425, 428, 424, 424, 350; $[\alpha]_D^{20}$ =−0.4° (benzene)	Sol in many organic solvents; insol in water	Fairly stable to heat decomposed by sunlight and alkali; can be stabilized
Prenylmenaquinone-6 Vitamin K_2 K_2 (30) Farnoquinone Menaquinone-6 2-Methyl-2-difarnesyl-1,-4-naphthoquinone Abbreviation: MQ-6	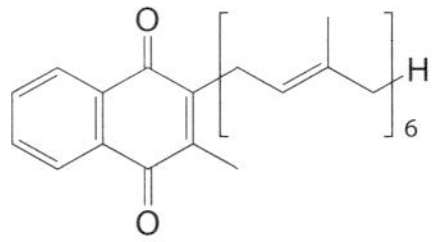 n=6 $C_{41}H_{56}O_2$ mol wt 580.89	Yellow crystals; mp 53.5 to 54.5°C; λ_{max}=243, 248, 261, 270, 325 nm (petroleum ether); $E_{1\,cm}^{1\%}$ =304, 320, 290, 292, 53	Slightly less sol than K_1	Similar to K_1
Prenylmenaquinone-7 K_2 (35) Menaquinone-7 2-Methyl-3-all *trans*-farnesylgeranylgeranyl-1,4-naphthoquinone Abbreviation: MQ-7	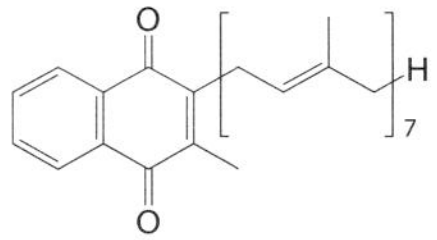 n=7 $C_{46}H_{64}O_2$ mol wt 649.02	Yellow crystals; mp 57°C; λ_{max}=243, 248, 261, 270, 325 nm (petroleum ether); $E_{1\,cm}^{1\%}$ =278, 295, 266, 267, 48	—	Similar to K_1
Menaquinone Vitamin K_3 Menadione Menaphthone 2-Methyl-1,4-naphthoquinone Abbreviation: MK	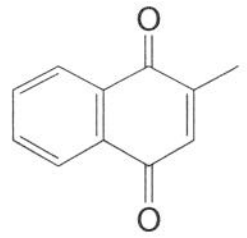 $C_{11}H_8O_2$ mol wt 172.19	Bright yellow crystals; mp 105 to 107°C; λ_{max}=244, 253, 264, 325, 328 nm (hexane); $E_{1\,cm}^{1\%}$ =1,150 (at 244 nm)	Sol in many organic solvents; insol in water; water sol forms are available	Fairly stable in air; decomposed by sun light

PROPERTIES OF VITAMINS (Continued)

VITAMIN K WATER SOLUBLE COMPOUNDS

Compound	Formula	Properties	Solubility (g/100 ml)	Stability
Menadione sodium bisulfite 2-Methyl-1,4-naphthoquinone sodium bisulfite Abbreviation: MSB	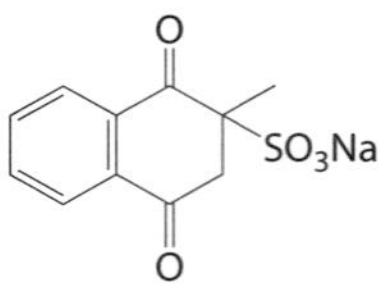 $C_{11}H_8O_2HaHSO_3 \cdot 3H_2O$ mol wt 330.29	White hygroscopic crystals; λ_{max}=229, 267 nm (aqueous solution); $E_{1\,cm}^{1\%}$ = 868, 290	50, water; sol in alcohol; almost insol in benzene and water	–
Menadiol diphosphate (tetrasodium salt) 2-Methyl-1,4-naphthalenediol diphosphoric acid ester tetrasodium salt	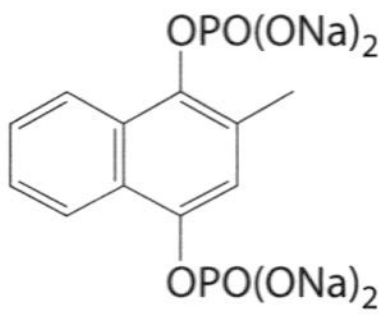 $C_{11}H_8Na_4O_8P_2$ mol wt 422.09 (anhydrous) 530.18 (hexahydrate)	Hexahydrate, white to pinkish crystals; hygroscopic	Very sol in water; insol in ether and acetane	–
Menadione dimethyl-pyrimidinol bisulphate Abbreviation: MPB	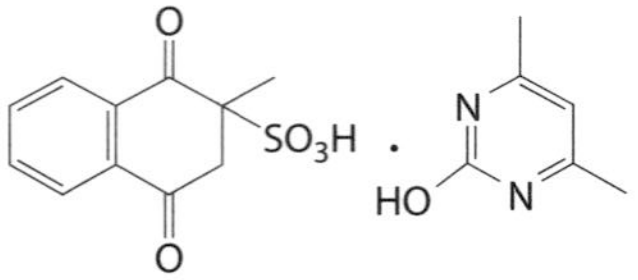 $C_{17}H_{18}O_6N_2S$ mol wt 378.41	White crystals; mp 215 to 217°C λ_{max}=229, 267, 297 nm (water); $E_{1\,cm}^{1\%}$ =810, 285, 234	1, water	–

VITAMIN C

Compound	Formula	Properties	Solubility (g/100 ml)	Stability
Ascorbic acid L-Ascorbic acid Vitamin C Antiscorbutic factor Cevitamic acid Hexuronic acid Antiskorbutin L-Threo-3-ketohexonic acid-enc-lactone L-Threo-2,3,4,5,6-pentahydroxy-2-hexene-γ-lactone L-3-Ketothreo-hexuronic acid lactone	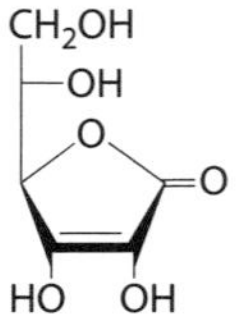 $C_6H_8O_6$ mol wt 176.13	Colorless or white crystals; mp 190 to 192°C; (decomposes); λ_{max}=245 nm (aqueous solution); $E_{1\,cm}^{1\%}$ =560; $[*]_D^{25}$ =+20.5 to 21.5° (aqueous solution); 1 mg = 20 IU or USP units; characteristic acid taste	33, water; 3.5, alcohol; 1, glycerol; insol in oils and most organic solvents	Stable to air when dry; oxidizes in solution; catalyzed by metals, copper and iron, and by alkali

PROPERTIES OF VITAMINS (Continued)

Compound	Formula	Properties	Solubility (g/100 ml)	Stability
		PABA		
p-Aminobenzoic acid PABA 4-Aminobenzoic acid Achromotrichia factor Antigray hair factor	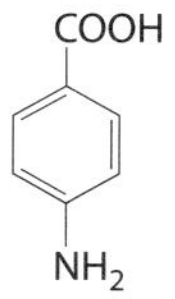 COOH–⬡–NH$_2$ $C_7H_7NO_2$ mol wt 137.13	Monoclinic prisms; mp 187°C; λ_{max}=266 nm (water); $E_{1\,cm}^{1\%}$ -1,070; optically inactive	0.5, water; very sol in alcohol, ether and glacial acetic acid; slightly sol in benzene	Stable in dry form; unstable in presence of ferric salts and oxidizing agents
		BIOTIN		
d-Biotin Vitamin H Coenzyme R Factor X Bios 11 or 11 B Egg white injury preventative or factor β-Hexahydro-2-oxo-1 H-thiene [3,4]-Imidazole-4-valeric acid 3,4-(2′-Ketoimidazolido)-2-(ω-carboxybutyl)-thiophane	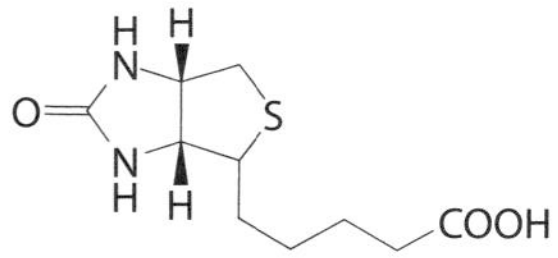 (biotin structure, COOH) $C_{10}H_{16}N_2O_3S$ mol wt 244.31	Colorless needle crystals; mp 230 to 233°C (decomposes); $[\alpha]_D^{21}$ =+91° (0.1 N NaOH)	0.020, water; 0.080 alcohol; more sol in hot water or dilute alkali	Stable, dry, in air and heat; stable in acid solution; less stable in alkali solution
		CHOLINE		
Choline Bilineurine, amantine, gossypine, vidine (β-Hydroxyethyl) trimethylammonium hydroxide	$(H_3C)_3N(OH)CH_2CH_2OH$ $C_5H_{15}NO_2$ mol wt 121.18	Viscous hydroscopic liquid or colorless crystals; optically inactive	Very sol in water and alcohol; insol in ether	Stable in aqueous solution; decomposed by hot alkali
Chloride salt Choline chloride (β-Hydroxyethyl) trimethylammonium chloride (2-Hydroxyethyl) trimethylammonium chloride	$[(H_3C)_3NCH_2CH_2OH]Cl^-$ $C_5H_{14}ClNO$ mol wt 139.63	Deliquescent hydroscopic white crystals; optically inactive	Like choline	Similar to choline

PROPERTIES OF VITAMINS (Continued)

Compound	Formula	Properties	Solubility (g/100 ml)	Stability
Cyanobalamin Vitamin B_{12} Antipernicious anemia factor *Lactobacillus lactis*-Dorner factor Animal protein factor Extrinsic factor Zoopherin 5,6-Dimethylbenzimidazolyl cyanocobamide	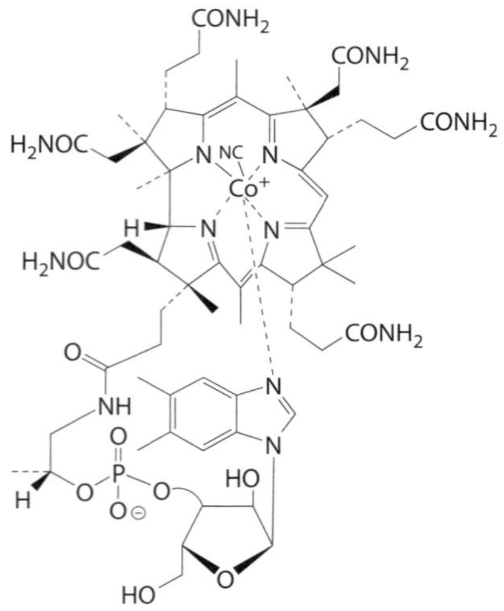 *VITAMIN B_{12}* $C_{63}H_{88}N_{14}O_{14}PCo$ mol wt 1,355.42	Dark red crystals; mp > 212°C (decomposes); λ_{max}=278, 361, 550 nm (aqueous solution); $E_{1cm}^{1\%}$ =115,204, 63; $[\alpha]_{656}^{25}$ =−59±9° (aqueous solution)	1.25, water; sol in alcohol, insol in chloroform, acetone, and ether	Most stable in aqueous solution at pH 4.5 to 5; decomposes slowly in weak acid or alkali; decomposition accelerated by oxidizing and reducing agents, ferrous salts, etc.
Pteroylglutamic acid Folic acid+ Folacin Vitamin M *Lactobacillus casei* factor Vitamin Bc, B_{10} or B_{11} Norite eluate factor Factor U or R Abbreviation: PGA+ or PteGlu[a] N-[4-([2-Amino-4-hydroxy-6-pteridyl)-methyl]amino)-benzoyl] glutamic acid	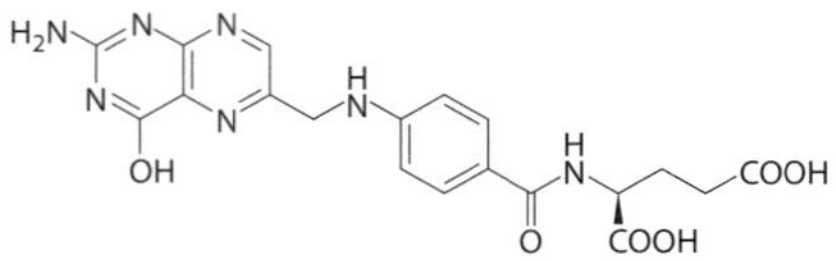 *FOLIC ACID[a] OR FOLACIN[b]* $C_{19}H_{19}N_7O_6$ mol wt 441.40	Yellowish-orange crystals; mp > 250°C (decomposes); λ_{max}=256, 283, 265 nm (alkaline solution); $E_{1cm}^{1\%}$ = 603, 600, 215	Sol in acetic acid, alkaline solution; very sol in acetone; insol in chloroform, ether and benzene	Relatively stable dry; unstable in acid medium; decomposed by sunlight and heat

PROPERTIES OF VITAMINS (Continued)

Compound	Formula	Properties	Solubility (g/100 ml)	Stability
FOLIC ACID^a OR FOLACIN^b (Continued)				
5-Formyltetrahydrofolic acid Folinic acid [N-[p-[(2-Amino-5-formyl-5,6,7,8-tetrahydro-4-hydroxy-6-pteridinyl)methyl] amino] benzoyl]glutamic acid Citrovorum factor Leucovorin Abbreviation: N^5-formyl THFA or N^5-F-PGAH$_4$ or 5-HCO-H$_4$PteGlu	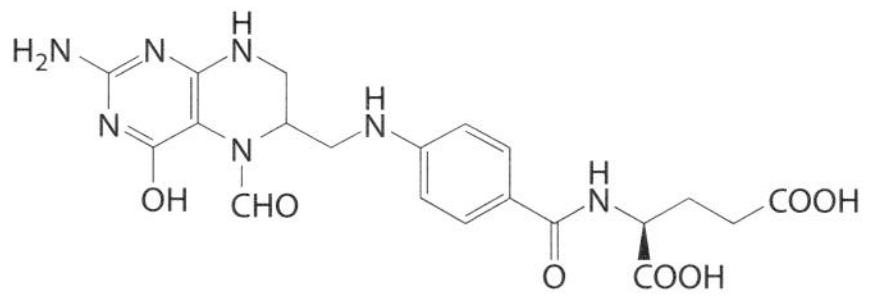 $C_{20}H_{23}N_7O_7$ mol wt 473.44	Colorless crystals; mp 240 to 250°C (decomposes); λ_{max}=282 nm (alkaline solution); = +14.26°	Very slightly sol in water	More stable at neutral or mild alkaline pH
INOSITOL				
Myo-inositol Inositol meso-Inositol i-Inositol Bios 1 Inosite Mouse antialopecia factor Rat antispectacle eye factor Hexahydroxycyclohexane Cyclohexanehexol, cyclohexitol	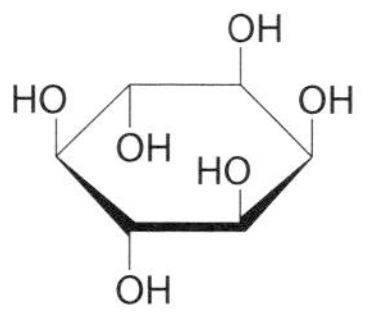 $C_6H_{12}O_6$ mol wt 180.16	White crystals; mp 225 to 227°C (anhydrous); mp 218°C (dihydrate); optically inactive; sweet taste	15, water; insol in absolute alcohol and ether	Relatively stable
NIACIN				
Nicotinic acid Niacin Antiblacktongue factor Pellegra preventive (PP) factor Vitamin PP Pyridine-3-carboxylic acid Pyridine-β-carboxylic acid	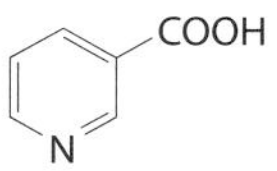 $C_6H_5NO_2$ mol wt 123.11	Colorless or white needles; mp 235 to 237°C; λ_{max}=261 nm (0.1 N HCl), 263 nm (pH 11); $E_{1cm}^{1\%}$=435, 260; optically inactive	1.6, water; 0.73 alcohol; insol in ether	Stable to air, light, and PH
Nicotinamide Niacinamide Nicotinic acid amide Vitamin PP Vitamin B$_3$ 3-Pyridinecarboxylic acid amide	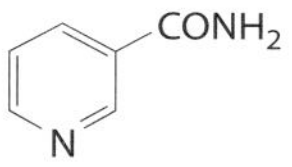 $C_6H_6N_2O$ mol wt 122.13	Colorless or white needles; mp129 to 131°C; λ_{max}=261 nm (0.1 N HCl), 262 nm (pH 11); $E_{1cm}^{1\%}$=432, 250; optically inactive	100, water; 66.6, alcohol; 10 glycerol; slightly sol in ether	Stable in air and heat, light, and pH (may hydrolyze to nicotinic acid)

PROPERTIES OF VITAMINS (Continued)

Compound	Formula	Properties	Solubility (g/100 ml)	Stability
		PANTOTHENIC ACID		
Pantothenic acid, D-(+) Chick antidermatitis factor Liver filtrate factor Antidermatosis vitamin Vitamin B_5 D(+)-N-(2,4-Dihydroxy-3,3-dimethylbutyryl)-β-alanine Pantoyl-β-alanine	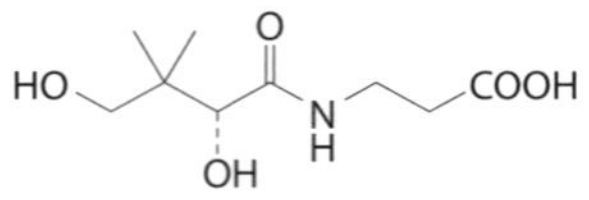 $C_9H_{17}NO_5$ mol wt 219.23	Colorless, hydroscopic, viscous oil; $[\alpha]_D^{25}=+37.5°$ (aqueous solution); only d-form is biologically active	Freely sol in water and acetic acid; moderately sol in alcohol; insol in benzene and chloroform	Stable at near neutral pH (5 to 7); unstable in acid or alkali; labile to prolonged heat
Calcium salt Calcium pantothenate	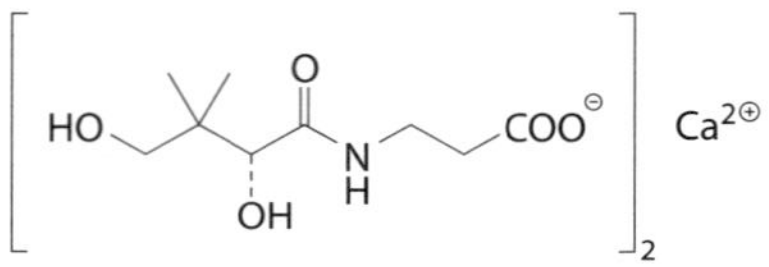 $C_{18}H_{32}CaN_2O_{10}$ mol wt 476.55	Colorless or white needles; mp 195 to 196°C (decomposes); $[\alpha]_D^{25}=+28.2°$ (aqueous solution); 1 g=70,000 to 75,000 chick units	35, water; very sol in glycerol and glacial acetic acid; slightly sol in ether, insol in benzene and chloroform	—
Pantothenyl alcohol Panthenol Provitamin for pantothenic acid N-(2,4-D-hydroxy-3,3-dimethylbutyryl-β-aminopropanol Pantoyl-β-aminopropanol	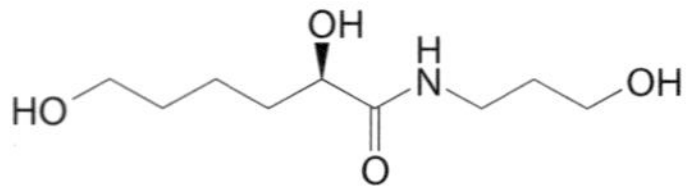 $C_9H_{19}NO_4$ mol wt 205.39	Viscous liquid (d); white crystals (dl); mp 64.5 to 67.5°C (dl); $[\alpha]_D^{20}=+29.7°$ (in water) (d)	Very sol in water and glycerol	More stable in solution than pantothenic acid or salts
		VITAMIN B_6		
Pyridoxine · HCl Pyridoxol hydrochloride Adermine hydrochloride Antiacrodynia factor Yeast eluate factor 5-Hydroxy-6-methyl-3,4-pyridinedimethanol hydrochloride 2-Methyl-3-hydroxy-4,5-bis(hydroxymethyl)pyridine hydrochloride	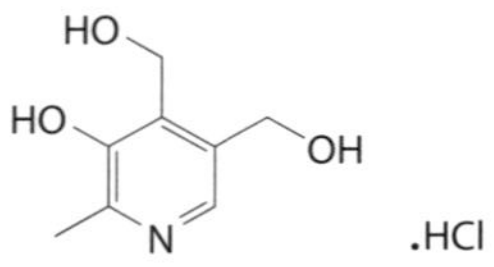 $C_8H_{11}NO_3 \cdot HCl$ mol wt 205.64	Platelets or rods, white crystals; mp 206°C (decomposes); $\lambda_{max}=291$ nm (0.17 N HCl); $E_{1cm}^{1\%}=422$; optically inactive	22, water; 1.1, ethanol; slightly sol in acetone; insol in ether	Stable, dry to air, light, and heat; sol in acid solution

PROPERTIES OF VITAMINS (Continued)

Compound	Formula	Properties	Solubility (g/100 ml)	Stability
VITAMIN B₆ (Continued)				
Pyridoxal 3-Hydroxy-5-(hydroxymethyl)-2-methylisonicotinaldehyde	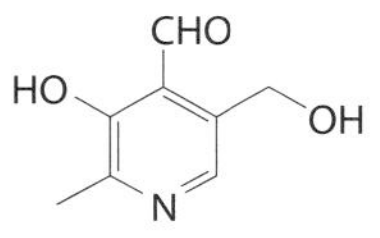 $C_8H_9NO_3$ mol wt 167.16	—	—	—
Pyridoxal · HCl	$C_8H_9NO_3 \cdot$ HCl mol wt 203.62	Colorless rhombic crystals; mp 165°C (decomposes); λ_{max}=292.5 nm; E_{mol}=7,600	—	—
Pyridoxamine Dihydrochloride 2-Methyl-3-hydroxy-4-aminomethyl-5-hydroxymethylpyridine dihydrochloride	$C_8H_{12}N_2O_2$ mol wt 168.19 $C_8H_{12}N_2O_2 \cdot$2HCl mol wt 241.12	Colorless crystals; mp 193 to 193.5°C Colorless platelets; mp 226 to 227°C (decomposes); λ_{max}=287.5 nm (at pH 1.94); E_{mol}=9,100	—	—
VITAMIN B₂				
Riboflavin Riboflavine Vitamin B₂ Vitamin G Lactoflavin Ovoflavin Hepatoflavin Lyochrome 7,8-Dimethyl-10-(D-ribo-2,3,4,5-tetrahydroxypentyl)isoalloxazine 7,8-Dimethyl-10-ribitylisoalloxazine	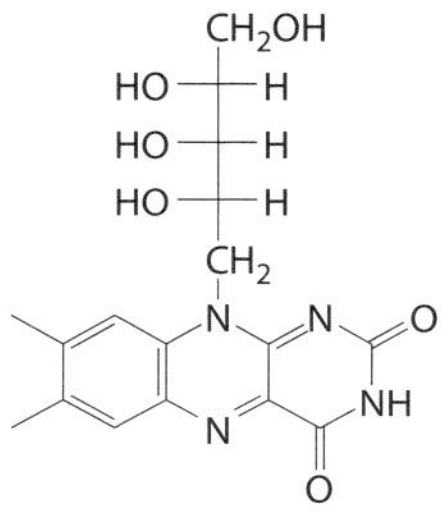 $C_{17}H_{20}N_4O_6$ mol wt 376.37	Yellow to orange-yellow; polymorphic crystals; mp 280 to 290°C (decomposes); λ_{max}=223, 266, 271, 444 nm(0.1HHCl); $E_{1\,cm}^{1\%}$ = 800, 870, 288, 310; $[\alpha]_D^{25}$ =−112 to 122° (dilute alcoholic NaOH); strong green fluorescence when irradiated by UV light	0.013, water; 0.040, alcohol; insol in ether, acetone, chloroform, and benzene; riboflavin phosphate, sodium salt quite water soluble	Unstable in alkali solution especially in light; stable in acid solution dark; reversibly reduced by sodium hydrosulfite and other reducing agents to dehydroriboflavin (leucoflavin); relatively stable in dry form
Vitamin B₂ phosphate sodium Riboflavin 5′-phosphate sodium Flavin mononucleotide Riboflavin 5′-phosphate ester monosodium salt	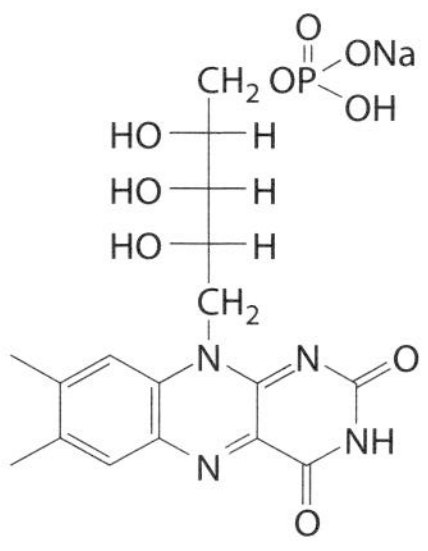 $C_{17}H_{20}N_4O_9PNa \cdot 2H_2O$ mol wt 514.37	Orange-yellow crystals; $[\alpha]_D^{20}$ =+38 to 42° (20% HCl)	4 to 11, water (depending on pH)	Similar to riboflavin

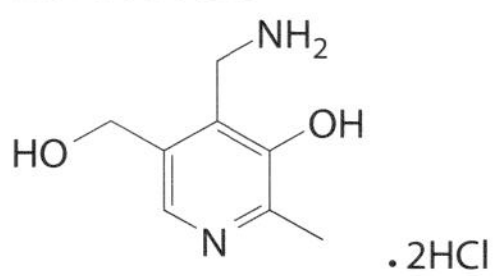

PROPERTIES of VITAMINS(Continued)

VITAMIN B_1

Compound	Formula	Properties	Solubility (g/100 ml)	Stability
Thiamine · HCl Thiamine chloride hydrochloride Vitamin B_1 hydrochloride Aneurine (hydrochloride) Oryzamin Antiberiberi vitamin 3-(4-Amino-2-methylpyrimidyl-5-methyl-4-methyl-5-(β-hydroxyethyl) thiazolium chloride hydrochloride	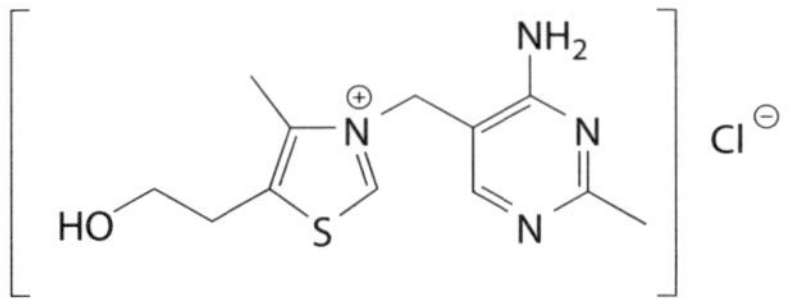 $C_{12}H_{17}ClN_4OS$ HCl mol wt 337.28	White monoclinic crystals; mp 246 to 250°C (decomposes); λ_{max} = 246 nm (0.1 N HCl); $E_{1cm}^{1\%}$ = 410; optically inactive; 1 mg=333 IU	100, water; 1, alcohol; insol in organic solvents	Stable when dry, stable in acid, unstable at alkaline pH, to prolonged heating, presence of bisulfite or thiaminase; very hygroscopic
Thiamine mononitrate Aneurine mononitrate Vitamin B_1 mononitrate	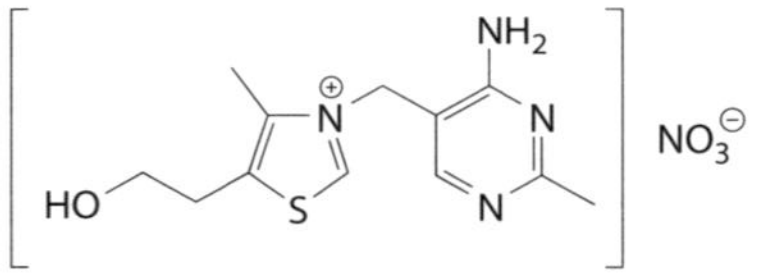 $C_{12}H_{17}N_5O_4S$ mol wt 327.36	White crystals; mp 196 to 200°C (decomposes); less hygroscopic than chloride hydrochloride; 1 mg=343 IU	2.7, water; insol in organic solvents	More stable than chloride salt in dry products; not hygroscopic

Compiled by J. C. Bauernfeind and E. De Ritter.

[a]Approved by IUPAC-IUB.
[b]Approved by IUNS.

BIOLOGICAL CHARACTERISTICS OF VITAMINS

Compound	Function	Deficiency Symptoms	Hyper-Use Symptoms	Coenzyme and Enzyme Involved	Remarks
Vitamin A	Primary role in vision as 11-cis retinal; synthesis of mucopolysaccharides; maintenance of mucous membranes and skin; bone development and growth; maintenance of cerebrospinal fluid pressure; production of corticosterone	Retarded growth, xerophthalmia, nyctalopia, hemeralopia, ataxia, tissue keratinization, cornification, desquamation, emaciation, lachrymation, impaired reproduction or hatchability (eggs), increased susceptibility to infection, optic nerve degeneration, odontoblast atrophy	Weight loss, bone abnormalities, inflammations, exfoliated epithelium, liver enlargement, pain, loss of hair, facial pigmentation	None identified	Other carotenes, α-, β-, γ-, and β-zeacarotene, cryptoxanthin, echinenone, certain apocarotenals, cis isomers of retinol, and dehydroretinol have fractional vitamin A activity, differing for various animal species; retinoic acid may be metabolically active form for certain functions
Vitamin D	Absorption and transport of calcium and phosphorus; synthesis of calcium protein carrier; interrelationship with parathyroid hormone; maintains alkaline phosphatase levels at bone site	Rickets, enlarged joints, softened bones, stilted gait, arched back, thin-shelled eggs, disturbed reproduction and hatchability, osteomalacia, faulty calcification of teeth, tetany, convulsions, raised plasma phosphatase, parturient paresis (dairy cattle)	Abnormal calcium deposition in bones and tissues, brittle or deformed bones, vomiting, abdominal discomfort, renal damage, weight loss	None identified	D_3 is the most effective form for the avian species; for man D_2 and D_3 are fully active; $1\alpha,25$ dihydroxycholecalciferol is the active metabolic form of D_3 and is considered by some, a hormone
Vitamin E	Biological antioxidant, interrelated with Se; metabolism of nucleic and sulfur amino acids; ubiquinone synthesis; detoxicant and oxidation-reduction action; stabilizes biological membranes against oxidative attack	Muscular dystrophy, encephalomalacia, hepatic necrosis, erythrocyte hemolysis, hock disorders, steatitis, reduced reproduction and hatchability, exudative diathesis, liver dystrophy, anemia, degeneration of testicular germinal epithelium, creatinuria	None identified; relatively nontoxic	None identified	Other tocols (β-, γ-, δ-tocopherol) and trienols (α-, β-, γ-, δ-tocotrienol) exist differing in ring substituents and in the side chain and having fractional vitamin activity; presence of unsaturated fat in the diet increases dietary vitamin E requirements
Vitamin K	Hepatic synthesis of prothrombin; synthesis of thromboplastin; needed in RNA formation and electron transport	Hemorrhage, impaired coagulation (low prothrombin levels), increased blood clotting time	Vomiting, albuminuria, porphyrinuria, polycythemia, splenomegaly, kidney and liver damage	None identified	K_1 is plant form of the vitamin; K_2, the microbiologically synthesized form; K_1 and K_2 are metabolically active forms; they also occur with longer or shorter side chain (isoprene) units; coumarin compounds and excess sulfa drugs are dietary stress agents for vitamin K
Ascorbic acid	Required for collagen formation; protects enzymes, hydrogen carriers, and adrenal steroids; functions in incorporation of iron into liver ferritin, folic acid into folinic acid; prevents scurvy, increases phagocytic activity	Scurvy, fragile capillaries, bleeding gums, loose teeth, anemia, follicular keratosis, sore muscles, weak bones, decreased egg shell strength, poor wound healing	None identified; relatively nontoxic	None identified	Dietary essential for man, monkey, guinea pig, fish, Indian pipistrel, Indian fruit bat, and flying fox; most other species synthesize it; glucoascorbic acid is an antagonist

BIOLOGICAL CHARACTERISTICS OF VITAMINS (Continued)

Compound	Function	Deficiency Symptoms	Hyper-Use Symptoms	Coenzyme and Enzyme Involved	Remarks
p-Aminobenzoic acid	Function or need not well understood or accepted; a microbial factor involved in melanin formation and pigmentation; inhibits oxidation of adrenaline; influences activity of tyrosinase; involved in microbial synthesis of folic acid	Nutritional achromotrichia (animals), retarded growth (chicks), disturbed lactation (mice)	Rash, nausea, fever, acidosis, vomiting, pruritis	None identified	Sulfa drugs, carbarzones, and others are antagonists; PAB and sulfa drugs have a common point of attack on certain enzyme systems; PAB has some chemotherapeutic uses
Biotin	As an enzyme component activating carbon dioxide and its transfer in amino acid, carbohydrate, and lipid metabolism; deamination of certain amino acids; synthesis of long-chain fatty acids; potassium metabolism; interrelationship with pyridoxine, cyanocobalamin, pantothenic acid, folic acid, and ascorbic acid	Dermatosis, perosis, scaly skin, loss of hair, spectacle alopecia, cracked foot pads or hoofs, impaired reproduction, retarded growth, anorexia, lassitude, sleeplessness, muscle pain, electrocardial changes	None identified	Carboxybiotin, carboxylases, transcarboxylases, carbamyl phosphate synthetase	Biocytin, a biotin bound form. is(+)-epsilon-*N*-biotinyl-L-lysine; mild oxidation converts biotin to sulfoxide, strong oxidation to the sulfone; biotin inactivated by avidin; antagonists are *α*-dehydro-biotin, biotin sulfone, avidin, and others
Choline	Source of methyl group, a methyl donor; for acetylcholine and phospholipid formation; essential for liver functioning	Fatty liver, hemorrhagic degeneration of kidneys, cirrhosis of liver, involution of thymus, enlarged spleen, retarded growth, impaired production (eggs) lactation and reproduction, perosis, muscle weakness or paralysis	Diarrhea and edema, erythrocyte formation inhibition	None identified	Choline occurs widespread in nature and is synthesized within the body to a limited extent; triethylcholine and others are choline antimetabolites
Cyanocoba.amin	Cofactor for methyl malonyl CoA isomerase; involved in isomerizations; dehydrogenations, methylations; interrelated in choline, folic acid, ascorbic acid, pantothenic acid, biotin, and *S*-amino acid metabolism; synthesis of nucleoproteins	Retarded growth, perosis, poor feathering, megaloblastic anemia, anorexia, degenerative changes in spinal cord, posterior incoordination, impaired hatchability	Polycythemia	Cyanocobalamin coenzyme	Other B_{12} molecule variations exist wherein cyanide is replaced by chlorine, bromine, hydroxylcyanate, nitrite thiocyanate, etc; 5,6-dichlorobenzimidazole is an antimetabolite

BIOLOGICAL CHARACTERISTICS OF VITAMINS (Continued)

Compound	Function	Deficiency Symptoms	Hyper-Use Symptoms	Coenzyme and Enzyme Involved	Remarks
Folic acid	Concerned with single carbon metabolism; for methyl, hydroxyl, and formyl transfers; for purine synthesis and normal histidine metabolism; interconversion of serine and glycine; a growth and hematopoietic factor; interrelationship with cyanocobalamin, ascorbic acid, iron, etc.	Retarded growth, sprue, diarrhea, macrocytic anemia, cervicular paralysis, reduction and abnormalities in white cells, dermatitis, impaired reproduction and lactation, perosis, poor feathering and lowered hatchability (poultry)	Obstruction of renal tubules	Tetrahydrofolic acid enzyme	The compound is a chelate, binding Co; certain molecular modifications of folic and tetrahydrofolic acid yield antagonists such as aminopterin, tetrahydroaminopterin, and others; the active forms of folic acid may have additional group such as formyl or methyl on the nitrogen in the molecule; thus, N^5-formyltetrahydrofolic acid is folinic acid and N^5-methyltetrahydrofolic acid is the form in blood
Inositol	Function or need not well understood or accepted; believed to be a lipotropic factor; a supply of methyl group functioning with cyanocobalamin; needed for acetyl choline production, functions in some microbial metabolic role	Fatty liver, hair loss, impaired reproduction and lactation, reduced growth (animal)	None identified; relatively nontoxic	None identified	Hexachlorocyclohexane (lindane) is an antimetabolite; while inositol can exist in eight cis-trans isomeric forms, only the optically inactive, *i*- or *meso*-inositol, is active; it occurs in animal tissues in free and phosphate ester form; inositol concentration is high in heart muscle, brain, and skeletal muscle
Nicotinic acid	Involved in enzyme mechanisms in carbohydrate, fat, and protein metabolism; functions as hydrogen transfer agent; interrelated with pyridoxine and tryptophan metabolism	Retarded growth, poor feathering or hair coat, black tongue (dog), pellagra, necrotic enteritis, impaired reproduction and hatchability, bowed legs (ducks and turkeys), enlarged hocks, perosis, stomatitis, diarrhea, headache, depression, paralysis, dermatitis	Vasodilation, flushing, tingling, pruritis, hyperhidrosis, nausea, abdominal cramps	NAD (DPN) NADP (TPN) dehydrogenases, coenzymes such as lactate dehydrogenase	In addition to vitamin need, niacin has pharmacological activity as a vasodilator; possessed to a markedly less degree by the amide; 3-acetyl-pyridine-6-aminonicotinamide and pyridine-3-sulfonic acid or amide are antagonists
Pantothenic acid	A component of coenzyme A; functions in acetylation (or 2 carbon) reactions in amino acid, carbohydrate, and fat metabolism; involved in biosynthesis of acetyl choline, steroids, triglycerides, phospholipids, and ascorbic acid; interrelationship with cyanocobalamin, folic acid, and biotin mechanisms	Retarded growth, dermatitis, anorexia, weakness, spastic abnormalities or gait, scours, achromotrichia (animals), adrenal hemorrhagic necrosis, burning sensation in hands and feet, impaired reproduction and hatchability	None identified; relatively nontoxic	Coenzyme A, acetylases	d-Form is the one in nature; the dl- and d-forms are also manufactured as calcium salts; ω-methyl pantothenic acid and other compounds are antagonists

BIOLOGICAL CHARACTERISTICS OF VITAMINS (Continued)

Compound	Function	Deficiency Symptoms	Hyper-Use Symptoms	Coenzyme and Enzyme Involved	Remarks
Pyridoxine	Functions in amino acid metabolism, decarboxylation, transamination, and desulfhydration; oxidation of amines and amino acid transport; phosphorylase activity of muscle; conversion of tryptophan to nicotinic acid and amino acids to biogenic amines	Retarded growth, hyperexcitability, myelin degeneration, convulsions, heart changes, spastic gait, nervousness, anorexia, insomnia, acrodynia, microcytic anemia, impaired production (eggs), reproduction and hatchability, tryptophan metabolites in urine	Convulsions and abnormal encephalograms	Pyridoxal phosphate, pyridoxamine phosphate, transaminases, amino acid decarboxylases	The term vitamin B_6 refers to the 3 compounds, pyridoxine (ol), pyridoxal, and pyridoxamine, the latter 2 being important metabolic forms; isonicotinic acid hydrazide, toxopyrimidine, deoxypyridoxine, and 1-amino-D-proline are antagonists; high tryptophan and/or methionine diets increase need for pyridoxine
Riboflavin	Functions as coenzyme; needed in cellular respiration, hydrogen and electron transfer; growth and tissue maintenance; role in visual mechanism	Retarded growth, ocular and orogenital disturbances, greasy scaling of nasolabial folds, cheeks, and chin, angular stomatitis, myelin degeneration, poor feathering or hair growth, impaired reproduction and hatchability, muscle weakness, curled toe paralysis, scours	Itching, paresthesia, anuria	Flavin mono- and dinucleotide, amino acid oxidase, cytochrome *c* reductase, succinic dehydrogenase, xanthine oxidase, others	Irradiation of alkaline solution produces lumiflavin of acid solution, lumichrome; riboflavin in solution is one of the most photosensitive compounds of the vitamin class; 5-deoxyriboflavin and several other compounds act as antimetabolites
Thiamine	Functions as a coenzyme; activation and transfer of active acetaldehyde, glycoaldehyde, and succinic semialdehyde; functions in carbohydrate metabolism	Polyneuritis, beriberi, convulsions, muscle paralysis, anorexia, bradycardia, heart dilation, myocardial lesions, retarded growth, edema, pyruvic acid accumulation in blood and tissues	Analgesic effect on peripheral nerves, vascular hypertension	Cocarboxylase, transketolase, carboxylases	Like most water-soluble vitamins, there is no significant tissue storage; amprolium, pyrithiamine, oxythiamine, and others are antimetabolites; the two important forms in production are the hydrochloride and the mononitrate

PROPERTIES FOR ASCORBIC ACID AND ASCORBATE-2-SULFATE

Compound

Ascorbic acid
Vitamin C
L-Ascorbic acid
L-Threoascorbic acid
L-Xyloascorbic acid (obsolete)
Cevatamic acid (obsolete)
Hexuronic acid (obsolete)
L-Threo-2,3,4,5,6-pentahydroxy-2-hexene-γ-lactone (also available as sodium ascorbate)

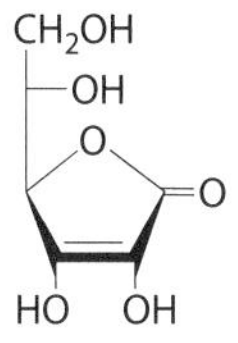

$C_6H_8O_6$ MW = 176.13

Properties

Colorless or white crystals; mp 191°, decomposes
λ_m (pH 5.2–10), 263.5 nm; ε_{263} 1.47 × 10^4
λ_m (acid soln), 243.5 nm; ε_{244}, 9,600
λ_m (isobestic), 252.8 nm
Characteristic acid taste

Solubility

33, water; 3.5, alcohol; 1, glycerol; sol, acetonitrile, dimethyl sulfoxide, dimethyl formamide, methanol, formic acid, acetic acid, etc., insol oils, and most organic solvents

Stability

Solid is stable in air in pure or tablet form unless hot and/or humid. Oxidizes autocatalytically in aqueous solution: Dilute (<0.1 mg/ml), very rapidly; concentrated (>1 mg/ml) moderate rate. Stable in solution if all O_2 excluded. Oxidation catalyzed by metals, especially Cu^{++} or Fe^{++}. Stabilized by acid solutions, especially metaphosphoric acid or trichloroacetic acid. Alkaline solutions unstable. Oxidation intermediates: Monodehydroascorbic acid (a free radical); dehydroascorbic acid; L-2,3-diketogulonate; plus many other products.

Compound

Ascorbate-2-sulfate
(Dipotassium salt)
Vitamin C_2
L-Ascorbic acid 2-sulfate
Potassium 2-O-sulfonato-L-ascorbate

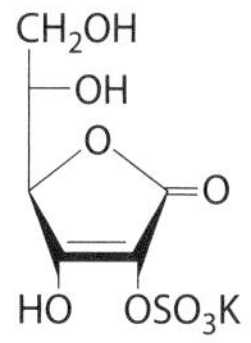

$C_6H_6O_9SK_2$ MW = 332.37

Properties

Colorless or white crystals, pKa, (sulfate group) ~2, pKa$_2$ (3-OH) ~3. λ_m(pH 4–10), 254 nm; ε_{254}, 17,700. λ_m (pH 2), 231 nm; ε_{231}, 11,000.

Solubility

Very sol, water; slightly sol, dimethyl formamide, dimethyl sulfoxide, acetonitrile; insoluble most organic solvents. Barium salt slightly sol in water.

Stability

Stable in dry solid form; moist solid subject to autocatalytic acid hydrolysis; stable in solution pH 5–9; hydrolyzes in solution < pH 5, rapidly below pH 3; not subject to air oxidation.

Compiled by Bert M. Tolbert.

BIOLOGICAL CHARACTERISTICS FOR ASCORBIC ACID

Compound

L-Ascorbic acid

Function

Prevents scurvy; required for normal collagen formation; involved in transferrition system; required for normal neurological function; a water soluble antioxidant and free radical scavenger; probably the functional cofactor for several hydroxylases; fundamental biochemical role(s) not known in animals nor in plants

Deficiency symptoms

Scurvy: Follicular hyperkeratosis, petechiae, ecchymosis, subconjunctional hemorrhage, joint effusions, dyspnea on exertion, swollen gums, neuropathy, edema, psychological impairment; terminal stage, heart blockage

Hyper-use symptoms

None substantiated; relatively nontoxic in oral form, probably due to limited gut transport and efficient urinary clearance

Coenzyme and enzymes involved

No proven coenzyme role; Cofactor roles implicated in prolyl hydroxylase, lysyl hydroxylase, p-hydroxyphenylpyruvate hydroxylase, homogentisic acid oxiginase, dopamine-β-hydroxylase and others

Subject to active transport through illium (in man), across blood-brain barrier and into the eye

Remarks

Probably present in all animals and plants; essential in the diet of man, monkey, baboon (probably all primates), guinea pig, trout, salmon, carp, catfish, several birds (especially insect or fruit eater, as swallows and redwing bulbul), Indian fruit bat, some insects; ascorbate-2-sulfate can substitute for ascorbate in trout and salmon but not in guinea pig; D-erythroascorbic acid (isoascorbic acid, erythorbic acid) is a common food additive, but has little nutritional value.

Compiled by Bert M. Tolbert.

VITAMERS

A vitamer is one of several chemical compounds, most often of closely related structure, which can fulfill the function of a given vitamin. One of the more prominent examples is Vitamin B_6 where the vitamers include pyridoxal, pyridoxine, 4-pyridoxic acid, and pyridoxal 5'-phosphate. Other examples include Vitamin B2 (flavin Vitamers), folic acid, vitamin K, biotin, Vitamin A, Vitamin D and vitamin E. Vitamers are not pseudovitamins. There are also phytonutrients which have been considered to have vitamin-like status. For example, the bioflavonoids are plant derivatives with antioxidant and anti-inflammatory properties; derived from citrus fruit rinds, berries, grains, and wines. Some are considered to have anticancer activity. Bioflavonoids/flavonoids are considered to be polyphenols (Albert, A., Manach, C., Morand, C., Rémésy, C., and Jimémez, L., Dietary polyphenols and the prevention of diseases, *Crit.Rev.Food Sci.Nutr.* 45, 287-306, 2005).

Vitamin A: β-Carotene and retinaldehyde and other retinol derivatives can be considered vitamers of vitamin A. Retinoic acid, although derived from β-Carotene via retinaldehyde or by the oxidation of retinol, is suggested to have a totally different function in growth and development unrelated to vision[1-6]. In addition, there is considerable interest in the use of vitamin A in cosmeceuticals/skin therapy[7-9].

Vitamin D: Vitamin D_2 and D_3 are derived from 7-dehydrocholesterol. Early work identified ergocalciferol (vitamin D_2) as an active component obtained from the irradiation of ergosterol. Later work established vitamin D_3 (cholecalciferol) as the major active component. Either must be converted to the 1α-hydroxylderivative (e.g. 1α,25-dihydroxyvitamin D_3 (1α,25-dihydroxyvitamin D)[10-13].

Vitamin E: There are a number of vitamin E (tocopherols/tocotrienols) vitamers[14-22] as well as the water-soluble derivative, Trolox[23-31] which is used as a standard for antioxidant measurements.

Vitamin K: Several vitamers[32-41] which have different activities in the support of γ-carboxylation reactions.

Thiamin (Vitamin B_1): Thiamine and several phosphate esters comprise the vitamers of thiamine[42-44] as well as some analogues[45-47].

Niacin (Vitamin B_3): Niacin is also referred to as nicotinic acid. Nicotinic acid and nicotinamide have equivalent activity and are incorporated into a coenzyme, nicotinamide adenine dinucleotide (NAD) and nicotinamide adenine dinucleotide phosphate (NADP). Nicotinic acid and nicotinamide are the major niacin vitamers[48-54].

Pyridoxine (Vitamin B_6): Multiple vitamers including pyridoxine, pyridoxal, pyridoxamine and the corresponding phosphates and precursors [55-61].

Biotin (Vitamin B_7; also Vitamin H): Biotin and precursors and precursors including 7-oxo-8-aminopelargonic acid[62-68].

Folic Acid (Vitamin B_9): Several vitamers[69-74] which are derivatives of the parent pteroyl glutamate.

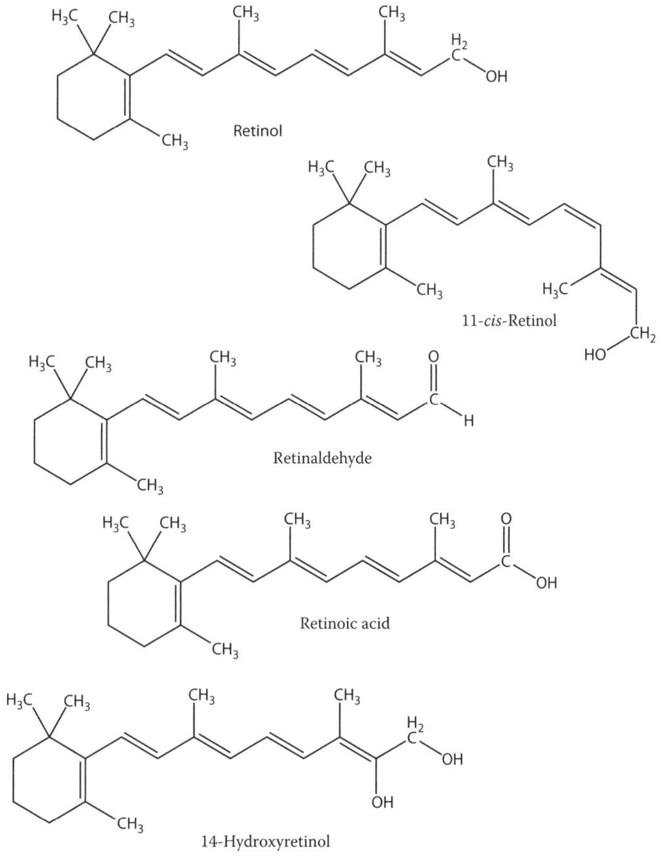

FIGURE 1 Structures of Vitamin A vitamers (retinol and retinol derivatives). See Sommer, A.., Vitamin a Deficiency and Clinical Diseases: An Historical Overview, *J.Nutr.* 138, 1835-1839, 2008; Dragsted, L.O., Biomarkers of Exposure to Vitamins A, C, and E their Relation to Lipid and Protein Oxidation Markers, *Eur.J.Nutr.* 47 (suppl 2), 3-18, 2008

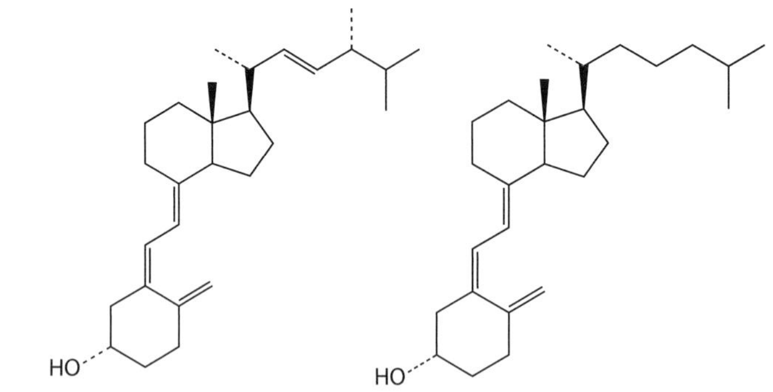

Vitamin D$_2$
Ergocalciferol

Vitamin D$_3$
Cholecalciferol

FIGURE 2   Structures of Vitamin D vitamers (Calciferol derivatives). See Holden, J.M., Lemar, L.E., and Exler, J., Vitamin D in Foods: Development of the US Department of Agriculture Database, *Am.J.Clin.Nutr.* 87, 1092S-1096S, 2008

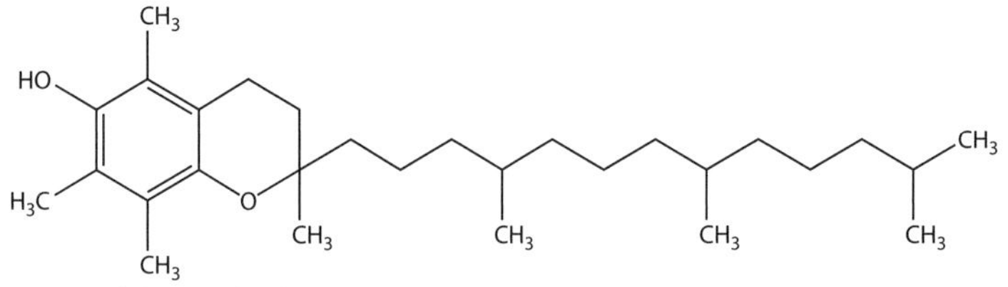

alpha-tocopherol

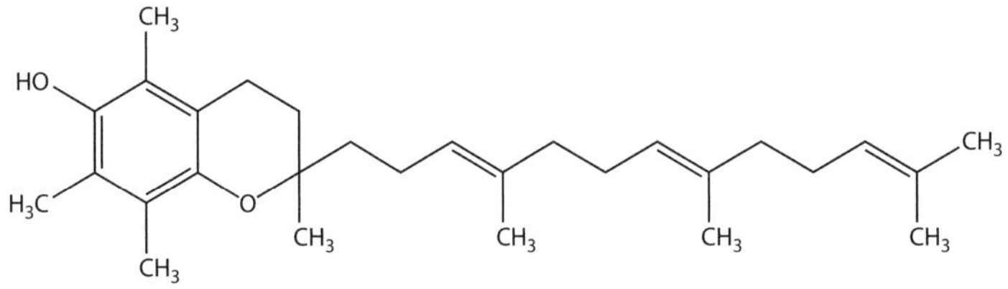

alpha-tocotrienol

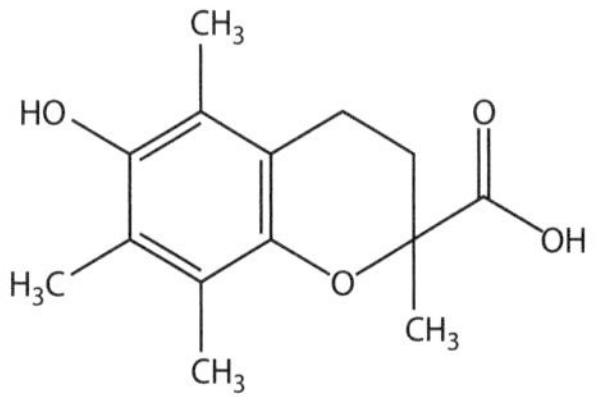

Trolox

FIGURE 3   Structures of Vitamin E vitamers (Tocopherol derivatives). See Yoshida, Y., Saito, Y., Jones, L.S., and Shigeri, Y., Chemical Reactivities and Physical Effects in Comparison between Tocopherols and Tocotrienols: Physiological Significance and Prospects as Antioxidants, *J.Biosci.Bioeng.* 104, 439-445, 2007; Clarke, M.W., Burnett, J.R., and Croft, K.D., Vitamin E in Human Health and Disease, *Crit.Rev.Clin.Lab.Sci.* 45, 417-450, 2008

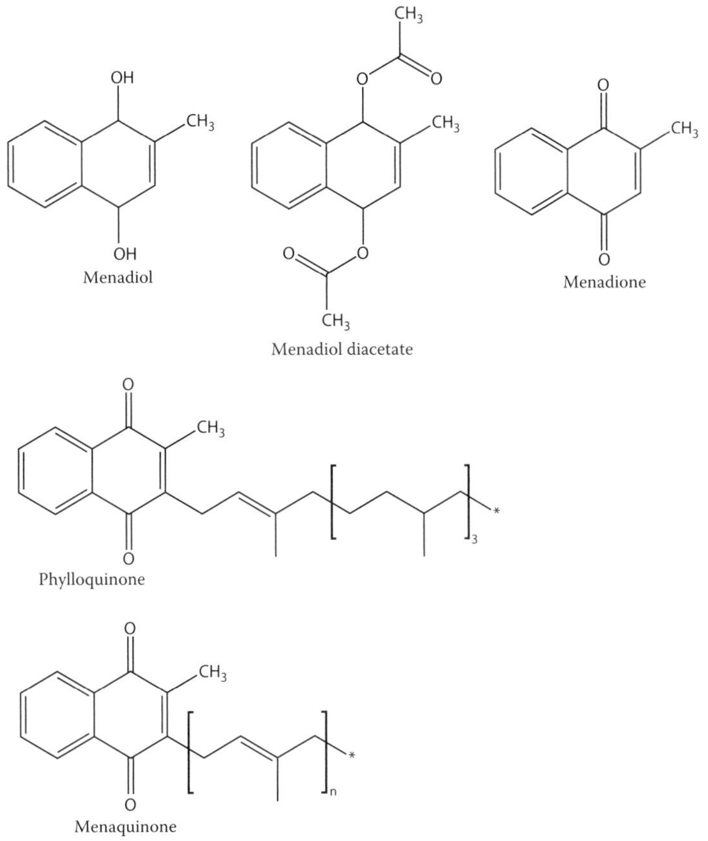

FIGURE 4 The structures of Vitamin K vitamers (Naphthoquinone derivatives). The K is derived from the German "koagulationvitamin." See Doisey, E.A., Brinkley, S.B., Thayer, S.A., and McKee, R.W., Vitamin K, *Science* 9, 58-62, 1940; Wolff, I.L. and Babior, B.M., Vitamin K and Warfarin. Metabolism, Function and Interaction, *Am.J.Med.* 53, 261-267, 1972; Sadowski, J.A. and Suttie, J.W., Mechanism of Action of Coumarins. Significance of Vitamin K epoxide, *Biochemistry* 13, 3696-3699, 1974; Yamada, Y., Inouye, G., Tahara, Y., and Kondo, K., The Structure of the Menaquinones with a Tetrahydrogenated Isoprenoid Side-Chain, *Biochim.Biophys.Acta* 488, 280-284, 1977; Bell, R.G., Vitamin K Activity and Metabolism of Vitamin K-1 Epoxide-1,4-diol, *J.Nutr.* 112, 287-292, 1982.

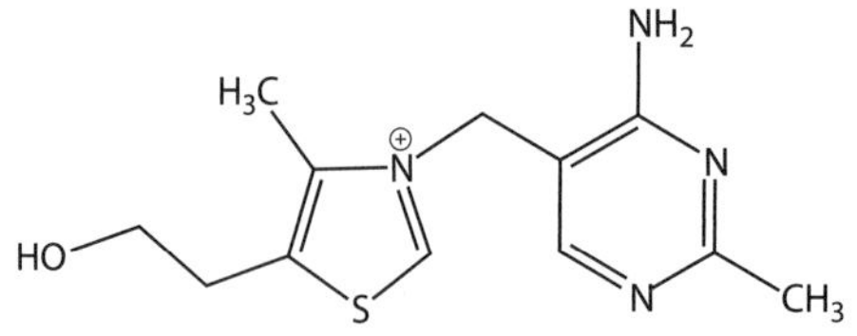

Thiamine

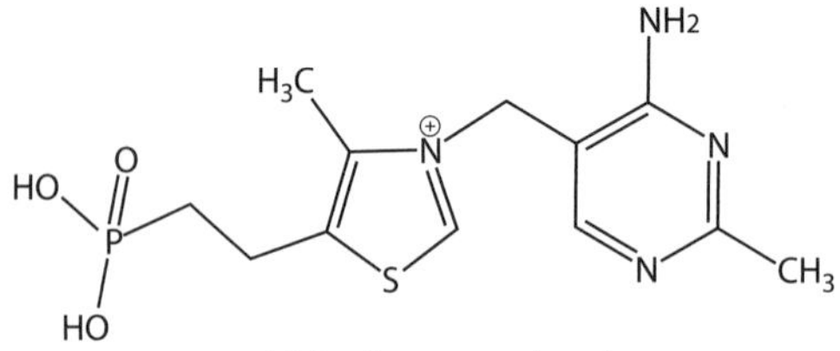

Thiamine monophosphate

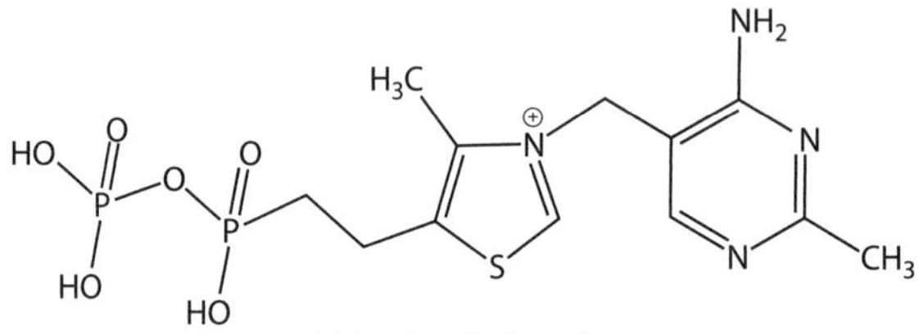

Thiamine Diphosphate

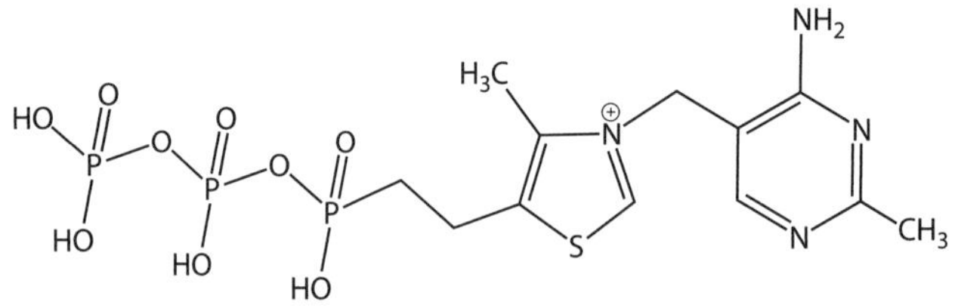

Thiamine Triphosphate

FIGURE 5 The structures of Vitamin B_1 vitamers (Thiamine and derivatives). See Maladrinos, G., Louloudi, M, and Hadjiliadis, N., Thiamine Models and Perspectives on the Mechanism of Action of Thiamine-Dependent Enzymes, *Chem.Soc.Rev.* 35, 684-692, 2006; Kowalska, E. and Kozik, A., The Genes and Enzymes Involved in the Biosynthesis of Thiamin and Thiamin Diphosphate in Yeasts, *Cell Mot.Biol.Lett.* 13, 271-282, 2008

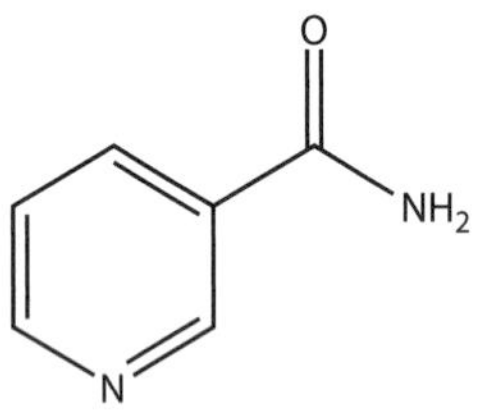

Nicotinamide

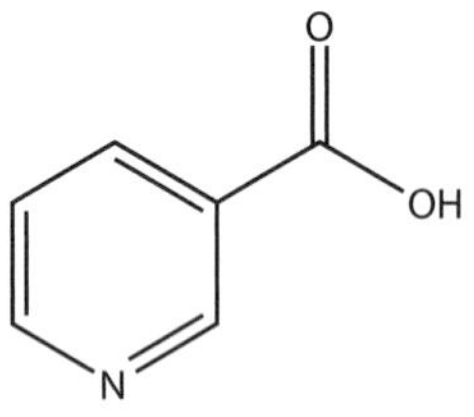

Nicotinic acid

FIGURE 6 The structures of Vitamin B_3 (Niacin) vitamers (Nicotinic acid and derivatives). See Skinner, P.J., Cherrier, M.C., Webb, P.J., *et al.*, 3-Nitro-4-Amino Benzoic Acids and 6-Amino Nicotinic Acids are Highly Selective Agonists for GPR109b, *Bioorg.Med.Chem. Lett.* 17, 6619-6622, 2007; Boovanahalli, S.K., Jin, X., Jin, Y. *et al.*, Synthesis of (Aryloxyacetylamino)-isonicotinic Acid Analogues as Potent Hypoxia-inducible factor (HIF)-1α Inhibitors, *Bioorg.Med.Chem.Lett.* 17, 6305-6310, 2007; Deng, Q., Frie, J.L., Marley, D.M. *et al.*, Molecular Modeling Aided Design of Nicotinic acid Receptor GPR109A Agonists, *Bioorg.Med.Chem.Lett.* 18, 4963-4967, 2008

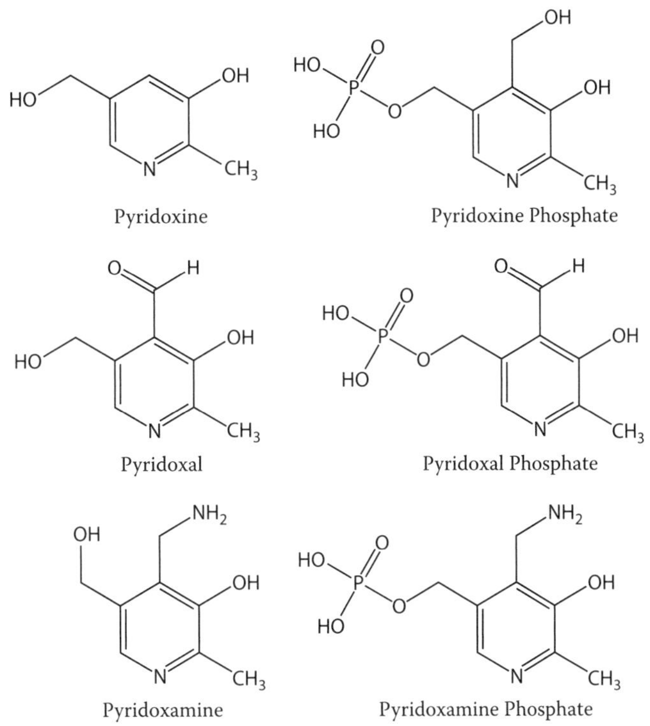

FIGURE 7 The structure of Pyridoxine (Vitamin B$_6$) vitamers. See Snell, E.E., Analogs of Pyridoxal or Pyridoxal Phosphate: Relation of Structure to Binding with Apoenzymes and to Catalytic Activity, *Vitam.Horm.* 28, 265-290, 1970; Drewke, C. and Leistner, E., Biosynthesis of Vitamin B$_6$ and Structurally Related Derivatives, *Vitam.Horm.* 61, 121-155, 2001; Garrido-Franco, M., Pyridoxine 5-Phosphate Synthase: *De novo* Synthesis of Vitamin B$_6$ and Beyond, *Biochim.Biophys.Acta* 1647, 92-97, 2003

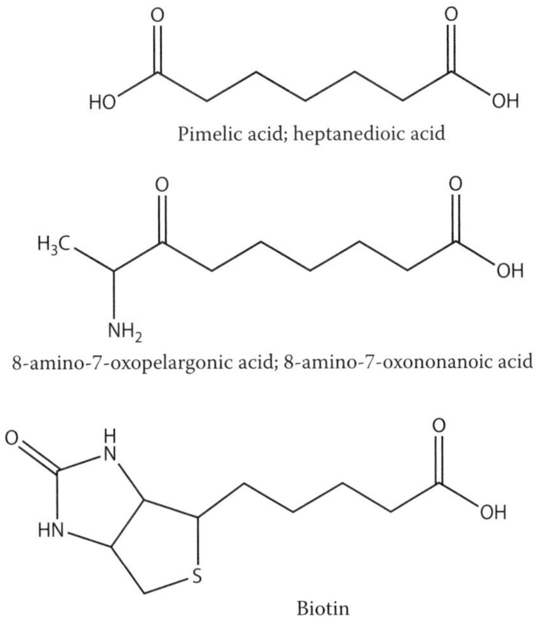

FIGURE 8 The structure of Biotin (Vitamin H) vitamers.

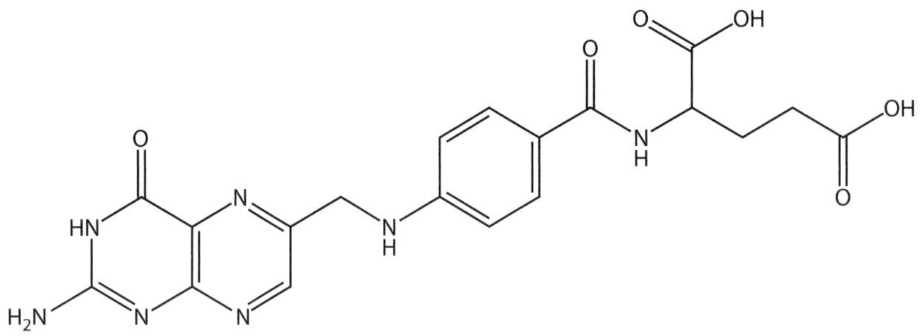

Folic acid (pteroylglutamic acid)

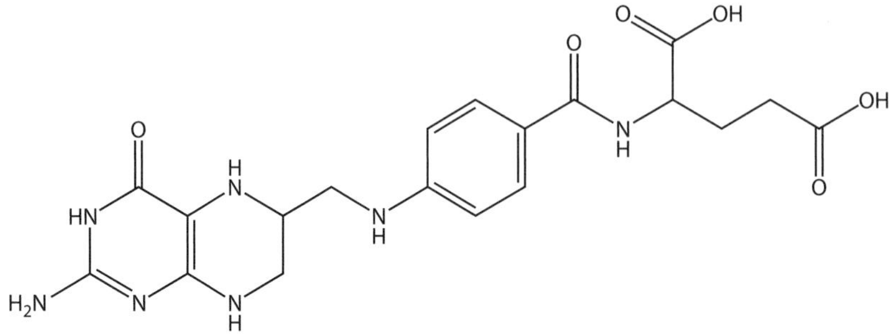

Tetrahydrofolic acid

FIGURE 9   The structure of Folic acid vitamers

References

1. Pitt, G.A., Chemical structure and the changing concept of vitamin A activity, *Proc.Nutr.Soc.* 42, 43-51, 1983

2. Wolf, G., The regulation of retinoic acid formation, *Nutr.Rev.* 54, 182-184, 1996

3. Clagett-Dame, M. and DeLuca, H.F., The role of vitamin A in mammalian reproduction and embryonic development, *Annu.Rev.Nutr.* 22, 347-381, 2002

4. Matt, N, Dupe, V., Garnier, J.M., Retinoic acid-dependent eye morphogenesis is orchestrated by neural crest cells, *Development* 132, 4789-4800, 2005

5. Moise, A.R., Isken, A., Dominguez, M., *et al.*, Specificity of zebrafish retinol saturase: formation of all-trans-13,14-dihydroretinol and all-trans-7,8-dihydroretinol, *Biochemistry* 46, 1811-1820, 2007

6. Reichrath, J., Lebmann, B., Carlberg, C., *et al.*, Vitamins as hormones, *Horm.Metab.Res.* 39, 71-84, 2007

7. Mayer, H., Bollag, W., Hanni, R., and Ruegg, R., Retinoids, a new class of compounds with prophylactic and therapeutic activities in oncology and dermatology, *Experientia* 34, 1105-1119, 1978

8. Zoubloulis, C.C., Retinoids—which dermatological indications will benefit in the near future?, *Skin Pharmacol.Appl.Skin Physiol.* 14, 303-315, 2001

9. Borg, O., Antille, C., Kaya, G., and Saurat, J.H., Retinoids in cosmeceuticals, *Dermatol.Ther.* 19, 289-296, 2006

10. Holick, M.F., The use and interpretation of assays for vitamin D and its metabolites, *J.Nutr.* 120, 1464-1469, 1990

11. Coburn, J.W., Tan, A.U., Jr., Levine, B.S., *et al.*, 1α-Hydroxy-vitamin D: a new look at an "old compound", *Nephrol.Dial.Transplant.* 11(supp 3), 153-157, 1996

12. Wikvall, K., Cytochrome P450 enzymes in the bioactivation of vitamin D to its hormonal form, *Int.J.Mol.Med.* 7, 201-209, 2001

13. Wu-Wong, J.R., Tian, J., and Goltzman, D., Vitamin D analogs as therapeutic agents: a clinical study update, *Curr.Opin.Investig.Drugs* 5, 32-326, 2004

14. Panfili, G., Fratianni, A., and Irano, M., Normal phase high-performance liquid chromatography method for the determination of tocopherols and tocotrienols in cereals, *J.Agric.Food Chem.* 51, 3940-3944, 2003

15. McCormick, C.C. and Parker, R.S., The cytotoxicity of vitamin E is both vitamer- and cell-specific and involves a selectable trait, *J.Nutr.* 124, 3335-3342, 2004

16. Sontag, T.J. and Parker, R.S., Vitamin E exhibits concentration- and vitamer-dependent impairment of microsomal enzyme activities, *Ann.N.Y.Acad.Sci.* 1031, 376-377, 2004

17. Amaral, J.S., Casal, S., Torres, D., *et al.*, Simultaneous determinations of tocopherols and tocotrienols in hazelnuts by a normal phase liquid chromatographic method, *Anal.Sci.* 21, 1545-1548, 2005

18. Cunha, S.C., Amaral, J.S. Ferandes, J.O., and Oliviera, B.P., Quantification of tocopherols and tocotrienols in Portuguese olive oils using HPLC with three different detection systems, *J.Agric.Food Chem.* 54, 3351-3356, 2006

19. Sookwong, P., Nakagawa, K., Murata, K., *et al.*, Quantitation of tocotrienols and tocopherol in various rice brans, *J.Agric.Food Chem.* 55, 461-466, 2007

20. Bustamante-Rangel, M. Delgado-Zamrreno, M.M., Sanchez-Perez A., and Carabias-Martinez, R., Determination of tocopherols and tocotrienols in cereals by pressurized liquid extraction-liquid chromatography-mass spectrometry, *Anal.Chim.Acta* 587, 216-221, 2007

21. Hunter, S.C. and Cahoon, E.B., Enhancing vitamin e in oilseeds: unraveling tocopherol and tocotrienol biosynthesis, *Lipids* 41, 97-108, 2007

22. Tsuzuki, W., Yunoki, R., and Yoshimura, H., Intestinal epithelial cell absorb γ-tocopherol faster than α-tocopherol, *Lipids* 42, 163-170, 2007

23. Kralli, A. and Moss, S.H., The sensitivity of an actinic reticuloid cell strain to near-ultraviolet radiation and its modification by trolox-C, a vitamin E analogue, *Br.J.Dermatol.* 116, 761-772, 1987

24. Nakamura, M., One-electron oxidation of Trolox C and vitamin E by peroxidase, *J.Biochem.* 110, 595-597, 1991

25. Miura, T., Muraoka, S., and Ogiso, T., Inhibition of hydroxyl radical-induced protein damages by trolox, *Biochem.Mol.Biol.Int.* 31, 125-133, 1993

26. Forrest, V.J., Kang, Y.H., McClain, D.E., *et al.*, Oxidative stress-induced apoptosis prevented by Trolox, *Free Radic.Biol.Med.* 16, 675-684, 1994

27. Albertini, R. and Abuja, P.M., Prooxidant and antioxidant properties of Trolox C, analogue of vitamin E, in oxidation of low-density lipoprotein, *Free Radic.Res.* 30, 181-188, 1999

28. Wang, C.C., Chu, C.Y., Chu, K.O., *et al.*, Trolox-equivalent antioxidant capacity assay versus oxygen radical absorbance capacity assay in plasma, *Clin.Chem.* 50, 952-954, 2004

29. Raspor, P., Plesnicar, S., Gazdag, Z., *et al.*, Prevention of intracellular oxidation in yeast: the role of vitamin E analogue, Trolox (6-hydroxy-2,5,7.8-tetramethylkorman-2-carboxyl acid), *Cell Biol.Int.* 29, 57-63, 2005

30. Abudu, N., Miller, J.J., and Levinson, S.S., Fibrinogen is a co-antioxidant that supplements the vitamin E analog trolox in a model system, *Free Radic Res.* 40, 321-331, 2005

31. Castro, I.A., Rogero, M.M., Junqueira, R.M., and Carrapeiro, M.M., Free radical scavenger and antioxidant capacity correlation of α-tocopherol and Trolox measured by three *in vitro* methodologies, *Int.J.Food Sci.Nutr.* 57, 75-82, 2006

32. Lefevere, M.F., De Lennheer, A.P., and Claeys, A.E., High-performance liquid chromatography assay of vitamin K in human serum, *J.Chromatog.* 186, 749-762, 1979

33. Lowenthan, J. and Vergel Rivera, G.M., Comparison of the activity of the *cis* and *trans* isomer of vitamin K1 in vitamin K-deficient and coumarin anticoagulant-pretreated rats, *J.Pharmacol.Exp.Ther.* 209, 330-333, 1979

34. Preusch, P.C. and Suttie, J.W., Stereospecificity of vitamin K-epoxide reductase, *J.Biol.Chem.* 258, 714-716, 1983

35. Hwang, S.M., Liquid chromatographic determination of vitamin K1 *trans-* and *cis-*isomers in infant formula, *J.Assoc.Off.Anal.Chem.* 68, 684-689, 1985

36. Will, B.H., Usui, Y., and Suttie, J.W., Comparative metabolism and requirement of vitamin K in chicks and rats, *J.Nutr.* 122, 2354-2360, 1992

37. Vermeer, C., Gijsbers, B.L., Craciun, A.M. *et al.*, Effects of vitamin K on bone mass and bone metabolism, *J.Nutr.* 126(4 suppl), 1187S-1191S, 1996

38. Gijsbers, B.L., Jie, K.S., and Vermeer, C., Effect of food composition on vitamin K absorption in human volunteers, *Br.J.Nutr.* 76, 223-229, 1996

39. Woolard, D.C., Indyk, H.E., Fong, B.Y., and Cook, K.K., Determination of vitamin K1 isomers in food by liquid chromatography with C_{30} bonded phase column, *J. AOAC Int.* 85, 682-691, 2002

40. Cook, K.K., Grundel, E., Jenkins, M.Y. and Mitchell, G.V., Measurement of *cis* and *trans* isomers of vitamin K1 in rat tissues by liquid chromatography with a C_{30} column, *J.AOAC Int.* 85, 832-840, 2002

41. Carrie, I., Pertoukalian, J., Vicaretti, R., *et al.*, Menequinone-4 concentration is correlated with sphingolipid concentration in rat brain, *J.Nutr.* 134, 167-172, 2004

42. Botticher, B. and Botticher, D., A new HPLC-method for the simultaneous determination of B_1-, B_2- and B_6-vitamers in serum and whole blood, *Int.J.Vitam.Nutr.Res.* 57, 273-278, 1987

43. Batifoulier, F., Verny, M.A., Bessom, C. *et al.*, Determination of thiamine and its phosphate esters in rat tissues analyzed as thiochromes on a RP-amide C_{16} column, *J.Chromatog.B.Analyt.Technol.Biomed. Life Sci.* 816, 67-72, 2005

44. Konings, E.J., Water-soluble vitamins, *J.AOAC Int.* 89, 285-288, 2006

45. Lowe, P.N., Leeper. F.J., and Perham, R.N., Stereoisomers of tetrahydrothiamine pyrophosphate, potent inhibitors of the pyruvate dehydrogenase multienzyme complex from *Escherichia coli*, *Biochemistry* 22, 150-157, 1983.

46. Klein, E., Nghiem, H.O., Valleix, A., *et al.*, Synthesis of stable analogues of thiamine di- and triphosphate as tools for probing a new phosphorylation pathway, *Chemistry* 8, 4649-4655, 2002

47. Erixon, K.M., Dabalos, C.L., and Leeper, F.J., Inhibition of pyruvate decarboxylase from *E.mobilis* by novel analogues of thiamine pyrophosphate: investigating pyrophosphate mimics, *Chem.Commun.* (9), 960-962, 2007

48. Sauberlich, H.E., Newer laboratory methods for assessing nutriculture of selected B-complex vitamins, *Annu.Rev.Nutr.* 4, 377-407, 1984

49. Stein, J., Hahn, A., and Rehner, G., High-performance liquid chromatographic determination of nicotinic acid and nicotinamide in biological samples applying post-column derivatization resulting in bathochrome absorption shifts, *J.Chromatog.B.Biomed.Appl.* 665, 71-78, 1995

50. Gillmor, H.A., Bolton, C.H., Hopton, M., *et al.*, Measurement of nicotinamide and *N*-methyl-2-pyridone-5-carboxamide in plasma by high performance liquid chromatography, *Biomed.Chromatog.* 13, 360-362, 1999

51. Khan, A.R., Khan, K.M, Perveen, S., and Butt, N., Determination of nicotinamide and 4-aminobenzoic acid in pharmaceutical preparations by LC, *J.Pharm.Biomed.Anal.* 29, 723-727, 2002

52. Chatzimichalakis, P.F., Samanidou, V.F., Verpoorte, R., and Papadoyannis, I.N., Development of a validated HPLC method for the determination of B-complex vitamins in pharmaceuticals and biological fluids after solid phase extraction, *J.Sep.Sci.* 27, 1181-1188, 2004

53. Hsieh, Y. and Chen, J., Simultaneous determination of nicotinic acid and its metabolites using hydrophilic interaction chromatography with tandem mass spectrometry, *Rapid Commun.Mass Spectrom.* 19, 3031-3036, 2005

54. Marszall, M.P., Markuszewski, M.J., and Kaliszan, R., Separation of nicotinic acid and its structural isomers using 1-ethyl-3-methylimidazolium ionic liquid as a buffer additive by capillary electrophoresis, *J.Pharm.Biomed.Anal.* 41, 329-332, 2006

55. Vanderslice, J.T., Maire, C.E., and Beecher, G.R., B_6 Vitamer analysis in human plasma by high performance liquid chromatography: a preliminary report, *Am.J.Clin.Nutr.* 34, 947-950, 1981

56. Hachey, D.L. Coburn, S.P., Brown, L.T., *et al.*, Quantitation of vitamin B6 in biological samples by isotope dilution mass spectrometry, *Anal. Biochem.* 151, 159-168, 1985

57. Driskell, J.A. and Chrisley, B.M., Plasma B-6 vitamer and plasma and urinary 4-pyridoxic acid concentrations in young women as determined using high performance liquid chromatography, *Biomed. Chromatogr.* 5, 198-201, 1991

58. Sharma, S.K. and Dakshinamurti, K., Determination of vitamin B6 vitamers and pyridoxic acid in biological samples, *J.Chromatogr.* 578, 45-51, 1992

59. Schaeffer, M.C., Gretz, D., Mahuren, J.D., and Coburn, S.P., Tissue B-6 vitamer concentrations in rats fed excess vitamin B-6, *J.Nutr.* 125, 2370-2378, 1995

60. Fu, T.F., di Salvo, M., and Schirch, V., Distribution of B6 vitamers in *Escherichia coli* as determined by enzymatic assay, *Anal.Biochem.* 298, 314-321, 2001

61. Bisp, M.R., Bor, M.V., Heinsvig, E.M., *et al.*, Determination of vitamin B6 vitamers and pyridoxic acid in plasma: development and evaluation of a high-performance liquid chromatography assay, *Anal. Biochem.* 305, 82-89, 2002

62. Eisenberg, M.A., The biosynthesis of biotin in growing yeast cells: The formation of biotin from an early intermediate, *Biochem.J.* 101, 598-600, 1966

63. Birnbaum, J., Pai, C.H., and Lichstein, H.C., Biosynthesis of biotin in microorganisms. V. Control of vitamer production, *J.Bacteriol.* 94, 1846-1853, 1967

64. Eisenberg, M.A. and Star, C., Synthesis of 7-oxo-8-aminopelargonic acid, a biotin vitamer, in cell-free extracts of *Escherichia coli* biotin auxotrophs, *J.Bacteriol.* 96, 1846-1843, 1967

65. Eisenberg, M.A. and Star, C., Synthesis of 7-oxo-8-aminopelargonic acid, a biotin vitamer, in cell-free extracts of *Escherichia coli* biotin autotrophs, *J.Bacteriol.* 96, 1291-1297, 1968

66. Ohsugi, M., Miyauchi, K., and Inoue, Y., Biosynthesis of biotin-vitamers from unsaturated higher fatty acids by bacteria, *J.Nutr.Sci. Vitaminol.* 31, 253-263, 1985

67. Sabatie, J., Speck, D., Reymund, J., *et al.*, Biotin formation by recombinant strains of *Escherichia coli*; influence of host physiology, *J.Biotechnol.* 20, 29-49, 1991

68. Phalip, V., Kuhn, I., Lemoine, Y., and Jeltsch, J.M., Characterization of the biotin biosynthesis pathway in *Saccharomyces cerevisiae* and evidence for a cluster containing B105, a novel gene involved in vitamer uptake, *Gene* 232, 43-51, 1999

69. Wegner, C., Trotz, M., and Nau, H., Direct determination of folate monoglutamates in plasma by high-performance liquid chromatography using an automatic precolumn-switching system as sample clean up procedure, *J.Chromatog.* 378, 55-65, 1986

70. Freisleben, A., Schieberle, P., and Rychlik, M., Syntheses of labeled vitamers of folic acid to be used as internal standards in stable isotope dilution assays, *J.Agric.Food Chem.* 50, 4760-4768, 2002

71. Freisleben, A., Schieberle, P., and Rychlik, M., Specific and sensitive quantification of folate vitamers in food by stable isotope dilution assay high-performance liquid chromatography-tandem mass spectrometry, *Anal.Bioanal.Chem.* 376, 149-156, 2003

72. Pfeiffer, C.M., Fazili, Z., McCoy, L., *et al.*, Determination of folate vitamers in human serum by stable-isotope-dilution tandem mass spectrometry and comparison with radioassay and microbiological assay, *Clin.Chem.* 50, 423-432, 2004

73. Smulders, Y.M., Smith, D.E, Kok, E.M., *et al.*, Cellular folate vitamer distribution during and after correction of vitamin B_{12} deficiency: a case for the methylfolate trap, *Br.J.Haematol.* 132, 623-629, 2006

74. Smith, D.E., Kok, R.M., Teerlink, T., *et al.*, Quantitative determination of erythrocyte folate vitamer distribution by liquid chromatography-tandem mass spectrometry, *Clin.Chem.Lab.Med.* 44, 450-459, 2006

VITAMIN NAMES DISCARDED

- Vitamins B_c, B_{10}, B_{11}, and B_x (these mostly have been used to refer to folic acid or folic acid precursors such *p*-aminobenzoic acid although it would appear that these terms were also used to refer to mixtures of vitamins).
- Vitamin M is a term that has recently been used to describe delta(Δ)-1-tetrahydrocannabinol Earlier the term vitamin M was used to describe a mixture of B-complex vitamins
- Vitamin B_4 (mostly used to describe adenine but occasionally for choline) Lecoq, R., The role of adenine (vitamin B4 in the metabolism of organic compounds and its repercussions on acid-base equilibrium, *J.Physiol.* 46, 406-410, 1954; Whelan, W.J., Vitamin B4, *IUBMB Life* 57, 125. 2005; Hartmann, J. and Getoff, N., Radiation-induced effect of adenine (vitamin B4) on mitomycin C activity. *In vitro* experiments, *Anticancer Res.* 26, 3005-3010, 2006
- Vitamin L (anthranilic acid; *o*-aminobenzoic acid)
- Vitamin Bc – folate although earlier used to describe the B complex vitamins
- Vitamin B_{10} –folate, precursors of folate such as *p*-aminobenzoic acid; R factor; also used to refer to vitamin A/retinoic acid; Wang, Y. and Okabe, N., Crystal structures and spectroscopic properties of Zinc(II) ternary complexes of Vitamin L, H' and their isomer *m*-aminobenzoic acid with bipyridine, *Chem.Pharm.Bull.* 53, 645-652, 2005

- Vitamin B_{11} – folic acid: Getoff, N., Transient absorption spectra and kinetics of folic acid (vitamin B11) and some kinetic data of folinic acid and methotrexate, *Oncol.Res.* 15, 295-300, 2005

Pseudovitamins (also described as "fake" vitamins; Young, V.R. and Newberne, P.M., Vitamins and cancer prevention: Issues and dilemmas, *Cancer* 47, 1226-1240, 1981)

Vitamin B_{17} (Laetrile®; amygdalin; 1-mandelonitrile-β-glucuronic acid)
Vitamin B_{15} (pangamic acid; not a chemically defined entity)
Vitamin B_{13} (Orotic acid)
H_3 (Gerovital)
U (methionine sulfonium salts)

General references for vitamins

Handbook of Vitamins, 2nd edn., ed. L.J. Machlin, Marcel Dekker, Inc., New York, NY, USA, 1991

Coumbs, G.F., *The Vitamins. Fundamental Aspects in Nutrition and Health*, 2nd edn., Academic Press, San Diego, CA, USA, 1998

Stipanuk, M.H., *Biochemical and Physical Aspects of Human Nutrition*, W.B. Saunders, Philadelphia, PA, USA, 2000

Bender, D.A., *Nutritional Biochemistry of the Vitamins*, 2nd Edn., Cambridge University Press, Cambridge, United Kingdom, 2003

Section IV
Nucleic Acids

UV SPECTRAL CHARACTERISTICS AND ACIDIC DISSOCIATION CONSTANTS OF 280 ALKYL BASES, NUCLEOSIDES, AND NUCLEOTIDES

B. Singer

The λ_{max} (nm), in those cases where more than one value has been reported, are either the most frequent value or an average of several values, the range being ± 1 to 2 nm for the λ_{max}. Since the λ_{min} is more sensitive than the λ_{max} to impurities in the sample, the values of λ_{min} in the table are generally the lowest reported. Values in parentheses are shoulders or inflexions. The cationic and anionic forms are either so stated by the authors or are arbitrarily taken at pH 1 and pH 13. Individual values are given for pK_a except when there are more than two values. In that case, a range is given.

Complete spectra representing a range of derivatives are shown in the figures, and reference to these is made in the table with an asterisk and number preceding the name of the compound.

All spectra were obtained in the author's laboratory from samples isolated from paper chromatograms. It is recognized that pH 1 or pH 13 is not ideal for obtaining the cationic or anionic forms when these pHs are close to a pK. Nevertheless, these conditions are useful for purposes of identification since the spectra are reproducible.

Additional data not quoted here are available in many of the references. These data include spectral characteristics in other solvents than H_2O and at other pH values, extinction coefficients, R_F values in various paper chromatographic systems, column chromatographic systems, methods of synthesis or preparation of alkyl derivatives, mass spectra, and NMR, optical rotatory dispersion and infrared spectra.

	Acidic		Basic			
	λ_{max}(nm)	λ_{min}(nm)	λ_{max}(nm)	λ_{min}(nm)	pK_a	References
*1ADENINE						
MONOALKYLATED						
*21-Methyl-	259	228	270	234	7.2	1–7
1-Ethyl-	260	233	271	242	6.9, 7.0	5, 8, 9
1-Isopropyl-	259		269			10
1-Benzyl-	260		271		7.0	11, 12
1-(2-Hydroxypropyl)-	259		271		7.2	13
1-(2-Hydroxyethylthioethyl)-	262		271		7.2	6
2-Methyl-	267	228	270 (280)	239	~5.1	3, 4
*33-Methyl-	274	235	273	244	6.1, 6.1	2, 5, 6, 13–15
3-Ethyl-	274	240	273	247	6.5	5, 8, 16
3-Isopropyl-	274		273			10
3-Benzyl-	275		272		5.1	17, 18
3-(2-Hydroxypropyl)-	274		273		6.0	13
3-(2-Diethyloaminoethyl)-	275	236	274	245		19
*4N^6-Methyl-	267	231	273 (280)	238	4.2, 4.2	1–4, 7, 15, 20, 21
N^6-Ethyl-	268	231	274(281)	241		8, 21
N^6-Butyl-	270		275			21
N^6-(2-Hydroxyethyl)-	272	233	273	236	3.7	72
N^6-(2-Diethylaminoethyl)-	275	233	274	239		19
*57-Methyl-	273	237	270(280)	230	~3.5, 3.6	8, 15, 18, 23, 24
7-Ethyl-	272	239	270 (280)	234		8, 16
7-Isopropyl-	272		272			10
9-Methyl-	261	230	262	228	3.9	5, 13, 15
9-Ethyl-	258	230	262	228	4.1	5, 13, 25
9-Isopropyl-	260		262			10
9-(2-Hydroxypropyl)-	259		261			13
9-(2-Diethylaminoethyl)	258	227	261	229		19
9-Benzyl-	259		261			18
DIALKYLATED						
1,N^6-Dimethyl-	261	230	273	245		1, 8, 20
1,7-Dimethyl-	270					26
1-Butyl, 7-methyl-	268					26
1,9-Dimethyl-	260	235	260(265)	235	9.08	13, 20, 27
1,9-Di(2-hydroxypropyl)-	260		260			13
1,9-Di(2-diethylaminoethyl)-	257	233	261	232		19

UV SPECTRAL CHARACTERISTICS AND ACIDIC DISSOCIATION CONSTANTS OF ALKYL BASES, NUCLEOSIDES, AND NUCLEOTIDES[a] (Continued)

| | Acidic | | Basic | | | |
	λ_{max}(nm)	λ_{min}(nm)	λ_{max}(nm)	λ_{min}(nm)	pK$_a$	References
[1]ADENINE (Continued)						
1-Ethyl-9-methyl-	261		261		9.16	27
1-Propyl-9-methyl-	261		261		9.15	27
3,N^6-Dimethyl-	281		287			28
3,N^6-Di(2-diethylaminoethyl)-	282	243	282	249		19
3,7-Dimethyl-	276	246	225, 280	221, 247	11	2, 13, 20
3,7-Dibenzyl-	278		281		9.6	18
3,7-Di(2-hydroxypropyl)-	278		281			13
3,7-Di(2-diethylaminoethyl)-	276	237	279	245		19
N^6,N^6-Dimethyl-	276	236	282	245		7, 21
N^6,N^6-Diethyl-	278		282			21, 29
N^6,7-Dimethyl-	279		275			24, 26
*[6]N^6-Methyl-7-ethyl-	277 (285)	241	276	244		8
N^6-Propyl-7-methyl-	281		277			24
N^6-Butyl-7-methyl-	279					26
N^6,9-Dimethyl-	265		268		4.02	27, 29
N^6,9-Di-2-diethylaminoethyl)-	266	229	270	233		19
N^6-Ethyl-9-methyl-	265		268		4.08	27, 29
N^6-Propyl-9-methyl-	266		270		4.14	27
N^6-Butyl-9-ethyl-	266		269			27
TRIALKYLATED						
1,N^6,N^6-Trimethyl-	221, 293	246	232, 301	262		30
3,$N^6,N^{6'}$-Trimethyl-	290	243	293	250		30
N^6,N^6,7-Trimethyl-	233, 293	250	291	246		30
N^6,N^6,9-Trimethyl-	269	234	276	237		29, 30
N^6,N^6-Dimethyl-9-ethyl-	270		277			25
3,N^6,7-Tribenzyl-	289		Unstable		9.4	18
[7]GUANINE						
MONOALKYLATED						
1-Methyl-	250 (270)	228	277 (262)	242	3.1	4, 7, 15, 23, 31–34
*[8]1-Ethyl-	251 (274)	229	278 (260)	243		35, 36
1-Isopropyl-	253					10
1-(2-Diethylaminoethyl)-	255, 275		257, 268			19
N^2-Methyl-[b]	251, 279	228	245–255, 278	238	3.3	4, 7, 23, 33, 34, 37, 38
*[9]N^2-Ethyl-	253 (280)	229	245, 279	263		37, 39
N^2-Isopropyl-	252		277			10
3-Methyl-	263 (244)	227	273	246		15, 33, 40, 41
*[10]3-Ethyl-	263 (244)	233	273	248		39
3-Isopropyl-	263		273			10
3-Benzyl-	263 (243)		274		4.00	42
O^6-Methyl-	286		246, 284			43, 44
*[11]O^6-Ethyl-	286	253	284 (246)	259		35, 44
O^6-Propyl-	286		246, 283			44
O^6-Isopropyl-	285 (230)		283 (245)			10, 45
O^6-Butyl-	286		246, 285			44
O^6-Isobutenyl-	286 (232)		283(245)			45
7-Methyl-	249 (272)	226	281 (240)	255	3.5	4, 23, 31, 32, 46
*[12]7-Ethyl-	249 (274)	233	280	258	3.7	5, 35
7-Isopropyl-	249 (274)		278 (240)			10, 45
7-Benzyl-	250		281		3.2	11
7-(2-Hydroxyethyl)-	250	229	281	261		47
7-(2-Hydroxypropyl)-	250, 272	229	280	257		13
7-(2-Diethylaminoethyl)-	253, 270		275 (250)			19
7-(β-Hydroxyethylthioethyl)-	250		281			48
8-Propyl-	249, 276		276			45
8-Isobutyl-	249, 278		276			45
8-(3-Methylbutyl)-	249, 276		275			45
9-Methyl	251, 276		268 (258)		2.9	33
*[13]9-Ethyl-	252, 277	230	253, 268	238		35
9-Isopropyl-	253, 276		256, 258			10, 45

UV SPECTRAL CHARACTERISTICS AND ACIDIC DISSOCIATION CONSTANTS OF ALKYL BASES, NUCLEOSIDES, AND NUCLEOTIDES[a] (Continued)

	Acidic		Basic			
	λ_{max}(nm)	λ_{min}(nm)	λ_{max}(nm)	λ_{min}(nm)	pK$_a$	References

[7]GUANINE (Continued)

DIALKYLATED

1,7-Dimethyl-	252 (272)	230	284(251)	262		23, 33, 35
[14]1,7-Diethyl-	252 (275)	232	285 (250)	263		49
1,9-Dimethyl-	254 (277)	229				33
N^2,N^2-Dimethyl-[b]	255 (289)	229	277–283			4, 34, 37, 38
7,8-Dimethyl-	249, 277		280 (235)		4.4	50
7,9-Dimethyl	254 (278)	229	c			28, 33
7,9-Di(2-diethylaminoethyl)-	257, 278		c			19
7-Methyl-9-ethyl-	254, 281		c		7.3	14
8,9-Dimethyl-	252, 277 (289)		280(252)		4.11	50

TRIALKYLATED

1,7,9-Trimethyl-	254, 280		c			33

[15]CYTOSINE

MONOALKYLATED

1-Methyl-	213, 283	241	274	250	4.55–4.61	51–53
O^2-Methyl-	260	241	270	246	5.41	53
3-Methyl-	273	240	294	251	7.4, 7.49	4, 53, 54
[16]3-Ethyl-	275	241	296	257		55
N^4-Methyl-	278	240	286 (230)	256	4.55	4, 56, 57
N^4-Ethyl-	277	244	284	253	4.58	55, 57
5-Methyl-	211, 284	242	288	254	4.6	58
6-Methyl-					5.13	59

DIALKYLATED

1,3-Dimethyl-	281	243	272	247	9.29–9.4	51, 53, 54
1,N^4-Dimethyl-	218, 285	244	274(235)	250	4.38–4.47	51, 53, 56
N^4,N^4-Dimethyl-	283	242	290(235)	259	4.15, 4.25	56, 57
1,5-Dimethyl-	291	244			4.76	60

TRIALKYLATED

1,3,N^4-Trimethyl-	212, 287	248	280	247	9.65	19, 51, 53
1,N^4,N^4-Trimethyl-	220, 288	248	283	242	4.2	56

5-HYDROXYMETHYLCYTOSINE

	Unmodified					
	279		284			7
3-Methyl-	278		296		7.1	14

[17]XANTHINE

[18]1-Methyl-[d]	260–265, (235)	239	283 (245)	257	1.3	35, 61, 62
3-Methyl-[d]	266		275 (232)		0.8	61, 62
[19]7-Methyl-[d]	267	233	289 (237)	255	0.8	32, 35, 61
9-Methyl-[d]	260		245, 278		2.0	61, 62
1,3-Dimethyl-	266		275		0.7	61, 62
1,7-Dimethyl-	~260		233, 289		0.5	61, 62
1,9-Dimethyl-	262		248, 277		2.5	61, 62
3,7-Dimethyl-	265		234, 274		0.3	61, 62
3,9-Dimethyl-	265		270 (240)		1.0	61, 62
7,9-Dimethyl-	239, 262		c			63
8,9-Dimethyl-	238, 265		245, 278			63
1,3,7-Trimethyl-	266				0.5	61
1,3,9-Trimethyl-	266				0.6	61
1.7,9-Trimethyl-	232, 262		c			63

[20]URACIL

1-Methyl-	208, 268	241	265	242	~1.8	58, 64
O^2-Ethyl-	218, 260		221, 265			58
[21]3-Methyl-	258	230	218, 283	245		4, 58, 65
O^4-Methyl-	267		276			65a
O^4-Ethyl-	269		220, 278			58
1,3-Dimethyl-	266	234	266	234		58
1,6-Dimethyl-	208, 268	234	266	241		65
O^2,3-Dimethyl-	213, 269					65 b
3,6-Dimethyl-	205, 259	231	281	245		65
5,6-Dimethyl-	267	236	275	245		65
1,5,6-Trimethyl-	206, 276	240	273	245		65

UV SPECTRAL CHARACTERISTICS AND ACIDIC DISSOCIATION CONSTANTS OF ALKYL BASES, NUCLEOSIDES, AND NUCLEOTIDES[a] (Continued)

	Acidic		Basic			
	λ_{max}(nm)	λ_{min}(nm)	λ_{max}(nm)	λ_{min}(nm)	pK$_a$	References
[*22]THYMINE						
1-Methyl-			269	244		66
1-(2-Diethylaminoethyl)-			265	248		19
3-Methyl-	266	237	290	248		31, 35, 66
[*23]3-Ethyl-	265	237	289	247		35
3-(2-Diethylaminoethyl)-			288	244		19
1,3-Di(2-diethylaminoethyl)-			269	245		19
1,O[4]-Di(2-diethylaminoethyl)-			274	245		19
O[2],3-Dimethyl-	217, 272					65b
HYPOXANTHINE						
1-Methyl-	249		260			15, 66a
3-Methyl-	253		265		2.61	15, 66a, 67
3-Ethyl-	254		266			68
3-Benzyl-	254		264(277)			69
3-(2-Diethylaminoethyl)-	260	234	262	243		19
O[6]-(3-Methyl-2-butenyl)-	247		262			45
7-Methyl-	250		262		2.12	15
9-Methyl-	250		254			15
1,7-Dimethyl-	252	232	267	237		70
1,7-Dibenzyl-	255		256			69
1,9-Dibenzyl-	263		259			69
3,7-Dimethyl-			267			20
3,7-Dibenzyl-	256		267			69
N[6],7-Dimethyl-	256		258			24
7,9-Dimethyl-	251		c			28
[*24]ADENOSINE						
[*25]1-Methyl-	257	231	258 (265)	233	8.8, 8.3	3, 4, 8, 16, 71
1-Methyldeoxy-	257	239	258	242		8, 72
[*26]1-Ethyl-	259	235	261, (268, 300)	237		8
1-Ethyldeoxy-	259	231	260 (268)	236		8
1-Benzyedeoxy-	259		259			11
2-Methyl-	258	230	264	227		3, 4, 73
[*27]N[6]-Methyl-	262	231	266	223	4.0	1–4, 8, 74, 74a
N[6]-Methyldeoxy-	262	231	266	226		4, 8, 72, 74
N[6]-Ethyl-	264	239	268	243		8
N[6]-Ethyldeoxy-	263	237	268	241		8
N[6]-Benzyl-	265	235	268	236		69, 75
N[6]-Benzyldeoxy-	264		268			11
N[6]-Butyl-	263		267			76
N[6]-(2-Hydroxyethyl)-	263	233	267	232	3.1	22
[*28]7-Ethyl-	268	239	c			8
[*29]1,N[6]-Dimethyl-	261	234	263 (300)	234		1, 8, 20
N[6],N[6]-Dimethyl-	268	233	276	237	4.5	3, 4, 30
[*30]N[6],7-Dimethyl-	276	241	c			8
N[6]-Methyl-7-ethyl-	276	242	c			8
[*31]GUANOSINE						
[*32]1-Methyl-	258 (280)	230	255 (270)	228	2.6	4, 7, 20, 34
1-Methyldeoxy-	257 (278)	232	255 (270)	229		20, 31, 77
1-Ethyl-	261 (272)	232	258 (270)	239	2.8	49
1-Ethyldeoxy-	256		257			77
1-Butyldeoxy-	258 (282)		257 (280)			77
N[2]-Methyl-[b]	251–258 (280–290)	222–234	248–258 (270–275)	227–238		4, 34, 38
O[6]-Methyl-	284 (243)	259	243, 277	239, 261	2.4	49
O[6]-Methyldeoxy-	284 (230)	252	243, 278	233, 261		31, 78
[*33]O[6]-Ethyl-	244, 286	239, 260	247, 278	233, 261	2.5	49
O[6]-Ethyldeoxy-	286	252	248, 280	261		77
O[6]-Butyldeoxy-	246, 287	260	248, 280	261		77
7-Methyl-	257 (275)	230	c		6.7–7.3	4, 14, 46, 49, 74

UV SPECTRAL CHARACTERISTICS AND ACIDIC DISSOCIATION CONSTANTS OF ALKYL BASES, NUCLEOSIDES, AND NUCLEOTIDES[a] (Continued)

	Acidic		Basic			
	λ_{max}(nm)	λ_{min}(nm)	λ_{max}(nm)	λ_{min}(nm)	pK$_a$	References
[31]*GUANOSINE (Continued)*						
7-Methyldeoxy-	256 (275)	229	c			31, 32, 74, 77
[34]*7-Ethyl-	258 (277)	238	c		7.2, 7.4	14, 49
7-Isopropyl-	256 (275)		c			45
7-Benzyl-	258		c		7.2	11
7-Butyldeoxy-	257 (280)		c			77
8-Methyl-	260 (273)		256		3.01	50
1,7-Dimethyl-	260 (270)	236	c			20, 49
[35]*1,7-Diethyl-	263 (270)	237	c			49
1-Methyl-7-ethyl-	259 (277)	233	c			49
1-Ethyl-7-methyl-	261 (275)	235	c			49
N^2,N^2-Dimethyl-	265 (290)	237	262 (283)	240		4, 7, 34, 38
N^2,O^6-Dimethyldeoxy-	288		249, 284			77
[36]*N^2,O^6-Diethyl-	246, 292	239, 267	252, 281	237, 268		35, 49
$N^2,N^2,7$-Trimethyl-	267, 300	239, 286	c			79
[37]*CYTIDINE						
O^2-Methyl-	233, 262	221, 243	Unstable		>8.6	35
[37a]*O^2-Ethyl-	233, 262	221, 243	Unstable		>8.6	35
3-Methyl-	278	243	225, 267	212, 244	8.3–8.9	4, 54–56, 71
[38]*3-Ethyl-	280	247	267	248	8.4	55
3-Ethyldeoxy-	280	245	268	247	8.6	55
3-Benzyl-	281		266		7.7	11
3-(2-Diethylaminoethyl)deoxy-	284	243	271			19
N^4-Methyl-	217, 281	207, 243	237, 273	250	3.85, 3.92	56, 57
N^4-Methyldeoxy-	282	242	236, 270	229, 248	4.01	57
[39]*N^4-Ethyl-	281	244	272	253	4.2	55
N^4-Ethyldeoxy-	279	247	272	253	4.2	55
$2'$-O N^4-Dimethyl-[e]	281	242	271	250		80
$2',3',5'$-Tri-O-methyl-N^4-methyl-[e]	281	242	271	247		81
5-Methyl-	288	245	278	255	4.28	60
5-Methyldeoxy-	287	246	278	255	4.40	60, 82
5-Ethyldeoxy-			278	255		83
6-Methyl	278	241	273	252	4.42	59, 84
6-Methyldeoxy-	278	241	273	252		84
$3,N^4$-Dimethyl-	286	249	277	249		55
[40]*$3,N^4$-Diethyl-	287	252	277	253		55
$3,N^4$Di(2-diethylaminoethyl)deoxy-	284	245				19
N^4, N^4-Dimethyl-	219, 285	245	279	238	3.7, 3.62	56, 57
N^4, N^4-Dimethyldeoxy-	287	245	278	238	3.79	57
[41]*N^4,N^4-Diethyl-	286	249	276	249		55
$2'$-O,N^4,N^4-Trimethyl-[e]	287	246	278	238		80
$N^4,5$-Dimethyl-			275 (234)	252		60
$N^4,5$-Dimethyldeoxy-	218, 287	246	275(235)		4.04	60
[42]*$3,N^4,N^4$-Triethyl-	287	252	289	253		55
[43]*URIDINE						
O^2-Methyl-	229, 251	213, 238				35, 86, 87
O^2-Ethyl-	228, 253	213, 237				87
3-Methyl-	263	233	262	233		4, 14, 55, 85
[44]*3-Ethyl-	262	235	264	237		35, 55
3-Benzyl-			264	235		83
3-(2-Hydroxyethyl)-	261	235	262			88
[44a]*O^4-Methyl-	271	235	274	239		80
6-Methyl-	261	230	264	242		65, 84
5,6-Dimethyl-	206, 269	236	270	248		65
[45]*THYMIDINE (DEOXY-)						
3-Methyl-	266	238	267	239		31, 35
[46]*3-Ethyl-	269	239	270	240		35
3-(2-Diethylaminoethyl)-			270	242		19
O^4-Methyl-(α)	279	245	279	243		89
O^4-Methyl-(β)	279	241	280	243		89

UV SPECTRAL CHARACTERISTICS AND ACIDIC DISSOCIATION CONSTANTS OF ALKYL BASES, NUCLEOSIDES, AND NUCLEOTIDES[a] (Continued)

	Acidic		Basic			
	λ_{max}(nm)	λ_{min}(nm)	λ_{max}(nm)	λ_{min}(nm)	pK$_a$	References
[*47]INOSINE						
1-Methyl-	250	223	249			4
1-Methyldeoxy-	250		250(265)			31
1-Benzyl-	251		249			69
1-(2-Hydroxyethyl)-	250	226	250	226		88
O^6-Methyl-	250		250			66a, 74a
[*48]7-Methyl-	252	221	c		6.4	35, 70
7-Ethyl-	252		c			68
1,7-Dimethyl-	265		c			70
XANTHOSINE						
7-Methyl-	262	237	c			32, 35
ADENYLIC ACID OR ADP[e,f]						
1-Methyl-	258	232	259 (268)	230		6, 35, 90
1-(2-Hydroxyethylthioethyl)-	261		261			6
2-Methyl-	259		263			73
N^6-Methyl-	261	231	265	229		6, 90
N^6-(2-Hydroxyethylthioethyl)-			268			6
N^6-(2-Hydroxyethyl)-	263	233	266	230		91
2,N^6-Dimethyl	263		269			92
GUANYLIC ACID OR GDP[e,f]						
1-Methyl-	258 (280)	230	256 (273)	230	<3	7, 93
7-Methyl-	259, 279	230	c		6.9–7.2	46, 94
O^6-Methyl-	245, 288	262	249, 281	263		95
N^2-Methyl-	263	237	263	240		96
N^2,N^2-Dimethyl-	265	232	263	237	~3	7, 96
$N^2,N^2,7$-Trimethyl-	262 (290)	237	c		7.4	96
CYTIDYLIC ACID OR CDP[e,f]						
3-Methyl-	276	242	223, 266	243	9.0, 9.2	97–99
N^4-Methyl-	217, 281	242	271	249	4.25	98, 99
N^4,N^4-Dimethyl-	219, 287	245	225, 278	238	4.0	98, 99
5-Methyl-	284		279			100
5-Methyldeoxy-	287	243	277	254	4.5	101
URIDYLIC ACID OR UDP[e,f]						
3-Methyl-	261	235	262	235		35, 91
3-Ethyl-	261		263			35
THYMIDYLIC ACID (DEOXY-)[f]						
[*49]3-Methyl-	267	240	268	241		35
3-Ethyl-	267	241	268	242		35

[a] Much of the data and some of the spectra were published in a review by B. Singer in *Prog. Nucleic Acids Res. Mol. Biol.*, 15, 219–284, 330–332 (1975).

[b] N²-Alkyl guanines and guanosines do not exhibit sharp maxima or minima, particularly in basic solution, as shown in the spectra published by Hall,[4] Smith and Dunn,[34] and Singer and Fraenkel-Conrat.[39] Therefore, some of the data are given as a range of values.

[c] All 7-alkyl purine nucleosides and nucleotides and 7,9-dialkyl purines are unstable in alkali and the imidazole ring opens at varying rates. For this reason spectral data obtained in alkaline solution do not represent the original compound and thus such data are omitted. The opening of the imidazole ring in alkali can be used as a means of identifying this class of alkyl compounds. Ring opening can lead to a number of derivatives.[39]

[d] Basic values are those of the dianion (pH 14).

[e] Alkylation of ribose does not cause any change in spectrum.

[f] Alkylated nucleoside diphosphates have the same spectral characteristics as alkylated nucleotides and the data are not separated. Alkylation of the phosphate group does not cause any change in spectrum.

References

1. Wacker and Ebert, *Z. Naturforsch.*, 14b, 709 (1959).
2. Brookes and Lawley, *J. Chem. Soc., (Lond.)*, p. 539 (1960).
3. Garrett and Mehta, *J. Am. Chem. Soc.*, 94, 8532 (1972).
4. Hall, *The Modified Nucleosides in Nucleic Acids.* Columbia University Press, New York, 1971.
5. Pal, *Biochemistry*, 1, 558 (1962).
6. Shooter, Edwards, and Lawley, *Biochem. J.*, 125, 829 (1971).
7. Venkstern and Baer, *Absorption Spectra of Minor Bases.*, Plenum Press, New York, 1965.
8. Singer, Sun, and Fraenkel-Conrat, *Biochemistry*, 13, 1913 (1974).
9. Ludlum, *Biochim. Biophys. Acta*, 174, 773 (1969).
10. Lawley, Orr, and Jarman, *Biochem. J.*, 145, 73 (1975).
11. Brookes, Dipple, and Lawley, *J. Chem. Soc. C*, p. 2026 (1968).
12. Leonard and Fujii, *Proc. Natl. Acad. Sci. U.S.A.*, 51, 73 (1964).

13. Lawley and Jarman, *Biochem. J.*, 126, 893 (1972).
14. Lawley and Brookes, *Biochem. J.*, 89, 127 (1963).
15. Elion, *J. Org. Chem.*, 27, 2478 (1962).
16. Lawley and Brookes, *Biochem. J.*, 92, 19c (1964).
17. Montgomery and Thomas, *J. Am. Chem. Soc.*, 85, 2672 (1963).
18. Montgomery and Thomas, *J. Heterocycl. Chem.*, l, 115 (1964).
19. Price, Gaucher, Koneru, Shibakawa, Sowa, and Yamaguchi, *Biochim. Biophys. Acta*, 166, 327 (1968).
20. Broom, Townsend, Jones, and Robins, *Biochemistry*, 3, 494 (1964).
21. Elion, Burgi, and Hitchings, *J. Am. Chem., Soc.*, 74, 411 (1952).
22. Windmueller and Kaplan, *J. Biol. Chem.*, 236, 2716 (1961).
23. Reiner and Zamenhof, *J. Biol Chem.*, 228, 475 (1957).
24. Prasad and Robins, *J. Am. Chem. Soc.*, 79, 6401 (1957).
25. Montgomery and Temple, *J. Am. Chem. Soc.*, 79, 5238 (1957).
26. Taylor and Loeffler, *J. Am. Chem. Soc.*, 82, 3147 (1960).
27. Itaya, Tanaka, and Fujii, *Tetrahedron*, 28, 535 (1972).
28. Jones and Robins, *J. Am. Chem. Soc.*, 84, 1914 (1962).
29. Robins and Lin, *J. Am. Chem. Soc.*, 79, 490 (1957).
30. Townsend, Robins, Loeppky, and Leonard, *J. Chem. Soc.* (Lond.), p. 5320 (1964).
31. Friedman, Mahapatra, Dash, and Stevenson, *Biochim. Biophys. Acta*, 103, 286 (1965).
32. Haines, Reese, and Todd, *J. Chem. Soc.* (Lond.), p. 5281 (1962).
33. Shapiro, *Prog. Nucleic Acid Res. Mol. Biol*, 8, 73 (1968).
34. Smith and Dunn, *Biochem. J.*, 72, 294 (1959).
35. Singer, unpublished.
36. Kriek and Emmelot, *Biochemistry*, 2, 733 (1963).
37. Elion, Lange, and Hitchings, *J. Am. Chem. Soc*, 78, 217 (1956).
38. Gerster and Robins, *J. Am. Chem. Soc.*, 87, 3752 (1965).
39. Singer and Fraenkel-Conrat, *Biochemistry*, 14, 772 (1975).
40. Lawley, Orr, and Shah, *Chem. Biol. Interact.*, 4, 431 (1971/72).
41. Townsend and Robins, *J. Chem. Soc.* (Lond.), p. 3008 (1962).
42. Miyaki and Shimizu, *Chem. Pharm. Bull.* (Tokyo), 18, 1446 (1970).
43. Lawley and Thatcher, *Biochem. J.*, 116, 693 (1970).
44. Balsiger and Montgomery, *J. Am. Chem. Soc.*, 25, 1573 (1960).
45. Leonard and Frihart, *J. Am. Chem. Soc.*, 96, 5894 (1974).
46. Hendler, Furer, and Srinivasan, *Biochemistry*, 9, 4141 (1970).
47. Brookes and Lawley, *J. Chem. Soc.* (Lond.), p. 3923 (1961).
48. Brookes and Lawley, *Biochem. J.*, 77, 478 (1960).
49. Singer, *Biochemistry*, 11, 3939 (1972).
50. Pfleiderer, Shanshal, and Eistetter, *Chem. Ber.*, 105, 1497 (1972).
51. Kenner, Reese, and Todd, *J. Chem. Soc.* (Lond.), p. 855 (1955).
52. Fox and Shugar, *Biochim. Biophys. Acta*, 9, 369 (1952).
53. Sukhorukov, Gukovskaya, Sukhoruchkina, and Lavrrenova, *Biophysica*, 17, 5 (1972).
54. Brookes and Lawley, *J. Chem. Soc.* (Lond.), p. 1348 (1962).
55. Sun and Singer, *Biochemistry*, 13, 1905 (1974).
56. Szer and Shugar, *Acta Biochim. Pol.*, 13, 177 (1966).
57. Wempen, Duschinsky, Kaplan, and Fox, *J. Am. Chem. Soc.*, 83, 4755 (1961).
58. Shugar and Fox, *Biochim. Biophys. Acta*, 9, 199 (1952).
59. Notari, Witiak, DeYoung, and Lin, *J. Med Chem.*, 15, 1207 (1972).
60. Fox, Praag, Wempen, Doerr, Cheong, Knoll, Eidinoff, Bendich, and Brown, *J. Am. Chem. Soc.*, 81, 178 (1959).
61. Lichtenberg, Bergmann, and Neiman, *J. Chem. Soc. C*, p. 1676 (1971).
62. Pfleiderer and Nubel, *Justus Liebigs Ann. Chem.*, 647, 155 (1961).
63. Pfleiderer, *Justus Liebigs Ann. Chem.*, 647, 161 (1961).
64. Brown, Hoerger, and Mason, *J. Chem. Soc.*, (Lond.), p. 211 (1955).
65. Wittenburg, *Collect. Czech. Chem. Commun.*, 36, 246 (1971).
65a. Wong and Fuchs, *J. Org. Chem.*, 35, 3786 (1970).
65b. Wong and Fuchs, *J. Org. Chem.*, 36, 848 (1971).
66. Wierzchowski, Litonska, and Shugar, *J. Am. Chem. Soc.*, 87, 4621 (1965).
66a. Miles, *J. Org. Chem.*, 26, 4761 (1961).
67. Bergmann, Levin, Kalmus, and Kwietny-Govrin, *J. Am. Chem. Soc.*, 26, 1504 (1961).
68. Rajabalee and Hanessian, *Can. J. Chem.*, 49, 1981 (1971).
69. Montgomery and Thomas, *J. Am. Chem. Soc.*, 28, 2304 (1963).
70. Michelson and Pochon, *Biochim. Biophys. Acta*, 114, 469 (1966).
71. Haines, Reese, and Todd, *J. Chem. Soc.* (Lond.), p. 1406 (1964).
72. Coddington, *Biochim. Biophys. Acta*, 59, 472 (1962).
73. Saneyoshi, Ohashi, Harada, and Nishimura, *Biochim. Biophys. Acta*, 262, 1 (1972).
74. Jones and Robins, *J. Am. Chem. Soc.*, 85, 193 (1963).
74a. Johnson, Thomas, and Schaeffer, *J. Am. Chem. Soc.*, 80, 699 (1958).
75. Kissman and Weiss, *J. Am. Chem. Soc.*, 21, 1053 (1956).
76. Fleysher, *J. Med. Chem.*, 15,187 (1972).
77. Fanner, Foster, Jarman, and Tisdale, *Biochem. J.*, 135, 203 (1973).
78. Loveless, *Nature*, 223, 206 (1969).
79. Saponara and Enger, *Nature*, 223, 1365 (1969).
80. Robins and Naik, *Biochemistry*, 10, 3591 (1971).
81. Kusmierek, Giziewica, and Shugar, *Biochemistry*, 12, 194 (1973).
82. Dekker and Elmore, *J. Chem. Soc.* (Lond.), p. 2864 (1951).
83. Imura, Tsuruo, and Ukita, *Chem. Pharm. Bull.*, 16, 1105 (1968).
84. Winkley and Robins, *J. Org. Chem.*, 33, 2822 (1968).
85. Miles, *Biochim. Biophys. Acta*, 22, 247 (1956).
86. Brown, Todd, and Varadarajan, *J. Chem. Soc.* (Lond.), p. 868 (1957).
87. Kimura, Fujisawa, Sawada, and Mitsunobu, *Chem. Lett.*, 691 (1974).
88. Holy, Bald, and Hong, *Collect. Czech. Chem. Commun.*, 36, 2658 (1971).
89. Lawley, Orr, Shah, Farmer, and Jarman, *Biochem. J.*, 135, 193 (1973).
90. Griffin and Reese, *Biochim. Biophys. Acta*, 68, 185 (1963).
91. Michelson and Grunberg-Manago, *Biochim. Biophys. Acta*, 91, 92 (1964).
92. Hattori, Ikehara, and Miles, *Biochim. Biophys. Acta*, 13, 2754 (1974).
93. Pochon and Michelson, *Biochim. Biophys. Acta*, 145, 321 (1967).
94. Lawley and Shah, *Biochem. J.*, 128, 117 (1972).
95. Gerchman, Dombrowski, and Ludlum, *Biochim. Biophys. Acta*, 272, 672 (1972).
96. Pochon and Michelson, *Biochim. Biophys. Acta*, 182, 17 (1969).
97. Ludlum and Wilhelm, *J. Biol Chem.*, 243, 2750 (1968).
98. Brimacombe and Reese, *J. Chem. Soc. C*, p. 588 (1966).
99. Brimacombe, *Biochim. Biophys. Acta*, 142, 24 (1967).
100. Szer, *Biochem. Biophys. Res. Commun.*, 20, 182 (1965).
101. Cohn, *J. Am. Chem. Soc.*, 73, 1539 (1951).

ULTRAVIOLET ABSORBANCE OF OLIGONUCLEOTIDES CONTAINING 2'-*O*-METHYLPENTOSE[a] RESIDUES

Compound	Min	Max	Absorbance ratios[b]				
			240	250	270	280	290
pH 7							
Am-Cp	225	256	0.46	0.88	0.63	0.17	0.02
Am-Gp	227	261	0.59	0.81	0.83	0.42	0.12
Am-Up	225	256	0.54	0.93	0.70	0.38	0.14
AM-Up	228	260	0.41	0.79	0.72	0.26	0.03
CM-Cp	249	268	0.95	0.85	1.17	0.87	0.32
Cm-Ap	227	261	0.62	0.83	0.81	0.42	0.13
Cm-Gp	224	254	0.87	1.04	0.92	0.74	0.33
Gm-Ap	225	256	0.57	0.93	0.72	0.40	0.16
Gm-Cp	225	255	0.82	1.00	0.93	0.69	0.28
Gm-Up	226	255	0.63	0.97	0.81	0.52	0.19
Um-Ap	228	258	0.46	0.83	0.71	0.25	0.03
Um-Gp	226	255	0.62	0.98	0.79	0.51	0.18
Um-Up	229	260	0.43	0.78	0.79	0.35	0.04
Am-Am-Up	228	258	0.44	0.83	0.70	0.25	0.04
Am-Gm-Cp	225	257	0.65	0.94	0.79	0.45	0.17
Am-Um-Gp	227	257	0.53	0.86	0.76	0.39	0.13
pH 2							
Am-Ap	229	257	0.44	0.84	0.70	0.23	0.05
Am-Cp	234	264	0.40	0.74	0.97	0.75	0.45
Am-Gp	227	256	0.54	0.93	0.71	0.42	0.23
Am-Up	229	258	0.44	0.82	0.72	0.27	0.05
Cm-Cp	239	278	0.28	0.47	1.67	1.96	1.39
Cm-Ap	234	264	0.39	0.73	0.98	0.76	0.48
Cm-Gp	232	276	0.48	0.79	1.06	1.08	0.73
Gm-Ap	227	256	0.54	0.92	0.70	0.42	0.22
Gm-Cp	230	274	0.53	0.83	1.07	1.04	0.67
Gm-Up	228	257	0.53	0.89	0.77	0.50	0.26
Um-Ap	229	257	0.48	0.86	0.72	0.27	0.05
Um-Gp	228	258	0.49	0.86	0.78	0.52	0.28
Um-Up	230	260	0.44	0.80	0.76	0.31	0.03
Am-Am-Up	230	257	0.48	0.86	0.71	0.27	0.05
Am-Gm-Cp	230	258	0.53	0.86	0.88	0.69	0.42
Am-Um-Gp	228	258	0.48	0.86	0.74	0.40	0.19
pH 12							
Am-Ap	227	257	0.42	0.84	0.68	0.19	0.03
Am-Cp	226	261	0.59	0.83	0.81	0.41	0.12
Am-Gp	228	258	0.50	0.86	0.76	0.36	0.06
Am-Up	230	260	0.55	0.83	0.69	0.18	0.03
Cm-Cp	249	268	0.94	0.86	1.15	0.88	0.34
Cm-Ap	228	262	0.60	0.81	0.83	0.43	0.13
Cm-Gp	230	268	0.71	0.87	1.03	0.73	0.22
Gm-Ap	228	259	0.50	0.85	0.80	0.38	0.06
Gm-Cp	230	268	0.74	0.89	1.02	0.71	0.20
Gm-Up	226	255	0.63	0.97	0.81	0.52	0.19
Um-Ap	230	258	0.56	0.84	0.69	0.21	0.04
Um-Gp	233	260	0.65	0.87	0.87	0.48	0.09
Um-Up	241	260	0.79	0.86	0.79	0.29	0.04
Am-Am-Up	229	258	0.53	0.86	0.71	0.24	0.04
Am-Gm-Cp	228	261	0.59	0.86	0.83	0.46	0.12
Am-Um-Gp	230	259	0.55	0.85	0.79	0.36	0.14

Compiled by A. R. Trim.

[a] The pentose is presumed to be 2'-*O*-methylribose since the dinucleotides were obtained from yeast ribonucleic acid. Evidence for the chemical constitution of the modified pentose is given in Howlett et al.[1] and Trim and Parker.[2]

[b] Absorbance ratios were calculated from optical densities at 240, 250, 270, 280, and 290 nm relative to that at 260 nm.

Values for trinucleotides are from Trim and Parker.[3] Data on dinucleotides are from Trim and Parker, *Biochem, J.,* 116, 589 (1970). With permission. Copyright by the Biochemical Society.

References

1. Howlett, Johnson, Trim, Eagles, and Self, *Anal. Biochem.*, 39, 429 (1971).
2. Trim and Parker, *Anal. Biochem.*, 46, 482 (1972).
3. Trim and Parker, unpublished data.

SPECTROPHOTOMETRIC CONSTANTS OF RIBONUCLEOTIDES

TABLE 1: Ultraviolet Absorbance of Mono- and Oligonucleotides

Compound	pH 7		Absorbance Ratios[a]					pH 1		Absorbance Ratios[a]					pH 12		Absorbance Ratios[a]					Slack[b] pct.
	λ_{min}	λ_{max}	240	250	270	280	290	λ_{min}	λ_{max}	240	250	270	280t	290	λ_{min}	λ_{max}	240	250	270	280	290	
Ap	226	259	0.41	0.79	0.65	0.14	0.00	229	257	0.46	0.87	0.67	0.22	0.00	227	259	0.40	0.79	0.66	0.14	0.00	cat.[c]
Cp	250	271	0.94	0.85	1.17	0.90	0.28	241	278	0.27	0.47	1.65	1.91	1.34	225	228	0.95	0.85	1.18	0.90	0.29	cat.
Gp	231	264	0.81	1.17	0.82	0.68	0.26	227	256	0.52	0.93	0.75	0.69	0.48	225	260	0.55	0.88	0.96	0.58	0.00	cat.
Up	230	261	0.39	0.75	0.82	0.33	0.00	229	261	0.41	0.78	0.79	0.30	0.00	242	261	0.75	0.83	0.79	0.27	0.00	cat.
pApA	227	258	0.41	0.83	0.66	0.19	0.00	229	256	0.45	0.86	0.67	0.21	0.00	229	258	0.42	0.82	0.66	0.18	0.00	1.0
ApCp	227	261	0.57	0.81	0.81	0.40	0.12	233	265	0.39	0.73	0.98	0.75	0.45	228	261	0.59	0.81	0.82	0.41	0.12	2.3
ApGp	225	256	0.58	0.97	0.71	0.37	0.13	228	257	0.47	0.88	0.71	0.42	0.24	229	259	0.46	0.83	0.79	0.35	0.00	0.9
ApUp	228	259	0.40	0.80	0.73	0.24	0.00	230	258	0.43	0.82	0.72	0.26	0.00	232	259	0.53	0.81	0.71	0.20	0.00	1.8
CpCp	250	269	0.94	0.85	1.17	0.90	0.30	241	278	0.24	0.44	1.69	1.96	1.37	250	270	0.95	0.85	1.19	0.92	0.33	1.3
CpGp	224	254	0.84	1.04	0.93	0.73	0.30	233	277	0.39	0.74	1.08	1.12	0.79	231	250	0.69	0.86	1.04	0.72	0.20	1.8
GpCp	224	255	0.83	1.01	0.94	0.70	0.27	233	276	0.43	0.76	1.08	1.10	0.76	231	267	0.71	0.88	1.04	0.70	0.17	1.7
GpUp	226	255	0.62	0.98	0.81	0.51	0.17	229	258	0.46	0.86	0.76	0.50	0.28	233	261	0.64	0.87	0.90	0.47	0.00	2.8
UpAp	228	259	0.42	0.81	0.70	0.24	0.00	229	257	0.45	0.85	0.72	0.27	0.00	231	259	0.54	0.83	0.70	0.21	0.00	2.9
UpGp	226	256	0.60	0.98	0.79	0.50	0.18	229	258	0.43	0.83	0.78	0.50	0.28	234	261	0.61	0.85	0.89	0.48	0.00	2.9
UpUp	229	260	0.41	0.78	0.80	0.33	0.00	230	260	0.42	0.79	0.78	0.31	0.00	242	260	0.79	0.85	0.78	0.29	0.00	0.4
pApApA	228	258	0.44	0.85	0.67	0.24	0.00	230	257	0.47	0.86	0.68	0.22	0.00	230	258	0.40	0.82	0.68	0.24	0.00	1.6
ApApCp	228	258	0.55	0.85	0.76	0.39	0.11	230	259	0.44	0.79	0.86	0.55	0.31	232	259	0.54	0.82	0.77	0.39	0.11	1.5
ApApGp	226	256	0.54	0.94	0.70	0.33	0.11	229	257	0.47	0.88	0.70	0.35	0.17	230	258	0.46	0.84	0.75	0.32	0.00	1.0
ApApUp	228	258	0.41	0.82	0.70	0.25	0.00	229	257	0.44	0.84	0.70	0.24	0.00	230	258	0.49	0.83	0.70	0.22	0.00	2.1
ApCpCp	229	262	0.66	0.82	0.89	0.52	0.18	236	270	0.37	0.67	1.16	1.04	0.69	234	263	0.67	0.81	0.91	0.54	0.20	2.6
ApCpGp	225	257	0.65	0.94	0.80	0.49	0.19	231	260	0.42	0.79	0.91	0.74	0.48	230	261	0.56	0.83	0.81	0.48	0.12	1.5
ApGpCp	226	257	0.65	0.94	0.81	0.47	0.18	231	260	0.43	0.79	0.89	0.70	0.45	230	261	0.56	0.83	0.87	0.47	0.12	1.8
ApGpUp	227	257	0.53	0.91	0.73	0.36	0.11	229	257	0.45	0.85	0.73	0.39	0.18	232	259	0.52	0.83	0.79	0.34	0.00	1.7
ApUpGp	227	257	0.51	0.90	0.73	0.36	0.11	229	258	0.43	0.83	0.74	0.40	0.19	231	259	0.51	0.82	0.80	0.34	0.00	1.9
CpApGp	226	257	0.67	0.93	0.82	0.53	0.21	231	260	0.42	0.79	0.91	0.74	0.48	230	260	0.58	0.84	0.87	0.51	0.12	0.9
CpCpGp	225	257	0.85	0.98	0.99	0.77	0.33	235	278	0.36	0.66	1.23	1.33	0.95	232	268	0.75	0.85	1.08	0.76	0.23	1.6
GpApCp	226	258	0.63	0.91	0.79	0.50	0.19	231	259	0.44	0.80	0.89	0.71	0.46	230	260	0.56	0.83	0.87	0.50	0.12	1.7
GpApUp	227	257	0.52	0.89	0.74	0.38	0.11	229	258	0.46	0.86	0.73	0.39	0.18	231	259	0.53	0.83	0.80	0.36	0.00	1.8
GpGpCp	224	254	0.79	1.06	0.87	0.65	0.27	230	260	0.45	0.82	0.92	0.87	0.61	231	266	0.61	0.87	1.00	0.65	0.14	2.6
GpGpUp	225	254	0.68	1.05	0.80	0.55	0.21	228	257	0.47	0.88	0.76	0.56	0.35	233	261	0.57	0.86	0.92	0.51	0.00	2.4
UpApGp	226	257	0.53	0.92	0.73	0.38	0.12	229	258	0.44	0.85	0.74	0.40	0.19	231	259	0.52	0.83	0.79	0.35	0.00	2.0
UpCpCp	230	264	0.72	0.83	0.99	0.65	0.20	236	273	0.32	0.59	1.28	1.22	0.79	248	266	0.92	0.88	1.02	0.67	0.22	2.3
UpUpGp	228	257	0.54	0.91	0.79	0.45	0.14	229	259	0.43	0.82	0.77	0.45	0.21	235	260	0.67	0.86	0.87	0.43	0.00	2.5

TABLE 1: Ultraviolet Absorbance of Mono- and Oligonucleotides (Continued)

Compound	pH 7 λ_{min}	λ_{max}	240	250	270	280	290	pH 1 λ_{min}	λ_{max}	240	250	270	280	290	pH 12 λ_{min}	λ_{max}	240	250	270	280	290	Slack[b] pct.
			Absorbance Ratios[a]							Absorbance Ratios[a]							Absorbance Ratios[a]					
pApApApA	229	257	0.43	0.85	0.68	0.25	0.00	229	257	0.45	0.86	0.69	0.22	0.00	230	257	0.42	0.84	0.67	0.25	0.00	0.7
ApApApCp	230	258	0.50	0.84	0.72	0.34	0.00	232	258	0.44	0.82	0.79	0.43	0.00	231	258	0.51	0.83	0.73	0.35	0.00	2.1
ApApApGp	227	256	0.51	0.92	0.69	0.32	0.00	229	257	0.46	0.87	0.69	0.31	0.13	230	258	0.45	0.86	0.74	0.31	0.00	0.5
CpCpCpGp	227	267	0.86	0.95	1.01	0.78	0.34	236	278	0.34	0.62	1.31	1.44	1.03	235	268	0.78	0.86	1.09	0.79	0.27	1.3
UpUpUpGp	228	258	0.49	0.85	0.80	0.41	0.00	230	260	0.41	0.80	0.80	0.43	0.16	237	260	0.69	0.85	0.84	0.38	0.00	2.3

Contributed by Jane N. Toal.

[a] Absorbance ratios were calculated from optical densities at 240, 250, 270, 280, and 290 mμ relative to that at 260 mμ. Ratios not calculated where optical density was less than 0.1.

[b] Slack = the absolute sum of the catalog mismatch at every wavelength.

[c] cat. = catalog.

From data in Toal, Rushizky, Pratt, and Sober, *Anal. Biochem.*, 23, 60 (1968). With permission of the copyright owners, Academic Press, New York.

TABLE 2: Hyperchromicity Ratios of Oligonucleotides at Different Wavelengths (mμ)[d]

Compound	pH 7						pH 1						pH 12					
	240	250	260	270	280	290	240	250	260	270	280	290	240	250	260	270	280	290
pApA	1.11	1.11	1.16	1.13	0.81	0.00	1.03	1.02	1.02	1.02	1.00	0.00	1.09	1.10	1.15	1.12	0.84	0.00
ApCp	1.10	1.09	1.08	1.11	1.05	0.88	1.02	1.02	1.02	1.02	1.03	1.01	1.09	1.10	1.08	1.10	1.04	0.91
ApGp	1.08	1.06	1.07	1.09	1.07	0.96	1.06	1.04	1.02	1.02	1.04	1.00	1.04	1.03	1.03	1.03	0.98	0.00
ApUp	1.08	1.09	1.09	1.07	0.97	0.00	1.06	1.05	1.04	1.03	1.01	0.00	1.01	1.03	1.04	1.03	0.95	0.00
CpCp	1.09	1.08	1.08	1.10	1.10	0.96	1.05	1.04	1.02	1.02	1.02	1.01	1.10	1.10	1.10	1.10	1.09	0.94
CpGp	1.08	1.06	1.05	1.09	1.10	0.94	1.08	1.04	1.02	1.02	1.03	1.01	1.07	1.05	1.04	1.05	1.03	0.88
GpCp	1.11	1.09	1.05	1.08	1.16	1.05	1.04	1.05	1.05	1.06	1.08	1.10	1.04	1.03	1.04	1.05	1.05	0.95
GpUp	1.07	1.05	1.05	1.07	1.08	0.99	1.12	1.09	1.08	1.10	1.12	1.09	1.02	1.02	1.02	1.02	1.01	0.00
UpAp	1.07	1.05	1.08	1.10	0.99	0.00	1.06	1.05	1.06	1.05	0.99	0.00	1.06	1.06	1.06	1.05	0.95	0.00
UpGp	1.05	1.02	1.02	1.07	1.06	0.91	1.08	1.05	1.01	1.01	1.03	1.00	1.01	1.01	1.00	1.01	0.97	0.00
UpUp	0.98	1.00	1.04	1.07	1.05	0.00	1.01	1.02	1.03	1.03	0.98	0.00	0.96	0.98	1.01	1.01	0.94	0.00
pApApA	1.20	1.20	1.27	1.23	0.78	0.00	1.05	1.05	1.04	1.03	1.01	0.00	1.33	1.25	1.29	1.24	0.79	0.00
ApApCp	1.20	1.19	1.22	1.22	0.99	0.85	1.04	1.04	1.04	1.04	1.04	1.02	1.21	1.21	1.23	1.22	1.00	0.90
ApApGp	1.11	1.11	1.15	1.16	1.01	0.85	1.01	1.01	1.00	1.00	1.01	0.98	1.07	1.07	1.11	1.10	0.92	0.00
ApApUp	1.18	1.16	1.21	1.18	0.88	0.00	1.06	1.05	1.05	1.04	0.99	0.00	1.14	1.14	1.18	1.14	0.86	0.00
ApCpCp	1.22	1.20	1.18	1.20	1.16	0.91	1.09	1.08	1.06	1.05	1.05	1.02	1.25	1.24	1.20	1.19	1.17	0.95
ApCpGp	1.14	1.11	1.12	1.16	1.13	0.90	1.06	1.06	1.05	1.05	1.05	1.01	1.11	1.10	1.09	1.10	1.05	0.85
ApGpCp	1.14	1.11	1.12	1.13	1.16	1.00	1.04	1.05	1.03	1.04	1.07	1.06	1.07	1.06	1.07	1.08	1.04	0.89
ApGpUp	1.10	1.08	1.10	1.12	1.09	0.93	1.09	1.07	1.05	1.05	1.05	1.01	1.05	1.04	1.04	1.04	0.97	0.00
ApUpGp	1.11	1.09	1.10	1.12	1.10	0.93	1.10	1.08	1.05	1.04	1.04	1.00	1.04	1.04	1.04	1.04	0.97	0.00
CpApGp	1.16	1.16	1.16	1.18	1.10	0.87	1.06	1.04	1.03	1.04	1.05	1.01	1.10	1.09	1.10	1.11	1.02	0.88
CpCpGp	1.15	1.12	1.10	1.15	1.15	0.89	1.05	1.05	1.03	1.03	1.04	1.01	1.09	1.08	1.07	1.08	1.08	0.89
GpApCp	1.19	1.17	1.13	1.18	1.11	0.94	1.04	1.06	1.05	1.06	1.08	1.04	1.11	1.11	1.10	1.11	1.02	0.90
GpApUp	1.14	1.13	1.12	1.13	1.05	0.92	1.07	1.07	1.07	1.07	1.09	1.06	1.04	1.06	1.06	1.05	0.96	0.00
GpGpCp	1.13	1.11	1.07	1.12	1.19	1.05	1.08	1.07	1.06	1.08	1.12	1.11	1.07	1.04	1.03	1.04	1.02	0.89
GpGpUp	1.09	1.08	1.07	1.11	1.11	0.97	1.12	1.09	1.08	1.09	1.11	1.07	1.07	1.03	1.02	1.01	0.99	0.00
UpApGp	1.13	1.10	1.12	1.15	1.08	0.94	1.10	1.06	1.05	1.05	1.05	1.01	1.08	1.07	1.07	1.07	1.00.	0.00
UpCpCp	1.09	1.07	1.09	1.14	1.14	0.91	0.99	1.02	1.03	1.04	1.05	1.02	1.02	1.03	1.07	1.11	1.10	0.90
UpUpGp	1.06	1.04	1.04	1.08	1.07	0.89	1.09	1.07	1.04	1.05	1.03	0.98	1.00	1.01	1.00	1.00	0.97	0.00
pApApApA	1.29	1.28	1.36	1.31	0.76	0.00	1.09	1.07	1.06	1.06	1.01	0.00	1.32	1.29	1.38	1.36	0.80	0.00
ApApApCp	1.23	1.23	1.07	1.25	0.92	0.00	1.02	1.03	1.03	1.03	1.02	0.00	1.22	1.25	1.29	1.26	0.96	0.00
ApApApGp	1.18	1.18	1.25	1.24	0.95	0.00	1.03	1.03	1.02	1.01	1.01	0.97	1.15	1.14	1.22	1.19	0.88	0.00
CpCpCpGp	1.21	1.16	1.15	1.20	1.22	0.93	1.07	1.06	1.04	1.04	1.05	1.02	1.16	1.12	1.12	1.14	1.14	0.92
UpUpUpGp	1.08	1.06	1.07	1.10	1.07	0.00	1.10	1.06	1.05	1.05	1.03	0.97	1.00	1.02	1.03	1.02	0.98	0.00

Contributed by Jane N. Toal.

[d] Hyperchromicity ratios were calculated as the optical density of a hydrolyzed compound divided by the optical density of the corresponding intact compound at the same wavelength. Ratios are calculated where optical density was less than 0.1

From data in Toal, Rushizky, Pratt, and Sober, *Anal. Biochem.*, 23, 60 (1968). With permission of the copyright owners, Academic Press, New York.

TABLE 3: Ultraviolet Absorbance of Mononucleotides in 7 M Urea

Compound	pH 7.0 (0.05 M Phosphate)							pH 1 (0.1 M HCl)							pH 12 (0.01 M NaOH)						
			Absorbance Ratios							Absorbance Ratios							Absorbance Ratios				
	λ_{min}	λ_{max}	240	250	270	280	290	λ_{min}	λ_{max}	240	250	270	280	290	λ_{min}	λ_{max}	240	250	270	280	290
pA	228	261	0.357	0.739	0.727	0.206	—	232	258	0.395	0.795	0.762	0.277	—	228	260	0.371	0.742	0.730	0.205	—
pC	251	272.5	0.998	0.881	1.240	1.068	0.424	243	281	0.342	0.456	1.795	2.333	1.949	251	272.5	1.004	0.877	1.237	1.078	0.438
pG	225	254	0.718	1.098	0.810	0.662	0.297	229	258	0.458	0.868	0.741	0.665	0.538	232	258	0.513	0.863	0.969	0.650	0.133
pU	231	263	0.353	0.706	0.898	0.438	—	231	262	0.361	0.718	0.879	0.423	—	242	262	0.704	0.801	0.855	0.355	—

TABLE 4: Ultraviolet Absorbance of Mononucleotides in 97% D_2O

Compound	pH 7.0 (0.05 M Phosphate)							pH 1 (0.1 M HCl)							pH 12 (0.01 M NaOH)						
			Absorbance Ratios							Absorbance Ratios							Absorbance Ratios				
	λ_{min}	λ_{max}	240	250	270	280	290	λ_{min}	λ_{max}	240	250	270	280	290	λ_{min}	λ_{max}	240	250	270	280	290
pA	226	258	0.414	0.817	0.618	0.136	—	229	257	0.473	0.875	0.650	0.206	—	227	258	0.418	0.814	0.608	0.135	—
pC	249	271	0.931	0.822	1.231	0.970	0.308	240	279	0.220	0.430	1.706	2.028	1.482	250	271	0.948	0.828	1.242	0.995	0.327
pG	223	252.5	0.589	1.181	0.829	0.664	0.248	227	256	0.535	0.957	0.740	0.705	0.505	230	264	0.543	0.883	0.951	0.549	—
pU	230	262	0.386	0.747	0.845	0.372	—	229	261	0.398	0.757	0.822	0.343	—	242	261	0.712	0.821	0.793	0.276	—

TABLE 5: Ultraviolet Absorbance of Mononucleotides in 90% V/V Ethylene Glycol

Compound	pH 7.0 (0.05 M Phosphate)							pH 1 (0.1 M HCl)							pH 12 (0.01 M NaOH)						
			Absorbance Ratios							Absorbance Ratios							Absorbance Ratios				
	λ_{min}	λ_{max}	240	250	270	280	290	λ_{min}	λ_{max}	240	250	270	280	290	λ_{min}	λ_{max}	240	250	270	280	290
pA	229	260	0.374	0.743	0.743	0.251	—	233	259	0.472	0.816	0.744	0.282	—	228	259	0.405	0.775	0.726	0.237	—
pC	253	279	1.082	0.937	1.253	1.162	0.518	243	274	0.372	0.443	1.838	2.480	2.210	252	274	1.040	0.909	1.267	1.165	0.560
pG	224	251	0.768	1.131	0.781	0.650	0.339	230	257.5	0.463	0.852	0.704	0.595	0.479	226	254	0.699	1.069	0.768	0.651	0.329
pU	232	262	0.344	0.687	0.932	0.486	—	232	262	0.366	0.712	0.897	0.445	—	232	263	0.353	0.698	0.931	0.481	—

From: data of Hoffman, J. L., and Bock, R. M., Mononucleoside 5'-phosphates obtained from P-L Biochemicals, Milwaukee, Wisconsin, were used. The spectra were determined using a Cary 15 spectrometer. These tables originally appeared in Sober, H., *Handbook of Biochemistry and selected data for Molecular Biology*, 2nd ed., Chemical Rubber Co., Cleveland, 1970.

PURINES, PYRIMIDINES, NUCLEOSIDES, AND NUCLEOTIDES: PHYSICAL CONSTANTS AND SPECTRAL PROPERTIES

David B. Dunn and Ross H. Hall

Data are included for most modified components of nucleic acids, for some naturally occurring purine and pyrimidine compounds, and for some related synthetic compounds. An index to the 246 compounds is provided (arranged as free bases, nucleosides, and nucleotides).

It is recommended that, where possible, compounds be referred to by the trivial names given in bold type. (Some of these, such as "wye", "wyosine" and "zeatosine", are new proposals made by W. E. Cohn and D. B. Dunn.) Systematic and other trivial names are given, particularly where these already occur in the literature. Compounds are arranged as: purines and pyrimidines, ribonucleosides (including arabinonucleosides), deoxyribonucleosides, ribonucleotides and deoxyribonucleotides. The principal compounds are arranged alphabetically, with derivatives grouped

after each, the position of the latter depending first on the locant of the substituent and second on its initial letter. Compounds with a modified, or more than one substituent group, follow the simpler compounds.

The symbols are in accord with the examples and principles set out by the IUPAC-IUB Commission on Biochemical Nomenclature. A summary of these rules is found elsewhere in the Handbook. The "3-letter" symbols are proposed for use in tables, figures, equations involving the monomeric units themselves, the "1-letter" symbols for sequences. For deoxyribonucleosides in sequences, d may precede the sequence and thus be eliminated from each residue. Symbols that have been proposed or used, but are not now recommended are marked by an asterisk. The following symbols for substituents are used:

Substituent	Structure	Symbols 3-Letter	1-Letter
acetyl-	CH_3CO-	Ac	ac
(-)amino-, imino-	NH_2-, $-NH-$, $NH=$	NH_2,NH	n
-α-aminobutyric acid (3-carboxy-3-aminopropyl-)	$HOOC-C(NH_2)-(CH_2)_2-$	(NH_2Bto) (NH_2CxPr)	nbt
arabinosyl-		Ara	a
		a	
butyl-	$CH_3(CH_2)_3-$	Bu	b
-butyramide (3-carbamoylpropyl-)	$NH_2CO(CH_2)_3-$	Btn NcPr	
-butyric acid (3-carboxypropyl-)	$HOOC(CH_2)_3-$	Bto CxPr	bt
dihydro		H_2	h
carbamoyl-	NH_2CO-	Nc	nc
carbamoylmethyl- (-acetamide)	NH_2COCH_2-	Ncm NcMe	ncm
-carbonyl-	$-CO-$	CO	c
carboxy- (-oxycarbonyl-)	$HOOC-$, $-OOC-$	Cx	c
carboxymethyl- (-acetic acid)	$HOOC-CH_2-$	Cm	cm
		CxMe	
cis		*cis*	
formamido-	$HCONH-$	Fn	fn
glycino-	$HOOC-CH_2NH-$	Gly	g
hydroperoxy-	$HOO-$	O_2	o_2
methoxy-	CH_3O-	MeO	mo
-oxy-	$-O-$	O	o
pentyl-	$CH_3(CH_2)_4-$	Pe	
putrescino- (aminobutylamino-)	$NH_2(CH_2)_4NH-$	Put (NH_2BuNH)	nbn
propyl-	$CH_3(CH_2)_2-$	Pr	
ribosyl-		Rib	r
(-)thio (mercapto-)	$-S-$, $S=$	S	s
threonino-	$CH_3-CHOH-C(COOH)NH-$	Thr	t
threoninocarbonyl-	$CH_3-CHOH-C(COOH)NH-CO-$	(ThrCO)	tc
trans		*tr*	
hydroxy-	$HO-$	HO	o
hydroxymethyl- (-methanol)	$HOCH_2-$	HOMe Hm	om hm*
isopropeno-methyl-		im*	
iso		*iso*	
isopentenyl- (3-methyl-2-butenyl-)	$(CH_3)_2C=CHCH_2-$	iPe Pe^i	i
methoxycarbonylmethyl- (-acetic acid methyl ester)	$CH_3O-CO-CH_2-$	MeCm	mcm
3-methoxycarbonyl-3- methoxyformamidopropyl-	$CH_3O-CO-C(NHCOOCH_3)-(CH_2)_2-$	Y (MeO)$_2$ FnBto	y m$_2$ fnbt
methyl-	CH_3-	Me	m

Data were taken where possible on chemically synthesized material.

The first reference cited in the origin and synthesis column gives the origin of the compound used to obtain the principal spectral data. C = chemical synthesis; E = prepared enzymically; R = isolated from RNA; D = isolated from DNA; N = isolated from natural product other than nucleic acids.

The melting points were taken from the first reference to chemical synthesis except where otherwise indicated by footnote (a); dec. signifies decomposition.

For $[a]^t_D$ the temperature is given as a superscript and the concentration and solvent used in obtaining the value in parenthesis; H_2O = in water; EtOH = in ethanol; MeOH = in methanol; Me_2SO = in dimethylsulfoxide.

pK values were taken from the first reference quoted except where values differed by only 0.2 pH units when a mean value was used. References that give values deviating from those quoted by more than ±0.1 pH units are marked with an asterisk (*). The pK values involved are similarly marked. The pK values for nucleotides are only those for the nucleoside moiety, not phosphate ionizations. Where pK values were determined from electrophoretic mobilities (footnote p) values were obtained from mobilities relative to the parent compounds and pK values of these given by Jordan (1955) in *The Nucleic Acids*, Chargaff and Davidson, Eds., Academic Press, New York I, p 447.

Where possible, all spectral values are given at pH values away from pK values; exceptions to this are indicated by a footnote. Data obtained at a pH value where the compound was unstable were obtained soon after subjecting the material to this pH and differ from those of the decomposition products. The reference giving the maximum amount of spectral data is cited first; other references giving additional data are marked by footnotes. (These, c to e, indicate that some but not necessarily all the data mentioned come from this reference.) The ratio columns give ratio of absorption at the wavelength given to that at 260 nm. In the pH column of spectral data. H_2O=in water; EtOH=in ethanol; MeOH=in methanol. References that give spectral data differing from the values quoted are marked with an asterisk as are also the values involved. For this, deviations were marked only where they were more than ±1 nm for λ_{max} or λ_{min}, ±5 per cent for ε_{max}, and ±10 per cent for spectral ratios. (The latter applies to all instances where spectral ratios are quoted in the additional reference but not necessarily to all references giving spectra.)

PURINES, PYRIMIDINES, NUCLEOSIDES AND NUCLEOTIDES: PHYSICAL CONSTANTS AND SPECTRAL PROPERTIES

PURINES AND PYRIMIDINES

No.	Compound	Symbol 3-Letter	Symbol 1-Letter	Structure	Formula (Mol Wt)	Melting Point °C	$[\alpha]_D$	pK Basic	pK Acidic
1	Adenine	Ade		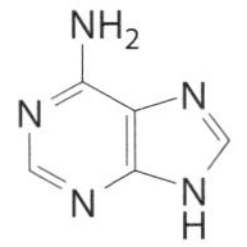	$C_5H_5N_5$ (135.13)	360° (dec) (sublimes 220°)	–	<1,4.15	9.8
2	1-(Δ²-Isopentenyl)adenine / 1-(γ,γ-Dimethylallyl)adenine	1iPeAde / 1PeⁱAde		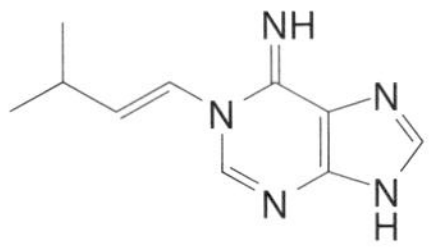	$C_{10}H_{13}N_5$ (203.24)	237–238°	–	7.1*	11.6ʰ
3	1-Methyladenine	1MeAde		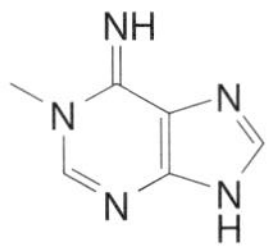	$C_6H_7N_5$ (149.16)	296–299° (dec)	–	7.2	11.0
4	1,N⁶-Dimethyladenine / 1-Methyl-6-methylaminopurine	1,6Me₂ Ade		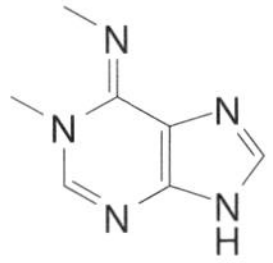	$C_7H_9N_5$ (163.18)	236° (picrate)	–	–	–

Acidic Spectral Data

No.	pH	λ_{max}	ε_{max} (×10⁻³)	λ_{min}	230	240	250	270	280	290
1	1	262.5	13.2	229	0.22	0.42	0.76	0.85	0.38	0.04
2	1	260	13.4	233	–	–	–	–	–	0.02*
3	4	259	11.7	228*	0.20	0.41	0.80	0.73	0.23*	–
4	1	261	12.9	230*	–	–	–	–	–	–

(Spectral Ratios: 230, 240, 250, 270, 280, 290)

Neutral Spectral Data

pH	λ_{max}	ε_{max} (×10⁻³)	λ_{min}	230	240	250	270	280	290
7	260.5	13.4	226	0.21	0.43	0.76	0.67	0.13	0.01
(EtOH)	273	12.3	246	–	–	–	–	–	–
8.8	270	11.9	242	1.43	0.39	0.52	1.34	0.87	0.14
–	–	–	–	–	–	–	–	–	–

(Spectral Ratios: 230, 240, 250, 270, 280, 290)

Alkaline Spectral Data

pH	λ_{max}	ε_{max} (×10⁻³)	λ_{min}	230	240	250	270	280	290
12	269	12.3	237	0.60	0.36	0.57	1.15	0.60	0.03
13	274*	15.2*	242	–	–	–	–	–	–
13	270*	14.4	239*	0.64	.29	0.55*	1.27	0.85	0.35*
11	274	12.7	245	–	–	–	–	–	–

(Spectral Ratios: 230, 240, 250, 270, 280, 290)

REFERENCES

No.	Origin and Synthesis	pK	$[\alpha]_D$	Spectral Data	Mass Spectra	Rf
1	C: 231,256ᵃ,257,337,357,358	56,22,220,51,336		232ᵇ,242ᵇᵉ,205,317ᵇ,358	–	258,21,18,242,291,292,402
2	C: 14,15	15,16*		14,15*	15	14,32,15
3	C: 8,10ᵃ,16,R: 9	8		8ᵇ,9ᵇ,10,11*,12*,265*ᵇ,317ᵇ	–	8–13,15
4	C: 24,12,30	–	–	24,12*	–	12,24

PURINES, PYRIMIDINES, NUCLEOSIDES AND NUCLEOTIDES: PHYSICAL CONSTANTS AND SPECTRAL PROPERTIES (Continued)

No.	Compound	Symbol (3-Letter)	Symbol (1-Letter)	Structure	Formula (Mol Wt)	Melting Point °C	$[\alpha]_D^t$	pK Basic	pK Acidic
5	2-Hydroxyadenine Isoguanine	2HOAde	isoGua*	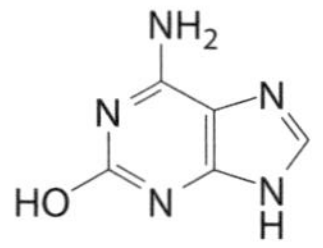	$C_5H_5N_5O$ (151.13)	>360°	—	4.5	9.0
6	2-Methyladenine	2MeAde		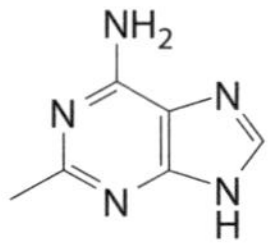	$C_6H_7N_5$ (149.16)	>340°	—	~5.1[f]	—
7	3-(Δ²-Isopentenyl]adenine Triacanthine	3iPeAde	3Pe^iAde	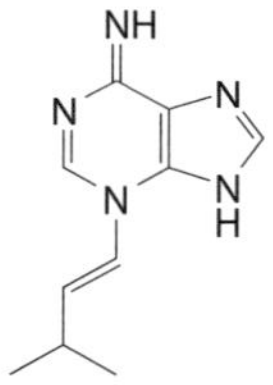	$C_{10}H_{13}N_5$ (203.24)	231–232°	—	—	5.4
8	N⁶-Glycinocarbonyladenine N-(Purin-6-ylcarbamoyl)glycine	6(Gly C O) Ade		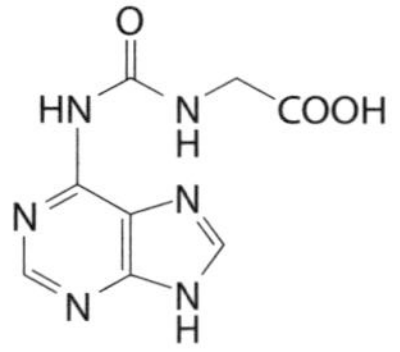	$C_8H_8N_6O_3$ (236.19)	233–234° (dec)	—	—	—

Acidic Spectral Data

No.	pH	λ_{max}	ε_{max} (×10⁻³)	λ_{min}	230	240	250	270	280	290
5	2	284*	11.7*	248	—	—	0.45	—	3.16	—
6	1	266	12.9	229	0.26	0.48	0.79	1.03	0.56	0.04
7	1	277	18.3	239	—	—	—	—	—	—
8	1.4	276.5	18.6	235	0.46	0.44	0.61	1.62	1.82	0.92

Neutral Spectral Data

No.	pH	λ_{max}	ε_{max} (×10⁻³)	λ_{min}	230	240	250	270	280	290
5	7	240, 286	7.8, 8.0*	210, 255	—	—	—	—	—	—
6	7	263	12.7	226	—	0.46	0.89	0.83	0.16	0.04
7	7	273	12.5	247	—	—	—	—	—	—
8	6.2	269	17.4	231	0.40	0.49	0.68	1.43	0.98	0.17

Alkaline Spectral Data

No.	pH	λ_{max}	ε_{max} (×10⁻³)	λ_{min}	230	240	250	270	280	290
5	12	284	12.3	253	—	—	0.80	—	3.47	—
6	13	271	10.7	238	0.77	0.40	0.61	1.28	0.84	0.04
7	—	—	—	—	—	—	—	—	—	—
8	12.3	278	16.2	240	1.34	0.37	0.54	1.76	2.10	1.23

REFERENCES

No.	Origin and Synthesis	Spectral Data	Mass Spectra	pK	$[\alpha]_D^t$	R_f
5	C: 282,340[a],358,111	28,283[abc],223[a]*,282[b],284[b],340,358*	—	22	—	340
6	C: 17[g],257[a],340,111 R: 18	18[b],317[bcde],20[d],21,257	—	19	—	18,21,291,340
7	C: 287,1	1	—	1	—	—
8	C: 408,R: 295	408	408	—	—	408

PURINES, PYRIMIDINES, NUCLEOSIDES AND NUCLEOTIDES: PHYSICAL CONSTANTS AND SPECTRAL PROPERTIES (Continued)

No.	Compound	Symbol 3-Letter	1-Letter	Structure	Formula (Mol Wt)	Melting Point °C	$[\alpha]_D^t$	pK Basic	pK Acidic
9	N^6-Threoninocarbonyladenine; N-(Purin-6-ylcarbamoyl)threonine	6(ThrCO)Ade		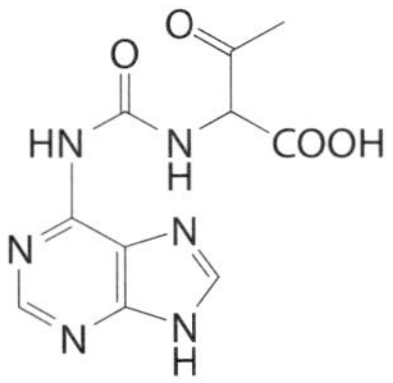 (structure)	$C_{10}H_{12}N_6O_4$ (280.24)	215–220°	$+30^{25}$ (0.4, H_2O)	<2[p]	~3
10	N^6-(Δ^2-Isopentenyl)adenine (N^6-($\gamma\gamma$-Dimethylallyl)adenine; 6-(3-Methyl-2-butenylamino)purine)	6iPeAde; 6Pe^{i}Ad		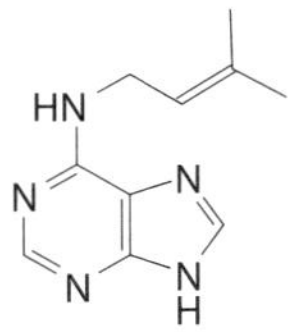 (structure)	$C_{10}H_{13}N_5$ (203.24)	213–215°	–	3.4*	10.4*
11	N^6-(Δ^2-Isopentenyl)-2-methylthioadenine; 6-(3-Methyl-2-butenylamino)-2-methylthiopurine	2MeS6Pe^{i}Ade; 2MeS6iPeAde		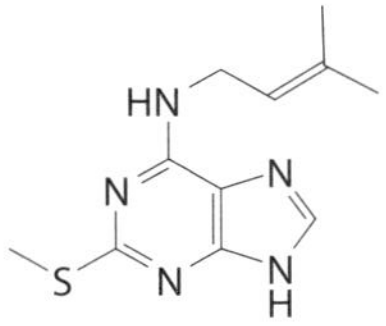 (structure)	$C_{11}H_{15}N_5S$ (249.32)	259–260°	–	–	–
12	N^6-Methyladenine; 6-Methylaminopurine	6MeAde		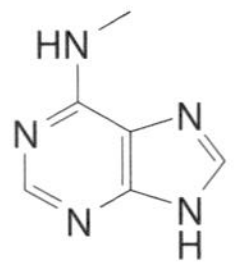 (structure)	$C_6H_7N_5$ (149.16)	319–320°	–	<1.4,2	10.0

Acidic Spectral Data

No.	pH	λ_{max}	ε_{max} (×10⁻³)	λ_{min}	230	240	250	270	280	290
9	1.6[4]	277*	20.6	234*	0.42	0.41	0.60	1.63	1.76*	0.86*
10	1	273[a]	18.6	235	0.14	0.22	0.57	1.27	1.11	0.64
11	1	253	21.7	217	–	–	–	–	–	–

Neutral Spectral Data

No.	pH	λ_{max}	ε_{max} (×10⁻³)	λ_{min}	230	240	250	270	280	290
9	5	276 / 269	16.9 / 19.2	232*	0.28	0.40	0.62	1.51*	1.13*	0.20*
10	7	276⁻ / 269	18.9 / 19.4	225	0.06	0.17	0.38	1.31	0.95	0.26
11	7	242	25.2	220	–	–	–	–	–	–

Alkaline Spectral Data

No.	pH	λ_{max}	ε_{max} (×10⁻³)	λ_{min}	230	240	250	270	280	290
	12	278	18.1	240	1.33*	0.32*	0.52	1.81	2.18*	1.32*
	13	275	18.1	240	0.86	0.28	0.49	1.70	1.60	0.54
	10	287	14.8	256	–	–	–	–	–	–

No.	pK	$[\alpha]_D^t$	Origin and Synthesis	Spectral Data	Mass Spectra	R_f
9	4,90	408	C: 408,290,409,4 R: 3	290[b],3*[b],408	4,404,408	290
10	16	–	C: 32,16[a],15	32,317[be],14*,15	15,406	32,15
11	–	–	C: 293,274	293	–	–
12	22	–	C: 22,23,24[a],12 D: 21 R: 18	21[b],18[b],25[bc],317*[be],28[d],8,23,12*,24*,265*[b]	–	9,10,12,13,18,21,22,24–27,291,292

REFERENCES

PURINES, PYRIMIDINES, NUCLEOSIDES AND NUCLEOTIDES: PHYSICAL CONSTANTS AND SPECTRAL PROPERTIES (Continued)

No.	Compound	Symbol 3-Letter	Symbol 1-Letter	Structure	Formula (Mol Wt)	Melting Point °C	$[\alpha]_D^t$	pK Basic	pK Acidic
13	N^6,N^6-Dimethyladenine 6-D-methylaminopurine	6Me2Ade		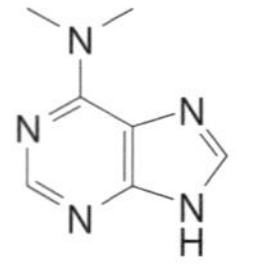	$C_7H_9N_5$ (163.18)	257–258°	–	<1,3.9	10.5
14	7-Methyladenine	7MeAde		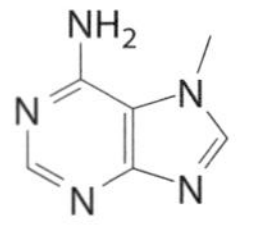	$C_6H_7N_5$ (149.16)	336° (dec)	–	4.2	–
15	Cytosine	Cyt		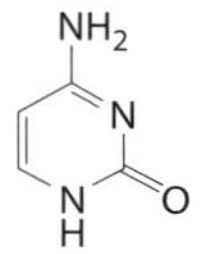	$C_4H_5N_3O$ (111.10)	312° (dec)	–	4.6*	12.2
16	3-Methylcytosine	3MeCyt		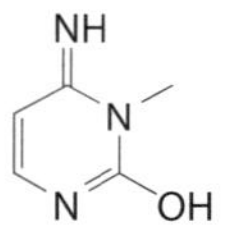	$C_5H_7N_3O$ (125.13)	242–245° (HCl salt)	–	7.4	>13

Acidic Spectral Data and Neutral Spectral Data

No.	Acidic pH	λ_{max}	ε_{max} (×10⁻³)	λ_{min}	230	240	250	270	280	290	Neutral pH	λ_{max}	ε_{max} (×10⁻³)	λ_{min}	230	240	250	270	280	290
12	(EtOH)	292	15.9	277	0.22	0.31	0.64	1.06	0.70	0.32*	(EtOH)	279	15.9	257	0.13	0.24	0.55	0.98	0.55	0.07
13	1	267	15.3*	232*	0.29	0.27	0.57	1.33	1.36	0.94	7	266	16.2	231*	–	–	–	–	–	–
14	1	277	15.6	236	0.58*	0.35	0.60	1.35	1.06*	0.21*	7	275	17.8	–	–	–	–	–	–	–
15	1	273*	14.0*	237	0.37	0.22	0.48	1.53	1.53	0.78	–	–	–	–	–	–	–	–	–	–
	1	276	10.0	239							7	267	6.1	247	1.13	0.86	0.78	1.05	0.58	0.08

Alkaline Spectral Data

No.	pH	λ_{max}	ε_{max} (×10⁻³)	λ_{min}	230	240	250	270	280	290
13	13 (EtOH)	273	15.9*	239*	0.77	0.39	0.55	1.48	1.19	0.25
14	13	281	17.8	245	1.62	0.54	0.53	1.86	2.61	2.09*
15	12	270*	10.6*	231	0.40	0.48	0.70	1.21	0.80	0.06
16	13q	281.5	7.1	251	2.26	1.12	0.66	1.68	2.13	1.4

REFERENCES

No.	Origin and Synthesis	pK	$[\alpha]_D^t$	Spectral Data
13	C: 23,338,29a,22,R: 18	22	–	18b,23d,28c,d,21,265*b,317b,338
14	C: 33,34	35	–	36b,33*d,35b,37*b
15	C: 64,233,352,396	59,66*,80	–	66b,242be,232b,317b,205,64
16	C: 57,30	58,57	–	57,58b,317b

No.	Mass Spectra	R_f
13	–	18,22,32
14	–	26,33,36
15	–	258,21,242,402
16	–	13,57

PURINES, PYRIMIDINES, NUCLEOSIDES AND NUCLEOTIDES: PHYSICAL CONSTANTS AND SPECTRAL PROPERTIES (Continued)

No.	Compound	Symbol 3-Letter	Symbol 1-Letter	Structure	Formula (Mol Wt)	Melting Point °C	$[\alpha]_D^t$	pK Basic	pK Acidic
17	N^6-Acetylcytosine / 4-Acetylamino-2-pyrimidinone	4AcCyt		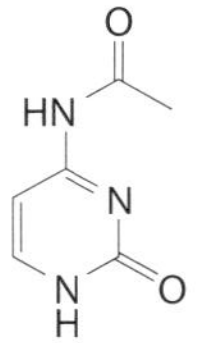 (structure)	$C_6H_7N_3O_2$ (153.14)	326–328°	–	–	–
18	N^4-Methylcytosine / 4-Methylamino-2-pyrimidinone	4MeCyt		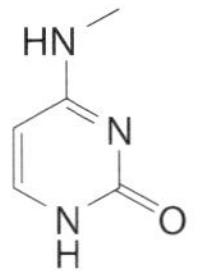 (structure)	$C_5H_7N_3O$ (125.13)	277–280° (dec)	–	4.5	12.7
19	5-Methylcytosine	5MeCyt		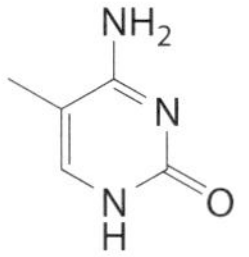 (structure)	$C_5H_7N_3O$ (125.13)	270°	–	4.6	12.4
20	5-Hydroxymethylcytosine	5HmCyt / 5HOMeCyt		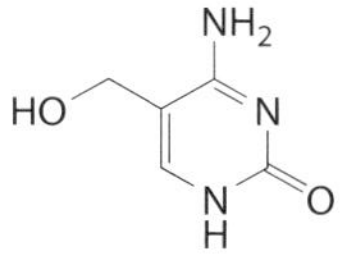 (structure)	$C_5H_7N_3O_2$ (141.13)	>200° (dec)	–	4.3	~13

Acidic Spectral Data

No.	pH	λ_{max}	ε_{max} (×10⁻³)	λ_{min}	230	240	250	270	280	290
16	4	274	9.4	240	0.52	0.27	0.50	1.47	1.33	0.56
17	–	–	–	–	–	–	–	–	–	–
18	1	277	10.5*	240	0.60	0.29	0.50	1.56	1.66	1.03

Neutral Spectral Data

No.	pH	λ_{max}	ε_{max} (×10⁻³)	λ_{min}	230	240	250	270	280	290
16	–	244.5	14.2	226	–	–	–	–	–	–
	7	293	4.9	270	–	–	–	–	–	–
18	7	267	7.2*	248	1.07	0.89	0.79	1.07	0.66	0.14

Alkaline Spectral Data

No.	pH	λ_{max}	ε_{max} (×10⁻³)	λ_{min}	230	240	250	270	280	290
16	14	282	7.9	251	5.10	1.60	0.60	2.60	3.28	2.6
	12	294	11.9	250	–	–	0.53	–	5.90	9.30
18	14	286	8.0*	256	3.18	2.02	0.96	2.02	3.27	3.23

REFERENCES

No.	Origin and Synthesis	$[\alpha]_D^t$	pK	Spectral Data	Mass Spectra	R_f
17	C: 280,281,396	–	–	281	–	–
18	C: 59,60a,61,62 R: 353	–	59,63	59b,60*,317b	–	59
19	C: 64,65a	–	66	66b,67b,64b,68b,265b,317b,	–	67,69,70
20	C: 71	–	72	73b,72b,71,74,265*b,317b	74	69,70,73,75,74,402

PURINES, PYRIMIDINES, NUCLEOSIDES AND NUCLEOTIDES: PHYSICAL CONSTANTS AND SPECTRAL PROPERTIES (Continued)

No.	Compound	Symbol 3-Letter	Symbol 1-Letter	Structure	Formula (Mol Wt)	Melting Point °C	$[\alpha]_D$	pK Basic	pK Acidic
21	6-Amino-5-N-methylformamidoisocytosine 2,6-Diamino-4-hydroxy-5-N-methylformamidopyrimidine	6NH₂5(M⁴eFn) isoCyt		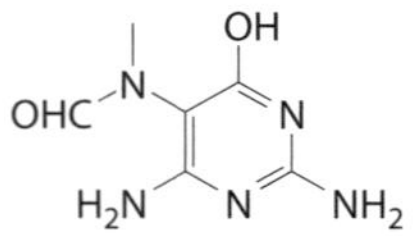	$C_6H_9N_5O_2$ (183.17)	–	–	3.8	9.9ᴷ
22	Guanine	Gua		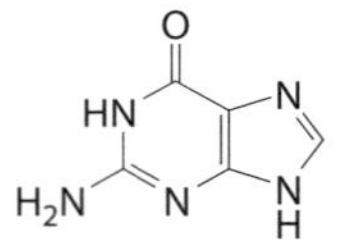	$C_5H_5N_5O$ (151.13)	>350°	–	<0, 3.2	9.6*, 12.4
23	1-Methylguanine	1MeGua		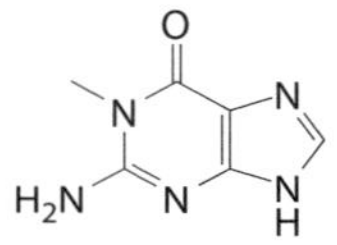	$C_6H_7N_5O$ (165.16)	None (dec)	–	ˉ0, 3.1	10.5
24	N²-Methylguanine 6-Hydroxy-2-methylaminopurine	2MeGua		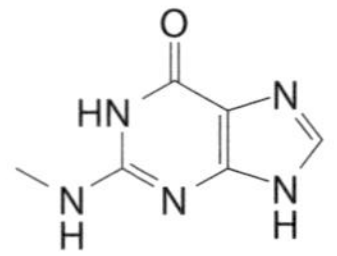	$C_6H_7N_5O$ (165.16)	–	–	3.3	8.9, 12.8
25	N²,N²-Dimethylguanine 2-Dimethylamino-6-hydroxypurine	2Me₂Gua		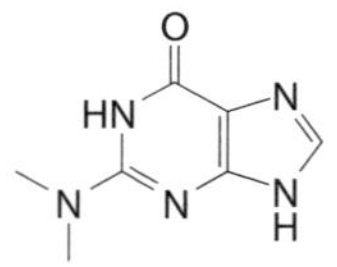	$C_7H_9N_5O$ (179.18)	–	–	–	–

Acidic Spectral Data

No.	pH	λ_{max}	ε_{max} (×10⁻³)	λ_{min}	Spectral Ratios 230	240	250	270	280	290
19	1	283	9.8	242	0.97	0.26	0.40	1.90	2.62	2.43
20	1	279	9.7	241	0.63	0.25	0.45	1.68	1.97	1.37
21	1	263	17.8	232*	0.37	0.39	0.59	0.92	0.34	0.04
22	1	248	11.4	224	0.71	1.18	1.37	0.85	0.84*	0.50*
		276	7.35	267						
23	1	250	10.2	227	0.47	0.93	1.28	0.87	0.81	0.50
		272	7.1	–						
24	1	250*	13.9*	228*	0.51	0.95	1.34*	0.62*	0.64	0.54*
		279	6.2*	–						
25	1	256*	19.0*	233*	0.36	0.48	0.92	0.52	0.37*	0.39

Neutral Spectral Data

No.	pH	λ_{max}	ε_{max} (×10⁻³)	λ_{min}	Spectral Ratios 230	240	250	270	280	290
19	7	273	6.2	252	1.60	1.10	0.80	1.35	1.21	0.54
20	7	269	5.7	251	0.59	0.93	0.80	1.14	0.80	0.15
21	7	264	13.8	242	0.73	–	–	–	–	–
22	7	246	10.7*	225	0.73	1.26	1.42	0.99	1.04	0.54
		276	8.15*	262						
23	7	248	10.0	227	0.54	1.01	1.24	0.96	0.93	0.46
		272	7.9	264						
24	7	249	14.1	227	0.56	0.86	1.18	0.72	0.72	0.49
		277	8.3	266						
25	7	250	17.0	229	0.55	0.87	1.20	0.61	0.62	0.59

Alkaline Spectral Data

No.	pH	λ_{max}	ε_{max} (×10⁻³)	λ_{min}	Spectral Ratios 230	240	250	270	280	290
19	14	289	8.1	254	5.05	1.97	0.84	2.02	3.64	4.71
20	13	283	7.6	254	3.91	2.59	0.83*	1.97	2.98	2.50
21	13	262	9.8	242	0.72	0.67	0.71	0.78	0.31	0.09
22	11	274	8.0	255	–	–	0.99	–	1.14	0.59
	14	274	9.9	238	–	–	0.81	–	1.24	0.61
23	13	277	8.7	241*	1.49	0.64	0.80	1.12	1.19	0.81
24	11	244	9.5*	263*	–	–	–	–	–	–
		278*	7.2	–						
25	12	282*	9.2*	265	–	1.26	1.15	0.95	1.05	0.95

REFERENCES

No.	Origin and Synthesis	$[\alpha]_D$	pK	Spectral Data	Mass Spectra	R_f
21	C: 41	–	41,76	90,41*ᶜ,ᵈ	–	41
22	C: 234,257ᵃ,354	–	170,56*,232	232ᵇ,242*ᵇᵉ,317ᵇ,205,257*	–	21,40,258,242,402
23	C: 33ᵍ,38ᵃ	–	39,27	40ᵇ,41ᶜᵈ,27ᵇᵉ,39ᵇᶜᵉ,25,33,42,265*ᵇ,317ᵇ	–	24,25,27,33,40,42
24	C: 43ᵍ,44,355	–	2	40ᵇ,44ᶜᵈ,317ᵇᵈᵉ,43*,2*,27*,265*ᵇ,25	–	25,27,33,40
25	C: 43ᵍ,44	–	–	317ᵇ,40ᵇᶜᵉ,44ᶜᵈ,265*ᵇ,43	–	40

PURINES, PYRIMIDINES, NUCLEOSIDES AND NUCLEOTIDES: PHYSICAL CONSTANTS AND SPECTRAL PROPERTIES (Continued)

No.	Compound	Symbol 3-Letter	Symbol 1-Letter	Structure	Formula (Mol Wt)	Melting Point °C	$[\alpha]_D^t$	pK Basic	pK Acidic
26	7-Methylguanine	7MeGua		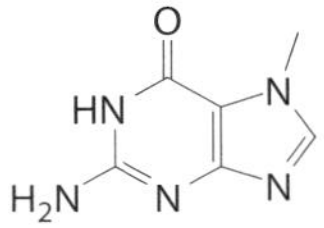	$C_6H_7N_5O$ (165.16)	>390° (dec)	—	⁻0, 3.5	9.9*
27	Hypoxanthine		Hyp	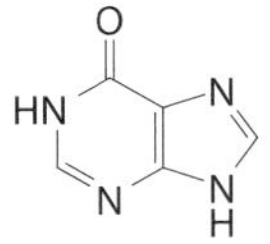	$C_5H_4N_4O$ (136.11)	>350° (dec)	—	2.0	8.9*, 12.1
28	1-Methylhypoxanthine	1MeHyp		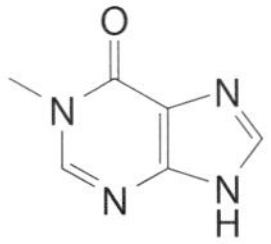	$C_6H_6N_4O$ (150.14)	311–312°	—	⁻2	8.9*, 13
29	3-Methylhypoxanthine	3MeHyp		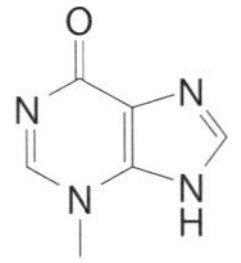	$C_6H_6N_4O$ (150.14)	>280° (dec)	—	2.6	8.3
30	7-Methylhypoxanthine	7MeHyp		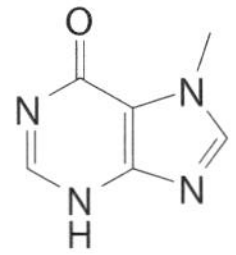	$C_6H_6N_4O$ (150.14)	355° (dec)	—	2.1	8.9

Acidic Spectral Data

No.	pH	λ_{max}	ε_{max} (×10⁻³)	λ_{min}	230	240	250	270	280	290
26	1	250, 272	10.6*, 6.9*	228*	0.55	0.99	1.30	0.84	0.79	0.52
27	0	248	10.8	215	—	—	1.45	—	0.04	0.00
28	1	249	9.4	219	0.59	1.11	1.37	0.43	0.10	0.01
29	0	253	11.0	—	—	—	—	—	—	—
30	0	250	10.2	224	—	—	—	—	—	—

Neutral Spectral Data

No.	pH	λ_{max}	ε_{max} (×10⁻³)	λ_{min}	230	240	250	270	280	290
26	7	283	9.3	235	1.54	1.42	1.46	1.35	1.87	1.73
27	6	248, 283, 249.5	5.7*, 7.4*, 10.7	261, 222	0.53	1.05	1.32	0.57	0.09*	0.01*
28	5	251	9.4	223	0.51	0.95	1.31	0.53	0.16	0.02
29	5	264	14.0	—	—	—	—	—	—	—
30	5	256	9.5	229	0.42	0.81	0.97	0.55	0.07	0.00

Alkaline Spectral Data

No.	pH	λ_{max}	ε_{max} (×10⁻³)	λ_{min}	230	240	250	270	280	290
26	12	280*	7.4*	257	1.92	1.50	1.013	1.50	1.89	1.47
27	11	259	11.1	232	—	—	0.84	0.84	0.12	0.01
27	14	263	11.5	233	0.48	0.46	0.71	—	0.19*	0.01*
28	11	260	9.7	242	—	—	0.97	0.53	0.16	0.02
29	11	265	10.9	—	—	—	—	—	—	—
30	11	262	10.6	230	0.36	0.49	0.76	0.83	0.21	0.00

REFERENCES

No.	Origin and Synthesis	pK	$[\alpha]_D^t$	Spectral Data	Mass Spectra	R_f
26	C: 45,10	39,27*	—	46[b],47[b],42*[b],35[b],39,41,48	—	10,13,24,27,41,42,46,334
27	C: 257,337,358,359,377	22,170,253,51*	—	232[b],242*[e],334*,317[b],205,257,358	—	258,242,278
28	C: 33,26,360	33,27*	—	49,33[d],317[bc],27,121,360	—	24,26,27,49,33,121
29	C: 26,50[a]	33	—	33,121	—	26,33,121
30	C: 45	33,51	—	52,33[c,d]	—	10,26,27,33

PURINES, PYRIMIDINES, NUCLEOSIDES AND NUCLEOTIDES: PHYSICAL CONSTANTS AND SPECTRAL PROPERTIES (Continued)

No.	Compound	Symbol 3-Letter	Symbol 1-Letter	Structure	Formula (Mol Wt)	Melting Point °C	$[\alpha]_D^t$	pK Basic	pK Acidic
31	Uracil	Ura		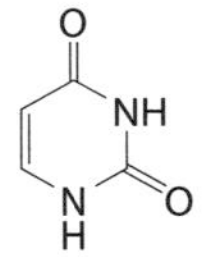	$C_4H_4N_2O_2$ (112.09)	315° (dec)	—	—	9.5, >13
32	1-Methyluracil	1MeUra		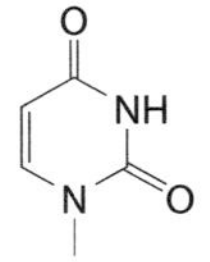	$C_5H_6N_2O_2$ (126.11)	232–233°	—	—	9.7
33	2-Thiouracil	2SUra 2Sra		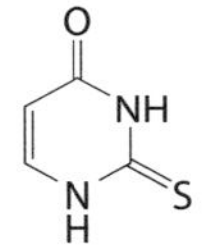	$C_4H_4N_2OS$ (128.15)	340° (dec)	—	—	—
34	2-Methylthiouracil / S-Methyl-2-thiouracil	2(MeS)Ura 2MeSra		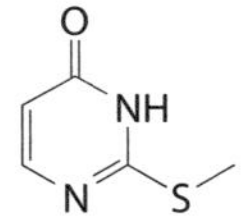	$C_5H_6N_2OS$ (142.18)	196–198°	—	—	—

Acidic Spectral Data

No.	pH	λ_{max}	ε_{max} (×10⁻³)	λ_{min}	230	240	250	270	280	290
31	0ᵍ	260	7.8	229	0.23	0.48	0.80	0.68	0.30	0.05
	4	259.5	8.2	227			0.84		0.17	0.01
32	2	272	—	—		—	0.60	—	0.64	0.08
33	1.5	270*	13.9*	241						
34	1 (MeOH)	221	10.0							
		248	7.9							

Neutral Spectral Data

No.	pH	λ_{max}	ε_{max} (×10⁻³)	λ_{min}	230	240	250	270	280	290
31	7	259.5	8.2	227	0.21	0.47	0.84	0.68	0.17	0.01
32	7	267	9.8	232	0.17	0.25	0.59	1.12	0.70	0.12
33	7.4	268*	11.8	240						
34	(MeOH)	228	7.9							
		286	7.9							

Alkaline Spectral Data

No.	pH	λ_{max}	ε_{max} (×10⁻³)	λ_{min}	230	240	250	270	280	290
31	12	284	6.2	241	1.09	0.56	0.71	1.25	1.40	1.27
32	12	265*	7.0	241	0.92	0.53	0.68	1.00	0.44	0.03
33	11	259	10.7	243*						
		307*	6.8	291*						
34	13 (MeOH)	222	15.9							
		243	7.9							

REFERENCES

No.	Origin and Synthesis	$[\alpha]_D^t$	pK	Spectral Data
31	C: 364,362ᵃ,84,395	—	66,80	66ᵇ,242ᵇᵉ,232ᵇᶜᵈᵉ,317ᵇ
32	C: 79,78,362,	—	66,80	66ᵇ,81 *ᵇ,ᶜ,ᵉ
33	C: 84	—	—	277ᵇ,331ᵇᶜ,86*ᵇᵈ
34	C: 301	—	—	301

No.	Mass Spectra	Rf
31	—	258,40,242,402
32	—	69,81
33	—	—
34	—	301

PURINES, PYRIMIDINES, NUCLEOSIDES AND NUCLEOTIDES: PHYSICAL CONSTANTS AND SPECTRAL PROPERTIES (Continued)

No.	Compound	Symbol (3-Letter)	Symbol (1-Letter)	Structure	Formula (Mol Wt)	Melting Point °C	$[\alpha]_D^t$	pK Basic	pK Acidic
35	5-Carboxymethyl-2-thiouracil / 2-Thiouracil-5-acetic acid	5Cm2SUra / 5Cm2Sra / 5(CxMe)2SUra		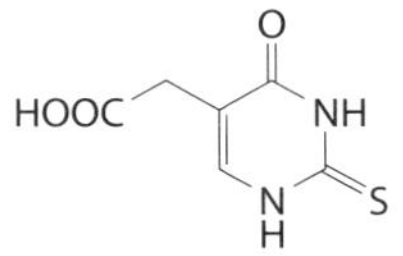 (HOOC–)	$C_6H_6N_2O_3S$ (186.19)	275–279° (dec)	—	—	4.15, 8.4, ~13.5
36	5-Carbamoylmethyl-2-thiouracil / 2-Thiouracil-5-acetamide	5Ncm2Sra / 5(NcMe)2SUra / 5(NcMe)2Sra / 5Ncm2SUra		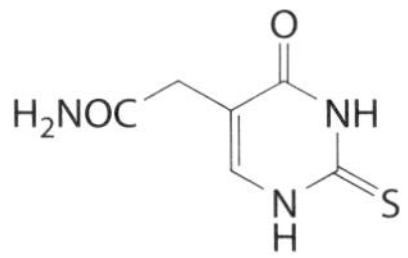 ($H_2NOC–$)	$C_6H_7N_3O_2S$ (185.21)	269–270° (dec)	—	—	—
37	5-(Methoxycarbonylmethyl)-2-thiouracil / 2-Thio-5-carboxymethyluracil methyl ester	5(MeCm)2Sura / 5MeCm2Sra / 5(MeCxMe)2Sra		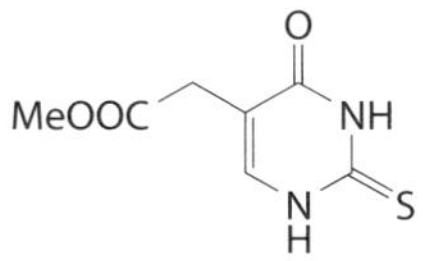 (MeOOC–)	$C_7H_8N_2O_3S$ (200.22)	218–220°	—	—	—

Acidic Spectral Data

No.	pH	λ_{max}	ε_{max} ($\times 10^{-3}$)	λ_{min}	230	240	250	270	280	290
35	1	276	15.6	242						
36	1	214 / 274	13.9 / 14.9	240						
37	1.5	214 / 276	15.6 / 16.6	240	0.49	0.31	0.47	1.48	1.52	1.44

Neutral Spectral Data

No.	pH	λ_{max}	ε_{max} ($\times 10^{-3}$)	λ_{min}	230	240	250	270	280	290
35	7	215 / 275	– / 16.6[a]	240	0.47	0.32	0.47	1.45	1.49	1.41

Alkaline Spectral Data

No.	pH	λ_{max}	ε_{max} ($\times 10^{-3}$)	λ_{min}	230	240	250	270	280	290
35	13[a]	236 / 259 / 313	9.9 / 11.1 / 8.3	244 / 288						
36	12	234 / 260 / 313	– / 12.0[a] / 8.6[a]	244 / 290	0.91	0.83	0.85	0.80	0.59	0.51

REFERENCES

No.	Origin and Synthesis	Spectral Data	$[\alpha]_D^t$	pK	Mass Spectra	R_f
35	C: 302	302	—	302	—	302
36	C: 302	302	—	—	—	302
37	C: 5,302	317[b],302[c,d],5[c]	—	—	—	302

PURINES, PYRIMIDINES, NUCLEOSIDES AND NUCLEOTIDES: PHYSICAL CONSTANTS AND SPECTRAL PROPERTIES (Continued)

No.	Compound	Symbol 3-Letter	Symbol 1-Letter	Structure	Formula (Mol Wt)	Melting Point °C	$[\alpha]_D^t$	pK Basic	pK Acidic
38	5-Methoxy-2-thiouracil / 4-Hydroxy-2-mercapto-5-methoxypyrimidine	5MeO2SUra 5MeO2Sra		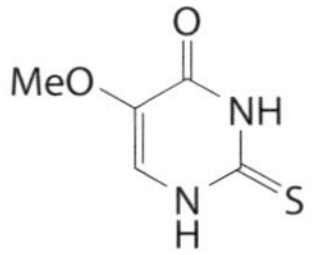	$C_5H_6N_2O_2S$ (158.18)	280–281° (dec)	–	–	–
39	2-Thiothymine / 5-Methyl-2-thiouracil	2Sthy 5Me2SUra 5Me2Sra		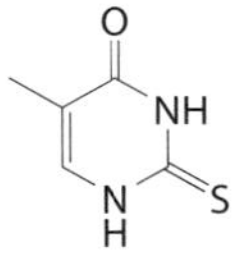	$C_5H_6N_2OS$ (142.18)	265–267°	–	–	–
40	3-Methyluracil	3MeUra		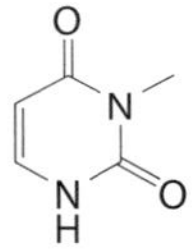	$C_5H_6N_2O_2$ (126.12)	179°	–	–	10.0
41	4-Thiouracil	4Sura 4Sra		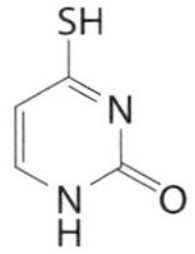	$C_4H_4N_2OS$ (128.15)	289–290° (dec)	–	–	–

Acidic Spectral Data

No.	pH	λ_{max}	ε_{max} (×10⁻³)	λ_{min}	230	240	250	270	280	290
38										
39	1	276	19.5	–	–	–	0.86	–	0.14	0.02
	3	259	7.3	–						
41	1	327	17.5	277						

Spectral Ratios

Neutral Spectral Data

No.	pH	λ_{max}	ε_{max} (×10⁻³)	λ_{min}	230	240	250	270	280	290
38										
39	7	274	16.9	230	0.29	0.46	0.84	0.66	0.14	0.02
	7	259	7.3	275						
41	7	328	16.6							

Spectral Ratios

Alkaline Spectral Data

No.	pH	λ_{max}	ε_{max} (×10⁻³)	λ_{min}	230	240	250	270	280	290
39	11	260	16.6							
	12	309	7.3	243	1.43	0.34	0.37	2.23	3.47	3.01
		283	10.7							
41	11	335	17.6	278						

Spectral Ratios

REFERENCES

No.	Origin and Synthesis	pK	$[\alpha]_D^t$	Spectral Data	Mass Spectra	R_f
38	C: 397	–	–	–	–	–
39	C: 301	–	–	86	–	–
40	C: 82,79ᵃ,83	66,80	–	66ᵇ,81 ᵇ,ᵉ	–	13,69,81
41	C: 85,84	–	–	86ᵇ,317ᵇ	–	–

PURINES, PYRIMIDINES, NUCLEOSIDES AND NUCLEOTIDES: PHYSICAL CONSTANTS AND SPECTRAL PROPERTIES (Continued)

No.	Compound	Symbol 3-Letter	Symbol 1-Letter	Structure	Formula (Mol Wt)	Melting Point °C	$[\alpha]_D$	pK Basic	pK Acidic
42	5-Carboxymethyluracil / Uracil-5-acetic acid	5CmUra / 5CxMeUra		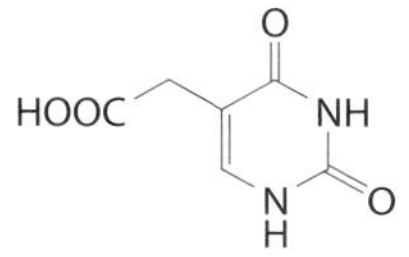	$C_6H_6N_2O_4$ (170.12)	316–318° (dec)	–	–	4.3, 10.0
43	5-(Methoxycarbonylmethyl)uracil / 5-Carboxymethyluracil methyl ester / Uracil-5-acetic acid methyl ester	5MeCmUra / 5(MeCxMe)Ura		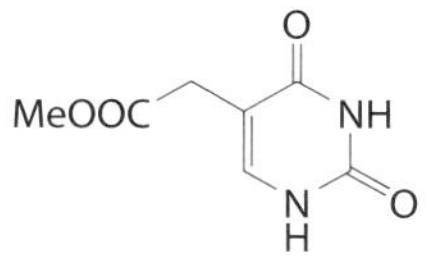	$C_7H_8N_2O_4$ (184.15)	236–237°	–	–	–
44	5-Hydroxyuracil	5HOUra		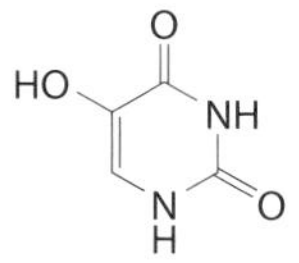	$C_4H_4N_2O_3$ (128.09)	>300° (dec)	–	–	8.0
45	5-Methoxyuracil / 2,4-Dihydroxy-5-methoxy-pyrimidine	5MeOUra		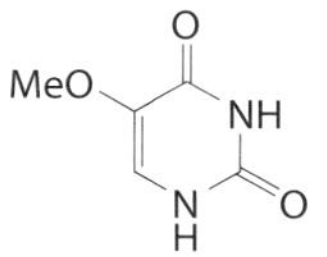	$C_5H_6N_2O_3$ (142.12)	341–345° (dec)	–	–	–

No.	Acidic Spectral Data pH	λ_{max}	ε_{max} (×10⁻³)	λ_{min}	Spectral Ratios 230	240	250	270	280	290	Neutral Spectral Data pH	λ_{max}	ε_{max} (×10⁻³)	λ_{min}	Spectral Ratios 230	240	250	270	280	290
42	1	262	8.1	231	0.23	0.37	0.71	0.89	0.42	0.05	7	264	7.6	234	0.33	0.33	0.66	0.98	0.56	0.13
43	1	262	8.2	232							6	278*	6.4	244	1.08	0.53	0.54	1.45	1.55	1.18
44	2	278*	6.4	244	1.16	0.58	0.56	1.52	1.63	1.26										
45	–																			

No.	Alkaline Spectral Data pH	λ_{max}	ε_{max} (×10⁻³)	λ_{min}	Spectral Ratios 230	240	250	270	280	290
42	13	290	5.5	246	1.95	0.78	0.63	1.40	1.77	2.00
43	12	239 / 305*	6.5* / 5.7	270	–	–	1.65	0.69	0.83	1.13

REFERENCES

No.	Origin and Synthesis	pK	$[\alpha]_D$	Spectral Data	Mass Spectra	R_f
42	C: 279,278,302ᵃ	302	–	278ᵇ,302ᵈ	278	278,302,402
43	C: 302	–	–	302	–	302
44	C: 87,88ᵃ,89	90	–	87ᵇ,317ᵇᶜᵈᵉ,91*ᵇ	–	–
45	C: 397	–	–	–	–	–

PURINES, PYRIMIDINES, NUCLEOSIDES AND NUCLEOTIDES: PHYSICAL CONSTANTS AND SPECTRAL PROPERTIES (Continued)

No.	Compound	Symbol 3-Letter	1-Letter	Structure	Formula (Mol Wt)	Melting Point °C	[α]b	pK Basic	pK Acidic
46	Thymine / 5-Methyluracil	Thy / 5MeUra		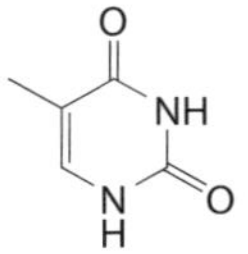 (structure)	$C_5H_6N_2O_2$ (126.11)	310° (dec)	—	—	9.9 > 13
47	5-(Putrescinomethyl)uracil / 5-(4-Aminobutylaminomethyl) uracil; N-T:yminylputrescine	5(PutMe)Ura / 5(NH₂BuNHMe)Ura / 5Put:Thy*		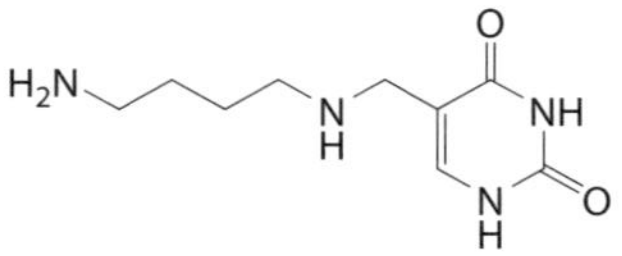 (structure)	$C_9H_{16}N_4O_2$ (212.25)	255° (dec) (HCl salt)	—	—	—
48	5-(Methylaminomethyl)uracil	5(MeNHMe)Ura		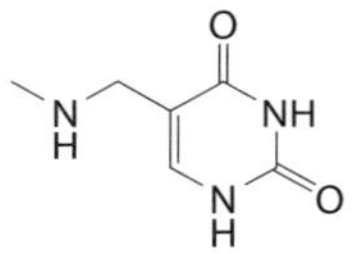 (structure)	$C_6H_9N_3O_2$ (155.16)	230–232°	—	—	—
49	5-Hydroxymethyluracil	5HmUra / 5HOMeUra		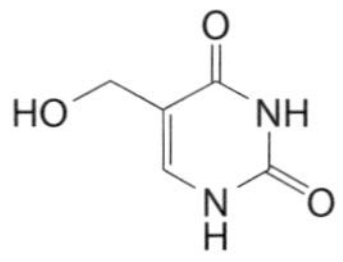 (structure)	$C_5H_6N_2O_3$ (142.11)	260–300° (dec)	—	—	9.4* *14

Acidic Spectral Data

No.	pH	λ_{max}	ε_{max} (×10⁻³)	λ_{min}	230	240	250	270	280	290
46	4	264.5	7.9	233	—	—	0.67	0.97	0.53	0.09
47	1	261	7.8	230	—	—	—	—	—	—
48	1	262	—	230	0.01	0.07	0.70	0.85	0.39	0.13
49	2	261	8.0	231	—	—	0.77	—	0.32	—
		206	9.55							

Neutral Spectral Data

No.	pH	λ_{max}	ε_{max} (×10⁻³)	λ_{min}	230	240	250	270	280	290
46	7	264.5	7.9	233	0.28	0.35	0.67	0.96	0.53	0.09
47	7	262	—	230.5	—	—	—	—	—	—
48	7.5	262	8.1	230	0.01	0.07	0.70	0.85	0.39	0.13
49	7	261	9.5	231	0.27	0.43	0.77	0.80	0.33	0.05
	H₂O	207								

Alkaline Spectral Data

No.	pH	λ_{max}	ε_{max} (×10⁻³)	λ_{min}	230	240	250	270	280	290
46	12	291	5.4	244	1.53	0.68	0.65	1.24	1.31	1.41
47	13	288.5	—	246	—	—	—	—	—	—
48	10.1	287	—	244	3.88	0.42	0.39	1.69	2.18	2.32
49	12	286	7.4	245	1.77	0.75	0.67	1.39	1.80	1.75

REFERENCES

No.	Origin and Synthesis	[α]tD	pK	Spectral Data	Mass Spectra	Rf
46	C: 263,361,254,101,348,356,395	—	66,80	66b,232b,317b	—	258,21,18,402
47	C: 402 D: 402	—	—	402	402	402
48	C: 270	—	—	270	270	270
49	C: 69,92,93 D: 96	—	72,94*	94b,69cde,317bd,72b,95b,74	—	69,96,97,74,402

PURINES, PYRIMIDINES, NUCLEOSIDES AND NUCLEOTIDES: PHYSICAL CONSTANTS AND SPECTRAL PROPERTIES (Continued)

No.	Compound	Symbol (3-Letter)	Symbol (1-Letter)	Structure	Formula (Mol Wt)	Melting Point °C	$[\alpha]_D^t$	pK Basic	pK Acidic
50	S(+)5-(4,5-Dihydroxpentyl)uracil	5(HO)$_2$PeUra		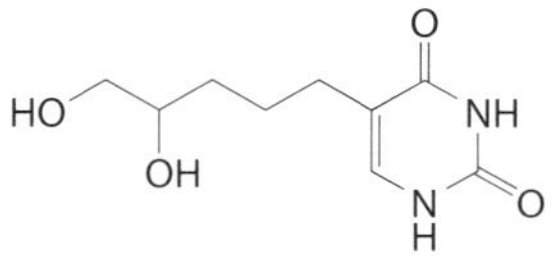	C$_9$H$_{14}$N$_2$O$_4$ (214.22)	255–226°	—	—	9.7
51	Dihydrouracil 5,6-Dihydrouracil	H$_2$Ura		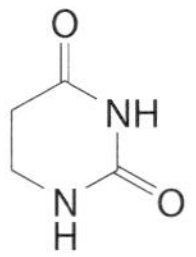	C$_4$H$_6$N$_2$O$_2$ (114.10)	275–276°	—	—	—
52	5,6-Dihydrothymine 5,6-Dihydro-5-methyluracil	H$_2$Thy		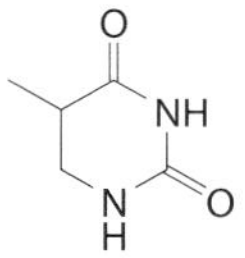	C$_5$H$_8$N$_2$O$_2$ (128.13)	264–265°	—	—	—
53	Orotic acid Uracil-6-carboxylic acid; 6-Carboxyuracil	Oro 6CxUra		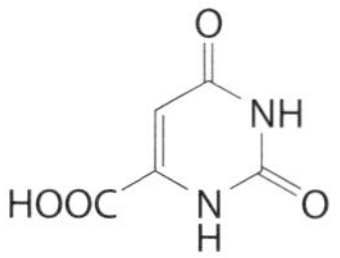	C$_5$H$_4$N$_2$O$_4$ (156.10)	345° (dec)	—	—	2.4, 9.5, >13

Acidic Spectral Data

No.	pH	λ_{max}	ε_{max} (×10⁻³)	λ_{min}	230	240	250	270	280	290
50	2.8	264	7.05	—	—	—	—	—	—	—
51	—	—	—	—	—	—	—	—	—	—
52	—	—	—	—	—	—	—	—	—	—
53	1	280	7.5	241	0.61	0.41	0.54	1.54	1.82	1.56
		227	36.0							

(Spectral Ratios: 230, 240, 250, 270, 280, 290)

Neutral Spectral Data

No.	pH	λ_{max}	ε_{max} (×10⁻³)	λ_{min}	230	240	250	270	280	290
50										
51										
52										
53	7	279	7.7	241	0.68	0.43	0.57	1.49	1.71	1.36
		230	31.5	249						

(Spectral Ratios: 230, 240, 250, 270, 280, 290)

Alkaline Spectral Data

No.	pH	λ_{max}	ε_{max} (×10⁻³)	λ_{min}	230	240	250	270	280	290
50	12	291	5.9	—	—	—	—	—	—	—
51	13°	230	8.2	224*	—	—	—	—	—	—
52	13°	230	8.1	—	—	—	—	—	—	—
53	12	286	6.0	244	1.36	0.80	0.80	1.38	1.71	1.72
		240	8.9	222						
		231	33.4	252	8.11	3.64	0.80	1.39	1.36	1.44

(Spectral Ratios: 230, 240, 250, 270, 280, 290)

REFERENCES

No.	Origin and Synthesis	pK	$[\alpha]_D^t$	Spectral Data	Mass Spectra	R$_f$
50	C: 326 D: 325,326	325	—	326,325	326,325	401,325
51	C: 98,85,99-102	—	—	103ᵇ,317ᵇᶜ,383*ᵇ	—	383
52	C: 101,99	—	—	103ᵇ,383	—	383
53	C: 104-106	66,105	—	66ᵇ,104ᵇ	—	69

PURINES, PYRIMIDINES, NUCLEOSIDES AND NUCLEOTIDES: PHYSICAL CONSTANTS AND SPECTRAL PROPERTIES (Continued)

No.	Compound	Symbol (3-Letter)	Symbol (1-Letter)	Structure	Formula (Mol Wt)	Melting Point °C	$[\alpha]_D^t$	pK Basic	pK Acidic
54	Wye²ʲ (Formerly "Yt base" or "Yt⁺") 4,9-Dihydro-4,6-dimethyl-9-oxo-1H-imidazo[1,2-a]purine; 1,N^2-isopropeno-3-methylguanine	Wye	ImGua*	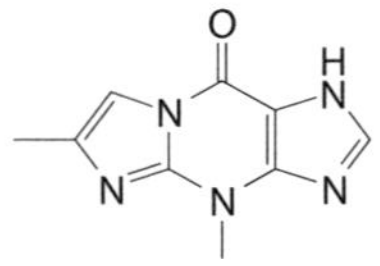 (structure)	$C_9H_9N_5O$ (203.20)	–	–	3.66	8.52
55	Wybutineʲ 7-[3-(Methoxycarbonyl)-3-(methoxyformamido)propyl] wyeʲ (Formerly "Y base" or "Yt*"); α-(Carboxyamino)-4,9-dihydro-4,6-dimethyl-9-oxo-1H-imidazo[1,2-a]purine-7-butyric acid dimethyl ester	Y-Wye (MeO)₂ FnBtoWyeʲ	Y-imGua*	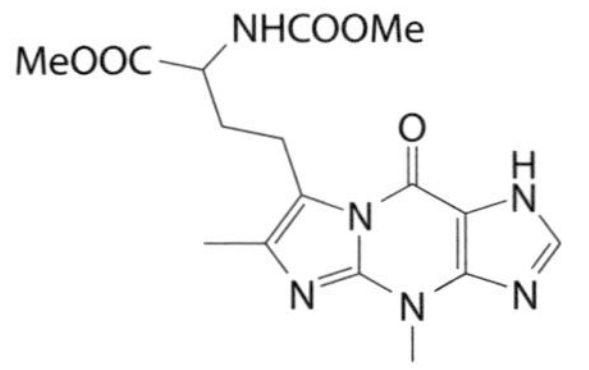 (structure)	$C_{16}H_{20}N_6O_5$ (376.38)	204–206°	–	3.7*	8.6*
56	Peroxywybutineʲ 7-[2-(Hydroperoxy)-3-(methoxycarbonyl)-3-(methoxyformamido)propyl] wyeʲ (Formerly "Yw base, Yr base, Peroxy Y base" or "Yw⁺") α-(Carboxyamino)-4,9-dihydro-β-hydroperoxy-4,6-dimethyl-9-oxo-1H-imidazo[1,2-a]purine-7-butyric acid dimethyl ester	O₂ Y-Wye O₂ (MeO)₂ FnBto Wyeʲ	O₂ Y-imGua*	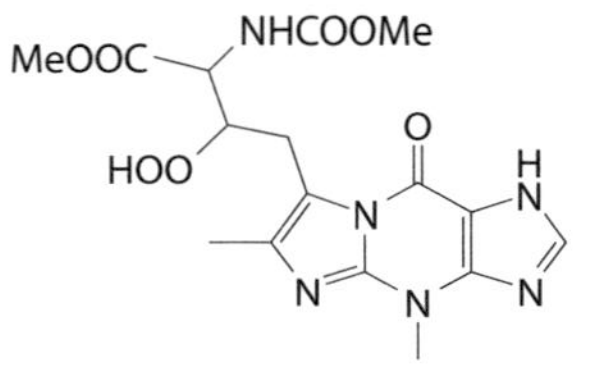 (structure)	$C_{16}H_{20}N_6O_7$ (408.37)	–	–	~3.3ᵖ	~9ᵖ

Acidic Spectral Data

No.	pH	λ_{max}	ε_{max} (×10⁻³)	λ_{min}	230	240	250	270	280	290
54	1.5	284	8.7	244	7.55	1.17	0.89	1.34	1.96	1.87
55	2	233 286	35.6 7.6	254*	–	3.45	1.04*	1.03	1.12*	1.12
56										

Neutral Spectral Data

No.	pH	λ_{max}	ε_{max} (×10⁻³)	λ_{min}	230	240	250	270	280	290
54	6.0	264 307	5.3 5.9	282	6.73	2.30	0.67	1.01	0.54	0.74
55	6.5	235 263 313	32.0 5.8 5.0	258* 288*	–	5.35	1.25	0.85	0.54*	0.50
56	H₂O	236 260*	32.0 6.0	211 256	5.02	4.07	1.16	0.81	0.56	0.69

Alkaline Spectral Data

No.	pH	λ_{max}	ε_{max} (×10⁻³)	λ_{min}	230	240	250	270	280	290
54	11.0	275 301	6.6 8.05	284						
55	10	236 265* 304	32.8 6.8 7.2	262* 287	–	4.95	1.35	1.01	0.91	0.86
56										

REFERENCES

No.	Origin and Synthesis	Spectral Data	pK	$[\alpha]_D$	Mass Spectra	R_f
54	C: 312 R: 312	312ᵇ,403ᵇ	312	–	312	312
55	C: 320 R: 316,310	320,310,316*ᵇ	320,310*	–	310,311	324,316,310,311
56	R: 327,321,324	327ᵇ,321*ᵇ	324	–	321,327	324,321,327

PURINES, PYRIMIDINES, NUCLEOSIDES AND NUCLEOTIDES: PHYSICAL CONSTANTS AND SPECTRAL PROPERTIES (Continued)

No.	Compound	Symbol 3-Letter	Symbol 1-Letter	Structure	Formula (Mol Wt)	Melting Point °C	$[\alpha]_D^t$	pK Basic	pK Acidic
57	Xanthine	Xan		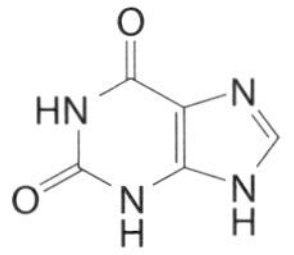 (structure)	$C_5H_4N_4O_2$ (152.11)	None (dec)	–	~0.8	7.5*, 11.1*
58	7-Methylxanthine	7MeXan		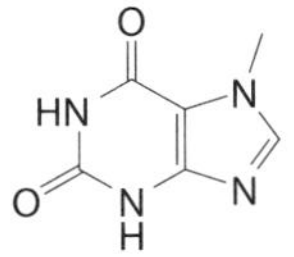 (structure)	$C_6H_6N_4O_2$ (166.14)	~380° (dec)	–	–	8.4, ~13
59	Zeatin 6-(trans-4-Hydroxy-3-methyl-2-butenylamino)purine; N^6-(trans-4-Hydroxy isopentenyl)adenine	Zea 6(tr HoiPe) Ade		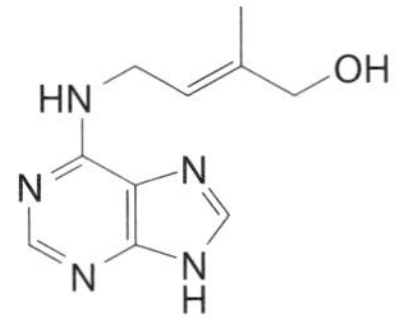 (structure)	$C_{10}H_{13}N_5O$ (219.24)	207–208°	–	–	–

Acidic Spectral Data

No.	pH	λ_{max}	ε_{max} (×10⁻³)	λ_{min}	230	240	250	270	280	290
57	0ᵍ	260	9.15	242	–	–	0.77	–	0.15	0.01
58	1ᵍ	268	9.3	241	0.53	0.42	0.66	0.94	0.40	0.04
59	2	207	14.5	235	–	0.38	–	–	0.75	–
	1	275	14.65		–	–	–	–	–	–

Neutral Spectral Data

No.	pH	λ_{max}	ε_{max} (×10⁻³)	λ_{min}	230	240	250	270	280	290
57	6	308	5.2	281	–	–	0.57	–	0.61	0.07
		267	10.25	239						
58	7ᵃ	–	–	240	0.45	0.42	0.62	1.16	0.69	0.14
59	6	269	10.0	240	–	–	–	–	–	–
	7	212	17.1	233	–	–	–	–	–	–
		270	16.2							

Alkaline Spectral Data

No.	pH	λ_{max}	ε_{max} (×10⁻³)	λ_{min}	230	240	250	270	280	290
57	10	240	8.9	222	–	–	1.29	–	1.71	0.92
		277	9.3	257						
58	13	–	–	257	1.81	1.33	1.12	1.63	2.28	2.07
59	14	284	9.4	257	–	–	1.11	–	2.39	2.27
	13	220	15.9	242	–	–	–	–	–	–
		276	14.65							

REFERENCES

No.	Origin and Synthesis	$[\alpha]_D^t$	pK	Spectral Data
57	C: 231,377	–	22,248,170*,53*,55*,253*	232ᵇ,242ᵉ,205
58	C: 45	–	53–56	53ᵇ,41ᶜ·ᵈ,27*,54*
59	C: 227 N: 228	–	227	227

No.	Mass Spectra	R_f
57	–	258,242
58	–	10,27
59	–	32,227

PURINES, PYRIMIDINES, NUCLEOSIDES AND NUCLEOTIDES: PHYSICAL CONSTANTS AND SPECTRAL PROPERTIES (Continued)

No.	Compound	Symbol (3-Letter)	Symbol (1-Letter)	Structure	Formula (Mol Wt)	Melting Point °C	$[\alpha]_D$	pK Basic	pK Acidic
				RIBONUCLEOSIDES					
60	Adenosine	Ado	A	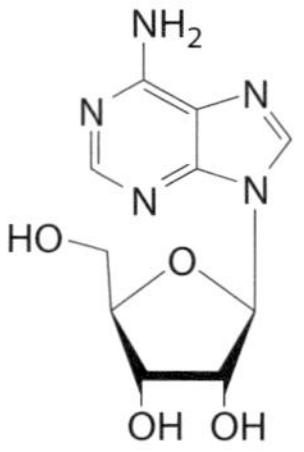	$C_{10}H_{13}N_5O_4$ (267.24)	235–236°	-61^{25} (1.0, H_2O)	3.5*	12.5
61	1-(Δ^2-Isopentenyl)adenosine 1-($\gamma\gamma$-Dimethylallyl)adenosine 1-(3-Methyl-2-butenyl)adenosine	1Pe^{i}Ado 1Pe^{i}A	i^{1}A	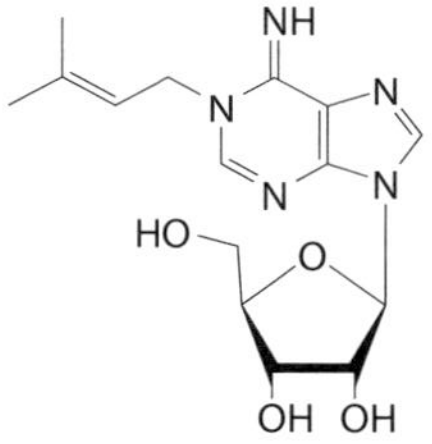	$C_{15}H_{21}N_5O_4$ (355.36)	131–133° (HBr salt)	—	8.5	—
62	1-Methyladenosine	1MeAdo	m^{1}A	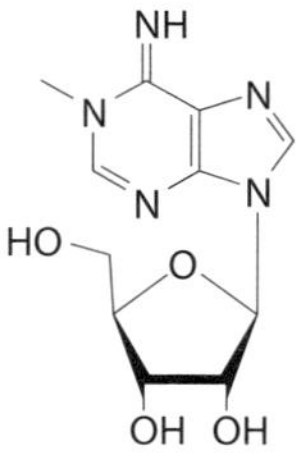	$C_{11}H_{15}N_5O_4$ (281.27)	214–217° (dec)	-59^{26} (2.0, H_2O)	8.8*	—

Acidic Spectral Data

No.	pH	λ_{max}	ε_{max} (×10⁻³)	λ_{min}	230	240	250	270	280	290
60	1	257	14.6	230	0.23	0.44	0.84	–	0.22	0.03
61	1	257.5	13.9	235						
62	2	258	13.7	232*	0.28*	0.41	0.81	0.70	0.23*	0.04

Neutral Spectral Data

No.	pH	λ_{max}	ε_{max} (×10⁻³)	λ_{min}	230	240	250	270	280	290
60	6	260	14.9	227	0.18	0.42	0.78	–	0.14	0.00
61	7	258	13.6	235						
62	7	258	13.9*	232*	0.29	0.41	0.81	0.70	0.23	0.05

Alkaline Spectral Data

No.	pH	λ_{max}	ε_{max} (×10⁻³)	λ_{min}	230	240	250	270	280	290
60	11	259	15.4	227	0.24	0.40	0.79	–	0.15	0.00
61	13°	259*	14.25*	236						0.31
62	10.5°	259*	14.6	231*	0.25	0.38	0.76	0.76	0.35	0.30

REFERENCES

No.	Origin and Synthesis	$[\alpha]_D$	pK	Spectral Data	Mass Spectra	R_f
60	C: 111,235,363,350,351,379	363,235,350	250,255*,336,371,51,220	232b,212cde,242e,183b,317b,205	315,341,317,405	258,110,18,242,291,292
61	C: 319,15ᵃ,14	–	15	319,15*	15	15,32,319
62	C: 10,30	10	15,318*	90,10*d,317*bd,318*b,109,12b	314,317	10,12,314,291

PURINES, PYRIMIDINES, NUCLEOSIDES AND NUCLEOTIDES: PHYSICAL CONSTANTS AND SPECTRAL PROPERTIES (Continued)

No.	Compound	Symbol (3-Letter)	Symbol (1-Letter)	Structure	Formula (Mol Wt)	Melting Point °C	$[\alpha]_D^t$	pK (Basic)	pK (Acidic)
63	1,N^6-Dimethyladenosine; 1-Methyl-6-methylamino-9-β-D-ribofuranosylpurine	1,6Me$_2$, Ado	$m_2^{1,6}A$; $m^1 m^6 A$	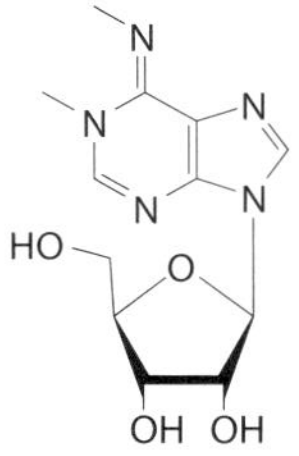	$C_{12}H_{17}N_5O_4$ (295.30)	206°	—	—	—
64	2-Hydroxyadenosine; Crotonoside; Isoguanosine	2HOAdo; isoGuo*	o^2A; isoG*	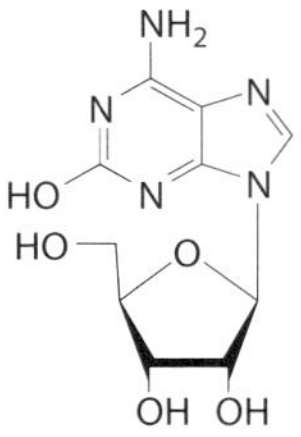	$C_{10}H_{13}N_5O_5$ (283.25)	237–252° (dec)	-71^{26} (1.06, 0.1N NaOH)	—	—
65	2-Methyladenosine	2MeAdo	m^2A	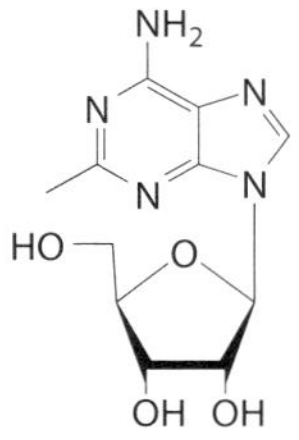	$C_{11}H_{15}N_5O_4$ (281.27)	>200° (dec) (picrate)	-66.6^{25} (1.0, H_2O)	—	—

Acidic Spectral Data

No.	pH	λ_{max}	ε_{max} (×10⁻³)	λ_{min}	230	240	250	270	280	290
63	1	261	14.2	234	0.26	0.31	0.68	0.81	0.47	0.18
64	1.2	235; 283*	6.14; 12.7*	—	—	—	—	—	—	—
65	1	258*	14.0	230	0.22	0.44	0.84	0.86	0.40	0.05

Neutral Spectral Data

No.	pH	λ_{max}	ε_{max} (×10⁻³)	λ_{min}	230	240	250	270	280	290
64	H_2O	247*; 293*	8.9; 11.1*	—	—	—	—	—	—	—
65	6	264	14.5	228	0.24	0.41	0.75	0.84	0.17	0.01

Alkaline Spectral Data

No.	pH	λ_{max}	ε_{max} (×10⁻³)	λ_{min}	230	240	250	270	280	290
63	14	262	14.9	234	—	—	—	—	—	—
64	12.8	251*; 285*	6.9*; 10.55*	—	—	—	—	—	—	—
65	13	263	15.2	230	0.24	0.41	0.75	0.84	0.17	0.01

REFERENCES

No.	Origin and Synthesis	$[\alpha]_D^t$	pK	Spectral Data	Mass Spectra	R_f
63	C: 24,12	—	—	24,90ᵉ 12	—	24,12
64	C: 269,N: 268	269	—	269ᵇ,268*ᵇ	—	269
65	E: 18 C: 111,342 R: 18,291	342	—	18ᵇ,342*ᶜᵈ,317ᵇᶜᵉ,291ᵇ	317	18,291

PURINES, PYRIMIDINES, NUCLEOSIDES AND NUCLEOTIDES: PHYSICAL CONSTANTS AND SPECTRAL PROPERTIES (Continued)

No.	Compound	Symbol 3-Letter	Symbol 1-Letter	Structure	Formula (Mol Wt)	Melting Point °C	$[\alpha]_D^t$	pK Basic	pK Acidic
66	2-Methylthioadenosine	2MeSAdo	ms^2 A	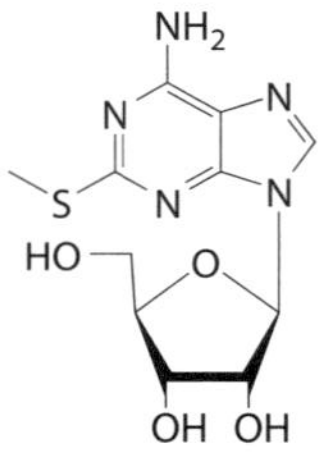	$C_{11}H_{15}N_5O_4 S$ (313.33)	227°	$+4^{29}$ (1.0, 0.1N HCl)	—	—
67	N^6-Glycinocarbonyladenosine N-[(9-β-D-Ribofuranosylpurin-6-yl)carbamoyl]glycine	6(GlyCO)Ado	gc^6 A	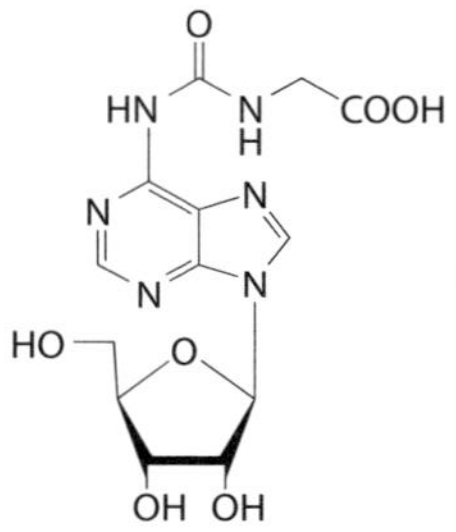	$C_{13}H_{14}N_6O_7$ (368.31)	214–216°	—	—	—
68	N^6-Methyl-N^6-glycinocarbonyladenosine N-[(9-β-D-Ribofuranosylpurin-6-yl)-N-methylcarbamoyl]glycine	6Me6(GlyCO) Ado	$m^6 gc^6$ A	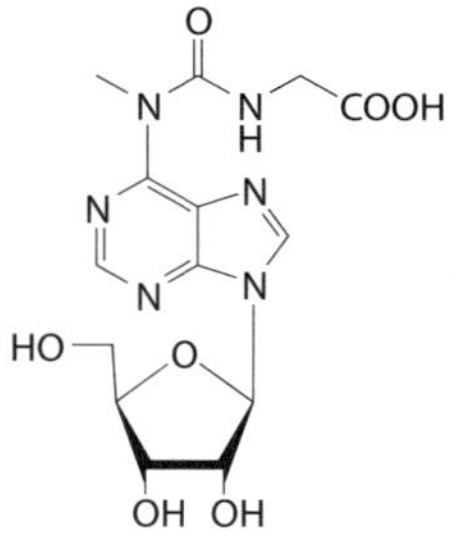	$C_{14}H_{18}N_6O_7$ (382.33)	173–174°	33.09^{23} (0.55 H_2O)	—	—

Acidic Spectral Data

No.	pH	λ_{max}	ε_{max} (×10⁻³)	λ_{min}	230	240	250	270	280	290
66	1	270	15.2							
67	1.2	271 276	18.2 19.1	238	0.58	0.37	0.59	1.58	1.58	0.57
68	1.0	283	16.7	239	0.75	0.44	0.62	1.70	2.39	1.83

Neutral Spectral Data

No.	pH	λ_{max}	ε_{max} (×10⁻³)	λ_{min}	230	240	250	270	280	290
66	7	235 274	17.7 13.5							
67	5.0	269 276	20.9 17.1	230	0.12	0.28	0.59	1.37	0.78	0.00
68	5.5	277 284	17.1 16.7	236	0.49	0.35	0.60	1.68	2.14	1.28

Alkaline Spectral Data

No.	pH	λ_{max}	ε_{max} (×10⁻³)	λ_{min}	230	240	250	270	280	290
66	13	235 274*	17.8* 13.5*							
67	12.1°	270ᵐ 277 298	13.5ᵐ 13.8 12.8	236	0.37	0.32	0.57	1.41	1.00	0.60
68	13°	277 284	15.4 15.8	245	1.36	0.79	0.76	1.38	1.85	1.33

REFERENCES

No.	Origin and Synthesis	$[\alpha]_D^t$	pK	Spectral Data	Mass Spectra	R_f
66	C: 399,111ᵃ	111	—	399,111*,299	—	299
67	C: 408,295 R: 295	—	—	408,295ᵇ,330	295,408,330	295,408,330
68	C: 330	330	—	330	—	330

PURINES, PYRIMIDINES, NUCLEOSIDES AND NUCLEOTIDES: PHYSICAL CONSTANTS AND SPECTRAL PROPERTIES (Continued)

No.	Compound	Symbol 3-Letter	Symbol 1-Letter	Structure	Formula (Mol Wt)	Melting Point °C	$[\alpha]_D$	pK Basic	pK Acidic
69	N^6-Threoninocarbonyladenosine N-[(9-β-D-Ribofuranosylpurin-6-yl)carbamoyl]-L-threonine; N-(Nebularin-6-ylcarbamoyl)-L-threonine	6(ThrCO)Ado	tc^{6}A	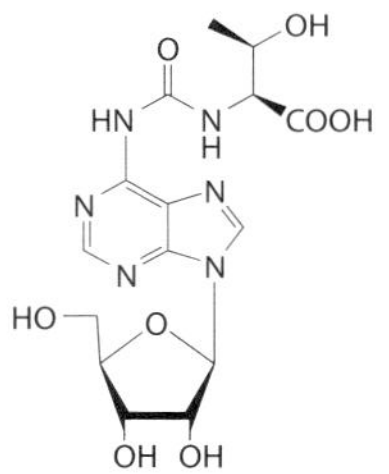	$C_{15}H_{20}N_6O_8$ (412.36)	204–207°	-13.9^{25} (1.0, Me$_2$SO)	<2^p	~3.0^p
70	N^6-Methyl-N^6-threoninocarbonyladenosine N-[(9-β-D-Ribofuranosylpurin-6-yl)-N-methylcarbamoyl]threonine	6Me6(ThrCO)Ado	m^{6}tc^{6}A	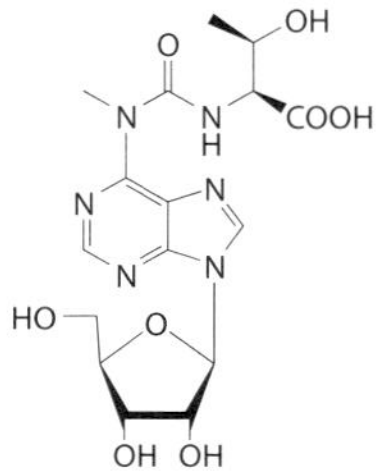	$C_{16}H_{22}N_6O_8$ (426.39)	159–160°	5.31^{23} (0.49, H$_2$O)	<2^p	~2.8^p
71	N^2-(Δ^2-Isopentenyl)adenosine; N^6-(3-Methyl-2-butenyl)adenosine; 6-($\gamma\gamma$-Dimethylallylamino)-9-β-D-ribofuranosyl purine	6iPeAdo 6PeiAdo	i^{6}A	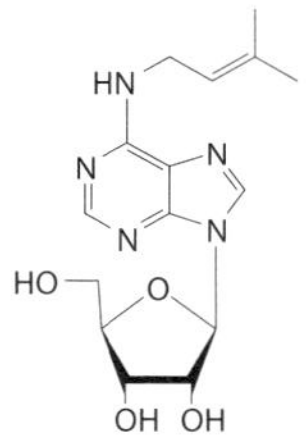	$C_{15}H_{21}N_5O_4$ (335.36)	145–147°	-103^{28} (0.14, EtOH)	3.8	—

Neutral Spectral Data

No.	pH	λ_{max}	ε_{max} ($\times 10^{-3}$)	λ_{min}	230	240	250	270	280	290
69	6.5	269 / 276	22.9 / 19.4	231	0.17	0.30	0.59	1.40	0.93	0.00
70	5.5	278 / 284	23.5 / 23.1	237	0.53	0.39	0.61	1.64	2.36	1.69
71	7	269	20.0	234*	0.24	0.29	0.52	1.17	0.94*	0.36*

Alkaline Spectral Data

pH	λ_{max}	ε_{max} ($\times 10^{-3}$)	λ_{min}	230	240	250	270	280	290
12.4°	270^m / 277 / 299	16.3^m / 16.0 / 11.1	236	0.40	0.34	0.59	1.44	1.16	0.42
13°	277 / 283	22.5 / 21.5	249	1.57	1.01	0.91	1.23	1.44	1.01
12	269	20.0	234			0.52*	1.17	0.94	0.36

Acidic Spectral Data

No.	pH	λ_{max}	ε_{max} ($\times 10^{-3}$)	λ_{min}	230	240	250	270	280	290
69	1.6^g	277	21.6	238	0.59	0.39	0.57	1.54	1.51	0.56
70	1.0^g	277 / 283	22.3 / 22.3	237	0.54	0.40	0.62	1.62	2.10	1.36
71	1	265	20.4	232	0.27	0.30	0.57	1.05	0.69*	0.18*

REFERENCES

No.	Origin and Synthesis	$[\alpha]_D^t$	pK	Spectral Data	Mass Spectra	R_f
69	C: 408,290,R: 4,296,407	408	407,90	290^b,6,296^b,297^b,317^b,313^b,330	408,330	290,295–297,407,408,330
70	C: 330 R: 297,407	330	407	300^b,297^b,313^b	297	297,407,330
71	C: 32,31,15,319,14 R: 32,116	319	15	32^b,116*^{b},317*^{b},14,31,15,265*^{b}	32,317,365,405	32,116,117,15,293,319

PURINES, PYRIMIDINES, NUCLEOSIDES AND NUCLEOTIDES: PHYSICAL CONSTANTS AND SPECTRAL PROPERTIES (Continued)

No.	Compound	Symbol 3-Letter	Symbol 1-Letter	Structure	Formula (Mol Wt)	Melting Point °C	$[\alpha]_D^t$	pK Basic	pK Acidic
72	N^6-'Δ^2-Isopentenyl)-2-methylthioadenosine 6-(3-Methyl-2-butenyl-amino)-2-methylthio-9-β-D-ribofuranosylpurine	2MeS6iPeAdo 2MeS6iPe^{i}Ado	ms^{2}i^{6}A	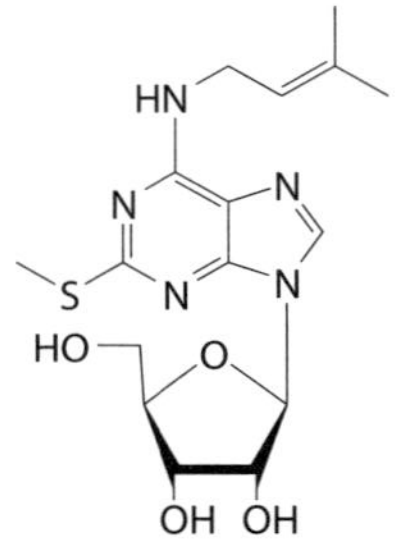	$C_{16}H_{23}N_5O_4S$ (381.147)	194–195°	—	—	—
73	N^6-(4-Hydroxyisopentenyl)adenosine N^6-(cis-4-Hydroxy-3-methyl-2-butenyl)adenosine cisZeatosine	6(HOiPe)Ado 6(HOPe^i)Ado 6(cisHOiPe)Ado	oi^{6}A	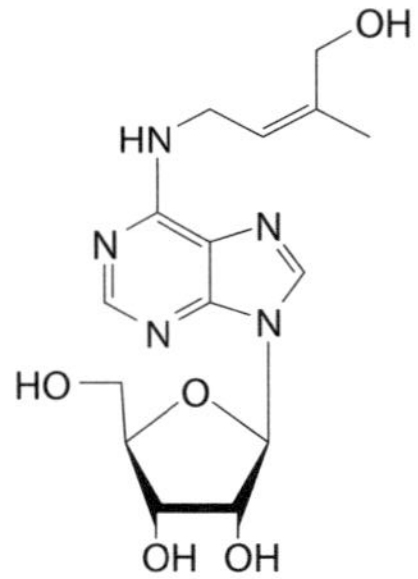	$C_{15}H_{21}N_5O_5$ (351.36)	206°	-98^{27}_{546} (0.02, H_2O)	—	—

Acidic Spectral Data

No.	pH	λ_{max}	ε_{max} ($\times10^{-3}$)	λ_{min}	230	240	250	270	280	290
72	1 (EtOH)	246 286 265	18.6 16.1 20.4	265	0.83	1.12	1.27	1.00	1.12	1.13
73	1	265	20.4	—	0.53	0.42	0.68	1.01	0.65	0.21

Neutral Spectral Data

No.	pH	λ_{max}	ε_{max} ($\times10^{-3}$)	λ_{min}	230	240	250	270	280	290
72	(EtOH)	244 283 268	25.3 18.1 20.0	258	1.39	2.39	1.78	1.39	1.85	1.67
73	7	268	20.0		0.36	0.27	0.53	1.16	0.79	0.27

Alkaline Spectral Data

No.	pH	λ_{max}	ε_{max} ($\times10^{-3}$)	λ_{min}	230	240	250	270	280	290
72	10 (EtOH)	243 283	24.9 18.0	259	1.39	2.40	1.79	1.40	1.85	1.68
73	12	268	20.0		0.43	0.29	0.50	1.23	0.85	0.29

REFERENCES

No.	Spectral Data	pK	$[\alpha]_D^t$
72	293[b],288[b],299[b],317[b],335[b]	—	—
73	117	—	117

No.	Origin and Synthesis	Mass Spectra	R_f
72	C: 293,274 R: 274,288,299,335	293,299,274,317,332	274,288,293,299,335
73	R: 117[a],332	117,332,317,366	117,332

PURINES, PYRIMIDINES, NUCLEOSIDES AND NUCLEOTIDES: PHYSICAL CONSTANTS AND SPECTRAL PROPERTIES (Continued)

No.	Compound	Symbol 3-Letter	Symbol 1-Letter	Structure	Formula (Mol Wt)	Melting Point °C	$[\alpha]_D^t$	pK Basic	pK Acidic
74	N^6-(4-Hydroxyisopentenyl)-2-methylthioadenosine N^6-(4-Hydroxy-3-methyl-2-butenyl)-2-methylthioadenosine Methylthio-*ciszeatosine*	2MeS6(HOiPe)Ado 2MeS6(HOPei)Ado	$ms^2\,oi^6\,A$	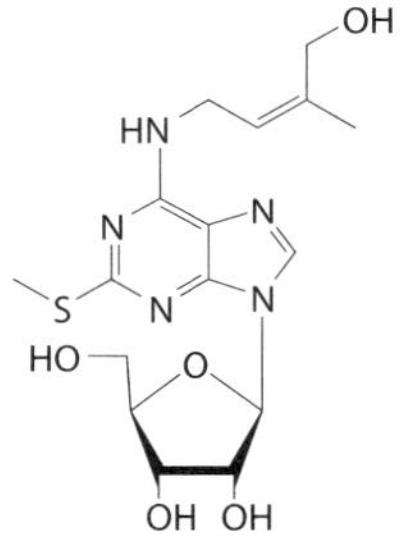 [structure]	$C_{16}H_{23}N_5O_5S$ (397.46)	155–156°	–	–	–
75	N^6-Methyladenosine 6-Methylaminopurine ribonucleoside	6MeAdo	m^6A	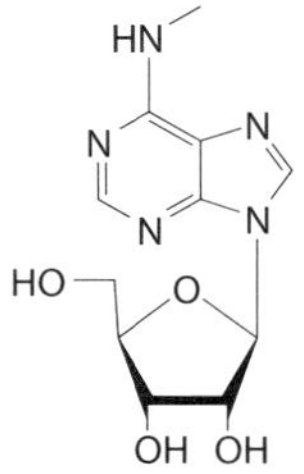 [structure]	$C_{11}H_{15}N_5O_4$ (281.27)	219–221°	-54^{26} (0.6, H_2O)	4.0	–
76	N^6-Methyl-2-methylthioadenosine	2MeS6MeAdo	ms^2m^6A	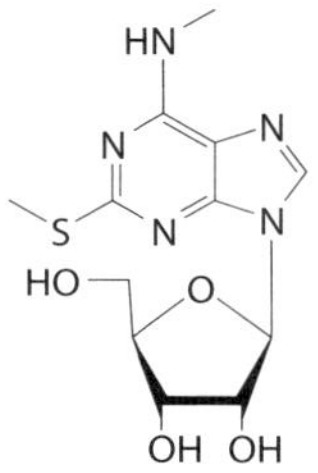 [structure]	$C_{12}H_{17}N_5O_4S$ (327.36)	–	–	–	–

Acidic Spectral Data

No.	pH	λ_{max}	ε_{max} (×10⁻³)	λ_{min}	230	240	250	270	280	290	$[\alpha]_D^t$	pK
74	1 (EtOH)	245 285	18.5 15.5	223 263	0.20	0.31	0.66	0.88	0.41	0.13	–	–
75	1	262	16.6	231							112	15
76											–	–

Neutral Spectral Data

No.	pH	λ_{max}	ε_{max} (×10⁻³)	λ_{min}	230	240	250	270	280	290	Spectral Data
74	7 (EtOH)	243 282	24.3 17.2	223 258	0.17	0.22	0.57	1.09	0.68	0.24	332
75	7	266	15.9	229							
76	7	241.5 280.5									

Alkaline Spectral Data

No.	pH	λ_{max}	ε_{max} (×10⁻³)	λ_{min}	230	240	250	270	280	290	Mass Spectra	R_f
74	12 (EtOH)	243 282	24.7 17.3	226 258	0.17	0.22	0.57	1.09	0.68	0.24	332	332
75	13	266	15.9*	232*							292,314,317,405	10,12,18,110,291,292,297,314 299,288,335
76											–	

REFERENCES

No.	Origin and Synthesis	Spectral Data
74	C: 332 R: 332,294	332
75	E: 18 C: 10,112,166,30 R: 18,292	18[b],112[c,d],10*[c],317[b,e],292[b],8,12
76	–	299

PURINES, PYRIMIDINES, NUCLEOSIDES AND NUCLEOTIDES: PHYSICAL CONSTANTS AND SPECTRAL PROPERTIES (Continued)

No.	Compound	Symbol (3-Letter)	Symbol (1-Letter)	Structure	Formula (Mol Wt)	Melting Point °C	$[\alpha]_b$	pK Basic	pK Acidic
77	N^6,N^6-Dimethyladenosine 6-Dimethylamino-9-β-D-ribofuranosylpurine	6Me₂Ado	m$^6_2$A	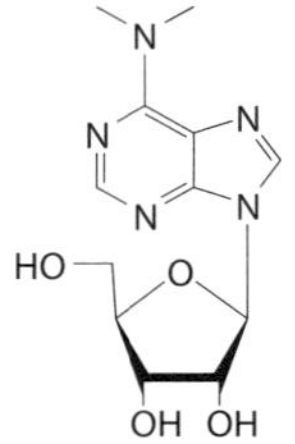	$C_{12}H_{17}N_5O_4$ (295.30)	183–184°	-62.6^{25} (2.6, H₂O)	4.5	–
78	$O^{2'}$-Methyladenosine 2'-C-Methyladenosine	2'MeAdo	Am	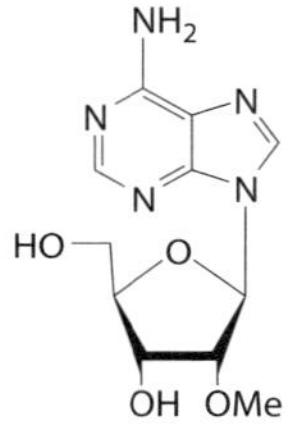	$C_{11}H_{15}N_5O_4$ (281.27)	201–202°	-57.9^{23} (1.0, H₂O)	–	–
79	$O^{2'}$-Ribosyladenosine 2'-O-Ribosyladenosine	ORibAdo		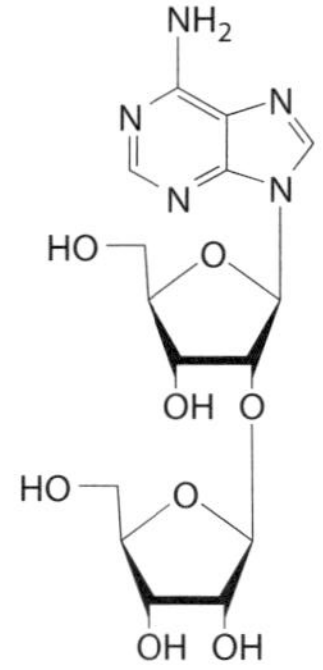	$C_{15}H_{21}N_5O_8$ (399.36)	–	–	–	–

Acidic Spectral Data

No.	pH	λ_{max}	ε_{max} (×10⁻³)	λ_{min}	230	240	250	270	280	290	$[\alpha]^t_D$
77	1	268	18.4	234	0.24	0.27	0.60	1.20	0.94	0.42	113
78	1	257	13.8	229	0.24	0.47	0.86	0.71	0.21	–	108,370
79	1.5	257	–	230	0.36	0.47	0.84	0.72	0.32	0.12	–

Neutral Spectral Data

pH	λ_{max}	ε_{max} (×10⁻³)	λ_{min}	230	240	250	270	280	290	pK
7	275	18.8	236	0.27	0.19	0.46	1.52	1.54	0.96	15
7	259	13.9*	228	0.17	0.40	0.80	0.70	0.15	–	–
7	259	–	227	0.35	0.47	0.82	0.67	0.21	0.04	–

Alkaline Spectral Data

pH	λ_{max}	ε_{max} (×10⁻³)	λ_{min}	230	240	250	270	280	290
13	276	19.2	237	0.17	0.09	0.37	1.57	1.68	1.01
12	259	13.9	227	0.20	0.39	0.79	0.70	0.15	–
11	259	–	227	0.34	0.47	0.79	0.66	0.20	0.05

REFERENCES

No.	Origin and Synthesis	Spectral Data	Mass Spectra	R_f
77	E: 18 C: 114,113,115,339	18[b],114[cd],317[be],113	314,405	18,32,117,291,314
78	C: 107,108[a],369,370	317[b],107[d],369*	315,317,367	107,109,110,369
79	C: 285 R: 110	110,317[b]	–	110

PURINES, PYRIMIDINES, NUCLEOSIDES AND NUCLEOTIDES: PHYSICAL CONSTANTS AND SPECTRAL PROPERTIES (Continued)

No.	Compound	Symbol 3-Letter	Symbol 1-Letter	Structure	Formula (Mol Wt)	Melting Point °C	$[\alpha]_D^t$	pK Basic	pK Acidic
80	7-α-D-Ribofuranosyladenine Pseudovitamin B$_{12}$ nucleoside	7αDAdo		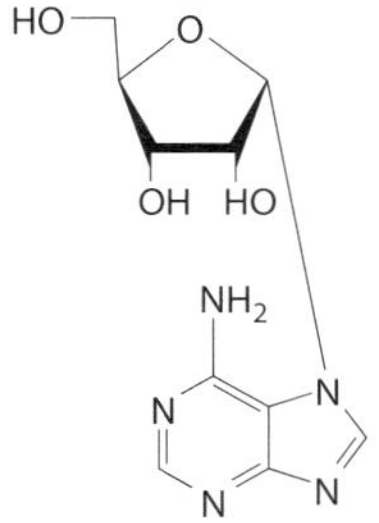	$C_{10}H_{13}N_5O_4$ (267.24)	220–222° (0.4, H$_2$O)	0^{25} (0.4, H$_2$O), 3.9		–
81	2-Methyl-7-α-ribofuranosyladenine Factor A nucleoside	2Me7αDAdo		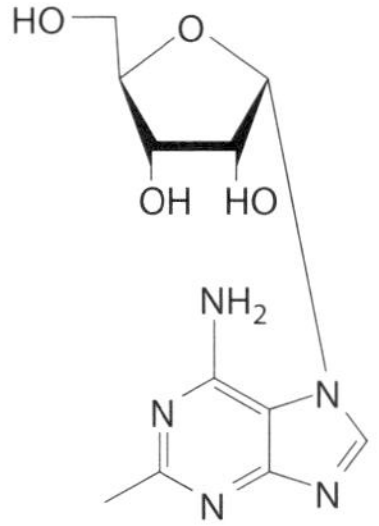	$C_{11}H_{15}N_5O_4$ (281.27)	219–220°	4.8		–
82	Cytidine	Cyd	C	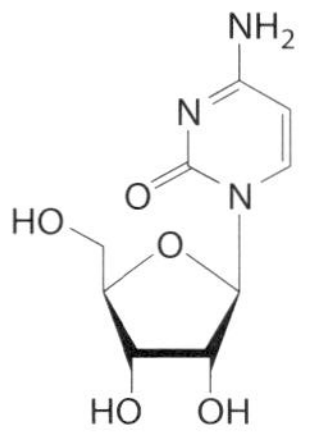	$C_9H_{13}N_3O_5$ (243.22)	224–225° (dec) (sulfate)	+34.2^{16} (2.0, H$_2$O), 4.15		12.5*

Acidic Spectral Data

No.	pH	λ_{max}	ε_{max} (×10^{-3})	λ_{min}	230	240	250	270	280	290
80	1	273	13.6	239	0.86	0.50	0.66	1.39	1.19	0.33
81	1	273	13.2	238	0.95	0.55	0.70	1.42	1.24	0.35
82	1	280	13.4	242	0.63	0.28	0.45	1.70	2.10	1.55

Neutral Spectral Data

No.	pH	λ_{max}	ε_{max} (×10^{-3})	λ_{min}	230	240	250	270	280	290
80	H$_2$O	271	9.8	233	0.59	0.65	0.78	1.26	0.97	0.25
81	7	276	9.1	233	0.85	0.90	0.89	1.40	1.35	0.51
82	7	229.5	8.3	226	0.98	0.82	0.86	1.28	0.93*	0.28*
		271	9.1	250						

Alkaline Spectral Data

No.	pH	λ_{max}	ε_{max} (×10^{-3})	λ_{min}	230	240	250	270	280	290
80	13	271	9.8	–	–	–	–	–	–	–
81	12	244	5.7	–	–	–	–	–	–	–
		276	9.1							
82	13^a	272.5	9.15	251	1.14	1.00	0.87	1.23	1.02	0.38
	14	273	9.2	252						

REFERENCES

No.	Origin and Synthesis	$[\alpha]_D^t$	pK	Spectral Data	Mass Spectra	R$_f$
80	N: 119 C: 120	120	119	119^b,120	–	119,120
81	N: 121^a	–	121	121^b	–	18
82	R: 236,181,C: 251,128,252,272,386,379	236,147	163*,231	163^b,232$^{b c}$,242^{*e},181^b,212,183^b,317^b,205	315,317	258,110,298,242

PURINES, PYRIMIDINES, NUCLEOSIDES AND NUCLEOTIDES: PHYSICAL CONSTANTS AND SPECTRAL PROPERTIES (Continued)

No.	Compound	Symbol 3-Letter	Symbol 1-Letter	Structure	Formula (Mol Wt)	Melting Point °C	$[\alpha]_D^t$	pK Basic	pK Acidic
83	2-Thiocytidine	2Syd 2SCyd	s^{2}C	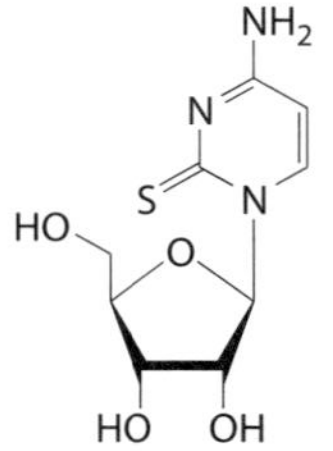	$C_9H_{13}N_3O_4S$ (259.28)	208–209°	+64.2[25] (1.8, H$_2$O)	–	–
84	3-Methylcytidine	3MeCyd	m^{3}C	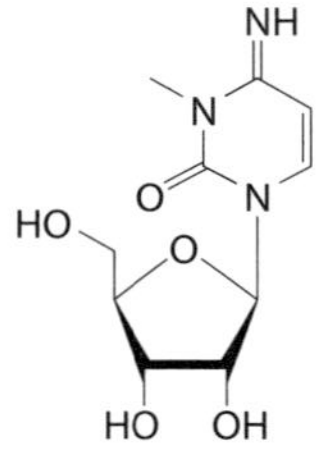	$C_{10}H_{15}N_3O_5$ (257.24)	193–194° (methosulfate)	–	8.7*	>12
85	N^4-Acetylcytidine 4-Acetylamino-1-β-D-ribofuranosyl-2-pyrimidinone	4AcCyd	ac^{4}C	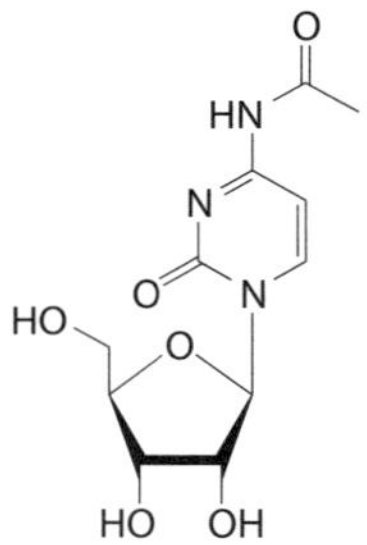	$C_{11}H_{15}N_3O_6$ (285.25)	208–209°	+60.1[23] (1.0, H$_2$O)	<1.5	–

Acidic Spectral Data

No.	pH	λ_{max}	ε_{max} (×10^{-3})	λ_{min}	230	240	250	270	280	290
83	0	229 276	17.0 17.4	213 250	1.44*	1.14	0.86	1.33*	1.30*	1.11
84	4	278* 241	11.8 12.4	243*	0.75	0.33	0.47	1.67	1.89*	1.21
85	1$^{0.4}$	308	13.8	267	2.4	2.2*	1.7*	0.78	1.4	2.4*

Neutral Spectral Data

No.	pH	λ_{max}	ε_{max} (×10^{-3})	λ_{min}	230	240	250	270	280	290
83	7	249*	22.3	220	0.59	0.95	1.08	0.78	0.50*	0.42
84	7	278 247*	11.6 15.2	243 226	0.55	0.36	0.48	1.63	1.82	1.18
85	7	297*	8.7	271*	0.77	1.65	1.80	0.46	0.61*	0.95

Alkaline Spectral Data

No.	pH	λ_{max}	ε_{max} (×10^{-3})	λ_{min}	230	240	250	270	280	290
83	13	252	–	229	0.68	0.70	1.04	0.91	0.72	0.39
84	12°	266 302	9.0 14.0	243* 241	1.12 0.88	0.68 0.97	0.73 0.72	1.01 1.34	0.69* 1.57	0.26 1.84
85	13°									

REFERENCES

No.	Origin and Synthesis	$[\alpha]_D^t$	pK	Spectral Data	Mass Spectra	R_f
83	C: 272,271*,285,301 R: 270,329	285,301	–	271*,270°,317^{bc},301*,329^b	–	272,329
84	C: 57,30,125	–	58,57,59*	57,317^{bcde},58^b,59*,125*,126	317	57,59,110,125
85	C: 127,244,245,328 R: 116,298	127	127	127,116*^{bce},317^{bd},244*,245,265*^{b},298^b	314,317,298	116,298,304,314

PURINES, PYRIMIDINES, NUCLEOSIDES AND NUCLEOTIDES: PHYSICAL CONSTANTS AND SPECTRAL PROPERTIES (Continued)

No.	Compound	Symbol 3-Letter	Symbol 1-Letter	Structure	Formula (Mol Wt)	Melting Point °C	$[\alpha]_D^t$	pK Basic	pK Acidic
86	N^4-Methylcytidine 4-Methylamino-1-β-D-ribofuranosyl-2-pyrimidinone	4MeCyd	m⁴C	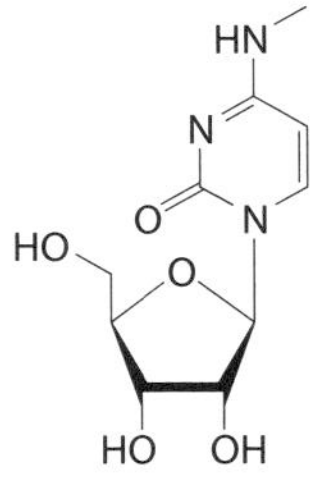 [structure]	$C_{10}H_{15}N_3O_5$ (257.24)	237° (dec)	–	3.9	–
87	N^4-Methyl-2-thiocytidine	4Me2Syd 4Me2Scyd	m⁴s²C	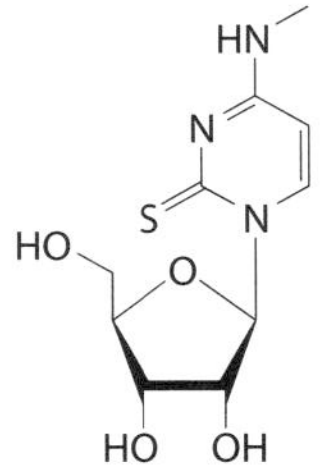 [structure]	$C_{10}H_{15}N_3O_4S$ (273.32)	189–190°	–	–	–
88	5-Methylcytidine	5MeCyd	m⁵C	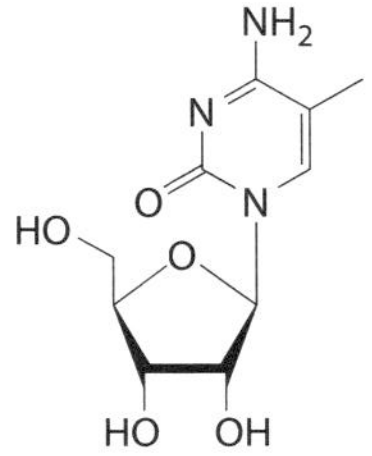 [structure]	$C_{10}H_{15}N_3O_5$ (257.24)	210–211° (dec)	-3^{23} (2.5, 1N NaOH)	4.3	>13

Acidic Spectral Data

No.	pH	λ_{max}	ε_{max} (×10⁻³)	λ_{min}	230	240	250	270	280	290
86	1	281	14.3	243	0.85	0.37	0.49	1.72	2.13	1.73
87	0	238 276	18.5 18.1	216 257	1.31	1.50	1.05	1.39	1.40	0.86
88	0	287	12.6	245	1.72	0.42	0.33	2.24	3.59	3.96

Neutral Spectral Data

No.	pH	λ_{max}	ε_{max} (×10⁻³)	λ_{min}	230	240	250	270	280	290
86	7	237	9.2	227	0.88	0.91	0.84	1.15	0.92	0.38
87	H₂O	271 224 262	11.6 14.9 27.4	250 236	0.49	0.51	0.64	0.92	0.65	0.37
88	7	277	8.9	255	1.53	1.34	1.03	1.33	1.45	0.96

Alkaline Spectral Data

No.	pH	λ_{max}	ε_{max} (×10⁻³)	λ_{min}	230	240	250	270	280	290
86	14	236	8.9	251	0.93	0.93	0.84	1.22	1.06	0.51
87	–	273	11.6	–	–	–	–	–	–	–
88	14ᵍ	279	9.0	256	1.58	1.35	1.05	1.37	1.58	1.19

REFERENCES

No.	Origin and Synthesis	pK	$[\alpha]_D^t$	Spectral Data	Mass Spectra	R_f
86	C: 59,128	128,59	–	59ᵇ,128,317ᵇ	–	59,129,130,353
87	C: 271	–	–	271	–	–
88	C: 128	128	128	128ᵇ,132ᵇ,317ᵇ	314,317	110,132,314

PURINES, PYRIMIDINES, NUCLEOSIDES AND NUCLEOTIDES: PHYSICAL CONSTANTS AND SPECTRAL PROPERTIES (Continued)

No.	Compound	Symbol 3-Letter	Symbol 1-Letter	Structure	Formula (Mol Wt)	Melting Point °C	$[\alpha]_D^t$	pK Basic	pK Acidic
89	$O^{2'}$-Methylcytidine / 2'-O-Methylcytidine	2'MeCyd	Cm	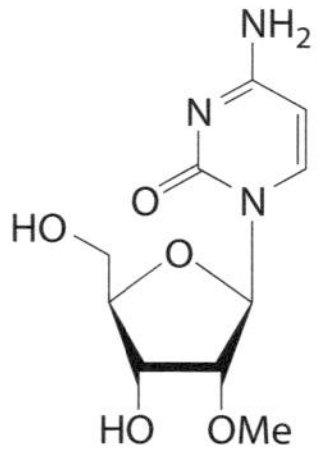 (structure)	$C_{10}H_{15}N_3O_5$ (257.24)	252–253°	$+54^{21}$ (1.1, H_2O)	4.2^p	–
90	$N^4,O^{2'}$-Dimethylcytidine	2',4Me$_2$Cyd	m^{4}Cm	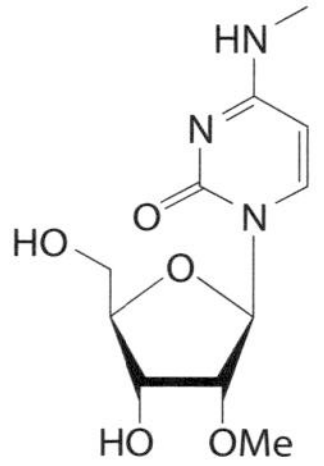 (structure)	$C_{11}H_{17}N_3O_5$ (271.27)	–	–	3.9	–
91	5-N-Methylformamido-6-ribosylamino isocytosine^t / 2-Amino-4-hydroxy-5-N-methylformamido-6-ribosylaminopyrimidine	5MeFn6(RibNH) isoCyt		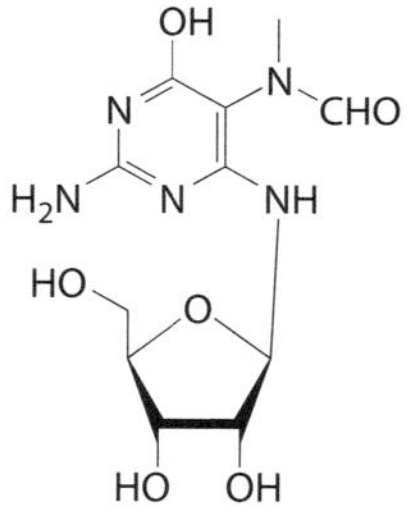 (structure)	$C_{11}H_{17}N_5O_6$ (315.29)	>180° (dec)	$+32.5^{25}$ (1, H_2O)	~0.6	9.6

Acidic Spectral Data

No.	pH	λ_{max}	ε_{max} (×10^{-3})	λ_{min}	230	240	250	270	280	290
89	1	281	12.9*	241	0.62	0.03	0.43	1.75	2.13	1.61
90	1	281	–	243	0.79	0.37	0.49	1.65	1.99	1.66
91	0.1°	270	25.1^n	239	0.49	0.44	0.54	1.41	0.85	0.14

Neutral Spectral Data

No.	pH	λ_{max}	ε_{max} (×10^{-3})	λ_{min}	230	240	250	270	280	290
89	(EtOH)	273 / 238	8.2 / –	252 / 227	0.87	0.88	0.83	1.15	0.92	0.34
90	7	270	–	250						
91	7	273	–	247	0.96	0.51	0.46	1.76	1.27	0.24

Alkaline Spectral Data

No.	pH	λ_{max}	ε_{max} (×10^{-3})	λ_{min}	230	240	250	270	280	290
89	11	272	8.9	251	1.46	0.96	0.85	1.18	0.94	0.31
90	–	–	–	–						
91	12	265	16.3	244	0.74	0.57	0.59	0.99	0.24	0.02

REFERENCES

No.	Origin and Synthesis	pK	$[\alpha]_D^t$	Spectral Data	Mass Spectra	R_f
89	C: 124,369^a	124	124	124,369*^{cd},317^b	315,317	109,110,124,369
90	R: 131	131	–	131^b	–	131
91	C: 137,41	52	137	52,137^d,13^b,41	–	41

PURINES, PYRIMIDINES, NUCLEOSIDES AND NUCLEOTIDES: PHYSICAL CONSTANTS AND SPECTRAL PROPERTIES (Continued)

No.	Compound	Symbol 3-Letter	Symbol 1-Letter	Structure	Formula (Mol Wt)	Melting Point °C	$[\alpha]_D^t$	pK Basic	pK Acidic
92	Guanosine	Guo	G	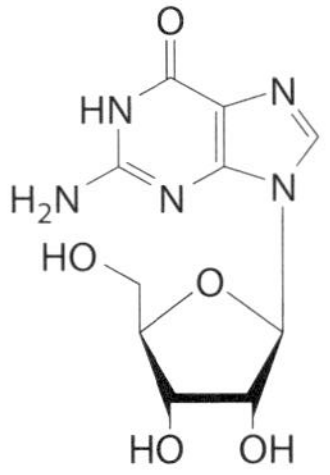	$C_{10}H_{13}N_5O_5$ (283.24)	>235° (dec)	-72^{26} (1.4, 0.1N NaOH)	1.6*	9.2*, 12.4
93	1-Methylguanosine	1MeGuo	m¹G	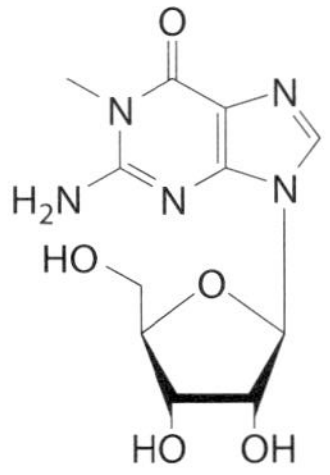	$C_{11}H_{15}N_5O_5$ (297.27)	225–227° (dec)	–	⁻2.4[l]	–
94	N^2-Methylguanosine 2-Methylamino-9-β-ribofuranosyl-purin-6-one	2MeGuo	m²G	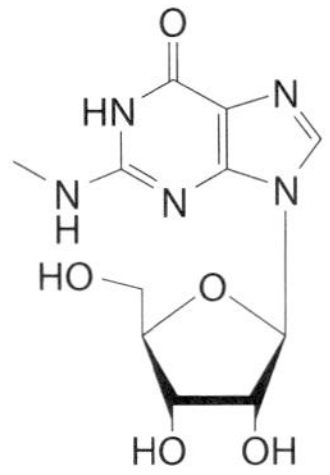	$C_{11}H_{15}N_5O_5$ (297.27)	>200° (dec)	-34.6^{26} (1.0 Me₂SO₄/EtOH)	2.3[p]	9.7[p]

Acidic Spectral Data

No.	pH	λ_{max}	ε_{max} (×10⁻³)	λ_{min}	230	240	250	270	280	290
92	1[aq]	271	22.3	246						
	0.7[q]	256	12.3	228	0.26	0.56	0.94	0.75	0.70	0.50
93	1	258	11.4*	232*	0.28	0.45	0.85	0.77	0.71	0.53
94	1	259	14.2	231	0.52	0.49	0.85	0.73	0.56	0.52

Neutral Spectral Data

No.	pH	λ_{max}	ε_{max} (×10⁻³)	λ_{min}	230	240	250	270	280	290
92	6	253	13.6	223	0.36	0.79	1.15	0.83	0.67	0.28*
93	6	256	13.1*	225	0.28	0.61	1.00	0.84	0.63	0.21
94	7	253	15.9	225	0.35	0.70	1.18	0.68	0.58	0.52

Alkaline Spectral Data

No.	pH	λ_{max}	ε_{max} (×10⁻³)	λ_{min}	230	240	250	270	280	290
92	11.3	256–266	11.3	230	–	–	0.89	–	0.61	0.13
93	13	–	–	–	0.43	0.56	0.88	0.97	0.60	0.09
94	13	256*	12.9*	231	0.39	0.61	0.99	0.83	0.63	0.22*
	13	258*	12.0*	237*	0.98	0.70	0.93	0.91	0.82	0.48

REFERENCES

No.	Origin and Synthesis	$[\alpha]_D^t$	pK
92	C: 237,231,351,372,373,111	237,351	231,170*,334*
93	E: 40 C: 24,400	–	–
94	E: 40 C: 44,372,381	44,381	40

No.	Spectral Data	Mass Spectra	R_f
92	232[b],242*[e],183[b],317[b],212,205,111	314,315,374,317	258,110,40,189,314,242
93	40[b],375[cd],317*[bce],24*,265*[b],400	314,317	24,40,110,314
94	40[b],375[cd],317*[bcde],381,44*	314,317	40,44,110,314

PURINES, PYRIMIDINES, NUCLEOSIDES AND NUCLEOTIDES: PHYSICAL CONSTANTS AND SPECTRAL PROPERTIES (Continued)

No.	Compound	Symbol (3-Letter)	Symbol (1-Letter)	Structure	Formula (Mol Wt)	Melting Point °C	$[\alpha]_D$	pK Basic	pK Acidic
95	N^2,N^2-2-Dimethylguanosine 2-Dimethylamino-9-β-ribofuranosyl-purin-6-one	$2Me_2Guo$	m_2^2G	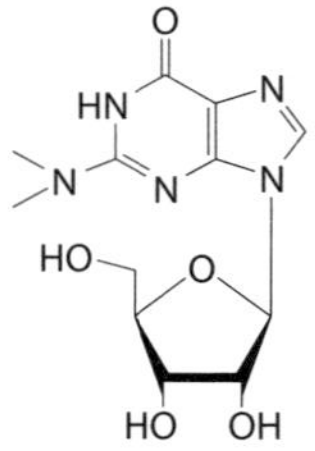	$C_{12}H_{17}N_5O_5$ (311.30)	242° (dec)	-35.6^{26} (1.1 Me_2SO_4/EtOH)	2.5[p]	9.7[p]
96	7-Methylguanosine	$7MeGuo$	m^7G	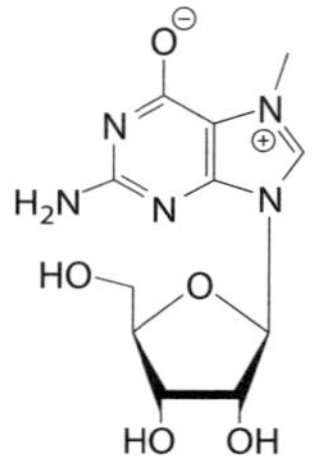	$C_{11}H_{15}N_5O_5$ (297.27)	165° (hemihydrate)	-35.5^{27} (0.4, H_2O)	r	7.0*
97	N^2,N^2-7-Trimethylguanosine 2-Dimethylamino-7-methyl-9-β-d-ribofuranosylpurin-6-or-e	$2,2,7Me_3Guo$	$M_2^{2,7}G$ $m_2^2 m^7G$	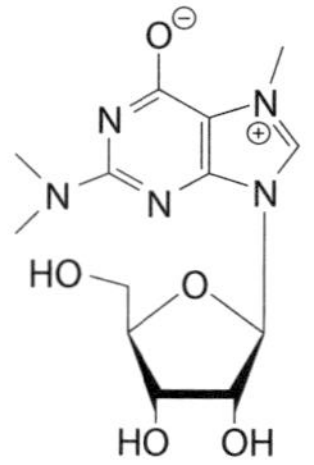	$C_{13}H_{19}N_5O_5$ (325.33)	—	—	r	—

Acidic Spectral Data

No.	pH	λ_{max}	ε_{max} (×10⁻³)	λ_{min}	230	240	250	270	280	290
95	1	265	17.7*	236*	0.45	0.32	0.63	0.92	0.56	0.48
96	3	257	10.7*	230	0.27	0.47	0.88	0.73	0.68	0.53
97	1	266 295	—	240	0.67	0.32	0.57	1.06	0.59	0.49

Neutral Spectral Data

No.	pH	λ_{max}	ε_{max} (×10⁻³)	λ_{min}	230	240	250	270	280	290
95	7	260	18.9[ms]	228	0.18	0.40	0.78	0.76	0.57	0.50
96	7[q]	258 281	8.5 7.4	238*	—	—	0.89	—	1.04	0.90
97	5	266	10.3	239	0.67	0.32	0.57	1.08	0.63	0.51

Alkaline Spectral Data

pH	λ_{max}	ε_{max} (×10⁻³)	λ_{min}	230	240	250	270	280	290
11	262	12.2	240	1.37	0.75	0.85	—	0.78	—
13	263	14.3*	242*	2.10	0.84	0.90	0.88	1.46	0.65
9°	282	8.0	242				1.16		1.30
10°	234 302	—	280	2.04	1.93	1.31	0.84	0.62	0.73

REFERENCES

No.	Origin and Synthesis	Spectral Data	pK	$[\alpha]_D$	Mass Spectra	R_f
95	E: 40 C: 44,372,381	40[p],375[cd],317*[bcde],44*[cd],265*[b],381*	40	44,381	314,317	40,44,110,314
96	C: 13,41,10	52,13[bd],41*[cd],10*[s],334	41,13,334*	10	314,317	10,41,314,334
97	C: 7,376	7[b]	—	—	376	—

PURINES, PYRIMIDINES, NUCLEOSIDES AND NUCLEOTIDES: PHYSICAL CONSTANTS AND SPECTRAL PROPERTIES (Continued)

No.	Compound	Symbol 3-Letter	Symbol 1-Letter	Structure	Formula (Mol Wt)	Melting Point °C	$[\alpha]_D^t$	pK Basic	pK Acidic
98	$O^{2'}$-Methylguanosine / 2'-O-Methyl guanosine	2'MeGuo	Gm	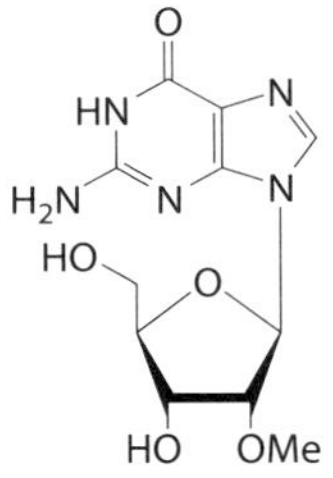	$C_{11}H_{15}N_5O_5$ (297.27)	218–220°	-38.4^{22} (0.6, H_2O)	–	–
99	Inosine	Ino	I	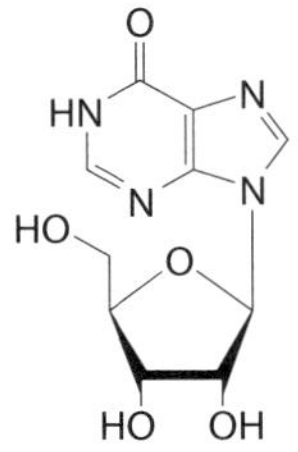	$C_{10}H_{12}N_4O_5$ (268.23)	218°	-58.8 (2.5, H_2O)	1.2	8.8,12.
100	1-Methylinosine	1MeIno	m^1I	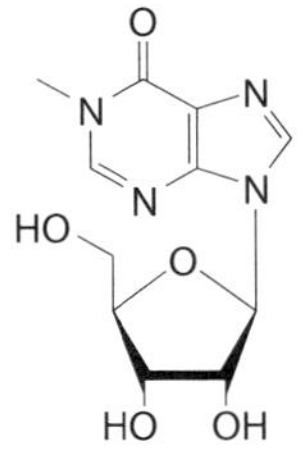	$C_{11}H_{14}N_4O_5$ (282.25)	210–212°	-49.2^{28} (0.5, H_2O)	–	–

Acidic Spectral Data

No.	pH	λ_{max}	ε_{max} (×10⁻³)	λ_{min}	230	240	250	270	280	290
98	1	256	10.7	–	–	–	–	–	0.11	0.00
99	0	251	10.9	221	–	–	1.21	–	0.25	0.03
	3	248	12.2	223	–	–	1.68	–	0.25	0.03
100	2ᵈ	250	10.4	223	0.59	1.14	1.42	0.57	0.23	0.03

Neutral Spectral Data

No.	pH	λ_{max}	ε_{max} (×10⁻³)	λ_{min}	230	240	250	270	280	290
99	6	248.5	12.3	223	–	–	1.68	–	0.25	0.03
100	6	249	10.4	223*	0.64	1.35	1.69	0.71	0.40	0.07

Alkaline Spectral Data

No.	pH	λ_{max}	ε_{max} (×10⁻³)	λ_{min}	230	240	250	270	280	290
98	11	258	9.8	224	–	–	1.05	–	0.18	0.01
99	11	253	13.1	–	–	–	1.6	0.67	0.35	0.07
100	12	249	10.7	–	0.86	1.28				0.07

REFERENCES

No.	Origin and Synthesis	pK	$[\alpha]_D^t$	Spectral Data	Mass Spectra	R_f
98	C: 108	–	108	108	315	109,110
99	R: 231ᵃ C: 377,378	170,231,51,250	255	232ᵇ,205,317ᵇ	314,317	110,314,378
100	C: 10,122	–	10	49,10*ᵈ,410*ᵉ,122	314,317	10,49,110,122,314

PURINES, PYRIMIDINES, NUCLEOSIDES AND NUCLEOTIDES: PHYSICAL CONSTANTS AND SPECTRAL PROPERTIES (Continued)

No.	Compound	Symbol 3-Letter	Symbol 1-Letter	Structure	Formula (Mol Wt)	Melting Point °C	$[\alpha]_D^t$	pK Basic	pK Acidic
101	2-Methylinosine	2MeIno	m²I	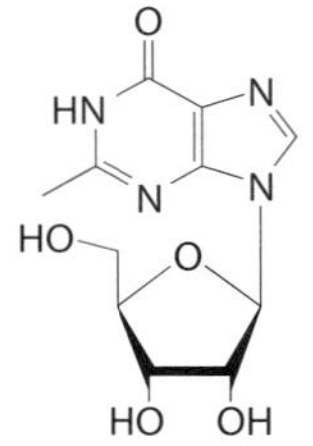 [structure]	$C_{11}H_{14}N_4O_5$ (282.25)	165–166°	-50.0^{26} (1.0, H_2O)	–	–
102	Nebularine 9-β-Ribofuranosylpurine	Neb		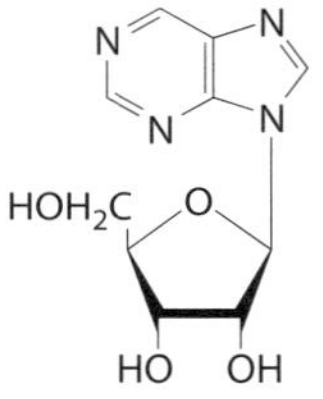 [structure]	$C_{10}H_{12}N_4O_4$ (252.33)	181–182°	-48.6^{25} (1.0, H_2O)	2.1	–
103	Uridine	Urd	U	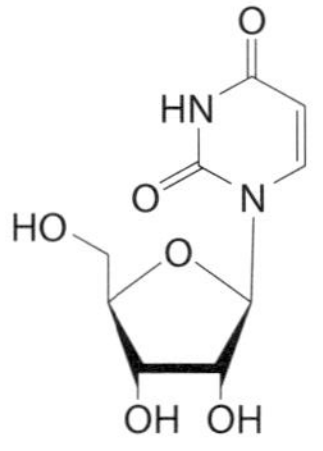 [structure]	$C_9H_{12}N_2O_6$ (244.20)	165–166°	$+9.6^{16}$ (2.0, H_2O)	–	9.2, 12.5

Acidic Spectral Data

No.	pH	λ_{max}	ε_{max} (×10⁻³)	λ_{min}	230	240	250	270	280	290
101	1	253	11.9	235	0.51	0.49	0.74	0.85	0.29	0.07
102	1	262	5.9	230	–	–	0.74	–	0.35	0.03
103	1	262	10.1	230	–	–	0.74	–	0.35	0.03

Neutral Spectral Data

No.	pH	λ_{max}	ε_{max} (×10⁻³)	λ_{min}	230	240	250	270	280	290
101	7	251.5	12.7	222	0.35	0.54	0.71	0.68	0.11	0.04
102	H_2O	262	7.1	230	–	–	0.74	–	0.35	0.03
103	7	262	10.1							

Alkaline Spectral Data

No.	pH	λ_{max}	ε_{max} (×10⁻³)	λ_{min}	230	240	250	270	280	290
101	13	258	13.1	234	0.45	0.48	0.71	0.75	0.22	0.11
102	13	262	7.1	243	–	–	0.83	–	0.29	0.02
103	12	262	7.45							

REFERENCES

No.	Origin and Synthesis	$[\alpha]_D^t$	pK	Spectral Data	Mass Spectra	R_f
101	C: 377	377	–	377	–	377
102	C: 123,363,391	123,363,391	123	123ᵇ,391	405	–
103	R: 236,181 C: 379,276ᵃ,380	236,276	163,231	163ᵇ,232*ᵇᵉ,181ᵇ,212,183ᵇ,	341,315,317	258,110,40

PURINES, PYRIMIDINES, NUCLEOSIDES AND NUCLEOTIDES: PHYSICAL CONSTANTS AND SPECTRAL PROPERTIES (Continued)

No.	Compound	Symbol 3-Letter	Symbol 1-Letter	Structure	Formula (Mol Wt)	Melting Point °C	$[\alpha]_D^t$	pK Basic	pK Acidic
104	2-Thiouridine	2Srd 2SUrd	s^2U	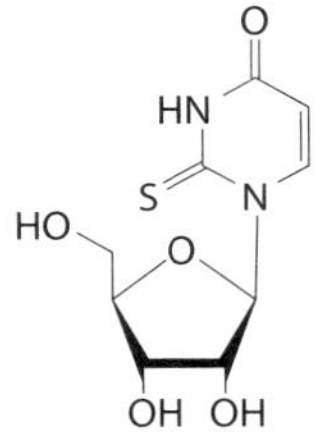	$C_9H_{12}N_2O_5S$ (260.26)	214°	$+39^{20}$ (1.2, H_2O)	–	8.8
105	2,4-Dithiouridine	2,4Srd 2,4S$_2$ Urd	$s^{2,4}U$ s^2s^4U	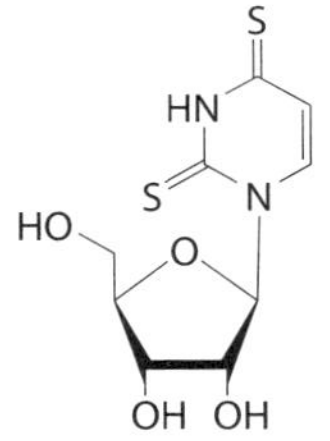	$C_9H_{12}N_2O_4S_2$ (276.33)	166–167°	–	–	7.4
106	5-Carboxymethyl-2-thiouridine 2-Thiouridine-5-acetic acid	5Cm2SUrd 5Cm2Srd 5(CxMe)2SUrd	cm^5s^2U cm^5S	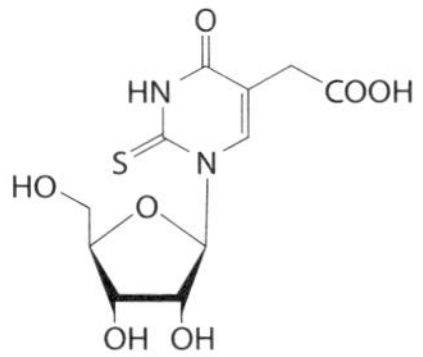	$C_{11}H_{14}N_2O_7S$ (318.31)	–	–	–	–

No.	Acidic Spectral Data pH	λ_{max}	ε_{max} (×10⁻³)	λ_{min}	230	240	250	270	280	290	Neutral Spectral Data pH	λ_{max}	ε_{max} (×10⁻³)	λ_{min}	230	240	250	270	280	290
104	2	279	16.4	247	1.32	0.63	0.53	1.66	1.86	1.66	7	218* 275*	16.2 13.6*	247	1.16	0.70	0.62	1.47	1.57	1.40
105	–	–	–	–	–	–	–	–	–	–	5.8	283	22.5	–	–	–	–	–	–	–
106	1	274.5									7	277.5								

No.	Alkaline Spectral Data pH	λ_{max}	ε_{max} (×10⁻³)	λ_{min}	230	240	250	270	280	290
104	14 9* 12	264.5 241 239 271	7.5 21.8 21.0 13.4	243 261	1.07	1.43	1.22	1.02	0.88	0.59
105	9	280 320	16.9 24.8	–	–	–	–	–	–	–
106										

REFERENCES

No.	Origin and Synthesis	$[\alpha]_D^t$	pK	Spectral Data	Mass Spectra	R_f
104	C: 276,285,301 E: 277	276,301	285	276*ᵇ,277ᵇ,301*ᶜᵈ,306ᵇ,305ᵇ	–	276,306,307,305
105	C: 271	–	271	271	–	271
106	R: 303	–	–	303	–	303

PURINES, PYRIMIDINES, NUCLEOSIDES AND NUCLEOTIDES: PHYSICAL CONSTANTS AND SPECTRAL PROPERTIES (Continued)

No.	Compound	Symbol (3-Letter)	Symbol (1-Letter)	Structure	Formula (Mol Wt)	Melting Point °C	$[\alpha]_D^t$	pK Basic	pK Acidic
107	5-Carbamoylmethyl-2-thiouridine / 2-Thiouridine-5-acetamide	5Ncm2Srd / 5Ncm2SUrd / 5(NcMe)2Srd	ncm5s2U / ncm5S	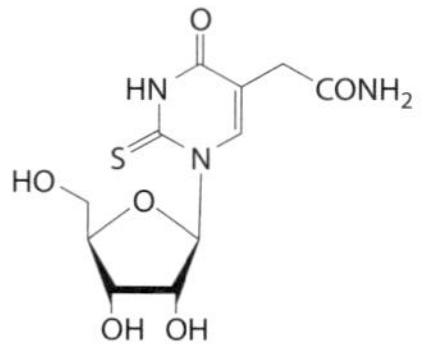 (structure, CONH₂)	$C_{11}H_{15}N_3O_6S$ (317.32)	217–218°	–	–	–
108	5-(Methoxycarbonylmethyl)-2-thiouridine / 5-Carboxymethyl-2-thiouridine methyl ester	5(MeCm)2SUrd / 5(MeCm)2Srd	mcm5s2U / mcm5S	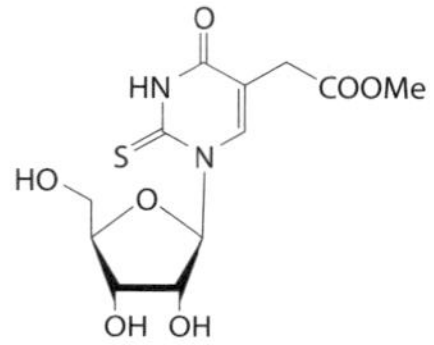 (structure, COOMe)	$C_{12}H_{16}N_2O_7S$ (332.33)	199°	$+19.8^{20}$ (0.5, H_2O)	–	–
109	5-Methoxy-2-thiouridine	5MeO2SUrd / 5MeO2Srd	mo5S / mo5s2U	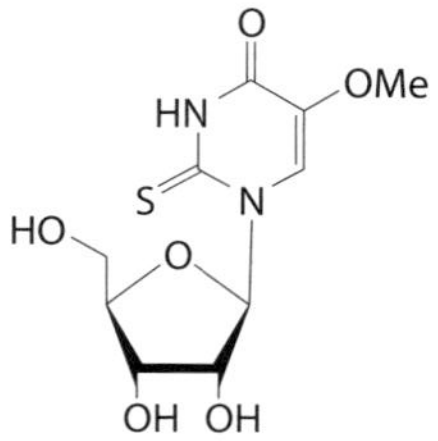 (structure, OMe)	$C_{10}H_{14}N_2O_6S$ (290.30)	221–222°	$+18.2^{24}$ (0.5, H_2O)	–	–

Acidic Spectral Data

No.	pH	λ_{max}	ε_{max} (×10⁻³)	λ_{min}	230	240	250	270	280	290
107										
108	1	277*	15.6	244	1.22	0.73	0.68	1.49	1.54	1.41
109										

Neutral Spectral Data

No.	pH	λ_{max}	ε_{max} (×10⁻³)	λ_{min}	230	240	250	270	280	290
107	(MeOH)	221 / 276	12.9 / 13.7							
108	7	220 / 277*	15.3 / 15.8	244	1.18	0.84	0.76	1.42	1.45	1.24
109	(MeOH)	227 / 285	10.0 / 12.6							

Alkaline Spectral Data

No.	pH	λ_{max}	ε_{max} (×10⁻³)	λ_{min}	230	240	250	270	280	290
107	13 (MeOH)	242 / 272	18.2 / 15.6							
108	12	242 / 271	22.4 / 15.8	261	1.06	1.37	1.16	1.02	0.86	0.46
109	13 (MeOH)	248 / 272	20.0 / 12.6							

REFERENCES

No.	Origin and Synthesis	$[\alpha]_D^t$	pK	Spectral Data	Mass Spectra	R_f
107	C: 301	–	–	301	–	301
108	C: 5,310 R: 275,306	301	–	5,301^cd,317^bc,275^*b	317.5,301,275	275,301,303,306,307
109	C: 301	301	–	301	–	–

PURINES, PYRIMIDINES, NUCLEOSIDES AND NUCLEOTIDES: PHYSICAL CONSTANTS AND SPECTRAL PROPERTIES (Continued)

No.	Compound	Symbol 3-Letter	Symbol 1-Letter	Structure	Formula (Mol Wt)	Melting Point °C	$[\alpha]_D^t$	pK Basic	pK Acidic
110	5-Methyl-2-thiouridine / 2-Thio-1-ribosylthymine	5Me2Surd / 5MeSrd	m5s2U / s2T	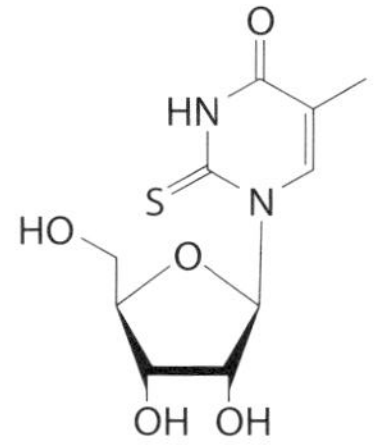	$C_{10}H_{14}N_2O_5S$ (274.30)	217°	$+31^{28}$ (1.23, H₂O)	–	–
111	5-(Methylaminomethyl)-2-thiouridine / 5-(N-Methylaminomethyl)-2-thiouridine	5(MeNHMe)Srd / 5(MeNHMe)2Surd	mnm5s2U / mnm5S	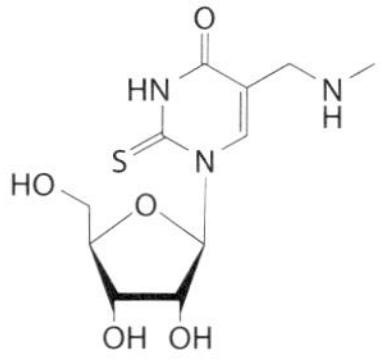	$C_{11}H_{17}N_3O_5S$ (303.33)	–		–	–
112	3-Methyluridine	3MeUrd	m3U	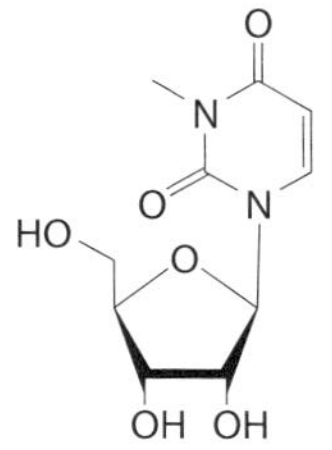	$C_{10}H_{14}N_2O_6$ (258.23)	119–120°	$+20.1^{26}$ (H₂O)	–	–

Acidic Spectral Data

No.	pH	λ_{max}	ε_{max} (×10⁻³)	λ_{min}	230	240	250	270	280	290
110	2	218 273	17.4 14.8	243	1.05	0.53	0.58	1.41	1.43	1.22
111	1	220 273	–	242	0.84	0.56	0.66	1.31	1.29	1.20
112	2ʸ	262*	9.5	232	0.31	0.39	0.74	0.90	0.35*	0.04*

Neutral Spectral Data

No.	pH	λ_{max}	ε_{max} (×10⁻³)	λ_{min}	230	240	250	270	280	290
110	7	219 272	16.2 14.1	247	1.06	0.77	0.69	1.40	1.39	1.15
111	7	220 273	–	242	0.84	0.56	0.66	1.33	1.31	1.22
112	7	262	9.5*	232	0.31	0.39	0.74	0.90	0.35*	0.04*

Alkaline Spectral Data

No.	pH	λ_{max}	ε_{max} (×10⁻³)	λ_{min}	230	240	250	270	280	290
110	9*	239	21.0	259	1.35	1.64	1.20	1.10	1.04	0.60
111	13	243	–	227	1.00	1.33	1.21	1.01	0.87	0.49
112	12	263*	9.4	232	0.32	0.38	0.70	0.91	0.46*	0.10

No.	Origin and Synthesis	$[\alpha]_D^t$	pK
110	C: 276,307 R: 307	276	–
111	R: 270,305	–	–
112	C: 139,125,140–142,81	139	–

REFERENCES

No.	Spectral Data	Mass Spectra	R_f
110	C: 276,307b	–	276,307,306
111	270,317b,305b	270,317	270,305
112	317b,142*,13*,81*,139,126*	314,317	13,57,81,110,125,126,142,143,173,314

PURINES, PYRIMIDINES, NUCLEOSIDES AND NUCLEOTIDES: PHYSICAL CONSTANTS AND SPECTRAL PROPERTIES (Continued)

No.	Compound	Symbol (3-Letter)	Symbol (1-Letter)	Structure	Formula (Mol Wt)	Melting Point °C	$[\alpha]_D$	pK Basic	pK Acidic
113	3-(3-Amino-3-carboxypropyl)uridine; Uricine-3-(α-aminobutyric acid)	$3NH_2$ BtoUrd; $3(NH_2\,CxPr)$Urd	nbt^3U	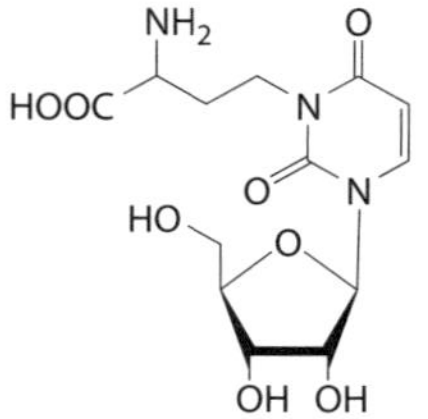 [structure]	$C_{13}H_{19}N_3O_8$ (345.31)	161–163° (2HCl salt)	–	–	–
114	4-Thiouridine	4Srd; 4SUrd	s^4U	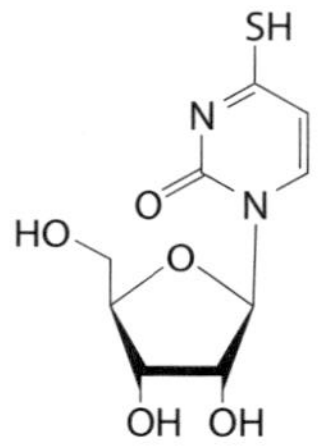 [structure]	$C_9H_{12}N_2O_5S$ (260.26)	135–138° (dec)	–	–	8.2
115	4-Thiouridine disulfide; *bis*(4-4′-Dithiouridine)	$(4SUrd)_2$		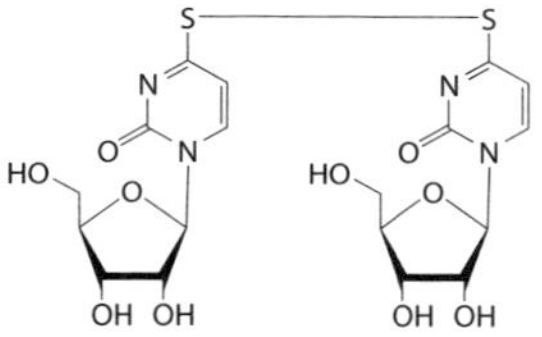 [structure]	$C_{18}H_{22}N_4O_{10}S_2$ (518.51)	188–190°	–	–	–

Acidic Spectral Data

No.	pH	λ_{max}	ε_{max} (×10⁻³)	λ_{min}	230	240	250	270	280	290
113	2	263	8.5	233	0.41	0.45	0.76	0.89	0.43	0.05
114	2	245; 331	5.2*; 17.0*	275*	–	–	–	–	–	–
115	–	–	–	–	–	–	–	–	–	–

Neutral Spectral Data

No.	pH	λ_{max}	ε_{max} (×10⁻³)	λ_{min}	230	240	250	270	280	290
113	6	263	8.5	233	0.41	0.45	0.76	0.89	0.43	0.05
114	6.5	245; 331	4.0*; 21.2*	225; 274	–	–	–	–	–	–
115	7	261; 309	–	236; 278	–	–	–	–	–	–

Alkaline Spectral Data

No.	pH	λ_{max}	ε_{max} (×10⁻³)	λ_{min}	230	240	250	270	280	290
113	12	263	8.5	234	0.46	0.52	0.77	0.89	0.44	0.08
114	12	316	19.7*	268*	–	–	–	–	–	–
115	–	–	–	–	–	–	–	–	–	–

REFERENCES

No.	Origin and Synthesis	pK	$[\alpha]_D$	Spectral Data	Mass Spectra	R_f
113	C: 300 R: 300,345	–	–	300[b]	300	300,345
114	C: 344,128 R: 144	144	–	344*,144[bcd],317*[b],128	–	144,344
115	C: 128	–	–	128,208[b]	–	144

PURINES, PYRIMIDINES, NUCLEOSIDES AND NUCLEOTIDES: PHYSICAL CONSTANTS AND SPECTRAL PROPERTIES (Continued)

No.	Compound	Symbol: 3-Letter	Symbol: 1-Letter	Structure	Formula (Mol Wt)	Melting Point °C	$[\alpha]^t_D$	pK: Basic	pK: Acidic
116	5-Carboxymethyluridine Uridine-5-acetic acid	5CmUrd 5(CxMe)Urd	cm⁵U	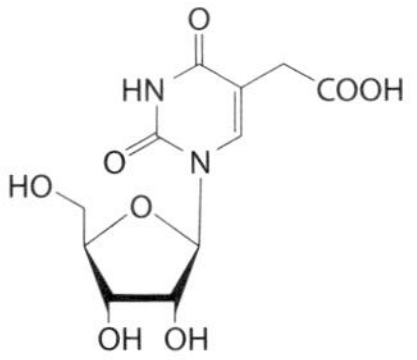	$C_{11}H_{14}N_2O_8$ (302.24)	242–244°	−24.3[25] (1*N* NaOH)	–	4.2, 9.8
117	5-Carbamoylmethyluridine Uridine-5-acetamide	5NcmUrd 5(NcMe)Urd	ncm⁵U	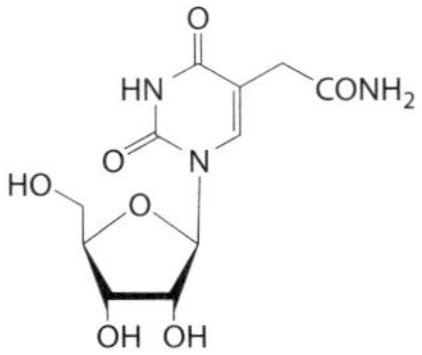	$C_{11}H_{15}N_3O_7$ (301.26)	227–230°	–	–	–
118	5-(Methoxycarbonylmethyl)uridine 5-Carboxymethyluridine methyl ester Uridine-5-acetic acid methyl ester	5MeCmUrd 5(MeCxMe)Urd	mcm⁵U	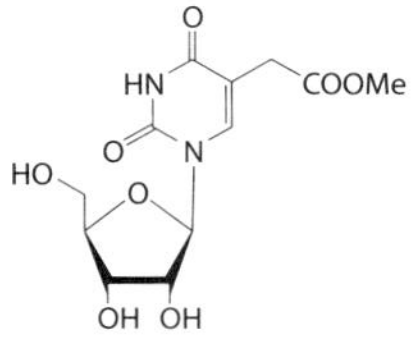	$C_{12}H_{16}N_2O_8$ (316.27)	163–165°	–	–	–
119	5-Hydroxyuridine	5HOUrd	o⁵U	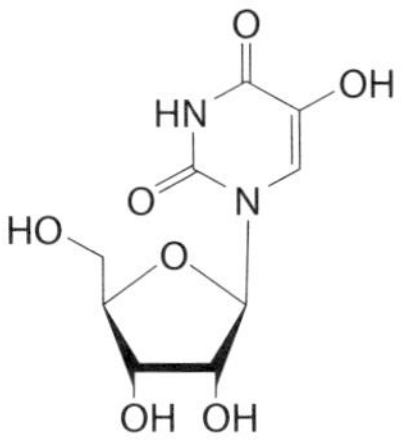	$C_9H_{12}N_2O_7$ (260.20)	242–245°	–	–	7.8

No.	Acidic Spectral Data										Neutral Spectral Data										Alkaline Spectral Data									
	pH	λ_{max}	ε_{max} (×10⁻³)	λ_{min}	Spectral Ratios 230	240	250	270	280	290	pH	λ_{max}	ε_{max} (×10⁻³)	λ_{min}	Spectral Ratios 230	240	250	270	280	290	pH	λ_{max}	ε_{max} (×10⁻³)	λ_{min}	Spectral Ratios 230	240	250	270	280	290
116	1	265	9.7	234*	0.39	0.40	0.69	1.02	0.64	0.19	7	266.5	–	236	0.50	0.41	0.68	1.07	0.75	0.27	13	266.5*	7.1	245	1.29	0.84	0.80	1.05	0.71	0.25
117	1	265	10.0	232*	–	–	–	–	–	–	H_2O	266	–	233	0.24	0.34	0.67	1.05	0.66	0.17	13	267*	7.0	244	1.10	0.78	0.79	0.98	0.54	0.11
118	1	265	5.2	232																										
119	2	280	–	245	0.92	0.50	0.56	1.52	1.76	1.46	7[q]	280*	8.2	–	–	–	–	–	–	–	12	306	–	267	–	1.60	1.29	0.70	0.89	1.16

REFERENCES

No.	Origin and Synthesis	$[\alpha]^t_D$	pK	Spectral Data	Mass Spectra	R_f
116	R: 278 C: 302,322[a]	322	302	278[b],302[*d],317[b]	–	278,302,303,304
117	C: 302 R: 398	–	–	302*,398[*ce]	–	302
118	C: 302,304 R: 304	–	–	302,414[b]	414	302,303,304
119	C: 145-147 R: 148	–	382	148[b],145[cd],146*	317	148

PURINES, PYRIMIDINES, NUCLEOSIDES AND NUCLEOTIDES: PHYSICAL CONSTANTS AND SPECTRAL PROPERTIES (Continued)

No.	Compound	Symbol 3-Letter	Symbol 1-Letter	Structure	Formula (Mol Wt)	Melting Point °C	$[\alpha]_D^t$	pK Basic	pK Acidic
120	5-Carboxymethoxyuridine / Uridine-5-oxyacetic acid	5CmOUrd / 5(CxMeO)Urd	cmo5U	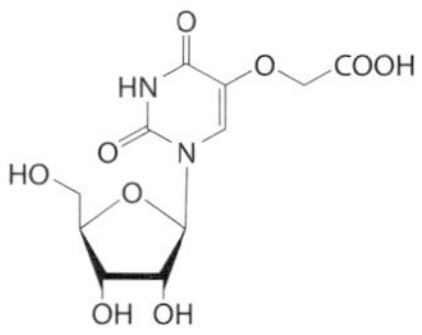	$C_{11}H_{14}N_2O_9$ (318.24)	–	–	–	2.9
121	5-Methyluridine / 1-β-Ribofuranosylthymine / Ribosylthymine	5MeUrd / r'Thd* / Thd	m5U / T	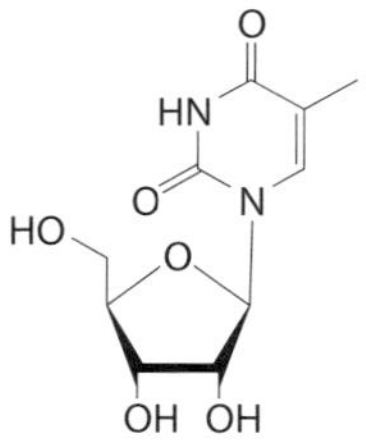	$C_{10}H_{14}N_2O_6$ (258.23)	183–185°	-10^{31} (2.0, H_2O)	–	9.7
122	5-Hydroxymethyluridine	5HmUrd / 5HOMeUrd	om5U / hm5U*	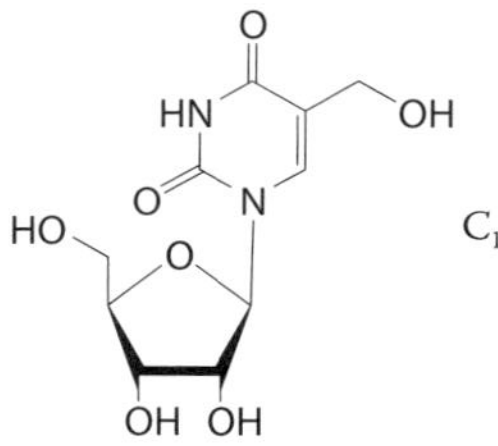	$C_{10}H_{14}N_2O_7$ (274.23)	167–168°	–	–	–

Acidic Spectral Data

No.	pH	λ_{max}	ε_{max} (×10⁻³)	λ_{min}	230	240	250	270	280	290
120	2ᵃ	277	8.4	243	0.44	0.39	0.67	1.07	0.74	0.27
121	1	267	9.9*	235*			0.70		0.52	
122	2	264	9.5	233						

Neutral Spectral Data

No.	pH	λ_{max}	ε_{max} (×10⁻³)	λ_{min}	230	240	250	270	280	290
120	7.5	280	7.6	247	0.44	0.39	0.67	1.07	0.74	0.27
121	7	267	9.8	236					0.53	
122	7	263	–	–						

Alkaline Spectral Data

No.	pH	λ_{max}	ε_{max} (×10⁻³)	λ_{min}	230	240	250	270	280	290
120	13	278	6.7	252	1.31	0.91	0.83	1.08	0.75	0.31*
121	13	268	7.5	246						
122	12	263	7.0	243			0.79		0.45	

REFERENCES

No.	Origin and Synthesis	$[\alpha]_D^t$	pK	Spectral Data	Mass Spectra	R_f
120	C: 308 R: 308,309	–	308	308,309[b],313[b]	308	308,309
121	E: 18,–49 C: 150,69,380,386,379	150,322,379	150	18[b],150[bd],149[bcd],317[be],69[*],265[*b],379	314,317	18,69,110,142,149,314,304
122	C: 69,151	–	–	69,94[ce]	–	69

PURINES, PYRIMIDINES, NUCLEOSIDES AND NUCLEOTIDES: PHYSICAL CONSTANTS AND SPECTRAL PROPERTIES (Continued)

No.	Compound	Symbol 3-Letter	Symbol 1-Letter	Structure	Formula (Mol Wt)	Melting Point °C	$[\alpha]_D$	pK Basic	pK Acidic
123	Dihydrouridine 5,6-Dihydrouridine	H₂Urd	hU D	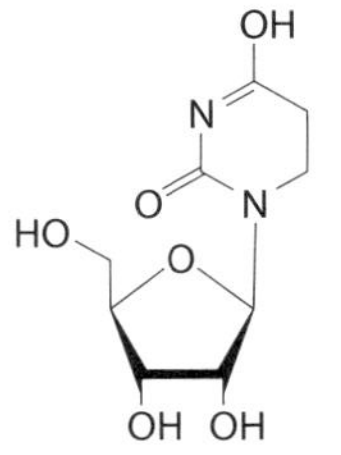	$C_9H_{14}N_2O_6$ (246.22)	106–108°	-36.8^{20} (2.1, H₂O)	–	–
124	5-Methyl-5,6-dihydrouridine 5,6-Dihydroribosylthymine	5MeH₂ Urd	m⁵D m⁵hU	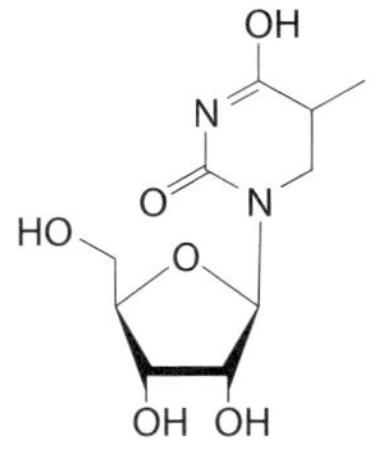	$C_{10}H_{16}N_2O_6$ (260.25)	–	–	–	–
125	O2′-Methyluridine 2′-O-Methyluridine	2′MeUrd	Um	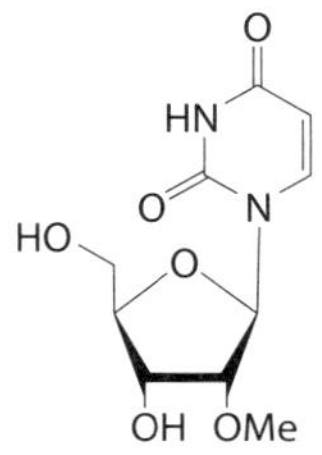	$C_{10}H_{14}N_2O_6$ (258.23)	159°	$+41^{20}$ (1.6, H₂O)	–	$^{-}9.3^p$

Acidic Spectral Data

No.	pH	λ_{max}	ε_{max} (×10⁻³)	λ_{min}	230	240	250	270	280	290
123	–	–	–	–	–	–	–	–	–	–
124										
125	2	263*	10.0	231	0.23	0.39	0.75	0.86	0.38	0.04

Neutral Spectral Data

No.	pH	λ_{max}	ε_{max} (×10⁻³)	λ_{min}	230	240	250	270	280	290
123	H₂O	208	6.6	–	–	–	–	–	–	–
124										
125	7	263	10.1	231	0.23	0.39	0.75	0.86	0.38	0.04

Alkaline Spectral Data

No.	pH	λ_{max}	ε_{max} (×10⁻³)	λ_{min}	230	240	250	270	280	290
123	13°	235*	10.1							
124										
125	12	262	7.4	243	0.98	0.76	0.83	0.82	0.3	0.03

REFERENCES

No.	Origin and Synthesis	$[\alpha]_D$	pK	Mass Spectra	Spectral Data	R_f
123	C: 266,384,99,147	384,412	–	314,317	266,283[cd],375*,265[b]	314,383,333
124	C: 333 R: 333	–	–	–	–	333
125	C: 124,369	124	124,138	315,317	124*,317[bcde],369	109,110,124,369

PURINES, PYRIMIDINES, NUCLEOSIDES AND NUCLEOTIDES: PHYSICAL CONSTANTS AND SPECTRAL PROPERTIES (Continued)

No.	Compound	Symbol (3-Letter)	Symbol (1-Letter)	Formula (Mol Wt)	Melting Point °C	$[\alpha]_D^t$	pK Basic	pK Acidic
126	5,O²'-Dimethyluridine / O²'-Methylribothymidine	2′,5Me₂ Urd	m⁵Um / Tm	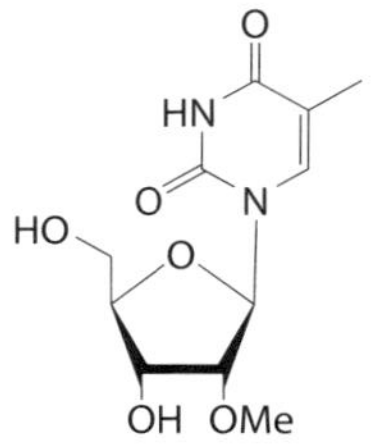 $C_{11}H_{16}N_2O_6$ (272.26)	–	–	–	–
127	Orotidine / Uridine-6-carboxylic acid / 6-Carboxyuridine	Ord / 6CxUrd	O	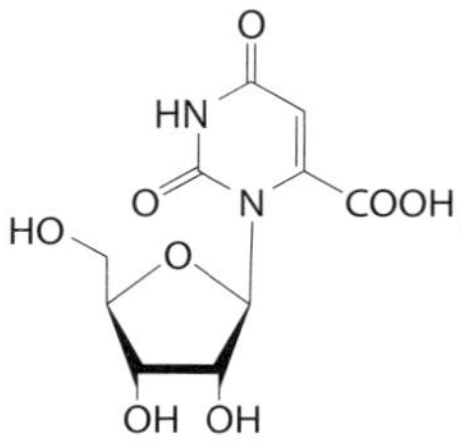 $C_{10}H_{12}N_2O_8$ (288.21)	183–184° (CHA salt[y])	–	–	–
128	Spongouridine / 1-β-D-Arabinofuranosyluracil	AraUrd / aUrd	aU	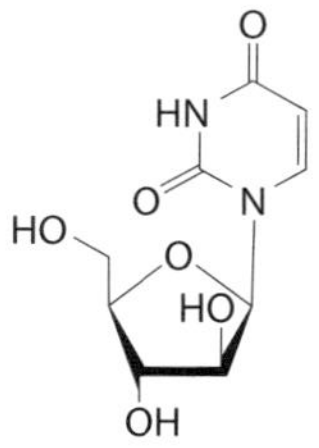 $C_9H_{12}N_2O_6$ (244.20)	222–224°	+131[20] (0.63, H₂O)	–	9.3

Acidic Spectral Data

No.	pH	λ_{max}	ε_{max} (×10⁻³)	λ_{min}	230	240	250	270	280	290
126	1	268	–	237	0.52	0.37	0.66	1.10	0.76	0.26
127	1	267	9.8	234	0.39	0.41	0.66	1.12	0.81	0.37
128	1	264	9.3	232	–	–	–	–	–	–

Neutral Spectral Data

No.	pH	λ_{max}	ε_{max} (×10⁻³)	λ_{min}	230	240	250	270	280	290
126	7	267	–	237	0.51	0.36	0.64	1.09	0.74	0.26
127	–	–	–	–	–	–	–	–	–	–
128	H₂O	263	10.5*	231	–	–	–	–	–	–

Alkaline Spectral Data

No.	pH	λ_{max}	ε_{max} (×10⁻³)	λ_{min}	230	240	250	270	280	290
126	13	267	–	247	–	0.88	0.78	1.05	0.67	0.15
127	13	266	7.8	245	1.07	0.83	0.83	1.04	0.71	0.29
128	11.5	263	7.2	242	–	–	–	–	–	–
128	14	265	7.9	241	–	–	–	–	–	–

REFERENCES

No.	Origin and Synthesis	$[\alpha]_D^t$	pK	Spectral Data	Mass Spectra	R_f
126	C: 325 R: 323	–	–	323	–	323
127	N: 160	–	–	160[b]	–	–
128	N: 393 C: 394,392	392	392	393*[b],392[cd]	–	392,393

PURINES, PYRIMIDINES, NUCLEOSIDES AND NUCLEOTIDES: PHYSICAL CONSTANTS AND SPECTRAL PROPERTIES (Continued)

No.	Compound	Symbol 3-Letter	Symbol 1-Letter	Structure	Formula (Mol Wt)	Melting Point °C	$[\alpha]^t_b$	pK Basic	pK Acidic
129	Spongothymidine; 1-β-D-Arabinofuranosylthymine	AraThd; aThd; 5MeaUrd	aT; m⁵aU	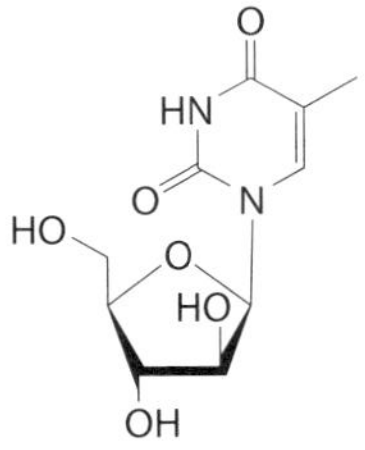	$C_{10}H_{14}N_2O_6$ (258.23)	238–242°	93^{24}_{589} (0.5, H₂O);	–	9.8
130	Pseudouridine β; f-Pseudouridine; Pseudouridine C; 5-β-D-Ribofuranosyluracil; 5-Ribosyluracil	Ψrd; βfΨrd; ΨrdC*	Ψ	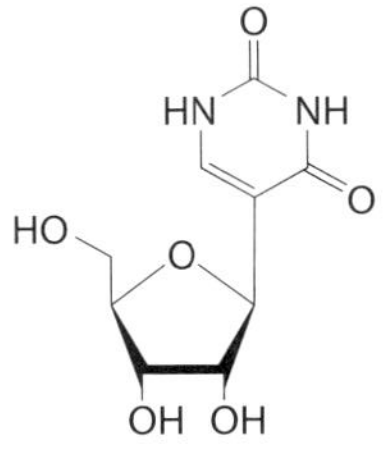	$C_9H_{12}N_2O_6$ (244.20)	223–224°	–3.0 (1.0, H₂O)	–	9.0*, >13

Acidic Spectral Data

No.	pH	λ_{max}	ε_{max} (×10⁻³)	λ_{min}	230	240	250	270	280	290
129	1	268	10.0	236	0.36	0.31	0.61	1.15	0.85	0.36
130	2ᶜ	262	7.9*	233	–	–	0.74	–	0.42*	0.06*

Neutral Spectral Data

No.	pH	λ_{max}	ε_{max} (×10⁻³)	λ_{min}	230	240	250	270	280	290
129	7	268	10.0	236	0.36	0.31	0.61	1.15	0.85	0.36
130	7	263	8.1*	233	0.33	0.42	0.74	0.90	0.44	0.08

Alkaline Spectral Data

No.	pH	λ_{max}	ε_{max} (×10⁻³)	λ_{min}	230	240	250	270	280	290
129	12	269	7.9	245	1.17	0.70	0.69	1.15	0.77	0.22
130	12	286	7.7*	245	2.06	0.73	0.62	1.51	2.06	2.16
130	14ⁿ	279	5.7ⁿ	248	2.31	1.11	0.61	1.67	2.09	1.51

REFERENCES

No.	Origin and Synthesis	pK	$[\alpha]^t_D$	Spectral Data	Mass Spectra	R_f
129	N: 161 C: 162	163	162	163ᵇ,161ᵇ	–	393
130	N: 153 C: 154,387 R: 388	155,156,94*,95*	95	155,94*bcde,154cd,95*bce,156b,157,265*b,317b	314,317	110,157–159,314

PURINES, PYRIMIDINES, NUCLEOSIDES AND NUCLEOTIDES: PHYSICAL CONSTANTS AND SPECTRAL PROPERTIES (Continued)

No.	Compound	Symbol (3-Letter)	Symbol (1-Letter)	Structure	Formula (Mol Wt)	Melting Point °C	$[\alpha]_D^t$	pK Basic	pK Acidic
131	α-f-Pseudouridine; Pseudouridine B5-α-D-Ribofuranosyluracil	αfΨrd; Ψrd B*	Ψ_B^*	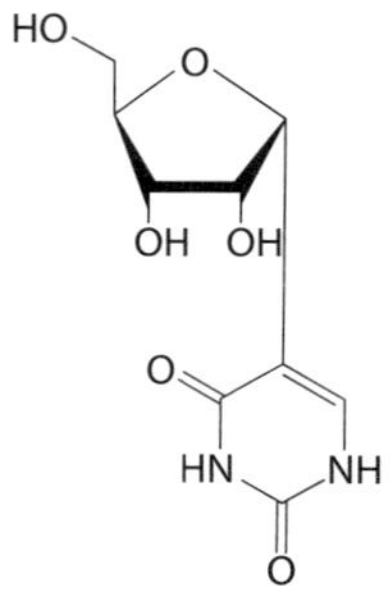 (structure)	$C_9H_{12}N_2O_6$ (244.20)	–	–	–	9.2*, >13
132	β-p-Pseudouridine; Pseudouridine A$_s$;5-β-D-Ribopyranosyluracil	βpΨrd; Ψrd A$_s^*$	Ψ_{AS}^*	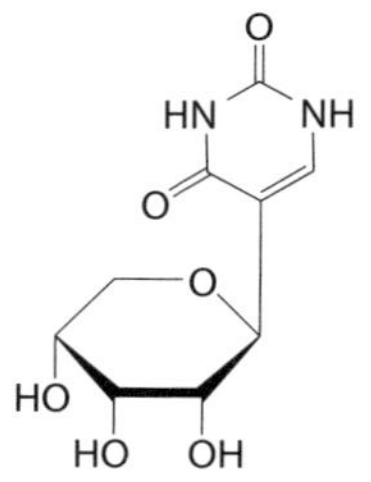 (structure)	$C_9H_{12}N_2O_6$ (244.20)	–	–	–	9.6, >13

No.	Acidic pH	Acidic λ_{max}	Acidic ε_{max} (×10⁻³)	Acidic λ_{min}	230	240	250	270	280	290	Neutral pH	Neutral λ_{max}	Neutral ε_{max} (×10⁻³)	Neutral λ_{min}	230	240	250	270	280	290
131	〃	–	–	–	–	–	–	–	–	–	7	264	–	234	0.33	0.38	0.70	0.95	0.51	0.09
132	〃	–	–	–	–	–	–	–	–	–	7	262	8.3	231	0.25	0.41	0.75	0.83	0.34	0.04

Alkaline Spectral Data

No.	pH	λ_{max}	ε_{max} (×10⁻³)	λ_{min}	230	240	250	270	280	290
131	12	288	–	245	1.88	0.76	0.70	1.26	1.46	1.56
131	14ª	279	–	248	2.20	1.02	0.61	1.56	1.81	1.37
132	12	286	9.2	244	2.10	0.60	0.56	1.69	2.50	2.66
132	14ª	281	7.5ª	247	2.24	0.81	0.54	1.73	2.28	1.93

REFERENCES

No.	Origin and Synthesis	$[\alpha]_D^t$	pK	Spectral Data	Mass Spectra	R_f
131	C: 154 R: 94,388	–	156,94*	155,94[bce],156[b]	–	158
132	R: 94,388 C: 154	–	94	155,94[bcde]	–	97

PURINES, PYRIMIDINES, NUCLEOSIDES AND NUCLEOTIDES: PHYSICAL CONSTANTS AND SPECTRAL PROPERTIES (Continued)

No.	Compound	Symbol 3-Letter	Symbol 1-Letter	Structure	Formula (Mol Wt)	Melting Point °C	$[\alpha]_D^t$	pK Basic	pK Acidic
133	α-p-Pseudouridine; Pseudouridine A_F; 5-α-D-Ribopyranosyluracil	αpΨrd; Ψrd A_F*	Ψ_{AF}*	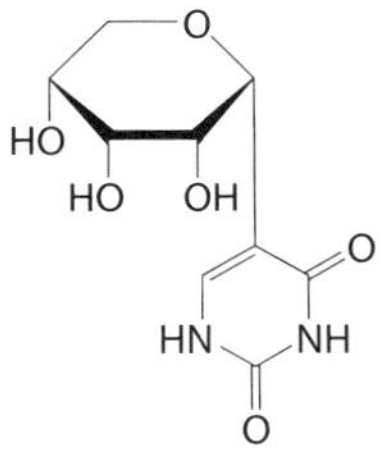	$C_9H_{12}N_2O_6$ (244.20)	—	—	—	9.6, >13
134	1-Methylpseudouridine; 1-Methyl-5-ribosyluracil	1MeΨrd	$m^1\psi$	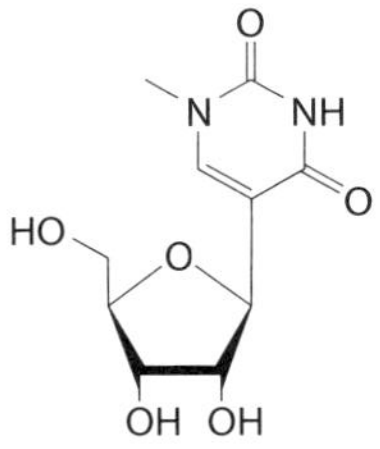	$C_{10}H_{14}N_2O_6$ (258.23)	—	—	—	—
135	O²'-Methylpseudouridine; 2'-O-Methylpseudouridine	2'MeΨrd	ψm	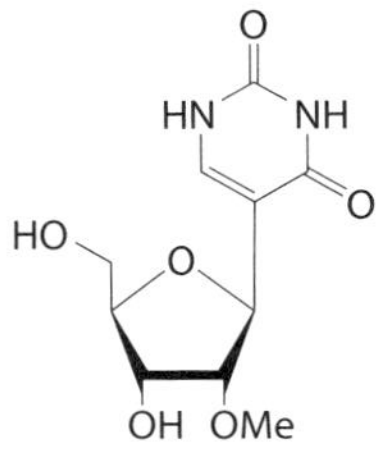	$C_{10}H_{14}N_2O_6$ (258.23)	—	—	—	—

No.		Acidic Spectral Data											Neutral Spectral Data									
	pH	λ_{max}	ε_{max} (×10⁻³)	λ_{min}	230	240	250	270	280	290	pH	λ_{max}	ε_{max} (×10⁻³)	λ_{min}	230	240	250	270	280	290
133	v	—	—	—	—	—	—	—	—	—	7	263	—	233	0.27	0.37	0.71	0.90	0.42*	0.05
134	2	—	—	—	—	—	0.54	—	1.03*	0.38	7	265	—	—	—	—	—	—	0.66	—
135	1	261	—	—	—	—	—	—	—	—	7	261	—	—	—	—	—	—	—	—

	Alkaline Spectral Data									
pH	λ_{max}	ε_{max} (×10⁻³)	λ_{min}	230	240	250	270	280	290
12	287	—	244*	1.59	0.62	0.67	1.26	1.49	1.58
14⁴	278	—	248	2.09	1.04	0.65	1.52	1.75	1.32
12	269	—	246	—	—	0.62	—	0.69*	0.08
14	272	—	—	—	—	—	—	—	—
13	281	—	—	—	—	—	—	1.05	—

REFERENCES

No.	Origin and Synthesis	$[\alpha]_D^t$	pK	Mass Spectra	Spectral Data	R_f
133	R: 94,388 C: 154	—	94	—	155,94^bce	97
134	C: 94,81	—	—	—	94*b,81^be,345^c	81,345
135	R: 118,415	—	—	—	118^b	118,415

PURINES, PYRIMIDINES, NUCLEOSIDES AND NUCLEOTIDES: PHYSICAL CONSTANTS AND SPECTRAL PROPERTIES (Continued)

No.	Compound	Symbol 3-Letter	Symbol 1-Letter	Structure	Formula (Mol Wt)	Melting Point °C	$[\alpha]_D^t$	pK Basic	pK Acidic
136	Wyosine[i] (Formerly "Yt") 1-N^2-Isopropeno-3-methylguanosine; 4,9-Dihydro-4,6-dimethyl-9-oxo-3-β-D-ribofuranosyl 1H-imidazo[1,2-a]purine	Wyo imGuo*	W	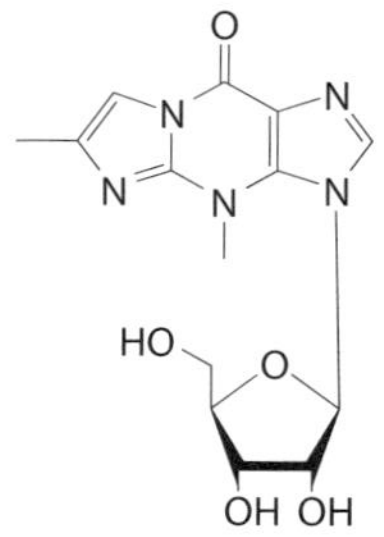	$C_{14}H_{17}N_5O_5$ (335.32)	–	–	–	–
137	Wybutosine[l] (Formerly "Y") 7-[3-(Methoxycarbonyl)-3-(methoxyformamido)propyl] wyosine[l], α-(Carboxyamino)-4,9-dihydro-4,6-dimethyl-9-oxo-3-β-D-ribofuranosyl-1H-imidazo[1,2-a]purine-7-butyric acid dimethyl ester	Y-Wyo (MeO)$_2$ FnBto-Wyo[j] Y-imGuo*	yW m$_2$ fnbtW	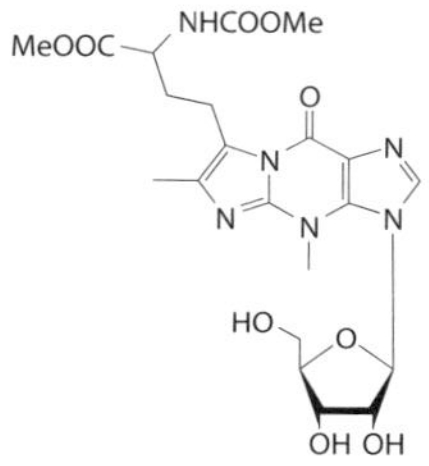	$C_{21}H_{28}N_6O_9$ (508.49)	–	–	–	–

No.		Acidic Spectral Data										Neutral Spectral Data											Alkaline Spectral Data										
	pH	λ_{max}	ε_{max} (×10⁻³)	λ_{min}	Spectral Ratios 230	240	250	270	280	290	pH	λ_{max}	ε_{max} (×10⁻³)	λ_{min}	Spectral Ratios 230	240	250	270	280	290	pH	λ_{max}	ε_{max} (×10⁻³)	λ_{min}	Spectral Ratios 230	240	250	270	280	290
136											8.5	236 295		257	4.14	4.63	1.31	1.01	1.14	1.37	12	236 295		257	4.14	4.63	1.31	1.01	1.14	1.37
137											7	240		213 283	3.58	4.70	2.40	0.93	0.88	0.89	13°	236 269		255						

REFERENCES

No.	Origin and Synthesis	$[\alpha]_D^t$	pK	Spectral Data	Mass Spectra	R_f
136	R: 403	–	–	403[b]	–	–
137	R: 346	–	–	346[b]	–	–

PURINES, PYRIMIDINES, NUCLEOSIDES AND NUCLEOTIDES: PHYSICAL CONSTANTS AND SPECTRAL PROPERTIES (Continued)

No.	Compound	Symbol 3-Letter	Symbol 1-Letter	Structure	Formula (Mol Wt)	Melting Point °C	$[\alpha]_D^t$	pK Basic	pK Acidic
138	Xanthosine	Xao	X	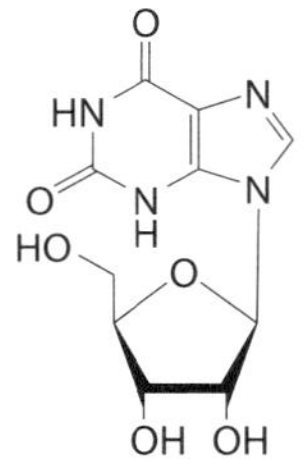	$C_{10}H_{12}N_4O_6$ (284.23)	—	$-51.2[30]$ (8, 0.3N NaOH)	<2.5	5.7* 13.0
139	Zeatosine; Ribosylzeatin; N6-(trans-4-Hydroxy-3-methyl-2-butenyl)adenosine; N6-(trans-4-Hydroxy isopentenyl)adenosine	Zeo 6(trHOiPe)Ado	Z	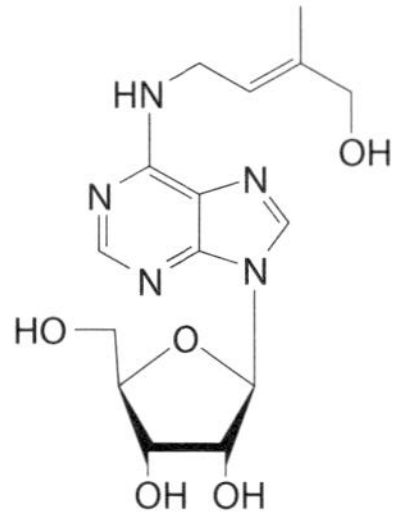	$C_{15}H_{21}N_5O_5$ (351.36)	180–182°	—	—	—

DEOXYRIBONUCLEOSIDES

No.	Compound	Symbol 3-Letter	Symbol 1-Letter	Structure	Formula (Mol Wt)	Melting Point °C	$[\alpha]_D^t$	pK Basic	pK Acidic
140	Deoxyadenosine	dAdo	dA	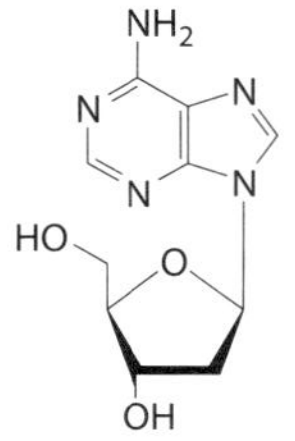	$C_{10}H_{13}N_5O_3$ (251.24)	187–189°	$-26.0[21]$ (1.0, H_2O)	3.8	—

Acidic Spectral Data

No.	pH	λ_{max}	ε_{max} (×10⁻³)	λ_{min}	230	240	250	270	280	290
138	3	235 263	8.4 8.95	248	—	—	0.75	—	0.28	0.03
139	1	208 266	19.8 18.5	235	—	—	—	—	—	—
140	2	258	14.5	228	—	—	0.83	—	0.24	—

Neutral Spectral Data

No.	pH	λ_{max}	ε_{max} (×10⁻³)	λ_{min}	230	240	250	270	280	290
138	8	295 248 278	10.2 8.9	223 264	—	—	1.30	—	1.13	0.61
139	7	211 270	19.3 17.8	233	—	—	—	—	—	—
140	7	260	15.2	225	—	—	0.79	—	0.15	<0.01

Alkaline Spectral Data

No.	pH	λ_{max}	ε_{max} (×10⁻³)	λ_{min}	230	240	250	270	280	290
138	14	303 252 276	8.6 9.3	288 230 262	—	—	1.12	—	1.16	0.59
139	11	215 270	18.1 18.3	235	—	—	—	—	—	—
140	13[d]	261	14.9	—	—	—	—	—	—	—

REFERENCES

No.	Origin and Synthesis	$[\alpha]_D^t$	pK	Spectral Data
138	C: 224,377	224	150,53*,55*,170	232[b],205
139	C: 277	—	—	227
140	D: 238 C: 249,267,390,389	239,249,389	246,223	246,389[c,d],390[d],317[b],267[b],223,240,249

No.	Mass Spectra	R_f
138	—	—
139	—	32,227
140	341,317	21,389

PURINES, PYRIMIDINES, NUCLEOSIDES AND NUCLEOTIDES: PHYSICAL CONSTANTS AND SPECTRAL PROPERTIES (Continued)

No.	Compound	Symbol (3-Letter)	Symbol (1-Letter)	Structure	Formula (Mol Wt)	Melting Point °C	$[\alpha]_D^t$	pK Basic	pK Acidic
141	N^6-Methyldeoxyadenosine / 6-Methylaminopurinedeoxyribonucleoside	6MedAdo d6MeAdo	m⁶dA	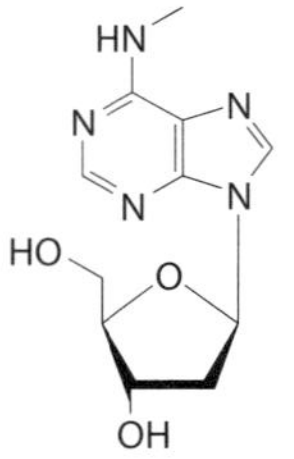	$C_{11}H_{15}N_5O_3$ (265.27)	206–208°	-23.5^{26} (1.0, H₂O)	–	–
142	Deoxycytidine	dCyd	dC	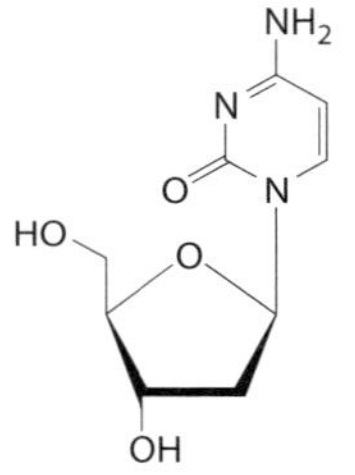	$C_9H_{13}N_3O_4$ (227.22)	200–201°	$+82.4^{19}$ (1.31, 1N NaOH)	4.3	>13
143	N^4-Methyldeoxycytidine / 1-β-2'-Deoxyribofuranosyl-4-methylamino-2-pyrimidinone	4MedCyd d4MeCyd	m⁴dC	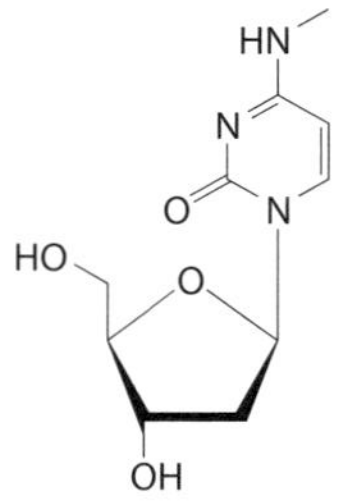	$C_{10}H_{15}N_3O_4$ (241.24)	191–193°	$+48^{28}$ (1.2, H₂O)	4.0	–

Acidic Spectral Data

No.	pH	λ_{max}	ε_{max} (×10⁻³)	λ_{min}	230	240	250	270	280	290
141	1.5	261	16.6*	232*	0.21	0.30	0.65	0.88	0.40	0.11
142	1	280	13.2	241	–	–	0.42	–	2.15	1.61
143	1	282	14.6	242	–	–	–	–	1.98	–

Neutral Spectral Data

No.	pH	λ_{max}	ε_{max} (×10⁻³)	λ_{min}	230	240	250	270	280	290
141	7	265	15.4	229*	0.29	0.35	0.61	1.07	0.65	0.22
142	7	271	9.0	250	–	–	0.83	–	0.97	0.31
143	7	236 / 270	9.1 / 11.7	229	–	–	–	–	–	–

Alkaline Spectral Data

No.	pH	λ_{max}	ε_{max} (×10⁻³)	λ_{min}	230	240	250	270	280	290
141	11	265	15.4	226*	0.17	0.30	0.58	1.08	0.63	0.11
142	11	271	9.0	250	–	–	0.83	–	0.97	0.31
143	13^q	271.5	9.1	250	–	–	–	–	–	–
143	12	236 / 270	9.1 / 11.7	229	–	–	–	–	–	–

REFERENCES

No.	Mass Spectra	R_f
141	–	10,18,21
142	–	201
143	–	129

No.	Origin and Synthesis	$[\alpha]_D^t$	pK	Spectral Data
141	C: 10 D: 21	10	–	10*,21*be,317*bcde
142	D: 238,241 C: 247,129	241	163	163b,232be,317b
143	C: 129	129	129	129

PURINES, PYRIMIDINES, NUCLEOSIDES AND NUCLEOTIDES: PHYSICAL CONSTANTS AND SPECTRAL PROPERTIES (Continued)

No.	Compound	Symbol 3-Letter	1-Letter	Structure	Formula (Mol Wt)	Melting Point °C	$[\alpha]_D^t$	pK Basic	pK Acidic
144	5-Methyldeoxycytidine	5MedCyd d5MeCyd	m5dC	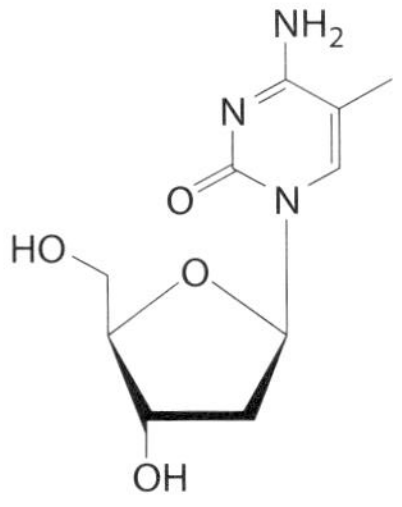	$C_{10}H_{15}N_3O_4$ (241.24)	211–212°	+43[22] (1.4, H_2O)	4.4	>13
145	5-Hydroxymethyldeoxycytidine	5HmdCyd d5HmCyd 5(HOMe)dCyd	om5dC hm5dC*	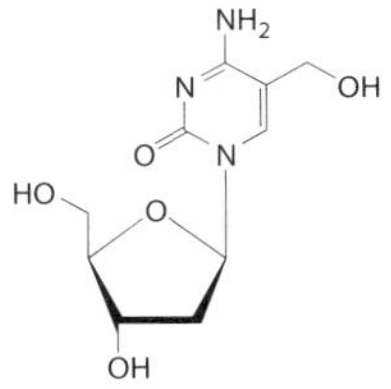	$C_{10}H_{15}N_3O_5$ (257.24)	203° (dec)	+51[20] (H_2O)	3.5	—

Acidic Spectral Data

No.	pH	λ_{max}	ε_{max} (×10⁻³)	λ_{min}	230	240	250	270	280	290
					Spectral Ratios					
144	1	287	12.4*	245	1.34	0.43	0.42	1.93	2.93	3.12
145	1	283	12.6	243	0.16	0.21*	0.64*	1.88*	2.47*	2.27*

Neutral Spectral Data

No.	pH	λ_{max}	ε_{max} (×10⁻³)	λ_{min}	230	240	250	270	280	290
					Spectral Ratios					
144	7	277	8.5	255	1.47	1.29	1.00	1.37	1.54	1.01
145	7	272	—	247	—	0.88	0.97	1.19	1.17	0.58*

Alkaline Spectral Data

No.	pH	λ_{max}	ε_{max} (×10⁻³)	λ_{min}	230	240	250	270	280	290
					Spectral Ratios					
144	14	279	8.8	255	1.57	1.27	0.98	1.43	1.67	1.13
145	13	274	—	252	—	1.21	0.97	1.26	1.17*	0.68*

REFERENCES

No.	Origin and Synthesis	$[\alpha]_D^t$	pK	Spectral Data	Mass Spectra	R_f
144	C: 128,391[a] D: 133,134	391	128	128[b],133*[b],134[b],391,317[b]	—	70
145	D: 75,135 C: 136,385	136	75	75[b],135*[bce]136*[d],317[b]	—	70,75

PURINES, PYRIMIDINES, NUCLEOSIDES AND NUCLEOTIDES: PHYSICAL CONSTANTS AND SPECTRAL PROPERTIES (Continued)

No.	Compound	Symbol (3-Letter)	Symbol (1-Letter)	Structure	Formula (Mol Wt)	Melting Point °C	$[\alpha]_D^t$	pK Basic	pK Acidic
146	Deoxyguanosine	dGuo	dG	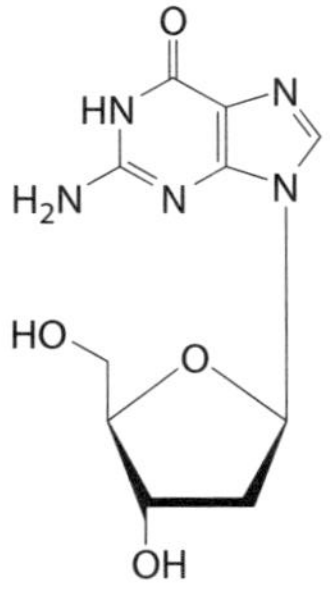	$C_{10}H_{13}N_5O_4$ (267.24)	250°	$-30.2^{23.5}$ (0.2, H_2O)	2.5	–
147	1-Methyldeoxyguanosine	1MedGuo / d1MeGuo	m¹dG	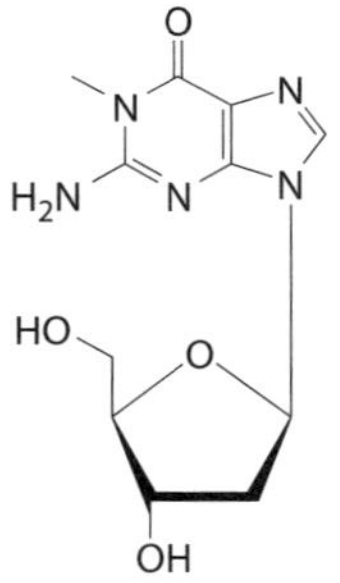	$C_{11}H_{15}N_5O_4$ (281.27)	249–250° (dec)	–	–	–

Acidic Spectral Data

No.	pH	λ_{max}	ε_{max} (×10⁻³)	λ_{min}	230	240	250	270	280	290
146	1°	255	12.1	232	0.26	0.60	1.0	0.84	0.69	0.47
147	1°	257	12.1	–	–	–	–	–	–	–

Neutral Spectral Data

No.	pH	λ_{max}	ε_{max} (×10⁻³)	λ_{min}	230	240	250	270	280	290
146	H_2O	254	13.0	223	0.38	0.81	1.16	0.75	0.68	0.27
147	–	–	–	–	–	–	–	–	–	–

Alkaline Spectral Data

No.	pH	λ_{max}	ε_{max} (×10⁻³)	λ_{min}	230	240	250	270	280	290
146	12	260	9.2	230	0.40	0.55	0.87*	0.98	0.61	0.09
147	11	254	13.6	–	–	–	–	–	–	–

REFERENCES

No.	Origin and Synthesis	$[\alpha]_D^t$	pK	Spectral Data	Mass Spectra	R_f
146	C: 267 D: 238	267	246	317ᵇ,242ᵉ,267*ᵇᶜᵈ	317	242
147	C: 24	–	–	24	–	24

PURINES, PYRIMIDINES, NUCLEOSIDES AND NUCLEOTIDES: PHYSICAL CONSTANTS AND SPECTRAL PROPERTIES (Continued)

No.	Compound	Symbol (3-Letter)	Symbol (1-Letter)	Structure	Formula (Mol Wt)	Melting Point °C	$[\alpha]_D^t$	pK Basic	pK Acidic
148	7-Methyldeoxyguanosine	7MedGuo / d7MeGuo	m7dG	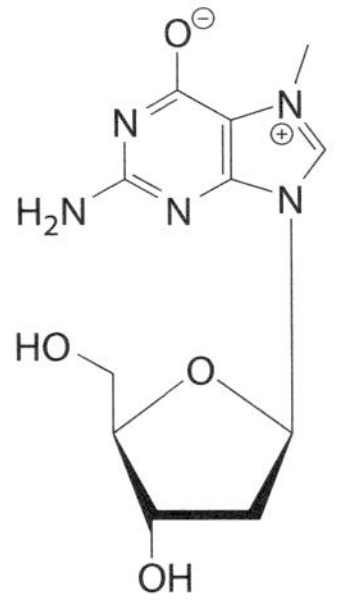	$C_{11}H_{15}N_5O_4$ (281.27)	None (dec)	–	–	–
149	Deoxyuridine	dUrd	dU	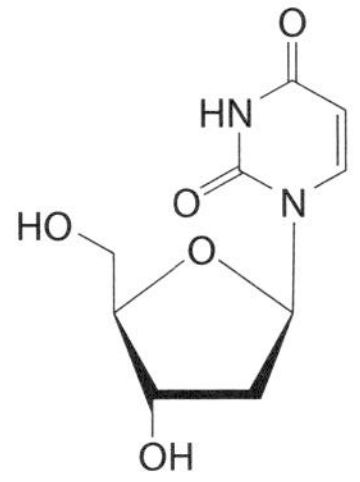	$C_9H_{12}N_2O_5$ (228.20)	163°	$+50.0^{22}$ (1.1 1N NaOH)	–	9.3, >13
150	Thymidine 5-Methyldeoxyuridine	dThd	dT	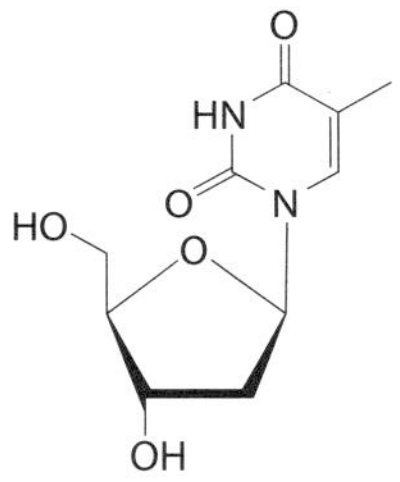	$C_{10}H_{14}N_2O_5$ (242.23)	183–184°	$+32.8^{16}$ (1.04, 1N NaOH)	–	9.8, >13

Acidic Spectral Data

No.	pH	λ_{max}	ε_{max} (×10⁻³)	λ_{min}	230	240	250	270	280	290
148	1°	256	10.8	229	–	0.20	0.74	0.83	0.32*	–
149	1	262	10.2	231	0.20	0.40	0.74	0.83	0.32*	–
150	1	267	9.65	235	0.33	0.34	0.65	1.06	0.70	0.22

Neutral Spectral Data

No.	pH	λ_{max}	ε_{max} (×10⁻³)	λ_{min}	230	240	250	270	280	290
148	6	257	10.2	235	–	0.20	0.74	0.83	–	–
149	7	262	10.2	231	0.20	0.40	0.74	0.83	0.32*	–
150	7	267	9.65	235	0.32	0.33	0.65	1.06	0.70	0.21*

Alkaline Spectral Data

No.	pH	λ_{max}	ε_{max} (×10⁻³)	λ_{min}	230	240	250	270	280	290
148	9°	–	–	–	–	–	–	–	–	–
149	12	262	7.6	242	0.95	0.70	0.80	0.80	0.27*	–
150	13	267	7.4	246	1.18	0.76	0.74	1.05	0.65	0.16

REFERENCES

No.	Origin and Synthesis	pK	Spectral Data	$[\alpha]_D^t$	Mass Spectra	R_f
148	C: 10,41	–	10ᶜ,41ᶜ	–	–	10
149	D: 238,243ᵃ	163	163ᵇ,317ᵇᵉ,232*	243	341,317	243
150	D: 238 C: 128,356,69	163	163ᵇ,242ᵉ,232*ᵇ,317ᵇ	241	317	21,18,69,242

PURINES, PYRIMIDINES, NUCLEOSIDES AND NUCLEOTIDES: PHYSICAL CONSTANTS AND SPECTRAL PROPERTIES (Continued)

No.	Compound	Symbol 3-Letter	Symbol 1-Letter	Structure	Formula (Mol Wt)	Melting Point °C	$[\alpha]_D^t$	pK Basic	pK Acidic
151	5-Hydroxymethyldeoxyuridine	5HmdUrd 5(HOMe)dUrd d5HmUrd	om⁵dU hm⁵dU*	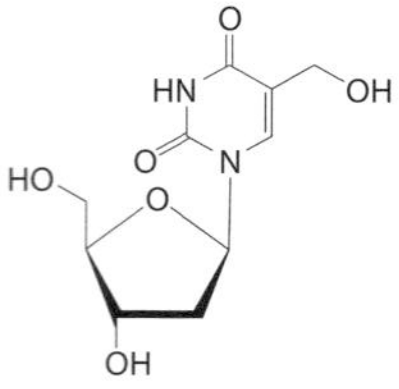	$C_{10}H_{14}N_2O_6$ (258.23)	180–182°	$+19^{20}$ (H_2O)	–	–
152	2-Thiothymidine 5-Methyl-2-thiodeoxyuridine	2SdThd	s²dT	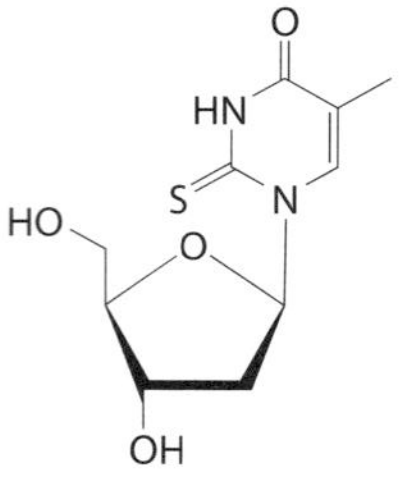	$C_{10}H_{14}N_2O_4S$ (258.30)	182–183°	$+16^{20}$ (0.5 MeOH)	–	–

Acidic Spectral Data

No.	pH	λ_{max}	ε_{max} ($\times 10^{-3}$)	λ_{min}	230	240	250	270	280	290
151	2	264	9.6	233	0.27	0.37	0.70	0.97	0.51	0.10

Neutral Spectral Data

No.	pH	λ_{max}	ε_{max} ($\times 10^{-3}$)	λ_{min}	230	240	250	270	280	290
151	7	264	9.6ⁿ	233	0.27	0.37	0.70	0.97	0.51	0.10
152	(MeOH)	221	15.2							

Alkaline Spectral Data

No.	pH	λ_{max}	ε_{max} ($\times 10^{-3}$)	λ_{min}	230	240	250	270	280	290
151	12	264	7.0	243	1.13	0.72	0.75	0.95	0.54*	0.18
152	13	242	22.8							

REFERENCES

No.	Origin and Synthesis	$[\alpha]_D^t$	pK	Spectral Data	Mass Spectra	R_f
151	C: 69,152ᵃ	152	–	317ᵇ, 69*ᶜᵈ	–	69
152	C: 301	301	–	301	–	–

PURINES, PYRIMIDINES, NUCLEOSIDES AND NUCLEOTIDES: PHYSICAL CONSTANTS AND SPECTRAL PROPERTIES (Continued)

No.	Compound	Symbol 3-Letter	Symbol 1-Letter	Structure	Formula (Mol Wt)	Melting Point °C	$[\alpha]_D$	pK Basic	pK Acidic
		RIBONUCLEOTIDES							
153	Adenosine 2'-phosphate	Ado-2'-P	2'-AMP	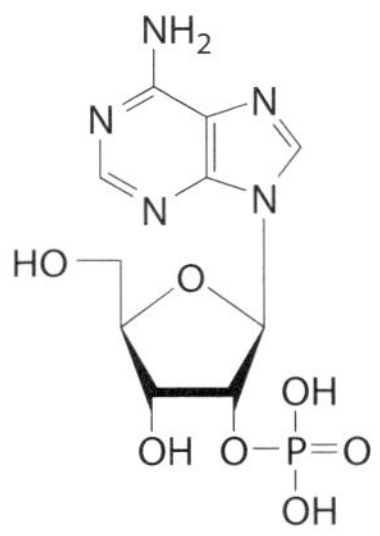	$C_{10}H_{14}N_5O_7P$ (347.22)	183° (dec)	-65.4^{22} (0.5, 0.5M, Na_2HPO_4)	3.8	—
154	Adenosine 3'-phosphate	Ado-3'-P	Ap A-	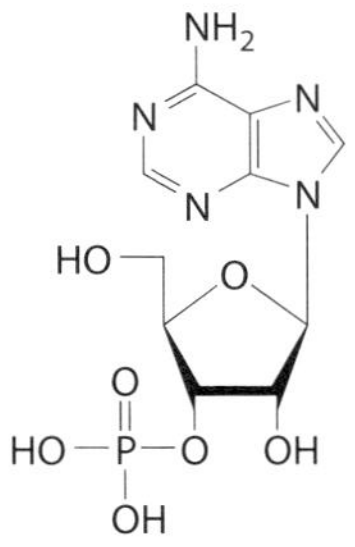	$C_{10}H_{14}N_5O_7P$ (347.22)	195° (dec)	-45.4^{22} (0.5, 0.5M Na_2HPO_4)	3.65	—
155	Adenosine 5'-phosphate	Ado-5'-P	PA -A	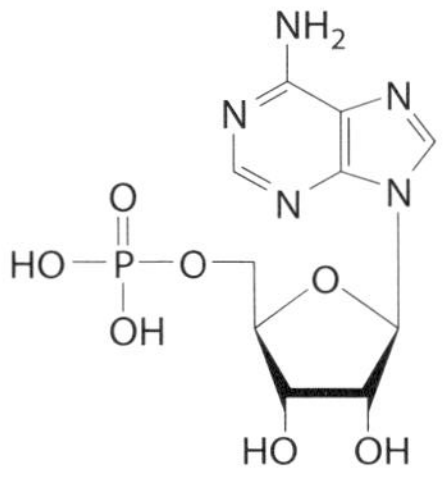	$C_{10}H_{14}N_5O_7P$ (347.22)	192° (dec)	-46.3^{24} (H_2O)	3.8	—

Acidic Spectral Data

No.	pH	λ_{max}	ε_{max} ($\times10^{-3}$)	λ_{min}	230	240	250	270	280	290
153	2	257*	14.4*	229*	-	-	0.85	0.71	0.23	0.04
154	1	257	15.1	230	-	-	0.85	0.71	0.22*	0.04*
155	2	257	15.0	230	0.23	0.43	0.84	0.68	0.22	0.44

Neutral Spectral Data

pH	λ_{max}	ε_{max} ($\times10^{-3}$)	λ_{min}	230	240	250	270	280	290
7	276	16.3	-	-	-	0.80	-	0.15	0.01
7	259^z	15.4^z	-	-	-	0.80	-	0.15	0.01
7	259	15.4	227*	0.18	0.39	0.79	0.66	0.16	0.01

Alkaline Spectral Data

pH	λ_{max}	ε_{max} ($\times10^{-3}$)	λ_{min}	230	240	250	270	280	290
12 (MeOH)	264	16.6	-	-	-	0.80	-	0.15	-
13	259^z	15.4^z	227	-	-	0.78	0.73	0.22	0.05
11	259	15.4	227	-	-	0.79	-	0.15	-

REFERENCES

No.	Origin and Synthesis	pK	$[\alpha]_D$	R_f	Spectral Data	Mass Spectra
153	C: 191,193,219 R: 182,221	220,170,218	221	258,219,263	179^e,223^z	-
154	C: 193,190,219 R: 182,221,194^z	220,170,218	221	258,219,263,292	265*^{b},179^e	-
155	C: 191,197,368 R: 195 N: 198	220,170,212	368	188,219,263,368	212,179^e,184^b,183^b,368*	374

PURINES, PYRIMIDINES, NUCLEOSIDES AND NUCLEOTIDES: PHYSICAL CONSTANTS AND SPECTRAL PROPERTIES (Continued)

No.	Compound	Symbol (3-Letter)	Symbol (1-Letter)	Structure	Formula (Mol Wt)	Melting Point °C	$[\alpha]_D^t$	pK Basic	pK Acidic
156	Adenosine 5'-diphosphate	Ado-5'-P$_2$ ADP	ppA	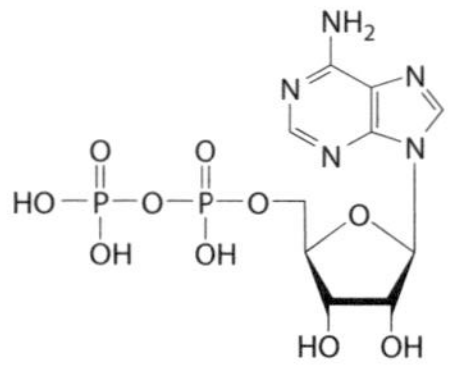	$C_{10}H_{15}N_5O_{10}P_2$ (427.21)	–	–	3.9	–
157	Adenosine 5'-triphosphate	Ado-5'-P$_3$ ATP	pppA	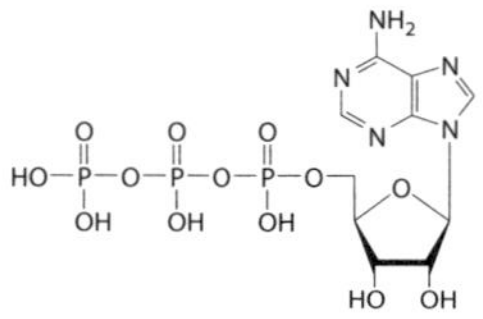	$C_{10}H_{16}N_5O_{13}P_3$ (507.19)	–	–	4.1	–
158	1-Methyladenosine 3'(2')-phosphate	1MeAdo-3'(2')-P	m^1 Ap m^1 A- for 3'	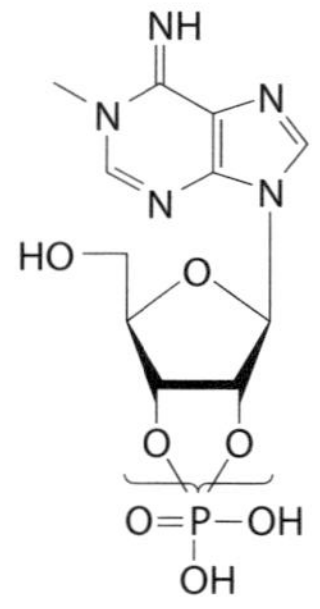	$C_{11}H_{16}N_5O_7P$ (361.25)	–	–	8.8[1]	–

Acidic Spectral Data

No.	pH	λ_{max}	ε_{max} (×10⁻³)	λ_{min}	230	240	250	270	280	290
156	2	257	15.0	230	–	–	0.85	–	0.21	–
157	2	257	14.7	230	–	–	0.85	–	0.22	–
158	2	258	13.2^w	230	0.24	0.44	0.83	0.67	0.26	0.07

Neutral Spectral Data

No.	pH	λ_{max}	ε_{max} (×10⁻³)	λ_{min}	230	240	250	270	280	290
156	7	259	15.4	227	–	–	0.78	–	0.16	–
157	7	259	15.4	227	–	–	0.80	–	0.15	–
158	–	–	–	–	–	–	–	–	–	–

Alkaline Spectral Data

No.	pH	λ_{max}	ε_{max} (×10⁻³)	λ_{min}	230	240	250	270	280	290
156	11	259	15.4	227	–	–	0.78	–	0.15	–
157	11	259	15.4	227	–	–	0.80	–	0.15	–
158	13^b	259	12.9^w	–	–	–	0.77	0.76	0.4	0.32

REFERENCES

No.	Origin and Synthesis	$[\alpha]_D^t$	pK	Spectral Data	Mass Spectra	R$_f$
156	C: 255,211,188,196 E: 226 N: 198	–	212	212,183^b,206	–	188
157	C: 211,214,188 N: 198	–	212	212^b,206,183^b	–	188
158	C: 90,292 R: 77	–	–	90	–	143,291,292

PURINES, PYRIMIDINES, NUCLEOSIDES AND NUCLEOTIDES: PHYSICAL CONSTANTS AND SPECTRAL PROPERTIES (Continued)

No.	Compound	Symbol 3-Letter	Symbol 1-Letter	Structure	Formula (Mol Wt)	Melting Point °C	$[\alpha]_D$	pK Basic	pK Acidic
159	1-Methyladenosine 5'-phosphate	1MeAdo-5'-P 1MeAMP	pm¹ A -m¹ A	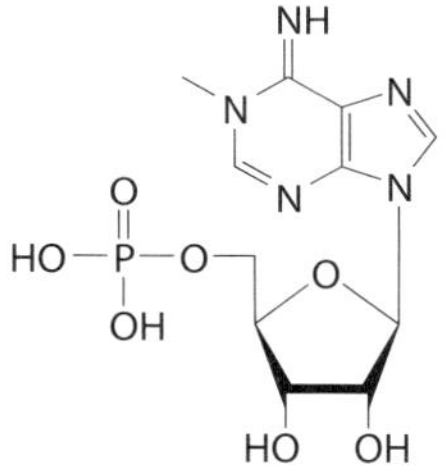	$C_{11}H_{16}N_5O_7P$ (361.25)	–	–	8.8¹	–
160	1-Methyladenosine 5'-diphosphate	1MeAdo-5'-P 1MeADP₂	ppm¹ A	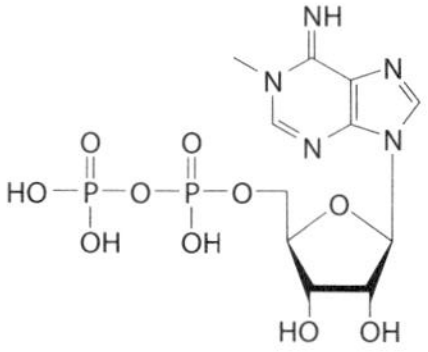	$C_{11}H_{17}N_5O_{10}P_2$ (441.23)	–	–	–	–
161	2-Methyladenosine 3'-phosphate	2MeAdo-3'-P	m² Ap	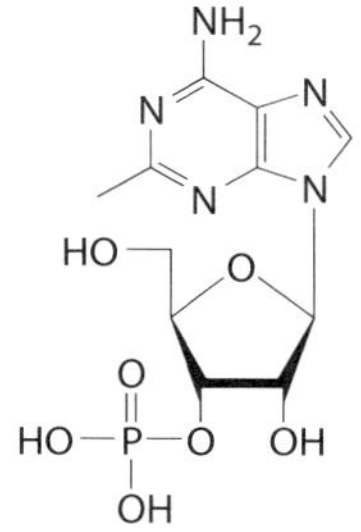	$C_{11}H_{16}N_5O_7P$ (361.25)	–	–	–	–

Acidic Spectral Data

No.	pH	λ_{max}	ε_{max} (×10⁻³)	λ_{min}	230	240	250	270	280	290
159	2	258	–	232	0.34	0.46	0.81	0.74	0.32	0.10
160	2	257	11.9	234	0.37	0.44	0.84	0.66	0.23	0.04
161	1	259	10.8	–	–	–	–	–	–	–

Neutral Spectral Data

No.	pH	λ_{max}	ε_{max} (×10⁻³)	λ_{min}	230	240	250	270	280	290
159	H₂O⁴	259	–	233	0.22	0.34	0.77	0.74	0.27	–
160	H₂O⁴	–	–	–	–	–	–	–	–	–
161	H₂O	264	12.9	–	–	–	–	–	–	–

Alkaline Spectral Data

No.	pH	λ_{max}	ε_{max} (×10⁻³)	λ_{min}	230	240	250	270	280	290
159	12⁰	259	13.1ʷ	230*	0.22*	0.36*	0.75*	0.71	0.36	0.3
160	12⁰	259	12.5	232	0.32	0.40	0.75	0.74	0.36	0.31
161	13	263	13.1	–	–	–	–	–	–	–

REFERENCES

No.	Origin and Synthesis	$[\alpha]_D$	pK	Spectral Data	Mass Spectra	R_f
159	C: 164,8	–	–	164,8*bd	–	8,164
160	C: 164	–	–	164ᵇ	–	164
161	C: 291 R: 291	–	–	291ᵇ	–	291

PURINES, PYRIMIDINES, NUCLEOSIDES AND NUCLEOTIDES: PHYSICAL CONSTANTS AND SPECTRAL PROPERTIES (Continued)

No.	Compound	Symbol (3-Letter)	Symbol (1-Letter)	Structure	Formula (Mol Wt)	Melting Point °C	$[\alpha]_D^t$	pK Basic	pK Acidic
162	2-Methyladenosine 5′-phosphate	2MeAdo-5′-P 2MeAMP	pm²A	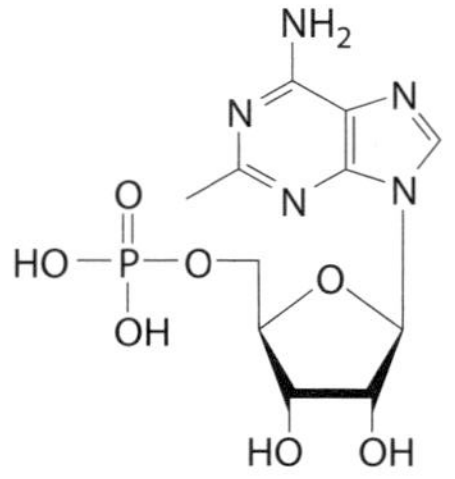	$C_{11}H_{16}N_5O_7P$ (361.25)	260° (dec) (Ba salt)	—	—	—
163	N^6-(Δ^2-Isopentenyl)adenosine 5′-phosphate: 6-($\gamma\gamma$-Dimethylallylamino)-9-β-D-ribofuranosylpurine 5′-phosphate	6Pe¹Ado-5′-P 6iPeAdo-5′-P 6iPeAMP	pi⁶A	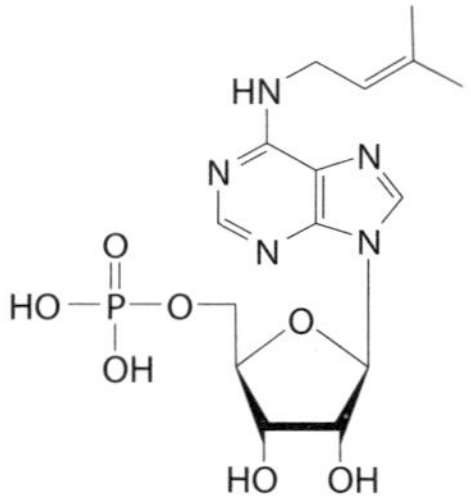	$C_{15}H_{22}N_5O_7P$ (415.35)	—	—	—	—
164	N^6-Methyladenosine 3′(2′)-phosphate	6MeAdo-3′(2′)-P	m6 Ap m6 A- for 3′	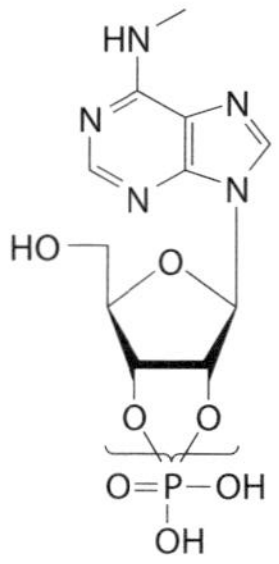	$C_{11}H_{16}N_5O_7P$ (361.25)	—	—	—	—

Acidic Spectral Data

No.	pH	λ_{max}	ε_{max} (×10⁻³)	λ_{min}	230	240	250	270	280	290
162	1	259	10.9	232.5						
163	1	264	20.9							
164	1	262	18.3	231*			0.64	0.91	0.45	0.14

Neutral Spectral Data

No.	pH	λ_{max}	ε_{max} (×10⁻³)	λ_{min}	230	240	250	270	280	290
162	6	264	13.2	233						
163	7	267	19.2							

Alkaline Spectral Data

No.	pH	λ_{max}	ε_{max} (×10⁻³)	λ_{min}	230	240	250	270	280	290
162	13	264	13.4	232						
163	13	268	19.0	230*						
164	12	266*		230*			0.58	1.05	0.67	0.22

REFERENCES

No.	Origin and Synthesis	$[\alpha]_D^t$	pK	Spectral Data	Mass Spectra	R_f
162	C: 342	—	—	342	—	—
163	C: 319	—	—	319	—	319
164	R: 165,292 C: 292	—	—	265*^b,292*^c,165*^c,90^e	—	143,165,291,292

PURINES, PYRIMIDINES, NUCLEOSIDES AND NUCLEOTIDES: PHYSICAL CONSTANTS AND SPECTRAL PROPERTIES (Continued)

No.	Compound	Symbol 3-Letter	Symbol 1-Letter	Structure	Formula (Mol Wt)	Melting Point °C	$[\alpha]_D^b$	pK Basic	pK Acidic
165	N^6-Methyladenosine 5′-phosphate	6MeAdo-5′-P 6MeAMP	pm⁶A -m⁶A	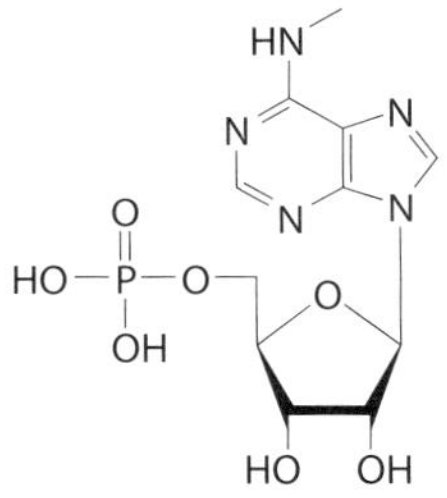	$C_{11}H_{16}N_5O_7P$ (361.25)	–	–	-3.7[p]	–
166	N^6-Methyladenosine 5′-diphosphate	6MeAdo-5′-P₂ 6MeADP	ppm⁶A	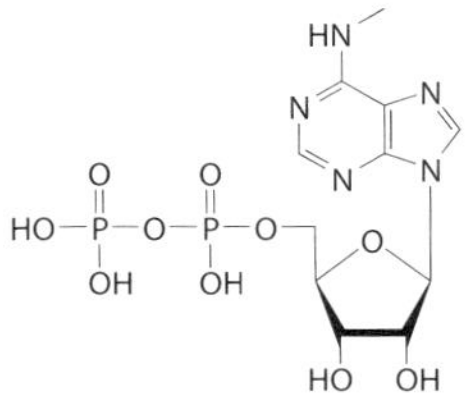	$C_{11}H_{17}N_5O_{10}P_2$ (441.23)	–	–	-3.7[p]	–
167	N^6,N^6-Dimethyladenosine 5′-phosphate	6Me₂Ado-5′-P 6Me₂AMP	pm₂⁶A -m₂⁶A	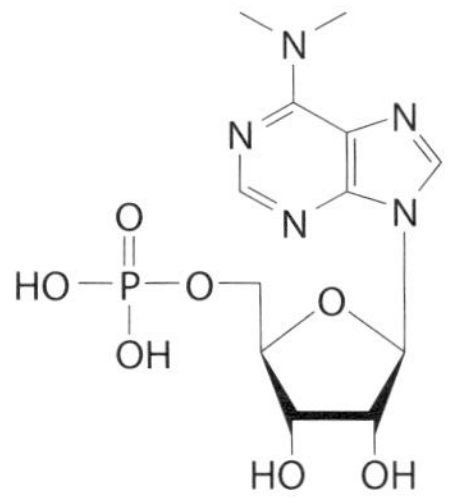	$C_{12}H_{18}N_5O_7P$ (375.28)	225° (dec)	-51^{20} (2.0, H₂O)	–	–

No.		Acidic Spectral Data											Neutral Spectral Data									
	pH	λ_{max}	ε_{max} (×10⁻³)	λ_{min}	230	240	250	270	280	290		pH	λ_{max}	ε_{max} (×10⁻³)	λ_{min}	230	240	250	270	280	290
165	2	261	16.3	231	0.28	0.39	0.73	0.85	0.36*	0.13		H₂O	264	13.4	229	0.17	0.29	0.63	0.97	0.56	0.17
166	2	262	15.7	231	0.18	0.32	0.69	0.84	0.29	0.08		–	–	–	–	–	–	–	–	–	–
167	H₂O	268	18.3	–	–	–	–	–	–	–		7	274	–	–	–	–	–	–	–	–

No.		Alkaline Spectral Data									
	pH	λ_{max}	ε_{max} (×10⁻³)	λ_{min}	230	240	250	270	280	290
165	12	266*	15.2[a]	231*	0.20	0.32*	0.60	1.08	0.66*	0.26*
166	12	265	15.4	229	0.14	0.26	0.60	0.99	0.57	0.18
167	–	–	–	–	–	–	–	–	–	–

REFERENCES

No.	Origin and Synthesis	pK	$[\alpha]_D^b$	Spectral Data	Mass Spectra	R_f
165	C: 164,166	164	–	164*,8*ᵇᶜᵈᵉ	–	164,166,407
166	C: 164	164	–	164ᵇ	–	164
167	C: 115,166	–	115	115,166ᶜ	–	115,166,167

PURINES, PYRIMIDINES, NUCLEOSIDES AND NUCLEOTIDES: PHYSICAL CONSTANTS AND SPECTRAL PROPERTIES (Continued)

No.	Compound	Symbol 3-Letter	Symbol 1-Letter	Structure	Formula (Mol Wt)	Melting Point °C	$[\alpha]_D$	pK Basic	pK Acidic
168	N^6,N^6-Dimethyladenosine 5′-diphosphate	6Me₂ Ado-5′P₂ / 6Me₂ ADP	ppm⁶ A	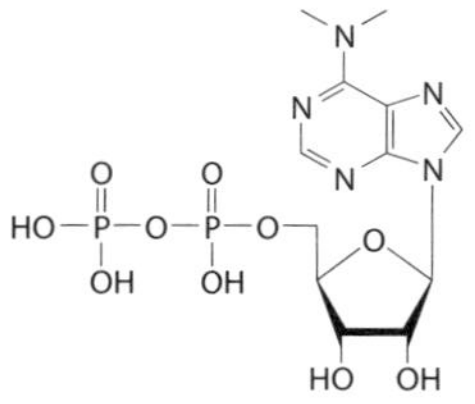	$C_{12}H_{19}N_5O_{10}P_2$ (455.26)	–	–	–	–
169	N^6-Threoninocarbonyladenosine 3′(2′)-phosphate; N-[(9-β-D-Ribofuranosylpurin-6-yl)- N-carbamoyl] threonine 3′(2′)-phosphate	6(ThrCO) Ado-3′(2′)-P	tc⁶ AP for 3′	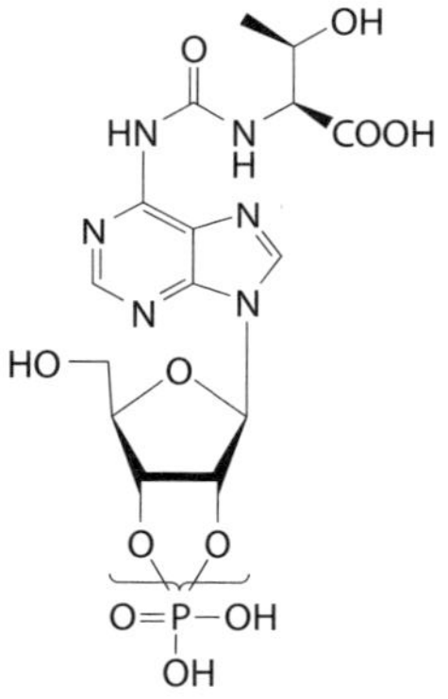	$C_{15}N_{21}N_6O_{11}P$ (492.34)	–	–	2.1ᵖ	–

Neutral Spectral Data

No.	pH	λ_{max}	ε_{max} (×10⁻³)	λ_{min}	230	240	250	270	280	290
168	–	–	–	–	–	–	–	–	–	–
169	5	269	–	231	0.2	0.33	0.62	1.38	0.87	0.03

Acidic Spectral Data

No.	pH	λ_{max}	ε_{max} (×10⁻³)	λ_{min}	230	240	250	270	280	290
168	–	–	–	–	–	–	–	–	–	–
169	–	–	–	–	–	–	–	–	–	–

Alkaline Spectral Data

No.	pH	λ_{max}	ε_{max} (×10⁻³)	λ_{min}	230	240	250	270	280	290
168	–	–	–	–	–	–	–	–	–	–
169	–	–	–	–	–	–	–	–	–	–

REFERENCES

No.	Origin and Synthesis	Spectral Data	pK	$[\alpha]_D$	Mass Spectra	R_f
168	C: 167	–	–	–	–	167
169	R: 90,410	90	90	–	–	–

PURINES, PYRIMIDINES, NUCLEOSIDES AND NUCLEOTIDES: PHYSICAL CONSTANTS AND SPECTRAL PROPERTIES (Continued)

No.	Compound	Symbol 3-Letter	Symbol 1-Letter	Structure	Formula (Mol Wt)	Melting Point °C	$[\alpha]_D^t$	pK Basic	pK Acidic
170	N^6-Threoninocarbonyladenosine 5′-phosphate N-[(9-β-D-Ribofuranosylpurin-6-yl)- N-carbamoyl] threonine 3′(2′)-phosphate	6(ThrCO)Ado-5′-P 6(ThrCO) AMP	ptc^6 A	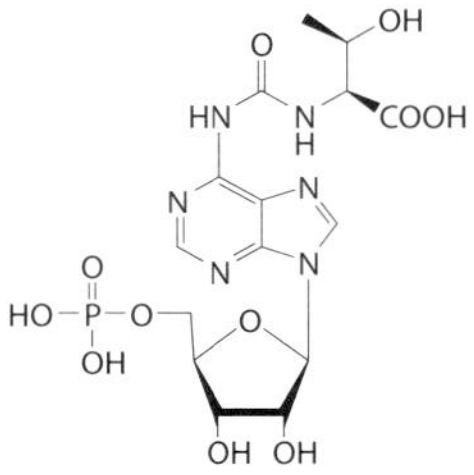	$C_{15}H_{21}N_6O_{11}P$ (492.34)	—	—	—	~3.0^p
171	N^6-Methyl-N^6-threoninocarbonyladenosine 5′-phosphate N-[(9-β-D-Ribofuranosylpurin-6-yl)- N-methylcarbamoyl] threonine 5′-phosphate	6Me6(ThrCO) Ado-5′-P 6Me6(ThrCO) AMP	pm^{6}tc^{6}A	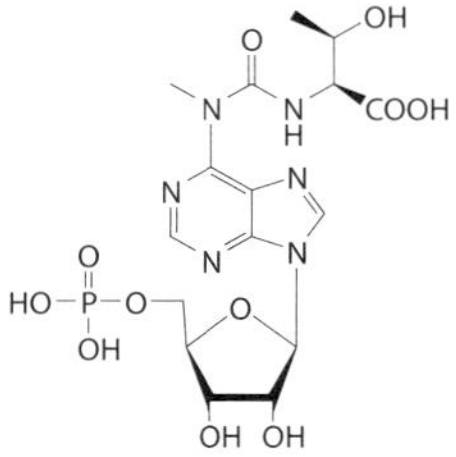	$C_{16}H_{23}N_6O_{11}P$ (506.37)	—	—	—	~3.0^p

No.		Acidic Spectral Data										Neutral Spectral Data								
	pH	λ_{max}	ε_{max} ($\times10^{-3}$)	λ_{min}	230	240	250	270	280	290	pH	λ_{max}	ε_{max} ($\times10^{-3}$)	λ_{min}	230	240	250	270	280	290
170	1	276		237	0.49	0.41	0.61	1.53	1.45	0.48	6.8	275 269 276		231	0.28	0.37	0.63	1.29	0.69	0.03
171	1	283		240	0.82	0.58	0.68	1.41	1.89	1.68	6.8	278		239	0.83	0.63	0.73	1.47	1.87	1.25

No.	Alkaline Spectral Data									
	pH	λ_{max}	ε_{max} ($\times10^{-3}$)	λ_{min}	230	240	250	270	280	290
170	13°	269 277 297		239 273 287	0.73	0.48	0.65	1.37	1.07	0.71
171	13	278		238	0.73	0.54	0.68	1.47	1.89	1.29

REFERENCES

No.	Origin and Synthesis	$[\alpha]_D^t$	pK	Spectral Data	Mass Spectra	R_f
170	R: 407	—	407	407^b	—	407
171	R: 407	—	407	407^b	—	407

PURINES, PYRIMIDINES, NUCLEOSIDES AND NUCLEOTIDES: PHYSICAL CONSTANTS AND SPECTRAL PROPERTIES (Continued)

No.	Compound	Symbol 3-Letter	Symbol 1-Letter	Structure	Formula (Mol Wt)	Melting Point °C	$[\alpha]_D^t$	pK Basic	pK Acidic
172	Cytidine 2′-phosphate	Cyd-2′-P	2′CMP	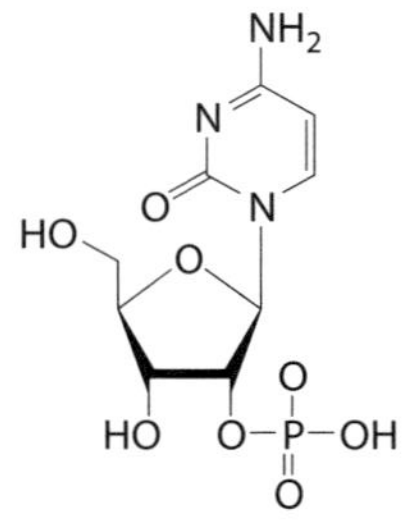	$C_9H_{14}N_3O_8P$ (323.21)	238–240° (dec)	$+20.7^{20}$ (1.0, H_2O)	4.4	—
173	Cytidine 3′-phosphate	Cyd-3′-P	Cp / C-	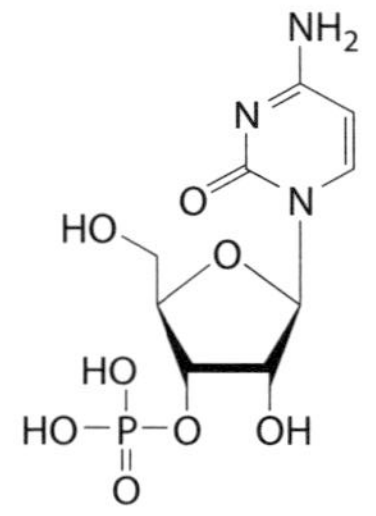	$C_9H_{14}N_3O_8P$ (323.21)	232–234° (dec)	$+49.4^{20}$ (1.0, H_2O)	4.3	—
174	Cytidine 5′-phosphate	Cyd-5′-P	pC / -C	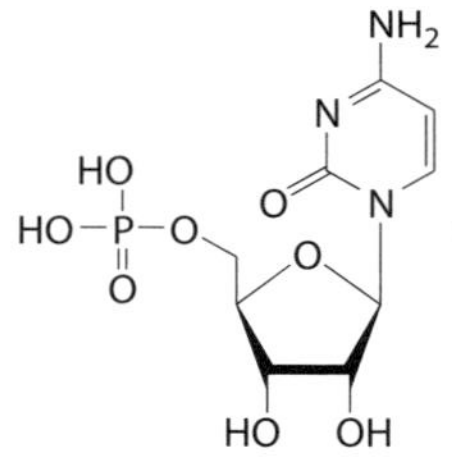	$C_9H_{14}N_3O_8P$ (323.21)	233° (dec)	$+27.1^{14}$ (0.54, H_2O)	4.5	—

Acidic Spectral Data

No.	pH	λ_{max}	ε_{max} (×10⁻³)	λ_{min}	230	240	250	270	280	290
172	2	278	12.7	240	—	—	0.48	—	1.80	1.22
173	2	279	13.0	240	—	—	0.45*	1.51	2.00*	1.43*
174	2	280*	13.2	241*	0.56	0.25	0.44	1.73	2.09	1.55

Neutral Spectral Data

No.	pH	λ_{max}	ε_{max} (×10⁻³)	λ_{min}	230	240	250	270	280	290
172	7	285	—	—	—	—	0.90	—	0.85	0.26
173	7	270ᶠ	9.0ᶠ	250ᶠ	—	—	0.86	—	0.93	0.30
174	7	271	9.1	249	1.07	0.92	0.84	1.21	0.98	0.33

Alkaline Spectral Data

No.	pH	λ_{max}	ε_{max} (×10⁻³)	λ_{min}	230	240	250	270	280	290
172	12	272	8.6	250	—	—	0.9	—	0.85	0.26
173	12	272	8.9	250	—	—	0.86	1.16	0.93	0.30*
174	11	271	9.1	249	—	—	0.84	—	0.98	0.33

REFERENCES

No.	Origin and Synthesis	$[\alpha]_D^t$	pK	Spectral Data
172	R:215ᵃ,192,178	215	218,170,192	223,179ᵇᵉ
173	R: 215ᵃ,192,178	215	218,170,192	223,179ᵇᵉ,181ᶻᵇᶜᵈ,251,265*ᵇᵉ
174	C: 190,196,368 R: 195 N: 198	190	212	212,183ᵇᶜ,184ᵇ,179,205,368*

No.	Mass Spectra	R_f
172	—	258
173	—	258
174	—	74

PURINES, PYRIMIDINES, NUCLEOSIDES AND NUCLEOTIDES: PHYSICAL CONSTANTS AND SPECTRAL PROPERTIES (Continued)

No.	Compound	Symbol 3-Letter	Symbol 1-Letter	Structure	Formula (Mol Wt)	Melting Point °C	$[\alpha]_D^t$	pK Basic	pK Acidic
175	Cytidine 5′-diphosphate	Cyd-5′-P₂ CDP	ppC	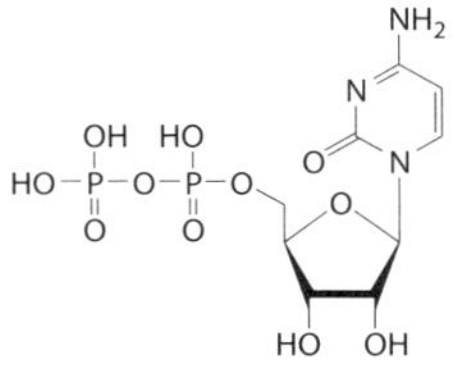	$C_9H_{15}N_3O_{11}P_2$ (403.18)	–	–	4.6	–
176	Cytidine 5′-triphosphate	Cyd-5′-P₃ CTP	pppC	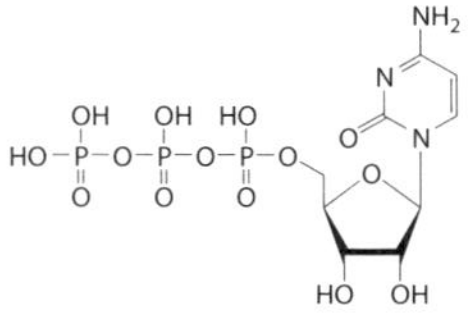	$C_9H_{16}N_3O_{14}P_3$ (483.16)	–	–	4.8	–
177	2-Thiocytidine 3′(2′)-phosphate	2Syd-3′(2′)-P 2SCyd-3′(2′)-P	S²Cp for 3′	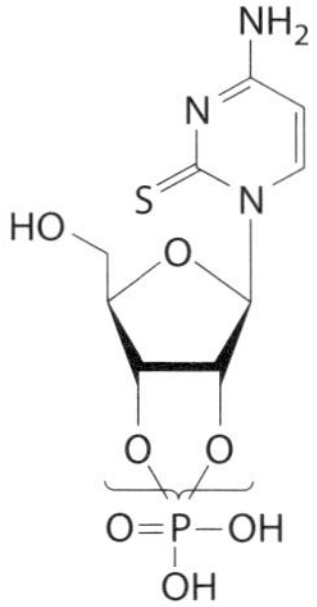	$C_9H_{14}N_3O_7SP$ (339.27)	–	–	3.6ᵖ	–

No.		Acidic Spectral Data									Neutral Spectral Data									Alkaline Spectral Data								
	pH	λ_{max}	ε_{max} (×10⁻³)	λ_{min}	Spectral Ratios 230	240	250	270	280	290	pH	λ_{max}	ε_{max} (×10⁻³)	λ_{min}	Spectral Ratios 230	240	250	270	280	290	pH	λ_{max}	ε_{max} (×10⁻³)	λ_{min}	Spectral Ratios 230	240	250	270	280	290
175	2	280	12.8	241	–	–	0.46	–	2.07	1.48	7	271	9.1	249	–	–	0.83	–	0.98	0.32	11	271	9.1	249	–	–	0.83	–	0.98	–
176	2	280	12.8	241	–	–	0.45	–	2.12	–	7	271	9.0	249	–	–	0.84	–	0.97	–	11	271	9.0	249	–	–	0.84	–	0.97	–
177	1	227 276	–	247	1.24	0.91	0.75	1.24	1.28	0.83	H_2O	248	–	220	0.64	0.94	1.08	0.88	0.69	0.39	13	249	–	228	0.71	0.93	1.08	0.89	0.71	0.4

REFERENCES

No.	Origin and Synthesis	$[\alpha]_D^t$	pK	Spectral Data	Mass Spectra	R_f
175	C: 196 N: 198	–	212	212,179ᵉ,183ᵇ	–	–
176	N: 198	–	212	212,183ᵇ	–	–
177	R: 270,329 C: 329	–	329	270ᵇ,329ᵇ	–	270,329

PURINES, PYRIMIDINES, NUCLEOSIDES AND NUCLEOTIDES: PHYSICAL CONSTANTS AND SPECTRAL PROPERTIES (Continued)

No.	Compound	Structure	Symbol 3-Letter	Symbol 1-Letter	Formula (Mol Wt)	Melting Point °C	$[\alpha]_D^t$	pK Basic	pK Acidic
178	3-Methylcytidine 3'(2')-phosphate		3MeCyd-3'(2')-P	m³CP m³C - for 3'	$C_{10}H_{16}N_3O_8P$ (337.22)	–	–	~9.0^p	–
179	3-Methylcytidine 5'-phosphate		3MeCyd-5'-P 3MeCMP	pm³C -m³C	$C_{10}H_{16}N_3O_8P$ (337.22)	–	–	–	–
180	3-Methylcytidine 5'-diphosphate		3MeCyd-5'-P₂ 3MeCDP	ppm³C	$C_{10}H_{17}N_3O_{11}P_2$ (417.21)	–	–	~9.0^p	–

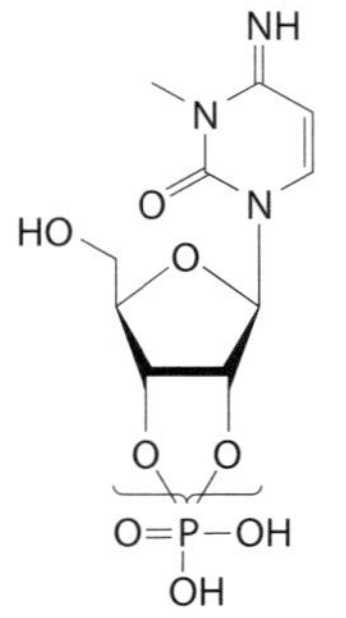

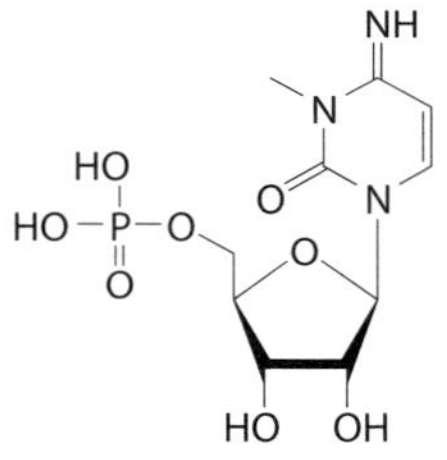

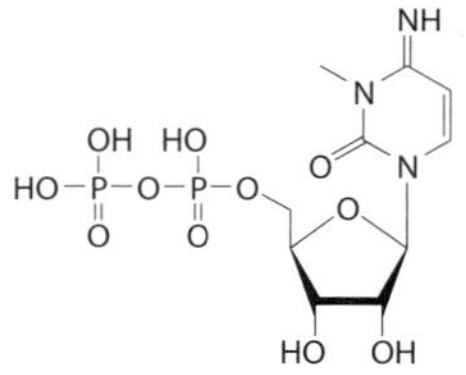

No.		Neutral Spectral Data										
	pH	λ_{max}	ε_{max} ($\times10^{-3}$)	λ_{min}	230	240	250	270	280	290		
178	7	276	11.2	242	0.66	0.34	0.50	1.56	1.64	0.98		
179	–	–	–	–	–	–	–	–	–	–		
180	7	277	11.0	241	–	–	–	–	–	–		

No.		Acidic Spectral Data									
	pH	λ_{max}	ε_{max} ($\times10^{-3}$)	λ_{min}	230	240	250	270	280	290
178	1	276	11.5	242	–	–	–	–	–	–
179	–	–	–	–	–	–	–	–	–	–
180	1	278	11.0	241	–	–	–	–	–	–

No.		Alkaline Spectral Data									
	pH	λ_{max} ($\times10^{-3}$)	ε_{max}	λ_{min}	230	240	250	270	280	290
178	–	–	–	–	–	–	–	–	–	290
179	–	–	–	–	–	–	–	–	–	–
180	–	–	–	–	–	–	–	–	–	–

No.	Origin and Synthesis	$[\alpha]_D^t$	pK	Spectral Data	Mass Spectra	R_f
178	C: 130 R: 413	–	130	130,413^e	–	13,130,143
179	C: 125	–	–	–	–	125
180	C: 130	–	130	130	–	130

REFERENCES

PURINES, PYRIMIDINES, NUCLEOSIDES AND NUCLEOTIDES: PHYSICAL CONSTANTS AND SPECTRAL PROPERTIES (Continued)

No.	Compound	Symbol 3-Letter	Symbol 1-Letter	Structure	Formula (Mol Wt)	Melting Point °C	$[\alpha]_D$	pK Basic	pK Acidic
181	N^4-Methylcytidine 3'(2')-phosphate	4MeCyd-3'(2')-P	m⁴Cp, m⁴C- for 3'	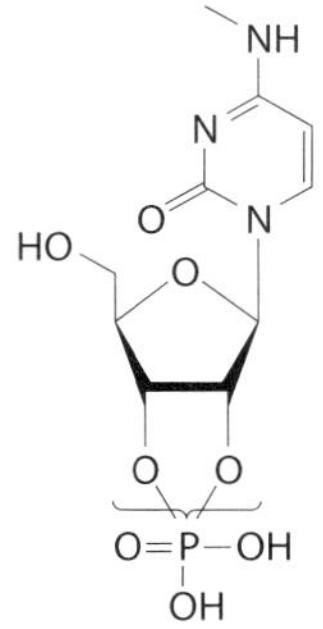	$C_{10}H_{16}N_3O_8P$ (337.22)	—	—	—	—
182	N^4-Methylcytidine 5'-phosphate	4MeCyd-5'-P, 4MeCMP	pm⁴C, -m⁴C	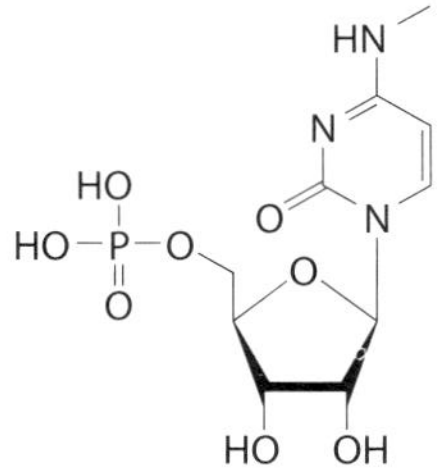	$C_{10}H_{16}N_3O_8P$ (337.22)	—	—	—	—
183	N^4-Methylcytidine 5'-diphosphate	4MeCyd-5'-P₂, 4MeCDP	ppm⁴C	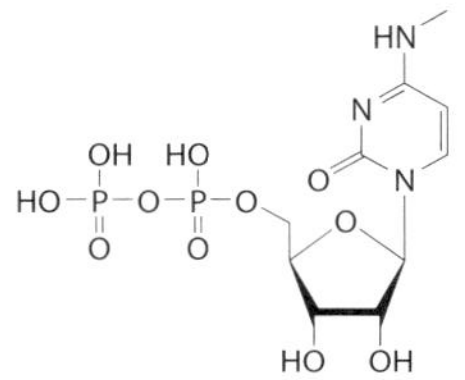	$C_{10}H_{17}N_3O_{11}P_2$ (417.21)	—	—	—	—

Acidic Spectral Data

No.	pH	λ_{max}	ε_{max} (×10⁻³)	λ_{min}	230	240	250	270	280	290
181	1	281	12.9	242	—	—	—	—	—	—
182	1	280	14.8	242	—	—	—	—	—	—
183	1	280	12.9	241	—	—	—	—	—	—

Neutral Spectral Data

No.	pH	λ_{max}	ε_{max} (×10⁻³)	λ_{min}	230	240	250	270	280	290
181	H₂O	237	—	227	—	—	—	—	—	—
182		272	—	248	—	—	—	—	—	—
183		—	—	—	—	—	—	—	—	—

Alkaline Spectral Data

No.	pH	λ_{max}	ε_{max} (×10⁻³)	λ_{min}	230	240	250	270	280	290
181	—	—	—	—	—	—	—	—	—	—
182	—	—	—	—	—	—	—	—	—	—
183	—	—	—	—	—	—	—	—	—	—

REFERENCES

No.	Origin and Synthesis	$[\alpha]_D$	pK	Spectral Data	Mass Spectra	R_f
181	C: 130	—	—	130	—	130
182	C: 130,59,168	—	—	130,168ᶜ	—	130,59
183	C: 130,59	—	—	130	—	130,59

PURINES, PYRIMIDINES, NUCLEOSIDES AND NUCLEOTIDES: PHYSICAL CONSTANTS AND SPECTRAL PROPERTIES (Continued)

No.	Compound	Symbol 3-Letter	Symbol 1-Letter	Structure	Formula (Mol Wt)	Melting Point °C	$[\alpha]_D^t$	pK Basic	pK Acidic
184	5-Methylcytidine 5′-phosphate	5MeCyd-5′-P, 5MeCMP	pm^5C, $-m^5C$	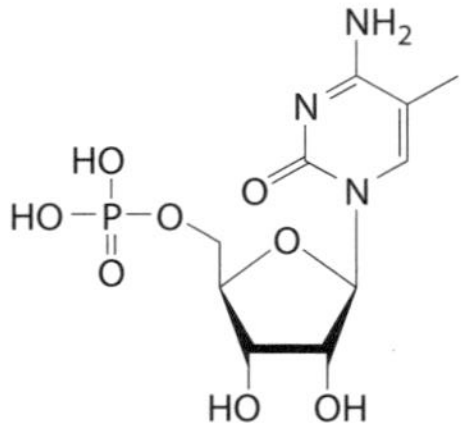	$C_{10}H_{16}N_3O_8P$ (337.22)	—	—	—	—
185	5-Methylcytidine 5′-diphosphate	5MeCyd-5′-P$_2$, 5MeCDP	ppm^5C	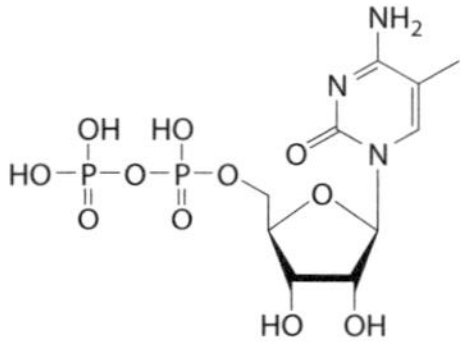	$C_{10}H_{17}N_3O_{11}P_2$ (417.21)	—	—	—	—
186	Guanosine 2′-phosphate	Guo-2′-P, 2′-GMP		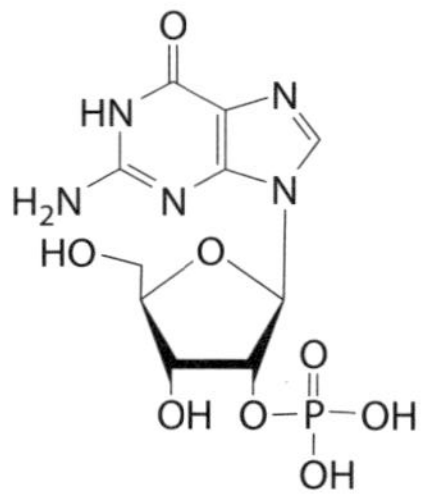	$C_{10}H_{14}N_5O_8P$ (363.22)	175–180° (dec)^z (dihydrate)	-57.0^{25z} (1.0, 2% NaOH)	—	—

No.		Acidic Spectral Data										Neutral Spectral Data									
	pH	λ_{max}	ε_{max} (×10⁻³)	λ_{min}	230	240	250	270	280	290	pH	λ_{max}	ε_{max} (×10⁻³)	λ_{min}	230	240	250	270	280	290
184	4^q	284	10.7	—	—	—	0.9	—	0.68	0.48	—	8	278	8.8	—	—	—	—	—	—
185	1	—	—	—	—	—	—	—	—	—	—	—	—	—	—	—	—	—	—	—
186											7	—	—	—	—	—	1.15	—	0.68	0.29

No.		Alkaline Spectral Data									
	pH	λ_{max}	ε_{max} (×10⁻³)	λ_{min}	230	240	250	270	280	290
184	—	—	—	—	—	—	—	—	—	—
185	—	—	—	—	—	—	—	—	—	—
186	12	—	—	—	—	—	0.89	—	0.60	0.11

REFERENCES

No.	Origin and Synthesis	$[\alpha]_D^t$	pK	Spectral Data	Mass Spectra	R_f
184	C: 59	—	—	169	—	59
185	C: 59	—	—	—	—	59
186	R: 222^{az},170	222^z	—	179	—	334

PURINES, PYRIMIDINES, NUCLEOSIDES AND NUCLEOTIDES: PHYSICAL CONSTANTS AND SPECTRAL PROPERTIES (Continued)

No.	Compound	Symbol 3-Letter	Symbol 1-Letter	Structure	Formula (Mol Wt)	Melting Point °C	$[\alpha]_D^t$	pK Basic	pK Acidic
187	Guanosine 3′-phosphate	Guo-3′-P, 3′-GMP	Gp, G-	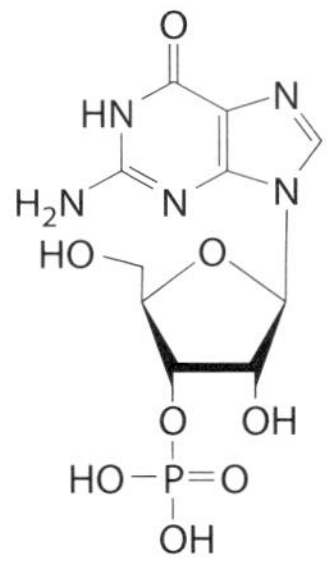	$C_{10}H_{14}N_5O_8P$ (363.22)	175–180° (dec)^z (dihydrate)	-57.0^{25z} (1.0, 2% NaOH)	2.3	9.7
188	Guanosine 5′-phosphate	Guo-5′-P, GMP	pG, -G	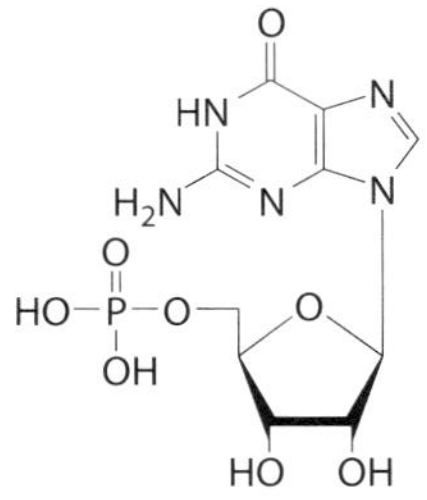	$C_{10}H_{14}N_5O_8P$ (363.22)	190–200° (dec)	—	2.4	9.4
189	Guanosine 5′-diphosphate	Guo-5′-P$_2$, GDP	ppG	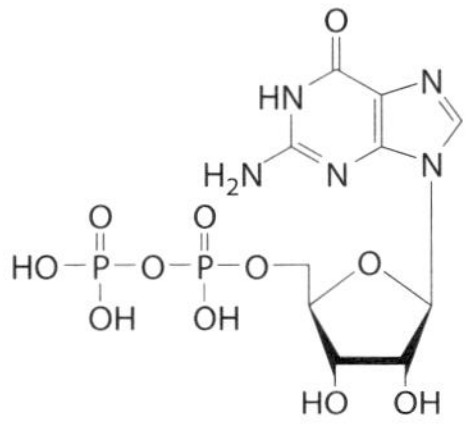	$C_{10}H_{15}N_5O_{11}P_2$ (443.21)	—	—	2.9	9.6

Acidic Spectral Data

No.	pH	λ_{max}	ε_{max} (×10^{-3})	λ_{min}	230	240	250	270	280	290	$[\alpha]_D^t$
187	1	257	12.2	228*	—	—	0.93	0.77	0.69	0.49	222^z
188	1	256	12.2	228	0.22	0.55	0.96	0.74	0.67	0.29	—
189	1	256	12.3	228	—	—	0.95	—	0.67	—	—

Neutral Spectral Data

No.	pH	λ_{max}	ε_{max} (×10^{-3})	λ_{min}	230	240	250	270	280	290	pK
187	7	252	13.4*	227*	—	—	1.15	0.86	0.68	0.29	231
188	7	252	13.7	224	0.36	0.81	1.16	0.81	0.66	0.29	212
189	7	253	13.7	224	—	—	1.15	—	0.66	—	212

Alkaline Spectral Data

No.	pH	λ_{max}	ε_{max} (×10^{-3})	λ_{min}	230	240	250	270	280	290	R_f
187	10.8^u	257	11.25	230*	—	—	0.92	1.00	0.64	0.15	258,334
188	11	258	11.6	230	0.38	0.82	0.90	0.97	0.61	0.29	189,213,334
189	11	258	11.7	230	—	—	0.91	—	0.61	—	213,334

REFERENCES

No.	Origin and Synthesis	Spectral Data	Mass Spectra
187	R: 222^{az},170	232^b,179^e,265^{be},400^{*be}	—
188	C: 190,199,189 N: 198 R: 195	212,183^b,184^b,198	—
189	C: 213,196 N: 198	212,183^b	—

PURINES, PYRIMIDINES, NUCLEOSIDES AND NUCLEOTIDES: PHYSICAL CONSTANTS AND SPECTRAL PROPERTIES (Continued)

No.	Compound	Symbol (3-Letter)	Symbol (1-Letter)	Structure	Formula (Mol Wt)	Melting Point °C	$[\alpha]_D^t$	pK Basic	pK Acidic
190	Guanosine 5′-triphosphate	Guo-5′-P$_3$ GTP	pppG	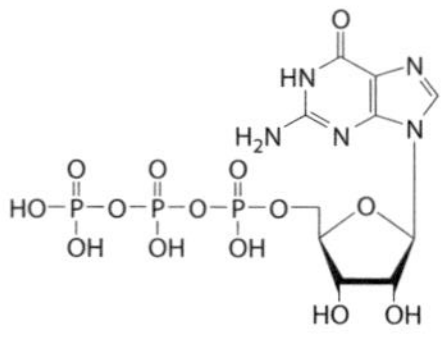	$C_{10}H_{16}N_5O_{14}P_3$ (523.19)	–	–	3.3	9.3
191	1-Methylguanosine 3′(2′)-phosphate	1MeGuo-3′(2′)-P	m^1 GP / m^1 G- for 3′	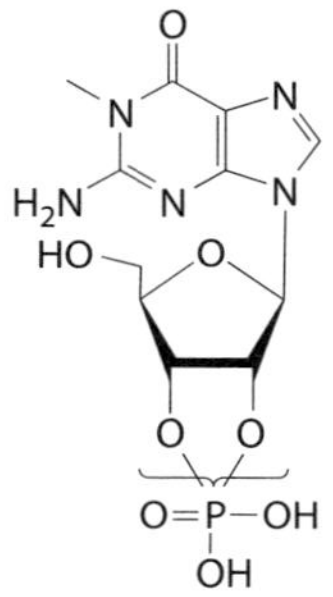	$C_{11}H_{16}N_5O_8P$ (377.25)	–	–	2.4^p	–

No.	pH	Acidic λ_{max}	ε_{max} (×10^{-3})	λ_{min}	230	240	250	270	280	290
190	1	256	12.4	228	–	–	0.96	–	0.67	0.51*
191	1	258	11.4^n	230*	0.21	0.45	0.86	0.8	0.72	–

No.	pH	Neutral λ_{max}	ε_{max} (×10^{-3})	λ_{min}	230	240	250	270	280	290
190	7	253	13.7	223	–	–	1.17	–	0.66	–
191	H$_2$O	255	12.4*	222	0.23	0.67	1.04	0.86	0.63	0.20

No.	pH	Alkaline λ_{max}	ε_{max} (×10^{-3})	λ_{min}	230	240	250	270	280	290
190	11	257	11.9	230	–	–	0.92	–	0.59	–
191	13	256	13.0^n	227*	0.3	0.66	1.02	0.84	0.63	0.20*

REFERENCES

No.	pK	$[\alpha]_D^t$	Origin and Synthesis	Spectral Data	Mass Spectra	R$_f$
190	212	–	C: 213 N: 198	212^b,183^b	–	213,334
191	40	–	R: -65,40	90,265*^{b},165	–	165

PURINES, PYRIMIDINES, NUCLEOSIDES AND NUCLEOTIDES: PHYSICAL CONSTANTS AND SPECTRAL PROPERTIES (Continued)

No.	Compound	Symbol 3-Letter	Symbol 1-Letter	Structure	Formula (Mol Wt)	Melting Point °C	$[\alpha]_D^t$	pK Basic	pK Acidic
192	N^2-Methylguanosine 3'(2')-phosphate	2MeGuo-3'(2')-P	m^2 Gp, m^2 G- for 3'	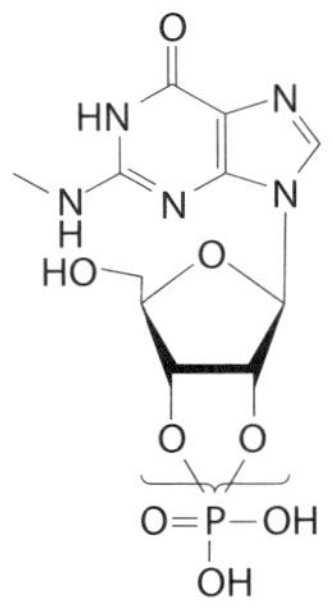	$C_{11}H_{16}N_5O_8P$ (377.25)	–	–	2.4^p	–
193	N^2,N^2-Dimethylguanosine 3'(2')-phosphate	2Me₂ Guo-3'(2')-P	m^2_2Gp, m^2_2G-; for 3'	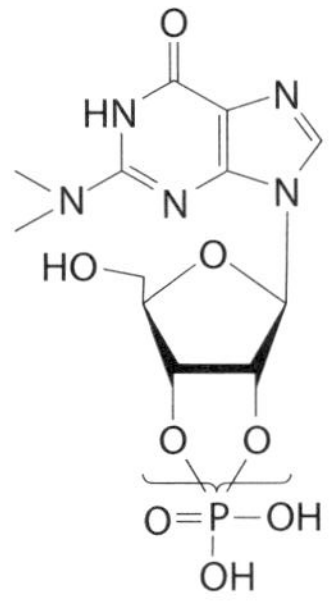	$C_{12}H_{18}N_5O_8P$ (391.28)	–	–	2.6^p	–

No.		Acidic Spectral Data									Neutral Spectral Data									
	pH	λ_{max}	ε_{max} (×10⁻³)	λ_{min}	230	240	250	270	280	290	pH	λ_{max}	ε_{max} (×10⁻³)	λ_{min}	230	240	250	270	280	290
192	1	259	14.2^n	232	0.29	0.44	0.81	0.77	0.60	0.53	H_2O	253	15.7^n	224	0.4	0.78	1.12	0.71	0.69	0.45
193	1	265	17.7^n	237*	0.42	0.29	0.62	0.97	0.58*	0.57	H_2O	259	19.2^n	228	0.25	0.47	0.84	0.72	0.58	0.50

	Alkaline Spectral Data								
pH	λ_{max}	ε_{max} (×10⁻³)	λ_{min}	230	240	250	270	280	290
13	258	13.3^n	236	0.72	0.59	0.89	0.91	0.78	0.41
13	263	14.9^n	241*	1.18	0.54	0.77	0.93	0.83	0.60*

REFERENCES

No.	Origin and Synthesis	$[\alpha]_D^t$	pK	Spectral Data	Mass Spectra	R_f
192	R: 40	–	40	90	–	–
193	R: 165,40	–	40	90,265*^{b},165	–	165

PURINES, PYRIMIDINES, NUCLEOSIDES AND NUCLEOTIDES: PHYSICAL CONSTANTS AND SPECTRAL PROPERTIES (Continued)

No.	Compound	Symbol 3-Letter	Symbol 1-Letter	Structure	Formula (Mol Wt)	Melting Point °C	$[\alpha]_D^t$	pK Basic	pK Acidic
194	7-Methylguanosine 2′-phosphate	7MeGuo-2′-P		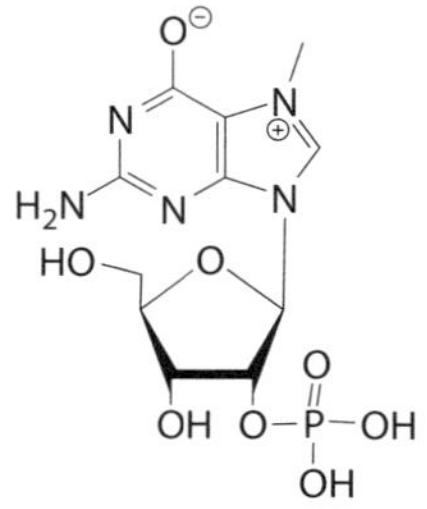 (structure)	$C_{11}H_{16}N_5O_8P$ (377.25)	–	–	r	7.0
195	7-Methylguanosine 3′-phosphate	7MeGuo-3′-P	m⁷Gp, m⁷G-	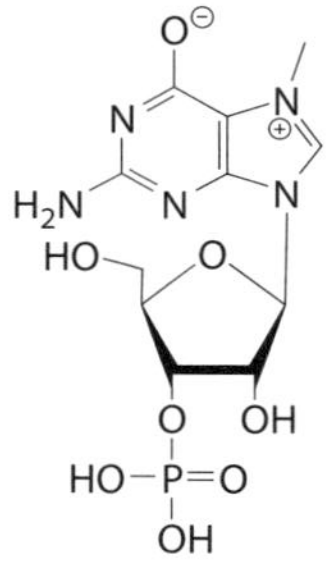 (structure)	$C_{11}H_{16}N_5O_8P$ (377.25)	–	–	r	6.9

Acidic Spectral Data

No.	pH	λ_{max}	ε_{max} (×10⁻³)	λ_{min}	230	240	250	270	280	290
194	2	257	12.6	230	–	–	–	–	–	–
195	2	257	13.2	230	0.26	0.51	0.89	0.74	0.68	0.52

Neutral Spectral Data

No.	pH	λ_{max}	ε_{max} (×10⁻³)	λ_{min}	230	240	250	270	280	290
194	7.4[q]	258, 280	9.6, 9.0	239, 271	–	–	–	–	–	–
195	7.4[q]	258, 282	9.8, 9.6	240, 270	–	–	–	–	–	–

Alkaline Spectral Data

No.	pH	λ_{max}	ε_{max} (×10⁻³)	λ_{min}	230	240	250	270	280	290
194	12[at]	268	9.6	245	–	–	0.92	1.09	1.4	–
195	8.9[b], 12[at]	258, 282	–9.9	241, 245	–	0.84	0.92	1.09	1.4	1.26

REFERENCES

No.	Origin and Synthesis	$[\alpha]_D^t$	pK	Spectral Data	Mass Spectra	R_f
194	C: 334 R: 77	–	334	334	–	334
195	C: 334,90 R: 77	–	334	334[b],90[e]	–	334

PURINES, PYRIMIDINES, NUCLEOSIDES AND NUCLEOTIDES: PHYSICAL CONSTANTS AND SPECTRAL PROPERTIES (Continued)

No.	Compound	Symbol (3-Letter)	Symbol (1-Letter)	Structure	Formula (Mol Wt)	Melting Point °C	$[\alpha]_D^t$	pK Basic	pK Acidic
196	7-Methylguanosine 5′-phosphate	7MeGuo-5′-P / 7MeGMP	pm^7 G / -m^7 G	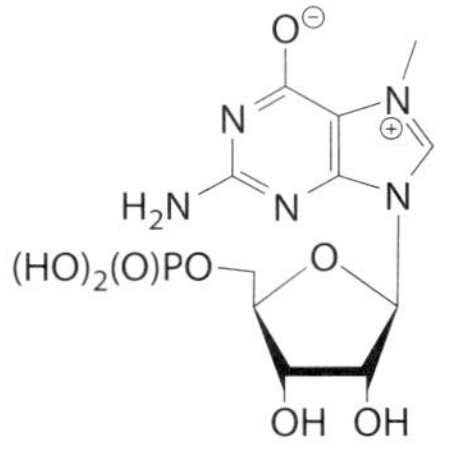	$C_{11}H_{16}N_5O_8P$ (377.25)	–	–	r	7.1
197	7-Methylguanosine 5′-diphosphate	7MeGuo-5′-P$_2$ / 7MeGDP	ppm^7 G	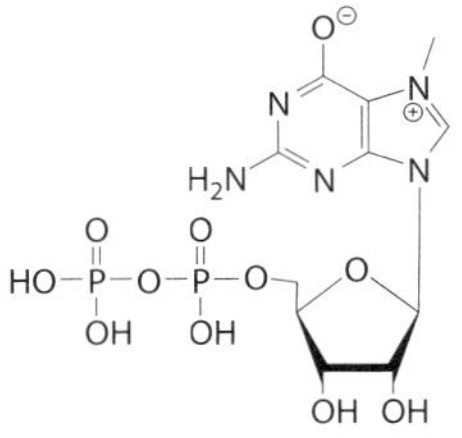	$C_{11}H_{17}N_5O_{11}P_2$ (457.23)	–	–	r	7.2
198	7-Methylguanosine 5′-triphosphate	7MeGuo-5′-P$_3$ / 7MeGTP	pppm^7 G	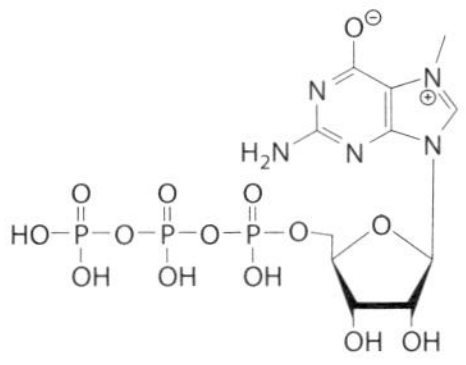	$C_{11}H_{18}N_5O_{14}P_3$ (537.21)	–	–	r	7.5

Acidic Spectral Data

No.	pH	λ_{max}	ε_{max} (×10^{-3})	λ_{min}	230	240	250	270	280	290
196	2	257	12	230	–	–	–	–	–	–
197	2	257	11.0	230	–	–	–	–	–	–
198	2	257	11.7	230	–	–	–	–	–	–

Neutral Spectral Data

No.	pH	λ_{max}	ε_{max} (×10^{-3})	λ_{min}	230	240	250	270	280	290
196	7.4^d	258 / 280	10.3 / 8.6	236 / 271	–	–	–	–	–	–
197	7.4^d	258 / 280	8.9 / 7.3	236 / 271	–	–	–	–	–	–
198	7.4^d	258	9.8	236	–	–	–	–	–	–

Alkaline Spectral Data

No.	pH	λ_{max}	ε_{max} (×10^{-3})	λ_{min}	230	240	250	270	280	290
196	8.9°	266	9.9	245	–	0.78	0.93	1.11	1.43	1.29
197	12^{at}	282	8.3	242	–	–	–	–	–	–
197	12as	268	7.0	244	–	–	–	–	–	–
198	12°	272 / 281	8.55	242 / 243	–	–	–	–	–	–

REFERENCES

No.	Origin and Synthesis	Spectral Data	pK	$[\alpha]_D^t$	Mass Spectra	R_f
196	C: 334,125	334^b	334	–	–	125,334
197	C: 334	334	334	–	–	334
198	C: 334	334	334	–	–	334

PURINES, PYRIMIDINES, NUCLEOSIDES AND NUCLEOTIDES: PHYSICAL CONSTANTS AND SPECTRAL PROPERTIES (Continued)

No.	Compound	Symbol (3-Letter)	Symbol (1-Letter)	Structure	Formula (Mol Wt)	Melting Point °C	$[\alpha]_D$	pK Basic	pK Acidic
199	Inosine 3'(2')-phosphate	Ino-3'(2')-P	Ip for 3'	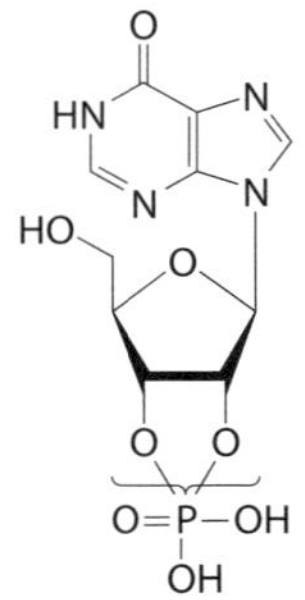	$C_{10}H_{13}N_4O_8P$ (348.21)	–	–	–	–
200	Inosine 5'-phosphate	IMP Ino-5'-P	pI	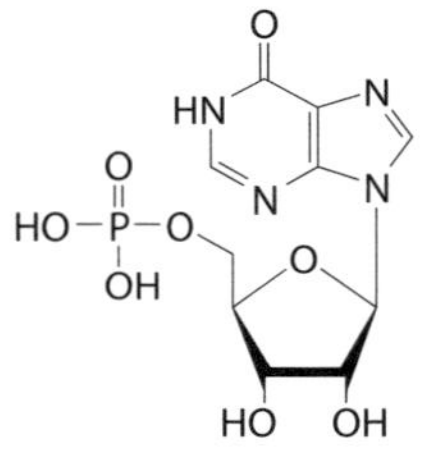	$C_{10}H_{13}N_4O_8P$ (348.21)	–	-18.4^{24} (0.9, 0.2 NHCl)	–	–
201	Inosine 5'-diphosphate Inosinic acid	IDP Ino-5'-P$_2$	ppI	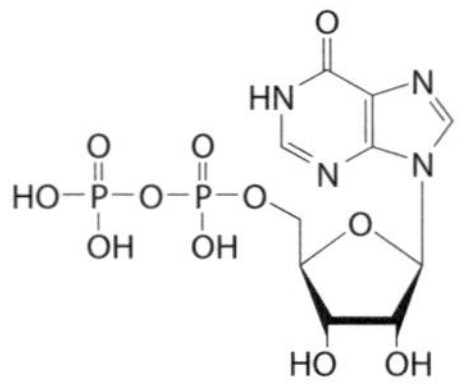	$C_{10}H_{14}N_4O_{11}P_2$ (428.19)	–	–	–	–

Neutral Spectral Data

No.	pH	λ_{max}	ε_{max} (×10⁻³)	λ_{min}	230	240	250	270	280	290
199	5	280	8.0	271	0.63	1.32	1.59	0.63	0.29	0.04
200	6	248	12.2	222	–	–	1.68	–	0.25	–
201	6	248.5	12.2	22.5	–	–	1.68	–	0.25	–

Acidic Spectral Data

No.	pH	λ_{max}	ε_{max} (×10⁻³)	λ_{min}	230	240	250	270	280	290
199	–	–	–	–	–	–	–	–	–	–
200	–	–	–	–	–	–	–	–	–	–
201	–	–	–	–	–	–	–	–	–	–

Alkaline Spectral Data

No.	pH	λ_{max}	ε_{max} (×10⁻³)	λ_{min}	230	240	250	270	280	290
199	–	–	–	–	–	–	–	–	–	–
200	–	–	–	–	–	–	–	–	–	–
201	–	–	–	–	–	–	–	–	–	–

REFERENCES

No.	Origin and Synthesis	$[\alpha]_D$	pK	Spectral Data	Mass Spectra	R_f
199	R: 90	–	–	90	–	–
200	C: 368	368	–	368,411^e	–	368,411
201	C: 411	–	–	411	–	411

PURINES, PYRIMIDINES, NUCLEOSIDES AND NUCLEOTIDES: PHYSICAL CONSTANTS AND SPECTRAL PROPERTIES (Continued)

No.	Compound	Symbol 3-Letter	Symbol 1-Letter	Structure	Formula (Mol Wt)	Melting Point °C	$[\alpha]_D^t$	pK Basic	pK Acidic
202	1-Methylinosine 3'(2')-phosphate	1MeIno-3'(2')-P	M^1Ip for 3'	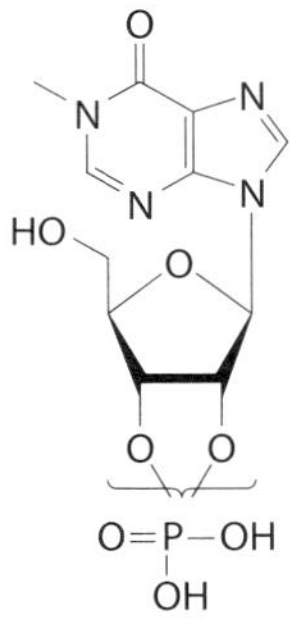	$C_{11}H_{15}N_4O_8P$ (362.24)	—	—	—	—
203	Uridine 2'-phosphate	Urd-2'-P 2'-UMP		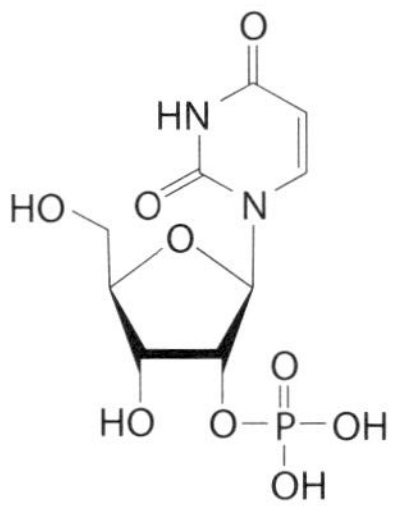	$C_9H_{13}N_2O_9P$ (324.18)	190–191° (dec)^z (Diammonium salt)	$+22.3^{22z}$ (2.0, H_2O)	—	—

No.	Acidic Spectral Data pH	λ_{max}	ε_{max} (×10⁻³)	λ_{min}	230	240	250	270	280	290
202	—	260*,x	9.9^x	—	—	—	—	—	—	—
203	2	260*,x	9.9^x	230^x	—	—	0.8	—	0.28	0.03

No.	Neutral Spectral Data pH	λ_{max}	ε_{max} (×10⁻³)	λ_{min}	230	240	250	270	280	290
202	4	249	—	233	0.66	1.35	1.65	0.72	0.41	0.07
203	7	260*,z	10.0^z	230^z	—	—	0.78	—	0.30	0.03

No.	Alkaline Spectral Data pH	λ_{max}	ε_{max} (×10⁻³)	λ_{min}	230	240	250	270	280	290
202	9.5	249.5	—	224	0.67	1.3	1.67	0.73	0.43	0.09
203	12	261^z	7.3^z	242^z	—	—	0.85	—	0.25	0.02

REFERENCES

No.	Origin and Synthesis	$[\alpha]_D^t$	pK	Spectral Data	Mass Spectra	R_f
202	R: 410	—	—	410	—	—
203	R: 178,170,216^{az}	216^z	—	181*zb,179e,90c	—	258

PURINES, PYRIMIDINES, NUCLEOSIDES AND NUCLEOTIDES: PHYSICAL CONSTANTS AND SPECTRAL PROPERTIES (Continued)

No.	Compound	Symbol 3-Letter	1-Letter	Structure	Formula (Mol Wt)	Melting Point °C	$[\alpha]_D^t$	pK Basic	pK Acidic
204	Uridine 3′-phosphate	Urd-3′-P 3′-UMP	Up U-	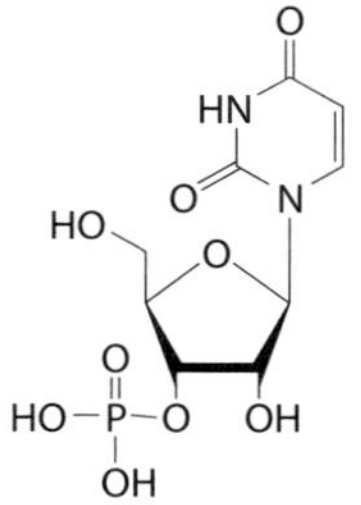	$C_9H_{13}N_2O_9P$ (324.18)	192°	$+22.3^{22z}$ (2.0, H_2O)	–	9.4
205	Uridine 5′-phosphate 5′-Uridylic acid	Urd-5′-P UMP	pU -U	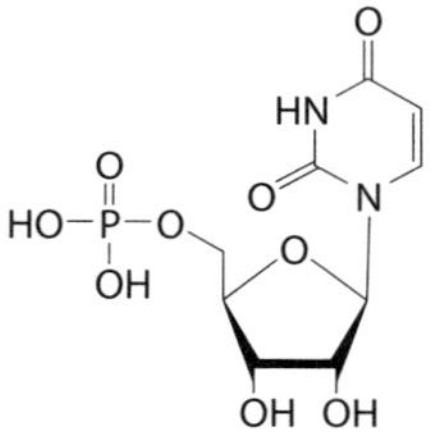	$C_9H_{13}N_2O_9P$ (324.18)	190–202° (dibrucine salt)	$+3.44^{28}$ (1.02, 10% HCl)	–	9.5
206	Uridine 5′-diphosphate	Urd-5′-P$_2$ UDP	ppU	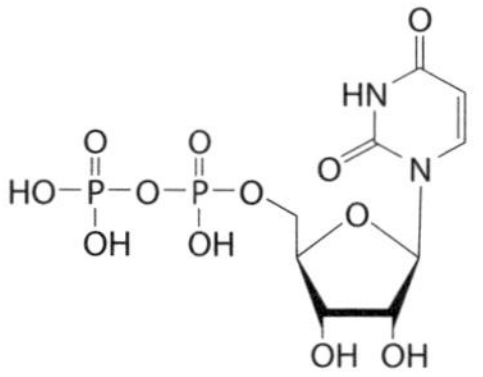	$C_9H_{14}N_2O_{12}P_2$ (404.16)	–	–	–	9.4

Neutral Spectral Data

No.	pH	λ_{min}	λ_{max}	ε_{max} (×10⁻³)	Spectral Ratios 230	240	250	270	280	290
204	7	230^z	262^z	10.0^z	–	0.21	0.73	0.87	0.35	0.03
205	7	230	262	10.0	0.21	0.38	0.73	0.87	0.39	0.03
206	7	230	262	10.0			0.73		0.39	–

Alkaline Spectral Data

No.	pH	λ_{min}	λ_{max}	ε_{max} (×10⁻³)	Spectral Ratios 230	240	250	270	280	290
204	13	241	261	7.8	0.79	–	0.83	0.85	0.28*	0.02*
205	11	241	261	7.8	0.79	0.5	0.8	–	0.31	0.02
206	11	241	261	7.9			0.8	–	0.32	–

Acidic Spectral Data

No.	pH	λ_{max}	ε_{max} (×10⁻³)	λ_{min}	Spectral Ratios 230	240	250	270	280	290
204	1	262	10	230	–	–	0.76	0.82	0.32*	0.03*
205	2	262	10.0	230	–	–	0.73	–	0.39*	0.03
206	2	262	10.0	230	–	–	0.73	–	0.39*	0.04

REFERENCES

No.	Origin and Synthesis	$[\alpha]_D^t$	pK	Spectral Data	Mass Spectra	R_f
204	C: 190 R: 178,170,216^z	216^z	223	265*[b],181[zbcd],179[e]	–	258
205	C: 264,190[a],368 R: 195 N: 198	217	212	21,183[be],184[b],179[e]	–	210,74
206	C: 210 N: 198	–	212	212,183[b],179*[e]	–	210

PURINES, PYRIMIDINES, NUCLEOSIDES AND NUCLEOTIDES: PHYSICAL CONSTANTS AND SPECTRAL PROPERTIES (Continued)

No.	Compound	Symbol (3-Letter)	Symbol (1-Letter)	Structure	Formula (Mol Wt)	Melting Point °C	$[\alpha]_D^t$	pK (Basic)	pK (Acidic)
207	Uridine 5′-triphosphate	Urd-5′-P$_3$ UTP	pppU	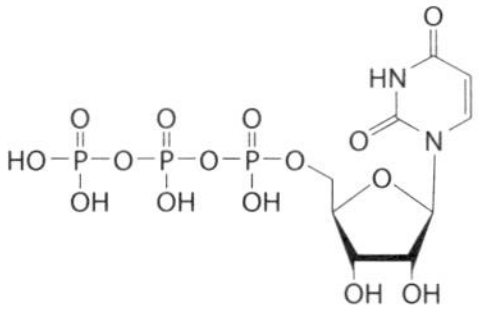 (structure)	C$_9$H$_{15}$N$_2$O$_{15}$P$_3$ (484.15)	–	–	–	9.6
208	2-Thiouridine 5′-phosphate	2SUrd-5′-P 2Srd-5′-P 2SUMP	ps^{2}U p^{2}S	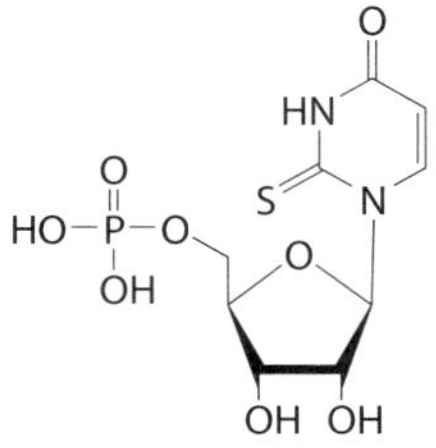 (structure)	C$_9$H$_{13}$N$_2$O$_8$PS (340.25)	–	–	–	–
209	5-(Methoxycarbonylmethyl)-2-thiouridine 3′-phosphate 5-Carboxymethyl-2-thiouridine methyl ester 3′-phosphate	5MeCm2SUrd-3′-P 5MeCm2SrD-3′-P	mcm^{5}s^{2}Up mcm^{5}Sp	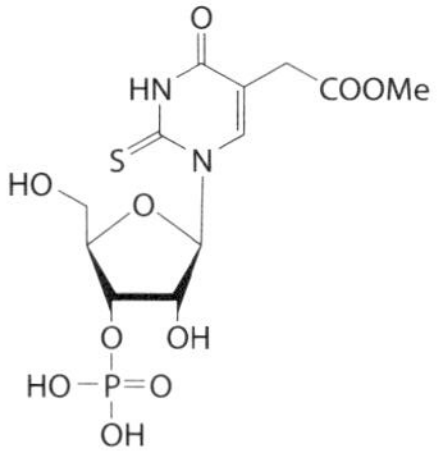 (structure)	C$_{12}$H$_{17}$N$_2$O$_{10}$PS (412.31)	–	–	–	–

Acidic Spectral Data

No.	pH	λ_{max}	ε_{max} (×10^{-3})	λ_{min}	230	240	250	270	280	290
207	2	262	10.0	230	0.21	0.37	0.75	0.88	0.38	–
208		275		243						
209	1									

Neutral Spectral Data

No.	pH	λ_{max}	ε_{max} (×10^{-3})	λ_{min}	230	250	270	280	290
207	7	262	10.0	230	0.21	0.75	0.86	0.38	–
208	H$_2$O	272		243					
209	7	275		243					

Alkaline Spectral Data

No.	pH	λ_{max}	ε_{max} (×10^{-3})	λ_{min}	230	240	250	270	280	290
207	11	261	8.1	239	0.79	0.65	0.81	0.78	0.31*	–
208	12°	241		213						
209										

REFERENCES

No.	Origin and Synthesis	$[\alpha]_D^t$	pK	R$_f$	Mass Spectra	Spectral Data
207	C: 210 N: 198,204	–	204,212	210	–	212[b],204[*e],183[b]
208	C: 344	–	–	344	–	344
209	R: 306	–	–	–	–	306[b]

PURINES, PYRIMIDINES, NUCLEOSIDES AND NUCLEOTIDES: PHYSICAL CONSTANTS AND SPECTRAL PROPERTIES (Continued)

No.	Compound	Symbol 3-Letter	Symbol 1-Letter	Structure	Formula (Mol Wt)	Melting Point °C	$[\alpha]_D^t$	pK Basic	pK Acidic
210	5-(Methylaminomethyl)-2-thiouridine 3'-phosphate	5(MeNHMe)2Srd-3'-P 5(MeNHMe)2SUrd-3'-P	mnm5s2Up mnm5Sp	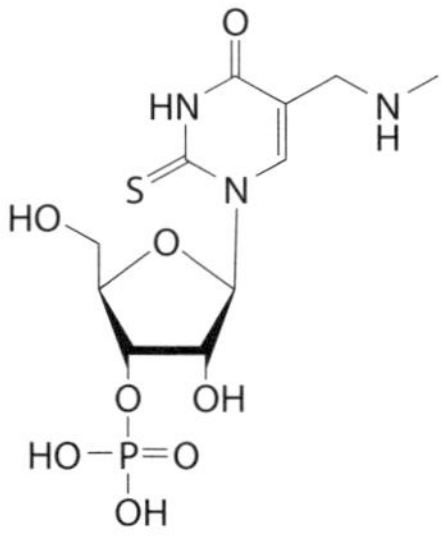 (structure)	$C_{11}H_{18}N_3O_8PS$ (383.32)	—	—	—	—
211	3-Methyluridine 3'(2')-phosphate	3MeUrd-3'(2')-P	m3UP m3U- for 3'	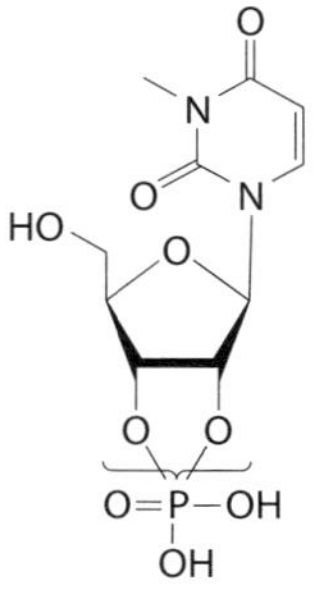 (structure)	$C_{10}H_{15}N_2O_9P$ (338.21)	—	—	—	—

Acidic Spectral Data

No.	pH	λ_{max}	ε_{max} ($\times 10^{-3}$)	λ_{min}	230	240	250	270	280	290
210										
211	2ᵛ	258	—	233	—	—	—	—	—	—

Neutral Spectral Data

No.	pH	λ_{max}	ε_{max} ($\times 10^{-3}$)	λ_{min}	230	240	250	270	280	290
210										
211	H₂O	262	8.8ᵃ	—	—	—	0.77	—	0.45	—

Alkaline Spectral Data

No.	pH	λ_{max}	ε_{max} ($\times 10^{-3}$)	λ_{min}	230	240	250	270	280	290
210										
211	11.6	260	9.3	233	—	—	—	—	—	—

REFERENCES

No.	Origin and Synthesis	$[\alpha]_D^t$	pK	Spectral Data	Mass Spectra	R_f
210	R: 305	—	—	—	—	305
211	C: 173,142,174	—	—	173,174ᶜ,142ᵃᶜᵈ	—	13,142,143,173

PURINES, PYRIMIDINES, NUCLEOSIDES AND NUCLEOTIDES: PHYSICAL CONSTANTS AND SPECTRAL PROPERTIES (Continued)

No.	Compound	Symbol 3-Letter	Symbol 1-Letter	Structure	Formula (Mol Wt)	Melting Point °C	$[\alpha]_D^t$	pK Basic	pK Acidic
212	3-Methyluridine 5′-phosphate	3MeUrd-5′-P 3MeUMP	Pm³U -m³U	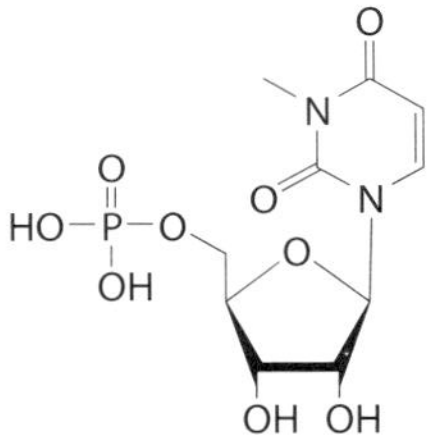	$C_{10}H_{15}N_2O_9P$ (338.21)	–	–	–	–
213	4-Thiouridine 3′(2′)-phosphate	4Srd-3′(2′)-P 4SUrd-3′(2′)-P	s⁴Up ⁴SP s⁴U- for 3′	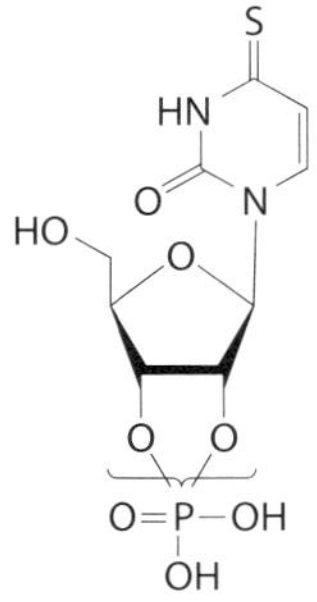	$C_9H_{13}N_2O_8PS$ (340.25)	–	–	–	–

Acidic Spectral Data

No.	pH	λ_{max}	ε_{max} (×10⁻³)	λ_{min}	Spectral Ratios 230	240	250	270	280	290
212	–	–	–	–	–	–	–	–	–	–
213	1	245 331	4.0 20.6*	225 276	–	–	–	–	–	–

Neutral Spectral Data

No.	pH	λ_{max}	ε_{max} (×10⁻³)	λ_{min}	Spectral Ratios 230	240	250	270	280	290
212	H₂O	262	8.8	232	–					
213	5.6	245 331	4.0 20.6	225 276						

Alkaline Spectral Data

No.	pH	λ_{max}	ε_{max} (×10⁻³)	λ_{min}	Spectral Ratios 230	240	250	270	280	290
213	13	315*	18.3*	257						

REFERENCES

No.	Origin and Synthesis	Spectral Data	$[\alpha]_D^t$	pK	Mass Spectra	R_f
212	C: 344	344	–	–	–	125,344
213	C: 343 R: 144	343,144*b	–	–	–	343

PURINES, PYRIMIDINES, NUCLEOSIDES AND NUCLEOTIDES: PHYSICAL CONSTANTS AND SPECTRAL PROPERTIES (Continued)

No.	Compound	Symbol 3-Letter	Symbol 1-Letter	Structure	Formula (Mol Wt)	Melting Point °C	$[\alpha]_D^t$	pK Basic	pK Acidic
214	4-Thiouridine 5′-phosphate	4Srd-5′-P 4SUrd-5′-P 4SUMP	ps^4U $-s^4U$ p^4S	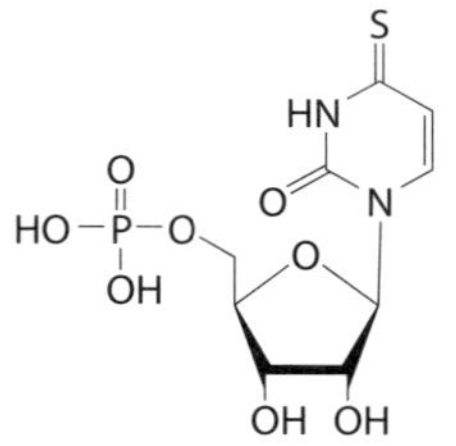	$C_9H_{13}N_2O_8PS$ (340.25)	–	–	–	–
215	5-Carboxymethyluridine 3′(2′)-phosphate Uridine-5-acetic acid 3′(2′)-phosphate	5CmUrd-3′(2′)-P 5CxMeUrd-3′(2′)-P	cm^5Up for 3′	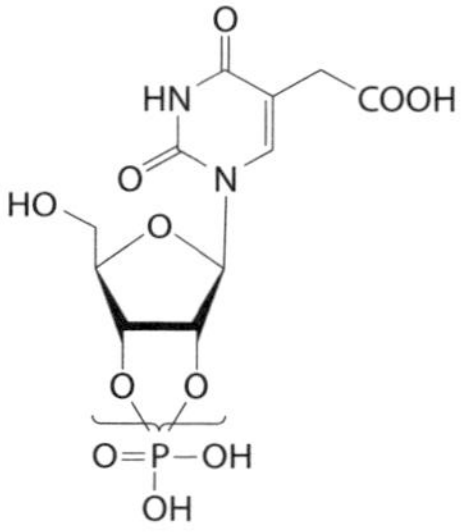	$C_{11}H_{15}N_2O_{11}P$ (382.22)	–	–	–	~4[P]

No.	Acidic pH	λ_{max}	ε_{max} (×10⁻³)	λ_{min}	230	240	250	270	280	290	Neutral pH	λ_{max}	ε_{max} (×10⁻³)	λ_{min}	230	240	250	270	280	290
214											H_2O	245 331	– 20.6	225 274						
215	2	265	9.7[n]	232	0.26	0.37	0.71	0.97	0.55	0.11	7	267	9.8[n]	232	0.23	0.28	0.62	1.06	0.69	0.14

No.	Alkaline pH	λ_{max}	ε_{max} (×10⁻³)	λ_{min}	230	240	250	270	280	290
214										
215	12.3	266	7.0[n]	242	1.13	0.74	0.77	0.98	0.55	0.07

REFERENCES

No.	Origin and Synthesis	$[\alpha]_D^t$	pK	Spectral Data	Mass Spectra	R_f
214	C: 344,343	–	–	344	–	343,344
215	R: 278,410	–	278	410	–	278

PURINES, PYRIMIDINES, NUCLEOSIDES AND NUCLEOTIDES: PHYSICAL CONSTANTS AND SPECTRAL PROPERTIES (Continued)

No.	Compound	Symbol 3-Letter	Symbol 1-Letter	Structure	Formula (Mol Wt)	Melting Point °C	$[\alpha]^t_D$	pK Basic	pK Acidic
216	5-Carbamoylmethyluridine 3'(2')-phosphate	5NcmUrd-3'(2')-P, 5NcMeUrd-3'(2')-P	ncm5Up for 3'	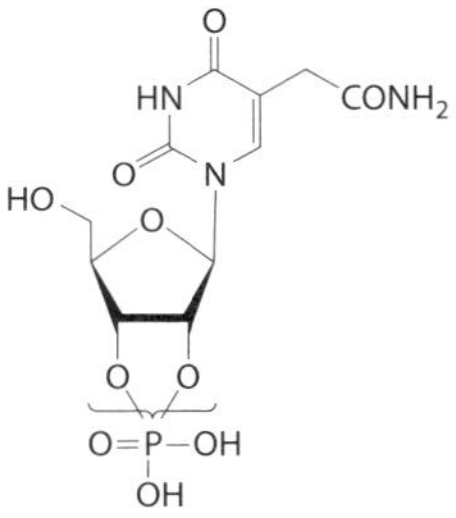	$C_{11}H_{16}N_3O_{10}P$ (381.24)	–	–	–	–
217	5-Hydroxyuridine 5'-phosphate	5HOUrd-5'-P, 5(HO)UMP	po5U, o5U	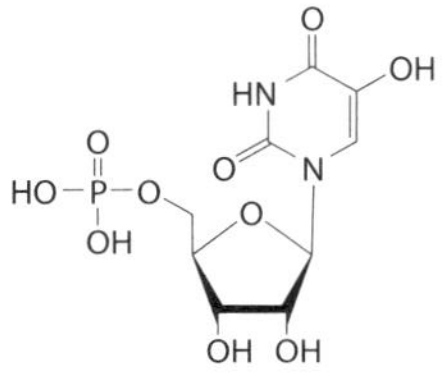	$C_9H_{13}N_2O_{10}P$ (340.18)	–	–	–	–

Acidic Spectral Data

No.	pH	λ_{max}	ε_{max} ($\times 10^{-3}$)	λ_{min}	230	240	250	270	280	290
216	2	266	10.0^n	232	0.31	0.38	0.71	0.98	0.55	0.11
217	–	–	–	–	–	–	–	–	–	–

Neutral Spectral Data

No.	pH	λ_{max}	ε_{max} ($\times 10^{-3}$)	λ_{min}	230	240	250	270	280	290
216	6	266	10.2^n	232	0.30	0.39	0.68	1.02	0.62	0.16
217	6	278	–	245	–	–	–	–	–	–

Alkaline Spectral Data

No.	pH	λ_{max}	ε_{max} ($\times 10^{-3}$)	λ_{min}	230	240	250	270	280	290
216	11.5	265	6.9^n	244	1.08	0.76	0.79	0.97	0.55	0.13
217	9	236, 300	–	268	–	–	–	–	–	–

REFERENCES

No.	Origin and Synthesis	Spectral Data	$[\alpha]^t_D$	pK	Mass Spectra	R_f
216	R: 398,410	410	–	–	–	–
217	C: 146,74	146,74	–	–	–	74

PURINES, PYRIMIDINES, NUCLEOSIDES AND NUCLEOTIDES: PHYSICAL CONSTANTS AND SPECTRAL PROPERTIES (Continued)

No.	Compound	Symbol: 3-Letter	Symbol: 1-Letter	Structure	Formula (Mol Wt)	Melting Point °C	$[\alpha]^t_D$	pK: Basic	pK: Acidic
218	5-Methyluridine 3′(2′)-phosphate Ribosylthymine 3′(2′)-phosphate	5MeUrd-3′(2′)-P Thd-3′(2′)-P	M5Upm5U- Tp T-for 3′	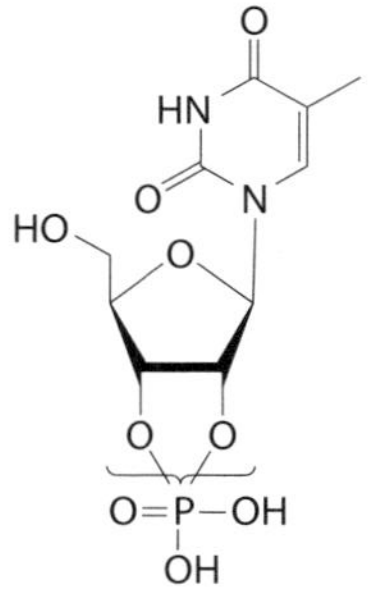	$C_{10}H_{15}N_2O_9P$ (338.21)	–	–	–	–
219	5-Methyluridine 5′-phosphate Ribosylthymine 5′-phosphate	Thd-5′-P TMP 5MeUMP	PT -T pm5U -m5U	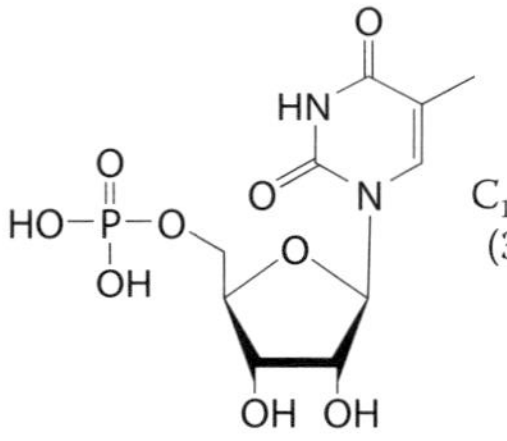	$C_{10}H_{15}N_2O_9P$ (338.21)	–	−12.3[26] (2.0, 0.1*N* HCl)	–	–

No.	Acidic Spectral Data										Neutral Spectral Data										Alkaline Spectral Data									
	pH	λ_{max}	ε_{max} (×10−3)	λ_{min}	Spectral Ratios						pH	λ_{max}	ε_{max} (×10−3)	λ_{min}	Spectral Ratios						pH	λ_{max}	ε_{max} (×10−3)	λ_{min}	Spectral Ratios					
					230	240	250	270	280	290					230	240	250	270	280	290					230	240	250	270	280	290
218	1	267*	9.8	235	–	–	0.68	1.05	0.66	0.23	–	–	–	–	–	–	–	–	–	–	13	268	–	247	–	–	0.79	1.04	0.69	0.23
219	2	267	8.8	–	–	–	–	–	–	–	–	–	–	–	–	–	–	–	–	–	–	–	–	–	–	–	–	–	–	–

REFERENCES

No.	Origin and Synthesis	$[\alpha]^t_D$	pK	Spectral Data	Mass Spectra	R_f
218	R: 165 C: 142	–	–	265[b],165*	–	165,142
219	C: 368	368	–	368	–	142

PURINES, PYRIMIDINES, NUCLEOSIDES AND NUCLEOTIDES: PHYSICAL CONSTANTS AND SPECTRAL PROPERTIES (Continued)

No.	Compound	Symbol 3-Letter	Symbol 1-Letter	Structure	Formula (Mol Wt)	Melting Point °C	$[\alpha]_D^t$	pK Basic	pK Acidic
220	5-Methyluridine 5′-diphosphate; Ribosylthymine 5′-diphosphate	Thd-5′-P₂ / TDP / 5MeUDP	ppm⁵U / ppT / ppT	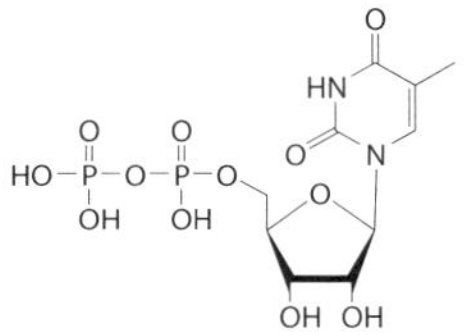	$C_{10}H_{16}N_2O_{12}P_2$ (418.18)	—	—	—	—
221	Pseudouridine 3′(2′) phosphate; β-f-Pseudouridine 3′(2′)-phosphate; 5-Ribosyluracil 3′(2′)-phosphate	ψrd-3′(2′)-P	Ψp / ψ- / For 3′	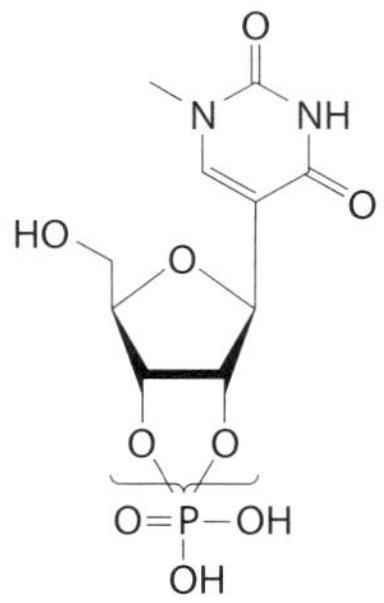	$C_9H_{13}N_2O_9P$ (324.18)	—	—	—	9.6
222	Pseudouridine 5′-phosphate; β-f-Pseudouridine 5′-phosphate; 5-Ribosyluracil 5′-phosphate	ψrd-5′-P / ψMP	Pψ / -ψ	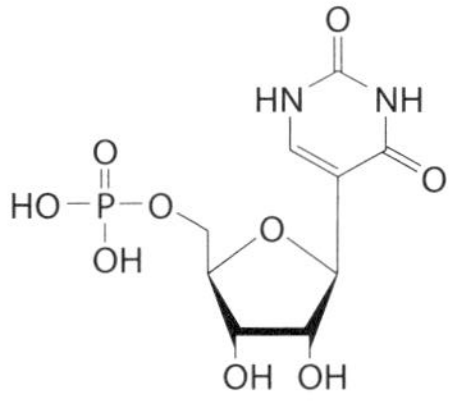	$C_9H_{13}N_2O_9P$ (324.18)	—	—	—	—

Acidic Spectral Data

No.	pH	λ_{min}	λ_{max}	ε_{max} (×10⁻³)	230	240	250	270	280	290
220	2	234	268	10.0	0.37	0.34	0.64	1.10	0.77	0.27
221	2*	233	263	8.4	0.30	0.41	0.75	0.86	0.40*	0.07*
222	—	—	—	—	—	—	—	—	—	—

Neutral Spectral Data

No.	pH	λ_{min}	λ_{max}	ε_{max} (×10⁻³)	230	240	250	270	280	290
220	—	—	—	—	—	—	—	—	—	—
221	7	233	263	—	0.28	0.39	0.74	0.85	0.40	0.07
222	—	—	—	—	—	—	—	—	—	—

Alkaline Spectral Data

No.	pH	λ_{min}	λ_{max}	ε_{max} (×10⁻³)	230	240	250	270	280	290
220	—	—	—	—	—	—	—	—	—	—
221	12	246	286	8.4	2.13	0.75	0.64*	1.54*	2.06*	2.14*
222	12	—	—	—	—	—	—	—	1.40	—

REFERENCES

No.	Origin and Synthesis	$[\alpha]_D^t$	pK	Spectral Data	Mass Spectra	R_f
220	C: 175	—	—	175	—	175
221	R: 157,94	—	157	157ᵇ,94,158,265*ᵇ	—	165
222	C: 158,176	—	—	158	—	158

PURINES, PYRIMIDINES, NUCLEOSIDES AND NUCLEOTIDES: PHYSICAL CONSTANTS AND SPECTRAL PROPERTIES (Continued)

No.	Compound	Symbol (3-Letter)	Symbol (1-Letter)	Structure	Formula (Mol Wt)	Melting Point °C	$[\alpha]_D$	pK Basic	pK Acidic
223	Pseucouridine 5'-diphosphate; β-ƒ-Pseudouridine 5'-diphosphate; 5-Ribosyluracil 5'-diphosphate	ψrd-5'-P₂, ψDP	—	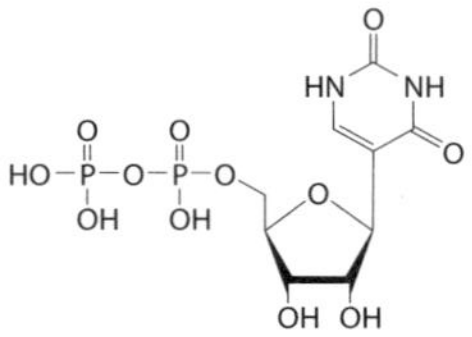	$C_9H_{14}N_2O_{12}P_2$ (404.16)	—	—	—	—
224	Orotidine 5'-phosphate; 6-Carboxyuridine 5'-phosphate	Ord-5'-P, OMP, 6CxUMP	pO, -O	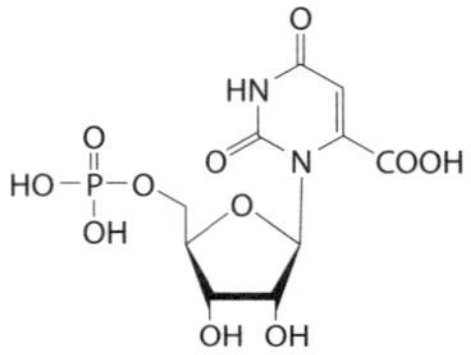	$C_{10}H_{13}N_2O_{11}P$ (368.19)	—	—	—	—
225	5-N-Methyl formamido-6-ribosylamino iso cystosine 3'(2')-phosphate‡; 2-Amino-4-hydroxy-5-N-methylformamido-6-ribosylaminopyrimidine 3'(2')-phosphate	5MeFn6(RibNH) isoCyt-3'(2')-P	—	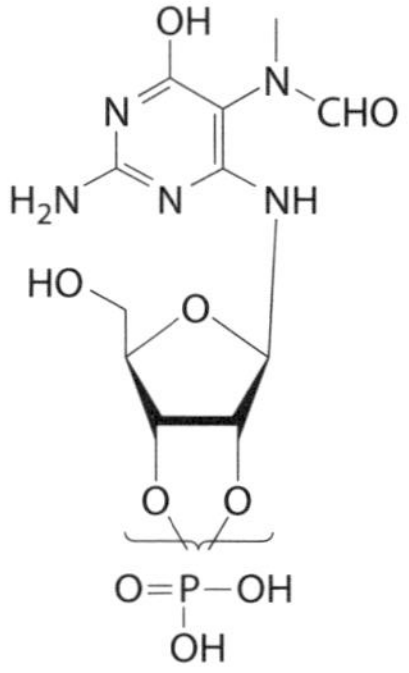	$C_{11}H_{18}N_5O_9P$ (395.27)	—	—	—	—

Acidic Spectral Data

No.	pH	λ_{max}	ε_{max} (×10⁻³)	λ_{min}	230	240	250	270	280	290
223	—	—	—	—	—	—	—	—	—	—
224	—	—	—	—	—	—	—	—	—	—
225	2	273	14.0ᵃ	247	1.00	0.50	0.47	1.81	1.41	0.31

Neutral Spectral Data

No.	pH	λ_{max}	ε_{max} (×10⁻³)	λ_{min}	230	240	250	270	280	290
223	—	—	—	—	—	—	—	—	—	—
224	7	266	—	—	—	—	—	—	0.66	—
225	—	—	—	—	—	—	—	—	—	—

Alkaline Spectral Data

No.	pH	λ_{max}	ε_{max} (×10⁻³)	λ_{min}	230	240	250	270	280	290
223	12	—	—	—	—	—	—	—	—	—
224	13	265	10.5ᵃ	244	0.77	0.51	0.56	1.03	1.30	0.26
225	—	—	—	—	—	—	—	—	—	0.05

REFERENCES

No.	Origin and Synthesis	$[\alpha]_D$	pK	Spectral Data	Mass Spectra	R_f
223	C: 158	—	—	158	—	158
224	E: 177	—	—	177	—	—
225	C: 90	—	—	90,334ᵇ	—	334

PURINES, PYRIMIDINES, NUCLEOSIDES AND NUCLEOTIDES: PHYSICAL CONSTANTS AND SPECTRAL PROPERTIES (Continued)

No.	Compound	Symbol (3-Letter)	Symbol (1-Letter)	Structure	Formula (Mol Wt)	Melting Point °C	$[\alpha]_D^t$	pK Basic	pK Acidic
				DEOXYRIBONUCLEOTIDES					
226	Deoxyadenosine 3′-phosphate	dAdo-3′P / 3′dAMP	dAp / dA-	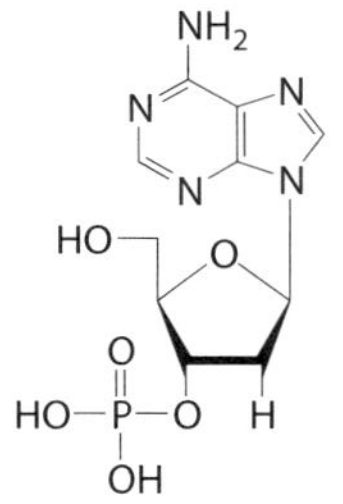 (structure)	$C_{10}H_{14}N_5O_6P$ (331.22)	—	—	—	—
227	Deoxyadenosine 5′-phosphate	dAdo-5′-P / dAMP	pdA / -dA	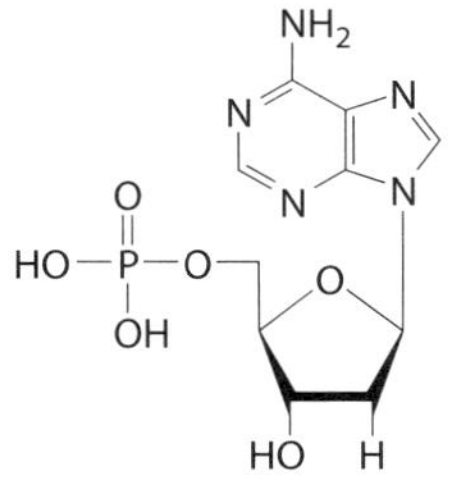 (structure)	$C_{10}H_{14}N_5O_6P$ (331.22)	142°	-38.0^{19} (0.23, H_2O)	~4.4	—
228	Deoxyadenosine 5′-triphosphate	dado-5′-P$_3$ / dATP	pppdA	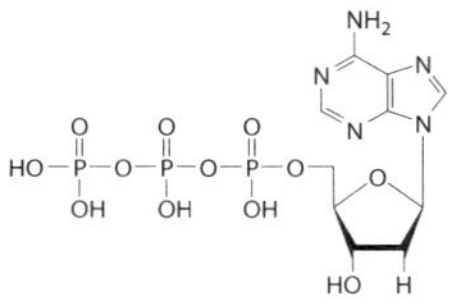 (structure)	$C_{10}H_{16}N_5O_{12}P_3$ (491.19)	—	—	—	—

Acidic Spectral Data

No.	pH	λ_{max}	ε_{max} (×10⁻³)	λ_{min}	Spectral Ratios 230	240	250	270	280	290
226	—	—	—	—	—	—	—	—	—	—
227	2	258	14.3*	230	—	—	0.82	—	0.23	0.04
228	—	—	—	—	—	—	—	—	—	—

Neutral Spectral Data

No.	pH	λ_{max}	ε_{max} (×10⁻³)	λ_{min}	Spectral Ratios 230	240	250	270	280	290
226	7	—	—	—	—	—	0.79	0.68	0.14	—
227	7	—	15.3	—	—	0.42	0.8	0.66	0.14	0.01
228	7	—	—	—	—	—	0.77	—	0.14	—

Alkaline Spectral Data

No.	pH	λ_{max}	ε_{max} (×10⁻³)	λ_{min}	Spectral Ratios 230	240	250	270	280	290
226	—	—	—	—	—	—	—	—	—	—
227	—	—	—	—	—	—	—	—	—	—
228	—	—	—	—	—	—	—	—	—	—

REFERENCES

No.	Origin and Synthesis	$[\alpha]_D^t$	pK	Spectral Data	Mass Spectra	R_f
226	D: 263 C: 203	—	—	263	—	263
227	D: 200a,260 C: 202	260	180	186,185*e,200,223cd	—	263
228	E: 209	—	—	209	—	—

PURINES, PYRIMIDINES, NUCLEOSIDES AND NUCLEOTIDES: PHYSICAL CONSTANTS AND SPECTRAL PROPERTIES (Continued)

No.	Compound	Symbol 3-Letter	Symbol 1-Letter	Structure	Formula (Mol Wt)	Melting Point °C	$[\alpha]_D^t$	pK Basic	pK Acidic
229	N^6-Methyldeoxyadenosine 5′-phosphate	6MedAdo-5′-P 6MedAMP d6MeAdo-5′-P	Pm6dA -m6dA	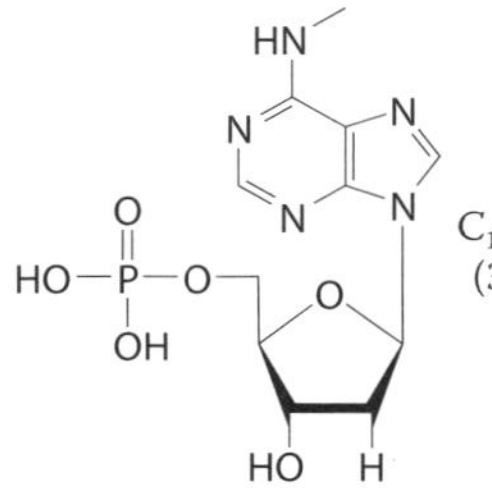 (structure)	$C_{11}H_{16}N_5O_6P$ (345.25)	—	—	3.6^p	—
230	Deoxycytidine 3′-phosphate	dCyd-3′-P 3′-dCMP	dCp dC-	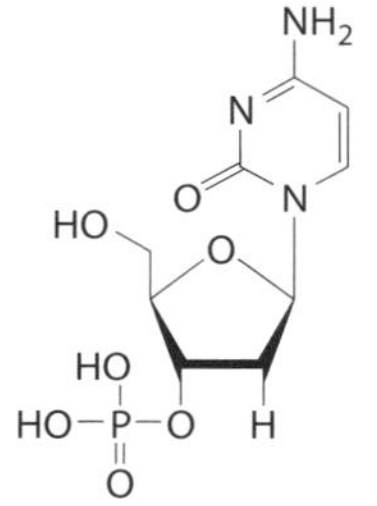 (structure)	$C_9H_{14}N_3O_7P$ (307.20)	196–197° (dec)	$+57.0^{17}$ (1.35, H_2O)	—	—
231	Deoxycytidine 5′-phosphate	dCyd-5′-P dCMP	pdC -dC	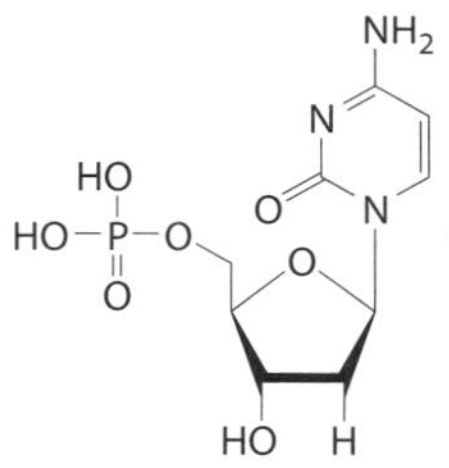 (structure)	$C_9H_{14}N_3O_7P$ (307.20)	183–184° (dec)	$+35.0^{21}$ (0.2, H_2O)	4.6	—

Acidic Spectral Data

No.	pH	λ_{max}	ε_{max} (×10⁻³)	λ_{min}	230	240	250	270	280	290
229	4^d	266	—	—	0.18	0.28	0.63	1.09	0.66	0.25
230	3	—	—	—	—	—	—	—	2.0	—
231	2	280	13.5	239	—	—	0.43	—	2.12	1.55

Neutral Spectral Data

No.	pH	λ_{max}	ε_{max} (×10⁻³)	λ_{min}	230	240	250	270	280	290
229	—	—	—	—	—	—	—	—	—	—
230	7	—	—	—	—	—	0.84	1.19	0.93	—
231	7	271	9.3	249	—	0.91	0.82	1.25	0.99	0.30

Alkaline Spectral Data

No.	pH	λ_{max}	ε_{max} (×10⁻³)	λ_{min}	230	240	250	270	280	290
229	—	—	—	—	—	—	—	—	—	—
230	13	266	—	234	0.18	0.24	0.57	1.08	0.63	0.21
231	12	—	—	—	—	—	0.82	—	0.99	0.30

REFERENCES

No.	Origin and Synthesis	Spectral Data	$[\alpha]_D^t$	pK	Mass Spectra	R_f
229	D: 21	21^b	—	21	—	—
230	D: 263 C: 201,203,202	$263,201^e$	201	—	—	203,201,263
231	C: 201 D: 200	$185^b,179^e,186^e,201$	207	180	—	185,201,263,74

PURINES, PYRIMIDINES, NUCLEOSIDES AND NUCLEOTIDES: PHYSICAL CONSTANTS AND SPECTRAL PROPERTIES (Continued)

No.	Compound	Symbol (3-Letter)	Symbol (1-Letter)	Structure	Formula (Mol Wt)	Melting Point °C	$[\alpha]_D^t$	pK Basic	pK Acidic
232	Deoxycytidine 5′-triphosphate	dCyd-5′-P$_3$ / dCTP	pppdC	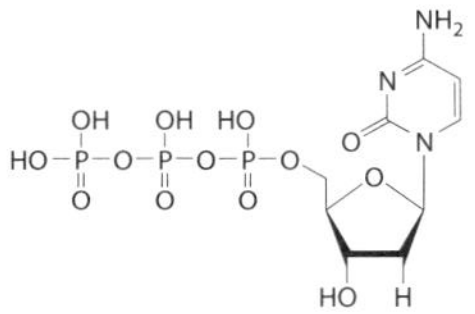	$C_9H_{16}N_3O_{13}P_3$ (467.17)	–	–	–	–
233	5-Methyldeoxycytidine 5′-phosphate	5MedCyd-5′-P / 5MedCMP / d5MeCyd-5′-P	pm^5 dC / -m^5 dC	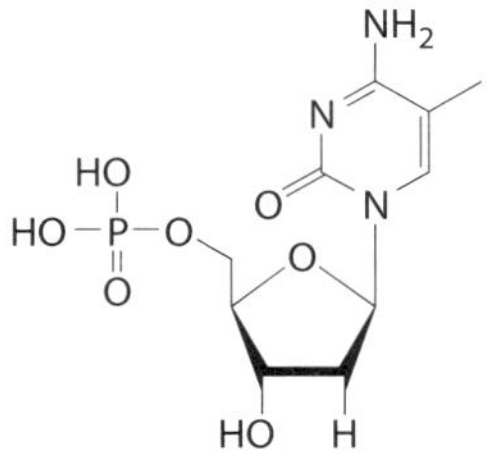	$C_{10}H_{16}N_3O_7P$ (321.22)	–	–	4.4	–
234	5-Hydroxymethyldeoxycytidine 5′-phosphate	5HmdCyd-5′-P / 5HmdCMP / 5HOMedCMP	pom^5 dC / -om^5 dC / phm^5 dC* / -hm^5 dC*	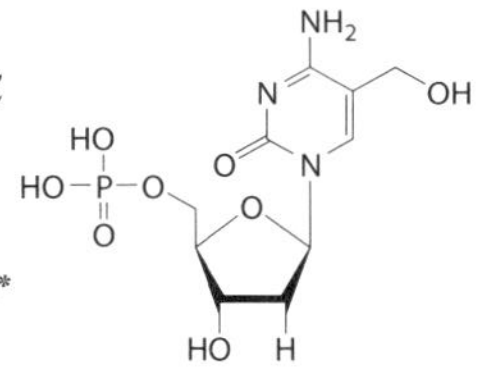	$C_{10}H_{16}N_3O_8P$ (337.22)	–	–	–	–

No.		Acidic Spectral Data									
	pH	λ_{max}	ε_{max} (×10^{-3})	λ_{min}	230	240	250	270	280	290	
232	2	–	–	–	–	–	0.44	2.10	2.14	3.44	
233	2	287	–	244	1.51	0.43	0.36	2.10	3.14	2.53	
234	1	284	12.5	245	1.12*	0.39	0.44	1.89	2.68		

No.	Neutral Spectral Data										Alkaline Spectral Data									
	pH	λ_{max}	ε_{max} (×10^{-3})	λ_{min}	230	240	250	270	280	290	pH	λ_{max}	ε_{max} (×10^{-3})	λ_{min}	230	240	250	270	280	290
232	7	–	–	–	–	–	–	–	–	–	–	–	–	–	–	–	–	–	–	–
233	7	278	–	254	1.52	1.29	0.95	1.40	1.52	1.01	12	278	–	–	–	–	–	–	–	–
234	7	275	7.7	254	1.50	1.10	0.90	1.35	1.33	0.71	12	275	7.7	254	1.40*	1.08	0.93	1.33	1.31	0.65

REFERENCES

No.	Origin and Synthesis	$[\alpha]_D^t$	pK	Spectral Data	Mass Spectra	R$_f$
232	E: 209 C: 214	–	–	209	–	–
233	D: 68	–	170	68^b,186	–	–
234	C: 74 E: 70 D: 171,172	–	–	74,70*^{bcde},171^d,172*	–	74

PURINES, PYRIMIDINES, NUCLEOSIDES AND NUCLEOTIDES: PHYSICAL CONSTANTS AND SPECTRAL PROPERTIES (Continued)

No.	Compound	Symbol (3-Letter)	Symbol (1-Letter)	Structure	Formula (Mol Wt)	Melting Point °C	$[\alpha]_D^t$	pK Basic	pK Acidic
235	Deoxyguanosine 3'-phosphate	dGuo-3'-P, 3'-dGMP	dGp, dG-	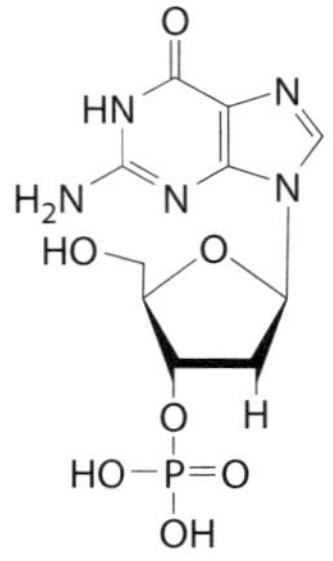	$C_{10}H_{14}N_5O_7P$ (347.23)	—	—	—	—
236	Deoxyguanosine 5'-phosphate	dGuo-5'-P, dGMP	pdG, -dG	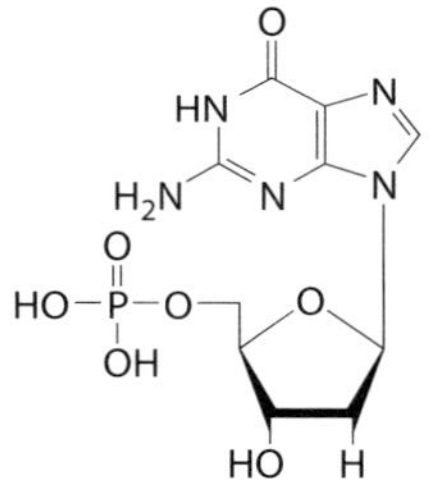	$C_{10}H_{14}N_8O_7P$ (347.23)	180–182°	-31^{19} (0.43, H_2O)	2.9	9.7

Acidic Spectral Data

No.	pH	λ_{max}	ε_{max} (×10⁻³)	λ_{min}	230	240	250	270	280	290
235	—	—	—	—	—	—	—	—	—	—
236	1°	255	11.8	228	—	—	1.02	—	0.70	—
	2^q	—	—	—	—	—	1.03	—	0.70	0.46

Neutral Spectral Data

No.	pH	λ_{max}	ε_{max} (×10⁻³)	λ_{min}	230	240	250	270	280	290
235	7	—	—	—	—	—	—	—	—	—
236	7	—	—	—	—	0.79	1.20	0.82	0.67	—
							1.13	0.81	0.67	0.27

Alkaline Spectral Data

No.	pH	λ_{max}	ε_{max} (×10⁻³)	λ_{min}	230	240	250	270	280	290
235	—	—	—	—	—	—	—	—	—	—
236	—	—	—	—	—	—	—	—	—	—

REFERENCES

No.	Origin and Synthesis	$[\alpha]_D^t$	pK	Spectral Data	Mass Spectra	R_f
235	D: 263 C: 203	—	—	263	—	263
236	D: 200,180[a],259	259	180	186,185[e],200,223[cde]	—	185,263

PURINES, PYRIMIDINES, NUCLEOSIDES AND NUCLEOTIDES: PHYSICAL CONSTANTS AND SPECTRAL PROPERTIES (Continued)

No.	Compound	Symbol (3-Letter)	Symbol (1-Letter)	Structure	Formula (Mol Wt)	Melting Point °C	$[\alpha]_D^t$	pK Basic	pK Acidic
237	Deoxyguanosine 5′-triphosphate	dGuo-5′-P₃ dGTP	pppdG	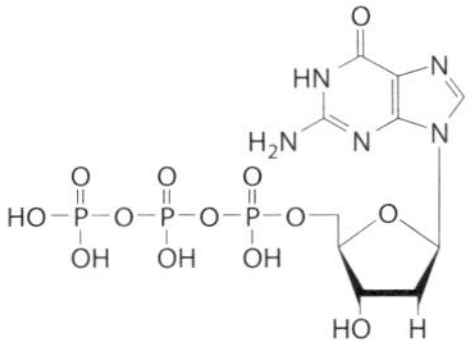 (structure)	$C_{10}H_{16}N_5O_{13}P_3$ (507.20)	—	—	—	—
238	7-Methyldeoxyguanosine 5′-phosphate	7MedGuo-5′-P 7MedGMP d7MeGuo-5′-P	pm⁷ dG -m⁷ dG	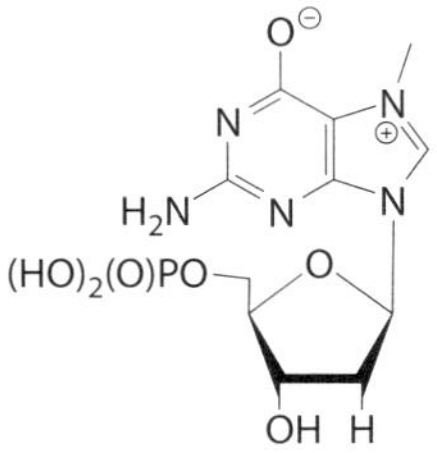 (structure)	$C_{11}H_{16}N_5O_7P$ (361.24)	—	—	r	—
239	7-Methyldeoxyguanosine 5′-triphosphate	7MedGuo-5′-P₃ 7MedGTP	pppm⁷ dG	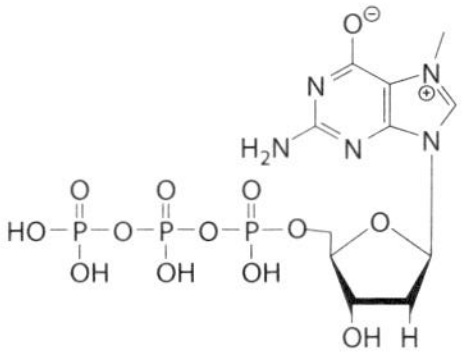 (structure)	$C_{11}H_{18}N_5O_{13}P_3$ (521.21)	—	—	r	7.5

Acidic Spectral Data

No.	pH	λ_{max}	ε_{max} (×10⁻³)	λ_{min}	230	240	250	270	280	290
237	—	—	—	—	—	—	—	—	—	—
238	—	257	10.6	230	—	—	—	—	—	—
239	2	—	—	—	—	—	—	—	—	—

Neutral Spectral Data

No.	pH	λ_{max}	ε_{max} (×10⁻³)	λ_{min}	230	240	250	280	290
237	7	256	9.8	—	—	—	1.14	0.66	—
238	7ᵠ	283	7.8	236	—	—	—	—	—
239	7.4ᵠ	258 / 280	8.9 / 7.25	271	—	—	—	—	—

Alkaline Spectral Data

No.	pH	λ_{max}	ε_{max} (×10⁻³)	λ_{min}	230	240	250	270	280	290
237	—	—	—	—	—	—	—	—	—	—
238	12ᵇ	281	7.9	243	—	—	—	—	—	—
239	—	—	—	—	—	—	—	—	—	—

REFERENCES

No.	Origin and Synthesis	$[\alpha]_D^t$	pK	Spectral Data	Mass Spectra	R_f
237	E: 209 C: 214	—	—	209	—	—
238	C: 46	—	—	46	—	46
239	C: 334	—	334	334	—	334

PURINES, PYRIMIDINES, NUCLEOSIDES AND NUCLEOTIDES: PHYSICAL CONSTANTS AND SPECTRAL PROPERTIES (Continued)

No.	Compound	Symbol (3-Letter)	Symbol (1-Letter)	Structure	Formula (Mol Wt)	Melting Point °C	$[\alpha]_D^t$	pK Basic	pK Acidic
240	Deoxyuridine 3'-phosphate	dUrd-3'-P / 3'-dUMP	dUp / dU-	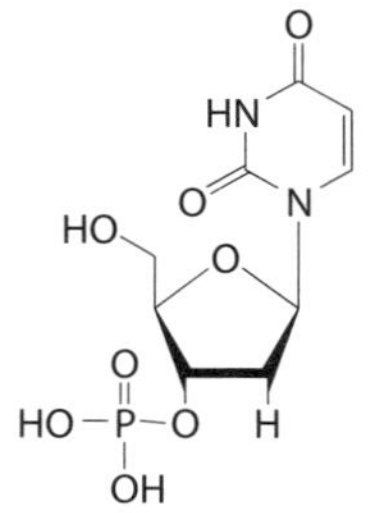	$C_9H_{13}N_2O_8P$ (308.18)	—	—	—	—
241	Deoxyuridine 5'-phosphate	dUrd-5'P / dUMP	pdU / -dU	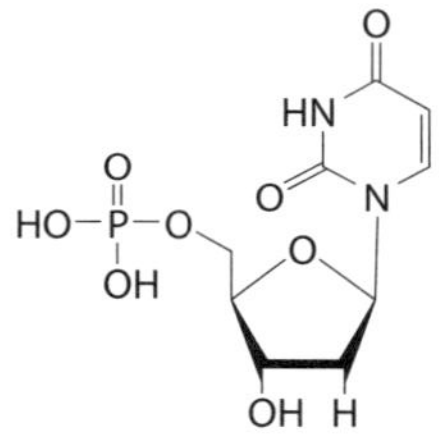	$C_8H_{13}N_2O_8P$ (308.18)	—	—	—	—
242	Deoxyuridine 5'-triphosphate	dUrd-5'-P₃ / dUTP	pppdU	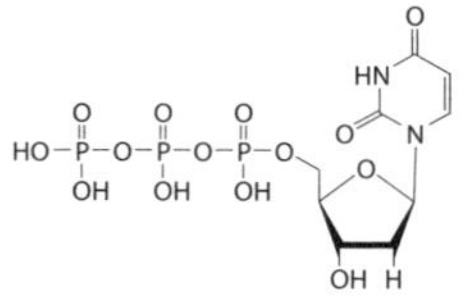	$C_9H_{15}N_2O_{14}P_3$ (468.15)	—	—	—	—

Acidic Spectral Data

No.	pH	λ_{max}	ε_{max} (×10⁻³)	λ_{min}	230	240	250	270	280	290
240	—	—	—	—	—	—	—	—	—	—
241	2	260	9.8	231	—	—	0.72	—	0.45	—
242	1	262	—	—	—	—	—	—	—	—

Neutral Spectral Data

No.	pH	λ_{max}	ε_{max} (×10⁻³)	λ_{min}	230	240	250	270	280	290
240	—	—	—	—	—	—	—	—	—	—
241	7	260	—	230	—	—	—	—	—	—
242	—	—	—	—	—	—	—	—	—	—

Alkaline Spectral Data

No.	pH	λ_{max}	ε_{max} (×10⁻³)	λ_{min}	230	240	250	270	280	290
240	—	—	—	—	—	—	—	—	—	—
241	12	261	7.6[a]	241	—	—	—	—	—	—
242	—	—	—	—	—	—	—	—	—	—

REFERENCES

No.	Origin and Synthesis	$[\alpha]_D^t$	pK	Mass Spectra	Spectral Data	R_f
240	—	—	—	—	—	—
241	E: 229	—	—	—	230[b]	74
242	C: 289	—	—	—	289	—

PURINES, PYRIMIDINES, NUCLEOSIDES AND NUCLEOTIDES: PHYSICAL CONSTANTS AND SPECTRAL PROPERTIES (Continued)

No.	Compound	Symbol (3-Letter)	Symbol (1-Letter)	Structure	Formula (Mol Wt)	Melting Point °C	$[\alpha]_D^t$	pK Basic	pK Acidic
243	Thymidine 3′-phosphate	dThd-3′-P / 3′dTMP	dTp / dT-	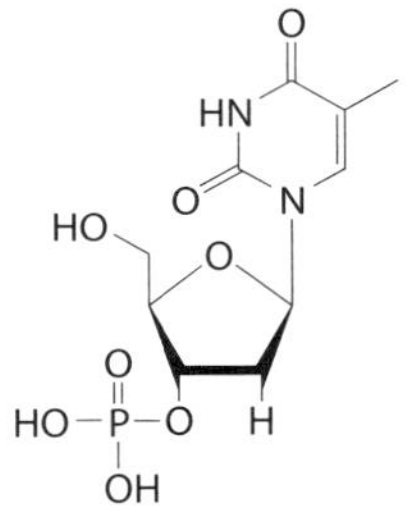	$C_{10}H_{15}N_2O_8P$ (322.21)	178°q (dibrucine salt)	$+7.3^{20}$ (1.5, H_2O)	—	—
244	Thymidine 5′-phosphate	dThd-5′-P / dTMP	pdT / -dT	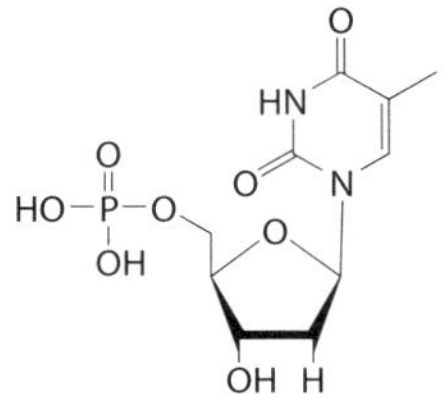	$C_{10}H_{15}N_2O_8P$ (322.21)	175° (dibrucine salt)	-4.4^{21} (0.4, H_2O)	—	10.0
245	Thymidine 5′-triphosphate	dThd-5′-P$_3$ / dTTP	pppdT	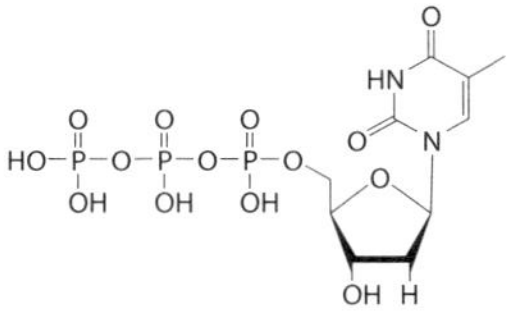	$C_{10}H_{17}N_2O_{14}P_3$ (482.18)	—	—	—	—

Acidic Spectral Data

No.	pH	λ_{max}	ε_{max} (×10⁻³)	λ_{min}	230	240	250	270	280	290
243	2	—	—	—	—	—	—	—	0.69	—
244	2	267	102	—	—	—	0.64	—	0.72	0.23
245	2	—	—	—	—	—	0.64	—	0.72	—

Neutral Spectral Data

No.	pH	λ_{max}	ε_{max} (×10⁻³)	λ_{min}	230	240	250	270	280	290
243	7	267	9.5	—	—	—	0.65	1.08	0.71	—
244	7	267	10.2	—	—	0.34	0.65	1.10	0.73	0.24
245	—	—	—	—	—	—	—	—	—	—

Alkaline Spectral Data

No.	pH	λ_{max}	ε_{max} (×10⁻³)	λ_{min}	230	240	250	270	280	290
243	—	—	—	—	—	—	—	—	—	—
244	12	—	—	—	—	—	0.74	—	0.67	0.17
245	—	—	—	—	—	—	—	—	—	—

REFERENCES

No.	Origin and Synthesis	pK	$[\alpha]_D^t$	Spectral Data	Mass Spectra	R_f
243	C: 262,202,187[a],203 D: 263	—	187	262,187[e],263[e]	—	187,262,263
244	C: 187,202 D: 200	180	207	185[b],186[e],179[e],223	—	187,185,263
245	E: 209 C: 209	—	—	209	—	—

PURINES, PYRIMIDINES, NUCLEOSIDES AND NUCLEOTIDES: PHYSICAL CONSTANTS AND SPECTRAL PROPERTIES (Continued)

No.	Compound	Symbol 3-Letter	1-Letter	Structure	Formula (Mol Wt)	Melting Point °C	$[\alpha]_D^t$	pK Basic	pK Acidic
246	5-Hydroxymethyldeoxyuridine 5'-phosphate	5HmdUrd-5'P 5HmdUMP 5(HOMe) dUMP	pom⁵ dU -om⁵ dU phm⁵ dU*	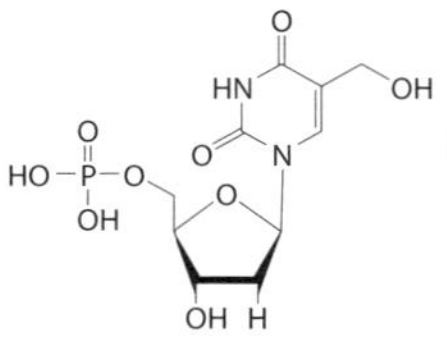 (see structure)	$C_{10}H_{15}N_2O_9P$ (338.21)	–	–	–	–

No.	pH	λ_{max}	ε_{max} (×10⁻³)	λ_{min}	230	240	250	270	280	290
					Acidic Spectral Data — Spectral Ratios					
246	2	264	10.2	234	0.32	0.37	0.69	0.97	0.56	0.11

pH	λ_{max}	ε_{max} (×10⁻³)	λ_{min}	230	240	250	270	280	290
			Neutral Spectral Data — Spectral Ratios						
–	–	–	–	–	–	–	–	–	–

pH	λ_{max}	ε_{max} (×10⁻³)	λ_{min}	230	240	250	270	280	290
			Alkaline Spectral Data — Spectral Ratios						
12	264		244*	1.15	0.75	0.80	0.95	0.48	0.09

No.	Origin and Synthesis	$[\alpha]_D^t$	pK	Spectral Data	Mass Spectra	R_f
246	C: 74 D: 96,171,349	–	–	74,171[d],96	–	74

REFERENCES

a Melting point from this reference.
b Full spectrum given.
c λ_{max} and/or λ_{min} from this reference.
d ε_{max} from this reference.
e Spectral ratios from this reference.
f pK of 2-methyl-6-methylaminoadenine. Compare the similar pK of adenine and N^6-methyladenine.[15]
g Spectral data taken on material synthesized this way then further purified by paper chromatography.
h In 50% dimethylformamide (HCONMe₂).
i In 50% dimethylsulfoxide/ethanol (Me₂ SO/EtOH).
j For an explanation of this nomenclature and abbreviations, see General Remarks on Wyosine in Natural Occurrence of Modified Nucleosides.
k pK of 6-amino-5-formamidoisocytosine.
l pK of nucleotide.
m λ_{max} and ε_{max} due to adenosine.
n ε calculated from spectral data using ε_{max} acid of nucleoside.
o Decomposes at this pH.
p Determined from electrophoretic mobility.
q Values very dependent on pH (near pK).
r Basic ionization at all pH values.
s Spectral data in water and pH 11 indicate decomposition.
t Alkaline degradation product of 7-methylguanosine or nucleotide.
v Spectra in acid and neutral similar.
w ε estimated from conversion to N^6-methyladenosine 3'(2')- or 5'-phosphate using ε of 15.2 × 10³ in alkali, for the N^6-isomers.
x Based on ε of 7-methylguanosine 3'(2')-phosphate assuming quantitative conversion in alkali. For a possible error in this value see General Remarks on 7-Methylguanosine in Natural Occurrence of Modified Nucleosides.
y Cyclohexylamine salt.
z Data on mixed 2' and 3' phosphates.

The authors are indebted to a number of authors who supplied unpublished data, provided original spectra for calculations of the values or gave advice on the selection of the most reliable data. They also wish particularly to thank Mr. I. H. Flack, Mr. R. Thedford and Miss L. Csonka for their assistance in the preparation of the table.

References

1. Leonard and Deyrup, *J. Am. Chem. Soc.*, 84, 2148 (1962).
2. Shapiro and Gordon, *Biochem. Biophys. Res. Commun.*, 17, 160 (1964).
3. Chheda, Hall, Magrath, Mozejko, Schweizer, Stasiuk, and Taylor, *Biochemistry*, 8, 3278 (1969).
4. Schweizer, Chheda, Baczynskyj, and Hall, *Biochemistry*, 8, 3283 (1969).
5. Baczynskyj, Biemann, Fleysher, and Hall, *Can. J. Biochem.*, 47, 1202 (1969).
6. Hall, *Biochemistry*, 3, 769 (1964).
7. Saponara and Enger, *Nature*, 223, 1365 (1969).
8. Brookes and Lawley, *J. Chem. Soc.*, 539 (1960) and unpublished.
9. Dunn, *Biochim. Biophys. Acta*, 46, 198 (1961).
10. Jones and Robins, *J. Am. Chem. Soc.*, 85, 193 (1963) and unpublished.
11. Mandel, Srinivasan, and Borek, *Nature*, 209, 586 (1966).
12. Wacker and Ebert, *Z. Naturforsch.*, 14B, 709 (1959).
13. Lawley and Brookes, *Biochem. J.*, 89, 127 (1963).
14. Leonard, Achmatowicz, Loeppky, Carraway, Grimm, Szweykowska, Hamzi, and Skoog, *Proc. Natl. Acad. Sci. USA*, 56, 709 (1966).
15. Martin and Reese, *J. Chem. Soc.*, 1731 (1968).
16. Leonard and Fujii, *Proc. Natl. Acad. Sci. USA*, 51, 73 (1964).
17. Baddiley, Lythgoe, McNeil, and Todd, *J. Chem. Soc.*, 383 (1943).
18. Littlefield and Dunn, *Biochem. J.*, 70, 642 (1958).
19. Lynch, Robins, and Cheng, *J. Chem. Soc.*, 2973 (1958).
20. Baddiley, Lythgoe, and Todd, *J. Chem. Soc.*, 318 (1944).
21. Dunn and Smith, *Biochem. J.*, 68, 627 (1958).
22. Albert and Brown, *J. Chem. Soc.*, 2060 (1954).
23. Elion, Burgi, and Hitchings, *J. Am. Chem. Soc.*, 74, 411 (1952).
24. Broom, Townsend, Jones, and Robins, *Biochemistry*, 3, 494 (1964).
25. Adler, Weissman, and Gutman, *J. Biol. Chem.*, 230, 717 (1958).
26. Elion, *The Chemistry and Biology of Purines* CIBA Symposium, Churchill, London, 1957, 39.
27. Weissman, Bromberg, and Gutman, *J. Biol. Chem.*, 224, 407 (1957).
28. Mason, *J. Chem. Soc.*, 2071 (1954).
29. Baker, Joseph, and Schaub, *J. Org. Chem.*, 19, 631 (1954).
30. Brederick, Haas, and Martini, *Chem. Ber.*, 81, 307 (1948).
31. Hall, Robins, Stasiuk, and Thedford, *J. Am. Chem. Soc.*, 88, 2614 (1966).
32. Robins, Hall, and Thedford, *Biochemistry*, 6, 1837 (1967).
33. Elion, *J. Org. Chem.*, 27, 2478 (1962).
34. Fischer, *Chem. Ber.*, 31, 104 (1898).
35. Pal, *Biochemistry*, 1, 558 (1962).
36. Lawley and Brookes, *Biochem. J.*, 92, 19c (1964).
37. Gulland and Holiday, *J. Chem Soc.*, 765 (1936).
38. Traube and Dudley, *Chem. Ber.*, 46, 3839 (1913).
39. Pfleiderer, *Annalen*, 647, 167 (1961).
40. Smith and Dunn, *Biochem. J.*, 72, 294 (1959).
41. Haines, Reese, and Todd, *J. Chem. Soc.*, 5281 (1962).
42. Reiner and Zamenhof, *J. Biol. Chem.*, 228, 475 (1957).
43. Elion, Lange, and Hitchings, *J. Am. Chem. Soc.*, 78, 217 (1956).
44. Gerster and Robins, *J. Am. Chem. Soc.*, 87, 3752 (1965).
45. Fischer, *Chem. Ber.*, 30, 2400 (1897).
46. Lawley, *Proc. Chem. Soc.*, 290 (1957).
47. Gulland and Story, *J. Chem. Soc.*, 692 (1938).
48. Brookes and Lawley, *J. Chem. Soc.*, 3923 (1961).
49. Hall, *Biochem. Biophys. Res. Commun.*, 13, 394 (1963) and unpublished.
50. Traube and Winter, *Arch. Pharm.*, 244, 11 (1906).
51. Ogston, *J. Chem. Soc.*, 1713 (1936).
52. Cohn, unpublished.
53. Cavalieri, Fox, Stone, and Chang, *J. Am. Chem. Soc.*, 76, 1119 (1954).
54. Pfleiderer and Nübel, *Annalen*, 647, 155 (1961).
55. Ogston, *J. Chem. Soc.*, 1376 (1935).
56. Taylor, *J. Chem. Soc.*, 765 (1948).
57. Brookes and Lawley, *J. Chem. Soc.*, 1348 (1962) and unpublished.
58. Ueda and Fox, *J. Am. Chem. Soc.*, 85, 4024 (1963).
59. Szer and Shugar, *Acta Biochim. Polon.*, 13, 177 (1966).
60. Ueda and Fox, *J. Org. Chem.*, 29, 1770 (1964).
61. Brown, *J. Appl. Chem.* 5, 358 (1955).

62. Johns, *J. Biol. Chem.*, 9, 161 (1911).
63. Brown, J. *Appl Chem.*, 9, 203 (1959).
64. Hitchings, Elion, Falco, and Russell, *J. Biol. Chem.*, 177, 357 (1949).
65. Wheeler and Johnson, *Am. Chem. J.*, 31, 591 (1904).
66. Shugar and Fox, *Biochim. Biophys. Acta*, 9, 199 (1952).
67. Wyatt, *Biochem. J.*, 48, 581 (1951).
68. Cohn, *J. Am. Chem. Soc.*, 73, 1539 (1951); also Beaven, Holiday, and Johnson, in *The Nucleic Acids, I*, Chargaff and Davidson, Eds., Academic Press, New York, 1955, 520.
69. Cline, Fink, and Fink, *J. Am. Chem. Soc.*, 81, 2521 (1959).
70. Flaks and Cohen, *J. Biol. Chem.*, 234, 1501 (1959).
71. Miller, *J. Am. Chem. Soc.*, 77, 752 (1955).
72. Fissekis, Myles, and Brown, *J. Org. Chem.*, 29, 2670 (1964).
73. Wyatt and Cohen, *Biochem. J.*, 55, 774 (1953).
74. Alegria, *Biochim Biophys. Acta*, 149, 317 (1967) and unpublished.
75. Loeb and Cohen, *J. Biol. Chem.*, 234, 364 (1959).
76. Pohland, Flynn, Jones, and Shive, *J. Am. Chem. Soc.*, 73, 3247 (1951).
77. Dunn, *Biochem. J.*, 86, 14P (1963).
78. Hilbert and Johnson, *J. Am. Chem. Soc.*, 52, 2001 (1930).
79. Brown, Hoerger, and Mason, *J. Chem. Soc.*, 211 (1955).
80. Levene, Bass, and Simms, *J. Biol. Chem.*, 70, 229 (1926).
81. Scannell, Crestfield, and Allen, *Biochim. Biophys. Acta*, 32, 406 (1959).
82. Whitehead, *J. Am. Chem. Soc.*, 74, 4267 (1952).
83. Johnson and Heyl, *Am. Chem. J.*, 37, 628 (1907).
84. Wheeler and Liddle, *Am Chem. J.*, 40, 547 (1908).
85. Fox and Van Praag, *J. Am. Chem. Soc.*, 82, 486 (1960).
86. Elion, Ide, and Hitchings, *J. Am. Chem. Soc.*, 68, 2137 (1946).
87. Wang, *J. Am. Chem. Soc.*, 81, 3786 (1959).
88. Johnson and McCollum, *J. Biol. Chem.*, 1, 437 (1906).
89. Behrend and Roosen, *Annalen*, 251, 235 (1889).
90. Dunn, D. B. and Flack, I. H., unpublished.
91. Stimson, *J. Am. Chem. Soc.*, 71, 1470 (1949).
92. Johnson and Litzinger, *J. Am. Chem. Soc.*, 58, 1940 (1936).
93. Dornow and Petsch, *Annalen*, 588, 45 (1954).
94. Cohn, *J. Biol. Chem.*, 235, 1488 (1960).
95. Yu and Allen, *Biochim. Biophys. Acta*, 32, 393 (1959).
96. Kallen, Simon, and Marmur, *J. Mol. Biol.*, 5, 248 (1962).
97. Chambers and Kurkov, *Biochemistry*, 3, 326 (1964).
98. di Carlo, Schultz, and Kent, *J. Biol Chem.*, 199, 333 (1952).
99. Green and Cohen, *J. Biol. Chem.*, 225, 397 (1957).
100. Brown and Johnson, *J. Am. Chem. Soc.*, 45, 2702 (1923).
101. Fischer and Roeder, *Chem. Ber.*, 34, 3751 (1901).
102. Lengfeld and Stieglitz, *Am. Chem. J.*, 15, 504 (1893).
103. Batt, Martin, Ploeser, and Murray, *J. Am. Chem. Soc.*, 76, 3663 (1954).
104. Mitchell and Nyc, *J. Am. Chem. Soc.*, 69, 674 and 1382 (1947).
105. Bachstetz, *Chem. Ber.*, 63b, 1000 (1930).
106. Johnson and Schroeder, *J. Am. Chem. Soc.*, 54, 2941 (1932).
107. Broom and Robins, *J. Am. Chem. Soc.*, 87, 1145 (1965).
108. Khwaja and Robins, *J. Am. Chem. Soc.*, 88, 3640 (1966).
109. Hall, *Biochim. Biophys. Acta*, 68, 278 (1963) and unpublished.
110. Hall, *Biochemistry*, 4, 661 (1965) and unpublished.
111. Davoll and Lowy, *J. Am. Chem. Soc.*, 74, 1563 (1952).
112. Johnson, Thomas, and Schaeffer, *J. Am. Chem. Soc.*, 80, 699 (1958).
113. Kissman, Pidacks, and Baker, *J. Am. Chem. Soc.*, 77, 18 (1955).
114. Townsend, Robins, Loeppky, and Leonard, *J. Am. Chem. Soc.*, 86, 5320 (1964).
115. Andrews and Barber, *J. Chem. Soc.*, 2768 (1958).
116. Feldmann, Dütting, and Zachau, *Hoppe-Seyler's Z. Physiol. Chem.*, 347, 236 (1966).
117. Hall, Csonka, David, and McLennan, *Science*, 156, 69 (1967) and unpublished.
118. Hall, *Biochemistry*, 3, 876 (1964).
119. Friedrich and Bernhauer, *Chem. Ber.*, 89, 2507 (1956).
120. Montgomery and Thomas, *J. Am. Chem. Soc.*, 85, 2672 (1963).
121. Friedrich and Bernhauer, *Chem. Ber.*, 90, 465 (1957).
122. Miles, *J. Org. Chem.*, 26, 4761 (1961).
123. Brown and Weliky, *J. Biol. Chem.*, 204, 1019 (1953).
124. Furukawa, Kobayaski, Kanai, and Honjo, *Chem. Pharm. Bull. Jap.*, 13, 1273 (1965) and unpublished.
125. Haines, Reese, and Todd, *J. Chem. Soc.*, 1406 (1964).
126. Hall, *Biochem. Biophys. Res. Commun.*, 12, 361 (1963).

127. Van Montagu and Stockx, *Arch. Intern. Physiol. Biochim*, 73, 158 (1965) and unpublished.
128. Fox, van Pragg, Wempen, Doerr, Cheong, Knoll, Eidinoff, Bendich, and Brown, *J. Am. Chem. Soc.*, 81, 178 (1959).
129. Wempen, Duschinsky, Kaplan, and Fox, *J. Am. Chem. Soc.*, 83, 4755 (1961).
130. Brimacombe and Reese, *J. Chem. Soc.*, 588C (1966).
131. Nichols and Lane, *Can. J. Biochem.*, 44, 1633 (1966).
132. Dunn, *Biochim. Biophys. Acta*, 38, 176 (1960).
133. Dekker and Elmore, *J. Chem. Soc.*, 2864 (1951).
134. Cohen and Barner, *J. Biol Chem.*, 226, 631 (1957).
135. Cohen, *Cold Spring Harbor Symp. Quant. Biol.*, 18, 221 (1953).
136. Brossmer and Röhm, *Angew. Chem. Int.*, 20, 742 (1963).
137. Townsend and Robins, *J. Am. Chem. Soc.*, 85, 242 (1963).
138. Smith and Dunn, *Biochim. Biophys. Acta*, 31, 573 (1959).
139. Miles, *Biochim. Biophys. Acta*, 22, 247 (1956).
140. Visser, Barron, and Beltz, *J. Am. Chem. Soc.*, 75, 2017 (1953).
141. Levene and Tipson, *J. Biol. Chem.*, 104, 385 (1934).
142. Thedford, Fleysher, and Hall, *J. Med. Chem.*, 8, 486 (1965).
143. Brimacombe, Griffin, Haines, Haslam, and Reese, *Biochemistry*, 4, 2452 (1965).
144. Lipsett, *J. Biol Chem.*, 240, 3975 (1965).
145. Roberts and Visser, *J. Am. Chem. Soc.*, 74, 668 (1952).
146. Ueda, *Chem. Pharm. Bull. Jap.*, 8, 455 (1960).
147. Levene and La Forge, *Chem. Ber.*, 45, 608 (1912).
148. Lis and Passarge, *Arch. Biochem. Biophys.*, 114, 593 (1966).
149. Reichard, *Acta Chem. Scand.*, 9, 1275 (1955).
150. Fox, Yung, Davoll, and Brown, *J. Am. Chem. Soc.*, 78, 2117 (1956).
151. Farkas and Sorm, *Coll. Czech. Chem. Commun.*, 28, 1620 (1963).
152. Brossmer and Röhm, *Angew. Chem. Int.*, 3, 66 (1964).
153. Cohn, Kurkov, and Chambers, *Biochem. Prep.*, 10, 135 (1963).
154. Shapiro and Chambers, *J. Am. Chem. Soc.*, 83, 3920 (1961).
155. Chambers, *Progr. Nucleic Acid Res. Mol. Biol.*, 5, 349 (1966) and Shapiro, R., Reeves, R. R., and Chambers, R. W., unpublished.
156. Ofengand and Schaefer, *Biochemistry*, 4, 2832 (1965).
157. Davis and Allen, *J. Biol. Chem.*, 227, 907 (1957).
158. Chambers, Kurkov, and Shapiro, *Biochemistry*, 2, 1192 (1963).
159. Michelson and Cohn, *Biochemistry*, 1, 490 (1962).
160. Michelson, Drell, and Mitchell, *Proc. Natl. Acad. Sci. USA*, 37, 396 (1951).
161. Bergmann and Feeney, *J. Org. Chem.*, 16, 981 (1951).
162. Fox, Yung, and Bendich, *J. Am. Chem. Soc.*, 79, 2775 (1957).
163. Fox and Shugar, *Biochim. Biophys. Acta*, 9, 369 (1952).
164. Griffin and Reese, *Biochim. Biophys. Acta*, 68, 185 (1963) and unpublished.
165. Davis, Carlucci, and Roubein, *J. Biol. Chem.*, 234, 1525 (1959).
166. Ikehara, Ohtsuka, and Ishikawa, *Chem. Pharm. Bull. Jap.*, 9, 173 (1961).
167. Griffin, Haslam, and Reese, *J. Mol. Biol.*, 10, 353 (1964).
168. Ikehara, Ueda, and Ikeda, *Chem. Pharm. Bull. Jap.*, 10, 767 (1962).
169. Szer, *Biochem. Biophys. Res. Commun.*, 20, 182 (1965).
170. Cohn, *in The Nucleic Acids* I, Chargaff, Davidson, Academic Press, New York, 1955, 211.
171. Kuno and Lehman, *J. Biol. Chem.*, 237, 1266 (1962).
172. Lehman and Pratt, *J. Biol. Chem.*, 235, 3254 (1960).
173. Szer and Shugar, *Acta Biochim. Polon.*, 7, 491 (1960).
174. Letters and Michelson, *J. Chem. Soc.*, 71 (1962).
175. Griffin, Todd, and Rich, *Proc. Natl. Acad. Sci. USA*, 44, 1123 (1958) and unpublished.
176. Goldberg and Rabinowitz, *Biochim. Biophys. Acta*, 54, 202 (1961).
177. Lieberman, Kornberg, and Simms, *J. Am. Chem. Soc.*, 76, 2844 (1954).
178. Cohn, *J. Am. Chem. Soc.*, 72, 2811 (1950).
179. Cohn, in *The Nucleic Acids* I, Chargaff, Davidson, Eds., Academic Press, New York, 1955, 513.
180. Hurst, Marko, and Butler, *J. Biol. Chem.*, 204, 847 (1953).
181. Ploeser and Loring, *J. Biol. Chem.*, 178, 431 (1949).
182. Cohn, *J. Cell Comp. Physiol.*, 38, supp, 1, 21 (1951).
183. Anon., Pabst Laboratories, Circular OR-10 (1956).
184. Steiner and Beers Jr., in *Polynucleotides*. Elsevier, New York, 1961, 155.
185. Shapiro and Chargaff, *Biochim. Biophys. Acta*, 26, 596 (1957).
186. Sinsheimer, *J. Biol. Chem.*, 208, 445 (1954).
187. Michelson and Todd, *J. Chem. Soc.*, 951 (1953).
188. Clark, Kirby, and Todd, *J. Chem. Soc.*, 1497 (1957).
189. Chambers, Moffatt, and Khorana, *J. Am. Chem. Soc.*, 79, 3747 (1957).
190. Michelson and Todd, *J. Chem. Soc.*, 2476 (1949).
191. Brown, Fasman, Magrath, and Todd, *J. Chem. Soc.*, 1448 (1954).
192. Loring, Bortner, Levy, and Hammell, *J. Biol. Chem.*, 196, 807 (1952).
193. Barker, *J. Chem. Soc.*, 3396 (1954).
194. Jones and Perkins, *J. Biol. Chem.*, 62, 557 (1925).
195. Cohn and Volkin, *Arch. Biochem. Biophys.*, 35, 465 (1952).
196. Chambers, Shapiro, and Kurkov, *J. Am. Chem. Soc.*, 82, 970 (1960).
197. Brown, Haynes, and Todd, *J. Chem. Soc.*, 3299 (1950).
198. Schmitz, Hurlbert, and Potter, *J. Biol. Chem.*, 209, 41 (1954).
199. Chambers, Moffatt, and Khorana, *J. Am. Chem. Soc.*, 77, 3416 (1955).
200. Volkin, Khym, and Cohn, *J. Am. Chem. Soc.*, 73, 1533 (1951).
201. Michelson and Todd, *J. Chem. Soc.*, 34 (1954).
202. Tener, *J. Am. Chem. Soc.*, 83, 159 (1961).
203. Schaller, Weimann, Lerch, and Khorana, *J. Am. Chem. Soc.*, 85, 3821 (1963).
204. Lipton, Morell, Frieden, and Bock, *J. Am. Chem. Soc.*, 75, 5449 (1953).
205. Volkin and Cohn in *Methods of Biochemical Analysis*, I, Glick, Ed., Interscience, New York, 1954. 287.
206. Morell and Bock, Am. Chem. Soc. 126th Meeting, New York, Div. of Biol., *Abstracts*, 1954, 44C.
207. Klein and Thannhauser, *Z. Physiol. Chem.*, 231, 96 (1935).
208. Lipsett, *Cold Spring Harbor Symp. Quant. Biol.*, 31, 449 (1966).
209. Lehman, Bessman, Simms, and Kornberg, *J. Biol. Chem.*, 233, 163 (1958).
210. Hall and Khorana, *J. Am. Chem. Soc.*, 76, 5056 (1954).
211. Khorana, *J. Am. Chem. Soc.*, 76, 3517 (1954).
212. Bock, Nan-Sing Ling, Morell, and Lipton, *Arch. Biochem. Biophys.*, 62, 253 (1956).
213. Chambers and Khorana, *J. Am. Chem. Soc.*, 79, 3752 (1957).
214. Smith and Khorana, *J. Am. Chem. Soc*, 80, 1141 (1958).
215. Loring and Luthy, *J. Am. Chem. Soc.*, 73, 4215 (1951).
216. Loring, Roll, and Pierce, *J. Biol. Chem.*, 174, 729 (1948).
217. Levene and Tipson, *J. Biol. Chem.*, 106, 113 (1934).
218. Cavalieri, *J. Am. Chem. Soc.*, 75, 5268 (1953).
219. Brown and Todd, *J. Chem. Soc.*, 44 (1952).
220. Alberty, Smith, and Bock, *J. Biol. Chem.*, 193, 425 (1951).
221. Reichard, Takenaka, and Loring, *J. Biol. Chem.*, 198, 599 (1952).
222. Levene, *J. Biol. Chem.*, 41, 483 (1920).
223. California Corporation for Biochemical Research, *Properties of Nucleic Acids Derivatives*, 4th Revision, 1961.
224. Levene and Jacobs, *Chem. Ber.*, 43, 3150 (1911).
225. Chambers, Moffatt, and Khorana, *J. Am. Chem. Soc.*, 79, 4240 (1957).
226. Le Page, *Biochem. Preps.*, 1, 1, (1949).
227. Shaw, Smallwood, and Wilson, *J. Chem. Soc.*, 921C (1969).
228. Carrington, Shaw, and Wilson, *J. Chem. Soc.*, 6864 (1965).
229. Scarano, *Boll. Soc. Ital. Biol. Sper.*, 34, 722 (1958).
230. Scarano, *Boll. Soc. Ital. Biol. Sper.*, 34, 727 (1958).
231. Levene and Bass, *in The Nucleic Acids*. Chemical Catalog Co, New York, 1931.
232. Beaven, Holiday, Johnson, (1955) in *The Nucleic Acids* I, Chargaff, Davidson, Eds., Academic Press, New York, 1955, 493.
233. Hilbert, Jansen, and Hendricks, *J. Am. Chem. Soc.*, 57, 552 (1935).
234. Traube, *Chem. Ber.*, 33, 1371 (1900).
235. Davoll, Lythgoe, and Todd, *J. Chem. Soc.*, 967 (1948).
236. Elmore, *J. Chem. Soc.*, 2084 (1950).
237. Davoll and Lowy, *J. Am. Chem. Soc.*, 73, 1650 (1951).
238. Andersen, Dekker, and Todd, *J. Chem. Soc.*, 2721 (1952).
239. Klein, *Z. Physiol. Chem.*, 224, 244 (1934).
240. Deutsch, unpublished data.
241. Schindler, *Helv. Chim. Acta*, 32, 979 (1949).
242. Hotchkiss, *J. Biol. Chem.*, 175, 315 (1948).
243. Dekker and Todd, *Nature*, 166, 557 (1950).
244. Watanabe and Fox, *Angew. Chem.*, 78, 589 (1966).
245. Mizuno, Itoh, and Tagawa, *Chem. Ind.*, 1498 (1965).
246. Anon., Schwarz Bioresearch, data (1966).

247. Hoffer, Duschinsky, Fox, and Yung, *J. Am. Chem. Soc.*, 81, 4112 (1959).
248. Wood, *J. Chem. Soc.*, 89, 1839 (1906).
249. Ness and Fletcher, *J. Am. Chem. Soc.*, 82, 3434 (1960).
250. Albert, *Biochem. J.*, 54, 646 (1953).
251. Fox, Yung, Wempen, and Doerr, *J. Am. Chem. Soc.*, 79, 5060 (1957).
252. Howard, Lythgoe, and Todd, *J. Chem. Soc.*, 1052 (1947).
253. Bergmann and Dikstein, *J. Am. Chem. Soc.*, 77, 691 (1955).
254. Johnson and Mackenzie, *Am. Chem. J.*, 42, 353 (1909).
255. Levene, Simms, and Bass, *J. Biol. Chem.*, 70, 243 (1926).
256. Baddiley, Lythgoe, and Todd, *J. Chem. Soc.*, 386 (1943).
257. Robins, Dille, Willits, and Christensen, *J. Am. Chem. Soc.*, 75, 263 (1953).
258. Carter, *J. Am. Chem. Soc.*, 72, 1466 (1950).
259. Klein and Thannhauser, *Z. Physiol. Chem.*, 218, 173 (1933).
260. Klein and Thannhauser, *Z. Physiol. Chem.*, 224, 252 (1934).
261. Embden and Schmidt, *Z. Physiol. Chem.*, 181, 130 (1929).
262. Turner and Khorana, *J. Am. Chem. Soc.*, 81, 4651 (1959).
263. Cunningham, *J. Am. Chem. Soc*, 80, 2546 (1958).
264. Hall and Khorana, *J. Am. Chem. Soc.*, 77, 1871 (1955).
265. Venkstern and Baev, in *Absorption Spectra of Minor Components and Some Oligonucleotides of Ribonucleic Acid*, Science Publishing House, Moscow, 1967; English transl., *Absorption Spectra of Minor Bases, Their Nucleosides, Nucleotides and Selected Oligoribonucleotides*, Plenum Press Data Division, New York, 1965.
266. Hanze, *J. Am. Chem. Soc.*, 89, 6720, (1967).
267. Venner, *Chem. Ber.*, 93, 140 (1960).
268. Falconer, Gulland, and Story, *J. Chem. Soc.*, 1784 (1939).
269. Davoll, *J. Am. Chem. Soc.*, 73, 3174 (1951).
270. Carbon, David, and Studier, *Science*, 161, 1146 (1968) and unpublished.
271. Ueda, Iida, Ikeda, and Mizuno, *Chem. Pharm. Bull Jap.*, 16, 1788 (1968) and unpublished.
272. Ueda and Nishino, *J. Am. Chem. Soc.*, 90, 1678 (1968).
273. Ueda, Iida, Ikeda, and Mizuno, *Chem. Pharm. Bull. Jap.*, 14, 666 (1966).
274. Burrows, Armstrong, Skoog, Hecht, Boyle, Leonard, and Occolowitz, *Science*, 161, 691 (1968) and unpublished.
275. Baczynskyj, Biemann, and Hall, *Science*, 159, 1481 (1968).
276. Shaw, Warrener, Maguire, and Ralph, *J. Chem. Soc.*, 2294 (1958).
277. Strominger and Friedkin, *J. Biol. Chem.*, 208, 663 (1954).
278. Gray and Lane, *Biochemistry*, 7, 3441 (1968).
279. Johnson and Speh, *Am. Chem. J.*, 38, 602 (1907).
280. Codington, Fecher, Maguire, Thomson, and Brown, *J. Am. Chem. Soc.*, 80, 5164 (1958).
281. Brown, Todd, and Varadarajan, *J. Chem. Soc.*, 2384 (1956).
282. Bendich, Tinker, and Brown, *J. Am. Chem. Soc.*, 70, 3109 (1948).
283. Cavalieri, Bendich, Tinker, and Brown, *J. Am. Chem. Soc.*, 70, 3875 (1948).
284. Wyngaarden and Dunn, *Arch. Biochem. Biophys.*, 70, 150 (1957).
285. Lee and Wigler, *Biochemistry*, 7, 1427 (1968).
286. Lis and Passarge, *Physiol. Chem. Phys.*, 1, 68 (1969).
287. Fujii and Leonard in *Synthetic Procedures in Nucleic Acid Chemistry.*, 1, Zorbach, Tipson, Eds., Interscience, New York, 1968, 13.
288. Nishimura, Yamada, and Ishikura, *Biochim. Biophys. Acta*, 179, 517 (1969).
289. Bessman, Lehman, Adler, Zimmerman, Simms, and Kornberg, *Proc. Natl. Acad. Sci. USA*, 44, 633 (1958).
290. Chheda, *Life Sci.*, 8, 979 (1969) and unpublished.
291. Saneyoshi, Ohashi, Harada, and Nishimura, *Biochim. Biophys. Acta*, 262, 1 (1972).
292. Saneyoshi, Harada, and Nishimura, *Biochim. Biophys. Acta*, 190, 264 (1969).
293. Burrows, Armstrong, Skoog, Hecht, Boyle, Leonard, and Occolowitz, *Biochemistry*, 8, 3071 (1969).
294. Hecht, Leonard, Burrows, Armstrong, Skoog, and Occolowitz, *Science*, 166, 1272 (1969).
295. Schweizer, McGrath, and Baczynskyj, *Biochem. Biophys. Res. Commun.*, 40, 1046 (1970).
296. Ishikura, Yamada, Murao, Saneyoshi, and Nishimura, *Biochem. Biophys. Res. Commun.*, 37, 990 (1969).
297. Kimura-Harada, von Minden, McCloskey, and Nishimura, *Biochemistry*, 11, 3910 (1972).
298. Ohashi, Murao, Yahagi, von Minden, McCloskey, and Nishimura, *Biochim. Biophys. Acta*, 262, 209 (1972).
299. Harada, Gross, Kimura, Chang, Nishimura, and RajBhandary, *Biochem. Biophys. Res. Commun.*, 33, 299 (1968).
300. Ohashi, Maeda, McCloskey, and Nishimura, *Biochemistry*, 13, 2620 (1974).
301. Vorbrüggen and Strehlke, *Chem. Ber.*, 106, 3039 (1973).
302. Fissekis and Sweet, *Biochemistry*, 9, 3136 (1970).
303. Kwong and Lane, *Biochim. Biophys. Acta*, 224, 405 (1970).
304. Tumaitis and Lane, *Biochim. Biophys. Acta*, 224, 391 (1970).
305. Ohashi, Saneyoshi, Harada, Hara, and Nishimura, *Biochem. Biophys. Res. Commun.*, 40, 866 (1970).
306. Yoshida, Takeishi, and Ukita, *Biochim. Biophys. Acta*, 228, 153 (1971).
307. Kimura-Harada, Saneyoshi, and Nishimura, *FEBS Lett.*, 13, 335 (1971).
308. Murao, Saneyoski, Harada, and Nishimura, *Biochem. Biophys. Res. Commun.*, 38, 657 (1970).
309. Ishikura, Yamada, and Nishimura, *Biochim. Biophys. Acta*, 228, 471 (1971).
310. Nakanishi, Furutachi, Funamizu, Grunberger, and Weinstein, *J. Am. Chem. Soc.*, 92, 7617 (1970).
311. Thiebe, Zachau, Baczynskyj, Biemann, and Sonnenbichler, *Biochim. Biophys. Acta*, 240, 163 (1971).
312. Kasai, Goto, Takemura, Goto, and Matsuura, *Tetrahedron Lett.*, 29, 2725 (1971).
313. Nishimura, *Progr. Nucl. Res. Mol. Biol.*, 12, 49 (1972).
314. Hecht, Gupta, and Leonard, *Anal. Biochem.*, 30, 249 (1969).
315. Howlett, Johnson, Trim, Eagles, and Self, *Anal. Biochem.*, 39, 429 (1971).
316. Thiebe and Zachau, *Eur. J. Biochem.*, 5, 546 (1968).
317. Hall, *The Modified Nucleosides in Nucleic Acids.*, Columbia University Press, New York, 1971, 27.
318. Macon and Wolfenden, *Biochemistry*, 7, 3453 (1968).
319. Grimm and Leonard, *Biochemistry*, 6, 3625 (1967).
320. Funamizu, Terahara, Feinberg, and Nakanishi, *J. Am. Chem. Soc.*, 93, 6706 (1971).
321. Blobstein, Grunberger, Weinstein, and Nakanishi, *Biochemistry*, 12, 188 (1973).
322. Ivanovics, Rousseau, and Robins, *Physiol. Chem. Phys.*, 3, 489 (1971).
323. Gross, Simsek, Raba, Limburg, Heckman, and RajBhandary, *Nucl. Acid Res.*, 1, 35 (1974).
324. Yoshikami and Keller, *Biochemistry*, 10, 2969 (1971).
325. Brandon, Gallop, Marmur, Hayashi, and Nakanishi, *Nature New Biol.*, 239, 70 (1972).
326. Hayashi, Nakanishi, Brandon, and Marmur, *J. Am. Chem. Soc.*, 95, 8749 (1973).
327. Feinberg, Nakanishi, Barciszewski, Rafalski, Augustyniak, and WiewiÓrowski, *J. Am. Chem. Soc.*, 96, 7797 (1974).
328. Sasaki and Mizuno, *Chem. Pharm. Bull. Jap.*, 15, 894 (1967).
329. Yamada, Saneyoshi, Nishimura, and Ishikura, *FEBS Lett.*, 7, 207 (1970).
330. Dutta, Hong, Murphy, Mittleman, Chheda, *Biochemistry*, 14, 3144 (1975) and unpublished
331. Miller, Robin, and Astwood, *J. Am. Chem. Soc.*, 67, 2201 (1945).
332. Burrows, Armstrong, Kaminek, Skoog, Bock, Hecht, Damman, Leonard, and Occolowitz, *Biochemistry*, 9, 1867 (1970).
333. Jacobson and Bonner, *Biochem. Biophys. Res. Commun.*, 33, 716 (1968).
334. Hendler, Fürer, and Srinivasan, *Biochemistry*, 4141 (1970).
335. Yamada, Nishimura, and Ishikura, *Biochim. Biophys. Acta*, 247, 170 (1971).
336. Harkins and Freiser, *J. Am. Chem. Soc.*, 80, 1132 (1958).
337. Richter, Loeffler, and Taylor, *J. Am. Chem. Soc.*, 82, 3144 (1960).
338. Breshears, Wang, Bechtolt, and Christensen, *J. Am. Chem. Soc.*, 81, 3789 (1959).
339. Zemlicka and Sorm, *Coll. Czech. Chem. Commun.*, 30, 1880 (1965).
340. Taylor, Vogl, and Cheng, *J. Am. Chem. Soc.*, 81, 2442 (1959).
341. Biemann and McCloskey, *J. Am. Chem. Soc.*, 84, 2005 (1962).

342. Yamazaki, Kumashiro, and Takenishi, *J. Org. Chem.*, 33, 2583 (1968).
343. Saneyoshi and Sawada, *Chem. Pharm. Bull Jap.*, 17, 181 (1969).
344. Kochetkov, Budowsky, Shebaev, Yeliseeva, Grachev, and Demushkin, *Tetrahedron*, 19, 1207 (1963).
345. Saponara and Enger, *Biochim. Biophys. Acta*, 349, 61 (1974).
346. RajBhandary, Faulkner, and Stuart, *J. Biol. Chem.*, 243, 575 (1968).
347. Fink, Lanks, Goto, and Weinstein, *J. Biol. Chem.*, 10, 1873 (1971).
348. Guyot and Mentzer, *Compt. Rend.*, 246, 436 (1958).
349. Roscoe and Tucker, *Virology*, 29, 157 (1966).
350. Kissman and Weiss, *J. Org. Chem.*, 21, 1053 (1956).
351. Furukawa and Honjo, *Chem. Pharm. Bull. Jap.*, 16, 1076 (1968).
352. Wempen, Brown, Ueda, and Fox, *Biochemistry*, 4, 54 (1965).
353. Ziff and Fresco, *Biochemistry*, 8, 3242 (1969).
354. Yamazaki, Kumashiro, and Takenishi, *J. Org. Chem.*, 32, 1825 (1967).
355. Shapiro, Cohen, Shiuey, and Maurer, *Biochemistry*, 8, 238 (1969).
356. Ulbricht, *Tetrahedron*, 6,225 (1959).
357. Ichikawa, Kato, and Takenishi, *J. Het. Chem.*, 2, 253 (1965).
358. Shaw, *J. Biol. Chem.*, 185, 439 (1950).
359. Taylor and Cheng, *Tetrahedron Lett.*, 12, 9 (1959).
360. Townsend and Robins, *J. Org. Chem.*, 27, 990 (1962).
361. Scherp, *J. Am. Chem. Soc.*, 68, 912 (1946).
362. Shaw and Warrener, *J. Chem. Soc.*, 157 (1958).
363. Fox, Wempen, Hampton, and Doerr, *J. Am. Chem. Soc.*, 80, 1669 (1958).
364. Davidson and Baudisch, *J. Am. Chem. Soc.*, 48, 2379 (1926).
365. Armstrong, Evans, Burrows, Skoog, Petit, Dahl, Steward, Strominger, Leonard, Hecht, and Occolowitz, *J. Biol. Chem.*, 245, 2922 (1970).
366. Babcock and Morris, *Biochemistry*, 9, 3701 (1970).
367. Shaw, Desiderio, Tsuboyama, and McCloskey, *J. Am. Chem. Soc.*, 92, 2510 (1970).
368. Shimizu, Asai, and Nishimura, *Chem. Pharm. Bull. Jap.*, 15, 1847 (1967).
369. Martin, Reese, and Stephenson, *Biochemistry*, 7, 1406 (1968).
370. Gin and Dekker, *Biochemistry*, 7, 1413 (1968).
371. Izatt, Hansen, Rytting, and Christensen, *J. Am. Chem. Soc.*, 87, 2760 (1965).
372. Yamazaki, Kumashiro, and Takenishi, *J. Org. Chem.*, 32, 3032 (1967).
373. Davoll, Lythgoe, and Todd, *J. Chem. Soc.*, 1685 (1948).
374. McCloskey, Lawson, Tsuboyama, Krueger, and Stillwell, *J. Am. Chem. Soc.*, 90, 4182 (1968).
375. Randerath, Yu, and Randerath, *Anal. Biochem.*, 48, 172 (1972).
376. Reddy, Ro-Choi, Henning, Shibata, Choi, and Busch, *J. Biol. Chem.*, 247, 7245 (1972).
377. Yamazaki, Kumashiro, and Takenishi, *J. Org. Chem.*, 32, 3258 (1967).
378. Baddiley, Buchanan, Hardy, and Stewart, *J. Chem. Soc.*, 2893 (1959).
379. Nishimura, Shimizu, and Iwai, *Chem. Pharm. Bull. (Jap.)*, 12, 1471 (1964).
380. Wempen and Fox, *Methods Enzymol*, 12A, 59 (1967).
381. Gerster and Robins, *J. Org. Chem.*, 31, 3258 (1966).
382. Visser, in *Synthetic Procedures in Nucleic Acid Chemistry vol. 1*, Zorbach, Tipson, Eds., Interscience John Wiley & Sons, New York, 1968. 428
383. Janion and Shugar, *Acta Biochim. Polon.*, 7, 309 (1960).
384. Cerutti, Kondo, Landis, and Witkop, *J. Am. Chem. Soc.*, 90, 771 (1968).
385. Prystaš and Šorm, *Coll. Czech. Chem. Commun.*, 31, 1053 (1966).
386. Prystaš and Šorm, *Coll. Czech. Chem. Commun.*, 31, 1035 (1966).
387. Brown, Burdon, and Slatcher, *J. Chem. Soc.*, 1051 (1968).
388. Cohn, *Methods Enzymol.*, 12A, 101 (1967).
389. Anderson, Goodman, and Baker, *J. Am. Chem. Soc.*, 81, 3967 (1959).
390. Pedersen and Fletcher, *J. Am. Chem. Soc.*, 82, 5210 (1960).
391. Wempen and Fox, *Methods Enzymol.*, 12A, 76 (1967).
392. Brown, Todd, and Varadarajan, *J. Chem. Soc.*, 2388 (1956).
393. Bergmann and Burk, *J. Org. Chem.*, 20, 1501 (1955).
394. Fox, Miller, and Wempen, *J. Med. Chem.*, 9, 101 (1966).
395. Wheeler and Merriam, *Am. Chem. J.*, 29, 478 (1903).
396. Wheeler and Johnson, *Am. Chem. J.*, 29, 492 (1903).
397. Chesterfield, McOmie, and Tute, *J. Chem. Soc.*, 4590 (1960).
398. Dunn and Trigg, *John Innes Institute Report*, 1972, 142 and unpublished.
399. Schaeffer and Thomas, *J. Am. Chem. Soc.*, 80, 3738 (1958).
400. Broude, Budowsky, and Kochekov, *Mol. Biol.*, 1, 214 (1967).
401. Marmur, Brandon, Neubort, Ehrlich, Mandel, and Konvicka, *Nature New Biol.*, 1, 239, 68 (1972).
402. Kropinski, Bose, and Warren, *Biochemistry*, 12, 151 (1973).
403. Takemura, Kasai, and Goto, *J. Biochem., (Japan)*, 75, 1169 (1974).
404. Hecht and McDonald, *Anal. Biochem.*, 47, 157 (1972).
405. McCloskey, Futrell, Elwood, Schram, Panzica, and Townsend, *J. Am. Chem. Soc.*, 95, 5762 (1973).
406. Hecht, *Anal. Biochem.*, 44, 262 (1971).
407. Cunningham and Gray, *Biochemistry*, 13, 543 (1974).
408. Chheda and Hong, *J. Med. Chem.*, 14, 748 (1971) and unpublished.
409. Dyson, Hall, Hong, Dutta, and Chheda, *Can. J. Biochem.*, 50, 237 (1972).
410. Dunn and Trigg, *Biochem. Soc. Trans.*, 3, 656 (1975) and unpublished.
411. P. L. Biochemicals, Biochemical reference guide & price list, (1973).
412. Cushley, Watanabe, and Fox, *J. Am. Chem. Soc.*, 89, 394 (1967).
413. Dunn and Flack, *John Innes Institute Report*, 1970, 76 and unpublished.
414. Kuntzel, Weissenbach, Wolff, Tumaitis-Kennedy, Lane, and Dirheimer, *Biochimie*, 57, 61 (1975).
415. Gray, *Biochemistry*, 13, 5453 (1974).

CHEMICAL MODIFICATION OF NUCLEIC ACIDS

The chemical modification of nucleic acids is not as complex as that of proteins since there are fewer monomer units and, for all practical purposes, only nitrogen as a nucleophilic reactive group; the nitrogen is reactive as a primary and secondary amine. Reaction at the primary amine groups of, for example, adenine, is referred to an exocyclic modification whereas reaction at the imine nitrogens of pyrimidines and purine rings is referred to as an endocyclic modification. There are also ring-opening reactions and cross-linking reactions.

TABLE 1: Chemical Modification of Nucleic Acids

Reagent	Base Modified	Product	Reference
Aldehydes[a]	Most data on pyrimidines with less on purines	Various adducts	1-10
Alkylation	Purine and pyrimidines	Various product resulting from environmental agents such as ethylene oxides and nitrogen mustards	11-20
Diethylpyrocarbonate	Purine and pyrimidines	Carboxethylation of the N-7 site on the purine ring followed by ring opening; pyrimidines react at primary amino groups to yield carbethoxy derivatives[b]	21-28
Dimethyl Sulfate (Methylation)	Purines and Pyrimidines	N^7-Methylguanidine with minor reaction at the N^1 and N^3-positions; reaction also occurs at N^3 in adenine and at N^3-position in cytidine; reactivity is controlled by polynucleotide structure[c,d]	29-41
Hydrazine	Pyrimidines	Ring cleavage yielding pyrazole derivatives[e] and the ribosyl backbone ; reaction used sequence analaysis and footprinting	42-47
Hydroxylamine	Purines and pyrimidines	Hydroxamate formation at "exocyclic" nitrogen; conversion of guanine to isoxazolone.	48-55
Nitrous Acid (HNO$_2$)	Purines and pyrimidines	Deamination; crosslinking at guanine, cytosine bases	56-64
Potassium Permanganate	Purines and Pyrmidines	Oxidation at double bonds; preferential reaction with thymidine	65-75
Sodium Bisulfate	Cytosine	Uracil (5-methylcytosine is converted to thymine)[f]	76-84

[a] e.g. formaldehyde, acetaldehyde, acrolein, crotonaldehyde; 4-hydroxy-2-nonenal (4-HNE)

[b] Diethylpyrocarbonate modification is used for DNA footprinting (Fox, K.R., Webster, R., Phelps, R.J., *et al.*, Sequence selective binding of bis-daunorubicin WP631 to DNA, *Eur.J.Biochem.* 271, 3556-3566, 2004)

[c] The N^7-position of guanine is always reactive for methylation (the N^7-position of guanine is the most reactive site in nucleic acids); methylation of guanine also occurs at the N^1-position of guanine at high concentrations of methylating agents. Methylation of the N^3-position of cytidine occurs in single-stranded DNA but is blocked in double-stranded DNA. Methylation (alkylation) also occurs at the N^1 and N^3 position of adenine but methylation is restricted in at the N^1 position in double-stranded DNA. *In vivo* methylation of DNA occurs at cytidine residues largely in CpG islands (Ehrlich, M. and Wang, R.Y., 5-Methylcytosine in eukaryotic DNA, *Science* 212, 1350-1357, 1981; Lewis, J. and Bird, A., DNA methylation and chromatin structure, *FEBS Lett.* 285, 155-159, 1991; Cheng, X., Structure and function of DNA methyltransferases, *Annu.Rev.Biophys.Biomol.Struct.* 24, 293-318, 1995; Scheule, R.K., The role of CpG motifs in immunostimulation and gene therapy, *Adv.Drug Deliv.Rev.* 44, 119-134, 2000).

[d] Base treatment of N^7-methylguanine results in opening of the imidazolium ring while N^3-methylcytidine undergoes base-catalyzed deamination to give N^3-methyluridine. Methylation of cytidine blocks conversion to thymidine by sodium bisulfite

[e] 4-methyl-5-pyrazolone with thymidylic acid; 3(5)-aminopyrazole with deoxycytidine

[f] Conversion to uracil does not occur with 5-methylcytosine; 5-methylcytosine is converted to thymine. The rate of reaction of 5-methylcytosine with sodium bisulfite is much slower than reaction of cytosine.

References for Table 1

1. Alegria, A.H., Hydroxymethylation of pyrimidine mononucleotides with formaldehyde, *Biochim.Biophys.Acta* 149, 317-324, 1967

2. Feldman, M.Y., Reactions of nucleic acids and nucleoproteins with formaldehyde, *Prog.Nucleic Acid Res.Mol.Biol.* 13, 1-49, 1973

3. McGhee, J.D. and von Hippel, P.H., Formaldehyde as a probe of DNA structure. I. Reaction with exocyclic amine groups of DNA bases, *Biochemistry* 25, 1281-1296, 1975

4. McGhee, J.D. and von Hippel, P.H., Formaldehyde as a probe of DNA structure. II. Reaction with endocyclic imine groups of DNA bases, *Biochemistry* 25, 1297-1303, 1975

5. Yamazaki, Y. and Suzuki, H., A new method of chemical modification of N^6-amino group in adenine nucleotides with formaldehyde and a thiol and its application to preparing immobilized ADP and ATP, *Eur.J.Biochem.* 92, 197-207, 1978

6. Chung, F.L., Young, R., and Hecht, S.S., Formation of cyclic 1, N^2-propanodeoxyguanosine adducts in DNA upon reaction with acrolein or crotonaldehyde, *Cancer Res.* 44, 990-995, 1984

7. Winter, C.K., Segall, H.J., and Haddon, W.F., Formation of cyclic adducts of deoxyguanosine with the aldehydes *trans*-4-hydroxy-2-hexenal and *trans*-4-hydroxynonenal *in vitro*, *Cancer Res.* 46, 5682-5686, 1986

8. Kennedy, G., Slaich, P.K., Golding, B.T., and Watson, W.P., Structure and mechanism of formation of a new adduct from formaldehyde and guanosine, *Chem.Biol.Interact.* 102, 93-100, 1996

9. Hecht, S.S., McIntee, E.J., and Wang, M., New DNA adducts of crotonaldehyde and acetaldehyde, *Toxicology* 166, 31-36, 2001

10. Kurtz, A.J. 1, N^2-deoxyguanosine adducts of acrolein, crotonaldehyde, and *trans*-4-hydroxynonenal cross-link to peptides via Schiff base linkage, *J.Biol.Chem.* 278, 5970-5976, 2003

11. Singer, B., The chemical effects of nucleic acid alkylation and their relation to mutagenesis and carcinogenesis, *Prog.Nucleic Res.Mol. Biol.* 15, 219-284, 1975

12. Lawley, P.D., DNA as a target of alkylating carcinogenesis, *Br.Med. Bull.* 36, 19-24, 1980

13. Coles, B., Effects of modifying structure on electrophilic reactions with biological nucleophiles, *Drug Metab.Rev.* 15, 1307-1334, 1984-1985

14. Wild, C.P., Antibodies to DNA alklylation adducts as analytical tools its chemical carcinogenesis, *Mutat.Res.* 233, 219-233, 1990

15. Lawley, P.D., Alkylation of DNA and its aftermath, *Bioessays* 17, 561-568, 1995

16. Bolt, H.M., Quantification of endogenous carcinogens. The ethylene oxide paradox, *Biochem.Pharmacol.* 52, 1-5, 1996

17. Rios-Blanco, M.N., Plna, K., Faller, T., *et al.*, Propylene oxide: mutagenesis, carcinogenesis and molecular dose, *Mutat.Res.* 380, 179-197, 1997

18. Wilson, D.S. and Szostak, J.W., In vitro selection of functional nucleic acids, *Annu.Rev.Biochem.* 68, 611-647, 1999

19. Denny, W.A., DNA minor groove alkylating agents, *Curr.Med.Chem.* 8, 533-544, 2001

20. Mishina, Y. and He, C., Oxidative dealkylation DNA repair mediated by the mononuclear non-heme iron AlkB proteins, *J.Inorg.Biochem.* 100, 670-678, 2006

21. Leonard, N.J., McDonald, J.J., Henderson, R.E.I., and Reichmann, M.E., Reaction of diethyl pyrocarbonate with nucleic acid components. Adenosine. *Biochemistry* 10, 335-3342, 1971

22. Solymosy, F., Hüvös, P., Gulyás, A., *et al.*, Diethyl pyrocarbonate, a new tool in the chemical modification of nucleic acids, *Biochim. Biophys.Acta* 238, 406-426, 1971

23. Vincze, A., Henderson, R.E.I., McDonald, J.J., and Leonard, N.J., Reaction of diethyl pyrocarbonate with nucleic acid components. Bases and nucleosides derived from guanine, cytosine, and uracil, *J.Amer.Chem.Soc.* 95, 2677-2683, 1973

24. Ehrenfeld, E., Interaction of Diethylpyrocarbonate with poliovirus double-stranded RNA, *Biochem.Biophys.Res.Commun.* 56, 214-219, 1974

25. Herr, W., Diethyl pyrocarbonate: A chemical probe for secondary structure in negatively supercoiled DNA, *Proc.Nat.Acad.Sci.USA* 82, 8009-8013, 1985

26. Johnston, B.H. and Rich, A., Chemical probes of DNA conformation: Detection of Z-DNA at nucleotide resolution, *Cell* 42, 713-724, 1985

27. Runkel, L. and Nordheim, A., Chemical footprinting of the interaction between left-handed *Z*-DNA and anti-*Z*-DNA antibodies by Diethylpyrocarbonate carboethoxylation, *J.Mol.Biol.* 189, 487-501, 1986

28. Buckle, M. and Buc, H., Fine mapping of DNA single-stranded regions using base-specific chemical probes: Study of an open complex formed between RNA polymerase and the *lac* UV₅ promoter, *Biochemistry* 28, 4388-4396, 1989

29. Jordan, D.O., The physical properties of nucleic acids, in *The Nucleic Acids. Chemistry and Biology*, Vol. 1, ed. E. Chargaff and J.N. Davidson, Academic Press, New York, NY, USA, Chapter 13, pps. 447-492, 1955

30. Kanduc, D., tRNA chemical modification In vitro and in vivo formation of 1,7-dimethylguanosine at high concentrations of methylating agents, *Biochem.Biophys.Acta* 653, 9-17, 1981

31. Singer, B., The chemical effects of nucleic acid alkylation and their relation to mutagenesis and carcinogenesis, *Prog.Nucl.Acids Res.Mol. Biol.* 15, 219-284, 1975

32. Behmoaras, T., Toulme, J.-J., and Relene, C., Specific recognition of apurinic sites in DNA by a tryptophan-containing peptide, *Proc.Nat. Acad.Sci.USA* 78, 926-930, 1981

33. Mhaskar, D.N., Chang, M.J.W., Hart, R.W., and D'Ambrosio, S.M., Analysis of alkylated sites at *N*-3 and *N*-7 positions of purines as an indicator for chemical carcinogens, *Cancer Res.* 41, 223-229, 1981

34. Kirkegaard, K., Buc, H., Spassky, A., and Wang, J.C., Mapping of single-stranded regions in duplex DNA at the sequence level: Single-strand-specific cytosine Methylation in RNA polymerase-promoter complexes, *Proc.Nat.Acad.Sci.USA* 80, 2544-2548, 1983

35. Potaman, V.N. and Sinden, R., Stabilization of triple-helical nucleic acids by basic oligopeptides, *Biochemistry* 34, 14885-14892, 1995

36. Lawley, P.D., Effects of some chemical mitogens and carcinogens on nucleic acids, *Prog.Nucl.Acid Res.Mol.Biol.* 5, 89-131, 1996

37. Dobner, T., Buchner, D., Zeller, T., *et al.*, Specific nucleoprotein complexes within adenovirus capsids, *Biol.Chem.* 382, 1373-1377, 2001

38. Hock, T.D., Nick, H.S., and Agarwal, A. Upstream stimulatory factors, USF1 and USF2, bind to the human haem oxygenase-1 proximal promoter *in vivo* and regulates its transcription, *Biochem.J.* 383, 209-218, 2004

39. Lagor, W.R., de Groh, E.D., and Ness, G.C., Diabetes alters the occupancy of the hepatic 5-hydroxy-3-methylglutaryl-CoA reductase promoter, *J.Biol.Chem.* 280, 36601-36608, 2005

40. Haugen, S.P., Berkmen, M.B., Ross, W., *et al.*, tRNA promoter regulates by nonoptimal binding of sigma region 1.2: an additional recognition element for RNA polymerase, *Cell* 125, 1069-1082, 2006

41. Temperli, A., Türler, H., Rüst, P. *et al.*, Studies on the nucleotide arrangement in deoxyribonucleic acids. IX. Selective degradation of pyrimidines deoxyribonucleotides, *Biochim.Biophys.Acta* 91, 462-476, 1964

42. Cashmore, A.R. and Peterson, G.B., The degradation of DNA by hydrazine: a critical study of the suitability of the reaction for the quantitative determination of purine nucleotide sequences, *Biochim. Biophys.Acta* 174, 591-603, 1969

43. Türler, H. and Chargaff, E., Studies on the nucleotide arrangement in deoxyribonucleic acids. XII. Apyrimidinic acid from calf-thymus deoxyribonucleic acid: preparation and properties, *Biochim.Biophys. Acta* 195, 446-455, 1969

44. Maxam, A.M. and Gilbert, W., A new method for sequencing DNA, *Proc.Nat.Acad.Sci.USA* 74, 560-564, 1977

45. Cashmore, A.R. and Petersen, G.B., The degradation of DNA by hydrazine: identification of 3-ureidopyrazole as a product of the hydrazinolysis of deoxycytidylic acid residues, *Nucleic Acids Res.* 5, 2485-2891, 1978

46. Peattie, D.A., Direct chemical method for sequencing RNA, *Proc.Nat. Acad.Sci.USA* 76, 1760-1764, 1979

47. Tolson, D.A. and Nicholson, N.H., Sequencing RNA by a combination of exonuclease digestion and uridine-specific chemical cleavage using MALDI-TOF, *Nucleic Acids Res.* 26, 446-451, 1998

48. Small, G.D. and Gordon, M.P., Reaction of hydroxylamine and methoxyamine with the ultraviolet-induced hydrate of cytidine, *J.Mol. Biol.* 14, 281-291, 1968

49. Brown, D.M. and Osborne, M.R., The reaction of adenosine with hydroxylamine, *Biochim.Biophys.Acta* 247, 514-518, 1971

50. Fraenkel-Conrat, H., and Singer, B., The chemical basis for the mutagenicity of hydroxylamine and methoxyamine, *Biochim.Biophys.Acta* 262, 264-268, 1972

51. Iida, S., Chung, K.C., and Hayatsu, H., The reaction of hydroxylamine with 4-thiouridine, *Biochim.Biophys.Acta* 308, 198-204, 1973

52. Kasai, H. and Nishimura, S., Hydroxylation of deoxyguanosine at the C-8 position by ascorbic acid and other reducing agents, *Nucleic Acids Res.* 12, 2137-2145, 1984

53. Johnston, B.H., Hydroxylamine and methoxyamine as probes of DNA structure, *Methods Enzymol.* 212, 180-194, 1992

54. Simandan, T., Sun, J., and Dix, T.A., Oxidation of DNA bases, deoxyribonucleosides and homopolymers by peroxyl radicals, *Biochem.J.* 335, 233-240, 1998

55. Tessman, I., Poddar, R.K., and Kumar, S., Identification of the altered bases in mutated single-stranded DNA. I. In vitro mutagenesis by hydroxylamine, ethyl methanesulfonate and nitrous acid, *J.Mol.Biol.* 93, 352-363, 1964

56. Stuy, J.H., Inactivation of transforming deoxyribonucleic acid by nitrous acid, *Biochem.Biophys.Res.Commun.* 6, 328-333, 1961

57. Horn, E.E. and Herriott, R.M., The mutagenic action of nitrous acid on "single-stranded" (denatured) *Hemophilus* transforming DNA, *Proc.Nat.Acad.Sci.USA* 48, 1409-1416, 1962

58. Kotaka, T. and Baldwin, R.L., Effects of nitrous acid on the DAT copolymer as a template for DNA polymerase, *J.Mol.Biol.* 93, 323-329, 1964

59. Carbon, J. and Curry, J.B., A change in the specificity of transfer RNA after partial deamination with nitrous acid, *Proc.Nat.Acad.Sci.USA* 59, 467-474, 1968

60. Shapiro, R. and Yamaguchi, H., Nucleic acid reactivity and conformation I. Deamination of cytosine by nitrous acid, *Biochim.Biophys.Acta.* 281, 501-506, 1972

61. Verly, W.G. and Lacroix, M. DNA and nitrous acid, *Biochim.Biophys.Acta* 414, 185-192, 1975

62. Dubelman, S. and Shapiro, R, A method for the isolation of crosslinked nucleosides from DNA: application to cross-links induced by nitrous acid, *Nucleic Acids Res.* 4, 1815-1827, 1977

63. Shapiro, R. Dubelman, S., Feinberg, A.M., *et al.*, Isolation and identification of cross-linked nucleosides from nitrous acid treated deoxyribonucleic acid, *J.Amer.Chem.Soc.* 99, 302-303, 1977

64. Edfeldt, N.B., Harwood, E.A., Sigurdsson, S.T. *et al.*, Solution structure of a nitrous acid induced DNA interstrand cross-link, *Nuc.Acids Res.* 32, 2785-2794, 2004

65. Darby, C.K., Jones, A.S., Tittensor, J.R., and Walker, R.T., Chemical degradation of DNA oxidized by permanganate, *Nature* 216, 793-794, 1967

66. Hayatsu, H. and Ukita, T., The selective degradation of pyrimidines in nucleic acids by permanganate oxidation, *Biochem.Biophys.Res.Commun.* 29, 556-561. 1967

67. Rubin, C.M. and Schmid, C.W., Pyrimidine-specific chemical reactions useful for DNA sequencing, *Nucleic Acids Res.* 8, 4613-4619, 1980

68. Fritzsche, E., Hayatsu, H., Igloi, G.L., *et al.*, The use of permanganate as a sequencing reagent for identification of 5-methylcytosine residues in DNA, *Nucleic Acisd Res.* 15, 5517-5528, 1987

69. Sasse-Dwight, S. and Gralla, J.D., KMnO₄ as a probe for *lac* promoter DNA melting and mechanism *in vivo*, *J.Biol.Chem.* 264, 8074-8081, 1989

70. Klysik, J., Rippe, K., and Jovin, T.M., Reactivity of parallel-stranded DNA to chemical modification reagents, *Biochemistry* 29, 9831-9839, 1990

71. Jiang, H., Zacharias, W., and Amirhaeri, S., Potassium permanganate as an *in situ* probe for B-Z and Z-Z junctions, *Nucleic Acids Res.* 19, 6943-6948, 1991

72. Nawamura, T., Negishi, K., and Hayatsu, H., 8-Hydroxyguanine is not produced by permanganate oxidation of DNA, *Arch.Biochem.Biophys.* 311, 523-524, 1994

73. Bailly, C. and Waring, M.J., Comparison of different footprinting methodologies for detecting binding sites for a small ligand on DNA, *J.Biomol.Struct.Dyn.* 12, 869-898, 1995

74. Kahl, B.F. and Paule, M.R., The use of Diethylpyrocarbonate and potassium permanganate as probes for strand separation and structural distortions in DNA, *Methods Mol.Biol.* 148, 63-75, 2001

75. Spicuglia, S., Kumar, S., Chasson, L., Potassium permanganate as a probe to map DNA-protein interactions *in vivo*, *J.Biochem.Biophys.Methods* 59, 189-194, 2004

76. Hayatsu, H., Wataya, Y., Kai, K., and Iida, S., Reaction of sodium bisulfite with uracil, cytosine, and their derivatives, *Biochemistry* 9, 2858-2865, 1970

77. Shapiro, R., Braverma, B., Louis, J.B., and Servis, R.E., Nucleic-acid reactivity and conformation 2. Reaction of cytosine and uracil with sodium bisulfite, *J.Biol.Chem.* 248, 4060-4064, 1973

78. Wang, R.Y.-H., Gehrke, C.W., and Ehrlich, M., Comparison of bisulfite of 5-methyldeoxycytidine and deoxycytidine residues, *Nucleic Acids Res.* 8, 4777-4790, 1980

79. Frommer, M., McDonald, L.E, Millar, D.S., *et al.*, A genomic sequencing protocol that yields a positive display of 5-methylcytosine residues in individual DNA strands, *Proc.Nat.Acad.Sci.USA* 89, 1827-1831, 1992

80. Chen, H. and Shaw, B.R., Kinetics of bisulfite-induced cytosine deamination in single-stranded DNA, *Biochemistry* 32, 3535-3539, 1993

81. Herman, J.G., Graff, J.R., Myöhänen, S., *et al.*, Methylation-specific PCR: A novel PCR assay for methylation status of CpG islands, *Proc.Nat.Acad.Sci.USA* 93, 9821-9826, 1996

82. Hong, K.-M., Yang, S.-H., Guo, M., *et al.*, Semiautomatic detection of DNA methylation at CpG islands, *BioTechniques* 38, 354-358, 2005

83. Ordway, J.M., Bedell, J.A, Citek, R.W. *et al.*, MethylMapper: a method for high-throughput, multilocus bisulfite sequence analysis and reporting, *BioTechniques* 39, 464-470, 2005

84. Zhou, D., Qiao, W., Yang, L., and Lu, Z., Bisulfite-modified target DNA array for aberrant methylation analysis, *Anal.Biochem.* 351, 26-35, 2006

There are modifications of ribose moiety in ribonucleotides such as periodate oxidation which was used in early structural analysis. Oxidation of the ribose ring has also been used to couple RNA to protein amino groups and to amino-containing matrices for affinity chromatography.

TABLE 2: Reaction of Periodate with Nucleic Acids

Reaction	Conditions	Reference
Coupling of periodate-oxidized RNA to hydrazide-agarose	140 μL 0.2 M NaIO₄ added to 500 – 3000 μg RNA in a volume of 1 mL 0.1 M sodium acetate, pH 5.0 for one hours at 23°C in the dark; reaction terminated with 80 μL ethylene glycol; after removal of reactants, the modified RNA was coupled to hydrazide matrix in the same solvent	1
Coupling of double-stranded RNA to agarose	12 μL 0.1 M NaIO₄ (sodium meta-periodate) /10 A₂₆₀ units of RNA(40 A₂₆₀ units/mL) in 0.1 M sodium acetate, pH 5.0, and incubated for 1 hr at 23°C. The reaction is terminated by precipitation of the RNA with ethanol. Coupling to agarose was accomplished in 0.1 M sodium acetate, pH 5.0	2

References for Table 2

1. Robberson, D.L. and Davidson, N., Covalent coupling of ribonucleic acid to agarose, *Biochemistry* 11, 533-537. 1972

2. Langland, J.O., Pettiford, S.M., and Jacobs, B.L., Nucleic acid affinity chromatography: Preparation and characterization of double-stranded RNA agarose, *Protein Exp.Purif.* 6, 25-32, 1995

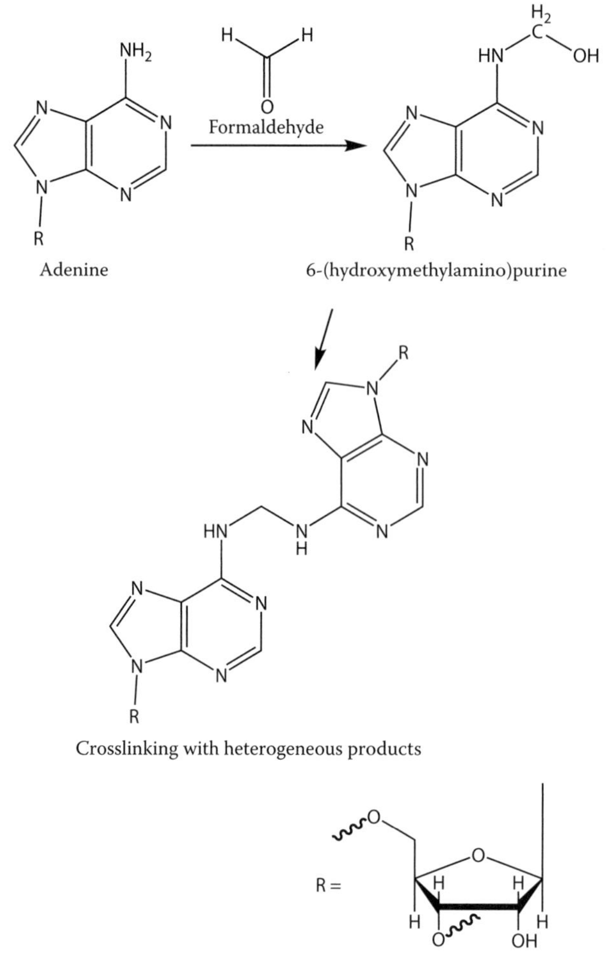

FIGURE 1 The reaction of formaldehyde with adenine in ribonucleic acid. (Adapted from *Nucleic Acids in Chemistry and Biology*, ed. G.M. Blackburn and M.J. Galt, Oxford University Press, Oxford, UK, 1996.)

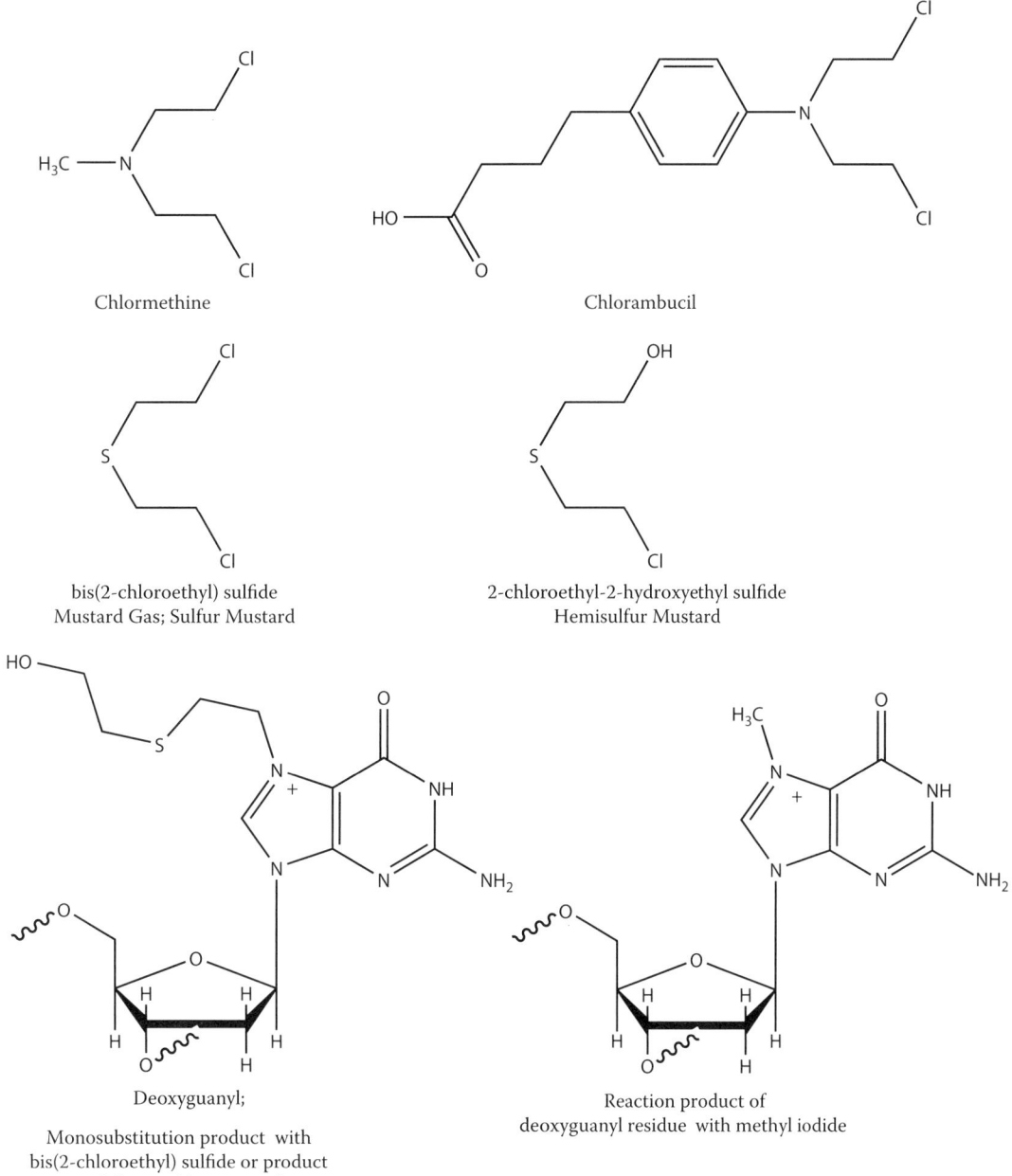

Chlormethine

Chlorambucil

bis(2-chloroethyl) sulfide
Mustard Gas; Sulfur Mustard

2-chloroethyl-2-hydroxyethyl sulfide
Hemisulfur Mustard

Deoxyguanyl;

Monosubstitution product with
bis(2-chloroethyl) sulfide or product
with hemisulfur mustard

Reaction product of
deoxyguanyl residue with methyl iodide

FIGURE 2   Some alkylating agents for the modification of DNA. Nitrogen mustards are described as a group of bis(2-chloroalkylamines). The original mustards were chloroalkyl disulfides. Also shown is the methylation of purine with methyl iodide. The N^7-position on the guanine ring is the most susceptible site for alkylation. See *Nucleic Acids in Chemistry and Biology*, ed. G.M. Blackburn and M.J. Galt, Oxford University Press, Oxford, UK, 1996; Denny, W.A., DNA minor groove alkylating agents, *Curr.Med. Chem.* 8, 533-544, 2001.

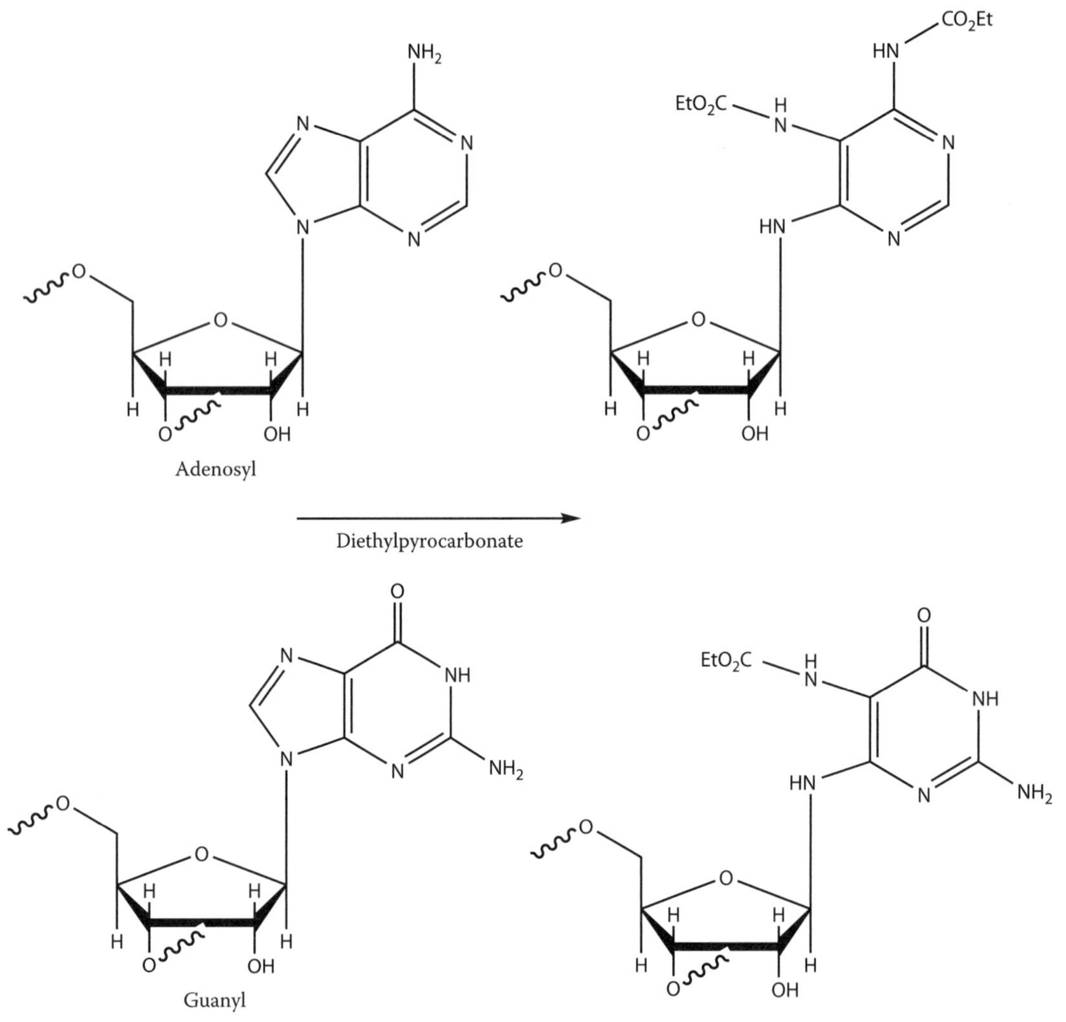

FIGURE 3 The reaction of diethylpyrocarbonate with adenyl and guanyl residues in ribonucleic acid.

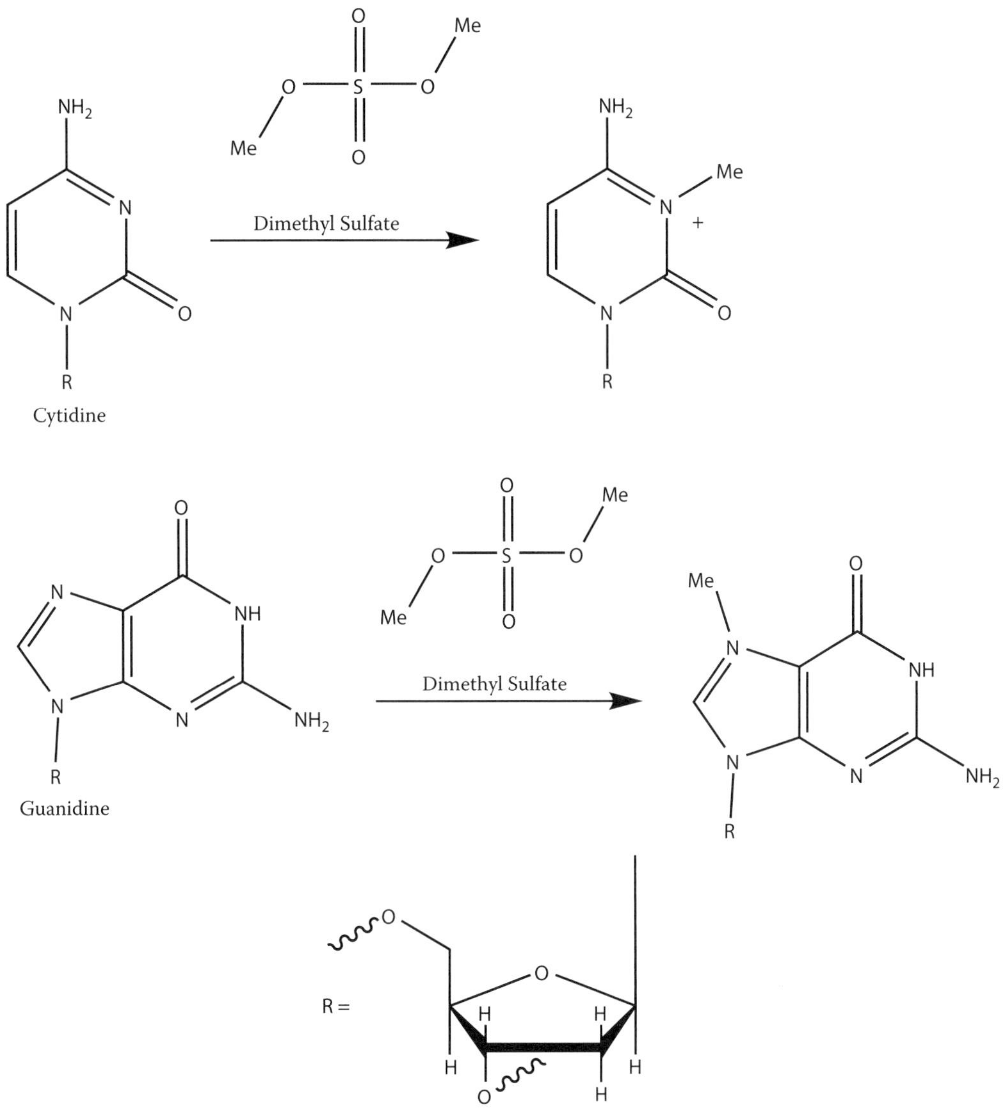

FIGURE 4 The modification of purines and pyrimidines with dimethyl sulfate.

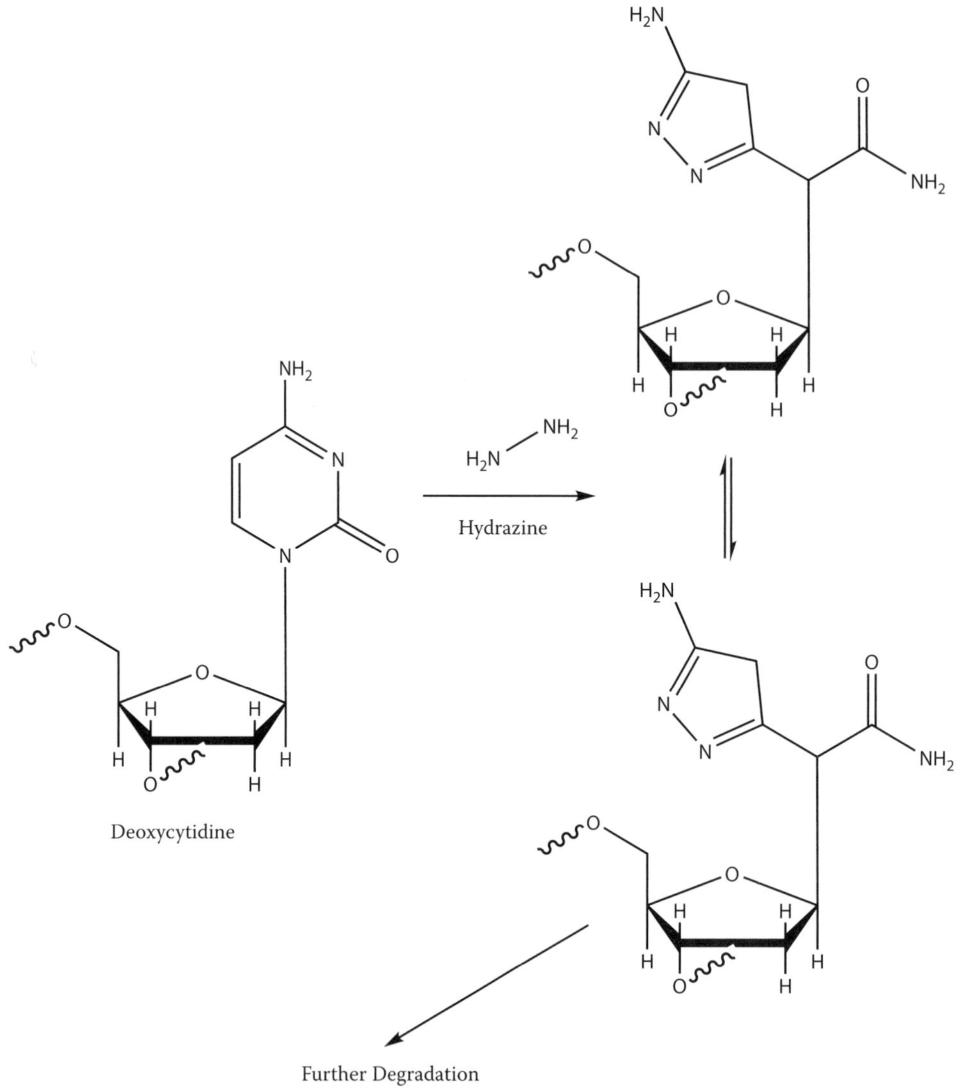

FIGURE 5 The degradation of cytidine with hydrazine. (Adapted from Cashmore, A.R. and Petersen, G.B., The degradation of DNA by hydrazine: identification of 3-ureidopyrazole as a product of the hydrazinolysis of deoxycytidylic acid residues, *Nucleic Acids Res.* 5, 2485-2491, 1978.)

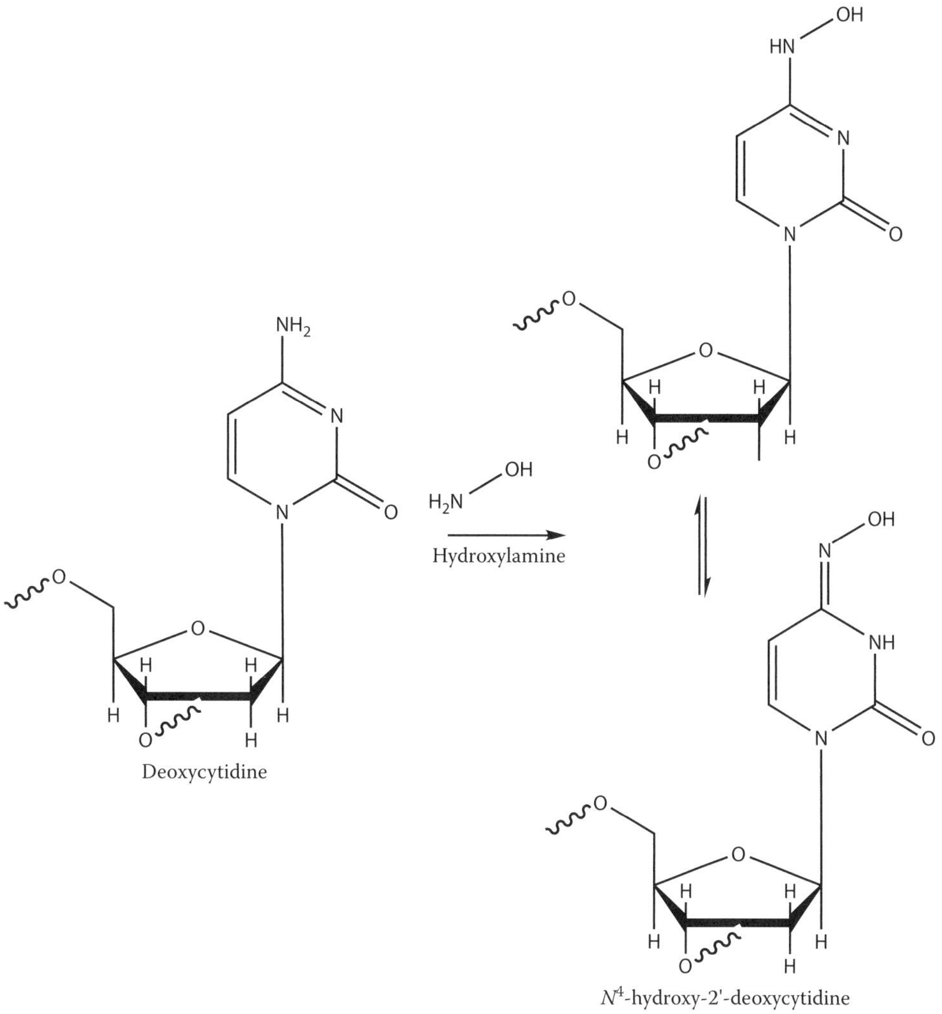

FIGURE 6 The reaction of hydroxylamine with cytosine. (Adapted from Blackburn, G.M., Jarvis, S., Ryder, M.C., *et al.*, Kinetics and mechanism of reaction of hydroxylamine with cytosine and its derivatives, *J.Chem.Soc.Perkins Trans* 1, 370-375, 1975 and *Nucleic Acids in Chemistry and Biology*, ed. G.M. Blackburn and M.J. Galt, Oxford University Press, Oxford, UK, 1996.)

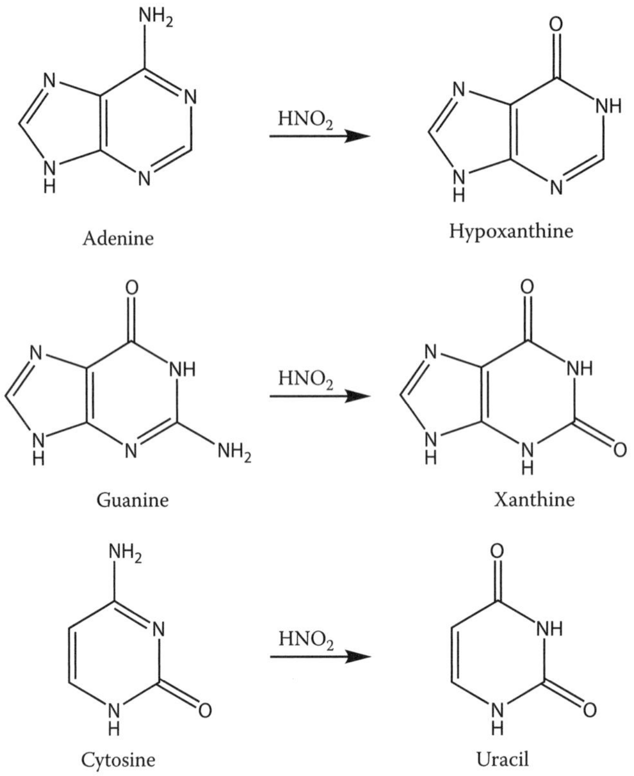

FIGURE 7 The reaction of purines and pyrimidines with nitrous acid resulting in deamination. The riboside derivative of hypoxanthine is inosine.

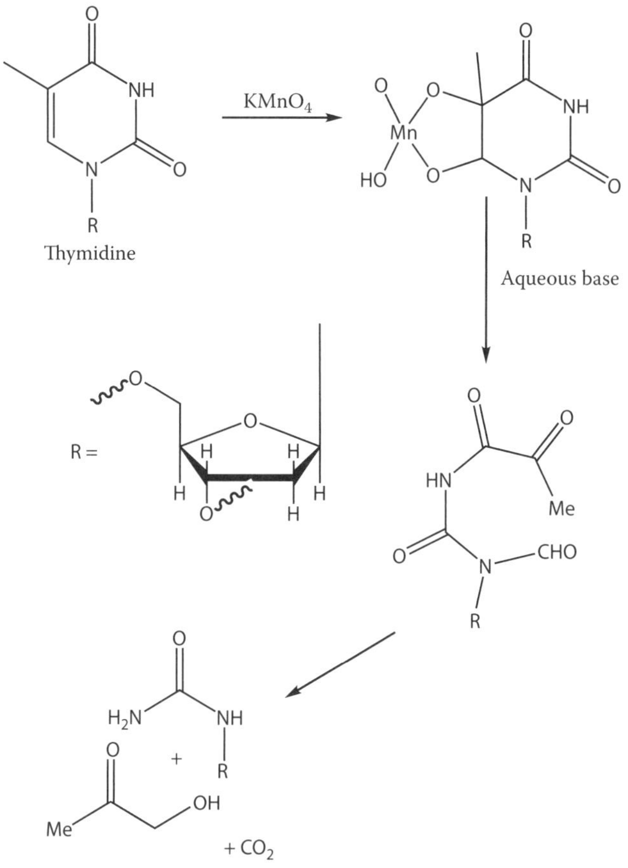

FIGURE 8 The oxidation of thymidine with potassium permanganate. Adapted from Bui, C.T., Rees, K., and Cotton, R.G.H., Permanganate oxidation reactions of DNA: Perspective in biological studies, *Nucleosides, Nucleotides, and Nucleic Acids* 22, 1835-1855, 2003.

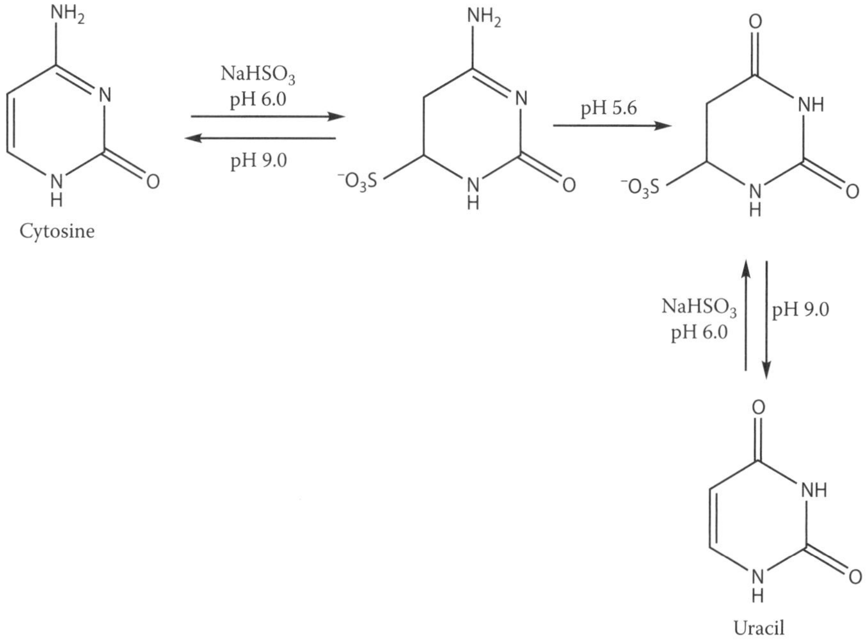

FIGURE 9 The modification of cytosine with sodium bisulfite results in the formation of uracil.

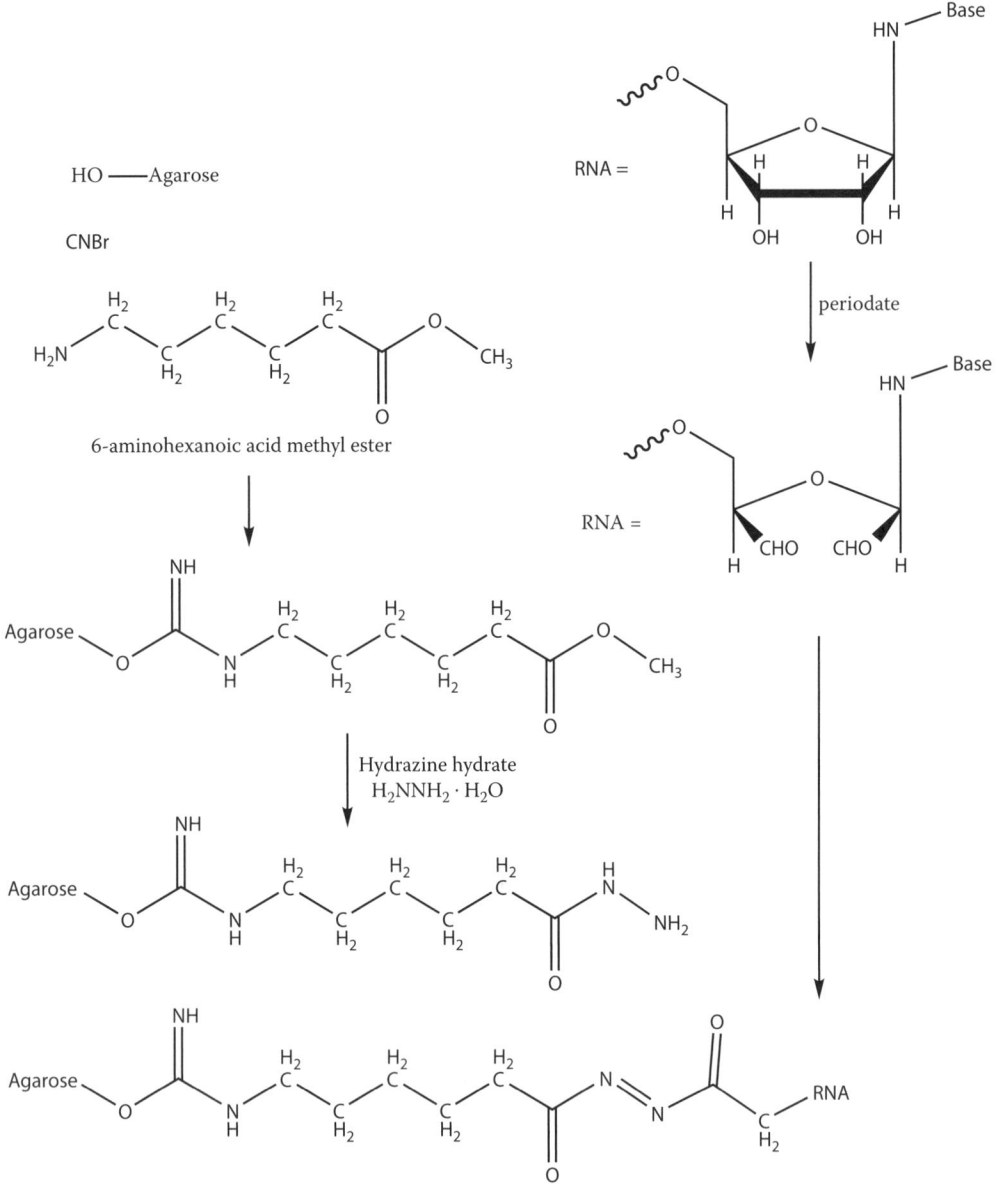

FIGURE 10 The covalent coupling of RNA to an agarose matrix. The RNA is oxidized with periodate and then coupled to an alkyl hydrazide derivative of agarose.

General references for the chemical modification of nucleic acids

Jordan, D.O., The reaction of nucleic acids with mustard gas, *Biochem.J.* 42, 308-316, 1948

Jordan, D.O., Nucleic acids, purines, and pyrimidines, *Annu.Rev.Biochem.* 21, 209-244, 1952

Jordan, D.O., The physical properties of nucleic acids, in *The Nucleic Acids. Chemistry and Biology*, Vol. 1., ed. E.Chargaff and J.N. Davidson, Academic Press, New York, New York, Chapter 13, pps 447-492, 1955.

Lawley, P.D., Effects of some chemical mutagens and carcinogens on nucleic acids, *Prog.Nucl.Acid.Res.Mol.Biol.* 5, 89-131, 1966

Lawley, P.D., Effects of some chemical mutagens and carcinogens on nucleic acids, *Prog.Nucleic Acid Res.Mol.Biol.* 5, 89-131, 1966

Singer, B. and Fraenkel-Conrat, H., The role of conformation in chemical modification, *Prog.Nucl.Acid Res.Mol.Biol.* 9, 1-29. 1969

Kochetkov, N.K. and Budowsky, E.T., The chemical modification of nucleic acids, *Prog.Nucl.Acid Res.Mol.Biol.* 9, 403-438, 1969

Steinschneider, A., Effect of methylamine on periodate-oxidized adenosine 5'-phosphate, *Biochemistry* 10,173-178, 1971

Solymosy, S., Hüvös, P., Gulyás, A., *et al.*, Diethyl pyrocarbonate, a new tool in the chemical modification of nucleic acids, *Biochim.Biophys. Acta.* 238, 406-416, 1971.

Lawley, P.D., Orr, D.J., and and Shah, S.A., Reaction of alkylating mutagens and carcinogens with nucleic acids: *N*-3 of guanine as a site of alkylation by *N*-methyl-*N*-nitrosourea and dimethyl sulphate, *Chem.Biol. Interact.* 4, 431-434, 1972

Uziel, M., Periodate oxidation and amine-catalyzed elimination of the terminal nucleoside from adenylate or ribonucleic acid. Products of overoxidation, *Biochemistry* 12, 938-942, 1973.

Singer, B., The chemical effects of nucleic acid alkylation and their relation to mutgenesis and carcinogensis, *Prog.Nucl.Acid Res.Mol.Biol.* 15, 219-284, 1975

Maxam, A.M. and Gilbert, W., A new method for sequencing DNA, *Proc. Nat.Acad.Sci.USA* 74, 560-564, 1977

Swenson, D.H. and Lawley, P.D., Alkylation of deoxyribonucleic acid by carcinogens dimethyl sulphate, ethylmethanesulphonate, *N*-ethyl-*N*-nitrosourea and *N*-methyl-*N*-nitrosourea. Relative reactivity of the phosphodiester site thymidylyl(3'-5')thymidine, *Biochem.J.* 171, 575-587, 1978

Erhesamann, C., Baudin, F., Mougel, M., *et al.,* Probing the structure of DNA in solution, *Nucleic Acids Res.* 15, 9109-9128

Chemistry of Nucleosides and Nucleotides, Volume 1, ed. L.B. Townsend, Plenum Press, New York, New York, 1988

Chemistry of Nucleosides and Nucleotides, Volume 2, ed. L.B. Townsend, Plenum Press, New York, New York, 1991

Oakley, E.J., DNA methylation analysis: A review of current methodologies, *Pharmacol.Therapeut.* 84, 389-400, 1991

Glennon, R.A. and Tejon-Butl, S., Mesoionic nucleosides and heterobases, in *Chemistry of Nucleosides and Nucleotides,* Volume 2, ed. L.B. Townsend, Plenum Press, New York, New York, Chapter 1, pps. 1-21, 1991

Adams, R.L.P., Knowler, J.T. and Leader, D.P., *The Biochemistry of the Nucleic Acids*, 11th Edn., Chapman & Hall, London, 1992

Brown, D.J.. Evans, R.E., Cowder, W.B., and Fenn, M.D., *The Pyrimidines*, Interscience/John Wiley, New York, New York, 1994

Chemistry of Nucleosides and Nucleotides, Vol. 3, ed. L.B. Townsend, Plenum Press, New York, New York, 1994

Shaw, G., The synthesis and chemistry of imidazole and benzimidazole nucleosides and nucleotides, in *Chemistry of Nucleosides and Nucleotides*, Vol. 3, ed. L.B. Townsend, Plenum Press, New York, New York, Chaper 4, pps. 263-420, 1994

Nucleic Acids in Chemistry and Biology, 2nd Edn., ed. G.M. Blackburn and M.J. Gait, Oxford University Press, Oxford, UK, 1996

Oakeley, E.J., DNA methylation analysis: a review of current methodologies, *Pharmacol. Therapeut.* 84, 389-400, 1999

Ordway, J.M., Bedell, J.A., Citek, R.W., *et al.*, MethylMapper: A method for high-throughput, multilocus bisulfite sequence analysis and reporting, *BioTechniques* 39, 464-470, 2005

Chen, X, Dudgeon, N., Shen, L., and Wang, J.H., Chemical modification of gene silencing oligonucleotides for drug discovery and development, *Drug Discov.Today* 10, 587-593. 2005

Zhang, W.-Y., Du, Q., Wahlestedt, C., and Liang, Z., RNA interference with chemically modified siRNA, *Curr.Top.Med.Chem.* 6, 893-900, 2006

TRANSFECTION TECHNOLOGIES

1 Overview of transfection technologies

Transfection refers to the process by which foreign DNA (transgene) is incorporated into and expressed by a eukaryotic cell[1]. Stable transfection describes the incorporation of the transgene into the host cellular genome such that it is transferred with other host genetic material to daughter cells. In general, transfected cells must be "selected" from transiently transfected cells to obtain a stable transfected type. Selection is accomplished by inclusion of a "selectable" trait such as drug resistance such as the inclusion of dihydrofolate reductase (DHFR) gene to provide for growth in the presence of methotrexate when a DHFR-deficient cell is used as substrate[2-6]. The work of Robert Schimke and colleagues on the development of this system was one the critical factors in the development of the current biotechnology industry. In transient transfection, the transgene persists for several days with a peak period of expression after the first day. Transformation is the term used to describe the incorporation of naked DNA into prokaryocytes while conjugation refers to bacterial-bacterial exchange of DNA and transduction refers to the phage-mediated transfer of DNA. The use of naked DNA is usually ineffective with eukaryotic cells and a carrier or process is required[7]. Electroporation appears to be the most popular method with cationic lipids used to a less extent. Calcium phosphate and polyethyleneimine are some less frequently used. DEAE-dextran is used the least. Other technologies such as the gene gun (biolistic labeling)[8-11] are also being developed.

References

1. Baum, C., Transfection, in *Encyclopedia of Molecular Biology*, ed. T.E. Creighton, John Wiley and Sons, Inc., New York, NY, USA, 1999
2. Schimke, R.T., Gene amplification and methotrexate resistance in cultured animal cells, *Harvey Lect.* 76, 1-25, 1980-1981.
3. Schimke, R.T., Gene amplification in cultured animal cells, *Cell* 37, 705-713, 1984
4. Assaraf, Y.G., Molina, A., and Schimke, R.T., Sequential amplification of dihydrofolate reductase and multidrug resistance genes in Chinese hamster ovary cells selected for stepwise resistance to the lipid-soluble antifolate trimetrexate, *J.Biol.Chem.* 264, 19326-18234, 1989
5. Sharma, R.C., and Schimke, R.T., The propensity for gene amplification: a comparison of protocols, cell lines, and selection agents, *Mutat.Res.* 304, 243-260, 1994
6. Jun, S.C., Kim, M.N., Baik, J.Y., Hwang, S.O. and Lee, G.M., Selection strategies for the establishment for the establishment of recombinant Chinese hamster ovary cell line with dihydrofolate reductase-mediated gene amplification, *Appl.Microbiol.Biotechnol.* 69, 162-169, 2005
7. Primrose, S.B. and Twyman, R.M., *Principles of Gene Manipulation and Genomics*, Blackwell, Malden, MA, USA, 2006
8. O'Brien, J.A., and Loomis, S.C., Biolistic transfection of neuronal cultures using a hand-held gene gun, *Nat.Protoc.* 1, 9787-981, 2006
9. O'Brien, J.A., and Loomis, S.C., Biolistic labeling of neuronal culture and intact tissue using a hand-held gene gun, *Nat.Protoc.* 1, 1517-1521, 2006
10. Zhang, M., Tao, W., and Pianetta, P.A., Dynamics modeling of biolistic gene guns, *Phys.Med.Biol.* 52, 1485-1493, 2007
11. Lain, W.H., Chang, C.H., Chen, Y.J., *et al.*, Intracellular delivery can be achieved by bombarding cells of tissues with accelerated molecules or bacteria without the need for carrier particles, *Exp.Cell Res.* 313, 53-64, 2007

2 Calcium phosphate transfection in eukaryotic cell culture

Calcium phosphate is used to transfect mammalian cells. It was first used by Graham and Van Der Eb in 1973[1] and despite considerable use over the past three decades, this technique is still poorly understood[2,3]. It seems likely the use of calcium phosphate evolved from the earlier use of calcium for in bacterial transformation[4-9]. Graham and Van der Eb[1] separated the calcium phosphate-mediated transfection into three steps:

(1) Calcium phosphate + DNA → DNA-CaPO$_4$
(2) DNA-CaPO$_4$ + Cells → DNA transport into Cell (DNA-Cells)
(3) DNA-Cells → Cell Growth

Each of these steps has been subsequently demonstrated to be critical and has critical process attributes and critical process parameters (See Table 1). Other observations which are important to calcium phosphate-mediated transfection include:

- Inclusion of a "carrier" DNA appears to enhance the efficiency of transfection[10-12].
- pH is critical with a <u>very</u> narrow pH optima between 7.0-7.2[1,3,13]. Tris buffer was used in the first study on calcium phosphate transfection[1] but was replaced by HEPES[1,14] to provide stronger buffering in the critical pH range. HEPES is a problematic buffer[15] and it is of interest that one of the basic studies on the use of calcium phosphate[13] used BES instead of HEPES.

References

1. Graham, F.L. and van der Eb, A.J., A new technique for the assay of infectivity of human adenovirus 5 DNA, *Virology* 52, 456-467, 1973; Graham, F.L. and van der Eb, *Virology* 54, 536-539, 1973.
2. Yang, Y.-W. and Yang, J.-C., Calcium phosphate as a gene carrier: electron microscopy, *Biomaterials* 18, 213-217, 1997
3. Jordan, M. and Wurm, F., Transfection of adherent and suspended cells by calcium phosphate, *Methods* 33, 136-143, 2004
3a. Chang, P.L., Calcium phosphate-mediated DNA transfection, in *Gene Therapeutics: Methods and Applications of Direct Gene Transfer*, ed. J.W. Wolff, Birkhauser, Boston, MA, USA, pps 157-179, 1994
3ab. Conn, K.J., Degterev, A., Fontanilla, M.R., *et al.*, Calcium phosphate transfection, *DNA Transfer to Cultured Cells*, ed. K. Ravid and R.I. Freshney, Wiley-Liss, New York, NY, USA, Chapter 6, pps. 111-124, 1996

TABLE 1: Critical Process Attributes and Critical Process Parameters for Calcium Phosphate Transfection

Calcium phosphate + DNA → DNA-CaPO₄	1. Source of Calcium ions[3ab, 16, 17](See note 1)
	2. DNA – Calcium -phosphate(order of mixing)[17]
	3. Source of orthophosphate[17](See note 1)
	4. pH of mixing and choice of buffer[1,2,3,13]
	5. Method of mixing[16,17,20]
	6. Temperature of mixing[1,2,13,18]
	7. DNA concentration[1,3,13,14,17,18]
	8. Mixing container[17]
DNA-CaPO₄ + Cells → DNA transport into Cell (DNA-Cells)	1. Cells growth status (S Stage)[1,20]
	2. Time of "storage" of DNA-CaPO₄ before addition to cells[1,3,13,18,19,20]
	3. Presence or absence of media during addition of DNA-CaPO₄[2]
	4. Presence or absence of serum during addition of DNA-CaPO₄[2] (See Note 2)
	5. Time of contact with cells[1,3ab,17,20]
	6. Volume of transfectant solution added to medium[21]
	7. Use of EDTA wash to remove reagents (See Note 3)
DNA-Cells → Cell Growth	1. Growth time
	2. Growth temperature

Note 1: Most investigators use a sterile source of calcium ions and phosphate solutions. Also, filtration immediately prior to mixing is a frequently cited practice.

Note 2: Fetal Calf Serum (FCS, 10% final concentration in medium, usually Dulbecco's minimal essential medium; DMEM) is usually included in the medium for culturing the cells. Protein has an effect on the organization of calcium phosphate crystals. An additional complication is provided by CO_2 which, as HCO_3^-/CO_3^{2-} will incorporate into calcium phosphate crystals[16]. The cell cultures are usually maintained under CO_2 and CO_2 levels have been shown to be critical during the incubation of DNA with cells[3,13].

Note 3: Calcium phosphate is toxic to some cells. Individual protocols differ on the washing procedure for the cells following the calcium phosphate step.

4. Tyeryar, F.J. and Lawton, W.D., Factors affecting transformation of *Pasturella novicida*, *J.Bacteriol.* 104, 1312-1317, 1970

5. Osowiecki H. and Skalinska, B.A., The conditions of transfection of *Escherichia coli* cells untreated with lysozyme. I. The effect of some factors on the efficiency of transfection with lambda phage DNA, *Mol. Gen.Genet.* 133, 335-343 1974.

6. Erhlich, M., Sarafyan, L.P., and Myers, D.J., Interaction of microbial DNA with cultured mammalian cells. Binding of the donor DNA to the cell surface, *Biochim.Biophys.Acta* 454 397-409, 1976

7. Kahmann, R., Kamp, D., and Zipser, D., Transfection of *Escherichia coli* by Mu DNA, *Mol.Gen.Genet.* 149, 323-328, 1976.

8. Norgard, M.V. and Imaeda, T., Physiological factors involved in the transformation of *Mycobacterium smegmatis J.Bacteriol.* 133 1254-1262, 1978

9. Dagert, M. and Ehrlich, S.D., Prolonged incubation in calcium chloride improves the competence of *Escherichia coli* cells, *Gene* 6 23-28, 1979.

10. Graham, F.L., van der Eb, A.J., and Heijneker, H.L., Size and location of the transforming region in human adenovirus type 5 DNA, *Nature* 251, 687-691, 1974.

11. Bacchetti, S. and Graham, F.L., Transfer of the gene for thymidine kinase to thymidine kinase-deficient human cells by purified herpes simplex viral DNA, *Proc.Nat.Acad.Sci.USA* 74, 1590-1594, 1977.

12. Wigler, M., Pellicer, A., Silverstein, S., and Axel, R., Biochemical transfer of single-copy eukaryotic genes using total cellular DNA as donor, *Cell* 14, 725-731, 1978

13. Chen, C. and Okayama, H., High-efficiency transformation of mammalian cells by plasmid DNA, *Mol.Cell.Biol.* 7, 2745-2752, 1987.

14. Graham, F.L. and van der Eb, A.J., Transformation of rat cells by DNA of human adenovirus 5, *Virology* 54, 536-539, 1974

15. Chirpich, T.P., The effect of different buffers on terminal deoxynucleotidyl transferase activity, *Biochim.Biophys.Acta* 518, 535-538, 1978; Tadolini, B., Iron autoxidation in Mops and Hepes buffers, *Free Radic. Res. Commun.* 4, 149-160, 1987; Simpson, J.A., Cheeseman, K.H., Smith, S.E., and Dean, R.T., Free-radical generation by copper ions and hydrogen peroxide. Stimulation by Hepes buffer, *Biochem.J.* 254, 519-523, 1988; Abas, L. and Guppy M., Acetate: a contaminant in Hepes buffer, *Anal.Biochem.* 229, 131-140, 1995; Schmidt, K., Pfeiffer, S., and Mayer, B., Reaction of peroxynitrite with HEPES or MOPS results in

the formation of nitric oxide donors, *Free Radic.Biol.Med.* 24, 859-862, 1998; Fulop, L., Szigeti, G., Magyar, J., *et al.*, Differences in electrophysiological and contractile properties of mammalian cardiac tissues bathed in bicarbonate – and HEPES-buffered solutions, *Acta Physiol.Scand.*178, 11-18, 2003; Mash, H.E., Chin, Y.P., Sigg, L., *et al.*, Complexation of copper by zwitterionic aminosulfonic (good) buffers, *Anal.Chem.* 75, 671-677, 2003 Sokolowska, M., and Bal, W., Cu(II) complexation by "non-coordinating" *N*-2-hydroxyethylpiperazine-*N'*-ethanesulfonic acid (HEPES buffer), *J.Inorg.Biochem.* 99, 1653-1660, 2005; Zhao, G. and Chasteen, N.D., Oxidation of Good's buffers by hydrogen peroxide, *Anal.Biochem.* 349, 262-267, 2006; Hartman, R.F. and Rose, S.D., Kinetics and mechanism of the addition of nucleophiles to alpha,beta-unsaturated thiol esters, *J.Org.Chem.* 71, 6342-6350, 2006.

16. Röszler, S., Sewing, A., Stözel, M., *et al.*, Electrochemically assisted deposition of thin calcium phosphate coating at near-physiological pH and temperature, *J.Biomed.Mater.Res.* 64A, 655-663, 2002.

17. Cosaro, C.M. and Pearson, M.L., Enhancing the efficiency of DNA-mediated gene transfer in mammalian cells, *Somat.Cell Genet.* 7, 603-616, 1981.

18. Jordan, M., Schallhorn, A., and Wurn, F.M., Transfecting mammalian cells: optimization of critical parameters affecting calcium-phosphate precipitate formation *Nucleic Acids Res.* 24, 596-601, 1996.

19. Coonrod, A., Li, F.-Q., and Horwitz, M., On the mechanism of DNA transfection: efficient gene transfer without viruses, *Gene Therapy* 4 1313-1321 1997

20. Watanabe, S.Y., Albsoul-Younes, A.M., Kawano, T., *et al.*, Calcium phosphate-mediated transfection of primary cultured brain neurons using GFP expression as a marker: application to single neuron physiology, *Gene* 270 61-68, 2001.

21. Fu, H., Hu, Y., McNelis, T., and Hollinger J.O., A calcium phosphate-based gene delivery system, *J.Biomed.Mater.Res.* 74A 40-48, 2005.

3 DEAE-Dextran and transfection in eukaryotic cell culture

The ability of DEAE-Dextran to promote nucleic acid uptake into eukaryotic cells was first noted by Vaheri and Pagano[1-2]. Dimethylsulfoxide (DMSO) was also observed to enhance nucleic acid uptake[3,4]. DMSO was subsequently demonstrated to enhance calcium phosphate transfection[5]. DMSO "osmotic shock" was later shown to markedly enhance DEAE-dextran transfection efficiency[6-8]. DEAE-dextran is used less than poly-ethyleneimine. It would appear that DEAE-dextran is used more with Cos cells or BHK cells than with CHO cells. It is also noted that polyamines such as DEAE-dextran have been demonstrated to enhance nucleic acid transfer in *Escherichia coli* spheroblast[9].

References

1. Vaheri, A. and Pagano, J.S., Infectious poliovirus RNA: A sensitive method of assay, *Virology* 27, 434-436, 1965
2. Pagano, J.S., McCutchan, J.H., and Vaheri, A., Factors influencing the enhancement of the infectivity of poliovirus ribonucleic acid by diethylaminoethyl-dextran, *J.Virol.* 1, 891-897, 1967
3. Tovell, D.R. and Colter, J.S., Observations on the assay of infectious viral ribonucleic acid: effects of DMSO and DEAE-dextran, *Virology* 32, 84-92, 1967
4. Tovell, D.R. and Colter, J.R., The interaction of tritium-labelled Mengo virus RNA and L cells: the effects of DMSO and DEAE-dextran, *Virology* 37, 624-631, 1969
5. Stow, N.D. and Wilkie, N.M., An improved technique for obtaining enhanced infectivity with herpes simplex virus type 1 DNA, *J.Gen. Virol.* 33, 447-458, 1976
6. Lopata, M.A., Cleveland, D.W., and Sollner-Webb, B., High level transient expression of a chloramphenicol acetyl transferase gene by DEAE-dextran mediated DNA transfection coupled with a dimethyl sulfoxide or glycerol shock treatment, *Nucleic Acids Res.* 12, 5707-5117, 1984
7. Kluxen, F.W. and Lubbert, H., Maximal expression of recombinant cDNAs in COS cells for use in expression cloning, *Anal.Biochem.* 208, 352-356, 1993
8. Schwartz, J.J. and Rosenberg, R., DEAE-dextran transfection, in *DNA Transfer to Cultured Cells*, ed. K. Ravid and R.I. Freshney, Wiley-Liss, New York, NY, USA, 1998
9. Henner, W.D., Kleber, I., and Benzinger, R., Transfection of *Escherichia coli* spheroblasts. 3. Facilitation of transfection and stabilization of spheroblasts by different basic polymers, *J.Virol.* 12, 741-747, 1973

TABLE 2: Some Examples of DEAE-Dextran Transfection in Cultured Eukaryotic Cells

Ehrlich, M., Sarafyan, L.P., and Myers, D.J., Interaction of microbial DNA with cultured mammalian cells. Binding of the donor DNA to the cell surface, *Biochim.Biophys.Acta* 454, 397-409, 1976

Evaluated the effect of various polyamines on the uptake of microbial DNA by cultured fibroblasts. In the presence of a polycation such as DEAE-dextran, 10-30% of DNA was bound to the cell; with DNA alone, 0.5-5% of the DNA was cell associated.

Gopal, T.V., Gene transfer method for transient gene expression, stable transformation, and cotransformation of suspension cell culture, *Mol.Cell Biol.* 5, 1188-1190, 1985

Cells (mouse myeloma, erythroleukemia) were bound to concanavalin A-coated tissue culture plates and treated with DEAE-dextran and plasmid DNA. Subsequent treatment with 40% polyethylene glycol facilitated the update of DNA by the substrate cells.

Takai, T. and Ohmori, H., DNA transfection of mouse lymphoid cells by the combination of DEAE-dextran-mediated DNA uptake and osmotic shock procedure, *Biochim.Biophys.Acta* 1048, 105-109. 1990

Mouse lymphoid cells are first treated with DEAE-dextran followed by osmotic shock (hypertonic Tris hydrochloride buffer with 0.5 M sucrose and 10% PEG followed by hypertonic RPMI 1640)

Pazzagli, M., Devine, J.H., Peterson, D.O., and Baldwin, T.O., Use of bacterial and firefly luciferase as reporter genes in DEAE-dextran-mediated transfection of mammalian cells, *Anal.Biochem.* 204, 315-323, 1992

Development of luciferase transgenes as reporter genes.

Gauss, G.H. and Lieber, M.R, DEAE-dextran enhances electroporation of mammalian cells, *Nuc.Acids Res.* 20, 6739-6740, 1992

DEAE-dextran increases transfection efficiency in several mammalian cell lines including human lymphoid cells and hamster fibroblast cells.

Kluxen, F.W. and Lubbert, H., Maximal expression of recombinant cDNAs in COS cells for use in expression cloning, *Anal.Biochem.* 208, 352-356, 1993

B-Galactosidase was used as reporter gene in COS cells. DEAE-dextran was found to be superior to electroporation or lipofection in transfection efficiency. COS-1 cells expressed more protein than COS-7 cells.

Gonzalez, A.L. and Joly, E., A simple procedure to increase efficiency of DEAE-dextran transfection of COS cells, *Trends Genet.* 11, 216-217, 1995

Technical improvements in DEAE-dextran transfection technology

Yang, Y.W. and Yang, J.C., Studies of DEAE-dextran-mediated gene transfer, *Biotechnol.Appl.Biochem.* 25, 47-51, 1997

Gene transfer into FR3T3 cells. Optimal ratio of DEAE-dextran to DNA of 50 to 1. It is suggested that the binding of the DEAE-dextran/DNA to the cell surface determines DNA transfection efficiency.

Mack, K.D., Wei, R., Elbagarri, A., *et al.*, A novel method of DEAE-dextran mediated transfection of adherent primary cultured human macrophages, *J.Immunol.Methods* 211, 79-86, 1998

DEAE-dextran concentration, DNA quantity, and incubation time were the three critical factors in the transfection of adherent human macrophages.

Schenborn, E.T. and Goiffon, V. DEAE-dextran transfection of mammalian cultured cells, *Methods Mol.Biol.* 130, 147-153, 2000

Technical review

Pari, G.S. and Xu, Y., Gene transfer into mammalian cells using calcium phosphate and DEAE-dextran, *Methods Mol.Biol.* 245, 25-32, 2004

Technical review comparing transfection technologies (calcium phosphate and DEAE-dextran)

Hermans, E., Generation of model cell lines expressing recombinant G-protein-coupled receptors, *Methods Mol.Biol.* 259, 137-153, 2004

Technical review comparing transfection technologies (calcium phosphate, DEAE-dextran, cationic lipids, and electroporation)

Escher, G., Hoang, A., Georges, S., *et al.*, Demethylation using epigenetic modifier, 5-azacytidine, increases the efficiency of transient transfection of macrophages, *J.Lipid Res.* 46, 356-365, 2005

Several transfection technologies were evaluated using a CMV-LacZ plasmid containing a bacterial β-galactosidase as a reporter gene. DEAE-dextran(with DMSO osmotic shock) was slightly more effective than a variety of cationic lipids.

4　Electroporation for the transfection of eukaryotic cells

Electroporation is a process for transfection where cells are suspended at relatively high density (10^6-10^7mL) and subjected to a short (milliseconds to microseconds) electric pulses[1-8]. While electroporation tends to be challenging to cells with cell losses up to 50%, it does see extensive use in commercial biotechnology where it is viewed as a robust process. Iontophoresis is an alternative drug delivery related to electroporation when a low-level electrical current is used to administer ionic drug to the skin[9-12]. Electroporation has been reported to promote protein uptake by cells[13-14]

References

1. Neumann, E. and Sowers, A. E., *Electroporation and Electrofusion in Cell Biology*, Plenum Press, New York, NY, USA, 1989
2. Chang, D.C., *Guide to Electroporation and Electrofusion*, Academic Press, San Diego, CA, USA, 1992
3. Nickoloff, J.A., *Animal Cell Electroporation and Electrofusion protocols*, Humana Press, Totowa, NJ, USA, 1995
4. Nickoloff, J.A., *Plant Cell Electroporation and Electrofusion Protocols*, Humana Press, Totowa, NJ, USA, 1995
5. Nickoloff, J.A., *Electroporation Protocols for Microorganisms*, Humana Press, Totowa, NJ, USA, 1995
6. Lynch, P.T. and Davey, M.R., *Electrical Manipulation of Cells*, Chapman & Hall, New York, NY, USA, 1996
7. Polk, C. and Postow, E., *Handbook of Biological Effects of Electromagnetic Fields*, CRC Press, Boca Raton, FL, USA, 1996
8. Peña, L., *Transgenic Plants: Methods and Protocols*, Human Press, Totowa, NJ, USA, 2005
9. Kanikkannan, N., Iontophoresis-based transdermal delivery systems, *BioDrugs* 16, 339-347, 2002
10. Gehl, J., Electroporation: theory and methods, perspectives for drug delivery, gene therapy and research, *Acta Physiol.Scand.* 177, 437-447, 2003
11. Nanda, A., Nanda, S. and Ghilzai, N.M., Current developments using emerging transdermal technologies in physical enhancement methods, *Curr.Drug Deliv.* 3, 233-242, 2006
12. Mayes, S. and Ferrone, M., Fentanyl HCl patient-controlled iontophoretic transdermal system for the management of acute postoperative pain, *Ann.Pharmacother.* 40, 2178-2186, 2006
13. Lambert, H, Pankov, R., Gauthier, J., and Hancock, R., Electroporation-mediated uptake of proteins into mammalian cells, *Biochem.Cell Biol.* 68, 729-734, 1990
14. Mohr, J.C., de Pablo, J.J., and Palecek, S.P., Electroporation of human embryonic stem cells: small and macromolecule loading and DNA transfection, *Biotechnol.Prog.* 22, 825-834, 2006

TABLE 3:　Some Examples of Electroporation for Transfection of Eukaryotic Cells

Maxwell, I.H. and Maxwell, F., Electroporation of mammalian cells with a firefly luciferase expression plasmid: kinetics of transient expression differ markedly among cells types, *DNA* 7, 557-5562, 1988

Transfection by electroporation was examined in a several mammalian cell types. In some cell lines, expression of the transgene was maximal at 12 hours followed by a rapid decline. It was concluded that the time course of transgene expression following electroporation must be determined for each cell line.

Andreason, G.L. and Evans, G.A., Optimization of electroporation for transfection of mammalian cell lines, *Anal.Biochem.* 180, 269-275, 1989

Evaluated transfection technologies (electroporation, calcium phosphate, and DEAE-dextran) in several rodent cells lines including PC12 and B50 (rat neuroblastoma) with firefly luciferase as the reporter gene. The relative effective of a specific transfection technology varied with the cell line. It was observed that electroporation was more effective at 23°C than at 4°C.

Gauss, G.H. and Lieber, M.R., DEAE-dextran enhances electroporation of mammalian cells, *Nucleic Acids Res.* 20, 6739-6740, 1992

The presence of DEAE-dextran increase transfection efficiency with electroporation.

Rols, M.P., Delteil, C, Serin, G., and Teissie, J. ,Temperature effects on electroporation of mammalian cells, *Nucleic Acids Res.* 22, 540, 1994

Experiments with CHO cells in 10 mm phosphate, pH 7.2, containing 250 mM sucrose and 1 mM $MgCl_2$ showed that there are different temperature optima for the steps in the transfection process. Incubation at 4°C prior to electroporation yielded the best results while incubation at 37°C after electroporation increased transfection efficiency.

Teifel, M., Heine, L.T. ,Milbredt, S., and Friedl, P., Optimization of transfection of human endothelial cells, *Endothelium* 5, 21-35, 1997

Electroporation is suggested to more useful than lipofection, DEAE-dextran, or calcium phosphate for the transfection of endothelial cells.

Nickoloff, J.A. and Reynolds, J.A., Electroporation-mediated gene transfer efficiency is reduced by linear plasmid carrier DNAs, *Anal.Biochem.* 205, 237-243, 1992

Linear DNA inhibited transfection by electroporation in CHO cells while circular plasmids enhanced transfection efficiency with electroporation.

Melkonyan, H., Sorg, C. and Klempt, M., Electroporation efficiency in mammalian cells is increased by dimethyl sulfoxide (DMSO), *Nucleic Acids Res.* 24, 4356-4367, 1996

DMSO improved transfection efficiency by electroporation in four cells lines (HL60, TR146, Cos-7, and L132). DMSO was present during electric pulse and in the following incubation.

Yang, T.A., Heiser, W.C., and Sedivy, J.M., Efficient *in situ* electroporation of mammalian cells grown on microporous membranes, *Nucleic Acids Res.* 23, 2803-2810, 1995

Electroporation with cells grown on microporous membranes such as polyethylene terephthalate or polyester.

Baum, C., Forster, P., Hegewisch-Becker, S. and Harbers, K., An optimized electroporation protocol applicable to a wide range of cell lines, *BioTechniques* 17, 1058-1062, 1994

Optimization of voltage for the electroporation step. With optimization, electroporation is superior to other methods of Transfection

Delteil, C., Teissie, J., and Rols, M.P., Effect of serum on *in vitro* electrically mediated gene delivery and expression in mammalian cells, *Biochim. Biophys.Acta* 1467, 362-368, 2000

Serum increased transfection efficiency mediated by electroporation in CHO cells.

Bodwell, J., Swift, F., and Richardson, J., Long duration electroporation for achieving high level expression of glucocorticoid receptor in mammalian cell lines, *J.Steroid Biochem.Mol.Biol.* 68, 77-82, 1999

Long duration electroporation (LDE) uses a lower voltage (440-500 V/cm) for a longer time (140 milliseconds). LDE allowed transient expression in Cos-7 cells at high levels usually seen only in stable cell lines. It was critical that the cells be log phase and less than 70% confluent in culture.

TABLE 3: Some Examples of Electroporation for Transfection of Eukaryotic Cells (Continued)

Golzio, M., Mora, M.P., Raynaud, C., *et al.*, Control by osmotic pressure of voltage-induced permeabilization and gene transfer in mammalian cells, *Biophys.J.* 74, 3015-3022, 1998

Electric pulsing at low osmolarity (10 mM phosphate-125 mM sucrose-1 mM MgCl₂, pH 7.4) increased electroporation efficiency

Muller, K.J., Horbaschek, M., Lucas, K., *et al.*, Electrotransfection of anchorage-dependent mammalian cells, *Exp. Cell Res.* 288, 344-353, 2003

Methods are developed for the *in situ* electroporation of anchorage-dependent cells.

Distler, J.H.W., Jüngel, A., Kurowska-Stolarska, M., *et al.*, Nucleofection: a new, highly efficient transfection method for primary human keratinocytes, *Exptl.Dermatol.* 14, 315-320, 2005

Nucleofection is an electroporation technique where DNA directly enters the nucleus.

Leclerre, P.G., Panjwani, A., Docherty, R., *et al.*, Effective gene delivery to adult neurons by a modified form of electroporation, *J.Neurosci.Methods* 142, 137-143, 2005

Used nucleofection for transfection of adult neurons

Barry, P.A., Efficient electroporation of mammalian cells in culture, *Methods Mol.Biol.* 245, 207-214, 2004

Review of electroporation technology

Buchser, W.J., Pardinas, J.R., Shi, Y., *et al.*, 96-Well electroporation method for transfection of mammalian central neurons, *BioTechniques* 41, 619-624, 2006

A 96-well electroporation platform has been developed.

---

5 Lipofection for transfection of eukaryotic cells in culture

Cationic or neutral lipids can be used to prepare liposomes to condense nucleic acids for transfer into cells[1-9]. These supramolecular structures are known as a lipoplexes. Lipoplexes are sometimes referred to as polyplexes but it would seem that the term polyplex is used more often to describe polyamine complexes with DNA such a PEI-DNA or DNA-DEAE-dextran complexes. The cationic lipids include *N*-[1-(2,3-dioleyl)propyl]-*N,N,N*-trimethylammonium chloride (DOTAP) and *N*-[1-(2,3-dioleyloxy) propyl]-*N,N,N*-trimethylammonium chloride (DOTMA)[10,11]. The inclusion of neutral lipids such as cholesterol or 1,2-dioleoyl-*sn*-glycero-3-phosphoethanolamine (DOPE) has been shown to enhance transfection[12-14]. It is thought that cationic lipids bind to the DNA condensing into a liposomal structure which can then fuse with the cell wall and undergo membrane fusion/endocytosis forming an intracellular endocytotic vesicle which is destabilized in the cytoplasm with the release of DNA. The DNA then enters the nucleus as postulated for the other polyplexes where the transgene may or may not be expressed in, most likely, a transient manner.

References

1. Clark, P.R. and Hersh, E.M., Cationic lipid-mediated gene transfer: current concepts, *Curr.Opin.Mol.Ther.* 1, 158-176, 1999
2. Ulrich, A.S., Biophysical aspects of using liposomes as delivery vehicles, *Biosci.Rep.* 22, 1291-150, 2002
3. Pedroso de Lima, M.C., Neves, S., Filipe, A., Duzgunes, N., and Simoes, S., Cationic liposomes for gene delivery: from biophysics to biological applications, *Curr.Med.Chem.* 10, 1221-1231, 2003
4. May, E. and Ben-Shaul, A., Modeling of cationic lipid-DNA complexes, *Curr.Med.Chem.* 11, 151-167, 2004
5. Zhang, S., Xu, Y., Wang, B., *et al.*, Cationic compounds used in lipoplexes and polyplexes for gene delivery, *J.Control.Release.* 100, 165-180, 2004
6. Simoes, S., Filipe, A., Faneca, H., *et al.*, Cationic liposomes for gene delivery, *Expert Opin.Drug Deliv.* 2, 237-254, 2005
7. Khalil, I.A., Kogure, K., Akita, S. and Harashima, H., Uptake pathways and subsequent intracellular trafficking in nonviral gene delivery, *Pharmacol.Rev.* 58, 32-45, 2006
8. Wasungu, L. and Hookstra, D., Cationic lipids, lipoplexes and intracellular delivery of genes, *J.Control.Release* 116, 2555-265, 2006
9. Huang, L. and Hung, M.C., *Non-viral vectors for Gene Therapy*, Elsevier/Academic Press, Amsterdam, Netherlands, 2005
10. Zhang, S., Xu, Y., Wang, B., *et al.*, Cationic compounds used in lipoplexes and polyplexes for gene delivery, *J.Control.Release* 100, 165-180, 2004
11. Wasungu, L. and Hookstra, D., Cationic lipids, lipoplexes and intracellular delivery of genes, *J.Control.Release* 116, 255-264, 2006

TABLE 4: Some Examples of the Use of Lipofection for Transfection of Mammalian Cell Culture

Liu, F., Yang, J., Huang, L., and Liu, D., New cationic lipid formulations for gene transfer, *Pharm.Res.* 13, 1856-1860, 1996

Tween was combined with lipid components for transfection studies

Keogh, M.C., Chen, D., Lupu, F., *et al.*, High efficiency reporter gene transfection of vascular tissue *in vitro* and *in vivo* using a cationic lipid-DNA complex, *Gene Ther.* 4, 162-171, 1997

The most important factors in cationic lipid transfection of a plasmid containing a luciferase reporter gene to rabbit or human arterial tissue was the ratio of lipid reagent to DNA, DNA concentration, transfection time and the presence or absence of serum. However, there was variation with cell line. Hep2 was the only cell line showing a positive effect of serum. The study did show the necessity of establishing optimal conditions for each cell line.

Zelphati, O., Nguyen, C., Ferrari, M., *et al.*, Stable and monodisperse lipoplex formulations for gene therapy, *Gene Ther.* 5, 1272-1282, 1998

A lipoplex formulation was developed which could be stored frozen without losing biological activity or physical stability. The critical parameters for formulation success are size of the cationic liposome, the rate and method of DNA and cationic lipid mixing, and ionic strength of suspension vehicle.

Ferrari, M.E, Ngyen, C.M., Zelphati, O., *et al.*, Analytical methods for the characterization of cationic lipid-nucleic acid complexes, *Hum.Gene Ther.* 9, 341-351, 1998

TABLE 4: Some Examples of the Use of Lipofection for Transfection of Mammalian Cell Culture (Continued)

Lipid recovery, total DNA, free DNA, nuclease sensitivity, and physical stability by filtration are proposed for the evaluation of formulation variables on the physical properties of lipoplexes.

Xu, Y., Hui, S.W., Frederik, P. and Szoka, F.C., Jr., Physicochemical characterization and purification of cationic lipoplexes, *Biophys.J.* 77, 341-353, 1999

Lipoplexes were prepared with varying excesses of DOTAP or DNA and separated into positively charged lipoplexes or negatively charged lipoplexes. Positively charged lipoplexes had high transfection activity and reduced toxicity. There were also structural differences between the two classes of lipoplexes.

Lin, A.J, Slack, N.L., Ahmad, A., *et al.*, Structure and structure-function studies of lipid/plasmid DNA complexes, *J.Drug.Target.* 8, 13-27, 2000

There appears to be a relationship between physical structure of lipoplexes and transfection efficiency in mouse fibroblast L-cells. One structure consisted of DNA between two ordered multilamellar structures. Lipids used included DOTAP, DOPE, and DOPC.

Nchinda, G., Uberla, K., and Zschornig, O., Characterization of cationic lipid DNA transfection complexes differing in susceptibility to serum inhibition, *BMC Biotechnol.* 2, 12, 2002

Evaluated the effect of serum on transfection efficiency.

Simberg, D., Weisman, S., Talmon, Y., and Barenholz, Y., DOTAP (and other cationic lipids): chemistry, biophysics, and transfection, *Crit.Rev.Ther. Drug Carrier Syst.* 21, 257-317, 2004

Evaluation of the properties of DOTAP in cell transfection.

Bengali, E., Pannier, A.K., Segura, T., *et al.*, Gene delivery through cell culture substrate adsorbed DNA complexes, *Biotechnol.Bioeng.* 90, 290-302, 2005

Pretreatment of tissue culture plates with serum enhanced transfection by lipoplexes by increasing the number of transfected cells with similar level of expression.

Decastro, M., Saijoh, Y., and Schoenwolf, O.C., Optimized cationic lipid-based gene delivery reagents for use in developing vertebrate embryos, *Dev. Dyn.* 235, 2210-2219, 2006

This study evaluated Lipofectamine, Lipofectamine 2000, and Lipofectamine with a disulfide-linked pegylated lipid for GFP transgene expression. Significant levels of ectopic gene expression were observed.

---

12. Rose, J.K., Buonocore, L., and Whitt, M.A., A new cationic liposome reagent mediating nearly quantitative transfection of animal cells, *Biotechniques* 10, 520-525, 1991
13. Felgner, J.H., Kumar, R., Sridhar, C.N., *et al.*, Enhanced gene delivery and mechanism studies with a novel series of cationic lipid formulations, *J.Biol.Chem.* 269, 2550-2561, 1994
14. Teifel, M., Heine, L.T., Milbredt, S., and Friedl, P., Optimization of transfection of human endothelial cells, *Endothelium* 5, 21-35, 1997

6   Polyethyleneimine and transfection in eukaryotic cell culture

Polyethyleneimine (PEI) is one of several vehicles used for non-viral transfection of cells and results in transient rather than stable transfection. PEI appears to be used more frequently that DEAE-dextran. Polyamines such as PEI function in transfection by condensing of DNA with the formation of a polyplex (the term polyplex refers to the DNA/carrier condensate). Branched forms and linear forms of PEI are used. The polyplex binds to the cell membrane, passes into the cell via endocytosis, is released from the endosome and passes into the nucleus[1-12]. Early studies by Pollard and coworkers[1] showed the polycation was not only required for effective endocytosis of DNA, polycation also stimulated nuclear uptake of the DNA[1]. These early studies have been extended by a number of workers including Breung and coworkers[2]. There have been a number studies on the endocytotic process but the mechanism remains elusive[8,12,]. Recent studies[13] have suggested that process efficiency is limited by passage of DNA into the nucleus and expression of all inserted DNA. PEI is also useful for transferring RNAi derivatives to the cytoplasm[14-18]. There is one report on the use of PEI as a transmembrane carrier for fluorescently labeled proteins[19].

References

1. Pollard, R., Remy, J.S., Loussouarn, G., *et al.*, Polyethyleneimine but not cationic lipids promote transgene delivery to the nucleus in mammalian cells, *J.Biol.Chem.* 273, 7507-511, 1998
2. Breunig, M., Lungwitz, U., Liebl, R. *et al.*, Mechanistic insights into linear polyethyleneimine-mediated gene transfer, *Biochim.Biophys. Acta* 1770, 196-205, 2007
3. Wagner, E., Effects of membrane-active agents in gene delivery, *J.Control.Release* 53, 155-158, 1998
4. Choosakookriang, S., Lobo, B.A., Koe, G.S., *et al.*, Biophysical characterization of PEI/DNA complexes, *J.Pharm.Sci.* 92, 1710-1722, 2003
5. Sonawane, N.D., Szoka, F.C., Jr., and Verkman, A.S., Chloride accumulation and swelling in endosomes enhances DNA transfer by polyamine-DNA polyplexes, *J.Biol.Chem.* 278, 33826-44832, 2003
6. Akinc, A., Thomas, M., Klibanov, A.M., and Langer, R., Exploring polyethyleneimine–mediated DNA transfection and the proton sponge hypothesis, *J.Gene Med.* 7, 657-663, 2005
7. Demeneix, B. and Behr, J.P., Polyethyleneimine (PEI), *Adv.Genet.* 53, 217-230, 2005
8. Heidel, J., Mishra, S., and Davis, M.E., Molecular conjugates, in *Gene Therapy and Gene Delivery Systems*, ed. D.V. Schoffer and W. Zhou, Springer-Verlag, Berlin, Germany. pps. 7-39, 2005
9. Nimesh, S., Goyal, A., Pawar, V., *et al.*, Polyethyleneimine nanoparticles as efficient transfecting agents for mammalian cells, *J.Control Release* 110, 457-468, 2006
10. Brissault, B., Leborgne, C., Guis, C., *et al.*, Linear topology confers *in vivo* gene transfer activity to polyethyleneimine, *Bioconjug.Chem.* 17, 759-765, 2006
11. Kodaoma, K., Katayama, Y., Shoji, Y., and Nakashima, H., The features and shortcomings for gene delivery of current non-viral carriers, *Curr.Med.Chem.* 13, 2155-2161, 2006
12. Rejman, J., Conese, M., and Hoekstra, D., Gene transfer by means of lipo- and polyplexes: role of clathrin and caveolae-mediated endocytosis, *J.Liposome Res.* 16, 237-247, 2006
13. Carpentier, E., Paris, S., Kamen, A.A., and Durocher, Y., Limiting factors governing protein expression following polyethyleneimine-mediated gene transfer in HEK293-EBNA1 cells, *J.Biotechnol.* 128, 268-280, 2007

TABLE 5: Some Examples of the Use of Polyethyleneimine for Gene Transfection in Cell Culture

Vancha, A.R., Govindaraju, S., Parsa, K.V., *et al.*, Use of polyethyleneimine polymer in cell culture as attachment factor and lipofection enhancer, *BMC Biotechnol.* 4, 23, 2004

PEI as attachment factor for weakly anchoring cell lines. PEI also appears to enhance lipofection efficiency. Experimental results with PC-12 and HEK-293 cells.

Schlatter, S., Stansfield, S.H., Dinnis, D.M., *et al.*, On the optimal ratio of heavy to light chain genes for efficient recombinant antibody production by CHO cells, *Biotechnol.Prog.* 21, 122-133, 2005

PEI transfection in CHO cells. It was determined that a low heavy chain/light chain ratio was optimal for transient expression

Mennesson, E., Erbacher, P., Piller, V., *et al.*, Transfection efficiency and uptake process of polyplexes in human lung endothelial cells: a comparative study non-polarized and polarized cells, *J.Gene Med.* 7, 729-738, 2005

Compared transfection efficiency of PEI polyplexes with histidylated polylysine (polylysine where approximately 50% of the ε-amino groups are substituted with a histidyl-residue-see Midoux, P. and Monsigny, M., Efficient gene transfer by histidylated polylysine/pDNA complexes, *Bioconjug. Chem.* 10, 406-411, 1999) with human lung microvascular endothelial cells in monolayer culture. Transfection was more effective with PEI polyplexes as judged with YOYO-labelled plasmids and luciferase activity.

Bengali, Z., Pannier, A.K., Segura, T., *et al.*, Gene delivery through cell culture substrate adsorbed DNA complexes, *Biotechnol.Bioeng.* 90, 290-302, 2005

Transfection efficiency increased when serum-coated tissue culture plate used. The polyplex complexes were immobilized on the serum-coated plate prior to the transfection step. Transfection efficiency was measured by luciferase(transgene) expression NIH/3T3 cells.

Breunig, M., Lungwitz, U., Liebl, R., *et al.*, Gen delivery with low molecular weight linear polyethyleneimines, *J.Gene Med.* 7, 1287-1298, 2005

Low molecular weight PEI polymers (5-9 kDa) may be better than higher molecular weight materials as a result of higher transfection efficiency and lower cytotoxicity. Transfection was measure by green fluorescent protein expression in either CHO cells or HeLa cells.

Banerjee, P., Weissleder, R. and Bogdanov, A., Jr., Linear polyethyleneimine grafted to a hyperbranched poly(ethylene glycol)-like core: a copolymer for gene delivery, *Bioconjug.Chem.* 17, 125-131, 2006

The block copolymer of branched PEI and hyperbranched PEG provided DNA condensates with longer life stability and much higher transfection efficiency that a block copolymer of branched PEI and PEG.

Arnold, A.S., Laporte, V., Dumont, S., *et al.*, Comparing reagents for efficient transfection of human primary myoblasts: FuGENE 6, Effectene and ExGen 500, *Fundam.Clin.Pharmacol.* 20, 81-89, 2006

Fugene 6 (a cationic lipid), Effectene (a cationic lipid), and ExGen500 (a linear PEI polymer) were compared with respect to their transfection efficiency in primary myoblast cells. Transfection was measured by cell viability (mitochondrial dehydrogenase activity). Fugene 6 was the most effective but was more dependent on the presence of serum than ExGen.

Braga, D., Laize, V, Tiago, D.M., and Cancela, M.L., Enhanced DNA transfer into fish bone cells using polyethyleneimine, *Mol.Biotechnol.* 34, 51-54, 2006

Transfection efficiency measured with green fluorescent protein and luciferase.

Galbraith, D.J., Tait, A.S., Racher, A.J., *et al.*, Control of culture environment for improved polyethyleneimine-mediated transient production of recombinant monoclonal antibodies by CHO cells, *Biotechnol.Prod.* 22, 753-762, 2006

Used a branched 25 kDa PEI with suspension-adapted CHO cells. The production of recombinant MABs was followed by ELISA and with alkaline phosphatase reporter gene in static culture and green fluorescent protein in shake cultures. The addition of growth factors to media improved production.

Breunig, M., Langwitz, U., Liebl, R., *et al.*, Mechanistic insights into linear polyethyleneimine-mediated gene transfer, *Biochim.Biophys.Acta* 1770, 196-205, 2007

Linear PEI (6.6 kDa) demonstrated high transfection efficiency (44%) with relatively low cytotoxicity. CHO cells were used in adherent culture. A plasmid with green fluorescent protein transgene was used as the reported gene.

14. Bologna, J.C., Dorn, G., Natt, F., and Weiler, J., Linear polyethyleneimine as a tool for comparative studies of antisense and short double-stranded RNA oliogonucleotides, *Nucleosides, Nucleotides Nucleic Acids* 22, 1729-1731, 2003

15. Urban-Klein, B,. Werth, S., Abuharbeid, S., *et al.*, RNAi-mediated gene targeting through systemic application of polyethyleneimine (PEI)-complexed siRNA *in vivo*, *Gene Ther.* 12, 461-466, 2005

16. Grayson, A.C., Doody, A.M., and Putman, D., Biophysical and structural characterization of polyethyleneimine-mediated siRNA delivery *in vitro*, *Pharm.Res.* 23, 1868-1876, 2006

17. Werth, S., Urban-Klein, G., Dai, L, *et al.*, A low molecular weight fraction of polyethyleneimine (PEI) displays increased transfection efficiency of DNA and siRNA in fresh or lyophilized complexes, *J.Control.Delivery* 112, 257-270, 2006

18. Putnam, D. and Doody, A., RNA-interference effectors and their delivery, *Crit.Rev.Ther.Drug.Carrier Syst.* 23, 137-164, 2006

19. Didenko, V.V., Ngo, R. and Baskin, D.S., Polyethyleneimine as a trans-membrane carrier of fluorescently labeled proteins and antibodies, *Anal.Biochem.* 344, 168-173, 2005

General references for transfection

Gene Therapy Technologies, Applications and Regulatory. From Laboratory to Clinic, ed. A. Meager, John Wiley & Sons, Ltd., Chichester, UK, 1995

DNA Transfer to Cultured Cells, ed. K.Ravid and R.I. Freshney, Wiley-Liss, New York, NY, USA, 1998

Understanding Gene Therapy, ed. N.R. Lemoine, Bios/Springer, Oxford, UK, 1999

Gene Therapy. Therapeutic Mechanisms and Strategies, ed. N.S. Templeton and D.D. Losic, Marcel Dekker, Inc., New York, NY, USA, 2000

Freshney, I.A., *Culture of Animal Cells. A Manual of Basic Techniques*, 4th edn., Wiley-Liss, New York, NY, 2000

Gene Therapy and Gene Delivery Systems, ed. D.V. Schoffer and W. Zhou, Springer-Verlag, Berlin, Germany, 2005

Section V
Carbohydrates

INTRODUCTION TO CARBOHYDRATES

Nomenclature for carbohydrates

The reader is directed to recommendations from the International Union of Pure and Applied Chemistry (IUPAC) (1,2). The IUPAC recommendations are contained in the excellent work on carbohydrate structure by Collins (3). There are also excellent discussions of nomenclature in several reference texts (4,5). The reader is also directed to several recent articles concerning specific nomenclature issues (6-17). There are also IUPAC guidelines for glycolipids (18) as well as a separate article on the nomenclature of inositol derivatives (19).

The following are some simple definitions which might be useful. The reader is directed to the formal IUPAC documents for detail.

- **Aldaric acids** are aldoses where the aldehyde and terminal primary alcohol function are replaced by carboxylic acid residues.
- **Alditols** are polyhydric alcohols where the aldehyde function in an aldolose or the ketone function in a ketose has been replaced by a hydroxyl function.
- **Aldonic acids** are monosaccharides which contain an aldehyde and a carboxylic acid function.
- **Aldoses** are monosaccharides with a terminal aldehyde function.
- **Aldosuloses** are monosaccharides which contain an aldehyde and a ketone function.
- An **Amino sugar** is a monosaccharide where a hydroxyl group has been replaced by an amino group such as with glucosamine.
- **Carbohydrates** includes monosaccharides, disaccharides, oligosaccharides, and oligosaccharides or glycans. The term **sugar** is usually confined to carbohydrates with lower molecular weights such as monosaccharides and disaccharides.
- **Deoxy** is a term describing a monosaccharide where an hydroxyl group has been replaced by a hydrogen.
- **Dialdoses** are dialdehyde monosaccharides.
- **Disaccharide** refers to a compound composed of two monosaccharide units connected by a glycosidic bond.
- **Furanose** (from furan) designates a five-membered ring.
- The term **glycan** is used to describe a oligosaccharide covalently bound to a protein. The **N-linked glycans** are frequently branched and are described a monoantennary, biantennary, triantennary, etc.
- A **glycerolipid** is a glycolipid with one or more glycerol groups.
- **Glycodiuloses** are diketo monosaccharides (5-keto-fructose).
- A **glycolipid** is a compound composed of one or more monosaccharide units bound by a glycosidic linkage to a hydrophobic group such as sphingoid or ceramide.
- A **glycoside (glycosidic) bond** joins monosaccharide units into disaccharides and oligosaccharide/polysaccharides. The glycoside bond joins the hydroxyl group on the anomeric carbon of a monosaccharide with any hydroxyl group on another monosaccharide through an acetal linkage.
- A **glycoside** is an acetal derivative of the cyclic form of a sugar where the hydrogen in the hemiacetal is replaced by an alkyl, aryl, or similar group.
- **Glycosonic acids** (**ketoaldonic acids**) are monosaccharides which contain a ketone function and a carboxylic acid function.
- A **glycosphingolipid** is a lipid containing at least one monosaccharide residue and either a sphingoid or ceramide.
- **Monosaccharide** denotes a single or monomer unit without glycosidic connection to another monosaccharide. The suffix **–ose** is used to designate a carbohydrate (**aldose**) such a glucose, pentose, hexose, galactose, etc. The suffix **–ulose** is used to designate a ketose, which is a carbohydrate containing a ketone function.
- The term **oligosaccharide** describes a compound composed of 10-20 monosaccharide units connected by glycosidic bonds. The length quoted here is arbitrary. Oligosaccharide is used to describe a oligomer of defined length rather than one, such as with starch or glycogen, of undefined length.
- The term **polysaccharide** is usually used to describe a large amorphous polymer consisting of monosaccharide units connected by glycosidic bonds.
- **Pyranose** (from pyran) designates a six-membered ring
- A **reducing sugar** contains an aldehyde or ketone function which reacts with alkaline ferricyanide or alkaline cupric tartrate (Fehling's solution).
- **Trisaccharide** describes a compound composed of three monosaccharides connected by glycosidic bonds.

1. McNaught, A.D., Nomenclature of carbohydrates, *Pure & Appl.Chem.* 68, 1919-2006, 1996
2. McNaught, A.D., Nomenclature of carbohydrates (recommendations 1996), *Adv.Carbohydr.Chem.Biochem.* 52, 43-177, 1996
3. *Dictionary of Carbohydrates*, 2nd edn., ed. P.M. Collins, Chapman and Hall/CRC, Boca Raton, FL, 2006
4. Collins, P.M. and Ferrier, R.J., *Monosaccharides Their Chemistry and Their Roles in Natural Products*, Wiley, Chichester, United Kingdom, 1995
5. *Glycoscience, Chemistry and Chemical Biology*, ed. B.O. Fraser-Reid, K. Tatsuta, and J. Thiem, Spinger-Verlag, Berlin, Germany, 2001
6. Hitchcock, P.J., Leive, L., Makela, P.H., *et al.*, Lipopolysaccharide nomenclature–past, present, and future, *J.Bacteriol.* 166, 699-705, 1986
7. Moynihan, P.J., Update on the nomenclature of carbohydrates and their dental effects, *J.Dent.* 26, 209-218, 1998
8. Berreau, U. and Stenutz, R., Web resources of the carbohydrate chemist, *Carbohydr.Res.* 339, 929-936, 2004
9. Murthy, P.P., Structure and nomenclature of inositol phosphates, phosphoinositides, and glycosylphosphatidylinositols, *Subcell. Biochem.* 39, 1-19, 2006
10. Chester, M.A., IUPAC-IUB joint commission on biochemical nomenclature (JCBN). Nomenclature of glycolipids–recommendations on 1997, *Eur.J.Biochem.* 257, 293-298, 1998
11. Roberts, M.C., Sutcliffe, J., Courvalin, P., *et al.*, Nomenclature for macrolide and macrolide-lincosamide-streptogramin B resistance determinants, *Antimicrob.Agents Chemother.* 43, 2823-2830, 1999
12. Henry, S. and Moulds, J.J., Preview 2000: Proposal for a new terminology to describe carbohydrate histo-blood group antigens/glycoproteins within the ISBT terminology framework, *Immunohematol.* 16, 49-56, 2000

13. Chai, W., Piskarev, V., and Lawson, A.M., Branching patter and sequence analysis of underivatized oligosaccharides by combined MS/MS of singly and doubly charged molecular ions in negative-ion, *J.Amer.Soc.Mass Spectrom.* 13, 670-679, 2002

14. Lohmann, K.K., and von der Lieth, C.W., GLYCO-FRAGMENT: A web tool to support the interpretation of mass spectra of complex carbohydrates, *Proteomics* 3, 2028-2035, 2003

15 Hilden, L. and Johansson, G., Recent developments on cellulases and carbohydrate-bindnig modules with cellulose affinity, *Biotechnol.Lett.* 26, 1683-1693, 2004

16. Morelle, W. and Michalski, J.C., Glycomics and mass spectrometry, *Curr.Pharm.Des.* 11, 2615-2645, 2005

17. Meitei, N.S., and Banerjee, S., Interpretation support for multistage MS: a mathematical method for theoretical generation of glycan fragments and calculation of their masses, *Proteomics* 7, 2530-2540, 2007

18. Chester, M.A., Nomenclature for glycolipids, *Adv.Carbohydr.Chem. Biochem.* 55, 312-326, 1999

19. Murthy, P.P., Structure and nomenclature of inositol phosphates, phosphoinositides, and glycosylphosphatidylinositols, *Subcell. Biochem.* 39, 1-19, 2006

General references to carbohydrate chemistry

- *The Amino Sugars. The Chemistry and Biology of Compounds Containing Amino Sugars*, ed. R.W. Jeanloz, Academic Press, New York, NY, USA, 1969
- *The Carbohydrates Chemistry and Biochemistry*, 2nd edn., ed. W. Pigman and D. Horton, Academic Press, New York, NY, USA, 1970
- *Carbohydrates*, ed. P.M. Collins, Chapman & Hall, London, United Kingdom, 1987
- *Carbohydrate Chemistry*, ed. J.F. Kennedy, Oxford, Clarendon Press, Oxford, United Kingdom, 1988
- *CRC Handbook of Chromatography, Vol II, Carbohydrates*, ed. S.C. Churms and J. Sherma, CRC Press, Boca Raton, FL, USA, 1991
- Collins, P.M. and Ferrier, R.J., *Monosaccharides. Their Chemistry and Their Role in Natural Products*, Wiley, Chichester, United Kingdom, 1995
- McNaught, A.D., Nomenclature of carbohydrates, *Pure & Appl.Chem.* 68, 1919-2006, 1996
- Lehman, J., *Carbohydrates Structure and Biology*, Thieme, Stuttgart, Germany, 1998
- *Carbohydrate Biotechnology Protocols*, ed. C. Burke, Humana Press, Totowa, NJ, USA, 1999
- *Carbohydrates Structure, Syntheses, and Dynamics*, ed. D. Finch, Kluwer Academic, Dordrecht, Netherlands, 1999
- Osborn, H. and Khan, T., *Oligosaccharides Their Synthesis and Biological Roles*, Oxford University Press, Oxford, UK, 2000
- *Carbohydrates in Chemistry and Biology*, ed. B. Ernst, G.A. Hart, and P. Sinay, Wiley-VCH, Weinheim, Germany, 2000
- *Glycoscience, Chemistry and Chemical Biology*, ed. B.O. Fraser-Reid, K. Tatsuta, and J. Thiem, Springer-Verlag, Berlin, Germany, 2001
- Vogel, P., De novo synthesis of monosaccharides, in *Glycoscience, Chemistry and Chemical Biology*, ed.B.O. Fraser-Reid, K. Tatsuta, and J. Thiem, Springer-Verlag, Berlin, Germany, 2001
- Fernandez-Bolaños, J.G. Al-Masaudi, N.A., and Mann, I., Sugar derivatives having a sulfur in the ring, *Adv. Carbohydr.Chem.Biochem.* 57, 21-98, 2001
- Casu, B. and Lindahl, U., Structure and biological interactions of heparin and heparin sulfate, *Adv.Carbohydr.Chem. Biochem.* 57, 159-206, 2001
- Ferrier, R.J. and Hoberg, J.O., Synthesis and reactions of unsaturated sugars, *Adv.Carbohydr.Chem.Biochem.* 58, 55-119, 2003
- Černý, M., Chemistry of anhydro sugars, *Adv.Carbohydr. Chem.Biochem.* 58, 121-198, 2003
- *NMR Spectroscopy of Glycoconjugates*, ed. J. Jiménez-Barbero and T. Peters, Wiley-VCH, Weinheim, Germany, 2003
- *Capillary Electrophoresis of Carbohydrates*, ed. P. Thibault and S. Honda, Humana Press, Totowa, NJ 2003
- *Chemistry and Biology of Hyaluron*, ed. H.C. Garg and C.A. Hales, Elsevier, Amsterdam, Netherlands, 2004
- Tomasik, P. and Schilling, C.H., Chemical modification of starch, *Adv.Carbohydr.Chem.Biochem.* 59, 1760-403, 2004
- de Lederkremer, R.M. and Gallo-Rodriguez, C., Naturally occurring monosaccharides: properties and synthesis, *Adv. Carbohydr.Chem.Biochem.* 59, 9-67, 2004
- *Handbook of Carbohydrate Engineering*, ed. K.J. Yarema, CRC/Taylor & Francis, Boca Raton, FL, USA, 2005
- *Chemistry of Polysaccharides*, ed. G.E. Zaikov, VSP Brill, Leiden, Netherlands, 2005
- *Polysaccharides for Drug Delivery and Pharmaceutical Applications*, ed. R.H. Marchessault, F. Ravenelle, and X.X. Zho, American Chemical Society, Washington, DC, USA, 2006
- *Dictionary of Carbohydrates*, 2nd edn., ed. P.M. Collins, Chapman and Hall/CRC, Boca Raton, FL, 2006

Recent references for selected monosaccharides

Structures for these carbohydrates are available on p. 390

Aldaric acids

- Fonseca, A., Utilization of tartaric acid and related compounds by yeasts: taxonomic implications, *Can.J.Microbiol.* 38, 1242-1251, 1992
- Talemaka. M., Yan, X., Ono, H., *et al.*, Caffeic acid derivatives in the roots of yacon (*Smallanthus sonchifolius*), *J.Agric.Food Chem.* 51, 793-796, 2003
- Lakatos, A., Bertani, R., Kiss, T., et al., Al(III) ion complexes of saccharic acid and mucic acid: a solution and solid-state study, *Chemistry* 10, 1281-1290, 2004
- Dornyei, A., Garribba, E., Jakusch, T., *et al.*, Vanadium (IV,V) complexes of D-saccharic and mucic acids in aqueous solution, *Dalton Trans.*June 21(12), 1882-1892, 2004
- Gonzalez, J.C., Daier, V., Garcia, S., *et al.*, Redox and complexation chemistry of the Cr(VI)/Cr(V)-D-galacturonic acid system, *Dalton Trans.* Augtust 7(15), 2288-2296, 2004

Aldonic acids

- Limberg, G., Klaffke, W., and Thiem, J., Conversion of aldonic acids to their corresponding 2-keto-3-deoxy analogs by the non-carbohydrate enzyme dihydroxy acid dehydratase (DHAD), *Bioorg.Med.Chem.* 3, 487, 494, 1995
- Yang, B.Y. and Montgomery, R., Oxidation of lactose with bromine, *Carbohydr.Res.* 340, 2698-2705, 2005

Allose altrose allulose

- Kemp, M.B. and Quayle, J.R., Microbial growth on C_1 compounds. Incorporation of C_1 units into allulose phosphate by extracts of *Pseudomonas methanica*, *Biochem.J.* 99, 41-48, 1966

- Hough, L. and Stacey, B.E., Biosynthesis of allitol and D-allulose in *Itea* plants–incorporation of $^{14}CO_2$, *Phytochemistry* 5, 215-222, 1966
- Cree, G.M. and Perlin, A.S., *O*-Isopropylidene derivatives of D-allulose (D-psicose) and D-erythro-hexopyranos-2,3-diulose, *Canad.J.Biochem.* 46, 765-770, 1968
- Poulson, T.S., Chang, Y.Y., and Hove-Jensen, B., D-Allose catabolism of *Escherichia coli*: Involvement of alsI and regulation of als regulon expression by allose and ribose, *J.Bacteriol.* 181, 7126-7130, 1999

Arabinose

- Quiocho, F.A., Molecular features and basic understanding of protein-carbohydrate interactions: the arabinose-binding protein-sugar complex, *Curr.Top.Microbiol.Immunol.* 139, 135-148, 1988
- Sultana, I., Mizanur, R.M., Takeshita, K., *et al.*, Direct production of D-arabinose from D-xylose by a coupling reaction using D-xylose isomerase, D-tagatose 3-epimerase and D-arabinose isomerase, *J.Biosci.Bioeng.* 95, 342-347, 2003
- Kim, P., Current studies on biological tagatose production using L-arabinose isomerase: a review and future perspective, *Appl.Microbiol.Biotechnol.* 65, 243-249, 2004
- Wang, H. and Ng, T.B., First report of an arabinose-specific fungal lectin, *Biochem.Biophys.Res.Commun.* 337, 621-625, 2005
- Pramod, S.N. and Venkatesh, Y.P., Utility of pentose colorimetric assay for the purification of potato lectin, an arabinose-rich glycoprotein, *Glycoconj.J.* 23, 481-488, 2006
- Bercier, A., Plantier-Royon, R., and Portella, C., Convenient conversion of wheat hemicelluloses pentoses (d-xylose and l-arabinose) into a common intermediate, *Carbohydr.Res.* 342, 2450-2455, 2007

Arabitol (arabinitol)

- de Repentigny, L. and Reiss, E., Current trends in immunodiagnosis of candidiasis and aspergillosis, *Rev.Infect.Dis.* 6, 301-312, 1984
- Christensson, B., Sigmundsdottir, G., and Larsson, L., D-Arabinitol–a marker for invasive candidiasis, *Med. Mycol.* 37, 391-396, 1999
- Mohan, S. and Pinto, B.M., Zwitterionic glycosidase inhibitors: salacinol and related analogues, *Carbohydr.Res.* 342, 1551-1580, 2007

Arabonic acid

- Jahn, M., Baynes, J.W., and Spiteller, G., The reation of hyaluronic acid and its monomers, glucuronic acid and *N*-acetylglucosamine, with reactive oxygen species, *Carbohydr.Res.* 321, 228-234, 1999
- Soroka, N.V., Kulminskaya, A.A., Eneyskaya, E.V., *et al.*, Synthesis of arabinitol 1-phosphate and its use for characterization of arabinitol-phosphate dehydrogenase, *Carbohydr.Res.* 340, 539-546, 2005
- Nunes, F.M., Reis, A., Domingues, M.R., and Coimbra, M.A., Characterization of galactomannan derivatives in roasted coffee beverages, *J.Agric.Food Chem.* 54, 3428-3439, 2006

Butanetetrol

- Sakamoto, I., Ichimura, K., and Ohrui, H., Synthesis of 2-C-methyl-D-erythritol and 2-C-methyl-L-threitol; determination of the absolute configuration of 2-C-methyl-1,2,3,4-butanetetrol isolated, *Biosci.Biotechnol.Biochem.* 64, 1915-1922, 2000
- Romero, C.M., Lozano, J.M., Sancho, J., and Giraldo, G.I., Thermal stability of β-lactoglobulin in the presence of aqueous solution of alcohols and polyols, *Int,J.Biol. Macromol.* 40, 423-428, 2007

Conduritol

- Quaronine, A., Gershon, E., and Semenza, G., Affinity labeling of the active sites in the sucrase-isomaltase complex from small intestine, *J.Biol.Chem.* 249, 6424-6433, 1974
- Kwon, Y.U. and Chung, S.K., Facile synthetic routes to all possible enantiomeric pairs of conduritol stereoisomers via efficient enzymatic resolution of conduritol B and C derivatives, *Org.Lett.* 3, 3013-3016, 2001
- Freeman, S., Hudlicky, T., New oligomers of conduritol-F and mucoinositol. Synthesis and biological evaluation as glycosidase inhibitors, *Bioorg.Med.Chem.Lett.* 14, 1209-1212, 2004
- Cere, V., Minzoni, M., Pollicino, S., et al., A general procedure of the synthesis of stereochemically pure conduritol derivatives practical also for solid-phase chemistry, *J.Comb.Chem.* 8, 74-78, 2006

Dialdoloses

- Thealander, O., Acids and other oxidation products, in *The Carbohydrates Chemistry and Biology.* 2nd edn., ed. W.Pigman and D. Horton, Academic Press, New York, NY, USA Chapter 23, pps. 1013-1100, 1980
- Green, J.W., Oxidative reactions and degradations, in *The Carbohydrates Chemistry and Biology.* 2nd edn., ed. W.Pigman and D. Horton, Academic Press, New York, NY, USA Chapter 24, pps. 1101-1166, 1980
- Avigad, G., Amorel, D., Asensio, C., and Horecker, B.L., The D-galactose oxidase of *Polyporus circinatus*, *J.Biol.Chem.* 237, 2736-2743, 1962

Fructose

- Benkovic, S.J. and deMaine, M.M., Mechanism of action of fructose 1,6-bisphosphatase, *Adv.Enzymol.Relat.Areas Mol.Biol.* 53, 45-82, 1982
- Hers, H.G., The discovery and the biological role of fructose 2,6-bisphosphate, *Biochem.Soc.Trans.* 12, 729-735, 1984
- Hanover, L.M. and White, J.S., Manufacturing, composition, and applications of fructose, *Am.J.Clin.Nutr.* 58 (5 suppl), 724S-732S, 1993
- Schalkwijk, C.G., Stehouwer, C.D., and van Hinsbergh, V.W., Fructose-mediated non-enzymatic glycation: sweet coupling or bad modification, *Diabetes Metab.Res.Rev.* 20, 369-382, 2004
- Reddy, M.R. and Erion, M.D., Computer-aided drug design strategies used in the discovery of fructose 1,6-bisphosphatase inhibitors, *Curr.Pharm.Des.* 11, 283-294, 2005

Galactose

- Mann, B.J., Mirelman, D., and Petri, W.A., Jr., The D-galactose-inhibitable lectin of *Entamoeba histolytica*, *Carbohydr.Res.* 213, 331-338, 1991
- McPherson, M.J., Stevens, C., Baron, A.J., *et al.*, Galactose oxidase: molecular analysis and mutagenesis studies, *Biochem.Soc.Trans.* 21, 752-756, 1993

- Halcrow, M., Phillips, S., and Knowles, P., Amine oxidases and galactose oxidase, *Subcell.Biochem.* 35, 183-231, 2000
- Whittaker, J.W., Galactose oxidase, *Adv.Protein Chem.* 60, 1-49, 2002
- Whittakcr, J.W., The radical chemistry of galactose oxidase, *Arch.Biochem.Biophys.* 433, 227-239, 2005
- Cho, C.S., Seo, S.J., Park, I.K., *et al.,* Galactose-carrying polymers as extracellular matrices for liver tissue engineering, *Biomaterials* 27, 576-585, 2006

Glucitol (sorbitol)

- Jeffery, J. and Jornvall, H., Sorbitol dehydrogenase, *Adv. Enzymol.Relat.Areas Mol.Biol.* 61, 47-106, 1988
- Suarez, G., Nonezymatic browning of proteins and the sorbitol pathway, *Prog.Clin.Biol.Res.* 304, 14-162, 1989
- Silveira, M.M. and Jonas, R., The biotechnological production of sorbitol, *Appl.Microbiol.Biotechnol.* 59, 400-408, 2002
- Dworacka, M., Winiarska, H., Szymanska, M., *et al.,* 1,5-anhydro-D-glucitol: a novel marker of glucose excursions, *Int.J.Clin.Pract.Suppl.I* July(129), 40-44, 2002
- El-Kabbani, O., Darmanin, C., and Chung, R.P., Sorbitol dehydrogenase: structure, function and ligand design, *Curr.Med.Chem.* 11, 365-476, 2004
- Jonas, R. and Silveira, M.M., Sorbitol can be produced not only chemically but also biotechnologically, *Appl.Biochem. Biotechnol.* 1991, 321-336, 2004
- Obrosova, I.G., Increased sorbitol pathway activity generates oxidative stress in tissue sites for diabetic complications, *Antioxid.Redox.Signal.* 7, 1543-1552, 2005
- Yardimci, H. and Leheny, R.L., Aging of the Johari-Goldstein relaxation in the glass-forming liquids sorbitol and xylitol, *J.Chem.Phys.* 124, 214503, 2006
- Pedruzzi, I., Malvessi, E., Mata, V.G., *et al.,* Quantification of lactobionic acid and sorbitol from enzymatic reaction of fructose and lactose by high-performance liquid chromatography, *J.Chromatog. A* 1145, 128-132, 2007
- Shah, P.P. and Roberts, C.J., Molecular solvation in water-methanol and water-sorbitol mixtures: the roles of preferential hydration, hydrophobicity, and the equation of state, *J.Phys.Chem.B.* 111, 4467-4476, 2007
- Jungo, C., Schenk, J., Pasquier, M., *et al.,* A quantitative analysis of the benefits of mixed feeds of sorbitol and methanol for the production of recombinant avidin with *Pichia pastoris, J.Biotechnol.* 131, 57-66, 2007

Glucose

- Fraser, C.G., Analytical goals for glucose analyses, *Ann. Clin.Biochem.* 23, 379-389, 1986
- Carruthers, A., Facilitated diffusion of glucose, *Physiol.Rev.* 70, 1135-1176, 1990
- James, D.E. Piper, R.C., and Slot, J.W., Targeting of mammalian glucose transporters, *J.Cell.Sci.* 104, 607-612, 1993
- Mueckler, M., Facilitative glucose transporters, *Eur. J. Biochem.* 219, 713-725, 1994
- Bhosale, S.H., Rao, M.B., and Deshpande, V.V., Molecular and industrial aspects of glucose isomerase, *Microbiol.Rev.* 60, 280-300, 1996
- Schmidt, K.C., Lucignani, G., and Sokoloff, L., Fluorine-18-fluorodeoxyglucose PET to determine regional cerebral glucose utilization: a re-examination, *J.Nucl.Med.* 37, 394-399, 1996

- Pischetsrieder, M., Chemistry of glucose and biochemical pathways for biological interest, *Perit.Dial.Int.* 20(suppl 2), S26-S30, 2000
- Bonnefont-Ruousselot, D., Glucose and reactive oxygen species, *Curr.Opin.Clin.Nutr.Metab.Care* 5, 510568, 2002
- Anthony, C., The quinoprotein dehydrogenases for methanol and glucose, *Arch.Biochem.Biophys.* 428, 2-9, 2004
- Fang, H., Kaur, G., and Wang, B., Progress in boronic acid-based fluorescent glucose sensors, *J.Fluoresc.* 14, 481-489, 2004

Glucose oxidation

- Merbouh, N., Francois Thaburet, J., Ibert, M., et al., Facile nitroxide-mediated oxidations of D-glucose to D-glucaric acid, *Carbohydr.Res.* 336, 75-78, 2001
- Ibert, M., Marsais, F., Merbouh, N., and Bruckner, C., Determination of the side-products formed during the nitroxide-mediated bleach oxidation of glucose to glucaric acid, *Carbohydr.Res.* 337, 1059-1063, 2002
- Fischer, K. and Bipp, H.P., Generation of organic acids and monosaccharides by hydrolytic and oxidative transformation of food processing residues, *Biosour.Technol.* 96, 831-842, 2005

Glucuronic acid

- *Glucuronic Acid Free and Combined, Chemistry, Biochemistry, Pharmacology, and Medicine,* ed. G.J. Dutton, Academic Press, New York, NY, USA, 1966
- Clarke, D.J.. and Buschell, B., The uridine diphosphate glucuronosyltransferase multigene family: Function and regulation, in Conjugation-Deconjugation Reactions in Drug Metabolism and Toxicity, ed. F.C. Kauffman, Spinger-Verlag, Berlin, Germany, 1994
- Bedford, C.G., Glucuronic acid conjugates, *J.Chromatogr. B.Biomed.Sci.Appl.* 717 313-326, 1998
- Durham, T.B. and Miller, M.J., Conversion of glucuronic acid glycosides to novel bicyclic beta-lactams, *Org.Lett.* 4, 135-138, 2002
- Pitt, N., Duane, R.M., O'Brien, A., et a., Synthesis of a glucuronic acid and glucose conjugate library and evaluation of effects on endothelial cell growth, *Carbohydr.Res.* 339, 1873-1887, 2004
- Engstrom, K.M., Daanen, J.F., Wagaw, S., and Stewart, A.O., Gram scale synthesis of the glucuronide metabolite of ABT-724, *J.Org.Chem.* 71, 8278-8383, 2006
- Sorich, M.J., McKinnon, R.A., Miners, J.O., and Smith, P.A., The importance of local chemical structure for chemical metabolism by human uridine-5'-diphosphate-glucuronosyltransferase, *J.Chem.Inf.Model.* 46, 2692-2697, 2006
- Jantti, S.E., Kiriazis, A., Reinila, R.R., *et al.,* Enzyme-assistance and characterization of glucuronide conjugates of neuroactive steroids, *Steroids* 72, 287-296, 2007

Glycodiuloses

- Avigad, G., and England, S., 5-Ketofructose I. Chemical characterization and analytical determination of the dicarbonylhexose produced by *Gluconobacter cerinus, J.Biol. Chem.* 240, 2290-2296, 1965
- Yamada, Y., Aida, K., Uemura, T., Enzymatic studies on oxidation of sugar and sugar alcohols. 2. Purification and properties of NADPH-linked 5-ketofructose reductase, *J.Biochem.* 61, 803-811, 1967

- Schrimsher, J.L., Wingfield, P.T., Bernard, A., et al., Purification and characterization of 5-ketofructose reductase from Erwinia citreus, Biochem.J. 253, 511-516, 1988
- White, R.H., and Xu, H.M., Methylglyoxal is an intermediate in the biosynthesis of 6-deoxy-5-ketofructose-1-phosphate: A precursor for aromatic amino acid biosynthesis in *Methanocaldococcus jannaschii*, *Biochemstry* 45, 12355-12379, 2006

Glycosonic Acids

- Truesdell, S.J., Sims, J.C., Boerman, P.A., et al., Pathways for metabolism of ketoaldonic acids in an *Erwinia* sp., *J.Bacteriol.* 173, 6651-6666, 1991

Glycosuloses

- Fry, S.C., Dumville, J.C., and Miller, J.G., Fingerprinting of polysaccharides attacked by hydroxyl radicals in vitro and in the cell wall of ripening pear fruit, *Biochem.J.* 357, 727-737, 2001
- Baker N, Egyud LG. 3-Deoxy-D-glucosulose in fed and fasted mouse livers. *Biochim. Biophys. Acta*, 1968 Sep 3;165(2):293-6

Heptoses and heptose derivatives

- Adams, G.A., Quadling, C., and Perry, M.B., D-*glycero*-D-*manno*-hepose as a component of lipopolysaccharides from gram negative bacteria, *Canad.J.Microbiol.* 13, 1605-1613, 1967
- Leshem, B., Sharoni, Y., and Dimant, E., The hyperglycemic effect of 1-deoxy-D-*manno*-heptulose. Inhibition of hexokinase, glucokinase, and insulin release *in vitro*, *Canad.J.Biochem.* 52, 1078-1081, 1974
- Shaw, P.E., Wilson, C.W., III, and Knight, R.J., Jr., High-performance liquid chromatographic analysis of D-*manno*-heptulose, perseitol, glucose, and fructose in avocado cultivers, *J.Agric.Food Chem.* 28, 379-382, 1980
- Penner, J.L., and Aspinall, G.O., Diversity of lipopolysaccharide saccharides from gram-negative bacteria, *J.Infect. Dis.* 176(Suppl 2), S135-S138, 1997
- Müller-Loennies, S., Brade, L., and Brade, H., Neutralizing and cross-reactive antibodies against enterobacterial lipopolysaccharides, *Int.J.Med.Microbiol.* 297, 321-340, 2007

- Johnson, L.N., Acharya, K.R., Jordon, M.D., and McLaughlin, P.J., Refined crystal structures of the phosphorylase-heptulose-2-phosphate-oligosaccharide-AMP-complex, *J.Mol.Biol.* 211, 645-661, 1990
- Gomez, R.V., Kolender, A.A., and Valero, O., Synthesis of polyhydroxyl amino acids based on D- an L-alanine from D-glycero-D-gulo-heptone-1,4 lactone, *Carbohydr.Res.*341, 1498-1504, 2006

Lactone formation

- de Lederkremer, R.M. and Marino, C., Acids and other products of the oxidation of sugars, *Adv.Carbohydr.Chem. Biochem.* 58, 199-306, 2003

Octuloses

- Weiser, W., Lehmann, J., Chiba, S., *et al.*, Steric course of the hydration of D-gluco-octenitol catalyzed by α-glucosidase and by trehalose, *Biochemistry* 27, 2294-2300, 1988
- Williams, J.F., Clark, M.G., and Arora, K.K., ^{14}C labelling of octulose bisphosphates by L-type pentose pathway reactions in liver *in situ* and *in vitro*, *Biochem.Int.* 11, 97-106, 1985
- Williams, J.F. and MacLeod, J.K., The metabolic significance of octulose phosphates in the photosynthetic carbon reduction cycle in spinach, *Photosynth.Res.* 90, 125-148, 2006

Shikimic acid

- Davis, B.D., Aromatic biosynthesis. I. The role of shikimic acid, *J.Biol.Chem.* 191, 315-325, 1951
- Yoshida, S., and Hasegawa, M., A micro-colorimetric method for the determination of shikimic acid, *Arch. Biochem.Biophys.* 70, 377-381, 1957
- Cox, G.B. and Gibson, F., The role of shikimic acid in the biosynthesis of vitamin K2, *Biochem.J.* 100, 1-6, 1966
- Stavric, B. and Stoltz, D.R., Shikimic acid, *Food Cosmet. Toxicol.* 14, 141-145, 1976
- Kramer, M., Bongaerts, J., Bovenberg, R., *et al.*, Metabolic engineering for microbial production of shikimic acid, *Metab.Eng.* 5, 277-283, 2003

Validatol

- Horri, S., Iwasa, T., Mizuta, E., and Kameda, Y., Studies on validamycins, new antibiotics. VI. Valdiamine, hydroxyvalidamine and validatol, new cyclitols, *J.Antibiot.* 24, 59-63, 1971

SOME TRIVIAL NAMES FOR MONOSACCHARIDES[a]

Allose (All)	*allo*-hexose	Mannosamine (ManN)	2-amino-2-deoxy-mannose
Altrose (Alt)	*altro*-hexose	Mannose (Man)	*manno*-hexose
Erythrose	*erythro*-tetrose	Psicose (Psi)	*ribo*-hex-2-ulose
Erythrulose	*glycero*-tetrulose	Quinovose (Qui)	6-deoxyglucose
Fructose (Fru)	*arabino*-hex-2-ulose	Rhamnosamine (RhaN)	2-amino-2,6-dideoxymannose
Galactosamine (GalN)	2-amino-2-deoxygalactose	Rhamnose (Rha)	6-deoxymannose
Galactose (Gal)	*galacto*-hexose	Ribose (Rib)	*ribo*-pentose
Glucose (Glu)	*gluco*-hexose	Ribulose (Rul)	*erythro*-pent-2-ulose
Glyceraldehyde	2,3-dihydroxy-propanal	Sorbose (Sor)	*xylo*-hex-2-ulose
Glycerol (Gro)	1,2,3-propanetriol	Tagatose (Tag)	*lyxo*-hex-2-ulose
Gulose (Gul)	*gulo*-hexose	Talose (Tal)	*talo*-hexose
Idose (Ido)	*ido*-hexose	Tartaric acid	Erythraric/Threaric Acid
Lyxose (Lyx)	*lyxo*-hexose	Xylose (Xyl)	*xylo*-pentose

[a] Adapted from McNaught, A.D., Nomenclature of carbohydrates, *IUPAC Pure & Applied Chemistry* 68, 1919-2008, 1996

Selected references for disaccharides

Structures for these carbohydrates are available on p.390.

Cellobiose

- Camevascini, G., A cellulose assay coupled to cellobiose dehydrogenase, *Anal.Biochem.* 147, 419-427, 1985
- Ichikawa, Y., Ichikawa, R., and Kuzuhara, H., Synthesis, from cellobiose, of a trisaccharide closely related to the GlNAc - - - -GlcA - - - - -GlcN segment of the antithrombin-binding sequence of heparin, *Carbohydr.Res.* 141, 273-282, 1985
- Ali, S.A., Eary, J.F., Warren, S.D., *et al.*, Synthesis and radio-iodination of tyramine cellobiose for labeling monoclonal antibodies, *Int.J.Rad.Appl.Instrum.B.* 15, 557-561, 1988
- Tewari, Y.B., and Goldberg, R.N., Thermodynamics of hydrolysis of disaccharides, cellobiose, gentiobiose, isomaltose, and maltose, *J.Biol.Chem.* 264, 3966-3971, 1989
- Reizer, J., Reizer, A., and Saier, M.H., Jr,. The cellobiose permease of *Escherichia coli* consists of three proteins and is homologous to the lactose permease of *Staphylococcus aureus*, *Res.Microbiol.* 141, 1061-1067, 1990
- Kremer, S.M. and Wood, P.M., Production of Fenton's reagent by cellobiose oxidase from cellulolytic cultures of *Phanerochaete chrysoporium*, *Eur.J.Biochem.* 208, 807-814, 1992
- Henriksson, G., Johansson, G., and Pettersson, G., A critical review of cellobiose dehydrogenases, *J.Biotechnol.* 78, 91-113, 2000
- Maeda, A., Kataoka, H., Adachi, S., and Matsuno, R., Transformation of cellubiose to 3-ketocellobiose by the EDTA-treated *Agrobacterium tumefaciens* cells, *J.Biosci. Bioeng.* 95, 608-611, 2003
- Mason, M.G., Nicholis, P., and Wilson, M.T., Rotting by radicals - the role of cellobiose oxidoreductase?, *Biochem. Soc.Trans.* 31, 1335-1336, 2003
- Koto, S., Hirooka, M., Tashiro, T., *et al.*, Simple preparations of alkyl and cycloalkyl α-glycosidases of maltose, cellobiose, and lactose, *Carbohydr.Res.* 339, 2415-2424, 2004
- Stoica, L., Ludwig, R., Haltrich, D., and Gorton, L., Third-generation biosensor for lactose based on newly discovered cellobiose dehydrogenase, *Anal.Chem.* 78, 393-398, 2006
- Zamocky, M., Ludwig, R., Peterbauer, C., *et al.*, Cellobiose dehydrogenase - a flavocytochrome from wood-degrading, phytopathogenic and saprotropic fungi, *Curr.Protein Pept. Sci.* 7. 255-280, 2006

Isomaltose

- Dowd, M.K., Reilly, P.J., and French, A.D., Relaxed-residue conformational mapping of the three linkage bonds of isomaltose and gentiobiose with MM3 (92), *Biopolymers* 34, 625-638, 1994
- Brand, D.A., Lium H.S., and Zhu, Z.S., The dependence of glucan conformation dynamics on linkage position and stereochemistry, *Carbohydr.Res.* 278, 11-26, 1995
- Vetere, A., Gamini, A., Campa, C., and Paoletti, S., Regiospecific transglycolytic synthesis and structural characterization of 6-*O*-glucopyranosyl-glucopyranose (isomaltose), *Biochem.Biophys.Res.Commun.* 274, 99-104, 2000
- Hendrix, D.L. and Salvucci, M.E., Isobemisiose: an unusual trisaccharide abundant in the silverleaf whitefly, *Bemisia argentifolii*, *J.Insect Physiol.* 47, 423-432, 2001
- Streigel, A.M., Anomeric configuration, glycosidic linkage, and the solution conformational entropy of *O*-linked disaccharides, *J.Am.Chem.Soc.* 125, 4146-4148, 2003

- Kanou, M., Nakanishi, K., Hashimoto, A., and Kameoka, T., Influence of monosaccharides and its glycosidic linkage on infrared spectral characteristics of disaccharides in aqueous solutions, *Appl.Spectrosc.* 59, 885-892, 2005
- Cai, Y., Liu, J., Shi, Y., *et al.*, Determination of several sugars in serum by high-performance anion-exchange chromatography with pulsed amperometric detection, *J.Chromatog.A.* 1085, 98-103, 2005
- Pereira, C.S., Kony, D., Baron, R., *et al.*, Conformational and dynamical properties of disaccharides in water: a molecular dynamics study, *Biophys.J.* 90, 4337-4344, 2006
- Mikami, B., Iwamoto, N., Malle, D., *et al.*, Crystal structure of pullulanase: evidence for parallel binding of oligosaccharides in the active site, *J.Mol.Biol.* 359, 690-707, 2006
- Maruta, K., Watanabe, H., Nishimoto, T., *et al.*, Acceptor specificity of trehalose phosphorylase from *Thermoanaerobacter brockii*: production of novel nonreducing trisaccharide, 6-*O*-α-galactopyranosyl trehalose, *J.Biosci.Bioeng.* 101, 385-390, 2006

Isomaltulose

- Bucke, C., Carbohydrate transformation by immobilized cells, *Biochem.Soc.Symp.* 48, 25-38, 1983
- Takazoe, I., Frostell, G., Ohta, K., *et al.*, Palatinose - a sucrose substitute. Pilot studies, *Swed.Dent.J.* 9, 81-87, 1985
- Thompson, J., Robrish, S.A., Pikis, A., *et al.*, Phosphorylation and metabolism of sucrose and its five linkage-isomeric α-D-glucosyl-D-fructoses by *Klebsiella pneumoniae*, *Carbohydr.Res.* 331, 149-161, 2001
- Lina, B.A., Jonker, D., and Kozianowski, G., Isomaltulose (Palatinose): a review of biological and toxicological studies, *Food Chem.Toxicol.* 40, 1375-1381, 2002
- Zhang, D. Li, N., Swaminathan, K., and Zhang, L.H., A motif rich in charged residues determines product specificity in isomaltulose synthase, *FEBS Lett.* 534, 151-155, 2003
- Ahn, S.J., Yoo, J.H., Lee, M.C., *et al.*, Enhanced conversion of sucrose to isomaltulose by a mutant of *Erwinia rhapontici*, *Biotechnol.Lett.* 25, 1179-1183, 2003
- Kawaguti, H.Y., Buzzato, M.F., and Sato, H.H., Isomaltulose production using free cells: optimization of a culture medium containing agricultural wastes and conversion in repeated-batch processes, *J.Ind.Microbiol.Biotechnol.* 34, 261-269, 2007
- Park, S.E., Cho, M.B., Lim, J.K., *et al.*, A new colorimetric method for determining the isomerization activity of sucrose isomerase, *Biosci.Biotechnol.Biochem.* 71, 583-586, 2007
- Achten, J., Jentjens, R.L., Brouns, F., and Jeukendrup, A.E., Exogenous oxidation of isomaltulose is lower than that of sucrose during exercise in men, *J.Nutr.* 137, 1143-1148, 2007

Lactose

- Yang, S.T. and Silva, E.M., Novel products and new technologies for use of a familiar carbohydrate, milk lactose, *J.Dairy Sci.* 78, 2541-2562, 1995
- Sahin-Toth, M., Dunten, R.L., and Kaback, H.R., The lactose permease of *Escherichia coli*: a paradigm for membrane transport proteins, *Soc.Gen.Physiol.Ser.* 48, 1-9, 1993
- Ramakrishnan, B., Boeggeman, E., and Qasba, P.K., β-1,4-Galactosyltransferase and lactose synthase: molecular mechanical devices, *Biochem.Biophys.Res.Commun.* 291, 1113-1118, 2002
- Adam, A.C., Rubio-Texeira, M., and Poaina, J., Lactose: the milk sugar from a biotechnological perspective, *Crit.Rev. Food Sci.Nutr.* 44, 553-557, 2004

- Guan, L. and Kaback, H.R., Lessons from lactose permease, *Annu.Rev.Biophys.Biomol.Struct.* 35, 67-91, 2006
- Pedruzzi, I., Malvessi, E., Mata, V.G., *et al.*, Quantification of lactobionic acid and sorbitol from enzymatic reaction of fructose and lactose by high-performance liquid chromatography, *J.Chromatog.A.* 1145, 128-132, 2007
- Jurs, S. and Thiem, J., From lactose towards a novel galactosylated cyclooctenone, *Carbohydr.Res.* 342, 1238-1243, 2007
- Meltretter, J., Seeber, S., Humeny, A., *et al.*, Site-specific formation of Maillard, oxidation, and condensation products from whey proteins during reaction with lactose, *J.Agric.Food Chem.* 55, 6096-6103, 2007
- Higl, B., Kurtmann, L., Carlsen, C.U., *et al.*, Impact of water activity, temperature, and physical state on the storage stability of *Lactobacillus paracasei* ssp. *paracasei* freeze-dried in a lactose matrix, *Biotechnol.Prog.* 23, 794-800, 2007

Maltose

- ¹H NMR studies of maltose, maltoheptaose, α-,β-, and γ-cyclodextrins, and complexes in aqueous solutions with hydroxyl protons as structural probes, *J.Org.Chem.* 68, 1671-1678, 2003
- Hernandez-Luis, F., Amado-Gonzalez, E., and Esteso, M.A., Activity coefficients of NaCl in trehalose-water and maltose-water at 298.15 K, *Carbohydr.Res.* 388, 1415-1424, 2003
- Kawakami, K. and Ida, Y., Direct observation of the enthalpy relaxation and the recovery processes of maltose-based amorphous formulation by isothermal microcalorimetry, *Pharm.Res.* 20, 1430-1436, 2003
- Lourdin, D., Colonna, P., and Ring, S.G., Volumetric behavior of maltose-water, maltose-glycerol and starch-sorbitol-water systems mixtures in relation to structural relaxation, *Carbohydr.Res.* 338, 2883-2887, 2003
- Haghight, K.S., Imura, Y., Oomori, T., *et al,*. Decomposition kinetics of maltose in subcritical water, *Biosci.Biotechnol. Biochem.* 68, 91-95, 2004
- Mundt, S. and Wedzicha, B.L., Comparative study of the composition of melanodins, from glucose and maltose, *J.Agric.Food Chem.* 52, 4256-4260, 2004
- Weise, S.E., Kim, K.S., Stewart, R.P., and Sharkey, T.D., β-Maltose is the metabolically active anomer of maltose during transitory starch degradation, *Plant Physiol.* 137, 756-761, 2005
- Mundt, S. and Wedzicha, B.L., Role of glucose in the Maillard browning of maltose and glycine: a radiochemical approach, *J.Agric.Food Chem.* 53, 6798-6803, 2005
- Rodriguez, S., Lona, L.M., and Franco, T.T., The effect of maltose on dextran yield and molecular weight distribution,. *Bioprocess Biosyst.Eng.* 28, 9-14, 2005
- Shirke, S. and Ludescher, R.D., Dynamic site heterogeneity in amorphous maltose and mannitol from spectral heterogeneity in erythrosine B phosphorescence, *Carbohydr.Res.* 340, 2661-2669, 2005
- Noel, T.R., Parker, R., Brownsely ,G.J., *et al.*, Physical aging of starch, maltodexin, and maltose, *J.Agric.Food Chem.* 53, 8580-8585, 2005
- Sanz, M.L., Cote, G.L., Gibson, G.R., and Rastall, R.A., Influence of glycosidic linkages and molecular weight on the fermentation of maltose-based oligosaccharides by human gut bacteria, *J.Agric.Food Chem.* 54, 9779-9784, 2006
- Jarusiewicz, J.A., Sherma, J., and Fried, B., Thin layer chromatographic analysis of glucose and maltose in estivated *Biomphalaria glabrata* snails and those infected with *Schistomsoma mansoni*, *Comp.Biochem.Physiol.B.Biochem. Mol.Biol.* 145, 346-349, 2006

Sucrose

- Kitts, D.D., Wu, C.H., Kopec, A. and Nagasawa, T. Chemistry and genotoxicity of caramelized sucrose, *Mol. Nutr.Food Res.* 50, 1180-1190, 2006
- Dangaran, K.L. and Krochta, J.M., Kinetics of sucrose crystallization in whey protein films, *J.Agric.Food Chem.* 54, 7152-7158, 2006
- Blanshard, J.M., Muhr, A.H., and Gough, A., Crystallization from concentrated sucrose solutions, *Adv.Exp.Med.Biol.* 302, 639-655, 1991
- Edye, L.A. and Clarke, M.A., Sucrose loss and color formation in sugar manufacture, *Adv.Exp.Med.Biol.* 434, 123-133, 1998
- Starzak, M., Peacock, S.D., and Mathiouthi, M., Hydration number and water activity models for the sucrose-water system: a critical review, *Crit.Rev.Food Sci.Nutr.* 40, 327-367, 2000
- Salerno, G.L. and Curatti, L., Origin of sucrose metabolism in higher plants: when, how and why?, *Trends Plant Sci.* 8, 63-69. 2003
- Queneau, Y., Jarosz, S., Lewandowski, B., and Fitremann, J., Sucrose chemistry and applications of sucrochemicals, *Adv.Carbohydr.Chem.Biochem.* 61, 217-292, 2007
- Jaradat, D.M., Mebs, S., Checinska, L., and Luger, P., Experimental charge density of sucrose at 20K: bond topological, atomic, and intermediate quantitative properties, *Carbohydr.Res.* 342, 1480-1489, 2007
- Bhugra, C., Ramphatla, S., Bakri, A., *et al.*, Prediction of the onset of crystallization of amorphous sucrose below the calorimetric glass transition temperature from correlations with mobility, *J.Pharm.Sci.* 96, 1258-1269, 2007
- Quintas, M., Brandao, T.R., Silva, C.L., and Cunha, R.L., Modeling viscosity temperature dependence of supercooled sucrose solutions - the random-walk approach, *J.Phys.Chem.B.* 111, 3192-3196, 2007
- Tombari, E., Salvetti, G., Ferrari, C., and Johari, G.P., Kinetics and thermodynamics of sucrose hydrolysis from real-time enthalpy and heat capacity measurements, *J.Phys. Chem.B.* 111, 496-501, 2007

Trehalose

- Lillford, P.J. and Holt, C.B., *In vitro* uses of biological cryoprotectants, *Philo.Trans.B.Soc.Lond.B.Biol.Sci.* 357, 945-951, 2002
- Kawai, K., Hagiwara, T., Takai, R., and Suzuki, T., Comparative investigation by two analytical approaches of enthalpy relaxation for glassy glucose, sucrose, maltose, and trehalose, *Pharm.Res.* 22, 490-495, 2005
- Cordone, L., Cottone, G., Giuffrida, S., *et al.*, Internal dynamics and protein-matrix coupling in trehalose-coated proteins, *Biochim.Biophys.Acta.* 1749, 252-281, 2005
- Lerbret, A., Bordat, P. Affouard, F., *et al.*, How homogeneous are the trehalose, maltose, and sucrose water solutions? An insight from molecular dynamics simulations, *J.Phys.Chem.B.* 109, 11046-11057, 2005
- Pereira, C.S. and Hunenberger, P.M., Interaction of the sugars trehalose, maltose and glucose with a phospholipid bilayer: a comparative molecular dynamics study, *J.Phys. Chem.B* 110, 11572-11581, 2006
- Empadinhas, N. and da Costa, M.S., Diversity and biosynthesis of compatible solutes in hyper/thermophiles, *Int. Microbiol.* 9, 199-206, 2006
- Lerbret, A., Bordat, P., Affouard, F., *et al.*, How do trehalose, maltose, and sucrose influence some structural and

dynamical properties of lysozyme? Insight from molecular dynamics simulations, *J.Phys.Chem.B.* 111, 9410-9420, 2007

- Crowe, J.H., Trehalose as a "chemical chaperone": fact and fantasy, *Adv.Exp.Med.Biol.* 594, 143-158, 2007

Trehalulose

- Ooshima, T., Izumitani, A., Minami, T., *et al.,* Trehalulose does not induce dental caries in rats infected with mutans streptococci, *Caries Res.* 25, 277-282, 1991
- Salvucci, M.E., Distinct sucrose isomerases catalyze trehalulose synthesis in whiteflies, *Bemisia argentifolii,* and *Erwinia rhaptici, Comp.Biochem.Physiol.B.Biochem.Mol. Biol.* 135, 385-395, 2003
- Ravaud, S., Watzlawick, H., Haser, R., *et al.,* Expression, purification, crystallization and preliminary x-ray crystallographic studies of the trehalulose synthase MutB from *Pseudomonas mesoacidophila* MX-45, *Acta Crystallogr. Sect.F. Struct.Biol.Cryst.Commun.* 61, 100-103, 2005
- Ravaud, S., Robert, X. Watzlawick, H., *et al.,* Trehalulose synthase native and carbohydrate complexed structures provide insights into sucrose isomerization, *J.Biol.Chem.* 282, 28126-28136, 2007

Glycosides

- The term glycoside refers to a compound composed of a carbohydrate and a non-carbohydrate moiety. The non-carbohydrate moiety is referred to as the aglycone. There are *O*-glycosides and *C*-glycosides. The carbohydrate portions generally contain between one and three monosaccharide units connected by a glycosidic bond.

- *O*-Glycosides are composed of saccharides where the hydroxyl group of a hemiacetal is replaced by a alkyloxy or aryloxy group. A thioglycoside is a glycoside where the hydroxyl group of a hemiacetal is replaced by a alkylthio or arylthio group. The replacement of the hydroxyl group of a hemiacetal with an alkylamine or arylamine function is called a glycosylamine. Replacement of the hydroxyl group with an alkyl or aryl group results in a *C*-glycoside. A *C*-glycoside is also referred to as an anhydroalditol. A *C*-glycoside is not cleaved by hydrolysis. The carbohydrate portion may be a five-membered ring (furanoside), a six-membered ring (pyranoside), or rarely a seven-membered ring (septanoside).

General reference for glycosides

- Waller, G.R. and Yamasaki, K., *Saponins used in traditional and modern medicine,* Plenum Press, New York, NY, USA, 1996
- Lehman, J., *Carbohydrate Structure and Biology,* Thieme, Stuttgart, Germany, 1998
- *Naturally Occurring Glycosides,* ed. R. Khan, John Wiley & Sons, Ltd., Chichester, United Kingdom, 1999
- *Saponins in Food, Feedstuffs and Medicinal Plants,* ed .W. Olezek and A. Mersten, Kluwer Academic, Dordrecht, Netherlands, 2000
- *Biologically Active Natural Products: Pharmaceuticals,* ed. S.J. Cutler and H.G. Cutler, CRC Press, Boca Raton, FL, 2000
- Garegg, P.J., Synthesis and reactions of glycosides, *Adv. Carbohydr.Chem.Biochem.* 59, 69-134, 2004
- Levy, D.E. and Fügedi, P., *The Organic Chemistry of Sugars,* Taylor & Francis, Boca Raton, FL, USA, 2006

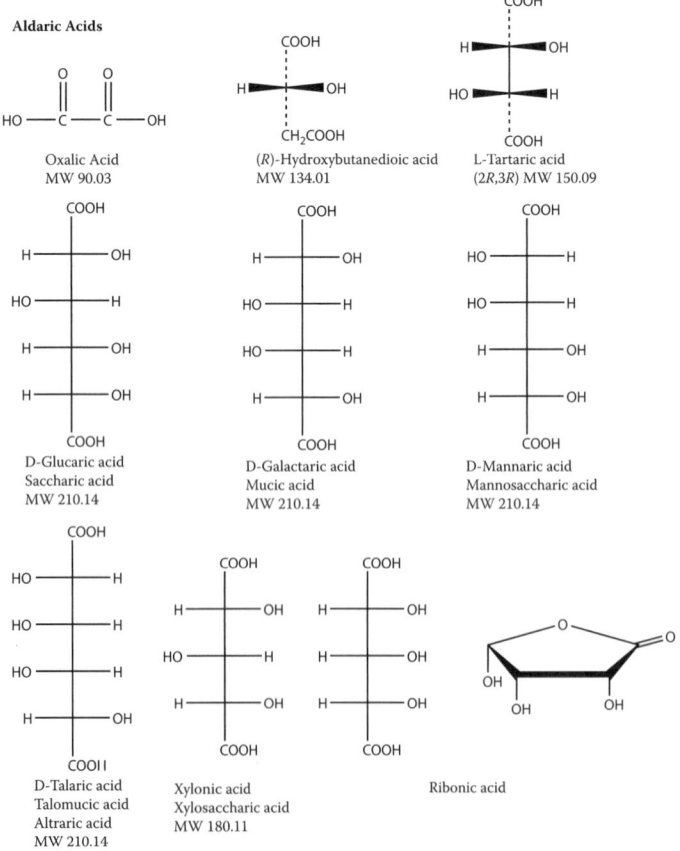

Aldaric Acids

Oxalic Acid
MW 90.03

(*R*)-Hydroxybutanedioic acid
MW 134.01

L-Tartaric acid
(2*R*,3*R*) MW 150.09

D-Glucaric acid
Saccharic acid
MW 210.14

D-Galactaric acid
Mucic acid
MW 210.14

D-Mannaric acid
Mannosaccharic acid
MW 210.14

D-Talaric acid
Talomucic acid
Altraric acid
MW 210.14

Xylonic acid
Xylosaccharic acid
MW 180.11

Ribonic acid

Aldaric acids are a group of carbohydrate dicarboxylic acids characterized by the general formula
$HOOC-(CHOH)_n-COOH$

Aldonic Acids

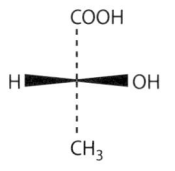

Glycollic acid
(hydroxyacetic acid)
MW 76.05

Glyceric acid
2,3-Dihydroxypropanoic acid
MS 106.08

D-Lactic acid
(2*R*)-Hydroxypropanoic acid

L-Lactic acid
(2*S*)-Hydroxypropanoic acid

Lactic acid
MW 90.08

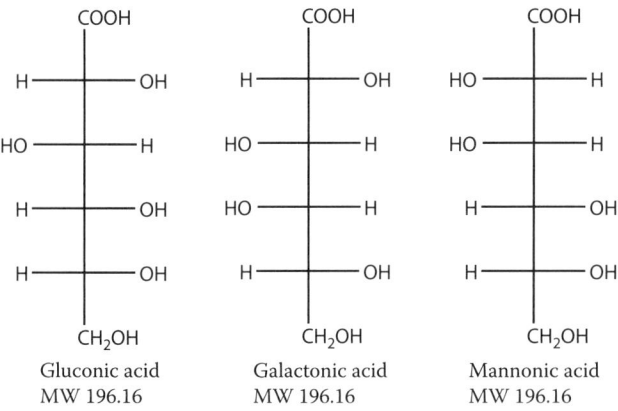

Gluconic acid
MW 196.16

Galactonic acid
MW 196.16

Mannonic acid
MW 196.16

Aldonic acids are carboxylic acids formed by
the oxidation of aldehyde functions

Allose and Altrose

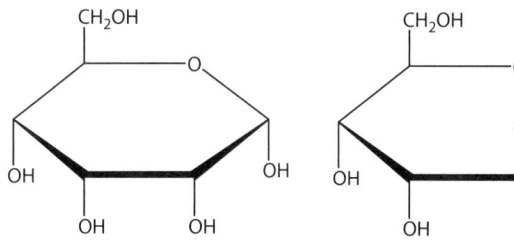

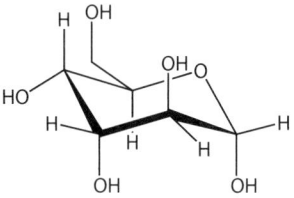

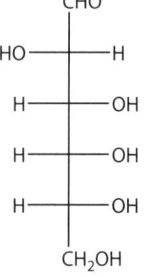

Allose
MW 181.16

Altrose
MW 189.16

Arabinose and Derivatives

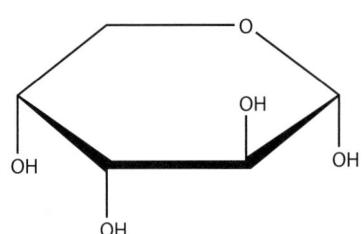

Arabinose (alpha-D-pyranose form)

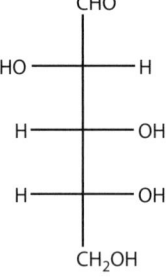

Arabinose
MW 15013

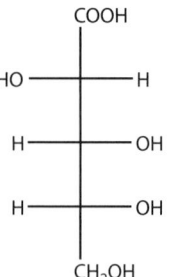

Arabonic acid
(Arabinonic acid
MW 166.13

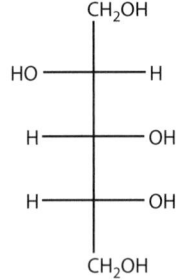

D-Arabitol   MW 152.15

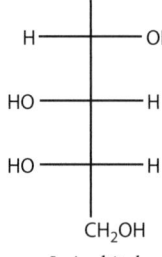

L-Arabitol

Conduritol

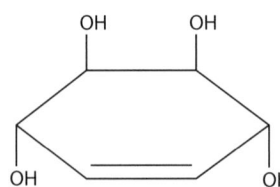

Conduritol A
(5-cyclohexene-1,2,3,4-tetrol)
MW 146.4

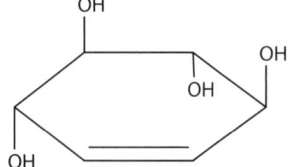

Conduritol B
MW 146.14

Butanetetrol

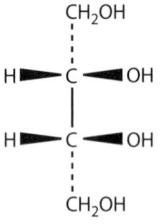

(*R**,*S**)-1,2,3,4-Butanetetrol
Erythritol
MW 122.12

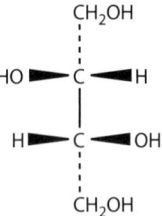

(*R**,*R**)-1,2,3,4-Butanetetrol
Threitol
MW 122.12

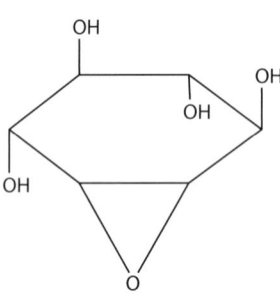

Conduritol B epoxide

Dialdoses

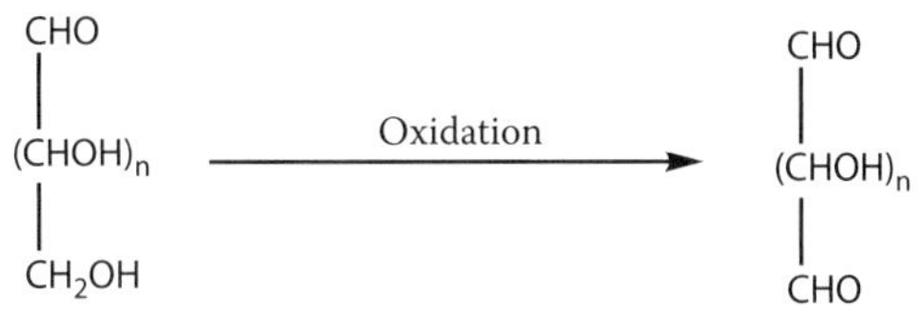

Aldose

Oxidation

Dialdose

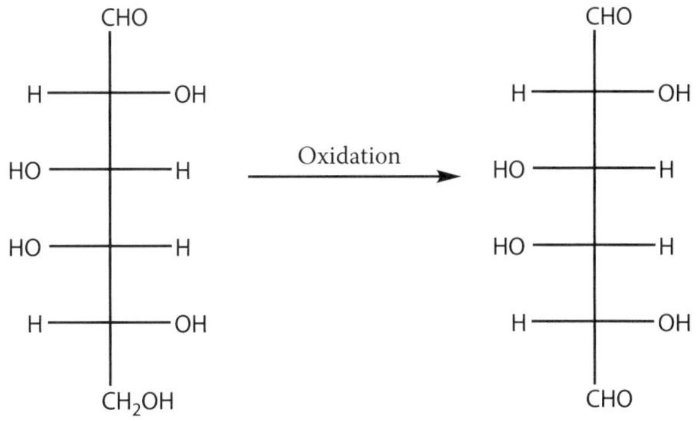

D-Galactose

Oxidation

D-*galacto*-Hexosedialdose

Fructose

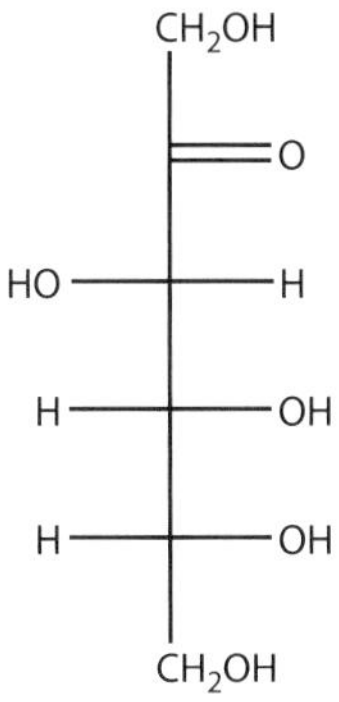

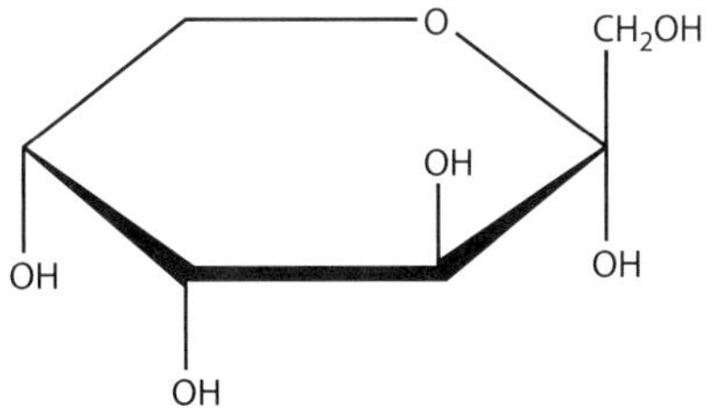

alpha-D-pyranose

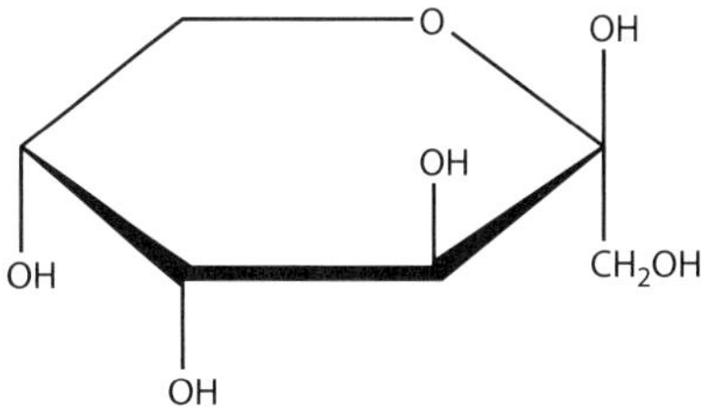

beta-D-pyranose

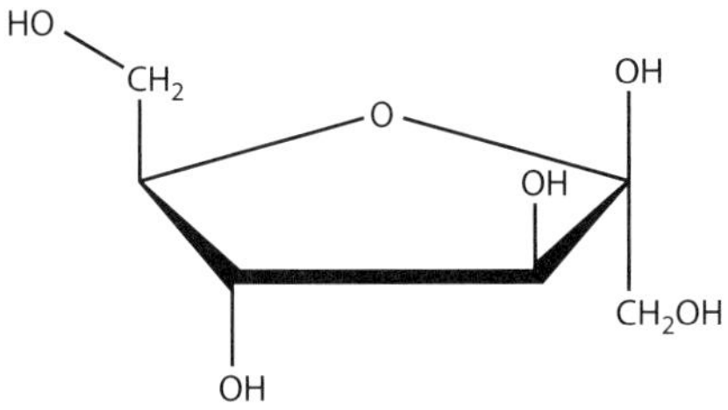

beta-D-furanose
D-Fructose
MW 180.16

Galactose and Galactose Derivatives

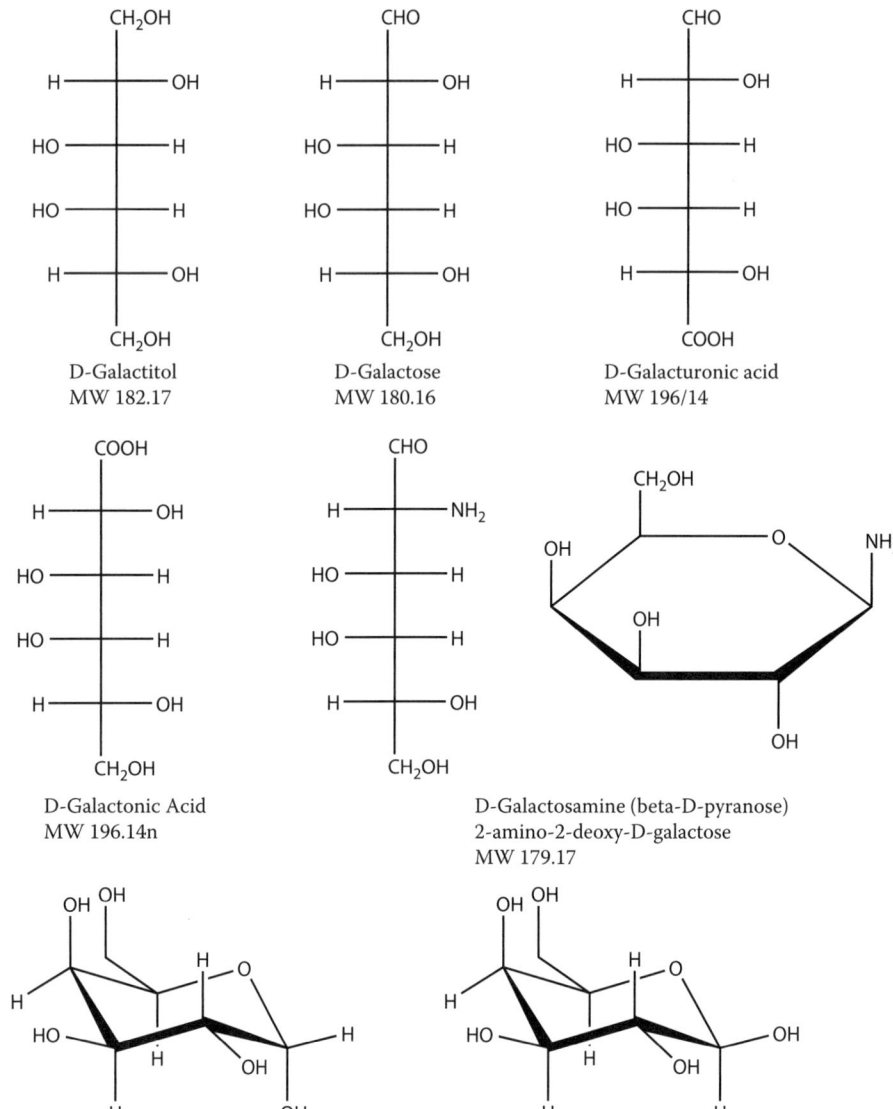

D-Galactitol
MW 182.17

D-Galactose
MW 180.16

D-Galacturonic acid
MW 196/14

D-Galactonic Acid
MW 196.14n

D-Galactosamine (beta-D-pyranose)
2-amino-2-deoxy-D-galactose
MW 179.17

D-galactose, alpha-pyranose form

D-galactose, beta-pyranose form

Glucitol (sorbitol)

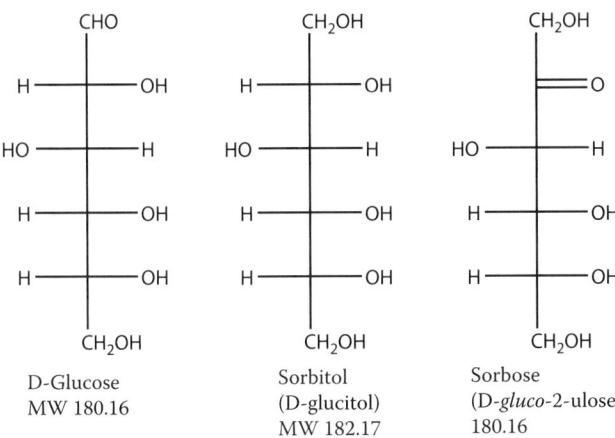

D-Glucose
MW 180.16

Sorbitol
(D-glucitol)
MW 182.17

Sorbose
(D-*gluco*-2-ulose)
180.16

Different Presentations of Glucose

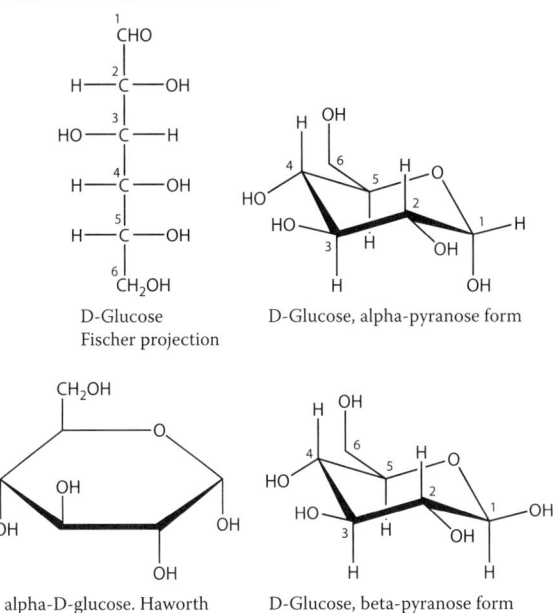

D-Glucose
Fischer projection

D-Glucose, alpha-pyranose form

alpha-D-glucose. Haworth
representation

D-Glucose, beta-pyranose form

Glucose and Glucose Derivatives

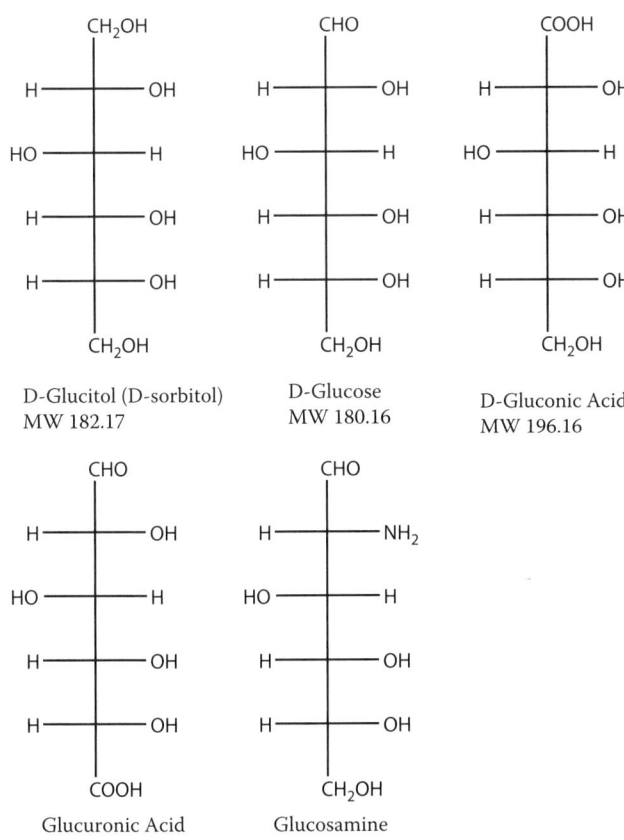

D-Glucitol (D-sorbitol)
MW 182.17

D-Glucose
MW 180.16

D-Gluconic Acid
MW 196.16

Glucuronic Acid
194.14

Glucosamine
2-Amino-2-deoxyglucose
MW 179.17

Glucose Oxidation

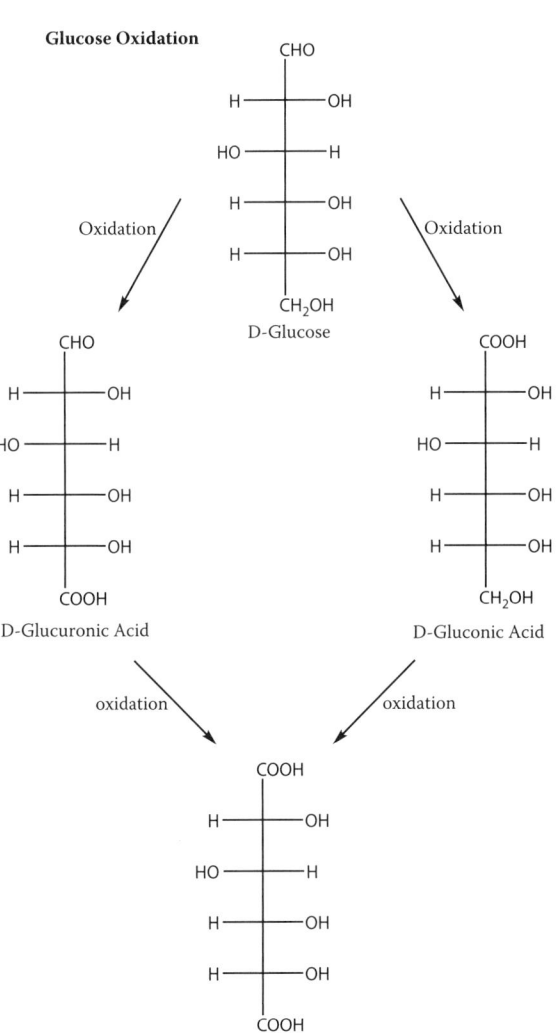

Oxidation

Oxidation

D-Glucose

D-Glucuronic Acid

D-Gluconic Acid

oxidation

oxidation

D-Glucaric Acid

Glucuronic Acid

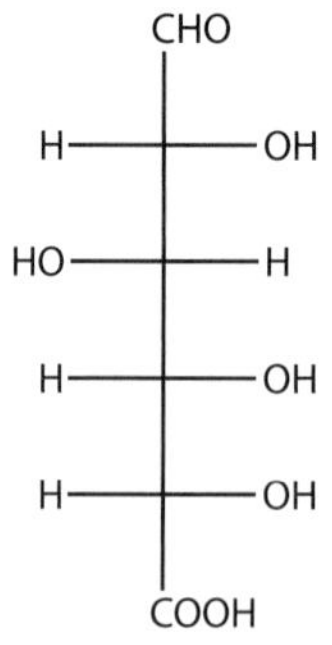

CHO

H———OH

HO———H

H———OH

H———OH

COOH

D-Glucuronic Acid

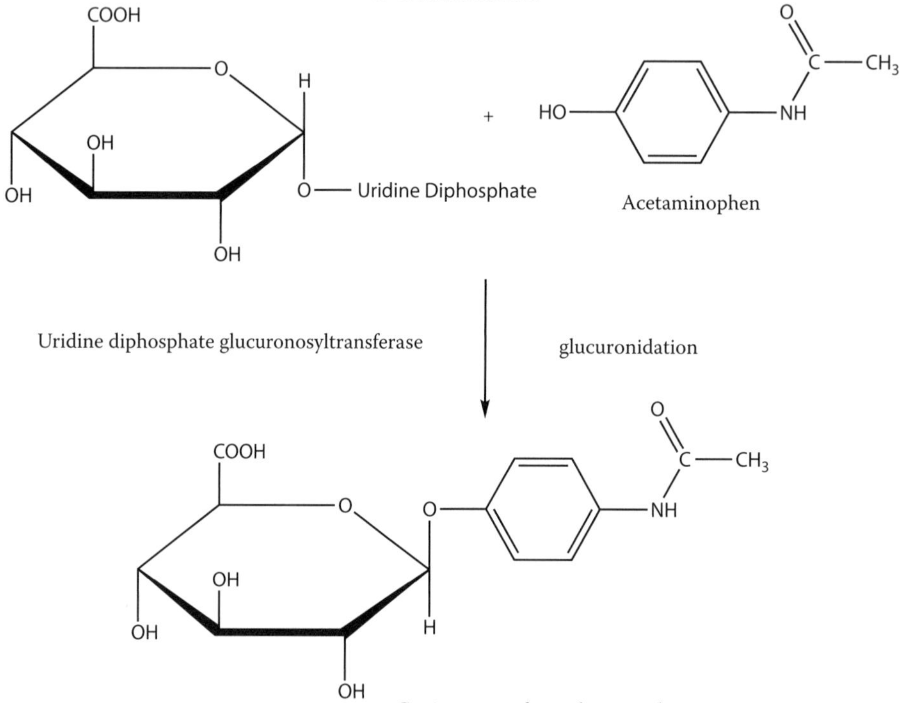

COOH

OH

OH

O

H

O——— Uridine Diphosphate

+

HO————————NH————C————CH₃
                         ‖
                         O

Acetaminophen

Uridine diphosphate glucuronosyltransferase   glucuronidation

COOH

OH

OH

O

O————————NH————C————CH₃
                        ‖
                        O

H

OH

Conjugate product, glucuronidate

Glycodiuloses

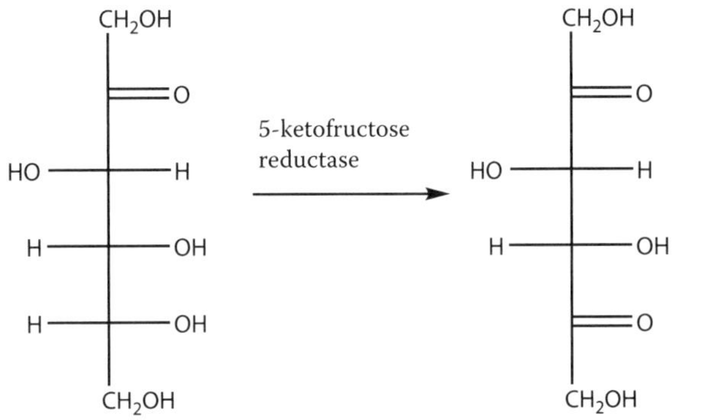

CH$_2$OH
‖
— O
HO — — H
H — — OH
H — — OH
|
CH$_2$OH

D-Fructose

→ 5-ketofructose reductase →

CH$_2$OH
‖
— O
HO — — H
H — — OH
‖
— O
|
CH$_2$OH

D-*threo*-2,5-hexodiulose
5-keto-D-fructose

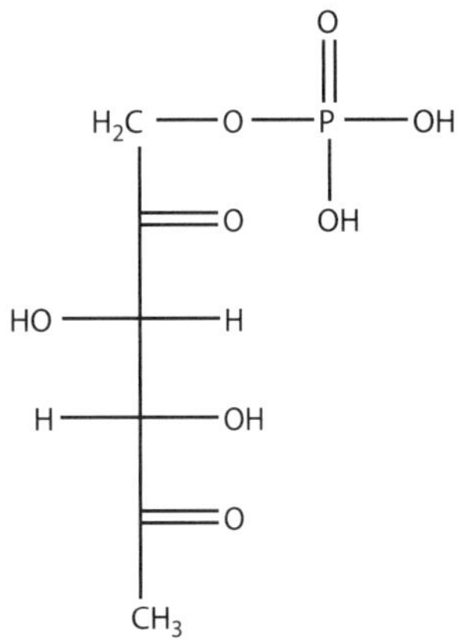

H$_2$C — O — P(=O) — OH
|
OH
‖
— O
HO — — H
H — — OH
‖
— O
|
CH$_3$

6-deoxy-5-ketofructose
1-phosphate

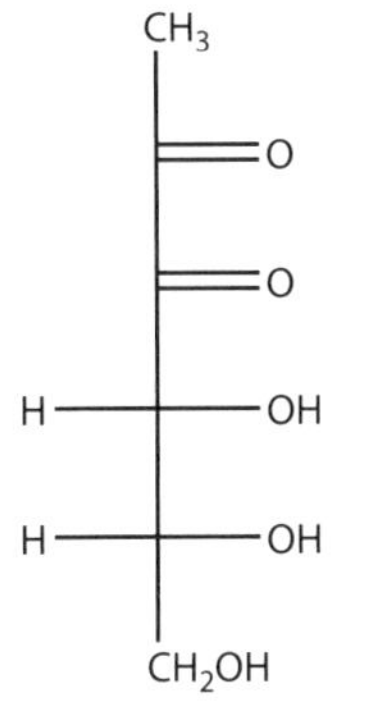

CH$_3$
‖
— O
‖
— O
H — — OH
H — — OH
|
CH$_2$OH

1-deoxy-2,3-hexodiulose

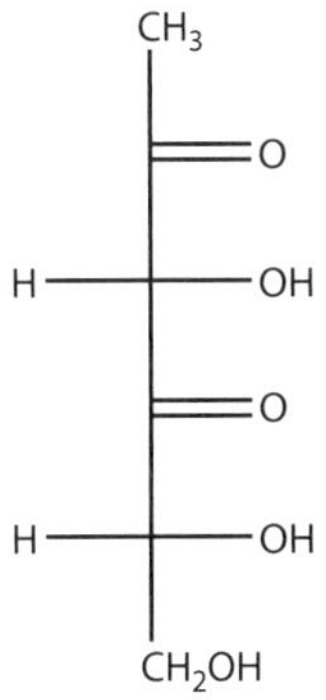

CH$_3$
‖
— O
H — — OH
‖
— O
H — — OH
|
CH$_2$OH

1-deoxy-2,4-hexodiulose

CH$_3$
H — — OH
‖
— O
‖
— O
H — — OH
|
CH$_2$OH

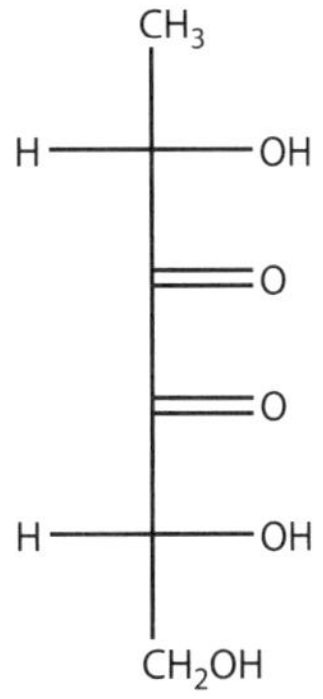

1-deoxy-3,4-hexodiulose

Glycosulose

Glycosuloses are obtained by the oxidation of a secondary hydroxyl groups in an aldose yielding a ketone

aldo-2-uloses

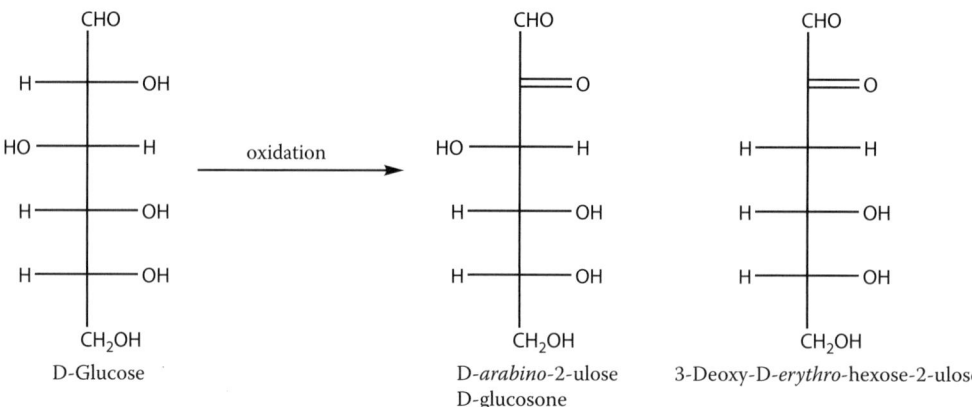

D-Glucose   D-*arabino*-2-ulose   3-Deoxy-D-*erythro*-hexose-2-ulose
                             D-glucosone

Glycosonic Acids (ketoaldonic acids) are obtained by the oxidation of secondary hydroxyl group in an aldonic acid yielding a ketone

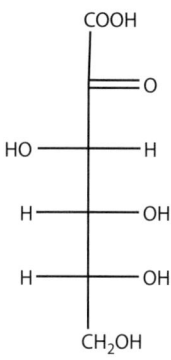

D-*arabino*-2-hexosonic acid

Heptose Derivatives

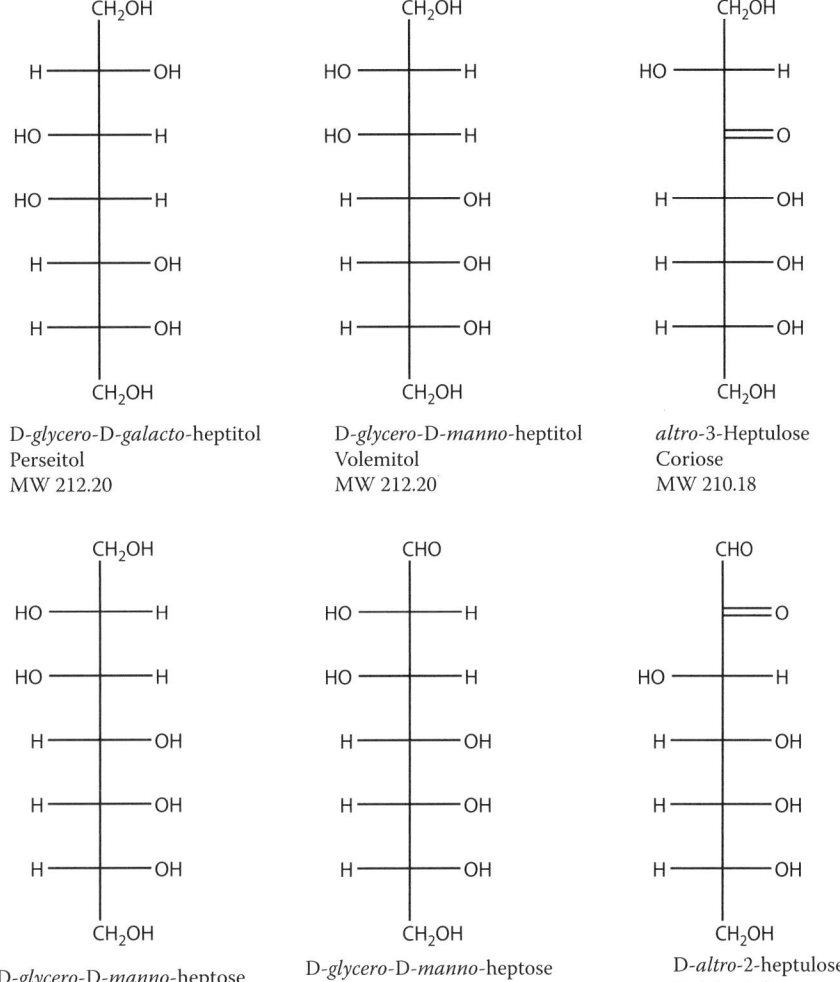

D-*glycero*-D-*galacto*-heptitol
Perseitol
MW 212.20

D-*glycero*-D-*manno*-heptitol
Volemitol
MW 212.20

altro-3-Heptulose
Coriose
MW 210.18

D-*glycero*-D-*manno*-heptose

D-*glycero*-D-*manno*-heptose

D-*altro*-2-heptulose
Sedoheptulose

Heptoses and heptonic acids

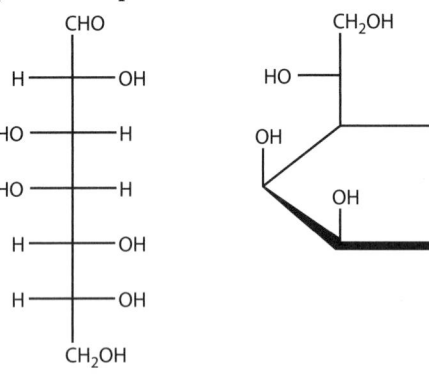

D-*glycero*-D-*galacto*-heptose
MW 210.18
alpha-D-pyranose form

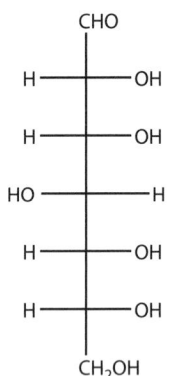

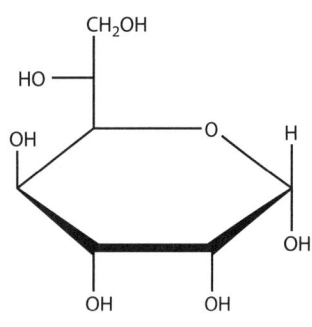

D-*glycero*-D-*gulo*-heptose
MW 210.18
alpha-D-pyranose form

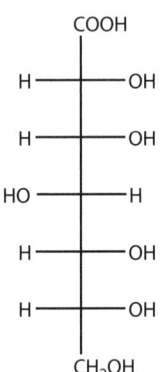

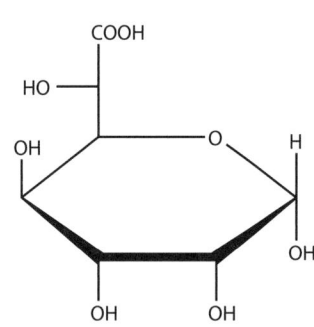

D-*glycero*-D-*gulo*-heptonic acid
MW 220.18

Lactone Formation

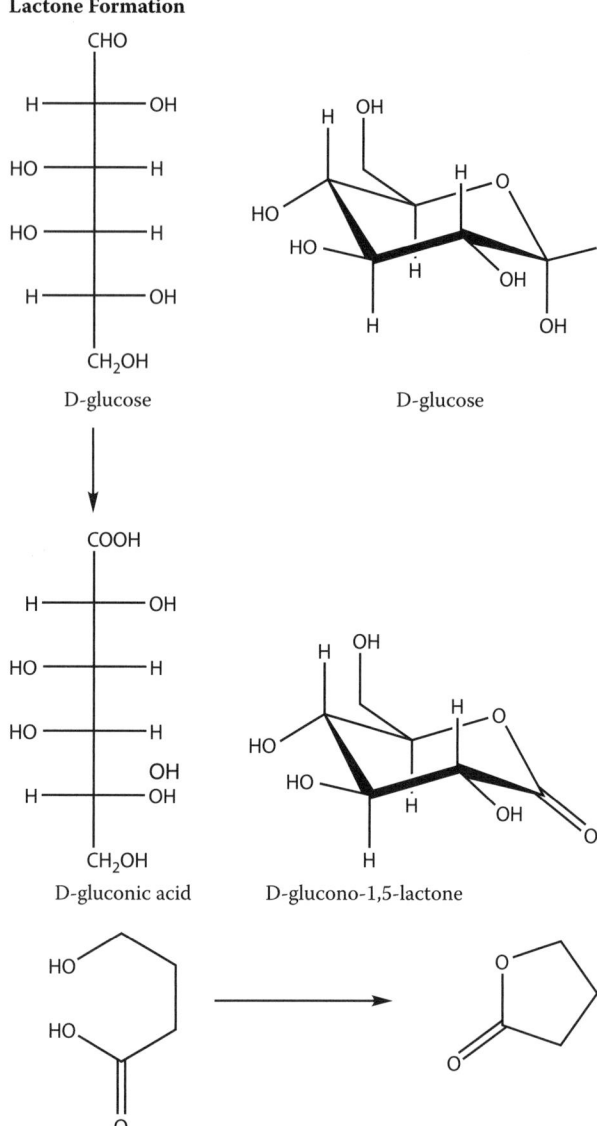

D-glucose   D-glucose

D-gluconic acid   D-glucono-1,5-lactone

gamma-hydroxybutyric acid   gamma-butyrolactone

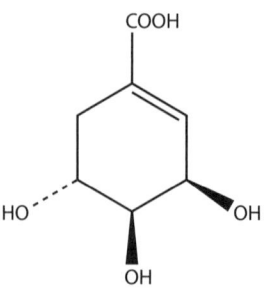

Shikimic acid
3,4,5-Trihydroxy-1-cyclohexene-1-carboxylic acid (*3R,4S,5R*)
MW 174.15

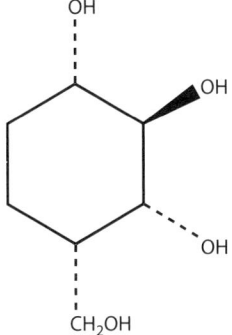

Validatol
4-Hydroxymethyl-1,2,3-hexanetriol, (1S,2R,3R,4S)
MW 162.19

Cellobiose; Lactose: Maltose

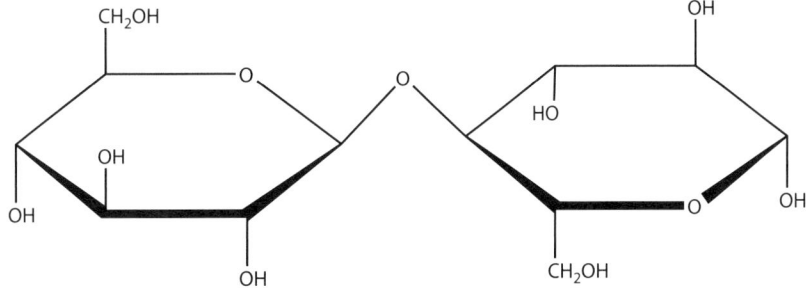

Cellobiose; MW 342.30
4-O-beta-D-glucopyranosyl-D-glucose

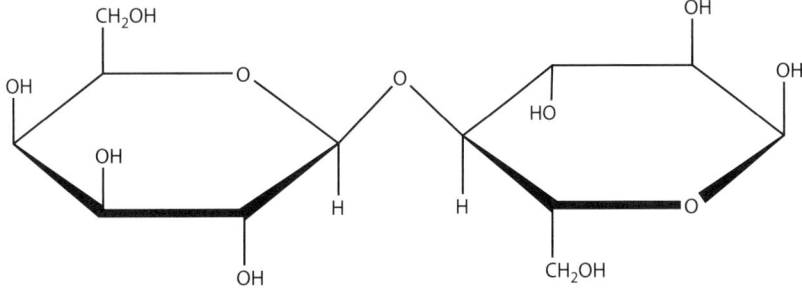

Lactose; MW 342,3
4-O-beta-D-galactopyranosyl-D-glucose

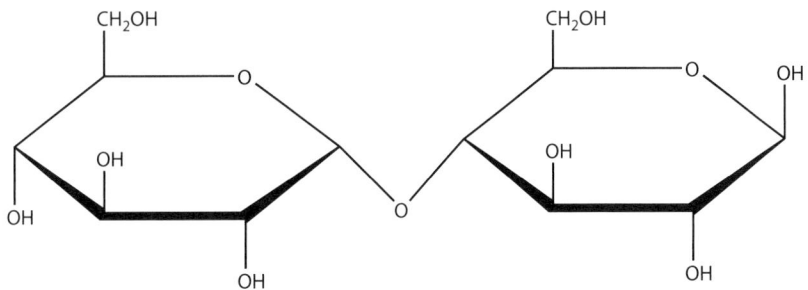

Maltose, MW 342.30
4-O-alpha-D-glucopyranosyl-D-glucose

Melibiose Trehalose Isomaltose

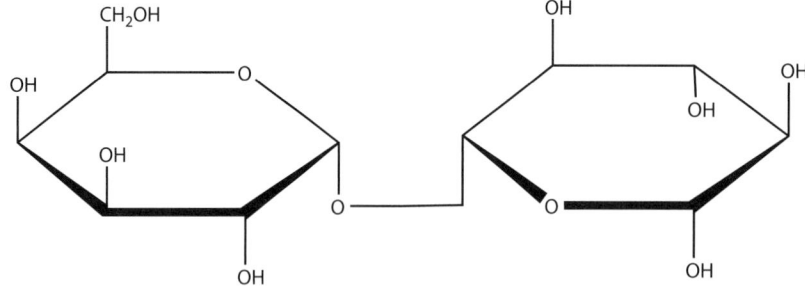

Melibiose MW 342.30
6-*O*-alpha-D-galactopyranosyl-D-glucose

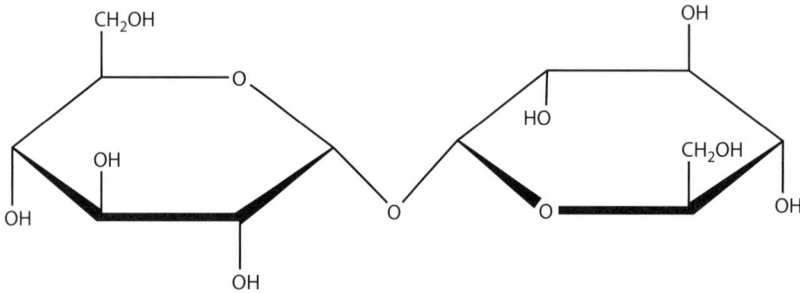

Trehalose MW 342.30
alpha-D-glucopyranosyl-alpha-D-glucopyranoside

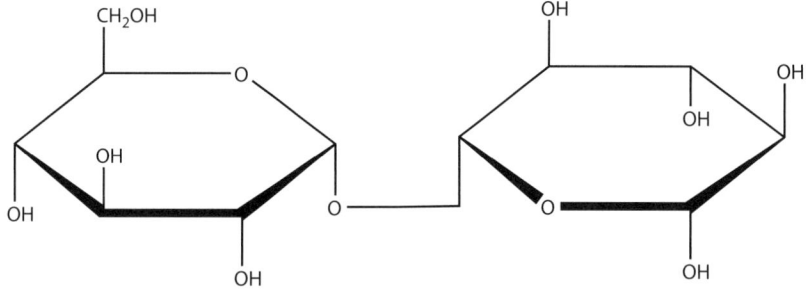

Isomaltose MW 342.30
Brachiose
6-*O*-alpha-D-glucopyranosyl-D-glucose

Sucrose Trehalulose Isomaltulose

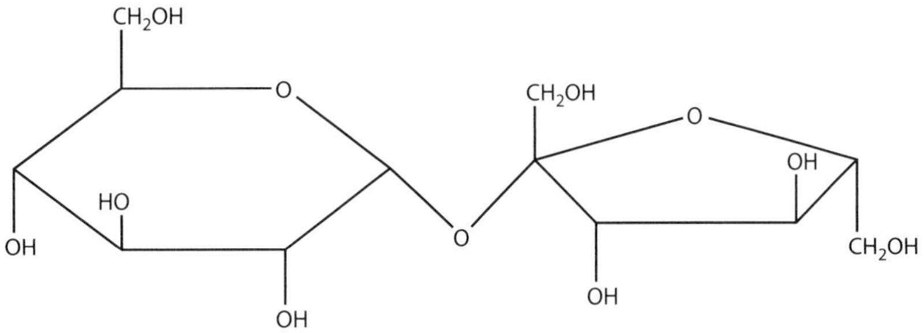

Sucrose MW 342.30
alpha-D-glucopyranosyl-beta-D-fructofuranoside

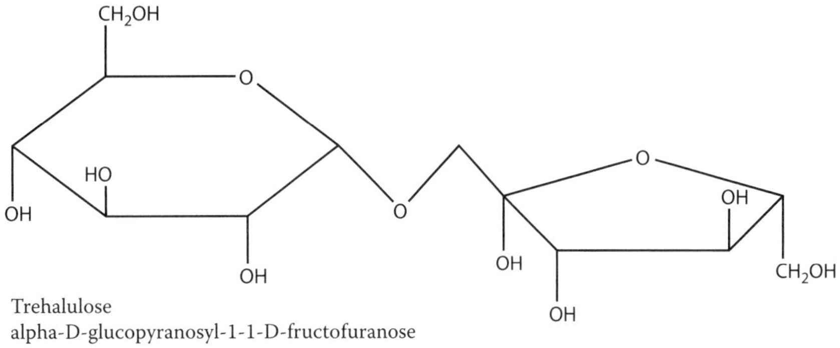

Trehalulose
alpha-D-glucopyranosyl-1-1-D-fructofuranose

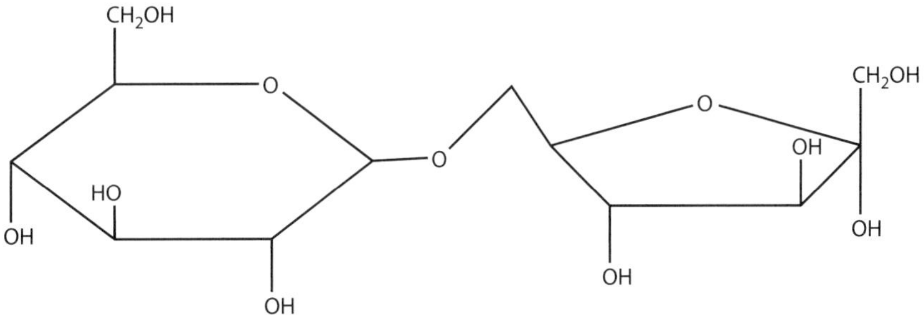

Isomaltulose' Palatinose
alpha-D-glucopyranosyl-1,6-D-fructofuranose

Chemical modification of carbohydrates

General Modification of Carbohydrates

- Hoover, R. and Sosulski, F.W., Composition, structure, functionality, and chemical modification of legume starches: a review, *Can.J.Physiol.Pharmacol.* 69, 79-92, 1991
- Stevens, C.V., Meriggi, A., and Booten, K., Chemical modification of inulin, a valuable renewable resource, and its industrial applications, *Biomacromolecules* 2, 1-16, 2001
- Lin, C.H. and Lin, C.C., Enzymatic and chemical approaches for the synthesis of sialyl glycoconjugates, *Adv. Exp.Med.Biol.* 491, 215-230, 2001
- Kellam, B., De Bank, P.A., and Shakesheff, K.M., Chemical modification of mammalian cell surfaces, *Chem.Soc.Rev.* 32, 327-337k, 2003
- Tomasik, P. and Schilling, C.H., Chemical modification of starch, *Adv.Carbohydr.Chem.Biochem.* 59, 175-403, 2004
- Jayakumuar, R., Nwe, N., Tokura, S., *et al,*. Sulfated chitin and chitosan as novel biomaterials, *Int.J.Biol.Macromol.* 40, 175-181, 2007

Glycol cleavage

- Salomies, H., Luukkanen, L., and Knuutila, R., Oxidation of beta-blocking agents - - VII. Periodate oxidation of labetalol, *J.Pharm.Biomed.Anal.* 7, 1447-1451, 1989
- Chai, W., Stoll, M.S., Cashmore, G.C., and Lawson, A.M., Specificity of mild periodate oxidation of oligosaccharide-alditols: relevance to the analysis of the core-branching pattern of *O*-linked glycoprotein oligosaccharides, *Carbohydr. Res.* 239, 107-115, 1993
- Brevnov, M.G., Gritsenko, O.M., Mikhailov, S.N., *et al.*, DNA duplexes with reactive dialdehyde groups as novel reagents for cross-linking to restriction-modification enzymes, *Nucleic Acids Res.* 25, 3302-3309, 1997
- Zhong, Y.L. and Shing, T.K., Efficient and facile glycol cleavage oxidation using improved silica gel-supported sodium metaperiodate, *J.Org.Chem.* 62, 2622-2624, 1997
- Bhavandandan, V.P.. Ringler, N.J., and Gowda, D.C., Identification of the glycosidically bound sialic acid in mucin glycoproteins that reacts like "free sialic acid" in the Warren assay, *Glycobiology* 8, 1077-1086, 1998
- Balakrishnan, B., Lesieur, S., Labarre, D., and Jayakrishnan, A., Periodate oxidation of sodium alginate in water and in

ethanol-water mixture: A comparative study, *Carbohydr. Res.* 340, 1425-1429, 2005
- Liu, B., Burdine, L., and Kodadek, T., Chemistry of periodate-mediated cross-linking of 3,4-dihydroxyphenylalanine-containing molecules to proteins, *J.Am.Chem.Soc.* 128, 15228-15235, 2006
- Perlin, A.S., Glycol-cleavage oxidation, *Adv.Carbohydr. Chem.Biochem.* 60, 184-210, 2006

Enzymatic oxidation of carbohydrates

- De Lederkremer, R.M. and Marino, C., Acids and other products of the oxidation of sugars, *Adv.Carbohydr.Chem. Biochem.* 58, 199-306, 2003
- Varela, O., Oxidative reactions and degradations of sugars and polysaccharides, *Adv.Carbohydr.Chem.Biochem.* 58, 308-369, 2003
- van Hellemond, E.W., Leferink, N.G.,H, Heuts, D.P.H.M., *et al.*, Occurrence and biocatalytic potential of carbohydrate oxidases, *Adv.Appd.Microbiol.* 60, 17-54, 2006
- Perlin, A.S., Glycol-cleavage oxidation, *Adv.Carb.Chem. Biochem.* 60, 184-250, 2006

Glucose oxidase

- Leskovac, V., Trivic, S., Wohlfahrt, G., *et al.*, Glucose oxidase from *Aspergillus niger*: the mechanism of action of oxygen, quinines, and one-electron acceptors, *Int.J.Biochem.Cell. Biol.* 37, 731-750, 2005
- Willner, I., Baron, R., and Willner,B., Integrated nanoparticle-biomolecule systems for biosensing and bioelectronics, *Biosens.Bioelectron.* 22, 1841-1852, 2007

Galactose oxidase

- Whittaker, J.W., The radical chemistry of galactose oxidase, *Arch.Biochem.Biophys.* 433, 227-239, 2005
- Kondakova, L., Yanishpolskii, V., Tertykh, V., and Bulglova, T., Galactose oxidase immobilized on silica in an analytical determination of galactose-containing carbohydrates, *Analyt.Sci.* 23, 97-101, 2007
- Alberton, D., de Olivera, L.S., Peralta, R.M. *et al.*, Production, purification, and characterization of a novel galactose oxidase from *Fusarium acumindatum*, *J.Basic Microbiol.* 47, 203-212, 2007

Glycol Cleavage

TABLE A: NATURAL ALDITOLS, INOSITOLS, INOSOSES, AND AMINO ALDITOLS AND INOSAMINES

Substance[a] (Synonym) Derivative	Chemical Formula	Melting Point °C	Specific Rotation[b] [α]$_D$	Reference[c]	Chromatography, R Value, and Reference[d]			
					ELC	GLC	PPC	TLC
(A)	(B)	(C)	(D)	(E)	(F)	(G)	(H)	(I)
ALDITOLS								
Glycerol (glycerine)	$C_3H_8O_3$	20	None (meso)	1	24rib (2)	4.69 (3)	32f (4)	62f (5)
Triacetate	$C_9H_{14}O_6$	4	None	1		3.60 (6)		
Trimethyl ether	$C_6H_{14}O_3$	Oil, bp 148	None	1		8xyll (7)		
Trimethylsilyl ether						5.94 (8)		67f (35t)
Tris(o-nitrobenzoate)	$C_{24}H_{17}N_3O_{12}$	190–192	None	9egp				
1-Deoxyglycerol[e] (1,2-propanediol, propylene glycol)	$C_3H_8O_3$	Oil, bp 188–189	Racemic	1,10	5rib (2)	3.00 (11)	35f (12)	80f (5)
Bis(p-nitrobenzoate)	$C_{17}H_{14}N_2O_8$	125–126		9egp				
Erythritol	$C_4H_{10}O_4$	118–120	None (meso)	13	53rib (2)	14.1 (3)	23f (4)	51f (5)
Tetraacetate	$C_{12}H_{18}O_8$	85	None	13		13.2 (6)		
Tetramethyl ether	$C_8H_{18}O_4$		None			26xyll (7)		
Trimethylsilyl ether						28.5 (14)		
Tetrakis(p-nitrobenzoate)	$C_{32}H_{22}N_4O_{16}$	251–252	None	9egp				
1,4-Dideoxyerythritol (2,3-butyleneglycol)	$C_4H_{10}O_2$	25,34	None (meso)	15	6rib (2)		55f (12)	
Dibenzoate	$C_{18}H_{18}O_4$	77		16				
1,4-Dideoxy-D-threitol	$C_4H_{10}O_2$	19	–13	15				
Diacetate	$C_8H_{14}O_4$	Oil, bp 192–194 (745 mm)	+1.4	17				
Bis(p-nitrobenzoate)	$C_{18}H_{16}N_2O_8$	142–144	–52.7±0.5 (c 4, CHCl$_3$)	18				
L-Threitol	$C_4H_{10}O_6$	88–89	–4.5	19,22p	96rib (2)		144glc^f (20)	
Tetraacetate	$C_{12}H_{18}O_{10}$	Oil, bp 145 (0.05 mm)	–32 (C$_2$H$_5$OH)		45xyll^f (21)			
Trimethylsilyl ether				22p		27.5^f (14)		
Di-O-benzylidene ether	$C_{18}H_{18}O_4$	218–220	+87.2 (c 0.4, acetone)	23				
Tetrakis(p-nitrobenzoate)	$C_{32}H_{22}N_4O_{16}$	219–221^f						
1,4-Dideoxy-L-threitol	$C_4H_{10}O_2$	Oil, bp ca 170	+12.4	18				
Bis(p-nitrobenzoate)	$C_{18}H_{16}N_2O_8$	141–143	+52 (CHCl$_3$)	18				
1,4-Dideoxy-DL-threitol	$C_4H_{10}O_2$	7.6	Racemic	24				
Diacetate	$C_8H_{14}O_4$	41–41.5	Racemic	24				
D-Arabinitol (D-arabitol)	$C_5H_{12}O_5$	103	+7.8 (c 8, borax solution)	25	124rib (2)	38.1 (3)	14f (4)	32f (26)
Pentaacetate	$C_{15}H_{22}O_{10}$	76	+37.2 (CHCl$_3$)	27				
L-Arabinitol	$C_5H_{12}O_5$	101–102	–7.2 (c 9, borax solution)	28p, 30cp		38.1 (3)		70f (31)
			–32 (c 0.4, 5% molybdate)	29				
Pentaacetate	$C_{15}H_{22}O_{10}$	72–73		30		44.4 (6)		
Pentamethyl ether	$C_{10}H_{22}O_5$					105xyll (7)		
Trimethylsilyl ether						40glcl (32)		
Ribitol (adonitol)	$C_5H_{12}O_5$	102	None (meso)	33	76rib (2)	39.9 (3)	14f (4)	37f (26)
Pentaacetate	$C_{15}H_{22}O_{10}$	51	None	34		40.0 (6)		
Pentamethyl ether	$C_{10}H_{22}O_5$		None			60xyll (7)		
Trimethylsilyl ether			None			32.8, 38.5 (35g)		92f (35t)

TABLE A: NATURAL ALDITOLS, INOSITOLS, INOSOSES, AND AMINO ALDITOLS AND INOSAMINES (Continued)

Substance[c] (Synonym) Derivative	Chemical Formula	Melting Point °C	Specific Rotation[b] $[\alpha]_D$	Reference[c]	Chromatography; R Value, and Reference[d]			
					ELC	GLC	PPC	TLC
(A)	(B)	(C)	(D)	(E)	(F)	(G)	(H)	(I)
ALDITOLS (Continued)								
Xylitol	$C_5H_{12}O_5$	61,92.5–93.5	None (meso)	28,36	155rib (2)	30.3 (3)	14f (4)	26f (26)
Pentaacetate	$C_{15}H_{22}O_{10}$	62.5–63	None	36		52.8 (6)		
Pentamethyl ether	$C_{10}H_{22}O_5$		None			100xyll (7)		
Trimethylsilyl ether			None			46myol (84)		
Galactitol (dulcitol)	$C_6H_{14}O_6$	186–187	None (meso)	37e	145rib (2)	144.4 (6)	7f (4)	24f (5)
Hexaacetate	$C_{18}H_{26}O_{12}$	167.5–168.5	None	37g		388xyll (7)		
Hexamethyl ether	$C_{12}H_{26}O_6$		None			12.47 (38)		
Trimethylsilyl ether	$C_{24}H_{62}O_6Si_6$	78	None	84				
D-Glucitol (sorbitol)	$C_6H_{14}O_6$	112	–1.8 (15°)	39	161rib (2)	144.4 (6)	8f (4)	22f (5)
Hexaacetate	$C_{18}H_{26}O_{12}$	99	+12.5 (c 0.8, $CHCl_3$)	40,41		246xyll (7)		
Hexamethyl ether	$C_{12}H_{26}O_6$					27 (32)		
Trimethylsilyl ether								
1,5-Anhydro-D-glucitol (polygalitol)	$C_6H_{12}O_5$	140–141	+42.4	42	173rib (2)		92sorl (44)	
Tetraacetate	$C_{14}H_{20}O_9$	73–74	+38.9 ($CHCl_2$)	42				
L-Iditol	$C_6H_{14}O_6$	73.5	–3.5 (c 10)	43		153xyll[f] (21)		
Hexaacetate	$C_{18}H_{26}O_{12}$	121.5	–25.7 ($CHCl_3$)	43,45	130rib (2)	38.1 (3)	8f (4)	27f (5)
D-Mannitol	$C_6H_{14}O_6$	166	–0.21	46				
Hexaacetate	$C_{18}H_{26}O_{12}$	126	+16 (5% molybdate)	29		127.2 (6)		83f (50a)
Hexamethyl ether	$C_{12}H_{26}O_6$		+18.8 (acetic acid)	47,48		284xyll (7)		
Trimethylsilyl ether						9.46 (8)		
1,5-Anhydro-D-mannitol (styrachitol)	$C_6H_{12}O_5$	157	–49.9	49				
Tetraacetate	$C_{14}H_{20}O_9$	66–67	–20.9 (C_2H_5OH)	51				
D-Mannitol 1-acetate	$C_8H_{16}O_7$	124–125	+4	52cp				
D-*glycero*-D-*galacto*-Heptitol (L-*glycero*-D-*manno*-heptitol, perseitol)	$C_7H_{16}O_7$	183–185,188	–1.1	53,54	140rib (2)		100perl (55a)	
Heptaacetate	$C_{21}H_{30}O_{14}$	119–120.5	+24.5 (5% molybdate)	29				
			–14 ($CHCl_3$)	53				
D-*glycero*-D-*gluco*-Heptitol (L-*glycero*-D-*talo*-heptitol, β-sedoheptitol)	$C_7H_{16}O_7$	131–132	+46 (5% molybdate)	55c	171rib (2)		>120suc (55b)	
Tri-O-methyl-ene-β-sedoheptitol	$C_{10}H_{16}O_7$	Sublimes 130, 276–278d	–23.3 (c 0.4, $CHCl_3$)	56				
D-*glycero*-D-*ido*-Heptitol	$C_7H_{16}O_7$		0.0	57c	168rib (2)		78–82 gal (57a)	
Heptaacetate	$C_{21}H_{30}O_{14}$	180–181	+24 ($CHCl_3$)	57		621aral (57b)		
Heptabenzcate	$C_{56}H_{44}O_{14}$			57				
L-*glycero*-L-*ido*-Heptitol	$C_7H_{16}O_7$	111–112	None (meso)	57c			74gal (57a)	
Heptaacetate	$C_{21}H_{30}O_{14}$	175–176	None	57		719aral (57b)		
D-*glycero*-D-*manno*-	$C_7H_{16}O_7$	153	+2.6	58				
Heptitol (D-*glycero*-D-*talo*-heptitol, volemitol)			+55 (5% molybdate)	29	140rib (2)			
Heptaacetate	$C_{21}H_{30}O_{14}$	62	+36.1 ($CHCl_3$)	58, 59				

TABLE A: NATURAL ALDITOLS, INOSITOLS, INOSOSES, AND AMINO ALDITOLS AND INOSAMINES (Continued)

Substance[a] (Synonym) Derivative (A)	Chemical Formula (B)	Melting Point °C (C)	Specific Rotation[b] $[\alpha]_D$ (D)	Reference[c] (E)	Chromatography, R Value, and Reference[d]			
					ELC (F)	GLC (G)	PPC (H)	TLC (I)
ALDITOLS (Continued)								
D-*erythro*-D-*galacto*- Octitol	$C_8H_{18}O_8 \cdot H_2O$	169–170	−11 (5% molybdate)	55c			83perl (55a)	
Octaacetate	$C_{24}H_{34}O_{16}$	99–100	+2 (CHCl$_3$)	55				
INOSITOLS								
Asteritol (an inositol monomethyl ether)	$C_7H_{14}O_6$	Sublimes, melts 164	+157 (c 0.01)	60			32f (60)	
Bettitol (a dideoxyinositol)	$C_6H_{12}O_4$	224		61				
D-Bornesitol (D-*myo*-inositol monomethyl ether)	$C_7H_{14}O_6$	201–202	+31.4	62	20rib (2)			
Pentaacetate	$C_{17}H_{24}O_{11}$	138–139	+11.8 (c 0.8, acetone)	62				
Trimethylsilyl ether						86myol (84)		
L-Bornesitol (1-O-methyl L-*myo*-inositol)	$C_7H_{14}O_6$	205–206	−32.1	63, 65cp	15glc (82)		19glc (66)	
Pentaacetate	$C_{17}H_{24}O_{11}$	142–143, 157	−11.2 (CHCl$_3$)	64		8.27 (93)		
Conduritol (a 2,3-dehydro-2,3-dideoxy-inositol)	$C_6H_{10}O_4$	142–143	None (meso)	67				
Dihydroconduritol	$C_6H_{12}O_4$	204	None	68				
Tetraacetate	$C_{14}H_{18}O_4$	bp 165 (0.6 mm)	None	68				
Dambonitol (1,3-di-O-methyl-*myo*-inositol)	$C_8H_{16}O_6$	206, 210	None (meso)	69, 70	0glc (82)		20f (70)	
Tetraacetate	$C_{16}H_{24}O_{10}$	202	None	71c				
D-Inositol [*d*-inositol, *chiro*- (+)-inositol,(+)-inositol, D-*chiro*-inositol]	$C_6H_{12}O_6$	246–247d	+60 +65	72c, 73	83glc (82)		17f 12f (72, 74)	
Hexaacetate	$C_{18}H_{24}O_{12}$	230–235	+68	74				
Hexabenzoate	$C_{48}H_{36}O_{12}$	215–220		74				
		252–253	+64.5	74,75				
L-Inositol [*l* inositol, *levo*-inositol, *chiro*-(−)-inositol]	$C_6H_{12}O_6$	247	−64.1	1, 76p 1	23rib (2)		49pinl (132)	
Hexaacetate	$C_{18}H_{24}O_{12}$	96						
Trimethylsilyl ether						82myol (84)		
D, L-Inositol	$C_6H_{12}O_6$	253	Racemic	77				
Hexaacetate	$C_{18}H_{24}O_{12}$	111	None	77		9.62 (93)		
Hexabenzoate	$C_{48}H_{36}O_{12}$	213	None	78cp		13.3 (79)		
Trimethylsilyl ether								
1-O-Methyl-(+)-inositol	$C_7H_{14}O_6$	207–208	+60.7	75	32glc[f] (82)			
Pentaacetate	$C_{17}H_{14}O_{11}$	110.5–111.5	+29.1 (CHCl$_3$)	75				
1-O-Methyl-*muco*-inositol	$C_7H_{14}O_6$	Gum		80cp,81	30glc (80)		109pinl (81)	
Pentaacetate	$C_{17}H_{14}O_6$	bp ca 200 (vac)		80				
Pentabenzoate	$C_{42}H_{34}O_{11}$	Amorphous 95–100		80				
myo-Inositol (*meso*-inositol)	$C_6H_{12}O_6$	225–227	None (meso)	72p	16rib (2)	13.42 (93)	2f(4)	27f (50b)
Hexaacetate	$C_{18}H_{24}O_{12}$	206–208	None	83cp				79f (50a)
		221–213		85				
Trimethylsilyl ether	$C_{24}H_{60}O_6Si_6$	118–119		84g		10.3, 24.3 (8, 79)		
neo-Inositol	$C_6H_{12}O_6$	314		85,86	43rib (2)			
Hexaacetate	$C_{18}H_{24}O_{12}$	251–253		86cpt		9.17 (93)		

TABLE A: NATURAL ALDITOLS, INOSITOLS, INOSOSES, AND AMINO ALDITOLS AND INOSAMINES (Continued)

Substance (Synonym) Derivative (A)	Chemical Formula (B)	Melting Point °C (C)	Specific Rotation $[\alpha]_D$ (D)	Reference (E)	Chromatography, R Value, and Reference			
					ELC (F)	GLC (G)	PPC (H)	TLC (I)
INOSITOLS (Continued)								
Laminitol (6-*C*-methyl-*myo* inositol)	C$_7$H$_{14}$O$_6$	226–269	–3	87c, 88cp	83glc (87a)		10f (87b)	
Hexaacetate	C$_{19}$H$_{26}$O$_{12}$	153	–19.6±1 (CHCl$_3$)	88			27f (87c)	
Leucanthemitol (a dehydro dideoxy inositol)	C$_6$H$_{10}$O$_4$	131–132	+101.5	89,90			85–80f (89)	
Dihydroleucanthemitol	C$_6$H$_{12}$O$_4$	161	–40	90				
Liniodendritol (1,4-di-*O*-methyl-*myo*-inositol)	C$_8$H$_{16}$O$_6$	224	–25	66,91			52glc (66)	
Tetraacetate	C$_{16}$H$_{24}$O$_{10}$	139	–24	91				
Mytilitol (a *C*-methyl-*scyllo*-inositol)	C$_7$H$_{14}$O$_6$	259	None (meso)	92	<10sorl (94)			
Hexaacetate	C$_{19}$H$_{26}$O$_{12}$	180–181	None	92				
d-Ononitol [(+)-ononitol, 4-*O*-methyl-L-*myo*-inositol]	C$_7$H$_{14}$O$_6$	172	+6.6	64	60glc (82)		32f (66)	
Pentaacetate	C$_{17}$H$_{24}$O$_{11}$	121	–11.1 (c 0.8, CHCl$_3$)	65cp				
Trimethylsilyl ether						80myol (84)		
d-Pinitol [(+)-pinitol, 3-*O*-methyl-(+)-inositol]	C$_7$H$_{14}$O$_6$	185–186	+65.5	95,96	35 rib (2)	7.07 (93)	9f (4)	
Pentaacetate	C$_{17}$H$_{24}$O$_{11}$	98	+8.6 (C$_2$H$_5$OH)	97				
l-Pinitol [(–)-pinitol, 3-*O*-methyl-(–)-inositol]	C$_7$H$_{14}$O$_6$	186	–65	98,99				
Trimethylsilyl ether						64myol (84)		
l-Quebrachitol (1L-2-*O*-methyl-*chiro*-inositol)	C$_7$H$_{14}$O$_6$	190–191	–80.2 (28°)	100–102p	27rib (2)	6.90 (93)	7f (4)	
Pentaacetate	C$_{17}$H$_{24}$O$_{11}$	96–98	–25.1 (29° CHCl$_3$)	100,101		58 (84)		
Trimethylsilyl ether								
allo-Quercitol (5-deoxy-*allo*-inositol)	C$_6$H$_{12}$O$_5$	262–263		103egt		11.9 (128)	140glc (103)	
Pentaacetate	C$_{16}$H$_{22}$O$_{10}$							
d-Quercitol [(+)-*proto*-quercitol, 4-deoxy-(+)-inositol]	C$_6$H$_{12}$O$_5$	235	+24.2	104,105	31glc (82)	8.4 (128)		
Pentaacetate	C$_{16}$H$_{22}$O$_{10}$							
Pentabenzoate	C$_{41}$H$_{32}$O$_{10}$	155	+61	106				
l-Quercitol [(–)-*proto*-quercitol]	C$_6$H$_{12}$O$_5$	238–239	–26	105,107				
Pentabenzoate	C$_{41}$H$_{32}$O$_{10}$	154–155	–62.8 (ethyl acetate)	105				
d-Quinic acid (a dideoxy carboxy-*dextro*-inositol)	C$_7$H$_{12}$O$_6$	164	+44 (c 10)	61				
l-Quinic acid	C$_7$H$_{12}$O$_6$	162	–42.1	108			45f (109c)	
Tetraacetate	C$_7$H$_{12}$O$_6$	130–136	–22.5 (C$_2$H$_5$OH)	110				
Trimethylsilyl ether	C$_{15}$H$_{20}$O$_{10}$					8.47,2.1 (8,113)		
5-Dehydroquinic acid	C$_7$H$_{10}$O$_6$	134–136,140–142	–82.4 (28°, C$_2$H$_5$OH)	111,112cp			30fg (109)	
Protocatechuic acid	C$_7$H$_6$O$_2$	201–202		111				
Trimethylsilyl ether						4.6 (113)		
Scyllitol (*scyllo*-inositol, cocositol)	C$_6$H$_{12}$O$_6$	352–353	None (meso)	114,115	7rib (2)		0f (4)	
Hexaacetate	C$_{18}$H$_{24}$O$_{12}$	299–300	None	114,115		22.25 (93)		
Trimethylsilyl ether	C$_{24}$H$_{60}$O$_6$Si$_6$	179–180	None	84		93myol (84) / 9.8,18.4 (8,79)		
O-Methyl-*scyllo*-inositol	C$_7$H$_{14}$O$_6$	243	None (meso)	116cp			65pinl (132)	
Pentaacetate	C$_{17}$H$_{24}$O$_{11}$	192–193	None	116t				

TABLE A: NATURAL ALDITOLS, INOSITOLS, INOSOSES, AND AMINO ALDITOLS AND INOSAMINES (Continued)

INOSITOLS (Continued)

Substance[a] (Synonym) Derivative (A)	Chemical Formula (B)	Melting Point °C (C)	Specific Rotation[b] $[\alpha]_D$ (D)	Reference[c] (E)	Chromatography, R Value, and Reference[d]			
					ELC (F)	GLC (G)	PPC (H)	TLC (I)
Sequoyitol (5-O-methyl *myo*-inositol)	$C_7H_{14}O_6$	234–235	None (meso)	117	25rib (2)		27f (72)	
Pentaacetate	$C_{17}H_{24}O_{11}$	198	None	117		9.76 (93)		
Shikimic acid (3,4-anhydroquinic acid)	$C_7H_{10}O_5$	183–184	−200 (16°)	118,119			60f (109)	
		190–191	−161 (c 0.6)	120				
Methyl shikimate	$C_8H_{12}O_5$	115	−136 (CH_3OH)	121				
Triacetate								
Trimethylsilyl ether	$C_{13}H_{16}O_8$	Syrup	−60	120		8.94,2.7 (8,113)		
5-Dehydroshikimic acid	$C_7H_8O_5$	150–152	−57.5 (28°, C_2H_5OH)	122			50fg (109)	
Methyl ester	$C_8H_{10}O_5$	124–126	−47.1±3 (c 0.2, C_2H_5OH)	122				
Trimethylsilyl ether						6.9 (113)		
Validitol[1S-(3,4,6/5)-3-hydroxymethyl-4,5,6-trihydroxycyclohexane]	$C_7H_{14}O_4$	119–121	−39	123				
Deoxyvalditol [1S-(4,6/5)-3-methyl-4,5,6-trihydroxycyclohexane]	$C_7H_{14}O_3$		−19.4	123				
Viburnitol [*l*-quercitol;[h] a deoxy *levo*-inositol, (−)-*vibo*-quercitol, 1-deoxy-D-*myo*-inositol]	$C_6H_{12}O_5$	174	−73.9	124	31glc (82)			
Pentaacetate	$C_{16}H_{22}O_{10}$	179–180	−49.5±1	125,126	<10sorl (94)	9.5 (128)	65pinl (132)	
		158–159		127				
		112–113		127				
		122–126		124–126				
Bioinose (*myo*-inosose-2, *scyllo*-inosose, a dehydro or keto inositol)	$C_6H_{10}O_6$	196–197	None (meso)	129			43pinl (132c)	
Pentaacetate	$C_{16}H_{20}O_{11}$	121–213	None	129				
Phenylhydrazone	$C_{12}H_{16}N_2O_5$	220–222	None	130				
Trimethylsilyl ether	$C_{21}H_{50}O_6Si_5$	98	None	84		75myol (84)		
myo-Inosose-1 (*vibo*-inosose, *d*-inosose, L-*myo*-inosose-1)	$C_6H_{10}O_6$	138–139	+19.6	131			27pinl (132c)	
Phenylhydrazone	$C_{12}H_{16}N_2O_5$	196–197	−55.3 (C_5H_5N-C_2H_5OH)	131				
Trimethylsilyl ether						9.53 (8)		
2,3-Didehydro-D-inositol	$C_6H_8O_6$	No constants known		133				
Bis-(phenylhydrazone)	$C_{18}H_{20}N_4O_4$	217d	−250→−222 (24 hr, c 0.5, C_5H_5, N-C_2H_5OH 1:1)	133				
2,3-Didehydro-L-inositol	$C_6H_8O_6$	No constants known		133				
Bis-(phenylhydrazone)	$C_{18}H_{20}N_4O_4$	217d	+240→+214 (24 hr c 0.8, C_5H_5N-C_2H_5 OH 1:1)	133				
2,3-Didehydro-4-deoxy-*epi*-inositol	$C_6H_8O_5$	No constants known		134				
Bis-(phenylhydrazone)	$C_{18}H_{20}N_4O_3$	198–199	+54.8 (C_5H_5 N- C_2H_5OH 1:1)	134			15f (134)	
2-Aminoethanol (ethanolamine)	C_2H_7NO	Oil, bp 171	None	1				
Hydrochloride	$C_2H_7NO{\cdot}HCl$	100	None	1				

TABLE A: NATURAL ALDITOLS, INOSITOLS, INOSOSES, AND AMINO ALDITOLS AND INOSAMINES (Continued)

Substance[a] (Synonym) Derivative (A)	Chemical Formula (B)	Melting Point °C (C)	Specific Rotation[b] $[\alpha]_D$ (D)	Reference[c] (E)	Chromatography, R Value, and Reference[d]			
					ELC (F)	GLC (G)	PPC (H)	TLC (I)
		INOSITOLS (Continued)						
Picrate	$C_8H_{10}N_4O_8$	158	None	1				
Trimethylsilyl ether			None			3.14 (8)		
A 2-amino-2,4-dideoxy-1-O-methyl-tetritol	$C_5H_{13}NO_2$	No constants known		135				
N-acetyl acetate	$C_9H_{17}NO_4$	Oil, bp 91 (0.1 μ)		135				
Actinamine (1,3-dideoxy-1,3-*N,N'*-dimethyl-*myo*-inositol)	$C_8H_{18}N_2O_4$	135–136		136,137c				
Dihydrochloride	$C_8H_{18}N_2O_4 \cdot$ HCl	>300		136				
N,N'-Diacetyl tetraacetate	$C_{20}H_{30}N_2O_{10}$	205–206		136				
Bluensidine (3-O-carbamoyl-1-deoxy-1-guanidino-*scyllo*-inositol)	$C_8H_{16}N_4O_6$	No constants known		138				
Hydrochloride	$C_8H_{16}N_4O_6 \cdot$ HCl	190–194d	+1±0.5	138				
N,N'-Diacetyl tetraacetate	$C_{20}H_{28}N_4O_{12}$	250–251d	+5 (c 0.9, CHCl$_3$)	138				
neo-Inosamine-2 (2-amino-2-deoxy-*neo*-inositol)	$C_6H_{13}NO_5$	239–241d	None (meso)	139,140				
Hydrochloride	$C_6H_{13}NO_5 \cdot$ HCl	217–221d	None	139,140				
N-Acetyl pentaacetate	$C_{18}H_{25}NO_{11}$	277–278	None	139,140				
scyllo-Inosamine	$C_6H_{13}NO_5 \cdot$ HCl	No constants known	141			47glcn (149)		
N-Acetyl pentaacetate	$C_{18}H_{25}NO_{11}$	209–301	None (meso)	142				
Streptamine (1,3-diamino-1,3-dideoxy-*scyllo*-inositol)	$C_6H_{14}N_2O_4$	>290 (205s)	None (meso)	143				
Dihydroiodide	$C_6H_{14}N_2O_4 \cdot$ 2HI	>280d	None	143				
N,N'-Diacetyl tetraacetate	$C_{18}H_{26}N_2O_{10}$	342–345d	None	143				
2-Deoxystreptamine	$C_6H_{14}N_2O_3$	221–223	None (meso)	144				
Dihydrobromide	$C_6H_{14}N_2O_3 \cdot$ 2HBr	283–286d	None	145				
N,N'-Diacetyl triacetate	$C_{16}H_{24}N_2O_8$	340–350	None	144				
Deoxy-N-methylstreptamine (hyosamine)	$C_7H_{16}N_2O_3$	183–186	+39.8	146				
		160–162	–31.1	146				
		130–133d	–17.8	147l				
N,N'-Diacetyl triacetate	$C_{17}H_{26}N_2O_8$	204	Racemic	144				
Dipicrate		238–240d		147				
Streptidine (1,3-dideoxy-1,3-diguanidino-*scyllo*-inositol)	$C_8H_{18}N_6O_4$		None (meso)	143				
N,N'-Diacetyl tetraacetate	$C_{20}H_{30}N_6O_{10}$	342–345	None	143				
Dipicrate	$C_{20}H_{24}N_{12}O_{18}$	284–285d	None	148				

TABLE A: NATURAL ALDITOLS, INOSITOLS, INOSOSES, AND AMINO ALDITOLS AND INOSAMINES (Continued)

Substance[a] (Synonym) Derivative	Chemical Formula	Melting Point °C	Specific Rotation[b] $[\alpha]_D$	Reference[c]	Chromatography, R Value, and Reference[d]			
					ELC	GLC	PPC	TLC
(A)	(B)	(C)	(D)	(E)	(F)	(G)	(H)	(I)
INOSITOLS (Continued)								
Validamine [1S-(1,2,4/3,5)-5-hydroxymethyl-2,3,4-trihydroxycyclohexylamine]	$C_7H_{15}NO_4$		+60.6	150				
Hydrochloride	$C_7H_{15}NO_4 \cdot$ HCl	229–232d	+57.4 (N HCl)	123				
Hydroxyvalidamine	$C_7H_{15}NO_5$	164–165	+80.7	123				
epi-Validamine	$C_7H_{15}NO_4$	210	+5.8	150				
Valienamine (1,6-dehydro-validamine)	$C_7H_{13}NO_4$	No constants known		150				
Hydrochloride	$C_7H_{13}NO_4 \cdot$ HCl		+68.6 (N HCl)	150				
N-Acetyl tetraacetate	$C_{17}H_{23}NO_9$	95	+30.2 (CHCl$_3$)	150				

Compiled by George G. Maher.

a In order of increasing carbon chain length in the parent compounds grouped in the classes – alditols, inositols, inososes, amino alditols, and inosamines.

b $[\alpha]_D$ for 1–5 g solute, *c*, per 100 ml aqueous solution at 20–25°C unless otherwise given.

c References for melting point and specific rotation data. Letter indicates the reference also has chromatographic data.

d R value times 100, given relative to that of the compound indicated by abbreviation: f = s solvent front, g = gas, p = paper, and t = thin-layer. R value times 100, given relative to that of the compound indicated by abbreviation: f = s solvent front, gal = galactose, glc = glucose, glcl = glucitol, glcn = glucosamine, myol = *myo*-inositol, perl = perseitol, pinl = pinitol, rib = ribose, sorl = sorbitol, suc = sucrose, xyll = xylitol, (as the pentaacetate or the pentamethyl ether, as pertains), and aral = arabinitol (as the pentaacetate). Under gas chromatography (Column GLC or G) numbers without code indication signify retention time in minutes. The conditions of the chromatography are correlated with the reference given in parentheses and are found in Table 5.

e Said to exist as a phosphate ester also.[10]

f Data given are for the enanthiomorphic isomer.

g The author names as 3-dehydroquinic acid, but it is actually 5-dehydroquinic.

h The early given name, *l*-quercitol, of this compound does not make it the enanthiomorph of *d*-quercitol; other isomeric relations are involved.

i This compound is isomeric with the previous one in regard to the *N*-methyl group position.

References

1. Pollock and Stevens, *Dictionary of Organic Compounds*, Oxford University Press, New York, 1965.
2. Frahn and Mills, *Aust. J. Chem.*, 12, 65 (1959).
3. Dooms, Declerck, and Verachtert, *J. Chromatogr.*, 42, 349 (1969).
4. Bourne, Lees, and Weigel, *J. Chromatogr.*, 11, 253 (1963).
5. de Simone and Vicedomini, *J. Chromatogr.*, 37, 538 (1968).
6. Sawardeker, Sloneker, and Jeanes, *Anal. Chem.*, 37, 1602 (1965).
7. Whyte, *J. Chromatogr.*, 87, 163 (1973).
8. Roberts, Johnston, and Fuhr, *Anal. Biochem.*, 10, 282 (1965).
9. Dutton and Unrau, *Can. J. Chem.*, 43, 924, 1738 (1965).
10. Lindberg, *Ark. Kemi. Mineral. Geol.*, 23, A 2 (1946–1947).
11. Weatherall, *J. Chromatogr.*, 26, 251 (1967).
12. Borecký and Gasparič, *Collect. Czech. Chem. Commun.*, 25, 1287 (1960).
13. Bamberger and Landsiedl, *Monatsh. Chem.*, 21, 571 (1900).
14. Dutton, Gibney, Jensen, and Reid, *J. Chromatogr.*, 36, 152 (1968).
15. Ward, Pettijohn, Lockwood, and Coghill, *J. Am. Chem. Soc.*, 66, 541 (1944).
16. Ciamician and Silber, *Ber. Dtsch. Chem. Ges.*, 44, 1280 (1911).
17. Morell and Auernheimer, *J. Am. Chem. Soc.*, 66, 792 (1944).
18. Rubin, Lardy, and Fischer, *J. Am. Chem. Soc.*, 74, 425 (1952).
19. Bertrand, *C. R. Acad. Sci.*, 130, 1472 (1900).
20. Batt, Dickens, and Williamson, *Biochem. J.*, 77, 272 (1960).
21. Oades, *J. Chromatogr.*, 28, 246 (1967).
22. Hu, McComb, and Rendig, *Arch. Biochem. Biophys.*, 110, 350 (1965).
23. Dutton and Unrau, *J. Chromatogr.*, 20, 78 (1965).
24. Wilson and Lucas, *J. Am. Chem. Soc.*, 58, 2396 (1936).
25. Asahina and Yanagita, *Ber. Dtsch. Chem. Ges.*, 67, 799 (1934).
26. Němec, Kefurt, and Jarý, *J. Chromatogr.*, 26, 116 (1967).
27. Frèrejacque, *C. R. Acad. Sci.*, 208, 1123 (1939).
28. Onishi and Suzuki, *Agric. Biol. Chem.*, 30, 1139 (1966).
29. Richtmyer and Hudson, *J. Am. Chem. Soc.*, 73, 2249 (1951).
30. Touster and Harwell, *J. Biol. Chem.*, 230, 1031 (1958).
31. Grasshof, *J. Chromatogr.*, 14, 513 (1964).
32. Dutton, Reid, Rowe, and Rowe, *J. Chromatogr.*, 47, 195 (1970).
33. Wessely and Wang, *Monatsh. Chem.*, 72, 168 (1938).
34. Binkley and Wolfrom, *J. Am. Chem. Soc.*, 70, 2809 (1948).
35. Gregory, *J. Chromatogr.*, 36, 342 (1968).
36. Wolfrom and Kohn, *J. Am. Chem. Soc.*, 64, 1739 (1942).
37. Wells, Pittman, and Egan, *J. Biol. Chem.*, 239, 3192 (1964).
38. Horowitz and Delman, *J. Chromatogr.*, 21, 302 (1966).
39. Von Lippmann, *Ber. Dtsch. Chem. Ges.*, 60, 161 (1927).
40. Haas and Hill, *Biochem. J.*, 26, 987 (1932).
41. Jeger, Norymberski, Szpilfogel, and Prelog, *Helv. Chim. Acta*, 29, 684 (1946).
42. Richtmyer, Carr, and Hudson, *J. Am. Chem. Soc.*, 65, 1477 (1943).
43. Bertrand, *Bull. Soc. Chim. Fr. Ser. 3*, 33, 166 (1905).
44. Britton, *Biochem. J.*, 85, 402 (1962).
45. Perlin, Mazurek, Jaques, and Kavanagh, *Carbohydr. Res.*, 7, 369 (1968).
46. Braham, *J. Am. Chem. Soc.*, 41, 1707 (1919).
47. Patterson and Todd, *J. Chem. Soc.* (Lond.), p. 2876 (1929).
48. Iwate, *Chem. Zentralbl.*, 2, 177 (1929).
49. Zervas, *Ber. Dtsch. Chem. Ges.*, 63, 1689 (1930).
50. Hay, Lewis, and Smith, *J. Chromatogr.*, 11, 479 (1963).
51. Asahina, *Ber. Dtsch. Chem. Ges.*, 45, 2363 (1912).
52. Lindberg, *Acta Chem. Scand.*, 7, 1119, 1123 (1953).
53. Jones and Wall, *Nature*, 189, 746 (1961).
54. Maquenne, *Ann. Chim. Phys. Ser. 6*, 19, 5 (1890).
55. Charlson and Richtmyer, *J. Am. Chem. Soc.*, 82, 3428 (1960).
56. Buck, Foster, Richtmyer, and Zissis, *J. Chem. Soc.* (Lond.), p. 3633 (1961).
57. Onishi and Perry, *Can. J. Microbiol.*, 11, 929 (1965).
58. Bougault and Aliard, *C.R. Acad. Sci.*, 135, 796 (1902).
59. Maclay, Hann, and Hudson, *J. Org. Chem.*, 9, 293 (1944).
60. Ackerman, *Hoppe-Seyler's Z. Physiol. Chem.*, 336, 1 (1964).
61. Von Lippmann, *Ber. Dtsch. Chem. Ges.*, 34, 1159 (1901).
62. King and Jurd, *J. Chem. Soc.* (Lond.), p. 1192 (1953).
63. Bien and Ginsburg, *J. Chem. Soc.* (Lond.), p. 3189 (1958).
64. Pilouvier, *C.R. Acad. Sci.*, 241, 983 (1955).
65. Post and Anderson, *J. Am. Chem. Soc.*, 84, 478 (1962).
66. Angyal and Bender, *J. Chem. Soc.* (Lond.), p. 4718 (1961).
67. Kübler, *Arch. Pharm.*, 246, 620 (1908).
68. Dangschat and Fischer, *Naturwissenschaften*, 27, 756 (1939).
69. DeJong, *Recl. Trav. Chim. Pays-Bas*, 27, 257 (1908).
70. Kiang and Loke, *J. Chem. Soc.* (Lond.), p. 480 (1956).
71. Angyal, Gilham, and MacDonald, *J. Chem. Soc.* (Lond.), p. 1417 (1957).
72. Ballou and Anderson, *J. Am. Chem. Soc.*, 75, 648 (1953).
73. Umezawa, Okami, Hashimoto, Suhara, Hamada, and Takeuchi, *J. Antibiot. Ser. A*, 18, 101 (1965).
74. Dzhumyrko and Shinkaxenko, *Chem. Nat. Compd.* (USSR), 7, 638 (1971).
75. Foxall and Morgan, *J. Chem. Soc.* (Lond.), p. 5573 (1963).
76. Smith, *Biochem. J.*, 57, 140 (1954).
77. Tanret, *C. R. Acad. Sci.*, 145, 1196 (1907).
78. Cosgrove, *Nature*, 194, 1265 (1962).
79. Lee and Ballou, *J. Chromatogr.*, 18, 147 (1965).
80. Adhikari, Bell, and Harvey, *J. Chem. Soc.* (Lond.), p. 2829 (1962).
81. Utkin, *Chem. Nat. Compd.* (USSR), 4, 234 (1968).
82. Angyal and McHugh, *J. Chem. Soc.* (Lond.), p. 1423 (1957).
83. Lindberg, *Acta Chem. Scand.*, 9, 1093 (1955).
84. Loewus, *Carbohydr. Res.*, 3, 130 (1966).
85. Allen, *J. Am. Chem. Soc.*, 84, 3128 (1962).
86. Cosgrove and Tate, *Nature*, 200, 568 (1963).
87. Lindberg and Wickberg, *Ark. Kemi*, 13, 447 (1959).
88. Posternak and Falbriard, *Helv. Chim. Acta*, 44, 2080 (1961).
89. Kindl, Kremlicka, and Hoffman-Ostenhof, *Monatsh. Chem.*, 97, 1783 (1966).
90. Plouvier, *C. R. Acad. Sci.*, 255, 360 (1962).
91. Plouvier, *C. R. Acad. Sci.*, 241, 765 (1955).
92. Ackermann, *Ber. Dtsch. Chem. Ges.*, 54, 1938 (1921).
93. Krzeminski and Angyal, *J. Chem. Soc.* (Lond.), p. 3251 (1962).
94. Bourne, Hutson, and Weigel, *J. Chem. Soc.* (Lond.), p. 4252 (1960).
95. Maquenne, *Ann. Chim. Phys. Ser. 6*, 22, 264 (1891).
96. Anderson, Fischer, and MacDonald, *J. Am. Chem. Soc.*, 74, 1479 (1952).
97. Pease, Reider, and Elderfield, *J. Org. Chem.*, 5, 198 (1940).
98. Plouvier, *C. R. Acad. Sci.*, 243, 1913 (1956).
99. Anderson, Takeda, Angyal, and McHugh, *Arch. Biochem. Biophys.*, 78, 518 (1958).
100. DeJong, *Recl. Trav. Chim. Pays-Bas*, 25, 48 (1906).
101. Adams, Pease, and Clark, *J. Am. Chem. Soc.*, 62, 2194 (1940).
102. Haustveit and Wold, *Carbohydr. Res.*, 29, 325 (1973).
103. Bourne, Percival, and Smestad, *Carbohydr. Res.*, 22, 75 (1972).
104. Prunier, *Ann. Chim. Phys. Ser. 5*, 15, 5 (1878).
105. McCasland, Naumann, and Durham, *Carbohydr.Res.*, 4, 516 (1967).
106. Bauer and Moll, *Arch. Pharm.*, 280, 37 (1942).
107. Plouvier, *C.R. Acad. Sci.*, 253, 3047 (1961).
108. Gorter, *Annchem*, 359, 221 (1908).
109. Haslam, Turner, Sargent, and Thompson, *J. Chem. Soc.* (Lond.), p. 1493 (1971).
110. Ervig and Koenigs, *Ber. Dtsch. Chem. Ges.*, 22, 1457 (1889).
111. Weiss, Davis, and Mingioli, *J. Am. Chem. Soc.*, 75, 5572 (1953).
112. Adlersberg and Sprinson, *Biochemistry*, 3, 1855 (1964).
113. Shyluk, Youngs, and Gamborg, *J. Chromatogr.*, 26, 268 (1967).
114. Muller, *J. Chem. Soc.* (Lond.), p. 1767 (1907).
115. Posternak, *Helv. Chim. Acta*, 25, 746 (1942).
116. Ueno, Hasegawa, and Tsuchiya, *Carbohydr. Res.*, 29, 520 (1973).
117. Sherrard and Kurth, *J. Am. Chem. Soc.*, 51, 3139 (1929).
118. Eijkman, *Ber. Dtsch. Chem. Ges.*, 24, 1278 (1891).
119. Eijkman, *Recl. Trav. Chim. Pays-Bas*, 4, 32 (1885).
120. McCrindle, Overton, and Raphael, *J. Chem. Soc.* (Lond.), p. 1560 (1960).
121. Grewe, Buttner, and Burmeister, *Angew. Chem.*, 69, 61 (1957).
122. Salamon and Davis, *J. Am. Chem. Soc.*, 75, 5567 (1953).
123. Horii, Iwasa, Mizuta, and Kameda, *J. Antibiot. Ser. A*, 24, 59 (1971).
124. Power and Tutin, *J. Chem. Soc.* (Lond.), p. 624 (1904).
125. Angyal, Gorin, and Pitman, *J. Chem. Soc.* (Lond.), p. 1807 (1965).
126. Posternak and Schopfer, *Helv. Chim. Acta*, 33, 343 (1950).

127. Nakajima and Kurihara, *Ber. Dtsch. Chem. Ges.*, 94, 515 (1961).

128. Angyal, Gorin, and Pitman, *J. Chem. Soc.* (Lond.), p. 1807 (1965).

129. Stanacev and Kates, *J. Org. Chem.*, 26, 912 (1961).

130. Posternak, *Helv. Chim. Acta*, 19, 1333 (1936).

131. Magasanik and Chargaff, *J. Biol. Chem.*, 175, 929 (1948).

132. Post and Anderson, *J. Am. Chem. Soc.*, 84, 471 (1962).

133. Magasanik and Chargaff, *J. Biol. Chem.*, 174, 173 (1948).

134. Berman and Magasanik, *J. Biol. Chem.*, 241, 800 (1966).

135. Stevens, Gillis, French, and Haskell, *J. Am. Chem. Soc.*, 80, 6088 (1958).

136. Johnson, Gourlay, Tarbell, and Autrey, *J. Org. Chem.*, 28, 300 (1963).

137. Nakajima, Kurihara, Hasegawa, and Kurokawa, *Justus Liebigs Ann. Chem.*, 689, 243 (1965).

138. Bannister and Argoudelis, *J. Am. Chem. Soc.*, 85, 119 (1963).

139. Allen, *J. Am. Chem. Soc.*, 78, 5691 (1956).

140. Patrick, Williams, Waller, and Hutchings, *J. Am. Chem. Soc.*, 78, 2652 (1956).

141. Walker and Walker, *Biochim. Biophys. Acta*, 170, 219 (1968).

142. Carter, Clark, Lytle, and McCasland, *J. Biol. Chem.*, 175, 683 (1948).

143. Peck, Hoffhine, Peel, Graber, Holly, Mozingo, and Folkers, *J. Am. Chem. Soc.*, 68, 776 (1946).

144. Nakajima, Hasegawa, and Kurihara, *Justus Liebigs Ann. Chem.*, 689, 235 (1965).

145. Maeda, Murase, Mawatari, and Umezawa, *J. Antibiot. Ser. A*, 11, 73 (1958).

146. Neuss, Koch, Malloy, Day, Huckstep, Dorman, and Roberts, *Helv. Chim. Acta*, 53, 2314 (1970).

147. Kondo, Sezaki, Koika, and Akita, *J. Antibiot. Ser. A*, 18, 192 (1965).

148. Peck, Graber, Walti, Peel, Hoffhine, and Folkers, *J. Am. Chem. Soc.*, 68, 29 (1946).

149. Nakajima, Kurihara, and Hasegawa, *Ber. Dtsch. Chem. Ges.*, 95, 141 (1962).

150. Horii and Kameda, *J. chem. Soc. D Chem. Commun.*, 746, 747 (1972).

TABLE B: NATURAL ACIDS OF CARBOHYDRATE DERIVATION

Substance[a] (Synonym) Derivative (A)	Chemical Formula (B)	Melting Point °C (C)	Specific Rotation[b] $[\alpha]_D$ (D)	Reference[c] (E)	Chromatography, R Value, and Reference[d] ELC (F)	GLC (G)	PPC (H)	TLC (I)
ALDONIC ACIDS								
Glycollic acid (hydroxyacetic acid)	$C_2H_4O_3$	80	None	1,5c	48Cl (2)			75f (4)
Acetate	$C_4H_6O_4$	66–68	None	1				
Ammonium salt	$C_2H_7NO_3$	102	None	1		10.45MU (3)		
Trimethylsilyl ester ether								
D-Glyceric acid	$C_3H_6O_4$	Gum	dextro	6			85f (8)	55f (4)
Amide	$C_3H_7NO_3$	99.5–100	−63.1 (CH₃OH)	7				
Calcium salt	$CaC_6H_{10}O_8$		+10.9	8p				
Methyl ester methyl ether	$C_6H_{12}O_4$					24xyll (11)		
L-Glyceric acid	$C_3H_6O_4$	Gum	levo	1				
Calcium salt	$CaC_6H_{10}O_8$	134–135	−12 (30°)	9				
Trimethylsilyl ester ether						16glc (10)		
D-Lactic acid (2-hydroxypropionic acid, 3-deoxy-D-glyceric acid)	$C_3H_6O_3$	26–27	−2.3	1,5c	42Cl (2)			
Acetate	$C_5H_8O_4$	Oil, bp 171–172	+54.3	1				
Amide	$C_3H_7NO_2$	49–51	$[\alpha]^{18}_{Hg}$ +22.2	1				
L-Lactic acid	$C_3H_6O_3$	25–26	+3.8 (15°)	1				
Methyl ester methyl ether	$C_5H_{10}O_3$	Oil, bp 45 (22 mm)	−95.5	1				
Trimethylsilyl ester ether						10.65MU (3)		
D,L-Lactic acid	$C_3H_6O_3$	18	Racemic	1				
Acetate	$C_5H_8O_4$	57–60	Racemic	1				
Amide	$C_3H_7NO_2$	75.5	Racemic	1				
3-Hydroxypropionic acid (2-deoxyglyceric acid)	$C_3H_6O_3$	Syrup	None	1				
Methyl ester methyl ether	$C_5H_{10}O_3$	Oil, bp 142–143	None	1				
Hydroxypyruvic acid (2-keto-glyceric acid, 2-triulosonic acid)	$C_3H_4O_4$		None	12				
p-Nitrophenylhydrazone	$C_9H_9N_3O_5$	260	None	1				
Pyruvic acid (2-keto-3 deoxyglyceric acid)	$C_3H_4O_3$	13.6	None	1	50Cl (2)			
p-Nitrophenylhydrazone	$C_9H_9N_3O_4$	220	None	1				
Methyl ester	$C_3H_6O_3$	Oil, bp 134–137	None	1				
Trimethylsilyl ester				1		10.91MU (3)		
D-Arabinonic acid (arabonic acid)	$C_5H_{10}O_6$	114–116	+10.5 (c6)	5c,13	8.6[e] (14)		20glc (15)	16f (16)
Phenylhydrazide	$C_{11}H_{16}N_2O_5$	208–209	−13	17				
Tetraacetate	$C_{13}H_{18}O_{10}$	135–136	+32.5	18				
Trimethylsilyl ester ether						55glc (10)		
L-Arabinonic acid	$C_5H_{10}O_6$	118–119	−9.6→−41.7[f]	19			16f (28)	35f (32)
Amide	$C_5H_{11}NO_5$	135–136	+37.2	20				
1,4-Lactone (γ-lactone)	$C_5H_8O_5$	95–98	−71.6	21,22cp			73f (28)	40f[g] (16)
Lactone trimethylsilyl ether						97glc (10)		75f[g] (23)

417

TABLE B: NATURAL ACIDS OF CARBOHYDRATE DERIVATION (Continued)

ALDONIC ACIDS (Continued)

Substance[a] (Synonym) Derivative (A)	Chemical Formula (B)	Melting Point °C (C)	Specific Rotation[b] $[\alpha]_D$ (D)	Reference[c] (E)	Chromatography, R Value, and Reference[d]			
					ELC (F)	GLC (G)	PPC (H)	TLC (I)
Methyl ester methyl ether	$C_{10}H_{20}O_6$					295xyll (11)		
Phenylhydrazide	$C_{11}H_{16}N_2O_5$	215		24				62f (32)
Pentulosonic acid, 3-deoxy- D-*glycero*-2-(2-keto-3-deoxy-D-arabonic acid)	$C_5H_8O_5$	No constants known	No constants known	29			61f (29)	
2,4-Dinitrophenylhydrazone	$C_{11}H_{12}N_4O_8$	163		29			14f (29)	
Pentulosonic acid, D-*threo*-4- (4-*keto*-D-arabonic acid)	$C_5H_8O_6$	154–155	−10.3	27				
Brucine salt	$C_{28}H_{34}N_2O_{10}$		−29.4	27				
Pentulosonic acid, 3-deoxy- L-*glycero*-2- (2-keto-3-deoxy-L-arabonic acid)	$C_5H_8O_5$	No constants known	No constants known	28			29f (28)	
2,4-Dinitrophenylhydrazone	$C_{11}H_{10}N_4O_7$[h]	220–223d	−22.7 (c 0.3, dioxane)	30				
1,4-Lactone	$C_5H_6O_4$						82f (28)	
L-Lyxonic acid	$C_5H_{10}O_6$	114[g]	+82.7[g]	5c,25,36[g] 45[g]				
1,4-Lactone	$C_5H_8O_5$	108–110[g]	+70[g]	46p				
Phenylhydrazide	$C_{11}H_{16}N_2O_5$	163	+13.7	26cp				
D-Ribonic acid	$C_5H_{10}O_6$	112–113	−17[f]	5c,31				
Amide	$C_5H_{11}NO_5$	136–137	+17	33				
1,4-Lactone	$C_5H_8O_5$	77	+17→+8 (13 days)	34,35p				
Methyl ester methyl ether	$C_{10}H_{20}O_6$					142xyll (11)		38f (32)
D-Xylonic acid	$C_5H_{10}O_6$	Syrup	−2.9→+20.1[f]	5c,36	9.1[e] (14)			73f (23)
Amide	$C_5H_{11}NO_5$	81–82	+44.5→+23.8	37				61f (32)
Brucine salt	$C_{28}H_{36}N_2O_{10}$	170–172	−37.4	38				
1,4-Lactone	$C_5H_8O_5$	98–101	+91.8→+86.7	21		126glc (10)		79f (23)
Lactone trimethylsilyl ether						269xyll (11)		
Methyl ester methyl ether	$C_{10}H_{20}O_6$					48glc (10)		
Trimethylsilyl ester ether								
L-Xylonic acid	$C_5H_{10}O_6$	No constants known	No constants known	26				
Brucine salt	$C_{28}H_{36}N_2O_{10}$	177–178	+24.3	26				
1,4-Lactone	$C_5H_8O_5$	97	−82.2	39				
Tetraacetate	$C_{13}H_{18}O_{10}$	86–88	−4.5 (c 2, C_2H_5OH)	40				
D-Altronic acid	$C_6H_{12}O_7$	110[g]	+8	5c,8c,47[g]				
1,4-Lactone	$C_6H_{10}O_6$	150–152	+35	8p				
Phenylhydrazide	$C_{12}H_{18}N_2O_6$		+18.4[g]	42,47[g]			132galn (41)	
D-Fuconic acid	$C_6H_{12}O_6$	No constants known	No constants known	43p				
1,4-Lactone	$C_6H_{10}O_5$	104[g]		48				
Methyl ester methyl ether	$C_{11}H_{22}O_6$					438xyll (11)		

TABLE B: Natural Acids of Carbohydrate Derivation

TABLE B: NATURAL ACIDS OF CARBOHYDRATE DERIVATION (Continued)

Substance[a] (Synonym) Derivative (A)	Chemical Formula (B)	Melting Point °C (C)	Specific Rotation[b] $[\alpha]_D$ (D)	Reference[c] (E)	Chromatography, R Value, and Reference[d]			
					ELC (F)	GLC (G)	PPC (H)	TLC (I)
ALDONIC ACIDS (Continued)								
Hexulosonic acid, 3,6-dideoxy- D-*threo*-2- (2-keto-3-deoxy-D-fuconic acid)	$C_6H_{10}O_5$	No constants known	No constants known	44				78f (44)
D-Galactonic acid	$C_6H_{12}O_7$	122	−11.2 → −57.6f	49,50	8.4e (14)			38f (32)
Amide	$C_6H_{13}NO_6$	148	−13.6 → −17	21				
1,4-Lactone	$C_6H_{10}O_6$	175	+31.5	51				
Methyl ester methyl ether	$C_{12}H_{24}O_7$	110–112, 132–133	−73 → +63.7	8,21,52		1185xyll (11)	100galn (41)	78f (23)
Pentaacetate	$C_{16}H_{22}O_{12}$	131–132	+12 (CHCl₃)	53				
Phenylhydrazide	$C_{12}H_{18}N_2O_6$	203	+10.4	24				
L-Galactonic acid	$C_6H_{12}O_7$	No constants known	No constants known	5c,8c				
Amide	$C_6H_{13}NO_6$	175	−30	54				
1,4-Lactone	$C_6H_{10}O_6$	110, 134–135	+77	55,56				
Lactone trimethylsilyl ether	$C_{15}H_{22}O_{12}$	Oil		57		66myo (57)	68f (71)	
Pentaacetate		132–133	−14 (28°, CHCl₃)	54				
Hexulosonic acid, D-*lyxo*-2- (2-keto-D-galactonic acid)	$C_6H_{10}O_7$	169	−5	58				
Brucine salt	$C_{29}H_{36}N_2O_{11}$	172	−22.5 (50% C₂H₅OH)	58				
Hexulosonic acid, 3-deoxy- D-*threo*-2 (2-keto-3-deoxy-D-galactonic acid)	$C_6H_{10}O_6$		+15	59cp,60			57f (61)	
Lactone phenylhydrazone	$C_{12}H_{14}N_2O_4$	213–214	−270 (c 0.5, C₅H₅N)	60				
Phenylhydrazone phenylhydrazide	$C_{18}H_{22}N_4O_4$	204–205	+13.9 (C₅H₅N)	60				
Potassium salt	$C_6H_9KO_6$	159–163d		61				
Hexulosonic acid, D-*arabino*-5- (5-keto- L-galactonic or D-tagaturonic acid)	$C_6H_{10}O_7$	108–109	−17g	62p			62galn (41)	
Brucine salt	$C_{29}H_{36}N_2O_{11}$	148–149, 189–190dg		62,64g				
Calcium salt · 5H₂O	$CaC_{12}H_{18}O_{14} \cdot 5H_2O$		−14	63				
D-Gluconic acid	$C_6H_{12}O_7$	120–121	−6.9 → +7.3	5c,21	25Cl (2)		10glc (15)	40f (32)
Amide	$C_6H_{13}NO_6$	143–144	+31.2	20,65				
1,4-Lactone	$C_6H_{10}O_6$	133–135	+68 → +17.7	8c,21			210glc (15)	75f (23)
Lactone trimethylsilyl ether						198glc (10)		
1,5-Lactone	$C_6H_{10}O_6$	150–152	+66 → +8.8	21				
Methyl ester methyl ether	$C_{12}H_{24}O_7$					708xyll (11)		
Pentaacetate	$C_{16}H_{22}O_{12}$	110–111	+11.5 (CHCl₃)	40				
Trimethylsilyl ester ether						100glc (10)		
D-Gluconic acid, 6-*O*-(*N,N*-dimethylglycyl)- (pangamic acid, vitamin B₁₅)	$C_{10}H_{19}NO_8$	No constants known	No constants known	66				
Amide hydrochloride	$C_{10}H_{20}N_2O_7 \cdot HCl$	92–95d	+20.9 (CH₃OH)	66				
Lactone hydrochloride	$C_{10}H_{17}NO_7 \cdot HCl$	69–73	+32.3 (CH₃OH)	66				

TABLE B: NATURAL ACIDS OF CARBOHYDRATE DERIVATION (Continued)

Substance (Synonym) Derivative (A)	Chemical Formula (B)	Melting Point °C (C)	Specific Rotation[b] $[\alpha]_D$ (D)	Reference[c] (E)	Chromatography, R Value, and Reference[d]			
					ELC (F)	GLC (G)	PPC (H)	TLC (I)
ALDONIC ACIDS (Continued)								
Hexonic acid, 2-deoxy-D-*arabino*- (2-deoxy-D-gluconic acid)	$C_6H_{12}O_6$	142–144	+2	67,68cpt				
1,4-Lactone	$C_6H_{10}O_5$	93–95	+68	68,69				61f (70)
Methyl ester methyl ether	$C_{11}H_{22}O_6$	156		69		353xyll (11)		
Phenylhydrazide	$C_{12}H_{18}N_2O_5$							
Hexulosonic acid, D-*arabino*-2- (2-keto-D-gluconic acid)	$C_6H_{10}O_7$		−99.6 (dil HCl)	5c,72	10.7[e] (14)		10glc (15)	18glc (73)
Brucine salt	$C_{29}H_{36}N_2O_{11}$	179–182	−59.4 (c 0.4)	74p				
Calcium salt · $3H_2O$	$CaC_{12}H_{18}O_{11} \cdot 3H_2O$	153d	−70.8	63,76p				
Methyl ester methyl ether	$C_{11}H_{20}O_7$			72		348xyll (11)		
Phenylhydrazone phenylhydrazide	$C_{18}H_{24}N_4O_6$	121	−36.1 (H_2O, C_5H_5N)					
Hexulosonic acid, 3-deoxy-D-*erythro*-2- (2-keto-3-deoxy-D-gluconic acid)	$C_6H_{10}O_6$	119.5–120	−45.2	60,75			116galn (62)	
Calcium salt · $1/2H_2O$	$CaC_{12}H_{18}O_{12} \cdot 1/2H_2O$	229d	−29.2 (c 6)	77p				
Phenylhydrazone	$C_{12}H_{16}N_2O_5$		+168 (C_5H_5N)	60				
Hexulosonic acid, D-*xylo*-5- (5-keto-D-gluconic acid)	$C_6H_{10}O_7$	174–175d	−14.5	78,79	9.3[e] (14)		24f (80eg)	
Brucine salt	$C_{29}H_{36}N_2O_{11}$		−24	64			79f (81)	
Calcium salt · $5H_2O$	$CaC_{12}H_{18}O_{14} \cdot 3H_2O$		−11.7 (dil HCl)	79				
Methyl ester methyl ether	$C_{11}H_{20}O_7$					538xyll (11)		
Hex-2,5-diulosonic acid, D-*threo*- (2,5-diketo-D-gluconic acid)	$C_6H_8O_7$	No constants known	No constants known	82p	100kgu (88)		59glc (83)	
Calcium salt · $3H_2O$	$CaC_{12}H_{14}O_{14} \cdot 3H_2O$	156–157d	−51 ± 5	82				
Bis(2,4-dinitrophenylhydrazone)	$C_{18}H_{16}N_4O_{13}$		+57.2 (C_5H_5N)	82				
L-Gluconic acid	$C_6H_{12}O_7$	120–121[g]	−6.9→+7.3[g]	5c,21,g8	25Cl[g] (2)		10glc[g] (15)	40f[g] (32)
Barium salt	$BaC_{12}H_{22}O_{14}$		−6.4 (c 8)	84				
Brucine salt	$C_{29}H_{38}N_2O_{11}$	181–182	−25.4	85				
Phenylhydrazide	$C_{12}H_{18}N_2O_6$	203–204	−11.7	85,86				
L-Gulonic acid	$C_6H_{12}O_7$	Syrup	0	5c,87	47kgu (88)		30f (89a)	31f[g] (32)
1,4-Lactone	$C_6H_{10}O_6$	183–185	+55	56			84galn (41)	53f[g] (32)
Lactone trimethylsilyl ether						70myo (57)		
Methyl ester methyl ether	$C_{12}H_{24}O_7$					632xyll (11)		
Hexulosonic acid, L-*xylo*-2- (2-keto-L-gulonic acid)	$C_6H_{10}O_7$	170–171	−48.8	63	100kgu (88)			35glc (73)
Brucine salt · H_2O	$C_{29}H_{36}N_2O_{11} \cdot H_2O$	114		90				
Sodium salt	$C_6H_9NaO_7$	145	−24.4	90				
Trimethylsilyl ester ether						164glc (91)		
Hexulosono-1,4-lactone, L-*xylo*-2- (2-keto-L-gulono-γ-lactone, L-ascorbic acid)[i]	$C_6H_8O_6$	190d	+24.7	96c	25Cl (2)		90f (89b)	57f (32)
Bis(phenylhydrazone)	$C_{18}H_{18}N_4O_4$	187		1				
Trimethylsily. ester ether						183glc		
Hexulosono-1,4-lactone 2-sulfate, L-*xylo*-2- (ascorbic acid sulfate)	$C_6H_8O_9S$		+98.5 (pH 8.6)	96c			40f (96)	39f (97p)
Ammonium salt	$C_6H_{10}NO_9S$		+202.8	96				

TABLE B: NATURAL ACIDS OF CARBOHYDRATE DERIVATION (Continued)

ALDONIC ACIDS (Continued)

Substance[a] (Synonym) Derivative (A)	Chemical Formula (B)	Melting Point °C (C)	Specific Rotation[b] $[\alpha]_D$ (D)	Reference[c] (E)	Chromatography, R Value, and Reference[d]			
					ELC (F)	GLC (G)	PPC (H)	TLC (I)
Hexulosonic acid, L-xylo-3- (3-keto-L-gulonic acid)	C6H10O7	No constants known	No constants known	92				
Hexulosonic acid, D-lyxo-5- (5-keto-L-gulonic acid or D-fructuronic acid)	C6H10O7	No constants known	No constants known	62,71	52kgu (88)		62f (41)	
Brucine salt	C29H36N2O11	195–197	−15.5	62,64			78f (81)	
Potassium salt	C6H9KO7	160–165	+11	93				
Hex-2,3-diulosonic acid, L-threo- (2,3-diketo-L-gulonic acid)	C6H8O7	No constants known	No constants known	94,95				
Bis(2,4-dinitrophenylhydrazone)	C18H16N4O13	281		95				
Hexulosonic acid, 4-deoxy-D-threo-5 (4-deoxy-5-keto-D-idonic acid)	C6H10O6	No constants known	No constants known	98,99			40f (98)	
Phenylosazone	C18H22N4O4	113d		98				
Sodium salt	C6H9NaO6	67–68	+5.5	98				
L-Idonic acid	C6H12O7	No constants known	No constants known	5c,100	47kgu (88)			
Brucine salt	C29H38N2O11	190–192	−17	8				
1,4-Lactone	C6H10O6	Syrup	+50,+4.5	8,101			111f (41)	
Phenylhydrazide	C12H18N2O6	115–117	+12.5	101				
Sodium salt	C6H11NaO7	179	+8	102				
D-Mannonic acid	C6H12O7	No constants known	−15.6	5c,103	47kgu (88)		4f (104)	
Brucine salt	C29H38N2O11	203	−27.8	105				
1,4-Lactone	C6H10O6	151–152	+51.5	8,21			112f (41)	62f8 (23)
1,5-Lactone	C6H10O6	158–160	+114→+30.3	21				
Methyl ester methyl ether	C12H24O7	68–70	+23 (CHCl3)	106				
Pentaacetate	C16H22O12					779xyll (11)		
Phenylhydrazide	C12H18N2O6	212–214	+15.8	42,107				
Hex-2,5-diulosonic acid, 3-deoxy-D-glycero- (3-deoxy-2,5-diketo-D-mannonic acid)	C6H8O6	No constants known	No constants known	108			17k (108)	
Hexulosonic acid, 2,3,6-trideoxy-D-glycero-4-(4-keto-2,3,6-trideoxy-D-mannonic acid, 5-hydroxy-4-ketohexanoic acid)	C6H10O4		−17.9	109c			82f (110c)	
L-Mannonic acid, 6-deoxy-(L-rhamnonic acid)	C6H12O6	141–142	+4→−32.7	5c, 36, 86				
Amide	C6H13NO5	149–151	+27.5	111				
1,4-Lactone	C6H10O5	172–182	−39.2,−51	8,21				
1,5-Lactone	C6H10O5		−100→−35.1	21				
Methyl ester methyl ether	C11H20O5					361xyll (11)		69f (23)
Heptulosonic acid, 3-deoxy-D-arabino-2-	C7H12O7	No constants known	No constants known	112c			23f (113)	
Ammonium salt	C7H15NO7	96	+42.4	114				
1,4-Lactone	C7H10O6	167	−5.8	113				
1,5-Lactone	C7H10O6		+33	115			39f (113)	

TABLE B: NATURAL ACIDS OF CARBOHYDRATE DERIVATION (Continued)

Substance[a] (Synonym) Derivative (A)	Chemical Formula (B)	Melting Point °C (C)	Specific Rotation[b] $[\alpha]_D$ (D)	Reference[c] (E)	ELC (F)	GLC (G)	PPC (H)	TLC (I)
ALDONIC ACIDS (Continued)								
Methyl ester glycoside	$C_9H_{14}O_6$	148	+78.2 (CH$_3$OH)	115				
7-Phosphate	$C_7H_{13}O_{10}P$						49f (112)	
Octulosonic acid, 3-deoxy- D-*manno*-2-	$C_8H_{14}O_8$	No constants known	No constants known	116c	100kdo (117)		52kdh (116)	
Ammonium salt · H$_2$O	$C_8H_{17}NO_8 \cdot H_2O$	125–126	+41.3	118c			25ara (116)	
1,4-Lactone	$C_8H_{12}O_7$	192–194	+31.8	118c				
Pentaacetate	$C_{18}H_{24}O_{13}$	98–103		118c				63f (118)
8-Phosphate	$C_8H_{15}O_{11}P$			118c			51f (119)	
URONIC ACIDS								
Glyoxylic acid	$C_2H_2O_3$	98	None	1,5c	57Cl (2)			30f (16)
Methyl ester	$C_3H_4O_3$	53	None	1				
Phenylhydrazone Trimethylsilyl ester	$C_8H_8N_2O_2$	144,137d	None	1				
Trimethylsilyl ester								
Malonic semi-aldehyde (2-deoxy-glyceruronic acid, formylacetic acid)	$C_3H_4O_3$	No constants known	No constants known	8		12.65MU (3)		
Semicarbazone	$C_4H_7N_3O_3$	116d		1				
Glyceruronic acid (tartonic semialdehyde)	$C_3H_4O_4$	No constants known	No constants known	8				
D-Lyxuronic acid	$C_5H_8O_6$	No constants known	No constants known	120				
Calcium salt · 2H$_2$O	$CaC_{10}H_{14}O_{12} \cdot 2H_2O$		−23→−53	120				
Methyl ester	$C_6H_{10}O_6$	140g	−37.7→−23g	121				
Phenylosazone salt	$C_{23}H_{28}N_6O_4$	164d		122				
α- D-Galacturonic acid · H$_2$O	$C_6H_{10}O_7 \cdot H_2O$	159–160 (110–115s)	+97.9→+50.9	123	7.9e (14)	85f (41)	32f (32)	
p-Bromophenylhydrazone salt	$Br_2C_{18}H_{24}N_4O_6$	145–146	+9 ± 2 (c 0.7, CH$_3$OH)	124				
2,3,4-Tri-O-methyl ether	$C_{12}H_{22}O_7$	160	+27→	123			67tmg (138a)	
β- D-Galacturonic acid	$C_6H_{10}O_7$	160	+27→+55.6	123			18glc (153.141c)	
p-Bromophenylhydrazone	$Br_2C_{12}H_{15}BrN_2O_6$	150–151	+11.5±2 (CH$_3$OH)	124				
Brucine salt	$C_{29}H_{36}N_2O_{11}$	180	−7.7	123				
Methyl ester β-methyl pyranoside	$C_8H_{14}O_7$	194–196	−49 (CH$_3$OH)	125pt				
β- D-Glucuronic acid	$C_6H_{10}O_7$	156	+11.7→ + 36.3	126,127	9.3e (14)		46f (41)	13f (4)
Brucine salt · H$_2$O	$C_{29}H_{36}N_2O_{11} \cdot H_2O$	156–157	−15.1	128				
Phenylhydrazone phenylhydrazide phenylhydrazide	$C_{18}H_{22}N_4O_5$	182		129				
2,3,4-Tri-O-methyl ether	$C_{12}H_{22}O_7$					372tmg (156)	84tmg (138a)	
Trimethylsilyl ether						7 (154)		

TABLE B: NATURAL ACIDS OF CARBOHYDRATE DERIVATION (Continued)

Substance[a] (Synonym) Derivative	Chemical Formula	Melting Point °C	Specific Rotation[b] [α]_D	Reference[c]	Chromatography, R Value, and Reference[d]			
(A)	(B)	(C)	(D)	(E)	ELC (F)	GLC (G)	PPC (H)	TLC (I)
URONIC ACIDS (Continued)								
D-Glucurono-3,6-lactone	$C_6H_8O_6$	163–165, 180	+18.6	130,131			125f (41)	58f (32)
D-Glucuronic acid, 3-O-methyl-	$C_7H_{12}O_7$	Syrup	+6	132,133				
p-Bromophenylosazone salt	$Br_3C_{25}H_{39}N_6O_5$	157	−104→−14	134				
Methyl ester α-methyl pyranoside	$C_9H_{16}O_7$	88.5–89	+150 (CH_3OH)	135ct				
D-Glucuronic acid, 4-O-methyl-	$C_7H_{12}O_7$	Syrup	+48,+82	136p,137p	9.2e (14)		145galu (138b)	
Amide α-methyl-pyranoside	$C_8H_{15}O_6$	232	+143 (c0.7)	139				
Methyl ester	$C_8H_{14}O_7$	123–124	+41					
Methyl ester α-methyl pyranoside	$C_9H_{16}O_7$	203–216	+145.5 (CH_3OH, 15°)	140				
L-Glucuronic acid	$C_6H_{10}O_7$	174–176	−33.3 (c 0.5)	141c			40asc (141a) 310glcu (141b)	
3,6-Lactone	$C_6H_8O_6$		−19	142pt				
Lactone 1,2,5-triacetate	$C_{12}H_{14}O_9$	195–196	−85.4 ($CHCl_3$)	142				
L-Guluronic acid	$C_6H_{10}O_7$	Syrup		143,147cp	81manu (144)	28glc (153,141c)	28glc (153,141c)	53glc (145)
3,6-Lactone	$C_6H_8O_6$	141–142	+81.7	146ct			377glcu (141b)	150glc (145)
Hexulosuronic acid, 4-deoxy-L-erythro-5-(4-deoxy-5-keto-D-mannuronic acid)	$C_6H_8O_6$	No constants known	No constants known	148c			26l (148)	
Hexulosuronic acid, 4-deoxy-L-threo-5-(4-deoxy-5-keto-L-iduronic acid)	$C_6H_8O_6$	No constants known	No constants known	108c			26l (1480)	
Methyl ester β-methyl glucoside	$C_8H_{12}O_6$		+192.5 (CH_3OH)	125			76f (149)	30f (125)
Methyl ester α-methyl glycoside	$C_8H_{12}O_6$		−67 (CH_3OH)	125				
L-Iduronic acid	$C_6H_{10}O_7$	131–133	+37→+33	150,151	79glcu (152)		65glc (131,141c)	
3,6-Lactone	$C_6H_8O_6$	Syrup	+30 (18°)	131			85glc (131,141c)	
Sodium salt	$C_6H_9NaO_7$						264glcu (141c)	
Trimethylsilyl ether						5 (154)		
ALDARIC ACIDS								
α-D-Mannuronic acid · H_2O	$C_6H_{10}O_7 \cdot H_2O$	120–130 (110s)	+16→−6.1 (c 6.8)	155	100manu (144)		22f (80)	36f (32)
3,6-Lactone	$C_6H_8O_6$	140–141	+89.3	155				
β-D-Mannuronic acid	$C_6H_{10}O_7$	165–167	−47.9→−23.9	155			52f (153)	53f (32)
p-Bromophenylhydrazone salt	$Br_2C_{18}H_{24}N_4O_6$	143–144d	+48.5 ± 1(CH_3OH)	124				
Oxalic acid	$C_2H_2O_4$	189–190	None	1	70Cl (2)			88f (4)
Dihydrate	$C_2H_2O_4 \cdot 2H_2O$	102, 150–160s	None	1				
Diamide	$C_2H_4N_2O_2$	320d	None	1				

TABLE B: NATURAL ACIDS OF CARBOHYDRATE DERIVATION (Continued)

ALDARIC ACIDS (Continued)

Substance[a] (Synonym) Derivative (A)	Chemical Formula (B)	Melting Point °C (C)	Specific Rotation[b] $[\alpha]_D$ (D)	Reference[c] (E)	Chromatography, R Value, and Reference[d]			
					ELC (F)	GLC (G)	PPC (H)	TLC (I)
Dimethyl ester	$C_4H_6O_4$	54	None	1				
Trimethylsilyl ester						11.14MU (3)		
Malonic acid (deoxytartonic acid)	$C_3H_4O_4$	135–136	None	1	63Cl (2)			
Diamide	$C_3H_6N_2O_2$	170	None	1				
Dimethyl ester	$C_5H_8O_4$	Bp 181	None	1				
Trimethylsilyl ester						12.MU (3)		
Tartronic acid	$C_3H_4O_5$	141–142	None (meso)	1	57Cl (2)			73f (4)
Diamide	$C_3H_6N_2O_3$	198	None	1				
Dimethyl ester	$C_5H_8O_5$	44–45	None	1				
Trimethylsilyl ester ether						28glc (10)		
D-Threaric acid (l-tartaric)	$C_4H_6O_6$	170	−15	157	59Cl (2)			58f (4)
Trimethylsilyl ester ether						16.8MU (3)		
L-Threaric acid (d-tartaric)	$C_4H_6O_6$	170	+15 (15°)	158,159				
Diamide	$C_4H_8N_2O_4$	195	+106.5	160				
Dimethyl ester	$C_6H_{10}O_6$	48, 61.5	+2.7	161–163				
L-Malic acid (hydroxysuccinic acid, deoxytartaric acid)	$C_4H_6O_5$	100	−2.3 (c 8.4)	164,165	57Cl (2)		56f (166a)	60f (166b)
Acetate	$C_4H_8O_6$	132	−37.9	167				
Diamide	$C_4H_8N_2O_3$	156–157	+6	168				
Dimethyl ester	$C_6H_{10}O_5$							
Trimethylsilyl ester ether						15MU (3)		
Oxaloacetic acid (hydroxymaleic acid, ketosuccinic acid)	$C_4H_4O_5$	No constants known	No constants known		48Cl (2)			
Diamide	$C_4H_6N_2O_3$	124d		1				
Dimethyl ester	$C_6H_8O_5$	77		1				
Trimethylsilyl ester ether						15.65MU (3)		
Allaric acid (allo-mucic acid)	$C_6H_{10}O_8$	188–192d		169				
D-Galactaric acid (mucic acid)	$C_6H_{10}O_8$	215d	None (meso)	170	200pa (171cp)		90mal (170)	
Dimethyl ester	$C_8H_{14}O_8$	184–186	None	171				
Bis(phenylhydrazide)	$C_{18}H_{22}N_4O_6$	242	None	170				
D-Glucaric acid (saccharic acid)	$C_6H_{10}O_8$	125–126	+6.9→+20.6	172	10.8e (14)		2f (89b)	59f (4)
Diamide	$C_6H_{12}N_2O_6$	172–173	+13.3	20				
1,4-Lactone · H_2O	$C_6H_8O_7 \cdot H_2O$	90–95		173			23f (173)	43f (32)
6,3-Lactone	$C_6H_8O_7$	143–145		173			30f (173)	85f (32)
Bis(phenylhydrazide)	$C_{18}H_{22}N_4O_6$	209–210		173				
D-Mannaric acid (D-manno-saccharic acid)	$C_6H_{10}O_8$	128.5	+3.5→+48.7	174				
Diamide	$C_6H_{12}N_2O_6$	188–189.5	−24.4	20				
Bis(phenylhydrazide)	$C_{18}H_{22}N_4O_6$	214–216d		175				

TABLE B: NATURAL ACIDS OF CARBOHYDRATE DERIVATION (Continued)

AMINO SUGAR ACIDS

Substance (Synonym) Derivative (A)	Chemical Formula (B)	Melting Point °C (C)	Specific Rotation[b] $[\alpha]_D$ (D)	Reference[c] (E)	Chromatography, R Value, and Reference[d]			
					ELC (F)	GLC (G)	PPC (H)	TLC (I)
Glycine (aminoacetic acid, amino deoxyglycollic acid)	$C_2H_5NO_2$	233–236d	None	176	150ala (177)		32f (178)	22f (179)
Amide	$C_2H_6N_2O$	65–67	None	1				
Hydrochloride	$C_2H_5NO_2 \cdot HCl$	185	None	1				
Methyl ester	$C_3H_7NO_2$	Bp 54 (50 mm)	None	1				
N,N-Trimethylsilyl amine ester						13.18MU (3)		
Sarcosine (N-methylaminoacetic acid)	$C_3H_7NO_2$	212–213d	None	1				
Hydrochloride	$C_3H_7NO_2 \cdot HCl$	168–170	None	1				
N-Trimethylsilyl amine ester						11.43MU (3)		
L-Serine (2-amino-2-deoxy-L-glyceric acid, 2-amino-3-hydroxypropionic acid)	$C_3H_7NO_3$	228d[g]	−7.3 (c 5.5, 26°)	1,180p	215ala (177)		28f (178)	47f (179)
Methyl ester hydrochloride	$C_4H_9NO_3 \cdot HCl$	167		1				
N-Trimethylsilyl amine ester ether						13.8MU (3)		
L-Alanine (2-amino-2,3-di-deoxy-L-glyceric acid, deoxy-L-serine, 2-aminopropionic acid)	$C_3H_7NO_2$	297d	+2.7	1	100ala (177)		49f (178)	26f (179)
Amide	$C_3H_8N_2O$	72		1				
Hydrochloride	$C_3H_7NO_2 \cdot HCl$	204	+10.4	1				
N-Trimethylsilyl amine ester						11.05MU (3)		
Acetoacetic acid, 2-amino-	$C_4H_7NO_3$	No constants known	No constants known	1				
Ethyl ester hydrochloride	$C_6H_{11}NO_3 \cdot HCl$	95d		1				
L-Xylonic acid, 2-amino-2-deoxy (polyoxamic acid)	$C_5H_{11}NO_5$	171–173d	+2.8	181				
N-Acetyl 1,4-lactone	$C_7H_{11}NO_5$	150–152		181				
L-Xylonic acid, 2-amino-2-deoxy-5-O-carbamoyl-	$C_6H_{12}N_2O_6$	226–232d	+1.3	181				
L-Xylonic acid, 2-amino-2,3-dideoxy-	$C_5H_{11}NO_4$	Syrup	+11	181				
L-Xylonic acid, 2-aminocarbamoyl-	$C_6H_{12}N_2O_5$	215–216d	+5.8	181				
N-Acetyl 1,4-lactone	$C_8H_{12}N_2O_5$	181–191		181				
D-Galactonic acid, 2-amino-2-deoxy- (D-galactosaminic acid, chondrosaminic acid)	$C_6H_{13}NO_6$	198–203d	−5 (c 0.6)	60				
N-Acetyl acid · H_2O	$C_8H_{15}NO_7 \cdot H_2O$	102–103	−33.8→ −16 (3 days)	182pt				
N-Acetyl 1, 4-lactone	$C_8H_{13}NO_6$	165		183				
D-Alluronic acid, 3-acetamido-3-deoxy-	$C_8H_{13}NO_7$	182–183		184				
3-Amino-3-deoxy- D-allose-hydrochloride	$C_6H_{13}NO_5 \cdot HCl$	157–160d	+25 (c 0.7)	199				
D-Alluronic acid, 5-amino-5-deoxy-	$C_6H_{11}NO_6$	No constants known	No constants known	181				
D-Galacturonic acid, 2-amino-2-deoxy- (D-galactosaminuronic acid)	$C_6H_{11}NO_6$	160d	+84 (pH 2, HCl)	186p	68gNUA (192a)		46f (185)	
α- D-Galactosamine · HCl	$C_6H_{13}NO_5 \cdot HCl$	185	+121→+80	187			69F (194)	

TABLE B: NATURAL ACIDS OF CARBOHYDRATE DERIVATION (Continued)

Substance (Synonym) Derivative (A)	Chemical Formula (B)	Melting Point °C (C)	Specific Rotation $[\alpha]_D$ (D)	Reference [c] (E)	Chromatography, R Value, and Reference [d]			
					ELC (F)	GLC (G)	PPC (H)	TLC (I)
AMINO SUGAR ACIDS (Continued)								
β-D-Galactosamine · HCl	$C_6H_{13}NO_5 \cdot HCl$	187	+44→	187				
D-Glucuronic acid, 2-amino-2-deoxy- (D-glucosaminuronic acid)	$C_6H_{11}NO_6$	120–172d	+55	188,189	7.2^m (188cp)		57F (185)	
α-D-Glucosamine	$C_6H_{13}NO_5$	88	+100→+47	190	9^m (188cp)		76f (194)	
β-D-Glucosamine	$C_6H_{13}NO_5$	110–111	+28→+47	190			90glcN (189)	
Methyl α-glycoside	$C_7H_{13}NO_6$	203–207 (196s)	+126.3	189			38f (191)	
N-Acetyl furanurono-1,4-lactone	$C_8H_{11}NO_6$	177–178	+43.6	191				
D-Guluronic acid, 2-amino-2-deoxy- (D-gulosaminuronic acid)	$C_6H_{11}NO_6$	No constants known	No constants known	193c	71gNUA (193a)		43glcN (193b)	
D-Gulosamine · HCl	$C_6H_{13}NO_5 \cdot HCl$	150–170d	+34→−19	200,201			77f (194)	
Hexuronic acid, 2-amino-2-deoxy-	$C_6H_{11}NO_6$	150–170d	+11.5	192c	40gNUA (192a)		27glcN (192b)	
N-Trimethylsilyl amine ester ether						128eic (192c)		
Hex-2-enuronic acid, 4-amino-2,3,4-trideoxy-D-*erythro*	$C_6H_9NO_4$	No constants known	No constants known	195				
Methyl α-pyranoside	$C_7H_{11}NO_4$	>270d	+30.5 (c 0.3)	196				
N-Acetyl methyl α-pyranoside methyl ester	$C_{10}H_{15}NO_5$	145–146	+87 (CHCl₃)	196				
D-Mannuronic acid, 2-amino-2-deoxy- (D-mannosaminuronic acid)	$C_6H_{11}NO_6$	92–94	−9.9 (c 0.6)	197,198	115gNUA (193a)		49glcN (193b)	
D-Mannosamine hydrochloride	$C_6H_{13}NO_5 \cdot HCl$	178–180d	−3	202			80f (194)	
Destomic acid	$C_7H_{15}NO_7$	207–209d	+1.9 −12.1→−30.6 (2 N HCl)	211 211				
Methyl ester · HCl	$C_8H_{17}NO_7 \cdot HCl$	150–151d		211				
Muramic acid (2-amino-3-O-(D-1-carboxyethyl)-2-deoxy-D-glucose)	$C_9H_{17}NO_7$	155	+165→+123 (3 h)	203cp,204cp	13glcN (210a)		76f (194)	
N-Acetylmuramic acid	$C_{11}H_{19}NO_8$	122–124	+59→+39 (6 h)	205	194gNAc (210b)		100glcN (205)	
N-Acetyl trimethylsilyl ether glycoside						152myot (206a)		
N-Acetyl muramic acid 6-acetate · 1/2 H₂O	$C_{13}H_{21}NO_9 \cdot 1/2\ H_2O$	176	+56 (c 0.6)	206,207			245gNAc (206b)	
N-Acetyl 6-acetate trimethylsilyl ether glycoside						175myot (206a)		
N-Glycolyl muramic acid · H₂O	$C_{11}H_{19}NO_9 \cdot H_2O$		+56 (50% C₂H₅OH)	208			130gNac (208a)	62gNAc (208b)
manno-Muramic acid [2-amino-3-O-(D-1-carboxyethyl)-2-deoxy-D-mannose]	$C_9H_{17}NO_7$		+21 (c 0.6, 79% C₂H₅OH)	209c	13glcN (210a)		180glcN (209a)	25f (209b)

TABLE B: NATURAL ACIDS OF CARBOHYDRATE DERIVATION (Continued)

Substance[a] (Synonym) Derivative	Chemical Formula	Melting Point °C	Specific Rotation[b] $[\alpha]_D$	Reference[c]	Chromatography, R Value, and Reference[d]			
					ELC	GLC	PPC	TLC
(A)	(B)	(C)	(D)	(E)	(F)	(G)	(H)	(I)
AMINO SUGAR ACIDS (Continued)								
N-Acetyl *manno*-muramic acid	$C_{11}H_{19}NO_8$	No constants known	No constants known		189gNAc (210b)		160gNAC (209a)	
N-Acetyl trimethylsilyl ether glycoside						19.6 (209c)		
Nonulosonic acid, 5-acetamino-3,5-dideoxy- D-*glycero*- D-*galacto*- (*N*-acetylneuraminic acid, gynaminic acid, lactaminic acid, sialic acid)	$C_{11}H_{19}NO_9$[n]	183–186d	−31.7	212,213			14f (212)	17f (214)
Quinoxaline deriv.	$C_{17}H_{23}N_3O_7$	204–205	−100 (*c* 0.3, 1:1 H_2O, CH_3SOCH_3)	215				
Methyl glycoside tetratrifluoroacetate						80–82 (216)		
N-Acetyl neuraminic acid 4-acetate	$C_{13}H_{21}NO_{10}$	200d	−62±1	217				
N-Acetyl neuraminic acid 7-acetate · H_2O	$C_{13}H_{21}NO_{10} \cdot H_2O$	138–140d	+6.2±2	217			233nana (217)	
N-Acetyl neuraminic acid 7,8(9)-diacetate · CH_3OH	$C_{15}H_{23}NO_{11} \cdot CH_3OH$	130–131d	+9.2±2	217			466nana (217)	
N-Glycoly lneuraminic acid	$C_{11}H_{19}NO_{10}$	189–191d	−33.6	218c, 220ep			30f (218)	47nana (219c)
Methyl glycoside tetratrifluoroacetate						86–88 (216)		
N-Acetoglycolyl-4-*C* methyl-4,9-dideoxyneuraminic acid	$C_{14}H_{23}NO_9$	No constants known	No constants known					144nana (219c)
N-Glycolyl sialic acid 4-acetate	$C_{13}H_{21}NO_{11}$	No constants known	No constants known				123nana (221ct)	
N-Glycolyl-8-*O*-methyl-neuraminic acid	$C_{12}H_{21}NO_{10}$	No constants known	No constants known				25–35f (222)	
Hf-Neuraminic acid	Unknown	No constants known	No constants known				31f (223cegt)	

Compiled by George G. Maher.

[a] In order of increasing carbon chain length in the parent compounds grouped in the classes-aldonic, uronic, aldaric, and amino sugar acids.
[b] $[\alpha]_D$ for 1–5 g solute, *c*, per 100 ml aqueous solution at 20–25°C, unless otherwise given.
[c] References for melting point and specific rotation data. Letter indicates the reference also has chromatographic data according to: c = column, e = electrophoresis, g = gas, p = paper, and t = thin-layer.
[d] R value times 100, given relative to that of the compound indicated by abbreviation: f = solvent front, ala = alanine, ara = arabinose, asa = ascorbic acid, Cl = chloride ion, eic = eicosane, galn = galactono-1,4-lactone, galU = galacturonic acid, glc = glucose, glcN = glucosamine, glcU = glucuronic acid, gNAc = *N*-acetyl-glucosamine, gNUA = glucosaminuronic acid, kdh = 3-deoxy-*erythro*-hexulosonic acid, kdo = 3-deoxy-*manno*-octulosonic acid, kgu = 2-keto-gulonic acid, mal = malonic acid, manU = mannuronic acid, myo = *myo*-inositol, myot = *myo*-inositol trimethylsilyl ether, MU = methylene standard hydrocarbon units, nana = *N*-acetyl-neuraminic acid, pa = picric acid, rha = rhamnose, tmg = 2,3,4,6-tetra-*O*-methyl glucose, and xyll = xylitol pentamethylether. Under gas chromatography (column Glc or G) numbers without code indications signify retention time in minutes. The conditions of the chromatography are correlated with the reference given in parentheses and are found in Table E.
[e] Value is in cm/h.
[f] Equilibrates with the lactone.
[g] Data given are for the enanthiomorphic isomer.
[h] The analytical elemental analysis indicates the compound is an anhydride.
[i] The enol form is L-ascorbic acid.
[j] Reference 99 terms this compound 2-deoxy-5-keto-D-gluconic acid; neither name nor structure seem definite.
[k] Value is in cm/9 h.
[l] Value is in cm/24 h.
[m] Value is in cm/1.5 h.
[n] Some workers relate the formula $C_{12}H_{21}NO_{10}$ whose elemental analysis is little different from that of $C_{11}H_{19}NO_9$.

Rererences

1. Pollock and Stevens, *Dictionary of Organic Compounds,* Oxford University Press, New York, 1965.
2. Gross, *Chem. Ind.,* p. 1219 (1959).
3. Butts, *Anal. Biochem.,* 46, 187 (1972).
4. Baraldi, *J. Chromatogr.,* 42, 125 (1969).
5. Carlsson and Samuelson, *Anal. Chim. Acta,* 49, 248 (1970).
6. Frankland and McGregor, *J. Chem. Soc.* (Lond.), p.513 (1893).
7. Frankland, Wharton, and Aston, *J. Chem. Soc.* (Lond.), p. 269 (1901).
8. Isherwood, Chen, and Mapson, *Biochem. J.,* 56, 1–15 (1954).
9. Wolfrom and DeWalt, *J. Am. Chem. Soc.,* 70, 3148 (1948).
10. Verhaar and de Wilt, *J. Chromatogr.,* 41, 168 (1969).
11. Whyte, *J. Chromatogr.,* 87, 163 (1973).
12. Hough and Jones, *J. Chem. Soc.* (Lond.), p. 4052 (1952).
13. Robbins and Upson, *J. Am. Chem. Soc.,* 62, 1074 (1940).
14. Theander, *Sven. Kem. Tidskr.,* 70, 393 (1958).
15. Bourne, Hutson, and Weigel, *J. Chem. Soc.* (Lond.), p. 5153 (1960).
16. Wolfrom, Patin, and Lederkremer, *J. Chromatogr.,* 17, 488 (1965).
17. Hardegger, Kreiss, and El Khadem, *Helv. Chim. Acta,* 35, 618 (1952).
18. Robbins and Upson, *J. Am. Chem. Soc.,* 62, 1074 (1940).
19. Rehorst, *Ber. Dtsch. Chem. Ges.,* 63, 2280 (1930).
20. Hudson and Komatsu, *J. Am. Chem. Soc.,* 41, 1141 (1919).
21. Isbell and Frush, *J. Res. Nat. Bur. Stand.,* 11, 649 (1933).
22. Assarson, Lindberg, and Vorbrueygen, *Acta Chem. Scand.,* 13, 1395 (1959).
23. Němec, Kefurt, and Jarý, *J. Chromatogr.,* 26, 116 (1967).
24. Bates, *Polarimetry, Saccharimetry and the Sugars,* Nat. Bur. Stand. Circ. C440, U.S. Govt. Print. Off., Washington, D.C., 1942, 790.
25. Gardner and Wenis, *J. Am. Chem. Soc.,* 73, 1855 (1951).
26. Kanfer, Ashwell, and Burns, *J. Biol. Chem.,* 235, 2518 (1960).
27. Liebster, Kulhanek, and Tadra, *Chem. Listy,* 47, 1075 (1953).
28. Weimberg, *J. Biol. Chem.,* 234, 727 (1959).
29. Palleroni and Doudoroff, *J. Biol. Chem.,* 223, 499 (1956).
30. Kurata and Sakurai, *Agric. Biol. Chem.,* 32, 1250 (1968).
31. Ladenberg, Tishler, Wellmann, and Babson, *J. Am. Chem. Soc.,* 66, 1217 (1944).
32. Hay, Lewis, and Smith, *J. Chromatogr.,* 11, 479 (1963).
33. Wolfrom, Bennett, and Crum, *J. Am. Chem. Soc.,* 80, 944 (1958).
34. Steiger, *Helv. Chim. Acta,* 19, 189 (1936).
35. Hough, Jones, and Mitchell, *Can. J. Chem.,* 36, 1720 (1958).
36. Rehorst, *Justus Liebigs Ann. Chem.,* 503, 143, 154 (1933).
37. Weerman, *Recl. Trav. Chim. Pays-Bas,* 37, 15, 40 (1917).
38. Menzinsky, *Ber. Dtsch. Chem. Ges.,* 68, 822 (1935).
39. Heyns and Stein, *Justus Liebigs Ann. Chem.,* 558, 194 (1947).
40. Major and Cook, *J. Am. Chem. Soc.,* 58, 2474, 2477 (1936).
41. Hickman and Ashwell, *J. Biol. Chem.,* 241, 1424 (1966).
42. Hickman and Ashwell, *J. Biol. Chem.,* 235, 1566 (1960).
43. Dahms and Anderson, *J. Biol. Chem.,* 247, 2222, 2228 (1972).
44. Dahms and Anderson, *J. Biol. Chem.,* 247, 2233 (1972).
45. Isbell, *J. Res. Nat. Bur. Stand.,* 29, 227 (1942).
46. Gorin and Perlin, *Can. J. Chem.,* 34, 693 (1956).
47. Humoller, McManus, and Austin, *J. Am. Chem. Soc.,* 58, 2479 (1936).
48. Mortensson-Egnund, Schöyen, Howe, Lee, and Harboe, *J. Bacteriol.,* 98, 924 (1969).
49. Kiliani, *Ber. Dtsch. Chem. Ges.,* 55, 75 (1922).
50. Pryde, *J. Chem. Soc.* (Lond.), p. 1808 (1923).
51. Glattfeld and MacMillan, *J. Am. Chem. Soc.,* 56, 2481 (1934).
52. Levene and Meyer, *J. Biol. Chem.,* 46, 307 (1921).
53. Hurd and Sowden, *J. Am. Chem. Soc.,* 60, 235 (1938).
54. Wolfrom, Berkebile, and Thompson, *J. Am. Chem. Soc.,* 71, 2360 (1949).
55. Fukunaga and Kubata, *Bull. Chem. Soc. Jap.,* 13, 272 (1938).
56. Wolfrom and Anno, *J. Am. Chem. Soc.,* 74, 5583 (1952).
57. Loewus, *Carbohydr. Res.,* 3, 130 (1966).
58. Ettel, Liebster, and Tadra, *Chem Listy,* 46, 45 (1952).
59. Claus, *Biochem. Biophys. Res. Commun.,* 20, 745 (1965).
60. Kuhn, Weiser, and Fischer, *Justus Liebigs Ann. Chem.,* 628, 207 (1959).
61. Ley and Doudoroff, *J. Biol. Chem.,* 227, 745 (1957).
62. Ashwell, Wahba, and Hickman, *J. Biol. Chem.,* 235, 1559 (1960).
63. Regna and Caldwell, *J. Am. Chem. Soc.,* 66, 243, 244, 246 (1944).
64. Hart and Everett, *J. Am. Chem. Soc.,* 61, 1822 (1939).
65. Wolfrom, Thompson, and Evans, *J. Am. Chem. Soc.,* 67, 1793 (1945).
66. Yurkevich, Vereikina, Dolgikh, and Preobrazheuskii, *J. Gen. Chem. USSR,* 37, 1201 (1967).
67. Hughes, Overend, and Stacey, *J. Chem. Soc.* (Lond.), p. 2846 (1949).
68. Bauer and Biely, *Collect. Czech. Chem. Commun.,* 33, 1165 (1968).
69. Fischer and Dangschat, *Helv. Chim. Acta,* 20, 705 (1937).
70. Williams and Egan, *J. Bacteriol.,* 77, 167 (1959).
71. Kilgore and Starr, *Biochim. Biophys. Acta,* 30, 652 (1958).
72. Ohle and Berend, *Ber. Dtsch. Chem. Ges.,* 60, 1159 (1927).
73. Waldi, *J. Chromatogr.,* 18, 417 (1965).
74. Cirelli and de Lederkremer, *Chem. Ind.,* 1139 (1971).
75. Paerels, *Recl. Trav. Chim. Pays-Bas,* 80, 985 (1961).
76. Henderson, *J. Am. Chem. Soc.,* 79, 5304 (1957).
77. Merrick and Roseman, *J. Biol. Chem.,* 235, 1274 (1960).
78. Boutroux, *Ann. Chim. Phys. Ser. 6,* 21, 565 (1890).
79. Barch, *J. Am. Chem. Soc.,* 55, 3656 (1933).
80. Strobel, *J. Biol. Chem.,* 245, 32 (1970).
81. Rosenthal, Spaner, and Brown, *J. Chromatogr.,* 13, 152 (1964).
82. Wakisaka, *Agric. Biol. Chem.,* 28, 819 (1964).
83. Katznelson, Tanenbaum, and Tatum, *J. Biol. Chem.,* 204, 43 (1953).
84. Hudson, *J. Am. Chem. Soc.,* 73, 4498 (1951).
85. Upson, Sands, and Whitnah, *J. Am. Chem. Soc.,* 50, 519 (1928).
86. Barber and Hassid, *Bull. Res. Counc. Isr. Sect. A.,* 11, 249 (1963).
87. Burns, *J. Am. Chem. Soc.,* 79, 1257 (1957).
88. Okazaki, Kanzaki, Sasajima, and Terada, *Agric. Biol. Chem.,* 33, 207 (1969).
89. Puhakainen and Hanninen, *Acta Chem. Scand.,* 26, 3599 (1972).
90. Heyns, *Justus Liebigs Ann. Chem.,* 558, 177 (1947).
91. deWilt, *J. Chromatogr.,* 63, 379 (1971).
92. Grollman and Lehninger, *Arch. Biochem. Biophys.,* 69, 458 (1957).
93. Okazaki, Kanzaki, Doi, Nara, and Motizuki, *Agric. Biol. Chem.,* 32, 1250 (1968).
94. Smiley and Ashwell, *J. Biol. Chem.,* 236, 357 (1961).
95. Penney and Zilva, *Biochem. J.,* 37, 403 (1943).
96. Mead and Finamore, *Biochemistry,* 8, 2652 (1969).
97. Mumma and Verlangieri, *Biochim. Biophys. Acta,* 273, 249 (1972).
98. Berman and Magasanik, *J. Biol. Chem.,* 241, 807 (1966).
99. Anderson and Magasanik, *J. Biol. Chem.,* 246, 5653, 5662 (1971).
100. Takagi, *Agric. Biol. Chem.,* 26, 717 (1962).
101. Hamilton and Smith, *J. Am. Chem. Soc.,* 76, 3543 (1954).
102. Kanzaki and Okazaki, *Agric. Biol. Chem.,* 34, 432 (1970).
103. Levene, *J. Biol. Chem.,* 59, 123 (1924).
104. Phillips and Criddle, *J. Chem. Soc.* (Lond.), p. 3404 (1960).
105. Pervozvanski, *Microbiology* (USSR), 8, 915 (1939).
106. Wolfrom, Konigsberg, and Weisblat, *J. Am. Chem. Soc.,* 61, 576 (1939).
107. Gakhokidge and Gvelukashvili, *J. Gen. Chem. USSR,* 22, 143 (1952).
108. Preiss and Ashwell, *J. Biol. Chem.,* 238, 1571, 1577 (1963).
109. Bloom and Westerfeld, *Biochemistry,* 5, 3204 (1966).
110. Hirabayashi and Harada, *Agric. Biol. Chem.,* 33, 276 (1969).
111. Kuhn, Bister, and Dafeldecker, *Justus Liebigs Ann. Chem.,* 617, 115 (1958).
112. Srinivasan and Sprinson, *J. Biol. Chem.,* 234, 716 (1959).
113. Paerels and Geluk, *Recl. Trav. Chim. Pays-Bas,* 89, 813 (1970).
114. Charon and Szabo, *J. Chem. Soc. (Lond.),* Perk I, 1175 (1973).
115. Adlersberg and Sprinson, *Biochemistry,* 3, 1855 (1964).
116. Ghalambor, Levine, and Heath, *J. Biol. Chem.,* 241, 3207 (1966).
117. Dröge, Lehmann, Lüderitz, and Westphal, *Eur. J. Biochem.,* 14, 175 (1970).
118. Hershberger, Davis, and Binkley, *J. Biol. Chem.,* 243, 1578, 1585 (1968).
119. Levin and Racker, *J. Biol. Chem.,* 234, 2532 (1959).
120. Ameyama and Kondo, *Bull. Agric. Chem. Soc. Jap.,* 22, 271, 380 (1958).
121. Hulyalkar and Perry, *Can. J. Chem.,* 43, 3241 (1965).
122. Bergmann, *Ber. Dtsch. Chem. Ges.,* 54, 1362 (1921).
123. Ehrlich and Schubert, *Ber. Dtsch. Chem. Ges.,* 62, 1987, 2022 (1929).
124. Niemann, Schoeffal, and Link, *J. Biol. Chem.,* 101, 337 (1933).
125. Kováč, Hirsch, and Kováčik, *Carbohydr. Res.,* 32, 360 (1974).
126. Winmann, *Ber. Dtsch. Chem. Ges.,* 62, 1637 (1929).
127. Ehrlich and Rehorst, *Ber. Dtsch. Chem. Ges.,* 58, 1989 (1925).
128. Ehrlich and Rehorst, *Ber. Dtsch. Chem. Ges.,* 62, 628 (1929).

129. Bergmann and Wolff, *Ber. Dtsch. Chem. Ges.*, 56, 1060 (1923).
130. Goebel and Babers, *J. Biol. Chem.*, 100, 573, 743 (1933).
131. Fischer and Schmidt, *Ber. Dtsch. Chem. Ges.*, 92, 2184 (1954).
132. Das Gupta and Sarkar, *Text. Res. J.*, 24, 705, 1071 (1954).
133. Marsh, *J. Chem. Soc.* (Lond.), p. 1578 (1952).
134. Levene and Meyer, *J. Biol. Chem.*, 60, 173 (1924).
135. Kovác, *Carbohydr. Res.*, 31, 323 (1973).
136. Currie and Timell, *Can. J. Chem.*, 37, 922 (1959).
137. Jones and Painter, *J. Chem. Soc.* (Lond.), p. 669 (1957).
138. Tyler, *J. Chem. Soc.* (Lond.), p. 5288, 5300 (1965).
139. Jones and Nunn, *J. Chem. Soc.* (Lond.), p. 3001 (1955).
140. Wacek, Leitinger, and Hochbahn, *Monatsh. Chem.*, 90, 555, 562 (1959).
141. Charalampous and Lyras, *J. Biol Chem.*, 228, 1 (1957).
142. Sowa, *Can. J. Chem.*, 47, 3931 (1969).
143. Sutter and Reichstein, *Helv. Chim. Acta*, 21, 1210 (1938).
144. Haug and Larsen, *Acta Chem. Scand.*, 15, 1395 (1961).
145. Gunther and Schweiger, *J. Chromatogr.*, 34, 498 (1968).
146. Fischer and Dörfel, *Hoppe-Seyler's Z. Physiol. Chem.*, 302, 186 (1955).
147. Whistler and Schweiger, *J. Am. Chem. Soc.*, 80, 5701 (1958).
148. Preiss and Ashwell, *J. Biol. Chem.*, 237, 309, 317 (1962).
149. Heim and Neukom, *Helv. Chim. Acta*, 45, 1737 (1962).
150. Cifonelli, Ludowieg and Dorfman, *J. Biol. Chem.*, 233, 541 (1958).
151. Shafizadeh and Wolfrom, *J. Am. Chem. Soc.*, 77, 2568 (1955).
152. St. Cyr, *J. Chromatogr.*, 47, 284 (1970).
153. Fischer and Dörfel, *Hoppe-Seyler's Z. Physiol. Chem.*, 301, 224 (1955).
154. Lehtonen, Kärkkäinen, and Haahti, *Anal. Biochem.*, 16, 526 (1966).
155. Schoeffel, Link, *J. Biol. Chem.*, 100, 397 (1933).
156. Stephen, Kaplan, Taylor, and Leisegang, *Tetrahedron Suppl.*, 7, 233 (1966).
157. Pasteur, *Ann. Chim. Phys. Ser.*, 3, 28, 71 (1850).
158. Walden, *Ber. Dtsch. Chem. Ges.*, 29, 1701 (1896).
159. Fandolt, *Ber. Dtsch. Chem. Ges.*, 6, 1075 (1873).
160. Frankland, and Slater, *J. Chem. Soc.* (Lond.), 83, 1354 (1903).
161. Anschütz, and Pictet, *Ber. Dtsch. Chem. Ges.*, 13, 1176 (1880).
162. Patterson, *J. Chem. Soc.* (Lond.), 85, 765 (1904).
163. Frankland and Wharton, *J. Chem. Soc.* (Lond.), 69, 1310 (1896).
164. Pasteur, *Justus Liebigs Ann. Chem.*, 82, 331 (1852).
165. Schneider, *Ber. Dtsch. Chem. Ges.*, 13, 620 (1880).
166. Walczyk and Burczyk, *Chem. Anal.* (Warsaw), 17, 404 (1972).
167. Anschütz and Bennert, *Justus Liebigs Ann. Chem.*, 254, 165 (1889).
168. Lutz, Dissertation, University of Rostock (1899); *Chem. Zentralbl.*, II, 1013 (1900).
169. Bond, D., *J. Chem. Soc. D*, p. 338 (1969).
170. Anet and Reynolds, *Nature*, 174, 930 (1954).
171. Kessler, Neufeld, Feingold, and Hassid, *J. Biol. Chem.*, 236, 308 (1961).
172. Rehorst, *Ber. Dtsch. Chem. Ges.*, 61, 163 (1928).
173. Marsh, *Biochem. J.*, 86, 77 1963; 87, 82 (1963); 89, 108 (1963).
174. Rehorst, *Ber. Dtsch. Chem. Ges.*, 65, 1476 (1932).
175. Matsui, Okada, and Ishidata, *J. Biochem. (Tokyo)*, 57, 715 (1965).
176. Tobie and Ayres, *J. Am. Chem. Soc.*, 64, 725 (1942).
177. Katz and Lewis, *Anal. Biochem.*, 17, 306 (1966).
178. Wright, Jr.Burton, and Berry, Jr. *Arch Biochem. Biophys.*, 86, 94 (1960).
179. Frei, Fukui, Lieu , T., and Frodyma, *Chemia* (Aarau), 20, 24 (1966).
180. Fusari, Haskell, Frohardt, and Bartz, *J. Am. Chem. Soc.*, 76, 2881 (1954).
181. Isono, Asahi, and Suzuki, *J. Am. Chem. Soc.*, 91, 7490 (1969).
182. Zissis, Diehl, and Fletcher, Jr. *Carbohydr. Res.*, 28, 327 (1973).
183. Karrer and Mayer, *Helv. Chim. Ada*, 20, 407 (1937).
184. Iwasaki, *Yakugaku Zasshi*, 82, 1380 (1962); *Chem. Abstr.*, 59, 758 (1963).
185. Heyns, Kiessling, Lindenberg, and Paulsen, *Ber. Dtsch. Chem. Ges.*, 92, 2435 (1959).
186. Heyns and Beck, *Ber. Dtsch. Chem. Ges.*, 90, 2443 (1957).
187. Levene, *J. Biol. Chem.*, 57, 337 (1923).
188. Williamson and Zamenhof, *J. Biol. Chem.*, 238, 2255 (1963).
189. Heyns, Paulsen, *Ber. Dtsch. Chem. Ges.*, 88, 188 (1955).
190. Westphal and Holzmann, *Ber. Dtsch. Chem. Ges.*, 75, 1274 (1942).
191. Weidmann, Fauland, Helbig, and Zimmerman, *Justus Liebigs Ann. Chem.*, 694, 183 (1966).
192. Romanowska and Reinhold, *Eur. J. Biochem.*, 36, 160 (1973).
193. Torii, Sakakibara, Kuroda, *Eur. J. Biochem.*, 37, 401 (1973).
194. Crumpton, *Biochem. J.*, 72, 479 (1959).
195. Ōtake, Takeuchi, Endō, and Yonehara, *Tetrahedron Lett.*, 1405 (1965).
196. Watanabe, Goody, Fox, *Tetrahedron*, 26, 3883 (1970).
197. Kundu, Crawford, Prajsnar, Reed, and Rosenthal, *Carbohydr. Res.*, 12, 225 (1970).
198. Perkins, *Biochem. J.*, 86, 475 (1963); 89, 104P (1963).
199. Koto, Kawakatsu, and Zen, *Bull. Chem. Soc. Jap.*, 46, 876 (1973).
200. Tarasiejska and Jeanloz, *J. Am. Chem. Soc.*, 79, 2660 (1957).
201. Kuhn and Bister, *Justus Liebigs Ann. Chem.*, 617, 92 (1958).
202. Lemieux and Nagabhushan, *Can. J. Chem.*, 46, 401 (1968).
203. Strange and Kent, *Biochem. J.*, 71, 333 (1959).
204. Lambert and Zilliken, *Ber. Dtsch. Chem. Ges*, 93, 2915 (1960).
205. Flowers and Jeanloz, *J. Org. Chem.*, 28, 1564, 2983 (1963).
206. Osawa, Sinay, Halford, and Jeanloz, *Biochemistry*, 8, 3369 (1969).
207. Ghuysen and Strominger, *Biochemistry*, 2, 1119 (1963).
208. Sinay, *Carbohydr. Res.*, 16, 113 (1971).
209. Sinay, Halford, Choudhary, Gross, and Jeanloz, *J. Biol. Chem.*, 247, 391 (1972).
210. Hoshino, Zehavi, Sinay, and Jeanloz, *J. Biol. Chem.*, 247, 381 (1972).
211. Kondo, Akita, and Sezaki, *J. Antibiot.* (Tokyo) Ser. A, 19, 137 (1966).
212. Faillard, *Hoppe-Seyler's Z. Physiol. Chem.*, 307, 62 (1957).
213. Zilliken and McGlick, *Naturwissenschaften*, 43, 536 (1956).
214. Khorlin and Privalova, *Chem. Nat. Compd.* (U S S R), 3, 159 (1967).
215. Kuhn and Baschang, *Justus Liebigs Ann. Chem.*, 659, 156 (1962).
216. Zanetta, Breckenridge, and Vincendon, *J. Chromatogr.*, 69, 291 (1972).
217. Blix and Lindberg, *Acta Chem. Scand.*, 14, 1809 (1960).
218. Faillard and Blohm, *Hoppe-Seyler's Z. Physiol. Chem.*, 341, 167 (1965).
219. Hotta, Kurokawa, and Isaka, *J. Biol. Chem.*, 245, 6307 (1970).
220. Brunetti, Jourdian, and Roseman, *J. Biol. Chem.*, 237, 2447 (1962).
221. Hakomori and Saito, *Biochemistry*, 8, 5082 (1969).
222. Warren, *Biochim. Biophys. Acta*, 83, 129 (1964).
223. Isemura, Zahn, and Schmid, *Biochem. J.*, 131, 509 (1973).

TABLE C: NATURAL ALDOSES

Substance[a] (Synonym) Derivative (A)	Chemical Formula (B)	Melting Point °C (C)	Specific Rotation[b] $[\alpha]_D$ (D)	Reference[c] (E)	Chromatography, R Value, and Reference[d] ELC (F)	GLC (G)	PPC (H)	TLC (I)
Acetaldehyde (deoxyglycolaldehyde)	C_2H_4O	Liquid, bp 21	None	1		0.6 (2)		
2,4-Dinitrophenylhydrazone	$C_8H_8N_4O_4$	146, 163.5–164.5	None	1				52f (3)
D-Glyceraldehyde (glycerose)	$C_3H_6O_3$	Syrup	+13.5±0.5	4	<10sor (5)		310f (6)	58f (7)
Dimethone	$C_{19}H_{28}O_6$	199–201	+197.5±0.5 (c 0.7, C_2H_5OH)	8				
2,4-Dinitrophenylhydrazone	$C_9H_{10}N_4O_6$	155–156		8				10f (3)
Trimethylsilyl ether				10		48glct (9)		
D-Glyceraldehyde, 3,3-bis-(C-hydroxymethyl)- (D-apiose)	$C_5H_{10}O_5$	138–139	−(29 (CH_3OH) +5.6 (c 10) (15°)	11,12				
Benzylphenylhydrazone	$C_{18}H_{22}N_2O_4$	137–138	$[\alpha]_{579}$ −78.5 (C_5H_5N)	11,12				
Di-O-iso-propylidene ether	$C_{11}H_{20}O_5$		+54±1 (c 0.55, C_2H_5OH)	13c,p				
2,3,4-Tri-O-methyl ether	$C_8H_{16}O_5$		−1.7 (CH_3OH)	14		234atemg (14a)	91temg (14b)	91temg (14c)
D-Glyceraldehyde, 3,3-bis-(C-hydroxymethyl)-3-deoxy- (cordycepose)	$C_5H_{10}O_4$	Syrup	−26 (c 0.6, C_2H_5OH)	15				
Cordyceponic acid phenyl hydrazide	$C_{11}H_{16}N_2O_4$	151	+26±3 (c 0.3, C_2H_5OH)	15				
β-D-Arabinose	$C_5H_{10}O_5$	155	−175→−103	16,411c	<10sor (5)		130f (6)	49f (17)
Benzylphenylhydrazone	$C_{18}H_{22}N_2O_4$	177–178	+14.4 (CH_3OH)(16°)	18				
2,3,5-Tri-O-methyl ether	$C_8H_{16}O_5$	Syrup	+40 (CH_3OH)	20		39etemg (19b); 33glct (9)	97temg (19a)	38,51f (40)e
Trimethylsilyl ether								
D-Arabinose, 2-O-methyl-	$C_6H_{12}O_5$	Syrup	−102±3	21,22	42temg (19c)		16f (19d)	56trma (23a)f
2-O-Methyl-D-arabinitol	$C_6H_{14}O_5$	98–99	−11 (CH_3OH)	24				32trmal (23a)f
Methyl β-pyranoside	$C_7H_{14}O_5$	113				56βtma (27a)		
Phenylhydrazone	$C_{12}H_{18}N_2O_4$			25				
α-L-Arabinose	$C_5H_{10}O_5$	Amorphous, 158	+55.4→+105	26				
β-L-Arabinose	$C_5H_{10}O_5$	160	+190.6→+104.5	27	30rib (28)		21f (29)	31f (30a)
p-Nitrophenylhydrazone	$C_{11}H_{15}N_3O_6$	181	+22.6 (1:1, C_5H_5N, C_2H_5OH)	31				
α-Pyranoside tetraacetate	$C_{13}H_{18}O_9$	97	+42.5 ($CHCl_3$)	32				
β-Pyranoside tetraacetate	$C_{13}H_{18}O_9$	86	+147.2 ($CHCl_3$)	32				
2,3,5-Tri-O-methyl ether	$C_8H_{16}O_5$	Syrup		33		7.07 (34)		167trma (23a)
Trimethylsilyl ether								
L-Arabinose 3-sulfate	$C_5H_{10}O_8S$		+75 (c 0.6)	35	118gals (35a)		83gal (35b)	
α-L-Lyxose	$C_5H_{10}O_5$	105	+5.8→+13.5	36,411c	42rib (28)f		25f (29)	46f (37)f
p-Bromophenylhydrazone	$BrC_{11}H_{15}N_2O_4$	157–158	−30.1→−10 (C_5H_5N)	38				
p-Nitrophenylhydrazone	$C_{11}H_{15}N_3O_6$	172		36				
α-Pyranoside tetraacetate Trimethylsilyl ether	$C_{13}H_{18}O_9$	93–94	+25 ($CHCl_3$)g	39		26glct (9)		
L-Lyxose, 5-deoxy-3-C-formyl- (streptose)	$C_6H_{10}O_5$	Syrup	−18	41				
Streptosonolactone	$C_6H_8O_5$	146–148	−37 (c 0.7)	42				

TABLE C: NATURAL ALDOSES (Continued)

Substance[a] (Synonym) Derivative (A)	Chemical Formula (B)	Melting Point °C (C)	Specific Rotation[b] $[\alpha]_D$ (D)	Reference[c] (E)	Chromatography, R Value, and Reference[d] ELC (F)	GLC (G)	PPC (H)	TLC (I)
L-Lyxose, 5-deoxy-3-C-hydroxymethyl- (dihydrostreptose)	$C_6H_{12}O_5$	Syrup	−24	41,43				
Dihydrostreptosonolactone	$C_6H_{10}O_5$	140.5–142.5	−32	41				
L-Lyxose, 3-C-formyl- (hydroxystreptose)	$C_6H_{10}O_6$	No constants known	No constants known	44				
3-Hydroxy-2-hydroxy-methyl-1,4-pyrone	$C_6H_6O_4$	152–153		44				
L-Lyxose, 2-O-methyl-	$C_6H_{12}O_5$	120–121	+6	45				
Pyranoside triacetate	$C_{12}H_{18}O_8$	Syrup	−10.5 ($CHCl_3$)	46				
2,3,4-Tri-O-methyl ether	$C_9H_{18}O_5$	Syrup	−21.6	46				
Pentose, 4,5-anhydro-5-deoxy-D-erythro-	$C_5H_8O_3$	No constants known	No constants known	47ce			75f (47)	
Pentose, 2-deoxy-D-erythro- (2-deoxy-D-ribose)	$C_5H_{10}O_4$	96–98	−91→−58	48	<10sor (5)		40f (49)	266glc (50)
p-Nitrophenylhydrazone	$C_{11}H_{15}N_3O_5$	160	−11.1 (c 0.1, C_2H_5OH)(14°)	51				
Pyranoside tr.acetate	$C_{11}H_{16}O_7$	98	−171.8 (c 0.5, $CHCl_3$)	51				
Tetra-O-acetyl-2-deoxy-D-ribitol	$C_{13}H_{18}O_8$					57xyll (52)		
Trimethylsilyl ether						6.53 (34)		
D-Ribose	$C_5H_{10}O_5$	87	−23.7	53,411c	40sor (5)		25f (49)	39f (30a)
p-Bromophenylhydrazone	$BrC_{11}H_{15}N_2O_4$	164–165	+10.3 (C_2H_5OH)	54				
α-Pyranoside tetraacetate	$C_{13}H_{18}O_9$	Syrup	+46.1 (CH_3OH)	55				
β-Pyranoside tetraacetate	$C_{13}H_{18}O_9$	110	−54.5 (CH_3OH)	55				
2,3,4-Tri-O-methyl ether	$C_8H_{16}O_5$	88.5–91		56			62f (56)	
Trimethylsilyl ether						32glct (9)		57f (59a)
D-Ribose, 2-C-hydroxymethyl- (D-hamamelose)	$C_6H_{12}O_6$	110–111	+7.7→−7.0	57			18f (57)[f]	
Ammonium D-hamamelonate	$C_6H_{15}NO_7$	152	$[\alpha]_{578}$ −3.9 (c 10)	57				
Hamamelonolactone	$C_6H_{10}O_6$	Syrup		57			30f (57)[f]	
p-Nitrophenylhydrazone	$C_{12}H_{17}N_3O_7$	165–166	$[\alpha]_{578}$ +144 (C_5H_5N)	57				
D-Ribose, 1′,5-di-O-galloyl-2-C-hydroxymethyl- (hamamelitannin)	$C_{20}H_{20}O_{14}$	145–147.5	+31.3	58t			26f (58)	
Methyl α-furanoside	$C_{21}H_{22}O_{14} \cdot H_2O$	147–150	+33 (C_2H_5OH)	58			49f (58)	
Methyl α-furanoside hexamethyl ether	$C_{27}H_{34}O_{14}$	94–95	+43 (c 0.5, C_2H_5OH)	58				
Methyl β-furanoside	$C_{21}H_{22}O_{14} \cdot H_2O$	206–208	−30.6 (C_2H_5OH)	58			54f (58)	
Methyl β-furanoside hexamethyl ether	$C_{27}H_{34}O_{14}$	162.5–163.5	−28.5 (acetone)	58				
Ribose, 5-deoxy-5-S-methyl-5-thio- (5-methylthioribose)	$C_6H_{12}O_4S$	Syrup	+41.9 (CH_3OH)(30°)	60e,61			71f (60a)	55f (60b)
Trimethylsilyl ether						11.75 (60c)		

TABLE C: NATURAL ALDOSES (Continued)

Substance[a] (Synonym) Derivative (A)	Chemical Formula (B)	Melting Point °C (C)	Specific Rotation[b] [α]$_D$ (D)	Reference[c] (E)	Chromatography, R Value, and Reference[d] ELC (F)	GLC (G)	PPC (H)	TLC (I)
α-D-Xylose	$C_5H_{10}O_5$	145	+93.6→+18.8	62,63,411c	17rib (28)		15f (49)	33f (30a)
p-Bromophenylhydrazone	$BrC_{11}H_{15}N_2O_4$	128	−20.7	64,65				84f (69)
α-Pyranoside tetraacetate	$C_{13}H_{18}O_9$	59	+89.3 ($CHCl_3$)	66				
β-Pyranoside tetraacetate	$C_{13}H_{18}O_9$	128	−24.7 ($CHCl_3$)	66				
2,3,4-Tri-O-methyl ether	$C_8H_{16}O_5$	84–86	+45.5→+22 (c 0.7)	68		44,59kβ temg (67)	97temg (68)	28f (69)
Trimethylsilyl ether						8.47 (34)		57f (59a)
D-Xylose, 5-deoxy-	$C_5H_{10}O_4$		+16	70				
p-Bromophenylhydrazone Triacetate	$BrC_{11}H_{15}N_2O_3$	69–70d	−26.1 (C_5H_5N)	71				
Bis-Trimethylsilyl ether	$C_{11}H_{16}O_7$		+60.9 ($CHCl_3$)	71		4.5 (72)e		
D-Xylose, 2-O-methyl-	$C_6H_{12}O_5$	132–133	+37±6 (c 0.27)	73	6.8h (77)		38temg (74)	
Methyl glycosides	$C_7H_{14}O_5$	111–112 (β-anomer)	−67.7 ($CHCl_3$)(β-anomer)	75		411,623kβtemg (80)		
2-O-Methyl xylitol tetraacetate	$C_{14}H_{22}O_9$					189dmxyl (76a)		
D-Xylose, 3-O-methyl-	$C_6H_{12}O_5$	95	+45→+19	77cp,78	13.6h (77)	189dmxyl (76a)	136xyl (82a)	
3-O-Methyl xylitol tetraacetate	$C_{14}H_{22}O_9$							
3-O-Methyl-D-xylono-lactone	$C_6H_{10}O_5$	90	+72→+40 (c 0.9)	79cep				
Methyl glycosides	$C_7H_{14}O_5$					355,557kβtemg (80)		
Phenylosazone	$C_{18}H_{22}N_4O_4$	172		83				
D-Xylose, 4-O-methyl-	$C_6H_{12}O_5$	Syrup	+9±2	81,82		233dmxyl (76b)	126xyl (82a)	
4-O-Methyl-D-xylose diethyl dithioacetal triacetate	$C_{14}H_{22}O_9$							
4-O-Methyl xylitol tetraacetate	$C_7H_{14}O_5$	95 (β-anomer)	−69 (β-anomer)	81		54xyll (82c)		
Methyl glycosides						435kβtemg (80)		
Phenylosazone	$C_{18}H_{22}N_4O_4$	158–158.5	+25→0 (C_2H_2OH-C_5H_5N)	84				
L-Xylose, 3-O-methyl-	$C_6H_{12}O_5$		−18	85		292temga (85c)	120rha (85a)	112rha (85b)
3-O-Methyl xylitol tetraacetate	$C_{14}H_{22}O_9$					14.61f (86)		
Methyl β-furanoside	$C_7H_{14}O_5$							
Aldgarose A (a dideoxy-C-hydroxyethylhexose carbonate)	$C_9H_{16}O_6$	No constants known	No constants known	87				
Methyl aldgaroside A	$C_{10}H_{18}O_6$	91–94		87				
Aldgarose B (a dideoxy-C-hydroxyethylhexose carbonate)	$C_9H_{16}O_6$	No constants known	No constants known	87				
Methyl aldgaroside B[i]	$C_{10}H_{18}O_6$	175–177	−41 (CH_3OH)	87,88				
D-Allose	$C_6H_{12}O_6$	130–132	+14.5±0.3	89,90,411c	75rib (28)			
Allitol	$C_6H_{14}O_6$	150–151		90	92rib (28)	88manl (91b)	110glc (91a)	
Allitol hexaacetate								
Allonolactone	$C_6H_{10}O_6$						109f (383)	
p-Bromophenylhydrazone	$BrC_{12}H_{17}N_2O_5$	145–147	−6.7 (C_2H_5OH)	92				
β-Pyranoside pentaacetate	$C_{16}H_{22}O_{11}$	93–93.5	−13.7 ($CHCl_3$)	93				
Trimethylsilyl ether						81glct (9)		

TABLE C: NATURAL ALDOSES (Continued)

Substance[a] (Synonym) Derivative (A)	Chemical Formula (B)	Melting Point °C (C)	Specific Rotation[b] $[\alpha]_D$ (D)	Reference[c] (E)	Chromatography, R Value, and Reference[d]			
					ELC (F)	GLC (G)	PPC (H)	TLC (I)
D-Allose, 6-deoxy-	$C_6H_{12}O_5$	132–135, 151–152	$-4\rightarrow+1.2$	94	127rha (96a)		100rha (95)	44rha (96b)
p-Bromophenylhydrazone	$BrC_{12}H_{17}N_2O_4$	138–140, 145–146	$-21.9\rightarrow-11.8$ (C_5H_5N)	96,97				
6-Deoxy-D-allitol pentaacetate Pyranoside tetraacetate	$C_{16}H_{24}O_{10}$ $C_{14}H_{20}O_9$	109–110	$+10.4$ ($CHCl_3$)	99,100		55aral (98p)		
D-Allose, 6-deoxy-2-O-methyl- (javose)	$C_7H_{14}O_5$	112–114	$-54\rightarrow-40,-8.2$	94,101	8i (94a)		50aco (94b)	28f (102 ep)
Methyl α-pyranoside	$C_8H_{16}O_5$	Syrup	$+90\pm3$ ($CHCl_3$)	101				
Methyl β-pyranoside	$C_8H_{16}O_5$	97–98	-82.8 ± 1 (CH_3OH)	102				41f (102)
D-Allose, 6-deoxy-3-O-methyl-	$C_7H_{14}O_5$	122–123	$+9$ (30°)	103	16.5i (94a)		52aco (94b)	129rha (96b, ep)
Methyl α-pyranoside	$C_8H_{16}O_5$	110–111	$+195\pm5$ (c 0.7, CH_3OH)	103t				
Methyl β-pyranoside	$C_8H_{16}O_5$	153–154	-37.1 ± 2 (CH_3OH)	104c				
D-Allose, 6-deoxy-2,3-di-O-methyl- (mycinose)	$C_8H_{16}O_5$	102–106	$-46\rightarrow-29$	105,411c				
Methyl α-pyranoside	$C_9H_{18}O_5$	88–88.5	$+140$ (c 0.7, $CHCl_3$)	106t				
Methyl β-pyranoside	$C_9H_{18}O_5$	101–103	-36 ($CHCl_3$)(27°)	105,107t				
β-Pyranoside diacetate	$C_{12}H_{20}O_7$		$+32$ (c 0.4, $CHCl_3$)	106				
D-Altrose, 6-deoxy-	$C_6H_{12}O_5$	Syrup	$+16.2$	108	164rha (96a)		110rha (95)	128rha (96b)
p-Bromophenylhydrazone	$BrC_{12}H_{17}N_2O_4$	155,177		109,110	104glcl (312)			
6-Deoxy-D-altritol (1-deoxy-D-talitol)	$C_6H_{14}O_5$							
6-Deoxy-D-altritol pentaacetate	$C_{16}H_{24}O_{10}$			109		61aral (98p)		
Methyl α-pyranoside	$C_7H_{14}O_5$	Syrup	$+118$ (CH_3OH)(16°)					
D-Altrose, 6-deoxy-3-O-methyl- (D-vallarose)	$C_7H_{14}O_5$	111–113	$+8.6\rightarrow+22.3$ (c 0.6)(18°)	111,112				
Methyl α-pyranoside	$C_8H_{16}O_5$		$+133\pm2$	112				
α-Pyranoside triacetate	$C_{13}H_{20}O_8$	112–113	$+14.8\pm2$ (15°) ($CHCl_3$)	112				
β-Pyranoside triacetate	$C_{13}H_{20}O_8$	79	-96.5 ± 2 (15°) ($CHCl_3$)	112				
D-Altrose, 6-deoxy-4-O-methyl- (sordarose)	$C_7H_{14}O_5$	Syrup	$+29$ (c 0.45)	113				139rha (113)
Methyl α-pyranoside	$C_8H_{16}O_5$		$+153$ (CH_3OH)	113t				
Methyl β-pyranoside	$C_8H_{16}O_5$		-12 (CH_3OH)	113t				
L-Altrose, 6-deoxy-	$C_6H_{12}O_5$	Gum	-16.1 ± 2	114pt				
L-Altrose, 6-deoxy-3-O-methyl- (L-vallarose)	$C_7H_{14}O_5$	106–110	-17.2 ± 2 (c 0.9)	115	20i (94a)		75aco (94b)	154rha (96b, ep)
α-Pyranoside triacetate	$C_{13}H_{20}O_8$	112–113s, 122–123	-12 ± 1 ($CHCl_3$)	116t				
Antiarose	$C_6H_{12}O_5$	Syrup	levo	117,118				
Antiaronolactone	$C_6H_{10}O_5$	Syrup	-30	117,118				
Antiaronic acid phenylhydrazide	$C_{12}H_{18}N_2O_5$	143–145		117,118				

TABLE C: NATURAL ALDOSES (Continued)

Substance[a] (Synonym) Derivative (A)	Chemical Formula (B)	Melting Point °C (C)	Specific Rotation[b] $[\alpha]_D$ (D)	Reference[c] (E)	Chromatography, R Value, and Reference[d] ELC (F)	GLC (G)	PPC (H)	TLC (I)
α-D-Galactose	$C_6H_{12}O_6$	167	+150.7→+80.2	119	28rip (28)		6f (49)	21f (30a)
Pyranose tetramethyl ether	$C_{10}H_{20}O_6$	71–73	+149.4→+116.9	120			92temg (68)	78f (30b)
α-Pyranoside pentaacetate	$C_{16}H_{22}O_{11}$	96	+106.7 ($CHCl_3$)	122				
Trimethylsilyl ether					8.94 (34)		66f (59)	
α-D-Galactopyranoside, ethyl-	$C_8H_{16}O_6$	143	+185	121			35f (150)	
β-D-Galactose	$C_6H_{12}O_6$	143–145	+52.8→+80.2	119,411c				
p-Nitrophenylhydrazone	$C_{12}H_{17}N_3O_7$	192	+70 (c 0.3, C_5H_5N-C_2H_5OH)	31				81f (69)
β-Pyranose pentaacetate	$C_{16}H_{22}O_{11}$	142	+25 ($CHCl_3$)	122				54f (59)
Dimethyl acetal			No constants known			108glct (9)		
d-Galactose, 3,6-anhydro-	$C_6H_{10}O_5$	153–155	+12,+21.3 (10°)	123–125				
Trimethylsilyl ether	$C_6H_{16}O_6$		+29 (17°)	125				
Diphenylhydrazone	$C_{18}H_{20}N_2O_4$	109, 139–140	+34.5→+23.6 (CH_3OH) (14°)	124			170rha (124)	
Methyl α-pyranoside	$C_7H_{12}O_5$		+80,+175 (10°)	123,124				
d-Galactose, 4,6-O-(1-carboxyethylidene)-	$C_9H_{14}O_8$	No constants known	No constants known	126				
Ammonium salt	$C_9H_{17}NO_8$		+51 (c 0.43)	127			20gal (127)	
Ethanolate	$C_9H_{14}O_8 \cdot C_2H_5OH$		+49	126				
4,6-O-(1-carboxyethylidene methyl ester)-d-galactitol	$C_{10}H_{18}O_8$	104–105	−18 (c 0.6, CH_3OH)	126				
Methyl α-pyranoside methyl ester	$C_{11}H_{18}O_8$	Syrup	+133 ($CHCl_3$)	128				
α-D-Galactose, 6-deoxy- (d-fucose, rhodeose)	$C_6H_{12}O_5$	140–145	+120→+76.3 (c 10)	129,412c			19f (49)	25f (130)
Benzylphenylhydrazone	$C_{19}H_{24}N_2O_4$	178–179	−14.9 (c 0.4, CH_3OH)	129				
6-Deoxy-D-galactitol	$C_6H_{14}O_5$				100sor (132)			
Fucitol pentaacetate	$C_{16}H_{24}O_{10}$							
α-Pyranoside tetraacetate	$C_{14}H_{20}O_9$	92–93	+129 ($CHCl_3$)	133		7.7 (131)		
Trimethylsilyl ether						6.2,7.1 (59b)		61,69f (59a)
d-Galactose, 6-deoxy-2-O-methyl-	$C_7H_{14}O_5$	155–161	+73→+87	134			180fuc (135)	
Methyl α-pyranoside	$C_8H_{16}O_5$	Syrup	+173.6	136			180mfuc (136)	
Methyl β-pyranoside	$C_8H_{16}O_5$	98.5–99.5	+3.5 (CH_3OH)	136				
d-Galactose, 6-deoxy-3-O-methyl- (digitalose)	$C_7H_{14}O_5$	106, 119	+106	137	12.5 (94a)		30aco (94b)	104rha (96b, ep)
Digitalonolactone	$C_7H_{14}O_6$	137–138	−83	137				
Methyl α-pyranoside	$C_8H_{16}O_5$	98.5–100	+198 (c 0.7, CH_3OH)	136			217mfuc (136)	
Methyl β-pyranoside	$C_8H_{16}O_5$	108–110	+9.9 (c 0.3, CH_3OH)	136			163mfuc (136)	
d-Galactose, 6-deoxy-4-O-methyl- (curacose)	$C_7H_{14}O_5$	131–132	+102.6→+80.6 (c 0.9)	38,138			142mfuc (136)	
Methyl α-pyranoside	$C_8H_{16}O_5$	144–145	−14.6 (c 0.9, CH_3OH)	138			52f (38)	
p-Tolylsulfonylhydrazone	$C_{14}H_{22}N_2O_6S$	134	−16→−3 (C_5H_5N)	138				
d-Galactose, 6-deoxy-2,3-di-O-methyl-	$C_8H_{16}O_5$	75–76	+73,+105	136,139			267fuc (135)	
Methyl α-pyranoside	$C_9H_{18}O_5$	Syrup	+190 (acetone)	135,136			85temg (136)	
Methyl β-pyranoside	$C_9H_{18}O_5$	Syrup	+0.7 (acetone)	135,136			97temg (136)	

TABLE C: NATURAL ALDOSES (Continued)

Substance[a] (Synonym) Derivative	Chemical Formula	Melting Point °C	Specific Rotation[b] $[\alpha]_D$	Reference[c]	Chromatography, R Value, and Reference[d]			
					ELC	GLC	PPC	TLC
(A)	(B)	(C)	(D)	(E)	(F)	(G)	(H)	(I)
Onic acid phenylhydrazide	$C_{14}H_{22}N_2O_5$	99–103	+21.5±3 (c 0.7, CH_3OH)	135				
D-Galactose,								
6-deoxy-2,4-di-O-methyl- (labilose)	$C_8H_{16}O_5$	129	+82 (c 0.5)(27°)	140cp			275fuc (135)	
Labilitol	$C_8H_{18}O_5$							
Methyl α-pyranoside	$C_9H_{18}O_5$	85	+37.4 (c 0.5, $CHCl_3$)(30°)	140				17f (140a)
Methyl β-pyranoside	$C_9H_{18}O_5$	111	+176 ($CHCl_3$)(30°)	140				13f (140b)
Methyl α-D-fucopyranoside trimethyl ether	$C_{12}H_{24}O_5$	96	−20.9 ($CHCl_3$)(30°)	140		4.6 (140c)		18f (140b)
Methyl β-D-fucopyranoside trimethyl ether	$C_{12}H_{24}O_5$	93–98	+213 (c 0.3)	140				28f (140b)
D-Galactose, 2-O-methyl-	$C_7H_{14}O_6$	148–149	+11.2	173			35temg (68)	
Methyl α-pyranoside	$C_9H_{18}O_5$	Syrup	+84.9 (c 0.5)(16°)	141				
Methyl β-pyranoside	$C_9H_{18}O_5$	131–132	+180 (CH_3OH)	142				
2-O-Methyl-N-phenyl-D-galactosylamine	$C_{13}H_{19}NO_5$	164–165	+1.69	142,143				
2-O-Methyl-D-galactitol trimethylsilyl ether				141		289temgt (144)		
D-Galactose, 3-O-methyl- (madurose)	$C_7H_{14}O_6$	144–147	+150→+108	145,146			73rib (145a)	2f (30b)
Methyl β-pyranoside	$C_8H_{16}O_6$	Syrup	+31.9	146				
3-O-Methyl-D-galactitol trimethylsilyl ether						294temgt (144)		
Trimethylsilyl ether						16 (145b)		
D-Galactose, 4-O-methyl-	$C_7H_{14}O_6$	207, 218	+62→+92	147–149			31f (150)	2f (30b)
4-O-Methyl-D-galactitol trimethylsilyl ether						294temgt (144)		
4-O-Methyl-N-phenyl-D-galactosylamine Trimethylsilyl ether	$C_{13}H_{19}NO_5$	167–168	−84→−39 (CH_3OH)	148		19,26 (149)		
D-Galactose, 6-O-methyl-	$C_7H_{14}O_6$	119–120	+76 (17°)	141			85rib (145a)	6f (69)
6-O-Methyl-D-galactitol trimethylsilyl ether						299temgt (144)		
Methyl α-pyranoside	$C_8H_{16}O_6$	137–138	+165	151				21,28f (40)
Methyl pyranoside tetramethyl ether	$C_{12}H_{22}O_6$			152		200βtemg (67)		
Phenylhydrazone	$C_{13}H_{20}N_2O_5$	181.5–182.5						
D-Galactose 2-sulfate	$C_6H_{12}O_9S$		+52 (as Ba salt)	153	100glcs (153a)		24gal (153b)	
D-Galactose 4-sulfate	$C_6H_{12}O_9S$		+64 (c 0.5)(as NH_4 salt); +58.4 (16°)(as Na salt)	154; 155	104glc (155a)		27f (155b)	
D-Galactose 6-sulfate	$C_6H_{12}O_9S$		+49 (c 0.33)(as NH_4 salt); +47 (16°)(as Na salt)	154; 155	132glc (155a)		13f (155b)	
L-Galactose	$C_6H_{12}O_6$	163–165	−78	156				
Phenylhydrazone	$C_{12}H_{18}N_2O_5$	158–160	+21.6	157				
L-Galactose, 3,6-anhydro-	$C_6H_{10}O_5$		−39.4→−25.2	158				

TABLE C: NATURAL ALDOSES (Continued)

Substance[a] (Synonym) Derivative	Chemical Formula	Melting Point °C	Specific Rotation[b] $[\alpha]_D$	Reference[c]	Chromatography, R Value, and Reference[d]			
(A)	(B)	(C)	(D)	(E)	ELC (F)	GLC (G)	PPC (H)	TLC (I)
Dimethyl acetal	$C_8H_{16}O_6$		-16.2	159cgp				50f (159a)
3,6-Anhydro-L-dulcitol	$C_6H_{12}O_5$	138.5–140	-10	160gpt				
Methyl α-pyranoside	$C_7H_{12}O_5$		-77 (c 0.9)(19°)	161				
Methyl β-pyranoside	$C_7H_{12}O_5$	118	-113.5	162				
L-Galactose, 3,6-anhydro-2-O-methyl-	$C_7H_{12}O_5$		-14.3 (c 0.6)(12°)	153,163				
Dimethyl acetal	$C_9H_{18}O_6$		-20.4	159cgp				60f (159a)
3,6-Anhydro-2-O-methyl-dulcitol	$C_7H_{14}O_5$		-13	160gpt			49f (160)	
3,6-Anhydro-2-O-methyl-L-galactonic acid	$C_7H_{12}O_6$	141–142	-70.3 (c 0.8)(12°)	158				
Methyl α-pyranoside	$C_8H_{14}O_5$							70f (159b)
Methyl β-pyranoside	$C_8H_{14}O_5$							80f (159b)
α-L-Galactose, 6-deoxy- (L-fucose)	$C_6H_{12}O_5$	145	-124.1→-76.4	163	22rib (28)		83rha (95)	49f (37)
6-Deoxy-L-galactitol (L-fucitol)	$C_6H_{14}O_5$	126.5–127.5		52		73xyll (52)	94rha (95)	
L-Fucitol pentaacetate	$C_{16}H_{24}O_{10}$							
L-Fucitol trimethylsilyl ether						8.33 (167)		
Methyl α-pyranoside	$C_7H_{14}O_5$	158–159	-191 (18°)	164			53temg (164)	
Methyl β-pyranoside	$C_7H_{14}O_5$	126–127	+10.5 (18°)	164			48temg (164)	
α-Pyranoside tetraacetate	$C_{14}H_{20}O_9$	92–93	-113 ($CHCl_3$)	165				
β-Pyranoside tetraacetate	$C_{14}H_{20}O_9$	172	-39 ($CHCl_3$)	166				
Trimethylsilyl ether					—	7.2 (34)		
L-Galactose, 6-deoxy-2-O-methyl-	$C_7H_{14}O_5$	149–150	-68→-85 (18°)	164,168			60temg (164)	
Methyl α-pyranoside	$C_8H_{16}O_5$	Syrup	-179 (26°) (CH_3OH)	136			180mfuc (136)	
Methyl β-pyranoside	$C_8H_{16}O_5$	98–99	+17.2 (c 0.5) (CH_3OH)	136			217mfuc (136)	
L-Galactose, 6-deoxy-3-O-methyl- (L-digitalose)	$C_7H_{14}O_5$	110	-97	164,170gpt			45glc (169)	
Methyl α-pyranoside	$C_8H_{16}O_5$	76–78,130–132	-200,-173 (c 0.4)	169,170pt				
Phenylosazone	$C_{19}H_{24}N_4O_3$	178–179d		169				
2,3,4,-Tri-O-methyl-L-fucose	$C_9H_{18}O_5$	36–37	-184→-128	172			92temg (164)	
Methyl α-L-fucopyranoside trimethyl ether	$C_{10}H_{16}O_5$	97–98	-209	164,172			72βtemg (80)	
Methyl β-L-fucopyranoside trimethyl ether	$C_{10}H_{16}O_5$	101.5–102.5	-21.1	172				
L-Galactose, 6-deoxy-4-sulfate	$C_6H_{12}O_8S$		-55	174g	100gals (175a)		81glc (175b)	
L-Galactose, 2-O-methyl-	$C_7H_{14}O_6$		-75 (c 0.5)(18°)	153ep				
2-O-Methyl-L-galactonolactone	$C_7H_{12}O_6$		+17 (18°)	153				
L-Galactose, 3-O-methyl-[1]	$C_7H_{14}O_6$			159p	74xyl (159c)			
L-Galactose, 4-O-methyl-	$C_7H_{14}O_6$	203–206	-84 (17°)	141gp				
4-O-Methyl-N-phenyl-L-galactopyranosyl amine	$C_{13}H_{19}NO_5$	167–168		176				
L-Galactose, 6-O-methyl-[1]	$C_7H_{14}O_6$			159	92xyl (159c)			
L-Galactose 3-sulfate	$C_6H_{12}O_9S$	Syrup	-32 (c 0.56)(as NH_4 salt)	127			75gal (127)	

TABLE C: NATURAL ALDOSES (Continued)

Substance[a] (Synonym) Derivative (A)	Chemical Formula (B)	Melting Point °C (C)	Specific Rotation[b] $[\alpha]_D$ (D)	Reference[c] (E)	Chromatography, R Value, and Reference[d] — ELC (F)	GLC (G)	PPC (H)	TLC (I)
L-Galactose 6-sulfate	$C_6H_{12}O_9S$	Syrup	−43 (c 0.56)(as NH_4 salt); −47 (c 0.2)(as Na salt)	127; 177			56gal (127)	
α-D-Glucopyranoside, ethyl- Tetra-p-nitrobenzoate	$C_8H_{16}O_6$; $C_{36}H_{28}N_4O_{18}$	113–114; 110–115	+150.3	1,180c; 1			60f (180a)	
Trifluoroacetate						6.2 (180b)		78f (69)
β-D-Glucopyranoside, ethyl- Tetraacetate Tetra-p-nitrobenzoate	$C_8H_{16}O_6$; $C_{16}H_{24}O_{10}$; $C_{36}H_{28}N_4O_{18}$	98–100; 101–108; 215–216	−37.9; +16.2 ($CHCl_3$); +28 (18°)(acetone)	1,180c; 1; 1				
Trifluoroacetate Trimethylsilyl ether						10.1 (180b); 195mmt (192)		
β-D-Glucopyranoside, methyl- Tetraacetate 2,3,4,6-Tetramethyl ether[k] Trimethylsilyl ether	$C_7H_{14}O_6$; $C_{15}H_{22}O_{10}$; $C_{11}H_{22}O_6$	102–104; 104–105; 40–41	−32; −18.7 ($CHCl_3$); −17.3	181; 182,183; 185	6rib (28)	358damg (184); 100, 143βtemg (80); 49, 107glct (9)	24f (150)	61f (30c); 76f (69); 46f (40)
α-D-Glucose Methyl α-pyranoside Methyl α-pyranoside tetramethyl ether Monohydrate	$C_6H_{12}O_6$; $C_7H_{14}O_6$; $C_{11}H_{22}O_6$; $C_6H_{12}O_6 \cdot H_2O$	146; 166; Syrup; 83	+112→+52.7; +158.9; +144 (acetone)	186,411c; 186; 194,195; 186	16rib (28); 9rib (28)	6.9 (189)	8f (49); 20f (49)	25f (30a); 29f (130); 29f (40)
Pentaacetate	$C_{16}H_{22}O_{11}$	114	+101.6 ($CHCl_3$)	182,187			27f (188a)	68f (188b)
2,3,4,6-Tetra-C-methyl-D-glucopyranose Trimethylsilyl ether	$C_{10}H_{20}O_6$	96	+92→+84	193	0 (5, 28)		82f (150)	86f (30b)
β-D-Glucose	$C_6H_{12}O_6$	148–150	+18.7→+52.7	186		100glct (9)		66f (59)
p-Nitrophenylhydrazone	$C_{12}H_{17}N_3O_7$	189	+21.5 (C_5H_5N, C_2H_5OH); −88 (c 0.5)	65,191; 31				
Pentaacetate Trimethylsilyl ether	$C_{16}H_{22}O_{11}$	135	+3.8 (c 7, $CHCl_3$)	182,190		157glct (9)	21f (188a)	81f (69)
D-Glucose 6-acetate Phenylhydrazone Tetrabenzoate	$C_8H_{14}O_7$; $C_{14}H_{20}N_2O_6$; $C_{35}H_{30}O_{11}$	133–135; 136; 183–184	+48; −13	196,197; 197; 198			28temg (197)	56f (59)
D-Glucose 6-acetate, 2,3,4-tri-O-[(+)-3-methylvaleryl]-	$C_{26}H_{44}O_{10}$	104–106	+30.2 ($CHCl_3$)	199c				
D-Glucose 1-benzoate (periplanetin) Tetraacetate	$C_{13}H_{16}O_7$; $C_{21}H_{24}O_{11}$	193; 140–141	−26.8	178p, 179; 178				
D-Glucose 6-benzoate Monohydrate β-Pyranoside tetraacetate	$C_{13}H_{16}O_7$; $C_{13}H_{16}O_7 \cdot H_2O$; $C_{21}H_{24}O_{11}$	Amorphous; 104–106; 132	+48 (C_2H_5OH); +32.9 ($CHCl_3$)	200; 178; 201				
D-Glucose 4-gallate[m] Heptaacetate	$C_{13}H_{16}O_{10}$; $C_{27}H_{30}O_{17}$	211–212; 125–126	−25.6 (18°); −24.3 (acetylene tetrachloride)	202[m]; 202				
D-Glucose 1-γ-hydroxy-α-methylenebutyrate (1-tuliposide A)	$C_{11}H_{18}O_8$		+64	203			312glc (203)	

TABLE C: NATURAL ALDOSES (Continued)

Substance[a] (Synonym) Derivative (A)	Chemical Formula (B)	Melting Point °C (C)	Specific Rotation[b] $[\alpha]_D$ (D)	Reference[c] (E)	Chromatography, R Value, and Reference[d]			
					ELC (F)	GLC (G)	PPC (H)	TLC (I)
D-Glucose 6-γ-hydroxy-α-methylenebutyrate (6-tuliposide A)	$C_{11}H_{18}O_8$	No constants known	No constants known	203				
D-Glucose 1-β,γ-di hydroxy-α-methylene-butyrate (1-tuliposide B)	$C_{11}H_{18}O_9$		+56	203			149glc (203)	
D-Glucose (tuliposide C)		No constants known	No constants known	203			100glc (203)	
D-Glucose 3-malonate	$C_9H_{14}O_9$	No constants known	No constants known	212p				
D-Glucose di-β-nitropropionate (endecaphyllin D)	$C_{12}H_{18}N_2O_{12}$	145–146		204t				
(endecaphyllin E)	$C_{12}H_{18}N_2O_{12}$	132–134, 138		204,205			27temg (205)	
D-Glucose tri-β-nitropropionate (endecaphyllin A, karakin)	$C_{15}H_{21}N_3O_{15}$	120–122	+4.5	204,205p				
Monoacetate	$C_{17}H_{23}N_3O_{16}$	125.5–126.5		204				
Diacetate	$C_{19}H_{25}N_3O_{17}$	103		205				
(endecaphyllin B)	$C_{15}H_{21}N_3O_{15}$	125–126.5		204				
(endecaphyllin C)	$C_{15}H_{21}N_3O_{15}$	150–152.5		204				
D-Glucose tetra-β-nitropropionate (endecaphyllin X, hiptagin)	$C_{18}H_{24}N_4O_{18}$	104–105.5		204,206				
D-Glucose, 4,6-O-(1'-carboxyethylidene)	$C_9H_{14}O_8$	No constants known	No constants known	207				
α-D-Glucose, 6-deoxy- (chinovose, epirhamnose, quinovose)	$C_6H_{12}O_5$	139–140	+73.3→+29.7 (c 8)	208,411c	148rha (96a)		96rha (96c)	99rha (96b)
6-Deoxy-D-glucitol (1-deoxy-L-gulitol)	$C_6H_{14}O_5$	151–152	+66.9→+5.4	209	94glcl (312)			
D-Glucomethylonic acid lactone	$C_6H_{10}O_5$							
Methyl α-pyranoside trimethylsilyl ether						3.4 (210a)		
Methyl β-pyranoside trimethylsilyl ether						3.6 (210a)		
Methyl α- or β-pyranoside	$C_7H_{14}O_5$							48f (210b)
Pyranoside tetraacetate	$C_{14}H_{20}O_9$	145	+23 (CHCl$_3$)	211				
α-D-Glucose, 6-deoxy-3-O-methyl- (D-thevetose)	$C_7H_{14}O_5$	116, 126	+84→+33	213,214 cp	15.5' (94a)		50aco (94b)	
Methyl α-pyranoside	$C_8H_{16}O_5$	86–87	+148±2	215				
Methyl α-pyranoside triacetate	$C_{13}H_{20}O_8$	105	+122 (acetone)	213				
Methyl β-pyranoside	$C_8H_{16}O_5$	116–117	-44±2	215				
Methyl β-pyranoside triacetate	$C_{13}H_{20}O_8$	121	+6 (acetone)	213				
D-Glucose, 6-deoxy-2,3-di-O-methyl-	$C_8H_{16}O_5$	Syrup	+40.4±2	216ep				
Methyl β-pyranoside	$C_9H_{18}O_5$	76–78	-49 (CHCl$_3$)	216				
D-Glucose, 6-deoxy-6-sulfonic acid (6-sulfoquinovose)	$C_6H_{12}O_8S$	No constants known	No constants known	217			39f (217)	
Allyl α-pyranoside cyclohexylamine salt	$C_{15}H_{29}NO_8S$	151.5–153	$[\alpha]$ Na$_{589}$ +86	217				

TABLE C: NATURAL ALDOSES (Continued)

Substance[a] (Synonym) Derivative (A)	Chemical Formula (B)	Melting Point °C (C)	Specific Rotation[b] [α]D (D)	Reference[c] (E)	Chromatography, R Value, and Reference[d]			
					ELC (F)	GLC (G)	PPC (H)	TLC (I)
Methyl α-pyranoside cyclohexylamine salt	$C_{13}H_{27}NO_8S$	173–174	+87	217				
α-D-Glucose, 3-O-methyl-	$C_7H_{14}O_6$	162–167	+98→+59.5 (c 0.4)	218c	13rib (28)		184glc (219a)	22f (219c)
Methyl α-pyranoside	$C_8H_{16}O_6 \cdot ½H_2O$	80–81	+164±2 (c 0.9)	220				22f (219c)
Methyl α-pyranoside triacetate	$C_{14}H_{22}O_9$					180damg (184) 3.46 (219b)		
Trimethylsilyl ether								
β-D-Glucose, 3-O-methyl-	$C_7H_{14}O_6$	130–132	+31.9→+55.1	221	90glc (223)			4f (30b)
3-O-Methyl-N-phenyl-D-glucopyrancside	$C_{13}H_{19}NO_5$	152–153	-108→-46±2 (c 0.5, CH₃OH)	223				
Methyl β-pyranoside	$C_8H_{16}O_6$	Syrup	-26 (c 5.5)	222				
Methyl β-pyranoside triacetate	$C_{14}H_{22}O_9$					266damg (184) 181amgl (184)		
Penta-O-acetyl-3-O-methyl-D-glucitol	$C_{17}H_{26}O_{11}$							
β-Pyranoside tetraacetate	$C_{15}H_{22}O_{10}$	95–96	-5.2 (CHCl₃)	224				
D-Glucose, 6-O-methyl-	$C_7H_{14}O_6$	139–141	+57.5	225	87glc (219d)		176glc (219a)	22f (219c)
Methyl α-py-anoside	$C_8H_{16}O_6$	Syrup	+127.9	226				
Methyl α-pyranoside triacetate	$C_{14}H_{22}O_9$					110damg (184)		
Methyl β-pyranoside	$C_8H_{16}O_6$	133–135	-27	227				
Methyl β-pyranoside triacetate	$C_{14}H_{22}O_9$					139damg (184)		
α-Pyranoside tetraacetate	$C_{15}H_{22}O_{10}$	119–120	+111.8 (CHCl₃)	228				
β-Pyranoside tetraacetate	$C_{15}H_{22}O_{10}$	91–93	+20.9 (CHCl₃)	228				
Trimethylsilyl ether						3.76 (219b)		
α-D-Glucose, 2,3-di-O-methyl-	$C_8H_{16}O_6$	85–87	+81.9→+48.3 (acetone)	135,221			211fuc (135)	13f (69)
Methyl α-pyranoside	$C_7H_{18}O_6$	80–82	+142.6	221				
Methyl α-pyranoside diacetate	$C_{11}H_{22}O_8$					15 (184)		
Tetra-O-acetyl-2,3-di-O-methyl-D-glucitol	$C_{16}H_{26}O_{10}$					29 (184)		
β-D-Glucose, 2,3-di-O-methyl-	$C_8H_{16}O_6$	108–110, 121	+5.9→+50.9 (acetone)	221,229	20.6glc (19c)		57temg (19a)	
2,3-Di-O-m-ethyl-N-phenyl-D-glucopyranosylamine	$C_{14}H_{21}NO_5$	134	-83 (CHCl₃)	230				
Methyl β-pyranoside	$C_7H_{18}O_6$	62–64	-36.6	222				
Methyl β-pyranoside diacetate	$C_{11}H_{22}O_8$					72.5damg (184)		
α-L-Glucose	$C_6H_{12}O_6$	141–143	-95.5→-51.4	231				
L-Glucose, 6-deoxy- (epi-rhamnose)	$C_6H_{12}O_5$	143–145	-84.7→-30.1	232,233			97rha (95)	
Diethyl mercaptal	$C_{10}H_{22}O_4S_2$	97–98	+47.1	233				
L-Glucose, 6-deoxy-3-O-methyl- (L-thevetose)	$C_7H_{14}O_5$	126–129	-36.9±2	234	99rha (96a)		247rha (96c)	153rha (96b)
α-Pyranos.de triacetate	$C_{13}H_{20}O_8$	103–104	-113 (CH₃OH)	235,236				
β-Pyranoside triacetate	$C_{13}H_{20}O_8$	118–119	-7.5±2 (acetone)	234				
D-Gulose, 6-deoxy- (antiarose)	$C_6H_{12}O_5$	130–131	-38	71,237	115rha (96a)	83aral (238a)	106rha (96c)	43rha (96b)
p-Bromophenylhydrazone	$BrC_{12}H_{17}N_2O_4$	135–136	-49→-34.7 (C₅H₅N)	71				
6-Deoxy-D-gulono-lactone	$C_6H_{10}O_5$	180–181		71				
6-Deoxy-D-gulitol pentaacetate	$C_{16}H_{24}O_{10}$							

TABLE C: NATURAL ALDOSES (Continued)

Substance[a] (Synonym) Derivative (A)	Chemical Formula (B)	Melting Point °C (C)	Specific Rotation[b] $[\alpha]_D$ (D)	Reference[c] (E)	Chromatography, R Value, and Reference[d]			
					ELC (F)	GLC (G)	PPC (H)	TLC (I)
Tetra-O-acetylglycoside	$C_{14}H_{20}O_9$	137–139	+5.2 (CHCl$_3$)	239			13f (49)	
L-Gulose	$C_6H_{12}O_6$							
L-Gulono-γ-lactone	$C_6H_{10}O_6$	183–185	+21.3	240,241,411c	53frib (28)			
Methyl α-pyranoside	$C_7H_{14}O_6$	77[f]	+55	242				
Methyl α-pyranoside tetraacetate	$C_{15}H_{22}O_{10}$	96–97	+109.4[f]	186				
Methyl β-pyranoside	$C_7H_{14}O_6$	176[f]	−96.5 (c 0.8, CHCl$_3$)	240				
Methyl β-pyranoside tetraacetate	$C_{15}H_{22}O_{10}$	64–66.5	−83.3[f]	186				
Methyl pyranoside tetramethyl ether	$C_{11}H_{22}O_6$		+33 (CHCl$_3$)	240				
Trimethylsilyl ether						1.37 (243); 66, 74glct (9)		
L-Gulose, 6-deoxy-	$C_6H_{12}O_5$	No constants known	No constants known	244				91rha (244a)
p-Bromophenylhydrazone	$BrC_{12}H_{17}N_2O_4$	135–137	+13±2 (C$_2$H$_5$OH)	244				
6-Deoxy-L-gulitol (1-deoxy-D-glucitol)	$C_6H_{14}O_5$			244	98glct (312)			
Methyl α-pyranoside triacetate	$C_{13}H_{20}O_8$	Syrup	+35 (CHCl$_3$)	244				73f (244b)
Methyl β-pyranoside triacetate	$C_{13}H_{20}O_8$	Syrup	−55 (CHCl$_3$)	244				62f (244b)
Methyl 6-deoxy-L-guloside	$C_7H_{14}O_5$					4 (244c)		
Hexodialdose, D-galacto-	$C_6H_{10}O_6$	Syrup	None (meso)	245			60f (245)	46f (246)
Methyl β-pyranoside	$C_7H_{12}O_6$							
Methyl β-pyranoside dimmer hexaacetate	$C_{26}H_{36}O_{18}$	206–108	+41.6 (CHCl$_3$)	246c				
Tetraacetate	$C_{14}H_{18}O_{10}$	184d		247				
Hexodialdo-1,5-pyranose,4-deoxy-4-ene-L-threo-	$C_6H_8O_5$	No constants known	No constants known	246				75f (246)
2,4-Dinitrophenylhydrazone	$C_{12}H_{12}N_4O_8$	148–152		246				
Methyl α-pyranoside diacetate	$C_{11}H_{14}O_7$		+32 (c 6.1, CH$_3$OH)	248t				
Methyl β-pyranoside diacetate	$C_{11}H_{14}O_7$		+309 (CHCl$_3$)	248t				
Hexopyranose, 1,1'-anhydro-2,6-dideoxy-4-C-(1'-hydroxyethyl)-	$C_8H_{14}O_4$	153–154	−144	249				
3,5-Dinitrobenzoate	$C_{15}H_{16}N_2O_9$	167–168		249				
Hexose, 2-deoxy-D-arabino- (2-deoxy-D-glucose)	$C_6H_{12}O_5$	146	+38.3→+45.9 (c 0.5) (18°)	250	<10sor (5)		97rha (95)	68f (37)
2-Deoxy-D-glucitol	$C_6H_{14}O_5$				100sor (132)		22f (49)	
2-Deoxy-D-glucitol pentaacetate	$C_{16}H_{24}O_{10}$					15 (131)		
2-Deoxy-N-phenyl-D-arabino-hexopyranosylamine	$C_{12}H_{17}NO_4$	193–194	−138→−106 (C$_5$H$_5$N)	250				
α-Pyranoside tetraacetate	$C_{14}H_{20}O_9$	91	+12.3 (c 0.3, C$_2$H$_5$OH)	250				35f (252)
		109.7–110.7	+107.7 (CHCl$_3$)	251				
β-Pyranoside tetraacetate	$C_{14}H_{20}O_9$	75–78	+30 (c 0.2, C$_2$H$_5$OH)	250				33f (252)
		92.2–93.2	−2.8 (CHCl$_3$)	251				
Trimethylsilyl ether						64glct (9)		

TABLE C: NATURAL ALDOSES (Continued)

Substance[a] (Synonym) Derivative (A)	Chemical Formula (B)	Melting Point °C (C)	Specific Rotation[b] $[\alpha]_D$ (D)	Reference[c] (E)	Chromatography, R Value, and Reference[d]			
					ELC (F)	GLC (G)	PPC (H)	TLC (I)
Hexose, 4-deoxy-D-*arabino*- (4-deoxy-D-altrose or idose)	$C_6H_{12}O_5$	No constants known	No constants known	253			168glc (253a)	127glc (253a)
Dibenzylmercaptal	$C_{20}H_{26}O_4S_2$	103–105	+146 (c 0.2, C_2H_5OH)	254				
4-Deoxy-D-*arabino*-hexitol			+97	253			128glc (253a)	
4-Deoxy-D-*arabino*-hexitol pentaacetate						190xyll (253b)		
Hexose, 2,6-dideoxy-D-*arabino*- (canarose, chromose C, olivose)	$C_6H_{12}O_4$	86–98	+19.6, +25 (H_2O)	256, 257			54f (258a)	14f (258b)
2,4-Dinitrophenylhydrazone	$C_{12}H_{16}N_4O_7$	100–103	+95.9, +110 (acetone); +31	256, 257; 258				
Methyl α-pyranoside	$C_7H_{14}O_4$	132–132.5	+131 (c 0.7, C_2H_5OH)	256			75f (268)	38f (258c)
Methyl β-pyranoside	$C_7H_{14}O_4$	84	−85 (C_2H_5OH)	258			70f (268)	27f (258c)
Hexose, 2,6-dideoxy-3-O-carbamoyl-D-*arabino*-	$C_7H_{13}NO_5$	No constants known	No constants known	258				
Methyl α-pyranoside	$C_8H_{15}NO_5$	146–149	+137 (acetone)	259				28f (259a)
Methyl α-pyranoside acetate	$C_{10}H_{17}NO_6$			259				
Olivose	$C_6H_{12}O_4$	100–110	+19.9 (H_2O), +60.8 (acetone)	259				30f (259b)
Hexose, 2,6-diceoxy-3-C-methyl-D-*arabino*- (D-3-epimycarose, evermicose)	$C_7H_{14}O_4$	108–112	+20.7 (24h)	260t			77f (262)	
1,4-Diacetate	$C_{11}H_{18}O_6$	73	+39.5 ($CHCl_3$)	260				
Hexose, 2,6-dideoxy-3-O-methyl-D-*arabino*- (D-oleandrose)	$C_7H_{14}O_4$	62–63	−12.5	261, 262			85cymo (263a)	44f (263b)
Hexonolactone	$C_7H_{14}O_4$	Syrup	+12.8±2 (acetone) (14°)	261				
Hexonic acid phenylhydrazide	$C_{13}H_{20}N_2O_4$	134–135	−20.6± (16°)(c 0.8, CH_3OH)	261				
Hexose, 3,6-dideoxy-L-*arabino*- (D-tyvelose)	$C_6H_{12}O_4$	95–99, 143–144	+24±2	264–266	122rha (273)		129rhaq (265)	33f (266)
3,6-Dideoxy-D-*arabino*-hexitol	$C_6H_{12}O_4$	113–115	−35±2 (c 0.7)	265			107rha (304)	40f (266)
Methyl α-pyranoside	$C_7H_{14}O_4$	84.5–85.5	$[\alpha]_{5461}$ +137±2 (CH_3OH)	267t				
Methyl β-pyranoside	$C_7H_{14}O_4$	Oil	$[\alpha]_{5461}$ −72±2 (CH_3OH)	267t				
Hexose, 2,6-dideoxy-3-C-methyl-L-*arabino*- (olivomycose)[j]	$C_7H_{14}O_4$	103–106	−13→−22 (26°) 1.5h	258,268			80f (258a)	
Methyl α-pyranoside	$C_8H_{16}O_4$	Syrup	−147 (C_2H_5OH)	268c			58f (268)	
Methyl α-pyranoside *iso*-butyrate[n]	$C_{12}H_{22}O_5$	Syrup	−123 (c 0.6, C_2H_5OH)	268			77f (268)	
Methyl β-pyranoside	$C_8H_{16}O_4$	93–94	+50 (C_2H_5OH)	268			73f (268)	
Methyl β-pyranoside *iso*-butyrate[n]	$C_{12}H_{22}O_5$	Syrup	+29 (C_2H_5OH)	268				
Hexose, 4-O-acetyl-2,6-dideoxy-3-C-methyl-L-*arabino*- (chromose B)	$C_9H_{16}O_5$	Syrup	−24	269p				

TABLE C: NATURAL ALDOSES (Continued)

Substance[a] (Synonym) Derivative (A)	Chemical Formula (B)	Melting Point °C (C)	Specific Rotation[b] $[\alpha]_D$ (D)	Reference[c] (E)	Chromatography, R Value, and Reference[d]			
					ELC (F)	GLC (G)	PPC (H)	TLC (I)
Hexose, 4-O-iso-butyryl-2,6-dideoxy-3-C-methyl-L-arabino-	$C_{11}H_{20}O_5$	Syrup	$-43 \to -33.5$ (c 0.5)	258			84f (258a)	
Hexose, 2,6-dideoxy-3-O-methyl-L-arabino- (L-oleandrose)	$C_7H_{14}O_4$	62–63	$+11.9\pm2.5$	270			14p (290)	
2,4-Dinitrophenylhydrazone	$C_{13}H_{18}N_4O_7$	155–160		271				
Hexonic acid phenylhydrazide	$C_{13}H_{20}N_2O_4$	135–136	$+21.1\pm3$ (c 0.8 CH$_3$OH)	270				
Methyl α-pyranoside	$C_8H_{16}O_4$	Syrup	-125.6 (C$_2$H$_5$OH)	272ct				
Methyl β-pyranoside	$C_8H_{16}O_4$	74–78	$+71.5$ (C$_2$H$_5$OH)	272ct				
Hexose, 3,6-dideoxy-L-arabino- (ascarylose, L-tyvelose)	$C_6H_{12}O_4$	Syrup	-25	265,273	122rha (273)	s	129rha (265)	
Ascarylitol	$C_6H_{14}O_4$	112–113	$+38\pm3$ (CH$_3$OH)	273			107rha (304)	
Hexose, 3-C-methyl-4-O-methyl-3-nitro-2,3,6-trideoxy-L-arabino- (evernitrose)	$C_8H_{15}NO_5$	88–92	$-4.9 \to -19,4$ (C$_2$H$_5$OH)	274				
Evernitronolactone	$C_8H_{13}NO_5$	63–64	-70	274				
Monoacetate	$C_{10}H_{17}NO_5$	58–59	-20.5 (C$_2$H$_5$OH)	274				
Hexose, 2,3,6-trideoxy-D-erythro- (erythro-amicitose)	$C_6H_{12}O_3$	Oil, bp 65–70	$+28.6$ (CHCl$_3$)	275				
2,4-Dinitrophenylhydrazone	$C_{12}H_{16}N_4O_6$	137.5–138, 152–153	-10 (c 0.9, C$_5$H$_5$N)	275,276	0.1 (276a)		79f (276b)	
Methyl α-pyranoside	$C_7H_{14}O_3$	Syrup	$+142\pm1$ (18°)	277				33f (277)
Methyl β-pyranoside	$C_7H_{14}O_3$	Syrup	-21.9 (c 0.8, CHCl$_3$)	278				32f (278)
Hexose, 2,6-dideoxy-D-lyxo- (2-deoxy-fucose, oliose)	$C_6H_{12}O_4$	Syrup	$+46, +53$	269c,279p			35f (269)	35f (280b)
3,4-Di-O-methyl-D-oliose	$C_6H_{12}O_4$	61–64	$+117$ (30°)(c 0.5, CHCl$_3$)	280				37f (280c)
Methyl α-pyranoside	$C_7H_{14}O_4$	70–72	$+122$ (16°)(CHCl$_3$)	281				
Methyl α-pyranoside 3,4-di-O-methyl ether	$C_9H_{18}O_4$	Syrup	$+115, +133$ (C$_2$H$_5$OH)	280				67f (280d)
Hexose, 3-O-acetyl-2,6-dideoxy-D-lyxo- (chromose D)	$C_8H_{14}O_5$	115–116.5	$+100 \to +78$ (29°)	279,281t			67f (279)	38f (280e)
Methyl α-pyranoside	$C_7H_{16}O_5$	Syrup	$+142$ (16°)(CHCl$_3$)	281				59f (281a)
Methyl α-pyranoside 4-O-methyl ether	$C_{10}H_{18}O_5$	Syrup	$+104$ (15°)(CHCl$_3$)	281				82f (281b)
Hexose, 2,6-dideoxy-3-O-methyl-D-lyxo- (diginose)	$C_7H_{14}O_4$	90–92	$+56\pm4$	283,284			11.4p (290)	
Diginonolactone	$C_7H_{12}O_4$	Syrup	-30 (14°)(acetone)	283			87cymo (263a)	50f (263b)
Hexose, 2,6-dideoxy-4-O-methyl-D-lyxo- (chromose A, olivomose)	$C_7H_{14}O_4$	158–162	$+98.5 \to +89$ (c 0.5)	268,285			65f (258a)	
2,4-Dinitrophenylhydrazone	$C_{13}H_{18}N_4O_7$	146–147	$+150$ (c 0.4, C$_2$H$_5$OH)(26°)	286				
Methyl α-pyranoside	$C_8H_{16}O_4$	98		268,286				61f (258c)
Methyl β-pyranoside	$C_8H_{16}O_4$	152–153	-37.4 (c 0.4, C$_2$H$_5$OH)(26°)	268,286				54f (258c)

TABLE C: NATURAL ALDOSES (Continued)

Substance[a] (Synonym) Derivative (A)	Chemical Formula (B)	Melting Point °C (C)	Specific Rotation[b] $[\alpha]_D$ (D)	Reference[c] (E)	ELC (F)	GLC (G)	PPC (H)	TLC (I)
						Chromatography, R Value, and Reference[d]		
Olivomose 3-acetate	$C_9H_{16}O_5$		+69 (c 0.4)	280				42f (280a)
Hexose, 2,6-dideoxy-L-*lyxo*-	$C_6H_{12}O_4$	103–106	−90.4→−61.6	287,288pt			10.7° (289)	
2,6-Dideoxy-L-*lyxo*-hexonolactone	$C_6H_{10}O_4$	167–169	+31.2±2 (acetone)	263			55drho (263c)	
Hexonic acid phenylhydrazide	$C_{12}H_{18}N_2O_4$	78–85	−8.5±2	263				
Hexose, 2,6-dideoxy-3-*O*-methyl-L-*lyxo*- (L-diginose)	$C_7H_{14}O_4$		−65	290			11.4° (290)	
Hexose, 2,6-dideoxy-D-*ribo*- (digitoxose)	$C_6H_{12}O_4$	110	+46.4	291			128rha (95)	94f (292)
2,6-Dideoxy-D-*ribo*-hexitol trimethylsilyl ether						46.5glcs (294)		
Digitoxonolactone	$C_6H_{10}O_4$	123–125	−29.5±2 (c 0.6, acetone)	263				
Digitoxonic acid phenylhydrazide	$C_{12}H_{18}N_2O_4$		−17.8±2 (c 0.4)	263			63drho (263c)	
Methyl α-pyranoside	$C_7H_{14}O_4$	Bp 98–100 (10⁻¹ torr)	+178.4 (CHCl₃)	293				
Phenylhydrazone	$C_{12}H_{18}N_2O_3$	204–209	+215 (C₂H₅OH, C₂H₅N)	307				
Trimethylsilyl ether						18glct (9)		
Hexose-2,6-dideoxy-3-*O*-methyl-D-*ribo*- (cymarose)	$C_7H_{14}O_4$	83, 90, 93	+55	295,296			83f (150)	100 (410)
Cymaronolactone	$C_7H_{12}O_4$	Syrup	−25	263			100 (263a)	54f (263b)
Cymaronic acid phenylhydrazide	$C_{13}H_{20}N_2O_4$	155–156	+1.4 (c 0.7, CH₃OH)(16°)	263,297				
Methyl α-pyranoside	$C_8H_{16}O_4$	34–36	+210 (14°)(CH₃OH)	298				
Hexose, 3,6-dideoxy-D-*ribo*- (paratose)	$C_6H_{12}O_4$	Syrup	+10±2 (c 0.9)	273,299p	138rha (273)		125rha (300)	
Methyl α-pyranoside	$C_7H_{14}O_4$	Syrup	$[\alpha]_{578}$ +170 (CHCl₃)	301				33f (301)
Methyl β-pyranoside	$C_7H_{14}O_4$	63–65	−60 (CH₃OH)	266p				
Paratitol	$C_6H_{14}O_4$	67–68	−18±2 (c 0.9)	299			71f (299)	
Hexose, 4,6-dideoxy-3-*O*-methyl-D-*ribo*- (chalcose, lancavose)	$C_7H_{14}O_4$	96–99	+120→+76	302,303t				
4,6-Dideoxy-D-*ribo*-hexose	$C_6H_{12}O_4$			305			71f (305a)	10f (305b)
Methyl α-pyranoside	$C_8H_{16}O_4$	101.5–102	+184.5 (CHCl₃)	302			65f (302)	
Methyl β-pyranoside	$C_8H_{16}O_4$		−21 (27°)(CHCl₃)	302				30f (305b)
Hexose, 2,6-dideoxy-3-*C*-methyl-L-*ribo*- (L-mycarose)[s]	$C_7H_{14}O_4$	128–129	−31.1	319	47glc (320a)		65f (320b)	
3,4-Di-*O*-methyl mycarose	$C_9H_{18}O_4$	83–86	−20 (CHCl₃)	321				
Methyl α-pyranoside[t]	$C_8H_{16}O_4$	Syrup	+22, +54 (CHCl₃)	319,324	48glc (320a)			
Methyl β-pyranoside[t]	$C_8H_{16}O_4$	62	−155 (CHCl₃)	319,324	37glc (320a)			
Mycaronolactone	$C_7H_{12}O_4$	108–109	−35	319				
1,3,4-Triacetate	$C_{13}H_{20}O_7$	133–135	−61.3 (26°)(C₂H₅OH)	325				
Mycarose 4-*O*-acetate	$C_9H_{16}O_5$	Bp 65 (1.5 mm)		327				
Methyl α-pyranoside	$C_{10}H_{18}O_5$	Bp 88–99 (2 mm)	−148 (CHCl₃)	326				
Methyl β-pyranoside	$C_{10}H_{18}O_5$	Bp 104–106 (1.5 mm)	+25 (CHCl₃)	326				

TABLE C: NATURAL ALDOSES (Continued)

(A) Substance[a] (Synonym) Derivative	(B) Chemical Formula	(C) Melting Point °C	(D) Specific Rotation[b] $[\alpha]_D$	(E) Reference[c]	Chromatography, R Value, and Reference[d]			
					(F) ELC	(G) GLC	(H) PPC	(I) TLC
Mycarose 4-O-n-butyrate	$C_{11}H_{20}O_5$	No constants known	No constants known	326				
Methyl α-pyranoside	$C_{12}H_{22}O_5$	Bp 112–122 (3 mm)	−137 ($CHCl_3$)	326				
Methyl β-pyranoside	$C_{12}H_{22}O_5$	Bp 118–119 (2 mm)	+16.9 ($CHCl_3$)	326				
Mycarose 4-O-iso-valerate	$C_{12}H_{22}O_5$	No constants known	No constants known	326				
Methyl α-pyranoside	$C_{13}H_{24}O_5$	Bp 115–116 (2 mm)	−135.5 ($CHCl_3$)	326				
Methyl β-pyranoside	$C_{13}H_{24}O_5$	Bp 117–118 (0.7 mm)	+13.5 ($CHCl_3$)	326				
Mycarose 4-O-n-propionate	$C_{10}H_{18}O_5$	No constants known	No constants known	326				
Methyl α-pyranoside	$C_{11}H_{20}O_5$	Bp 98–100 (2 mm)	−145 ($CHCl_3$)	326				
Methyl β-pyranoside	$C_{11}H_{20}O_5$	Bp 109–110 (2 mm)	+20 ($CHCl_3$)	326				
Hexose, 2,6-dideoxy-3-C-methyl-3-O-methyl-L-ribo- (cladinose)[s]	$C_8H_{16}O_4$	Bp 120–132 (0.25 mm)	−23.1	328	0glc (320a)		74f (320b)	
Cladinitol	$C_8H_{18}O_4$	66–67	−25 (27°)(95% C_2H_5OH)	329			160ole (329)	
Cladinose diacetate	$C_{12}H_{20}O_6$	123–125	−36 (CH_3OH)	323				
Cladinonolactone 3,5-dinitrobenzoate	$C_{15}H_{16}N_2O_9$	27.5–28.5[f]		330				
Methyl α-pyranoside	$C_9H_{18}O_4$	87–91	−6.9	322,328				
Hexose, 2,6-dideoxy-3-O-methyl-L-ribo- (L-cymarose)	$C_7H_{14}O_4$		−53.6±2	295				
L-Cymaronic acid phenylhydrazide	$C_{13}H_{20}N_2O_4$	153–154	0.3±3 (c 0.7, CH_3OH)	295				
Hexose, 2,3,6-trideoxy-D-threo- (threo-amicitose)	$C_6H_{12}O_3$	Syrup	+10.2 (acetone)	306	2.3 (276a)			116digx (306)
2,4-Dinitrophenylhydrazone	$C_{12}H_{16}N_4O_6$	105–106, 121–122	+13.7 (c 0.9, C_5H_5N)	276,306				
Hexose, 2,3,6-trideoxy-L-threo- (L-threo-amicitose, rhodinose)	$C_6H_{12}O_3$	Syrup	−11±6	307,308[c]			71f (307a)	73f (307b)
2,4-Dinitrophenylhydrazone	$C_{12}H_{16}N_4O_6$	121–122	−14.9 (c 0.5, C_5H_5N)	306,309				
Hexose, 2-deoxy-D-xylo- (2-deoxy-D-gulose or idose)	$C_6H_{12}O_5$	Syrup	+12±2	310,311			47digx (310)	
2-Deoxy-D-xylo-hexonolactone	$C_6H_{10}O_5$	Syrup	−56.6±2 (acetone)	311				
2-Deoxy-D-xylo-hexonic acid phenylhydrazide	$C_{12}H_{18}N_2O_5$	124–126	−8.1±2 (CH_3OH)	311				
2-Deoxy-xylo-hexitol	$C_6H_{14}O_5$		−3.9→+3.9±2 (18°)	313	107glcl[f] (312)			
Hexose, 2,6-dideoxy-D-xylo- (boivinose)	$C_6H_{12}O_4$	96–98	−14 (acetone)	238			176digx (310) 522 gal (238b)	

TABLE C: NATURAL ALDOSES (Continued)

Substance[a] (Synonym) Derivative (A)	Chemical Formula (B)	Melting Point °C (C)	Specific Rotation[b] $[\alpha]_D$ (D)	Reference[c] (E)	Chromatography, R Value, and Reference[d]			
					ELC (F)	GLC (G)	PPC (H)	TLC (I)
Methyl α-pyranoside	$C_7H_{14}O_4$	Syrup	+108.7±2 (CH_3OH)	313			75cym (315)	
Hexose, 2,6-dideoxy-3-O-methyl-D-*xylo*- (sarmentose)	$C_7H_{14}O_4$	78–79	+12→+15.8	314				
Methyl α-pyranoside	$C_8H_{16}O_4$	33, 36	+156±1 (acetone)	316				
Methyl β-pyranoside	$C_8H_{16}O_4$	46–45	−39.4±1.5 (acetone)	316				
Sarmentonolactone	$C_7H_{12}O_4$						66cymo (263a)	42f (263b)
Hexose, 3,6-dideoxy-D-*xylo*- (abequose)	$C_6H_{12}O_4$	138–139	−3.2±0.6	264,265			117rha (95)q	
3,6-Dideoxy-D-*xylo*- hexitol	$C_6H_{14}O_4$	92–93	+51±2	265				
Methyl α-pyranoside	$C_7H_{14}O_4$	Syrup	$[\alpha]_{5461}$ +102±5 (CH_3OH)	267				
Methyl β-pyranoside	$C_7H_{14}O_4$	Syrup	$[\alpha]_{5461}$ −90±3 (CH_3OH)	267				
Hexose, 2,6-dideoxy-4-C-[1'-hydroxyethyl]-L-*xylo*-	$C_8H_{16}O_5$	No constants known	No constants known	317				
Methyl α-pyranoside (?) (glycoside A2)	$C_9H_{18}O_5$		+15 (c 0.5, $CHCl_3$)	317			31f (317a)	25f (317b)
Methyl β-pyranoside (?) (glycoside A1)	$C_9H_{18}O_5$		−104 (c 0.7, $CHCl_3$)	317				48f (317b)
Hexose, 2,6-dideoxy-3-C-methyl-L-*xylo*- (axenose)	$C_7H_{14}O_4$	111–112	−28.5	318				
Methyl α-pyranoside	$C_8H_{16}O_4$	101–103	−142	318				
Methyl β-pyranoside	$C_8H_{16}O_4$	122–123	+38	318				
Hexose, 2,6-dideoxy-3-C-methyl-3-O-methyl-L-*xylo*- (arcancse)	$C_8H_{16}O_4$	96–98	−20.9 (C_2H_5OH)	303				93f (303)
Arcanitol	$C_8H_{18}O_4$	bp 110 (0.05 torr)	−2.0 (C_2H_5OH)	303				
Arcanose 4-acetate	$C_{10}H_{18}O_5$	Oil	−52.3 (C_2H_5OH)	303				
Methyl pyranoside acetate	$C_{11}H_{20}O_5$	Oil	−24.3 (c 6.5, C_2H_5OH)	303				
Hexose, 2,6-dideoxy-4-C-[1'-oxoethyl]-L-*xylo*-	$C_8H_{14}O_5$	No constants known	No constants known	317				
Methyl α-pyranoside (?) (glycoside B2)	$C_9H_{16}O_5$		+46 (c 0.18, $CHCl_3$)	317			70f (317a)	57f (317c)
Methyl β-pyranoside (?) (glycoside B1)	$C_9H_{16}O_5$		−60 (c 0.12, $CHCl_3$)	317				70f (317c)
Hexose, 3,6-dideoxy-L-*xylo*- (colitose)	$C_6H_{12}O_4$	Syrup	+4	265,331			116rha (265)	
Colitito	$C_6H_{14}O_4$	92–94	−51±2	265				
p-Nitrophenylsulfonylhydrazone	$C_{18}H_{21}N_3O_7S$	141		331				
Everninose (a deoxy-O-methyl-hexose)	$C_7H_{14}O_5$	186–188	−69	255			56f (255)	
p-Tolylsulfonylhydrazone	$C_{14}H_{22}N_2O_6S$	135–137	+36.2 (C_5H_5N)	255				
Variose (a dideoxy-C-methyl-hexose)	$C_7H_{14}O_4$		+54 (c 0.5)	333c				45f (333)
Vinelose (a 6-deoxy-3-C-methyl-2-O-methyl-hexose)	$C_8H_{16}O_5$	Oil	$[\alpha]_{546}$ +12 (14.5°)	334	62glc (334c)		454rha (334a)	
Diacetate	$C_{12}H_{20}O_7$					120erya (282)		
Vinelitol	$C_8H_{18}O_5$	Oil	$[\alpha]_{546}$ −6.4 (19.8°)	334			346rha (334a)	

TABLE C: NATURAL ALDOSES (Continued)

Substance[a] (Synonym) Derivative (A)	Chemical Formula (B)	Melting Point °C (C)	Specific Rotation[b] $[α]_D$ (D)	Reference[c] (E)	Chromatography, R Value, and Reference[d]			
					ELC (F)	GLC (G)	PPC (H)	TLC (I)
Vinelitol triacetate	$C_{14}H_{24}O_8$	Oil	$[α]_{546}$ −45.7 (CHCl$_3$)	334				
Vinelose, O-[2'-O-methyl-glycolyl]-	$C_{11}H_{20}O_7$	Oil	−26.5 (c 0.9, CHCl$_3$)	334	0glc (334c)		76f (334b)	
D-Idose[v]	$C_6H_{12}O_6$	Syrup	+15.8±1	335,336			9f (49)	
D-Iditol	$C_6H_{14}O_6$	See Table 1 also	See Table 1 also	334			80glc (91a)	
Methyl α-pyranoside	$C_7H_{14}O_6$	67–68	+99.8±1	336				
Methyl β-pyranoside	$C_7H_{14}O_6$	Syrup	−81.1±1	336				
Pentaacetate	$C_{16}H_{22}O_{11}$	91–92	+54.3±2 (CHCl$_3$)	336				
Trimethylsilyl ether						50glct (9)		
L-Idose	$C_6H_{12}O_6$	Syrup	−17.4	339p	115rib^f (28)		172glc (337)	
L-Iditol	$C_6H_{14}O_6$	See Table 1 also	See Table 1 also		100glcl^f (312)			
Methyl α-pyranoside	$C_7H_{14}O_6$	Syrup	−98 (CH$_3$OH)	338		75glct (9)	57f (338)	
Methyl α-pyranoside trimethylsilyl ether								
Methyl β-pyranoside	$C_7H_{14}O_6$	Syrup	+61 (27°)(CH$_3$OH)	338		64glct (9)	63f (338)	
Methyl β-pyranoside trimethylsilyl ether								
L-Idopyranose, 1,6-anhydro-	$C_6H_{10}O_5$	128–129	+113 (acetone)	337			265glc (337)	
β-Triacetate	$C_{12}H_{16}O_8$	85–86	+75.5 (c 0.7, (CHCl$_3$)	337c				
β-Trimethyl ether	$C_9H_{16}O_5$	39–40	+88 (19°)(CHCl$_3$)	338c				
α-D-Mannose	$C_6H_{12}O_6$	133	+29.3→+14.5	340	35rib (28)		8f (49)	30f (30)
Methyl α-pyranoside	$C_7H_{14}O_6$	193–194	+79.2	186	17rib (28)		42f (49)	2f (69)
Methyl α-pyranoside tetramethyl ether	$C_{11}H_{22}O_6$	38–39	+69 (CH$_3$OH)	341gt		154atemg (67)	98temg (341)	54f (69)
Pentaacetate	$C_{16}H_{22}O_{11}$	64	+55 (CHCl$_3$)	182				
2,3,4,6-Tetra-O-methyl-D-mannose	$C_{10}H_{20}O_6$	50–52	+7.4→+2.4	186			100temg (342)	
Trimethylsilyl ether						70glct (9)		56f (59)
β-D-Mannose	$C_6H_{12}O_6$	132	−17→+14.6	343,411c				
Methyl β-pyranoside	$C_7H_{14}O_6$							
p-Nitrophenylhydrazone	$C_{12}H_{17}N_3O_7$	202–203	+56 (C$_5$H$_5$N, C$_2$H$_5$OH)	64,65,191,344				
Pentaacetate	$C_{16}H_{22}O_{11}$	117–118	−25.3 (CHCl$_3$)	182,343				
Trimethylsilyl ether						108glct (9)		54f (59)
D-Mannose, 3-O-carbamoyl-	$C_7H_{13}NO_7$	No constants known	No constants known	240				
Methyl α-pyranoside	$C_8H_{15}NO_7$	Amorphous	+49 (CH$_3$OH)	345cg				
Methyl α-pyranoside triacetate	$C_{14}H_{19}NO_{10}$	142.5	+35.8 (CHCl$_3$)	345t				40f (244 b,g)
Methyl α-pyranoside tribenzoate	$C_{29}H_{27}NO_{10}$	Glass	−19.5 (c 0.9, CHCl$_3$)	345				33f (345)
D-Mannose, 4,6-O-(1-carboxy-ethylidene)-	$C_9H_{14}O_8$	No constants known	No constants known	346gpt				
D-Mannose, 6-deoxy- (D-rhamnose)	$C_6H_{12}O_5$	86–90	−7.0	347,411c	100glcl (312)		100rha (347)	37f (130)
D-Rhamnitol	$C_6H_{14}O_5$						97rha^f (95)	38f (197)
Rhamnitol pentaacetate	$C_{16}H_{24}O_{10}$					7.1 (131)		
Rhamnitol pentamethyl ether	$C_{11}H_{24}O_5$					7.1 (353)		
D-Mannose, 6-deoxy-3-C-methyl- (D-evalose)	$C_7H_{14}O_5$	Glass	−4.7→−5.2	348				

TABLE C: NATURAL ALDOSES (Continued)

Substance[a] (Synonym) Derivative (A)	Chemical Formula (B)	Melting Point °C (C)	Specific Rotation[b] $[\alpha]_D$ (D)	Reference[c] (E)	Chromatography, R Value, and Reference[d]			
					ELC (F)	GLC (G)	PPC (H)	TLC (I)
2,3,4-Tri-O-methyl-D-evalose (D-nogalose)[u]	$C_{10}H_{20}O_5$	115–120	+18.3→+6.3 (CH_3OH) 24h	348				
D-Mannose, 6-deoxy-2-O-methyl-	$C_7H_{14}O_5$		−22	350cg	32glc (350a)		137rha (350b)	
D-Mannose, 6-deoxy-3-O-methyl- (D-acofriose)	$C_7H_{14}O_5$		−27 (c 0.9)	350	56glc (350a)		132rha (350b)	
D-Mannose, 6-deoxy-2,3-di-O-methyl-	$C_8H_{16}O_5$		W	351			192rha (352e)	
D-Mannose, 6-deoxy-3,4-di-O-methyl-	$C_8H_{16}O_5$	86–88	W	351			198rha (352e)	
2,3,4-Tri-O-methyl-D-rhamnose	$C_9H_{18}O_5$		W				232rha (352)	
Methyl 2,3,4-tri-O-methyl-α pyranoside	$C_{10}H_{20}O_5$		W	350		31 α temg (350c)		
2,3,4-Tri-O-methyl-D-rhamnonic acid phenylhydrazide	$C_{16}H_{25}N_2O_5$	181–183	−36.3 (c 0.14, C_2H_5OH)	350				
D-Mannose, 3-O-methyl-	$C_7H_{14}O_6$	133–134	+14→+3 (c 0.6)	354,355	48glc (354a)		94rha (354b)	176glc (356p)
D-Mannose, 2,6-di-O-methyl- (D-curamicose)	$C_8H_{16}O_6$	Syrup	+10.3, +22.4	357,358ep	0 (357c)		72glc (357a)	
Curamicitol (1,5-di-O-methyl-L-mannitol)	$C_8H_{18}O_6$	Syrup	+18.5 (c 0.5)	357				
Curamiconolactone	$C_8H_{14}O_6$	Syrup	+58	357		368 α temg (357b)		
2,6-Di-O-methyl-D-mannonic acid phenylhydrazide	$C_{14}H_{22}N_2O_6$	130	−40 (c 0.5)	357				
Methyl α-pyranoside	$C_9H_{18}O_6$					324 α temg (357b)		
Trimethylsilyl ether						77,96 α temg (357b)		
α-L-Mannose, 6-deoxy- (L-rhamnose)	$C_6H_{12}O_5 \cdot H_2O$	93–94	−8.6→+8.2	63,359	100rha (96a)		100rha (96c)	100rha (96b)
Methyl α-pyranoside	$C_7H_{14}O_5$	109–110	−62.5	362			76f (360a)	
Methyl α-pyranoside trimethyl ether	$C_{10}H_{20}O_5$	Syrup	−15.1	361		44 β temg (67)		
Rhamnitol trimethylsilyl ether						66glcs (294)		
2,3,4-Tri-O-methyl-L-rhamnose	$C_9H_{18}O_5$	Syrup	+26	361			102temg (68)	
Triemthylsilyl ether						30glct (9)		
β-L-Mannose, 6-deoxy-	$C_6H_{12}O_5$	123–125	+38.4→+8.9	363	32rib (28)		53digx (310)	46f (30)
Methyl β-pyranoside	$C_7H_{14}O_5$	138–140	+95.4	364				
p-Nitrophenylhydrazone	$C_{12}H_{17}N_3O_6$	190–191	−50→−8.5 (C_5H_5N, C_2H_5OH)	64				
L-Rhamnonolactone	$C_6H_{10}O_5$	98–99	+13.9 (c 15, $C_2H_5Cl_2$)	364				69f (17)
Tetraacetate	$C_{14}H_{20}O_9$	115–121	+15.5 (H_2O)					
L-Mannose, 6-deoxy-3-C-methyl-2,3,4-tri-O-methyl- (nogalose)	$C_{10}H_{20}O_5$		−10.6 (CH_3OH)	332				
Methyl pyranoside	$C_{11}H_{22}O_5$	41–43	−48.4 (CH_3OH)	349				

TABLE C: NATURAL ALDOSES (Continued)

Substance[a] (Synonym) Derivative (A)	Chemical Formula (B)	Melting Point °C (C)	Specific Rotation[b] $[\alpha]_D$ (D)	Reference[c] (E)	Chromatography, R Value, and Reference[d] — ELC (F)	GLC (G)	PPC (H)	TLC (I)
Nogalitol	$C_{10}H_{22}O_5$	Oil	-13 (CH_3OH)	349				
Nogalonolactone	$C_{10}H_{18}O_5$	Bp 76 (0.1 mm)	+15.9 ($CHCl_3$); +6.7 (CH_3OH)	349; 332				
L-Mannose,6-deoxy-5-C-methyl-4-O-methyl- (noviose,5,5-di-C-methyl-4-O-methyl-L-lyxose)	$C_8H_{16}O_5$	133–134	+22.6 (50% C_2H_5OH)	365				
Methyl α-pyranoside	$C_9H_{18}O_5$	68–70	-62±2 (C_2H_5OH)	366				
Methyl β-pyranoside	$C_9H_{18}O_5$	61–68	+113.8	367				
Novono-γ-lactone	$C_8H_{14}O_5$	111–113	-35 (0.1 N HCl)	368				
Noviose, 3-O-carbamoyl-	$C_9H_{17}NO_6$	124–126	+45.3 (C_2H_5OH)	367				
Methyl α-pyranoside	$C_{10}H_{19}NO_6$	194–195	-24.7 (C_2H_5OH)	367				
Methyl β-pyranoside	$C_{10}H_{19}NO_6$	117–118	+124±4 (C_2H_5OH)	366				
L-Mannose, 6-deoxy-2-O-methyl-	$C_7H_{14}O_5$	113–114	+31 (27°)	134,352,369			148rha (352)	
6-Deoxy-2-O-methyl-N-phenyl-L-mannopyranosylamine	$C_{13}H_{19}NO_4$	152	+43 (C_5H_5N)	370				
6-Deoxy-2-O-methyl-L-mannonolactone	$C_7H_{12}O_5$	116–117	-62	370				
Methyl pyranoside	$C_8H_{16}O_5$	139–140		134				
L-Mannose, 6-deoxy-3-O-methyl- (L-acofriose)	$C_7H_{14}O_5$	112–116	+37.3±2	115	72rha (96a)		212rha (96c)	135rha (96b)
Acofrionolactone	$C_7H_{12}O_5$	Syrup	-20 (15°)	371				
Methyl glycoside	$C_8H_{16}O_5$	128–130	+57 ($C_5H_5N \cdot C_2H_5OH$)	372		486βtemg (67)		
Phenylosazone	$C_{19}H_{26}N_4O_3$							
L-Mannose, 6-deoxy-2,3-di-O-methyl-	$C_8H_{16}O_5$	Syrup	+47.6	372,373	2glc (375a)		83glc (373)	
6-Deoxy-2,3-di-O-methyl-N-phenyl-L-mannopyranosylamine	$C_{14}H_{21}NO_4$	136–137	+147.8→+42.8 (c 0.4, C_2H_5OH) 70h	374,357				
2,4-Dinitrophenyl-hydrazone	$C_{14}H_{20}N_4O_8$	168d	+45.4 (c 0.6, di-oxane)	375				
Methyl α-pyranoside	$C_9H_{18}O_5$	Syrup	-6, -14	373,374				
L-Mannose, 6-deoxy-2,4- di-O-methyl-	$C_8H_{16}O_5$	82–91, 93	+42 (CH_3OH); -19 (16°), +10.6	377cp; 352,375,376	2glc (375a)		87glc (375b)	
6-Deoxy-2,4-di-O-methyl-N-phenyl-L-mannopyranosylamine	$C_{14}H_{21}NO_4$	133–134	+137 (c 0.7, CH_3OH)	377p				
6-Deoxy-2,4-di-O-methyl-L-mannonolactone	$C_8H_{14}O_5$	141–142	+110→+7 (c 0.4, C_2H_5OH)	376				
2,4-Dinitrophenylhydrazone	$C_{14}H_{20}N_4O_8$	Syrup	+47 (15°)(c 0.9)	376				
Methyl α-pyranoside	$C_9H_{18}O_5$	164–165d	+39 (dioxane)	375				
Methyl β-pyranoside	$C_9H_{18}O_5$	Syrup	-68 (CH_3OH)	375				
L-Mannose, 6-deoxy-3,4-di-O-methyl-	$C_8H_{16}O_5$	98–100	+85 (c 0.5, CH_3OH)	375	40glc (375a)		88glc (375b)	
2,4-Dinitrophenyl-hydrazone	$C_{14}H_{20}N_4O_8$	98–99	+18.5 (c 0.5)	378,379				
Methyl glycoside	$C_9H_{18}O_5$	170	-75.6 (dioxane)	375		110βtemg (67)		
D-Talose	$C_6H_{12}O_6$	128–132	+16.9	380,411c	70sor (5)		19glc (379)	
Methylphenylhydrazone	$C_{13}H_{20}N_2O_5$	143–144		380				36f (17)

TABLE C: NATURAL ALDOSES (Continued)

Substance (Synonym) Derivative (A)	Chemical Formula (B)	Melting Point °C (C)	Specific Rotation [α]D (D)	Reference (E)	Chromatography; R Value, and Reference			
					ELC (F)	GLC (G)	PPC (H)	TLC (I)
α-Pyranoside pentaacetate	C16H22O11	106–107	+70.2 (CHCl3)	381				
Talitol	C6H14O6				138rib (28)		110sor (382)	31f (17)
Talitol hexaacetate	C18H26O12	129–131				102manl (91b)	107f (383)	68f (17)
Talonolactone	C6H10O6							
Trimethylsilyl ether						86glct (9)		
D-Talose, 6-deoxy- (D-talomethylose)	C6H12O5		+20.6	347	98glc (312)		124rha (351)	
6-Deoxy-D-talitol (1-deoxy-D-altritol)	C6H14O5							
6-Deoxy-D-talitol pentaacetate	C16H24O10	91–91.5	+76 (26°)(CH3OH)	384				
Methyl α-pyranoside triacetate	C13H20O8		+53 (C5H5N:C2H5OH, 2:3)	347		41xyll (82c)		
Phenylosazore	C18H24N4O3	176–178						
D-Talose, 6-deoxy-3-O-methyl- (D-acovenose)	C7H14O5	Syrup	+16.5	82	106rha (82b)		163xyl (82a)	
6-Deoxy-3-O-methyl-D-talitol tetraacetate	C15H24O9					31xyll (82c)		
L-Talose, 6-deoxy- (L-talomethylose)	C6H12O5	126–127	−20.5±1.4	385cp 360ep		113rha (96a)	189rha (96c)	49rha (96b)
p-Bromophenylhydrazone	BrC12H17N2O4	145–147	−10→+4 (16°)(c 0.8, C2H5OH)	386				
6-Deoxy-L-talonolactone	C6H10O5	134–135	+33±2 (18°)	386				
Methylphenylhydrazone	C13H20N2O4	136–137	−12 (17°)(c 0.8, C2H5OH)	386				
Methyl α-pyranoside	C7H14O5	63–65	−104	360	39glc (360b)		81f (360a)	
Methyl α-pyranoside triacetate	C13H20O8	91–92	−73.3 (c 1.2, CH3OH)	360				
L-Talose, 6-deoxy-3-O-methyl- (L-acovenose)	C7H14O5	Amorphous	−19.4	351,387	103rha (96a)		385rha (96c)	94rha (96b)
L-Acovenonolactone	C7H12O5	167–168	+29.4±2 (16°) (CH3OH)	387,388pt				
Heptose, D-glycero-D-galacto-	C7H14O7	139–140	+47→+64 (c 0.5)	389ep, 392cp	40sor (5)		15f (391)	
2,5-Dichlorcphenylhydrazone	C13Cl2H18N2O6 See Table 1 for this compound	203–204		389				
Heptitol, D-glycero-D-galacto-	See Table 1 for this compound							
β-Hexaacetate	C19H26O13	109–110	+30.4 (CHCl3)	390				
Trimethylsilyl ether						427×glct (9)		
Heptose, D-glycero-D-gluco- Heptitol, D-glycero-D-gluco-	C7H14O7 See Table 1 for this compound	156–157	+17→+46.2 (6h)	394cp	20'sor (5)		80glc (393)	
α-Hexaacetate	C19H26O13	180–182	+105 (CCl2H2)	394				
β-Hexaacetate	C19H26O13	133–134	+19.6 (CHCl3)	394				
Heptose, D-glycero-D-manno-	C7H14O7		+21 (CH3OH)	395, 397 egpt	80'sor (5)		85glc (398a)	
Heptonolactone, D-glycero-D-manno Heptitol, D-glycero-D-manno-	C7H12O7 See Table 1 for this compound	164–165	+48 (c 0.2)	395cp			100hep (395)	
D-glycero-D-manno-heptitol heptaacetate	C21H30O14	Syrup	+34 (c 0.4, CHCl3)	397g		198glca (398b)		

TABLE C: NATURAL ALDOSES (Continued)

Substance[a] (Synonym) Derivative	Chemical Formula	Melting Point °C	Specific Rotation[b] $[\alpha]_D$	Reference[c]	Chromatography, R Value, and Reference[d]			
					ELC	GLC	PPC	TLC
(A)	(B)	(C)	(D)	(E)	(F)	(G)	(H)	(I)
α-Hexaacetate	$C_{19}H_{26}O_{13}$	139–140	+65 ($CHCl_3$)	396cp				
Methyl pyranoside	$C_8H_{16}O_7$	Syrup	+47 (CH_3OH)	396				
p-Nitrophenylhydrazone	$C_{13}H_{19}N_3O_8$	176–177		396				
Trimethylsilyl ether						338mant (398b)		
Heptose, L-*glycero*-D-*manno*-	$C_7H_{14}O_7{\cdot}H_2O$	179–181	+14 (*c* 0.9)	397, 399	80glc(397)		64glc (398a)	
L-*glycero*-D-*manno*-heptitol heptaacetate	$C_{21}H_{30}O_{14}$	116	−11 ($CHCl_3$)	397g		229glca (398b)		
Heptose diethyl dithioacetal	$C_{11}H_{24}O_6S_2$	201–202	+9.9 (C_5H_5N)	400				
Hexabenzoate	$C_{49}H_{38}O_{13}$	100	−32 ($CHCl_3$)	397				
Trimethylsilyl ether						461mant (398b)		
Heptose, 7-deoxy-L-*glycero*-D-*manno*-	$C_7H_{14}O_6$	No constants known	No constants known	401				
Heptononitrile acetate						16.9 (401a)		
Potassium heptonate trimethylsilyl ether						14.2 (401b)		
Heptose, 6-deoxy-D-*manno*-	$C_7H_{14}O_6$		+30±5	402,403			ca. 10 (402a)	
6-Deoxy-D-*manno*-heptitol hexaacetate	$C_{19}H_{28}O_{12}$					125glca (402b)		
Methyl α-pyranoside	$C_8H_{16}O_6$		+80 (*c* 0.5)	403 egp				
Methyl α-pyranoside tetraacetate	$C_{16}H_{24}O_{10}$	77–78	+62 (*c* 0.4, $CHCl_3$)	403				
Heptose, unidentified		No constants known	No constants known	404–407[z]			68glc (406a) 99glc (406b)	
Octose, 6-amino-6,8-dideoxy-7-*O*-methyl-D-*erythro*-D-*galacto*- (celestose)	$C_9H_{19}NO_6$	No constants known	No constants known	408				
Pentaacetate	$C_{19}H_{29}NO_{11}$	215, 216t, 234, 234.5		409				

Compiled by George C. Maher.

[a] In alphabetical order of parent sugar names within groups of increasing carbon chain length in the parent compounds.

[b] $[\alpha]_D$ for 1, 5 g solute, *c*, per 100 ml aqueous solution at 20, 25°C, unless otherwise given.

[c] References for melting point and specific rotation data. Letter indicates that the reference also has chromatographic data as follows: c = column, e = electrophoresis, g = gas, p = paper, and t = thin layer.

[d] R value times 100, given relative to that of the compound indicated by abbreviation: f = solvent front, aco = acovenose, amgl = tetraacetyl dimethyl glucitol, aral = arabinitol pentaacetate, cym = cymarose, cymo = cymaronolactone, damg = methyl 4,6-di-*O*-acetyl-2,3-di-*O*-methyl-α-D-glucopyranoside, digx = digitoxose, dmxyl = 2,3-di-*O*-methyl-xylitol triacetate, drho = 2-deoxy-rhamnonolactone, erya = erythritol tetraacetate, fuc = fucose, gal = galactose, gals = galactose 6-sulfate, glc = glucose, glca = glucitol hexaacetate, glcl = glucitol, glcs = glucitol trimethylsilyl ether, glct = glucose trimethylsilyl ether, hep = D-*glycero*-D-*manno*-heptonolactone, manl = mannitol hexaacetate, mant = mannose trimethylsilyl ether, mfuc = methyl α-fucopyranoside, mmt = methyl α-mannopyranose trimethylsilyl ether, ole = oleandrose, rha = rhamnose, rib = ribose, sor = sorbitol, temg = 2,3,4,6-tetra-*O*-methyl-glucose, αtemg = methyl 2,3,4,5-tetra-*O*-methyl-α-glucoside, βtemg = methyl 2,3,4,5-tetra-*O*-methyl-β-glucoside, temga = 2,3,4,6-tetra-*O*-methyl-glucitol diacetate, temgt = 2,3,4,6-tetra-*O*-methyl-galactose trimethylsilyl ether, βtma = methyl 2,3,4-tri-*O*-methyl-β-D-arabinoside, trma = 2,3,4-tri-*O*-methyl-L-arabinose, trmal = 2,3,4-tri-*O*-methyl-L-arabinitol, xyl = xylose, and xyll = xylitol pentaacetate. Under gas chromatography (column GLC or G), numbers without code indication signify retention time in min. The conditions of the chromatography are correlated with the reference given in parentheses and are found in Table E.

[e] As the methyl furanoside.

[f] Data are for the enanthiomorphic isomer.

[g] By inference from the enanthiomorphic isomer.

[h] Value is in cm/3.75 hr.

[i] Not, however, anomeric with aldgaroside A. Reference 88 assigns a 4-deoxy-β-D-*ribo*-hexopyranose or 4-deoxy-D-allose structure.

[j] Value is in cm/5 hr.

TABLE C: NATURAL ALDOSES (Continued)

k As the methyl pyranoside.

l The reference is not clear as to D or L configuration.

m Reference 202 structures this as a glycoside, not an ester; but it may be an ester in view of the recent comparative isolates of Reference 58 under D-ribose, C-hydroxymethyl-.

n The inference is that olivomycose *iso*-butyrate exists in the parent antibiotic.

o Value is in cm/±7 hr.

p Value is in cm/23 hr.

q Reference 95 has the definitive names for tyvelose and abequose reversed from those of References 265 and 300.

r Although rhodinose is the L-isomer, it is not clear whether the sugar form in ydiginic acid is D or L.

s Early work had supported an L-*xylo*-configuration,[333] but additional study has established the L-*ribo*-.[320, 322]

t A recent publication gives the reverse sign to the glycoside's optical rotation.

u A more recent publication has tentatively assigned the L-rhamnose configuration to the natural compound and called it nogalose.[349]

v Though never proved, some evidence exists for this sugar, or for L-altrose, in the polysaccharide varianose.

w As data here are meager, see those for the enanthiomorphic isomer.

x Column was at 210° instead of 140°.

y Value is for the D-*glycero*-L-*hexo*- isomer.

z These may well be D-*glycero*-D-*manno*-, D-*glycero*-L-*manno*-, D-*glycero*-L-*manno*-, and L-*glycero*-D-*manno*- for References 404, 405, and 406, respectively; Reference 407 remains uncertain.

References

1. Pollock and Stevens, *Dictionary of Organic Compounds.* Oxford University Press, New York, 1965.
2. Williams and Tucknott, *J. Sci. Food Agric.*, 22, 264 (1971).
3. Bloem, *J. Chromatogr.*, 35, 108 (1968).
4. Wohl and Momber, *Ber. Dtsch. Chem. Ges.*, 50, 456 (1917).
5. Bourne, Hutson, and Weigel, *J. Chem. Soc.* (Lond.), p. 4252 (1960).
6. Bourne, Hutson, and Weigel, *J. Chem. Soc.* (Lond.), p. 5153 (1960).
7. Bancher, Scherz, and Kaindl, *Mikrochim. Acta,*,p. 1043 (1964).
8. Fischer and Baer, *Helv. Chim. Acta.*, 17, 622 (1934).
9. Sweeley, Bentley, Makita, and Wells, *J. Am. Chem. Soc.*, 85, 2497 (1963).
10. Williams and Jones, *Can. J. Chem.*, 42, 69 (1964).
11. Vongerichten, *Justus Liebigs Ann. Chem.*, 321, 71 (1902).
12. Schmidt, *Justus Liebigs Ann. Chem.*, 483, 115 (1930).
13. Duff, *Biochem. J.*, 94, 768 (1965).
14. Hulyalkar, Jones, and Perry, *Can. J. Chem.*, 43, 2085 (1965).
15. Bentley, Cunningham, and Spring, *J. Chem. Soc.* (Lond.), p. 2301 (1951).
16. Hockett and Hudson, *J. Am. Chem. Soc.*, 56, 1632 (1934).
17. Němec, Kefurt, and Jarý, *J. Chromatogr.*, 26, 116 (1967).
18. Fischer, Bergmann, and Schotts, *Ber. Dtsch. Chem. Ges.*, 53, 522 (1920).
19. Misaki and Yukawa, *J. Biochem.* (Tokyo), 59, 511 (1966).
20. Haworth, Peat, and Whetstone, *J. Chem. Soc.* (Lond.), p. 1975 (1938).
21. Halliburton and McIlroy, *J. Chem. Soc.* (Lond.), 299 (1949).
22. Lynch, Olney, and Wright, *J. Sci. Food Agric.* 56 (1958).
23. Williams and Jones, *Can. J. Chem.*, 45, 275 (1967).
24. Sowden, Oftedahl, and Kirkland, *J. Org. Chem.*, 27, 1791 (1962).
25. Jones, Kent, and Stacey, *J. Chem. Soc.* (Lond.), p. 1341 (1947).
26. Vogel, *Helv. Chim. Acta*, 11, 1210 (1928).
27. Montgomery and Hudson, *J. Am. Chem. Soc.*, 56, 2074 (1934).
28. Frahn and Mills, *Aust. J. Chem.*, 12, 65 (1959).
29. Phillips and Criddlc, *J. Chem. Soc.* (Lond.), p. 3404 (1960).
30. Wolfrom, Patin, and de Lederkremer, *J. Chromatogr.*, 17, 488 (1965).
31. Whistler and Kirby, *J. Am. Chem. Soc.*, 78, 1755 (1956).
32. Hudson and Dale, *J. Am. Chem. Soc.*, 40, 995 (1918).
33. Jones, *J. Chem. Soc.* (Lond.), p. 1055 (1947).
34. Roberts, Johnston, and Fuhr, *Anal. Biochem.*, 10, 282 (1965).
35. Mackie and Percival, *Biochem. J.*, 91, 5P (1964).
36. Alberda van Ekenstein, *Chem. Weekbl.*, 11, 189 (1914).
37. Adachi, *J. Chromatogr.*, 17, 295 (1965).
38. Galmarini and Deulofeu, *Tetrahedron*, 15, 76 (1961).
39. Levene and Wolfrom, *J. Biol. Chem.*, 78, 525 (1928).
40. Gee, *Anal. Chem.*, 35, 354 (1963).
41. Dyer, McGonigal, and Rice, *J. Am. Chem. Soc.*, 87, 654 (1965).
42. Kuehl, Jr., Flynn, Brink, and Folkers, *J. Am. Chem. Soc.*, 68, 2679 (1946).
43. Tatsuoka, Kusaka, Miyake, Inone, Hitomi, Shiraishi, Iwasaki, and Imanishi, *Pharm. Bull.*, 343 (1957).
44. Stodola, Shotwell, Borud, Benedict, and Riley, Jr., *J. Am. Chem. Soc.*, 73, 2290, 5912 (1951).
45. Brimacombe and Mofti, *J. Chem. Soc. D Chem. Commun.*, 241 (1971).
46. Ganguly, Sarre, and Morton, *J. Chem. Soc. D Chem. Commun.*, 1488 (1969).
47. Hogenkamp and Barker, *J. Biol. Chem.*, 236, 3097 (1961).
48. Deriaz, Overend, Stacey, Teece, and Wiggins, *J. Chem. Soc.* (Lond.), p. 1879 (1949).
49. Bourne, Lees, and Weigel, *J. Chromatogr.*, 11, 253 (1963).
50. Lombard, *J. Chromatogr.*, 26, 283 (1967).
51. Allerton and Overend, *J. Chem. Soc.* (Lond.), p. 1480 (1951).
52. Oades, *J. Chromatogr.*, 28, 246 (1967).
53. Phelps, Isbell, and Pigman, *J. Am. Chem. Soc.*, 56, 747 (1934).
54. Levene and Tipson, *J. Biol. Chem.*, 115, 731 (1936).
55. Zinner, *Ber. Dtsch. Chem. Ges.*, 86, 817 (1953).
56. Barker and Smith, *J. Chem. Soc.* (Lond.), p. 1323 (1955).
57. Burton, Overend, and Williams, *J. Chem. Soc.* (Lond.), p. 3433, 3446 (1965).
58. Ezekial, Overend, and Williams, *Carbohydr. Res.*, 11, 233 (1969).
59. Kärkkainen, Haohti, and Lehtonen, *Anal. Chem.*, 38, 1316 (1966).
60. Schroeder, Barnes, Bohinski, Mumma, and Mallette, *Biochim. Biophys. Acta*, 273, 254 (1972).
61. Levene and Sobotka, *J. Biol. Chem.*, 65, 55 (1925).
62. Hudson and Yanovsky, *J. Am. Chem. Soc.*, 39, 1013 (1917).
63. Isbell and Pigman, *J. Res. Natl. Bur. Stand.*, 18, 141 (1937).
64. Alberda van Ekenstein and Blanksma, *Recl. Trav. Chim. Pays-Bas*, 22, 434 (1903).
65. Reclaire, *Ber. Dtsch. Chem. Ges.*, 41, 3665 (1908).
66. Hudson and Johnson, *J. Am.Chem. Soc.* 37, 2748 (1915).
67. Stephen, Kaplan, Taylor, and Leisegang, *Tetrahedron*, Suppl. 7, 233 (1966).
68. Tyler, *J. Chem. Soc.* (Lond.), pp. 5288, 5300 (1965).
69. Hay, Lewis, and Smith, *J. Chromatogr.*, 11, 479 (1963).
70. Gorin, Hough, and Jones, *J. Chem. Soc.* (Lond.), p. 2140 (1953).
71. Levene and Compton, *J. Biol. Chem.*, 111, 325 (1935).
72. Ryan, Arzoumanian, Acton, and Goodman, *J. Am. Chem. Soc.*, 86, 2497 (1964).
73. Andrews and Hough, *Chem. Ind*, p.1278 (1956).
74. Alam and McIlroy, *J. Chem. Soc. Sect. C Org. Chem.*, p. 1579 (1967).
75. Robertson and Speedie, *J. Chem. Soc.* (Lond.), p. 824 (1934).
76. Lance and Jones, *Can. J. Chem.*, 45, 1995 (1967).
77. Laidlaw, *J. Chem. Soc.* (Lond.), p. 752 (1954).
78. Aspinall and McKay, *J. Chem. Soc.* (Lond.), p. 1059 (1958).
79. Laidlaw and Percival, *J. Chem. Soc.* (Lond.), p. 528 (1950).
80. Aspinall, *J. Chem. Soc.* (Lond.), p. 1676 (1963).
81. Hough and Jones, *J. Chem. Soc.* (Lond.), p. 4349 (1952).
82. Weckesser, Mayer, and Fromme, *Biochem. J.*, 135, 293 (1973).
83. Percival and Willox, *J. Chem. Soc.* (Lond.), p. 1608 (1949).
84. Wintersteiner and Klingsberg, *J. Am. Chem. Soc.*, 71, 939 (1949).
85. Weckesser, Rosenfelder, Mayer, and Luderitz, *Eur. J. Biochem.*, 24, 112 (1971).
86. Anderle, Kováč, and Anderlová, *J. Chromatogr.*, 64, 368 (1972).
87. Kunstmann, Mitscher, and Bohonos, *Tetrahedron Lett.*, 839 (1966).
88. Paulsen and Redlich, *Angew. Chem. Int. Ed.*, 11, 1021 (1972).
89. Beylis, Howard, and Perold, *J. Chem. Soc. D Chem. Commun.*, p. 597 (1971).
90. Steiger and Reichstein, *Helv. Chim. Acta*, 19, 184 (1936).
91. Scher and Ginsburg, *J. Biol. Chem.*, 243, 2385 (1968).
92. Levene and Jacobs, *Ber. Dtsch. Chem. Ges.*, 43, 3141 (1910).
93. Lerner and Kohn, *J. Med. Chem.*, 7, 655 (1964).
94. Muhlradt, Weiss, and Reichstein, *Justus Liebigs Ann. Chem.*, 685, 253 (1965).
95. MacLennan and Randall, *Anal. Chem.*, 31, 2020 (1959).
96. Kaufmann, Mühlradt, and Reichstein, *Helv. Chim. Acta*, 50, 2287 (1967).
97. Keller and Reichstein, *Helv. Chim. Acta*, 32, 1607 (1949).
98. Perry and Daoust, *Carbohydr. Res.*, 31, 131 (1973).
99. Levene and Compton, *J. Biol. Chem.*, 117, 37 (1937).
100. Iselin and Reichstein, *Helv. Chim. Acta*, 27, 1203 (1944).
101. Brimacombe and Husain, *Chem. Commun.*, 630 (1966).
102. Hoffman, Weiss, and Reichstein, *Helv. Chim. Acta*, 49, 2209 (1966).
103. Brimacombe and Portsmouth, *J. Chem. Soc. Sect. C Org. Chem.*, p. 499 (1966).
104. Krasso and Weiss, *Helv. Chim. Acta*, 49, 1113 (1966).
105. Dion, Woo, and Bartz, *J. Am. Chem. Soc.*, 84, 880 (1962).
106. Brimacombe, Ching, and Stacey, *J. Chem. Soc. Sect. C Org. Chem.*, 197 (1969).
107. Brimacombe, Stacey, and Tucker, *J. Chem. Soc.* (Lond.), p. 5391 (1964).
108. Jäger, *Dissertation (Basel)*, (1959).
109. Gut and Prins, *Helv. Chim. Acta*, 29, 1555 (1946).
110. Iwadare, *Bull. Chem. Soc. Jap.*, 17, 296 (1942).
111. Krauss, *Dissertation (Basel)*, (1959).
112. Grob and Prins, *Helv. Chim. Acta*, 28, 840 (1945).
113. Hauser and Sigg, *Helv. Chim. Acta*, 54, 1178 (1971).
114. Ellwood and Kirk, *Biochem. J.*, 122, 14P (1971).
115. Kaufmann, *Helv. Chim. Acta*, 48, 83 (1965).
116. Brimacombe, Da'aboul, and Tucker, *J. Chem. Soc. Sec. C Org. Chem.*, p. 3762 (1971).
117. Kiliani, *Ber. Dtsch. Chem. Ges.*, 46, 667 (1913).
118. Kiliani,*Arch. Pharm.*, 234, 449 (1896); *Chem. Zentralbl.*, 67, II, 591 (1896).

119. Ruber, Minsaas, and Lyche, *J. Chem. Soc.* (Lond.), p. 2173 (1929).
120. Charlton, Haworth, and Hickinbottom, *J. Chem. Soc.* (Lond), p. 1527 (1927).
121. Nottbohm and Mayer, *Vorratspflege Lebensmitrelforsch.*, 243 (1938).
122. Hudson and Parker, *J. Am. Chem. Soc.*, 37, 1589 (1915).
123. O'Neill, *J. Am. Chem. Soc*, 77, 2837 (1955).
124. Araki and Hirase, *Bull. Chem. Soc. Jap.*, 29, 770 (1956).
125. Clingman and Nunn, *J. Chem. Soc.* (Lond.), p. 493 (1959).
126. Gorin and Spencer, *Can. J. Chem.*, 42, 1230 (1964).
127. Nunn, Parolis, and Russell, *Carbohydr. Res.*, 29, 281 (1973).
128. Gorin and Ishikawa, *Can. J. Chem.*, 45, 521 (1967).
129. Votoček and Valentin, *Collect. Czech. Chem. Commun.*, 2, 36 (1930).
130. Lato, Brunelli, Ciuffini, and Mezzetti, *J. Chromatogr.*, 34, 26 (1968).
131. Shaw and Moss, *J. Chromatogr.*, 41, 350 (1969).
132. Bourne, Hutson, and Weigel, *J. Chem. Soc.* (Lond.), p. 35 (1961).
133. Levvy and McAllan, *Biochem. J.*, 80, 433 (1961).
134. MacPhillamy and Elderfield, *J. Org. Chem.*, 4, 150 (1939).
135. Khare, Schindler, and Reichstein, *Helv. Chim. Acta*, 45, 1534 (1962).
136. Springer, Desai, and Kolechi, *Biochemistry*, 3, 1076 (1964).
137. Lamb and Smith, *J. Chem. Soc.* (Lond.), p. 422 (1936).
138. Gros, *Carbohydr. Res.*, 2, 56 (1966).
139. Schmidt and Wernicke, *Justus Liebigs Ann. Chem.*, 556, 179 (1944).
140. Akita, Maeda, and Umezawa, *J. Antibiot.* (Tokyo) *Ser. A*, 71, 200 (1964).
141. Nunn and Parolis, *Carbohydr. Res.*, 6, 1, (1968); 8, 361 (1968)
142. Bell and Williamson, *J. Chem. Soc.* (Lond.), p. 1196 (1938).
143. Oldham and Bell, *J. Am. Chem. Soc.*, 60, 323 (1938).
144. Freeman, Stephan, and Van der Bijl, *J. Chromatogr.*, 73, 29 (1972).
145. Lechevalier and Gerber, *Carbohydr. Res.*, 13, 451 (1970).
146. Reber and Reichstein, *Helv. Chim. Acta*, 28, 1164 (1945).
147. Hirst and Jones, *J. Chem. Soc* (Lond.), p. 506 (1946).
148. Jeanloz, *J. Am. Chem. Soc.*, 76, 5684 (1954).
149. Itasaka, *J. Biochem.* (Tokyo), 60, 52 (1966).
150. Kocourek, Tichá, and Koštiv, *J. Chromatogr.*, 24, 117 (1966).
151. Goldstein, Hamilton, and Smith, *J. Am. Chem. Soc.*, 79, 1190 (1957).
152. Hassid and Su, *Biochemistry*, 1, 468 (1962).
153. Bowker and Turvey, *J. Chem. Soc.* (Lond.), p. 983, 989 (1968).
154. Love and Percival, *J. Chem. Soc.* (Lond.), 3338 (1964).
155. Turvey and Williams, *J. Chem. Soc.* (Lond.), p. 2119 (1962); p. 2242 (1963).
156. Anderson, *J. Biol. Chem.*, 100, 249 (1933).
157. Fischer and Hertz, *Ber. Dtsch. Chem. Ges.*, 25, 1247 (1892).
158. Araki and Hirase, *Bull. Chem. Soc. Jap.*, 26, 463 (1953); 33, 291 (1960).
159. Kochetkov, Usov, and Miroshnikova, *J. Gen. Chem. USSR Engl. Ed.*, 40, 2457, 2461 (1970).
160. Usov, Lotov, and Kochetkov, *J. Gen. Chem. USSR Engl Ed.*, 41, 1156 (1971).
161. Nunn and von Holdt, *J. Chem. Soc.* (Lond.), p. 1094 (1957).
162. Duff and Percival, *J. Chem. Soc.* (Lond.), p. 830 (1941).
163. Minsaas, *Recl. Trav. Chim. Pays-Bas*, 50, 424 (1933).
164. Gardiner and Percival, *J. Chem. Soc.* (Lond.), p. 1414 (1958).
165. Levvy and McAllan, *Biochem. J.*, 80, 433 (1961).
166. Westphal and Feier, *Ber. Dtsch. Chem. Ges.*, 89, 582 (1956).
167. Horowitz and Delman, *J. Chromatogr.*, 21, 302 (1966).
168. Anderson, Andrews, and Hough, *Chem. Ind.*, 1453 (1957).
169. Conchie and Percival, *J. Chem. Soc.* (Lond.), p. 827 (1950).
170. Percival and Young, *Carbohydr. Res.*, 32, 195 (1974).
171. Dejter-Juszynski and Flowers, *Carbohydr. Res.*, 28, 61 (1973).
172. Schmidt, Mayer, and Distelmaier, *Justus Liebigs Ann. Chem.*, 555, 26 (1943).
173. James and Smith, *J. Chem. Soc.* (Lond.), p. 739, 746 (1945).
174. Katzman and Jeanloz, *J. Biol. Chem.*, 248, 50 (1973).
175. Anno, Seno, and Ota, *Carbohydr. Res.*, 13, 167 (1970).
176. Araki, Arai, and Hirasi, *Bull. Chem. Soc. Jap.*, 40, 959 (1967).
177. Turvey and Rees, *Nature*, 189, 831 (1961).
178. Quilico, Piozzi, Pavan, and Mantia, *Tetrahedron*, 5, 10 (1959).
179. Zervas, *Ber. Dtsch. Chem. Ges.*, 64, 2289 (1931).
180. Imanari and Tamura, *Agric. Biol. Chem.*, 35, 321 (1971).
181. Plouvier, *C.R. Acad. Sci.*, 256, 1397 (1963).
182. Hudson and Dale, *J. Am. Chem. Soc.*, 37, 1264, 1280 (1915).
183. Harris, Hirst, and Wood, *J. Chem. Soc.* (Lond.), p. 2108 (1932).
184. Jones and Jones, *Can. J. Chem.*, 47, 3269 (1969).
185. Purdie and Irvine, *J. Chem. Soc.* (Lond.), p. 1049 (1904).
186. Bates, Polarimetry, Saccharimetry and the Sugars: National Bureau of Standards Circular C440. U.S. Gov. Print. Off., Washington, D.C., 1942.
187. Georg, *Helv. Chim. Acta*, 12, 261 (1929).
188. Micheel and Berendes, *Mikrochim. Acta*, 519 (1963).
189. Brennan, *J. Chromatogr.*, 59, 231 (1971).
190. Brigl and Schreyer, *Hoppe Seyler's Z. Physiol. Chem.*, 160, 214 (1926).
191. Alberda van Ekenstein and Blanksma, *Recl. Trav. Chim. Pays-Bas*, 24, 33 (1905).
192. Yoshida, Honda, Iino, and Kato, *Carbohydr. Res.*, 10, 333 (1969).
193. Irvine and Oldham, *J. Chem. Soc.* (Lond.), p. 1744 (1921).
194. Purdie and Irvine, *J. Chem. Soc.* (Lond.), p. 1049 (1904).
195. Irvine and Moodie, *J. Chem. Soc.* (Lond.), p. 1578 (1906).
196. Duff, *J. Chem. Soc.* (Lond.), p. 4730 (1957).
197. Duff, Webley, and Farmer, *Biochem. J.*, 65, 21P (1957).
198. Josephson, *Ber. Dtsch. Chem. Ges.*, 62, 317 (1929).
199. Schumacher, *Carbohydr. Res.*, 13, 1 (1970).
200. Ohle, *Biochem. Z.*, 131, 611 (1922).
201. Brigl and Grüner, *Justus Liebigs Ann. Chem.*, 495, 60 (1932).
202. Fischer and Bergmann, *Ber. Dtsch. Chem. Ges.*, 51, 1760, 1804 (1918).
203. Tschesche, Kämmerer, and Wulff, *Tetrahedron Lett.*, 701 (1968).
204. Finnegan, Mueller, and Morris, *Proc. Chem. Soc. London*, 182 (1963).
205. Carter, *J. Sci. Food Agric.*, 2, 54 (1951).
206. Finnegan and Stephani, *J. Pharm. Sci.*, 57, 353 (1968).
207. Sloneker and Orentas, *Can. J. Chem.*, 40, 2188 (1962).
208. Fischer and Lieberman, *Ber. Dtsch. Chem. Ges.*, 26, 2415 (1893).
209. Fischer and Zach, *Ber. Dtsch. Chem. Ges.*, 45, 3761 (1902).
210. Evans, Long, Jr., and Parrish, *J. Chromatogr.*, 32, 602 (1968).
211. Staněk and Tajmr, *Chem. Listy*, 52, 551 (1958).
212. Ebert and Zenk, *Arch. Mikrobiol.*, 54, 276 (1966).
213. Frèrejacque, *C.R. Acad. Sci.*, 230, 127 (1950).
214. Korte, *Ber. Dtsch. Chem. Ges.*, 88, 1527 (1955).
215. Reyle and Reichstein, *Helv. Chim. Acta*, 35, 195 (1956).
216. Allgeier, Weiss, and Reichstein, *Helv. Chim. Acta*, 50, 456 (1967).
217. Miyano and Benson, *J. Am. Chem. Soc.*, 84, 59 (1962).
218. Chanley, Ledeen, Wax, Nigrelli, and Sobotka, *J. Am. Chem. Soc.*, 81, 5180 (1959).
219. Saier Jr. and Ballou, *J. Biol. Chem.*, 243, 992 (1968).
220. Jeanloz and Gut, *J. Am. Chem. Soc.*, 76, 5793 (1954).
221. Irvine and Scott, *J. Chem. Soc.* (Lond.), p. 582, 571, 575 (1913).
222. Oldham, *J. Am. Chem. Soc.*, 56, 1360 (1934).
223. Jeanloz, Rapin, and Hakomori, *J. Org. Chem.*, 26, 3939 (1961).
224. Levene and Raymond, *J. Biol. Chem.*, 88, 513 (1930).
225. Lee and Ballou, *J. Biol. Chem.*, 239, 3602 (1964).
226. Helferich, Klein, and Schafer, *Ber. Dtsch. Chem. Ges.*, 59, 79 (1926).
227. Helferich and Himmen, *Ber. Dtsch. Chem. Ges.*, 62, 2136, 2141 (1929).
228. Helferich and Günther, *Ber. Dtsch. Chem. Ges.*, 64, 1276 (1931).
229. White and Rao, *J. Am. Chem. Soc.*, 75, 2617 (1953).
230. Christensen and Smith, *J. Am. Chem. Soc.*, 79, 4492 (1957).
231. Fischer, *Ber. Dtsch. Chem. Ges.*, 23, 2618 (1890).
232. Makarevich and Kolesnikov, *Chem. Nat. Compd.*, (*USSR*), 164 (1969).
233. Zissis, Richtmyer, and Hudson, *J. Am. Chem. Soc.*, 73, 4714 (1951).
234. Blindenbacher and Reichstein, *Helv. Chim. Acta*, 31, 1669 (1948).
235. Frèrejacque and Hasenfratz, *C.R. Acad. Sci.*, 222, 815 (1946).
236. Frèrejacque and Durgeat, *C.R. Acad. Sci.*, 228, 1310 (1949).
237. Doebel, Schlittler, and Reichstein, *Helv. Chim. Acta*, 31, 688 (1948).
238. Perry and Daoust, *Can. J. Chem.*, 51, 3039 (1973).
239. Capek, Tikal, Jarý, and Masojidková, *Collect. Czech. Chem. Commun.*, 36, 1973 (1971).
240. Takita, Maeda, Umezawa, Omoto, and Umezawa, *J. Antibiot.* (Tokyo) *Ser. A*, 22, 237 (1969).
241. Evans and Parrish, *Carbohydr. Res.*, 28, 359 (1973).
242. Wolfrom and Anno, *J. Am. Chem. Soc.*, 74, 5583 (1952).
243. Cooke and Percival, *Carbohydr. Res.*, 32, 383 (1974).
244. Ohashi, Kawabe, Kono, and Ito, *Agric. Biol. Chem.*, 37, 2379 (1973).
245. Avigad, Amaral, Asensio, and Horecker, *J. Biol. Chem.*, 237, 2736 (1962).
246. Maradufer and Perlin, *Carbohydr. Res.*, 32, 127 (1974).
247. Wolfrom and Usdin, *J. Am. Chem. Soc.*, 75, 4318 (1953).

248. Perlin, Mackie, and Dietrich, *Carbohydr. Res.*, 18, 185 (1971).
249. Webb, Broschard, Cosulich, Mowat, and Lancaster, *J. Am. Chem. Soc.*, 84, 3183 (1962).
250. Overend, Stacey, and Stanék, *J. Chem. Soc.* (Lond.), p. 2841 (1949).
251. Bonner, *J. Org. Chem.*, 26, 908 (1961).
252. Wirz and Hatdegger, *Helv. Chim. Acta*, 54, 2017 (1971).
253. Keleti, Mayer, Fromme, and Lüderitz, *Eur. J. Biochem.*, 16, 284 (1970).
254. Černý, Pacak, and Stanék, *Chem. Ind.*, p. 945 (1961).
255. Herzog, Meseck, Delorenzo, Murawski, Charney, and Rosselet, *Appl. Microbiol.*, 13, 515 (1965).
256. Zorbach and Ciaudelli, *J. Org. Chem.*, 30, 451 (1965).
257. Studer, Panavaram, Gavilanes, Linde, and Meyer, *Helv. Chim. Acta*, 46, 23 (1963).
258. Berlin, Esipov, Kiseleva, and Kolosov, *Chem. Nat. Compd.*, (*USSR*), 3, 280 (1967).
259. Brufani, Keller-Schierlein, Löffler, Mansperger, and Zähner, *Helv. Chim. Acta*, 51, 1293 (1968).
260. Ganguly and Sarre, *J. Chem. Soc. D Chem. Commun.*, p. 1149 (1969).
261. Vischer and Reichstein, *Helv. Chim. Acta*, 27, 1332 (1944).
262. Tschesehe and Buschauer, *Justus Liebigs Ann. Chem.*, 603, 59 (1957).
263. Allgeier, *Helv. Chim. Acta*, 51, 311, 668 (1968).
264. Westphal, Lüderitz, Fromme, and Joseph, *Angew. Chem.*, 65, 555 (1953).
265. Fouquey, Lederer, Lüderitz, Polonsky, Staub, Stirm, Tirelli, and Westphal, *C.R. Acad. Sci.*, 246, 2417 (1958).
266. Williams, Szarek, and Jones, *Can. J. Chem.*, 49, 796 (1971).
267. Stirm, Lüderitz, and Westphal, *Justus Liebigs Ann. Chem.*, 696, 180 (1966).
268. Berlin, Esipov, Kolosov, Shemyakin, and Brazhnikova, *Tetrahedron Lett.*, p. 1323 (1964).
269. Miyamoto, Kawamatsu, Shinohara, Nakadaira, and Nakanishi, *Tetrahedron*, 22, 2785 (1966).
270. Blindenbacher and Reichstein, *Helv. Chim. Acta*, 31, 2061 (1948).
271. Hesse, *Ber. Dtsch. Chem. Ges.*, 70, 2264 (1937).
272. Celmer and Hobbs, *Carbohydr. Res.*, 1, 137 (1965).
273. Davies, *Nature*, 191, 43 (1961).
274. Ganguly, Sarre, and Reimann, *J. Am. Chem. Soc.*, 90, 7129 (1968).
275. Stevens, Nagarajan, and Haskell, *J. Org. Chem.*, 27, 2991 (1962).
276. Stevens, Cross, and Toda, *J. Org. Chem.*, 28, 1283 (1963).
277. Albano and Horton, *J. Org. Chem.*, 34, 3519 (1969).
278. Williams, Szarek, and Jones, *Carbohydr. Res.*, 20, 49 (1971).
279. Berlin, Esipov, Kolosov, and Shemyakin, *Tetrahedron Lett.*, p. 1431 (1966).
280. Berlin, Borisova, Esipov, Kolosov, and Kirvoruchko, *Chem. Nat. Compd.*, (*USSR*), 5, 89, 94 (1969).
281. Brimacombe and Portsmouth, Carbohydr. Res., 1, 128 (1965); *Chem. Ind.*, p. 468 (1965).
282. Howarth, Szarek, and Jones, *Can. J. Chem.*, p. 46, 3375 (1968).
283. Shoppe and Reichstein, *Helv. Chim. Acta*, 25, 1611 (1942).
284. Tamm and Reichstein, *Helv. Chim. Acta*, 31, 1630 (1948).
285. Brimacombe, Portsmouth, and Stacey, *J. Chem. Soc.* (Lond.), p. 5614 (1965).
286. Miyamoto, Kawamatsu, Shinohara, Asahi, Nakedaira, Kakisawa, Nakanishi, and Bhacca, *Tetrahedron Lett.*, p.693 (1963).
287. Iselin and Reichstein, *Helv. Chim. Acta*, 27, 1200 (1944).
288. Brockmann and Waehneldt, *Naturwissenschaften*, 48, 717 (1961).
289. Wyss, Jager, and Schindler, *Helv. Chim. Acta*, 43, 664 (1960).
290. Renkonen, Schindler, and Reichstein*Helv. Chim. Acta*, 42, 182 (1959); 39, 1490 (1956).
291. Kilani, *Arch. Pharm.*, 234, 486 (1896).
292. Stahl and Kaltenbach, *J. Chromatogr.*, 5, 351 (1961).
293. Haga, Chonan, and Tejima, *Carbohydr. Res.*, 16, 486 (1971).
294. El-Dash and Hodge, *Carbohydr. Res.*, 18, 259 (1971).
295. Krasso, Weiss, and Reichstein, *Helv. Chim. Acta*, 46, 1691 (1963).
296. Jacobs, *J. Biol. Chem.*, 88, 519 (1930).
297. Bolliger and Ulrich, *Helv. Chim. Acta*, 35, 93 (1952).
298. Prins, *Helv. Chim. Acta*, 29, 378 (1949).
299. Fouquey, Polonsky, Lederer, Westphal, and Lüderitz, *Nature*, 182, 944 (1958).
300. Davies, Staub, Fromme, Lüderitz, and Westphal, *Nature*, 181, 822 (1958).
301. Ekborg and Svensson, *Acta Chem. Scand.*, 27, 1437 (1973).
302. Woo, Dion, and Bartz, *J. Am. Chem. Soc.*, 83, 3352 (1961).
303. Keller-Schierlein and Roncari, *Helv. Chim. Acta*, 45, 138 (1962); 49, 705 (1966).
304. Westphal and Lüderitz, *Angew. Chem.*, 72, 881 (1960).
305. Kochetkov and Usov, *Bull. Acad. Sci. USSR Engl. Ed.*, p.471 (1965).
306. Stevens, Blumbergs, and Wood, *J. Am. Chem. Soc.*, 86, 3592 (1964).
307. Brockmann and Waehneldt, *Naturwissenschaften*, 50, 43 (1963).
308. Jr.Rinehart and Borders, *J. Am. Chem. Soc.*, 85, 4037 (1963).
309. Haines, *Carbohydr. Res.*, 21, 99 (1972).
310. Kowalewski, Schindler, Jäger, and Reichstein, *Helv. Chim. Acta*, 43, 1214, 1280 (1960).
311. Golab and Reichstein, *Helv. Chim. Acta*, 44, 616 (1961).
312. Angus, Bourne, and Weigel, *J. Chem. Soc.* (Lond.), p. 22 (1965).
313. Bolliger and Reichstein, *Helv. Chim. Acta*, 36, 302 (1953).
314. Jacobs and Bigelow, *J. Biol. Chem.*, 96, 355 (1932).
315. Abisch, Tamm, and Reichstein, *Helv. Chim. Acta*, 42, 1014 (1959).
316. Hauenstein and Reichstein, *Helv. Chim. Acta*, 33, 446 (1950).
317. Matern, Grisebach, Karl, and Achenbach, *Eur. J. Biochem.*, 29, 1, 5 (1972).
318. Arcamone, Barbieri, Franceschi, Penco, and Vigevani, *J. Am. Chem. Soc.*, 95, 2008 (1973).
319. Regna, Hochstein, Wagner, and Woodward, *J. Am. Chem. Soc.*, 75, 4625 (1953).
320. Hofheinz, Grisebach, and Friebolin, *Tetrahedron*, 18, 1265 (1962).
321. Lemal, Pacht, and Woodward, *Tetrahedron*, 18, 1275 (1962).
322. Flaherty, Overend, and Williams, *J. Chem. Soc, Sect. C, Org. Chem.*, p. 398 (1966).
323. Foster, Inch, Lehmann, Thomas, Webber, and Wyer, *Proc. Chem. Soc.* London, p. 254 (1962); *Chem. Ind.*, p. 1619 (1962).
324. Paul and Tchelitcheff, *Bull. Soc. Chim. Fr.*, p. 443 (1957).
325. Jaret, Mallams, and Reimann, *J. Chem. Soc.* (Lond.) *Perk. I.*, 1374 (1973).
326. Omura, Katagiri, and Hata, *J. Antibiot.* (Tokyo), 21, 272 (1968).
327. Watanabe, Fujii, and Satake, *J. Biochem.* (Tokyo), 50, 197 (1961).
328. Flynn, Sigal, Wiley, and Gerzon, *J. Am. Chem. Soc.*, 76, 3121 (1954).
329. Corcoran, *J. Biol. Chem.*, 236, PC27 (1961).
330. Wiley and Weaver,*J. Am. Chem. Soc.*, 77, 3422 (1955); 78, 808 (1956).
331. Lüderitz, Staub, Stirm, and Westphal, *Biochem. Z.*, 330, 193 (1958).
332. Wiley, Mackellar, Carron, and Kelly, *Tetrahedron Lett.*, p. 663 (1968).
333. Zhdanovich, Lokshin, Kuzovkov, and Rudaya, *Chem. Nat. Compd.*, (*USSR*), 625 (1971).
334. Okuda, Suzuki, Suzuki, *J. Biol. Chem.*, 242, 958 (1967); 243, 6353 (1968).
335. Haworth, Raistrick, and Stacey, *Biochem. J*, 29, 2668 (1935).
336. Sorkin and Reichstein, *Helv. Chim. Acta*, 662 (1945).
337. Stoffyn and Jeanloz, *J. Biol. Chem.*, 235, 2507 (1960).
338. Baggett, Stoffyn, and Jeanloz, *J. Org. Chem.*, 28, 1041 (1963).
339. Vargha, *Ber Dtsch. Chem. Ges.*, 87, 1351 (1954).
340. Levene *J. Biol. Chem.* 57, 329 (1923); 59, 129(1924)
341. Bishop, Perry, Blank, and Cooper, *Can. J. Chem.*, 43, 30 (1965).
342. Hamilton, Partlow, and Thompson, *J. Am. Chem. Soc.*, 82, 451 (1960).
343. Levine, Hansen, and Sell, *Carbohydr. Res.*, 382 (1968).
344. Butler and Cretcher, *J. Am. Chem. Soc.*, 53, 4358, 4363 (1931).
345. Omoto, Takita, Maeda, and Umezawa, *Carbohydr. Res.*, 30, 239 (1973).
346. Dutton and Yang, *Can. J. Chem.*, 50, 2382 (1972).
347. Markovitz, *J. Biol. Chem.*, 237, 1767 (1962).
348. Ganguly and Saksena, *J. Chim. Soc. D Chem. Commun.*, p. 531 (1973).
349. Wiley, Duchamp, Hsiung, and Chidestet, *J. Org. Chem.*, 36, 2670 (1971).
350. Morrison, Young, Perry, and Adams, *Can. J. Chem.*, 45, 1987 (1967).
351. MacLennan, *Biochem. J.*, 82, 394 (1962).
352. MacLennan, Smith and Randell, *Biochem. J.*, 74, 3P (1960); 80, 309 (1961)
353. Ovodov and Evtushenko, *J. Chromatogr.*, 31, 527 (1967).
354. Caudy and Baddiley, *Biochem. J.*, 98, 15 (1966).
355. Aspinall and Zweifel, *J. Chem. Soc.* (Lond.), p. 2271 (1957).
356. Scheer, Terai, Kulkami, Conant, Wheat, and Plowe, *J. Bacteriol.*, 103, 525 (1970).
357. Perry and Webb, *Can. J. Chem.*, 47, 31 (1969).

358. Gros, Deulofeu, Galmarini, and Frydman, *Experientia*, 24, 323 (1968).
359. Behrend, *Ber. Dtsch. Chem. Ges.*, 11, 1353 (1878).
360. Collins and Overend, *J. Chem. Soc.* (Lond.), p. 1912 (1965).
361. Purdie and Young, *J. Chem. Soc.* (Lond.), p. 89, 1194 (1906).
362. Fischer, *Ber. Dtsch. Chem. Ges.*, 28, 1158 (1895).
363. Fischer, *Ber. Dtsch. Chem. Ges.*, 29, 324 (1896).
364. Fischer, Bergmann, and Rabe, *Ber. Dtsch. Chem. Ges.*, 53, 2362 (1920).
365. Vaterlaus, Kiss, and Spieglberg, *Helv. Chim. Acta*, 47, 381 (1964).
366. Barker, Homer, Keith, and Thomas, *J. Chem. Soc.* (Lond.), p. 1538 (1963).
367. Hinman, Caron, and Hoeksema, *J. Am. Chem. Soc.*, 79, 3789 (1957).
368. Walton, Rodin, Stammer, Holly, and Folkers, *J. Am. Chem. Soc.*, 80, 5168 (1958).
369. Young and Elderfield, *J. Org. Chem.*, 241 (1942).
370. Andrews, Hough, and Jones, *J. Am. Chem. Soc.*, 77, 125 (1955).
371. Hirst, Percival, and Williams, *J. Chem. Soc.* (Lond.), p. 1942 (1958).
372. Schmidt, Plankenhorn, and Kübler, *Ber. Dtsch. Chem. Ges.*, 75, 579 (1942).
373. Brown, Hough, and Jones, *J. Chem. Soc.* (Lond.), p. 1125 (1950).
374. Percival and Percival, *J. Chem. Soc.* (Lond.), p. 690 (1950).
375. Butler, Lloyd, and Stacey, *J. Chem. Soc.* (Lond.), p. 1531, 1537 (1955).
376. Charalambous and Percival, *J. Chem. Soc.* (Lond.), p. 2443 (1954).
377. Geerdes and Smith, *J. Am. Chem. Soc.*, 77, 3572 (1955).
378. Chaput, Michel, and Lederer, *Experientia*, 17, 107 (1961).
379. Hirst, Hough, and Jones, *J. Chem. Soc.* (Lond.), p. 928, 3145 (1949).
380. Wiley and Sigal, *J. Am. Chem. Soc.*, 80, 1010 (1958).
381. Pigman and Isbell, *J. Res. Natl. Bur. Stand.*, 19, 189 (1937).
382. Britton, *Biochem. J.*, 85, 402 (1962).
383. Hickman and Ashwell, *J. Biol. Chem.*, 241, 1424 (1966).
384. Stevens, Glinski, and Taylor, *J. Org. Chem.*, 33, 1586 (1968).
385. MacLennon, *Biochim. Biophys. Acta*, 48, 600 (1961).
386. Schmutz, *Helv. Chim. Acta*, 31, 1719 (1948).
387. Von Euw and Reichstein, *Helv. Chim. Acta*, 33, 485 (1950).
388. Kapur and Allgeier, *Helv. Chim. Acta*, 51, 89 (1968).
389. Sephton and Richtmyer, *J. Org. Chem.*, 28, 1691 (1963).
390. Strobach and Szabo, *J. Chem. Soc.* (Lond.), p. 3970 (1963).
391. Isherwood and Jermyn, *Biochem. J.*, 48, 515 (1951).
392. MacLennon and Davies, *Biochem. J.*, 66, 562 (1957).
393. Davies, *Biochem. J.*, 67, 253 (1957).
394. Begbie and Richtmyer, *Carbohydr. Res.*, 272 (1966).
395. Richtmyer and Charlson, *J. Am. Chem. Soc.*, 82, 3428 (1960).
396. Hulyalkar, Jones, and Perry, *Can. J. Chem.*, 41, 1490 (1963).
397. Young and Adams, *Can. J. Chem.*, 43, 2929 (1965).
398. Adams, Quadling, and Perry, *Can. J. Microbiol.*, 13, 1605 (1967).
399. Teuber, Bevill, and Osborn, *Biochemistry*, 3303 (1969).
400. Weidell and Hoppe-Seyler's, *Z. Physiol. Chem.*, 299, 253 (1955).
401. Varma, Varma, Allen, and Wardi, *Carbohydr. Res.*, 32, 386 (1974).
402. Hellerqvist, Lindberg, Samuelson, and Brubaker, *Acta Chem. Scand.*, 26, 1389 (1972).
403. Boren, Eklind, Garegg, Lindberg, and Pilotti, *Acta Chem. Scand.*, 26, 4143 (1972).
404. Davies, *Nature*, 180, 1129 (1957).
405. Missale, Colajacomo, and Bologna, *Boll. Soc. Ital. Biol. Sper.*, 36, 1885 (I960); *Chem. Abstr.*, 55, 24869 (1961).
406. Kuriki and Kurahashi, *J. Biochem.* (Tokyo), 58, 308 (1965).
407. Fraenkel, Osborn, Horecker, and Smith, *Biochem. Biophys. Res. Commun.*, 11, 423 (1963).
408. Hoeksema, *J. Am. Chem. Soc*, 90, 755 (1968).
409. Hoeksema and Hinman, *J. Am. Chem. Soc.*, 86, 4979 (1964).
410. Tschesche and Kohl, *Tetrahedron*, 24, 4359 (1968).
411. Martinsson and Samuelson, *J. Chromatogr.*, 50, 429 (1970).
412. Walborg Jr. and Kondo, *Anal. Biochem.*, 37, 323 (1970).

TABLE D: NATURAL KETOSES

Substance[a] (Synonym) Derivative (A)	Chemical Formula (B)	Melting Point °C (C)	Specific Rotation[b] $[\alpha]_D$ (D)	Reference[c] (E)	ELC (F)	GLC (G)	PPC (H)	TLC (I)
					Chromatography, R Value, and Reference[d]			
Triosulose, 3-deoxy- (3-deoxy-2-keto-glyceraldehyde, methyl glyoxal, pyruvic aldehyde)[e]	$C_3H_4O_2$	Oil, bp 72	None	1		6.3, 26.6 (3)		75f (2)
bis-2,4-Dinitrophenylhydrazone	$C_{15}H_{12}N_8O_8$	308–309	None	1			88f (7)	113 for (17)
p-Nitrophenylhydrazone	$C_9H_9N_3O_3$	217	None	1				
Triulose (2-keto-glyceritol, dihydroxyacetone)	$C_3H_6O_3$	80 (dimer)	None	1			55f (4)	40f (2)
Diacetate	$C_7H_{10}O_6$	46–47	None	1			0f (8)	
bis-2,4-Dinitrophenylhydrazone	$C_{15}H_{12}N_8O_9$	277–278	None	1				
p-Nitrophenylhydrazone	$C_9H_{11}N_3O_4$	160	None	1				
Trimethylsilyl ether	Dimer			1		82glct (5)		
Triulose, 1-amino-1,3-dideoxy- (aminoacetone)	C_3H_7NO	No constants known	No constants known	1				
Hydrochloride	$C_3H_7NO\cdot HCl$	75	None	1				
Triulose, dideoxy (acetone)	C_3H_6O	Liquid, bp 56	None	1		1.6 (6)		73f (14)
2,4-Dinitrophenylhydrazone	$C_9H_{10}N_4O_4$	128	None	1			93f (7)	
p-Nitrophenylhydrazone	$C_9H_{11}N_3O_2$	149	None	1				
Tetradiulose, 1,4-dideoxy- (2,3-butanedione, diacetyl, dimethylglyoxal)	$C_4H_6O_2$	Liquid, bp 88	None	None	1, 6		4.1 (3)	82f (9)
bis-2,4-Dinitrophenylhydrazone	$C_{16}H_{14}N_8O_8$	252–254	None	1			0f (8)	
erythro-Butane-2,3-diol	$C_4H_{10}O_2$	23.4	None (meso)	1	6rib (10)	See also Table 1		
D,L-threo-Butane-2,3-diol	$C_4H_{10}O_2$	7.6	None (racemic)	1	33rib (10)	See also Table 1		
Tetrulose, L-glycero- (L-erythrulose, keto-erythritol, L-treulose)	$C_4H_8O_4$	Syrup	+12	11,12		225glct (13)		
o-Nitrophenylhydrazone	$C_{10}H_{13}N_3O_8$	152–153	+48 (C_2H_5OH) (18°)	15				
Tetrulose, 1,4-dideoxy-D-glycero- (acetoin, 3-hydroxybutan-2-one)	$C_4H_8O_2$	Liquid, bp 143	–1.4 (neat), –105 (H_2O)	1,6,16		24 (3)	91f (9)	
2,4-Dinitrophenylhydrazone	$C_{10}H_{12}N_2O_5$	114–116	–12 ($CHCl_3$)	1				
Pentodiulose, 1,3,5-trideoxy-2,4- (acetyl-acetone, 2,4-pentane-dione)	$C_5H_8O_2$	Liquid bp 139 (746 mm)	None	1		17.3 (3)		
2,4-Dinitrophenylhydrazone	$C_{11}H_{12}N_4O_5$	122	None	22				
o-Nitrophenylhydrazone	$C_{11}H_{13}N_3O_3$	100s, 135m	None	23				
erythro-Pentane-2,4-diol	$C_5H_{12}O_2$				0rib (10)			
threo-Pentane-2,4-diol	$C_5H_{12}O_2$				0rib (10)			
Pentosulose, 3-deoxy-D- (3-deoxy-D-pentosone)	$C_5H_8O_4$	Oil	+7	17–10	250van (10a)		430glc (19b)	
bis-2,4-Dinitrophenylhydrazone	$C_{17}H_{16}N_8O_{10}$	259	+294 (dioxane)	18, 19ct				
3-Deoxy-pentitol acetates	$C_{13}H_{16}O_8$					16.5, 18.5 (19c)		66 for (17)
Pentosulose, D-threo-[f] (xylosone, xylosulose)	$C_5H_8O_5$	No constants known	No constants known	17				

457

TABLE D: NATURAL KETOSES (Continued)

Substance[a] (Synonym) Derivative (A)	Chemical Formula (B)	Melting Point °C (C)	Specific Rotation[b] $[\alpha]_D$ (D)	Reference[c] (E)	Chromatography, R Value, and Reference[d]			
					ELC (F)	GLC (G)	PPC (H)	TLC (I)
bis-2,4-Dinitrophenylhydrazone	$C_{17}H_{17}N_8O_{11}$	231	+187 (c 0.36, dioxane)	17	100van (20a)			
Methyl β-pyranoside	$C_6H_{10}O_5$	No constants known	No constants known				162glc (20b)	
Pentulose, 4-C-methyl 1,3,5-trideoxy- (diacetone alcohol, 4-hydroxy-4-methyl-2-pentanone)	$C_6H_{12}O_2$	Liquid, bp 164	None	1		37.8 (3)		
2,4-Dinitrophenylhydrazone	$C_{12}H_{16}N_4O_5$	198–199	None	1				
Pentulose, D-erythro- (adonose, D-ribulose)	$C_5H_{10}O_5$	Syrup	−15	24,25cp	209rib (10)		38f (26)	
o-Nitrophenylhydrazone	$C_{11}H_{15}N_3O_6$	165–166.5	−52 ±5 (CH_3OH)	27				
Trimethylsilyl ether						25, 33, 35glct (28)		
Pentulose, L-erythro- (L-ribulose)	$C_5H_{10}O_5$	Syrup	+16.6	27p, 29			68f (30c)	
o-Nitrophenylhydrazone	$C_{11}N_{15}N_3O_6$	162–163	+47.4 (c 0.3, CH_3OH)	30				
Pentulose, erythro-3-	$C_5H_{10}O_5$	126–127	None (meso)	31cp	9.1[g] (31)			
2,5-Dichlorophenylhydrazone	$C_{11}Cl_2H_{14}N_2O_4$		None	32cp				
Pentulose, D-threo- (D-xylulose)	$C_5H_{10}O_5$	Syrup	−33	24p	194rib (10)	43f (26)		
p-Bromophenylhydrazone	$BrC_{11}H_{15}N_2O_4$	126–128	+24.1→−31 (C_5H_5N) 7d	24, 33				
2,4-Dinitro-zenylhydrazone	$C_{11}H_{14}N_4O_8$	175–176		32				
Pentulose, 5-deoxy-D-threo- (5-deoxy-D-xylulose)	$C_5H_{10}O_4$	−5 ±1 (CH_3OH) 34c		34		153rha (34)		
1-Deoxy-D-arabinitol	$C_5H_{12}O_4$				103glcl (43)		45f (44)	
1-Deoxy-L-xylitol	$C_5H_{12}O_4$				96glcl (43)[h]			
Phenylosazone	$C_{17}H_{20}N_4O_2$	174–175	+74→+7 (C_5H_5N-C_2H_5OH)	34				
Pentulose, L-threo- (L-xylulose, L-xylulose, xyloketose)	$C_5H_{10}O_5$	Syrup	+33.3	25p,35,36			65f (30c)	117 xyl (36)
p-Bromophenylhydrazone	$BrC_{11}H_{15}N_2O_4$	128–129	−26→+31.9 (C_5H_5N)	35				
Phenylosazone	$C_{17}H_{20}N_4O_3$	162–164	No constants known	25				
Hexodiulose, 6-deoxy-D-erythro-2,5-	$C_6H_{10}O_5$	No constants known	No constants known	37				
1-Deoxy-L-altritol	$C_6H_{14}O_6$	106–108	−2.6 (18°)	58	98glcl (43)[h]		36f (44)[h]	
1-Deoxy-D-galactitol	$C_6H_{14}O_5$				100glcl (43)[h]		31f (44)	
1-Deoxy-D-talitol	$C_6H_{14}O_5$				104glcl (43)			
Hexodiulose, D-threo-2,5- (5-keto-fructose)	$C_6H_{10}O_6$	157–159, 172–174	−85	38,39p			70fru (38)	
bis-Phenylhydrazone	$C_{18}H_{22}N_4O_4$	133–135, 141	−164 (C_5H_5N)	38,39				
Hexos-2,3-diulose, 4,6-dideoxy- (actinospectose)	$C_6H_8O_4$	No constants known	No constants known	40				
Hexosulose, D-arabino- (D-glucosone)	$C_6H_{10}O_6$	Syrup	−10.6→+7.9 (c 8.5) (15°)	41,42			25glc (45)	10 for (17)
2,4-Dinitrophenylosazone	$C_{18}H_{18}N_8O_{12}$	253d		46				
Phenylosazone	$C_{18}H_{22}N_4O_4$	206–208	−75→−41 (c 0.7, C_5H_5N-C_2H_5OH, 2:3)	45				

TABLE D: NATURAL KETOSES (Continued)

Substance[a] (Synonym) Derivative (A)	Chemical Formula (B)	Melting Point °C (C)	Specific Rotation[b] $[\alpha]_D$ (D)	Reference[c] (E)	Chromatography, R Value, and Reference[d]			
					ELC (F)	GLC (G)	PPC (H)	TLC (I)
Tetraacetate·H₂O	$C_{14}H_{18}O_{10} \cdot H_2O$	112	+14.7→+53.7 (20% C₂H₅OH) 96h	47				
Hexos-5-ulose, 6-deoxy-D-*arabino*-	$C_6H_{10}O_5$		-4.3 (CH₃OH) (12°)	48,49			38f (49)	
bis(*p*-Nitrophenyl-)	$C_{18}H_{20}N_6O_8$	211d	+1.1 (c 0.6, C₅H₅N) (15°)	49				
See also 1-Deoxy-D-galactitol and -D-talitol under Hexodiulose, 6-deoxy-D-*erythro*-2,5- above								
Hexosulose, 3-deoxy-D-*erythro*-	$C_6H_{10}O_5$		-2.5→+1.5 (c 6) (27°)	17,46,50			200–270, 137 106 glc (51)	37 for (17)
2,4-Dinitrophenylosazone	$C_{18}H_{18}N_8O_{11}$	251d, 265d	+86 (c 0.09, DMSO)	46,51,52				
Hexos-4-ulose, 3,6-dideoxy-D-*erythro*-	$C_6H_{10}O_4$	No constants known	No constants known	53t				
iso-Propylidene ether	$C_9H_{14}O_4$	Oil, bp 47–49 (0.3 mm)	+166.3 (27°)	53				
Hexos-4-ulose, 3,6-dideoxy-L-*erythro*-	$C_6H_{10}O_4$	No constants known	No constants known	54				
Hexos-4-ulose, 2,3,6-trideoxy-L-*glycero*- (cinerulose A)	$C_6H_{10}O_3$			55				
Methyl α-pyranoside	$C_7H_{12}O_3$		+310±2 (CHCl₃)[h]	56				
Methyl α-pyranoside *p*-nitrophenylhydrazone	$C_{13}H_{17}N_3O_4$	158–159	+347±2 (c 0.6, CHCl₃)[h]	56				70f (56)[h]
Hexos-5-ulose, D-*lyxo*-	$C_6H_{10}O_6$	157–158	-86.6	57p				
bis(*p*-Nitrophenylhydrazone)	$C_{18}H_{20}N_6O_9$	173–174		57				
bis-Phenylhydrazone	$C_{18}H_{22}N_4O_4$	123–124	-138.5 (c 0.25, C₂H₅N)	57				
Hexosulose, 6-deoxy-L-*lyxo*- (angustose, 2-keto-fucose)	$C_6H_{10}O_5$	115–116	+18 (C₂H₅OH)	58c			36f (58)	
Methyl pyranoside dimethyl acetal	$C_9H_{18}O_6$		-19.3 (CH₃OH)	58				
Methyl pyranoside	$C_{11}H_{22}O_6$	77–78	-53.3 (C₂H₅OH) (18°)	58				
6-Deoxy-L-talitol (1-deoxy-L-altritol)	$C_6H_{14}O_5$	106–108	-2.6 (18°)	58	98glcl (43)[h]		36f (44)[h]	
Hexos-3-ulose, D-*ribo*- (3-keto-D-glucose)	$C_6H_{10}O_6 \cdot 5H_2O$	58–60	+14.8 (26°)	59			116fru (59)	
Methyl α-pyranoside trimethyl ether	$C_{10}H_{18}O_6$	82.5–83.5	$[\alpha]_{578}$+164 (c 0.9, CHCl₃)	60t				
Methyl β-pyranoside trimethyl ether	$C_{10}H_{18}O_6$	117.5–119.5	$[\alpha]_{578}$-24 (CHCl₃)	60t				
Hexos-3-ulose, 6-deoxy-L-*ribo*-, 2-acetate	$C_8H_{12}O_6$	No constants known	No constants known	61				
Hexulose, β-D-*arabino*-β-D-fructose, levulose)[i]	$C_6H_{12}O_6$	102–104	-133.5→-92	62,63	75rib (10)		38f (4)	
β-Furanoside tetra-benzoate	$C_{34}H_{28}O_{10}$	174–175	-165 (CHCl₃)	65				28f (64)
Methyl α-pyranoside tetramethyl ether	$C_{11}H_{22}O_6$							87f (66)
Methyl β-pyranoside tetramethyl ether	$C_{11}H_{22}O_6$							23f (67)
p-Nitrophenylhydrazone	$C_{12}H_{17}N_3O_7$	176		68				16f (67)
β-Pyranoside tetra-acetate	$C_{14}H_{20}O_{10}$	132–132	-91.6 (CHCl₃)	69,70				
Trimethylsilyl ether	$C_{21}H_{52}O_6Si_5$		-73.8 (hexane)	106		69glct (28)		
Hexulose, 6-deoxy-D-*arabino*- (D-rhamnulose)	$C_6H_{12}O_5$	Syrup	-6±1, -13±2	73,74			120rha (74)	
1-Deoxy-L-gulitol	$C_6H_{14}O_5$				94glcl (43)			
1-Deoxy-D-mannitol	$C_6H_{14}O_5$				100glcl (43)			

TABLE D: NATURAL KETOSES (Continued)

Substance[a] (Synonym) Derivative	Chemical Formula	Melting Point °C	Specific Rotation[b] $[\alpha]_D$	Reference[c]	Chromatography, R Value, and Reference[d]			
					ELC	GLC	PPC	TLC
(A)	(B)	(C)	(D)	(E)	(F)	(G)	(H)	(I)
o-Nitrophenylhydrazone	$C_{12}H_{17}N_3O_6$	136–137	+40±3 (C_2H_5OH)	73	103rib (10)		38f (4)	46f (64)
Hexulose, D-lyxo- (D-tagatose)	$C_6H_{12}O_6$	131–132	+2.7 → −4	75				
Phenylosazone	$C_{18}H_{22}N_4O_4$	186–187		76				
Pyranose pentaacetate	$C_{16}H_{22}O_{11}$	132	$[\alpha]_{578}$ +30.2 ($CHCl_3$)	77,78				
Tagaturonic acid	See Table 2							
Trimethylsilyl ether						varied 60–159 glct (79)		
Hexulose, 6-deoxy-L-lyxo- (L-fuculose)	$C_6H_{12}O_5$	68–69	+3.4±1	80				
o-Nitrophenyl hydrazone	$C_{12}H_{17}N_3O_6$	162–163		81			50f (81)	
See also 1-Deoxy-D-galacitol and -L-altritol under Hexodiulose, 6-deoxy-D-erythro-2,5-, above								
Hexulose, D-ribo- (D-allulose, D-psicose)	$C_6H_{12}O_6$	Amorphous	+3.2, +4.7	82,83	188rib (10)		145glc (84)	
Methyl glycoside tetramethyl ether	$C_{11}H_{22}O_6$	104 (2 mm)	−28.3 (C_2H_5OH) (18°)	82				
Phenylosazone	$C_{18}H_{22}N_4O_4$	159–163	−74 → −68 (C_5H_5N)	85p				
Pyranose pentaacetate	$C_{16}H_{22}O_{11}$	63–65	−21.5 ($CHCl_3$) (29°)	86c				
Hexulose, L-xylo- (L-sorbose)	$C_6H_{12}O_6$	159–161	−43.1	87	73rib (10)		33f (4)	43f (64)
α-Pyranoside pentaacetate	$C_{16}H_{22}O_{11}$	97	−56.5 ($CHCl_3$)	88				
β-Pyranoside pentaacetate	$C_{16}H_{22}O_{11}$	113.8	+74.4 ($CHCl_3$)	88				
Trimethylsilyl ether	$C_{21}H_{52}O_6Si_5$		−16.4 (hexane)	106		85glct (28)		
Hexulose, 6-deoxy-L-xylo- (6-deoxy-L-sorbose)	$C_6H_{12}O_5$	88	−25±2 (c 0.7)	74	98glct (43)		134rha (74)	
1-Deoxy-D-glucitol	$C_6H_{14}O_5$							
Phenylosazone	$C_{18}H_{22}N_4O_3$	184–185		89				
Pyran-4-one, 3,5-dihydroxy-2-hydroxymethyl- (iso-kojic acid, hydroxy-kojic acid, oxykojic acid)	$C_6H_6O_5$	187	None	90			46f (91)	
3,5-Di-o-methyl ether	$C_8H_{10}O_5$	115.5–117	None	90				
2,4-Dinitrophenylhydrazone	$C_{12}H_{10}N_4O_8$	118.5–119.5	None	90				
Pyran-4-one, 3,5-dihydroxy-2-methyl- (5-hydroxymaltol, oxymaltol)	$C_6H_6O_4$	156–156.5, 184–184.5	None	90,92		67.5 (92a)	72f (91)	43f (92b)
3,5-Di-o-methyl ether	$C_8H_{10}O_5$	98	None	90				
Pyran-4-one, 2,3-dihydro-3,5-dihydroxy-6-methyl-(dihydrohydroxy-maltol)	$C_6H_8O_4$	67–70	None	92		65 (92a)	61f (91)	38f (92b)
Pyran-4-one, 5-hydroxy-2-hydroxymethyl- (kojic acid)	$C_6H_6O_4$	150–152	None	91				
Diacetate	$C_{10}H_{10}O_6$	101–103	None	93				
2,4-Dinitrophenylhydrazone	$C_{18}H_{14}N_8O_{10}$	221–224	None	94				
Phenylosazone	$C_{18}H_{18}N_8O_2$	169–170	None	92				
Pyran-4-one, 3-hydroxy-2-methyl- (maltol)	$C_6H_6O_3$	161–162	None	95,96,98p		44 (92a)		45f (92b)
Benzoate	$C_{12}H_{10}O_4$	114–115	None	95				
Methyl ether	$C_7H_8O_3$	Liquid, bp 78–79 (4 mm)						
Phenylurethane	$C_{13}H_{11}NO_4$	152–153	None	95				
Trimethylsilyl ether						4.6 (100)		

TABLE D: NATURAL KETOSES (Continued)

Substance[a] (Synonym) Derivative (A)	Chemical Formula (B)	Melting Point °C (C)	Specific Rotation[b] $[\alpha]_D$ (D)	Reference[c] (E)	Chromatography, R Value, and Reference[d] — ELC (F)	GLC (G)	PPC (H)	TLC (I)
Pyran-4-one, 3-hydroxy-5-methyl-[k] (iso-maltol)	$C_6H_6O_3$	101±2	None	97				75f (92b)
Benzoate	$C_{13}H_{10}O_4$	100–101	None	97				
2,4-Dinitrophenylhydrazone	$C_{12}H_{10}N_4O_6$	216	None	97				
Methyl ether	$C_7H_8O_3$	101.5–103	None	92,97		21 (92a)		50f (92b)
Heptulose, D-allo-	$C_7H_{14}O_7$	128–130	+52.8 (c 0.2)	101c, 122	100rib (10)		99fru (101)	
D-glycero-D-allo-Heptitol	$C_7H_{16}O_7$	144.5–146	None (meso)	122	144rib (10)			
D-glycero-D-altro-Heptitol	$C_7H_{16}O_7$	125–128	−0.3±0.4	107,122				
Phenylosazone	$C_{19}H_{26}N_4O_5$	164–167		101			41f (103)	
Heptulose, D-altro- (sedoheptose, sedoheptulose)	$C_7H_{14}O_7$	Amorphous	+2.5 (c 10)	101,102				
See D-glycero-D-gluco-and D-glycero-D-talo-Heptitols in Table A								
2,7-Anhydro-β-pyranose (sedoheptulosan)	$C_7H_{12}O_6$	155–156	−145	101			73f (103)	
Pyranose hexaacetate	$C_{19}H_{26}O_{13}$	98–99.5	+59 (CHCl₃)	105				
Sedoheptulosan hydrate	$C_7H_{12}O_6 \cdot H_2O$	101–102 (91s)	−132	104				
Sedoheptulosan trimethylsilyl ether				106		112glct (28)		
Trimethylsilyl ether	$C_{25}H_{62}O_7Si_6$		+18 (hexane)			166glct (106)		
Heptulose, D-altro-3- (coriose)	$C_7H_{14}O_7$	169–171	+20	101,107gp			99fru (101)	
See D-glycero-D-altro-Heptitol above and D-glycero-D-talo-Heptitol in Table A								
Pentabenzoate	$C_{42}H_{34}O_{12}$	100–102		107				
Heptulose, L-galacto- (perseulose)	$C_7H_{14}O_2 \cdot \tfrac{1}{2}H_2O$	100–115	−80, −90	108,109			88manh (110)	
See L-glycero-manno-Heptitol (D-glycero-D-galacto-Heptitol) in Table A								
L-glycero-D-gluco-Heptitol	$C_7H_{16}O_7$	198–199	−90→−45	110	176rib (10)[h]			
Phenylosazone	$C_{19}H_{26}N_4O_5$	112	−113 (CHCl₃)	111				
Pyranose hexaaceate	$C_{19}H_{26}O_{13}$							
Heptulose, L-gluco-	$C_7H_{14}O_7$	171–172	−68	110	100sor (116)[h]		93manh (110)	
L-glycero-L-gulo-Heptitol	$C_7H_{16}O_7$	129–130	None (meso)	3	160rib (10)[h]			
L-glycero-L-ido-Heptitol	$C_7H_{16}O_7$		−0.8	114	168rib (10)[h]			
Phenylosazone	$C_{19}H_{26}N_4O_5$	181–182d	+6→−35.3 (C₅H₅N-C₂H₅OH,2:3) 96h	114				
Trimethylsilyl ether	$C_{25}H_{62}O_7Si_6$		+36 (hexane)[h]	106		206glct (106)[h]		
Heptulose, L-gulo-	$C_7H_{14}O_7$	Syrup	−28	112			53f (118)	
See L-glycero-D-gluco-Heptitol above and D-glycero-D-gluco-Heptitol in Table A								
2,7-Anhydro-L-gulo-heptulose (guloheptulosan)	$C_7H_{12}O_6$	113–115[l]	−39.7	112			65f (118)	
Phenylosazone	$C_{19}H_{26}N_4O_5$	197–200d	+111→+65 (C₅H₅N) 47h	112				
Heptulose, D-ido-	$C_7H_{14}O_7$		−20	104			50f (118)	
See D-glycero-D-ido-Heptitol in Table A								
D-glycero-L-ido-Heptitol	$C_7H_{16}O_7$		None (meso)		182rib (10)			

TABLE D: NATURAL KETOSES (Continued)

Substance[a] (Synonym) Derivative (A)	Chemical Formula (B)	Melting Point °C (C)	Specific Rotation[b] $[\alpha]_D$ (D)	Reference[c] (E)	Chromatography, R Value, and Reference[d]			
					ELC (F)	GLC (G)	PPC (H)	TLC (I)
Anhydro-D-*ido*-heptulose (idoheptulosan)	$C_7H_{12}O_6$	172	−34±8 (c 0.3)	113			86rib (113)	
Phenylosazone	$C_{19}H_{26}N_4O_5$	178–179	+11.6→−43.4 (C_5H_5N) 72h	104				
Heptulose, D-*manno*-	$C_7H_{14}O_7$	152	+29.4	101c,115	40sor (116)		102glc (117)	
See D-*glycero*-D-*galacto*- and D-*glycero*-D-*talo*-Heptitols in Table 1								
Phenylosazone	$C_{19}H_{26}N_4O_5$	200	+74→+35	110,119				
Pyranose hexaacetate	$C_{19}H_{26}O_{13}$	110	+39 ($CHCl_3$)	120				
Trimethylsilyl ether	$C_{25}H_{62}O_7Si_6$		+30.4 (hexane)	106		194glct (106)		
Heptulose, D-*talo*-	$C_7H_{14}O_7$	135–137	+47.4→+12.9 (6h)	121,122			138manh (110)	
See D-*glycero*-D-*altro*-Heptitol above								
D-*glycero*-L-*altro*-Heptitol	$C_7H_{16}O_7$		None (meso)	3				
Octulose, D-*glycero*-L-*galacto*	$C_8H_{16}O_8$	Syrup	−57, −43.4→−13.4 (c 0.6)	123cp, 124cp			42sed (101)	
2,5-Dichlorophenyl-hydrazone	$C_{14}Cl_2H_{20}N_2O_7$	178–180		123				
Octulose, D-*glycero*-D-*manno*-	$C_8H_{16}O_8$	Syrup	+20, +25 (CH_3OH)	101c,121p			46sed (101) 80glc (117)	
2,5-Dichlorophenyl-hydrazone	$C_{14}Cl_2H_{20}N_2O_7$	169–170		121				
Phenylosazone	$C_{20}H_{26}N_4O_6$	188–189d	−20	121				
Heptulose, D-*ido*-	$C_7H_{14}O_7$			104			50f (118)	
See D-*glycero*-D-*ido*-Heptitol in Table A								
D-*glycero*-L-*ido*-Heptitol	$C_7H_{16}O_7$		None (meso)					
Anhydro-D-*ido*-heptulose (idoheptulosan)	$C_7H_{12}O_6$	172	−34±8 (c 0.3)	113	182rib (10)		86rib (113)	
Phenylosazone	$C_{19}H_{26}N_4O_5$	178–179	+11.6→−43.4 (C_5H_5N) 72h	104				
Heptulose, D-*manno*-	$C_7H_{14}O_7$	152	+29.4	101c,115	40sor (116)		102glc (117)	
See D-*glycero*-D-*galacto*- and D-*glycero*-D-*talo*-Heptitols in Table A								
Phenylosazone	$C_{19}H_{26}N_4O_5$	200	+74→+35	110,119				
Pyranose hexaacetate	$C_{19}H_{26}O_{13}$	110	+39 ($CHCl_3$)	120				
Trimethylsilyl ether	$C_{25}H_{62}O_7Si_6$		+30.4 (hexane)	106		194glct (106)		
Heptulose, D-*talo*-	$C_7H_{14}O_7$	135–137	+47.4→+12.9 (6h)	121,122			138manh (110)	
See D-*glycero*-D-*altro*-Heptitol above								
D-*glycero*-L-*altro*-Heptitol	$C_7H_{16}O_7$		None (meso)	3				
Octulose, D-*glycero*-L-*galacto*	$C_8H_{16}O_8$	Syrup	−57, −43.4→−13.4 (c 0.6)	123cp, 124cp			42sed (101)	
2,5-Dichlorophenyl-hydrazone	$C_{14}Cl_2H_{20}N_2O_7$	178–180		123				
Octulose, D-*glycero*-D-*manno*-	$C_8H_{16}O_8$	Syrup	+20, +25 (CH_3OH)	101c, 121p			46sed (101) 80 glc (117)	

TABLE D: NATURAL KETOSES (Continued)

Substance[a] (Synonym) Derivative (A)	Chemical Formula (B)	Melting Point °C (C)	Specific Rotation[b] $[\alpha]_D$ (D)	Reference[c] (E)	Chromatography, R Value, and Reference[d] ELC (F)	GLC (G)	PPC (H)	TLC (I)
2,5-Dichlorophenyl-hydrazone	$C_{14}Cl_2H_{20}N_2O_7$	169–170		121				
Phenylosazone	$C_{20}H_{26}N_4O_6$	188–189d		121				
See D-erythro-D-galacto-Octitol in Table A								
D-erythro-D-talo-Octitol	$C_8H_{18}O_8$		None (meso)	3				
Nonulose, D-erythro-L-galacto-	$C_9H_{18}O_9$	Syrup	−9.7, −37 (H₂O) −36.2 (95% CH₃OH)	101cp, 125cp				
2,5-Dichlorophenyl-osazone	$C_{21}Cl_4H_{24}N_4O_7$	238–240		125				
D-arabino-D-gluco-Nonitol	$C_9H_{20}O_9$	180–181		125				
D-arabino-D-manno-Nonitol	$C_9H_{20}O_9$	192–193		125				
Trimethylsilyl ether						18.2 (125)		
Nonulose, D-erythro-L-gluco-	$C_9H_{18}O_9$	Syrup	−40 (c 0.6)	126cp				
2,5-Dichlorophenylosazone	$C_{21}Cl_4H_{24}N_4O_7$	248–250d		126				

Compiled by George G. Maher.

a In alphabetical order by parent sugar names within groups of increasing carbon chain length in the parent compounds.

b $[\alpha]_D$ for 1–5 g solute, c, per 100 ml aqueous solution at 20–25°C, unless otherwise given.

c References for m.p. and specific rotation data.

d R value times 100, given relative to that of the compound indicated by abbreviation: f = solvent front, for = formaldehyde 2,4-dinitrophenylhydrazone, fru = fructose, glc = glucose, glcl = glucitol, glct = glucose trimethylsilyl ether, manh = manno-heptulose, rha = rhamnose, rib = ribose, sed = sedoheptulose, sor = sorbitol, van = vanillin, xyl = xylose. Under gas chromatography (column GLC or G), numbers without code indication signify retention time in minutes. The conditions of the chromatography are correlated with the reference given in parentheses and are found in Table E.

e A possible finding of glyoxal as a component in ethanol distillery streams exists, retention time of 29.9 min, but the identification needs other evidence.³

f There is no clear evidence that the configuration is D- or L.

g Value is in cm³/3 hr.

h Data are for the enantiomorphic isomer.

i The 1/2H₂O and 2H₂O forms also exist.

j From cured tobacco and animal products fructose and amino acid combinations, which are probably rearranged glycoside compounds and hence not included here, have been isolated; see References 71 and 72.

k This is an early structural name. The modern, preferred one is 3-hydroxy-2-furyl methyl ketone.⁹⁸ While perhaps not carbohydrates in a strict sense, this group of pyranose compounds is included because of their intimate relationship.

l A hemihydrate of lower melting point forms from the anhydrous anhydro sugar upon aging.

References

1. Pollock and Stevens, *Dictionary of Organic Compounds*, Oxford University Press, New York, 1965.
2. Bancher, Scherz, and Kaindl, *Mikrochim. Acta*, 1043 (1964).
3. Maher, unpublished data.
4. Adachi, *Anal. Biochem.*, 224 (1964).
5. Verhaar and de Wilt, *J. Chromatogr.*, 41, 168 (1969).
6. Williams and Tucknott, *J. Sci. Food Agric.*, 22, 264 (1971).
7. Bush and Hockaday, *J. Chromatogr.*, 433 (1962).
8. Gasparic and Vecera, *Collect. Czech. Chem. Commun.*, 22, 1426 (1957).
9. Reio, *J. Chromatogr.*, 338 (1958).
10. Frahn and Mills, *Aust. J. Chem.*, 12, 65 (1959).
11. Bertrand, *Bull. Soc. Chim. Fr. Ser. 3*, 23, 681 (1904).
12. Hu, McComb, and Rendig, *Arch. Biochem. Biophys.*, 110, 350 (1965).
13. Batt, Dickens, and Williamson, *Biochem. J.*, 77, 272 (1960).
14. Bloehm, *J. Chromatogr.*, 35, 108 (1968).
15. Müller, Montigel, and Reichstein, *Helv. Chim. Acta*, 20, 1468 (1937).
16. Blom, *J. Am. Chem. Soc.*, 67, 494 (1945).
17. Kato, Tsusaka, and Fujimaki, *Agric. Biol. Chem.*, 34, 1541 (1970).
18. Kurata and Sakurai, *Agric. Biol. Chem.*, 31, 170,177 (1967).
19. Humphries and Theander, *Acta Chem. Scand.*, 25, 883 (1971).
20. Brimacombe, Brimacombe, and Lindberg, *Acta Chem. Scand.*, 14, 2236 (1960).
21. Theander, *Acta Chem. Scand.*, 11, 717 (1957).
22. Spence and Degering, *J. Am. Chem. Soc.*, 66, 1624 (1944).
23. Schöpf and Ross, *Justus Liebigs Ann. Chem.*, 546, 30 (1941).
24. Hickman and Ashwell, *J. Am. Chem. Soc.*, 78, 6209 (1956).
25. Futterman and Roe, *J. Biol. Chem.*, 215, 257 (1955).
26. Uehara and Takeda, *J. Biochem. (Tokyo)*, 56, 42 (1964).
27. Horecker, Smyrniotis, and Seegmiller, *J. Biol. Chem.*, 193, 383 (1951).
28. Sweeley, Bentley, Makita, and Wells, *J. Am. Chem. Soc.*, 85, 2497 (1963).
29. Reichstein, *Helv. Chim. Acta*, 17, 996 (1934).
30. Simpson, Wolin, and Wood, *J. Biol. Chem.*, 230, 457 (1958).
31. Ashwell and Hickman, *J. Am. Chem. Soc.*, 77, 1062 (1955).
32. Stanković, Linek, and Fedoroňko, *Carbohydr. Res.*, 10, 579 (1969).
33. Ashwell and Hickman, *J. Biol. Chem.*, 226, 65 (1957).
34. Gorin, Hough, and Jones, *J. Chem. Soc. (Lond.)*, 2140 (1953).
35. Levene and LaForge, *J. Biol. Chem.*, 18, 319 (1914).
36. Wolfrom and Bennett, *J. Org. Chem.*, 30, 458 (1965).
37. Chassy, Sugimori, and Suhadolnik, *Biochim. Biophys. Acta*, 130, 12 (1966).
38. Avigad and England, *J. Biol. Chem.*, 240, 2290, 2297, 2302 (1965).
39. Whiting and Coggins, *Chem. Ind.*, 1925 (1963).
40. Hoeksema, Argoudelis, and Wiley, *J. Am. Chem. Soc.*, 84, 3212 (1962).
41. Bean and Hassid, *Science*, 124, 171 (1956).
42. Bayne, Collie, and Fewster, *J. Chem. Soc. (Lond.)*, 2766 (1952).
43. Angus, Bourne, and Weigel, *J. Chem. Soc. (Lond.)*, 22 (1965).
44. Bourne, Lees, and Weigel, *J. Chromatogr.*, 11, 253 (1963).
45. Walton, *Can. J. Chem.*, 47, 3483 (1969).
46. Kato, *Agric. Biol. Chem.*, 27, 461 (1963).
47. Maurer, *Ber. Dtsch. Chem. Ges.*, 63, 25 (1930).
48. Mann and Woolf, *J. Am. Chem. Soc.*, 79, 120 (1957).
49. Takahashi and Nakajima, *Tetrahedron Lett.*, 2285 (1967).
50. Együd, *Carbohydr. Res.*, 23, 307 (1972).
51. Fodor, Sachetto, Szent-Györgyi, and Együd, *Proc. Natl. Acad. Sci. U.S.A.*, 57, 1644 (1967).
52. El Khadem, Horton, Meshreki, and Nashed, *Carbohydr. Res.*, 17, 183 (1971).
53. Stevens, Schultze, Smith, Pillai, Rubenstein, and Strominger, *J. Am. Chem. Soc.*, 95, 5767 (1973).
54. Richle, Winkler, Hawley, Dobler, and Keller-Schierlein, *Helv. Chim. Acta*, 55, 467 (1972).
55. Keller-Schierlein and Richle, *Chimia*, 24, 35 (1970).
56. Albano and Horton, *Carbohydr. Res.*, 11, 485 (1969).
57. Weidenhagen and Bernsee, *Angew. Chem.*, 72, 109 (1960).
58. Yunsten, *J. Antibiot. Ser. A.*, 77, 233 (1958).
59. Fukui and Hochster, *J. Am. Chem. Soc.*, 85, 1697 (1963).
60. Kenne, Larm, and Svensson, *Acta Chem. Scand.*, 26, 2473 (1972).
61. Kupchan, Sigel, Guttman, Restivo, and Bryan, *J. Am. Chem. Soc.*, 94, 1353 (1972).
62. Hudson and Brauns, *J. Am. Chem. Soc.*, 38, 1216 (1916).
63. Hudson and Yanovsky, *J. Am. Chem. Soc.*, 39, 1025 (1917).
64. Adachi, *J. Chromatogr.*, 17, 295 (1965).
65. Brigl and Schinle, *Ber. Dtsch. Chem. Ges.*, 66, 325 (1933).
66. Hay, Lewis, and Smith, *J. Chromatogr.*, 11, 479 (1963).
67. Gee, *Anal. Chem.*, 35, 350 (1963).
68. Reclaire, *Ber. Dtsch. Chem. Ges.*, 41, 3665 (1908).
69. Pacsu and Rich, *J. Am. Chem. Soc.*, 55, 3018 (1933).
70. Barry and Honeyman, *Adv. Carbohydr. Chem.*, 60,85 (1952).
71. Tomita, Noguchi, and Tamaki, *Agric. Biol. Chem.*, 29, 515, 959 (1965).
72. Yamamoto and Noguchi, *Agric. Biol. Chem.*, 37, 2185 (1973).
73. Morgan and Reichstein, *Helv. Chim. Acta*, 21, 1023 (1938).
74. Hough and Jones, *J. Chem. Soc.* (Lond.), 4052 (1952).
75. Reichstein and Bosshard, *Helv. Chim. Acta*, 17, 753 (1934).
76. Adachi, *J. Agric. Chem. Soc. Jap.*, 32, 309 (1958).
77. Khouvine and Tomoda, *C.R. Acad. Sci.*, 205, 736 (1937).
78. Khouvine, Arragon, and Tomoda, *Bull. Soc. Chim. Fr. Ser. 5*, 354 (1939).
79. Tesarik, *J. Chromatogr.*, 65, 295 (1972).
80. Barnett and Reichstein, *Helv. Chim. Acta*, 21, 913 (1938).
81. Green and Cohen, *J. Biol. Chem.*, 219, 557 (1956).
82. Yüngsten, *J. Antibiot. (Tokyo) Ser. A*, 11, 244 (1958).
83. Wolfrom, Thompson, and Evans, *J. Am. Chem. Soc.*, 67, 1793 (1945).
84. Bourne, Percival, and Smestad, *C.R. Acad. Sci.*, 260, 999 (1965).
85. Strecker, Gouret, and Montreuil, *Compt. Rend.*, 260, 999 (1965).
86. Binkley and Wolfrom, *J. Am. Chem. Soc.*, 70, 3940 (1948).
87. Schlubach and Vorwerk, *Ber. Dtsch. Chem. Ges.*, 66, 1251 (1933).
88. Schlubach and Graefe, *Justus Liebigs Ann. Chem.*, 532, 211 (1937).
89. Muller and Reichstein, *Helv. Chim. Acta*, 21, 263 (1938).
90. Terada, Suzuki, and Kinoshita, *Agric. Biol. Chem.*, 25, 802, 939 (1961).
91. Sato, Yamada, Aida, and Uemura, *Agric. Biol. Chem.*, 33, 1606 (1969).
92. Shaw, Tatum, and Berry, *Carbohydr. Res.*, 16, 207 (1971).
93. Chittenden, *Carbohydr. Res.*, 11, 424 (1969).
94. Corbett, *J. Chem. Soc. (Lond.)*, 3213 (1959).
95. Schenck and Spielman, *J. Am. Chem. Soc.*, 67, 2276 (1945).
96. Spielman and Freifelder, *J. Am. Chem. Soc.*, 69, 2908 (1947).
97. Hodge and Nelson, *Cereal Chem.*, 38, 207 (1961).
98. Potter and Patton, *J. Dairy Sci.*, 39, 978 (1956).
99. Fisher and Hodge, *J. Org. Chem.*, 29, 776 (1964).
100. Gunner, Hand, and Sahasrabudhe, *J. Assoc. Of. Anal. Chem.*, 51, 959 (1968).
101. Begbie and Richtmyer, *Carbohydr. Res.*, 272 (1966).
102. LaForge and Hudson, *J. Biol. Chem.*, 30, 61 (1917).
103. Wood, *J. Chromatogr.*, 35, 352 (1968).
104. Pratt, Richtmyer, and Hudson, *J. Am. Chem. Soc.*, 74, 2203, 2210 (1952).
105. Richtmyer and Pratt, *J. Am. Chem. Soc.*, 78, 4717 (1956).
106. Okuda and Konishi, *J. Chem. Soc. D Chem. Commun.*, 796 (1969).
107. Okuda and Konishi, *Tetrahedron*, 24, 6907 (1968).
108. Bertrand, *Bull. Soc. Chim. Fr. Ser. 4*, 51, 629 (1909).
109. Hann and Hudson, *J. Am. Chem. Soc.*, 61, 336 (1939).
110. McComb and Rendig, *Arch. Biochem. Biophys.*, 95, 316 (1961); 97, 562 (1962).
111. Khouvine and Arragon, *C.R. Acad. Sci.*, 206, 917 (1938).
112. Stewart, Richtmyer, and Hudson, *J. Am. Chem. Soc.*, 74, 2206 (1952).
113. Gorin and Jones, *J. Chem. Soc.* (Lond.), 1537 (1953).
114. Maclay, Hann, and Hudson, *J. Am. Chem. Soc.*, 64, 1606 (1942).
115. Bevenne, White, Secor, and Williams, *J. Assoc. Of. Agric. Chem.*, 44, 265 (1961).
116. Bourne, Hutson, and Weigel, *J. Chem. Soc.* (Lond.), 4252 (1960).
117. Rendig, McComb, and Hu, *J. Agric. Food Chem.*, 12, 421 (1964).
118. Noggle, *Arch. Biochem. Biophys.*, 43, 238 (1953).
119. LaForge, *J. Biol. Chem.*, 28, 511 (1917).
120. Montgomery and Hudson, *J. Am. Chem. Soc.*, 61, 1654 (1939).
121. Charlson and Richtmyer, *J. Am. Chem. Soc.*, 82, 3428 (1960).
122. Pratt and Richtmyer, *J. Am. Chem. Soc.*, 77, 6326 (1955).
123. Sephton and Richtmyer, *J. Org. Chem.*, 28, 1691 (1963).
124. Jones and Sephton, *Can. J. Chem.*, 38, 753 (1960).
125. Sephton and Richtmyer, *Carbohydr. Res.*, 289 (1966).
126. Sephton and Richtmyer, *J. Org. Chem.*, 28, 2388 (1963).

Table E: CHROMATOGRAPHIC CONDITIONS FOR CHROMATOGRAPHY DATA IN TABLES A–D

Reference[a]	Conditions[b]
(A)	**(B)**

FOR TABLE A

(2) Whatman® No. 4 paper, pH 9.6 $Na_2 HAsO_3$ (as 19.8 g As_2O_3/1 and NaOH added), 90-min runs at 20–25 V/cm, 20–25° using 18–20° cooling water.

(3) Polypak 1 (120–200 mesh), 0.5 or 1 m by 6 mm O.D. glass column, N_2 gas at 32.5 ml/min, 250° isothermic or programmed 150–250° at 4°/min.

(4) Whatman No. 1 paper, EtAc-HAc-H_2O (9:2:2 v/v), front moves 30 cm in 3–4 h, descending.

(5) Silica Gel H, 0.25 mm thick, 30 min 110° activation, EtOH-32% NH_3-H_2O (21:2:3.5).

(6) Chromosorb® W (80–100 mesh) with 10% Carbowax® 20 *M* terminated with terephthalic acid, 4 ft by 0.25 in. O.D. copper column, He gas at 85 ml/min, 190°.

(7) Gas-Chrom® Q with 3% QF-1 (w/w), 180 cm by 0.45 cm O.D. coiled hardened aluminum column, N_2 gas at 60 ml/min, 110°.

(8) Chromosorb W (60–80 mesh) with 20% diethylene glycol succinate, 6 ft by 0.25 in. O.D. aluminum column, He gas at 55 ml/min, programmed 70–220° at 15°/min.

(11) Silanized Celite® (60–80 mesh) with 15% LAC-2R-446 diethylene glycol adipate cross-linked with pentaerythritol, 20 ft by 0.37 in. O.D. copper column, H_2 gas at 300 ml/min, 175°.

(12) Whatman No. 3 paper, water-saturated, EtAc descending.

(14) Diatoport® S (60–80 mesh) with 20% SF96, 8 ft by 0.25 in. O.D. coiled copper column, He gas at 88 ml/min, 130° for 6 min and programmed to 220° at 3°/min.

(20) Whatman No. 1 paper, EtAc-Pyr-H_2O (12:5:4), descending.

(21) Chromosorb W coated by hexamethyldisilazane (60–80 mesh) with 10% LAC-1R-296 diethylene glycol adipate, 1.84 m by 0.32 cm I.D. glass column, N, gas at 30–50 ml/min, programmed 170–220° at 0.8°/min.

(26) Silica Gel G (240–270 μm), 2 mm thick, dried at room temperature and 55–60° relative humidity 1–3 days, MeC(O)Et-HAc-H_2O (60:20:20).

(31) Magnesium silicate, air-dried plates with no chamber saturation, PrOH-H_2O (5:5 v/v).

(32) Diatoport S (80–100 mesh) with 10% SF-96, 8 ft by 0.25 in. O.D. copper column, He gas, 190°.

(35g) Silanized Diatomate C (85–100 mesh) with 3.5% SE 52, 5 ft glass column, gas (not specified) flow rate at 45 ml/min.

(35t) Merck® Silica Gel F 254, C_6H_6.

(38) Celite AW (80–100 mesh) with 10% butanediol succinate polyester, 6 ft by 0.25 in. O.D. glass column, He gas at 28 ml/min, programmed 125–215° at 4°/min.

(44) Whatman No. 1 paper, AmOH-Pyr-H_2O (4:3:2 v/v).

(50a) Silica Gel G, dried overnight at 135°, upper phase of C_6H_6 -EtOH-H_2O-0.8 sp gr NH_4OH (200:47:15:1 v/v).

(50b) Same as (50a) but solvent is BuOH-HAc-EtOEt-H_2O (9:6:3:1 v/v).

(55a) Whatman No. 1 paper, BuOH-EtOH-H_2O (40:11:19), descending.

(55b) Whatman No. 1 paper, EtAc-HAc-HC(O)OH-H_2O (18:3:1:4), descending.

(57p) Whatman No. 1 paper, EtAc-Pyr-H_2O (10:4:3 v/v), descending.

(57g) Chromosorb W (80–100 mesh) with 10% neopentylglycol sebacate polyester, 120 cm by 0.5 cm I.D. glass column, Ar gas at 150 ml/min, 205°.

(60) Whatman No. 1 paper, BuOH-HAc-H_2O (4:1:1), descending (?).

(66) Whatman No. 1 paper, BuOH-Pyr-H_2O (6:4:3).

(70) Whatman No. 1 paper, BuOH-EtOH-H_2O (4:1:5 v/v) upper phase, descending.

(72) Whatman No. 1 paper, MeC(O)Me-H_2O (95:5 v/v), descending.

(74) Paper not described, BuOH-MeC(O)Me-H_2O (4:1:5).

(79) Anakrom® ABS (100–110 mesh) with 5% QF-1, 5 ft by 0.13 in. O.D. column (material of construction not given), N_2 gas at 20 ml/min, 150°.

(80) Whatman No. 3 paper, 0.05 *M* $Na_2B_4O_7$, 15 h run at 10 V/cm, room temperature.

(81) Whatman No. 1 paper, MeC(O)Me-BuOH-H_2O (3:1:1), ascending.

(82) Whatman No. 1 paper, 0.15 *M* $Na_2B_4 O_7$, 10 V/cm, room temperature.

(84) Gas-Chrom Q (100–120 mesh) with 3% silicone polymer JXR, 1.8 m by 4 mm O.D. coiled glass column, N_2 gas at 60 ml/min, 140° for 2 min and programmed to 210° at 3°/min.

(87a) pH 10 Borate buffer solvent, only detail given.

(87b) Whatman No. 1 paper, MeC(O)Me-aq 10% HAc (4:1).

(87c) Whatman No. 1 paper impregnated with DMSO; *iso*-propylether solvent.

(89) Breite, Macherey, Nagel, and Company No. MN847 paper, BuOH-Pyr-H_2O (10:3:3 v/v).

(93) Embacel kieselguhr (60–100 mesh) with 1.5% polyester resin LAQ l-R-296, 115 cm by 4.5 mm I.D. glass column, N_2 gas at 30 ml/min, 215°.

(94) Whatman No. 3 MM paper, aq $Na_2 MoO_4 \cdot 2H_2O$ (25 g/1,200 ml) made to pH 5 with H_2SO_4, 2-h run at 30–60 V/cm, cooled.

(103) Whatman No. 1 paper, EtAc-HAc-HC(O)OH-H_2O (18:3:1:4), descending.

(109) Whatman No. 1 paper, *sec*-BuOH-HAc-H_2O (14:1:5).

(113) Anachrom ABS with 10% XE-60, 24 in. by 0.19 in. O.D. stainless steel column, He gas at 70 ml/min, 170°.

(128) Acid-alkali-washed Celite with 1.5% LAC l-R-296 polyester, N_2 gas at 30–40 ml/min, 210°.

Table E: CHROMATOGRAPHIC CONDITIONS FOR CHROMATOGRAPHY DATA IN TABLES A–D (Continued)

Reference[a] (A)	Conditions[b] (B)

FOR TABLE A (Continued)

(132)   Whatman No. 1 paper, MeC(O)Me-H$_2$O (9:1 v/v), descending.
(134)   Whatman No. 3 MM paper, EtAc-HAc-H$_2$O (3:1:3), descending.
(145)   Toyo No. 50 paper, BuOH-HAc-H$_2$O (4:2:1 v/v).
(149)   Whatman No. 1 paper, Pyr-EtAc-HAc-H$_2$O (5:5:1:3), descending (?).

FOR TABLE B

(2)   Whatman No. 3 MM paper, ca. 0.1 M (NH$_4$)$_2$CO$_3$ (7.9 g/l) at pH 8.9, 25-min run at 80 V/cm, 8°.
(3)   Chromosorb W(HP) (80–100 mesh) with 3% OV-1, 1.9 m by 4 mm I.D. glass column, He gas at 80 ml/min, programmed 70–325° at 10°/min.
(4)   Cellulose MN300HR, 350 μm thick, dried 24 h at room temperature, *iso*-PrOH-EtAc-H$_2$O (23.5:65:11.5).
(8)   Whatman No. 1 paper, PrOH-conc NH$_3$ (7:3 v/v), descending.
(10)   Chromosorb G-AW-DMCS (60–80 mesh) with 3% polypenyl ether-5 ring, 4.5 m by 0.13 in. O.D. A1S1 321 stainless steel coiled column, Ar gas at 20 ml/min, 200°.
(11)   Gas-Chrom Q with 3% QF-1 (w/w), 180 cm, by 0.45 cm O.D. coiled hardened aluminum column, N$_2$ gas at 60 ml/min, 110°.
(14)   Whatman No. 1 paper, 0.05 M acetate buffer at pH 4, 1-h run at 35 V/cm.
(15)   Whatman No. 1 paper, BuOH-EtOH-H$_2$O (4:1:5 v/v) upper phase, descending.
(16)   Avirin* microcrystalline cellulose, dried 30–60 min at 80°, EtAc-HAc-HC(O)OH-H$_2$O (18:3:1:4).
(23)   Silica Gel G (240–270 μm), 2 mm thick, dried at room temperature and 55–60° relative humidity 1–3 days MeC(O)Et-HAc-H$_2$O (60:20:20).
(28)   Whatman No. 1 paper, BuOH-Pyr-H$_2$O (6:4:3), ascending.
(29)   Paper not described, PrOH-HC(O)OH-H$_2$O (6:3:1) for the acid, BuOH saturated with aq 3% conc NH$_3$ for the hydrazone.
(32)   Silica Gel G, dried overnight at 135°, BuOH-HAc-H$_2$O (2:1:1 v/v).
(41)   Whatman No. 1 paper, EtAc-HAc-H$_2$O (3:1:3), upper phase (?).
(44)   Whatman No. 1 paper, BuOH-Pyr-H$_2$O (6:4:3).
(57)   Gas-Chrom Q (100–200 mesh) with 3% silicone polymer JXR, 1.8 m by 4 mm O.D. coiled glass column, N$_2$ gas at 60 ml/min, 140° for 2 min and programmed at 210° at 3°/min.
(61)   Whatman No. 1 paper, acid-washed, PrOH-HC(O)OH-H$_2$O (6:3:1), descending.
(62)   Whatman No. 1 paper, EtAc-HAc-H$_2$O (3:1:3), upper phase (?).
(70)   Whatman No. 1 paper, BuOH-HAc-H$_2$O (4:1:1), descending (?).
(71)   Whatman No. 1 paper, Pyr-EtAc-HAc-H$_2$O (5:5:1:3), descending (?).
(73)   Kieselgur G, phosphate-impregnated, air-dried overnight, BuOH-MeC(O)Me-phosphate buffer (40:50:10).
(80)   Whatman No. 1 paper, BuOH-HAc-H$_2$O (4:1:5), descending.
(81)   Whatman No. 1 paper, BuOH-HAc-H$_2$O (4.4:1.6:4.0), descending.
(83)   Whatman No. 1 paper, water-saturated isobutyric acid, descending.
(88)   Paper not described, 0.1 M formate buffer at pH 3, 45-min run at 50 V/cm.
(89a)   Whatman P20 phosphocellulose paper, PrOH-EtAc-25% NH$_3$-H$_2$O (5:1:1:3), descending.
(89b)   Whatman No. 1 paper, EtAc-Pyr-H$_2$O (80:34:12), descending.
(91)   Chromosorb G (HP) (80–100 mesh) with 1% OV-17, 3 m by 3 mm I.D. stainless steel column, He gas at 35 ml/min, 130–190°.
(96)   Whatman No. 1 paper, BuOH-HAc-H$_2$O (5:2:3 v/v), descending.
(97)   Silica Gel Supelcosil 12B, 0.5 mm thick, CHCl$_3$-MeOH-HAc-H$_2$O (65:50:2:12 v/v).
(98)   Whatman No. 3HR paper, EtAc-HAc-H$_2$O (3:1:3), descending.
(104)   Whatman No. 1 paper, BuOH-HAc-H$_2$O (4:1:5), descending.
(108)   Whatman No. 1 paper, EtAc-HAc-HC(O)OH-H$_2$O (18:3:1:4), descending.
(110)   Toyo-Roshi No. 52 paper, BuOH-HAc-H$_2$O (4:1:5 v/v), descending.
(112)   Whatman No. 1 paper, acid-washed, *tert*-AmOH-98% HC(O)OH-H$_2$O (3:3:1), descending.
(113)   Schleicher & Schull No. 2043 paper, washed 2 days with developing solvent, BuOH-HAc-H$_2$O (4:1:1 v/v).
(116)   Whatman No. 1 paper, BuOH-Pyr-0.1 N HCL (5:3:2), descending.
(117)   Pyr-HAc-H$_2$O (100:40:860 v/v, pH 5.3), only details given.
(118)   Whatman Chromedia SG-41, 0.20 mm thick, MeOH-CHCl$_3$ (2:98), ascending.
(119)   Whatman No. 3 paper, EDTA-washed, EtOH-HAc-H$_2$O (70:1:29), ascending.
(125)   Silica Gel G, CHCl$_3$-MeC(O)Me (4:1).
(138a)   Whatman No. 1 paper, BuOH-HAc-H$_2$O (4:1:5), descending.
(138b)   Whatman No. 1 paper, Pyr-EtAc-HAc-H$_2$O (5:5:1:3), descending (?).
(141a)   Whatman No. 1 paper, BuOH-HAc-H$_2$O (100:21:50), ascending.
(141b)   Whatman No. 1 paper, EtAc-Pyr-H$_2$O (40:11:6), descending.
(141c)   Whatman No. 1 paper, Pyr-EtAc-HAc-H$_2$O (5:5:1:3), descending (?).
(144)   Schleicher & Schüll No. 2043b paper, 0.01 M borax sol 0.007 M with CaCl$_2$ pH 9.2, 0.5 mA/cm.
(145)   Cellulose Pulver MN300HR, 0.5 mm thick, *iso*-PrOH-Pyr-HAc-H$_2$O (8:8:1:4).
(148)   Whatman No. 1 paper, BuOH-HAc-H$_2$O (50:12:25), descending.
(149)   Whatman No. 1 paper, EtAc-HAc- HC(O)OH-H$_2$O (18:3:1:4), descending.
(152)   Whatman No. 1 paper, 0.1 M cadmium acetate, 90-min run at 20 V/cm.
(153)   Whatman No. 1 paper, EtAc-Pyr-H$_2$O (40:11:6), descending.
(154)   Siliconized Gas-Chrom P (100–140 mesh) with 1% SE-30, column not detailed, N$_2$ gas at 30 cc/min, 140°.

Table E: CHROMATOGRAPHIC CONDITIONS FOR CHROMATOGRAPHY DATA IN TABLES A–D (Continued)

Reference[a] (A)	Conditions[b] (B)

FOR TABLE B (Continued)

(156)	Chromosorb W (80–100 mesh) with 14% ethylene glycol succinate polyester, 3 ft column, He gas at 23 psi, 155°.
(166a)	Whatman No. 1 paper, BuOH-HAc-H_2O (4:1:1), descending (?).
(166b)	Silica Gel (Merck), BuOH-HAc-H_2O (4:1:1), thickness not given.
(170)	Whatman No. 1 paper, EtOH-0.88d. NH_3-H_2O (80:5:15), descending.
(171)	0.2 M NH_4 Ac buffer, pH 5.8, only detail given.
(173)	Whatman No. 1 paper, BuOH-HAc-H_2O (4:1:5), descending.
(177)	Chromedia® CC 41 binder free cellulose, 0.01 M NH_3 and 0.0033 M HAc, pH 10.2, 16-min run at 4.5 kV.
(178)	Whatman No. 1 paper, sec-BuOH-90% HC(O)OH-H_2O (15:3:2), ascending.
(179)	Silica Gel G (Merck), BuOH-HAc-H_2O (3:1:1).
(185)	Paper not described, DMF-iso-PrOH-MeC(O)Et-H_2O (10:25:45:20).
(188)	Whatman No. 1 paper, 0.05 M acetate pH 4.5 buffer, 90-min run at 20 V/cm.
(189)	Paper not described, BuOH-HAc-H_2O (7:7:2.3).
(191)	Whatman No. 1 paper, BuAc-HAc-EtOH-H_2O (3:2:1:1).
(192a)	Whatman No. 3 MM paper, aq $Na_2MoO_4 \cdot 2H_2O$ (25 g/1,200 ml) made to pH 5 with H_2SO_4, 2 h-run at 30–60 V/cm, cooled.
(192b)	Whatman No. 1 paper, BuOH-Pyr-H_2O (6:4:3).
(192c)	Glass beads (120–140 mesh) with 0.05% OV-11, 150 cm by 0.3 cm I.D. stainless steel column, gas not described, programmed from 80° at 10°/min.
(193a)	Toyo No. 51A paper with the buffer of (192a) above.
(193b)	Toyo No. 51A paper, EtAc-Pyr-HAc-H_2O (5:5:1:3 v/v), descending.
(194)	Whatman No. 1 paper, phenol-aq NH_3, only detail given.
(205)	Whatman No. 1 paper, BuOH-Pyr-H_2O (6:4:3).
(206a)	Gas-Chrom A (60–80 mesh) with 3% OV-17, 300 cm by 0.3 cm I.D. stainless steel column, gas not described, programmed from 120° at 5°/min.
(206b)	Whatman No. 1 paper, BuOH-HAc-H_2O (6:1:1 v/v), descending.
(208a)	Cellulose MN 300, 0.25 mm thick, BuOH-Pyr-H_2O (6:4:3).
(208b)	Whatman No. 1 paper, BuOH-HAc-H_2O (50:12:25), descending.
(209a)	Whatman No. 1 paper, BuOH-HAc-H_2O (5:2:2 v/v), descending.
(209b)	Cellulose Sigmacell, BuOH-HAc-H_2O (5:1:2 v/v), thickness not given.
(209c)	Gas-Chrom A (60–80 mesh) with 3% OV-17, 300 cm by 0.3 cm I.D. stainless steel column, gas not described, programmed from 120° at 5°/min.
(210a)	Whatman No. 3 MM paper, water-washed, BuOH-Pyr-HAc-H_2O (40:5:1:954), pH 5.8, 75-min run at 35 V/cm.
(210b)	Whatman No. 1 paper, 0.2 M pyridine acetate buffer pH 6.5, time not given, 57 V/cm.
(212)	Schleicher & Schüll No. 2043b paper, BuOH-HAc-H_2O (4:1:5).
(214)	KSK Silica Gel (150–200 mesh), PrOH-H_2O (7:3), thickness not given.
(216)	Varaport 30 with 5% OV-210, 2 m by 2 mm I.D. Pyrex® U glass column, N_2 gas at 7.5 ml/min, programmed from 90° at 1°/min.
(217)	Whatman No. 1 paper, BuOH-Pyr-H_2O (6:4:3).
(218)	Schleicher & Schüll No. 2043b paper, BuOH-PrOH-0.1 N HCl (1:2:1).
(219)	Silica Gel F_{245} (Merck) impregnated with 0.2 M NaH_2PO_4, thickness not given, BuOH-EtOH-H_2O (2:1:1).
(221)	Whatman No. 3 MM paper, BuOH-HAc-H_2O (4:1:5).
(222)	Whatman No. 17 paper, EtOH-H_2O (70:30 v/v).
(223)	Whatman No. 3 MM paper, BuOH-PrOH-0.1 N HCl (1:2:1 v/v).

FOR TABLE C

(2)	Chromosorb W (60–80 mesh) with 10% Carbowax 20M, 12 ft by 0.125 in. I.D. column, N_2 gas at 40 ml/min, programmed 65–210° at 8°/min.
(3)	Silica Gel G (Merck 7731) with 0.3 parts Carbowax 4000 and 0.0035 parts Tinopol WG (w/w), 0.3 mm thick, dried 30 min at 80°, benzene-heptane (65:35 v/v).
(5)	Whatman No. 3 MM paper, aq $Na_2MoO_4 \cdot 2H_2O$ (25 g/1,200 ml) made to pH 5 with H_2SO_4, 2-h run at 30–60 V/cm, cooled.
(6)	Whatman No. 1 paper, BuOH-EtOH-H_2O (4:1:5 v/v) organic phase.
(7)	Kieselgel G (Merck), BuOH-HAc-H_2O (4:1:5).
(9)	Acid-washed and silanized Chromosorb W (80–100 mesh) with 3% SE-52, 6 ft by 0.25 in. O.D. coiled stainless steel column, unspecified gas at 75 ml/min, 140°.
(14a)	Chromosorb W (80–100 mesh) with 10% Carbowax 6000, 120 cm by 0.5 cm I.D. glass column, no gas or temperature data.
(14b)	Whatman No. 1 paper, BuOH-EtOH-H_2O (10:3:3 v/v), descending.
(14c)	Silica Gel G, dried 10 h at 110°, solvent as in (14b).
(17)	Silica Gel G (240–270 μm), 2 mm thick, dried at room temperature and 55–60° relative humidity 1–3 days, MeC(O)Et-HAc-H_2O (60:20:20).
(19a)	Whatman No. 1 paper, BuOH-EtOH-H_2O (4:1:5 v/v) upper phase, descending.
(19b)	Neosorb NC with 15% butanediol succinate polyester, 100 cm by 0.4 cm I.D. stainless steel column, He gas at 40 ml/min, 200°.
(19c)	Whatman No. 1 paper, 0.1 M borate buffer at pH 9.6, 400 V, 3.5 h.
(19d)	Whatman No. 1 paper, MeC(O)Et-H_2O azeotrope.
(23a)	Silica Gel G, 0.25 mm thick, MeC(O)Et-H_2O azeotrope (85:7 v/v).
(23b)	Chromosorb W (100–120 mesh) with 15% LAC-4R-886, 120 cm by 0.5 cm ID. glass column, unspecified gas at 20 ml/min, 200°.

Table E: CHROMATOGRAPHIC CONDITIONS FOR CHROMATOGRAPHY DATA IN TABLES A–D (Continued)

Reference[a] (A)	Conditions[b] (B)

FOR TABLE C (Continued)

(28)	Whatman No. 4 paper, pH 9.6 Na$_2$ HAsO$_3$ (as 19.8 g As$_2$O$_3$/1 and NaOH added), 90-min runs at 20–25 V/cm, 20–25° using 18–20° cooling water.
(29)	Whatman No. 1 paper (?), BuOH-HAc-H$_2$O (4:1:5), descending.
(30a)	Avirin microcrystalline cellulose, dried 30–60 min at 80°, BuOH-HAc-H$_2$O (3:1:1).
(30b)	Avirin microcrystalline cellulose, dried 30–60 min at 80°, MeC(O)Et-water azeotrope.
(30c)	Avirin microcrystalline cellulose, dried 30–60 min at 80°, Pyr-EtAc-HAc-H$_2$O (5:5:1:3).
(34)	Chromosorb W (60–80 mesh) with 20% diethylene glycol succinate, 6 ft by 0.25 in. O.D. aluminum column, He gas at 55 ml/min programmed 70–220° at 15°/min.
(35a)	Whatman 3 MM paper, 0.05 M Pyr-HAc buffer, pH 6.
(35b)	Whatman 3 MM paper, EtAc-Pyr-H$_2$O (10:4:3 v/v).
(37)	"Kieselgel nach Stahl" (Merck), with 0.1 M sodium bisulfite, 0.25 mm thick, dried 1 h at 110–120°, EtAc-HAc-MeOH-H$_2$O (6:1.5:1.5:1).
(38)	Whatman No. 1 paper, BuOH-EtOH-H$_2$O (4:1:5 v/v) upper phase, descending.
(40)	Silica Gel G (Merck), 0.25 mm thick, dried 0.5 h at 100°, EtOEt-toluene (2:1 v/v).
(47)	Whatman No. 1 paper, BuOH-EtOH-H$_2$O (52.5:32:15.5), descending.
(49)	Whatman No. 1 paper, EtAc-HAc-H$_2$O (9:2:2 v/v), front moves 30 cm in 3–4 h, descending.
(50)	Kieselgel G (Merck) with 0.15 M phosphate buffer, pH 8, 0.25 mm thick, dried overnight, phenol-H$_2$O (75:25 w/v), ascending.
(52)	Chromosorb W coated by hexamethyldisilazane (60–80 mesh) with 10% LAC-IR-296 diethylene glycol adipate, 1.84 m by 0.32 cm I.D. glass column, N$_2$ gas at 30–50 ml/min, programmed 170–220° at 0.8°/min.
(56)	Whatman No. 1 paper, water-saturated BuOH.
(57)	Whatman No. 1 paper, BuOH-EtOH-H$_2$O (4:1:5 v/v) upper phase, descending.
(58)	Whatman No. 1 paper, BuOH-EtOH-H$_2$O (4:1:5 v/v) upper phase, descending.
(59a)	Kieselgel G (Merck), 0.25 mm thick, dried 2 h at 120°, activated 15 min at 120°, benzene.
(59b)	Acid-washed, siliconized Gas-Chrom P (100–140 mesh) with 1% SE-30, 6 ft by 4 mm glass U column treated with hexamethyldisilazane, N$_2$ gas at 40–60 ml/min, 150°.
(60a)	Whatman No. 1 paper, EtAc-HAc-H$_2$O (3:1:1 v/v), descending.
(60b)	Silica Gel G, 0.5 mm thick, CHCl$_3$-MeOH-H$_2$O (65:25:4 v/v).
(60c)	Unspecified support with 3% SE-30 (methyl silicone), 5 ft by 0.125 in. unspecified column, He gas at 25 ml/min, 150°.
(67)	Chromosorb W (80–100 mesh) with 14% ethylene glycol succinate polyester, 3 ft column, He gas at 23 psi, 155°.
(68)	Whatman No. 1 paper, BuOH-EtOH-H$_2$O (4:1:5 v/v) upper phase, descending.
(69)	Silica Gel G, dried overnight at 135°, upper phase of C$_6$H$_6$-EtOH-H$_2$O-0.8 sp gr NH$_4$OH (200:47:15:1 v/v).
(72)	Acid-washed Chromosorb W (80–100 mesh) with 20% butanediol succinate, 1.5 m by 0.375 cm unspecified column, He gas at 120 ml/min, 145°.
(74)	Whatman No. 1 paper, BuOH-EtOH-H$_2$O (4:1:5 v/v) upper phase, descending.
(76a)	Acid-washed Chromosorb W (100–120 mesh) with 2% octylphenoxypoly (oxyethylene) ethanol, 115 cm by 3.8 mm unspecified U column purged 1 d at 180°, He gas at 50 ml/min, 165°.
(76b)	Support of (76a) with 5% LAC-4R-886 polyester wax, purged 1 d at 200°, He gas at 100 ml/min, 190°.
(77)	Whatman No. 1 paper, borate buffer, pH 10, 400 V.
(80)	Acid-washed Celite (80–100 mesh) with 15% butanediol succinate, 120 cm by 0.5 cm unspecified column, unspecified gas at 80–100 ml/min, 175°.
(82a)	Whatman No. 1 paper, BuOH-Pyr-H$_2$O (6:4:3).
(82b)	Whatman No. 3 paper, borate buffer, pH 10.4, 26.3 V/cm, 4.25-h run.
(82c)	ECNSS-M liquid phase in a glass column at 155–165° only data given.
(85a)	Whatman No. 1 paper, BuOH-Pyr-H$_2$O (6:4:3).
(85b)	Substrate not given, BuOH-Pyr-H$_2$O (6:4:3 v/v).
(85c)	Glass column at 155–165° only detail, see *Angew. Chem.*, 82, 643 (1970).
(86)	Embacel AW (60–70 mesh) with 5% XE-60, 6 ft by 0.25 in. O.D. unspecified column, N$_2$ gas at 39 ml/min, programmed 100–220° at 10°/min.
(91a)	Whatman No. 1 paper, EtAc-Pyr-H$_2$O (3.6:1:1.15), descending.
(91b)	Unspecified support with 3% ECNSS-M (ethylene glycol succinate combined with cyanoethylsilicone), 10 ft by 0.25 in. unspecified column, N$_2$ gas at 60 ml/min, 175°.
(94a)	Whatman No. 3 paper, borate buffer, pH 10.4, 30 V/cm, 5-h run.
(94b)	Whatman No. 1 paper, toluol-BuOH (1:2)/H$_2$O.
(95)	Whatman No. 1 paper, BuOH-Pyr-H$_2$O (6:4:3).
(96a)	Whatman No. 3 paper, borate buffer, pH 10.4, 26.3 V/cm, 4.25-h run.
(96b)	Kieselgel G (Merck), activated 1 h at 120°, EtAc-*iso*-PrOH-MeOH (70:15:15).
(96c)	Whatman No. 1 paper, toluol-BuOH (1:2)/H$_2$O.
(98)	Gas-Chtom Q (100–120 mesh) with 3% ECNSSM, 1.3 m by 3 mm I.D. glass U column, gas and temperature.
(102)	Kieselgel G (Merck), activated 1 h at 120°, EtAc-*iso*-PrOH-MeOH (70:15:15).
(113)	Kieselgel G (Merck), activated 1 h at 120°, EtAc-*iso*-PrOH-MeOH (70:15:15).
(124)	Whatman No. 1 paper, BuOH-EtOH-H$_2$O (40:11:19).
(127)	Whatman No. 1 paper, EtAc-Pyr-H$_2$O (10:4:3 v/v), descending.
(130)	TIC grade silica gel (Fluka DO) with 0.03 M H$_3$ BO$_3$, 0.3 mm thick, dried at room temperature 24 h, activated 1 h at 110°, BuOH-HAc-H$_2$O (4:1:5).

Table E: CHROMATOGRAPHIC CONDITIONS FOR CHROMATOGRAPHY DATA IN TABLES A–D (Continued)

Reference[a] (A)	Conditions[b] (B)

FOR TABLE C (Continued)

(131)   Chromosorb W (60–80 mesh) with 2% EGSS-X, 2.74 mm by 3.16 mm I.D. glass column, N_2 gas at 40 ml/min, 199°.

(135)   Whatman No. 1 paper, BuOH-H_2O (1:1?), descending.

(136)   Whatman No. 1 paper, MeC(O)Et-H_2O (4:1).

(140a)   Silica Gel, C_6H_6-MeC(O)Me (1:1).

(140b)   Silica Gel, C_6H_6-MeC(O)Me (5:1).

(140c)   Shimazer Gaschromatograph apparatus, NGS, 136°, only details given.

(144)   Chromosorb W (60–80 mesh) with 3% OV-101, 1.8 m by 6.25 mm O.D. glass column, He gas at 60 ml/min, 140°.

(145a)   Whatman No. 1 paper, EtAc-HAc-H_2O (3:1:3), upper phase (?).

(145b)   Diatoport W (60–80 mesh) with 10% SE-30, 6 ft by 0.25 in. column, He gas at 46 ml/min, 190°.

(149)   Chromosorb W (HMDS) with 5% Ucon LB 55OX, 2 m by 3 mm stainless steel column, undescribed gas and temperature.

(150)   Whatman No. 3 paper, BuOH-HAc-H_2O (10:1:3), descending.

(153a)   Whatman 3 MM paper, 0.05 M Pyr-HAc buffer, pH 6, 40 V/cm.

(153b)   Whatman No. 54 paper, EtAc-HAc-H_2O (6:3:2).

(155a)   Whatman 3 MM paper, 0.1 M borate buffer, pH 10, 40 V/cm., 1-h run.

(155b)   Whatman No. 1 paper, BuOH-EtOH-H_2O (5:1:4), descending.

(159a)   Silica Gel (KSK), $CHCl_3$-MeC(O)Me (6:4).

(159b)   Silica Gel (KSK), water-saturated cyclohexanol.

(159c)   Volodarskii M paper, 0.2 M borate buffer, pH 9.2, 8.5 V/cm, 5-h run.

(160)   Volodarskii C paper, EtAc-HAc-HC(O)OH-MeC(O)Et-H_2O (17:3:1:15:5), descending.

(164)   Whatman No. 1 paper, BuOH-EtOH-H_2O (4:1:5 v/v) upper phase, descending.

(167)   Celite AW (80–100 mesh) with 10% butanediol succinate polyester, 6 ft by 0.25 in. O.D. glass column, He gas at 28 ml/min, programmed 125–215° at 4°/min.

(169)   Whatman No. 1 paper, BuOH-EtOH-H_2O (4:1:5 v/v) upper phase, descending.

(175a)   Toyo No. 51 paper, 0.1 M HAc-Pyr buffer, pH 6.5, 3000 V, 30-min run.

(175b)   Toyo No. 51 paper, BuOH-HAc-H_2O (50:12:25) descending.

(180a)   Toyo-Roshi No. 51A paper, BuOH-Pyr-H_2O (6:4:3), descending.

(180b)   Gas-Chrom P with 2% XF-1105, 1.8 m by 4 mm I.D. glass column, N_2 gas at 70 ml/min 140°.

(184)   Chromosorb W (60–80 mesh) 15% LAC-4R-886, 120 cm by 0.5 cm I.D. glass column, unspecified gas at 190 ml/min and 190°.

(188a)   Ederol 208 paper impregnated with 40% dimethylformamide in acetone, cyclohexane-benzene-dimethyl-formamide (7:2:1 v/v).

(188b)   Kieselgel G (Merck), 0.25 mm thick, cyclohexane-diisopropyl ether-Pyr (4:4:2 v/v).

(189)   Chromosorb W (100–125 mesh) with 10% diethylene glycol succinate, column not given, N_2 gas at 45 ml/min, 175°.

(192)   Chromosorb W (60–80 mesh) with 1.5% SE-30, 1.5 m by 3 mm I.D. stainless steel column, N_2 gas at undisclosed flow, 170°.

(197)   Whatman No. 1 paper, water-saturated BuOH.

(203a)   Whatman No. 1 paper, EtAc-Pyr-H_2O (3.6:1:1.15), descending, (?).

(203b)   Kieselgel G (Merck), C_6H_6-MeC(O)Me (2:1).

(205)   BuOH-HAc-H_2O only detail given.

(210a)   Chromosorb W (80–100 mesh) with 10% Carbowax 6000, 6 ft by 0.25 in. unspecified column, N_2 gas at 120 ml/min, 145°.

(210b)   Silica Gel G (nach Stahl), EtAc-EtOH-H_2O (15:2:1 v/v).

(217)   Whatman No. 4 paper, phenol-H_2O, 20°.

(219a)   Whatman No. 1 paper, BuOH-Pyr-H_2O (10:3:3 v/v), descending.

(219b)   Aeropak 30 with 10% Carbowax, 5 ft by 0.125 in. stainless steel column, N_2 gas at 15–20 ml/min, 170°.

(219c)   Silica Gel H, EtAc-HAc-HC(O)OH-H_2O (18:3:1:4 v/v), ascending.

(219d)   Whatman No. 1 paper, borate buffer, pH 10, 1,250 V, 2-h run.

(223)   Whatman No. 3 paper, borate buffer, pH 10, 1,500 V.

(238a)   Gas-Chrom Q (100–200 mesh) with 3% ECNSSM, 1.3 m by 3 mm I.D. glass U column, gas and temperature not given.

(238b)   Whatman No. 1 paper, Pyr-EtAc-H_2O (2:5:5: v/v), upper layer, descending.

(241)   Whatman No. 4 paper, 0.05 M sodium tetraborate, 90-min runs at 20–25 V/cm, 20–25° using 18–20° cooled water.

(243)   Acid-washed Celite (80–100 mesh) with 10% m-bis(m-phenoxyphenoxy) benzene, 120 cm by 0.5 cm unspecified column, unspecified gas at 80–100 ml/min, 175°.

(244a)   Silica Gel G (Merck), MeC(O)Et saturated with H_2O.

(244b)   Silica Gel G (Merck), EtAc-toluene (2:1).

(244c)   Chromosorb W with SE-30, 150 cm by 0.3 cm stainless steel column, N_2 gas at 70 ml/min, programmed 110–170° at 4° /min.

(245)   Whatman No. 1 paper, PrOH-HAc-H_2O (6:1:2 v/v), descending.

(246)   Silica Gel G, PrOH-EtAc-H_2O (3:2:1 v/v).

(252)   Kieselgel G (Merck), activated at 140°, EtOEt-cyclohexane (5:1).

(253a)   Cellulose MN 300, 0.25 mm thick, BuOH-Pyr-H_2O (6:4:3), (?).

(253b)   Chromosorb G (80–100 mesh) with 3% ECNSS-M, 200 cm by 0.318 cm column, only details given.

(255)   Whatman No. 1 paper, BuOH-EtOH-H_2O (4:1:5 v/v) upper phase, descending.

(258a)   Whatman No. 2 paper, BuOH-EtOH-H_2O (4:1:5 v/v) upper phase, descending.

(258b)   Silica gel as aqueous silicic acid (activity grade IV); <150 mesh, 0.5 mm thick, C_6H_6-MeC(O)Me(1:1).

(258c)   Neutral alumina (Al_2O_3, activity grade V), 0.5 mm thick, C_6H_6-MeC(O)Me (1:1).

(259a)   Kieselgel F254 Fertigplatten (Merck), activated at 140°; EtAc.

(259b)   EtAc-MeOH (8:2), only detail.

Table E: CHROMATOGRAPHIC CONDITIONS FOR CHROMATOGRAPHY DATA IN TABLES A–D (Continued)

Reference[a] (A)	Conditions[b] (B)

FOR TABLE C (Continued)

(262)   Schliecher and Schüll No. 2043b paper, EtAc-Pyr-H$_2$O (2:1:2).

(263a)   Whatman No. 1 paper, toluene-BuOH (9:1), 6-h run.

(263b)   Kieselgel G (Merck), EtAc-MeOH (9:1), 45-min run.

(263c)   Whatman No. 1, paper toluene-BuOH-Pyr-H$_2$O, 18-h run.

(265)   Whatman No. 1 paper, BuOH-Pyr-H$_2$O (6:4:3).

(266)   Silica Gel G, EtAc-MeOH (19:1) or (9:1).

(268)   Whatman No. 2 paper, BuOH-EtOH-H$_2$O (4:1:5 v/v) upper phase, descending.

(269)   Whatman No. 1 paper, water-saturated BuOH, ascending.

(273)   Borate buffer in the method of Foster, *Chem. Ind.*, 828, 1050 (1952).

(276a)   Spinco No. 300–846 or Schleicher and Schüll No. 2043 A gl paper, 0.083 M borax at pH 9.2, 380 V, 2 h and 10 min run.

(276b)   Paper not given, MeOH saturated with heptane.

(277)   Silica Gel G, 0.25 mm thick, activated at 120°, CH$_2$Cl$_2$-EtOEt (1:1).

(278)   Silica Gel G, EtAc-petroleum ether (3:2 v/v).

(279)   Whatman No. 2 paper, BuOH-EtOH-H$_2$O (4:1:5 v/v) upper phase, descending.

(280a)   Neutral alumina (Al$_2$O$_3$, activity grade V), 0.5 mm thick, C$_6$H$_6$ MeC(O)Me (3:1).

(280b)   Silica Gel as aqueous silica acid (activity grade IV); <150 mesh, 0.5 mm thick, CHCl$_3$-MeC(O)Me (1:2).

(280c)   Neutral alumina (AlCO$_3$, activity grade V); 0.5 mm thick, C$_6$H$_6$ -MeC(O)Me (2:1).

(280d)   Silica Gel as aqueous silica acid (activity grade IV); <150 mesh, 0.5 mm thick, EtAc-heptane (3:1).

(280e)   Silica Gel as aqueous silicic acid (activity grade IV); <150 mesh, 0.5 mm thick, C$_6$H$_6$ -MeC(O)Me (1:1).

(281a)   Silica Gel (Merck), EtAc.

(281b)   Silica Gel (Merck), EtAc-CHCl$_3$ (1:1 v/v).

(289)   Whatman No. 1 paper, water-saturated BuOH, 17 h.

(290)   Whatman No. 1 paper, toluene-BuOH (4:1)/H$_2$O, 18-h run.

(292)   Kieselgur G, EtAc-65% *iso*-PrOH (65:35 v/v).

(294)   Chromosorb W (80–100 mesh) with 3% SE-52 silicone gum, 6 ft by 0.125 in. O.D. stainless steel column, unspecified gas, programmed from 120° at 2°/min.

(299)   Whatman No. 3 paper, BuOH-Pyr-H$_2$O (6:4:3).

(300)   Whatman No. 3 paper, BuOH-Pyr-H$_2$O (6:4:3).

(301)   Kieselgel F254 Fertigplatten (Merck), activated at 140°; EtAc, (?).

(302)   *tert*-BuOH-HAc-H$_2$O (2:2:1), only detail given.

(303)   Silica Gel (Merck), EtAc.

(304)   Whatman No. 1 paper, BuOH-Pyr-H$_2$O (6:4:3).

(305a)   Lenigrad factory N$_2$ "M" paper, BuOH-HAc-H$_2$O (4:1:1), ascending.

(305b)   Alumina with Brockmann activity of II-III, C$_6$H$_6$ -MeOH (9:1).

(306)   Silica Gel G, CHCl$_3$-MeC(O)Me (1:7).

(307a)   Silica Gel G, CHCl$_3$-MeC(O)Me (1:7).

(307b)   Whatman No. 1 paper, BuOH-HAc-H$_2$O (4:1:1), descending (?).

(310)   Whatman No. 1 paper, BuOH-MeC(O)Et (1:1)/borate buffer 50%, descending.

(312)   Whatman No. 3 MM paper, aq Na$_2$MoO$_4$ · 2H$_2$O (25 g/1,200 ml) made to pH 5 with H$_2$SO$_4$, 2-h run at 30–60 V/cm, cooled.

(315)   Toluene-MeC(O)Et (1:1)/H$_2$O, descending, 23-h run.

(317a)   Schleicher and Schüll No. 2043b paper, BuOH-borate buffer.

(317b)   Silica thin-layer foil (Merck, Kieselgel-Fertigfolien), CH$_2$Cl$_2$ -MeOH (100:5 v/v).

(317c)   Silica thin-layer foil (Merck, Kieselgel-Fertigfolien), CH$_2$Cl$_2$ -MeOH (100:3 v/v).

(320a)   Paper not given, 0.1 M sodium borate buffer, pH 10, 1,600 V, 0°.

(320b)   Paper not given, BuOH-H$_2$O, [see *J. Chromatogr.*, 3, 63 (1960)].

(329)   Whatman No. 1 paper, solvent as in (315) above.

(333)   Silica Gel as aqueous silicic acid (activity grade IV); <150 mesh, 0.5 mm thick, C$_6$H$_6$ -MeC(O)Me (1:1).

(334a)   Toyo No. 51A paper, toluene-BuOH-H$_2$O (1:1:2), descending.

(334b)   Toyo No. 51A paper, EtAc-HAc-H$_2$O (3:1:3), descending.

(334c)   Toyo No. 51A paper, 0.1 M sodium borate pH 9.5, 30 V/cm, 70-min run.

(337)   Whatman No. 1 paper, BuOH-EtOH-H$_2$O (4:1:1).

(338)   Schleicher and Schüll No. 589 green paper, *teri*-AmOH-PrOH-H$_2$O (3:1:1), descending.

(341)   Whatman No. 1 paper, MeC(O)Et saturated with H$_2$O, descending.

(342)   Whatman No. 1 paper, MeC(O)Et-H$_2$O (10:1 v/v).

(345)   Silica Gel, C$_6$H$_6$-EtAc (7:1).

(347)   Whatman No. 1 paper, MeC(O)Et saturated with H$_2$O, descending.

(350a)   Whatman No. 3 MM paper, 0.05 M sodium tetraborate, 25 V/cm, 2–3-h run.

(350b)   Whatman No. 1 paper, BuOH-Pyr-H$_2$O (6:4:3).

(350c)   Chromosorb W (80–100 mesh) with 10% neopentylglycol sebacate polyester, 120 cm by 0.5 cm I.D. glass column, Ar gas at 150 ml/min, 205°.

(352)   Whatman No. 1 paper, BuOH-EtOH-H$_2$O (4:1:5 v/v) upper phase, descending.

(354a)   0.05 M Sodium tetraborate, only detail given.

(354b)   Whatman No. 1 paper, BuOH-EtOH-H$_2$O (4:1:5 v/v) upper phase, descending.

Table E: CHROMATOGRAPHIC CONDITIONS FOR CHROMATOGRAPHY DATA IN TABLES A–D (Continued)

Reference[a] (A)	Conditions[b] (B)

FOR TABLE C (Continued)

(356) Cellulose Sigmacell, BuOH-HAc-H$_2$O (5:1:2 v/v), thickness not given.

(357a) Whatman No. 1 paper, Pyr-EtAc-H$_2$O (2:5:5 v/v), upper layer, descending.

(357b) Chromosorb W (80–100 mesh) with 10% neopentylglycol sebacate polyester, 120 cm by 0.5 cm I.D. glass column, Ar gas at 150 ml/min, 205°.

(357c) 0.05 M Sodium tetraborate, only detail given.

(360a) Whatman No. 1 paper, EtAc-PrOH-H$_2$O (5:3:2).

(360b) Whatman No. 4 paper, pH 9.6 Na$_2$HAsO$_3$ (as 19.8 g As$_2$O$_3$/l and NaOH added), 90-min runs at 20–25 V/cm, 20–25° using 18–20° cooling water.

(373) Whatman No. 1 paper, BuOH-EtOH-H$_2$O-NH$_3$ (40:10:49:1).

(375a) Borate buffer in the method of Foster, *Chem. Ind.*, 828, 1050 (1952).

(375b) Whatman No. 1 paper, BuOH-EtOH-H$_2$O (4:1:5 v/v) upper phase, descending.

(379) Whatman No. 1 paper, BuOH-EtOH-H$_2$O-NH$_3$ (40:10:49:1).

(382) Whatman No. 1 paper, AmOH-Pyr-H$_2$O (4:3:2 v/v).

(383) Whatman No. 1 paper, EtAc-HAc-H$_2$O (3:1:3), upper phase (?).

(391) Whatman No. 1 paper, EtAc-Pyr-H$_2$O (2:1:2) water-poor phase, descending.

(393) Whatman No. 1 paper, BuOH-Pyr-H$_2$O (6:4:3).

(395) Whatman No. 1 paper, EtAc-HAc-HC(O)OH-H$_2$O (18:3:1:4), descending.

(397) Whatman No. 3 MM paper, 0.1 M sodium tetraborate, 25 V/cm, 2–3-h run.

(398a) Whatman No. 1 paper, MeC(O)Me-H$_2$O (95:5 v/v), descending.

(398b) Chromosorb W (80–100 mesh) with 10% neopentylglycol sebacate polyester, 120 cm by 0.5 cm I.D. glass column, Ar gas at 150 ml/min, 204° for acetates and ·165° for ethers.

(401a) Acid-washed Chromosorb W (100–200 mesh) with 10% LAC-4R-886 polyester wax, 5 ft by 0.125 in. coiled metal column, N$_2$ gas at 75 ml/min, 190°.

(401b) Acid-washed, siliconized Gas-Chrom P (100–400 mesh) with 15% SE-30, same column as (401a), N$_2$ gas at 60 ml/min, 180°.

(402a) Whatman No. 1 paper, EtAc-HAc-H$_2$O (9:2:2 v/v), front moves 30 cm in 3–4 h, descending.

(402b) ECNSS-M, only data given; glass column at 155–165° assumed.

(406a) Whatman No. 1 paper, MeC(O)Me-H$_2$O (95:5 v/v), descending.

(406b) Whatman No. 1 paper, *iso*-BuOH-Pyr-H$_2$O (10:3:3 v/v).

(410) Kieselgel G (Merck), C$_6$H$_6$ -MeC(O)Me (2:1).

FOR TABLE D

(2) Kieselgel G (Merck), BuOH-HAc-H$_2$O (4:1:5).

(3) Gas-Chrom Z (60–80 mesh) with 20% Carbowax 20M, 7 ft by 0.25 in. O.D. stainless steel U column, He gas at 80 ml/min, 90°.

(4) Toyo No. 51 paper, *n*-PrOH-EtAc-H$_2$O (7:1:2), ascending.

(5) Chromosorb G-AW-DMCS (60–80 mesh) with 3% polypenyl ether-5 ring, 4.5 m by 0.13 in. O.D. A1S1 321 stainless steel coiled column, Ar gas at 20 ml/min, 200°.

(6) Chromosorb W (60–80 mesh) with 10% Carbowax 20M, 12 ft by 0.125 in. I.D. column, N$_2$ gas at 40 ml/min, programmed 65–210° at 8°/min.

(7) Whatman No. 3 MM paper, *n*-BuOH-0.5 M NH$_3$ in H$_2$O (1:1 v/v), descending.

(8) Whatman No. 4 paper dipped in 25% solution of DMF in EtOH, dried 10–15 min, cyclohexane.

(9) Whatman No. 1 paper, *iso*-BuC(O)Me-HC(O)OH-H$_2$O (1,000:4:96 v/v), descending.

(10) Whatman No. 4 paper, pH 9.6 Na$_2$HAsO$_3$ (as 19.8 g As$_2$O$_3$/1 and NaOH added), 90-min runs at 20–25 V/cm, 20–25° using 18–20° cooling water.

(13) Whatman No. 1 paper, EtAc-Pyr-H$_2$O (12:5:4), descending.

(14) Silica Gel G (Merck 7731) with 0.3 parts Carbowax 4000 and 0.0035 parts Tinopol WG (w/w), 0.3 mm thick, dried 30 min at 80°, benzene-heptane (65:35 v/v).

(17) Silica Gel G (Merck), after Stahl, 0.5 mm thick, toluene-EtAc (2:1 v/v).

(19a) Whatman No. 3 paper, bisulphite buffer, pH 4.7, 12–18 V/cm, 50°.

(19b) Whatman No. 1 paper, BuOH-EtOH-H$_2$O (4:1:5 v/v) upper phase, descending.

(19c) 3% ECNSS containing column, 150°, only details given.

(20a) Whatman No. 3 paper, bisulphite buffer, pH 4.7, 12–18 V/cm, 50°, (see also Reference 21).

(20b) Whatman No. 1 paper, EtAc-HAc-H$_2$O (3:1:3), upper phase (?).

(26) Toyo-Roshi No. 50 paper, *iso*-PrOH-H$_2$O (6:1), ascending.

(28) Acid-washed and silanized Chromosorb W (80–100 mesh) with 3% SE-52, 6 ft by 0.25 in. O.D. coiled stainless steel column, unspecified gas at 75 ml/min, 140°.

(30) Whatman No. 4 paper, water-saturated phenol, descending.

(31) Paper not given, 0.05 M borate buffer, pH 10, 500 V, 3-h run.

(34) Whatman No. 1 paper, BuOH-EtOH-H$_2$O (40:11:19).

(36) Microcrystalline cellulose (Avirin), BuOH-EtOH-H$_2$O (40:11:19 v/v), ascending.

(38) Whatman No. 1 paper, BuOH-EtOH-H$_2$O (5:2:2 v/v), descending.

(43) Whatman No. 3 MM paper, aq Na$_2$MoO$_4$·2H$_2$O (25 g/1,200 ml) made to pH5 with H$_2$SO$_4$, 2-h run at 30–60 V/cm, cooled.

(44) Whatman No. 1 paper, EtAc-HAc-H$_2$O (9:2:2: v/v), front moves 30 cm in 3–4 h, descending.

(45) Paper not given, phenol-H$_2$O (4:1).

Table E: CHROMATOGRAPHIC CONDITIONS FOR CHROMATOGRAPHY DATA IN TABLES A–D (Continued)

Reference[a] (A)	Conditions[b] (B)

FOR TABLE D (Continued)

(49)	Whatman No. 1 paper, water-saturated BuOH.
(51)	Whatman No. 1 paper, BuOH-HAc-H_2O (4:1:1), descending (?).
(56)	Silica Gel G, 0.25 mm thick, activated at 120°, CH_2Cl_2-EtOEt (1:1).
(58)	Paper not given, BuOH-HAc-H_2O (4:5:1).
(59)	Whatman No. 1 paper, MeC(O)Me-aq 10% HAc (4:1).
(64)	"Kieselgel nach Stahl" (Merck), with 0.1 M sodium bisulfite, 0.25 mm thick, dried 1 h at 110–120°. EtAc-HAc-MeOH-H_2O (6:1.5:1.5:1).
(66)	Silica Gel G, dried overnight at 135°, upper phase of C_6H_6-EtOH-H_2O-0.8 sp gr NH_4OH (200:47:15:1 v/v).
(67)	Silica Gel G (Merck), 0.25 mm thick, dried 0.5 h at 100°, EtOEt-toluene (2:1 v/v).
(74)	Whatman No. 1 paper, BuOH-EtOH-H_2O (40:11:19).
(79)	3–10% XE-60 on silanized glass capillary, 28 m by 0.17 mm I.D., N_2 gas flow not given, 130–170°.
(81)	Whatman No. 1 paper, MeC(O)Et-EtC(O)OH-H_2O (75:25:30 v/v), ascending.
(84)	Whatman No. 1 paper, BuOH-EtOH-H_2O (40:11:19).
(91)	Toyo-Roshi No. 50 paper, PrOH-H_2O (4:1 v/v).
(92a)	Gas-Chrom P (60–80 mesh) with 20% Carbowax 20M, 9 ft by 0.25 in O.D. stainless steel column, He gas at1 80 ml/min, 80°, programmed irregularly at 215° in 76 min.
(92b)	Bio Sil A, dried overnight at 135°, upper phase of C_6H_6-EtOH-H_2O-0.8 sp gr NH_4OH (200:47:15:1 v/v).
(100)	Diatoport A (80–100 mesh) with 10% UCW 98, 6 ft by 0.125 in. O.D. stainless steel column, He gas at 120 ml/min, 130°.
(101)	Whatman No. 1 paper, EtAc-HAc-HC(O)OH-H_2O (18:3:1:4), descending.
(103)	Whatman No. 1 paper, phenol-H_2O (80:10 w/v).
(106)	Chromosorb W with 1.5% SE-30, 225 cm by 4 mm stainless steel column.
(110)	Whatman No. 1 paper, EtAc-Pyr-H_2O (8:2:1 v/v), descending.
(113)	Whatman No. 1 paper, EtAc-HAc-H_2O (9:2:2 v/v), front moves 30 cm in 3–4 h, descending.
(116)	Whatman No. 3 MM paper, aq Na_2 $MoO_4 \cdot 2H_2O$ (25g/1,200 ml) made to pH5 with H_2SO_4, 2-h run at 30–60 V/cm, cooled.
(117)	Whatman No. 1 paper, EtAc-Pyr-H_2O (8:2:1 v/v), descending.
(118)	Whatman No. 1 paper, water saturated phenol, pH 5.5, descending.
(125)	Gas-Chrom A with 3% SE-52, 183 cm by 0.6 cm column, N_2 gas at 100 ml/min, programmed 75–280 at 11°/min.

Compiled by George G. Maher.

[a] The reference number correlates with the same reference number in the respective preceding tables.

[b] Abbreviations used in developing solvent compositions: Ac = CH_3 C(O)O, Am = *n*-amyl, Bu = *n*-butyl, DMF = dimethylformamide, DMSO = dimethylsulfoxide, Et = ethyl, Me = methyl, Pr = *n*-propyl, Pyr = pyridine. For trade names of support and liquid phase materials, one is referred to an adequate chemical supply house catalog of chiomatographic materials, or to *Handbook of Chromatography*, Vol. 2, Zweig, G. and Sherma, J., Eds., CRC Press, Cleveland, Ohio, 1972, 255.

CARBOHYDRATE PHOSPHATE ESTERS

The next two tables are taken directly from 3rd Edition of *The Handbook of Biochemistry and Molecular Biology* which was edited by late Gerald Fasman. This material appeared in the original edition of *The Handbook of Biochemistry and Molecular Biology* which was edited by Herbert Sober. These tables have been retained in the current edition without modification to ensure the retention of data. It is the intent of the editor to ensure that this material is revised for subsequent editions.

Substance[a] (Synonym) (A)	Hydrolysis Constant[b] $k \times 10^3$ (B)	Hydrolysis Temperature °C (C)	Hydrolysis Medium (D)	Ester Group Form[c] (E)	Specific Rotation[d] (F)	Concentration[e] Solvent (G)	Melting Point, °C (H)	Reference (I)
2-Aminoethanol 1-phosphate	—	—	—	FA	—	—	236–238	1
D-Glycerol 1-phosphate (α-L-glycerophosphate, L-glycerin 3-phosphate)	0.15	80	Water pH 6.3	Ag	+1.0	6.5	—	2–4
D-Erythritol 4-phosphate	—	—	—	FA	-2.6, +2.6	—	—	5, 6
	—	—	—	dcha	-2.3	—	183–186	6
L-Erythritol 4-phosphate	—	—	—	dcha	+2.3	—	186–190	6
L-Ribitol 1-phosphate (D-ribitol 5-phosphate)	<5[f]	100	N HCl	FA	—	—	—	7
D-*myo*-Inositol 1-phosphate (*myo*-inositol 3-phosphate)	—	—	—	dcha	+9.3	pH 2.0	210–215d	8, 9, 10
	—	—	—	dcha	-3.2	pH 9.0	—	8
	—	—	—	B	—	—	238	11
L-*myo*-Inositol 1-phosphate (*myo*-inositol 3-phosphate)	0.99	100	Water pH 2.0	FA	-9.8	pH 2.0	—	12
	—	—	—	dcha	+3.4	pH 9.0	—	12
L-*myo*-Inositol 2-phosphate	—	—	—	FA	0 (meso)	—	194–203	12
	—	—	—	B	—	—	244–247	13
	—	—	—	cha	—	—	202–205	13
	—	—	—	dcha	—	—	215–220d	13
myo-Inositol 1,4-diphosphate	—	—	—	trcha	—	—	186–197d	9
myo-Inositol 4,5-diphosphate	—	—	—	trcha	—	—	182–183d	14
myo-Inositol hexaphosphate (phytic acid)	—	—	—	d CH₃ ester	—	—	247–249	14
D-Mannitol 1-phosphate	<0.5[f]	100	N HCl	FA	—	—	—	125
Shikimic acid 5-phosphate	—	—	—	K · H₂O	-107.6	(29°)	—	15
Bis(2,3-dihydroxypropyl) hydrogen phosphate phosphate (α,α-diglycerophosphate)	150[f]	100	N HNO₃	FA	—	—	—	16
2,3-Dihydroxypropyl *myo*-inositol hydrogen phosphate (1-O-glycerophosphoryl-*myo*-inositol)	15[f]	100	N HCl	FA	—	—	—	17
D-Glyceric acid 2-phosphate	—	—	—	cha	-14	6, pH 3.5	—	18
	—	—	—	FA	+24.3, +13	N HCl	115–117d	18, 126
D-Glyceric acid 3-phosphate	1.8[f]	125	N HCl	—	-68, +5	(NH₄)₂MoO₄	—	19, 20
	—	—	—	Ba	-14.5, +14	N HCl	—	19, 21
	—	—	—	Ba	-725	MoO₄	—	19, 20
D-Glyceric acid 1,3-diphosphate	26	38	Water	—	+5	(NH₄)₂ MoO₄	—	21
D-Glyceric acid 2,3-diphosphate	—	—	—	FA	very small	—	—	19
	—	—	—	Ba	-2, -4	—	—	23
3-Hydroxypyruvic acid phosphate (2-deoxy-2-keto-glyceric acid phosphate)	15[f]	90	N HCl	Na	-4, +4.6	6-17, N HNO₃	—	24, 25
	—	—	—	—	—	6-28	—	24–26
	—	—	—	—	—	—	—	27

CARBOHYDRATE PHOSPHATE ESTERS (Continued)

Substance[a] (Synonym)	Hydrolysis Constant[b] k × 10³	Hydrolysis Temperature °C	Hydrolysis Medium	Ester Group Form[c]	Specific Rotation[d]	Concentration[e] Solvent	Melting Point °C	Reference
(A)	(B)	(C)	(D)	(E)	(F)	(G)	(H)	(I)
D-Erythronic acid 4-phosphate	—	—	—	trcha	−20	—	—	28, 29
D-Galacturonic acid 1-phosphate methyl ester triacetate	—	—	—	d Benzyl ester	+12.2	CHCl₃	84.5–86	127, 128
D-Gluconic acid 6-phosphate	0.21	100	N HCl	FA	+0.2 (5461)	—	—	30, 31
	—	—	—	FA lactone	+18 (5461)	—	—	31
D-arabino-Hexulosonic acid 6-phosphate (2-keto-D-gluconic acid)	3[f]	94	$N\,H_2SO_4$	K	+3.3	—	—	32
3-Deoxy-D-erythro-hexulosonic acid 6-phosphate (3-deoxy-2-keto-D-gluconic acid)	5–6[f]	100	N HCl	FA	—	—	—	33
D-Glucuronic acid 1-phosphate	—	—	—	K	+53.6	(19°)	—	34
L-xylo-Hexulosonic acid diphosphate (2-keto-L-gulonic acid)	No constants	—	—	—	—	—	—	35
D-Glyceraldehyde 3-phosphate	37.5	100	N HCl	FA	+12	—	—	36–39
D-Erythrose 4-phosphate	15	100	NH_2SO_4	FA	0	—	—	40
D-Arabinose 5-phosphate	3[f]	100	NH_2SO_4	FA	—	—	—	41, 42
L-Arabinofuranosyl phosphate	—	—	—	FA	+16.9	—	—	22
L-Arabinopyranosyl phosphate	—	—	—	Ba	+48.2	—	—	22
	—	—	—	dcha (α-anomer)	+30.8	(26°)	144–150	44
	—	—	—	dcha (β-anomer)	91	(26°)	155–161	44
α-D-Ribofuranosyl phosphate	1.25[f]	20	0.01 N HCl	FA	—	—	—	43
	—	—	—	dcha	−40.3	—	—	45
β-D-Ribofuranosyl phosphate	0.63[f]	20	0.01 N HCl	Ba	−9.3	—	—	43
	—	—	—	dcha	−13.6	Ethanol	—	45
D-Ribopyranosyl phosphate	1.25[f]	20	0.1 N HCl	Ba	−47.1	5% Acetic acid	—	43
D-Ribose 2-phosphate	—	—	—	Ba	−6.8	—	—	46
	—	—	—	dB	−27.5	H₂O:C₅H₅N, 1:1	112–114d	46
D-Ribose 3-phosphate	—	—	—	Ba	−6.8	—	—	46
	—	—	—	dB	−35, −28	H₂O:C₅H₅N, 1:1	114–117d	46, 47
D-Ribose 5-phosphate	—	—	—	Na	−9.7	—	—	48, 49
	4.5	100	0.25 N HCl	FA	18 ± 2	0.2, 1 N HCl	—	50, 51
	0.5	100	0.25 N HCl	Ba	+6	—	—	48, 52
α-D-Ribofuranose 1,5-diphosphate	1.66	70	0.01 N HCl	FA	—	—	—	53, 54
D-Ribofuranose 5-phosphate 1-pyro-phosphate	—	—	—	tcha	+20.8	0.43	—	53
	30[g]	65	Acetate pH 4 buffer	—	—	—	—	55
2-Deoxy-α-D-erythro-pentosyl phosphate (2-deoxy-D-ribosyl phosphate)	—	—	—	cha	−34.5	—	—	56
2-Deoxy-β-D-erythro-pentosyl phosphate (2-deoxy-β-D-ribosyl phosphate)	13–17[f]	—	Acetate pH 4.5 buffer	FA	—	—	—	16, 57
	—	—	—	cha	−15.8	—	—	56

CARBOHYDRATE PHOSPHATE ESTERS (Continued)

Substance[a] (Synonym) (A)	Hydrolysis — Constant[b] $k \times 10^3$ (B)	Hydrolysis — Temperature °C (C)	Hydrolysis — Medium (D)	Ester Group Form[c] (E)	Specific Rotation[d] (F)	Concentration[e] Solvent (G)	Melting Point, °C (H)	Reference (I)
2-Deoxy-D-erythro-pentofuranose 5-phosphate (2-deoxy-D-ribose 5-phosphate)	50[f]	100	N HCl	FA	+19	0.47	—	58, 59
2-Deoxy-D-erythro-pentofuranose 1,5-diphosphate (2-deoxy-D-ribose)	>3	—	—	Ba	+10.8, 16.5	0.52	—	60, 61
	—	100	Water pH 4	FA	—	—	—	62
D-Xylose 5-phosphate	<5[h]	100	N HCl	FA	—	—	—	62
	4	100	N HCl	FA	+8	—	—	63
	1.1	100	6 N H$_2$SO$_4$	FA	+3.2	—	—	64
	—	—	—	Na	+5	—	—	63
	—	—	—	Ba	—	—	—	63
L-Fucose 1-phosphate	100	80	Water pH 2	FA	+148	—	—	129
α-D-Galactosyl phosphate	5.9	37	0.25 N HCl	FA	+108	0.2 N HCl	—	65
	—	—	—	K	+92	—	—	65
	—	—	—	Ba	+78.5	—	—	65
β-D-Galactosyl phosphate	5.6	37	0.25 N HCl	dcha	+31.3	—	147–153	44
	—	—	—	Ba	+21	—	—	66
	—	—	—	dcha	+25.2	(16°)	145–151	44
D-Galactose 6-phosphate	—	—	—	Ba	−11.9	0.5	—	67, 68
	—	—	—	CH$_3$ β-D-glycoside of dcha	—	—	138–144d	69
3,6-Dideoxy-D-xylo-hexosyl phosphate (abequose 1-phosphate)	—	—	—	FA	+1.5 (α-anomer)	0.5	—	70
	—	—	—	FA	−3.8 (β-anomer)	0.5	—	70
α-D-Glucopyranosyl phosphate	1.3, 2.99	37	0.25 N HCl	FA	+118	—	—	37, 71, 72
	—	—	—	Ba	+75.5	—	—	37
	—	—	—	dcha	+64	(26°)	163–169	44
	—	—	—	K	+78	—	—	73
β-D-Glucopyranosyl phosphate	5	33	N HCl	dB	+0.5	—	—	74
	15	33	N HCl	dB	−20	—	—	74
D-Glucose 6-phosphate	—	—	—	dcha	+7.3	(26°)	137–143	44
	0.23	100	N HCl	FA	+35.7	—	—	31, 76
	—	—	—	Ba	+18	—	157d	31, 76, 77
	—	—	—	K	+21.2	—	95–97d	77, 78
	—	—	—	CH$_3$ α-D-glycoside of dcha	+61	—	157–159d	69
α-D-Glucose 1,6-diphosphate	0.78	30	N H$_2$SO$_4$	FA	+83 ± 4	0.2	—	79
β-D-Glucose 1,6-diphosphate	3.15	30	N H$_2$SO$_4$	FA	−19 ± 2	0.2	—	79
2-Deoxy-D-arabino-hexose 6-phosphate (2-deoxy-D-glucose)	>10[f]	100	N HCl	—	—	—	—	80, 81, 123
D-Mannosyl phosphate	—	—	—	FA	+58	—	—	75
	—	—	—	Ba	+36	—	—	75

CARBOHYDRATE PHOSPHATE ESTERS (Continued)

Substance[a] (Synonym)	Hydrolysis Constant[b] k × 10³	Temperature °C	Medium	Ester Group Form[c]	Specific Rotation[d]	Concentration[e] Solvent	Melting Point, °C	Reference
(A)	(B)	(C)	(D)	(E)	(F)	(G)	(H)	(I)
D-Mannose 6-phosphate	—	—	—	dcha	+28.7	—	—	83
	0.29	100	N HCl	FA	15.1 (5461 Å)	—	—	76
D-Mannose 1,6-diphosphate	No constants	—	—	Ba	3.5 (5461 Å)	0.7	—	82
		—	—	—	—	—	—	84
Dihydroxyacetone phosphate	33.7	100	N HCl	—	—	—	—	37
L-glycero-Tetrulose phosphate (L-erythrulose phosphate)	10[f]	100	N HCl	—	—	—	—	85, 100
D-erythro-Pentulosyl 5-phosphate (D-ribulose phosphate)	5	100	N H₂SO₄	FA	−29.6	0.2 N HCl	—	86, 87
L-erythro-Pentulosyl 5-phosphate (L-ribulose phosphate)	5–12	90	N H₂SO₄	FA	+28	0.26, 0.2 N HBr	—	88
D-erythro-Pentulose 1,5-diphosphate (D-ribulose diphosphate)	15[i]	100	N H₂SO₄	—	—	—	—	89
D-threo-Pentulose phosphate (D-xylulose phosphate)	86	100	N HCl	—	—	—	—	90, 91
D-Fructose 1-phosphate	70	100	N HCl	FA	−64.2 (5461 Å)	11.3	—	92
	—	—	—	Ba	−(39 (5461 Å)	6.1	—	92
	—	—	—	B	−52.1 (5461 Å)	—	—	92
	—	—	—	Ba	−30.4	(26°)	—	93
D-Fructose 6-phosphate	4.4	100	N HCl	Ba	3.6	10	—	76, 94
D-Fructose 1,6-diphosphate	52	100	N HCl	FA	+4.1	13.6	—	94, 95
D-threo-2,5-Hexodiulose phosphate (5-keto-D-fructose phosphate)	70	98	N HCl	trcha	—	—	163–166d	96
	—	—	—	FA	—	—	—	130
L-Fuculosyl phosphate	60	100	N HCl	Ba	−2.3	—	—	97
α-D-ribo-Hexos-3-ulose 1-phosphate (3-keto-D-glucose 1-phosphate)	100	100	N H₂SO₄	FA	+68(15°)	0.7	—	131
6-Deoxy-L-arabino-hexosyl phosphate (L-rhamnulose phosphate)	46[f]	100	N HCl	Ba	+8.9	(30°) 5.2, pH 4 Acetic acid	—	98
L-Sorbose 1-phosphate	60[f]	100	N HCl	dcha	−16.5	—	171–173d	99
	—	—	—	mono K	−7.2	0.1 N HCl	—	101
Sedoheptulose 7-phosphate (D-altro-2-heptulose 7-phosphate)	0.28[f]	100	N H₂SO₄	Ba	—	—	—	102
Sedoheptulose 1 7-diphosphate	20	100	N H₂SO₄	FA	—	—	—	103
	0.28[h]	100	N H₂SO₄	FA	—	—	—	103
D-arabino-3-Heptulose phosphate	No constants	—	—	—	—	—	—	104
D-manno-Heptulose phosphate	No constants	—	—	—	—	—	—	105
A heptulose monophosphate	4	100	N HCl	Ba	+8 (5461 Å)	—	—	106
D-glycero-D-altro-Octulose 1-phosphate	20[f]	100	N HCl	FA	—	—	—	132

CARBOHYDRATE PHOSPHATE ESTERS (Continued)

Substance[a] (Synonym)	Hydrolysis Constant[b] $k \times 10^3$	Hydrolysis Temperature °C	Hydrolysis Medium	Ester Group Form[c]	Specific Rotation[d]	Concentration[e] Solvent	Melting Point, °C	Reference
(A)	(B)	(C)	(D)	(E)	(F)	(G)	(H)	(I)
D-glycero-D-altro-Octulose 8-phosphate	$<2^f$	100	N HCl	FA	—	—	—	132
D-glycero-D-altro-Octulose 1,8-diphosphate	ca 10	100	N HCl	FA	—	—	—	132
2-Amino-2-deoxy-D-galactosyl phosphate (glucosamine-1-phosphate)	—	—	—	FA N-acetyl der	+178	—	—	107
	—	—	—	K N-acetyl der	+112.4	—	—	108
	—	—	—	FA	+142.6	—	—	108
2-Amino-2-deoxy-D-galactose 6-phosphate	0.8^f	110	6 N HCl	FA	+57.8	0.1	—	109, 11
	—	—	—	N-acetyl der	+48.4	0.05 M Na acetate	—	110
2-Amino-2-deoxy-α-D-glucosyl phosphate (glucosamine-1-phosphate)	60^f	100	N HClO$_4$	FA	—	—	—	111
2-Amino-2-deoxy-β-D-glucosyl phosphate	230^f	100	N HCl	FA	+100	—	—	112
	3.7^f	37	N H$_2$SO$_4$	K N-acetyl der	+79	—	—	112
	—	—	—	FA	−43 (calcd)	—	—	112
	—	—	—	K N-acetyl der	(40 (calcd))	—	—	112
2-Amino-2-deoxy-D-glucose 6-phosphate	86^f	26	N H$_2$SO$_4$	dNA N-acetyl der	−1.7	—	170–171d	113, 114
	0.06^f	100	N HCl	FA	+54	0.5	170–180d	111, 115
	—	—	—	Ba	+53	(18°)0.3, pH 2.5 8, 0.5 M Na acetate	—	115
	—	—	—	FA N-acetyl der	+29.5	—	—	110
2-Amino-2-deoxy-D-gluconic acid 6-phosphate	1.24	100	N H$_2$SO$_4$	FA	−6.8	—	—	116
2-Amino-3-O-(2-carboxyethyl)-2-deoxy-D-glucose phosphate (muramic acid phosphate)	0.8^f	100	6 N HCl	—	—	—	—	109
α-Lactosyl phosphate	2^f	37	N HCl	Ba	+73.3	—	—	117, 118
β-Lactosyl phosphate	6^f	37	N HCl	Ba	+24.8	—	—	117, 118
2-O-α-D-Mannosyl-L-myo-inositol 1-phosphate	No constants	—	—	—	—	—	—	83
Sucrose 1-phosphate	5.9^f	100	N H$_2$SO$_4$	FA	—	—	—	119
Maltose-1-phosphate	—	—	—	—	+147	—	—	124
Trehalose phosphate	0.16	—	—	FA	+185	—	—	120
	—	—	—	Ba	+132 (5461 Å)	0.1 N HCl	—	121
Trehalose 6,6′-diphosphate	1.6^f	100	3 N HCl	FA	+99.3	0.7	—	122

a In order of increasing carbon chain-length in the parent compounds, grouped in the classes: alditols, acids, aldoses, ketoses, and disaccharides. The term phosphate ester denotes a carbohydrate dihydrogen phosphate wherein the two acidic groups of the acid ester are combined with suitable cations.

b For the ester group that is farthest in the carbon chain structure from the asymmetric center which determines the D or L configuration of the parent compound.

c FA = free acid, Ag = silver salt, K = potassium salt, Na = sodium salt, B = brucine salt, Ba = barium salt, cha = cyclohexylammonium salt, d = di; tr = tri, and t = tetra.

d $[\alpha]_D$ at the sodium D line, 5876 Å, unless indicated otherwise in parentheses.

e In grams per 100 ml of solution at 20–25°C. Other temperatures are in parentheses. Unless otherwise indicated, the concentration is 1–5 g in water.

f Calculated by the contributors from the data of the reference cited, using $k = 0.30$/time in min for 50% hydrolysis of the ester linkage.

g For the pyrophosphate group.

h For the second ester group.

i Both ester linkages hydrolyze equally.

j D-Mannonic acid 6-phosphate and 2-deoxy-D-arabino-hexose 6-phosphate have been reported as skin metabolites in unnatural environments (see Reference 123).

References

1. Grollman and Osborn, *Biochemistry*, 1571 (1964).
2. Baer and Kates, *J. Am. Chem. Soc.*, 70, 1394 (1948).
3. Kiessling and Schuster, *Ber. Dtsch. Chem. Ges.*, 71, 123 (1938).
4. Weil-Malherbe and Green, *Biochem. J.*, 49, 286 (1951).
5. Shetter, *J. Am. Chem. Soc.*, 78, 3722 (1956).
6. MacDonald, Fischer, and Ballou, *J. Am. Chem. Soc.*, 78, 3720 (1956).
7. Baddiley, Buchanan, Carss, and Mathias, *J. Chem. Soc.* (Lond.), p. 4583 (1956).
8. Ballou and Pizer, *J. Am. Chem. Soc.*, 81, 4745 (1959).
9. Posternak *Helv. Chim. Acta*, 41, 1891 (1958); 1959 42, 390 (1959)
10. Eisenberg, Jr. and Bolden, *Biochem. Biophys. Res. Commun.*, 21, 100 (1965).
11. Woolley, *J. Biol. Chem.*, 147, 581 (1943).
12. Pizer and Ballou, *J. Am. Chem. Soc.*, 81, 915 (1959).
13. Brown and Hall, *J. Chem. Soc.* (Lond.), p. 357 (1959).
14. Angyal and Tate, *J. Chem. Soc.* (Lond.), p. 4122 (1961).
15. Wolff and Kaplan, *J. Biol. Chem.*, 218, 849 (1957).
16. Weiss and Mingioli, *J. Am. Chem. Soc.*, 78, 2894 (1956).
17. Maruo and Benson, *J. Am. Chem. Soc.*, 79, 4564 (1957).
18. Lepage, Mumma, and Bensen, *J. Am. Chem. Soc.*, 82, 3713 (1960).
19. Ballou and Fischer, *J. Am. Chem. Soc.*, 76, 3188 (1954).
20. Kiessling, *Ber. Dtsch. Chem. Ges.*, 68, 243 (1935).
21. Meyerhof and Oesper, *J. Biol. Chem.*, 179, 1371–1381 (1949).
22. Wright and Khorana, *J. Am. Chem. Soc.*, 80, (1994) (1958).
23. Negelein and Brömel, *Biochem. Z.*, 303, 132 (1939).
24. Baer, *J. Biol. Chem.*, 185, 763 (1950).
25. Greenwald, *J. Biol. Chem.*, 63, 339 (1925).
26. Sutherland, Posternak, and Cori, *J. Biol. Chem.*, 181, 153 (1949).
27. Ballou and Hesse, *J. Am. Chem. Soc.*, 78, 3718 (1956).
28. Barker and Wold, *J. Org. Chem.*, 28, 1847 (1963).
29. Ishii, Hashimoto, Tachibana, and Yoshikawa, *Biochem. Biophys. Res. Commun.*, 10, 19 (1963).
30. Patwardhan, *Biochem. J.*, 28, 1854 (1934).
31. Robison and King, *Biochem. J.*, 25, 323 (1931).
32. Ciferri, Blakley, and Simpson, *Can. J. Microbiol.*, 277 (1959).
33. MacGee and Doudoroff, *J. Biol. Chem.*, 210, 617 (1954).
34. Barker, Bourne, Fleetwood, and Stacey, *J. Chem. Soc.*, (Lond.), p. 4128 (1958).
35. Moses, Ferrier, and Calvin, *Proc. Natl. Acad. Sci. (U.S.A.)*, 48, 1644 (1962).
36. Fischer and Baer, *Ber. Dtsch. Chem. Ges.*, 337, 1040 (1932)65.
37. Kiessling, *Ber. Dtsch. Chem. Ges.*, 67, 869 (1934).
38. Meyerhof and Junowicz-Kocholaty, *J. Biol. Chem.*, 71, 149 (1943).
39. Ballou and Fischer, *J. Am. Chem. Soc.*, 77, 3329 (1955).
40. Ballou, Fischer, and MacDonald, *J. Am. Chem. Soc.*, 77, 5967 (1955).
41. Volk, *J. Biol. Chem.*, 234, 1931 (1959).
42. Volk, *Biochim. Biophys. Acta*, 37, 365 (1960).
43. Wright and Khorana, *J. Am. Chem. Soc.*, 78, 811 (1956).
44. Putman and Hassid, *J. Am. Chem. Soc.*, 79, 5057 (1957).
45. Tener, Wright, and Khorana, *J. Am. Chem. Soc.*, 79, 441 (1957).
46. Khym, Doherty, and Cohn, *J. Am. Chem. Soc.*, 76, 5523 (1954).
47. Loring, Moss, Levy, and Hain, *Arch. Biochem. Biophys.*, 65, 578 (1956).
48. Albaum and Umbreit, *J. Biol. Chem.*, 167, 369 (1947).
49. Levene and Harris, *J. Biol. Chem.*, 101, 419 (1933).
50. Horecker and Smyrniotis, *Arch. Biochem. Biophys.*, 29, 232 (1950).
51. Michelson and Todd, *J. Chem. Soc.* (Lond.), p. 2476 (1949).
52. Levene and Stiller, *J. Biol. Chem.*, 104, 299 (1934).
53. Tener and Khorana, *J. Am. Chem. Soc.*, 80, 1999 (1958).
54. Klenow, *Arch. Biochem. Biophys.*, 46, 186 (1953).
55. Kornberg, Lieberman, and Simms, *J. Biol. Chem.*, 215, 389 (1955).
56. MacDonald and Fletcher Jr., *J. Am. Chem. Soc.*, 84, 1262 (1962).
57. Friedkin, *J. Biol. Chem.*, 184, 449 (1950).
58. Racker, *J. Biol. Chem.*, 196, 347 (1952).
59. MacDonald and Fletcher Jr., *J. Am. Chem. Soc.*, 81, 3719 (1959).
60. Szabó and Szabó, *J. Chem. Soc.*, (Lond.), p. 5139 (1964).
61. Ukita and Nagasawa, *Chem. Pharm. Bull.* (Tokyo), 7, 655 (1959).
62. Tarr, *Chem. Ind.*, 562 (1957).
63. Levene and Raymond, *J. Biol. Chem.*, 102, 347 (1933).
64. Gorin, Hough, and Jones, *J. Chem. Soc.* (Lond.), p. 582 (1955).
65. Kosterlitz, *Biochem. J.* 33, 1087 (1939); 37, 318 (1943).
66. Reithel, *J. Am. Chem. Soc.*, 67, 1056 (1945).
67. Inouye, Tannenbaum, and Hsia, *Nature*, 193, 67 (1962).
68. Tanaka, *Yakugaku Zasshi*, 81, 797 (1961); *Chem. Abstr.*, 55, 27064 (1961).
69. Szabó and Szabó, *J. Chem. Soc.* (Lond.), p. 3762 (1960).
70. Antonakis, *Bull. Soc. Chim. Fr.*, 2112 (1965).
71. Cori, Colowick, and Cori, *J. Biol. Chem.*, 121, 465 (1937).
72. Meagher and Hassid, *J. Am. Chem. Soc.*, 68, 2135 (1946).
73. Wolfrom and Pletcher, *J. Am. Chem. Soc.*, 63, 1050 (1941).
74. Wolfrom, Smith, Pletcher, and Brown, *J. Am. Chem. Soc.*, 64, 23 (1942).
75. Colowick, *J. Biol. Chem.*, 124, 557 (1938).
76. Robison, *Biochem. J.*, 26, 2191 (1932).
77. Saito and Noguchi, *Nippon Kagaku Zasshi*. 82, 469 (1961); *Chem. Abstr.*, 56. 11678 (1962).
78. Lardy and Fischer, *J. Biol. Chem.*, 164, 513 (1946).
79. Posternak, *J. Biol. Chem.*, 180, 1269 (1949).
80. Crane and Sols, *J. Biol. Chem.*, 210, 597 (1954).
81. DeMoss and Happel, *J. Bacteriol.*, 70, 104 (1955).
82. Leloir, *Fortschr. Chem. Org. Naturst.*, 47 (1951).
83. Hill and Ballou, *J. Biol. Chem.*, 241, 895 (1966).
84. Leloir, McElroy, Glass, Phosphorus Metabolism. Johns Hopkins Press, Baltimore, 1951, 67.
85. Charalampous and Mueller, *J. Biol. Chem.*, 201, 161 (1953).
86. Horecker, Smyrniotis, and Seegmiller, *J. Biol. Chem.*, 193, 383 (1951).
87. Hurwitz, Weissbach, Horecker, and Smyrniotis, *J. Biol. Chem.*, 218, 769 (1956).
88. Simpson and Wood, *J. Am. Chem. Soc.*, 78, 5452 (1956); *J. Biol. Chem.*, 230, 473 (1958).
89. Horecker, Hurwitz, and Weissbach, *J. Biol. Chem.*, 218, 785 (1956).
90. Glock, *Biochem. J.*, 52, 575 (1952).
91. Stumpf and Horecker, *J. Biol. Chem.*, 218, 753 (1956).
92. Tanko and Robison, *Biochem. J.*, 29, 961 (1935).
93. Pogell, *J. Biol. Chem.*, 201, 645 (1953).
94. Neuberg, Lustig, and Rothenberg, *Arch. Biochem. Biophys.*, 33 (1944).
95. MacLeod and Robison, *Biochem. J.*, 27, 286 (1933).
96. McGilvery, *J. Biol. Chem.*, 200, 835 (1953).
97. Heath and Ghalambar, *J. Biol. Chem.*, 237, 2423 (1962).
98. Chiu and Feingold, *Biochem. Biophys. Acta*, 92, 489 (1964).
99. Chiu, Otto, Power, and Feingold, *Biochim. Biophys. Acta*, 127, 249 (1966).
100. Gillett and Ballou, *Biochemistry*, 547 (1963).
101. Hers, *Biochim. Biophys. Acta*, 416 (1952).
102. Mann and Lardy, *J. Biol. Chem.*, 187, 339 (1950).
103. Horecker, Smyrniotis, Hiatt, and Marks, *J. Biol. Chem.*, 212, 827 (1955).
104. Sie, Nigam, and Fishman, *J. Am. Chem. Soc.*, 81, 6083 (1959); 82, 1007 (1960).
105. Nordahl and Benson, *J. Am. Chem. Soc.*, 76, 5054 (1954).
106. Robison, Macfarlane, and Tazelaar, *Nature*, 142, 114 (1938).
107. Cardini and Leloir, *J. Biol. Chem.*, 225, 318 (1957).
108. Carlson, Swanson, and Roseman, *Biochemistry*, 402 (1964).
109. Liu and Gotschlich, *J. Biol. Chem.*, 238, 1928 (1963).
110. Distler, Merrick, and Roseman, *J. Biol. Chem.*, 230, 497 (1958).
111. Brown, *J. Biol. Chem.*, 204, 877 (1953).
112. Maley, Maley, and Lardy, *J. Am. Chem. Soc.*, 78, 5303 (1956).
113. O'Brien, *Biochim. Biophys. Acta*, 86, 628 (1964).
114. Baluja, Chase, Kenner, and Todd, *J. Chem. Soc.* (Lond.), p. 4678 (1960).
115. Anderson and Percival, *J. Chem. Soc.* (Lond.), p. 814 (1956).
116. Grieling and Kisters, *Hoppe-Seyler' Z. Physiol. Chem.*, 346, 77 (1966).
117. Gander, Petersen, and Boyer, *Arch. Biochem. Biophys.*, 69, 85 (1957).
118. Sasaki and Taniguchi. *Nippon Nogeikagaku Kasihi*, 33. 183 (1959); *Chem. Abstr.*, 54. 308 (1960).
119. Leloir and Cardini, *J. Biol. Chem.*, 214, 157 (1955).
120. Cabib and Leloir, *J. Biol. Chem.*, 231, 259 (1958).
121. Robison and Morgan, *Biochem. J.*, 22, 1277 (1928).

122. Narumi and Tsumita, *J. Biol. Chem.*, 240, 2271 (1965).
123. Brooks, Lawrence, and Ricketts, *Biochem. J.*, 73, 566 (1959); *Nature*, 187, 1028 (1960).
124. Narumi and Tsumita, *J. Biol. Chem.*, 242, 2233 (1967).
125. Angval and Russell, *Aust. J. Chem.*, 22, 383 (1969).
126. Seamark, Tate, and Smeaton, *J. Biol. Chem.*, 243, 2424 (1968).
127. Volk, *J. Bacteriol.*, 95, 782 (1968).
128. Pippen and McCready, *J. Org. Chem.*, 16, 262 (1951).
129. Ishihara, Massaro, and Heath, *J. Biol. Chem.*, 243, 1103 (1968).
130. Avigad and Englard, *J. Biol. Chem.*, 243, 1511 (1968).
131. Fukui, *J. Bacterial.*, 97, 793 (1969).
132. Bartlett and Bucolo, *Biochem, Biophys. Res. Commun.*, 474 (1960).

CARBOHYDRATE PHOSPHATE ESTERS (Continued)

Substance[a] (Synonym)	Constant[b] $k \times 10^3$	Temp °C	Medium	Ester Group Form[c]	Specific Rotation	Concentration[e] Solvent	Melting Point, °C	Reference
TRIOSES								
D-Glycerol 1-phosphate (α-1-glycerophosphate, 1-glycerol 3-phosphate)	0.15	80	Water	Ag	+1.0	6.5	—	1–3
Bis (2,3-dihydroxypropyl) hydrogen phosphate (α,α-diglycerophosphate)	350[f]	100	pH 6.3	FA	-1.45	10	—	4
			N HNO_3	FA	—	—	—	
D-Glyceraldehyde 3-phosphate	86	100	N HCl	FA	+14	—	—	5–8
Dihydroxyacetone phosphate	77	100	N HCl	—	—	—	—	6
D-Glyceric acid 2-phosphate	0.40[f]	125	N HCl	FA	+13	N HCl	—	9,10
					+5	$(NH_4)_2MoO_4$	—	9
D-Glyceric acid 3-phosphate	0.40[f]	125	N HCl	Ba	-14.5	N HCl	—	9,10,11
				Ba	-745	$(NH_4)_2MoO_4$	—	9
D-Glyceric acid 1,3-diphosphate	26	38	Water	FA	-2.3	—	—	12
D-Glyceric acid 2,3-diphosphate	—	—	—	Ba	-2, -4	6–17, N HNO_3	—	13,14
			—	Na	-4, +4.6	6–28	—	13–15
3-Hydroxypropionic acid phosphate (2-deoxyglyceric acid phosphate)	35[f]	90	N HCl	—	—	—	—	16
TETROSES								
D-Erythritol 4-phosphate	—	—	—	FA	+2.6	—	—	17
L-Erythritol 4-phosphate	—	—	—	dcha	-2.3, -2.6	—	183–186	17,18
D-Erythrose 4-phosphate	86	100	N H_2SO_4	dcha	+2.3	—	186–190	17
D-glycero-Tetrulose 1-phosphate	100	100	1 N HCl	FA	0	—	—	19
D-glycero-Tetrulose 4-phosphate (D-erythrulose phosphate)	46	100	1 N HCl	—	—	—	—	20
				—	-1.4	—	—	21,22
D-glycero-Tetrulose 1,4-diphosphate	23[f]	100	N HCl	FA	-11.7 (4000)	—	—	20
D-Erythronic acid 4-phosphate	—	—	—	FA	-20	—	—	23
				trcha	—	—	—	23
D-Erythronolactone 2-phosphate	—	—	—	cha	-55.0	N HCl	—	24,25
4-Deoxy-D-erythronic acid 2-phosphate	—	—	—	FA	+15	N HCl	—	24
4-Deoxy-D-erythronic acid 3-phosphate	—	—	—	FA	-14.5	N HCl	—	26
				FA	-737	$(NH_4)_2MoO_4$	—	26
PENTOSES								
L-Ribitol 1-phosphate (D-ribitol 5-phosphate)	No constants	100	N HCl	FA	—	—	—	27
Ribitol 1,5-diphosphate	—	—	—	—	—	—	—	28
D-Xylitol 5-phosphate	—	—	—	Ba	+1.27	—	—	29
2-Deoxy-D-erythro-pentitol 5-phosphate	—	—	—	Ba	-16.8	—	—	30
				cha	-10.0	—	—	31
α-D-Apio-D-furanosyl phosphate	35	26	0.25 N H_2SO_4	—	—	—	—	32
α-D-Apio-L-furanosyl phosphate	83	26	0.25 N H_2SO_4	—	—	—	—	32
L-Arabinofuranosyl phosphate	—	—	—	FA	+16.9	—	—	33

CARBOHYDRATE PHOSPHATE ESTERS (Continued)

Substance[a] (Synonym)	Hydrolysis			Ester Group Form[c]	Specific Rotation	Concentration[e] Solvent	Melting Point, °C	Reference
	Constant[b] $k \times 10^3$	Temp °C	Medium					
PENTOSES (Continued)								
L-Arabinopyranosyl phosphate	—	—	—	Ba	+48.2	—	—	33
(α-Anomer)	—	—	—	dcha	+30.8	(26°)	144–150	34
(β-Anomer)	—	—	—	dcha	+91	(26°)	155–161	34
α-D-Arabinofuranosyl phosphate	—	—	—	Ba	+6.4	—	—	33
α-D-Arabinopyranosyl phosphate	—	—	—	cha	-39.1	—	—	33
D-Arabinose 5-phosphate	3[f]	100	$N\,H_2SO_4$	FA	-18.8	—	—	35, 36
	—	—	—	Ba	-48.6	50% Pyridine	—	37
	—	—	—	B	—	—	—	37
α-D-Ribofuranosyl phosphate	2.9	20	0.01 N HCl	FA	+40.3	—	—	38
	—	—	—	dcha	—	—	—	39
	275	25	0.5 N HCl	—	—	—	—	40
	6[f]	25	0.01 N HCl	—	—	—	—	41
β-D-Ribofuranosyl phosphate	5	25	0.01 N HCl	Ba	-9.3	—	—	41
	1.5[f]	20	0.01 N HCl	dcha	-13.6	Ethanol	—	38
	—	—	—	dcha	—	—	—	39
β-D-Ribopyranosyl phosphate	0.4	25	0.01 N HCl	—	-47.1	5% Acetic acid	—	41
D-Ribopyranose 2-phosphate	2.9[f]	20	0.01 N HCl	Ba	-6.8	—	—	38
D-Ribose 2-phosphate	—	—	—	Ba	-27.5	$H_2O{:}C_5H_5N$, 1:1	112–114d	42
	—	—	—	dB	-6.8	—	—	42
D-Ribose 3-phosphate	—	—	—	Ba	-35;-28	$H_2O{:}C_5H_5N$, 1:1	114–117d	42, 43
	—	—	—	dB	-97	—	—	44, 45
	13	100	$0.25\,N\,H_2SO_4$	Na	+38	Half-saturated boric acid	—	45
D-Ribose 5-phosphate	0.5	100	0.25 N HCl	FA	18±2	0.2, 1 N HCl	—	46, 47
	—	—	—	Ba	+6	—	—	44, 48
α-D-Ribofuranose 1,5-diphosphate	1.66	70	0.01 N HCl	FA	+20.8	0.43	—	49, 50
	—	—	—	tcha	+33.5	—	—	49
D-Ribofuranose 5-phosphate 1-pyrophosphate	69[g]	65	Acetate pH 4 buffer	FA	—	—	—	51
	—	—	—	—	—	—	—	52
α-D-Xylopyranosyl phosphate	6.2	36	0.38 N HCl	Ba	+65	5% Acetic	—	53
	—	—	—	Ba	+70.9	—	—	53
	—	—	—	K	+76	—	—	53, 54
β-D-Xylopyranosyl phosphate	—	—	—	cha	+58	—	—	34
	—	—	—	cha	+0.8	—	—	34
D-Xylose 3-phosphate	—	—	—	Ba	-13.3	5% acetic	—	54
D-Xylose 5-phosphate	—	—	—	Ba	-1.27	—	—	29
	9	100	N HCl	FA	—	—	—	55
	1.1	100	$6\,N\,H_2SO_4$	FA	+8	—	—	56

CARBOHYDRATE PHOSPHATE ESTERS (Continued)

Substance[a] (Synonym)	Constant[b] $k \times 10^3$	Temp °C	Medium	Ester Group Form[c]	Specific Rotation	Concentration[e] Solvent	Melting Point, °C	Reference
PENTOSES (Continued)								
2-Deoxy-α-D-*erythro*-pentofuranosyl phosphate (2-deoxy-α-D-ribosyl phosphate)	—	—	—	Na	+3.2	$H_2O{:}C_5H_5N$, 1:1	—	55
	—	—	—	Ba	+5	—	—	55
2-Deoxy-β-D-*erythro*-pentofuranosyl phosphate (2-deoxy-β-D-ribosyl phosphate)	57[f]	21	Acetate pH 4.5 buffer	cha	+34.5, +38.8	—	—	57–59
2-Deoxy-β-D-*erythro*-pentofuranosyl phosphate (2-deoxy-β-D-ribosyl phosphate)	—	—	—	cha	-15.8	—	—	58
2-Deoxy-D-*erythro*-pentofuranose 5-phosphate (2-deoxy-D-ribose 5-phosphate)	110[f]	100	N HCl	FA	+19	0.47	—	60, 61
2-Deoxy-D-*erythro*-pentofuranose 1,5-diphosphate (2-deoxy-D-ribose 1,5-diphosphate)	33	20	Water pH 4	Ba	+10.8, 16.5	0.52	—	30, 31
	—	—	—	FA	—	—	—	62
2-Deoxy-D-*threo*-pentose 5-phosphate	65[h]	100	N HCl	FA	-35	—	—	62
3-Deoxy-D-*erythro*-pentose 5-phosphate	—	—	—	Ba	-10.6	—	—	63
D-*erythro*-Pentulose 1-phosphate	No constants	—	—	Ba		—	—	64
D-*erythro*-Pentulose 5-phosphate (D-ribulose phosphate)	—	—	—	Ba		—	—	65
D-*erythro*-Pentulose 5-phosphate (D-ribulose phosphate)	11[f]	100	$N\,H_2SO_4$	FA	-29.6	$0.2N$ HCl	—	66, 67
D-*erythro*-Pentulose 1,5-diphosphate (D-ribulose diphosphate)	20[i]	100	$N\,H_2SO_4$	Ba	-40	$0.02N$ HCl	—	68
	—	—	—	—		—	—	69
L-*erythro*-Pentulose 5-phosphate (L-ribulose phosphate)	57[i]	100	$0.1\,N$ HCl	—		—	—	70
	30[f]	90	$N\,H_2SO_4$	FA	+28	0.26, $0.2\,N$ HBr	—	71
D-*threo*-Pentulose 5-phosphate (D-xylulose phosphate)	200[f]	100	N HCl	B	-37.8	—	—	72, 73
	—	—	—	—		—	—	29, 74
2-Deoxy-D-*erythro*-pentonic acid 5-phosphate	—	—	—	Ba	+1.95	—	—	30
HEXOSES								
D-Mannitol 1-phosphate	—	100	N HCl	FA	-20.5	—	—	75
2-Amino-2-deoxy-D-glucitol 3-phosphate	—	—	—	FA	+16.7	—	—	76
β-D-Galactofuranosyl phosphate	—	—	—	Ba	+148	—	—	77
α-D-Galactopyranosyl phosphate	14	37	$0.25\,N$ HCl	FA	+98	$0.2\,N$ HCl	—	78
	—	—	—	K	+92	—	—	78
	—	—	—	Ba	+78.5	—	—	78
	—	—	—	dcha		—	147–153	78
β-D-Galactopyranosyl phosphate	12	37	$0.25\,N$ HCl	Ba	+31.3	—	—	34
	—	—	—	dcha	+21	—	—	79
	—	—	—	dcha		—	145–151	34
D-Galactose 3-phosphate	—	—	—	K	+25.2	—	—	80
D-Galactose 6-phosphate	—	—	—	FA	+36.5	—	—	80
α-D-Galactopyranose 1,6-diphosphate	—	—	—	Ba	+25.2	(16°)	—	81, 82
	—	—	—	FA	+111	—	—	51
α-D-Glucopyranosyl phosphate	3.0	37	$0.25\,N$ HCl	FA	+118	—	—	83, 53

CARBOHYDRATE PHOSPHATE ESTERS (Continued)

HEXOSES (Continued)

Substance[a] (Synonym)	Hydrolysis Constant[b] $k \times 10^3$	Temp °C	Medium	Ester Group Form[c]	Specific Rotation	Concentration[e] Solvent	Melting Point, °C	Reference
	—	—	—	Ba	+75.5	—	—	83
	—	—	—	dcha	+64	(26°)	163–169	34
	—	—	—	K	+78	—	—	84
β-D-Glucopyranosyl phosphate	11.5	33	N HCl	dB	+0.5	—	—	85
	35	33	N HCl	dB	−20	—	—	85
α-L-Glucopyranosyl phosphate	—	—	—	dcha	+7.3	(26°)	137–143	34
	—	—	—	Ba	−73.2	—	—	86
	—	—	—	K	−78.2	—	—	86
D-Glucose 2-phosphate	5	100	0.1 N HCl	K	+15	—	—	87
	—	—	—	FA	+35	—	—	87
D-Glucose 3-phosphate	—	—	—	FA	+39 (5461)	—	—	88
	—	—	—	Ba	+26.5	—	—	89
D-Glucose 4-phosphate	—	—	—	B	−14.5	50% Pyridine	—	90
D-Glucose 5-phosphate	—	—	—	B	−45.3	Pyridine	—	91
D-Glucose 6-phosphate	—	—	—	Ba	+15	—	—	92
	0.5	100	N HCl	FA	+35.7	—	157d	93, 94
	—	—	—	Ba	+18	—	95–97d	93–95
	—	—	—	K	+21.2	—	—	95, 96
3-O-Methyl-D-glucose 6-phosphate	—	—	—	cha	+22	—	—	97
α-D-Glucose 1,6-diphosphate	0.78	30	N H₂SO₄	FA	+83±4	0.2	—	98
	—	—	—	dcha	+31.0	0.5	—	99
β-D-Glucose 1,6-diphosphate	3.15	30	N H₂SO₄	FA	−19±2	0.2	—	98
α-L-Idopyranosyl phosphate	—	—	—	Li	−27.5	—	—	100
	—	—	—	cha	−32	0.4	—	100
α-D-Mannopyranosyl phosphate	1.9	30	0.95 N H₂SO₄	FA	+58	—	—	101, 102
	—	—	—	Ba	+36	—	—	101
	—	—	—	dcha	+28.7	—	—	103
β-D-Mannopyranosyl phosphate	—	—	—	Li	+46.3	—	—	100
	—	—	—	cha	−6.5	—	180	104
D-Mannose 6-phosphate	0.67	100	N HCl	FA	+15.1 (5461)	0.7	—	94
	—	—	—	Ba	+3.5 (5461)	—	—	70
α-D-Mannopyranose 1,6-diphosphate	0.48	30	0.95 N HCl	K	+29.9	—	—	102
2-Acetamido-2-deoxy-α-D-galactopyranosyl phosphate	22	37	1 N HCl	FA	28.5	—	—	51
	—	—	—	—		—	—	105
	—	—	—	Li	+189 (5780)	—	—	106
	—	—	—	Li	+197 (5780)	—	—	107
	—	—	—	K	+112	—	—	108
	—	—	—	FA	+178	—	—	109
2-Acetamido-2-deoxy-α-D-galactopyranosyl phosphate 6-sulfate	—	—	—	Ba	+71.5	—	—	110

CARBOHYDRATE PHOSPHATE ESTERS (Continued)

HEXOSES (Continued)

Substance[a] (Synonym)	Hydrolysis Constant[b] $k \times 10^3$	Temp °C	Medium	Ester Group Form[c]	Specific Rotation	Concentration[e] Solvent	Melting Point, °C	Reference
2-Amino-2-deoxy-α-D-galactopyranosyl phosphate (galactosamine 1-phosphate)	170	100	1 N HCl	—	—	—	—	111
2-Amino-2-deoxy-D-galactose 6-phosphate	—	—	—	FA	+143	—	—	108
	0.8[f]	—	6 N HCl	FA	+57.8	0.1	—	112, 113
	—	110	—	N-Acetyl der	+48.4	0.5 M Na acetate	—	113
6-Deoxy-6-fluoro-α-D-galactopyranosyl phosphate	3.1	37	0.25 N HCl	K	+81	0.2 (18°)	—	114
	69	60	0.25 N HCl	—	—	—	—	114
2-Acetamido-2-deoxy-α-D-glucopyranosyl phosphate	3.7	37	1 N HCl	K	+79	—	—	115
2-Acetamido-2-deoxy-β-D-glucopyranosyl phosphate	4.1	37	1 N HCl	Ca	+107	—	—	116
	1.2	26	1.33 N HCl	K	+76	—	—	117
	—	—	—	Li	+144	—	—	107
	86	26	1.33 N HCl	Na	−1.7	—	170–171d	117, 118
2-Amino-2-deoxy-α-D-glucopyranosyl phosphate (glucosamine 1-phosphate)	—	—	—	K	—	−40 (Calcd)	—	115
	230	100	1 N HCl	K	+100	—	—	115
2-Amino-2-deoxy-β-D-glucopyranosyl phosphate	140[f]	100	N HClO₄	FA	−20 (Calcd)	—	—	119
	—	—	—	FA	+70	—	178–179d	115, 118
	—	—	—	FA	+79	—	—	120
	—	—	—	FA	—	—	—	76
	—	—	—	FA	+29.5	8, 0.5 M Na acetate	—	113
2-Acetamido-2-deoxy-D-glucose 6-phosphate	0.16	100	N HCl	FA	+54	0.5 (18°)	170–180d	119, 121
	—	—	—	Ba	+53	0.3, pH 2.5	—	121
2-Amino-2-deoxy-D-glucose 6-phosphate	—	—	—	FA	+73	—	—	51
2-Acetamido-2-deoxy-D-glucose 1,6-diphosphate	No constants	—	—	FA	+84	—	—	51
2-Amino-2-deoxy-D-glucose 1,6-diphosphate	0.8[f]	100	6 N HCl	—	+79	—	—	112
2-Amino-3-O-(2-carboxyethyl) 2-deoxy-D-glucose 6-phosphate (muramic acid phosphate)	No constants	—	—	FA	—	—	—	122
N-Acetylmuramic acid 1-phosphate	—	—	—	cha	+60.5	—	158–162	123
3-Deoxy-3-fluoro-α-D-glucopyranosyl phosphate	—	—	—	—	—	—	—	124
3-Deoxy-3-fluoro-D-glucose-6-phosphate	—	—	—	FA	+60.5	—	—	124
4-Thio-α-D-glucopyranosyl phosphate	—	—	—	cha	+136.6	0.29	—	125
5-Thio-α-D-glucopyranosyl phosphate	—	—	—	cha	+21.3	—	—	126
2-Acetamido-2-deoxy-α-D-mannopyranosyl phosphate	—	—	—	cha	+45	—	159–163	127
2-Deoxy-α-D-arabino-hexopyranosyl phosphate (2-deoxy-α-D-glucose 1-phosphate)	—	—	—	cha	—	0.5	136–141	128
2-Deoxy-D-arabino-hexose 6-phosphate (2-deoxy-D-glucose 6-phosphate)	>10[f]	100	—	—	—	—	—	129–131

CARBOHYDRATE PHOSPHATE ESTERS (Continued)

Substance[a] (Synonym)	Hydrolysis — Constant[b] k×10³	Hydrolysis — Temp °C	Hydrolysis — Medium	Ester Group Form[c]	Specific Rotation	Concentration[e] Solvent	Melting Point, °C	Reference
HEXOSES (Continued)								
2-Deoxy-D-lyxo-hexose 3-phosphate (2-deoxy-D-galactose 3-phosphate)	—	—	—	FA	+25	—	—	80
2-Deoxy-D-lyxo-hexose 6-phosphate (2-deoxy-D-galactose 6-phosphate)	—	—	—	FA	+41	—	—	80
3-Deoxy-α-D-ribo-hexopyranosyl phosphate (3-deoxy-α-D-glucose 1-phosphate)	—	—	—	Et$_3$N	+61.2	0.47	—	132
3-Deoxy-β-D-ribo-hexopyranosyl phosphate (3-deoxy-β-D-glucose 1-phosphate)	—	—	—	Et$_3$N	+0.7	0.42	—	132
3-Deoxy-D-ribo-hexose 6-phosphate (3-deoxy-D-glucose 6-phosphate)	—	—	—	Ba	+6.6	—	—	30
3-Deoxy-α-D-xylo-hexopyranosyl phosphate (3-deoxy-α-D-galactose 1-phosphate)	—	—	—	Ba	+3.8	—	—	133
	—	—	—	B	-24.2	—	—	133
3-Deoxy-β-D-xylo-hexopyranosyl phosphate (3-deoxy-β-D-galactose 1-phosphate)	—	—	—	Ba	+4	—	—	134
	—	—	—	Ba	-96	—	—	134
4-Deoxy-α-D-xylo-hexopyranosyl phosphate (4-deoxy-α-D-glucose 1-phosphate)	—	—	—	Et$_3$N	+72.5	0.43	—	132
4-Deoxy-β-D-xylo-hexopyranosyl phosphate (4-deoxy-β-D-glucose 1-phosphate)	—	—	—	Et$_3$N	+3.3	0.21	—	132
3,6-Dideoxy-D-xylo hexosyl phosphate (abequose 1-phosphate)	—	—	—	FA	1.5 (α-Anomer)	0.5	—	135
	—	—	—	FA	-3.8 (β-Anomer)	0.5	—	135
6-Deoxy-α-D-glucopyranosyl phosphate	—	—	—	Et$_3$N	+80.1	—	—	132
6-Deoxy-β-D-glucopyranosyl phosphate	—	—	—	Et$_3$N	+1.8	—	—	132
α-L-Fucopyranosyl phosphate	—	—	—	cha	-77.8	0.11	—	136
β-L-Fucopyranosyl phosphate	230	80	Water pH 2	FA	—	—	—	137
α-L-Rhamnopyranosyl phosphate	—	—	—	cha	-20.5	—	—	138
β-L-Rhamnopyranosyl phosphate	25	37	pH 4	cha	-21.5	—	—	139
β-D-Fructopyranosyl phosphate	34	37	pH 4	cha	+11.9	—	—	138
β-D-Fructofuranosyl phosphate	—	—	—	Ba	-83.3	—	—	140
	—	—	—	cha	-77.9	—	—	141
	86	37	pH 4	Na	-53.6	—	—	140
D-Fructose 1-phosphate	21	100	0.1 N HCl	FA	—	—	—	142
	160	100	N HCl	Ba	-64.2 (5461)	11.3	—	142
	—	—	—	B	-39 (5461)	6.1	—	142
	—	—	—	Ba	-52.1 (5461)	—	—	142
	—	—	—	Ba	-30.4	(26°)	—	143
D-Fructose 6-phosphate	10	100	N HCl	Ba	+3.6	10	—	94, 144
D-Fructose 1,6-diphosphate	112	100	N HCl	FA	+4.0	13.6	—	144,145
	—	—	—	trcha	—	—	163–166d	146
L-Fuculose 1-phosphate	140	100	N HCl	Ba	-2.3	—	—	147

CARBOHYDRATE PHOSPHATE ESTERS (Continued)

Hydrolysis spans the columns Constant, Temp, and Medium.

Substance[a] (Synonym)	Constant[b] $k \times 10^3$	Temp °C	Medium	Ester Group Form[c]	Specific Rotation	Concentration[e] Solvent	Melting Point, °C	Reference
HEXOSES (Continued)								
L-Rhamnulose 1-phosphate	106[f]	100	N HCl	Ba	+8.9 (30°)	5.2 pH 4 Acetic acid	—	148
L-Sorbose 1-phosphate	90	—	—	dcha	—	—	171–173d	149
(L-Sorbose 1-phosphate)	—	99	1 N HCl	K	-16.5	0.1 N HCl	—	150
L-Sorbose 6-phosphate	—	—	—	Ba	-7.2	—	—	150
D-Tagatose 6-phosphate	11	99	1 N HCl	Ba	-12.0	—	—	150
(D-Tagatose 6-phosphate)	—	—	—	Ba	+5.6	—	—	151
α-D-*ribo*-Hexopyranosyl-3-ulose phosphate (3-keto-D-glucose 1-phosphate)	—	100	N H$_2$SO$_4$	FA	+68 (15°)	0.7	—	152
D-*xylo*-Hexos-5-ulose-6-phosphate (5-ketoglucose 6-phosphate)	No constants	—	—	—	—	—	—	153
D-*threo*-2,5-Hexodiulose phosphate (5-keto-D-fructose phosphate)	125	98	N HCl	FA	—	—	—	154
D-Gluconic acid 6-phosphate	0.48	100	N HCl	FA	+0.2 (5461)	—	—	93, 155
(D-Gluconic acid 6-phosphate)	—	—	—	FA lactone	+21 (5461)	—	—	93
D-Mannono-1,4-lactone 6-phosphate	0.117	100	N HCl	—	+54.1 (5461)	—	—	155
D-Mannono-1,5-lactone 6-phosphate	0.117	100	N HCl	—	+60.6 (5461)	—	—	155
2-Amino-2-deoxy-D-gluconic acid 6-phosphate	1.24	100	N H$_2$SO$_4$	FA	-6.8	—	—	156
2-Deoxy-D-*arabino*-hexonic acid 6-phosphate	—	—	—	cha	+6	Ethanol	—	157
3-Deoxy-D-*ribo*-hexonic acid 6-phosphate	—	—	—	B	-20.2	—	—	133
D-Glucosaccharinic acid 6-phosphate	—	—	—	—	+62	—	—	158
α,β-D-Glucometasaccharinic acid 6-phosphate	—	—	—	Ba	-5.5	—	—	97
α,β-D-Glucometasaccharinic acid 5-phosphate	—	—	—	Ba	-6.7	—	—	97
β-D-Galactopyranosyluronic acid phosphate (β-D-galacturonic acid 1-phosphate)	—	—	—	Benzylamine	-14	—	—	159
α-D-Glucopyranosyluronic acid phosphate (α-D-glucuroric acid 1-phosphate)	0.23	61	0.01 N HCl	K	+53.6	—	—	160
α-L-Idopyranosyluronic acid phosphate (α-Liduronic acid phosphate)	—	—	—	K	+51	—	135–138	161
(α-L-Idopyranosyluronic acid phosphate)	—	—	—	cha	-15.8	—	—	100
α-D-Mannopyranosyluronic acid phosphate (α-D-mannuronic acid 1-phosphate)	—	—	—	Li	+19.1	—	—	100
D-*arabino*-Hexulosonic acid 6-phosphate (2-keto-D-gluconic acid 6-phosphate)	3[f]	94	N H$_2$SO$_4$	K	+3.3	—	—	162
3-Deoxy-D-*erythro*-hexulosonic acid 6-phosphate (3-deoxy-2-keto-D-gluconic acid 6-phosphate)	16	100	N HCl	FA	—	—	—	163
L-*xylo*-Hexulosonic acid diphosphate (2-keto-L-gluconic acid diphosphate)	No constants	—	—	—	—	—	—	164
α-D-*xylo*-Hexopyranos-4-ulosyluronic acid phosphate (4-keto-α-D-glucuronic acid 1-phosphate)	—	—	—	—	+30	—	—	165

CARBOHYDRATE PHOSPHATE ESTERS (Continued)

Substance[a] (Synonym)	Hydrolysis			Ester Group Form[c]	Specific Rotation	Concentration[e] Solvent	Melting Point, °C	Reference
	Constant[b] $k \times 10^3$	Temp °C	Medium					
HEPTOSES, OCTOSES								
D-*glycero*-D-*galacto*-Heptose phosphate	—	—	—	Ba	+26.8	—	—	166
L-*glycero*-α-D-*manno*-Heptopyranosyl phosphate	—	—	—	cha	+32	—	—	167
Sedoheptulose 7-phosphate (D-*altro*-2-heptulose 7-phosphate)	9	100	N H₂SO₄	FA	—	—	—	168
Sedoheptulose 1,7-diphosphate	9	100	1 N HCl	Ba	+8 (5461)	—	—	169, 170
	17ᶠ	100	N H₂SO₄	FA	—	—	—	168
	9	100	N H₂SO₄	FA	—	—	—	168
D-*gluco*-3-Heptulose phosphate	No constants	—	—	—	—	—	—	171
D-*manno*-Heptulose phosphate	No constants	—	—	—	—	—	—	172
A heptulose monophosphate	9–2	100	N HCl	Ba	+8 (5461)	—	—	170
3-Deoxy-D-*gluco*-heptonic acid 7-phosphate	—	—	—	cha	+9.2	—	—	173
3-Deoxy-D-*arabino*-Heptulosonic acid 7-phosphate	—	—	—	FA	+42	—	—	173
	—	—	—	K	+15.7	—	—	173
D-*glycero*-D-*altro*-Octulose 1-phosphate	46ᶠ	100	N HCl	FA	—	—	—	174
D-*glycero*-D-*altro*-Octulose 8-phosphate	<2ᶠ	100	N HCl	FA	—	—	—	174
D-*glycero*-D-*altro*-Octulose 1,8-diphosphate	ca. 23	100	N HCl	FA	—	—	—	174
CYCLITOLS								
D-*myo*-Inositol 1-phosphate (*myo*-inositol 3-phosphate)	—	—	—	dcha	+9.3	pH 2.0	210–215d	175–177
	—	—	—	dcha	-3.2	pH 9.0	—	175
	—	—	—	B	—	—	238	178
L-*myo*-Inositol 1-phosphate (*myo*-inositol 3-phosphate)	2ᶠ	100	Water pH 2.0	dcha	-9.8	pH 2.0	—	179
	—	—	—	—	+3.4	pH 9.0	—	179
myo-Inositol 2-phosphate	—	—	—	FA	0 (meso)	—	196–198	180
	—	—	—	B	—	—	244–247	180
	—	—	—	cha	—	—	203–205	180
	—	—	—	dcha	—	—	210–212	180
myo-Inositol 1,4-diphosphate	—	—	—	trcha	—	—	186–197d	181
myo-Inositol 4,5-diphosphate	—	—	—	trcha	—	—	182–183d	181
myo-Inositol hexaphosphate (phytic acid)	—	—	—	d CH₃ ester	—	—	247–249	182
(-)-Inositol 3-phosphate	—	—	—	FA	-25.6	—	>250	183
3-O-Methyl-(+)-inositol 4-phosphate (pinitol 4-phosphate)	—	—	—	FA	+20.5	—	>250	183
2,3-Dihydroxypropyl *myo*-inositol hydrogen phosphate (1-O-glycerophosphoryl-*myo*-inositol)	35ᶠ	100	N HCl	FA	—	—	—	184
2-O-α-D-Mannosyl-L-*myo*-inositol 1-phosphate	No constants	—	—	cha	-14	6, pH 3.5	115–117d	184–185
	—	—	—	—	—	—	—	103
Shikimic acid 5-phosphate	—	—	—	K H₂O	-107.6	(29°)	—	186

CARBOHYDRATE PHOSPHATE ESTERS (Continued)

Substance[a] (Synonym)	Hydrolysis			Ester Group Form[c]	Specific Rotation	Concentration[e] Solvent	Melting Point, °C	Reference
	Constant[b] $k \times 10^3$	Temp °C	Medium					
DISACCHARIDES								
α-Cellobiosyl phosphate	—	—	—	cha	—	—	—	187
N-Acetyl α-chondrosinyl phosphate	—	—	—	Bu$_3$N	+66.4	—	—	188
α-Lactosyl phosphate	4[f]	37	N HCl	Ba	+73.3	—	—	189,190
	—	—	—	FA	+99.5	—	—	191
β-Lactosyl phosphate	13[f]	37	N HCl	Ba	+24.8	—	—	189,190
	—	—	—	FA	+31.5	—	—	191
α-Maltosyl phosphate	3.2	36	0.38 N HCl	Ba	+107	—	—	53
	140	100	N HCl	—	+147	—	—	192
Sucrose 6-phosphate	—	—	—	Ba	+35.4	—	—	193
	—	—	—	K	+34	—	—	193
Trehalose 6-phosphate	13[f]	100	N H$_2$SO$_4$	—	—	—	—	194
	—	—	—	B	+31 (5461)	—	—	195
	—	—	—	Ba	+99	—	—	196
	0.6	—	—	FA	+185	0.1 N HCl	—	197
Trehalose 6,6′-diphosphate	—	—	—	Ba	+132 (5461)	—	—	195
	—	—	—	cha	+62	—	—	196
	3.8[f]	100	3 N HCl	FA	+99.3	0.7	—	198

Compiled by Donald L. MacDonald, with acknowledgments to George C. Maher and the late Melville L. Wolfrom, compilers of the corresponding tables in earlier editions.

[a] Compounds are grouped in order of increasing chain length. Within each group, they are arranged in the order alcohols, aldoses, ketoses, and acids.

[b] For the ester group at the lowest numbered carbon atom. The k values are in min^{-1} (natural logarithms).

[c] FA = free acid, Ag = silver salt, B = brucine salt, Ba = barium salt, cha = cyclohexylammonium salt, K = potassium salt, Na = sodium salt, d = di, tr = tri, t = tetra.

[d] $[\alpha]$, at the sodium D line, 5890 Å, unless otherwise indicated in parenthesis.

[e] Concentrations are 1–5 g/100 ml of solution at 20–25°; solvents other than water are indicated, and other temperatures are given in parentheses.

[f] Calculated by the contributor from the data of the reference cited; k = 0.69/time in minutes for 50% hydrolysis of the ester linkage.

[g] For the pyrophosphate group.

[h] For the second ester group.

[i] Both ester groups hydrolyze with equal ease.

References

1. Kiessling and Schuster, *Ber. Dtsch. Chem. Ges.*, 71, 123 (1938).
2. Weil-Malherbe and Green, *Biochem. J.*, 49, 286 (1951).
3. Baer and Fischer, *J. Biol. Chem.*, 128, 491 (1939).
4. Maruo and Benson, *J. Am. Chem. Soc.*, 79, 4564 (1957).
5. Fischer and Baer, *Ber. Dtsch. Chem. Ges.*, 337, 1040 (1932)65.
6. Kiessling, *Ber. Dtsch. Chem. Ges.*, 67, 869 (1934).
7. Meyerhof and Junowicz-Kocholaty, *J. Biol. Chem.*, 71, 149 (1943).
8. Ballou and Fischer, *J. Am. Chem. Soc.*, 77, 3329 (1955).
9. Ballou and Fischer, *J. Am. Chem. Soc.*, 76, 3188 (1954).
10. Kiessling, *Ber. Dtsch. Chem. Ges.*, 68, 243 (1935).
11. Meyerhof and Schultz, *Biochem. Z.*, 297, 60 (1938).
12. Negelein and Brömel, *Biochem. Z.*, 303, 132 (1939).
13. Baer, *J. Biol. Chem.*, 185, 763 (1950).
14. Greenwald, *J. Biol. Chem.*, 63, 339 (1925).
15. Sutherland, Posternak, and Cori, *J. Biol. Chem.*, 181, 153 (1949).
16. Ballou and Hesse, *J. Am. Chem. Soc.*, 78, 3718 (1956).
17. MacDonald, Fischer, and Ballou, *J. Am. Chem. Soc.*, 78, 3720 (1956).
18. Shetter, *J. Am. Chem. Soc.*, 78, 3722 (1956).
19. MacDonald, Ballou, and Fischer, *J. Am. Chem. Soc.*, 77, 5967 (1955).
20. Chu and Ballou, *J. Am. Chem. Soc.*, 83, 1711 (1961).
21. Charalampous and Mueller, *J. Biol. Chem.*, 201, 161 (1953).
22. Gillett and Ballou, *Biochemistry*, 547 (1963).
23. Taylor and Ballou, *Biochemistry*, 553 (1963).
24. Barker and Wold, *J. Org. Chem.*, 28, 1847 (1963).
25. Ishii, Hashimoto, Tachibana, and Yoshikawa, *Biochem. Biophys. Res. Commun.*, 10, 19 (1963).
26. Ballou, *J. Am. Chem. Soc.*, 79, 984 (1957).
27. Baddiley, Buchanan, Carss, and Mathias, *J. Chem. Soc.* (Lond.), p. 4583 (1956).
28. Applegarth, Buchanan, and Baddiley, *J. Chem. Soc.* (Lond.), 1213 (1965).
29. Moffatt and Khorana, *J. Am. Chem. Soc.*, 79, 1194 (1957).
30. Szabó and Szabó, *J. Chem. Soc.* (Lond.), 7, 5139 (1964).
31. Ukita and Nagasawa, *Chem. Pharm. Bull.* (Tokyo), p. 655 (1959).
32. Mendicino and Hanna, *J. Biol. Chem.*, 245, 6113 (1970).
33. Wright and Khorana, *J. Am. Chem. Soc.*, 80, 1994 (1958).
34. Putman and Hassid, *J. Am. Chem. Soc.*, 79, 5057 (1957).
35. Volk, *J. Biol. Chem.*, 234, 1931 (1959).
36. Volk, *Biochim. Biophys. Acta*, 37, 365 (1960).
37. Levene and Christman, *J. Biol. Chem.*, 123, 607 (1938).
38. Wright and Khorana, *J. Am. Chem. Soc.*, 78, 811 (1956).
39. Tener, Wright, and Khorana, *J. Am. Chem. Soc.*, 79, 441 (1957).
40. Kalckar, *J. Biol. Chem.*, 167, 477 (1947); Plesner and Klenow, *Methods Enzymol.*, 3 181 (1957).
41. Halmann, Sanchez, and Orgel, *J. Org. Chem.*, 34, 3702 (1969).
42. Khym, Doherty, and Cohn, *J. Am. Chem. Soc.*, 76, 5523 (1954).
43. Loring, Moss, Levy, and Hain, *Arch. Biochem. Biophys.*, 65, 578 (1956).
44. Albaum and Umbreit, *J. Biol. Chem.*, 167, 369 (1947).
45. Levene and Harris, *J. Biol. Chem.*, 101, 419 (1933).
46. Horecker and Smyrniotis, *Arch. Biochem. Biophys.*, 29, 232 (1950).
47. Michelson and Todd, *J. Chem. Soc. (Lond.)*, 2476 (1949).
48. Levene and Stiller, *J. Biol. Chem.*, 104, 299 (1934).
49. Tener and Khorana, *J. Am. Chem. Soc.*, 80, (1958) (1999).
50. Klenow, *Arch. Biochem. Biophys.*, 46, 186 (1953).
51. Hanna and Mendicino, *J. Biol. Chem.*, 245, 4031 (1970).
52. Kornberg, Lieberman, and Simms, *J. Biol. Chem.*, 215, 389 (1955).
53. Meagher and Hassid, *J. Am. Chem. Soc.*, 68, 2135 (1946).
54. Antia and Watson, *J. Am. Chem. Soc.*, 80, 6134 (1958).
55. Levene and Raymond, *J. Biol. Chem.*, 102, 347 (1933).
56. Gorin, Hough, and Jones, *J. Chem. Soc.* (Lond.), p. 582 (1955).
57. Friedkin, *J. Biol. Chem.*, 184, 499 (1950).
58. MacDonald and Fletcher Jr., *J. Am. Chem. Soc.*, 84, 1262 (1962).
59. Tarr, *Can. J. Biochem. Physiol.*, 36, 517 (1958).
60. Racker, *J. Biol. Chem.*, 196, 347 (1952).
61. MacDonald and Fletcher Jr., *J. Am. Chem. Soc.*, 81, 3719 (1959).
62. Tarr, *Chem. Ind.* (Lond.), p. 562 (1957).
63. Antonakis, Dowgiallo, and Szabo, *Bull. Soc. Chim. Fr.*, 1355 (1962).
64. Szabó and Szabó, *J. Chem. Soc.* (Lond.), p. 2944 (1965).
65. Stewart and Ballou, *J. Org. Chem.*, 32, 1065 (1967).
66. Horecker, Smyrniotis, and Seegmiller, *J. Biol. Chem.*, 193, 383 (1951).
67. Hurwitz, Weissbach, Horecker, and Smyrniotis, *J. Biol. Chem.*, 218, 769 (1956).
68. Kornberg quoted in Horecker, *Methods Enzymol.*, 3, 190 1957
69. Horecker, Hurwitz, and Weissbach, *J. Biol. Chem.*, 218, 785 (1956).
70. Leloir, *Fortschr. Chem. Org. Naturst.*, 47 (1951).
71. Simpson and Wood, *J. Am. Chem. Soc.*, 78, 5452 (1956); *J. Biol. Chem.*, 230, 473 (1958).
72. Glock, *Biochem. J.*, 52, 575 (1952).
73. Stumpf and Horecker, *J. Biol. Chem.*, 218, 753 (1956).
74. Barnwell, Saunders, and Watson, *Can. J. Chem.*, 33, 711 (1955).
75. Wolff and Kaplan, *J. Biol. Chem.*, 218, 849 (1957).
76. Lambert and Zilliken, *Chem. Ber.*, 96, 2350 (1963).
77. Chittenden, *Carbohydr. Res.*, 25, 35 (1972).
78. Kosterlitz *Biochem. J* 33, 1087 (1939); 37, 318 (1973)
79. Reithel, *J. Am. Chem. Soc.*, 67, 1056 (1945).
80. Foster, Overend, and Stacey, *J. Chem. Soc.* (Lond.), p. 980 (1951).
81. Inouye, Tannenbaum, and Hsia, *Nature*, 193, 67 (1962).
82. Tanaka, *Yakugaku Zasshi*, 81, 797 (1961); *Chem. Abstr.*, 55, 27064 (1961).
83. Cori, Colowick, and Cori, *J. Biol. Chem.*, 121, 465 (1937).
84. Wolfrom and Pletcher, *J. Am. Chem. Soc.*, 63, 1050 (1941).
85. Wolfrom, Smith, Pletcher, and Brown, *J. Am. Chem. Soc.*, 64, 23 (1942).
86. Potter, Sowden, Hassid, and Doudoroff, *J. Am. Chem. Soc.*, 70, 1751 (1948).
87. Farrar, *J. Chem. Soc. (Lond.)*, 3131 (1949).
88. Josephson and Proffe, *Ann.*, 481, 91 (1930).
89. Levene and Raymond, *J. Biol. Chem.*, 89, 479 (1930).
90. Levene and Raymond, *J. Biol. Chem.*, 91, 751 (1931).
91. Raymond, *J. Biol. Chem.*, 113, 375 (1936).
92. Josephson and Proffe, *Biochem. Z.*, 258, 147 (1933).
93. Robison and King, *Biochem. J.*, 25, 323 (1931).
94. Robison, *Biochem. J.*, 26, 2191 (1932).
95. Saito and Noguchi, *Nippon Kagaku Zasshi*, 82, 469 (1961); *Chem. Abstr.*, 56, 11678 (1962).
96. Lardy and Fischer, *J. Biol. Chem.*, 164, 513 (1946).
97. Lewak and Szabó, *J. Chem. Soc. (Lond.)*, 3975 (1963).
98. Posternak, *J. Biol. Chem.*, 180, 1269 (1949).
99. Buck, *Carbohydr. Res.*, 247 (1968).
100. Perchimlides, Osawa, Davidson, and Jeanloz, *Carbohydr. Res.*, 463 (1967).
101. Colowick, *J. Biol. Chem.*, 124, 557 (1938).
102. Posternak and Rosselet, *Helv. Chim. Acta*, 36, 1614 (1953).
103. Hill and Ballou, *J. Biol. Chem.*, 241, 895 (1966).
104. Pridhar and Behrman, *Carbohydr. Res.*, 23, 456 (1972).
105. Leloir, Cardini, and Olavarria, *Arch. Biochem. Biophys.*, 74, 84 (1958).
106. Davidson and Wheat, *Biochim. Biophys. Acta*, 72, 112 (1963).
107. Kim and Davidson, *J. Org. Chem.*, 28, 2475 (1963).
108. Carlson, Swanson, and Roseman, *Biochemistry*, 402 (1964).
109. Cardini and Leloir, *J. Biol. Chem.*, 225, 317 (1958).
110. Olavesen and Davidson, *Biochim. Biophys. Acta*, 101, 245 (1965).
111. Cardini and Leloir, *Arch. Biochem. Biophys.*, 45, 55 (1953).
112. Liu and Gotschlich, *J. Biol. Chem.*, 238, 1928 (1963).
113. Distler, Merrick, and Roseman, *J. Biol. Chem.*, 230, 497 (1958).
114. Kent and Wright, *Carbohydr. Res.*, 22, 193 (1972).
115. Maley, Maley, and Lardy, *J. Am. Chem. Soc.*, 78, 5503 (1956).
116. Leloir and Cardini, *Biochim. Biophys. Acta*, 20, 33 (1956).
117. O'Brien, *Biochim. Biophys. Acta*, 86, 628 (1964).
118. Baluja, Chase, Kenner, and Todd, *J. Chem. Soc.* (Lond.), p. 4678 (1960).
119. Brown, *J. Biol. Chem.*, 204, 877 (1953).
120. Westphal and Stadler, *Angew. Chem.*, 75, 452 (1963).
121. Anderson and Percival, *J. Chem. Soc.* (Lond.), 814 (1956).
122. Jeanloz, Konami, and Osawa, *Biochemistry*, 10, 192 (1971).
123. Heymann, Turdiu, Lee, and Barkulis, *Biochemistry*, 1393 (1968).
124. Wright, Taylor, Brunt, Brownsley, *J. Chem. Soc. Chem. Commun.*, 691 (1972); Wright, Taylor, *Carbohydr. Res.*, 32, 366 (1974)
125. Kochetkov, Shibaev, Kusov, and Troitskii, *Izv. Akad. Nauk SSSR, Ser. Khim.*, 425 (1973); *Bull. Acad. Sci. USSR, Div. Chem. Sci.*, 22, 408 (1973).
126. Whistler and Stark, *Carbohydr. Res.*, 13, 15 (1970).

127. Salo and FletcherJr., *Biochemistry*, 878 (1970).
128. Shibaev, Kusov, Kuchar, and Kochetkov, *Izv. Akad. Nauk SSSR, Ser. Khim.*, 992 (1973); *Bull. Acad. Sci. USSR, Div. Chem. Sci.*, 22, 886 (1973).
129. Crane and Sols, *J. Biol. Chem.*, 210, 597 (1954).
130. DeMoss and Happel, *J. Bacteriol.*, 70, 104 (1955).
131. Brooks, Lawrence, and Ricketts, *Biochem. J.*, 73, 566 (1959); *Nature*, 187, 1028 (1960).
132. Shibaev, Kusov, Kuchar, and Kochetkov, *Izv. Akad. Naud SSSR, Ser. Khim.*, 430 (1973); *Bull. Acad. Sci. USSR, Div. Chem. Sci.*, 22, 408 (1973).
133. Dahlgard and Kaufmann, *J. Org. Chem.*, 25, 781 (1960).
134. Antonakis, *Compt. Rend.*, 258, 3511 (1964).
135. Antonakis, *Bull. Soc. Chim. Fr.*, 2112 (1965).
136. Schanbacher and Wilken, *Biochim. Biophys. Acta*, 141, 646 (1967); Leaback, Heath, and Roseman, *Biochemistry*, 8, 1351 (1969)
137. Ishihara, Massaro, and Heath, *J. Biol. Chem.*, 243, 1103 (1968).
138. Pridhar and Behrman, *Biochemistry*, 12, 997 (1973).
139. Barber, *Biochim. Biophys Acta.*, 141, 174 (1967); Chatterjee and MacDonald, *Carbohydr. Res.*, 6, 253 (1968).
140. Pontis and Fischer, *Biochem. J.*, 89, 452 (1963).
141. MacDonald, *J. Org. Chem.*, 31, 513 (1966).
142. Tako and Robison, *Biochem. J.*, 29, 961 (1935).
143. Pogell, *J. Biol. Chem.*, 201, 645 (1953).
144. Neuberg, Lustig, and Rothenberg, *Arch Biochem. Biophys.*, 33 (1944).
145. MacLeod and Robison, *Biochem. J.*, 27, 286 (1933).
146. McGilvery, *J. Biol. Chem.*, 200, 835 (1953).
147. Heath and Ghalambar, *J. Biol. Chem.*, 237, 2423 (1962).
148. Chiu and Feingold, *Biochim. Biophys. Acta*, 92, 489 (1964).
149. Chiu, Otto, Power, and Feingold, *Biochim. Biophys. Acta*, 127, 249 (1966).
150. Mann and Lardy, *J. Biol. Chem.*, 187, 339 (1950).
151. Totton and Lardy, *J. Biol. Chem.*, 181, 701 (1949).
152. Fukui, *J. Bacteriol.*, 97, 793 (1969).
153. Kiely and FletcherJr., *J. Org. Chem.*, 33, 3723 (1968).
154. Avigad and England, *J. Biol. Chem.*, 243, 1511 (1968).
155. Patwardhan, *Biochem. J.*, 28, 1854 (1934).
156. Greiling, Kisters, and Hoppe-, *Seyler's Z. Physiol. Chem.*, 346, 77 (1966).
157. Wolfrom and Franks, *J. Org. Chem.*, 29, 3645 (1964).
158. Lee, *J. Org. Chem.*, 28, 2473 (1963).
159. Touster and Reynolds, *J. Biol. Chem.*, 197, 863 (1952).
160. Barker, Bourne, Fleetwood, and Stacey, *J. Chem. Soc.* (Lond.), 4128 (1958).
161. Marsh, *J. Chem. Soc.* (Lond.), 1578 (1952).
162. Ciferri, Blakley, and Simpson, *Can. J. Microbiol.*, 277 (1959).
163. MacGee and Doudoroff, *J. Biol. Chem.*, 210, 617 (1954).
164. Moses, Ferrier, and Calvin, *Proc. Natl. Acad. Sci. U.S.A.*, 48, 1644 (1962).
165. Stroud and Hassid, *Biochem. Biophys. Res. Commun.*, 15, 65 (1964).
166. Strobach and Szabó, *J. Chem. Soc.* (Lond.), 3970 (1963).
167. Teuber, Bevill, and Osborn, *Biochemistry*, 3303 (1968).
168. Horecker, Smyrniotis, Hiatt, and Marks, *J. Biol. Chem.*, 212, 827 (1955).
169. Benson, Paech, Tracy, Modern Methods of Plant Analysis. Springer-Verlag, Berlin, 1955, 113.
170. Robison, Macfarlane, and Tazelaar, *Nature*, 142, 114 (1938).
171. Sie, Nigam, and Fishman, *J. Am. Chem. Soc.*, 82, 1007 (1960).
172. Nordahl and Benson, *J. Am. Chem. Soc.*, 76, 5054 (1954).
173. Sprinson, Rothschild, and Sprecher, *J. Biol. Chem.*, 238, 3170 (1963).
174. Bartlett and Bucolo, *Biochem. Biophys. Res. Commun.*, 474 (1960).
175. Ballou and Pizer, *J. Am. Chem. Soc.*, 82, 3333 (1960).
176. Posternak, *Helv. Chim. Acta*, 42, 390 (1959).
177. EisenbergJr. and Bolden, *Biochem. Biophys. Res. Commun.*, 21, 100 (1965).
178. Woolley, *J. Biol. Chem.*, 147, 581 (1943).
179. Pizer and Ballou, *J. Am. Chem. Soc.*, 81, 915 (1959).
180. Brown and Hall, *J. Chem. Soc.* (Lond.), 357 (1959).
181. Angyal and Tate, *J. Chem. Soc.* (Lond.), 4122 (1961).
182. Angyal and Russell, *Aust. J. Chem.*, 22, 383 (1969).
183. Kilgour and Ballou, *J. Am. Chem. Soc.*, 80, 3956 (1958).
184. Lepage, Mumma, and Bensen, *J. Am. Chem. Soc.*, 82, 3713 (1960).
185. Seamark, Tate, and Smeaton, *J. Biol. Chem.*, 243, 2424 (1968).
186. Weiss and Mingioli, *J. Am. Chem. Soc.*, 78, 2894 (1956).
187. Shibaev, Kusov, Troitskii, and Kochetkov, *Izv. Akad. Nauk SSSR, Ser. Khim.*, 182 (l914); *Bull. Acad. Sci. USSR, Div. Chem. Sci.*, 23, 171 (1974).
188. Olavesen and Davidson, *J. Biol. Chem.*, 240, 992 (1965).
189. Gander, Petersen, and Boyer, *Arch. Biochem. Biophys.*, 69, 85 (1957).
190. Sasaki and Taniguchi, *Nippon Nogeikagaku Kasihi*, 33, 183 (1959); *Chem. Abstr.*, 54, 308 (1960).
191. Reithel and Young, *J. Am. Chem. Soc.*, 74, 4210 (1952).
192. Narumi and Tsumita, *J. Biol. Chem.*, 242, 2233 (1967).
193. Buchanan, Cummerson, and Turner, *Carbohydr. Res.*, 21, 283 (1972).
194. Leloir and Cardini, *J. Biol. Chem.*, 214, 157 (1955).
195. Robison and Morgan, *Biochem. J.*, 22, 1277 (1928).
196. MacDonald and Wong, *Biochim. Biophys. Acta*, 86, 390 (1964).
197. Cabib and Leloir, *J. Biol. Chem.*, 231, 259 (1958).
198. Narumi and Tsumita, *J. Biol. Chem.*, 240, 2271 (1965).

THE NATURALLY OCCURRING AMINO SUGARS

2-AMINO SUGARS

Compound and Formula	Source	Physical Constants[a] MP	Physical Constants[a] $[\alpha]_D$	R_{GlcN} Values on Paper Solvent Systems[b] A	B	C	D	E	R_{GlcN} on Ion Exchange System[c] I	II	Reference
Glucosamine (chitosamine): 2-amino-2-deoxy-D-glucose	Polysaccharides of bacteria, fungi, invertebrates, chitin, antibiotics, higher plants, vertebrates, UDP-complexes	88; 190–210; 110–111	(α) FB +100 → +47.5 (α) HCl; +100 → +72 (β) FB; +28 → +47.5 (β) HCl; +25 → +72.6	1.00	1.00	1.00	1.00	1.00	1.00	1.00	1–4
Galactosamine (chondrosamine): 2-amino-2-deoxy-D-galactose	Polysaccharides of bacteria, fungi, invertebrates, antibiotics, vertebrates, UDP-complexes	185	(α) HCl:W–HCl +121 → +80 (β) HCl:W–HCl +44 → +80	0.90; 0.80[Δ]	0.90	1.04	1.05	0.94	1.17; 1.20	1.03	1–4
Mannosamine: 2-amino-2-deoxymannose	Pneumococcus type XIX polysaccharide E. coli, Salmonella, animal metabolite (N-acetyl-D-isomer)	178–180	HCl – 3	1.05; 1.13[Δ]; 1.17[Δ]	—	1.19	1.05	—	1.06	1.12	1–4, 6[e], 7[e], 8
D-Gulosamine: 2-amino-2-deoxy-D-gulose	Antibiotics: streptolin B, streptothricin	150–170	(α) HCl +6.1 → –17.9 (β) HCl	1.00; 1.01	—	—	1.05	—	1.21	1.20	1–4
2-Aminoheptose: D-glycero-2-deoxy-2-amino-gulo-(or ido-) heptose[j]	Anacystis nidulans cell wall	—	—	—	—	—	—	—	2.19	—	46
6-O-Methyl-D-Glucosamine: 6-O-methyl-2-amino-2-deoxy-D-glucose	Lipopolysaccharide of Rhodo-pseudomonas palustris	—	—	—	—	—	—	—	—	—	54
D-Quinovosamine: 2-amino-2,6-dideoxy-D-glucose	Achromobacter georgio politanum Salmonella. Proteus vulgaris Arizona, Neurospora crassa, Vibrio cholera, Brucella species	165–170	HCl + 55.5	1.85	2.23	2.5	—	1.40	1.43	1.26	3, 4, 9[d], 10, 45, 47, 55, 56
L-Quinovosamine: 2-amino-2,6-dideoxy-L-glucose	Lipopolysaccharide of Shigella boydii	—	—	—	—	—	—	—	—	—	57
D-Fucosamine: 2-amino-2,6-dideoxy-D-galactose	C. violaceum; B. licheniformis; B. cereus; Erysipelothrix insidiosa; Pseudomonas aeruginosa	155 chars; 170–175 decomp.	HCl +91 ± 2	1.32	1.94	2.4	1.31	1.20	1.73; 1.75; 1.95	1.24	1, 2, 4; 10, 11, 12; 13[d], 14[d]; 48[d], 49[d], 53[d]
L-Fucosamine: 2-amino-2,6-dideoxy-L-galactose	Polysaccharide of pneumococcus type V Citrobacter freundii 05: H30 mucopolysaccharide	155 chars	HCl –93.4 ± 2	—	—	—	—	—	—	—	2, 15[d], 16[d]

THE NATURALLY OCCURRING AMINO SUGARS (Continued)

Compound and Formula	Source	Physical Constants[a] MP	$[\alpha]_D$	R_{GlcN} Values on Paper Solvent Systems[b] A	B	C	D	E	R_{GlcN} on Ion Exchange System[c] II	II	Reference
2-AMINO SUGARS (Continued)											
Pneumosamine: 2-amino-2,6-dideoxy-L-talose	Pneumococcus type V	162–163	HCl +6.9 → +10.4	—	—	—	—	—	—	1.35	2, 4, 16[d]
L-Rhamnosamine: 2-amino-2,6-dideoxy-L-rhamnose	E. coli 03:K2ab(L):H2	—	HCl +22.5	—	—	—	—	—	—	1.48	50[d]
3-AMINO SUGARS											
Kanosamine: 3-amino-3-deoxy-D-glucose	Antibiotics: kanamycin group	—	—	—	—	—	—	—	—	—	2[c], 17[d] 18, 21
3-Amino-3,6-dideoxy-D-glucose	Lipopolysaccharides, Citrobacter freundii, Salmonella, E. coli	Syrup	—	1.6	2.0 2.2	—	—	1.36	1.34 1.37	129	10[e], 19[d] 20[e], 21[e]
3-Amino-3,6-dideoxy-D-galactose	Lipopolysaccharide, Xanthomonas campestris. Salmonella, E. coli Arizona	—	—	1.2	1.6 1.8	—	—	—	1.54	120	20[e], 21[e] 22[d]
Mycosamine: 3-amino-3,6-dideoxy-D-mannose	Antibiotics: amphotericin B, nystatin, pimaricin	—	—	1.4 18	1.8 2.2	—	—	—	1.28	1.26	2, 17[d] 20, 21
3-Aminoribose: 3-amino-3-deoxy-D-ribose	3'-amino-3'-deoxy-adenosine from Helminthosporium sp. and Cordyceps militaris, antibiotic: puromycin	154–155 157–158 161	HCl -37 → -24.0 Ac -25	—	—	—	—	—	—	—	2, 17, 23, 24
4-AMINO SUGARS											
Viosamine: 4-amino-4,6-dideoxy-D-glucose	Lipopolysaccharide Chromobacterium violaceum. E. coli, TDP-nucleotides in Escherichia and Salmonella	132–138	HCl -9 → +21	1.47	1.45	1.3	—	—	1.43	—	2, 3, 11[d] 12, 25[d], 51

R_rhamnose solvent systems (for 4-amino sugars):

Compound	J	K	L	M	N	Reference
Viosamine	1.38	0.42	0.94	0.5 0.7	1.0 1.5	26[d]
4-Amino-4,6-dideoxy-D-galactose (Source: TDP-nucleotide in Escherichia, Salmonella and Pasteurella, Lipopolysaccharide of E. coli.)	1.25	0.88	1.16	0.5 0.7	1.0 1.5	26[d], 51

Compound and Formula	Source	MP	$[\alpha]_D$	A	B	C	D	E	II	II	Reference
6-AMINO SUGARS											
6-Amino-6-deoxy-D-glucose	Antibiotics: kanamycin A	161–162	HCl +23.0 → +50.1	—	—	—	—	—	—	—	2, 17[d], 18[d]

THE NATURALLY OCCURRING AMINO SUGARS (Continued)

DIAMINO-SUGARS

Compound and Formula	Source	MP	$[\alpha]_D$	R_GlcN Paper: A	B	C	D	E	Ion Exch: II	II	Reference
2,3-Diamino-2,3,6-trideoxyhexose:											58
Neosamine B: 2,6-diamino-2,6-dideoxy-L-idose	Lipopolysaccharides of *Rhodospirillaceae*; Antibiotics: neomycin, paronomycin. zygomycin	—	HCl +17	—	—	—	—	—	—	—	2, 27, 28
Neosamine C: 2,6-diamino-2,6-dideoxy-D-glucose	Antibiotics: neomycin B, neomycin A, zygomycin A	—	HCl +67	—	—	—	—	—	—	4.7[g]	2, 27, 28
				N	S						
2,4-Diamino-2,4-trideoxy-hexose	Polysaccharide from *Bacillus lichenformis*	—	—	0.92	1.09	—	—	—	—	—	2, 29[d], 30
2,4-Diamino-2,4-trideoxy-hexose	UDP-nucleotide synthesized by *D. pneumoniae* Type XIV	—	—	—	—	—	—	—	—	—	31

N-ACETYLATED[h] AND N-METHYLATED AMINO SUGARS

Compound and Formula	Source	MP	$[\alpha]_D$	R_GlcN: A	B	C	D	E	Reference
2-Acetamido-D-glucose	—	205 / 182–184	(α) +64 → +40.9 ; (β) −21.5 → +40.9	1.71 / 1.84[i] / 0.96[Δ]	—	—	1.10	—	1, 3
2-Acetamido-D-galactose	—	172–173	(α) +115 → +86	1.59 / 1.70[i] / 0.75[Δ]	—	—	1.19	—	—
2-Acetamido-D-mannose	—	105–108	(β) −21 → +10	1.79 / 1.84[i] / 1.05[Δ]	—	—	1.12	—	1, 2, 3
2-Acetamido-D-gulose	—	—	−55 → −59	1.87	—	—	1.18	—	1[e], 3
2-Acetamido-2,6-dideoxy-D-glucose	—	—	—	—	—	—	—	—	—
2-Acetamido-2,6-dideoxy-D-galactose	—	—	—	2.12	—	—	1.35	—	3
2-Acetamido-2,6-dideoxy-L-galactose	—	195–198	−79	—	—	—	—	—	16[d]
2-Acetamido-2,6-dideoxy-L-talose	—	—	—	—	—	—	—	—	—

Compound and Formula	Source	MP	$[\alpha]_D$	R_fucose: A	B	F	G	H	I	Reference
3-Acetamido-3-deoxy-D-glucose	—	199–202	−43	—	—	0.71	1.13	0.90	0.92	18, 32
3-Acetamido-3,6-dideoxy-D-glucose	—	—	—	1.2	1.3	—	—	—	—	20, 21

THE NATURALLY OCCURRING AMINO SUGARS (Continued)

N-ACETYLATED[h] AND N-METHYLATED AMINO SUGARS (Continued)

Compound and Formula	Source	MP	$[\alpha]_D$	A	B	C	D	E	II	II	Reference
3-Acetamido-3,6-dideoxy-D-galactose	—	174–176	+114	1.1 1.2	1.4	1.20	1.31	1.24	1.18	—	20, 21 22[d], 32
3-Acetamido-3,6-dideoxy-D-mannose	—	191–192	−46E	1.4	1.6	1.88	1.31	1.37	1.37	—	20, 21, 32, 33
3-Acetamido-3-deoxy-D-ribose	—	—	—	—	—	1.71	1.33	1.28	1.29	—	32
				A	**J**	**G**	**K**	**L**	**M**		$R_{rhamnose}$
4-Acetamido-4,6-dideoxy-D-glucose	—	—	—	0.97	1.20	1.27	0.61	1.02	1.09	—	26[d]
4-Acetamido-4,6-dideoxy-D-galactose	—	—	—	0.94	1.30	1.39	0.56	0.99	1.07	—	26[d]
				A	**N**	**Q**	**R**	**S**			R_{GlcN}
6-Acetamido-6-deoxy-D-glucose	—	196–198	+44.0 → +34.9	—	—	—	—	—	—	—	18[d]
2-Amino-4-acetamido-2,4,6-trideoxy-L-altrose·HCl	—	216–219	+115 → +94	1.70	2.33	1.38	1.8	1.38	—	—	23, 30
2,4-Diacetamido-2,4,6-trideoxyhexose	—	262–264	+67 W–E	—	7.50	—	—	—	—	—	29, 30
N-methyl-L-glucosamine: 2-deoxy-2-methylamino-L-glucose	Antibiotics: Streptomycin group	160–163 Gum 165–166	HCl −103 → −88 FB −65 M N-acetyl −51	—	—	—	—	—	—	—	2, 17, 34[d]
Desosamine: 3-dimethylamino-3,4,6-trideoxy-D-xylohexose	Antibiotics: erythromycin, griseomycin, methymycin. narbomycin, neomethymycin, oleandomycin, picromycin, plicacetin	189–191	HCl +49.5 +53 4 E	—	—	—	—	—	—	—	2, 17[d] 35
Mycaminose: 3-dimethyl-amino-3,6-dideoxy-D-glucose	Antibiotics: carbomycin, spiromycin, leucomycin	115–116	HCl +31	—	—	—	—	—	—	—	2, 17[d], 36
Rhodosamine: 3-dimethyl-amino-2,3,6-trideoxy-L-lyxohexose	Antibiotics: cinerubins. pyrromycin rhodomycins	152–153	HCl −65.2	—	—	—	—	—	—	—	2, 37
Amosamine: 4-dimethylamino-4,6-dideoxy-D-glucose	Antibiotics: amicetin	192–193	HCl +45.5	—	—	—	—	—	—	—	2, 17[d], 38
4-Dimethylamino-2,3,4,6-tetradeoxyhexose	Antibiotics: spiramycins	75	+62.6 +83.9 M	—	—	—	—	—	—	—	2, 39[d]

Physical Constants[a]; *R_GlcN Values on Paper Solvent Systems[b]*; *R_GlcN on Ion Exchange System[c]*

THE NATURALLY OCCURRING AMINO SUGARS (Continued)

Compound and Formula	Source	Physical Constants[a]		R_GlcN Values on Paper Solvent Systems[b]					R_GlcN on Ion Exchange System[c]		Reference
		MP	$[\alpha]_D$	A	B	C	D	E	I	II	
ACIDIC AMINO SUGARS											
Glucosaminuronic acid: 2-amino-2-deoxy-D-glucuronic acid	Polysaccharide of *Haemophilus influenza* type d. *Staphylococcus*	172	+55	0.30 0.35	—	—	0.40	0.46	0.70	—	1[f], 2, 3 4[e], 40[d]
D-Galactosaminuronic acid: 2-amino-2-deoxy-D-galacturonic acid	Vi antigens: *E. coli, Paracolobacterium ballerup* and *S. typhosa*	160	W-HCl, pH2 +84.5	0.10	0.83	0.58	—	0.54	1.00	—	2–4
Mannosaminuronic acid: 2-amino-2-deoxymannuronic acid	Polysaccharide of *Micrococcus lysodeikticus*, K7 antigen of *E. coli* (D-configuration)	—	—	—	—	—	—	—	—	—	2
Gulosaminuronic acid: 2-amino-2-deoxy guluronic acid	Polysaccharide of *Vibrio parahaemolyticus*	—	—	0.42	—	—	—	—	—	—	52
		—	—	—	—	—	—	0.48	—	—	59
4-amino-4-deoxy-D-hexuronic acid	Antibiotics: gougeroutin, blastocidins	—	—	—	—	—	—	—	—	—	60
Muramic acid	Bacterial cell walls, spores, nucleotide	—	+109	1.00	—	—	—	—	—	—	
	complexes, cell walls of blue–green algae	—	—	1.02	1.86	2.10	0.83	1.11	1.1 1.2	—	2, 3, 41[d]
Neuraminic acid: 5-amino-3,5-dideoxy-D-glycero-D-galacto nonulosonic acid. Occurs in Nature as:	Polysaccharide of bacteria, invertebrates, vertebrates	—	—	—	—	—	—	—	—	—	2
N-acetylneuraminic acid		185–187	−31 ± 2	0.48	—	—	—	—	—	—	—
N-glycolylneuraminic acid		185–187	−31 ± 2	0.33	—	—	—	—	—	—	8, 42
N-4-O-diacetylneuraminic acid		200	−61 ± 1	—	—	—	—	—	—	—	—
N-7-O-diacetylneuraminic acid		138–140	+6 ± 2	—	—	—	—	—	—	—	—
Bovine N-acetyl-O-diacetylneuraminic acid		130–131	+9 ± 2	—	—	—	—	—	—	—	—
KETO-AMINO SUGARS											
2-acetamido-4-keto 2,6-dideoxyhexose	UDP-nucleotide made by enzyme of *D. pneumoniae* type XIV and *Citrobacter freundii* ATCC 10053	—	—	—	—	—	—	—	—	—	31, 43

Compiled by Rudolf A. Raff and Robert W. Wheat.

THE NATURALLY OCCURRING AMINO SUGARS (Continued)

a   MP. melting po.nt; $[\alpha]_D$ – Rotation solvent; water, unless otherwise indicated in which case W-E = water:ethanol (1:1); W-HCl = water-HCl; E = ethanol; M = methanol. The abbreviations (α) or (β) indicate the anomer, FB = free base; HCl indicates the hydrochloride: Ac the acetate.

b   Solvent system:

A.   n-butanol-pyridine-water (6:4:3)
A$^\Delta$.   solvent system A on paper treated with 0.1 M BaCl$_2$ or BaAc$_2$
B.   n-butanol glacial acetic acid-water (5:1:2)
C.   phenol-water (70:30)
D.   phenol-water (80:20), ammonia atmosphere
E.   ethylacetate-pyridine-water-acetic acid (5:5:3:1)
F.   pyridine-ethyl acetate-water (10:36:11.5)
G.   phenol-water (80:20)
H.   n-butanol-acetic acid-water (5:1.2:2.5)
I.   ethylacetate-pyridine-n-butanol-butyric acid water (10:10:5:1:5)

J.   isobutyric acid-N ammonia (10:0.6)
K.   pyridine-ethyl acetate-water (10:3.0:1.5 upper layer)
L.   n-butanol-ethanol-water (13:8:4)
M.   n-butanol-acetic acid-water (3:1:1)
N.   n-butanol-ethanol-water (4:1:1)
O.   n-butanol-pyridine-0.1 N HCl (5:3:2)
P.   n-butanol-acetic acid-water (25:6:25)
Q.   ethylacetate-pyridine-water (2:1:2)
R.   n-butanol-ethanol-water (5:1:4)
S.   n-propanol-1 per cent ammonia (7:3)

c   System I: Dowex 50 H$^+$ column 1 × 50 cm packed according to Gardell and eluted with 0.33 N HCl.[44]

System II: Technicon amino acid analyzer modified by Brendel et al.4 and eluted with 0.133 M pyridine-acetic acid (0.82 M) buffer pH 3.85.

d   Denotes the reference describing the characterization of the compound in question.

e   Denotes chromatographic identification.

f   Indicates that the recorded physical properties are for the synthetic compound.

g   Brendel et al. system using 3.1 M pyridine-acetic buffer at pH 4.5.[4]

h   Not all of the N-acetamido compounds have necessarily been found in nature, but are listed here for convenience.

i N-acetylmannosamine and N-acetylgalactosomine have an R$_{N'}$-acetylglucosamine of 0.4–0.5 when run on Borate-treated paper with solvent A.

j   The c-2 carbon configuration is in doubt.

References

1. Horton, *Advanced Carbohyd. Chem.*, 15, 59 (1960).
2. Sharon, in *The Amino Sugars*, Balazo and Jeanloz, Eds., Vol. 2A, Academic Press, New York, 1965, 1.
3. Wheat, *Method Enzymol.*, 8, 60 (1966).
4. Brendel, Roszel, Wheat, and Davidson, *Anal. Biochem.*, 18, 147 (1967); Brendel, Steele, Wheat, and Davidson, *Anal. Biochem.*, 18, 161 (1967).
5. Sharbarova, Buchanan, and Baddiley, *Biochim. Biophys., Acta*, 57, 146 (1962).
6. Luderitz, Jann, and Wheat, *Comp. Biochem.*, 26, in press.
7. Rude and Goebel, *J. Expt. Med.*, 116, 73 (1962).
8. Neuberger, Marshall, and Gottschalk, in *Glycoproteins*, Gottschalk, Ed., Elsevier, Amsterdam, 1966, 158.
9. Smith, *Biochem. Biophys. Res. Commun.*, 15, 593 (1964); Colwell, Smith, and Chapman, *Can. J. Microbiol.*, 14, 165 (1968).
10. Raff and Wheat, *J. Biol. Chem.*, 242, 4610 (1967).
11. Wheat, Rollins, and Leatherwood, *Biochem. Biophys. Res. Commun.*, 9, 120 (1962).
12. Smith, Leatherwood, and Wheat, *J. Bacteriol.*, 84, 100 (1962).
13. Crumpton and Davies, *Biochem. J.*, 70, 729 (1958).
14. Wheat, Rollins, and Leatherwood, *Nature*, 202, 492 (1965).
15. Barry and Roark, *Nature*, 202, 493 (1965).
16. Barker, Brimacombe, Horn, and Stacey, *Nature*, 189, 303 (1961).
17. Dutcher, *Advance Carbohyd. Chem.*, 18, 259 (1965).
18. Cron, Forbig, Johnson, Schmitz, Whitehead, Hooper, and Lemieux, *J. Amer. Chem. Soc.*, 80, 2342 (1958).
19. Raff and Wheat, *Fed. Proc.*, 26, 281 (1967).
20. Lüderitz, Ruschmann, Westphal, Raff, and Wheat, *J. Bacteriol.*, 94, 5 (1967).
21. Jann, Jann, and Mueller-Seitz, *Nature*, 215, 170 (1967).
22. Ashwell and Volk, *J. Biol. Chem.*, 240, 4549 (1965).
23. Baker, Schaub, and Kissman, *J. Amer. Chem. Soc.*, 77, 5911 (1955).
24. Baer and Fischer, *J. Amer. Chem. Soc.*, 81, 5184 (1959).
25. Stevens, Blumberg, Daniker, Wheat, Kujomoto, and Rollins, *J. Amer. Chem. Soc.*, 85, 3061 (1963).
26. Matsuhashi and Strominger, *J. Biol. Chem.*, 239, 2454 (1964).
27. Rinehart, Jr., Woo, and Argoudelis, *J. Amer. Chem. Soc.*, 80, 6461 (1958).
28. Hichens and Rinehart, Jr., *J. Amer. Chem. Soc.*, 85, 1547 (1963).
29. Zehavi and Sharon, *Israel J. Chem.*, 2, 322 (1964).
30. Sharon and Jeanloz, *J. Biol. Chem.*, 235, 1 (1960).
31. Distker, Kauffman, and Roseman, *Arch. Biochem. Biophys.*, 116, 466 (1966).
32. Ashwell, Brown, and Volk, *Arch. Biochem. Biophys.*, 112, 648 (1965).
33. Walters, Dutcher, and Wintersteiner, *J. Amer. Chem. Soc.*, 79, 5076 (1957).
34. Kuehl, Jr., Flynn, Holly, Mozingo, and Folkers, *J. Amer. Chem. Soc.*, 68, 536 (1946); *J. Amer. Chem. Soc.*, 69, 3032 (1947).
35. Brockman, Konig, and Oster, *Chem. Ber.*, 87, 856 (1954).
36. Hochstein and Regna, *J. Amer. Chem. Soc.*, 77, 3353 (1955).
37. Brockmann, Spohler, and Waehneldt, *Chem. Ber.*, 96, 2925 (1963).
38. Stevens, Gasser, Mukherjee, and Haskell, *J. Amer. Chem. Soc.*, 78, 6212 (1956).
39. Paul and Tchelitcheff, *Bull. Soc. Chem. (France)*, 734 (1957).
40. Hanession and Haskell, *J. Biol. Chem.*, 239, 2758 (1964).
41. Strange and Kent, *Biochem. J*, 11, 333 (1959).
42. Bourillon and Michon, *Bull. Soc. Chim. Biol.*, 41, 267 (1959).
43. Raff and Wheat, *Fed. Proc.*, 24, 478 (1965).
44. Gardell, *Acta Chem. Scand.*, 7, 207 (1953).
45. Luderitz, Gmeiner, Kickhofen, Mayer, Westphal, and Wheat, *J. Bacteriol.*, 95, 490 (1968).
46. Weiss, Drews, Jann, and Jann, personal communication from Dr. Drews (1969).
47. Livington, *J. Bacteriol.*, 99, 85 (1969).
48. Erler, *Arch. Exp. Veteriarmed*, 22, (6), 1155 (1968).
49. Suziki, *Biochim. Biophys. Acta*, 177, 371 (1969).
50. Jann and Jann, *Eur. J. Biochem.*, 5, 173 (1967).
51. Jann and Jann, *Eur. J. Biochem.*, 2, 26 (1967).
52. Mayer, *Eur. J. Biochem.*, 8, 139 (1969).
53. Sharon, Shif, and Zehavi, *Biochem. J.*, 93, 210 (1964).
54. Mayer and Framberg, *Europ. J. Biochem.*, 44, 181 (1974).
55. Jackson, G. D., *University of South Wales, Australia*, personal communication, 1974.
56. Bowser, Wheat, Foster, and Leong, *Inf. Imm.* 9, 722 (1974).
57. Dmitriev, Backinowsky, Kochetkov, and Khomenko, *Europ. J. Biochem.*, 34, 513 (1973).
58. Weckesser, Drews, Fromme, Mayer, *Arch. Mikrobiol.*, 92, 123 (1973).
59. Jann and Jann, *Europ. J. Biochem.*, 37, 401 (1973).
60. Kotick, Klein, Watanabe, and Fox, *Carb. Res.*, 11, 369 (1969).

OLIGOSACCHARIDES (INCLUDING DISACCHARIDES)

The next two tables are taken directly from 3rd Edition of *The Handbook of Biochemistry and Molecular Biology* which was edited by the late Gerald Fasman. This material appeared in the original edition of *The Handbook of Biochemistry and Molecular Biology* which was edited by Herbert Sober. These tables have been retained in the current edition without modification to assure the retention of data. It is the intent of the Editor to assure that this material is revised for subsequent editions.

Substance[a] (Synonym) (A)	Derivative (B)	Chemical Formula (C)	Melting Point °C (D)	Specific Rotation[b] $[\alpha]_D$ (E)	Reference (F)
DISACCHARIDES					
O-6-Deoxy-3-O-methyl-β-D-Allp-(1 → 4)-D-digitoxose (Drebyssobiose)		$C_{13}H_{24}O_8$	108–110	+25.8	324
	Drebyssobionic acid lactone	$C_{13}H_{22}O_8$	125–131	+29.6 (c 0.7, CHCl$_3$)	324
O-(4-O-Benzoyl)-D-apiosyl-(1 → 2)-D-glucopyranosyl benzoate		$C_{25}H_{26}O_{12}$	148–150	−106 (CH$_3$OH)	1, 2
	Pentaacetate	$C_{35}H_{36}O_{17}$	203–204	−35 (CHCl$_3$)	1
O-D-Fruf-(2 → 2)-D-Fruf-anhydride (Alliuminoside)		$C_{12}H_{20}O_{10}$	92–93	−23.8	3
	Hexaacetate	$C_{24}H_{32}O_{16}$	98–99	−29.3	3
O-β-D-Fruf-(2 → 1)-D-Fru (Inulobiose)		$C_{12}H_{22}O_{11}$	—	−32.5, −72.4	4, 5
	Octaacetate	$C_{28}H_{38}O_{19}$	—	−6.5, −14.2 (CHCl$_3$)	4, 5
O-D-Fruf-(2 → 1)-D-Fru		$C_{12}H_{22}O_{11}$	—	−26.3	6
	Octaacetate	$C_{28}H_{38}O_{19}$	—	+14.2 (CHCl$_3$)	6
O-D-Fruf-(2 → 6)-D-Fruf (Levanbiose)		$C_{12}H_{22}O_{11}$	—	−20.8 (17°)	322, 323
O-D-Fruf-(2 → ?)-D-Fru (Sogdianose)		$C_{12}H_{22}O_{11}$	156–158	−16.4	3
	Octaacetate	$C_{28}H_{38}O_{19}$	94–95	−28.7 (CHCl$_3$)	3
O-β-D-Fruf-(2 → 1)-α-D-Glcp (Sucrose)		$C_{12}H_{22}O_{11}$	188, 170[c]	+66.5	7
	Octaacetate	$C_{28}H_{38}O_{19}$	69, 75[c]	+59.6 (CHCl$_3$)	8, 9
O-β-D-Fruf-(2 → ?)-D Glc (Ceratose)		$C_{12}H_{22}O_{11}$	—	+21	10
O-β-D-Fruf-(2 → 6)-D Glc		$C_{12}H_{22}O_{11}$	—	+5	11
O-β-D-Fruf-(2 → 6)-D-2dGlc		$C_{12}H_{22}O_{10}$	s 56, 85	+62 → +26.8	12
O-β-D-Galp-(1 → 4)-D-Allp		$C_{12}H_{22}O_{11}$	112–113	+29	13
O-β-D-Galp-(1 → 2)-D-arabinitol (Umbilicin)		$C_{11}H_{22}O_{10}$	138–139	−81	14
	Octaacetate	$C_{27}H_{38}O_{18}$	84–85	−20 (CHCl$_3$)	14
O-β-D-Galp-(1 → 4)-D-arabinitol		$C_{11}H_{22}O_{10}$	178–181	—	13
O-α-D-Galp-(1 → 3)-L-Ara		$C_{11}H_{20}O_{10}$	192–194	+80 → +65	15, 16
	Heptamethyl ether	$C_{18}H_{34}O_{10}$	89	+164	17, 18
O-β-D-Galp-(1 → 3)-L-Ara		$C_{11}H_{20}O_{10}$	177–178	+30 → +45	13
O-β-D-Galp-(1 → 4)-L-Arap		$C_{11}H_{20}O_{10}$	No constants known	—	13
	Methyl β-glycoside	$C_{11}H_{22}O_{10}$	216–218	+163	13
O-β-D-Galp-(1 → 5)-L-Araf		$C_{11}H_{20}O_{10}$	—	−13, −18, +85	13, 19, 20
	Heptamethyl ether	$C_{18}H_{34}O_{10}$	—	−36, −45 (CH$_3$OH)	19, 20
O-β-D-Galp-(1 → 2)-D-erythritol		$C_{10}H_{20}O_9$	182–184	+7	13, 21
	Heptaacetate	$C_{24}H_{34}O_{16}$	114	+109 (c 0.6, acetone)	22
O-α-D-Galp-(1 → 2)-D-Fruf		$C_{12}H_{22}O_{11}$	s 170, 179	+81.5	23
O-β-D-Galp-(1 → 2)-D-Fruf	Octaacetate	$C_{12}H_{22}O_{11}$	—	−33	24
		$C_{28}H_{38}O_{19}$	169–170	+41.3 (CHCl$_3$)	25
O-β-D-Galp-(1 → 3)-D-Fucp		$C_{12}H_{22}H_{10}$	245–246	+78	13
O-α-D-Galp-(1 → 3)-D-Galp		$C_{12}H_{22}O_{11}$	—	+161	317
	Octaacetate	$C_{28}H_{38}O_{19}$	157.5–158.5	+110.2 (c 0.5, CHCl$_3$)	318
O-β-D-Galp-(1 → 3)-D-Galp		$C_{12}H_{22}O_{11}$	151–152	+69 → +55	16
			163–170	+75 → +60	26
			204–206	+56	13
O-α-D-Galp-(1 → 4)-D-Galp		$C_{12}H_{22}O_{11}$	210–211	+172	209, 317
	O-α-D-Galp-(1→4)-D-galactitol	$C_{12}H_{24}O_{11}$	—	+119	317
O-β-D-Galp-(1 → 4)-D Galp		$C_{12}H_{22}O_{11}$	195–198	+85 → +67	13
	Octaacetate	$C_{28}H_{38}O_{19}$	172–173	+57.3 (CHCl$_3$)	27

OLIGOSACCHARIDES (INCLUDING DISACCHARIDES) (Continued)

Substance[a] (Synonym) (A)	Derivative (B)	Chemical Formula (C)	Melting Point °C (D)	Specific Rotation[b] $[\alpha]_D$ (E)	Reference (F)
DISACCHARIDES (Continued)					
O-α-D-Gal$p$-(1 → 5)-D-Gal$f$		$C_{12}H_{22}O_{11}$	—	+133	317
	O-α-D-Gal$p$-(1 → 5)-D-galactitol	$C_{12}I_{24}O_{11}$	—	+122	317
O-α-D-Gal$p$-(1 → 6)-D-Gal$p$ (Swietenose)		$C_{12}H_{22}O_{11}$	—	+149	28, 29
	Octaacetate	$C_{28}H_{38}O_{19}$	223–227	+186 (c 0.5, CHCl$_3$)	29
O-β-D-Gal$p$-(1 → 6)-D-Gal$p$		$C_{12}H_{22}O_{11}$	97–100 (· CH$_3$OH)	+26	24, 30
			168–170	+28	31
	Octamethyl ether	$C_{20}H_{38}O_{11}$	68–70	−5.7 (CH$_3$OH)	31
O-β-D-Gal$p$-(1 → 4)-3,6-anhydro-L-Gal (Agarobiose)		$C_{12}H_{20}O_{10}$	—	−21.5 → −16.4	32
			—	+19	33
	Dimethyl acetal	$C_{14}H_{24}O_{10}$	163–166	−36 (MeOH)	34
O-D-Gal$p$-(1 → 6)-D-GalNAcp		$C_{14}H_{25}NO_{11}$	—	+142	35
O-α-D-Gal$p$-(1 → 2)-D-Glc		$C_{12}H_{22}O_{11}$	118–120	+145	36
			155d	+161 → +153	37
	Octaacetate	$C_{28}H_{38}O_{19}$	176–178	+153 (CHCl$_3$)	37
O-β-D-Gal$p$-(1 → 2)-D-Glc		$C_{12}H_{22}O_{11}$ · H$_2$O	175d	+48 → +39	38
O-β-D-Gal$p$-(1 → 3)-D-Glc		$C_{12}H_{22}O_{11}$	167–169	+37 (c 0.8)	13
	Monohydrate	$C_{12}H_{22}O_{11}$ · H$_2$O	202–204	+77 → +41	39
	Phenylosazone	$C_{24}H_{34}N_4O_9$	184–185	—	39
	Methyl α-glycoside	$C_{13}H_{24}O_{11}$	148.5–150	+18.3 (c 0.5)	325
O-β-D-Gal$p$-(1 → 4)-α-D-Glc$p$		$C_{12}H_{22}O_{11}$ · H$_2$O	202	+83.5 → +52.6	40, 41
	Octaacetate	$C_{28}H_{38}O_{19}$	152	+53.6 (c 10, CHCl$_3$)	42
O-β-D-Gal$p$-(1 → 4)-β-D-Glc$p$ (Lactose)		$C_{12}H_{22}O_{11}$	252	+34.2 → 53.6	43
	Octaacetate	$C_{28}H_{38}O_{19}$	90	−4.7 (c 10, CHCl$_3$)	42
O-α-D-Gal$p$-(1 → 6)-α-D-Glc		$C_{12}H_{22}O_{11}$	183–184	+166 → 142	44
O-α-D-Gal$p$-(1 → 6)-β-D-Glc$p$ (Melibiose)		$C_{12}H_{22}O_{11}$ · 2H$_2$O	85–86	+123 → +143	7, 44
	Melibiitol	$C_{12}H_{24}O_{11}$	173–175	+111	45
	Octaacetate	$C_{28}H_{38}O_{19}$	177	+102.5	46
O-β-D-Gal$p$-(1 → 6)-D-Glc$p$ (Allolactose, Lactobiose)		$C_{12}H_{22}O_{11}$	165, 174–176	+25 +37.5	27, 47–49
	Octaacetate	$C_{28}H_{38}O_{19}$	165	−0.5 (CHCl$_3$)	50
O-D-Gal-(? → ?)-D-Glc		$C_{12}H_{22}O_{11}$	205	−27	49
O-β-D-Gal$p$-(1 → 3)-D-6d-Glc$p$		$C_{12}H_{22}O_{10}$	246–248	+25	13
O-β-D-Gal$p$-(1 → 3)-D-GlcNAc (Lacto-N-biose I)		$C_{14}H_{25}NO_{11}$	166–167	+32 → +14	51, 52
O-β-D-Gal$p$-(1 → 4)-D-GlcNAcp (Lactosamine, N-acetyl)		$C_{14}H_{25}O_{11}$	No constants known	—	53
	Monomethanolate	$C_{14}H_{25}NO_{11}$ · CH$_3$OH	172	+51.2 → +27.8	53
	Heptaacetate	$C_{28}H_{39}NO_{18}$	222–223	+61.5 (30°) (CHCl$_3$)	53
	Methyl β-glycoside	$C_{15}H_{27}NO_{11}$	243–245d	−23.1	54
O-α-D-Gal$p$-(1 → 6)-D-GlcNAcp		$C_{14}H_{25}NO_{11}$	138d	+118	35, 317
	Heptaacetate	$C_{28}H_{39}NO_{18}$	141–142	+50.2 (CHCl$_3$)	317
O-β-D-Gal$p$-(1 → 6)-D-GlcNAcp		$C_{14}H_{25}O_{11}$	157–159	+32.1 → +27.3	51, 53
O-α-D-Gal$p$-(1 → 1)-glycerol		$C_9H_{18}O_8$	150–152	+155	55
O-β-D-Gal$p$-(1 → 1)-glycerol		$C_9H_{18}O_8$	139–140	+3.8, −7, −73	55–57
	Hexabenzoate	$C_{51}H_{42}O_{14}$	133–134	−6 (CHCl$_3$)	57
O-α-D-Gal$p$-(1 → 2)-glycerol (Floridoside)		$C_9H_{18}O_8$	86–87, 129–130	+151, +163	22, 58, 59
	Hexaacetate	$C_{21}H_{30}O_{14}$	101	+114 (acetone)	60
O-α-D-Gal$p$-(1 → 1)-*myo*-inositol (Galactinol)		$C_{12}H_{22}O_{11}$ · 2H$_2$O	220–222	+135.6	61
	Nonamethyl ether	$C_{21}H_{40}O_{11}$	96.5–98	+119	62
O-β-D-Gal$p$-(1 → 5)-*myo*-inositol		$C_{12}H_{22}O_{11}$	248–252	0	63
O-β-D-Gal$p$-(1 → 3)-D-mannitol (Peltigeroside)		$C_{12}H_{24}O_{11}$	161–163	−55.5, −61	64, 65
O-D-Gal$p$-(1 → ?)-D-Man		$C_{12}H_{22}O_{11}$	No constants given	—	13
	Methyl glycoside	$C_{13}H_{24}O_{11}$	168–170	+51 (c 0.7)	13
O-β-D-Gal$p$-(1 → 3)-α-L-Rha		$C_{12}H_{22}O_{10}$	200–202	+2 → +15	15
O-β-D-Gal$p$-(1 → 3)-D-Rib		$C_{11}H_{20}O_{10}$	No constants given	—	13
O-β-D-Gal$p$-(1 → 5)-D-Rib		$C_{11}H_{20}O_{10}$	No constants given	—	13

OLIGOSACCHARIDES (INCLUDING DISACCHARIDES) (Continued)

Substance[a] (Synonym) (A)	Derivative (B)	Chemical Formula (C)	Melting Point °C (D)	Specific Rotation[b] $[\alpha]_D$ (E)	Reference (F)
DISACCHARIDES (Continued)					
O-β-D-Gal*p*-(1 → 2)-D-Xyl		$C_{11}H_{20}O_{10}$	159–160	+25	13
	Heptamethyl ether	$C_{18}H_{34}O_{10}$	—	–5.9 (CH_3OH)	66
O-β-D-Gal*p*-(1 → 3)-D-Xyl		$C_{11}H_{20}O_{10}$	196–200	+27 → +10	13
O-3,6-anhydro-α-D-Gal*p*-(1 → 3)-D-Gal*p* 4-sulfate		$C_{12}H_{20}O_{13}S$	No constants known	—	326
O-3,6-anhydro-L-Gal*p*-(1 → 3)-D-Gal Neoagarobiose)		$C_{12}H_{20}O_{10}$	207–208	+34.4 → +20.3	67
	Hexaacetate	$C_{24}H_{32}O_{16}$	112	+1.6 (27°) ($CHCl_3$)	67
O-α-D-GalUA*p*-(1 → 4)-D-GalUA (Digalacturonic acid)		$C_{12}H_{18}O_{13}$	125–135d	+154	68, 69
	Methyl α-glycoside dimethyl ester	$C_{15}H_{24}O_{13}$	120–122	+162.6	70
O-4,5-anhydro-α-D-GalUA*p*-(1 → 4)-D-GalUA		$C_{12}H_{16}O_{12}$	—	+177.8	71
O-β-D-Glc*p*-(1 → 3)-L-Ara		$C_{11}H_{20}O_{10}$	199–201	+70	13
O-β-D-Glc*p*-(1 → 4)-D-Ara		$C_{11}H_{20}O_{10}$	135	–104	327
	Heptaacetate	$C_{25}H_{34}O_{17}$	214	–62 ($CHCl_3$)	327
O-α-D-Glc*p*-(1 → 1)-D-Fru*f*		$C_{12}H_{22}O_{11}$	—	+49	72
O-α-D-Glc*p*-(1 → 3)-D-Fru (Turanose)		$C_{12}H_{22}O_{11}$	157	+22 → +75.3	45, 73
	Octaacetate I	$C_{28}H_{38}O_{19}$	216–217	+20.5 ($CHCl_3$)	74
	Octaacetate II	$C_{28}H_{38}O_{19}$	158	+107 ($CHCl_3$)	74
O-α-D-Glc*p*-(1 → 4)-D-Fru (Maltulose)		$C_{12}H_{22}O_{11}$	113–115d	+58 → +64	75
O-α-D-Glc*p*-(1 → 5)-D-Fru (Leucrose)		$C_{12}H_{22}O_{11}$	161–163	–8.8 → –6.8	76
	Heptaacetate	$C_{26}H_{36}O_{18}$	150–151	—	76
O-α-D-Glc*p*-(1 → 6)-D-	Phenylosazone	$C_{24}H_{34}O_4O_9$	186–188	—	77
		$C_{12}H_{22}O_{11}$	—	+97.2	72, 78
Fru*f* (palatinose, Isomaltulose)	Phenylosazone	$C_{24}H_{34}N_4O_9$	173–175	—	72, 77
O-β-D-Glc*p*-(1 → 3)-D-Fuc		$C_{12}H_{22}O_{10}$	250–252	+45	13
O-α-D-Glc*p*-(1 → 1)-D-GAL*f*		$C_{12}H_{22}O_{11}$	No constants given	—	79
O-β-D-Glc*p*-(1 → 6)-D-Gal		$C_{12}H_{22}O_{11}$	128–130	+15 ±1	80, 328
	Octaacetate	$C_{28}H_{38}O_{19}$	—	+1 ($CHCl_3$)	81
O-α-Glc*p*-(1 → 1)-α-D-Glc*p* (α,α-Trehalose)		$C_{12}H_{22}O_{11}$	214–216	+199	7, 82
	Dihydrate	$C_{12}H_{22}O_{11} \cdot 2H_2O$	94–100	+180	82
	Octaacetate	$C_{28}H_{38}O_{19}$	98	+162.3 (*c* 10) ($CHCl_3$)	8
O-β-Glc*p*-(1 → 1)-β-D-Glc*p* (β,β-Trehalose)		$C_{12}H_{22}O_{11}$	130–135	–41.5	83, 84
	Octaacetate	$C_{28}H_{38}O_{19}$	181	–19 ($CHCl_3$)	85
O-α-D-Glc*p*-(1 → 1)-β-D-Glc*p*		$C_{12}H_{22}O_{11}$	80d, 150–153	+70 (*c* 0.2)	84, 86
	Octaacetate	$C_{28}H_{38}O_{19}$	120, 140	+67, +82 ($CHCl_3$)	84–86
O-α-D-Glc*p*-(1 → 2)-D-Glc (Kojibiose)		$C_{12}H_{22}O_{11}$	175	+135	87, 88
	α-Octaacetate	$C_{28}H_{38}O_{19}$	166	+153 ($CHCl_3$)	89
	β-Octaacetate	$C_{28}H_{38}O_{19}$	118	+112 ($CHCl_3$)	90
O-β-D-Glc*p*-(1 → 2)-D-Glc (Sophorose)		$C_{12}H_{22}O_{11}$	195–196	+20	83, 91, 92
	β-Octaacetate	$C_{28}H_{38}O_{19}$	191–192	–8 ($CHCl_3$)	91–93
	α-Octaacetate	$C_{28}H_{38}O_{19}$	111	+45 ($CHCl_3$)	94
O-α-D-Glc*p*-(1 → 3)-D-Glc (Nigerose, Sakebiose)		$C_{12}H_{22}O_{11}$	—	+135, +145	89, 95, 96
	Octaacetate	$C_{28}H_{38}O_{19}$	151–153	+80 ($CHCl_3$)	97
O-β-D-Glc*p*-(1 → 3)-α-D-Glc (α-Laminaribiose)		$C_{12}H_{22}O_{11}$	202–205	+25.5 → +17.5	98
	α-Octaacetate ethanolate	$C_{28}H_{38}O_{19} \cdot C_2H_5OH$	77–78	+20 ($CHCl_3$)	99
O-β-D-Glc*p*-(1 → 3)-β-D-Glc (β-Laminaribiose)		$C_{12}H_{22}O_{11}$	188–192	+7.5 → +20.8	100
	β-Octaacetate	$C_{28}H_{38}O_{19}$	160–161	–28.6 (17°) ($CHCl_3$)	101
O-β-D-Glc*p*-(1 → 4)-β-D-Glc*p* (β-Cellobiose)		$C_{12}H_{22}O_{11}$	225	+14.2 → +34.6 (*c* 8)	83, 102
	β-Octaacetate	$C_{28}H_{38}O_{19}$	202	–14.7 ($CHCl_3$)	42
	α-Octaacetate	$C_{28}H_{38}O_{19}$	229	+41.0 ($CHCl_3$)	42
O-α-D-Glc*p*-(1 → 4)-α-D-Glc*p* (α-Maltose)		$C_{12}H_{22}O_{11}$	108	+173	103, 104
	α-Octaacetate	$C_{28}H_{39}O_{19}$	125	+123 ($CHCl_3$)	42
O-α-D-Glc*p*-(1 → 4)-β-D-Glc*p* (β-Maltose)	Monohydrate	$C_{12}H_{22}O_{11} \cdot H_2O$	102–103	+112 → +130	103, 104
	β-Octaacetate	$C_{28}H_{38}O_{19}$	159–160	+62.6 ($CHCl_3$)	42
O-α-D-Glc*p*-(1 → 6)-Glc*p* (Isomaltose)		$C_{12}H_{22}O_{11}$	Amorph	+103.2, +122	105, 106
	β-Octaacetate	$C_{28}H_{38}O_{19}$	144–145	+96.9 ($CHCl_3$)	105
	α-Octaacetate				
O-β-D-Glc*p*-(1 → 6)-α-D-Glc*p* (α-Gentiobiose)	Dimethanolate	$C_{12}H_{22}O_{11} \cdot 2CH_3 OH$	85–86	+31 → +9.6	7

OLIGOSACCHARIDES (INCLUDING DISACCHARIDES) (Continued)

Substance[a] (Synonym) (A)	Derivative (B)	Chemical Formula (C)	Melting Point °C (D)	Specific Rotation[b] $[\alpha]_D$ (E)	Reference (F)
DISACCHARIDES (Continued)					
		$C_{28}H_{38}O_{19}$	189	+52.4 (CHCl$_3$)	107, 108
O-β-D-Glcp-(1 → 6)-β-D-Glcp (β-Gentiobiose)		$C_{12}H_{22}O_{11}$	190	–3 → +10.5	83, 109
	β-Octaacetate	$C_{28}H_{38}O_{19}$	193	–5.4 (c 6, CHCl$_3$)	107, 108
O-D-Glcp-(1 → 4)-D-2d-Glcp		$C_{12}H_{22}O_{10}$	204–205	+23,+123	110, 327
O-D-Glcp-(1 → ?)-3-O-methyl-D-6d-Glcp (Kondurangobiose)		$C_{13}H_{24}O_{10}$	202–204d	+20	111
	Hexaacetate	$C_{25}H_{36}O_{16}$	188	—	112
O-D-Glcp-(1 → ?)-3-O-methyl-D-6d-Glcp (Thevetobioside)		$C_{13}H_{24}O_{10}$	233	–64.4 (CH$_3$OH)	113
	Hexaacetate	$C_{25}H_{36}O_{16}$	168–170	–70°(18°) (CHCl$_3$)	113
O-α-D-Glcp-(1 → 1)-α-D-GlcNp (Trehalosamine)	Hydrochloride	$C_{12}H_{23}NO_{10} \cdot HCl$	197d	+176 (c 0.02)	114, 330
	N-Acetyl heptaacetate	$C_{28}H_{39}NO_{18}$	100–102	+152 (CHCl$_3$)	114, 330
O-D-Glcp-(1 → 4)-D-GlcNp		$C_{12}H_{23}NO_{10}$	180–185d	+147,+100 → +81	110, 115
	N-Acetyl derivative	$C_{14}H_{25}NO_{11}$	144.5–146	+85 → +39	115
O-β-D-Glcp-(1 → 4)-D-GlcNAcp		$C_{14}H_{25}O_{11}$	160	+23	327
O-D-Glcp-(1 → 2)-D-GlcUA		$C_{12}H_{20}O_{12}$	—	+89	116
O-α-D-Glcp-(1 → 1)-glycerol		$C_9H_{18}O_8$	—	+128 (27°) (c 0.7)	117
	Hexaacetate	$C_{21}H_{30}O_{14}$	92	+110 (27°) (acetone)	117
O-β-D-Glcp-(1 → 1)-glycerol		$C_9H_{18}O_8$	—	+12,–32 (c 0.5)	118, 331
	Hexa-p-nitrobenzoate	$C_{51}H_{36}N_6O_{26}$	105–108	—	119
	Hexaacetate	$C_{21}H_{30}O_{14}$	144	–4.4 ± 1 (c 0.4, CHCl$_3$)	331
O-D-Glcp-(1 → 1)-myo-inositol		$C_{12}H_{22}O_{11}$	256–260	–17.5	63
	Nonamethyl ether	$C_{21}H_{40}O_{11}$	95–98	–10 (C$_2$H$_5$OH)	63
O-D-Glcp-(1 → 5)-myo-inositol		$C_{12}H_{22}O_{11}$	158–162	–20	63
O-β-D-Glcp-(1 → 1)-mannitol		$C_{12}H_{24}O_{11}$	140–141	–18	120
	Nonabenzoate	$C_{75}H_{60}O_{20}$	88–94	+39.7 (18°) (CHCl$_3$)	101
O-β-D-Glcp-(1 → 3)-mannitol		$C_{12}H_{24}O_{11}$	97–100	–6	121
O-β-D-Glcp-(1 → 4)-D-Manp		$C_{12}H_{22}O_{11}$	179–182	+6	122
	Monohydrate	$C_{12}H_{22}O_{11} \cdot H_2O$	133–135	+5.5	123
	Octaacetate	$C_{28}H_{38}O_{19}$	201	+34 (CHCl$_3$)	124
O-D-Glc-(1 → 2)-L-Rha (Bryobioside)		$C_{12}H_{22}O_{10}$	175, 188–190	+61.5,+ 66.1	125, 126
	Heptaacetate	$C_{26}H_{36}O_{17}$	127	+67.2 (CHCl$_3$)	126
O-α-D-Glcp-(1 → 1)-L-Sor		$C_{12}H_{22}O_{11}$	178–180	+33	127
	Octaacetate	$C_{28}H_{38}O_{19}$	—	+38 (CHCl$_3$)	127
O-α-D-Glcp-(1 → 2)-β-D-threo-pentuloside (Glucoxyluloside)		$C_{11}H_{20}O_{10}$	156–157	+43	128
	Heptaacetate	$C_{25}H_{34}O_{17}$	180–181	+22 (CHCl$_3$)	128
O-β-D-Glcp-(1 → 2)-D-Xyl		$C_{11}H_{20}O_{10}$	202–206	+7	13
	Phenylosazone	$C_{23}H_{32}N_4O_8$	213–215	—	129
O-α-D-Glcp-(1 → 3)-D-Xyl		$C_{11}H_{20}O_{10}$	—	+87.5	130
O-β-D-Glcp-(1 → 3)-D-Xyl		$C_{11}H_{20}O_{10}$	120	–6.4	129, 336
	Phenylosazone	$C_{23}H_{32}N_4O_8$	204–206	—	332
O-α-D-Glcp-(1 → 4)-D-Xyl	Dihydrate	$C_{11}H_{20}O_{10} \cdot 2H_2O$	78, s 58	+97.5	131
O-β-D-Glcp-(1 → 4)-D-Xyl (Securidabiose)		$C_{11}H_{20}O_{10}$	—	–7, –27.3	327, 329, 333
	Heptaacetate	$C_{25}H_{34}O_{17}$	154–157	–18.2 (CHCl$_3$)	333
O-3-O-Methyl-D-Glcp-(1 → 4)-D-Glc		$C_{13}H_{24}O_{11}$	No constants known	—	334
O-β-D-GlcNAcp-(1 → 3)-D-Gal (Lacto-N-biose II)		$C_{14}H_{25}NO_{11}$	131–133	+45.5 → +35.7	132
	Heptaacetate	$C_{28}H_{39}NO_{18}$	212–213	+79.5	335
O-α-D-GlcNAcp-(1 → 6)-D-Galp		$C_{14}H_{25}NO_{11}$	—	+125.6 (c 0.2)	133
O-β-D-GlcNAcp-(1 → 6)-D-Galp		$C_{14}H_{25}NO_{11}$	—	+9.2	132, 133
	Heptaacetate	$C_{28}H_{39}NO_{18}$	197–198	+6.3 (CHCl$_3$)	134
O-β-D-GlcNAcp-(1 → 4)-3-O-(D-1-carboxyethyl)-D-GlcNac		$C_{19}H_{32}N_2O_{13}$	—	+10 → +4.5	135
	Pentaacetate methyl ester	$C_{30}H_{44}N_2O_{18}$	235–236	+40 (c 0.2, CHCl$_3$)	135
O-β-D-GlcNAcp-(1 → 4)-6-O-acetyl-3-O-(D-1-carboxyethyl)-D-GlcNAc		$C_{21}H_{34}N_2O_{14}$	—	+23.5	136

OLIGOSACCHARIDES (INCLUDING DISACCHARIDES) (Continued)

Substance[a] (Synonym) (A)	Derivative (B)	Chemical Formula (C)	Melting Point °C (D)	Specific Rotation[b] $[\alpha]_D$ (E)	Reference (F)
DISACCHARIDES (Continued)					
O-β-D-GlcNAcp-(1 → 6)-3-O-(D-1-carboxyethyl)-D-GlcNAc		$C_{19}H_{32}N_2O_{13}$	—	+16 → +14	135
	Pentaacetate methyl ester	$C_{30}H_{44}N_2O_{18}$	240–241	+40 (c 0.4, CHCl₃)	135
	Methyl α-glycoside methyl ester tetraacetate	$C_{29}H_{44}N_2O_{17}$	288–289	+54 (CHCl₃)	137
O-β-D-GlcNAcp-(1 → 4)-α-D-GlcNAcp (N,N′-Diacetylchitobiose)		$C_{16}H_{28}N_2O_{11}$	260–262d	+16.6	138
	Hexaacetate	$C_{28}H_{40}N_2O_{17}$	308–309d	+55 (18°) (c 0.5, CH₃COOH)	139
O-D-GlcNAc-(1 → 4)-D-GlcUA		$C_{14}H_{23}NO_{12}$	—	+106	140, 141
O-3-O-(D-1-carboxyethyl)-β-D-GlcNAcp-(1 → 4)-D-GlcNAcp (N-acetylmuramyl-D-glucose)		$C_{19}H_{32}N_2O_{13}$	—	+12.4 (c 0.5)	142
O-6-O-acetyl-3-O-(D-1-carboxyethyl)-β-D-GlcNAcp-(1 → 4)-D-GlcNAcp		$C_{21}H_{34}N_2O_{14}$	—	+15.4 (c 0.5)	142
O-β-D-GlcUAp-(1 → 3)-D-GlcNac (N-acetylchondrosine)		$C_{14}H_{23}NO_{12}$	—	+35	143, 144
	Heptaacetate	$C_{28}H_{37}NO_{19}$	221–222	–2.1 (CHCl₃)	144
	Methyl ester hydrochloride	$C_{13}H_{23}NO_{11} \cdot HCl$	155–156	+39 (CH₃OH)	145
O-β-D-GlcUAp-(1 → 3)-D-GalNAc 6-sulfate	Barium salt	$BaC_{14}H_{22}NO_{14}S$	—	–4.4 (c 0.2)	143, 337
O-β-D-GlcUAp-(1 → 3)-D-GlcNAc (N-acetylhyalobiuronic acid)		$C_{14}H_{23}NO_{12}$	—	–32 (28°)	146
	Methyl α-glycoside methyl ester pentaacetate	$C_{26}H_{37}NO_{17}$	236–238	+30 (c 0.7, CHCl₃)	147
O-4,5-Dideoxy-β-L-*threo*-hex-4-enopyranosyluronic acid-(1 → 3)-D-GlcNAc		$C_{14}H_{21}NO_{11}$	—	–20	148
	Pentaacetate	$C_{24}H_{31}NO_{16}$	190–192	—	148
O-2,6-Diamino-2,6-dideoxy-D-*arabino*-hexopyranosyl-(1 → ?)-deoxystreptamine (Neamine, Neomycin A)		$C_{12}H_{26}N_4O_6$	256d	+123 (c 0.5)	149
	Hydrochloride	$C_{12}H_{26}N_4O_6 \cdot 4HCl$	233d	+82	150, 338
	N-Salicylidene derivative	$C_{40}H_{42}N_4O_{10}$	198–201	—	151
O- {2-Amino-4-[(1-carboxyformimidoyl)amino]-2,3,4,6-tetradeoxy-D-hexopyranosyl}-(1 → ?)-(+)-inositol (Kasugamycin)		$C_{14}H_{25}N_3O_{10}$	206–210d	+125	152
O-α-D-*ribo*-Hexopyranosyl-3-ulose-(1 → 2)-β-D-Fruf-(3-Keto-sucrose)	Trihydrate	$C_{12}H_{20}O_{11} \cdot 3H_2O$	82–83	+40 (c 7.5)	153
O-α-D-*ribo*-Hexopyranosyl-3-ulose-(1 → 1)-α-D-Glcp (3-Keto-trehalose)		$C_{12}H_{20}O_{11}$	114	+151.1	154
O-α-D-*ribo*-Hexopyranosyl-3-ulose-(1 → 4)-D-Glcp (3-Keto-maltose)		$C_{12}H_{20}O_{11}$	107	+87.2	154
O-β-D-*ribo*-Hexopyranosyl-3-ulose-(1 → 4)-D-Glcp (3-Keto-cellobiose)		$C_{12}H_{20}O_{11}$	135	+19.9	339
O-β-D-*xylo*-Hexopyranosyl-3-ulose-(1 → 4)-D-Glcp (3-Keto-lactose)		$C_{12}H_{20}O_{11}$	118–120	+39.2	154
O-β-D-Manp-(1 → 4)-*meso*-erythritol		$C_{10}H_{20}O_9$	160–162	–38	155, 156
O-D-Manp-(1 → 1)-D-GlcNp		$C_{12}H_{23}NO_{10}$	No constants known	—	340
	Hydrochloride	$C_{12}H_{23}NO_{10} \cdot HCl$	b 230	+91.3	340
	N-Acetyl octaacetate	$C_{28}H_{39}NO_{18}$	91.5–93	+250 (CHCl₃)	340
O-β-D-Manp-(1 → 4)-α-D-Glcp		$C_{12}H_{22}O_{11}$	210–212	+12,+18	123, 157
	Octaacetate	$C_{28}H_{38}O_{19}$	162–168	—	158
O-α-D-Manp-(1 → ?)-glyceric acid		$C_9H_{16}O_9$	88–89	+105 (15°)	159
	Sodium salt	$C_9H_{15}O_9Na$	270d	—	159
O-β-D-Manp-(1 → 4)-α-D-Manp		$C_{12}H_{22}O_{11}$	204	–5 → –8	160
	Octamethyl ether	$C_{20}H_{38}O_{11}$	—	–12 (CHCl₃)	161
O-β-D-Manp-(1 → 4)-β-D-Manp		$C_{12}H_{22}O_{11}$	193–194	–7.7 → –2.2	162
	Monohydrate	$C_{12}H_{22}O_{11} \cdot H_2O$	122–124	—	123
O-4,5-Anhydro-β-D-ManUAp-(1 → 4)-D-ManUA		$C_{12}H_{16}O_{12}$	135–136.5d	–8	163
O-β-L-Rhap-(1 → 6)-D-Galp (Robinobiose)		$C_{12}H_{22}O_{10}$	Amorph	+2.70	164
	Heptaacetate	$C_{26}H_{36}O_{17}$	113	–9.9, –19.2 (CHCl₃)	165

OLIGOSACCHARIDES (INCLUDING DISACCHARIDES) (Continued)

Substance[a] (Synonym) (A)	Derivative (B)	Chemical Formula (C)	Melting Point °C (D)	Specific Rotation[b] [α]$_D$ (E)	Reference (F)
DISACCHARIDES (Continued)					
O-β-L-Rhap-(1 → 6)-D-Glcp (Rutinose)		$C_{12}H_{22}O_{10}$	189 192d, Amorph	+3.2 → −0.8	166, 167
	Heptaacetate	$C_{26}H_{36}O_{17}$	168–169	−29.7 (CHCl₃)	168
O-α-D-Xylp-(1 → 2)-β-D-Fruf		$C_{11}H_{20}O_{10}$	—	+62	169
O-D-Xylp-(1 → 1)-D-Glcp dibenzoate		$C_{25}H_{28}O_{12}$	147–148	−106.7 (CH₃OH)	1, 2, 170
	Pentaacetate	$C_{35}H_{38}O_{17}$	203	—	1, 170
O-D-Xylp-(1 → 4)-D-Glcp		$C_{11}H_{20}O_{10}$	—	+70	123, 171
O-α-D-Xylp-(1 → 6)-D-Glc		$C_{11}H_{20}O_{10}$	—	+122	341
O-β-D-Xylp-(1 → 6)-D-Glc (Primeverose)		$C_{11}H_{20}O_{10}$	208	+24.1 → −3.3	10, 66, 172
	Heptaacetate	$C_{25}H_{34}O_{17}$	216	−23.5 (CHCl₃)	172
O-β-D-Xylp-(1 → 3)-D-Xyl (Rhodymenabiose)		$C_{10}H_{18}O_9$	192–193	−35 → −22	173, 174
O-β-D-Xylp-(1 → 4)-D-Xylp (Xylobiose)		$C_{10}H_{18}O_9$	195–197	−40 → −27	173, 175
	Hexaacetate	$C_{22}H_{30}O_{15}$	155–156	−75 (c 10, CHCl₃)	176
TRISACCHARIDES					
O-L-Araf-(1 → 3)-O-D-Xylp-(1 → 4)-D-Xylp		$C_{15}H_{26}O_{13}$	—	−19.3	177
	Octamethyl ether	$C_{23}H_{42}O_{13}$	—	−13.3 (CH₃OH)	177, 178
O-D-Fruf-(2 → 1)-O-D-Fruf-(2 → 1)-D-Fru (Inulotriose)		$C_{18}H_{32}O_{16}$	—	No constants known	179
O-D-Fruf-(2 → ?)-O-D-Fruf-(2 → 2)-D-Fruf-anhydride (Polygontin)		$C_{18}H_{30}O_{15}$	207–208	−52.9	3
	Nonaacetate	$C_{36}H_{48}O_{24}$	84–85	−38.4 (CHCl₃)	3
O-D-Fruf-(2 → ?)-O-D-Fruf-(2 → 2)-D-Fruf- (Trifructan)		$C_{18}H_{32}O_{16}$	—	−22.3	6
	Decaacetate	$C_{38}H_{52}O_{26}$	—	+8.5	6
O-D-Fruf-(2 → ?)-O-DFruf-(2 → 1)-dFru		$C_{18}H_{32}O_{16}$	—	−10.2	180
O-D-Fruf-(2 → ?)-O-D-Fruf-(2 → ?)-D-Gal (Labiose)		$C_{18}H_{32}O_{16}$	s 126–128, 205d	+136.7	181
	Hendecaacetate	$C_{40}H_{54}O_{27}$	88	+122.5 (CHCl₃)	181
O-D-Fruf-(2 → 1)-O-D-Fruf-(2 → 1)-D-Glcp		$C_{18}H_{32}O_{16}$	193	+28, +33	11, 182–184
O-D-Fruf-(2 → 6)-O-D-Fruf-(2 → 1)-D-Glcp		$C_{18}H_{32}O_{16}$	143	+22	11, 185, 342
O-β-D-Fruf-(2 → 6)-O-α-D-Glcp-(1 → 2)-β-D-Fruf (Neokestose)		$C_{18}H_{32}O_{16}$	—	+15, +22.2	185–187
	Hendecamethyl ether	$C_{29}H_{54}O_{16}$	—	−28 (c 10)	187
O-α-D-Fucp-(1 → 2)-O-β-D-Galp-(1 → 4)-D-Galp (Fucisido-lactose)		$C_{18}H_{32}O_{15}$	230–213d	−53.5 → −57.5 (c 0.2)	51
	p-Tolylsulfonylhydrazone	$C_{25}H_{40}N_2O_{17}S$	205–206	−73 (C₅H₅N:H₂O)	51
O-α-D-Galp-(1 → 1)-O-β-D-Fruf-(2 → 1)-D-Glcp		$C_{18}H_{32}O_{16}$	—	+131	188
O-α-D-Galp-(1 → 3)-O-β-D-Fruf-(2 → 1)-D-Glcp		$C_{18}H_{32}O_{16}$	—	+98	188
O-β-D-Galp-(1 → 4)-O-β-D-Fruf-(2 → 1)-D-Glcp		$C_{18}H_{32}O_{16}$	—	+44.1	189
O-α-D-Galp-(1 → 6)-O-β-D-Fruf-(2 → 1)-D-Glcp (Planteose)	Dihydrate	$C_{18}H_{32}O_{16} \cdot 2H_2O$	123–124	+125.2	190
	Hendecaacetate	$C_{40}H_{54}O_{27}$	135	+97 (CHCl₃)	190
O-β-D-Galp-(1 → 3)-O-[α-L-Fucp-(1 → 4)]-D-GlcNAcp		$C_{20}H_{35}NO_{15}$	—	−44 ± 3 (c 0.3)	191
O-β-D-Galp-(1 → 6)-O-β-D-Galp-(1 → ?)-D-Fru		$C_{18}H_{32}O_{16}$	—	−28	24
O-D-Galp-(1 → ?)-O-D-Galp-(1 → ?)-D-Glc (Lactotriose)		$C_{18}H_{32}O_{16}$	No constants known	—	192
	Hendecaacetate	$C_{40}H_{54}O_{27}$	120–122	—	192
O-β-D-Galp-(1 → 3)-O-β-D-Galp-(1 → 4)-α-D-Glcp	Trihydrate	$C_{18}H_{32}O_{16} \cdot 3H_2O$	197–200	+56 → +43	15, 24
	Hendecaacetate	$C_{40}H_{54}O_{27}$	108–110	+17.2 (c 0.6, CHCl₃)	343
O-β-D-Galp-(1 → 4)-O-β-D-Galp-(1 → 4)-D-Glcp		$C_{18}H_{32}O_{16}$	228–231	+68 → +45	13
O-β-D-Galp-(1 → 6)-O-β-D-Gal-(1 → 4)-D-Glcp	Dihydrate	$C_{18}H_{32}O_{16} \cdot 2H_2O$	187, s 167	+34	24, 193
O-α-D-Galp-(1 → 6)-O-α-D-Galp-(1 → 6)-D-Glc (Manninotriose)		$C_{18}H_{32}O_{16}$	Amorph 150	+167	194
	Hendecaacetate	$C_{40}H_{54}O_{27}$	s 105	+135 (C₂H₅OH)	194
O-β-D-Galp-(1 → 6)-O-β-D-Galp-(1 → 6)-D-Glc		$C_{18}H_{32}O_{16}$	No constants known	—	195

OLIGOSACCHARIDES (INCLUDING DISACCHARIDES) (Continued)

Substance[a] (Synonym) (A)	Derivative (B)	Chemical Formula (C)	Melting Point °C (D)	Specific Rotation[b] $[\alpha]_D$ (E)	Reference (F)
TRISACCHARIDES (Continued)					
O-α-D-Galp-(1 → 6)-O-β-D-Galp-(1 → 1)-glycerol		$C_{15}H_{28}O_{13}$	188–189	—	196
			196–198	+90	55
	Nonamethyl ether	$C_{24}H_{46}O_{13}$	Oil	+65 (CHCl$_3$)	344
O-D-Galp-(1 → 6)-O-D-Galp-(1 → 1)-*myo*-inositol		$C_{18}H_{32}O_{16}$	—	+145.1	197
O-α-D-Galp-(1 → 2)-O-α-D-Glcp-(1 → 2)-β-D-Fruf (Umbelliferose)		$C_{18}H_{32}O_{16}$	—	+125	198
O-α-D-Galp-(1 → 3)-O-α-D-Glcp-(1 → 2)-β-D-Fruf		$C_{18}H_{32}O_{16}$	No constants given	—	199
O-β-D-Galp-(1 → 4)-O-α-D-Glcp-(1 → 2)-β-D-Fruf (Lactosucrose)	Pentahydrate	$C_{18}H_{32}O_{16}\cdot 5H_2O$	181, s 150	+59	200
	Hendecaacetate	$C_{40}H_{54}O_{27}$	131	+44 (CHCl$_3$)	201
O-α-D-Galp-(1 → 6)-O-α-D-Glcp-(1 → 2)-β-D-Fruf (Raffinose)	Pentahydrate	$C_{18}H_{32}O_{16}\cdot 5H_2O$	77–78, 118	+101, +123	202
	Hendecaacetate	$C_{40}H_{54}O_{27}$	99–101	+92, +100 (c 8, C$_2$H$_5$OH)	203, 204
O-D-Galp-(1 → 4)-O-D-Glcp-(1 → ?)-L-Fuc		$C_{18}H_{32}O_{15}$	No constants known	—	205
O-β-D-Galp-(1 → 4)-O-[α-D-Glcp-(1 → 2)]-D-Glcp		$C_{18}H_{32}O_{16}$	—	+103	206, 207
O-α-D-Galp-(1 → 2)-O-α-D-Glcp-(1 → 1)-glycerol		$C_{15}H_{28}O_{13}$	171	+170 ± 3	208
	Nonaacetate	$C_{33}H_{46}O_{22}$	97	+145 ± 3(c 0.5, CHCl$_3$)	208
O-α-D-GalUAp-(1 → 4)-O-α-D-GalUAp-(1 → 4)-O-D-GalUAp (Trigalacturonic acid		$C_{18}H_{26}O_{19}$	135–142d	+154, + 187	69, 209
O-α-D-Glcp(1 → 2)-O-β-D-Fruf-(1 → 2)-β-D-Fruf (Isokestose, 1-Kestose)		$C_{18}H_{32}O_{16}$	82–88d, 90–92, 105–110	+25	184, 210
			148	+29.3	211
			200–201	+28.9	212
	Hendecamethyl ether	$C_{29}H_{54}O_{16}$	—	+27.9	211–214
O-α-D-Glcp(1 → 2)-O-β-D-Fruf-(6 → 2)-β-D-Fruf (Kestose)		$C_{18}H_{32}O_{16}$	145	+28	212, 215
	Hendecamethyl ether	$C_{29}H_{54}O_{16}$	—	+25.8 (18°)	215
O-α-D-Glcp-(1 → 2)-O-β-D-Fruf-(3 → 1)-α-D-Galp		$C_{18}H_{32}O_{16}$	No constants known	—	216
O-α-D-Glcp-(1 → 3)-O-β-D-Fruf-(2 → 1)-α-D-Glcp (Melezitose)	Dihydrate	$C_{18}H_{32}O_{16}\cdot 2H_2O$	153–154	+88.2	217, 218
	Hendecaacetate	$C_{40}H_{54}O_{27}$	117	+103.6 (CHCl$_3$)	219
O-α-D-Glcp-(1 → 4)-O-α-D-Glcp-(1 → 2)-β-D-Fruf (Erlose)		$C_{18}G_{32}O_{16}$	—	+121.8	220, 221
	Hendecaacetate	$C_{40}H_{54}O_{27}$	68–73	+92.7 (15°) (CHCl$_3$)	222
O-α-D-Glcp-(1 → 6)-O-α-D-Glcp-(1 → 2)-β-D-Fruf		$C_{18}H_{32}O_{16}$	118–120	+102.5 (18°)	223, 224
O-β-D-Glcp-(1 → 6)-O-α-D-Glcp-(1 → 2)-β-D-Fruf (Gentianose)		$C_{18}H_{32}O_{16}$	210	+33.4	225, 226
O-α-D-Glcp-(1 → 6)-O-α-D-Glcp(1 → 5)-D-Fruf (5-O-α-Isomaltosylfructose)		$C_{18}H_{32}O_{16}$	No constants known	—	77
O-α-D-Glcp-(1 → 6)-O-α-D-Glcp(1 → 6)-D-Fruf (Isomaltotriulose)		$C_{18}H_{32}O_{16}$	—	+118	72
O-α-D-Glcp-(1 → 3)-O-β-D-Glcp-(1 → 1)-α-D-Glcp		$C_{18}H_{32}O_{16}$	—	+139	227
	Hendecaacetate	$C_{40}H_{54}O_{27}$	183–184	+110.5 (CHCl$_3$)	345
O-β-D-Glcp-(1 → 3)-O-β-D-Glcp-(1 → 3)-β-D-Glcp(Laminaritriose)		$C_{18}O_{32}O_{16}$	—	+2.4 (17°)	101, 228
	Hendecaacetate	$C_{40}H_{54}O_{27}$	120–121	−40 (CHCl$_3$)	101
O-α-D-Glcp-(1 → 3)-O-α-D-Glcp-(1 → 4)-D-Glcp		$C_{18}H_{32}O_{16}$	—	+169.5	231, 232
O-β-D-Glcp-(1 → 3)-O-β-D-Glcp-(1 → 4)-α-D-Glcp		$C_{18}H_{32}O_{16}$	229–231	+18.7 → +13	229, 230
	Hendecaacetate	$C_{40}H_{54}O_{27}$	121–123	−22.2 (CHCl$_3$)	230

OLIGOSACCHARIDES (INCLUDING DISACCHARIDES) (Continued)

Substance[a] (Synonym) (A)	Derivative (B)	Chemical Formula (C)	Melting Point °C (D)	Specific Rotation[b] $[\alpha]_D$ (E)	Reference (F)
TRISACCHARIDES (Continued)					
O-α-D-Glc$p$-(1 → 2)-O[α-D-Glc$p$-(1 → 4)]-D-Glc$p$		$C_{18}H_{32}O_{16}$	—	+140 (27°)	346
O-α-D-Glc$p$-(1 → 2)-O-[β-D-Glc$p$-(1 → 4)]-D-Glc$p$		$C_{18}H_{32}O_{16}$	—	+93	88
O-α-D-Glc$p$-(1 → 4)-O-α-D-Glc$p$-(1 → 3)]-D-Glc		$C_{18}H_{32}O_{16}$	No constants known	—	231, 232
O-β-D-Glc$p$-(1 → 4)-O-β-D-Glc$p$-(1 → 3)]-D-Glc		$C_{18}H_{32}O_{16}$	236–239	+16.5 → +11.7	229
	Hendecaacetate I	$C_{40}H_{54}O_{27}$	108–110	−8.3 (CHCl$_3$)	230
	Hendecaacetate II	$C_{40}H_{54}O_{27}$	186–188	−24 (CHCl$_3$)	233
O-α-D-Glc$p$-(1 → 4)-O-α-Glc$p$-(1 → 4)]-D-Glc$p$ (Maltotriose)		$C_{18}H_{32}O_{16}$	150, Amorph	+160	234, 235
	Hendecaacetate	$C_{40}H_{54}O_{27}$	134–136	+86 (CHCl$_3$)	234, 236
O-β-D-Glc$p$-(1 → 4)-O-β-D-Glc$p$-(1 → 4)-D-Glc$p$ (Cellotriose)		$C_{18}H_{32}O_{16}$	—	+25	230, 237
	α-Hendecaacetate	$C_{40}H_{54}O_{27}$	220–222	+22.6 (CHCl$_3$)	238
	β-Hendecaacetate	$C_{40}H_{54}O_{27}$	199.5–200.5	−10.8 (CHCl$_3$)	230
O-α-D-Glc$p$-(1 → 4)-O-α-D-Glc$p$-(1 → 6)-D-Glc		$C_{18}H_{32}O_{16}$	—	+159	239
	Hendecaacetate	$C_{40}H_{54}O_{27}$	139–141	+112 (CHCl$_3$)	239
O-α-D-Glc$p$-(1 → 6)-O-α-D-Glc$p$-(1 → 4)-D-Glc$p$ (Panose)		$C_{18}H_{32}O_{16}$	222–224	+154	222, 240
	O-α-Isomalto-pyranosyl-(1→4)-D-glucitol dodecaacetate	$C_{42}H_{58}O_{28}$	147–149	+120 (0.5% C_2H_5OH in CHCl$_3$)	241
O-α-D-Glc$p$-(1 → 6)-O-α-D-Glc$p$-(1 → 6)-D-Glc (Isomaltotriose, Dextrantriose)		$C_{18}H_{32}O_{16}$	—	+145	106
	Hendecabenzoate	$C_{95}H_{87}O_{27}$	226–227	+131 (CHCl$_3$)	242
O-β-D-Glc$p$-(1 → 6)-O-β-D-Glc$p$-(1 → 6)-D-Glc (Gentiotriose, Luteose)		$C_{18}H_{32}O_{16}$	—	−10.3	243, 244
	Hendecaacetate	$C_{40}H_{54}O_{27}$	214–215	−9.4 (CHCl$_3$)	224
O-α-D-Glc$p$-(1 → 6)-O-α-D-Glc$p$-(1 → 2)-D-GlcUA		$C_{18}H_{30}O_{17}$	—	+110 (c 0.07)	116
O-α-D-Glc$p$-(1 → 2)-O-α-D-Glc$p$-(1 → 1)-glycerol		$C_{15}H_{28}O_{13}$	—	+172 ± 2	347
O-α-D-Glc$p$-(1 → 6)-O-α-D-Glc$p$-(1 → 1)-glycerol		$C_{15}H_{28}O_{13}$	—	+148	117
O-β-D-Glc$p$-(1 → 1)-O-[β-D-Glc$p$-(1 → 6)]-D-mannitol		$C_{18}H_{34}O_{16}$	—	−14	120
	Dodecaacetate	$C_{42}H_{58}O_{28}$	136–138	−7.7 (CHCl$_3$)	245
O-β-D-Glc$p$-(1 → 6)-O-β-D-Glc$p$-(1 → 2)-L-Rha (Bryodulcoside)		$C_{18}H_{32}O_{15}$	No constants known	—	126
	Nonaacetate	$C_{35}H_{30}O_{24}$	172–173	+54 (CHCl$_3$)	126
O-β-D-Glc$p$-(1 → 4)-O-β-D-Man$p$-(1 → 4)-D-Man$p$		$C_{18}H_{32}O_{16}$	—	−9	123, 246
O-3-O-Methyl-D-Glc$p$-(1 → 4)-O-D-Glc$p$-(1 → 4)-D-Glc$p$		$C_{19}H_{34}O_{14}$	No constants known	—	334
O-(N-methyl-α-L-GlcNp)-(1 → 2)-O-(3-C-formyl-α-L-5d-Lyx$f$)-(1 → 1)-2,4-dideoxy-2,4-diguanidino-*scyllo*-inositol (Streptomycin)	Trihydrochloride	$C_{21}H_{37}N_7O_{12} \cdot 3HCl$	—	−86.7	247
O-(N-Methyl-α-L-GlcNp)-(1 → 2)-O-(3-C-hydroxymethyl-β-L-5d-Lyx$f$)-(1 → 1)-(2-O-carbamoyl-4-deoxy-4-guanidino-*scyllo*-inositol (Bluensomycin)		$C_{21}H_{39}N_5O_{14}$	No constants known	—	248
O-β-D-GlcNAcp-(1 → 4)-O-β-D-GlcNAcp-(1 → 4)-D-GlcNAcp (Chitotriose)		$C_{24}H_{41}N_3O_{16}$	304–306d	+3.8 → +2.2	139, 140
		$C_{40}H_{57}N_3O_{24}$	315	+33 (c 0.2, CH$_3$COOH)	249
O-β-D-GlcNAcp-(1 → ?)-O-β-D-GlcNAcp-(1 → 3)-D-GlcNAc	Octaacetate	$C_{22}H_{36}N_2O_{17}$	—	−16	250
O-β-D-GlcNAcp-(1 → 4)-O-N-acetylmuramic acid-(1 → 4)-D-GlcNAcp		$C_{27}H_{45}N_3O_{18}$	No constants known	—	348

OLIGOSACCHARIDES (INCLUDING DISACCHARIDES) (Continued)

Substance[a] (Synonym) (A)	Derivative (B)	Chemical Formula (C)	Melting Point °C (D)	Specific Rotation[b] $[\alpha]_D$ (E)	Reference (F)
TRISACCHARIDES (Continued)					
O-α-3-Amino-3-deoxy-D-Glc*p*-(1 → 6)-O-[6-amino-6-deoxy-α-D-Glc*p*-type (1 → 4)]-1,3-diamino-1,2,3-trideoxy-*scyllo*-inositol (Kanamycin A)		$C_{18}H_{36}N_4O_{11}$	200	+99, +149 (H_2O)	349, 350
				+146 (0.1 $N\ H_2SO_4$)	252
	Tetra-*N*-acetyl kanomycin	$C_{26}H_{44}N_4O_{15}$	250–255d, 280–282d	+115	252, 350
	Tetra-*N*-2,4-di nitrophenylheptaacetate	$C_{56}H_{58}N_{12}O_{34}$	210–213d	+52 (acetone)	349
O-3-Amino-3-deoxy-α-D-Glc*p*-(1 → 6)-O-[α-D-GlcN*p*-(1 → 4)]-1,3-diamino-1,2,3-trideoxy-*scyllo*-inositol (Kanamycin C)		$C_{18}H_{36}N_4O_{11}$	270d	+126, +139	351, 352
	Tetra-*N*-2,4-di-nitrophenylheptaacetate	$C_{56}H_{58}N_{12}O_{34}$	208–211d	+299 (*c* 0.8, acetone)	352
O-3-Amino-3-deoxy-α-D-Glc*p*-(1 → 6)-O-[2,6-diamino-2,6-dideoxy-α-D-Glc*p*-(1 → 4)]-1,3-diamino-1,2,3-trideoxy-*scyllo*-inositol (Kanamycin B)		$C_{18}H_{37}N_5O_{10}$	178–182d	+130 (*c* 0.5)	353
	Penta-*N*-acetyl-kanamycin B	$C_{28}H_{47}N_5O_{15}$	250d	+110	353
	Penta-*N*-2,4-di nitrophenylhexaacetate	$C_{60}H_{59}N_{15}O_{36}$	217–218d	+240 (*c* 0.4, acetone)	354
Destomycin A (an aminoheptosidohexosido-*scyllo*-diamino-dideoxyinositol)		$C_{20}H_{39}N_3O_{14}$	180–190d	+7	253
	Tri-*N*-acetyl destomycin A	$C_{26}H_{45}N_3O_{17}$	240–260d	—	253, 254
		$C_{20}H_{39}N_3O_{14}$	140–200d	+6	253
Destomycin B	Tri-*N*-acetyl destomycin B	$C_{26}H_{45}N_3O_{17}$	220–240d	—	253
O-*N*-Acetylmuramic acid-(1 → 4)-O-D-GlcNAc*p*-(1 → 4)-*N*-acetylmuramic acid		$C_{30}H_{49}N_3O_{20}$	No constants known	—	355
O-(*N*-Acetylneuraminic acid)-(2 → 3)-O-β-D-Gal*p*-(1 → 4)-β-D-Glc*p* (Neuraminyl lactose, Lactaminyl lactose, Sialyl lactose)		$C_{23}H_{39}NO_{19}$	—	+16.8	255
	—		—	+(6 ($CH_3\ SOCH_3$)	256
O-(*N*-Acetylneuraminic acid)-(2 → 3)-O-β-D-Gal*p* 6-sulfate)-(1 → 4)-β-D-Glc*p* (Neuraminyl lactose sulfate)		$C_{23}H_{39}NO_{22}S$	No constants known	—	257
O-(*N*-Acetylneuraminic acid)-(2 → 6)-O-β-D-Gal*p*-(1 → 4)-β-D-Glc*p*-(6-Neuraminyl lactose)		$C_{23}H_{39}NO_{19}$	—	+27.9	255
O-α-D-Man*p*-(1 → 3)-O-α-D-Gla*p*-(1 → 2)-glycerol (Mannosyl floridoside)		$C_{15}H_{28}O_{13}$	Amorph	—	258
	Nonaacetate	$C_{33}H_{46}O_{22}$	153–154	+103 ($CHCl_3$)	258
O-β-D-Man*p*-(1 → 4)-O-β-D-Glc*p*-(1 → 4)-D-Glc*p*		$C_{18}H_{32}O_{16}$	—	+9.5 (28°)	123
O-β-D-Man*p*-(1 → 4)-O-β-D-Man*p*-(1 → 4)-D-Glc*p*		$C_{18}H_{32}O_{16}$	—	–12 ± 3	123, 259
O-D-Man*p*-(1 → 6)-O-D-Man*p*-(1 → 6)-D-Glc*p* (Laevidulinose)		$C_{18}H_{32}O_{16}$	—	–11.5, –15	260, 261
	Hendecaacetate	$C_{40}H_{54}O_{37}$	95–100	+18	260, 261
O-β-D-Man*p*-(1 → 4)-O-β-D-Man*p*-(1 → 4)-α-D-Man*p*		$C_{18}H_{32}O_{16}$	214–216	–23	123
	Monohydrate	$C_{18}H_{32}O_{16}H_2O$	166–167	—	123
	Mannobiosyl mannitol dodecaacetate	$C_{42}H_{58}O_{28}$	112	–20 ± 5($CHCl_3$)	160
Rhamninose		$C_{18}H_{32}O_{14}$	135–140d, Amorph	–41	319
	Hexaacetate	$C_{30}H_{44}O_{20}$	95–100	–30.9 ($C_2H_5\ OH$)	319
O-α-D-Xyl*p*-(1 → 6)-O-β-D-Glc*p*-(1 → 4)-D-Glc*p*		$C_{17}H_{30}O_{15}$	147–150	+150	123
O-D-Xyl*p*-(1 → 3)-O-D-Xyl*p*-(1 → 4)-D-Xyl*p*		$C_{15}H_{26}O_{13}$	225	–52 → –47	173
O-β-D-Xyl*p*-(1 → 4)-O-β-D-Xyl*p*-(1 → 4)-β-D-Xyl*p* (Xylotriose)		$C_{15}H_{26}O_{13}$	215–216	–48.1	262, 263
	Octaacetate	$C_{31}H_{42}O_{21}$	108–109.5	–83 ($CHCl_3$)	263

Handbook of Biochemistry and Molecular Biology

OLIGOSACCHARIDES (INCLUDING DISACCHARIDES) (Continued)

Substance[a] (Synonym) (A)	Derivative (B)	Chemical Formula (C)	Melting Point °C (D)	Specific Rotation[b] $[\alpha]_D$ (E)	Reference (F)
TETRASACCHARIDES					
D-Fructose tetraose (Veronicin) (non-reducing)		$C_{24}H_{42}O_{21}$	170, 188	−29.4	264
	Tetradecaacetate	$C_{52}H_{70}O_{35}$	92	−21.1 (CHCl₃)	264
D-Fructose tetraose		$C_{24}H_{42}O_{21}$	—	−17.3	180
O-D-Fruf-(2 → 1)-O-[d-Fruf-(2 → 2)]-O-[d-Fruf-(2 → 6)]-D-Glcp (Neobifurcose)		$C_{24}H_{42}O_{21}$	—	+14.4, + 16.7	184, 265
	Tridecamethyl ether	$C_{37}H_{68}O_{21}$	—	−35.4	278
[O-D-Fruf-(2 → 1)]₃ d-Glcp (Inulotriosyl glucose)		$C_{24}H_{42}O_{21}$	—	−2	266
O-L-Fucp-(1 → 2)-O-D-Galp-(1 → 4)-O-[l-Fucp-(1 → 3)]-D-Glcp (Lactodifucotetraose)		$C_{24}H_{42}O_{19}$	—	−17.1, −106	205, 267
O-D-Galp-(1 → ?)-O d-Galp-(1 → ?)-O-D-Fruf-(2 → 1)-D-Glcp (Sesamose)		$C_{24}H_{42}O_{21}$	No constants given	—	268
[O-β-D-Galp-(1 → 4)-]₃ d-Glcp		$C_{24}H_{42}O_{21}$	—	+43	13
[O-α-D-Galcp-(1 → 6)-]₂ O-β-D-Galcp-(1 → 1)-glycerol		$C_{21}H_{38}O_{18}$	—	+114	344
	Dodecamethyl ether	$C_{33}H_{62}O_{18}$	—	+79.8 (CHCl₃)	344
[O-α-D-Galp-(1 → 6)-]₂ α-D-Glcp-(1 → 2)-D-Fruf (Stachyose, Manneotetrose)		$C_{24}H_{42}O_{21}$	140s, 170	+131, +146	194, 269
	Tetradecaacetate	$C_{52}H_{70}O_{35}$	95–96, Amorph	+120 (C₂H₅OH)	270
	Tetradecamethyl ether	$C_{38}H_{70}O_{21}$	—	+130 (CHCl₃)	271
O-D-Galp-(1 → 4)-O-D-Galp-(1 → 6)-O-α-D-Glcp-(1 → 2)-D-Fruf		$C_{24}H_{42}O_{21}$	—	+143	272
d-Galactose-3,6-anhydro-L-galactose tetrasaccharide (Neoagarotetrose)	Dihydrate	$C_{24}H_{38}O_{19} \cdot 2H_2O$	104–107	−2.8	273
	Decaacetate	$C_{44}H_{58}O_{29}\,2H_2$	121	−15.8 (CHCl₃)	273
O-α-D-Galp-(1 → 6)-O-α-D-Glcp-(1 → 2)-O-β-D-Fruf-(1 → 1) α-D-Glcp (Lychnose)		$C_{24}H_{42}O_{21}$	—	+155	272
O-α-D-Galp-(1 → 6)-O-α-D-Glcp-(1 → 2)-O-β-D-Fruf-(3 → 1)-α-D-Galp (Isolychnose)		$C_{24}H_{42}O_{21}$	No constants given	—	274
O-β-D-Galp-(1 → 3)-O-β-D-GlcNAcp-(1 → 3)-O-β-D-Galp-(1 → 4)-D-Glcp (Lacto-N-tetraose)		$C_{16}H_{45}NO_{21}$	205d	+25.2	51
O-α-D-GalUAp-(1 → 4)-O-α-D-GalUAp-(1 → 4)-O-α-D-GalUAp-(1 → 4)-D-GalUAp (Tetra-galacturonic acid)	Trihydrate	$C_{14}H_{34}O_{25} \cdot 3H_2O$	160–170d	—	275
O-α-D-Glcp-(1 → 2)-[O-β-D-Fruf-(1 → 2)-]₂ O-β-D-Fruf (Nystose)		$C_{24}H_{42}O_{21}$	131–132	+10.6	276
			110–115	+17.9	277, 278
O-α-D-Glcp-(1 → 2)-O-[β-D-Fruf-(2 → 6)]-O-β-D-Fruf-(1 → 2)-D-Fruf (Bifurcose)		$C_{24}H_{42}O_{21}$	156	+8	211
	Tetradecamethyl ether	$C_{38}H_{70}O_{21}$	—	+3 (CHCl₃)	211
O-α-D-Glcp-(1 → 2)-[O-β-D-Fruf-(6 → 2)-]₂ β-D-Fruf		$C_{24}H_{42}O_{21}$	—	−7	279
[O-α-D-Glcp-(1 → 4)-]₂ O-D-Glcp-(1 → 2)-D-Fruf (Maltosylsucrose)		$C_{24}H_{42}O_{21}$	No constants given	—	221
[O-β-D-Glcp-(1 → 3)-]₃ β-D-Glcp (Laminaritetraose)		$C_{24}H_{42}O_{21}$	—	−5.9	101, 228
	Tetradecaacetate	$C_{52}H_{70}O_{35}$	122–123	−46.2 (CHCl₃)	101
O-α-D-Glcp-(1 → 3)-O-α-D-Glcp-(1 → 4)-O-α-D-Glcp-(1 → 3)-D-Glcp		$C_{24}H_{42}O_{21}$	—	+181 (c 0.7)	356
O-β-D-Glcp-(1 → 3)-[O-β-D-Glcp-(1 → 4)-]₂ D-Glcp	α-Anomer	$C_{24}H_{42}O_{21}$	221–224d	+12.9 → +10.6	280
	β-Anomer · 2H₂O	$C_{24}H_{42}O_{21} \cdot 2H_2O$	180–181	+7 → +10	280
O-β-D-Glcp-(1 → 4)-O-β-D-Glcp-(1 → 3)-O-β-D-Glcp-(1 → 4)-D-Glcp		$C_{24}H_{42}O_{21}$	223–226	+19.8 (90% CH₃COOH)	280, 281
[O-β-D-Glcp-(1 → 4)-]₂ O-β-D-Glcp-(1 → 3)-β-D-Glcp		$C_{24}H_{42}O_{21}$	241–245d	+11.4 → +8.4	280
	Tetradecaacetate	$C_{52}H_{70}O_{35}$	118–121	—	282
[O-α-D-Glcp-(1 → 4)-]₃-D-Glcp (Maltotetraose)		$C_{24}H_{42}O_{21}$	—	+166	235, 283
	Methyl glycoside	$C_{25}H_{44}O_{21}$	—	+213	235
[O-β-D-Glcp-(1 → 4)-]₃ D-Glcp (Cellotetraose)		$C_{24}H_{42}O_{21}$	251	+11.3 → +17	237, 301
	α-Tetradecaacetate	$C_{52}H_{70}O_{35}$	226–227	+12.5 (CHCl₃)	284
	β-Tetradecaacetate	$C_{52}H_{70}O_{35}$	223–225	−18.2 (CHCl₃)	284

OLIGOSACCHARIDES (INCLUDING DISACCHARIDES) (Continued)

Substance[a] (Synonym) (A)	Derivative (B)	Chemical Formula (C)	Melting Point °C (D)	Specific Rotation[b] $[\alpha]_D$ (E)	Reference (F)
TETRASACCHARIDES (Continued)					
O-α-D-Glc$p$-(1 → 6)-[O-α-D-Glc$p$-(1 → 4)-]$_2$ D-Glc$p$		$C_{24}H_{42}O_{21}$	—	+177 (15°) (c 0.5)	285
[O-α-D-Glc$p$-(1 → 6)-]2 O-α-D-Glc$p$-(1 → 4)-D-Glc$p$		$C_{24}H_{42}O_{21}$	—	+164	240
	Methyl glycoside monomethanolate	$C_{25}H_{44}O_{21}$ CH$_3$OH	192–193	+189	286
Glucose tetrasaccharide, cyclic?		$C_{24}H_{40}O_{20}$	—	+168	287
	Dodecaacetate	$C_{48}H_{64}O_{32}$	—	+157	287
[O-α-D-Glc$p$-(1 → 6)-]$_3$ 3-O-methyl-D-Glc		$C_{25}H_{44}O_{21}$	—	+129	357
[O-α-D-Glc$p$-(1 → 6)-]$_2$ O-α-D-Glc$p$-(1 → 1)-glycerol		$C_{21}H_{38}O_{18}$	—	+166 (27°) (c 0.5)	117
O-3-O-methyl-D-Glc$p$-(1 → 4)-[O-D-Glc$p$-(1 → 4)-]$_2$ D-Glc		$C_{25}H_{44}O_{21}$	No constants known	—	334
O-D-GlcNAcp-(1 → 4)-O-N-acetylmuramic acid-(1 → 4)-O-D-GlcNAcp-(1 → 4)-N-acetylmuramic acid		$C_{38}H_{62}N_4O_{35}$	No constants known	—	355
O-β-D-GlcNAcp-(1 → 6)-O-β-N-acetyl-neuraminic acid-(1 → 4)-O-β-D-GlcNAcp-(1 → 6)-N-acetylneuraminic acid		$C_{38}H_{62}N_4O_{25}$	No constants given	—	288
O-β-D-GlcUAp-(1 → 3)-O-D-GlcNAcp (1 → ?)-O-β-D-GlcUAp-(1 → 3)-D-GlcNAc		$C_{28}H_{44}N_2O_{23}$	200d	−41 → −53 (27°)	289
O-4-O-Methyl-α-D-GlcUA-(1 → 2)-[O-β-D-Xyl$p$-(1 → 4)-]$_2$ D-Xyl$p$		$C_{22}H_{36}O_{19}$	—	+23.4	263
O-(2,6-Diamino-2,6-dideoxy-α-arabino-hexopyranosyl)-(1 → 3)-O-β-D-Rib$f$-(1 → 5)-O-[2,6-diamino-2,6-dideoxy-α-D-arabino-hexopyranosyl-(1 → 4)]- 1,3-diamino-1,2,3-trideoxy-$scyllo$-inositol (Neomycin C, Streptothricin B II, Framycetin)		$C_{23}H_{46}N_6O_{13}$	No constants given	—	290
	Hexa-N-acetyl neomycin C	$C_{35}H_{58}N_6O_{19}$	—	+94.5 (28°)	291
O-(2,6-Diamino-2,6-dideoxy-α-D-arabino-hexopyranosyl)-(1 → 3)-O-β-D-Rib$f$-(1 → 5)-O-[α-D-GlcNp-(1 → 4)]-1,3-diamino-1,2,3-trideoxy-$scyllo$-inositol (Paromomycin II)	Penta-N-acetyl	$C_{23}H_{45}N_5O_{14}$	No constants given	—	290
	Paromomycin II	$C_{33}H_{55}N_5O_{19}$	—	+64 (27°)	292
O-(2,6-Diamino-2,6-dideoxy-α-L-xylo-hexopyranosyl)-(1 → 3)-O-β-D-Rib$f$-(1 → 5)-O-[2,6-diamino-2,6-dideoxy-α-D-arabino-hexopyranosyl-(1 → 4)]-1,3-diamino-1,2,3-trideoxy-$scyllo$-inositol (Neomycin B, Streptothricin B I)		$C_{23}H_{46}N_6O_{13}$	No constants given	—	290
	Hexa-N-acetyl neomycin B	$C_{35}H_{58}N_6O_{19}$	—	+47.8 (28°)	291
O-(2,6-Diamino-2,6-dideoxy-α-L-xylo-hexopyranosyl)-(1 → 3)-O-β-D-Rib$f$-(1 → 5)-O-[α-D-GlcNp-(1 → 4)]-1,3-diamino-1,2,3-trideoxy-$scyllo$-inositol (Paromomycin I)		$C_{23}H_{45}N_5O_{14}$	No constants given	—	290
O-α-D-Man$p$-(1 → 4)-O-N-methyl-α-L-GlcNp-(1 → 2)-O-3-C-formyl-β-L-5d-Lyx$f$)-(1 → 1)-2,4-dideoxy-2,4-di-guanidino-$scyllo$-inositol (Streptomycin B)	Trihydrochloride	$C_{27}H_{47}N_7O_{17}$ · 3HCl	179–182d	−47	293
O-β-D-Xyl$p$-(1 → 4)-O-[α-L-Ara$f$-(1 → 3)]-O-β-D-Xyl$p$-(1 → 4)-D-Xyl		$C_{20}H_{34}O_{17}$	—	−75	294
	Decaacetate	$C_{40}H_{54}O_{27}$	179–180	−85 (CHCl$_3$)	294
O-D-Xyl-(1 → 4)-O-D-Xyl-(1 → 3)-O-D-Xyl-(1 → 4)-D-Xyl$p$		$C_{20}H_{34}O_{17}$	—	−56.7	173
[O-β-D-Xyl$p$-(1 → 4)-]$_3$ D-Xyl$p$ (Xylotetraose)		$C_{20}H_{34}O_{17}$	224–226	−61.9	262, 263
	Decaacetate	$C_{40}H_{54}O_{27}$	200–201	−92.4 (CHCl$_3$)	263
Scorodose		$C_{24}H_{42}O_{21}$	200, Amorph	−41.5	295
	Acetate	Unknown	85–90	−28.5 (CHCl$_3$)	295
Pentasaccharides					
Fructose pentasaccharide		$C_{30}H_{52}O_{26}$	—	+8, −23	180, 279
O-Fru$f$-(2 → ?)-stachyose		$C_{30}H_{52}O_{26}$	No constants given	—	179

OLIGOSACCHARIDES (INCLUDING DISACCHARIDES) (Continued)

Substance[a] (Synonym) (A)	Derivative (B)	Chemical Formula (C)	Melting Point °C (D)	Specific Rotation[b] $[\alpha]_D$ (E)	Reference (F)
TETRASACCHARIDES(Continued)					
O-α-L-Fuc$p$-(1 → 2)-(Lacto-N-tetraose) (Lacto-N-fucopentaose I)		$C_{32}H_{55}O_{25}$	216	$-11 \to -16.3$	51
O-β-D-Gal$p$-(1 → 3)-O-[α-L-Fuc$p$-(1 → 4)]-O-β-D-GlcNAcp-(1 → 3)-O-β-D-Gal$p$-(1 → 4)-D-Glc$p$ (Lacto-N-fucopentaose II)		$C_{32}H_{55}O_{25}$	213–215	$-28 \to +30.4$	297
Lactopentaose B		$C_{37}H_{62}N_2O_{28}$	—	+15	298
Lactopentaose C		$C_{37}H_{62}N_2O_{28}$	—	+13	298
[O-α-D-Gal$p$-(1 → 6)-]$_3$ O-α-D-Glc$p$-(1 → 2)-β-D-Fru$f$ (Verbascose)		$C_{30}H_{52}O_{26}$	219–220, 253	+169.9	299, 300
	Heptadecaacetate	$C_{64}H_{86}O_{43}$	132	+130.4	299, 300
O-D-Glc$p$-(1 → 2)-O-[D-Fru$f$-(2 → 6)]-O-D-Fru$f$(1 → 2)-O-D-Fru$f$-(1 → 2)-D-Fru$f$		$C_{30}H_{52}O_{26}$	—	-3.5	211
	Heptadecamethyl ether	$C_{47}H_{86}O_{26}$	—	-10.5 $(CHCl_3)$	211
O-α-D-Glc$p$-(1 → 2)-[O-β-D-Fru$f$-(6 → 2)-]$_3$ β-D-Fru$f$		$C_{30}H_{52}O_{26}$	—	-11.2	279, 296
[O-α-D-Glc$p$-(1 → 4)-]$_4$ -D-Glc$p$ (Maltopentaose)		$C_{30}H_{52}O_{26}$	—	+178	302, 320
[O-β-D-Glc$p$-(1 → 4)-]$_4$ -D-Glc$p$ (Cellopentaose)		$C_{30}H_{52}O_{26}$	No constants given	—	237
	α-Heptadecaacetate	$C_{64}H_{86}O_{43}$	246–249	—	303
	β-Heptadecaacetate	$C_{64}H_{86}O_{43}$	238.5–239	-18.5 $(CHCl_3)$	284
[O-α-D-Glc$p$-(1 → 6)-]$_3$ O-α-D-Glc$p$-(1 → 4)-D-Glc$p$		$C_{30}H_{52}O_{26}$	—	+167 (c 0.7)	240
	Methyl α-glycoside monoethanolate	$C_{31}H_{54}O_{26}$	161–164	+188.7	286
		C_2H_5OH			
O-4-O-Methyl-α-D-GlcUA-(1 → 2)-[O-β-D-Xyl$p$-(1 → 4)-]$_3$ D-Xyl$p$		$C_{27}H_{44}O_{23}$	—	+0.6	263
[O-β-D-Xyl$p$-(1 → 4)-]$_4$ β-D-Xyl$p$ (Xylopentaose)		$C_{25}H_{42}O_{21}$	240–242	-72.9	262, 263
	Dodecaacetate	$C_{49}H_{66}O_{33}$	249–250	-98 $(CHCl_3)$	263
HEXASACCHARIDES					
Frutose hexasaccharides (Arctose) Nonreducing)		$C_{36}H_{62}O_{31}$	178	-41	264
	Eicosaacetate	$C_{76}H_{102}O_{51}$	108	-36.5 (15°) $(CHCl_3)$	264
(Campanulin) (Non-reducing)		$C_{36}H_{62}O_{31}$	170	-23	264
	Eicosaacetate	$C_{76}H_{102}O_{51}$	73	-0.2 $(CH_3 COCH_3)$	264
O-α-L-Fuc$p$-(1 → 4)-O-[β-D-Gal$p$-(1 → 3)-O-β-D-GlcNAcp-(1 → 3)-O-β-D-Gal$p$-(1 → 4)-[O-α-lFuc$p$-(1 → 3)]-D-Glc$p$ (Lacto-N-difucohexaose II)		$C_{38}H_{65}O_{29}$	218–220d	-68.8	191, 304
(O-D-Gal$p$-)$_4$ O-D-Gal$p$-(1 → 2)-D-Fru$f$ (Lycopose) (Non-reducing)		$C_{36}H_{62}O_{31}$	270	+187	305
	Eicosaacetate	$C_{76}H_{102}O_{51}$	150	+174.5 (9°) (CH_3COCH_3)	305
[O-α-D-Gal$p$-(1 → 6)-]$_4$ O-α-D-Glc$p$-(1 → 2)-β-D-Fru$f$ (Ajugose)	Hexahydrate	$C_{36}H_{62}O_{31} \cdot 6H_2O$	204–205	+163	306
O-α-D-Glu$p$-(1 → 2)-[O-β-D-Fru$f$-(6 → 2)-]$_4$ β-D-Fru$f$		$C_{36}H_{62}O_{31}$	—	-19	279
O-D-Glc$p$-(1 → ?)-[O-D-Fru$f$-(? → 2)-]4D-Fru$f$		$C_{36}H_{62}O_{31}$	—	-5.3	278
	Eicosamethyl ether	$C_{56}H_{102}O_{31}$	—	-37	278
[O-α-D-Glc$p$(1 → 4)-]$_5$ D-Glc$p$ (Maltohexaose)		$C_{36}H_{62}O_{31}$	—	+180	302, 320
[O-β-D-Glc$p$(1 → 4)-]$_5$ D-Glc$p$ (Cellohexaose)		$C_{36}H_{62}O_{31}$	229–231	+168 (15°)	237, 307
	α-Eicosaacetate	$C_{76}H_{102}O_{51}$	252–255	—	303
	β-Eicosaacetate	$C_{76}H_{102}O_{51}$	241–243	-18.9 $(CHCl_3)$	284
[O-α-D-Glc$p$-(1 → 4)-]$_6$ (Schardinger α-dextrin, cyclohexaamylose)		$C_{36}H_{60}O_{30}$	—	+151	308
	Octadecaacetate	$C_{72}H_{96}O_{48}$	—	+107 $(CHCl_3)$	309
[O-α,β-D-Glc$p$-(1 → 5)-]$_6$	Dodecahydrate	$C_{36}H_{60}O_{30} \cdot 12H_2O$	290–300	+152	310
	Octadecamethyl ether	$C_{54}H_{96}O_{30}$	98–103	+161	310
O-4-O-Methyl-α-D-GluUA-(1 → 2)-[O-β-D-Xyl$p$-(1 → 4)-]$_4$ D-Xyl$p$		$C_{32}H_{52}O_{27}$	—	-11.8	263

OLIGOSACCHARIDES (INCLUDING DISACCHARIDES) (Continued)

Substance[a] (Synonym) (A)	Derivative (B)	Chemical Formula (C)	Melting Point °C (D)	Specific Rotation[b] $[\alpha]_D$ (E)	Reference (F)
HEXASACCHARIDES (Continued)					
[O-D-Xylp-(1 → 4-)$_2$ O-D-Xylp-(1 → 3)-[O-D-Xylp-(1 → 4)-]$_2$ D-Xylp		$C_{30}H_{50}O_{25}$	169–173	–154 $[\alpha]_{436}$	311
[O-β-D-Xylp-(1 → 4)-]$_5$ D-Xylp (Xylohexaose)	Dihydrate	$C_{30}H_{50}O_{25} \cdot 2H_2O$	236–237	+72.8	262, 312
	Tetradecaacetate	$C_{58}H_{78}O_{39}$	260–261	—	312
HEPTASACCHARIDES					
Fructose heptasaccharide (Asparagose)		$C_{42}H_{72}O_{36}$	—	–35.7	313
[O-α-D-Galp-(1 → 6)-]$_5$ O-α-D-Glcp-(1 → 2)-β-D-Fruf	Tetrahydrate	$C_{42}H_{72}O_{36} \cdot 4H_2O$	246–248	+168	306, 321
[O-α-D-Galp-(1 → 6)-]$_4$ O-α-D-Glcp-(1 → 2)-O-β-D-Fruf-(3 → 1)-α-D-Galp		$C_{42}H_{72}O_{36}$	No constants given	—	314
[O-α-D-Glcp-(1 → 4)-]$_7$ (Schardinger β dextrin, cycloheptaamylose)		$C_{42}H_{70}O_{35}$	—	+162	308
	Heneicosaacetate	$C_{84}H_{112}O_{56}$	—	+121 (CHCl$_3$)	309
O-4-O-Methyl-α-D-GlcUA-(1 → 2)-[O-β-D-Xylp-(1 → 4)-]$_5$ d-Xylp		$C_{37}H_{60}O_{31}$	—	–20.8	263
[O-β-D-Xylp-(1 → 4)-]$_6$ d-Xylp		$C_{35}H_{58}O_{29}$	232–234	–74	262, 312
OCTASACCHARIDES					
[O-α-D-Galp-(1 → 6)-]$_6$ O-α-D-Glcp-(1 → 2)-β-D-Fruf	Tetrahydrate	$C_{48}H_{82}O_{41} \cdot 4H_2O$	267–268	+168	306
Di(lacto-N-tetraose)		$C_{52}H_{88}N_2O_{41}$	No constants given	—	315
[O-α-D-Glcp-(1 → 4)-]$_3$ (Schardinger γ-dextrin, cyclooctaamylose)		$C_{48}H_{80}O_{40}$	—	+180	316
	Tetracosaacetate	$C_{96}H_{128}O_{64}$	—	+137 (CHCl$_3$)	309
O-4-O-Methyl-α-D-GlcUA-(1 → 2)-[O-β-D-Xylp-(1 → 4)-]$_6$ d-Xylp		$C_{42}H_{68}O_{35}$	—	–25.7	263
[O-β-D-Xylp-(1 → 4)-]$_7$ d-Xylp (Xylooctaose)		$C_{40}H_{66}O_{33}$	No constants given	—	262

Compiled by George G. Maher.

[a] In alphabetical order by the sequence (starting at the nonreducing end) of the component monosaccharide glycosyl units constituting the oligosaccharide arranged within the groups—disaccharides, trisaccharides, etc. The oligosaccharides entered are only those which have been as a naturally existing entity or have been found to derive from the known reaction of a known natural enzyme on a known natural carbohydrate substrate under conditions not foreign to natural biological systems.

[b] $[\alpha]_b$ for 1–5 g solute, c per 100 ml aqueous solution at 20–25°C unless otherwise given.

[c] Crystallizes in one of two forms, depending on solvent used.

References

1. Hansson, Johansson, and Lindberg, *Acta Chem. Scand.*, 20, 2358 (1966).
2. Hemming and Ollis, *Chem. Ind.*, p. 85 (1953).
3. Strepkov, *Zh. Obshch. Khim.*, 28, 3143 (1958).
4. Pazur and Gordon, *J. Am. Chem. Soc.*, 75, 3458 (1953).
5. Schlubach and Scheffler, *Justus Liebigs Ann. Chem.*, 588, 192 (1954).
6. Strepkov, *Dokl. Akad. Nauk SSSR.*, 124, 1344 (1959).
7. Bates, *Polarimetry, Saccharimetry and the Sugars*, Nat. Bur. Stand. Circ. C440, U.S. Gov. Print. Off., Washington, D.C., (1942).
8. Hudson and Johnson, *J. Am. Chem. Soc.*, 37, 2748 (1915).
9. Brigl and Scheyer, *Hoppe-Seyler's Z. Physiol. Chem.*, 160, 214 (1926).
10. Wallenfels and Lehmann, *Ber. Dtsch. Chem. Ges.*, 90, 1000 (1957).
11. Bacon, *Biochem. J.*, 57, 320 (1954).
12. Barber, *J. Am. Chem. Soc.*, 81, 3722 (1959).
13. Gorin, Spencer and Phaff, *Can. J. Chem.*, 42, 1341, 2307 (1964).
14. Lindberg, Wachmeister, and Wickberg, *Acta Chem. Scand.*, 6, 1052 (1952).
15. Gorin, Haskins, and Westlake, *Can. J. Chem.*, 44, 2083 (1966).
16. Aspinall, Auret, and Hirst, *J. Chem. Soc.* (Lond.), p. 4408 (1958).
17. Charlson, Nunn, and Stephan, *J. Chem. Soc.* (Lond), p. 269 (1955).
18. Shaw, Stephan, and Fuller, *J. Chem. Soc.* (Lond.), p. 2287 (1965).
19. Goldstein, Smith, and Srivastava, *J. Am. Chem. Soc.*, 79, 3858 (1957).
20. Srivastava and Smith, *J. Am. Chem. Soc.*, 79, 982 (1957).
21. Charlson, Gorin, and Perlin, *Can. J. Chem.*, 34, 1811 (1956).
22. Austin, Hardy, Buchanan, and Baddiley, *J. Chem. Soc.* (Lond), p. 1419 (1965).
23. Feingold, Avigad, and Hestrin, *J. Biol. Chem.*, 224, 295 (1957).
24. Ballio and Russi, *J. Chromatogr.*, 4, 117 (1960).
25. Helferich and Steinpreis, *Ber. Dtsch. Chem. Ges.*, 91, 1794 (1958).
26. Ball and Jones, *J. Chem. Soc.* (Lond.), p. 905 (1958).
27. Masamune and Kamiyama, *Jap. J. Exp. Med.*, 66, 43 (1957).
28. Ingle and Bhide, *J. Indian Chem. Soc.*, 35, 516 (1958).
29. Turton, Bebbington, Dixon, and Pacsu, *J. Am. Chem. Soc.*, 77, 2565 (1955).
30. Meier, *Acta Chem. Scand.*, 16, 2275 (1962).
31. Haq and Adams, *Can. J. Chem.*, 39, 1563 (1961).
32. Hirase and Araki, *Bull. Chem. Soc. Jap.*, 27, 105 (1953).
33. Yoshikawa and Watanabe, *Hyogo Noka Daigaku Kenkyu Hokoku*, 3, 53 (1957); *Chem. Abstr.*, 52, 19198 (1958).
34. Clingman, Nunn, and Stephan, *J. Chem. Soc.* (Lond.), p. 197 (1957).
35. Watkins, *Nature*, 181, 117 (1958).
36. Wickstrom, *Acta Chem. Scand.*, 11, 1473 (1957).
37. Lehmann and Beck, *Justus Liebigs Ann. Chem.*, 630, 56 (1960).
38. Beck and Wallenfels, *Justus Liebigs Ann. Chem.*, 655, 173 (1962).
39. Kuhn and Baer, *Ber. Dtsch. Chem. Ges.*, 87, 1560 (1954).
40. Trey, *Z. Phys. Chem.*, 46, 620 (1903).
41. Gillis, *Recl. Trav. Chim. Pays-Bas*, 39, 88, 677 (1920).
42. Hudson and Johnson, *J. Am. Chem. Soc.*, 37, 1270, 1276 (1915).

43. Tanret, *Z. Phys. Chem.*, 53, 692 (1905).
44. Fletcher and Diehl, *J. Am. Chem. Soc.*, 74, 5774 (1952).
45. Assarson and Theander, *Acta Chem. Scand.*, 12, 1319 (1958).
46. Hudson and Johnson, *J. Am. Chem. Soc.*, 37, 2752 (1915).
47. Pazur, Tipton, Budovich, and Marsh, *J. Am. Chem. Soc.*, 80, 119 (1958).
48. Helferich and Sparmberg, *Ber. Dtsch. Chem. Ges.*, 66, 806 (1933).
49. Polonovski and Lespangol, *C. R. Acad. Sci.*, 192, 1319 (1931); 195, 465 (1932).
50. Bredereck, Wagner, Geissel, Gross, Hutten, and Ott, *Ber. Dtsch. Chem. Ges.*, 95, 3056 (1962).
51. Kuhn, Gauhe, and Baer, *Ber. Dtsch. Chem. Ges.*, 87, 289, 1553 (1954); 88, 1135, 1713 (1955); 89, 2513, 2514 (1956).
52. Glick, Chen, and Zillikin, *J. Biol. Chem.*, 237, 981 (1962).
53. Zilliken, Smith, Rose, and Gyorgy, *J. Biol. Chem.*, 208, 299 (1954); 217, 79 (1955).
54. Kuhn and Kirschenlohr, *Justus Liebigs Ann. Chem.*, 600, 135 (1956).
55. Wickberg, *Acta Chem. Scand.*, 12, 1183, 1187 (1959).
56. Carter, McCluer, and Slifer, *J. Am. Chem. Soc.*, 78, 3735 (1956).
57. Reeves, Latour, and Lousteau, *Biochemistry*, 3, 1248 (1964).
58. Colin, *Bull. Soc. Chim. Fr. Ser. 5*, 4, 277 (1937).
59. Su and Hassid, *Biochemistry*, 1, 468 (1962).
60. Putman and Hassid, *J. Am. Chem. Soc.*, 76, 2221 (1954).
61. Brown and Serro, *J. Am. Chem. Soc.*, 75, 1040 (1953).
62. Kabat, MacDonald, Ballou, and Fischer, *J. Am. Chem. Soc.*, 75, 4507 (1953).
63. Gorin, Horitsu and Spencer, *Can. J. Chem.*, 43, 2259 (1965).
64. Pueyo, *C. R. Acad. Sci.*, 248, 2788 (1959).
65. Lindberg, Silvander, and Wachtmeister, *Acta Chem. Scand.*, 18, 213 (1964).
66. Kooiman, *Recl. Trav. Chim. Pays-Bas*, 80, 849 (1961).
67. Araki and Arai, *Bull. Chem. Soc. Jap.*, 29, 339 (1956).
68. McCready, McComb, and Black, *J. Am. Chem. Soc.*, 76, 3035 (1954).
69. Bhattacharjee and Timell, *Can. J. Chem.*, 43, 758 (1965).
70. Gee, Jones, and McCready, *J. Org. Chem.*, 23, 620 (1958).
71. Nagel and Vaughn, *Arch. Biochem. Biophys.*, 94, 328 (1961).
72. Avigad, *Biochem. J.*, 73, 587 (1959).
73. Hudson and Pacsu, *J. Am. Chem. Soc.*, 52, 2519 (1930).
74. Pacsu, *J. Am. Chem. Soc.*, 54, 3649 (1932).
75. Hough, Jones, and Richards, *J. Chem. Soc.* (Lond.), p. 2005 (1953).
76. Stodola, Koepsell, and Sharpe, *J. Am. Chem. Soc.*, 74, 3202 (1952); 78, 2514 (1956).
77. Bourne, Hutson, and Weigel, *Biochem. J.*, 79, 549 (1961).
78. Weidenhagen and Lorenz, *Angew. Chem.*, 69, 641 (1957).
79. Bourne, Hartigan, and Weigel, *J. Chem. Soc.* (Lond.), 1088 (1961).
80. Knox, *Biochem. J.*, 94, 534 (1965).
81. Goldstein and Whelan, *J. Chem. Soc.* (Lond.), p. 4264 (1963).
82. Birch, *J. Chem. Soc.* (Lond.), p. 3489 (1965).
83. Peat, Whelan, and Hinson, *Nature*, 170, 1056 (1952).
84. Sharp and Stacey, *J. Chem. Soc.* (Lond.), p. 285 (1951).
85. Micheel and Hagel, *Ber. Dtsch. Chem. Ges.*, 85, 1087 (1952).
86. Matsuda, *J. Agric. Chem. Soc. Jap.*, 30, 119 (1956).
87. Takiura and Koizuma, *Yakugaku Zasshi*, 82, 852 (1962).
88. Bailey, Barker, Bourne, Grant, and Stacey, *J. Chem. Soc.* (Lond.), p. 1895 (1958).
89. Peat, Whelan, and Hinson, *Chem. Ind.*, 385 (1955).
90. Sato and Aso, *Nature*, 180, 984 (1957).
91. Freudenberg, Knauber, and Cramer, *Ber. Dtsch. Chem. Ges.*, 84, 114 (1951).
92. Finan and Warren, *J. Chem. Soc.* (Lond), p. 5229 (1963).
93. Vis and Fletcher, *J. Am. Chem. Soc.*, 78, 4709 (1956).
94. Rabate, *Bull. Soc. Chim. Fr. Ser. 5*, 7, 565 (1940).
95. Haq and Whelan, *J. Chem. Soc.* (Lond.), p. 1342 (1958).
96. Watanabe and Aso, *J. Agr. Res.* (Tokyo), 11, 109 (1960).
97. Wolfrom and Thompson, *J. Am. Chem. Soc.*, 77, 6403 (1955); 78, 4116 (1956).
98. Weismann and Meyer, *J. Am. Chem. Soc.*, 76, 1753 (1954).
99. Freudenberg and Oertzen, *Justus Liebigs Ann. Chem.*, 574, 37 (1951).
100. Connell, Hirst, and Percival, *J. Chem. Soc.* (Lond.), p. 3494 (1950).
101. Peat, Whelan, and Lawley, *J. Chem. Soc.* (Lond.), p. 724, 729 (1958).
102. Peterson and Spencer, *J. Am. Chem. Soc.*, 49, 2822 (1927).
103. Gillis, *Natuurwet. Tijdschr.*, 12, 193 (1930); *Chem. Zentralbl.*, 1, 256 (1931).
104. Hudson and Yanovsky, *J. Am. Chem. Soc.*, 39, 1013 (1917).
105. Wolfrom, Georges, and Miller, *J. Am. Chem. Soc.*, 71, 125 (1949).
106. Jeanes, Wilham, Jones, Tsuchiya, and Rist, *J. Am. Chem. Soc.*, 75, 5911 (1953).
107. Zemplén, *Hoppe-Seyler's Z. Physiol. Chem.*, 85, 399 (1913).
108. Hudson and Johnson, *J. Am. Chem. Soc.*, 39, 1272 (1917).
109. Thompson and Wolfrom, *J. Am. Chem. Soc.*, 75, 3605 (1953).
110. Selinger and Schramm, *J. Biol. Chem.*, 236, 2183 (1961).
111. Baytop, Tanher, Tekman, and Oner, *Folia Pharm.* (Istanbul), 4, 464 (1960).
112. Korte, *Ber. Dtsch. Chem. Ges.*, 88, 1527 (1955).
113. Frérejacque, *C. R. Acad. Sci.*, 246, 459 (1958).
114. Arcamone and Bizioli, *Gazz. Chim. Ital.*, 87, 896 (1957).
115. Wolfrom, Vercellotti, and Horton, *J. Org. Chem.*, 27, 705 (1962); 28, 278 (1963).
116. Barker, Gomez-Sanchez, and Stacey, *J. Chem. Soc.* (Lond.), p. 3264 (1959).
117. Sawai and Hehre, *J. Biol. Chem.*, 237, 2047 (1962).
118. Jermyn, *Aust. J. Biol. Sci.*, 11, 114 (1958).
119. Dutton and Unrau, *Can. J. Chem.*, 42, 2048 (1964).
120. Lindbergh, *Acta Chem. Scand.*, 7, 1119 (1953).
121. Lindberg, Silvander, and Wachtmeister, *Acta Chem. Scand.*, 17, 1348 (1963).
122. Meier, *Acta Chem. Scand.*, 14, 749 (1960).
123. Perila and Bishop, *Can. J. Chem.*, 39, 815 (1961).
124. Gyaw and Timell, *Can. J. Chem.*, 38, 1957 (1960).
125. Palleroni and Doudoroff, *J. Biol. Chem.*, 219, 957 (1956).
126. Tunmann and Scheherer, *Arch. Pharm.*, 292, 745 (1959).
127. Hassid, Doudoroff, Barker, and Dore, *J. Am. Chem. Soc.*, 67, 1394 (1945).
128. Hassid, Doudoroff, Barker, and Dore, *J. Am. Chem. Soc.*, 68, 1465 (1946).
129. Barker, Bourne, Hewitt, and Stacey, *J. Chem. Soc.* (Lond.), p. 3541 (1957).
130. Barker, Stackey, and Stroud, *Nature*, 189, 138 (1961).
131. Putman, Litt, and Hassid, *J. Am. Chem. Soc.*, 77, 4351 (1955).
132. Yosizawa, *Biochim. Biophys, Acta*, 52, 588 (1961).
133. Lloyd and Roberts, *J. Chem. Soc.* (Lond.), p. 6910 (1965).
134. Kuhn and Kirschenlohr, *Ber. Dtsch. Chem. Ges.*, 87, 384 (1954).
135. Sharon, Osawa, Flowers, and Jeanloz, *J. Biol. Chem.*, 241, 223 (1966).
136. Tipper, Ghuysen, and Strominger, *Biochemistry*, 4, 468 (1965).
137. Flowers and Jeanloz, *J. Org. Chem.*, 28, 1564 (1963).
138. Osawa and Nakazawa, *Biochim. Biophys. Acta*, 130, 56 (1966).
139. Barker, Foster, Stacey, and Webber, *Chem. Ind.*, p. 208 (1957); *J. Chem. Soc.* (Lond.) 2218 (1958).
140. Barker, Foster, Khmelnitski, and Webber, *Bull. Soc. Chim. Biol.*, 42, 1799 (1960).
141. Danishefsky and Steiner, *Biochim. Biophys. Acta*, 101, 37 (1965).
142. Tipper and Strominger, *Biochem. Biophys. Res. Commun.*, 22, 48 (1966).
143. Martinez, Wolfe, and Nakada, *J. Bacteriol.*, 78, 217 (1959).
144. Olavesen and Davidson, *J. Biol. Chem.*, 240, 992 (1965).
145. Wolfrom, Madison, and Cron, *J. Am. Chem. Soc.*, 74, 1491 (1952).
146. Weissmann and Meyer, *J. Am. Chem. Soc.*, 74, 4729 (1952); 76, 1753 (1954).
147. Jeanloz and Jeanloz, *Biochemistry*, 3, 121 (1964).
148. Linker, Meyer, and Hoffman, *J. Biol. Chem.*, 219, 13 (1956).
149. Leach and Teeters, *J. Am. Chem. Soc.*, 73, 2794 (1951); 74, 3187 (1952).
150. Carter, Dyer, Shaw, Rinehart, and Hichens, *J. Am. Chem. Soc.*, 83, 3723 (1961).
151. Ito, Nishio, and Ogawa, *J. Antibiot. Ser. A*, 17, 189 (1964).
152. Suhara, Maeda, Umezawa, and Ohno, *Tetrahedron Lett.*, 1239 (1966).
153. Fukui, Hochster, Durbin, Grebner, and Feingold, *Bull. Res. Counc Isr.*, 11A, 262 (1963).
154. Fukui and Hochster, *Can. J. Biochem.*, 41, 2363 (1963).
155. Gorin, Haskins, and Spencer, *Can. J. Biochem. Physiol.*, 38, 165 (1960).
156. Gorin and Perlin, *Can. J. Chem.*, 39, 2474 (1961).
157. Tyminski and Timell, *J. Am. Chem. Soc.*, 82, 2823 (1960).
158. Merler and Wise, *Tappi*, 41, 80 (1958).
159. Colin and Angier, *C. R. Acad. Sci.*, 208, 1450 (1939).
160. Jones and Painter, *J. Chem. Soc.* (Lond.), p. 669 (1959).
161. Jones and Nicholson, *J. Chem. Soc.* (Lond.), p. 27 (1958).

162. Whistler and Stein, *J. Am. Chem. Soc.*, 73, 4187 (1951).
163. Tsujino, *Agric. Biol. Chem.*, 27, 236 (1963).
164. Zemplén and Gerecs, *Ber. Dtsch. Chem. Ges.*, 68, 2054 (1935).
165. Zemplén, Gerecs, and Flesch, *Ber. Dtsch. Chem. Ges.*, 71, 774 (1938).
166. Charaux, *C. R. Acad. Sci.*, 178, 1312 (1924); 180, 1419 (1925).
167. Gorin and Perlin, *Can. J. Chem.*, 37, 1930 (1959).
168. Zemplén and Gerecs, *Ber Dtsch. Chem. Ges.*, 71, 2520 (1938).
169. Hestrin, Feingold, and Avigad, *J. Am. Chem. Soc.*, 77, 6710 (1955).
170. Power and Salway, *J. Chem. Soc.* (Lond.), p. 1062 (1914).
171. Amanmuradov and Abubakirov, *Chem. Nat. Compd.* (USSR), 1, 292 (1965).
172. Helferich and Rouch, *Justus Liebigs Ann. Chem.*, 455, 168 (1927).
173. Howard, *Biochem. J.*, 67, 643 (1957).
174. Curtis and Jones, *Can. J. Chem.*, 38, 1305 (1960).
175. Ball and Jones, *J. Chem. Soc.* (Lond.), p. 33 (1958).
176. Whistler, Bachrack, and Tu, *J. Am. Chem. Soc.*, 74, 3059 (1952).
177. Bishop, *J. Am. Chem. Soc.*, 78, 2840 (1956).
178. Banerji and Rao, *Aust. J. Chem.*, 17, 1059 (1964).
179. Pazur, *J. Am. Chem. Soc.*, 75, 6323 (1953).
180. Schlubach and Berndt, *Justus Liebigs Ann. Chem.*, 677, 172 (1964).
181. Strepkov, *Zh. Obshch. Khim.*, 9, 1489 (1939).
182. Bacon, *Nature*, 184, 1957 (1959).
183. Pridham, *Biochem. J.*, 76, 13 (1960).
184. Barker and Carrington, *J. Chem. Soc.* (Lond.), p. 3588 (1953).
185. Haq and Adams, *Can. J. Chem.*, 39, 1165 (1961).
186. Allen and Bacon, *Biochem. J.*, 63, 200 (1956).
187. Gross, Blanchard, and Bell, *J. Chem. Soc.* (Lond.), p. 1727 (1954).
188. Davy and Courtois, *C. R. Acad. Sci.*, 261, 3483 (1965).
189. Suzuki and Hehre, *Arch. Biochem. Biophys.*, 105, 339 (1964).
190. French, Wild, Young, and James, *J. Am. Chem. Soc.*, 709, 3664 (1953).
191. Rege, Painter, Watkins, and Morgan, *Nature*, 204, 740 (1964).
192. Wallenfels, Bernt, and Limberg, *Justus Liebigs Ann. Chem.*, 579, 113 (1952).
193. Ballio and Russi, *Tetrahedron*, 9, 125 (1960).
194. Tanret, *Bull. Soc. Chim. Fr. Ser. 3*, 27, 947 (1902).
195. Pazur, Marsh, and Tipton, *J. Am. Chem. Soc.*, 80, 1433 (1958).
196. Sastry and Kates, *Biochemistry*, 3, 1271 (1964).
197. Petek, Villarroya, and Courtois, *C. R. Acad. Sci.*, 265D, 195 (1966).
198. Wickstrom and Baerheim-Svendsen, *Acta Chem. Scand.*, 10, 1199 (1956).
199. MacLeod and McCorquodale, *Nature*, 182, 815 (1958).
200. Avigad, *J. Biol. Chem.*, 229, 121 (1957).
201. Aso and Yamauchi, *Agric. Biol. Chem.*, 25, 10 (1961).
202. Haworth, Hirst, and Ruell, *J. Chem. Soc. (Lond.)*, 3125 (1923).
203. Scheibler and Mittelmeier, *Ber. Dtsch. Chem. Ges.*, 23, 1438 (1890).
204. Tanret, *Bull. Soc. Chim. Fr. Ser. 3*, 13, 261 (1895).
205. Montreuil, *C. R. Acad. Sci.*, 242, 192, 828 (1956).
206. Bailey, Barker, Bourne, and Stacey, *Nature*, 176, 1164 (1955).
207. Yamauchi and Aso, *Nature*, 189, 753 (1961).
208. Brandish, Shaw, and Baddiley, *J. Chem. Soc. Sect. C Org. Chem.*, 521 (1966).
209. Jones and Reid, *J. Chem. Soc.* (Lond.) 1361 (1954); p. 1890 (1955).
210. Kurasawa, Yamamoto, Igaue, and Nakamura, *Nippon Nogei-Kagaku Kaishi*, 30, 624 (1956).
211. Schlubach and Koehn, *Justus Liebigs Ann. Chem.*, 614, 126 (1958).
212. Binkley, *Int. Sugar J.*, 66, 46 (1964).
213. Barker, Bourne, and Carrington, *J. Chem. Soc.* (Lond.), p. 2125 (1954).
214. Bacon and Bell, *J. Chem. Soc.* (Lond.), p. 2528 (1953).
215. Albon, Bell, Blanchard, Gross, and Rundell, *J. Chem. Soc.* (Lond), p. 24 (1953).
216. Courtois, LeDizet, and Petek, *Bull. Soc. Chim. Biol.*, 41, 1261 (1959).
217. Kuhn and von Grundherr, *Ber. Dtsch. Chem. Ges.*, 59, 1655 (1926).
218. Von Lippmann, *Ber. Dtsch. Chem. Ges.*, 60, 161 (1927).
219. Hudson and Sherwood, *J. Am. Chem. Soc.*, 40, 1456 (1918).
220. White and Maher, *J. Am. Chem. Soc.*, 75, 1259 (1953).
221. Wolf and Ewart, *Arch. Biochem. Biophys.*, 58, 365 (1955).
222. Takiura and Nakagawa, *Yakugaku Zasshi*, 83, 301, 305 (1963).
223. Barker, Bourne, and Theander, *J. Chem. Soc.* (Lond.), p. 2064 (1957).
224. Baron and Guthrie, *Ann. Entomol. Soc. Am.*, 53, 220 (1960).
225. Meyer, *Hoppe-Seyler's Z. Physiol. Chem.*, 6, 135 (1882).
226. Binaghi and Falqui, *Chem. Zentralbl.*, 2, 44 (1926).
227. Krieglstein and Fischer, *Hoppe-Seyler's Z. Physiol. Chem.*, 344, 209 (1966).
228. Feingold, Neufeld, and Hassid, *J. Biol. Chem.*, 233, 783 (1958).
229. Perlin and Suzuki, *Can. J. Chem.*, 40, 50 (1962).
230. Peat, Whelan, and Roberts, *J. Chem. Soc.* (Lond.), p. 3916 (1957).
231. Barker, Bourne, O'Mant, and Stacey, *J. Chem. Soc.* (Lond.), p. 2448 (1957).
232. Reese and Mandels, *Can. J. Microbiol.*, 10, 103 (1964).
233. Moscatelli, Ham, and Rickes, *J. Biol. Chem.*, 236, 2858 (1961).
234. Wolfrom, Georges, Thompson, and Miller, *J. Am. Chem. Soc.*, 71, 2873 (1949).
235. Peat, Whelan, and Jones, *J. Chem. Soc.* (Lond.), p. 2490 (1957).
236. Sugihara and Wolfrom, *J. Am. Chem. Soc.*, 71, 3357 (1949).
237. Walker and Wright, *Arch. Biochem. Biophys.*, 69, 362 (1957).
238. Wolfrom and Fields, *Tappi*, 41, 204 (1958).
239. French, Taylor, and Whelan, *Biochem. J.*, 90, 616 (1964).
240. Bailey, Barker, Bourne, and Stacey, *J. Chem. Soc.* (Lond.), p. 3536 (1957).
241. Wolfrom, Thompson, and Galkowski, *J. Am. Chem. Soc.*, 73, 4093 (1951).
242. Turvey and Whelan, *Biochem. J.*, 67, 49 (1957).
243. Conchie, Moreno, and Cardini, *Arch. Biochem. Biophys.*, 94, 342 (1961).
244. Haq and Whelan, *J. Chem. Soc.* (Lond.), p. 4543 (1956).
245. Peat, Whelan, and Evans, *J. Chem. Soc.* (Lond.), p. 175 (1960).
246. Schwarz and Timell, *Can. J. Chem.*, 41, 1381 (1963).
247. Kuehl, Peck, Hoffhine, Graber, and Folkers, *J. Am. Chem. Soc.*, 68, 1460 (1946).
248. Bannister and Argoudelis, *J. Am. Chem. Soc.*, 85, 119, 234 (1963).
249. Zechmeister and Toth, *Ber. Dtsch. Chem. Ges.*, 65, 161 (1932).
250. Linker, Meyer, and Weissmann, *J. Biol. Chem.*, 213, 237 (1955).
251. Umezawa, Ueda, Maeda, Yagashita, Kondo, Okami, Utahara, Osato, Nitta, and Takeushi, *J. Antibiot. Ser. A.*, 23, 298 (1957).
252. Cron, Johnson, Palermiti, Perron, Taylor, Whitehead, and Hooper, *J. Am. Chem. Soc.*, 80, 752 (1958).
253. Kondo, Sezaki, Koike, Shimura, Akita, Satoh, and Hara, *J. Antibiot. Ser. A*, 18, 38 (1965).
254. Kondo, Akita, and Koike, *J. Antibiot. Ser. A*, 19, 139 (1966).
255. Schneir and Rafelson, *Biochim. Biophys. Acta*, 130, 1 (1966).
256. Kuhn and Brossmer, *Ber. Dtsch. Chem. Ges.*, 92, 1667 (1959).
257. Ryan, Carubelli, Caputto, and Trucco, *Biochim. Biophys. Acta*, 101, 252 (1965); *J. Biol. Chem.*, 236, 2381 (1961).
258. Lindberg, *Acta Chem. Scand.*, 9, 1093, 1097 (1955).
259. Aspinall, Begbie, and McKay, *J. Chem. Soc.* (Lond.), p. 214 (1962).
260. Mayeda, *J. Biochem.* (Tokyo), 1, 131 (1922).
261. Ohtsuki, *Acta Phytochim.* (Japan), 4, 1 (1928).
262. Whistler and Masak, *J. Am. Chem. Soc.*, 77, 1241 (1955).
263. Marchessault and Timell, *J. Polymer. Sci. C.*, 2, 49 (1963).
264. Murakami, *Acta Phytochim.*, 14, 101 (1944); 15, 105, 109 (1949).
265. Schlubach and Berndt, *Justus Liebigs Ann. Chem.*, 647, 41 (1961).
266. Pazur, *J. Biol. Chem.*, 199, 217 (1952).
267. Kuhn and Gauhe, *Justus Liebigs Ann. Chem.*, 611, 249 (1958).
268. Hatanaka, *Arch. Biochem. Biophys.*, 82, 188 (1959).
269. French, Wild, and James, *J. Am. Chem. Soc.*, 75, 3664 (1953).
270. Onuki, *Sci. Pap. Inst. Phys. Chem. Res.* (Tokyo), 20, 201 (1933).
271. Laidlaw and Wylam, *J. Chem. Soc.* (Lond.), p. 567 (1953).
272. Davy and Courtois, *C. R. Acad. Sci.*, 261, 3483 (1965).
273. Araki and Arai, *Bull. Chem. Soc. Jap.*, 30, 287 (1957).
274. Courtois and Ariyoshi, *Bull. Soc. Chim. Biol.* (Paris), 42, 737 (1960).
275. Demain and Phaff, *Arch. Biochem. Biophys.*, 51, 114 (1954).
276. Binkley and Altenburg, *Int. Sugar J.*, 67, 110 (1965).
277. Kurasawa, Yamamoto, Igaue, and Nakamura, *Nippon Nogei Kagaku Kaishi*, 30, 696 (1956).
278. Strepkov, *Dokl. Akad. Nauk SSSR*, 125, 216 (1959).
279. Schlubach and Koehn, *Justus Liebigs Ann. Chem.*, 606, 130 (1957).
280. Parrish, Perlin, and Reese, *Can. J. Chem.*, 38, 2094 (1960).
281. Perlin and Suzuki, *Can. J. Chem.*, 40, 50 (1962).
282. Igarashi, Igoshi, and Sakurai, *Agric Biol. Chem.*, 30, 1254 (1966).
283. Whistler and Hickson, *J. Am. Chem. Soc.*, 76, 1671 (1954).
284. Wolfrom and Fields, *Tappi*, 41, 204 (1958); 40, 335 (1957).
285. Duncan and Manners, *Biochem. J.*, 69, 343 (1958).
286. Jones, Jeanes, Stringer, and Tsuchiya, *J. Am. Chem. Soc.*, 78, 2499 (1956).

287. Akiya, *J. Pharm. Soc. Jap.*, 58, 71 (1938).
288. Salton and Ghuysen, *Biochim. Biophys. Acta*, 45, 355 (1960).
289. Weissman, Meyer, Sampson, and Linker, *J. Biol. Chem.*, 208, 417 (1954).
290. Rinehart, Hichens, Argoudelis, Chilton, Carter, Georgiadis, Schaffner, and Schillings, *J. Am. Chem. Soc.*, 84, 3218 (1962).
291. Rinehart, Argoudelis, Goss, Sohler, and Schaffner, *J. Am. Chem. Soc.*, 82, 3938 (1960).
292. Haskell, French, and Bartz, *J. Am. Chem. Soc.*, 81, 3482 (1959).
293. Fried and Titus, *J. Biol. Chem.*, 168, 391 (1947).
294. Goldschmid and Perlin, *Can. J. Chem.*, 41, 2272 (1963).
295. Kihara, *Proc. Imp. Acad.* (Tokyo), 5, 349 (1929); 11, 552 (1935); *J. Agric. Chem. Soc.* (Japan), 12, 1044 (1936).
296. Schlubach, Berndt, and Chiemprasert, *Justus Liebigs Ann. Chem.*, 665, 191 (1963).
297. Kuhn, Baer, and Gauhe, *Ber. Dtsch. Chem. Ges.*, 91, 364 (1958).
298. Kuhn and Gauhe, *Ber. Dtsch. Chem. Ges.*, 95, 513 (1962).
299. Bourquelot and Bridel, *C. R. Acad. Sci.*, 151, 760 (1910).
300. Murakami, *Proc. Imp. Acad.* (Tokyo), 16, 14 (1940).
301. Zechmeister and Toth, *Ber. Dtsch. Chem. Ges.*, 64, 854 (1931).
302. Hoover, Nelson, Milner, and Wei, *J. Food Sci.*, 30, 253 (1965).
303. Wolfrom, Dacons, and Fields, *Tappi*, 39, 803 (1956).
304. Kuhn and Gauhe, *Ber. Dtsch. Chem. Ges.*, 93, 647 (1960).
305. Murakami, *Acta Phytochim.*, 13, 37 (1942).
306. Herissey, Fleury, Wickstrom, Courtois, and Dizet, *C.R. Acad. Sci.*, 239, 824 (1954); *Bull. Soc. Chim. Biol.* (Paris), 36, 1507 (1954).
307. Akiya and Tomoda, *J. Pharm. Soc. Jap.*, 76, 571 (1956).
308. French and Rundle, *J. Am. Chem. Soc.*, 64, 1651 (1942).
309. Fruedenberg and Jacobi, *Justus Liebigs Ann. Chem.*, 518, 102 (1935).
310. Akiya and Watanabe, *J. Pharm. Soc. Jap.*, 70, 576 (1950).
311. Bjorndel, Eriksson, Garegg, Lindberg, and Swan, *Acta Chem. Scand.*, 19, 2309 (1965).
312. Whistler and Tu, *J. Am. Chem. Soc.*, 74, 3609 (1952); 75, 645 (1954).
313. Murakami, *Acta Phytochim.*, 10, 43 (1937).
314. Courtois, Dizet, and Wickstrom, *Bull. Soc. Chim. Biol.* (Paris), 40, 1059 (1958).
315. Malpress and Hytten, *Biochem. J.*, 68, 708 (1958).
316. French, Knapp, and Pazur, *J. Am. Chem. Soc.*, 72, 5150 (1950).
317. Clancy and Whelan, *Arch. Biochem. Biophys.*, 118, 724, 730 (1967).
318. Morgan and O'Neill, *Can. J. Chem.*, 37, 1201 (1959).
319. Tanret, *Bull. Soc. Chim. Fr. Ser. 3*, 21, 1065 (1899).
320. Maher, unpublished results (1957).
321. Herissey, Fleury, Wickstrom, Courtois, and Dizet, *Bull. Soc. Chim. Biol.* (Paris), 36, 1507 (1964).
322. Aspinall and Telfer, *J. Chem. Soc.* (Lond.), p. 1106 (1955).
323. Anderson, *Acta. Chem. Scand.*, 21, 828 (1967).
324. Allgeier, *Helv. Chim. Acta*, 51, 668 (1968).

325. Glaudemans, *Carbohydr. Res.*, 10, 213 (1969).
326. Anderson, Dolan, and Rees, *J. Chem. Soc. Sect. C Org. Chem.*, 596 (1968).
327. Alexander, *Arch. Biochem. Biophys.*, 123, 240 (1968).
328. Levene and Tipson, *J. Biol. Chem.*, 125, 355 (1938).
329. Manners and Stark, *Carbohydr. Res.*, 3, 102 (1966).
330. Umezawa, Tatsuta, and Muto, *J. Antibiot. Ser. A*, 20, 388 (1967).
331. Brundish and Baddiley, *Carbohydr. Res.*, 8, 308 (1968).
332. Ferrier and Prasad, *J. Chem. Soc.* (Lond.), p. 7429 (1965).
333. Zatula and Kolesnikov, *Chem. Natur. Compd.* (USSR), 3, 138 (1967).
334. Saier and Ballou, *J. Biol. Chem.*, 243, 992 (1968).
335. Shapiro, Acher, and Rachaman, *J. Org. Chem.*, 32, 3767 (1967).
336. Duncan, Manners, and Thompson, *Biochem. J.*, 73, 295 (1959).
337. Olavesen and Davidson, *Biochim. Biophys. Acta*, 101, 245 (1965).
338. Umezawa, Tatsuta, Tsuchiya, and Kitazawa, *J. Antibiot. Ser. A*, 20, 53 (1967).
339. Hayano and Fukui, *J. Biochem.* (Tokyo), 64, 901 (1968).
340. Uramoto, Otake, and Yonehara, *J. Antibiot. Antibiot. Ser. A*, 20, 236 (1967).
341. Srivastava and Singh, *Carbohydr. Res.*, 4, 326 (1967).
342. Bollman, Hirschmüller, and Schmidt-Berg-Lorenz, *Int. Sugar J.*, 67, 143 (1965).
343. Beith-Halahmi and Flowers, *Carbohydr. Res.*, 81, 340 (1968).
344. Urbas, *Can. J. Chem.*, 46, 49 (1968).
345. Fischer and Krieglstein, *Hoppe-Seyler's Z. Physiol. Chem.*, 348, 1252 (1967).
346. Siddiqui and Furgala, *Carbohydr. Res.*, 6, 250 (1968).
347. Fischer and Seyferth, *Hoppe-Seyler's Z. Physiol. Chem.*, 349, 1662 (1968).
348. Pollock, Chipman, and Sharon, *Arch. Biochem. Biophys.*, 120, 235 (1967).
349. Umezawa, Tatsuta, and Koto, *J. Antibiot. Set. A*, 21, 367 (1968); *Bull. Chem. Soc. Jap.*, 42, 533 (1969).
350. Hasegawa, Kurihara, Nishimura, and Nakajima, *Agric. Biol. Chem.*, 32, 1130 (1968).
351. Murase, *J. Antibiot.* (Tokyo) *Ser. A*, 14, 156, 367 (1961).
352. Umezawa, Koto, Tatsuta, and Tsumura, *Bull. Chem. Soc. Jap.*, 42, 529 (1969).
353. Ito, Nishio, and Ogawa, *J. Antibiot. Ser. A*, 17, 189 (1964).
354. Umezawa, Koto, Tatsuta, Hineno, Nishimura, and Tsumura, *J. Antibiot. Ser. A*, 21, 424 (1968).
355. Leyh-Bouille, Ghuysen, Tipper, and Strominger, *Biochemistry*, 5, 3079 (1966).
356. Tung and Nordin, *Biochim. Biophys. Acta*, 158, 154 (1968).
357. Barker, Bourne, Grant, and Stacey, *J. Chem. Soc.* (Lond.), p. 601 (1958).

OLIGOSACCHARIDES (INCLUDING DISACCHARIDES)

Substance[a] (Synonym) (A)	Derivative (B)	Chemical Formula (C)	Melting Point °C (D)	Specific Rotation[b] $[\alpha]_D$ (E)	Reference (F)
DISACCHARIDES					
O-α-L-Ara$f$-(1 → 3)-L-Ara		$C_{10}H_{18}O_9$	—	0	4
O-β-L-Ara$f$-(1 → 3)-L-Ara$f$		$C_{10}H_{18}O_9$	—	+89,+94	1, 2
	Phenylosazone	$C_{22}H_{30}N_4O_7$	200	—	1, 2
O-β-L-Ara$p$-(1 → 3)-L-Ara$f$		$C_{10}H_{18}O_9$	—	—	3
O-α-L-Ara$f$-(1 → 5)-L-Ara$f$		$C_{10}H_{18}O_9$	—	−72, −87	2, 4
	Phenylosazone	$C_{22}H_{30}N_4O_7$	177	—	2, 4
O-β-L-Ara$p$-(1 → 4)-L-Ara		$C_{10}H_{18}O_9$	—	+193	5
	Hexaacetate	$C_{22}H_{30}O_{15}$	167	—	5
O-α-L-Ara$p$-(1 → 5)-L-Ara		$C_{10}H_{18}O_9$	143	−14 → −18	4
O-L-Ara$f$-(1 ⟩ 3) D Xyl		$C_{10}H_{10}O_9$	—	—	6
O-β-L-Ara$p$-(1 → 2)-D-Glc		$C_{11}H_{20}O_{10}$	210–220	+151	7
O-α-L-Ara$p$-(1 → 3)-D-Glc		$C_{11}H_{20}O_{10}$	176–178	+54.4	8
O-α-L-Ara$p$-(1 → 4)-D-Glc		$C_{11}H_{20}O_{10}$	—	+41.9	8
O-α-L-Ara$f$-(1 → 6)-D-Glc	Heptaacetate	$C_{25}H_{34}O_{17}$	108	−20	9
O-α-L-Ara$p$-(1 → 6)-D-Glc		$C_{11}H_{20}O_{10}$	210	+37.2, +56	8, 10

OLIGOSACCHARIDES (INCLUDING DISACCHARIDES) (Continued)

Substance[a] (Synonym) (A)	Derivative (B)	Chemical Formula (C)	Melting Point °C (D)	Specific Rotation[b] $[\alpha]_D$ (E)	Reference (F)
DISACCHARIDES (Continued)					
O-β-L-Ara$f$-(1 → 6)-D-Glc		$C_{11}H_{20}O_{10}$	—	+73	9
O-L-Ara$f$-(1 → 6)-D-Gal		$C_{11}H_{20}O_{10}$	—	—	11
O-L-Ara-(1 →)-D-Man	Heptaacetate	$C_{25}H_{34}O_{17}$	147–149	—	12
O-β-D-Xyl$p$-(1 → 2)-L-Ara		$C_{10}H_{18}O_9$	167–168	+32 → +33	13
O-α-D-Xyl$p$-(1 → 3)-L-Ara		$C_{10}H_{18}O_9$	117–119	+173 → +182	14
	Hexaacetate	$C_{22}H_{30}O_{15}$	168–170	+106	14
O-α-Xyl-(1 → 5)-L-Ara		$C_{10}H_{18}O_9$	—	+74	15
O-β-D-Xyl$p$-(1 → 5)-L-Ara		$C_{10}H_{18}O_9$	—	−41 → −47	16, 17
O-α-D-Xyl$p$-(1 → 2)-D-Xyl		$C_{10}H_{18}O_9$	—	—	18
O-α-D-Xyl$p$-(1 → 3)-D-Xyl		$C_{10}H_{18}O_9$	178	+118	19
O-β-D-Xyl$p$-(1 → 3)-D-Xyl (Rhodymenabiose)		$C_{10}H_{18}O_9$	192–193	−18.4 → −22	20
	Phenylosazone	$C_{22}H_{30}N_4O_7$	194–196	+47	20
O-α-D-Xyl$p$-(1 → 4)-D-Xyl		$C_{10}H_{18}O_9$	—	—	19
O-β-D-Xyl$p$-(1 → 4)-D-Xyl (Xylobiose)		$C_{10}H_{18}O_9$	185–190	−20 → −30	21, 22
	Hexaacetate	$C_{22}H_{30}O_{15}$	155–156	−74 → −75	23
O-β-D-Xyl-(1 → 6)-D-Gal		$C_{11}H_{20}O_{10}$	194–196	−3.6	24
O-β-D-Xyl$p$-(1 → 3)-D-Glc		$C_{11}H_{20}O_{10}$	—	—	25
O-D-Xyl$p$-(1 → 4)-D-Glc$p$		$C_{11}H_{20}O_{10}$	—	+70	284
O-α-D-Xyl$p$-(1 → 6)-D-Glc (Isoprimeverose)		$C_{11}H_{20}O_{10}$	205	+121 → +127	26
	Methylglycoside pentaacetate	$C_{21}H_{30}O_{14}$	123–124	+66	26
O-β-D-Xyl$p$-(1 → 6)-D-Glc (Primeverose)		$C_{11}H_{20}O_{10}$	215	−23	27, 28
	Phenylosazone	$C_{23}H_{32}N_4O_8$	224–226	—	26, 29
O-α-D-Rib$f$-(1 → 6)-D-Glc		$C_{11}H_{20}O_{10}$	—	+77	30
O-β-D-Rib$f$-(1 → 6)-D-Glc		$C_{11}H_{20}O_{10}$	—	0	30
	Heptaacetate	$C_{25}H_{34}O_{17}$	108–110	+3	30
O-α-D-Xyl$p$-(1 → 2)-β-D-Fru$f$		$C_{11}H_{20}O_{10}$	—	+62	283
O-D-Gal$p$-(1 → 2)-D-Ara		$C_{11}H_{20}O_{10}$	143–144	+34.4	31
	Heptaacetate	$C_{25}H_{34}O_{17}$	139–142	+40.6(CHCl$_3$)	31
O-β-D-Gal$p$-(1 → 3)-D-Ara		$C_{11}H_{20}O_{10}$	166–168	−55 → −65	32
	Heptaacetate	$C_{25}H_{34}O_{17}$	154	−80.8(CHCl$_3$)	32
O-α-D-Gal$p$-(1 → 3)-L-Ara$f$	Heptamethyl ether	$C_{18}H_{34}O_{10}$	—	+102(CHCl$_3$)	33
O-β-D-Gal$p$-(1 → 4)-L-Ara$p$		$C_{11}H_{20}O_{10}$	177–178	+30 → +45	34
	Methyl β-glycoside	$C_{12}H_{22}O_{10}$	216–218	+163	34
O-β-D-Gal$p$-(1 → 5)-L-Ara$f$		$C_{11}H_{20}O_{10}$	—	−13, −18	35, 36
O-β-D-Gal$f$-(1 → 2)-D-Arabinitol (Umbilicin)		$C_{11}H_{22}O_{10}$	138–139	−81	37
	Octaacetate	$C_{27}H_{38}O_{18}$	84–85	−20(CHCl$_3$)	37
O-β-D-Gal$p$-(1 → 4)-D-Arabinitol		$C_{11}H_{22}O_{10}$	178–181	—	34
O-β-D-Gal$p$-(1 → 2)-D-Xyl		$C_{11}H_{20}O_{10}$	159–160	+25, +30	34, 38
O-β-D-Gal$p$-(1 → 3)-D-Xyl		$C_{11}H_{20}O_{10}$	196–200	+27–+10	34
O-β-D-Gal$p$-(1 → 4)-D-Xyl		$C_{11}H_{20}O_{10}$	201–211	+15	38
O-β-D-Gal$p$-(1 → 3)-D-Rib		$C_{11}H_{20}O_{10}$	200–202	+2 → +15	39
O-β-D-Gal$p$-(1 → 5)-D-Rib		$C_{11}H_{20}O_{10}$	—	—	34
O-α-D-Gal$p$-(1 → 3)-D-Gal		$C_{11}H_{20}O_{11}$	—	+161	40
	Octaacetate	$C_{28}H_{38}O_{19}$	157.5–158.5	+1 10.2 (c 0.5 CHCl$_3$)	41
O-β-D-Gal$p$-(1 → 3)-D-Gal		$C_{12}H_{22}O_{11}$	151–152	+69 → +55	42
			163–170	+75 → +60	43
			204–206	+56	34
O-α-D-Gal$p$-(1 → 4)-D-Gal		$C_{12}H_{22}O_{11}$	210–211	+172	40, 44
O-β-D-Gal$p$-(1 → 4)-D-Gal		$C_{12}H_{22}O_{11}$	195–198	+85 → +67	34
	Octaacetate	$C_{28}H_{38}O_{19}$	172–173	+57.3(CHCl$_3$)	45
O-α-D-Gal$f$-(1 → 5)-D-Gal		$C_{12}H_{22}O_{11}$	—	+133	40
O-β-D-Gal$f$-(1 → 5)-D-Gal		$C_{12}H_{22}O_{11}$	—	−65	46
O-β-D-Gal$f$-(1 → 5)-D-Galactitol		$C_{12}H_{24}O_{11}$	149–151	−65	46
O-α-D-Gal$p$-(1 → 6)-D-Gal (Swietenose)		$C_{12}H_{22}O_{11}$	—	+149	47, 48
	Octaacetate	$C_{28}H_{38}O_{19}$	223–227	+189(c 0.5 CHCl$_3$)	48
O-β-D-Gal$p$-(1 → 6)-D-Gal		$C_{12}H_{22}O_{11}$	97–100	+26	49, 50
			168–170	+28	51
	Octamethyl ether	$C_{20}H_{38}O_{11}$	68–70	−5.7(CH$_3$OH)	51
O-α-D-Gal$p$-(1 → 2)-D-Glc		$C_{12}H_{22}O_{11}$	118–120	+145	52
			155d	+161 → +153	53
	Octaacetate	$C_{28}H_{38}O_{19}$	176–178	+153(CHCl$_3$)	53

OLIGOSACCHARIDES (INCLUDING DISACCHARIDES) (Continued)

Substance[a] (Synonym) (A)	Derivative (B)	Chemical Formula (C)	Melting Point °C (D)	Specific Rotation[b] $[\alpha]_D$ (E)	Reference (F)
DISACCHARIDES (Continued)					
O-β-D-Galp-(1 → 2)-D-Glc		$C_{12}H_{22}O_{11}$	175	+48 → +39	54
O-α-D-Galp-(1 → 3)-D-Glc		$C_{12}H_{22}O_{11}$	169–170	+64.5	55
O-β-D-Galp-(1 → 3)-D-Glc		$C_{12}H_{22}O_{11}$	167–169	+37(c 0.8)	34
	Monohydrate	$C_{12}H_{24}O_{12}$	202–204	+77 → +41	56
O-β-D-Galp-(1 → 4)-α-D-Glcp (α-lactose)	Monohydrate Octaacetate	$C_{12}H_{24}O_{12}$	202	+83.5 → +52.6	57, 58
		$C_{28}H_{38}O_{19}$	152	+5 3.6 (c 10, CHCl₃)	59
O-β-D-Galp-(1 → 4)-β-D-Glcp (β-lactose)		$C_{12}H_{22}O_{11}$	252	+34.2 → +53.6	60
	Octaacetate	$C_{28}H_{38}O_{19}$	90	−4.7 (c 10, CHCl₃)	59
O-α-D-Galp-(1 → 6)-α-D-Glc (α-Melibiose)		$C_{12}H_{22}O_{11}$	183–184	+166 → +142	61
O-α-D-Galp-(1 → 6)-β-D-Glc (β-Melibiose)		$C_{12}H_{22}O_{11}$	85–86	+123 → +143	61
	Octaacetate	$C_{28}H_{38}O_{19}$	177	+102.5	62
O-β-D-Galp-(1 → 6)-D-Glc (Allolactose)		$C_{12}H_{22}O_{11}$	165, 174–176	+25,+37.5	15, 63–65
	Octaacetate	$C_{18}H_{38}O_{19}$	165	−0.5(CHCl₃)	66
O-β-D-Galp-(1 → 4)-α-D-Man (α-Epilactose)	Monohydrate	$C_{12}H_{24}O_{12}$	150–160	}+30 → +38	67
O-β-D-Galp-(1 → 4)-β-D-Man (β-Epilaetose)		$C_{12}H_{22}O_{11}$	196–197		67
	Octaacetate	$C_{28}H_{38}O_{19}$	96–97		67
O-α-D-Galp-(1 → 6)-D-Man (Epimolibiose)		$C_{12}H_{22}O_{11}$	201–202	+123 → +124	68
O-α-D-Galp-(1 → 2)-D-Fruf		$C_{12}H_{22}O_{11}$	170	+81.5	85
O-β-D-Galp-(1 → 2)-D-Fruf		$C_{12}H_{22}O_{11}$	—	−33	49
	Octaacetate	$C_{28}H_{38}O_{19}$	169–170	+41.3(CHCl₃)	69
O-β-D-Galp-(1 → 3)-D-Fruf		$C_{12}H_{22}O_{11}$	245–246	+78	34
O-β-D-Galp-(1 → 3)-D-GlcNAc (Lacto-N-biose I)		$C_{14}H_{25}NO_{11}$	166–170	+14	70, 71
O-β-D-Galp-(1 → 4)-D-GlcNAc (N-Acetyllactosamine)		$C_{14}H_{25}NO_{11}$	168–170	+27 → +28	72, 73
	Heptaacetate	$C_{28}H_{39}NO_{18}$	222–223	+61.5	72, 73
O-α-D-Galp-(1 → 6)-D-GlcNAcp		$C_{14}H_{25}NO_{11}$	138	+118	74
O-β-D-Galp-(1 → 6)-D-GlcNAcp (N-Acetylallolactosamine)		$C_{14}H_{25}NO_{11}$	157–159	+32.1 → +27.3	75, 76
O-α-D-Galp-(1 → 6)-D-GalNAc		$C_{14}H_{25}NO_{11}$	—	+142	74
O-β-D-Galp-(1 → 4)-D-ManNAc		$C_{14}H_{25}NO_{11}$	233	+38.5	77
O-β-D-Galp-(1 → 4)-3,6-anhydro-L-Gal (Agarobiose)		$C_{12}H_{20}O_{10}$	—	−21.5 → −16.4	78, 79
	Dimethylacetal	$C_{14}H_{24}O_{10}$	163–166	−36(CH₃OH)	78, 79
O-β-D-Galp-(1 → 3)-D-6dGlc		$C_{12}H_{22}O_{10}$	246–248	+25	34
O-α-D-Glcp-(1 → 2)-D-Ara	Heptaacetate	$C_{25}H_{34}O_{17}$	138–140	+46	80
O-β-D-Glcp-(1 → 2)-D-Ara	Heptaacetate	$C_{25}H_{34}O_{17}$	199–200	−46 → −47	81
O-α-D-Glcp-(1 → 3)-D-Ara	Monohydrate	$C_{11}H_{22}O_{11}$	119–121	+47	82
O-β-D-Glcp-(1 → 3)-D-Ara		$C_{11}H_{20}O_{10}$	161	−90 → −94	83, 84
O-α-D-Glcp-(1 → 3)-L-Ara	Dihydrate	$C_{11}H_{24}O_{12}$	—	+156	86
O-β-D-Glcp-(1 → 3)-L-Ara		$C_{11}H_{20}O_{10}$	199–201	+70	34
O-α-D-Glcp-(1 → 4)-L-Ara		$C_{11}H_{20}O_{10}$	—	—	87
O-α-D-Glcp-(1 → 2)-D-Xyl	Heptamethyl ether	$C_{18}H_{34}O_4$	—	+109 → +112	88, 89
O-β-D-Glcp-(1 → 2)-D-Xyl		$C_{11}H_{20}O_{10}$	202–206	+7	34
	Phenylosazone	$C_{23}H_{32}N_4O_8$	213–215	—	90
O-α-D-Glcp-(1 → 3)-D-Xyl		$C_{11}H_{20}O_{10}$	—	+87.5	91
O-β-D-Glcp-(1 → 3)-D-Xyl		$C_{11}H_{20}O_{10}$	—	−0.6(c 0.4)	90
O-α-D-Glcp-(1 → 4)-D-Xyl	Dihydrate	$C_{11}H_{24}O_{10}$	78	+97.5	92
O-α-D-Glcp-(1 → 1)-D-Galf		$C_{12}H_{22}O_{11}$	—	—	72
O-β-D-Glcp-(1 → 2)-D-Gal		$C_{12}H_{22}O_{11}$	171–172	+42.6	93
O-α-D-Glcp-(1 → 3)-D-Gal		$C_{12}H_{22}O_{11}$	—	+138	94
O-β-D-Glcp-(1 → 3)-D-Gal (Solabiose)		$C_{12}H_{22}O_{11}$	200	+35 → +40	95
	Octaacetate	$C_{28}H_{38}O_{19}$	75	+27	95
O-α-D-Glcp-(1 → 4)-D-Gal		$C_{12}H_{22}O_{11}$		+140	96
O-β-D-Glcp-(1 → 4)-D-Gal		$C_{12}H_{22}O_{11}$	246–247	+41.5	97
	Octaacetate	$C_{28}H_{38}O_{19}$	165–166	+26.8	97
O-α-D-Glcp-(1 → 6)-D-Gal		$C_{12}H_{22}O_{11}$	—	—	98
O-β-D-Glcp-(1 → 6)-D-Gal		$C_{12}H_{22}O_{11}$	—	+10 → +20	98, 99
	Octaacetate	$C_{28}H_{38}O_{19}$	149–150	—	98, 99
O-α-D-Glcp-(1 → 1)-α-D-Glcp (αα′ Trehalose)		$C_{12}H_{22}O_{11}$	214–216	+199	103
	Octaacetate	$C_{28}H_{38}O_{19}$	98	+162.3(c 10, CHCl₃)	104
O-β-D-Glcp-(1 → 1)-β-D-Glcp ββ′-Trehalose)		$C_{12}H_{22}O_{11}$	130–135	−41.5	105, 106
	Octaacetate	$C_{28}H_{38}O_{19}$	181	−19(CHCl₃)	107
O-α-D-Glcp-(1 → 1)-β-D-Glcp (αβ-Trehalose)		$C_{12}H_{22}O_{11}$	150–153	+70(c 0.2)	106, 108
	Octaacetate	$C_{28}H_{38}O_{19}$	120,140	+67,+82(CHCl₃,)	106, 108

OLIGOSACCHARIDES (INCLUDING DISACCHARIDES) (Continued)

Substance[a] (Synonym) (A)	Derivative (B)	Chemical Formula (C)	Melting Point °C (D)	Specific Rotation[b] $[\alpha]_D$ (E)	Reference (F)
DISACCHARIDES (Continued)					
O-α-D-Glc$p$-(1 → 2)-D-Glc (Kojibiose)		$C_{12}H_{22}O_{11}$	188	+137	100, 101
	α-Octaacetate	$C_{28}H_{38}O_{19}$	166	+152($CHCl_3$)	113
	β-Octaacetate	$C_{28}H_{38}O_{19}$	118	+112($CHCl_3$)	114
O-β-D-Glc$p$-(1 → 2)-D-Glc		$C_{12}H_{22}O_{11}$	195–196	+20	105, 109, 110
(Sophorose)	α-Octaacetate	$C_{28}H_{38}O_{19}$	111	+45($CHCl_3$)	109, 110, 112
	β-Octaacetate	$C_{28}H_{38}O_{19}$	191–192	−8($CHCl_3$)	111
O-α-D-Glc$p$-(1 → 3)-D-Glc		$C_{12}H_{22}O_{11}$	—	+135,+145	113, 115, 116
(Nigerose)	Octaacetate	$C_{28}H_{38}O_{19}$	151–153	+80($CHCl_3$)	117
O-β-D-Glc$p$-(1 → 3)-α-D-Glc (α-Laminaribiose)		$C_{12}H_{22}O_{11}$	202–205	+25.5 → +17.5	118
	α-Octaacetate	$C_{28}H_{38}O_{19}$	77–78	+20($CHCl_3$)	119
O-β-D-Glc$p$-(1 → 3)-β-D-Glc (β-Laminaribiose)		$C_{12}H_{22}O_{11}$	188–192	+7.5 → +20.8	120
	β-Octaacetate	$C_{28}H_{38}O_{19}$	160–161	−28.6(17°) ($CHCl_3$)	121
O-α-D-Glc$p$-(1 → 4)-α-D-Glc$p$ (α-Maltose)		$C_{12}H_{22}O_{11}$	108	+173	122, 123
	α-Octaacetate	$C_{28}H_{38}O_{19}$	125	+123($CHCl_3$)	59
O-β-D-Glc$p$-(1 → 4)-β-D-Glc$p$ (β-Maltose)	Monohydrate	$C_{12}H_{24}O_{12}$	102–103	+112 → +130	122, 123
	β-Octaacetate	$C_{28}H_{38}O_{19}$	159–160	+62.6($CHCl_3$)	59
O-β-D-Glc$p$-(1 → 4)-β-D-Glc$p$ (β-Celiobiose)		$C_{12}H_{22}O_{11}$	225	+14.2 → +34.6(c 8)	105, 124
	α-Octaacetate	$C_{28}H_{38}O_{19}$	229	+41.0($CHCl_3$)	59
	β-Octaacetate	$C_{28}H_{38}O_{19}$	202	−14.7($CHCl_3$)	59
O-α-D-Glc$p$-(1 → 6)-D-Glc$p$ (Isomaltose)		$C_{12}H_{22}O_{11}$	—	+103.2,+122	125, 126
	β-Octaacetate	$C_{28}H_{38}O_{19}$	144–145	+96.9($CHCl_3$)	125
O-β-D-Glc$p$-(1 → 6)β-D-Glc$p$ (α-Gentiobiose)	Dimethanolate	$C_{14}H_{30}O_{13}$	85–86	+31 → +9.6	127
	α-Octaacetate	$C_{28}H_{38}O_{19}$	189	+52.4($CHCl_3$)	128, 129
O-β-D-Glc$p$-(1 → 6)-β-D-Glc$p$ (β-Gentiobiose)		$C_{12}H_{22}O_{11}$	190	−3 → +10.5	105, 130
	β-Octaacetate	$C_{28}H_{38}O_{19}$	193	−5.4(c 6,$CHCl_3$)	128, 129
O-β-D-Glc$p$-(1 → 1)-Mannitol		$C_{12}H_{24}O_{11}$	140–141	−18	161
O-D-Glc-(1 → 3)-D-Man		$C_{12}H_{22}O_{11}$	165	+27.9	131
	Octaacetate	$C_{28}H_{38}O_{19}$	142–143	+35.6($CHCl_3$)	131
O-β-D-Glc$p$-(1 → 3)-Mannitol		$C_{12}H_{24}O_{11}$	97–100	−6	162
O-β-D-Glc$p$-(1 → 3)-Man (Epimaltose)		$C_{12}H_{22}O_{11}$	213–215	+115	132, 133
	Octaacetate	$C_{28}H_{38}O_{19}$	157	+117($CHCl_3$)	132, 133
O-β-D-Glc$p$-(1 → 4)-α-D-Man (α-Epicellobiose)		$C_{12}H_{22}O_{11}$	135–137	+5.8,+12.5	134, 135
	α-Octaacetate	$C_{28}H_{38}O_{19}$	202–204	+36($CHCl_3$)	135, 136
O-β-D-Glc$p$-(1 → 4)-β-D-Man (β-Epicellobiose)		$C_{12}H_{22}O_{11}$	—	−6.5 → +5.8($CHCl_3$)	135, 136
	β-Octaacetate	$C_{28}H_{38}O_{19}$	165	−13	135, 136
O-β-D-Glc$p$-(1 → 6)-D-Man (Epigentiobiose)	Monohydrate	$C_{12}H_{22}O_{11}$	137–138	−11	137, 138, 139
	α-Octaacetate	$C_{28}H_{38}O_{19}$	110–112	+26 → +29($CHCl_3$)	137, 138, 139
	β-Octaacetate	$C_{28}H_{38}O_{19}$	132–140	−20.6	137, 138, 139
O-α-D-Glc$p$-(1 → 1)-D-Fru		$C_{12}H_{22}O_{11}$	—	+49	140
O-β-D-Glc$p$-(1 → 1)-D-Fru	Dihydrate	$C_{12}H_{26}O_{13}$	132–135	−59.2	141, 142
	Octaacetate	$C_{28}H_{38}O_{19}$	128–129	−14	143
O-α-D-Glc$p$-(1 → 3)-D-Fru (Turanose)		$C_{12}H_{22}O_{11}$	157	+22 → +75.3	144
	Octaacetate I	$C_{28}H_{38}O_{19}$	216–217	+20.5($CHCl_3$)	145
	Octaacetate II	$C_{28}H_{38}O_{19}$	158	+107($CHCl_3$)	145
O-α-D-Glc$p$-(1 → 4)-D-Fru (Maltulose)		$C_{12}H_{22}O_{11}$	113–115	+58 → +64	146
O-β-D-Glc$p$-(1 → 4)-D-Fru (Cellobiulose)		$C_{12}H_{22}O_{11}$	—	−60.1	149
O-α-D-Glc$p$-(1 → 5)-D-Fru (Leucrose)		$C_{12}H_{22}O_{11}$	161–163	−8.8 → 6.8	147
	Heptaacetate	$C_{26}H_{36}O_{18}$	150–151	—	147
	Phenylosazone	$C_{24}H_{34}N_4O_9$	186–188	—	148
O-α-D-Glc$p$-(1 → 6)-D-Fru (Palatinose, Isomaltulose)		$C_{12}H_{22}O_{11}$	—	+97.2	140, 150
	Phenylosazone	$C_{24}H_{34}N_4O_9$	173–175	—	140, 148
O-α-D-Glc$p$-(1 → 1)-α-D-GlcNp (Trehalosamine)	Hydrochloride	$C_{12}H_{24}N_{10}Cl$	—	+176(c 0.02)	151
	N-Acetylhepta-acetate	$C_{28}H_{39}NO_{18}$	100–102	—	151
O-α-D-Glc$p$-(1 → 4)-D-GlcN	Hydrochloride	$C_{12}H_{24}N_{10}Cl$	180–185	+81	153

OLIGOSACCHARIDES (INCLUDING DISACCHARIDES) (Continued)

Substance[a] (Synonym) (A)	Derivative (B)	Chemical Formula (C)	Melting Point °C (D)	Specific Rotation[b] $[\alpha]_D$ (E)	Reference (F)
DISACCHARIDES (Continued)					
O-α-D-Glcp-(1 → 4)-D-GlcNAc		$C_{14}H_{26}NO_{11}$	144–146	+39	153
O-β-D-Glcp-(1 → 4)-D-GlcNAc		$C_{14}H_{26}NO_{11}$	—	—	154
O-β-D-Glcp-(1 → 3)-D-GalNAc	Dihydrate	$C_{14}H_{29}NO_{13}$	155–157	+19	152
O-β-D-Glcp-(1 → 4)-L-Fuc		$C_{12}H_{22}O_{10}$	—	–71	155
	Heptaacetate	$C_{26}H_{36}O_{17}$	228–230	–59	155
O-β-D-Glcp-(1 → 4)-L-Rha (Scillabiose)		$C_{12}H_{22}O_{10}$	—	–24.8	156, 157
	Heptaacetate	$C_{26}H_{36}O_{17}$	96–97	–50.4	156, 157
O-D-Glc-(1 → 2)-L-Rha (Bryobioside)		$C_{12}H_{22}O_{10}$	175,188–190	+61.5,+66.1	159, 160
	Heptaacetate	$C_{26}H_{36}O_{17}$	127	$+67.2(CHCl_3)$	160
O-α-D-Glcp-(1 → 2)-D-GlcUA		$C_{12}H_{20}O_{12}$	—	+89	158
O-D-Glcp-(1 → 4)-D-2d-Glcp		$C_{12}H_{22}O_{10}$	—	+123	163
O-D-Glcp-(1 →)-3-O-Methyl-6d-D-Glcp (Kondurangobiose)		$C_{13}H_{24}O_{10}$	202–204	+20	164
	Hexaacetate	$C_{25}H_{36}O_{16}$	188	—	165
O-D-Glcp-(1 →)-3-O-Methyl-6d-D-Glc (Thevetobioside)		$C_{13}H_{24}O_{10}$	233	$–66.4 (CH_3 OH)$	166
	Hexaacetate	$C_{26}H_{36}O_{16}$	168–170	$–70(18°)(CHCl_3)$	166
O-α-D-Manp-(1 → 3)-D-Gal		$C_{12}H_{22}O_{11}$	—		167
O-D-Manp-(1 → 6)-D-Gal		$C_{12}H_{22}O_{11}$	—	+134	168
O-β-D-Manp-(1 → 4)-α-D-Glc		$C_{12}H_{22}O_{11}$	210–212	+12,+18	169, 170
	Octaacetate	$C_{28}H_{38}O_{19}$	162–168	—	171
O-α-D-Manp-(1 → 6)-D-Glc		$C_{12}H_{22}O_{11}$	—	+73	172
	Octaacetate	$C_{28}H_{38}O_{19}$	83–87	+33	172
O-β-D-Manp-(1 → 6)-D-Glc		$C_{12}H_{22}O_{11}$	209–210	–5	173
O-α-D-Manp-(1 → 2)-D-Man		$C_{12}H_{22}O_{11}$	—	+42	480
O-α-D-Manp-(1 → 3)-D-Man		$C_{12}H_{22}O_{11}$	—	+50	174
O-α-D-Manp-(1 → 4)-D-Man		$C_{12}H_{22}O_{11}$	—	+49	175
O-β-D-Manp-(1 → 4)-α-D-Manp		$C_{12}H_{22}O_{11}$	204	–5 → –8	176
	Octamethyl ether	$C_{20}H_{38}O_{11}$	—	–12	177
O-β-D-Manp-(1 → 4)-β-D-Manp		$C_{12}H_{22}O_{11}$	193–194	–7.7 → –2.2	178
O-α-D-Manp-(1 → 6)-D-Manp		$C_{12}H_{22}O_{11}$	196–197	+52, +62	177, 179
	Octaacetate	$C_{28}H_{38}O_{19}$	152–153	+19.6	180
O-β-D-Manp-(1 → 6)-D-Manp		$C_{12}H_{22}O_{11}$	—	–12.4	177
O-α-D-Manp-(1 → 3)-D-GlcNAc		$C_{14}H_{25}NO_{11}$	129–130	+61 → +58	181
O-α-D-Manp-(1 → 4)-D-GlcNAc		$C_{14}H_{25}NO_{11}$	154–156	+77 → +66 $(50\% CH_3 OH)$	182
	Heptaacetate	$C_{28}H_{34}O_{18}$	113–114	$+1.5(CHCl_3)$	182
O-α-D-Manp-(1 → 6)-D-GlcNAc		$C_{14}H_{25}NO_{11}$	142–144	+38 → +35	183
	Heptaacetate	$C_{28}H_{34}NO_{18}$	75–77	$+68(CHCl_3)$	183
O-β-D-Manp-(1 → 6)-D-GlcN	Hydrochloride	$C_{14}H_{25}NO_{11}$	—	+55	184
O-β-D-Fru f-(2 → 1)-α-D-Glcp (Sucrose)		$C_{12}H_{22}O_{11}$	188,170	+66.5	127
	Octaacetate	$C_{28}H_{38}O_{19}$	69, 75	$+59.6(CHCl_3)$	104
O-β-D-Fruf-(2 →)-D-Glc (Ceratose)		$C_{12}H_{22}O_{11}$	—	+21	191
O-β-D-Fruf-(2 → 6)-D-Glc		$C_{12}H_{22}O_{11}$	—	+5	192
O-D-Fruf-(2 → 2)D-Fruf-anhydride (Alliuminoside)		$C_{12}H_{20}O_{10}$	92–93	–23.8	185
	Hexaacetate	$C_{24}H_{32}O_{16}$	98–99	–29.3	185
O-β-D-Fruf-(2 → 1)-D-Fru (Inulobiose)		$C_{12}H_{22}O_{11}$	—	–32.5,72.4	186, 187
	Octaacetate	$C_{28}H_{38}O_{19}$	—	$–6.5,–14.2 (CHCl_3)$	186, 187
O-D-Fruf-(2 → 6)-D-Fru (Levanbiose)		$C_{12}H_{22}O_{11}$	—	–20.8(17°)	188, 189
O-D-Fruf-(2 →)-D-Fru (Sogdianose)		$C_{12}H_{22}O_{11}$	156–158	–16.4	185
	Octaacetate	$C_{28}H_{38}O_{19}$	94–95	$–28.7(CHCl_3)$	185
O-β-D-Fruf-(2 → 6)-D-2dGlc		$C_{12}H_{22}O_{10}$	56, 85	+62 → +26.8	193
O-α-L-Fucp-(1 → 2)-D-Gal		$C_{12}H_{22}O_{10}$	—	–56.7	194, 195
	Benzylglycoside	$C_{19}H_{29}O_{11}$	205–207	–97.8(c 0.92)	194
O-L-Fuc-(1 →)-D-Glc		$C_{12}H_{22}O_{10}$	—	—	196
O-α-D-Fucp-(1 → 6)-D-Glc		$C_{12}H_{22}O_{10}$	—	+125	197
O-α-L-Fucp-(1 → 2)-L-Fuc		$C_{12}H_{22}O_9$	185–190	–169	198, 199
O-α-L-Fucp-(1 → 3)-L-Fuc		$C_{12}H_{22}O_9$	198–200	–191	198
O-α-L-Fucp-(1 → 4)-L-Fuc		$C_{12}H_{22}O_9$	—	–170	198
O-α-L-Fucp-(1 → 4)-D-GlcNAc		$C_{14}H_{25}NO_{10}$	128–129	–24 → –25	200
	Hexaacetate	$C_{26}H_{37}O_{16}$	94–96	$(c\ 0.8,50\% CH_3OH)$ $–10°(CHCl_3)$	200

OLIGOSACCHARIDES (INCLUDING DISACCHARIDES) (Continued)

Substance[a] (Synonym) (A)	Derivative (B)	Chemical Formula (C)	Melting Point °C (D)	Specific Rotation[b] [α]$_D$ (E)	Reference (F)
DISACCHARIDES (Continued)					
O-α-L-Fuc*p*-(1 → 6)-D-GlcNAc		$C_{14}H_{25}NO_{10}$	—	−51	201
O-α-L-Fuc*p*-(1 → 2)-D-Tal		$C_{12}H_{22}NO_{10}$	—	−120	195
O-α-L-Rha*p*-(1 → 2)-D-Gal		$C_{12}H_{22}O_{10}$	75–80	−3.5	202
O-α-L-Rha*p*-(1 → 6)-D-Gal (Robinobiose)		$C_{12}H_{22}O_{10}$	—	0 → +2.7	203, 204
	Heptaacetate	$C_{26}H_{36}O_{17}$	113	−19	203
O-β-L-Rha*p*-(1 → 2 or 4)-D-Glc (Neohesperidose)		$C_{12}H_{22}O_{10}$	—	—	205
O-α-L-Rha*p*-(1 → 6)-D-Glc (Rutinose)		$C_{12}H_{22}O_{10}$	189–192	+0.8 → +3.2	206, 204
	β-Heptaacetate	$C_{26}H_{36}O_{17}$	167	−28	206
O-α-D-GalNAc*p*-(1 → 3)-D-Gal		$C_{14}H_{25}NO_{11}$	179–181	+150,+200.7	212, 213
O-α-D-GalNAc*p*-(1 → 4)-D-Gal		$C_{14}H_{25}NO_{11}$	186–187	+203	214
O-β-D-GlcNAc*p*-(1 → 3)-D-Gal (Lacto-*N*-biose II)		$C_{14}H_{25}NO_{11}$	131–133	+45.5 → +35.7	207
O-β-D-GlcNAc*p*-(1 → 6)-α-D-Gal*p*		$C_{14}H_{25}NO_{11}$	—	+125.6(*c* 0.2)	208
O-β-D-GacNAc*p*-(1 → 6)-β-D-Glc		$C_{14}H_{25}NO_{11}$	—	+3.7	209
	Heptaacetate	$C_{28}H_{39}NO_{18}$	218–219	−9.5	209
O-β-D-GlcNAc*p*-(1 → 4)-D-GlcNAc*p* (*N,N′*-Diacetyl-chitobiose)		$C_{16}H_{28}N_2O_{11}$	260–262	+16.6	268
	Hexaacetate	$C_{28}H_{40}N_2O_{17}$	308–309	+55(18°) (*c* 0.5,CH_3 COOH)	269
O-α-D-GlcNAc*p*-(1 → 6)-D-GlcNAc		$C_{16}H_{28}N_2O_{11}$	215	+125	210
O-β-D-GlcNAc*p*-(1 → 6)-D-GlcNAc		$C_{16}H_{28}N_2O_{11}$	200	+6	211
O-D-GlcNAc-(1 → 4)-D-GlcUA		$C_{14}H_{20}NO_{12}$	—	+106	270, 271
O-D-GalUA-(1 →)-L-Ara		$C_{11}H_{18}O_{11}$	—	+58.2	215
O-α-D-GalUA*p*-(1 → 4)-D-Xyl*p*		$C_{11}H_{18}O_{11}$	—	+67	216
O-D-GalUA*p*-(1 → 3)-D-Gal		$C_{12}H_{20}O_{12}$	—	—	217
O-D-GalUA*p*-(1 → 4)-D-Gal		$C_{12}H_{20}O_{12}$	—	—	218
O-D-GalUA*p*-(1 → 6)-D-Gal		$C_{12}H_{20}O_{12}$	—	—	219
O-GalUA-(1 →)-Fuc		$C_{12}H_{20}O_{11}$	—	—	220
O-α-D-GalUA*p*-(1 → 2)-L-Rha		$C_{12}H_{20}O_{11}$	—	+65 → +69	221, 222
O-α-D-GalUA*p*-(1 → 4)-D-GalUA		$C_{12}H_{18}O_{13}$	125–135	+154	223, 224
	Methyl α-glyco-side dimethyl ester	$C_{15}H_{24}O_{13}$	120–122	+162.6	225
O-α-D-GlcUA*p*-(1 → 2)-D-Xyl		$C_{11}H_{18}O_{11}$	—	+88 → +98	233
O-β-D-GlcUA*p*-(1 → 2)-D-Xyl		$C_{11}H_{18}O_{11}$	—	+5.7	234
O-α-D-GlcUA*p*-(1 → 3)-D-Xyl*p*		$C_{11}H_{18}O_{11}$	—	+18 → +57	235
O-β-D-GlcUA*p*-(1 → 3)-D-Xyl		$C_{11}H_{18}O_{11}$	—	+3.7	236
O-α-D-GlcUA*p*-(1 → 4)-D-Xyl		$C_{11}H_{18}O_{11}$	—	—	237
O-β-GlcUA-(1 → 2)-Lyx		$C_{11}H_{18}O_{11}$	—	—	238
O-α-D-GlcUA*p*-(1 → 3)-D-Gal		$C_{12}H_{20}O_{12}$	—	—	226
O-α-D-GlcUA-(1 → 4)-D-Gal	Ba-salt	$C_{12}H_{19}O_{12}Ba$	—	+67	227
O-β-D-GlcU A*p*-(1 → 4)-D-Gal		$C_{12}H_{20}O_{12}$	—	+15	228
O-β-D-GlcUA*p*-(1 → 6)-D-Gal		$C_{12}H_{20}O_{12}$	116–120	−3	229
O-β-D-GlcUA*p*-(1 → 4)-D-Glc (Cellobiouronic acid)		$C_{12}H_{20}O_{12}$	189	+7.6	230, 231
	Heptaacetate	$C_{26}H_{34}O_{19}$	239	+32.9	230
O-β-D-GlcUA*p*-(1 → 6)-D-Glc	α-Heptaacetate methyl ester	$C_{27}H_{36}O_{19}$	201–202	+48.4	232
O-β-D-GlcUA*p*-(1 → 2)-D-Man*p*		$C_{12}H_{20}O_{12}$	—	−32 → −33	239, 240
O-α-D-GlcUA*p*-(1 → 4)-D-Man*p*		$C_{12}H_{20}O_{12}$	—	—	241
O-α-D-GlcUA*p*-(1 → 2)-L-Rha		$C_{12}H_{20}O_{11}$	—	+63	242
O-β-D-GlcUA*p*-(1 → 4)-L-Rha		$C_{14}H_{20}O_{11}$	—	−6 to −22	243, 244
O-β-D-GlcUA*p*-(1 → 3)-D-GalNAc (*N*-Acetylchondrosine)		$C_{14}H_{23}NO_{12}$	—	+35	276, 277
O-β-D-GlcUA*p*-(1 → 3)-D-GlcNAc (*N*-Acetylhyalobiuronic acid)		$C_{14}H_{23}NO_{12}$	—	−32(28°)	278
O-D-GlcUA*p*-(1 → 6)-D-GlcN		$C_{12}H_{21}NO_{11}$	—	—	245
O-β-D-GlcUA*p*-(1 → 2)-D-GlcUA	Ba-salt	$C_{12}H_{16}O_{13}Ba_2$	—	−5.2	246
O-α-NANA-(2 → 6)-D-GalNAc		$C_{19}H_{32}N_2O_{14}$	—	—	247
O-α-NANA-(2 → 6)-D-GlcNAc		$C_{19}H_{32}N_2O_{14}$	—	—	248
O-α-D-Gal*p*-(1 → 1)-glycerol		$C_9H_{18}O_8$	150–152	+155	249
O-β-D-Gal*p*-(1 → 1)-glycerol		$C_9H_{18}O_8$	139–140	+3.8,−7,−73	* 249–251
	Hexabenzoate	$C_{51}H_{42}O_{14}$	133–134	−6($CHCl_3$)	251
O-α-D-Gal*p*-(1 → 2)-glycerol (Floridoside)		$C_9H_{18}O_8$	86–87 129–130	+151,+163	252, 253

OLIGOSACCHARIDES (INCLUDING DISACCHARIDES) (Continued)

Substance[a] (Synonym) (A)	Derivative (B)	Chemical Formula (C)	Melting Point °C (D)	Specific Rotation[b] $[\alpha]_D$ (E)	Reference (F)
DISACCHARIDES (Continued)					
	Hexaacetate	$C_{21}H_{30}O_{14}$	101	+114(acetone)	255
O-β-D-Galp-(1 → 2)-D-erythritol		$C_{10}H_{20}O_9$	182–184	+7	275
	Heptaacetate	$C_{24}H_{34}O_{11}$	114	+109(c 0.6 acetone)	252
O-α-D-Galp-(1 → 1)-myo-inositol (Galactinol)	Dihydrate	$C_{12}H_{26}O_{13}$	220–222	+135.6	256
	Nonamethyl ether	$C_{21}H_{40}O_{11}$	96.5–98	+119	257
O-β-D-Galp-(1 → 5)-myo-inositol		$C_{12}H_{22}O_{11}$	248–252	0	258
O-β-D-Galp-(1 → 3)-D-mannitol (Peltigeroside)		$C_{12}H_{24}O_{11}$	161–163	−55.5,−61	259, 260
O-3,6-Anhydro-L-Galp-(1 → 3)-D-Gal (Neoagarobiose)		$C_{12}H_{20}O_{10}$	207–208	+34.4 → +203	261
	Hexaacetate	$C_{24}H_{32}O_{16}$	112	+1.6(27°) (CHCl$_3$)	261
O-(4,5-Anhydro-α-D-GalUAp)-(1 → 4)-D-GalUA		$C_{12}H_{16}O_{12}$	—	+177.8	262
O-α-D-Glcp-(1 → 1)-glycerol		$C_9H_{18}O_8$	—	+128(27°) (c0.7)	263
	Hexaacetate	$C_{21}H_{30}O_{14}$	92	+110(27°) (acetone)	263
O-β-D-Glcp-(1 → 1)-glycerol		$C_9H_{18}O_8$	—	+12	264, 265
	Hexa-P-nitro-benzoate	$C_{51}H_{36}N_6O_{26}$	105–108	—	265
O-D-Glcp-(1 → 1)-myoinositol		$C_{12}H_{22}O_{11}$	256–260	−17.5	258
	Nonamethylether	$C_{21}H_{40}O_{11}$	95–98	−10(C$_2$H$_5$OH)	258
O-D-Glcp-(1 → 5)-myoinositol		$C_{12}H_{22}O_{11}$	158–162	−20	258
O-α-D-Glcp-(1 → 2)-L-Sor		$C_{12}H_{22}O_{11}$	178–180	+33	266
	Octaacetate	$C_{28}H_{38}O_{19}$	—	+38(CHCl$_3$)	266
O-α-D-Glcp-(1 → 2)-β-D-*threo*-pentuloside (Glucoxyluloside)		$C_{11}H_{20}O_{10}$	156–157	+43	267
	Heptaacetate	$C_{25}H_{34}O_{17}$	180–181	+22(CHCl$_3$)	267
O-β-D-Manp-(1 → 4)-meso-erythritol		$C_{10}H_{20}O_9$	160–162	−38	272, 273
O-α-D-Manp-(1 → ?)-glyceric acid		$C_9H_{16}O_9$	88–89	+105(15°)	274
O-(4,5-Dideoxy-α-L-threo-hex-4-enopyranosyluronic acid)-(1 → 3)-D-GlcNAc		$C_{14}H_{21}NO_{11}$	—	−20	279
	Pentaacetate	$C_{24}H_{31}NO_{16}$	190–192	—	279
O-α-D-*ribo*-Hexopyranosyl-3-ulose-(1 → 2)-β-D-Fruf (3-Ketosucrose)	Trihydrate	$C_{12}H_{26}O_{14}$	82–83	+40(c 7.5)	280
O-α-D-*ribo*-Hexopyranosyl-3-ulose-(1 → 1)-α-D-Glcp (3-Ketotrehalose)		$C_{12}H_{20}O_{11}$	114	+151.1	281
O-α-D-*ribo*-Hexopyranosyl-3-ulose-(1 → 4)-D-Glcp (3-Ketomaltose)		$C_{12}H_{20}O_{11}$	107	+87.2	281
O-β-D-*xylo*-Hexopyranosyl-3-ulose-(1 → 4)-D-Glcp (Keto-lactose)		$C_{12}H_{20}O_{11}$	118–120	+39.2	281
O-4,5-Anhydro-β-D-ManUAp-(1 → 4)-D-ManUA		$C_{12}H_{16}O_{12}$	135–136.5	−8	282
TRISACCHARIDES					
O-α-L-Arap-(1 → 5)-O-α-L-Arap-(1 → 5)-L-Ara		$C_{15}H_{26}O_{13}$	—	−35	285
O-α-L-Araf-(1 → 3)-O-D-Xylp-(1 → 4)-D-Xylp		$C_{15}H_{26}O_{13}$	—	−19.3	286
	Octamethyl ether	$C_{23}H_{42}O_{13}$	—	−13.3(CH$_3$OH)	286, 287
O-D-Xylp-(1 → 3)-O-D-Xylp-(1 → 4)-D-Xylp		$C_{15}H_{26}O_{13}$	225	−52 → −47	288
O-β-D-Xylp-(1 → 4)-O-β-D-Xylp-(1 → 4)-β-D-Xylp (Xylotriose)		$C_{15}H_{26}O_{13}$	215–216	−48.1	289, 290
	Octaacetate	$C_{31}H_{42}O_{21}$	108–109.5	−83(CHCl$_3$)	290
O-α-D-Xylp-(1 → 6)-O-β-D-Glcp-(1 → 4)-D-Glcp		$C_{17}H_{30}O_{15}$	147–150	+150	169
O-L-Gal-(1 → 4)-Xyl-(1 → 2)-L-Ara		$C_{16}H_{28}O_{14}$	217–219	−61	291
O-α-D-Galp-(1 → 3)-O-α-D-Galp-(1 → 3)-D-Gal		$C_{18}H_{32}O_{16}$	237–239	+146	292
O-β-D-Galp-(1 → 3)-O-α-D-Galp-(1 → 3)-D-Gal		$C_{18}H_{32}O_{16}$	240–245	+51	293
O-D-Gal-(1 → 6)-O-D-Gal-(1 → 3)-D-Gal		$C_{18}H_{32}O_{16}$	—	+20.5,+36	294, 295
O-α-D-Gal-(1 → 6)-O-α-D-Gal-(1 → 6)-D-Gal		$C_{18}H_{32}O_{16}$	—	—	296
O-β-D-Galp-(1 → 3)-O-β-D-Galp-(1 → 4)-D-Glc	Trihydrate	$C_{18}H_{38}O_{19}$	197–200	+56 → +43	49
	Undecaacetate	$C_{40}H_{54}O_{27}$	108–110	+17.2(c 0.61 CHCl$_3$)	297
O-β-D-Galp-(1 → 4)-O-β-D-Galp-(1 → 4)-D-Glcp		$C_{18}H_{32}O_{16}$	228–231	+68 → +4S	34
O-α-D-Galp-(1 → 6)-O-β-D-Galp-(1 → 4)-D-Glc		$C_{18}H_{32}O_{16}$	187, 167	+34	49, 298, 299
O-α-D-Galp-(1 → 6)-O-α-D-Galp-(1 → 6)-D-Glc		$C_{18}H_{32}O_{16}$	150	+167	300
	Undecaacetate	$C_{40}H_{54}O_{27}$	105	+135(C$_2$H$_5$OH)	300
O-β-D-Galp-(1 → 6)-O-β-D-Galp-(1 → 6)-D-Glc		$C_{18}H_{32}O_{16}$	—	—	301
O-D-Galp-(1 →)-O-D-Galp-(1 →)-D-Glc (Lactotriose)		$C_{18}H_{32}O_{16}$	—	—	302
	Undecaacetate	$C_{40}H_{54}O_{27}$	120–122	—	302
O-β-D-Galp-(1 → 6)-O-β-D-Galp-(1 →)-D-Fru		$C_{18}H_{32}O_{16}$	—	−28	49
O-α-D-Galp-(1 → 6)-O-α-D-Galp-(1 → 6)-D-Manp		$C_{18}H_{32}O_{16}$	—	+131	303

OLIGOSACCHARIDES (INCLUDING DISACCHARIDES) (Continued)

Substance[a] (Synonym)	Derivative	Chemical Formula	Melting Point °C	Specific Rotation[b] $[\alpha]_D$	Reference
(A)	(B)	(C)	(D)	(E)	(F)

TRISACCHARIDES (Continued)

Substance[a] (Synonym)	Derivative	Chemical Formula	Melting Point °C	Specific Rotation[b] $[\alpha]_D$	Reference
O-β-D-Galp-(1 → 4)-O-β-D-Glcp-(1 → 4)-D-Gal		$C_{18}H_{32}O_{16}$	—	+22.2	304
	Phenylosazone	$C_{30}H_{44}N_4O_{14}$	211	—	304
O-β-D-Glcp-(1 → 4)-O-β-D-Glcp-(1 → 6)-D-Glcp		$C_{18}H_{32}O_{16}$	257	+22	305
	Phenylosazone	$C_{30}H_{44}N_4O_{14}$	233	−50.5	305
O-α-D-Galp-(1 → 2)-O-α-D-Glcp-(1 → 2)-β-D-Fruf (Umbelliferose)		$C_{18}H_{32}O_{16}$	—	+125	306
O-α-D-Galp-(1 → 3)-O-α-D-Glcp-(1 → 2)-β-D-Fruf		$C_{18}H_{32}O_{16}$	—	+90.5	307
O-β-D-Galp-(1 → 4)-O-α-D-Glcp-(1 → 2)-β-D-Fruf (Lactosylsucrose)	Pentahydrate	$C_{18}H_{42}O_{21}$	181	+59	308
	Undecaacetate	$C_{40}H_{54}O_{27}$	131	+44(CHCl_3,)	309
O-α-D-Galp-(1 → 6)-O-α-D-Glcp-(1 → 2)-β-D-Fruf (Raffinose)	Pentahydrate	$C_{18}H_{42}O_{21}$	77–78,118	+101,+123	310
	Undecaacetate	$C_{40}H_{54}O_{27}$	99–101	+92,+100(c 8, C_2H_5OH)	311, 312
O-α-D-Galp-(1 → 6)-O-β-D-Manp-(1 → 4)-D-Man		$C_{18}H_{32}O_{16}$	228–229	+98.4	313
O-α-D-Galp-(1 → 1)-O-β-D-Fruf-(2 → 1)-D-Glc		$C_{18}H_{32}O_{16}$	—	+131	314
O-α-D-Galp-(1 → 3)-O-β-D-Fruf-(2 → 1)-D-Glcp		$C_{18}H_{32}O_{16}$	—	+98	314
O-β-D-Galp-(1 → 4)-O-β-D-Fruf-(2 → 1)-D-Glcp		$C_{18}H_{32}O_{16}$	—	+44.1	315
O-α-D-Galp-(1 → 6)-O-β-D-Fruf-(2 → 1)-D-Glcp (Planteose)	Dihydrate	$C_{18}H_{36}O_{18}$	123–124	+125.2	316
	Undecaacetate	$C_{40}H_{54}O_{27}$	135	+97(CHCl_3)	316
O-β-D-Galp-(1 → 6)-O-β-D-Fruf (2 → 1)-α-D-Glcp		$C_{18}H_{32}O_{16}$	—	—	317
O-D-Galp-(1 → 3)-O-[β-D-Glcp-(1 → 4)]-L-Fuc		$C_{18}H_{32}O_{15}$	—	+23	318
O-β-D-Galp-(1 → 4)-O-[α-L-Fucp-(1 → 3)]-D-Glc (3-Fucosyllaetose)		$C_{18}H_{32}O_{15}$	—	—	319
O-β-D-Galp-(1 → 4)-O-[α-D-Glcp-(1 → 2)]-D-Glcp		$C_{18}H_{32}O_{16}$	—	+103	320, 321
O-β-D-Galp-(1 → 3)-O-β-D-GlcNAcp-(1 → 3)-D-Gal (Lacto-N-triose I)	Dihydrate	$C_{20}H_{39}NO_{18}$	202	+40.7	322, 323
	Phenylosazone	$C_{32}H_{47}N_5O_{14}$	230	—	324, 325
O-β-D-Galp-(1 → 4)-O-β-D-GlcNAp-(1 → 3)-D-Gal		$C_{20}H_{35}O_{16}$	—	—	326
O-D-Gal-(1 →)-GalUA-(1 →)-L-Rha		$C_{18}H_{30}O_{17}$	—	+5.2	327
O-β-D-Glcp-(1 → 2)-O-β-D-Glcp-(1 → 4)-D-Galp		$C_{18}H_{32}O_{16}$	250–260	+13.1	328, 329
O-β-D-Glcp-(1 → 4)-O-β-D-Glcp-(1 → 6)-D-Gal		$C_{18}H_{32}O_{16}$	—	+9.5	330
	Phenylosazone		207	—	330
O-α-D-Glcp-(1 → 2)-O-[β-D-Galp-(1 → 4)]-D-Glcp		$C_{18}H_{32}O_{16}$	—	+103	331
O-α-D-Glcp-(1 → 2)-O-α-D-Glcp-(1 → 2)-D-Glc (Kojitriose)		$C_{18}H_{32}O_{16}$	—	—	358
O-α-D-Glcp-(1 → 3)-O-β-D-Glcp-(1 → 1)-α-D-Glcp		$C_{18}H_{32}O_{16}$	—	+139	367
O-α-D-Glcp-(1 → 3)-O-α-D-Glcp-(1 → 3)-D-Glc		$C_{18}H_{32}O_{16}$	—	—	365
O-β-D-Glcp-(1 → 3)-O-β-D-Glcp-(1 → 3)-β-D-Glcp (Laminaritriose)		$C_{18}H_{32}O_{16}$	—	+2.4(17°)	368
O-α-D-Glcp-(1 → 3)-O-β-D-Glcp-(1 → 4)-D-Glcp		$C_{18}H_{32}O_{16}$	—	+169.5	337, 338
O-β-D-Glcp-(1 → 3)-O-β-D-Glcp-(1 → 4)-D-Glcp		$C_{18}H_{32}O_{16}$	229–231	+11.6+13.6	332, 333
	Undecaacetate	$C_{40}H_{54}O_{27}$	120–122	−20	332, 333
O-β-D-Glcp-(1 → 3)-O-β-D-Glcp-(1 → 6)-D-Glc		$C_{18}H_{32}O_{16}$	—	−3.5 → −6	353
O-α-D-Glcp-(1 → 4)-O-α-D-Glcp-(1 → 3)-D-Glc		$C_{18}H_{32}O_{16}$	—	—	337, 338
O-β-D-Glcp-(1 → 4)-O-β-D-Glcp-(1 → 3)-D-Glc		$C_{18}H_{32}O_{16}$	236–239	+16.5 → +11.7	339
	Undecaacetate I	$C_{40}H_{54}O_{27}$	108–110	−8.3(CHCl_3)	340
	Undecaacetate II	$C_{40}H_{54}O_{27}$	186–188	−24(CHCl_3)	341
O-α-D-Glcp-(1 → 4)-O-α-D-Glcp-(1 → 4)-D-Glcp (Maltotriose)		$C_{18}H_{32}O_{16}$	150	+160	342, 343
	Undecaacetate	$C_{40}H_{54}O_{27}$	134–136	+86(CHCl_3)	342, 344
O-β-D-Glcp-(1 → 4)-O-β-D-Glcp-(1 → 4)-D-Glcp (Cellotriose)		$C_{18}H_{32}O_{16}$	—	+25	340, 345
	α-Undecaacetate	$C_{40}H_{54}O_{27}$	220–222	+226(CHCl_3)	346
	β-Undecaacetate	$C_{40}H_{54}O_{27}$	199.5–200.5	−10.8(CHCl_3)	340
O-α-D-Glcp-(1 → 4)-O-α-D-Glcp-(1 → 6)-D-Glc		$C_{18}H_{32}O_{16}$	—	+159	347
	Undecaacetate	$C_{40}H_{54}O_{27}$	139–141	+112(CHCl_3)	347
O-α-D-Glcp-(1 → 4)-O-β-D-Glcp-(1 → 6)-D-Glc	α-Undecaacetate	$C_{40}H_{54}O_{27}$	174–176	+80	354
	β-Undecaacetate	$C_{40}H_{54}O_{27}$	233	+42	355
O-β-D-Glcp-(1 → 4)-O-α-D-Glcp-(1 → 6)-D-Glc		$C_{18}H_{32}O_{16}$	—	—	357
O-β-D-Glcp-(1 → 4)-O-β-D-Glcp-(1 → 6)-D-Glc		$C_{18}H_{32}O_{14}$	247–252	+8.4	356
O-α-D-Glcp-(1 → 6)-O-α-D-Glcp-(1 → 3)-D-Glc		$C_{18}H_{32}O_{16}$	—	+150	366
	Undecaacetate	$C_{40}H_{54}O_{27}$	117–119	+117 → +121	366

OLIGOSACCHARIDES (INCLUDING DISACCHARIDES) (Continued)

Substance[a] (Synonym) (A)	Derivative (B)	Chemical Formula (C)	Melting Point °C (D)	Specific Rotation[b] $[\alpha]_D$ (E)	Reference (F)
		TRISACCHARIDES (Continued)			
O-α-D-Glc$p$-(1 → 6)-O-β-D-Glc$p$-(1 → 3)-D-Glc		$C_{18}H_{32}O_{16}$	—	+67	359
O-β-D-Glc$p$-(1 → 6)-O-[β-D-Glc$p$-(1 → 3)]-D-Glc		$C_{18}H_{32}O_{16}$	—	+14.4	360
O-β-D-Glc$p$-(1 → 6)-O-β-D-Glc$p$-(1 → 3)-D-Glc		$C_{18}H_{32}O_{16}$	—	−3.2 → −4.2	360, 361
O-α-D-Glc$p$-(1 → 6)-O-[α-D-Glc$p$-(1 → 4)]-D-Glc$p$		$C_{18}H_{32}O_{16}$	—	—	362
O-β-D-Glc$p$-(1 → 6)-O-[α-D-Glc$p$-(1 → 4)]-D-Glc$p$		$C_{18}H_{32}O_{16}$	—	+84	363, 364
O-β-D-Glc$p$-(1 → 6)-O-β-D-Glc$p$-(1 → 4)-D-Glc$p$		$C_{18}H_{32}O_{16}$	—	+9.2 → +10.2	335, 336
	Undecaacetate	$C_{40}H_{54}O_{27}$	205	−13	335
O-α-D-Glc$p$-(1 → 6)-O-α-D-Glc$p$-(1 → 4)-D-Glc$p$ (Panose)		$C_{18}H_{32}O_{16}$	222–224	+154	348, 349
O-β-D-Glc$p$-(1 → 6)-O-[β-D-Glc$p$-(1 → 4)]-D-Glc$p$		$C_{18}H_{32}O_{16}$	—	—	334
O-α-D-Glc$p$-(1 → 6)-O-α-D-Glc$p$-(1 → 6)-D-Glc (Isomaltotriose)		$C_{18}H_{32}O_{16}$	—	+145	126
	Undecabenzoate	$C_{95}H_{87}O_{27}$	226–227	+131(CHCl$_3$)	350
O-β-D-Glc$p$-(1 → 6)-O-β-D-Glc$p$-(1 → 6)-D-Glc (Gentiotriose)		$C_{18}H_{32}O_{16}$	—	−10.3	351, 352
	Undecaacetate	$C_{40}H_{54}O_{27}$	214–215	−9.4(CHCl$_3$)	352
O-α-D-Glc$p$-(1 → 4)-O-α-D-Glc$p$-(1 → 2)-β-D-Fru$f$ (Erlose)		$C_{18}H_{32}O_{16}$	—	+121.8	369, 370
	Undecaacetate	$C_{40}H_{54}O_{27}$	68–73	+92.7(15°)	348
O-α-D-Glc$p$-(1 → 6)-O-α-D-Glc$p$-(1 → 2)-β-D-Fru$f$		$C_{18}H_{32}O_{16}$	118–120	+102.5(18°)	371, 372
O-β-D-Glc$p$-(1 → 6)-O-α-D-Glc$p$ (1 → 2)-β-D-Fru$f$ (Gentianose)		$C_{18}H_{32}O_{16}$	210	+33.4	373, 374
O-α-D-Glc$p$-(1 → 6)-O-α-D-Glc$p$-(1 → 5)-D-Fru$f$		$C_{18}H_{32}O_{16}$	—	—	148
O-α-D-Glc$p$-(1 → 6)-O-α-D-Glc$p$-(1 → 6)-D-Fru$f$ (Isomaltotriulose)		$C_{18}H_{32}O_{16}$	—	+118	140
O-α-D-Glc$p$-(1 → 6)-O-α-D-Glc$p$-(1 → 2)-D-GlcUA		$C_{18}H_{30}O_{17}$	—	+110(c0.07)	389
O-β-D-Glc$p$-(1 → 6)-O-β-D-Glc$p$-(1 → 2)-L-Rha		$C_{18}H_{32}O_{15}$	—	—	160
	Nonaacetate	$C_{36}H_{50}O_{24}$	172–173	+54(CHCl$_3$)	160
O-β-D-Glc$p$-(1 → 4)-O-β-D-Man$p$-(1 → 4)-D-Man$p$		$C_{18}H_{32}O_{16}$	—	−9	388
O-α-D-Glc$p$-(1 → 2)-O-β-D-Fru$f$-(1 → 2)-O-β-D-Fru$f$ (Isokestose)		$C_{18}H_{32}O_{16}$	200-201	+28.9	375, 390
	Undecamethyl ether	$C_{29}H_{54}O_{16}$	—	+27.9	376–378
O-α-D-Glc$p$-(1 → 2)-O-β-D-Fru$f$-(1 → 2)-β-D-Fru$f$ (Kestose)		$C_{18}H_{32}O_{16}$	145	+28	375, 379, 391, 392
	Undecamethyl ether	$C_{29}H_{54}O_{16}$	—	+25.8(18°)	379
O-α-D-Glc$p$-(1–3)-O-β-D-Fru$f$-(2 → l)-α-D-Glc$p$ (Melezitose)	Dihydrate	$C_{18}H_{36}O_{18}$	153–154	+88.2	380, 381
	Undecaacetate	$C_{40}H_{54}O_{27}$	117	+103.6(CHCl$_3$)	382
O-β-D-Man$p$-(1 → 4)-O-α-D-Glc$p$-(1 → 4)-D-Glc$p$		$C_{18}H_{32}O_{16}$	—	+9.5(28°)	169
O-β-D-Man$p$-(1 → 4)-O-α-D-Man$p$-(1 → 4)-D-Glc$p$		$C_{18}H_{32}O_{16}$	—	−12±3	169, 383
O-D-Man$p$-(1 → 6)-O-D-Man$p$-(1 → 6)-D-Glc$p$ (Laevidulinose)		$C_{18}H_{32}O_{16}$	—	−11.5,−15	384, 385
	Undecaacetate	$C_{40}H_{54}O_{27}$	95–100	+18	384, 385
O-α-D-Man$p$-(1 → 2)-O-α-D-Man$p$-(1 → 2)-D-Man		$C_{18}H_{32}O_{16}$	—	+63.0	480
O-β-D-Man$p$-(1 → 4)-O-β-D-Man$p$-(1 → 4)-α-D-Man$p$		$C_{18}H_{32}O_{16}$	214–216	−23	176
O-α-D-Man$p$-(1 → 6)-O-α-D-Man$p$-(1 → 6)-D-Man		$C_{18}H_{32}O_{16}$	—	+71.4	480
O-α-D-Man$p$-(1 → 6)-O-β-D-Glc$p$-(1 → 6)-D-Glc	Undecaacetate	$C_{40}H_{54}O_{27}$	118–119	+20.2(CHCl$_3$)	386
O-α-D-Man$p$-(1 → 6)-O-β-D-GlcNAcp-(1 → 4)-D-GlcNAc		$C_{22}H_{38}N_2O_{16}$	186–188	+36 → +34	387
	Nonaacetate	$C_{40}H_{56}N_2O_{16}$	134–135	+37°(CHCl$_3$)	387
O-β-D-Fru$f$-(2 → 6)-O-α-D-Glc$p$-(1–2)-β-D-Fru$f$ (Neokestose)		$C_{18}H_{32}O_{16}$	—	+21 → +22	390
O-D-Fru-(2 → ?)-O-D-Gal-(1 → ?)-D-Fru (Labiose)	Trihydrate	$C_{18}H_{38}O_{19}$	126–128	+136	394
	Undecaacetate	$C_{40}H_{54}O_{19}$	88	+122.5(CHCl$_3$)	394
O-D-Fru$f$-(2 → 1)-O-D-Fru$f$-(2 → 1)-D-Fru (Inulotriose)		$C_{18}H_{32}O_{16}$	—	—	401
O D Fru$f$ (2 → ?) O D Fru$f$ (2 → 2) D Fru$f$ anhydride (Polygontin)		$C_{18}H_{30}O_{15}$	207–208	−52.9	185
	Nonaacetate	$C_{36}H_{48}O_{24}$	84–85	−38.4(CHCl$_3$)	185
O-D-Fru$f$-(2 →)-O-D-Fru$f$-(2 → 2)-D-Fru$f$ (Trifructan)		$C_{18}H_{32}O_{16}$	—	−22.3	402
	Decaacetate	$C_{36}H_{52}O_{26}$	—	+8.5	402
O-α-L-Fuc$p$-(1 → 2)-O-β-D-Gal$p$-(1 → 4)-D-Glc (2′-Fucosyllactose)		$C_{18}H_{32}O_{15}$	230-231	−53.5 → −57.5 (c 0.2)	395

OLIGOSACCHARIDES (INCLUDING DISACCHARIDES) (Continued)

Substance[a] (Synonym) (A)	Derivative (B)	Chemical Formula (C)	Melting Point °C (D)	Specific Rotation[b] $[\alpha]_D$ (E)	Reference (F)
	TRISACCHARIDES (Continued)				
	p-Tolylsulfonyl hydrazone	$C_{25}H_{40}N_2O_{17}S$	205–206	$-73(C_5H_5N:H_2O)$	395
O-L-Fuc*p*-(1 → 4)-*O*-D-Glc*p*-(1 → 4)-D-Glc		$C_{18}H_{32}O_{15}$	—	—	396
O-α-D-GalNAc*p*-(1 → 3)-*O*-β-D-Gal*p*-(1 → 3)-D-GlcNAc		$C_{22}H_{38}N_2O_{16}$	—	+110,+136	397, 398
O-α-D-GalNAc*p*-(1 → 3)-*O*-β-D-Gal*p*-(1 → 4)-D-GlcNAc		$C_{22}H_{38}N_2O_{16}$	—	+147	398
O-α-D-GalNAc*p*-(1 → 4)-*O*-β-D-Gal*p*-(1 → 4)-D-GlcNAc		$C_{22}H_{38}N_2O_{16}$	273–275	+140	400
O-β-D-GlcNAc*p*-(1 → 3)-*O*-β-D-Gal*p*-(1 → 3)-D-GlcNAc*p*		$C_{22}H_{38}N_2O_{16}$	—	+19.5	403
O-β-D-GlcN Ac*p*-(1 → 6)-*O*-β-D-Gal*p*-(1 → 3)-D-GlcNAc*p*		$C_{22}H_{38}N_2O_{16}$	—	+51.6	403
O-β-D-GlcNAc*p*-(1 → 3)-*O*-[β-D-GlcNAc*p*-(1 → 6)]-D-Gal*p*		$C_{22}H_{38}N_2O_{16}$	—	+6.5	404
O-β-D-GlcNAc*p*-(1 → 3)-*O*-β-D-Gal*p*-(1 → 4)-D-Glc (Lacto-*N*-triose II)		$C_{20}H_{35}NO_{16}$	201–202	+40.7	405–407
	Phenylosazone	$C_{32}H_{47}N_5O_{14}$	230	—	405–407
O-β-D-GlcNAc*p*-(1 → 4)-*O*-β-D-GlcNAc*p*-(1 → 4)-D-GlcNAc*p* (Chitotriose)		$C_{24}H_{41}N_3O_{16}$	304–306	+3.8 → +2.2	269, 270
	Octaacetate	$C_{40}H_{57}N_3O_{24}$	315	+33(*c* 0.2, CH_3COOH)	408
O-β-D-GlcNAc*p*-(1 →)-*O*-β-D-GlcUA*p*-(1 → 3)-D-GlcNAc		$C_{22}H_{36}N_2O_{17}$	—	−16	409
O-D-GalUA-(1 →)-*O*-L-Ara-(1 →)-Xyl		$C_{16}H_{26}O_{15}$	—	—	410
O-α-D-GalUA-(1 → 2)-*O*-L-Rha-(1 → 4)-D-Gal		$C_{17}H_{28}O_{15}$	—	—	411
O-GlcUA-(1 → ?)-*O*-Xyl-(1 → ?)-Gal		$C_{17}H_{28}O_{16}$	—	+40	412
O-β-D-GlcUA*p*-(1 → 2)-*O*-D-Man-(1 → ?)-D-Glc		$C_{18}H_{30}O_{17}$	—	—	413
O-β-D-GlcUA*p*-(1 → 4)-*O*-β-D-Xyl*p*-(1 → 4)-D-Xyl*p*		$C_{16}H_{26}O_{15}$	—	+38	414
O-α-D-GalNAc*p*-(1 → 3)-[*O*-α-L-Fuc*p*-(1 → 2)]-D-Gal		$C_{20}H_{35}O_{15}$	—	—	484
O-α-D-Gal*p*-(1 → 3)-[*O*-α-L-Fuc*p*-(1 → 2)]-D-Gal		$C_{18}H_{32}O_{15}$	—	—	484
O-α-NANA-(2 → 3)-*O*-β-D-Gal*p*-(1 → 3)-D-GalNAc		$C_{25}H_{42}N_2O_{19}$			485
O-α-NANA-(2 → 3)-*O*-β-D-Gal*p*-(1 → 4)-D-Glc (3′-Sialyllactose)		$C_{23}H_{39}NO_{19}$	—	+16.8	415
		$C_{23}H_{39}NO_{19}$	—	+6 (DMSO)	416
O-α-NGNA-(2 → 3)-*O*-β-D-Gal*p*-(1 → 4)-D-Glc		$C_{23}H_{39}N_{20}S$	—	—	462
O-α-NANA-(2 → 3)-*O*-β-D-Gal*p*-6-Sulfate-(1 → 4)-D-Glc		$C_{23}H_{39}NO_{22}S$	—	—	417
O-α-NANA-(2 → 6)-*O*-β-D-Gal*p*-(1 → 4)-D-Glc (6′-Sialyllactose)		$C_{23}H_{39}NO_{19}$	—	+27.9	415
O-α-D-Gal*p*-(1 → 2)-*O*-α-D-Glc*p*-(1 → 1)-glycerol		$C_{15}H_{28}O_{13}$	171	+170±3	418
	Nonaacetate	$C_{33}H_{40}O_{22}$	97	+145±3 (*c* 0.5, CHCl₃)	418
O-α-D-Glc*p*-(1 → 6)-*O*-α-D-Glc*p*-(1 → 1)-glycerol		$C_{15}H_{28}O_{13}$	—	+148	263
O-β-D-Glc*p*-(1 → 1)-*O*-[β-D-Glc*p*-(1 → 6)]-D-Mannitol		$C_{18}H_{34}O_{16}$	—	−14	161
	Dodecaacetate	$C_{42}H_{58}O_{28}$	136–138	−7.7 (CHCl₃)	419
O-α-D-Man*p*-(1 → 3)-*O*-α-D-Gal*p*-(1 → 2)-glycerol		$C_{15}H_{28}O_{13}$	—	—	420
	Nonaacetate	$C_{33}H_{46}O_{22}$	153–154	+103 (CHCl₃)	420
O-α-D-GalUA*p*-(1 → 4)-*O*-α-D-GalUA*p*-(1 → 4)-D-GalUAp (Trigalacturonic acid)		$C_{18}H_{26}O_{19}$	135–142	+154, +187	224, 421
O-β-D-Xyl*p*-(1 → 4)-*O*-[α-L-Ara*f*-(1 → 3)]-*O*-β-D-Xyl*p*-(1 → 4)-D-Xyl		$C_{20}H_{34}O_{17}$	—	−75	422
	Decaacetate	$C_{40}H_{54}O_{27}$	179–180	−85 (CHCl₃)	422
O-D-Xyl-(1 → 4)-*O*-D-Xyl-(1 → 3)-*O*-D-Xyl-(1 → 4)-D-Xyl*p*		$C_{20}H_{34}O_{17}$	—	−56.7	288
O-β-D-Xyl*p*-(1 → 4)-*O*-β-D-Xyl*p*-(1 → 4)-*O*-β-D-Xyl*p*-(1 → 4)-D-Xyl*p* (Xylotetraose)		$C_{20}H_{34}O_{17}$	224–226	−61.9	289, 290
	Decaacetate	$C_{40}H_{54}O_{27}$	200–201	−92.4 (CHCl₃)	290
O-D-Gal*p*-(1 → ?)-*O*-D-Gal*p*-(1 → ?)-*O*-D-Fru*f*-(2 → l)-D-Glc*p* (Sesamose)		$C_{24}H_{42}O_{21}$	—		423
O-β-D-Gal*p*-(1 → 4)-*O*-β-D-Gal*p*-(1 → 4)-*O*-β-D-Gal*p*-(1 → 4)-D-Glc*p*		$C_{24}H_{42}O_{21}$	—	+43	34
O-D-Gal*p*-(1 → 4)-*O*-D-Gal*p*-(1 → 6)-*O*-α-D-Glc*p*-(1 → 2)-D-Fru*f*		$C_{24}H_{42}O_{21}$	—	+143	424
O-α-D-Gal*p*-(1 → 6)-*O*-α-D-Gal*p*-(1 → 6)-*O*-α-D-Glc*p*-(1 → 2)-D-Fru*f* (Stachyose)		$C_{24}H_{42}O_{21}$	140, 170	+131 ,+146	300, 425
	Tetradecaacetate	$C_{52}H_{70}O_{35}$	95–96	+120 (C₂H₅OH)	426
O-α-D-Gal*p*-(1 → 6)-*O*-α-D-Glc*p*-(1 → 2)-*O*-β-D-Fru*f*-(1 → 1)-α-D-Gal*p* (Lychnose)		$C_{24}H_{42}O_{21}$	—	+155	424
O-α-D-Gal*p*-(1 → 6)-*O*-α-D-Glc*p*-(1 → 2)-*O*-β-D-Fru*f*-(3 → 1)-α-D-Gal (Isolychnose)		$C_{24}H_{42}O_{21}$	—	—	427

OLIGOSACCHARIDES (INCLUDING DISACCHARIDES) (Continued)

Substance[a] (Synonym) (A)	Derivative (B)	Chemical Formula (C)	Melting Point °C (D)	Specific Rotation[b] $[\alpha]_D$ (E)	Reference (F)
TETRASACCHARIDES (Continued)					
O-D-Galp-(1 → 6)-O-D-Galp-(1 → 3)-O-D-Galp-(1 → 3)-D-Gal		$C_{24}H_{42}O_{21}$	—	—	428
O-D-Galp-(1 → 6)-O-D-Galp-(1 → 6)-O-D-Galp-(1 → 3)-D-Gal		$C_{24}H_{42}O_{21}$	—	—	428
O-β-D-Galp-(1 → 3)-O-β-D-GlcNAcp-(1 → 3)-O-β-D-Galp-(1 → 4)-D-Glc (Lacto-N-tetraose)		$C_{26}H_{45}O_{21}$	205	+25.2	70
O-β-D-Galp-(1 → 4)-O-β-D-GlcNAcp-(1 → 3)-O-β-D-Galp-(1–4)-D-Glc (Lacto-N-neotetraose)	Trihydrate	$C_{26}H_{51}NO_{24}$	214–218	+27	429
O-β-D-Glcp(1 → 2)-O-[β-D-Xylp-(1 → 3)]-O-β-D-Glcp-(1 → 4)-D-Galp		$C_{23}H_{40}O_{20}$	188	+2	430
O-β-D-Glcp-(1 → 4)-O-β-D-Glcp-(1 → 4)-O-β-D-Galp-(1 → 4)-D-GlcUA			—	—	431
O-β-D-Glcp-(1 → 3)-O-β-D-Glcp-(1 → 4)-O-β-D-Glcp-(1 → 4)-D-Glcp	α-Anomer	$C_{24}H_{42}O_{21}$	221–224	+12.9 → +10.6	432
	β-Anomer dihydrate	$C_{24}H_{46}O_{23}$	180–181	+7 → +10	432
O-β-D-Glcp-(1 → 4)-O-β-D-Glcp-(1 → 3)-O-β-D-Glcp-(1 → 4)-D-Glcp		$C_{24}H_{42}O_{21}$	223–226	+19.8 (90% CH$_3$COOH)	432, 433
O-β-D-Glcp-(1 → 4)-O-β-D-Glcp-(1 → 4)-O-β-D-Glcp-(1 → 3)-β-D-Glcp		$C_{24}H_{42}O_{21}$	241–245	+11.4 → +8.4	432
	Tetradecaacetate	$C_{52}H_{70}O_{35}$	118–121	—	434
O-β-D-Glcp-(1 → 4)-O-α-D-Glcp-(1 → 4)-O-α-D-Glcp-(1 → 4)-D-Glcp (Maltotetraose)		$C_{24}H_{42}O_{21}$	—	+166	435
	Methyl glycoside	$C_{25}H_{44}O_{21}$	—	+213	435
O-β-D-Glcp-(1 → 4)-O-β-D-Glcp-(1 → 4)-O-β-D-Glcp-(1 → 4)-D-Glcp		$C_{24}H_{42}O_{21}$	251	+11.3 → +17	436
	α-Tetradecaacetate	$C_{52}H_{70}O_{35}$	226–227	+12.5 (CHCl$_3$)	437
	β-Tetradecaacetate	$C_{52}H_{70}O_{35}$	223–225	−18.2 (CHCl$_3$)	437
O-α-D-Glcp-(1 → 6)-O-α-D-Glcp-(1 → 4)-O-α-D-Glcp-(1 → 4)-D-Glcp		$C_{24}H_{42}O_{21}$	—	+177 (15°) (c 0.5)	438
O-α-D-Glcp-(1 → 6)-O-α-D-Glcp-(1 → 6)-O-α-D-Glcp-(1 → 4)-D-Glcp		$C_{24}H_{42}O_{21}$	—	+164	439
	Methyl glycoside monomethanolate	$C_{26}H_{48}O_{22}$	192–193	+189	439
O-α-D-Glcp-(1 → 4)-O-α-D-Glcp-(1 → 6)-O-α-D-Glcp-(1 → 4)-D-Glcp		$C_{24}H_{42}O_{21}$	—	—	440
O-β-D-Glcp-(1 → 4)-O-β-D-Glcp-(1 → 6)-O-β-D-Glcp-(1 → 6)-D-Glcp	Tetradecaacetate	$C_{52}H_{70}O_{35}$	239–240	−19.6	441
O-α-D-Glcp-(1 → 6)-O-[α-D-Glcp-(1 → 3)]-O-α-D-Glcp (1 → 6) D Glcp		$C_{24}H_{42}O_{21}$	—	—	442
O-α-D-Glcp-(1 → 2)-O-β-D-Fruf-(1 → 2)-O-β-D-Fruf-(1 → 2)-β-D-Fruf (Nystose)		$C_{24}H_{42}O_{21}$	131–132	+10.6	443
			110–115	+17.9	444, 445
O-α-D-Glcp-(1 → 4)-O-α-D-Glcp-(1 → 4)-O-α-D-Glcp-(1 → 2)-D-Fruf		$C_{24}H_{42}O_{21}$	—	—	370
O-β-D-Glcp-(1 → 3)-O-β-D-Glcp-(1 → 3)-O-β-D-Glcp-(1 → 3)-β-D-Glcp		$C_{24}H_{42}O_{21}$	—	−5.9	119, 368
	Tetradecaacetate	$C_{52}H_{70}O_{35}$	122–123	−46.2 (CHCl$_3$)	119
O-α-D-Manp-(1 → 3)-O-α-D-Manp-(1 → 2)-O-α-D-Manp-(1 → 2)-D-Man		$C_{24}H_{42}O_{21}$	—	+65.7	481
O-α-D-Manp-(1 → 6)-O-α-D-Manp-(1 → 6)-O-α-D-Manp-(1 → 6)-D-Man		$C_{24}H_{42}O_{21}$	—	+60.8	481
O-α-D-Manp-(1 → 2)-O-α-D-Manp-(1 → 3)-O-α-D-Manp-(1 → 2)-D-Man		$C_{24}H_{42}O_{21}$	—	—	482
D-Fructose tetraose (Veronicin)(nonreducing)		$C_{24}H_{42}O_{21}$	170,188	−29.4	446
	Tetradecaacetate	$C_{52}H_{70}O_{35}$	92	−21.1 (CHCl$_3$)	446
O-D-Fruf-(2 → 1)-O-[D-Fruf-(2 → 2)]-O-[D-Fruf-(2 → 6)]-D-Glc (Neobifurcose)		$C_{24}H_{42}O_{21}$	—	+14.4,+16.7	447, 448
	Tridecamethyl ether	$C_{37}H_{68}O_{21}$	—	−35.4	445
O-D-Fruf-(2 → 1)-O-D-Fruf-(2 → 1)-O-D-Fruf-(2 → 1)-D-Glcp (Inulotriosylglucose)		$C_{24}H_{42}O_{21}$	—	−2	449
O-α-L-Fucp-(1 → 2)-O-β-D-Galp-(1 → 4)-O-[α-L-Fucp-(1 → 3)]-D-Glc (Lactodifucotetraose)		$C_{24}H_{42}O_{19}$		−17.1,-106	450

OLIGOSACCHARIDES (INCLUDING DISACCHARIDES) (Continued)

Substance[a] (Synonym) (A)	Derivative (B)	Chemical Formula (C)	Melting Point °C (D)	Specific Rotation[b] $[\alpha]_D$ (E)	Reference (F)
TETRASACCHARIDES (Continued)					
O-α-D-GalUAp-(1 → 4)-O-α-D-GalUAp-(1 → 4)-O-α-D-GalUAp-(1 → 4)-D-GalUAp	Trihydrate	$C_{24}H_{40}O_{28}$	160–170	—	451
O-α-D-GalNAcp-(1 → 3)-[O-α-L-Fuc$p$-(1 → 2)]-O-β-D-Gal$p$-(1 → 4)-D-Glc		$C_{26}H_{45}NO_{20}$	—	—	486
O-α-NANA-(2 → 8)-O-α-NANA-(2 → 3)-O-β-D-Gal$p$-(1 → 3)-D-GalNAc		$C_{36}H_{59}N_3O_{27}$	—	—	485
O-β-D-GlcUAp-(1 → 3)-O-D-GlcNAcp-(1 →)-O-β-D-GlcUAp-(1 → 3)-D-GlcNAc		$C_{28}H_{44}N_2O_{23}$	200	−41 → −53(27°)	452
O-α-NANA-(2 → 8)-O-α-NANA-(2 → 3)-O-β-D-Gal$p$-(1 → 4)-D-Glc		$C_{34}H_{56}N_2O_{27}$	—	—	462
O-α-D-Glc$p$-(1 → 6)-O-α-D-Glc$p$-(1 → 6)-O-α-D-Glc$p$-(1 → 1)-glycerol		$C_{21}H_{38}O_{18}$	—	+166 (27°)(c 0.5)	263
PENTASACCHARIDES					
[O-β-D-Xyl$p$-(1 → 4)-]$_4$-β-D-Xyl$p$ (Xylopentaose)		$C_{25}H_{42}O_{21}$	240–242	−72.9	289, 290
	Dodecaacetate	$C_{49}H_{76}O_{33}$	249–250	−98 (CHCl$_3$)	290
[O-α-D-Gal$p$-(1 → 6)-]$_3$-O-α-D-Glc$p$-(1 → 2)-β-D-Fru$f$ (Verbascose)		$C_{30}H_{52}O_{26}$	219–220,253	+169.9	453, 454
	Heptadecaacetate	$C_{64}H_{86}O_{43}$	132	+130.4	453, 454
O-β-D-Gal$p$-(1 → 3)-O-[α-L-Fuc$p$-(1 → 4)]-O-β-D-GlcNAcp-(1 → 3)-O-β-D-Gal$p$-(1 → 4)-D-Glc$p$ (Lacto-N-fucopentaose II)		$C_{32}H_{55}NO_{25}$	213–215	−28 → +30.4	455
O-D-Glc$p$-(1 → 2)-O-[D-Fru$f$-(2 → 6)]-O-D-Fru$f$-(1 → 2)-O-D-Fru$f$-(1 → 2)-D-Fru$f$		$C_{30}H_{52}O_{26}$	—	−3.5	456
	Heptadeca methylether	$C_{47}H_{86}O_{26}$	—	−10.5 (CHCl$_3$)	456
[O-α-D-Glc$p$-(1 → 6)-]$_3$-O-α-D-Glc$p$-(1 → 4)-D-Glc$p$		$C_{30}H_{52}O_{26}$	—	+167 (c 0.7)	349
	Methyl-α-glycoside monoethanolate	$C_{33}H_{60}O_{27}$	161–164	+188.7	437
[O-β-D-Glc$p$-(1 → 4)-]$_4$-D-Glc$p$ (Cellopentaose)		$C_{30}H_{52}O_{26}$	—	—	345
	α-Heptaacetate	$C_{64}H_{86}O_{43}$	246–249	—	457
	β-Heptaacetate	$C_{64}H_{86}O_{43}$	238.5–239	−18.5 (CHCl$_3$)	437
[O-α-D-Glc$p$-(1 → 4)-]$_4$-D-Glc$p$ (Maltopentaose)		$C_{30}H_{52}O_{26}$	—	+178	458
[O-α-D-Glc$p$-(1 → 2)-[O-β-D-Fru$f$-(6 → 2)-]$_3$-β-D-Fru$f$		$C_{30}H_{52}O_{26}$	—	−11.2	459, 460
O-β-D-Gal$p$-(1 → 4)-O-[α-L-Fuc$p$-(1 → 3)]-O-β-D-GlcNAcp-(1 → 3)-O-β-D-Gal$p$-(1 → 4)-D-Glc$p$ (Lacto-N-fucopentaose III)		$C_{32}H_{55}NO_{25}$	275–277	+7.35	461
[O-α-D-Man$p$-(1 → 2)-]$_4$-D-Man		$C_{30}H_{52}O_{26}$	—	—	483
O-α-L-Fuc$p$-(1 → 2)-O-β-D-Gal$p$-(1 → 3)-O-β-D-GlcNAcp-(1 → 3)-O-β-D-Gal$p$-(1 → 4)-D-Glc$p$ (Lacto-N-fucopentaose I)		$C_{32}H_{55}NO_{25}$	216	−11 → −16.3	70
O-α-D-GalNAcp-(1 → 3)-[O-α-L-Fuc$p$-(1 → 2)]-O-β-D-Gal$p$-(1 → 4)-[O-α-L-Fuc$p$-(1 → 3)]-D-Glc		$C_{32}H_{55}NO_{24}$	—	—	484
O-α-D-Gal$p$-(1 → 3)-[O-α-L-Fuc$p$-(1 → 2)]-O-β-D-Gal$p$-(1 → 4)-[O-α-L-Fuc$p$-(1 → 3)]-D-Glc		$C_{30}H_{52}O_{24}$	—	—	484
O-α-NANA-(2 → 3)-O-β-D-Gal$p$-(1 → 3)-O-β-D-GlcNAcp-(1 → 3)-O-β-D-Gal$p$-(1 → 4)-D-Glc (LST-a)		$C_{37}H_{62}N_2O_{29}$	—	—	462
O-β-D-Gal$p$-(1 → 3)-[O-α-NANA-(2 → 6)-]-O-β-D-GlcNAcp-(1 → 3)-O-β-D-Gal$p$-(1 → 4)-D-Glc (LST-b)		$C_{37}H_{62}N_2O_{29}$	—	+14.5 → +15.0	462
O-α-NANA-(2 → 6)-O-β-D-Gal$p$-(1 → 4)-O-β-D-GlcNAcp-(1 → 3)-O-β-D-Gal$p$-(1 → 4)-D-Glc (LST-c)		$C_{37}H_{62}N_2O_{29}$	—	+13	462
O-α-NGNA-(2 → 3)-O-β-D-Gal$p$-(1 → 4)-O-β-D-GlcNAcp-(1 → 3)-O-β-D-Gal$p$-(1 → 4)-D-Glc (G$_{LNnT}$ I NGNA)		$C_{37}H_{62}N_2O_{30}$	—	—	463
HEXASACCHARIDES					
[O-β-D-Xyl$p$-(1 → 4)-]$_5$-D-Xyl$p$ (Xylohexaose)	Dihydrate	$C_{30}H_{54}O_{27}$	236–237	+72.8	289, 464
	Tetradecaacetate	$C_{58}H_{78}O_{39}$	260–261	—	464
[O-D-Xyl$p$-(1 → 4)-]$_2$-O-D-Xyl$p$-(1 → 3)-[O-D-Xyl$p$-(1 → 4)-)$_2$-D-Xyl$p$		$C_{30}H_{50}O_{25}$	169–173	−154[α]$_{436}$	465
(O-D-Gal$p$-)$_4$-O-D-Gal$p$-n-(1 → 2)-D-Fru$f$ (Lycopose)		$C_{36}H_{62}O_{31}$	270	+187	466
	Eicosa acetate	$C_{76}H_{102}O_{51}$	150	+174.5(9°) (CH$_3$COCH$_3$)	466

Substance[a] (Synonym) (A)	Derivative (B)	Chemical Formula (C)	Melting Point °C (D)	Specific Rotation[b] $[\alpha]_D$ (E)	Reference (F)
HEXASACCHARIDES (Continued)					
[O-α-D-Galp-(1 → 6)-]₄-O-α-D-Glcp-(1 → 2)-β-D-Fruf (Ajugose)	Hexahydrate	$C_{36}H_{74}O_{37}$	204–205	+163	467
O-α-D-Glcp-(1 → 2)-(O-β-D-Fruf-(6 → 2)-]₄-β-D-Fruf		$C_{36}H_{62}O_{31}$	—	–19	479
O-D-Glcp-(1 → ?)-[O-D-Fruf-(? → 2)-]₄-O-Fruf		$C_{36}H_{62}O_{31}$	—	–5.3	445
	Eicosamethyl ether	$C_{56}H_{102}O_{31}$	—	–37	445
[O-α-D-Glcp-(1 → 4)-]₅-D-Glc (Maltohexaose)		$C_{36}H_{62}O_{31}$	—	+180	458
O-β-D-Glcp-(1 → 4)-]₅-D-Glc (Cellohexaose)		$C_{36}H_{62}O_{31}$	229–231	+168 (15°)	345, 468
	α-Eicosaacetate	$C_{76}H_{102}O_{51}$	252–255	—	457
	β-Eicosaacetate	$C_{76}H_{102}O_{51}$	241–243	–18.9 (CHCl₃)	437
[O-α-D-Glcp-(1 → 4)-]₆ (Cyclohexaamylose)		$C_{36}H_{60}O_{30}$	—	+151	469
	Octaacetate	$C_{72}H_{96}O_{48}$	—	+107 (CHCl₃)	470
[O-α-D-Manp-(1 → 2)-]₅-D-Man		$C_{36}H_{62}O_{31}$	—	—	483
O-β-D-Galp-(1 → 3)-O-β-D-GlcNAcp-(1 → 3)-[O-β-D-Galp-(1 → 4)-O-β-D-GlcNAcp-(1 → 6)-]-O-β-D-Galp-(1 → 4)-D-Glc (Lacto-N-hexaose)		$C_{40}H_{68}N_2O_{31}$	—	—	471
O-β-D-Galp-(1 → 4)-O-β-D-GlcNAcp-(1 → 3)-[O-β-D-Galp-(1 → 4)-O-β-D-GlcNAcp-(1 → 6)]-O-β-D-Galp-(1 → 4)-D-Glc (Lacto-N-neohexaose)		$C_{40}H_{68}N_2O_{31}$	—	—	472
O-α-L-Fucp-(1 → 2)-O-β-D-Galp-(1 → 3)-[O-α-L-Fucp-(1 → 4)]-O-β-D-GlcNAcp-(1 → 3)-O-β-D-Galp-(1 → 4)-D-Glc (Lacto-N-difucohexaose I)		$C_{38}H_{65}NO_{29}$	—	—	473
O-β-D-Galp-(1 → 3)-[O-α-L-Fucp-(1 → 4)]-O-β-D-GlcNAcp-(1 → 3)-O-β-D-Galp-(1 → 4)-[O-α-L-Fucp-(1 → 3)]-D-Glc (Lacto-N-difucohexaose II)		$C_{38}H_{65}NO_{29}$	218–220	–68.8	474
O-α-NANA-(2 → 3)-O-β-D-Galp(1 → 3)-[O-α-NANA-(2 → 6)]-O-β-D-GlcNAcp-(1 → 3)-O-β-D-Galp-(1 → 4)-D-Glc (Disialyl-lacto-N-tetraose)		$C_{48}H_{79}N_3O_{37}$	—	—	475
O-α-D-GalNAcp-(1 → 3)-[O-α-L-Fucp-(1 → 2)]-O-β-D-Galp-(1 → 3)-O-β-D-GlcNAcp-(1 → 3)-O-β-D-Galp-(1 → 4)-D-Glc		$C_{40}H_{68}N_2O_{30}$	—	—	486
O-α-D-4-O-methyl-GlcUA-(1 → 2)-[O-β-D-Xylp-(1 → 4)-]₄-D-Xylp		$C_{32}H_{52}O_{27}$	—	–11.8	290
HEPTASACCHARIDES					
Fructose heptasaccharide (Asparagose)		$C_{42}H_{72}O_{36}$	—	–35.7	476
[O-β-D-Xylp-(1 → 4)-]₆-D-Xylp (Xyloheptaose)		$C_{35}H_{58}O_{29}$	232–234	–74	289, 464
[O-α-D-Galp-(1 → 6)-]₅-O-α-D-Glcp-(1 → 2)-β-D-Fruf	Tetrahydrate	$C_{42}H_{80}O_{40}$	246–248	+168	467
[O-α-D-Galp-(1 → 6)-]₄-O-α-D-Glcp-(1 → 2)-O-β-D-Fruf-(3 → 1)-α-D-Galp		$C_{42}H_{72}O_{36}$	—	—	477
[O-α-D-Glcp-(1 → 4)-]₇ (Cycloheptaamylose)		$C_{42}H_{70}O_{35}$	—	+162	469
	Heneicosa acetate	$C_{84}H_{112}O_{56}$	—	+121 (CHCl₃)	470
[O-α-D-Manp-(1 → 2)-]₆-D-Man		$C_{42}H_{72}O_{36}$	—	—	483
Fucosyllacto-N-hexaose		$C_{46}H_{78}N_2O_{35}$	—	—	472
Fucosyllacto-N-neohexaose		$C_{46}H_{78}N_2O_{35}$	—	—	472
O-α-NANA-(2 → 6)-O-β-D-Galp-(1 → 4)-O-β-D-GlcNAcp-(1 → 6)-[O-β-D-Galp-(1 → 3)-O-β-D-GlcNAcp-(1 → 3)-]O-β-D-Galp-(1 → 4)-D-Glc		$C_{51}H_{85}N_3O_{39}$	—	—	472
O-α-NANA-(2 → 6)-O-β-D-Galp(1 → 4)-O-β-D-GlcNAcp-(1 → 6)-[O-β-D-Galp(1 → 4)-O-β-D-GlcNAcp-(1 → 3)-]-O-β-D-Galp-(1 → 4)-D-Glc		$C_{51}H_{85}N_3O_{39}$	—	—	472
O-α-D-4-O-methyl-GlcUA-(1 → 2)-[O-β-D-Xylp-(1 → 4)-]₅-D-Xylp		$C_{37}H_{60}O_{31}$	—	–20.8	290
OCTASACCHARIDES					
[O-β-D-Xylp-(1 → 4)-]₇-D-Xylp (Xylooctaose)		$C_{40}H_{66}O_{33}$	—	—	289
[O-α-D-Galp-(1 → 6)-]₆-O-α-D-Glcp-(1 → 2)-β-D-Fruf	Tetrahydrate	$C_{48}H_{90}O_{45}$	267–268	+168	467
[O-α-D-Glcp-(1 → 4)-]₈ (Cyclooctaamylose)		$C_{40}H_{80}O_{40}$	—	+180	478
	Tetracosaacetate	$C_{96}H_{128}O_{64}$	—	+137 (CHCl₃)	470
Difucosyl-lacto-N-hexaose			—	—	472
Difucosyl-lacto-N-neohexaose			—	—	472

OLIGOSACCHARIDES (INCLUDING DISACCHARIDES) (Continued)

Substance[a] (Synonym) (A)	Derivative (B)	Chemical Formula (C)	Melting Point °C (D)	Specific Rotation[b] $[\alpha]_D$ (E)	Reference (F)
OCTASACCHARIDES (Continued)					
Sialyl-fucosyl-lacto-*N*-neohexaose		$C_{57}H_{95}N_3O_{43}$	—	—	472
O-α-D-4-*O*-methyl-GlcUA-(1 → 2)-[*O*-β-D-Xyl*p*-(1 → 4)-]$_6$-D-Xyl*p*		$C_{42}H_{68}O_{25}$	—	−25.7	290
O-β-Gal*p*-(1 → 4)-*O*-β-GlcNAc*p*-(1 → 2)-*O*-α-Man*p*-(1 → 3)-[*O*-β-Gal*p*-(1 → 4)-*O*-β-GlcNAc*p*-(1 → 2)-*O*-α-Man*p*-(1 → 6)]-*O*-β-Man*p*-(1 → 4)-GlcNAc		$C_{54}H_{91}N_3O_{41}$	—	—	393

Compiled by Akira Kobata.

[a] In the order of pentose, hexose, methylpentose, hexosamine, hexuronic acid, and sialic acid at the nonreducing end of the oligosaccharide, arranged within the groups-disaccharides, trisaccharides, etc. Oligosaccharides which have other component than sugar are entered at the end of each group.

[b] $[\alpha]_D$ for 1–5 g solute, *c*, per 100 ml aqueous solution at 20–25°C unless otherwise noted.

References

1. Aspinall, Hirst, and Nicolson, *J. Chem. Soc.* (Lond.), p. 1697 (1959).
2. Andrews, Hough, and Powell, *Chem. Ind.*, 658 (1956).
3. Haq and Adams, *Can. J. Chem.*, 39, 1563 (1961).
4. Smith and Stephen, *J. Chem. Soc.* (Lond.), p. 4892 (1961).
5. Jones and Nicholson, *J. Chem. Soc.* (Lond.), p. 27 (1958).
6. Aspinall and Cairncross, *J. Chem. Soc.* (Lond.), p. 3998 (1960).
7. Lehmann and Beck, *Justus Liebigs Ann. Chem.*, 630, 56 (1960).
8. Wallenfels and Beck, *Justus Liebigs Ann. Chem.*, 630, 46 (1960).
9. Gorin, *Can. J. Chem.*, 40, 275 (1962).
10. Helferich and Bredreck, *Justus Liebigs Ann. Chem.*, 465, 166 (1928).
11. Aspinall and Nicholson, *J. Chem. Soc.* (Lond.), p. 2503 (1960).
12. Gakhokidze and Kutidze, *J. Gen. Chem. U.S.S.R.*, 22, 247 (1952).
13. Aspinall and Ferrier, *J. Chem. Soc.* (Lond.), p. 4188 (1957).
14. Montgomery, Smith, and Srivastava, *J. Am. Chem. Soc.*, 79, 698 (1957).
15. Erskine and Jones, *Can. J. Chem.*, 35, 1174 (1957).
16. Ball and Jones, *J. Chem. Soc.* (Lond.), p. 4871 (1957).
17. Andrews, Ball, and Jones, *J. Chem. Soc.* (Lond.), p. 4090 (1953).
18. Rosenberg and Zamenhof, *J. Biol. Chem.*, 237, 1040 (1962).
19. Ball and Jones, *J. Chem. Soc.* (Lond.), p. 33 (1958).
20. Curtis and Jones, *Can. J. Chem.*, 38, 1305 (1960).
21. Myhre and Smith, *J. Org. Chem.*, 26, 4609 (1961).
22. Aspinall and Ross, *J. Chem. Soc.* (Lond.), p. 3674 (1961).
23. Whistler, Bacchrach, and Tu, *J. Am. Chem. Soc.*, 74, 3059 (1952).
24. Ball and Jones, *J. Chem. Soc.* (Lond.), p. 4871 (1957).
25. Kuhn, Low, and Trischmann, *Angew. Chem.*, 68, 212 (1956).
26. Zemplén and Bognár, *Chem. Ber.*, 72, 1160 (1939).
27. Wallenfels and Lehmann, *Chem. Ber.*, 90, 1000 (1957).
28. Bridel and Charaux, *C. R. Acad. Sci.*, 180, 1219 (1925).
29. Helferich and Steinpreis, *Chem. Ber.*, 91, 1794 (1958).
30. Gorin, *Can. J. Chem.*, 40, 275 (1962).
31. Gakhokide and Kobiashvili, *J. Gen. Chem. U.S.S.R.*, 22, 244, 247 (1952).
32. Whistler and Yagi, *J. Org. Chem.*, 26, 1050 (1961).
33. Smith, *J. Chem. Soc.* (Lond.), p. 744 (1939).
34. Gorin, Spencer, and Phaff, *Can. J. Chem.*, 42, 1341, 2307 (1964).
35. Goldstein, Smith, and Srivastava, *J. Am. Chem. Soc.*, 79, 3858 (1957).
36. Srivastava and Smith, *J. Am. Chem. Soc.*, 79, 982 (1957).
37. Lindberg, Wachmeister, and Wickberg, *Acta Chem. Scand.*, 6, 1052 (1952).
38. Montgomery, Smith, and Srivastava, *J. Am. Chem. Soc.*, 79, 698 (1957).
39. Gorin, Haskins, and Westlake, *Can. J. Chem.*, 44, 2083 (1966).
40. Clancy and Whelan, *Arch. Biochem. Biophys.*, 118, 724 (1967).
41. Morgan and O'Neil, *Can. J. Chem.*, 37, 1201 (1959).
42. Aspinall, Auret, and Hirst, *J. Chem. Soc.* (Lond.), p. 4408 (1958).
43. Ball and Jones, *J. Chem. Soc.* (Lond.), p. 905 (1958).
44. Jones and Reid, *J. Chem. Soc.* (Lond.), p. 1361 (1954); 1890 (1955).
45. Masamune and Kamiyama, *J. Exp. Med.* (Tokyo), 66, 43 (1957).
46. Gorin and Spencer, *Can. J. Chem.*, 37, 499 (1959).
47. Ingle and Bhide, *J. Indian Chem. Soc.*, 35, 516 (1958).
48. Turton, Bebbington, Dixon, and Pacsu, *J. Am. Chem. Soc.*, 77, 2565 (1955).
49. Ballio and Russi, *J. Chromatogr.*, 4, 117 (1960).
50. Meier, *Acta Chem. Scand.*, 16, 2275 (1962).
51. Haq and Adams, *Can. J. Chem.*, 39, 1563 (1961).
52. Wickstrom, *Acta Chem. Scand.*, 11, 1473 (1957).
53. Lehmann and Beck, *Justus Liebigs Ann. Chem.*, 630, 56 (1960).
54. Beck and Wallenfels, *Justus Liebigs Ann. Chem.*, 655, 173 (1962).
55. Gakhokide, *Chem. Abstr.*, 47, 6875 (1953).
56. Kuhn and Baer, *Chem. Ber.*, 87, 1560 (1954).
57. Trey, *Z. Phys. Chem.*, 46, 620 (1903).
58. Gillis, *Rec. Trav. Chim. Pay-Bas*, 39, 88, 677 (1920).
59. Hudson and Johnson, *J. Am. Chem. Soc.*, 37, 1270, 1276 (1915).
60. Tanret, *Hoppe Seyler's Z. Physiol. Chem.*, 53, 692 (1905).
61. Fletcher and Diehl, *J. Am. Chem. Soc.*, 74, 5774 (1952).
62. Hudson and Johnson, *J. Am. Chem. Soc.*, 37, 2752 (1915).
63. Pazur, Tipton, Budovieh, and Marsh, *J. Am. Chem. Soc.*, 80, 119 (1958).
64. Helferich and Sparmberg, *Chem. Ber.*, 66, 806 (1933).
65. Polonovski and Lespagnol, *C. R. Acad. Sci.*, 192, 1319 (1931).
66. Bredereck, Wagner, Geissel, Gross, Hutten, and Ott, *Chem. Ber.*, 95, 3056 (1962).
67. Haskins, Hann, and Hudson, *J. Am. Chem. Soc.*, 64, 1852 (1942).
68. Jones, Hough, and Richards, *J. Chem. Soc.* (Lond.), p. 295 (1954).
69. Helfrich and Steinpreis, *Chem. Ber.*, 91, 1794 (1958).
70. Kuhn and Gauhe, *Chem. Ber.*, 93, 647 (1960).
71. Alessandrini, Schmidt, Zilliken, and György, *J. Biol. Chem.*, 220, 71 (1956).
72. Kuhn and Kirschenlohr, *Justus Liebigs Ann. Chem.*, 600, 135 (1956).
73. Okuyama, *Tohoku J. Exp. Med.*, 68, 313 (1958).
74. Watkins, *Nature*, 181, 117 (1958).
75. Kuhn, Baer, and Gauhe, *Chem. Ber.*, 88, 1713 (1955).
76. Zilliken, Smith, Rose, and Gyorgy, *J. Biol. Chem.*, 208, 299 (1954).
77. Kuhn and Gauhe, *Chem. Ber.*, 94, 842 (1961).
78. Hirase and Araki, *Bull. Chem. Soc. Jap.*, 27, 105 (1953).
79. Clingman, Nunn, and Stephan, *J. Chem. Soc* (Lond.), p. 197 (1957).
80. Gakhokidze, *J. Gen. Chem. U.S.S.R.*, 16, 1923 (1946).
81. Weissmann and Meyer, *J. Am. Chem. Soc.*, 76, 1753 (1954).
82. Lindberg and Wickberg, *Acta Chem. Scand.*, 8, 821 (1954).
83. Zemplén, *Chem. Ber.*, 59, 1254 (1926).
84. Gakhokidze, *J. Gen. Chem. U.S.S.R.*, 16, 1914 (1946).
85. Feingold, Avigad, and Hestrin, *J. Biol. Chem.*, 224, 295 (1957).
86. Hassid, Doudoroff, Potter, and Barker, *J. Am. Chem. Soc.*, 70, 306 (1948).
87. Andrews and Jones, *J. Chem. Soc.* (Lond.), p. 1724 (1954).
88. Gupta, *J. Chem. Soc.*, (Lond.), p.5262 (1961).
89. Gorrod and Jones, *J. Chem. Soc.* (Lond.), p. 2522 (1954).
90. Barker, Boume, Hewitt, and Stacey, *J. Chem. Soc.* (Lond.), p. 3541 (1957).
91. Barker, Stacey, and Stroud, *Nature*, 189, 138 (1961).
92. Putman, Lilt, and Hassid, *J. Am. Chem. Soc.*, 77, 4351 (1955).

93. Gakhokidze and Kutidze, *J. Gen. Chem. U.S.S.R.*, 22, 139 (1952).
94. Flowers, *Carbohydr. Res.*, 18, 211 (1971).
95. Kuhn, Low, and Trischmann, *Chem. Ber.*, 88, 1492 (1955).
96. Jones and Perry, *J. Am. Chem. Soc.*, 79, 2787 (1957).
97. Kuhn and Low, *Chem. Ber.*, 86, 1027 (1953).
98. Lloyd and Roberts, *Proc Chem. Soc.*, 250 (1960).
99. Freudenberg, Noe, and Knopf, *Chem. Ber.*, 60, 238 (1927).
100. Matsuda, *Nature*, 180, 985 (1957).
101. Haq and Whelan, *Nature*, 178, 1225 (1956).
102. Bourne, Hartigan, and Weigel, *J. Chem. Soc.* (Lond.), p. 1088 (1961).
103. Birch, *J. Chem. Soc.* (Lond.), p. 3489 (1965).
104. Hudson and Johnson, *J. Am. Chem. Soc.*, 37, 2748 (1915).
105. Peat, Whelan, and Hinson, *Nature*, 170, 1056 (1952).
106. Sharp and Stacey, *J. Chem. Soc.* (Lond.), p. 285 (1951).
107. Micheel and Hagel, *Chem. Ber.*, 85, 1087 (1952).
108. Matsuda, *J. Agric. Chem. Soc. Jap.*, 30, 119 (1956).
109. Freudenberg, Knauber, and Cramer, *Chem. Ber.*, 84, 144 (1951).
110. Finan and Warren, *J. Chem. Soc.* (Lond.), p. 5229 (1963).
111. Rabat, *Bull. Soc. Chim. Fr. Ser. 5*, 5, 7, 565 (1940).
112. Vis and Fletcher, *J. Am. Chem. Soc.*, 78, 4709 (1956).
113. Peat, Whelan, and Hinson, *Chem. Ind.*, 385 (1955).
114. Sato and Aso, *Nature*, 180, 984 (1957).
115. Haq and Whelan, *J. Chem. Soc.* (Lond.), p. 1342 (1958).
116. Watanabe and Aso, *J. Agric. Res.* (Tokyo), 11, 109 (1960).
117. Wolfrom and Thompson, *J. Am. Chem. Soc.*, 77, 6403 (1955).
118. Weismann and Meyer, *J. Am. Chem. Soc.*, 76, 1753 (1954).
119. Freudenberg and Oertzen, *Justus Liebigs Ann. Chem.*, 574, 37 (1951).
120. Connel, Hirst, and Percival, *J. Chem. Soc.* (Lond.), p. 3494 (1950).
121. Peat, Whelan, and Lawley, *J. Chem. Soc.* (Lond.), p. 724, 729 (1958).
122. Gillis, *Natuurwet. Tijdschr.* (Ghent), 12, 193 (1930); *Chem. Zentralbl.*, 1, 256 (1931).
123. Hudson and Yanovski, *J. Am. Chem. Soc.*, 39, 1013 (1917).
124. Peterson and Spencer, *J. Am. Chem. Soc.*, 49, 2822 (1927).
125. Wolfrom, Georges, and Miller, *J. Am. Chem. Soc.*, 71, 125 (1949).
126. Jeanes, Wilham, Jonones, Tsuchiya, and Rist, *J. Am. Chem. Soc.*, 75, 5911 (1953).
127. Bates, *Polarimetry, Saccharimetry and the Sugars*, National Bureau of Standards Circ. C440, U.S. Gov. Print. Off., Washington, D.C, 1942.
128. Zemplèn, *Hoppe-Seyler's Z. Physiol. Chem.*, 85, 399 (1913).
129. Hudson and Johnson, *J. Am. Chem. Soc.*, 39, 1272 (1917).
130. Thompson and Wolfrom, *J. Am. Chem. Soc.*, 75, 3605 (1953).
131. Gakhokide and Gvelukashvli, *J. Gen. Chem. U.S.S.R.*, 22, 143 (1952).
132. Hudson, *J. Org. Chem.*, 9, 470 (1944).
133. Haworth, Hirst, and Reynolds, *J. Chem. Soc* (Lond.), p. 302 (1934).
134. Brauns, *J. Am. Chem. Soc.*, 48, 2776 (1926).
135. Haworth, Hirst, Streight, Thomas, and Welb, *J. Chem. Soc.* (Lond.), p. 2636 (1930).
136. Haskins, Hann, and Hudson, *J. A m. Chem. Soc.*, 63, 1724 (1941).
137. Lindberg, *Acta Chem. Scand.*, 7, 1218 (1953).
138. Bredereck, Wagner, Kuhn, and Ott, *Chem. Ber.*, 93, 1201 (1960).
139. Peat, Whelan, and Evans, *J. Chem. Soc.* (Lond.), p. 175 (1960).
140. Avigad, *Biochem. J.*, 73, 587 (1959).
141. Gakhokidze and Gvelukashvli, *J. Gen. Chem. U.S.S.R.*, 22, 143 (1952).
142. Haworth, Hirst, and Reynolds, *J. Chem. Soc.* (Lond.), p. 302 (1934).
143. Brauns, *J. Am. Chem. Soc.*, 48, 2776 (1926).
144. Hudson and Pacsu, *J. Am. Chem. Soc.*, 52, 2519 (1930).
145. Pacsu, *J. Am. Chem. Soc.*, 54, 3649 (1932).
146. Hough, Jones, and Richards, *J. Chem. Soc.* (Lond.), p.2005 (1953).
147. Stodola, Koepsell, and Sharpe, *J. Am. Chem. Soc.*, 74, 3202 (1952); 78, 2514 (1956).
148. Bourne, Hutson, and Weigel, *Biochem. J.*, 79, 549 (1961).
149. Corbett and Kenner, *J. Chem. Soc.* (Lond), p. 1431 (1955).
150. Weidenhagen and Lotenz, *Angew. Chem.*, 69, 641 (1957).
151. Arcamone and Bizioli, *Gazz. Chim. Ital.*, 87, 896 (1957).
152. Wolfrom and Juliano, *J. Am. Chem. Soc.*, 82, 1673 (1960).
153. Selinger and Schramm, *J. Biol. Chem.*, 236, 2183 (1961).
154. Barker, Heidelberger, Stacey, and Tipper, *J. Chem. Soc.* (Lond), p. 3468 (1958).
155. Gorin and Spencer, *Can. J. Chem.*, 39, 2275 (1961).
156. Zemplén, *Chem. Abstr.*, 33, 4202 (1939).
157. Stoll, Keis, and von Wartburg, *Helv. Chim. Acta*, 35, 2495 (1952).
158. Barker, Gómez-Sánchez, and Stacey, *J. Chem. Soc.* (Lond.), p. 3264 (1959).
159. Palleroni and Doudoroff, *J. Biol. Chem.*, 219, 957 (1956).
160. Tunmann and Scheherer, *Arch. Pharm.*, 292, 745 (1959).
161. Lindberg, *Acta Chem. Scand.*, 7, 1119 (1953).
162. Lindberg, Silvander, and Wachtmeister, *Acta Chem. Scand.*, 17, 1348 (1963).
163. Selinger and Schramm, *J. Biol. Chem.*, 236, 2183 (1961).
164. Baytop, Tanher, Tekman, and Oner, *Folia Pharm.* (Istanbul), 4, 464 (1960).
165. Korte, *Chem. Ber.*, 88, 1527 (1955).
166. Frérejacque, *C. R. Acad. Sci.*, 246, 459 (1958).
167. Lindberg, *Acta Chem. Scand.*, 8, 869 (1954); 9, 1093, 1097 (1954).
168. Freudenberg, Wolf, Knopf, and Zaheer, *Chem. Ber.*, 61, 1743 (1928).
169. Perila and Bishop, *Can. J. Chem.*, 39, 815 (1961).
170. Tyminski and Timell, *J. Am. Chem. Soc.*, 82, 2823 (1960).
171. Merler and Wise, *Tappi (Tech. Assoc. Pulp Pap. Ind.)*, 41, 80 (1958).
172. Gorin and Perlin, *Can. J. Chem.*, 37, 1930 (1959).
173. Gorin and Perlin, *Can. J. Chem.*, 39, 2474 (1961).
174. Jones and Nicholson, *J. Chem. Soc.* (Lond.), p. 27 (1958).
175. Aspinall, Rashbrook, and Kessler, *J. Chem. Soc,* (Lond.), p. 215 (1958).
176. Jones and Painter, *J. Chem. Soc.* (Lond.), p. 669 (1959).
177. Jones and Nicholson, *J. Chem. Soc.* (Lond.), p. 27 (1958).
178. Wistler and Stein, *J. Am. Chem. Soc.*, 73, 4187 (1951).
179. Peat, Turvey, and Doyle, *J. Chem. Soc.* (Lond.), p. 3918 (1961).
180. Talley, Reynolds, and Evans, *J. Am. Chem. Soc.*, 65, 575 (1943).
181. Shaban and Jeanloz, *Carbohydr. Res.*, 17, 193 (1971).
182. Shaban and Jeanloz, *Carbohydr. Res.*, 20, 17 (1971).
183. Shaban and Jeanloz, *Carbohydr. Res.*, 17, 411 (1971).
184. Barker, Murray, Stacey, and Stroud, *Nature*, 191, 143 (1961).
185. Strepkov, *Zh. Obshch. Khim.*, 28, 3143 (1958).
186. Pazur and Gordon, *J. Am. Chem. Soc.*, 75, 3458 (1953).
187. Schlubach and Scheffler, *Justus Liebigs Ann. Chem.*, 588, 192 (1954).
188. Aspinall and Telfer, *J. Chem. Soc.* (Lond.), p. 1106 (1955).
189. Anderson, *Acta Chem. Scand.*, 21, 828 (1967).
190. Brigl and Scheyer, *Hoppe-Seyler's Z. Physiol. Chem.*, 160, 214 (1926).
191. Wallenfels and Lehmann, *Chem. Ber.*, 90, 1000 (1957).
192. Bacon, *Biochem. J.*, 57, 320 (1954).
193. Barber, *J. Am. Chem. Soc.*, 81, 3722 (1959).
194. Levy, Flowers, and Sharon, *Carbohydr. Res.*, 4, 305 (1967).
195. Kuhn, Baer, and Gauhe, *Justus Liebigs Ann. Chem.*, 611, 242 (1958).
196. Eagon and Dedonder, *C. R. Acad. Sci.*, 241, 579 (1955).
197. Gorin and Perlin, *Can. J. Chem.*, 37, 1930 (1959).
198. Coté, *J. Chem. Soc.* (Lond.), p. 2248 (1959).
199. O'Neill, *J. Am. Chem. Soc.*, 76, 5074 (1954).
200. Shaban and Jeanloz, *Carbohydr. Res.*, 20, 399 (1971).
201. Dejter-Juszynski and Flowers, *Carbohydr. Res.*, 23, 41 (1972).
202. Kuhn, Low, and Trischmann, *Chem. Ber.*, 88, 1492 (1955).
203. Zemplén, Gerecs, and Flesch, *Chem. Ber.* 71, 2511 (1938).
204. Gorin and Perlin, *Can. J. Chem.*, 37, 1930 (1959).
205. Zemplén, Tettamanti, and Farago, *Chem. Ber.*, 71, 2511 (1938).
206. Zemplén and Gerecs, *Chem. Ber.*, 67, 2049 (1934).
207. Yoshizawa, *Biochim. Biophys. Acta*, 52, 588 (1961).
208. Lloyd and Roberts, *J. Chem. Soc.* (Lond.), p. 6910 (1965).
209. Kuhn and Kirschenlohr, *Chem. Ber.*, 87, 384 (1954).
210. Foster and Horton, *J. Chem. Soc.* (Lond.), p. 1890 (1958).
211. Wang and Tai, *Acta Chem. Sinicia*, 25, 50 (1959).
212. Schiffman, Rabat, and Leskowitz, *J. Am. Chem. Soc.*, 84, 73 (1962).
213. Yoshizawa, *J. Biochem.* (Tokyo), 51, 1 (1962).
214. Shinohara, *Tohoku J. Exp. Med.*, 67, 141 (1958).
215. Anderson and Fireman, *J. Biol. Chem.*, 109, 437 (1935).
216. Roudier and Eberhard, *Bull. Soc. Chim. Fr.*, 28, 2074 (1958).
217. Dhar and Mukherjee, *J. Sci. ind. Res.* (India), 18B, 219 (1959).
218. Hirst and Dunstan, *J. Chem. Soc.* (Lond.), p. 2332 (1953).
219. Aspinall and Nicholson, *J. Chem. Soc.* (Lond.), p. 2503 (1960).
220. Aspinall and Fanshawe, *J. Chem. Soc.* (Lond.), p. 4215 (1961).
221. Anderson and Crowder, *J. Am. Chem. Soc.*, 52, 3711 (1930).
222. Tipson, Christman, and Levene, *J. Biol. Chem.*, 128, 609 (1939).
223. McCready, McComb, and Black, *J. Am. Chem. Soc.*, 76, 3035 (1954).

224. Bhattacharjee and Timll, *Can. J. Chem.*, 43, 758 (1965).
225. Gee, Jones, and McCready, *J. Org. Chem.*, 23, 620 (1958).
226. Mathur and Mukherjee, *J. Sci. lnd. Res.* (India), 13B, 452 (1954).
227. Mukherjee and Strivastava, *J. Am. Chem. Soc.*, 77, 422 (1955).
228. Gorin and Spencer, *Can. J. Chem.*, 39, 2282 (1961).
229. Hotchkiss and Goebel, *J. Biol. Chem.*, 115, 285 (1936).
230. Jayne and Demmig, *Chem. Ber.*, 95, 356 (1960).
231. Jones and Perry, *J. Am. Chem. Soc.*, 79, 2787 (1957).
232. Helferich and Beiger, *Chem. Ber.*, 90, 2492 (1957).
233. Hamilton, Spriesterbach, and Smith, *J. Chem. Soc.*, 79, 443 (1957).
234. Bowering and Timell, *J. Am. Chem. Soc.*, 82, 2827 (1960).
235. Bishop, *Can. J. Chem.*, 33, 1521 (1955).
236. Bishop, *Can. J. Chem.*, 31, 134 (1953).
237. Whistler and Hough, *J. Am. Chem. Soc.*, 75, 4919 (1953).
238. Davidson and Meyer, *J. Am. Chem. Soc.*, 77, 4796 (1955).
239. Drummond and Percival, *J. Chem. Soc.*, (Lond.), p. 3908 (1961).
240. Smith and Stephen, *J. Chem. Soc.* (Lond.), p. 4892 (1961).
241. Barker, Foster, Siddiqui, and Stacey, *J. Chem. Soc.* (Lond.), p. 2358 (1958).
242. Hirst, Percival, and Williams, *J. Chem. Soc.* (Lond.), p. 1942 (1958).
243. O'Donnell and Perciva, *J. Chem. Soc.* (Lond.), p. 2168 (1959).
244. McKinnell and Percival, *J. Chem. Soc.* (Lond.), p. 2082 (1962).
245. Danishefsky, Eiber, and Langholtz, *Biochem. Biophys. Res. Commun.*, 3, 571 (1960).
246. Voss and Pfirschke, *Chem. Ber.*, 70, 132 (1937).
247. Graham and Gottschalk, *Biochim. Biophys. Acta*, 38, 513 (1960).
248. Gottschalk and Graham, *Biochim. Biophys. Acta*, 34, 380 (1959).
249. Wickberg, *Acta Chem. Scand.*, 1183, 1187 (1959).
250. Carter, McCluer, and Slifer, *J. Am. Chem. Soc.*, 78, 3735 (1956).
251. Reeves, Latour, and Lousteau, *Biochemistry*, 3, 1248 (1964).
252. Austin, Hardy, Buchanan, and Baddiley, *J. Chem. Soc.* (Lond.), 1419 (1965).
253. Colin, *Bull. Soc. Chim. Fr. Ser. 5*, 4, 277 (1937).
254. Su and Hassid, *Biochemistry*, 1, 468 (1962).
255. Putman and Hassid, *J. Am. Chem. Soc.*, 76, 2221 (1954).
256. Brown and Serro, *J. Am. Chem. Soc.*, 75, 1040 (1953).
257. Kabat, MacDonald, Ballou, and Fischer, *J. Am. Chem. Soc.*, 75, 4507 (1953).
258. Gorin, Horitsu, and Spencer, *Can. J. Chem.*, 43, 2259 (1965).
259. Pueyo, *C. R. Acad. Sci.*, 248, 2788 (1959).
260. Lindberg, Silvander, and Wachtmeister, *Acta Chem. Scand.*, 18, 213 (1964).
261. Araki and Arai, *Bull. Chem. Soc. Jap.*, 29, 339 (1956).
262. Nagel and Vaughn, *Arch. Biochem. Biophys.*, 94, 328 (1961).
263. Sawai and Hehre, *J. Biol. Chem.*, 237, 2047 (1962).
264. Jermyn, *Aust. J. Biol. Sci.*, 11, 114 (1958).
265. Dutton and Unrau, *Can. J. Chem.*, 42, 2048 (1964).
266. Hassid, Doudoroff, Barker, and Dore, *J. Am. Chem. Soc.*, 67, 1394 (1945).
267. Hassid, Doudoroff, Barker, and Dore, *J. Am. Chem. Soc.*, 68, 1465 (1946).
268. Osawa and Nakazawa, *Biochim. Biophys. Acta*, 130, 56 (1966).
269. Barker, Foster, Stacey, and Webber, *Chem. Ind.*, 208 (1957).
270. Barker, Foster, Khmelnitski, and Webber, *Bull. Soc. Chim. Biol.*, 42, 1799 (1960).
271. Danishefski and Steiner, *Biochim. Biophys. Acta*, 101, 37 (1965).
272. Gorin, Haskins, and Spencer, *Can. J. Biochem. Physiol.*, 38, 165 (1960).
273. Gorin and Perlin, *Can. J. Chem.*, 39, 2474 (1941).
274. Colin and Angier, *C. R. Acad. Sci.*, 208, 1450 (1939).
275. Charlson, Gorin, and Perlin, *Can. J. Chem.*, 34, 1811 (1956).
276. Martinez, Wolfe, and Nakada, *J. Bacteriol.*, 78, 217 (1959).
277. Olavesen and Davidson, *J. Biol. Chem.*, 240, 992 (1965).
278. Weissmann and Meyer, *J. Am. Chem. Soc.*, 74, 4729 (1952); 76, 1753 (1954).
279. Linker, Meyer, and Hoffman, *J. Biol. Chem.*, 219, 13 (1956).
280. Fukui, Hochster, Durbin, Grebner, and Feingold, *Bull. Res. Counc. Isr.*, 11A, 262 (1963).
281. Fukui and Hochster, *Can. J. Biochem.*, 41, 2363 (1963).
282. Tsujino, *Agric. Biol. Chem.*, 27, 236 (1963).
283. Hestrin, Feingold, (and Avigad, *J. Am. Chem. Soc.*, 77, 6710 (1955).

284. Amanmuradov and Abubakirov, *Chem. Nat. Compd.* (USSR), 1, 292 (1965).
285. Smith and Stephen, *J. Chem. Soc.* (Lond.), p. 4892 (1961).
286. Bishop, *J. Am. Chem. Soc.*, 78, 2840 (1956).
287. Banerji and Rao, *Aust. J. Chem.*, 17, 1059 (1964).
288. Howard, *Biochem. J.*, 67, 643 (1957).
289. Whistler and Masak, *J. Am. Chem. Soc.*, 77, 1241 (1955).
290. Marchessault and Timell, *J. Polym. Sci.*, C2, 49 (1963).
291. Whistler and Corbett, *J. Am. Chem. Soc.*, 77, 6328 (1955).
292. Morgan and O'Neil, *Can. J. Chem.*, 37, 1201 (1959).
293. Aspinall, Hirst, and Ramstad, *J. Chem. Soc.* (Lond.), p. 593 (1958).
294. Haq and Adams, *Can. J. Chem.*, 39, 1563 (1961).
295. Smith and Stephen, *J. Chem. Soc.* (Lond.), p. 4892 (1961).
296. Courtois, *Carbohydrate Chemistry of Substances of Biological Interest, 4th Int. Congr. Biochem.*, Vol. 1, Wolfrom Pergamon Press, London. . Vol. 1, 146 (1959).
297. Halahmi, Flowers, and Shapiro, *Carbohydr. Res.*, 5, 25 (1967).
298. Ballio and Russi, *Tetrahedron*, 9, 125 (1960).
299. Yamashita and Kobata, *Arch. Biochem. Biophys.*, 161, 164 (1974).
300. Tanret, *Bull. Soc. Chim. Ser.*, 27, 947 (1902).
301. Pazur, Marsh, and Timpton, *J. Am. Chem. Soc.*, 80, 1433 (1958).
302. Wallenfels, Bernt, and Limberg, *Justus Liebigs Ann. Chem.*, 579, 113 (1952).
303. Chaudun, Courtois, and Dizet, *Bull. Soc. Chim. Biol.*, 42, 227 (1960).
304. Freudenberg, Wolf, Knopf, and Zaheer, *Chem. Ber.*, 61, 1743 (1928).
305. Helferich and Schafer, *Justus Liebigs Ann. Chem.*, 450, 229 (1926).
306. Wickstrom and Svendsen, *Acta Chem. Scand.*, 10, 1199 (1956).
307. Macleod and McCorquodale, *Nature*, 815, (1958).
308. Avigad, *J. Biol. Chem.*, 229, 121 (1957).
309. Aso and Yamauchi, *Agric. Biol. Chem.*, 25, 10 (1961).
310. Haworth, Hirst, and Ruell, *J. Chem. Soc.* (Lond.), p. 3125 (1923).
311. Scheibler and Mittelmeier, *Chem. Ber.*, 23, 1438 (1890).
312. Tarnet, *Bull. Soc. Chim. Fr. Ser.*, 13, 261 (1895).
313. Whistler and Durso, *J. Am. Chem. Soc.*, 74, 5140 (1952).
314. Davy and Courtois, *C. R. Acad. Sci.*, 261, 3483 (1965).
315. Suzuki, and Hehre, *Arch. Biochem. Biophys.*, 105, 339 (1964).
316. French, Wild, Young, and James, *J. Am. Chem. Soc.*, 75, 709 (1953).
317. Pazur, Marsh, and Tipton, *J. Biol. Chem.*, 233, 277 (1958).
318. Gorin and Spencer, *Can. J. Chem.*, 39, 2275 (1961).
319. Montreuil, *C. R. Acad. Sci.*, 242, 192 (1956).
320. Bailey, Barker, Bourne, and Stacey, *Nature*, 176, 1164 (1955).
321. Yamauchi and Aso, *Nature*, 189, 753 (1961).
322. Kuhn, Baer, and Gauhe, *Chem. Ber.*, 89, 2514 (1956).
323. Kuhn, Baer, and Gauhe, *Justus Liebigs Ann. Chem.*, 611, 242 (1958).
324. Kuhn, Baer, and Gauhe, *Chem. Ber.*, 91, 364 (1958).
325. Kuhn and Gauhe, *Chem. Ber.*, 93, 647 (1960).
326. Kobata and Ginsburg, *J. Biol. Chem.*, 244, 5496 (1969).
327. Whistler and Conrad, *J. Am. Chem. Soc.*, 76, 3544 (1954).
328. Kuhn and Löw, *Chem. Ber.*, 86, 1027 (1953).
329. Kuhn, Löw, and Trischmann, *Chem. Ber.*, 90, 203 (1957); *Angew. Chem.*, 68, 212 (1958).
330. Freudenberg, Wolf, Knopf, and Zaheer, *Chem. Ber.*, 61, 1743 (1928).
331. Bailey, Barker, Bourne, Grant, and Stacey, *J. Chem. Soc.* (Lond.), p. 1895 (1958).
332. Peat, Whelan, and Roberts, *J. Chem. Soc.* (Lond.), p. 3916 (1957).
333. Parrish, Perlin, and Reese, *Can. J. Chem.*, 38, 2094 (1960).
334. Klemer, *Chem. Ber.*, 89, 2583 (1956).
335. Crook and Stone, *Biochem. J.*, 65, 1 (1957).
336. Berger and Eberhart, *Biochem. Biophys. Res. Comm.*, 6, 62 (1961).
337. Barker, Boume, O'mant, and Stacey, *J. Chem. Soc.* (Lond.), p.2448 (1957).
338. Reese and Mandels, *Can. J. Microbiol.*, 10, 103 (1964).
339. Perlin and Suzuki, *Can. J. Chem.*, 40, 50 (1962).
340. Peat, Whelan, and Roberts, *J. Chem. Soc.* (Lond.), p. 3916 (1957).
341. Moscatelli, Ham, and Rickes, *J. Biol. Chem.*, 236, 2858 (1961).
342. Wolfrom, Georges, Thompson, and Miller, *J. Am. Chem. Soc.*, 71, 2873 (1949).
343. Peat, Whelan, and Jones, *J. Chem. Soc.* (Lond.), p. 2490 (1957).
344. Sugihara and Wolfrom, *J. Am. Chem. Soc.*, 71, 3357 (1949).

345. Walker and Wright, *Arch. Biochem. Biophys.*, 69, 362 (1957).
346. Wolfrom and Fields, *Tappi (Tech. Assoc. Pulp Pap. Ind.)*, 41, 204 (1958).
347. French, Taylor, and Whelan, *Biochem. J.*, 90, 616 (1964).
348. Takiura and Nakagawa, *Yakugaku Zasshi*, 83, 301, 305 (1963).
349. Bailey, Barker, Bourne, and Atacey, *J. Chem. Soc.* (Lond.), p. 3536 (1957).
350. Turvey and Whelan, *Biochem. J.*, 67, 49 (1957).
351. Conchie, Moreno, and Cardini, *Arch. Biochem. Biophys.*, 94, 342 (1961).
352. Haq and Whelan, *J. Chem. Soc.* (Lond.), p. 4543 (1956).
353. Peat, Whelan, and Evans, *J. Chem. Soc.* (Lond.), p. 175 (1960).
354. Thompson and Wolfrom, *J. Am. Chem. Soc.*, 77, 3567 (1955).
355. Asp and Lindberg, *Acta Chem. Scand.*, 5, 665 (1951).
356. Zemplén and Gerecs, *Chem. Ber.*, 64, 1545 (1931).
357. Zemplén, Bruckner, and Gerecs, *Chem. Ber.*, 64, 744 (1931).
358. Shibasaki, *Tohoku J. Agric. Res.*, 6, 171 (1955).
359. Peat, Whelan, and Lawley, *J. Chem. Soc.* (Lond.), p. 729 (1958).
360. Turvey and Evans, *J. Chem. Soc.* (Lond.), p. 2366 (1960).
361. Handa and Nishizawa, *Nature*, 192, 1078 (1961).
362. Peat, Turvey, and Evans, *J. Chem. Soc.* (Lond.), p. 3223 (1959).
363. Klemer, *Chem. Ber.*, 92, 218 (1959).
364. Goldstein and Lindberg, *Acta Chem. Scand.*, 16, 383 (1962).
365. Peat, Whelan, Turvey, and Morgan, *J. Chem. Soc.* (Lond.), p. 623 (1961).
366. Abdullah, Goldstein, and Whelan, *J. Chem. Soc.* (Lond.), p. 176 (1962).
367. Krieglstein and Fischer, *Hoppe-Seyler's Z. Physiol. Chem.*, 344, 209 (1966).
368. Feingold, Neufeld, and Hassid, *J. Biol. Chem.*, 233, 783 (1958).
369. White and Maher, *J. Am. Chem. Soc.*, 75, 1259 (1953).
370. Wolf and Ewart, *Arch. Biochem. Biophys.*, 58, 365 (1955).
371. Barker, Bourne, and Theander, *J. Chem. Soc.* (Lond.), p. 2064 (1957).
372. Baron and Guthrie, *Ann. Entomol. Soc. Am.*, 53, 220 (1960).
373. Meyer, *Hoppe-Seyler's Z. Physiol. Chem.*, 6, 135 (1882).
374. Binaghi and Falqui, *Chem. Zentralbl.*, 2, 44 (1926).
375. Binkley, *Int. Sugar J.*, 66, 46 (1964).
376. Schlubach and Koehm, *Justus Liebigs Ann. Chem.*, 614, 126 (1958).
377. Barker, Bourne, and Carrington, *J. Chem. Soc.* (Lond.), p. 2125 (1954).
378. Bacon and Bell, *J. Chem. Soc.* (Lond.), p. 2528 (1953).
379. Albon, Bell, Blanchard, Gross, and Rundell, *J. Chem. Soc.* (Lond.), 24 (1953).
380. Kuhn and von Grundherr, *Chem. Ber.*, 59, 1655 (1926).
381. Von Lippmann, *Chem. Ber.*, 60, 161 (1927).
382. Hudson and Sherwood, *J. Am. Chem. Soc.*, 40, 1456 (1918).
383. Aspinall, Begbie, and Mckay, *J. Chem. Soc.* (Lond.), p. 214 (1962).
384. Mayeda, *J. Biochem.* (Tokyo), 1, 131 (1922).
385. Ohtsuki, *Acta Phytochim.*, 4, 1 (1928).
386. Talley and Evans, *J. Am. Chem. Soc.*, 573, 575 (1943).
387. Shaban and Jeanloz, *Carbohydr. Res.*, 19, 311 (1971).
388. Schwarz and Timell, *Can. J. Chem.*, 41, 1381 (1963).
389. Barker, Gomez-Sanchez, and Stacey, *J. Chem. Soc.* (Lond.), p. 3264 (1959).
390. Allen and Bacon, *Biochem. J.*, 63, 200 (1956).
391. Bacon and Bell, *J. Chem. Soc.* (Lond.), p. 2528 (1953).
392. Bacon, *Biochem. J.*, 57, 320 (1954).
393. Wolfe, Senior, and Kin, *J. Biol. Chem.*, 249, 1828 (1974).
394. Strepkov, *J. Gen. Chem. U.S.S.R.*, 9, 1489 (1939).
395. Kuhn, Gauhe, and Baer, *Chem. Ber.*, 87, 289, 1553 (1954); 89, 2513, 2514 (1956).
396. Nakazawa, *J. Biochem.* (Tokyo), 46, 1579 (1959).
397. Schiffman, Kabat, and Leskowitz, *J. Am. Chem. Soc.*, 84, 73 (1962).
398. Lister-Cheese, and Morgan, *Nature*, 191, 149 (1961).
399. Masamune, Yoshizawa, and Haga, *Tohoku J. Exp. Med.*, 64, 257 (1956).
400. Masamune and Shinohara, *Tohoku J. Exp. Med.*, 69, 65 (1958).
401. Pazur, *J. Am. Chem. Soc.*, 75, 6323 (1953).
402. Strepkov, *Dokl. Akad. Nauk SSSR*, 124, 1344 (1959).
403. Okuyama, *Seikagaku*, 33, 134, 821 (1961).
404. Yoshizawa, *J. Biochem.* (Tokyo), 51, 145 (1962).
405. Kuhn, Baer, and Gauhe, *Justus Liebigs Ann. Chem.*, 611, 242 (1958).
406. Kuhn and Gauhe, *Chem. Ber.*, 93, 647 (1960).
407. Kuhn, Baer, and Gauhe, *Chem. Ber.*, 91, 364 (1958).
408. Zechmeister and Toth, *Chem. Ber.*, 65, 161 (1932).
409. Linker, Meyer, and Weissmann, *J. Biol. Chem.*, 213, 237 (1955).
410. Anderson, Gillette, and Seely, *J. Biol. Chem.*, 140, 569 (1941).
411. Buchi and Deuel, *Helv. Chim. Acta*, 37, 1392 (1954).
412. Falconer and Adams, *Can. J. Chem.*, 34, 338 (1956).
413. Sloneker and Jeanes, *Can. J. Chem.*, 40, 2066 (1962).
414. Whistler and McGilvray, *J. Am. Chem. Soc.*, 77, 2212 (1955).
415. Schneir and Rafelson, *Biochem. Biophys. Acta*, 130, 1 (1966).
416. Kuhn and Brossmer, *Chem. Ber.*, 92, 1667 (1959).
417. Ryan, Carubelli, Caputto, and Trucco, *Biochim. Biophys. Acta*, 101, 252 (1965).
418. Brandish, Shaw, and Baddiley, *J. Chem. Soc.i* (Lond.), p. C521 (1966).
419. Peat, Whelan, and Evans, *J. Chem. Soc.* (Lond.), p. 175 (1960).
420. Lindberg, *Acta Chem. Scand.*, 9, 1093, 1097 (1955).
421. Jones and Reid, *J. Chem. Soc.* (Lond.), p. 1361 (1954); p. 1890 (1955).
422. Goldschmid and Perlin, *Can. J. Chem.*, 41, 2272 (1963).
423. Hatanaka, *Arch. Biochem. Biophys.*, 82, 188 (1959).
424. Davy and Courtois, *C. R. Acad. Sci.*, 261, 3483 (1965).
425. French, Wild, and James, *J. Am. Chem. Soc.*, 75, 3664 (1953).
426. Onuki, *Sci. Pap. Inst. Phys. Chem. Res.* (Japan), 20, 201 (1933).
427. Courtois and Ariyoshi, *Bull. Soc. Chim. Biol.*, 42, 737 (1960).
428. Haq and Adams, *Can. J. Chem.*, 39, 1563 (1961).
429. Kuhn, Gauhe, and Baer, *Chem. Ber.*, 95, 513, 518 (1962).
430. Kuhn, Löw, and Trischmann, *Chem. Ber.*, 90, 203 (1957).
431. Torriani and Pappenheimer, *J. Biol. Chem.*, 237, 1 (1962).
432. Parrish, Perlin, and Reese, *Can. J. Chem.*, 38, 2094 (1960).
433. Perlin and Suzuki, *Can. J. Chem.*, 40, 50 (1962).
434. Igarashi, Igoshi, and Sakurai, *Agric. Biol. Chem.*, 30, 1254 (1966).
435. Whistler and Hickson, *J. Am. Chem. Soc.*, 76, 1671 (1954).
436. Zechmeister and Toth, *Chem. Ber.*, 64, 854 (1931).
437. Wolfrom and Fields, *Tappi (Tech. Assoc. Pulp Pap. Ind.)*, 41, 204 (1958); 40, 335 (1957).
438. Duncan and Manners, *Biochem. J.*, 69, 343 (1958).
439. Jones, Jeanes, Stringer, and Tsuchiya, *J. Am. Chem. Soc.*, 78, 2499 (1956).
440. Pazur and Ando, *J. Biol. Chem.*, 235, 297 (1960).
441. Helferich, Schafer, and Bauerlein, *Justus Liebigs Ann. Chem.*, 465, 166 (1928).
442. Bailey, Hutson, and Weigel, *Biochem. J.*, 80, 514 (1961).
443. Binkley and Altenburg, *Int. Sugar J.*, 67, 110 (1965).
444. Kurosawa, Yamamoto, Igaue, and Nakamura, *Nippon Nogei Kagaku Kaishi*, 30, 696 (1956).
445. Strepkov, *Dokl. Akad. Nauk SSSR*, 125, 216 (1959).
446. Murakami, *Acta Phytochim.*, 14, 101 (1944); 15, 105, 109 (1949).
447. Barker and Carrington, *J. Chem. Soc.* (Lond.), p. 3588 (1953).
448. Schlubach and Berndt, *Justus Liebigs Ann. Chem.*, 647, 41 (1961).
449. Pazur, *J. Biol. Chem.*, 199, 217 (1952).
450. Kuhn and Gauhe, *Justus Liebigs Ann. Chem.*, 611, 249 (1958).
451. Demain and Phaff, *Arch. Biochem. Biophys.*, 51, 114 (1954).
452. Wiessman, Meyer, Sampson, and Linker, *J. Biol. Chem.*, 208, 417 (1954).
453. Bourquelot and Bridel, *C. R. Acad. Sci.*, 151, 760 (1910).
454. Murakami, *Proc. Imp. Acad.* (Tokyo), 16, 14 (1940).
455. Kuhn, Baer, and Gauhe, *Chem. Ber.*, 91, 364 (1958).
456. Schlubach and Koehn, *Justus Liebigs Ann. Chem.*, 614, 126 (1958).
457. Wolfrom, Dacons, and Fields, *Tappi (Tech. Assoc. Pulp Pap. Ind.)*, 39, 803 (1956).
458. Hoover, Nelson, Milnei, and Wei, *J. Food Sci.*, 30, 253 (1965).
459. Schlubach and Koehn, *Justus Liebigs Ann. Chem.*, 606, 130 (1957).
460. Schlubach, Berndt, and Chiemprasert, *Justus Liebigs. Ann. Chem.*, 665, 191 (1963).
461. Kobata and Ginsburg, *J. Biol. Chem.*, 244, 5496 (1969).
462. Kuhn, Gauhe, and Baer, *Chem. Ber.*, 86, 827 (1953).
463. Wiegandt and Schulze, *Z. Naturforsch.*, 24b, 945 (1969).
464. Whistler and Tu, *J. Am. Chem. Soc.*, 74, 3609 (1952); 75, 645 (1954).
465. Bjorndel, Eriksson, Garegg, Lindberg, and Swan, *Acta Chem. Scand.*, 19, 2309 (1965).
466. Murakami, *Acta Phytochim.*, 13, 37 (1942).
467. Herissey, Fleury, Wickstrom, Courtois, and Dizet, *Compt. Rend.*, 239, 824 (1954).

468. Akiya and Tomoda, *J. Pharm. Soc. Jap.*, 76, 571 (1956).
469. French and Rundle, *J. Am. Chem. Soc.*, 64, 1651 (1942).
470. Freudenberg and Jacobi, *Justus Liebigs Ann. Chem.*, 518, 102 (1935).
471. Kobata and Ginsburg, *J. Biol. Chem.*, 247, 1525 (1972).
472. Kobata and Ginsburg, *Arch. Biochem. Biophys.*, 150, 273 (1972).
473. Kuhn, Baer, and Gauhe, *Justus Liebigs Ann. Chem.*, 611, 242 (1958).
474. Kuhn and Gauhe, *Chem. Ber.*, 93, 647 (1960).
475. Grimmonprez and Montreuil, *Bull. Soc. Chim. Biol.*, 50, 843 (1968).
476. Murakami, *Acta Phytochim.*, 10, 43 (1937).
477. Courtois, Dizet, and Wickstrom, *Bull. Soc. Chim. Biol.*, 40, 1059 (1958).
478. French, Knapp, and Pazur, *J. Am. Chem. Soc.*, 72, 5150 (1950).
479. Schlubach and Koehn, *Justus Liebigs Ann. Chem.*, 606, 130 (1957).
480. Suzuki and Sunayama, *Jap. J. Microbiol.*, 13, 95 (1969).
481. Suzuki, Sunayama, and Saito, *Jap. J. Microbiol.*, 12, 19 (1968).
482. Suzuki and Sunayama, *Jap. J. Microbiol.*, 12, 413 (1968).
483. Sunayama, *Jap. J. Microbiol.*, 14, 27 (1970).
484. Lundblad, Hallgren, Rudmark, and Svensson, *Biochemistry*, 12, 3341 (1973).
485. Huttunen and Miettinen, *Acta Chem. Scand.*, 19, 1486 (1965).
486. Kobata and Ginsbuig, *J. Biol. Chem.*, 245, 1484 (1970).

MUCOPOLYSACCHARIDES (GLYCOSAMINOGLYCANS)

Name[a]	Repeating Unit[b]	$[\alpha]_D$, Degrees[c]	Infrared, Cm^{-1},[d]	Intrinsic Viscosity (21)	Molecular Weight	Occurrence
Chitin	$(1 \rightarrow 4)$-O-2-Acetamido-2-deoxy-β-D-glucopyranose (1)	−14 to +56[i] (1)	884–890 (1)	—	—	Skeletal substance of arthropods, molluscs, and annelids, cell wall of many fungi, green algae
Chondroitin	$(1 \rightarrow 4)$-O-β-D-Glucopyranosyluronic acid-$(1 \rightarrow 3)$-2-acetamido-2-deoxy-β-D-galactopyranose (2, 3)	−21 (3)	—	—	—	Bovine cornea
Chondroitin 4-sulfate (Chondroitin sulfate A)	$(1 \rightarrow 4)$-O-β-D-glucopyranosyluronic acid-$(1 \rightarrow 3)$-2-acetamido-2-deoxy-4-O-sulfo-β-D-galactopyranose (4)	−26 to −30 (4)	724, 851, 930 (17, 18)	0.2–1.0	5×10^4 (22)	Bone cornea, cartilage, notochord, skin
Chondroitin 6-sulfate (Chondroitin sulfate C)	$(1 \rightarrow 4)$-O-Glucopyranosyluronic acid-$(1 \rightarrow 3)$-2-acetamido-2-deoxy-6-O-sulfo-β-D-galactopyranose (5)	−12 to −22 (5)	775, 820, 1,000 (17,18)	0.2–1.3	5×10^3– 50×10^3 (21)	Cartilage, aorta, skin, umbilical cord
Dermatan sulfate (Chondroitin sulfate B, β-heparin)	$(1 \rightarrow 4)$-O-α-L-Idopyranosyluronic acid-$(1 \rightarrow 3)$-2-acetamido-2-deoxy-4-O-sulfo-β-D-galactopyranose (6)[e]	−55 to −63 (13)	724, 851, 930 (18)	0.5–1.0	1.5×10^4– 4×10^4 (21)	Aorta, skin, umbilical cord, tendon
Heparan sulfate (Heparitin sulfate)	Glucopyranosyluronic acid-$(1 \rightarrow 4)$-2-amino-2-deoxy-O-sulfo-D-glucopyranose (7)[f]	+39 to +69 (14)	920, 1,050 (19)	—	—	Aorta, lung
Heparin	$(1 \rightarrow 4)$-O-α-D-Glucopyranosyluronic acid-$(1 \rightarrow 4)$-2-sulfoamino-2-deoxy-6-O-sulfo-α-D-glucopyranose and $(1 \rightarrow 4)$-O-α-L-idopyranosyluronic acid-2-sulfate-$(1 \rightarrow 4)$-2-sulfoamino-2-deoxy-6-O-sulfo-α-D-glucopyranose (8, 9)[g]	+48 (15)	890, 940 (14)	0.1–0.2	8×10^3– 20×10^3 (23)	Liver, lung, skin, mast cells
Hyaluronic acid	$(1 \rightarrow 4)$-O-β-D-Glucopyranosyluronic acid-$(1 \rightarrow 3)$-2-acetamido-2-deoxy-β-D-glucopyranose (10)	−68 (16)	900, 950 (16)	2.0–48	2×10^5– 10×10^5 (24)	Synovial fluid, vitreous humor, umbilical cord, skin
Keratan sulfate (Keratosulfate)	$(1 \rightarrow 3)$-O-β-D-Galactopyranose-$(1 \rightarrow 4)$-2-acetamido-2-deoxy-6-O-sulfo-β-D-glucopyranose (7, 11)[h]	+4.5	775, 820, 998 (20)	0.2–0.5	8×10^3– 12×10^3 (21)	Aorta, cornea, cartilage, nucleus pulposus

Compiled by I. Danishefsky.

[a] Certain mucopolysaccharides have been given different names by various investigators. In such cases, alternative names are given in parentheses.

[b] The repeating unit, which is indicated, is the one that is most prevalent. In many mucopolysaccharides there may be some variations in parts of the chain (microheterogeneity), especially with regard to sulfation. The units involved in the linkage to the proteins are not given here.

[c] Many of the values for specific rotation and intrinsic viscosity are given as ranges because these depend on the method of isolation and the original tissue source. In these cases, the references cited are review articles which include the original research papers.

[d] Fingerprint region.

[e] Also contains small but significant amounts of glucuronic acid.[25,26]

[f] Some amino groups are acetylated and others are sulfated. There are also variations in the position of O-sulfate. Part of the uronic acid is iduronic acid.

[g] There are O-sulfates in about one third of the uronic acid residues. Although both iduronate and glucuronate are present their relative proportion is not completely defined.[27]

[h] Some of the galactose is sulfated on position-6.

[i] In HCl; change in rotation occurs as a result of hydrolysis.

References

1. Foster, Webber, *Adv. Carbohydr. Chem.*, 15, 371 (1960).
2. Meyer, Linker, Davidson, and Weissmann, *J. Biol. Chem.*, 205, 611 (1953).
3. Davidson and Meyer, *J. Biol. Chem.*, 211, 605 (1954).
4. Jeanloz, *Methods Carbohydr. Chem.*, 5, 110 (1965).
5. Jeanloz, *Methods Carbohydr. Chem.*, 5, 113 (1965).
6. Jeanloz, *Methods Carbohydr. Chem.*, 5, 114 (1965).
7. Cifonelli, *Carbohydr. Res.*, 8, 233 (1968).
8. Danishefsky, Steiner, Bella, and Friedlander, *J. Biol. Chem.*, 244, 1741 (1969).
9. Wolfrom, Honda, and Wang, *Carbohydr. Res.*, 10, 259 (1969).
10. Brimacombe and Webber, *Mucopolysaccharides.* Elsevier, Amsterdam. 43 (1964).
11. Bhavanandam and Meyer, *Science*, 151, 1404 (1966).
12. Irvine, *J. Chem. Soc.*, 95, 564 (1909).
13. Brimacombe and Webber, *Mucopolysaccharides.* Elsevier, Amsterdam. 82 (1964).
14. Linker, Hoffman, Sampon, and Meyer, *Biochim. Biophys. Acta*, 29, 443 (1958).

15. Danishefsky, Eiber, and Carr, *Arch. Biochem. Biophys.*, 90, 114 (1960).
16. Danishefsky and Bella, *J. Biol. Chem.*, 241, 143 (1966).
17. Orr, *Biochim. Biophys. Acta*, 14, 173 (1954).
18. Mathews, *Nature*, 181, 421 (1958).
19. Schiller, *Biochim. Biophys. Acta*, 32, 315 (1959).
20. Brimacombe and Webber, *Mucopolysaccharides,*. Elsevier, Amsterdam. 138 (1964).
21. Mathews, *Clin. Orthop.*, 48, 267 (1966).

22. Mathews, *Arch. Biochem. Biophys.*, 61, 367 (1956).
23. Barlow, Sanderson, and McNeil, *Arch. Biochem. Biophys.*, 94, 518 (1961).
24. Silpananta, Dunston, and Ogston, *Biochem. J.*, 109, 43 (1968).
25. Hoffman, Linker, and Meyer, *Arch. Biochem. Biophys.*, 69, 435 (1957).
26. Framsson and Roden, *J. Biol. Chem.*, 242, 4161 (1967).
27. Helting and Lindahl, *J. Biol Chem.*, 246, 5442 (1972).

Section VI
Physical and Chemical Data

IUBMB-IUPAC JOINT COMMISSION ON BIOCHEMICAL NOMENCLATURE (JCBN) RECOMMENDATIONS FOR NOMENCLATURE AND TABLES IN BIOCHEMICAL THERMODYNAMICS

(IUPAC Recommendations 1994)
http://www.chem.qmul.ac.uk/iubmb/thermod/

Robert A. Alberty

Department of Chemistry, Massachusetts Institute of Technology, Cambridge, Massachusetts 02139, USA

*Membership of the Panel on Biochemical Thermodynamics during the preparation of this report (1991–1993) was as follows:

Convener: R. A. Alberty (USA); A. Cornish-Bowden (France); Q. H. Gibson (USA); R. N. Goldberg (USA); G. G. Hammes (USA); W. Jencks (USA); K. F. Tipton (Ireland); R. Veech (USA); H. V. Westerhoff (Netherlands); E. C. Webb (Australia).

†Membership of the IUBMB–IUPAC Joint Commission on Biochemical Momenclature (JCBN):

Chairman: J. F. G Vliegenthart (Netherlands); *Secretary*: A. J. Barrett (UK); A. Chester (Sweden); D. Coucouvanis (USA); C Liébecq (Belgium); K. Tipton (Ireland); P. Venetianer (Hungary); *Associate Members*: H. B. F. Dixon (UK); J. C. Rigg (Netherlands).

Chemical equations are written in terms of specific ionic and elemental species and balance elements and charge, whereas biochemical equations are written in terms of reactants that often consist of species in equilibrium with each other and do not balance elements that are assumed fixed, such as hydrogen at constant pH. Both kinds of reaction equations are needed in biochemistry. When the pH and the free concentrations of certain metal ions are specified, the apparent equilibrium constant K' for a biochemical reaction is written in terms of sums of species and can be used to calculate a standard transformed Gibbs energy of reaction $\Delta_r G'^\circ$. Transformed thermodynamic properties can be calculated directly from conventional thermodynamic properties of species. Calorimetry or the dependence of K' on temperature can be used to obtain the standard transformed enthalpy of reaction $\Delta_r H'^\circ$. Standard transformed Gibbs energies of formation $\Delta_f G'^\circ(i)$ and standard transformed enthalpies of formation $\Delta_f H'^\circ(i)$ for reactants (sums of species) can be calculated at various T, pH, pMg, and ionic strength (I) if sufficient information about the chemical reactions involved is available. These quantities can also/be calculated from measurement of K' for a number of reactions under the desired conditions. Tables can be used to calculate $\Delta_r G'^\circ$ and $\Delta_r H'^\circ$ for many more reactions.

Contents

1. Preamble

In 1976 an Interunion Commission on Biothermodynamics (IUPAC, IUB, IUPAB) published *Recommendations for Measurement and Presentation of Biochemical Equilibrium Data* (ref. 1). This report recommended symbols, units, and terminology for biochemical equilibrium data and standard conditions for equilibrium measurements. These recommendations have served biochemistry well, but subsequent developments indicate that new recommendations and an expanded nomenclature are needed. In 1985 the Interunion Commission on Biothermodynamics published *Recommendations for the Presentation of Thermodynamic and Related Data in Biology (1985)* (ref. 2).

Reproduced from:
Pure & Appl. Chem., Vol. 66, No.8, pp. 1641–1666, 1994.
© 1994 IUPAC

Before discussing the new recommendations, some of the basic recommendations of 1976 are reviewed and the recommended changes in these basic matters are given.

2. Basic 1976 recommendations on symbols and nomenclature

In the 1976 *Recommendations* (ref. 1), the overall reaction for the hydrolysis of ATP to ADP was written as

$$\text{total ATP} + \text{H}_2\text{O} = \text{total ADP} + \text{total P}_\text{i} \tag{1}$$

and the expressions for the apparent equilibrium constant K' and the apparent standard Gibbs energy change $\Delta G^{\circ\prime}$ were written as*

$$K' = \frac{[\text{total ADP}][\text{total P}_\text{i}]}{[\text{total ATP}]} \tag{2}$$

$$\Delta G^{0\prime} = -RT \ln \frac{[\text{total ADP}][\text{total P}_\text{i}]}{[\text{total ATP}]} \tag{3}$$

where these are equilibrium concentrations, recommended to be molar concentrations. The 1976 Recommendations further recommended that information about the experimental conditions could be indicated by writing $K_c'(\text{pH} = x, \text{etc.})$ and $\Delta G_c^{\circ\prime}(\text{pH} = x,$ etc.), where the subscript c indicates that molar concentrations are used. The 1976 Recommendations pointed out that the hydrolysis of ATP can also be formulated in terms of particular species of reactants and products. For example, at high pH and in the absence of magnesium ion

$$\text{ATP}^{4-} + \text{H}_2\text{O} = \text{ADP}^{3-} + \text{P}_\text{i}^{2-} + \text{H}^+ \tag{4}$$

leads to the equilibrium constant expression

$$K_{\text{ATP}^{4-}} = \frac{[\text{ADP}^3][\text{P}_\text{i}^2][\text{H}^+]}{[\text{ATP}^{4-}]} \tag{5}$$

where the equilibrium constant $K_{\text{ATP}}4-$ is independent of pH. The 1976 *Recommendations* went on to show how K' is related to $K_{\text{ATP}}4-$.

3. Corresponding new recommendations

The new recommendation is that reaction 1 be should be written as

$$\text{ATP} + \text{H}_2\text{O} = \text{ADP} + \text{P}_\text{i} \tag{6}$$

where ATP refers to an equilibrium mixture of ATP^{4-}, HATP^{3-}, $\text{H}_2\text{ATP}^{2-}$, MgATP^{2-}, MgHATP^-, and Mg_2ATP at the specified pH and pMg. This is referred to as a *biochemical equation* to emphasize that it describes the reaction that occurs at specified pH and pMg. The apparent equilibrium constant K' is made up of the equilibrium concentrations of the reactants relative to the standard state concentration c°, which is 1 M; note that M is an abbreviation for mol L^{-1}.

$$K' = \frac{([\text{ADP}]/c^\circ)[\text{P}_\text{i}]/c^\circ)}{([\text{ATP}]/c^\circ)} = \frac{[\text{ADP}][\text{P}_\text{i}]}{[\text{ATP}]c^\circ} \tag{7}$$

The term c° arises in the derivation of this equilibrium constant expression from the fundamental equation of thermodynamics and makes the equilibrium constant dimensionless. The logarithm of K' can only be taken if it is dimensionless (ref. 3). The standard state concentration used is an absolutely essential piece of information for the interpretation of the numerical value of an equilibrium constant. The apparent equilibrium constant K' is a function of T, P, pH, pMg, and I (ionic strength). Various metal ions may be involved, but Mg^{2+} is used as an example. As described below,

$$\Delta_\text{r} G^{\prime\circ} = -RT \ln K' \tag{8}$$

where $\Delta_\text{r} G^{\prime\circ}$ is the standard transformed Gibbs energy of reaction. The important point is that when the pH, and sometimes the free concentrations of certain metal ions, are specified, the criterion of equilibrium is the transformed Gibbs energy G' (ref. 4, 5). The reason for this name is discussed later in the section on transformed thermodynamic properties. Since the apparent equilibrium constant K' yields the standard value (the change from the initial state with the separated reactants at c° to the final state with separated products at c°) of the change in the transformed Gibbs energy G', the superscript $^\circ$ comes after the prime in $\Delta_\text{r} G^{\prime\circ}$. The subscript r (recommended in ref. 3) refers to a reaction and is not necessary, but it is useful in distinguishing the standard transformed Gibbs energy of reaction from the standard transformed Gibbs energy of formation $\Delta_\text{f} G^{\prime\circ}(i)$ of reactant i, which is discussed below.

The hydrolysis of ATP can also be described by means of a *chemical equation* such as

$$\text{ATP}^{4-} + \text{H}_2\text{O} = \text{ADP}^{3-} + \text{HPO}_4^{2-} + \text{H}^+ \tag{9}$$

A *chemical equation* balances atoms and charge, but a *biochemical equation* does not balance H if the pH is specified or Mg if pMg is specified, and therefore does not balance charge. Equation 9 differs from equation 4 in one way that is significant but not of major importance. Writing HPO_4^{2-}, rather than P_i^{2-} is a move in the direction of showing that atoms and charge balance in equation 9. Strictly speaking ATP^{4-} ought to be written $\text{C}_{10}\text{H}_{12}\text{O}_{13}\text{N}_5\text{P}_3^{4-}$. That is not necessary or advocated here, but we will see later that the atomic composition of a biochemical species is used in calculating standard transformed thermodynamic properties. Chemical equation 9 leads to the following equilibrium constant expression

$$K = \frac{[\text{ADP}^{3-}][\text{HPO}_4^{2-}][\text{H}^4]}{[\text{ATP}^{4-}](c^\circ)^2} \tag{10}$$

where $c^\circ = 1$ M. The equilibrium constant K is a function of T, P, and I. This equilibrium constant expression does not completely describe the equilibrium that is reached except at high pH and in the absence of Mg^{2+}. Chemical equations like equation 4 are useful in analyzing biochemical reactions and are often referred to as reference equations; thus, the corresponding equilibrium constants may be represented by K_ref. The effect of Mg^{2+} is discussed here, but this should be taken as only an example because the effects of other metal ions can be handled in the same manner.

---

* *Note* Abbreviations used in this document are: AMP, adenosine 5′-monophosphate; ADP, adenosine 5′-diphosphate; ATP, adenosine 5′-triphosphate; Glc, glucose; Glc-6-P, glucose 6-phosphate; P$_\text{i}$, orthophosphate. The designator (aq) is understood as being appended to all species that exist in aqueous solution.

4. Additional new recommendations

4.1 Recommendations concerning chemical reactions

The thermodynamics of reactions of species in aqueous solution is discussed in every textbook on physical chemistry, but this section is included to contrast the nomenclature with that of the next section and to respond to the special needs of biochemistry. As mentioned in section 3, equilibrium constants of chemical reactions that are used in biochemistry are taken to be functions of T, P, and I. Therefore, the standard thermodynamic properties are also functions of T, P, and I. The standard Gibbs energy of reaction $\Delta_r G°$ for reaction 9 is calculated using

$$\Delta_r G° = -RT \ln K \tag{11}$$

and there are corresponding values of $\Delta_r H°$ and $\Delta_r S°$ that are related by

$$\Delta_r G° = \Delta_r H° - T\Delta_r S° \tag{12}$$

The standard enthalpy of reaction is given by

$$\Delta_r H° = RT^2 \left[\frac{\partial \ln K}{\partial T} \right]_{P,I} \tag{13}$$

If $\Delta_r H°$ is independent of temperature in the range considered, it can be calculated using

$$\Delta_r H° = [RT_1 T_2/(T_2 - T_1)] \ln (K_2/K_1) \tag{14}$$

If the standard molar heat capacity change $\Delta_r CP°$ is not equal to zero and is independent of temperature, the standard molar enthalpy of reaction varies with temperature according to

$$\Delta_r H°(T) = \Delta_r H°(T^*) + \Delta_r C_P°(T - T^*) \tag{15}$$

The reference temperature T^* is usually taken as 298.15 K. In this case, $\Delta_r G°$ and K vary with temperature according to (ref. 6)

$$\Delta_r G°(T) = -RT \ln K(T)$$
$$= \Delta_r H°(T^*) + \Delta_r C_P°(T - T^*) + T\{\Delta_r G°(T^*)$$
$$- \Delta_r H°(T^*)\}/T^* - T\Delta_r C_P° \ln (T/T^*) \tag{16}$$

Additional terms containing $(\partial \Delta_r C_P°/\partial T)_P$ and higher order derivatives may be needed for extremely accurate data or for a very wide temperature range.

Equations 13 and 14 are exact only when the equilibrium constants are based on a molality standard state. If the equilibrium constants were determined with a standard state based on molarity, these equilibrium constants should be converted to a molality basis prior to using equation 13. For dilute aqueous solutions, $m_i = c_i/\rho$ where m_i and c_i are, respectively, the molality and the molarity of substance i and ρ is the mass density of water in kg L^{-1}. If this conversion is not made, there is an error of $RT^2(\partial \ln \rho/\partial T)_{P,I}$ for each unsymmetrical term in the equilibrium constant. This quantity is equal to 0.187 kJ mol^{-1} for dilute aqueous solutions at 298.15 K. Similar statements pertain to equations 27 and 28, which are given later in this document.

Since the standard thermodynamic properties $\Delta_r G°$ and $\Delta_r H°$ apply to the change from the initial state with the separated reactants at $c°$ to the final state with separated products at $c°$, it is of interest to calculate the changes in the thermodynamic properties under conditions where the reactants and products have specified concentrations other than $c°$. The change in Gibbs energy $\Delta_r G$ in an isothermal reaction in which the reactants and products are not all in their standard states, that is, not all at 1 M, is given by

$$\Delta_r G = \Delta_r G° + RT \ln Q \tag{17}$$

where Q is the reaction quotient of specified concentrations of species. The reaction quotient has the same form as the equilibrium constant expression, but the concentrations are arbitrary, rather than being equilibrium concentrations. Ideal solutions are assumed. The change in Gibbs energy $\Delta_r G$ in an isothermal reaction is related to the change in enthalpy $\Delta_r H$ and change in entropy $\Delta_r S$ by

$$\Delta_r G = \Delta_r H - T\Delta_r S \tag{18}$$

The corresponding changes in entropy and enthalpy are given by

$$\Delta_r S = \Delta_r S° - R \ln Q \tag{19}$$

$$\Delta_r H = \Delta_r H° \tag{20}$$

The standard reaction entropy can be calculated from the standard molar entropies of the reacting species: $\Delta_r S° = \Sigma v_i°(i)$, where v_i is the stoichiometric number (positive for products and negative for reactants) of species i.

The electromotive force E of an electrochemical cell is proportional to the $\Delta_r G$ for the cell reaction.

$$\Delta_r G = -|v_e|FE \tag{21}$$

where $|v_e|$ is the number of electrons transferred in the cell reaction and F is the Faraday constant (96 485.31 C mol^{-1}). Substituting equation 17 yields

$$E = E° - \frac{RT}{|v_e|F} \ln Q \tag{22}$$

where $E° = -\Delta_r G°/|ve|F$ is the standard electromotive force, that is the electromotive force when all of the species are in their standard states, but at the ionic strength specified for $\Delta_r G°$. The electromotive force for a cell is equal to the difference in the electromotive forces of the half cells.

The standard Gibbs energy and enthalpy of reaction can be calculated from the formation properties of the species.

$$\Delta_r G° = \Sigma v_i \Delta_f G°(i) \tag{23}$$

$$\Delta_r H° = \Sigma v_i \Delta_f H°(i) \tag{24}$$

where the v_i is the stoichiometric numbers of species i. The standard entropy of formation of species i can be calculated using

$$\Delta_f S°(i) = [\Delta_f H°(i) - \Delta_f G°(i)]/T \tag{25}$$

Two special needs of biochemistry are illustrated by considering the seven species in Table I. The first part of Table I gives the standard thermodynamic properties as they are found in the standard thermodynamic tables (see Appendix). The standard

Table I: Standard Formation Properties of Aqueous Species at 298.15 K.

	$\Delta_f H°$/kJ mol^{-1}	$\Delta_f G°$/kJ mol^{-1}
$I = 0$ M		
H_2O	−285.83	−237.19
H^+	0.00	0.00
Mg^{2+}	−467.00	−455.30
HPO_4^{2-}	−1299.00	−1096.10
$H_2PO_4^-$	−1302.60	−1137.30
$MgHPO_4$	−1753.80	−1566.87
Glucose	−1262.19	−915.90
$I = 0.25$ M		
H_2O	−285.83	−237.19
H^+	0.41	-0.81
Mg^{2+}	−465.36	−458.54
HPO_4^{2-}	−1297.36	−1099.34
$H_2PO_4^-$	−1302.19	−1138.11
$MgHPO_4$	−1753.80	−1566.87
Glucose	−1262.19	−915.90

thermodynamic tables give the standard formation properties for the standard state, which is the state in a hypothetical ideal solution with a concentration of 1 M but the properties of an infinitely dilute solution and the activity of the solvent equal to unity. This means that the tabulated thermodynamic properties apply at $I = 0$. Since many biochemical reactions are studied at about $I = 0.25$ M, the tabulated values of $\Delta_f G°(i)$ and $\Delta_f H°(i)$ in *The NBS Tables of Chemical Thermodynamic Properties* and the *CODATA Key Values for Thermodynamics* have to be corrected to ionic strength 0.25 M, as described in Section 5.3. The values at $I = 0.25$ M are given in the second part of Table I. No adjustments are made for H_2O, $MgHPO_4$, and glucose because the ionic strength adjustment is negligible for neutral species. We will see in Section 5.1 that the transformed Gibbs energy G' is the criterion of equilibrium at specified pH and pMg. In Section 5.4 we will see that the calculation of transformed thermodynamic properties involves the adjustment of the standard formation properties of species to the desired pH and pMg by use of formation reactions involving H^+ and Mg^{2+}. The standard transformed formation properties of species can be calculated at any given pH and pMg in the range for which the acid dissociation constants and magnesium complex dissociation constants are known. However, for the purpose of making tables, it is necessary to choose a pH and pMg that is of general interest. For the tables given here, pH = 7 and pMg = 3 are used because they are close to the values in many living cells. Table II shows the result of these calculations for the species in Table I. The ions H^+ and Mg^{2+} do not appear in this table because it applies at pH = 7 and pMg = 3. These methods have been applied to calculate the transformed formation properties

Table II: Standard Transformed Formation Properties of Species at 298.15 K, pH = 7, pMg = 3, and I = 0.25 M.

	$\Delta_f H'°$/kJ mol^{-1}	$\Delta_f G'°$/kJ mol^{-1}
H_2O	−286.65	−155.66
HPO_4^{2-}	−1297.77	−1058.57
$H_2PO_4^-$	−1303.01	−1056.58
$MgHPO_4$	−1288.85	−1050.44
Glucose	−1267.11	−426.70

of P_i (ref. 4); glucose 6-phosphate (ref. 5); adenosine, AMP, ADP, and ATP (ref. 7).

Tables I and II can be extended by use of measured equilibrium constants and enthalpies of reaction for enzyme-catalyzed reactions. For example, the species of glucose 6-phosphate can be added to these two tables because the equilibrium constant for the hydrolysis of glucose 6-phosphate has been measured at several temperatures, and because the acid dissociation constant and magnesium complex dissociation constant for glucose 6-phosphate are known at more than one temperature. However, as the standard thermodynamic properties are not known for any species of adenosine, AMP, ADP, or ATP, it is necessary to adopt the convention that $\Delta_f G° = \Delta_f H° = 0$ for adenosine in dilute aqueous solution at each temperature. This convention was introduced for H^+ a long time ago. This method has been used to calculate the standard enthalpies and standard Gibbs energies of formation of adenosine phosphate species relative to H_2ADP^- at 298.15 K (ref. 8). When this convention is used, it is not possible to calculate the enthalpy of combustion of adenosine, but it is possible to calculate $\Delta_r G°$ and $\Delta_r H°$ for reactions of adenosine that do not reduce it to CO_2, H_2O, and N_2. If the standard thermodynamic properties of all of the species of a reactant are known, $\Delta_r G'°$ and $\Delta_r H'°$ can be calculated at any specified pH and pMg, as described in the next section. When $\Delta_f G°$ and $\Delta_f H°$ are eventually determined for adenosine in dilute aqueous solution, the values of $\Delta_f G°$ and $\Delta_f H°$ of the other species in the ATP series can be calculated, but this will not alter the equilibrium constants and enthalpies of reaction that can be calculated using the tables calculated using the assumption that $\Delta_f G° = \Delta_f H° = 0$ for adenosine.

In making these calculations, the pH has been defined by pH $= -\log_{10}([H^+]/c°)$, rather than in terms of the activity, as it is in more precise measurements. The reason for doing this is that approximations are involved in the interpretation of equilibrium experiments on biochemical reactions at the electrolyte concentrations of living cells. Even Na^+ and K^+ ions are bound weakly by highly charged species of biochemical reactants, like ATP. As an approximation the acid dissociation constants and magnesium complex dissociation constants are taken to be functions of the ionic strength and the different effects of Na^+ and K^+ are ignored. These approximations can be avoided in more precise work, but only at the cost of a large increase in the number of parameters and the amount of experimental work required.

4.2 Recommendations concerning biochemical reactions

When pH and pMg are specified, a whole new set of transformed thermodynamic properties come into play (ref. 4, 5). These properties are different from the usual Gibbs energy G, enthalpy H, entropy S, and heat capacity at constant pressure C_p, and they are referred to as the transformed Gibbs energy G', transformed enthalpy H', transformed entropy S', and transformed heat capacity at constant pressure Cp'. The standard transformed Gibbs energy of reaction $\Delta_r G'°$ is made up of contributions from the standard transformed enthalpy of reaction $\Delta_r H'°$ and the standard transformed entropy of reaction $\Delta_r S'°$.

$$\Delta_r G'° = \Delta_r H'° - T\Delta_r S'° \tag{26}$$

The standard transformed enthalpy of reaction is given by

$$\Delta_r H'° = RT^2 \left(\frac{\partial \ln K'}{\partial T} \right)_{P, pH, pMg, I} \tag{27}$$

If $\Delta_r H'^\circ$ is independent of temperature in the range considered, it can be calculated using

$$\Delta_r H'^\circ = [RT_1 T_2 / (T_2 - T_1)] \ln (K_2'/K_1') \qquad (28)$$

where K_2' and K_1' are measured at the same P, pH, pMg, and I. If $\Delta_r H'^\circ$ is dependent on temperature, a more complicated equation involving an additional parameter $\Delta_r C_P'^\circ$, the standard transformed heat capacity of reaction at constant pressure, can be used with the assumption that

$$\Delta_r H'^\circ(T) = \Delta_r H'^\circ(298.15 \text{ K}) + (T - 298.15 \text{ K})\Delta_r C_P'^\circ \qquad (29)$$

This more complicated equation is analogous to equation 16. The standard transformed reaction entropy can be calculated from the standard transformed molar entropies of the reacting species: $\Delta_r S^\circ = \Sigma v_i^i \bar{S}^\circ(i)$, where the v_i' are the apparent stoichiometric numbers (positive for products and negative for reactants) of the reactants i in a biochemical reaction written in terms of reactants (sums of species) (for example, reaction 6).

The standard transformed enthalpy of reaction $\Delta_r H'^\circ$ can also be calculated from calorimetric measurements. When that is done it is necessary to make corrections for the enthalpies of reaction caused by the change $\Delta_r N(H^+)$ in the binding of H^+ and in the change $\Delta_r N(Mg^{2+})$ in the binding of Mg^{2+} in the reaction (see equation 38 below). The change in binding of an ion in a biochemical reaction is equal to the number of ions bound by the products at the specified pH and pMg minus the number of the ions bound by the reactants. Note that $\Delta_r N(H^+)$ and $\Delta_r N(Mg^{2+})$ are dimensionless.

The change in binding of H^+ and Mg^{2+} in a biochemical reaction can be calculated if the acid dissociation constants and magnesium complex dissociation constants for the reactants are known; the equilibrium constant for the biochemical reaction itself does not have to be known for this calculation. Earlier calculations of the production of H^+ and Mg^{2+} used a different sign convention (ref. 9). The changes in binding $\Delta_r N(H^+)$ and $\Delta_r N(Mg^{2+})$ are given by

$$\Delta_r N(H^+) = \Sigma v_i \overline{N}_H(i) \qquad (30)$$

$$\Delta_r N(Mg^{2+}) = \Sigma v_i \overline{N}_{Mg}(i) \qquad (31)$$

where the v_i' are the apparent stoichiometric numbers of the reactants i in a biochemical reaction. $\overline{N}_H(i)$ is the number of H bound by an average molecule of reactant i at T, P, pH, pMg, and I. $\overline{N}_H(i)$ is calculated from $\Sigma r_i N_H(i)$ where r_i is the mole fraction of species i in the equilibrium mixture of the species of the reactant at the specified pH and pMg. $N_H(i)$ is the number of hydrogen atoms in species i. The average numbers $\overline{N}_H(i)$ and $\overline{N}_{Mg}(i)$ can be included in tables of transformed thermodynamic properties at specified pH and pMg so that $\Delta_r N(H^+)$ and $\Delta_r N(Mg^{2+})$ can be readily calculated for biochemical reactions. The sign and magnitude of $\Delta_r N(H^+)$ and $\Delta_r N(Mg^{2+})$ are important because they determine the effect of pH and pMg on the apparent equilibrium constant K'. It can be shown (ref. 9, 10) that

$$\Delta_r N(H^+) = -\left(\frac{\partial \log_{10} K'}{\partial \text{pH}}\right)_{T, P, \text{pMg}, I} \qquad (32)$$

$$\Delta_r N(Mg^{2+}) = -\left(\frac{\partial \log_{10} K'}{\partial \text{pMg}}\right)_{T, P, \text{pH}, I} \qquad (33)$$

where the logarithms are $\log_{10}$. A pHstat can be used to measure $\Delta_r N(H^+)$ directly.

Since the standard transformed thermodynamic properties $\Delta_r G'^\circ$ and $\Delta_r H'^\circ$ apply to the change from the initial state with the separated reactants at c° to the final state with separated products at c°, it is of interest to calculate the changes in the transformed thermodynamic properties under conditions where the reactants and products have the concentrations they do in a living cell. The change in transformed Gibbs energy $\Delta_r G'$ in an isothermal reaction in which the reactants and products are not all in their standard states, that is, not all at 1 M, is given by

$$\Delta_r G' = \Delta_r G'^\circ + RT \ln Q' \qquad (34)$$

where Q' is the apparent reaction quotient of specified concentrations of reactants (sums of species). The change in transformed Gibbs energy $\Delta_r G'$ in an isothermal reaction at specified pH and pMg is related to the change in transformed enthalpy $\Delta_r H'$ and change in transformed entropy $\Delta_r S'$ by

$$\Delta_r G' = \Delta_r H' - T\Delta_r S' \qquad (35)$$

The corresponding changes in transformed entropy and transformed enthalpy are given by

$$\Delta_r S' = \Delta_r S'^\circ - R \ln Q' \qquad (36)$$

$$\Delta_r H' = \Delta_r H'^\circ \qquad (37)$$

The calorimetrically determined enthalpy of reaction $\Delta_r H(\text{cal})$ includes the enthalpies of reaction of H^+ and Mg^{2+} (consumed or produced) with the buffer at the specified T, P, pH, pMg, and I. The standard transformed enthalpy of reaction $\Delta_r H'^\circ$ can be calculated using (ref. 11)

$$\Delta_r H'^\circ = \Delta_r H(\text{cal}) - \Delta_r N(H^+)\Delta_r H^\circ(\text{HBuff}) - \Delta_r N(Mg^{2+})\Delta_r H^\circ(\text{MgBuff}) \qquad (38)$$

$\Delta_r H^\circ(\text{HBuff})$ is the standard enthalpy for the acid dissociation of the buffer, and $\Delta_r H^\circ(\text{MgBuff})$ is the standard enthalpy for the dissociation of the magnesium complex formed with the buffer. The values of $\Delta_r N(H^+)$ and $\Delta_r N(Mg^{2+})$ can be determined experimentally using equations 32 and 33, or they can be calculated if sufficient data on acid and magnesium complex dissociation constants are available.

When the pH is specified, the electromotive force of an electrochemical cell can be discussed in terms of the concentrations of reactants (sums of species) rather than species. When this is done the electromotive force of the cell or of a half cell is referred to as the apparent electromotive force E'. When this is done equation 34 becomes

$$E' = E'^\circ - \frac{RT}{|v_e|F} \ln Q' \qquad (39)$$

where $E'^\circ = -\Delta_r G'^\circ / |v_e| F$ is the standard apparent electromotive force at that pH. The symbol E'° is also used for the standard apparent reduction potential for an electrode reaction.

Biochemists have not had the advantage of having tables of standard formation properties of reactants at some standard set of conditions involving pH and pMg. Currently, information on biochemical reactions is tabulated as standard transformed Gibbs energies of reaction $\Delta_r G'^\circ$ and, in some cases, standard transformed enthalpies of reaction $\Delta_r H'^\circ$ at specified, T, P, pH, pMg, and I. Standard transformed formation properties have not been calculated because of lack of thermodynamic information to connect reactants in aqueous solution with the elements in their standard states and because of lack of knowledge as to how to calculate standard thermodynamic properties for a reactant like ATP that is made up of an equilibrium mixture of species at a given pH and pMg. The solution to the first problem is to assign zeros to a minimum number of species. This is what is done with $H^+(aq)$ a long time ago. The solution to the second problem is that when pH and pMg are specified, the various species ATP^{4-}, $HATP^{3-}$, $MgATP^{2-}$, etc. of ATP become pseudoisomers. That is, the relative concentrations of the various species are then a function of temperature only. At a given T, P, pH, pMg, and I, the relative concentrations can be calculated, and the standard transformed thermodynamic properties of ATP (sum of species) can be calculated. The equations for doing this are given in Section 5.2. Thus, ATP at a given pH and pMg can be treated like a single species with the properties $\Delta_f G'^\circ$, $\Delta_f H'^\circ$, and $\Delta_f S'^\circ$.

When pH and pMg are specified, the transformed formation properties (indicated by a subscript f) of reactants are defined by (ref. 4)

$$\Delta_r G'^\circ = \Sigma \, \nu_i' \Delta_f G'^\circ(i) \qquad (40)$$

$$\Delta_r H'^\circ = \Sigma \, \nu_i' \Delta_f H'^\circ(i) \qquad (41)$$

where the ν_i' are the apparent stoichiometric numbers of the reactants i in a biochemical reaction written in terms of reactants. These formation properties apply to reactants like ATP (that is, sums of species) at a specified T, P, pH, pMg, and I. The corresponding standard transformed entropy of formation of a reactant like ATP can be calculated using

$$\Delta_f G'^\circ(i) = \Delta_f H'^\circ(i) - T \Delta_f S'^\circ(i) \qquad (42)$$

Table III gives standard transformed enthalpies of formation and standard transformed Gibbs energies of formation that have been calculated at 298.15 K, pH = 7, pMg = 3, and $I = 0.25$ M (ref. 7). The values for creatine phosphate are based on the recent work of Teague and Dobson (ref. 12). The adjustment of standard formation properties of species to standard transformed formation properties at the desired pH and pMg has been mentioned in connection with Table II. When a reactant exists as a single species at pH = 7 and pMg = 3, the transformed formation properties of the species in Table II go directly into Table III. Water has to be

Table III: Standard Transformed Formation Properties of Reactants (sums of species) at 298.15 K, pH = 7, pMg = 3, and I = 0.25 M. This Table Uses the Convention that $D_f G^\circ = D_f H^\circ = 0$ for Adenosine in Dilute Aqueous Solution

	$\Delta_f H'^\circ$/kJ mol^{-1}	$\Delta_f G'^\circ$/kJ mol^{-1}
ATP	−2981.79	−2102.88
ADP	−2000.19	−1231.48
AMP	−1016.59	−360.38
A (adenosine)	−5.34	529.96
Glc-6-P	−2279.09	−1318.99
Glc (glucose)	−1267.11	−426.70
CrP	−1509.75	−750.37
Cr (creatine)	−540.08	107.69
P_i	−1299.13	−1059.55
$H_2O(l)$	−286.65	−155.66

included in this table because its formation properties must be used in equations 40 and 41, even though it does not appear in the expression for the apparent equilibrium constant. When a reactant exists at pH = 7 and pMg = 3 as an equilibrium mixture of species, isomer group thermodynamics (see Section 5.2) has to be used to calculate standard transformed formation properties (Table III) for that reactant.

The large number of significant figures in Table I might appear to indicate that these thermodynamic properties are known very accurately, but this is misleading. The values in such a table are used only by subtracting them from other values in the table, and so the only things that are important are differences between values. The values have to be given in the table with enough significant figures so that thermodynamic information in the differences is not lost. The following examples illustrate uses of this table.

This kind of table will make it easier to make thermodynamic calculations on systems of enzymatic reactions, like glycolysis. Currently, to calculate $\Delta_r G'^\circ$ for the net reaction of glycolysis, 10 reactions must be added and $\Delta_r G'^\circ$ must be multiplied by 2 for some of them.

Example 1. Calculate $\Delta_r G'^\circ$, $\Delta_r H'^\circ$, $\Delta_r S'^\circ$, and K' for the glucokinase reaction (EC 2.7.1.2) (ref. 13) at 298.15 K, pH = 7, pMg = 3, and $I = 0.25$ M.

ATP + Glc = ADP + Glc-6-P
$\Delta_r G'^\circ = (-1231.48 - 1318.99 + 2102.88 + 426.70)$ kJ mol^{-1}
     $= -20.89$ kJ mol^{-1}
$\Delta_r H'^\circ = (-2000.19 - 2279.09 + 2981.79 + 1267.11)$ kJ mol^{-1}
     $= -30.38$ kJ mol^{-1}
$\Delta_r S'^\circ = (-30.38 + 20.89) \times 10^3$ J mol^{-1}/298.15 K $= -31.83$ J K^{-1} mol^{-1}

$$K' = \exp\left(\frac{20\,890\,\mathrm{J\,mol^{-1}}}{8.3145\,\mathrm{J K^{-1}} \times 298.15\mathrm{K}}\right) = 4.57 \times 10^3$$

Example 2. Calculate $\Delta_r G'$, $\Delta_r H'$, and $\Delta_r S'$ for the glucokinase reaction at 298.15 K, pH = 7, pMg = 3, and $I = 0.25$ M when the reactant concentrations are [ATP] = 10^{-5} M, [ADP] = 10^{-3} M, [Glc] = 10^{-4} M, and [Glc-6-P] = 10^{-2} M.

$Q' = (10^{-3})(10^{-2})/(10^{-5})(10^{-4}) = 10^4$
$\Delta_r G' = \Delta_r G'^\circ + RT\ln Q' = -20.89 + (8.3145 \times 10^{-3})(298.15)\ln 10^4$
     $= 1.94$ kJ mol^{-1}

Since $\Delta_r G'$ is positive, the glucokinase reaction cannot occur in the forward direction under these conditions.

$\Delta_r H' = \Delta_r H'^\circ = -30.38$ kJ mol^{-1}
$\Delta_r S' = \Delta_r S'^\circ - R\ln Q' = -108.41$ J K^{-1} mol^{-1}

Note that

$\Delta_r G' = \Delta_r H' - T\Delta_r S' = -30.38 - (298.15)(-0.10841) = 1.94$ kJ mol^{-1}

4.3 The importance of distinguishing between chemical equations and biochemical equations

Both types of equations are needed in biochemistry. Chemical equations are needed when it is important to keep track of all of the atoms and charges in a reaction, as in discussing the mechanism of chemical change. Biochemical reactions are needed to answer the question as to whether a reaction goes in the forward or backward direction at specified T, P, pH, pMg, and I, or for calculating the equilibrium extent of such a reaction. Therefore, it is essential to be able to distinguish between these two types of equations on sight. The reaction equations in *Enzyme Nomenclature* [ref. 13] are almost exclusively biochemical equations. In the case of the hydrolysis of adenosine triphosphate to adenosine diphosphate and inorganic phosphate, it is clear that equation 9 is a chemical equation and equation 6 is a biochemical equation. Equation 6 does not indicate that hydrogen ions or magnesium ions are conserved, but it is meant to indicate that C, O, N, and P are conserved. Equation 6 indicates the form of the expression for the apparent equilibrium constant K' at specified T, P, pH, pMg, and I. Equation 9 indicates that electric charge is conserved, and the abbreviations ATP^{4-} and ADP^{3-} can be replaced by the atomic compositions of these ions to show that C, H, O, N, and P are conserved. Equation 9 indicates the form of the expression for the equilibrium constant K at specified T, P, and I. Currently, the hydrolysis of ATP is often represented by $ATP + H_2O = ADP + P_i + H^+$ in textbooks and research papers, but this is a hybrid of a chemical equation and a biochemical equation and does not have an equilibrium constant. Furthermore, this "equation" does not give the correct stoichiometry. The correct stoichiometry with respect to H^+ is obtained by use of equation 32 and is $\Delta_r N(H^+) = 0.62$ at 298.15 K, 1 bar, pH = 7, pMg = 3, and $I = 0.25$ M. The convention is that H_2O is omitted in the equilibrium expression for K or K' when reactions in dilute aqueous solutions are considered.

In writing biochemical equations, words are often used to avoid the implication that hydrogen atoms and charge are being balanced, but it is important to understand that all other atoms are balanced. For example,

$$\text{pyruvate} + \text{carbonate} + \text{ATP} = \text{oxaloacetate} + \text{ADP} + P_i \quad (43)$$

is a biochemical equation and

$$C_3H_3O_3^- + HCO_3^- + ATP^{4-} = C_4H_2O_5^{2-} + ADP^{3-}$$
$$+ HPO_4^{2-} + H^+ . \quad (44)$$

is a chemical reaction. There is no unique way to write a chemical reaction; for example, this equation could be written with $H_2PO_4^-$ and no H^+ on the right hand side. It can also be written with CO_2 on the left-hand side, but then it is necessary to be clear about whether this CO_2 is in the solution or gas phase. In equation 43 the word carbonate refers to the sum of the species CO_2, H_2CO_3, HCO_3^-, and CO_3^{2-} in aqueous solutions.

It is important to realize that $K' = K$ for reactions where the reactants are nonelectrolytes (ref. 14). An example is the hydrolysis of sucrose to glucose and fructose. Of course sugars do have ionizable groups, but we are usually not interested in the dissociations that occur above pH = 12. For racemases, $K' = K$. There are other reactions for which K' is approximately equal to K because a product has very nearly the same acid dissociation constant as a reactant.

The need to clearly distinguish between biochemical equations and chemical equations raises problems with some abbreviations that are widely used. For example, the use of NAD^+ in a biochemical equation makes it look like this charge should be balanced. NAD^+ is also not a suitable abbreviation for use in a chemical equation because it is actually a negative ion.

Chemical equations and biochemical equations should not be added or subtracted from each other because their sum or difference does not lead to an equation that has an equilibrium constant. On the other hand, chemical equations can be added to chemical equations, and biochemical equations can be added to biochemical equations.

The net equation for a system of biochemical reactions can also be written as a chemical equation or a biochemical equation, but the equilibrium constants are, of course, in general different. Net equations in the form of biochemical equations are especially useful for determining whether the system of reactions goes in the forward or backward direction at specified T, P, pH, pMg, and I.

These recommendations apply also to reactions catalyzed by RNA enzymes (ref. 15), catalytic antibodies (ref. 16), and synthetic enzymes (ref. 17) (sometimes called ribozymes, abzymes, and synzymes, respectively). The reactions catalyzed have apparent equilibrium constants K' that are functions of pH and free concentratioins of certain metal ions. Both biochemical equations and chemical equations can be written for these reactions if the reactants are weak acids or bind metal ions.

4.4 Experimental matters

In reporting results on equilibrium measurements on biochemical reactions it is extremely important to give enough information to specify T, P, pH, pMg (or free concentration of any other cation that is bound by reactants), and I at equilibrium. The most difficult of these variables are pMg and I. The calculation of pMg in principle requires information on the composition of the solution in terms of species, and this requires information on the dissociation constants of all of the weak acids and magnesium complex ions. However, if the metal ion binding constants of the buffer are known and the reactants are at low concentrations compared with the buffer, the concentrations of free metal ions can be calculated approximately. It is important to specify the composition of the solution and calculate the ionic strength, even if it can only be done approximately. Other important issues are the purities of materials, the methods of analysis, the question as to whether the same value of apparent equilibrium constant was obtained from both directions, and assignment of uncertainties. For calorimetric measurements, it is important to measure the extent of reaction. An important part of any thermodynamic investigation is the clear specification of the substances used and the reaction(s) studied. It is very helpful to readers to give *Enzyme Nomenclature* (ref. 13) identification numbers of enzymes and Chemical Abstracts Services registry numbers for reactants. IUPAC has published a "Guide to the Procedures for the Publication of Thermodynamic Data" (ref. 18), and CODATA has published a "Guide for the Presentation in the Primary Literature of Numerical Data Derived from Experiments" (ref. 19).

It is recommended that equilibrium and calorimetric measurements on biochemical reactions be carried out over as wide a range of temperature, pH, pMg, and I as is practical. For the study of biochemical reactions under "near physiological conditions," the following set of conditions is recommended: $T = 310.15$ K, pH = 7.0, pMg = 3.0, and $I = 0.25$ M. It is also recognized that there is no unique set of physiological conditions and that for many purposes it will be necessary and desirable to study biochemical reactions under different sets of conditions.

For the purpose of relating results obtained on biochemical reactions to the main body of thermodynamic data (NBS Tables and other tables listed in the Appendix) the results of experiments should be treated so as to yield results for a chemical (reference) reaction at $T = 298.15$ K and $I = 0$. If this calculation is done, the method of data reduction and all auxiliary data used should be reported. It is also recognized that while a standard state based upon the concentration scale has been widely used in biochemistry, the molality scale has significant advantages for many purposes and can also be used for the study of biochemical reactions and for the calculation of thermodynamic properties.

5. Thermodynamic background

5.1 Transformed thermodynamic properties

The definition of a transformed Gibbs energy is a continuation of a process that starts with the first and second laws of thermodynamics, but is not always discussed in terms of Legendre transforms. The combined first and second law for a closed system involving only pressure-volume work is

$$dU = TdS - PdV \tag{45}$$

where U is the internal energy and S is the entropy. The criterion for spontaneous change at specified S and V is $(dU)_{S,V} \leq 0$. That is, if S and V are held constant, U can only decrease and is at a minimum at equilibrium. To obtain a criterion at specified S and P, the enthalpy was defined by the Legendre transform $H = U + PV$ so that $(dH)_{S,P} \leq 0$. To obtain a criterion at specified T and V, the Helmholtz energy was defined by the Legendre transform $A = U - TS$ so that $(dA)_{T,V} \leq 0$. To obtain a criterion at specified T and P, the Gibbs energy was defined by the Legendre transform $G = H - TS$ so that $(dG)_{T,P} \leq 0$. The Gibbs energy is especially useful because it provides the criterion for equilibrium at specified T and P. Two Legendre transforms can be combined. For example, the internal energy can be transformed directly to G by use of $G = U + PV - TS$. Alberty and Oppenheim (ref. 20, 21) used a Legendre transform to develop a criterion for equilibrium for the alkylation of benzene by ethylene at a specified partial pressure of ethylene. Wyman and Gill (ref. 22) have described the use of transformed Gibbs energies in describing macromolecular components in solution.

In 1992, Alberty (refs. 4, 5) used the Legendre transform

$$G' = G - n'(H^+)\mu(H^+) - n'(Mg^{2+})\mu(Mg^{2+}) \tag{46}$$

to define a transformed Gibbs energy G' in terms of the Gibbs energy G. Here $n'(H^+)$ is the total amount of H^+ in the system (bound and unbound) and $\mu(H^+)$ is the specified chemical potential for H^+, which is given for an ideal solution by

$$\mu(H^+) = \mu(H^+)^\circ + RT \ln ([H^+]/c^\circ) \tag{47}$$

where $\mu(H^+)^\circ$ is the chemical potential of H^+ at 1 M in an ideal solution at specified T, P, and I. The transformed Gibbs energy G' is defined in order to obtain a criterion of spontaneous change at T, P, pH, and pMg. It can be shown that $(dG')_{T,P,pH,pMg} \leq 0$, so that G' is at a minimum when T, P, pH, and pMg are held constant. This is the fundamental justification for the use of G' in biochemistry. Under the appropriate circumstances, the magnesium term can be left out or be replaced by a term in another metal ion. The reaction is generally an enzyme-catalyzed reaction, but these concepts apply to any reaction involving a weak acid or metal ion complex when the pH and concentration of free metal ion at equilibrium are specified.

A consequence of equation 46 is that the chemical potential μ_i of each species in the system is replaced by the transformed chemical potential μ_i' given by

$$\mu_i' = \mu_i - N_H(i)\mu(H^+) - N_{Mg}(i)\mu(Mg^{2+}) \tag{48}$$

where $N_H(i)$ is the number of hydrogen atoms in species i and $N_{Mg}(i)$ is the number of magnesium atoms in species i.

Although thermodynamic derivations are carried out using the chemical potential, in actual calculations, the chemical potential μ_i of species i is replaced by the Gibbs energy of formation $\Delta_f G_i$

and the transformed chemical potential μ_i' of species i is replaced by the transformed Gibbs energy of formation $\Delta_f G_i'$, where

$$\Delta_f G_i = \Delta_f G_i^\circ + RT \ln ([i]/c^\circ) \tag{49}$$

$$\Delta_f G_i' = \Delta_f G_i'^\circ + RT \ln ([i]/c^\circ) \tag{50}$$

for ideal solutions. The calculation of $\Delta_f G_i'^\circ$ for a species is discussed in Section 5.4 and the calculation of $\Delta_f G_i'^\circ$ for a reactant is discussed in Section 5.5.

Once the $\Delta_f G_i'^\circ$ for the species ($H_2PO_4^-$, HPO_4^{2-}, $MgHPO_4$) of P_i, for example, have been calculated, the next question is how can these values be combined to obtain the value of $\Delta_f G_i'^\circ$ for P_i? The equations for this calculation are given in the next section.

5.2 Isomer group thermodynamics

A problem that has to be faced in biochemical thermodynamics at specified pH and pMg is that a reactant may consist of various species in equilibrium at the specified pH and pMg. Fortunately, a group of isomers (or pseudoisomers) in equilibrium with each other have thermodynamic properties just like a species does, but we refer to the properties of a pseudoisomer group as transformed properties. The problem of calculating a standard transformed Gibbs energy of formation of a reactant like ATP also arises when a reactant exists in isomeric forms (or hydrated and unhydrated forms), even if it is not a weak acid and does not complex with metal ions, so first we discuss a simple isomerization. The thing that characterizes an isomer group in ideal solutions is that the distribution within the isomer group is a function of temperature only. For such solutions, the standard Gibbs energy of formation of an isomer group $\Delta_f G^\circ(iso)$ can be calculated from the standard Gibbs energies of formation $\Delta_f G_i^\circ$ of the various isomers using (ref. 23)

$$\Delta_f G^\circ(iso) = -RT \ln \sum_{i=1}^{N_1} \exp\left(-\Delta_f G_i^\circ / RT\right) \tag{51}$$

where N_1 is the number of isomers in the isomer group. The standard enthalpy of formation $\Delta_f H^\circ(iso)$ of the isomer group can be calculated using (ref. 24)

$$\Delta_f H^\circ(iso) = \sum_{i=1}^{N_1} r_i \Delta_f H_i^\circ \tag{52}$$

where r_i is the equilibrium mole fraction of the ith species within the isomer group that is given by

$$r_i = \exp\{[\Delta_f G^\circ(iso) - \Delta_f G_i^\circ]/RT\} \tag{53}$$

The standard entropy of formation of the isomer group $\Delta_f S^\circ(iso)$ is given by

$$\Delta_f S^\circ(iso) = \sum_{i=1}^{N_1} r_i \Delta_f S_i^\circ - R \sum_{i=1}^{N_1} r_2 \ln r_i \tag{54}$$

These equations can be used for pseudoisomer groups (for example, the species of ATP at specified pH and pMg) by using the transformed thermodynamic properties of the species.

For pseudoisomer groups, equations 51, 52, and 53 become

$$\Delta_f G^{i\circ}(reactant) = -RT \ln \sum_{i=1}^{N_1} \exp(-\Delta_f G_i'^\circ/RT) \tag{55}$$

$$\Delta_f H'^\circ (\text{reactant}) = \sum_{i=1}^{N_1} r_i \Delta_f H_i'^\circ \qquad (56)$$

$$r_i = \exp\{[\Delta_f G'^\circ (\text{reactant}) - \Delta_f G_i'^\circ]/RT\} \qquad (57)$$

where i refers to a species at specified pH and specified free concentrations of metal ions that are bound.

5.3 Adjustment for ionic strength

The ionic strength has a significant effect on the thermodynamic properties of ions, and the extended Debye-Huckel theory can be used to adjust the standard Gibbs energy of formation and the standard enthalpy of formation of ion i to the desired ionic strength (ref. 25-28). At 298.15 K these adjustments can be approximated by

$$\Delta_f G_i^\circ (I) = \Delta_f G_i^\circ (I=0) - 2.91482 z_i^2 I^{1/2}/(1 + BI^{1/2}) \qquad (58)$$

$$\Delta_f H_i^\circ (I) = \Delta_f H_i^\circ (I=0) + 1.4775 z_i^2 I^{1/2}/(1 + BI^{1/2}) \qquad (59)$$

where kJ mol^{-1} are used, z_i is the charge on ion i, and $B = 1.6$ L$^{1/2}$ mol$^{-1/2}$. Since for H$^+$, $\Delta_f G^\circ = 0$ and $\Delta_f H^\circ = 0$ at each temperature at $I = 0$, $\Delta_f G^\circ$(H$^+$, 298.15 K, $I = 0.25$ M) = -0.81 kJ mol^{-1} and $\Delta_f H^\circ$(H$^+$, 298.15 K, $I = 0.25$ M) = 0.41 kJ mol^{-1}. For the purpose of these recommendations, pH $= -\log_{10}([\text{H}^+]/c^\circ)$ and pMg $= -\log_{10}([\text{Mg}^{2+}]/c^\circ)$, as discussed above Section 4.1.

The adjustment of thermodynamic quantities from one solution composition to another using ionic strength effects alone is an approximation that works well at low ionic strengths (< 0.1 M) but it can fail at higher ionic strengths. Rigorous treatments require the use of interaction parameters (ref. 29) and a knowledge of the composition of the solution. While a substantial body of information on these parameters exists for aqueous inorganic solutions, there is very little of this type of data available for biochemical substances. Therefore, it is important that complete information on the compositions of the solutions used in equilibrium and calorimetric measurements be reported so that when values of these interaction parameters eventually become available, the results can be treated in a more rigorous manner. Specific ion effects are especially important when nucleic acids, proteins, and other polyelectrolytes are involved (refs. 30, 31).

5.4 Adjustment of standard thermodynamic properties of species to the desired pH and pMg

When pH and pMg are specified, the various species of ATP, for example, become pseudoisomers; that is their relative concentrations are a function of temperature only. The procedure for calculating the transformed chemical potential μ_i' of a species has been indicated in equation 48. For actual calculations the chemical potentials μ_i of species are replaced with $\Delta_f G_i$ (see equation 49), and the transformed chemical potentials μ_i' of species are replaced with $\Delta_f G_i'$ (see equation 50). Thus equation 48 for a species can be written

$$\Delta_f G_i'^\circ = \Delta_f G_i^\circ - N_H(i)[\Delta_f G^\circ (\text{H}^+) + RT\ln([\text{H}^+]/c^\circ)] \\ - N_{Mg}(i)[\Delta_f G^\circ (\text{Mg}^{2+}) + RT\ln([\text{Mg}^{2+}]/c^\circ)] \qquad (60)$$

where $N_H(i)$ is the number of hydrogen atoms in species i. The corresponding equation for the standard transformed enthalpy of formation of species i is

$$\Delta_f H_i'^\circ = \Delta_f H_i^\circ - NH(i)\Delta_f H^\circ (\text{H}^+) - NMg(i)\Delta_f H^\circ (\text{Mg}^{2+}) \qquad (61)$$

since the enthalpy of an ion in an ideal solution is independent of its concentration.

In adjusting standard Gibbs energies of formation to a specified pH, there is the question as to whether to count all of the hydrogens or only those involved in the reaction under consideration. However, the recommendation here is to adjust for all of the hydrogens in a species because all of them may be ultimately removed in biochemical reactions. This has been done in Tables II and III.

There is a simple way to look at the standard transformed Gibbs energy of formation $\Delta_f G_i'^\circ$ and the standard transformed enthalpy of formation $\Delta_f H_i'^\circ$ of species i, and that is that they are the changes in formation reactions of the species with H$^+$ at the specified pH and Mg^{2+} at the specified pMg on the left-hand side of the formation reaction (ref. 32). For H$_2$PO$_4^-$,

$$\text{P(s)} + 2\text{O}_2(\text{g}) + 2\text{H}^+(\text{pH} = 7) + 3\text{e}^- = \text{H}_2\text{PO}_4^- \qquad (62)$$

$$\Delta_f G'^\circ(\text{H}_2\text{PO}_4^-) = \Delta_f G^\circ(\text{H}_2\text{PO}_4^-) - 2\{\Delta_f G^\circ(\text{H}^+) + RT\ln 10^{-\text{pH}}\} \qquad (63)$$

$$\Delta_f H'^\circ(\text{H}_2\text{PO}_4^-) = \Delta_f H^\circ(\text{H}_2\text{PO}_4^-) - 2\Delta_f H^\circ(\text{H}^+) \qquad (64)$$

The quantities $\Delta_f G^\circ$(H$^+$) and $\Delta_f H^\circ$(H$^+$) are included because they are equal to zero only at zero ionic strength. The electrons required to balance the formation reaction are assigned $\Delta_f G^\circ$(e$^-$) = $\Delta_f H^\circ$(e$^-$) = 0. This calculation can be made with either the standard thermodynamic properties at $I = 0$ or at some specified ionic strength.

The calculation of $\Delta_f G'^\circ$ and $\Delta_f H'^\circ$ for HPO$_4^{2-}$ and MgHPO$_4$ follow this same pattern, with Mg^{2+}(pMg = 3) also on the left-hand side of the formation reaction of MgHPO$_4$.

For a pseudoisomer group in which $\Delta_f G^\circ$ and $\Delta_f H^\circ$ are not known for any species, zero values have to be assigned to one of the species, as described in Section 4.1.

5.5 Calculation of the standard formation properties of a pseudoisomer group at specified pH and pMg

H$_2$PO$_4^-$, HPO$_4^{2-}$ and MgHPO$_4$ form a pseudoisomer group when pH and pMg are specified. Therefore, equations 55-57 can be used to calculate $\Delta_f G'^\circ(\text{P}_i)$ and $\Delta_f H'^\circ(\text{P}_i)$ for inorganic phosphate at the desired pH and pMg. These calculations have been made for inorganic phosphate and glucose 6-phosphate at pH = 7 and pMg = 3 by Alberty (ref. 4), and for adenosine, AMP, ADP, and ATP by Alberty and Goldberg (ref. 7) using the convention that $\Delta_f G^\circ = \Delta_f H^\circ = 0$ for neutral adenosine.

For less common and more complicated reactants, the acid dissociation constants and magnesium complex dissociation constants may not be known. The $\Delta_f G_i'^\circ$ values of the reactants at pH = 7 and pMg = 3 can, however, be calculated if K' has been measured at pH = 7 and pMg = 3 for a reaction in which $\Delta_f G_i'^\circ$ is known for the other reactants. For example, this approach can be used to calculate $\Delta_f G_i'^\circ$ for the reactants in glycolysis.

5.6 The actual experiment and thought experiments

In the laboratory, a biochemical equilibrium experiment is actually carried out at specified T and P, and the pH is measured at equilibrium. Buffers are used to hold the pH constant, but there may be a change in the pH if the catalyzed reaction produces or consumes acid. pMg at equilibrium has to be calculated, and this can be done accurately only if the acid dissociation constants and magnesium complex dissociation constants are known for all of the reactants and buffer components (ref. 12, 33). In the absence of this information pMg can be calculated approximately if the buffer binds H$^+$ and Mg^{2+}, these dissociation constants are known, and the concentrations of the reactants are much smaller than the concentration of the buffer components that are primarily responsible for the binding of Mg^{2+}. We can hope that some day there will be a pMg electrode as convenient as the pH electrode.

When we interpret the thermodynamics of a biochemical equilibrium experiment, we use an idealized thought experiment that is equivalent to the laboratory experiment. In the laboratory experiment, the buffer determines the approximate pH, but the pH will drift if H^+ is produced or consumed. The pH should be measured at equilibrium because the composition and $\Delta_r G'^\circ$ and $\Delta_r H'^\circ$ depend on this pH. Since the experimental results depend on the final pH, we can imagine that the experiment was carried out in a reaction vessel with a semipermeable membrane (permeable to H^+ and an anion, and impermeable to other reactants) with a pH reservoir on the other side. If the binding of H^+ by the products is greater than that of the reactants, H^+ will diffuse in from the pH reservoir as the reaction proceeds. If the binding of H^+ by the reactants is greater, H^+ will diffuse out of the reaction vessel as the reaction proceeds. Thus hydrogen ion is not conserved in the reaction vessel in this idealized thought experiment. Similar statements can be made about Mg^{2+}. In calorimetric experiments, corrections have to be made for the enthalpies of reaction due to the production of H^+ and Mg^{2+} to obtain $\Delta_r H'^\circ$, as mentioned earlier.

The thermodynamic interpretation of the apparent equilibrium constant K' uses $\Delta_f G'^\circ$ and $\Delta_r H'^\circ$. These quantities correspond with another thought experiment in which the separated reactants, each at 1 M at the specified T, P, final pH, final pMg, and I react to form the separated products, each at 1 M at the specified T, P, final pH, final pMg, and I.

5.7 Linear algebra

It is generally understood that chemical equations conserve atoms and charge, but it is not generally known how the conservation equations for a chemical reaction system can be calculated from a set of chemical equations or how an independent set of chemical equations can be calculated from the conservation equations for the system. Nor is it well known that conservation equations in addition to atom and charge balances may arise from the mechanism of reaction. The quantitative treatment of conservation equations and chemical reactions requires the use of matrices and matrix operations (ref. 10, 23). When the equilibrium concentrations of species such as H^+ and Mg^{2+} are specified, these species and electric charge are not conserved, and so a biochemical equation should not indicate that they are conserved. The current practice of using words like acetate and symbols like ATP and P_i is satisfactory provided that people understand the reason for using these words and symbols. It should be possible to distinguish between chemical equations and biochemical equations on sight, and this means that different symbols should be used for the reactants in these two types of equations.

A set of simple chemical equations has been discussed from the viewpoint of linear algebra (ref. 34). The hydrolysis of ATP to ADP and P_i at specified pH has also been discussed from the viewpoint of linear algebra which shows why the 4 chemical equations reduce down to a single biochemical equation (ref. 35).

The conservation matrix for a biochemical reaction is especially useful for the identification of the constraints in addition to element balances (ref. 36).

6. Recommendations on thermodynamic tables

The papers by Alberty (ref. 5) and Alberty and Goldberg (ref. 7) show four types of tables of thermodynamic properties of biochemical reactants: (1) $\Delta_f G^\circ$ and $\Delta_f H^\circ$ for species at 298.15 K, 1 bar (0.1 MPa), $I = 0$. (2) $\Delta_f G^\circ$ and $\Delta_f H^\circ$ for species at 298.15 K, 1 bar, $I = 0.25$ M. (3) $\Delta_f G'^\circ$ and $\Delta_f H'^\circ$ for species at 298.15 K, 1 bar, pH = 7, pMg = 3, and $I = 0.25$ M. (4) $\Delta_f G'^\circ$ and $\Delta_f H'^\circ$ for reactants (sum of species) at 298.15 K, 1 bar, pH = 7, pMg = 3, and $I = 0.25$ M. Table 1 contains the most basic information for calculating $\Delta_f G^\circ$ and $\Delta_f H^\circ$ for reference reactions at $I = 0$ and corresponds with the NBS and CODATA Tables. Table 4 is the most convenient for calculating $\Delta_r G'^\circ$ and $\Delta_r H'^\circ$ under normal experimental conditions of 298.15 K, 1 bar, pH = 7, pMg = 3, and $I = 0.25$ M. Currently, thermodynamic information in biochemistry is stored as K' and $\Delta_r H'^\circ$ when pH and pMg are specified and as K and $\Delta_r H^\circ$ for reactions in terms of species. In order to calculate K' and $\Delta_r H'^\circ$ or K and $\Delta_r H^\circ$ for a reaction that has not been studied, it is currently necessary to add and subtract known reactions. It would be more convenient to be able to look up reactants in a table and add and subtract their formation properties to calculate K' and $\Delta_r H'^\circ$ or K and $\Delta_r H^\circ$, as is usually done for chemical reactions. One reactant can be involved in hundreds of reactions, and so it is more economical to focus on the reactants. The usefulness of such a table increases rapidly with its length. As mentioned in Section 4.3 columns for $\overline{N}_H(i)$ and $\overline{N}_{Mg}(i)$ can be included so that the change in binding of H and Mg at specified T, P, pH, pMg, and I can be readily calculated by use of equations 30 and 31.

The choice of 298.15 K, 1 bar, pH = 7, pMg = 3, and $I = 0.25$ M is arbitrary, but these conditions are often used. Tables can be constructed for other conditions (T, pH, pMg, and I) if sufficient information is available. H_2O has to be included in this table because its $\Delta_f G'^\circ(H_2O)$ and $\Delta_f H'^\circ(H_2O)$ have to be included in the summations in equations 40 and 41 when it is a reactant, even though H_2O is omitted in the expression for the apparent equilibrium constant.

The most basic principle is that thermodynamic tables on biochemical reactants at pH = 7 and pMg = 3 should be consistent with the usual thermodynamic tables to as great an extent as possible. A great deal is already known about the thermodynamics of reactions in aqueous solution, and this is all of potential value in biochemistry. The standard transformed formation properties of inorganic phosphate and glucose 6-phosphate at pH = 7 and pMg = 3 can be calculated since the standard formation properties of inorganic phosphate and glucose are well known and the properties of glucose 6-phosphate can be calculated from $\Delta_r G'^\circ$ and $\Delta_r H'^\circ$ for the glucose 6-phosphatase reaction. This is true for many other biochemical reactants. Sometimes it is necessary to use the convention that $\Delta_f G^\circ = \Delta_f H^\circ = 0$ for a reference species, as described in the discussion of the ATP series. It is difficult and expensive to obtain these missing data because biochemical reactants are often rather large molecules and contain a large number of elements. The methods described here make it possible to calculate $\Delta_f G'^\circ$ and $\Delta_f H'^\circ$ for P_i, glucose 6-phosphate, adenosine, AMP, ADP, and ATP at temperatures in the approximate range 273-320 K, pH in the approximate range 3-10, pMg in the range above about 2, and ionic strengths in the approximate range 0-0.35 M. Since the choice of reference species is arbitrary to a certain extent, it is desirable to have international agreement on these choices. This agreement is required so that thermodynamic properties in different tables of this type can be used together.

For many biochemical reactants, the acid and magnesium complex dissociation constants have not been measured, but this does not mean that these reactants cannot be included in a table of $\Delta_f G'^\circ$ and $\Delta_f H'^\circ$ at 298.15 K, pH = 7, pMg = 3, and I = 0.25 M. What is required is that the apparent equilibrium constant K' for a reaction involving this reactant (or pair of reactants) with reactants with known properties has been determined at pH = 7, pMg = 3, and I = 0.25 M at more than one temperature. If the pKs of a reactant are unknown, there is a problem in calculating the equilibrium pMg, but this uncertainty may not be large if the concentration of Mg^{2+} is controlled by a buffer with known binding properties and the equilibrium concentration of the reactants is low.

The fact that biochemical reactions are often organized in series will facilitate the construction of thermodynamic tables. When a series starts or ends with a reactant with known formation properties, knowlege of the apparent equilibrium constants in the series makes it possible to calculate $\Delta_f G'^\circ$ for reactants in the series at pH = 7. For example, consider glycolysis for which the apparent equilibrium constants for the 10 reactions have been known for some time. Since the standard transformed thermodynamic properties of glucose in aqueous solution are known, the $\Delta_f G'^\circ$ values for the 17 reactants at pH = 7, including pyruvate, can be calculated with just one problem. Since $\Delta_f G^\circ$ is not known for NAD or NADH, one of the species has to be assigned $\Delta_f G^\circ = \Delta_f H^\circ = 0$. Since the thermodynamic properties of pyruvate are known, this provides a check on the calculation of $\Delta_f G_i'^\circ$ of the reactants in glycolysis.

The following conventions are recommended:

1. When a reactant exists only in an electrically neutral form at pH = 7 and pMg = 3 and $\Delta_f G^\circ$ and $\Delta_f H^\circ$ for that form in dilute aqueous solution are known, the values of $\Delta_f G'^\circ$ and $\Delta_f H'^\circ$ are calculated by adjusting for the content of H. An example is glucose.

2. When a reactant exists in a single ionized form in the neighborhood of pH = 7 and pMg = 3, the values of $\Delta_f G^\circ$ and $\Delta_f H^\circ$ for that form in the usual tables (which apply at I = 0) have to be adjusted to I = 0.25 M with the extended Debye-Hückel theory and adjusted for H to obtain the entry to the table of $\Delta_f G'^\circ$ and $\Delta_f H'^\circ$ values. Obviously, thermodynamic properties of H^+ and Mg^{2+} will not be found in the table of $\Delta_f G'^\circ$ values. Also, ions like Ca^{2+} which bind significantly with multiply charged negative species of biochemical reactants cannot be put in the table because they require treatment like Mg^{2+}. Values of $\Delta_f G^\circ$ and $\Delta_f H^\circ$ for Krebs cycle intermediates that exist at pH = 7 and pMg = 3 in a single ionic form calculated by Miller and Smith-Magowan (Appendix, ref. 8) can be adjusted for H and Mg and used in the proposed table after the values have been corrected to I = 0.25 M. An example is succinate.

3. When a reactant exists in several ionized or complexed forms that are at equilibrium at pH = 7 and pMg = 3 and the standard thermodynamic properties of all of the ionized and complexed forms are known, the values of $\Delta_f G'^\circ$ and $\Delta_f H'^\circ$ of reactants at pH = 7, pMg = 3 , I = 0.25 M can be calculated using isomer group thermodynamics. Examples are inorganic phosphate (P_i), pyrophosphate, carbonate, citrate, and glucose 6-phosphate.

4. When a reactant exists in several ionized or complexed forms with known dissociation constants and $\Delta_f G^\circ$ and $\Delta_f H^\circ$ are not known for any species of the reactant, $\Delta_f G'^\circ$ and $\Delta_f H'^\circ$ for the species can only be calculated by assigning one of them $\Delta_f G^\circ = \Delta_f H^\circ = 0$ in dilute aqueous solution. The $\Delta_f G^\circ$ and $\Delta_f H^\circ$ values of the various species have to be adjusted to an ionic strength of 0.25 M and adjusted for H and Mg, so that $\Delta_f G'^\circ$ and $\Delta_f H'^\circ$ can be calculated for

the reactant (sum of species) by use of isomer group thermodynamics, as illustrated here for the ATP series. NAD and NADH also provide an example, which is a little different because the acid and magnesium complex dissociation constants are believed to be identical.

5. If acid dissociation and magnesium dissociation constants are not known for a reactant, it can still be put into a table at pH = 7 and pMg = 3 if apparent equilibrium constants have been measured under these conditions for a reaction involving this reactant with other reactants whose transformed thermodynamic properties are known. Examples are the many reactants in glycolysis other than glucose, ATP, ADP, Pi, NAD, NADH, and H_2O.

6. It is not necessary to have columns in tables for $\Delta_f S'^\circ$ and $\Delta_f S^\circ$ because these can be treated as dependent properties and can be calculated from equations 25 and 42.

7. Nomenclature

Symbol	Name	Unit
A	extensive Helmholtz energy of a system	kJ
B	parameter in the extended Debye-Hückel theory	$L^{-1/2}\,mol^{-1/2}$
c_i	concentration of species i	$mol\,L^{-1}$
c°	standard state concentration (1 M)	$mol\,L^{-1}$
$\Delta_r C_P^\circ$	standard heat capacity at constant pressure of reaction at T, P, and I	$J\,K^{-1}\,mol^{-1}$
$\Delta_r C_P'^\circ$	standard transformed heat capacity of reaction at constant T, P, pH, pMg, and I	$J\,K^{-1}\,mol^{-1}$
E	electromotive force	V
E°	standard electromotive force of a cell or half cell	V
E'	apparent electromotive force at specified pH	V
E'°	standard apparent electromotive force of a cell or half cell at specified pH	V
F	Faraday constant (96 485.31 C mol^{-1})	$C\,mol^{-1}$
G	extensive Gibbs energy of a system	kJ
G'	extensive transformed Gibbs energy of a system	kJ
$\Delta_r G$	reaction Gibbs energy for specified concentrations of species at specified T, P, and I	$kJ\,mol^{-1}$
$\Delta_r G^\circ$	standard reaction Gibbs energy of a specified reaction in terms of species at specified T, P, and I	$kJ\,mol^{-1}$
$\Delta_r G'$	transformed reaction Gibbs energy in terms of reactants (sums of species) for specified concentrations of reactants and products at specified T, P, pH, pMg, and I	$kJ\,mol^{-1}$
$\Delta_r G'^\circ$	standard transformed reaction Gibbs energy of a specified reaction in terms of reactants (sums of species) at specified T, P, pH, pMg and I	$kJ\,mol^{-1}$
$\Delta_f G(i)$	Gibbs energy of formation of species i at a specified concentration of i and specified T, P, and I	$kJ\,mol^{-1}$
$\Delta_f G^\circ(i)$	standard Gibbs energy of formation of species i at specified T, P, and I	$kJ\,mol^{-1}$

Symbol	Name	Unit
$\Delta_f G'(i)$	transformed Gibbs energy of formation of species i or reactant i (sum of species) at specified concentration and specified T, P, pH, pMg, and I	kJ mol⁻¹
$\Delta_f G'^\circ(i)$	standard transformed Gibbs energy of formation of species i or reactant i (sum of species) at specified T, P, pH, pMg, and I	kJ mol⁻¹
H	extensive enthalpy of a system	kJ
H'	extensive transformed enthalpy of a system	kJ
$\Delta_r H(\text{cal})$	calorimetrically determined enthalpy of reaction that includes the enthalpies of reaction of H⁺ and Mg²⁺ (consumed or produced) with any buffer in solution	kJ mol⁻¹
$\Delta_r H$	enthalpy of reaction of a specified reaction in terms of species at specified T, P, and I	kJ mol⁻¹
$\Delta_r H^\circ$	standard enthalpy of reaction of a specified reaction in terms of species at specified T, P, and I	kJ mol⁻¹
$\Delta_r H'$	transformed enthalpy of reaction of a specified reaction in terms of reactants (sums of species) for specified concentrations of reactants and products at specified T, P, pH, pMg, and I	kJ mol⁻¹
$\Delta_r H'^\circ$	standard transformed enthalpy of a specified reaction in terms of reactants (sums of species) at specified T, P, pH, pMg and I	kJ mol⁻¹
$\Delta_f H(i)$	enthalpy of formation of species i at specified T, P, and I	kJ mol⁻¹
$\Delta_f H^\circ(i)$	standard enthalpy of formation of species i at specified T, P, and I	kJ mol⁻¹
$\Delta_f H'(i)$	transformed enthalpy of formation of species i or reactant i (sum of species) at specified T, P, pH, pMg, and I	kJ mol⁻¹
$\Delta_f H'^\circ(i)$	standard transformed enthalpy of formation of species i or reactant i (sum of species) at specified T, P, pH, pMg, and I	kJ mol⁻¹
I	ionic strength	mol L⁻¹
K	equilibrium constant for a specified reaction written in terms of concentrations of species at specified T, P, and I (omitting H₂O when it is a reactant)	dimensionless
K'	apparent equilibrium constant for a specified reaction written in terms of concentrations of reactants (sums of species) at specified T, P, pH, pMg, and I (omitting H₂O when it is a reactant)	dimensionless
m_i	molality of i	mol kg⁻¹
n_i or $n(i)$	amount of species i	mol
$n'(i)$	amount of species (bound and unbound) or amount of reactant i (that is, sum of species)	mol
$N_H(i)$	number of H atoms in species i	dimensionless
$N_{Mg}(i)$	number of Mg atoms in species i	dimensionless
$\overline{N}_H(i)$	average number of H atoms in reactant i at specified T, P, pH, pMg, and I	dimensionless
$\Delta_r N(\text{H}^+)$	change in binding of H⁺ in a biochemical reaction at specified T, P, pH, pMg, and I	dimensionless
$\Delta_r N(\text{Mg}^{2+})$	change in binding of Mg⁺² in a biochemical reaction at specified T, P, pH, pMg, and I	dimensionless
N_I	number of isomers in an isomer group	dimensionless
pH	$-\log_{10}([\text{H}^+]/c^\circ)$	dimensionless
pMg	$-\log_{10}([\text{Mg}^{2+}]/c^\circ)$	dimensionless
pX	$-\log_{10}([\text{X}]/c^\circ)$	dimensionless
P	pressure	bar
Q	reaction quotient of specified concentrations of species in the same form as the equilibrium constant expression	dimensionless
Q'	apparent reaction quotient of specified concentrations of reactants and products (sum of species) in the same form as the apparent equilibrium constant expression	dimensionless
R	gas constant (8.31451 J K⁻¹ mol⁻¹)	J K⁻¹ mol⁻¹
r_i or $r(i)$	equilibrium mole fraction of i within a specified class of molecules	dimensionless
S	extensive entropy of a system	J K⁻¹
S'	extensive transformed entropy of a system	J K⁻¹
$\overline{S}^\circ(i)$	standard molar entropy of species i at specified T, P, and I	J K⁻¹ mol⁻¹
$\overline{S}'^\circ(i)$	standard molar transformed entropy of species i or reactant i at specified T, P, pH, pMg, and I	J K⁻¹ mol⁻¹
$\Delta_r S$	entropy of reaction of a specified reaction in terms of species at specified T, P, and I	J K⁻¹ mol⁻¹
$\Delta_r S^\circ$	standard entropy of reaction of a specified reaction in terms of ionic species at specified T, P, and I	J K⁻¹ mol⁻¹
$\Delta_r S'$	transformed entropy of reaction of a specified reaction in terms of reactants (sums of species) for specified concentrations of reactants and products at specified T, P, pH, pMg, and I	J K⁻¹ mol⁻¹
$\Delta_r S'^\circ$	standard transformed entropy of a specified reaction in terms of sums of species at specified T, P, pH, pMg and I	J K⁻¹ mol⁻¹
$\Delta_f S^\circ(i)$	standard entropy of formation of species i at specified T, P, and I	J K⁻¹ mol⁻¹
$\Delta_f S'^\circ(i)$	standard transformed entropy of formation of species i or reactant i (sum of species) at specified T, P, pH, pMg, and I	J K⁻¹ mol⁻¹
T	temperature	K
U	extensive internal energy of a system	kJ
V	volume	L
z_i	charge of ion i with sign	dimensionless
ρ	density	kg m⁻³
$\mu(i)$	chemical potential of species i at specified T, P, and I	kJ mol⁻¹
$\mu'(i)$	transformed chemical potential of species i or reactant (sum of species) at specified T, P, pH, pMg, and I [can be replaced by $\Delta_f G'(i)$]i	kJ mol⁻¹

Symbol	Name	Unit
$\mu^{\circ}(i)$	standard chemical potential of species i at specified T, P, and I [can be replaced by $\Delta_f G^{\circ}(i)$]	kJ mol^{-1}
ν_e	number of electrons in a cell reaction	dimensionless
ν_i or $\nu(i)$	stoichiometric number of species i in a specified chemical reaction	dimensionless
$\nu'(i)$	apparent stoichiometric number of reactant i in a specified biochemical reaction	dimensionless

8. References

1. Wadsö, I., Gutfreund, H., Privlov, P., Edsall, J. T., Jencks, W. P., Strong, G. T., and Biltonen, R. L. (1976) Recommendations for Measurement and Presentation of Biochemical Equilibrium Data, *J. Biol. Chem. 251*, 6879-6885; (1976) *Q. Rev. Biophys. 9*, 439-456.

2. Wadsö, I., and Biltonen, R. L.(1985) Recommendations for the Presentation of Thermodynamic Data and Related Data in Biology, *Eur. J. Biochem. 153*, 429-434.

3. Mills, I., Cvitas, T., Homann, K., Kallay, N., and Kuchitsu, K. (1988 and 1993) *Quantities, Units and Symbols in Physical Chemistry*, Blackwell Scientific Publications, Oxford.

4. Alberty, R. A. (1992) *Biophys. Chem. 42*, 117-131.

5. Alberty, R. A. (1992) *Biophys. Chem. 43*, 239-254.

6. Clarke, E. C. W., and D. N. Glew, D. N. (1966) *Trans. Faraday Soc. 62*, 539-547.

7. Alberty, R. A., and Goldberg, R. N. (1992) *Biochemistry 31*, 10610-10615.

8. Wilhoit, R. C. (1969) *Thermodynamic Properties of Biochemical Substances, in Biochemical Microcalorimetry*, H. D. Brown, ed., Academic Press, New York.

9. Alberty, R. A. (1969) *J. Biol. Chem. 244*, 3290-3302.

10. Alberty, R. A. (1992) *J. Phys. Chem. 96*, 9614-9621.

11. Alberty, R. A., and Goldberg, R. N. (1993) *Biophys. Chem. 47*, 213-223.

12. Teague, W. E., and Dobson, G. P. (1992) *J. Biol. Chem. 267*, 14084-14093.

13. Webb, E. C. (1992) *Enzyme Nomenclature*, Academic Press, San Diego.

14. Alberty, R. A., and Cornish-Bowden, A. (1993) *Trends Biochem. Sci. 18*, 288-291.

15. Cech, T. R., Herschlag, D., Piccirilli, J. A., and Pyle, J. A. (1992) J. Biol. Chem. 256, 17479-82.

16. Blackburn, G. M., Kang, A. S., Kingsbury, G. A., and Burton, D. R. (1989) *Biochem. J. 262*, 381-391.

17. Pike, V. W. (1987) in *Biotechnology* (H.-J. Rehm and G. Reed, eds.), vol. 7a, 466-485, Verlag-Chemie.

18. "A Guide to the Procedures for the Publication of Thermodynamic Data", (1972) *PureAppl. Chem. 289*, 399-408. (Prepared by the IUPAC Commission on Thermodynamics and Thermochemistry.)

19. "Guide for the Presentation in the Primary Literature of Numerical Data Derived from Experiments". (February 1974) Prepared by a CODATA Task Group. Published in *National Standard Reference Data System News*.

20. Alberty, R. A., and Oppenheim, I. (1988) *J. Chem. Phys. 89*, 3689-3693.

21. Alberty, R. A., and Oppenheim, I. (1992) *J. Chem. Phys. 96*, 9050-9054.

22. Wyman, J., and Gill, S. J. (1990) *Binding and Linkage*, University Science Books, Mill Valley, CA.

23. Smith, W. R., and Missen, R. W. (1982) *Chemical Reaction Equilibrium Analysis: Theory and Algorithms*, Wiley-Interscience, New York.

24. Alberty, R. A. (1983) *I & EC Fund. 22*, 318-321.

25. Goldberg, R. N., and Tewari, Y. B. (1989) *J. Phys. Chem. Ref. Data 18*, 809-880.

26. Larson, J. W., Tewari, Y. B., and Goldberg, R. N. (1993) *J. Chem. Thermodyn. 25*, 73-90.

27. Goldberg, R. N., and Tewari, Y. B. (1991) *Biophys. Chem. 40*, 241-261.

28. Clarke, E. C. W., and Glew, D. N. (1980) *J. Chem. Soc., Faraday Trans. 1 76*, 1911-1916.

29. Pitzer, K. S. (1991) Ion Interaction Approach: Theory and Data Correlation, in *Activity Coefficients in Electrolyte Solutions,* 2nd Edition, K. S. Pitzer, editor, CRC Press, Boca Raton, Fla.

30. Record, M. T., Anderson, C. F., and Lohman, T. M. (1978) *Q. Rev. Biophys. 11*, 2.

31. Anderson, C. F., and Record, M. T. (1993) *J. Phys. Chem. 97*, 7116-7126.

32. Alberty, R. A. (1993) *Pure Appl. Chem. 65*, 883-888.

33. Guynn, R. W., and Veech, R. L. (1973) *J. Biol. Chem. 248*, 6966-6972.

34. Alberty, R. A. (1991) *J. Chem. Educ. 68*, 984.

35. Alberty, R. A. (1992) *J. Chem. Educ. 69*, 493.

36. Alberty, R. A. (1994) *Biophys. Chem. 49*, 251-261.

9. Appendix: Survey of current biochemical thermodynamic tables

(The reader is cautioned on distinguishing chemical reactions from biochemical reactions.)

1. Burton, K., Appendix in Krebs, H. A., and Kornberg, H. L. (1957) *Energy Transformations in Living Matter*, Springer-Verlag, Berlin.

2. Atkinson, M. R., and R. K. Morton, R. K. (1960) in *Comparative Biochemistry*, Volume II, *Free Energy and Biological Function*, Florkin, M., and Mason, H. (eds.), Academic Press, New York.

3. Wilhoit, R. C. (1969) *Thermodynamic Properties of Biochemical Substances, in Biochemical Microcalorimetry*, H. D. Brown (ed.), Academic Press, New York. This article gives standard thermodynamic properties of a large number of species at zero ionic strength. In a separate table standard enthalpies and standard Gibbs energies of formation of adenosine phosphate species are given relative to H_2ADP^- at 298.15 K.

4. Thauer, R. K., Jungermann, K., and Decker, K. (1977) *Bacteriological Reviews 41*, 100-179. Standard Gibbs energies of formation of many species of biochemical interest at 298.15 K. Table of standard Gibbs energies of reaction corrected to pH 7 by adding $m\Delta_f G^{\circ}(H^+)$, where m is the net number of protons in the reaction.

5. Goldberg, R. N. (1984) *Compiled thermodynamic data sources for aqueous and biochemical systems: An annotated bibliography (1930-1983)*, National Bureau of Standards Special Publication 685, U. S. Government Printing Office, Washington, D. C. A general and relatively complete guide to compilations of thermodynamic data on biochemical and aqueous systems.

6. Rekharsky, M. V., Galchenko, G. L., Egorov, A. M., and Berezin, I. V. (1986) Thermodynamics of Enzymatic Reactions, in *Thermodynamic Data for Biochemistry and Biotechnology*, H.-J. Hinz (ed.), Springer-Verlag, Berlin. Tables of $\Delta_r H^{\circ}$, $\Delta_r G^{\circ}$, and $\Delta_r S^{\circ}$ at pH 7 and 298.15 K, but the reactions are written in terms of ionic species so that there is a question about the interpretation of the parameters.

7. Goldberg, R. N., and Tewari, Y. B. (1989) Thermodynamic and Transport Properties of Carbohydrates and their Monophosphates: The Pentoses and Hexoses, *J. Phys. Chem. Ref. Data 18*, 809-880 . Values on a very large number of reactions at 298.15 K carefully extrapolated to zero ionic strength. $\Delta_r H^{\circ}$ and $\Delta_f G^{\circ}$ for a large number of sugars and their phosphate esters.

8. Miller, S. L., and Smith-Magowan, D. (1990) The Thermodynamics of the Krebs Cycle and Related Compounds, *J. Phys. Chem. Ref. Data 19*, 1049-1073. A critical evaluation for a large number of reactions and properties of substances at 298.15 K.

9. Goldberg, R. N., and Tewari, Y. B. (1991) Thermodynamics of the Disproportionation of adenosine 5'-diphosphate to adenosine 5'-triphosphate and Adenosine 5'-monophosphate, *Biophys. Chem. 40*, 241-261. Very complete survey of data on this reaction and on the acid dissociation and magnesium complex dissociations involved.

10. Goldberg, R. N., Tewari, Y. B., Bell, D., Fazio, K., and Anderson, E. (1993) Thermodynamics of Enzyme-Catalyzed Reactions; Part 1. Oxidoreductases, *J. Phys. Chem. Ref. Data, 22,* 515-582. This review contains tables of apparent equilibrium constants and standard transformed molar enthalpies for the biochemical reactions catalyzed by the oxidoreductases.
11. Goldberg, R. N., and Tewari, Y. B. (1994) Thermodynamics of Enzyme-Catalyzed Reactions: Part 2. Transferases, *J. Phys. Chem. Ref. Data 23,* 547-617.

Standard thermodynamic tables

1. Wagman, D. D., Evans, W. H., Parker, V. B., Schumm, R. H., Halow, I., Bailey, S. M., Churney, K. L., and Nutall, R. L. (1982) *The NBS Tables of Chemical Thermodynamic Properties, J. Phys. Chem. Ref. Data, 11,* Suppl. 2.
2. Cox, J. D., Wagman, D. D., and Medvedev, M. V. (1989) *CODATA Key Values for Thermodynamics,* Hemisphere, Washington, D. C.

STANDARD TRANSFORMED GIBBS ENERGIES OF FORMATION FOR BIOCHEMICAL REACTANTS

Robert N. Goldberg and Robert A. Alberty

This table contains values of the standard transformed Gibbs energies of formation $\Delta_f G'^\circ$ for 130 biochemical reactants. Values of $\Delta_f G'^\circ$ are given at pH 7.0, the temperature 298.15 K, and the pressure 100 kPa for three ionic strengths: $I = 0$, $I = 0.1$ mol/L and $I = 0.25$ mol/L. The table can be used for calculating apparent equilibrium constants K' and standard apparent reduction potentials E'° for biochemical reactions. Such a listing is more compact than tabulating the actual apparent equilibrium constants or standard apparent reduction potentials, which would require a very large number of reactant–product combinations. In the table, all reactants are in aqueous solution unless indicated otherwise.

A biochemical reactant is a sum of species. For example, ATP consists of an equilibrium mixture of the aqueous species ATP^{4-}, $HATP^{3-}$, H_2ATP^{2-}, $MgATP^{2-}$, etc. Similarly, phosphate refers to the equilibrium mixture of the aqueous species PO_4^{3-}, HPO_4^{2-}, $H_2PO_4^-$, H_3PO_4, $MgHPO_4$, etc. Biochemical reactions are written using biochemical reactants in terms of an apparent equilibrium constant K', which is distinct from the standard equilibrium constant K. This subject is discussed in an IUPAC report (see Reference 1 below).

The apparent equilibrium constant K' and the standard transformed Gibbs energy change $\Delta_r G'^\circ$ for a biochemical reaction can be calculated from the $\Delta_f G'^\circ$ values by using the relationship

$$-RT \ln K' = \Delta_r G'^\circ = \Sigma v'_i \, \Delta_f G'^\circ,$$

where the summation is over all of the biochemical reactants. The quantity v'_i is the stoichiometric number of reactant i (v'_i is positive for reactants on the right side of the equation and negative for reactants on the left side); R is the gas constant. As an example, the hydrolysis reaction of ATP is

$$ATP + H_2O(l) = ADP + phosphate.$$

At pH 7.00 and $I = 0.25$ M, $\Delta_r G'^\circ$ and K' are calculated as follows:

$$\Delta_r G'^\circ = \{-1424.70 - 1059.49 - (-2292.50 - 155.66)\} \cdot (kJ \; mol^{-1})$$
$$= -36.03 \; kJ \; mol^{-1}$$

$$K' = \exp[-(-36030 \; J \; mol^{-1})/\{(8.3145 \; J \; mol^{-1} \; K^{-1}) \cdot (298.15 \; K)\}$$
$$= 2.05 \cdot 10^6$$

An example involving a biochemical half-cell reaction is

$$acetaldehyde(aq) + 2 \; e^- = ethanol(aq).$$

At 298.15 K, pH 7.00, and $I = 0$, the standard apparent reduction potential E'° can be calculated as follows

$$E'^\circ = -(1/nF) \cdot \{\Delta_f G'^\circ(ethanol) - \Delta_f G'^\circ(acetaldehyde)\},$$

where n is the number of electrons in the half-cell reaction and F is the Faraday constant. Then,

$$E'^\circ = [-1/(2 \cdot 9.6485 \cdot 10^4 \; C \; mol^{-1})] \cdot (58.10 \cdot 10^3 \; J \; mol^{-1}$$
$$- 20.83 \cdot 10^3 \; J \; mol^{-1}) = -0.193 \; V$$

References

1. Alberty, R.A., Cornish-Bowden, A., Gibson, Q.H., Goldberg, R.N., Hammes, G., Jencks, W., Tipton, K.F, Veech, R., Westerhoff, H.V., and Webb, E.C. *Pure Appl. Chem.* 66, 1641-1666, 1994.
2. Alberty, R.A., *Arch. Biochem. Biophys.*, 353, 116-130, 1998; 358, 25-39, 1998.
3. Alberty, R.A., *Thermodynamics of Biochemical Reactions*, Wiley-Interscience, New York, 2003.
4. Alberty, R.A., *BasicBiochemData2: Data and Programs for Biochemical Thermodynamics*, <http://library.wolfram.com/infocenter/MathSource/797>.

Reactant	$\Delta_f G'^\circ$(I = 0) kJ mol^{-1}	$\Delta_f G'^\circ$(I = 0.1 M) kJ mol^{-1}	$\Delta_f G'^\circ$(I = 0.25 M) kJ mol^{-1}
Acetaldehyde	20.83	23.27	24.06
Acetate	−249.46	−248.23	−247.83
Acetone	80.04	83.71	84.90
Acetyl Coenzyme A	−60.49	−58.65	−58.06
Acetylphosphate	−1109.34	−1107.57	−1107.02
cis-Aconitate	−797.26	−800.93	−802.12
Adenine	510.45	513.51	514.50
Adenosine	324.93	332.89	335.46
Adenosine 5′-diphosphate (ADP)	−1428.93	−1425.55	−1424.70
Adenosine 5′-monophosphate (AMP)	−562.04	−556.53	−554.83
Adenosine 5′-triphosphate (ATP)	−2292.61	−2292.16	−2292.50
D-Alanine	−91.31	−87.02	−85.64
Ammonia	80.50	82.34	82.93
D-Arabinose	−342.67	−336.55	−334.57
L-Asparagine	−206.28	−201.38	−199.80

STANDARD TRANSFORMED GIBBS ENERGIES OF FORMATION FOR BIOCHEMICAL REACTANTS (Continued)

Reactant	$\Delta_f G'^\circ (I = 0)$ kJ mol^{-1}	$\Delta_f G'^\circ (I = 0.1$ M) kJ mol^{-1}	$\Delta_f G'^\circ (I = 0.25$ M) kJ mol^{-1}
L-Aspartate	−456.14	−453.08	−452.09
1,3-Biphosphoglycerate	−2202.06	−2205.69	−2207.30
Butanoate	−72.94	−69.26	−68.08
1-Butanol	227.72	233.84	235.82
Citrate	−963.46	−965.49	−966.23
Isocitrate	−956.82	−958.84	−959.58
Coenzyme A (CoA)	−7.98	−7.43	−7.26
CO(aq)	−119.90	−119.90	−119.90
CO(g)	−137.17	−137.17	−137.17
CO_2(aq)[total]	−547.33	−547.15	−547.10
CO_2(g)	−394.36	−394.36	−394.36
Creatine	100.41	105.92	107.69
Creatinine	256.55	260.84	262.22
L-Cysteine	−59.23	−55.01	−53.65
L-Cystine	−187.03	−179.69	−177.32
Cytochrome c [oxidized]	0.00	−5.51	−7.29
Cytochrome c [reduced]	−24.51	−26.96	−27.75
Dihydroxyacetone phosphate	−1096.60	−1095.91	−1095.70
Ethanol	58.10	61.77	62.96
Ethyl acetate	−18.00	−13.10	−11.52
Ferredoxin [oxidized]	0.00	−0.61	−0.81
Ferredoxin [reduced]	38.07	38.07	38.07
Flavine adenine dinucleotide (FAD) [oxidized]	1238.65	1255.17	1260.51
Flavine adenine dinucleotide (FAD) [reduced]	1279.68	1297.43	1303.16
Flavin adenine dinucleotide-enzyme (FADenz) [oxidized]	1238.65	1255.17	1260.51
Flavin adenine dinucleotide-enzyme (FADenz) [reduced]	1229.96	1247.71	1253.44
Flavin mononucleotide (FMN) [oxidized]	759.17	768.35	771.32
Flavin mononucleotide (FMN) [reduced]	800.20	810.61	813.97
Formate	−311.04	−311.04	−311.04
D-Fructose	−436.03	−428.69	−426.32
D-Fructose 1,6-diphosphate	−2202.84	−2205.66	−2206.78
D-Fructose 6-phosphate	−1321.71	−1317.16	−1315.74
Fumarate	−521.97	−523.19	−523.58
D-Galactose	−429.45	−422.11	−419.74
α-D-Galactose 1-phosphate	−1317.50	−1313.01	−1311.60
D-Glucose	−436.42	−429.08	−426.71
α-D-Glucose 1-phosphate	−1318.03	−1313.34	−1311.89
D-Glucose 6-phosphate	−1325.00	−1320.37	−1318.92
Glutamate	−377.82	−373.54	−372.16
D-Glutamine	−128.46	−122.34	−120.36
Glutathione [oxidized]	1198.69	1214.60	1219.74
Glutathione [reduced]	625.75	634.76	637.62
Glutathione-coenzyme A	563.49	572.06	574.83
D-Glyceraldehyde 3-phosphate	−1088.94	−1088.25	−1088.04
Glycerol	−177.83	−172.93	−171.35
sn-Glycerol 3-phosphate	−1080.22	−1077.83	−1077.13
Glycine	−180.13	−177.07	−176.08
Glycolate	−411.08	−409.86	−409.46
Glycylglycine	−200.55	−195.65	−194.07
Glyoxylate	−428.64	−428.64	−428.64
H_2(aq)	97.51	98.74	99.13
H_2(g)	79.91	81.14	81.53
H_2O(l)	−157.28	−156.05	−155.66
H_2O_2(aq)	−54.12	−52.89	−52.50
3-Hydroxypropanoate	−318.62	−316.17	−315.38
Hypoxanthine	249.33	251.77	252.56
Indole	503.49	507.78	509.16
Lactate	−316.94	−314.49	−313.70
Lactose	−688.29	−674.83	−670.48
L-Leucine	167.18	175.14	177.71
L-Isoleucine	175.53	183.49	186.06
D-Lyxose	−349.58	−343.46	−341.48
Malate	−682.88	−682.85	−682.85
Maltose	−695.65	−682.19	−677.84
D-Mannitol	−383.22	−374.65	−371.89

STANDARD TRANSFORMED GIBBS ENERGIES OF FORMATION FOR BIOCHEMICAL REACTANTS (Continued)

Reactant	$\Delta_f G'^\circ (I = 0)$ kJ mol^{-1}	$\Delta_f G'^\circ (I = 0.1$ M) kJ mol^{-1}	$\Delta_f G'^\circ (I = 0.25$ M) kJ mol^{-1}
Mannose	−430.52	−423.18	−420.81
Methane(aq)	125.50	127.94	128.73
Methane(g)	109.11	111.55	112.34
Methanol	−15.48	−13.04	−12.25
L-Methionine	−63.40	−56.67	−54.49
N_2(aq)	18.70	18.70	18.70
N_2(g)	0.00	0.00	0.00
Nicotinamide Adenine Dinucleotide (NAD) [oxidized]	1038.86	1054.17	1059.11
Nicotinamide Adenine Dinucleotide (NAD) [reduced]	1101.47	1115.55	1120.09
Nicotinamide Adenine Dinucleotide Phosphate (NADP) [oxidized]	163.73	173.52	176.68
Nicotinamide Adenine Dinucleotide Phosphate (NADP) [reduced]	229.67	235.79	237.77
O_2(aq)	16.40	16.40	16.40
O_2(g)	0.00	0.00	0.00
Oxalate	−673.90	−676.35	−677.14
Oxaloacetate	−713.38	−714.60	−715.00
Oxalosuccinate	−979.05	−979.05	−979.05
2-Oxoglutarate	−633.58	−633.58	−633.58
Palmitate	979.25	997.61	1003.54
L-Phenylalanine	232.42	239.15	241.33
Phosphate	−1058.56	−1059.17	−1059.49
2-Phospho-D-glycerate	−1340.72	−1341.32	−1341.79
3-Phospho-D-glycerate	−1346.38	−1347.19	−1347.73
Phosphoenolpyruvate	−1185.46	−1188.53	−1189.73
1-Propanol	143.84	148.74	150.32
2-Propanol	134.42	139.32	140.90
Pyrophosphate	−1934.95	−1939.13	−1940.66
Pyruvate	−352.40	−351.18	−350.78
Retinal	1118.78	1135.91	1141.45
Retinol	1170.78	1189.14	1195.07
Ribose	−339.23	−333.11	−331.13
Ribose 1-phosphate	−1215.87	−1212.24	−1211.14
Ribose 5-phosphate	−1223.95	−1220.32	−1219.22
Ribulose	−336.38	−330.26	−328.28
L-Serine	−231.18	−226.89	−225.51
Sorbose	−432.47	−425.13	−422.76
Succinate	−530.72	−530.65	−530.64
Succinyl Coenzyme A	−349.90	−348.06	−347.47
Sucrose	−685.66	−672.20	−667.85
Thioredoxin [oxidized]	0.00	0.00	0.00
Thioredoxin [reduced]	54.32	55.41	55.74
L-Tryptophan	364.78	372.12	374.49
L-Tyrosine	68.82	75.55	77.73
Ubiquinone [oxidized]	3596.07	3651.15	3668.94
Ubiquinone [reduced]	3586.06	3642.37	3660.55
Urate	−206.03	−204.81	−204.41
Urea	−42.97	−40.53	−39.74
Uric acid	−197.07	−194.63	−193.84
L-Valine	80.87	87.60	89.78
D-Xylose	−350.93	−344.81	−342.83
D-Xylulose	−346.59	−340.47	−338.49

ENTHALPY, ENTROPY, AND FREE ENERGY VALUES
FOR BIOCHEMICAL REDOX REACTIONS

(Data are reported for pH 7.0 and 298 K)

Oxidation Half Reaction	$E°'$–Volts	ΔG kJ Mole^{-1}	ΔH kJ Mole^{-1}	ΔS J Mole^{-1} Deg^{-1}	References (Enthalpy Data)
Non protein reactions					
$2\,Cys \rightleftharpoons (Cys)_2 + 2H^+ + 2e^-$	0.32	−61.5	40.2	341	1, 2
$2GSH \rightleftharpoons (GS)_2 + 2H^+ + 2e^-$	0.23	−44.4	25.1	233	1, 2
$2HOC_2H_4SH \rightleftharpoons (HOC_2H_4S)_2 + 2H^+ + 2e^-$	—	—	38.1	—	1, 2
$L(+)Lactate \rightleftharpoons pyruvate + 2H^+ + 2e^-$	0.18	−34.7	78.2	379	1, 2
$H_4Folate \rightleftharpoons H_2\,folate + 2H^+ + 2e^-$	0.18	−34.7	211.7	828	3
$Ascorbate \rightleftharpoons dehydroascorbate + 2H^+ + 2e^-$	−0.06	11.7	77.8	222	—
$FMNH_2 \rightleftharpoons FMN + 2H^+ + 2e^-$	−0.22	42.3	56.1	46.0	4
$Hydroquinone \rightleftharpoons benzoquinone + 2H^+ + 2e^-$	−0.29	55.6	177.4	408	1, 2
$NADH \rightleftharpoons NAD^+ + H^+ + 2e^-$	−0.32	61.6	29.2	−108	2, 5, 6
$NADPH \rightleftharpoons NADP^+ + H^+ + 2e^-$	−0.324	62.4	25.3	−124	2, 5, 7
$Fe(CN)_6 \rightleftharpoons Fe(CN)_6^{-3} + e^-$	−0.36	69.4	111.7	142	8[a]
Protein reactions					
$Fe(II)hemerythrin \rightleftharpoons Fe(III)hemerythrin + e^-$	—	—	102.5	—	9
$Ferrocytochrome\ c \rightleftharpoons Ferrocytochrome\ c + e^-$ (mammalian)	−0.26	25	59	114	1
$Ferrocytochrome\ c \rightleftharpoons Ferrocytochrome\ c + e^-$ (bacterial)	—	—	79.5	—	1

Compiled by Neal Langerman.

[a] This reaction is the commonly used reference reaction.

References

1. Watt and Sturtevant, personal communication.
2. Schott and Sturtevant, personal communication.
3. Rothman, Kisliuk, and Langerman, *J. Biol. Chem.*, 248, 7845 (1973).
4. Beaudette and Langerman, *Arch. Biochem. Biophys.*, 161, 125 (1974).
5. Burton, *Biochem. J.*, 143, 365 (1974).
6. Poe, Gutfreund, and Estabrook, *Arch. Biochem. Biophys.*, 122, 204 (1967).
7. Engel and Dalziel, *Biochem. J.*, 105, 691 (1967).
8. Hanania, Irvine, Eaton, and George, *J. Phys. Chem.*, 71, 2022 (1967).
9. Langerman and Sturtevant, *Biochemistry*, 10, 2809 (1971).

OXIDATION-REDUCTION POTENTIALS, ABSORBANCE BANDS AND MOLAR ABSORBANCE OF COMPOUNDS USED IN BIOCHEMICAL STUDIES

Paul A. Loach

In addition to the references[1-5,7,8] in the table, other generally used sources of oxidation-reduction data are: *Biochemist's Handbook*, D. Van Nostrand Co., Princeton, N.J. (1961); *The Encyclopedia of Electrochemistry*, C. A. Hampel, Ed., Reinhold Publishing Corp. New York (1964); *Oxidation-Reduction Potentials in Bacteriology and Biochemistry*, sixth ed., L. F. Hewitt, McCorquodale and Co. Ltd., London, (1950); *Biochemisches Taschenbuch* part II, Springer-Verlag, New York, (1964).

The oxidation-reduction couples are listed according to decreasing values of E° or E°′. When both values are available, the order is according to E°′. Unless otherwise indicated, E°′ is the mid-point potential for a particular couple at pH7. Temperatures are not listed; most of the data are relevant to room temperature (20°C to 30°C). When more exact conditions are desired (ionic strength, concentration, temperature, nature of data used to derive E° or E°′) the reader should consult the reference listed.

	System	$E°$	$E°′$	λ_{max}	E_{mM}	Reference
1.	F_2(gas)/F^-	2.87	—	—	—	1
2.	$H_2N_2O_2/N_2$ (gas)	2.65	—	—	—	1
3.	$S_2O_8^{2-}/SO_4^{2-}$	2.0	—	—	—	1
4.	H_2O_2/H_2O	1.77	—	—	—	1
5.	MnO_4^-/MnO_2	1.69	—	—	—	1
6.	$HClO_2/HClO$	1.64	—	—	—	1
7.	H_5IO_6/IO_3^-	1.6	—	—	—	1
8.	MnO_4^-/Mn^{2+}	1.51	—	—	—	1
9.	Mn^{3+}/Mn^{2+}	1.4	—	—	—	2
10.	Cl_2(gas)/Cl^-	1.359	—	—	—	1
11.	ClO_2(gas)/$HClO_2$	1.27	—	—	—	1
12.	MnO_2/Mn^{2+}	1.23	—	—	—	1
13.	$[Mn^{3+}(PO_4)_2]^{3-}/[Mn^{2+}(PO_4)_2]^{4-}$	1.22	—	—	—	2
14.	Pt^{2+}/Pt	1.2	—	—	—	1
15.	IO_3^-/I_2	1.19	—	—	—	1
16.	ClO_4^-/ClO_3^-	1.19	—	—	—	1
17.	ClO_3^-/ClO_2(gas)	1.15	—	—	—	1
18.	$[Cu^{3+}(IO_6)_2]^{7-}/[Cu^{2+}(IO_6)_2]^{8-}$, pH 8	—	1.1	—	—	2
	pH 12		0.7			
19.	Br_2/Br^-	1.087	—	—	—	1
20.	N_2O_4(gas)/HNO_2	1.07		—	—	1
21.	Fe^{3+}/Fe^{2+} O-phenanthroline	1.06	—	—	—	3
22.	$[IrCl_6]^{2-}/[IrCl_6]^{3-}$	1.05	—	—	—	1
23.	VO_3^-/VO^{2+}	1.0	—	—	—	2
24.	IO_4^-/IO_3^-	1.375	0.96	—	—	2
25.	HNO_2/NO(gas)	0.99	—	—	—	1
26.	*p*-Toluenesulfochloramide, Na salt (Chloramine-T)	1.52	0.90	—	—	2
27.	Nitrosoguanidine/Nitroguanidine	0.85	—	—	—	4
28.	O_2/H_2O	1.229	0.816	—	—	1
29.	NO_3^-/N_2O_4(gas)	0.80	—	—	—	1
30.	Ag^+/Ag	0.7994	—	—	—	1
31.	1,2-Benzoquinone	0.792	—	—	—	5
32.	Hg_2^{2+}/Hg	0.792	—	—	—	1
33.	Zn Octaethylporphyrin (methanol)	—	0.78	—	—	71
34.	Fe^{3+}/Fe^{2+}	0.771	—	—	—	1
35.	$[Mo^{3+}(CN)_6]^{3-}/[Mo^{2+}(CN)_6]^{4-}$	0.73	—	—	—	5
36.	Porphyrexide	—	0.725	—	—	5
37.	Pyrogallol	0.713	—	—	—	5
38.	NO(gas)/$H_2N_2O_2$	0.71	—	—	—	1
39.	Hg^{2+}/Hg_2^{2+}	0.91	—	—	—	1
40.	1,2-Naphthoquinone-4-sulfonate	0.628	—	—	—	5
41.	Mn^{4+}/Mn^{3+} Hematoporphyrin IX, pH 9.9	—	0.626	400 (ox)	70	6
42.	MnO_4^-/MnO_4^{2-}	0.6	—	—	—	1
43.	$S_2O_6^{2-}/H_2SO_3$	0.6	—	—	—	1
44.	$[W(CN)_8]^{3-}/[W(CN)_8]^{4-}$	0.57	—	—	—	5
45.	Porphyrindin	—	0.565	—	—	5
46.	NH_2OH/NH_4	—	0.562	—	—	7

557

OXIDATION-REDUCTION POTENTIALS, ABSORBANCE BANDS AND MOLAR ABSORBANCE
OF COMPOUNDS USED IN BIOCHEMICAL STUDIES (Continued)

	System	$E°$	$E°'$	λ_{max}	E_{mM}	Reference
47.	$H_3AsO_4/HAsO_2$	0.56	—	—	—	1
48.	o-Tolidine	—	0.55	—	—	5
49.	Cu^{2+}/Cu^+ Hemocyanin	—	0.540	350	—	8, 9
				600		
50.	I_2/I^-	0.536	—	—	—	1
51.	Cu^+/Cu	0.521	—	—	—	1
52.	Bacteriochlorophyll a (methanol)	—	0.52	—	—	71
53.	$S_2O_3^{2-}/S$	0.5	—	—	—	1
54.	$S_2O_4^{2-}/S_2O_3^{2-}$	1.03	0.484	—	—	1,7
55.	MoO_2^{2+}/MoO^{3+}	0.48	—	—	—	1
56.	Phenylhydrazine sulfonate	0.437	—	—	—	5
57.	2-Methyl-1,4-naphthoquinone (Menadione-Vitamin K_3)	0.422	—	—	—	5
58.	P_{700}	—	0.43	—	—	10
59.	$P_{890}(P_{0.44})$	—	0.44	—	—	11, 12 13, 14
60.	NO_3^-/NO_2^-	0.94	0.421	—	—	1, 7
61.	$H_2SO_3/S_2O_3^{2-}$	0.40	—	—	—	1
62.	2,5-Dihydroxy-1,4-benzoquinone	—	0.38	—	—	5
63.	Adrenalin	0.809	0.380	—	—	4, 5
64.	p-Aminodimethylaniline	—	0.38	—	—	5
65.	Fe^{3+}/Fe^{2+} Cytochrome f	—	0.365	413 (ox)	—	66
				423(red)		
				525(red)		
				555(red)		
66.	$[Fe(CN)_6]^{3-}/[Fe(CN)_6]^{4-}$	—	0.36	—	—	5
				420(ox)	1.000	15
67.	o-Quinone/Diphenol		0.35	—		5
68.	Fe^{3+}/Fe^{2+} Cytochrome c_{550} (R. rubrum)	—	0.338	409(ox)	—	16
				416(red)		
				521(red)		
				550(red)		
69.	Cu^{2+}/Cu	0.337	—	—	—	1
70.	Fe^{3+}/Fe^{2+} Acetate, pH 5	—	0.34	—	—	5
71.	Fe^{3+}/Fe^{2+} Cytochrome c_5 (Azotobacter)	—	0.32	420(red)	—	17
				526(red)		
				555(red)		
72.	$As5^+/As^{3+}$	—	0.316	—	—	2
73.	p-Aminophenol	0.779	0.314	—	—	5
74.	$O_2(gas)/H_2O_2$	0.69	0.295	—	—	7
75.	Fe^{3+}/Fe^{2+} Cyt. c_4 (Azotobacter)	—	0.30	411(ox)	115.8	17
				416(red)	157.2	
				522(red)	17.6	
				551(red)	23.8	
76.	Fe^{3+}/Fe^{2+} Cyt c_{552} (Pseudomonas)	—	0.300	409(ox)	—	36
				416(red)		
				520(red)		
				552(red)		
77.	1,4-Benzoquinone	0.699	0.293	—	—	5
78.	Fe^{3+}/Fe^{2+} Cyt a	—	0.29	—	—	18
79.	p-Quinone/Hydroquinone	—	0.28	—	—	5
80.	2,6-Dibromo-2'-SO_3H indophenol	—	0.273	—	—	5
81.	Fe^{3+}/Fe^{2+} Malonate, pH 4	—	0.26	—	—	5
82.	2,5-Dihydroxyphenylacetic acid (Homogentisic acid)	0.687	0.260	—	—	5
83.	Fe^{3+}/Fe^{2+} Salicylate, pH 4	—	0.26	—	—	5
84.	Fe^{3+}/Fe^{2+}Cyt c	—	0.254	407(ox)	—	19, 20
				415(red)	125	
				521(red)	15.9	
				550(red)	27.7	
85.	2,6,2'-Trichloroindophenol	—	0.254	—	—	5
86.	Fe^{3+}/Fe^{2+} Chlorocruorin(pyridine)$_2$	—	0.246	434(red)		29
				544(red)		
				562(red)		
87.	Indophenol	—	0.228	—	—	5
88.	o-Toluidine Blue	0.677	0.224	—	—	5
89.	Phenol Blue	—	0.224	—	—	5
90.	Fe^{3+}/Fe^{2+} Cyt c_1	—	0.22	410(ox)		37

OXIDATION-REDUCTION POTENTIALS, ABSORBANCE BANDS AND MOLAR ABSORBANCE OF COMPOUNDS USED IN BIOCHEMICAL STUDIES (Continued)

	System	$E°$	$E°'$	λ_{max}	E_{mM}	Reference
				418(red)	116	
				524(red)	11.6	
				554(red)	24.1	
91.	Fe^{3+}/Fe^{2+} Mesoporphyrin poly D, L-(lysine-phenylalanine), pH 4	—	0.22	—		21
92.	Fe^{3+}/Fe^{2+} Cyt b_2 (yeast)	—	0.219	—	—	22
93.	2,6-Dichlorophenolindophenol (DCPIP)	—	0.217	600	20.6	5, 23
94.	2,6-Dibromoindophenol	—	0.216	—	—	5
95.	Janus Green	—	0.21	—	—	5
96.	3-Aminothiazine	—	0.208	—	—	5
97.	Butyryl-Co A dehydrogenase $FAD^+/FADH_2$ (Cu present)	—	0.187	—	—	24
98.	Fe^{3+}/Fe^{2+} Hemoglobin (H 6.0),	—	0.17	500(ox)	9.0	25, 26
				630(ox)	4.0	27
	pH 7.0	—	0.144	—	—	26, 28
99.	SO_4^{2-}/H_2SO_3	0.17	—	—	—	1
100.	2,6-Dibromo-2′-methoxyindophenol	—	0.161	—	—	5
101.	Sn^{4+}/Sn^{2+}	—	0.15	—	—	2
102.	Adrenodoxin	—	0.15	414(ox)	5.7	9
103.	2,6-Dimethylindophenol	—	0.148	—	—	5
104.	1,2-Naphthoquinone	0.547	0.143	-	—	4
105.	Fe^{3+}/Fe^{2+} PPIX(pyridine)$_2$,	—	0.137	419(red)	192	29
				525(red)	17.5	30
				557(red)	34.4	
	pH9	—	0.09	—	—	5
106.	l-Naphthol-2-sulfonate indophenol	—	0.123	—	—	5
107.	Fe^{3+}/Fe^{2+} Cyt$_{553}$ (*R. spheroides*)	—	0.120	412(ox)	—	31
				418(red)		
				523(red)		
				553(red)		
108.	Toluylene Blue	—	0.115	—	—	5
109.	Fe^{3+}/Fe^{2+} Cyt$_{552}$ (*Chromatium*)	—	0.100	410(ox)	—	16
				417(red)		
				525(red)		
				552(red)		
110.	Ubiquinone/Ubihydroquinone (in 95% ethanol)	0.542	0.10	275	15	32
111.	TiO^{2+}/Ti^{3+}	—	0.10	—	—	3
112.	$S_4O_6^{2-}/S_2O_3^{2-}$	0.08	—	—	—	1
113.	Dehydroascorbic acid/ascorbic acid,	—	0.058	—	—	5
	pH 4	—	0.166	—	—	7
	pH 8.7	—	−0.012	—	—	2
114.	*N*-Methylphenazinium methosulfate (PMS)	—	0.08	387(ox)	23.8	33
				388(semi-		5
				450 (quinone)		
115.	Fe^{3+}/Fe^{2+} Cyt b (mitochondrial)	—	0.077	—	—	34
			0.050	429	114	
				532		
				561	21	
			−0.040			18
116.	$[W^{5+}(OH^-)_4(CN^-)_4]^{3-}/[W^{4+}/OH^-)_4(CN^-)_4]^{4-}$	—	0.07	—	—	35
117.	Thionine	0.563	0.064	—	—	5
118.	Thioindigo-tetrasulfonate	0.409	0.063	—	—	5
119.	Phenazine ethosulfate	—	0.055	—	—	5
120.	Cresyl Blue	0.583	0.047	632	—	5
121.	Fe^{3+}/Fe^{2+} Myoglobin	—	0.046	500(ox)	9.1	28, 27
				630(ox)	3.5	
122.	Fe^{3+}/Fe^{2+} Cyt b_3 (plants)	—	0.040	560(red)	—	39
				529(red)	—	
123.	1,4-Naphthoquinone	0.470	0.036	—	—	5
124.	Toluidine Blue	0.534	0.034	—	—	5
125.	Fumaric/Succinate	—	0.031	—	—	7
126.	$[Ni(C_{10}H_{10})]^+/Ni(C_{10}H_{10})$	—	0.03	—	—	40
127.	Thiazine Blue	—	0.027	—	—	5
128.	Gallocyanine	—	0.021	—	—	5
129.	Fe^{3+}/Fe^{2+} Cyt b_5 (microsomal)	—	0.02	413(ox)	117	41

OXIDATION-REDUCTION POTENTIALS, ABSORBANCE BANDS AND MOLAR ABSORBANCE
OF COMPOUNDS USED IN BIOCHEMICAL STUDIES (Continued)

	System	$E°$	$E°'$	λ_{max}	E_{mM}	Reference
				423(red)	170	
				526(red)	13	
				555(red)	26	
130.	Thioindigo disulfonate	0.347	0.014	—	—	5
131.	Methylene Blue	0.532	0.011	688(ox)	—	5
132.	Fe^{3+}/Fe^{2+} Methylated heme undecapeptide of Cyt c (pyridine)	—	0.008	—	—	68
133.	Fe^{3+}/Fe^{2+} Hematoporphyrin (pyridine)$_2$	—	+0.004	519(red)	—	8, 29
				545(red)		
134.	Fe^{3+}/Fe^{2+} Oxalate	—	0.002	—	—	5
135.	3-Methyl-9-phenyl isoalloxazine	—	−0.002	—	—	5
136.	Fe^{3+}/Fe^{2+} Cytochromoid c (*Chromatium*)	—	−0.040	406(ox)	—	42
				418(red)		
				525(red)		
				552(red)		
137.	Fe^{3+}/Fe^{2+} Cytochromoid c (*R. rubrum*)	—	−0.008	390(ox)	—	70
				424(red)		
				568(red)		
138.	Crotonyl-CoA/Butyryl-CoA	—	−0.015	—	—	38
139.	Pyocyanine	0.235	−0.034	690(ox)	4.5	33
			−0.038	370		5
140.	Indigo-tetrasulfonate	0.365	−0.046	—	—	5
141.	2-Methyl-3-phytyl-1,4-naphthoquinone (Vitamin K$_1$/ Dihydro-Vitamin K$_1$)	0.363	−0.05	—	—	5
142.	Luciferin	—	−0.05	—	—	4
				490(ox)	8.85	
				380(ox)	10.8	
143.	Fe^{3+}/Fe^{2+} Rubredoxin		−0.057	333(ox)	6.3	69
				311(red)	10.8	
144.	Fe^{3+}/Fe^{2+} Cyt b$_6$ (Chloroplasts)	—	−0.06	563(red)	—	43
145.	Methyl Capri Blue	0.477	−0.061	—	—	5
146.	Fe^{3+}/Fe^{2+} Mesoporphyrin (pyridine)$_2$	—	−0.063	—	—	8
147.	H_2O_3/HS_2O_4	−0.08	—	—	—	1
148.	Fe^{3+}/Fe^{2+} Mesoporphyrin poly-D, L-(glu- phe), pH 9	—	−0.07	—	—	21
149.	Xanthine oxidase	—	−0.08	—	—	5
150.	Indigo-trisulfonate	0.332	−0.081	—	—	5
151.	Fe^{3+}/Fe^{2+}1, 3, 5, 8-Tetramethyl porphyrin-6,7-dipropionic acid methyl ester-2,4-disulfonic acid	—	−0.09	—	—	44
152.	Thiohistidine	—	−0.09	—	—	4
153.	$[V(C_{10}H_{10})]^{2+}/[V(C_{10}H_{10})]^+$	—	−0.08	—	—	40
154.	Glyoxylate/Glycollate	—	−0.090	—	—	7
155.	Fe^{3+}/Fe^{2+} Heme undecapeptide from Cyt c (pyridine)	—	−0.092	403(ox)	117	68
				413(red)	155	
				521(red)		
				551(red)		
156.	6,8,9-Trimethyl isoalloxazine	—	−0.109	—	—	5
157.	Chloraphine	0.274	−0.115	—	—	5
158.	CO_2(gas)/CO(gas)	−0.12	—	—	—	1
159.	Yellow enzyme FMN/FMNH$_2$	—	−0.122	—	—	45
160.	Indigo-disulfonate	0.291	−0.125	—	—	5
161.	9-Phenyl isoalloxazine	—	−0.126	—	—	4
162.	Vitamin K reductase	—	−0.127	—	—	46
163.	Fe^{3+}/Fe^{2+} PPIX (histidine)$_2$, pH9.5	—	−0.138	—	—	5
164.	2-OH-1,4-Naphthoquinone	—	−0.139	—	—	5
165.	Thioglycolic acid	—	−0.14	—	—	4
166.	Fe^{3+}/Fe^{2+} (Pyrophosphate)	—	−0.14	—	—	5
167.	2-Amino-*N*-methyl phenazine methosulfate	—	−0.145	—	—	5
168.	Indigo-monosulfonate	0.262	−0.157	—	—	5
169.	Hydroxypyruvate/Glycerate	—	−0.158	—	—	7
170.	Oxaloacetate/Malate	—	−0.166	—	—	47
171.	Brilliant Alizarin Blue	—	−0.173	—	—	5
172.	Alloxazine	—	−0.170	—	—	5
173.	Mn^{3+}/Mn^{2+} Methyl pheophorbide *a*	—	−0.180	370(ox)	39	48
				425(ox)	31	

OXIDATION-REDUCTION POTENTIALS, ABSORBANCE BANDS AND MOLAR ABSORBANCE
OF COMPOUNDS USED IN BIOCHEMICAL STUDIES (Continued)

	System	$E°$	$E°'$	λ_{max}	E_{mM}	Reference
				475(ox)	13	
				665(ox)	17	
				418(red)	120	
				647(red)	24	
174.	2-Methyl-3-hydroxy-1,4-naphthoquinone (Phthiocol)	—	−0.180	—	—	5
		—	—	—	—	4
175.	9-Methylisoalloxazine	—	−0.183	—	—	4
176.	Fe^{3+}/Fe^{2+} PPIX $(CN^-)_2$, pH9.9	—	−0.183	—	—	5
177.	Anthraquinone-2,6-disulfonate	0.228	−0.184	—	—	5
178.	Pyruvate/Lactate	—	−0.185	—	—	4
			−0.190			7
179.	Fe^{3+}/Fe^{2+} Protoporphyrin IX (borate buffer), pH 8.2	—	−0.188	—	—	4, 49
180.	Neutral Blue	0.17	−0.19	—	—	5
181.	Dihydroxy acetone-P/	—	−0.19	—	—	4
	α-Glycero-P		−0.192	—	—	7
182.	Acetaldehyde/Ethanol	—	−0.197	—	—	4, 7
183.	$[Ti(C_{10}H_{10})]^{2+}/[Ti(C_{10}H_{10})]^+$	—	−0.20	—	—	40
184.	$SO_4^{2-}/S_2O_6^{2-}$	−0.2	—	—	—	1
185.	Fe^{3+}/Fe^{2+} Heme undecapeptide, Cyt c (imidazole)	—	−0.201	—	—	68
186.	Riboflavin	—	−0.208	260	27.7	4
				375(ox)	10.6	
				450(ox)	12.2	50
187.	Fe^{3+}/Fe^{2+} Cyt c_3 (*Desulforibro desulfuricans*)	—	−0.205	410(ox)	—	51
				419(red)		
				525(red)		
				553(red)	4.2	
188.	Fe^{3+}/Fe^{2+} Heme octapeptide from Cyt c	—	−0.205	397(ox)	140	52
				414(red)	128	
				520(red)	6	
				550(red)	10	
189.	$[Ru(NH_3)_6]^{3+}/[Ru(NH_3)_6]^{2+}$	—	−0.214	—	—	53
190.	Fe^{3+}/Fe^{2+} Heme octapeptide from Cyt c (imidazole)	—	−0.217	405(ox)	122	68
				416(red)	162	
				520(red)		
				550(red)		
191.	Anthraquinone-1-sulfonate	0.195	−0.218	—	—	5
192.	$FMN/FMNH_2$, pH 7.09	—	−0.219	260	27.1	54
			−0.211	375(ox)	10.4	50
				450(ox)	12.2	
193.	$FAD/FADH_2$	—	−0.219	260	37	54
				375(ox)	9.3	
				450(ox)	11.3	
194.	6,7,9-Trimethyl-isoalloxazine (Lumiflavin)	—	−0.223	—	—	54
195.	Janus Green B	—	−0.225	—	—	5
196.	Fe^{3+}/Fe^{2+} Protoporphyrin IX (phosphate buffer), pH 8.2	—	−0.226	395(ox)	55	4
				495(ox)	7	30
				620(ox)	6	
197.	Glutathione	—	−0.23	—	—	7, 5
			−0.34			
198.	Acetoacetyl CoA/B-OH-Butyryl CoA	—	−0.238	—	—	7
199.	S(rhombic)/H_2S	0.14	−0.243	—	—	1, 7
200.	Acetylmethyl carbinol/butane-2,3-diol	—	−0.244	—	—	7
201.	Fe^{3+}/Fe^{2+} Copoporphyrin $(CN^-)_2$, pH 9.6	—	−0.247	—	—	5
202.	3-Acetylpyridine-NAD	—	−0.248	—	—	55
203.	Phenosafranine	0.280	−0.252	—	—	5
204.	V^{3+}/V^{2+}	−0.255	—	—	—	5
205.	Co^{3+}/Co^{2+} Mesoporphyrin (pyridine)$_2$	—	−0.265	—	—	5
206.	Mn^{3+}/Mn^{2+}Hematoporphyrin IX dimethyl ester	—	−0.268	—	—	6
207.	Fe^{3+}/Fe^{2+} Peroxidase (horseradish)	—	−0.271	415(ox)	60	56, 27, 57
				500(ox)	10.0	
				640(ox)	3.0	
208.	Fruotose-sorbitol	—	−0.272	—	—	5
209.	H_3PO_4/H_3PO_3	−0.276	—	—	—	1
210.	Rosindulin 2G	0.139	−0.281	—	—	5

OXIDATION-REDUCTION POTENTIALS, ABSORBANCE BANDS AND MOLAR ABSORBANCE
OF COMPOUNDS USED IN BIOCHEMICAL STUDIES (Continued)

	System	$E°$	$E°'$	λ_{max}	E_{mM}	Reference
211.	Thionicotinamide-NAD	—	−0.285	400(red)	—	58
212.	Acetone/Isopropanol	—	−0.281	—	—	5
			−0.286			
213.	Safranine T	0.235	−0.289	—	—	5
214.	Lipoic acid	—	−0.29	—	—	5
215.	Indulin Scarlet	0.047	−0.299	—	—	5
216.	Thiophenol	—	−0.30	—	—	4
217.	4-Aminoacridine	—	−0.301	—	—	59
218.	Acridine	—	−0.313	—	—	59
219.	$NAD^+/NADH$	−0.105	−0.320	259(ox)	18	7, 50
				259(red)	15	
				339(red)	6.2	
220.	$NADP^+/NADPH$	—	−0.324	259(ox)	18	7, 50
				259(red)	15	
				339(red)	6.2	
221.	Neutral Red	0.240	−0.325	—	—	5
222.	Cystine/Cysteine	—	−0.340	240(ox) (shoulder)	0.050	7
223.	Lipoyl dehydrogenase	—	−0.34	—	—	60
224.	NAD^+/α-NADH	—	−0.341	259(ox)	17	61
				346(red)		
225.	Mn^{3+}/Mn^{2+} Hematoporphyrin IX	—	−0.342	370(ox)	79	6
				460(ox)	50	
				545(ox)	12	
				770(ox)	1.3	
				416(red)	175	
				545(red)	18	
226.	Acetoacetate/β-hydroxybutyrate	—	−0.346	—	—	7
227.	Uric acid/Xanthine	—	−0.36	—	—	7
228.	Benzyl viologen	—	−0.36	—	—	5
229.	Gluconolactone/Glucose	—	−0.364	—	—	7
230.	3-Aminoacridine	—	−0.369	—	—	59
231.	Xanthine/Hypoxanthine	—	−0.371	248.5(ox)	10.2	7, 50
				278(ox)	8.9	
232.	Mn^{3+}/Mn^{2+} Mesoporphyrin (pyridine)$_2$	—	−0.387	—	—	5
233.	1-Aminoacridine	—	−0.394	—	—	59
234.	Cr^{3+}/Cr^{2+}	−0.40	—	—	—	1, 2
235.	N-Methyl nicotinamide	—	−0.419	—	—	5
236.	CO_2/Formate	−0.20	−0.42	—	—	1
237.	Fe^{3+}/Fe^{2+} Ferredoxin (*Clostridium*)	—	−0.413	300(ox)		65
				390(ox)	6	
238.	H^+/H_2	0.000	−0.421	—	—	5
239.	Fe^{3+}/Fe^{2+} Ferredoxin (spinach)	—	−0.432	325	—	65
				420(ox)		
				463(ox)		
240.	Methyl violgoen	—	−0.44	—	—	5
241.	Xanthine oxidase	—	−0.45	—	—	63
242.	SO_4^{2-}/SO_3^{2-}	—	−0.454	—	—	7
243.	Gluconate/Glucose	—	−0.44	—	—	62
			−0.47			7
244.	2-Aminoacridine	—	−0.486		—	59
245.	Oxalate/Glyoxalate	—	−0.50	—	—	7
246.	H_3PO_3/H_3PO_2	−0.50	—	—	—	1
247.	$SO_3^{2-}/S_2O_4^{2-}$	—	−0.527	—	—	7
			−0.471			
248.	Acetate/acetaldehyde	—	−0.581	—	—	7
			−0.589			
249.	2,8-Diaminoacridine	—	−0.731	—	—	59
250.	SiO_2/Si	−0.86	—	—	—	1
251.	5-Aminoacridine	—	−0.916	—	—	59
252.	N_2(gas)/H_3NOH^+	−1.87	—	—	—	1
253.	Formamidine sulfinic acid	—	−1.5	—	—	64

Compiled by Paul A. Loach.

References

1. Latimer, in *The Oxidation States of the Elements and Their Potentials in Aqueous Solution*, 2nd ed., Prentice-Hall, New York, 1952.
2. Berka, Vulterin, and Zyka, in *Newer Redox Titrants*, Pergamon Press, New York, 1965.
3. Koltoff, Belcher, Stenger, and Matsuyama, in *Volumetric Analysis*, Vol. III. Interscience, New York, 1957.
4. Lardy, in *Respiratory Enzymes*, Burgess, Minneapolis, 1949.
5. Clark, in *Oxidation-Reduction Potentials of Organic Systems*, Williams & Wilkins, Baltimore, 1960.
6. Loach and Calvin, *Biochemistry*, 2, 361 (1963).
7. Burton, *Ergeb. Physiol.*, 49, 275 (1957).
8. Martell and Calvin, in *Chemistry of Metal Chelate Complexes*, Prentice-Hall, New York, 1958.
9. Klotz and Klotz, *Science*, 121, 477 (1955).
10. Kok, *Biochim. Biophys. Acta*, 48, 527 (1961).
11. Goodheer, *Biochim. Biophys. Acta*, 38, 389 (1960).
12. Clayton, *Photochem. Photobiol.*, 1, 201 (1962).
13. Loach, Androes, Maksim, and Calvin, *Photochem. Photobiol*, 2, 443 (1963).
14. Kuntz, Loach, and Calvin, *Biophys. J.*, 4, 277 (1964).
15. Minakami, Ringler, and Singer, *J. Biol. Chem.*, 237, 569 (1962).
16. Kamen and Vernon, *Biochim. Biophys. Acta*, 17, 10 (1955).
17. Tissieres, *Biochim. J.*, 64, 582 (1956).
18. Ball, *Biochem. Z.*, 295, 262 (1938).
19. Rodkey and Ball, *J. Biol. Chem.*, 182, 17 (1950).
20. Theorell and Åkeson, *J. Am. Chem. Soc.*, 63, 1804 (1941).
21. Lautsch, Brouer, and Becker, *Z. Electrochem.*, 61, 174 (1957).
22. Cutolo, *Arzneimittelforsch*, 8, 581 (1958).
23. Armstrong, *Biochim. Biophys. Acta*, 86, 194 (1964).
24. Green, Mii, Mahler, and Bock, *J. Biol. Chem.*, 206, 1 (1954).
25. Havemann, *Biochem. Z.*, 314, 118 (1943).
26. Taylor and Hastings, *J. Biol. Chem.*, 131, 649 (1939).
27. George, Beetleston, and Griffith, in *Symposia on Hematin Enzymes*, Canberra, , 1959.
28. Taylor and Morgan, *J. Biol. Chem.*, 144, 15 (1942).
29. Falk, in *Porphyrins and Metalloporphyrins*, Elsevier, New York, 1964.
30. Shack and Clark, *J. Biol. Chem.*, 171, 143 (1947).
31. Orlando, *Biochim. Biophys. Acta*, 57, 373 (1962).
32. Morton, Gloor, Schindler, Wilson, Chopard-dit-Jean, Hemming, Isler, Leat, Pennock, Ruegg, Schwieter, and Wiss, *Helv. Chim. Acta*, 41, 2343 (1958).
33. Jagendorf and Marguiliea, *Arch. Biochem. Biophys.*, 90, 184 (1960).
34. Holton and Colpa-Boonstra, *Biochem. J.*, 76, 179 (1960).
35. Mikhalevich and Litvinchuk, *Zh. Neorgan. Khim.*, 9, 2391 (1964).
36. Kamen and Lakeda, *Biochim. Biophys. Acta*, 21, 518 (1956).
37. Green, Jarnefelt, and Tisdale, *Biochim. Biophys. Acta*, 31, 34 (1959).
38. Hauge, *J. Am. Chem. Soc.*, 78, 5266 (1956).
39. Hartree, *Advanc. Enzymol.*, 18, 1 (1957).
40. Pauson, *Quart. Rev.*, 9, 391 (1955).
41. Velick and Strittmatter, *J. Biol. Chem.*, 221, 265 (1956).
42. Newton and Kamen, *Biochim. Biophys. Acta*, 21, 71 (1956).
43. Hill, *Nature*, 174, 501 (1954).
44. Walter, *J. Biol. Chem.*, 196, 151 (1952).
45. Vestling, *Acta Chem. Scand.*, 9, 1600 (1955).
46. Martius and Marki, *Biochem. Z.*, 333, 111 (1960).
47. Burton and Wilson, *Biochem. J.*, 54, 86 (1953).
48. Loach and Calvin, *Nature*, 22, 343 (1964).
49. Cowgill and Clark, *J. Biol. Chem.*, 198, 33 (1952).
50. Weber, in *Biochemist's Handbook*, Long, Ed., Spon, London, 1961., 81.
51. Postgate, *Biochim. Biophys. Acta*, 18, 427 (1955).
52. Harbury and Loach, *J. Biol. Chem.*, 235, 3640 (1960).
53. Endicott, and Taube, *Inorg. Chem.*, 4, 437 (1965).
54. Lowe and Clark, *J. Biol. Chem.*, 221, 983 (1956).
55. Rodkey, *J. Biol. Chem.*, 234, 188 (1959).
56. Harbury, *J. Biol. Chem.*, 225, 1009 (1957).
57. Theotell, *Enzymologia*, 10, 3 (1942).
58. Anderson and Kaplan, *J. Biol. Chem.*, 234, 1226 (1959).
59. Breyer, Buchanan, and Duewell, *J. Chem. Soc.*, 360, 000 (1944).
60. Searls and Sanadi, *Proc. Natl. Acad. Sci. (USA)*, 45, 697 (1957).
61. Kaplan, in *The Enzymes*, 2nd, ed., Vol. 3, Boyer, Lardy, and Myrbäck, Eds., Academic Press, New York, 1960, 105.
62. Strecker and Korkes, *J. Biol. Chem.*, 196, 769 (1952).
63. Mackler, Mahler, and Green, *J. Biol. Chem.*, 210, 149 (1954).
64. Shashova, *Biochemistry*, 3, 1719 (1964).
65. Tagawa and Arnon, *Nature*, 195, 537 (1962).
66. Davenport and Hill, *Proc. R. Soc. B (England)*, 139, 327 (1952).
67. Wateri and Kimura, *Biochem. Biophys. Res. Commun.*, 24, 106 (1966).
68. Harbury and Loach, *J. Biol. Chem.*, 235, 3646 (1960).
69. Lovenberg and Sobel, *Proc. Natl. Acad. Sci. (USA)*, 54, 193 (1965).
70. Bartsch and Kamen, *J. Biol. Chem.*, 230, 41 (1958).
71. Fuhrhop and Mauzerall, *J. Am. Chem. Soc.*, 91, 4174 (1969).

CALORIMETRIC ΔH VALUES ACCOMPANYING CONFORMATIONAL CHANGES OF MACROMOLECULES IN SOLUTION

Macromolecule	Mol	$S_{20,w}$	Solvent	pH[a]	Concentration[a]	Temperature[b] °C	Type of Transition	Type of Measurement	ΔH[e] kcal/mol	Ref.
Pepsin	3.5×10^5	—	0.05 M phosphate and about 0.15 M KCl	7.16	0.2–0.5%	15	Denaturation	Heat of mixing	22[c]	1, 2
				6.41		35			69[c]	
Trypsin	2.0×10^4	—	0.1 M NaCl	1.4–2.5	0.2–0.5%	25	Denaturation	Heat of mixing	8.0	3
Fibrin	3.3×10^5	—	1.0 M NaBr-acetate, phosphate	6.08	2.91%	25	Polymerization	Heat of mixing	−19	4
				6.88			Clotting		−44.5	
Fibrinogen	3.3×10^5	—	Phosphate	6–8.5	5 g/l	25	Clotting	Heat of mixing	−44	5
Mercaptalbumin	6.7×10^4	—	0.1 M NaCl	2.8–4.7	2–3.5%	25	Denaturation	Heat of mixing	1.5–3.4[d]	3, 6
						15			1.5[d]	6
Ferrihemoglobin	6.8×10^4	—	0.02 M sodium formate	3.2–3.8	0.76%	25	Denaturation	Heat of mixing	10 ± 0.3	7
					0.61, 1.17%	15			−76 ± 1.6	
Horse serum albumin	6.9×10^4	—	0.1 M glycine	7.0	2%	55	Denaturation	Heat capacity	90 ± 15	8
						68			75 ± 10	
						76			55 ± 7	
Myoglobin	1.78×10^4	—	0.15 M KCl	4.5	3 g/l	30	Denaturation	Heat of mixing	40	9
sperm whale	1.76×10^4		0.1 M glycine	9.5	3.0 g/l	85		Heat capacity	200	41, 42
				10.6		78			178	
				11.0		72			134	
				11.5		63			100	
				12.25		50			73	
Ribonuclease A	1.37×10^4	—	0.15 M KCl	2.8	1.385 and 2.69%	43	Denaturation	Heat capacity	70 ± 1	10
			0.1 M KCl	2.2	0.5–1.0 g/l	45		Heat of mixing	109 ± 5	11
			0.15 M KCl	2.8	1.5%	44		Heat capacity	86.5 ± 4.4	12
			Water	7.80	3.41–7.22%	60			99 ± 8	75
			1.5 M urea		7.14%	55			87 ± 4	76
			2 M urea		4.86–7.43%	53.5			83 ± 5	75
			2.5 M urea		7.26%	52			81 ± 13	75
			3 M urea		5.73%	48.5			71 ± 4	76
			4 M urea		7.02%	46			68 ± 6	75
			1 M guanidine HCl		4.61%	50			79 ± 1	76
			2 M guanidine HCl		4.31%	37.5			55 ± 3	76
			1 M hexa-methylene-tetramine		9.5	61			99 ± 2	75

CALORIMETRIC ΔH VALUES ACCOMPANYING CONFORMATIONAL CHANGES OF MACROMOLECULES IN SOLUTION (Continued)

Macromolecule	Mol	$S_{20,w}$	Solvent	pH[a]	Concentration[a]	Temperature[b] °C	Type of Transition	Type of Measurement	ΔH[c] kcal/mol	Ref.
Ribonuclease A (Continued)			2 M hexamethylenetetramine		4.7–14.4%	60			105 ± 10	
			15 mM cacodylate	9.24	5.16–6.63%	61			99 ± 2	
			0.1 M glycine, acetate	2.4	0.1–1.0%	36			52	65
				3.3		47			66	
				3.7		50			73	
				4.44		54			77	
				6.0		59			89	
Ribonuclease—bovine pancreatic	1.37×10^4	—	0.04 M glycine	5.5	2.0 g/l	69.0	Denaturation	Heat capacity	115.0	43, 44
				4.0		57.0			108.0	
				3.3		49.0			97.0	
				2.75		42.0			91.5	
			HCl	0.36	0.5%	31.5			61	64
				1.05		29.9			59	
			0.2 M glycine	2.02	0.1–2.7%	31.2			66	
				2.80		40.6			88	
				3.28	0.5%	45.8			105	
			0.2 M acetate	4.04		52.3			126	
				5.00		57.8			151	
			0.2 M NaCl	6.23		60.8			155	
				7.00		61.3			168	
				7.80		61.2			178	
Ribonuclease S'	—		15 mM cacodylate	7.0	5.16–6.63%	47.1	Denaturation	Heat capacity	111	75
Ribonuclease S protein	—		15 mM cacodylate	7.0	5.16–6.63%	37.6	Denaturation	Heat capacity	55	75
Ribonuclease S	—		15 mM cacodylate	7.0	5.16–6.63%	47.7	Denaturation	Heat capacity	107	75
			0.3 M NaCl		75μM S-protein	5	S-protein+S-peptide=RNase S'	Heat of mixing	−23.6	55
						10			−25.2	
						15			−28.4	
						20			−33.3	
						25			−39.8	
						30			−47.9	
						35			−57.5	
						40			−68.8	
						0	S-protein+Met (O₂)-S-13-Peptide=Met (O₂)-13-RNase S'		−18.9	
						5			−19.8	
						10			−21.9	
						15			−25.0	
						20			−29.2	

CALORIMETRIC ΔH VALUES ACCOMPANYING CONFORMATIONAL CHANGES OF MACROMOLECULES IN SOLUTION (Continued)

Macromolecule	Mol	$S_{20,w}$	Solvent	pH^a	Concentration[a]	Temperature[b] °C	Type of Transition	Type of Measurement	ΔH[c] kcal/mol	Ref.
Tropocollagen Rat skin	3.6×10^5	—	Acetic acid, no salt	3.5	0.1–0.4 g/l	25			−34.4	
						30			−40.6	
						35			−47.8	
						40			−56.0	
Pike skin	3.6×10^5	—	Acetic acid, no salt	3.5	0.1–0.4 g/l	40.8	Denaturation	Heat capacity	1.53 residue	39, 40
Merlang skin	3.6×10^5	—	Acetic acid, no salt	3.5	0.1–0.4 g/l	30.6	Denaturation	Heat capacity	1.24 residue	39, 40
Cod skin	3.6×10^5	—	Acetic acid, no salt	3.5	0.1–0.4 g/l	21.5	Denaturation	Heat capacity	0.88 residue	39, 40
						20.0	Denaturation	Heat capacity	0.75 residue	39, 40
Lysozyme — egg white[h]	1.45×10^4	—	0.1 M phosphate	5.37	5.34–6.16%	76.5	Denaturation	Heat capacity	138 ± 7	62
			4 M urea		5.23–9.67%	65.5			103 ± 7	
			7 M urea		5.92%	55.0			80 ± 3	
			1 M guanidine HCl		4.40%	67.5			103 ± 6	
			2 M guanidine HCl		5.22%	58			85 ± 4	
			1 M hexa-methylene-tetramine		4.72–5.53%	—			121 ± 9	
			2 M hexa-methylene-tetramine		5.72%				121 ± 6	
	1.43×10^4		0.04 M glycine	4.5	10–5.0 g/l	78.5			141	44
				4.0		77.0			134	
				3.0		74.5			133	
				2.6		69.0			125	
				2.5		66.0			119	
				2.0		56.0			106	
				1.5		48.0			91	
			Water-HCl	1.0	2.5%	46		Heat of mixing	56 ± 8	13
Ovalbumin	4.5×10^4	—	3 M guanidine HCl	1.25	4.47–22.4 g/l	25			30 ± 3	61
			0.1 M glycine	10.0	~2%	77.5	Denaturation	Heat capacity	210 ± 13	60
				9		73			172 ± 13	
				5		68			119 ± 13	
				4.5		62			95 ± 13	
				4		57			84 ± 13	
				3		52			45 ± 13	
Chymotrypsin	—	—	0.01 M KCl	2.0 ± 0.08	2.0–10.0 g/l	25	Denaturation	Heat of mixing	50 ± 10	50
Chymotrypsinogen	—	—	0.01 M KCl	2.0 ± 0.08	2.0–10.0 g/l	40	Denaturation	Heat of mixing	110	50
						50			123	
	2.57×10^4	—	Water HCl	1.95	0.21–0.26%	40.6			103	51
				2.03		42.0			102	
				2.08		42.0			99	
			0.05 M glycine HCl	2.59		48.0			126	

CALORIMETRIC ΔH VALUES ACCOMPANYING CONFORMATIONAL CHANGES OF MACROMOLECULES IN SOLUTION (Continued)

Macromolecule	Mol	$S_{20,w}$	Solvent	pH^a	Concentration^a	Temperature^b °C	Type of Transition	Type of Measurement	ΔH^e kcal/mol	Ref.
				2.99		53.9			145	
				3.02		54.2			135	
Me₂ SO-chymotrypsin	—	—	0.01 M KCl	2.0 ± 0.08	2.0–10.0 g/l	25	Denaturation	Heat of mixing	32	50
						40			73	
Chymotrypsin-bovine	2.52×10^4	—	0.04 M glycine	4.0	1.0–5.0 g/l	56.4	Denaturation	Heat capacity	162	44, 46
				3.4		55.4			155	
				3.1		52.0			149	
				2.8		48.6			142	
				2.6		44.8			132	
				2.2		38.2			108	
Chymotrypsinogen A	—	—	0.1 M NaCl, 0.1 M hydro-cinnamate	7.4	0.4 mM	25	Activation to π chymotrypsin	Heat of mixing	0 ± 0.5	66
	2.51×10^4		HCl	3	5.92%	56	Denaturation	Heat capacity	154 ± 8	76
				2	6.77%	42.5			112 ± 3	
π Chymotrypsin	—	—	Various NaCl, CaCl₂, and buffers	7.4	0.06–0.7 mM	25	Conversion to δ-chymotrypsin	Heat of mixing	−2 ± 1	66
α-Chymotrypsinogen	2.45×10^4	—	0.01 M glycine acetate	2.3	0.1–1.0%	43	Denaturation	Heat capacity	78	41
				2.6		49			102	
				2.8		51			110	
				3.4		58			130	
				4.0		61			140	
				5.0		62			148	
α-Chymotrypsin	—	—	0.05 M phosphate, 0.2 M KCl	7.8	0.1–0.4 mM	25	Dimerization	Heat of mixing	−17.1 ± 1.2	67
Cytochrome c bovine heart	1.24×10^4	—	0.04 M Glycine	4.8	1.0–5.0 g/l	78.0	Denaturation	Heat capacity	107	44, 47
				4.5		77.0			103	
				3.9		72.0			96	
				3.7		70.0			93	
				3.4		66.0			84	
				3.2		62.0			77	
				3.0		59.0			70	
				2.8		52.5			60	
Poly (β-benzyl-L-aspartate)	—	—	5.2 mol % CHCl₂CO₂H, 94.8 mol % CHCl₂CHCl₂	—	0–1%	−0.8	Coil-helix	Heat of mixing	0.358 residue	13
			5.7 mol % CHCl₂CO₂H, 94.3 mol % CHCl₂CHCl₂			7.4			0.334 residue	
			6.0 mol % CHCl₂CO₂H, 94.0 mol % CHCl₂CHCl₂			17.8			0.298 residue	
			6.4 mol % CHCl₂CO₂H, 93.6 mol % CHCl₂CHCl₂			28.8			0.229 residue	
			93.3 mol % CHCl₂CO₂H, 6.7 mol % CHCl₂CHCl₂			38.0			0.169 residue	

CALORIMETRIC ΔH VALUES ACCOMPANYING CONFORMATIONAL CHANGES OF MACROMOLECULES IN SOLUTION (Continued)

Macromolecule	Mol	$S_{20,w}$	Solvent	$_pH^a$	Concentration[a]	Temperature[b] °C	Type of Transition	Type of Measurement	ΔH[e] kcal/mol	Ref.
Poly (γ-benzyl-L-glutamate)	5×10^5	—	47 mol % $CHCl_2CO_2H$, 53 mol % $CHCl_3$	—	2–3 wt/vol%	0	Coil-helix	Heat capacity	0.75 residue	73
			51 mol % $CHCl_2CO_2H$, 49 mol % $CHCl_3$			2			0.71 residue	
			56 mol % $CHCl_2CO_2H$, 44 mol % $CHCl_3$			9			0.68 residue	
			62 mol % $CHCl_2CO_2H$, 38 mol % $CHCl_3$			15			0.61 residue	
			78 mol % $CHCl_2CO_2H$, 22 mol % $CHCl_3$			41			0.40 residue	
			46 mol % $CHCl_2CO_2H$, 54 mol % $CHCl_2CHCl_2$			-21			0.84 residue	
			52 mol % $CHCl_2CO_2H$, 48 mol % $CHCl_2CHCl_2$			-15			0.80 residue	
			65 mol % $CHCl_2CO_2H$, 35 mol % $CHCl_2CHCl_2$			3			0.72 residue	
			74 mol % $CHCl_2CO_2H$, 26 mol % $CHCl_2CHCl_2$			14			0.57 residue	
			79 mol % $CHCl_2CO_2H$, 21 mol % $CHCl_2CHCl_2$			21			0.84 residue	
			81 mol % $CHCl_2CO_2H$, 19 mol % $CHCl_2CHCl_2$			39			0.30 residue	
			37 mol % CH_2ClCH_2Cl, 63 mol % $CHCl_2CO_2H$			-24			0.86 residue	
			47 mol % $CHCl_2CO_2H$, 53 mol % CH_2ClCH_2Cl			-10			0.81 residue	
			56 mol % CH_2ClCH_2Cl, 44 mol % $CHCl_2CO_2H$			2			0.73 residue	
			63 mol % $CHCl_2CO_2H$, 37 mol % CH_2ClCH_2Cl			13			0.66 residue	
			69 mol % $CHCl_2CO_2H$, 31 mol % CH_2ClCH_2Cl			23			0.56 residue	
			74 mol % $CHCl_2CO_2H$, 26 mol % CH_2ClCH_2Cl			31			0.49 residue	
			77 mol % $CHCl_2CO_2H$, 23 mol % CH_2ClCH_2Cl			40			0.36 residue	
	2.35×10^5		19 wt % CH_2ClCH_2Cl, 81 wt % $CHCl_2CO_2H$		0.257 m residue	32			0.43 residue	14, 15
					0.132 m residue				0.68 residue	
					0.068 m residue C → O				0.81 residue	
									0.95 ± 0.03 residue	
	2.7×10^5		25 vol % CH_2ClCH_2Cl, 75 vol % $CHCl_2CO_2H$		0.097 m residue	26			0.525 ± 0.08 residue	16

CALORIMETRIC ΔH VALUES ACCOMPANYING CONFORMATIONAL CHANGES OF MACROMOLECULES IN SOLUTION (Continued)

Macromolecule	Mol	$S_{20,w}$	Solvent	pH[a]	Concentration[a]	Temperature[b] °C	Type of Transition	Type of Measurement	ΔH[e] kcal/mol	Ref.
Poly (γ-benzyl-L-glutamate) (Continued)	1.6×10^5		$(CHCl_2CO_2H\text{-}CH_2ClCH_2Cl) \rightarrow 100\%$ $CHCl_2CO_2H$		7 mM residue	30		Heat of solution	0.70 ± 0.05 residue	17
	3.5×10^5		$CH_2ClCH_2Cl\text{-}CHCl_2CO_2H$ 82 wt % $CHCl_2CO_2H$		0.25 m residue	37		Heat capacity	0.38 residue	18
					0.07 m residue	43			0.79 residue	
			83 wt % $CHCl_2CO_2H$		0.25 m residue	40			0.32 residue	
					0.13 m residue	44			0.62 residue	
					0.07 m residue	46			0.78 residue	
			85 wt % $CHCl_2CO_2H$		0.25 m residue	46			0.29 residue	
					0.13 m residue	50			0.58 residue	
					0.07 m residue	53			0.76 residue	
			88 wt % $CHCl_2CO_2H$		0.25 m residue	—			0.26 residue	
					0.13 m residue				0.55 residue	
					0.07 m residue				0.74 residue	
	2.9×10^5		25 vol % $CHCl_3$, 75 vol % $CHCl_2CO_2H$		0.139–0.082 m residue	25		Heat of mixing	1.0 ± 0.1 residue	31
	1.6×10^5		20 vol % CH_2ClCH_2Cl, 80 vol % $CHCl_2CO_2H$		2 g/l	30		Heat of solution	0.65 ± 0.3 residue	32
			25 vol % $CHCl_3$, 75 vol % $CHCl_2CO_2H$						0.65 ± 0.3 residue	
	2.0×10^5		25 vol % CH_2ClCH_2Cl, 75 vol % $CHCl_2CO_2H$		0.123–0.7 m residue	25		Heat of mixing	0.75 ± 0.2 residue	33
	3.5×10^4		75 vol % $CHCl_2CO_2H$, 25 vol % CH_2ClCH_2Cl		3%	—		Heat capacity	0.3 ± 0.8 residue	58
	4.5×10^4								0.38 ± 0.8 residue	
	9.9×10^4								0.49 ± 0.5 residue	
	29.0×10^4								0.615 ± 0.8 residue	
	33.5×10^4								0.59 ± 0.8 residue	
	43.5×10^4								0.60 ± 0.5 residue	
	55.0×10^4								0.515 ± 0.6 residue	
Poly-γ-benzyl-L-glutamate (deuterated)	2.7×10^5	—	34 vol % CH_2ClCH_2Cl, 66 vol % $CHCl_2CO_2H$	—	3%	8.5	Coil-helix	Heat capacity	0.67 ± 0.05 residue	19
			18 vol % CH_2ClCH_2Cl, 82 vol % $CHCl_2CO_2H$			40			0.38 ± 0.05 residue	
Poly (N-γ-carbobenzoxy-L-α-γ-diaminobutyric acid)	—	—	$CH_2ClCH_2Cl/CHCl_2CO_2H$	—	~2 g/l	30	Solvation	Heat of solution	−0.6 residue	68

CALORIMETRIC ΔH VALUES ACCOMPANYING CONFORMATIONAL CHANGES OF MACROMOLECULES IN SOLUTION (Continued)

Macromolecule	Mol	$S_{20,w}$	Solvent	pH[a]	Concentration[a]	Temperature[b] °C	Type of Transition	Type of Measurement	ΔH[c] kcal/mol	Ref.
Poly (N-δ-carbobenzoxy-L-ornithine)	—	—	CH₂ClCH₂Cl/ CHCl₂CO₂H	—	~2 g/l	30	Order-disorder		0.255 ± .025 residue	
							Order-disorder	Heat of solution	−0.65 residue	
Poly(L-lysine)	1.1×10^5	—	0.1 *M* KCl	6.0	0.25%	15, 25	Coil-helix	Heat of mixing	1.2 residue	63
							α-β		0 residue	
Poly (ε-carbobenzoxy-L-lysine)	7.5×10^5	—	CH₂ClCH₂Cl/ CHCl₂CO₂H	—	0.1%	15, 25	Coil-helix	Heat of mixing	0.62 ± 0.04 residue	59
	1.5×10^5, 2.75×10^5		37 vol % CHCl₂CO₂H, 63 vol % CHCl₃		3%	26		Heat capacity	0.21 ± 0.06 residue	20
Poly (L-glutamic acid)	$0.4–1.0 \times 10^5$	—	0.1 *M* KCl	4.5–5.5	0.5 g/l	30	Coil-helix	Heat of mixing	−1.1 ± 0.2 residue	21
Poly (γ-ethyl-L-glutamate)	1.3×10^5	—	40 vol % CH₂ClCH₂Cl, 60 vol % CHCl₂CO₂H	—	2 g/l	30	Coil-helix	Heat of solution	0.65 ± 0.3 residue	32
	4.0×10^5		35 vol % CH₂ClCH₂Cl, 65 vol % CHCl₂CO₂H						0.65 ± 0.3 base pair	
Salmon DNA	—	21.7	0.1 *M* NaCl	6.0	0.15–0.6 g/l	25	Acid denaturation	Heat of mixing	8.31 base pair	23, 24
Herring spermatozoa DNA	—	—	0.015 *M* NaCl, 1.5 m*M* citrate	—	1%	75	Thermal denaturation	Heat capacity	5 base pair	15
Ps. fluorescens DNA	—	20.1	0.1 *M* NaCl	6.0	0.15–0.6 g/l	25	Acid denaturation	Heat of mixing	7.83 base pair	24
S. marcescens DNA	—	17.4	0.1 *M* NaCl	6.0	0.15–0.6 g/l	25	Acid denaturation	Heat of mixing	7.83 base pair	24
Sea urchin DNA	—	23.3	0.1 *M* NaCl	6.0	0.15–0.6 g/l	25	Acid denaturation	Heat of mixing	8.03 base pair	24
Calf thymus DNA	>10^6	—	0.015 *M* NaCl, 1.5 m*M* citrate	6.0	10 g/l	72	Thermal denaturation	Heat capacity	7.0 ± 0.5 base pair	25
			0.15 *M* Phosphate	11.3	~4 m*M* base pair	34	Denaturation		8.3 ± 0.5 base pair	52
				11.15		44.2			9.5 ± 0.5 base pair	
				11.00		47.2			9.5 ± 0.5 base pair	
				10.90		50.6			9.3 ± 0.5 base pair	
				10.70		56.6			10.0 ± 0.7 base pair	
				10.60		58.8			9.1 ± 0.5 base pair	
				10.45		63.5			9.2 ± 0.7 base pair	
				10.30		68.8			10.4 ± 0.7 base pair	

CALORIMETRIC ΔH VALUES ACCOMPANYING CONFORMATIONAL CHANGES OF MACROMOLECULES IN SOLUTION (Continued)

Macromolecule	Mol	$S_{20,w}$	Solvent	$_pH^a$	Concentration[a]	Temperature[b] °C	Type of Transition	Type of Measurement	ΔH^e kcal/mol	Ref.
Calf thymus DNA (Continued)			1 mM phosphate, 1.5 mM Na$^+$	7.0		58.05			6.4 ± 0.3 base pair	
			1 mM phosphate, 6.5 mM Na$^+$			64.5			6.8 ± 0.3 base pair	
			1 mM phosphate, 11.2 mM Na$^+$			68.8			6.9 ± 0.3 base pair	
			1 mM phosphate, 51 mM Na$^+$			77.0			7.2 ± 0.3 base pair	
Cl. perfrigens DNA	—	—	1.0 mM KCl, 1.5 mM sodium citrate	7.0	5–6 g/l	55	Denaturation	Heat capacity	7.73 base pair	53
M. lysodeikticus DNA	—	—	1.0 mM KCl, 1.5 mM sodium citrate	7.0	5–6 g/l	79	Denaturation	Heat capacity	8.52 base pair	53
T$_2$ phage DNA	—	—	3 mM phosphate, 0.2 M NaCl	7.0	0.5 g/l	84.8	Denaturation	Heat capacity	9.65 base pair	37
			3 mM phosphate, 0.115 M NaCl			81.2			9.42 base pair	
			3 mM phosphate, 0.057 M NaCl			75.0			9.28 base pair	
			3 mM phosphate, 0.036 M NaCl			71.5			9.14 base pair	
			3 mM phosphate, 0.014 M NaCl			66.0			9.15 base pair	
			3 mM phosphate, 0.009 M NaCl			69.0			8.90 base pair	
			3 mM phosphate, 0.2 M NaCl	8.5		82.5			8.94 base pair	
			3 mM phosphate, glycine	8.9		76.5			8.03 base pair	
			3 mM glycine, 0.2 M NaCl	9.3		71.8			7.78 base pair	
				9.6		66.3			7.14 base pair	
			3 mM citrate, phosphate, 0.20 M NaCl	5.4		84.0			9.40 base pair	37, 38
			3 mM citrate, 0.20 M NaCl	4.8		82.3			8.57 base pair	
				4.3		76.0			6.60 base pair	
				4.0		71.5			5.43 base pair	
				3.8		68.0			5.00 base pair	
				3.5		64.0			4.70 base pair	
				3.2		55.0			7.00 base pair	
Salmon sperm DNA	—	—	1.0 mM HCl, 1.5 mM sodium citrate	7.0	5–6 g/l	60.6	Denaturation	Heat capacity	7.84 base pair	53
M$_4$ coliphage DNA	4×10^7	—	0.015 M HCl, 0.1 M KCl, 2.2 M urea, citrate	3.25	0.045–0.066 g/l	27	Denaturation	Heat of mixing	9.5 ± 1.5 base pair	69
tRNA^{Phe}-yeast	—	—	0.01 M tris, 50 µM Mg^{2+}	7.0	~0.12%	66.5	Unfolding	Heat capacity	140	48
			0.01 M tris, 0.1 mM Mg^{2+}			70			156	
			0.01 M tris, 0.25 mM Mg^{2+}			73			216	
			0.01 M tris, 5 mM Mg^{2+}			80.5			248	

CALORIMETRIC ΔH VALUES ACCOMPANYING CONFORMATIONAL CHANGES OF MACROMOLECULES IN SOLUTION (Continued)

Macromolecule	Mol	$S_{20,w}$	Solvent	$_pH^a$	Concentration[a]	Temperature[b] °C	Type of Transition	Type of Measurement	ΔH^e kcal/mol	Ref.
tRNA^{Phe}-yeast (Continued)			5 mM phosphate, 0.1 M NaCl 0.2 mM Mg^{2+}			68			175	
			5 mM NaCl, 1 mM MgCl$_2$	7.2	~10 μM	57		Heat of mixing	123 ± 25	49
			5 mM citrate, 1 mM MgSO$_4$	6.5	0.06–0.08 mM	49		Heat capacity	200 ± 30	71
			5 mM citrate, 5 mM MgSO$_4$			70			250 ± 20	
			5 mM citrate, 0.08 M MgSO$_4$, 0.5 M NaCl			76.5			240 ± 20	
			5 mM citrate, 8 mM MgSO$_4$, 0.1 M NaCl			76.5			240 ± 20	
			5 mM citrate, 0.08 M MgSO$_4$, 0.5 M NaCl			76.5			240 ± 20	
			5 mM citrate, 0.02 M MgSO$_4$			76.5 / 79			220 ± 20 / 230 ± 20	
			5 mM citrate 10 mM MgSO$_4$	6.5	1.25–2.5 g/l	60		Heat of mixing	310	71
Poly(A·U)	—	4.5–10.0 Poly(U), 8.0–12.0 Poly(A)	0.1 M KCl, 0.01 M cacodylate	6.6	20 mM–50 μM nucleotide	25	Poly (A)+ Poly (U)=Poly(A·U)	Heat of mixing	−5.9 ± 0.2 base pair	26
			0.5 M KCl, 0.01 M cacodylate						−5.9 ± 0.2 base pair	
			1.0 M KCl, 0.01 M cacodylate						−4.75 ± 0.3 base pair	
		2.1–12.2 Poly(A)	0.1 M KCl, 0.01 M cacodylate						−5.95 ± 0.1 base pair	
		6.1 Poly(A), 7.2 Poly(U)	0.1 M KCl, 0.01 M cacodylate	7.0	35 mM nucleotide	10			−6.29 ± 0.19 base pair[f]	2
						25			−6.97 ± 0.17 base pair[f]	
						40			−7.72 ± 0.29 base pair[f]	
			0.1 M NaCl, 0.01 M cacodylate	6.8	80 mM nucleotide	24			−5.95 ± 0.1 base pair[f]	
						37			−6.50 ± 0.1 base pair	
			0.5 M NaCl, 0.01 M cacodylate			37			−6.69 ± 0.1 base pair	
	~10^5	—	0.01 M citrate, 0.057 M NaCl		8.5 mM base pair	49		Heat capacity	−6.7 base pair	29
			0.01 M citrate, 0.10 M NaCl			54.8			−7.2 base pair	
			0.01 M citrate, 15 M NaCl			58.4 / 85–90		Extrapolated	−7.7 base pair / −8.5 ± 0.5 base pair	

CALORIMETRIC ΔH VALUES ACCOMPANYING CONFORMATIONAL CHANGES OF MACROMOLECULES IN SOLUTION (Continued)

Macromolecule	Mol	$S_{20,w}$	Solvent	pH^a	Concentration^a	Temperature^b °C	Type of Transition	Type of Measurement	ΔH^c kcal/mol	Ref.
Poly(A·U) (Continued)	—	9.53 Poly(A), 6.15 Poly(U)	0.018 M NaCl, 5 mM cacodylate	6.9–7.0	5.0 mM nucleotide	44.5		Heat capacity	−7.38 ± 0.08 base pair	34
			0.043 M NaCl, 5 mM cacodylate			51.3			−7.95 ± 0.07 base pair	
			0.103 M NaCl, 0.01 M cacodylate		6.04 mM nucleotide	58.3			−8.20 ± 0.2 base pair	
			0.104 M NaCl, 0.01 M cacodylate		5.0 mM nucleotide	58.2			−8.20 ± 0.24 base pair	
		7.56 Poly(A), 5.62 Poly(U)	0.011 M KCl, 5 mM cacodylate			35.9			−6.44 ± 0.22 base pair	
			0.012 M KCl, 5 mM cacodylate			36.2			−6.44 ± 0.22 base pair	
			0.040 M KCl, 5 mM cacodylate		2.28 mM nucleotide	47			−6.83 ± 0.33 base pair	
			0.054 M KCl, 5mM cacodylate		5.0 mM nucleotide	48.7			−6.85 ± 0.11 base pair	
			0.055 M KCl, 5 mM cacodylate			48.8			−6.85 ± 0.11 base pair	
	~10^5		0.06 M cations, 3.3 mM citrate	6.5	3.76 mM base pair	49.4			6.8 ± 0.4 base pair	3
			0.063 M cations, 3.3 mM citrate		1.88 mM base pair	51.2			−6.9 ± 0.4 base pair	
			0.06 M cations, 3.3 mM citrate		0.984 mM base pair	51			−6.9 ± 0.4 base pair	3
			0.46 M cations, 0.01 M citrate	6.8	Not given	56.1			−8.2 base pair	36
			0.50 M cations, 0.01 M citrate			70.0			−8.8 base pair	
						54.1			−8.1 base pair	
			0.57 M cations, 0.01 M citrate			71.6			−8.7 base pair	
						53.5			−8.0 base pair	
						74.5		Extrapolated	−8.9 base pair	
						95			−9.5 ± 0.5 base pair	
			5 mM NaCl		8.5 mM nucleotide	45.8	Poly(A) + Poly(U)= Poly(A·U)	Heat capacity	−6.6 base pair	70
			5 mM NaCl, D$_2$O			47.7			−6.6 base pair	
			0.1 M NaCl, 0.01 M cacodylate			24	Poly(A·U)+Poly(U)= Poly(A·2U)	Heat of mixing	−3.82 ± 0.1(A·2U) residue	28
						37			−3.5 ± 0.5 (A·2U) residue	
			0.5 M NaCl, 0.01 M cacodylate			24			−3.80 ± 01 (A·2U) residue	

CALORIMETRIC ΔH VALUES ACCOMPANYING CONFORMATIONAL CHANGES OF MACROMOLECULES IN SOLUTION (Continued)

Macromolecule	Mol	$S_{20,w}$	Solvent	pH[a]	Concentration[a]	Temperature[b] °C	Type of Transition	Type of Measurement	ΔH[c] kcal/mol	Ref.
	—	7.56 Poly(A), 5.62 Poly(U)	0.015 M KCl, 5mM cacodylate		7.72 mM nucleotide	37			-4.09 ± 0.1 (A·2U) residue	36
					7.72 mM nucleotide	28.4		Heat capacity	-1.24 ± 0.1 (A·2U) residue	
					7.50mM nucleotide	28.6			-1.24 ± 0.15 (A·2U) residue	
	—	9.53 Poly(A), 6.15 Poly(U)	0.018 M NaCl, 5 mM cacodylate	6.9–7.0	5.0 mM nucleotide	32.6		Heat capacity	-1.29 ± 0.16 (A·2U) residue	34
			0.019 M NaCl, 5 mM cacodylate			31.5			-1.29 ± 0.16 (A·2U) residue	
	~10⁵	—	0.46 M cations, 0.01 M citrate	6.8	Not given	56.1			-4.1 ± 2 (A·2U) residue	
			0.50 M cacodylate 0.01 M citrate			54.1			-4.3 ± 2 (A·2U) residue	
			0.57 M cations, 0.01 M cacodylate			53.1			-4.2 ± 2 (A·2U) residue	
	~10⁵		0.01 M citrate, 0.5 M NaCl			54.3	2 Poly(A·U)=Poly(A·2U)+Poly(A)		3.2 (A·2U) residue	29
						85–90		Extrapolated	4.5 ± 0.5 (A·2U) residue	
	—	9.53 Poly(A), 6.15 Poly(U)	0.263 M NaCl, 0.01 M cacodylate	6.9–7.0	5.0 mM nucleotide	57.5		Heat capacity	2.76 ± 0.1 (A·2U) residue	34
			0.46 M cations, .01 M citrate	6.8	Not given	56.1			4.1 (A·2U) residue	36
			0.50 M cations, .0 M citrate			54.1			3.8 (A·2U) residue	
			0.57 M cations, .01 M citrate			53.5			3.8 (A·2U) residue	
						95	Poly(A·2U)=Poly(A)+ 2 Poly(U)	Extrapolated	5.1 ± 0.5 (A·2U) residue	
			0.01 M citrate, 0.50 M cations			72.1			11.9 (A·2U) residue	
						85–90			12.5 ± 0.5 (A·2U) residue	
		7.56 Poly A, 5.62 Poly U	268 M NaCl, 0.01 M cacodylate	6.9–7.0	7.0 mM nucleotide	67.9		Heat capacity	12.7 ± 0.13 (A·2U) residue	34
			5.5 mM KCl, 5mM cacodylate		6.0 mM nucleotide	49.1			10.0 ± 0.25 (A·2U) residue	
			5.6 mM KCl, 5mM cacodylate			49.3			10.0 ± 0.25 (A·2U) residue	
	~10⁵	—	0.46 M cations, 0.01 M citrate	6.8	Not given	70.0			12.9 (A·2U) residue	36
			0.50 M cations, 0.01 M cacodylate			71.6			13.0 (A·2U) residue	
			0.57 M cations, 0.01 M cacodylate			74.5			13.1 (A·2U) residue	

CALORIMETRIC ΔH VALUES ACCOMPANYING CONFORMATIONAL CHANGES OF MACROMOLECULES IN SOLUTION (Continued)

Macromolecule	Mol	$S_{20,w}$	Solvent	pH[a]	Concentration[a]	Temperature[b] °C	Type of Transition	Type of Measurement	ΔH[e] kcal/mol	Ref.
	—	9.53 Poly(A), 6.15 Poly(U)	1.8 mM NaCl, 5 mM cacodylate	6.9–7.0	7.5 mM nucleotide	95	Poly(A)+2 Poly(U)=Poly(A·U) +Poly(U)	Extrapolated	13.5 ± 0.5 (A·2U) residue	34
						45.5		Heat capacity	-8.38 ± 0.14 (A·U) residue	
			1.9 mM NaCl, 5 mM cacodylate			45.0			-8.38 ± 0.14 (A·U) residue	
		7.56 Poly(A), 5.62 Poly(U)	1.5 mM KCl, 5 mM cacodylate		7.72 mM nucleotide	38.8			-7.49 ± 0.23 (A·U) residue	
					7.50 mM nucleotide	38.6			-7.49 ± 0.23 (A·U) residue	
Poly A	~10⁵	4.23	0.1 M NaCl, 0.01 M tris	7.30	C → O from 0.37–1.34 g/l	35	Helix-coil	Heat capacity	9.4 residue	3
	—		Various salt concentrations	6.8	7.8 mM base pair	90–95	Double helix-coil	Extrapolated	4.5 ± 2 residue	2
			0.20 M citrate, 0.15 M NaCl, HCl	5.50	0.0132 m nucleotide	31.5		Heat capacity	3.36 base pair	
				5.30		39.2			4.09 base pair	
				5.06		47.1			4.67 base pair	
				4.89		56.6			5.13 base pair	
				4.70		65.5			5.57 base pair	
				4.20		85.5			5.90 base pair	
	—	6.1	0.1 M KCl, 0.01 M cacodylate	4.0		10		Heat of mixing	1.80 ± 0.25[g] residue	27
						25			2.74 ± 0.20[g] residue	
Poly(dA-dT)	—	—	5 mM NaCl, 1 mM cacodylate, 1 mM citrate	7.0	5.9–11.6 mM base pair	40	Helix-coil	Heat capacity	7.9 ± 0.14 base pair	56
Poly(I·C)	5 × 10⁵	—	.01 M citrate, .063 m Na⁺	6.9 ± 0.1	1.8 mm base pair	54.1	Poly(I)+Poly(C)= Poly(I·C)	Heat capacity	-6.5 ± 0.4 base pair	54
			.01 M citrate, 0.104 m Na⁺		1.78 mm base pair	60.8			6.8 ± 0.4 base pair	
			.01 M citrate, 0.303 m Na⁺			67.6			-7.6 ± 0.4 base pair	
			.01 M citrate, 0.503 m Na⁺			70.7			-7.9 ± 0.4 base pair	
			.01 M citrate, 1.003 m Na⁺			73.9			-8.0 ± 0.4 base pair	
	—		0.02–0.2 M NaCl	8.0	70 mM base pair	20		Heat of mixing	-5.59 ± 0.02 base pair	57
			0.1–0.4 M NaCl			37			-5.59 ± 0.01 base pair	
Poly(I)	—	—	0.1 M citrate, 1.0 m Na⁺	6.9 ± 0.1	1.79 mM base pair	43.6	Triple helix-coil	Heat capacity	-1.9 ± 0.4 residue	54

CALORIMETRIC ΔH VALUES ACCOMPANYING CONFORMATIONAL CHANGES OF MACROMOLECULES IN SOLUTION (Continued)

Macromolecule	Mol	$S_{20,w}$	Solvent	pH[a]	Concentration[a]	Temperature[b] °C	Type of Transition	Type of Measurement	ΔH[e] kcal/mol	Ref.
PolyC	—	—	1 mM acetate, 0.01 M Na+	3.68	~2.8 g/l	40	Double helix-coil	Heat capacity	4.06 base pair	72
				4.33		63			5.20 base pair	
				4.48		75			5.25 base pair	
				4.55		74			5.27 base pair	
				4.85		72			5.12 base pair	
				5.20		61			4.95 base pair	
				5.50		54			4.34 base pair	
				5.76		47			3.62 base pair	

Compiled by Gordon C. Kresheck.

a Final value in mixing experiments; m is used for molal concentration; M for molar.
b Transition temperature for heat capacity experiments.
c These values depend upon the choice of expressing pepsin.
d Value depended upon commercial source of protein.
e The manner of treating ionization changes and baseline shifts varies from worker to worker and may introduce differences between the results reported by different laboratories. The latter is discussed in some detail in Reference 48 for heat capacity measurements.
f Heat change corrected for unfolding poly A before reaction.
g These calorimetric data were recalculated by Stevens and Felsenfeld[77] to yield values of 6.5 and 8.5 kcal mol nucleotide for the single helix-coil transition.
h Data are also found in Reference 74 for this protein, although complete experimental conditions were not given.

References

1. Buzzell and Sturtevant, *J. Am. Chem. Soc.*, 74, 1983 (1952).
2. Sturtevant, *J. Phys. Chem.*, 58, 97 (1954).
3. Gutfreund and Sturtevant, *J. Am. Chem. Soc.*, 75, 5447 (1953).
4. Sturtevant, Laskowski, Donnelly, and Scheraga, *J. Am. Chem. Soc.*, 77, 6163 (1955).
5. Laki and Kitzinger, *Nature* (Lond.), 178, 985 (1956).
6. Bro and Sturtevant, *J. Am. Chem. Soc.*, 80, 1789 (1958).
7. Forrest and Sturtevant, *J. Am. Chem. Soc.*, 82, 585 (1960).
8. Privalov and Monaselidze, *Biofizika*, 8, 420 (1963).
9. Hermans and Rialdi, *Biochemistry*, 4, 1277 (1965).
10. Beck, Gill, and Downing, *J. Am. Chem. Soc.*, 87, 901 (1965).
11. Kresheck and Scheraga, *J. Am. Chem. Soc.*, 88, 4588 (1966).
12. Danforth, Krakauer, and Sturtevant, *Rev. Sci. Instrum.*, 38, 484 (1967).
13. McKnight, Ph.D. thesis, University of Massachusetts, 1974.
14. Ackermann and Ruterjans, *Z. Phys. Chem.*, 41, 116 (1964).
15. Ackermann and Ruterjans, *Ber Bunsenges Phys. Chem.*, 68, 850 (1964).
16. Karasz, O'Reilly, and Bair, *Nature* (Lond.), 202, 693 (1964).
17. Giacometti and Turolla, *Z. Phys. Chem.*, 51, 108 (1966).
18. Ackermann and Neumann, *Biopolymers*, 5, 649 (1967).
19. Karasz and O'Reilly, *Biopolymers*, 4, 1015 (1966).
20. Karasz, O'Reilly, and Bair, *Biopolymers*, 3, 241 (1965).
21. Rialdi and Hermans, *J. Am. Chem. Soc.*, 88, 5719 (1966).
22. Klump, Neumann, and Ackermann, *Biopolymers*, 7, 423 (1969).
23. Sturtevant and Geiduschek, *J. Am. Chem. Soc.*, 80, 2911 (1958).
24. Bunville, Geiduschek, Rawitscher, and Sturtevant, *Biopolymers*, 3, 213 (1965).
25. Ruterjans, thesis, University of Munster, Germany, 1965.
26. Steiner and Kitzinger, *Nature (Lond.)*, 194, 1172 (1962).
27. Rawitscher, Ross, and Sturtevant, *J. Am. Chem. Soc.*, 85, 1915 (1963).
28. Ross and Scruggs, *Biopolymers*, 3, 491 (1965).
29. Neumann and Ackermann, *J. Phys. Chem.*, 71, 2377 (1967).
30. Epand and Scheraga, *J. Am. Chem. Soc.*, 89, 3888 (1967).
31. Kagemoto and Fugishiro, *Makromol. Chem.*, 114, 139 (1968).
32. Giacometti, Turolla, and Boni, *Biopolymers*, 6, 441 (1968).
33. Kagemoto and Jujishiro, *Biopolymers*, 6, 1753 (1968).
34. Krakauer and Sturtevant, *Biopolymers*, 6, 491 (1968).
35. Hinz, Schmitz, and Ackermann, *Biopolymers*, 7, 611 (1969).
36. Neumann and Ackermann, *J. Phys. Chem.*, 73, 2170 (1969).
37. Privalov, *Mol. Biol.* (Mosc.), 3, 690 (1969).
38. Privalov, Ptitsyn, and Birstein, *Biopolymers*, 8, 559 (1969).
39. Privalov, *Biofizika*, 13, 955 (1968).
40. Privalov and Tiktopulo, *Biopolymers*, 9, 127 (1970).
41. Privalov, Khechinashvili, and Atanasov, *Biopolymers*, 10, 1865 (1971).
42. Atanasov, Khechinashvili, and Privalov, *Mol. Biol.* (Mosc.), 6, 33 (1972).
43. Privalov, Tiktopulo, and Khechinashvili, *Int. J. Protein Peptide Res.*, 5, 229 (1973).
44. Privalov and Khechinashvili, *J. Mol. Biol.*, 86, 665 (1974).
45. Khcchinashvili, Privalov, and Tiktopulo, *FEBS Lett.*, 30, 57 (1973).
46. Tischenko, Tiltopulo, and Privalov, *Biofizika*, 19, 400 (1974).
47. Khechinashvili and Privalov, *Biofizika*, 19, 14 (1974).
48. Brandts, Jackson, and Ting, *Biochemistry*, 13, 3595 (1974).
49. Levy, Rialdi, and Biltonen, *Biochemistry*, 11, 4138 (1972).
50. Biltonen, Schwartz, and Wadsö, *Biochemistry*, 10, 3417 (1971).
51. Jackson and Brandts, *Biochemistry*, 9, 2294 (1970).
52. Shiao and Sturtevant, *Biopolymers*, 12, 1829 (1973).
53. Klump and Ackermann, *Biopolymers*, 10, 513 (1971).
54. Hinz, Haar, and Ackermann, *Biopolymers*, 9, 923 (1970).
55. Hearn, Richards, Sturtevant, and Watt, *Biochemistry*, 10, 806 (1971).
56. Scheffler and Sturtevant, *J. Mol. Biol.*, 42, 577 (1969).
57. Ross and Scruggs, *J. Mol. Biol.*, 45, 567 (1969).
58. Kagemoto and Karasz, *Analytical Calorimetry*. Vol. 2, Plenum Press, New York, 1970, 147.
59. Giacometti, Turolla, and Boni, *Biopolymers*, 9, 979 (1970).
60. Privalov, *Biofizika*, 3, 308 (1963).
61. Atha and Ackers, *J. Biol. Chem.*, 246, 5845 (1971).
62. Delben and Crescenzi, *Biochim. Biophys. Acta*, 194, 615 (1969).
63. Chou and Scheraga, *Biopolymers*, 10, 657 (1971).
64. Tsong, Hearn, Warthall, and Sturtevant, *Biochemistry*, 9, 2666 (1970).
65. Gerassimov and Mikhailov, *Soobshch. Akad. Nauk Gruz. SSSR*, 64, 185 (1971).
66. Sturtevant and Beres, *Biochemistry*, 10, 2120 (1971).
67. Shiao and Sturtevant, *Biochemistry*, 8, 4910 (1969).
68. Giacometti, Turolla, and Verdini, *J. Am. Chem. Soc.*, 93, 3092 (1971).
69. Rialdi and Profumo, *Biopolymers*, 6, 899 (1968).
70. Klump, *Biopolymers*, 11, 2331 (1972).
71. Bode, Schernau, and Ackermann, *Biophys. Chem.*, 1, 214 (1974).
72. Klump, in press.
73. Simon and Karasz, *Thermochim. Acta*, 8, 97 (1974).
74. McKnight and Karasz, *Thermochim. Acta*, 5, 339 (1973).
75. Delben, Crescenzi, and Quadrifoglio, *Int. J. Protein Peptide Res.*, 3, 57 (1971).
76. Crescenzi and Delben, *Int. J. Protein Peptide Res.*, 3, 57 (1971).
77. Stevens and Felsenfeld, *Biopolymers*, 2, 293 (1964).

FREE ENERGIES OF HYDROLYSIS AND DECARBOXYLATION

William P. Jencks

One of the reasons that there has been so much confusion and disagreement regarding the free energies of hydrolysis of "energy-rich" compounds of biochemical interest is that it is uncommon for any two workers to express their results according to the same nomenclature and conventions. The following summary may be helpful in making use of these tables of free energies of hydrolysis and decarboxylation.

The equilibrium constant, K_I, for the hydrolysis of an ester may be expressed according to the Equation 1 and the free energy of hydrolysis according to Equation 2, using the convention that the concentration of water is expressed in the same units as the other reactants and pure water is 55.5 M. For glycine ethyl ester the values of $K_I = 0.43$ and $\Delta G° = +500$ cal/mol at 39° reflect

$$K_I = \frac{[RCOOH][HOR']}{[RCOOR'][HOH]} \quad (1)$$

$$\Delta G_I° = -RT \ln K_I \quad (2)$$

the fact that −OH and −OC₂H₅ have approximately the same affinity for the carbonyl group.

For biochemical reactions, which usually take place in dilute aqueous solution, it is generally more convenient to take the activity of pure water as 1.0, and this convention will be adopted here. For glycine ethyl ester the values of $K_I = 24$ and $\Delta G_I° = -1,970$ cal/mol according to this convention reflect the fact that the driving force toward hydrolysis which results from the high concentration of water compared to the other reactants is hidden in the equilibrium expression by the convention that the activity of liquid water is 1.0. This extra driving force amounts to −RT in $55.5 \cong -2,400$ cal/mol and is one reason that free energies of hydrolysis expressed according to this convention are unlikely to be equal to heats of hydrolysis. The standard states of the other reactants according to this convention are ideal 1 M solutions of the non-ionized species. This commonly leads to difficulty for a compound such as glycine which does not exist in a non-ionized form in appreciable concentration and the standard state is commonly modified, as in the case of the values for glycine ethyl ester given here, to refer to a species in which the *reacting* groups are non-ionized; i.e., to H⁺₃ NCH₂COOH. instead of H₂NCH₂COOH. This convention, which we shall call convention I, gives a single value of $\Delta G°$ which is true regardless of the pH. Its use requires that only the actual concentrations (or activities) of the *particular ionic species* which are given in the equilibrium expression be included in calculations. For example, in order to calculate the free energy of hydrolysis of glycine ethyl ester from the results of an experiment carried out at pH 3.0, it is necessary to insert the concentration of H⁺₃ NCH₂COOH which is present at equilibrium at this pH, not the stoichiometric concentration of total glycine.

It is often convenient, especially when the ionization constants of the reactants are not accurately known, to use convention II in which the concentrations (or activities) of the reactants are given in terms of some convenient ionic species that is present under the conditions of the experiments and any hydrogen ions present in the equilibrium expression are included. For glycine ethyl ester this is shown in Equations 3 and 4 and the value of $\Delta G_{II}°$ is + 1,440 cal/mol. As in the case of convention I, this convention refers only to the concentrations of the particular

$$K_{II} = \frac{[RCOO^-][H^+][HOR']}{[RCOOR'][HOH]} \quad (3)$$

$$\Delta G_{II}° = -RT \ln K_{II} \quad (4)$$

ionic species given in the equilibrium expression which may be present in a given solution. The $G_{II}°$ value of $\Delta G_{II}°$ is independent of pH.

To interconvert these pH-independent free energies with free energies which hold for stoichiometric concentrations of reactants and products at a given pH, which would be found experimentally, it is only necessary to substitute in Equation 5 the actual concentrations present at the desired pH of the particular ionic species of the reactants which are present at that pH,

$$\Delta G' = \Delta G° + RT \ln \frac{[products]}{[reactants]} \quad (5)$$

including any hydrogen ions given off or taken up in the reaction. Equation 5 is the basic equation which relates concentrations (or activities) to free energies. When a reaction is at equilibrium $\Delta G' = 0$ and the standard free energy is then a logarithmic function of the *equilibrium* concentrations of the reactants and products (Equation 6). When the reactants and products are all in the standard state of activity 1.0 the concentration term drops out and the free energy of the system is equal to the standard free energy of the reaction (Equation 7).

$$\Delta G° = -RT \ln \frac{[C]_{eq}[D]_{eq}}{[A]_{eq}[B]_{eq}} = -RT \ln K \quad (6)$$

$$\Delta G' = \Delta G° + RT \ln \frac{1 \times 1}{1 \times 1} = \Delta G° \quad (7)$$

Thus, the standard free energy is the difference in free energy between a system in which all the reactants and products are in the standard states ($\Delta G' = \Delta G°$) and the same system at equilibrium ($\Delta G° = 0$).

A useful special case of Equation 5 is the situation in which the total concentrations of all reactants and products, except hydrogen ion, are 1.0 M at a given pH. This gives a value of $\Delta G°'$ which refers to 1.0 M total concentrations of all the ionic species of the reactants and products and is *true only at the specified pH*. This convention III is the convention which refers most directly to experimental results and is most useful in making comparisons of free energies of hydrolysis under physiological conditions. $\Delta G°'$

is also sometimes referred to as $\Delta G'$, ΔG°_{anal} ΔG°_{exp} and (unfortunately) as ΔG°.

For example, the value of $\Delta G^{\circ'}$ for glycine ethyl ester is calculated from the ΔG° of convention I by inserting the fraction of the 1 M total glycine that is present as the free acid at pH 7.0, as shown in Equation 8.

$$\Delta G^{\circ'} = \Delta G^{\circ}_I + RT \ln \frac{[RCOOH][HOR']}{[RCOOR'][HOH]}$$

$$= +500 + 1,420 \log \frac{(1 \times 10^{-4,-6})(1)}{(0.85)(1)} \qquad (8)$$

$$= -8,400 \text{ cal/mol}$$

The fact that glycine ethyl ester is only 85% in the protonated form at pH 7.0 introduces a further small correction. The same value of $\Delta G^{\circ'}$ may be calculated from the ΔG° of convention II by inserting the hydrogen ion activity at pH 7.0 into the equilibrium expression (Equation 9), because at pH 7 glycine is entirely in the form of the carboxylate anion, which is the form which is used in the equilibrium expression according to this convention.

$$\Delta G^{\circ'} = \Delta G^{\circ}_{II} + RT \ln \frac{[RCOOH^-][H^+][HOR']}{[RCOOR'][HOH]}$$

$$= 1,440 + 1,420 \log \frac{(1)(10^{-7})(1)}{(0.85)(1)} \qquad (9)$$

$$= -8400 \text{ cal/mol}$$

These interconversions between pH-independent and pH-dependent free energies may generally be carried out without difficulty if the following two simple rules are followed:

1. The equilibrium expression for the pH-independent equilibrium constant and free energy of hydrolysis may include any desired ionic species of the reactants, but must be based on a *balanced equation* for the reaction which includes any *hydrogen ions* which are given off or taken up.
2. The actual concentrations (or activities) of the *particular ionic species* given in this expression and which are present at a given pH value must be substituted in the expression for the pH-independent equilibrium constant or free energy.

A final convention gives the molar free energy of hydrolysis under conditions in which the reactants are at concentrations other than 1.0 M. For glycine ethyl ester at pH 7.0 under conditions in which the reactants and products, except for water and hydrogen ion, are present at a concentration of $10^{-3} M$, the free energy of hydrolysis is –12,660 cal/mol, as shown in Equation 10. The importance of specifying and understanding the particular convention that is being used is illustrated by the range of values from +1,440 to –12,660 calories/mole for the free energy of hydrolysis of glycine ethyl ester according to the different conventions. At pH 7 and under physiological conditions, glycine ethyl ester has a free energy of hydrolysis which clearly places it in the category of "high-energy" or "energy-rich" compounds. All of these conventions are correct and are useful for different purposes.

$$\Delta G' = \Delta G^{\circ'}_{pH7} + RT \ln \frac{[gly]_{tot}[HOEt]_{tot}}{[glyOEt]_{tot}[HOH]}$$

$$= -8,400 + 1,420 \log \frac{(10^{-3})(10^{-3})}{(10^{-3})(1)} \qquad (10)$$

$$-12,660 \text{ cal/mol}$$

Complexation of the compounds which are involved in an equilibrium with other compounds which may be present in the solution is a common cause of difficulty in the determination of equilibria and free energies of hydrolysis. The most important example of this in biochemical reactions is the binding of magnesium and other ions to phosphate and polyphosphates. This problem may be dealt with in several ways:

1. The reaction may be carried out under conditions in which the complexing ions are present in negligible concentrations compared to the compounds involved in the equilibrium under study.
2. The concentrations of the free and complexed species of the reactants and ions may be calculated from equilibrium constants for complex formation, which must be determined in separate experiments or be obtained from the literature. There is still some disagreement in the literature as to the correct values for these complexing constants for many compounds and ions of biochemical importance.
3. The reaction may be carried out in the presence of an excess of ions under conditions in which most of the reactants exist in the form of the complex. The equilibrium constant and free energy are then obtained for this particular set of experimental conditions or may be expressed in terms of reactions of the complexed species. For example, the affinity of Mg^{2+} toward ATP^{4-} and toward ATP^{3-} is much larger than that toward HPO_4^{2-}, so that the equilibrium in the presence of excess Mg^{2+} may be expressed according to Equation 11. Most of the free energies of hydrolysis of polyphosphate compounds which are given in these tables refer to conditions in which Mg^{2+} is present in excess and most or all of the polyphosphates exist as the magnesium complexes. There is still no general agreement regarding these values.

$$K = \frac{[Mg \cdot ADP^-][HPO_4^=][H^+]}{[Mg \cdot ATP^=][H_2O]} \qquad (11)$$

The free energies in these tables generally refer to concentrations rather than activities of the reactants. Thermodynamic values extrapolated to zero ionic strength are of theoretical interest, but have not often been obtained for reactions of biochemical importance.

It is worth noting that these equilibria refer only to aqueous solutions, and that a large fraction of a cell or cell particle is not aqueous. The perturbation of equilibria that may occur in nonaqueous systems is illustrated by the fact that esters of long chain fatty acids can be formed at equilibrium in a nonaqueous phase which is in contact with neutral buffer, although the equilibrium in the buffer solution is far toward hydrolysis.

It cannot be pointed out too often that thermodynamic measurements and conventions say nothing about the *pathway* by which a reaction takes place; i.e., the equilibrium state of a system under a given set of experimental conditions is the same regardless of the pathway by which equilibrium is attained. Thus, the common practice of calling a particular ionic species the "reactive" species from an observed change in the stoichiometric equilibrium position of a reaction with changing pH is incorrect. It is this independence of reaction pathway that makes it equally legitimate to specify the equilibrium constant of a reaction according to any of a number of equations which contain different ionic species of reactants and products; the only requirement is that the equations balance. Different equilibrium constants will be obtained, of course, from the different equations.

Further information regarding the methods for dealing with free energies of hydrolysis may be found in References 1–4. Carpenter[5] has prepared a useful summary of the dependence on pH of $\Delta G^{\circ\prime}$ for several classes of compounds of biochemical interest.

FREE ENERGIES OF HYDROLYSIS OF ESTERS OF ACETIC ACID AND RELATED COMPOUNDS AT 25°

Compound	$-\Delta G^{\circ a}$	$-G^{\circ\prime}_{PH7}{}^b$	Reference
Acetic anhydride	15,700	21,800	6
p-Nitrophenyl acetate	9,430	13,010	3
m-Nitrophenyl acetate	8,550	11,610	3
p-Chlorophenyl acetate	7,590	10,650	3
Phenyl acetate	7,390	10,450	3
p-Methylphenyl acetate	6,890	9,950	3
p-Methoxyphenyl acetate	6,590	9,650	3
Acetyl hypochlorite	ca. 5,950	ca. 9,214	13, 14[f]
Acetyl phosphate	6,690[c]	10,300	7, 3
N,O-Diacetyl-N-methylhydroxylamine[d]	6,190	9,250	3
4-Pyridinealdoxime acetate[c] (37°)	5,670	8,730	8
Glycine ethyl ester (39°)	1,970	8,400	9
Valyl RNA (30°)	2,000[g]	8,400[g]	10
Trifluoroethyl acetate	4,970	8,030	4
Acetylcarnitine (35°)	4,150	7,210	4, 11
Acetylcholine	2,940	6,000	4
Chloroethyl acetate	2,840	5,900	4
Methoxyethyl acetate	2,180	5,240	4
Ethyl acetate	1,660	4,720	4

Compiled by William P. Jencks.

[a] Standard free energy of hydrolysis based on a standard state of 1 *M* concentrations of the *uncharged* reactants and products and an activity of pure water of 1.0 (convention I).
[b] Standard free energy of hydrolysis at pH 7.0 based on a standard state of 1 *M* total stoichiometric concentration of reactants and products, except hydrogen ion, and on an activity of pure water of 1.0 (convention III). Values for derivatives of acetic acid are based on a thermodynamic pK$_a$ of 4.76 for acetic acid and a ΔG for ionization of acetic acid at pH 7.0 of 3,060 cal/mol. Values for $\Delta G^{\circ\prime}_{pH7}{}^b$ for acetate derivatives based on a pK$'_a$ of 4.63 ± 0.02 at ionic strength 0.2 to 1.0[12] are 180 cal/mol more negative.
[c] For the dianions of acetyl phosphate and phosphate.
[d] For hydrolysis of the ester.
[e] Based on (closely similar) equilibrium constants with several thiol esters and the ΔG° for N,S-dicetyl-β-mercaptoethylamine.
[f] From the data of De la Mare in acetic acid containing traces of water[13] and an ionization constant of 4.1×10^{-8} for hypochlorous acid.[14]
[g] Based on Reference 10. $\Delta G^{\circ\prime}_{pH7.0}$ =−7,700 for ATP (→ PP and AMP), and pK$_a$ valine = 2.32.

FREE ENERGIES OF HYDROLYSIS OF THIOL ESTERS

Compound	$-\Delta G^{\circ a}$	$-G^{\circ\prime}_{PH7}{}^b$	Reference
N,S-Diacetyl-β-mercaptoethylamine	4,460	7,520	4
S-Acetylmercaptoacetate	4,140	7,200	4
S-Acetyhnercaptopropanol	4,400	7,460	9
2-Diethylaminoethane thioacetate	—	7,470[c]	8
S-Acetylthiophenol	—	7,450[c]	8
2-Di*iso*sopropylaminoethane thioacetate	—	6,720[c]	8
S-Acetylglutathione	—	7,500[c]	8
	—	7,830[d]	7
Acetyl coenzyme A,	—[c]	7,520[e]	4
pH 7.2	—	7,100[f]	15
	—	7,370[g]	7, 16

Compiled by William P. Jencks.

[a] Standard free energy of hydrolysis based on a standard state of 1 *M* concentrations of the *uncharged* reactants and products and an activity of pure water of 1.0 (convention I).
[b] Standard free energy of hydrolysis at pH 7.0 based on a standard state of 1 *M* total stoichiometric concentration of reactants and products, except hydrogen ion, and on an activity of pure water of 1.0 (convention III). Values for derivatives of acetic acid are based on a thermodynamic pK$_a$ of 4.76 for acetic acid and a ΔG for ionization of acetic acid at pH 7.0 of 3,060 cal/mol. Values for $\Delta G^{\circ\prime}_{pH7}{}^b$ for acetate derivatives based on a pK$'_a$ of 4.63 ± 0.02 at ionic strength 0.2 to 1.0[12] are 180 cal/mol more negative.
[c] Based on Reference 8 and 4-pyridinealdoxime acetate (see previous table).
[d] Based on Reference 7 and acetylimidazole (see following table).
[e] Based on N,S-diacetyl-β-mercaptoethylamine.
[f] Based on the equilibria for the condensation of acetate and acetyl coenzyme A with oxaloacetate to give citrate with the correction for citrate ionization recalculated as described in the text above.
[g] Based on equilibria with acetyl phosphate and acetylimidazole.[7,16]

FREE ENERGIES OF HYDROLYSIS OF AMIDES

Compound	$-\Delta G^{\circ\prime}_{PH7}{}^a$	Reference
Acetylimidazole	12,970	3
10-Formyltetrahydrofolic acid (pH 7.7, 37°)	5,830[b]	17
Asparagine	3,600	18
Glutamine	3,400	18
N-Dimethylpropionamide	2,100[c]	19
Propionamide	2,100[c]	19
Hippurylanilide (pH 5.0, 39°)	1,470	20
Benzoyltyrosyl-glycylanilide (pH 6.5, 23°)	1,360	21
Benzoyltyrosyl-glycinamide	420[d]	22
N-Acetyltyrosine hydroxamic acid	1,870	23
Acetohydroxamic acid	−200	23
N-Methylpropionamide	−300[c]	19

Compiled by William P. Jencks.

[a] Standard free energy of hydrolysis at pH 7.0 based on a standard state of 1 *M* total stoichiometric concentration of reactants and products, except hydrogen ion, and on an activity of pure water of 1.0 (convention III).
[b] Based on Reference 17 and $\Delta G^{\circ\prime}_{pH7}$ 7.7 = −8,030 cal/mol for ATP.
[c] Data for other simple amides at elevated temperatures are given in Reference 19.
[d] For cleavage of the peptide bond, to fully ionized products.

FREE ENERGIES OF HYDROLYSIS OF PHOSPHATES[a]

Compound	$-\Delta G^{\circ\prime b}$	Reference
Phosphoenolpyruvate, pH 7.0	14,800	24
pH 7.4–8.4	12,800[b]	2
β-Aspartyl phosphate, pH 8.0, 15°	13,000[c]	25
Carbamyl phosphate, pH 9.5	ca. 12,300[d]	26
3-Phosphoglyceroyl phosphate, pH 6.9	11,800	2
Acetyl phosphate, pH 7.0	10,300	7, 3
Creatine phosphate, pH 7.0, 7.5, 37°	10,300	27
Phosphoarginine, pH 8.0, excess Mg^{++}	7,700[e]	28
Uridine diphosphate glucose (glycoside cleavage), pH 7.6	7,300	29
Adenosine triphosphate ($\rightarrow$ AMP, PP), pH 7.0, excess Mg^{++}	7,700[f]	4, 9
pH 7.5, excess Mg^{++}	10,300[g]	30
Adenosine triphosphate ($\rightarrow$ ADP, Pi), 37°, pH 7.0, excess Mg^{++}	7,300	31–33
25°, pH 7.4, $10^{-3}M$ Mg^{++}	8,800	44
25°, pH 7.4, 0 Mg^{++}	9,600	44
Pyrophosphate, pH 7.0	8,000	24
pH 7.0, 0.005 M Mg^{++}	4,500	24
Cytidine-2′-3′-phosphate (cyclic)$\rightarrow$ 3′-phosphate, pH 7.0	5,000	34
Glucose-1-phosphate, 25°, pH 7.0	5,000	35
N-Acetylethanolamine phosphate, pH 7.0	2,900	36
Glucose-6-phosphate, 25°, pH 7.0	3,300	35
α-Glycerophosphate, 38°, pH 8.5	2,200	37
pH 5.8	2,600	3
Hexose-6-phosphates, 38°, pH 8.5	2,800 ± 200	37
pH 5.8	3,200 ± 200	37

Compiled by William P. Jencks.

[a] For a more detailed compilation, see Reference 2.

[b] Standard free energy of hydrolysis based on a standard state of 1 M total stoichiometric concentration of reactants and products, except hydrogen ion, and on an activity of pure water of 1.0 (convention III).

[c] Based on Reference 25 and $-\Delta G^{\circ\prime}_{pH\,8.0} = -8,400$ cal/mol for ATP.

[d] Based on Reference 26 and $\Delta G^{\circ\prime}_{pH\,9.5} = -10,440$ cal/mol for ATP.

[e] Based on Reference 28 and $\Delta G^{\circ\prime}_{pH\,8.0} = -8,400$ cal/mol for ATP.

[f] Based on acetyl coenzyme A and the ATP-activated synthesis of acetyl coenzyme A.

[g] Based on a series of equilibria and the assumption of the similarity of the terminal phosphates of ATP and ADP. This value is not consistent with the preceding estimate.

FREE ENERGIES OF HYDROLYSIS OF GLYCOSIDES

Compound	$-\Delta G^{\circ\prime}_{PH7}{}^{a}$	Reference
Uridine diphosphoglucose (pH 7.6)	7,300[b,c,d]	29
Sucrose	7,000[b]	38
Levan (fructofuranoside 2-6-fructose)	5,000[b]	39
Glucose-1-phosphate	5,000	35
Maltose	4,000	39
Glycogen	4,000	39
Amylose (α(l-4)glucosidic)	3,400	39

Compiled by William P. Jencks.

[a] Standard free energy of hydrolysis at pH 7.0 based on a standard state of 1 M total stoichiometric concentration of reactants and products, except hydrogen ion, and on an activity of pure water of 1.0 (convention III).

[b] Based on a corrected value for sucrose hydrolysis in Reference 38.

[c] Recalculated from K = 1.6, $\Delta G^{\circ\prime} = -300$ from the data of Reference 29.

[d] The values for thymidine diphosphoglucose and adenosine diphosphoglucose are very similar.[45,46]

FREE ENERGIES OF DECARBOXYLATION

Compound	$-\Delta G^{\circ a}$	Reference
Oxaloacetate$^=$ $\leftrightarrows$ Pyruvate$^-$ + HCO_3^-	6,200	24
Methylmalonyl-CoA $\leftrightarrows$ Propionyl-CoA + HCO_3^-		
30°	6,200[h]	24,40
28°	7,510[c]	41
Enzyme-biotin-CO_2 $\leftrightarrows$ Enzyrne-biotin + HCO_3^- (pH 7.0, 0°)	4,700	43

Compiled by William P. Jencks.

[a] Based on the indicated ionic species of reactants and products. Units are cal/mole.

[b] Based on oxaloacetate decarboxylation and K for carbon dioxide transfer; K = [pyruvate] [D-methyl-malonyl-CoA] / [propionyl-CoA] =1.0.[47]

[c] Based on ATP-coupled carboxylation and $-\Delta G^{\circ\prime}_{pH8.1} = -8.540$ for ATP hydrolysis. The corresponding value at 37° from the data of Reference 42 is -11.500 cal/mol.[42]

References

1. Johnson, in *The Enzymes*, Vol. 3, 2nd ed., Boyer, Lardy, Myrbäck, Eds., Academic, New York, 1960, chap, 21, 407.
2. Atkinson and Morton, in *Comparative Biochemistry*, Vol. 2,. Florkin and Mason, Eds., Academic, New York, 1960, chap. 1, 1.
3. Gerstein and Jeacks, *J. Am. Chem. Soc.*, 86, 4655 (1964).
4. Jencks and Gilchrist, *J. Am. Chem. Soc.*, 86, 4651 (1964).
5. Carpenter, *J. Am. Chem. Soc.*, 82, 1111 (1960).
6. Jencks, Barley, Barnett, and Gilchrist, *J. Am. Chem. Soc.*, 88, 4464 (1966).
7. Stadtman, in *The Mechanism of Enzyme Action*, McElroy and Glass, Eds., Johns Hopkins, Baltimore, 1954, 581
8. O'Neill, Kohl, and Epstein, *Biochem. Pharm.*, 8, 399 (1961).
9. Jencks, Cordes, and Carriuolo, *J. Biol. Chem.*, 235, 3608 (1960).
10. Berg, Bergmann, Ofengand, and Dieckmann, *J. Biol. Chem.*, 236, 1726 (1961).
11. Fritz, Schultz, and Srere, *J. Biol. Chem.*, 238, 2509 (1963).
12. Bjerrum, Schwarzenbach, and Sillén, Stability Constants. Chemical Society, London, 1957.
13. De la Mare, Hilton, and Vernon, *J. Chem. Soc.*, 4039 (1960).
14. Mauger and Soper, *J. Chem. Soc.*, p. 71 (1946).
15. Tate and Datta, *Biochem. J.*, 94, 470 (1965).
16. Sly and Stadtman, *J. Biol. Chem.*, 238, 2639 (1963).
17. Himes and Rabinowitz, *J. Biol. Chem.*, 237, 2903 (1962).
18. Benzinger, Kitzinger, Hems, and Burton, *Biochem. J.*, 71, 400 (1959).
19. Morawetz and Otaki, *J. Am. Chem. Soc.*, 85, 463 (1963).
20. Carty and Kirschenbaum, *Biochim. Biophys. Acta*, 110, 399 (1965).
21. Gawron, Glaid, Boyle, and Odstrchel, *Arch. Biochem. Biophys.*, 95, 203 (1961).
22. Dobry, Fruton, and Sturtevant, *J. Biol. Chem.*, 195, 149 (1952).
23. Jencks, Caplow, Gilchrist, and Kallen, *Biochemistry*, 1313 (1963).
24. Wood, Davis, and Lochmüller, *J. Biol. Chem.*, 241, 5692 (1966).
25. Black and Wright, *J. Biol. Chem.*, 213, 27 (1955).
26. Jones and Lipmann, *Proc. Natl. Acad. Sci. U.S.A.*, 46, 1194 (1960).
27. Kuby and Noltmann in *The Enzymes* Vol. 4, 2nd. ed., Boyer, Lardy and Myrbäck, Eds., Academic, New York, 1962, chap 31, 515
28. Uhr, Marcus, and Morrison, *J. Biol. Chem.*, 241, 5428 (1966).
29. Avigad, *J. Biol. Chem.*, 239, 3613 (1964).
30. Schuegraf, Ratner, and Warner, *J. Biol. Chem.*, 235, 3597 (1960).
31. Atkinson, Johnson, and Morton, *Nature*, 184, 1925 (1959).
32. Robbins and Boyer, *J. Biol. Chem.*, 224, 121 (1957).
33. Benzinger, Kitzinger, Hems, and Burton, *Biochem. J.*, 71, 400 (1959).
34. Bahr, Cathou, and Hammes, *J. Biol. Chem.*, 240, 3372 (1965).
35. Atkinson, Johnson, and Morton, *Biochem. J.*, 79, 12 (1961).

36. Dayan and Wilson, *Biochim. Biophys. Acta*, 77, 446 (1963).

37. Meyerhof and Green, *J. Biol. Chem.*, 178, 655 (1949).

38. Neufeld and Hassid, *Adv Carbohyd. Chem.*, 18, 309 (1963) (footnote 166, p.329).

39. Dedonder, *Ann. Rev. Biochem.*, 30, 347 (1961).

40. Wood and Stjernholm, *Proc. Natl. Acad. Sci. U.S.A.*, 47, 289 (1961).

41. Kaziro, Grossman, and Ochoa, *J. Biol. Chem.*, 240, 64 (1965).

42. Halenz, Feng, Hegre, and Lane, *J. Biol. Chem.*, 237, 2140 (1962).

43. Wood, Lochmüller, Riepertinger, and Lynen, *Biochem. Z.*, 337, 247 (1963).

44 Alberty, R. A., personal communication.

45. Avigad and Milnetr, *Meth. Enzymol.*, 8, 341 (1966).

46. Murata, Sugiyama, Minimikawa, and Akazawa, *Arch. Biochem. Biophys.*, 113, 34 (1966).

47. Wood, H., personal communication.

This section originally appeared in Sober, Ed., *Handbook of Biochemistry and selected data for Molecular Biology*, 2nd ed., Chemical Rubber Co., Cleveland, 1970.

DECI-NORMAL SOLUTIONS OF OXIDATION AND REDUCTION REAGENTS

Atomic and molecular weights in the following table are based upon the 1965 atomic weight scale and the isotope C-12. The weight in grams of the compound in 1 cc of the following deci-normal solutions is found by dividing the H equivalent in the last column by 1,000.

Name	Formula	Atomic or Molecular Weight	Hydrogen Equivalent	0.1 Hydrogen Equivalent in g
Antimony	Sb	121.75	$\frac{1}{2}$ Sb	6.0875
Arsenic	As	74.9216	$\frac{1}{2}$ As	3.7461
Arsenic trisulfide	As_2S_3	246.0352	$\frac{1}{2}$ As_2S_3	6.1509
Arsenous oxide	As_2O_3	197.8414	$\frac{1}{4}$ As_2O_3	4.9460
Barium peroxide	BaO_2	169.3388	$\frac{1}{2}$ BaO_2	8.4669
Barium peroxide hydrate	$BaO_2 \cdot 8H_2O$	313.4615	$\frac{1}{2}$ $BaO_2\,8H_2O$	15.6730
Calcium	Ca	40.08	$\frac{1}{2}$ Ca	2.004
Calcium carbonate	$CaCO_3$	100.0894	$\frac{1}{2}$ $CaCO_3$	5.0045
Calcium hypochlorite	$Ca(OCl)_2$	142.9848	$\frac{1}{4}$ $Ca(OCl)_2$	3.5746
Calcium oxide	CaO	56.0794	$\frac{1}{2}$ CaO	2.8040
Chlorine	Cl	35.453	Cl	3.5453
Chromium trioxide	CrO_3	99.9942	$\frac{1}{3}$ CrO_3	3.3331
Ferrous ammonium sulfate	$FeSO_4(NH_4)SO_4 \cdot 6H_2O$	392.0764	$FeSO_4(NH_4)_2SO_4 \cdot 6H_2O$	39.2076
Hydroferrocyanic acid	$H_4Fe(CN)_6$	215.9860	$H_4Fe(CN)_6$	21.5986
Hydrogen peroxide	H_2O_2	34.0147	$\frac{1}{2}$ H_2O_2	1.7007
Hydrogen sulfide	H_2S	34.0799	$\frac{1}{2}$ H_2S	1.7040
Iodine	I	126.9044	I	12.6904
Iron	Fe	55.847	Fe	5.5847
Iron oxide (ferrous)	FeO	71.8464	FeO	7.1846
Iron oxide (ferric)	Fe_2O_3	159.6922	$\frac{1}{2}$ Fe_2O_3	7.9846
Lead peroxide	PbO_2	239.1888	$\frac{1}{2}$ PbO_2	11.9594
Manganese dioxide	MnO_2	86.9368	$\frac{1}{2}$ MnO_2	4.3468
Nitric acid	HNO_3	63.0129	$\frac{1}{3}$ HNO_3	2.1004
Nitrogen trioxide	N_2O_3	76.0116	$\frac{1}{4}$ N_2O_3	1.9002
Nitrogen pentoxide	N_2O_5	108.0104	$\frac{1}{5}$ N_2O_5	1.8001
Oxalic acid	$C_2H_2O_4$	90.0358	$\frac{1}{2}$ $C_2H_2O_4$	4.5018
Oxalic acid hydrate	$C_2H_2O_4 \cdot 2H_2O$	126.0665	$\frac{1}{2}$ $C_2H_2O_4 \cdot 2H_2O$	6.3033
Oxygen	O	15.9994	$\frac{1}{2}$ O	0.8000
Potassium dichromate	$K_2Cr_2O_7$	294.1918	$\frac{1}{6}$ $K_2Cr_2O_7$	4.9032
Potassium chlorate	$KClO_3$	122.5532	$\frac{1}{6}$ $KClO_3$	2.0425
Potassium chromate	K_2CrO_4	194.1076	$\frac{1}{3}$ K_2CrO_4	6.4733
Potassium ferrocyanide	$K_4Fe(CN)_6$	368.3621	$K_4Fe(CN)_6$	36.8362
Potassium ferrocyanide	$K_4Fe(CN)_6 \cdot 3H_2O$	422.4081	$K_4Fe(CN)_6 \cdot 3H_2O$	42.2408
Potassium iodide	KI	166.0064	KI	16.6006
Potassium nitrate	KNO_3	101.1069	$\frac{1}{3}$ KNO_3	3.3702
Potassium perchlorate	$KClO_4$	138.5526	$\frac{1}{8}$ $KClO_4$	1.7319
Potassium permanganate	$KMnO_4$	158.0376	$\frac{1}{5}$ $KMnO_4$	3.1608
Sodium chlorate	$NaClO_3$	106.4410	$\frac{1}{6}$ $NaClO_3$	1.7740
Sodium nitrate	$NaNO_3$	84.9947	$\frac{1}{3}$ $NaNO_3$	2.8332
Sodium thiosulfate	$Na_2S_2O_3 \cdot 5H_2O$	248.1825	$Na_2S_2O_3 \cdot 5H_2O$	24.8183

DECI-NORMAL SOLUTIONS OF OXIDATION AND REDUCTION REAGENTS (Continued)

Name	Formula	Atomic or Molecular Weight	Hydrogen Equivalent	0.1 Hydrogen Equivalent in g
Stannous chloride	$SnCl_2$	189.5960	$\frac{1}{2} SnCl_2$	9.4798
Stannous oxide	SnO	134.6894	$\frac{1}{2} SnO$	6.7345
Sulfur dioxide	SO_2	64.0628	$\frac{1}{2} SO_2$	3.2031
Tin	Sn	118.69	$\frac{1}{2} Sn$	5.935

This table originally appeared in Sober, Ed., *Handbook of Biochemistry and Selected Data for Molecular Biology*, 2nd ed., Chemical Rubber Co., Cleveland, 1970.

GUIDELINES FOR POTENTIOMETRIC MEASUREMENTS IN SUSPENSIONS PART A. THE SUSPENSION EFFECT

(IUPAC Technical Report)

Srecko F. Oman[1,‡], M. Filomena Camões[2], Kipton J. Powell[3], Raj Rajagopalan[4], and Petra Spitzer[5]

[1]*Faculty of Chemistry and Chemical Technology, University of Ljubljana, Aškerčeva 5,1000 Ljubljana, Slovenia;*
[2]*Departamento de Química e Bioquímica, University of Lisbon (CECUL/DQB), Faculdade de Sciências da Universidade de Lisboa, Edifício C8, Pt-1749-016, Lisboa, Portugal;*
[3]*Department of Chemistry, University of Canterbury, Christchurch, New Zealand;*
[4]*Department of Chemical and Biomolecular Engineering, National University of Singapore, 117576, The Republic of Singapore;*
[5]*Physikalisch-Technische Bundesanstalt (PTB), Postfach 3345, D-38023, Braunschweig, Germany*

*Membership of the Analytical Chemistry Division during the final preparation of this report:

President: R. Lobinski (France); *Titular Members:* K. J. Powell (New Zealand); A. Fajgelj (Slovenia); R. M. Smith (UK); M. Bonardi (Italy); P. De Bièvre (Belgium); B. Hibbert (Australia); J.-Å. Jönsson (Sweden); J. Labuda (Slovakia); W. Lund (Norway); *Associate Members:* Z. Chai (China); H. Gamsjäger (Austria); U. Karst (Germany); D. W. Kutner (Poland); P. Minkkinen (Finland); K. Murray (USA); *National Representatives:* C. Balarew (Bulgaria); E. Dominguez (Spain); S. Kocaoba (Turkey); Z. Mester (Canada); B. Spivakov (Russia); W. Wang (China); E. Zagatto (Brazil); *Provisional Member:* N. Torto (Botswana).

‡Corresponding author: E-mail: srecko.oman@fkkt.uni-lj.si

Abstract: An explanation of the origin and interpretation of the suspension effect (SE) is presented in accordance with "pH Measurement: IUPAC Recommendations 2002" [*Pure Appl. Chem.* **74**, 2169 (2002)]. It is based on an analysis of detailed schemes of suspension potentiometric cells and confirmed with experimental results. Historically, the term "suspension effect" evolved during attempts to determine electrochemically the thermodynamically defined activity of H+ (aq) in suspensions. The experimental SE arises also in determining other pIon values, analogous to pH values.

The SE relates to the observation that for the potential generated when a pair of electrodes (e.g., reference electrode, RE, and glass electrode) is placed in a suspension, the measured cell voltage is different from that measured when they are both placed in the separate equilibrium solution (eqs). The SE is defined here as the sum of: (1) the difference between the mixed potential of the indicator electrode (IE) in a suspension and the IE potential placed in the separated eqs; and (2) the anomalous liquid junction potential of the RE placed in the suspension. It is not the consequence of a boundary potential between the sediment and its eqs in the suspension potentiometric cells as is stated in the current definition of the SE.

Keywords: operational definition of suspension effect; suspension effect; pH; suspension potentiometric cell; IUPAC Analytical Chemistry Division; pIon; boundary potential; mixed potential; soil pH; anomalous liquid junction potential.

1. The suspension effect explained on the basis of analysis of potentiometric cells

1.1 Introduction

Potentiometry is an electroanalytical technique based on the measurement of the potential of an electrochemical cell, composed of a measuring and a reference electrode (RE), both immersed in the measuring solution to be measured.

In homogeneous solutions, direct potentiometry is used for the estimation of ion activities (e.g., pH) and potentiometric titrations for determination of the amount concentration of ionic species. These measuring techniques are also applied to suspensions or sols of different materials (containing positively or negatively charged particles) in aqueous dispersion media. Although pH measurement in soil suspensions is highly relevant to this work, ion exchanger suspensions were chosen preferentially as models due to their simplicity.

The most frequently applied direct potentiometric method is the measurement of pH. Therefore, the determination of pH is selected to explain the essential procedures and experimental set-up for the potentiometric techniques applied to suspensions or sols.

The most recent definitions, procedures, and terminology relating to pH measurements in dilute aqueous solutions in the temperature range 5–50 °C are given in the IUPAC Recommendations 2002 [1]. In this reference, the glass electrode cell V is proposed for practical pH measurements [1, p. 2187]:

$$\text{reference electrode} \,|\, \text{KCl} \,(c \geq 3.5 \text{ mol dm}^{-3})$$
$$\text{solution pH(X)} \,|\, \text{glass electrode (cell V)}$$

Typically, the galvanic cells used for practical pH measurements conform to the characteristics of cell V; therefore, the results obtained in these practical pH measurements approximate results obtained by cell V.

Although "the quantity pH is intended to be a measure of the activity of hydrogen ions in [homogeneous] solutions" [1], and measurements using cell V include an unknown liquid junction potential, cell V is also used for practical measurement of pH in suspensions with the electrodes usually positioned in different phases.

When pH is measured in (i) a suspension (or its sediment) or (ii) in its equilibrium solution (eqs), the measured pH value in

Reproduced from:
Pure & Appl. Chem., Vol. 79, No.1, pp. 67–79, 2007.
doi: 10.1351/pac200779010067

each of these constituent parts is different, even though the total system is in equilibrium. None of these pH values represents the (thermodynamically) true H^+ activity in a suspension. This observation has caused serious problems for the theory and practice of pH measurements, problems which remain unresolved.

1.2 Consideration of the "glass electrode cell" containing a suspension

The term "suspension" should mean a *uniform equilibrated multiphase system*. It can be separated into the eqs and the sediment. If, when separated, the eqs and the sediment remain in physical and electrical contact, they represent a *combined suspension system*. The separated supernatant, obtained by sedimentation, centrifugation, or filtration, does not necessarily give absolutely equivalent solutions, yet they can be considered eqs, because the differences between them can be neglected with respect to the characteristics of the measured values.

In this work, the term "suspension" means the dispersion of electrically charged solid particles in water or in an aqueous solution. The origin of the charges can be adsorption or ionization, or as a property of the ion exchanger beads. The positively or negatively charged particles of different sizes found in soils provide another example. However, for this document (and in the literature that it relies upon) ion exchanger particles (which, depending on solution pH lower than 7 will be mostly in the H^+ form) were chosen as a representative example for the study of pH measurements in suspension. These particles reduce the experimental effort and make a simple approach possible. Experiments showed essentially the same results when particles of other types were used [2].

The bulk liquid (of any electrolyte concentration) in the suspension will be identical to the supernatant of this suspension, when it is separated in whatever manner into two parts. It is different from that in the diffusion layers of individual particles, which are responsible for the mixed electrode potential when they are in contact with the glass electrode part of the pH electrode (Section 1.4). The diffusion layer of individual particles contributes to the anomalous liquid junction potential observed in pH measurements in suspensions.

The two separated parts are (i) the *sediment*, which can be considered the most concentrated suspension possible, and (ii) the clear (homogeneous, non-turbid) solution above it. This solution is called the eqs, if the suspension is equilibrated before separation. It is proposed to call the combination of an eqs and sediment, which are in physical and electrical contact, a *combined suspension system*.

For pH measurement in a combined suspension system, the following specific positions for the glass and REs are possible:

1. both electrodes are positioned in the eqs;
2. both electrodes are in the sediment;
3. the glass electrode is in the sediment and the RE is in the eqs; or
4. the electrodes are in the reverse position from that in 3.
   In addition, it is possible to measure the pH in each *separated suspension component*, which means that:
5. both electrodes are in the separated "eqs" or
6. both electrodes are in the separated sediment.
   The pH measurement is possible also with
7. both electrodes in the original, (nonseparated) equilibrated suspension, the concentration of which should (ideally) be practically constant during the measurement.

In a suspension in equilibrium, the electrochemical potential $\tilde{\mu}_{H^+}$ is equal throughout the system, therefore, the different electrode arrangements 1 to 7 could be expected to give the same pH values. Nevertheless, the electrode combinations 1 and 2 and the analogous pair 5 and 6 show large (and nearly equal) pH differences, as do the combinations 3 and 4. These pH differences were named the *suspension effect* (SE) for the first time in 1930 [5,6]. Subsequently, the nature of this effect has been studied intensively by many authors; a list of references may be found in reviews, e.g., [4,7–9]. However, there has been no consensus on the origin of, or explanation for, the SE.

An acceptable explanation of this phenomenon follows from a detailed analysis of the suspension cells (combinations 1 and 2, or 5 and 6) and from their cell potential differences, ΔE, from which the corresponding pH differences can be calculated; this explanation is supported by recent experimental observations [3].

1.3 Detailed schemes for potentiometric cells used in suspensions

Scheme 1 shows both electrodes in the eqs (combined with sediment, system 1, or separated, system 5).

$$[A] \quad Ag \mid AgCl \mid KCl \ \vdots \ \mid KCl \ \vdots \quad eqs\ (H^+) \ \blacksquare \ HCl \mid AgCl \mid Ag \qquad E_A$$
$$E_{ref1} \qquad E_j(S_A) \quad E_j(A) \qquad E_g(A) \qquad E_{ref2} \quad (\equiv E_{soln})$$

Scheme 1

Here, S indicates the separator (membrane) of the salt bridge and G the glass membrane of the glass electrode. The symbol $\vdots$ represents the region of the KCl solution in the separator and that in the contact range of eqs with KCl solution diffused from the RE. (The extent of the KCl layer in eqs is exaggerated in the scheme.) E_{ref} is the potential of the RE, E_g is the potential of the glass electrode. $E_j(S_A)$ and $E_j(A)$ are the liquid junction potentials; these are of negligible magnitude due to the approximately equal transport numbers of K^+ and Cl^-, as established by potentiometric measurements [3,13].

This scheme is equivalent to that for cell V (above), but considers the interphases in detail.

Scheme 2 shows both electrodes in the sediment (combined with eqs, or separated, systems 2 and 6, respectively).

$$[B] \ Ag \mid AgCl \mid KCl \ \vdots \ \mid KCl,HX \ \vdots \ \mid eqs\ (H^+),\ HX \ \#\hspace{-3pt}\vert\ HCl \mid AgCl \mid Ag \qquad E_B$$
$$E_{ref1} \qquad E_j(S_B) \qquad E_{j\ anomal}(B) \qquad E_{g\ mix}(B) \qquad E_{ref} \quad (\equiv E_{susp})$$

Scheme 2

Again, S, G, and $\vdots$ have the same meaning as in scheme [A]. The symbol ▓ represents the ion exchanger X (sediment of X in H^+ form) bathed in KCl solution which diffuses from the RE, and the symbol $\vdots$ means the ion exchanger X (in H^+ form) dispersed in eqs. The sign $\#\hspace{-3pt}\vert$ represents the glass electrode in intimate contact with the suspension particles (HX) and entrained eqs; this evokes the mixed electrode potential. $E_{j,anomal}$ is the anomalous liquid junction potential and $E_{g,mix}$ is the mixed potential of the glass electrode. $E_j(S_B)$ is negligible [3], and E_{ref} is defined as in [A].

When comparing cells [A] and [B], the first component of the potential difference arises from junction potentials, viz. $E_j(A) + E_j(S_A) - E_{j,anomal}(B) - E_j(S_B)$, which can be approximated to $E_j(A) - E_{j,anomal}(B)$, as $E_j(S_A)$ and $E_j(S_B)$ are negligible [3]. This difference occurs because in cell [B] the filling solution, which diffuses from the salt bridge into the suspension, exchanges K+ for the H+ counter-ions of the particles, which changes the ion arrangement in the suspension and most importantly affects the approximate equality of ion transport numbers in the KCl diffusion front. This effect is termed an anomalous liquid junction potential and represents the *suspension effect of the second kind*, SE 2, as defined in [2,3,10]. The second component of the potential difference is $E_{g',mix}(B) - E_g(A)$; this results from the small suspension particles making intimate contact with the electrode surface, and is called the *suspension effect of the first kind*, SE 1 [3,11]. This arises because the electrode is in contact with the (true) eqs and at the same time in intimate contact with charged particles. In the latter contact regions, an overlapping of the double layers of the particles and the electrode occurs and causes a different H+ activity in comparison with the activity existing in contacts of eqs with the electrode. This gives rise to a mixed potential [2,3,12], as discussed in Section 1.4.

It is evident that the cell potentials E_A and E_B will differ in two component potentials: (1) $E_j(A)$ and $E_{j,anomal}(B)$ and (2) $E_g(A)$ and $E_{g',mix}(B)$. As both electrodes are in the same phase there can be no boundary potential component $E_{boundary}$, either in E_A or in E_B.

Scheme 3, System 3, shows the glass electrode in the sediment and RE in the eqs of a "combined suspension system" in equilibrium.

$$
\begin{array}{ccccc}
& S & & G & \\
[C]\ Ag\ |\ AgCl\ |\ KCl\ \|\ KCl\ &\ eqs\ \|\ eqs,\ HX\ &\ HCl\ |\ AgCl\ |\ Ag & E_C \\
E_{ref1} & E_j(S_C)\quad E_j(C)\quad E_b & & E_{g\,mix}(C) & E_{ref2}
\end{array}
$$

Scheme 3

The boundary potential is represented by E_b (= $E_{boundary}$); all other symbols have the analogous meanings as above. The cell potential E_C differs from the potential E_A (Scheme 1) only in the potential of the glass electrode because, as discussed below, E_b is negligible [3]. In [A] the potential of the glass electrode $E_g(A)$ is a single potential, because the electrode is in contact with a homogeneous solution. However, in [C] it is a multiple or mixed electrode potential, as in [B].

The potential E_b (also known as a Donnan potential) at the eqs/sediment boundary is often considered, without foundation, as arising from an effective semipermeable membrane. It has been established experimentally [2,3] that when, for example, the movable electrode penetrates the sediment phase ("perforates" the "fictitious membrane") no measurable step-change of the electrode potential occurs. Thus, $E_b = 0$ and can be neglected. However, the electrode potential changes proportionately with progressive immersion of the electrode in the sediment. This is in accordance with the above interpretation that a mixed potential forms.

A Donnan potential exists at the solid–solution interface around individual particles (because the fixed ions inside the particles cannot cross the interfaces), but it does not exist where the bulk eqs is constricted in the interstitial eqs channels between the particles.

1.4 Analysis of the schemes and findings
Potential at the eqs/sediment boundary.

Analysis of the above cell schemes shows that the effect of suspended sediment material on two electrochemical processes is responsible for the SE. The SE is not a result of a hypothetical membrane and the corresponding potential, which might be ascribed to the boundary between the eqs and the sediment. It has been established experimentally that $E_b \approx 0$ [2,3] both in control experiments, which included an appropriate agitation of the RE in the suspension, and in experiments in which a restrained flow of the solution filling the salt bridge was used or the direction of the flow was reversed.

Liquid junction potential.

The liquid junction potential formed at the contact of the RE salt bridge with the sediment can show a much greater value than when in contact with the eqs.

The experiments carried out using a "movable electrode" to establish the existence of the "hypothetical membrane" between eqs and the sediment [10] showed that the *cause* of this change in liquid junction potential is, in fact, ion exchange between the sediment particles and the electrolyte solution flowing from the salt bridge [10]. This change in cell potential begins even before the tip of the salt bridge of the movable RE penetrates the interface [10]. This explanation is also accepted in Galster's monograph on pH measurement [13].

The ions of the diffused filling solution may exchange with counterions (e.g., H+) in the (colloid) particle double layers and change the solution composition in the particle environment, which will affect *the approximately equal ion transport numbers of the diffusing electrolyte solution*. This is the fundamental reason for the development and maintenance of the *anomalous liquid junction potential*, which can be regarded as the *nature* of the changed potential. This potential can be considered as a systematic error of measurement and can be eliminated (as described in Section 2.6). The magnitude of this potential is usually of the order of some tens of mV, but it can attain more than 100 mV [2,3,7,8].

Indicator electrode (IE) potential and its duration.

The change of the IE potential when the electrode comes into intimate contact with the charged particles can also be followed by means of the above-mentioned movable electrode. Experiments confirm the interpretation that the overlapping of the electrode double layer with the double layers of particles is the *cause* of the potential change of the IE when introduced into a suspension [3,4,11,12]. If in the combined suspension system, a movable IE perforates the fictitious membrane, where a phase boundary potential difference between supernatant and the slurry phase should exist, an instantaneous electrode potential change would occur, but it does not! By a step-by-step movement of the electrode into the bulk of the suspension, a progressive increase of the electrode surface in contact with particles occurs and the electrode potential changes in parallel. This leads to the interpretation that the electrode potential change in the suspension depends on surface processes at the electrode and not on effects associated with a membrane. With further penetration of the electrode into the suspension, the contact regions on the electrode surface increase and with this the influence on the value of the mixed potential.

After the introduction of the electrode into the suspension and establishment of contacts with the particles, the electrode potential becomes an *irreversible mixed potential* [12], because two electrochemical reactions proceed simultaneously on the same electrode surface. This potential remains essentially constant for a period of time which exceeds the time required for a potentiometric measurement. This mixed potential can be regarded as the *nature* of the changed IE potential in suspensions, which cannot be eliminated from any measurement. The mixed potential in cells [B] or [C] depends on the species, smoothness of the electrode and the particles, the ionic strength of the solution, and the particle charge and size [11,12]. Its value is usually not greater than some tens of mV.

The potentiometric cell, represented schematically by cell [C], is generally adopted as the most suitable for soil pH measurements. The analysis of its scheme shows that $E_j(B)$ (required in cell [B]) is replaced by $E_j(C)$ (which is ≈ 0) and E_b (which is = 0). The cell potential E_C changes measurably only when the mixed potential of the IE $E_g(C_{mix})$ changes. The systematic RE error is eliminated from the cell potential because the filling solution of the RE does not flow into the suspension. From the steady-state potential E_C, a useful approximation of the pH of a suspension can be calculated, because E_C depends on the contribution to the H+ activity from both the particles and the eqs. If the eqs is not completely free from colloidal particles, this may represent (at most) a small uncertainty which must be taken into account.

The interpretation of the experimental results obtained in the study of hydrogen ion activity in suspensions [12] is applicable in general to a potentiometric estimation of ion activities in suspensions measured with different ion-selective electrodes (ISEs) in combination with the RE. For the latter measurements, the symbol pH used in this work should be replaced by the symbol pIon.

1.5 Conclusions

In the publication "Measurement of pH: IUPAC Recommendations 2002" [1], cell V is recommended for practical pH measurement *in solutions*. Because this cell is identical to the pH cells most frequently used in laboratory measurements in the past, the results from both are equivalent and in accordance with the recommendations. Cell V and some other cells are used also for pH measurements *in suspensions*, with the electrodes positioned usually separately, the glass electrode in the sediment, and the RE in its eqs. The pH values measured separately in the suspension (or in its sediment) or in the eqs are different, even though the suspension and solution are in equilibrium. This pH difference, which can be expressed in terms of the corresponding differences in the cell potentials, ΔE, is called the "suspension effect".

An analysis of the detailed schemes for the potentiometric cells used in such suspension measurements provides an acceptable explanation of the SE. The SE is the sum of two galvanic potential changes, which occur when the electrodes are transferred from the eqs to the suspension (or sediment):

1. The change in potential of the IE, which changes to *an irreversible mixed potential probe* (a consequence of the overlapping of the diffuse double layers of the electrode with the double layers of particles, when the electrode makes intimate contact with them).

2. The change in the liquid junction potential that exists between the salt-bridge solution of the RE and either the eqs or the suspension. In the latter case, contact of the flowing electrolyte from the salt bridge with the suspension particles gives rise to *an anomalous junction potential*.

Each of these phenomena has been confirmed with experiments [2–4, 10–12].

Measurements on the suspension potentiometric cell [B] shows no evidence for a potential boundary (as is also the case in cell [A]), characterized as a "semipermeable membrane". Experiments have established that there is *no measurable boundary potential* existing between the eqs and the sediment. Therefore, the SE does not include a measurable boundary potential; the SE cannot be interpreted as a boundary or Donnan potential.

2. Guidelines for practical pH measurements in soil suspensions

2.1 Introduction

The revised view of potentiometric measurements in suspensions (Section 1), which takes into account the currently presented definition of the SE and its interpretation in "Guidelines for potentiometric measurements in suspensions: Part B. Guidelines for practical pH measurements in soil suspensions (IUPAC Recommendations 2007)" [*Pure Appl. Chem.* **79**, 81 (2007)], provides an explanation for the results obtained by different potentiometric measurement techniques when applied to suspensions. This is important especially in the determination of soil pH.

Each of the different methods used provides a soil pH value, which is often neither clearly defined nor understood. The results can involve a large uncertainty and may only approximate the actual pH value. Only one experimental method is considered to provide a pH value with acceptable uncertainty in regard to the influence of soil solution components on a plant. However, a comparison of results obtained by several potentiometric methods can give a meaningful insight to the true H+ activity in the suspensions.

The SE that contributes to the measurement value should not be considered a very significant characteristic of a suspension, but rather it is a troublesome difference between two cell potentials, both of which affect the determination of the *actual pH value of the suspension*. The thermodynamically defined H+ ion activity in a suspension cannot be equated with any potentiometrically determined pH value.

The methods for pH measurement in dilute aqueous solutions, as in IUPAC Recommendations 2002 [6], are taken as the basis for pH measurements of suspensions. From the definition and interpretation of the SE presented here, a revised view of pH measurements in suspensions becomes possible. The effects, which occur in the measurement system due to the suspension characteristics, are analyzed below for five different measurement protocols applied to cells which contain the eqs or the suspension. These effects influence the potential difference measured in suspension potentiometric cells. Each of these measurement protocols is applicable to soil pH measurements. Advice is given on the reasonable choice and use of electrodes in suspension measurements.

To codify the different expressions of the suspension pH (soil pH), measured by cell [C] or cell V, some expressions can be proposed that specify the technique used.

The term "direct suspension pH" is used when the original sample is measured directly by cell V [1], with both electrodes in the original suspension (or soil), analogous to the pH measurement in solutions (noted as, e.g., "direct soil pH"). The term "modified direct pH" is used when the suspension is modified in any way before measurement (e.g., with water or electrolyte solution added to the original sample); this must be explicitly noted (e.g., "modified direct soil pH (1:2 w)") and the notation explained.

The term "effective suspension pH" can be used in the case where the original suspension is separated into two parts (combined suspension system) and the pH is measured with the IE in the sediment and the RE in the eqs of the cell [C], either without any prior modification of the suspension ("effective soil pH") or with a modification of the suspension ("modified effective" suspension pH). In this case, any water or solution added to the original sample must be noted explicitly (e.g., "modified effective soil pH (1:5 KCl)"). Each of these measurements is considered to give an approximation to the true pH in soil solution, that is the pH to which an object immersed in this suspension (e.g., a root) could be exposed.

The *true* pH means the pH value measured in the clear eqs separated from the equilibrated original suspension (e.g., "true soil pH"). Analogous to the above, the term "modified true" is used when the sample was modified before measurements (e.g., "modified true soil pH (1:2 CaCl₂)"). These values, in combination with the direct pH values, are used for determination of the SE.

For routine work, the corresponding abbreviations are proposed:

D soil pH, MD soil pH (1:2 w)
E soil pH, ME soil pH (2:5 KCl)
T soil pH, MT soil pH (1:5 CaCl₂), etc.

The values should be valid for measurements at 20 °C; "w" means distilled water, "KCl" 1 mol kg⁻¹ solution and "CaCl₂" 0.01 mol kg⁻¹ solution of salts, if not indicated otherwise.

2.2 Nature of suspensions and their relation to the pH electrode potential

In this report, the aqueous suspensions considered are defined as *charged solid particles* of not strictly determined sizes, which are dispersed in an aqueous *dispersing medium* (water or aqueous solution). This medium surrounds the particles permanently, even when they are settled and form a suspension *sediment*. In a suspension of, for example, ion exchanger beads (declared to be in H⁺-form), which are in equilibrium with the surrounding eqs, the particles together with their double layers may contain a larger or smaller concentration of H⁺ than that existing in the bulk solution. These charged particles could be regarded as reservoirs of ions that are blocked from the eqs by an equilibrium Donnan potential. Thus, a suspension contains at least two phases of either similar or different activity of H⁺.

The electrochemical potential $\bar{\mu}_{H^+}$ is the same throughout the whole equilibrated suspension system. Therefore, *the pH of the eqs* can be considered to be the true pH value of the whole interstitial solution in a suspension or sediment (which is not disturbed by the measurement). The pH of the eqs can be measured practically by means of cell V, defined in IUPAC Recommendations 2002 [1]. In the case of an equilibrated soil suspension, it could be considered as the true soil pH value.

When the pH electrode is placed into the suspension, the particles do not influence its electrode potential during the measurement [7,10] until the reservoirs come into intimate contact with the electrode, resulting in an overlapping of the double layers of both. As a consequence of contact regions, the number depending on the particle size, the IE potential changes [12]. The change is proportional to the ratio of contact surfaces to the total electrode surface and to the double-layer thicknesses.

Whereas the potential of the pH glass electrode positioned in the eqs follows the Nernst equation, it changes to an *irreversible mixed (or corrosion) potential* when the electrode is transferred to the *corresponding suspension or sediment*, as described in Section 1.4. Different electrodes may show different mixed potentials in the same suspension, and the same electrode may show different mixed potentials when it contacts particles of different sizes in the same suspension.

The contents of reservoirs can be estimated approximately by selected methods given in Section 2.5.

2.3 Nature of suspensions and their relation to the reference electrode potential

The *heterogeneous character of suspensions* also influences the potential of the reference part of the potentiometric cell (represented by the RE connected with the salt bridge), which is immersed in the suspension. The liquid junction potential between the RE and its salt bridge remains unchanged during the cell potential measurement. The liquid junction potential between the filling solution of the salt bridge (containing cations and anions of approximately equal transport numbers) and the eqs can practically be neglected. In contrast, the liquid junction potential between the diffused filling solution of the salt bridge and the suspension particles can be significant. It is called the *anomalous liquid junction potential*. In regard to the measurement technique, it represents a systematic error of measurement, and can be eliminated only by avoiding the salt-bridge filling solution from coming in contact with the suspension particles.

2.4 Relationship between the suspension–equilibrium solution–sediment and the positioning of the electrodes

Because the origin and cause of the SEs have not been clarified since the beginning of their study (in 1930), different modified techniques were introduced into routine determination of soil pH which give different and not clearly explained, nevertheless useful, results.

With regard to the electrode positioning in a uniform or in a combined system, the following classification of potentiometric techniques is possible: both electrodes placed (1) in the original suspension system, (2) in the eqs of the suspension, (3) in the suspension sediment, or (4) the IE in the sediment and RE in the eqs of the suspension, and (5) in the reverse mode to (4).

In these methods, the above-mentioned relations between the electrodes and the measured medium must be considered, and for 4 and 5 also the possible influence of the boundary between the sediment and the eqs on the measured cell potential. As described in Section 1.5 no measurable (Donnan) boundary potential exists at this interface; thus, it is not a "virtual continuous semipermeable membrane", as has been shown in control experiments [3,12]. Nevertheless, in spite of this fact, in some recent publications it is erroneously assumed that a boundary potential between the eqs and the sediment is the main contributor to the SE [16–18].

2.5 Discussion of modified methods of pH measurements in soil suspensions

This part provides guidelines for pH measurement. With the aid of the proposed definition and interpretation of the SE in "Guidelines for potentiometric measurements in suspensions: Part B. Guidelines for practical pH measurements in soil suspensions (IUPAC Recommendations 2007)" [*Pure Appl. Chem.* **79**, 81 (2007)], processes and techniques are discussed and the significance of the results obtained is explained.

The in situ *"soil pH"* can be measured only in wet soil if it contains enough water so that the *water activity a* ≈ 1. If this is not the case, deionized or rain water is added to the soil to obtain a homogenized wet *soil paste*, similar to the original wet soil. In these cases, the suspension is not separated into the eqs and the sediment. Measurements of pH with both electrodes in nearly dry soils are meaningless from a sheer physicochemical point of view.

In *routine pH measurements*, a greater amount of water is added to the soil to form a diluted aqueous suspension. This must *be equilibrated and separated into sediment and the corresponding eqs*. In laboratory measurements, a complete separation is performed by centrifugation, otherwise the separation is obtained by sedimentation, in which case the imperfect separation must be taken into account in assessing the uncertainty of the result.

For better-defined results, the air-dried pulverized soil is sieved, mixed with deionized water in known proportions by mass, and the pH is measured in the eqs after separation. The result for a 1:2 soil/water system can be given as "soil pH (1:2 w)". Different soil/water proportions show different pH values, which need a suitable interpretation to give useful information. Protocols for the sampling of soil populations are described in "Terminology in soil sampling (IUPAC Recommendations 2005)" [19].

Method 1 (cell potential E1)

The *direct pH measurement of the original suspension* by means of cell V (analogous to IUPAC Recommendations 2002, which is recommended for measurement in homogeneous solutions), with *both electrodes placed in the soil suspension*, gives a result which is different from those obtained with other methods. This pH_1 value has *no reasonable pH meaning*, because it is calculated from a cell potential E_1, which contains the unknown mixed potential of the IE and the anomalous liquid junction potential, the latter representing a systematic error of the measurement (Scheme 2 in Section 1.3). These two potentials are responsible for the SE. Nevertheless, pH_1 can be used in comparison with other pH_n values as a repeatable suspension characteristic. Also, the *pH of a soil paste* can be considered as a result of Method 1 with the same significance.

Method 2 (cell potential E2)

In this method, a known amount of water or of salt solution is added to the soil sample and the equilibrated suspension separated into the eqs and the sediment. For pH measurement, *both electrodes are positioned in the eqs*, which may, or may not, be in contact with the sediment. The cell potential E_2 is equivalent to the difference of the electrode potentials E_A of the cell [A] in Section 1.3. The value pH_2 calculated from E_2 is not influenced by the diffuse layer of the suspended particles. This pH value can be adopted as the pH of the whole suspension system if the suspension is not disturbed by the measurement.

The amount of water added to the original suspension must be reported with the results. The air-dried soil-to-liquid mass ratio of the suspension should be reproducible with acceptable precision as it determines the measured cell potential. The ratio should be expressed explicitly; e.g., for the ratio 1:2 as E_2(1:2 w) for water, or E_2(1:2 KCl) for KCl solution and E_2(1:2 CaCl$_2$) for CaCl$_2$ solution, whichever is used as the dispersing medium. The comparison of E_2 values, obtained in water-eqs and solution-eqs, respectively, allows an estimation of the amount of H$^+$ set free from particles for different soil-to-liquid ratios after these were exchanged by K$^+$ or Ca^{2+} [7]. In routine work, the measured E_2 values are expressed as corresponding pH_2 values. These can be regarded as the best defined, "true pH" value measured for a suspension.

Method 3 (cell potential E3)

The measurement is performed with *both electrodes in the separated sediment* of a suspension and is *equivalent to that in Method 1*, except for the fact that the particle concentration is the maximum possible. The sediment may, or may not, be in contact with the separated part of eqs. The cell scheme is given in Section 1.3, Scheme 2. Both E_j and E_g contribute to pH_3, and it cannot be used for pH evaluation of a suspension, but it is used for the determination of the total SE as described in the definition of the SE.

Method 4 (cell potential E4)

4(a) The electrode position in Method 4 is obtained by transferring *the IE* from the eqs, as it is positioned in Method 2, *into the sediment, while the RE remains in eqs*. As can be seen from the cell [C] (Scheme 3 in Section 1.3), the IE potential changes to a mixed

potential, the value of which depends on the pH of the eqs and on the H$^+$ activity of the diffuse layer of the contacting suspension particles. Because the RE potential and the diffusion potential remain unchanged, the measured cell potential E_4 differs from E_2 by the potential difference ΔE known as SE 1 (Section 1.3.) For an equilibrated suspension soil/water ratio of ½, this is given by ΔE_{4-2} (1:2 w) $= E_4 - E_2$. The pH_4 values obtained with the same IE in different (soil) suspensions allow an *approximate comparison* of the H$^+$ activity to which a (charged) surface similar to that of the IE (e.g., of a root in the measured soil) could be exposed, when coming in contact with the particles of these soils. Any change in pH_4 indicates a change of the electrode mixed potential, which depends on the particle contacts with the electrode surface.

Except for a method only applicable to the laboratory, where the filling solution of the salt bridge is exposed to a negative pressure [2], three other variations of Method 4 are used in routine practice, 4(b), (c), and (d). In these, the amount of the separated eqs is minimal and contact of the salt-bridge filling solution with the particles is avoided.

4(b) In this method, only a small amount of the eqs is needed for a measurement. It employs an RE connected with the suspension by two salt bridges in series (double salt bridge), of which the second one is filled with eqs.

4(c) In this modification, a strip of filter paper wetted with eqs is used for the electrolytic connection between the suspension and the salt bridge. When brought into contact with the suspension [9], a minimal amount of the clear eqs diffuses along the strip to the salt bridge.

4(d) In this method, a special combination electrode is used which, during the measurement, has only the pH sensing element in contact with the suspension. The eqs "climbs up" the specially prepared surface of the electrode stem to form the contact with the salt-bridge solution. In this case, the combination electrode does not *measure pH without the SE*, as it is often declared, because the presence of the SE 1 is unavoidable.

Method 4 is used very frequently in routine work, because the measured pH_4 values, though not absolutely repeatable, depend on the sum of H$^+$ activities contributed from the eqs and from the particles.

Method 5 (cell potential E5)

In this method, the IE is placed in the eqs and the RE in the sediment. It is used solely when the anomalous liquid junction potential (i.e., the systematic error of measurement), equal to the cell potential difference $\Delta E_{5-2} = E_5 - E_2$, is to be determined. The derived pH_5 value is not very relevant in routine work. From the methods discussed above, the most appropriate one can be used to obtain the information of interest. An illustration of the above methods applied to soil pH measurement is presented in [7]. It is seen that different soils show different pH_2 values (and pH differences), which can be used for the characterization of these soils in agronomy. The treatise relating to soil pH measurement can be applied—*cum grano salis*—to the general potentiometric pIon measurement in soils and in other suspensions. It should be emphasized once more that, by measurement of the voltage of any suspension galvanic cell no thermodynamically defined quantity can be obtained.

2.6 Devices and their application in practical "soil pH" measurements

Because these guidelines are based on the IUPAC Recommendations 2002 [1], the definitions given in that Glossary for pH measurement in real solutions, are also valid when applied to pH measurement in suspensions. Nevertheless, some additional points should be noted.

The *electrodes* used may be "single" or "combination", but combination ones are suitable only in some cases.

Single IEs used in suspensions should be glass or other solid-state ISE, having smooth surfaces and providing fast responses and reproducible results. Electrodes of the second kind (e.g., Ag/AgCl, Sb/Sb$_2$O$_3$) do not have smooth surfaces, and for this and other reasons they show a greater or unexpected contribution to the SE [12].

The *single RE* may be constructed with one salt bridge which is filled with the same filling solution as the electrode ("half bridge"). Two salt bridges in series (a double salt bridge) are also feasible. The term "double salt bridge" in this case is more appropriate than the term "double junction" electrode. Both kinds of salt bridge are sealed with a separator (capillary, porous ceramic plug, frit, ground glass sleeve, or other). From the separator of the single salt bridge, which is in contact with the measured medium, its filling solution (e.g., saturated KCl solution of the RE half-cell) always flows or diffuses, even if the filling solution is gel-stabilized [10,14,15]. This can give rise to an anomalous liquid junction potential when it contacts the suspension particles. This can be minimized if the final half of the double salt bridge and the separator are filled with, for example, the eqs of the measured suspension for both the test solution measurements and the electrode calibration. It must be emphasized that the filling solution which flows from the separator to the sediment boundary can cause large systematic errors, even if the separator is placed in eqs near this boundary [10]. This can be minimized with a shielding tube, which is pulled onto the salt bridge and perforated by a small side-aperture (about 1.5 cm above the bottom of the tube), providing liquid and electrical contact between the two sides [11].

Combination electrodes are often used for measurements in suspensions. They are not suitable for measurements in *combined suspension systems*, except if the electrode is placed so that its indicator half-cell is connected with the sediment and the reference half-cell with the eqs so that SE 2 is minimized. The cell voltages measured with combination electrodes, the IE of which is immersed in a suspension, always include SE 1, notwithstanding that the electrodes are often declared to "measure soil pH without suspension effect", as found in advertisements. Only some combination electrodes of special construction could possibly eliminate SE 2. *ISFET (combined) electrodes* also cannot avoid the SE in

suspension measurements as the experiments showed; the results are therefore equivalent to those obtained with method 1.

For measurements in suspensions, the electrodes must be *placed* in the suspension in such a way that any differentiation in particle sizes around the sensing element of the electrode is avoided and a stable position of the electrodes is assured.

The instruments for voltage measurement should have a high input resistance (as pH meters generally have). The potential differences of the suspension potentiometric cells, which have relatively small ohmic inner resistances (e.g., cells with the metal or solid membrane used as a halogenide IE), can be measured with voltmeters of smaller input resistance, but the readings are not stable and are difficult to interpret.

References

1. R. P. Buck, S. Rondinini, A. K. Covington, F. G. K. Baucke, C. M. A. Brett, M. F. Camoes, M. J. T. Milton, T. Mussini, R. Naumann, K. W. Pratt, R Spitzer G. S. Wilson. *Pure Appl. Chem.* **74**, 2169 (2002).
2. S. F. Oman, I. Lipar. *Electrochim. Acta* **42**, 15 (1997).
3. S. Oman. *Talanta* **51**, 21 (2000).
4. J. Th. G. Overbeek. J. *Colloid Sci.* **8**, 593 (1953).
5. H. Pallmann. *Kolloid Beih.* **30**, 334 (1930).
6. G. Wiegner. *Kolloid-Z.* **51**, 49 (1930).
7. S. F. Oman. *Acta Chim. Slov.* **47**, 519 (2000).
8. Yu. M. Chernoberezhskii. "The Suspension Effect" in *Surface and Colloid Science*. 2nd ed., Vol. 12, E. Matijevic (Ed.), pp. 359–453, Plenum Press, New York (1982).
9. T. R. Yu. *Ion-Selective Electrode Rev.* **7**, 165 (1985).
10. Lehrwerk Chemie, Vol. 5, G. Ackermann et al. *Elektrolytgleichgewichte und Elektrochemie*, 4th ed., p. 70, VEB Deutscher Verlag, Leipzig (1985).
11. S. Oman, A. Godec. *Electrochim. Acta* **36**, 59 (1991).
12. S. F. Oman. *Acta Chim. Slov.* **51**, 189 (2004).
13. H. Galster. *pH Measurement*, Chaps. 3.2, 3.4, 6.2, VCH, Weinheim (1991).
14. A. K. Covington. *Ion-Selective Electrode Methodology*, Vol. I, p. 60, CRC Press, Boca Raton (1980).
15. R. E. Dohner, D. Wegmann, W. E. Morf, W. Simon. *Anal. Chem.* **58**, 2585 (1986).
16. R. P. Buck, E. Lindner. *Pure Appl. Chem.* **66**, 2533 (1994).
17. D. H. Everett, L. K. Koopal. *Chem. Int.* **25**, 18 (2003) and refs. therein.
18. R. J. Hunter. *Colloid Science*, 2nd ed., p. 354, Oxford University Press, New York (2001).
19. P. de Zorzi, S. Barbizzi, M. Belli, G. Ciceri, A. Fajgelj, D. Moore, U. Sansone, M. Van der Perk. *Pure Appl. Chem.* **77**, 827 (2005).

IONIZATION CONSTANTS OF ACIDS AND BASES

W. P. Jencks and J. Regenstein

These pK_a' values were taken from the original literature and from several extensive compilations of such data, of which the most important are

> Albert, *Ionization Constants of Acids and Bases*, Methuen, London, 1962.
>
> Bell, *The Proton in Chemistry*, 2nd ed., Cornell, Ithaca, New York, 1973.
>
> Brown, McDaniel, and Häfliger, in Braude and Nachod, *Determination of Organic*
>
> *Structures by Physical Methods*, Academic Press, New York, 1955.
>
> Kortum, Vogel, and Andrussow, *Dissociation Constants of Organic Acids in Aqueous Solution*, Butterworths, London, 1961.
>
> Perrin, *Dissociation Constants of Organic Bases in Aqueous Solution*, Butterworths,
>
> London, 1965.
>
> Yukawa, Ed., *Handbook of Organic Structural Analysis*, Benjamin, New York, 1965.

A particularly valuable source of dissociation constants obtained under a variety of experimental conditions is provided by Sillen, L. G. and Martell, A. E., Eds., *Stability Constants*, Special Publications No. 17 and 25, Chemical Society, London, 1964 and 1971. This compilation also lists association constants of metals for a variety of inorganic and organic ligands.

The compounds selected were those which were thought most likely to be useful to biochemists and chemists and these compilations should be consulted for information on compounds which are not included here.

All values are reported as $pK_a' = -\log K_a' = 14 - pK_b'$. K_a' is the ionization constant

$$\frac{[H^+][A^-]}{[HA]} \text{ or } \frac{[H^+][B]}{[HB^+]} \text{ or } \frac{[A^{n-1}][H^+]}{[HA^n]}$$

Temperatures are not indicated because variations of pK_a' with temperature are generally smaller than the variations of the data from different sources for other reasons, but most of the data were obtained at or near 25°. Ionization constants which are reported as thermodynamic values at 25° are indicated with an asterisk, *, but some of these may only represent values measured at low ionic strength.

These pK_a' values and a measured pH should not be used to obtain an *exact* measure of the ratio of acid to base in a given solution. Ionic strength and specific salt effects, as well as possible errors in the reported pK_a' values, are likely to make such estimates inaccurate. It should be kept in mind that the effect of increasing ionic strength is generally to decrease the apparent pK_a' of neutral and anionic acids and to increase the pK_a of cationic acids. These effects are particularly large for polyanions, such as phosphates.

There is some intentional redundancy in the tables to facilitate the location of listings for compounds that might be listed in several sections. The pK_a' values for amines refer to the ionization of the conjugate acids of the amines except for a few nitrogen acids, which undergo an acidic ionization.

The pH of a solution at a given ionic strength and temperature is given by

$$pH = pK_a' + \log \frac{(base)}{(acid)}$$

in which the pK_a' is measured under the same experimental conditions. The following relationships are useful to have readily available to estimate the ratio of acid to base at a given pH or to estimate the buffer ration of acid required to give a given pH; the compiler keeps a copy of these numbers on his desk.

For graphical plots of a large number of substituted phosphorus compounds see Ref. (83). For complex chelating agents of aliphatic amines, see also Reference 77.

Fraction Base or Acid	pH
5% or 95%	$pK_a' \pm 1.25$
10% or 90%	$pK_a' \pm 0.95$
15% or 85%	$pK_a' \pm 0.75$
20% or 80%	$pK_a' \pm 0.60$
25% or 75%	$pK_a' \pm 0.48$
30% or 70%	$pK_a' \pm 0.37$
35% or 65%	$pK_a' \pm 0.27$
40% or 60%	$pK_a' \pm 0.18$
45% or 55%	$pK_a' \pm 0.09$
50% or 50%	$pK_a' \pm 0$

INORGANIC ACIDS

Compound	pK'_a	Reference	Compound	pK'_a	Reference
AgOH	3.96	4	$H_3P_2O_7^-$	2.36*	77
$Al(OH)_3$	11.2	28	$H_2P_2O_7^{2-}$	6.60*	77
$As(OH)_3$	9.22	28	$HP_2O_7^{3-}$	9.25*	77
H_3AsO_4	2.22, 7.0, 13.0	28	$HReO_4$	−1.25	30
$H_2AsO_4^-$	6.98*	77	HSCN	0.85	77
$HAsO_4^{2-}$	11.53*	77	H_3SiO_3	10.0	34
H_3AsO_3	9.22*	—	H_2S	7.00*	77
H_3AuO_3	13.3, 16.0	78	HS^-	12.92*	77
H_3BO_3	9.23	28	H_2SO_3	1.9, 7.0, 1.76*	28, 77
$H_2B_4O_7$	4.00	34	H_2SO_4	1.9	28
$HB_4O_7^-$	9.00	34	HSO_3^-	7.21*	77
$Be(OH)_2$	3.7	4	HSO_4^-	1.99*	77
HBr	−9.00	31	$H_2S_2O_3$	0.60,* 1.72*	77
HOBr	8.7	28	$H_2S_2O_4$	1.9	29
HOCl	7.53, 7.46	28, 33	H_2Se	3.89*	77
$HClO_2$	2.0	28	HSe^-	11.00*	77
$HClO_3$	−1.00	28	H_2SeO_3	2.6, 8.3; 2.62*	28
HCN	9.40	34	$HSeO_3^-$	8.32	77
H_2CO_3	6.37, 6.35,* 3.77*	34, 23	H_2SeO_4	Strong, 2.0	28
HCO_3^-	10.33*	—	$HSeO_4^-$	2.00	34
H_2CrO_4	−0.98, 0.74	30, 77	$HSbO_2$	11.0	34
$HCrO_4^-$	6.50*	2, 30	HTe	5.00	34
HOCN	3.92	34	H_2Te	2.64, 11.0	34, 78
HF	3.17*	77	H_2TeO_3	2.7, 8.0	28
H_3GaO_3	10.32, 11.7	78	$Te(OH)_6$	6.2, 8.8	28
H_2GeO_3	8.59, 12.72	34, 78	H_2VO_4	8.95	30
$Ge(OH)_4$	8.68, 12.7	28	HVO_4^{2-}	14.4	30
HI	−10.0	31	$H_4V_6O_{17}$	1.96	78
HOI	11.0	28	Cacodylic acid	1.57,* 6.27*	99
HIO_3	0.8	28	$(CH_3)_2As(O)OH$		
$H_4IO_6^-$	6.00	34			
H_5IO_6	1.64, 1.55, 8.27	34, 28, 78	*SUBSTITUTED AsO_3H_2*		
	3.29, 6.70, 15.0	—	CH_3-	3.61,* 8.18*	97
$HMnO_4$	−2.25	30	CH_3CH_2-	3.89,* 8.35*	—
NH_3OH^+	5.98*	12	$CH_3(CH_2)_4$-	4.14,* 10.07*	—
NH_4^+	9.24*	77	$CH_3(CH_2)_5$-	4.16,* 9.19*	—
HN_3	4.72*	77	$COOH(CH_2)$-	2.94,* 4.67,* 7.68*	—
H_3N	33	153	$COOH(CH_2)_4$-	2.00,* 4.89,* 7.74*	—
HNCS	~−2.0	143	$o\text{-}CH_3C_6H_4$-	3.82,* 8.85*	—
HNO_2	3.29	28	$m\text{-}CH_3C_6H_4$-	3.82,* 8.60*	—
HNO_3	−1.3	28	$p\text{-}CH_3C_6H_4$-	3.70,* 8.68*	—
$N_2H_5^+$	7.99*	77	$o\text{-}NH_2C_6H_4$-	3.77,* 8.66*	—
$H_2N_2O_2$	7.05	34	$m\text{-}NH_2C_6H_4$-	4.05,* 8.62*	—
$H_2N_2O_2^-$	11.0	34	$p\text{-}NH_2C_6H_4$-	4.05,* 8.92*	—
H_2NSO_3H	1.0	80			
H_2OsO_5	12.1	34	*HYDRATED METAL IONS*		
H_2O	15.7	—	Ti^{3+}	1.15	98
H_3O^+	−1.7	—	Bi^{+3}	1.58	—
$Pb(OH)_2$	6.48 (10.92)	4, 78	Fe^{+3}	2.80	—
PH_3	27	156	Hg^{+2}	2.60, 3.70	—
H_3PO_2	2.0, 2.23,* 1.07	28, 77	Sn^{+2}	4.00	—
H_3PO_4	2.12*	77	Cr^{+3}	3.80	—
$H_2PO_4^-$	7.21*	77	Al^{+3}	4.96	—
HPO_4^{2-}	12.32*	77	Sc^{+3}	4.96	—
H_3PO_5	1.12, 5.51, 12.80	102	Fe^{+2}	8.30	—
H_3PO_3	2.0, 1.07	28, 77	Cu^{+2}	8.30	—
$H_2PO_3^-$	6.58*	77	Ni^{+2}	9.30	—
$H_4P_2O_7$	1.52*	77	Zn^{+2}	9.60, 10.84	—

*Thermodynamic value.

PHOSPHATES AND PHOSPHONATES

Compound	pK_a'	Reference
PHOSPHATES		
Phosphate	212,* 7.21,* 12.32*	77
Glyceric acid 2-phosphate	3.6, 7.1	53
Enolpyruvic acid	3.5, 6.4	53
Methyl-	1.54, 6.31	55
Ethyl-	1.60, 6.62	55
n-Propyl-	1.88, 6.67	55
n-Butyl-	1.80, 6.84	55
Dimethyl-	1.29	55
Diethyl-	1.39	55
Di-*n*-propyl-	1.59	55
Di-*n*-butyl-	1.72	55
Glucose-3-	0.84, 5.67	56
Glucose-4-	0.84, 5.67	56
α-Glycero-	1.40, 6.44	54
β-Glycero-	1.37, 6.34	54
3-Phosphoglyceric acid	1.42, 3.42	54
2-Phosphoglyceric acid	1.42, 3.55, 7.1	—
Peroxymonophosphoric acid	4.85	69
Diphosphoglyceric acid	7.40, 7.99	54
Glyceraldehyde-	2.10, 6.75	54
Dioxyacetone-	1.77, 6.45	54
Hexose di-	1.52, 6.31	54
Fructose-6-	0.97, 6.11	54
Glucose-6-	0.94, 6.11	54
Glucose-1-	1.10, 6.13	54
Pyrophosphoric acid	0.9, 2.0, 6.6, 9.4	54
Phosphopyruvic acid	3.5, 6.38	54
DL-Phosphoserine	6.19	145
Creatine phosphate	2.7, 4.5	54
Arginine phosphate	2.8, 4.5, 9.6, 11.2	54
Amino phosphate	(−0.9), 2.8, 8.2	54
Trimetaphosphate	2.05	77
Trimethyl phosphine	8.80*	99
Triphosphate	8.90, 6.26, 2.30	77
Tetrametaphosphate	2.74	77
Fluorophosphate	0.55, 4.8	56
See also under *Nucleic Acid Derivatives*		
PHOSPHONATES		
$H_2O_3P(CH_2)_4PO_3H_2$	< 2, 2.75, 7.54, 8.38	57
$H_2O_3P(CH_2)_3PO_3H_2$	< 2, 2.65, 7.34, 8.35	57
$H_2O_3PCH_2CH(CH_3)\text{-}PO_3H_2$	< 2, 2.6, 7.00, 9.27	57
$H_2O_3PCH_2PO_3H_2$	< 2, 2.57, 6.87, 10.33	57
Methyl-	2.35, 7.1*	57, 97
Ethyl-	2.43, 7.85*	57, 97
n-Propyl-	2.45, 8.18*	57, 97
Isopropyl-	2.55, 7.75	57
n-Butyl-	2.59, 8.19	57
Isobutyl-	2.70, 8.43	57
s-Butyl-	2.74, 8.48	57
t-Butyl-	2.79, 8.88	57
Neopentyl-	2.84, 8.65	57
1,1-Dimethylpropyl-	2.88, 8.96	57
n-Hexyl-	2.6, 7.9	57
n-Dodecyl-	8.25	57
$CH_3(CH_2)_5CH(COOH)\text{-}$	1	57
$CF_3\text{-}$	1.16, 3.93	57
$CCl_3\text{-}$	1.63, 4.81	57
$NH_3^+CH_2\text{-}$	2.35, 5.9	57
$(^-OOCCH_2)_2NH^+CH_2\text{-}$	5.57	57
$CHCl_2\text{-}$	1.14, 5.61	57
$CH_2Cl\text{-}$	1.40, 6.30	57
$CH_2Br\text{-}$	1.14, 6.52	57

Compound	pK_a'	Reference
PHOSPHONATES (Continued)		
$(^-OOCCH_2)_2NH^+\text{-}(CH_2)_2\text{-}$	6.54	57
$CH_2I\text{-}$	1.30, 6.72	57
	2.45, 7.00	57
$C_6H_5CH=CH\text{-}$	2.00, 7.1	57
$HOCH_2\text{-}$	1.91, 7.15	57
	2.1	57
$C_6H_5NH(CH_2)_3\text{-}$	7.17	57
$Br(CH_2)_2\text{-}$	2.25, 7.3	57
$CH_3(CH_2)_5CH(COO^-)\text{-}$	7.5	57
$C_6H_5CH_2\text{-}$	2.3, 7.55	57
$NH_3^+(CH_2)_4\text{-}$	2.55, 7.55	57
$NH_3^+(CH_2)_5\text{-}$	2.6, 7.6	57
$NH_3^+(CH_2)_{10}\text{-}$	8.00	57
$^-OOC(CH_2)_{10}\text{-}$	8.25	57
$(CH_3)_3SiCH_2\text{-}$	3.22, 8.70	57
$C_6H_5CH_2\text{-}$	3.3†, 8.4†	57
$(C_6H_5)_3C\text{-}$	3.85†, 9.00†	57
ARYLPHOSPHONIC ACIDS		

$2X\text{-}RC_6H_3PO_3H_2$		—	57
X	*R*		
Cl	4-O_2N	1.12, 6.14	—
Br	5-O_2N	6.14	—
Cl	5-Cl	6.63	—
Cl	H	1.63, 6.98	—
Br	H	1.64, 7.00	—
Br	5-CH_3	1.81, 7.15	—
Cl	4-NH_2	7.33	—
CH_3O	4-O_2N	1.53, 6.96	—
CH_3O	H	2.16, 7.77	—
CH_3O	4-NH_2	8.22	—
HO	4-O_2N	1.22, 5.39	—
O_2N	H	1.45, 6.74	—
F	H	1.64, 6.80	—
I	H	1.74, 7.06	—
NH_2	H	7.29	—
CH_3	H	2.10, 7.68	—
C_6H_5	H	8.13	—
HOOC	H	1.71, 9.17	—

SUBSTITUTED-PO_2H_2		
$CH_3\text{-}$	3.08*	97
$CH_3CH_2\text{-}$	3.29*	97
$CH_3(CH_2)_2\text{-}$	3.46*	97
$(CH_3)_2CH\text{-}$	3.56*	97
$CH_3(CH_2)_3\text{-}$	3.41*	97
$(CH_3)_3C\text{-}$	4.24*	97
$C_6H_5\text{-}$	2.1*	97
$p\text{-}BrC_6H_4\text{-}$	2.1*	97
$p\text{-}CH_3OC_6H_4\text{-}$	2.35*	97
$p\text{-}(CH_3)_2N\text{-}$	2.1*, 4.1*	97

X=	pK_a'				Ref. 2
	−H		−NH_3^+		
$X(CH_2)PO_3H_2$	2.35	7.1	1.85	5.35	—
$X(CH_2)_2PO_3H_2$	2.45	7.85	2.45	7.00	—
$X(CH_2)_4PO_3H_2$	—	—	2.55	7.55	—
$X(CH_2)_5PO_3H_2$	—	—	2.6	7.65	—
$X(CH_2)_6PO_3H_2$	2.6	7.9	—	—	—
$X(CH_2)_{10}PO_3H_2$	—	—	—	8.00	—

* Thermodynamic value.

† These values were obtained in 50 per cent ethanol.

For graphical plots of a large number of substituted phosphorus compounds see Ref. 83

CARBOXYLIC ACIDS

Compound	pK_a'	Reference
ALIPHATIC		
Acetic Acids, Substituted		
H-	4.76*	?
O$_2$N-	1.68*	2
(CH$_3$)$_3$N$^+$-	1.83*	2
(CH$_3$)$_2$NH$^+$-	1.95	2
CH$_3$NH$_2$$^+$-	2.16*	2
NH$_3$$^+$-	2.31*	2
CH$_3$SO$_2$-	2.36*	2
NC-	2.43*	2
C$_6$H$_5$SO$_2$-	2.44	2
HO$_2$C-	2.83*	2
C$_6$H$_5$SO-	2.66	2
F-	2.66	2
Cl-	2.86*	2
Br-	2.86	2
Cl$_2$-	1.29	2
F$_2$-	1.24	2
Br$_2$	1.48	142
Br$_3$-	0.66	2
Cl$_3$-	0.65	2
F$_3$-	0.23, (−0.26)	2
HON-	3.01	2
F$_3$C-	3.07*	2
ClF$_2$	0.46	159
N$_3$-	3.03	2
I-	3.12	2
C$_6$H$_5$O-	3.12	2
C$_2$H$_5$O$_2$C-	3.35	2
C$_6$H$_5$S-	3.52*	2
CH$_3$O-	3.53	2
NCS-	3.58	2
CH$_3$CO-	3.58*	2
C$_2$H$_5$O-	3.60	2
n-C$_3$H$_7$O-	3.65	2
n-C$_4$H$_9$O-	3.66	2
Sec. -C$_4$H$_9$O-	3.67	2
HS-	3.67*	2
i-C$_3$H$_7$O-	3.69*	2
CH$_3$S-	3.72*	2
i-C$_3$H$_7$S-	3.72*	2
C$_6$H$_5$CH$_2$S-	3.73*	2
C$_2$H$_5$S-	3.74*	2
n-C$_3$H$_7$S-	3.77*	2
n-C$_4$H$_9$S-	3.81*	2
HO-	3.83*	2
$^-$O$_3$S-	4.05	2
(C$_6$H$_5$)$_3$CS-	4.30*	2
C$_6$H$_5$-	4.31*	2
CH$_2$=CH-	4.35*	2
CH$_3$-	4.88*	2
$^-$O$_2$Se-	5.43	2
$^-$O$_2$C-	5.69*	2
(CH$_3$)$_2$-	4.86	2
(C$_6$H$_5$CH$_2$)$_2$-	4.57	2
(C$_6$H$_5$)$_2$	3.96	2
(CH$_3$)$_3$-	5.01	2
CH$_3$CHOH-	3.9	2
(CH$_3$CH$_2$)$_2$-	4.74	
(CH$_3$)$_2$ (CN)-	2.43	
(CH$_3$)$_2$ C(CN)-	2.40	
HC≡C-	1.84	142
CH$_3$C≡C-	2.60	142
C$_6$H$_5$C≡C-	2.23	142
CH$_3$CH=CH	4.69	142
C$_6$H$_5$CH=CH-	4.44	142
3,5-Di-NO$_2$C$_6$H$_5$-	2.82	142
OHC-	3.32	142
C$_6$H$_5$CO-	1.32	142

*Thermodynamic value.

Substituent	Propionic α	Propionic β	Butyric α	Butyric β	Butyric γ	Valeric α	Valeric δ
			STRAIGHT-CHAIN, SUBSTITUTED				
H-	4.88	—	4.82*	—	—	4.86*	—
O$_2$N-	—	3.81	—	—	—	—	—
(CH$_3$)$_3$N$^+$-	—	—	—	—	—	—	—
H$_3$N$^+$-	2.34	3.60	—	—	4.23	—	4.27*
NC-	2.43	3.99*	—	—	4.44*	—	—
HO$_2$C-	—	4.19*	—	—	4.34*	—	4.42*
Cl-	2.80	4.08	2.84	4.06	4.52	—	4.70
Br-	2.98	4.02	2.99	—	4.58	—	4.72
HON=	3.32	4.01	3.15	—	—	3.19	4.64
F$_3$C-	—	4.18*	—	—	4.49	—	—
I–	3.12	4.06	—	—	4.64	—	4.77
C$_6$H$_5$O-	3.11	4.27	3.17	—	—	—	—
C$_2$H$_5$O$_2$C-	—	4.52	—	—	—	—	4.60
CH$_3$O-	3.52	4.46	—	—	4.68	—	4.72
CH$_3$CO-	—	4.60	—	—	4.67	—	4.72
C$_2$H$_5$O-	3.61	4.50	—	—	4.70	—	—
HS-	3.70*	4.34*	—	—	—	—	—
HO-	3.86*	4.51	4.22	4.52	4.72	3.89	—
$^-$O$_3$S-	4.22	—	—	—	—	—	—
C$_6$H$_5$-	4.31	4.66*	—	—	4.76*	—	—
H$_2$CCH-	—	4.68*	—	—	4.72*	—	—
CH$_3$-	4.86*	4.82*	4.78	4.78*	4.86*	—	4.88*
$^-$O$_2$Se-	5.48	6.00	5.48	—	—	5.48	—
$^-$O$_2$C-	—	5.48*	—	—	5.42*	—	5.41*

Compound	pK_a'	Reference
GENERAL ALIPHATIC		
Acetoacetic	3.58	6
Acetopyruvic	2.61, 7.85 (enol)	6
Aconitic, *trans*-	2.80, 4.46	6
Adipamic	4.37	101
Aminomalonic	3.32, 9.83	77
Betaine	1.84	6
α-Bromobutyric	2.97	77
N-Butylaminoacetic	2.29, 10.07	77
Caproic	4.88	101
Caprylic	4.89	101
N-(Carbamoylmethyl)- iminodiacetic	2.30, 6.60	77
β-Carboxymethylamino- propionic	3.61, 9.46	77
2-Carboxyethyliminodiacetic	2.06, 3.69, 9.66	77
Citric	3.09, 4.75, 5.41	6
Crotonic	4.69	6
Cyanomethyliminodiacetic	3.06, 4.34	77
Cyclohexane carboxylic	4.90	153
Cyclopentane carboxylic	4.99	153
Cyclopropane carboxylic	4.83	153
α-Diaminobutyric	1.85, 8.24, 10.44	77
α,β-Diaminopropionic	1.23, 6.69	77
Di-(carboxymethyl)- aminomethylphosphonic	2.25, 5.57, 10.76	77
Diethylaminoacetic	2.04, 10.47	77
Dihydroxyfumaric	1.14	6
α,β-Dimercaptosuccinic	2.40, 3.46, 9.44, 11.82	77
Dimethylaminoacetic	2.08, 9.80	77
2,2-Dimethylbutanoic	4.93	131
2,2-Dimethylpropionic	5.03	131
2-Ethylbutanoic	4.75	131
α-Ethylbutyric	4.74	130
Ethyl hydrogen adipate	4.60	101
Ethyl hydrogen diethyl-malonate	3.64	101
Ethyl hydrogen dimethyl-malonate	3.52	101
Ethyl hydrogen ethyl-malonate	3.40	101
Ethyl hydrogen malonate	3.35	101

CARBOXYLIC ACIDS (Continued)

Compound	pK'_a	Reference
GENERAL ALIPHATIC (Continued)		
Ethyl hydrogen methyl-malonate	3.41	101
Ethyl hydrogen sebacate	4.84	101
Ethyl hydrogen suberate	4.84	101
Ethyl hydrogen succinate	4.52	101
N-Ethylaminoacetic	2.30, 10.10	77
Ethylenediaminetetraacetic	2.00, 2.67, 0.26, 6.16, 10.26, 0.96	6, 94
Ethylenediamine-N,N-diacetic	5.58, 11.05	77
Formic	3.77*	2
Fumaric	3.03, 4.54	6
Gluconic	3.86*	77
Glutaramic	4.40	101
Glyceric	3.55	6
Glycolic	3.82	6
Glyoxylic	3.32	6
Heptanoic	4.89	131
Hexanoic	4.86	129
Homogentisic	4.40	6
α-Hydroxybutyric	3.65	77
β-Hydroxybutyric	4.39	77
N-2-Hydroxyethyliminodiacetic	2.2, 8.73	77
β-Hydroxypropionic	3.73	77
3-Hydroxypropyliminodiacetic	2.06, 9.24	77
Iminodiacetic	2.98*, 9.89*	77
Iminodipropionic	4.11, 9.61	77
β-Iodopropionic	4.04*	77
Isobutyric	4.86*	77
Isocaproic	4.85	130
Isohexanoic	4.85	129
Isovaleric	4.78	129
N-Isopropylaminoacetic	2.36, 10.06	77
α-Keto-β-methyl valeric	2.3	6
Lactic	3.86	6
Maleic	1.93, 6.58	6
Malic	3.40, 5.2	6
Malonamic	3.64	101
Mandelic	3.41	77
α-Mercaptobutyric	3.53	77
2-Mercaptoethyliminodiacetic	−2.14, 8.17, 10.79	77
2-Methoxyethyliminodiacetic	2.2, 8.96	77
N-Methylaminoacetic	2.24, 10.01	77
Methyl hydrogen succinate	4.49	101
Methyliminodiacetic	2.81, 10.18	77
2-Methylthioethyliminodiacetic	2.1, 8.91	77
Nitrilotriacetic	3.03, 3.07, 10.70	77
Octanoic	4.90	131
Oenanthylic	4.89	77
Oxalic	1.25, 4.14	77
Oxalacetic (trans-enol)	2.56, 4.37	6, 97
(cis-enol)	2.15, 4.06	6
Pelargonic	4.95	101
Pentanoic	4.84	131
2-Methyl-	4.78	131
3-Methyl-	4.77	131
4-Methyl-	4.85	131
2,2-Dimethyl-	4.97	131
2-Phosphonoethyliminodiacetic	1.95, 2.45, 6.54, 10.46	77
Pivalic	5.05	153
N-n-Propylaminoacetic	2.25, 10.03	77
Protocatechuic	4.48	6
Pyruvic	2.50	6
Succinamic	4.54	101
N-2-Sulfoethyliminodiacetic	1.92, 2.28, 8.16	77
Tartaric D or L	2.89, 4.16	6
meso-	3.22, 4.85	6
Thiophene-2-carboxylic	3.53	129
Vinylacetic	4.42	6

Compound	pK'_a	Reference
GENERAL ALIPHATIC (Continued)		
$CH_3CH_2OCH_2COOH$	3.65*	97
o-$CH_3C_6H_4OCH_2COOH$	3.23*	97
m-$CH_3C_6H_4OCH_2COOH$	3.20*	97
p-$CH_3C_6H_4OCH_2COOH$	3.22*	97
2,6-$(CH_3)_2C_6H_3OCH_2$-COOH	3.36*	97
o-$CH_3OC_6H_4OCH_2COOH$	3.23*	97
m-$CH_3OC_6H_4OCH_2$-COOH	3.14*	97
p-$CH_3OC_6H_4OCH_2COOH$	3.21	97
$CH_3COCH_2COCOOH$	2.58*, 8.50*	97
$CH_3CH(OH)COOCH$-(CH_3) COOH	2.95*	97
$CH_2ClCH(OH)COOH$	3.12*	97
$CH_3CH(OH)CHClCOOH$	2.59*	97
$CH_3CHClCH(OH)COOH$	3.08*	97
$CH_2ClC(CH_3)(OH)COOH$	3.20*	97
$COOHCHClCH(OH)$-COOH	2.32*	97
$CH_3COOC(CH_2COOH)_2$-COOH	2.49*	97

SULFUR CONTAINING CARBOXYLIC ACIDS

Compound	pK'_a	Reference
$HOOCH_2SCH_2COOH$	3.30*, 4.50*	97
$HOOCCH_2SCH_2CH_2$-COOH	3.31*, 4.34*	97
$HOOCCH_2S(CH_2)_5SCH_2$-COOH	3.49*, 4.41*	97
$CH_3SCH(CH_3)SCH_2$-COOH	3.77*	97
$(CH_3)_2CHSCH(CH_3)$-COOH	3.78*	97

$$\begin{array}{l} CH_3\text{-}CH\text{-}COOH \\ \quad | \\ \quad S \\ \quad \backslash \\ HOOC\text{-}CH\text{-}CH_3 \end{array} \qquad 4.62^* \qquad 97$$

$$\begin{array}{l} CH_3\text{-}CH\text{-}COOH \\ \quad | \\ \quad S \\ \quad | \\ CH_3\text{-}CH\text{-}COOH \end{array} \qquad 4.57^* \qquad 97$$

$$\begin{array}{l} CH_3\text{-}CH\text{-}COOH \\ \quad | \\ \quad S\text{-}S \\ \quad \diagup \\ CH_3\text{-}CH\text{-}COOH \end{array} \qquad 3.14^* \qquad 97$$

$$\begin{array}{l} CH_3\text{-}CH\text{-}COOH \\ \quad | \\ \quad S\text{-}S \\ \quad \diagup \\ COOH\text{-}CH\text{-}CH_3 \end{array} \qquad 3.15^* \qquad 97$$

Compound	pK'_a	Reference
$HOOC(CH_2)_9S(CH_2)_2$-NH_2	4.00*, 8.30*	97
$HOOC(CH_2)_{10}S(CH_2)_2$-NH_2	2.6*, 9.6*	97
$CH_3SO_2CH(CH_3)COOH$	2.44*	97

UNSATURATED ACIDS, CIS AND TRANS

$$R_1CH{=}CR_2COOH$$

cis-Acid

trans-Acid

R_1	R_2	cis-Acid	trans-Acid	Reference
H-	H-	4.25*	4.25*	2
CH_3-	H-	4.44*	4.69*	2
Cl-	H-	3.32	3.65	2
C_6H_5-	H-	3.88*	4.44*	2
o-ClC_6H_4-	H-	3.91	4.41	2
o-BrC_6H_4-	H-	4.02	4.41	2
CH_3-	CH_3-	4.30	5.02	2
C_6H_5-	H-	5.26†	5.58†	2
2,4,6-$(CH_3)_3$-C_6H_2-	H-	6.12†	5.70†	2
C_6H_5-	CH_3-	4.98†	5.98†	2

*Thermodynamic value.

CARBOXYLIC ACIDS (Continued)

Compound	pK'$_a$	Reference	Compound	pK'$_a$	Reference
*UNSATURATED DICARBOXYLIC ACIDS**			*DICARBOXYLIC ACIDS AND DERIVATIVES* (Continued)*		
Acetylenediacrboxylic	1.73, 4.40	2	Diisopropyl-	2.12, 8.85	136
Bromofumaric	1.46, 3.57	2	Ethyl-*n*-propyl-	2.15, 7.43	2
Bromomaleic	1.45, 4.62	2	Di-*n*-propyl-	2.07, 7.51	2
Chlorofumaric	1.78, 3.81	2	Glutaric	4.34, 5.42	2
Chloromaleic	1.72, 3.86	2	β-Methyl-	4.25, 6.22	2
Citraconic (Dimethylmaleic acid)	2.29, 6.15	2	β-Ethyl-	4.29, 6.33	2
Fumaric	3.02, 4.38	2	β-Isopropyl-	4.30, 5.51	129
Itaconic (l-Propene-2,3-dicarboxylic acid)	3.85, 5.45	2	β-*n*-Propyl-	4.31. 6.39	2
Maleic	1.92, 6.23	2	β,β-Dimethyl-	3.70, 6.29	2
Mesaconic (Dimethylfumaric acid)	3.09, 4.75	2	β,β -Methylethyl-	3.62, 6.70	2
Phthalic	2.95, 5.41	2	β,β-Diethyl-	3.62, 7.12	2
Δ^1-Tetrahydrophthalic	3.01, 5.34	2	β,β-Di-*n*-propyl-	3.69, 7.31	2
			β,β-Pentamethylene-	3.49, 6.96	129
ALICYCLIC DICARBOXYLIC ACIDS			Succinic	4.19, 5.48	2
cis-Caronic (1,1-dimethylcyclopropane-2,3-dicarboxylic acid)	2.34*, 8.31*	2	Methyl-	4.07, 5.64	101
trans-Caronic	3.83*, 5.32*	2	Ethyl-	4.07, 5.89	101
1,2-(*trans*-Cyclopropanedicarboxylic	3.65*, 5.13*	2	Tetramethyl-	3.50, 7.28	2
1,2-*cis*-Cyclopropanedicarboxylic	3.33*, 6.47*	2	DL-1:2-Dichloro-	1.68, 3.18	20
trans-Ethyleneoxide-dicarboxylic	1.93, 3.25	2	*meso*-1:2-Dichloro-	1.74, 3.24	20
1,2-*trans*-Cyclobutanedicarboxylic	3.94, 5.55	132	DL-1:2-Dibromo-	1.48	20
1,3-*trans*-Cyclobutanedicarboxylic	3.81, 5.28	2	*meso*-1:2-Dibromo-	1.42, 2.97	20
1,2-*trans*-Cyclopentanedicarboxylic	3.89, 5.91	2	DL-1:2-Dimethyl-	3.93, 6.00	20
1,3-*trans*-Cyclopentanedicarboxylic	4.40, 5.45	2	*meso*-1:2-Dimethyl-	3.77, 5.36	20
1,2-*trans*-Cyclohexanedicarboxylic	4.18, 5.93	2	D-Tartaric	3.03, 4.45	20
1,3-*trans*-Cyclohexanedicarboxylic	4.31, 5.73	2	*meso*-Tartaric	3.29, 4.92	20
1,4-*trans*-Cyclohexanedicarboxylic	4.18, 5.42	2	Adipic	4.42, 5.41	2
cis-Ethyleneoxidedicarboxylic	1.94, 3.92	2	Pimelic	4.48, 5.42	2
1,2-*cis*-Cyclobutanedicarboxylic	4.16, 6.23	132	Suberic	4.52, 5.40	2
1,3-*cis*-Cyclobutanedicarboxylic	4.03, 5.31	2	Azelaic	4.55, 5.41	2
1,2-*cis*-Cyclopentanedicarboxylic	4.37, 6.51	2			
1,3-*cis*-Cyclopentanedicarboxylic	4.23, 5.53	2	*LYSERGIC ACID AND DERIVATIVES*		
1,2-*cis*-Cyclohexanedicarboxylic	4.34, 6.76	2	Ergometrine	6.8	2
1,3-*cis*-Cyclohexanedicarboxylic	4.10, 5.46	2	Ergometrinine	7.3	2
1,4-*cis*-Cyclohexanedicarboxylic	4.44, 5.79	2	Dihydroergometrine	7.4	2
			α-Dihydrolysergol	8.3	2
HYDROXYCYCLOHEXANECARBOXYLIC ACIDS			β-Dihydrolysergol	8.2	2
Cyclohexanecarboxylic	4.90	2	6-Methylergoline	8.85	2
cis-1,2-	4.80	2	Lysergic acid	7.8, 3.3	2
cis-1,3-	4.60	2	Isolysergic acid	8.4, 3.4	2
cis-1,4-	4.84	2	α-Dihydrolysergic	8.3, 3.6	2
trans-1,2-	4.68	2	γ-Dihydrolysergic	8.6, 3.6	2
trans-1,3-	4.82	2			
trans-1,4-	4.68	2	For complex chelating agents, see also Ref. (84).		
BICYCLO(2.2.2)OCTANE-1-CARBOXYLIC ACIDS, 4-SUBSTITUTED			*AROMATIC*		
H-	6.75	2	Anthracene-1-COOH	3.69	2
C$_2$H$_5$O$_2$C-	6.31	2	Anthracene-9-COOH	3.65	2
NC-	5.90	2	2-Furan-COOH	3.16	153
HO-	6.33	2	3-Furan-COOH	3.95	153
Br-	6.08	2	Naphthalene-2-COOH	4.17	2
			Naphthalene-1-COOH	3.69	2
*DICARBOXYLIC ACIDS AND DERIVATIVES**			Naphthol-1-COOH	9.85	153
Oxalic	1.23, 4.19	2	Naphthol-2-COOH	9.63	153
Malonic	2.83, 5.69	2	1-Phenyl-5-methyl-1,2,3-triazole-4-COOH	3.73	126
Methyl-	3.05, 5.76	2	1-Phenyl-1,2,3-triazole-4-COOH	2.88	126
Ethyl-	2.99, 5.83	2	1-Phenyl-1,2,3-triazole-4,5-(COOH)$_2$	2.13, 4.93	126
Ethylisoamyl-	2.50, 7.31	129	2-Pyrrole-COOH	4.45	153
n-Propyl-	3.00, 5.84	2	2-Thiophen-COOH	3.53	153
i-Propyl-	2.94, 5.88	2	3-Thiophen-COOH	4.10	153
Dimethyl-	3.17, 6.06	2	1,2,3-Triazole-4-COOH	3.22, 8.73	126
Methylethyl-	2.86, 6.41	2	1,2,3-Triazole-4,5-(COOH)$_2$	1.86, 5.90, 9.30	126
Diethyl-	2.21, 7.29	2			

*Thermodynamic value.

CARBOXYLIC ACIDS (Continued)

Benzoic Acid	ortho	meta	para	Reference	Compound		pK$_a'$		Reference

AROMATIC (Continued) | ORTHO-SUBSTITUTED BENZOIC ACIDS

Benzoic Acid	ortho	meta	para	Reference	Compound	pK$_a'$	Reference
					2-CH$_3$-	3.91*	2
					2-t-C$_4$H$_9$-	3.46	2
Substituted Benzoic Acids			X—⬡—COOH	2, 97, 100, 101	2,6-(CH$_3$)$_2$-	3.21	2
					2,3,4,6-(CH$_3$)$_4$-	4.00	2
H-	4.20*	—	—	—	2,3,5,6-(CH$_3$)$_4$-	3.52	2
O$_2$N-	2.17*	3.45*	3.44	—	2-C$_2$H$_5$-	3.77	2
CH$_3$CO-	4.14	3.83	3.70	—	2-C$_6$H$_5$-	3.46*	2
CH$_3$SO$_2$-	—	3.64*	3.52*	—	2,4,6-(CH$_3$)$_3$-	3.43	2
CH$_3$S-	—	5.53	5.74	—	2,3,4,5-(CH$_3$)$_4$-	4.22	2
HS-	5.02	5.42	5.56	—	2,4-OH-	3.22*	97
Br-	2.85*	3.81*	4.00*	—	2,6-OH-	1.22*	97
F-	3.27*	3.87*	4.14*	—			
CH$_3$O-	4.09*	4.09*	4.47*	—			
n-C$_3$H$_7$O-	4.24*	4.20*	4.46*	—			
n-C$_4$H$_9$O-	—	4.25*	4.53*	—			
C$_6$H$_5$O-	3.53*	3.95*	4.52*	—			
CH$_3$-	3.91*	4.24*	4.34*	—			
(CH$_3$)$_2$CH-	—	—	4.35*	—			
(CH$_3$)$_3$N$^+$-	1.37	3.45	3.43	—			
NC-	3.14*	3.60*	3.55*	—			
HO$_2$C-	2.95*	3.54	3.51	—			
F$_3$C-	—	3.79	—	—			
HO-	2.98*	4.08*	4.58*	—			
I–	2.86*	3.86*	3.93	—			
Cl–	2.94*	3.83*	3.99*	—			
(CH$_3$)$_3$Si-	—	4.24*	4.27*	—			
C$_2$H$_5$O-	4.21*	4.17*	4.45*	—			
i-C$_3$H$_7$O-	4.24*	4.15*	4.68*	—			
n-C$_5$H$_{11}$O-	—	—	4.55*	—			
C$_6$H$_5$-	3.46*	—	—	—			
CH$_3$CH$_2$-	3.77	—	4.35*	—			
(CH$_3$)$_3$C-	3.46	4.28	4.40*	—			
NH$_2$-	2.05*	3.07*	2.38*	—			
	4.95*	4.73*	4.89*	—			
SO$_2$NH$_2$-	—	3.54	3.47	—			
CH$_3$CO$_2$-	3.48	4.00	4.38	—			
CH$_3$CONH-	3.63	4.07	4.28	—			
$^-$HO$_3$P-	3.78	4.03	3.95	—			
$^-$O$_3$S-	—	4.15	4.11	—			
(CH$_3$)$_2$N-	8.42	5.10	5.03	—			
$^-$HO$_3$As-	—	—	4.22	—			
$^-$O$_2$C-	5.41	4.60	4.82	—			
CH$_3$NH-	5.33	5.10	5.04	—			

Acid	Position of Carboxyl	pK^I	pK^{II}	pK^{III}	pK^{IV}	pK^V	pK^{VI}

BENZENE POLYCARBOXYLIC ACIDS (2)

Acid	Position of Carboxyl	pK^I	pK^{II}	pK^{III}	pK^{IV}	pK^V	pK^{VI}
Benzoic	1	4.17*	—	—	—	—	—
Phthalic	1, 2	2.98*	5.28*	—	—	—	—
Isophthalic	1, 3	3.46*	4.46*	—	—	—	—
Terephthalic	1, 4	3.51*	4.82*	—	—	—	—
Hemimellitic	1, 2, 3	2.80*	4.20*	5.87*	—	—	—
Trimellitic	1, 2, 4	2.52*	3.84*	5.20*	—	—	—
Trimesic	1, 3, 5	3.12*	3.89*	4.70*	—	—	—
Mellophanic	1, 2, 3, 4	2.06*	3.25*	4.73*	6.21*	—	—
Prehnitic	1, 2, 3, 5	2.38*	3.51*	4.44*	5.81*	—	—
Pyromellitic	1, 2, 4, 5	1.92*	2.87*	4.49*	5.63*	—	—
Benzenepentacarboxylic	1, 2, 3, 4, 5	1.80*	2.73*	3.97*	5.25*	6.46*	—
Mellitic	1, 2, 3, 4, 5, 6	1.40*	2.19*	3.31*	4.78*	5.89*	6.96*

*Thermodynamic value.

PHENOLS

Phenol	*ortho*	*meta*	*para*	Reference
H-	9.95*	—	—	52, 97, 100, 153
$(CH_3)_3N^+$-	7.42	8	8	—
CH_3SO_2-	—	8.40	7.83	—
CH_3CO-	—	9.19	8.05	—
$C_2H_5O_2C$-	—	—	8.50*	—
$C_3H_5CH_2O_2C$-	—	—	8.41*	—
Br-	8.42*	9.11*	9.34	—
F-	8.81*	9.28*	9.95*	—
HO-	9.48	9.44	9.96	—
CH_3-	10.28*	10.08	10.19*	—
CH_3O-	9.93	9.65	10.20	—
^-O_2C-	13.82	9.94*	9.39*	153
$^{--}O_3P$-	—	10.2	9.9	—
C_6H_5-	9.93	9.59	9.51	—
O_2N-	7.23*	8.35*	7.14*	—
OCH-	6.79	8.00	7.66	—
NC-	—	8.61	7.95	—
CH_3O_2C-	—	—	8.47*	—
n-$C_4H_9O_2C$-	—	—	8.47*	—
I-	8.51	9.17*	9.31	—
Cl-	8.48*	9.02*	9.38*	—
CH_3S-	—	9.53	9.53	—
$HOCH_2$-	9.92*	9.83*	9.82*	—
C_2H_5-	10.2	9.9	10.0	—
H_2N-	9.71	9.87	10.30	—
^-O_3S-	—	9.29	9.03	—
$^{--}O_3As$-	—	—	8.37	—
NO-	—	—	6.35	—
H_2NCO-	8.37*	—	—	—

Name	pK$'_a$	Reference
POLYSUBSTITUTED PHENOLS		
2,3-Dimethyl-	10.54	101
2,4-Dimethyl-	10.60	101
2,5-Dimethyl-	10.41	101
2,6-Dimethyl-	10.63	101
2,4,5-Trimethyl-	10.88	101
2,3,5-Trimethyl-	10.69	101

Name	pK$'_a$	Reference
POLYSUBSTITUTED PHENOLS (Continued)		
3,4-Dimethyl-	10.36	101
3,5-Dimethyl-	10.19	101
2,4-Dichloro-	7.85	101
3-Chloro-2-nitro-	6.75	101
3-Chloro-4-nitro-	6.80	101
3-Bromo-2-nitro-	6.78	101
3-Bromo-4-nitro-	6.84	101
3-Iodo-2-nitro-	6.89	101
3-Iodo-4-nitro-	6.94	101
2,4-Dinitro-	4.11	101
2,5-Dinitro-	5.22	101
2,6-Dinitro-	5.23	101
3,4-Dinitro-	5.42	101
2,4,6-Trinitro-	0.96	101
2-Chloro-4-nitro-	5.45	79
2-Nitro-4-chloro-	6.46	79
2-OCH_3-4-$CH_2CH=CH_2$-	10.00*	97
2-OCH_3-4-OHC-	7.40*	97
2-OCH_3-6-OHC-	7.91*	97
2-OCH_3-5-OHC-	8.89*	97
Chromotropic acid	5.36, 15.6	6
2-Amino-4,5-dimethylphenol hydrochloride	10.4, 5.28	51
4,5-Dihydroxybenzene-1,3-disulfonic acid	7.66, 12.6	77
Kojic acid	9.40	77
Resorcinol	9.15 (30°)	50
3-Hydroxyanthranilic acid	10.09, 5.20	51
2-Aminophenol hydrochloride	9.99, 4.86	51
SUBSTITUTED CATECHOLS		
3-Nitro-	6.66	101
4-Nitro-	6.89	101
3,4-Dinitro-	4.39	101
4-Formyl-	7.36	101
4-Hydroxyiminomethyl-	8.68	101
3-Methyl-	9.28	101
3-Methoxy-	9.28	101
4-Benzoyl-	7.74	101
4-Cyano-	7.72	101

*Thermodynamic value.

ALCOHOLS AND OTHER OXYGEN ACIDS

Compound	pK'_a	Reference
ALCOHOLS, SIMPLE		
Choline	13.9	6
CF_3CH_2OH	12.43	63
$CF_3CH(OH)CH_3$	11.8	63
$C_3F_7CH_2OH$	11.4†	63
$(C_3F_7)_2CHOH$	10.6†	63
$CH\equiv CCH_2OH$	13.55	64
$C(CH_2OH)_4$	14.1	64
$CH_2OHCHOHCH_2OH$	14.4	64
CH_2OHCH_2OH	14.77	64
$CH_3OCH_2CH_2OH$	14.82	64
CH_3OH	15.54	64
$CH_2=CHCH_2OH$	~15.52	64
H_2O	15.74	64
CH_3CH_2OH	16	64
CCl_3CH_2OH	12.24	64
$CHF_2CH_2CH_2OH$	12.74	64
$CHCl_2CH_2OH$	12.89	64
CH_2ClCH_2OH	14.31	64
$CF_3C(CH_3)_2OH$	11.6	64
$HOCH_2CF_2CF_2CH_2OH$	11	64
$C_3F_7CH(C_2F_5)OH$	10.48	65
$(C_3F_7)_2CHOH$	10.52	65
$(CF_3)_2CHOH$	9.3	108
$(CF_3)_3COH$	5.4	108
$(CF_3)C(OH)CF_2NO$	3.9	108
$(CF_3)_2C(CH_3)OH$	9.6	122
$(CF_3)_2CHOH$	9.3	122
$(CF_3)_2C(CClF_2)OH$	5.3	122
$(CF_3)_2C(CCl_3)OH$	5.1	122
F_2CHCH_2OH	13.11	142
FCH_2CH_2OH	14.20	142
Br_3CCH_2OH	12.70	142
Br_2CHCH_2OH	13.29	142
$BrCH_2CH_2OH$	14.38	142
ICH_2CH_2OH	14.56	142
$NCCH_2CH_2OH$	14.03	142
$C_2H_5OCH_2CH_2OH$	14.98	142
$C_6H_5OCH_2CH_2OH$	14.60	142
$C_6H_5CH_2CH_2OH$	15.48	142
$HOCH_2CH_2OH$	15.11	142
$C_2H_5CH_2OH$	15.92	142
$C_3H_7CH_2OH$	15.87	142
$i\text{-}C_3H_7CH_2OH$	15.91	142
$(CH_3)_3CCH_2OH$	16.04	142
$CH_3C\equiv CCH_2OH$	14.16	142
$C_6H_5C\equiv CCH_2OH$	13.87	142
$CH_2=CHCH_2OH$	15.48	142
$CH_3CH=CHCH_2OH$	15.80	142
$C_6H_5CH=CHCH_2OH$	15.62	142
$C_6H_5CH_2OH$	15.44	142
3,5-Di-$NO_2C_6H_3CH_2OH$	14.43	142
$OHCCH_2OH$	14.80	142
CH_3COCH_2OH	14.19	142
$C_6H_5COCH_2OH$	13.33	142

Substituted Triphenylmethanols in	H_2SO_4	$HClO_4$	HNO_3	Ref. 66
CARBONIUM IONS				
4,4′,4″-Trimethoxy-	0.82	0.82	0.80	
4,4′-Dimethoxy-	−1.24	−1.14	−1.11	
4-Methoxy-	−3.40	−3.56	−3.41	
4-Methyl-	−5.41	−5.67	—	
4-Trideuteriomethyl-	−5.43	−5.67	—	
3,3′,3″-Trimethyl-	−6.35	−5.95	—	
Unsubstituted triphenylmethanol	−6.63	−6.89	−6.60	
4,4′,4″-Trichloro-	−7.74	−8.01	—	
4-Nitro-	−9.15	−9.76	—	

*Thermodynamic value.
†50% aqueous methanol.

Compound	pK'_a	Reference
HYDROXAMIC ACIDS		
Aceto-	9.40	68
o-Aminobenzo-	9.17	93
p-Aminobenzo-	9.32	93
Benzo-	8.88	68
n-Butyro-	9.48	68
Chloroaceto-	8.40	93
p-Chlorobenzo-	9.59	68
p-Chlorophenoxyaceto-	8.75	93
Cyclohexano-	9.75	93
Formo-	8.65	93
Furo-	8.45	72
Glycine-	7.40	72
Hexano-	9.75	93
Hippuro-	8.80	72
p-Hydroxybenzo-	8.93	93
N-Hydroxyphthalimide	7.00, 6.10	71, 72
Indole-3-aceto-	9.58	93
L-Lacto-	9.35	93
D-Lysine-	7.93	93
L-Lysine-	7.93	93
p-Methylbenzo-	8.90	72
p-Methoxybenzo-	9.00	68
α-Naphtho-	~7.7	68
Nicotin-	8.30	72
*iso*Nicotin-	7.85	72
Nicotin-methiodide	6.46	72
m-Nitrobenzo-	8.07	72
p-Nitrobenzo-	8.01	93
Phenylaceto	9.19	68
N-Phenylbenzo-	9.15	93
N-Phenylnicotino-	8.00	93
Phthalo-	9.48	93
Picolin-	8.50	72
Propiono-	9.46	68
Pyrimidine-2-carbox-	7.88	72
Salicyl-	7.43	72
Tropo-	9.09	72
L-Tyrosine	9.20	93
OXIMES		
Acet-	12.42	18
Acetophenone	11.48	18
Benzophenone	11.3	18
Benzoquinoline mon-	6.25	93
1,2,3-Cyclohexanetrionetri-	8.0	76
Diethyl ket-	12.6	18
Isonitrosoacetone (INA)	8.3	76
Isonitrosoacetylacetone (INAA)	7.4	76
5-Methyl-1,2,3-cyclohexanetrione-1,3-di-	8.3	76
5-Methyl-1,2,3-cyclohexanetrionetri-	8.0	76
Phenylglyoxald-	8.30	93
Pyridine-2-ald-	3.56*, 10.17*	99
Pyridine-3-ald-	3.94*, 10.32*	99
Pyridine-4-ald-	4.58*, 9.91*	99
Pyridine-4-aldoxime dodeciodide	8.50	93
Pyridine-2-aldoxime heptiodide	8.00	93
Pyridine-2-aldoxime methiodide	8.00	93
Pyridine-3-aldoxime methiodide	9.20	93
Pyridine-4-aldoxime methiodide	8.50	93
Pyridine-4-aldoxime pentiodide	8.50	93
3-Pyridine-1,2-ethanedione-2-oxime methiodide	7.20	93
4-Pyridine-1,2-ethanedione-2-oxime methiodide	7.10	93

ALCOHOLS AND OTHER OXYGEN ACIDS (Continued)

PEROXIDES, ROOH

R=	H	CH_3	C_2H_5	$iso\text{-}C_3H_7$	$tert\text{-}C_4H_9$	$iso\text{-}C_4H_9$	Reference
	11.6	11.5	11.8	12.1	12.8	12.8	70

Compound	pK_a'	Reference
PEROXY ACIDS		
Acetic	8.2	70
n-Butyric	8.2	70
Formic	7.1	70
Peroxydiphosphoric	5.18, 7.68	85
Peroxymonophosphoric	4.85	90
Peroxymonosulfuric	9.4	69
Propionic	8.1	70

X	2—	3—	4—
PYRIDINE 1-OXIDES AND DERIVATIVES (REF. 99, 47, 67)			
H-	0.79	—	
CH₃CONH-	−0.42*	0.99*	1.59*
H₂N-	2.67*	1.47*	3.59*
C₆H₅CH₂S-	−0.23*	—	2.09*
HOOC-	—	0.09	−0.48
	—	2.73*	2.86*
(CH₃)₂N-	2.27*	—	3.88*
HO-	5.97*	—	5.76*
CH₃O-	1.23*	—	2.05*
CH₃NH-	2.61*	—	3.85*
C₆H₅-	0.77*	0.74*	0.83*
CH₃-	2.61*	1.08*	1.29*
C₂H₅O-	1.18*	—	—
C₆H₅CH₂O-	—	—	1.99
NO₂-	—	—	−1.7
COOC₄H₉-	—	0.03	—

Compound	pK_a'	Reference
PYRIDINE 1-OXIDES AND DERIVATIVES		
2-Amino-1-methoxypyridinium perchlorate	12.4	67
4-Amino-1-methoxypyridinium perchlorate	>11	67
1-Benzyloxypyrid-2-one	−1.7	67
1-Benzyloxypyrid-4-one	2.58	67
4-Dimethylamino-1-methoxypyridinium perchlorate	>11	67
2-Methylamino-1-methoxypyridinium toluene-*p*-sulfonate	>11	67
1-Methoxypyrid-2-one	−1.3	67
1-Methoxypyrid-4-one	2.57	67
3-R-Pyrazine-1-Oxides		
R=CN–	−1.12	121
Cl–	−1.05	121
CH₃O–	−0.45	121

*Thermodynamic value.

Compound	pK_a'	Reference
PYRIDINE 1-OXIDES AND DERIVATIVES (Continued)		
NH₂–	−1.92, 1.50	121
H–	0.05	121
CH₃–	0.46	121
(CH₃)₂N–	−1.77, 1.34	121
	−1.80, 1.34	121
SULFINIC ACIDS		
Benzene-	1.84, 2.16	73
p-Bromobenzene-	1.89	73
p-Chlorobenzene-	1.81	73
m-Nitrobenzene-	1.88	73
p-Nitrobenzene-	1.86	73
p-Toluene-	1.99	73
OTHER OXYGEN ACIDS		
CF₃CH₂NHOH	11.3	108
(CF₃)₂CHNHOH	8.5	108
(CF₃)₃CNHOH	5.9	108
CF₃CHNOH	8.9	108
(CF₃)₂CNOH	6.0	108
Glutaconic dialdehyde	5.75	153
Hydroxylamine	13.7	133
Mannitol	13.5	100
Sucrose	12.7	100
Phenylboric acid	8.86	100
Pyridine-4-aldehyde	12.20	153
β-Phenylethylboric acid	10	100
Lyxose	12.11	25
Ribose	12.11	25
2-Deoxyribose	12.61	25
Xylose	12.15	25
Arabinose	12.34	25
Fructose	12.03	25
2-Deoxyglucose	12.52	25
Galactose	12.35	25
(CF₃)₂C(OH)₂	6.58	108
(CF₂Cl)₂C(OH)₂	6.67	108
(CF₂H)₂C(OH)₂	8.79	108
CF₂ClCF₂HC(OH)₂	7.90	108
Trimethylamine-*N*-oxide	4.6	18
Triethylamine-*N*-oxide	5.13*	99
Acetaldehyde hydrate	13.48	91
Formaldehyde hydrate	13.29	91
Glucose	12.43*	97
Mannose	13.50*	97
Sorbose	13.57	97
Acetamide-H⁺	−0.025	149, 150
Chloroacetamide-H⁺	−0.26	149, 150
Dichloroacetamide-H	−0.26	149, 150
N-Methylacetamide-H⁺	0.26	149, 150
N,N-Dimethylacetamide-H⁺	0.62	149, 150
Biotin-H⁺	−1.13	149, 150
Desthiobiotin-H⁺	−0.97	149, 150
Dimethylurea-H⁺	−0.20	149, 150
Formamide-H⁺	0.12	149, 150
N,N-Diethylformamide-H⁺	0.36	149, 150
N,N-Dimethylformamide-H⁺	0.18	149, 150
N-Methylformamide-H⁺	0.52	149, 150
Imidazolidone-H⁺	−1.05	149, 150
(CH₃)₂CHCH(OH)₂	13.77	160
CH₃CH(OH)₂	13.57	160
CH₂(OH)₂	13.27	160
CF₃CH(OH)₂	10.20	160
CCl₃CH(OH)₂	10.04	160
C₆H₅C(OH)₂CF₃	10.00	160

AMINO ACIDS

Compound	pK'_a	Reference	Compound	pK'_a	Reference
Alanine	2.34, 9.69	6	**4-Aminophenylacetic acid**	3.60, 5.26	99
N-Acetyl-	3.72	97	**2-Aminophenylarsonic acid**	3.77, 8.66	99
Amide	8.02*	99	**2-Aminophenylboric acid**	4.53, 9.31	99
3-(2-Aminoethyldithio)-	8.28, 9.30	99	β-**Aminopropionic acid**	3.55*, 10.23*	97
Carbamyl-	3.89	99	**4-Aminosalicylic acid**	1.78, 3.63	99
N-Ethyl-	2.22, 10.22	99	α-**Aminotricarballylic acid**	2.10, 3.60, 4.60, 9.82	99
N-Methyl-	2.22, 10.19	99	α-**Aminovaleric acid**	4.20	130
N-n-Propyl-	2.21, 10.19	99	**2-Anilinoethylsulfonic acid**	3.80	99
β-(2-Pyridyl)-	1.37, 4.02, 9.22	99	**Arginine**	12.48, 2.17, 9.04	6
β-(3-Pyridyl)-	1.77, 4.64, 9.10	99	**Argininosuccinic acid**	>12, 1.62, 9.58, 2.70,	—
β-(4-Pyridyl)-	4.85	99		4.26	
β-Alanine	3.60, 10.19	6	**Asparagine**	2.02, 8.8	6
N-Acetyl-	4.44	129	α-Hydroxy-	2.28, 7.20	99
Carbamyl-	4.49	129	β-Hydroxy-	2.09, 8.29	99
Allothreonine	2.11, 9.01	99	**Aspartic acid**	2.09, 3.86, 9.82	99
O-Methyl-	1.92, 8.90	99	Diamide	7.00	99
γ-Aminoacetoacetic acid	2.9, 8.3	99	Hydroxy-	1.91, 3.51, 9.11	99
α-**Aminoadipic acid**	2.14, 4.21	101	**Azaserine**	8.55	101
2-Aminobenzoic acid	2.19, 4.95	99	m-Benzbetaine	3.22	99
N,N-Dimethyl-	1.4, 8.49	99	f-Benzbetaine	3.25	99
3-Hydroxy-	5.19, 10.12	99	Betaine	1.84	99
N-Methyl-	1.97, 5.34	99	γ-**Butyrobetaine**	3.94	99
3-Aminobenzoic acid	3.29, 5.10	99	**Canaline**	2.40-, 3.70, 9.20	99
4-Aminobenzoic acid	2.50, 4.87	99	**Canavanine**	2.50, 6.60, 9.25	99
4-Aminobutylphosphonic acid	2.55, 7.55, 10.9	99	**L-Citrulline**	2.43, 9.41	99
4-Aminobutylsulphonic acid	10.65	99	**Creatine**	2.67, 11.02	6
α-**Aminobutyric acid**	2.55, 9.60	6	**Creatinine**	4.84, 9.2	6
Carbamoyl-α-amino-n-butyric	3.89	129	**Cycloserine**	4.4, 7.4	101
γ-**Aminobutyric acid**	4.23, 10.43	6	**Cysteine**	10.78, 1.71, 8.33	6
Carbamyl-	4.68	129	Ethyl ester	6.69, 9.17	99
2-Aminobutyric acid	2.27, 9.68	99	Methyl ester	6.56, 8.99	99
α-**Amino-n-caproic acid**	2.33	129	S-Ethyl-	1.94, 8.69	99
ε-**Aminocaproic acid**	4.37	129	S-Methyl-	8.75	99
10-Aminodecylphosphonic acid	8.0, 11.25	99	**Cystine**	1.65, 7.85	6
10-Aminodecylsulphonic acid	11.35	99	**L-Cystine diamide**	5.93, 6.90	99
10-Amino-n-dodecanoic acid	4.648	99	**2,4-Diaminobutyric acid**	1.85, 8.28, 10.50	99
Aminoethylphosphoric acid	2.45, 7.0, 10.8	99	**2,3-Diaminopropionic acid**	1.23, 6.73, 9.56	99
2-Aminoethylsulphoric acid	8.95	99	**2,7-Diaminosuberic acid**	1.84, 2.64, 9.23, 9.89	99
ω-**Aminoheptanoic acid**	4.50	136	**3-Dimethylaminopropionic acid**	9.85	99
6-Aminohexanoic acid	4.37, 10.81	99	**Formamidinoglutaric acid**	2.7, 4.4, 11.3	99
α-**Aminoisobutyric acid**	2.36, 10.21	6	**Formamidinoacetic acid**	2.6, 11.5	99
Carbamyl-	4.46	129	**Glutamic acid**	2.19, 4.25, 9.67	6
α-**Aminoisocaproic acid**	2.33	129	Diethyl ester	7.04	99
α-**Aminoisovaleric acid**	2.29	129	γ-Monobenzyl ester	2.17, 9.00	99
δ-**Aminolaevulinic acid**	4.05, 8.90	99	α-Monoethyl ester	3.85, 7.84	99
Aminomethylphosphonic acid	2.35, 5.9	99	γ-Monoethyl ester	2.15, 9.19	99
Aminomethylsulfonic acid	5.75	99	**Glutamine**	2.17, 9.13	6
α-**Amino-β-methyl-n-valeric acid**	2.32	129	**Glycine**	2.34, 9.6	6
1-Aminonaphthalene-2-sulfonic acid	1.71	99	N-Acetyl-	3.67	99
2-Aminonaphthalene-1-sulfonic acid	2.35	99	N,N-Bis(2-hydroxyethyl)-	2.50, 8.11	99
3-Amino-1-naphthoic acid	2.61, 4.39	99	N-n-Butyl-	2.35, 10.25	99
4-Aminopentanoic acid	3.97, 10.46	99	Carbamyl-	3.88*	97
5-Aminopentylsulfonic acid	10.95	99	Chloroacetyl-	3.38*	97

*Thermodynamic value.

AMINO ACIDS (Continued)

Compound	pK_a'	Reference	Compound	pK_a'	Reference
N,N-Diethyl-	2.04, 10.47	99	*o*-Fluoro-	2.12*, 9.01*	97
Dihydroxyethyl-	8.08*	97	*m*-Fluoro-	2.10*, 8.98*	97
N,N-Dimethyl-	2.08–, 9.80	99	*p*-Fluoro-	2.13*, 9.05*	97
N-Ethyl-	2.34*, 10.23	99	*p*-HOSO$_2$NH-	1.99*, 8.64*, 10.26*	97
Ethyl ester	7.83	99	α-Methyl-	9.57	99
Formyl-	3.43*	97	Methyl ester	7.00	99
N-Isobutyl-	2.35, 10.12	99	**Phenylglycine**	1.83, 4.39	99
Methyl ester	7.73	99	*m*-Chloro-	1.05, 3.93	99
Histamine	5.0, 9.7	6	*p*-Chloro-	1.46, 4.04	99
Histidine	6.0, 1.82, 9.17	6	*m*-Cyano-	0.28, 3.78	99
Amide	5.78, 7.64	99	*m*-Methyl-	1.89, 4.60	99
2-Mercapto-	1.84, 8.47, 11.4	99	*p*-Methyl-	1.97, 4.85	99
1-Methyl-	6.58, 8.60	99	**Proline**	1.99, 10.60	6
2-Methyl-	1.7, 7.2, 9.5	99	**2-Pyrrolidone-5-carboxylic acid**	3.32	6
Methyl ester	7.33, 5.38	99	(glutamic acid)		
Homocysteine	2.22, 8.87	101	**Sarcosine**	2.23, 10.01	6
β-**Hydroxyglutamate**	2.27, 4.29, 9.66	99	Amide	8.31*	99
N-Hydroxyethylethylenediamine-			*N*-Dimethylamide	8.82*	99
triacetic acid			*N*-Methylamide	8.24*	99
Hydroxylysine	2.13, 8.62, 9.67	6	**Serine**	2.21, 9.15	6
Hydroxyproline	1.92, 9.73	6	Amide	7.30	99
Imidazolelactic acid	2.96, 7.35	99	Methyl ester	7.10	99
Isoasparagine	2.97, 8.02	99	**Taurine**	1.5, 8.74	6
N-Acetyl	3.99	151	**Thiolhistidine**	<1.5, 11.4, 1.84, 8.47	6
N-Carbobenzoxy-	4.05	151	**Threonine**	2.63, 10.43	6
Isocreatine	2.84	99	*O*-Methyl-	2.02, 9.00	101
Isoglutamine	3.81, 7.88	99	**5,5,5-Trifluoroleucine**	2.05, 8.92	111
N-Acetyl-	4.34	151	**6,6,6-Trifluoronorleucine**	2.164, 9.46	111
N-Carbobenzoxy-	4.39	151	**4,4,4-Trifluorothreonine**	1.55, 7.82	111
Isoleucine	2.36, 9.68	6	**4,4,4-Trifluorovaline**	1.54, 8.10	111
Isoserine	2.72, 9.33	99	**Tryptophan**	2.38, 9.39	6
Leucine	2.36, 9.60	6	Amide	7.5	99
Amide	7.80	99	**Tyrosine**	10.07, 2.20, 9.11	6
Ethyl ester	7.57	99	Amide	7.48, 9.89	99
Lombricine	8.9	99	3,5-Dibromo-	2.17, 6.45, 7.60	99
Lysine	2.18, 8.95	6	3,5-Dichloro-	2.22, 6.47, 7.62	99
Hydroxy-	2.13, 8.62, 9.67	99	Diiodo-	6.48, 2.12, 7.82	6
Methionine	2.28, 9.21	6	Ethyl ester	7.33, 9.80	22
Amide	7.53	99	*O*-Methyl-	9.27	21
N-Methylaminodiacetic acid	2.15	129	*O*-Methyl, ethyl ester	7.31	22
Nitrilotriacetic acid	1.88, 2.48, 4.28	129	*N*-Trimethyl-	9.75	21
Norleucine	2.39, 9.76	6	**Urocanic acid**	5.8, 3.5	—
Norvaline	2.30, 9.78	99	**Valine**	2.32, 9.62	6
Octopine	1.40, 2.30, 8.72, 11.34	99	Amide	8.00	99
Ornithine	1.71, 8.69	6	Hydroxy-	2.55, 9.77	99
Phenylalanine	1.83, 9.13	6	β-Mercapto-	2.0, 8.0, 10.5	99
Amide	7.22	99	CH$_3$CH$_2$OCONHCH$_2$COOH	3.66*	97
o-Chloro-	2.23, 8.94	99	CH$_3$CONH(CH$_2$)$_2$COOH	4.45	97
m-Chloro-	2.17, 8.91	99	NH$_2$CONH(CH$_2$)$_2$COOH	4.49	97
p-Chloro-	2.08, 8.96	99	DL-CH$_3$CH$_2$CH(NH$_2$)COOH	2.29, 9.83	97
3,4-Dihydroxy-	2.32*, 8.68*, 9.88*	97	CH$_3$CONHCH(CH$_2$CH$_3$)COOH	3.72	97
2,4-Diiodo-3-hydroxy-	2.12*, 6.48*, 7.82*	97	NH$_2$CONHCH(CH$_2$CH$_3$)COOH	3.89	97
			NH$_2$CONH(CH$_2$)$_3$COOH	4.68	97
			(CH$_3$)$_2$C(NH$_2$)COOH	2.36*, 10.205*	97

*Thermodynamic value.

AMINO ACIDS (Continued)

Compound	pK'_a	Reference	Compound	pK'_a	Reference
NH$_2$CONHC(CH$_3$)$_2$COOH	4.46	97	(COOHCH$_2$CH$_2$)$_2$N(CH$_2$)$_2$N(CH$_2$CH$_2$C OOH)	3.00*, 3.43*, 6.77*, 9.60*	97
DL-CH$_3$(CH$_2$)$_2$CH(NH$_2$)COOH	4.36*, 9.72*	97	NH$_2$COCH$_2$COOH	3.64*	97
CH$_3$CH$_2$CH(NH$_2$)CH$_2$COOH	4.02*, 10.40*	97	NH$_2$CO(CH$_2$)$_2$COOH	4.54*	97
NH$_2$(CH$_2$)$_4$COOH	4.20*, 10.69*	97	NH$_2$CO(CH$_2$)$_3$COOH	4.60*	97
DL-(CH$_3$)$_2$CHCH(NH$_2$)COOH	2.29, 9.74	97	NH$_2$CO(CH$_2$)$_4$COOH	4.63*	97
NH$_2$(CH$_2$)$_5$COOH	4.43*, 10.75*	97			
NH$_2$(CH$_2$)$_{11}$COOH	4.65*	97	Oxyproline	1.92*, 9.73*	97
NH$_2$(CH$_2$)$_3$CH(NH$_2$)COOH	1.94*, 8.65*, 10.76*	97			
NH$_2$CONH(CH$_2$)$_3$CH(NH$_2$)COOH	2.43*, 9.41*	97			
COOH(CH$_2$)$_2$CH(NH$_2$)COOCH$_2$CH$_3$	3.85*, 7.84*	97	COOHCH$_2$CH(OH)CH(NH$_2$)COOH	2.32*, 4.23*, 9.56*	97
CH$_3$CH$_2$OCO(CH$_2$)$_2$CH(NH$_2$)COOH	2.15*, 9.19*	97	CH$_3$CH$_2$SCH$_2$CH(NH$_2$)COOH	2.03*, 8.60*	97
COOHCH$_2$NHCH$_2$COOH	2.54*, 9.12*	97	CF$_3$(CH$_2$)$_3$CH(NH$_2$)COOH	2.16*, 9.46*	97
COOHCH$_2$N(CH$_3$)CH$_2$COOH	2.15*, 10.09*	97	CF$_3$CH(CH$_3$)CH$_2$CH(NH$_2$)COOH	2.05*, 8.94*	97
N(CH$_2$COOH)$_3$	2.96*, 10.23	97	CF$_3$(CH$_2$)$_2$CH(NH$_2$)COOH	2.04*, 8.92*	97
COOH(CH$_2$)$_2$NHCH$_2$COOH	3.61*, 9.46*	97	CF$_3$CH(CH$_3$)CH(NH$_2$)COOH	1.54*, 8.10*	97
COOH(CH$_2$)$_2$NH(CH$_2$)$_2$COOH	4.11*, 9.61*	97	CF$_3$CH$_2$CH(NH$_2$)COOH	1.60*, 8.17*	97
COOHCH$_2$NH(CH$_2$)$_2$NHCH$_2$COOH	6.42*, 9.46*	97	CF$_3$CH(OH)CH(NH$_2$)COOH	1.55*, 7.82*	97
(CH$_3$)$_2$N(CH$_2$)$_2$N(CH$_2$COOH)$_2$	6.05*, 10.07*	97	CF$_3$CH(NH$_2$)CH$_2$COOH	2.76*, 5.82*	97
(COOHCH$_2$)$_2$N(CH$_2$)$_2$N(CH$_2$COOH)$_2$	6.27*, 10.95*	97			
COOH(CH$_2$)$_2$N(CH$_2$COOH) (CH$_2$)$_2$N(CH$_2$COOH(CH$_2$)$_2$COOH	3.00*, 3.79*, 5.98*, 9.83*	97			
COOH(CH$_2$)$_2$NH(CH$_2$)$_2$NH(CH$_2$)$_2$ COOH	6.87*, 9.60*	97			

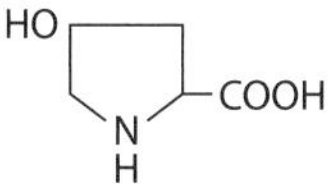

*Thermodynamic value.

PEPTIDES

Compound	pK_a'	Reference	Compound	pK_a'	Reference
Ala-Ala-(LD)	3.12, 8.30	27	Gly-Ala-Ala-Gly	3.30, 7.93	99
Ala-Ala-(LL)	3.30, 8.14	27	Gly-Asp	2.81, 4.45, 8.60	99
Ala-Ala-Ala-(3D)	3.39, 8.06	27	Gly-asparagine	2.82, 7.20	99
Ala-Ala-Ala-(DLL)	3.37, 8.06	27	Gly-Gly	3.06, 8.13	6
Ala-Ala-Ala-Ala-(DLLL)	3.42, 7.99	27	Gly-Gly-cystine	2.71, 7.94	99
Ala-Ala-Ala-(3L)	3.39, 8.03	27	Gly-Gly-Gly	3.26, 7.91	23
Ala-Ala-Ala-Ala-(4L)	3.42, 7.94	27	Gly-His	6.79, 8.20	99
Ala-Ala-Ala-(LDL)	3.31, 8.13	27	Gly-Leu	3.10, 8.41	99
Ala-Ala-Ala-Ala-(LDLL)	3.22, 7.99	27	Gly-Pro	2.81, 8.65	99
Ala-Ala-Ala-(LLD)	3.37, 8.05	27	Gly-sarcosine	2.98, 8.57	99
Ala-Ala-Ala-Ala-(LLDL)	3.24, 7.93	27	Gly-Ser	2.92, 8.10	99
Ala-Gly	3.16, 8.24	27	Gly-Ser-Gly	3.23, 7.99	99
Ala-Gly-Gly	3.19, 8.15	99	Gly-Trp	8.06	99
Ala-Lys-Ala-(3L)	3.15, 7.65, 10.30	27	Gly-Tyr	2.93, 8.45, 10.49	99
Ala-Lys-Ala-(LDL)	3.33, 7.97, 10.36	27	Gly-Val	3.15, 8.18	99
Ala-Lys-Ala-(LDLL)	3.32, 8.01, 10.37	27	His-Gly	2.36, 6.27, 8.57	99
Ala-Lys-Ala-(LLD)	3.29, 7.84, 10.49	27	His-His	5.54, 6.80, 7.82	99
Ala-Lys-Ala-Ala-(4L)	3.58, 8.01, 10.58	27	Leu-asparagine	2.83, 8.23	99
Ala-Lys-Ala-Ala-Ala-(5L)	3.53, 7.75, 10.35	27	Leu-Tyr	2.87, 8.36, 10.28	99
Ala-Lys-Ala-Ala-Ala-(LDLLL)	3.30, 7.85, 10.29	27	Lys-Ala-(LD)	3.00, 7.74, 10.63	27
β-Ala-1-methylhistidine	2.64, 7.04, 9.49	99	Lys-Ala-(LL)	3.22, 7.62, 10.70	27
Ala-Pro	3.04, 8.38	99	Lys-Glu	2.98, 4.47, 8.45, 11.30	99
β-Ala-Bis	2.73, 6.87, 9.73	99	Lys-Lys-(LD)	2.85, 7.53, 9.92, 10.98	27
Anserine	7.0, 2.65, 9.5	6	Lys-Lys-(LL)	3.01, 7.53, 10.05, 11.01	27
Asparaginyl-Gly	2.90, 7.25	99	Lys-Lys-Lys-(3L)	3.08, 7.34, 9.80, 10.54, 11.32	27
Asp-Asp	2.70, 3.40, 4.70, 8.26	99	Lys-Lys-Lys-(LDD)	2.94, 7.14, 9.60, 10.38, 11.09	27
α-Aspartyl-histidine	2.45, 3.02, 6.82, 7.98	99	Lys-Lys-Lys-(LDL)	2.91, 7.29, 9.79, 10.54, 11.42	27
β-Aspartyl-histidine	1.93, 2.95, 6.93, 8.72	99	Met-Met	2.22, 9.27	99
Asp-Gly	2.10, 4.53, 9.07	99	Methyl-Leu-Gly	3.29, 7.82	99
Asp-Tyr	2.13, 3.57, 8.92, 10.23	99	Phe-Ala-Arg	2.60, 7.54, 12.43	99
Carnosine	6.83, 9.51	6	Phe-Gly	3.13, 7.62	99
Cys-Cys	2.65, 7.27, 9.35, 10.85	99	Phenylalanylglycine amide	6.72	99
Cys-Gly-Gly	3.13, 6.36, 6.95	99	Pro-Gly	3.19, 8.97	99
Cys-Gly-Gly-Gly-Gly	3.21, 6.01, 6.87	99	Sarcosyl-Gly	3.14, 8.66	99
L-Cystinylcystine	1.87, 2.94, 6.53, 7.66	99	Sarcosyl-Leu	3.15, 8.67	99
N,N-Dimethylglycyl-glycine	3.11, 8.09	99	Sarcosylsarcosine	2.89, 9.18	99
N,N-Dimethyl-leucyl-glycine	7.78	99	Ser-Gly	3.10, 7.33	99
Glutaminyl-glutamic acid	3.14, 4.38, 7.62	99	Ser-Leu	3.08, 7.45	99
Glutaminyl-glycine	3.15, 7.52	99	Tyr-Tyr	3.52, 7.68, 9.80, 10.26	99
Glutathione	3.59, 8.75, 9.65	77	Val-Gly	3.23, 8.00	99
Glutathione, oxidized	3.15, 4.03, 8.57, 9.54	77			
Gly-Ala (L), (D)	3.17, 8.23	27			
Gly-Ala-Ala (LD)	3.30, 8.17	27			
Gly-Ala-Ala (LL)	3.38, 8.10	27			

NITROGEN COMPOUNDS

X	XNH_3^+	$XCH_2NH_3^+$	$X(CH_2)_2NH_3^+$	$X(CH_2)_3NH_3^+$	$X(CH_2)_4NH_3^+$	$X(CH_2)_5NH_3^+$	Reference

ALIPHATIC AMINES, SIMPLE

Primary Amines

X	XNH_3^+	$XCH_2NH_3^+$	$X(CH_2)_2NH_3^+$	$X(CH_2)_3NH_3^+$	$X(CH_2)_4NH_3^+$	$X(CH_2)_5NH_3^+$	Reference
H–	9.25*	10.64*	10.67*	10.58*	10.61*	10.63	2
HF_2C–	—	7.52	—	—	—	—	—
RO_2C–	—	7.75	9.13	9.71	10.15*	10.37	2
HO–	5.96*	—	9.50*	—	—	—	—
C_6H_5–	4.58*	9.37*	9.83*	10.20*	10.39*	10.49*	2
H_2N–	8.12*	—	9.98	10.65*	10.84*	11.05*	2
$H_2C=CH$–	—	9.69	—	—	—	—	—
CH_3–	10.64*	10.67*	10.58*	10.61*	10.63*	10.64*	2

X	Me_2N	–H	$–NH_3^+$	$–CO_2^-$	$–SO_3^-$	$–PO_3^-$	Reference
$X-NH_3^+$		9.25*	-0.88	—	1	10.25	2
$X(CH_2)NH_3^+$		10.64	—	9.77	5.75	10.8	2
$X(CH_2)_2NH_3^+$	5.98, 9.30	10.67	—	10.19	9.20	10.8	2, 118
$X(CH_2)_3NH_3^+$	9.91, 7.67	10.58	8.59	10.43	10.05	—	2, 118
$X(CH_2)_4NH_3^+$	8.44, 10.17	10.61	9.31	10.77	10.65	10.9	2, 118
$X(CH_2)_5NH_3^+$	9.07, 10.44	10.63	9.74	10.75	10.95	11.0	2, 118
$X(CH_2)_8NH_3^+$		10.65	10.10	—	—	—	2
$X(C_2)_{10}NH_3^+$		10.64	—	—	11.35	11.25	2

For complex chelating agents of aliphatic amines, see also Ref. 7

Compound	pK_a'	Reference	Compound	pK_a'	Reference
PRIMARY AMINES			**PRIMARY AMINES (Continued)**		
1-Acetamido-2-aminoethane	9.05*	99	2-Amino-3-hydroxyindan	8.13*	99
1-Acetoxy-2-aminoethane	9.1	99	1-Amino-2-hydroxy-2-methylpropane	9.25*	99
Acetylhydrazine	3.24*	99	2-Amino-1-hydroxy-2-methylpropane	9.71*	99
β-Alanine ester	9.13	1	1-Amino-5-hydroxypentane	10.46*	99
Allylamine	9.49	1	1-Amino-3-hydroxypropane	9.96*	99
1-Amino-2-benzamidoethane	9.13*	99	2-Amino-1-hydroxypropane	9.43*	99
1-Amino-2-benzylaminoethane	6.48*, 9.41*	99	1-Amino-2-(2-hydroxypropyl) aminoethane	6.94*, 9.86*	99
1-Amino-2-bromoethane	8.49*	99			
1-Amino-3-bromopropane	8.93*	99	1-Amino-2-(3-hydroxypropyl) aminoethane	6.78*, 9.67*	99
1-Amino-2-butylaminoethane	7.53*, 10.30*	99			
(3-Amino)butylbenzene	9.79*	99	2-Aminoindan	9.57*	99
(4-Amino)butylbenzene	10.36*	99	5-Aminoindan	5.31*	99
γ-Amino-n-butyric acid ester	9.71	1	1-Amino-2-isopropylaminoethane	7.70*, 10.62*	99
1-Amino-2-diethylaminoethane	7.07*, 10.02*	99	Aminomalonic acid	3.32, 9.83	77
1-Amino-2-dimethylaminoethane	6.63*, 9.53*	99	1-Amino-2-mercaptoethane	8.27*, 10.53*	99
1-Amino-2,2-dimethylpropane	10.24*	99	1-Amino-2-methylbutane	10.64	99
1-Amino-2-ethylaminoethane	7.63*, 10.56*	99	1-Amino-3-methylbutane	10.60*	99
1-(2-Aminoethyl)piperidine	6.38*, 9.89*	99	2-Amino-2-methylbutane	10.72*	99
2-Aminoethylsulfonic acid	9.08	77	1-Amino-3-methylcyclohexane	10.56* cis, 10.61* trans	99
(2-Aminoethyl)trimethylammonium chloride	7.1	99			
			1-Amino-2-methylpropane	10.72*	99
1-Aminofluorene	3.87*	99	2-Amino-2-methylpropane	10.68*	99
2-Aminofluorene	4.64*	99	1-Amino-2-methylaminoethane	6.86*, 10.15*	99
3-Aminofluorene	4.82*	99	1-Amino-2-methylthioethane	9.49*	99
4-Aminofluorene	3.39*	99	1-Aminononane	10.64*	99
1-Amino-2-furfurylaminoethane	6.20*, 9.72*	99	1-Aminooctane	10.65*	99
1-Aminoheptane	10.66*	99	2-Aminooctane	10.49*	99
2-Aminoheptane	10.67*	99	1-Aminopentane	10.63*	99
1-Aminohexane	10.56*	99	3-Aminopentane	10.42*	99
1-Amino-4-hydroxybutane	10.35*	99	1-Amino-2-propylaminoethane	8.24*, 11.04*	99
2-Amino-1-hydroxybutane	9.52*	99	1-Aminoprop-2-ene	9.49*	99
1-Amino-2-hydroxycycloheptane	9.25*	99	1-Aminoprop-2-yne	8.15*	99
1-Amino-2-hydroxycyclopentane	9.70* cis, 9.28 trans	99	1-Aminotetradecane	10.62*	99
2-Amino-2'-hydroxydiethyl sulfide	9.04, 9.41	77, 99	1-Amino-3,3,3-trichloropropane	9.65*	99
1-Amino-2-(2-hydroxyethyl) aminoethane	6.83*, 9.82*	99	Benzamide	-1.85	99
			Benzoylhydrazine	2.97	114

*Thermodynamic value.

NITROGEN COMPOUNDS (Continued)

Compound	pK_a'	Reference	Compound	pK_a'	Reference
PRIMARY AMINES (Continued)			*PRIMARY AMINES (Continued)*		
Benzyl-	9.34	1	Methyl benzimidate	5.8*	99
bis(2-aminoethyl)disulfide	8.82*, 9.16*	99	*N*-Methylethylenedi-	7.56, 10.40	77
1,2-bis(2-aminoethyl)thioethane	8.69*, 9.62*	99	2-Methylthioethyl-	9.18	77
1,2-bisglycylamidoethane	7.63*, 8.35*	99	Octyl-	10.65	153
n-Butyl-	10.59	1	*neo*-Pentyl-	10.21	1
t-Butyl-	10.55	1	Phenylamyl-	10.49	2
N-n-Butylethylenedi-	7.53, 10.30	77	(δ-Phenylbutyl-	10.40	2
Carbamylmethyl-	7.93	153	β-Phenylethyl-	9.83	1
N-(Carbamoylmethyl)iminodiacetic acid	2.30, 6.60	77	Phenylmethyl-	9.34	153
Cyanoethyl-	7.7	153	γ-Phenylpropyl-	10.20	1
Cyanomethyl-	5.34	153	Propionamide	−0.49*	99
Ethoxycarbonylethyl-	9.13	153	*n*-Propyl-	10.53	1
Cyclohexyl-	10.64	1	*N-n*-Propylethylenedi-	7.54, 10.34	77
Cyclohexylmethyl	10.49	1	*sec*-Butyl-	10.56	1
1,4-Diaminobutane-	9.24, 10.72	146	Semicarbazide	3.65*	99
2,3-Diaminobutane, *meso*	6.92, 9.97	77	1,2,3-Triaminopropane	3.72, 7.95, 9.59	77
2,3-Diaminobutane, *racemic*	6.91, 10.00	77	Triaminotriethyl-	8.56, 9.59, 10.29	77
2,2′-Diaminodiethyl-	3.58, 8.86, 9.65	77	β,β′,β″-Triaminotriethyl-	8.42, 9.44, 10.13	87
2,2′-Diaminodiethyl sulfide	8.84, 9.64	77	2,2,2-Trichloroethyl-	5.47*	99
2,3-Diamino-2,3-dimethylbutane	6.56, 10.13	77	2-Trienylmethyl-	8.92	77
1,3-Diamino-2,2-dimethylpropane	8.18, 10.22	77	Triethylenedi-	8.8*	—
3,3′-Diaminodi-*n*-propyl-	8.02, 9.70, 10.70	77	Trifluoroethyl-	5.7	10
N,N′-Di-(2-aminoethyl)ethylenedi-	3.32, 6.67, 9.20, 9.92	77	Trimethylsilylmethyl-	10.96	1
1,2-Di-(2-aminoethylthio)ethane	8.42, 9.32	77	Thioacetamide	−1.76*	99
1,3-Diamino-2-hydroxypropane	7.93*, 9.69*	99	Tris-(hydroxymethyl)aminomethane	8.10	77
1,2-Diamino-2-methylpropane	6.79, 10.00	77	Undecyl-	10.63	153
1,3-Diaminopropan-2-ol	8.23, 9.68	77	Vinylmethyl-	9.69	153
1,2-Diaminopropane	7.13*, 10.00*	99	$CF_3SO_2^-$	5.8	128
1,3-Diaminopropane	8.64*, 10.62*	99			
β-Difluoroethyl-	7.52	1	*SECONDARY AMINES*		
N,N′-Diglycylethylenedi-	7.63, 8.35	77	*N*-(2-Acetamido)-2-aminoethane-sulfonic acid	6.9	108
2,3-Dimethoxybenzyl-	9.41*	99	Acetamidoglycine	7.7	108
3,4-Dimethoxybenzyl-	9.39*	99	Acetanilide	0.61	4
N,N′-Diethylethylenedi-	7.07, 10.02	77	1-Acetylpiperazine	7.94*	99
N,N′-Dimethylethylenedi-	6.63, 9.53	77	Allylmethyl-	10.11	1
4-4′-Diaminostilbene	3.9*, 5.2*	99	*N*-2-Aminoethylpiperazine	8.51*, 9.63*	99
1-Dimethylamino-3-hydrazinobutane	5.90*, 9.23*	99	Aminomethylcyclohexane	10.59*	99
Ethanol-	9.50	1	1-4-Benzoquinoneimine	3.9*	99
Ethyl-	10.63	1	*N*-Benzoylpiperazine	7.78	1
Ethylenedi-	9.98, 7.52	1, 77	Benzylethyl-	9.68	1
Ethylenediamine-*N,N*-diacetic acid	7.63, 8.35	77	Benzylmethyl-	9.58	1
N-Ethylethylenedi-	7.63, 10.56	77	α-Benzylpyrrolidine	10.36	2
Furfuryl-	8.89	77	α-Benzylpyrroline-	7.08	2
D-Glucos-	7.75*	99	1,2-Bisethylaminoethane	7.70*, 10.46*	99
Glycine ester	7.75	1	1,2-Bisfurfurylaminoethane	5.74*, 8.61*	99
Hexadecyl-	10.61	153	1,2-Bisisopropylaminoethane	7.59*, 10.40*	99
Hydrazine	8.10	1	1,2-Bismethylaminoethane	7.40*, 10.16*	99
N-(2-Hydroxyethyl)ethylenedi-	6.83, 9.82	77	1,2-Bispropylaminoethane	7.53*, 10.27*	99
Hydroxyl-	5.97	1	*N*-Butylaminoacetic acid	2.29, 10.07	77
2-Hydroxy-3-methoxybenzylamine	8.70, 11.06*	99	*t*-Butylcyclohexyl-	11.23	1
3-Hydroxy-2-methoxybenzylamine	8.89*, 10.54*	99	*N*-Carbethoxypiperazine	8.28	1
4-Hydroxy-3-methoxybenzylamine	8.94*, 10.52*	99	β-Carboxymethylaminopropionic acid	3.61, 9.46	77
2-(2-Hydroxypropylamino)ethyl-	6.94, 9.86	77	*cis*-2,6-Dimethylpiperidine	10.92	3
2-(3-Hydroxypropylamino)ethyl-	6.78, 9.76	77	α-Cyclohexylpyrrolidine	10.80	2
N-Isopropylethylenedi-	7.70, 10.62	77	α-Cyclohexylpyrroline	7.95	2
Isopropyl-	10.63	1	Diallyl-	9.29	1
Mercaptoethyl-	8.27, 10.53	77	Di-*n*-butyl-	11.25	1
Methoxy-	4.60*	12	Diethyl-	10.98	1
Methoxycarbonylmethyl-	7.66	153	*N,N′*-Diethylethylenedi-	7.70, 10.46	77
2-Methoxyethyl-	9.20	77	Di-(trimethylsilylmethyl)-	11.40	1
Methyl-	10.62	1	Di-(hydroxyethyl)amine	8.88*	99
N-Methylaminoacetic acid	2.24, 10.01	77	Diisobutyl-	10.50	1
Methyl-α-amino-β-mercaptopropionate	6.56, 8.99	77	Diisopropyl-	11.05	1
			Dimethyl-	10.64	1

*Thermodynamic value.

NITROGEN COMPOUNDS (Continued)

Compound	pK'_a	Reference	Compound	pK'_a	Reference
SECONDARY AMINES (Continued)			**Tertiary Amines (Continued)**		
N,N'-Dimethylethylenedi-	7.40, 10.16	77	1-Diethylaminobutan-(4)-	10.1	5
N,O-Dimethylhydroxyl-	4.75*	12	1-Diethylaminohexan-(6)-	10.1	5
1-Diphenylmethoxy-2-methylaminoethane	9.12*	99	1-Diethylaminohexanethiol-(6)-	10.1	5
			1-Diethylaminopropan-(3)-	8.0, 10.5	5
Di-n-propyl-	11.00	1	N-Diethylcysteamine	7.8, 10.75	5
N,N'-Di-n-propylethylenedi-	8.14, 10.97	77	Di-(2-hydroxyethyl)aminoacetic acid	8.08	77
Di-sec-butyl-	11.01	1	Dimethylaminoacetic acid	2.08, 9.80	77
N-Ethylaminoacetic acid	2.30, 10.10	77	1-Dimethylamino-2-hydroxyethane	9.31*	99
Ethylenediamine-N,N'-diacetic acid	6.42, 9.46	77	1-Dimethylaminoprop-2-ene	8.64*	99
α-Ethylpyrrolidine	10.43	2	1-Dimethylaminoprop-2-yne	6.97*	99
α-Ethylpyrroline	7.43	2	Dimethyl-n-butyl-	10.02	1
Iminodiacetic acid	2.98, 9.89	77	Dimethyl-t-butyl-	10.52	1
Iminodipropionic acid	4.11, 9.61	77	N-Dimethylcysteamine	7.95, 10.7	5
N-Isopropylaminoacetic acid	2.36, 10.06	77	Dimethylethyl-	9.99	1
Methylaminocyclopentane	10.85*	99	N-Dimethylhydroxylamine	5.20*	12
1-Methylaminoprop-2-ene	10.11*	99	Dimethylisobutyl-	9.91	1
N-Methylglucamine	9.62*	99	Dimethylisopropyl-	10.30	1
N-Methylhydroxyl-	5.96*	12	N,N-Dimethylmethoxy-	3.65	1
N-Methylmethoxy-	4.75	1	1,2-Dimethylpiperidine	10.26	2
2-Methylpiperidine	11.08*	99	Dimethyl-n-propyl	9.99	1
3-Methylpiperidine	11.07*	99	1,2-Dimethylpyrrolidine	10.26	2
Methyltrifluorethyl-	6.05	10	1,2-Dimethyl-Δ²-pyrroline	11.94	2
Morpholine	8.36	1	Dimethyl-sec-butyl-	10.40	1
Piperazine	5.68, 9.82	77	1,2-Dimethyl-Δ²-tetrahydropyridine	11.57	2
Piperidine	11.22	1	Dimethyltrifluoroethyl-	4.75	10
N-n-Propylaminoacetic acid	2.28, 10.03	77	l-Dipropylaminopropane	10.26*	99
1-Propylaminopropane	11.00	11	N-Dipropylcysteamine	8.00, 10.8	5
Pyrrolidine	11.27	1	1-Ethoxycarbonyl-4-methylpiperazine	7.31*	99
α-(p-Tolyl)-pyrrolidine	10.01	2	N-Ethyl-cis-2,3-iminobutane	8.56	7
α-(p-Tolyl)pyrroline	7.59	2	N-Ethyl-1,2-iminobutane	8.18	7
1-Tosylpiperazine	7.39	3	1-Ethyl-2-methylpiperidine	10.70	2
Trimethyleneimine	11.29	1	l-Ethyl-2-methyl-Δ²-pyrroline	11.92	2
N-Tris(hydroxymethyl)methylglycine	8.15	108	1-Ethyl-2-methylpyrrolidine	10.64	2
			1-Ethyl-2-methyl-Δ²-tetrahydropyridine	11.57	2
TERTIARY AMINES			N-Ethylmorpholine	7.70	1
N-(2-Acetamido)iminodiacetic acid	6.6	99	N-Ethylpiperidine	10.40	1
1-Acetyl-2-diethylaminoethane	9.04*	99	N-Ethyl-trans-2,3-iminobutane	9.47	7
Allyldimethyl-	8.73	1	Hexamethylenetetra-	5.13	77
N-Allylmorpholine	7.05	1	N-2-Hydroxyethyliminodiacetic acid	2.2, 8.73	77
N-Allylpiperidine	9.68	1	N-2-Hydroxyethylpiperazine-N'-2-ethanesulfonic acid	7.55	99
4-(2-Aminoethyl)morpholine	4.84, 9.45	77			
1-Benzoyl-4-methylpiperazine	6.78*	99	1-Hydroxy-2-(2-hydroxyethylmethyl)aminoethane	8.52*	99
1-Benzylcarbonyl-2-diethylaminoethane	9.40*	99			
			1,2-Iminoethane	7.93	7
1-Acetyl-2-dimethylaminoethane	8.37*	99	1-Methyl-2-n-butylpyrrolidine	10.24	2
1-Benzylcarbonyl-2-dimethylaminoethane	8.30*	99	1-Methyl-2-n-butyl-Δ²-pyrroline	11.90	2
			N-β-Mercaptoethylmorpholine	6.65, 9.8	5
Benzyldiethyl-	9.48	1	N-β-Mercaptoetnylpiperidine	7.95, 11.05	5
Benzyldimethyl-	8.93	1	Methyl-β-diethylaminoethylsulfide	9.8	5
Bis(2-chloroethyl)aminoethane	6.55*	99	Methyldiethyl-	10.29	1
1-Bis(2-chloroethyl)amino-2-methoxyethane	5.45*	99	Methyliminodiacetic acid	2.81, 10.18	77
			1-Methyl-4-nitrosopiperazine	5.93*	99
1,2-Bisdimethylaminopropane	5.40*, 9.49*	99	N-Methylmorpholine	7.41	1
1,3-Bisdimethylaminopropane	7.7*, 9.8*	99	N-Methylpiperidine	10.08	1
N,N-Bis(2-hydroxyethyl)-2-aminoethanesulfonic acid	7.15	99	N-Methylpyrrolidine	10.46	1
			N-Methyltrimethyleneimine	10.40	1
N,N-Bis(2-hydroxyethyl)glycine	8.35	99	2-(N-Morpholino)ethanesulfonic acid	6.15	99
1-n-Butylpiperidine	10.47*	99	Piperazine-N,N'-bis(2-ethanesulfonic acid	6.8	99
1-n-Butyl-2-methyl-Δ²-pyrroline	11.90	2			
1-Chloro-2-diethylaminoethane	8.80*	99	Propargyldimethyl-	7.05	1
N-Chloro-N-ethylaminoethane	1.02*	99	Propargylethyldimethyl-	8.88	1
N-Chloro-N-methylaminomethane	0.46*	99	Propargylmethyldimethyl-	8.33	1
1-Cyanomethylpiperidine	4.55*	99	1-n-Propylpiperidine	10.48	2
Diallylmethyl-	8.79	1	Triallyl-	8.31	1
Diethylaminoacetic acid	2.04, 10.47	77	Tri-n-butyl-	10.89	1

*Thermodynamic value.

NITROGEN COMPOUNDS (Continued)

Compound	pK_a'	Reference
TERTIARY AMINES (Continued)		
Triethanol-	7.77	1
Triethyl-	10.65	1
Triethylenedi-	4.18, 8.19,	77
Trimethyl-	9.76	1
	2.97, 8.82	116
Trimethylhydroxylamine	3.65	12
Tri-n-propyl-	10.65	1
N-Tris(hydroxymethyl)-methyl-2-aminoethanesulfonic acid	7.5	99
$(CH_3)_2NCH_3$	9.76	119
$(CH_3CH_2)_2NCH_3$	10.29	119
$(CH_2)_2\,NCH_3$	7.86	119
N-Methylaziridine		
$(CH_2)_3\,NCH_3$	10.40	119
N-Methylazetidine		
$(CH_2)_4\,NCH_3$	10.46	119
N-Methylpyrrolidine		
$(CH_2)_5\,NCH_3$	10.08	119
N-Methylpiperidine		

BENZYLAMINES, MONOSUBSTITUTED

	2	3	4	99
Chloro-	5.20*	—	—	—
Methoxy-	9.70*	9.15*	9.47*	—
Methyl-	9.19*	9.33*	9.36*	—
Sulfamoyl-	8.53*	8.55*	8.52*	—
	10.11*	10.14*	10.08*	—

Compound	pK_a'	Reference
CYANOAMINES		
N-Piperidine-CH_2CN		8
Et_2NCN	−2.0	8
$Et_2N(CH_2)_2CN$	7.65	8
$Et_2N(CH_2)_4CN$	10.08	8
$Et_2NC(CH_3)_2CN$	9.13	8
$EtN(CH_2CN)_2$	−0.6	8
$EtN(CH_2CH_2CN)_2$	4.55	8
H_2NCH_2CN	5.34	8
N-Amphetamine-$(CH_2)_2$-CN	7.23	8
N-Norcodeine-$(CH_2)_2CN$	5.68	8
Dimethylcyanimide	1.2	9
Diethylcyanimide	1.2	9
Aminoacetonitrile	5.3	9
Diethylaminoacetonitrile	4.5	9
2-Amino-2-cyanopropane	5.3	9
β-Isopropylaminopropionitrile	8.0	9
β-Diethylaminopropionitrile	7.6	9
1-Amino-2-cyanoethane	7.7*	99
Et_2NCH_2CN	4.55	8
$Et_2N(CH_2)_3CN$	9.29	8
$Et_2N(CH_2)_5CN$	10.46	8
$HN(CH_2CN)_2$	0.2	8
$HN(CH_2CH_2CN)_2$	5.26	8
$N(CH_2CH_2CN)_3$	1.1	8
N-Piperidine-$C(CH_3)_2CN$	9.22	8
N-Methamphetamine-$(CH_2)_2CN$	6.95	8
Methyl cyanamide	1.2	9
Ethyl cyanamide	1.2	9
Cyanamide	1.1	9
Dimethylaminoacetonitrile	4.2	9

*Thermodynamic value.

Compound	pK_a'	Reference
CYANOAMINES (Continued)		
β-Aminopropionitrile	7.7	9
β-Dimethylaminopropionitrile	7.0	9
β,β''-Dicyanodiethylamine	5.2	9
CYCLIC AMINES		
1,2-Iminoethane	7.98	7
cis-2,3-Iminobutane	8.72	7
1,2-Imino-2-methylpropane	8.61	7
1,2-Iminobutane	8.29	7
$trans$-2,3-Iminobutane	8.69	7
In 80 percent methyl cellosolve:		
Pentamethylene-	9.99	2
Hexamethylene-	10.00	2
Heptamethylene-	9.77	2
Octamethylene-	9.39	2
Nonamethylene-	9.14	2
Decamethylene-	9.04	2
Undecamethylene-	9.31	2
Dodecamethylene-	9.31	2
Tridecamethylene-	9.35	2
Tetradecamethylene-	9.35	2
Hexadecamethylene-	9.29	2
Heptadecamethylene-	9.27	2
Cyclohexyl-	9.82	2
Cycloheptyl-	9.99	2
Cycloöctyl-	10.01	2
Cyclononyl-	9.95	2
Cyclodecyl-	9.85	2
Cycloundecyl-	9.71	2
Cyclododecyl-	9.62	2
Cyclotridecyl-	9.63	2
Cyclotetradecyl-	9.54	2
Cyclopentadecyl-	9.54	2
Cycloheptadecyl-	9.57	2
Cyclooctadecyl-	9.54	2

PHENYLETHYLAMINES

Compound	pK_a'	Reference
2-Phenylethylamine	9.78	11
N-Methyl-2-(3,4-dihydroxyphenyl)-ethylamine	8.78	11
N-Methyl-2-phenyl-	10.31	11
Epinephrine	8.55	11
Arterenol	8.55	11

R_1	R_2	R_3	R_4	pK_a'
H	H	H	H	9.78
H	H	OH	H	8.90
H	OH	OH	H	8.81
OH	H	OH	H	8.67
H	OH	H	H	9.22
OH	OH	H	H	8.93
OH	OH	OH	H	8.58
H	H	H	CH_3	10.31
H	H	OH	CH_3	9.31
H	OH	OH	CH_3	8.62
OH	H	OH	CH_3	8.89
H	OH	H	CH_3	9.36
OH	OH	H	CH_3	8.78
OH	OH	OH	CH_3	8.55

NITROGEN COMPOUNDS (Continued)

Compound	pK$_a'$	Reference	Compound	pK$_a'$	Reference
ALKALOIDS AND DERIVATIVES			*ALKALOIDS AND DERIVATIVES (Continued)*		
Acetylscopolamine	7.35*	99	10-Hydroxycodeine	7.12*	99
Aconitine	8.35*	99	Hyoscyamine	9.68*	99
Alypine	3.8*, 9.5*	99	Isolysergic acid	3.33*, 8.46*	99
Anhydroplatynecine	9.40*	99	Isopilocarpine	7.18*	99
Apomorphine	7.20*, 8.92*	99	Isoretronecanol	10.83*	99
Aposcopolamine	7.72*	99	Lysergic acid	3.32*, 7.82*	99
Arecaidine	9.07*	99	*N*-Methyl-1-benzoylecgonine	8.65*	99
Arecaidine methyl ester	7.64*	99	6-Methyl ergoline	8.87*	99
Arecoline	7.41*	99	Morphine	8.07*, 9.85*	99
Aspidospermine	7.63*	99	Myosmine	5.26*	99
Atropine	9.85*	99	Narceine	3.5*, 9.3*	99
Benzoylecgonine	11.80*	99	Narcotine	5.86*	99
Benzoylecgonine methyl ester	8.74*	99	Nicotine	3.13*, 8.02*	99
N-Benzyltriacanthine	5.94*	99	Nicotine dimethohydroxide	7.88*, 10.23*	99
Berberine	11.73*	99	Nicotine isomethohydroxide	5.35*, 11.72*	99
Brucine	2.50*, 8.16*	99	Nicotine methohydroxide	8.54*, 12.04*	99
N-Butylveratramine	7.20*	99	Nicotine monomethobromide	3.09*	99
Cevadine	9.05*	99	Nicotine oxide	5.00*	99
Cinchonidine	4.17*, 8.40*	99	Nicotyrine	4.76*	99
Cinchonine	4.28*, 8.35*	99	Norcodeine	5.68*	99
Cocaine	8.39*	99	Norcurarine	8.5*	99
Codeine	8.21*	99	Norhyoscyamine	10.28*	99
Colchicine	1.85*	99	Nornicotyrine	4.35*	99
Cupreine	7.63*	99	Optochine	4.05*, 8.5*	99
Cytisine	1.20*, 8.12*	99	Papaverine	6.40*	99
Deacetylaspidospermine	2.70*, 8.45*	99	Pelletierine	9.45*	99
Desoxyretronecine	9.51*	99	*N*-Pentylveratramine	7.28*	99
Dicodide	7.95*	99	Perlolidine	4.01*, 11.39*	99
Dihydroarecaidine	9.70*	99	Perloline	8.54*	99
Dihydroarecaidine, methyl ester	8.39*	99	Physostigmine	1.96*, 8.08*	99
Dihydrocodeine	8.75	99	Pilocarpate	7.47*	99
α-Dihydrolysergic acid	3.57*, 8.45*	99	Pilocarpine	1.63*, 7.05*	99
γ-Dihydrolysergic acid	3.60*, 8.71*	99	Piperine	1.98*	99
Dihydromorphine	9.35*	99	Platynecine	10.20*	99
Dihydroergonovine	7.38*	99	*N*-Propylveratramine	7.20*	
α-Dihydrolysergol	8.30*	99	Pseudoecgonine	9.70*	99
β-Dihydrolysergol	8.23*	99	Pseudoecgonine methyl ester	8.15*	99
Dihydronicotyrine	7.07*	99	Pseudotropine	9.86*	99
Dilaudide	7.8*	99	Quinidine	4.2*, 8.77*	99
Ecgonine	10.91	99	Quinine	4.32*, 8.4*	99
Ecgonine methyl ester	9.16*	99	Retronecanol	10.88*	99
Emetine	7.56*, 8.43*	99	Retronecine	8.88*	99
Ergobasine	6.79*	99	Scopolamine	7.55*	99
Ergobasinine	7.43*	99	Scopoline	8.20*	99
Ergometrinine	7.32*	99	Sempervirine	10.6*	99
Ergonovine	6.73*	99	Solanine	7.54*	99
Ethylmorphine	8.08*	99	Sparteine	4.80*, 11.96*	99
N-Ethylveratramine	7.40*	99	Stovaine	7.9*	99
β-Eucaine	9.35*	99	Strychnine	8.26*	99
Eucodal	8.6*	99	Tetradehydroyohimbine	10.69*	99
Gelsemine	9.79*	99	Tetrahydro-α-morphimethine	8.65*	99
Harmine	7.61*	99	Tetrahydroserpentine	10.55*	99
Harmol	7.86*, 9.51*	99	Thebaine	8.15*	99
Heliotridane	11.40*	99	Theobromine	10.00*	99
Heliotridene	10.55*	99	Theophylline	8.6*	99
Heroin	7.6*	99	Triacanthine	6.0*	99
Homatropine	9.7*	99	Tropacocaine	9.88*	99
Hydrastine	6.63*	99	Tropine	10.33*	99
Hydrastinine	11.58*	99	Veratramine	7.49*	99
Hydroquinine	8.87*	99	Yohimbine	3*, 7.45*	99

*Thermodynamic value.

NITROGEN COMPOUNDS (Continued)

ANILINES (2, 99)

Monosubstituted

Substituent	ortho	meta	para
H-	4.62*	4.64*	4.58*
$(CH_3)_3N^+$-	—	2.26	2.51
CH_3O_2C-	2.16	3.56	2.30
CH_3SO_2-	—	2.68*	1.48
CH_3S-	—	4.05	4.40
Br-	2.60*	3.51*	3.91*
F-	2.96*	3.38*	4.52*
CH_3O-	4.49*	4.20*	5.29*
C_6H_5-	3.78*	4.18	4.27*
$(CH_3)_3C$-	3.78	—	—
^-O_3S-	—	3.80	3.32
H_3N^+-	1.3	2.65	3.29
O_2N-	−0.28*	2.45*	0.98*, 1.11*
HO_2C-	2.04	3.05	2.32
$C_2H_5O_2C$-	2.10	—	2.38
F_3C-	—	3.49*	2.57*
HO-	4.72	4.17	5.50
Cl-	2.62*	3.32*	3.81*
$(CH_3)_3Si$-	—	4.64*	4.36*
C_2H_5O-	4.47*	4.17*	5.25*
CH_3-	4.38*	4.67*	5.07*
$^-HO_3As$-	3.77	4.05	4.05
H_2N-	4.47	4.88	6.08
CH_3CO-	2.22*	3.59*	2.19*
CN-	0.95*	2.75	1.74
C_2H_5-	4.37*	4.70*	—
C_6H_5CO-	—	—	2.17*
n-Butyl	4.26*	—	—
t-Butyl	5.03*	4.66*	4.95*
HCO-	—	—	1.76*
I-	2.60*	3.61*	3.78*
Isopropyl-	4.42*	4.67*	—
HS-	3.00*	—	—
	6.59*	—	—
Sulfamoyl	1.0*	2.90*	2.02*

Compound	pK_a'	Reference
3-Amino-2,6-dihydroxyaniline	2.9*, 5.6*, 9.3*, 11.5*	99
3-Amino-4,6-dihydroxyaniline	3.8*, 6.0*, 9.8*, 12.0*	99
3-Amino-2-hydroxyaniline	2.7*, 5.5*, 10.5*	99

Compound	pK_a'	Reference
3-Amino-4-hydroxyaniline	3.1*, 5.7*, 10.5*	99
4-Bromo-2,6-dimethylaniline	3.54*	99
3-Bromo-4-methylaniline	3.98*	99
4-Bromo-2-methylaniline	3.58*	99
3,5-di-t-Butylaniline	4.97	88
3-Chloro-5-methoxyaniline	3.10	88
3-Chloro-4-methylaniline	4.05*	88
2,4-Diaminoaniline	3.7*, 6.1*	99
2,4-Dibromoaniline	2.3*	99
2,6-Dibromoaniline	0.38*	99
3,5-Dibromoaniline	2.34*	99
2,4-Dichloroaniline	2.05*	99
2,5-Dichloroaniline	1.57*	99
3,5-Dichloroaniline	2.37	138
2,4-Dihydroxyaniline	5.7*, 9.3*, 11.3*	99
2,6-Dihydroxyaniline	5.1*, 9.3*, 11.6*	99
3,5-Diiodoaniline	2.37	138
3,5-Dimethoxyaniline	3.82	88
3-Methoxy-5-nitroaniline	2.11	88
2,3-Dimethylaniline	4.70*	99
2,4-Dimethylaniline	4.84*	99
2,5-Dimethylaniline	4.57*	99
2,6-Dimethylaniline	3.89*	99
3,4-Dimethylaniline	5.17*	99
3,5-Dimethylaniline	4.91*	99
2,4-Dinitroaniline	−4.27	120
3,5-Dinitroaniline	0.23	138
2,3,5,6-Tetramethylaniline	4.30*	99
2,4,6-Trimethylaniline	4.38*	99
3,4,5-Trimethylaniline	5.12*	99

Compound	pK_1	pK_2	Ref.
3-Amino-5-nitrobenzoic acid	1.55	3.55	147
methyl ester	1.47	—	147
ethyl ester	1.52	—	147
Ethyl N-(m-carboxyphenyl)glycinate	1.15	4.30	147
Ethyl N-(m-methoxycarbonylphenyl)glycinate	1.06	—	147
Ethyl N-(m-ethoxycarbonylphenyl)glycinate	1.11	—	147
Ethyl N-phenylglycinate	2.08	—	147
Methyl N-phenylglycinate	2.07	—	147
Ethyl N-(p-carboxyphenyl)glycinate	—	4.88	147

N-SUBSTITUTED ANILINES

R	C_6H_5NHR	$C_6H_5N(CH_3)R$	$C_6H_5NR_2$	$2\text{-}CH_3C_6H_4NHR$	$2\text{-}CH_3C_6H_4NR_2$
				(2)	
H-	4.58	4.85	4.58	4.39	4.39
CH_3-	4.85	5.06	5.06	4.59	5.86
C_2H_5-	5.11	5.98	6.56	4.92	7.18
n-C_3H_7-	5.02	—	5.59	—	—
n-C_4H_9-	4.95	—	~5.7	—	—
i-C_4H_9-	—	5.20	—	—	—
sec-C_4H_9-	—	6.03	—	—	—
t-C_6H_{12}-	6.30	—	—	—	—
Cyclopentyl-	5.30	6.71	—	5.07	—
Cyclohexyl-	5.60	6.35	—	5.34	—
t-C_4H_9	6.95	7.52	—	6.49	—

* Thermodynamic value.

NITROGEN COMPOUNDS (Continued)

N-SUBSTITUTED ANILINES[119]

R=	$4\text{-}NO_2C_6H_4R$	$4\text{-}HOOCC_6H_4R$	C_6H_5R
$(CH_3)_2N\text{-}$	0.65	1.40	4.22
			4.39
$CH_3CH_2N\text{-}$	1.75	2.45	5.71
			5.85
$(CH_2)_3\text{-}N\text{-}$ Azetidine	0.34		5.59
			4.08
$(CH_2)_4\text{-}N\text{-}$ Pyrrolidine	−0.42	0.39	3.71
			3.45
			3.24
$(CH_2)_5\text{-}N\text{-}$ Piperidine	2.46	2.67	4.60
			5.22
			4.93
$(CH_2)_6\text{-}N\text{-}$ Hexahydroazepine	−0.15		

Compound	pK'_a	Reference	Compound	pK'_a	Reference
N,N-DIETHYL-			***ORTHO-SUBSTITUTED, IN 50% ETHANOL***		
2,4-Dinitro-	0.18*	99	H-	4.25	2
2-Methyl-	7.23*	99	$2\text{-}CH_3\text{-}$	3.98, 4.09	2
3-Methyl-	7.12*	99	$2,3\text{-}(CH_3)_2\text{-}$	4.42	2
4-Methyl-	7.13*	99	$2,4\text{-}(CH_3)_2\text{-}$	4.61	2
4-Nitroso-	4.11*	99	$2,5\text{-}(CH_3)_2\text{-}$	4.17, 4.23	2
N,N-DIMETHYL-			$2,6\text{-}(CH_3)_2\text{-}$	3.42, 3.49	2
H–	5.07	52	$3,5\text{-}(CH_3)_2\text{-}$	4.48	2
$m\text{-}NO_2\text{-}$	2.63	52	$2\text{-}CH_3\text{-}$	4.09	2
$m\text{-}CN\text{-}$	2.97	52	$2\text{-}(CH_3)_2CH\text{-}$	4.06	2
$p\text{-}NO_2$	0.61	52	$2\text{-}(CH_3)_3C\text{-}$	3.38	2
$p\text{-}CN\text{-}$	1.78	52	$2,6\text{-}(CH_3)_2\text{-}4\text{-}(CH_3)_3C\text{-}$	3.88	2
$p\text{-}NO\text{-}$	4.54	52	$2,4\text{-}(CH_3)_2\text{-}6\text{-}(CH_3)_3C\text{-}$	3.43	2
N-DIMETHYL-, IN 50% ETHANOL			$2\text{-}CH_3\text{-}4,6\text{-}(CH_3)_3C\text{-}$	3.31	2
H-	4.21, 4.09	2	$2,4,6\text{-}[(CH_3)_3C]_3\text{-}$	< 2	2
$m\text{-}CH_3\text{-}$	4.66	2	***2-NITROANILINE***		
$p\text{-}C_2H_5\text{-}$	4.69	2	R=H-	−0.29	120
$o\text{-}(CH_3)_2CH\text{-}$	5.05	2	$4\text{-}CH_3O\text{-}$	0.77	120
$p\text{-}CH_3CH_2CH_2CH_2\text{-}$	4.62	2	$4\text{-}CH_3\text{-}$	0.43	120
$o\text{-}(CH_3)_3C\text{-}$	4.26	2	4-F-	−0.44	120
$p\text{-}I\text{-}$	3.43, 2.73	2	4-Cl-	−1.03	120
$p\text{-}Br\text{-}$	3.52, 2.82	2	4-Br-	−1.05	120
$p\text{-}Cl\text{-}$	3.33	2	$4\text{-}CF_3\text{-}$	−2.25	120
$m\text{-}(CH_3)_3Si\text{-}$	4.41	2	$4\text{-}CH_3OCO$	−2.61	120
$o\text{-}CH_3O\text{-}$	5.49	2	$4\text{-}NO_2\text{-}$	−4.27	120
$o\text{-}CH_3\text{-}$	5.15, 5.07	2	$4\text{-}CH_3CO\text{-}$	−2.85	120
$p\text{-}CH_3\text{-}$	4.94	2	4-HO-	1.20	120
$p\text{-}CH_3CH_2CH_2\text{-}$	4.43	2	$3\text{-}CH_3\text{-}$	−0.09	120
$p\text{-}(CH_3)_2CH\text{-}$	4.77	2	$3\text{-}CH,O\text{-}$	−0.72	120
$p\text{-}(CH_3)_2CHCH_2\text{-}$	4.19	2	3-Cl-	−1.48	120
$p\text{-}(CH_3)_3C\text{-}$	4.65	2	3-Br-	−1.48	120
$m\text{-}Br\text{-}$	3.08	2	$3\text{-}NO_2\text{-}$	−2.49	120
$m\text{-}Cl\text{-}$	3.09	2	3-HO-	−0.55	120
$p\text{-}F\text{-}$	4.01	2	***6-NITROANILINE***		
$p\text{-}(CH_3)_3Si\text{-}$	3.99	2	R=2-Cl-	−2.41	120
$p\text{-}CH_3O\text{-}$	5.14, 5.16	2	$2\text{-}NO_2\text{-}$	−5.56	120
N-METHYL			$2,4\text{-}Cl_2\text{-}$	−3.16	120
4-Chloro-	3.9*	99	$4\text{-}CH_3\text{-}2\text{-}NO_2\text{-}$	−4.45	120
4-Chloro, 2-nitro-	−1.49*	99	$2\text{-}4\text{-}(NO_2)_2\text{-}$	−10.23	120
2-Methyl-	4.62*	99	***OTHER ANILINE DERIVATIVES, IN 50% ETHANOL***		
3-Methyl-	5.00*	99	**Unhindered aniline**	4.19	40
4-Methyl-	5.36*	99	$p\text{-}Aminodiphenyl$	3.81	40
2-Methoxycarbonyl-	3.53*	99	2-Naphthylamine	3.77	40
4-Methoxycarbonyl-	2.32*	99			

*Thermodynamic value.

NITROGEN COMPOUNDS (Continued)

Compound	pK'a	Reference
3-Phenanthrylamine	3.59	40
Hindered-o-aminodiphenyl	3.03	40
peri		
1-Naphthylamine	3.40	40
9-Phenanthrylamine	3.19	40
3-Aminopyrene	2.91	40
meso		
9-Anthrylamine	2.7	40
m-Aminodiphenyl	3.82	40
2-Aminofluorene	4.21	40
2-Phenanthrylamine	3.60	40
2-Anthrylamine	3.40	40
1-Phenanthrylamine	3.23	40
1-Anthrylamine	3.22	40

o-AMINOPHENOLS

Compound	pK'a	Reference
3-Hydroxyanthranilic acid	10.09, 5.20	51
2-Aminophenol hydrochloride	9.99, 4.86	51
2-Amino-4,5-dimethyl-phenolhydrochloride	10.40, 5.28	51

INDICATORS

Compound	pK'a	Reference
p-Aminoazobenzene	2.82, 2.76	
4-Chloro-2-nitroaniline	−1.02, −1.03	60
4,6 Dichloro-2-nitroaniline	−3.61, −3.32	60
6-Bromo-2,4-dinitroaniline	−6.64, −6.71	60
N,N-Dimethyl-2,4- dinitroaniline	−1.00	60
p-Nitrodiphenylamine	−2.4 to −2.9, −2.50	60
4-Methyl-2,6-dinitroaniline	−3.96, −4.44	60

SUBSTITUTED NAPHTHYLAMINES

Compound		pK'a	Reference
$1\text{-}NH_2\text{-}$		3.92*	2
$1\text{-}NH_2\text{-}2\text{-}NO_2\text{-}$		−1.6	2
$1\text{-}NH_2\text{-}3\text{-}NO_2\text{-}$		2.22	2
$1\text{-}NH_2\text{-}4\text{-}NO_2\text{-}$		0.54	2
$1\text{-}NH_2\text{-}5\text{-}NO_2\text{-}$		2.80	2
$1\text{-}NH_2\text{-}6\text{-}NO_2\text{-}$		3.15	2
$1\text{-}NH_2\text{-}7\text{-}NO_2\text{-}$		2.83	2
$1\text{-}NH_2\text{-}8\text{-}NO_2\text{-}$		2.79	2
$1\text{-}NH_2\text{-}2\text{-}SO_3\text{-}$		1.71	2
$1\text{-}NH_2\text{-}3\text{-}SO_3\text{-}$		3.20*	2
$1\text{-}NH_2\text{-}4\text{-}SO_3\text{-}$		2.81*	2
$1\text{-}NH_2\text{-}5\text{-}SO_3\text{-}$		3.69*	2
$1\text{-}NH_2\text{-}6\text{-}SO_3\text{-}$		3.80*	2
$1\text{-}NH_2\text{-}7\text{-}SO_3\text{-}$		3.66*	2
$1\text{-}NH_2\text{-}8\text{-}SO_3\text{-}$		5.03*	2
$2\text{-}NH_2\text{-}$		4.11*	2
$2\text{-}NH_2\text{-}1\text{-}NO_2\text{-}$		−1.0	2
$2\text{-}NH_2\text{-}3\text{-}NO_2\text{-}$		2.93	2
$2\text{-}NH_2\text{-}4\text{-}NO_2\text{-}$		2.63	2
$2\text{-}NH_2\text{-}5\text{-}NO_2\text{-}$		3.16	2
$2\text{-}NH_2\text{-}6\text{-}NO_2\text{-}$		2.75	2
$2\text{-}NH_2\text{-}7\text{-}NO_2\text{-}$		3.13	2
$2\text{-}NH_2\text{-}8\text{-}NO_2\text{-}$		2.86	2
$2\text{-}NH_2\text{-}1\text{-}SO_3\text{-}$		2.35	2
$2\text{-}NH_2\text{-}4\text{-}SO_3\text{-}$		3.70	2
$2\text{-}NH_2\text{-}5\text{-}SO_3\text{-}$		3.96*	2
$2\text{-}NH_2\text{-}6\text{-}SO_3\text{-}$		3.74*	2
$2\text{-}NH_2\text{-}7\text{-}SO_3\text{-}$		3.95*	2
$2\text{-}NH_2\text{-}8\text{-}SO_3\text{-}$		3.89*	2
2-Naphthylamine	X	4.16	88
$1\text{-}NH_2,3\text{-}X$	NO_2	2.07	88
	CN	2.26	88
	Cl	2.66	88
	Br	2.67	88
	I	2.82	88
	$COOCH_3$	3.12	88
	OCH_3	3.26	88

*Thermodynamic value.

SUBSTITUTED NAPHTHYLAMINES

Compound		pK'a	Reference
	OH	3.30	88
	CH_3	3.96	88
	Cl	2.71	88
$2\text{-}NH_2,5\text{-}X$	NO_2	3.01	88
	OH	4.07	88
$1\text{-}NH_2,5\text{-}X$	NO_2	2.73	88
	OH	3.96	88
	Cl	3.34	88
	NH_2	4.21	88
$1\text{-}NH_2,7\text{-}X$	NO_2	2.55	88
	Cl	3.48	88
	OCH_3	4.07	88
	OH	4.20	88
$1\text{-}NH_2,2\text{-}X$	NO_2	−1.74	88
$1\text{-}X,2\text{-}NH_2$	NO_2	−0.85	88
$1\text{-}NH_2,8\text{-}X$	NO_2	2.79	88
$2\text{-}NH_2,4\text{-}X$	NO_2	2.43	88
	CN	2.66	88
	Cl	3.38	88
	Br	3.40	88
	I	3.41	88
	$COOCH_3$	3.38	88
	OCH_3	4.05	88
$1\text{-}NH_2,6\text{-}X$	NO_2	2.89	88
X	Cl	3.48	88
	OCH_3	3.90	88
	OH	3.97	88
$2\text{-}NH_2, 7\text{-}X$	NO_2	3.10	88
	Cl	3.71	88
	OCH_3	4.19	88
	OH	4.25	88
	NH_2	4.66	88
$2\text{-}NH_2,6\text{-}X$	NO_2	2.62	88
	OCH_3	4.64	88
$2\text{-}NH_2,8\text{-}X$	NO_2	2.73	88
$1\text{-}NH_2,4\text{-}X$	NO_2	0.54	88
	Br	3.21	88
$2\text{-}NH_2,3\text{-}X$	NO_2	1.48	88

HETEROCYCLIC COMPOUNDS

Compound	pK'a	Reference
Adenine	4.15, 9.80	6
Adenine deoxyriboside-5'-phosphoric acid	4.4, 6.4	99
Adenosine	3.63, 12.5	6, 99
1-Oxide	2.25, 12.86	99
ADP	3.95, 6.3, (7.20*)	36, 113
5-Amino-1(β-D-ribosyluronic acid) uracil	3.06	99
2'-AMP	3.81, 6.17	6
3'-AMP	3.74, 5.92	6
5'-AMP	3.74, 6.2–6.4	6
5-Aminouridine	3.11	99
1-D-Arabinosyl-5-methylcytosine	4.1	99
ATP	4.00 (4.1), 6.5 (7.68*)	36, 113
Barbital	7.85, 12.7	37
Barbituric acid	3.9, 12.5	37
N-n-Butyl-5-fluoro-2'-deoxycytidine	2.21	99
CDP	4.44, (7.18*)	6, 113
CDP (deoxy)	4.8, 6.6	6
2'-CMP	4.3–4.4, 6.19[a]	6
3'-CMP	4.16–4.31, 6.04	6
5'-CMP	4.5, 6.3	6
CTP	4.6, 6.4, (7.65*)	6, 113
Cytidine	4.22, 12.5	35
Cytosine	4.45, 12.2	6

NITROGEN COMPOUNDS (Continued)

NUCLEOSIDES, NUCLEOTIDES, AND RELATED COMPOUNDS (Continued)

Compound	pK'_a	Reference
Cytosine (deoxy)	4.25, ~13	6, 101
Deoxycytidine-5′-phosphoric acid	4.6, 6.6	99
2,6-Diaminopurine	5.09, 10.77	6
N,N-Dimethylcytidine	3.58	99
N,N-Dimethyl-2′-deoxycytidine	3.75	99
N-Ethyl-5-fluoro-2′-deoxycytidine	2.21	99
5-Fluorocytidine	2.22	99
5-Fluoro-2′-deoxycytidine	2.39	99
5-Fluoro-*N,N*-dimethyl-2′-deoxycytidine	1.89	99
5-Fluoro-*N*-methyl-2′-deoxycytidine	2.14	99
GDP	2.9, 9.6, 6.3, (7.19*)	6, 113
1-D-Glucopyranosylcytosine	3.78	99
GMP (2′+3′)	2.3, 9.36, 0.7, 5.9	6
5′-GMP	2.4, 9.4, 6.1	6
5′-GMP (deoxy)	2.9, 9.7, 6.4	6
GTP	3.3, 9.3, 6.5, (7.65*)	6, 113
Guanine	3.3, 9.2, 12.3	6
Guanine deoxyriboside-3′-phosphoric acid	2.9, 6.4, 9.7	99
Guanosine	1.6, 9.16, 12.5	35
Guanosine-3′-phosphoric acid	0.7, 2.3, 5.92, 9.38	99
Guanosine (deoxy)	1.6–2.2, 9.16–9.5	6
Hypoxanthine	1.98, 8.94, 12.10	6
5′-IMP	8.9, 1.54, 6.04	6
Inosine	1.2, 8.75, 12.5	6, 35
ITP	7.68*	113
IDP	7.18*	113
N-Methylcytidine	3.88	99
1-Methylcytidine	8.7	99

NUCLEOSIDES, NUCLEOTIDES, AND RELATED COMPOUNDS (Continued)

Compound	pK'_a	Reference
5-Methylcytidine	4.21	99
5-Methylcytidylic acid	4.4	99
5-Methyicytosine	4.6, 12.4	6
5-Methylcytosine deoxyriboside	4.5, 13.0	6
5-Methylcytosine deoxyriboside 5-phosphate	4.4, 13	6, 100
N-Methyl-2′-deoxycytidine	3.97	99
5-Methyl-2′-deoxycytidine	4.33	99
1-Methyluracil	9.95	37
3-Methyluracil	9.75	37
1-Methylxanthine	7.7, 12.05	38
3-Methylxanthine	8.5, (8.1), 11.3	38
7-Methylxanthine	8.5, (8.3)	38
9-Methylxanthine	6.3	38
Orotic acid	2.8, 9.45, 13	6
Purine	2.52, 8.90	37
Pyrimidine	1.30	37
Thymidine	9.8	6
Thymine	0, 9.9, >13.0	6
5′-TMP	10.0, 1.6, 6.5	6
UDP	9.4, 6.5, (7.16*)	6, 113
UMP(2′+3′)	9.43, 1.02, 5.88	6
5′-UMP	9.5, 6.4	6
Uracil	0.5, 9.5, 13.0	6
Uracil deoxyriboside	9.3, >13	6, 101
Uric acid	5.4, 10.3	6
Uridine	9.17, 12.5	35
UTP	9.5, 6.6, (7.58*)	6, 113
Xanthine	0.8, 7.44, 11.12	6
Xanthosine	0, 5.5, 13.0	6

HETEROCYCLIC BASES

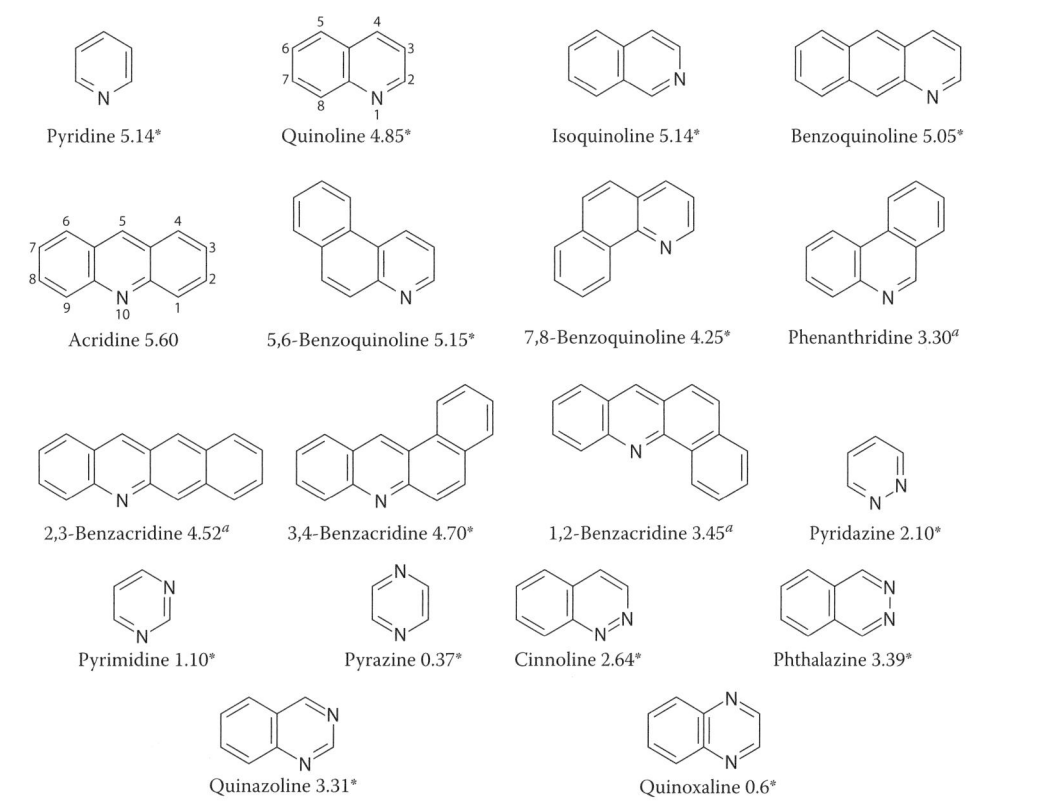

Pyridine 5.14*

Quinoline 4.85*

Isoquinoline 5.14*

Benzoquinoline 5.05*

Acridine 5.60

5,6-Benzoquinoline 5.15*

7,8-Benzoquinoline 4.25*

Phenanthridine 3.30[a]

2,3-Benzacridine 4.52[a]

3,4-Benzacridine 4.70*

1,2-Benzacridine 3.45[a]

Pyridazine 2.10*

Pyrimidine 1.10*

Pyrazine 0.37*

Cinnoline 2.64*

Phthalazine 3.39*

Quinazoline 3.31*

Quinoxaline 0.6*

*Thermodynamic value.

[a] In 50% EtOH

NITROGEN COMPOUNDS (Continued)

Acridine	1-	2-	3-	4-	5-	9-
H-	5.60*	4.11[a]	—	—	—	—
H$_2$N-	4.40*	8.04*	5.88*	6.04*	9.99*	9.95*
	3.59[a]	7.61[a]	5.03[a]	5.50[a]	9.45[a]	—
HO	4.18[a]	4.86[a]	5.52	4.45[a]	−0.32	—
	10.7[a]	9.9[a]	8.81	9.4[a]	>12	—
CH$_3$-	3.95[a]	—	4.60[a]	—	4.70[a]	—
H$_2$N-(1-CH$_3$-)-	—	—	—	4.79[a]	9.73[a]	3.22[a]
1,9-(CH$_3$)$_2$-	2.88[a]	—	—	—	—	—
COOH-	—	5.26*	—	7.76*	—	5.0*

Reference 3, 39, 99

Compound	pK$_a'$	Reference
HETEROCYCLICS (Continued)		
Acridine	5.62	39
3-Amino-7-carboxy-	2.3*, 9.0*	99
3-Amino-6-chloro-	7.22*	99
3-Amino-7-chloro-	6.91*	99
9-Amino-1-hydroxy-	5.57*	99
9-Amino-2-hydroxy-	7.67*	99
9-Amino-3-hydroxy-	6.59*	99
9-Amino-4-hydroxy-	7.01*	99
3-Amino-7-sulfo-	7.6*	99
2,7-Diamino-	6.14*	99
3,6-Diamino-	9.65*	99
3,7-Diamino-	8.11	99
3,9-Diamino-	8.11	99
5-Methoxy-	7	39
9-Methoxy-	−0.32*	99
2-Sulpho-	4.78*	99
Aureomycin	3.30, 7.44, 9.27	77
Azacycloundecane 1-Methyl-6-hydroxy-7-oxo-	9.1*	99
Azaindole		
4-	6.9	154
5-	8.3	154
6-	8.0	154
7-	4.6	154
3,4-Di-	4.0, 11.1	154
3,5-Di-	6.1, 10.9	154
2,5, 7-Tri-	2.8, 9.5	154
4-Amino-2, 5, 7-tri-	4.6, 10.8	154
Azepine Hexahydro-	11.07*	99
Azetidine	11.29*	99
Aziridine	8.01*	99
1:2-Benzacridine	3.45[a]	19
5-Amino-	8.13[a]	19
7-Amino-	4.05[a]	19
8-Amino-	6.72, 5.97[a]	19
4′,5-Diamino-	8.44[a]	19
2:3-Benzacridine	4.52[a]	19
5-Acetamido-	4.56[a]	19
5-Amino-	9.72[a]	19
7-Amino-	5.38[a]	19
5-Amino-6, 7, 8, 9-tetrahydro-	9.66[a]	19
3:4-Benzacridine	4.70, 4.16[a]	19
8-Acetamido-	4.48[a]	19
5-Amino-	8.4l[a]	19
7-Amino-	5.03[a]	19
8-Amino-	7.42, (651)[a]	19
8-Dimethylamino-	7.31, 6.99	19

Compound	pK$_a'$	Reference
HETEROCYCLICS (Continued)		
Benzimidazole	5.4, 12.78	43, 86, 99 107
2-Amino-	7.51*	—
5-Amino-	3.04*, 6.07*	—
6-Amino-	3.0, 6.0	—
2-Aminomethyl-	7.69*, 3.46*	—
1-α-D-Arabopyranosyl-	4.19*	—
1-α-L-Arabopyranosyl-	4.06*	—
1-α-L-Arabopyranosyl-5,6-dimethyl-	4.56*	—
1-α-L-Arabopyranosyl-5-methyl-	4.30	—
1-Ethyl-	5.59*	—
2-Ethyl-	6.18*	—
1-β-D-Glucopyranosyl-	3.97*	—
1-β-D-Glucopyranosyl-5, 6-dimethyl-	4.60*	—
1-β-D-Glucopyranosyl-5-methyl-	4.29*	—
4-Hydroxy-	5.3, 9.5	—
4-Hydroxy-6-amino-	5.9	—
1-Hydroxymethyl-	5.41*	—
2-Hydroxymethyl-	5.40*, 11.55*	—
4-Hydroxy-6-nitro-	3.05	—
4-Methoxy-	5.1	—
1-Methyl-	5.54*	—
2-Methyl-	5.58	—
4-Methyl-	5.65*	—
5-Methyl-	5.78*	—
2-Methyl-4-hydroxy-6-amino-	6.65	—
1-Methyl-2-hydroxymethyl-	5.55, 11.45	—
2-Methyl-4-hydroxy-6-nitro-	3.9	—
4-Nitro-	3.33*	—
5-Nitro-	3.48*	—
6-Nitro-	3.05, 10.6	—
5-F-	1.67*, 4.92*	123, 124
5-Br-	1.98*, 4.66*	123, 124
5-CF$_3$-	2.28*, 4.22*	123, 124
5-Cl-	1.94*, 4.70*	123, 124
Benzo[c]cinnoline	2.20*	99
6,7-Benzoquinazoline	5.2*	99
5,6-Benzoquinoline	5.15, 3.90[a]	19
4-Amino-	7.99[a]	19
2-Amino-4-methyl-	7.14, 6.51[a]	19
4-Amino-2-methyl-	8.45[a]	19
1′-Amino-	5.03	19
3′-Amino-	4.02[a]	19
4′-Amino-	5.20, 4.10[a]	19
2′,4′-Diamino-	4.91[a]	19
2-Methyl-	4.44[a]	19
6,7-Benzoquinoline	5.05, 3.84[a]	19
3-Amino-	4.78, 3.73[a]	19
4-Amino-	8.75[a]	19
4-Amino-2-methyl-	9.45[a]	19
4-Amino-2-methyl-8-chloro-	5.95[a]	19
8-Chloro-	2.5[a]	19
3,4-Diamino-	8.15[a]	19
7,8-Benzoquinoline	4.25, 3.15[a]	19
4-Amino-	7.68[a]	19
2-Amino-4-methyl-	6.74, 6.02[a]	19
4-Amino-2-methyl-	7.96[a]	19
6-Amino-2-methyl-	5.23[a]	19
1′-Amino-2-methyl-	4.75[a]	19
Benzoxazole	(decomp.)	19
2-Amino-	3.73	19
Benzthiazole	1.2, 0.1[a]	19
2-Amino-	4.51	19
Benztriazole	1.6, 8.64*	19
Bispidine	10.25*	99

*Thermodynamic value.

[a]In 50% alcohol

NITROGEN COMPOUNDS (Continued)

Compound	pK'_a	Reference
HETEROCYCLICS (Continued)		
Caffeine	0.61	4
Cinchonine	7.2	4
Cinnoline	2.70, 2.29	19, 39
4-Amino-	6.84	19
3-Hydroxy-	8.64, 0.21	39
4-Hydroxy-	9.27, 0.35	39
5-Hydroxy-	7.40, 1.92	39
6-Hydroxy-	7.52, 3.65	39
7-Hydroxy-	7.56, 3.31	39
8-Hydroxy-	8.20, 2.74	39
4-Methoxy-	3.21	39
4-Methylthio-	3.09*	99
5,6,7,8-Tetrahydro-	4.30*	99
α,α'-Dipyridyl	4.43	6
4.5-Diazaindan	4.12*	99
1,4-Diazaindene	6.92*	99
1,5-Diazaindene	8.24*	99
1,6-Diazaindene	7.93*	99
1,7-Diazaindene	4.57*	99
Flavone	−1.2	154
Δ^2-Dihydro-2-methyl-	11.1	154
Furan, 2-(2-aminoethylaminomethyl)-	6.54*, 9.87*	99
Furan, 3-(2-aminoethylaminomethyl)-	6.70*, 9.86*	99
Furan, 2-aminomethyl-	8.89	99
Gramine	16.00	152
5-Benzyloxy-	16.90	152
Histamine	6.0	43
Histidine methyl ester	5.2, 7.1	43
4-Hydroxymethyl-	6.39	99
2-(2-Imidazolyl)-	4.53*	99
4-(3-Methoxycarbonylpropyl)-	7.3	99
1-Methyl-	6.95*	99
2-Methyl-	7.85	99
4-Methyl-	7.51	99
1-Methyl-4-chloro-	3.10	106
1-Methyl-5-chloro-	4.75	106
1-Methyl-4-nitro-	−0.53*	99
1-Methyl-5-nitro-	2.13*	99
1-Methyl-4-phenyl-	5.78*	99
2-Nitro-	7.15	106
4-Nitro-	−0.05*	99
5-Nitro-	9.20	106
4-Nitro-5-chloro-	5.85	106
5-Nitro-4-chloro-	5.85	106
2-Phenyl-	6.48*, 13.32*	99
4-Phenyl-	6.10*, 13.42*	99
5-Phenyl-	6.10, 13.42	107
4-(2-Pyridyl)-	5.42*	99
2,4,5-Trimethyl-	8.92*	99
Hydantoin	9.16	42
5:5-Dimethyl-2-thio-	8.71	42
5:5-Diphenyl-2-thio-	7.69	42
5-Isopropyl-2-thio-	8.70	42
1-Methyl-5,5-pentamethylene-2-thio-	9.25	42
3-Methyl-5,5-pentamethylene-2-thio-	11.23	42
5:5-Pentamethylene-2-thio-	8.79	42
3:5:5-Trimethyl-2-thio-	10.80	42
Imidazoles	7.05, 14.52	107
4-(2-Acetoxyethyl)-	6.97*	99
N-Acetyl-	3.6	99
N-Acetylhistidine	7.05	43
4-Aminomethyl-	9.37*, 4.71*	99
5-Amino-4-(*N*-methylcarboxamidino)-	9.5*	99
4-Bromo-	3.7	43
4-Carbamoyl-	3.7*, 11.8*	99
4-(3-Carbamoylpropyl)-	6.52*	99
Carbobenzoxy-l-histidyl-l-tyrosine ethyl ester	6.25	43
4-(2′,4′-Dihydroxyphenyl)-	6.45	43
4-Chloro-1-methyl-	6.23*	99
2,4-Dimethyl-	8.36*	99
2,4-Dinitro-	2.85	106
2,5-Dinitro-	2.85	106
2,4-Diphenyl-	5.64*, 12.53*	99
2,5-Diphenyl-	5.64, 12.53	107
4,5-Diphenyl-	5.90*, 12.80*	99
1-Ethyl-	7.30*	99
2-Ethyl-	8.00	99
1*H*-Imidazo[4,5-*b*]pyridine	3.92*, 11.11*	99
1*H*-Imidazo[4, 5-*c*]pyridine	6.10*, 10.88*	99
Imidazolidines		
2-Imino-1-methyl-4-oxo-	4.80*	99
2-Imino-4-oxo-	4.76*	99
n-Nitrimino-	−1.36*	99
2-Imidazoline		
2-(*N*-Benzylanilinomethyl)-	2.45*, 10.13*	99
2-(3-Diethylamino-1-phenyl)propyl-	8.41*, 10.09*	99
2-(3-Dimethylamino-1-phenyl)propyl-	7.98*, 9.99*	99
2-Diphenylmethyl-	9.78*	99
Imidazoline		
2-(3-(2-Hydroxynaphthyl))-	7.01*, 10.85*	99
2-(2-Hydroxyphenyl)-	6.63*, 12.58*	99
4-Methyl-5-carboxylic acid-	2.49, 7.02	144
1-Methyl-2-carboxylic acid-	1.26, 7.25	144
1*H*-Indazole	1.22*, 14*	99
3-Amino-	3.12*	99
4-Amino-	3.26*	99
5-Amino-	5.12*	99
6-Amino-	3.99*	99
7-Amino-	3.02*	99
Indole	−2.4*	99
Indole	−3.6, 16.97	154, 152
2-Amino-	8.15*	99
2-Amino-1-methyl-	9.60*	99
1,2-Dimethyl-	0.34*	99
2,3-Dimethyl-	−1.10*	99
1-Methyl-	−1.80	99
2-Methyl-	−0.10*	99
3-Methyl-	−3.35*	99
3-Formyl-	12.36	152
3-Acetyl-	12.99	152
5-Nitro-	14.75	152
5-Cyano-	15.24	152
5-Bromo-	16.13	152
5-Fluoro-	16.30	152
4-Fluoro-	16.30	152

*Thermodynamic value.

NITROGEN COMPOUNDS (Continued)

Compound	pK_a'	Reference	Compound	pK_a'	Reference
HETEROCYCLICS (Continued)			*HETEROCYCLICS (Continued)*		
2-Carboxylate-	17.13	152	5-NO₂-2-Carboxylate	14.91	152
3-Carboxylate-	15.59	152	5-Br-2-Carboxylate	16.10	152
5-Carboxylate-	16.92	152	5-MeO-2-Carboxylate	17.03	152
3-Acetic acid-	16.90	152	Isoalloxazine, 7,8 dichloro-		
3-Carbinol-	16.50	152	10-(3-Dibutylaminopropyl)-	8.0*	99
L-Tryptophanol	16.91	152	10-(4-Diethylaminobutyl)-	9.7*	99
L-Tryptophan	16.82	152	10-(2-Diethylaminoethyl)-	7.7*	99
4-Methyltryptophan	16.90	152	10-(5-Diethylaminopropyl)-	9.1*	99
5-Hydroxytryptophan	19.20	152	10-(3-Piperidinopropyl)-	9.0*	99
6-Methoxytryptophan	16.70	152	Isoxazole	1.3*	27
Tryptamine	16.60	152	3,5-Dimethyl-	−2*	27
Serotonin	19.50	152	4,5-Dimethyl-	0*	27
Gramine	16.00	152	3-Methyl-	−1*	27
5-Benzyloxydramine	16.90	152	5-Methyl-	2.3*	27
Skatole	16.60	152	3,4,5-Trimethyl-	−1*	27

	1	3	4	5	6	7	8	Reference
Isoquinoline	5.46	5.14	—	—	—	—	—	19, 44, 99
OH-	−1.2	2.18*	4.78*	5.40	5.85	5.68*	5.66	—
	—	9.62*	8.70*	8.45	9.15	8.90*	8.40	—
NH₂-	7.59*	5.05	6.26*	5.59	7.16*	6.20*	6.04*	—
CH₃-	—	5.64*	—	—	—	—	—	—
Br-	—	—	3.31*	—	—	—	—	—
SH-	−1.9*	0.39*	—	—	—	—	—	—
	10.86*	8.62*	—	—	—	—	—	—
CH₃O-	3.01*	—	—	—	—	—	—	—
CH₃S-	3.89*	3.37*	—	—	—	—	—	—
NO₂-	—	—	1.35*	3.49*	3.43*	3.57*	3.55*	—

Compound	pK_a'	Reference	Compound	pK_a'	Reference
HETEROCYCLICS (Continued)			*HETEROCYCLICS (Continued)*		
Isoquinoline			2-Ethyl-	6.13*, 12.9*	99
1,2,3,4-Tetrahydro-	9.4	154	2-Hydroxy-	11.66*	99
Isoquinoline-*N*-oxide	1.01	47	2-Methyl-	6.10*, 12.9*	99
Morpholine	8.39*	99	1,5-Naphthyridine	2.84*	99
N-(3-Acetyl-3,3-diphenyl)propyl-	6.83*	99	1,6-Naphthyridine	3.76*	99
N-(2-Acetyl-2-phenyl)ethyl-	6.23*	99	1,7-Naphthyridine	3.61*	99
N-Allyl-	7.02*	99	1,8-Naphthyridine	3.36*	99
N-(2-Amino)ethyl-	4.84*, 9.45*	99	8-Hydroxy-6-methyl-1,6-naphthyridinium chloride	4.34	44
N-(2-Benzylcarbonyl-2-phenyl)ethyl-	6.17*	99	8-Hydroxy-1,6-naphthyridine	4.08	44
N-(2-Bis-2-hydroxypropyl)aminoethyl-	7.9*	99	Oxazoline		
N-(3-Cyano-3,3-diphenyl)propyl-	6.04*	99	4-Carbamoyl-2-phenyl-	2.9	96
N-(3-Cyano-1-methyl-3,3-diphenyl)propyl-	5.5*, 8.4*	99	2-Methyl-Δ²-	5.5	96
N-(2-Diphenylmethyl-carbonyl)ethyl-	6.39*	99	2-Phenyl-Δ²-	4.4	96
N-(3,3-Diphenyl)propyl-	7.20*	99	4-Methyl-	1.07	144
N-(3,3-Diphenyl-3-propyl-carbonyl)propyl-	7.17*	99	Ethyl-4-Methyl-5-carboxylate-	0.83	144
N-Ethyl-	7.67*	99	4-Methyl-5-carboxylic acid-	0.95, 2.88	144
N-(3-Ethylcarbonyl-3,3-diphenyl)propyl-	6.95*	99	Perimidine	6.35*	99
N-(3-Ethylcarbonyl-1-methyl-3,3-diphenyl) propyl	6.68*	99	3,4-Pentamethylene-5,6,7,8-tetrahydrocinnoline	6.03*	99
N-(3-Ethylcarbonyl-2-methyl-3,3-diphenyl)propyl-	7.12*	99	1,5-Phenanthroline	0.75*, 4.10	99
N-(2-Hydroxy-3-morpholino)propyl-	5.00*, 6.98*	99	1,7-Phenanthroline	1,4	99
N-Methyl-	7.38*	99	1,10-Phenanthroline	4.84*	99
N-(1-Methyl-3,3-diphenyl)propyl-	6.85*	99	4-Bromo-	4.01*	99
N-(2-Morpholino)ethyl-	3.63*, 6.65*	99	3-Chloro-	3.97*	99
N-(3-Morpholino)propyl-	6.25*, 7.25*	99	4-Chloro-	4.30*	99
1(3)*H*-Naphth[l,2-d]imidazole	5.27*	99	5-Chloro-	4.24*	99
1*H*-Naphth[2,3-d]imidazole	5.21*. 12.58*	99	4-Cyano-	3.56*	99
2-Amino-	6.99*	99	3-Ethyl-	4.96*	99
2-Carboxymethylthio-	1.9*, 4.72*	99			

*Thermodynamic value.

NITROGEN COMPOUNDS (Continued)

Compound	pK_a'	Reference	Compound	pK_a'	Reference
HETEROCYCLICS (Continued)			*HETEROCYCLICS (Continued)*		
4-Ethyl-	5.42*	99	2,4-di-OH-	< 1.0, 7.91	101
4-Hydroxy-	2.17*	99	2,6-di-OH-	6.7, 11.6	101
2-Methyl-	4.98*	99	2,7-di-OH-	5.83, 10.07	101
3-Methyl-	4.98*	99	4,6-di-OH-	6.08, 9.73	101
5-Methyl-	5.26*	99	4,7-di-OH-	6.82, 10.02	101
5-Nitro-	3.55*	99	6,7-di-OH-	6.87, 10.0	101
2-Phenyl-	4.88*	99	2,4,6-tri-OH-	5.73, 9.41	101
3-Phenyl-	4.80*	99	3,4,7-tri-OH-	3.61	101
4-Phenyl-	4.88*	99	4-OH-6-Me-	8.19	101
5-Phenyl-	4.70*	99	4-OH-7-Me-	8.09	101
o-Phenanthroline	4.27[a], 5.2	19	6-OH-5-Me-	3.73, 10.6	101
p-Phenanthroline	3.12[a]	19	7-OH-8-Me-	1.1	101
1:10-Diamino-3,8-dimethyl-	8.76[a], 6.31[a]	19	2-OH-1-Me-	<1, 11.43	101
6-*m*-Phenanthroline	3.11[a]	19	2-OH-3-Me-	1, 11.01	101
1-Amino-	ca. 7.3, 7.29[a]	19	2-OH-3,6,7-tri-Me-	< 2, 11.36	101
Phenazine	1.23	39	2-OH-6,7,8-tri-Me-	< 2, 10.26	101
1-Amino-	2.6[a]	19	4-OH-1-Me-	1.25	101
2-Amino-	4.75, 3.46[a]	19	4-OH-3-Me-	−0.47	101
1,3-Diamino-	5.64[a]		4-OH-6,7-di-Me-	8.39	101
2,3-Diamino-	4.74	19	4-OH-6,7,8-tri-Me-	4.70, 9.46	101
2:7-Diamino-	4.63, 3.9[a]	19	7,8-Dihydro-6-OH-	4.78, 10.54	101
1-Hydroxy-	1.61*, 8.33*		5,6-Dihydro-7-OH-	3.36, 9.94	101
2-Hydroxy-	7.5, 2.6	99	5,6-Dihydro-6,7-di-OH-5-Me-	2.91, 9.33	101
Phenanthridine	4.65	39	5,6-Dihydro-4,7-di-OH-6-Me-	8.43, 11.40	101
2-Amino-9-methyl-	5.66[a]	44	4,7-di-OH-6-CHO-	5.93, 9.31	101
6-Amino-	6.88	19	4,7-di-OH-6-COOH-	ca. 3, 6.69, 10.15	101
7-Amino-9-methyl-	5.23[a]	40	2-MeO-	2.13	101
9-Amino-	7.31, 6.75[a]	19	4-MeO-	1.04	101
2:7-Diamino-9-methyl-	6.26[a]	19	6-MeO-	3.60	101
2-Hydroxy-	8.79, 4.82	19	7-MeO-	1.64	101
6-Hydroxy-	8.43, 5.35	44	2-NH_2-	4.29	101
7-Hydroxy-	4.38, 8.68	44	4-NH_2-	3.51	101
9-Hydroxy (phenanthridone)	<−1.5	44	6-NH_2-	4.15	101
9-Methoxy-	2.38	44	7-NH_2-	2.96	101
2,7,9-Triamino-	8.06[a]	44	2,4-di-NH_2-	5.32	101
Phenothiazine		19	4,6-di-NH_2-	4.37	101
10-(2-Diethylaminoethyl)-	9.06*	99	4,7-di-NH_2-	4.97	101
10-(2-Dimethylaminobutyl)-	9.02*	99	2,4,7-tri-NH_2-	6.03	101
10-(2-Dimethylaminoethyl)-	8.66*	99	4,6,7-tri-NH_2-	5.57	101
3,7-Diamino-	5.3, 4.4	154	2,4,6,7-tetra-NH_2-	6.86	101
Phthalazine	3.47	19	2-MeNH-	3.64	101
1-Amino-	6.60	19	4-MeNH-	4.33	101
1-Hydroxy-	11.99, −2	39	2-Me_2N-	3.03	101
1-Mercapto-	−3.43*, 9.99*	99	4-Me_2N-	4.33	101
1-Methoxy-	3.73*	99	6-Me_2N-	4.31	101
1-Methyl-	4.37*	99	7-Me_2N-	2.56	101
1-Methylthio-	3.44*	99	2-NH_2-7-OH-	1.5*, 7.50*	99
1-Phenyl-	3.51*	99	2-NH_2-4-CH_3O-	3.44*	99
Picolinic acid	5.52	4	2-NH_2-4-CH_3-	2.81*	99
Trimethyl(2,6-di-tert-butyl-4-picolyl) ammonium	3.51	125	2-NH_2-6-CH_3-	4.03*	99
			2-NH_2-7-CH_3-	3.73*	99
Pteridines	4.12	101	4-NH_2-2-CH_3-	4.28*	99
6-Cl-	3.68*	99	2-NH_2-4-OH-	2.31, 7.92	101
2-Me-	4.87	101	4-NH_2-2-OH-	3.21, 9.97	101
4-Me-	2.94	101	2-NH_2-4,6-di-OH-	1.6, 6.3, 9.23	101
7-Me-	3.49	101	2-MeCONH-	2.67	101
6,7-di-Me-	2.93	101	4-MeCONH-	1.21	101
2,6,7-tri-Me-	3.76	101	4-H_2NNH-	4.00	101
2-OH-	< 2, 11.13	101	2-SH-	9.98	101
4-OH-	< 1.5, 7.98	101	4-SH-	6.81	101
6-OH-	3.67, 6.7	101	7-SH-	5.5	101
7-OH-	1.2, 6.41	101	2-MeS-	2.2	101
			4-MeS-	2.59	101

NITROGEN COMPOUNDS (Continued)

Compound	pK$_a'$	Reference	Compound	pK$_a'$	Reference
HETEROCYCLICS (Continued)			*HETEROCYCLICS (Continued)*		
7-MeS-	2.49	101	2,6,8-tri-NH$_2$-	2.41, 6.23,	101
4-MeS-7-Me-	< 2	101		10.96	
4-SH-7-Me-	7.02	101	6-MeNH-	1, 4.18, 9.99	101
3,4-Dihydro-2-OH-	0*, 12.6*	99	8-MeNH-	4.78, 9.56	101
3,4-Dihydro-4-OH-	4.75*, 11.25*	99	2-Me$_2$N-	4.02, 10.22	101
5,6-Dihydro-4-OH-	2.94*, 10.24*	99	6-Me$_2$N-	<1, 3.84, 10.5	101
7,8-Dihydro-2-OH-	3.46*	99	8-Me$_2$N-	1, 4.80, 9.73	101
7,8-Dihydro-4-OH-	0.32*, 12.13*	99	2-C$_6$H$_5$NH-	4.2, 10.1	101
6-OH-2-CH$_3$-	4.65*, 6.33*	99	2-Me-6-MeNH-	5.08	101
6-OH-4-CH$_3$-	4.08*, 6.41*	99	2-NH$_2$-6-OH-(Guanine)	3.3, 9.2, 12.3	101
6-OH-7-CH$_3$-	3.69*, 7.20*		6-NH$_2$-2-OH-	4.51, 8.99	101
7,8-Dihydro-4,6-dimethyl-	6.0	154	2-NH$_2$-6-SH-	8.2, 11.6	101
5,6,7,8-Tetrahydro-	6.6	154	6-CHO-	2.4, 8.8	101
5,6,7,8-Tetrahydro-5-formyl-	5.0	154	6-HONH-	3.88, 9.88, >12	101
5,6,7,8-Tetrahydro-4-methyl-	6.7	154	6-NH$_2$CONH-	2.35, 9.95	101
5,6,7,8-Tetrahydro-4-hydroxy-	3.9, 10.1	154	6-CN-	ca. 0.3, 6.88	101
			6-CF$_3$-	0, 7.35	101
Pteridines			8-CF$_3$-	1.0, 5.12	101
7-OH-2-CH$_3$-	1.68*, 6.70*	99	6-Me-	2.6, 9.02	101
3-Methyl-4-pteridone	−0.47	44	8-Me-	2.85, 9.37	101
1-Methyl-4-pteridone	1.25	44	9-Me-	2.36	101
			8-Ph-	2.68, 8.09	101
Pteroylglutamic acid	8.26	77	2-OH-	8.43, 11.90	101
			6-OH- (Hypoxanthine)	8.94, 12.10	101
Purine	2.39, 8.93	101	8-OH-	8.24, >12	101
2-Amino-6,8-bistrifluoro-methyl-	0.3*, 5.02*	99	2,8-di-OH-	7.65, 9.7	101
6-Amino-9-cyclohexyl-amino-	4.19*	99	2,6-di-OH- (Xanthine)	7.44, 11.12	101
2-Amino-8-phenyl-	3.97*, 9.21*	99	1-Me-2,6-di-OH-	7.7, 12.5	101
2-Amino-6-trifluoromethyl-	1.85*, 8.87*	99	3-Me-2,6-di-OH-	8.33, 11.9	101
8-Carboxy-	0*, 2.93*, 9.41*	99	7-Me-2,6-di-OH-	8.7, 10.7	191
6-Chloro-	7.85*	99	1,3-di-Me-2,6-di-OH-	8.81	101
6-Cyclohexylamino-	4.2*, 10.2*	99	1,7-di Me-2,6-di-OH-	8.71	101
9-Cyclohexyl-6-cyclohexyl-amino-	4.4*	99	3,7-di Me-2,6-di-OH-	9.97	101
2,6-Diamino-8-trifluoro-methyl-	3.68*, 7.55*	99	2,6,8-tri-OH- (Uric acid)	5.75, 10.3	101
1,6-Dihydro-1,7-dimethyl-6-oxo-	2.13*	99	2,6,8-tri-OH-3-Me-	5.75, >12	101
7,8-Dihydro-7,9-dimethyl-8-oxo-	2.8*	99	2,6,8-tri-OH-1-Me-	5.75, 10.6	101
1,2-Dihydro-8-hydroxy-1-methyl-2-oxo-	−0.5*, 7.0*, 13.0*	99	2,6,8-tri-OH-1,3-di-Me-	5.75	101
			2,6,8-tri-OH-3,7-di-Me-	5.5, 12	101
1,6-Dihydro-8-hydroxy-1-methyl-6-oxo-	8.54*, 11.87*	99	2,6,8-tri-OH-1,3,7-tri-Me-	6.0	101
2,3-Dihydro-8-hydroxy-3-methyl-2-oxo-	1.25*, 8.0*, 13.0*	99	2,6,8-tri-OH-3,7,9-tri-Me-	8.35	101
			6-OH-9-Me-	1.86, 9.32	101
1,6-Dihydro-6-imino-1-methyl-	11.0*, 7.0*	99	8-OH-7-Me-	2.69, 8.20	101
1,2-Dihydro-1-methyl-2-oxo-	1.8*, 8.80*	99	8-OH-9-Me-	2.80, 9.05	101
1,2-Dihydro-2-oxo-1-β-d-ribofuranosyl-	1.5*, 8.55*	99	2-MeO-	2.44, 9.2	101
6-Dimethylcarbamoyl-	0*, 7.9*	99	6-MeO-	1.98, 8.94	101
6-(Ethoxycarbonyl)amino-	2.4*, 9.63*, 12.2*	99	8-MeO-	3.14, 7.73	101
6-Ethylamino-2-methyl-	5.01*	99	Xanthine	7.53, 11.63	101
6-Hydrazonomethyl-	2.8*, 9.2*	99	1-Me-	7.7, 12.05	101
8-Hydroxymethyl-	2.58*, 8.83*	99	3-Me-	8.10, 11.3	101
6-Hydroxy-2-trifluoro-methyl-	1.1*, 5.1*, 11.2*	99	7-Me-	8.30	101
			9-Me-	6.25	101
7-Methyl-	2.25*	99	Hypoxanthine	8.8, 12.0	101
8-Methylthio-	2.92*, 7.70*	99	Uric acid	5.78, 5.85	101
Purine 1-oxide, 6-amino-	2.69*, 8.845*, 15.4*	99	Pyrazine	1.1 (0.6)	49, 39
			3-Acetyl-2-aminomethylene-amino-	5.49*	99
2-SH-	7.15*, 10.4	101	2-Amidino-3-methylamino-	8.96*	99
6-SH-	7.77, 10.84	101	2-Amino-	3.14	19
8-SH-	6.64, 11.64	101	2-Amino-3-carboxy-	3.70*	99
2-MeS-	8.91	101	2-Carbomoyl-	−0.5*	99
6-MeS-	8.74	101	2-Carbamoyl-3-methyl-amino-	2.09*	99
2-NH$_2$-	−0.28, 3.80, 9.93	101	2,3-Dicarboxy-	0.9*, 3.57*	99
6-NH$_2$- (adenine)	<1, 4.22, 9.8	101	2,5-Dimethyl-	2.1	49
8-NH$_2$-	4.68, 9.36	101	2,6-Dimethyl-	2.5*	99
2,6-di-NH$_2$-	<1, 5.09, 10.7	101			

*Thermodynamic value.

NITROGEN COMPOUNDS (Continued)

HETEROCYCLICS (Continued)

Compound	pK$_a'$	Reference
2-Dimethylamino-	3.24*	99
2-Hydroxy-	−0.1*, 8.25*	99
2-Mercapto	−0.73*, 6.34*	99
2-Methoxy-	0.75	39
2-Methyl-	−5.25*, 1.45*	99
1-Methyl-2-pyrazone	−0.04	39
2-Methylamino-	3.39*	99
2-Methylthio-	0.48*	99
2-Sulphanilamido-	6.04*	99
2,3,5,6-Tetramethyl-	2.8	49
Trimethyl-	−0.35*, 2.8*	99
Pyrazole	2.48	99
3-(2-Aminoethyl)-	2.02*, 9.61*	99
1,3-Dimethyl-	3.11*	99
3,5-Dimethyl-	4.38*	99
1-Methyl-	2.04*	99
3-Methyl-	3.56*	99
Pyrazolo[4,5-4,5]pyrimidine 6-Amino-	4.96*, 10.19*	99
Pyrazolo[5′,4′-4,5]pyrimidine	2.80*, 9.58*	99
6-Amino-	4.55*, 10.88*	99
6-Amino-1′-melhyl-	4.28*	99
6-Amino-2′-methyl-	5.37*, 11.34*	99
6-Amino-3′-methyl-	4.57*, 11.15*	99
6-Anilino-	3.88*	99
6-Benzylamino-	4.12*, 10.97*	99
2,6-Diamino-	4.63*, 11.25*	99
6-Diethylamino-	4.67*	99
6-Dimethylamino-	4.49*, 11	99
6-Ethylamino-	4.56*, 10.94*	99
6-Furfurylamino-	3.97*	99
6-Isopropylamino-	4.58*, 11.03*	99
6-Methylamino-	4.49*, 10.59*	99
1-Methyl-	2.46*	99
6-Methylthio-	1.0*, 9.69*	99
6-Phenethylamino-	4.34*	99
Pyridazine	2.33	19
3-Amino-	5.19	19
4-Amino-	6.65*	99
3-Amino-6-methyl-	5.29*	99
3-n-Butyl-	3.49*	99
3-Carbamoyl-	1.0*	99
4-Carbamoyl-	1.0*	99
3,5-Dihydroxy-	−2.2*	99
3,6-Dihydroxy-	5.67, −2.2, 13	39
3,6-Dimethoxy-	1.61	39
3,4-Dimethyl-	4.10*	99
3,5-Dimethyl-	4.11*	99
3,6-Dimethyl-	3.99*	99
4,5-Dimethyl-	4.13*	99
3-Hydroxy-	10.46, −1.8	39
4-Hydroxy-	8.68, 1.07	39
4-Hydroxy-2-methylpyridazinium chloride	1.74	44
3-Mercapto-	−2.68*. 8.32*	99
4-Mercapto-	−0.75*, 6.55*	99
3-Methoxy-	2.52	39
4-Methoxy-	3.70	39
3-Methyl-	3.46*	99
4-Methyl-	3.53*	99
3-Methylthio-	2.26*	99
4-Methylthio-	3.24*	99
3-Sulfanilamido-	7.06*	99

Compound	pK$_a'$ 2-	3-	4-
Pyridine (2, 46–48, 88, 99, 105, 117, 140)			
H-	5.17*	—	—
Cl-	0.72*	2.84*	3.88
I-	1.82*	3.25*	4.02*
CH$_3$CH$_2$-	5.97*	5.70*	6.02*
(CH$_3$)$_3$C-	5.76*	5.82*	5.99*
HO-	0.75	4.86	3.27
	11.62	8.72	11.09
NO$_2$-	−2.06	0.81	1.23
SO$_3$-	—	2.9	—
CH$_3$O-	3.28	4.88	6.62
C$_2$H$_5$O-	—	—	6.67
F-	−0.44*	2.97*	—
Br-	0.90*	2.84*	3.82
CH$_3$-	5.97*	5.68*	6.02*
(CH$_3$)$_2$CH-	5.83*	5.72*	6.02*
CH$_3$CO-	—	3.18*	—
CONH$_2$-	2.10	3.40	3.61
NC-	−0.26*	1.45	1.90*
CH$_3$CONH-	4.09*	4.46	5.87
EtOOC-	—	3.35	3.45
NH$_2$-	6.71*	6.03*	9.11
	−7.6	−1.5	−6.3
C$_6$H$_5$CONH-	3.33*	3.80*	5.32*
COOH-	0.99*	2.00*	1.77*
	5.39*	4.83*	4.84*
HCO-	3.80*	3.80*	4.77*
	12.80*	13.10*	12.20*
H$_2$NNHCO-	—	1.86*	1.82*
	3.86*	3.29*	3.52*
	12.27*	11.47*	10.79*
HS-	−1.07*	2.26*	1.43*
	10.00*	7.03*	8.86*
CH$_3$OCO-	2.21*	3.13*	3.26*
CH$_3$NH-	—	—	9.66
CH$_3$S-	3.59*	4.42*	5.94*
C$_6$H$_5$-	4.48*	4.80*	5.55*
CH$_2$=CH-	4.98*	—	5.62*
Benzyl-	5.13*	—	—
Benzylthio-	3.23*	—	5.41*
t-Butyl-	5.76*	5.82*	5.99*
Dimethylaminoethyl-	3.46*	4.30*	4.66*
	8.75*	8.86*	8.70*
Dimethylaminomethyl-	2.58*	3.17*	3.39*
	8.12*	8.00*	7.66*
Hexyl-	5.95*	—	—
Methanesulfonamido-	1.10*	3.43*	3.64*
	8.02*	7.02*	9.07*
N-Methylacetamido-	2.01*	3.52*	4.62*
N-Methylbenzamido-	1.44*	3.66*	4.68*
N-Methylmethanesulfonamido-	1.73*	3.94*	5.14*
Piperidineoethyl-	3.59*	4.25*	4.68*
	9.29*	8.81*	9.06*
Piperidinomethyl-	2.61*	3.16*	3.90*
	8.51*	8.30*	7.88*
2-Pyridyl-	4.44*	—	—
3-Pyridyl-	4.42*	4.60*	—
	1.52*	3.0*	—
4-Pyridyl-	4.77*	4.85*	4.82*
	1.19*	3.0*	

*Thermodynamic value.

NITROGEN COMPOUNDS (Continued)

Compound	pK$'_a$	Reference	Compound	pK$'_a$	Reference
HETEROCYCLICS (Continued)			*HETEROCYCLICS (Continued)*		
Pyridine-*N*-oxides: see oxygen acids			4,5-Diamino-2-chloro-	4.79, 0.08	105
Pyridines			2,3-Diamino-6-chloro-	3.02, −0.91	105
2,3-Me$_2$-	6.60	48	2,4-Dichloro-5-amino-	0.73	105
2,4-Me$_2$-	6.72	48	2:4-Dihydroxy-	13, 1.37, 6.50	39
2,5-Me$_2$-	6.47	48	4-Ethoxy-3-nitro-	2.67	105
2,6-Me$_2$-	6.77	48	2-Hydroxy-3-nitro-	−4.00, 8.52	105
3,4-Me$_2$-	6.52	48	4-Hydroxy-3-nitro-	−0.70, 7.65	105
3,5-Me$_2$-	6.14	48	1-Methyl-2-pyridone	0.32	39
2,4,6-Me$_3$-	7.48	48	1-Methyl-4-pyridone	3.33	39
2-Me,5-Et-	6.51	48	4-Methylamino-3-nitro-	5.19	105
2-Amino-3-nitro-	2.38	105	1-Methylpyrid-2-one acetylimine	7.12	46
2-Amino-3-nitro-	2.42, −12.4	141	1-Methylpyrid-4-one acetylimine	11.03	46
2-Amino-5-nitro-	2.80, −12.1	141	1-Methylpyrid-4-one benzylimine	9.89	46
3-Amino-2-nitro-	0.02, −9.07	141	2,4,6-Trihydroxy-	4.6, 9.0, 13.00	39
4-Amino-3-nitro-	5.04	105	Trimethyl(2,6-di-*tert*-butyl-4-pyridyl)ammonium ion	1.65	125
3-Amino-4-methylamino-	0.38, 9.57	105	Trimethyl(2-*tert*-butyl-6-pyridyl) ammonium ion	<−1	125
4-Amino-3-methylamino-	0.12, 9.37	105	Dimethyl(2,6-di-tert-butyl-4-pyridyl)ammonium ion	1.6	125
2-Amyl-	6.00*	45	Δ′-Tetrahydro-2-methyl-	9.6	154
2-Benzamido-	3.33	46	Δ′-Tetrahydro-1,2-dimethyl-	11.4	154
2-Benzyl-	5.13	45	1,4-Dihydro-1,4,4-trimethyl-	7.4	154
2-Chloro-3-nitro-	−2.6	105	Pyrimidines		
2,3-Diamino-	7.00, −0.01	105			
3,4-Diamino-	9.14, 0.49	105			

	pK$_a$	3,4-(NH$_2$)$_2$	3-NH$_2$	4-NH$_2$	5-NH$_2$	6-NH$_2$	2-NH$_2$
2-OH[140,141]		−0.87	2.78	−5.14	−0.61	−6.12	
		4.16		2.65	3.77	2.32	
		13.43		13.54	11.65	11.38	
2-OMe		5.68, 1.06	3.35	−5.86			
				7.05	4.28	4.64	
1-Me-2=O	0.32		2.94		0.05	2.09	
1-Me-4=O	3.33		3.88				
4-OH			0.04				−6.58
			3.84				5.04
			11.38				10.69
4-OMe			7.30				
2-NH$_2$			−12.1		−10.7	−3.73	
			0.38		1.97	6.00	7.62
			6.73		6.46		
3-NH$_2$				−10.7			
				0.80			
				9.19			

Monosubstituted(99)	2	4	5	Compound	pK$'_a$	Reference
Amino-	3.45*	5.69*	2.51*	Cytidine	4.08, 12.5	134
n-Butylamino-	4.14*	—	—	Cytosine	4.60, 12.16	101
Carboxy-	−1.13*	—	—	5-Me-cytosine	4.6, 12.4	101
	2.85*	—	—	Isocytosine	4.01, 9.42	101
Dimethylamino-	3.93*	6.32*	—	Thymine	9.90	134
Ethylamino-	3.89*	—	—	Thymidine	9.79	134
Hydroxy-	2.15*	1.66*	1.85*	Uracil	9.45	101
	9.20*	8.63*	6.80*	Uridine	9.30, 12.59	134
Mercapto-	1.35*	0.68*	—	1-Me-	9.99	101
	7.10*	6.94*	—	3-Me-	9.71	101
Methoxy-	<1	2.5*	—	5-Me-	9.94	101
Methyl-	—	1.91*	—	1,3-di-Me-	None	101
Methylamino-	3.79*	6.09*	—	Orotic acid	2.40, 9.45	101
Methylthio-	0.59*	—	—			
Sulfanilamido-	6.34*	6.17*	6.22*			
	2.0*	—	—			

*Thermodynamic value.

NITROGEN COMPOUNDS (Continued)

Compound	pK′$_a$	Reference	Compound	pK′$_a$	Reference
HETEROCYCLICS (Continued)			*HETEROCYCLICS (Continued)*		
Pyrimidines, substituted			4-Amino-	7.23*	—
4,6-Bisdimethylamino-	6.34*	99	5-Amino-	0.99*, 3.93*	99
4,5-Bismethylamino-	6.77*	99	4-Amino-6-chloro-	3.55*	99
4,6-Bismethylamino-	6.32*	99	5-Amino-4,6-dihydroxy-	3.6*, 8.9*	99
5-Bromo-2,4-dihydroxy-	−7.25*, 7.83*	99	6-Amino-5-formylamino-4-hydroxy-	2.5*, 9.9*	99
5-Bromo-2-methylamino-	2.09*	99	4-Amino-6-hydroxy-	3.30*, 10.81*	99
4-Butylamino-2-hydroxy-	4.67*	99	4-Amino-6-hydroxy-5-methyl-	3.58*, 11.10*	99
4-Butylamino-2-mercapto-	3.2*, 11.15*	99	4-Amino-6-methyl-	7.7*	99
4-Carboxy-6-hydroxy-	2.8*, 8.4*	99	5-Amino-6-methylthio-	5.44*	99
2-Chloro-4-methylamino-	2.90*	99	4,5-Diamino-	2.50*, 7.60*	99
4-Chloro-2-methylamino-	2.59*	99	4,6-Diamino-	6.81*	99
4-Chloro-6-dimethylamino-	2.42*	99	Pyrimidines, 4-amino substituted		
4-Chloro-6-methylamino-	2.24*	99	5-Aminomethyl-2-methyl-	4.0*, 7.1*	99
1,2-Dihydro-2-imino-1-methyl-	10.71*	99	5-Carboxymethylamino-	2.99*, 6.70*	99
1,4-Dihydro-4-imino-1-methyl-	12.18	99	6-Chloro-	2.10*	—
l,2-Dihydro-1-methyl-2-oxo-	2.45*	99	6-Chloro-2-methylamino-	3.79*	99
1,4-Dihydro-1-methyl-4-oxo-	2.02*	99	1,2-Dihydro-1,5-dimethyl-2-oxo	4.69*	99
1,6-Dihydro-1-methyl-6-oxo-	1.79*	99	1,6-Dihydro-6-imino-1-methyl-	11.94*	99
1,2-Dihydro-1-methyl-2-thio-	1.66*	99	1,2-Dihydro-1-methyl-2-oxo-	4.57*	99
1,4-Dihydro-1-methyl-6-thio-	0.56*	99	1,6-Dihydro-1-methyl-6-oxo-	0.98*	99
2,4-Dihydroxy-	−3.38*, 9.45*	99	2,3-Dihydro-3-methyl-2-oxo-	7.4*	99
2,4-Dihydroxy-5-amino-	3.20, 8.52	140	5-Fluoro-2-hydroxy-	2.90*	99
2,6-Dihydroxy-4-amino-	0.00, 8.69, 15.32	140	2-Hydroxy-	4.60*, 12.16*	99
2,6-Dihydroxy-4,5-diamino-	4.56	140	6-Hydroxy-	1.36*, 10.08*	99
2,6-(=0)-1,3-Methyl-4,5-diamino-	4.44	140	2-Hydroxy-5-methyl-	4.6*, 12.4*	99
4,5-Dihydroxy-	1.99*, 7.52*, 11.69*	99	6-Hydroxy-2-methylamino-	3.20*, 11.06*	99
4,6-Dihydroxy-	5.4	39	2-Hydroxy-5-nitro-	7.40*	99
4,6-Dimethyl-	2.7*	99	2-Mercapto-	3.29*, 10.66*	99
4-Dimethylamino-2-hydroxy-	4.21*	99	2-Methoxy-	5.3*	99
4-Dimethylamino-6-hydroxy-	1.22*, 10.49*	99	6-Methoxy-	4.00*	99
4,6-Dimethyl-2-methyl-amino-	5.23*	99	6-Methyl-	6.16*	99
4-Dimethylamino-2-methoxy-	6.13*	99	2-Methylamino-	7.53*	99
4-Dimethylamino-6-methoxy-	4.27*	99	6-Methylamino-	6.30*	99
4-Ethoxy-2-hydroxy-	1.00*, 10.7*	99	6-Methylamino-5-nitro-	2.73*	99
4-Ethylamino-2-mercapto-	3.08*, 11.15*	99	5-Trifluoracetamido-	3.91*	99
4-Ethylamino-2-hydroxy-	4.56*	99	5-Amino-	6.00*	99
5-Ethyl-2-methylamino-	4.29*	99	6-Amino-	5.99*	99
4-Hydroxy-5-methoxy-	1.75*, 8.64*	99	6-Amino-5-bromo-	4.20*	99
2-Hydroxy-4-methyl-	3.06*, 9.9*	99	5-Amino-2-chloro-	2.63*	99
4-Hydroxy-6-methyl-	2.06*, 9.1*	99	5-Amino-2-chloro-6-ethoxycarbonyl-	1.27*	99
2-Mercapto-4-methyl-	2.1*, 8.1*	99	5-Amino-6-dimethylamino-6-ethoxycarbonyl-	−2.03*, 6.47*	99
4-Mercapto-6-methyl-	1.8*, 7.3*	99	5-Amino-6-ethoxycarbonyl-2-hydroxy-	−3.21*, 3.22*	99
2-Mercapto-4-methyl-amino-	3.07*, 11.12*	99	5-Amino-6-ethoxycarbonyl-2-mercapto-	2.11*	99
4-Methyl-2-methylthio-	1.86*	99	5-Amino-6-hydroxy-	1.28*, 3.54*, 9.89*	99
4-Methyl-6-methylthio-	3.16*	99	5-Amino-2-hydroxy-	4.34*, 11.48*	99
4-Methyl-2-sulphanilamido-	7.06*	99	6-Amino-2-hydroxy-	6.47*, 11.98*	99
1,4,5,6-Tetrahydro-	13.0	99	5-Amino-2-mercapto-	2.93*, 10.42*	99
1,4,5,6-Tetrahydro-2-amino-	14.1	154	5-Amino-2-methylthio-	5.03*	99
2,4,6-Trihydroxy- (Barbituric acid)	3.9, 12.5	39	5,6-Diamino-	1.41*, 5.75*	99
2,4,5-Trihydroxy- (Bobarbituric acid)	8.11, 11.48	39	Pyrimidines, 5-amino substituted		
Pyrimidines, 2-amino substituted			4-Carboxy-6-carboxy-methylamino-2-chloro-	3.05*, 4.48*	99
5-Bromo-	1.95*	99	4-Carboxy-6-carboxy-methylamino-2-dimethylamino-	9.85*	99
4-Chloro-1,6-dihydro-6-imino-1-methyl-	9.87*	99	4-(1-Carboxyethylidene)imino-	3.04*, 7.10*	99
4-Diethylaminoethylamino-5,6-dimethyl-	7.9*, 9.7*	99	4-Carboxymethylamino-	2.9*, 6.59*	99
4-Diethylaminoethylamino-6-methyl-	7.5*, 9.55*	99	2-Ethoxy-4-ethoxycarbonyl-6-ethoxycarbonylmethyl-amino-	4.58*	99
1,4-Dihydro-4-imino-1-methyl-	12.9*	99			
4,5-Dimethyl-	5.0*	99	4-Methyl-	3.06*	99
4,6-Dimethyl-	4.85	39	1-Methyl-2-pyrimidone	2.50	39
4-Dimethylamino-	7.94*	99	1-Methyl-4-pyrimidone	1.8	39
4,6-Diphenyl-	3.74*	99	3-Methyl-4-pyrimidone	1.84	39
4-Hydroxy-	3.91*, 9.54*	99			
4-Methyl-	4.11*	99			
5-Nitro-	0.35	99			
1,4,5,6-Tetrahydro-	14.11*	99			

*Thermodynamic value.

NITROGEN COMPOUNDS (Continued)

Compound	pK'_a	Reference	Compound	pK'_a	Reference
4-Pyrone	0.1	154	2-Benzyl-	10.31*	99
2,6-Dimethyl-	0.4	154	1-n-Butyl-2-methyl-	10.61*	99
3-Hydroxy-	7.9	154	2-n-Butyl-1-methyl-	10.20*	99
Pyrrole	17.51	152	2-Carbamoyl-	8.82*	99
2,4-Dimethyl-	2.55*	99	1-Cyanomethyl-	4.8*	99
2,5-Dimethyl-	−0.71*	99	2-Cyclohexyl-	10.76*	99
3,4-Dimethyl-	0.66*	99	1,2-Dimethyl-	10.20*	99
3-Ethyl-2,4-dimethyl-	3.54*	99	1-Ethyl-2-methyl-	10.56*	99
1-Methyl-	−2.90*	99	1-Methyl-	10.32*	99
2-Methyl-	−0.21	99	2-(p-Tolyl)-	9.95	99
3-Methyl-	−1.00	99	2-Pyrrolines		
2,3,4,5-Tetramethyl-	3.77	99	2-Benzyl-	7.06*	99
2,3,4-Trimethyl-	3.94	99	2-n-Butyl-1-methyl-	11.84*	99
2,3,5-Trimethyl-	2.00	99	2-Cyclohexyl-	7.91*	99
Δ'-Dihydro-l,2-dimethyl-	11.9	154	1,2-Dimethyl-	11.90*	99
Δ'-Dihydro-2-ethyl-	7.9	154	1-Ethyl-2-methyl-	11.84*	99
Δ³-Dihydro-1-methyl-	9.9	154	2-Phenyl-	6.7	99
Pyrrolidine	11.27*	99			
3-Amino-2,5-dioxo-	5.9*, 9.0*	99			
1-(2-Aminoethyl)-	6.56*. 9.74*	99			

Quinazoline(19,39,99)	2	4	5	6	7	8
NH_2-	4.43	5.73	3.56*	3.2[a]	4.59*	2.4[a]
OH-	1.30	2.12	3.62*	3.12	3.20*	3.41
CH_3O-	1.31	3.13	3.39*	2.83*	2.87*	3.49*
Cl-	−1.6	—	3.73*	3.53*	3.25*	3.28*
SH-	0.26*	1.51*	—	—	—	—
	8.18*	8.47*	—	—	—	—
CH_3-	4.50*	2.44*	3.61*	3.39*	3.15*	3.18*
CH_3S-	1.60*	2.97*	—	—	—	—
NO_2-	—	—	3.73*	4.16*	4.03*	3.98*

Compound	pK'_a	Reference
3,4-Dihydro-	1.47, 9.2	99, 154
3,4-Dihydro-2-methyl-	10.2	154
3,4-Dihydro-3-methyl-	9.2	154
3,4-Dihydro-3-methyl-4-hydroxy	7.6	154
2:4-Dihydroxy-	9.78, 2.5	39
2,4-Dimethyl-	3.58*	99
3-Methiodide	7.26	39
3-Oxide	1.47*	99
1,2,3,4-Tetrahydro-	10.2	154

Quinoline(2,44,99)		3	4	5	6	7	8
H–	4.85*	4.80	4.69*	—	—	—	—
H_2N-	7.25*	4.86*	9.08*	5.37*	5.54*	6.56*	3.90*
HO-	−0.36	4.30	2.27	5.20	5.17	5.48	5.13
	11.74	8.06	11.25	8.54	8.88	8.85	9.89
CH_3	5.42	5.14	5.20	4.62	4.92	5.08	4.60
	5.8	—	5.6	—	—	—	—
F–	—	2.36*	—	3.68*	4.00*	4.04*	3.08*
HO_2C-	4.96*	4.62*	4.53*	4.81*	4.98*	4.97*	7.20*
Br-	1.05*	2.75*	—	3.62*	3.91*	3.87*	3.33*
Cl-	—	2.63*	3.72*	3.65*	3.99*	3.85*	3.12*
HS-	1.44*	2.29*	0.77*	—	—	—	—
	10.25*	6.17*	8.87*	—	—	—	—
CH_3O-	3.16*	—	6.45*	—	5.03*	—	—
CH_3S-	3.67*	3.84*	5.85*	4.46*	4.71*	—	3.46*
O_2N-	—	1.03	—	2.69*	2.72*	2.40*	2.55*

*Thermodynamic value.
[a] In 50% ethanol on methanol

NITROGEN COMPOUNDS (Continued)

Compound	pK_a'	Reference	Compound	pK_a'	Reference
HETEROCYCLICS (Continued)			*HETEROCYCLICS (Continued)*		
Quinoline			4-COO-	10.55	158
4-Amino-7-chloro-	8.23*	99	4-COOCH₃-	9.46	158
4-Amino-8-hydroxy-	6.91*, 10.71*	99	4-COOC₂H₅-	9.44	158
5-Amino-8-hydroxy-	5.67*, 11.24*	99	4-CO CH₃-	9.45	158
4-Amino-6-methoxy-	8.93*	99	4CONH₂ -	9.38	158
8-Amino-6-methoxy-	3.38*	99	4-CN-	8.07	158
6-Bromo-4-chloro-	2.83*	99	4-OH-	9.44	158
4-Chloro-6-ethoxy-	3.82*	99	4-OCH₃-	9.31	158
4-Chloro-6-fluoro-	2.95*	99	4-OCOCH₃-	8.99	158
4-Chloro-6-methoxy-	3.93	99	4-Cl-	8.62	158
4-Chloro-6-methyl-	3.96	99	4-Br-	8.49	158
4,6-Dichloro-	2.81	99	4-I-	8.70	158
2:4-Diamino-	9.45	19	4-NH₂-	10.10	158
1,4-Dihydro-4-imino-1-methyl-	12.4*	99	4-NHCH₃-	10.28	158
1,2-Dihydro-1-methyl-2-oxo-	−0.71*	99	4-N(CH₃)₂-	10.11	158
1,2-Dihydro-1-methyl-2-thio-	−1.6	99	4-NHCOCH₃-	9.54	158
2:4-Dihydroxy-	5.86, 0.76	39	4-NHCOOC₂H₅-	9.57	158
2,3-Dimethyl-	4.94*	99	4-NO₂-	7.65	158
2,4-Dimethyl-	5.12*	99	Riboflavin	9.93	77
2,6-Dimethyl-	6.1*	99	Serotonine	19.50	152
2,7-Dimethyl-	5.02*	99	Sulfadiazine	6.48	6
2,8-Dimethyl-	4.11*	99	Sulfapyridine	8.43	6
1-Methyl-2-quinolone	−0.71	39	Sulfaguanidine	11.25	6
1-Methyl-4-quinolone	2.46	39	Sulfathiazole	7.12	6
1,2,3,4-Tetrahydro-	5.0	154	Terramycin	3.10, 7.26, 9.11	77
Quinoxaline	0.8, 0.56	19, 39	1,4,5,8-Tetraazanaphthalene	2.47*	99
2-Amino-	3.96	19	Tetramethylenediamine	10.7	4
5-Amino-	2.62	19	1,2,4,5-Tetrazine		
6-Amino-	2.95	19	3,6-Diethyl-1,4-dihydro-	4.23*	99
2-Carbamoyl-	−0.4*	99	1,4-Dihydro-	2.25*	99
2:3-Diamino-	4.70	19	Tetrazole	4.9	51
2:3-Dihydroxy-	9.52	39	5-Chloro-	2.1	51
2-Hydroxy-	9.08, −1.37	39	5-Amino-	1.8,6.0	51
4-Hydroxy-	10.01, 2.85	39	1,2,4-Thiadiazole		
5-Hydroxy-	8.65, 0.9	39	3-Amino-5-phenyl-	0.1*	99
6-Hydroxy-	7.92, 1.40	39	5-Amino-3-phenyl-	1.4*	99
5-Hydroxy-1-methyl-quinoxalinium chloride	5.74	44	1,3,4-Thiadiazole		
			2-Amino-5-phenyl-	2.9*	99
1,2,3,4-Tetrahydro-	2.1, 4.8	154	2-Benzylamino-5-phenyl-	2.5*	99
2-Mercapto-	−1.24*, 7.20*	99	2-Ethylamino-5-phenyl-	3.05*	99
1-Methiodide-	5.74	39	2-Methylamino-5-phenyl-	2.8*	99
2-Methoxy-	0.28*	99	Thiazine		
2-Methyl-	0.95*	99	Δ²-Dihydro-2-methyl-	7.6	154
2-Methylamino-	4.07*	99	1,4-Thiazine		
2-Methylthio-	0.29*	99	Tetrahydro-	8.40*	99
1,5-Naphthyridine	2.91	39	Thiazole	2.44*	99
Quinuclidine	10.95*	99	2-Amino-	5.36, 5.39	99, 41
3-Carbamoyl-	9.67*	99	5-Carbamoyl-	0.6*	99
3-Cyano-	7.81*	99	4-Methyl-5-carboxylic acid-	3.51	144
3-Phenyl-	10.23*	99	4-Methyl-2-carboxylic acid-	1.20, 3.18	144
4-CH₃-	10.88	158	Δ²-Dihydro-2-methyl-	5.2	154
4-CH₂CH₂-	10.95	158	1,3,4-Thiazole	−2.5*	99
4-CH(CH₂)₃-	11.02	158	2-Nitramino-		
4-C(CH₂)₂-	11.07	158	Thiazolidine	6.22*, 6.31	99, 95
4-CH=CH₂-	10.60	158	4-Carboxy-	1.42*, 6.30*	99
4-C₆H₆-	10.20	158	4-Carboxylate methyl ester	4.00	95
4-CH₂OH-	10.45	158	4-Methoxycarbonyl-	3.91*	99
4-CH₂ OCH₃-	10.50	158	Thiazolo[5,4-*d*]pyrimidine		
4-CH₂OCOCH₃-	10.27	158	7-Amino-	2.74*	99
4-CH₂OTs-	9.87	158	7-Methylamino-	2.81*	99
4-CH₂Cl-	10.19	158	Thiophene		
4-CH₂Br-	10.13	158			
4-CH₂I-	10.12	158			
4-CH(OH)₂-	9.90	158			

*Thermodynamic value.

NITROGEN COMPOUNDS (Continued)

Compound	pK'_a	Reference	Compound	pK'_a	Reference
HETEROCYCLICS (Continued)			*HETEROCYCLICS (Continued)*		
2-(2-Aminoethylaminomethyl)-	6.29*, 9.77	99	4,5 Dibromo-	5.37	126
2-Aminomethyl-	8.92	99	1-Phenyl-5-methyl-4-carboxylic acid-	3.73	126
1,2,4-Triazanaphthalene	−0.82*	99	1-Phenyl-4-carboxylic acid-	2.88	126
1,3,5-Triazanaphthalene			1-Phenyl-4,5-dicarboxylic acid-	2.13, 4.93	126
4-Hydroxy-	8.98*	99	4-Carboxylic acid-	3.22, 8.73	126
1,3,8-Triazanaphthalene			4,5-Dicarboxylic acid-	1.86 5.90, 9.30	126
2-Hydroxy-	1.81*, 10.06	99	1,2,3-Triazole	1.17*, 9.51*	99
1,4,5-Triazanaphthalene	1.20	39	1-Methyl-	1.25*	99
8-Hydroxy-	8.76, 0.60	39	1,2,4-Triazole	10.3, 2.3	154
1,4,6-Triazanaphthalene	3.05*	99	3,5-Dimethyl-	3.8	154
5-Hydroxy-	11.05, 0.78	39	3,5-Diamino-	12.1, 4.4	154
1,2,4-Triazine			3-Methyl-	10.7, 3.3	154
3-Amino-	3.09*	99	3-Amino-	11.1, 4.0	154
1,3,5-Triazine			3-Chloro-	8.1	154
2-Amino-4,6-bisethylamino-	6.18*	99	3,5-Dichloro-	5.2	154
2-Amino-4,6-dimethyl-	3.56*	99	1,2,3-Triazolo(5′,4′-4,5)pyrimidine	2.03*, 4.96*	99
2,4-Diamino-	5.88*	99	Tryptamine	16.60	152
2,4-Diamino-6-guanidino-	9.4*	99	Tryptophan	16.82	152
2:4-Dihydroxy-	6.5	39	4-Methyl-	16.90	152
2,4,6-Triamino-	5.1*	99	5-Hydroxy-	19.20	152
2,4,6-Trisdi(2-hydroxy)ethylamino-	4.70*	99			
2,4,6-Trisguanidino-	4.6*, 7.6*, 10.3*	99			
2,4,6-Trishydroxymethylamino-	4.0*	99			
Triazole					
Benzo-	8.38	126			

*Thermodynamic value.

SPECIAL NITROGEN COMPOUNDS

Compound	pK$_a'$	Reference	Compound	pK$_a'$	Reference
AMIDINES AND GUANIDINES			***C-SUBSTITUTED-N-PHENYLAMIDINIUM IONS***		
Acetamidine	12.52	19	EtO-	7.71	114
			H$_2$N-	10.77	114
C-SUBSTITUTED AMIDINIUM IONS			MeO-	7.41	114
BuO-	10.15	114	MeS-	7.14	114
CH$_2$CHCH$_2$O-	9.70	114	PhNH-	10.42	144
EtO-	10.02	114			
Me-	12.41	114	***AMIDOXIMES***		
Me$_2$N-	13.4	114			
MeO-	9.72	114	Benz-	4.99	17
MeS-	9.83	114	Malon-	~4.77	17
NH$_2$-	13.86	114	Ox-	3.02	17
Ph-	11.6	114	α-Phenylacet-	5.24	17
PrO-	10.16	114	Succin-	3.11, 5.97	17
1-Chloro-4-N^1-methylguanidino benzene	12.6*	99	o-Tolu-	4.03	17
1-Chloro-4-N^3-methylguanidino benzene	10.85*		p-Tolu-	5.14	17
			HYDRAZINES (30°)		
Diguanide	3.07, 13.25	99	Hydrazine	8.07	13
Ethyl-	3.08, 11.47	99	Acet-	3.24	15
Ethylene-	11.34, 1.74, 2.88, 11.76	77	N,N-Diethyl-	7.71	13
Methyl-	3.00, 11.42	99	N,N'-Diethyl-	7.78	13
Phenyl-	2.16, 10.71	77	N,N-Dimethyl-	7.21	13
			N,N'-Dimethyl-	7.52	13
Guanidine	13.6	99	Ethyl-	7.99	13
Acetyl-	8.26	99	Glycylhydrazide	2.38, 7.69	15
2-Anthryl-	11.0	99	Isonicotinhydrazide	1.85, 3.54, 10.77	77
Carbamoyl-	3.76	99	Methyl-	7.87	13
N,N'-Dimethyl-	13.4	99	Phenyl-	5.21 (15°)	14
N,N'-Dimethyl-	13.6	99	Tetramethyl-	6.30	13
N,N'-Diphenyl-	10.12	99	Trimethyl-	6.56	13
N-Methyl-	13.4	99	HOCH$_2$CH$_2$-	7.12	148
N-Methyl-N'-nitro-	12.40	99	C$_6$H$_5$-CH$_2$-	6.83	148
2-Naphthyl-	10.7	99	C$_6$H$_5$-O(CH$_2$)$_2$-	6.80	148
Nitro-	−0.55, 12.20	99	C$_2$H$_5$OOC-CH$_2$-	5.97	148
Nitroamino-	−1, 10.60	99	NC-(CH$_2$)$_2$-	5.91	148
Nitroso-	2.13, 11.70	99	CHF$_2$−CF$_2$−CH(CH$_3$)-	5.59	148
Pentamethyl-	13.8	99	HC≡C-CH$_2$-	5.46	148
N-Phenyl-	10.77	99	CF$_3$−CH$_2$-	5.38	148
N,N,N'-Trimethyl-	13.6	99	CHF$_2$(CF$_2$)$_3$ −CH$_2$-	5.34	148
N,N',N''-Trimethyl-	13.9	99	C$_6$H$_5$−CH(CF$_3$)-	4.88	148
Triphenyl-	9.10	99			
			HYDRAZONES OF		
1-SUBSTITUTED GUANIDINIUM IONS			Benzophenone	3.85	
CN-	−0.4	114	p-Chloro-	3.53	16
H$_2$NCO-	7.85	114	p, p'-Dimethoxy-	4.38	16
MeO-	7.46	114	p,p'-Dichloro-	3.13	16
$^-$O$_2$CCH$_2$O-	7.51	114	p-Methoxyacetophenone	4.94	16
O-Methylisourea	9.80	20	Phenyl-2-thienyl ketone	3.80	16
N-Phenyl-	7.3	20			
S-Methylisothiourea	9.83	20	***SEMICARBAZONES OF***		
N-Phenyl-	7.14	20	Acetone	1.33	14
			Acetaldehyde	1.10	14
1-SUBSTITUTED 3-NITROGUANIDINES			Benzaldehyde	0.96	14
Bz-	8.10	114	Furfural	1.44	14
EtO$_2$CCH$_2$-	11.20	114	Pyruvic acid	0.59	14
H$_2$NCO-	7.50	114	Semicarbazide	3.66	14
NCCH$_2$-	9.30	114			
NH$_2$-	10.60	114	***NITROGEN ACIDS***		
Ph-	10.50	114	Dimedone	5.23	18
			Diphenylthiocarbazone	4.5	6
			Nitrourea	4.57	18
			Nitrourethane	3.28	18
			Phthalimide	8.30	18

*Thermodynamic value.

SPECIAL NITROGEN COMPOUNDS (Continued)

Compound	pK'_a	Reference	Compound	pK'_a	Reference
			Diphenylthiocarbazone	4.5	6
			Thiourea	−0.96	4
OTHER			Urea	0.18	4,99
Acetamide	−0.51	4	O-Allyl-	9.70	99
Azobenzene	−2.48*	99	O-n-Butyl-	10.15	99
4-Amino-3′-methyl-	−2.88*	99	O-Cyclohexyl-	10.19	99
4-Amino-4′-methyl-	3.04	99	O-Isobutyl-	10.30	99
4-Dimethylamino-	−1.3*, 3.226*	99	O-Isopentyl-	10.11	99
4-Hydroxy-	−0.93*, 8.2	99	O-Methyl	0.9	99
Benzamide			O-Methyl-	9.72	99
3,5-Dinitro-4-methyl-	−2.77	110	O-Phenethyl-	10.03	99
4-Methoxy-	−1.46	110	Phenyl-	−0.3	99
3-Nitro-	−2.25	110	N-Phenyl-O-methyl-	7.3	20
2,3,6-Trichloro-	−3.10	110	N-Propyl-	10.16	99
3,4,5-Trimethoxy-	−1.86	110			

*Thermodynamic value.

THIOLS

Compound	pK_a'	Reference	Compound	pK_a'	Reference
N-Acetylcysteine	9.52	112	2-Mercaptopropionic acid	4.32, 10.30	153
N-Acetyl-β-mercaptoisoleucine	10.30	112	Methyl cysteine	6.5, (7.5)	81
N-Acetylpenicillamine	9.90	112	Methyl [β-diethylaminoethyl] sulfide	9.8	5
O-Aminothiophenol	6.59	81			
p-Chlorothiophenol	7.50	81	Methyl thioglycolate	7.8	23
Cysteine	1.8, 8.3, 10.8	23	p-Nitrobenzenethiol	5.1	58
Cysteine ethyl ester	6.53, 9.05	112	Penicillamine	7.90, 10.42	112
Cysteinylcysteine	2.65, 7.27, 9.35, 10.85	23	Thiocyanic acid	−1.84	104
1-Diethylaminobutane-(4)	10.1	5	Thioglycolic acid	3.67, 10.31	23
1-Diethylaminohexane-(6)	10.1	5	Thiophenol	7.8, 6.52	59, 81, 82
1-Diethylaminopropane-(3)	8.0, 10.5	5			
N-Diethylcysteamine	7.8, 10.75	5	Pentafluoro-	2.68	155
N-Dimethylcysteamine	7.95, 10.7	5	p-Me-	6.82	157
N-Dipropylcysteamine	8.00, 10.8	5	p-OMe-	6.77	157
Ethyl mercaptan	10.50	81	m-Me-	6.66	157
Glutathione	2.12, 3.59, 8.75, 9.65	23	m-OMe-	6.38	157
DL-Homocysteine	8.70, 10.46	112	p-Cl-	6.13	157
2-Mercaptoethanesulfonate	7.53 (9.1)	81	p-Br-	6.02	157
Mercaptoethanol	9.5	23	m-Cl-	5.78	157
Mercaptoethylamine	8.6, 10.75	23	p-COMe-	5.33	157
N-β-Mercaptoethylmorpholine	6.65, 9.8	5	m-NO$_2$-	5.24	157
N-β-Mercaptoethylpiperidine	7.95, 11.05	5	p-NO$_2$	4.71, 4.50	157
β-Mercaptoisoleucine	8.10, 10.6	112	l-Thio-D-sorbitol	9.35	81
o-Mercaptophenylacetic acid	4.28, 7.67	59	N-Trimethyl cysteine	8.6	23

X=	−H	−S$^-$	−SH	X=	−H	−S$^-$	−SH
X(CH$_2$)$_2$SH	12.0	13.96	10.75	X(CH$_2$)$_3$SH	—	13.24	11.14
X(CH$_2$)$_4$SH	12.4	13.25	11.50	X(CH$_2$)$_5$SH	—	13.27	11.82

Compound	pK_a'	Reference	Compound	pK_a'	Reference
Mercaptans, RSH			n-C$_3$H$_7$-	10.65	82
			t-C$_4$H$_9$-	11.05	82
R			(CH$_3$)$_2$CH-	10.86*	103
C$_6$H$_5$CH$_2$-	9.43	82	(CH$_3$)$_3$C-	11.22*	103
HOCH$_2$CH(OH)CH$_2$-	9.51	82	HOCH$_2$CH$_2$-	9.72	103
CH$_2$=CHCH$_2$-	9.96	82	CH$_3$CONHCH$_2$CH$_2$-	9.92	103
n-C$_4$H$_9$-	10.66	82	$^-$OCOCH$_2$-	10.68*	103
t-C$_5$H$_{11}$-	11.21	82	$^-$OCOCH$_2$CH$_2$-	10.84*	103
C$_2$H$_5$OCOCH$_2$-	7.95	82	o-$^-$OCOC$_6$H$_4$-	8.88*	103
C$_2$H$_5$OCH$_2$CH$_2$-	9.38	82	p-$^-$OCOC$_6$H$_4$-	5.80*	103
HOCH$_2$CH(OH)CH$_2$-	9.66	82	CH$_3$CO-	3.62*	103

*Thermodynamic value.

CARBON ACIDS

Compound	pK'_a	Reference	Compound	pK'_a	Reference
Acetone	c. 20	24	$CH_3COCH_2COCH_3$	9	74
Acetonitrile	c. 25	24	$CH_3COCHBrCOCH_3$	7	74
Acetylacetone	8.95	24	$CH_3COCH_2COCF_3$	4.7	74
Benzoylacetone (enol)	8.23	24	$C_6H_5COCH_2NC_5H_5$	10.51	74
Diacetylacetone	7.42	153	$CH(COCH_3)_3$	5.85	74
Dihydroresorcinol	5.26	153	$CH_3SO_2CH_3$	c. 23	74
Ethyl acetoacetate	10.68	153	$CH(SO_2CH_3)_3$	Strong	74
Dimethylsulfone	14	24	$C_2H_5O_2CCH_2CN$	9	74
Hydrocyanic acid	9.21	25	$CH_3CO_2C_2H_5$	c. 24.5	74
Nitroethane	8.44	153	$CHC_2H_5(CO_2C_2H_5)_2$	15	74
Nitromethane	10.21	153	CH_3CONH_2	c. 25	74
2-Nitropropane	7.74	18	$CH_2(CO_2C_2H_5)_2$	13.3	74
Saccharin	1.6	18	CH_3CO_2H	c. 24	74
Triacetylmethane	5.81	153	thiophene–$\overset{O}{\overset{\parallel}{C}}CH_2\overset{O}{\overset{\parallel}{C}}CF_3$	6.10	74
CH_4	40	127			
$CH(NO_2)_3$	0	127	cyclohexanone-$CO_2C_2H_5$	10.96	74
CH_3CN	25	127			
$CH_2(CN)_3$	12	127			
$CH(CN)_3$	0	127	cyclopentanone-$COCH_3$	7.82	74
$CH_3CHClNO_2$	7	74			
$CH_3COCH_2NO_2$	5.1	74			
$CH(NO_2)_3$	Strong	74	cyclopentanone-$CO_2C_2H_5$	10.5	74
$CH_3COCHCl_2$	15	74			
$CH_3COCHC_2H_5CO_2C_2H_5$	12.7	74			
$CH_3COCHCH_3COCH_3$	11	74	cyclohexanone-$CCH_3(=O)$	10.1	74
$CH_3COCH_2COC_6H_5$	9.4	74			
$C_6H_5COCH_2COCF_3$	6.82	74			
CH_3COCH_2CHO	5.92	74			
$CH_3COCH_2CO_2CH_3$	10	74			
$CH_3SO_2CH_2SO_2CH_3$	14	74			
$CH_3SO_2CH(COCH_3)_2$	4.3	74			
$C_2H_5O_2CCH_2NO_2$	5.82	74			
$CH_2(NO_2)_2$	3.57	74			
CH_3COCH_2Cl	c. 16.5	74			
$CH_3COCH_2CO_2C_2H_5$	10.68	74	$CH_2(CHO)_2$	5	74

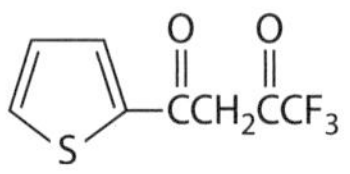

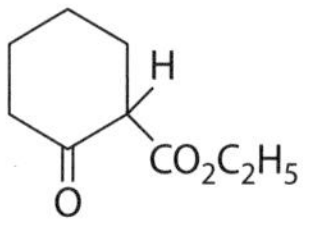

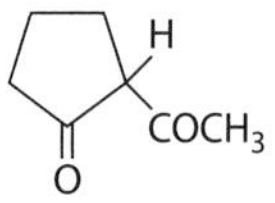

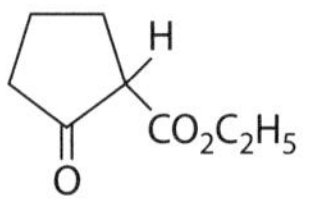

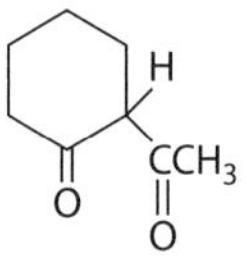

$RC(NO_2)_2H$ R=	pK'_a	References
CH_3-	5.13	127
CH_3CH_2-	5.49	127
$CH_3CH_2CH_2$-	5.35	127
$(CH_3)_2CHCH_2$-	5.36	127
$CH_3(CH_2)_2CH_2$-	5.34	127
$CH_3(CH_2)_3CH_2$-	5.37	127
$CH_3(CH_2)_4CH_2$-	5.46	127
$CH_3(CH_2)_5CH_2$-	5.46	127
$CH_3(CH_2)_6CH_2$-	5.46	127
$CH_3(CH_2)_7CH_2$-	5.45	127
$CH_2=CHCH_2$-	4.95	127
$HOCH_2CH_2$-	4.44	127
H-	3.57	127
C_6H_5-	3.71	127
$CH(NO_2)_2CH_2$-	1.09	127
$CH_3C(NO_2)_2CH_2$-	1.35	127
$CH_3OOCCH_2CH_2$-	4.34	127
CH_3OCH_2-	3.48	127
$N{\equiv}CCH_2CH_2$-	3.45	127
$O_2NCH_2CH_2$	3.24	127
CH_3OOCCH_2-	3.08	127
$(CH_3)_3NCH_2$-	-1.87	127
$N{\equiv}CCH_2$-	2.27	127
$(CH_3)_2CH$-	6.71	127

MISCELLANEOUS

Compound	pK'_a	Reference	Compound	pK'_a	Reference
ANTIBIOTICS AND VITAMINS			*INDICATORS AND DYES (Continued)*		
Chlorotetracycline	3.30, 7.44, 9.27	99	Methyl yellow	3.25	28
5-Desoxypyridoxal	4.17, 8.14	99	Neutral red	7.4	28
Dimethyloxytetracycline	7.5, 9.4	99	Nile blue A	2.4	99
Isochlorotetracycline	3.1, 6.7, 8.3	99	Phenol blue	−6.5, 4.8	99
O-Methylpyridoxal	4.75	99	Phenolindophenol	−5.3, 0.95, 8.1	99
Oxytetracycline	3.27, 7.32, 9.11	99	Phenol red	8.03	97
Pyridoxal	4.20, 8.66, 13	99	Pinachrom (M)	7.31	99
Pyridoxal 5-phosphate	4.14, 6.20 8.69	99	*N*-Propylanilinesulfonephthalein	1.57, 13.11	99
Pyridoxamine	3.31, 7.90, 10.4	99	Propylhelianthin	3.95	99
Pyridoxamine 5-phosphate	2.5, 3.69, 5.76, 8.61, 10.92	99	Pyronine B	7.7	99
Pyridoxine	5.00, 8.96	99	Quinaldine red	2.63	99
Riboflavin	10.02	99	Rhodamine B	3.2	99
Tetracycline	3.30, 7.68, 9.69	99	Safranin O	6.4	99
			Thioflavine T	2.7	99
INDICATORS AND DYES			Thionine	6.9	99
Acridine red	3.1	99	Thymol blue (1)	1.65	28
Anilinesulfonephthalein	1.59, 12.26	99	Thymol blue (2)	9.2	28
N-Benzylanilinesulfonephthalein	0.30, 12.76	99	Toluidine blue 0	7.5	99
Bindschedler's green	−2.5	99	Tropeoline 00	2.0	28
Bismarck brown Y	5.0	99			
Brilliant cresyl blue	3.2	99	*PORPHYRINS, BILE PIGMENTS AND STEROIDS*		
Bromocresol green	4.9	28	Biliverdin	3	99
Bromocresol purple	6.46	97	Chlorin e$_6$	1.9	99
Bromophenol blue	4.1	28	Chlorin p$_6$ trimethyl ester	1.4	99
Bromothymol blue	7.3	28	Coproporphyrin I	7.13, 4.2	99
Butylhelianthin	4.01	99	Deuteroporphyrin IX dimethyl ester disulfonic acid	0.3, 4.7	99
Chlorophenol blue	4.43	97	Dipyrrylmethene	8.50	99
Chlorophenol red	7.96	97	Mesobiliviolin	4.0	99
Chrysoidin Y	5.3	99	*N*-Methyl coproporphyrin I	0.7, 11.3	99
Congo red	4.19	97	*N*-Methyl coproporphyrin I methyl esterI	0.7, 8.3	99
m-Cresol purple	1.70	97	Methylphaeophorbid a	0.2	99
Cyanine	8.62	99	Methylphaeophorbid b	−0.1, 1.9	99
N-Ethylanilinesulfonephthalein	1.73, 13.20	99	Phaeopurpurin 18 methyl ester	−0.2, 2.1	99
Ethylhelianthin	4.34	99	Phyllochlorin	2.1, 4.6	99
Hexylhelianthin	3.71	99	Phylloporphyrin	2.5, 5	99
Iodophenol blue	2.19	97	Pyrrochlorin	2.0, 4.5	99
N-Methylanilinesulfonephthalein	1.36, 12.94	99	Pyrroporphyrin	2.0, 4.5	99
Methylene blue	3.8	99	Rhodin g$_7$	1.6	99
Methylene green	3.2	99	Rhodochlorin dimethyl ester	0.9	99
Methylhelianthin	3.76	99	*b*-Rhodochlorin dimethyl ester	0.2, 2.8	99
Methyl orange	3.45	28	Rhodoporphyrin	1.2, 3.7	99
Methyl red (1)	2.3	28	Stercobilin	7.60	99
Methyl red (2)	5.0	28	*d*-Urobilin	7.20	99
			i-Urobilin	7.40	99

References

1. Hall, *J. Am. Chem. Soc.*, 79, 5441 (1957).
2. Brown, McDaniel, Häfliger, in *Determination of Organic Structures by Physical Methods.*,Braude, Nachod, Eds., Academic Press, New York, 1955, P. 567.
3. Hall, *J. Am. Chem. Soc.*, 79, 5439 (1957).
4. Hodgman, Ed., *Handbook of Chemistry and Physics*, The Chemical Rubber Co., Cleveland, 1951. p.1636
5. Franzen, *Chem. Ber.*, 90, 623 (1957).
6. Dawson, Elliott, Elliott, and Jones, *Data for Biochemical Research*, Clarendon, Oxford, 1959.
7. Buist and Lucas, *J. Am. Chem. Soc.*, 79, 6157 (1957).
8. Stevenson and Williamson, *J. Am. Chem. Soc.*, 80, 5943 (1958).
9. Soloway and Lipschitz, *J. Org. Chem.*, 23, 613 (1958).
10. Bissell and Finger, *J. Org. Chem.*, 24, 1256 (1959).
11. Tuckerman, Mayer, and Nachod, *J. Am. Chem. Soc.*, 81, 92 (1959).
12. Bissot, Parry, and Campbell, *J. Am. Chem. Soc.*, 79, 796 (1957).
13. Hinman, *J. Org. Chem.*, 23, 1587 (1958).
14. Conant and Bartlett, *J. Am. Chem. Soc.*, 54, 2881 (1932).
15. Lindegreen and Niemann, *J. Am. Chem. Soc.*, 71, 1504 (1949).
16. Harnsberger, Cochran, and Szmant, *J. Am. Chem. Soc.*, 77, 5048 (1955).
17. Pearse and Pflaum, *J. Am. Chem. Soc.*, 81, 6505 (1959).
18. Bell and Higginson, *Proc. Roy. Soc.*, 197A, 141 (1949).
19. Albert, Goldacre, and Phillips, *J. Chem. Soc.*, 2240 (1948).
20. Dippy, Hughes, and Rozanski, *J. Chem. Soc.*, 2492 (1959).
21. Edsall, Martin, Bruce, and Hollingworth, *Proc. Natl. Acad. Sci. U.S.*, 44, 505 (1958).
22. Martin, Edsall, Wetlaufer, and Hollingworth, *J. Biol. Chem.*, 233, 1429 (1958).
23. Edsall and Wyman, *Biophysical Chemistry*, Academic Press, New York, 1958.

24. Pearson and Dillon, *J. Am. Chem. Soc.*, 75, 2439 (1953).
25. Ang, *J. Chem. Soc.*, 3822 (1959).
26. Martin and Fernelius, *J. Am. Chem. Soc.*, 81, 1509 (1959).
27. Ellenbogen, *J. Am. Chem. Soc.*, 74, 5198 (1952).
28. Kolthoff and Elving, *Treatise on Analytical Chemistry.*, Interscience Encyclopedia, New York, 1959,1.
29. Edwards, *J. Am. Chem. Soc.*, 76, 1540 (1954).
30. Bailey, Carrington, Lott, and Symons, *J. Chem. Soc.*, 290 (1960).
31. Brownstein and Stillman, *J. Phys. Chem.*, 63, 2061 (1959).
32. Meier and Schwarzenbach, *Helv. Chim. Acta*, 40, 907 (1957).
33. Ingham and Morrison, *J. Chem. Soc.*, 1200 (1933).
34. Hildebrand, *Principles of Chemistry*, Macmillan, New York, 1940.
35. Baddiley, in *The Nucleic Acids*, Chargaff and Davidson, Eds., Academic Press, New York, I, 1955, 137.
36. Circular OR-18, Pabst Laboratories, Milwaukee, Wisc, April 1961.
37. Bendich, in *The Nucleic Acids*, Chargaff and Davidson, Eds., Academic Press, New York, I, 1955, 81.
38. Jordan, *in The Nucleic Acids*, Chargaff and Davidson, Eds., Academic Press, New York, I, 1955, 447.
39. Albert and Phillips, *J. Chem. Soc.*, 1294 (1956).
40. Elliott and Mason, *J. Chem. Soc.*, 2352 (1959).
41. Angyal and Angyal, *J. Chem. Soc.*, 1461 (1952).
42. Edward and Nielsen, *J. Chem. Soc.*, 5075 (1957).
43. Bruice and Schmir, *J. Am. Chem. Soc.*, 80, 148 (1958).
44. Mason, *J. Chem. Soc.*, 674 (1958).
45. Linnell, *J. Org. Chem.*, 25, 290 (1960).
46. Jones and Katrizky, *J. Chem. Soc.*, 1317 (1959).
47. Jaffee and Doak, *J. Am. Chem. Soc.*, 77, 4441 (1955).
48. Clarke and Rothwell, *J. Chem. Soc.*, 1885 (1960).
49. Keyworth, *J. Org. Chem.*, 24, 1355 (1959).
50. Gawron, Duggan, and Grelecki, *Anal. Chem.*, 24, 969 (1952).
51. Sims, *J. Chem. Soc.*, 3648 (1959).
52. Fickling, Fischer, Mann, Packer, and Vaughan, *J. Am. Chem. Soc.*, 81, 4226 (1959).
53. Wold and Ballou, *J. Biol. Chem.*, 227, 301 (1957).
54. McElroy and Glass, *Phosphorus Metabolism.* Johns Hopkins, Baltimore, I, 1951.
55. Kumler and Eiler, *J. Am. Chem. Soc.*, 65, 2355 (1943).
56. Van Wazer, *Phosphorus and its Compounds*, Inter-Science, New York, I, 1958.
57. Freedman and Doak, *Chem. Rev.*, 57, 479 (1957).
58. Ellman, *Arch. Biochem. Biophys.*, 74, 443 (1958).
59. Pascal and Tarbell, *J. Am. Chem. Soc.*, 79, 6015 (1957).
60. Bascombe and Bell, *J. Chem. Soc.*, 1096 (1959).
61. Gawron and Draus, *J. Am. Chem. Soc.*, 80, 5392 (1958).
62. Mukherjee and Grunwald, *J. Phys. Chem.*, 62, 1311 (1958).
63. Ballinger and Long, *J. Am. Chem. Soc.*, 81, 1050 (1959).
64. Ballinger and Long, *J. Am. Chem. Soc.*, 82, 795 (1960).
65. Haszeldine, *J. Chem. Soc.*, 1757 (1953).
66. Deno, Berkheimer, Evans, and Peterson, *J. Am. Chem. Soc.*, 81, 2344 (1959).
67. Gardner and Katrizky, *J. Chem. Soc.*, 4375 (1957).
68. Wise and Brandt, *J. Am. Chem. Soc.*, 77, 1058 (1955).
69. Fortnum, Battaglia, Cohen, and Edwards, *J. Am. Chem. Soc.*, 82, 778 (1960).
70. Everett and Minkoff, *Trans. Faraday. Soc.*, 49, 410 (1953).
71. Bauer and Miarka, *J. Am. Chem. Soc.*, 79, 1983 (1957).
72. Green, Sainsbury, Saville, and Stansfield, *J. Chem. Soc.*, 1583 (1958).
73. Burkhard, Sellers, DeCou, and Lambert, *J. Org. Chem.*, 24, 767 (1959).
74. Bell, *The Proton in Chemistry*, Cornell, Ithaca, 1959.
75. Stewart and Maeser, *J. Am. Chem. Soc.*, 46, 2583 (1924).
76. Jencks and Carriuolo, *J. Am. Chem. Soc.*, 82, 1778 (1960).
77. Bjerrum, Schwarzenbach, and Sillen, *Stability Constants of Metal-Ion Complexes, Part I, IInorganic Ligands*, Chemical Society, London, 1957.
78. Parsons, *Handbook of Electrochemical Constants*, Butterworths, London, 1959.
79. Bower and Robinson, *J. Phys. Chem.*, 64, 1078 (1960).
80. Candlin and Wilkins, *J. Chem. Soc.*, 4236 (1960).
81. Danehy and Noel, *J. Am. Chem. Soc.*, 82, 2511 (1960).
82. Kreevoy, Harper, Duvall, Wilgus, and Ditsch, *J. Am. Chem. Soc.*, 82, 4899 (1960).
83. Kabachnik, Mastrukova, Shipov, and Melentyeva, *Tetrahedron.*, 9, 10 (1960).
84. Bjerrum, Schwarzenbach, and Sillen, *Stability Constants of Metal-Ion Complexes, Part I, Organic Ligands*, Chemical Society, London, 1957.
85. Crutchfield and Edwards, *J. Am. Chem. Soc.*, 82, 3533 (1960).
86. Lane and Quinlan, *J. Am. Chem. Soc.*, 82, 2994, 2997 (1960).
87. Moeller and Ferrús, *J. Phys. Chem.*, 64, 1083 (1960).
88. Bryson, *J. Am. Chem. Soc.*, 82, 4858, 4862, 4871 (1960).
89. Henderson and Streuli, *J. Am. Chem. Soc.*, 82, 5791 (1960).
90. Fortnum, Battaglia, Cohen, and Edwards, *J. Am. Chem. Soc.*, 82, 778 (1960).
91. Bell and McTigue, *J. Chem. Soc.*, 2983 (1960).
92. Li, Miller, Solony, and Gillis, *J. Am. Chem. Soc.*, 82, 3737 (1960).
93. Cohen and Erlanger, *J. Am. Chem. Soc.*, 82, 3928 (1960).
94. Olson and Margerum, *J. Am. Chem. Soc.*, 82, 5602 (1960).
95. Ratner and Clarke, *J. Am. Chem. Soc.*, 59, 200 (1937).
96. Porter, Rydon, and Schofield, *Nature*, 182, 927 (1958).
97. Kortum, Vogel, and Andrussow, *Dissociation Constants of Organic Acids and Aqueous Solution*, Butterworths, London, 1961.
98. King, *Qualitative Analysis and Electrolytic Solutions*, Harcourt Brace, New York, 1959.
99. Perrin, *Dissociation Constants of Organic Bases in Aqueous Solution*, Butterworths, London, 1965.
100. Albert and Serjeant, *Ionization Constants of Acids and Bases.*, John Wiley & Sons, New York, 1962.
101. Yukawa, Eds., *Handbook of Organic Structural Analysis.* Benjamin, New York, 1965, 584.
102. Battaglia and Edwards, *Inorg. Chem.*, 4, 552 (1965).
103. Irving, Nelander, and Wadsö, *Acta Chem. Scand.*, 18, 769 (1964).
104. Morgan, Stedman, and Whincup, *J. Chem. Soc.*, 4813 (1965).
105. Barlin, *J. Chem. Soc.*, 2150 (1964).
106. Gallo, Pasqualucci, Radaell, and Lancini, *J. Org. Chem.*, 29, 862 (1964).
107. Walba and Isensee, *J. Org. Chem.*, 26, 2789 (1961).
108. D'Yatkin, Mochalina, and Knunyants, *Tetrahedron.*, 21, 2991 (1965).
109. Bell and McTigue, *J. Chem. Soc.*, 2985 (1960).
110. Yates and Riordan, *Can. J. Chem.*, 43, 2328 (1965).
111. *Tables for Identification of Organic Compounds*, Chemical Rubber Co., 3rd ed., 1967.
112. Friedman, Cavins, and Wall, *J. Am. Chem. Soc.*, 87, 3672 (1965).
113. Phillips, Eisenberg, George, and Rutman, *J. Biol. Chem.*, 240, 4393 (1965).
114. Charton, *J. Org. Chem.*, 30, 969 (1965).
115. Good, Winget, Winter, Connolly, Izawa, and Singh, *Biochemistry*, 5, 467 (1966).
116. Paelotti, Stern, and Vacca, *J. Phys. Chem.*, 69, 3759 (1965).
117. Spinner, *J. Chem. Soc.*, 3855 (1963).
118. Hine, Via, and Jensen, *J. Org. Chem.*, 36, 2926 (1971).
119. Eastes, Aldridge, Minesinger, and Kamlet, *J. Org. Chem.*, 36, 3847 (1971).
120. Kamlet and Minesinger, *J. Org. Chem.*, 36, 610 (1971).
121. Paulder and Humphrey, *J. Org. Chem.*, 35, 3467 (1970).
122. Filler and Schure, *J. Org. Chem.*, 32, 1217 (1967).
123. Walba, Stiggall, and Coutts, *J. Org. Chem.*, 32, 1954 (1967).
124. Walba and Ruiz-Velasco, *J. Org. Chem.*, 34, 3315 (1969).
125. Deutsch and Cheung, *J. Org. Chem.*, 38, 1124 (1973).
126. Hansen, West, Baca, and Blank, *J. Am. Chem. Soc.*, 90, 6588 (1971).
127. Sitzman, Adolph, and Kamlet, *J. Am. Chem. Soc.*, 90, 2815 (1968).
128. Hendrickson, Bergeron, Giga, and Sternbach, *J. Am. Chem. Soc.*, 95, 3412 (1973).
129. Christensen, Izatt, and Hansen, *J. Am. Chem. Soc.*, 89, 213 (1967).
130. Christensen, Oscarson, and Izatt, *J. Am. Chem. Soc.*, 90, 5949 (1968).
131. Christensen, Slade, Smith, Izatt, and Tsang, *J. Am. Chem. Soc.*, 92, 4164 (1970).
132. Bloomfield and Fuchs, *J. Chem. Soc. B.*, 363 (1970).
133. Hughes, Nicklin, and Shrimanker, *J. Chem. Soc.*, B, 3485 (1971).
134. Christensen, Rytting, and Izatt, *J. Chem. Soc.*, B, 1643 (1970).
135. Christensen, Rytting, and Izatt, *J. Chem. Soc.*, B, 1646 (1970).

136. Ives and Prasad, *J. Chem. Soc., B*, 1652 (1970).

137. Ives and Mosely, *J. Chem. Soc., B*, 1655 (1970).

138. Bolton and Hall, *J. Chem. Soc., B*, 1247 (1970).

139. Chuchani and Frohlich, *J Chem. Soc., B*, 1417 (1970).

140. Barlin and Pfleinderer, *J. Chem. Soc., B*, 1425 (1971).

141. Bellobono and Favini, *J. Chem. Soc., B*, 2034 (1971).

142. Takahashi, Cohen, Miller, and Peake, *J. Org. Chem.*, 36, 1205 (1971).

143. Crowell and Hankins, *J. Phys. Chem.*, 73, 1380 (1969).

144. Haake and Bausher, *J. Phys. Chem.*, 72, 2213 (1968).

145. Mäkitie and Mirttinen, *Suomen Kemistilehti, B*, 44, 155 (1971).

146. Koskinen and Nikkilä, *Suomen Kemistilehti, B*, 45, 89 (1971).

147. Serjeant, *Aust. J. Chem.*, 22, 1189 (1969).

148. Pollet and VandenEynde, *Bull. Soc. Chim. Belges*, 77, 341 (1968).

149. Wada and Takenaka, *Bull. Chem. Soc. Jap.*, 44, 2877 (1971).

150. Caplow, *Biochemistry*, 8, 2656 (1969).

151. Nozaki and Tanford, *J. Biol. Chem.*, 242, 4731 (1967).

152. Yagil, *Tetrahedron*, 23, 2855 (1967).

153. Albert, *Ionization Constants of Acids and Bases*, Methuen and Co. Ltd, London, 1962.

154. Albert, *Heterocyclic Chemistry*, Athalone Press, London, 1968.

155. Jencks and Salvesen, *J. Am. Chem. Soc.*, 93, 4433 (1971).

156. Bell, *The Proton in Chemistry*, 2nd ed., Cornell, Ithaca, N.Y., 1973.

157. DeMaria, Fini, and Hall, *J. Chem. Soc. Perkin II*, 1969 (1973).

158. Ceppi, Eckhardt, and Grob, *Tetrahedron Lett.*, 37, 3627 (1973).

159. Kurz and Farrar, *J. Am. Chem. Soc.*, 91, 6057 (1969).

160. Hine and Koser, *J. Org. Chem.*, 36, 1348 (1971).

GUIDELINES FOR NMR MEASUREMENTS FOR DETERMINATION OF HIGH AND LOW pK$_a$ VALUES

(IUPAC Technical Report)

Konstantin Popov[1,2‡], Hannu Rönkkömäki[3], and Lauri H. J. Lajunen[4]

[1]*Institute of Reagents and High Purity Substances (IREA), Bogorodsky val-3, 107258, Moscow, Russia;* [2]*Moscow State University of Food Production, Volokolamskoye Sh. 11, 125080 Moscow, Russia;* [3]*Finnish Institute of Occupational Health, Oulu Regional Institute of Occupational Health, Laboratory of Chemistry, Aapistie 1 FIN-90220 Oulu, Finland;* [4]*Department of Chemistry, University of Oulu, P.O. Box 3000, FIN-90014 Oulu, Finland*

*Membership of the Analytical Chemistry Division during the final preparation of this report was as follows:

President: K. J. Powell (New Zealand); *Titular Members:* D. Moore (USA); R. Lobinski (France); R. M. Smith (UK); M. Bonardi (Italy); A. Fajgelj (Slovenia); B. Hibbert (Australia); J.-Å. Jönsson (Sweden); K. Matsumoto (Japan); E. A. G. Zagatto (Brazil); *Associate Members:* Z. Chai (China); H. Gamsjäger (Austria); D. W. Kutner (Poland); K. Murray (USA); Y. Umezawa (Japan); Y Vlasov (Russia); *National Representatives:* J. Arunachalam (India); C. Balarew (Bulgaria); D. A. Batistoni (Argentina); K. Danzer (Germany); E. Domínguez (Spain); W. Lund (Norway); Z. Mester (Canada); *Provisional Member:* N. Torto (Botswana).

‡Corresponding author: E-mail: ki-popov@mtu-net.ru

Abstract: Factors affecting the NMR titration procedures for the determination of pK$_a$ values in strongly basic and strongly acidic aqueous solutions ($2 \geq$ pH ≥ 0 and $14 \geq$ pH ≥ 12) are analyzed. Guidelines for experimental procedure and publication protocols are formulated. These include: calculation of the equilibrium H$^+$ concentration in a sample; avoidance of measurement with glass electrode in highly acidic (basic) solutions; exclusion of D$_2$O as a solvent; use of an individual sample isolated from air for each pH value; use of external reference and lock compounds; use of a medium of constant ionic strength with clear indication of the supporting electrolyte and of the way the contribution of any ligand to the ionic strength of the medium is accounted for; use of the NMR technique in a way that eliminates sample heating to facilitate better sample temperature control (e.g., ^{1}H-coupled NMR for nuclei other than protons, GD-mode, CPD-mode, etc.); use of Me$_4$NCl/Me$_4$NOH or KCl/KOH as a supporting electrolyte in basic solution rather than sodium salts in order to eliminate errors arising from NaOH association; verification of the independence of the NMR chemical shift from background electrolyte composition and concentration; use of extrapolation procedures.

Keywords: NMR titration; dissociation constants; acidity constants; chemical shift dependence on medium; high and low pK measurement; IUPAC Analytical Chemistry Division.

Introduction

Numerical data for acid–base equilibria (lg K_a values) have contributed significantly to the theoretical foundation of modern organic and inorganic chemistry [1,2]. In particular, the ligand acid dissociation constants (pK$_a$) correlate strongly with complex stability for many classes of ligands [3]. The related linear Gibbs energy relationships may be used for prediction of metal complex stability constants K_{ML} in cases where their direct experimental measurement is difficult or impossible [2,4,5].

Many important acid–base equilibria take place in highly basic or highly acidic aqueous solutions. For strongly acidic aquametal ions (e.g., Tl^{III}, Bi^{III}, Ti^{IV}, Th^{IV}, Be^{II}, Pd^{II}), the measurement of stability constants frequently requires solutions of low pH (pH ≥ 2) [6], while complex formation frequently involves ligands with very small pK$_a$ values. By contrast, many technologies and complexation reactions require pH ≥ 12 [7] and ligands that are strongly basic (e.g., phosphonates, anionic forms of sugars, hydroxybenzoates, polyamines). In both cases, the application of glass electrode-based potentiometry does not give reliable results [8].

In recent years, a variety of new techniques have been developed as alternatives to the classical potentiometric titration procedure. Among these is nuclear magnetic resonance (NMR), which has a unique application for microscopic acid dissociation constant measurements [9] as well as for work in highly basic and highly acidic media [1,6–8]. Although early reports on the use of the NMR technique were not promising [1], later work revealed good concurrence with potentiometric results for compounds with pK$_a$ values in the range $11 \geq$ pK$_a \geq 3$ [10–11]. Recently, fully automated pH-NMR titration equipment for protonation studies has been reported [10a,10c,12–14]. However, the pK$_a$ values estimated from NMR measurements in strongly basic (acidic) solutions often differ significantly from those obtained by potentiometry. The higher reliability of equilibrium data based on NMR measurements in the ranges $2 \geq$ pH and pH ≥ 12 is widely recognized [7,8b,8c,8d,12,13].

At the same time, diverse experimental conditions have been used for protonation and stability constant measurements by NMR. This affects the reliability and the comparison of the resulting equilibrium constants. Further, many authors have not used a standard approach to the chemical shift reference application, preparation of samples, pD/pH corrections, ionic strength control, etc. [11,14–17]. This in turn has resulted in a considerable disparity among the calculated constants. The present report is therefore focused on general recommendations for the application of NMR spectroscopy to the determination of protonation (dissociation) constants in aqueous solution, with an emphasis on titration procedures in highly acidic or highly basic media ($2 \geq$ pH ≥ 0 and $14 \geq$ pH ≥ 12). At the same time, it provides some guidelines for the critical treatment of the NMR-based pK$_a$ values published earlier.

Reproduced From:
Pure Appl. Chem., Vol. 78, N0.3, pp. 663–675, 2006.
doi:10.1351/pac200678030663
© 2006 IUPAC

637

Factors affecting the accuracy of NMR titrations

Acid dissociation constants can be expressed in terms of activity (thermodynamic constants) or concentrations (concentration, conditional constants). In the former case, the activity constant $K_a = a_H a_L / a_{HL} a^\circ$, or a mixed activity-concentration constant $K_a = a_H [L]/[HL]c^\circ$ are considered, where $c^\circ = 1$ mol dm^{-3} is the standard amount concentration; a° is the corresponding activity; a_H, a_L, a_{HL} represent activities; and [H], [L], [HL] amount concentrations of H$^+$, L$^-$, and HL species, respectively. IUPAC recommends for solution equilibrium studies the determination of concentration-based constants $K_a = $ [H] [L]/[HL]c° [18, 19]. In the present paper, the term pK_a always indicates the concentration constant valid for a particular ionic strength I and temperature, while pH corresponds to the concentration p[H] scale (p[H$^+$]); i.e., we define p[H] = $-$lg {[H$^+$]/c°} unless otherwise is stated. In a similar way, p[D] should correspond to $-$lg {[D$^+$]/c°}. This requires either calibration of a pH meter by solutions with known [D$^+$] at a particular I, or the direct calculation of [D$^+$] in strongly basic (acidic) solutions when the concentration of L can be neglected. However, this ideal condition is seldom if ever fulfilled, and the common practice is based on the "pH meter readings" in D$_2$O solutions after the pH meter was calibrated in H$_2$O buffer solution [8a] (see eqs. 5–7 and further discussion). Obviously, this approach gives some value of pD as unclear function of activity a_D and cannot be recommended for work in concentrated (>0.1 mol dm^{-3}) solutions of bases (acids).

For the dissociation equilibrium of the protonated ligand HL (charge numbers are neglected):

$$HL \rightleftharpoons L + H \qquad (1)$$

the acidity constant K_a is defined at a particular ionic strength I as $K_a = $ [L][H]/[HL]c°. Then p$K_a = -$lg $K_a = $ p[H] $+$ lg ([HL]/[L]) and at half-neutralization p[H] becomes a reasonable estimate of pK_a as [HL] = [L] and lg ([HL]/[L]) = 0.

However, many research groups use the NMR technique for D$_2$O solutions and therefore operate with measurements of pD in terms of activity as indicated earlier. The corresponding mixed activity-concentration constant is denoted here as K_a(D^{2}O) = a_D[L]/[DL]c°. Then the K_a(D$_2$O) values are recalculated by means of some empirical and very arbitrary equations (see further discussion) into some equilibrium activity-concentration constant

K_a(H$_2$O), which is supposed to indicate $K_a = a_H$[L]/[HL]c° for H$_2$O solutions, although there is no rigorous background for that supposition.

From the p[H] dependence of the chemical shift, the pK_a can be determined, using ^{1}H, ^{13}C, ^{14}N, ^{15}N, ^{31}P, ^{19}F, or any other NMR-active nucleus in a ligand [9a]. Since proton dissociation from HL changes the electron density, species HL and L reveal different chemical shifts, denoted as δ_{HL} and δ_L. Most acids in aqueous solutions are characterized by rapid proton-transfer reactions on the NMR time-scale. Thus, the observed chemical shift of any one nucleus represents the single concentration-weighted average δ_{obs} of the chemical shifts for the nucleus of each chemical species in the equilibrium:

$$\delta_{obs} = x_{HL}\delta_{HL} + x_L\delta_L \qquad (2)$$

where x_{HL} and x_L denote the mole fractions of equilibrium species HL and L. The dynamically averaged chemical shift δ_{obs} provides a good measure for the degree of ionization (proton dissociation):

$$x_L = (\delta_{obs} - \delta_{HL})/(\delta_L - \delta_{HL}) \qquad (3)$$

The mole fractions can be expressed in terms of p[H] and pK_a [1, 13]:

$$pK_a = p[H] + lg[(\delta_L - \delta_{obs})/(\delta_{obs} - \delta_{HL})] \qquad (4)$$

It is easy to demonstrate from eq. 4 that: (a) for an acid HL, a plot of δ_{obs} vs. p[H] has the shape of a titration curve lying between the asymptotes δ_L and δ_{HL}, with a point of inflection at p[H] = pK_a, $\delta_{obs} = (\delta_L + \delta_{HL})/2$; (b) the titration curve is symmetrical about the inflection point (Fig. 1), which gives a possible simple method of estimating pK_a, δ_L, and δ_{HL}.

The normal procedure for a NMR titration is based on the dependence of the chemical shift δ_{obs} on p[H], with subsequent treatment of experimental data via routine software. Therefore, three constants are to be found from the δ_{obs} vs. p[H] data by computer analysis of this nonlinear equation, and preliminary values can often be found directly from the plot. A significant advantage of the NMR technique is associated with the possibility of titrating a mixture of ligands, including impurities, if the total concentration of the ligands (and therefore of impurities

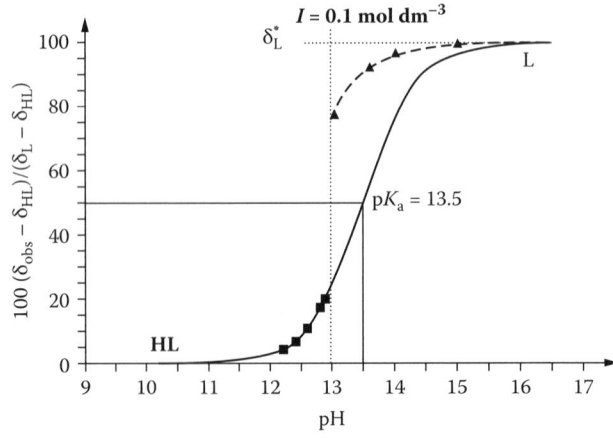

FIGURE 1: Simulated NMR titration curve for the hypothetical 0·001 mol dm^{-3} acid HL with pK_a = 13.5 at I = 0.1 mol dm^{-3} (solid line), plotted by SPECIES [35]. Squares refer to hypothetical experimental NMR-titration points at I = 0.1 mol dm^{-3}. Their range is limited by the ionic strength I (high pH limit) and by the requirement [OH$^-$] >> [HL] (low pH limit). Dashed line and triangles refer to a simulated NMR titration of the same acid at I = 1.0 mol dm^{-3}, which provides the value δ_L^* directly or via extrapolation.

associated with the ligands) is much less then the base (acid) concentration.

It is important to stress that for classical potentiometry with a glass electrode the inflection point can be observed only if pK_a is close to (pK_w)$^{1/2}$ [1]. For a NMR titration, the situation is completely different. As far as only mole fractions, instead of the total acid concentration, are involved in the data evaluation, NMR facilitates pK_a measurement outside the range of potentiometry if high or low p[H] are determined by means other than glass electrode readings [13]. Thus, the main sources of errors in NMR-based pK_a determinations are the accuracy and precision of δ_{obs} and p[H] values.

Chemical shifts

General conventions for chemical shifts are comprehensively considered in recent IUPAC recommendations [20]. In the present paper, we will focus only on specific problems associated with NMR-based pK_a determinations, bearing in mind that many research groups involved in solution chemistry equilibria still do not have modern NMR equipment and have to work with routine spectrometers. It is essential that the chemical shift measurement being made for each datum point is reliable. Another important requirement is to obtain from the set of chemical shifts such a pK_a value that can pass comparison with other equilibrium constants.

As described in ref. [20], there are three types of referencing method that could reasonably be applied in titrations: *internal referencing*, *substitution method*, and *external referencing*. These methods all have various advantages and disadvantages in relevance to NMR titration.

Internal referencing may lead to intermolecular interactions between ligand, solvent, and reference compound. Further, in many spectrometers the sample must normally include a deuterium-containing molecule for magnetic field stabilization ("lock"). For many purposes, all these interactions can be safely ignored, but for NMR-based titrations at high and low p[H] a considerable caution is needed. The use of D_2O (as the "lock") instead of H_2O as a solvent, and the addition of uncontrolled "small" amounts of a reference compound like sodium 3-(trimethylsilyl)propane-1-sulfonate (DSS), dimethyl sulfoxide (DMSO), *tert*-butylalcohol or 1, 4-dioxane inside a sample (internally), became common practice for ^{1}H and ^{13}C NMR [12,15,16]. In some cases (^{13}C NMR), the added reference compound is itself deuterated (e.g., (^{2}H$_6$)DMSO or DSS, deuterated at the CH$_2$ positions), thus providing the lock signal as well. Modern NMR techniques give the possibility to work with very low concentrations of DSS. Therefore, it gives a negligible contribution to ionic strength and to solution properties. It is demonstrated to be effective at p[H] 0–1 [12]. At the same time, little is known about the properties of internal references at elevated p[H]. Nevertheless, any internal substance can potentially participate in association processes either with the cation under complex formation study or with the background electrolyte and is therefore generally undesirable from the point of view of equilibrium studies.

For ^{1}H NMR titrations, the use of D_2O as a solvent instead of H_2O is a common procedure. This is usually done to eliminate masking of a substrate peak by the H_2O resonance [15–17]. The use of D_2O internally raises the problems of how to effect pD measurement with a standard glass electrode, as well as the relationship between pK_a(H_2O) and pK_a(D_2O). The proposed simple empirical eq. 5 derived for ionic strength $I = 0.001$–0.01 mol dm^{-3} and 25°C [21] to obtain values on the conventional pD scale from glass electrode readings is widely accepted, although it is frequently used far outside of the originally intended ionic strength

and temperature limits:

$$pD = pH\text{-meter reading}^* + 0.40 \qquad (5)$$

Some authors, however, use eqs. 6 or 7 [22,23]:

$$pD = pH\text{-meter reading} + 0.44 \, (22°C, I = 0.01 \text{ mol dm}^{-3}) \qquad (6)$$

$$pD = pH\text{-meter reading} + 0.50 \, (22°C, I = 0.1 \text{ mol dm}^{-3}) \qquad (7)$$

Although the difference between pD values calculated by different equations is not large, it is a substantial contribution to systematic error, even for low ionic strength and room temperature, but particularly for high ionic strengths and high temperatures.

There is an even greater diversity of relationships between pK_a(H_2O) and pK_a(D_2O). Both quantities are ionic strength-dependent. The proposed empirical equations yield significantly different results and seem very arbitrary: relationships depend on the nature and number of compounds studied [24–26]. It is observed that the activity coefficient products undergo significant changes when one goes from light to heavy water [27]. It is obvious that at present a correct extrapolation of pK_a(D_2O) to an aqueous phase pK_a(H_2O) is not possible, and that the systematic errors for calculated values are outside the accepted uncertainty for pK_a(H_2O) values derived directly from NMR measurements with external D_2O. Besides, pK_a(D_2O) values can hardly pass comparison with other equilibrium constants measured in H_2O, and their use for complex formation equilibria in H_2O is very doubtful. Assuming the above difficulties, the use of internal D_2O is not recommended.

For ^{31}P NMR, the use of internal referencing at high and low p[H] is difficult, and external reference application is widely used and recommended [7c,7d,8b,8c,11a,11b,16,17a,17b].

Substitution method uses measurement of sample and reference spectra in two separate experiments. It became feasible due to implementation of stable, internally locked spectrometers. In this procedure, the sample and reference materials are not mixed. This benefits the equilibrium study. If locking is not used, the magnet should not be reshimmed between running the sample and reference solution, since this changes the applied magnetic field [20]. This can become a disadvantage for time-consuming ^{13}C NMR-based equilibrium experiments because the ligand concentrations have to be small.

External referencing involves sample and reference contained separately in coaxial cylindrical tubes. A single spectrum is recorded, which includes signals from both the sample and the reference. It is also an ideal situation for equilibrium study as far as both reference and "lock" substances are separated from the ligand solution. The external reference procedure generally requires corrections arising from differences in bulk magnetic susceptibility between sample and reference [20]. This is important for precise chemical shift measurements, but for the relative change of δ_{obs} between δ_L and δ_{HL} for a series of nearly identical aqueous solutions in a narrow pH range (either p[H] 0–2 or 12–14) with constant ionic strength and a constant sample volume it is insignificant. Numerous measurements of ^{31}P NMR-based pK_a values revealed no influence from this factor [7c,7d,8b, 8c,11a,11b,16,17a,11b]. Alternatively, magic-angle spinning could be used. Therefore, such a technique seems to be the preferable choice.

---

* pH-meter reading for solutions in D_2O when the pH electrodes are calibrated with standard aqueous buffers.

p[H] Values and titration procedure

An important source of error in NMR-based pK_a determinations is the accuracy and precision of the p[H] values. Determining extreme values of p[H] requires special attention, since glass electrodes cannot be used reliably [8a,8c,12,13]. Therefore, the traditional single-sample NMR titration is recommended [8c,13,27,29,30]. A set of individual samples with constant monoprotic acid HL (or ligand L) concentration (e.g., 0.01 mol dm^{-3}), constant ionic strength (e.g., 1 mol dm^{-3}) and varying p[H] value are prepared one-by-one ("constant volume titration") in such a way that a strong acid or a strong base added for desired p[H] adjustment is taken in a significant excess over HL or L (e.g., 0.1–1.0 mol dm^{-3}). This permits the equilibrium p[H] value to be calculated reliably as it is equated to the total amount of a strong acid (strong base) added to the sample [27]. Since each sample is prepared individually from stock solutions, the ionic strength can be very precisely controlled [12]. The use of "lock" and reference substances externally excludes their undesirable influence on the equilibrium system. Alternatively, in an approach developed by Hägele [13], the glass electrode can be completely avoided by adding an indicator molecule to the sample for in situ p[H] monitoring. However, this method is primarily based on the procedure stated above.

The proposed method is equally valid for both strongly acidic (pH < 2) and strongly basic (pH > 12) solutions, although some peculiarities do exist in the latter. For the acidic medium, p[H] is directly derived from the total strong acid concentration. In the case of highly basic solutions, the initially calculated p[OH$^-$] values have to be converted to the p[H] scale, using appropriate pK_w values to allow calculation of the corresponding pK_a values. Some important issues that restrict the application of the above method and influence the data quality should also be considered.

Titration procedure and titration curve treatment

A full-scale NMR titration for a single proton equilibrium 1 will provide values of δ_{HL}, δ_L, and some intermediate chemical shift values applicable to a particular pH at a constant I. Ideally, a titration spans over 4 pH units with the half-neutralization point in the middle of this pH range. For extremely high or low pK_a, this condition is not achievable: for $pK_a = 13.5$, the value for δ_L has to be measured at pH 15.5, while for $pK_a = 0.5$ a direct observation of δ_{HL} requires pH = −1.5.

If the ionic strength is 1.0 mol dm^{-3} (NaCl/NaOH), then the highest pH attainable at 25°C (pH$_{max}$) is less than 13.72 (pK_w for 1 mol dm^{-3} NaOH*), while for I = 0.1 mol dm^{-3} NaCl/NaOH pH$_{max}$ < 12.78 (limitation due to pK_w and I). Thus for $pK_a = 13.5$ at $I = 1.0$ mol dm^{-3} (NaCl/NaOH) only about 80 % of the titration curve is accessible, providing a value for δ_{HL} and a half-neutralization point, but not for δ_L. In the case of $I = 0.1$ mol dm^{-3} (NaCl/NaOH*), about 30 % of the full curve can be obtained experimentally, but excluding the half-neutralization point and δ_L, Fig. 1 (square points). Although comprehensive software (SigmaPlot, WinEQNMR) permits calculation of pK_a and δ_L values for very weak acids on the basis of data at different pH values below that for the half-neutralization point, the corresponding constants have a large error. But in some cases,

the programs fail to produce results and experimental measurement of high pK_a at low ionic strength becomes impossible. This can be illustrated by the last dissociation step of nitrilotris (methylenephosphonic acid) (NTPH, H$_6$ntph) and ethylenediaminetetra (methylenephosphonic acid) (EDTPH, H$_8$edtph)**, Table 1.

On the other hand, if the initial δ_{HL} experimental value and the subsequent 30–40 % of a complete titration curve are supported by at least one final high pH titration point to provide δ_L, then the precision of the pK_a calculation becomes sufficiently high and the measurement becomes feasible.

For those nuclei with chemical shift poorly dependent on the ionic strength and nature of the supporting electrolyte (the case of ^{31}P and ^{13}C), δ_L can be obtained by titration of the same system at a higher or even uncontrolled ionic strength until the "plateau" is reached (triangle points, Fig. 1). The resultant δ_L^* is very close to δ_L, e.g., δ_L ($I = 0.1$ mol dm^{-3}) ~ δ_L^* ($I = 1.0$ mol dm^{-3}). Then the following two-step procedure is recommended. The first step involves the titration of a ligand at a sufficiently high ionic strength, e.g., 1.0 mol dm^{-3} NaCl (triangles in Fig. 1) or 1.5 mol dm^{-3} NaCl, etc., rather than in 0.1 mol dm^{-3} NaCl. This gives two advantages. The first is that the pH$_{max}$ is shifted from 12.78 to ca. 14. The second derives from the fact that sodium ion forms weak complexes with L (e.g., phosphonic acids). Therefore, the whole titration curve is shifted to a lower pH range as the total sodium concentration is increased. Both of these factors facilitated the direct observation of a "plateau" corresponding to δ_L^* ($I = 1$ mol dm^{-3}).

Due to the fact that δ_L (0.1 mol dm^{-3} NaCl) is practically equal to δ_L^* (1 mol dm^{-3} NaCl), then the δ_L^*-value could be used instead of δ_L along with experimental points obtained for $I = 0.1$ mol dm^{-3} (square points, Fig. 1). Therefore, within the second step, δ_L^* is assigned to a conventional pH = 16 or pH = 17, where the titration curve definitely has a plateau. A titration is repeated for 0.1 mol dm^{-3} NaCl solutions, a δ_L^* point is added to the experimental data set, and a pK_a^* value is calculated. The subsequent treatment of the united data reveals a significant increase in the accuracy of pK_a. This can be seen from Table 1, where both constants pK_a (calculated without δ_L^*) and pK_a^* (calculated with a δ_L^* point) are represented. If δ_L is significantly dependent on ionic strength, then the extrapolation procedure proposed by Popov, Lajunen, and Rönkkömäki [28,29] could be applied.

Ligand concentration

Calculation of p[H] from the acid stoichiometry requires a low ligand concentration: for a monoprotic acid $C_{HL} < 0.01$ I. Recent developments of the NMR technique make it possible to now work with very dilute solutions. In case of the organophosphonates, concentrations C_L ~ 0.001 mol dm^{-3} are quite suitable for ^{31}P NMR titrations [28,29].

By contrast, for ^{13}C NMR titrations, the ligand concentration has to be rather high (about 0.1 mol dm^{-3}) in order to perform the titration in a reasonable time. Therefore, the equilibrium [OH$^-$] cannot be equal to the total [OH$^-$] added to the system. For this case (e.g., 0.1 mol dm^{-3} HL), another two-step procedure reported for sucrose dissociation constant measurements [31] is recommended. In the first step, the equilibrium [OH$^-$] is taken as equal to the total [OH$^-$] added, and the full titration curve is plotted, mathematically treated, and the pK_a, δ_L, and δ_{HL} values are calculated. The difference between δ_L and δ_{HL} chemical shifts defines the linear scale of OH$^-$ consumption by the ligand: 0 mol dm^{-3}

---

* Reliable values for pK_w are measured only for some common supporting electrolytes, e.g., NaCl, NaClO$_4$, KNO$_3$, etc. For 1 mol dm^{-3} NaOH, the pK_w value found for 1 mol dm^{-3} NaCl is valid as far as the difference in corresponding activity coefficients is negligible. The same situation is observed for the 0.1 mol dm^{-3} NaCl/NaOH system. However, it is not the case for a complete substitution of 1 mol dm^{-3} NO$_3^-$ for OH$^-$ or of 1 mol dm^{-3} K$^+$ (Na$^+$) for H$^+$ (acidic solutions).

** The PINs (preferred IUPAC names) for NTPH and EDTPH are: [nitrilo-tris(methylene)]tris(phosphonic acid) and ethane-1, 2-diyldinitrilotetra-kis(methylene)tetrakis(phosphonic acid).

TABLE 1: Dissociation Constants pK_a for Hedtph[-7] and Hntph[-5] Derived from ^{31}P NMR Measurements by SigmaPlot Data Treatment.[a]

Ligand	I/(mol dm^{-3})	t/°C	pK_a	pK_a*	Reference
Hedtph[-7]	0.1 (KNO$_3$)	25	Calculation failed	13.29 ± 0.07	[28]
	0.15 (NaCl)	37	13 ± 1	12.86 ± 0.07	[28]
Hntph[-5]	0.1 (KNO$_3$)	25	12.2 ± 0.3	12.9 ± 0.1	[29]

[a] pK_a and pK_a* represent constants calculated without δ_L* and with δ_L* values, respectively; see text for other explanations.

TABLE 2: Dissociation Constant of 0.1 mol dm^{-3} Sucrose (HL) from ^{13}C NMR Titration at 60°C in 1 mol dm^{-3} NaCl/NaOH [31].

Procedure	δ_L/ppm	δ_{HL}/ppm	pK_a	R
One-step data treatment	103.00	101.98	12.40 ± 0.05	0.999
Two-step data treatment	102.96	101.98	12.30 ± 0.05	0.999

(δ_{HL}) and 0.1 mol dm^{-3} (δ_L) for a 0.1 mol dm^{-3} solution of L. Within the second step, all the experimental values δ_{obs} are treated again with redefined values of p[OH], and an improved value of pK_a is calculated. As indicated in Table 2, the correction due to the second step reveals a systematic error of 0.1 in pK_a.

Background electrolyte and ionic strength

To date, the background electrolyte effect on chemical shifts has been inadequately studied. Equilibrium concentration products are ionic strength-dependent, yet numerous NMR titration experiments have been performed without ionic strength control [14c,16c], and have produced pK_a values in reasonable agreement with potentiometric results. In part, this arises from the fact that the chemical shifts depend on concentrations, rather than the activities of various species in solution [32]. The best agreement has been demonstrated for systems studied by ^{13}C and ^{31}P NMR [28,29,33]. For the ^{13}C and ^{31}P NMR resonances in alkylcarboxylic and alkylphosphonic acids, the chemical shifts correlate linearly with the background electrolyte concentration. However, this effect is normally negligible in comparison with that associated with a ligand dissociation or complex formation. This fact offers a unique possibility to use ^{13}C and ^{31}P NMR chemical shifts, δ_L, of a ligand, measured at high pH and high ionic strength, for calculations of pK_a at low ionic strength [28,29]. General observation reveals that the chemical shift depends on both the nature of the nucleus and its position in the ligand. The ^{31}P nuclei in phosphonic (–PO$_3$H$_2$, –PO$_3$H$^-$, –PO$_3$$^{2-}$) as well as [13]C nuclei in carboxylate or methylenic groups (–CO$_2$$^-$,–CH$_2$–, –CH$_3$) are relatively isolated

from solution by oxygen or hydrogen atoms. Thus, their chemical shifts are mostly sensitive to the substrate intramolecular processes (deprotonation/protonation, complex formation), while the solvent changes give the least contribution. On the other hand, the nuclei that contact the solvent directly, e.g., [133]Cs$^+$, [35]Cl$^-$, are more affected by medium effects. Therefore, a NMR titration under variable ionic strength is not desirable, unless the independence of chemical shift δ on ionic strength I is demonstrated.

Among the supporting electrolytes for 14 ≥ pH ≥ 12, the use of 1.0 mol dm^{-3} Me$_4$NCl/Me$_4$NOH is recommended as there is no reported evidence for Me$_4$NOH self-association. In the case of KOH and NaOH, corrections for base self-association could be needed. The uncertainty is associated with imprecise knowledge of the MOH stability constants. Table 3 represents the estimation of errors if the MOH stability constants recommended by Baes and Mesmer [34] are used. Table 3 also demonstrates that for NaOH solutions the pH scale has to be corrected, while for KOH no correction is needed. However, it should be mentioned that Martell [35] gives significantly higher stability constants for MOH ion pairing. Thus, the corresponding corrections could be larger.

Another important issue for NMR titration is the need for a clear indication as to whether the contribution of the ligand to the total ionic strength is considered or not. For monobasic acids, this contribution could be negligible, but it is not the case for polyprotic substrates such as EDTPH. In basic 0.01 mol dm^{-3} solutions of EDTPH, the ligand contribution to the total ionic strength constitutes 0.25 mol dm^{-3} for Hedtph[-7] and 0.33 for edtph[-8].

Special care should be taken over supporting electrolyte purity. Indeed, in 1 mol dm^{-3} Me$_4$NCl/Me$_4$NOH medium, the concentration of Ca^{2+} impurities in the supporting electrolyte can be comparable with the ligand content in the system [8c].

Another important peculiarity of the titration procedure at high and low pH arises from a complete substitution of either cation or anion. Indeed, within the constant background electrolyte concentration, e.g., 1 mol dm^{-3} at 25°C, the ionic strength can change significantly. For example, the complete substitution of 1 mol dm^{-3} KNO$_3$ for 1 mol dm^{-3} KOH induces the change of

TABLE 3: Calculated –lg {[H$^+$]/mol dm^{-3}} for MOH Solutions in 0.1 and 1.0 mol dm^{-3} MCl/MOH.[a]

MOH	Total [OH$^-$], mol dm^{-3}	Free[a] [OH$^-$], mol dm^{-3}	pH Calculated without correction for MOH Association	pH Corrected for MOH Association[a]	ΔpH
NaOH	0.1000	0.0947	12.75	12.73	0.02
	1.00	0.69	13.75	13.54	0.21
KOH	0.1000	0.0998	13.15	13.15	0.00
	1.00	0.91	14.15	14.11	0.04

[a] Free [OH$^-$] is calculated with the SPECIES software [36] using MOH stability constants lg K_1 from [34] (for ionic strength 1.0 mol dm^{-3} lg K_1 = –0.5 for NaOH and –0.8 for KOH), [H$^+$] is calculated from [OH$^-$] using pK_w = 13.75 for 1 mol dm^{-3} NaCl and 14.16 for 1 mol dm^{-3} KCl [37].

mean activity coefficient from 0.444 to 0.733. In the same way, a substitution of 1 mol dm^{-3} KNO$_3$ for 1 mol dm^{-3} HNO$_3$ results in a change of activity coefficient from 0.444 to 0.730. At the same time for 1 mol dm^{-3} NaCl/NaOH system, the corresponding change is negligible (0.657 and 0.674)*. Therefore, a proper choice of supporting electrolyte, or clear indication of corresponding corrections, is needed.

Temperature

Dissociation constants, as well as pK_w, are temperature-dependent [35]. A temperature variation of 20–30°C can result in a change of 0.2–0.3 in pK_a (or more). Especially critical are the high pK_a – values. For example, for Hntph^{-5} dissociation $\Delta H = -38.8$ kJ mol^{-1} for $I = 0.1$ mol dm^{-3} and 25°C ([36], Mini Database). Therefore, pK_a = 13.30 at 25°C and 12.98 at 40°C. The difference in 0.1 pK_a unit per 5°C is significant for dissociation constant. The temperature dependence of pK_w additionally affects all the measurements in basic solutions. For example, in 0.51 mol dm^{-3} NaCl solutions, pK_w changes from 13.71 (25°C) to 12.96 (50°C). In this respect, the noise associated with ^{1}H-decoupling widely used in early NMR measurements might have led to some errors in pK_a values due to significant energy dissipation and therefore to a sample heating. Although modern multipulse decoupling methods (GD-mode, CPD-mode) dissipate less energy, some caution is needed to control the process. In some cases, ^{1}H-coupled spectra are the better choice.

Guidelines

Recommendations for NMR titrations in solutions of high and low pH (2 ≥ pH ≥ 0 and 14 ≥ pH ≥ 12) are intended to be a supplement to the IUPAC guidelines for the determination of stability constants [19] and to a standard format for the publication of stability constant measurements [38] considering the peculiarities of NMR spectroscopy mentioned above. Some of these requirements are also valid for the range 12 ≥ pH ≥ 2.

1. Within the NMR titration procedure at high and low pH solutions (2 ≥ pH ≥ 0 and 14 ≥ pH ≥ 12), the equilibrium H$^+$ concentration should be calculated from solution stoichiometry, not measured with a glass electrode. For this reason, the ligand concentration has to be ≤ 0.001 mol dm^{-3}. For higher ligand concentrations, the titration is possible, but corrections for strong base (strong acid) consumption by a ligand are necessary.
2. Arrangement of a titration procedure. Sets of samples should be prepared in such a way that the concentration of the ligand and the total ionic strength remain constant, while the supporting electrolyte composition is varied from sample to sample to provide different concentrations of OH$^-$ or H$^+$. For highly basic solutions, the total concentration of added base should be much greater than the ligand concentration ([OH$^-$] >> [L]). For highly acidic media, the same requirement applies to [H$^+$] ([H$^+$] >> [L]). Thus, the total concentration of added base (acid) can be treated as the equilibrium concentration (i.e., the OH$^-$ or H$^+$ consumption by the substrate can be neglected). This circumvents the problems associated with pH measurements with

the glass electrode. An additional advantage of such an approach is that the protonation constants are derived in terms of concentration, not activity.

3. A medium of constant ionic strength should be used, with clear indication of the supporting electrolyte and of the way the contribution of the deprotonated ligand and of the change in a background electrolyte composition (e.g., change of [Cl$^-$] for [OH$^-$] or [Cl$^-$] for [H$^+$]) to medium ionic strength is taken into account.
4. The supporting electrolytes Me$_4$NCl/Me$_4$NOH or KCl/KOH should be used in basic solution rather than NaCl/NaOH of LiCl/LiOH in order to eliminate errors arising from NaOH and LiOH association. A clear indication of the pK_w used is necessary.
5. External reference and "lock" compounds should be used to eliminate any possible interactions with the ligand and additional changes of the medium.
6. Water should be used as a solvent rather then D$_2$O or H$_2$O/D$_2$O mixtures. This eliminates the need for pD/pH corrections and makes the pK_a values obtained comparable and compatible with values derived from potentiometric measurements performed in H$_2$O.
7. An NMR procedure should be selected, and described clearly, that will minimize possible sample heating (e.g., ^{1}H-coupled NMR, GD-technique, etc.) and provide confidence in temperature control.
8. The calculation of pK_a requires the chemical shift value for the free ligand L (δ_L) and for the protonated species HL (δ_{HL}) along with a number of intermediate experimental values. This is seldom possible for high (low) pH range. In those cases where the δ_{HL} or δ_L value is not available due to ionic strength (and pH) limitations, it should be derived either directly from higher ionic strength measurements (for ionic strength-independent resonance) or by an extrapolation of high ionic strength values to the lower I used in the experiment (for ionic strength-dependent resonance).

Acknowledgments

The authors are grateful to the IUPAC Analytical Chemistry Division for support within the Grant 2001-038-2-500 as well as to the Finnish Academy of Science, which supported preparation of the present paper in part. We are also thankful to K. J. Powell, P. M. May, E. D. Becker, and R. K. Harris for valuable comments and suggestions.

References

1. R. F. Cookson. *Chem. Rev.* **74**, 5 (1974).
2. H. Irving, H. S. Rossotti. *Acta Chem. Scand.* **10**, 72 (1956).
3. (a) H. H. Jensen, L. Lyngbye, M. Bols. *Angew. Chem.* **40**, 3447 (2001); (b) C. M. Chang, M. K. Wung. *TheoChem.* **417**, 237 (1997); (c) G. Thirot. *Bull. Soc. Chim. Fr.* 3559 (1967).
4. (a) T. Shi, L. I. Elding. *Inorg. Chem.* **36**, 528 (1997); (b) G. Anderegg. *Inorg. Chim. Acta* **180**, 69 (1991); (c) P. R. Wells. *Linear Free Energy Relationships*, Academic Press, London (1968).
5. R. M. Smith, A. E. Martell, R. J. Motekaitis. *Inorg. Chim. Acta* **99**, 207 (1985).
6. (a) S. Nakamura, K. Yamashita. *Phosphorus Res. Bull.* **11**, 1 (2000); (b) P. Coupe, D. Williams, H. Lyn. *J. Chem. Soc., Perkin Trans. 2* 1595 (2001); (c) V. B. Fainerman, D. Vollhardt, R. Johann. *Langmuir* **16**, 7731 (2000); (d) A. S. Goldstein. U.S. Pat. 5929008 (1999); (e) S. R. Chen, M. G. F. Thomas. Eur. Pat. EP 564232 (1993) and Eur. Pat. EP 564248; (f) P. G. Yohannes, K. Bowman-James. *Inorg. Chim. Acta* **209**, 115 (1993); (j) K. M. Thompson, W. P. Griffith, M. Spiro. *J. Chem.*

* Mean activity coefficients are taken from *CRC Handbook of Chemistry and Physics*, 82nd ed., R. Lide (Ed.), CRC Press, Boca Raton, FL (2001–2003).

Soc., Faraday Trans. **89**, 1203 (1993); (h) E. Okutsu, Y. Kudo, S. Hori, K. Hasanuma. Japan Pat. JP 67073147 (1986); (i) R. R. Dague, J. N. Veenstra, T. W. McKim. J. *Water Pollut. Control Fed.* **52**, 2204 (1980).

7. (a) E. Matczak-Jon, B. Kurzak, W. Sawka-Dobrowolska, P. Kafarski, B. Lejczak. J. *Chem. Soc., Dalton Trans.* 3455 (1996); (b) L. Alderighi, A. Bianchi, L. Biondi, L. Calabi, M. De Miranda, P. Gans, S. Ghelli, R Losi, L. Paleari, A. Sabatini, A. Vacca. J. *Chem. Soc., Perkin Trans.* 2 2741 (1999); (c) J. Rohovec, M. Kyvala, P. Vojtisek, P. Hermann. I. Lukes. *Eur. J. Inorg. Chem. Soc.* 195 (2000); (d) I. Lukes, L. Blaha, F. Kesner, J. Rohovec, R Hermann. J. *Chem. Soc., Dalton Trans.* 2629 (1997).

8. (a) R. G. Bates. *Determination of pH: Theory and Practice*, 2nd ed., John Wiley, New York (1973); (b) I. Lukes, K. Bazakas, R Hermann, R Vojtisek. J. *Chem. Soc., Dalton Trans.* 939 (1992); (c) K. Popov, E. Niskanen, H. Rönkkömäki, L. H. J. Lajunen. *New J. Chem.* **23**, 1209 (1999).

9. (a) D. L. Rabenstein, S. P. Hari, A. Kaerner. *Anal. Chem.* **69**, 4310 (1997); (b) D. L. Rabenstein, T. L. Soyer. *Anal. Chem.* **48**, 1141 (1976).

10. (a) J. Glaser, U. Henriksson, T. Klason. *Acta Chem. Scand.* **A40**, 344 (1986); (b) D. T. Major, A. Laxer, B. Fisher. J. *Org. Chem.* **67**, 790 (2002); (c) F. Reneiro, C. Guillou, C. Frassinetti, S. Ghelli. *Anal. Biochem.* **319**, 179 (2003); (d) C. Frassineti, S. Ghelli, P. Gans, A. Sabatini, M. S. Moruzzi, A. Vacca. *Anal. Biochem.* **231**, 374 (1995).

11. (a) H. Rönkkömäki, J. Jokisaari, L. H. J. Lajunen. *Acta Chem. Scand.* **47**, 331 (1993); (b) K. Sawada, T. Miyagawa, T. Sakaguchi, K. Doi. J. *Chem. Soc., Dalton Trans.* 3777 (1993).

12. Z. Szakacs, G. Hägele. *Talanta* **62**, 819 (2004).

13. Z. Szakacs, G. Hägele, R. Tyka. *Anal. Chim. Acta* **522**, 247 (2004).

14. (a) M. Peters, L. Siegfried, T. A. Kaden. J. *Chem. Soc., Dalton Trans.* 1603 (1999); (b) J. Ollig, G. Haegele. *Comput. Chem.* **19**, 287 (1995); (c) H.-Z. Cai, T. A. Kaden. *Helv. Chim. Acta* **77**, 383 (1994).

15. R. Delgado, L. C. Siegfried, T. Kaden. *Helv. Chim. Acta* **73**, 140 (1990).

16. (a) T. G. Appleton, J. R. Hall, A. D. Harris, H. A. Kimlin, I. J. McMahon. *Austr. J. Chem.* **37**, 1833 (1984); (b) T. G. Appleton, J. R. Hall, I. J. McMahon. *Inorg. Chem.* **25**, 726 (1986); (c) T. G. Appleton, J. R. Hall, S. F. Ralph, C. S. M. Thompson. *Inorg. Chem.* **28**, 1989 (1989).

17. (a) I. N. Marov, L. V. Ruzaikina, V. A. Ryabukhin, P. A. Korovaikov, N. M. Dyatlova. *Koord. Khim. (Russ. J. Coord. Chem.)* **3**, 1334 (1977); (b) I. N. Marov, L. V. Ruzaikina, V. A. Ryabukhin, R A. Korovaikov, A. V. Sokolov. *Koord. Khim. (Russ. J. Coord. Chem.)* **6**, 375 (1980); (c) B. Song, J. Reuber, C. Ochs, F. E. Hahn, C. Luegger, C. Orvig. *Inorg. Chem.* **40**, 1527 (2001).

18. A. Braibanti, G. Ostacoli, P. Paoletti, L. D. Pettit, S. Sammartano. *Pure Appl. Chem.* **59**, 1721 (1987).

19. G. H. Nancollas, M. B. Tomson. *Pure Appl. Chem.* **54**, 2675 (1982).

20. R. K. Harris, E. D. Becker, S. M. Cabral de Menezes, R. Goodfellow, P. Granger. *Pure Appl. Chem.* **73**, 1795 (2001).

21. R K. Glasoe, F. A. Long. J. *Phys. Chem.* **64**, 188 (1960).

22. K. Mikkelsen, S. O. Nielsen. J. *Phys. Chem.* **64**, 632 (1960).

23. C. F. G. C. Geraldes, A. M. Urbano, M. C. Apoim, A. D. Sherry, K.-T. Kuan, R. Rajagopalan, F. Maton, R. N. Muller. *Magn. Reson. Imaging* **13**, 401 (1995).

24. S. R Dagnall, D. N. Hague, M. E. McAdam, A. D. Moreton. J. *Chem. Soc., Faraday Trans. 1* **81**, 1483 (1985).

25. R. Delgado, J. J. R. Frausto Da Silva, M. T. S. Amorim, M. F. Cabral, S. Chaves, J. Costa. *Anal. Chim. Acta* **245**, 271 (1991).

26. C. A. Blindauer, A. Holy, H. Dvorakova, H. Sigel. J. *Chem. Soc., Perkin Trans. 2* 2353 (1997).

27. R Salomaa, A. Vesala, S. Vesala. *Acta Chem. Scand.* **23**, 2107 (1969).

28. K. Popov, A. Popov, H. Rönkkömäki, A. Vendilo, L. H. J. Lajunen. J. *Solution Chem.* **31**, 511 (2002).

29. A. Popov, H. Rönkkömäki, L. H. J. Lajunen, A. Vendilo, K. Popov. *Inorg. Chim. Acta* **353**, 1 (2003).

30. G. Grossmann, K. A. Burkov, G. Hägele, L. A. Myund, C. Verwey, S. Hermens, S. M. Arat-ool. *Inorg. Chim. Acta* **357**, 797 (2004).

31. K. Popov, N. Sultanova, H. Rönkkömäki, M. Hannu-Kuure, L. H. J. Lajunen, I. F. Bugaenko, V. I. Tuzhilkin. *Food Chem.* **96**, 248 (2006).

32. C. A. Eckert, M. M. McNiel, B. A. Scott, L. A. Halas. *AIChE J.* **32**, 820 (1986).

33. K. Popov, N. Sultanova, H. Rönkkömäki, M. Hannu-Kuure, L. H. J. Lajunen. Unpublished data on acetate ion ^{13}C chemical shift dependence on background electrolyte concentration.

34. C. F. Baes Jr., R. E. Mesmer. *The Hydrolysis of Cations*, John Wiley, New York (1976).

35. NIST Standard Reference Database 46. NIST Critically Selected Stability Constants of Metal Complexes, Version 4.0, compiled by A. E. Martell, R. M. Smith, R. J. Motekaitis, Texas A&M University (1997).

36. IUPAC Stability Constants Database (for Windows 95/98), Version 4.06, compiled by L. D. Pettit, K. J. Powell, Academic Software and K. J. Powell, Sourby Old Farm, Timble, UK (1999); available from <www.acadsoft.co.uk>.

37. I. Kron, S. L. Marshall, P. M. May, G. T. Hefter, E. Königsberger. *Monatsh. Chem.* **126**, 819 (1995).

38. D. Tuck. *Pure. Appl. Chem.* **61**, 1161 (1989).

MEASUREMENT AND INTERPRETATION OF ELECTROKINETIC PHENOMENA

(IUPAC Technical Report)

A. V. Delgado[1,‡], F. González-Caballero[1], R. J. Hunter[2], L. K. Koopal[3], and J. Lyklema[3]

[1]*University of Granada, Granada, Spain;* [2]*University of Sydney, Sydney, Australia;*
[3]*Wageningen University, Wageningen, The Netherlands*

With contributions from (in alphabetical order): S. Alkafeef, College of Technological Studies, Hadyia, Kuwait; E. Chibowski, Maria Curie Sklodowska University, Lublin, Poland; C. Grosse, Universidad Nacional de Tucumán, Tucumán, Argentina; A. S. Dukhin, Dispersion Technology, Inc., New York, USA; S. S. Dukhin, Institute of Water Chemistry, National Academy of Science, Kiev, Ukraine; K. Furusawa, University of Tsukuba, Tsukuba, Japan; R. Jack, Malvern Instruments, Ltd., Worcestershire, UK; N. Kallay, University of Zagreb, Zagreb, Croatia; M. Kaszuba, Malvern Instruments, Ltd., Worcestershire, UK; M. Kosmulski, Technical University of Lublin, Lublin, Poland; R. Nöremberg, BASF AG, Ludwigshafen, Germany; R. W. O'Brien, Colloidal Dynamics, Inc., Sydney, Australia; V. Ribitsch, University of Graz, Graz, Austria; V. N. Shilov, Institute of Biocolloid Chemistry, National Academy of Science, Kiev, Ukraine; F. Simon, Institut für Polymerforschung, Dresden, Germany; C. Werner; Institut für Polymerforschung, Dresden, Germany; A. Zhukov, University of St. Petersburg, Russia; R. Zimmermann, Institut für Polymerforschung, Dresden, Germany

‡Membership of the Division Committee during preparation of this report (2004–2005) was as follows:

President: R. D. Weir (Canada); *Vice President:* C. M. A. Brett (Portugal); *Secretary:* M. J. Rossi (Switzerland); *Titular Members:* G. H. Atkinson (USA); W. Baumeister (Germany); R. Fernández-Prini (Argentina); J. G. Frey (UK); R. M. Lynden-Bell (UK); J. Maier (Germany); Z.-Q. Tian (China); *Associate Members:* S. Califano (Italy); S. Cabral de Menezes (Brazil); A. J. McQuillan (New Zealand); D. Platikanov (Bulgaria); C. A. Royer (France); *National Representatives:* J. Ralston (Australia); M. Oivanen (Finland); J. W. Park (Korea); S. Aldoshin (Russia); G. Vesnaver (Slovenia); E. L. J. Breet (South Africa).

‡Corresponding author

Abstract: In this report, the status quo and recent progress in electrokinetics are reviewed. Practical rules are recommended for performing electrokinetic measurements and interpreting their results in terms of well-defined quantities, the most familiar being the ζ-potential or electrokinetic potential. This potential is a property of charged interfaces, and it should be independent of the technique used for its determination. However, often the ζ-potential is not the only property electrokinetically characterizing the electrical state of the interfacial region; the excess conductivity of the stagnant layer is an additional parameter. The requirement to obtain the ζ-potential is that electrokinetic theories be correctly used and applied within their range of validity. Basic theories and their application ranges are discussed. A thorough description of the main electrokinetic methods is given; special attention is paid to their ranges of applicability as well as to the validity of the underlying theoretical models. Electrokinetic consistency tests are proposed in order to assess the validity of the ζ-potentials obtained. The recommendations given in the report apply mainly to smooth and homogeneous solid particles and plugs in aqueous systems; some attention is paid to nonaqueous media and less ideal surfaces.

Keywords: Electrokinetics; zeta potential; conductivity; aqueous systems; surface conductivity; IUPAC Physical and Biophysical Chemistry Division.

Reproduced From:
Pure Appl. Chem., Vol. 77, No. 10, pp. 1753–1805, 2005.
DOI:10.1351/pac200577101753
© 2005 IUPAC

1. Introduction

1.1 Electrokinetic phenomena

Electrokinetic phenomena (EKP) can be loosely defined as all those phenomena involving tangential fluid motion adjacent to a charged surface. They are manifestations of the electrical properties of interfaces under steady-state and isothermal conditions. In practice, they are often the only source of information available on those properties. For this reason, their study constitutes one of the classical branches of colloid science, *electrokinetics*, which has been developed in close connection with the theories of the electrical double layer (EDL) and of electrostatic surface forces [1–4].

From the point of view of nonequilibrium thermodynamics, EKP are typically cross-phenomena, because thermodynamic forces of a certain kind create fluxes of another type. For instance, in *electro-osmosis* and *electrophoresis*, an electric force leads to a mechanical motion, and in *streaming current (potential)*, an applied mechanical force produces an electric current (potential). First-order phenomena may also provide valuable information about the electrical state of the interface: for instance, an external electric field causes the appearance of a *surface current*, which flows along the interfacial region and is controlled by the *surface conductivity* of the latter. If the applied field is alternating, the *electric permittivity* of the system as a function of frequency will display one or more relaxations. The characteristic frequency and amplitude of these relaxations may yield additional information about the electrical state of the interface. We consider these first-order phenomena as *closely related* to EKP.

1.2 Definitions

Here follows a brief description of the main and related EKP [1–9].

- *Electrophoresis* is the movement of charged colloidal particles or polyelectrolytes, immersed in a liquid, under the influence of an external electric field. The *electrophoretic velocity*, v_e (m s^{-1}), is the velocity during electrophoresis. The *electrophoretic mobility*, u_e (m^2 V^{-1} s^{-1}), is the magnitude of the velocity divided by the magnitude of the electric field strength. The mobility is counted positive if the particles move toward lower potential (negative electrode) and negative in the opposite case.

- *Electro-osmosis* is the motion of a liquid through an immobilized set of particles, a porous plug, a capillary, or a membrane, in response to an applied electric field. It is the result of the force exerted by the field on the counter-charge in the liquid inside the charged capillaries, pores, etc. The moving ions drag the liquid in which they are embedded along. The *electro-osmotic velocity*, v_{eo} (m s^{-1}), is the uniform velocity of the liquid far from the charged interface. Usually, the measured quantity is the volume flow rate of liquid (m^3 s^{-1}) through the capillary, plug, or membrane, divided by the electric field strength, $Q_{to\ eo}$ (m^4 V^{-1} s^{-1}), or divided by the electric current, $Q_{to\ eo}$ (m^3 C^{-1}). A related concept is the *electro-osmotic counter-pressure*, Δp_{eo} (Pa), the pressure difference that must be applied across the system to stop the electro-osmotic volume flow. The value Δp_{eo} is considered to be positive if the high pressure is on the higher electric potential side.

- *Streaming potential (difference)*, U_{str} (V), is the potential difference at zero electric current, caused by the flow of liquid under a pressure gradient through a capillary, plug, diaphragm, or membrane. The difference is measured across the plug or between the ends of the capillary. Streaming potentials are created by charge accumulation caused by the flow of counter-charges inside capillaries or pores.

- *Streaming current*, I_{str} (A), is the current through the plug when the two electrodes are relaxed and short-circuited. The *streaming current density*, j_{str} (A m^{-2}), is the streaming current per area.

- *Dielectric dispersion* is the change of the electric permittivity of a suspension of colloidal particles with the frequency of an applied alternating current (ac) field. For low and middle frequencies, this change is connected with the polarization of the ionic atmosphere. Often, only the low-frequency dielectric dispersion (LFDD) is investigated.

- *Sedimentation potential*, U_{sed} (V), is the potential difference sensed by two identical electrodes placed some vertical distance L apart in a suspension in which particles are sedimenting under the effect of gravity. The electric field generated, U_{sed}/L, is known as the *sedimentation field*, E_{sed} (V m^{-1}). When the sedimentation is produced by a centrifugal field, the phenomenon is called *centrifugation potential.*

- *Colloid vibration potential*, U_{CV} (V), measures the ac potential difference generated between two identical relaxed electrodes, placed in the dispersion, if the latter is subjected to an (ultra)sonic field. When a sound wave travels through a colloidal suspension of particles whose density differs from that of the surrounding medium, inertial forces induced by the vibration of the suspension give rise to a motion of the charged particles relative to the liquid, causing an alternating electromotive force. The manifestations of this electromotive force may be measured, depending on the relation between the impedance of the suspension and that of the measuring instrument, either as U_{CV} or as *colloid vibration current*, I_{CV} (A).

- *Electrokinetic sonic amplitude* (ESA) method provides the amplitude, A_{ESA} (Pa), of the (ultra)sonic field created by an ac electric field in a dispersion; it is the counterpart of the colloid vibration potential method.

- *Surface conduction* is the excess electrical conduction tangential to a charged surface. It will be represented by the *surface conductivity*, K^σ (S), and its magnitude with respect to the bulk conductivity is frequently accounted for by the *Dukhin number*, Du (see eq. 12 below).

1.3 Model of charges and potentials in the vicinity of a surface

1.3.1 Charges

The electrical state of a charged surface is determined by the spatial distribution of ions around it. Such a distribution of charges has traditionally been called EDL, although it is often more complex than just two layers, and some authors have proposed the term "electrical interfacial layer". We propose here to keep the traditional terminology, which is used widely in the field. The simplest picture of the EDL is a physical model in which one layer of the EDL is envisaged as a fixed charge, the surface or titratable charge, firmly bound to the particle or solid surface, while the other layer is distributed more or less diffusely within the solution in contact with the surface. This layer contains an excess of counterions (ions opposite in sign to the fixed charge), and has a deficit of co-ions (ions of the same sign as the fixed charge).

For most purposes, a more elaborate model is necessary [3, 10]: the uncharged region between the surface and the locus of hydrated counterions is called the *Stern layer*, whereas ions beyond it form the *diffuse layer* or *Gouy layer* (also, Gouy–Chapman layer). In some cases, the separation of the EDL into a charge-free Stern layer and a diffuse layer is not sufficient to interpret experiments. The Stern layer is then subdivided into an *inner Helmholtz layer* (IHL), bounded by the surface and the *inner Helmholtz plane* (IHP) and an *outer Helmholtz layer* (OHL), located between the IHP and the *outer Helmholtz plane* (OHP). This situation is shown in Fig. 1 for a simple case. The necessity of this subdivision may occur when some ion types (possessing a chemical affinity for the surface in addition to purely Coulombic interactions), are specifically adsorbed on the surface, whereas other ion types interact with the surface charge only through electrostatic forces. The IHP is the locus of the former ions, and the OHP determines the beginning of the diffuse layer, which is the generic part of the EDL (i.e., the part governed by purely electrostatic forces). The fixed surface-charge density is denoted σ^0, the charge density at the IHP σ^i, and that in the diffuse layer σ^d. As the system is electroneutral.

$$\sigma^0 + \sigma^i + \sigma^d = 0 \qquad (1)$$

1.3.2 Potentials

As isolated particles cannot be linked directly to an external circuit, it is not possible to change their *surface potential* at will

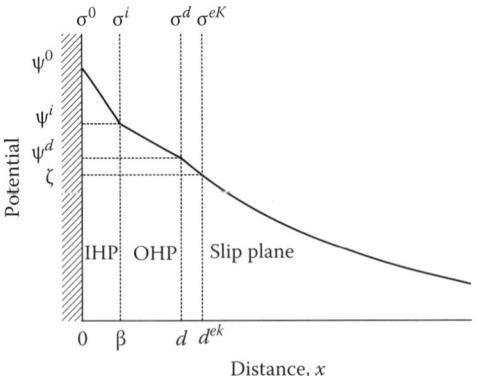

FIGURE 1: Schematic representation of the charges and potentials at a positively charged interface. The region between the surface (electric potential ψ^0; charge density σ^0) and the IHP (distance β from the surface) is free of charge. The IHP (electric potential ψ^i; charge density σ^i) is the locus of specifically adsorbed ions. The diffuse layer starts at $x = d$ (OHP), with potential ψ^d and charge density σ^d. The *slip plane* or *shear plane* is located at $x = d^{ek}$. The potential at the slip plane is the *electrokinetic* or *zeta-potential*, ζ; the *electrokinetic charge density* is σ^{ek}.

by applying an external field. Contrary to mercury and other electrodes, the surface potential, ψ^0, of a solid is therefore not capable of operational definition, meaning that it cannot be unambiguously measured without making model assumptions. As a consequence, for disperse systems it is the surface charge that is the primary parameter, rather than the surface potential. The potential at the OHP, at distance d from the surface, is called the *diffuse-layer potential*, ψ^d (sometimes also known as *Stern potential*): it is the potential at the beginning of the diffuse part of the double layer. The potential at the IHP, located at distance β ($0 \le \beta \le d$) from the surface, the *IHP potential*, is given the symbol ψ^i. All potentials are defined with respect to the potential in bulk solution.

Concerning the ions in the EDL, some further comments can be of interest. Usually, a distinction is made between *indifferent* and *specifically adsorbing* ions. Indifferent ions adsorb through Coulomb forces only; hence, they are repelled by surfaces of like sign, attracted by surfaces of opposite sign, and do not preferentially adsorb on an uncharged surface. Specifically adsorbing ions possess a chemical or specific affinity for the surface in addition to the Coulomb interaction, where chemical or specific is a collective adjective, embracing all interactions other than those purely Coulombic. It was recommended in [10], and is now commonly in use to restrict the notion of *surface ions* to those that are constituents of the solid, and hence are present on the surface, and to proton and hydroxyl ions. The former are covalently adsorbed. The latter are included because they are always present in aqueous solutions, their adsorption can be measured (e.g., by potentiometric titration) and they have, for many surfaces, a particularly high affinity. The term *specifically adsorbed* then applies to the sorption of all other ions having a specific affinity to the surface in addition to the generic Coulombic contribution. Specifically adsorbed charges are located within the Stern layer.

1.4 Plane of shear, electrokinetic potential, and electrokinetic charge density

Tangential liquid flow along a charged solid surface can be caused by an external electric field (*electrophoresis, electro-osmosis*) or by an applied mechanical force (*streaming potential,*

current). Experience and recent molecular dynamic simulations [11] have shown that in such tangential motion usually a very thin layer of fluid adheres to the surface: it is called the *hydrodynamically stagnant layer*, which extends from the surface to some specified distance, d^{ek}, where a so-called hydrodynamic slip plane is assumed to exist. For distances to the wall, $x < d^{ek}$, one has the *stagnant layer* in which no hydrodynamic flows can develop. Thus, we can speak of a distance-dependent viscosity with roughly a step-function dependence [12]. The space charge for $x > d^{ek}$ is hydrodynamically mobile and electro-kinetically active, and a particle (if spherical) behaves hydrodynamically as if it had a radius $a + d^{ek}$. The space charge for $x < d^{ek}$ is hydrodynamically immobile, but can still be electrically conducting. The potential at the plane where slip with respect to bulk solution is postulated to occur is identified as the *electrokinetic* or *zeta-potential*, ζ. The diffuse charge at the solution side of the slip plane equals the negative of the *electrokinetic (particle) charge*, σ^{ek}.

General experience indicates that the plane of shear is located very close to the OHP. Both planes are abstractions of reality. The OHP is interpreted as a sharp boundary between the diffuse and the non diffuse parts of the EDL, but it is very difficult to locate it exactly. Likewise, the slip plane is interpreted as a sharp boundary between the hydrodynamically mobile and immobile fluid. In reality, none of these transitions is sharp. However, liquid motion may be hindered in the region where ions experience strong interactions with the surface. Therefore, it is feasible that the immobilization of the fluid extends further out of the surface than the beginning of the diffuse part of the EDL. This means that, in practice, the ζ-potential is equal to or lower in magnitude than the diffuse-layer potential, ψ^d. In the latter case, the difference between ψ^d and ζ is a function of the ionic strength: at low ionic strength, the decay of the potential as a function of distance is small and $\zeta = \psi^d$; at high ionic strength, the decay is steeper and $|\zeta| \le |\psi^d|$. A similar reasoning applies to the electrokinetic charge, as compared to the diffuse charge.

1.5 Basic problem: Evaluation of ζ-potentials

The notion of slip plane is generally accepted in spite of the fact that there is no unambiguous way of locating it. It is also accepted that ζ is fully defined by the nature of the surface, its charge (often determined by pH), the electrolyte concentration in the solution, and the nature of the electrolyte and of the solvent. It can be said that for any interface with all these parameters fixed, ζ is a well-defined property.

Experience demonstrates that different researchers often find different ζ-potentials for supposedly identical interfaces. Sometimes, the surfaces are not in fact identical: the high specific surface area and surface reactivity of colloidal systems make ζ very sensitive to even minor amounts of impurities in solution. This can partly explain variations in electrokinetic determinations from one laboratory to another. Alternatively, since ζ is not a directly measurable property, it may be that an inappropriate model has been used to convert the electrokinetic signal into a ζ-potential. The level of sophistication required (for the model) depends on the situation and on the particular phenomena investigated. The choice of measuring technique and of the theory used depends to a large extent on the purpose of the electrokinetic investigation.

There are instances in which the use of simple models can be justified, even if they do not yield the correct ζ-potential. For example, if electrokinetic measurements are used as a sort of quality-control tool, one is interested in rapidly (online) detecting modifications in the electrical state of the interface rather than in obtaining accurate ζ-potentials. On the other hand, when the purpose is to compare the calculated values of ζ of a system under given conditions using different electrokinetic techniques, it may be essential to find a true ζ-potential. The same applies to those cases in which ζ will be used to perform calculations of other physical quantities, such as the Gibbs interaction energy between particles. Furthermore, there may be situations in which the use of simple theories may be misleading even for simple quality control. For example, there are ranges of ζ-potential and double-layer thickness for which the electrophoretic mobility does not depend linearly on ζ, as assumed in the simple models. Two samples might have the same true ζ-potential and quite different mobilities because of their different sizes. The simple theory would lead us to believe that their electrical surface characteristics are different when they are not.

An important complicating factor in the reliable estimation of ζ is the possibility that charges behind the plane of shear may contribute to the excess conductivity of the double layer (stagnant-layer or inner-layer conductivity.) If it is assumed that charges located between the surface and the plane of shear are electrokinetically inactive, then the ζ-potential will be the only interfacial quantity explaining the observed electrokinetic signal.

Otherwise, a correct quantitative explanation of EKP will require the additional estimation of the stagnant-layer conductivity (SLC). This requires more elaborate treatments [2, 3, 13–17] than standard or classical theories, in which only conduction at the solution side of the plane of shear is considered.

It should be noted that there are a number of situations where electrokinetic measurements, without further interpretation, provide extremely useful and unequivocal information, of great value for technological purposes. The most important of these situations are

- Identification of the isoelectric point (or point of zero z-potential) in titrations with a potential determining ion (e.g., pH titration).
- Identification of the isoelectric point in titrations with other ionic reagents such as surfactants or polyelectrolytes.
- Identification of a plateau in the adsorption of an ionic species indicating optimum dosage for a dispersing agent.

In these cases, the complications and digressions, which are discussed below, are essentially irrelevant. The electrokinetic property (or the estimated ζ-potential) is then zero or constant and that fact alone is of value.

1.6 Purpose of the document

The present document is intended to deal mainly with the following issues, related to the role of the different EKP as tools for surface chemistry research. Specifically, its aims are:

- Describe and codify the main and related EKP and the quantities involved in their definitions.
- Give a general overview of the main experimental techniques that are available for electrokinetic characterization.
- Discuss the models for the conversion of the experimental signal into ζ-potential and, where appropriate, other double-layer characteristics.
- Identify the validity range of such models, and the way in which they should be applied to any particular experimental situation.

The report first discusses the most widely used EKP and techniques, such as electrophoresis, streaming-potential, streaming current, or electro-osmosis. Attention is also paid to the rapidly growing techniques based on dielectric dispersion and electro-acoustics.

2. Elementary theory of electrokinetic phenomena

All electrokinetic effects originate from two generic phenomena, namely, the *electro-osmotic flow* and the *convective electric surface current* within the EDL. For nonconducting solids, Smoluchowski [18] derived equations for these generic phenomena, which allowed an extension of the theory to all other specific EKP. Smoluchowski's theory is valid for any shape of a particle or pores inside a solid, provided the (local) curvature radius a largely exceeds the Debye length κ^{-1}:

$$\kappa a \gg 1 \tag{2}$$

where κ is defined as

$$\kappa = \left\{ \frac{\sum_{i=1}^{N} e^2 z_i^2 n_i}{\varepsilon_{rs} \varepsilon_0 kT} \right\}^{1/2} \tag{3}$$

with e the elementary charge, z_i, n_i the charge number and number concentration of ion i (the solution contains N ionic species), ε_{rs} the relative permittivity of the electrolyte solution, ε_0 the electric permittivity of vacuum, k the Boltzmann constant, and T the thermodynamic temperature. Note that under condition (2), a curved surface can be considered as flat for any small section of the double layer. This condition is traditionally called the "thin double-layer approximation", but we do not recommend this language, and we rather refer to this as the "large κa limit". Many aqueous dispersions satisfy this condition, but not those for very small particles in low ionic strength media.

Electro-osmotic flow is the liquid flow along any section of the double layer under the action of the tangential component E_t of an external field E. In Smoluchowski's theory, this field is considered to be independent of the presence of the double layer, i.e., the distortion of the latter is ignored*. Also, because the EDL is assumed to be very thin compared to the particle radius, the hydrodynamic and electric field lines are parallel for large κa. Under these conditions, it can be shown [3] that at a large distance from the surface the liquid velocity (electro-osmotic velocity), v_{eo}, is given by

$$v_{eo} = -\frac{\varepsilon_{rs}\varepsilon_0 \zeta}{\eta} E \tag{4}$$

where η is the dynamic viscosity of the liquid. This is the *Smoluchowski equation for the electro-osmotic slip velocity*. From this, the electro-osmotic flow rate of liquid per current, $Q_{eo,I}$ ($m^3 \ s^{-1} \ A^{-1}$), can be derived

$$Q_{eo,I} = \frac{Q_{eo}}{I} = -\frac{\varepsilon_{rs}\varepsilon_0 \zeta}{\eta K_L} \tag{5}$$

K_L being the bulk liquid conductivity ($S \ m^{-1}$) and I the electric current (A).

It is impossible to quantify the distribution of the electric field and the velocity in pores with unknown or complex geometry.

---

* The approximation that the structure of the double layer is not affected by the applied field is one of the most restrictive assumptions of the elementary theory of EKP.

However, this fundamental difficulty is avoided for $\kappa a \gg 1$, when eqs. 4 and 5 are valid [3].

Electrophoresis is the counterpart of *electro-osmosis*. In the latter, the liquid moves with respect to a solid body when an electric field is applied, whereas during electrophoresis the liquid as a whole is at rest, while the particle moves with respect to the liquid under the influence of the electric field. In both phenomena, such influence on the double layer controls the relative motions of the liquid and the solid body. Hence, the results obtained in considering electro-osmosis can be readily applied for obtaining the corresponding formula for electrophoresis. The expression for the electrophoretic velocity, that is, the velocity of the particle with respect to a medium at rest, becomes, after changing the sign in eq. 4

$$v_e = \frac{\varepsilon_{rs}\varepsilon_0 \zeta}{\eta} E \qquad (6)$$

and the electrophoretic mobility, u_e

$$u_e = \frac{\varepsilon_{rs}\varepsilon_0 \zeta}{\eta} \qquad (7)$$

This equation is known as the *Helmholtz–Smoluchowski* (HS) equation for electrophoresis.

Let us consider a capillary with circular cross-section of radius a and length L with charged walls. A pressure difference between the two ends of the capillary, Δp, is produced externally to drive the liquid through the capillary. Since the fluid near the interface carries an excess of charge equal to σ^{ek}, its motion will produce an electric current known as *streaming current*, I_{str}:

$$I_{str} = -\frac{\varepsilon_{rs}\varepsilon_0\, a^2}{\eta}\frac{\Delta p}{L}\zeta \qquad (8)$$

The observation of this current is only possible if the extremes of the capillary are connected through a low-resistance external circuit (short-circuit conditions). If this resistance is high (open circuit), transport of ions by this current leads to the accumulation of charges of opposite signs between the two ends of the capillary and, consequently, to the appearance of a potential difference across the length of the capillary, the *streaming-potential*, U_{str}. This gives rise to a *conduction current*, I_c:

$$I_c = K_L\, a^2 \frac{U_{str}}{L} \qquad (9)$$

The value of the *streaming-potential* is obtained by the condition of equality of the conduction and streaming currents (the net current vanishes)

$$\frac{U_{str}}{\Delta p} = \frac{\varepsilon_{rs}\varepsilon_0 \zeta}{\eta K_L} \qquad (10)$$

For large κa, eq. 10 is also valid for porous bodies.

As described, the theory is incomplete in mainly three aspects: (i) it does not include the treatment of strongly curved surfaces (i.e., surfaces for which the condition $\kappa a \gg 1$ does not apply); (ii) it neglects the effect of *surface conduction* both in the diffuse and the inner part of the EDL; and (iii) it neglects EDL polarization. Concerning the first point, the theoretical analysis described above is based on the assumption that the interface is flat or that its radius of curvature at any point is much larger than the

double-layer thickness. When this condition is not fulfilled, the Smoluchowski theory ceases to be valid, no matter the existence or not of surface conduction of any kind. However, theoretical treatments have been devised to deal with these surface curvature effects. Roughly, in order to check if such corrections are needed, one should simply calculate the product κa, where a is a characteristic radius of curvature (e.g., particle radius, pore or capillary radius). When describing the methods below, we will give details about analytical or numerical procedures that can be used to account for this effect.

With respect to surface conductivity, a detailed account is given in Section 3 and mention will be made to it where necessary in the description of the methods. Here, it may suffice to say that it may be important when the ζ-potential is moderately large (>50 mV, say.)

Finally, the *polarization* of the double layer implies accumulation of excess charge on one side of the colloidal particle and depletion on the other. The resulting induced dipole is the source of an electric field distribution that is superimposed on the applied field and affects the relative solid/liquid motion. The extent of polarization depends on surface conductivity, and its role in electrokinetics will be discussed together with the methodologies.

3. Surface conductivity and electrokinetic phenomena

Surface conduction is the name given to the excess electric conduction that takes place in dispersed systems owing to the presence of the electric double layers. Excess charges in them may move under the influence of electric fields applied tangentially to the surface. The phenomenon is quantified in terms of the *surface conductivity*, K^σ, which is the surface equivalent to the bulk conductivity, K_L. K^σ is a surface excess quantity just as the surface concentration Γ_i of a certain species i. Whatever the charge distribution, K^σ can always be defined through the two-dimensional analog of Ohm's law

$$j^\sigma = K^\sigma E \qquad (11)$$

where j^σ is the (excess) *surface current density* (A m^{-1}).

A measure of the relative importance of surface conductivity is given by the dimensionless Dukhin number, Du, relating surface (K^σ) and bulk (K_L) conductivities

$$Du = \frac{K^\sigma}{K_L a} \qquad (12)$$

where a is the local curvature radius of the surface. For a colloidal system, the total conductivity, K, can be expressed as the sum of a solution contribution and a surface contribution. For instance, for a cylindrical capillary, the following expression results:

$$K = \left(K_L + 2K^\sigma/a\right) = K_L(1 + 2Du) \qquad (13)$$

The factor 2 in eq. 13 applies for cylindrical geometry. For other geometries, its value may be different.

As mentioned, HS theory does not consider surface conduction, and only the solution conductivity, K_L, is taken into account to derive the tangential electric field within the double layer. Thus, in addition to eq. 2, the applicability of the HS theory requires

$$Du \ll 1 \qquad (14)$$

The surface conductivity can have contributions owing to the diffuse-layer charge outside the plane of shear, $K^{\sigma d}$, and to the stagnant layer $K^{\sigma i}$:

$$K^{\sigma} = K^{\sigma d} + K^{\sigma i} \qquad (15)$$

Accordingly, Du can be written as

$$Du = \frac{K^{\sigma d}}{K_L a} + \frac{K^{\sigma i}}{K_L a} = Du^{d} + Du^{i} \qquad (16)$$

The $K^{\sigma d}$ contribution is called the *Bikerman surface conductivity* after Bikerman, who found a simple equation for $K^{\sigma d}$ (Eq. 17). The SLC may include a contribution due to the specifically adsorbed charge and another one due to the part of the diffuse-layer charge that may reside behind the plane of shear. The charge on the solid surface is generally assumed to be immobile; it does not contribute to K^{σ}.

The conductivity in the diffuse double layer outside the plane of shear, $K^{\sigma d}$, consists of two parts: a migration contribution, caused by the movement of charges with respect to the liquid; and a convective contribution, due to the electro-osmotic liquid flow beyond the shear plane, which gives rise to an additional mobility of the charges and hence leads to an extra contribution to $K^{\sigma d}$. For the calculation of $K^{\sigma d}$, the Bikerman equation (eq. 17, see below) can be used. This equation expresses $K^{\sigma d}$ as a function of the electrolyte and double-layer parameters. For a symmetrical z-z electrolyte, a convenient expression is

$$K^{\sigma d} = \frac{2e^2 N_A z^2 c}{kT\kappa} \left[D_{+}(e^{-ze\zeta/2kT} - 1)\left(1 + \frac{3m_{+}}{z^2}\right) \right.$$
$$\left. + D_{-}(e^{ze\zeta/2kT} - 1)\left[1 + \frac{3m_{-}}{z^2}\right] \right] \qquad (17)$$

where c is the electrolyte amount concentration (mol m^{-3}), N_A is the Avogadro constant (mol^{-1}), and m_{+} (m_{-}) is the dimensionless mobility of the cations (anions)

$$m_{\pm} = \frac{2}{3}\left(\frac{kT}{e}\right)^2 \frac{\varepsilon_{rs}\varepsilon_0}{\eta D_{\pm}} \qquad (18)$$

where $D_{\pm}$ (m^2 s^{-1}) are the ionic diffusion coefficients. The parameters $m_{\pm}$ indicate the relative contribution of electro-osmosis to the surface conductivity.

The extent to which K^{σ} influences the electrokinetic behavior of the systems depends on the value of Du. For the Bikerman part of the conductivity, Du^{d} can be written explicitly. For a symmetrical z-z electrolyte and identical cation and anion diffusion coefficients so that $m_{+} = m_{-} = m$:

$$Du^{d} \equiv \frac{K^{\sigma d}}{K_L a} = \frac{2}{\kappa a}\left(1 + \frac{3m}{z^2}\right)\left[\cosh\left(\frac{ze\zeta}{2kT}\right) - 1\right] \qquad (19)$$

From this equation, it follows that Du^{d} is small if $\kappa a \gg 1$, and ζ is small. Substitution of this expression for Du^{d} in eq. 16 yields

$$Du = \frac{2}{\kappa a}\left(1 + \frac{3m}{z^2}\right)\left[\cosh\left(\frac{ze\zeta}{2kT}\right) - 1\right]\left(1 + \frac{K^{\sigma i}}{K^{\sigma d}}\right) \qquad (20)$$

This equation shows that, in general, Du is dependent on the ζ-potential, the ion mobility in bulk solution, and $K^{\sigma i}/K^{\sigma d}$. Now, the condition $Du \ll 1$ required for application of the HS theory is achieved for $\kappa a \gg 1$, rather low values of ζ, and $K^{\sigma i}/K^{\sigma d} < 1$.

4. Methods

4.1 Electrophoresis

4.1.1 Operational definitions; recommended symbols and terminology; relationship between the measured quantity and ζ-potential

Electrophoresis is the translation of a colloidal particle or polyelectrolyte, immersed in a liquid, under the action of an externally applied field, E, constant in time and position-independent. For uniform and not very strong electric fields, a linear relationship exists between the steady-state *electrophoretic velocity*, v_e (attained by the particle roughly a few milliseconds after application of the field) and the applied field

$$v_e = u_e E \qquad (21)$$

where u_e is the quantity of interest, the *electrophoretic mobility*.

4.1.2 How and under which conditions the electrophoretic mobility can be converted into ζ-potential

As discussed above, it is not always possible to rigorously obtain the ζ-potential from measurements of electrophoretic mobility only. We give here some guidelines to check whether the system under study can be described with the standard electrokinetic models:

a. Calculate κa for the suspension.
b. If $\kappa a \gg 1$ ($\kappa a > 20$, say), we are in the large κa regime, and simple analytical models are available.
   b.1 Obtain the mobility u_e for a range of indifferent electrolyte concentrations. If u_e decreases with increasing electrolyte concentration, use the HS formula, eq. 7, to obtain ζ.
      b.1.1 If the ζ value obtained is low ($\zeta \leq 50$ mV, say), concentration polarization is negligible, and one can trust the value of ζ.
      b.1.2 If ζ is rather high ($\zeta > 50$ mV, say), then HS theory is not applicable. One has to use more elaborate models. The possibilities are: (i) the numerical calculations of O'Brien and White [22]; (ii) the equation derived by Dukhin and Semenikhin [5] for symmetrical z-z electrolytes

$$\frac{3}{2}\frac{\eta e}{\varepsilon_{rs}\varepsilon_0 kT}u_e = \frac{3y^{ek}}{2} - 6$$

$$\times \left(\frac{y^{ek}(1 + 3m/z^2)\sinh^2(zy^{ek}/4) + [2z^{-1}\sinh(zy^{ek}/2) - 3my^{ek}]\ln\cosh(zy^{ek}/4)}{\kappa a + 8(1 + 3m/z^2)\sinh^2(zy^{ek}/4) - (24m/z^2)\ln\cosh(zy^{ek}/4)}\right) \qquad (22)$$

where m was defined in eq. 18 and

$$y^{ek} = \frac{e\zeta}{kT} \qquad (23)$$

is the dimensionless ζ-potential. For aqueous solutions, m is about 0.15. O'Brien [4] found that eq. 22 can be simplified by neglecting terms of order $(\kappa a)^{-1}$ as follows:

$$\frac{3}{2}\frac{\eta e}{\varepsilon_{rs}\varepsilon_0 kT}u_e$$

$$=\frac{3}{2}y^{ek}-\frac{6\left[\dfrac{y^{ek}}{2}-\dfrac{\ln 2}{z}\left\{1-\exp(-zy^{ek})\right\}\right]}{2+\dfrac{\kappa a}{1+3m/z^2}\exp\left(-\dfrac{zy^{ek}}{2}\right)} \quad (24)$$

Note that both the numerical calculations and eqs. 22 and 24 automatically account for diffuse-layer conductivity.

b.2 If a maximum in u_e (or in the apparent ζ-potential deduced from the HS formula) vs. electrolyte concentration is found, the effect of stagnant-layer conduction is likely significant. For low ζ-potentials, when concentration polarization is negligible, the following expression can be used [23]:

$$u_e=\frac{\varepsilon_{rs}\varepsilon_0}{\eta}\zeta\left[1+\frac{Du}{1+Du}\left(\frac{kT}{e|\zeta|}\frac{2\ln 2}{z}-1\right)\right] \quad (25)$$

with Du including both the stagnant- and diffuse-layer conductivities. This requires additional information about the value of $K^{\sigma i}$ (see Section 3).

c. If κa is low, the O'Brien and White [22] numerical calculations remain valid, but there are also several analytical approximations. For $\kappa a < 1$, the *Hückel–Onsager* (HO) equation applies [4]:

$$u_e=\frac{2}{3}\frac{\varepsilon_{rs}\varepsilon_0}{\eta}\zeta \quad (26)$$

d. For the transition range between low and high κa, Henry's formula can be applied if ζ is presumed to be low (<50 mV; in such conditions, surface conductivity and concentration polarization are negligible). For a nonconducting sphere, Henry derived the following expression:

$$u_e=\frac{2}{3}\frac{\varepsilon_{rs}\varepsilon_0}{\eta}\zeta f_1(\kappa a) \quad (27)$$

where the function f_1 varies smoothly from 1.0, for low values of κa, to 1.5 as κa approaches infinity. Henry [24] gave two series expansions for the function f_1 one for small κa and one for large κa. Ohshima [25] has provided an approximate analytical expression which duplicates the Henry expansion almost exactly. Ohshima's relation is

$$f_1(\kappa a)=1+\frac{1}{2}\left[1+\left(\frac{2.5}{\kappa a[1+2\exp(-\kappa a)]}\right)\right]^{-3} \quad (28)$$

Equation 27 can be used in the calculation of the electrophoretic mobility of particles with nonzero bulk

conductivity, K_p. With that aim, it can be modified to read [2]

$$u_e=\frac{2}{3}\frac{\varepsilon_{rs}\varepsilon_0}{\eta}\zeta\times F_1(\kappa a,K_p) \quad (29)$$

$$F_1(\kappa a,K_p)=1+\frac{2-2K_{rel}}{1+K_{rel}}[f_1(\kappa a)-1] \quad (30)$$

with

$$K_{rel}=\frac{K_p}{K_L} \quad (31)$$

e. If SLC is likely for a system with low κa, then as discussed before, ζ ceases to be the only parameter needed for a full characterization of the EDL, and additional information on K^σ is required (see Section 3). Numerical calculations like those of Zukoski and Saville [9] or Mangelsdorf and White [10–12] can be used.

Figure 2a allows a comparison to be established between the predictions of the different models mentioned, for the case of spheres. For the κa chosen, the curvature is enough for the HS theory to be in error, the more so the higher $|\zeta|$. According to Henry's treatment, the electrophoretic mobility is lower than predicted by the simpler HS equation. Note also that Henry's theory fails for low-to-moderate ζ-potentials; this is a consequence of its neglecting concentration polarization. The full O'Brien and White theory demonstrates that as ζ increases, the mobility is lower than predicted by either Henry's or HS calculations. The existence of surface conduction can account for this. In addition, for sufficiently high ζ-potential, the effect of concentration polarization is a further reduction of the mobility, which goes through a maximum and eventually decreases with the increase of ζ-potential.

The effect of κa on the $u_e(\zeta)$ relationship is depicted in Fig. 2b. Note that the maximum is more pronounced with the larger κa, and that the electrophoretic mobility increases (in the range of κa shown) with the former. Finally, Fig. 2c demonstrates the drastic change that can occur in the mobility-ζ-potential trends if SLC is present. This quantity always tends to decrease u_e, as the total surface conductivity is increased, as compared to the case of diffuse-layer conductivity alone.

4.1.3 Experimental techniques available: Samples

i. Earlier techniques, at present seldom used in colloid science:

• *Moving boundary* [26]. In this method, a boundary is mechanically produced between the suspension and its equilibrium serum. When the electric field is applied, the migration of the solid particles provokes a displacement of the solid/liquid separation whose velocity is in fact proportional to v_e. The traditional moving-boundary method contributed to a great extent to the knowledge of proteins and polyelectrolytes as well as of colloids. It inspired gel electrophoresis, presently essential in such important fields as genetic analysis.

• *Mass transport* electrophoresis [27]. The mass transport method is based on the fact the application of a

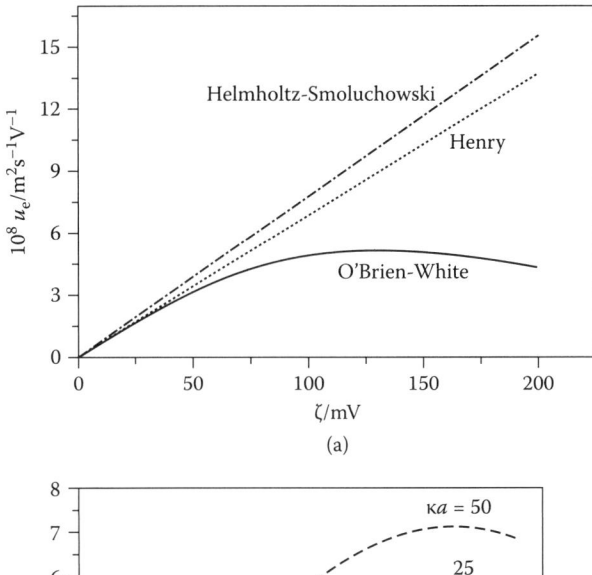

(a)

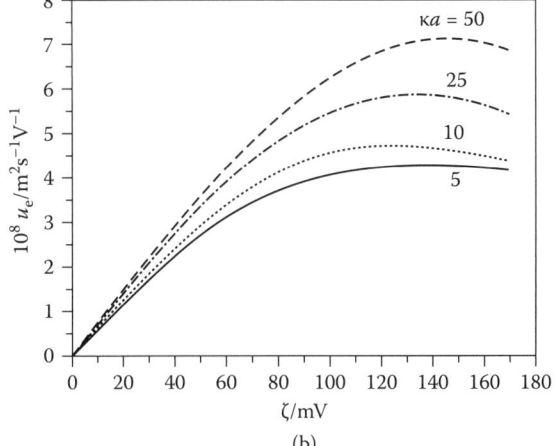

(b)

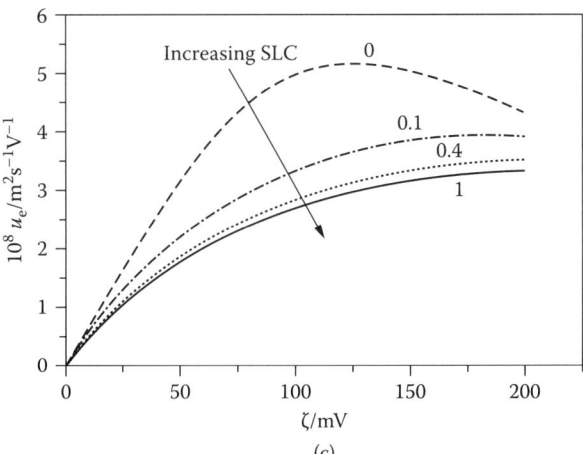

(c)

FIGURE 2: (a) Electrophoretic mobility u_e plotted as a function of the ζ-potential according to different theoretical treatments, all neglecting SLC: HS, O'Brien–White (full theory), Henry (no surface conductance), for $\kappa a = 15$. (b) Role of κa on the mobility-ζ relationship (O'Brien–White theory). (c) Effect of SLC on the electrophoretic mobility-ζ-potential relationship for the same suspensions as in part (a). The ratios between the diffusion coefficients of counterions in the stagnant layer and in the bulk electrolyte are indicated (the upper curve corresponds to zero SLC).

known potential difference to the suspension causes the particles to migrate from a reservoir to a detachable collection chamber. The electrophoretic mobility is deduced from data on the amount of particles moved after a certain time, which can be determined by simply weighing the collection chamber or otherwise analyzing its contents.

ii. Microscopic (visual) microelectrophoresis
Probably the most widespread method until the 1980s, microscopic (visual) microelectrophoresis is based on the direct observation, with a suitable magnifying optics, of individual particles in their electrophoretic motion. In fact, it is not the particle that is seen, but a bright dot on a dark background, due to the Tyndall effect, that is the strong lateral light scattering of colloidal particles.

Size range of samples
The ultramicroscope is necessary for particles smaller than 0.1 μm. Particles about 0.5 μm can be directly observed using a travelling microscope illuminated with a strong (cold) light source.

Advantages and prerequisites of the technique
- The particles are directly observed in their medium.
- The suspensions to be studied should be stable and dilute; if they are not, individual particles cannot be identified under the microscope. However, in dilute systems; the aggregation times are very long, even in the worst conditions, so that velocities can likely be measured.

Problems involved in the technique and proposed actions to solve them
- Its main limitations are the bias and subjectivity of the observer, who can easily select only a narrow range of velocities, which might be little representative of the true average value of the suspension. Furthermore, measurements usually take a fairly long time, and this can bring about additional problems such as Joule heating, pH changes, and so on. Hence, some manufacturers of commercial apparatus have modified their designs to include automatic tracking by digital image processing.
- Recall that electrophoresis is the movement of the particles with respect to the fluid, which is assumed to be at rest. However, the observed velocity is in fact relative to the instrument, and this is a source of error, as an electro-osmotic flow of liquid is also induced by the external field if the cell walls are charged, which is often the case. If the cell is open, the velocity over its section would be constant and equal to its value at the outer double-layer boundary. However, in almost all experimental set-ups, the measuring cell is closed, and the electro-osmotic counter-pressure provokes a liquid flow of Poiseuille type. The resulting velocity profile for the case of a cylindrical channel is given by [4]

$$v_L = v_{eo}\left(2\frac{r^2}{a^2} - 1\right) \qquad (32)$$

where v_{eo} is the electro-osmotic liquid velocity in the channel, a is the capillary radius, and r is the radial distance from the cylinder axis. From eq. 32, it is clear that $v_L = 0$ if $r = a/\sqrt{2}$, so that the true electrophoretic velocity will be displayed only by particles moving in a cylindrical shell placed at 0.292 a from the channel wall. It is easy to estimate the uncertainties associated with errors in the measuring position: if $a \sim 2$ mm and the microscope has a focus depth of $\sim$50 μm, then an error of 2% in the velocity will be always present. A more accurate, although time-consuming method, consists in measuring the whole parabolic velocity profile to check for absence of systematic errors. These arguments also apply to electrophoresis cells with rectangular or square cross-sections.

Some authors (see, e.g., [28]) have suggested that a procedure to avoid this problem would be to cover the cell walls, whatever their geometry, with a layer of uncharged chemical species, for instance, polyacrylamide. However, it is possible that after some usage, the layer gets detached from the walls, and this would mask the electrophoretic velocity measured at an arbitrary depth, with an electro-osmotic contribution, the absence of which can only be ascertained by measuring u_e of standard, stable particles, which in turn remains an open problem in electrokinetics.

A more recent suggestion [29] is to perform the electrophoresis measurements in an alternating field with frequency much larger than the reciprocal of the characteristic time τ for steady electro-osmosis ($\tau \sim 1$ s), but smaller than that of steady electrophoresis ($\tau \sim 10^{-4}$ s). Under such conditions, no electro-osmotic flow can develop and hence the velocity of the particle is independent of the position in the cell.

Another way of overcoming the electro-osmosis problem is to place both electrodes providing the external field inside the cell, completely surrounded by the suspension; since no net external field acts on the charged layer close to the cell walls, the associated electro-osmotic flow will not exist [30].

iii. Electrophoretic light scattering (ELS)
These are automated methods based on the analysis of the (laser) light scattered by moving particles [31–34]. They have different principles of operation [35]. The most frequently used method, known as laser Doppler velocimetry, is based on the analysis of the intensity autocorrelation function of the scattered light. The method of phase analysis light scattering (PALS) [36–38] has the adavantage of being suited for particles moving very slowly, for instance, close to their isoelectric point. The method is capable of detecting electrophoretic mobilities as low as 10^{-12} m^2 V^{-1} s^{-1}, that is, 10^{-4} μm s^{-1}/V cm^{-1} in practical mobility units (note that mobilities typically measurable with standard techniques must be above $\sim$10^{-9} m^2 V^{-1} s^{-1}). These techniques are rapid, and measurements can be made in a few seconds. The results obtained are very reproducible, with typical standard deviations less than 2 %. A small amount of sample is required for analysis, often a few millilitres of a suitable dispersion. However, dilution of the sample may be required, and therefore the sample preparation technique becomes very important.

Samples that can be studied
a. Sample composition
Measurements can be made of any colloidal dispersion where the continuous phase is a transparent liquid and the dispersed phase has a refractive index which differs from that of the continuous phase.

b. Size range of samples
The lower size limit is dependent upon the sample concentration, the refractive index difference between disperse and continuous phase, and the quality of the optics and performance of the instrument. Particle sizes down to 5 nm can be measured under optimum conditions.

The upper size limit is dependent upon the rate of sedimentation of particles (which is related to particle size and density). ELS methods are inherently directional in their measurement plane. Hence, for a horizontal field, samples can be measured while they are sedimenting. Measurement is possible so long as there are particles present in the detection volume. Typically, measurements are possible for particles with diameters below 30 μm.

c. Sample conductivity
The conductivity of samples that can be measured ranges from that of particles dispersed in deionized water up to media containing greater than physiological saline. In high salt concentration, the Joule heating of the sample will affect the particle mobility, and thermostating of the cell is not at all easy. Reduction of the applied voltage decreases this effect, but will also reduce the resolution obtainable from the measurement.

The presence of some ions in the medium is recommended (e.g., 10^{-4} mol/L NaCl) as this will stabilize the field in the cell and will improve the repeatability of measurements. Furthermore, some salt is always needed anyway because otherwise the double layer becomes ill-defined.

d. Sample viscosity
There is no particular limit as to the viscosity range of samples that can be measured. But it must be emphasized that increasing the viscosity of the medium will reduce the mobility of the particles and may require longer observation times, with the subsequent increased risk of Joule heating.

e. Permittivity
Measurements in a large variety of solvents are possible, depending on the instrument configuration.

f. Fluorescence
Sample fluorescence results in a reduction in the signal-to-noise ratio of the measurement. In severe cases, this may completely inhibit measurements.

Sample preparation
Many samples will be too concentrated for direct measurement and will require dilution. How this dilution is carried out is critical. The aim of sample preparation is to preserve the existing state of the particle surface during the process of dilution. One way to ensure this is by filtering or gently centrifuging some clear liquid from the original sample and using this to dilute the original concentrated sample. In this way, the equilibrium between surface and liquid is perfectly maintained. If extraction of a supernatant is not possible, then just letting a sample naturally sediment and using the fine particles left in the supernatant is a good alternative method. The possibility also exists of dialyzing the concentrate against a solution of the desired ionic composition. Another method is to imitate the original medium as closely as possible. This should be done with regard to pH, concentration of each ionic species in the medium, and concentration of any other additive that might be present.

However, attention must be paid to the possible modification of the surface compositon upon dilution, particularly when polymers or polyelectrolytes are in solution [39]. Also, if the particles are positively charged, care must be taken to avoid long storage in glass containers, as dissolution of glass can lead to adsorption of negatively charged species on the particles. For emulsion systems, dilution is always problematic, because changing the phase volume ratio may alter the surface properties due to differential solubility effects.

Ranges of electrolyte and particle concentration that can be investigated

Microelectrophoresis is a technique where samples must be dilute enough for particles not to interfere with each other. For any system under investigation, it is recommended that an experiment should be done to check the effect of concentration on the mobility. The concentration range which can be studied will depend upon the suitability of the sample (e.g., size, refractive index) and the optics of the instrument. By way of example, a 200-nm polystyrene latex standard (particle refractive index 1.59, particle absorbance 0.001) dispersed in water (refractive index 1.33) can be measured at a solids concentration ranging from 2×10^{-3} to 1×10^{-6} g/cm^3.

Standard samples for checking correct instrument operation

Microelectrophoresis ELS instruments are constructed from basic physical principles and as such need not be calibrated. The correctness of their operation can only be verified by measuring a sample of which the ζ-potential is known. A pioneering study in this direction was performed in 1970 by a group of Japanese surface and colloid chemists, forming a committee under the Division of Surface Chemistry in the Japan Oil Chemists Society [6, 39]. This group measured and compared ζ-potentials of samples of titanium dioxide, silver iodide, silica, microcapsules, and some polymer latexes. The study involved different devices in nine laboratories, and concluded that the negatively charged PSSNa (polystyrene-sodium *p*-vinylbenzenesulfonate copolymer) particles prepared as described in [40] could be a very useful standard, providing reliable and reproducible mobility data. Currently, there is no negative ζ-potential standard available from the U.S. National Institute of Standards and Technology (NIST).

A positively charged sample available from NIST is Standard Reference Material (SRM) 1980. It contains a 500 mg/L goethite (α–FeOOH) suspension saturated with 100 μmol/g phosphate in a 5×10^{-2} mol/L sodium perchlorate electrolyte solution at a pH of 2.5. When prepared according to the procedure supplied by NIST, the certified value and uncertainty for the positive electrophoretic mobility of SRM1980 is 2.53 ± 0.12 μm s^{-1}/V cm^{-1}. This will give a ζ-potential of $+32.0 \pm 1.5$ mV if the HS equation (eq. 7) is used.

4.2 Streaming current and streaming potential

4.2.1 *Operational definitions; recommended symbols and terminology; conversion of the measured quantities into ζ-potential*

The phenomena of streaming current and streaming potential occur in capillaries and plugs and are caused by the charge displacement in the EDL as a result of an applied pressure inducing the liquid phase to move tangentially to the solid. The *streaming current* can be detected directly by measuring the electric current between two positions, one upstream and the other downstream. This can be carried out via nonpolarizable electrodes, connected to an electrometer of sufficiently low internal resistance.

4.2.1.1 Streaming current

The first quantity of interest is the *streaming current per pressure drop*, $I_{str}/\Delta p$ (SI units: A Pa^{-1}), where I_{str} is the measured current, and Δp the pressure drop. The relation between $I_{str}/\Delta p$ and ζ-potential has been found for a number of cases:

a. If $\kappa a \gg 1$ (a is the capillary radius), the HS formula can be used

$$\frac{I_{str}}{\Delta p} = -\frac{\varepsilon_{rs}\varepsilon_0 \zeta}{\eta} \frac{A_c}{L} \tag{33}$$

where A_c is the capillary cross-section, and L its length. If instead of a single capillary, the experimental system is a porous plug or a membrane, eq. 33 remains approximately valid, provided that $\kappa a \gg 1$ everywhere in the pore walls. In the case of porous plugs, attention has to be paid to the fact that a plug is not a system of straight parallel capillaries, but a random distribution of particles with a resulting porosity and tortuosity, for which an equivalent capillary length and cross-section is just a simplified model. In addition, the use of eq. 33 requires that the conduction current in the system is determined solely by the bulk conductivity of the supporting solution. It often happens that surface conductivity is important, and, besides that, the ions in the plug behave with a lower mobility than in solution.

A_c/L can be estimated experimentally as follows [40, 41]. Measure the resistance, R_∞, of the plug or capillary wetted by a highly concentrated (above 10^{-2} mol/L, say) electrolyte solution, with conductivity K^∞_L. Since for such a high ionic strength the double-layer contribution to the overall conductivity is negligible, we may write

$$\frac{A_c}{L} = \frac{1}{K^\infty_L R_\infty} \tag{34}$$

In addition, theoretical or semi-empirical models exist that relate the apparent values of A_c and L (external dimensions of the plug) to the volume fraction, ϕ, of solids in the plug. For instance, according to [42]

$$\frac{A_c}{L} = \frac{A_c^{ap}}{L^{ap}} \exp(B\phi) \tag{35}$$

where B is an empirical constant that can be experimentally determined by measuring the electro-osmotic volume flow for different plug porosities. In eq. 35, L^{ap} and A_c^{ap} are the apparent (externally measured) length and cross-sectional area of the plug, respectively. An alternative expression was proposed in [43]:

$$\frac{A_c}{L} = \frac{A_c^{ap}}{L^{ap}} \phi_L^{-5/2} \tag{36}$$

where ϕ_L is the volume fraction of liquid in the plug (or void volume fraction). Other estimates of A_c/L can be found in [44–46].

For the case of a close packing of spheres, theoretical treatments are available involving the calculation of streaming current using cell models. No simple expressions can be given in this case; see [3, 47–52] for details.

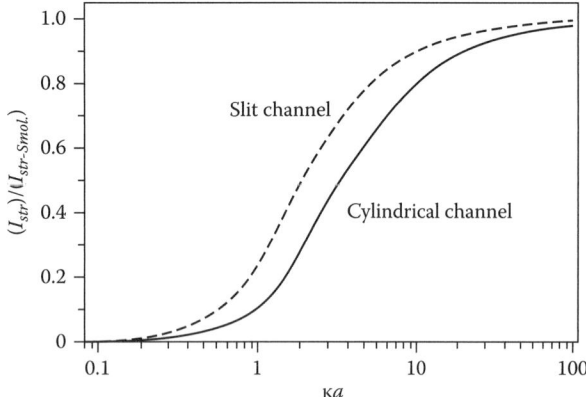

FIGURE 3: Streaming current divided by the applied pressure difference, eqs. 37 and 39, relative to the Smoluchowski formula, eq. 33, plotted as a function of the product κa (a: capillary radius, or slit half-width) for slit- and cylindrical-shaped capillaries.

b. If κa is intermediate ($\kappa a \approx 1$–10, say), the HS equation is not valid. For low ζ, curvature effects can be corrected by means of the Burgen and Nakache theory [4, 49]:

$$\frac{I_{str}}{\Delta p} = \frac{\varepsilon_{rs}\varepsilon_0\zeta}{\eta}\frac{A_c}{L}[1 - G(\kappa a)] \tag{37}$$

where

$$G(\kappa a) = \frac{\tanh(\kappa a)}{\kappa a} \tag{38}$$

for slit-shaped capillaries ($2a$ corresponds in this case to the separation of the parallel solid walls). In the case of cylindrical capillaries of radius a, the calculattion was first carried out by Rice and Whitehead [50]. They found that the function $G(\kappa a)$ in eq. 37 reads

$$G(\kappa a) = \frac{2I_1(\kappa a)}{\kappa a I_0(\kappa a)} \tag{39}$$

where I_0 and I_1 are the zeroth- and first-order modified Bessel functions of the first kind, respectively. Fig. 3 illustrates the importance of this curvature correction.

c. If the ζ-potential is not low and κa is small, no simple expression for I_{str} can be given, and only numerical procedures are available [52].

4.2.1.2 Streaming potential

The *streaming potential difference* (briefly, streaming potential) U_{str} can be measured between two electrodes, upstream and downstream in the liquid flow, connected via a high-input impedance voltmeter. The quantity of interest is, in this case, the ratio between the streaming potential and the pressure drop, $U_{str}/\Delta p$ (V Pa^{-1}). The conversion into ζ-potentials can be realized in a number of cases.

a. If $\kappa a \gg 1$ and surface conduction can be neglected, the HS formula can be used:

$$\frac{U_{str}}{\Delta p} = \frac{\varepsilon_{rs}\varepsilon_0\zeta}{\eta}\frac{1}{K_L} \tag{40}$$

b. The most frequent case (except for high ionic strengths, or high K_L) is that surface conductance, K^σ, is significant. Then the following equation should be used:

$$\frac{U_{str}}{\Delta p} = \frac{\varepsilon_{rs}\varepsilon_0\zeta}{\eta}\frac{1}{K_L(1+2Du)} \tag{41}$$

where Du is given by eqs. 12 and 20.

An empirical way of taking into account the existence of surface conductivity is to measure the resistance R_∞ of the plug or capillary in a highly concentrated electrolyte solution of conductivity K_L^∞. As for such a solution, Du is negligible, one can write

$$K_L^\infty R_\infty = \left(K_L + \frac{2K^\sigma}{a}\right)R_s \tag{42}$$

where R_s is the resistance of the plug in the solution under study, of conductivity K_L. Now, eq. 41 can be approximated by

$$\frac{U_{str}}{\Delta p} = \frac{\varepsilon_{rs}\varepsilon_0\zeta}{\eta}\frac{R_S}{K_L^\infty R_\infty} \tag{43}$$

c. If κa is intermediate ($\kappa a \sim 1...10$) and the ζ-potential is low, Rice and Whitehead's corrections are needed [50]. For a cylindrical capillary, the result is

$$\frac{U_{str}}{\Delta p} = \frac{\varepsilon_{rs}\varepsilon_0\zeta}{\eta}\frac{R_S}{K_L^\infty R_\infty}\frac{1 - \dfrac{2I_1(\kappa a)}{\kappa a I_0(\kappa a)}}{1 - \beta\left[1 - \dfrac{2I_1(\kappa a)}{\kappa a I_0(\kappa a)} - \dfrac{I_1^2(\kappa a)}{I_0^2(\kappa a)}\right]} \tag{44}$$

where

$$\beta = \frac{\left(\varepsilon_{rs}\varepsilon_0\kappa\zeta\right)^2}{\eta}\frac{R_S}{K_L^\infty R_\infty} \tag{45}$$

Figure 4 illustrates some results that can be obtained by using eq. 44.

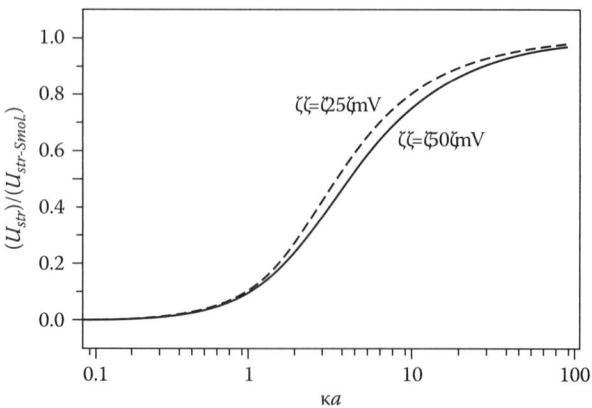

FIGURE 4: Streaming potential divided by the applied pressure difference, eq. 44, relative to its Smoluchowski value, eq. 40, as a function of the product κa (a: capillary radius), for the ζ-potentials indicated. Surface conductance is neglected.

d. As in the case of streaming current, for high ζ-potentials, only numerical methods are available (see, e.g., [53] for details).

In practice, instead of potential or current measurements for just one driving pressure, the streaming potential and streaming current are mostly measured at various pressure differences applied in both directions across the capillary system, and the slopes of the functions $U_{str} = U_{str}(\Delta p)$ and $I_{str} = I_{str}(\Delta p)$ are used to calculate the ζ-potential. This makes it possible to detect electrode asymmetry effects and correct for them. It is also advisable to verify that the Δp dependencies are linear and pass through the origin.

4.2.2 Samples that can be studied

Streaming potential/current measurements can be applied to study macroscopic interfaces of materials of different shape. Single capillaries made of flat sample surfaces (rectangular capillaries) and cylindrical capillaries can be used to produce microchannels for streaming potential/current measurements. Further, parallel capillaries and irregular capillary systems such as fiber bundles, membranes, and particle plugs can also be studied. Recall, however, the precautions already mentioned in connection with the interpretation of results in the case of plugs of particles. Other effects, including temperature gradients, Donnan potentials, or membrane potential can contribute to the observed streaming potential or electro-osmotic flow. An additional condition is the constancy of the capillary geometry during the course of the experiment. Reversibility of the signal upon variations in the sign and magnitude of Δp is a criterion for such constancy.

Most of the materials studied so far by streaming potential/current measurements, including synthetic polymers and inorganic non-metals, are insulating. Either bulk materials or thin films on top of carriers can be characterized. In addition, in some cases, semiconductors [54] and even bulk metals [55] have been studied, proving the general feasibility of the experiment.

Note that streaming potential/current measurements on samples of different geometries (flat plates, particle plugs, fiber bundles, cylindrical capillaries,...) each require their own set-up.

4.2.3 Sample preparation

The samples to be studied by streaming potential/current measurements have to be mechanically and chemically stable in the aqueous solutions used for the experiment. First, the geometry of the plug must be consolidated in the measuring cell. This can be checked by rinsing with the equilibrium liquid through repeatedly applying Δp in both directions until finding a constant signal. Another issue to consider is the necessity that the solid has reached chemical equilibrium with the permeating liquid; this may require making the plug from a suspension of the correct composition, followed by rinsing. Checking that the experimental signal does not change during the course of measurement may be a good practice. The presence or formation of air bubbles in the capillary system has to be avoided.

4.2.3.1 Standard samples

No standard samples have been developed specifically so far for streaming potential/current measurements, although several materials have been frequently analyzed and may, therefore, serve as potential reference samples [56, 57].

4.2.3.2 Range of electrolyte concentrations

From the operational standpoint, there is no lower limit to the ionic strength of the systems to be investigated by these methods, although in the case of narrow channels, very low ionic strengths

require effective electrical insulation of the set-up in order to prevent short-circuiting. However, such low ionic strength values can only be attained if the solid sample is extremely pure and insoluble. The upper value of electrolyte concentration depends on the sensitivity of the electrometer and on the applied pressure difference; usually, solutions above 10^{-1} mol/L of 1-1 charge-type electrolyte are difficult to measure by the present techniques.

4.3 Electro-osmosis
4.3.1 Operational definitions; recommended symbols and terminology; conversion of the measured quantities into ζ-potential

In electro-osmosis, a flow of liquid is produced when an electric field E is applied to a charged capillary or porous plug immersed in an electrolyte solution. If $\kappa a \gg 1$ everywhere at the solid/liquid interface, far from that interface the liquid will attain a constant (i.e., independent of the position in the channel) velocity (the electro-osmotic velocity) v_{eo}, given by eq. 4. If such a velocity cannot be measured, the convenient physical quantity becomes the *electro-osmotic flow rate*, Q_{eo} (m³ s⁻¹), given by

$$Q_{eo} = \iint_{A_c} v_{eo} \cdot \mathbf{dS} \tag{46}$$

where dS is the elementary surface vector at the location in the channel where the fluid velocity is v_{eo}. The counterparts of Q_{eo} are $Q_{eo,E}$ (flow rate divided by electric field) and $Q_{eo,I}$ (flow rate divided by current). These are the quantities that can be related to the ζ-potential. As before, several cases can be distinguished:

a. If $\kappa a \gg 1$ and there is no surface conduction:

$$Q_{eo,E} \equiv \frac{Q_{eo}}{E} = -\frac{\varepsilon_{rs}\varepsilon_0 \zeta}{\eta} A_c$$
$$Q_{eo,I} \equiv \frac{Q_{eo}}{I} = -\frac{\varepsilon_{rs}\varepsilon_0 \zeta}{\eta} \frac{1}{K_L} \tag{47}$$

b. With surface conduction, the expression for $Q_{eo,E}$ is as in eq. 47, and that for $Q_{eo,I}$ is

$$Q_{eo,I} = -\frac{\varepsilon_{rs}\varepsilon_0 \zeta}{\eta} \frac{1}{K_L(1+2Du)} \tag{48}$$

In eq. 48, the empirical approach for the estimation of Du can be followed:

$$Q_{eo,I} = -\frac{\varepsilon_{rs}\varepsilon_0 \zeta}{\eta} \frac{R_S}{K_L^\infty R_\infty} \tag{49}$$

c. Low ζ-potential, finite surface conduction, and arbitrary capillary radius [46]:

$$Q_{eo,E} = -\frac{\varepsilon_{rs}\varepsilon_0 \zeta}{\eta} A_c[1-G(\kappa a)]$$
$$Q_{eo,I} = -\frac{\varepsilon_{rs}\varepsilon_0 \zeta}{\eta} \frac{[1-G(\kappa a)]}{K_L(1+2Du)} \tag{50}$$
$$\cong -\frac{\varepsilon_{rs}\varepsilon_0 \zeta}{\eta} \frac{R_S[1-G(\kappa a)]}{K_L^\infty R_\infty}$$

where the function $G(\kappa a)$ is given by eq. 38.

d. When ζ is high and the condition $\kappa a \gg 1$ is not fulfilled, no simple expression can be given for Q_{eo}.

As in the case of streaming potential and current, the procedures described can be also applied to either plugs or membranes. If the electric field E is the independent variable (see eq. 47), then A_c must be estimated. In that situation, the recommendations suggested in Section 4.2.1 can be used, since eq. 47 can be written as

$$\frac{Q_{eo,E}}{\Delta V_{ext}} = -\frac{\varepsilon_{rs}\varepsilon_0 \zeta}{\eta} \frac{A_c}{L} \qquad (51)$$

where ΔV_{ext} is the applied potential difference.

4.3.2 Samples that can be studied
The same samples as with streaming current/potential, see Section 4.2.2.

4.3.3 Sample preparation and standard samples
See Section 4.2.3 referring to streaming potential/current determination.

4.4 Experimental determination of surface conductivity
Surface conductivities are excess quantities and cannot be directly measured. There are, in principle, three methods to estimate them.

i. In the case of plugs, measure the plug conductivity K_{plug} as a function of K_L. The latter can be changed by adjusting the electrolyte concentration. The plot of K_{plug} vs. K_L has a large linear range which can be extrapolated to $K_L = 0$ where the intercept represents K^σ. This method requires a plug and seems relatively straightforward.

ii. For capillaries, deduce K^σ from the radius dependence of the streaming potential, using eq. 41 and the definition of Du (eq. 12). This method is rather direct, but requires a range of capillaries with different radii, but identical surface properties [58, 59].

iii. Utilize the observation that, when surface conductivity is not properly accounted for, different electrokinetic techniques may give different values for the ζ-potential of the same material under the same solution conditions. Correct the theories by inclusion of the appropriate surface conductivity, and find in this way the value of K^σ that harmonizes the ζ-potential. This method requires insight into the theoretical backgrounds [60, 61], and it works best if the two electrokinetic techniques have a rather different sensitivity for surface conduction (such as electrophoresis and LFDD).

In many cases, it is found that the surface conductivity obtained in one of these ways exceeds $K^{\sigma d}$, sometimes by orders of magnitude. This means that $K^{\sigma i}$ is substantial. The procedure for obtaining $K^{\sigma i}$ consists of subtracting $K^{\sigma d}$ from K^σ. For $K^{\sigma d}$, Bikerman's equation (eq. 17) can be used. The method is not direct because this evaluation requires the ζ-potential, which is one of the unknowns; hence, iteration is required.

4.5 Dielectric dispersion
4.5.1 Operational definitions; recommended symbols and terminology; conversion of the measured quantities into ζ-potential
The phenomenon of dielectric dispersion in colloidal suspensions involves the study of the dependence on the frequency of the applied electric field of the electric permittivity and/or the electric conductivity of disperse systems. When dealing with such heterogeneous systems as colloidal dispersions, these quantities are defined as the electric permittivity and conductivity of a sample of homogeneous material, that when placed between the electrodes of the measuring cell, would have the same resistance and capacitance as the suspension. The dielectric investigation of dispersed systems involves determinations of their *complex permittivity, $\varepsilon^*(\omega)$* (F m^{-1}) and *complex conductivity $K^*(\omega)$* (S m^{-1}) as a function of the frequency ω (rad s^{-1}) of the applied ac field. These quantities are related to the volume, surface, and geometrical characteristics of the dispersed particles, the nature of the dispersion medium, and also to the concentration of particles, expressed either in terms of volume fraction, ϕ (dimensionless) or number concentration N (m^{-3}).

It is common to use the relative permittivity $\varepsilon_r^*(\omega)$ (dimensionless), instead of the permittivity

$$\varepsilon^*(\omega) = \varepsilon_r^*(\omega)\varepsilon_0 \qquad (52)$$

ε_0 being the permittivity of vacuum. $K^*(\omega)$ and $\varepsilon^*(\omega)$ are not independent quantities:

$$K^*(\omega) = K_{DC} - i\omega\varepsilon^*(\omega) = K_{DC} - i\omega\varepsilon_0\varepsilon_r^*(\omega) \qquad (53)$$

or, equivalently,

$$\begin{aligned} \text{Re}[K(\omega)] &= K_{DC} + \omega\varepsilon_0 \text{Im}[\varepsilon_r^*(\omega)] \\ \text{Im}[K^*(\omega)] &= -\omega\varepsilon_0 \text{Re}[\varepsilon_r^*(\omega)] \end{aligned} \qquad (54)$$

where K_{DC} is the direct-current (zero frequency) conductivity of the system.

The complex conductivity K^* of the suspension can be expressed as

$$K^*(\omega) = K_L + \delta K^*(\omega) \qquad (55)$$

where $\delta K^*(\omega)$ is usually called *conductivity increment* of the suspension. Similarly, the complex dielectric constant of the suspension can be written in terms of a *relative permittivity increment* or, briefly, *dielectric increment $\delta\varepsilon_r^*(\omega)$*:

$$\varepsilon_r^*(\omega) = \varepsilon_{rs} + \delta\varepsilon_r^*(\omega) \qquad (56)$$

As in homogeneous materials, the electric permittivity is the macroscopic manifestation of the electrical polarizability of the suspension components. Mostly, more than one relaxation frequency is observed, each associated with one of the various mechanisms contributing to the system's polarization. Hence, the investigation of the frequency dependence of the electric permittivity or conductivity allows us to obtain information about the characteristics of the disperse system that are responsible for the polarization of the particles.

The frequency range over which the dielectric dispersion of suspensions in electrolyte solutions is usually measured extends between 0.1 kHz and several hundred MHz. In order to define in this frame the *low-frequency* and *high-frequency* ranges, it is convenient to introduce an important concept dealing with a point in the frequency scale. This frequency corresponds to the *reciprocal of the Maxwell–Wagner–O'Konski relaxation time τ_{MWO}*

$$\omega_{MWO} \equiv \frac{1}{\tau_{MWO}} \equiv \frac{(1-\phi)K_p + (2+\phi)K_L}{\varepsilon_0\left[(1-\phi)\varepsilon_{rp} + (2+\phi)\varepsilon_{rs}\right]\varepsilon_{rs}} \qquad (57)$$

and it is called the Maxwell–Wagner–O'Konski relaxation frequency. In eq. 57, ε_{rp} is the relative permittivity of the dispersed particles. For low volume fractions and low permittivity of the particles ($\varepsilon_{rp} \ll \varepsilon_{rs}$), this expression reduces to

$$\tau_{MWO} \approx \frac{\varepsilon_{rs}\varepsilon_0}{K_L} \approx \frac{1}{2D\kappa^2} \quad (58)$$

where D is the mean diffusion coefficient of ions in solution. The last term in eq. 58 suggests that τ_{MWO} roughly corresponds in this case to the time needed for ions to diffuse a distance of the order of one Debye length. In fact, in such conditions τ_{MWO} equals τ_{el}, the so-called *relaxation time of the electrolyte solution*. It is a measure of the time required for charges in the electrolyte solution to recover their equilibrium distribution after ceasing an external perturbation. Its role in the time domain is similar to the role of κ^{-1} in assessing the double-layer thickness.

The low-frequency range can be defined by the inequality

$$\omega < \omega_{MWO} \quad (59)$$

For these frequencies, the characteristic value of the *conduction current density* in the electrolyte solution significantly exceeds the *displacement current density*, and the spatial distribution of the local electric fields in the disperse system is mainly determined by the distribution of ionic currents. The frequency dependence shown by the permittivity of colloidal systems in this frequency range is known as low-frequency dielectric dispersion or LFDD.

In the high-frequency range, determined by the inequality

$$\omega > \omega_{MWO} \quad (60)$$

the characteristic value of the displacement current density exceeds that of conduction currents, and the space distribution of the local electric fields is determined by polarization of the molecular dipoles, rather than by the distribution of ions.

4.6 Dielectric dispersion and ζ-potential: Models

a. Middle-frequency range: Maxwell–Wagner–O'Konski relaxation

There are various mechanisms for the polarization of a heterogeneous material, each of which is always associated with some property that differs between the solid, the liquid, and their interface. The most widely known mechanism of dielectric dispersion, the *Maxwell–Wagner dispersion*, occurs when the two contacting phases have different conductivities and electric permittivities. If the ratio ε_{rp}/K_p is different from that of the dispersion medium, i.e., if

$$\frac{\varepsilon_{rp}}{K_p} \neq \frac{\varepsilon_{rs}}{K_L} \quad (61)$$

the conditions of continuity of the normal components of the current density and the electrostatic induction on both sides of the surface are inconsistent with each other. This results in the formation of free ionic charges near the surface. The finite time needed for the formation of such a free charge is in fact responsible for the Maxwell–Wagner dielectric dispersion.

In the Maxwell–Wagner model, no specific properties are assumed for the surface, which is simply considered as a geometrical boundary between homogeneous phases. The Maxwell–Wagner model was generalized by O'Konski [61], who first took the surface conductivity K^σ explicitly into account. In his treatment, the conductivity of the particle is modified to include the contributions of both the solid material and the excess surface conductivity. This effective conductivity will be called K_{pef}:

$$K_{pef} = K_p + \frac{2K^\sigma}{a} \quad (62)$$

Both the conductivity and the dielectric constant can be considered as parts of a complex electric permittivity of any of the system's components. Thus, for the dispersion medium

$$\varepsilon_{rs}^* = \varepsilon_{rs} - i\frac{K_L}{\omega\varepsilon_0} \quad (63)$$

and for the particle

$$\varepsilon_{rp}^* = \varepsilon_{rp} - i\frac{K_{pef}}{\omega\varepsilon_0} \quad (64)$$

In terms of these quantities, the Maxwell–Wagner–O'Konski theory gives the following expression for the complex dielectric constant of the suspension:

$$\varepsilon_r^* = \varepsilon_{rs}^* \frac{\varepsilon_{rp}^* + 2\varepsilon_{rs}^* + 2\phi(\varepsilon_{rp}^* - \varepsilon_{rs}^*)}{\varepsilon_{rp}^* + 2\varepsilon_{rs}^* - \phi(\varepsilon_{rp}^* - \varepsilon_{rs}^*)} \quad (65)$$

b. Low-frequency range: dilute suspensions of nonconducting spherical particles with $\kappa a \gg 1$, and negligible $K^{\sigma i}$

At moderate or high ζ-potentials, mobile counterions are more abundant than coions in the EDL. Therefore, the contribution of the counterions and the coions to surface currents in the EDL differs from their contribution to currents in the bulk solution. Such difference gives rise to the existence of a field-induced perturbation of the electrolyte concentration, $\delta c(r)$, in the vicinity of the polarized particle. The ionic diffusion caused by $\delta c(r)$ provokes a low-frequency dependence of the particle's dipole coefficient, C_0^* (see below). This is the origin of the LFDD (α-dispersion) displayed by colloidal suspensions. Recall that the dipole coefficient relates the dipole moment d^* to the applied field E. For the case of a spherical particle of radius a, the dipole coefficient is defined through the relation

$$d^* = 4\pi\varepsilon_{rs}\varepsilon_0 a^3 C_0^* E \quad (66)$$

The calculation of this quantity proves to be essential for evaluation of the dielectric dispersion of the suspension [63–65]. A model for the calculation of the low-frequency conductivity increment $\delta K^*(\omega)$ and relative permittivity increment $\delta\varepsilon_r^*(\omega)$ from the dipole coefficient C_0^* when $\kappa a \gg 1$ is described in Appendix I. There it is shown that, in the absence of SLC, the only parameter of the solid/liquid interface that is needed to account for LFDD is the ζ-potential.

The overall behavior is illustrated in Fig. 5 for a dilute dispersion of spherical nonconducting particles ($a = 100$ nm, $\varepsilon_{rp} = 2$) in a 10^{-3} mol/L KCl solution ($\varepsilon_{rs} = 78.5$), and with negligible ionic conduction in the stagnant layer. In this figure, we plot the variation of the real and imaginary parts

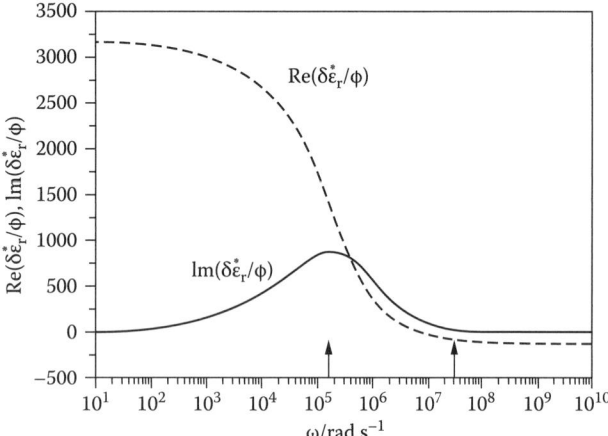

FIGURE 5:   Real and imaginary parts of the dielectric increment $\delta\varepsilon_r^*$ (divided by the volume fraction ϕ) for dilute suspensions of 100-nm particles in KCl solution, as a function of the frequency of the applied field for $\kappa a = 10$ and $\zeta = 100$ mV. The arrows indicate the approximate location of the α (low frequency) and Maxwell–Wagner–O' Konski relaxations.

of $\delta\varepsilon_r^*$ with frequency. Note the very significant effect of the double-layer polarization on the low-frequency dielectric constant of a suspension. The variation with frequency is also noticeable, and both the α- (around 10^5 s^{-1}) and Maxwell–Wagner ($\sim 2 \times 10^7$ s^{-1}) relaxations are observed, the amplitude of the latter being much smaller than that of the former for the conditions chosen. No other electrokinetic technique can provide such a clear account of the double-layer relaxation processes. The effect of ζ-potential on the frequency dependence of the dielectric constant is plotted in Fig. 6: the dielectric increment always increases with ζ, as a consequence of the larger concentration of counterions in the EDL: all processes responsible for LFDD are amplified for this reason.

The procedure for obtaining ζ from LFDD measurements is somewhat involved. It is advisable to determine experimentally the dielectric constant (or, equivalently, the conductivity) of the suspension over a wide frequency

range, and use eqs. I.2 and I.3 (see Appendix I) to estimate the LFDD curve that best fits the data. A simpler routine is to measure only the low-frequency values, and deduce ζ using the same equations, but substituting $\omega = 0$. However, the main experimental problems occur at low frequencies (see Section 4.5.3).

c. Dilute suspensions of nonconducting spherical particles with arbitrary κa, and negligible $K^{\sigma i}$

In this situation, there are no analytical expressions relating LFDD measurements to ζ. Instead, numerical calculations based on DeLacey and White's treatment [66] are recommended. As before, the computing routine should be constructed to perform the calculations a number of times with different ζ-potentials as inputs, until agreement between theory and experiment is obtained over a wide frequency range (or, at least, at low frequencies).

d. Dilute suspensions of nonconducting spherical particles with $\kappa a \gg 1$ and SLC

The problem of generalizing the theory by taking into account surface conduction caused by ions situated in the hydrodynamically stagnant layer has been dealt with in [60, 61, 64]. In theoretical treatments, SLC is equated to conduction within the Stern layer.

According to these models, the dielectric dispersion is determined by both ζ and $K^{\sigma i}$. This means that, as discussed before, additional information on $K^{\sigma i}$ (see the methods described in Section 4.4) must accompany the dielectric dispersion measurements. Using dielectric dispersion data alone can only yield information about the total surface conductivity [66].

e. Dilute suspensions of nonconducting spherical particles with arbitrary κa and SLC.

Only numerical methods are available if this is the physical nature of the system under study. The reader is referred to [13–16, 61, 62, 67–69]. Figure 7 illustrates how important the effect of SLC onRe $[\delta\varepsilon_r^*(\omega)]$ can be for the same conditions as in Fig. 5. Roughly, the possibility of increased ionic mobilities in the stagnant layer brings about a systematically larger dielectric increment of the suspension: surface currents are larger for a conducting stagnant layer, and hence the electrolyte concentration gradients, ultimately responsible for the dielectric dispersion, will also be increased.

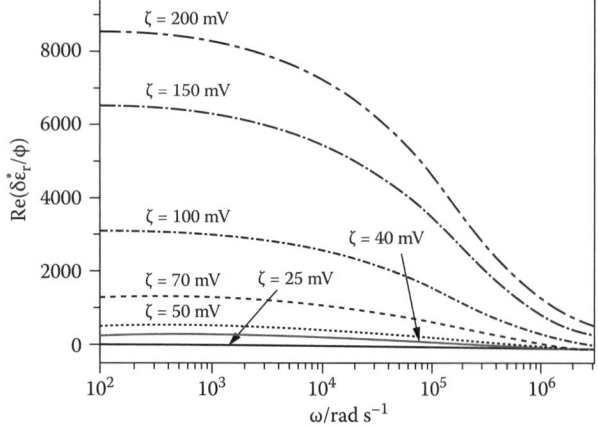

FIGURE 6:   Real part of the dielectric increment $\delta\varepsilon_r^*$ (per volume fraction) as a function of the frequency of the applied field for dilute suspensions of 100-nm particles in KCl solution. The ζ-potentials are indicated, and in all cases, $\kappa a = 10$.

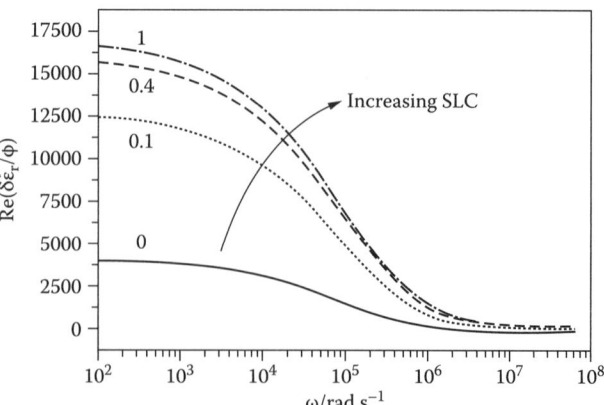

FIGURE 7:   Real part of the dielectric increment (per volume fraction) of dilute suspensions as in Fig. 5. The curves correspond to increasing importance of SLC; the ratios between the diffusion coefficients of counterions in the stagnant layer and in the bulk electrolyte are indicated (the lower curve corresponds to zero SLC).

f. Nondilute suspensions of nonconducting particles with large κa and negligible $K^{\sigma i}$

Considering that many suspensions of technological interest are rather concentrated, the possibilities of LFDD in the characterization of moderately concentrated suspensions have also received attention. This requires establishing a theoretical basis relating the dielectric or conductivity increments of such systems to the concentration of particles [70–72]. Here, we focus on a simplified model [73] that allows the calculation of the volume-fraction dependence of both the low-frequency value of the real part of $\delta\varepsilon_r^*$, and of the characteristic time τ_α of the α-relaxation. The starting point is the assumption that L_D, the length scale over which ionic diffusion takes place around the particle, can be averaged in the following way between the values of a very dilute ($L_D \approx a$) and a very concentrated ($L_D \approx b - a$; b is half the average distance between the centers of neighboring particles) dispersion:

$$L_D = \left(\frac{1}{a^2} + \frac{1}{(b-a)^2}\right)^{-1/2} \quad (67)$$

or, in terms of the particle volume fraction

$$L_D = a\left(1 + \frac{1}{(\phi^{-1/3}-1)^2}\right)^{-1/2} \quad (68)$$

From these expressions, the simplified model allows us to obtain the dielectric increment at low frequency as follows. Let us call

$$\Delta\varepsilon(0) \equiv \frac{\text{Re}[\delta\varepsilon_r^*(0)]}{\phi} \quad (69)$$

the *specific* (i.e., per unit volume fraction) *dielectric increment* (for $\omega \to 0$). In the case of dilute suspensions, this quantity is a constant (independent of ϕ), that we denote $\Delta\varepsilon_d(0)$:

$$\Delta\varepsilon(0) \equiv \frac{\text{Re}[\delta\varepsilon_r^*(0)]}{\phi}\bigg|_{\phi\to0} \quad (70)$$

The model allows us to relate eq. 69 with eq. 70 through the volume-fraction dependence of L_D:

$$\Delta\varepsilon(0) = \Delta\varepsilon_d(0)\left(1 + \frac{1}{(\phi^{-1/3}-1)^2}\right)^{-3/2} \quad (71)$$

A similar relationship can be established between dilute and concentrated suspensions in the case of the relaxation frequency $\omega_\alpha = 1/\tau_\alpha$:

$$\omega_\alpha = \omega_{\alpha d}\left(1 + \frac{1}{(\phi^{-1/3}-1)^2}\right) \quad (72)$$

where $\omega_{\alpha d}$ is the reciprocal of the relaxation time for a dilute suspension. Using this model, the dielectric increment and characteristic frequency of a concentrated suspension can be related to those corresponding to the dilute case which, in turn, can be related, as discussed above, to the ζ-potential and other double-layer parameters. A general treatment of the problem, valid for arbitrary values of κa and ζ can be found in [74].

Summing up, we can say that the dielectric dispersion of suspensions is an interesting physical phenomenon, extremely sensitive to the characteristics of the particles, the solution, and their interface. It can provide invaluable information on the dynamics of the EDL and the processes through which it is altered by the application of an external field. Because of the experimental difficulties involved in its determination, it is unlikely that dielectric dispersion measurements alone can be useful as a tool to obtain the ζ-potential of the dispersed particles.

4.6.1 Experimental techniques available

One of the most usual techniques for measuring the dielectric permittivity and/or the conductivity of suspensions as a function of the frequency of the applied field, is based on the use of a conductivity cell connected to an impedance analyzer. This technique has been widely employed since it was first proposed by Fricke and Curtis [75]. In most modern set-ups, the distance between electrodes can be changed (see, e.g., [69, 76–79]). The need for variable electrode separation stems from the problem of electrode polarization at low frequencies, since at sufficiently low frequencies the electrode impedance dominates over that of the sample. The method makes use of the assumption that electrode polarization does not depend on their distance. A so-called *quadrupole method* has been recently introduced [80] in which the correction for electrode polarization is optimally carried out by proper calibration. Furthermore, the method based on the evaluation of the logarithmic derivative of the imaginary part of raw $\varepsilon^*(\omega)$ data also seems to be promising [81].

These are not, however, the only possible procedures. A four-electrode method has also been employed with success [60, 61, 68] in this case, since the sensing and current-supplying electrodes are different, polarization is not the main problem, but the electronics of the experimental set-up is rather complicated.

4.6.2 Samples for LFDD measurements

There are no particular restrictions to the kind of colloidal particles that can be studied with the LFDD technique. The obvious precautions involve avoiding sedimentation of the particles during measurement, and control of the stability of the suspensions. LFDD quantities are most sensitive to particle size, particle concentration, and temperature. Hence, the constancy of the latter is essential. Another important concern deals with the effect of electrode polarization. Particularly at low frequencies, electrode polarization can be substantial and completely invalidate the data. This fact imposes severe limitations on the electrolyte concentrations that can be studied; it is very hard to obtain data for ionic strengths in excess of 1 to 5 mmol L^{-1}.

4.7 Electroacoustics
4.7.1 Operational definitions; recommended symbols and terminology; experimentally available quantities
Terminology

The term "electroacoustics" refers to two kinds of closely related phenomena:

- *Colloid vibration current* (I_{CV}) and *colloid vibration potential* (U_{CV}) are two phenomena in which a sound wave is passed through a colloidal dispersion and, as a result, electrical currents and fields arise in the suspension. When the wave travels through a dispersion of particles whose

density differs from that of the surrounding medium, inertial forces induced by the vibration of the suspension give
rise to a motion of the charged particles relative to the liquid, causing an alternating electromotive force. The manifestations of this electromotive force may be measured in
a way depending on the relation between the impedance of
the suspension and the properties of the measuring instrument, either as I_{CV} (for small impedance of the meter) or as
U_{CV} (for large one).

- The reciprocal effect of the above two phenomena is the
electrokinetic sonic amplitude (ESA), in which an alternating electric field is applied to a suspension and a sound
wave arises as a result of the motion of the particles caused
by their ac electrophoresis.

Colloid vibration potential/current may be considered as the
ac analog of sedimentation potential/current. Similarly, ESA
may be considered as the ac analog of classical electrophoresis. The relationships between electroacoustics and classical
EKP may be used for testing modern electroacoustic theories,
which should at least provide the correct limiting transitions
to the well-known and well-established results of the classical
electrokinetic theory. A very important advantage of the electroacoustic techniques is the possibility they offer to be applied
to concentrated dispersions.

4.7.1.1 Experimentally available quantities
4.7.1.1.1 Colloid vibration potential (U_{CV}) If a standing
sound wave is established in a suspension, a voltage difference
can be measured between two different points in the standing
wave. If measured at zero current flow, it is referred to as *colloid
vibration* potential. The measured voltage is due to the motion of
the particles: it alternates at the frequency of the sound wave and
is proportional to another measured value, ΔP, which is the pressure difference between the two points of the wave. The kinetic
coefficient, equal to the ratio

$$\frac{U_{CV}}{\Delta p} = \frac{U_{CV}}{\Delta p}\left(\omega, \phi, \Delta\rho/\rho, K^{\sigma i}, \zeta, \eta, a\right) \quad (73)$$

characterizes the suspension. In eq. 73, $\Delta\rho$ is the difference in density between the particles and the suspending fluid, of density ρ.

4.7.1.1.2 Colloid vibration current (I_{CV}) If the measurements of the electric signal caused by the sound wave in the
suspension, are carried out under zero U_{CV} conditions (short-circuited), an ac I_{CV} can be measured. Its value, measured between
two different points in the standing wave, is also proportional to
the pressure difference between those two points, and the kinetic
coefficient $I_{CV}/\Delta p$ characterizes the suspension and is closely
related to $U_{CV}/\Delta p$:

$$\frac{I_{CV}}{\Delta p} = \frac{I_{CV}}{\Delta p}\left(\omega, \phi, \Delta\rho/\rho, K^{\sigma i}, \zeta, \eta, a\right) = K^* \frac{U_{CV}}{\Delta p} \quad (74)$$

4.7.1.1.3 Electrokinetic sonic amplitude (ESA) This
refers to the measurement of the sound wave amplitude, which is
caused by the application of an alternating electric field to a suspension of particles of which the density is different from that of
the suspending medium. The ESA signal (i.e., the amplitude A_{ESA}
of the sound pressure wave generated by the applied electric field)
is proportional to the field strength E, and the kinetic coefficient

A_{ESA}/E can be expressed as a function of the characteristics of the
suspension

$$\frac{A_{ESA}}{E} = \frac{A_{ESA}}{E}\left(\omega, \phi, \Delta\rho/\rho, K^{\sigma i}, \zeta\right) \quad (75)$$

Measurement of I_{CV} or A_{ESA} rather than U_{CV} has the operational
advantage that it enables measurement of the kinetic characteristics, $I_{CV}/\Delta p$ and A_{ESA}/E, which are independent of the complex
conductivity K^* of the suspension, and thus knowledge of K^* is
not a prerequisite for the extraction of the ζ-potential from the
interpretation of the electroacoustic measurements. Note, however, that if SLC is significant, as with the other EKP, additional
measurements will be needed, as both ζ and $K^{\sigma i}$ are required to
fully characterize the interface.

4.7.2 Estimation of the ζ-potential from U_{CV}, I_{CV}, or A_{ESA}
There are two recent methods for the theoretical interpretation of the data of electroacoustic measurements and extracting
from them a value for the ζ-potential. One is based on the symmetry relation proposed in [82, 83] to express both kinds of electroacoustic phenomena (colloid vibration potential/current and
ESA) in terms of the same quantity, namely the *dynamic electrophoretic mobility, u_d^**, which is the complex, frequency-dependent
analog of the normal direct current (dc) electrophoretic mobility.
The second method is based on the direct evaluation of I_{CV} without using the symmetry relations, and hence it is not necessarily
based on the concept of dynamic electrophoretic mobility. Both
methods for ζ-potential determination from electroacoustic measurements are briefly described below.

Using the dynamic mobility method has some advantages:
(i) the zero frequency limiting value of u_d^* is the normal electrophoretic mobility, and (ii) the frequency dependence of u_d^*
can be used to estimate not only the ζ-potential, but also (for
particle radius > –40 nm) the particle size distribution. Since
the calculation of the ζ-potential in the general case requires a
knowledge of κa it is helpful to have available the most appropriate estimate of the average particle size (most colloidal dispersions are polydisperse, and there are many possible "average"
sizes which might be chosen). Although the calculation of the
ζ-potential from the experimental measurements would be a
rather laborious procedure, the necessary software for effecting
the conversion is provided as an integral part of the available
measuring systems, for both dilute and moderately concentrated sols. The effects of SLC can also be eliminated in some
cases, without access to alternative measuring devices, simply
by undertaking a titration with an indifferent electrolyte.

In the case of methods based on the direct evaluation of I_{CV},
the use of different frequencies, if available, or of acoustic attenuation measurements, allows the determination of particle size
distributions.

4.8 Method based on the concept of dynamic electrophoretic mobility
The symmetry relations lead to the following expressions,
relating the different electroacoustic phenomena to the dynamic
electrophoretic mobility, u_d^* [82, 83]:

$$\frac{U_{CV}}{\Delta p} \propto \phi \frac{\Delta\rho}{\rho} \frac{u_d^*}{K^*} \quad (76)$$

$$\frac{I_{CV}}{\Delta p} \propto \phi \frac{\Delta \rho}{\rho} u_d^* \qquad (77)$$

$$\frac{A_{ESA}}{E} \propto \phi \frac{\Delta \rho}{\rho} u_d^* \qquad (78)$$

Although the kinetic coefficients on the right-hand side of these relations (both magnitude and phase) can readily be measured at any particle concentration, there is some difficulty (see below) in the conversion of u_d^* to a ζ-potential, except for the simplest case of spherical particles in fairly dilute suspensions (up to a few vol %). In this respect, the situation is similar to that for the more conventional electrophoretic procedures. There are, though, some offsetting advantages. In particular, the ability to operate on concentrated systems obviates the problems of contamination which beset some other procedures. It also makes possible meaningful measurements on real systems without the need for extensive dilution, which can compromise the estimation of ζ-potential, especially in emulsions systems.

4.9 Dilute systems (up to ~4 vol %)

1. For a dilute suspension of spherical particles with $\kappa a \gg 1$ and arbitrary ζ-potential, the following equation can be used [84], which relates the dynamic mobility with the ζ-potential and other particle properties:

$$u_d^* = \left(\frac{2\varepsilon_{rs}\varepsilon_0 \zeta}{3\eta} \right)(1+f)G(\alpha) \qquad (79)$$

The restriction concerning double-layer thickness requires in practice that $\kappa a > ~20$ for reliable results, although the error is usually tolerable down to about $\kappa a = 13$.

The function f is a measure of the tangential electric field around the particle surface. It is a complex quantity, given by

$$f = \frac{1+i\omega' -[2Du+i\omega'(\varepsilon_{rp}/\varepsilon_{rs})]}{2(1+i\omega')+[2Du+i\omega'(\varepsilon_{rp}/\varepsilon_{rs})]} \qquad (80)$$

where $\omega' \equiv \omega\varepsilon_{rs}\varepsilon 0/K_L$ is the ratio of the measurement frequency, ω, to the Maxwell–Wagner relaxation frequency of the electrolyte. If it can be assumed that the tangential current around the particle is carried essentially by ions outside the shear surface, then Du is given by the Du^d (eqs. 16–19); see also [85]*.

The function $G(\alpha)$ is also complex and given by

$$G(\alpha) = \frac{1+(1+i)\sqrt{\alpha/2}}{1+(1+i)\sqrt{\alpha/2}+i\frac{\alpha}{9}(3+2\Delta\rho/\rho)} \qquad (81)$$

It is a direct measure of the inertia effect. The dimensionless parameter α is defined as

$$\alpha = \frac{\omega a^2 \rho}{\eta} \qquad (82)$$

so G is strongly dependent on the particle size. G varies monotonically from a value of unity, with zero phase angle, when a is small (less than 0.1 μm typically) to a minimum of zero and a phase lag of $\pi/4$ when a is large (say, $a > 10$ μm).

Equations 80–82 are, as mentioned above, applicable to systems of arbitrary ζ-potential, even when conduction occurs in the stagnant layer, in which case Du in eq. 80 must be properly evaluated. They have been amply confirmed by measurements on model suspensions [86], and have proved to be of particular value in the estimation of high ζ-potentials [87]. The maximum that appears in the plot of dc mobility against ζ-potential ([22], see also Fig. 2) gives rise to an ambiguity in the assignment of the ζ-potential, which does not occur when the full dynamic mobility spectrum is available.

2. For double layers with $\kappa a < ~20$, there are approximate analytical expressions [88, 89] for dilute suspensions of spheres, but they are valid only for low ζ-potentials. They have been checked against the numerical solutions for general κa and ζ [90, 91], and those numerical calculations have been subjected to an experimental check in [92].

3. The effect of particle shape has been studied in [93,94], again in the limit of dilute systems with thin double layers. This analysis has been extended in [95] to derive formulae for cylindrical particles with zero permittivity and low ζ-potentials, but for arbitrary κa. The results are consistent with those of [93, 94].

4.10 Concentrated systems

The problem of considering the effect on the electroacoustic signal of hydrodynamic or electrostatic interactions between particles in concentrated suspensions was first theoretically tackled [96] by using the cell model of Levine and Neale [97, 98] to provide a solution that was claimed to be valid for U_{CV} measurements on concentrated systems (see [87] for a discussion on the validity of such approach in the high-frequency range).

It is possible to deal with concentrated systems at high frequency without using cell models in the case of near neutrally buoyant systems (where the relative density compared to water is in the range 0.9–1.5) using a procedure developed by O'Brien [99, 100]. In that case, the interparticle interactions can be treated as pairwise additive and only nearest-neighbor interactions need to be taken into account. An alternative approach is to estimate the effects of particle concentration considering in detail the behavior of a pair of particles in all possible orientations [101, 102]. Empirical relations have been developed that appear to represent the interaction effects for more concentrated suspensions up to volume fractions of 30% at least. In [103], an example can be found where the dynamic mobility was analyzed assuming the system to be dilute. The resulting value of the ζ-potential (ζ_{app}) was then corrected for concentration using the semi-empirical relation:

$$\zeta_{corr} = \zeta_{app} \exp\{2\phi[1+s(\phi)]\}, \quad \text{with } s(\phi) = \frac{1}{1+(0.1/\phi)^4} \qquad (83)$$

Finally, O'Brien et al. [104] have recently developed an effective analytical solution to the concentration problem for the dynamic mobility.

4.11 Method based on the direct evaluation of the I_{CV}

This approach [105] applies a "coupled phase model" [106–109] for describing the speed of the particle relative to the liquid. The

---

* If this assumption breaks down, Du must be estimated by reference to a suitable surface conductance model or, preferably, by direct measurement of the conductivity over the frequency range involved in the measurement (normally from about 1 to 40 MHz) [86]. Another procedure involved the analysis of the results of a salt titration (see previous subsection).

Kuwabara cell model [110] yields the required hydrodynamic parameters, such as the drag coefficient, whereas the Shilov–Zharkikh cell model [111] was used for the generalization of the Kuwabara cell model to the electrokinetic part of the problem.

The method allows the study of polydisperse systems without using any superposition assumption. It is important in concentrated systems, where superposition does not work because of the interactions between particles.

An independent exact expression for I_{CV} in the quasi-stationary case of low frequency, using Onsager's relationship and the HS equation, and neglecting the surface conductivity effect ($Du \ll 1$), is [105]

$$I_{CV}\big|_{\omega \to 0} = \frac{\varepsilon_{rs}\varepsilon_0 \zeta K_{DC}\phi}{\eta K_L}\frac{(\rho_P - \rho_S)}{\rho_S}\frac{dp}{dx} \qquad (84)$$

where x is the coordinate in the direction of propagation of the pressure wave, ρ_S is the density of the suspension (not of the dispersion medium: this is essential for concentrated suspensions, see [112]). Equation 84 is the analog of the HS equation for stationary electrophoretic mobility. It has been claimed to be valid over the same wide range of real dispersions, with particles of any shape and size, and any concentration.

4.11.1 Experimental procedures

In the basic experimental procedures, an ac voltage is applied to a transducer that produces a sound wave which travels down a delay line and passes into the suspension. This acoustic excitation causes a very small periodic displacement of the fluid. Although the particles tend to stay at rest because of their inertia, the ionic cloud surrounding them tends to move with the fluid to create a small oscillating dipole moment. The dipole moment from each particle adds up to create an electric field that can be sensed by the receiving transducer. The voltage difference that then appears between the electrodes (measured at zero current flow) is proportional to U_{CV}. The current, measured under short-circuit conditions, is a measure of I_{CV}. Alternatively, an ac electric field can be applied across the electrodes. A sound wave is thereby generated near the electrodes in the suspension, and this wave travels along the delay line to the transducer. The transducer output is then a measure of the ESA effect. For both techniques, measurements can be performed for just one frequency or for a set of frequencies ranging between 1 and 100 MHz, depending on the instrument.

Recently, the term *electroacoustic spectrometry* has been coined to refer to the measurement of the dynamic mobility as a function of frequency (see Fig. 8 for some examples of the kind of spectra that can be obtained). The plot of magnitude and phase of u_d^* over a range of (MHz) frequencies may yield information about the particle size as well as the ζ-potential [86, 113], provided the particles are in a suitable size range (roughly 0.05 μm < a < 5 μm). For smaller particles, the signal provides a measure of the ζ-potential, but an independent estimate of size is needed for quantitative results. That can be provided by, for instance, ultrasonic attenuation measurements with the same device.

4.11.2 Samples for calibration

The original validation of the ESA technique was achieved by comparing the theoretical dynamic mobility spectrum, as given by eqs. 80–82, with the experimentally measured values. That comparison has been done for a number of silica, alumina, titania, and goethite samples, and some pharmaceutical emulsion samples

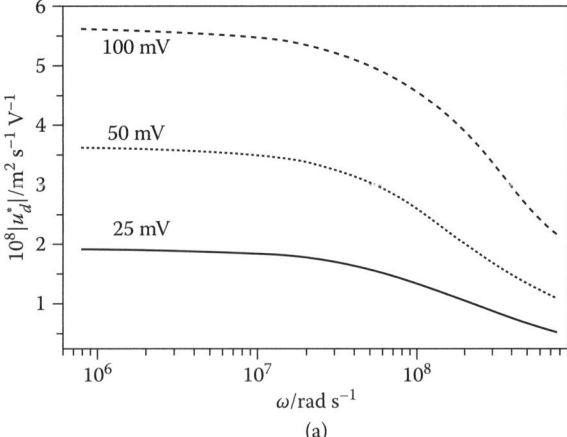

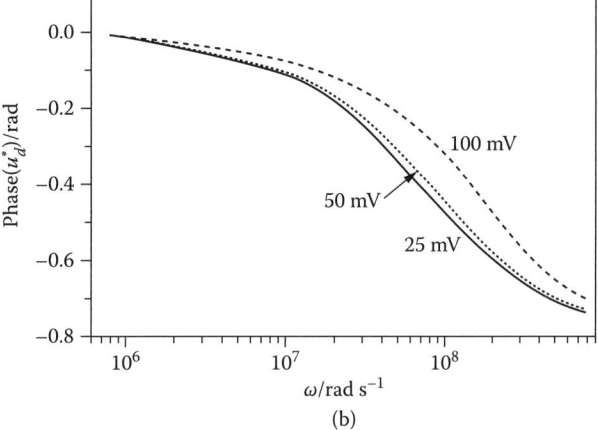

FIGURE 8: Modulus (a) and phase angle (b) of the dynamic mobility of spherical particles in a KCl solution with $\kappa a = 20$ as a function of frequency for different ζ-potentials; cf., eqs. 80–82. Parameters: particle radius 100 nm; dielectric constant of the particles (dispersion medium): 2 (78.54); density of the particles (dispersion medium): 5×10^3 (1×10^3) kg m^{-3}.

[86]. Presently, the calibration standard is a salt solution (potassium dodecatungstosilicate). This salt has a significant ESA signal that can be calculated from its known transport and equilibrium thermodynamic properties (namely, the transport numbers and partial molar volumes of the ions) [85]. Once calibrated, the instrument allows a direct measure of both the size and ζ-potential of the particles. When applied to monodisperse suspensions of spherical particles, the instrument gives results that agree, for both size and ζ-potential, with more conventional methods.

In I_{CV} measurements, calibration is often carried out using a colloid with known ζ-potential, such as Ludox™ (commercially available colloidal silica), which is diluted to a mass concentration of 0.1 g/cm³ with 10^{-2} mol L^{-1} KCl.

5. Electrokinetics in nonaqueous systems

5.1 Difference with aqueous systems: Permittivity

The majority of the investigations of EKP are devoted to aqueous systems, for which the main charging mechanisms of solid/water interfaces have been established and the EDL properties are fairly well known. All general theories of EKP assume that the equations relating them to the ζ-potential are applicable to any liquid medium characterized by two bulk properties: *electric permittivity* $\varepsilon_{rs}\varepsilon_0$ and *viscosity* η. The value of ε_{rs} is an important parameter of the liquid phase, because it determines the dissociation of the electrolytes embedded in it. Most nonaqueous solvents are less polar than water, and hence ε_{rs} is lower. All liquids can be classified roughly as nonpolar ($\varepsilon_{rs} \leq 5$), weakly polar ($5 < \varepsilon_{rs} \leq 12$), moderately polar ($12 < \varepsilon_{rs} \leq 40$), and polar ($\varepsilon_{rs} > 40$).

For $\varepsilon_{rs} > 40$, dissociation of most dissolved electrolytes is complete, and all equations concerning EDL or EKP remain unmodified, except that a lower value for ε_{rs} has to be used. For moderately polar solvents, the electrolyte dissociation is incomplete, which means that the concentration of charged species (i.e., the species that play a role in EDL or EKP) may be lower than the corresponding electrolyte concentration. Moreover, the charge number of the charge carriers can become lower. For instance, in solutions of $Ba(NO_3)_2$, we may find not (only) Ba^{2+} and NO_3^- ions, but (also) $Ba(NO_3)^+$ complexes. The category of weakly polar solvents is a transition to the class of nonpolar liquids, for which the relative permittivity (at infinite frequency) equals the square of the index of refraction. Such liquids exhibit no self-dissociation, and the notion of electrolytes almost completely dissociated loses its meaning. However, even in such media, special types of molecules may dissociate to some extent and give rise to the formation of EDL and EKP. It is this group that we shall now emphasize.

Unlike in aqueous media, dissociation in these solvents occurs only for potential electrolytes that contain ions of widely different sizes. Once dissociation occurs, the tendency for re-dissociation is relatively small because the charge on the larger ion is distributed over a large area. However, the concentration of ions (c_+, c_-) is very small. An indication of the magnitude of c_+ and c_- can be obtained from conductivity measurements.

Particles embedded in such a liquid can become charged when one type of ion can be preferentially adsorbed. Typically, the resulting surface charges, σ^0, are orders of magnitude lower than in aqueous systems. However, because of the very low EDL capacitance, the resulting surface potentials, ψ^0, are of the same order of magnitude as in aqueous systems.

The very slow decay of the potential with distance has two consequences. First, as the decay between surface and slip plane is negligible, $\psi^0 \approx \zeta$. This simplifies the analysis. Second, the slow decay implies that colloid particles "feel" each other's presence at long range. So, even dilute sols may behave physically as "concentrated" and formation of small clusters of coagulated particles is typical.

Interestingly, electrophoresis can help in making a distinction between concentrated and dilute systems by studying the dependence of the electrophoretic mobility on the concentration of dispersed particles. If there is no dependence, the behavior is that of a dilute system. In this case, equations devised for dilute systems can be used. Otherwise, the behavior is effectively that of a concentrated system, and ESA or U_{CV} measurements are more appropriate [86].

5.2 Experimental requirements of electrokinetic techniques

The methods described are also applicable to electrokinetic measurements in nonaqueous systems, but some precautions need to be taken. Microelectrophoresis cells are often designed for aqueous and water-like media whose conductivity is high relative to that of the material from which the cell is made. A homogeneous electric field between the electrodes of such cells filled with well- or moderately conducting liquids is readily achieved. However, when the cell is filled with a low-conductivity liquid, the homogeneity of the electric field can be disturbed by the

more conducting cell walls and/or by the surface conduction due to adsorption of traces of water dissolved from the solvents of low polarity on the more hydrophilic cell walls. Special precautions (coating the walls with hydrophobic layers) are necessary to improve the field characteristics. The electric field in cells of regular geometrical shape is calculated from the measured current density and the conductivity of the nonaqueous liquid. Because of the low ionic strength of the latter, electrode polarization may occur and it can sometimes be observed as bubble formation. Hence, an additional pair of electrodes is often used for the measurements of the voltage gradient across the cell.

For the correct measurement of the *streaming potential* in nonaqueous systems, attention must be paid to ensure that the resistance of the capillary, plug, or diaphragm filled with liquid is at least 100 times less than the input impedance of the electrical measuring device. The usual practice is to use millivoltmeter-electrometers with input resistance higher than 10^{14} Ω and platinum gauze electrodes placed in contact with the ends of the capillary or plug for both resistance and streaming potential measurements. The resistance is usually measured with an ac bridge in the case of polar solvents (typical frequencies used are around a few kHz) and by dc methods in the case of nonpolar or weakly polar liquids. The use of data recording on the electrometer output is common practice to check the rate of attainment of equilibrium and possible polarization effects.

5.3 Conversion of electrokinetic data into ζ-potentials

The first step in interpreting the electrokinetic behavior of nonaqueous systems must be to check whether the system behaves as a dilute or as a concentrated system (see Section 5.2). In the dilute regime, all theories described for aqueous systems can be used, provided one can find the right values of the essential parameters κa, K_p, and K^σ.

The calculation of κ requires knowledge of the ionic strength of the solution; this, in turn, can be estimated from the measurement of the dialysate conductivity, K_L, and knowledge of the mobilities and valences of the ionic species.

The effective conductivity of the solid particle, K_{pef}, including its surface conductivity* can be calculated from the experimental values of conductivities of the liquid K_L and of a dilute suspension of particles, K, with volume fraction ϕ using either Street's equation

$$\frac{K_{pef}}{K^L} = 1 + \frac{2}{3}\frac{\frac{K}{K_L}-1}{\phi} \quad (85)$$

or Dukhin's equation

$$\frac{K_{pef}}{K^L} = 2\frac{(1-\phi)-\frac{K}{K_L}(1+\phi/2)}{(1-\phi)\frac{K}{K_L}-(1+2\phi)} \quad (86)$$

which accounts for interfacial polarization.

---

* The low conductivity of nonaqueous media with respect to aqueous solutions is the main reason why often the finite bulk conductivity of the dielectric solids cannot be neglected. In contrast, if one can assume that no water is adsorbed at the interface, any SLC effect can be neglected. Furthermore, the joint adsorption of the ionic species of both signs or (and) the adsorption of such polar species as water at the solid/liquid interface can produce an abnormally high surface conduction. This is not due to the excess of free charges in the EDL, commonly taken into account in the parameter K^σ of the generalized theories of EKP. Rather, it is conditioned by the presence of thin, highly conducting adsorption layers. Therefore, the ratio K_{pef}/K_L is an important parameter that has to be estimated for nonaqueous systems.

For the estimation of the ζ-potential in the case of electrophoresis, considering that low κa values are not rare, it is suggested to use Henry's theory, after substitution of K_{pef} estimated as described above, for the particle conductivity. In formulas, it is suggested to employ eqs. 29 and 30.

Concerning streaming potential/current or electro-osmotic flow measurements, all the above-mentioned features of nonaqueous systems must be taken into account to obtain correct values of ζ-potential from this sort of data in either single capillaries, porous plugs, or diaphragms. In view of the low κa values normally attained by nonaqueous systems, the Rice and Whitehead curvature corrections are recommended [50]; see eqs. 37–39, 44, and 50.

Finally, electroacoustic characterization of ζ-potential in nonaqueous suspensions requires subtraction of the background arising from the equilibrium dispersion medium. This is imperative because the electroacoustic signal generated by particles in low- or nonpolar media is very weak. It is recommended to make measurements at several volume fractions in order to ensure that the signal comes, in fact, from the particles.

Note that there is one more issue that complicates the formulation of a rigorous electrokinetic theory in concentrated nonaqueous systems, namely, the conductivity of a dispersion of particles in such media becomes position-dependent, as the double layers of the particles may occupy most of the volume of the suspension. This is a significant theoretical obstacle in the elaboration of the theory. In the case of EKP based on the application of nonstationary fields (dielectric dispersion, electroacoustics), this problem can be overcome. This is possible because one of the peculiarities of low- and nonpolar liquids compared to aqueous systems is the very low value of the Maxwell–Wagner–O'Konski frequency, eqs. 57 and 58. This means that in the modern electroacoustic methods based on the application of electric or ultrasound fields in the MHz frequency region, all effects related to conductivity can be neglected, because that frequency range is well above the Maxwell–Wagner characteristic frequency of the liquid. This makes electroacoustic techniques most suitable for the electrokinetic characterization of suspensions in nonaqueous media.

6. Remarks on non-ideal surfaces

6.1 General comments

The general theory of EKP described so far strictly applies to *ideal*, *nonporous*, and *rigid* surfaces or particles. By *ideal*, we mean smooth (down to the scale of molecular diameters) and chemically homogeneous, and we use *rigid* to describe those particles and surfaces that do not deform under shear. We will briefly indicate interfaces that are nonporous and rigid as *hard interfaces*. For hard particles and surfaces, there is a sharp change of density in the transition between the particle or surface and the surrounding medium. Hard particles effectively lead to stacking of water (solvent, in general) molecules; that is, only for hard surfaces the notion of a stagnant layer close to the real surface is conceptually simple. Not many surfaces fulfill these conditions. For instance, polystyrenesulfonate latex colloid has, in fact, a heterogeneous interface: the hydrophilic sulfate end-groups of its polymer chains occupy some 10 % of the total surface of the particles, the remainder being hydrophobic styrene. Moreover, the sulfate groups will protrude somewhat into the solution. Generally speaking, many interfaces are far from molecularly smooth, as they may contain pores or can be somewhat deformable. Such interfaces can briefly be indicated as *soft interfaces*. For soft interfaces, such as, for instance, rigid particles with "soft" or

"hairy" polymer layers, gel-type penetrable particles and water–air or water–oil interfaces, the molecular densities vary gradually across the phase boundary. A main problem in such cases is the description of the hydrodynamics, and in some cases it is even questionable if a discrete slip plane can be defined operationally.

The difficulties encountered when interpreting experimental results obtained for non-ideal interfaces depend on the type and magnitude of the non-idealities and on the aim of the measurements. In practice, one can always measure some quantity (like u_e), apply some equation (like HS) to compute what we can call an "effective" ζ-potential, but the physical interpretation of such a ζ-potential is ambiguous. We must keep in mind that the obtained value has only a relative meaning and should not be confused with an actual electrostatic potential close to the surface. Nevertheless, such an "effective ζ" can help us in practice, because it may give some feeling for the sign and magnitude of the part of the double-layer charge that controls the interaction between particles. When the purpose of the measurement is to obtain a realistic value of the ζ-potential, there is no general recipe. It may be appropriate to use more than one electrokinetic method and to take into account the specific details of the non-ideality as well as possible in each model for the calculation of the ζ-potential. If the ζ-potentials resulting from both methods are similar, physical meaning can be assigned to this value.

Below, we will discuss different forms of non-ideality in somewhat more detail. We will mainly point out what the difficulties are, how these can be dealt with and where relevant literature can be found.

6.2 Hard surfaces

Some typical examples of non-ideal particles that still can be considered as hard are discussed below. Attention is paid to size and shape effects, surface roughness, and surface heterogeneity. For hard non-ideal surfaces, both the stagnant-layer concept and the ζ-potential remain locally defined and experimental data provide some average electrokinetic quantity that will lead to an average ζ-potential. The kind of averaging may depend on the electrokinetic method used, therefore, different methods may give different average ζ-potentials.

6.2.1 Size effects

For rigid particles that are spherical with a homogeneous charge density, but differ in size, a rigorous value of the ζ-potential can be found if the electrokinetic quantity measured is independent of the particle radius, a. In general, this will be the case for $\kappa a \gg 1$ (HS limit). In the case of electrophoresis, the mobility is also independent of a for $\kappa a < 1$ (HO limit). For other cases, the particle size will come into play and most of the time an average radius has to be used to calculate an average ζ-potential [3, 114]. The type of average radius used will, in principle, affect the average ζ-potential. Furthermore, different EKP may require different averages. For instance, in [115] it was found that a simple number average is suitable for analysis of electrophoresis in polydisperse systems; a volume average, on the other hand, was found to be the best choice when discussing dielectric dispersion. In principle, the best solution would be found if the full size distribution is measured, but even in such a case the signal-processing procedure must be taken into account. For instance, in ELS methods, the scattering of light by particles of different sizes is determinant for the average u_e value found by the instrument. In this context, measuring the dynamic mobility spectrum may be useful in providing information on both the size and the ζ-potential from the same signal.

6.2.2 Shape effects

For nonspherical particles (cylinders, discs, rods, etc.) of homogeneous ζ-potential, the induced dipole moment is different for different orientations with respect to the applied field (i.e., it acquires a tensorial character). Only when the systems are sufficiently simple (cylinders, rods) is it sufficient to distinguish between a parallel and a normal orientation of the dipoles to the applied field [116]. In spite of these additional difficulties, some approximate approaches to either the electrophoresis [117–120] or the permittivity [121, 122] of suspensions of nonspherical particles are available.

A more complicated situation arises if the particles are polydisperse in shape and size. Except if $\kappa a \gg 1$ (HS equation valid) or if $\kappa a \ll 1$ (HO limit), the only (approximate) approach is to define an "equivalent spherical particle" (for instance, one having the same volume as the particle under study) and use theories developed for spheres. In that case, the "average" ζ-potential that is obtained depends on the type(s) of polydispersity of the sample and the definition used for the "equivalent sphere".

6.2.3 Surface roughness

Most theories described so far assume that the interface is smooth down to a molecular scale. However, even the surfaces prepared with the strictest precautions show some degree of roughness, which will be characterized by R, the typical dimension of the mountains or valleys present on the surface. However, surface roughness (with R as a measure of the roughness size) affects the position of the plane of shear except in conditions where $\kappa R \gg 1$ (HS limit) or $\kappa R \ll 1$ (roughness not seen). If it is assumed that the outer parts of the asperities determine the position of the slip plane, there will be diffusely bound or even free ions in the valleys. This will inevitably lead to large surface conductivities behind the apparent slip plane, and to an additional mechanism of stagnancy [6, 116, 123]. Due care must be taken to measure this conductivity and to take it into account in evaluating the ζ-potentials. Experiments with surfaces of well-defined roughness are required to gain further insight in the complications.

We will not proceed with a more thorough description of these highly specialized—and mostly unsolved—topics. As a rule of thumb, we recommned that the reader deal with an interface of known, high roughness to use a simple approach based on the calculation of an effective ζ-potential obtained under the assumption that the interface is smooth, but to refrain from using this effective ζ for further calculations. This is particularly true if the estimation of interaction energies between the particles is sought, as it has been shown [124] that asperities have a considerable effect on both electrostatic and van der Waals forces between colloid particles.

6.2.4 Chemical surface heterogeneity

In general, chemical surface heterogeneity will also lead to surface charge heterogeneity. When the charges do not protrude into the solution, the position of the slip plane is unaffected. With charge heterogeneity, often two extreme cases are considered: (1) a *random* or *regular* heterogeneity that is assumed to lead to one (smeared-out) surface charge density and also one (averaged) electrokinetic potential and electrokinetic charge density at given solution conditions; and (2) a *patchwise* heterogeneity with large patches. The patches will lead to an inhomogenous charge distribution. In this case, it is usually assumed that each patch has its own smeared-out charge density at given solution conditions. The characteristic size of the patches should be at least of the order of the Debye length, otherwise the surface may be considered as regularly heterogeneous with one smeared-out charge density over the entire surface. Particles with random heterogeneity can be treated in the same way as particles with a homogeneous charge density, so it does not present additional problems.

The patchwise heterogeneity case applies, for instance, to surfaces with different crystal faces. The electrokinetic theory for this type of surface has been considered by several authors [118, 119, 125, 126]. Anderson et al. [127–129] have also performed experimental checks on the validity of some theoretical predictions. In the case of spherical particles, it has been demonstrated [130] that their motion depends on the first, second, and third moments of the ζ-potential distribution along the surface of the particle. It is important to note that the second moment (dipole moment) brings about a rotational motion superimposed on the translational (when present) electrophoretic motion. Clay particles [126, 131], with anionic charges at the plate surfaces and pH-dependent charges at the edges, are typical examples of systems with patchwise heterogeneous surfaces.

A rather specific situation may appear in the case of sparse polyelectrolyte adsorption onto oppositely charged surfaces. This may lead to a mosaic-type charge distribution [132, 133]. Not only the very concepts of slip plane and hence ζ-potential are doubtful here, but also electrokinetic data alone can lead us to erroneous conclusions. For instance, one can find a high $|u_e|$ value for particles with attached polyelectrolytes and from this predict a high colloidal stability. This might not be found, since the patchwise nature of the charges might induce flocculation even in systems that have considerable average ζ-potential. Such instability will be due to attraction between oppositely charged patches (corresponding to regions with and without adsorbed polyelectrolyte).

6.3 Soft particles

Some familiar examples in which the interface must be considered as soft are discussed below. The examples refer to two different groups. The first consists of hard particles with hairy, grafted, or adsorbed layers and of particles that are (partially) penetrable. The hydrodynamic permeability and the conductivity in the permeable layer make the interpretation of the data complicated. The second group are the water–oil or water–air interfaces. Droplets and bubbles comprise a specific class of "soft" particles, for which the definition of the slip plane is an academic question. When and how this issue can be solved depends on the surface active components present at the interface.

6.3.1 Charged particles with a soft uncharged layer

For a good understanding of charged hard particles covered by nonionic surfactants or uncharged polymers, the position of the slip plane and conduction behind the slip plane must be considered. It is very useful to also investigate the bare particles. Neutral adsorbed layers reduce the tangential fluid flow and the tangential electric current near a particle surface, but to a different extent [134]. Both reductions have to be considered for a correct interpretation of the electrokinetic results. By doing this, the comparison of the results between bare and covered particles may give information about the net particle potential and effective particle charge, the surface charge adjustment, and the adsorbed-layer thickness [135]. Let us mention that electroacoustic studies of the dynamic mobility spectrum of particles coated with adsorbed neutral polymer can give information on the thickness of the adsorbed layer and the shift of the slip plane due to the polymer [136].

6.3.2 Uncharged particles with a soft charged layer

The surface of many latex particles can be considered as being itself uncharged, but with charged groups at the end of oligomeric or polymeric "hairs" that protrude into the solution. The extension of the "hairs" from the surface into the liquid determines the position of the apparent slip plane. Due to the ion penetrability of the stagnant layer, the ionic strength will affect the distance between the surface and this (apparent) slip plane, and the electric conduction in the stagnant layer and this will lead to a more complex ionic strength dependence of the ζ-potential than with rigid particles. Some studies account for these effects through the introduction of a "softness factor" [137, 138] as a fitting parameter. Stein [139] discusses the electrokinetics of polystyrene latex particles in more detail. Hidalgo-Álvarez et al. [140] discuss the anomalous electrokinetic behavior of polymer colloids in general.

6.3.3 Charged particles with a soft charged layer

Charged particles with adsorbed or grafted polyelectrolyte or ionic surfactant layers fall in this class. The complications arising with these systems are similar to those mentioned for the uncharged surfaces with charged layers. Adsorbed layers impede tangential fluid flow; therefore, in the presence of the bound layer the hydrodynamic particle radius increases and the apparent slip plane is moved outwards. This affects the tangential electric current. In most cases, the particle surface and the bound layers will have opposite charges, and, therefore, the electrostatic potential profile is rather complicated. The EKP will be dominated by conduction within the slip plane and the potential at the hydrodynamic boundary relatively far away from the surface [114, 141, 142]. To unravel the situation, a systematic investigation is required that considers the electrokinetic behavior of both uncovered and covered particles, the shift of the apparent slip plane and the conduction behind the slip plane. When qualitative information is required with respect to the net charge of the particle plus adsorbed layer, the sign of the ζ-potential is important, as it will indicate whether the particle charge is overcompensated or not (within the plane of shear).

6.3.4 Ion-penetrable or partially penetrable particles

Some proteins, many biological cells, and other natural particles are penetrable for water and ions. The most important complications for the description of the electrokinetic behavior of such particles are associated with the conductivity, the dielectric constant, and the liquid transport inside the particles. An additional complication occurs when the particles are able to swell depending on the solution conditions.

When the particles are only partially penetrable, we may consider them as hard with a gel-like corona. This situation is very similar to hard particles coated with a polyelectrolyte layer. In the limit of a very small particle with a very thick corona, the penetrable particle limit results. The simplest models assume that the electrical potential inside the gel layer is the Donnan potential, whereas the hindered motion of liquid in it is represented by a friction parameter incorporated in the Navier–Stokes equation [143–146]. Ohshima's theory of electrophoresis of soft spheres basically ranges from hard particles to penetrable particles. References to his work can be found in his review on EKP [140].

Systems with biological or medical relevance that have received some systematic attention are protein-coated latex particles [147], electrophoresis of biological cells [148], liposomes [149], and bacteria [150].

6.3.5 Liquid droplets and gas bubbles in a liquid

The electrophoresis of uncontaminated liquid droplets and gas bubbles in a liquid show quite different behavior from that of rigid particles. The main reason is that flow may also occur within the droplets or bubbles because there is momentum transfer across the interface. The classical notion of a slip plane loses its meaning. Due to the flow inside the droplet or bubble, the tangential velocity of the liquid surrounding the droplet does not have to become zero at the surface of the particle. As a result, the electrophoretic mobility is higher than for a corresponding rigid body. However, to model this situation and to arrive at a conversion of the mobility to a ζ-potential is not a trivial task. As there is no slip plane, even the HS model cannot be applied.

In general, however, droplets or bubbles are not stable without an adsorbed layer of a surface active component (surfactant, polymer, poly electrolyte, protein). Most layers will make the surface inextensible (rigid) if (nearly) completely covered. In this case, the surface behaves as rigid, and it is possible to use the treatments for rigid particles [3]. Even the presence of the double layer itself, through the primary electroviscous effect, makes momentum transfer to the liquid drop very slight [151].

For partially covered bubbles (as in flotation), the situation is more complex; surface inhomogeneities may occur under the influence of shear, and Marangoni effects become important. Problems arise with regard to the definition of the slip plane, its location, and the charge inhomogeneity [152].

7. Discussion and recommendations

From the analyses given above, it will be clear that the computation of the ζ-potential from experimental data is not always a trivial task. For each method, the fundamental requirement for the conversion of experimental data is that models should be correctly used within their range of applicability. The bottom line is that the ζ-potential does exist as a characteristic of a charged surface under fixed conditions. This is fortunate, because otherwise, how could two authors compare their results with the same material using two electrokinetic techniques, or even two different versions of the same technique?

This means that we must focus on the correct use of the existing theories, and in their improvement, if necessary. The sometimes-asked question: "Is the computed ζ-potential independent of the technique used?" must be answered with *yes*, because ζ-potentials are unique characteristics for the charge state of interfaces under given conditions. Whether by different techniques identical ζ-potentials are indeed observed depends on the quality of both the technique and the interpretation. Measuring different electrokinetic properties for a given material in a given liquid medium, and checking the equality of ζ-potentials using appropriate models, is what Lyklema calls an *electrokinetic consistency test* [3], and what Dukhin and Derjaguin called *integrated electrosurface investigation* [5].

Another related question that could be asked is "Is the ζ-potential the only property characterizing the electrical equilibrium state of the interfacial region?" The answer is *most often it is not*. Most recent models and experimental results demonstrate that in the majority of cases the conductivity behind the slip plane must also be taken into account. This implies that for a correct evaluation, the *Du* number should also be measured. Using *Du* only to characterize the interface has, in general, no practical interest either. One experimental electrokinetic technique suffices only if $K^{\sigma i}/K^{\sigma d}$ is small because then $Du(\zeta,K^{\sigma i}) \approx Du(\zeta)$. The problem now is how does one know that $K^{\sigma i}$ is in fact that small? We have some possibilities:

- If we have also access to the value of the potentiometrically measured surface charge, σ0, the values of σek (calculated from the value of the z-potential obtained by neglecting conductance behind the slip plane) and d can be compared. This comparison allows us to reach a first estimate of σi. From σi Ksi could be estimated assuming—as appears to be the case, at least for monovalent counterions—that the mobilities of ions in the inner layer are comparable to those in the bulk. When Du(z,Ksi) ≈ Du(z,0) = Du(z), the calculated value of z is correct, otherwise, it has to be recalculated taking the inner-layer conduction into account.

- In the case of capillaries or plugs, the total conductivity and the streaming potential can be measured. An initial value of Ksd can be obtained from z using Bikerman's equation, and Ksi can be calculated. The improved value of z can now be obtained using a suitable model including inner-layer conduction, and again Kσd and Ksi can be obtained. If the difference between z computed without and with finite Ksi is high, we should go back to the calculation of z and iterate till the difference between two steps is small.

If these approaches are not possible, there is no other way but employing different electrokinetic techniques on the same sample and performing the consistency test. In this respect, the best procedure would be using one technique in which neglecting $K^{\sigma i}$ underestimates ζ (electrophoresis, streaming potential, e.g.), and another in which ζ is overestimated (dc conductivity, dielectric dispersion) [3].

For practical reasons—not every worker has available the numerical routines required for nonanalytical theories—another question emerges: "When can we use with confidence HS equations for the different types of electrokinetic data?"

Although most published data on ζ-potentials are based on the various versions of the HS equation for EKP, let us stress that this approach is correct only if (eq. 2) $\kappa a \gg 1$, where a is a characteristic dimension of the system (curvature radius of the solid particle, capillary radius, equivalent pore radius of the plug,...), and furthermore, the surface conductance of any kind must be low, i.e., $Du(\zeta,K^{\sigma i}) \ll 1$. Thus, in the absence of independent information about $K^{\sigma i}$, additional electrokinetic determinations can only be avoided for sufficiently large particles and high electrolyte concentrations.

Another caveat can be given, even if the previously mentioned conditions on dimensions and Dukhin number are met. For concentrated systems, the possibility of the overlap of the double layers of neighboring particles cannot be neglected if the concentration of the dispersed solids in a suspension or a plug is high. In such cases, the validity of the HS equation is also doubtful and cell models for either electrophoresis, streaming potential, or electroosmosis are required. Use of the latter two kinds of experiments or of electroacoustic or LFDD measurements is recommended. In all cases, a proper model accounting for interparticle interaction must be available.

8. Appendix I. Calculation of the low-frequency dielectric dispersion of suspensions

Neglecting SLC, the complex conductivity increment is related to the dipole coefficient as follows [62–64]:

$$\delta K^*(\omega) = \frac{3\phi}{a^3} K_L C_0^* = \delta K^* \Big|_{\omega \to 0}$$
$$+ 9\phi K_L \frac{(R^+ - R^-)H}{2AS} \frac{i\omega\tau_\alpha}{1+\sqrt{S}\sqrt{i\omega\tau_\alpha}+i\omega\tau_\alpha} \quad (I.1)$$

Its low-frequency value is a real quantity:

$$\delta K^* \Big|_{\omega \to 0} = 3\phi K_L \left(\frac{2Du(\zeta)-1}{2Du(\zeta)+2} - \frac{3(R^+-R^-)H}{2B} \right) \quad (I.2)$$

The dielectric increment of the suspension can be calculated from $\delta K^* \Big|_{\omega \to 0}$ as follows:

$$\delta \varepsilon_r^*(\omega) = -\frac{1}{\omega\varepsilon_0}\left(\delta K^* - \delta K^* \Big|_{\omega=0}\right)$$
$$= \frac{9}{2}\phi\varepsilon_{rs}' \frac{\tau_\alpha}{\tau_{el}} \frac{(R^+-R^-)H}{AS} \frac{1}{1+\sqrt{S}\sqrt{i\omega\tau_\alpha}+i\omega\tau_\alpha} \quad (I.3)$$

Here:

$$R^{\pm} = 4\frac{\exp\left(\mp\frac{zy^{ek}}{2}\right)-1}{\kappa a} + 6m^{\pm}\left[\frac{\exp\left(\mp\frac{zy^{ek}}{2}\right)-1}{\kappa a} \pm z\frac{zy^{ek}}{\kappa a}\right] \quad (I.4)$$

and

$$\tau_\alpha = \frac{a^2}{2D_{ef}}\frac{1}{S} \quad (I.5)$$

is the value of the relaxation time of the low-frequency dispersion. It is assumed that the dispersion medium is an aqueous solution of an electrolyte of z-z charge type. The definitions of the other quantities appearing in eqs. I.1–5 and the ζ-potential are as follows:

$$D_{ef} = \frac{2D^+D^-}{D^++D^-} \quad (I.6)$$

$$A = 4Du(\zeta)+4 \quad (I.7)$$

$$B = (R^++2)(R^-+2)-U^+-U^--(U^+R^-+U^-R^+)/2 \quad (I.8)$$

$$S = \frac{B}{A} \quad (I.9)$$

$$H = \frac{(R^+-R^-)(1-z^2\Delta^2)-U^++U^-+z\Delta(U^++U^-)}{A} \quad (I.10)$$

$$U^{\pm} = \frac{48 m^{\pm}}{\kappa a} \ln\left[\cosh\frac{z y_{\zeta}}{4}\right] \qquad (I.11)$$

$$\Delta = \frac{D^- - D^+}{z\left(D^- + D^+\right)} \qquad (I.12)$$

The factor $1/S$ is of the order of unity, and comparison of eqs. 58 and I.5 leads to the important conclusion that

$$\frac{\tau_{\alpha}}{\tau_{\mathrm{MWO}}} \approx (\kappa a)^2 \qquad (I.13)$$

This means that for the case of $\kappa a \gg 1$, the characteristic frequency for the α-dispersion is $(\kappa a)^2$ times lower than that of the Maxwell–Wagner dispersion.

9. Acknowledgments

Financial assistance from IUPAC is gratefully acknowledged. The Coordinators of the MIEP working party wish to thank all members for their effort in preparing their contributions, suggestions, and remarks.

10. References

1. S. S. Dukhin. *Adv. Colloid Interface Sci.* **61,** 17 (1995).
2. R. J. Hunter. *Foundations of Colloid Science,* Chap. 8, Oxford University Press, Oxford (2001).
3. J. Lyklema. *Fundamentals of Interfaces and Colloid Science,* Vol. II, Chaps. 3, 4, Academic Press, New York (1995).
4. R. J. Hunter. *Zeta Potential in Colloid Science,* Academic Press, New York (1981).
5. S. S. Dukhin and B. V. Derjaguin. "Electrokinetic phenomena", in *Surface and Colloid Science,* Vol. 7, Chap. 2, E. Matijević (Ed.), John Wiley, New York (1974).
6. A. V. Delgado (Ed.). *Interfacial Electrokinetics and Electrophoresis,* Marcel Dekker, New York (2001).
7. Manual of symbols and terminology for physicochemical uantities and units, Appendix II: Definitions, terminology and symbols in colloid and surface chemistry. Part I. *Pure Appl. Chem.* **31,** 577 (1972).
8. I. Mills, T. Cvitaš, K. Homann, N. Kallay, K. Kuchitsu. *Quantities, Units and Symbols in Physical Chemistry,* 2nd ed., Sect. 2.14, International Union of Pure and Applied Chemistry, Blackwell Science, Oxford (1993).
9. A. D. McNaught and A. Wilkinson. *Compendium of Chemical Terminology. IUPAC Recommendations,* 2nd ed., International Union of Pure and Applied Chemistry, Blackwell Science, Oxford (1997).
10. J. Lyklema. *Pure Appl. Chem.* **63,** 885 (1995).
11. J. Lyklema, S. Rovillard, J. de Coninck. *Langmuir* **14,** 5659 (1998).
12. J. W. Lorimer. *J. Membr. Sci.* **14,** 275 (1983).
13. C. F. Zukoski and D. A. Saville. *J. Colloid Interface Sci.* **114,** 32 (1986).
14. C. S. Mangelsdorf and L. R. White. *J. Chem. Soc., Faraday Trans.* **86,** 2859 (1990).
15. C. S. Mangelsdorf and L. R. White. *J. Chem. Soc., Faraday Trans.* **94,** 2441 (1998).
16. C. S. Mangelsdorf and L. R. White. *J. Chem. Soc., Faraday Trans.* **94,** 2583 (1998).
17. J. Lyklema and M. Minor. *Colloids Surf., A* **140,** 33 (1998).
18. M. von Smoluchowski. In *Handbuch der Electrizität und des Magnetismus (Graetz),* Vol. II, p. 366, Barth, Leipzig (1921).
19. J. J. Bikerman. *Z. Physik. Chem.* **A163,** 378 (1933).
20. J. J. Bikerman. *Kolloid Z.* 72, 100 (1935).
21. J. J. Bikerman. *Trans. Faraday Soc.* **36,** 154 (1940).
22. R. W. O'Brien and L. R. White. J. *Chem. Soc., Faraday Trans. II* **74,** 1607 (1978).
23. R. W. O'Brien and R. J. Hunter. *Can. J. Chem.* **59,** 1878 (1981).
24. D. C. Henry. *Proc. R. Soc. London* **A133,** 106 (1931).
25. H. Ohshima. *J. Colloid Interface Sci.* **168,** 269 (1994).
26. R. P. Tison. *J. Colloid Interface Sci.* **60,** 519 (1977).
27. A. Homola and A. A. Robertson. *J. Colloid Interface Sci.* **51,** 202 (1975).
28. K. Furusawa, Y. Kimura, T. Tagawa. In *Polymer Adsorption and Dispersion Stability,* E. D. Goddard and B. Vincent (Eds.), ACS Symposium Series Vol. 240, p. 131, American Chemical Society, Washington, DC (1984).
29. M. Minor, A. J. van der Linde, H. P. van Leeuwen, J. Lyklema. *J. Colloid Interface Sci.* **189,** 370 (1997).
30. E. E. Uzgiris. *Rev. Sci. Instrum.* **45,** 74 (1974).
31. R. Ware and W. H. Flygare. *J. Colloid Interface Sci.* **39,** 670 (1972).
32. E. E. Uzgiris. *Optics Comm.* **6,** 55 (1972).
33. E. Malher, D. Martin, C. Duvivier. *Studia Biophys.* **90,** 33 (1982).
34. E. Malher, D. Martin, C. Duvivier, B. Volochine, J. F. Stolz. *Biorheology* **19,** 647 (1982).
35. K. Oka and K. Furusawa. In *Electrical Phenomena at Interfaces,* H. Ohshima and K. Furusawa (Eds.), Chap. 8, Marcel Dekker, New York (1998).
36. J. F. Miller, K. Schätzel, B. Vincent. *J. Colloid Interface Sci.* **143,** 532 (1991).
37. F. Manerwatson, W. Tscharnuter, J. Miller. *Colloids Surfaces A* **140,** 53 (1988).
38. J. F. Miller, O. Velev, S. C. C. Wu, H. J. Ploehn. *J. Colloid Interface Sci.* **174,** 490 (1995).
39. Japanese Surface and Colloid Chemical Group (Japan Oil Chemists Soc.). *Yukagaku* **25,** 239 (1975).
40. F. Fairbrother and H. Mastin. *J. Chem. Soc.* 2319 (1924).
41. D. R. Briggs. *J. Phys. Chem.* **32,** 641 (1928).
42. M. Chang and A. A. Robertson. *Can. J. Chem. Eng.* **45,** 66 (1967).
43. G. J. Biefer and S. G. Mason. *Trans. Faraday Soc.* **55,** 1234 (1959).
44. J. Happel and P. A. Ast. *Chem. Eng. Sci.* **11,** 286 (1960).
45. J. Schurz and G. Erk. *Progr. Colloid Polym. Sci.* **71,** 44 (1985).
46. S. Levine, G. Neale, N. Epstein. *J. Colloid Interface Sci.* **57,** 424 (1976).
47. J. L. Anderson and W. H. Koh. *J. Colloid Interface Sci.* **59,** 149 (1977).
48. S. Levine, J. R. Marriot, G. Neale, N. Epstein. *J. Colloid Interface Sci.* **52,** 136 (1975).
49. D. Burgreen and F. R. Nakache. *J. Phys. Chem.* **68,** 1084 (1964).
50. C. L. Rice and P. Whitehead. *J. Phys. Chem.* **69,** 4017 (1965).
51. R. W. O'Brien and W. T. Perrins. *J. Colloid Interface Sci.* **99,** 20 (1984).
52. D. Erickson and D. Li. *J. Colloid Interface Sci.* **237,** 283 (2001).
53. A. Szymczyk, B. Aoubiza, R Fievet, J. Pagetti. *J. Colloid Interface Sci.* **192,** 105 (2001).
54. N. Spanos and P. G. Koutsoukos. *J. Colloid Interface Sci.* **214,** 85 (1999).
55. J. M. Gierbers, J. M. Kleijn, M. A. Cohen-Stuart. *J. Colloid Interface Sci.* **248,** 88 (2002).
56. O. El-Gholabzouri, M. A. Cabrerizo, R. Hidalgo-Álvarez. *J. Colloid Interface Sci.* **261,** 386 (2003).
57. A. Pettersson and J. B. Rosenholm. *Langmuir* **18,** 8447 (2002).
58. J. Th. G. Overbeek. In *Colloid Science,* Vol. I, H. R. Kruyt (Ed.), Chap. V, Elsevier, Amsterdam (1952).
59. C. Werner, R. Zimmermann, T. Kratzmüller. *Colloids Surf., A* **192,** 205 (2001).
60. J. Kijlstra, H. P. van Leeuwen, J. Lyklema. *Langmuir* **9,** 1625 (1993).
61. J. Kijlstra, H. P. van Leeuwen, J. Lyklema. *J. Chem. Soc., Faraday Trans.* **88,** 3441 (1992).
62. C. T. O'Konski. *J. Phys. Chem.* **64,** 605 (1960).
63. S. S. Dukhin and V. N. Shilov. *Dielectric Phenomena and the Double Layer in Disperse Systems and Poly electrolytes,* John Wiley, Jerusalem (1974).
64. C. Grosse and V. N. Shilov. *J. Phys. Chem.* **100,** 1771 (1996).
65. B. Nettleblad and G. A. Niklasson. *J. Colloid Interface Sci.* **181,** 165 (1996).
66. E. H. B. DeLacey and L. R. White. *J. Chem. Soc. Faraday Trans. II* **77,** 2007 (1981).
67. V. N. Shilov, A. V. Delgado, F. González-Caballero, C. Grosse. *Colloids Surf., A* **192,** 253 (2001).

68. L. A. Rosen, J. C. Baygents, D. A. Saville. *J. Chem. Phys.* **98,** 4183 (1993).

69. F. J. Arroyo, F. Carrique, T. Bellini, A. V. Delgado. *J. Colloid Interface Sci.* **210,** 194 (1999).

70. E. Vogel and H. Pauli. *J. Chem. Phys.* **89,** 3830 (1988).

71. Yu. B. Borkovskaja and V. N. Shilov. *Colloid J.* **54,** 173 (1992).

72. V. N. Shilov and Yu. B. Borkovskaja. *Colloid J.* **56,** 647 (1994).

73. A. V. Delgado, F. J. Arroyo, F. Gonzalez-Caballero, V. N. Shilov, Y. Borkovskaja. *Colloids Surf., A* **140,** 139 (1998).

74. F. Carrique, F. J. Arroyo, M. L. Jiménez, A. V. Delgado. *J. Chem. Phys.* **118,** 1945 (2003).

75. H. Fricke and H. J. Curtis. *J. Phys. Chem.* **41,** 729 (1937).

76. F. J. Arroyo, F. Carrique, A. V. Delgado. *J. Colloid Interface Sci.* **217,** 411 (1999).

77. M. M. Springer, A. Korteweg, J. Lyklema. *J. Electroanal. Chem.* **153,** 55 (1983).

78. K. Lim and E. I. Frances. *J. Colloid Interface Sci.* **110,** 201 (1986).

79. C. Grosse, A. J. Hill, K. R. Foster. *J. Colloid Interface Sci.* **127,** 167 (1989).

80. C. Grosse and M. C. Tirado. *Mater. Res. Soc. Symp. Proc.* **430,** 287 (1996).

81. M. L. Jiménez, F. J. Arroyo, J. van Turnhout, A. V. Delgado. *J. Colloid Interface Sci.* **249,** 327 (2002).

82. R. W O'Brien. *J. Fluid Mech.* **212,** 81 (1990).

83. R. W. O'Brien, B. R. Midmore, A. Lamb, R. J. Hunter. *Faraday Discuss. Chem Soc.* **90,** 301 (1990).

84. R. W. O'Brien. *J. Fluid Mech.* **190,** 71 (1988).

85. R. W. O'Brien, D. Cannon, W. N. Rowlands. *J. Colloid Interface Sci.* **173,** 406 (1995).

86. R. J. Hunter. *Colloids Surf., A* **141,** 37 (1998).

87. R. J. Hunter and R. W. O'Brien. *Colloids Surf., A* **126,** 123 (1997).

88. R. W. O'Brien. U.S. Patent 5,059,909 (1991).

89. R. P. Sawatsky and A. J. Babchin. *J. Fluid Mech.* **246,** 321 (1993).

90. C. S. Mangelsdorf and L. R. White. *J. Colloid Interface Sci.* **160,** 275 (1993).

91. C. S. Mangelsdorf and L. R. White. *J. Chem. Soc., Faraday Trans.* **88,** 3567 (1992).

92. S. E. Gibb and R. J. Hunter. *J. Colloid Interface Sci.* **224,** 99 (2000).

93. M. Loewenberg and R. W. O'Brien. *J. Colloid Interface Sci.* **150,** 158 (1992).

94. M. Loewenberg. *J. Fluid Mech.* **278,** 149 (1994).

95. H. Ohshima. *J. Colloid Interface Sci.* **185,** 131 (1997).

96. B. J. Marlow, D. Fairhurst, H. P. Pendse. *Langmuir* **4,** 611 (1988).

97. S. Levine and G. Neale. *J. Colloid Interface Sci.* **47,** 520 (1974).

98. S. Levine and G. Neale. *J. Colloid Interface Sci.* **49,** (1974).

99. R. W. O'Brien, W. N. Rowlands, R. J. Hunter. In *Electro-acoustics for Characterization of Particulates and Suspensions,* S. B. Malghan (Ed.), NIST Special Publication 856, pp. 1–22, National Institute of Standards and Technology, Washington, DC (1993).

100. R. J. Hunter. "Electrokinetic characterization of emulsions", in *Encyclopaedic Handbook of Emulsion Technology,* J. Sjoblom (Ed.), Chap. 7, Marcel Dekker, New York (2001).

101. P. Rider and R. W. O'Brien. *J. Fluid Mech.* **257,** 607 (1993).

102. J. Ennis, A. A. Shugai, S. L. Carnie. *J. Colloid Interface Sci.* **223,** 21 (2000).

103. S. B. Johnson, A. S. Russell, P. J. Scales. *Colloids Surf., A* **141,** 119 (1998).

104. R. W. O'Brien, A. Jones, W. N. Rowlands. *Colloids Surf., A* **218,** 89 (2003).

105. A. S. Dukhin, V. N. Shilov, H. Ohshima, P. J. Goetz. *Langmuir* **15,** 6692 (1999).

106. A. H. Harker and J. A. G. Temple. *J. Phys. D.: Appl. Phys.* **21,** 1576 (1988).

107. R. L. Gibson and M. N. Toksoz. *J. Acoust. Soc. Amer.* **85,** 1925 (1989).

108. A. S. Dukhin and P. J. Goetz. *Langmuir* **12,** 4987 (1996).

109. A. S. Ahuja. *J. Appl. Phys.* **44,** 4863 (1973).

110. S. Kuwabara. *J. Phys. Soc. Jpn.* **14,** 527 (1959).

111. V. N. Shilov, N. I. Zharkikh, Y. B. Borkovskaya. *Colloid J.* **43,** 434 (1981).

112. A. S. Dukhin, V. N. Shilov, H. Ohshima, P. J. Goetz. In *Interfacial Electrokinetics and Electrophoresis,* A. V. Delgado (Ed.), Chap. 17, Marcel Dekker, New York (2001).

113. R. J. Hunter and R. W. O'Brien. In *Encyclopedia of Colloid and Surface Science,* A. Hubbard (Ed.), p. 1722, Marcel Dekker, New York (2002).

114. H. Ohshima. In *Electrical Phenomena at Interfaces. Fundamentals, Measurements and Applications,* H. Ohshima and K. Furusawa (Eds.), Chap. 2, Marcel Dekker, New York (2002).

115. F. Carrique, F. J. Arroyo, A. V. Delgado. *J. Colloid Interface Sci.* **206,** 206 (1998).

116. S. S. Dukhin. *Adv. Colloid Interface Sci.* **44,** 1 (1993).

117. R. W. O'Brien and D. N. Ward. *J. Colloid Interface Sci.* **121,** 402 (1988).

118. M. C. Fair and J. L. Anderson. *J. Colloid Interface Sci.* **127,** 388 (1989).

119. D. Velegol, J. L. Anderson, Y. Solomentsev. In *Interfacial Electrokinetics and Electrophoresis,* A. V. Delgado (Ed.), Chap. 6, Marcel Dekker, New York (2002).

120. J. Y. Kim and B. J. Yoon. In *Interfacial Electrokinetics and Electrophoresis,* A. V. Delgado (Ed.), Chap. 7, Marcel Dekker, New York (2002).

121. C. Grosse and V. N. Shilov. *J. Colloid Interface Sci.* **193,** 178 (1997).

122. C. Grosse, S. Pedrosa, V. N. Shilov. *J. Colloid Interface Sci.* **220,** 31 (1999).

123. S. S. Dukhin, R. Zimmermann, C. Werner. *Colloids Surf., A* **195,** 103 (2001).

124. J. Y. Walz. *Adv. Colloid Interface Sci.* **74,** 119 (1998).

125. S. A. Allison. *Macromolecules* **29,** 7391 (1996).

126. M. Teubner. *J. Chem. Phys.* **76,** 5564 (1982).

127. M. C. Fair and J. L. Anderson. *Langmuir* **8,** 2850 (1992).

128. D. Velegol, J. L. Anderson, S. Garoff. *Langmuir* **12,** 675 (1996).

129. J. L. Anderson, D. Velegol, S. Garoff. *Langmuir* **16,** 3372 (2000).

130. J. L. Anderson. *J. Colloid Interface Sci.* **105,** 45 (1985).

131. H. van Olphen. *An Introduction to Clay Colloid Chemistry,* John Wiley, New York (1977).

132. S. Akari, W. Schrepp, D. Horn. *Langmuir* **12,** 857 (1996).

133. K. E. Bremmell, G. J. Jameson, S. Biggs. *Colloids Surf., A* **139,** 199 (1998).

134. M. A. Cohen-Stuart, F. H. W. H. Waajen, S. S. Dukhin. *Colloid Polym. Sci.* **262,** 423 (1984).

135. L. K. Koopal, V. Hlady, J. Lyklema. *J. Colloid Interface Sci.* **121,** 49 (1988).

136. L. Kong, J. K. Beattie, R. J. Hunter. *Phys. Chem. Chem. Phys.* **3,** 87 (2001).

137. H. Ohshima. *J. Colloid Interface Sci.* **163,** 474 (1995).

138. H. Ohshima and K. Makino. *Colloids Surf., A* **109,** 71 (1996).

139. H. N. Stein. In *Interfacial Electrokinetics and Electrophoresis,* A. V. Delgado (Ed.), Chap. 21, Marcel Dekker, New York (2002).

140. R. Hidalgo-Álvarez, A. Martín, A. Fernández, D. Bastos, F. Martínez, F. J. de las Nieves. *Adv. Colloid Interface Sci.* **67,** 1 (1996).

141. H. Ohshima. In *Interfacial Electrokinetics and Electrophoresis,* A. V. Delgado, (Ed.), Chap. 5, Marcel Dekker, New York (2002).

142. H. J. Keh. In *Interfacial Electrokinetics and Electrophoresis,* A. V. Delgado (Ed.), Chap. 15, Marcel Dekker, New York (2002).

143. C. Tanford and J. A. Reynolds. *Biochim. Biophys. Acta* **457,** 133 (1976).

144. D. Stigter. *Cell Biophys.* **11,** 139 (1987).

145. D. Stigter. *J. Phys. Chem. B* **104,** 3402 (2000).

146. R. J. Hill, D. A. Saville, W. B. Russel. *J. Colloid Interface Sci.* **258,** 56 (2003).

147. A. Martín-Rodríguez, J. L. Ortega-Vinuesa, R. Hidalgo-Álvarez. In *Interfacial Electrokinetics and Electrophoresis,* A. V. Delgado (Ed.), Chap. 22, Marcel Dekker, New York (2002).

148. E. Lee, F. Y. Yen, J. P. Hsu. In *Interfacial Electrokinetics and Electrophoresis,* A. V. Delgado (Ed.), Chap. 23, Marcel Dekker, New York (2002).

149. R. Barchini, H. P. van Leewen, J. Lyklema. *Langmuir* **16,** 8238 (2000).

150. A. Van der Wal, M. Minor, W. Norde, A. J. B. Zehnder, J. Lyklema. *Langmuir* **13,** 165 (1997).

151. S. A. Nespolo, M. A. Bevan, D. Y. C. Chan, F. Grieser, G. W. Stevens. *Langmuir* **17,** 7210 (2001).

152. C. Yang, T. Dabros, D. Li, J. Czarnecki, J. H. Masliyah. *J. Colloid Interface Sci.* **243,** 128 (2001).

11. List of symbols

Note: SI base (or derived) units are given in parentheses for all quantities, except dimensionless ones.

a (m)	particle radius, local curvature radius, capillary radius
A_c (m^2)	capillary cross-section
A_c^{ap} (m^2)	apparent (externally measured) capillary cross-section
A_{ESA} (Pa)	electrokinetic sonic amplitude
b (m)	half distance between neighboring particles
c (mol m^{-3})	electrolyte concentration
$c_+ (c_-)$ (mol m^{-3})	concentration of cations (anions)
C_0^*	dipole coefficient of particles
d (m)	distance between the surface and the outer Helmholtz plane
d^* (C m)	complex dipole moment
d^{ek} (m)	distance between the surface and the slip plane
D (m^2 s^{-1})	diffusion coefficient of counterions (or average diffusion coefficient of ions)
$D+ (D-)$ (m^2 s^{-1})	diffusion coefficient of cations (anions)
Du	Dukhin number
Du^d	Dukhin number associated with diffuse-layer conductivity
Du^i	Dukhin number associated with stagnant-layer conductivity
e (C)	elementary charge
E (V m^{-1})	applied electric field
E_{sed} (V m^{-1})	sedimentation field
E_t (V m^{-1})	tangential component of external field
F (C mol^{-1})	Faraday constant
$f_1(\kappa a)$, $F_1(\kappa a, Kp)$	Henry's functions
I (A)	electric current intensity
$I_0\ I_1$	zeroth- (first-) order modified Bessel functions of the first kind
I_c (A)	conduction current
I_{CV} (A)	colloid vibration current
Istr (A)	streaming current
j^σ (A m^{-1})	surface current density
j_{str} (A m^{-2})	streaming current density
k (J K^{-1})	Boltzmann constant
K (S m^{-1})	total conductivity of a colloidal system
K_{DC} (S m^{-1})	direct current conductivity of a suspension
K_L (S m^{-1})	conductivity of dispersion medium
K_L^∞ (S m^{-1})	conductivity of a highly concentrated ionic solution
K_p (S m^{-1})	conductivity of particles
K_{plug} (S m^{-1})	conductivity of a plug of particles
K_{pef} (S m^{-1})	effective conductivity of particles
K_{rel}	ratio between particle and liquid conductivities
K^* (S m^{-1})	complex conductivity of a suspension
K^σ (S)	surface conductivity
$K^{\sigma d}$ (S)	diffuse-layer surface conductivity
$K^{\sigma i}$ (S)	stagnant-layer surface conductivity
L (m)	capillary length, characteristic dimension
L^{ap} (m)	apparent (externally measured) capillary cross-section

L_D (m)	ionic diffusion length
m	dimensionless ionic mobility of counterions
$m+ (m-)$	dimensionless ionic mobility of cations (anions)
n (m^{-3})	number concentration of particles
N_A (mol^{-1})	Avogadro constant
n_i (m^{-3})	number concentration of type i ions
Q_{eo} (m^3 s^{-1})	electro-osmotic flow rate
$Q_{eo,E}$ (m^4 s^{-1} V^{-1})	electro-osmotic flow rate per electric field
$Q_{eo,I}$ (m^3 C^{-1})	electro-osmotic flow rate per current
r (m)	spherical or cylindrical radial coordinate
R (m)	roughness of a surface
R_s (Ω)	electrical resistance of a capillary or porous plug in an arbitrary solution
R_∞ (Ω)	electrical resistance of a capillary or porous plug in a concentrated ionic solution
T (K)	thermodynamic temperature
U_d^r (m^2 s^{-1} V^{-1})	dynamic electrophoretic mobility
U_{CV} (V)	colloid vibration potential
v_e (m^2 s^{-1} V^{-1})	electrophoretic mobility
U_{sed} (V)	sedimentation potential
U_{str} (V)	streaming potential
v_e (m s^{-1})	electrophoretic velocity
v_{eo} (m s^{-1})	electro-osmotic velocity
v_L (m s^{-1})	liquid velocity in electrophoresis cell
y^{ek}	dimensionless ζ-potential
z	common charge number of ions in a symmetrical electrolyte
z_i	charge number of type i ions
α	relaxation of double-layer polarization, degree of electrolyte dissociation, dimensionless parameter used in electroacoustics
β (m)	distance between the solid surface and the inner Hemholtz plane (see also eq. 45 for another use of this symbol)
Γ_i (m^{-2})	surface concentration of type i ions
δ_c (mol m^{-3})	field-induced perturbation of electrolyte amount concentration
δK^* (S m^{-1})	conductivity increment of a suspension
$\delta_{\varepsilon r}$	relative dielectric increment of a suspension
Δ_p (Pa)	applied pressure difference
Δ_{peo} (Pa)	electro-osmotic counter-pressure
$\Delta Vext$ (V)	applied potential difference
$\Delta_\varepsilon(0)$	low-frequency dielectric increment per volume fraction
$\Delta_{\varepsilon d}(0)$	value of $\Delta\varepsilon(0)$ for suspensions with low volume fractions
$\Delta\rho$ (kg m^{-3})	density difference between particles and dispersion medium
ε^* (F m^{-1})	complex electric permittivity of a suspension
ε_r^*	complex relative permittivity of a suspension
ε_{rp}	relative permittivity of the particle
ε_{rp}^*	complex relative permittivity of a particle
ε_{rs}	relative permittivity of the dispersion medium

ε_{rs}^{*}	complex relative permittivity of the dispersion medium	τ_{MWO} (s)	characteristic time of the Maxwell–Wagner–O'Konski relaxation
ε_0 (F m^{-1})	electric permittivity of vacuum	τ_{α} (s)	relaxation time of the low-frequency dispersion
ζ (V)	electrokinetic or ζ-potential		
ζ_{app} (V)	electrokinetic or ζ-potential not corrected for the effect of particle concentration	ϕ	volume fraction of solids
		ϕ_L	volume fraction of liquid in a plug
		ψ^d (V)	diffuse-layer potential
η (Pa s)	dynamic viscosity	ψ^i (V)	inner Helmholtz plane potential
κ (m^{-1})	reciprocal Debye length	ψ^0 (V)	surface potential
ρ (kg m^{-3})	density of dispersion medium	ω (s^{-1})	angular frequency of an ac electric field
ρ_p (kg m^{-3})	density of particles		
ρ_s (kg m^{-3})	density of a suspension	ω_{MWO} (s^{-1})	Maxwell–Wagner–O'Konski characteristic frequency
σ^d (C m^{-2})	diffuse charge density		
σ^{ek} (C m^{-2})	electrokinetic charge density	ω_{α} (s^{-1})	characteristic frequency of the α-relaxation
σ^i (C m^{-2})	surface charge density at the inner Helmholtz plane		
		$\omega_{\alpha d}$ (s^{-1})	characteristic frequency of the α-relaxation for a dilute suspension
σ^0 (C m^{-2})	titratable surface charge density		

MEASUREMENT OF pH DEFINITION, STANDARDS, AND PROCEDURES

(IUPAC Recommendations 2002)

Working Party on pH

R. P. Buck (Chairman)[1], S. Rondinini (Secretary)[2,†], A. K. Covington (Editor)[3], F. G. K. Baucke[4], C. M. A. Brett[5], M. F. Camões[6], M. J. T. Milton[7], T. Mussini[8], R. Naumann[9], K. W. Pratt[10], P. Spitzer[11], and G. S. Wilson[12]

[1]*101 Creekview Circle, Carrboro, NC 27510, USA;* [2]*Dipartimento di Chimica Fisica ed Elettrochimica, Università di Milano, Via Golgi 19, I-20133 Milano, Italy;* [3]*Department of Chemistry, The University, Bedson Building, Newcastle Upon Tyne, NE1 7RU, UK;* [4]*Schott Glasswerke, P.O. Box 2480, D-55014 Mainz, Germany;* [5]*Departamento de Química, Universidade de Coimbra, P-3004-535 Coimbra, Portugal;* [6]*Departamento de Química e Bioquimica, University of Lisbon (SPQ/DQBFCUL), Faculdade de Ciencias, Edificio CI-5 Piso, P-1700 Lisboa, Portugal;* [7]*National Physical Laboratory, Centre for Optical and Environmental Metrology, Queen's Road, Teddington, Middlesex TW11 0LW, UK;* [8]*Dipartimento di Chimica Fisica ed Elettrochimica, Università di Milano, Via Golgi 19, I-20133 Milano, Italy;* [9]*MPI for Polymer Research, Ackermannweg 10, D-55128 Mainz, Germany;* [10]*Chemistry B324, Stop 8393, National Institute of Standards and Technology, 100 Bureau Drive, ACSL, Room A349, Gaithersburg, MD 20899-8393, USA;* [11]*Physikalisch-Technische Bundesanstalt (PTB), Postfach 33 45, D-38023 Braunschweig, Germany;* [12]*Department of Chemistry, University of Kansas, Lawrence, KS 66045, USA*

†*Corresponding author*

Abstract: The definition of a "primary method of measurement" [1] has permitted a full consideration of the definition of primary standards for pH, determined by a primary method (cell without transference, Harned cell), of the definition of secondary standards by secondary methods, and of the question whether pH, as a conventional quantity, can be incorporated within the internationally accepted system of measurement, the International System of Units (SI, Système International d'Unités). This approach has enabled resolution of the previous compromise IUPAC 1985 Recommendations [2]. Furthermore, incorporation of the uncertainties for the primary method, and for all subsequent measurements, permits the uncertainties for all procedures to be linked to the primary standards by an unbroken chain of comparisons. Thus, a rational choice can be made by the analyst of the appropriate procedure to achieve the target uncertainty of sample pH. Accordingly, this document explains IUPAC recommended definitions, procedures, and terminology relating to pH measurements in dilute aqueous solutions in the temperature range 5–50°C. Details are given of the primary and secondary methods for measuring pH and the rationale for the assignment of pH values with appropriate uncertainties to selected primary and secondary substances.

Contents

Reproduced from:
Pure Appl. Chem., Vol. 74, No. 11, pp. 2169–2200, 2002.
© 2002 IUPAC

Abbreviations used

BIPM	Bureau International des Poids et Mesures, France
CRMs	certified reference materials
EUROMET	European Collaboration in Metrology (Measurement Standards)
NBS	National Bureau of Standards, USA, now NIST
NIST	National Institute of Science and Technology, USA
NMIs	national metrological institutes
PS	primary standard
LJP	liquid junction potential
RLJP	residual liquid junction potential
SS	secondary standard

1. Introduction and scope

1.1 pH, a single ion quantity

The concept of pH is unique among the commonly encountered physicochemical quantities listed in the IUPAC Green Book [3] in that, in terms of its definition [4],

$$pH = -\lg a_H$$

it involves a single ion quantity, the activity of the hydrogen ion, which is immeasurable by any thermodynamically valid method and requires a convention for its evaluation.

1.2 Cells without transference, Harned cells

As will be shown in Section 4, primary pH standard values can be determined from electrochemical data from the cell without transference using the hydrogen gas electrode, known as the Harned cell. These primary standards have *good* reproducibility and *low* uncertainty. Cells involving glass electrodes and liquid junctions have considerably *higher* uncertainties, as will be discussed later (Sections 5.1, 10.1). Using evaluated uncertainties, it is possible to rank reference materials as primary or secondary in terms of the methods used for assigning pH values to them. This ranking of primary (PS) or secondary (SS) standards is consistent with the metrological requirement that measurements are traceable with stated uncertainties to national, or international, standards by an unbroken chain of comparisons each with its own stated uncertainty. The accepted definition of traceability is given in Section 12.4. If the uncertainty of such measurements is calculated to include the hydrogen ion activity convention (Section 4.6), then the result can also be traceable to the internationally accepted SI system of units.

1.3 Primary pH standards

In Section 4 of this document, the procedure used to assign primary standard [pH(PS)] values to primary standards is described. The only method that meets the stringent criteria of a primary method of measurement for measuring pH is based on the Harned cell (Cell I). This method, extensively developed by R. G. Bates [5] and collaborators at NBS (later NIST), is now adopted in national metrological institutes (NMIs) worldwide, and the procedure is approved in this document with slight modifications (Section 3.2) to comply with the requirements of a primary method.

1.4 Secondary standards derived from measurements on the Harned cell (Cell I)

Values assigned by Harned cell measurements to substances that do not entirely fulfill the criteria for primary standard status are secondary standards (SS), with pH(SS) values, and are discussed in Section 8.1.

1.5 Secondary standards derived from primary standards by measuring differences in pH

Methods that can be used to obtain the difference in pH between buffer solutions are discussed in Sections 8.2–8.5 of these Recommendations. These methods involve cells that are practically more convenient than the Harned cell, but have greater uncertainties associated with the results. They enable the pH of other buffers to be compared with primary standard buffers that have been measured with a Harned cell. It is recommended that these are secondary methods, and buffers measured in this way are secondary standards (SS), with pH(SS) values.

1.6 Traceability

This hierarchical approach to primary and secondary measurements facilitates the availability of traceable buffers for laboratory calibrations. Recommended procedures for carrying out these calibrations to achieve specified uncertainties are given in Section 11.

1.7 Scope

The recommendations in this Report relate to analytical laboratory determinations of pH of dilute aqueous solutions (≤ 0.1 mol kg^{-1}). Systems including partially aqueous mixed solvents, biological measurements, heavy water solvent, natural waters, and high-temperature measurements are excluded from this Report.

1.8 Uncertainty estimates

The Annex (Section 13) includes typical uncertainty estimates for the use of the cells and measurements described.

2. Activity and the definition of pH

2.1 Hydrogen ion activity

pH was originally defined by Sørensen in 1909 [6] in terms of the concentration of hydrogen ions (in modern nomenclature) as pH = $-\lg(c_H/c^\circ)$ where c_H is the hydrogen ion concentration in mol dm^{-3}, and $c^\circ = 1$ mol dm^{-3} is the standard amount concentration. Subsequently [4], it has been accepted that it is more satisfactory to define pH in terms of the relative activity of hydrogen ions in solution

$$pH = -\lg a_H = -\lg(m_H \gamma_H / m^\circ) \qquad (1)$$

where a_H is the relative (molality basis) activity and γ_H is the molal activity coefficient of the hydrogen ion H$^+$ at the molality m_H, and m° is the standard molality. The quantity pH is intended to be a measure of the activity of hydrogen ions in solution. However, since it is defined in terms of a quantity that cannot be measured by a thermodynamically valid method, eq. 1 can be only a *notional definition* of pH.

3. Traceability and primary methods of measurement

3.1 Relation to SI

Since pH, a single ion quantity, is not determinable in terms of a fundamental (or base) unit of any measurement system, there was some difficulty previously in providing a proper basis for the traceability of pH measurements. A satisfactory approach is now available in that pH determinations can be incorporated into the SI if they can be traced to measurements made using a method that fulfills the definition of a "Primary method of measurement" [1].

3.2 Primary method of measurement

The accepted definition of a primary method of measurement is given in Section 12.1. The essential feature of such a method is that it must operate according to a well-defined measurement equation in which all of the variables can be determined experimentally in terms of SI units. Any limitation in the determination of the experimental variables, or in the theory, must be included within the estimated uncertainty of the method if traceability to the SI is to be established. If a convention is used without an estimate of its uncertainty, true traceability to the SI would not be established. In the following section, it is shown that the Harned cell fulfills the definition of a primary method for the measurement of the acidity function, $p(a_H\gamma_{Cl})$, and subsequently of the pH of buffer solutions.

4. Harned cell as a primary method for the absolute measurement of pH

4.1 Harned cell

The cell without transference defined by

$$Pt \mid H_2 \mid buffer\ S,\ Cl^- \mid AgCl \mid Ag \qquad\qquad Cell\ I$$

known as the Harned cell [7], and containing standard buffer, S, and chloride ions, in the form of potassium or sodium chloride, which are added in order to use the silver–silver chloride electrode. The application of the Nernst equation to the spontaneous cell reaction:

$$^{1/2}H_2 + AgCl \rightarrow Ag(s) + H^+ + Cl^-$$

yields the potential difference E_1 of the cell [corrected to 1 atm (101.325 kPa), the partial pressure of hydrogen gas used in electrochemistry in preference to 100 kPa] as

$$E_1 = E^\circ - [(RT/F)\ln 10]\ \lg[(m_H\gamma_H/m^\circ)(m_{Cl}\gamma_{Cl}/m^\circ)] \qquad (2)$$

which can be rearranged, since $a_H = m_H\gamma_H/m^\circ$, to give the acidity function

$$p(a_H\gamma_{Cl}) = -\lg(a_H\gamma_{Cl}) = (E_1 - E^\circ)/[(RT/F)\ln 10] + \lg(m_{Cl}/m^\circ) \qquad (2')$$

where E° is the standard potential difference of the cell, and hence of the silver–silver chloride electrode, and γ_{Cl} is the activity coefficient of the chloride ion.

> *Note 1*: The sign of the standard electrode potential of an electrochemical reaction is that displayed on a high-impedance voltmeter when the lead attached to standard hydrogen electrode is connected to the minus pole of the voltmeter.

The steps in the use of the cell are summarized in Fig. 1 and described in the following paragraphs.

The standard potential difference of the silver–silver chloride electrode, E°, is determined from a Harned cell in which only HCl is present at a fixed molality (e.g., $m = 0.01$ mol kg^{-1}). The application of the Nernst equation to the HCl cell

$$Pt \mid H2 \mid HCl(m) \mid AgCl \mid Ag \qquad\qquad Cell\ Ia$$

gives

$$E_{Ia} = E^\circ - [(2RT/F)\ln 10]\ \lg[(m_{HCl}/m^\circ)(\gamma_{\pm HCl})] \qquad (3)$$

where E_{Ia} has been corrected to 1 atmosphere partial pressure of hydrogen gas (101.325 kPa) and $\gamma_{\pm HCl}$ is the mean ionic activity coefficient of HCl.

4.2 Activity coefficient of HCl

The values of the activity coefficient ($\gamma_{\pm HCl}$) at molality 0.01 mol kg^{-1} and various temperatures are given by Bates and Robinson [8]. The standard potential difference depends in some not entirely

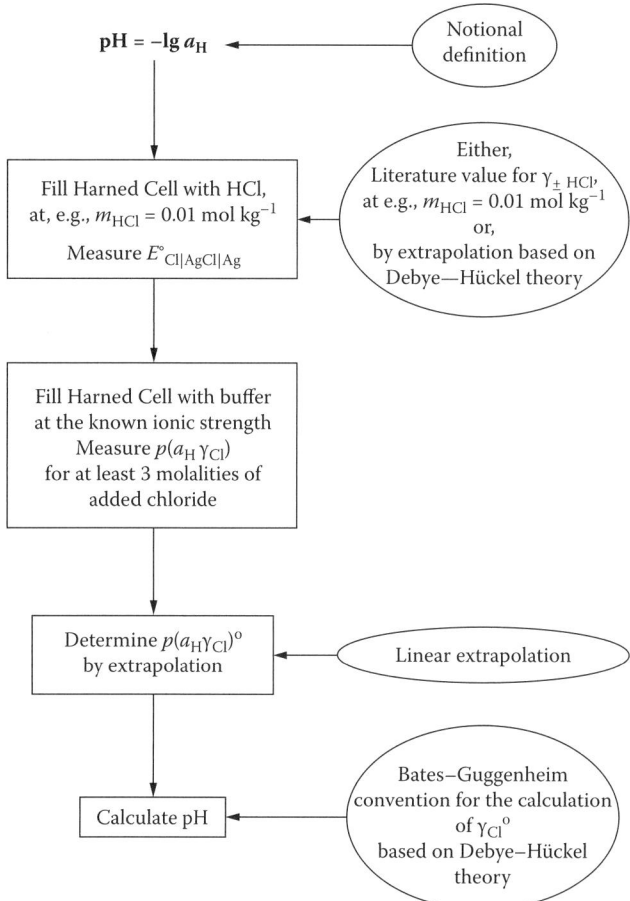

FIGURE 1   Operation of the Harned cell as a primary method for the measurement of absolute pH.

understood way on the method of preparation of the electrodes, but individual determinations of the activity coefficient of HCl at 0.01 mol kg^{-1} are more uniform than values of E°. Hence, the practical determination of the potential difference of the cell with HCl at 0.01 mol kg^{-1} is recommended at 298.15 K at which the mean ionic activity coefficient is 0.904. Dickson [9] concluded that it is not necessary to repeat the measurement of E° at other temperatures, but that it is satisfactory to correct published smoothed values by the observed difference in E° at 298.15 K.

4.3 Acidity function

In NMIs, measurements of Cells I and Ia are often done simultaneously in a thermostat bath. Subtracting eq. 3 from eq. 2 gives

$$\Delta E = E_1 - E_{Ia} = -[(RT/F)\ln 10]\{\lg[(m_H\gamma_H/m^\circ)(m_{Cl}\gamma_{Cl}/m^\circ)] \\ - \lg[(m_{HCl}/m^\circ)^2\gamma^2_{\pm HCl}]\} \qquad (4)$$

which is independent of the standard potential difference. Therefore, the subsequently calculated pH does not depend on the standard potential difference and hence does not depend on the assumption that the standard potential of the hydrogen electrode, $E^\circ(H^+\mid H_2) = 0$ at all temperatures. Therefore, the Harned cell can give an exact comparison between hydrogen ion activities at two different temperatures (in contrast to statements found elsewhere, see, for example, ref. [5]).

The quantity $p(a_H\gamma_{Cl}) = -\lg(a_H\gamma_{Cl})$, on the left-hand side of eq. 2', is called the acidity function [5]. To obtain the quantity pH (according to eq. 1), from the acidity function, it is necessary to evaluate $\lg \gamma_{Cl}$ by independent means. This is done in two steps:

(i) the value of $\lg(a_H\gamma_{Cl})$ at zero chloride molality, $\lg(a_H\gamma_{Cl})^\circ$, is evaluated and (ii) a value for the activity of the chloride ion γ°_{Cl}, at zero chloride molality (sometimes referred to as the limiting or "trace" activity coefficient [9]) is calculated using the Bates–Guggenheim convention [10]. These two steps are described in the following paragraphs.

4.4 Extrapolation of acidity function to zero chloride molality

The value of $\lg(a_H\gamma_{Cl})^\circ$ corresponding to zero chloride molality is determined by linear extrapolation of measurements using Harned cells with at least three added molalities of sodium or potassium chloride ($I < 0.1$ mol kg^{-1}, see Sections 4.5 and 12.6)

$$-\lg(a_H\gamma_{Cl}) = -\lg(a_H\gamma_{Cl})^\circ + Sm_{Cl} \tag{5}$$

where S is an empirical, temperature-dependent constant. The extrapolation is linear, which is expected from Brønsted's observations [11] that specific ion interactions between oppositely charged ions are dominant in mixed strong electrolyte systems at constant molality or ionic strength. However, these acidity function measurements are made on mixtures of weak and strong electrolytes at constant buffer molality, but not constant total molality. It can be shown [12] that provided the change in ionic strength on addition of chloride is less than 20 %, the extrapolation will be linear without detectable curvature. If the latter, less-convenient method of preparation of constant total molality solutions is used, Bates [5] has reported that, for equimolal phosphate buffer, the two methods extrapolate to the same intercept. In an alternative procedure, often useful for partially aqueous mixed solvents where the above extrapolation appears to be curved, multiple application of the Bates–Guggenheim convention to each solution composition gives identical results within the estimated uncertainty of the two intercepts.

4.5 Bates–Guggenheim convention

The activity coefficient of chloride (like the activity coefficient of the hydrogen ion) is an immeasurable quantity. However, in solutions of low ionic strength ($I < 0.1$ mol kg^{-1}), it is possible to calculate the activity coefficient of chloride ion using the Debye–Hückel theory. This is done by adopting the Bates–Guggenheim convention, which assumes the trace activity coefficient of the chloride ion γ°_{Cl} is given by the expression [10]

$$\lg \gamma^\circ_{Cl} = -AI^{1/2}/(1 + Ba\,I^{1/2}) \tag{6}$$

where A is the Debye–Hückel temperature-dependent constant (limiting slope), a is the *mean* distance of closest approach of the ions (ion size parameter), Ba is set equal to 1.5 (mol kg^{-1})$^{-1/2}$ at all temperatures in the range 5–50 °C, and I is the ionic strength of the buffer (which, for its evaluation requires knowledge of appropriate acid dissociation constants). Values of A as a function of temperature can be found in Table A.9 and of B, which is effectively unaffected by revision of dielectric constant data, in Bates [5]. When the numerical value of $Ba = 1.5$ (i.e., without units) is introduced into eq. 6 it should be written as

$$\lg \gamma^\circ_{Cl} = -AI^{1/2}/[1 + 1.5\,(I/m^\circ)^{1/2}] \tag{6'}$$

The various stages in the assignment of primary standard pH values are combined in eq. 7, which is derived from eqs. 2', 5, 6',

$$\begin{aligned}\mathrm{pH(PS)} = \lim_{m_{Cl}\to 0} &\{(E_I - E^\circ)/[(RT/F)\ln 10] + \lg(m_{Cl}/m^\circ)\} \\ &- AI^{1/2}/[1 + 1.5\,(I/m^\circ)^{1/2}],\end{aligned} \tag{7}$$

and the steps are summarized schematically in Fig. 1.

5. Sources of uncertainty in the use of the harned cell

5.1 Potential primary method and uncertainty evaluation

The presentation of the procedure in Section 4 highlights the fact that assumptions based on electrolyte theories [7] are used at three points in the method:

   i. The Debye–Hückel theory is the basis of the extrapolation procedure to calculate the value for the standard potential of the silver–silver chloride electrode, even though it is a published value of $\gamma_{\pm HCl}$ at, e.g., $m = 0.01$ mol kg^{-1}, that is recommended (Section 4.2) to facilitate E° determination.
   ii. Specific ion interaction theory is the basis for using a linear extrapolation to zero chloride (but the change in ionic strength produced by addition of chloride should be restricted to no more than 20 %).
   iii. The Debye–Hückel theory is the basis for the Bates–Guggenheim convention used for the calculation of the trace activity coefficient, γ°_{Cl}.

In the first two cases, the inadequacies of electrolyte theories are sources of uncertainty that limit the extent to which the measured pH is a true representation of $\lg a_H$. In the third case, the use of eq. 6 or 7 is a convention, since the value for Ba is not directly determinable experimentally. Previous recommendations have not included the uncertainty in Ba explicitly within the calculation of the uncertainty of the measurement.

Since eq. 2 is derived from the Nernst equation applied to the thermodynamically well-behaved platinum–hydrogen and silver–silver chloride electrodes, it is recommended that, when used to measure $-\lg(a_H\gamma_{Cl})$ in aqueous solutions, the Harned cell *potentially* meets the agreed definition of a primary method for the measurement. The word "potentially" has been included to emphasize that the method can only achieve primary status if it is operated with the highest metrological qualities (see Sections 6.1–6.2). Additionally, if the Bates–Guggenheim convention is used for the calculation of $\lg \gamma^\circ_{Cl}$, the Harned cell *potentially* meets the agreed definition of a primary method for the measurement of pH, subject to this convention if a realistic estimate of its uncertainty is included. The uncertainty budget for the primary method of measurement by the Harned cell (Cell I) is given in the Annex, Section 13.

Note 2: The experimental uncertainty for a typical primary pH(PS) measurement is of the order of 0.004 (see Table 4).

5.2 Evaluation of uncertainty of the Bates–Guggenheim convention

In order for a measurement of pH made with a Harned cell to be traceable to the SI system, an estimate of the uncertainty of each step must be included in the result. Hence, it is recommended that an estimate of the uncertainty of 0.01 (95% confidence interval) in pH associated with the Bates–Guggenheim convention is used. The extent to which the Bates–Guggenheim convention represents the "true" (but immeasurable) activity coefficient of the chloride ion can be calculated by varying the coefficient Ba between 1.0 and 2.0 (mol kg^{-1})$^{1/2}$. This corresponds to varying the ion-size parameter between 0.3 and 0.6 nm, yielding a range of ± 0.012 (at $I = 0.1$ mol kg^{-1}) and ± 0.007 (at $I = 0.05$ mol kg^{-1}) for γ°_{Cl} calculated using equation [7]. Hence, an uncertainty of 0.01 should cover the full extent of variation. This must be included in the uncertainty of pH values that are to be regarded as traceable to the SI. pH values stated without this contribution to their uncertainty cannot be considered to be traceable to the SI.

5.3 Hydrogen ion concentration

It is rarely required to calculate hydrogen ion concentration from measured pH. Should such a calculation be required, the only consistent, logical way of doing it is to assume $\gamma_H = \gamma_{Cl}$ and set the latter to the appropriate Bates–Guggenheim conventional value. The uncertainties are then those derived from the Bates–Guggenheim convention.

5.4 Possible future approaches

Any model of electrolyte solutions that takes into account both electrostatic and specific interactions for individual solutions would be an improvement over use of the Bates–Guggenheim convention. It is hardly reasonable that a fixed value of the ion-size parameter should be appropriate for a diversity of selected buffer solutions. It is hoped that the Pitzer model of electrolytes [13], which uses a virial equation approach, will provide such an improvement, but data in the literature are insufficiently extensive to make these calculations at the present time. From limited work at 25 °C done on phosphate and carbonate buffers, it seems that changes to Bates–Guggenheim recommended values will be small [14]. It is possible that some anomalies attributed to liquid junction potentials (LJPs) may be resolved.

6. Primary buffer solutions and their required properties

6.1 Requisites for highest metrological quality

In the previous sections, it has been shown that the Harned cell provides a primary method for the determination of pH. In order for a particular buffer solution to be considered a primary buffer solution, it must be of the "highest metrological" quality [15] in accordance with the definition of a primary standard. It is recommended that it have the following attributes [5: p. 95;16,17]:

- High buffer value in the range 0.016–0.07 (mol OH⁻)/pH
- Small dilution value at half concentration (change in pH with change in buffer concentration) in the range 0.01–0.20
- Small dependence of pH on temperature less than ±0.01 K⁻¹
- Low residual LJP <0.01 in pH (see Section 7)
- Ionic strength ≤0.1 mol kg⁻¹ to permit applicability of the Bates–Guggenheim convention
- NMI certificate for specific batch
- Reproducible purity of preparation (lot-to-lot differences of |ΔpH(PS)| < 0.003)
- Long-term stability of stored solid material

Values for the above and other important parameters for the selected primary buffer materials (see Section 6.2) are given in Table 1.

Note 3: The long-term stability of the solid compounds (>5 years) is a requirement not met by borax [16]. There are also doubts about the extent of polyborate formation in 0.05 mol kg⁻¹ borax solutions, and hence this solution is not accorded primary status.

6.2 Primary standard buffers

Since there can be significant variations in the purity of samples of a buffer of the same nominal chemical composition, it is essential that the primary buffer material used has been certified with values that have been measured with Cell I. The Harned cell has been used by many NMIs for accurate measurements of pH of

buffer solutions. Comparisons of such measurements have been carried out under EUROMET collaboration [18], which have demonstrated the high comparability of measurements (0.005 in pH) in different laboratories of samples from the same batch of buffer material. Typical values of the pH(PS) of the seven solutions from the six accepted primary standard reference buffers, which meet the conditions stated in Section 6.1, are listed in Table 2. These listed pH(PS) values have been derived from certificates issued by NBS/NIST over the past 35 years. Batch-to-batch variations in purity can result in changes in the pH value of samples of at most 0.003. The typical values in Table 2 should not be used in place of the certified value (from a Harned cell measurement) for a specific batch of buffer material.

The required attributes listed in Section 6.1 effectively limit the range of primary buffers available to between pH 3 and 10 (at 25 °C). Calcium hydroxide and potassium tetroxalate have been excluded because the contribution of hydroxide or hydrogen ions to the ionic strength is significant. Also excluded are the nitrogen bases of the type BH⁺ [such as tris(hydroxymethyl)aminomethane and piperazine phosphate] and the zwitterionic buffers (e.g., HEPES and MOPS [19]). These do not comply because either the Bates–Guggenheim convention is not applicable, or the LJPs are high. This means the choice of primary standards is restricted to buffers derived from oxy-carbon, -phosphorus, -boron, and mono-, di-, and tri-protic carboxylic acids. In the future, other buffer systems may fulfill the requirements listed in Section 6.1.

7. Consistency of primary buffer solutions

7.1 Consistency and the liquid junction potential

Primary methods of measurement are made with cells without transference as described in Sections 1–6. Less-complex, secondary methods use cells with transference, which contain liquid junctions. A single LJP is immeasurable, but differences in LJP can be estimated. LJPs vary with the composition of the solutions forming the junction and the geometry of the junction.

Equation 7 for Cell I applied successively to two primary standard buffers, PS₁, PS₂, gives

$$\Delta pH_I = pH_I(PS_2) - pH_I(PS_1) = \lim m_{Cl\to0}\{E_I(PS_2)/k - E_I(PS_1)/k\}$$
$$- A\{I_{(2)}^{1/2}/[1 + 1.5\,(I_{(2)}/m°)^{1/2}] - I_{(1)}^{1/2}/[1 + 1.5\,(I_{(1)}/m°)^{1/2}]\} \quad (8)$$

where $k = (RT/F)\ln 10$ and the last term is the ratio of trace chloride activity coefficients $\lg[\gamma°_{Cl(2)}/\gamma°_{Cl(1)}]$, conventionally evaluated via B-G eq. 6′.

Note 4: Since the convention may unevenly affect the $\gamma°_{Cl(2)}$ and $\gamma°_{Cl(1)}$ estimations, ΔpH_I differs from the true value by the unknown contribution: $\lg[\gamma°_{Cl(2)}/\gamma°_{Cl(1)}] - A\{I_{(1)}^{1/2}/[1 + 1.5(I_{(1)}/m°)^{1/2}] - I_{(2)}^{1/2}/[1 + 1.5(I_{(2)}/m°)^{1/2}]\}$.

A second method of comparison is by measurement of Cell II in which there is a salt bridge with two free-diffusion liquid junctions

Pt | H₂ | PS₂ ⋮ KCl (≥3.5 mol dm⁻³) ⋮ PS₁ | H₂ | Pt   Cell II

for which the spontaneous cell reaction is a dilution,

$$H^+(PS_1) \to H^+(PS_2)$$

which gives the pH difference from Cell II as

$$\Delta pH_{II} = pH_{II}(PS_2) - pH_{II}(PS_1) = E_{II}/k - [(E_{j2} - E_{j1})/k] \quad (9)$$

where the subscript II is used to indicate that the pH difference between the same two buffer solutions is now obtained from Cell II. ΔpH_{II} differs from ΔpH_I (and both differ from the true value ΔpH_I) since it depends on unknown quantity, the residual LJP,

TABLE 1   Summary of Useful Properties of Some Primary and Secondary Standard Buffer Substances and Solutions [5]

Salt or Solid Substance	Molecular Formula	Molality/ $mol\ kg^{-1}$	Molar Mass/ $g\ mol^{-1}$	Density/ $g\ dm^{-3}$	Amount Conc. at 20 °C/ $mol\ dm^{-3}$	Mass/g to Make 1 dm^3	Dilution Value $\Delta pH_{1/2}$	Buffer Value (β)/ $mol\ OH^-\ dm^{-3}$	pH Temperature Coefficient/ K^{-1}
Potassium tetroxalate dihydrate	$KH_3C_4O_8 \cdot 2H_2O$	0.1	254.191	1.0091	0.09875	25.101			
Potassium tetroxalate dihydrate	$KH_3C_4O_8 \cdot 2H_2O$	0.05	254.191	1.0032	0.04965	12.620	0.186	0.070	0.001
Potassium hydrogen tartrate (sat. at 25 °C)	$KHC_4H_4O_6$	0.0341	188.18	1.0036	0.034	6.4	0.049	0.027	−0.0014
Potassium dihydrogen citrate	$KH_2C_6H_5O_7$	0.05	230.22	1.0029	0.04958	11.41	0.024	0.034	−0.022
Potassium hydrogen phthalate	$KHC_8H_4O_4$	0.05	204.44	1.0017	0.04958	10.12	0.052	0.016	0.00012
Disodium hydrogen orthophosphate + potassium dihydrogen orthophosphate	Na_2HPO_4	0.025	141.958	1.0038	0.02492	3.5379	0.080	0.029	−0.0028
	KH_2PO_4	0.025	136.085			3.3912			
Disodium hydrogen orthophosphate + potassium dihydrogen orthophosphate	Na_2HPO_4	0.03043	141.959	1.0020	0.08665	4.302	0.07	0.016	−0.0028
	KH_2PO_4	0.00869	136.085		0.03032	1.179			
Disodium tetraborate decahydrate	$Na_2B_4O_7 \cdot 10H_2O$	0.05	381.367	1.0075	0.04985	19.012			
Disodium tetraborate decahydrate	$Na_2B_4O_7 \cdot 10H_2O$	0.01	381.367	1.0001	0.00998	3.806	0.01	0.020	−0.0082
Sodium hydrogen carbonate + sodium carbonate	$NaHCO_3$	0.025	84.01	1.0013	0.02492	2.092	0.079	0.029	−0.0096
	Na_2CO_3	0.025	105.99			2.640			
Calcium hydroxide (sat. at 25°C)	$Ca(OH)_2$	0.0203	74.09	0.9991	0.02025	1.5	−0.28	0.09	−0.033

TABLE 2 Typical Values of pH(PS) for Primary Standards at 0–50 °C (see Section 6.2)

Primary Standards (PS)	Temp./°C										
	0	5	10	15	20	25	30	35	37	40	50
Sat. potassium hydrogen tartrate (at 25 °C)						3.557	3.552	3.549	3.548	3.547	3.549
0.05 mol kg⁻¹ potassium dihydrogen citrate	3.863	3.840	3.820	3.802	3.788	3.776	3.766	3.759	3.756	3.754	3.749
0.05 mol kg⁻¹ potassium hydrogen phthalate	4.000	3.998	3.997	3.998	4.000	4.005	4.011	4.018	4.022	4.027	4.050
0.025 mol kg⁻¹ disodium hydrogen phosphate + 0.025 mol kg⁻¹ potassium dihydrogen phosphate	6.984	6.951	6.923	6.900	6.881	6.865	6.853	6.844	6.841	6.838	6.833
0.03043 mol kg⁻¹ disodium hydrogen phosphate + 0.008695 mol kg⁻¹ potassium dihydrogen phosphate	7.534	7.500	7.472	7.448	7.429	7.413	7.400	7.389	7.386	7.380	7.367
0.01 mol kg⁻¹ disodium tetraborate	9.464	9.395	9.332	9.276	9.225	9.180	9.139	9.102	9.088	9.068	9.011
0.025 mol kg⁻¹ sodium hydrogen carbonate + 0.025 mol kg⁻¹ sodium carbonate	10.317	10.245	10.179	10.118	10.062	10.012	9.966	9.926	9.910	9.889	9.828

$RLJP = (E_{j2} - E_{j1})$, whose exact value could be determined if the true ΔpH were known.

Note 5: The subject of liquid junction effects in ion-selective electrode potentiometry has been comprehensively reviewed [20]. Harper [21] and Bagg [22] have made computer calculations of LJPs for simple three-ion junctions (such as HCl + KCl), the only ones for which mobility and activity coefficient data are available. Breer, Ratkje, and Olsen [23] have thoroughly examined the possible errors arising from the commonly made approximations in calculating LJPs for three-ion junctions. They concluded that the assumption of linear concentration profiles has less-severe consequences (~0.1–1.0 mV) than the other two assumptions of the Henderson treatment, namely constant mobilities and neglect of activity coefficients, which can lead to errors in the order of 10 mV. Breer et al. concluded that their calculations supported an earlier statement [24] that in ion-selective electrode potentiometry, the theoretical Nernst slope, even for dilute sample solutions, could never be attained because of liquid junction effects.

Note 6: According to IUPAC recommendations on nomenclature and symbols [3], a single vertical bar (|) is used to represent a phase boundary, a dashed vertical bar (¦) represents a liquid–liquid junction between two electrolyte solutions (across which a potential difference will occur), and a double dashed vertical bar (¦¦) represents a similar liquid junction, in which the LJP is assumed to be effectively zero (~1 % of cell potential difference). Hence, terms such as that in square brackets on the right-hand side of eq. 9 are usually ignored, and the liquid junction is represented by ¦¦. However, in the Annex, the symbol ¦ is used because the error associated with the liquid junction is included in the analysis. For ease of comparison, numbers of related equations in the main text and in the Annex are indicated.

Note 7: The polarity of Cell II will be negative on the left, i.e., − | +, when pH(PS₂) > pH(PS₁). The LJP E_j of a single liquid junction is defined as the difference in (Galvani) potential contributions to the total cell potential difference arising at the interface from the buffer solution less that from the KCl solution. For instance, in Cell II, $E_{j1} = E(S_1) − E(KCl)$ and $E_{j2} = E(S_2) − E(KCl)$. It is negative when the buffer solution of interest is acidic and positive when it is alkaline, provided that E_j is principally caused by the hydrogen, or hydroxide, ion content of the solution of interest (and only to a smaller degree by its alkali ions or anions). The residual liquid junction potential (RLJP), the difference E_j(right) − E_j(left), depends on the relative magnitudes of the individual E_j values and has the opposite polarity to the potential difference E of the cell. Hence, in Cell II the RLJP, $E_{j1}(PS_1) − E_{j2}(PS_2)$, has a polarity + | − when pH(S₂) > pH(S₁).

Notwithstanding the foregoing, comparison of pH$_{II}$ values from the Cell II with two liquid junctions (eq. 9) with the assigned pH$_I$(PS) values for the same two primary buffers measured with Cell I (eq. 8) makes an estimation of RLJPs possible [5]:

$$[pH_I(PS_2) − pH_{II}(PS_2)] − [pH_I(PS_1) − pH_{II}(PS_1)] = (E_{j2} − E_{j1})/k = RLJP \qquad (10)$$

With the value of RLJP set equal to zero for equimolal phosphate buffer (taken as PS₁) then [pH$_I$(PS₂) − pH$_{II}$(PS₂)] is plotted against pH(PS). Results for free-diffusion liquid junctions formed in a capillary tube with cylindrical symmetry at 25 °C are shown in Fig. 2 [25, and refs. cited therein].

Note 8: For 0.05 mol kg⁻¹ tetroxalate, the published values [26] for Cell II with free-diffusion junctions are wrong [27,28].

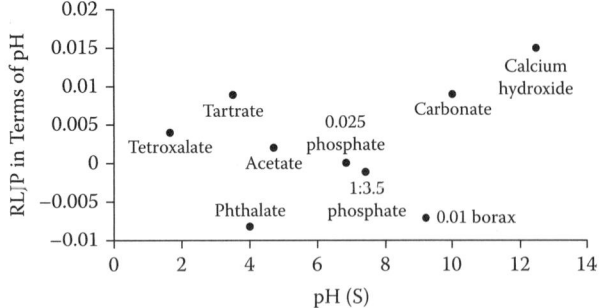

FIGURE 2 Some values of residual LJPs in terms of pH with reference to the value for 0.025 mol kg^{-1} Na$_2$HPO$_4$ + 0.025 mol kg^{-1} KH$_2$PO$_4$ (0.025 phosphate buffer) taken as zero [25].

Values such as those shown in Fig. 2 give an indication of the extent of possible systematic uncertainties for primary standard buffers arising from three sources:

i. Experimental uncertainties, including any variations in the chemical purity of primary buffer materials (or variations in the preparation of the solutions) if measurements of Cells I and II were not made in the same laboratory at the same occasion.

ii. Variation in RLJPs between primary buffers.

iii. Inconsistencies resulting from the application of the Bates–Guggenheim convention to chemically different buffer solutions of ionic strengths less than 0.1 mol kg^{-1}.

It may be concluded from examination of the results in Fig. 2, that a consistency no better than 0.01 can be ascribed to the primary pH standard solutions of Table 2 in the pH range 3–10. This value will be greater for less reproducibly formed liquid junctions than the free-diffusion type with cylindrical symmetry.

Note 9: Considering the conventional nature of eq. 10, and that the irreproducibility of formation of geometry-dependent devices exceeds possible bias between carefully formed junctions of known geometry, the RLJP contribution, which is included in the difference between measured potential differences of cells with transference, is treated as a statistical, and not a systematic error.

Note 10: Values of RLJP depend on the Bates–Guggenheim convention through the last term in eq. 8 and would be different if another convention were chosen. This interdependence of the single ion activity coefficient and the LJP may be emphasized by noting that it would be possible *arbitrarily* to reduce RLJP values to zero for each buffer by adjusting the ion-size parameter in eq. 6.

7.2 Computational approach to consistency

The consistency between conventionally assigned pH values can also be assessed by a computational approach. The pH values of standard buffer solutions have been calculated from literature values of acid dissociation constants by an iterative process. The arbitrary extension of the Bates–Guggenheim convention for chloride ion, to all ions, leads to the calculation of ionic activity coefficients of all ionic species, ionic strength, buffer capacity, and calculated pH values. The consistency of these values with primary pH values obtained using Cell I was 0.01 or lower between 10 and 40 °C [29,30].

8. Secondary standards and secondary methods of measurement

8.1 Secondary standards derived from Harned cell measurements

Substances that do not fulfill all the criteria for primary standards but to which pH values can be assigned using Cell I are considered to be secondary standards. Reasons for their exclusion as primary standards include, inter alia:

i. Difficulties in achieving consistent, suitable chemical quality (e.g., acetic acid is a liquid).

ii. High LJP, or inappropriateness of the Bates–Guggenheim convention (e.g., other charge-type buffers).

Therefore, they do not comply with the stringent criterion for a primary measurement of being of the highest metrological quality. Nevertheless, their pH(SS) values can be determined. Their consistency with the primary standards should be checked with the method described in Section 7. The primary and secondary standard materials should be accompanied by certificates from NMIs in order for them to be described as certified reference materials (CRMs). Some illustrative pH(SS) values for secondary standard materials [5,17,25,31,32] are given in Table 3.

8.2 Secondary standards derived from primary standards

In most applications, the use of a high-accuracy primary standard for pH measurements is not justified, if a traceable secondary standard of sufficient accuracy is available. Several designs of cells are available for comparing the pH values of two buffer solutions. However, there is no primary method for measuring the *difference* in pH between two buffer solutions for reasons given in Section 8.6. Such measurements could involve either using a cell successively with two buffers, or a single measurement with a cell containing two buffer solutions separated by one or two liquid junctions.

8.3 Secondary standards derived from primary standards of the same nominal composition using cells without salt bridge

The most direct way of comparing pH(PS) and pH(SS) is by means of the single-junction Cell III [33].

$$\text{Pt} \mid \text{H}_2 \mid \text{buffer S}_2 \mathbin{\vert\vert} \text{buffer S}_1 \mid \text{H}_2 \mid \text{Pt} \qquad \text{Cell III}$$

The cell reaction for the spontaneous dilution reaction is the same as for Cell II, and the pH difference is given, see Note 6, by

$$\text{pH(S}_2) - \text{pH(S}_1) = E_{III}/k \qquad \text{(11) cf.(A-7)}$$

The buffer solutions containing identical Pt | H$_2$ electrodes with an identical hydrogen pressure are in direct contact via a vertical sintered glass disk of a suitable porosity (40 μm). The LJP formed between the two standards of nominally the same composition will be particularly small and is estimated to be in the μV range. It will, therefore, be less than 10 % of the potential difference measured if the pH(S) values of the standard solutions are in the range $3 \le \text{pH(S)} \le 11$ and the difference in their pH(S) values is not larger than 0.02. Under these conditions, the LJP is not dominated by the hydrogen and hydroxyl ions but by the other ions (anions, alkali metal ions). The proper functioning of the cell can

TABLE 3 Values of pH(SS) of Some Secondary Standards from Harned Cell I Measurements

Secondary Standards	Temp./°C									
	0	5	10	15	20	25	30	37	40	50
0.05 mol kg^{-1} potassium tetroxalate[a] [5,17]		1.67	1.67	1.67	1.68	1.68	1.68	1.69	1.69	1.71
0.05 mol kg^{-1} sodium hydrogen diglycolate[b] [31]		3.47	3.47	3.48	3.48	3.49	3.50	3.52	3.53	3.56
0.1 mol dm^{-3} acetic acid + 0.1 mol dm^{-3} sodium acetate [25]	4.68	4.67	4.67	4.66	4.66	4.65	4.65	4.66	4.66	4.68
0.1 mol dm^{-3} acetic acid + 0.1 mol dm^{-3} sodium acetate [25]	4.74	4.73	4.73	4.72	4.72	4.72	4.72	4.73	4.73	4.75
0.02 mol kg^{-1} piperazine phosphate[c] [32]	6.58	6.51	6.45	6.39	6.34	6.29	6.24	6.16	6.14	6.06
0.05 mol kg^{-1} tris hydrochloride + 0.01667 mol kg^{-1} tris[c] [5]	8.47	8.30	8.14	7.99	7.84	7.70	7.56	7.38	7.31	7.07
0.05 mol kg^{-1} disodium tetraborate	9.51	9.43	9.36	9.30	9.25	9.19	9.15	9.09	9.07	9.01
Saturated (at 25°C) calcium hydroxide [5]	13.42	13.21	13.00	12.81	12.63	12.45	12.29	12.07	11.98	11.71

[a] potassium trihydrogen dioxalate ($KH_3C_4O_8$)

[b] sodium hydrogen 2,2′-oxydiacetate

[c] 2-amino-2-(hydroxymethyl)-1,3 propanediol or tris(hydroxymethyl)aminomethane

be checked by measuring the potential difference when both sides of the cell contain the same solution.

8.4 Secondary standards derived from primary standards using cells with salt bridge

The cell that includes a hydrogen electrode [corrected to 1 atm (101.325 kPa) partial pressure of hydrogen] and a reference electrode, the filling solution of which is a saturated or high concentration of the almost equitransferent electrolyte, potassium chloride, hence minimizing the LJP, is, see Note 6:

Ag|AgC|KCl (≥3.5 mol dm^{-3}) ⋮⋮ buffer S | H$_2$ | Pt   Cell IV

Note 11: Other electrolytes, e.g., rubidium or cesium chloride, are more equitransferent [34].

Note 12: Cell IV is written in the direction: *reference| indicator*

i. for conformity of treatment of all hydrogen ion-responsive electrodes and ion-selective electrodes with various choices of reference electrode, and partly,

ii for the practical reason that pH meters usually have one low impedance socket for the reference electrode, assumed negative, and a high-impedance terminal with a different plug, usually for a glass electrode.

With this convention, whatever the form of hydrogen ion-responsive electrode used (e.g., glass or quinhydrone), or whatever the reference electrode, the potential of the hydrogen-ion responsive electrode always decreases (becomes more negative) with increasing pH (see Fig. 3).

This convention was used in the 1985 document [2] and is also consistent with the treatment of ion-selective electrodes [35]. In effect, it focuses attention on the indicator electrode, for which the potential is then given by the Nernst equation for the single-electrode potential, written as a reduction process, in accord with the Stockholm convention [36]:

$$\text{For Ox} + ne^- \rightarrow \text{Red}, \quad E = E° - (k/n) \lg(a_{red}/a_{ox})$$

(where *a* is activity), or, for the hydrogen gas electrode at 1 atm partial pressure of hydrogen gas:

$$H^+ + e^- \rightarrow {}^{1/2}H_2 \quad E = E° + k \lg a_{H^+} = E° - k\text{pH}$$

The equation for Cell IV is, therefore:

$$\text{pH(S)} = -[E_{IV}(S) - E_{IV}°']/k \qquad (12)$$

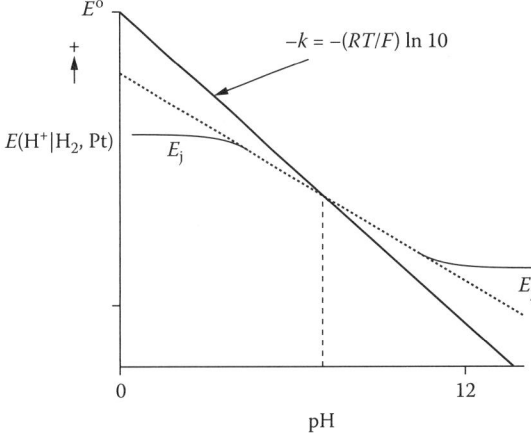

FIGURE 3 Schematic plot of the variation of potential difference (----) for the cell $^-$Ag|AgCl|KCl H$^+$(buffer)|H$_2$|Pt$^+$ with pH and illustrating the choice of sign convention. The effect of LJP is indicated (----) with its variation of pH as given by the Henderson equation (see, e.g., ref. [5]). The approximate linearity (----) in the middle pH region should be noted. Both lines have been grossly exaggerated in their deviation from the Nernst line since otherwise they would be indistinguishable from each other and the Nernst line. For the calomel electrode Hg|Hg$_2$Cl$_2$|KCl and the thallium amalgam|thallium(I) chloride electrode Hg|Tl(Hg)|TlCl|KCl, or any other constant potential reference electrode, the diagram is the same.

in which $E_{IV}^{o\prime}$ is the standard potential, which includes the term lg $a_{Cl}/m°$, and E_j is the LJP.

Note 13: Mercury–mercury(I) chloride (calomel) and thallium amalgam–thallium (I) chloride reference electrodes are alternative choices to the silver–silver chloride electrode in Cell IV.

The consecutive use of two such cells containing buffers S_1 and S_2 gives the pH difference of the solutions

$$pH(S_2) - pH(S_1) = -[E_{IV}(S_2) - E_{IV}(S_1)]/k \qquad (13) \text{ cf. (A-8)}$$

Note 14: Experimentally, a three-limb electrode vessel allowing simultaneous measurement of two Cell IIs may be used [25] with the advantage that the stability with time of the electrodes and of the liquid junctions can be checked. The measurement of cells of type II, which has a salt bridge with two liquid junctions, has been discussed in Section 7.

Cells II and IV may also be used to measure the value of secondary buffer standards that are not compatible with the silver–silver chloride electrode used in Cell I. Since the LJPs in Cells II and IV are minimized by the use of an equitransferent salt, these cells are suitable for use with secondary buffers that have a different concentration and/or an ionic strength greater than the limit ($I \le 0.1$ mol kg^{-1}) imposed by the Bates–Guggenheim convention. They may, however, also be used for comparing solutions of the same nominal composition.

8.5 Secondary standards from glass electrode cells

Measurements cannot be made with a hydrogen electrode in Cell IV, for example, if the buffer is reduced by hydrogen gas at the platinum (or palladium-coated platinum) electrode. Cell V involving a glass electrode and silver–silver chloride reference electrode may be used instead in consecutive measurements, with two buffers S_1, S_2 (see Section 11 for details).

8.6 Secondary methods

The equations given for Cells II to V show that these cannot be considered primary (ratio) methods for measuring pH difference [1], (see also Section 12.1) because the cell reactions involve transference, or the irreversible inter-diffusion of ions, and hence an LJP contribution to the measured potential difference. The value of this potential difference depends on the ionic constituents, their concentrations and the geometry of the liquid junction between the solutions. Hence, the measurement equations contain terms that, although small, are not quantifiable, and the methods are secondary not primary.

9. Consistency of secondary standard buffer solutions established with respect to primary standards

9.1 Summary of procedures for establishing secondary standards

The following procedures may be distinguished for establishing secondary standards (SS) with respect to primary standards:

    i. For SS of the same nominal composition as PS, use Cells III or II.
    ii. For SS of different composition, use Cells IV or II.
    iii. For SS not compatible with platinum hydrogen electrode, use Cell V (see Section 11.1).

Although any of Cells II to V could be used for certification of secondary standards with stated uncertainty, employing different procedures would lead to inconsistencies. It would be difficult to define specific terminology to distinguish each of these procedures or to define any rigorous hierarchy for them. Hence, the methods should include estimates of the typical uncertainty for each. The choice between methods should be made according to the uncertainty required for the application (see Section 10 and Table 4).

9.2 Secondary standard evaluation from primary standards of the same composition

It is strongly recommended that the preferred method for assigning secondary standards should be a procedure in which measurements are made with respect to the primary buffer of nominally the same chemical composition. All secondary standards should be accompanied by a certificate relating to that

TABLE 4 Summary of Recommended Target Uncertainties

	U(pH) (For coverage factor 2)	Comments
PRIMARY STANDARDS		
Uncertainty of PS measured (by an NMI) with Harned Cell I	0.004	
Repeatability of PS measured (by an NMI) with Harned Cell I	0.0015	
Reproducibility of measurements in comparisons with Harned Cell I	0.003	EUROMET comparisons
Typical variations between batches of PS buffers	0.003	
SECONDARY STANDARDS		
Value of SS compared with same PS material with Cell III	0.004	increase in uncertainty is negligible relative to PS used e.g., biological buffers
Value of SS measured in Harned Cell I	0.01	
Value of SS labeled against different PS with Cell II or IV	0.015	
Value of SS (not compatible with Pt | H$_2$) measured with Cell V	0.02	example based on phthalate
ELECTRODE CALIBRATION		
Multipoint (5 point) calibration	0.01–0.03	
Calibration (2-point) by bracketing	0.02–0.03	
Calibration (1-point), ΔpH = 3 and assumed slope	0.3	

Note: None of the above includes the uncertainty associated with the Bates–Guggenheim convention so the results cannot be considered to be traceable to SI (see Section 5.2).

particular batch of reference material as significant batch-to-batch variations are likely to occur. Some secondary standards are disseminated in solution form. The uncertainty of the pH values of such solutions may be larger than those for material disseminated in solid form.

9.3 Secondary standard evaluation when there is no primary standard of the same composition

It may sometimes be necessary to set up a secondary standard when there is no primary standard of the same chemical composition available. It will, therefore, be necessary to use either Cells II, IV, or V, and a primary or secondary standard buffer of different chemical composition. Buffers measured in this way will have a different status from those measured with respect to primary standards because they are not directly traceable to a primary standard of the same chemical composition. This different status should be reflected in the, usually larger, uncertainty quoted for such a buffer. Since this situation will only occur for buffers when a primary standard is not available, no special nomenclature is recommended to distinguish the different routes to secondary standards. Secondary buffers of a composition different from those of primary standards can also be derived from measurements on Cell I, provided the buffer is compatible with Cell I. However, the uncertainty of such standards should reflect the limitations of the secondary standard (see Table 4).

10. Target uncertainties for the measurement of secondary buffer solutions

10.1 Uncertainties of secondary standards derived from primary standards

Cells II to IV (and occasionally Cell V) are used to measure secondary standards with respect to primary standards. In each case, the limitations associated with the measurement method will result in a greater uncertainty for the secondary standard than the primary standard from which it was derived.

Target uncertainties are listed in Table 4. However, these uncertainties do not take into account the uncertainty contribution arising from the adoption of the Bates–Guggenheim convention to achieve traceability to SI units.

10.2 Uncertainty evaluation [37]

Summaries of typical uncertainty calculations for Cells I–V are given in the Annex (Section 13).

11. Calibration of pH meter-electrode assemblies and target uncertainties for unknowns

11.1 Glass electrode cells

Practical pH measurements are carried out by means of Cell V reference electrode | KCl (c ≥ 3.5 mol dm^{-3}) ¦¦ solution[pH(S) or pH(X)] | glass electrode Cell V and pH(X) is obtained, see Note 6, from eq. 14

$$pH(X) = pH(S) - [E_V(X) - E_V(S)] \qquad (14)$$

This is a one-point calibration (see Section 11.3).

These cells often use glass electrodes in the form of single probes or combination electrodes (glass and reference electrodes fashioned into a single probe, a so-called "combination electrode").

The potential difference of Cell V is made up of contributions arising from the potentials of the glass and reference electrodes and the liquid junction (see Section 7.1).

Various random and systematic effects must be noted when using these cells for pH measurements:

i. Glass electrodes may exhibit a slope of the E vs. pH function smaller than the theoretical value $[k = (RT/F)\ln 10]$, often called a sub-Nernstian response or practical slope k', which is experimentally determinable. A theoretical explanation for the sub-Nernstian response of pH glass electrodes in terms of the dissociation of functional groups at the glass surface has been given [38].

ii. The response of the glass electrode may vary with time, history of use, and memory effects. It is recommended that the response time and the drift of the electrodes be taken into account [39].

iii. The potential of the glass electrode is strongly temperature-dependent, as to a lesser extent are the other two terms. Calibrations and measurements should, therefore, be carried out under temperature-controlled conditions.

iv. The LJP varies with the composition of the solutions forming the junction, e.g., with pH (see Fig. 2). Hence, it will change if one solution [pH(S) or pH(X) in Cell V is replaced by another. It is also affected by the geometry of the liquid junction device. Hence, it may be different if a free-diffusion type junction, such as that used to measure the RLJP (see Section 7.1), is replaced by another type, such as a sleeve, ceramic diaphragm, fiber, or platinum junction [39,40].

v. Liquid junction devices, particularly some commercial designs, may suffer from memory and clogging effects.

vi. The LJP may be subject to hydrodynamic effects, e.g., stirring.

Since these effects introduce errors of unknown magnitude, the measurement of an unknown sample requires a suitable calibration procedure. Three procedures are in common use based on calibrations at one point (one-point calibration), two points (two-point calibration or bracketing), and a series of points (multipoint calibration).

11.2 Target uncertainties for unknowns

Uncertainties in pH(X) are obtained, as shown below, by several procedures involving different numbers of experiments. Numerical values of these uncertainties obtained from the different calibration procedures are, therefore, not directly comparable. It is, therefore, not possible at the present time to make a universal recommendation of the best procedure to adopt for all applications. Hence, the target uncertainty for the unknown is given, which the operator of a pH meter electrode assembly may reasonably seek to achieve. Values are given for each of the three techniques (see Table 4), but the uncertainties attainable experimentally are critically dependent on the factors listed in Section 11.1 above, on the quality of the electrodes, and on the experimental technique for changing solutions.

In order to obtain the overall uncertainty of the measurement, uncertainties of the respective pH(PS) or pH(SS) values must be taken into account (see Table 4). Target uncertainties given below, and in Table 4, refer to calibrations performed by the use of standard buffer solutions with an uncertainty U[pH(PS)] or U[pH(SS)] d 0.01. The overall uncertainty becomes higher if standards with higher uncertainties are used.

11.3 One-point calibration

A single-point calibration is insufficient to determine both slope and one-point parameters. The theoretical value for the slope can be assumed, but the practical slope may be up to 5% lower. Alternatively, a value for the practical slope can be assumed from the manufacturer's prior calibration. The one-point calibration, therefore, yields only an estimate of pH(X). Since both parameters may change with age of the electrodes, this is not a reliable procedure. Based on a measurement for which $\Delta pH = |pH(X) - pH(S)| = 5$, the expanded uncertainty would be $U = 0.5$ in pH(X) for $k' = 0.95k$, but assumed theoretical, or $U = 0.3$ in pH(X) for $\Delta pH = |pH(X) - pH(S)| = 3$ (see Table 4). This approach could be satisfactory for certain applications. The uncertainty will decrease with decreasing difference pH(X) – pH(S) and be smaller if k' is known from prior calibration.

11.4 Two-point calibration {target uncertainty, $U[pH(X)] = 0.02$–0.03 at $25\,°C$}

In the majority of practical applications, glass electrodes cells (Cell V) are calibrated by two-point calibration, or bracketing, procedure using two standard buffer solutions, with pH values $pH(S_1)$ and $pH(S_2)$, bracketing the unknown pH(X). Bracketing is often taken to mean that the $pH(S_1)$ and $pH(S_2)$ buffers selected should be those that are immediately above and below pH(X). This may not be appropriate in all situations and choice of a wider range may be better.

If the respective potential differences measured are $E_V(S_1)$, $E_V(S_2)$, and $E_V(X)$, the pH value of the unknown, pH(X), is obtained from eq. 15

$$pH(X) = pH(S_1) - [E_V(X) - E_V(S_1)]/k' \qquad (15)\ cf.\ (A\text{-}10)$$

where the practical slope factor (k') is given by

$$k' = [EV(S_1) - E_V(S_2)]/[pH(S_2) - pH(S_1)] \qquad (16)$$

An example is given in the Annex, Section 13.

11.5 Multipoint calibration {target uncertainty: $U[pH(X)] = 0.01$–0.03 at $25\,°C$}

Multipoint calibration is carried out using up to five standard buffers [39,40]. The use of more than five points does not yield any significant improvement in the statistical information obtainable.

The calibration function of Cell V is given by eq. 17

$$E_V(S) = E_V° - k'pH(S) \qquad (17)\ cf.\ (A\text{-}11)$$

where $E_V(S)$ is the measured potential difference when the solution of pH(S) in Cell V is a primary or secondary standard buffer. The intercept, or "standard potential", $E_V°$ and k', the practical slope are determined by linear regression of eq. 17 [39–41].

pH(X) of an unknown solution is then obtained from the potential difference, $E_V(X)$, by

$$pH(X) = [E_V° - E_V(X)]/k' \qquad (18)\ cf.\ (A\text{-}12)$$

Additional quantities obtainable from the regression procedure applied to eq. 17 are the uncertainties $u(k')$ and $u(E_V°)$ [40]. Multipoint calibration is recommended when minimum uncertainty and maximum consistency are required over a wide range of pH(X) values. This applies, however, only to that range of pH values in which the calibration function is truly linear. In nonlinear regions of the calibration function, the two-point method has clear advantages provided that $pH(S_1)$ and $pH(S_2)$ are selected to be as close to pH(X) as possible.

Details of the uncertainty computations for the multipoint calibration have been given [40], and an example is given in the Annex. The uncertainties are recommended as a means of checking the performance characteristics of pH meter-electrode assemblies [40]. By careful selection of electrodes for multipoint calibration, uncertainties of the unknown pH(X) can be kept as low as $U[pH(X)] = 0.01$.

In modern microprocessor pH meters, potential differences are often transformed automatically into pH values. Details of the calculations involved in such transformations, including the uncertainties, are available [41].

12. Glossary [2,15,44]

12.1 Primary method of measurement

A primary method of measurement is a method having the highest metrological qualities, whose operation can be completely described and understood, for which a complete uncertainty statement can be written down in terms of SI units.

A primary direct method measures the value of an unknown without reference to a standard of the same quantity.

A primary ratio method measures the value of a ratio of an unknown to a standard of the same quantity; its operation must be completely described by a measurement equation.

12.2 Primary standard

Standard that is designated or widely acknowledged as having the highest metrological qualities and whose value is accepted without reference to other standards of the same quantity.

12.3 Secondary standard

Standard whose value is assigned by comparison with a primary standard of the same quantity.

12.4 Traceability

Property of the result of a measurement or the value of a standard whereby it can be related to stated references, usually national or international standards, through an unbroken chain of comparisons all having stated uncertainties. The concept is often expressed by the adjective traceable. The unbroken chain of comparisons is called a traceability chain.

12.5 Primary pH standards

Aqueous solutions of selected reference buffer solutions to which pH(PS) values have been assigned over the temperature range 0–50 °C from measurements on cells without transference, called Harned cells, by use of the Bates–Guggenheim convention.

12.6 Bates–Guggenheim convention

A convention based on a form of the Debye–Hückel equation that approximates the logarithm of the single ion activity coefficient of chloride and uses a fixed value of 1.5 for the product Ba in the denominator at all temperatures in the range 0–50 °C (see eqs. 4, 5) and ionic strength of the buffer < 0.1 mol kg^{-1}.

12.7 Secondary pH standards

Values that may be assigned to secondary standard pH(SS) solutions at each temperature:

i. with reference to [pH(PS)] values of a primary standard of the same nominal composition by Cell III,

ii with reference to [pH(PS)] values of a primary standard of different composition by Cells II, IV or V, or

iii by use of Cell I.

Note 15: This is an exception to the usual definition, see Section 12.3.

12.8 pH glass electrode

Hydrogen-ion responsive electrode usually consisting of a bulb, or other suitable form, of special glass attached to a stem of high-resistance glass complete with internal reference electrode and internal filling solution system. Other geometrical forms may be appropriate for special applications, e.g., capillary electrode for measurement of blood pH.

12.9 Glass electrode error

Deviation of a glass electrode from the hydrogen-ion response function. An example often encountered is the error due to sodium ions at alkaline pH values, which by convention is regarded as positive.

12.10 Hydrogen gas electrode

A thin foil of platinum electrolytically coated with a finely divided deposit of platinum or (in the case of a reducible substance) palladium metal, which catalyzes the electrode reaction: $H^+ + e \rightarrow {}^1/_2 H_2$ in solutions saturated with hydrogen gas. It is customary to correct measured values to standard 1 atm (101.325 kPa) partial pressure of hydrogen gas.

12.11 Reference electrode

External electrode system that comprises an inner element, usually silver–silver chloride, mercury–mercury(I) chloride (calomel), or thallium amalgam–thallium(I) chloride, a chamber containing the appropriate filling solution (see 12.14), and a device for forming a liquid junction (e.g., capillary) ceramic plug, frit, or ground glass sleeve.

12.12 Liquid junction

Any junction between two electrolyte solutions of different composition. Across such a junction there arises a potential difference, called the liquid junction potential. In Cells II, IV, and V, the junction is between the pH standard or unknown solution and the filling solution, or the bridge solution (q.v.), of the reference electrode.

12.13 Residual liquid junction potential error

Error arising from breakdown in the assumption that the LJPs cancel in Cell II when solution X is substituted for solution S in Cell V.

12.14 Filling solution (of a reference electrode)

Solution containing the anion to which the reference electrode of Cells IV and V is reversible, e.g., chloride for silver–silver chloride electrode. In the absence of a bridge solution (q.v.), a high concentration of filling solution comprising almost equitransferent cations and anions is employed as a means of maintaining the LJP small and approximately constant on substitution of unknown solution for standard solution(s).

12.15 Bridge (or salt bridge) solution (of a double junction reference electrode)

Solution of high concentration of inert salt, preferably comprising cations and anions of equal mobility, optionally interposed between the reference electrode filling and both the unknown and standard solution, when the test solution and filling solution are chemically incompatible. This procedure introduces into the cell a second liquid junction formed, usually, in a similar way to the first.

12.16 Calibration

Set of operations that establish, under specified conditions, the relationship between values of quantities indicated by a measuring instrument, or measuring system, or values represented by a material measure or a reference material, and the corresponding values realized by standards.

12.17 Uncertainty (of a measurement)

Parameter, associated with the result of a measurement, which characterizes the dispersion of the values that could reasonably be attributed to the measurand.

12.18 Standard uncertainty, u_x

Uncertainty of the result of a measurement expressed as a standard deviation.

12.19 Combined standard uncertainty, $u_c(y)$

Standard uncertainty of the result of a measurement when that result is obtained from the values of a number of other quantities, equal to the positive square root of a sum of terms, the terms being the variances, or covariances of these other quantities, weighted according to how the measurement result varies with changes in these quantities.

12.20 Expanded uncertainty, U

Quantity defining an interval about the result of a measurement that may be expected to encompass a large fraction of the distribution of values that could reasonably be attributed to the measurand.

Note 16: The fraction may be viewed as the coverage probability or level of confidence of the interval.

Note 17: To associate a specific level of confidence with the interval defined by the expanded uncertainty requires explicit or implicit assumptions regarding the probability distribution characterized by the measurement result and its combined standard uncertainty. The level of confidence that may be attributed to this interval can be known only to the extent to which such assumptions may be justified.

Note 18: Expanded uncertainty is sometimes termed overall uncertainty.

12.21 Coverage factor

Numerical factor used as a multiplier of the combined standard uncertainty in order to obtain an expanded uncertainty

Note 19: A coverage factor is typically in the range 2 to 3. The value 2 is used throughout in the Annex.

13. Annex: measurement uncertainty

Examples are given of uncertainty budgets for pH measurements at the primary, secondary, and working level. The calculations are done in accordance with published procedures [15,37].

When a measurement (y) results from the values of a number of other quantities, $y = f(x_1, x_2, \ldots x_i)$, the combined standard uncertainty of the measurement is obtained from the square root of the expression

$$u_c^2(y) = \sum_{i=1}^{n} \left(\frac{\partial f}{\partial x_i} \right) \cdot u^2(x_i)$$

where $\frac{\partial f}{\partial x_i}$ is called the sensitivity coefficient (c_i). This equation holds for uncorrelated quantities. The ∂x_i equation for correlated quantities is more complex.

The uncertainty stated is the expanded uncertainty, U, obtained by multiplying the standard uncertainty, $u_c(y)$, by an appropriate coverage factor. When the result has a large number of degrees of freedom, the use of a value of 2 leads to approximately 95% confidence that the true value lies in the range $\pm U$. The value of 2 will be used throughout this Annex.

The following sections give illustrative examples of the uncertainty calculations for Cells I–V.

After the assessment of uncertainties, there should be a reappraisal of experimental design factors and statistical treatment of the data, with due regard for economic factors before the adoption of more elaborate procedures.

A.1 Uncertainty budget for the primary method of measurement using Cell I

Experimental details have been published [42–45].

A-1.1 Measurement equations

The primary method for the determination of pH(PS) values consists of the following steps (Section 4.1):

1. Determination of the standard potential of the Ag | AgCl electrode from the acid-filled cell (Cell Ia)

$$E^\circ = E_a + 2k \lg(m_{HCl}/m^\circ) + 2k \lg \gamma_{HCl} - (k/2) \lg(p^\circ/p_{H_2})$$
$$\text{(A-2) cf. (3)}$$

where $E_{Ia} = E_a - (k/2) \lg(p^\circ/p_{H_2})$, $k = (RT/F)\ln 10$, p_{H_2} is the partial pressure of hydrogen in Cell Ia, and p° is the standard pressure.

2. Determination of the acidity function, $p(a_H\gamma_{Cl})$, in the buffer-filled cell (Cell I)

$$-\lg(a_H\gamma_{Cl}) = (E_b - E^\circ)/k + \lg(m_{Cl}/m^\circ)$$
$$-(1/2) \lg(p^\circ/p_{H_2}), \qquad \text{(A-3) cf. (2)}$$

where $E_I = E_b - (k/2) \lg(p^\circ/p_{H_2})$, p_{H_2} is the partial pressure of hydrogen in Cell I, and p° the standard pressure.

3. Extrapolation of the acidity function to zero chloride concentration

$$-\lg(a_H\gamma_{Cl}) = -\lg(a_H\gamma_{Cl})^\circ + Sm_{Cl} \qquad \text{(A-4) cf. (5)}$$

4. pH Determination

$$pH(PS) = -\lg(a_H\gamma_{Cl})^\circ + \lg \gamma^\circ_{Cl} \qquad \text{(A-5)}$$

where $\lg \gamma^\circ_{Cl}$ is calculated from the Bates–Guggenheim convention (see eq. 6). Values of the Debye–Hückel limiting law slope for 0 to 50 °C are given in Table A.9 [46].

A-1.2 Uncertainty budget

Example: PS = 0.025 mol kg^{-1} disodium hydrogen phosphate + 0.025 mol kg^{-1} potassium dihydrogen phosphate.

TABLE A.1 Calculation of Standard Uncertainty of the Standard Potential of the Silver–Silver Chloride Electrode (E°) from Measurements in m_{HCl} = 0.01 mol kg^{-1}

Quantity	Estimate x_i	Standard Uncertainty $u(x_i)$	Sensitivity Coefficient $|c_i|$	Uncertainty Contribution $u_i(y)$
E/V	0.464	2×10^{-5}	1	2×10^{-5}
T/K	298.15	8×10^{-3}	8.1×10^{-4}	6.7×10^{-6}
m_{HCl}/mol kg^{-1}	0.01	1×10^{-5}	5.14	5.1×10^{-5}
p_{H_2}/kPa	101.000	0.003	1.3×10^{-7}	4.2×10^{-7}
ΔE(Ag/AgCl)/V Bias potential	3.5×10^{-5}	3.5×10^{-5}	1	3.5×10^{-5}
$\gamma_\pm$	0.9042	9.3×10^{-4}	0.0568	5.2×10^{-6}

$u_c(E^\circ) = 6.5 \times 10^{-5}$ V

Note 20: The uncertainty of method used for the determination of hydrochloric acid concentration is critical. The uncertainty quoted here is for potentiometric silver chloride titration. The uncertainty for coulometry is about 10 times lower.

TABLE A.2 Calculation of the Standard Uncertainty of the Acidity Function $\lg(a_H\gamma_{Cl})$ for m_{Cl} = 0.005 mol kg^{-1}

Quantity	Estimate x_i	Standard Uncertainty $u(x_i)$	Sensitivity Coefficient $|c_i|$	Uncertainty Contribution $u_i(y)$
E/V	0.770	2×10^{-5}	16.9	3.4×10^{-4}
E°/V	0.222	6.5×10^{-5}	16.9	1.1×10^{-3}
T/K	298.15	8×10^{-3}	0.031	2.5×10^{-4}
m_{Cl}/mol kg^{-1}	0.005	2.2×10^{-6}	86.86	1.9×10^{-4}
p_{H_2}/kPa	101.000	0.003	2.2×10^{-6}	7×10^{-6}
ΔE(Ag/AgCl)/V	3.5×10^{-5}	3.5×10^{-5}	16.9	5.9×10^{-4}

$u_c[\lg(a_H\gamma_{Cl})] = 0.0013$

Note 21: If, as is usual practice in some NMIs [42–44], acid and buffer cells are measured at the same time, then the pressure measuring instrument uncertainty quoted above (0.003 kPa) cancels, but there remains the possibility of a much smaller bubbler depth variation between cells.

TABLE A.3 S_1 = **Primary Buffer, pH(PS) = 4.005, u(pH) = 0.003; S_2 = Secondary Buffer, pH(SS) = 6.86. Free-Diffusion Junctions with Cylindrical Symmetry Formed in Vertical Tubes Were Used [25]**

Quantity	Estimate x_i	Standard Uncertainty $u(x_i)$	Sensitivity Coefficient $|c_i|$	Uncertainty Contribution $u_i(y)$
pH(S_1)	4.005	0.003	1	0.003
E_{II}/V	0.2	1×10^{-5}	16.9	1.7×10^{-4}
$(E_{j2} - E_{j1})$/V	3.5×10^{-4}	3.5×10^{-4}	16.9	6×10^{-3}
T/K	298.15	0.1	1.2×10^{-5}	1.2×10^{-6}

$u_c[\text{pH}(S_2)] = 0.007$

Note 22: The error in E_{II} is estimated as the scatter from 3 measurements. The RLJP contribution is estimated from Fig. 2 as 0.006 in pH; it is the principal contribution to the uncertainty.

The standard uncertainty due to the extrapolation to zero added chloride concentration (Section 4.4) depends in detail on the number of data points available and the concentration range. Consequently, it is not discussed in detail here. This calculation may increase the expanded uncertainty (of the acidity function at zero concentration) to $U = 0.004$.

As discussed in Section 5.2, the uncertainty due to the use of the Bates–Guggenheim convention includes two components:

  i. The uncertainty of the convention itself, and this is estimated to be approximately 0.01. This contribution to the uncertainty is required if the result is to be traceable to SI, but will not be included in the uncertainty of "conventional" pH values.

  ii The contribution to the uncertainty from the value of the ionic strength should be calculated for each individual case.

The typical uncertainty for Cell I is between $U = 0.003$ and $U = 0.004$.

A.2 Uncertainty budget for secondary pH buffer using Cell II

$$\text{Pt} \mid \text{H}_2 \mid S_2 \mathrel{\vdots} \text{KCl} (\geq 3.5 \text{ mol dm}^{-3}) \mathrel{\vdots} S_1 \mid \text{H}_2 \mid \text{Pt} \qquad \text{Cell II}$$

where S_1 and S_2 are different buffers.

A.2.1 Measurement equations

1. Determination of pH(S_2) $\text{pH}_{II}(S_2)$

$$- \text{pH}_{II}(S_1) = E_{II}/k - (E_{j2} - E_{j1})/k \qquad \text{(A-6) cf. (9)}$$

2. Theoretical slope, $k = (RT/F)\ln 10$

A.2.2 Uncertainty budget

Therefore, $U[\text{pH}(S_2)] = 0.014$.

A.3 Uncertainty budget for secondary pH buffer using Cell III

$$\text{Pt} \mid \text{H}_2 \mid \text{Buffer } S_2 \mathrel{\vdots} \text{Buffer } S_1 \mid \text{H}_2 \mid \text{Pt} \qquad \text{Cell III}$$

A.3.1 Measurement equations

1. $\text{pH}(S_2) - \text{pH}(S_1) = (E_{III} + E_j)/k$ (A-7) cf. (11)
2. $k = (RT/F)\ln 10$

For experimental details, see refs. [16,33,38].

Therefore, $U[\text{pH}(S_2)] = 0.004$. The uncertainty is no more than that of the primary standard PS_1.

A.4 Uncertainty budget for secondary pH buffer using Cell IV

$$\text{Ag} \mid \text{AgCl} \mid \text{KCl} (\geq 3.5 \text{ mol dm}^{-3}) \mathrel{\vdots} \text{buffer}$$
$$S_1 \text{or } S_2 \mid \text{H}_2 \mid \text{Pt} \qquad \text{Cell IV}$$

A.4.1 Measurement equations

1. Determination of pH(S_2)

$$\text{pH}_{IV}(S_2) - \text{pH}_{IV}(S_1) = -[E_{IV}(S_2) - E_{IV}(S_1)]/k$$
$$- (E_{j2} - E_{j1})/k \qquad \text{(A-8) cf. (13)}$$

2. Theoretical slope, $k = (RT/F)\ln 10$

TABLE A.4 pH (S_2) Determination. S_1 = Primary Standard (PS) and S_2 = Secondary Standard (SS) are of the Same Nominal Composition. Example: 0.025 mol kg^{-1} Disodium Hydrogen Phosphate + 0.025 mol kg^{-1} Potassium Dihydrogen Phosphate, PS_1 = 6.865, u(pH) = 0.002

Quantity	Estimate x_i	Standard Uncertainty $u(x_i)$	Sensitivity Coefficient $|c_i|$	Uncertainty Contribution $u_i(y)$
pH(PS_1)	6.865	2×10^{-3}	1	2×10^{-3}
$[E(S_2) - E(S_1)]$/V	1×10^{-4}	1×10^{-6}	16.9	16.9×10^{-6}
$[E_{id}(S_2) - E_{id}(S_1)]$/V	1×10^{-6}	1×10^{-6}	16.9	1.7×10^{-5}
Ej/V	1×10^{-5}	1×10^{-5}	16.9	16.9×10^{-5}
T/K	298.15	2×10^{-3}	5×10^{-6}	1×10^{-8}

$u_c[\text{pH}(S_2)] = 0.002$

Note 23: $[E_{id}(S_2) - E_{id}(S_1)]$ is the difference in cell potential when both compartments are filled with solution made up from the same sample of buffer material. The estimate of E_j comes from the observations made of the result of perturbing the pH of samples by small additions of strong acid or alkali, and supported by Henderson equation considerations, that E_j contributes about 10 % to the total cell potential difference [33].

TABLE A.5 Example from the Work of Paabo and Bates [5] Supplemented by Private Communication from Bates to Covington. S_1 = 0.05 mol kg^{-1} Equimolal Phosphate; S_2 = 0.05 mol kg^{-1} Potassium Hydrogen Phthalate. KCl = 3.5 mol dm^{-3}. S_1 = Primary Buffer PS$_1$, pH = 6.86, u(pH) = 0.003, S_2 = Secondary Buffer SS$_2$, pH = 4.01

Quantity	Estimate x_i	Standard Uncertainty $u(x_i)$	Sensitivity Coefficient $|c_i|$	Uncertainty Contribution $u_i(y)$
pH(S$_1$)	6.86	0.003	1	0.003
$\Delta E_{IV}/V$	0.2	2.5×10^{-4}	16.9	4×10^{-3}
$(E_{j2} - E_{j1})/V$	3.5×10^{-4}	3.5×10^{-4}	16.9	6×10^{-3}
T/K	298.15	0.1	1.78×10^{-3}	1.78×10^{-4}

$u_c[\text{pH}(S_2)] = 0.008$

Note 24: The estimate of the error in ΔE_{IV} comes from an investigation of several 3.5 mol dm^{-3} KCl calomel electrodes in phosphate solutions. The RLJP contribution for free-diffusion junctions is estimated from Fig. 2 as 0.006 in pH.

A.4.2 Uncertainty budget

Therefore, U[pH(S$_2$)] = 0.016.

A.5 Uncertainty budget for unknown pH(X) buffer determination using Cell V

Ag | AgCl | KCl (≥3.5 mol dm^{-3}) ¦ Buffer pH(S) or pH(X) | glass electrode   Cell V

A.5.1 Measurement equations: 2-point calibration (bracketing)

1. Determination of the practical slope (k')

$$k' = [(E_V(S_2) - E_V(S_1)]/[\text{pH}(S_2) - \text{pH}(S_1)] \quad \text{(A-9) cf. (16)}$$

2. Measurement of unknown solution (X)

$$(X) \text{ pH}(X) = \text{pH}(S_1)$$
$$-[E_V(X) - E_V(S_1)]/k' - (E_{j2} - E_{j1})/k' \quad \text{(A-10) cf. (15)}$$

A.5.2 Uncertainty budget

Example of two-point calibration (bracketing) with a pH combination electrode [47].

A.5.3 Measurement equations for multipoint calibration

$$E_V(S) = E_V° - k'\text{pH}(S) \quad \text{(A-11) cf. (17)}$$
$$\text{pH}(X) = [E_V° - E_V(X)]/k' \quad \text{(A-12) cf. (18)}$$

Uncertainty budget:

Example: Standard buffers pH(S$_1$) = 3.557, pH(S$_2$) = 4.008, pH(S$_3$) = 6.865, pH(S$_4$) = 7.416, pH(S$_5$) = 9.182; pH(X) was a "ready-to-use" buffer solution with a nominal pH of 7.

A combination electrode with capillary liquid junction was used. For experimental details, see ref. [41]; and for details of the calculations, see ref. [45].

The uncertainty will be different arising from the RLJPs if an alternative selection of the five standard buffers was used. The uncertainty attained will be dependent on the design and quality of the commercial electrodes selected.

Therefore, $U[\text{pH}(X)] = 0.01$.

TABLE A.6 Primary Buffers PS$_1$, pH = 7.4, u(pH) = 0.003; PS$_2$, pH = 4.01, u(pH) = 0.003. Practical Slope (k′) Determination

Quantity	Estimate x_i	Standard Uncertainty $u(x_i)$	Sensitivity Coefficient $|c_i|$	Uncertainty Contribution $u_i(y)$
$\Delta E/V$	0.2	5×10^{-4}	2.95×10^{-1}	1.5×10^{-4}
T/K	298.15	0.1	1.98×10^{-4}	1.98×10^{-5}
$(E_{j2} - E_{j1})/V$	6×10^{-4}	6×10^{-4}	2.95×10^{-1}	1.8×10^{-4}
ΔpH	3.39	4.24×10^{-3}	1.75×10^{-2}	7.40×10^{-5}

$u_c(k') = 2.3 \times 10^{-4}$

TABLE A.7 pH(X) Determination

Quantity	Estimate x_i	Standard Uncertainty $u(x_i)$	Sensitivity Coefficient $|c_i|$	Uncertainty Contribution $u_i(y)$
pH(S$_1$)	7.4	0.003	1	0.003
$\Delta E/V$	0.03	1.40×10^{-5}	16.95	2.37×10^{-4}
$(E_{j2} - E_{j1})/V$	6.00×10^{-4}	6.00×10^{-4}	16.95	1.01×10^{-2}
k'/V	0.059	2.3×10^{-4}	9.01	2.1×10^{-3}

$u_c[\text{pH}(X)] = 1.6 \times 10^{-2}$

Note 25: The estimated error in ΔE comes from replicates. The RLJP is estimated as 0.6 mV. Therefore, $U[\text{pH}(X)] = 0.021$.

TABLE A.8

Quantity	Estimate x_i	Standard Uncertainty $u(x_i)$	Sensitivity Coefficient $|c_i|$	Uncertainty Contribution $u_i(y)$
$E°/V$	−0.427	5×10^{-4}	16.96	0.0085
T/K	298.15	0.058	1.98×10^{-4}	1.15×10^{-5}
$E(X)/V$	0.016	2×10^{-4}	16.9	0.0034
k'/V	0.059	0.076×10^{-3}	67.6	0.0051

$u_c[\text{pH}(X)] = 0.005$

Note 26: There is no explicit RLJP error assessment as it is assessed statistically by regression analysis.

TABLE A.9 Values of the Relative Permittivity of Water [46] and the Debye–Hückel limiting Law Slope for Activity Coefficients as lg γ in eq. 6. Values are for 100.000 kPa, but the Difference from 101.325 kPa (1 atm) is Negligible

$t/°C$	Relative Permittivity	$A/\text{mol}^{-\frac{1}{2}}\,\text{kg}^{\frac{1}{2}}$
0	87.90	0.4904
5	85.90	0.4941
10	83.96	0.4978
15	82.06	0.5017
20	80.20	0.5058
25	78.38	0.5100
30	76.60	0.5145
35	74.86	0.5192
40	73.17	0.5241
45	71.50	0.5292
50	69.88	0.5345

14. Summary of recommendations

- IUPAC recommended definitions, procedures, and terminology are described relating to pH measurements in dilute aqueous solutions in the temperature range 0–50 °C.
- The recent definition of *primary method of measurement* permits the definition of primary standards for pH, determined by a primary method (cell without transference, called the Harned cell) and of secondary standards for pH.
- pH is a conventional quantity and values are based on the Bates–Guggenheim convention. The assigned uncertainty of the Bates–Guggenheim convention is 0.01 in pH. By accepting this value, pH becomes traceable to the internationally accepted SI system of measurement.
- The required attributes (listed in Section 6.1) for primary standard materials effectively limit the number of primary substances to six, from which seven primary standards are defined in the pH range 3–10 (at 25 °C). Values of pH(PS) from 0–50 °C are given in Table 2.
- Methods that can be used to obtain the difference in pH between buffer solutions are discussed in Section 8. These methods include the use of cells with transference that are practically more convenient to use than the Harned cell, but have greater uncertainties associated with the results.
- Incorporation of the uncertainties for the primary method, and for all subsequent measurements, permits the uncertainties for all procedures to be linked to the primary standards by an unbroken chain of comparisons. Despite its conventional basis, the definition of pH, the establishment of pH standards, and the procedures for pH determination

are self-consistent within the confidence limits determined by the uncertainty budgets.

- Comparison of values from the cell with liquid junction with the assigned pH(PS) values of the same primary buffers measured with Cell I makes the estimation of values of the RLJPs possible (Section 7), and the consistency of the seven primary standards can be estimated.
- The Annex (Section 13) to this document includes typical uncertainty estimates for the five cells and measurements described, which are summarized in Table 4.
- The hierarchical approach to primary and secondary measurements facilitates the availability of recommended procedures for carrying out laboratory calibrations with traceable buffers grouped to achieve specified target uncertainties of unknowns (Section 11). The three calibration procedures in common use, one-point, two-point (bracketing), and multi-point, are described in terms of target uncertainties.

15. References

1. BIPM. *Com. Cons. Quantité de Matière* **4** (1998). See also: M. J. T. Milton and T. J. Quinn. *Metrologia* **38**, 289 (2001).
2. A. K. Covington, R. G. Bates, R. A. Durst. *Pure Appl. Chem.* **57**, 531 (1985).
3. IUPAC. *Quantities, Units and Symbols in Physical Chemistry,* 2nd ed., Blackwell Scientific, Oxford (1993).
4. S. P. L. Sørensen and K. L. Linderstrøm-Lang. *C. R. Trav. Lab. Carlsberg* **15**, 6 (1924).
5. R. G. Bates. *Determination of pH*, Wiley, New York (1973).
6. S. P. L. Sørensen. *C. R. Trav. Lab. Carlsberg* **8**, 1 (1909).
7. H. S. Harned and B. B. Owen. *The Physical Chemistry of Electrolytic Solutions*, Chap. 14, Reinhold, New York (1958).
8. R. G. Bates and R. A. Robinson. *J. Solution Chem.* **9**, 455 (1980).
9. A. G. Dickson. *J. Chem. Thermodyn.* **19**, 993 (1987).
10. R. G. Bates and E. A. Guggenheim. *Pure Appl. Chem.* **1**, 163 (1960).
11. J. N. Brønsted. *J. Am. Chem. Soc.* **42**, 761 (1920); **44**, 877, 938 (1922); **45**, 2898 (1923).
12. A. K. Covington. Unpublished.
13. K. S. Pitzer. In K. S. Pitzer (Ed.), *Activity Coefficients in Electrolyte Solutions*, 2nd ed., p. 91, CRC Press, Boca Raton, FL (1991).
14. A. K. Covington and M. I. A. Ferra. *J. Solution Chem.* **23**, 1 (1994).
15. *International Vocabulary of Basic and General Terms in Metrology* (VIM), 2nd ed., Beuth Verlag, Berlin (1994).
16. R. Naumann, Ch. Alexander-Weber, F. G. K. Baucke. *Fresenius' J. Anal. Chem.* **349**, 603 (1994).
17. R. G. Bates. *J. Res. Natl. Bur. Stand., Phys. Chem.* **66A** (2), 179 (1962).
18. P. Spitzer. *Metrologia* **33**, 95 (1996); **34**, 375 (1997).
19. N. E. Good, G. D. Wright, W. Winter, T. N. Connolly, S. Isawa, K. M. M. Singh. *Biochem. J.* **5**, 467 (1966).
20. A. K. Covington and M. J. F. Rebelo. *Ion-Sel. Electrode Rev.* **5**, 93 (1983).
21. H. W. Harper. *J. Phys. Chem.* **89**, 1659 (1985).
22. J. Bagg. *Electrochim. Acta* **35**, 361, 367 (1990); **37**, 719 (1992).
23. J. Breer, S. K. Ratkje, G.-F. Olsen. *Z. Phys. Chem.* **174**, 179 (1991).
24. A. K. Covington. *Anal. Chim. Acta* **127**, 1 (1981).

25. A. K. Covington and M. J. F. Rebelo. *Anal. Chim. Acta* **200**, 245 (1987).
26. D. J. Alner, J. J. Greczek, A. G. Smeeth. *J. Chem. Soc. A* 1205 (1967).
27. F. G. K. Baucke. *Electrochim. Acta* **24**, 95 (1979).
28. A. K. Clark and A. K. Covington. Unpublished.
29. M. J. G. H. M. Lito, M. F. G. F. C. Camoes, M. I. A. Ferra, A. K. Covington. *Anal. Chim. Acta* **239**, 129 (1990).
30. M. F. G. F. C. Camoes, M. J. G. H. M. Lito, M. I. A. Ferra, A. K. Covington. *Pure Appl. Chem.* **69**, 1325 (1997).
31. A. K. Covington and J. Cairns. *J. Solution Chem.* **9**, 517 (1980).
32. H. B. Hetzer, R. A. Robinson, R. G. Bates. *Anal. Chem.* **40**, 634 (1968).
33. F. G. K. Baucke. *Electroanal. Chem.* **368**, 67 (1994).
34. P. R. Mussini, A. Galli, S. Rondinini. *J. Appl. Electrochem.* **20**, 651 (1990); C. Buizza, P. R. Mussini, T. Mussini, S. Rondinini. *J. Appl. Electrochem.* **26**, 337 (1996).
35. R. P. Buck and E. Lindner. *Pure Appl. Chem.* **66**, 2527 (1994).
36. J. Christensen. *J. Am. Chem. Soc.* **82**, 5517 (1960).
37. *Guide to the Expression of Uncertainty* (GUM), BIPM, IEC, IFCC, ISO, IUPAC, IUPAP, OIML (1993).
38. F. G. K. Baucke. *Anal. Chem.* **66**, 4519 (1994).
39. F. G. K. Baucke, R. Naumann, C. Alexander-Weber. *Anal. Chem.* **65**, 3244 (1993).
40. R. Naumann, F. G. K. Baucke, P. Spitzer. In *PTB-Report W-68*, P. Spitzer (Ed.), pp. 38–51, Physikalisch-Technische Bundesanstalt, Braunschweig (1997).
41. S. Ebel. In *PTB-Report W-68*, P. Spitzer (Ed.), pp. 57–73, Physikalisch-Technische Bundesanstalt, Braunschweig (1997).
42. H. B. Kristensen, A. Salomon, G. Kokholm. *Anal. Chem.* **63**, 885 (1991).
43. P. Spitzer, R. Eberhadt, I. Schmidt, U. Sudmeier. *Fresenius' J. Anal. Chem.* **356**, 178 (1996).
44. Y. Ch. Wu, W. F. Koch, R. A. Durst. *NBS Special Publication*, 260, p. 53, Washington, DC (1988).
45. BSI/ISO 11095 *Linear calibration using reference materials* (1996).
46. D. A. Archer and P. Wang. *J. Phys. Chem. Ref. Data* **19**, 371 (1990). See also D. J. Bradley and K. S. Pitzer. *J. Phys. Chem.* **83**, 1599 and errata 3799 (1979); D. P. Fernandez, A. R. H. Goodwin, E. W. Lemmon, J. M. H. Levelt Sengers, R. C. Williams. *J. Phys. Chem. Ref. Data* **26**, 1125 (1997).
47. A. K. Covington, R. Kataky, R. A. Lampitt. Unpublished.

GENERAL COMMENTS ON BUFFERS

The major factor in biological pH control in eukaryotic cells is the carbon dioxide-bicarbonate-carbonate buffer (Scheme I) system[1-4]. There other biological buffers such as bulk protein and phosphate anions which can provide some buffering effect, metabolites such as lactic acid which can lower pH and tris(hydroxymethylaminomethyl) methane, (THAM®) has been used to treat acid base disorders[5-7]. pH control in prokaryotic cells is mediated by membrane transport of various ions including hydrogen, potassium and sodium[8-10].

Scheme I

$$CO_2 + H_2O = H_2CO_3; \ H_2CO_3 + H_2O = HCO_3^{-1} + H_3O^+$$
$$\times \ (pKa \ 6.15); \ HCO_3^{-1} + H_2O = CO_3^{-2-} \ (pKa \ 10.3)$$

See Jungas, R.L., Best literature values for the pK of carbonic and phosphoric acid under physiological conditions, *Anal. Biochem.* 349, 1–15, 2006

In the laboratory, the bicarbonate/carbonate buffer system can only be used in the far alkaline range (pH 9-11) and unless "fixed" by a suitable cation such as sodium, can be volatile. A variety of buffers, most notably the "Good" buffers which were developed by Norman Good and colleagues[10a], have been developed over the years to provide pH control in *in vitro* experiments. While effective in controlling pH, the numerous non-buffer effects that buffer salts have on experimental systems are somewhat less appreciated. Some effects, such as observed with phosphate buffers, are based on biologically significant interactions with proteins and, as such, demonstrate specificity. Other effects, such as metal ion chelation, can be considered general. There are some effects where the stability of a reagent is dependent on both pH and buffer species. One example is provided by the stability of phenylmethylsulfonyl fluoride (PMSF)[11]. PMSF was less stable in Tris buffer than in either HEPES or phosphate buffer; PMSF is less stable in HEPES than in phosphate buffer. Activity was measured by the ability of PMSF to inhibit chymotrypsin; all activity was lost in Tris (10 mM; pH 7.5) after one hour at 25°C while activity was fully retained in phosphate (10 mM, pH 7.5). This is likely a reflection of the nucleophilic property of Tris[12,13] which appears to be enhanced in the presence of divalent cations such as zinc[14]. The loss of activity, presumably the result of the hydrolysis of the fluoride to hydroxyl function, is more marked at more alkaline pH. Tris can also function as phosphoacceptor in assays for alkaline phosphatase but was not as effective as 2-amino-2-methyl-1,3-propanediol[15]. The various nitrogen-based buffers such as Tris, HEPES, CAP, and BICINE influence colorimetric protein assays [16-18].

Other specific examples are presented in Table 1.

References

1. Lubman, R.L. and Crandall, E.D., Regulation of intracellular pH in alveolar epithelial cells, *Amer.J.Physiol.* 262, L1-L14, 1992
2. Lyall, V. and Biber, T.O.L., Potential-induced changes in intracellular pH, *Amer.J.Physiol.* 266, F685-F696, 1994
3. Palmer, L.G., Intracellular pH as a regulator of Na+ transport, *J.Membrane Biol.* 184, 305-311, 2001
4. Vaughn-Jones, R.D. and Spitzer, K.W., Role of bicarbonate in the regulation of intracellular pH in the mammalian ventricular myocyte, *Biochem.Cell Biol.* 80, 579-596, 2002
5. Henschler, D., Trispuffer(TAHM) als therapeuticum, *Deutsch.Med. Wochenschr.* 88, 1328-1331, 1963
6. Nahas, G.G., Sutin, K.M., Fermon, C., *et al*, Guidelines for the treatment of academia with THAM, *Drugs* 55, 191-224, 1998
7. Rehm, M. and Finsterer, U., Treating intraoperative hypercholoremic acidosis with sodium bicarbonate or tris-hydroxymethyl amino methane, *Anesthes.Analg.* 96, 1201-1208, 2003
8. Kashket, E.R. and Wong, P.T., The intracellular pH of *Escherichia coli*, *Biochim.Biophys.Acta* 193, 212-214, 1969
9. Padan, E. and Schuldiner, S., Intracellular pH regulation in bacterial cells, *Methods Enzymol.* 125, 327-352, 1986
10. Booth, I.R., The regulation of intracellular pH in bacteria, *Novartis Found.Symp.* 221, 19-28, 1999
10a. Good, N.E., Winget, G.D., Winter, W., *et al.*, Hydrogen ion buffers for biological research, *Biochemistry* 5, 467-477, 1966
11. James, G.T., Inactivation of the protease inhibitor phenylmethylsulfonyl fluoride in buffers, *Anal.Biochem.* 86, 574-579, 1978
12. Acharya, A.S., Roy, R.P., and Dorai, B., Aldimine to ketoamine isomerization (Amadori rearrangement) potential at the individual nonenzymic glycation sites of hemoglobin A: preferential inhibition of glycation by nucleophiles at sites of low isomerization potential, *J.Protein Chem.* 10, 345-358, 1991
13. Mattson, A., Boutelje, J., Csoregh, I., *et al.*, Enhanced stereoselectivity in pig liver esterase catalyzed diester hydrolysis. The role of a competitive inhibitor, *Bioorg.Med.Chem.* 2, 501-508, 1994
14. Tomida, H. and Schwartz, M.A., Further studies on the catalysis of hydrolysis and aminolysis of benzylpenicillin by metal chelates, *J.Pharm.Sci.* 72, 331-335, 1983
15. Stinson, R.A., Kinetic parameters for the cleaved substrate, and enzyme and substrate stability, vary with the phosphoacceptor in alkaline phosphatase catalysis, *Clin.Chem.* 39, 2293-2297, 1993
16. Kaushal, V. and Barnes, L.D., Effect of zwitterionic buffers on measurement of small masses of protein with bicinchoninic acid, *Anal. Biochem.* 157, 291-294, 1986
17. Lleu, P.L. and Rebel, G., Interference of Good's buffers other biological buffers with protein determination, *Anal.Biochem.* 192, 215-218, 1991
18. Sapan, C.V., Lundblad, R.L., and Price, N.C., Colorimetric protein assay techniques, *Biotechnol.Appl.Biochem.* 29, 99-108, 1999

TABLE 1: Effects of Buffers

Buffer	Observation
ACES	Competitive inhibitor of γ-aminobutyric acid receptor binding[1].
ADA	Competitive inhibitor of γ-aminobutyric acid receptor binding[1]; chelation of calcium ions[2].
BES	Interacts with DNA yielding distortion of DNA electrophoretograms[3].
BICINE	Chelation of calcium ions[2]; protects liver alcohol dehydrogenase from inactivation by iodoacetic acid[4].
Borate	Anomalous complex formation with nucleic acids[5]; complex formation with carbohydrates[6,7]; participant in the modification of arginine residues by 1,2-cyclohexanedione[8].
Cacodylic Acid	Reaction with sulfhydryl compounds[9].
Carbonate	Enhances rate of reaction of phenylglyoxal with arginine residues in proteins[10]; modulation of peroxynitrite reactions with proteins[11,12]; modulation of Cu^{2+} oxidation reactions[13-15].
Citrate	Chelation of calcium ions[2].
HEPES	Free radical generation[16,17] and complexation of copper ions[18]; reported adverse effects in tissue culture[19,20].
MES	Complexes copper ions[21].
MOPS	Adverse effect on smooth muscle contraction[22]; Oxidation of metal ions[22]; formation of nitric oxide donors on incubation with peroxynitrite[24]; slow reaction with hydrogen peroxide[25].
Phosphate	Catalysis of the racemization of 5-phenylhydantoins[26,27].
PIPES	Binding to bile salt-stimulated lipase[28]; variation in physiological response based on vendor source[29]; inhibition of a K^+-activated phosphatase[30].
TES	Interaction with extracellular matrices[31]; inhibition of the interaction of proteoglycans with type 1 collagen[32].
Tricine	Chelating agent[2]; tricine radicals have been reported in the presence of peroxide-forming enzymes[33].
Tris	Nucleophile[34,35] and enzyme inhibitor[36].

References to Table 1

1. Tunnicliff, G. and Smith, J.A., Competitive inhibition of gamma-aminobutyric acid receptor binding by N-hydroxyethylepiperazine-N-2-ethanesulfonic acid and related buffers, *J.Neurochem.* 36, 1122-1126, 1981

2. Durham, A.C., A survey of readily available chelators for buffering calcium ion concentrations in physiological solutions, *Cell Calcium* 4, 33-46, 1983

3. Stellwagen, N.C., Bossi, A., Gelfi, C. and Righetti, P.G., DNA and buffers: Are there any noninteracting neutral pH buffers?, *Anal. Biochem.* 287, 167-175, 2000

4. Syvertsen, C. and McKinley-McKee, J.S., Affinity labelling of liver alcohol dehydrogenase. Effect of pH and buffers on affinity labelling with iodoacetic acid and (R,S)-2- bromo-3-(5-imidazolyl)propionic acid, *Eur.J.Biochem.* 117, 165-170, 1981

5. Biyani, M. and Nishigaki, K., Sequence-specific and nonspecific mobilities of single-stranded oligonucleotides observed by changing the borate buffer concentration, *Electrophoresis* 24, 628-633, 2003

6. Zittle, Z.A., Reaction of borate with substances of biological interest, *Adv.Enzymol.Relat.Sub.Biochem.* 12, 493-527, 1951

7. Weitzman, S., Scott, V., and Keegstra, K., Analysis of glycoproteins as borate complexes by polyacrylamide gel electrophoresis, *Anal. Biochem.* 438-449, 1979

8. Patthy, L. and Smith, E.L., Reversible modification of arginine residues. Application to sequence studies by restriction of tryptic hydrolysis to lysine residues, *J.Biol.Chem.* 250, 557-564, 1975

9. Jacobson, K.B., Murphey, J.B., and Sarma, B.D., Reaction of cacodylic acid with organic thiols, *FEBS Lett.* 22, 80-82, 1972

10. Cheung, S.T. and Fonda, M.L., Reaction of phenylglyoxal with arginine. The effect of buffers and pH, *Biochem.Biophys.Res.Commun.* 90, 940-947, 1979

11. Uppu, R.M., Squadrito, G.L., and Pryor, W.A., Acceleration of peroxynitrite oxidations by carbon dioxide, *Arch.Biochem.Biophys.* 327, 335-343, 1996

12. Denicola, A., Freeman, B.A., Trujillo, M., and Radi, R., Peroxynitrite reaction with carbon dioxide/bicarbonate: kinetics and influence on peroxynitrite-mediated oxidations, *Arch.Biochem.Biophys.* 333, 49-58, 1996

13. Munday, R., Munday, C.M. and Winterbourn, C.C., Inhibition of copper-catalyzed cysteine oxidation by nanomolar concentrations of iron salts, *Free Rad.Biol.Med.* 36, 757-764, 2004

14. Jansson, P.J., Del Castillo, U., Lindqvist, C., and Nordstrom, T., Effects of iron on vitamin C/copper-induced hydroxyl radical generation in bicarbonate-rich water, *Free Rad.Res.* 39, 565-570, 2005

15. Ramirez, D.C., Mejiba, S.E. and Mason, R.P., Copper-catalyzed protein oxidation and its modulation by carbon dioxide: enhancement of protein radicals in cells, *J.Biol.Chem.* 280, 27402-27411, 2005

16. Tadolini, B., Iron autoxidation in Mops and Hepes buffers, *Free Radic.Res.Commun.* 4, 149-160, 1987

17. Simpson, J.A., Cheeseman, K.H., Smith, S.E., and Dean, R.T., Free-radical generation by copper ions and hydrogen peroxide. Stimulation by Hepes buffer, *Biochem.J.* 254, 519-523, 1988

18. Sokolowska, M. and Bal, W., Cu(II) complexation by "non-coordinating" N-2-hydroxyethylpiperazine-N'-ethanesulfonic acid (HEPES buffer), *J.Inorg.Biochem.* 99, 1653-1660, 2005

19. Bowman, C.M., Berger, E.M., Butler, E.N. *et al.*, HEPES may stimulate cultured endothelial-cells to make growth-retarding oxygen metabolites, *In Vitro Cell.Devel.Biol.* 21, 140-142, 1985

20. Magonet, E., Briffeuil, E., Polimay, Y., and Ronveaux, M.F., Adverse-effects of HEPES on human-endothelial cells in culture, *Anticancer Res.* 7, 901, 1987

21. Mash, H.E., Chin, Y.P., Sigg, L., *et al.*, Complexation of copper by zwitterionic aminosulfonic (Good) buffers, *Anal.Chem.* 75, 671-677, 2003

22. Altura, B.M., Carella, A., and Altura, B.T., Adverse effects of Tris, HEPES, and MOPS buffers on contractile responses of arterial and venous smooth muscle induced by prostaglandins, *Prostaglandins Med.* 5, 123-130, 1980

23. Tadolini, B., and Sechi, A.M., Iron oxidation in Mops and Hepes buffers, *Free Radic.Res.Commun.* 4, 149-160, 1987

24. Schmidt, K., Pfeiffer, S., and Meyer, B., Reaction of peroxynitrite with HEPES or MOPS results in the formation of nitric oxide donors, *Free Radic.Biol.Med.* 24, 859-862, 1998

25. Zhao, G. and Chasteen, J.D., Oxidation of Good's buffers by hydrogen peroxide, *Anal.Biochem.* 349, 262-267, 2006

26. Dudley, K.H. and Bius, D.L., Buffer catalysis of the racemization reaction of some 5-phenylhydantoins and its relation to in vivo metabolism of ethotoin, *Drug.Metab.Dispos.* 4, 340-348, 1976

27. Lazarus, R.A., Chemical racemization of 5-benzylhydantoin, *J.Org. Chem.* 55, 4755-4757, 1990

28. Moore, S.A., Kingston, R.L., Loomes, K.M., *et al.*, The structure of truncated recombinant human bile salt-stimulated lipase reveals bile salt-independent conformational flexibility at the active-site loop and provides insight into heparin binding, *J.Mol.Biol.* 312, 511-523, 2001

29. Schmidt, J., Mangold, C., and Deitmer, J., Membrane responses evoked by organic buffers in identified leech neurones, *J.Exp.Biol.* 199, 327-335, 1996

30. Robinson, J.D. and Davis, R.L., Buffer, pH, and ionic strength effects on the ($Na^+ + K^+$)-ATPase, *Biochim.Biophys.Acta* 912, 343-347, 1987

31. Poole, C.A., Reilly, H.C., and Flint, M.H., The adverse effects of HEPES, TES, and BES zwitterionic buffers on the ultrastructure of cultured chick embryo epiphyseal chondrocytes, *In Vitro* 18, 755-765, 1982

32. Pogány, G., Hernandez, D.J., and Vogel, K.G., The *in Vitro* interaction of proteoglycans with type I collagen is modulated by phosphate, *Archs.Biochem.Biophys.* 313, 102-111, 1994

33. Grande, H.J. and Van der Ploeg, K.R., Tricine radicals as formed in the presence of peroxide producing enzymes, *FEBS Lett.* 95, 352-356, 1978

34. Oliver, R.W. and Viswanatha, T., Reaction of tris(hydroxymethyl) aminomethane with cinnamoyl imidazole and cinnamoyltrypsin, *Biochim.Biophys.Acta* 156, 422-425, 1968

35. Ray, T., Mills, A., and Dyson, P., Tris-dependent oxidative DNA strand scission during electrophoresis, *Electrophoresis* 16, 888-894, 1995

36. Qi, Z., Li, X., Sun, D., *et al.*, Effect of Tris on catalytic activity of MP-11, *Bioelectrochemistry* 68, 40-47, 2006

LIST OF BUFFERS

Common Name	Chemical Name	M.W	Properties and Comment
ACES	2-[2-amino-2-oxoethyl)amino] ethanesulfonic acid	182.20	One of the several "Good" buffers

Good, N.E., Winget, G.D., Winter, W., *et al.*, Hydrogen ion buffers for biological research, *Biochemistry* 5, 467-477, 1966; Tunnicliff, G. and Smith, J.A, Competitive inhibition of gamma-aminobutyric acid receptor binding by N-hydroxyethylpiperazine-N'-2-ethanesulfonic acid and related buffers, *J.Neurochem.* 36, 1122-1126, 1981; Chappel, D.J., N-[(carbamoylmethyl)amino] ethanesulfonic acid improves phenotyping of α-1-antitrypsin by isoelectric focusing on agarose gel, *Clin.Chem.* 31, 1384-1386, 1985; Liu, Q., Li, X., and Sommer, S.S., pk-Matched running buffers for gel electrophoresis, *Anal.Biochem.* 270, 112-122, 1999; Taha, M., Buffers for the physiological pH range: acidic dissociation constants of zwitterionic compounds in various hydroorganic media, *Ann.Chim.* 95, 105-109, 2005.

Acetic Acid/Sodium Acetate	acetic acid (usually with sodium hydroxide to provide sodium acetate	60.0/ 82.0	Frequently used in chromatography with cation exchange matrices below pH 6.0; therapeutic use for acid-base disorders; buffer for peritoneal dialysis. It is an organic acid and a natural product (as is the case with citrate and phosphate)

Chohan, I.S., Vermylen, J., Singh, I. *et al.*, Sodium acetate buffer: a diluent of choice in the clot lysis time technique, *Thromb.Diath.Haemorrh.* 33, 226-229, 1975; Lim, C.K. and Peters, T.J., Ammonium acetate: a general purpose buffer for clinical applications of high-performance liquid chromatography, *J.Chromatog.* 3126, 397-406, 1984; Kodama, C., Kodama, T., and Yosizawa, Z., Methods for analysis of urinary glycosaminoglycans, *J.Chromatog.* 429, 293-313, 1988; Stegmann, S., Norgren, R.B., Jr., and Lehman, M.N., Citric acid-ammonium acetate buffer, *Biotech.Histochem.* 1, 27-28,1991; Cuvelier, A., Bourguignon, J., Muir, J.F., *et al.*, Substitution of carbonate by acetate buffer for IgG coating in sandwich ELISA, *J.Immunoassay* 17, 371-382, 1996; Urbansky, E.T., Cooper, B.T., and Margerum, D.W., Disproportionation kinetics of hypoiodous acid as catalyzed and suppressed by acetic acid-acetate buffer, *Inorg.Chem.* 36, 1338-1344, 1997; Watanabe, N., Shirakami, Y., Tomiyoshi, K., *et al.*, Direct labeling of macroaggregated albumin with indium-111-chloride using acetate buffer, *J.Nucl.Med.* 38, 1590-1592, 1997; Righetti, P.G. and Gelfi, C., Capillary electrophoresis of DNA in the 20-500 bp range: recent developments, *J.Biochem.Biophys.Methods* 41, 75-90, 1999; Sen Gupta, K.K., Pal, B., and Begum, B.A., Reactivity of some sugars and sugar phosphates toward gold (III) in sodium acetate-acetic acid buffer medium, *Carbohydr.Res.* 330, 115-123, 2001. Citations for clinical use: Man, N.K., Itakura, Y., Chauveau, P., and Yamauchi, T., Acetate-free biofiltration: state of the art, *Contrib. Nephrol.* 108, 87-93, 1994; Maiorca, R., Cancarini, G.C., Zubani, R., *et al.*, Differing dialysis treatment strategies and outcome, *Nephrol.Dial. Transplant.* 11(Suppl 2), 134-139, 1996; Naka, T. and Bellomo, R., Bench-to-bedside review: Treating acid-base abnormalities in the intensive care unit – the role of renal replacement therapy, *Crit.Care* 8, 108-114, 2004; Khanna, A. and Kurtzman, N.A., Metabolic alkalosis, *J.Nephrol.* 19(Suppl 9), S86-S96, 2006.

ADA	N-(2-amino-2-oxoethyl)-N-(carboxymethyl)glycine N-(2-acetamido)iminodiacetic acid	190.2	A "Good" buffer

LIST OF BUFFERS (Continued)

Common Name	Chemical Name	M.W	Properties and Comment

Good, N.E., Winget, G.D., Winter, W., *et al.*, Hydrogen ion buffers for biological research, *Biochemistry* 5, 467-477, 1966; Tunnicliff, G. and Smith, J.A., Competitive inhibition of gamma-aminobutyric acid receptor binding by *N*-2-hydroxyethylpiperazine-*N'*-2-*e*-ethanesulfonic acid and related buffers, *J.Neurochem.* 36, 1122-1126, 1981; Durham, A.C., A survey of readily available chelators for buffering calcium ion concentrations in physiological solutions, *Cell Calcium* 4, 33-46, 1983; Kaushal, V. and Barnes, L.D., Effect of zwitterionic buffers on measurement of small masses of protein with bicinchoninic acid, *Anal.Biochem.* 157, 291-294, 1986; Robinson, J.D. and Davis, R.L., Buffer, pH, and ionic strength effects on the (Na$^+$, + K$^+$)-ATPase, *Biochim.Biophys.Acta* 912, 343-347, 1987; Pietrzkowski, E and Korohoda, W., Extracellular ATP and ADA-buffer enable chick embryo fibroblasts to grow in secondary culture in protein-free, hormone-free, extracellular growth factor-free media, *Folia Histochem.Cytobiol.* 26, 143-152, 1988; Righetti, P.G., Chiari, M., and Gelfi, C., Immobilized pH gradients: effect of salts, added carrier ampholytes and voltage gradients on protein patterns, *Electrophoresis* 9, 65-73, 1988; Bers, D.M., Hryshko, L.V., Harrison, S.M., and Dawson, D.D., Citrate decreases contraction and Ca current in cardiac muscle independent of its buffering action, *Am.J.Physiol.* 260, C900-C909, 1991; Delaney, J.P., Kimm, G.E., and Bonsack, M.E., The influence of luminal pH on the severity of acute radiation enteritis, *Int.J.Radiat.Biol.* 61, 381-386, 1992; Taha, M., Buffers for the physiological pH range: acidic dissociation constants of zwitterionic compounds in various hydroorganic media, *Ann.Chim.* 95, 105-109, 2005.

BES   *N,N*-bis(2-hydroxyethyl)-   213.3   A "Good" buffer, not frequently used, similar to MES,
                       2-aminoethanesulfonic acid;   HEPES
                       N,N-bis(2-hydroxyethyl)taurine

Good, N.E., Winget, G.D., Winter, W., *et al.*, Hydrogen ion buffers for biological research, *Biochemistry* 5, 467-477, 1966; Kaushal, V. and Barnes, L.D., Effect of zwitterionic buffers on the measurement of small masses of protein with bicinchoninic acid, *Anal.Biochem.* 157, 291-294, 1986; MacKerrow, S.D., Merry, J.M., and Hoeprich, P.D., Effects of buffers on testing of Candida species susceptibility to flucytosine, *J.Clin.Microbiol.* 25, 885-888, 1987; Tuli, R.K. and Holtz, W., The effect of zwitterionic buffers on the feasibility of Boer goat semen, *Theriogenology* 37, 947-951, 1992; Stellwagen, N.C., Bossi, A., Gelfi, C., and Righetti, P.G., DNA and buffers: are there any noninteracting, neutral pH buffers, *Anal.Biochem.* 287, 167-175, 2000; Hosse, M. and Wilkinson, K.J., Determination of electrophoretic mobilities and hydrodynamic radii of three humic substances as a function of pH and ionic strength, *Environ.Sci.Technol.* 35, 4301-4306, 2002; Taha, M., Buffers for the physiological pH range: acidic dissociation contstants of zwitterionic compounds in various hydroorganic media, *Ann.Chim.* 95, 105-109, 2005.

Bicine   *N,N*-bis-(2-hydroxyethyl)glycine;   163.2
                       N,N-bis(2-hydroxyethyl)
                       amino-acetic acid

Kanfer, J.N., Base exchange reactions of the phospholipids in rat brain particles, *J.Lipid Res.* 13, 468-476, 1972; Williams-Smith, D.L., Bray, R.C., Barber, M.J., *et al.*, Changes in apparent pH on freezing aqueous buffer solutions and their relevance to biochemical electron-paramagnetic-resonance spectroscopy, *Biochem.J.* 167, 593-600, 1977; Syvertsen, C. and McKinley-McKee, J.S., Affinity labeling of liver alcohol dehydrogenase. Effects of pH and buffers on affinity labeling with iodoacetic acid and (R, S)-2-bromo-3-(5-imidazoyl)propionic acid, *Eur.J.Biochem.* 117, 165-170, 1981; Ito, S., Takaoka, T., Mori, H., and Teruo, A., A sensitive new method for measurement of guanase with 8-azaguanine in bicine bis-hydroxy ethyl glycine buffer as substrate, *Clin.Chim.Acta* 115, 135-144, 1981; Nakon, R., Krishnamoorthy, C.R., Free-metal ion depletion by "Good's" buffers, *Science* 221, 749-750, 1983; Ito, S., Xu, Y., Keyser, A.J., and Peters, R.L., Histochemical demonstration of guanase in human liver with guanine in bicine buffer as substrate, *Histochem.J.* 16, 489-499, 1984; Roy, R.N., Gibbons, J.J, Baker, G., and Bates, R.G., Standard electromotive force of the H$_2$-AgCL:Ag cell in 30, 40, and 50 mass% dimethyl sulfoxide/water from -20 to 25°; pK$_2$ and pH values for a standard "Bicine" buffer solution at subzero temperatures, *Cryobiology* 21, 672-681, 1984; Vaidya, N.R., Gothoskar, B.P., and Banerji, A.P., Column isoelectric focusing in nature pH gradients generated by biological buffers, *Electrophoresis* 11, 156-161, 1990; Wiltfang, J., Arold, N., and Neuhoff, V., A new multiphasic buffer system for sodium sulfate-polyacrylamide gel electrophoresis of proteins and peptides with molecular masses 100,000-1000, and their detection with picomolar sensitivity, *Electrophoresis* 12, 352-366, 1991; Rabilloud, T., Vuillard, L., Gilly, C., and Lawrence, J.J., Silver-staining of proteins in polyacrylamide gels: a general overview, *Cell.Mol. Biol.* 40, 57-75, 1994; Gordon-Weeks, R., Koren'kov, V.D., Steele, S.H., and Leigh, R.A., Tris is a competitive inhibitor of K+ activation of the vacuolar H+-pumping pyrophosphatase, *Plant Physiol.* 114, 901-905, 1997; Luo, Q., Andrade, J.D., and Caldwell, K.D., Thin-layer ion-exchange

LIST OF BUFFERS (Continued)

Common Name	Chemical Name	M.W	Properties and Comment

chromatography of proteins, *J.Chromatog. A* 816, 97-105, 1998; Churchill, T.A. and Kneteman, N.M., Investigation of a primary requirement of organ preservation solutions: supplemental buffering agents improve hepatic energy production during cold storage, *Transplanation* 65, 551-559, 1998; Taha, M., Thermodynamic study of the second-stage dissociation of *N,N*-bis-(2-hydroxyethyl)glycine (bicine) in water at different ionic strength and different solvent mixtures, *Ann.Chim.* 94, 971-978, 2004; Taha, M., Buffers for the physiological pH range: acidic dissociation constants of zwitterionic compounds in various hydroorganic media, *Ann.Chim.* 95, 105-109, 2005; Williams, T.I., Combs, J.C., Thakur, A.P., *et al.*, A novel Bicine running buffer system for doubled sodium dodecyl sulfate – polyacrylamide gel electrophoresis of membrane proteins, *Electrophoresis* 27, 2984-2995, 2006.

Common Name	Chemical Name	M.W	Properties and Comment
Borate: Sodium Borate (sodium tetraborate/Boric Acid	$Na_2B_4O_7/H_3BO_3$ Sodium borate decahydrate is borax)	61.8/ 201.2	Borate buffers have long history of use; borate well-known for interaction with carbohydrates; participates in the reversible modification of arginine residues by 1,2-cyclohexanedione

$$B(OH)_3 + 2H_2O \rightleftharpoons B(OH)_4^- + H_3O^+$$

Adjutantis, G., Electrophoretic separation of filter paper of the soluble liver-cell proteins of the rat using borate buffer, *Nature* 173, 539-540, 1954; Consden, R. and Powell, M.N., The use of borate buffer in paper electrophoresis of serum, *J.Clin.Pathol.* 8, 150-152, 1955; Cooper, D.R., Effect of borate buffer on the electrophoresis of serum, *Nature* 181, 713-714, 1958; Cooper, D.R., Effect of borate buffer on the electrophoresis of serum, *Nature* 181, 713-714, 1958; Poduslo, J.F., Glycoprotein molecular-weight estimation using sodium dodecyl sulfate-pore gradient electrophoresis: Comparison of Tris-glycine and Tris-borate-EDTA buffer systems, *Anal.Biochem.* 114, 131-139, 1981; Shukun, S.A. and Zav'yalov, V.P., Peculiar features of application of pH gradients formed in borate buffer with a polyhydroxy compound for separation of proteins in a free-flow electrophoretic apparatus, *J.Chromatog.* 496, 121-128, 1989; Patton, W.F., Chung-Welch, N., Lopez, M.F., *et al.*, Tris-tricine and Tris-borate buffer systems provide better estimates of huma mesothelial cell intermediate filament protein molecular weights than the standard Tris-glycine system, *Anal.Biochem.* 197, 25-33, 1991; Roden, L., Yu, H., Jin, J., and Greenshields, J., Separation of *N*-acetylglucosamine and *N*-acetylmannosamine by chromatography on Sephadex in borate buffer, *Anal.Biochem.* 209, 188-191, 1993; Yokota, H., van den Engh, G., Mostert, M., and Trask, B.J., Treatment of cells with alkaline borate buffer extends the capability of interphase FISH mapping, *Genomics* 25, 485-491, 1995; Biyani, M. and Nishigaki, K., Sequence-specific and nonspecific mobilities of single-stranded oliogonucleotides observed by changing the borate buffer concentration, *Electrophoresis* 24, 628-633, 2003; Zhao, Y., Yang, X., Jiang, R. *et al.*, Chiral separation of synthetic vicinal diol compounds by capillary zone electrophoresis with borate buffer and β-cyclodextrin as buffer additives, *Anal.Sci.* 22, 747-751, 2006. Articles focusing on the interaction of borate with carbohydrates and other polyols include: Zittle, C.A., Reaction of borate with substances of biological interest, *Adv.Enzymol.Relat.Sub.Biochem.* 12, 493-527, 1951; Larsson, U.B. and Samuelson, O. Anion exchange separation of organic acids in borate medium: influence of the temperature, *J.Chromatog.* 19, 404-411, 1965; Lin, F.M. and Pomeranz, Y., Effect of borate on colorimetric determinations of carbohydrates by the phenol-sulfuric acid method, *Anal.Biochem.* 24, 128-131, 1968; Haug, A., The influence of borate and calcium on the gel formation of a sulfated polysaccharide from *Ulva lactuca*, *Acta Chem.Scand. B.* 30, 562-566, 1976; Weitzman, S., Scott, V., and Keegstra, K., Analysis of glycoproteins as borate complexes by polyacrylamide gel electrophoresis, *Anal.Biochem.* 97, 438-449, 1979; Honda, S., Takahashi, M., Kakehi, K. and Ganno, S., Rapid, automated analysis of monosaccharides by high-performance anion-exchange chromatograpy of borate complexes with fluorometric detection using 2-cyanoacetamide, *Anal.Biochem.* 113, 130-138, 1981; Rothman, R.J. and Warren, L., Analysis of IgG glycopeptides by alkaline borate gel filtration chromatography, *Biochim.Biophys.Acta* 955, 143-153, 1988; Todd, P. and Elsasser, W., Nonamphometric isoelectric focusing: II. Stablity of borate-glycerol pH gradients in recycling isoelectric focusing, *Electrophoresis* 11, 947-952, 1990. Selected studies on the effect of borate on the modification of arginine with 1,2-cyclohexanedione include: Patthy, L. and Smith, E.L., Reversible modification of arginine residues. Application to sequence studies by restriction of tryptic hydrolysis to lysine residues, *J.Biol.Chem.* 250, 557-564, 1975; Patthy, L. and Smith, E.L., Identification of functional arginine residues in ribonuclease A and lysozyme, *J.Biol.Chem.* 250, 565-569,1975; Menegatti, E., Ferroni, R., Benassi, C.A., and Rocchi, R., Arginine modification in Kunitz bovine trypsin inhibitor through 1,2-cyclohexanedione, *Int.J.Pept.Protein Res.* 10, 146-152, 1977; Kozik, A., Guevara, I. and Zak, Z., 1,2-Cyclohexanedione modification of arginine residues in egg-white riboflavin-binding protein, *Int.J.Biochem.* 20, 707-711, 1988.

Common Name	Chemical Name	M.W	Properties and Comment
Cacodylic Acid	Dimethylarsinic Acid	138.10	Buffer salt in neutral pH range; largely replaced because of toxicity.

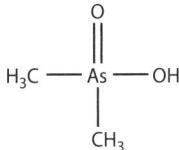

McAlpine, J.C., Histochemical demonstration of the activation of rat acetylcholinesterase by sodium cacodylate and cacodylic acid using the thioacetic acid method, *J.R.Microsc.Soc.* 82, 95-106, 1963; Jacobson, K.B., Murphy, J.B., and Das Sarma, B., Reaction of cacodylic acid with organic thiols, *FEBS Lett.* 22, 80-82, 1972; Travers, F., Douzou, P., Pederson, T., and Gunsalus. I.C., Ternary solvents to investigate proteins at sub-zero temperature, *Biochimie* 57, 43-48, 1975; Young, C.W., Dessources, C., Hodas, S., and Bittar, E.S., Use of cationic disc electrophoresis near neutral pH in the evaluation of trace proteins in human plasma, *Cancer Res.* 35, 1991-1995, 1975; Chirpich, T.P., The effect of different buffers on terminal deoxynucleotidyl transferase activity, *Biochim.Biophys.Acta* 518, 535-538, 1978; Nunes, J.F., Aguas, A.P., and Soares, J.O., Growth of fungi in cacodylate buffer, *Stain Technol.* 55, 191-192, 1980; Caswell, A.H. and Bruschwig, J.P., Identification and extraction of proteins that compose the triad junction of skeletal muscle, *J.Cell Biol.* 99, 929-939, 1984; Parks, J.C. and Cohen, G.M., Glutaraldehyde fixatives for preserving the chick's inner ear, *Acta Otolaryngol.* 98, 72-80, 1984; Song, A.H. and Asher, S.A., Internal intensity standards for heme protein UV resonance Raman studies: Excitation profiles of cacodylic acid and sodium selenate, *Biochemistry* 30, 1199-1205, 1991; Henney, P.J., Johnson, E.L., and Cothran, E.G., A new buffer system for acid PAGE typing of equine protease inhibitor, *Anim.Genet.* 25, 363-364, 1994; Jezewska, M.J., Rajendran, S., and Bujalowski, W., Interactions of the 8-kDa domain of rat DNA polymerase beta with DNA, *Biochemistry* 40, 3295-3307, 2001; Kenyon, E.M. and Hughes, M.F., A concise review of the toxicity and carcinogenicity of dimethylarsinic acid, *Toxicology* 160, 227-236, 2001; Cohen, S.M., Arnold, L.L., Eldan, M., *et al.*, Methylated arsenicals: the implications of metabolism and carcinogenicity studies in rodents to human risk management, *Crit.Rev.Toxicol.* 99-133, 2006.

LIST OF BUFFERS (Continued)

Common Name	Chemical Name	M.W	Properties and Comment
CAPS	3-(cyclohexylamino)-1-propanesulfonic acid	221.3	A zwitterionic buffer similar to a "Good" buffer

Lad, P.J. and Leffert, H.L., Rat liver alcohol dehydrogenase. I. Purification and characterization, *Anal.Biochem.* 133, 350-361, 1983; Kaushal, V. and Barnes, L.D., Effect of zwitterionic buffers on measurement of small masses of protein with bicinchoninic acid, *Anal.Biochem.* 157, 291-294, 1986; Himmel, H.M. and Heller, W., Studies on the interference of selected substances with two modifications of the Lowry protein determination, *J.Clin. Chem.Clin.Biochem.* 25, 909-913, 1987; Nguyen, A.L., Luong, J.H., and Masson, C., Determination of nucleotides in fish tissues using capillary electrophoresis, *Anal.Chem.* 62, 2490-2493, 1990; Jin, Y. and Cerletti, N., Western blotting of transforming growth factor β2. Optimization of the electrophoretic transfer, *Appl.Theor.Electrophor.* 3, 85-90, 1992; Ng, L.T., Selwyn, M.J., and Choo, H.L., Effect of buffers and osmolality on anion uniport across the mitochondrial inner membrane, *Biochim.Biophys.Acta* 1143, 29-37, 1993; Venosa, R.A., Kotsias, B.A., and Horowicz, P., Frog striated muscle is permeable to hydroxide and buffer anions, *J.Membr.Biol.* 139, 57-74, 1994; Righetti, P.G., Bossi, A. and Gelfi, C., Capillary isoelectric focusing and isoelectric buffers: an evolving scenario, *J.Capillary Electrophor.* 4, 47-59, 1997; Bienvenut, W.V., Deon, C., Sanchez, J.C., and Hochstrasser, D.F., *Anal.Biochem.* 307, 297-303, 2002; Zaitseva, J., Holland, I.B., and Schmitt, L., The role of CAPS buffer in expanding the crystallization space of the nucleotide-binding domain of the ABC transporter haemolysin B from *Escherichia coli*, *Acta Crystallogr.D.Biol.Crystallogr.* 60, 1076-1084, 2004; Kannamkumarath, S.S., Wuilloud, R.G., and Caruso, J.A., Studies of various elements of nutritional and toxicological interest associated with different molecular weight fractions in Brazil nuts, *J.Agric.Food Chem.* 52, 5773-5780, 2004; Hautala, J.T., Wiedmer, S.K., and Riekkola, M.L., Influence of pH on formation and stability of phosphatidylcholine/phosphatidylserine coatings in fused-silica capillaries, *Electrophoresis* 26, 176-186, 2005; Taha, M., Buffers for the physiological pH range: acidic dissociation constants of zwitterionic compounds in various hydroorganic media, *Ann.Chim.* 95, 105-109, 2005; Tu, J., Halsall, H.B., Seliskar, C.J. *et al.*, Estimation of logP(ow) values for neutral and basic compounds by microchip microemulsion electrokinetic chromatography with indirect fluorometric detection (muMEEKC-IFD), *J.Pharm.Biomed.Anal.* 38, 1-7, 2005.

CAPSO	3-(Cyclohexylamino)-2-hydroxy-1-propanesulfonic acid	237.3	A zwitterionic buffer similar to a "Good" buffer

Delaney, J.P., Kimm, G.E., and Bonsack, M.E., The influence of lumenal pH on the severity of acute radiation enteritis, *Int.J.Radiat.Biol.* 61, 381-386, 1992; McGregor, D.P., Forster, S., Steven, J. *et al.*, Simultaneous detection of microorganisms in soil suspension based on PCR amplification of bacterial 16S rRNA fragments, *BioTechniques* 21, 463-466, 1996; Liu, Q. Li, X., and Somer, S.S. pK-Matched running buffers for gel electrophoresis, *Anal. Biochem.* 270, 112-122, 1999; Quiros, M., Parker, M.C., and Turner, N.J., Tuning lipase enantioselectivity in organic media using solid-state buffers, *J.Org.Chem.* 66, 5074-5079, 2001; Okuda, M., Iwahori, K., Yamashita, I., and Yoshimura, H., Fabrication of nickel and chromium nanoparticles using the protein cage of apoferritin, *Biotechnol.Bioeng.* 84, 187-194, 2003; Vespalec, R., Vlckova, M., and Horakova, H., Aggregation and other intermolecular interactions of biological buffers observed by capillary electrophoresis and UV photometry, *J.Chromatog.A* 1051, 75-84, 2004; Taha, M., Buffers for the physiological pH range: acidic dissociation constants of zwitterionic compounds in various hydroorganic media, *Ann.Chim.* 95, 105-109, 2005.

Carbonate	Sodium bicarbonate Sodium Carbonate Ammonium bicarbonate ammonium carbonate;		The ammonium salt system is a volatile buffer. The carbonate buffer is considered to be a physiological buffer. Bicarbonate buffers are used in renal dialysis.

Nagasawa, K. and Uchiyama, H., Preparation and properties of biologically active fluorescent heparins, *Biochim.Biophys.Acta* 544, 430-440, 1978; Horejsi, V. and Hilgert, I., Simple polyacrylamide gel electrophoresis in continuous carbonate buffer system suitable for the analysis of ascites fluids of hybridoma bearing mice, *J.Immunol.Methods* 86, 103-105, 1986; Chang, G.G. and Shiao, S.L., Possible kinetic mechanism of human placental alkaline phosphatase in vivo as implemented in reverse micelles, *Eur.J. Biochem.* 220, 861-870, 1994; Steinitz, M. and Tamir, S., An improved method to create nitrocellulose particles suitable for the immobilization of antigen and antibody, *J.Immunol.Methods* 187, 171-177, 1995; Wang, Z., Gurel, O., Baatz, J.E. and Notter, R.H., Acylation of pulmonary surfactant protein-C is required for its optimal surface active interactions with phospholipids, *J.Biol.Chem.* 271, 19104-19109, 1996; Petersen, A. and Steckhan, E., Continuous indirect electrochemical regeneration of galactose oxidase, *Bioorg.Med.Chem.* 7, 2203-2208, 1999; Medda, R., Padiglia, A., Messana, T., *et al.*, Separation of diadenosine polyphosphates by capillary electrophoresis, *Electrophoresis* 21, 2412-2416, 2000; Bartzatt, R., Fluorescent labeling of drugs and simple organic compounds containing amine functional groups, utilizing dansyl chloride in Na$_2$CO$_3$ buffer, *J.Pharmacol.Toxicol.Methods* 45, 247-253, 2001; Bruno, F., Curini, R., Di Corcia, A., *et al.*, Determination of surfactants and some of their metabolites in untreated and anaerobically digested sewage sludge by subcritical water extraction followed by liquid chromatography-mass spectrometry, *Environ.Sci.Technol.* 36, 4156-4161, 2002; Chen, X.L., Sun, C.Y., Zhang, Y.Z., and Gao, P.J., Effects of different buffers on the thermostability and autolysis of a cold-adapted proteases MCP-01, *J.Protein Chem.* 21,523-527, 2002; Duman, M., Saber, R., and Piskin, E., A new approach for immobilization of

LIST OF BUFFERS (Continued)

Common Name	Chemical Name	M.W	Properties and Comment

oligonucleotides onto piezoelectric quartz crystal for preparation of a nucleic acid sensor following hybridization, *Biosens.Bioelectron.* 18, 1355-1363, 2003; Talu, G.F. and Diyamandoglu, V., Formate ion decomposition in water under UV irradiation at 253.7 nm, *Environ.Sci.Technol.* 38, 3984-3993, 2004; Dwight, S.J., Gaylord, B.S., Hong, J.W., and Bazan, G.C., Perturbation of fluorescence by nonspecific interactions between anionic poly(phenylenevinylene)s and proteins. Implications for biosensors, *J.Am.Chem.Soc.* 126, 16850-16859, 2004; Willems, A.V., Deforce, D.L., Van Peteghem, C.H., and Van Bocxlaer, J.F., Development of a quality control method for the characterization of oligonucleotides by capillary zone electrophoresis-electrospray ionization-quadrupole time of flight-mass spectrometry, *Electrophoresis* 26, 1412-1423, 2005; Asberg, P.,Bjork, P., Hook, F., and Inganas, O., Hydrogels from a water-soluble zwitterionic polythiophene: dynamics under pH change and biomolecular interactions observed using quartz crystal microbalance with dissipation monitoring, *Langmuir* 21, 7292-7298, 2005; Shah, M., Meija, J., Cabovska, B.,, and Caruso, J.A., Determination of phosphoric acid triesters in human plasma using solid-phase microextraction and gas chromatography coupled to inductively coupled plasma mass spectrometry, *J.Chromatog.A.* 1103, 329-336, 2006; Di Pasqua, A.J., Goodisman, J., Kerwood, D.J. *et al.*, Activation of carboplatin by carbonate, *Chem.Res.Toxicol.* 18, 139-149, 2006; Ormond, D.R. and Kral, T.A., Washing methogenic cells with the liquid fraction from a Mars soil stimulant and water mixture, *J.Microbiol.Methods* 67, 603-605, 2006; Binter, A., Goodisman, J., and Dabrowiak, J.C., Formation of monofunctional cisplatin-DNA adducts in carbonate buffer, *J.Inorg.Biochem.* 100, 1219-1224, 2006. Alkaline carbonate buffers have been used as the medium for proteins for application to microplates for immunoassays such as ELISA assays (Rote, N.S., Taylor, N.L. Shigeoka, A.O., *et al.*, Enzyme-linked immunosorbent assay for group B streptococcal antibodies, *Infect. Immun.* 27, 118-123, 1980; Hubschle, O.J., Lorenz, R.J., and Matheka, H.D., Enzyme-linked immunosorbent assay for detection of bluetongue virus antibodies, *Am.J.Vet.Res.* 42, 61-65, 1981; Solling, H., and Dinesen, B., The development of a rapid ELISA for IgE utilizing commercially available reagents, *Clin.Chim.Acta* 130, 71-83, 1983; Mowat, W.P. and Dawson, S., Detection and identification of plant viruses by ELISA using crude sap extracts and unfractionated antisera, *J.Virol.Methods* 15, 233-247, 1987; Ferris, N.P., Powell, H., and Donaldson, A.I., Use of pre-coated immunoplates and freeze-dried reagents for the diagnosis of foot-and-mouth disease and swine vesicular disease by enzyme-linked immunosorbent assay [ELISA], *J.Virol.Methods* 19,197-206, 1988; Cutler, S.J. and Wright, D.J., Comparison of immunofluorescence and enzyme linked immunosorbent assays for diagnosing Lyme disease, *J.Clin.Pathol.* 42, 869-871, 1989; Oshima, M. and Atassi, M.Z., Comparison of peptide-coating conditions in solid phase assays for detection of anti-peptide antibodies, *Immunol.Invest.* 18, 841-851, 1989; Martin, R.R., Relationships among luteoviruses based on nucleic acid hybridization and serological studies, *Intervirology* 31,23-30, 1990; Houen, G. and Koch, C., A non-denaturing enzyme linked immunosorbent assay with protein preadsorbed onto aluminum hydroxide, *J.Immunol.Methods* 200, 99-105, 1997; Shrivastav, T.G., Basu, A., and Kariya, K.P., Substitution of carbonate buffer by water for IgG immobilization in enzyme linked immunosorbent assay, *J.Immunoassay Immunochem.* 24, 191-203, 2003). Bicarbonate buffers also have an effect on the reaction of phenylglyoxal with proteins (Cheung, S.T. and Fonda, M.L., Reaction of phenylglyoxal with arginine. The effect of buffers and pH, *Biochem.Biophys.Res.Commun.* 90, 940-947, 1979). Bicarbonate also enhances the binding of iron to transferrin(Matinaho, S., Karhumäki, P., and Parkkinen, J., Bicarbonate inhibits the growth of *Staphylococcus epidermidis* in platelet concentrates by lowering the level of non-transferrin-bound iron, *Transfusion* 45, 1768-173, 2005

Cholamine (2-aminoethyl) trimethyl-ammonium chloride hydrochloride

Blasie, C.A. and Berg, J.M., Structure-based thermodynamic analysis of a coupled metal binding-protein folding reaction involving a zinc finger peptide, *Biochemistry* 41, 15068-15073, 2002; Zwiorek, K., Kloeckner, J., Wagner, E., and Coester, C., Gelatin nanoparticles as a new and simple gene delivery system, *J.Pharm.Pharm.Sci.* 7, 22-28, 2005.

| **Citric Acid** | 2-hydroxy-1,2,3-propanetricarboxylic acid | 192.1 | Compounds found in a variety of biological tissues and cells; involved in energy metabolism (citric acid cycle; Krebs cycle; Krebs, H.A., The citric acid cycle and the Szent-Gyorgyi cycle in pigeon breast muscle, *Biochem.J.* 34, 775-779, 1940). Also used as biological buffer. |

Citric acid has three carboxylic acid functions which permits buffering capacity from pH 2.0 to pH 12. Citric acid also chelate divalent cations and is used an anticoagulant for the collection of blood based its ability to chelate calcium ions. Chelation of metal ions is responsible for the observed inhibition of many enzymes. For early observations, see Smith, E.G., Dipeptidases, *Methods Enzymol.* 2, 93-114, 1955; McDonald, M.R., Deoxyribonucleases, *Methods Enzymol.* 2, 437-447, 1955; Kornberg, A., Adenosine phosphokinase, *Methods Enzymol.* 2, 497-500, 1955; Koshland, D.E., Jr., Preparation and properties of acetyl phosphatase, *Methods Enzymol.* 2, 556-556, 1955. The ability to chelate calcium serves as the basis for use as a decalcification agent. The polyvalent nature of citrate provide some unique characteristics such as the differentiation of muscle fiber types for histochemistry (Matoba, H., and Gollnick, P.D., Influence of ionic composition, buffering agent, and pH on the histochemical demonstration of myofibrillar actomyosin ATPase, *Histochemistry* 80, 609-614, 1984) and the activation of "prothrombin"(Seegers, W.H., McClaughery, R.I., and Fahey, J.L., Some properties of purified prothrombin and its activation with sodium citrate, *Blood* 5, 421-433, 1950; Lanchantin, G.F., Friedman, J.A., and Hart, D.W., The conversion of human prothrombin to thrombin by sodium citrate. Analysis of the reaction mixture, *J.Biol.Chem.* 240, 3276-3282, 1965; Aronson, D.L. and Mustafa, A.J., The activation of human factor X in sodium citrate: the role of factor VII, *Thromb.Haemostas.* 36, 104-114, 1976). Citrate is a polyvalent anion and like other polyvalent anions such as phosphate and sulfate, citrate can cause a "salting-out" phenomena (Hegardt, F.G. and Pie, A., Sodium citrate salting-out of the human blood serum proteins, *Rev.Esp.Fisiol.* 24, 161-168, 1968; Carrea, G., Pasta, P., and Vecchio, G., Effect of the lyotropic series of anions on denaturation and renaturation of 20-β-hydroxysteroid dehydrogenase, *Biochim.Biophys.Acta* 784, 16-23,

Common Name	Chemical Name	M.W	Properties and Comment

1984; Nakano, T., Yuasa, H., and Kanaya, Y., Suppression of agglomeration in fluidized bed coating. III. Hofmeister series in suppression of particle agglomeration, *Pharm.Res.* 16, 1616-1620, 1999; Nakano, T. and Yuasa, H., Suppression of agglomeration in fluidized bed coating. IV. Effects of sodium citrate concentration on the suppression of particle agglomeration and the physical properties of HPMC film, *Int.J.Pharm.* 215, 3-12, 2001; Mani, N. and Jun, H.W., Microencapsulation of a hydrophilic drug into a hydrophobic matrix using a salting-out procedure. I: Development and optimization of the process using factorial design, *J.Microencapsulation* 21, 125-135, 2004). Citrate also has an effect on partitioning in aqueous two-phase systems (Andrews, B.A., Schmidt, A.S., and Asenjo, J.A., Correlation for the partition behavior of proteins in aqueous two-phase systems: effect of surface hydrophobicity and charge, *Biotechnol.Bioeng.* 90, 380-390, 2005). Citrate has proved useful in the solubilization of proteins, usually, but not always, from mineralized/calcified matrices (Faludi, E. and Harsanyi, V., The effect of Na$_3$-citrate on the solubility of cryoprecipitate [citrate effect of cryoprecipitate], *Haematologia* 14, 207-214, 1981; Myllyla, R., Preparation of antibodies to chick-embryo galactosylhydroxylysyl glucosyltransferase and their use for an immunological characterization of the enzyme of collagen synthesis, *Biochim.Biophys.Acta* 658, 299-307, 1981; Guy, O., Robles-Diaz, G., Adrich, Z., *et al.,* Protein content of precipitates present in pancreatic juice of alcoholic subjects and patients with chronic calcifying pancreatitis, *Gastroenterology* 84, 102-107, 1983; Collingwood, T.N., Shanmugam, M., Daniel, R.M., and Langdon, A.G., M[III]-facilitated recovery and concentration of enzymes from mesophilic and thermophilic organisms, *J.Biochem. Biophys.Methods* 19, 281-286, 1989). Citrate is useful for dissociating protein complexes in some situations by binding to specific anion binding sites; the ability of citrate to function as a buffering at low pH is an advantage [Kuo, T.T., Chow, T.Y., Lin, X.T., *et al.,* Specific dissociation of phage Xp12 by sodium citrate, *J.Gen.Virol.* 10,199-202, 1971(in this case, the dissociation is reflection of metal ion binding; the dissociation is associated with the loss of biological activity – see Lark, K.G. and Adams, M.H., The stability of phage as a function of the ionic environment, *Cold Spring Harbor Symposium on Quantitative Biology*, 18, 171-183, 1953); Sheffery, M. and Newton, A., Reconstitution and purification of flagellar filaments from *Caulobacter crescentus*, *J.Bacteriol.* 132, 1027-1030, 1977; Brooks, S.P. and Nicholls, P., Anion and ionic strength effects upon the oxidation of cytochrome c by cytochrome c oxidase, *Biochim.Biophys.Acta* 680, 33-43, 1982; Berliner, L.J., Sugawara, Y., and Fenton, J.W., 2nd, Human alpha-thrombin binding to nonpolymerized fibrin-Sepharose: evidence for an anionic binding region, *Biochemistry* 24, 7005-7009, 1985; Kella, N.K. and Kinsella, J.E., Structural stability of beta-lactoglobulin in the presence of kosmotropic salts. A kinetic and thermodynamic study, *Int.J.Pept.Protein Res.* 32, 396-405, 1988; Oe, H., Takahashi, N., Doi. E., and Hirose, M., Effects of anion binding on the conformations of the two domains of ovotransferrin, *J.Biochem.* 106, 858-863, 1989; Polakova, K., Karpatova, M., and Russ, G., Dissociation of β-2-microglobulin is responsible for selective reduction of HLA class I antigenicity following acid treatment of cells, *Mol.Immunol.* 30, 1223-1230, 1993; Lecker, D.N. and Khan, A., Model for inactivation of α-amylase in the presence of salts: theoretical and experimental studies, *Biotechnol.Prog.* 14, 621-625, 1998; Rabiller-Baudry, M. and Chaufer, B., Small molecular ion adsorption on proteins and DNAs revealed by separation techniques, *J.Chromatog.B.Analyt.Technol.Biomed.Life.Sci.* 797, 331-345, 2003; Raibekas, A.A., Bures, E.J., Siska, C.C., *et al.,* Anion binding and controlled aggregation of human interleukin-1 receptor antagonist, *Biochemistry* 44, 9871-9879, 2005). A special application of citrate dissociation of protein complexes is the isolation and dissociation of antigen-antibody complexes (Woodroffe, A.J. and Wilson, C.B., An evaluation of elution techniques in the study of immune complex glomerulonephritis, *J.Immunol.* 118, 1788-1794, 1977; Ehrlich, R. and Witz, I.P., The elution of antibodies from viable murine tumor cells, *J.Immunol.Methods* 26, 345-353, 1979; McIntosh, R.M., Garcia, R., Rubio, L., *et al.,* Evidence of an autologous immune complex pathogenic mechanism in acute poststreptococcal glomerulonephritis, *Kidney Int.* 14, 501-510, 1978; Theofilopoulos, A.N., Eisenberg, R.A., and Dixon, F.J., Isolation of circulating immune complexes using Raji cells. Separation of antigens from immune complexes and production of antiserum, *J.Clin.Invest.* 61, 1570-1581, 1978; Tomino, Y., Sakai, H., Endoh, M., *et al.,* Cross-reacivity of eluted antibodies from renal tissues of patients with Henoch-Schonlein purpura nephritis and IgA nephropathy, *Am.J.Nephrol.* 3, 315-318, 1983). A more complex and poorly understood application of citrate buffers is in epitope retrieval (Shi, S.R., Chaiwun, B., Young, L., *et al.,* Antigen retrieval techniques utilizing citrate buffer or urea solution for immunohistochemical demonstration of androgen receptor in formalin-fixed paraffin sections, *J.Histochem.Cytochem.* 41, 1599-1604, 1993; Langlois, N.E., King, G., Herriot, R., and Thompson, W.D., Non-enzymatic retrieval of antigen permits staining of follicle centre cells by the rabbit polyclonal antibody to protein gene product 9.5, *J.Pathol.* 173, 249-253, 1994; Leong, A.S., Microwaves in diagnostic immunohistochemistry, *Eur.J.Morphol.* 34, 381-383, 1996; Lucas, D.R., al-Abbadi, M., Teabaczka, P., *et al.,* c-Kit expression in desmoid fibroblastosis. Comparative immunohistochemical evaluation of two commercial antibodies, *Am.J.Clin.Pathol.* 119, 339-345, 2003). Additional work has indicated that citrate is useful but not unique for epitope retrieval (Imam, S.A., Young, L., Chaiwun, B., and Taylor, C.B., Comparison of two microwave based antigen-retrieval solutions in unmasking epitopes in formalin-fixed tissues for immunostaining, *Anticancer Res.* 15, 1153-1158, 1995; Pileri, S.A., Roncador, G., Ceccarelli, C., *et al.,* Antigen retrieval techniques in immunohistochemistry: comparison of different methods, *J.Pathol.* 183, 116-123, 1997; Rocken, C. and Roessner, A., An evaluation of antigen retrieval procedures for immunoelectron microscopic classification of amyloid deposits, *J.Histochem.Cytochem.* 47, 1385-1394, 1999). Citrate buffer has been useful in affinity chromatography (Ishikawa, K. and Iwai, K., Affinity chromatography of cysteine-containing histone, *J.Biochem.* 77, 391-398, 1975; Chadha, K.C., Grob, P.M., Mikulski, A.J., *et al.,* Copper chelate affinity chromatography of human fibroblast and leucocyte interferons, *J.Gen.Virol.* 43, 701-706, 1979; Tanaka, H., Sasaki, I., Yamashita, K. *et al.,* Affinity chromatography of porcine pancreas deoxyribonuclease I on DNA-binding Sepharose under non-digestive conditions, using its substrate-binding site, *J.Biochem.* 88, 797-806, 1980 Smith, R.L. and Griffin, C.A., Separation of plasma fibronectin from associated hemagglutinating activity by elution from gelatin-agarose at pH 5.5, *Thromb. Res.* 37, 91-101, 1985). Citrate is also used for immunoaffinity chromatography including chromatography on Protein A (Martin, L.N., Separation of guinea pig IgG subclasses by affinity chromatography on protein A-Sepharose, *J.Immunol.Methods* 52, 205-212, 1982; Compton, B.J., Lewis, M.A., Whigham, F., *et al.,* Analtyical potential of protein A for affinity chromatography of polyclonal and monoclonal antibodies, *Anal.Chem.* 61, 1314-1317, 1989; Giraudi, G. and Baggiani, C. Strategy for fractionating high-affinity antibodies to steroid hormones by affinity chromatography, *Analyst* 121, 939-944, 1996; Arakawa, T., Philo, J.S., Tsumoto, K., *et al.,* Elution of antibodies from a Protein-A column by aqueous arginine solutions, *Protein Expr. Purif.* 36, 244-248, 2004; Ghose, S., McNerney, T., and Hubbard, B., Protein A affinity chromatography for capture and purification of monoclonal antibody and Fc-fusion protein: Practical considerations for process development, in *Process Scale Bioseparations for the Biopharmaceutical Industry*, ed. A.A. Shukla, M.R. Etzel, and S. Gadam, , ed. A.A. Shukla, M.R. Etzel, and S. Gadam, CRC/Taylor & Francis, Boca Raton, FL., Chapter 16, pps. 462-489, 2007).

LIST OF BUFFERS (Continued)

Common Name	Chemical Name	M.W	Properties and Comment
HEPES	4-(2-hydroxyethyl)-1-piperizineethanesulfonic acid	238.3	a "Good" buffer; reagent purity has been an issue; metal ion binding must be considered; there are buffer-specific effects which are poorly understood; component of tissue fixing technique

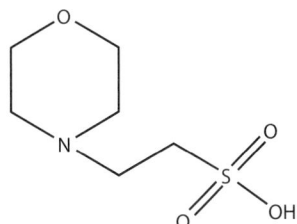

Good, N.E., Winget, G.D., Winter, W., *et al.*, Hydrogen ion buffers for biological research, *Biochemistry* 5, 467-477, 1966; Turner, L.V. and Manchester, K.L., Interference of HEPES with the Lowry method, *Science* 170, 649, 1970; Chirpich, T.P., The effect of different buffers on terminal deoxynucleotidyl transferase activity, *Biochim.Biophys.Acta* 518, 535-538, 1978; Tadolini, B., Iron autoxidation in Mops and Hepes buffers, *Free Radic.Res. Commun.* 4, 149-160, 1987; Simpson, J.A., Cheeseman, K.H., Smith, S.E., and Dean, R.T., Free-radical generation by copper ions and hydrogen peroxide. Stimulation by Hepes buffer, *Biochem.J.* 254, 519-523, 1988; Abas, L. and Guppy M., Acetate: a contaminant in Hepes buffer, *Anal.Biochem.* 229, 131-140, 1995; Schmidt, K., Pfeiffer, S., and Mayer, B., Reaction of peroxynitrite with HEPES or MOPS results in the formation of nitric oxide donors, *Free Radic.Biol.Med.* 24, 859-862, 1998; Wiedorn, K.H., Olert, J., Stacy, R.A., *et al.*, HOPE – a new fixing technique enables preservation and extraction of high molecular weight DNA and RNA of >20 kb from paraffin-embedded tissues. Hepes-glutamic acid buffer mediated Organic solvent Protection Effect, *Pathol.Res.Pract.* 198, 735-740, 2002; Fulop, L., Szigeti, G., Magyar, J., *et al.*, Differences in electrophysiological and contractile properties of mammalian cardiac tissues bathed in bicarbonate – and HEPES-buffered solutions, *Acta Physiol.Scand.* 178, 11-18, 2003; Mash, H.E., Chin, Y.P., Sigg, L., *et al.*, Complexation of copper by zwitterionic aminosulfonic (good) buffers, *Anal.Chem.* 75, 671-677, 2003 Sokolowska, M., and Bal, W., Cu(II) complexation by "non-coordinating" *N*-2-hydroxyethylpiperazine-*N'*-ethanesulfonic acid (HEPES buffer), *J.Inorg.Biochem.* 99, 1653-1660, 2005; ; Zhao, G. and Chasteen, N.D., Oxidation of Good's buffers by hydrogen peroxide, *Anal.Biochem.* 349, 262-267, 2006; Hartman, R.F. and Rose, S.D., Kinetics and mechanism of the addition of nucleophiles to alpha,beta-unsaturated thiol esters, *J.Org.Chem.* 71, 6342-6350, 2006.

MES	1-morpholineethanesulfonic acid; 2-(4-morpholino)ethanesulfonate	198.2	A "Good" buffer

Good, N.E., Winget, G.D., Winter, W., *et al.*, Hydrogen ion buffers for biological research, *Biochemistry* 5, 467-477, 1966; Bugbee, B.G. and Salisbury, F.B., An evaluation of MES (2(*N*-morpholino)ethanesulfonic acid) and Amberlite 1RC-50 as pH buffers for nutrient growth studies, *J.Plant Nutr.* 8, 567-583, 1985; Kaushal, V. and Barnes, L.D., Effect of zwitterionic buffers on measurement of small masses of proiten with bicinchoninic acid, *Anal. Biochem.* 157, 291-294, 1986; Grady, J.K., Chasteen, N.D., and Harris, D.C., Radicals from "Good's" buffers, *Anal.Biochem.* 173, 111-115, 1988; Le Hir, M., Impurity in buffer substances mimics the effect of ATP on soluble 5'-nucleotidase, *Enzyme* 45, 194-199, 1991; Pedrotti, B., Soffientini, A., and Islam, K., Sulphonate buffers affect the recovery of microtubule-associated proteins MAP1 and MAP2: evidence that MAP1A promotes microtubule assembly, *Cell Motil.Cytoskeleton* 25, 234-242, 1993; Vasseur, M., Frangne, R., and Alvarado, F., Buffer-dependent pH sensitivity of the fluorescent chloride-indicator dye SPQ, *Am.J.Physiol.* 264, C27-C31, 1993; Frick, J. and Mitchell, C.A., Stabilization of pH in solid-matrix hydroponic systems, *HortScience* 28, 981-984, 1993; Yu, Q., Kandegedara, A., Xu, Y., and Rorabacher, D.B., Avoiding interferences from Good's buffers: A continguous series of noncomplexing tertiary amine buffers covering the entire range of pH 3-11, *Anal.Biochem.* 253, 50-56, 1997; Gelfi, C., Vigano, A., Curcio, M., *et al.*, Single-strand conformation polymorphism analysis by capillary zone electrophoresis in neutral pH buffer, *Electrophoresis* 21, 785-791, 2000; Walsh, M.K., Wang, X., and Weimer, B.C., Optimizing the immobilization of single-stranded DNA onto glass beads, *J.Biochem.Biophys.Methods* 47, 221-231, 2001; Hosse, M. and Wilkinson, K.J., Determination of electrophoretic mobilities and hydrodynamic radii of three humic substances as a function of

LIST OF BUFFERS (Continued)

Common Name	Chemical Name	M.W	Properties and Comment

pH and ionic strength, *Environ.Sci.Technol.* 35, 4301-4306, 2001; Mash, H.E., Chin, Y.P., Sigg, L., *et al.*, Complexation of copper by zwitterionic aminosulfonic (good) buffers, *Anal.Chem.* 75, 671-677, 2003; Ozkara, S., Akgol, S., Canak, Y., and Denizli, A., A novel magnetic adsorbent for immunoglobulin-g purification in a magnetically stabilized fluidized bed, *Biotechnol.Prog.* 20, 1169-1175, 2004; Hachmann, J.P. and Amshey, J.W., Models of protein modification in Tris-glycine and neutral pH Bis-Tris gels during electrophoresis: effect of pH, *Anal.Biochem.* 342, 237-345, 2005; Krajewska, B. and Ciurli, S., Jack bean (*Canavalia ensiformis*) urease. Probing acid-base groups of the active site by pH variation, *Plant Physiol. Biochem.* 43, 651-658, 2005; Zhao, G. and Chasteen, N.D., Oxidation of Good's buffers by hydrogen peroxide, *Anal.Biochem.* 349, 262-267, 2006.

MOPS	3-(*N*-morpholino)propanesulfonic acid; 4-morpholinepropanesulfonic acid	209.3	A "Good" buffer

Good, N.E., Winget, G.D., Winter, W., *et al.*, Hydrogen ion buffers for biological research, *Biochemistry* 5, 467-477, 1966; Altura, B.M., Altura, B.M., Carella, A. and Altura, B.T., Adverse effects of Tris, HEPES and MOPS buffers on contractile responses of arterial and venous smooth muscle induced by prostaglandins, *Prostaglandins Med.* 5, 123-130, 1980; Tadolini, B., Iron autoxidation in Mops and Hepes buffers, *Free Radic.Res.Commun.* 4, 149-160, 1987; Tadolini, B. and Sechi, A.M., Iron oxidation in Mops buffer. Effect of phosphorus containing compounds, *Free Radic.Res.Commun.* 4, 161-172, 1987; Tadolini, B., Iron oxidation in Mops buffer. Effect of EDTA,. hydrogen peroxide and FeCl$_3$, *Free Radic.Res.Commun.* 4, 172-182, 1987; Ishihara, H. and Welsh, M.J., Block by MOPS reveals a conformation change in the CFTR pore produced by ATP hydrolysis, *Am.J.Physiol.* 273, C1278-C1289, 1997; Schmidt, K., Pfeiffer, S., and Meyer, B., Reaction of peroxynitrite with HEPES or MOPS results in the formation of nitric oxide donors, *Free Radic.Biol.Med.* 24, 859-862, 1998; Hodges, G.R. and Ingold, K.U., Superoxide, amine buffers and tetranitromethane: a novel free radical chain reaction, *Free Radic.Res.* 33, 547-550, 2000; Corona-Izquierdo, F.P. and Membrillo-Hernandez, J., Biofilm formation in *Escherichia coli* is affected by 3-(*N*-morpholino)propane sulfonate (MOPS), *Res.Microbiol.* 153, 181-185, 2002; Mash, H.E., Chin, Y.P., Sigg, L., *et al.*, Complexation of copper by zwitterionic aminosulfonic (good) buffers, *Anal.Chem.* 75, 671-677, 2003; Denizli, A., Alkan, M., Garipcan, B., *et al.*, Novel metal-chelate affinity adsorbent for purification of immunoglobulin-G from human plasma, *J.Chromatog.B.Analyt.Technol.Biomed.Life.Sci.* 795, 93-103, 2003; Emir, S., Say, R., Yavuz, H., and Denizli, A., A new metal chelate affinity adsorbent for cytochrome C, *Biotechnol.Prog.* 20, 223-228, 2004; Cvetkovic, A., Zomerdijk, M., Straathof, A.J., *et al.*, Adsorption of fluorescein by protein crystals, *Biotechnol.Bioeng.* 87, 658-668, 2004; Zhao, G. and Chasteen, J.D., Oxidation of Good's buffers by hydrogen peroxide, *Anal.Biochem.* 349, 2620267, 2006; Vrakas, D., Giaginis, C. and Tsantili-Kakoulidou, A., Different retention behavior of structurally diverse basic and neutral drugs in immobilized artificial membrane and reversed-phase high performance liquid chromatography: comparison with octanol-water partitioning, *J.Chromatog.A.* 1116, 158-164, 2006; de Carmen Candia-Plata, M., Garcia, J., Guzman, R., *et al.*, Isolation of human serum immunoglobulins with a new salt-promoted adsorbent, *J.Chromatog.A.* 1118, 211-217, 2006.

Phosphate

Phosphate buffers are among the most common buffers used for biological studies. It is noted that the use of phosphate solutions in early transfusion medicine lead to the discovery of the importance of calcium ions in blood coagulation (Hutchin, P., History of blood transfusion: A tercentennial look, *Surgery* 64, 685-700, 1968). Phosphate-buffer saline (PBS; generally 0.01 M sodium phosphate – 0.14 M NaCl, pH 7.2 – Note, an incredible variation in PBS exists so it is necessary to verify composition – the only common factor that this writer finds is 0.01 M (10 mM) phosphate) is extensively used. Sodium phosphate buffers are the most common but there is extensive use of potassium phosphate buffers and mixtures of sodium and potassium. Unfortunately many investigators simply refer to phosphate buffer without respect to counter ion. Also, investigators will prepare a stock solution of sodium phosphate[usually sodium dihydrogen phosphate (sodium phosphate, monobasic) or disodium hydrogen phosphate (sodium phosphate, dibasic) and adjust pH as required with (usually) hydrochloric acid and/or sodium hydrogen. This is not preferable and, if used, must be described in the text to permit other investigators to repeat the experiment. pH changes in phosphate buffers during freezing can be dramatic due to precipitation of phosphate buffer salts (van den Berg, L. and Rose, D., Effect of freezing on the pH and composition of sodium and potassium phosphate solutions: The reciprocal system KH$_2$PO$_4$-Na$_2$PO$_4$-H$_2$O, *Arch.Biochem.Biophys.* 81, 319-329, 1959; Murase, N. and Franks, F., Salt precipitation during the freeze-concentration of phosphate buffer solutions, *Biophys.Chem.* 34, 393-300, 1989; Pikal-Cleland, K.A. and Carpenter, J.F., Lyophilization-induced protein denaturation in phosphate buffer systems: monomeric and tetrameric beta-galactosidase, *J.Pharm.Sci.* 90, 1255-1268, 2001; Gomez, G., Pikal, M., and Rodriguez-Hornedo, N., Effect of initial buffer composition on pH changes during far-from-equilibrium freezing of sodium phosphate buffer solutions, *Pharm.Res.* 18, 90-97, 2001; Pikal-Cleland, K.A., Cleland, J.L., Anchorodoquy, T.J. and Carpenter, J.F., Effect of glycine on pH changes and protein stability during freeze-thawing in phosphate buffer systems, *J.Pharm .Sci.* 91, 1969-1979, 2002). Phosphate bind divalent cations in solutions

LIST OF BUFFERS (Continued)

Common Name	Chemical Name	M.W	Properties and Comment

and can form insoluble salts. Phosphate influences biological reactions by binding cations such as calcium, platinum and iron (Staum, M.M., Incompatibility of phosphate buffer in 99^m Tc-sulfur colloid containing aluminum ion, *J.Nucl.Med.* 13, 386-387, 1972; Frank, G.B., Antagonism by phosphate buffer of the twitch ions in isolated muscle fibers produced by calcium-free solutions, *Can.J.Physiol.Pharmacol.* 56, 523-526, 1978; Hasegawa, K., Hashi, K., and Okada, R., Physicochemical stability of pharmaceutical phosphate buffer solutions. I. Complexation behavior of Ca(II) with additives in phosphate buffer solutions, *J.Parenter.Sci.Technol.* 36, 128-133, 1982; Abe, K., Kogure, K., Arai, H., and Nakano, M., Ascorbate induced lipid peroxidation results in loss of receptor binding in tris, but not in phosphate, buffer. Implications for the involvement of metal ions, *Biochem.Int.* 11, 341-348, 1985; Pedersen, H.B., Josephsen, J., and Keerszan, G., Phosphate buffer and salt medium concentrations affect the inactivation of T4 phage by platinum(II) complexes, *Chem.Biol.Interact.* 54, 1-8, 1985; Kuzuya, M., Yamada, K., Hayashi, T., *et al.*, Oxidation of low-density lipoprotein by copper and iron in phosphate buffer, *Biochim.Biophys.Acta* 1084, 198-201, 1991. Also see Wolf, W.J., and Sly, D.A., Effects of buffer cations on chromatography of proteins on hydroxylapatite, *J.Chromatog.* 15, 247-250, 1964; Taborsky, G., Oxidative modification of proteins in the presence of ferrous ion and air. Effect of ionic constituents of the reaction medium on the nature of the oxidation products, *Biochemistry* 12, 1341-1348, 1973; Millsap, K.W., Reid, G., van der Mei, H.C., and Busscher, H.J., Adhesion of *Lactobacillus* species in urine and phosphate buffer to silicone rubber and glass under flow, *Biomaterials* 18, 87-91, 1997; Gebauer, P. and Bocek, P., New aspects of buffering with multivalent weak acids in capillary zone electrophoresis: pros and cons of the phosphate buffer, *Electrophoresis* 21, 2809-2813, 2000; Gebauer, P., Pantuikova, P. and Bocek, P., Capillary zone electrophoresis in phosphate buffer – known or unknown?, *J.Chromatog.A* 894, 89-93, 2000; Buchanan, D.D., Jameson, E.E., Perlette, J., *et al.*, Effect of buffer, electric field, and separation time on detection of aptamers-ligand complexes for affinity probe capillary electrophoresis, *Electrophoresis* 24, 1375-1382, 2003; Ahmad, I., Fasihullah, Z. and Vaid, F.H., Effect of phosphate buffer on photodegradation reactions of riboflavin in aqueous solution, *J.Photochem.Photobiol.B* 78, 229-234, 2005.

| **PIPES** | piperazine-*N,N*'-bis(2-ethanesulfonic acid) 1,4-piperazinediethane sulfonic acid | 302.4 | A "Good" buffer |

Good, N.E., Winget, G.D., Winter, W., *et al.*, Hydrogen ion buffers for biological research, *Biochemistry* 5, 467-477, 1966; Olmsted, J.B. and Borisy, G.G., Ionic and nucleotide requirements for microtubule polymerization in vitro, *Biochemistry* 14, 2996-3005, 1975; Baur, P.S. and Stacey, T.R., The use of PIPES buffer in the fixation of mammalian and marine tissues for electron microscopy, *J.Micros.* 109, 315-327, 1977; Schiff, R.I. and Gennaro, J.F., Jr., The influence of the buffer on maintenance of tissue liquid in specimens for scanning electron microscopy, *Scan.Electron Microsc.* (3), 449-458, 1979; Altura, B.M., Altura, B.T., Carella, A., and Turlapty, P.D., Adverse effects of artificial buffers on contractile responses of arterial and venous smooth muscles, *Br.J.Pharmacol.* 69, 207-214, 1980; Syvertsen, C. and McKinley-McKee, J.S., Affinity labeling of liver alcohol dehydrogenase. Effects of pH and buffers on affinity labelling with iodoacetic acid and (*R,S*-2-bromo-3-(5-imidazolyl)propionic acid, *Eur.J.Biochem.* 117, 165-170, 1981; Roy, R.N., Gibbons, J.J., Padron, J.L., *et al.*, Revised values of the paH of monosodium 1,4-piperazinediethanesulfonate ("Pipes") in water other buffers in isotonic saline at various temperatures, *Clin.Chem.* 27, 1787-1788, 1981; Waxman, P.G., del Campo, A.A., Lowe, M.C., and Hamel, E., Induction of polymerization of purified tubulin by sulfonate buffers. Marked differences between 4-morpholineethananesulfonate (Mes) and 1,4-piperazineethanes ulfonate(Pipes), *Eur.J.Biochem.* 129, 129-136, 1981; Yamamoto, K. and Ogawa, K., Effects of NaOH-PIPES buffer used in aldehyde fixative on alkaline phosphatase activity in rat hepatocytes, *Histochemistry* 77, 339-351, 1983; Haviernick, S., Lalague, E.D., Corvellec, M.R., *et al.*, The use of Hanks'—pipes buffers in the preparation of human, normal leukocytes for TEM observation, *J.Microsc.* 135, 83-88, 1984; Simpson, J.A., Cheeseman, K.H., Smith, S.E., and Dean, R.T., Free-radical generation by copper ions and hydrogen peroxide. Stimulation by Hepes buffers, *Biochem.J.* 254, 519-523, 1988; Prutz, W.A. The interaction between hydrogen peroxide and the DNA-Cu(I) complex: effects of pH and buffers, *Z.Naturforsch.* 45, 1197-1206, 1990; Le Hir, M., Impurity in buffer substances mimics the effects of ATP on soluble 5'-nucleotidase, *Enzyme* 45, 194-199, 1991; Lee, B.H. and Nowak, T., Influence of pH on the Mn^{2+} activation of and binding to yeast enolase: a functional study, *Biochemistry* 31, 2165-2171, 1992; Tedokon, M., Suzuki, K., Kayamori, Y., *et al.*, Enzymatic assay of inorganic phosphate with the use of sucrose phosphorylase and phosphoglucomutase, *Clin.Chem.* 38, 512-515, 1992; Correla, J.J., Lipscomb, L.D., Dabrowiak, J.C., *et al.*, Cleavage of tubulin by vandate ion, *Arch.Biochem.Biophys.* 309, 94-104, 1994; Schmidt, J., Mangold, C., and Deitmer, J., Membrane responses evoked by organic buffers in identified leech neurones, *J.Exp.Biol.* 199,327-335, 1996; Yu, Q., Kandegedara, A., Xu, Y., and Rorabacher, D.B., Avoiding interferences from Good's buffers: A contiguous series of noncomplexing tertiary amine buffers covering the entire pH range of pH 3-11, *Anal.Biochem.* 253, 50-56, 1997; Rover Junior, L., Fernandes, J.C., de Oliveira Neto, G., *et al.*, Study of NADH stability using ultraviolet-visible spectrophotometric analysis and factorial design, *Anal.Biochem.* 260, 50-55, 1998; Moore, S.A., Kingston, R.L., Loomes, K.M., *et al.*, The structure of truncated recombinant human bile salt-stimulated lipase reveals bile salt-independent conformational flexibility at the active-site loop and provides insights into heparin binding, *J.Mol.Biol.* 3 12, 511-523, 2001; Sani, R.K., Peyton, B.M., and Dohnalkova, A., Toxic effects of uranium on *Desulfovibrio desulfuricans* G20, *Environ.Toxicol.Chem.* 25, 1231-1238, 2006.

LIST OF BUFFERS (Continued)

Common Name	Chemical Name	M.W	Properties and Comment
TES	N-tris(hydroxymethyl) methyl-2-aminoethane-sulfonic acid	229.3	A "Good" buffer.

Good, N.E., Winget, G.D., Winter, W., *et al.*, Hydrogen ion buffers for biological research, *Biochemistry* 5, 467-477, 1966; Itagaki, A. and Kimura, G., Tes and HEPES buffers in mammalian cell cultures and viral studies: problem of carbon dioxide requirement, *Exp.Cell Res.* 83, 351-361, 1974; Bridges, S. and Ward, B., Effect of hydrogen ion buffers on photosynthetic oxygen evolution in the blue-green alga, *Agmenellum quadruplicatum*, *Microbios* 15, 49-56, 1976; Bailyes, E.M., Luzio, J.P., and Newby, A.C., The use of a zwitterionic detergent in the solubilization and purification of the intrinsic membrane protein 5'-nucleotidase, *Biochem.Soc.Trans.* 9, 140-141, 1981; Poole, C.A., Reilly, H.C., and Flint, M.H., The adverse effects of HEPES, TES, and BES zwitterionic buffers on the ultrastructure of cultured chick embryo epiphyseal chondrocytes, *In Vitro* 18, 755-765, 1982; Nakon, R. and Krishnamoorthy, C.R., Free-metal ion depletion by "Good's" buffers, *Science* 221, 749-750, 1983; del Castillo, J., Escalona de Motta, G., Eterovic, V.A., and Ferchmin, P.A., Succinyl derivatives of N-tris (hydroxylmethyl) methyl-2-aminoethane sulphonic acid: their effects on the frog neuromuscular junction, *Br.J.Pharmacol.* 84, 275-288, 1985; Kaushal, V. and Varnes, L.D., Effect of zwitterionic buffers on measurement of small masses of protein with bicinchoninic acid, *Anal.Biochem.* 157, 291-294, 1986; Bhattacharyya, A. and Yanagimachi, R., Synthetic organic pH buffers can support fertilization of guinea pig eggs, but not as efficiently as bicarbonate buffer, *Gamete Res.* 19, 123-129, 1988; Veeck, L.L., TES and Tris (TEST)-yolk buffer systems, sperm function testing, and in vitro fertilization, *Fertil.Steril.* 58, 484-486, 1992; Kragh-Hansen, U. and Vorum, H., Quantitative analyses of the interaction between calcium ions and human serum albumin, *Clin.Chem.* 39, 202-208, 1993; Jacobs, B.R., Caulfield, J., and Boldt, J., Analysis of TEST (TES and Tris) yolk buffer effects of human sperm, *Fertil.Steril.* 63, 1064-1070, 1995; Stellwagne, N.C., Bossi, A., Gelfi, C., and Righetti, P.G., DNA and buffers: are there any noninteracting, neutral pH buffers?, *Anal.Biochem.* 287, 167-175, 2000; Taylor, J., Hamilton, K.L., and Butt, A.G., HCO_3^- potentiates the cAMP-dependent secretory response of the human distal colon through a DIDS-sensitive pathway, *Pflugers Arch.* 442, 256-262, 2001; Taha, M., Buffers for the physiological pH range: acidic dissociation constants of zwitterionic compounds in various hydroorganic media, *Ann. Chim.* 95, 105-109, 2005.

| **Tricine** | N-[tris(hydroxymethyl) methyl] glycine; N-[2-hydroxy-1,1-bis-(hydroxymethyl)ethyl] glycine | 179.2 | A "Good" buffer which is also used as a chelating agent, useful for cupric ions. Tricine is also used to complex technetium-99(99mTc) in cancer therapy. |

Garder, R.S., The use of tricine buffer in animal tissue cultures, *J.Cell Biol.* 42, 320-321, 1969; Spendlove, R.S., Crosbie, R.B., Hayes, S.F., and Keeler, R.F., TRICINE-buffered tissue culture media for control of mycoplasma contaminants, *Proc.Soc.Exptl.Biol.Med.* 137, 258-263, 1971; Bates, R.G., Roy, R.N., and Robinson, R.A., Buffer standards of tris(hydroxymethyl)methylglycine ("tricine") for the physiological range pH 7.2 to 8.5, *Anal.Chem.* 45, 1663-1666, 1973; Roy, R.N., Robinson, R.A., and Bates, R.G., Thermodynamics of the two dissociation steps of N-tris(hydroxymethyl)methylglycine ("tricine") in water from 5 to 50 degrees, *J.Amer.Chem.Soc.* 95, 8231-8235, 1973; Grande, H.J. and van der Ploeg, K.R., Tricine radicals as formed in the presence of peroxide producing enzymes, *FEBS Lett.* 95, 352-356, 1978; Roy, R.N., Gibbons, J.J., and Baker, G.E., Acid dissociation constants and pH values for standard "bes" and "tricine" buffer solutions in 30, 40, and 50 mass% dimethyl sulfoxide/water between 25 and -25°C, *Cryobiology* 22, 589-600, 1985; Hall, M.S. and Leach, F.R., Stability of firefly luciferase in tricine buffer and in a commercial enzyme stabilizer, *J.Biolumin.Chemilumin.* 2, 41-44, 1988; Patton, W.F., Chung-Welch, N., Lopez, M.F., *et al,*, Tris-tricine and tris-borate buffer systems provide better estimates of human mesothelial cell intermediate filament protein molecular weights than the standard Tris-glycine system, *Anal.Biochem.* 197, 25-33, 1991; [99mTc] tricine: a useful precursor complex for the radiolabeling of hydrazinonicotinate protein conjugates, *Bioconjugate Chem.* 6, 635-638, 1995; Wisdom, G.B., Molecular weight determinations using polyacrylamide gel electrophoresis with tris-tricine buffers, *Methods Mol.Biol.* 73, 97-100, 1997; Barrett, J.A., Crocker, A.C., Damphousee, D.J.. Biological evaluation of thrombus imaging agents utilizing water soluble phospines and tricine as coligands when used to label a hydrazinonicotinamide-modified cyclic glycoprotein IIb/IIIa receptor antagonist with 99mTc, *Bioconjug.Chem.* 8, 155-160, 1997; Bangard, M., Behe, M., Guhlke, S., *et al.*, Detection of somatostatin receptor-positive tumours using the new 99mTc-tricine-HYNIC-D-Phe1-Tyr3-octreotide: first results in patients and comparison with 111In-D-Phe1-octreotide, *Eur.J.Nucl.Med.* 27, 628-637, 2000; Ramos silva, M., Paixao, J.A. , Matos Beja, A., and Alte da Veiga, L., Conformational flexibility of tricine as a chelating agent in catena-poly-[[(tricinato)copper(II)]-mu-chloro], *Acta Crystallogr.C.* 57, 9-11, 2001; Silva, M.R., Paixo, J.A., Beja, A., and Alte da Veiga, L., N-[Tris(hydroxymethyl)methyl]glycine(tricine), *Acta Crystallogr.C.*

LIST OF BUFFERS (Continued)

Common Name	Chemical Name	M.W	Properties and Comment

57, 421-422, 2001; Su, Z.F., He, J., Rusckowski, M., and Hnatowich, D.J., In vitro cell studies of technetium-99m-labeled RGD-HYNIC peptide, a comparison of tricine and EDDA as co-ligands, *Nucl.Med.Biol.* 30, 141-149, 2003; Le, Q.T. and Katunuma, N., Detection of protease inhibitors by a reverse zymography method, performed in a tris(hydroxylmethyl)aminomethane-Tricine buffer system, *Anal.Biochem.* 324, 237-240, 2004.

Triethanolamine — tris(2-hydroxyethyl) amine — 149.2 — Buffer; transdermal transfer reagent

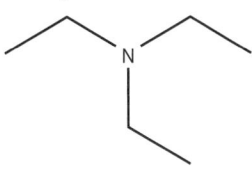

Fitzgerald, J.W., The tris-catalyzed isomerization of potassium D-glucose 6-*O*-sulfate, *Can.J.Biochem.* 53, 906-910, 1975; Buhl, S.N., Jackson, K.Y., and Graffunder, B., Optimal reaction conditions for assaying human lactate dehydrogenase pyruvate-to-lactate at 25, 30, and 37 degrees C, *Clin.Chem.* 24, 261-266, 1978; Myohanen, T.A., Bouriotas, V., and Dean, P.D. Affinity chromatography of yeast alpha-glucosidase using ligand-mediated chromatography on immobilized phenylboronic acids, *Biochem.J.* 197, 683-688, 1981; Shinomiya, Y., Kato, N., Imazawa, M., and Miyamoto, K., Enzyme immunoassay of the myelin basic protein, *J.Neurochem.* 39, 1291-1296, 1982; Arita, M., Iwamori, M., Higuchi, T., and Nagai, Y., 1,1,3,3-tetramethylurea and triethanolaminme as a new useful matrix for fast atom bombardment mass spectrometry of gangliosides and neutral glycosphingolipids, *J.Biochem.* 93, 319-322, 1983; Cao, H. and Preiss, J., Evidence for essential arginine residues at the active site of maize branching enzymes, *J.Protein Chem.* 15, 291-304, 1996; Knaak, J.B., Leung, H.W., Stott, W.T., *et al.*, Toxicology of mono-, di-, and triethanolamine, *Rev.Environ. Contim.Toxicol.* 149, 1-86, 1997; Liu, Q., Li, X., and Sommer, S.S., pK-matched running buffers for gel electrophoresis, *Anal.Biochem.* 270, 112-122, 1999; Sanger-van de Griend, C.E., Enantiomeric separation of glycyl dipeptides by capillary electrophoresis with cyclodextrins as chiral selectors, *Electrophoresis* 20, 3417-3424, 1999; Fang, L., Kobayashi, Y., Numajiri, S., *et al.*, The enhancing effect of a triethanolamine-ethanol-isopropyl myristate mixed system on the skin permeation of acidic drugs, *Biol.Pharm.Bull.* 25, 1339-1344, 2002; Musial, W. and Kubis, A., Effect of some anionic polymers of pH of triethanolamine aqueous solutions, *Polim.Med.* 34, 21-29, 2004.

Triethylamine — *N,N*-diethylethanamine — 101.2 — ion-pair reagent; buffer

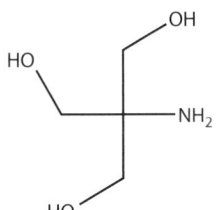

Brind, J.L., Kuo, S.W., Chervinsky, K., and Orentreich, N., A new reversed phase, paired-ion thin-layer chromatographic method for steroid sulfate separations, *Steroids* 52, 561-570, 1988; Koves, E.M., Use of high-performance liquid chromatography-diode array detection in forensic toxicology, *J.Chromatog.A* 692, 103-119, 1995; Cole, S.R. and Dorsey, J.G., Cyclohexylamine additives for enhanced peptide separations in reversed phase liquid chromatography, *Biomed.Chromatog.* 11, 167-171, 1997; Gilar, M., and Bouvier, E.S.P., Purification of crude DNA oligonucleotides by solid-phase extraction and reversed-phase high-performance liquid chromatography, *J.Chromatog.A* 890, 167-177, 2000; Loos, R. and Barcelo, D., Determination of haloacetic acids in aqueous environments by solid-phase extraction followed by ion-pair liquid chromatography-electrospray ionization mass spectrometric detection, *J.Chromatog.A.* 938, 45-55, 2001; Gilar, M., Fountain, K.J., Budman, Y., *et al.*, Ion-pair reversed phase high-performance liquid chromatography analysis of oligonucleotides: retention prediction, *J.Chromatog.A.* 958, 167-182, 2002; El-dawy, M.A., Mabrouk, M.M., and El-Barbary, F.A., Liquid chromatographic determination of fluoxetine, *J.Pharm.Biomed.Anal.* 30, 561-571, 2002; Yang, X., Zhang, X., Li, A., *et al.*, Comprehensive two-dimensional separations based on capillary high-performance liquid chromatography and microchip electrophoresis, *Electrophoresis* 24, 1451-1457, 2003; Murphey, A.T., Brown-Augsburger, P., Yu, R.Z., *et al.*, Development of an ion-pair reverse-phase liquid chromatographic/tandem mass spectrometry method for the determination of an 18-mer phosphorothioate oligonucleotide in mouse liver tissue, *Eur.J.Mass Spectrom.* 11, 209-215, 2005; Xie, G., Sueishi, Y., and Yamamoto, S., Analysis of the effects of protic, aprotic, and multi-component solvents on the fluorescence emission of naphthalene and its exciplex with triethylamine, *J.Fluoresc.* 15, 475-483, 2005.

Tris — tris(hydroxymethyl) aminomethylmethane — 121.14 — Buffer

LIST OF BUFFERS (Continued)

Common Name	Chemical Name	M.W	Properties and Comment

Bernhard, S.A., Ionization constants and heats of tris(hydroxymethyl)aminomethane and phosphate buffers, *J.Biol.Chem.* 218, 961-969, 1956; Rapp, R.D. and Memminger, M.M., Tris (hydroxymethyl)aminomethane as an electrophoresis buffer, *Am.J.Clin.Pathol.* 31, 400-403, 1959; Rodkey, F.L., Tris(hydroxymethyl)aminomethane as a standard for Kjeldahl nitrogen analysis, *Clin.Chem.* 10, 606-610, 1964; Oliver, R.W. and Viswanatha, T., Reaction of tris(hydroxymethyl)aminomethane with cinnamoyl imidazole and cinnamoyltrypsin, *Biochim.Biophys.Acta* 156, 422-425, 1968; Douzou, P., Enzymology at sub-zero temperatures, *Mol.Cell.Biochem.* 1, 15-27, 1973; The tris-catalyzed isomerization of potassium D-glucose 6-*O*-sulfate, *Can.J.Biochem.* 53, 906-910, 1975; Visconti, M.A. and Castrucci, A.M., Tris buffer effects on melanophore aggregating responses, *Comp.Biochem. Physiol.C* 82, 501-503 1985; Stambler, B.S., Grant, A.O., Broughton, A., and Strauss, H.C., Influences of buffers on dV/dtmax recovery kinetics with lidocaine in myocardium, *Am.J.Physiol.* 249, H663-H671, 1985; Nakano, M. and Tauchi, H., Difference in activation by Tris(hydroxymethyl) aminomethane of Ca,Mg-ATPase activity between young and old rat skeletal muscles, *Mech.Aging.Dev.* 36, 287-294, 1986; Oliveira, L., Araujo-Viel, M.S., Juliano, L., and Prado, E.S., Substrate activation of porcine kallikrein *N-α* derivatives of arginine 4-nitroanilides, *Biochemistry* 26, 5032-5035, 1987; Ashworth, C.D. and Nelson, D.R., Antimicrobial potentiation of irrigation solutions containing tris-[hydroxymethyl] aminomethane-EDTA, *J.Am.Vet.Med.Assoc.* 197, 1513-1514, 1990; Schacker, M., Foth, H., Schluter, J., and Kahl, R., Oxidation of tris to one-carbon compounds in a radical-producing model system, in microsomes, in hepatocytes and in rats, *Free Radic.Res.Commun.* 11, 339-347, 1991; Weber, R.E., Use of ionic and zwitterionic (Tris/BisTris and HEPES) buffers in studies on hemoglobin function, *J.Appl.Physiol.* 72, 1611-1615, 1992; Veeck, L.L., TES and Tris (TEST)-yolk buffer systems, sperm function testing, and in vitro fertilization, *Fertil.Steril.* 58, 484-486, 1992; Shiraishi, H., Kataoka, M., Morita, Y., and Umemoto, J., Interaction of hydroxyl radicals with tris (hydroxymethyl) aminomethane and Good's buffers containing hydroxymethyl or hydroxyethyl residues produce formaldehyde, *Free Radic.Res.Commun.* 19, 315-321, 1993; Vasseur, M., Frangne, R., and Alvarado, F., Buffer-dependent pH sensitivity of the fluorescent chloride-indicator dye SPQ, *Am.J.Physiol.* 264, C27-C31, 1993; Niedernhofer, L.J., Riley, M., Schnez-Boutand, N., *et al.*, Temperature dependent formation of a conjugate between tris(hydroxymethyl)aminomethane buffer and the malondialdehyde-DNA adduct pyrimidopurinone, *Chem.Res.Toxicol.* 10, 556-561, 1997; Trivic, S., Leskovac, V., Zeremski, J., *et al.*, Influence of Tris(hydroxymethyl)aminomethane on kinetic mechanism of yeast alcohol dehydrogenase, *J.Enzyme Inhib.* 13, 57-68, 1998; Afifi, N.N., Using difference spectrophotometry to study the influence of different ions and buffer systems on drug protein binding, *Drug Dev.Ind.Pharm.* 25, 735-743, 1999; AbouHaider, M.G. and Ivanov, I.G., Non-enzymatic RNA hydrolysis promotedby the combined catalytic activity of buffers and magnesium ions, *Z.Naturforsch.* 54, 542-548, 1999; Shihabi, Z.K., Stacking of discontinuous buffers in capillary zone electrophoresis, *Electrophoresis* 21, 2872-2878, 2000; Stellwagen, N.C, Bossi, A., Gelfi, C., and Righetti, P.G., DNA and buffers: are theire any nointeracting, neutral pH buffers?, *Anal.Biochem.* 287, 167-175, 2000;Burcham, P.C., Fontaine, F.R., Petersen, D.R., and Pyke, S.M., Reactivity of Tris(hydroxymethyl) aminomethane confounds immunodetection of acrolein-adducted proteins, *Chem. Res.Toxicol.* 16, 1196-1201, 2003; Koval, D., Kasicka, V., and Zuskova, I., Investigation of the effect of ionic strength of Tris-acetate background electrolyte on electrophoretic mobilities of mono-, di-, and trivalent organic anions by capillary electrophoresis, *Electrophoresis* 26, 3221-3231, 2005; Kinoshita, T., Yamaguchi, A., and Tada, T., Tris(hydroxymethyl)aminomethane induced conformational change and crystal-packing contraction of porcine pancreatic elastase, *Acta Crystallograph.Sect.F .Struct. Biol. Cryst.Commun.* 62, 623-626, 2006; Qi, Z., Li, X., Sun, D., *et al,*, Effect of Tris on catalytic activity of MP-11, *Bioelectrochemistry* 68, 40-47, 2006.

BRØNSTED ACIDITIES

Definition: A molecule (conjugate acid) which liberates a proton in solution; the residual molecular ion is a conjugate base

$$EH_x \rightarrow EH_{x-1}$$

- Acid strength depends on the strength of the E-H bond.
- Electronegativity of E influences the polarity of the E-H bond
- Energy of solvation of $[EH_{x-1}]^-$ – small anions have more favorable solvation energies

pKa values of EH_x

CH_4	−58
NH_3	39
OH_2	14
SH_2	7
SeH_2	4
TeH_2	3
FH	3

Mingo, D.M.P., *Essential Trends of Inorganic Chemistry*, Oxford University Press, Oxford, United Kingdom, 1998

Lewis acids

Definition: A molecule which can accept electrons.

$$AlCl_3 + Cl^- \rightarrow [AlCl_4]^-$$

Lewis Base is an electron donor, for example PR_3—which becomes a "stronger" base as the electron donating properties of the R group increases as long as the Lewis acid is a simple electron acceptor and steric effects are not important.

References

Blackwell, J.A. and Carr, S.W., The role of Lewis Acid-Base processes in ligand-exchange chromatography of benzoic acid derivatives on zirconium oxide, *Anal.Chem.* 64, 853-862, 1992.

Hancock, R.D., Bartolotti, L.J., and Kaltsoyannis, N., Density functional theory-based prediction of some aqueous-phase chemistry of superheavy element 111. Roentgenium(I) is the 'softest' metal ion, *Inorg. Chem.* 45, 10780-10785, 2006.

MEASUREMENT OF pH

Roger G. Bates and Maya Paabo

Definition of pH

The following definition of pH has received the endorsement of the International Union of Pure and Applied Chemistry.

1. *Operational definition.* In all existing national standards the definition of pH is an operational one. The electromotive force E_x of the cell:

Pt, H_2 |solution X|concentrated KCl solution| reference electrode

is measured and likewise the electromotive force E_s of the cell:

Pt, H_2 |solution S|concentrated KCl solution| reference electrode

both cells being at the same temperature throughout and the reference electrodes and bridge solutions being identical in the two cells. The pH of the solution X, denoted by pH(X), is then related to the pH of the solution S, denoted by pH(S), by the definition:

$$pH(X) = pH(S) + \frac{E_x - E_s}{(RT \ln 10)/F}$$

where R denotes the gas constant, T the thermodynamic temperature, and F the faraday constant. Thus defined the quantity pH is dimensionless.

To a good approximation, the hydrogen electrodes in both cells may be replaced by other hydrogen ion responsive electrodes, e.g., glass or quinhydrone. The two bridge solutions may be of any molality not less than 3.5 mol kg^{-1}, provided they are the same (see *Pure Appl. Chem.*, 1, 163, 1960).

2. *Standards.* The difference between the pH of two solutions having been defined as above, the definition of pH can be completed by assigning a value of pH at each temperature to one or more chosen solutions designated as standards. A series of pH(S) values for seven suitable standard reference solutions is given in Table 1. The constants for calculating pH(S) values over the temperature range for 0 to 95°C are given in Table 2.

If the definition of pH given above is adhered to strictly, then the pH of a solution may be slightly dependent on which standard solution is used. These unavoidable deviations are caused not only by imperfections in the response of the hydrogen ion electrodes but also by variations in the liquid-junction potentials resulting from the different ionic compositions and mobilities of the several standards and from differences in the structure of the liquid–liquid boundary. In fact such variations in measured pH are usually too small to be of practical significance. Moreover, the acceptance of several standards allows the use of the following alternative definition of pH.

The electromotive force E_x is measured, and likewise the electromotive forces E_1 and E_2, of two similar cells with the solution X replaced by the standard solutions S_1 and S_2 such that E_1 and E_2 values are on either side of, and as near as possible to, E_x. The pH of solution X is then obtained by assuming linearity between pH and E, that is to say

$$\frac{pH(X) - pH(S_1)}{pH(S_2) - pH(S_1)} = \frac{E_x - E_1}{E_2 - E_1}$$

This procedure is especially recommended when the hydrogen-ion-responsive electrode is a glass electrode.

Standard solutions

The pH meter or other electrometric pH assembly does not, strictly speaking, measure the pH but rather indicates a difference between the pH of an unknown solution (X) and a standard solution (S), both of which are at the same temperature. The pH meter should always be standardized routinely with two reference solutions of assigned pH, chosen if possible to bracket the pH of the test solution. These standards are prepared as indicated in Table 3. For convenience, air weights of the buffer salts are given. A good grade of distilled or de-ionized water should be used; for the four solutions of highest pH, the water should be freed of dissolved carbon dioxide by boiling or purging. For a detailed discussion of the properties of the primary standard buffer solutions, the reader is referred to chapter 4 of R. G. Bates, *Determination of pH*, 2nd ed., (John Wiley and Sons, Inc., New York, 1973).

Highly pure buffer materials should be used. These materials are obtainable commercially; they are also distributed as certified standard reference materials by the National Bureau of Standards. It should be noted that individual lots show slight variations; hence, the values certified for a particular lot may differ slightly from those given in Table 1.

The use of two or more standard reference solutions may disclose small inconsistencies in the standardization of the pH meter, depending on which standards are chosen. When this is the case, the best results are often obtained by assuming linearity between E and pH between the two calibrating points bracketing the pH of the unknown.

Electrodes

Although the hydrogen electrode is the ultimate standard on which the pH scale is based, in practice the convenient and versatile glass electrode is favored for the vast majority of pH measurements. New glass electrodes, or those that have been allowed to dry out, should be conditioned by soaking in water for several hours before use and after exposure to nonaqueous or dehydrating media. Some glass electrodes are designed especially for use at high temperatures, while others are best suited to low-temperature use. Special "high pH" electrodes are also available. For optimum results, careful attention should be paid to selection of the proper electrode for the problem at hand.

Glass electrodes of small dimensions are of great utility when sample volumes are limited. The pH-sensitive glasses are, however, moderately soluble, and small amounts of alkali are dissolved from the glass surface by the solutions in which the electrode is immersed. For this reason, the most accurate results are obtained when the ratio of the electrode area to sample volume is small.

The concentrated solution of potassium chloride that joins the reference electrode with the unknown or standard solution is reasonably effective in reducing the liquid-junction potential to small, fairly constant, values. It is important to assure that the flow of bridge solution into the test solution is neither excessive nor completely interrupted by crystallization of salt in the aperture where liquid-liquid contact is established.

Temperature gradients within the pH cell are a common source of difficulty, marked by variability and inaccuracy in the

TABLE 1   Values of pH(S) for Seven Primary Standard Solutions

$t/°C$	A	B	C	D	E	F	G
0	—	3.863	4.003	6.984	7.534	9.464	10.317
5	—	3.840	3.999	6.951	7.500	9.395	10.245
10	—	3.820	3.998	6.923	7.472	9.332	10.179
15	—	3.802	3.999	6.900	7.448	9.276	10.118
20	—	3.788	4.002	6.881	7.429	9.225	10.062
25	3.557	3.776	4.008	6.865	7.413	9.180	10.012
30	3.552	3.766	4.015	6.853	7.400	9.139	9.966
35	3.549	3.759	4.024	6.844	7.389	9.102	9.925
38	3.548	3.755	4.030	6.840	7.384	9.081	9.903
40	3.547	3.753	4.035	6.838	7.380	9.068	9.889
45	3.547	3.750	4.047	6.834	7.373	9.038	9.856
50	3.549	3.749	4.060	6.833	7.367	9.011	9.828
55	3.554	—	4.075	6.834	—	8.985	—
60	3.560	—	4.091	6.836	—	8.962	—
70	3.580	—	4.126	6.845	—	8.921	—
80	3.609	—	4.164	6.859	—	8.885	—
90	3.650	—	4.205	6.877	—	8.850	—
95	3.674	—	4.227	6.886	—	8.833	—

The compositions of the standard solutions are:

A: KH tartrate (saturated at 25°C)
B: KH_2 citrate, $m = 0.05$ mol kg^{-1}
C: KH phthalate, $m = 0.05$ mol kg^{-1}
D: KH_2PO_4, $m = 0.025$ mol kg^{-1}; Na_2HPO_4, $m = 0.025$ mol kg^{-1}
E: KH_2PO_4, $m = 0.008695$ mol kg^{-1}; Na_2HPO_4, $m = 0.03043$ mol kg^{-1}
F: $Na_2B_4O_7$, $m = 0.01$ mol kg^{-1}
G: $NaHCO_3$, $m = 0.025$ mol kg^{-1}; Na_2CO_3, $m = 0.025$ mol kg^{-1} where m denotes molality.

TABLE 2   Values of The Constants of The Equation: $pH(S) = \frac{A}{T} + B + CT + DT^2$ For Seven Primary Standard Buffer Solutions From 0 To 95°C

Solution	Temperature Range °C	A	B	C	10^5 D	Standard Deviation of The Fitted Curves
A. Tartrate	25 to 95	−1727.96	23.7406	−0.075947	9.2873	0.0016
B. Citrate	0 to 50	1280.4	−4.1650	0.012230	0	0.0010
C. Phthalate	0 to 95	1678.30	−9.8357	0.034946	−2.4804	0.0027
D. Phosphate	0 to 95	3459.39	−21.0574	0.073301	−6.2266	0.0017
E. Phosphate	0 to 50	5706.61	−43.9428	0.154785	−15.6745	0.0011
F. Borax	0 to 95	5259.02	−33.1064	0.114826	−10.7860	0.0025
G. Carbonate	0 to 50	2557.1	−4.2846	0.019185	0	0.0026

TABLE 3   Preparation of Primary Standard Buffer Solutions

Standard Solution	NBS SRM No.[a]	Buffer Substance	Weight In Air[b] (g)
A. Tartrate	188	$KHC_4H_4O_6$	(Satd. at 25°C)
B. Citrate	190	$KH_2C_6H_5O_7$	11.41
C. Phthalate	185d	$KHC_8H_4O_4$	10.12
D. Phosphate	186Ic	KH_2PO_4	3.388
	186IIb	Na_2HPO_4	3.533
E. Phosphate	186Ic	KH_2PO_4	1.179
	186IIb	Na_2HPO_4	4.302
F. Borax	187a	$Na_2B_4O_7 \cdot 10H_2O$	3.80
G. Carbonate	191	$NaHCO_3$	2.092
	192	Na_2CO_3	2.640

[a] These materials may be ordered from the Office of Standard Reference Materials, National Bureau of Standards, Washington, D.C. 20234.

[b] This weight of salt to be dissolved in water and diluted to 1 liter at 25°C to provide concentrations indicated in Table 1.

reading. Both of the electrodes, and the standard and test solutions as well, should be within a few degrees Celsius of the same temperature. For results of the highest reliability, temperature control should be provided. It is the function of the temperature compensator of the pH meter to adjust the pH-e.m.f. slope in such a manner that a difference of e.m.f. (in volts) is correctly converted to a difference of pH. This adjustment cannot compensate for inequalities of temperature through the cell or for differences between the temperature of the standard and test solutions.

Techniques

Electrodes and sample cups should be washed carefully with distilled or de-ionized water and gently dried with clean absorbent tissue. The electrodes are immersed in the first standard solution and the temperature compensator of the measuring instrument is set at the temperature of the solutions whose pH is to be measured. The standardization control of the instrument is adjusted until the meter is balanced at the known pH of the standard, as given in Table 1. This procedure is repeated with successive portions of the same standard until replacement causes no change in the position of balance. The electrodes are then washed once more and dried.

A second standard solution is selected and the measurement repeated without altering the position of the standardization control. The pH reading of this second solution is noted and the sample replaced with a second portion of the same solution. This replacement is continued until successive readings agree within 0.02 pH unit, when the electrodes and meter may be judged to be functioning properly. It is advisable to make a final check with one of the buffers at the conclusion of a series of measurements.

After the instrument is properly standardized, a portion of the test solution is placed in the sample cup and the pH reading noted. Successive portions are again used until two measurements agree within the limits imposed by the reproducibility of the measuring instrument and the temperature control. With the best meters, measurements on buffered solutions should be reproducible to 0.01 unit or even better. With water or poorly buffered solutions, values agreeing to 0.1 unit may have to be accepted. Some improvement will result if poorly buffered solutions are protected from carbon dioxide of the atmosphere during the period of the measurements.

Interpretation of pH numbers

The standard values of pH given in the table of an earlier section are based on hydrogen electrode potentials as measured in cells without a liquid junction. The uncertainty of the standard values is estimated at 0.005 unit. The accuracy of the results furnished by a given pH assembly adjusted with these primary standards is, however, further limited by inconsistencies which have their origin in defects of the glass electrode response and variations in the liquid-junction potential. For these reasons, the accuracy of experimental pH numbers can be considered to be better than 0.01 unit only under unusually favorable conditions.

The operational definition of pH fulfills adequately the need for an experimental scale capable of furnishing reproducible pH numbers. The interpretation of these numbers may be of secondary importance and should only be attempted when the standard and unknown solutions are matched so closely in composition that there is good reason to believe that the liquid-junction potential remains fairly constant when the standard is replaced by the unknown. In general, this will be the case when the unknowns are aqueous solutions of simple salts of total concentration not in excess of 0.2 M with pH values between 2.5 and 11.5.

When these "ideal" conditions prevail, the experimental pH can be considered to approach $-\log a_H$, where a_H is the conventional hydrogen ion activity defined in a manner consistent with the convention on which the standard values of pH(S) were based Bates, R. G., *J. Res. Natl. Bur. Standards*, 66A, 179 (1962). All quantitative applications of pH measurements, when justifiable, should therefore be based on the approximation $pH(X) \approx -\log a_H = -\log m_{H}\gamma_H$, where m is molality and γ is the activity coefficient.

Indicator methods

Acid-base indicators have the property of altering the color of a solution in the region 1 to 2 pH units as the pH changes. They are therefore useful for pH measurements, although in general the accuracy is inferior to that obtainable by electrometric procedures. A list of suitable indicators, their pH ranges and color changes, is given in Table 4.

Equal concentrations of the same indicator are added to the test solution and to each of a series of buffer solutions of known pH selected to bracket the pH of the test solution. Color comparisons are made with a colorimeter or spectrophotometer, and solutions of equal color are assumed to have the same pH. The pH of a series of suitable reference solutions can be determined in advance by electrometric methods. Alternative ly, tables of pH as a function of composition can be utilized. The compositions and pH values of a set of useful solutions covering the range pH 1 to 13 are summarized in Table 5.

References

1. Paabo, Bates, and Robinson, *J. Res. Natl. Bur. Standards*, 67A, 573 (1963).
2. Hetzer, Robinson, and Bates, *Anal. Chem.*, 40, 634 (1968).
3. Paabo and Bates, *J. Phys. Chem.*, 74, 702 (1970).
4. Hetzer, Bates, and Robinson, *J. Phys. Chem.*, 70, 2869 (1966).
5. Bates and Robinson, *Anal. Chem.*, 45, 420 (1973).
6. Durst and Staples, *Clin. Chem.*, 18, 206 (1972).
7. Bates, Roy, and Robinson, *Anal. Chem.*, 45, 1663 (1973).

TABLE 4   Acid-Base Indicators

Indicator	pH Range	Color Change	Indicator	pH Range	Color Change
Acid cresol red	0.2–1.8	Red–yellow	Metacresol purple	7.6–9.2	Yellow–purple
Acid metacresol purple	1.2–2.8	Red–yellow	Thymol blue	8.0–9.6	Yellow–blue
Acid thymol blue	1.2–2.8	Red–yellow	Phthalein red	8.6–10.2	Yellow–red
Bromophenol blue	3.0–4.6	Yellow–blue	Tolyl red	10.0–11.6	Red–yellow
Bromocresol green	3.8–5.4	Yellow–blue	Acyl red	10.0–11.6	Red–yellow
Methyl red	4.4–6.0	Red–yellow	Parazo orange	11.0–12.6	Yellow–orange
Chlorophenol red	5.2–6.8	Yellow–red	Acyl blue	12.0–13.6	Red–blue
Bromocresol purple	5.2–6.8	Yellow–purple	Benzo yellow	2.4–4.0	Red–yellow
Bromothymol blue	6.0–7.6	Yellow–blue	Benzo red	4.4–7.6	Red–blue
Phenol red	6.8–8.4	Yellow–red	Thymol red	8.0–11.2	Yellow–red
Cresol red	7.2–8.8	Yellow–red			

Courtesy of W. A. Taylor and Co.

TABLE 5   Buffer Solutions for Indicator Measurements and pH Control

25 ml 0.2 M KCl, x ml 0.2 M HCl, DILUTED TO 100 ml

pH	x	pH	x
1.00	67.0	1.50	20.7
1.10	52.8	1.60	16.2
1.20	42.5	1.70	13.0
1.30	33.6	1.80	10.2
1.40	26.6	1.90	8.1
—	—	2.00	6.5
—	—	2.10	5.1
—	—	2.20	3.9

50 ml 0.1 M KH PHTHALATE, x ml 0.1 M NaOH, DILUTED TO 100 ml

pH	x	pH	x
4.10	1.3	5.10	25.5
4.20	3.0	5.20	28.8
4.30	4.7	5.30	31.6
4.40	6.6	5.40	34.1
4.50	8.7	5.50	36.6
4.60	11.1	5.60	38.8
4.70	13.6	5.70	40.6
4.80	16.5	5.80	42.3
4.90	19.4	5.90	43.7
5.00	22.6	—	—

50 ml OF A MIXTURE 0.1 M WITH RESPECT TO BOTH KCl AND H₃BO₃, x ml 0.1 M NaOH, DILUTED TO 100 ml

pH	x	pH	x
8.00	3.9	9.00	20.8
8.10	4.9	9.10	23.6
8.20	6.0	9.20	26.4
8.30	7.2	9.30	29.3
8.40	8.6	9.40	32.1
8.50	10.1	9.50	34.6
8.60	11.8	9.60	36.9
8.70	13.7	9.70	38.9
8.80	15.8	9.80	40.6
8.90	18.1	9.90	42.2
—	—	10.00	43.7
—	—	10.10	45.0
—	—	10.20	46.2

50 ml 0.025 M BORAX, x ml 0.1 M HCl, DILUTED TO 100 ml

pH	x	pH	x
8.00	20.5	8.50	15.2
8.10	19.7	8.60	13.5
8.20	18.8	8.70	11.6
8.30	17.7	8.80	9.4
8.40	16.6	8.90	7.1
—	—	9.00	4.6
—	—	9.10	2.0

50 ml 0.1 M KH PHTHALATE, x ml 0.1 M HCl, DILUTED TO 100 ml

pH	x	pH	x
2.20	49.5	3.20	15.7
2.30	45.8	3.30	12.9
2.40	42.2	3.40	10.4
2.50	38.8	3.50	8.2
2.60	35.4	3.60	6.3
2.70	32.1	3.70	4.5
2.80	28.9	3.80	2.9
2.90	25.7	3.90	1.4
3.00	22.3	4.00	0.1
3.10	18.8	—	—

50 ml 0.1 M, KH2PO4 x ml 0.1 M NaOH, DILUTED TO 100 ml

pH	x	pH	x
5.80	3.6	6.80	22.4
5.90	4.6	6.90	25.9
6.00	5.6	7.00	29.1
6.10	6.8	7.10	32.1
6.20	8.1	7.20	34.7
6.30	9.7	7.30	37.0
6.40	11.6	7.40	39.1
6.50	13.9	7.50	41.1
6.60	16.4	7.60	42.8
6.70	19.3	7.70	44.2
—	—	7.80	45.3
—	—	7.90	46.1
—	—	8.00	46.7

50 ml 0.1 M TRIS(HYDROXMETHYL)AMINOMETHANE, x ml 0.1 M HCl, DILUTED TO 100 ml

pH	x	pH	x
7.00	46.6	8.00	29.2
7.10	45.7	8.10	26.2
7.20	44.7	8.20	22.9
7.30	43.4	8.30	19.9
7.40	42.0	8.40	17.2
7.50	40.3	8.50	14.7
7.60	38.5	8.60	12.4
7.70	36.6	8.70	10.3
7.80	34.5	8.80	8.5
7.90	32.0	8.90	7.0
—	—	9.00	5.7

50 ml 0.025 M BORAX, x ml 0.1 M NaOH, DILUTED TO 100 ml

pH	x	pH	x
9.20	0.9	10.20	20.5
9.30	3.6	10.30	21.3
9.40	6.2	10.40	22.1
9.50	8.8	10.50	22.7
9.60	11.1	10.60	23.3
9.70	13.1	10.70	23.80
9.80	15.0	10.80	24.25
9.90	16.7	—	—
10.00	18.3	—	—
10.10	19.5	—	—

TABLE 5 Buffer Solutions for Indicator Measurements and pH Control (Continued)

50 ml 0.05 M $NaHCO_3$, x ml 0.1 M NaOH, DILUTED TO 100 ml

pH	*x*	pH	*x*
9.60	5.0	10.60	19.1
9.70	6.2	10.70	20.2
9.80	7.6	10.80	21.2
9.90	9.1	10.90	22.0
10.00	10.7	11.00	22.7
10.10	12.2	—	—
10.20	13.8	—	—
10.30	15.2	—	—
10.40	16.5	—	—
10.50	17.8	—	—

50 ml 0.05 M Na_2HPO_4, x 1ml 0.1 M NaOH, DILUTED TO 100 ml

pH	*x*	pH	*x*
10.90	3.3	11.40	9.1
11.00	4.1	11.50	11.1
11.10	5.1	11.60	13.5
11.20	6.3	11.70	16.2
11.30	7.6	11.80	19.4
—	—	11.90	23.0
—	—	12.00	26.9

25 ml 0.2 M KCl, x ml 0.2 M NaOH, DILUTED TO 100 ml

pH	*x*	pH	*x*
12.00	6.0	12.50	20.4
12.10	8.0	12.60	25.6
12.20	10.2	12.70	32.2
12.30	12.8	12.80	41.2
12.40	16.2	12.90	53.0
—	—	13.00	66.0

Source: Bower and Bates, *J. Res. Natl. Bur. Standards*, 55, 197 (1955); Bates and Bower, *Anal. Chem.*, 28, 1322 (1956).

TABLE 6 pH Values For Miscellaneous Buffer Solutions Over A Range of Temperature

		t/°C										
Composition of the Buffer Solution	*m*[a]	0	5	10	15	20	25	30	35	40	45	50
	0.005	—	—	—	—	—	6.251	—	—	—	—	—
Potassium dihydrogen phosphate (*m*)	0.015	—	—	—	—	—	6.162	—	—	—	—	—
Sodium succinate (*m*) (1)	0.025	—	—	—	—	—	6.109	—	—	—	—	—
	0.02	6.580	6.515	6.453	6.394	6.338	6.284	6.234	6.185	6.140	6.097	6.058
Piperazine phosphate (*m*) (2)	0.05	6.589	6.525	6.463	6.404	6.348	6.294	6.243	6.195	6.149	6.106	6.066
	0.02	7.000	6.905	6.812	6.722	6.635	6.551	6.469	6.390	6.312	6.237	6.165
2,2-Bis(hydroxymethyl)-2,2′-2″-nitrilotriethanol (2*m*)	0.04	7.029	6.932	6.839	6.748	6.662	6.577	6.495	6.415	6.336	6.262	6.190
	0.06	7.050	6.953	6.859	6.767	6.681	6.595	6.513	6.434	6.353	6.280	6.208
	0.08	7.067	6.969	6.876	6.783	6.696	6.610	6.528	6.448	6.367	6.294	6.222
Hydrochloric acid (*m*) (3)	0.10	7.082	6.983	6.889	6.796	6.710	6.623	6.540	6.460	6.378	6.306	6.235
Morpholine (1.5*m*) Hydrochloric acid (*m*) (4)	0.10	8.963	8.828	8.702	8.579	8.458	8.343	8.231	8.120	8.013	7.908	7.806
Tris(hydroxymethyl)amino-methane ("Tris") (*m*), Tris. HCl (*m*) (5)	0.05	8.946	8.774	8.614	8.461	8.313	8.173	8.036	7.904[b]	7.777	7.654	7.537
Tris (*m*), Tris.HCl (3*m*) (6)	0.01667	8.471	8.303	8.142	7.988	7.840	7.698	7.563	7.433	7.307	7.186	7.070
N-Tris(hydroxymethyl)methyl-glycine ("Tricine") (*m*), Na Tricinate (*m*) (7)	0.05	—	8.485	8.375	8.271	8.175	8.079	7.988	7.902	7.817	7.740	7.663
("Tricine") (3*m*), Na Tricinate (*m*) (7)	0.02	—	8.023	7.916	7.813	7.713	7.621	7.527	7.437[c]	7.355	7.275	7.197

Compiled by Roger G. Bates and Maya Paabo
Contribution from the National Bureau of Standards , not subjected to copyright.

[a] mol kg^{-1}
[b] 7.851 at 37°C
[c] 7.407 at 37°C

General references for pH measurement

1. Spitzer, P. and Meinrath, G., Importance of traceable pH measurement, *Anal. Bioanal. Chem.*, 374, 765–766, 2002.
2. Baucke, F.G.K., New IUPAC recommendations on the measurement of pH — background and essentials, *Anal. Bioanal. Chem.*, 374, 772–777, 2002.
3. Spitzer, P. and Werner, B., Improved reliability of pH measurements, *Anal. Bioanal. Chem.*, 374, 787–795, 2002.
4. Covington, A.K., Bates, R.G., and Durst, R.A., Definition of pH scales, standard reference values, measurement of pH and related terminology, *Pure & Appl. Chem.*, 57, 531–542, 1985.

BUFFER SOLUTIONS

No.	Name	Range of pH Value	Temperature (°C)	ΔpH/K
GENERAL BUFFERS				
1	KCl/HCl (Clark and Lubs)[2]	1.0–2.2	Room	0
2	Glycine/HCl (Sørensen)[3]	1.2–3.4	Room	0
3	Na citrate/HCl (Sørensen)[3]	1.2–5.0	Room	0
4	K biphthalate/HCl (Clark and Lubs)[2]	2.4–4.0	20	+ 0.001
5	K biphthalate/NaOH (Clark and Lubs)[2]	4.2–6.2	20	
6	Na citrate/NaOH (Sørensen)[3]	5.2–6.6	20	+ 0.004
7	Phosphate (Sørensen).[3]	5.0–8.0	20	− 0.003
8	Barbital-Na/HCl (Michaelis)[4]	7.0–9.0	18	
9	Na borate/HCl (Sørensen)[3]	7.8–9.2	20	− 0.005
10	Glycine/NaOH (Sørensen)[3]	8.6–12.8	20	− 0.025
11	Na borate/NaOH (Sørensen)[3]	9.4–10.6	20	− 0.01
UNIVERSAL BUFFERS				
12	Citric acid/phosphate (McIlvaine)[5]	2.2–7.8	21	
13	Citrate-phosphate-borate/HCl (Teorell and Stenhagen)[6]	2.0–12.0	20	
14	Britton-Robinson[7]	2.6–11.8	25	At low pH:0 At high pH: − 0.02
BUFFERS FOR BIOLOGICAL MEDIA				
15	Acetate (Walpole)[8–10]	3.8–5.6	25	
16	Dimethylglutaric acid/NaOH[11]	3.2–7.6	21	
17	Piperazine/HCl[12,13]	4.6–6.4 8.8–10.6	20	
18	Tetraethylethylenediamine[a,13]	5.0–6.8 8.2–10.0	20	
19	Tris maleate[9,14]	5.2–8.6	23	
20	Dimethylaminoethylamine[a,13]	5.6–7.4 8.6–10.4	20	
21	Imidazole/HCl[15]	6.2–7.8	25	
22	Triethanolamine/HCl[16]	7.9–8.8	25	
23	N-Dimethylaminoleucylglycine/NaOH[17]	7.0–8.8	23	− 0.015
24	Tris/HCl[9]	7.2–9.0	23	− 0.02
25	2-Amino-2-methylpropane-1,3-diol/HCl[9,14]	7.8–10.0	23	
26	Carbonate (Delory and King)[9,8]	9.2–10.8	20	

[a] Can be combined with tris buffer to give a cationic universal buffer (see Semenza et al.[13]).

BUFFER SOLUTIONS (Continued)

pH	1	2	3	4	5	6	7	8	9	10	11	12	13	14	15	16a	16b	17	18	19	20	21	22	23	24	25	26
1.0	54.2	—	—	—	—	—	—	—	—	—	—	—	—	—	—	—	—	—	—	—	—	—	—	—	—	—	—
1.2	36.0	11.1	9.0	—	—	—	—	—	—	—	—	—	—	—	—	—	—	—	—	—	—	—	—	—	—	—	—
1.4	23.2	26.4	17.9	—	—	—	—	—	—	—	—	—	—	—	—	—	—	—	—	—	—	—	—	—	—	—	—
1.6	14.7	36.2	23.6	—	—	—	—	—	—	—	—	—	—	—	—	—	—	—	—	—	—	—	—	—	—	—	—
1.8	9.3	43.9	27.6	—	—	—	—	—	—	—	—	—	—	—	—	—	—	—	—	—	—	—	—	—	—	—	—
2.0	5.9	50.7	30.2	—	—	—	—	—	—	—	—	98.8	74.4	—	—	—	—	—	—	—	—	—	—	—	—	—	—
2.2	3.8	56.5	32.2	—	—	—	—	—	—	—	—	94.5	68.8	—	—	—	—	—	—	—	—	—	—	—	—	—	—
2.4	—	62.3	34.1	41.0	—	—	—	—	—	—	—	90.0	64.6	—	—	—	—	—	—	—	—	—	—	—	—	—	—
2.6	—	68.4	36.0	34.3	—	—	—	—	—	—	—	85.1	63.3	1.6	—	—	—	—	—	—	—	—	—	—	—	—	—
2.8	—	74.7	37.9	27.8	—	—	—	—	—	—	—	80.3	58.9	3.6	—	—	—	—	—	—	—	—	—	—	—	—	—
3.0	—	81.0	39.9	21.6	—	—	—	—	—	—	—	76.0	56.9	5.7	—	—	—	—	—	—	—	—	—	—	—	—	—
3.2	—	86.2	42.1	15.9	—	—	—	—	—	—	—	72.0	55.2	7.8	—	7.0	14.4	—	—	—	—	—	—	—	—	—	—
3.4	—	90.3	44.8	10.9	—	—	—	—	—	—	—	68.4	53.9	9.9	—	13.3	20.9	—	—	—	—	—	—	—	—	—	—
3.6	—	—	47.8	6.7	—	—	—	—	—	—	—	65.1	52.9	11.7	—	20.7	26.8	—	—	—	—	—	—	—	—	—	—
3.8	—	—	51.2	3.3	—	—	—	—	—	—	—	62.0	51.8	13.5	—	26.3	32.4	—	—	—	—	—	—	—	—	—	—
4.0	—	—	55.1	0.0	3.0	—	—	—	—	—	—	59.1	50.7	15.3	—	32.4	36.6	—	—	—	—	—	—	—	—	—	—
4.2	—	—	60.0	—	6.7	—	—	—	—	—	—	56.4	49.7	17.5	10.9	36.2	40.3	—	—	—	—	—	—	—	—	—	—
4.4	—	—	66.4	—	11.1	—	—	—	—	—	—	53.7	48.6	19.7	16.6	39.3	43.1	—	—	—	—	—	—	—	—	—	—
4.6	—	—	74.9	—	16.5	—	—	—	—	—	—	51.2	47.5	21.9	23.9	41.3	45.7	94.3	—	—	—	—	—	—	—	—	—
4.8	—	—	85.6	—	22.6	—	—	—	—	—	—	49.0	46.4	24.1	33.5	43.5	48.3	91.5	94.3	—	—	—	—	—	—	—	—
5.0	—	—	100.0	—	28.8	87.1	99.2	—	—	—	—	46.9	45.4	26.3	44.9	45.7	51.5	87.8	91.5	—	—	—	—	—	—	—	—
5.2	—	—	—	—	34.4	78.0	98.4	—	—	—	—	44.7	44.3	28.6	56.6	48.4	53.6	83.6	87.8	3.2	—	—	—	—	—	—	—
5.4	—	—	—	—	39.1	70.3	97.3	—	—	—	—	42.4	43.2	31.0	67.8	51.3	58.2	77.6	83.1	5.0	—	—	—	—	—	—	—
5.6	—	—	—	—	42.4	64.5	95.5	—	—	—	—	40.0	42.0	33.4	76.8	55.0	63.6	71.8	77.6	7.3	94.3	—	—	—	—	—	—
5.8	—	—	—	—	45.0	60.3	92.8	—	—	—	—	37.4	40.8	35.8	84.0	58.8	68.7	66.5	71.7	9.7	91.7	—	—	—	—	—	—
6.0	—	—	—	—	46.7	57.2	88.9	—	—	—	—	34.5	39.7	38.3	89.3	63.9	73.6	61.8	66.4	12.4	88.0	—	—	—	—	—	—
6.2	—	—	—	—	—	54.8	83.0	—	—	—	—	31.4	38.4	40.8	—	69.5	78.5	58.2	61.7	15.2	83.3	43.4	—	—	—	—	—
6.4	—	—	—	—	—	53.2	75.4	—	—	—	—	27.9	37.0	43.3	—	74.1	83.3	55.5	58.0	17.9	77.9	40.4	—	—	—	—	—
6.6	—	—	—	—	—	—	65.3	—	—	—	—	23.5	35.6	45.8	—	83.5	87.4	—	55.3	20.8	72.0	36.5	—	—	—	—	—
6.8	—	—	—	—	—	—	53.4	53.3	—	—	—	19.0	34.2	48.3	—	87.4	91.0	—	—	22.2	66.6	31.4	86.2	86.4	—	—	—
7.0	—	—	—	—	—	—	41.3	55.0	—	—	—	13.8	32.9	50.9	—	90.0	93.2	—	—	23.7	61.9	25.4	79.6	80.6	—	—	—
7.2	—	—	—	—	—	—	29.6	57.6	—	—	—	9.8	31.7	53.4	—	91.8	94.9	—	—	25.2	58.1	19.6	73.1	72.8	44.7	—	—
7.4	—	—	—	—	—	—	19.7	60.8	—	—	—	6.8	30.6	55.8	—	93.0	95.8	—	—	26.7	55.3	14.6	62.0	63.2	42.0	—	—
7.6	—	—	—	—	—	—	12.8	65.2	53.0	—	—	4.6	29.6	58.2	—	93.8	96.8	—	—	28.6	—	10.2	52.0	52.1	39.3	—	—
7.8	—	—	—	—	—	—	7.4	70.6	55.4	—	—	—	28.8	60.5	—	—	—	—	—	31.2	—	6.6	42.0	41.1	33.7	43.9	—
8.0	—	—	—	—	—	—	3.7	75.9	58.0	—	—	—	28.1	62.8	—	—	—	—	46.4	33.9	—	—	31.9	31.4	27.9	41.6	—
8.2	—	—	—	—	—	—	—	81.2	62.1	—	—	—	27.6	65.0	—	—	—	—	43.9	36.9	—	—	22.5	23.0	22.9	38.4	—
8.4	—	—	—	—	—	—	—	86.2	66.9	—	—	—	27.0	67.2	—	—	—	—	40.9	39.9	—	—	16.0	15.9	17.3	34.8	—
8.6	—	—	—	—	—	—	—	90.1	73.6	94.7	—	—	26.3	69.3	—	—	—	45.5	36.8	42.7	45.4	—	11.7	10.3	13.0	30.7	—
8.8	—	—	—	—	—	—	—	93.2	83.5	92.0	—	—	25.2	71.3	—	—	—	43.2	31.8	—	42.8	—	—	—	8.8	23.3	—
9.0	—	—	—	—	—	—	—	—	95.6	88.4	—	—	24.0	73.2	—	—	—	40.0	26.2	—	39.2	—	—	—	5.3	17.7	—
9.2	—	—	—	—	—	—	—	—	—	84.0	—	—	22.6	75.1	—	—	—	35.8	20.4	—	34.7	—	—	—	—	13.3	10.0
9.4	—	—	—	—	—	—	—	—	—	78.9	87.0	—	21.4	77.0	—	—	—	30.8	15.2	—	29.3	—	—	—	—	9.2	18.4
9.6	—	—	—	—	—	—	—	—	—	73.2	75.5	—	20.2	78.8	—	—	—	25.0	10.8	—	23.6	—	—	—	—	5.2	29.3
9.8	—	—	—	—	—	—	—	—	—	67.2	65.1	—	19.0	80.4	—	—	—	—	—	—	19.0	—	—	—	—	4.1	42.0

BUFFER SOLUTIONS (Continued)

pH	1	2	3	4	5	6	7	8	9	10	11	12	13	14	15	16a	16b	17	18	19	20	21	22	23	24	25	26	pH
10.0	—	—	—	—	—	—	—	—	—	62.5	59.6	—	18.1	81.8	—	—	—	19.4	7.4	—	13.1	—	—	—	—	2.3	53.4	10.0
10.2	—	—	—	—	—	—	—	—	—	58.8	56.4	—	17.1	83.1	—	—	—	14.3	—	—	9.2	—	—	—	—	—	63.7	10.2
10.4	—	—	—	—	—	—	—	—	—	55.7	54.1	—	16.5	84.3	—	—	—	10.0	—	—	6.2	—	—	—	—	—	73.1	10.4
10.6	—	—	—	—	—	—	—	—	—	53.6	52.3	—	16.0	85.4	—	—	—	6.9	—	—	—	—	—	—	—	—	81.2	10.6
10.8	—	—	—	—	—	—	—	—	—	52.2	—	—	15.5	86.5	—	—	—	—	—	—	—	—	—	—	—	—	87.9	10.8
11.0	—	—	—	—	—	—	—	—	—	51.2	—	—	14.7	87.8	—	—	—	—	—	—	—	—	—	—	—	—	—	11.0
11.2	—	—	—	—	—	—	—	—	—	50.4	—	—	13.5	89.3	—	—	—	—	—	—	—	—	—	—	—	—	—	11.2
11.4	—	—	—	—	—	—	—	—	—	49.5	—	—	11.7	91.3	—	—	—	—	—	—	—	—	—	—	—	—	—	11.4
11.6	—	—	—	—	—	—	—	—	—	48.7	—	—	9.1	94.5	—	—	—	—	—	—	—	—	—	—	—	—	—	11.6
11.8	—	—	—	—	—	—	—	—	—	47.6	—	—	5.5	99.0	—	—	—	—	—	—	—	—	—	—	—	—	—	11.8
12.0	—	—	—	—	—	—	—	—	—	46.0	—	—	1.3	—	—	—	—	—	—	—	—	—	—	—	—	—	—	12.0
12.2	—	—	—	—	—	—	—	—	—	43.2	—	—	—	—	—	—	—	—	—	—	—	—	—	—	—	—	—	12.2
12.4	—	—	—	—	—	—	—	—	—	39.1	—	—	—	—	—	—	—	—	—	—	—	—	—	—	—	—	—	12.4
12.6	—	—	—	—	—	—	—	—	—	31.8	—	—	—	—	—	—	—	—	—	—	—	—	—	—	—	—	—	12.6
12.8	—	—	—	—	—	—	—	—	—	21.4	—	—	—	—	—	—	—	—	—	—	—	—	—	—	—	—	—	12.8

Note: The table gives the volumes x (in ml) of the stock solutions listed that are required to make up a buffer solution of the desired pH value.

BUFFER SOLUTIONS (Continued)

Stock solutions and their amount of substance concentrations or mass and/or volume contents of the solutes

No	A	B	Composition of the Buffer
1	KCl 0.2 mol l^{-1} (14.91 g l^{-1})	HCl 0.2 mol l^{-1}	25 ml A + x ml B made up to 100 ml
2	Glycine 0.1 mol l^{-1} + NaCl 0.1 mol l^{-1} (1 l solution contains 7.507 g glycine + 5.844 g NaCl)	HCl 0.1 mol l^{-1}	x ml A + (100 − x) ml B
3	Disodium citrate 0.1 mol l^{-1} (1 l solution contains 21.01 g citric acid monohydrate + 200 ml NaOH 1 mol l^{-1})	HCl 0.1 mol l^{-1}	x ml A + (100 − x) ml B
4	Potassium biphthalate 0.1 mol l^{-1} (20.42 g l^{-1})	HCl 0.1 mol l^{-1}	50 ml A + x ml B made up to 100 ml
5	As No. 4	NaOH 0.1 mol l^{-1}	50 ml A + x ml B made up to 100 ml
6	As No. 3	NaOH 0.1 mol l^{-1}	x ml A + (100 − x) ml B
7	Potassium dihydrogen phosphate 1/15 mol l^{-1} (9.073 g l^{-1})	Disodium phosphate 1/15 mol l^{-1} (Na$_2$HPO$_4 \cdot$ 2 H$_2$O, 11.87 g l^{-1})	x ml A + (100 − x) ml B
8	Barbital-Na 0.1 mol l^{-1} (20.62 g l^{-1})	HCl 0.1 l^{-1}	x ml A + (100 − x) ml B
9	Boric acid, half-neutralized, 0.2 mol l^{-1} (corresponds to 0.05 mol l^{-1} borax solution; 1 l solution contains 12.37 g boric acid 100 ml NaOH 1 mol l^{-1})	HCl 0.1 mol l^{-1}	x ml A + (100 − x) ml B
10	As No. 2	NaOH 0.1 mol l^{-1}	x ml A + (100 − x) ml B
11	As No. 9	NaOH 0.1 mol l^{-1}	x ml A + (100 − x) ml B
12	Citric acid 0.1 mol l^{-1} (citric acid monohydrate 21.01 g l^{-1})	Disodium phosphate 0.2 mol l^{-1} (Na$_2$HPO$_4 \cdot$ 2 H$_2$O, 35.60 g l^{-1})	x ml A + (100 − x) ml B
13	To 100 ml citric acid and 100 ml phosphoric acid solution, each equivalent to 100 ml NaOH 1 mol l^{-1}, add 3.54 g boric acid and 343 ml NaOH 1 mol l^{-1}and make up to 1 l of solution	HCl 0.1 l^{-1}	20 ml A + x ml B made up to 100 l
14	Citric acid, potassium hydrogen phosphate, barbital, and boric acid, all 0.02857 mol l^{-1} (1 l solution contains 6.004 g citric acid monohydrate, 3.888 g potassium hydrogen phosphate, 5.263 g barbital, 1.767 g boric acid)	NaOH 0.2 mol l^{-1}	100 ml A + x ml B
15	Sodium acetate 0.1 mol l^{-1} (1 l solution contains 8.204 g C$_2$H$_3$O$_2$Na or 13.61 g C$_2$H$_3$O$_2$Na $\cdot$ 3 H$_2$O)	Acetic acid 0.1 mol l^{-1} (6.005 g l^{-1})	x ml A + (100 − x) ml B
16a	Dimethylglutaric acid 0.1 mol l^{-1} (16.02 g l^{-1})	NaOH 0.2 mol l^{-1}	(a) 100 ml A + x ml B made up to 1000 ml
16b	Dimethylglutaric acid 0.1 mol l^{-1} (16.02 g l^{-1})	NaOH 0.2 mol l^{-1}	(b) 100 ml A + x ml B + 5.844 g NaCl made up to 1000 ml NaCl − 0.1 mol l^{-1}
17	Piperazine 1 mol l^{-1} (86.14 g l^{-1})	HCl 0.1 mol l^{-1}	5 ml A + x ml B made up to 100 ml
18	Tetraethylethylenediamine 1 mol l^{-1} (172.32 g l^{-1})	HCl 0.1 mol l^{-1}	5 ml A + x ml B made up to 100 ml
19	Tris acid maleate 0.2 mol l^{-1} [1 l solution contains 24.23 g tris(hydroxymethyl)aminomethane + 23.21 g maleic acid or 19.61 g maleic annydride]	NaOH 0.2 mol l^{-1}	25 ml A + x ml B made up to 100 ml
20	Dimethylaminoethylamine 1 mol l^{-1} (88 g l^{-1})	HCl 0.1 mol l^{-1}	5 ml A + x ml B made up to 100 ml
21	Imidazole 0.2 mol l^{-1} (13.62 g l^{-1})	HCl 0.1 mol l^{-1}	25 ml A + x ml B made up to 100 ml
22	Triethanolamine 0.5 mol l^{-1} + ethylenediamine-tetraacetic acid disodium salt (1 l solution contains 74.60 g C$_6$H$_{15}$O$_3$N + 20 g C$_{10}$H$_{14}$O$_8$N$_2 \cdot$ Na$_2 \cdot$ 2 H$_2$O)	HCl 0.05 mol l^{-1}	10 ml A + x ml B made up to 100 ml
23	N-Dimethylaminoleucylglycine 0.1 mol l^{-1} + NaCl 0.2 mol l^{-1} (1 l solution contains 24.33 g C$_{10}$H$_{20}$O$_3$N$_2 \cdot$ ½ H$_2$O + 11.69 g NaCl)	NaOH 1 mol l^{-1} 100 ml made up to 1 l with solution A	x ml A + (100−x) ml B
24	Tris 0.2 mol l^{-1} [tris(hydroxymethyl)aminomethane 24.23 g l^{-1}]	HCl 0.1 mol l^{-1}	25 ml A + x ml B made up to 100 ml
25	2-Amino-2-methylpropane-1,3-diol 0.1 mol l^{-1} (10.51 g l^{-1})	HCl 0.1 mol l^{-1}	50 ml A + x ml B made up to 100 ml
26	Sodium carbonate anhydrous 0.1 mol l^{-1} (10.60 g l^{-1})	Sodium bicarbonate 0.1 mol l^{-1} (8.401 g l^{-1})	x ml A + (100−x) ml B

Note: When not otherwise specified, both stock and buffer solutions should be made up with distilled water free of CO$_2$. Only standard reagents should be used. If there is any doubt as to the purity or water content of solutions, their amount of substance concentration must be checked by titration. The volumes x (in ml) of stock solutions required to make up a buffer solution of the desired pH value are given in the table on the next page.

From Lenter, C, Ed., *Geigy Scientific Tables*, 8th ed., volume 3, Ciba-Geigy, Basel, 1984, pages 58–60. With permission.

AMINE BUFFERS USEFUL FOR BIOLOGICAL RESEARCH

Norman Good

All of these amines are highly polar, water-soluble substances. Their advantages and disadvantages must be determined empirically for each biological reaction system. For best buffering performance they should be used at pH's close to the pKa, preferably within ±0.5 pH units of the pKa and never more than ±1.0 unit from the pKa. Note that the pKa's, and therefore the pH's of buffered solutions, change with temperature in the manner indicated.

Chemical Name	Trivial Name or Acronym	Structure	pKa at 20°C	ΔpKa/°C
2-(N-Morpholino)ethanesulfonic acid	MES		6.15	−0.011
Bis(2-hydroxyethyl)imino-tris-(hydroxymethyl)methane	Bistris	$(HOCH_2CH_2)_2{=}N{-}C{\equiv}(CH_2OH)_3$	6.5	—
N-(2-Acetamido)iminodiacetic acid	ADA[a]		6.6	−0.011
Piperazine-N,N'-bis(2-ethanesulfonic acid)	PIPES		6.8	−0.0085
1,3-Bis[tris(hydroxymethyl)methylamino] propane	Bistrispropane	$(HOCH_2)_3{\equiv}C{-}NH(CH_2)_3\ NH{-}C{\equiv}(CH_2\ OH)_3$	6.8 (9.0)	—
N-(Acetamido)-2-aminoethanesulfonic acid	ACES	$H_2NCOCH_2\ N^+H_2CH_2CH_2SO_3^-$	6.9	−0.020
3-(N-Morpholino)propanesulfonic acid	MOPS		7.15	−0.013
N,N'-Bis(2-hydroxyethyl)-2-amino-ethanesulfonic acid	BES	$(HOCH_2CH_2)_2{=}N^+HCH_2CH_2SO_3^-$	7.15	−0.016
N-Tris(hydroxymethyl)methyl-2-aminoethanesulfonic acid	TES	$(HOCH_2)_3{\equiv}C{-}N^+H_2CH_2CH_2SO_3^-$	7.5	−0.020
N-2-Hydroxyethylpiperazine-N'-ethanesulfonic acid	HEPES[b]		7.55	−0.014
N-2-Hydroxyethylpiperazine-N'-propanesulfonic acid	HEPPS[b]		8.1	−0.015
N-Tris(hydroxymethyl)methylglycine	Tricine[a]	$(HOCH_2)_3{\equiv}C{-}N^+H_2CH_2COO^-$	8.15	−0.021
Tris(hydroxymethyl)aminomethane	Tris	$(HOCH_2)_3{\equiv}CNH_2$	8.3	−0.031
N,N-Bis(2-hydroxyethyl)glycine	Bicine[a]	$(HOCH_2CH_2)_2{=}N^+HCH_2COO^-$	8.35	−0.018
Glycylglycine	Glycylglycine[a]	$H_3N^+CH_2CONHCH_2COO^-$	8.4	−0.028
N-Tris(hydroxymethyl)methyl-3-amino-propanesulfonic acid	TAPS	$(HOCH_2)_3{\equiv}C{-}N^+H_2(CH_2)_3SO_3^-$	8.55	−0.027
1,3-Bis[tris(hydroxymethyl)-methylamino]propane	Bistrispropane	$(HOCH_2)_3{\equiv}C{-}NH(CH_2)_3NH{-}C{\equiv}(CH_2OH)_3$	9.0 (6.8)	—
Glycine	Glycine[a]	$H_3N^+CH_2COO^-$	9.9	—

Compiled by Norman Good.

[a] These substances may bind certain di- and polyvalent cations and therefore they may sometimes be useful for providing constant, low level concentrations of free heavy metal ions (heavy metal buffering).

[b] These substances interfere with and preclude the Folin protein assay.

For further information on these and other buffers, see Good and Izawa, in *Methods in Enzymology, Part B*, Vol. 24, Pietro, Ed., Academic Press, New York, 1972, 53.

PREPARATION OF BUFFERS FOR USE IN ENZYME STUDIES*

G. Gomori

The buffers described in this section are suitable for use either in enzymatic or histochemical studies. The accuracy of the tables is within ±0.05 pH at 23°. In most cases the pH values will not be off by more than ±0.12 pH even at 37° and at molarities slightly different from those given (usually 0.05 M).

The methods of preparation described are not necessarily identical with those of the original authors. The titration curves of the majority of the buffers recommended have been redetermined by the writer. The buffers are arranged in the order of ascending pH range. For more complete data on phosphate and acetate buffers over a wide range of concentrations, see Vol. I [10].*

*From Gomori, in *Methods in Enzymology*, Vol. 1, Colowick and Kaplan, Eds., Academic Press, New York, 1955, 138. With permission.

TABLE 1   Hydrochloric Acid-Potassium Chloride Buffer*

x	pH
97.0	1.0
78.0	1.1
64.5	1.2
51.0	1.3
41.5	1.4
33.3	1.5
26.3	1.6
20.6	1.7
16.6	1.8
13.2	1.9
10.6	2.0
8.4	2.1
6.7	2.2

* Stock solutions

A: 0.2 M solution of KCl (14.91 g in 1,000 ml)
B: 0.2 M HCl 50 ml of A + x ml of B, diluted to a total of 200 ml

Reference

1. Clark and Lubs, *J. Bacteriol.*, 2, 1 (1917).

TABLE 2   Glycine-HCL Buffer*

x	pH	x	pH
5.0	3.6	16.8	2.8
6.4	3.4	24.2	2.6
8.2	3.2	32.4	2.4
11.4	3.0	44.0	2.2

* Stock solutions

A: 0.2 M solution of glycine (15.01 g in 1,000 ml)
B: 0.2 M HCl 50 ml of A + x ml of B, diluted to a total of 200 ml

Reference

1. Sørensen, *Biochem. Z.*, 21, 131 (1909); 22, 352 (1909).

TABLE 3   Phthalate-Hydrochloric Acid Buffer*

x	pH	x	pH
46.7	2.2	14.7	3.2
39.6	2.4	9.9	3.4
33.0	2.6	6.0	3.6
26.4	2.8	2.63	3.8
20.3	3.0		

* Stock solutions

A: 0.2 M solution of potassium acid phthalate (40.84 g in 1,000 ml)
B: 0.2 M HCl 50 ml of A + x ml of B, diluted to a total of 200 ml

Reference

1. Clark and Lubs, *J. Bacteriol.*, 2, 1 (1917).

TABLE 4   Aconitate Buffer*

x	pH	x	pH
15.0	2.5	83.0	4.3
21.0	2.7	90.0	4.5
28.0	2.9	97.0	4.7
36.0	3.1	103.0	4.9
44.0	3.3	108.0	5.1
52.0	3.5	113.0	5.3
60.0	3.7	119.0	5.5
68.0	3.9	126.0	5.7
76.0	4.1		

* Stock solutions

A: 0.5 M solution of aconitic acid (87.05 g in 1,000 ml)
B: 0.2 M NaOH 20 ml of A + x ml of B, diluted to a total of 200 ml

Reference

1. Gomori, unpublished data.

TABLE 5   Citrate Buffer*

x	y	pH
46.5	3.5	3.0
43.7	6.3	3.2
40.0	10.0	3.4
37.0	13.0	3.6
35.0	15.0	3.8
33.0	17.0	4.0
31.5	18.5	4.2
28.0	22.0	4.4
25.5	24.5	4.6
23.0	27.0	4.8
20.5	29.5	5.0
18.0	32.0	5.2
16.0	34.0	5.4
13.7	36.3	5.6
11.8	38.2	5.8
9.5	41.5	6.0
7.2	42.8	6.2

* Stock solutions

A: 0.1 M solution of citric acid (21.01 g in 1,000 ml)

B: 0.1 M solution of sodium citrate (29.41 g $C_6H_5O_7Na_3 \cdot 2H_2O$ in 1,000 ml; the use of the salt with 5½ H_2O is not recommended). x ml of A + y ml of B, diluted to a total of 100 ml

Reference

1. Lillie, *Histopathologic Technique*, Blakiston, Philadelphia and Toronto, 1948.

TABLE 6   Acetate Buffer*

x	y	pH
46.3	3.7	3.6
44.0	6.0	3.8
41.0	9.0	4.0
36.8	13.2	4.2
30.5	19.5	4.4
25.5	24.5	4.6
20.0	30.0	4.8
14.8	35.2	5.0
10.5	39.5	5.2
8.8	41.2	5.4
4.8	45.2	5.6

* Stock solutions

A: 0.2 M solution of acetic acid (11.55 ml in 1,000 ml)

B: 0.2 M solution of sodium acetate (16.4 g of $C_2H_3O_2Na$ or 27.2 g of $C_2H_3O_2Na \cdot 3H_2O$ in 1,000 ml) x ml of A + y ml of B, diluted to a total of 100 ml

Reference

1. Walpole, *J. Chem. Soc.*, 105, 2501 (1914).

TABLE 7   Citrate-Phosphate Buffer*

x	y	pH
44.6	5.4	2.6
42.2	7.8	2.8
39.8	10.2	3.0
37.7	12.3	3.2
35.9	14.1	3.4
33.9	16.1	3.6
32.3	17.7	3.8
30.7	19.3	4.0
29.4	20.6	4.2
27.8	22.2	4.4
26.7	23.3	4.6
25.2	24.8	4.8
24.3	25.7	5.0
23.3	26.7	5.2
22.2	27.8	5.4
21.0	29.0	5.6
19.7	30.3	5.8
17.9	32.1	6.0
16.9	33.1	6.2
15.4	34.6	6.4
13.6	36.4	6.6
9.1	40.9	6.8
6.5	43.6	7.0

* Stock solutions

A: 0.1 M solution of citric acid (19.21 g in 1,000 ml)

B: 0.2 M solution of dibasic sodium phosphate (53.65 g of $Na_2HPO_4 \cdot 7H_2O$ or 71.7 g of $Na_2HPO_4 \cdot 12H_2O$ in 1,000 ml) x ml of A + y ml of B, diluted to a total of 100 ml

Reference

1. McIlvaine, *J. Biol. Chem.*, 49, 183 (1921).

TABLE 8   Succinate Buffer*

x	pH	x	pH
7.5	3.8	26.7	5.0
10.0	4.0	30.3	5.2
13.3	4.2	34.2	5.4
16.7	4.4	37.5	5.6
20.0	4.6	40.7	5.8
23.5	4.8	43.5	6.0

* Stock solutions

A: 0.2 M solution of succinic acid (23.6 g in 1,000 ml)

B: 0.2 M NaOH 25 ml of A + x ml of B, diluted to a total of 100 ml

Reference

1. Gomori, unpublished, data.

TABLE 9   Phthalate-Sodium Hydroxide Buffer*

x	pH	x	pH
3.7	4.2	30.0	5.2
7.5	4.4	35.5	5.4
12.2	4.6	39.8	5.6
17.7	4.8	43.0	5.8
23.9	5.0	45.5	6.0

* Stock solutions

A: 0.2 M solution of potassium acid phthalate (40.84 g in 100 ml)

B: 0.2 M NaOH 50 ml of A + x ml of B, diluted to a total of 200 ml

Reference

1. Clark and Lubs, *J. Bacteriol.*, 2, 1 (1917).

TABLE 10   Maleate Buffer*

x	pH	x	pH
7.2	5.2	33.0	6.2
10.5	5.4	38.0	6.4
15.3	5.6	41.6	6.6
20.8	5.8	44.4	6.8
26.9	6.0		

* Stock solutions

A: 0.2 M solution of acid sodium maleate (8 g of NaOH + 23.2 g of maleic acid or 19.6 g of maleic anhydride in 1,000 ml)

B: 0.2 M NaOH 50 ml of A + x ml of B, diluted to a total of 200 ml

Reference

1. Temple, *J. Am. Chem. Soc.*, 51, 1754 (1929).

TABLE 11   Cacodylate Buffer*

x	pH	x	pH
2.7	7.4	29.6	6.0
4.2	7.2	34.8	5.8
6.3	7.0	39.2	5.6
9.3	6.8	43.0	5.4
13.3	6.6	45.0	5.2
18.3	6.4	47.0	5.0
23.8	6.2		

* Stock solutions

A: 0.2 M solution of sodium cacodylate (42.8 g of Na(CH$_3$)$_2$ AsO$_2 \cdot$ 3H$_2$O in 1,000 ml)

B: 0.2 M HCl 50 ml of A + x ml of B, diluted to a total of 200 ml

Reference

1. Plumel, *Bull. Soc. Chim. Biol.*, 30, 129 (1949).

TABLE 12   Phosphate Buffer*

x	y	pH	x	y	pH
93.5	6.5	5.7	45.0	55.0	6.9
92.0	8.0	5.8	39.0	61.0	7.0
90.0	10.0	5.9	33.0	67.0	7.1
87.7	12.3	6.0	28.0	72.0	7.2
85.0	15.0	6.1	23.0	77.0	7.3
81.5	18.5	6.2	19.0	81.0	7.4
77.5	22.5	6.3	16.0	84.0	7.5
73.5	26.5	6.4	13.0	87.0	7.6
68.5	31.5	6.5	10.5	90.5	7.7
62.5	37.5	6.6	8.5	91.5	7.8
56.5	43.5	6.7	7.0	93.0	7.9
51.0	49.0	6.8	5.3	94.7	8.0

* Stock solutions

A: 0.2 M solution of monobasic sodium phosphate (27.8 g in 1,000 ml)

B: 0.2 M solution of dibasic sodium phosphate (53.65 g of Na$_2$HPO$_4 \cdot$ 7H$_2$O or 71.7 g of Na$_2$HPO$_4 \cdot$ 12H$_2$O in 1,000 ml) x ml of A + y ml of B, diluted to a total of 200 ml

Reference

1. Sørensen, *Biochem. Z.*, 21, 131 (1909); 22, 352 (1909).

TABLE 13   Tris(Hydroxymethyl)Aminomethane-Maleate (Tris-Maleate) Buffer*†

x	pH	x	pH
7.0	5.2	48.0	7.0
10.8	5.4	51.0	7.2
15.5	5.6	54.0	7.4
20.5	5.8	58.0	7.6
26.0	6.0	63.5	7.8
31.5	6.2	69.0	8.0
37.0	6.4	75.0	8.2
42.5	6.6	81.0	8.4
45.0	6.8	86.5	8.6

* Stock solutions

A: 0.2 M solution of Tris acid maleate (24.2 g. of tris(hydroxymethyl)aminomethane + 23.2 g of maleic acid or 19.6 g of maleic anhydride in 1,000 ml)

B: 0.2 M NaOH 50 ml of A + x ml of B, diluted to a total of 200 ml

†A buffer-grade Tris can be obtained from the Sigma Chemical Co., St. Louis, MO., or From Matheson Coleman & Bell, East Rutherford, NJ.

Reference

1. Gomori, *Proc. Soc. Exp. Biol. Med.*, 68, 354 (1948).

TABLE 14   Barbital Buffer*†

x	pH
1.5	9.2
2.5	9.0
4.0	8.8
6.0	8.6
9.0	8.4
12.7	8.2
17.5	8.0
22.5	7.8
27.5	7.6
32.5	7.4
39.0	7.2
43.0	7.0
45.0	6.8

* Stock solutions

A: 0.2 M solution of sodium barbital (veronal) (41.2 g in 1,000 ml)

B: 0.2 M HCl 50 ml of A + x ml of B, diluted to a total of 200 ml

†Solutions more concentrated than 0.05 M may crystallize on standing, especially in the cold.

Reference

1. Michaelis, *J. Biol. Chem.*, 87, 33 (1930).

TABLE 15   Tris(Hydroxymethyl)-Aminomethane (Tris) Buffer*†

x	pH
5.0	9.0
8.1	8.8
12.2	8.6
16.5	8.4
21.9	8.2
26.8	8.0
32.5	7.8
38.4	7.6
41.4	7.4
44.2	7.2

* Stock solutions

A: 0.2 M solution of tris(hydroxymethyl)aminomethane (24.2 g in 1,000 ml)

B: 0.2 M HCl 50 ml of A + x ml of B, diluted to a total of 200 ml

†A buffer-grade Tris can be obtained from the Sigma Chemical Co., St. Louis, MO., or from Matheson Coleman & Bell, East Rutherford, NJ.

TABLE 16   Boric Acid-Borax Buffer*

x	pH	x	pH
2.0	7.6	22.5	8.7
3.1	7.8	30.0	8.8
4.9	8.0	42.5	8.9
7.3	8.2	59.0	9.0
11.5	8.4	83.0	9.1
17.5	8.6	115.0	9.2

* Stock solutions

A: 0.2 M solution of boric acid (12.4 g in 1,000 ml)

B: 0.05 M solution of borax (19.05 g in 1,000 ml; 0.2 M in terms of sodium borate) 50 ml of A + x ml of B, diluted to a total of 200 ml

Reference

1. Holmes, *Anat. Rec.*, 86, 163 (1943).

TABLE 17   2-Amino-2-Methyl-1,3-Propanediol (Ammediol) Buffer*

x	pH	x	pH
2.0	10.0	22.0	8.8
3.7	9.8	29.5	8.6
5.7	9.6	34.0	8.4
8.5	9.4	37.7	8.2
12.5	9.2	41.0	8.0
16.7	9.0	43.5	7.8

* Stock solutions

A: 0.2 M solution of 2-amino-2-methyl-l,3-propanediol (21.03 g in 1,000 ml)

B: 0.2 M HCl 50 ml of A + x ml of B, diluted to a total of 200 ml

Reference

1. Gomori, *Proc. Soc. Exp. Biol. Med.*, 62, 33 (1946).

TABLE 18   Glycine-NaOH Buffer*

x	pH	x	pH
4.0	8.6	22.4	9.6
6.0	8.8	27.2	9.8
8.8	9.0	32.0	10.0
12.0	9.2	38.6	10.4
16.8	9.4	45.5	10.6

* Stock solutions

A: 0.2 M solution of glycine (15.01 g in 1,000 ml)

B: 0.2 M NaOH 50 ml of A + x ml of B, diluted to a total of 200 ml.

Reference

1. Sørensen, *Biochem. Z.*, 21, 131 (1909); 22, 352 (1909).

TABLE 19   Borax-NaOH Buffer*

x	pH
0.0	9.28
7.0	9.35
11.0	9.4
17.6	9.5
23.0	9.6
29.0	9.7
34.0	9.8
38.6	9.9
43.0	10.0
46.0	10.1

* Stock solutions

A: 0.05 M solution of borax (19.05 g in 1,000 ml; 0.02 M in terms of sodium borate)

B: 0.2 M NaOH 50 ml of A + x ml of B, diluted to a total of 200 ml

Reference

1. Clark and Lubs, *J. Bacteriol.*, 2, 1 (1917).

TABLE 20   Carbonate-Bicarbonate Buffer*

x	y	pH
4.0	46.0	9.2
7.5	42.5	9.3
9.5	40.5	9.4
13.0	37.0	9.5
16.0	34.0	9.6
19.5	30.5	9.7
22.0	28.0	9.8
25.0	25.0	9.9
27.5	22.5	10.0
30.0	20.0	10.1
33.0	17.0	10.2
35.5	14.5	10.3
38.5	11.5	10.4
40.5	9.5	10.5
42.5	7.5	10.6
45.0	5.0	10.7

* Stock solutions

A: 0.2 M solution of anhydrous sodium carbonate (21.2 g in 1,000 ml)

B: 0.2 M solution of sodium bicarbonate (16.8 g in 1,000 ml) x ml of A + y ml of B, diluted to a total of 200

Reference

1. Delory and King, *Biochem. J.*, 39, 245 (1945).

BUFFER FOR ACRYLAMIDE GELS (SINGLE-GEL SYSTEMS)

Buffer Solutions								
In Gel			In Electrode Vessels					
Buffer Ions	M	pH	Buffer Ions	M	pH	Characteristics of System	Apparatus Type[a]	Examples of Materials Investigated
Veronal	0.09	8.6	Same as in gel			—	VP	Serum[1]
Tris	0.070							
EDTA	0.007	8.7	Same as in gel			Albumin band too narrow for acurate quantitation	VP	Serum[1]
Borate[b]	0.010							
Tris	0.070		Same as in gel			Mobilities are low in this buffer	VP	
EDTA	0.007	8.65						
Ca lactate	0.0016							
Tris	0.006						VP	Serum[1]
Veronal	0.022	8.5						
Na lactate	0.017							
Acetate	0.05	5.7				Rapid; cooled in petroleum ether	FP	Lactic dehydrogenase isoenzymes[3]
Tris	0.068							
Citrate	0.025	7.2	Borate[b]	0.05	8.6	12 M urea in gel	VP	Myosin subunits[4]
Phenol:acetic acid: water 2:1:1			Same as in gel			—	FP	Ribosomal proteins[5]
Tris	0.010		Tris	0.04		Sample unstabilized	D	Serum[6]
Glycine	0.005	9.2	Glycine	0.022	9.2			
Tris borate	0.1	8.7	Same as in gel			Preliminary soaking of gel possible	VP	Serum[7]
Potassium acetate	0.02	2.9	Valine (anode)	0.3	4.0 (acetic)	Sucrose used to stabilize sample	D	Histones[8]
			Glycine (cathode)	0.3				
Tris glycine	0.37	9.5	Tris glycine	0.37	9.5	Sucrose use to stabilize sample	D	Serum[12]
Veronal	0.05	8.4	—	—	—	—	FP	Lactic dehydrogenase isoenzymes[3]
Tris	0.083	8.9	Same as in gel			7 M urea in gel	D	α-Crystallin[14]
Tris	0.37	8.8	Borate[b]	0.05	9.2	Sample concentrates as it enters gel	D	Serum[15]

[a] D, Disc; FP, flat plate, in the horizontal; VP, vertical plate.

[b] Molarity based on weight of orthoboric acid, (H_4BO_3) taken.

From *Electrophoresis of Proteins in Polyacrylamide and Starch Gels*. Gordon, A. H., Ed., (*Laboratory Techniques in Biochemistry and Molecular Biology*, Work, T. S. and Work, E., Gen. Eds.), North-Holland, New York, 1971, 40. With permission of Elsevier Science Publishers.

BUFFER FOR ACRYLAMIDE GELS WITH MORE THAN ONE LAYER

Gel Layer	Buffer Solutions — In Gel			Buffer Solutions — In Electrode Vessels			Characteristics of System	Apparatus Type[a]	Ref.
	Buffer Ions	M	pH	Buffer Ions	M	pH			
Sample (large pore)	Tris	0.062	6.7	Tris	0.005	8.3	The sample is set in the upper-the most gel and concentrates in spacer gel; the actual separation occurs in the bottom gel	D	9
	Chloride	0.06		Glycine	0.038				
Spacer (large pore)	Tris	0.062	6.7						
	Chloride	0.06							
Small-pore gel	Tris	0.37	8.9						
	Chloride and buffers	0.06							
Same gel layers as above				Tris	0.05	8.3	Same as Ref. 9 except that improved resolution is obtained if as above carried out at 4-5°C	D	10
				Glycine	0.38				
Spacer (large pore)	Tris	0.075	6.7	Tris	0.05	8.3	Same as Ref. 9 except that sample gel is omitted; the sample is stabilized with sucrose	D	2
	Chloride	0.075		Glycine	0.38				
Small-pore gel	Tris	0.046	8.7						
	Chloride	0.075							
Spacer (large pore)	Tris	0.049	6.7	Tris	0.025	8.4	Same as Ref. 9 except that sample gel is omitted; the sample is stabilized with acrylamide monomer	VP	11
	H_3PO_4	0.026		Glycine	0.19				
Small-pore gel	Tris	0.37	8.9						
	Chloride	0.06							
Sample (large pore)	Acetic acid	0.062	6.8	Acetic acid	0.14	4.5	An acid gel suitable for separation of bases	D	13
	KOH	0.06		β-Alanine	0.35				
Spacer (large pore)	Acetic acid	0.062	6.8						
	KOH	0.06							
Small-pore gel	Acetic acid	0.75	4.3						
	KOH	0.12							

[a] D: disc; FP: flat plate, in the horizontal); VP: vertical plate.

[b] Molarity based on weight of orthoboric acid, (H_4BO_3) taken. For additional buffers see Starch Gels. All buffers recommended for starch gel can be used in acrylamide gels.

From *Electrophoresis of Proteins in Polyacrylamide and Starch Gels*, Gordon, A. H., Ed., (*Laboratory Techniques in Biochemistry and Molecular Biology*, Work, T. S. and Work, E., Gen. Eds.), North-Holland, New York, 1971. 41. With permission of Elsevier Science Publishers.

References For Acrylamide Gel Tables

1. Ferris, T. G., Easterling, R. E., and Budd, R. E., *Anal. Biochem. 8*, 477, 1964.
2. Pun, J. Y. and Lombarozo, L., *Anal. Biochem.*, 9, 9, 1964.
3. Jensen, K., *Scand. J. Clin. Lab. Invest.*, 17, 192, 1965.
4. Small, P. A., Harington, W. F., and Keilley, W. W., *Biochim. Biophys. Acta.*, 49, 462, 1961.
5. Work, T. S., *J. Mol. Biol.*, 10, 544, 1964.
6. Matson, C. F., *Anal. Biochem.*, 13, 294, 1965.
7. Lorber, A., *J. Lab. Clin. Med.*, 64, 133, 1964.
8. Shepherd, G. R. and Gurley, L. R., *Anal. Biochem.*, 14, 356, 1966.
9. Davies, B. J., *Ann. N.Y. Acad. Sci.*, 121, 404, 1964.
10. Pastewka, J. V., Neass, A. T., and Peacock, A. C., *Clin. Chim. Acta*, 14, 219, 1966).
11. Ritchie, R. F., Harter, J. G., and Bayles, T. B., *J. Lab, Clin. Med.*, 68, 842, 1966.
12. Hjerten S., Jerstedt S., and Tiselius A., *Anal. Biochem.*, 11, 219, 1965.
13. Reisfeld R. A., Lewis U. J., and Williams D. E., *Nature*, 195, 281, 1962.
14. Bloemendal H., Bout W. S., Jongkind J. F., and Wisse J. H., *Exp. Eye Res.*, 1,300, 1962.
15. Gordon A. H. and Louis L. N., *Anal. Biochem.*, 21, 190, 1967.

STARCH GELS

TABLE 1   Starch Gels: Buffer Solutions

Buffer Ions	M	pH	Buffer Ions	M	pH	Materials Investigated (Examples Only)	Results
	In Gel			In Electrode Vessels			
Borate[a,b]	0.025	8.5	Borate	0.3	8.5	Serum[1]	
Tris	0.076	8.6	Borate	0.3	8.5	Serum,[2,3] diphtheria toxin,[2] horseradish	Citrate borate boundary is seen as a brown band
Citrate[a]	0.005					peroxidase,[2] haptoglobins,[5] and transferins[6]	The electrophoresis is over in less time compared to borate only
						Hemoglobins,[4] human and animal sera,[7] staphylococcal enterotoxin B[16]	
Tris	0.045		Tris	0.013			
Borate[a]	0.025	8.4	Borate	0.075	8.4	Hemoglobins[12-14]	Hb A and F are separated
EDTA	0.002		EDTA	0.006		Serum[17]	
Tris	0.1		Tris	0.01			
Malate[a]	0.01	7.4	Malate	0.1	7.4	Hemolysates invested for phosphoglucomutase[20]	
EDTA	0.001		EDTA	0.01			
$MgCl_2$	0.001		$MgCl_2$	0.01			
Glycine	0.05	8.9	Borate	0.3	8.9	Myeloma γ-globulin[8]	Five bands
Citrate[a]	0.0027		—	—			
Borate[a]	0.0076	8.0	Borate	0.38		Anterior pituitary hormones[15]	
Tris	0.0144		—	—			
Lithium	0.0020		Lithium	0.10			
Tris	0.03	8.4	Tris	0.05		Tissue dehydrogenases[21]	
HCl	—		HCl	—			
Acetate	0.1	5.0	—	—		Pepsin and gastricsin[23]	The enzymes were recovered from the gel

[a] Molarities based on weight of orthoboric (H_4BO_3) or other acid taken. When no cation is given, the acids have been adjusted to the pH shown by addition of NaOH.

[b] For optimum results, use borate concentration recommended for each individual batch of starch.

From Gordon, A. H., Ed., *Electrophoresis of Proteins in Polyacrylamide and Starch Gels* (in *Laboratory Techniques in Biochemistry and Molecular Biology*, Work, T. S. and Work, E., Gen. Eds.), North-Holland, New York, 1971, 100. With permission of Elsevier Science Publishers.

TABLE 2   Starch Gels Containing Urea Buffer Solutions

Buffer Ions	M	pH	Urea M	Buffer Ions	M	pH	Materials Investigated (Examples Only)	Results
		Buffer Solutions						
	In Gel				In Electrode Vessels			
Formate	0.05	3.0	8	Formate	0.2	2—3	γ-Myeloma, heavy chains,[10] antibodies[18]	Slightly sharper bands were obtained if alkylation was done as well as reduction
Aluminum lactate	0.05	3.1	3	Aluminum lactate			Gliadin[11]	
Glycine	0.035	7—8	8	Borate	0.3	8.2	γ-Globulin heavy and light chains[9]	Light chains well separated
Borate[a]	0.025	8.5	8	Borate	0.3	8.5	Haptoglobins[19]	Without prior reduction urea has no effect
Tris	0.076	8.6	7	Borate	0.3	8.6	α-and β-casein[22,25]	Urea treatment of α-crystallin, without reduction, leads to formation of many bands
Citrate	0.005							
Acetate	0.0365	5.6	6	—	—	—	E. coli ribosomal proteins[24]	At least 14 bands

[a] Molarities based on weight of orthoboric (H_4BO_3) or other acid taken. When no cation is given, the acids have been adjusted to the pH shown by addition of NaOH.

[b] For optimum results, use borate concentration recommended for each individual batch of starch.

From Gordon, A. H., Ed., *Electrophoresis of Proteins in Polyacrylamide and Starch Gels* (in *Laboratory Techniques in Biochemistry and Molecular Biology*, Work, T. S. and Work, E., Gen. Eds.), North-Holland, New York, 1971. 101. with permission of Elsevier Science Publishers.

References to Tables 1 and 2

1. Smithies, O., *Biophys. J.*, 61, 629, 1955.

2. Poulik, M.D., *Nature*, 180, 1477, 1957.

3. Poulik, M. D., *J. Immunol.*, 82, 502, 1959.

4. De Grouchy, J., *Rev. Fr. Etud. Clin. Biol.*, 3, 877, 1958.

5. Giblett, E. R., Motulsky, A. G., and Frazer, G. R., *Am. J. Human Genet.*, 18, 553, 1966.

6. Harris, H., Robson, E. B., and Siniscalco, M., *Nature*, 182, 452, 1958.

7. Krotski, W. A., Benjamin, D. C., and Weimer, H. E., *Can. J. Biochem.*, 44, 545, 1966.

8. Askonas, I., *Biochem. J.*, 79, 33, 1961.

9. Cohen, S. and Porter, R. R., *Biochem. J.*, 90, 278, 1964.

10. Edelman, G. M. and Poulik, M. D., *J. Exp. Med.*, 113, 861, 1961.

11. Woychick, J. H., Boundy, J. A., and Dimler, R. J., *Arch. Biochem. Biophys.*, 94, 477, 1961.

12. Winterhalter, K. H. and Huehns, E. R., *J Biol. Chem.*, 239, 3699, 1964.

13. Huehns, E. R., Dance, N., Beaven, G. H., Keil, J. V., Hecht, F., and Motulsky, A. G., *Nature*, 201, 1095, 1964.

14. Chernoff, A. I. and Pettit, N. M., *Blood*, 25, 646, 1965.

15. Ferguson, K. A. and Wallace, A. L. C., *Nature*, 190, 629, 1961.

16. Baird-Parker, A. C. and Jospeh, R. L., *Nature*, 202, 570, 1964

17. Poulik, M. D., in *Methods of Biochemical Analysis*, Vol. 14, Glick, D., Ed., Interscience, 1966, 455.

18. Edelman, G. M., Benacerraf, B., and Ovary, Z., *J. Exp. Med.*, 118, 229, 1963.

19. Smithies, O. and Connell, G. E., in *Biochemistry of Human Genetics*, Churchill, London, 1959, 178.

20. Spencer, N., Hopkinson, D. A., and Harris, H., *Nature*, 204, 742, 1964.

21. Tsao, M. U., *Arch. Biochem. Biophys.*, 90, 234, 1960.

22. Wake, R. G. and Baldwin, R. L., *Biochem. Biophys. Acta*, 47, 227, 1961.

23. Tang, J., Wolf, S., Caputto, R., and Trucco, R. E., *J. Biol. Chem.*, 234, 1174, 1959.

24. Waller, J. P. and Harris, J. I., *Proc. Nat. Acad. Sci. U.S.A.*, 47, 18, 1961.

25. Bloemendal, H., Bout, W. S., Jonkind, J. F., and Wisse, J. H., *Nature*, 193, 437, 1962.

INDICATORS FOR VOLUMETRIC WORK AND pH DETERMINATIONS

Indicator	Chemical Name	Acid Color	pH Range	Basic Color	Preparation
Methyl violet 6B	Tetra and pentamethylated *p*-rosaniline hydrochloride	Y	0.1–1.5	B	pH: 0.25% water
Metacresol purple (acid range)	*m*-Cresolsulfonphthalein	R	0.5–2.5	Y	pH: 0.10 g. in 13.6 ml 0.02 *N* NaOH, diluted to 250 ml. with water
Metanil yellow	4-Phenylamino-azobenzene-3′-sulfonic acid	R	1.2–2.3	Y	pH: 0.25% in ethanol
p-Xylenol blue (acid range)	1,4-Dimethyl-5-hydroxybenzenesulfonphthalein	R	1.2–2.8	Y	pH: 0.04% in ethanol
Thymol blue (acid range)	Thymolsulfonphthalein	R	1.2–2.8	Y	pH: 0.1 g. in 10.75 ml. 0.02 *N* NaOH, diluted to 250 ml. with water
Tropaeolin OO	Sodium *p*-diphenylamino-azobenzenesulfonate	R	1.4–2.6	Y	pH: 0.1% in water Vol.: 1% in water
Quinaldine red	2-(*p*-Dimethylaminostyryl)quinoline ethiodide	C	1.4–3.2	R	Vol.: 0.1% in ethanol
Benzopurpurine 4B	Ditolyl-diazo-bis-*α*-naphthyl-amine-4-sulfonic acid	B-V	1.3–4.0	R	pH, vol.: 0.1% in water
Methyl violet 6B	Tetra and pentamethylated *p*-rosaniline hydrochloride	B	1.5–3.2	V	pH, vol.: 0.25% in water
2,4-Dinitrophenol		C	2.6–4.0	Y	pH, vol.: 0.1 g. in 5 ml. ethanol, diluted to 100 ml. with water
Methyl yellow	*p*-Dimethylaminoazobenzene	R	2.9–4.0	Y	pH, vol.: 0.05% in ethanol
Bromphenol blue	Tetrabromophenolsulfonphthalein	Y	3.0–4.6	B	pH: 0.1 g. in 7.45 ml. 0.02 *N* NaOH, diluted to 250 ml. with water
Tetrabromophenol blue	Tetrabromophenol-tetrabromosulfon-phthalein	Y	3.0–4.6	B	pH: 0.1 g. in 5.00 ml. 0.02 *N* NaOH, diluted to 250 ml. with water
Direct purple	Disodium 4,4′-bis(2-amino-1-naphthylazo)-2,2′-stilbenedisulfonate	B-P	3.0–4.6	R	Vol.: 0.1 g. in 7.35 ml. 0.02 *N* NaOH, diluted to 100 ml. with water
Congo red	Diphenyl-diazo-bis-1-naphthylamine-4-sodium sulfonate	B	3.0–5.2	R	pH: 0.1% in water
Methyl orange	4′-Dimethylaminoazobenzene-4-sodium sulfonate	R	3.1–4.4	Y	Vol.: 0.1% in water
Brom-chlorphenol blue	Dibromodichlorophenolsulfonphthalein	Y	3.2–4.8	B	pH: 0.1 g. in 8.6 ml. 0.02 *N* NaOH, diluted to 250 ml. with water Vol.: 0.04% in ethanol
p-Ethoxychrysoidine	4′-Ethoxy-2,4-diaminoazobenzene	R	3.5–5.5	Y	Vol.: 0.1% in ethanol
α-Naphthyl red		R	3.7–5.0	Y	Vol.: 0.1% in ethanol
Sodium alizarinsulfonate	Dihydroxyanthraquinone sodium sulfonate	Y	3.7–5.2	V	pH, vol.: 1% in water
Bromcresol green	Tetrabromo-*m*-cresolsulfonphthalein	Y	3.8–5.4	B	pH: 0.10 g. in 7.15 ml. 0.02 *N* NaOH, diluted to 250 ml. with water
2,5-Dinitrophenol		C	4.0–5.8	Y	pH, vol.: 0.10 g. in 20 ml. Ethanol, then dilute to 100 ml. with water
Methyl red	4′-Dimethylaminoazobenzene-2-carboxylic acid	R	4.2–6.2	Y	pH: 0.10 g. in 18.6 ml. 0.02 *N* NaOH, diluted to 250 ml. with water Vol.: 0.1% in ethanol
Lacmoid		R	4.4–6.2	B	Vol.: 0.5% in ethanol
Azolitmin		R	4.5–8.3	B	Vol.: 0.5% in water
Litmus		R	4.5–8.3	B	Vol.: 0.5% in water
Cochineal	Complex hydroxyanthraquinone derivative	R	4.8–6.2	V	Vol.: Triturate 1 g. with 20 ml. Ethanol and 60 ml. water, let stand 4 days, and filter
Hematoxylin		Y	5.0–6.0	V	Vol.: 0.5% in ethanol.
Chlorphenol red	Dichlorophenolsulfonphthalein	Y	5.0–6.6	R	pH: 0.1 g. in 11.8 ml. 0.02 *N* NaOH, diluted to 250 ml. with water Vol.: 0.04% in ethanol
Bromcresol purple	Dibromo-*o*-cresolsulfonphthalein	Y	5.2–6.8	Pu	pH: 0.1 g. in 9.25 ml. 0.02 *N* NaOH, diluted to 250 ml with water Vol.: 0.02% in ethanol
Bromphenol red	Dibromophenolsulfonphthalein	Y	5.2–7.0	R	pH: 0.l g. in 9.75 ml. 0.02 *N* NaOH, diluted to 250 ml. with water Vol.: 0.04% in ethanol
Alizarin	1,2-Dihydroxyanthraquinone	Y	5.5–6.8	R	Vol.: 0.1% in ethanol

INDICATORS FOR VOLUMETRIC WORK AND pH DETERMINATIONS (Continued)

Indicator	Chemical Name	Acid Color	pH Range	Basic Color	Preparation
Dibromophenoltetrabromo-phenolsulfonphthalein		Y	5.6–7.2	Pu	pH: 0.1 g. in 1.21 ml. 0.1 N NaOH, diluted to 250 ml. with water
p-Nitrophenol		C	5.6–7.6	Y	pH, vol.: 0.25% in water
Bromothymol blue	Dibromothymolsulfonphthalein	Y	6.0–7.6	B	pH: 0.1 g. in 8 ml. 0.02 N NaOH, diluted to 250 ml. with water Vol.: 0.1% in 50% ethanol
Indo-oxine	5,8-Quinolinequinone-8-hydroxy-5-quinolyl-5-imide	R	6.0–8.0	B	Vol.: 0.05% in ethanol
Cucumin		Y	6.0–8.0	Br-R	Vol: saturated aq. soln.
Quinoline blue	Cyanine	C	6.6–8.6	B	Vol.: 1% in ethanol
Phenol red	Phenolsulfonphthalein	Y	6.8–8.4	R	pH: 0.1 g. in 14.20 ml. 0.02 N NaOH, diluted to 250 ml. with water Vol.: 0.1% in ethanol
Neutral red	2-Methyl-3-amino-6-dimethylaminophenazine	R	6.8–8.0	Y	pH, vol.: 0.1 g. in 70 ml. ethanol, diluted to 100 ml. with water
Rosolic acid aurin; corallin		Y	6.8–8.2	R	pH, vol.: 1% in 50% ethanol
Cresol red	o-Cresolsulfonphthalein	Y	7.2–8.8	R	pH: 0.1 g. in 13.1 ml. 0.02 N NaOH, diluted to 250 ml. with water Vol.: 0.1% in ethanol
α-Naphtholphthalein		P	7.3–8.7	G	pH. vol.: 0.1% in 50% ethanol
Metacresol purple (alkaline range)	m-Cresolsulfonphthalein	Y	7.4–9.0	P	pH: 0.1 g. in 13.1 ml. 0.02 N NaOH, diluted to 250 ml. with water Vol.: 0.1% in ethanol
Ethylbis-2,4-dinitrophenylacetate		C	7.5–9.1	B	Vol.: saturated soln. in equal volumes of acetone and ethanol
Tropaeolin OOO No. 1	Sodium α-naphtholazobenzene-sulfonate	Y	7.6–8.9	R	Vol.: 0.1% in water
Thymol blue (alkaline range)	Thymolsulfonphthalein	Y	8.0–9.6	B	pH: 0.1 g. in 10.75 ml. 0.02 N NaOH, diluted to 250 ml. with water Vol.: 0.1% in ethanol
p-Xylenol blue	1,4-Dimethyl-5-hydroxybenzenesulfonphthalein	Y	8.0–9.6	B	pH, vol.: 0.04% in elhanol
o-Cresolphthalein		C	8.2–9.8	R	pH, vol.: 0.04% in ethanol
α-Naphtholbenzein		Y	8.5–9.8	G	pH, vol.: 1% in ethanol
Phenolphthalein	3,3-Bis(p-hydroxyphenyl)-phthalide	C	8.2–10	R	Vol.: 1% in ethanol
Thymolphthalein		C	9.3–10.5	B	pH, vol.: 0.1% in ethanol
Nile blue A	Aminonaphthodiethylaminophenoxazine sulfate	B	10–11	P	Vol.: 0.1% in water
Alizarin yellow GG	3-Carboxy-4-hydroxy-3'-nitroazobenzene	Y	10–12	L	pH, vol.: 0.1% in 50% ethanol
Alizarin yellow R	3-Carboxy-4-hydroxy-4'-nitroazobenzene sodium salt	Y	10.2–12.0	R	pH, vol.: 0.1% in water
Poirrer's blue C4B		B	11–13	R	pH: 0.2% in water
Tropaeolin O	p-Benzenesulfonic acid-azoresorcinol	Y	11–13	O	pH: 0.1% in water
Nitramine	Picrylnitromethylamine	C	10.8–13	Br	pH: 0.1% in 70% ethanol
1,3,5-Trinitrobenzene		C	11.5–14	O	pH: 0.1% in ethanol
Indigo carmine	Sodium indigodisulfonate	B	11.6–14	Y	pH: 0.25% in 50% ethanol

Note: The indicator colors are abbreviated as follows: B, blue; Br, brown; C, colorless; G, green; L, lilac; O, orange; P, pink; Pu, purple; R, red; V, violet; and Y, yellow.

Mixed Indicators

Composition	Solvent	Transition pH	Acid Color	Transition Color	Basic Color
Dimethyl yellow, 0.05% + Methylene blue, 0.05%	alc.	3.2	Blue–violet	—	Green
Methyl orange, 0.02% + Xylene cyanole FF, 0.28%	50% alc.	3.9	Red	Gray	Green
Methyl yellow, 0.08% + Methylene blue, 0.004%	alc.	3.9	Pink	Straw–pink	Yellow–green
Methyl orange, 0.1% + Indigocarmine, 0.25%	aq.	4.1	Violet	Gray	Yellow–green
Bromcresol green, 0.1% + Methyl orange, 0.02%	aq.	4.3	Orange	Light green	Dark green
Bromcresol green, 0.075% + Methyl red, 0.05%	alc.	5.1	Wine–red	—	Green
Methyl red, 0.1% + Methylene blue, 0.05%	alc.	5.4	Red–violet	Dirty blue	Green
Bromcresol green, 0.05% + Chlorphenol red, 0.05%	aq.	6.1	Yellow–green	—	Blue–violet
Bromcresol purple, 0.05% + Bromthymol blue, 0.05%	aq.	6.7	Yellow	Violet	Violet–blue
Neutral red, 0.05% + Methylene blue, 0.05%	alc.	7.0	Violet–blue	Violet–blue	Green
Bromthymol blue, 0.05% + Phenol red, 0.05%	aq.	7.5	Yellow	Violet	Dark violet
Cresol red, 0.025% + Thymol blue, 0.15%	aq.	8.3	Yellow	Rose	Violet
Phenolphthalein, 0.033% + Methyl green, 0.067%	alc.	8.9	Green	Gray–blue	Violet
Phenolphthalein, 0.075% + Thymol blue, 0.025%	50% alc.	9.0	Yellow	Green	Violet
Phenolphthalein, 0.067% + Naphtholphthalein, 0.033%	50% alc.	9.6	Pale rose	—	Violet
Phenolphthalein, 0.033% + Nile blue, 0.133%	alc.	10.0	Blue	Violet	Red
Alizarin yellow, 0.033% + Nile blue, 0.133%	alc.	10.8	Green	—	Red–brown

ACID AND BASE INDICATORS

The following is a brief list of some acid-base indicators (a more comprehensive listing is available in the *CRC Handbook of Chemistry and Physics*, CRC Press, Boca Raton, FL, USA, Section 8). There is extensive use of acid-base indicators in the measurement of intracellular pH, as indicators for enzyme-catalyzed reactions, and in the measurement of material transfer across membranes and detection of changes in solid matrices.

Some Acid-Base Indicators (pH Indicators)

Indicator Dye	Acid Color	Basic Color	pKa	pH Range
Cresol Red (I) (*o*-cresolsulfonephthalein)	Red	Yellow		0.2 – 1.8
Crystal Violet	Green	Blue		0.0 – 2.0
Thymol Blue (I) (thymolsulfonephthalein)	Red	Yellow	1.6	1.2 – 2.8
Cresol Purple (metacresol purple; *m*-cresolpurple)	Red	Yellow	1.5	1.2 – 2.8
Bromophenol Blue	Yellow	Blue	4.1	3.0 – 4.6
Congo Red	Blue/Violet	Red		3.0 – 5.2
Methyl Red	Red	Yellow	5.1	4.4 – 6.3
Neutral Red	Red	Yellow	7.4	6.8 – 8.0
Phenol Red (phenolsulfonphthalein)	Yellow	Red	8.0	6.8 – 8.4
Cresol Red (II)	Yellow	Purple	8.4	7.3 – 8.8
Cresol Purple (II)	Yellow	Purple	8.3	7.4 – 9.0
Thymol Blue (II)	Yellow	Blue	9.2	8.0 – 9.6
Phenolphthalein	Colorless	Red	9.6	8.3 – 10.0
Nile Blue	Blue	Red	10.0	9.0 – 10.4
Nitramine (picrymethylnitramine)	Colorless	Orange-Brown		10.8 – 12.8

Acid-Base Indicators in Organic Solvents

Indicator	pKa		
	H$_2$O	Dimethylformamide	2-Propanol
Thymol Blue (I)	1.7		5.0
Cresol Red(I)			4.3
Bromophenol Blue	4.1		8.8
Neutral Red	7.4		7.2
Phenol Red	8.0	15.4	15.4
Cresol Purple	8.3	15.2	

References for the use
of acid-base indicators

Hammett, L.P. and Deyrup, A.J., A series of simple basic indicators. I. The acidity functions of mixtures of sulfuric and perchloric acids with water, *J.Amer.Chem.Soc.* 54, 2721-2739, 1932

Gardner, K.J., Use of acid-base indicator for quantitative paper chromatography of sugars, *Nature* 176, 929-930, 1955

Meikle, R.W., Paper chromatography of 2-halogenated carboxylic acids: *N,N*-dimethyl-*p*-phenylazoaniline as an acid-base indicator reagent, *Nature* 196, 61, 1962

Kolthoff, I.M., Bhowmik, S., and Chantooni, M.K., Acid-base indicator properties of sulfonephthaleins and benzeins in acetonitrile, *Proc. Nat.Acad.Sci.USA* 56, 13701376, 1966

Chance, B. and Scarpa, A., Acid-base indicator for the measurement of rapid changes in hydrogen ion concentration, *Methods Enzymol.* 24, 336-342, 1972

Benkovic, P.A., Hegazi, M., Cunningham, B.A., and Benkovic, S.J., Investigation of the pre-steady state kinetics of fructose bisphosphatase by employment of an indicator method, *Biochemistry* 18, 830-860, 1979

Smith, M.A. and Thompson, R.A., A method for the estimation of the activity of the inhibitor of the first component of complement, *J.Clin. Pathol.* 33, 167-170, 1980

Kogure, K., Alonso, O.F., and Martinez, E., A topographical measurement of brain pH, *Brain Res.* 195, 95-109, 1980

Kiernan, J.A, Chromoxane cyanine R. I. Physical and chemical properties of the dye and of some its iron complexes, *J.Microsc.* 143, 13-23, 1984

Paradiso, A.M., Tsien, R.Y., and Machen, T.E., Na^+-H^+ exchange in gastric glands as measured with a cytoplasmic-trapped, fluorescent pH indicator, *Proc.Nat.Acad.Sci.USA* 81, 7436-7440, 1984

Mera, S.L. and Davies, J.D., Differential Congo red staining: the effects of pH, non-aqueous solvents and the substrate, *Histochem.J.* 16, 195-210, 1984

Horie, K., Hagihara, H., Wada, A., and Fukutome, H., A highly sensitive photometric method for proton release or uptake: difference photometry, *Anal.Biochem.* 137, 80-87, 1984

Krchnak, V., Vagner, J., and Lebl, M., Noninvasive continuous monitoring of solid-phase peptide synthesis by acid-base indicator, *Int.J.Pept. Protein Res.* 32, 415-416, 1988

Rosenberg, R.M., Herreid, R.M., Piazza, G.J., and O'Leary, M.N., Indicator assay for amino acid decarboxylases, *Anal.Biochem.* 181, 55-65, 1989

Weiner, I.D. and Hamm, L.L., Use of fluorescent dye BCECF to measure intracellular pH in cortical collecting tubule, *Am.J.Physiol(Renal, Fluid, Electrolyte Physiol.),* 256, F957-F964, 1989

Bassnett, S., Reinisch, L., and Beebe, D.C., Intracellular pH measurement using single excitation-dual emission fluorescence ratios, *Am.J.Physiol.(Cell Physiol.)* 258, C171-C178, 1990

Anderson, R.E, Bjorkman, D. and McGreavy, J.M., Alteration of gastric surface cell pH regulation by sodium taurocholate, *J.Surg.Res.* 50, 65-71, 1991

Tortorello, M.L., Trotter, K.M., Angelos, S.M., *et al.*, Microtiter plate assays for the measurement of phage adsorption and infection in *Lactococcus* and *Enterococcus*, *Anal.Biochem.* 192, 362-366, 1991

Raley-Sussman, K.M., Sapolsky, R.M., and Kopito, R.R., $Cl^- \cdot HCO_3^-$-exchange function differs in adult and fetal rat hippocampal neurons, *Brain Res.* 614, 308-314, 1993

Reusch, H.P., Reusch, R., Rosskopf, D., *et al.*, Na^+/H^+ exchange in human lymphocytes and platelets in chronic and subacute metabolic acidosis, *J.Clin.Invest.* 92, 858-865, 1993

Mehta, V.D., Kulkarni, P.V., Mason, R.P., *et al.*, 6-Fluoropyridoxal: a novel probe of cellular pH using ^{19}F NMR spectroscopy, *FEBS Lett.* 349, 234-238, 1994

Optiz, N., Merten, E., and Acker, H., Evidence for redistribution-associated intracellular pH shifts of the pH-sensitive fluoroprobe carboxy SNARF-1, *Pflugers Arch.* 427, 332-342, 1994

Webb, B., Frame, J., Zhao, Z., *et al.*, Molecular entrapment of small molecules within the interior of horse spleen ferritin, *Arch.Biochem. Biophys.* 309, 178-183, 1994

Zhou, Y., Marcus, E.M., Haugland, R.P., and Opas, M., Use of a new fluorescent probe, seminaphthofluorescein-calcein, for determination of intracellular pH by simulataneous dual-emission imaging laser scanning confocal microscopy, *J.Cell Physiol.* 164, 9-16, 1995

Scheef, C.A., Oelkrug, D., and Schmidt, P.C., Surface acidity of solid pharmaceutical excipients III. Excipients for solid dosage forms, *Eur.J.Pharm.Biopharm.* 46, 209-213, 1998

Shao, P.G. and Bailey, L.C., Porcine insulin biodegradable polyester microspheres: stability and in vitro release characteristics, *Pharm.Dev. Technol.* 5, 1-9, 2000

Silver, R.B., Breton, S. and Brown, D., Potassium depletion increases proton pump (H(+)-ATPase) activity in intercalated cells of cortical collecting duct, *Am.J.Physiol.Renal Physiol.* 279, F195-F202, 2000

Chu, Y.I., Penland, R.L., and Wilhemus, K.R., Colorimetric indicators of microbial contamination in corneal preservation, *Cornea* 19, 517-520, 2000

Jayaraman, S., Song, Y., and Verkman, A.S., Airway surface liquid pH in well-differentiated airway epithelial cell cultures and mouse trachea, *Am.J.Physiol.Cell Physiol.* 281, C1504-155, 2001

Yu, E., Pan, J., and Zhou, H.M., A direct continuous pH-spectrophotometric assay for arginine kinase activity, *Protein Pept.Lett.* 9, 545-552, 2002

Hur, O., Niks, D., Casino, P., and Dunn, M.F., Proton transfer in the β-reaction catalyzed by tryptophan synthase, *Biochemistry* 41, 9991-10001, 2002

Sun, C. and Berg, J.C, A review of the different techniques for solid surface acid-base characterization, *Adv.Colloid Interface Sci.* 105, 1510175, 2003

Li, J., Chatterjee, K., Medek, A., *et al.*, Acid-base characterization of bromophenol blue-citrate buffer systems in the amorphous state, *J.Pharm.Sci.* 93, 697-712, 2004

Balderas-Hernandez, P, Rojas-Hernandez, A., Galvan, M., and Ramirez-Silva, M.T., Spectrophotometric study of the system Hg(II)-thymol blue-H_2O and its evidence through electrochemical means, *Spectrochimica Acta A Mol.Biomol.Spectrosc.* 60, 569-577, 2004

Gilman, J.B. and Vaida, V., Permeability of acetic acid through organic films at the air-aqueous interface, *J.Phys.Chem.A Mol.Spectros.Kinet. Environ.Gen.Theory* 110, 7581-7587, 2006

Sanchez-Armass, S., Sennoune, S.R., Maiti, D., *et al.*, Spectral imaging microscopy demonstrates cytoplasmic pH oscillations in glial cells, *Am.J.Physiol.Cell Physiol.* 290, C524-C538, 2006

General acid-base indicators

Widmer, M., Titrimety, in *Encyclopedia of Analytical Chemistry*, ed. R.A. Meyers, John Wiley & Sons, Ltd., Chichester, UK, pps. 13624-13636, 2002

Encyclopedia of Analytical Sciences, ed. A. Townshend, Academic Press, London, 1995

Butler, J.N., *Ionic Equilibrium. Solubility and pH Calculations*, John Wiley & Sons, New York, NY, USA, 1998

Westcott, G.C., *pH Measurements*, Academic Press, New York, NY, USA, 1978

Kotyk, A. and Slavík, J., *Intracellular pH and Its Measurement*, CRC Press, Boca Raton, FL, USA, 1989

Britton, H.T.S., *Hydrogen Ions. Their Determination and Importance in Pure and Industrial Chemistry*, D.Van Nostrand, New York, NY, USA, 1932

Kolthoff, I.M.(trans. Rosenblum, C.), *Acid-Base Indicators(Säure-Basen Indicatoren)*, MacMillan Company, New York, NY, USA, 1937

Kolthoff, I.M. and Laitinen, H.A., *pH and Electro Titration. The Colorimetric and Potentiometric Determination of pH, Potentiometry, Conductometry, and Voltometry(Polarography). Outline of Electrometric Titration*, John Wiley & Sons, Inc., New York, NY, USA, 1941

Webber, R.B., *The Book of pH*, George Newnes Ltd., London, UK, 1957

Fritz, J.S., *Acid-Base Titrations in Nonaqueous Solvents*, G.Frederick Smith Chemical Company, Columbus, Ohio, 1952

Clark, W.M. *The Determination of Hydrogen Ions*, 2nd edn., Williams & Wilkins, Baltimore, MD, USA, 1927

Radhuraman, B., Gustavson, G, van Hal, R.E.G., *et al.*, Extended-range spectroscopic pH measurement using optimized mixtures of dyes, *Appl.Spectrosc.* 60, 1461-1468, 2006

Non-aqueous titration

Kolade, Y.T., Adegbolagun, O.M., Idowu, O.S. *et al.*, Comparative determination of halofantrine tablets by titrimetry, spectrophotometry and liquid chromatography, *Afr.J.Med.Med.Sci.* 35, 79-84, 2006

Mera, S.L. and Davies, J.D., Differential congo red staining: the effects of pH, non-aqueous solvents and the substrate, *Histochem.J.* 16, 195-210, 1984

Cresol Purple

Schindler, J.F., Naranjo, P.A., Honaberger, D.A., *et al.*, Haloalkane dehalogenase: Steady-state kinetics and halide inhibition, *Biochemistry* 38, 5772-5778, 1999

Phenolphthalein

King, B.F., Liu, M., Townsend-Nicholson, A., *et al.*, Antagonism of ATP responses at P2X receptor subtypes by the pH indicator dye, phenol red, *Br.J.Pharmacol.* 145, 313-322, 2005

Riccio, M.L., Rossolini, G.M., Lombardi, G., *et al.*, Expression cloning of different bacterial phosphatase-encoding genes by Histochemical screening of genomic libraries onto an indicator medium containing phenolphthalein diphosphate and methyl green, *J.Appl.Microbiol.* 82, 177185, 1997

Gerber, H., Colorimetric determination of alkaline phosphatase as indicator of mammalian feces in corn meal: collaborative study, *J.Assoc.Off. Anal.Chem.* 69, 496-498, 1986

50a. Khalifab, R.G., The carbon dioxide hydration activity of carbonic anhydrase. I. Stop-flow kinetic studies on the native human isoenzymes B and C, *J.Biol.Chem.* 246, 2561-2573, 1971

Cresol Red

Borucki, B., Davanthan, S., Otto, H., *et al.*, Kinetics of proton uptake and dye binding by photoactive yellow protein in wild type and the E46Q and E46A mutants, *Biochemistry* 41, 10026-10037, 2002

Actis, L.A., Smoot, J.C., Baracin, C.E., and Findlay, R.H., Comparison of differential plating media and two chromatographic techniques for the detection of histamine production in bacteria, *J.Microbiol. Methods* 39, 79-90, 1999

Jeronimo, P.C., Araujo, A.N., Montenegro, M.C., *et al.*, Flow-through sol-gel optical biosensor for the colorimetric determination of acetazolamide, *Analyst* 130, 1190-1197, 2005

Nakamura, N. and Amao, Y., Optical sensor for carbon dioxide combining colorimetric change of a pH indicator and a reference luminescent dye, *Anal.Bioanal.Chem.* 376, 642-646, 2003

Yu, Z., Pan, J., and Zhou, H.M., A direct continuous pH-spectrophotometric assay for arginine kinase activity, *Protein Pept.Lett.* 9, 545-552, 2002

58a. Caselli, M., Mangone, A., Paoliollo, P., and Traini, A., Determination of the acid dissociation constant of bromocresol green and cresol red in water/AOT/isooctane reverse micelles by multiple linear regression and extended principal component analysis, *Ann.Chim.* 92, 501-512, 2002

Actis, L.A., Smoot, J.C, Barancin, C.E., and Findlay, R.H., Comparison of differential plating media and two chromatographic techniques for the detection of histamine production in bacteria, *J.Microbiol. Methods* 39, 79-90, 1999

Grabner, R., Influence of cationic amphiphilic drugs on the phosophatidylcholine hydrolysis by phospholipase A2, *Biochem.Pharmacol.* 36, 1063-1067, 1987

Horie, K., Hagihara, H., Wada, A., and Fukutome, H., A highly sensitive photometric method for proton release or uptake: difference photometry, *Anal.Biochem.* 137, 80-87, 1984

Velthuys, B.R., A third site of proton translocation in green plant photosynthetic electron transport, *Proc.Nat.Acad.Sci.USA* 75, 6031-6034, 1978

Crystal Violet

Kolade, Y.T., Adegboagun, O.M. Iodwu, O.S., *et al.*, Comparative determination of halofantrine tablets by titrimetry, spectrophometry and liquid chromatography, *Afr.J.Med.Med.Sci.* 35, 79-84, 2006

Bornscheuer, U.T., Altenbuchner, J., and Meyer, J.H., Directed evolution of an esterase for the stereoselective resolution of a key intermediate in the synthesis of epothilones, *Biotechnol.Bioeng.* 58, 554-559, 1998

Nakayasu, H., Crystal violet as an indicator dye for nonequilibrium pH gradient

Congo Red

Mera, S.L., and Davies, J.D., Differential Congo red staining: the effects of pH, non-aqueous solvents and the substrate, *Histochem.J.* 16, 195-210, 1984

Schneider, R.L., Chung, E.B., Leffall, L.D., Jr., and Syphax, B., Delineation of the canine gastric antrum with pH probe and dye indicator, *J.Natl. Med.Assoc.* 63, 202-204, 1971

Xu, S., Kramer, M., and Haag, R., pH-Responsive dendritic core-shell architectures as amphiphilic nanocarriers for polar drugs, *J.Drug Target.* 14, 367-374, 2006

Sekine, H., Iijima, K., Koike, T., *et al.*, Regional differences in the recovery of gastric acid secretion after *Helicobacter pylori* eradication: evaluations with Congo red chromoendoscopy, *Gastrointest.Endosc.* 64, 686-690, 2006

Parrish, N.M., Ko, C.G., Dick, J.D., *et al.*, Growth, Congo Red agar colony morphotypes and antibiotic susceptibility testing of *Mycobacterium avium* subspecies paratuberculosis, *Clin.Med.Res.* 2, 107-114, 2004

Thymol Blue

Balderas-Hernandez, P., Rojas-Hernandez, A., Galvan, M., and Ramirez-Silva, M.T., Spectrophotometric study of the system Hg(II)-thymol blue-H_2O and its evidence through electrochemical means, *Spectrochim.Acta A Mol.Biomol.Spectrosc.* 60, 569-577, 2004

Nakamura, N. and Amao, Y., Optical sensor for carbon dioxide combining colorimetric change of a pH indicator and a reference luminescent dye, *Anal.Bioanal.Chem.* 376, 642-646, 2003

Dowding, C.E., Borda, M.J., Fey, M.V., and Sparks, D.L., A new method for gaining insight into the chemistry of drying mineral surfaces using ATR-FTIR, *J.Colloid Interface Sci.* 292, 148-151, 2005

Saika, P.M., Bora, M., and Dutta, R.K., Acid-base equilibrium of anionic dyes partially bound to micelles of nonionic surfactants, *J.Colloid Interface Sci.* 285, 382-387, 2005

Bromophenol Blue

Govindarajan, R., Chatterjee, K., Gatlin, L., *et al.*, Impact of freeze-drying on ionization of sulfonphthalein probe molecule in trehalose-citrate systems, *J.Pharm.Sci.* 95, 1498-1510, 2006

Suzuki, Y., Theoretical analysis concerning the characterization of a dye-binding method for determining serum protein based on protein error of pH indicator: effect of buffer concentration of the color reagent on the color development, *Anal.Sci.* 21, 83-88, 2005

Li, J., Chatterjee, K., Medek, A. *et al.*, Acid-base characteristics of bromophenol blue-citrate buffer system in the amorphous state, *J.Pharm. Sci.* 93, 697-712, 2004

Shao, P.G. and Bailey, L.C., Porcine insulin biodegradable polyester microspheres: stability and in vitro release characteristics, *Pharm.Dev. Technol.* 5, 1-9, 2000

Koren, R. and Hammes, G.G., A kinetic study of protein-protein interactions, *Biochemistry* 15, 1165-1171, 1976

Methyl Red

Katsuda, T., Ooshima, H., Azuma, M., and Kato, J., New detection method for hydrogen gas for screening hydrogen-producing microorganisms using water-soluble Wilkinson's catalyst derivative, *J.Biosci.Bioeng.* 102, 220-226, 2006

Benedict, J.B., Cohen, D.E., Lovell, S., *et al.,* What is syncrystallization? States of the pH indicator methyl red in crystals of phthalic acid, *J.Am.Chem.Soc.* 128, 5548-5559, 2006

Pelechova, J., Petrova, L., Ujcova, E. and Martinkova, L., Selection of a hyperproducing strain of *Aspergillus niger* for biosynthesis of citric acid on unusual carbon substrates, *Folia Microbiol.* 35, 138-142, 1990

Phenol Red

Govindarajan, R., Chatterjee, K., Gatlin, L., *et al.,* Impact of freeze-drying on ionization of sulfonphthalein probe molecule in trehalose-citrate systems, *J.Pharm.Sci.* 95, 1498-1510, 2006

Chu, A., Morris, K., Greenberg, R. and Zhou, D. Stimulus induced pH changes in retinal implants, *Conf.Proc.IEEE Eng.Med.Biol.Soc.* 6, 4160-4162, 2004

Deng, C. and Chen ,R.R., A pH-sensitive assy for galactosyltransferase, *Anal.Biochem.* 330, 219-226, 2004

Still, K., Reading, L., and Scutt, A., Effects of phenol red on CRU-f differentiation and formation, *Calcif.Tissue Int.* 73, 173-179, 2003

Hur, O., Nik, D., Casino, P., and Dunn, M.F., Proton transfers in the β-reaction catalyzed by tryptophan synthase, *Biochemistry* 41, 9991-10001, 2002

Oh, K.H., Nam, S.H., and Kim, H.S., Directed evolution of N-carbamyl-D-amino acid amidohydrolase for simultaneous improvement of oxidative and thermal stability, *Biotechnol.Prog.* 18, 413-417, 2002

Girard, P., Jordan, M., Tsao, M. and Wurm, F.M., Small-scale bioreactor system for process development and optimization, *Biochem.Eng.J.* 7, 117-119, 2001

Jarrett, J.T, Choi, C.Y. and Matthews, R.G., Changes in protonation associated with substrate binding and Cob(I)alamin formation in cobalamin-dependent methionine synthase, *Biochemistry* 36, 15739-15748, 1997

Ahmed, Z. and Connor, J.A., Intracellular pH changes induced by calcium influx during electrical activity in molluscan neurons, *J.Gen.Physiol.* 75, 403-426, 1980

Connor, J.A. and Ahmed, Z., Diffusion of ions and indicator dyes in neural cytoplasm, *Cell.Mol.Neurobiol.* 4, 53-66, 1984

Clark, A.M. and Perrin, D.D., A re-investigation of the question of activators of carbonic anhydrase, *Biochem.J.* 48, 495-502, 1951

Nile Blue

Lie, C.-W., Shulok, J.R., Wong, Y.-K., *et al.,* Photosensitization, uptake, and retention of phenoxazine Nile Blue derivatives in human bladder carcinoma cells, *Cancer Res.* 51, 1109-1114, 1991

SPECIFIC GRAVITY OF LIQUIDS

Specific gravity and density are not identical although the abbreviation "d" is frequently used to designate specific gravity. Specific gravity and density are numerically equal when water is the standard of reference for specific gravity and g/ml is the unit designation for density.

The numerical value for specific gravity is usually written with a superscript (indicating the temperature of the liquid) and a subscript (indicating the temperature of the liquid to which it is referred), thus $d_4^{25} 1.724$ or sp. gr. 1.724_4^{25}. When

these are omitted in this table, the specific gravity at 20°C referred to water at 4°C is intended. When the standard of reference is not specified, for liquids and solids, it is understood to be water.

Water is most dense at 4°C, hence the sp. gr. of a liquid with reference to water will be higher at all other temperatures than it is at 4°C. To obtain the sp. gr. with reference to water at the same temperature as the liquid, multiply the sp. gr. of $\frac{15}{4}$, $\frac{20}{4}$, or $\frac{25}{4}$ by 1.001, 1.002, or 1.003, respectively.

(Items listed in the order of increasing specific gravities)

Liquid	Specific Gravity	Liquid	Specific Gravity
n-Pentane	0.626	Propionitrile	0.783
n-Hexane	0.660	Acetonitrile	0.783_{25}^{25}
1-Butyne	0.668_4^0	n-Butyl ether	0.784_4^0
Dimethylamine	0.680_4^0	Isopropyl alcohol	0.785
Isoprene	0.681	Isovaleronitrile	0.788
n-Heptane	0.684	Butyl alcohol, tertiary	0.789
2-Butyne	0.688^{25}	Methanol, anhydrous	0.791
1,5-Hexadiene	0.688	Acetone	0.792
Isopropylamine	0.694_4^{15}	Isobutyraldehyde	0.794
Butylamine, tertiary	0.696	Acrylonitrile	0.797
Triethylboron	0.696^{23}	Ethyl alcohol, anhydrous	$0.798_{15.56}^{15.56}$
Ethylamine	0.706_4^0	Valeronitrile	0.801
Diethylamine	0.711_4^{18}	Isovaleraldehyde	0.803_4^{17}
2,4-Hexadiene	0.711	n-Propyl alcohol	0.804
Diethyl ether	0.713	Allyl ether	0.805_0^{18}
n-Nonane	0.716	Ethyl methyl ketone	0.805
Triethylamine	0.723_4^{25}	Isobutyl alcohol	0.806_4^{15}
Butylamine, secondary	0.724	Propionaldehyde	0.807
Isopropyl ether	0.726	Butyl alcohol, secondary	0.808
Ethyl methyl ether	0.726_4^0	Amyl alcohol, tertiary	0.809
2,4-Heptadiene	$0.733_4^{21.5}$	Methyl propyl ketone	0.809
Isobutylamine	0.724_4^{25}	n-Butyl alcohol	0.810
Propyl ether	0.736	Cycloheptane	0.810
Methyl propyl ether	0.738	Cyclohexene	0.810
Dipropylamine	0.738	Isoamyl alcohol	0.813_4^{15}
Ethyl n-propyl ether	0.739	Ethyl propyl ketone	$0.813_4^{21.8}$
n-Butylamine	0.740	pri-n-Amyl alcohol	0.814
Undecane	0.741	Heptyl ether	0.815_4^0
N,N-Dimethylamylamine	0.743	Diethyl ketone	0.816_4^{19}
Ethyl isopropyl ether	0.745_4^0	Ethyl alcohol, 95 per cent	$0.816_{15.56}^{15.56}$
Isoamylamine	0.751	Butyraldehyde	0.817
Cyclopentane	0.751	Dipropyl ketone	0.817
Butyl ethyl ether	0.752	Ethyl butyl ketone	0.818
Isohexylamine	0.758_4^{25}	n-Hexyl methyl ketone	0.819
Isobutyl ether	0.761_4^{15}	1-Hexanol	0.819
Allylamine	0.761	3-Hexanol	0.819
n-Amylamine	0.761	Isoamyl alcohol, secondary	0.819
Butyl methyl ether	0.764_4^0	Pinacolin	0.821_4^0
Allyl ether ether	0.765	Amyl methyl ketone	0.822_4^{15}
n-Dodecane	0.766_4^0	Cycloheptene	0.823
Dibutylamine	0.767	n-Octyl alcohol	0.825
n-Butyl ether	0.769_{20}^{20}	2-Undecanone	0.826
Cyclopentene	0.774	Light Liquid Petrolatum	$0.828-0.880_{25}^{25}$
n-Heptylamine	0.777	2-Hexanol	0.829_4^0
Cyclohexane	0.778	n-Decyl alcohol	0.829
n-Octylamine	0.779_{20}^{20}	n-Undecylaldehyde	0.830
Isoamyl ether	0.781_{15}^{15}	Butyl methyl ketone	0.830_4^0

SPECIFIC GRAVITY OF LIQUIDS (Continued)

Liquid	Specific Gravity	Liquid	Specific Gravity
1-Undecanol	0.833^{23}_{4}	Cubeb oil	$0.905-0.925^{25}_{25}$
Acrolein	0.841	Eucalyptus oil	$0.905-0.925^{25}_{25}$
Orange oil	$0.842-0.846^{25}_{25}$	Diethyl Carbitol	0.907
Bitter orange oil	$0.845-0.851^{25}_{25}$	Styrene	0.907
Butyl chloride, tertiary	0.847^{15}_{4}	Undecylenic acid	0.908^{25}_{4}
Rose oil	$0.848-0.863^{30}_{15}$	Olive oil	$0.910-0.915^{25}_{25}$
Lemon oil	$0.849-0855^{25}_{25}$	Expressed almond oil	$0.910-0.915^{25}_{25}$
Amyl ether ketone	0.850^{0}_{4}	Persic oil	$0.910-0.923^{25}_{25}$
n-Amyl nitrite	0.853	Thyme oil	$0.910-0.935^{25}_{25}$
Rectified turpentine oil	$0.853-0.862^{25}_{25}$	n-Butyl nitrite	0.911^{0}_{4}
Dwarf pine needle oil	$0.853-0.871^{25}_{25}$	Peanut oil	$0.912-0.920^{25}_{25}$
Allyl alcohol	0.854	Mustard oil	$0.914-0.916^{15}_{15}$
Mesityl oxide	0.854	Corn oil	$0.914-0.921^{25}_{25}$
Myristica oil	$0.854-0.910^{25}_{25}$	Methyl propionate	0.915
p-Cymene	0.857	Glycerin trioleate	0.915
dl-Pinene	0.858	Cottonseed oil	$0.915-0.921^{25}_{25}$
Isopropyl chloride	0.859	Sesame oil	$0.916-0.921^{25}_{25}$
2-Diethylaminoethanol	0.860^{25}_{25}	Spearmint oil	$0.917-0.934^{25}_{25}$
Liquid Petrolatum	$0.860-0.905^{25}_{25}$	Cardamom oil	$0.917-0.947^{25}_{25}$
Piperidine	0.861	Coconut oil	$0.918-0.923^{25}_{25}$
Cumene	0.863	Cod liver oil	$0.918-0.927^{25}_{25}$
Coriander oil	$0.863-0.875^{25}_{25}$	Halibut liver oil	$0.920-0.930^{25}_{25}$
Orange flower oil	$0.863-0.880^{25}_{25}$	Eucalyptol	$0.921-0.923^{25}_{25}$
Phytol	0.864^{0}_{4}	Ethyl format	0.924^{25}_{4}
m-Xylene	0.864	Soya oil	$0.924-0.927^{15}_{15}$
Toluene	0.866	Linseed oil	$0.925-0.935^{25}_{25}$
Ethyl benzene	0.867	Pine oil	$0.927-0.940^{25}_{25}$
m-Cymene	0.870	Methyl acetate	0.928
Isoamyl acetate	0.870^{25}_{4}	Cellosolve	0.930
Isopropyl acetate	0.870	Ionone	$0.933-0.937^{25}_{25}$
Isobutyl nitrite	0.870^{20}_{20}	N,N-Diethylaniline	0.935
Butyl chloride, secondary	0.871	Furan	0.937
Octyl acetate	0.873^{20}_{20}	Allyl chloride	0.938
Isobutyl acetate	0.875	Valeric acid	0.942
Isoamyl nitrite	0.875^{25}_{25}	Castor oil	$0.945-0.965^{25}_{25}$
Bergamot oil	$0.875-0.880^{25}_{25}$	Cyclohexanone	0.948
Lavender oil	$0.875-0.888^{25}_{25}$	Pyrrole	0.948
o-Cymene	0.876	Cyclopentanone	0.948
Benzene	0.879^{15}_{4}	Cyclopentanol	0.949
Amyl acetate	0.879^{20}_{20}	Isobutyric acid	0.949
Geraniol	0.881^{16}_{4}	2-Picoline	0.950^{15}_{4}
n-Amyl chloride	0.883	Chenopodium oil	$0.950-0.980^{25}_{25}$
Isobutyl chloride	0.883^{15}	Myrcia oil	$0.950-0.990^{25}_{25}$
n-Butyl chloride	0.884	Cycloheptanone	0.951
Pine needle oil	$0.884-0.886^{15}_{15}$	Fennel oil	$0.953-0.973^{25}_{25}$
Citronella oil	$0.885-0.912^{25}_{25}$	Dimethylaniline	0.956
2-Dimethylaminoethanol	0.887	4-Picoline	0.957^{15}_{4}
n-Propyl acetate	0.887	3-Picoline	0.961^{15}_{4}
1-Menthol	0.890^{15}_{15}	Indan	0.965
Propyl chloride	0.890^{20}_{20}	Methyl cellosolve	0.966
Isoamyl chloride	0.893	Phenetole	0.967
Rosemary oil	$0.894-0.912^{25}_{25}$	Vitamin K_1	0.967^{25}_{25}
Oleic acid	0.895^{18}_{4}	Tetralin	0.970
Isodurene	0.896^{0}_{4}	Carvacrol	0.976
Peppermint oil	$0.896-0.908^{25}_{25}$	Pyridine	0.978^{25}_{4}
o-Xylene	0.897	Anise oil	$0.978-0.988^{25}_{25}$
Ethyl nitrite	0.900^{15}_{15}	Ethyl urethan	0.981
Caraway oil	$0.900-0.910^{25}_{25}$	Benzylamine	0.983^{19}_{4}
Ethyl acetate	0.902	Benzyl acetone	0.989^{23}_{17}
Linoleic acid	0.903	m-Toluidine	0.989

SPECIFIC GRAVITY OF LIQUIDS (Continued)

Liquid	Specific Gravity	Liquid	Specific Gravity
Carbitol	0.990	Diethanolamine	1.097
Dimethyl glyoxal	0.990^{15}_{15}	Benzyl chloride	1.103^{18}_{4}
Isoamyl benzoate	0.993^{19}_{4}	Aldol	1.103
Paraldehyde	0.994	Acetyl chloride	1.105
Anisole	0.995	Ethyl nitrate	1.105
Isoamyl nitrate	0.996^{22}_{4}	Chlorobenzene	1.107
Morpholine	0.999	Polyethylene Glycol 400	$1.110–1.140^{25}_{25}$
Water	0.9970^{0}_{4}	Cinnamaldehyde	1.112^{15}_{4}
Water	0.9999^{20}_{20}	Benzyl benzoate	1.118^{25}_{4}
Water	$1.0000^{4.08}_{4.08}$	Diethylene glycol	1.118^{20}_{20}
Isobutyl benzoate	1.002^{15}_{4}	Anisaldehyde	1.123
o-Toluidine	1.004	Diethyl phthalate	1.123^{25}_{4}
Indene	1.006	Triethanolamine	1.124
Nicotine	1.009	Polyethylene Glycol 300	$1.124–1.130^{25}_{25}$
Benzonitrile	1.010^{15}_{15}	Furfuryl alcohol	1.130
Hydrazine	1.011^{15}_{4}	Nitromethane	1.130
Ethanolamine	1.018	Formamide	1.134
Pimenta oil	$1.018–1.048^{25}_{25}$	Ethyl salicylate	1.136^{15}_{4}
Aniline	1.022	m-Nitrotoluene	1.157
Sparteine	1.023	Ethyl chloroacetate	1.159
Phenylethyl alcohol	1.024^{15}_{4}	Furfural	1.160
Dibenzylamine	1.026	Glycerol triacetate	1.161
Chloroacetal	1.026^{16}_{4}	o-Nitrotoluene	1.163
Acetophenone	1.033^{15}_{15}	Salicylaldehyde	1.167
1,4-Dioxane	1.034	Methyl salicylate	1.184^{25}_{25}
m-Cresol	1.034	Dimethyl phthalate	1.189^{25}_{25}
Glycerol tributyrate	1.035	Nitrobenzene	1.205^{15}_{4}
Propylene glycol	1.036^{25}_{4}	Isoamyl bromide	1.210^{15}_{4}
Phlorol	1.037^{12}	Benzoyl chloride	1.219^{15}_{15}
Butyl nitrate, secondary	1.038^{0}_{4}	sym.-Dichloroethyl ether	1.222
Bitter almond oil	$1.038–1.060^{25}_{25}$	Butyl bromide, tertiary	1.222
Clove oil	$1.038–1.060^{25}_{25}$	Formic acid	1.226^{15}_{4}
Ethyl succinate	1.040	Methyl chloroacetate	1.238^{20}_{20}
Benzyl ether	1.043	Amyl bromide	1.246^{0}_{4}
Benzyl alcohol	1.045^{25}_{4}	Lactic acid (dl)	1.249^{15}_{4}
Cinnamon oil	$1.045–1.063^{25}_{25}$	.uns.-Ethylene dichloride	1.252
o-Cresol	1.047	Butyl bromide, secondary	1.258
n-Butyl phthalate	1.047	Glycerol	1.260
n-Butyl nitrate	1.048^{0}_{4}	Carbon disulfide	1.263
Acetic acid (glacial)	1.049^{25}_{25}	n-Butyl bromide	1.269^{25}_{4}
Benzaldehyde	1.050^{15}_{4}	Isobutyl bromide	1.272^{15}_{4}
Ethyl benzoate	1.051^{15}_{4}	sym.-Dichloroethylene	1.291^{15}_{4}
Ethyl malonate	1.055	o-Dichlorobenzene	1.307^{20}_{20}
Benzyl acetate	1.057^{16}_{4}	Isopropyl bromide	1.310
Allyl benzoate	1.058^{15}_{15}	Ethylsulfuric acid	1.316^{17}_{4}
n-Propyl nitrate	1.058	Methylene chloride	1.335^{15}_{4}
Succinaldehyde	1.064	n-Propyl bromide	1.353
Thiophene	1.064	m-Xylyl bromide	1.371^{23}_{4}
Methyl carbonate	1.065^{17}_{4}	Benzotrichloride	1.380^{15}_{4}
Eugenol	1.066	Ethyl trichloroacetate	1.383
p-Chlorotoluene	1.070	Allyl bromide	1.398
m-Chlorotoluene	1.072	Ethyl bromide	1.430
Diethyl maleate	1.074^{15}_{15}	Benzyl bromide	1.438^{22}_{0}
Benzofuran	1.078^{15}_{15}	Hydrogen peroxide, anhydrous	1.465^{0}_{4}
o-Chlorotoluene	1.082	Trichloroethylene	1.465
Acetic anhydride	1.087^{15}_{4}	Chloroform	1.498^{15}_{15}
o-Anisidine	1.092	Bromobenzene	1.499^{15}_{15}
Methyl benzoate	1.094^{15}_{4}	Chloral	1.512
Quinoline	1.095	Trichloroethanol	1.550^{20}_{20}
m-Anisidine	1.096	Dichloroacetic acid	1.563

SPECIFIC GRAVITY OF LIQUIDS (Continued)

Liquid	Specific Gravity	Liquid	Specific Gravity
Benzoyl bromide	1.570_4^{15}	Ethyl iodide	1.933
Glycerophosphoric acid	1.590_4^{19}	Ethylene bromide	2.170_4^{25}
Nitroglycerin	1.592_4^{25}	Ethylene dibromide	2.172_{25}^{25}
Carbon tetrachloride	1.595	Methyl iodide	2.251
Tetrachloroethane	1.600	Bromal	2.300_4^{15}
Tetrachloroethylene	1.631_4^{15}	Methylene bromide	2.495
Chloropicrin	1.651	Bromoform	2.890
Diphosgene	1.653_4^{14}	Tetrabromoethane	2.964
Thionyl chloride	$1.655^{10.4}$	Methylene iodide	3.325
Acetyl bromide	1.663_4^{16}	Mercury	13.546
Isopropyl iodide	1.703		

Reprinted from *The Merck Index* (1960), 7th ed., Merck and Co., Rahway, N.J., pp. 1532–1535, with permission of the copyright owner.

VISCOSITY AND DENSITY TABLES

SUCROSE IN WATER, 0.0°C

Sucrose %	Density[a] g/ml	Viscosity[b] cP	Sucrose %	Density[a] g/ml	Viscosity[b] cP	Sucrose %	Density[a] g/ml	Viscosity[b] cP
0	1.0004	1.780	24	1.1037	4.646	48	1.2269	34.57
1	1.0043	1.830	25	1.1085	4.912	49	1.2324	39.23
2	1.0082	1.884	26	1.1133	5.202	50	1.2380	44.74
3	1.0122	1.941	27	1.1181	5.519	51	1.2436	51.29
4	1.0162	2.002	28	1.1229	5.866	52	1.2493	59.11
5	1.0203	2.066	29	1.1278	6.246	53	1.2550	68.52
6	1.0244	2.135	30	1.1327	6.665	54	1.2607	79.92
7	1.0285	2.208	31	1.1376	7.126	55	1.2665	93.85
8	1.0326	2.286	32	1.1426	7.635	56	1.2723	111.0
9	1.0368	2.369	33	1.1476	8.201	57	1.2781	132.3
10	1.0411	2.458	34	1.1527	8.829	58	1.2840	158.9
11	1.0453	2.552	35	1.1578	9.530	59	1.2899	192.6
12	1.0496	2.653	36	1.1629	10.31	60	1.2958	235.7
13	1.0539	2.761	37	1.1680	11.20	61	1.3018	291.4
14	1.0583	2.877	38	1.1732	12.19	62	1.3078	364.2
15	1.0627	3.001	39	1.1784	13.31	63	1.3138	460.6
16	1.0671	3.134	40	1.1836	14.58	64	1.3199	589.9
17	1.0716	3.277	41	1.1889	16.03	65	1.3260	766.0
18	1.0760	3.430	42	1.1942	17.69	66	1.3321	1010.
19	1.0806	3.596	43	1.1996	19.59	67	1.3383	1352.
20	1.0852	3.774	44	1.2050	21.78	68	1.3445	1842.
21	1.0898	3.967	45	1.2104	24.31	69	1.3507	2556.
22	1.0944	4.175	46	1.2159	27.24	70	1.3570	3621.
23	1.0991	4.401	47	1.2213	30.66			

SUCROSE IN WATER, 5.0°C

Sucrosé %	Density[a] g/ml	Viscosity[b] cP	Sucrosé %	Density[a] g/ml	Viscosity[b] cP	Sucrosé %	Density[a] g/ml	Viscosity[b] cP
0	1.0004	1.516	24	1.1027	3.831	48	1.2250	25.97
1	1.0043	1.558	25	1.1074	4.042	49	1.2306	29.28
2	1.0082	1.603	26	1.1122	4.272	50	1.2361	33.16
3	1.0121	1.650	27	1.1169	4.523	51	1.2417	37.73
4	1.0161	1.700	28	1.1218	4.796	52	1.2474	43.16
5	1.0201	1.753	29	1.1266	5.094	53	1.2530	49.62
6	1.0241	1.809	30	1.1315	5.422	54	1.2587	57.39
7	1.0282	1.869	31	1.1364	5.781	55	1.2645	66.79
8	1.0323	1.933	32	1.1413	6.177	56	1.2702	78.24
9	1.0365	2.001	33	1.1463	6.614	57	1.2760	92.30
10	1.0406	2.073	34	1.1513	7.099	58	1.2819	109.7
11	1.0448	2.150	35	1.1563	7.637	59	1.2877	131.4
12	1.0491	2.232	36	1.1614	8.236	60	1.2936	158.9
13	1.0534	2.319	37	1.1665	8.905	61	1.2996	194.0
14	1.0577	2.413	38	1.1717	9.656	62	1.3056	239.1
15	1.0620	2.513	39	1.1768	10.50	63	1.3116	297.9
16	1.0664	2.621	40	1.1820	11.45	64	1.3176	375.6
17	1.0708	2.736	41	1.1873	12.53	65	1.3237	479.4
18	1.0753	2.859	42	1.1926	13.76	66	1.3298	620.3
19	1.0798	2.992	43	1.1979	15.16	67	1.3360	814.3
20	1.0843	3.135	44	1.2033	16.76	68	1.3422	1086.
21	1.0889	3.290	45	1.2087	18.60	69	1.3484	1473.
22	1.0934	3.456	46	1.2140	20.73	70	1.3546	2034.
23	1.0981	3.636	47	1.2195	23.18			

VISCOSITY AND DENSITY TABLES (Continued)

SUCROSE IN WATER, 10.0°C

Sucrosé %	Density[a] g/ml	Viscosity[b] cP	Sucrosé %	Density[a] g/ml	Viscosity[b] cP	Sucrosé %	Density[a] g/ml	Viscosity[b] cP
0	1.0002	1.308	24	1.1016	3.206	48	1.2231	19.96
1	1.0040	1.343	25	1.1062	3.377	49	1.2286	22.37
2	1.0079	1.380	26	1.1109	3.562	50	1.2341	25.17
3	1.0118	1.420	27	1.1157	3.763	51	1.2397	28.45
4	1.0157	1.462	28	1.1204	3.982	52	1.2453	32.32
5	1.0196	1.506	29	1.1252	4.220	53	1.2510	36.89
6	1.0236	1.553	30	1.1300	4.481	54	1.2566	42.34
7	1.0277	1.603	31	1.1349	4.767	55	1.2623	48.87
8	1.0317	1.655	32	1.1398	5.080	56	1.2681	56.75
9	1.0358	1.711	33	1.1448	5.424	57	1.2739	66.35
10	1.0400	1.771	34	1.1498	5.805	58	1.2797	78.11
11	1.0442	1.835	35	1.1548	6.225	59	1.2855	92.65
12	1.0484	1.902	36	1.1598	6.692	60	1.2914	110.8
13	1.0526	1.974	37	1.1649	7.211	61	1.2973	133.6
14	1.0569	2.051	38	1.1700	7.790	62	1.3033	162.7
15	1.0612	2.134	39	1.1752	8.438	63	1.3093	200.0
16	1.0655	2.222	40	1.1803	9.167	64	1.3153	248.6
17	1.0699	2.316	41	1.1856	9.988	65	1.3214	312.5
18	1.0743	2.417	42	1.1908	10.92	66	1.3275	397.7
19	1.0788	2.525	43	1.1961	11.97	67	1.3336	512.9
20	1.0833	2.642	44	1.2014	13.17	68	1.3398	671.1
21	1.0878	2.767	45	1.2068	14.54	69	1.3460	891.7
22	1.0924	2.903	46	1.2122	16.11	70	1.3522	1205.
23	1.0969	3.049	47	1.2176	17.92			

SUCROSE IN WATER, 15.0°C

Sucrosé %	Density[a] g/ml	Viscosity[b] cP	Sucrosé %	Density[a] g/ml	Viscosity[b] cP	Sucrosé %	Density[a] g/ml	Viscosity[b] cP
0	0.9996	1.140	24	1.1002	2.719	48	1.2211	15.65
1	1.0034	1.170	25	1.1048	2.859	49	1.2265	17.44
2	1.0073	1.202	26	1.1095	3.010	50	1.2320	19.52
3	1.0111	1.235	27	1.1142	3.174	51	1.2376	21.93
4	1.0150	1.271	28	1.1189	3.352	52	1.2432	24.75
5	1.0189	1.308	29	1.1237	3.546	53	1.2488	28.06
6	1.0229	1.348	30	1.1285	3.757	54	1.2544	31.98
7	1.0269	1.390	31	1.1334	3.987	55	1.2601	36.64
8	1.0309	1.434	32	1.1382	4.239	56	1.2658	42.22
9	1.0350	1.481	33	1.1432	4.515	57	1.2716	48.95
10	1.0391	1.531	34	1.1481	4.818	58	1.2774	57.12
11	1.0432	1.584	35	1.1531	5.153	59	1.2832	67.12
12	1.0474	1.640	36	1.1581	5.522	60	1.2891	79.48
13	1.0516	1.701	37	1.1632	5.932	61	1.2950	94.85
14	1.0558	1.765	38	1.1682	6.386	62	1.3009	114.2
15	1.0601	1.833	39	1.1734	6.894	63	1.3069	138.7
16	1.0644	1.906	40	1.1785	7.461	64	1.3129	170.2
17	1.0688	1.985	41	1.1837	8.097	65	1.3189	211.0
18	1.0732	2.068	42	1.1889	8.813	66	1.3250	264.6
19	1.0776	2.158	43	1.1942	9.621	67	1.3311	336.0
20	1.0820	2.255	44	1.1995	10.54	68	1.3373	432.2
21	1.0865	2.358	45	1.2048	11.58	69	1.3434	563.8
22	1.0910	2.470	46	1.2102	12.77	70	1.3497	746.7
23	1.0956	2.590	47	1.2156	14.13			

VISCOSITY AND DENSITY TABLES (Continued)

SUCROSE IN WATER, 20.0°C

Sucrosé %	Densityª g/ml	Viscosityᵇ cP	Sucrosé %	Densityª g/ml	Viscosityᵇ cP	Sucrosé %	Densityª g/ml	Viscosityᵇ cP
0	0.9988	1.004	24	1.0987	2.333	48	1.2189	12.50
1	1.0026	1.030	25	1.1033	2.449	49	1.2244	13.86
2	1.0064	1.057	26	1.1079	2.575	50	1.2299	15.42
3	1.0102	1.086	27	1.1126	2.710	51	1.2354	17.23
4	1.0140	1.116	28	1.1173	2.857	52	1.2409	19.33
5	1.0179	1.148	29	1.1220	3.016	53	1.2465	21.79
6	1.0219	1.181	30	1.1268	3.189	54	1.2522	24.67
7	1.0258	1.217	31	1.1316	3.378	55	1.2578	28.07
8	1.0298	1.255	32	1.1365	3.583	56	1.2635	32.11
9	1.0339	1.295	33	1.1414	3.808	57	1.2693	36.96
10	1.0380	1.337	34	1.1463	4.053	58	1.2750	42.78
11	1.0421	1.382	35	1.1513	4.323	59	1.2808	49.85
12	1.0462	1.430	36	1.1563	4.621	60	1.2867	58.50
13	1.0504	1.480	37	1.1613	4.948	61	1.2926	69.15
14	1.0546	1.535	38	1.1663	5.311	62	1.2985	82.39
15	1.0588	1.592	39	1.1714	5.714	63	1.3044	99.01
16	1.0631	1.654	40	1.1766	6.163	64	1.3104	120.1
17	1.0674	1.720	41	1.1817	6.664	65	1.3164	147.0
18	1.0718	1.790	42	1.1870	7.226	66	1.3225	182.0
19	1.0762	1.865	43	1.1922	7.857	67	1.3286	227.8
20	1.0806	1.946	44	1.1975	8.570	68	1.3347	288.5
21	1.0851	2.032	45	1.2028	9.376	69	1.3408	370.2
22	1.0896	2.125	46	1.2081	10.29	70	1.3470	481.8
23	1.0941	2.225	47	1.2135	11.33			

SUCROSE IN WATER, 25.0° C

Sucrosé %	Densityª g/ml	Viscosityᵇ cP	Sucrosé %	Densityª g/ml	Viscosityᵇ cP	Sucrosé %	Densityª g/ml	Viscosityᵇ cP
0	0.9977	0.8913	24	1.0970	2.023	48	1.2167	10.14
1	1.0014	0.9139	25	1.1016	2.121	49	1.2221	11.19
2	1.0052	0.9376	26	1.1062	2.226	50	1.2276	12.39
3	1.0090	0.9625	27	1.1108	2.339	51	1.2331	13.77
4	1.0128	0.9886	28	1.1155	2.462	52	1.2386	15.37
5	1.0167	1.016	29	1.1202	2.595	53	1.2442	17.22
6	1.0206	1.045	30	1.1250	2.739	54	1.2498	19.39
7	1.0246	1.076	31	1.1298	2.895	55	1.2554	21.93
8	1.0285	1.108	32	1.1346	3.064	56	1.2611	24.92
9	1.0325	1.142	33	1.1395	3.249	57	1.2668	28.48
10	1.0366	1.179	34	1.1444	3.451	58	1.2726	32.73
11	1.0407	1.217	35	1.1493	3.672	59	1.2784	37.85
12	1.0448	1.258	36	1.1543	3.914	60	1.2842	44.04
13	1.0489	1.301	37	1.1593	4.181	61	1.2901	51.61
14	1.0531	1.347	38	1.1643	4.475	62	1.2959	60.93
15	1.0574	1.396	39	1.1694	4.799	63	1.3019	72.50
16	1.0616	1.449	40	1.1745	5.160	64	1.3078	86.99
17	1.0659	1.505	41	1.1797	5.560	65	1.3138	105.3
18	1.0702	1.564	42	1.1848	6.008	66	1.3199	128.8
19	1.0746	1.628	43	1.1901	6.509	67	1.3259	159.1
20	1.0790	1.696	44	1.1953	7.072	68	1.3320	198.8
21	1.0835	1.770	45	1.2006	7.706	69	1.3382	251.4
22	1.0879	1.848	46	1.2059	8.423	70	1.3444	322.0
23	1.0924	1.933	47	1.2113	9.236			

VISCOSITY AND DENSITY TABLES (Continued)

SUCROSE IN WATER, 30.0°C

Sucrosé %	Density[a] g/ml	Viscosity[b] cP	Sucrosé %	Density[a] g/ml	Viscosity[b] cP	Sucrosé %	Density[a] g/ml	Viscosity[b] cP
0	0.9963	0.7978	24	1.0951	1.771	48	1.2144	8.344
1	1.0000	0.8176	25	1.0997	1.854	49	1.2198	9.168
2	1.0038	0.8384	26	1.1043	1.943	50	1.2252	10.10
3	1.0075	0.8601	27	1.1089	2.039	51	1.2307	11.18
4	1.0113	0.8830	28	1.1136	2.143	52	1.2362	12.42
5	1.0152	0.9069	29	1.1183	2.255	53	1.2418	13.84
6	1.0191	0.9322	30	1.1230	2.376	54	1.2474	15.50
7	1.0230	0.9588	31	1.1278	2.506	55	1.2530	17.43
8	1.0270	0.9868	32	1.1326	2.648	56	1.2586	19.69
9	1.0310	1.016	33	1.1374	2.802	57	1.2643	22.36
10	1.0350	1.048	34	1.1423	2.970	58	1.2701	25.52
11	1.0391	1.081	35	1.1472	3.153	59	1.2758	29.30
12	1.0432	1.116	36	1.1522	3.353	60	1.2816	33.84
13	1.0473	1.154	37	1.1572	3.572	61	1.2875	39.34
14	1.0515	1.193	38	1.1622	3.813	62	1.2933	46.05
15	1.0557	1.235	39	1.1672	4.079	63	1.2992	54.30
16	1.0599	1.280	40	1.1723	4.372	64	1.3052	64.53
17	1.0642	1.328	41	1.1775	4.697	65	1.3112	77.35
18	1.0685	1.380	42	1.1826	5.058	66	1.3172	93.54
19	1.0728	1.434	43	1.1878	5.461	67	1.3232	114.2
20	1.0772	1.493	44	1.1931	5.912	68	1.3293	140.9
21	1.0816	1.555	45	1.1983	6.418	69	1.3354	175.8
22	1.0861	1.622	46	1.2036	6.988	70	1.3416	222.0
23	1.0906	1.694	47	1.2090	7.632			

Compiled by Norman G. Anderson based on equations developed by E. J. Barber in *J. Nat. Cancer Inst. Monograph*, **21**, 219 (**1966**).

[a] Original data were stated to a precision of about 1 part in 10,000. Maximum deviation from original data is 7 parts in 10,000.

[b] Precision of original data was between 1 part in 1,000 and 1 part in 10,000. Maximum deviation from original data is 4 parts in 1,000 in the range covered in this set of tables.

A LISTING OF LOG P VALUES, WATER SOLUBILITY, AND MOLECULAR WEIGHT FOR SOME SELECTED CHEMICALS[a]

Compound	M.W.	Log P[b]	Water Solubility(gm/L)[c]
Acetamide	59.07	−1.26	2.25×10^3
Acetic acid	60.05	−0.17	10×10^3
Acetic anhydride	102.09	−0.58	1.2×10^2
Acetoacetic acid	102.1	−0.98	1×10^3
Acetoin	88.11	−0.36	1×10^3
Acetone	58.08	−0.24	1×10^3
Acetophenone	120.15	1.58	6.13
N-Acetylcysteinamide	162.21	−0.29	5.8
N-Acetylcysteine		−0.64	
N-Aceylmethionine		−0.49	
Acetylsalicylic acid	180.16	1.19	4.6
Acridine	179.22	3.40	0.03
Acrolein	56.06	−0.01	2.13×10^2
Acrylamide	71.08	−0.67	6.4×10^2
Adenine	135.13	−0.09	1.0
Adenosine	267.25	−1.05	8.2
Alanine	89.09	−2.96	1.7×10^2
Aldosterone		1.08	
9-Aminoacridine	194.23	2.74	0.02
4-Aminobenzoic acid (p-aminobenzoic acid; PABA)	151.17	1.03	9.89
4-Aminobutyric acid (γ-aminobutyric acid; GABA)	103.12	−3.17	1.3×10^3
6-Aminohexanoic acid (ε-aminocaproic acid)	131.18	−2.95	5.05×10^2
Ammonium picrate	246.14	−1.40	1.6×10^2
Aniline		0.9	
Anisole		2.11	
ANS (1-amino-2-naphthalenesulfonic acid)	222.25	−0.97	2.23
Anthracene		4.45	
Arabinose	150.13	−3.02	1×10^3
Arginine	174.20	−4.20	1.82×10^2
Ascorbic acid	176.13	−1.64	1×10^3
Asparagine	132.12	−3.82	29.4
Aspartic acid	133.10	−3.89	5.0
Barbital (5,5-diethylbarbituric acid)	184.20	0.65	7
Barbituric acid	128.1	−1.47	
Benzamide	121.14	0.64	13.5
Benzamidine	120.16	0.65	27.9
Benzene	78.11	2.13	0.002
Benzoic acid	122.12	1.87	3.4
Betaine	117.15	−4.93	6.11×10^2
Biuret (imidodicarbonic acid)	103.08		1.5
Bromoacetic acid	138.95	0.41	93
2-Bromopropionic acid	152.98	0.92`	29.9
2,3-Butanediol	90.12	−0.36	7.6×10^2
2,3-Butanedione	86.09	−1.34	2×10^2
Butyl urea	116.16	0.41	46.3
3-Butyl hydroxy urea	132.16	0.32	23.5
Cacodylic acid	138.00	0.36	2×10^3
Carbon tetrachloride	153.82	2.83	0.8
Cholesterol	386.67	8.74	0.9
Chloroacetamide	93.51	−0.53	90
Chloroacetic anhydride	170.98	−0.07	68
Chloroacetyl chloride	112.94	−0.22	1.6×10^2
Chloroform	119.38	1.97	8
6-Chloroindole	151.60	3.25	0.1
p-Chloromercuribenzoic acid	357.16	1.48	0.3
Chlorosuccinic acid	152.54	−0.57	1.8×10^2
Cholic acid	405.58	2.02	0.2
Citric acid	192.13	−1.72	5.92×10^2
Congo red	696.68	2.63	1.2×10^2

**A LISTING OF LOG P VALUES, WATER SOLUBILITY, AND MOLECULAR WEIGHT
FOR SOME SELECTED CHEMICALS[a] (Continued)**

Compound	M.W.	Log P[b]	Water Solubility(gm/L)[c]
Corticosterone		1.94	
Cortisone		2.88	
Creatine	132.14	−3.72	13.3
Creatinine	113.12	−1.76	80
Crotonaldehyde (2-butenal)	70.09	0.60	1.8×10^2
Cyanoacetic acid	85.06	−0.76	7.7×10^2
Cyanogen	52.04	0.07	1.2×10^2
Cyanuric acid	129.08	0.61	2
Cyclohexanone		0.81	
Cysteine	121.16	−2.49	1.1×10^2
Cystine	240.30	−5.08	0.2
Cytidine	243.22	−2.51	1.8×10^2
Cytosine	111.10	−1.73	8
Deoxycholic acid	392.58	3.50	0.04
Deoxycorticosterone		2.88	
Dexamethasone		2.01	
Diazomethane	42.04	2.00	2
Dichloromethane		1.2	
Dicumarol	336.30	2.07	0.1
Diethyl ether (ethyl ether; ether)	74.1	0.9	
Diethylsuberate	230.31	3.35	0.7
Diethylsulfone	122.19	−0.59	1.4×10^2
N,N-Diethyl urea	116.2	0.1	4
Dihydroxyacetone	88.11	−0.49	16.2
Diketene	84.08	−0.39	5.3×10^2
Dimethylformamide		−1.04	
Dimethylguanidine	87.13	−0.95	1.6
Dimethylsulfoxide	78.13	−1.35	1×10^3
Dimethylphthalate		1.56	
1,4-Dinitrobenzene		1.47	
2,4-Dinitrophenol		1.55	
EDTA	292.25	−3.86	1
EDTA, sodium salt	360.17	−13.17	1×10^3
Ethanol (ethyl alcohol)	46.07	−0.31	$1 \times 10+3$
N-Hydroxy-1-ethylurea	104.11	−0.10	7
N-Ethylnicotinamide	150.18	0.31	41.2
N-Hydroxyurea	104.11	−0.76	
Estradiol		2.69	
N-Ethylthiourea	104.17	−0.21	24
Ethylurea	88.11	−0.74	26.4
Ethylene glycol	2.07	−1.36	1×10^3
Ethylene oxide	44.05	−0.30	1×10^3
Fluorescein	333.32	3.35	0.05
Fluoroacetone	76.07	−0.39	286
Folic acid	441.41	−2.00	0.002
Formaldehyde	30.03	0.35	400
Formic acid	48.03	−0.54	1×10^3
Galactose	180.16	−2.43	683
Glucose	180.16	−1.88	1.2×10^3
Glutamic acid	147.10	−3.69	8.6
Glutamine	146.15	−3.64	41
Glycerol	92.10	−1.76	1×10^3
Glycine	75.10	−3.21	2.5×10^2
Glyoxal	58.04	−1.66	1×10^3
Glyoxylic acid	74.04	−1.40	1×10^3
Guanidine	59.07	−1.63	1.8
Guanine	151.13	−0.91	2.1
Guanosine	283.25	−1.90	0.7
Hexanal	100.16	1.78	6
Hydroxyproline	131.13	−3.17	395

A LISTING OF LOG P VALUES, WATER SOLUBILITY, AND MOLECULAR WEIGHT
FOR SOME SELECTED CHEMICALS[a] (Continued)

Compound	M.W.	Log P[b]	Water Solubility(gm/L)[c]
Hydroxyurea	76.06	−1.80	224
Imidazole	68.08	−0.08	160
Indole	117.15	2.14	4
Inositol	180.16	−2.08	143
Iodoacetamide	184.96	−0.19	76
Isoleucine	131.18	−1.70	34
Isopropanol	60.10	0.05	1×10^3
Lactic acid	90.08	−0.72	1×10^3
Lactose	342.30	−5.43	195
Leucine	131.18	−1.52	22
Linoleic acid	280.45	7.05	0.00004
Lysine	146.19	−3.05	1×10^3
Maleic anhydride	98.06	1.62	5
Maltose	342.30	−5.43	780
Mannitol	182.17	−3.10	216
Mercaptoacetic acid	92.12	0.09	1×10^3
2-Mercaptobenzoic acid	154.19	2.39	0.7
Methane	16.04	1.09	0.002
Methanol	32.04	−0.77	1×10^3
Methionine	149.21	−1.87	57
Methotrexate	454.45	−1.85	2.6
Methylene blue	319.86	5.85	44
N-Methyl glycine	89.09	−2.78	300
5-Methylindole	131.18	2.68	0.5
Methyl isocyanate	57.05	0.79	29
Methylmalonic acid	118.09	−0.83	680
Methyl methacrylate	86.09	0.80	49
Methylmethane sulfonate	110.13	−0.66	1×10^3
Methyl thiocyanate	73.12	0.73	32
N-Methyl thiourea	119.21	−0.69	240
Methyl urea	74.08	−1.40	100
Naphthalene	128.17	3.29	220
Nicotinic acid	123.11	0.36	18
Ornithine	132.16	−4.22	1×10^3
Orotic acid	156.10	−0.83	2
Oxalic acid	90.06	−2.22	
Oxindole	133.15	1.16	9
Palmitic acid	256.43	7.17	0.0008
Paraldehyde	132.16	0.67	112
Pentobarbital	226.28	2.10	0.7
Phenol	94.11	1.46	83
Phenylalanine	165.19	−1.52	22
Phosgene	98.02	−0.71	475
Proline	115.13	−2.54	131
Prostaglandin E2	352.48	2.82	0.006
Propylamine	59.11	0.48	1×10^3
Propylene oxide	58.08	0.03	595
Pyridine	79.10	0.65	1×10^3
Pyridoxal	203.63	−3.32	500
Pyridoxal-5-phosphate	247.15	0.37	20
Pyridoxine	169.18	−0.77	282
Pyruvic acid	88.06	−1.24	1×10^3
Ribose	150.13	−2.32	
Sarin	140.10	0.72	1×10^3
Serine	105.09	−3.07	425
Sorbic acid	112.13	1.33	2
Sorbitol	182.17	−2.20	3×10^3
Stearic acid	284.49	8.23	0.03
Succinic anhydride	100.07	0.81	24
Succinimide	99.09	−0.85	196

A LISTING OF LOG P VALUES, WATER SOLUBILITY, AND MOLECULAR WEIGHT
FOR SOME SELECTED CHEMICALS[a] (Continued)

Compound	M.W.	Log P[b]	Water Solubility(gm/L)[c]
Sucrose	342.30	−3.70	2.12×10^3
Testosterone	288.43	3.32	0.03
Tetrahydrofuran	72.11	0.46	1×10^3
Threonine	119.12	−2.94	97
Toluene	92.14	2.73	0.5
2,4,6-Trinitrobenzene	257.12	0.23	21
Tryptophan	204.23	−1.06	12
Urea	60.06	−2.11	545
Valine	117.15	−2.26	60

[a] Adapted from *Handbook of Physical Properties of Organic Chemicals*, ed. P.H. Howard and W.M. Meylan, CRC Press, Boca Raton, FL, 1997

[b] $\text{Log P} = \log \frac{[\text{Concentration in 1-octanol}]}{[\text{concentration in water}]}$

See above Howard and Meylan and following for discussion of log P (log of partitioning coefficient for a substance between 1-octanol and water.

[c] Solubility values taken from various literature sources and in some cases are approximations

General references

Chuman, H., Mori, A., and Tanaka, H., Prediction of the 1-octanol/H_2O partition coefficient, Log *P*, by *Ab Initio* calculations: hydrogen-bonding effect of organic solutes on Log *P*, *Analyt.Sci.* 18, 1015-1020, 2002.

Hansch, C. and Leo, A., *Exploring QSAR. Fundamentals and Applications in Chemistry and Biology*, American Chemical Society, Washington, DC, 1995

Uttamsingh, V., Keller, D.A., and Anders, M.W., Acylase I-catalyzed deacetylation of *N*-acetyl-L-cysteine and *S*-Alkyl-*N*-acetyl-L-cysteines, *Chem.Res.Toxicol.* 11, 800-809, 1998

Yalkowsky, S.H. and He, Y., *Handbook of Aqueous Solubility Data*, CRC Press, Boca Raton, Florida, 2003

Halling, P.J., Thermodynamic predictions for biocatalysis in nonconventional media: theory, tests, and recommendations for experimental design and analysis, *Enzyme Microb.Technol.* 16, 178-206, 1994

Abrahams, M.H., Du, C.M., and Platts, J.A., Lipophilicity of the nitrophenols, *J.Org.Chem.* 65, 7114-7718, 2000

Lipinski, C.A., Lombardo, F., Dominy, B.W., and Feeney, P.J., Experimental and computational approaches to estimate solubility and permeability in drug discovery and development settings, *Adv.Drug.Deliv.Rev.* 46, 3-26, 2001

Valko, K., Du, C.M., Bevan, C., Reynolds, D.P., and Abraham, M.H., Rapid method for the estimation of octanol/water partition coefficient (Log P_{oct}) from gradient RP-HPLC retention and a hydrogen bond acidity term ($\Sigma \alpha_2^H$), *Curr.Medicin.Chem.* 8, 1137-1146, 2001

Avdeef, A., Physicochemical profiling (solubility, permeability and charge state), *Curr.Top.Med.Chem.* 1, 277-351, 2001

CHEMICALS COMMONLY USED
IN BIOCHEMISTRY AND MOLECULAR BIOLOGY AND THEIR PROPERTIES

Common Name	Chemical Name	M.W.	Properties and Comment
Acetaldehyde	Acetaldehyde, Ethanal	44.05	Manufacturing intermediate; modification of amino groups; toxic chemical; first product in detoxification of ethanol.

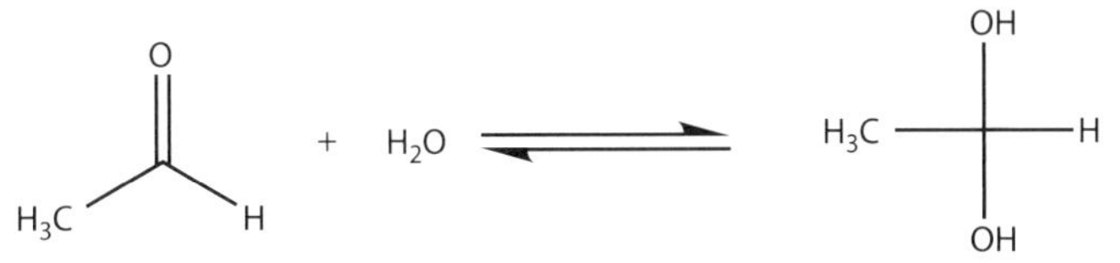

Acetaldehyde   *gem*-diol form (approximately 60%)

Burton, R.M. and Stadtman, E.R., The oxidation of acetaldehyde to acetyl coenzyme A, *J. Biol. Chem.* 202, 873–890, 1953; Gruber, M. and Wesselius, J.C., Nature of the inhibition of yeast carboxylase by acetaldehyde, *Biochim. Biophys. Acta* 57, 171–173, 1962; Holzer, H., da Fonseca-Wollheim, F., Kohlhaw, G., and Woenckhaus, C.W., Active forms of acetaldehyde, pyruvate, and glycolic aldehyde, *Ann. N.Y. Acad. Sci.* 98, 453–465, 1962; Brooks, P.J. and Theruvathu, J.A., DNA adducts from acetaldehyde: implications for alcohol-related carcinogenesis, *Alcohol* 35, 187–193, 2005; Tyulina, O.V., Prokopieva, V.D., Boldyrev, A.A., and Johnson, P., Erthyrocyte and plasma protein modification in alcoholism: a possible role of acetaldehyde, *Biochim. Biophys. Acta* 1762, 558–563, 2006; Pluskota-Karwatka, D., Pawlowicz, A.J., and Kronberg, L., Formation of malonaldehyde-acetaldehyde conjugate adducts in calf thymus DNA, *Chem. Res. Toxicol.* 19, 921–926, 2006.

Common Name	Chemical Name	M.W.	Properties and Comment
Acetic Acid	Acetic Acid, Glacial	60.05	Solvent (particular use in the extraction of collagen from tissue), buffer component (used in urea-acetic acid electrophoresis). Use in endoscopy as mucous-resolving agent.

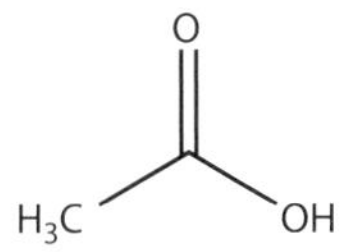

Banfield, A.G., Age changes in the acetic acid-soluble collagen in human skin, *Arch. Pathol.* 68, 680–684, 1959; Steven, F.S. and Tristram, G.R., The denaturation of acetic acid-soluble calf-skin collagen. Changes in optical rotation, viscosity, and susceptibility towards enzymes during serial denaturation in solutions of urea, *Biochem. J.* 85, 207–210, 1962; Neumark, T. and Marot, I., The formation of acetic-acid soluble collagen under polarization and electron microscrope, *Acta Histochem.* 23, 71–79, 1966; Valfleteren, J.R., Sequential two-dimensional and acetic acid/urea/Triton X-100 gel electrophoresis of proteins, *Anal. Biochem.* 177, 388–391, 1989; Smith, B.J., Acetic acid-urea polyacrylamide gel electrophoresis of proteins, *Methods Mol.Biol.* 32, 39–47, 1994; Banfield, W.G., MacKay, C.M., and Brindley, D.C., Quantitative changes in acetic acid-extractable collagen of hamster skin related to anatomical site and age, *Gerontologia* 12, 231–236, 1996; Lian, J.B., Morris, S., Faris, B. et al., The effects of acetic acid and pepsin on the crosslinkages and ultrastructure of corneal collagen, *Biochim. Biophys. Acta.* 328, 193–204, 1973; Canto, M.I., Chromoendoscopy and magnifying endoscopy for Barrett's esophagus, *Clin.Gastroenterol.Hepatol.* 3 (7 Suppl. 1), S12–S15, 2005; Sionkowska, A., Flash photolysis and pulse radiolysis studies on collagen Type I in acetic acid solution, *J. Photochem. Photobiol. B* 84, 38–45, 2006.

CHEMICALS COMMONLY USED IN BIOCHEMISTRY AND MOLECULAR BIOLOGY AND THEIR PROPERTIES (Continued)

Common Name	Chemical Name	M.W.	Properties and Comment
Acetic Anhydride	Acetic Anhydride	102.07	Protein modification (trace labeling of amino groups); modification of amino groups and hydroxyl groups.

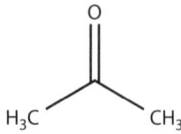

Jencks, W.P., Barley, F., Barnett, R., and Gilchrest, M., The free energy of hydrolysis of acetic anhydride, *J. Am. Chem. Soc.* 88, 4464–4467, 1966; Cromwell, L.D. and Stark, G.D., Determination of the carboxyl termini of proteins with ammonium thiocyanate and acetic anhydride, with direct identification of the thiohydantoins, *Biochemistry* 8, 4735–4740, 1969; Montelaro, R.C. and Rueckert, R.R., Radiolabeling of proteins and viruses *in vitro* by acetylation with radioactive acetic anhydride, *J. Biol. Chem.* 250, 1413–1421, 1975; Valente, A.J. and Walton, K.W., The binding of acetic anhydride- and citraconic anhydride-modified human low-density lipoprotein to mouse peritoneal macrophages. The evidence for separate binding sites, *Biochim. Biophys. Acta* 792, 16–24, 1984; Fojo, A.T., Reuben, P.M., Whitney, P.L., and Awad, W.M., Jr., Effect of glycerol on protein acetylation by acetic anhydride, *Arch. Biochem. Biophys.* 240, 43–50, 1985; Buechler, J.A., Vedvick, T.A., and Taylor, S.S., Differential labeling of the catalytic subunit of cAMP-dependent protein kinase with acetic anhydride: substrate-induced conformational changes, *Biochemistry* 28, 3018–3024, 1989; Baker, G.B., Coutts, R.T., and Holt, A., Derivatization with acetic anhydride: applications to the analysis of biogenic amines and psychiatric drugs by gas chromatography and mass spectrometry, *J. Pharmacol. Toxicol. Methods* 31, 141–148, 1994; Ohta, H., Ruan, F., Hakomori, S., and Igarashi, Y., Quantification of free Sphingosine in cultured cells by acetylation with radioactive acetic anhydride, *Anal. Biochem.* 222, 489–494, 1994; Yadav, S.P., Brew, K., and Puett, D., Holoprotein formation of human chorionic gonadotropin: differential trace labeling with acetic anhydride, *Mol. Endocrinol.* 8, 1547–1558, 1994; Miyazaki, K. and Tsugita, A., C-terminal sequencing method for peptides and proteins by the reaction with a vapor of perfluoric acid in acetic anhydride, *Proteomics* 4, 11–19, 2004.

Acetone	Dimethyl Ketone; 2-propanone	58.08	Solvent, protein purification (acetone powders); rare reaction with amino groups.

La Du, B., Jr. and Greenberg, D.M., The tyrosine oxidation system of liver. I. Extracts of rat liver acetone powder, *J. Biol. Chem.* 190, 245–255, 1951; Korn, E.D. and Payza, A.N., The degradation of heparin by bacterial enzymes. II. Acetone powder extracts, *J. Biol. Chem.* 223, 859–864, 1956; Ohtsuki, K., Taguchi, K., Sato, K., and Kawabata, M., Purification of ginger proteases by DEAE-Sepharose and isoelectric focusing, *Biochim. Biophys. Acta* 1243, 181–184, 1995; Selden, L.A., Kinosian, H.J., Estes, J.E., and Gershman, L.C., Crosslinked dimers with nucleating activity in actin prepared from muscle acetone powder, *Biochemistry* 39, 64–74, 2000; Abadir, W.F., Nakhla, V., and Chong, F., Removal of superglue from the external ear using acetone: case report and literature review, *J. Laryngol. Otol.* 109, 1219–1221, 1995; Jones, A.W., Elimination half-life of acetone in humans: case reports and review of the literature, *J. Anal. Toxicol.* 24, 8–10, 2000; Huang, L.P. and Guo, P., Use of acetone to attain highly active and soluble DNA packaging protein Gp16 of Phi29 for ATPase assay, *Virology* 312, 449–457, 2003; Paska, C., Bogi, K., Szilak, L. et al., Effect of formalin, acetone, and RNAlater fixatives on tissue preservation and different size amplicons by real-time PCR from paraffin-embedded tissues, *Diagn. Mol. Pathol.* 13, 234–240, 2004; Kuksis, A., Ravandi, A., and Schneider, M., Covalent binding of acetone to aminophospholipids *in vitro* and *in vivo*, *Ann. N.Y. Acad. Sci.* 1043, 417–439, 2005; Perera, A., Sokolic, F., Almasy, L. et al., On the evaluation of the Kirkwood–Buff integrals of aqueous acetone mixtures, *J. Chem. Physics* 123, 23503, 2005; Zhou, J., Tao, G., Liu, Q. et al., Equilibrium yields of mono- and di-lauroyl mannoses through lipase-catalyzed condensation in acetone in the presence of molecular sieves, *Biotechnol. Lett.* 28, 395–400, 2006.

Acetonitrile	Ethenenitrile, Methyl Cyanide	41.05	Chromatography solvent, general solvent.

CHEMICALS COMMONLY USED IN BIOCHEMISTRY AND MOLECULAR
BIOLOGY AND THEIR PROPERTIES (Continued)

Common Name	Chemical Name	M.W.	Properties and Comment

Hodgkinson, S.C. and Lowry, P.J., Hydrophobic-interaction chromatography and anion-exchange chromatography in the presence of acetonitrile. A two-step purification method for human prolactin, *Biochem. J.* 199, 619–627, 1981; Wolf-Coporda, A., Plavsic, F., and Vrhovac, B., Determination of biological equivalence of two atenolol preparations, *Int. J. Clin. Pharmacol. Ther. Toxicol.* 25, 567–571, 1987; Fischer, U., Zeitschel, U., and Jakubke, H.D., Chymotrypsin-catalyzed peptide synthesis in an acetonitrile-water-system: studies on the efficiency of nucleophiles, *Biomed. Biochim. Acta* 50, S131–S135, 1991; Haas, R. and Rosenberry, T.L., Protein denaturation by addition and removal of acetonitrile: application to tryptic digestion of acetylcholinesterase, *Anal. Biochem.* 224, 425–427, 1995; Joansson, A., Mosbach, K., and Mansson, M.O., Horse liver alcohol dehydrogenase can accept NADP+ as coenzyme in high concentrations of acetonitrile, *Eur. J. Biochem.* 227, 551–555, 1995; Barbosa, J., Sanz-Nebot, V., and Toro, I., Solvatochromic parameter values and pH in acetonitrile-water mixtures. Optimization of mobile phase for the separation of peptides by high-performance liquid chromatography, *J. Chromatog. A* 725, 249–260, 1996; Barbosa, J., Hernandez-Cassou, S., Sanz-Nebot, V., and Toro, I., Variation of acidity constants of peptides in acetonitrile-water mixtures with solvent composition: effect of preferential salvation, *J. Pept. Res.* 50, 14–24, 1997; Badock, V., Steinhusen, U., Bommert, K., and Otto, A., Prefractionation of protein samples for proteome analysis using reversed-phase high-performance liquid chromatography, *Electrophoresis* 22, 2856–2864, 2001; Yoshida, T., Peptide separation by hydrophilic-interaction chromatography: a review, *J. Biochem. Biophys. Methods* 60, 265–280, 2004: Kamau, P. and Jordan, R.B., Complex formation constants for the aqueous copper(I)-acetonitrile system by a simple general method, *Inorg. Chem.* 40, 3879–3883, 2001; Nagy, P.I. and Erhardt, P.W., Monte Carlo simulations of the solution structure of simple alcohols in water-acetonitrile mixtures, *J. Phys. Chem. B Condens. Matter Mater. Surf. Interfaces Biophys.* 109, 5855–5872, 2005; Kutt, A., Leito, I., Kaljurand, I. et al., A comprehensive self-consistent spectrophotometric acidity scale of neutral Bronstad acids in acetonitrile, *J. Org. Chem.* 71, 2829–2938, 2006.

Acetyl Chloride	Ethanoyl Chloride	78.50	Acetylating agent.

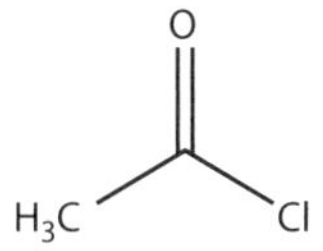

Hallaq, Y., Becker, T.C., Manno, C.S., and Laposata, M., Use of acetyl chloride/methanol for assumed selective methylation of plasma nonesterified fatty acids results in significant methylation of esterified fatty acids, *Lipids* 28, 355–360, 1993; Shenoy, N.R., Shively, J.E., and Bailey, J.M., Studies in C-terminal sequencing: new reagents for the synthesis of peptidylthiohydantoins, *J. Protein Chem.* 12, 195–205, 1993; Bosscher, G., Meetsma, A., and van De Grampel, J.C., Novel organo-substituted cyclophosphazenes via reaction of a monohydro cyclophosphazene and acetyl chloride, *Inorg. Chem.* 35, 6646–6650, 1996; Mo, B., Li, J., and Liang, S., A method for preparation of amino acid thiohydantoins from free amino acids activated by acetyl chloride for development of protein C-terminal sequencing, *Anal. Biochem.* 249, 207–211, 1997; Studer, J., Purdie, N., and Krouse, J.A., Friedel–Crafts acylation as a quality control assay for steroids, *Appl. Spectros.* 57, 791–796, 2003.

Acetylcysteine	*N*-acetyl-L-cysteine	163.2	Mild reducing agent for clinical chemistry (creatine kinase); therapeutic use for aminoacetophen intoxication; some other claimed indications.

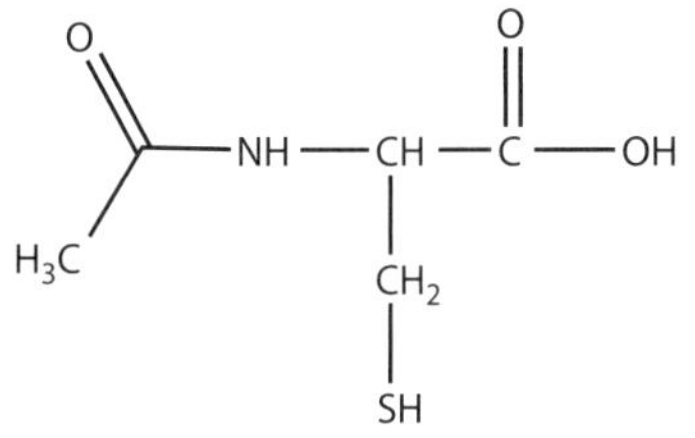

Szasz, G., Gruber, W., and Bernt, E., Creatine kinase in serum. I. Determination of optimum reaction conditions, *Clin. Chem.* 22, 650–656, 1976; Holdiness, M.R., Clinical pharmacokinetics of *N*-acetylcysteine, *Clin. Pharmacokinet.* 20, 123–134, 1991; Kelley, G.S., Clinical applications of *N*-acetylcysteine, *Altern. Med. Rev.* 3, 114–127, 1998; Schumann, G., Bonora, R., Ceriotti, F. et al., IFCC primary reference procedures for the measurement of catalytic activity concentrations of enzymes at 37°C. Part 2. Reference procedure for the measurement of catalytic concentration of creatine kinase, *Clin. Chem. Lab. Med.* 40, 635–642, 2002; Zafarullah, M., Li, W.Q., Sylvester, J., and Ahmad, M., Molecular mechanisms of *N*-acetylcysteine actions, *Cell. Mol. Life Sci.* 60, 6–20, 2003; Marzullo, L., An update of *N*-acetylcysteine treatment for acute aminoacetophen toxicity in children, *Curr. Opin. Pediatr.* 17, 239–245, 2005; Aitio, M.L., *N*-acetylcysteine — passé-partout or much ado about nothing? *Br. J. Clin. Pharmacol.* 61, 5–15, 2006.

CHEMICALS COMMONLY USED IN BIOCHEMISTRY AND MOLECULAR
BIOLOGY AND THEIR PROPERTIES (Continued)

Common Name	Chemical Name	M.W.	Properties and Comment
N-Acetylimidazole	1-acetyl-1*H*-imidazole	110.12	Reagent for modification of tyrosyl residues in proteins.

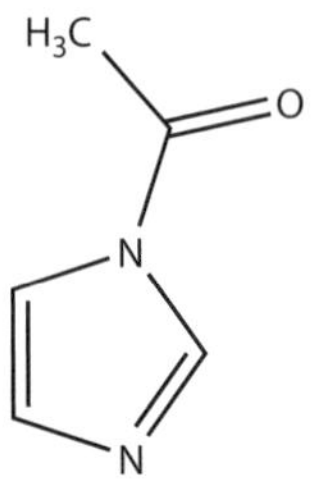

Lundblad, R.L., *Chemical Reagents for Protein Modification*, CRC Press, Boca Raton, FL, 2004; Gorbunoff, M.J., Exposure of tyrosine residues in proteins. 3. The reaction of cyanuric fluoride and *N*-acetylimidazole with ovalbumin, chymotrypsinogen, and trypsinogen, *Biochemistry* 44, 719–725, 1969; Houston, L.L. and Walsh, K.A., The transient inactivation of trypsin by mild acetylation with *N*-acetylimidazole, *Biochemistry* 9, 156–166, 1970; Shifrin, S. and Solis, B.G., Reaction of *N*-acetylimidazole with L-asparaginase, *Mol. Pharmacol.* 8, 561–564, 1972; Ota, Y., Nakamura, H., and Samejima, T., The change of stability and activity of thermolysin by acetylation with *N*-acetylimidazole, *J. Biochem.* 72, 521–527, 1972; Kasai, H., Takahashi, K., and Ando, T., Chemical modification of tyrosine residues in ribonuclease T1 with *N*-acetylimidazole and *p*-diazobenzenesulfonic acid, *J. Biochem.* 81, 1751–1758, 1977; Zhao, X., Gorewit, R.C., and Currie, W.B., Effects of *N*-acetylimidazole on oxytocin binding in bovine mammary tissue, *J. Recept. Res.* 10, 287–298, 1990; Wells, I. and Marnett, L.J., Acetylation of prostaglandin endoperoxide synthase by *N*-acetylimidazole: comparison to acetylation by aspirin, *Biochemistry* 31, 9520–9525, 1992; Cymes, G.D., Iglesias, M.M., and Wolfenstein-Todel, C., Chemical modification of ovine prolactin with *N*-acetylimidazole, *Int. J. Pept. Protein Res.* 42, 33–38, 1993; Zhang, F., Gao, J., Weng, J. et al., Structural and functional differences of three groups of tyrosine residues by acetylation of *N*-acetylimidazole in manganese-stabilizing protein, *Biochemistry* 44, 719–725, 2005.

| Acetylsalicylic Acid | 2-(acetoxy)benzoic Acid; Aspirin | 180.16 | Analgesic, anti-inflammatory; mild acetylating agent. |

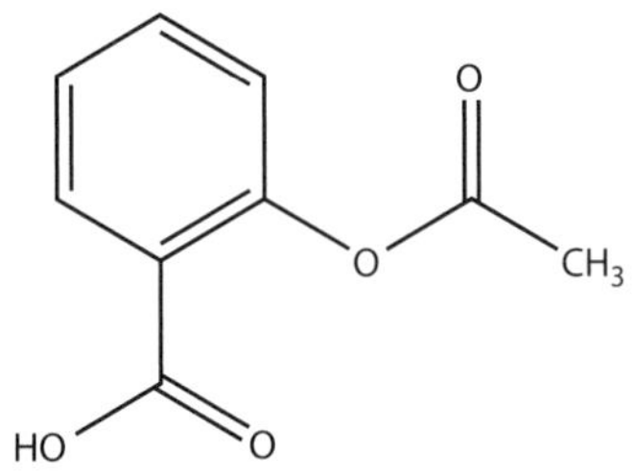

Hawkins, D., Pinckard, R.N., and Farr, R.S., Acetylation of human serum albumin by acetylsalicylic acid, *Science* 160, 780–781, 1968; Kalatzis, E., Reactions of aminoacetophen in pharmaceutical dosage forms: its proposed acetylation by acetylsalicylic acid, *J. Pharm. Sci.* 59, 193–196, 1970; Pinckard, R.N., Hawkins, D., and Farr, R.S., The inhibitory effect of salicylate on the actylation of human albumin by acetylsalicylic acid, *Arthritis Rheum.* 13, 361–368, 1970; Van Der Ouderaa, F.J., Buytenhek, M., Nugteren, D.H., and Van Dorp, D.A., Acetylation of prostaglandin endoperoxide synthetase with acetylsalicylic acid, *Eur. J. Biochem.* 109, 1–8, 1980; Rainsford, K.D., Schweitzer, A., and Brune, K., Distribution of the acetyl compared with the salicyl moiety of acetylsalicylic acid. Acetylation of macromolecules in organs wherein side effects are manifest, *Biochem. Pharmacol.* 32, 1301–1308, 1983; Liu, L.R. and Parrott, E.L., Solid-state reaction between sulfadiazine and acetylsalicyclic acid, *J. Pharm. Sci.* 80, 564–566, 1991; Minchin, R.F., Ilett, K.F., Teitel, C.H. et al., Direct *O*-acetylation of *N*-hydroxy arylamines by acetylsalicylic acid to form carcinogen-DNA adducts, *Carcinogenesis* 13, 663–667, 1992.

| Acrylamide | 2-propenamide | 71.08 | Monomer unit of polyacrylamide in gels, hydrogels, hard polymers; environmental carcinogen; fluorescence quencher. |

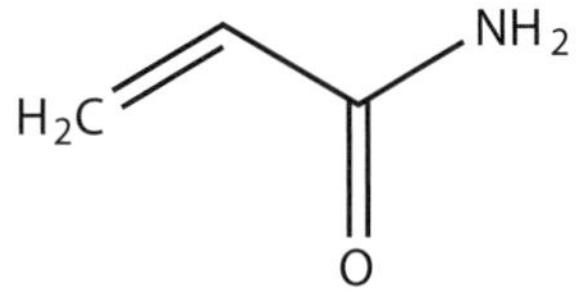

**CHEMICALS COMMONLY USED IN BIOCHEMISTRY AND MOLECULAR
BIOLOGY AND THEIR PROPERTIES (Continued)**

Common Name	Chemical Name	M.W.	Properties and Comment

Eftink, M.R. and Ghiron, C.A., Fluorescence quenching studies with proteins, *Anal. Biochem.* 114, 199–227, 1981; Dearfield, K.L., Abernathy, C.O., Ottley, M.S. et al., Acrylamide: its metabolism, developmental and reproductive effects, *Mutat. Res.* 195, 45–77, 1988; Williams, L.R., Staining nucleic acids and proteins in electrophoresis gels, *Biotech. Histochem.* 76, 127–132, 2001; Hamden, M., Bordini, E., Galvani, M., and Righetti, P.G., Protein alkylation by acrylamide, its *N*-substituted derivatives and crosslinkers and its relevance to proteomics: a matrix-assisted laser desorption/ionization-time of flight-mass spectrometry study, *Electrophoresis* 22, 1633–1644, 2001; Cioni, P. and Strambini, G.B., Tryptophan phosphorescence and pressure effects on protein structure, *Biochim. Biophys. Acta* 1595, 116–130, 2002; Taeymans, D., Wood, J., Ashby, P. et al., A review of acrylamide: an industry perspective on research, analysis, formation, and control, *Crit. Rev. Food Sci. Nutr.* 44, 323–347, 2004; Rice, J.M., The carcinogenicity of acrylamide, *Mutat. Res.* 580, 3–20, 2005; Besaratinia, A. and Pfeifer, G.P., DNA adduction and mutagenic properties of acrylamide, *Mutat. Res.* 580, 31–40, 2005; Hoenicke, K. and Gaterman, R., Studies on the stability of acrylamide in food during storage, *J. AOAC Int.* 88, 268–273, 2005; Castle, L. and Ericksson, S., Analytical methods used to measure acrylamide concentrations in foods, *J. AOAC Int.* 88, 274–284, 2005; Stadler, R.H., Acrylamide formation in different foods and potential strategies for reduction, *Adv. Exp. Med. Biol.* 561, 157–169, 2005; Lopachin, R.M. and Decaprio, A.P., Protein adduct formation as a molecular mechanism in neurotoxicity, *Toxicol. Sci.* 86, 214–225, 2005.

Common Name	Chemical Name	M.W.	Properties and Comment
Gamma (γ)-aminobutyric Acid (GABA)	4-aminobutanoic acid	103.12	Neurotransmitter.

Mandel, P. and DeFeudis, F.V., Eds., *GABA—Biochemistry and CNS Functions*, Plenum Press, New York, 1979; Costa, E. and Di Chiara, G., *GABA and Benzodiazepine Receptors*, Raven Press, New York, 1981; Racagni, G. and Donoso, A.O., *GABA and Endocrine Function*, Raven Press, New York, 1986; Squires, R.F., *GABA and Benzodiazepine Receptors*, CRC Press, Boca Raton, FL, 1988; Martin, D.L. and Olsen, R.W., *GABA in the Nervous System: The View at Fifty Years*, Lippincott, Williams & Wilkins, Philadelphia, PA, 2000.

Common Name	Chemical Name	M.W.	Properties and Comment
Amiloride	3,5-diamino-*N*-(amino-iminomethyl)-6-chloropyrazine-carboxamide	229.63	Sodium ion channel blocker.

Benos, D.J., A molecular probe of sodium transport in tissues and cells, *Am. J. Physiol.* 242, C131–C145, 1982; Garty, H., Molecular properties of epithelial, amiloride-blockable Na⁺ channels, *FASEB J.* 8, 522–528, 1994; Barbry, P. and Lazdunski, M., Structure and regulation of the amiloride-sensitive epithelial sodium channel, *Ion Channels* 4, 115–167, 1996; Kleyman, T.R., Sheng, S., Kosari, F., and Kieber-Emmons, T., Mechanism of action of amiloride: a molecular perspective, *Semin. Nephrol.* 19, 524–532, 1999; Alvarez de la Rosa, D., Canessa, C.M., Fyfe, G.K., and Zhang, P., Structure and regulation of amiloride-sensitive sodium channels, *Annu. Rev. Physiol.* 62, 573–594, 2000; Haddad, J.J., Amiloride and the regulation of NF-κ β: an unsung crosstalk and missing link between fluid dynamics and oxidative stress-related inflammation — controversy or pseudo-controversy, *Biochem. Biophys. Res. Commun.* 327, 373–381, 2005.

Common Name	Chemical Name	M.W.	Properties and Comment
2-Aminopyridine	α-aminopyridine	94.12	Precursor for synthesis of pharmaceuticals and reagents; used to derivatize carbohydrates for analysis; blocker of K+ channels.

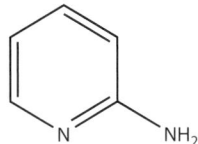

CHEMICALS COMMONLY USED IN BIOCHEMISTRY AND MOLECULAR
BIOLOGY AND THEIR PROPERTIES (Continued)

Common Name	Chemical Name	M.W.	Properties and Comment

Hase, S., Hara, S., and Matsushima, Y., Tagging of sugars with a fluorescent compound, 2-aminopyridine, *J. Biochem.* 85, 217–220, 1979; Hase, S., Ibuki, T., and Ikenaka, T., Reexamination of the pyridylamination used for fluorescence labeling of oligosaccharides and its application to glycoproteins, *J. Biochem.* 95, 197–203, 1984; Chen, C. and Zheng, X., Development of the new antimalarial drug pyronaridine: a review, *Biomed. Environ. Sci.* 5, 149–160, 1992; Hase, S., Analysis of sugar chains by pyridylamination, *Methods Mol. Biol.* 14, 69–80, 1993; Oefner, P.J. and Chiesa, C., Capillary electrophoresis of carbohydrates, *Glycobiology* 4, 397–412, 1994; Dyukova, V.I., Shilova, N.V., Galanina, O.E. et al., Design of carbohydrate multiarrays, *Biochim. Biophys. Acta* 1760, 603–609, 2006; Takegawa, Y., Deguchi, K., Keira, T. et al., Separation of isomeric 2-aminopyridine derivatized *N*-glycans and *N*-glycopeptides of human serum immunoglobulin G by using a zwitterionic type of hydrophilic-interaction chromatography, *J. Chromatog. A* 1113, 177–181, 2006; Suzuki, S., Fujimori, T., and Yodoshi, M., Recovery of free oligosaccharides from derivatives labeled by reductive amination, *Anal. Biochem.* 354, 94–103, 2006; Caballero, N.A., Melendez, F.J., Munoz-Caro, C., and Nino, A., Theoretical prediction of relative and absolute pK(a) values of aminopyridine, *Biophys. Chem.*, 124, 155–160, 2006.

Ammonium Bicarbonate   Acid Ammonium Carbonate   79.06   Volatile buffer salt.

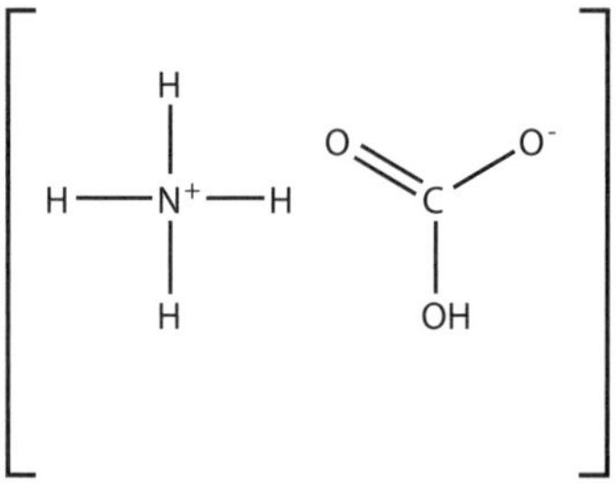

Gibbons, G.R., Page, J.D., and Chaney, S.G., Treatment of DNA with ammonium bicarbonate or thiourea can lead to underestimation of platinum-DNA monoadducts, *Cancer Chemother. Pharmacol.* 29, 112–116, 1991; Sorenson, S.B., Sorenson, T.L., and Breddam, K., Fragmentation of protein by *S. aureus* strain V8 protease. Ammonium bicarbonate strongly inhibits the enzyme but does not improve the selectivity for glutamic acid, *FEBS Lett.* 294, 195–197, 1991; Fichtinger-Schepman, A.M., van Dijk-Knijnenburg, H.C., Dijt, F.J. et al., Effects of thiourea and ammonium bicarbonate on the formation and stability of bifunctional cisplatinin-DNA adducts: consequences for the accurate quantification of adducts in (cellular) DNA, *J. Inorg. Biochem.* 58, 177–191, 1995; Overcashier, D.E., Brooks, D.A., Costantino, H.R., and Hus, C.C., Preparation of excipient-free recombinant human tissue-type plasminogen activator by lyophilization from ammonium bicarbonate solution: an investigation of the two-stage sublimation process, *J. Pharm. Sci.* 86, 455–459, 1997.

ANS   1-anilino-8-   299.4   Fluorescent probe for protein conformation;
                                               naphthalenenesulfonate   considered a hydrophobic probe; study of
                                                                 molten globules.

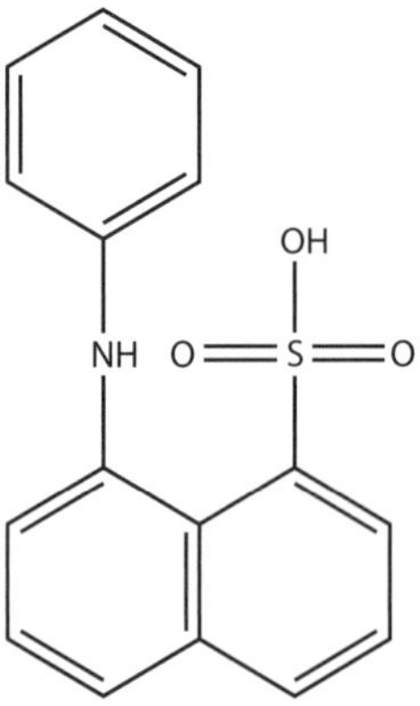

Ferguson, R.N., Edelhoch, H., Saroff, H.A. et al., Negative cooperativity in the binding of thyroxine to human serum prealbumin. Preparation of tritium-labeled 8-anilino-1-naphthalenesulfonic acid, *Biochemistry* 14, 282–289, 1975; Ogasahara, K., Koike, K., Hamada, M., and Hiraoka, T., Interaction of hydrophobic probes with the apoenzyme of pig heart lipoamide dehydrogenase, *J. Biochem.* 79, 967–975, 1976; De Campos Vidal, B., The use of the fluorescence probe 8-anilinonaphthalene sulfate (ANS) for collagen and elastin histochemistry, *J. Histochem. Cytochem.* 26, 196–201, 1978; Royer, C.A., Fluorescence spectroscopy, *Methods Mol. Biol.* 40, 65–89, 1995; Celej, M.S., Dassie, S.A., Freire, E. et al., Ligand-induced thermostability in proteins: thermodynamic analysis of ANS-albumin interaction, *Biochim. Biophys. Acta* 1750, 122–133, 2005; Banerjee, T. and Kishore, N., Binding of 8-anilinonaphthalene sulfonate to dimeric and tetrameric concanavalin A: energetics and its implications on saccharide binding studied by isothermal titration calorimetry and spectroscopy, *J. Phys. Chem. B Condens. Matter Mater. Surf. Interfaces Biophys.* 110, 7022–7028, 2006; Sahu, K., Mondal, S.K., Ghosh, S. et al., Temperature dependence of salvation dynamics and anisotropy decay in a protein: ANS in bovine serum albumin, *J. Chem. Phys.* 124, 124909, 2006; Wang, G., Gao, Y., and Geng, M.L., Analysis of heterogeneous fluorescence decays in proteins. Using fluorescence lifetime of 8-anilino-1-naphthalenesulfonate to probe apomyoglobin unfolding at equilibrium, *Biochim. Biophys. Acta* 1760, 1125–1137, 2006; Greene, L.H., Wijesinha-Bettoni, R., and Redfield, C., Characterization of the molten globule of human serum retinol-binding protein using NMR spectroscopy, *Biochemistry* 45, 9475–9484, 2006.

Common Name	Chemical Name	M.W.	Properties and Comment
Arachidonic Acid	5,8,11,14(all *cis*)-eicosatetraenoic acid	304.5	Essential fatty acid; precursor of prostaglandins, thromboxanes, and leukotrienes.

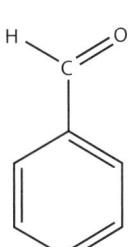

Moncada, S. and Vane, J.R., Interaction between anti-inflammatory drugs and inflammatory mediators. A reference to products of arachidonic acid metabolism, *Agents Actions Suppl.* 3, 141–149, 1977; Moncada, S. and Higgs, E.A., Metabolism of arachidonic acid, *Ann. N.Y. Acad. Sci.* 522, 454–463, 1988; Piomelli, D., Arachidonic acid in cell signaling, *Curr. Opin. Cell Biol.* 5, 274–280, 1993; Janssen-Timmen, U., Tomic, I., Specht, E. et al., The arachidonic acid cascade, eicosanoids, and signal transduction, *Ann. N.Y. Acad. Sci.* 733, 325–334, 1994; Wang, X. and Stocco, D.M., Cyclic AMP and arachidonic acid: a tale of two pathways, *Mol. Cell. Endocrinol.* 158, 7–12, 1999; Brash, A.R., Arachidonic acid as a bioactive molecule, *J. Clin. Invest.* 107, 1339–1345, 2001; Luo, M., Flamand, N., and Brock, T.G., Metabolism of arachidonic acid to eicosanoids within the nucleus, *Biochim. Biophys. Acta* 1761, 618–625, 2006; Balboa, M.A. and Balsinde, J., Oxidative stress and arachidonic acid mobilization, *Biochim. Biophys. Acta* 1761, 385–391, 2006.

| **Ascorbic Acid** | Vitamin C; 3-oxo-L-gulofuranolactone | 176.13 | Nutrition, antioxidant (reducing agent); possible antimicrobial function. |

Ascorbic acid   Dehydroascorbic acid

Barnes, M.J. and Kodicek, E., Biological hydroxylations and ascorbic acid with special regard to collagen metabolism, *Vitam. Horm.* 30, 1–43, 1972; Leibovitz, B. and Siegel, B.V., Ascorbic acid and the immune response, *Adv. Exp. Med. Biol.* 135, 1–25, 1981; Englard, S. and Seifter, S., The biochemical functions of ascorbic acid, *Annu. Rev. Nutr.* 6, 365–406, 1986; Levine, M. and Hartzell, W., Ascorbic acid: the concept of optimum requirements, *Ann. N.Y. Acad. Sci.* 498, 424–444, 1987; Padh, H., Cellular functions of ascorbic acid, *Biochem. Cell Biol.* 68, 1166–1173, 1990; Meister, A., On the antioxidant effects of ascorbic acid and glutathione, *Biochem. Pharmacol.* 44, 1905–1915, 1992; Wolf, G., Uptake of ascorbic acid by human neutrophils, *Nutr. Rev.* 51, 337–338, 1993; Kimoto, E., Terada, S., and Yamaguchi, T., Analysis of ascorbic acid, dehydroascorbic acid, and transformation products by ion-pairing high-performance liquid chromatography with multiwavelength ultraviolet and electrochemical detection, *Methods Enzymol.* 279, 3–12, 1997; May, J.M., How does ascorbic acid prevent endothelial dysfunction? *Free Rad. Biol. Med.* 28, 1421–1429, 2000; Smirnoff, N. and Wheeler, G.L., Ascorbic acid in plants: biosynthesis and function, *Crit. Rev. Biochem. Mol. Biol.* 35, 291–314, 2000; Arrigoni, O. and De Tullio, M.C., Ascorbic acid: much more than just an antioxidant, *Biochim. Biophys. Acta* 1569, 1–9, 2002; Akyon, Y., Effect of antioxidant on the immune response of *Helicobacter pyrlori*, *Clin. Microbiol. Infect.* 8, 438–441, 2002; Takanaga, H., MacKenzie, B., and Hediger, M.A., Sodium-dependent ascorbic acid transporter family SLC23, *Pflügers Arch.* 447, 677–682, 2004.

| **Benzaldehyde** | Benzoic Aldehyde; Essential Oil of Almond | 106.12 | Intermediate in manufacture of pharmaceuticals, flavors; reacts with amino groups, semicarbidizide. |

CHEMICALS COMMONLY USED IN BIOCHEMISTRY AND MOLECULAR
BIOLOGY AND THEIR PROPERTIES (Continued)

Common Name	Chemical Name	M.W.	Properties and Comment

Chalmers, R.M., Keen, J.N., and Fewson, C.A., Comparison of benzyl alcohol dehydrogenases and benzaldehyde dehydrogenases from the benzyl alcohol and mandelate pathways in *Acinetobacter calcoaceticus* and the TOL-plasmid-encoded toluene pathway in *Pseudomonas putida*. *N*-terminal amino acid sequences, amino acid composition, and immunological cross-reactions, *Biochem. J.* 273, 99–107, 1991; Pettersen, E.O., Larsen, R.O., Borretzen, B. et al., Increased effect of benzaldehyde by exchanging the hydrogen in the formyl group with deuterium, *Anticancer Res.* 11, 369–373, 1991; Nierop Groot, M.N. and de Bont, J.A.M., Conversion of phenylalanine to benzaldehyde initiated by an amino-transferase in *Lactobacillus plantarum*, *Appl. Environ. Microbiol.* 64, 3009–3013, 1998; Podyminogin, M.A., Lukhtanov, E.A., and Reed, M.W., Attachment of benzaldehyde-modified oligodeoxynucleotide probes to semicarbazide-coated glass, *Nucleic Acids Res.* 29, 5090–5098, 2001; Kurchan, A.N. and Kutateladze, A.G., Amino acid-based dithiazines: synthesis and photofragmentation of their benzaldehyde adducts, *Org. Lett.* 4, 4129–4131, 2002; Kneen, M.M., Pogozheva, I.D., Kenyon, G.L., and McLeish, M.J., Exploring the active site of benzaldehyde lyase by modeling and mutagenesis, *Biochim. Biophys. Acta* 1753, 263–271, 2005; Mosbacher, T.G., Mueller, M., and Schultz, G.E., Structure and mechanism of the ThDP-dependent benzaldehyde lyase from *Pseudomonas fluorescens*, *FEBS J.* 272, 6067–6076, 2005; Sudareva, N.N. and Chubarova, E.V., Time-dependent conversion of benzyl alcohol to benzaldehyde and benzoic acid in aqueous solution, *J. Pharm. Biomed. Anal.* 41, 1380–1385, 2006.

Benzamidine HCl 156.61 Inhibitor of trypticlike serine proteases.

Ensinck, J.W., Shepard, C., Dudl, R.J., and Williams, R.H., Use of benzamidine as a proteolytic inhibitor in the radio-immunoassay of glucagon in plasma, *J. Clin. Endocrinol. Metab.* 35, 463–467, 1972; Bode, W. and Schwager, P., The refined crystal structure of bovine beta-trypsin at 1.8 Å resolution. II. Crystallographic refinement, calcium-binding site, benzamidine-binding site and active site at pH 7.0., *J. Mol. Biol.* 98, 693–717, 1975; Nastruzzi, C., Feriotto, G., Barbieri, R. et al., Differential effects of benzamidine derivatives on the expression of *c-myc* and HLA-DR alpha genes in a human B-lymphoid tumor cell line, *Cancer Lett.* 38, 297–305, 1988; Clement, B., Schmitt, S., and Zimmerman, M., Enzymatic reduction of benzamidoxime to benzamidine, *Arch. Pharm.* 321, 955–956, 1988; Clement, B., Immel, M., Schmitt, S., and Steinman, U., Biotransformation of benzamidine and benzamidoxime *in vivo*, *Arch. Pharm.* 326, 807–812, 1993; Renatus, M., Bode, W., Huber, R. et al., Structural and functional analysis of benzamidine-based inhibitors in complex with trypsin: implications for the inhibition of factor Xa, tPA, and urokinase, *J. Med. Chem.* 41, 5445–5456, 1998; Henriques, R.S., Fonseca, N., and Ramos, M.J., On the modeling of snake venom serine proteinase interactions with benzamidine-based thrombin inhibitors, *Protein Sci.* 13, 2355–2369, 2004; Gustavsson, J., Farenmark, J., and Johansson, B.L., Quantitative determination of the ligand content in benzamidine Sepharose® 4 Fast Flow media with ion-pair chromatography, *J. Chromatog. A* 1070, 103–109, 2005.

Benzene Benzene 78.11 Solvent; a xenobiotic.

Lovley, D.R., Anaerobic benzene degradation, *Biodegradation* 11, 107–116, 2000; Snyder, R., Xenobiotic metabolism and the mechanism(s) of benzene toxicity, *Drug Metab. Rev.* 36, 531–547, 2004; Rana, S.V. and Verma, Y., Biochemical toxicity of benzene, *J. Environ. Biol.* 26, 157–168, 2005; Lin, Y.S., McKelvey, W., Waidyanatha, S., and Rappaport, S.M., Variability of albumin adducts of 1,4-benzoquinone, a toxic metabolite of benzene, in human volunteers, *Biomarkers* 11, 14–27, 2006; Baron, M. and Kowalewski, V.J., The liquid water-benzene system, *J. Phys. Chem. A Mol. Spectrosc. Kinet. Environ. Gen. Theory* 100, 7122–7129, 2006; Chambers, D.M., McElprang, D.O., Waterhouse, M.G., and Blount, B.C., An improved approach for accurate quantitation of benzene, toluene, ethylbenzene, xylene, and styrene in blood, *Anal. Chem.* 78, 5375–5383, 2006.

Benzidine *p*-benzidine; 184.24 Precursor for azo dyes; mutagenic agent;
  (1,1′-biphenyl)-4,4′-diamine   forensic analysis for bloodstains based on
    reactivity with hemoglobin.

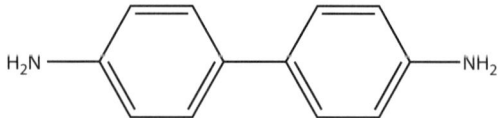

**CHEMICALS COMMONLY USED IN BIOCHEMISTRY AND MOLECULAR
BIOLOGY AND THEIR PROPERTIES (Continued)**

Common Name	Chemical Name	M.W.	Properties and Comment

Ahlquist, D.A. and Schwartz, S., Use of leuco-dyes in the quantitative colorimetric microdetermination of hemoglobin and other heme compounds, *Clin. Chem.* 21, 362–369, 1975; Josephy, P.D., Benzidine: mechanisms of oxidative activation and mutagensis, *Fed. Proc.* 45, 2465–2470, 1986; Choudhary, G., Human health perspectives on environmental exposure to benzidine: a review, *Chemosphere* 32, 267–291, 1996; Madeira, P., Nunes, M.R., Borges, C. et al., Benzidine photodegradation: a mass spectrometry and UV spectroscopy combined study, *Rapid Commun. Mass Spectrom.* 19, 2015–2020, 2005; Saitoh, T., Yoshida, S., and Ichikawa, J., Naphthalene-1,8-diylbis(diphenylmethylium) as an organic two-electron oxidant: benzidine synthesis via oxidative self-coupling of *N,N*-dialkylanilines, *J. Org. Chem.* 71, 6414–6419, 2006.

BIG CHAP/Deoxy BIG CHAP — *N,N*-bis(3-d-gluconamido-propyl) cholamide/*N,N*-bis(3-d-gluconamido-propyl) deoxycholamide — 878.1/ 862.1 — Nonionic detergents; protein solubilization, adenovirus gene transfer enhancement.

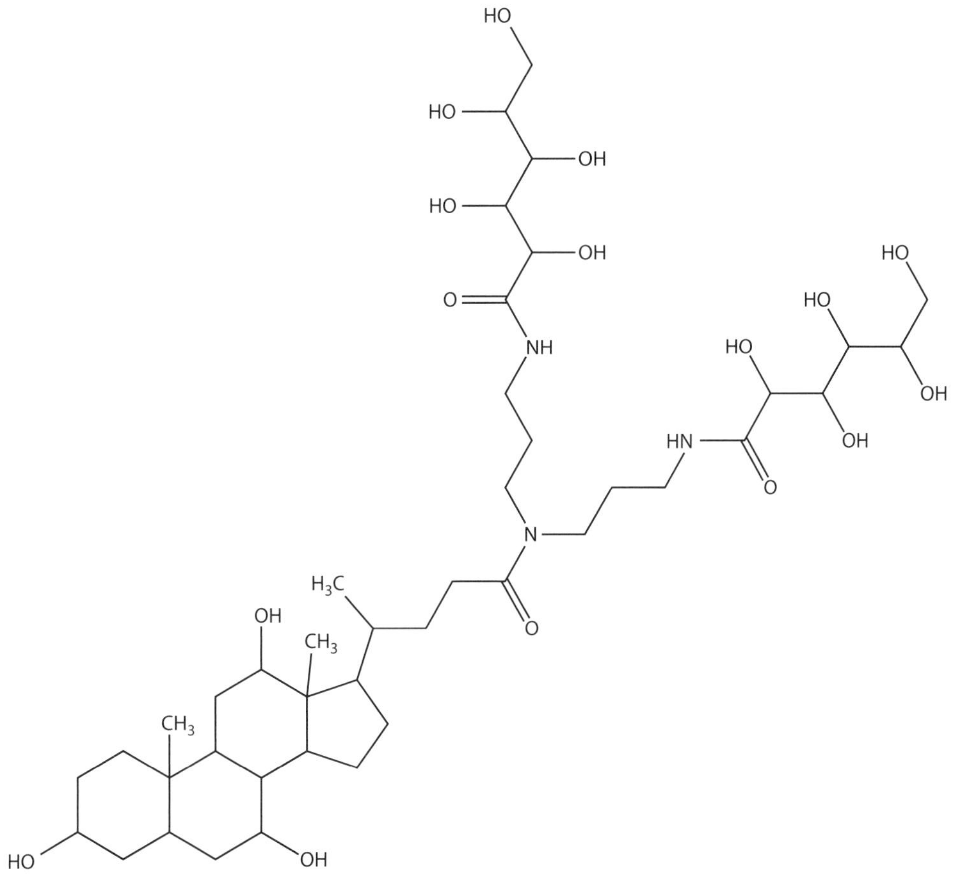

Bonelli, F.S. and Jonas, A., Reaction of lecithin: cholesterol acyltransferase with a water-soluble substrate: effects of surfactants, *Biochim. Biophys. Acta* 1166, 92–98, 1993; Aigner, A., Jager, M., Pasternack, R. et al., Purification and characterization of cysteine-*S*-conjugate *N*-acetyltransferase from pig kidney, *Biochem. J.* 317, 213–218, 1996; Mechref, Y. and Eirassi, Z., Micellar electrokinetic capillary chromatography with *in-situ* charged micelles. 4. Evaluation of novel chiral micelles consisting of steroidal glycoside surfactant borate complexes, *J. Chromatog. A* 724, 285–296, 1996; Abe, S., Kunii, S., Fujita, T., and Hiraiwa, K., Detection of human seminal gamma-glutamyl transpeptidase in stains using sandwich ELISA, *Forensic Sci. Int.* 91, 19–28, 1998; Akutsu, Y., Nakajima-Kambe, T., Nomura, N., and Nakahara, T., Purification and properties of a polyester polyurethane-degrading enzyme form *Comamonas acidovorans* TB-35, *Appl. Environ. Microbiol.* 64, 62–67, 1998: Connor, R.J., Engler, H., Machemer, T. et al., Identification of polyamides that enhance adenovirus-mediated gene expression in the urothelium, *Gene Therapy* 8, 41–48, 2001; Vajdos, F.F., Ultsch, M., Schaffer, M.L. et al., Crystal structure of human insulin-like growth factor-1: detergent binding inhibits binding protein interactions, *Biochemistry* 40, 11022–11029, 2001; Kuball, J., Wen, S.F., Leissner, J. et al., Successful adenovirus-mediated wild-type p53 gene transfer in patients with bladder cancer by intravesical vector instillation, *J. Clin. Oncol.* 20, 957–965, 2002; Susasara, K.M., Xia, F., Gronke, R.S., and Cramer, S.M., Application of hydrophobic interaction displacement chromatography for an industrial protein purification, *Biotechnol. Bioeng.* 82, 330–339, 2003; Ishibashi, A. and Nakashima, N., Individual dissolution of single-walled carbon nanotubes in aqueous solutions of steroid or sugar compounds and their Raman and near-IR spectral properties, *Chemistry*, 12, 7595–7602, 2006.

CHEMICALS COMMONLY USED IN BIOCHEMISTRY AND MOLECULAR
BIOLOGY AND THEIR PROPERTIES (Continued)

Common Name	Chemical Name	M.W.	Properties and Comment
Biotin	Coenzyme R	244.31	Coenzyme function in carboxylation reactions; growth factor; tight binding to avidin used for affinity interactions.

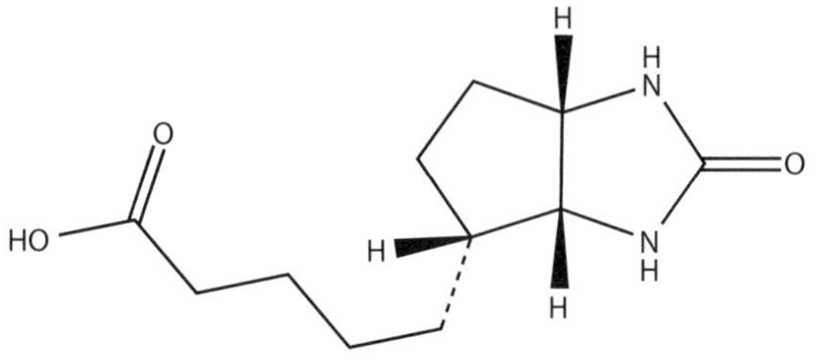

Biotin

Knappe, J., Mechanism of biotin action, *Annu. Rev. Biochem.* 39, 757–776, 1970; Dunn, M.J., Detection of proteins on blots using the avidin-biotin system, *Methods Mol. Biol.* 32, 227–232, 1994; Wisdom, G.B., Enzyme and biotin labeling of antibody, *Methods Mol. Biol.* 32, 433–440, 1994; Wilbur, D.S., Pathare, P.M, Hamlin, D.K. et al., Development of new biotin/streptavidin reagents for pretargeting, *Biomol. Eng.* 16, 113–118, 1999; Jitrapakdee, S. and Wallace, J.C., The biotin enzyme family: conserved structural motifs and domain rearrangements, *Curr. Protein Pept. Sci.* 4, 217–229, 2003; Nikolau, B.J., Ohlrogge, J.B., and Wurtels, E.S., Plant biotin-containing carboxylases, *Arch. Biochem. Biophys.* 414, 211–222, 2003; Fernandez-Mejia, C., Pharmacological effects of biotin, *J. Nutri. Biochem.* 16, 424–427, 2005; Wilchek, M., Bayer, E.A., and Livnah, O., Essentials of biorecognition: the (strept)avidin-biotin system as a model for protein–protein and protein–ligand interactions, *Immunol. Lett.* 103, 27–32, 2006; Furuyama, T. and Henikoff, S., Biotin-tag affinity purification of a centromeric nucleosome assembly complex, *Cell Cycle* 5, 1269–1274, 2006; Streaker, E.D. and Beckett, D., Nonenzymatic biotinylation of a biotin carboxyl carrier protein: unusual reactivity of the physiological target lysine, *Protein Sci.* 15, 1928–1935, 2006; Raichur, A.M., Voros, J., Textor, M., and Fery, A., Adhesion of polyelectrolyte microcapsules through biotin-streptavidin specific interaction, *Biomacromolecules* 7, 2331–2336, 2006. For biotin switch assay, see Martinez-Ruiz, A. and Lamas, S., Detection and identification of *S*-nitrosylated proteins in endothelial cells, *Methods Enzymol.* 396, 131–139, 2005; Huang, B. and Chen, C., An ascorbate-dependent artifact that interferes with the interpretation of the biotin switch assay, *Free Radic. Biol. Med.* 41, 562–567, 2006; Gladwin, M.T., Wang, X., and Hogg, N., Methodological vexation about thiol oxidation versus *S*-nitrosation — a commentary on "An ascorbate-dependent artifact that interferes with the interpretation of the biotin-switch assay," *Free Radic. Biol. Med.* 41, 557–561, 2006.

Biuret	Imidodicarbonic Diamide	103.08	Prepared by heating urea, reaction with cupric ions in base yields red-purple (the biuret reaction); nonprotein nitrogen (NPN) nutritional source.

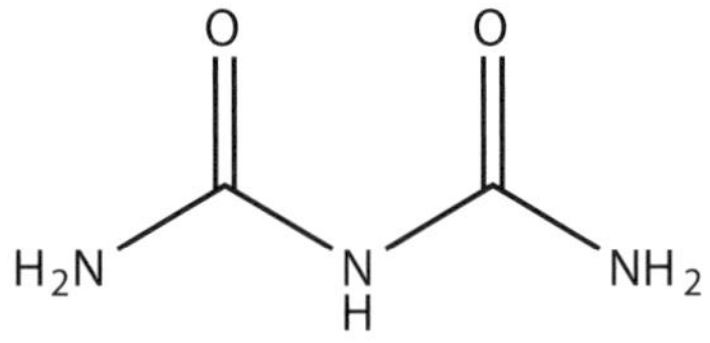

Jensen, H.L. and Schroder, M., Urea and biuret as nitrogen sources for *Rhizobium* spp., *J. Appl. Bacteriol.* 28, 473–478, 1965; Ronca, G., Competitive inhibition of adenosine deaminase by urea, guanidine, biuret, and guanylurea, *Biochim. Biophys. Acta* 132, 214–216, 1967; Oltjen, R.R., Slyter, L.L., Kozak, A.S., and Williams, E.E., Jr., Evaluation of urea, biuret, urea phosphate, and uric acid as NPN sources for cattle, *J. Nutr.* 94, 193–202, 1968; Tsai, H.Y. and Weber, S.G., Electrochemical detection of oligopeptides through the precolumn formation of biuret complexes, *J. Chromatog.* 542, 345–350, 1991; Gawron, A.J. and Lunte, S.M., Optimization of the conditions for biuret complex formation for the determination of peptides by capillary electrophoresis with ultraviolet detection, *Clin. Chem.* 51, 1411–1419, 2000; Roth, J., O'Leary, D.J., Wade, C.G. et al., Conformational analysis of alkylated biuret and triuret: evidence for helicity and helical inversion in oligoisocyates, *Org. Lett.* 2, 3063–3066, 2000; Hortin, G.L., and Mellinger, B., Cross-reactivity of amino acids and other compounds in the biuret reaction: interference with urinary peptide measurements, *Clin. Chem.* 51, 1411–1419, 2005.

Boric Acid	*o*-boric Acid	61.83	Buffer salt, manufacturing; complexes with carbohydrates and other polyhydroxyl compounds; therapeutic use as a topic antibacterial/antifungal agent.

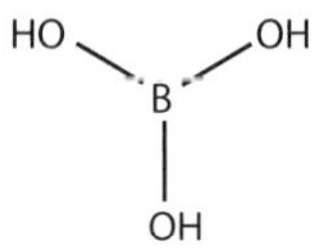

$$B(OH)_3 + 2H_2O \rightleftharpoons B(OH)_4^- + H_3O^+$$

**CHEMICALS COMMONLY USED IN BIOCHEMISTRY AND MOLECULAR
BIOLOGY AND THEIR PROPERTIES (Continued)**

Common Name	Chemical Name	M.W.	Properties and Comment

Sciarra, J.J. and Monte Bovi, A.J., Study of the boric acid–glycerin complex. II. Formation of the complex at elevated temperature, *J. Pharm. Sci.* 51, 238–242, 1962; Walborg, E.F., Jr. and Lantz, R.S., Separation and quantitation of saccharides by ion-exchange chromatography utilizing boric acid–glycerol buffers, *Anal. Biochem.* 22, 123–133, 1968; Lerch, B. and Stegemann, H., Gel electrophoresis of proteins in borate buffer. Influence of some compounds complexing with boric acid, *Anal. Biochem.* 29, 76–83, 1969; Walborg, E.F., Jr., Ray, D.B., and Ohrberg, L.E., Ion-exchange chromatography of saccharides: an improved system utilizing boric acid/2,3-butanediol buffers, *Anal. Biochem.* 29, 433–440, 1969; Chen, F.T. and Sternberg, J.C., Characterization of proteins by capillary electrophoresis in fused-silica columns: review on serum protein anlaysis and application to immunoassays, *Electrophoresis* 15, 13–21, 1994; Allen, R.C. and Doktycz, M.J., Discontinuous electrophoresis revisited: a review of the process, *Appl. Theor. Electrophor.* 6, 1–9, 1996; Manoravi, P., Joseph, M., Sivakumar, N., and Balasubramanian, H., Determination of isotopic ratio of boron in boric acid using laser mass spectrometry, *Anal. Sci.* 21, 1453–1455, 2005; De Muynck, C., Beauprez, J., Soetaert, W., and Vandamme, E.J., Boric acid as a mobile phase additive for high-performance liquid chromatography separation of ribose, arabinose, and ribulose, *J. Chromatog. A* 1101, 115–121, 2006; Herrmannova, M., Kirvankova, L., Bartos, M., and Vytras, K., Direct simultaneous determination of eight sweeteners in foods by capillary isotachophoresis, *J. Sep. Sci.* 29, 1132–1137, 2006; Alencar de Queiroz, A.A., Abraham, G.A., Pires Camillo, M.A. et al., Physicochemical and antimicrobial properties of boron-complexed polyglycerol-chitosan dendrimers, *J. Biomater. Sci. Polym. Ed.* 17, 689–707, 2006; Ringdahl, E.N., Recurrent vulvovaginal candidiasis, *Mol. Med.* 103, 165–168, 2006.

Common Name	Chemical Name	M.W.	Properties and Comment
BNPS-Skatole	(2-[2′-nitrophenyl-sulfenyl]-3-methyl-3′-bromoindolenine	363.23	Tryptophan modification, peptide-bond cleavage; derived from skatole, which is also known as boar taint.

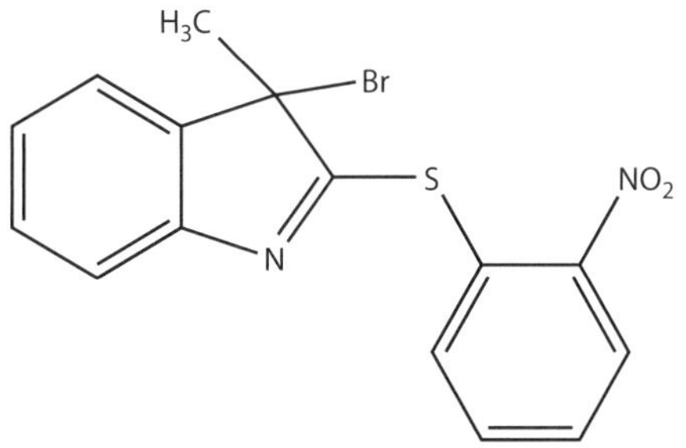

Boulanger, P., Lemay, P., Blair, G.E., and Russell, W.C., Characterization of adenovirus protein IX, *J. Gen. Virol.* 44, 783–800, 1979; Russell, J., Kathendler, J., Kowalski, K. et al., The single tryptophan residue of human placental lactogen. Effects of modification and cleavage on biological activity and protein conformation, *J. Biol. Chem.* 256, 304–307, 1981; Moskaitis, J.E. and Campagnoni, A.T., A comparison of the dodecyl sulfate-induced precipitation of the myelin basic protein with other water-soluble proteins, *Neurochem. Res.* 11, 299–315, 1986; Mahboub, S., Richard, C., Delacourte, A., and Han, K.K., Applications of chemical cleavage procedures to the peptide mapping of neurofilament triplet protein bands in sodium dodecyl sulfate-polyacrylamide gel electrophoresis, *Anal. Biochem.* 154, 171–182, 1986; Rahali, V. and Gueguen, J., Chemical cleavage of bovine beta-lactoglobulin by BPNS-skatole for preparative purposes: comparative study of hydrolytic procedure and peptide characterization, *J. Protein Chem.* 18, 1–12, 1999; Swamy, N., Addo, J., Vskokovic, M.R., and Ray, R., Probing the vitamin D sterol-binding pocket of human vitamin D-binding protein with bromoacetate affinity-labeling reagents containing the affinity probe at C-3, C-6, C-11, and C-19 positions of parent vitamin D sterols, *Arch. Biochem. Biophys.* 373, 471–478, 2000; Celestina, F. and Suryanarayana, T., Biochemical characterization and helix-stabilizing properties of HSNP-C′ from the thermophilic archaeon *Sulfolobus acidocaldarius*, *Biochem. Biophys. Res. Commun.* 267, 614–618, 2000; Kibbey, M.M., Jameson, M.J., Eaton, E.M., and Rosenzweig, S.A., Insulinlike growth factor binding protein-2: contributions of the C-terminal domain to insulinlike growth factor-1 binding, *Mol. Pharmacol.* 69, 833–845, 2006.

Common Name	Chemical Name	M.W.	Properties and Comment
***p*-Bromophenacyl Bromide**	2-bromo-1-(4-bromophenyl) ethanone; 4-bromophenacyl bromide	277.04	Modification of various residues in proteins: reagent for identification of carboxylic acids; phospholipase A2 inhibitor.

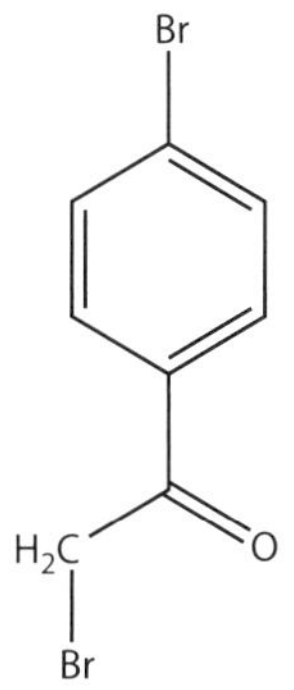

CHEMICALS COMMONLY USED IN BIOCHEMISTRY AND MOLECULAR
BIOLOGY AND THEIR PROPERTIES (Continued)

Common Name	Chemical Name	M.W.	Properties and Comment

Erlanger, B.F., Vratrsanos, S.M., Wasserman, N., and Cooper, A.G., A chemical investigation of the active center of pepsin, *Biochem. Biophys. Res. Commun.* 23, 243–245, 1966; Yang, C.C. and King, K., Chemical modification of the histidine residue in basic phospholipase A2 from the venom of *Naja nigricollis*, *Biochim. Biophys. Acta.* 614, 373–388, 1980; Darke, P.L., Jarvis, A.A., Deems, R.A., and Dennis, E.A., Further characterization and *N*-terminal sequence of cobra venom phospholipase A2, *Biochim. Biophys. Acta* 626, 154–161, 1980; Ackerman, S.K., Matter, L., and Douglas, S.D., Effects of acid proteinase inhibitors on human neutrophil chemotaxis and lysosomal enzyme release. II. Bromophenacyl bromide and 1,2-epoxy-3-(*p*-nitrophenoxy)propane, *Clin. Immunol. Immunopathol.* 26, 213–222, 1983; Carine, K. and Hudig, D., Assessment of a role for phospholipase A2 and arachidonic acid metabolism in human lymphocyte natural cytotoxicity, *Cell Immunol.* 87, 270–283, 1984; Duque, R.E., Fantone, J.C., Kramer, C. et al., Inhibition of neutrophil activation by *p*-bromophenacyl bromide and its effects on phospholipase A2, *Br. J. Pharmacol.* 88, 463–472, 1986; Zhukova, A., Gogvadze, G., and Gogvadze, V., *p*-bromophenacyl bromide prevents cumene hydroperoxide-induced mitochondrial permeability transition by inhibiting pyridine nucleotide oxidation, *Redox Rep.* 9, 117–121, 2004; Thommesen, L. and Laegreid, A., Distinct differences between TNF receptor 1- and TNR receptor 2-mediated activation of NF-κβ® *J. Biochem. Mol. Biol.* 38, 281–289, 2005; Yue, H.Y., Fujita, T., and Kumamoto, E., Phospholipase A2 activation by melittin enhances spontaneous glutamatergic excitatory transmission in rat substantia gelatinosa neurons, *Neuroscience* 135, 485–495, 2005; Costa-Junior, H.M., Hamaty, F.C., de Silva Farias, R. et al., Apoptosis-inducing factor of a cytotoxic T-cell line: involvement of a secretory phospholipase A(2), *Cell Tissue Res.* 324, 255–266, 2006; Marchi-Salvador, D.P., Fernandes, C.A., Amui, S.F. et al., Crystallization and preliminary X-ray diffraction analysis of a myotoxic Lys49-PLA2 from *Bothrops jararacussu* venom complexed with *p*-bromophenacyl bromide, *Acta Crystallograph. Sect. F Struct. Biol. Cryst. Commun.* 62, 600–603, 2006.

Calcium Chloride	$CaCl_2$	110.98	Anhydrous form as drying agent for organic solvents, variety of manufacturing uses; meat quality enhancement; therapeutic use in electrolyte replacement and bone cements; source of calcium ions for biological assays.

Barratt, J.O., Thrombin and calcium chloride in relation to coagulation, *Biochem. J.* 9, 511–543, 1915; Van der Meer, C., Effect of calcium chloride on choline esterase, *Nature* 171, 78–79, 1952; Bhat, R. and Ahluwalia, J.C., Effect of calcium chloride on the conformation of proteins. Thermodynamic studies of some model compounds, *Int. J. Pept. Protein Res.* 30, 145–152, 1987; Furihata, C., Sudo, K., and Matsushima, T., Calcium chloride inhibits stimulation of replicative DNA synthesis by sodium chloride in the pyloric mucosa of rat stomach, *Carcinogenesis* 10, 2135–2137, 1989; Ishikawa, K., Ueyama, Y., Mano, T. et al., Self-setting barrier membrane for guided tissue regeneration method: initial evaluation of alginate membrane made with sodium alginate and calcium chloride aqueous solutions, *J. Biomed. Mater. Res.* 47, 111–115, 1999; Vujevic, M., Vidakovic-Cifrek, Z., Tkalec, M. et al., Calcium chloride and calcium bromide aqueous solutions of technical and analytical grade in Lemna bioassay, *Chemosphere* 41, 1535–1542, 2000; Miyazaki, T., Ohtsuki, C., Kyomoto, M. et al., Bioactive PMMA bone cement prepared by modification with methacryloxypropyltrimethoxysilane and calcium chloride, *J. Biomed. Mater. Res. A* 67, 1417–1423, 2003; Harris, S.E., Huff-Lonegan, E., Lonergan, S.M. et al., Antioxidant status affects color stability and tenderness of calcium chloride-injected beef, *J. Anim. Sci.* 79, 666–677, 2001; Behrends, J.M., Goodson, K.J., Koohmaraie, M. et al., Beef customer satisfaction: factors affecting consumer evaluations of calcium chloride-injected top sirloin steaks when given instructions for preparation, *J. Anim. Sci.* 83, 2869–2875, 2005.

Cetyl Pyridinium Chloride	1-hexadecylpyridinium chloride	350.01	Cationic detergent; precipitating agent and staining agent for glycosaminoglycans; antimicrobial agent.

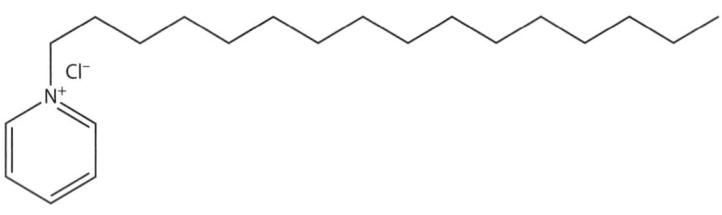

Laurent, T.C. and Scott, J.E., Molecular weight fractionation of polyanions by cetylpyridinium chloride in salt solutions, *Nature* 202, 661–662, 1964; Kiss, A., Linss, W., and Geyer, G., CPC-PTA section staining of acid glycans, *Acta Histochem.* 64, 183–186, 1979; Khan, M.Y. and Newman, S.A., An assay for heparin by decrease in color yield (DECOY) of a protein-dye-binding reaction, *Anal. Biochem.* 187, 124–128, 1990; Chardin, H., Septier, D., and Goldberg, M., Visualization of glycosaminoglycans in rat incisor predentin and dentin with cetylpyridinium chloride-glutaraldehyde as fixative, *J. Histochem. Cytochem.* 38, 885–894, 1990; Chardin, H., Gokani, J.P., Septier, D. et al., Structural variations of different oral basement membranes revealed by cationic dyes and detergent added to aldehyde fixative solution, *Histochem. J.* 24, 375–382, 1992; Agren, U.M., Tammi, R., and Tammi, M., A dot-blot assay of metabolically radiolabeled hyaluronan, *Anal. Biochem.* 217, 311–315, 1994; Maccari, F. and Volpi, N., Glycosaminoglycan blotting on nitrocellulose membranes treated with cetylpyridinium chloride afer agarose-gel electrophoretic separation, *Electrophoresis* 23, 3270–3277, 2002; Maccari, F. and Volpi, N., Direct and specific recognition of glycosaminoglycans by antibodies after their separation by agarose gel electrophoresis and blotting on cetylpyridinium chloride-treated nitrocellulose membranes, *Electrophoresis* 24, 1347–1352, 2003.

CHEMICALS COMMONLY USED IN BIOCHEMISTRY AND MOLECULAR BIOLOGY AND THEIR PROPERTIES (Continued)

Common Name	Chemical Name	M.W.	Properties and Comment
CHAPS	3-[(3-cholamidopropyl)-dimethylammonio]-1-propanesulfonate	614.89	Detergent, solubilizing agent; extensive use for the solubilization of membrane proteins.

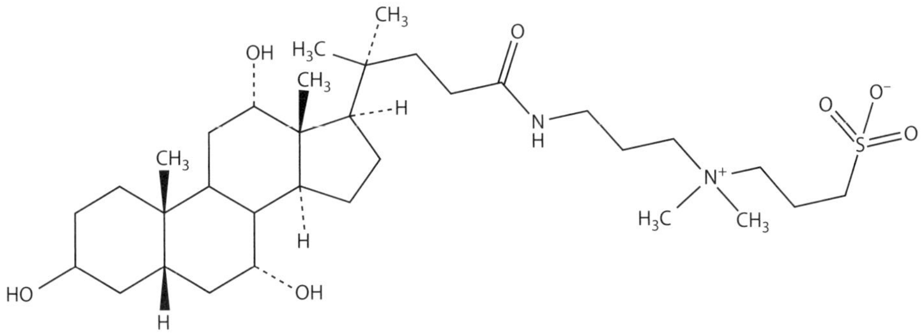

Hjelmeland, L.M., A nondenaturing zwitterionic detergent for membrane biochemistry: design and synthesis, *Proc. Natl. Acad. Sci. USA* 77, 6368–6370, 1980; Giradot, J.M. and Johnson, B.C., A new detergent for the solubilization of the vitamin K–dependent carboxylation system from liver microsomes: comparison with triton X-100, *Anal. Biochem.* 121, 315–320, 1982; Liscia, D.S., Alhadi, T., and Vonderhaar, B.K., Solubilization of active prolactin receptors by a nondenaturing zwitterionic detergent, *J. Biol. Chem.* 257, 9401–9405, 1982; Womack, M.D., Kendall, D.A., and MacDonald, R.C., Detergent effects on enzyme activity and solubilization of lipid bilayer membranes, *Biochim. Biophys. Acta* 733, 210–215, 1983; Klaerke, D.A. and Jorgensen, P.L., Role of Ca^{2+}-activated K^+ channel in regulation of NaCl reabsorption in thick ascending limb of Henle's loop, *Comp. Biochem. Physiol. A* 90, 757–765, 1988; Kuriyama, K., Nakayasu, H., Mizutani, H. et al., Cerebral GABAB receptor: proposed mechanisms of action and purification procedures, *Neurochem. Res.* 18, 377–383, 1993; Koumanov, K.S., Wolf, C., and Quinn, P.J., Lipid composition of membrane domains, *Subcell. Biochem.* 37, 153–163, 2004.

Chloroform	Trichloromethane	177.38	Used for extraction of lipids, usually in combination with methanol.

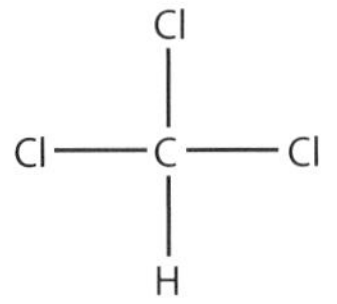

Stevan, M.A. and Lyman, R.L., Investigations on extraction of rat plasma phospholipids, *Proc. Soc. Exp. Biol. Med.* 114, 16–20, 1963; Wells, M.A. and Dittmer, J.C., A microanalytical technique for the quantitative determination of twenty-four classes of brain lipids, *Biochemistry* 5, 3405–3418, 1966; Colacicco, G. and Rapaport, M.M., A simplified preparation of phosphatidyl inositol, *J. Lipid. Res.* 8, 513–515, 1967; Curtis, P.J., Solubility of mitochondrial membrane proteins in acidic organic solvents, *Biochim. Biophys. Acta* 183, 239–241, 1969; Privett, O.S., Dougherty, K.A., and Castell, J.D., Quantitative analysis of lipid classes, *Am. J. Clin. Nutr.* 24, 1265–1275, 1971; Claire, M., Jacotot, B., and Robert, L., Characterization of lipids associated with macromolecules of the intercellular matrix of human aorta, *Connect. Tissue Res.* 4, 61–71, 1976; St. John, L.C. and Bell, F.P., Extraction and fractionation of lipids from biological tissues, cells, organelles, and fluids, *Biotechniques* 7, 476–481, 1989; Dean, N.M. and Beaven, M.A., Methods for the analysis of inositol phosphates, *Anal. Biochem.* 183, 199–209, 1989; Singh, A.K. and Jiang, Y., Quantitative chromatographic analysis of inositol phospholipids and related compounds, *J. Chromatog. B Biomed. Appl.* 671, 255–280, 1995.

CHEMICALS COMMONLY USED IN BIOCHEMISTRY AND MOLECULAR
BIOLOGY AND THEIR PROPERTIES (Continued)

Common Name	Chemical Name	M.W.	Properties and Comment
Cholesterol		386.66	The most common sterol in man and other higher animals. Cholesterol is essential for the synthesis of a variety of compounds including estrogens and vitamin D; also membrane component.

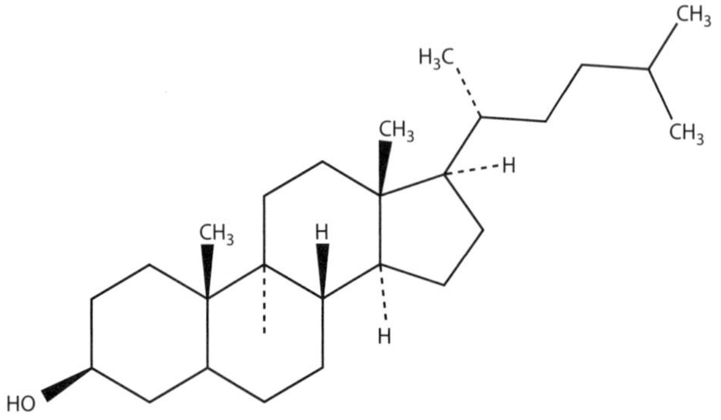

Cholesterol

Doree, C., The occurrence and distribution of cholesterol and allied bodies in the animal kingdom, *Biochem. J.* 4, 72–106, 1909; Heilbron, I.M., Kamm, E.D., and Morton, R.A., The absorption spectrum of cholesterol and its biological significance with reference to vitamin D. Part I: Preliminary observations, *Biochem. J.* 21, 78–85, 1927; Cook, R.P., Ed., *Cholesterol: Chemistry, Biochemistry, and Pathology*, Academic Press, New York, 1958; Vahouny, G.V. and Treadwell, C.R., Enzymatic synthesis and hydrolysis of cholesterol esters, *Methods Biochem. Anal.* 16, 219–272, 1968; Heftmann, E., *Steroid Biochemistry*, Academic Press, New York, 1970; Nestel, P.J., Cholesterol turnover in man, *Adv. Lipid Res.* 8, 1–39, 1970; Dennick, R.G., The intracellular organization of cholesterol biosynthesis. A review, *Steroids Lipids Res.* 3, 236–256, 1972; J. Polonovski, Ed., *Cholesterol Metabolism and Lipolytic Enzymes*, Masson Publications, New York, 1977; Gibbons, G.F., Mitrooulos, K.A., and Myant, N.B., *Biochemistry of Cholesterol*, Elsevier, Amsterdam, 1982; Bittman, R., *Cholesterol: Its Functions and Metabolism in Biology and Medicine*, Plenum Press, New York, 1997; Oram, J.P. and Heinecke, J.W., ATP-binding cassette transporter A1: a cell cholesterol exporter that protects against cardiovascular disease, *Physiol. Rev.* 85, 1343–1372, 2005; Holtta-Vuori, M. and Ikonen, E., Endosomal cholesterol traffic: vesicular and non-vesicular mechanisms meet, *Biochem. Soc. Trans.* 34, 392–394, 2006; Cuchel, M. and Rader, D.J., Macrophage reverse cholesterol transport: key to the regression of atherosclerosis? *Circulation* 113, 2548–2555, 2006.

Cholic Acid		408.57	Component of bile; detergent.

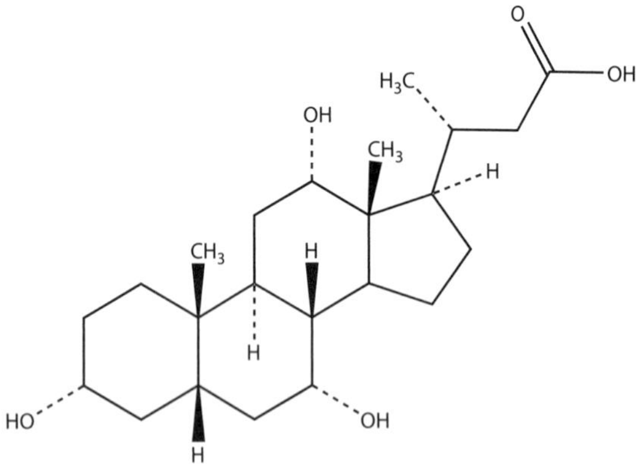

Schreiber, A.J. and Simon, F.R., Overview of clinical aspects of bile salt physiology, *J. Pediatr. Gastroenterol. Nutr.* 2, 337–345, 1983; Chiang, J.Y., Regulation of bile acid synthesis, *Front. Biosci.* 3, dl176–dl193, 1998; Cybulsky, M.I., Lichtman, A.H., Hajra, L., and Iiyama, K., Leukocyte adhesion molecules in atherogenesis, *Clin. Chim. Acta* 286, 207–218, 1999.

**CHEMICALS COMMONLY USED IN BIOCHEMISTRY AND MOLECULAR
BIOLOGY AND THEIR PROPERTIES (Continued)**

Common Name	Chemical Name	M.W.	Properties and Comment
Citraconic Anhydride	Methylmaleic Anhydride	112.1	Reversible modification of amino groups.

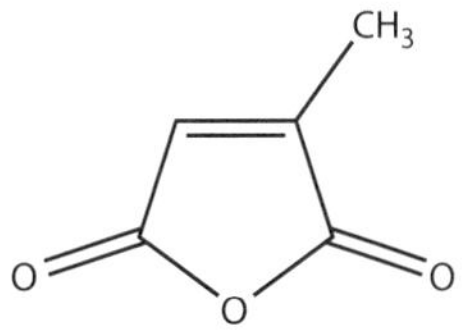

Dixon, H.B. and Perham, R.N., Reversible blocking of amino groups with citraconic anhydride, *Biochem. J.* 109, 312–314, 1968; Gibbons, I. and Perham, R.N., The reaction of aldolase with 2-methylmaleic anhydride, *Biochem. J.* 116, 843–849, 1970; Yankeelov, J.A., Jr. and Acree, D., Methylmaleic anhydride as a reversible blocking agent during specific arginine modification, *Biochem. Biophys. Res. Commun.* 42, 886–891, 1971; Takahashi, K., Specific modification of arginine residues in proteins with ninhydrin, *J. Biochem.* 80, 1173–1176, 1976; Brinegar, A.C. and Kinsella, J.E., Reversible modification of lysine in soybean proteins, using citraconic anhydride: characterization of physical and chemical changes in soy protein isolate, the 7S globulin, and lipoxygenase, *J. Agric. Food Chem.* 28, 818–824, 1980; Shetty, J.K. and Kensella, J.F., Ready separation of proteins from nucleoprotein complexes by reversible modification of lysine residues, *Biochem. J.* 191, 269–272, 1980; Yang, H. and Frey, P.A., Dimeric cluster with a single reactive amino group, *Biochemistry* 23, 3863–3868, 1984; Bindels, J.G., Misdom, L.W., and Hoenders, H.J., The reaction of citraconic anhydride with bovine alpha-crystallin lysine residues. Surface probing and dissociation-reassociation studies, *Biochim. Biophys. Acta* 828, 255–260, 1985; Al jamal, J.A., Characterization of different reactive lysines in bovine heart mitochondrial porin, *Biol. Chem.* 383, 1967–1970, 2002; Kadlik, V., Strohalm, M., and Kodicek, M., Citraconylation — a simple method for high protein sequence coverage in MALDI-TOF mass spectrometry, *Biochem. Biophys. Res. Commun.* 305, 1091–1093, 2003.

| **Coomassie Brilliant Blue G-250** | CI Acid Blue 90 | 854 | Most often used for the colorimetric determination of protein. |

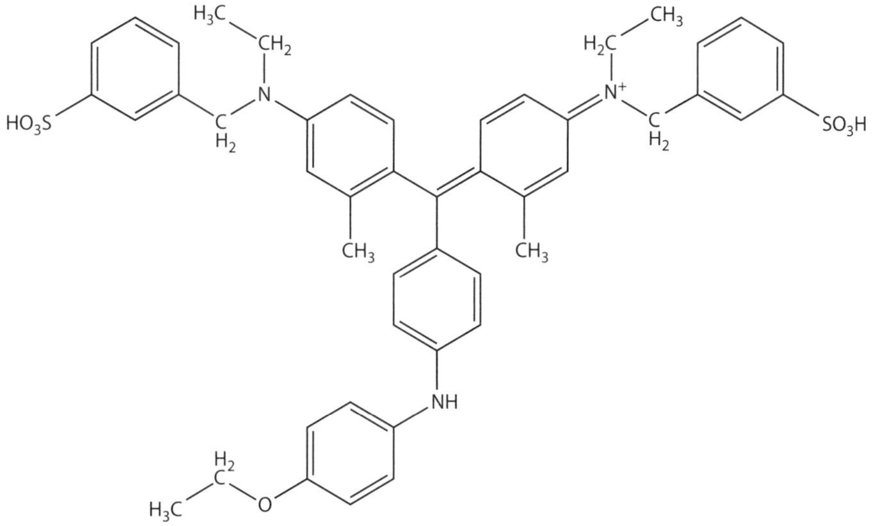

Bradford, M.M., A rapid and sensitive method for the quantitation of microgram quantities of protein utilizing the principle of protein-dye binding *Anal. Biochem.* 72, 248–254, 1976; Saleemuddin, M., Ahmad, H., and Husain, A., A simple, rapid, and sensitive procedure for the assay of endoproteases using Coomassie Brilliant Blue G-250, *Anal. Biochem.* 105, 202–206, 1980; van Wilgenburg, M.G., Werkman, E.M., van Gorkom, W.H., and Soons, J.B., Criticism of the use of Coomassie Brilliant Blue G-250 for the quantitative determination of proteins, *J.Clin. Chem.Clin. Biochem.* 19, 301–304, 1981; Mattoo, R.L., Ishaq, M., and Saleemuddin, M., Protein assay by Coomassie Brilliant Blue G-250-binding method is unsuitable for plant tissues rich in phenols and phenolases, *Anal. Biochem.* 163, 376–384, 1987; Lott, J.A., Stephan, V.A., and Pritchard, K.A., Jr., Evaluation of the Coomassie Brilliant Blue G-250 method for urinary proteins, *Clin. Chem.* 29, 1946–1950, 1983; Fanger, B.O., Adaptation of the Bradford protein assay to membrane-bound proteins by solubilizing in glucopyranoside detergents, *Anal. Biochem.* 162, 11–17, 1987; Marshall, T. and Williams, K.M., Recovery of proteins by Coomassie Brilliant Blue precipitation prior to electrophoresis, *Electrophoresis* 13, 887–888, 1992; Sapan, C.V., Lundblad, R.L., and Price, N.C., Colorimetric protein assay techniques, *Biotechnol. Appl. Biochem.* 29, 99–108, 1999.

CHEMICALS COMMONLY USED IN BIOCHEMISTRY AND MOLECULAR BIOLOGY AND THEIR PROPERTIES (Continued)

Common Name	Chemical Name	M.W.	Properties and Comment
Coomassie Brilliant Blue R-250	CI Acid Blue 83	826	Most often used for the detection of proteins on solid matrices such as polyacrylamide gels.

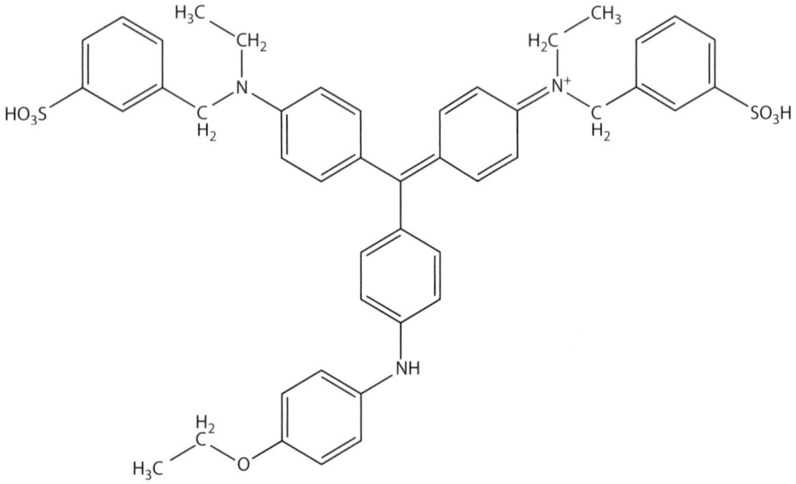

Vesterberg, O., Hansen, L., and Sjosten, A., Staining of proteins after isoelectric focusing in gels by a new procedure, *Biochim. Biophys. Acta* 491, 160–166, 1977; Micko, S. and Schlaepfer, W.W., Metachromasy of peripheral nerve collagen on polyacrylamide gels stained with Coomassie Brilliant Blue R-250, *Anal. Biochem.* 88, 566–572, 1978; Osset, M., Pinol, M., Fallon, M.J. et al., Interference of the carbohydrate moiety in Coomassie Brilliant Blue R-250 protein staining, *Electrophoresis* 10, 271–273, 1989; Pryor, J.L., Xu, W., and Hamilton, D.W., Immunodetection after complete destaining of Coomassie blue-stained proteins on immobilon-PVDF, *Anal. Biochem.* 202, 100–104, 1992; Metkar, S.S., Mahajan, S.K., and Sainis, J.K., Modified procedure for nonspecific protein staining on nitrocellulose paper using Coomassie Brilliant Blue R-250, *Anal. Biochem.* 227, 389–391, 1995; Kundu, S.K., Robey, W.G., Nabors, P. et al., Purification of commercial Coomassie Brilliant Blue R-250 and characterization of the chromogenic fractions, *Anal. Biochem.* 235, 134–140, 1996; Choi, J.K., Yoon, S.H., Hong, H.Y. et al., A modified Coomassie blue staining of proteins in polyacrylamide gels with Bismark brown R, *Anal. Biochem.* 236, 82–84, 1996; Moritz, R.L., Eddes, J.S., Reid, G.E., and Simpson, R.J., *S*- pyridylethylation of intact polyacrylamide gels and *in situ* digestion of electrophoretically separated proteins: a rapid mass spectrometric method for identifying cysteine-containing peptides, *Electrophoresis* 17, 907–917, 1996; Choi, J.K. and Yoo, G.S., Fast protein staining in sodium dodecyl sulfate polyacrylamide gel using counter ion-dyes, Coomassie Brilliant Blue R-250, and neutral red, *Arch. Pharm. Res.* 25, 704–708, 2002; Bonar, E., Dubin, A., Bierczynska-Krzysik, A. et al., Identification of major cellular proteins synthesized in response to interleukin-1 and interleukin-6 in human hepatoma HepG2 cells, *Cytokine* 33, 111–117, 2006.

Cy 2		714	Fluorescent label used in proteomics and gene expression; use for internal standard.

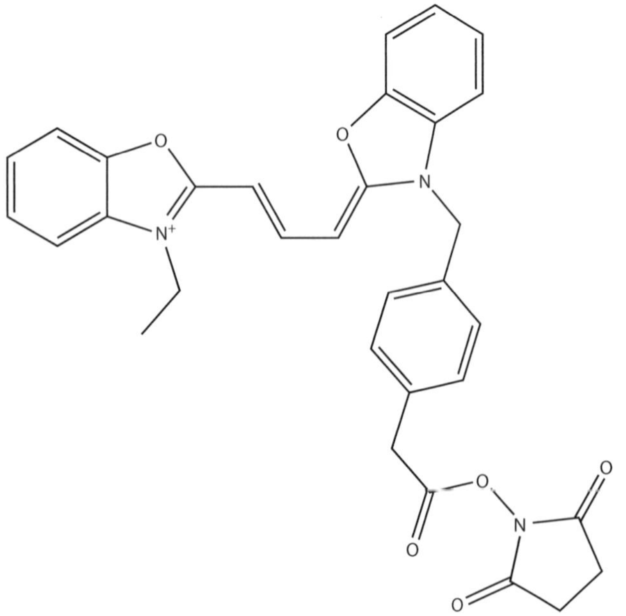

CHEMICALS COMMONLY USED IN BIOCHEMISTRY AND MOLECULAR
BIOLOGY AND THEIR PROPERTIES (Continued)

Common Name	Chemical Name	M.W.	Properties and Comment

Tonge, R., Shaw, J., Middleton, B. et al., Validation and development of fluorescence two-dimensional differential gel electrophoresis proteomics technology, *Proteomics* 1, 377–396, 2001; Chan, H.L., Gharbi, S., Gaffney, P.R. et al., Proteomic analysis of redox- and ErbB2-dependent changes in mammary luminal epithelial cells using cysteine- and lysine-labeling two-dimensional difference gel electrophoresis, *Proteomics* 5, 2908–2926, 2005; Misek, D.E., Kuick, R., Wang, H. et al., A wide range of protein isoforms in serum and plasma uncovered by a quantitative intact protein analysis system, *Proteomics* 5, 3343–3352, 2005; Doutette, P., Navet, R., Gerkens, P. et al., Steatosis-induced proteomic changes in liver mitochondria evidenced by two-dimensional differential in-gel electrophoresis, *J. Proteome Res.* 4, 2024–2031, 2005.

| **Cy 3** | | 911.0 | Fluorescent label used in proteomics and gene expression; in combination with Cy 5 is used for FRET-based assays. |

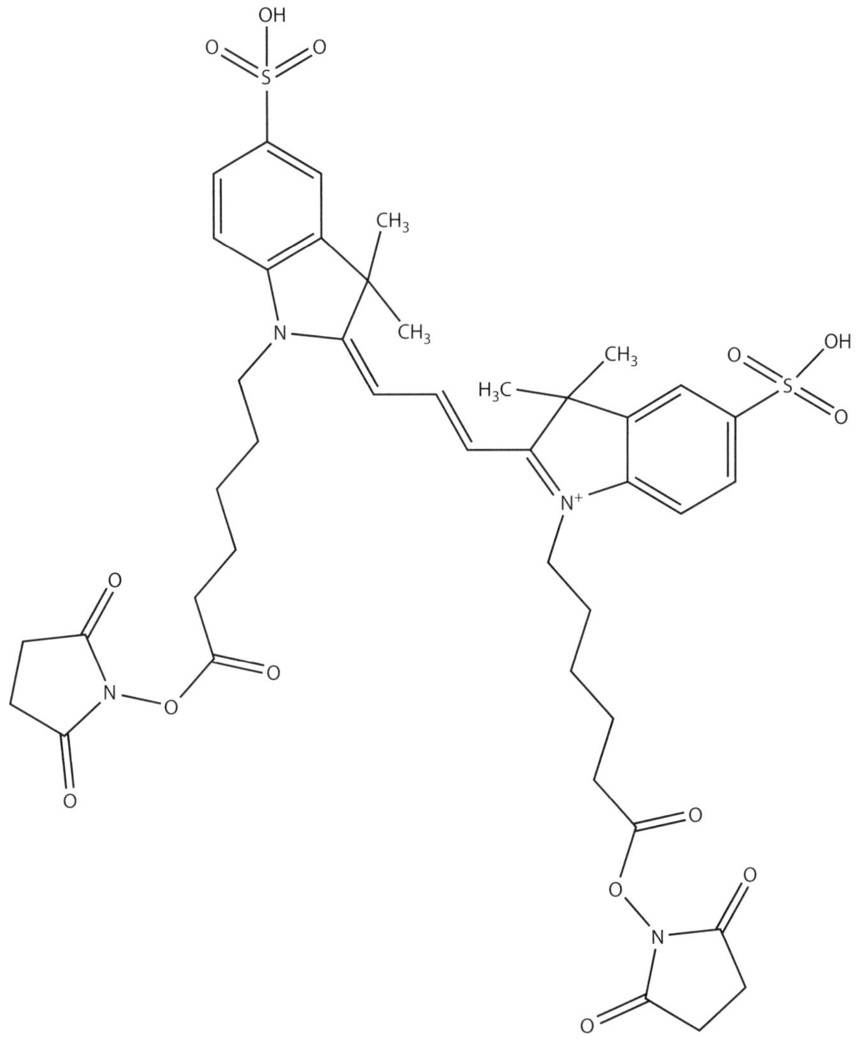

Brismar, H. and Ulfake, B., Fluorescence lifetime measurements in confocal microscopy of neurons labeled with multiple fluorophores, *Nat. Biotechnol.* 15, 373–377, 1997; Strohmaier, A.R., Porwol, T., Acker, H., and Spiess, E., Tomography of cells by confocal laser scanning microscopy and computer-assisted three-dimensional image reconstruction: localization of cathepsin B in tumor cells penetrating collagen gels *in vitro*, *J. Histochem. Cytochem.* 45, 975–983, 1997; Alexandre, I., Hamels, S., Dufour, S. et al., Colorimetric silver detection of DNA microarrays, *Anal. Biochem.* 295, 1–8, 2001; Shaw, J., Rowlinson, R., Nickson, J. et al., Evaluation of saturation labeling two-dimensional difference gel electrophoresis fluorescent dyes, *Proteomics* 3, 1181–1195, 2003.

**CHEMICALS COMMONLY USED IN BIOCHEMISTRY AND MOLECULAR
BIOLOGY AND THEIR PROPERTIES (Continued)**

Common Name	Chemical Name	M.W.	Properties and Comment
Cy 5		937.1	Fluorescent label used in proteomics and gene expression; also used in histochemistry.

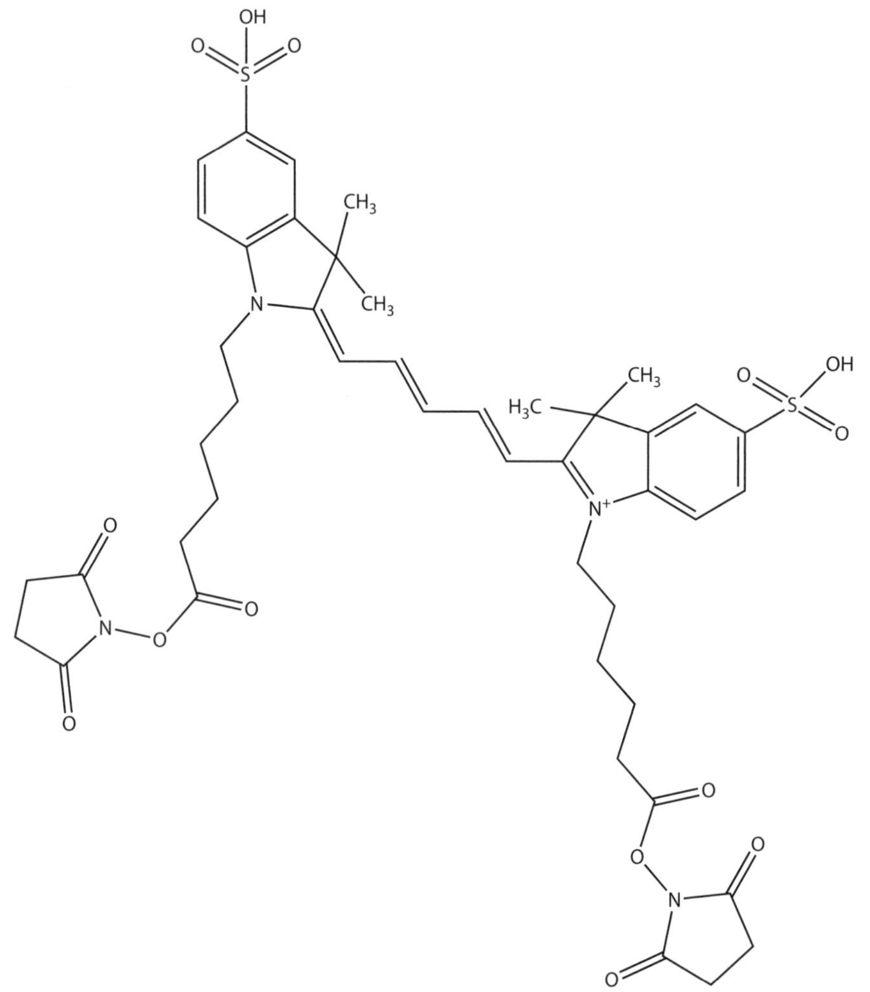

Uchihara, T., Nakamura, A., Nagaoka, U. et al., Dual enhancement of double immunofluorescent signals by CARD: participation of ubiquitin during formation of neurofibrillary tangles, *Histochem. Cell Biol.* 114, 447–451, 2000; Duthie, R.S., Kalve, I.M., Samols, S.B. et al., Novel cyanine dye-based dideoxynucleoside triphosphates for DNA sequencing, *Bioconjug. Chem.* 13, 699–706, 2002; Graves, E.E., Yessayan., D., Turner, G. et al., Validation of *in vivo* fluorochrome concentrations measured using fluorescence molecular tomography, *J. Biomed. Opt.* 10, 44019, 2005; Lapeyre, M., Leprince, J., Massonneau, M. et al., Aryldithioethyloxycarbonyl (Ardec): a new family of amine-protecting groups removable under mild reducing conditions and their applications to peptide synthesis, *Chemistry* 12, 3655–3671, 2006; Tang, X., Morris, S.L., Langone, J.J., and Bockstahler, L.E., Simple and effective method for generating single-stranded DNA targets and probes, *Biotechniques* 40, 759–763, 2006.

| **α-Cyano-4-hydroxycinnamic Acid** | 4-HCCA; Cinnamate | 189.2 | Used as matrix substance for MALDI; transport inhibitor and enzyme inhibitor. |

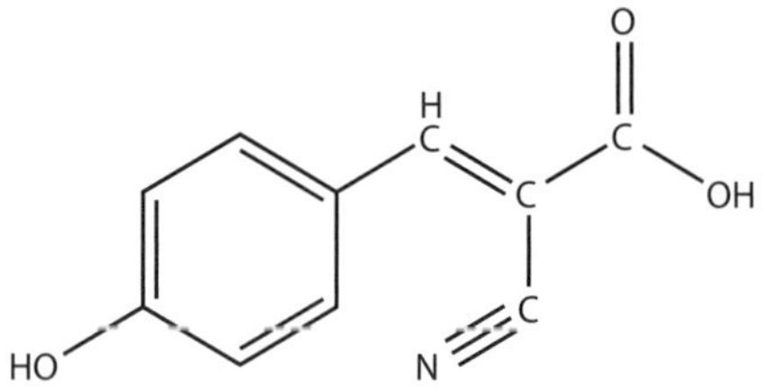

Common Name	Chemical Name	M.W.	Properties and Comment

Gobom, J., Schuerenberg, M., Mueller, M. et al., α-cyano-4-hydroxycinnamic acid affinity sample preparation. A protocol for MALDI-MS peptide analysis in proteomics, *Anal. Chem.* 73, 434–438, 2001; Zhu, X. and Papayannopoulos, I.A., Improvement in the detection of low concentration protein digests on a MALDI TOF/TOF workstation by reducing α-cyano-4-hydroxycinnamic acid adduct ions, *J. Biomol. Tech.* 14, 298–307, 2003; Neubert, H., Halket, J.M., Fernandez Ocana, M., and Patel, R.K., MALDI post-source decay and LIFT-TOF/TOF investigation of α-cyano-4-hydroxycinnamic acid cluster interferences, *J. Am. Soc. Mass Spectrom.* 15, 336–343, 2004; Kobayashi, T., Kawai, H., Suzuki, T. et al., Improved sensitivity for insulin in matrix-assisted laser desorption/ionization time-of-flight mass spectrometry by premixing α-cyano-4-hydroxycinnamic acid with transferrin, *Rapid Commun. Mass Spectrom.* 18, 1156–1160, 2004; Pshenichnyuk, S.A. and Asfandiarov, N.L., The role of free electrons in MALDI: electron capture by molecules of α-cyano-4-hydroxycinnamic acid, *Eur. J. Mass Spectrom.* 10, 477–486, 2004; Bogan, M.J., Bakhoum, S.F., and Agnes, G.R., Promotion of α-cyano-4-hydroxycinnamic acid and peptide cocrystallization within levitated droplets with net charge, *J. Am. Soc. Mass Spectrom.* 16, 254–262, 2005. As enzyme inhibitor: Clarke, P.D., Clift, D.L., Dooledeniya, M. et al., Effects of α-cyano-4-hydroxycinnamic acid on fatigue and recovery of isolated mouse muscle, *J. Muscle Res. Cell Motil.* 16, 611–617, 1995; Del Prete, E., Lutz, T.A., and Scharrer, E., Inhibition of glucose oxidation by α-cyano-4-hydroxycinnamic acid stimulates feeding in rats, *Physiol. Behav.* 80, 489–498, 2004; Briski, K.P. and Patil, G.D., Induction of Fox immunoreactivity labeling in rat forebrain metabolic loci by caudal fourth ventricular infusion of the monocarboxylate transporter inhibitor, α-cyano-4-hydroxycinnamic acid, *Neuroendocrinology* 82, 49–57, 2005.

Common Name	Chemical Name	M.W.	Properties and Comment
Cyanogen	C_2N_2; Ethanedinitrile	53.03	Protein crosslinking at salt bridges.

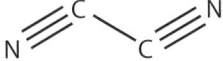

Ghenbot, G., Emge, T., and Day, R.A., Identification of the sites of modification of bovine carbonic anhydrase II (BCA II) by the salt bridge reagent cyanogen, C_2N_2, *Biochim. Biophys. Acta* 1161, 59–65, 1993; Karagozler, A.A., Ghenbot, G., and Day, R.A., Cyanogen as a selective probe for carbonic anhydrase hydrolase, *Biopolymers* 33, 687–692, 1993; Winters, M.S. and Day, R.A., Identification of amino acid residues participating in intermolecular salt bridges between self-associating proteins, *Anal. Biochem.* 309, 48–59, 2002; Winters, M.S. and Day, R.A., Detecting protein–protein interactions in the intact cell of *Bacillus subtilis* (ATCC 6633), *J. Bacteriol.* 185, 4268–4275, 2003.

Common Name	Chemical Name	M.W.	Properties and Comment
Cyanogen Bromide	CNBr; Bromide Cyanide	105.9	Protein modification; cleavage of peptide bonds; coupled nucleophiles to polyhydroxyl matrices; environmental toxicon derived from monobromamine and cyanide.

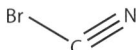

Hofmann, T., The purification and properties of fragments of trypsinogen obtained by cyanogen bromide cleavage, *Biochemistry* 3, 356–364, 1964; Chu, R.C. and Yasunobu, K.T., The reaction of cyanogen bromide and *N*-bromosuccinimide with some cytochromes C, *Biochim. Biophys. Acta* 89, 148–149, 1964; Inglis, A.S. and Edman, P., Mechanism of cyanogen bromide reaction with methionine in peptides and proteins. I. Formation of imidate and methyl thiocyanate, *Anal. Biochem.* 37, 73–80, 1970; Kagedal, L. and Akerstrom, S., Binding of covalent proteins to polysaccharides by cyanogen bromide and organic cyanates. I. Preparation of soluble glycine-, insulin- and ampicillin-dextran, *Acta Chem. Scand.* 25, 1855–1899, 1971; Sipe, J.D. and Schaefer, F.V., Preparation of solid-phase immunosorbents by coupling human serum proteins to cyanogen bromide–activated agarose, *Appl. Microbiol.* 25, 880–884, 1973; March, S.C., Parikh, I., and Cuatrecasas, P., A simplified method for cyanogen bromide activation of agarose for affinity chromatography, *Anal. Biochem.* 60, 149–152, 1974; Boulware, D.W., Goldsworthy, P.D., Nardella, F.A., and Mannik, M., Cyanogen bromide cleaves Fc fragments of pooled human IgG at both methionine and tryptophan residues, *Mol. Immunol.* 22, 1317–1322, 1985; Jaggi, K.S. and Gangal, S.V., Monitoring of active groups of cyanogen bromide-activated paper discs used as allergosorbent, *Int. Arch. Allergy Appl. Immunol.* 89, 311–313, 1989; Villa, S., De Fazio, G., and Canosi, U., Cyanogen bromide cleavage at methionine residues of polypeptides containing disulfide bonds, *Anal. Biochem.* 177, 161–164, 1989; Luo, K.X., Hurley, T.R., and Sefton, B.M., Cyanogen bromide cleavage and proteolytic peptide mapping of proteins immobilized to membranes, *Methods Enzymol.* 201, 149–152, 1991; Jennissen, H.P., Cyanogen bromide and tresyl chloride chemistry revisited: the special reactivity of agarose as a chromatographic and biomaterial support for immobilizing novel chemical groups, *J. Mol. Recognit.* 8, 116–124, 1995; Kaiser, R. and Metzka, L., Enhancement of cyanogen bromide cleavage yields for methionyl-serine and methionyl-threonine peptide bonds, *Anal. Biochem.* 266, 1–8, 1999; Kraft, P., Mills, J., and Dratz, E., Mass spectrometric analysis of cyanogen bromide fragments of integral membrane proteins at the picomole level: application to rhodopsin, *Anal. Biochem.* 292, 76–86, 2001; Kuhn, K., Thompson, A., Prinz, T. et al., Isolation of *N*-terminal protein sequence tags from cyanogen bromide-cleaved proteins as a novel approach to investigate hydrophobic proteins, *J. Proteome Res.* 2, 598–609, 2003; Macmillan, D. and Arham, L., Cyanogen bromide cleavage generates fragments suitable for expressed protein and glycoprotein ligation, *J. Am. Chem. Soc.* 126, 9530–9531, 2004; Lei, H., Minear, R.A., and Marinas, B.J., Cyanogen bromide formation from the reactions of monobromamine and dibromamine with cyanide ions, *Environ. Sci. Technol.* 40, 2559–2564, 2006.

**CHEMICALS COMMONLY USED IN BIOCHEMISTRY AND MOLECULAR
BIOLOGY AND THEIR PROPERTIES (Continued)**

Common Name	Chemical Name	M.W.	Properties and Comment
Cyanuric Chloride	2,4,6-trichloro-1,3,5-triazine	184.41	Coupling of carbohydrates to proteins; more recently for coupling of nucleic acid to microarray platforms.

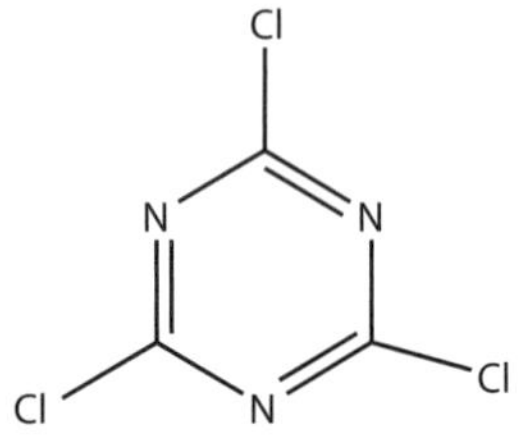

Gray, B.M., ELISA methodology for polysaccharide antigens: protein coupling of polysaccharides for adsorption to plastic tubes, *J. Immunol. Methods* 28, 187–192, 1979: Horak, D., Rittich, B., Safar, J. et al., Properties of RNase A immobilized on magnetic poly(2-hydroxyethyl methacrylate) microspheres, *Biotechnol. Prog.* 17, 447–452, 2001; Lee, P.H., Sawan, S.P., Modrusan, Z. et al., An efficient binding chemistry for glass polynucleotide microarrays, *Bioconjug. Chem.* 13, 97–103, 2002; Steinberg, G., Stromsborg, K., Thomas, L. et al., Strategies for covalent attachment of DNA to beads, *Biopolymers* 73, 597–605, 2004; Abuknesha, R.A., Luk, C.Y., Griffith, H.H. et al., Efficient labeling of antibodies with horseradish peroxidase using cyanuric chloride, *J. Immunol. Methods* 306, 211–217, 2005.

1,2-Cyclohexylenedinitrilotetraacetic acid	CDTA		Chelating agent suggested to have specificity for manganese ions; weaker for other metal ions such as ferric.

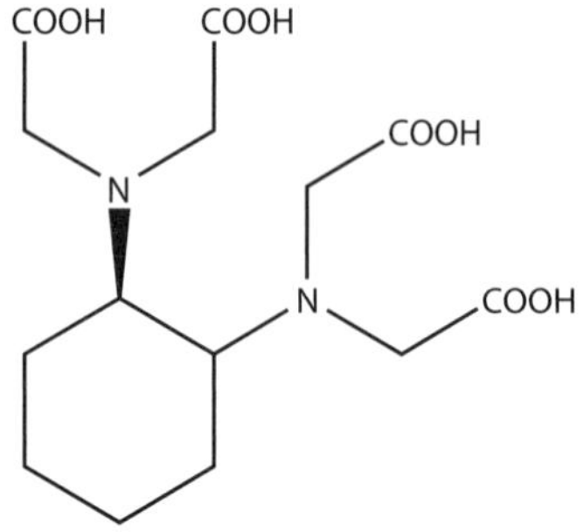

Tandon, S.K. and Singh, J., Removal of manganese by chelating agents from brain and liver of manganese, *Toxicology* 5, 237–241, 1975; Hazell, A.S., Normandin, L., Norenberg, M.D., Kennedy, G., and Yi, J.H., Alzheimer type II astrocyte changes following sub-acute exposure to manganese, *Neurosci. Lett.*, 396, 167–171, 2006; Hassler, C.S. and Twiss, M.R., Bioavailability of iron sensed by a phytoplanktonic Fe-bioreporter, *Environ. Sci. Tech.* 40, 2544–2551, 2006.

Dansyl Chloride	5-(dimethylamino)-1-naphthalenesulfonyl chloride	269.8	Fluorescent label for proteins; amino acid analysis.

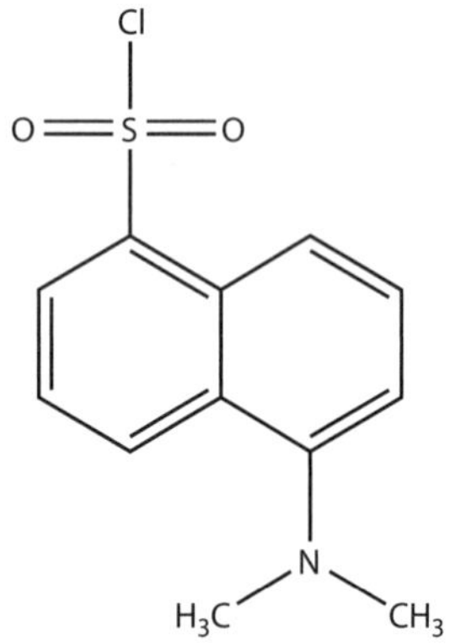

Hill, R.D. and Laing, R.R., Specific reaction of dansyl chloride with one lysine residue in rennin, *Biochim. Biophys. Acta* 132, 188–190, 1967; Chen, R.F., Fluorescent protein-dye conjugates. I. Heterogeneity of sites on serum albumin labeled by dansyl chloride, *Arch. Biochem. Biophys.* 128, 163–175, 1968; Chen, R.F., Dansyl-labeled protein modified with dansyl chloride: activity effects and fluorescence properties, *Anal. Biochem.* 25, 412–416, 1968; Brown, C.S. and Cunningham, L.W., Reaction of reactive sulfydryl groups of creatine kinase with dansyl chloride, *Biochemistry* 9, 3878–3885, 1970; Hsieh, W.T. and Matthews, K.S., Lactose repressor protein modified with dansyl chloride: activity effects and fluorescence properties, *Biochemistry* 34, 3043–3049, 1985;

**CHEMICALS COMMONLY USED IN BIOCHEMISTRY AND MOLECULAR
BIOLOGY AND THEIR PROPERTIES (Continued)**

Common Name	Chemical Name	M.W.	Properties and Comment

Scouten, W.H., van den Tweel, W., Kranenburg, H., and Dekker, M., Colored sulfonyl chloride as an activated agent for hydroxylic matrices, *Methods Enzymol.* 135, 79–84, 1987; Martin, M.A., Lin, B., Del Castillo, B., The use of fluorescent probes in pharmaceutical analysis, *J. Pharm. Biomed. Anal.* 6, 573–583, 1988; Walker, J.M., The dansyl method for identifying *N*-terminal amino acids, *Methods Mol. Biol.* 32, 321–328, 1994; Walker, J.M., The dansyl-Edman method for peptide sequencing, *Methods Mol. Biol.* 32, 329–334, 1994; Pin, S. and Royer, C.A., High-pressure fluorescence methods for observing subunit dissociation in hemoglobin, *Methods Enzymol.* 323, 42–55, 1994; Rangarajan, B., Coons, L.S., and Scarnton, A.B., Characterization of hydrogels using luminescence spectroscopy, *Biomaterials* 17, 649–661, 1996; Kang, X., Xiao, J., Huang, X., and Gu, X., Optimization of dansyl derivatization and chromatographic conditions in the determination of neuroactive amino acids of biological samples, *Clin. Chim. Acta* 366, 352–356, 2006.

| **DCC** | *N,N'*-dicyclohexyl-carbodiimide | 206.33 | Activates carboxyl groups to react with hydroxyl groups to form esters and with amines to form an amide bond; used to modify ion-transporting ATPases. Lack of water solubility has presented challenges. |

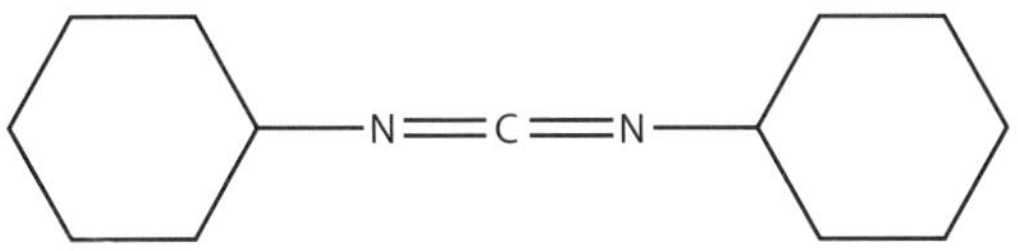

Chau, A.S. and Terry, K., Analysis of pesticides by chemical derivatization. I. A new procedure for the formation of 2-chloroethyl esters of ten herbicidal acids, *J. Assoc. Off. Anal. Chem.* 58, 1294–1301, 1975; Patel, L. and Kaback, H.R., The role of the carbodiimide-reactive component of the adenosine-5′-triphosphatase complex in the proton permeability of *Escherichia coli* membrane vesicles, *Biochemistry* 15, 2741–2746, 1976; Esch, F.S., Bohlen, P., Otsuka, A.S. et al., Inactivation of the bovine mitochondrial F1-ATPase with dicyclohexyl[14C]carbodiimide leads to the modification of a specific glutamic acid residue in the beta subunit, *J. Biol. Chem.* 256, 9084–9089, 1981; Hsu, C.M. and Rosen, B.P., Characterization of the catalytic subunit of an anion pump, *J. Biol. Chem.* 264, 17349–17354, 1989; Gurdag, S., Khandare, J., Stapels, S. et al., Activity of dendrimer-methotrexate conjugates on methotrexate-sensitive and -resistant cell lines, *Bioconjug. Chem.* 17, 275–283, 2006; Vgenopoulou, I., Gemperli, A.C., and Steuber, J., Specific modification of a Na+ binding site in NADH: quinone oxidoreductase from *Klebsiella pneumoniae* with dicyclohexylcarbodiimide, *J. Bacteriol.* 188, 3264–3272, 2006; Ferguson, S.A., Keis, S., and Cook, G.M., Biochemical and molecular characterization of a Na+-translocating F1Fo-ATPase from the thermophilic bacterium *Clostridium paradoxum*, *J. Bacteriol.* 188, 5045–5054, 2006.

| **Deoxycholic Acid** | Desoxycholic Acid | 392.57 | Detergent, nanoparticles. |

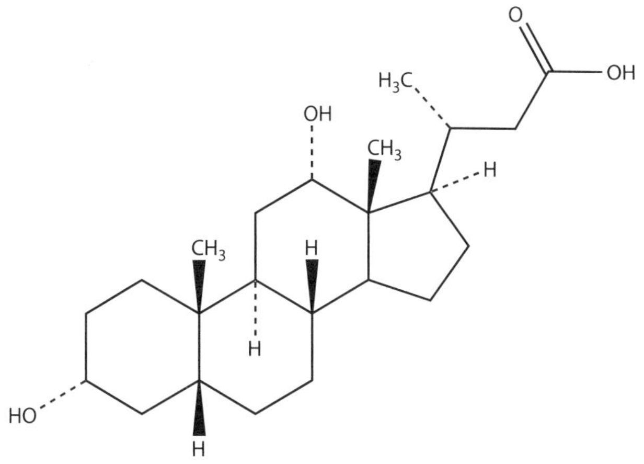

Akare, S. and Martinez, J.D., Bile acid-induced hydrophobicity-dependent membrane alterations, *Biochim. Biophys. Acta* 1735, 59–67, 2005; Chae, S.Y., Son, S., Lee, M. et al., Deoxycholic acid-conjugated chitosan oligosaccharide nanoparticles for efficient gene carrier, *J. Control. Release* 109, 330–344, 2005; Dall'Agnol, M., Bernstein, C., Bernstein, H. et al., Identification of S-nitrosylated proteins after chronic exposure of colon epithelial cells to deoxycholate, *Proteomics* 6, 1654–1662, 2006; Dotis, J., Simitsopoulou, M., Dalakiouridou, M. et al. Effects of lipid formulations of amphotericin B on activity of human monocytes against *Aspirgillus fumigatus*, *Antimicrob. Agents Chemother.* 128, 3490–3491, 2006; Darragh, J., Hunter, M., Pohler, E. et al., The calcium-binding domain of the stress protein SEP53 is required for survival in response to deoxycholic acid-mediated injury, *FEBS J.* 273, 1930–1947, 2006.

CHEMICALS COMMONLY USED IN BIOCHEMISTRY AND MOLECULAR BIOLOGY AND THEIR PROPERTIES (Continued)

Common Name	Chemical Name	M.W.	Properties and Comment
Deuterium Oxide D$_2$O	"Heavy Water"	20.03	Structural studies in proteins, enzyme kinetics; *in vivo* studies of metabolic flux.

Cohen, A.H., Wilkinson, R.R., and Fisher, H.F., Location of deuterium oxide solvent isotope effects in the glutamate dehydrogenase reaction, *J. Biol. Chem.* 250, 5343–5246, 1975; Rosenberry, T.L., Catalysis by acetylcholinesterase: evidence that the rate-limiting step for acylation with certain substrates precedes general acid-base catalysis, *Proc. Natl. Acad. Sci. USA* 72, 3834–3838, 1975; Viggiano, G., Ho, N.T., and Ho, C., Proton nuclear magnetic resonance and biochemical studies of oxygenation of human adult hemoglobin in deuterium oxide, *Biochemistry* 18, 5238–5247, 1979; Bonnete, F., Madern, D., and Zaccai, G., Stability against denaturation mechanisms in halophilic malate dehydrogenase "adapt" to solvent conditions, *J. Mol. Biol.* 244, 436–447, 1994; Thompson, J.F., Bush, K.J., and Nance, S.L., Pancreatic lipase activity in deuterium oxide, *Proc. Soc. Exp. Biol. Med.* 122, 502–505, 1996; Dufner, D. and Previs, S.F., Measuring *in vivo* metabolism using heavy water, *Curr. Opin. Clin. Nutr. Metab. Care* 6, 511–517, 2003; O'Donnell, A.H., Yao, X., and Byers, L.D., Solvent isotope effects on alpha-glucosidase, *Biochem. Biophys. Acta* 1703, 63–67, 2004; Hellerstein, M.K. and Murphy, E., Stable isotope-mass spectrometric measurements of molecular fluxes *in vivo*: emerging applications in drug development, *Curr. Opin. Mol. Ther.* 6, 249–264, 2004; Mazon, H., Marcillat, O., Forest, E., and Vial, C., Local dynamics measured by hydrogen/deuterium exchange and mass spectrometry of the creatine kinase digested by two proteases, *Biochimie* 87, 1101–1110, 2005; Carmieli, R., Papo, N., Zimmerman, H. et al., Utilizing ESEEM spectrscopy to locate the position of specific regions of membrane-active peptides within model membranes, *Biophys. J.* 90, 492–505, 2006.

| DFP | Diisopropylphosphoro-fluoridate; Isofluorophate | 184.15 | Classic cholinesterase inhibitor; inhibitor of serine proteases, some nonspecific reaction tyrosine. |

Baker, B.R., Factors in the design of active-site-directed irreversible inhibitors, *J. Pharm. Sci.* 53, 347–364, 1964; Dixon, G.H. and Schachter, H., The chemical modification of chymotrypsin, *Can. J. Biochem. Physiol.* 42, 695–714, 1964; Singer, S.J., Covalent labeling active site, *Adv. Protein Chem.* 22, 1–54, 1967; Kassell, B. and Kay, J., Zymogens of proteolytic enzymes, *Science* 180, 1022–1027, 1973; Fujino, T., Watanabe, K., Beppu, M. et al., Identification of oxidized protein hydrolase of human erythrocytes as acylpeptide hydrolase, *Biochim. Biophys. Acta* 1478, 102–112, 2000; Manco, G., Camardello, L., Febbraio, F. et al., Homology modeling and identification of serine 160 as nucleophile as the active site in a thermostable carboxylesterase from the archeon *Archaeoglobus fulgidus, Protein Eng.* 13, 197–200, 2000; Gopal, S., Rastogi, V., Ashman, W., and Mulbry, W., Mutagenesis of organophosphorous hydrolase to enhance hydrolysis of the nerve agent VX, *Biochem. Biophys. Res. Commun.* 279, 516–519, 2000; Yeung, D.T., Lenz, D.E., and Cerasoli, D.M., Analysis of active-site amino acid residues of human serum paraoxanse using competitive substrates, *FEBS J.* 272, 2225–2230, 2005; D'Souza, C.A., Wood, D.D., She, Y.M., and Moscarello, M.A., Autocatalytic cleavage of myelin basic protein: an alternative to molecular mimicry, *Biochemistry* 44, 12905–12913, 2005.

| Dichloromethane | Methylene Chloride | 84.9 | Lipid solvent; isolation of sterols, frequently used in combination with methanol. |

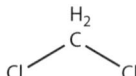

Bouillon, R., Kerkhove, P.V., and De Moor, P., Measurement of 25-hydroxyvitamin D3 in serum, *Clin. Chem.* 22, 364–368, 1976; Redhwi, A.A., Anderson, D.C., and Smith, G.N., A simple method for the isolation of vitamin D metabolites from plasma extracts, *Steroids* 39, 149–154, 1982; Scholtz, R., Wackett, L.P., Egli, C. et al., Dichloromethane dehalogenase with improved catalytic activity isolated form a fast-growing dichloromethane-utilizing bacterium, *J. Bacteriol.* 170, 5698–5704, 1988; Russo, M.V., Goretti, G., and Liberti, A., Direct headspace gas chromatographic determination of dichloromethane in decaffeinated green and roasted coffee, *J. Chromatog.* 465, 429–433, 1989; Shimizu, M., Kamchi, S., Nishii, Y., and Yamada, S., Synthesis of a reagent for fluorescence-labeling of vitamin D and its use in assaying vitamin D metabolites, *Anal. Biochem.* 194, 77–81, 1991; Rodriguez-Palmero, M., de la Presa-Owens, S., Castellote-Bargallo, A.I. et al., Determination of sterol content in different food samples by capillary gas chromatography, *J. Chromatog. A* 672, 267–272, 1994; Raghuvanshi, R.S., Goyal, S., Singh, O., and Panda, A.K., Stabilization of dichloromethane-induced protein denaturation during microencapsulation, *Pharm. Dev. Technol.* 3, 269–276, 1998; El Jaber-Vazdekis, N., Gutierrez-Nicolas, F., Ravelo, A.G., and Zarate, R., Studies on tropane alkaloid extraction by volatile organic solvents: dichloromethane vs. chloroform, *Phytochem. Anal.* 17, 107–113, 2006.

Common Name	Chemical Name	M.W.	Properties and Comment
Diethyldithiocarbamate	Ditiocarb; Dithiocarb; DTC	171.3 (Na)	Chelating agent with particular affinity for Pb, Cu, Zn, Ni; colorimetric determination of Cu.

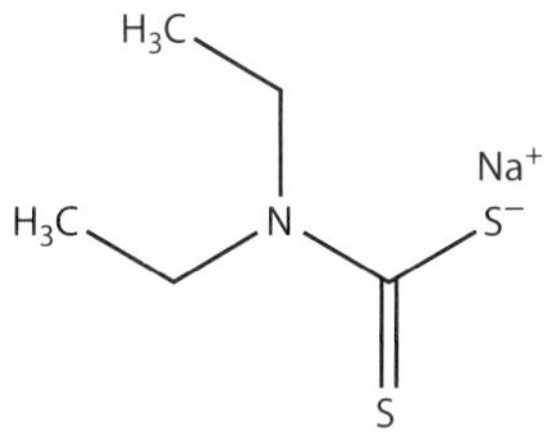

Matsuba, Y. and Takahashi, Y., Spectrophotometric determination of copper with *N,N,N',N'*-tetraethylthiuram disulfide and an application of this method for studies on subcellular distribution of copper in rat brains, *Anal. Biochem.* 36, 182–191, 1970; Koutensky, J., Eybl, V., Koutenska, M. et al., Influence of sodium diethyldithiocarbamate on the toxicity and distribution of copper in mice, *Eur. J. Pharmacol.* 14, 389–392, 1971; Xu, H. and Mitchell, C.L., Chelation of zinc by diethyldithiocarbamate facilitates bursting induced by mixed antidromic plus orthodromic activation of mossy fibers in hippocampal slices, *Brain Res.* 624, 162–170, 1993; Liu, J., Shigenaga, M.K., Yan, L.J. et al., Antioxidant activity of diethyldithiocarbamate, *Free Radic. Res.* 24, 461–472, 1996; Zhang, Y., Wade, K.L., Prestera, T., and Talalav, P., Quantitative determination of isothiocyanates, dithiocarbamates, carbon disulfide, and related thiocarbonyl compounds by cyclocondensation with 1,2-benzenedithiol, *Anal. Biochem.* 239, 160–167, 1996; Shoener, D.F., Olsen, M.A., Cummings, P.G., and Basic, C., Electrospray ionization of neutral metal dithiocarbamate complexes using in-source oxidation, *J. Mass Spectrom.* 34, 1069–1078, 1999; Turner, B.J., Lopes, E.C., and Cheema, S.S., Inducible superoxide dismutase 1 aggregation in transgenic amyotrophic lateral sclerosis mouse fibroblasts, *J. Cell Biochem.* 91, 1074–1084, 2004; Xu, K.Y. and Kuppusamy, P., Dual effects of copper-zinc superoxide dismutase, *Biochem. Biophys. Res. Commun.* 336, 1190–1193, 2005; Jiang, X., Sun, S., Liang, A. et al., Luminescence properties of metal(II)-diethyldithiocarbamate chelate complex particles and its analytical application, *J. Fluoresc.* 15, 859–864, 2005; Wang, J.S. and Chiu, K.H., Mass balance of metal species in supercritical fluid extraction using sodium diethyldithio-carbamate and dibuylammonium dibutyldithiocarbamate, *Anal. Sci.* 22, 363–369, 2006.

Diethylpyrocarbonate (DEPC)	Ethoxyformic Anhydride	162.1	Reagent for modification of proteins and DNA; used as a sterilizing agent; RNAse inhibitor for RNA purification; preservative for wine and fruit fluids.

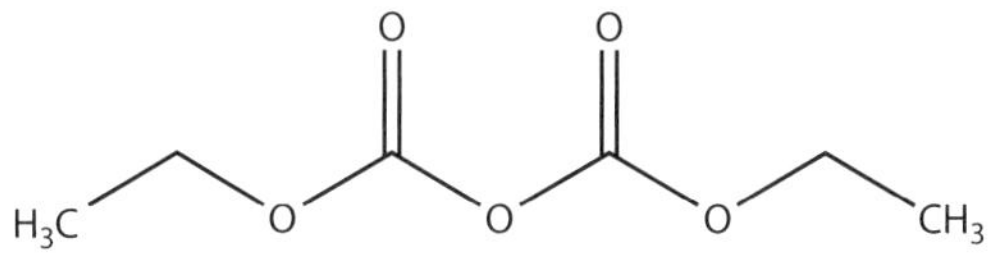

Wolf, B., Lesnaw, J.A., and Reichmann, M.E., A mechanism of the irreversible inactivation of bovine pancreatic ribonuclease by diethylpyrocarbonate. A general reaction of diethylpyrocarbonate with proteins, *Eur. J. Biochem.* 13, 519–525, 1970; Splittstoesser, D.F. and Wilkison, M., Some factors affecting the activity of diethylpyrocarbonate as a sterilant, *Appl. Microbiol.* 25, 853–857, 1973; Fedorcsak, I., Ehrenberg, L., and Solymosy, F., Diethylpyrocarbonate does not degrade RNA, *Biochem. Biophys. Res. Commun.* 65, 490–496, 1975; Berger, S.L., Diethylpyrocarbonate: an examination of its properties in buffered solutions with a new assay technique, *Anal. Biochem.* 67, 428–437, 1975; Lloyd, A.G. and Drake, J.J., Problems posed by essential food preservatives, *Br. Med. Bull.* 31, 214–219, 1975; Ehrenberg, L., Fedorcsak, I., and Solymosy, F., Diethylpyrocarbonate in nucleic acid research, *Prog. Nucleic Acid Res. Mol. Biol.* 16, 189–262, 1976; Saluz, H.P. and Jost, J.P., Approaches to characterize protein–DNA interactions *in vivo*, *Crit. Rev. Eurkaryot. Gene Expr.* 3, 1–29, 1993; Bailly, C. and Waring, M.J., Diethylpyrocarbonate and osmium tetroxide as probes for drug-induced changes in DNA conformation *in vitro*, *Methods Mol. Biol.* 90, 51–59, 1997; Mabic, S. and Kano, I., Impact of purified water quality on molecular biology experiments, *Clin. Chem. Lab. Med.* 41, 486–491, 2003; Colleluori, D.M., Reczkowski, R.S., Emig, F.A. et al., Probing the role of the hyper-reactive histidine residue of argininase, *Arch. Biochem. Biophys.* 444, 15–26, 2005; Wu, S.N. and Chang, H.D., Diethylpyrocarbonate, a histidine-modifying agent, directly stimulates activity of ATP-sensitive potassium channels in pituitary GH(3) cells, *Biochem. Pharmacol.* 71, 615–623, 2006.

Dimedone	5,5-dimethyl-1,3-cyclohexanedione	140.18	Originally described as reagent for assay of aldehydes; used as a specific modifier of sulfenic acid.

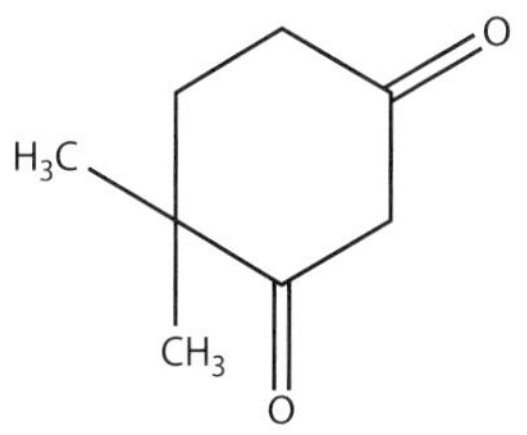

**CHEMICALS COMMONLY USED IN BIOCHEMISTRY AND MOLECULAR
BIOLOGY AND THEIR PROPERTIES (Continued)**

Common Name	Chemical Name	M.W.	Properties and Comment

Bulmer, D., Dimedone as an aldehyde-blocking reagent to facilitate the histochemical determination of glycogen, *Stain Technol.* 34, 95–98, 1959; Sawicki, E. and Carnes, R.A., Spectrophotofluorimetric determination of aldehydes with dimedone and other reagents, *Mikrochim. Acta* 1, 95–98, 1968; Benitez, L.V. and Allison, W.S., The inactivation of the acyl phosphatase activity catalyzed by the sulfenic acid form of glyceraldehyde 3-phosphate dehydrogenase by dimedone and olefins, *J. Biol. Chem.* 249, 6234–6243, 1974; Huszti, Z. and Tyihak, E., Formation of formaldehyde from *S*-adenosyl-L-[methyl-³H]methionine during enzymic transmethylation of histamine, *FEBS Lett.* 209, 362–366, 1986; Sardi, E. and Tyihak, E., Sample determination of formaldehyde in dimedone adduct form in biological samples by high-performance liquid chromatography, *Biomed. Chromatog.* 8, 313–314, 1994; Demaster, A.G., Quast, B.J., Redfern, B., and Nagasawa, H.T., Reaction of nitric oxide with the free sulfhydryl group of human serum albumin yields a sulfenic acid and nitrous oxide, *Biochemistry* 34, 14494–14949, 1995; Rozylo, T.K., Siembida, R., and Tyihak, E., Measurement of formaldehyde as dimedone adduct and potential formaldehyde precursors in hard tissues of human teeth by overpressurized layer chromatography, *Biomed. Chromatog.* 13, 513–515, 1999; Percival, M.D., Ouellet, M., Campagnolo, C. et al., Inhibition of cathepsin K by nitric oxide donors: evidence for the formation of mixed disulfides and a sulfenic acid, *Biochemistry* 38, 13574–13583, 1999; Carballal, S., Radi, R., Kirk, M.C. et al., Sulfenic acid formation in human serum albumin by hydrogen peroxide and peroxynitrite, *Biochemistry* 42, 9906–9914, 2003; Poole, L.B., Zeng, B.-B., Knaggs, S.A., Yakuba, M., and King, S.B., Synthesis of chemical probes to map sulfenic acid modifications on proteins, *Bioconjugate Chem.* 16, 1624–1628, 2005; Kaiserov, K., Srivastava, S., Hoetker, J.D. et al., Redox activation of aldose reductase in the ischemic heart, *J. Biol. Chem.* 281, 15110–15120, 2006.

Dimethylformamide (DMF)   *N,N*-dimethylformamide   73.09   Solvent.

Eliezer, N. and Silberberg, A., Structure of branched poly-alpha-amino acids in dimethylformamide. I. Light scattering, *Biopolymers* 5, 95–104, 1967; Bonner, O.D., Bednarek, J.M., and Arisman, R.K., Heat capacities of ureas and water in water and dimethylformamide, *J. Am. Chem. Soc.* 99, 2898–2902, 1977; Sasson, S. and Notides, A.C., The effects of dimethylformamide on the interaction of the estrogen receptor with estradiol, *J. Steroid Biochem.* 29, 491–495, 1988; Jeffers, R.J., Feng, R.Q., Fowlkes, J.B. et al., Dimethylformamide as an enhancer of cavitation-induced cell lysis *in vitro*, *J. Acoust. Soc. Am.* 97, 669–676, 1995; You, L. and Arnold, F.H., Directed evolution of subtilisin E in *Bacillus subtilis* to enhance total activity in aqueous dimethylformamide, *Protein Eng.* 9, 77–83, 1996; Szabo, P.T. and Kele, Z., Electrospray mass spectrometry of hydrophobic compounds using dimethyl sulfoxide and dimethylformamide, *Rapid Commun. Mass Spectrom.* 15, 2415–2419, 2001; Nishida, Y., Shingu, Y., Dohi, H., and Kobayashi, K., One-pot alpha-glycosylation method using Appel agents in *N,N*-dimethylformamide, *Org. Lett.* 5, 2377–2380, 2003; Shingu, Y., Miyachi, A., Miura, Y. et al., One-pot alpha-glycosylation pathway via the generation *in situ* of alpha-glycopyranosyl imidates I *N,N*-dimethylformamide, *Carbohydr. Res.* 340, 2236–2244, 2005; Porras, S.P. and Kenndler, E., Capillary electrophoresis in *N,N*-dimethylformamide, *Electrophoresis* 26, 3279–3291, 2005; Wei, Q., Zhang, H., Duan, C. et al., High sensitive fluorophotometric determination of nucleic acids with pyronine G sensitized by *N,N*-dimethylformamide, *Ann. Chim.* 96, 273–284, 2006.

Dimethyl Suberimidate (DMS)   Crosslinking agent.

Davies, G.E. and Stark, G.R., Use of dimethyl suberimidate, a crosslinking reagent, in studying the subunit structure of oligomeric proteins, *Proc. Natl. Acad. Sci. USA* 66, 651–656, 1970; Hassell, J. and Hand, A.R., Tissue fixation with diimidoesters as an alternative to aldehydes. I. Comparison of crosslinking and ultrastructure obtained with dimethylsuberimidate and glutaraldehyde, *J. Histochem. Cytochem.* 22, 223–229, 1974; Thomas, J.O., Chemical crosslinking of histones, *Methods Enzymol.* 170, 549–571, 1989; Roth, M.R., Avery, R.B., and Welti, R., Crosslinking of phosphatidylethanolamine neighbors with dimethylsuberimidate is sensitive to the lipid phase, *Biochim. Biophys. Acta* 986, 217–224, 1989; Redl, B., Walleczek, J., Soffler-Meilicke, M., and Stoffler, G., Immunoblotting analysis of protein–protein crosslinks within the 50S ribosomal subunit of *Escherichia coli*. A study using dimethylsuberimidate as crosslinking reagent, *Eur. J. Biochem.* 181, 351–256, 1989; Konig, S., Hubner, G., and Schellenberger, A., Crosslinking of pyruvate decarboxylase-characterization of the native and substrate-activated enzyme states, *Biomed. Biochim. Acta* 49, 465–471, 1990; Chen, J.C., von Lintig, F.C., Jones, S.B. et al., High-efficiency solid-phase capture using glass beads bonded to microcentrifuge tubes: immunoprecipitation of proteins from cell extracts and assessment of ras activation, *Anal. Biochem.* 302, 298–304, 2002; Dufes, C., Muller, J.M., Couet, W. et al., Anticancer drug delivery with transferrin-targeted polymeric chitosan vesicles, *Pharm. Res.* 21, 101–107, 2004; Levchenko, V. and Jackson, V., Histone release during transcription: NAP1 forms a complex with H2A and H2B and facilitates a topologically dependent release of H3 and H4 from the nucleosome, *Biochemistry* 43, 2358–2372, 2004; Jastrzebska, M., Barwinski, B., Mroz, I. et al., Atomic force microscopy investigation of chemically stabilized pericardium tissue, *Eur. Phys. J. E* 16, 381–388, 2005.

**CHEMICALS COMMONLY USED IN BIOCHEMISTRY AND MOLECULAR
BIOLOGY AND THEIR PROPERTIES (Continued)**

Common Name	Chemical Name	M.W.	Properties and Comment
Dimethyl Sulfate		126.1	Methylating agent; methylation of nucleic acids; used for a process called footprinting to identify sites of protein–nucleic acid interaction.

Nielsen, P.E., *In vivo* footprinting: studies of protein–DNA interactions in gene regulation, *Bioessay* 11, 152–155, 1989; Saluz, H.P. and Jost, J.P., Approaches to characterize protein–DNA interactions *in vivo*, *Crit. Rev. Eurkaryot. Gene Expr.* 3, 1–29, 1993; Saluz, H.P. and Jost, J.P., *In vivo* DNA footprinting by linear amplification, *Methods Mol. Biol.* 31, 317–329, 1994; Paul, A.L. and Ferl, R.J., *In vivo* footprinting of protein–DNA interactions, *Methods Cell Biol.* 49, 391–400, 1995; Gregory, P.D., Barbaric, S., and Horz, W., Analyzing chromatin structure and transcription factor binding in yeast, *Methods* 15, 295–302, 1998; Simpson, R.T., *In vivo* to analyze chromatin structrure, *Curr. Opin. Genet. Dev.* 9, 225–229, 1999; Nawrocki, A.R., Goldring, C.E., Kostadinova, R.M. et al., *In vivo* footprinting of the human 11β-hydroxysteroid dehydro-genase type 2 promoter: evidence for cell-specific regulation by Sp1 and Sp3, *J. Biol. Chem.* 277, 14647–14656, 2002; McGarry, K.C., Ryan, V.T., Grimwade, J.E., and Leonard, A.C., Two discriminatory binding sites in the *Escherichia coli* replication origin are required for DNA stand opening by initiator DnaA-ATP, *Proc. Natl. Acad. Sci. USA* 101, 2811–2816, 2004; Kellersberger, K.A., Yu, E., Kruppa, G.H. et al., Two-down characterization of nucleic acids modified by structural probes using high-resolution tandem mass spectrometry and automated data interpretation, *Anal. Chem.* 76, 2438–2445, 2004; Matthews, D.H., Disney, M.D., Childs, J.L. et al., Incorporating chemical modification constraints into a dynamic programming algorithm for prediction of RNA secondary structure, *Proc. Natl. Acad. Sci. USA* 101, 7287–7292, 2004; Forstemann, K. and Lingner, J., Telomerase limits the extent of base pairing between template RNA and temomeric DNA, *EMBO Rep.* 6, 361–366, 2005; Kore, A.R. and Parmar, G., An industrial process for selective synthesis of 7-methyl guanosine 5′-diphosphate: versatile synthon for synthesis of mRNA cap analogues, *Nucleosides Nucleotides Nucleic Acids* 25, 337–340, 2006.

Dioxane	1,4-diethylene Dioxide	88.1	Solvent.

Sideri, C.N. and Osol, A., A note on the purification of dioxane for use in preparing nonaqueous titrants, *J. Am. Pharm. Am. Pharm. Assoc.* 42, 586, 1953; Martel, R.W. and Kraus, C.A., The association of ions in dioxane-water mixtures at 25 degrees, *Proc. Natl. Acad. Sci. USA* 41, 9–20, 1955; Mercier, P.L. and Kraus, C.A, The ion-pair equilibrium of electrolyte solutions in dioxane-water mixtures, *Proc. Natl. Acad. Sci. USA* 41, 1033–1041, 1995; Inagami, T., and Sturtevant, J.M., The trypsin-catalyzed hydrolysis of benzoyl-L-arginine ethyl ester. I. The kinetics in dioxane-water mixtures, *Biochim. Biophys. Acta* 38, 64–79, 1980; Zaeklj, A. and Gros, M., Electrophoresis of lipoprotein, prestained with Sudan Black B, dissolved in a mixture of dioxane and ethylene glycol, *Clin. Chim. Acta* 5, 947, 1960; Krasner, J. and McMenamy, R.H., The binding of indole compounds to bovine plasma albumin. Effects of potassium chloride, urea, dioxane, and glycine, *J. Biol. Chem.* 241, 4186–4196, 1966; Smith, R.R. and Canady, W.J., Solvation effects upon the thermodynamic substrate activity: correlation with the kinetics of enzyme-catalyzed reactions. II. More complex interactions of alpha-chymotrypsin with dioxane and acetone which are also competitive inhibitors, *Biophys. Chem.* 43, 189–195, 1992; Forti, F.L., Goissis, G., and Plepis, A.M., Modifications on collagen structures promoted by 1,4-dioxane improve thermal and biological properties of bovine pericardium as a biomaterial, *J. Biomater. Appl.* 20, 267–285, 2006.

Dithiothreitol	1,4-dithiothreitol; DTT; Cleland's Reagent; *threo*-2,3-dihydroxy-1,4-dithiolbutane	154.3	Reducing agent.

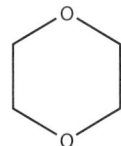

**CHEMICALS COMMONLY USED IN BIOCHEMISTRY AND MOLECULAR
BIOLOGY AND THEIR PROPERTIES (Continued)**

Common Name	Chemical Name	M.W.	Properties and Comment

Cleland, W.W., Dithiothreitol, a new protective reagent for SH groups, *Biochemistry* 3, 480–482, 1964; Gorin, G., Fulford, R., and Deonier, R.C., Reaction of lysozyme with dithiothreitol and other mercaptans, *Experientia* 24, 26–27, 1968; Stanton, M. and Viswantha, T., Reduction of chymotryptin A by dithiothreitol, *Can. J. Biochem.* 49, 1233–1235, 1971; Warren, W.A., Activation of serum creatine kinase by dithiothreitol, *Clin. Chem.* 18, 473–475, 1972; Hase, S. and Walter, R., Symmetrical disulfide bonds as *S*-protecting groups and their cleavage by dithiothreitol: synthesis of oxytocin with high biological activity, *Int. J. Pept. Protein Res.* 5, 283–288, 1973; Fleisch, J.H., Krzan, M.C., and Titus, E., Alterations in pharmacologic receptor activity by dithiothreitol, *Am. J. Physiol.* 227, 1243–1248, 1974; Olsen, J. and Davis, L., The oxidation of dithiothreitol by peroxidases and oxygen, *Biochim. Biophys. Acta.* 445, 324–329, 1976; Chao, L.P., Spectrophotometric determination of choline acetyltransferase in the presence of dithiothreitol, *Anal. Biochem.* 85, 20–24, 1978; Fukada, H. and Takahashi, K., Calorimetric study of the oxidation of dithiothreitol, *J. Biochem.* 87, 1105–1110, 1980; Alliegro, M.C., Effects of dithiothreitol on protein activity unrelated to thiol-disulfide exchange: for consideration in the analysis of protein function with Cleland's reagent, *Anal. Biochem.* 282, 102–106, 2000; Rhee, S.S. and Burke, D.H., Tris(2-carboxyethyl)phosphine stabilization of RNA: comparison with dithiothreitol for use with nucleic acid and thiophosphoryl chemistry, *Anal. Biochem.* 325, 137–143, 2004; Pan, J.C., Cheng, Y., Hui, E.F., and Zhou, H.M., Implications of the role of reactive cysteine in arginine kinase: reactivation kinetics of 5,5′-dithiobis-(2-nitrobenzoic acid)-modified arginine kinase reactivated by dithiothreitol, *Biochem. Biophys. Res. Commun.* 317, 539–544, 2004; Thaxton, C.S., Hill, H.D., Georganopoulou, D.G. et al., A bio-barcode assay based upon dithiothreitol-induced oligonucleotide release, *Anal. Chem.* 77, 8174–8178, 2005.

DMSO	Dimethylsulfoxide	78.13	Solvent; suggested therapeutic use; effect on cellular function; cyropreservative.

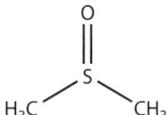

Huggins, C.E., Reversible agglomeration used to remove dimethylsulfoxide from large volumes of frozen blood, *Science* 139, 504–505, 1963; Yehle, A.V. and Doe, R.H., Stabilization of *Bacillus subtilis* phage with dimethylsulfoxide, *Can. J. Microbiol.* 11, 745–746, 1965; Fowler, A.V. and Zabin, I., Effects of dimethylsulfoxide on the lactose operon of *Escherichia coli*, *J. Bacteriol.* 92, 353–357, 1966; Williams, A.E. and Vinograd, J., The buoyant behavior of RNA and DNA in cesium sulfate solutions containing dimethylsulfoxide, *Biochim. Biophys. Acta* 228, 423–439, 1971; Levine, W.G., The effect of dimethylsulfoxide on the binding of 3-methylcholanthrene to rat liver fractions, *Res. Commun. Chem. Pathol. Pharmacol.* 4, 511–518, 1972; Fink, A.L, The trypsin-catalyzed hydrolysis of *N*-alpha-benzoyl-L-lysine *p*-nitrophenyl ester in dimethylsulfoxide at subzero temperatures, *J. Biol. Chem.* 249, 5072–5932, 1974; Hutton, J.R. and Wetmur, J.G., Activity of endonuclease S1 in denaturing solvents: dimethylsulfoxide, dimethylformamide, formamide, and formaldehyde, *Biochem. Biophys. Res. Commun.* 66, 942–948, 1975; Gal, A., De Groot, N., and Hochberg, A.A., The effect of dimethylsulfoxide on ribosomal fractions from rat liver, *FEBS Lett.* 94, 25–27, 1978; Barnett, R.E., The effects of dimethylsulfoxide and glycerol on Na⁺, K⁺-ATPase, and membrane structure, *Cryobiology* 15, 227–229, 1978; Borzini, P., Assali, G., Riva, M.R. et al., Platelet cryopreservation using dimethylsulfoxide/polyethylene glycol/sugar mixture as cryopreserving solution, *Vox Sang.* 64, 248–249, 1993; West, R.T., Garza, L.A., II, Winchester, W.R., and Walmsley, J.A., Conformation, hydrogen bonding, and aggregate formation of guanosine 5′-monophosphate and guanosine in dimethylsulfoxide, *Nucleic Acids Res.* 22, 5128–5134, 1994; Bhattacharjya, S. and Balarma, P., Effects of organic solvents on protein structures; observation of a structured helical core in hen egg-white lysozyme in aqueous dimethylsulfoxide, *Proteins* 29, 492–507, 1997; Simala-Grant, J.L. and Weiner, J.H., Modulation of the substrate specificity of *Escherichia coli* dimethylsulfoxide reductase, *Eur. J. Biochem.* 251, 510–515, 1998; Tsuzuki, W., Ue, A., and Kitamura, Y., Effect of dimethylsulfoxide on hydrolysis of lipase, *Biosci. Biotechnol. Biochem.* 65, 2078–2082, 2001; Pedersen, N.R., Halling, P.J., Pedersen, L.H. et al., Efficient transesterification of sucrose catalyzed by the metalloprotease thermolysin in dimethylsulfoxide, *FEBS Lett.* 519, 181–184, 2002; Fan, C., Lu, J., Zhang, W., and Li, G., Enhanced electron-transfer reactivity of cytochrome b5 by dimethylsulfoxide and *N,N′*-dimethylformamide, *Anal. Sci.* 18, 1031–1033, 2002; Tait, M.A. and Hik, D.S., Is dimethylsulfoxide a reliable solvent for extracting chlorophyll under field conditions? *Photosynth. Res.* 78, 87–91, 2003; Malinin, G.I. and Malinin, T.I., Effects of dimethylsulfoxide on the ultrastructure of fixed cells, *Biotech. Histochem.* 79, 65–69, 2004; Clapisson, G., Salinas, C., Malacher, P. et al., Cryopreservation with hydroxyethylstarch (HES) + dimethylsulfoxide (DMSO) gives better results than DMSO alone, *Bull. Cancer* 91, E97–E102, 2004.

EDC	1-ethyl-(3-dimethylamino propyl)carbodiimide; *N*-(3-dimethylamino-propyl)-*N*′-ethyl-carbodiimide	191.7 (HCl)	Water-soluble carbodiimide for the modification of carboxyl groups in proteins; zero-length crosslinking proteins; activation of carboxyl groups for amidation reactions, as for the coupling of amino-nucleotides to matrices for DNA microarrays.

CHEMICALS COMMONLY USED IN BIOCHEMISTRY AND MOLECULAR BIOLOGY AND THEIR PROPERTIES (Continued)

Common Name	Chemical Name	M.W.	Properties and Comment

Lin, T.Y. and Koshland, D.E., Jr., Carboxyl group modification and the activity of lysozyme, *J. Biol. Chem.* 244, 505–508, 1969; Carraway, K.L., Spoerl, P., and Koshland, D.E., Jr., Carboxyl group modification in chymotrypsin and chymotrypsinogen, *J. Mol. Biol.* 42, 133–137, 1969; Yamada, H., Imoto, T., Fujita, K. et al., Selective modification of aspartic acid-101 in lysozyme by carbodiimide reaction, *Biochemistry* 20, 4836–4842, 1981; Buisson, M. and Reboud, A.M., Carbodiimide-induced protein-RNA crosslinking in mammalian subunits, *FEBS Lett.* 148, 247–250, 1982; Millett, F., Darley-Usmar, V., and Capaldi, R.A., Cytochrome c is crosslinked to subunit II of cytochrome c oxidase by a water-soluble carbodiimide, *Biochemistry* 21, 3857–3862, 1982; Chen, S.C., Fluorometric determination of carbodiimides with trans-aconitic acid, *Anal. Biochem.* 132, 272–275, 1983; Davis, L.E., Roth, S.A., and Anderson, B., Antisera specificities to 1-ethyl-3-(3-dimethylaminopropyl) carbodiimide adducts of proteins, *Immunology* 53, 435–441, 1984; Ueda, T., Yamada, H., and Imoto, T., Highly controlled carbodiimide reaction for the modification of lysozyme. Modification of Leu129 or As119, *Protein Eng.* 1, 189–193, 1987; Ghosh, M.K., Kildsig, D.O., and Mitra, A.K., Preparation and characterization of methotrexate-immunoglobulin conjugates, *Drug. Des. Deliv.* 4, 13–25, 1989; Grabarek, Z. and Gergely, J., Zero-length crosslinking procedure with the use of active esters, *Anal. Biochem.* 185, 131–135, 1990; Gilles, M.A., Hudson, A.Q., and Borders, C.L., Jr., Stability of water-soluble carbodiimides in aqueous solutions, *Anal. Biochem.* 184, 244–248, 1990; Soinila, S., Mpitsos, G.J., and Soinila, J., Immunohistochemistry of enkephalins: model studies on hapten-carrier conjugates and fixation methods, *J. Histochem. Cytochem.* 40, 231–239, 1992; Soper, S.A., Hashimoto, M., Situma, C. et al., Fabrication of DNA microarrays onto polymer substrates using UV modification protocols with integration into microfluidic platforms for the sensing of low-abundant DNA point mutations, *Methods* 37, 103–113, 2005.

Common Name	Chemical Name	M.W.	Properties and Comment
EDTA	Ethylenediaminetetraacetic acid	292.24	Chelating agent; some metal ion-EDTA complexes (i.e., Fe^{2+}-EDTA) function as chemical nucleases.

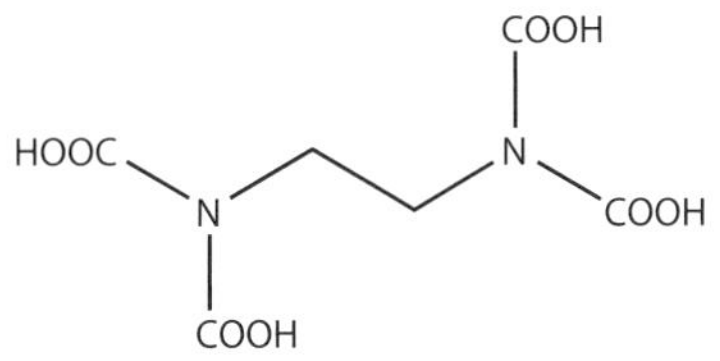

Flaschka, H.A., *EDTA Titrations: An Introduction to Theory and Practice*, Pergammon Press, Oxford, UK, 1964; West, T.S., *Complexometry with EDTA and Related Reagents*, BDH Chemicals Ltd., Poole (Dorset), UK, 1969; Pribil, R., *Analytical Applications of EDTA and Related Compounds*, Pergammon Press, Oxford, UK, 1972; Papavassiliou, A.G., Chemical nucleases as probes for studying DNA–protein interactions, *Biochem. J.* 305, 345–357, 1995; Martell, A.E., and Hancock, R.D., *Metal Complexes in Aqueous Solutions*, Plenum Press, New York, 1996; Loizos, N. and Darst, S.A, Mapping protein–ligand interactions by footprinting, a radical idea, *Structure* 6, 691–695, 1998; Franklin, S.J., Lanthanide-mediated DNA hydrolysis, *Curr. Opin. Chem. Biol.* 5, 201–208, 2001; Heyduk, T., Baichoo, N., and Henduk, E., Hydroxyl radical footprinting of proteins using metal ion complexes, *Met. Ions Biol. Syst.* 38, 255–287, 2001; Orlikowsky, T.W., Neunhoeffer, F., Goelz, R. et al., Evaluation of IL-8-concentrations in plasma and lyszed EDTA-blood in healthy neonates and those with suspected early onset bacterial infection, *Pediatr. Res.* 56, 804–809, 2004; Matt, T., Martinez-Yamout, M.A., Dyson, H.J., and Wright, P.E., The CBP/p300 TAZ1 domain in its native state is not a binding partner of MDM2, *Biochem. J.* 381, 685–691, 2004; Nyborg, J.K. and Peersen, O.B., That zincing feeling: the effects of EDTA on the behavior of zinc-binding transcriptional regulators, *Biochem. J.* 381, e3–e4, 2004; Haberz, P., Rodriguez-Castanada, F., Junker, J. et al., Two new chiral EDTA-based metal chelates for weak alignment of proteins in solution, *Org. Lett.* 8, 1275–1278, 2006.

Common Name	Chemical Name	M.W.	Properties and Comment
Ellman's Reagent	5,5′-dithiobis[2-nitro-benzoic] acid	396.35	Reagent for determination of sulfydryl groups/ disulfide bonds.

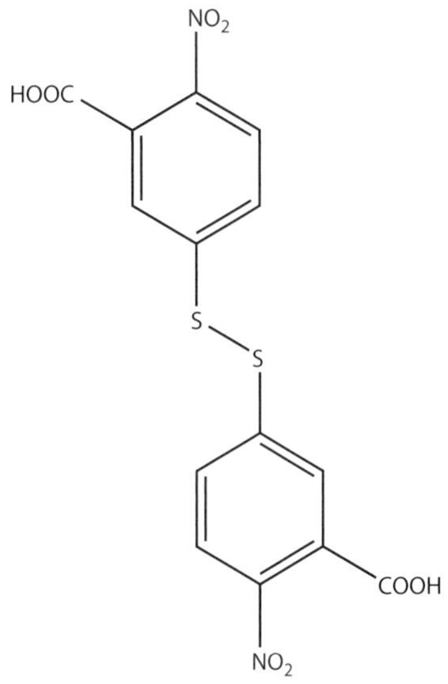

CHEMICALS COMMONLY USED IN BIOCHEMISTRY AND MOLECULAR
BIOLOGY AND THEIR PROPERTIES (Continued)

Common Name	Chemical Name	M.W.	Properties and Comment

Ellman, G.L., Tissue sulfydryl groups, *Arch. Biochem. Biophys.* 82, 70–77, 1959; Boyne, A.F. and Ellman, G.L., A methodology for analysis of tissue sulfydryl components, *Anal. Biochem.* 46, 639–653, 1972; Brocklehurst, K., Kierstan, M., and Little, G., The reaction of papain with Ellman's reagent (5,5′-dithiobis-(2-nitrobenzoate), *Biochem. J.* 128, 811–816, 1972; Weitzman, P.D., A critical reexamination of the reaction of sulfite with DTNB, *Anal. Biochem.* 64, 628–630, 1975; Hull, H.H., Chang, R., and Kaplan, L.J., On the location of the sulfhydryl group in bovine plasma albumin, *Biochim. Biophys. Acta* 400, 132–136, 1975; Banas, T., Banas, B., and Wolny, M., Kinetic studies of the reactivity of the sulfydryl groups of glyceraldehyde-3-phosphate dehydrogenase, *Eur. J. Biochem.* 68, 313–319, 1976; der Terrossian, E. and Kassab, R., Preparation and properties of *S*-cyano derivatives of creatine kinase, *Eur. J. Biochem.* 70, 623–628, 1976; Riddles, P.W., Blakeley, R.L., and Zerner, B., Ellman's reagent: 5,5′-dithiobis(2-nitrobenzoic acid) — a reexamination, *Anal. Biochem.* 94, 75–81, 1979; Luthra, N.P., Dunlap, R.B., and Odom, J.D., Characterization of a new sulfydryl group reagent: 6, 6′- diselenobis-(3-nitrobenzoic acid), a selenium analog of Ellman's reagent, *Anal. Biochem.* 117, 94–102, 1981; Di Simplicio, P., Tiezzi, A., Moscatelli, A. et al., The SH-SS exchange reaction between the Ellman's reagent and protein-containing SH groups as a method for determining conformational states: tubulin, *Ital. J. Biochem.* 38, 83–90, 1989; Woodward, J., Tate, J., Herrmann, P.C., and Evans, B.R., Comparison of Ellman's reagent with *N*-(1-pyrenyl)maleimide for the determination of free sulfydryl groups in reduced cellobiohydrolase I from *Trichoderma reesei, J. Biochem. Biophys. Methods* 26, 121–129, 1993; Berlich, M., Menge, S., Bruns, I. et al., Coumarins give misleading absorbance with Ellman's reagent suggestive of thiol conjugates, *Analyst* 127, 333–336, 2002; Riener, C.K., Kada, G., and Gruber, H.J., Quick measurement of protein sulfhydryls of Ellman's reagents and with 4,4′-dithiopyridine, *Anal. Bio. Anal. Chem.* 373, 266–276, 2002; Zhu, J., Dhimitruka, I., and Pei, D., 5-(2-aminoethyl)dithio-2-nitrobenzoate as a more base-stable alternative to Ellman's reagent, *Org. Lett.* 6, 3809–3812, 2004; Owusu-Apenten, R., Colorimetric analysis of protein sulfhydryl groups in milk: applications and processing effects, *Crit. Rev. Food Sci. Nutr.* 45, 1–23, 2005.

Ethanolamine	Glycinol	61.08	Buffer component; component of a phospholipid (phosphatidyl ethanolamine, PE).

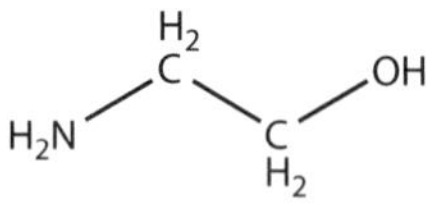

Vance, D.E. and Ridgway, N.D., The methylation of phosophatidylethanolamine, *Prog. Lipid Res.* 27, 61–79, 1988; Louwagie, M., Rabilloud, T., and Garin, J., Use of ethanolamine for sample stacking in capillary electrophoresis, *Electrophoresis* 19, 2440–2444, 1998; de Nogales, V., Ruiz, R., Roses. M. et al., Background electrolytes in 50% methanol/water for the determination of acidity constants of basic drugs by capillary zone electrophoresis, *J. Chromatog. A* 1123, 113–120, 2006.

Ethidium Bromide		394.31	

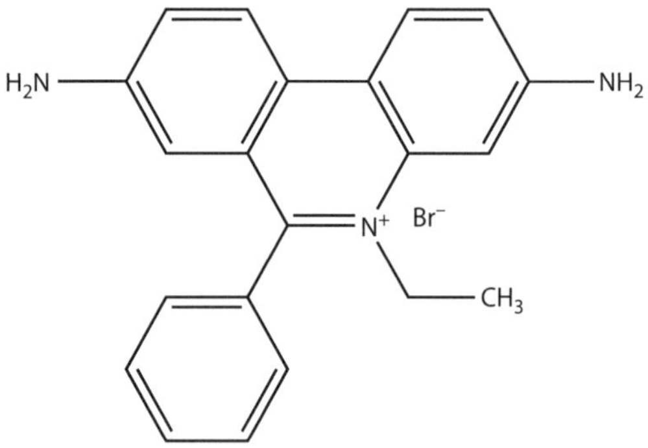

Sela, I., Fluorescence of nucleic acids with ethidium bromide: an indication of the configurative state of nucleic acids, *Biochim. Biophys. Acta* 190, 216–219, 1969; Le Pecq, J.B., Use of ethidium bromide for separation and determination of nucleic acids of various conformational forms and measurement of their associated enzymes, *Methods Biochem. Anal.* 20, 41–86, 1971; Borst, P., Ethidium DNA agarose gel electrophoresis: how it started, *IUBMB Life* 57, 745–747, 2005.

Ethyl Alcohol	Ethanol	46.07	Solvent; used to adjust solvent polarity; use in plasma protein fractionation.

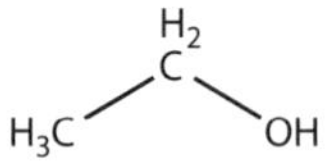

CHEMICALS COMMONLY USED IN BIOCHEMISTRY AND MOLECULAR
BIOLOGY AND THEIR PROPERTIES (Continued)

Common Name	Chemical Name	M.W.	Properties and Comment

Dufour, E., Bertrand-Harb, C., and Haertle, T., Reversible effects of medium dielectric constant on structural transformation of beta-lactoglobulin and its retinol binding, *Biopolymers* 33, 589–598, 1993; Escalera, J.B., Bustamante, P., and Martin, A., Predicting the solubility of drugs in solvent mixtures: multiple solubility maxima and the chameleonic effect, *J. Pharm. Pharmcol.* 46, 172–176, 1994; Gratzer, P.F., Pereira, C.A., and Lee, J.M., Solvent environment modulates effects of glutaraldehyde crosslinking on tissue-derived biomaterials, *J. Biomed. Mater. Res.* 31, 533–543, 1996; Sepulveda, M.R. and Mata, A.M., The interaction of ethanol with reconstituted synaptosomal plasma membrane Ca^{2+}, *Biochim. Biophys. Acta* 1665, 75–80, 2004; Ramos, A.S. and Techert, S., Influence of the water structure on the acetylcholinesterase efficiency, *Biophys. J.* 89, 1990–2003, 2005; Wehbi, Z., Perez, M.D., and Dalgalarrondo, M., Study of ethanol-induced conformation changes of holo and apo alpha-lactalbumin by spectroscopy anad limited proteolysis, *Mol. Nutr. Food Res.* 50, 34–43, 2006; Sasahara, K. and Nitta, K., Effect of ethanol on folding of hen egg-white lysozyme under acidic condition, *Proteins* 63, 127–135, 2006; Perham, M., Liao, J., and Wittung-Stafshede, P., Differential effects of alcohol on conformational switchovers in alpha-helical and beta-sheet protein models, *Biochemistry* 45, 7740–7749, 2006; Pena, M.A., Reillo, A., Escalera, B., and Bustamante, P., Solubility parameter of drugs for predicting the solubility profile type within a wide polarity range in solvent mixtures, *Int. J. Pharm.* 321, 155–161, 2006; Jenke, D., Odufu, A., and Poss, M., The effect of solvent polarity on the accumulation of leachables from pharmaceutical product containers, *Eur. J. Pharm. Sci.* 27, 133–142, 2006.

| **Ethylene Glycol** | 1,2-ethanediol | 62.07 | Solvent/cosolvent; increases viscosity (visogenic osmolyte); perturbant; cryopreservative. |

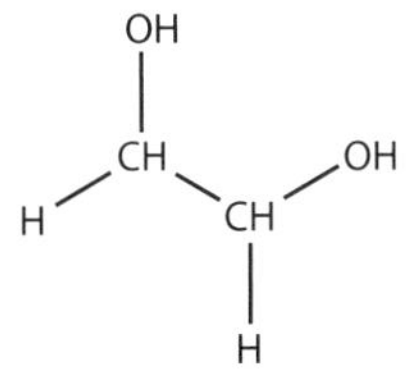

Tanford, C., Buckley, C.E., III, De, P.K., and Lively, E.P., Effect of ethylene glycol on the conformation of gamma-globulin and beta-lactoglobulin, *J. Biol. Chem.* 237, 1168–1171, 1962; Kay, C.M. and Brahms, J., The influence of ethylene glycol on the enzymatic adenosine triphosphatase activity and molecular conformation of fibrous muscle proteins, *J. Biol. Chem.* 238, 2945–2949, 1963; Narayan, K.A., The interaction of ethylene glycol with rat-serum lipoproteins, *Biochim. Biophys. Acta* 137, 22–30, 1968; Bello, J., The state of the tyrosines of bovine pancreatic ribonuclease in ethylene glycol and glycerol, *Biochemistry* 8, 4535–4541, 1969; Lowe, C.R. and Mosbach, K., Biospecific affinity chromatography in aqueous-organic cosolvent mixtures. The effect of ethylene glycol on the binding of lactate dehydrogenase to an immobilized-AMP analogue, *Eur. J. Biochem.* 52, 99–105, 1975; Ghrunyk, B.A. and Matthews, C.R., Role of diffusion in the folding of the alpha subunit of tryptophan synthase from *Escherichia coli*, *Biochemistry* 29, 2149–2154, 1990; Silow, M. and Oliveberg, M., High concentrations of viscogens decrease the protein folding rate constant by prematurely collapsing the coil, *J. Mol. Biol.* 326, 263–271, 2003; Naseem, F. and Khan, R.H., Effect of ethylene glycol and polyethylene glycol on the acid-unfolded state of trypsinogen, *J. Protein Chem.* 22, 677–682, 2003; Hubalek, Z., Protectants used in the cyropreservation of microorganisms, *Cryobiology* 46, 205–229, 2003; Menezo, Y.J., Blastocyst freezing, *Eur. J. Obstet. Gynecol. Reprod. Biol.* 155 (Suppl. 1), S12–S15, 2004; Khodarahmi, R. and Yazdanparast, R., Refolding of chemically denatured alpha-amylase in dilution additive mode, *Biochim. Biophys. Acta.* 1674, 175–181, 2004; Zheng, M., Li, Z., and Huang, X., Ethylene glycol monolayer protected nanoparticles: synthesis, characterization, and interactions with biological molecules, *Langmuir* 20, 4226–4235, 2004; Bonincontro, A., Cinelli, S., Onori, G., and Stravato, A., Dielectric behavior of lysozyme and ferricytochrome-c in water/ethylene-glycol solutions, *Biophys. J.* 86, 1118–1123, 2004; Kozer, N. and Schreiber, G., Effect of crowding on protein–protein association rates: fundamental differences between low and high mass crowding agents, *J. Mol. Biol.* 336, 763–774, 2004; Levin, I., Meiri, G., Peretz, M. et al., The ternary complex of *Pseudomonas aeruginosa* dehydrogenase with NADH and ethylene glycol, *Protein Sci.* 13, 1547–1556, 2004; Stupishina, E.A., Khamidullin, R.N., Vylegzhanina, N.N. et al., Ethylene glycol and the thermostability of trypsin in a reverse micelle system, *Biochemistry* 71, 533–537, 2006; Nordstrom, L.J., Clark, C.A., Andersen, B. et al., Effect of ethylene glycol, urea, and *N*-methylated glycines on DNA thermal stability: the role of DNA base pair composition and hydration, *Biochemistry* 45, 9604–9614, 2006.

| **Ethyleneimine** | Aziridine | 43.07 | Modification of sulfhydryl groups to produce amine functions; alkylating agent; reacts with carboxyl groups at acid pH; monomer unit for polyethylene amine, a versatile polymer. |

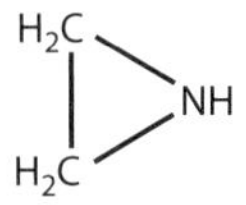

CHEMICALS COMMONLY USED IN BIOCHEMISTRY AND MOLECULAR
BIOLOGY AND THEIR PROPERTIES (Continued)

Common Name	Chemical Name	M.W.	Properties and Comment

Raftery, M.A. and Cole, R.D., On the aminoethylation of proteins, *J. Biol. Chem.* 241, 3457–3461, 1966; Fishbein, L., Detection and thin-layer chromatography of derivatives of ethyleneimine. I. *N*-carbamoyl and aziridines, *J. Chromatog.* 26, 522–526, 1967; Yamada, H., Imoto, T., and Noshita, S., Modification of catalytic groups in lysozyme with ethyleneimine, *Biochemistry* 21, 2187–2192, 1982; Okazaki, K., Yamada, H., and Imoto, T., A convenient *S*-2-aminoethylation of cysteinyl residues in reduced proteins, *Anal. Biochem.* 149, 516–520, 1985; Hemminki, K., Reactions of ethyleneimine with guanosine and deoxyguanosine, *Chem. Biol. Interact.* 48, 249–260, 1984; Whitney, P.L., Powell, J.T., and Sanford, G.L., Oxidation and chemical modification of lung beta-galactosidase-specific lectin, *Biochem. J.* 238, 683–689, 1986; Simpson, D.M., Elliston, J.F., and Katzenellenbogen, J.A., Desmethylnafoxidine aziridine: an electrophilic affinity label for the estrogen receptor with high efficiency and selectivity, *J. Steroid Biochem.* 28, 233–245, 1987; Musser, S.M., Pan, S.S., Egorin, M.J. et al., Alkylation of DNA with aziridine produced during the hydrolysis of *N,N′,N″*-triethylenethiophosphoramide, *Chem. Res. Toxicol.* 5, 95–99, 1992; Thorwirth, S., Muller, H.S., and Winnewisser, G., The millimeter- and submillimeter-wave spectrum and the dipole moment of ethyleneimine, *J. Mol. Spectroso.* 199, 116–123, 2000; Burrage, T., Kramer, E., and Brown, F., Inactivation of viruses by aziridines, *Dev. Biol.* (Basel) 102, 131–139, 2000; Brown, F., Inactivation of viruses by aziridines, *Vaccine* 20, 322–327, 2001; Sasaki, S., Active oligonucleotides incorporating alkylating agent as potential sequence- and base-selective modifier of gene expression, *Eur. J. Pharm. Sci.* 13, 43–51, 2001; Hou, X.L., Fan, R.H., and Dai, L.X., Tributylphosphine: a remarkable promoting reagent for the ring-opening reaction of aziridines, *J. Org. Chem.* 67, 5295–5300, 2002; Thevis, M., Loo, R.R.O., and Loo, J.A., In-gel derivatization of proteins for cysteine-specific cleavages and their analysis by mass spectrometry, *J. Proteome Res.* 2, 163–172, 2003; Sasaki, M., Dalili, S., and Yudin, A.K., *N*-arylation of aziridines, *J. Org. Chem.* 68, 2045–2047, 2003; Gao, G.Y., Harden, J.D., and Zhang, J.P., Cobalt-catalyzed efficient aziridination of alkenes, *Org. Lett.* 7, 3191–3193, 2005; Hopkins, C.E., Hernandez, G., Lee, J.P., and Tolan, D.R., Aminoethylation in model peptides reveals conditions for maximizing thiol specificity, *Arch. Biochem. Biophys.* 443, 1–10, 2005; Li, C. and Gershon, P.D., pK(a) of the mRNA cap-specific 2′-*O*-methyltransferase catalytic lysine by HSQC NMR detection of a two-carbon probe, *Biochemistry* 45, 907–917, 2006; Vicik R., Helten, H., Schirmeister, T., and Engels, B., Rational design of aziridine-containing cysteine protease inhibitors with improved potency: studies on inhibition mechanism, *ChemMedChem*, 1, 1021–1028, 2006.

| **Ethylene Oxide** | Oxirane | 44.05 | Sterilizing agent; starting material for ethylene glycol and other products such as nonionic surfactants. |

Windmueller, H.G., Ackerman, C.J., and Engel, R.W., Reaction of ethylene oxide with histidine, methionine, and cysteine, *J. Biol. Chem.* 234, 895–899, 1959; Starbuck, W.C. and Busch, H., Hydroxyethylation of amino acids in plasma albumin with ethylene oxide, *Biochim. Biophys. Acta* 78, 594–605, 1963; Guengerich, F.P., Geiger, L.E., Hogy, L.L., and Wright,. P.L., *In vitro* metabolism of acrylonitrile to 2-cyanoethylene oxide, reaction with glutathione, and irreversible binding to proteins and nucleic acids, *Cancer Res.* 41, 4925–4933, 1981; Peter, H., Schwarz, M., Mathiasch, B. et al., A note on synthesis and reactivity towards DNA of glycidonitrile, the epoxide of acrylonitrile, *Carcinogenesis* 4, 235–237, 1983; Grammer, L.C. and Patterson, R., IgE against ethylene oxide-altered human serum albumin (ETO-HAS) as an etiologic agent in allergic reactions of hemodialysis patients, *Artif. Organs* 11, 97–99, 1987; Bolt, H.M., Peter, H., and Fost, U., Analysis of macromolecular ethylene oxide adducts, *Int. Arch. Occup. Environ. Health* 60, 141–144, 1988; Young, T.L., Habraken, Y., Ludlum, D.B., and Santella, R.M., Development of monoclonal antibodies recognizing 7-(2-hydroxyethyl) guanine and imidazole ring-opened 7-(2-hydroxyethyl) guanine, *Carcinogenesis* 11, 1685–1689, 1990; Walker, V.E., Fennell, T.R., Boucheron, J.A. et al., Macromolecular adducts of ethylene oxide: a literature review and a time-course study on the formation of 7-(2-hydroxyethyl)guanine following exposure of rats by inhalation, *Mutat. Res.* 233, 151–164, 1990; Framer, P.B., Bailey, E., Naylor, S. et al., Identification of endogenous electrophiles by means of mass spectrometric determination of protein and DNA adducts, *Environ. Health Perspect.* 99, 19–24, 1993; Tornqvist, M. and Kautianinen, A., Adducted proteins for identification of endogenous electrophiles, *Environ. Health Perspect.* 99, 39–44, 1993; Galaev, I. Yu. and Mattiasson, B., Thermoreactive water-soluble polymers, nonionic surfactants, and hydrogels as reagents in biotechnology, *Enzyme Microb. Technol.* 15, 354–366, 1993; Segerback, D., DNA alkylation by ethylene oxide and mono-substituted epoxides, *IARC Sci. Publ.* 125, 37–47, 1994; Phillips, D.H. and Farmer, P.B., Evidence for DNA and protein binding by styrene and styrene oxide, *Crit. Rev. Toxicol.* 24 (Suppl.), S35–S46, 1994; Marczynski, B., Marek, W., and Baur, X., Ethylene oxide as a major factor in DNA and RNA evolution, *Med. Hypotheses* 44, 97–100, 1995; Mosely, G.A. and Gillis, J.R., Factors affecting tailing in ethylene oxide sterilization part 1: when tailing is an artifact… and scientific deficiencies in ISO 11135 and EN 550, *PDA J. Pharm. Sci. Technol.* 58, 81–95, 2004.

| *N*-**Ethylmaleimide** | 1-ethyl-1*H*-pyrrole-2,5-dione | 125.13 | Modification of sulfhydryl groups; basic building block for a number of reagents. Mechanism different from alkylating agent in that reaction involves a Michael addition. |

CHEMICALS COMMONLY USED IN BIOCHEMISTRY AND MOLECULAR BIOLOGY AND THEIR PROPERTIES (Continued)

Common Name	Chemical Name	M.W.	Properties and Comment

Lundblad, R.L., *Chemical Reagent for Protein Modification*, 3rd ed., CRC Press, Boca Raton, FL, 2004; Bowes, T.J. and Gupta, R.S., Induction of mitochondrial fusion by cysteine-alkylators ethyacrynic acid and *N*-ethylmaleimide, *J. Cell Physiol.* 202, 796–804, 2005; Engberts, J.B., Fernandez, E., Garcia-Rio, L., and Leis, J.R., Water in oil microemulsions as reaction media for a Diels–Alder reaction between *N*-ethylmaleimide and cyclopentadiene, *J. Org. Chem.* 71, 4111–4117, 2006; Engberts, J.B., Fernandez, E., Garcia-Rio, L., and Leis, J.R, AOT-based microemulsions accelerate the 1,3-cycloaddition of benzonitrile oxide to *N*-ethylmaleimide, *J. Org. Chem.* 71, 6118–6123, 2006; de Jong, K. and Kuypers, F.A., Sulphydryl modifications alter scramblase activity in murine sickle cell disease, *Br. J. Haematol.* 133, 427–432, 2006; Martin, H.G., Henley, J.M., and Meyer, G., Novel putative targets of *N*-ethylmaleimide sensitive fusion proteins (NSF) and alpha/beta soluble NSF attachment proteins (SNAPs) include the Pak-binding nucleotide exchange factor betaPIX, *J. Cell. Biochem.*, 99, 1203–1215, 2006; Carrasco, M.R., Silva, O., Rawls, K.A. et al., Chemoselective alkylation of *N*-alkylaminooxy-containing peptides, *Org. Lett.* 8, 3529–3532, 2006; Pobbati, A.V., Stein, A., and Fasshauer, D., N- to C-terminal SNARE complex assembly promotes rapid membrane fusion, *Science* 313, 673–676, 2006; Mollinedo, F., Calafat, J., Janssen, H. et al., Combinatorial SNARE complexes modulate the secretion of cytoplasmic granules in human neutrophils, *J. Immunol.* 177, 2831–2841, 2006.

| **Formaldehyde** | Methanal | 30.03 | Tissue fixation; protein modification; zero-length crosslinking; protein–nucleic acid interactions. |

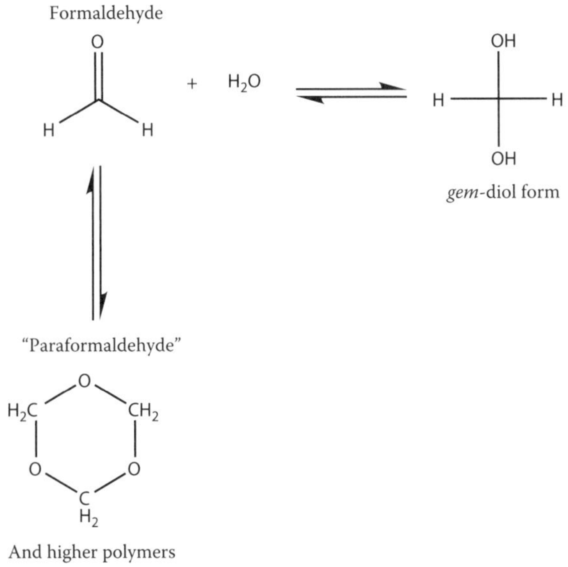

Formaldehyde

O
‖
H—C—H + H₂O ⇌ gem-diol form

OH
|
H—C—H
|
OH

gem-diol form

"Paraformaldehyde"

And higher polymers

Feldman, M.Y., Reactions of nucleic acids and nucleoproteins with formaldehyde, *Prog. Nucleic Acid Res. Mol. Biol.* 13, 1–49, 1973; Russell, A.D. and Hopwood, D., The biological uses and importance of glutaraldehyde, *Prog. Med. Chem.* 13, 271–301, 1976; Means, G.E., Reductive alkylation of amino groups, *Methods Enzymol.* 47, 469–478, 1977; Winkelhake, J.L., Effects of chemical modification of antibodies on their clearance for the circulation. Addition of simple aliphatic compounds by reductive alkylation and carbodiimide-promoted amide formation, *J. Biol. Chem.* 252, 1865–1868, 1977; Yamazaki, Y. and Suzuki, H., A new method of chemical modification of N^6-amino group in adenine nucleotides with formaldehyde and a thiol and its application to preparing immobilized ADP and ATP, *Eur. J. Biochem.* 92, 197–207, 1978; Geoghegan, K.F., Cabacungan, J.C., Dixon, H.B., and Feeney, R.E., Alternative reducing agents for reductive methylation of amino groups in proteins, *Int. J. Pept. Protein Res.* 17, 345–352, 1981; Kunkel, G.R., Mehradian, M., and Martinson, H.G., Contact-site crosslinking agents, *Mol. Cell. Biochem.* 34, 3–13, 1981; Fox, C.H., Johnson, F.B., Whiting, J., and Roller, P.P., Formaldehyde fixation, *J. Histochem. Cytochem.* 33, 845–853, 1985; Conaway, C.C., Whysner, J., Verna, L.K., and Williams, G.M., Formaldehyde mechanistic data and risk assessment: endogenous protection from DNA adduct formation, *Pharmacol. Ther.* 71, 29–55, 1996; Masuda, N., Ohnishi, T., Kawamoto, S. et al., Analysis of chemical modifications of RNA from formalin-fixed samples and optimization of molecular biology applications for such samples, *Nucleic Acids Res.* 27, 4436–4443, 1999; Micard, V., Belamri, R., Morel, M., and Guilbert, S., Properties of chemically and physically treated wheat gluten films, *J. Agric. Food Chem.* 48, 2948–2953, 2000; Taylor, I.A. and Webb, M., Chemical modification of lysine by reductive methylation. A probe for residues involved in DNA binding, *Methods Mol. Biol.* 148, 301–314, 2001; Perzyna, A., Marty, C., Facopre, M. et al., Formaldehyde-induced DNA crosslink of indolizino[1,2-b]quinolines derived from the A-D rings of camptothecin, *J. Med. Chem.* 45, 5809–5812, 2002; Yurimoto, H., Hirai, R., Matsuno, N. et al., HxlR, a member of the DUF24 protein family, is a DNA-binding protein that acts as a positive regulator of the formaldehyde-inducible hx1AB operon in *Bacillus subtilis*, *Mol. Microbiol.* 57, 511–519, 2005.

| **Formic Acid** | Methanoic Acid | 46.03 | Solvent; buffer component. |

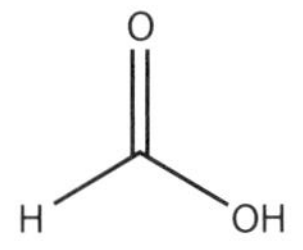

CHEMICALS COMMONLY USED IN BIOCHEMISTRY AND MOLECULAR
BIOLOGY AND THEIR PROPERTIES (Continued)

Common Name	Chemical Name	M.W.	Properties and Comment

Sarkar, P.B., Decomposition of formic acid by periodate, *Nature* 168, 122–123, 1951; Hass, P., Reactions of formic acid and its salts, *Nature* 167, 325, 1951; Smillie, L.B. and Neurath, H., Reversible inactivation of trypsin by anhydrous formic acid, *J. Biol. Chem* 234, 355–359, 1959; Hynninen, P.H. and Ellfolk, N., Use of the aqueous formic acid-chloroform-dimethylformamide solvent system for the purification of porphyrins and hemins, *Acta Chem. Scand.* 27, 1795–1806, 1973; Heukeshoven, J. and Dernick, R., Reversed-phase high-performance liquid chromatography of virus proteins and other large hydrophobic proteins in formic acid-containing solvents, *J. Chromatog.* 252, 241–254, 1982; Tarr, G.E. and Crabb, J.W., Reverse-phase high-performance liquid chromatography of hydrophobic proteins and fragments thereof, *Anal. Biochem.* 131, 99–107, 1983; Heukeshoven, J. and Dernick, R., Characterization of a solvent system for separation of water-insoluble poliovirus proteins by reversed-phase high-performance liquid chromatography, *J. Chromatog.* 326, 91–101, 1985; De Caballos, M.L., Taylor, M.D., and Jenner, P., Isocratic reverse-phase HPLC separation and RIA used in the analysis of neuropeptides in brain tissue, *Neuropeptides* 20, 201–209, 1991; Poll, D.J. and Harding, D.R., Formic acid as a milder alternative to trifluoroacetic acid and phosphoric acid in two-dimensional peptide mapping, *J. Chromatog.* 469, 231–239, 1989; Klunk W.E. and Pettegrew, J.W., Alzheimer's beta-amyloid protein is covalently modified when dissolved in formic acid, *J. Neurochem.* 54, 2050–2056, 1990; Erdjument-Bromage, H., Lui, M., Lacomis, L. et al., Examination of the micro-tip reversed phase liquid chromatographic extraction of peptide pools for mass spectrometric analysis, *J. Chromatog. A* 826, 167–181, 1998; Duewel, H.S. and Honek, J.F., CNBr/formic acid reactions of methionine- and trifluoromethionine-containing lambda lysozyme: probing chemical and positional reactivity and formylation side reactions of mass spectrometry, *J. Protein Chem.* 17, 337–350, 1998; Kaiser, R. and Metzka, L., Enhancement of cyanogen bromide cleavage yields for methionyl-serine and methionyl-threonine peptide bonds, *Anal. Biochem.* 266, 1–8, 1999; Rodriguez, J.C., Wong, L., and Jennings, P.A., The solvent in CNBr cleavage reactions determines the fragmentation efficiency of ketosteroid isomerase fusion proteins used in the production of recombinant peptides, *Protein Expr. Purif.* 28, 224–231, 2003; Zu, Y., Zhao, C., Li, C., and Zhang, L., A rapid and sensitive LC-MS/MS method for determination of coenzyme Q10 in tobacco (*Nicotiana tabacum* L.) leaves, *J. Sep. Sci.* 29, 1607–1612, 2006; Kalovidouris, M., Michalea, S., Robola, N. et al., Ultra-performance liquid chromatography/tandem mass spectrometry method for the determination of lercaidipine in human plasma, *Rapid Commun. Mass Spectrom.*, 20, 2939–2946, 2006; Wang, P.G., Wei, J.S., Kim, G. et al., Validation and application of a high-performance liquid chromatography-tandem mass spectrometric method for simultaneous quantification of lopinavir and ritonavir in human plasma using semi-automated 96-well liquid–liquid chromatography, *J. Chromatog. A*, 1130, 302–307, 2006.

Glutaraldehyde	Pentanedial	100.12	Protein modification; tissue fixation; sterilization agent approved by regulatory agencies; use with albumin as surgical sealant.

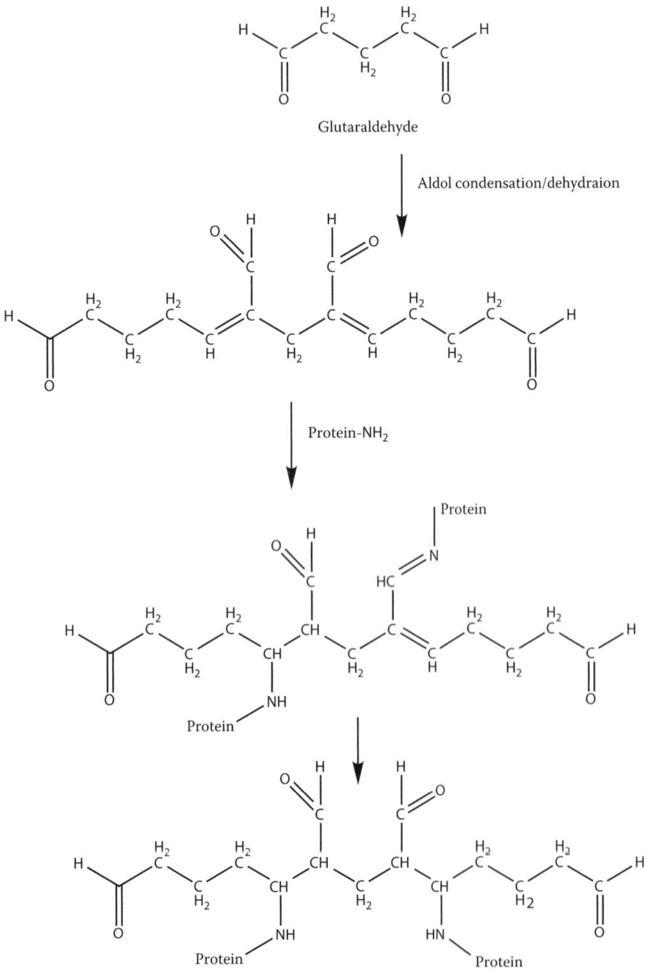

Glutaraldehyde

Aldol condensation/dehydraion

Protein-NH₂

**CHEMICALS COMMONLY USED IN BIOCHEMISTRY AND MOLECULAR
BIOLOGY AND THEIR PROPERTIES (Continued)**

Common Name	Chemical Name	M.W.	Properties and Comment

Hopwood, D., Theoretical and practical aspects of glutaraldehyde fixation, *Histochem. J.*, 4, 267–303, 1972; Hassell, J. and Hand, A.R., Tissue fixation with diimidoesters as an alternative to aldehydes. I. Comparison of crosslinking and ultrastructure obtained with dimethylsubserimidate and glutaraldehyde, *J. Histochem. Cytochem.* 22, 223–229, 1974; Russell, A.D. and Hopwood, D., The biological uses and importance of glutaraldehyde, *Prog. Med. Chem.* 13, 271–301, 1976; Woodroof, E.A., Use of glutaraldehyde and formaldehyde to process tissue heart valves, *J. Bioeng.* 2, 1–9, 1978; Heumann, H.G., Microwave-stimulated glutaraldehyde and osmium tetroxide fixation of plant tissue: ultrastructural preservation in seconds, *Histochemistry* 97, 341–347, 1992; Abbott, L., The use and effects of glutaraldehyde: a review, *Occup. Health* 47, 238–239, 1995; Jayakrishnan, A. and Jameela, S.R., Glutaraldehyde as a fixative in bioprosthesis and drug delivery matrices, *Biomaterials* 17, 471–484, 1996; Tagliaferro, P., Tandler, C.J., Ramos, A.J. et al., Immunofluorescence and glutaraldehyde fixation. A new procedure base on the Schiff-quenching method, *J. Neurosci. Methods* 77, 191–197, 1997; Cohen, R.J., Beales, M.P., and McNeal, J.E., Prostate secretory granules in normal and neoplastic prostate glands: a diagnostic aid to needle biopsy, *Hum. Pathol.* 31, 1515–1519, 2000; Chae, H.J., Kim, E.Y., and In, M., Improved immobilization yields by addition of protecting agents in glutaraldehyde-induced immobilization of protease, *J. Biosci. Bioeng.* 89, 377–379, 2000; Nimni, M.E., Glutaraldehyde fixation revisited, *J. Long Term Eff. Med. Implants* 11, 151–161, 2001; Fujiwara, K., Tanabe, T., Yabuchi, M. et al., A monoclonal antibody against the glutaraldehyde-conjugated polyamine, putrescine: application to immunocytochemistry, *Histochem. Cell Biol.* 115, 471–477, 2001; Chao, H.H. and Torchiana, D.F., Bioglue: albumin/glutaraldehyde sealant in cardiac surgery, *J. Card. Surg.* 18, 500–503, 2003; Migneault, I., Dartiguenave, C., Bertrand, M.J., and Waldron, K.C., Glutaraldehyde: behavior in aqueous solution, reaction with proteins, and application to enzyme crosslinking, *Biotechniques* 37, 790–796, 2004; Jearanaikoon, S. and Abraham-Peskir, J.V., An x-ray microscopy perspective on the effect of glutaraldehyde fixation on cells, *J. Microsc.* 218, 185–192, 2005; Buehler, P.W., Boykins, R.A., Jia, Y. et al., Structural and functional characterization of glutaraldehyde-polymerized bovine hemoglobin and its isolated fractions, *Anal. Chem.* 77, 3466–3478, 2005; Kim, S.S., Lim, S.H., Cho, S.W. et al., Tissue engineering of heart valves by recellularization of glutaraldehyde-fixed porcine values using bone marrow-derived cells, *Exp. Mol. Med.* 38, 273–283, 2006.

Glutathione	γ-GluCysGly	307.32	Reducing agent; intermediate in phase II detoxification of xenobiotics.

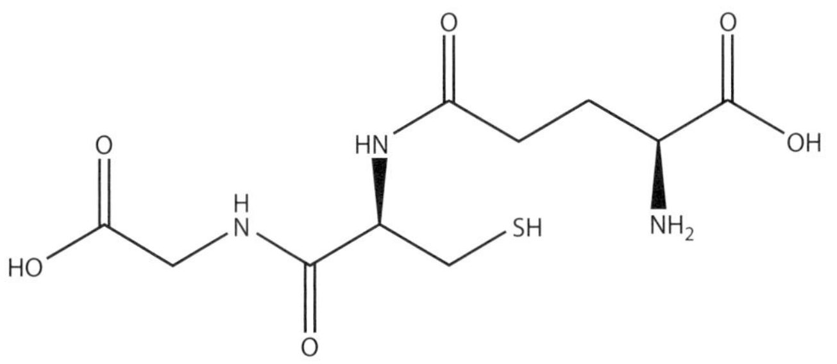

Arias, I.M. and Jakoby, W.B., *Glutathione, Metabolism and Function,* Raven Press, New York, 1976; Meister, A., *Glutamate, Glutamine, Glutathione, and Related Compounds,* Academic Press, Orlando, FL, 1985; Sies, H. and Ketterer, B., *Glutathione Conjugation: Mechanisms and Biological Significance,* Academic Press, London, UK, 1988; Tsumoto, K., Shinoki, K., Kondo, H. et al., Highly efficient recovery of functional single-chain Fv fragments from inclusion bodies overexpressed in *Escherichia coli* by controlled introduction of oxidizing reagent — application to a human single-chain Fv fragment, *J. Immunol. Methods* 219, 119–129, 1998; Jiang, X., Ookubo, Y., Fujii, I. et al., Expression of Fab fragment of catalytic antibody 6D9 in an *Escherichia coli in vitro* coupled transcription/translation system, *FEBS Lett.* 514, 290–294, 2002; Sun, X.X., Vinci, C., Makmura, L. et al., Formation of disulfide bond in p53 correlates with inhibition of DNA binding and tetramerization, *Antioxid. Redox Signal.* 5, 655–665, 2003; Sies, H. and Packer, L., Eds., *Glutathione Transferases and Gamma-Glutamyl Transpeptidases,* Elsevier, Amsterdam, 2005; Smith, A.D. and Dawson, H., Glutathione is required for efficient production of infectious picornativur virions, *Virology,* 353, 258–267, 2006.

Glycine	Aminoacetic Acid	75.07	Buffer component; protein-precipitating agent, excipient for pharmaceutical formulation.

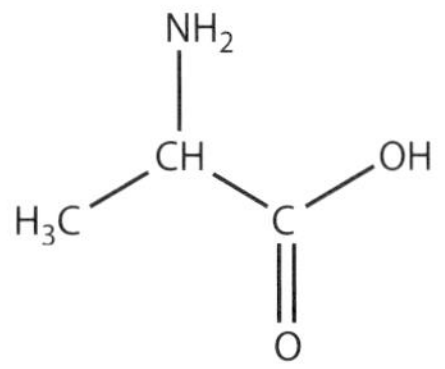

CHEMICALS COMMONLY USED IN BIOCHEMISTRY AND MOLECULAR
BIOLOGY AND THEIR PROPERTIES (Continued)

Common Name	Chemical Name	M.W.	Properties and Comment

Sarquis, J.L. and Adams, E.T., Jr., The temperature-dependent self-association of beta-lactoglobulin C in glycine buffers, *Arch. Biochem. Biophys.* 163, 442–452, 1974; Poduslo, J.F., Glycoprotein molecular-weight estimation using sodium dodecyl suflate-pore gradient electrophoresis: comparison of Tris-glycine and Tris-borate-EDTA buffer systems, *Anal. Biochem.* 114, 131–139, 1981; Patton, W.F., Chung-Welch, N., Lopez, M.F. et al., Tris-tricine and Tris-borate buffer systems provide better estimates of human mesothelial cell intermediate filament protein molecular weights than the standard Tris-glycine system, *Anal. Biochem.* 197, 25–33, 1991; Trasltas, G. and Ford, C.H., Cell membrane antigen-antibody complex dissociation by the widely used glycine-HC1 method: an unreliable procedure for studying antibody internalization, *Immunol. Invest.* 22, 1–12, 1993; Nail, S.L., Jiang, S., Chongprasert, S., and Knopp, S.A., Fundamentals of freeze-drying, *Pharm. Biotechnol.* 14, 281–360, 2002; Pyne, A., Chatterjee, K., and Suryanarayanan, R., Solute crystallization in mannitol-glycine systems — implications on protein stabilization in freeze-dried formulations, *J. Pharm. Sci.* 92, 2272–2283, 2003; Hasui, K., Takatsuka, T., Sakamoto, R. et al., Double immunostaining with glycine treatment, *J. Histochem. Cytochem.* 51, 1169–1176, 2003; Hachmann, J.P. and Amshey, J.W., Models of protein modification in Tris-glycine and neutral pH Bis-Tris gels during electrophoresis: effect of gel pH, *Anal. Biochem.* 342, 237–245, 2005.

Glyoxal   Ethanedial   58.04   Modification of proteins and nucleic acids; model for glycation reaction; fluorescent derivates formed with tryptophan.

Nakaya, K., Takenaka, O., Horinishi, H., and Shibata, K., Reactions of glyoxal with nucleic acids. Nucleotides and their component bases, *Biochim. Biophys. Acta* 161, 23–31, 1968; Canella, M. and Sodini, G., The reaction of horse-liver alcohol dehydrogenase with glyoxal, *Eur. J. Biochem.* 59, 119–125, 1975; Kai, M., Kojima, E., Okhura, Y., and Iwaski, M., High-performance liquid chromatography of N-terminal tryptophan-containing peptides with precolumn fluorescence derivatization with glyoxal, *J. Chromatog. A.* 653, 235–250, 1993; Murata-Kamiya, N., Kamiya, H., Kayi, H., and Kasai, H., Glyoxal, a major product of DNA oxidation, induces mutations at G:C sites on a shuttle vector plasmid replicated in mammalian cells, *Nucleic Acids Res.* 25, 1897–1902, 1997; Leng, F., Graves, D., and Chaires, J.B., Chemical crosslinking of ethidium to DNA by glyoxal, *Biochim. Biophys. Acta* 1442, 71–81, 1998; Thrornalley, P.J., Langborg, A., and Minhas, H.S., Formation of glyoxal, methylglyoxal, and 3-deoxyglucosone in the glycation of proteins by glucose, *Biochem. J.* 344, 109–116, 1999; Sady, C., Jiang, C.L., Chellan, P. et al., Maillard reactions by alpha-oxoaldehydes: detection of glyoxal-modified proteins, *Biochim. Biophys. Acta* 1481, 255–264, 2000; Olsen, R., Molander P., Ovrebo, S. et al., Reaction of glyoxal with 2'-deoxyguanosine, 2'-deoxyadenosine, 2'-deoxycytidine, cytidine, thymidine, and calf thymus DNA: identification of the DNA adducts, *Chem. Res. Toxicol.* 18, 730–739, 2005; Manini, P., La Pietra, P., Panzella, L. et al., Glyoxal formation by Fenton-induced degradation of carbohydrates and related compounds, *Carbohydr. Res.* 341, 1828–1833, 2006.

Guanidine	Aminomethanamidine	59.07	Chaotropic agent; guanidine hydrochloride
Guanidine Hydrochloride (GuCl)		95.53	use for study of protein denaturation; GTIC
Guanidine Thiocyanate (GTIC)		118.16	is considered to be more effective than GuCl; GTIC used for nucleic acid extraction.

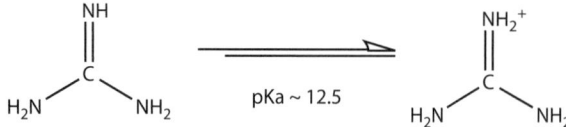

Hill, R.L., Schwartz, H.C., and Smith, E.L., The effect of urea and guanidine hydrochloride on activity and optical rotation of crystalline papain, *J. Biol. Chem.* 234, 572–576, 1959; Appella, E. and Markert, C.L., Dissociation of lactate dehydrogenase into subunits with guanidine hydrochloride, *Biochim. Biophys. Res. Commun.* 6, 171–176, 1961; von Hippel, P.H. and Wong, K.-Y., On the conformational stability of globular proteins. The effects of various electrolytes and nonelectrolytes on the thermal transition ribonuclease transition, *J. Biol. Chem.* 240, 3909–3923, 1965; Katz, S., Partial molar volume and conformational changes produced by the denaturation of albumin by guanidine hydrochloride, *Biochim. Biophys. Acta* 154, 468–477, 1968; Shortle, D., Guanidine hydrochloride denaturation studies of mutant forms of staphylococcal nuclease, *J. Cell Biochem.* 30, 281–289, 1986; Lippke, J.A., Strzempko, M.N., Rai, F.F. et al., Isolation of intact high-molecular-weight DNA by using guanidine isothiocyanate, *Appl. Environ. Microbiol.* 53, 2588–2589, 1987; Alberti, S. and Fornaro, M., Higher transfection efficiency of genomic DNA purified with a guanidinium thiocyanate–based procedure, *Nucleic Acids Res.* 18, 351–353, 1990; Shirley, B.A., Urea and guanidine hydrochloride denaturation curves, *Methods Mol. Biol.* 40, 177–190, 1995; Cota, E. and Clarke, J., Folding of beta-sandwich proteins: three-state transition of a fibronectin type III module, *Protein Sci.* 9, 112–120, 2000; Kok, T., Wati, S., Bayly, B. et al., Comparison of six nucleic

**CHEMICALS COMMONLY USED IN BIOCHEMISTRY AND MOLECULAR
BIOLOGY AND THEIR PROPERTIES (Continued)**

Common Name	Chemical Name	M.W.	Properties and Comment

acid extraction methods for detection of viral DNA or RNA sequences in four different non-serum specimen types, *J. Clin. Virol.* 16, 59–63, 2000; Salamanca, S., Villegas, V., Vendrell, J. et al., The unfolding pathway of leech carboxypeptidase inhibitor, *J. Biol. Chem.* 277, 17538–17543, 2002; Bhuyan, A.K., Protein stabilization by urea and guanidine hydrochloride, *Biochemistry* 41, 13386–13394, 2002; Jankowska, E., Wiczk, W., and Grzonka, Z., Thermal and guanidine hydrochloride-induced denaturation of human cystatin C, *Eur. Biophys. J.* 33, 454–461, 2004; Fuertes, M.A., Perez, J.M., and Alonso, C., Small amounts of urea and guanidine hydrochloride can be detected by a far-UV spectrophotometric method in dialyzed protein solutions, *J. Biochem. Biophys. Methods* 59, 209–216, 2004; Berlinck, R.G., Natural guanidine derivatives, *Nat. Prod. Rep.* 22, 516–550, 2005; Rashid, F., Sharma, S., and Bano, B., Comparison of guanidine hydrochloride (GdnHCl) and urea denaturation on inactivation and unfolding of human placental cystatin (HPC), *Biophys. J.* 91, 686–693, 2006; Nolan, R.L. and Teller, J.K., Diethylamine extraction of proteins and peptides isolated with a mono-phasic solution of phenol and guanidine isothiocyanate, *J. Biochem. Biophys. Methods* 68, 127–131, 2006.

Common Name	Chemical Name	M.W.	Properties and Comment
Hydrazine	N_2H_4	32.05	Reducing agent; modification of aldehydes and carbohydrates; hydrazinolysis used for release of carbohydrates from protein; derivatives such as dinitrophenyl-hydrazine used for analysis of carbonyl groups in oxidized proteins; detection of acetyl and formyl groups in proteins.

Schmer, G. and Kreil, G., Micro method for detection of formyl and acetyl groups in proteins, *Anal. Biochem.* 29, 186–192, 1969; Gershoni, J.M., Bayer, E.A., and Wilchek, M., Blot analyses of glycoconjugates: enzyme-hydrazine — a novel reagent for the detection of aldehydes, *Anal. Biochem.* 146, 59–63, 1985; O'Neill, R.A., Enzymatic release of oligosaccharides from glycoproteins for chromatographic and electrophoretic analysis, *J. Chromatog. A* 720, 201–215, 1996; Routier, F.H., Hounsell, E.F., and Rudd, P.M., Quantitation of the oligosaccharides of human serum IgG from patients with rheumatoid arthritis: a critical evaluation of different methods, *J. Immunol. Methods* 213, 113–130, 1998; Robinson, C.E., Keshavarzian, A., Pasco, D.S. et al., Determination of protein carbonyl groups by immunoblotting, *Anal. Biochem.* 266, 48–57, 1999; Merry, A.H., Neville, D.C., Royle, L. et al., Recovery of intact 2-aminobenzamide-labeled O-glycans released from glycoproteins by hydrazinolysis, *Anal. Biochem.* 304, 91–99, 2002; Vinograd, E., Lindner, B., and Seltmann, G., Lipopolysaccharides from *Serratia maracescens* possess one or two 4-amino-4-deoxy-L-arabinopyranose 1-phosphate residues in the lipid A and D-*glycero*-D-*talo*-Oct-ulopyranosonic acid in the inner core region, *Chemistry* 12, 6692–6700, 2006.

Common Name	Chemical Name	M.W.	Properties and Comment
Hydrogen Peroxide	H_2O_2	34.02	Oxidizing agent; bacteriocidal agent.
Hydroxylamine	H_3NO	33.03	
8-Hydroxyquinoline	8-quinolinol	145.16	Metal chelator.

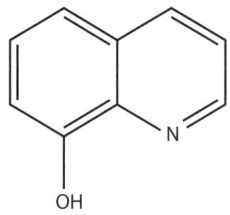

Common Name	Chemical Name	M.W.	Properties and Comment
Imidazole	1,3-diazole	69.08	Buffer component.

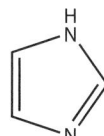

Common Name	Chemical Name	M.W.	Properties and Comment
2-Iminothiolane	Traut's Reagent (earlier as methyl-4-mercaptobutyrim-idate)	137.63	Introduction of sulfhydryl group by modification of amino group; sulfhydryl groups could then be oxidized to form cystine, which served as cleavable protein crosslink.

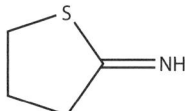

CHEMICALS COMMONLY USED IN BIOCHEMISTRY AND MOLECULAR BIOLOGY AND THEIR PROPERTIES (Continued)

Common Name	Chemical Name	M.W.	Properties and Comment

Traut, R.R., Bollen, A., Sun, T.-T. et al., Methyl-4-mercaptobutyrimidate as a cleavable crosslinking reagent and its application to the *Escherichia coli* 30S ribosome, *Biochemistry* 12, 3266–3273, 1973; Schram, H.J. and Dulffer, T., The use of 2-iminothiolane as a protein crosslinking reagent, *Hoppe Seylers Z. Physiol.Chem.* 358, 137–139, 1977; Jue, R., Lambert, J.M., Pierce, L.R., and Traut, R.R., Addition of sulfhydryl groups *Escherichia coli* ribosomes by protein modification with 2-iminothiolane (methyl 4-mercaptobutyrimidate), *Biochemistry* 17, 5399–5406, 1978; Lambert, J.M., Jue, R., and Traut, R.R., Disulfide crosslinking of *Escherichia coli* ribosomal proteins with 2-iminothiolane (methyl 4-mercaptobutyrimidate): evidence that the crosslinked protein pairs are formed in the intact ribosomal subunit, *Biochemistry* 17, 5406–5416, 1978; Alagon, A.C. and King, T.P., Activation of polysaccharides with 2-iminothiolane and its use, *Biochemistry* 19, 4341–4345, 1980; Tolan, D.R. and Traut, R.R., Protein topography of the 40 S ribosomal subunit from rabbit reticulocytes shown by crosslinking with 2-iminothiolane, *J. Biol. Chem.* 256, 10129–10136, 1981; Boileau, G., Butler, P., Hershey, J.W., and Traut, R.R., Direct crosslinks between initiation factors 1, 2, and 3 and ribosomal proteins promoted by 2-iminothiolane, *Biochemistry* 22, 3162–3170, 1983; Kyriatsoulis, A., Maly, P., Greuer, B. et al., RNA-protein crosslinking in *Escherichia coli* ribosomal subunits: localization of sites on 16S RNA which are crosslinked to proteins S17 and S21 by treatment with 2-iminothiolane, *Nucleic Acids Res.* 14, 1171–1186, 1986; Uchiumi, T., Kikuchi, M., and Ogata, K., Crosslinking study on protein neighborhoods at the subunit interface of rat liver ribosomes with 2-iminothiolane, *J. Biol. Chem.* 261, 9663–9667, 1986; McCall, M.J., Diril, H., and Meares, C.F., Simplified method for conjugating macrocyclic bifunctional chelating agents to antibodies via 2-iminothiolane, *Bioconjug. Chem.* 1, 222–226, 1990; Tarentino, A.L., Phelan, A.W., and Plummer, T.H., Jr., 2-iminothiolane: a reagent for the introduction of sulphydryl groups into oligosaccharides derived from asaparagine-linked glycans, *Glycobiology* 3, 279–285, 1993; Singh, R., Kats, L., Blattler, W.A., and Lambert, J.M., Formation of *N*-substituted 2-iminothiolanes when amino groups in proteins and peptides are modified by 2-iminothiolanes, *Anal. Biochem.* 236, 114–125, 1996; Hosono, M.N., Hosono, M., Mishra, A.K. et al., Rhenium-188-labeled anti-neural cell adhesion molecule antibodies with 2-iminothiolane modification for targeting small-cell lung cancer, *Ann. Nucl. Med.* 14, 173–179, 2000; Mokotoff, M., Mocarski, Y.M., Gentsch, B.L. et al., Caution in the use of 2-iminothiolane (Traut's reagent) as a crosslinking agent for peptides. The formation of *N*-peptidyl-2-iminothiolanes with bombesin (BN) antagonists (D-trp[6]-leu13-ψ[CH$_2$NH]-Phe[14]BN$_{6-14}$ and D-trp- gln-trp-NH$_2$, *J. Pept. Res.* 57, 383–389, 2001; Kuzuhara, A., Protein structural changes in keratin fibers induced by chemical modification using 2-iminothiolane hydrochloride: a Raman spectroscopic investigation, *Biopolymers* 79, 173–184, 2005.

| **Indole** | 2,3-benzopyrrole | 117.15 | |

| **Indole-3-acetic Acid** | Indoleacetic Acid; Heteroauxin | 175.19 | Plant growth regulator. |

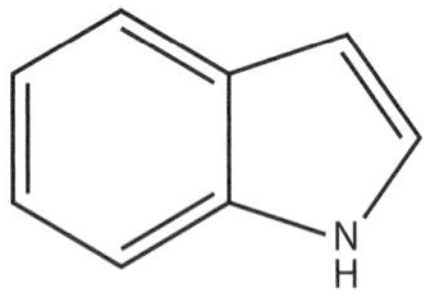

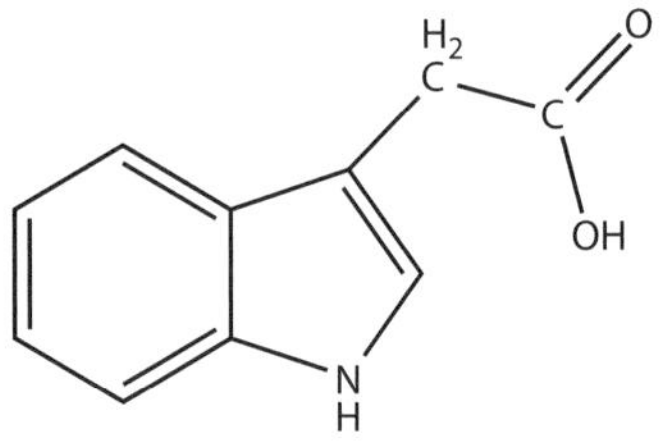

Kawaguchi, M. and Syono, K., The excessive production of indole-3-acetic and its significance in studies of the biosynthesis of this regulator of plant growth and development, *Plant Cell Physiol.* 37, 1043–1048, 1996; Normanly, J. and Bartel, B., Redundancy as a way of life-IAA metabolism, *Curr. Opin. Plant Biol.* 2, 207–213, 1999; Leyser, O., Auxin signaling: the beginning, the middle, and the end, *Curr. Opin. Plant Biol.* 4, 382–386, 2001; Ljung, K., Hull, A.K., Kowalczyk, M. et al., Biosynthesis, conjugation, catabolism, and homeostasis of indole-3-acetic acid in *Arabidopsis thaliana*, *Plant Mol. Biol.* 49, 249–272, 2002; Kawano, T. Roles of the reactive oxygen species-generating peroxidase reactions in plant defense and growth induction, *Plant Cell Rep.* 21, 829–837, 2003; Aloni, R., Aloni, E., Langhans, M., and Ullrich, C.I., Role of cytokine and auxin in shaping root architecture: regulating vascular differentiation, laterial root initiation, root apical dominance, and root gravitropism, *Ann. Bot.* 97, 882–893, 2006.

| **Iodoacetamide** | 2-iodoacetamide | 184.96 | Alkylating agents that react with a variety of nucleophiles in proteins and nucleic acids. Reaction is more rapid than the bromo or chloro derivatives. |

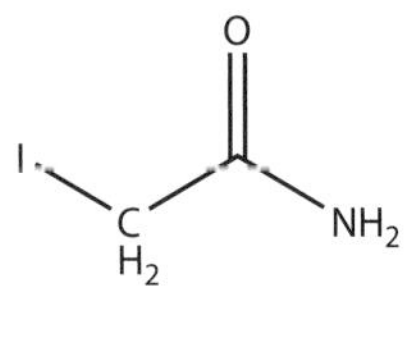

CHEMICALS COMMONLY USED IN BIOCHEMISTRY AND MOLECULAR BIOLOGY AND THEIR PROPERTIES (Continued)

Common Name	Chemical Name	M.W.	Properties and Comment
Iodoacetic Acid		185.95	

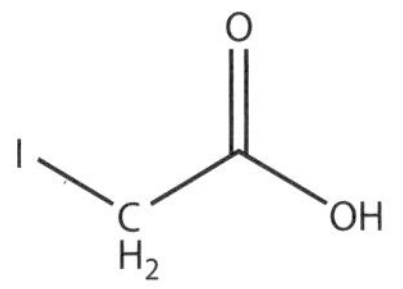

The amide is neutral and is not susceptible to either positive or negative influence from locally charged groups; iodoacetamide is frequently used to modify sulfhydryl groups as part of reduction and carboxymethylation prior to structural analysis. Crestfield, A.M., Moore, S., and Stein, W.H., The preparation and enzymatic hydrolysis of reduced and *S*-carboxymethylated proteins, *J. Biol. Chem.* 238, 622–627, 1963; Watts, D.C., Rabin, B.R., and Crook, E.M., The reaction of iodoacetate and iodoacetamide with proteins as determined with a silver/silver iodide electrode, *Biochim. Biophys. Acta* 48, 380–388, 1961; Inagami, T., The alkylation of the active site of trypsin with iodoacetamide in the presence of alkylguanidines, *J. Biol. Chem.* 240, PC3453–PC3455, 1965; Fruchter, R.G. and Crestfield, A.M., The specific alkylation by iodoacetamide of histidine-12 in the active site of ribonuclease, *J. Biol. Chem.* 242, 5807–5812, 1967; Takahashi, K., The structure and function of ribonuclease T. X. Reactions of iodoacetate, iodoacetamide, and related alkylating reagents with ribonuclease T, *J. Biochem.* 68, 517–527, 1970; Whitney, P.L., Inhibition and modification of human carbonic anhydrase B with bromoacetate and iodoacetate, *Eur. J. Biochem.* 16, 126–135, 1970; Harada, M. and Irie, M., Alkylation of ribonuclease from *Aspirgillus saitoi* with iodoacetate and iodoacetamide, *J. Biochem.* 73, 705–716, 1973; Halasz, P. and Polgar, L., Effect of the immediate microenvironment on the reactivity of the essential SH group of papain, *Eur. J. Biochem.* 71, 571–575, 1976; Franzen, J.S., Ishman, P., and Feingold, D.S., Half-of-the-sites reactivity of bovine liver uridine diphosphoglucose dehydrogenase toward iodoacetate and iodoacetamide, *Biochemistry* 15, 5665–5671, 1976; David, M., Rasched, I.R., and Sund, H., Studies of glutamate dehydrogenase. Methionione-169: the preferentially carboxymethylated residue, *Eur. J. Biochem.* 74, 379–385, 1977; Ohgi, K., Watanabe, H., Emman, K. et al., Alkylation of a ribonuclease from *Streptomyces erthreus* with iodoacetate and iodoacetamide, *J. Biochem.* 90, 113–123, 1981; Dahl, K.H. and McKinley-McKee, J.S., Enzymatic catalysis in the affinity labeling of liver alcohol dehydrogenase with haloacids, *Eur. J. Biochem.* 118, 507–513, 1981; Syvertsen, C. and McKinley-McKee, J.S., Binding of ligands to the catalytic zinc ion in horse liver alcohol dehydrogenase, *Arch. Biochem. Biophys.* 228, 159–169, 1984; Communi, D. and Erneux, C., Identification of an active site cysteine residue in type Ins(1,4,5)P³5-phosphatase by chemical modification and site-directed mutagenesis, *Biochem. J.* 320, 181–186, 1996; Sarkany, Z., Skern, T., and Polgar, L., Characterization of the active site thiol group of rhinovirus 21 proteinase, *FEBS Lett.* 481, 289–292, 2000; Lundblad, R.L., *Chemical Reagents for Protein Modification*, CRC Press, Boca Raton, FL, 2004.

| **Isatoic Anhydride** | 3,1-benzoxazine-2,4(1*H*)-dione | 163.13 | Fluorescent reagents for amines and sulfydryl groups; amine scavenger. |

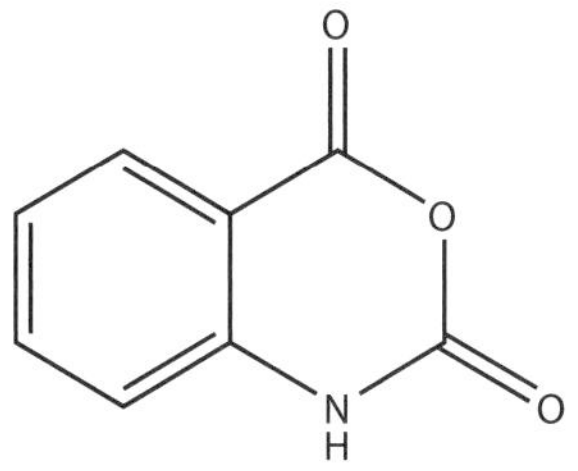

Gelb, M.H. and Abeles, R.H., Substituted isatoic anhydrides: selective inactivators of trypsinlike serine proteases, *J. Med. Chem.* 29, 585–589, 1986; Gravett, P.S., Viljoen, C.C., and Oosthuizen, M.M., Inactivation of arginine esterase E-1 of *Bitis gabonica* venom by irreversible inhibitors including a water-soluble carbodiimide, a chloromethyl ketone, and isatoic anhydride, *Int. J. Biochem.* 23, 1101–1110, 1991; Servillo, L., Balestrieri, C., Quagliuolo, L. et al., tRNA fluorescent labeling at 3′ end including an aminoacyl-tRNA-like behavior, *Eur. J. Biochem.* 213, 583–589, 1993; Churchich, J.E., Fluorescence properties of *o*-aminobenzoyl-labeled proteins, *Anal. Biochem.* 213, 229–233, 1993; Brown, A.D. and Powers, J.C., Rates of thrombin acylation and deacylation upon reaction with low molecular weight acylating agents, carbamylating agents, and carbonylating agents, *Bioorg. Med. Chem.* 3, 1091–1097, 1995; Matos, M.A., Miranda, M.S., Morais, V.M., and Liebman, J.F., Are isatin and isatoic anhydride antiaromatic and aromatic, respectively? A combined experimental and theoretic investigation, *Org. Biomol. Chem.* 1, 2566–2571, 2003; Matos, M.A., Miranda, M.S., Morais, V.M., and Liebman, J.F., The energetics of isomeric benzoxazine diones: isatoic anhydride revisited, *Org. Biomol. Chem.* 2, 1647–1650, 2004; Raturi, A., Vascratsis, P.O., Seslija, D. et al., A direct, continuous, sensitive assay for protein disulphide-isomerase based on fluorescence self-quenching, *Biochem. J.* 391, 351–357, 2005; Zhang, W., Lu, Y., and Nagashima, T., Plate-to-plate fluorous solid-phase extraction for solution-phase parallel synthesis, *J. Comb. Chem.* 7, 893–897, 2005.

| **Isoamyl Alcohol** | Isopentyl Alcohol; 3-methyl-1-butanol | 88.15 | Solvent. |

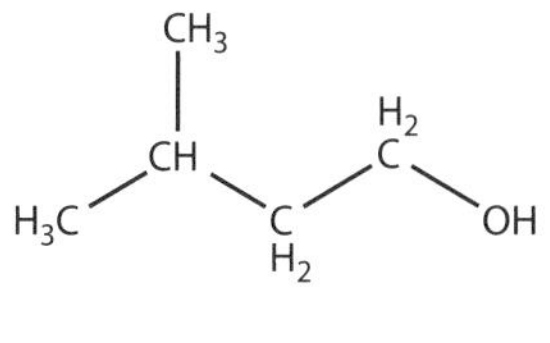

**CHEMICALS COMMONLY USED IN BIOCHEMISTRY AND MOLECULAR
BIOLOGY AND THEIR PROPERTIES (Continued)**

Common Name	Chemical Name	M.W.	Properties and Comment
Isopropanol	2-propanol	60.10	Solvent; precipitation agent for purification of plasmid DNA; reagent in stability test for identification of abnormal hemoglobins.

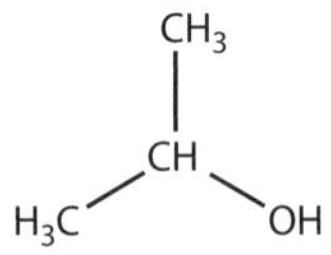

Brosious, E.M., Morrison, B.Y., and Schmidt, R.M., Effects of hemoglobin F levels, KCN, and storage on the isopropanol precipitation test for unstable hemoglobins, *Am. J. Clin. Pathol.* 66, 878–882, 1976; Bensinger, T.A. and Beutler, E., Instability of the oxy form of sickle hemoglobin and of methemoglobin in isopropanol, *Am. J. Clin. Pathol.* 67, 180–183, 1977; Acree, W.E., Jr. and Bertrand, G.L., A cholesterol-isopropanol gel, *Nature* 269, 450, 1977; Naoum, P.C. Teixeira, U.A., de Abreu Machado, P.E., and Michelin, O.C., The denaturation of human oxyhemoglobin A, A2, and S by isopropanol/buffer method, *Rev. Bras. Pesqui. Med. Biol.* 11, 241–244, 1978; Ali, M.A., Quinlan, A., and Wong, S.C., Identification of hemoglobin E by the isopropanol solubility test, *Clin. Biochem.* 13, 146–148, 1980; Horer, O.L. and Enache, C., 2-propanol dependent RNA absorbances, *Virologie* 34, 257–272, 1983; De Venditis, E., Masullo, M., and Bocchini, V., The elongation factor G carries a catalytic site for GTP hydrolysis, which is revealed by using 2-propanol in the absence of ribosomes, *J. Biol. Chem.* 261, 4445–4450, 1986; Wang, L., Hirayasu, K., Ishizawa, M., and Kobayashi, Y., Purification of genomic DNA from human whole blood by isopropanol-fractionation with concentrated NaI and SDS, *Nucleic Acids Res.* 22, 1774–1775, 1994; Dalhus, B. and Gorbitz, C.H., Glycyl-L-leucyl-L-tyrosine dehydrate 2-propanol solvate, *Acta Crystallogr. C* 52, 2087–2090, 1996; Freitas, S.S., Santos, J.A., and Prazeres, D.M., Optimization of isopropanol and ammonium sulfate precipitation steps in the purification of plasmid DNA, *Biotechnol. Prog.* 22, 1179–1186, 2006; Halano, B., Kubo, D., and Tagaya, H., Study on the reactivity of diarylmethane derivatives in supercritical alcohols media: reduction of diarylmethanols and diaryl ketones to diarylmethanes using supercritical 2-propanol, *Chem. Pharm. Bull.* 54, 1304–1307, 2006.

Isopropyl-β-D-thiogalactoside	IPTG, Isopropyl-β-D-thiogalactopyroa-noside	238.3	"Gratuitous" inducer of the *lac* operon.

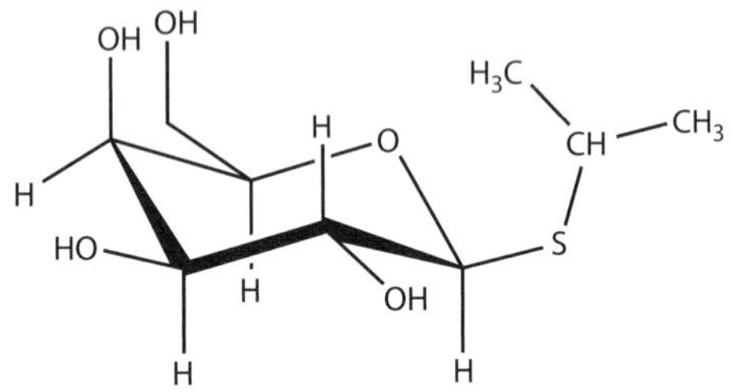

Cho, S., Scharpf, S., Franko, M., and Vermeulen, C.W., Effect of isopropyl-β-D-galactoside concentration on the level of *lac*-operon induction in steady state *Escherichia coli*, *Biochem. Biophys. Res. Commun.* 128, 1268–1273, 1985; Carlsson, U., Ferskgard, P.O., and Svensson, S.C., A simple and efficient synthesis of the induced IPTG made for inexpensive heterologous protein production using the *lac*-promoter, *Protein Eng.* 4, 1019–1020, 1991; Donovan, R.S., Robinson, C.W., and Glick, B.R., Review: optimizing inducer and culture conditions for expression of foreign proteins under control of the *lac* promoter, *J. Ind. Microbiol.* 16, 145–154, 1996; Hansen, L.H., Knudsen, S., and Sorensen, S.J., The effect of the lacy gene on the induction of IPTG-inducible promoters, studied in *Escherichia coli* and *Pseudomonas fluorescens*, *Curr. Microbiol.* 36, 341–347, 1998; Teich, A., Lin, H.Y., Andersson, L. et al., Amplification of ColE1 related plasmids in recombinant cultures of *Escherichia coli* after IPTG induction, *J. Biotechnol.* 64, 197–210, 1998; Ren, A. and Schaefer, T.S., Isopropyl-β-D-thiogalactoside (IPTG)-inducible tyrosine phosphorylation of proteins in *E. coli*, *Biotechniques* 31, 1254–1258, 2001; Ko, K.S., Kruse, J., and Pohl, N.L., Synthesis of isobutryl-C-galactoside (IBCG) as an isopropylthiogalactoside (IPTG) substitute for increased induction of protein expression, *Org. Lett.* 5, 1781–1783, 2003; Intasai, N., Arooncharus, P., Kasinrerk, W., and Tayapiwatana, C., Construction of high-density display of CD147 ectodomain on VCSM13 phage via gpVIII: effects of temperature, IPTG, and helper phage infection-period, *Protein Expr. Purif.* 32, 323–331, 2003; Faulkner, E., Barrett, M., Okor, S. et al., Use of fed-batch cultivation for achieving high cell densities for the pilot-scale production of a recombinant protein (phenylalanine dehydrogenase) in *Escherichia coli*, *Biotechnol. Prog.* 22, 889–897, 2006; Gardete, S., de Laencastre, H., and Tomasz, A., A link in transcription between the native pbpG and the acquired mecA gene in a strain of *Staphylococcus aureus*, *Microbiology* 152, 2549–2558, 2006; Hewitt, C.J., Onyeaka, H., Lewis, G. et al., A comparison of high cell density fed-batch fermentations involving both induced and noninduced recombinant *Escherichia coli* under well-mixed small-scale and simulated poorly mixed large-scale conditions, *Biotechnol. Bioeng.*, in press, 2006; Picaud, S., Olsson, M.E., and Brodelius, P.E., Improved conditions for production of recombinant plant sesquiterpene synthases in *Escherichia coli*, *Protein Expr. Purif.*, in press, 2006.

CHEMICALS COMMONLY USED IN BIOCHEMISTRY AND MOLECULAR BIOLOGY AND THEIR PROPERTIES (Continued)

Common Name	Chemical Name	M.W.	Properties and Comment
Maleic Anhydride	2,5-furandione	98.06	Modification of amino groups in proteins. The dimethyl derivative (dimethylmaleic anhydride) is used for ribosome dissociation; monomer for polymer.

Giese, R.W. and Vallee, B.L., Metallocenes. A novel class of reagents for protein modification. I. Maleic anhydride-iron tetracarbonyl, *J. Am. Chem. Soc.* 94, 6199–6200, 1972; Cantrell, M. and Craven, G.R., Chemical inactivation of *Escherichia coli* 30 S ribosomes with maleic anhydride: identification of the proteins involved in polyuridylic acid binding, *J. Mol. Biol.* 115, 389–402, 1977; Jordano, J., Montero, F., and Palacian, E., Relaxation of chromatin structure upon removal of histones H2A and H2B, *FEBS Lett.* 172, 70–74, 1984; Jordano, J., Montero, F., and Palacian, E., Rearrangement of nucleosomal components by modification of histone amino groups. Structural role of lysine residues, *Biochemistry* 23, 4280–4284, 1984; Palacian, E., Gonzalez, P.J., Pineiro, M., and Hernandez, F., Dicarboxylic acid anhydrides as dissociating agents of protein-containing structures, *Mol. Cell. Biochem.* 97, 101–111, 1990; Paetzel, M., Strynadka, N.C., Tschantz, W.R. et al., Use of site-directed chemical modification to study an essential lysine in *Escherichia coli* leader peptidase, *J. Biol. Chem.* 272, 9994–10003, 1997; Wink, M.R., Buffon, A., Bonan, C.D. et al., Effect of protein-modifying reagents on ecto-apyrase from rat brain, *Int. J. Biochem. Cell Biol.* 32, 105–113, 2000.

2-Mercaptoethanol	β-mercaptoethanol	78.13	Reducing agent; used frequently in the reduction and alkylation of proteins for structural analysis and for preservation of oxidation-sensitive enzymes.

Geren, C.R., Olomon, C.M., Jones, T.T., and Ebner, D.E., 2-mercaptoethanol as a substrate for liver alcohol dehydrogenase, *Arch. Biochem. Biophys.* 179, 415–419, 1977; Opitz, H.G., Lemke, H, and Hewlett, G., Activation of T-cells by a macrophage or 2-mercaptoethanol-activated serum factor is essential for induction of a primary immune response to heterologous red cells *in vitro*, *Immunol. Rev.* 40, 53–77, 1978; Burger, M., An absolute requirement for 2-mercaptoethanol in the *in vitro* primary immune response in the absence of serum, *Immunology* 37, 669–671, 1979; Nealon, D.A., Pettit, S.M., and Henderson, A.R., Diluent pH and the stability of the thiol group in monothioglycerol, *N*-acetyl-L-cysteine, and 2-mercaptoethanol, *Clin. Chem.* 27, 505–506, 1981; Dahl, K.H. and McKinley-McKee, J.S., Enzymatic catalysis in the affinity labeling of liver alcohol dehydrogenase with haloacids, *Eur. J. Biochem.* 118, 507–513, 1981; Righetti, P.G., Tudor, G., and Glanazza, E., Effect of 2-mercaptoethanol on pH gradients in isoelectric focusing, *J. Biochem. Biophys. Methods* 6, 219–227, 1982; Soderberg, L.S. and Yeh, N.H., T-cells and the anti-trinitrophenyl antibody response to fetal calf serum and 2-mercaptoethanol, *Proc. Soc. Exp. Biol. Med.* 174, 107–113, 1983; Ochs, D., Protein contaminants of sodium dodecyl sulfate-polyacrylamide gels, *Anal. Biochem.* 135, 470–474, 1983; Schaefer, W.H., Harris, T.M., and Guengerich, F.P., Reaction of the model thiol 2-mercaptoethanol and glutathione with methylvinylmaleimide, a Michael acceptor with extended conjugation, *Arch. Biochem. Biophys.* 257, 186–193, 1987; Obiri, N. and Pruett, S.B., The role of thiols in lymphocyte responses: effect of 2-mercaptoethanol on interleukin 2 production, *Immunobiology* 176, 440–449, 1988; Gourgerot-Pocidalo, M.A., Fay, M., Roche, Y., and Chollet-Martin, S., Mechanisms by which oxidative injury inhibits the proliferative response of human lymphocytes to PHA. Effect of the thiol compound 2-mercaptoethanol, *Immunology* 64, 281–288, 1988; Fong, T.C. and Makinodan, T., Preferential enhancement by 2-mercaptoethanol of IL-2 responsiveness of T blast cells from old over young mice is associated with potentiated protein kinase C translocation, *Immunol. Lett.* 20, 149–154, 1989; De Graan, P.N., Moritz, A., de Wit, M., and Gispen, W.H., Purification of B-50 by 2-mercaptoethanol extraction from rat brain synaptosomal plasma membranes, *Neurochem. Res.* 18, 875–881, 1993; Carrithers, S.L. and Hoffman, J.L., Sequential methylation of 2-mercaptoenthanol to the dimethyl sulfonium ion, 2-(dimethylthio)ethanol, *in vivo* and *in vitro*, *Biochem. Pharmacol.* 48, 1017–1024, 1994; Paul-Pretzer, K. and Parness, J., Elimination of keratin contaminant from 2-mercaptoethanol, *Anal. Biochem.* 289, 98–99, 2001; Adebiyi, A.P., Jin, D.H, Ogawa, T., and Muramoto, K., Acid hydrolysis of protein in a microcapillary tube for the recovery of tryptophan, *Biosci. Biotechnol. Biochem.* 69, 255–257, 2005; Adams, B., Lowpetch, K., Throndycroft, F. et al., Stereochemistry of reactions of the inhibitor/substrates L- and D-β-chloroalanine with β-mercaptoethanol catalyzed by L-aspartate aminotransferase and D-amino acid amino-transferase, respectively, *Org. Biomol. Chem.* 3, 3357–3364, 2005; Layeyre, M., Leprince, J., Massonneau, M. et al., Aryldithioethyloxycarbonyl (Ardec): a new family of amine-protecting groups removable under mild reducing conditions and their applications to peptide synthesis, *Chemistry* 12, 3655–3671, 2006; Okun, I., Malarchuk, S., Dubrovskaya, E. et al., Screening for caspace-3 inhibitors: effect of a reducing agent on the identified hit chemotypes, *J. Biomol. Screen.* 11, 694–703, 2006; Aminian, M., Sivam, S., Lee, C.W. et al., Expression and purification of a trivalent pertussis toxin-diphtheria toxin-tetanus toxin fusion protein in *Escherichia coli*, *Protein Expr. Purif.* 51, 170–178, 2006.

CHEMICALS COMMONLY USED IN BIOCHEMISTRY AND MOLECULAR
BIOLOGY AND THEIR PROPERTIES (Continued)

Common Name	Chemical Name	M.W.	Properties and Comment
(3-Mercaptopropyl)trimethoxysilane	3-(trimethoxysilyl)-1-propanethiol	196.34	Introduction of reactive sulfhydryl onto glass (silane) surface.

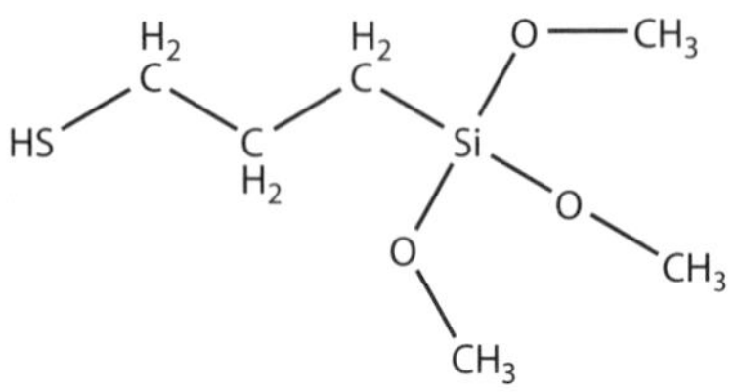

Jung, S.K. and Wilson, G.S., Polymeric mercaptosilane-modified platinum electrodes for elimination of interferants in glucose biosensors, *Anal. Chem.* 68, 591–596, 1996; Mansur, H.S., Lobato, Z.P., Orefice, R.L. et al., Surface functionalization of porous glass networks: effects on bovine serum albumin and porcine insulin immobilization, *Biomacromolecules* 1, 479–497, 2000; Kumar, A., Larsson, O., Parodi, D., and Liang, Z., Silanized nucleic acids: a general platform for DNA immobilization, *Nucleic Acids Res.* 28, E71, 2000; Zhang, F., Kang, E.T., Neoh, K.G. et al., Surface modification of stainless steel by grafting of poly(ethylene glycol) for reduction in protein adsorption, *Biomaterials* 22, 1541–1548, 2001; Jia, J., Wang, B., Wu, A. et al., A method to construct a third-generation horseradish peroxidase biosensor: self-assembling gold nanoparticles to three-dimensional sol-gel network, *Anal. Chem.* 74, 2217–2223, 2002; Abdelghani-Jacquin, C., Abdelghani, A., Chmel, G. et al., Decorated surfaces by biofunctionalized gold beads: application to cell adhesion studies, *Eur. Biophys. J.* 31, 102–110, 2002; Ganesan, V. and Walcarius, A., Surfactant templated sulfonic acid functionalized silica microspheres as new efficient ion exchangers and electrode modifiers, *Langmuir* 20, 3632–3640, 2004; Crudden, C.M., Sateesh, M., and Lewis, R., Mercaptopropyl-modified mesoporous silica: a remarkable support for the preparation of a reusable, heterogeneous palladium catalyst for coupling to reactions, *J. Am. Chem. Soc.* 127, 10045–10050, 2005; Yang, L., Guihen, E., and Glennon, J.D., Alkylthiol gold nanoparticles in sol-gel-based open tabular capillary electrochromatography, *J. Sep. Sci.* 28, 757–766, 2005.

Methanesulfonic Acid		96.11	Protein hydrolysis for amino acid analysis; deprotection during peptide synthesis; hydrolysis of protein substituents such as fatty acids.

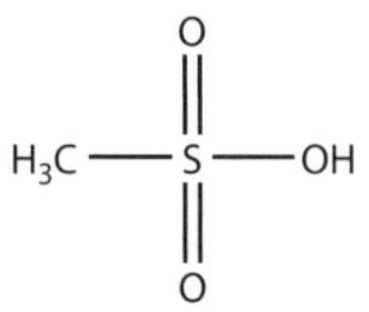

Simpson, R.J., Neuberger, M.R., and Liu, T.Y., Complete amino acid analysis of proteins from a single hydrolyzate, *J. Biol. Chem.* 251, 1936–1940, 1976; Kubota, M., Hirayama, T., Nagase, O., and Yajima, H., Synthesis of two peptides corresponding to an alpha-endophin and gamma-endorphin by the methanesulfonic acid deprotecting procedures, *Chem. Pharm. Bull.* 27, 1050–1054, 1979; Yajima, H., Akaji, K., Saito, H. et al., Studies on peptides. LXXXII. Synthesis of [4-Gln]-neurotensin by the methanesulfonic acid deprotecting procedure, *Chem. Pharm. Bull.* 27, 2238–2242, 1979; Sakuri, J. and Nagahama, M. Tryptophan content of *Clostridium perfringens* epsilon toxin, *Infect. Immun.* 47, 260–263, 1985; Malmer, M.F. and Schroeder, L.A., Amino acid analysis by high-performance liquid chromatography with methanesulfonic acid hydrolysis and 9-fluorenylmethyl-chloroformate derivatization, *J. Chromatog.* 514, 227–239, 1990; Weiss, M., Manneberg, M., Juranville, J.F. et al., Effect of the hydrolysis method on the determination of the amino acid composition of proteins, *J. Chromatog. A* 795, 263–275, 1998; Okimura, K., Ohki, K., Nagai, S., and Sakura, N., HPLC analysis of fatty acyl-glycine in the aqueous methanesulfonic acid hydrolysates of N-terminally fatty acylated peptides, *Biol. Pharm. Bull.* 26, 1166–1169, 2003; Wrobel, K., Kannamkumarath, S.S., Wrobel, K., and Caruso, J.A., Hydrolysis of proteins with methanesulfonic acid for improved HPLC-ICP-MS determination of seleno-methionine in yeast and nuts, *Anal. BioAnal. Chem.* 375, 133–138, 2003.

Methanol	Methyl Alcohol	32.04	Solvent.
Methylethyl Ketone (MEK)	2-butanal; 2-butanone	72.11	Solvent; with acid for cleavage of heme moiety of hemeproteins for preparation of apoproteins.

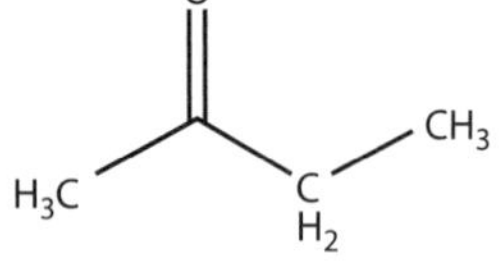

Teale, F.W., Cleavage of haem-protein link by acid methylethylketone, *Biochim. Biophys. Acta* 35, 543, 1959; Tran, C.D. and Darwent, J.R., Characterization of tetrapyridylporphyrinatozinc (II) apomyoglobin complexes as a potential photosynthetic model, *J. Chem. Soc. Faraday Trans. II*, 82, 2315–2322, 1986.

CHEMICALS COMMONLY USED IN BIOCHEMISTRY AND MOLECULAR
BIOLOGY AND THEIR PROPERTIES (Continued)

Common Name	Chemical Name	M.W.	Properties and Comment
Methylglyoxal	Pyruvaldehyde; 2-oxopropanal	72.06	Derived from oxidative modification of triose phosphate during glucose metabolism; model for glycation of proteins; reacts with amino groups in proteins and nucleic acids; involved in advanced glycation endproducts.

Szabo, G., Kertesz, J.C., and Laki, K., Interaction of methylglyoxal with poly-L-lysine, *Biomaterials* 1, 27–29, 1980; McLaughlin, J.A., Pethig, R., and Szent-Gyorgyi, A., Spectroscopic studies of the protein-methylglyoxal adduct, *Proc. Natl. Acad. Sci. USA* 77, 949–951, 1980; Cooper, R.A., Metabolism of methylglyoxal in microorganisms, *Annu. Rev. Microbiol.* 38, 49–68, 1984; Richard, J.P., Mechanism for the formation of methylglyoxal from triosephosphates, *Biochem. Soc. Trans.* 21, 549–553, 1993; Riley, M.L. and Harding, J.J., The reaction of methylglyoxal with human and bovine lens proteins, *Biochim. Biophys. Acta* 1270, 36–43, 1995; Thornalley, P.J., Pharmacology of methylglyoxal: formation, modification of proteins and nucleic acids, and enzymatic detoxification — a role in pathogenesis and antiproliferative chemotherapy, *Gen. Pharmacol.* 27, 565–573, 1996; Nagaraj, R.H., Shipanova, I.N., and Faust, F.M., Protein crosslinking by the Maillard reaction. Isolation, characterization, and *in vivo* detection of a lysine–lysine crosslink derived from methylglyoxal, *J. Biol. Chem.* 271, 19338–19345, 1996; Shipanova, I.N., Glomb, M.A., and Nagaraj, R.H., Protein modification by methylglyoxal: chemical nature and synthetic mechanism of a major fluorescent adduct, *Arch. Biochem. Biophys.* 344, 29–34, 1997; Uchida, K., Khor, O.T., Oya, T. et al., Protein modification by a Maillard reaction intermediate methylglyoxal. Immunochemical detection of fluorescent 5-methylimidazolone derivatives *in vivo*, *FEBS Lett.* 410, 313–318, 1997; Degenhardt, T.P., Thorpe, S.R., and Baynes, J.W., Chemical modification of proteins by methylglyoxal, *Cell. Mol. Biol.* 44, 1139–1145, 1998; Izaguirre, G., Kikonyogo, A., and Pietruszko, R., Methylglyoxal as substrate and inhibitor of human aldehyde dehydrogenase: comparison of kinetic properties among the three isozymes, *Comp. Biochem. Physiol. B Biochem. Mol. Biol.* 119, 747–754, 1998; Lederer, M.O. and Klaiber, R.G., Crosslinking of proteins by Maillard processes: characterization and detection of lysine–arginine crosslinks derived from glyoxal and methylglyoxal, *Bioorg. Med. Chem.* 7, 2499–2507, 1999; Kalapos, M.P., Methylglyoxal in living organisms: chemistry, biochemistry, toxicology, and biological implications, *Toxicol. Lett.* 110, 145–175, 1999; Thornalley, P.J., Landborg, A., and Minhas, H.S., Formation of glyoxal, methylglyoxal, and 3-deoxyglucose in the glycation of proteins by glucose, *Biochem. J.* 344, 109–116, 1999; Nagai, R., Araki, T., Hayashi, C.M. et al., Identification of *N*-epsilon-(carboxyethyl)lysine, one of the methylglyoxal-derived AGE structures, in glucose-modified protein: mechanism for protein modification by reactive aldehydes, *J. Chromatog. B Analyt. Technol. Biomed. Life Sci.* 788, 75–84, 2003.

Methyl Methanethiosulfonate (MMTS)	*S*-methyl Methanethiosulfonate	126.2	Modification of sulfhydryl groups.

Smith, D.J., Maggio, E.T., and Kenyon, G.L., Simple alkanethiol groups for temporary sulfhydryl groups of enzymes, *Biochemistry* 14, 766–771, 1975; Nishimura, J.S., Kenyon, G.L., and Smith, D.J., Reversible modification of the sulfhydryl groups of *Escherichia coli* succinic thiokinase with methanethiolating reagents, 5,5′-dithio-bis(2-nitrobenzoic acid), *p*-hydroxymercuribenzoate, and ethylmercurithiosalicylate, *Arch. Biochem. Biophys.* 170, 407–430, 1977; Bloxham, D.P., The chemical reactivity of the histidine-195 residue in lactate dehydrogenase thiomethylated at the cysteine-165 residue, *Biochem. J.* 193, 93–97, 1981; Gavilanes, F., Peterson, D., and Schirch, L., Methyl methanethiosulfate as an active site probe of serine hydroxymethyltransferase, *J. Biol. Chem.* 257, 11431–11436, 1982; Daly, T.J., Olson, J.S., and Matthews, K.S., Formation of mixed disulfide adducts as cysteine-281 of the lactose repressor protein affects operator- and inducer-binding parameters, *Biochemistry* 25, 5468–5474, 1986; Salam, W.H. and Bloxham, D.P., Identification of subsidiary catalytic groups at the active site of β-ketoacyl-CoA thiolase by covalent modification of the protein, *Biochim. Biophys. Acta* 873, 321–330, 1986; Stancato, L.F., Hutchison, K.A., Chakraborti, P.K. et al., Differential effects of the reversible thiol-reactive agents arsenite and methyl methanethiosulfonate on steroid binding by the glucocorticoid receptor, *Biochemistry* 32, 3739–3736, 1993; Hou, L.X. and Vollmer, S., The activity of *S*-thiolated modified creatine kinase is due to the regeneration of free thiol at the active site, *Biochim. Biophys. Acta* 1205, 83–88, 1994; Jensen, P.E., Shanbhag, V.P., and Stigbrand, T., Methanethiolation of the liberated cysteine residues of human α-2-macroglobulin treated with methylamine generates a derivative with similar functional characteristics as native β-2-macroglobulin, *Eur. J. Biochem.* 227, 612–616, 1995; Trimboli, A.J., Quinn, G.B., Smith, E.T., and Barber, M.J., Thiol modification and site-directed mutagenesis of the flavin domain of spinach NADH: nitrate reductase, *Arch. Biochem. Biophys.* 331, 117–126, 1996; Quinn, K.E. and Ehrlich, B.E., Methanethiosulfonate derivatives inhibits current through the rynodine receptor/channel, *J. Gen. Physiol.* 109, 225–264, 1997;

CHEMICALS COMMONLY USED IN BIOCHEMISTRY AND MOLECULAR
BIOLOGY AND THEIR PROPERTIES (Continued)

Common Name	Chemical Name	M.W.	Properties and Comment

Hashimoto, M., Majima, E., Hatanaka, T. et al., Irreversible extrusion of the first loop facing the matrix of the bovine heart mitochondrial ADP/ATP carrier by labeling the Cys(56) residue with the SH-reagent methyl methanethiosulfonate, *J. Biochem.* 127, 443–449, 2000; Spelta, V., Jiang, L.H., Bailey, R.J. et al., Interaction between cysteines introduced into each transmembrane domain of the rat P2X2 receptor, *Br. J. Pharmacol.* 138, 131–136, 2003; Britto, P.J., Knipling, L., McPhie, P., and Wolff, J., Thiol-disulphide interchange in tubulin: kinetics and the effect on polymerization, *Biochem. J.* 389, 549–558, 2005; Miller, C.M., Szegedi, S.S., and Garrow, T.A., Conformation-dependent inactivation of human betaine-homocysteine *S*-methyltransferase by hydrogen peroxide *in vitro*, *Biochem. J.* 392, 443–448, 2005.

| *N*-Methylpyrrolidone | 1-methyl-2-pyrrolidone | 99.13 | Polar solvent; transdermal transport of drugs. |

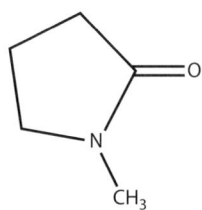

Barry, B.W. and Bennett, S.L., Effect of penetration enhancers on the permeation of mannitol, hydrocortisone, and progesterone through human skin, *J. Pharm. Pharmacol.* 39, 535–546, 1987; Forest, M. and Fournier, A., BOP reagent for the coupling of pGlu and Boc-His(Tos) in solid phase peptide synthesis, *Int. J. Pept. Protein Res.* 35, 89–94, 1990; Sasaki, H., Kojima, M., Nakamura, J., and Shibasaki, J., Enhancing effect of combining two pyrrolidone vehicles on transdermal drug delivery, *J. Pharm. Pharmacol.* 42, 196–199, 1990; Uch, A.S., Hesse, U., and Dressman, J.B., Use of 1-methyl-pyrrolidone as a solubilizing agent for determining the uptake of poorly soluble drugs, *Pharm. Res.* 16, 968–971, 1999; Zhao, F. Bhanage, B.M., Shirai, M., and Arai, M., Heck reactions of iodobenzene and methyl acrylate with conventional supported palladium catalysts in the presence of organic and/or inorganic bases without ligands, *Chemistry* 6, 843–848, 2000; Lee, P.J., Langer, R., and Shastri, V.P., Role of *n*-methyl pyrrolidone in the enhancement of aqueous phase transdermal transport, *J. Pharm. Sci.* 94, 912–917, 2005; Tae, G., Kornfield, J.A., and Hubbell, J.A., Sustained release of human growth hormone from *in situ* forming hydrogels using self-assembly of fluoroalkyl-ended poly(ethylene glycol), *Biomaterials* 26, 5259–5266, 2005; Babu, R.J. and Pandit, J.K., Effect of penetration enhancers on the transdermal delivery of bupranolol through rat skin, *Drug Deliv.* 12, 165–169, 2005; Luan, X. and Bodmeier, R., *In situ* forming microparticle system for controlled delivery of leupolide acetate: influence of the formulation and processing parameters, *Eur. J. Pharm. Sci.* 27, 143–149, 2006; Lee, P.J., Ahmad, N., Langer, R. et al., Evaluation of chemical enhancers in the transdermal delivery of lidocaine, *Int. J. Pharm.* 308, 33–39, 2006; Ruble, G.R., Giardino, O.X., Fossceco, S.L. et al., *J. Am. Assoc. Lab. Anim. Sci.* 45, 25–29, 2006.

| NBS | *N*-bromosuccinimide; 1-bromo-2,5-pyrrolidinedione | 178 | Protein modification reagent; bromination of olefins; analysis of a variety of other compounds. |

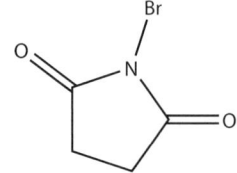

Sinn, H.J., Schrenk, H.H., Friedrich, E.A. et al., Radioiodination of proteins and lipoproteins using *N*-bromosuccinimide as oxidizing agent, *Anal. Biochem.* 170, 186–192, 1988; Tanemura, K., Suzuki, T., Nishida, Y. et al., A mild and efficient procedure for α-bromination of ketones using *N*-bromosuccinimide catalyzed by ammonium acetate, *Chem. Commun.* 3, 470–471, 2004; Lundblad, R.L., *Chemical Reagents for Protein Modification*, 3rd ed., CRC Press, Boca Raton, FL, 2004; Edens, G.J., Redox titration of antioxidant mixtures with *N*-bromosuccinimide as titrant: analysis by nonlinear least-squares with novel weighting function, *Anal. Sci.* 21, 1349–1354, 2005; Abdel-Wadood, H.M., Mohamed, H.A., and Mohamed, F.A., Spectrofluorometric determination of acetaminophen with *N*-bromosuccinimide, *J. AOAC Int.* 88, 1626–1630, 2005; Krebs, A., Starczyewska, B., Purzanowska-Tarasiewicz, H., and Sledz, J., Spectrophotometric determination of olanzapine by its oxidation with *N*-bromosuccinimide and cerium(IV) sulfate, *Anal. Sci.* 22, 829–833, 2006; Braddock, D.C., Cansell, G., Hermitage, S.A., and White, A.J., Bromoiodinanes with a I(III)-Br bond: preparation, X-ray crystallography, and reactivity as electrophilic brominating agents, *Chem. Commun.* 13, 1442–1444, 2006; Chen, G., Sasaki, M., Li, X., and Yudin, A.K., Strained enamines as versatile intermediates for stereocontrolled construction of nitrogen heterocycles, *J. Org. Chem.* 71, 6067–6073, 2006; Braddock D.C., Cansell, G., and Hermitage, S.A., Ortho-substituted iodobenzenes as novel organocatalysts for the transfer of electrophilic bromine from *N*-bromosuccinimide to alkenes, *Chem. Commun.* 23, 2483–2485, 2006.

CHEMICALS COMMONLY USED IN BIOCHEMISTRY AND MOLECULAR BIOLOGY AND THEIR PROPERTIES (Continued)

Common Name	Chemical Name	M.W.	Properties and Comment
NHS	N-hydroxysuccinimide; 1-hydroxy-2,5-pyrrolidinedione	111.1	Use in preparation of active esters for modification of amino groups (with carbodiimide); structural basis for reagents for amino group modification.

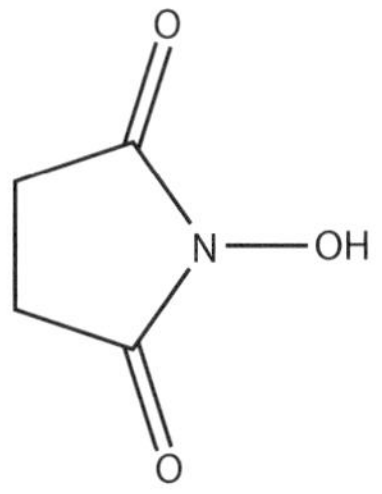

Anderson, G.W., Callahan, F.M., and Zimmerman, J.E., Synthesis of N-hydroxysuccinimide esters of acyl peptides by the mixed anhydride method, *J. Am. Chem. Soc.* 89, 178, 1967; Lapidot, Y., Rappoport, S., and Wolman, Y., Use of esters of N-hydroxysuccinimide in the synthesis of N-acylamino acids, *J. Lipid Res.* 8, 142–145, 1967; Holmquist, B., Blumberg, S., and Vallee, B.L., Superactivation of neutral proteases: acylation with N-hydroxysuccinimide esters, *Biochemistry* 15, 4675–4680, 1976; 't Hoen, P.A., de Kort, F., van Ommen, G.J., and den Dunnen, J.T., Fluorescent labeling of cRNA for microarray applications, *Nucleic Acids Res.* 31, e20, 2003; Vogel, C.W., Preparation of immunoconjugates using antibody oligosaccharide moieties, *Methods Mol. Biol.* 283, 87–108, 2004; Cooper, M., Ebner, A., Briggs, M. et al., Cy3B: improving the performance of cyanine dyes, *J. Fluoresc.* 14, 145–150, 2004; Lundblad, R.L., *Chemical Reagents for Protein Modification*, 3rd ed., CRC Press, Boca Raton, FL, 2004; Zhang, R., Tang, M., Bowyer, A. et al., A novel pH- and ionic-strength-sensitive carboxy methyl dextran hydrogel, *Biomaterials* 26, 4677–4683, 2005; Tyan, Y.C., Jong, S.B., Liao, J.D. et al., Proteomic profiling of erythrocyte proteins by proteolytic digestion chip and identification using two-dimensional electrospray ionization tandem mass spectrometry, *J. Proteome Res.* 4, 748–757, 2005; Lovrinovic, M., Spengler, M., Deutsch, C., and Niemeyer, C.M., Synthesis of covalent DNA-protein conjugates by expressed protein ligation, *Mol. Biosyst.* 1, 64–69, 2005; Smith, G.P., Kinetics of amine modification of proteins, *Bioconjug. Chem.* 17, 501–506, 2006; Yang, W.C., Mirzael, H., Liu, X., and Regnier, F.E., Enhancement of amino acid detection and quantitation by electrospray ionization mass spectrometry, *Anal. Chem.* 78, 4702–4708, 2006; Yu, G., Liang, J., He, Z., and Sun, M., Quantum dot-mediated detection of gamma-aminobutyric acid binding sites on the surface of living pollen protoplasts in tobacco, *Chem. Biol.* 13, 723–731, 2006; Adden, N., Gamble, L.J., Castner, D.G. et al., Phosphonic acid monolayers for binding of bioactive molecules to titanium surfaces, *Langmuir* 22, 8197–8204, 2006.

Ninhydrin	1-*H*-indene-1,2,3-trione Monohydrate	178.14	Reagent for amino acid analysis; reagent for modification of arginine residues in proteins; reaction with amino groups and other nucleophiles such as sulfhydryl groups.

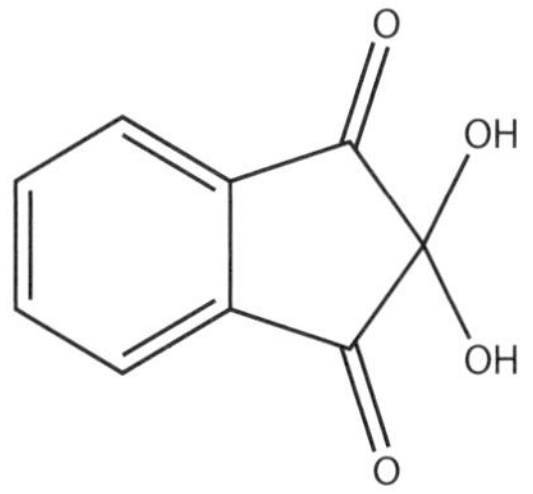

Duliere, W.L., The amino-groups of the proteins of human serum. Action of formaldehyde and ninhydrin, *Biochem. J.* 30, 770–772, 1936; Schwartz, T.B. and Engel, F.L., A photometric ninhydrin method for the measurement of proteolysis, *J. Biol. Chem.* 184, 197–202, 1950; Troll, W. and Cannan, R.K., A modified photometric ninhydrin method for the analysis of amino and imino acids, *J. Biol. Chem.* 200, 803–811, 1953; Moore, S. and Stein, W.H., A modified ninhydrin reagent for the photometric determination of amino acids and related compounds, *J. Biol. Chem.* 211, 907–913, 1954; Rosen, H., A modified ninhydrin colorimetric analysis for amino acids, *Arch. Biochem. Biophys.* 67, 10–15, 1957; Meyer, H., The ninhydrin reactions and its analytical applications, *Biochem. J.* 67, 333–340, 1957; Whitaker, J.R., Ninhydrin assay in the presence of thiol compounds, *Nature* 189, 662–663, 1961; Grant, D.R., Reagent stability in Rosen's ninhydrin method for analysis of amino acids, *Anal. Biochem.* 6, 109–110, 1963; Shapiro, R. and Agarwal, S.C., Reaction of ninhydrin with cytosine derivatives, *J. Am. Chem. Soc.* 90, 474–478, 1968; Moore, S., Amino acid analysis: aqueous dimethylsulfoxide as solvent for the ninhydrin reaction, *J. Biol. Chem.* 243, 6281–6283, 1968; McGrath, R., Protein measurement by ninhydrin determination of amino acids released by alkaline hydrolysis, *Anal. Biochem.* 49, 95–102, 1972; Lamothe, P.J. and McCormick, P.G., Role of hydrindantin in the determination of amino acids using ninhydrin, *Anal. Chem.* 45, 1906–1911, 1973; Quinn, J.R., Boisvert, J.G., and Wood, I., Semi-automated ninhydrin assay of Kjeldahl nitrogen, *Anal. Biochem.* 58, 609–614, 1974; Chaplin, M.R., The use of ninhydrin as a reagent for the reversible modification of arginine residues in proteins, *Biochem. J.* 155, 457–459, 1976; Takahashi, K., Specific modification of arginine residues in proteins with ninhydrin, *J. Biochem.* 80, 1173–1176, 1976; Yu, P.H. and Davis, B.A., Deuterium isotope effects in the ninhydrin reaction of primary amines, *Experientia* 38, 299–300, 1982; D'Aniello, A., D'Onofrio, G., Pischetola, M., and Strazzulo, L., Effect of various substances on the

CHEMICALS COMMONLY USED IN BIOCHEMISTRY AND MOLECULAR BIOLOGY AND THEIR PROPERTIES (Continued)

Common Name	Chemical Name	M.W.	Properties and Comment

colorimetric amino acid–ninhydrin reaction, *Anal. Biochem.* 144, 610–611, 1985; Macchi, F.D., Shen, F.J., Keck, R.G., and Harris, R.J., Amino acid analysis, using postcolumn ninhydrin detection, in a biotechnology laboratory, *Methods Mol. Biol.* 159, 9–30, 2000; Moulin, M., Deleu, C., Larher, F.R., and Bouchereau, A., High-performance liquid chromatography determination of pipecolic acid after precolumn derivatization using domestic microwave, *Anal. Biochem.* 308, 320–327, 2002; Pool, C.T., Boyd, J.G., and Tam, J.P., Ninhydrin as a reversible protecting group of amino-terminal cysteine, *J. Pept. Res.* 63, 223–234, 2004; Schulz, M.M., Wehner, H.D., Reichert, W., and Graw, M., Ninhydrin-dyed latent fingerprints as a DNA source in a murder case, *J. Clin. Forensic Med.* 11, 202–204, 2004; Buchberger, W. and Ferdig, M., Improved high-performance liquid chromatographic determination of guanidine compounds by precolumn derivatization with ninhydrin and fluorescence detection, *J. Sep. Sci.* 27, 1309–1312, 2004; Hansen, D.B., and Joullie, M.M., The development of novel ninhydrin analogues, *Chem. Soc. Rev.* 34, 408–417, 2005.

Common Name	Chemical Name	M.W.	Properties and Comment
Nitric Acid	HNO_3	63.01	Strong acid.
***p*-Nitroaniline (PNA)**	4-nitroaniline	138.13	Signal from cleavage of chromogenic substrate.

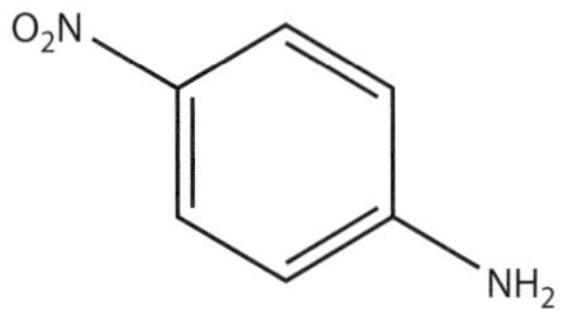

Common Name	Chemical Name	M.W.	Properties and Comment
2-Nitrobenzylsulfenyl Chloride	*o*-nitrophenylsulfenyl Chloride	189.6	Modification of tryptophan in proteins.

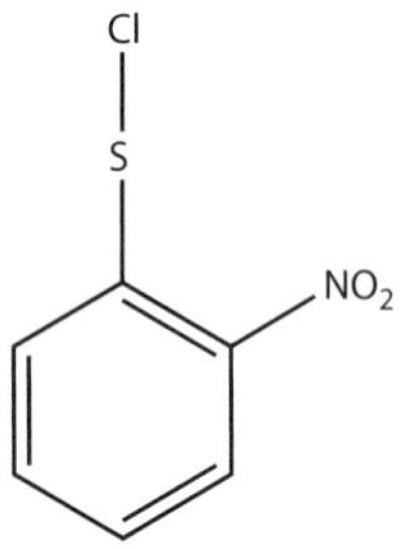

Fontana, A. and Scofone, E., Sulfenyl halides as modifying reagents for peptides and proteins, *Methods Enzymol.* 25B, 482–494, 1972; Sanda, A. and Irie, M., Chemical modification of tryptophan residues in ribonuclease form a *Rhizopus* sp., *J. Biochem.* 87, 1079–1087, 1980; De Wolf, M.J., Fridkin, M., Epstein, M., and Kohn, L.D., Structure-function studies of cholera toxin and its A and B protomers. Modification of tryptophan residues, *J. Biol. Chem.* 256, 5481–5488, 1981; Mollier, P., Chwetzoff, S., Bouet, F. et al., Tryptophan 110, a residue involved in the toxic activity but in the enzymatic activity of notexin, *Eur. J. Biochem.* 185, 263–270, 1989; Cymes, C.D., Iglesias, M.M., and Wolfenstein-Todel, C., Selective modification of tryptophan-150 in ovine placental lactogen, *Comp. Biochem. Physiol. B* 106, 743–746, 1993; Kuyama, H., Watanabe, M., Toda, C. et al., An approach to quantitate proteome analysis by labeling tryptophan residues, *Rapid Commun. Mass Spectrom.* 17, 1642–1650, 2003; Lundblad, R.L., *Chemical Reagents for Protein Modification*, 3rd ed., CRC Press, Boca Raton, FL, 2004; Matsuo, E., Toda, C., Watanabe, M., et al., Selective detection of 2-nitrobenzensulfenyl-labeled peptides by matrix-assisted laser desorption/ionization-time-of-flight mass spectrometry using a novel matrix, *Proteomics* 6, 2042–2049, 2006; Ou, K., Kesuma, D., Ganesan, K. et al., Quantitative labeling of drug-assisted proteomic alterations by combined 2-nitrobenzenesulfenyl chloride (NBS) isotope labeling and 2DE/MS identification, *J. Proteome Res.* 5, 2194–2206, 2006.

Common Name	Chemical Name	M.W.	Properties and Comment
***p*-Nitrophenol**	4-nitrophenol	139.11	Popular signal from indicator enzymes such as alkaline phosphatase.
***n*-Octanol**	1-octanol; Caprylic Alcohol	130.23	Partitioning between octanol and water is used to determine lipophilicity; a factor in QSAR studies.

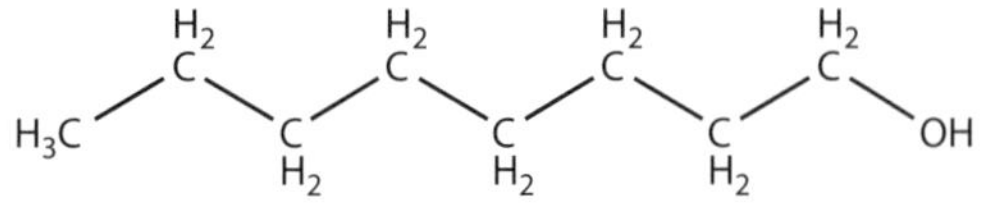

CHEMICALS COMMONLY USED IN BIOCHEMISTRY AND MOLECULAR BIOLOGY AND THEIR PROPERTIES (Continued)

Common Name	Chemical Name	M.W.	Properties and Comment

Marland, J.S. and Mulley, B.A., A phase-rule study of multiple-phase formation in a model emulsion system containing water, *n*-octanol, *n*-dodecane, and a non-ionic surface-active agent at 10 and 25 degrees, *J. Pharm. Pharmacol.* 23, 561–572, 1971; Dorsey, J.G. and Khaledi, M.G., Hydrophobicity estimations by reversed-phase liquid chromatography. Implications for biological partitioning processes, *J. Chromatog.* 656, 485–499, 1993; Vailaya, A. and Horvath, C., Retention in reversed-phase chromatography: partition or adsorption? *J. Chromatog.* 829, 1–27, 1998; Kellogg, G.E. and Abraham, D.J., Hydrophobicity: is logP(o/w) more than the sum of its parts? *Eur. J. Med. Chem.* 35, 651–661, 2000; van de Waterbeemd,H., Smith, D.A., and Jones, B.C., Lipophilicity in PK design: methyl, ethyl, futile, *J. Comput. Aided Mol. Des.* 15, 273–286, 2001; Bethod, A. and Carda-Broch, S., Determination of liquid–liquid partition coefficients by separation methods, *J. Chromatog. A* 1037, 3–14, 2004.

Octoxynol	Triton X-100™; Igepal CA-630™		Nonionic detergent; surfactant.

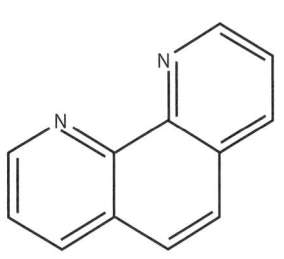

Octoxynol, n = 5–15

Peroxynitrite			
Petroleum Ether	Mixture of Pentanes and Hexanes	N/A	
Perchloric Acid	HClO$_4$	100.5	Oxidizing agent.
1,10-Phenanthroline Monohydrate	*o*-phenanthroline Hydrate	198.21	Chelating agent; inhibitor for metalloproteinases; use in design of synthetic nucleases and proteases.

CHEMICALS COMMONLY USED IN BIOCHEMISTRY AND MOLECULAR
BIOLOGY AND THEIR PROPERTIES (Continued)

Common Name	Chemical Name	M.W.	Properties and Comment

Hoch, F.L., Willams, R.J., and Vallee, B.L., The role of zinc in alcohol dehydrogenases. II. The kinetics of the instantaneous reversible inactivation of yeast alcohol dehydrogenase by 1,10-phenanthroline, *J. Biol. Chem.* 232, 453–464, 1958; Sigman, D.S. and Chen, C.H., Chemical nucleases: new reagents in molecular biology, *Annu. Rev. Biochem.* 59, 207–236, 1990; Pan, C.Q., Landgraf, R., and Sigman, D.S., DNA-binding proteins as site-specific nucleases, *Mol. Microbiol.* 12, 335–342, 1994; Galis, Z.S., Sukhova, G.K., and Libby, P., Microscopic localization of active proteases by *in situ* zymography: detection of matrix metalloproteinase activity in vascular tissue, *FASEB J.* 9, 974–980, 1995; Papavassiliou, A.G., Chemical nucleases as probes for studying DNA–protein interactions, *Biochem. J.* 305, 345–357, 1995; Perrin, D.M., Mazumder, A., and Sigman, D.S., Oxidative chemical nucleases, *Prog. Nucleic Acid Res. Mol. Biol.* 52, 123–151, 1996; Sigman, D.S., Landgraf, R., Perrin, D.M., and Pearson, L., Nucleic acid chemistry of the cuprous complexes of 1,10-phenanthroline and derivatives, *Met. Ions Biol. Syst.* 33, 485–513, 1996; Cha, J., Pedersen, M.V., and Auld, D.S., Metal and pH dependence of heptapeptide catalysis by human matrilysin, *Biochemistry* 35, 15831–15838, 1996; Kidani, Y. and Hirose, J., Coordination chemical studies on metalloenzymes. II. Kinetic behavior of various types of chelating agents towards bovine carbonic anhydrase, *J. Biochem.* 81, 1383–1391, 1997; Marini, I., Bucchioni, L., Borella, P. et al., Sorbitol dehydrogenase from bovine lens: purification and properties, *Arch. Biochem. Biophys.* 340, 383–391, 1997; Dri, P., Gasparini, C., Menegazzi, R. et al., TNF-induced shedding of TNF receptors in human polymorphonuclear leukocytes: role of the 55-kDa TNF receptor and involvement of a membrane-bound and non-matrix metalloproteinase, *J. Immunol.* 165, 2165–2172, 2000; Kito, M. and Urade, R., Protease activity of 1,10-phenanthroline-copper systems, *Met. Ions Biol. Syst.* 38, 187–196, 2001; Winberg, J.O., Berg, E., Kolset, S.O. et al., Calcium-induced activation and truncation of promatrix metalloproteinase-9 linked to the core protein of chondroitin sulfate proteoglycans, *Eur. J. Biochem.* 270, 3996–4007, 2003; Butler, G.S., Tam, E.M., and Overall, C.M., The canonical methionine 392 of matrix metalloproteinase 2 (gelatinase A) is not required for catalytic efficiency or structural integrity: probing the role of the methionine-turn in the metzincin metalloprotease superfamily, *J. Biol. Chem.* 279, 15615– 15620, 2004; Vauquelin, G. and Vanderheyden, P.M., Metal ion modulation of cystinyl aminopeptidase, *Biochem. J.* 390, 351–357, 2005; Schilling, S., Cynis, H., von Bohlen, A. et al., Isolation, catalytic properties, and competitive inhibitors of the zinc-dependent murine glutaminyl cyclase, *Biochemistry* 44, 13415–13424, 2005; Vik, S.B. and Ishmukhametov, R.R., Structure and function of subunit a of the ATP synthase of *Escherichia coli*, *J. Bioenerg. Biomembr.* 37, 445–449, 2005.

Phenol	Hydroxybenzene; Phenyl Hydroxide	94.11	Solvent; nucleic acid purification.

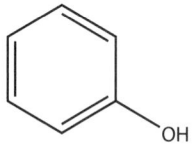

Braun, W., Burrous, J.W., and Phillips, J.H., Jr., A phenol-extracted bacterial deoxyribonucleic acid, *Nature* 180, 1356–1357, 1957; Habermann, V., Evidence for peptides in RNA prepared by phenol extraction, *Biochim. Biophys. Acta* 32, 297–298, 1959; Colter, J.S., Brown, R.A., and Ellem, K.A., Observations on the use of phenol for the isolation of deoxyribonucleic acid, *Biochim. Biophys. Acta* 55, 31–39, 1962; Lust, J. and Richards, V., Influence of buffers on the phenol extraction of liver microsomal ribonucleic acids, *Anal. Biochem.* 20, 65–76, 1967; Yamaguchi, M., Dieffenbach, C.W., Connolly, R. et al., Effect of different laboratory techniques for guanidinium-phenol-chloroform RNA extraction on A260/A280 and on accuracy of mRNA quantitation by reverse transcriptase-PCR, *PCR Methods Appl.* 1, 286–290, 1992; Pitera, R., Pitera, J.E., Mufti, G.J., Salisbury, J.R., and Nickoloff, J.A., Sepharose spin column chromatography. A fast, nontoxic replacement for phenol: chloroform extraction/ethanol precipitation, *Mol. Biotechnol.* 1, 105–108, 1994; Finnegan, M.T., Herbert, K.E., Evans, M.D., and Lunec, J., Phenol isolation of DNA yields higher levels of 8-deoxodeoxyguanosine compared to pronase E isolation, *Biochem. Soc. Trans.* 23, 430S, 1995; Beaulieux, F., See, D.M., Leparc-Goffart, I. et al., Use of magnetic beads versus guanidium thiocyanate-phenol-chloroform RNA extraction followed by polymerase chain reaction for the rapid, sensitive detection of enterovirus RNA, *Res. Virol.* 148, 11–15, 1997; Fanson, B.G., Osmack, P., and Di Bisceglie, A.M., A comparison between the phenol-chloroform method of RNA extraction and the QIAamp viral RNA kit in the extraction of hepatitis C and GB virus-C/hepatitis G viral RNA from serum, *J. Virol. Methods* 89, 23–27, 2000; Kochl, S., Niederstratter, N., and Parson, W., DNA extraction and quantitation of forensic samples using the phenol-chloroform method and real-time PCR, *Methods Mol. Biol.* 297, 13–30, 2005; Izzo, V., Notomista, E., Picardi, A. et al., The thermophilic archaeon *Sulfolobus solfatarius* is able to grow on phenol, *Res. Microbiol.* 156, 677–689, 2005; Robertson, N. and Leek, R., Isolation of RNA from tumor samples: single-step guanidinium acid-phenol method, *Methods Mol. Biol.* 120, 55–59, 2006.

Phenoxyethanol	2-phenoxyethanol	138.16	Biochemical preservative; preservative in personal care products.

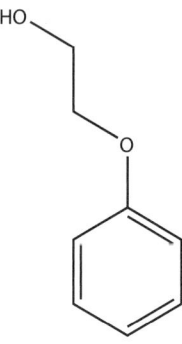

**CHEMICALS COMMONLY USED IN BIOCHEMISTRY AND MOLECULAR
BIOLOGY AND THEIR PROPERTIES (Continued)**

Common Name	Chemical Name	M.W.	Properties and Comment

Nakahishi, M., Wilson, A.C., and Nolan, R.A., Phenoxyethanol: protein preservative for taxonomists, *Science* 163, 681–683, 1969; Frolich, K.W., Anderson, L.M., Knutsen, A., and Flood, P.R., Phenoxyethanol as a nontoxic substitute for formaldehyde in long-term preservation of human anatomical specimens for dissection and demonstration purposes, *Anat. Rec.* 208, 271–278, 1984.

| **Phenylglyoxal** | Phenylglyoxal Hydrate | 134.13 | Modification of arginine residues. |

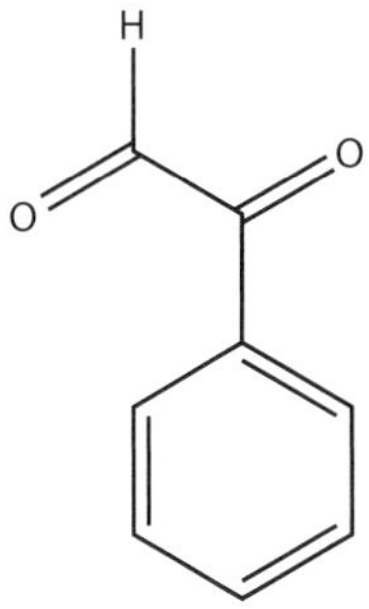

Takahashi, K., The reaction of phenylglyoxal with arginine residues in proteins, *J. Biol. Chem.* 243, 6171–6179, 1968; Bunzli, H.F. and Bosshard, H.R., Modification of the single arginine residue in insulin with phenylglyoxal, *Hoppe Seylers Z. Physiol. Chem.* 352, 1180–1182, 1971; Cheung, S.T. and Fonda, M.L., Reaction of phenylglyoxal with arginine. The effect of buffers and pH, *Biochem. Biophys. Res. Commun.* 90, 940–947, 1979; Srivastava, A. and Modak, M.J., Phenylglyoxal as a template site-specific reagent for DNA or RNA polymerases. Selective inhibition of initiation, *J. Biol. Chem.* 255, 917–921, 1980; Communi, D., Lecocq, R., Vanweyenberg, V., and Erneux, C., Active site labeling of inositol 1,4,5-triphosphate 3-kinase A by phenylglyoxal, *Biochem. J.* 310, 109–115, 1995; Eriksson, O., Fontaine, E., and Bernardi, P., Chemical modification of arginines by 2,3-butanedione and phenylglyoxal causes closure of the mitochondrial permeability transition pore, *J. Biol. Chem.* 273, 12669–12674, 1998; Redowicz, M.J., Phenylglyoxal reveals phosphorylation-dependent difference in the conformation of *Acanthamoeba* myosin II active site, *Arch. Biochem. Biophys.* 384, 413–417, 2000; Kucera, I., Inhibition by phenylglyoxal of nitrate transport in *Paracoccus denitrificans*; a comparison with the effect of a protonophorous uncoupler, *Arch. Biochem. Biophys.* 409, 327–334, 2003; Johans, M., Milanesi, E., Frank, M. et al., Modification of permeability transition pore arginine(s) by phenylglyoxal derivatives in isolated mitochondria and mammalian cells. Structure-function relationship of arginine ligands, *J. Biol. Chem.* 280, 12130–12136, 2005.

| **Phosgene** | Carbonyl Chloride; Carbon Oxychloride | 98.92 | Reagent for organic synthesis; preparation of derivatives for analysis. |

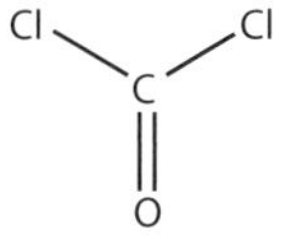

Wilchek, M., Ariely, S., and Patchornik, A., The reaction of asparagine, glutamine, and derivatives with phosgene, *J. Org. Chem.* 33, 1258–1259, 1968; Hamilton, R.D. and Lyman, D.J., Preparation of *N*-carboxy-α-amino acid anhydrides by the reaction of copper(II)-amino acid complexes with phosgene, *J. Org. Chem.* 34, 243–244, 1969; Pohl, L.R., Bhooshan, B., Whittaker, N.F., and Krishna, G., Phosgene: a metabolite of chloroform, *Biochem. Biophys. Res. Commun.* 79, 684–691, 1977; Gyllenhaal, O., Derivatization of 2-amino alcohols with phosgene in aqueous media: limitations of the reaction selectivity as found in the presence of *O*-glucuronides of alprenolol in urine, *J. Chromatog.* 413, 270–276, 1987; Gyllenhaal, O. and Vessman, J., Phosgene as a derivatizing reagent prior to gas and liquid chromatography, *J. Chromatog.* 435, 259–269, 1988; Noort, D., Hulst, A.G., Fidder, A., et al. *In vitro* adduct formation of phosgene with albumin and hemoglobin in human blood, *Chem. Res. Toxicol.* 13, 719–726, 2000; Lemoucheux, L. Rouden, J., Ibazizene, M. et al., Debenylation of tertiary amies using phosgene or triphosgen: an efficient and rapid procedure for the preparation of carbamoyl chlorides and unsymmetrical ureas. Application in carbon-11 chemistry, *J. Org. Chem.* 68, 7289–7297, 2003.

| **Picric Acid** | 2,4,6-trinitrophenol | 229.1 | Analytical reagent. |

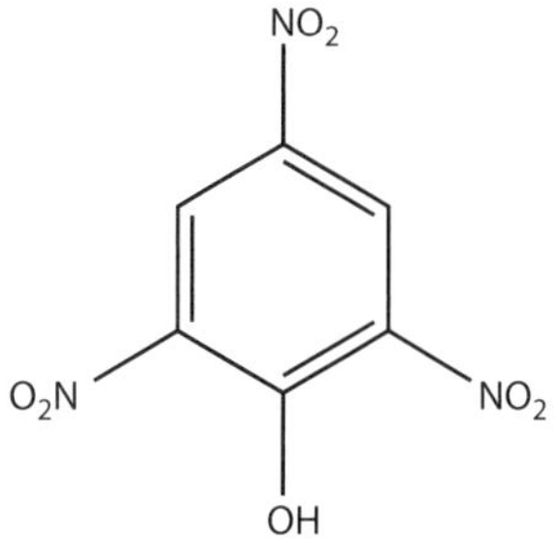

CHEMICALS COMMONLY USED IN BIOCHEMISTRY AND MOLECULAR BIOLOGY AND THEIR PROPERTIES (Continued)

Common Name	Chemical Name	M.W.	Properties and Comment

De Wesselow, O.L., The picric acid method for the estimation of sugar in blood and a comparison of this method with that of MacLean, *Biochem. J.* 13, 148–152, 1919; Newcomb, C., The error due to impure picric acid in creatinine estimations, *Biochem. J.* 18, 291–293, 1924; Davidsen, O., Fixation of proteins after agarose gel electrophoresis by means of picric acid, *Clin. Chim. Acta* 21, 205–209, 1968; Gisin, B.F., The monitoring of reactions in solid-phase peptide synthesis with picric acid, *Anal. Chim. Acta* 58, 248–249, 1972; Hancock, W.S., Battersby, J.E., and Harding, D.R., The use of picric acid as a simple monitoring procedure for automated peptide synthesis, *Anal. Biochem.* 69, 497–503, 1975; Vasiliades, J., Reaction of alkaline sodium picrate with creatinine: I. Kinetics and mechanism of formation of the mono-creatinine picric acid complex, *Clin. Chem.* 22, 1664–1671, 1976; Somogyi, P. and Takagi, H., A note on the use of picric acid-formaldehyde-glutaraldehyde fixative for correlated light and electron microscopic immunocytochemistry, *Neuroscience* 7, 1779–1783, 1982; Meyer, M.H., Meyer, R.A., Jr., Gray, R.W., and Irwin, R.L., Picric acid methods greatly overestimate serum creatinine in mice: more accurate results with high-performance liquid chromatography, *Anal. Biochem.* 144, 285–290, 1985; Knisley, K.A. and Rodkey, L.S., Direct detection of carrier ampholytes in immobilized pH gradients using picric acid precipitation, *Electrophoresis* 13, 220–224, 1992; Massoomi, F., Mathews, H.G., III, and Destache, C.J., Effect of seven fluoroquinolines on the determination of serum creatinine by the picric acid and enzymatic methods, *Ann. Pharmacother.* 27, 586–588, 1993.

Polysorbate   Tween 20   Nonionic detergent; surfactant.

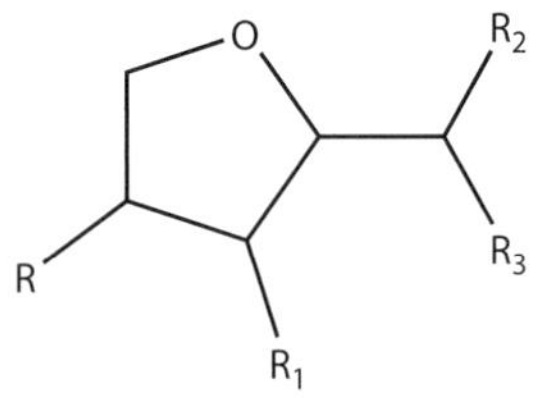

Polysorbates

Polyvinylpyrrolidone (PVP)   Povidone   N/A   Pharmaceutical; excipient; phosphate analysis.

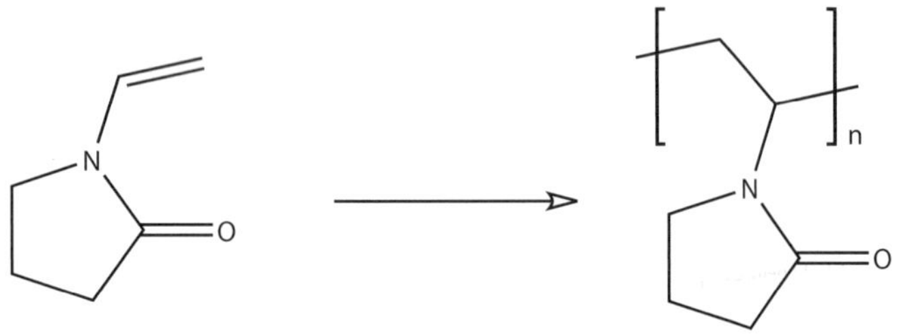

Morin, L.G. and Prox, J., New and rapid procedure for serum phosphorus using *o*-phenylenediamine as reductant, *Clin. Chim. Acta.* 46, 113–117, 1973; Ohnishi, S.T. and Gall, R.S., Characterization of the catalyzed phosphate assay, *Anal. Biochem.* 88, 347–356, 1978; Steige, H. and Jones, J.D., Determination of serum inorganic phosphorus using a discrete analyzer, *Clin. Chim. Acta.* 103, 123–127, 1980, Plaizier-Vercammen, J.A. and De Neve, R.E., Interaction of povidone with aromatic compounds. II: evaluation of ionic strength, buffer concentration, temperature, and pH by factorial analysis, *J. Pharm. Sci.* 70, 1252–1256, 1981; van Zanten, A.P. and Weber, J.A., Direct kinetic method for the determination of phosphate, *J. Clin. Chem. Clin. Biochem.* 25, 515–517, 1987; Barlow, I.M., Harrison, S.P., and Hogg, G.L., Evaluation of the Technicon Chem-1, *Clin. Chem.* 34, 2340–2344, 1988; Giulliano, K.A., Aqueous two-phase protein partitioning using textile dyes as affinity ligands, *Anal. Biochem.* 197, 333–339, 1991; Goldenheim, P.D., An appraisal of povidone-iodine and wound healing, *Postgrad. Med. J.*, 69 (Suppl. 3), S97–S105, 1993; Vemuri, S., Yu, C.D., and Roosdorp, N., Effect of cryoprotectants on freezing, lyophilization, and storage of lyophilized recombinant alpha 1-antitrypsin formulations, *PDA J. Pharm. Sci. Technol.* 48, 241–246, 1994; Anchordoquy, T.J. and Carpenter, J.F., Polymers protect lactate dehydrogenase during freeze-drying by inhibiting dissociation in the frozen state, *Arch. Biochem. Biophys.* 332, 231–238, 1996; Fleisher, W., and Reimer, K., Povidone-iodine in antisepsis — state of the art, *Dermatology* 195 (Suppl. 2), 3–9, 1997; Fernandes, S., Kim, H.S., and Hatti-Kaul, R., Affinity extraction of dye- and metal ion-binding proteins in polyvinalypyrrolidone-based aqueous two-phase system, *Protein Expr. Purif.* 24, 460–469, 2002; D'Souza, A.J., Schowen, R.L., Borchardt, R.T. et al., Reaction of a peptide with polyvinylpyrrolidone in the solid state, *J. Pharm. Sci.* 92, 585–593, 2003; Kaneda, Y., Tsutsumi, Y., Yoshioka, Y. et al., The use of PVP as a polymeric carrier to improve the plasma half-life of drugs, *Biomaterials* 25, 3259–3266, 2004; Art, G., Combination povidone-iodine and alcohol formulations more effective, more convenient versus formulations containing either iodine or alcohol alone: a review of the literature, *J. Infus. Nurs.* 28, 314–320, 2005; Yoshioka, S., Aso, Y., and Miyazaki, T., Negligible contribution of molecular mobility to the degradation of insulin lyophilized with poly(vinylpyrrolidone), *J. Pharm. Sci.* 95, 939–943, 2006.

**CHEMICALS COMMONLY USED IN BIOCHEMISTRY AND MOLECULAR
BIOLOGY AND THEIR PROPERTIES (Continued)**

Common Name	Chemical Name	M.W.	Properties and Comment
Pyridine	Azine	79.10	Solvent.

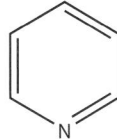

Klingsberg, E. and Newkome, G.R., Eds., *Pyridine and Its Derivatives,* Interscience, New York, 1960; Schoefield, K., *Hetero-aromatic Nitrogen Compounds; Pyrroles and Pyridines,* Butterworths, London, 1967; Hurst, D.T., *An Introduction to the Chemistry and Biochemistry and Pyrimidines, Purines, and Pteridines,* J. Wiley, Chichester, UK, 1980; Plunkett, A.O., Pyrrole, pyrrolidine, pyridine, piperidine, and azepine alkaloids, *Nat. Prod. Rep.* 11, 581–590, 1994; Kaiser, J.P., Feng, Y., and Bollag, J.M., Microbial metabolism of pyridine, quinoline, acridine, and their derivatives under aerobic and anaerobic conditions, *Microbiol. Rev.* 60, 483–498, 1996.

Common Name	Chemical Name	M.W.	Properties and Comment
Pyridoxal-5-phosphate (PLP)	Pyridoxal-5-(dihydrogen phosphate)	247.14	Selective modification of amino groups in proteins; affinity label for certain sites based on phosphate group.

Hughes, R.C., Jenkins, W.T., and Fischer, E.H., The site of binding of pyridoxal-5′-phosphate to heart glutamic-aspartic transaminase, *Proc. Natl. Acad. Sci. USA* 48, 1615–1618, 1962; Finseth, R. and Sizer, I.W., Complexes of pyridoxal phosphate with amino acids, peptides, polylysine, and apotransaminase, *Biochem. Biophys. Res. Commun.* 26, 625–630, 1967; Pages, R.C., Benditt, E.P., and Kirkwood, C.R., Schiff base formation by the lysyl and hydroxylysyl side chains of collagen, *Biochem. Biophys. Res. Commun.* 33, 752–757, 1968; Whitman, W.B., Martin, M.N., and Tabita, F.R., Activation and regulation of ribulose bisphosphate carboxylase-oxygenase in the absence of small subunits, *J. Biol. Chem.* 254, 10184–10189, 1979; Howell, E.E. and Schray, K.J., Comparative inactivation and inhibition of the anomerase and isomerase activities of phosphoglucose isomerase, *Mol. Cell. Biochem.* 37, 101–107, 1981; Colanduoni, J. and Villafranca, J.J., Labeling of specific lysine residues at the active site of glutamine synthetase, *J. Biol. Chem.* 260, 15042–15050, 1985; Peterson, C.B., Noyes, C.M., Pecon, J.M. et al., Identification of a lysyl residue in antithrombin which is essential for heparin binding, *J. Biol. Chem.* 262, 8061–8065, 1987; Diffley, J.F., Affinity labeling the DNA polymerase alpha complex. Identification of subunits containing the DNA polymerase active site and an important regulatory nucleotide-binding site, *J. Biol. Chem.* 263, 19126–19131, 1988; Perez-Ramirez, B. and Martinez-Carrion, M., Pyridoxal phosphate as a probe of the cytoplasmic domains of transmembrane proteins: application to the nicotinic acetylcholine receptor, *Biochemistry* 28, 5034–5040, 1989; Valinger, Z., Engel, P.C., and Metzler, D.E., Is pyridoxal-5′-phosphate an affinity label for phosphate-binding sites in proteins? The case of bovine glutamate dehydrogenase, *Biochem. J.* 294, 835–839, 1993; Illy, C., Thielens, N.M., and Arlaud, G.J., Chemical characterization and location of ionic interactions involved in the assembly of the C1 complex of human complement, *J. Protein Chem.* 12, 771–781, 1993; Hountondji, C., Gillet, S., Schmitter, J.M. et al., Affinity labeling of *Escherichia coli* lysyl-tRNA synthetase with pyridoxal mono- and diphosphate, *J. Biochem.* 116, 502–507, 1994; Brody, S., Andersen, J.S., Kannangara, C.G. et al., Characterization of the different spectral forms of glutamate-1-semialdehyde aminotransferase by mass spectrometry, *Biochemistry* 34, 15918–15924, 1995; Kossekova, G., Miteva, M., and Atanasov, B., Characterization of pyridoxal phosphate as an optical label for measuring electrostatic potentials in proteins, *J. Photochem. Photobiol. B* 32, 71–79, 1996; Kim S.W., Lee, J., Song, M.S. et al., Essential active-site lysine of brain glutamate dehydrogenase isoproteins, *J. Neurochem.* 69, 418–422, 1997; Martin, D.L., Liu, H., Martin, S.B., and Wu, S.J., Structural features and regulatory properties of the brain glutamate decarboxylase, *Neurochem. Int.* 37, 111–119, 2000; Jaffe, M. and Bubis, J., Affinity labeling of the guanine nucleotide binding site of transducin by pyridoxal 5′-phosphate, *J. Protein Chem.* 21, 339–359, 2002.

Common Name	Chemical Name	M.W.	Properties and Comment
Sodium Borohydride	$NaBH_4$	37.83	Reducing agent for Schiff bases; reduction of aldehydes; other chemical reductions.

CHEMICALS COMMONLY USED IN BIOCHEMISTRY AND MOLECULAR BIOLOGY AND THEIR PROPERTIES (Continued)

Common Name	Chemical Name	M.W.	Properties and Comment

Chaykin, S., King, L., and Watson, J.G., The reduction of DPN+ and TPN+ with sodium borohydride, *Biochim. Biophys. Acta* 124, 13–25, 1966; Cerutti, P. and Miller, N., Selective reduction of yeast transfer ribonucleic acid with sodium borohydride, *J. Mol. Biol.* 26, 55–66, 1967; Tanzer, M.L., Collagen reduction by sodium borohydride: effects of reconstitution, maturation, and lathyrism, *Biochem. Biophys. Res. Commun.* 32, 885–892, 1968; Phillips, T.M., Kosicki, G.W., and Schmidt, D.E., Jr., Sodium borohydride reduction of pyruvate by sodium borohydride catalyzed by pyruvate kinase, *Biochim. Biophys. Acta* 293, 125–133, 1973; Craig, A.S., Sodium borohydride as an aldehyde-blocking reagent for electron microscope histochemistry, *Histochemistry* 42, 141–144, 1974; Miles, E.W., Houck, D.R., and Floss, H.G., Stereochemistry of sodium borohydride reduction of tryptophan synthase of *Escherichia coli* and its amino acid Schiff's bases, *J. Biol. Chem.* 257, 14203–14210, 1982; Kumar, A., Rao, P., and Pattabiraman, T.N., A colorimetric method for the estimation of serum glycated proteins based on differential reduction of free and bound glucose by sodium borohydride, *Biochem. Med. Metab. Biol.* 39, 296–304, 1988; Lenz, A.G., Costabel, U., Shaltiel, S., and Levine, R.L., Determination of carbonyl groups in oxidatively modified proteins by reduction with tritiated sodium borohydride, *Anal. Biochem.* 177, 419–425, 1989; Yan, L.J. and Sohal, R.S., Gel electrophoresis quantiation of protein carbonyls derivatized with tritiated sodium borohydride, *Anal. Biochem.* 265, 176–182, 1998; Azzam, T., Eliyahu, H., Shapira, L. et al., Polysaccharide-oligoamine-based conjugates for gene delivery, *J. Med. Chem.* 45, 1817–1824, 2002; Purich, D.L., Use of sodium borohydride to detect acyl-phosphate linkages in enzyme reactions, *Methods Enzymol.* 354, 168–177, 2002; Bald, E., Chwatko, S., Glowacki, R., and Kusmierek, K., Analysis of plasma thiols by high-performance liquid chromatography with ultraviolet detection, *J. Chromatog. A* 1032, 109–115, 2004; Eike, J.H. and Palmer, A.F., Effect of $NABH_4$ concentration and reaction time on physical properties of glutaraldehyde-polymerized hemoglobin, *Biotechnol. Prog.* 20, 946–952, 2004; Zhang, Z., Edwards, P.J., Roeske, R.W., and Guo, L., Synthesis and self-alkylation of isotope-coded affinity tag reagents, *Bioconjug. Chem.* 16, 458–464, 2005; Studelski, D.R., Giljum, K., McDowell, L.M., and Zhang, L., Quantitation of glycosaminoglycans by reversed-phase HPLC separation of fluorescent isoindole derivatives, *Glycobiology* 16, 65–72, 2006; Floor, E., Maples, A.M., Rankin, C.A. et al., A one-carbon modification of protein lysine associated with elevated oxidative stress in human substantia nigra, *J. Neurochem.* 97, 504–514, 2006; Kusmierek, K., Glowacki, R., and Bald, E., Analysis of urine for cysteine, cysteinylglycine, and homocysteine by high-performance liquid chromatography, *Anal. BioAnal. Chem.* 385, 855–860, 2006.

Sodium Chloride	Salt; NaCl	58.44	Ionic strength; physiological saline.
Sodium Cholate		430.55	Detergent.

Lindstrom, J., Anholt, R., Einarson, B. et al., Purification of acetylcholine receptors, reconstitution into lipid vesicles, and study of agonist-induced channel regulation, *J. Biol. Chem.* 255, 8340–8350, 1980; Gullick, W.J., Tzartos, S., and Lindstrom, J., Monoclonal antibodies as probes of acetylcholine receptor structure. 1. Peptide mapping, *Biochemistry* 20, 2173–2180, 1981; Henselman, R.A. and Cusanovich, M.A., The characterization of sodium cholate solubilized rhodopsin, *Biochemistry* 13, 5199–5203, 1974; Ninomiya, R., Masuoka, K., and Moroi, Y., Micelle formation of sodium chenodeoxycholate and solublization into the micelles: comparison with other unconjugated bile salts, *Biochim. Biophys. Acta* 1634, 116–125, 2003; Simoes, S.I., Marques, C.M., Cruz, M.E. et al., The effect of cholate on solubilization and permeability of simple and protein-loaded phosphatidylcholine/sodium cholate-mixed aggregates designed to mediate transdermal delivery of macromolecules, *Eur. J. Pharm. Biopharm.* 58, 509–519, 2004; Reis, S., Moutinho, C.G., Matos, C. et al., Noninvasive methods to determine the critical micelle concentration of some bile acid salts, *Anal. Biochem.* 334, 117–126, 2004; Nohara, D., Kajiura, T., and Takeda, K., Determination of micelle mass by electrospray ionization mass spectrometry, *J. Mass Spectrom.* 40, 489–493, 2005; Guo, J., Wu., T., Ping, Q. et al., Solublization and pharmacokinetic behaviors of sodium cholate/lecithin-mixed micelles containing cyclosporine A, *Drug Deliv.* 12, 35–39, 2005; Bottari, E., Buonfigli, A., and Festa, M.R., Composition of sodium cholate micellar solutions, *Ann. Chim.* 95, 479–490, 2005; Schweitzer, B., Felippe, A.C., Dal Bo, A. et al., Sodium dodecyl sulfate promoting a cooperative association process of sodium cholate with bovine serum albumin, *J. Colloid Interface Sci.* 298, 457–466, 2006; Burton, M.I., Herman, M.D., Alcain, F.J., and Villalba, J.M., Stimulation of polyprenyl 4-hydroxybenzoate transferase activity by sodium cholate and 3- [(cholamidopropyl)dimethylammonio]-1-propanesulfonate, *Anal. Biochem.* 353, 15–21, 2006; Ishibashi, A. and Nakashima, N., Individual dissolution of single-walled carbon nanotubes in aqueous solutions of steroid of sugar compounds and their Raman and near-IR spectral properties, *Chemistry*, 12, 7595–7602, 2006.

Sodium Cyanoborohydride	$NaBH_3(CN)$	62.84	Reducing agent; considered more selective than $NaBH_4$.

Rosen, G.M., Use of sodium cyanoborohydride in the preparation of biologically active nitroxides, *J. Med. Chem.* 17, 358–360, 1974; Chauffe, L. and Friedman, M., Factors affecting cyanoborohydride reduction of aromatic Schiff's bases in proteins, *Adv. Exp. Med. Biol.* 86A, 415–424, 1977; Baues, R.J. and Gray, G.R., Lectin purification on affinity columns containing reductively aminated disaccharides, *J. Biol. Chem.* 252, 57–60, 1977; Jentoft, N. and Dearborn, D.G., Labeling of proteins by reductive methylation using sodium cyanoborohydride, *J. Biol. Chem.* 254, 4359–4365, 1979; Jentoft, N., and Dearborn, D.G., Protein labeling by reductive methylation with sodium cyanoborohydride: effect of cyanide and metal ions on the reaction, *Anal. Biochem.* 106, 186–190, 1980; Bunn, H.F. and Higgins, P.T., Reaction of monosaccharides with proteins: possible evolutionary significance, *Science* 213, 222–224, 1981; Geoghegan, K.F., Cabacungan, J.C., Dixon, H.B., and Feeney, R.E., Alternative reducing agents for reductive methylation of amino groups in proteins, *Int. J. Pept. Protein Res.* 17, 345–352, 1981; Habeeb, A.F., Comparative studies on radiolabeling of lysozyme by iodination and reductive methylation, *J. Immunol. Methods* 65, 27–39, 1983; Prakash, C. and Vijay, I.K., A new fluorescent tag for labeling of saccharides, *Anal. Biochem.* 128, 41–46, 1983; Acharya, A.S. and Sussman, L.G., The reversibility of the ketoamine linkages of aldoses with proteins, *J. Biol. Chem.* 259, 4372–4378, 1984; Climent, I., Tsai, L., and Levine, R.L., Derivatization of gamma-glutamyl semialdehyde residues in oxidized proteins by fluorescamine, *Anal. Biochem.* 182, 226–232, 1989; Hartmann, C. and Klinman, J.P., Reductive trapping of substrate to methylamine oxidase from *Arthrobacter* P1, *FEBS Lett.* 261, 441–444, 1990; Meunier, F. and Wilkinson, K.J., Nonperturbing fluorescent labeling of polysaccharides, *Biomacromolecules* 3, 858–864, 2002; Webb, M.E., Stephens, E., Smith, A.G., and Abell, C., Rapid screening by MALDI-TOF mass spectrometry to probe binding specificity at enzyme active sites, *Chem. Commun.* 19, 2416–2417, 2003; Sando, S., Matsui, K., Niinomi, Y. et al., Facile preparation of DNA-tagged carbohydrates, *Bioorg. Med. Chem. Lett.* 13, 2633–2636, 2003; Peelen, D. and Smith, L.M., Immobilization of anine-modified oligonucleotides on aldehyde-terminated alkanethiol monolayers on gold, *Langmuir* 21, 266 271, 2005; Mirzaei, H. and Regnier, F., Enrichment of carbonylated peptides using Girard P reagent and strong cation exchange chromatography, *Anal. Chem.* 78, 770–778, 2006.

CHEMICALS COMMONLY USED IN BIOCHEMISTRY AND MOLECULAR BIOLOGY AND THEIR PROPERTIES (Continued)

Common Name	Chemical Name	M.W.	Properties and Comment
Sodium Deoxycholate	Desoxycholic Acid, Sodium Salt	414.55	Detergent; potential therapeutic use with adipose tissue.

Bril, C., van der Horst, D.J., Poort, S.R., and Thomas, J.B., Fractionation of spinach chloroplasts with sodium deoxycholate, *Biochim. Biophys. Acta* 172, 345–348, 1969; Smart, J.E. and Bonner, J., Selective dissociation of histones from chromatin by sodium deoxycholate, *J. Mol. Biol.* 58, 651–659, 1971; Part, M., Tarone, G., and Comoglio, P.M., Antigenic and immunogenic properties of membrane proteins solubilized by sodium desoxycholate, papain digestion, or high ionic strength, *Immunochemistry* 12, 9–17, 1975; Johansson, K.E. and Wbolewski, H., Crossed immunoelectrophoresis, in the presence of tween 20 or sodium deoxycholate, or purified membrane proteins from *Acholeplasma laidlawii, J. Bacteriol.* 136, 324–330, 1978; Lehnert, T. and Berlet, H.H., Selective inactivation of lactate dehydrogenase of rat tissues by sodium deoxycholate, *Biochem. J.* 177, 813–818, 1979; Suzuki, N., Kawashima, S., Deguchi, K., and Ueta, N., Low-density lipoproteins form human ascites plasma. Characterization and degradation by sodium deoxycholate, *J. Biochem.* 87, 1253–1256, 1980; Robern, H., The application of sodium deoxycholate and Sephacryl S-200 for the delipidation and separation of high-density lipoprotein, *Experientia* 38, 437–439, 1982; Nedivi, E. and Schramm, M., The beta-adrenergic receptor survives solubilization in deoxycholate while forming a stable association with the agonist, *J. Biol. Chem.* 259, 5803–5808, 1984; McKernan, R.M., Castro, S., Poat, J.A., and Wong, E.H., Solubilization of the *N*-methyl-D-aspartate receptor channel complex from rat and porcine brain, *J. Neurochem.* 52, 777–785, 1989; Carter, H.R. Wallace, M.A., and Fain, J.N., Activation of phospholipase C in rabbit brain membranes by carbachol in the presence of GTP gamma S: effects of biological detergents, *Biochim. Biophys. Acta* 1054, 129–134, 1990; Shivanna, B.D. and Rowe, E.S., Preservation of the native structure and function of Ca2+-ATPase from sarcoplasmic reticulum: solubilization and reconstitution by new short-chain phospholipid detergent 1,2-diheptanoyl-*sn*-phosphatidylcholine, *Biochem. J.* 325, 533–542, 1997; Arnold, U. and Ulbrich-Hofmann, R., Quantitative protein precipitation from guandine hydrochloride-containing solutions by sodium deoxycholate/trichloroacetic acid, *Anal. Biochem.* 271, 197–199, 1999; Haque, M.E., Das, A.R., and Moulik, S.P., Mixed micelles for sodium deoxycholate and polyoxyethylene sobitan monooleate (Tween 80), *J. Colloid Interface Sci.* 217, 1–7, 1999; Srivastava, O.P. and Srivastava, K., Characterization of a sodium deoxycholate-activable proteinase activity associated with betaA3/A1-crystallin of human lenses, *Biochim. Biophys. Acta* 1434, 331–346, 1999; Rotunda, A.M., Suzuki, H., Moy, R.L., and Kolodney, M.S., Detergent effects of sodium deoxycholate are a major feature of an injectable phosphatidylcholine formulation used for localized fat dissolution, *Dermatol. Surg.* 30, 1001–1008, 2004; Asmann, Y.W., Dong, M., and Miller, L.J., Functional characterization and purification of the secretin receptor expressed in baculovirus- infected insect cells, *Regul. Pept.* 123, 217–223, 2004; Ranganathan, R., Tcacenco, C.M., Rosseto, R., and Hajdu, J., Characterization of the kinetics of phospholipase C activity toward mixed micelles of sodium deoxycholate and dimyristoyl-phophatidylcholine, *Biophys. Chem.* 122, 79–89, 2006.

Sodium Dodecylsulfate	Sodium Lauryl Sulfate, SDS	288.38	Detergent.

Sodium dodecylsulfate, SDS, lauryl sulfate, sodium salt

Shapiro, A.L., Vinuela, E., and Maizel, J.V., Jr., Molecular weight estimation of polypeptide chains by electrophoresis in SDS-polyacrylamide gels, *Biochem. Biophys. Res. Commun.* 28, 815–820, 1967; Shapiro, A.L., and Maizel, J.V., Jr., Molecular weight estimation of polypeptides by SDS-polyacrylamide gel electrophoresis: further data concerning resolving power and general considerations, *Anal. Biochem.* 29, 505–514, 1969; Weber, K. and Osborn, M., The reliability of molecular weight determinations of dodecyl sulfate-polyacryalmide gel electrophoresis, *J. Biol. Chem.* 244, 4406–4412, 1969; Weber, K. and Kuter, D.J., Reversible denaturation of enzymes by sodium dodecyl sulfate, *J. Biol. Chem.* 246, 4504–4509, 1971; de Haen, C., Molecular weight standards for calibration of gel filtration and sodium dodecyl sulfate-polyacrylamide gel electrophoresis: ferritin and apoferritin, *Anal. Biochem.* 166, 235–245, 1987; Smith, B.J., SDS polyacrylamide gel electrophoresis of proteins, *Methods Mol. Biol.* 32, 23–34, 1994; Guttman, A., Capillary sodium dodecyl sulfate-gel electrophoresis of proteins, *Electrophoresis* 17, 1333–1341, 1996; Bischoff, K.M., Shi, L., and Kennelly, P.J., The detection of enzyme activity following sodium dodecyl sulfate-polyacryalamide gel electrophoresis, *Anal. Biochem.* 260, 1–17, 1998; Maizel, J.V., SDS polyacrylamide gel electrophoresis, *Trends Biochem. Sci.* 35, 590–592, 2000; Robinson, J.M. and Vandre, D.D., Antigen retrieval in cells and tissues: enhancement with sodium dodecyl sulfate, *Histochem. Cell Biol.* 116, 119–130, 2001; Todorov, P.D., Kralchevsky, P.A., Denkov, N.D. et al., Kinetics of solublization of *n*-decane and benzene by micellar solutions of sodium dodecyl sulfate, *J. Colloid Interface Sci.* 245, 371–382, 2002; Zhdanov, S.A., Starov, V.M., Sobolev, V.D., and Velarde, M.G., Spreading of aqueous SDS solutions over nitrocellulose membranes, *J. Colloid Interface Sci.* 264, 481–489, 2003; Santos, S.F., Zanette, D., Fischer, H., and Itri, R., A systematic study of bovine serum albumin (BSA) and sodium dodecyl sulfate (SDS) interactions by surface tension and small angle X-ray scattering, *J. Colloid Interface Sci.* 262, 400–408, 2003; Biswas, A. and Das, K.P., SDS-induced structural changes in alpha-crystallin and its effect on refolding, *Protein J.* 23, 529–538, 2004; Jing, P., Kaneta, T., and Imasaka, T., On-line concentration of a protein using denaturation by sodium dodecyl sulfate, *Anal. Sci.* 21, 37–42, 2005; Choi, N.S., Hahm, J.H., Maeng, P.J., and Kim, S.H., Comparative study of enzyme activity and stability of bovine and human plasmins in electrophoretic reagents, β-mercaptoethanol, DTT, SDS, Triton X-100, and urea, *J. Biochem. Mol. Biol.* 38, 177–181, 2005; Miles, A.P. and Saul, A., Quantifying recombinant proteins and their degradation products using SDS-PAGE and scanning laser densitometry, *Methods Mol. Biol.* 308, 349–356, 2005; Thongngam, M. and McClements, D.J., Influence of pH, ionic strength, and temperature on self-association and interactions of sodium dodecyl sulfate in the absence and presence of chitosan, *Langmuir* 21, 79–86, 2005; Romani, A.P., Gehlen, M.H., and Itri, R., Surfactant-polymer aggregates formed by sodium dodecyl sulfate, poly(*N*-vinyl-2-pyrrolidone), and poly(ethylene glycol), *Langmuir* 21, 1271–1233, 2005; Gudiksen, K.L., Gitlin, I., and Whitesides, G.M., Differentiation of proteins based on characteristic patterns of association and denaturation in solutions of SDS, *Proc. Natl. Acad. Sci. USA* 103, 7968–7972, 2006; Freitas, A.A., Paulo, L., Macanita, A.L., and Quina, F.H., Acid-base equilibria and dynamics in sodium dodecyl sulfate micelles: geminate recombination and effect of charge stabilization, *Langmuir* 22, 7986–7893, 2006.

**CHEMICALS COMMONLY USED IN BIOCHEMISTRY AND MOLECULAR
BIOLOGY AND THEIR PROPERTIES (Continued)**

Common Name	Chemical Name	M.W.	Properties and Comment
Sodium Metabisulfite	Sodium Bisulfite	190.1	Mild reducing agent; converts unmethylated cytosine residues to uracil residues (DNA methylation).

Miller, R.F., Small, G., and Norris, L.C., Studies on the effect of sodium bisulfite on the stability of vitamin E, *J. Nutr.* 55, 81–95, 1955; Hayatsu, H., Wataya, Y., Kai, K., and Iida, S., Reaction of sodium bisulfite with uracil, cytosine, and their derivatives, *Biochemistry* 9, 2858–2865, 1970; Seno, T., Conversion of *Escherichia coli* tRNATrp to glutamine-accepting tRNA by chemical modification with sodium bisulfite, *FEBS Lett.* 51, 325–329, 1975; Tasheva, B. and Dessev, G., Artifacts in sodium dodecyl sulfate-polyacrylamide gel electrophoresis due to 2-mercaptoethanol, *Anal. Biochem.* 129, 98–102, 1983; Draper, D.E., Attachment of reporter groups to specific, selected cytidine residues in RNA using a bisulfite-catalyzed transamination reaction, *Nucleic Acids Res.* 12, 989–1002, 1984; Oakeley, E.J., DNA methylation analysis: a review of current methodologies, *Pharmacol. Ther.* 84, 389–400, 1999; Geisler, J.P., Manahan, K.J., and Geisler, H.E., Evaluation of DNA methylation in the human genome: why examine it and what method to use, *Eur. J. Gynaecol. Oncol.* 25, 19–24, 2004; Thomassin, H., Kress, C., and Grange, T., MethylQuant: a sensitive method for quantifying methylation of specific cytosines within the genome, *Nucleic Acids Res.* 32, e168, 2004; Derks, S., Lentjes, M.H., Mellebrekers, D.M. et al., Methylation-specific PCR unraveled, *Cell. Oncol.* 26, 291–299, 2004; Galm, O. and Herman, J.G., Methylation-specific polymerase chain reaction, *Methods Mol. Biol.* 113, 279–291, 2005; Ogino, S., Kawasaki, T., Brahmandam, M. et al., Precision and performance characteristics of bisulfite conversion and real-time PCR (MethyLight) for quantitative DNA methylation analysis, *J. Mol. Diagn.* 8, 209–217, 2006; Yang, I., Park, I.Y., Jang, S.M. et al., Rapid quantitation of DNA methylation through dNMP analysis following bisulfite PCR, *Nucleic Acids Res.* 34, e61, 2006; Wischnewski, F., Pantel, K., and Schwazenbach, H., Promoter demethylation and histone acetylation mediate gene expression of MAGE-A1, -A2, -A3, and -A12 in human cancer cells, *Mol. Cancer Res.* 4, 339–349, 2006; Zhou, Y., Lum, J.M., Yeo, G.H. et al., Simplified molecular diagnosis of fragile X syndrome by fluorescent methylation-specific PCR and GeneScan analysis, *Clin. Chem.* 52, 1492–1500, 2006.

Succinic Anhydride	Butanedioic Anhydride; 2,5-diketotetra-hydrofuran	100.1	Protein modification; dissociation of protein complexes.

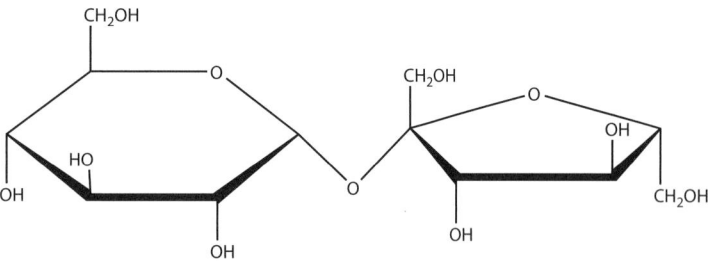

Habeeb, A.F., Cassidy, H.G., and Singer, S.J., Molecular structural effects produced in proteins by reaction with succinic anhydride, *Biochim. Biophys. Acta* 29, 587–593, 1958; Hass, L.F., Aldolase dissociation into subunits by reaction with succinic anhydride, *Biochemistry* 3, 535–541, 1964; Scanu, A., Pollard, H., and Reader, W., Properties of human serum low-density lipoproteins after modification by succinic anhydride, *J. Lipid Res.* 9, 342–349, 1968; Vasilets, I.M., Moshkov, K.A., and Kushner, V.P., Dissociation of human ceruloplasmin into subunits under the action of alkali and succinic anhydride, *Mol. Biol.* 6, 193–199, 1972; Tedeschi, H., Kinnally, K.W., and Mannella, C.A., Properties of channels in mitochondrial outer membrane, *J. Bioenerg. Biomembr.* 21, 451–459, 1989; Palacian, E., Gonzalez, P.J., Pineiro, M., and Hernandez, F., Dicarboxylic acid anhydrides as dissociating agents of protein-containing structures, *Mol. Cell. Biochem.* 97, 101–111, 1990; Pavliakova, D., Chu, C., Bystricky, S. et al., Treatment with succinic anhydride improves the immunogenicity of *Shigella flexneri* type 2a O-specific polysaccharide-protein conjugates in mice, *Infect. Immun.* 67, 5526–5529, 1999; Ferretti, V., Gilli, P., and Gavezzotti, A., X-ray diffraction and molecular simulation study of the crystalline and liquid states of succinic anhydride, *Chemistry* 8, 1710–1718, 2002.

Sucrose		342.30	Osmolyte; density gradient centrifugation.

Cann, J.R., Coombs, R.O., Howlett, G.J. et al., Effects of molecular crowding on protein self-association: a potential source of error in sedimentation coefficients obtained by zonal ultracentrifugation in a sucrose gradient, *Biochemistry* 33, 10185–10190, 1994; Camacho-Vanegas, O., Lorein, F., and Amaldi, F., Flat absorbance background for sucrose gradients, *Anal. Biochem.* 228, 172–173, 1995; Ben-Zeev, O. and Doolittle, M.H., Determining lipase subunit structure by sucrose gradient centrifugation, *Methods Mol. Biol.* 109, 257–266, 1999; Lustig, A., Engel, A., Tsiotis, G. et al., Molecular weight determination of membrane proteins by sedimentation equilibrium at the sucrose of nycodenz-adjusted density of the hydrated detergent micelle, *Biochim. Biophys. Acta* 1464, 199–206, 2000; Kim, Y.S., Jones, L.A., Dong, A. et al., Effects of sucrose on conformational equilibria and fluctuations within the native-state ensemble of proteins, *Protein Sci.* 12, 1252–1261, 2003; Srinivas, K.A., Chandresekar, G., Srivastava, R., and Puvanakrishna, R., A novel protocol for the subcellular fractionation of C3A hepatoma cells using sucrose-density gradient centrifugation, *J. Biochem. Biophys. Methods* 60, 23–27, 2004; Richter, W., Determining the subunit structure of phosphodiesterase using gel filtration and sucrose-density gradient centrifugation, *Methods Mol. Biol.* 307, 167–180, 2005; Cioni, P., Bramanti, E., and Strambini, G.B., Effects of sucrose on the internal dynamics of azurin, *Biophys. J.* 88, 4213–4222, 2005; Desplats, P., Folco, E. and Salerno, G.L., Sucrose may play an additional role to that of an osmolyte in *Synechocystis* sp. PCC 6803 salt-shocked cells, *Plant Physiol. Biochem.* 43, 133–138, 2005; Chen, L., Ferreira, J.A., Costa, S.M. et al., Compaction of ribosomal protein S6 by sucrose occurs only under native conditions, *Biochemistry* 21, 2189–2199, 2006.

**CHEMICALS COMMONLY USED IN BIOCHEMISTRY AND MOLECULAR
BIOLOGY AND THEIR PROPERTIES (Continued)**

Common Name	Chemical Name	M.W.	Properties and Comment
Sulfuric Acid	H_2SO_4	98.1	Strong acid; component of piranha solution with hydrogen peroxide.
Tetrabutylammonium Chloride		277.9	Ion-pair reagent for extraction and HPLC.

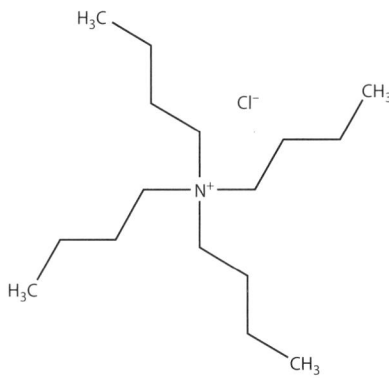

Walseth, T.F., Graff, G., Moos, M.C., Jr., and Goldberg, N.D., Separation of 5′-ribonucleoside monophosphates by ion-pair reverse-phase high-performance liquid chromatography, *Anal. Biochem.* 107, 240–245, 1980; Ozkul, A. and Oztunc, A., Determination of naprotiline hydrochloride in tables by ion-pair extraction using bromthymol blue, *Pharmzie* 55, 321–322, 2000; Cecchi, T., Extended thermodynamic approach to ion interaction chromatography. Influence of the chain length of the solute ion; a chromatographic method for the determination of ion-pairing constants, *J. Sep. Sci* 28, 549–554, 2005; Pistos, C., Tsantili-Kakoulidou, A., and Koupparis, M., Investigation of the retention/pH profile of zwitterionic fluoroquinolones in reversed-phase and ion-interaction high-performance liquid chromatography, *J. Pharm. Biomed. Anal.* 39, 438–443, 2005; Choi, M.M., Douglas, A.D., and Murray, R.W., Ion-pair chromatographic separation of water-soluble gold monolayer-protected clusters, *Anal. Chem.* 78, 2779–2785, 2006; Saradhi, U.V., Prarbhakar, S., Reddy, T.J., and Vairamani, M., Ion-pair solid-phase extraction and gas chromatography mass spectrometric determination of acidic hydrolysis products of chemical warfare agents from aqueous samples, *J. Chromatog. A*, 1129, 9–13, 2006.

Tetrahydrofuran	Trimethylene Oxide	72.1	Solvent; template for combinatorial chemistry.

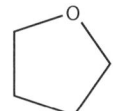

Leuty, S.J., Rapid dehydration of plant tissues for paraffin embedding; tetrahydrofuran vs. t-butanol, *Stain Technol.* 44, 103–104, 1969; Tandler, C.J. and Fiszer de Plazas, S., The use of tetrahydrofuran for delipidation and water solubilization of brain proteolipid proteins, *Life Sci.* 17, 1407–1410, 1975; Dressman, J.B., Himmelstein, K.J., and Higuchi, T., Diffusion of phenol in the presence of a complexing agent, tetrahydrofuran, *J. Pharm. Sci.* 72, 12–17, 1983; Diaz, R.S., Regueiro, P., Monreal, J., and Tandler, C.J., Selective extraction, solubilization, and reversed-phase high-performance liquid chromatography separation of the main proteins from myelin using tetrahydrofuran/water mixtures, *J. Neurosci. Res.* 29, 114–120, 1991; Santa, T., Koga, D., and Imai, K., Reversed-phase high-performance liquid chromatography of fullerenes with tetrahydrofuran-water as a mobile phase and sensitive ultraviolet or electrochemical detection, *Biomed. Chromatogr.* 9, 110–111, 1995; Lee, J., Kang, J.H., Lee, S.Y. et al., Protein kinase C ligands based on tetrahydrofuran templates containing a new set of phorbol ester pharmacophores, *J. Med. Chem.* 42, 4129–4139, 1999; Edwards, A.A., Ichihara, O., Murfin, S. et al., Tetrahydrofuran-based amino acids as library scaffolds, *J. Comb. Chem.* 6, 230–238, 2004; Baron, C.P., Refsgaard, H.H., Skibsted, L.H., and Andersen, M.L., Oxidation of bovine serum albumin initiated by the Fenton reaction — effect of EDTA, tert-butylhydroperoxide, and tetrahydrofuran, *Free Radic. Res.* 40, 409–417, 2006; Bowron, D.T., Finney, J.L., and Soper, A.K., The structure of liquid tetrahydrofuran, *J. Am. Chem. Soc.* 128, 5119–5126, 2006; Hermida, S.A., Possari, E.P., Souza, D.B. et al., 2′-deoxyguanosine, 2′-deoxycytidine, and 2′-deoxyadenosine adducts resulting from the reaction of tetrahydrofuran with DNA bases, *Chem. Res. Toxicol.* 19, 927–936, 2006; Li, A.C., Li, Y., Guirguis, M.S., Advantages of using tetrahydrofuran-water as mobile phases in the quantitation of cyclosporine A in monkey and rat plasma by liquid chromatography-tandem mass spectrometry, *J. Pharm. Biomed. Anal.* 43, 277–284, 2007.

CHEMICALS COMMONLY USED IN BIOCHEMISTRY AND MOLECULAR
BIOLOGY AND THEIR PROPERTIES (Continued)

Common Name	Chemical Name	M.W.	Properties and Comment
Tetraphenylphosphonium Bromide		419.3	Membrane-permeable probe; determination of metal ions.

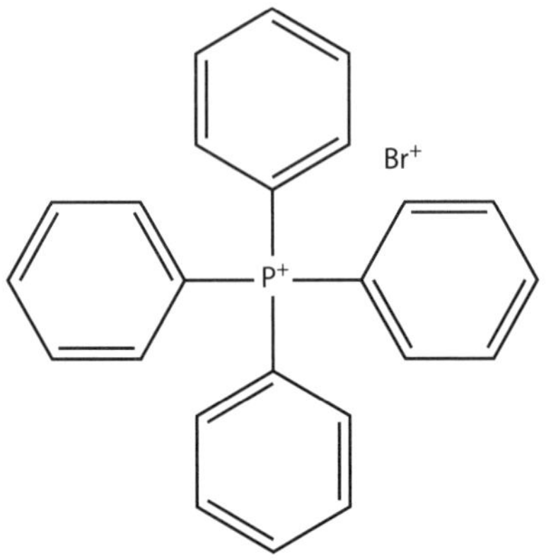

Boxman, A.W., Barts, P.W., and Borst-Pauwels, G.W., Some characteristics of tetraphenylphosphonium uptake into *Saccharomyces cerevisiae*, *Biochim. Biophys. Acta* 686, 13–18, 1982; Flewelling, R.F. and Hubbell, W.L., Hydrophobic ion interactions with membranes. Thermodynamic analysis of tetraphenylphosphonium binding to vesicles, *Biophys. J.* 49, 531–540, 1986; Prasad, R. and Hofer, M., Tetraphenylphosphonium is an indicator of negative membrane potential in *Candida albicans*, *Biochim. Biophys. Acta* 861, 377–380, 1986; Aiuchi, T., Matsunada, M., Nakaya, K., and Nakamura, Y., Calculation of membrane potential in synaptosomes with use of a lipophilic cation (tetraphenylphosphonium), *Chem. Pharm. Bull.* 37, 3333–3337, 1989; Nhujak T. and Goodall, D.M., Comparison of binding of tetraphenylborate and tetraphenylphosphonium ion to cyclodextrins studied by capillary electrophoresis, *Electrophoresis* 22, 117–122, 2001; Yasuda, K., Ohmizo, C., and Katsu, T., Potassium and tetraphenylphosphonium ion-selective electrodes for monitoring changes in the permeability of bacterial outer and cytoplasmic membranes, *J. Microbiol. Methods* 54, 111–115, 2003; Min, J.J., Biswal, S., Deroose, C., and Gambhir, S.S., Tetraphenylphosphonium as a novel molecular probe for imaging tumors, *J. Nucl. Med.* 45, 636–643, 2004.

Thionyl Chloride	Sulfurous Oxychloride	118.97	Preparation of acyl chlorides.

Rodin, R.L. and Gershon, H., Photochemical alpha-chlorination of fatty acid chlorides by thionyl chloride, *J. Org. Chem.* 38, 3919–3921, 1973; DuVal, G., Swaisgood, H.E., and Horton, H.R, Preparation and characterization of thionyl chloride-activated succinamidopropyl-glass as a covalent immobilization matrix, *J. Appl. Biochem.* 6, 240–250, 1984; Molnar-Perl, I., Pinter-Szakacs, M., and Fabian-Vonsik, V., Esterification of amino acids with thionyl chloride acidified butanols for their gas chromatographic analysis, *J. Chromatog.* 390, 434–438, 1987; Stabel, T.J., Casele, E.S., Swaisgood, H.E., and Horton, H.R., Anti-IgG immobilized controlled pore glass. Thionyl chloride-activated succinamidopropyl-gas as a covalent immobization matrix, *Appl. Biochem. Biotechnol.* 36, 87–96, 1992; Chamoulaud, G. and Belanger, D., Chemical modification of the surface of a sulfonated membrane by formation of a sulfonamide bond, *Langmuir* 20, 4989–4895, 2004; Porjazoska, A.,Yilmaz, O.K., Baysal, K. et al., Synthesis and characterization of poly(ethylene glycol)-poly(D,L-lactide-co-glycolide) poly(ethylene glycol) tri-block co-polymers modified with collagen: a model surface suitable for cell interaction, *J. Biomater. Sci. Polym. Ed.* 17, 323–340, 2006; Gao, C., Jin, Z.Q., Kong, H. et al., Polyurea-functionalized multiwalled carbon nanotubes: synthesis, morphology, and Ramam spectroscopy, *J. Phys. Chem. B* 109, 11925–11932, 2005; Chen, G.X., Kim, H.S., Park, B.H., and Yoon, J.S., Controlled functionalization of multiwalled carbon nanotubes with various molecular-weight poly(L-lactic acid), *J. Phys. Chem. B* 109, 22237–22243, 2005.

Thiophosgene	$CSCl_2$	115	
Thiourea	Thiocarbamide	76.12	Chaotropic agent; useful for membrane proteins; will react with haloacetyl derivatives such as iodoacetamide; protease inhibitor.

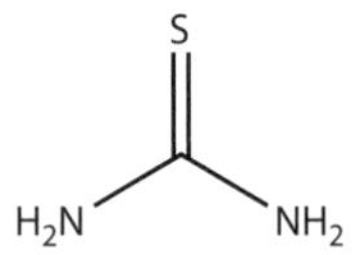

CHEMICALS COMMONLY USED IN BIOCHEMISTRY AND MOLECULAR
BIOLOGY AND THEIR PROPERTIES (Continued)

Common Name	Chemical Name	M.W.	Properties and Comment

Maloof, F. and Soodak, M., Cleavage of disulfide bonds in thyroid tissue by thiourea, *J. Biol. Chem.* 236, 1689–1692, 1961; Gerfast, J.A., Automated analysis for thiourea and its derivatives in biological fluids, *Anal. Biochem.* 15, 358–360, 1966; Lippe, C., Urea and thiourea permeabilities of phospholipid and cholesterol bilayer membranes, *J. Mol. Biol.* 39, 588–590, 1966; Carlsson, J., Kierstan, M.P., and Brocklehurst, K., Reactions of L-ergothioneine and some other aminothiones with 2,2′- and 4,4′-dipyridyl disulphides and of L-ergothioneine with iodoacetamide, 2-mercaptoimidazoles, and 4-thiopyridones, thiourea, and thioacetamide as highly reactive neutral sulphur nucleophiles, *Biochem. J.* 139, 221–235, 1974; Filipski, J., Kohn K.W., Prather, R., and Bonner, W.M., Thiourea reverses crosslinks and restores biological activity in DNA treated with dichlorodiaminoplatinum (II), *Science* 204, 181–183, 1979; Wasil, M., Halliwell, B., Grootveld, M. et al., The specificity of thiourea, dimethylthiourea, and dimethyl sulphoxide as scavengers of hydroxyl radicals. Their protection of alpha-1-antiproteinase against inactivation by hypochlorous acid, *Biochem. J.* 243, 867–870, 1987; Doona, C.J. and Stanbury, D.M., Equilibrium and redox kinetics of copper(II)–thiourea complexes, *Inorg. Chem.* 35, 3210–3216, 1996; Rabilloud, T., Use of thiourea to increase the solubility of membrane proteins in two-dimensional electrophoresis, *Electrophoresis* 19, 758–760, 1998; Musante, L., Candiano, G., and Ghiggeri, G.M., Resolution of fibronectin and other uncharacterized proteins by two-dimensional polyacrylamide electrophoresis with thiourea, *J. Chromatog. B* 705, 351–356, 1998; Nagy, E., Mihalik, R., Hrabak, A. et al., Apoptosis inhibitory effect of the isothiourea compound, tri-(2-thioureido-*S*-ethyl)-amine, *Immunopharmacology* 47, 25–33, 2000; Galvani, M., Rovatti, L., Hamdan, M. et al., Protein alkylation in the presence/absence of thiourea in proteome analysis: a matrix-assisted laser desorption/ionization-time-of-flight-mass spectrometry investigation, *Electrophoresis* 22, 2066–2074, 2001; Castellanos-Serra, L. and Paz-Lago, D., Inhibition of unwanted proteolysis during sample preparation: evaluation of its efficiency in challenge experiments, *Electrophoresis* 23, 1745–1753, 2002; Tyagarajan, K., Pretzer, E., and Wiktorowicz, J.E., Thiol-reactive dyes for fluorescence labeling of proteomic samples, *Electrophoresis* 24, 2348–2358, 2003; Fuerst, D.E., and Jacosen, E.N., Thiourea-catalyzed enantioselective cyanosilylation of ketones, *J. Am. Chem. Soc.* 127, 8964–8965, 2005; Gomez, D.E., Fabbrizzi, L., Licchelli, M., and Monzani, E., Urea vs. thiourea in anion recognition, *Org. Biomol. Chem.* 3, 1495–1500, 2005; George, M., Tan, G., John, V.T., and Weiss, R.G., Urea and thiourea derivatives as low molecular-mass organochelators, *Chemistry* 11, 3243–3254, 2005; Limbut, W., Kanatharana, P., Mattiasson, B. et al., A comparative study of capacitive immunosensors based on self-assembled monolayers formed from thiourea, thioctic acid, and 3-mercaptopropionic acid, *Biosens. Bioelectron.* 22, 233–240, 2006.

TNBS	Trinitrobenzenesulfonic Acid	293.2	Reagent for the determination of amino groups in proteins; also reacts with sulfydryl groups and hydrazides; used to induce animal model of colitis.

Habeeb, A.F., Determination of free amino groups in proteins by trinitrobenzenesulfonic acid, *Anal. Biochem.* 14, 328–336, 1966; Goldfarb, A.R., A kinetic study of the reactions of amino acids and peptides with trinitrobenzenesulfonic acid, *Biochemistry* 5, 2570–2574, 1966; Scheele, R.B. and Lauffer, M.A., Restricted reactivity of the epsilon-amino groups of tobacco mosaic virus protein toward trinitrobenzenesulfonic acid, *Biochemistry* 8, 3597–3603, 1969; Godin, D.V. and Ng, T.W., Trinitrobenzenesulfonic acid: a possible chemical probe to investigate lipid–protein interactions in biological membranes, *Mol. Pharmacol.* 8, 426–437, 1972; Bubnis, W.A. and Ofner, C.M., III, The determination of epsilon-amino groups in soluble and poorly soluble proteinaceous materials by a spectrophotometric method using trinitrobenzenesulfonic acid, *Anal. Biochem.* 207, 129–133, 1992; Cayot, P. and Tainturier, G., The quantification of protein amino groups by the trinitrobenzenesulfonic acid method: a reexamination, *Anal. Biochem.* 249, 184–200, 1997; Neurath, M., Fuss, I., and Strober, W., TNBS-colitis, *Int. Rev. Immunol.* 19, 51–62, 2000; Lindsay, J., Van Montfrans, C., Brennen, F. et al., IL-10 gene therapy prevents TNBS-induced colitis, *Gene Ther.* 9, 1715–1721, 2002; Whittle, B.J., Cavicchi, M., and Lamarque, D., Assessment of anticolitic drugs in the trinitrobenzenesulfonic acid (TNBS) rat model of inflammatory bowel disease, *Methods Mol. Biol.* 225, 209–222, 2003; Necefli, A., Tulumoglu, B., Giris, M. et al., The effects of melatonin on TNBS-induced colitis, *Dig. Dis. Sci.* 51, 1538–1545, 2006.

TNM	Tetranitromethane	196.03	Modification of tyrosine residues in proteins; crosslinking a side reaction as a reaction with cysteine; antibacterial and antiviral agent.

CHEMICALS COMMONLY USED IN BIOCHEMISTRY AND MOLECULAR
BIOLOGY AND THEIR PROPERTIES (Continued)

Common Name	Chemical Name	M.W.	Properties and Comment

Sokolovsky, M., Riordan, J.F., and Vallee, B.L., Tetranitromethane. A reagent for the nitration of tyrosyl residues in proteins, *Biochemistry* 5, 3582–3589, 1966; Nishikimi, M. and Yagi, K., Reaction of reduced flavins with tetranitromethane, *Biochem. Biophys. Res. Commun.* 45, 1042–1048, 1971; Kunkel, G.R., Mehrabian, M., and Martinson, H.G., Contact-site crosslinking agents, *Mol. Cell. Biochem.* 34, 3–13, 1981; Rial, E. and Nicholls, D.G., Chemical modification of the brown-fat-mitochondrial uncoupling protein with tetranitromethane and *N*-ethylmaleimide. A cysteine residue is implicated in the nucleotide regulation of anion permeability, *Eur. J. Biochem.* 161, 689–694, 1986; Prozorovski, V., Krook, M., Atrian, S. et al., Identification of reactive tyrosine residues in cysteine-reactive dehydrogenases. Differences between liver sorbitol, liver alcohol, and *Drosophila* alcohol dehydrogenase, *FEBS Lett.* 304, 46–50, 1992; Gadda, G., Banerjee, A., and Fitzpatrick, P.F., Identification of an essential tyrosine residue in nitroalkane oxidase by modification with tetranitromethane, *Biochemistry* 39, 1162–1168, 2000; Hodges, G.R. and Ingold, K.U., Superoxide, amine buffers, and tetranitro-methane: a novel free radical chain reaction, *Free Radic. Res.* 33, 547–550, 2000; Capeillere-Blandin, C., Gausson, V., Descamps-Latscha, B., and Witko-Sarsat, V., Biochemical and spectrophotometric significance of advanced oxidation protein products, *Biochim. Biophys. Acta* 1689, 91–102, 2004; Lundblad, R.L., *Chemical Reagents for Protein Modification*, CRC Press, Boca Raton, FL, 2004; Negrerie, M., Martin, J.L., and Nghiem, H.O., Functionality of nitrated acetylcholine receptor: the two-step formation of nitrotyrosines reveals their differential role in effectors binding, *FEBS Lett.* 579, 2643–2647, 2005; Carven, G.J. and Stern, L.J., Probing the ligand-induced conformational change in HLA-DR1 by selective chemical modification and mass spectrometry mapping, *Biochemistry* 44, 13625–13637, 2005.

Common Name	Chemical Name	M.W.	Properties and Comment
Trehalose	α-D-glucopyrano-glucopyranosyl-1,1-α-D-glucopyranoside; Mycose	342.3	A nonreducing sugar that is found in a variety of organisms where it is thought to protect against stress such as dehydration; there is considerable interest in the use of trehalose as a stabilizer in biopharmaceutical proteins.

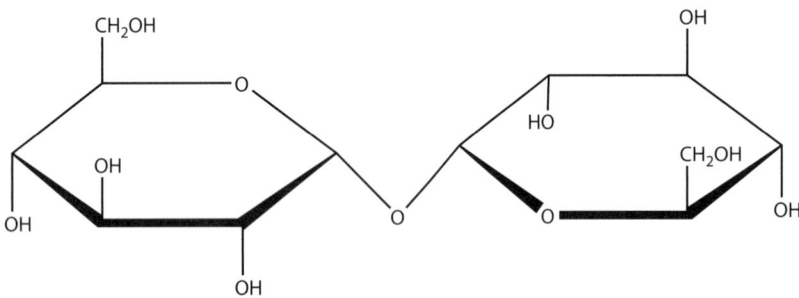

Elbein, A.D., The metabolism of alpha, alpha-trehalose, *Adv. Carbohydr. Chem. Biochem.* 30, 227–256, 1974; Wiemken, A., Trehalose in yeast, stress protectant rather than reserve carbohydrate, *Antonie Van Leeuwenhoek*, 58, 209–217, 1990; Newman, Y.M., Ring, S.G., and Colaco, C., The role of trehalose and other carbohydrates in biopreservation, *Biotechnol. Genet. Eng. Rev.* 11, 263–294, 1993; Panek, A.D., Trehalose metabolism — new horizons in technological applications, *Braz. J. Med. Biol. Res.* 28, 169–181, 1995; Schiraldi, C., Di Lernia, I., and De Rosa, M., Trehalose production: exploiting novel approaches, *Trends Biotechnol.* 20, 420–425, 2002; Elbein, A.D., Pan, Y.T., Pastuszak, I., and Carroll, D., New insights on trehalose: a multifunctional molecule, *Glycobiology* 13, 17R–27R, 2003; Gancedo, C. and Flores, C.L., The importance of a functional trehalose biosynthetic pathway for the life of yeasts and fungi, *FEMS Yeast Res.* 4, 351–359, 2004; Cordone, L., Cottone, G., Giuffrida, S. et al., Internal dynamics and protein-matrix coupling in trehalose-coated proteins, *Biochim. Biophys. Acta* 1749, 252–281, 2005.

Common Name	Chemical Name	M.W.	Properties and Comment
Trichloroacetic Acid		163.4	Protein precipitant.

Common Name	Chemical Name	M.W.	Properties and Comment

Chang, Y.C., Efficient precipitation and accurate quantitation of detergent-solubilized membrane proteins, *Anal. Biochem.* 205, 22–26, 1992; Sivaraman, T., Kumar, T.K., Jayaraman, G., and Yu. C., The mechanism of 2,2,2-trichloroacetic acid-induced protein precipitation, *J. Protein Chem.* 16, 291–297, 1997; Arnold, U. and Ulbrich-Hoffman, R., Quantitative protein precipation form guandine hydrochloride-containing solutions by sodium deoxycholate/trichloroacetic acid, *Anal. Biochem.* 271, 197–199, 1999; Jacobs, D.I., van Rijssen, M.S., van der Heijden, R., and Verpoorte, R., Sequential solubilization of proteins precipitated with trichloroacetic acid in acetone from cultured *Catharanthus roseus* cells yields 52% more spots after two-dimensional electrophoresis, *Proteomics* 1, 1345–1350, 2001; Garcia-Rodriguez, S., Castilla, S.A., Machado, A., and Ayala, A., Comparison of methods for sample preparation of individual rat cerebrospinal fluid samples prior to two-dimensional polyacrylamide gel electrophoresis, *Biotechnol. Lett.* 25, 1899–1903, 2003; Chen, Y.Y., Lin, S.Y., Yeh, Y.Y. et al., A modified protein precipitation procedure for efficient removal of albumin from serum, *Electrophoresis* 26, 2117–2127, 2005; Zellner, M., Winkler, W., Hayden, H. et al., Quantitative validation of different protein precipitation methods in proteome analysis of blood platelets, *Electrophoresis* 26, 2481–2489, 2005; Carpentier, S.C., Witters, E., Laukens, K. et al., Preparation of protein extracts from recalcitrant plant tissues: an evaluation of different methods for two-dimensional gel electrophoresis analysis, *Proteomics* 5, 2497–2507, 2005; Manadas, B.J., Vougas, K., Fountoulakis, M., and Duarte, C.B., Sample sonication after trichloroacetic acid precipitation increases protein recovery from cultured hippocampal neurons, and improves resolution and reproducibility in two-dimensional gel electrophoresis, *Electrophoresis* 27, 1825–1831, 2006; Wang, A., Wu, C.J., and Chen, S.H., Gold nanoparticle-assisted protein enrichment and electroelution for biological samples containing low protein concentration — a prelude of gel electrophoresis, *J. Proteome Res.* 5, 1488–1492, 2006.

| **Triethanolamine** | Tris(2-hydroxyethyl)amine | 149.2 | Buffer; transdermal transfer reagent. |

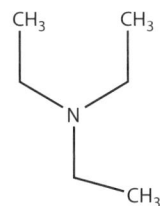

pKa approx. 9.5

Triethanolamine   Triethanolamine hydrochloride

Fitzgerald, J.W., The Tris-catalyzed isomerization of potassium D-glucose 6-*O*-sulfate, *Can. J. Biochem.* 53, 906–910, 1975; Buhl, S.N., Jackson, K.Y., and Graffunder, B., Optimal reaction conditions for assaying human lactate dehydrogenase pyruvate-to-lactate at 25, 30, and 37 degrees C, *Clin. Chem.* 24, 261–266, 1978; Myohanen, T.A., Bouriotas, V., and Dean, P.D., Affinity chromatography of yeast alpha-glucosidase using ligand-mediated chromatography on immobilized phenylboronic acids, *Biochem. J.* 197, 683–688, 1981; Shinomiya, Y., Kato, N., Imazawa, M., and Miyamoto, K., Enzyme immunoassay of the myelin basic protein, *J. Neurochem.* 39, 1291–1296, 1982; Arita, M., Iwamori, M., Higuchi, T., and Nagai, Y., 1,1,3,3-tetramethylurea and triethanolaminme as a new useful matrix for fast atom bombardment mass spectrometry of gangliosides and neutral glycosphingolipids, *J. Biochem.* 93, 319–322, 1983; Cao, H. and Preiss, J., Evidence for essential arginine residues at the active site of maize branching enzymes, *J. Protein Chem.* 15, 291–304, 1996; Knaak, J.B., Leung, H.W., Stott, W.T. et al., Toxicology of mono-, di-, and triethanolamine, *Rev. Environ. Contim. Toxicol.* 149, 1–86, 1997; Liu, Q., Li, X., and Sommer, S.S., pK-matched running buffers for gel electrophoresis, *Anal. Biochem.* 270, 112–122, 1999; Sanger-van de Griend, C.E., Enantiomeric separation of glycyl dipeptides by capillary electrophoresis with cyclodextrins as chiral selectors, *Electrophoresis* 20, 3417–3424, 1999; Fang, L., Kobayashi, Y., Numajiri, S. et al., The enhancing effect of a triethanolamine-ethanol-isopropyl myristate mixed system on the skin permeation of acidic drugs, *Biol. Pharm. Bull.* 25, 1339–1344, 2002; Musial, W. and Kubis, A., Effect of some anionic polymers of pH of triethanolamine aqueous solutions, *Polim. Med.* 34, 21–29, 2004.

| **Triethylamine** | *N,N*-diethylethanamine | 101.2 | Ion-pair reagent; buffer. |

CHEMICALS COMMONLY USED IN BIOCHEMISTRY AND MOLECULAR
BIOLOGY AND THEIR PROPERTIES (Continued)

Common Name	Chemical Name	M.W.	Properties and Comment

Brind, J.L., Kuo, S.W., Chervinsky, K., and Orentreich, N., A new reversed-phase, paired-ion thin-layer chromatographic method for steroid sulfate separations, *Steroids* 52, 561–570, 1988; Koves, E.M., Use of high-performance liquid chromatography-diode array detection in forensic toxicology, *J. Chromatog. A* 692, 103–119, 1995; Cole, S.R. and Dorsey, J.G., Cyclohexylamine additives for enhanced peptide separations in reversed-phase liquid chromatography, *Biomed. Chromatog.* 11, 167–171, 1997; Gilar, M. and Bouvier, E.S.P., Purification of crude DNA oligonucleotides by solid-phase extraction and reversed-phase high-performance liquid chromatography, *J. Chromatog. A* 890, 167–177, 2000; Loos, R. and Barcelo, D., Determination of haloacetic acids in aqueous environments by solid-phase extraction followed by ion-pair liquid chromatography-electrospray ionization mass spectrometric detection, *J. Chromatog. A* 938, 45–55, 2001; Gilar, M., Fountain, K.J., Budman, Y. et al., Ion-pair reversed-phase high-performance liquid chromatography analysis of oligonucleotides: retention prediction, *J. Chromatog. A* 958, 167–182, 2002; El-dawy, M.A., Mabrouk, M.M., and El-Barbary, F.A., Liquid chromatographic determination of fluoxetine, *J. Pharm. Biomed. Anal.* 30, 561–571, 2002; Yang, X., Zhang, X., Li, A. et al., Comprehensive two-dimensional separations based on capillary high-performance liquid chromatography and microchip electrophoresis, *Electrophoresis* 24, 1451–1457, 2003; Murphey, A.T., Brown-Augsburger, P., Yu, R.Z. et al., Development of an ion-pair reverse-phase liquid chromatographic/tandem mass spectrometry method for the determination of an 18-mer phosphorothioate oligonucleotide in mouse liver tissue, *Eur. J. Mass Spectrom.* 11, 209–215, 2005; Xie, G., Sueishi, Y., and Yamamoto, S., Analysis of the effects of protic, aprotic, and multi-component solvents on the fluorescence emission of naphthalene and its exciplex with triethylamine, *J. Fluoresc.* 15, 475–483, 2005.

Trifluoroacetic Acid   114.0   Ion-pair reagent; HLPC; peptide synthesis.

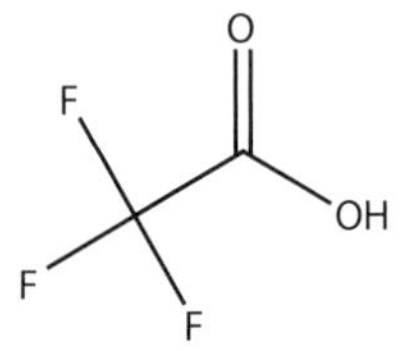

Rosbash, D.O. and Leavitt, D., Decalcification of bone with trifluoroacetic acid, *Am. J. Clin. Pathol.* 22, 914–915, 1952; Katz, J.J., Anhydrous trifluoroacetic acid as a solvent for proteins, *Nature* 174, 509, 1954; Uphaus, R.A., Grossweiner, L.I., Katz, J.J., and Kopple, K.D., Fluorescence of tryptophan derivatives in trifluoroacetic acid, *Science* 129, 641–643, 1959; Acharya, A.S., di Donato, A., Manjula, B.N. et al., Influence of trifluoroacetic acid on retention times of histidine-containing tryptic peptides in reverse phase HPLC, *Int. J. Pept. Protein Res.* 22, 78–82, 1983; Tsugita, A., Uchida, T., Mewes, H.W., and Ataka, T., A rapid vapor-phase acid (hydrochloric and trifluoroacetic acid) hydrolysis of peptide and protein, *J. Biochem.* 102, 1593–1597, 1987; Hulmes, J.D. and Pan, Y.C., Selective cleavage of polypeptides with trifluoroacetic acid: applications for microsequencing, *Anal. Biochem.* 197, 368–376, 1991; Eshragi, J. and Chowdhury, S.K., Factors affecting electrospray ionization of effluents containing trifluoroacetic acid for high-performance liquid chromatography/mass spectrometry, *Anal. Chem.* 65, 3528–3533, 1993; Apffel, A., Fischer, S., Goldberg, G. et al., Enhanced sensitivity for peptide mapping with electrospray liquid chromatography-mass spectrometry in the presence of signal suppression due to trifluoroacetic acid- containing mobiles phases, *J. Chromatog. A* 712, 177–190, 1995; Guy, C.A. and Fields, G.B., Trifluoroacetic acid cleavage and deprotection of resin-bound peptides following synthesis by Fmoc chemistry, *Methods Enzymol.* 289, 67–83, 1997; Morrison, I.M. and Stewart, D., Plant cell wall fragments released on solubilization in trifluoroacetic acid, *Phytochemistry* 49, 1555–1563, 1998; Yan, B., Nguyen, N., Liu, L. et al., Kinetic comparison of trifluoroacetic acid cleavage reactions of resin-bound carbamates, ureas, secondary amides, and sulfonamides from benzyl-, benzhyderyl-, and indole-based linkers, *J. Comb. Chem.* 2, 66–74, 2000; Ahmad, A., Madhusudanan, K.P., and Bhakuni, V., Trichloroacetic acid- and trifluoroacetic acid-induced unfolding of cytochrome C: stabilization of a nativelike fold intermediate(1), *Biochim. Biophys. Acta* 1480, 201–210, 2000; Chen, Y., Mehok, A.R., Mant, C.T. et al., Optimum concentration of trifluoroacetic acid for reversed-phase liquid chromatography of peptide revisited, *J. Chromatog. A* 1043, 9–18, 2004.

Tris(2-carboxyethyl) phosphine   TCEP   250.2   Reducing agent.

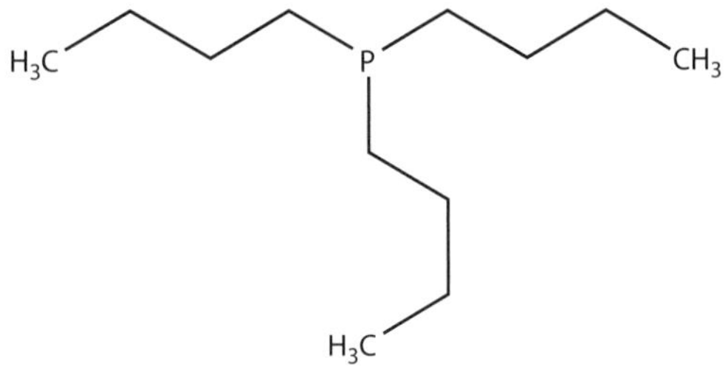

CHEMICALS COMMONLY USED IN BIOCHEMISTRY AND MOLECULAR
BIOLOGY AND THEIR PROPERTIES (Continued)

Common Name	Chemical Name	M.W.	Properties and Comment

Gray, W.R., Disulfide structures of highly bridged peptides: a new strategy for analysis, *Protein Sci.* 2, 1732–1748, 1993; Gray, W.R., Echistatin disulfide bridges: selective reduction and linkage assignment, *Protein Sci.* 2, 1749–1755, 1993; Han, J.C. and Han, G.Y., A procedure for quantitative determination of Tris(2-carboxyethyl)phosphine, an odorless reducing agent more stable and effective than dithiothreitol, *Anal. Biochem.* 220, 5–10, 1994; Wu, J., Gage, D.A., and Watson, J.T., A strategy to locate cysteine residues in proteins by specific chemical cleavage followed by matrix-assisted laser desorption/ionization-time-of-flight mass spectrometry, *Anal. Biochem.* 235, 161–174, 1996; Han, J., Yen. S., Han, G., and Han, F., Quantitation of hydrogen peroxide using Tris(2-carboxyethyl) phosphine, *Anal. Biochem.* 234, 107–109, 1996; Han, J., Clark, C., Han, G. et al., Preparation of 2-nitro-5-thiobenzoic acid using immobilized Tris(2-carboxyethyl) phosphine, *Anal. Biochem.* 268, 404–407, 1999; Anderson, M.T., Trudell, J.R., Voehringer, D.W. et al., An improved monobromobimane assay for glutathione utilizing Tris-(2-carboxyethyl)phosphine as the reductant, *Anal. Biochem.* 272, 107–109, 1999; Shafer, D.E., Inman, J.K. and Lees, A. Reaction of Tris(2-carboxyethyl)phosphine (TCEP) with maleimide and alpha-haloacyl groups: anomalous elution of TCEP by gel filtration, *Anal. Biochem.* 282, 161–164, 2000; Rhee, S.S. and Burke, D.H., Tris(2-carboxyethyl)phosphine stabilization of RNA: comparison with dithiothreitol for use with nucleic acid and thiophosphoryl chemistry, *Anal. Biochem.* 325, 137–143, 2004; Legros, C., Celerier, M.L., and Guette, C., An unusual cleavage reaction of a peptide observed during dithiothreitol and Tris(2-carboxyethyl)phosphine reduction: application to sequencing of HpTx2 spider toxin using nanospray tandem mass spectrometry, *Rapid Commun. Mass Spectrom.* 19, 1317–1323, 2004; Xu, G., Kiselar, J., He, Q., and Chance, M.R., Secondary reactions and strategies to improve quantitative protein footprinting, *Anal. Chem.* 77, 3029–3037, 2005; Valcu, C.M. and Schlink, K., Reduction of proteins during sample preparation and two-dimensional gel electrophoresis of woody plant samples, *Proteomics* 6, 1599–1605, 2006; Scales, C.W., Convertine, A.J., and McCormick, C.L., Fluorescent labeling of RAFT-generated poly(*N*-isopropylacrylamide) via a facile maleimide-thiol coupling reaction, *Biomacromolecules* 7, 1389–1392, 2006.

Urea	Carbamide	60.1	Chaotropic agent.

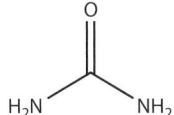

Edelhoch, H., The effect of urea analogues and metals on the rate of pepsin denaturation, *Biochim. Biophys. Acta* 22, 401–402, 1956; Steven, F.S. and Tristram, G.R., The denaturation of ovalbumin. Changes in optical rotation, extinction, and viscosity during serial denaturation in solution of urea, *Biochem. J.* 73, 86–90, 1959; Nelson, C.A. and Hummel, J.P., Reversible denaturation of pancreatic ribonuclease by urea, *J. Biol. Chem.* 237, 1567–1574, 1962; Herskovits, T.T., Nonaqueous solutions of DNA; denaturation by urea and its methyl derivatives, *Biochemistry* 2, 335–340, 1963; Subramanian, S., Sarma, T.S., Balasubramanian, D., and Ahluwalia, J.C., Effects of the urea–guanidinium class of protein denaturation on water structure: heats of solution and proton chemical shift studies, *J. Phys. Chem.* 75, 815–820, 1971; Strachan, A.F., Shephard, E.G., Bellstedt, D.U. et al., Human serum amyloid A protein. Behavior in aqueous and urea-containing solutions and antibody production, *Biochem. J.* 263, 365–370, 1989; Gervais, V., Guy, A., Teoule, R., and Fazakerley, G.V., Solution conformation of an oligonucleotide containing a urea deoxyribose residue in front of a thymine, *Nucleic Acids Res.* 20, 6455–6460, 1992; Smith, B.J., Acetic acid-urea polyacrylamide gel electrophoresis of proteins, *Methods Mol. Biol.* 32, 39–47, 1994; Buck, M., Radford, S.E., and Dobson, C.M., Amide hydrogen exchange in a highly denatured state. Hen egg-white lysozyme in urea, *J. Mol. Biol.* 237, 247–254, 1994; Shirley, B.A., Urea and guanidine hydrochloride denaturation curve, *Methods Mol. Biol.* 40, 177–190, 1995; Bennion, B.J. and Daggett, V., The molecular basis for the chemical denaturation of proteins by urea, *Proc. Natl. Acad. Sci. USA* 100, 5142–5147, 2003; Soper, A.K., Castner, E.W., and Luzar, A., Impact of urea on water structure: a clue to its properties as a denaturant? *Biophys.Chem.* 105, 649–666, 2003; Smith, L.J., Jones, R.M., and van Gunsteren, W.F., Characterization of the denaturation of human alpha-1-lactalbumin in urea by molecule dynamics simulation, *Proteins* 58, 439–449, 2005; Idrissi, A., Molecular structure and dynamics of liquids: aqueous urea solutions, *Spectrochim. Acta A Mol. Biomol. Spectrosc.* 61, 1–17, 2005; Chow, C., Kurt, N., Murphey, R.M., and Cavagnero, S., Structural characterization of apomyoglobin self-associated species in aqueous buffer and urea solution, *Biophys. J.* 90, 298–309, 2006.

Vinyl Pyridine	4-vinylpyridine	105.1	Modification of cysteine residues in protein.

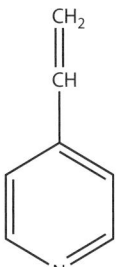

CHEMICALS COMMONLY USED IN BIOCHEMISTRY AND MOLECULAR
BIOLOGY AND THEIR PROPERTIES (Continued)

Common Name	Chemical Name	M.W.	Properties and Comment
Water	Hydrogen Oxide	18.0	Solvent.

Lumry, R. and Rajender, S., Enthalpy-entropy compensation phenomena in water solutions of proteins and small molecules: a ubiquitous property of water, *Biopolymers* 9, 1125–1227, 1970; Cooke, R. and Kuntz, I.D., The properties of water in biological systems, *Annu. Rev. Biophys. Bioeng.* 3, 95–126, 1974; Fettiplace, R. and Haydon, D.A., Water permeability of lipid membranes, *Physiol. Rev.* 60, 510–550, 1980; Lewis, C.A. and Wolfenden, R., Antiproteolytic aldehydes and ketones: substituent and secondary deuterium isotope effects on equilibrium addition of water and other nucleophiles, *Biochemistry* 16, 4886–4890, 1977; Wolfenden, R.V., Cullis, P.M., and Southgate, C.C., Water, protein folding, and the genetic code, *Science* 206, 575–577, 1979; Wolfenden, R., Andersson, L., Cullis, P.M., and Southgate, C.C., Affinities of amino acid side chains for solvent water, *Biochemistry* 20, 849–855, 1981; Cullis, P.M. and Wolfenden, R., Affinity of nucleic acid bases for solvent water, *Biochemistry* 20, 3024–3028, 1981; Radzicka, A., Pedersen, L., and Wolfenden, R., Influences of solvent water on protein folding: free energies of salvation of *cis* and *trans* peptides are nearly identical, *Biochemistry* 27, 4538–4541, 1988; Dzingeleski, G.D. and Wolfenden, R., Hypersensitivity of an enzyme reaction to solvent water, *Biochemistry* 32, 9143–9147, 1993; Timasheff, S.N., The control of protein stability and association by weak interactions with water: how do solvents affect these processes? *Annu. Rev. Biophys. Biomol. Struct.* 22, 67–97, 1993; Wolfenden, R. and Radzcika, A., On the probability of finding a water molecule in a nonpolar cavity, *Science* 265, 936–937, 1994; Jayaram, B. and Jain, T., The role of water in protein–DNA recognition, *Annu. Rev. Biophys. Biomol. Struct.* 33, 343–361, 2004; Pace, C.N., Trevino, S., Prabhakaran, E., and Scholtz, J.M., Protein structure, stability, and solubility in water and other solvents, *Philos. Trans. R. Soc. Lond. B Biol. Sci.* 359, 1225–1234, 2004; Rand, R.P., Probing the role of water in protein conformation and function, *Philos. Trans. R. Soc. Lond. B Biol. Sci.* 359, 1277–1284, 2004; Bagchi, B., Water dynamics in the hydration layer around proteins and micelles, *Chem. Rev.* 105, 3179–3219, 2005; Raschke, T.M., Water structure and interactions with protein surfaces, *Curr. Opin. Struct. Biol.* 16, 152–159, 2006; Levy, Y. and Onuchic, J.N., Water mediation in protein folding and molecular recognition, *Annu. Rev. Biophys. Biomol. Struct.* 35, 389–415, 2006; Wolfenden, R., Degrees of difficulty of water-consuming reactions in the absence of enzymes, *Chem. Rev.* 106, 3379–3396, 2006.

COMMON DETERGENTS USED IN BIOCHEMICAL RESEARCH

Name	Type	Composition	Form	CMC*	Cloud Point	Solubility	Applications
Brij 35	Nonionic	Polyoxyethylene lauryl ether (E_{23})	Solid white wax; density 1.18–1.22 (6)	0.058 g/l (4)	100°C (4)	Soluble in H_2O and (6) most organic solvents insoluble in oils	Column chromatography (7)
Tergitol TMN	Nonionic	Polyoxyethylene trimethyl nonanol (E_6)	Pale yellow liquid, 90%; density, 1.024 (6)	—	36°C (4)	Completely soluble in H_2O (6)	Membrane solubilization (2)
Triton X-45	Nonionic	Polyoxyethylene octylphenol (E_5)	Amber liquid, 100% (6)			Soluble in most organic solvents; insoluble in H_2O (6)	Membrane solubilization (2)
Triton X-100	Nonionic	Polyoxyethylene tert-octylphenol (E_{10})	Amber liquid (6)	0.16 g/l (4)	64°C (5)	Soluble in H_2O and (6) alcohols; slightly soluble in aromatic solvents	Membrane solubilization (2) gel electrophoresis (9)
Tween 20 (Polysorbate 20)	Nonionic	Polyoxyethylene sorbitan monolaurate (E_{22})	Yellow oily liquid (6)	0.14 g/l	95°C (4)	Soluble in H_2O and organic solvents (6)	Solubilizer (1, 4)
Tween 80 (Polysorbate 80)	Nonionic	Polyoxyethylene sorbitan monooleate (E_{25})	Amber liquid; density 1.05–1.10 (6)		93°C (4)	Soluble in H_2O and most organic solvents; insoluble in oils (6)	Solubilizer (1, 4)
CTAB	Cationic	Cetyltrimethyl-ammonium bromide	Creamy-white voluminous powder (13)	0.33 g/l (10)		Soluble in H_2O and organic solvents (13)	Membrane solubilization (2)
SDS	Anionic	Sodium dodecyl (lauryl) sulfate	White crystals, flakes, or powder (13)	2.3 g/l (10)		Soluble in H_2O	Protein solubilization (11), gel electrophoresis (12)
Cholate	Anionic	3,7,12-Trihydroxy-5 cholanate	White Powder			Partially soluble in H_2O; soluble in most organic solvents (13)	Membrane solubilization (2)
Deoxycholate	Anionic	3,12-Dihydroxy-5 cholanate	White Powder			Partially soluble in H_2O; soluble in most organic solvents (13)	Membrane solubilization (2)
Lubrol W	Nonionic	A fatty alcohol ethylene oxide condensate	Fawn colored waxy solids (6)		H_2O, 25°C	Soluble in H_2O, vegetable oils and fatty acids (6)	Membrane solubilization (14)
Atlas G	Nonionic	Polyoxyethylene (sorbitol) hexanolate					Membrane solubilization (14)
Span 20	Nonionic	Sorbitan mono-laurate					Membrane solubilization (14)

Compiled by A. Fulmer.

* Critical micelle concentration in H_2O, 25°C.

References

1. Elworthy, Florence, and Macfarlane, *Solubilization by Surface-Active Agents*, Chapman and Hall Ltd., London, 1968.
2. Foxx and Keith, *Membrane Molecular Biology*, Sinaur Associates, Stamford, Conn., 1972.
3. Swanson, Bradford, and McIlwain, *Biochem. J.*, 92, 235 (1964).
4. Schönfeldt, *Surface-Active Ethylene Oxide Adducts*, Pergamon Press, New York, 1969.
5. Shinoda, Nakagawa, Tamamuchi, and Isemura, *Colloidal Surfactants*, Academic Press, New York, 1963.
6. Sisley and Wood, *Encyclopedia of Surface-Active Agents*, Chemical Publishing, New York, 1964.
7. Morris and Morris, *Separation Methods in Biochemistry*, Pitman Publishing, New York, 1963.
8. Roodyn, *Biochem J.*, 85,177 (1962).
9. Alfageme, Zweider, Mahowald, and Cohen, *J. Biol. Chem.*, 249, 3729 (1974).
10. Mukerjee and Mysels, *Critical Micelle Concentrations of Aqueous Surfactants Systems*, Natl. Stand. Ref. Data Ser., Natl. Bureau Stand., 1971.
11. Reynolds and Tanford, *Proc. Natl. Acad. Sci. U.S.A.*, 66, 1002, (1970).
12. Dunder and Reukert, *J. Biol. Chem.*, 244, 5074 (1969).
13. Stecher, Ed., *The Merck Index of Chemicals and Drugs*, 7th ed., Merck & Co., Rahway, N.J., 1960.
14. Umbreit and Strominger, *Proc. Natl. Acad. Sci. U.S.A.*, 70, 2997 (1973).

General references for surfactants and detergents

Siskey, J.P.(trans. P.J. Wood), *Encyclopedia of Surface-Active Agents*, Chemical Publishing Company, New York, NY, USA, 1952

Ferguson, L.N., On the water solubilities of ethers, *J.Amer.Chem.Soc.* 77, 5288-5289, 1955

Nonionic Surfactants, ed. M.J. Schick, Marcel Dekker, New York, NY, USA, 1966

Jungermann, E., *Cationic Surfactants*, Marcel Dekker, New York, NY, USA, 1970

Cabane, B., Structure of some polymer-detergent aggregates in water, *J.Phys.Chem.* 81, 1639-1645, 1977

Helenius, A., McCaslin, D.R., Fries, E., and Tanford, C., Properties of detergents, *Methods Enzymol.* 56, 734-749, 1979

Membranes Detergents, and Receptor Solubilization, ed. J.C. Venter and L.C. Harrison, A.R.Liss, New York, NY, USA, 1984

Industrial Applications of Surfactants II, ed. D.R. Karsa, Royal Society of Chemistry, Cambridge, UK, 1990

Neugebauer, J.M., Detergents: an overview, *Methods Enzymol.* 182, 239-253, 1990

Industrial Applications of Surfactants III, ed. D.R. Karsa, Royal Society of Chemistry, Cambridge, UK, 1992

Surfactants in Lipid Chemistry: Recent Synthetic, Physical and Biodegradation Studies, ed. J.H.P. Tyman, Royal Society of Chemistry, Cambridge, UK, 1992

Lawrence, M.J., Surfactant systems: Their use in drug delivery, *Chem.Soc. Rev.* 23, 417-223, 1994

Porter, M.R., *Handbook of Surfactants*, 2nd edn., Blackie Academic & Professional, Glasgow, UK, 1994

Structure and Flow in Surfactant Solutions, ed. C.A. Herb and R.K. Prud'homme, American Chemical Society, Washington, DC, 1994
  - Chapter 23 (pps 320-336), Li, Y. and Dubin, P.L., Polymer-surfactant complexes
  - Chapter 26 (pps. 370-379), Smith, B.C., Chou, L.-C., Lu, B., and Zakin, J.L., Effect of counterion structure in flow birefringence and dray reduction behavior of quaternary ammonium salt cationic surfactants
  - Chapter 27 (pps. 380-393), Burgess, D.J. and Sahin, N.O., Interfacial rheology of β-casein solutions

Stevens, L., Solutions used in enzymology, in *Enzymology LabFax*, ed. P.C. Engel, Bios Scientific Publishers, Oxford, UK, Chaper 9, pps. 269-289, 1996

Cogdell, R.J. and Lindsay, J.G., Integral Membrane Proteins, in *Protein LabFax*, ed. N.C. Price, Bios Scientific Publishers, Oxford, UK, Chapter 10, pps. 101-107, 1996

Jönsson, B., Lindman, B., Holmberg, K., and Kronberg, B., *Surfactants and Polymers in Aqueous Solution*, John Wiley & Sons, Chichester, UK, 1998

Tsujii, K., *Surface Activity Principles, Phenomena, and Applications*, Academic Press, San Diego, CA, USA, 1998

Holmberg, K., *Novel Surfactants Preparations, Applications and Biodegradability*, Marcel Dekker, New York, NY, USA, 1998

Hill, R.M., *Silicone Surfactants*, Marcel Dekker, New York, NY, 1999

Industrial Applications of Surfactants IV, ed. D.R. Karsa, Royal Society of Chemistry, Cambridge, UK, 1999

Specialist Surfactants, ed. I.D. Robb, Blackie Academic & Professional, London, UK, 1997

Liposomes Rational Design, ed. A.S. Janoff, Marcel Dekker, New York, NY, USA, 1999

Hummel, D.O., *Handbook of Surfactant Analysis. Chemical, Physico-Chemical and Physical Methods*, John Wiley & Sons, Ltd., Chichester, UK, 2000

Corrigan, O.I. and Healy, A.M., Surfactants in pharmaceutical products and systems, in *Encyclopedia of Pharmaceutical Technology*, ed. J. Swarbrick and J.C. Boylan, Marcel Dekker, New York, NY, Volume 3, pp 2639-2653, 2002

Encyclopedia of Surface and Colloid Science, ed. A.T. Hubbard, Marcel Dekker, Inc, New York, NY, USA, 2002
  - Hirata, H. Surfactant molecular complexes, Volume 4, pps 5178-5204
  - Imamura, T., Surfactant-protein interactions, Volume 4, pps 5230-5243

Rosen, M.J., *Surfactants and Interfacial Phenomena*, Wiley-Interscience, Hoboken, NJ, USA, 2004

Goodwin, J.W., *Colloids and Interfaces with Surfactants and Polymers: An Introduction*, John Wiley, Hoboken, NJ, USA, 2004

Tadros, T.F., *Applied Surfactants. Principles and Applications*, Wiley-VCH, Weinheim, Germany, 2005

Handbook of Functional Lipids, ed. C.C. Akoh, CRC/Taylor & Francis, Boca Raton, FL, USA, 2006

SOME PROPERTIES OF DETERGENTS AND SURFACTANTS USED IN BIOCHEMISTRY AND MOLECULAR BIOLOGY

Structures for these compounds may be found on p. 816–818.

Detergent/Surfactant	Molecular Weight	Classification	Key References
Tween 20[a]		nonionic	1-5
Tween 80[b]	1,300	nonionic	6-10
Triton X-100[c]	650	nonionic	11-15
Nonidet P-40[d]	650	nonionic	16-20
Brij Deterents[e] (polyoxyethylene derivatives)		nonionic	21-25
Lubrol[f]	582	nonionic	26-30
Sodium dodecyl dulfate (SDS)[g]	288.4	anionic	31-35
Sodium deoxycholate[h]	432	anionic	36-40
Cetylpyridinium chloride[i]	340.0	cationic	41-45
Cetyltrimethylammonium bromide[j]	364.5	cationic	46-50
Tetradecyltrimethyl-ammonium bromide[k]	336	cationic	51-55
Betainesulfonate,[l] sulfobetaine, alkylsulfobetaine derivatives		zwitterionic	56-60
CHAPS[m]	614.9	zwitterionic	61-65
CHAPSO[n]	630.9	zwitterionic	66-70

Footnotes

[a] Tween 20 is a polysorbate surfactant (polyoxyethylene sorbitan fatty acid ester) useful in immunoassays and pharmaceutical development. Tween 20 reduced protein interaction and protein binding to surfaces. It is related to Tween 80. The Tween surfactants are condensation products of mono-substituted fatty acyl derivatives of sorbitan (SPAN®; a registered trademark of Atlas Chemical) and ethylene oxide. Span® 20 is sorbitan monolaurate and Tween® 20 (polyethoxyethylene (20) sorbitan monolaurate) is derived from Span® 20. Tween®20 has a viscosity of 200-400 cp at 25°C.

[b] Tween® 80 (polysorbate 80) is a polysorbate surfactant (polyoxyethylene sorbitan mono-oleate) is used in pharmaceutical formulation where it stabilizes proteins and in various diagnostics tests. Tween® 80 was added early to bacterial culture media and was also demonstrated to enhance the anti-bacterial activity of certain antibiotics. Tween® 80 has also been used for protein membrane studies and were originally manufactured by Atlas Power Company. Tween 80 has a viscosity of 600-800 cp at 25°C.

[c] Triton X-100 is a nonionic detergent known as octoxynol, octylphenoxy polyethoxyethanol. The Triton family of surfactants which is manufactured by the condensation of ethylene oxide with alkylphenols and contain between 5 and 15 ethylene oxide units; with Triton X-100, there is 9 or 10 ethylene oxide units while Triton X-30 contains three ethyloxide groups. Triton X-100 is also known as Igepal CA or Polydetergent G. These alkylaryl polyether alcohols were first manufacture by I.G. Farbenindustrie as Igepal products and subsequently manufactured and marketed in the United States as Triton. Triton X-100 is used for the lysis of cells and preparation of subcellular fractions and in the study of membrane proteins.

[d] Nonidet® is a registered trademark of the Shell Oil Company; however, it is not clear that Shell Oil Company is still associated with the manufacture and/or distribution of this material. A search of the Shell Oil website did not a result for Nonidet. Nonidet® P-40 [octylphenolpoly(ethyleneglycolether)] is a popular alkylphenyl ethoxylate nonionic detergent; also referred to as nonylphenylpolyethylene glycol; polyethyleneglycol-p-isooctyl-phenyl ether; octylphenoxy polyethoxy ethanol; Igepal CO 630. There is question as to the relation of nonidet P-40 to NP-40. Related to the Triton surfactants. It is useful to assure the provenance of a product labeled Nonidet P40 or NP-40.

[e] Brij® detergents are a series of polyoxyethylene ethers such as lauryl polyoxyethylene (dodecyl polyoxyethylate). This class of surfactants is referred to as poloxamers and includes polyoxypropylene. The number of ethylene oxide units determines the solubility of the dodecyl alcohol derivative:

Moles Ethylene Oxide/ Dodecyl Alcohol	Approximate Solubility of Product
0	Insoluble
2	Insoluble
4	Somewhat miscible
6	Slightly soluble
7	Soluble

[f] Lubrol® is a trademark of ICI referring to a series of nonionic surfactants similar to the Brij® surfactants in being a series of polyoxyethylene alkyl ethers. Lubrol® surfactants are used for membrane solubilization and more recently in the study of lipid rafts. While there are similarities between the various nonionic surfactants, there are also distinct differences.

[g] Sodium Dodecyl Sulfate; also known as sodium lauryl sulfate. An anionic detergent used for membrane solubilization and protein denaturation. Used in a popular electrophoretic procedure, SDS-gel electrophoresis where separation is presumed to occur on the basis of molecular weight; there are some exceptions to this assumption (Noel, D., Nikaido, K., and Ames, G.F., A single amino acid substitution in a histidine-transport protein drastically alters its mobility in sodium dodecyl sulfate-polyacrylamide gel electrophoresis, *Biochemistry* 18, 4159-4165, 1979; Briggs, M.M., Klevit, R.E., and Schachat, F.H., Heterogeneity of contractile proteins-purification and characterization of 2 species of troponin-T from rabbit fast skeletal-muscle, *J.Biol.Chem.* 259, 10369-10375, 1984; Sakakura, Y., Hirabayashi, J.,. Oda, Y. *et al.*, Structure of chicken 16-kDa β-galactoside-binding lectin-complete amino-acid-sequence, cloning of cDNA, and production of recombinant lectin, *J.Biol.Chem.* 265, 21573-21579, 1990; Okumura, N., Terasawa, F., Fujita, K. *et al.*, Difference in electrophoretic mobility and plasmic digestion profile between four recombinant fibrinogens, gamma 308K, gamma 3081, gamma 308A, and wild type (gamma 308N), *Electrophoresis* 21, 2309-2315, 2000). The

counterion to the lauryl sulfate moiety can also influence the interaction with proteins (Kubo, K. and Takagi, T., Modulation of the behavior of a protein in polyacrylamide gel electrophoresis in the presence of dodecyl sulfate by varying the cations, *Anal.Biochem.* 224, 572-579, 1995; Kubo, K., Effect of incubation of solutions of proteins containing sodium dodecyl sulfate on the cleavage of peptide bonds by boiling, *Anal.Biochem.* 225, 351-353, 1995).

^{h} Deoxycholic acid (3,12-dihydroxycholan-24-oic acid (3α, 5β, 12α), MW 392.6) is relatively insoluble, the sodium salt is more soluble. Deoxychlolate (desoxycholate) is used for membrane solubilization. A natural constituent of bile secretions in man and other mammals.

^{i} Cetylpyridinium chloride (1-hexadecylpyridinium chloride) is a cationic detergents which has pharmaceutical use as a preservative and a topical disinfectant. There has been significant use of this material as an active ingredient in mouthwashes. The interaction of cetylpyridinium chloride with glycosaminoglycans such as heparin is useful for characterization. Cetylpyridinium chloride has been used for the isolation of glycosaminoglycans and for the histochemical staining of glycosaminoglycans.

^{j} Cetyltrimethylammonium bromide (N,N,N-trimethylhexadecaaminium bromide; CTAB; cetrimonium bromide) was developed as a disinfectant and antiseptic. It has been used as detergent in a manner similar to SDS in the determination of protein molecular weight. CTAB is also used for the isolation and assay of nucleic acids.

^{k} Tetradecyltrimethyl-ammonium bromide (TTABr). TTABr is one of the components of Cetrimide®, a disinfectant. TTABr has seen occasional use for membrane protein solubilization and in the preparation of mixed micelles.

^{l} The "parent" compound for this class of surfactants is betaine sulfonate (sulfobetaine). There are non-detergent sulfobetaines (non detergent sulfobetaines such as 3-(1-pyridino)-1-propanesulfonate; NDSB 201)) which have been useful in preventing unwanted protein-protein aggregation (Vuillard, L., Braun-Breton, C., and Rabilloud, T., Non-detergent sulfobetaines: a new class of mild solubilization agents for protein purification, *Biochem.J.* 305, 337-343, 1995; Collins, T., D'Amico, S., Georlette, D., *et al.*, A nondetergent sulfobetaine prevents protein aggregation in microcalorimetric studies, *Anal.Biochem.* 352, 299-301,2006). Detergent sulfobetaines such Zwittergent® 3-12 (*n*-Dodecyl-N,N-dimethyl-3-ammonio-1-propanesulfonate) contain a long chain alkyl function (Wollstadt, K.H., Karkhanis, Y.D., Gnozzio, M.J. *et al.*, Potential of the sulfobetaine detergent Zwittergent 3-12 as a desorbing agent in biospecific and bioselective affinity chromatography, *J.Chromatog.* 497, 87-100, 1989). The CHAPS class of detergents are derivatives of sulfobetaine which contain a cholamide function.

^{m} 3-[(3-cholamidopropyl)dimethylamino]propanesulfonic acid

^{n} 3-[(3-cholamidopropyl)dimethylammonio]-2-hydroxy-1-propanesulfonate

References for table

1. Liljas, L., Lundahl, P., and Hjerten, S., Selective solubilization with Tween 20 of proteins from water-extracted human erythrocyte membranes. Analysis by gel electrophoresis in dodecylsulfate and in Tween 20, *Biochim.Biophys.Acta.* 352, 327-337, 1974.
2. Hoffman, W.L., and Jump, A.A., Tween 20 removes antibodies and other proteins from nitrocellulose, *J.Immunol.Methods* 94, 191-196, 1986.
3. Vandenberg, E.T. and Krull, U.J., The prevention of adsorption of interferents to radiolabeled protein by Tween 20, *J.Biochem.Biophys. Methods* 22, 269-277, 1991.
4. Feng, M., Morales, A.B., Poot, A., *et al.*, Effects of Tween 20 on the desorption of protein from polymer surfaces, *J.Biomater.Sci.Polym. Ed.* 7, 415-424, 1995
5. Kreilgaard, L., Jones, L.S., Randolph, T.W., *et al.*, Effect of Tween 20 on freeze-thawing- and agitation-induced aggregation of recombinant human factor XIII, *J.Pharm.Sci.* 87, 1597-1603, 1998.
6. Youmans, A.S. and Youmans, G.P., The effect of "tween 80" in vitro on the bacteriostatic activity of twenty compounds for *Mycobacterium tuberculosis*, *J.Bacteriol.* 56, 245-252, 1948.
7. Young, M., Dinda, M., and Singer, M., Effect of Tween 80 on lipid vesicle permeability, *Biochim. Biophys.Acta.* 735, 429-432, 1983.
8. Kerwin, B.A., Heller, M.C., Levin, S.H., and Randolph, T.W., Effects of Tween 80 and sucrose on acute short-term stability and long-term storage at -20°C of a recombinant hemoglobin, *J.Pharm.Sci.* 87, 1062-1068, 1998.
9. Arakawa, T. and Kita, Y., Protection of bovine serum albumin from aggregation by Tween 80, *J.Pharm.Sci.* 89, 646-651, 2000.
10. Hillgren, A., Lindgren, J., and Alden, M., Protection mechanism of Tween 80 during freeze-thawing of a model protein, LDH, *Int.J.Pharm.* 237, 57-69, 2002.
11. De Duve, C. and Wattiaux, R., Tissue fractionation studies. VII. Release of bound hydrolases by means of triton X-100, *Biochem.J.* 63, 606-608, 1956.
12. Ashani, Y. and Catravas, G.N., Highly reactive impurities in Triton X-100 and Brij 35: partial characterization and removal, *Anal. Biochem.* 109, 55-62, 1980.
13. Labeta, M.O., Fernandez, N., and Festenstein, H., Solubilization effect of Nonidet P-40, Triton X-100 and CHAPS in the detection of MHC-like glycoproteins, *J.Immunol.Methods* 112, 133-138, 1988.
14. Partearroyo, M.A., Urbaneja, M.A., and Goni, F.M., Effective detergent/lipid ratios in the solubilization of phosphatidylcholine vesicles by Triton X-100, *FEBS Lett.* 302, 138-140, 1992.
15. Blonder, J., Yu, L.R., Radeva, G., *et al.*, Combined chemical and enzymatic stable isotope labeling for quantitative profiling of detergent-insoluble membrane proteins isolated membrane proteins isolated using Triton X-100 and Brij-96, *J.Proteome Res.* 5, 349-360, 2006.
16. Schwartz, B.D. and Nathenson, S.G., Isolation of H-2 alloantigens solubilized by the detergent NP-40, *J.Immunol.* 107, 1363-1367, 1971.
17. Hosaka, Y. and Shimizu, Y.K., Artificial assembly of envelope particles of HVJ (Sendai virus). I. Asssembly of hemolytic and fusion factors from envelopes solubilized by Nonidet P40, *Virology* 49, 627-639, 1972.
18. Hart, D.A., Studies on nonidet P40 lysis of murine lymphoid cells. I. Use of cholera toxin and cell surface Ig to determine degree of dissociation of the plasma membrane, *J.Immunol.* 115, 871-875, 1975.
19. Soloski, M.J., Cabrera, C.V., Esteban, M., and Holowczak, J.A., Studies concerning the structure of and organization of the vaccinia virus nucleod. I. Isolation and characterization of subviral particles prepared by treating virions with guanidine-HCl, nonidet-P40, and 2-mercaptoethanol, *Virology* 99, 209-217, 1979.
20. Lanuti, P., Marchisio, M., Cantilena, S., *et al.*, A flow cytometry procedure for simultaneous characterization of cell DNA content and expression of intracellular protein kinase C-zeta, *J.Immunol.Methods* 315, 37-48, 2006.
21. Godson, G.N. and Sinsheimer, R.L., Use of Brij as a general method to prepare polyribosomes from *Escherichia coli.*, *Biochim.Biophys. Acta* 149, 489-495, 1967.
22. Ashani, Y. and Catravas, G.N., Highly reactive impurities in Triton X-100 and Brij 35: partial characterization and removal, *Anal. Biochem.* 109, 55-62, 1980
23. Krause, M., Rudolph, R., and Schwartz, E., The non-ionic detergent Brij 58P mimics chaperone effects, *FEBS Lett.* 532, 253-255, 2002.
24. Lee, Y.C., Simamora, P., and Yalkowsky, S.H., Effect of Brij-78 on systemic delivery of insulin from an ocular device, *J.Pharm.Sci.* 86, 430-433, 1997.
25. Chakraborty, T., Ghosh, S., and Moulik, S.P., Micellization and related behavior of binary and ternary surfactant mixtures in aqueous medium: cetyl pyridinium chloride (CPC) cetyl trimethyl ammonium bromide (CTAB), and polyoxythelene (10) cetyl ether (Brij-56) derived system,

26. Flawia, M.M. and Torres, H.N., Adenylate cyclase activity in lubrol-treated membranes from *Neurospora crassa, Biochim.Biophys.Acta* 289, 428-432, 1972

27. Young, J.L. and Stansfield, D.A., Solubilization of bovine corpus-luteum adenylate cyclase in Lubrol-PX, triton X-100 or digitonin and the stabilizing effect of sodium fluoride present in the solubilization medium, *Biochem.J.* 173, 919-924, 1978.

28. Hommes, F.A., Eller, A.G., Evans, B.A., and Carter, A.L., Reconstitution of ornithine transport in liposomes with Lubrol extracts of mitochondria, *FEBS Lett.* 170, 131-134, 1984.

29. Chamberlain, L.H., Detergents as tools for the purification and classification of lipid rafts, *FEBS Lett.* 559, 1-5, 2004.

30. Gil, C., Cubi, R., Blasi, J., and Aguilera, J., Synaptic proteins associate with a sub-set of lipid rafts when isolated from nerve endings at physiological temperature, *Biochem.Biophys.Res.Commun.* 348, 1334-1342, 2006

31. Emerson, M.F. and Holtzer, A., The hydrophobic bond in micellar systems. Effects of various additives on the stability of micelles of sodium dodecyl sulfate and of *n*-dodecyltrimethylammonium bromide, *J.Phys.Chem.* 71, 3320-3330, 1967.

32. Fish, W.W., Reynolds, J.A., and Tanford, C., Gel chromatography of proteins in denaturing solvents. Comparison between sodium dodecyl sulfate and guanidine hydrochloride as denaturants, *J.Biol.Chem.* 245, 5166-5168, 1970

33. Weber, K. and Kuter, D.J., Reversible denaturation of enzymes by sodium dodecyl sulfate, *J.Biol.Chem.* 246, 4504-4509, 1971.

34. Dai, S. and Tam, K.C., Effects of cosolvents on the binding interaction between poly(ethylene oxide) and sodium dodecyl sulfate, *J.Phys. Chem.B Condens Mater Surf Interfaces Biophys.* 110, 20794-20800, 2006.

35. Keller, S., Heerklotz, H., Jahnke, N., and Blume, A., Thermodyanamics of lipid membrane solubilization by sodium dodecyl sulfate, *Biophys.J.* 90, 4509-4521, 2006.

36. Benzonana, G., Study of bile salts micelles: properties of mixed oleate-deoxycholate solutions at pH 9.0, *Biochim.Biophys.Acta.* 176, 836-848, 1969

37. Olsenes, S., Removal of structural proteins from ribosomes by treatment with sodium dexoycholate I the presence of EDTA, *FEBS Lett.* 7, 211-213, 1970.

38. Ehrhart, J.C., and Chaveau, J., Differential solubilization of proteins, phospholipids, free and esterified cholesterol of rat liver cellular membranes by sodium deoxycholate, *Biochim.Biophys.Acta* 375, 434-445, 1975.

39. Robinson, N.C. and Tanford, C., The binding of deoxycholate, Triton X-100, sodium dodecyl sulfate, and phosphatidylcholine vesicles to cytochrome b5, *Biochemistry* 14, 369-378, 1975.

40. Ranganathan, R., Tcacenco, C.M., Rossetto, R., and Hajdu, J., Characterization of the kinetics of Phospholipase C activity toward mixed micelles of sodium deoxycholate and dimyristoylphosphatidylcholine, *Biophys.Chem.* 122 79-89, 2006

41. Malchiodi Albedi, F., Cassano, A.M., Ciaralli, F., *et al.*, Influence of cetylpyridinium chloride on the ultrastructural appearance of sulphated glycosaminoglycans in human colonic mucosa, *Histochemistry* 89, 397-401, 1988

42. Chardin, H., Septier, D., and Goldberg, M., Visualization of glycosaminoglycans in rat incisor predentin and dentin with cetylpyridinium chloride-glutaraldehyde as fixative, *J.Histochem.Cytochem.* 38, 885-894, 1900

43. Savolainen, H., Isolation and separation of proteoglycans, *J.Chromatogr.B Biomed. Sci.Appl.* 722, 255-262, 1999

44. Benamor, M., Aguersif, N., and Draa, M.T., Spectrophotometric determination of cetylpyridinium chloride in pharmaceutical products, *J.Pharm.Biomed.Anal.* 26, 151-154, 2001

45. Arrigler, V., Kogej, K., Majhenc, J., and Svetina, S., Interaction of cetylpyridinium chloride with giant vesicles, *Langmuir* 21, 7653-7661, 2005

46. Davies, G.E., Quaternary ammonium compounds: a new technique for the study of their bactericidal action and the results obtained with cetavlon (cetyltrimethylammonium bromide), *J.Hyg.*(Lond), 47, 271-277, 1949

47. Akin, D.T., Shapira, R., and Kinkade, J.M., Jr., The determination of molecular weights of biologically active proteins by cetyltrimethylammonium bromide-polyacrylamide gel electrophoresis, *Anal. Biochem.* 145, 170-176, 1985

48. Jost, J.P., Jiricny, J. and Saluz, H., Quantitative precipitation of short oligonucleotides with low concentrations of cetyltrimethylammonium bromide, *Nucleic Acid Res.* 17, 2143, 1989

49. Li, Y.F., Shu, W.Q., Feng, P., *et al.*, Determination of DNA with cetyltrimethylammonium bromide by the measurement of resonance light scattering, *Anal.Sci.* 17, 693-696, 2001

50. Carra, A., Gambino, G., and Schubert, A., A cetyltrimethylammonium bromide-based method to extract low-molecular-weight RNA from polysaccharide-rich plant tissues, *Anal.Biochem.* 360, 318-320, 2007.

51. Hayakawa, K., Santerre, J.P., and Kwak, J.C, The binding of cationic surfactants by DNA, *Biophys.Chem.* 17, 175-181, 1983

52. Castedo, A., Castillo, J.L.D., Suarez-Filloys, M.J., and Rodriguez, J.R., Effect of temperature on the mixed micellar tetradecyltrimethylammonium bromide-butanol system, *J.Colloid Interface Sci.* 196, 148-156, 1997

53. Medrzycka, K. and Zwierzykowski, W., Adsorption of alkyltrimethylammonium bromides at the various interfaces, *J.Colloid Interface Sci.* 230, 67-72, 2000

54. Stodghill, S.P., Smith, A.E., and O'haver, J.H., Thermodynamics of micellization and adsorption of three alkyltrimethylammonium bromides using isothermal titration calorimetry, *Langmuir* 20, 11387-11392, 2004

55. Rasmussen, C.D., Nielsen, H.B., and Andersen, J.E., Analysis of the purity of cetrimide by titrations, *PDA J. Pharm.Sci.Technol.* 60, 104-110, 2006.

56. Sims, N.R, Horvath, L.B., and Carnegie, P.R., Detergent solubilization and solubilization of 2':3'-cyclic nucleotide 3'-phosphodiesterase from isolated myelin and c6 cells, *Biochem.J.* 181, 367-375, 1979

57. Wong, R.K., Nichol, C.P., Sekar, M.C., and Roufogalis, B.D., The efficiency of various detergents for extraction and stabilization of acetylcholinesterase from bovine erythrocytes, *Biochem.Cell Biol.* 65, 8-18, 1987

58. Wydro, P. and Paluch, M., A study of the interaction of dodecyl sulfobetaine with cationic and anionic surfactant in mixed micelles and monolayers at the air/water interface, *J.Colloid Interface Sci.* 286, 387-391, 2005

59. Nyuta, K., Yoshimura, T., and Esumi, K., Surface tension and micellization of heterogemini surfactants containing quaternary ammonium salt and sulfobetaine moiety, *J.Colloid Interface Sci.* 301, 267-273, 2006.

60. Zanna, L. and Haeuw, J.F., Separation and quantitative analysis of alkyl sulfobetaine-type detergents by high-performance liquid chromatography and light-scattering detection, *J.Chromatog.B Analyt. Technol.Biomed.Life Sci.*, in press, 2007.

61. Hjelmeland, L.M., A nondenaturing zwitterionic detergent for membrane biochemistry: Design and synthesis, *Proc.Nat.Acad.Sci.USA* 77, 6368-6370, 1980

62. Ray, J.P., Mernoff, S.T., Sangameswaran, L., and de Blas, A.L., The Stokes Radius of the CHAPS-solubilized benzodiazepine receptor complex, *Neurochem.Res.* 10, 1221-1229, 1985

63. Labeta, M.O., Fernandez, N., and Festenstein, H., Solubilisation effect of Nonidet P-40, Triton X-100 and CHAPS in the detection of MHC-like glycoprotein, *J.Immunol.Methods* 112, 133-138, 1988

64. Banerjee, P., Buse, J.T., and Dawson, G., Asymmetric extraction of membrane lipids by CHAPS, *Biochim.Biophys.Acta* 1044, 305-314, 1990

65. Rouvinski, A., Gahali-Sass, I., Stav, I., *et al.*, Both raft- and non-raft proteins associate with CHAPS-insoluble complexes: some APP in large complexes, *Biochem.Biophys.Res.Commun.* 308, 750-758, 2003

66. Womack, M.D., Kendall, D.A., and MacDonald, R.C., Detergent effects on enzyme activity and solubilization of lipid bilayer membranes, *Biochim.Biophys.Acta.* 733, 210-215, 1983.

67. Saunders, C.R, 2nd and Prestegard, J.H., Magnetically orientable phospholipid bilayers containing small amounts of a bile salt analogue CHAPSO, *Biophys.J.* 58, 447-460, 1990

68. Gartner, W., Ullrich, D. and Vogt, K., Quantum yield of CHAPSO-solubilized rhodopsin and 3-hydroxy retinal containing bovine opsin, *Photochem.Photobiol.* 54, 1047-1055, 1991

69. Banerjee, P., Joo, J.B., Buse, J.T., and Dawson, G., Differential solubilization of lipids along with membrane proteins by different classes of detergents, *Chem.Phys.Lipids* 77, 65-78, 1995

70. Gehrig-Burger, K., Kohout, L., and Gimpl, G., CHAPSETEROL. A novel cholesterol-based detergent, *FEBS J.* 272, 800-812, 2005

Tween

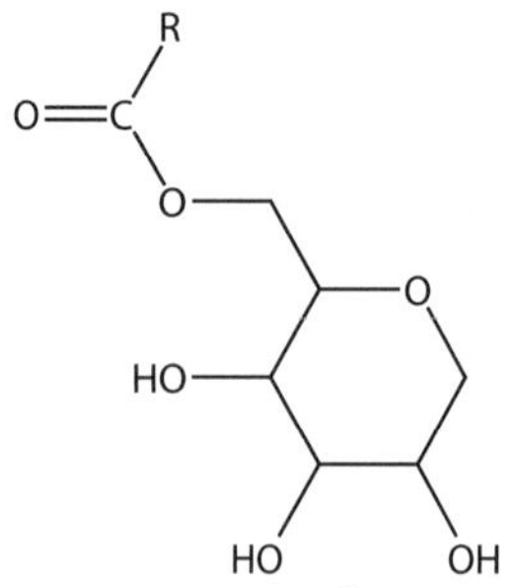

Where R = fatty acid

Sorbitan Fatty Acid Ester, SPAN®

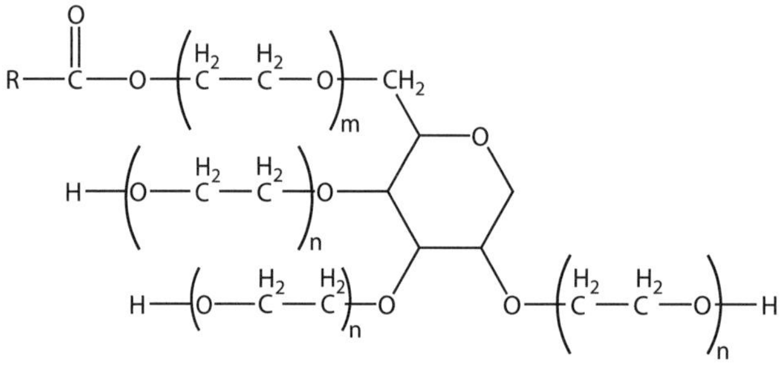

Polysorbate (Tween)

Triton Detergents

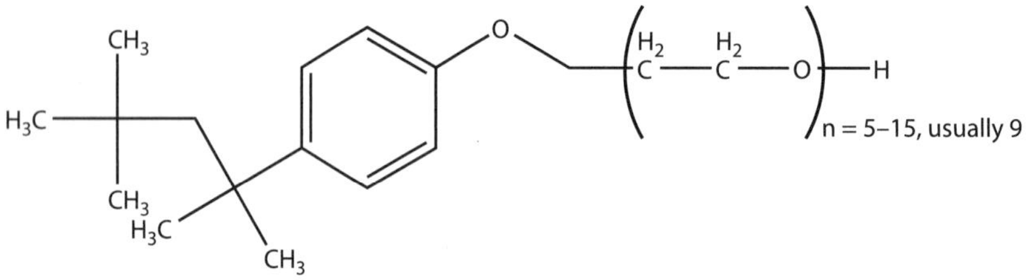

$n = 5$–15, usually 9

Octoxynol, Triton X®, Igepal® CA

Alkylphenoxy ethoxylate nonionic detergent

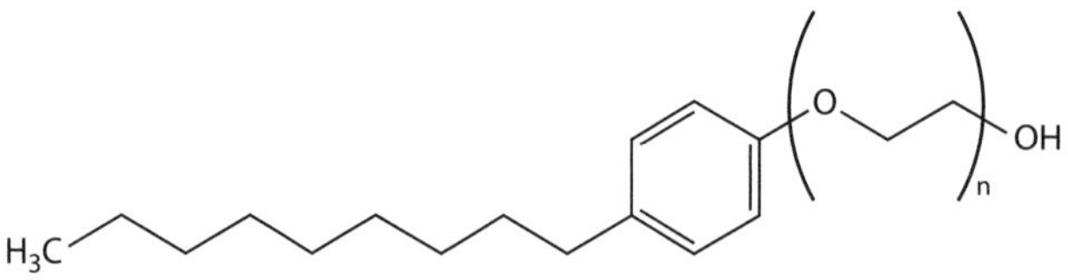

nonoxynol; non-ionic detergent

Polyoxyethyleneglycol Ethers (Polyoxyethylene ethers)

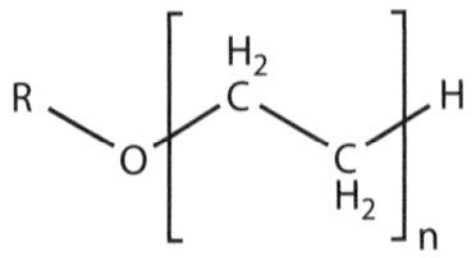

R=alkyl such as cetyl, dodecyl

Sodium dodecylsulfate, SDS, lauryl sulfate, sodium salt

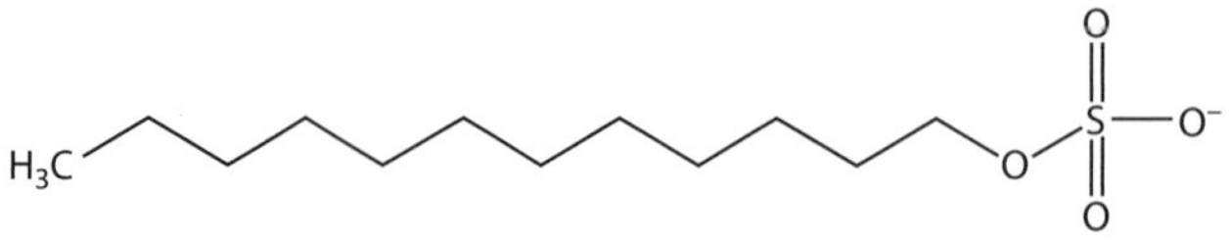

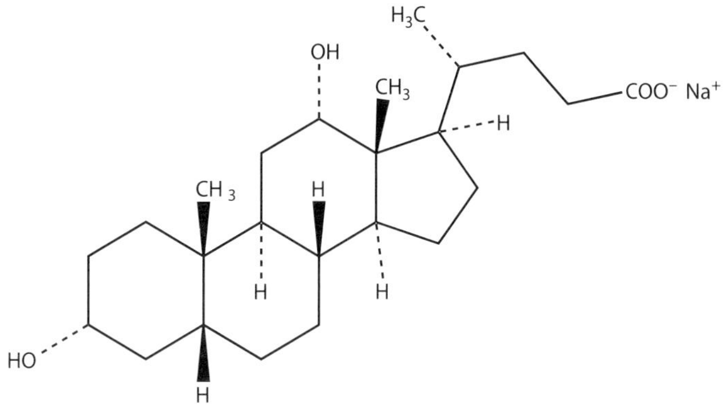

Cetylpyridinium Chloride

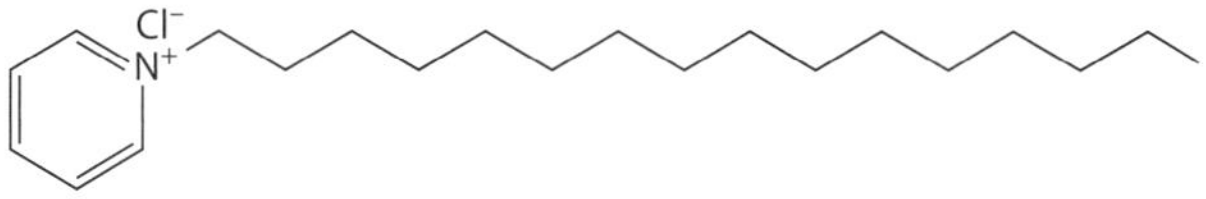

Betaine

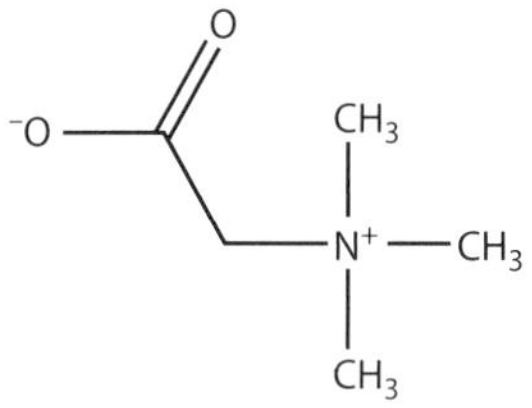

Betaine

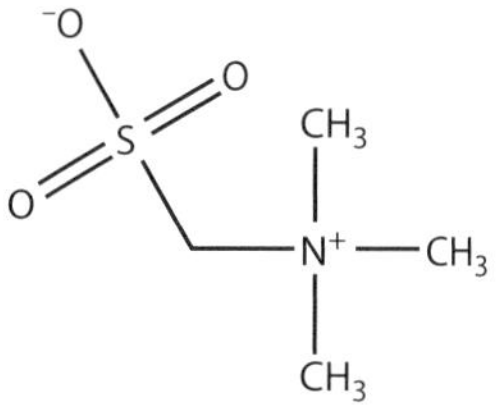

Sulfobetaine

Sulfobetaine Derivatives

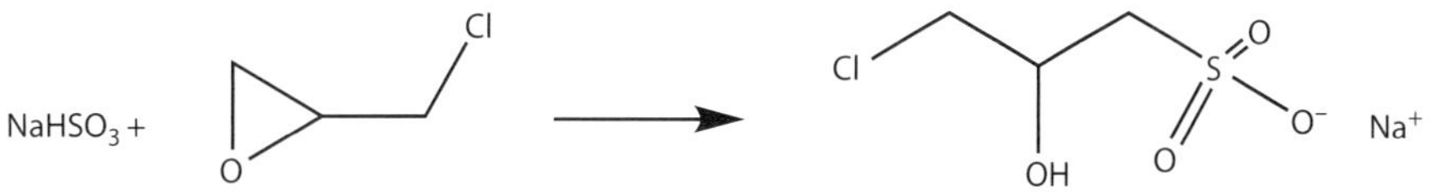

$NaHSO_3$ +

Epichlorohydrin

3-chloro-2-hydroxypropanesulfonate, sodium salt

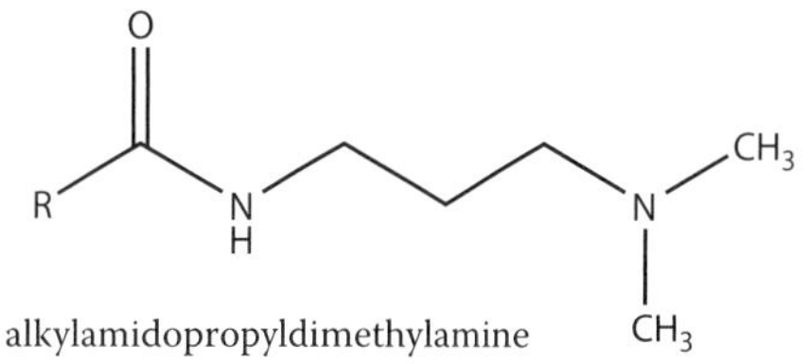

alkylamidopropyldimethylamine

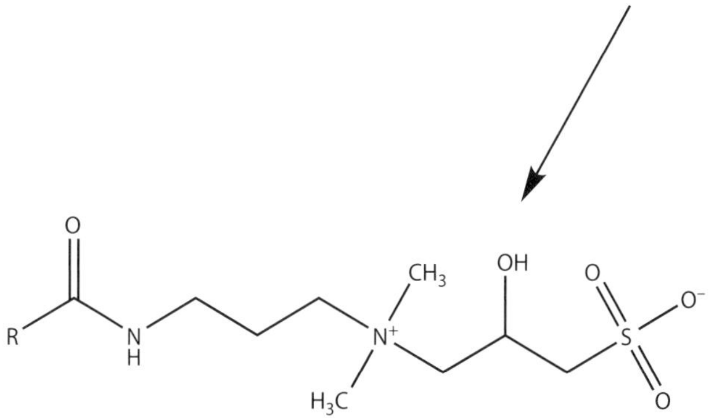

alkylsulfobetaine derivative

3-[(3-cholamidopropyl)dimethylammonio]-1-propanesulfonate (CHAPS)

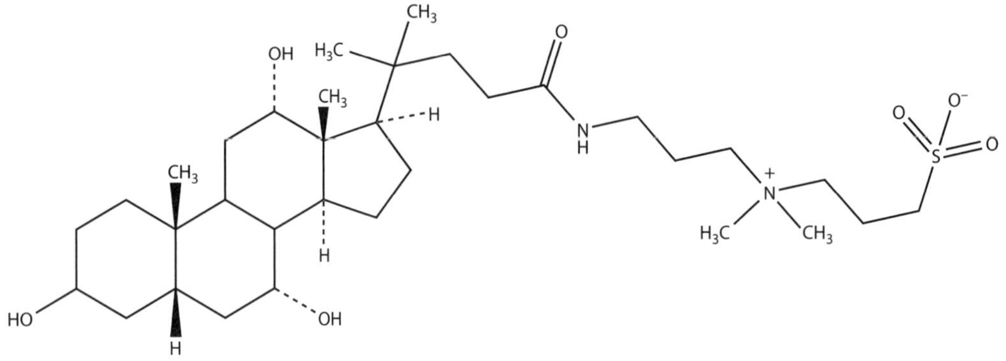

BigChap
N,N-bis(3-gluconamidopropyl)cholamide

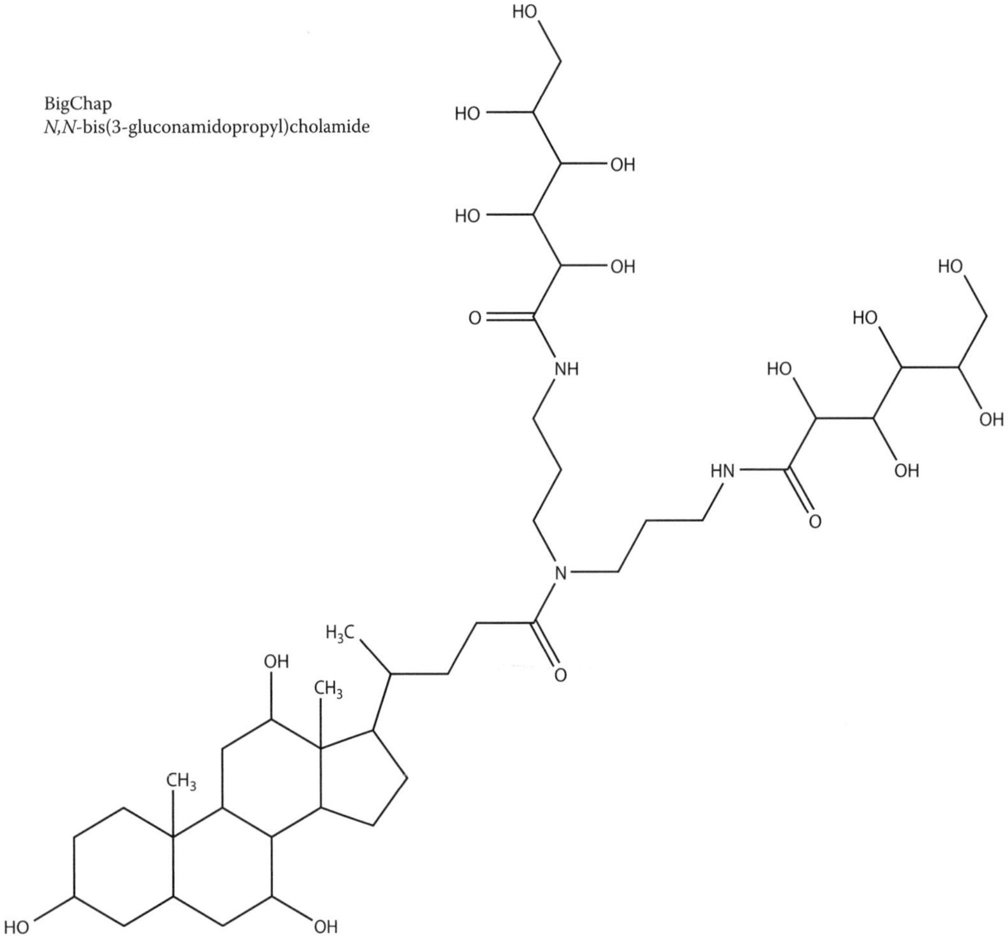

CHAPSO

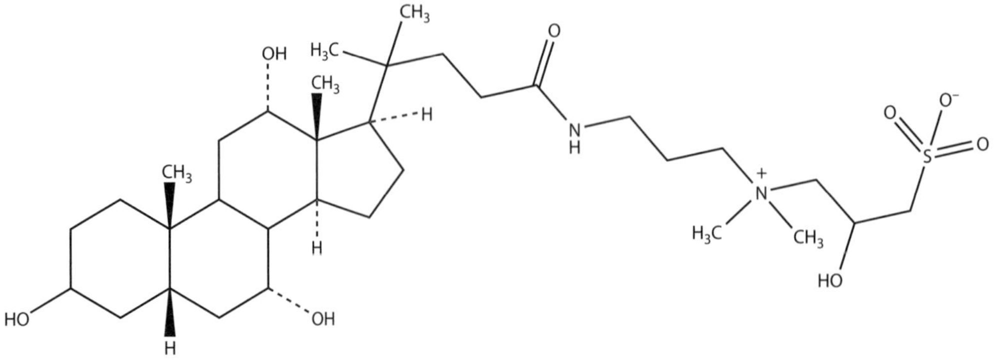

3-[(3-cholamidopropyl)dimethylammonio]-2-hydroxy-1-propanesulfonate (CHAPSO)

SOME BIOLOGICAL STAINS AND DYES

Name	Description

Acridine Orange MW 320 as the chloride hydrate

Acridine dyes are strongly yellow fluorescent dyes; stains for nucleic acids and is used for identification of the malaria parasite. Acridine orange is weakly basic, is permeable to membranes and tends to accumulate in intracellular acidic regions. Some use in photodynamic therapy for tumors. The binding of acridine orange to nucleic acids has been extensively studied.

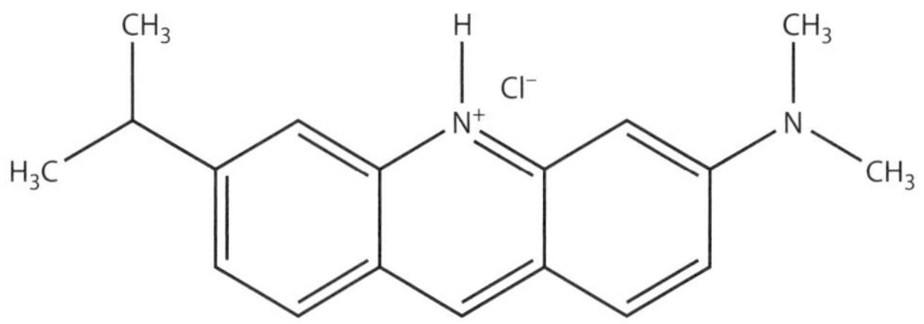

Steiner, R.F. and Beers, R.F., Jr., Spectral changes accompanying binding of acridine orange by polyadenylic acid, *Science* 127, 335-336, 1958; Mayor, H.D. and Hill, N.O., Acridine orange staining of a single-stranded DNA bacteriophage, *Virology* 14, 264-266, 1961; Boyle, R.E., Nelson, S.S., Dollish, F.R., and Olsen, M.J., The interaction of deoxyribonucleic acid and acridine orange, *Arch.Biochem.Biophys.* 96, 47-50, 1962; Leith, J.D., Jr., Acridine orange and acriflavine inhibit deoxyribonuclease action, *Biochim.Biophys.Acta.* 72, 643-644, 1963; Morgan, R.S. and Rhoads, D.G., Binding of acridine orange to yeast ribosomes, *Biochim.Biophys.Acta* 102, 311-313, 1965; Yamabe, S., A spectrophotometric study on binding of acridine orange with DNA, *Mol.Pharmacol.* 3, 556-560, 1967; Stewart, C.R., Broadening by acridine orange of the thermal transition of DNA, *Biopolymers* 6, 1737-1743, 1968; Clerc, S. and Barenholz, Y., A quantitative model for using acridine orange as a transmembrane pH gradient probe, *Anal.Biochem.* 259, 104-111, 1998; Zoccarto, F., Cavallini, L., and Alexandre, A., The pH-sensitive dye acridine orange as a tool to monitor exocytosis/endocytosis, *J.Neurochem.* 72, 625-633, 1999; Lyles, M.B., Cameron, I.L., and Rawls, H.R., Structural basis for the binding efficiency of xanthines with the DNA intercalator acridine orange, *J.Med.Chem.* 44, 4650-4660, 2001; Lyles, M.B. and Cameron, I.L., Interactions of the DNA intercalator acridine orange, with itself, with caffeine, and the double stranded DNA, *Biophys.Chem.* 96, 53-76, 2002; Keiser, J., Utzinger, J., Premji, Z. *et al.*, Acridine orange for malaria diagnosis: its diagnostic performance, its promotion and implementation in Tanzania, and the implications for malaria control, *Ann.Trop.Med.Paristol.* 96, 643-654, 2002; Lauretti, F., Lucas de Mel, F., Benati, F.J., *et al.*, Use of acridine orange staining for the detection of rotavirus RNA in polyacrylamide gels, *J.Virol.Methods* 114, 29-35, 2003; Ueda, H.., Murata, H., Takeshita, H., *et al.*, Unfiltered xenon light is useful for photodynamic therapy with acridine orange, *Anticancer Res.* 25, 3979-3983, 2005; Wang, F., Yang, J., Wu, X., *et al.*, Improvement of the acridine orange-protein-surfactant system for protein estimation based on aromatic ring stacking effect of sodium dodecyl benzene sulphonate, *Luminescence* 21, 186-194, 2006; Hiruma, H., Katakura, T., Takenami, T., *et al.*, Vesicle disruption, plasma membrane bleb formation, and acute cell death caused by illumination with blue light in acridine orange-loaded malignant melanoma cells, *J.Photochem.Photobiol.B* 86, 1-8, 2007.

Alizarin Blue MW 292   a carbonyl dye; used as a pH indicator and a stain for copper

Meloan, S.N. and Puchtler, H., Iron alizarin blue S stain for nuclei, *Stain Technol.* 49, 301-304, 1974; Rosenthal, A.R. and Appleton, B., Histochemical localization of intraocular copper foreign bodies, *Am.J.Ophthalmol.* 79, 613-625, 1975; Rao, N.A., Tso, M.O., and Rosenthal, A.R., Chalcosis in the human eye. A clinicopathologic study, *Arch Ophthalmol.* 94, 1379-1384, 1976; Amin, A.S. and Dessouki, H.A., Facile colorimetric methods for the quantitative determination of tetramisole hydrochloride, *Spectrochim.Acta A. Mol.Biomol.Spectros.* 58, 2541-2546, 2002

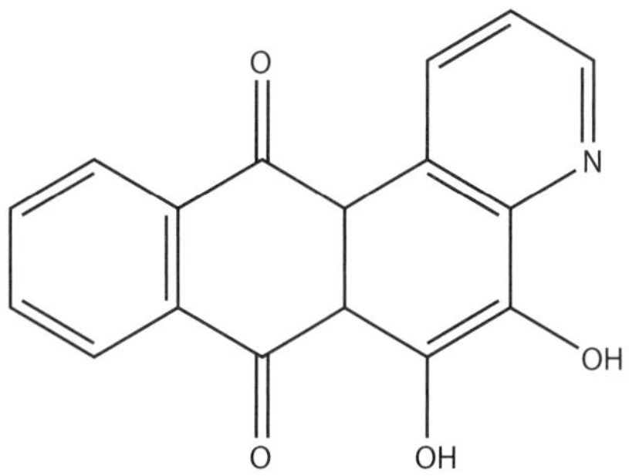

SOME BIOLOGICAL STAINS AND DYES (Continued)

Name	Description

Amido Black 10B(Naphthal Blue
Black; Amido Schwarz) MW 617
as sodium salt

Protein staining; originally developed as stain for collagen; protein determination.

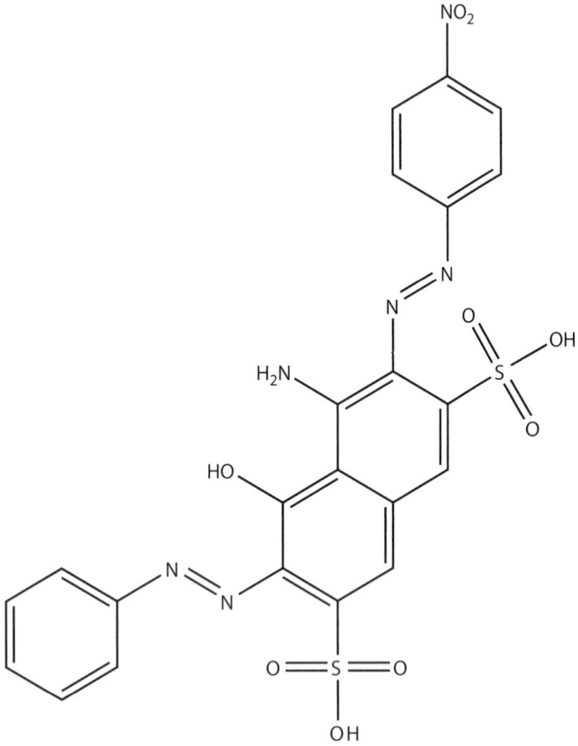

Mundkar, B. and Brauer, B., Selective localization of nucleolar protein with amido black 10B, *J.Histochem.Cytochem.* 14, 94-103, 1966; Mundkar, B. and Greenwood, H., Amido black 10B as a nucleolar stain for lymph nodes in Hodgkin's disease, *Acta Cytol.* 12, 218-226, 1968; Schaffner, W. and Weissmann, C., A rapid, sensitive, and specific method for the determination of protein in dilute solution, *Anal.Biochem.* 56, 502-514, 1973; Kolakowski, E., Determination of peptides in fish and fish products. Part 1. Application of amido black 10B for determination of peptides in trichloroacetic acid extracts of fish meat, *Nahrung* 18, 371-383, 1974; Kruski, A.W. and Narayan, K.A., Some quantitative aspects of the disc electrophoresis of ovalbumin using amido black 10B stain, *Anal.Biochem.* 60, 431-440, 1974; Wilson, C.M., Studies and critique of Amido Black 10B, Coomassie Blue R, and Fast Green FCT as stains for protein after polyacrylamide gel electrophoresis, *Anal.Biochem.* 96, 263-278, 1979; Kaplan, R.S. and Pedersen, P.L., Determination of microgram quantities of protein in the presence of milligram levels of lipid with amido black 10B, *Anal.Biochem.* 150, 97-104, 1985; Nettleton, G.S., Johnson, L.R., and Sehlinger, T.E., Thin layer chromatography of commercial samples of amido black 10B, *Stain Technol.* 61, 329-336, 1986; Tumakov, S.A., Elanskaia, L.N., Esin, M.S., and Drozdova, N.I., Quantitative determination of protein in small volumes of biological substances using amido black 10B(article in Russian), *Lab.Delo.* (5), 54-56, 1988; Schulz, J., Dettlaff, S., Fritzsche, U., *et al.*, The amido black assay: a simple and quantitative multipurpose test of adhesion, proliferation, and cytotoxicity in microplate cultures of keratinocytes (HaCaT) and other cell types growing adherently or in suspension, *J.Immunol.Methods* 167, 1-13, 1994; Gentile, F., Bali, E., and Pignalosa, G., Sensitivity and applications of the nondenaturing staining of proteins on polyvinylidene difluoride membranes with Amido Black 10B in water followed by destaining in water, *Anal.Biochem.* 245, 260-262, 1997; Plekhanov, A.Y., Rapid staining of lipids on thin-layer chromatograms with amido black 10B and other water-soluble stains, *Anal.Biochem.* 271, 186-187, 1999; Butler, P.J.G., Ubarretxene-Belandia, I., Warne, T., and Tate, C.G., The *Escherichia coli* multidrug transporter EmrE is a dimer in the detergent-solubilised state, *J.Mol.Biol.* 340, 797-808,. 2004

Azure Dyes (Azure A, Azure B or
azure blue)

Cationic dyes which are used to stain nucleic acid and sulfated glycosaminoglycans such as heparin. The sensitivity for staining sulfated glycosaminoglycans is increased with the presence of silver.

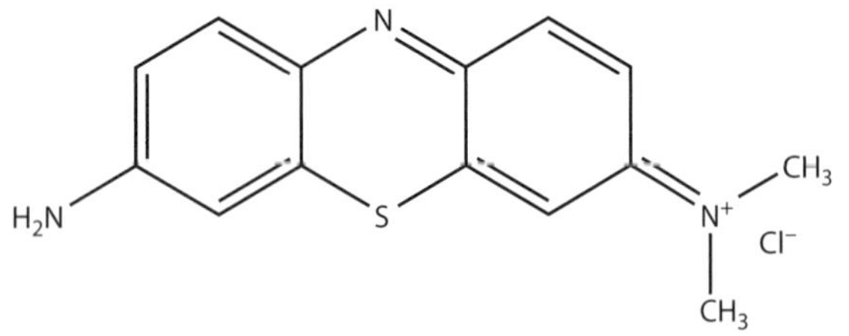

SOME BIOLOGICAL STAINS AND DYES (Continued)

Name   **Description**

Klein, F. and Szirmai, J.A., Quantitative studies on the interaction of azure A with deoxyribonucleic acid and deoxyribonucleoprotein, *Biochim.Biophys. Acta* 72, 48-61, 1963; Goldstein, D.J., A further note on the measurement of the affinity of a dye (Azure A) for histological substrates, *Q.J.Microsc.Sci.* 106, 299-306, 1965; Wollin, A. and Jaques, L.B., Analysis of heparin—azure A metachromasy in agarose gel, *Can.J.Physiol.Pharmacol.* 50, 65-71, 1972; Lohr, W., Sohmer, I., and Wittekind, D., The azure dyes: their purification and physicochemical properties. I. Purification of azure A, *Stain Technol.* 49, 359-366, 1974; Bennion, P.J., Horobin, R.W., and Murgatroyd, L.B., The use of a basic dye (azure A or toluidine blue) plus a cationic surfactant for selective staining of RNA: a technical and mechanistic study, *Stain Technol.* 50, 307-313, 1975; Dutt, M.K., Staining of DNA-phosphate groups with a mixture of azure A and acridine orange, *Microsc.Acta* 82, 285-289, 1979; Tadano-Aritomi, K., and Ishizuka, I., Determination of peracetylated sulfoglycolipids using the azure A method, *J.Lipid Res.* 24, 1368-1375, 1983; Gundry, S.R., Klein, M.D,. Drongowski, R.A., and Kirsh, M.M., Clinical evaluation of a new rapid heparin assay using the dye azure A, *Am.J.Surg.* 148, 191-194, 1984; Lyon, M. and Gallagher, J.T., A general method for the detection and mapping of submicrogram quantities of glycosaminoglycan oligosaccharides on polyacrylamide gels by sequential staining with azure A and ammoniacal silver, *Anal.Biochem.* 185, 63-70, 1990; van de Lest, C.H., Versteeg, E.M., Veerkamp, J.H. and van Kuppevelt, T.H., Quantification and characterization of glycosaminoglycans at the nanogram level by a combined azure A-silver staining in agarose gels, *Anal.Biochem.* 221, 356-361, 1994; Wang, L., Malsch, P. and Harenberg, J., Heparins, low-molecular weight heparins, and other glycosaminoglycans analyzed by agarose gel electrophoresis and azure A-silver staining, *Semin.Throm.Hemost.* 23, 11-16, 1997.

Biebrich Scarlet (Ponceau B), MW   Anionic diazo dye used as cytoplasmic stain and a stain for basic proteins. Biebrich
556 as disodium salt   scarlet also binds specifically to lysozyme and chymotrypsin in a specific manner and inhibits
                                                                      enzyme activity

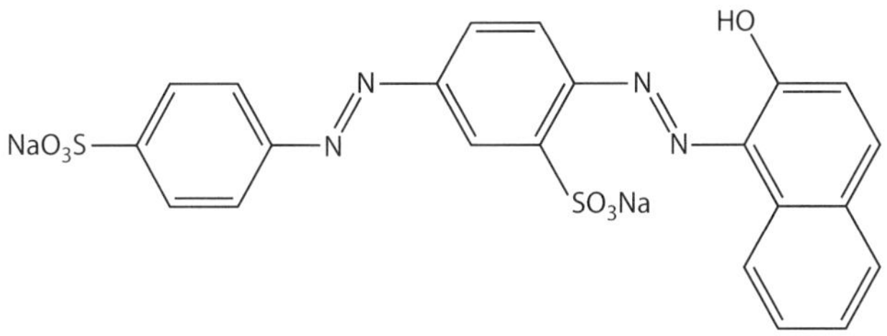

Douglas, S.D., Spicer, S.S., and Bartels, P.H., Microspectrophotometric analysis of basic protein rich sites stained with Biebrich scarlet, *J.Histochem. Cytochem.* 14, 352-360, 1966; Winkelman, J.W. and Bradley, D.F., Binding of dyes to polycations. I. Biebrich scarlet and histone interaction parameters, *Biochim.Biophys.Acta* 126, 536-539, 1966; Saint-Blancard, J., Allary, M., and Jolles, P., Influence of Biebrich scarlet on the lysis kinetics of *Micrococcus lysodeikticus* by several lysozymes, *Biochemie* 54, 1375-1376, 1972; Holler, E., Rupley, J.A., and Hess, G.P., Productive and unproductive lysozyme-chitosaccharide complexes. Equilibrium measurements, *Biochemistry* 14, 1088-1094, 1975; Giannini, I. and Grasselli, P., Proton transfer to a charged dye bound to the alpha-chymotrypsin active site studied by laser photolysis, *Biochim.Biophys.Acta* 445, 420-425, 1976; Clark, G. and Spicer, S.S., The assessing of acidophilia with Biebrich scarlet, ponceau de zylidine and woodstain scarlet, *Stain Techol.* 54, 13-16, 1979; Smith-Gill, S.J., Wilson, A.C., Potter, M., *et al.*, Mapping the antigenic epitope for a monoclonal antibody against lysozyme, *J.Immunol.* 128, 314-322, 1982; Mlynek, M.L., Comparative investigations on the specificity of Adams' reaction and the Biebrich scarlet stain for the demonstration of eosinophilic granules, *Klin. Wochenschr.* 63, 646-647, 1985; Garvey, W., Fathi, A., Bigelow, F., *et al.*, A combined elastic, fibrin and collagen stain, *Stain Technol.* 62, 365-368, 1987; Allcock, H.R. and Ambrosio, A.M., Synthesis and characterization of pH-sensitive poly(organophosphazene)hydrogels, *Biomaterials* 17, 2295-2302, 1996; Ma, F., Koike, K., Higuchi, T., *et al.*, Establishment of a GM-CSF-dependent megakaryoblastic cell line with the potential to differentiate into an eosinophilic linage in response to retinoic acids, *Br.J.Haematol.* 100, 427-435, 1998; Tan, K., Li, Y., and Huang, C., Flow-injection resonance light scattering detection of proteins at the nanogram level, *Luminescence* 20, 176-180, 2005.

SOME BIOLOGICAL STAINS AND DYES (Continued)

Name	Description
Blue tetrazolium; (Tetrazolium blue, bimethoxyneotetrazolium), MW 728 as the dichloride salt. Also nitro blue tetrazolium and tetranitro blue tetrazolium	A relatively large hydrophobic cation; histochemical stain, used for oxidoreductases. Forms a blue color in the presence of reducing agents which provided the basis for the early use of sulfydryl groups and other reducing compounds. Also used to measure free radicals, superoxide, and Amadori glycation products.

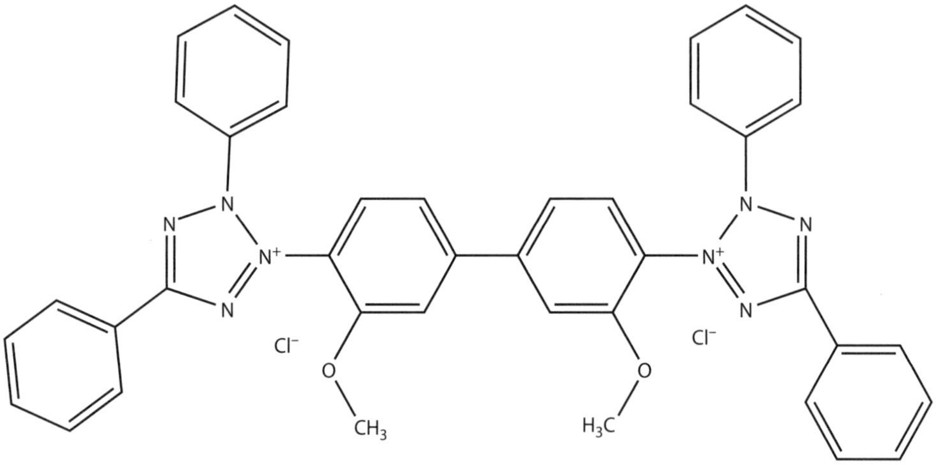

Tetrazolium Blue

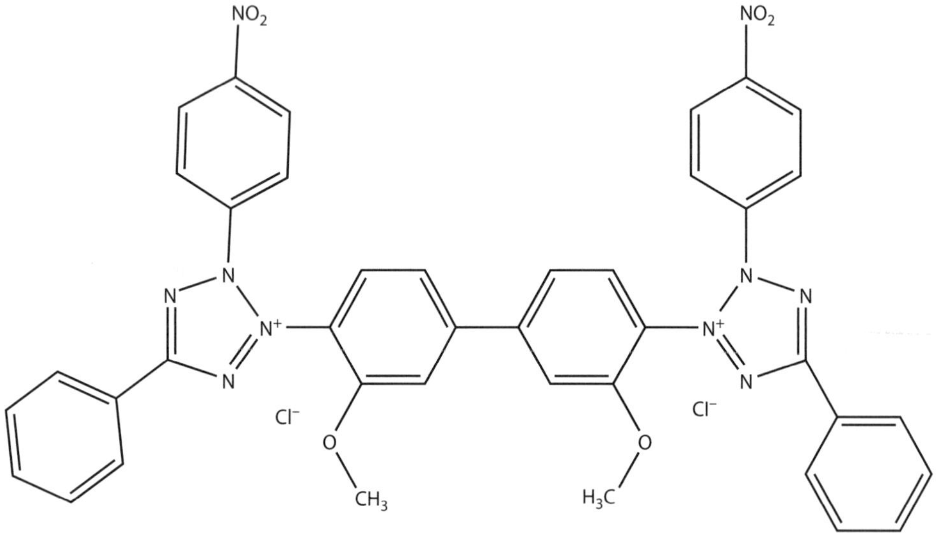

Nitro Blue Tetrazolium

Sulkowitch, H., Rutenburg, A.M., Lesses, M.F., *et al.*, Estimation of urinary reducing corticosteroids with blue tetrazolium, *N.Engl.J.Med.* 252, 1070-1075, 1955; Litteria, M. and Recknagel, R.O., A simplified blue tetrazolium reaction, *J.Lab.Clin.Med.* 48, 463-468, 1956; Leene, W. and van Iterson, W., Tetranitro—blue tetrazolium reduction in *Bacillus subtilis*, *J.Cell Biol.* 27, 237-241, 1965; Sedar, A.W. and Burde, R.M., The demonstration of the succinic dehydrogenase system in *Bacillus subtilis* using tetranitro—blue tetrazolium combined with techniques of electron microscopy, *J.Cell Biol.* 27, 53-66, 1965; Bhatnagar, R.S. and Liu, T.Z., Evidence for free radical involvement in the hydroxylation of proline: inhibition by nitro blue tetrazolium, *FEBS Lett.* 26, 32-34, 1972; Graham, R.E., Biehl, E.R., Kenner, C.T., *et al.*, Reduction of blue tetrazolium by corticosteroids, *J.Pharm.Sci.* 64, 226-230, 1975; DeChatelet, L.R. and Shirley, P.S., Effect of nitro blue tetrazolium dye on the hexose monophosphate shunt activity of human polymorphonuclear leukocytes, *Biochem.Med.* 14, 391-398, 1975; Oteiza, R.M., Wooten, R.S., Kenner, C.T., *et al.*, Kinetics and mechanism of blue tetrazolium reaction with corticosteroids, *J.Pharm.Sci.* 66, 1385-1388, 1977; DeBari, V.A. and Needle, M.A., Mechanism for transport of nitro-blue tetrazolium into viable and non-viable leukocytes, *Histochemistry* 56, 155-163, 1978; Biehl, E.R., Wooten, R., Kenner, C.T., and Graham, R.E., Kinetic and mechanistic studies of blue tetrazolium reaction with phenylhydrazines, *J.Pharm.Sci.* 67, 927-930, 1978; Schopf, R.E., Mattar, J., Meyenburg, W., *et al.*, Measurement of the respiratory burst in human monocytes and polymorphonuclear leukocytes by nitro blue tetrazolium reduction and chemiluminescence, *J.Immunol.Methods* 67, 109-117, 1984; Jue, C.K. and Lipke, P.N., Determination of reducing sugars in the nanomole range with tetrazolium blue, *J.Biochem.Biophys.Methods* 11, 109-115, 1985; Walker, S.M., Howie, A.F., and Smith, A.F., The measurement of glycosylated albumin

SOME BIOLOGICAL STAINS AND DYES (Continued)

Name	Description

by reduction of alkaline nitro-blue tetrazolium, *Clin.Chim.Acta* 156, 197-206, 1986; Ghiggeri, G.M., Candiano, G., Ginevri, F., *et al.*, Spectrophotometric determination of browning products of glycation of protein amino groups based on their reactivity with nitro blue tetrazolium salts, *Analyst* 113, 1101-1104, 1988; Brenan, M. and Bath, M.L., Indoxyl-tetranitro blue tetrazolium method for detection of alkaline phosphatase in immunohistochemistry, *J.Histochem.Cytochem.* 37, 1299-1301, 1989; Issopoulos, P.B., Sensitive colorimetric assay of cardidopa and methyldopa using tetrazolium blue chloride in pharmaceutical products, *Pharm.Weekbl.Sci.* 11, 213-217, 1989; Fattorossi, A., Nisini, R., Le Moli, S., *et al.*, Flow cytometric evaluation of nitro blue tetrazolium (NBT) reduction in human polymorphonuclear leukocytes, *Cytometry* 11, 907-912, 1990; Albiach, M.R., Guerri, J., and Moreno, P., Multiple use of blotted polyvinylidene difluoride membranes immunostained with nitro blue tetrazolium, *Anal. Biochem.* 221, 25-28, 1994; Chanine, R., Huet, M.P., Oliva, L., and Nadeau, R., Free radicals generated by electrolysis reduces nitro blue tetrazolium in isolated rat heart, *Exp.Toxicol.Pathol.* 49, 91-95, 1997.

Bromophenol Blue (bromphenol blue); tetrabromophenolsulfonphthalein, MW sultone, 670; sodium salt of sulfonic acid, 692

A vital stain; used for determination of protein, pH indicator. There are reports of specific binding to sites on proteins.

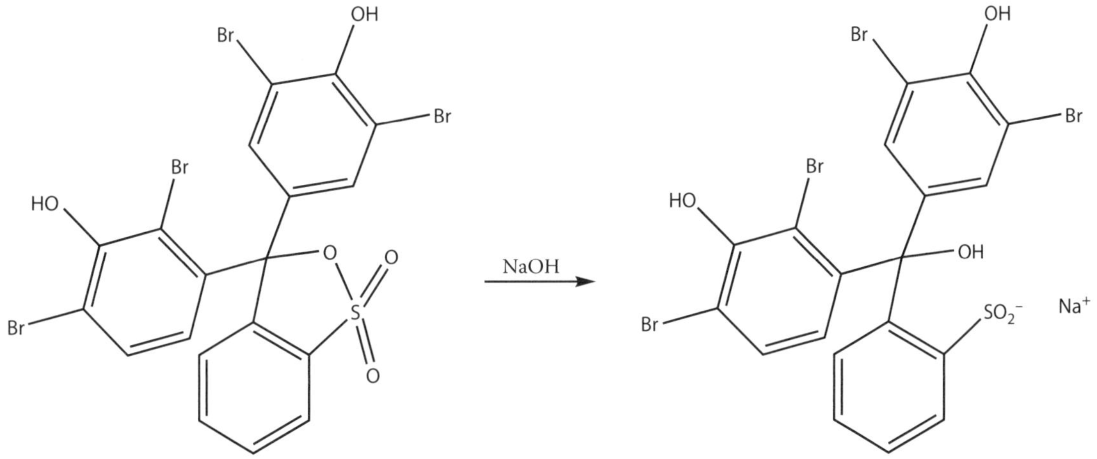

Bromophenol Blue (sultone)

Bjerrum, O.J., Interaction of bromophenol blue and bilirubin with bovine and human serum albumin determined by gel filtration, *Scand.J.Clin.Invest.* 22, 41-48, 1968; Ramalingam, K. and Ravidranath, M.H., An evaluation of the metachromasia of bromophenol blue, *Stain Technol.* 47, 179-184, 1972; Harruff, R.C. and Jenkins, W.T., The binding of bromophenol blue to aspartate aminotransferase, *Arch.Biochem.Biophys.* 176, 206-213,1976; Krishnamoorthy, G. and Prabhananda, B.S., Binding site of the dye in bromophenol blue-lysozyme complex. Protein magnetic resonance study in aqueous solutions, *Biochim.Biophys.Acta* 709, 53-57, 1982; Ahmad, H. and Saleemuddin, M., Bromophenol blue protein assay: improvement in buffer tolerance and adaptation for the measurement of proteolytic activity, *J.Biochem.Biophys.Methods* 7, 335-343, 1983; Subrahanian, M., Sheshadri, B.S., and Venkatappa, M.P., Interaction of lysozyme with dyes. II. Binding of bromophenol blue, *J.Biochem.* 96, 245-252, 1984; Ma, C.Q., Li, K.A., and Tong, S.Y., Microdetermination of proteins by resonance light scattering spectroscopy with bromophenol blue, *Anal.Biochem.* 239, 86-91, 1996; Cathey, J.C., Schmidt, C.A., and DeWoody, J.A., Incorporation of bromophenol blue enhances visibility of polyacrylamide gels, *BioTechniques* 22, 222, 1997; Trivedi, V.D., On the role of lysine residues in the bromophenol blue-albumin interaction, *Ital.J.Biochem.* 46, 67-73, 1997; Bertsch, M., Mayburd, A.L., and Kassner, R.J., The identification of hydrophobic sites on the surface of proteins using absorption difference spectroscopy of bromophenol blue, *Anal.Biochem.* 313, 187-195, 2003; Li, J., Chatterjee, K., Medek, A., *et al.*, Acid-base characteristics of bromophenol blue-citrate buffer systems in the amorphous state, *J.Pharm.Sci.* 93, 697-712, 2004; Sarma, S. and Dutta, R.K., Electronic spectral behavior of bromophenol blue in oil in water microemulsions stabilized by sodium dodecyl sulfate and *n*-butanol, *Spectrochim.Acta A Mol.Biomol. Spectrosc.* 64, 623-627, 2006; You, L., Wu, Z., Kim, T., and Lee, K., Kinetics and thermodynamics of bromophenol blue adsorption by a mesoporous hybrid gel derived from tetraethoxysilane and bis(trimethoxysilyl)hexane, *J.Colloid Interface Sci.* 300, 526-535, 2006; Zeroual, Y., Kim, B.S., Kim, C.S., *et al.*, A comparative study on biosorption characteristics of certain fungi for bromophenol blue dye, *Appl.Biochem.Biotechnol.* 134, 51-60, 2006

SOME BIOLOGICAL STAINS AND DYES (Continued)

Name	Description
Bromothymol Blue or Bromthymol Blue; dibromothymolsulfonphthalein, MW as sultone, 624; as sodium salt, 646	A lipophilic dye; serves as a vital stain to trace fluid movements; used as an ion pair reagent for the measurement of certain drugs; extensive use as pH indicator as there is a change from yellow to blue as pH changes from 6 to 8.0

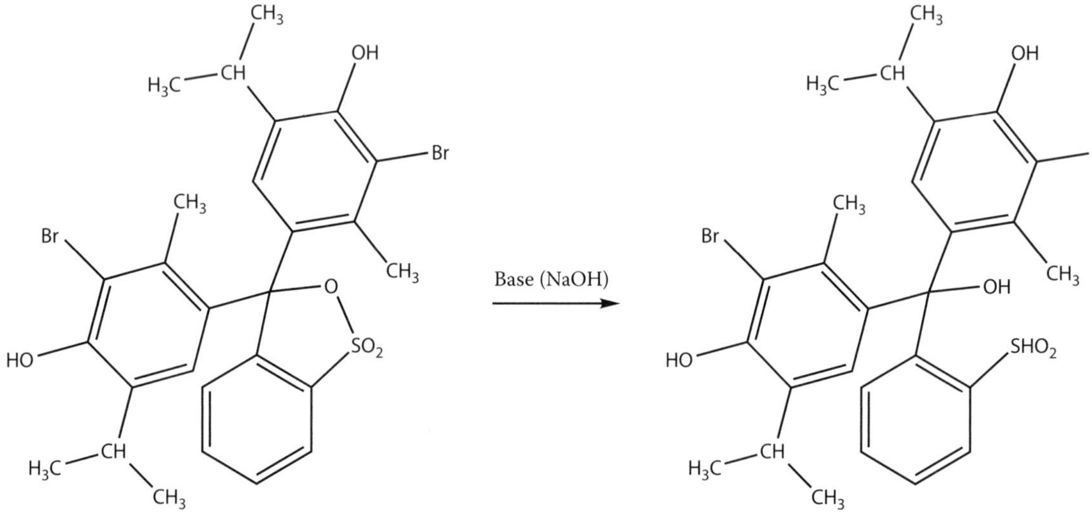

Azzone, G.F., Piemonte, G., and Massari, S., Intramembrane pH changes and bromthymol blue translocation in liver mitochondria, *Eur.J.Biochem.* 6, 207-212, 1968; Mitchell, P., Moyle, J., and Smith, L., Bromthymol blue as a pH indicator in mitochondrial suspensions, *Eur.J.Biochem.* 4, 9-19, 1968; Das Gupta, V. and Cadwallader, D.E., Determination of first pKa' value and partition coefficients of bromthymol blue, *J.Pharm.Sci.* 57, 2140-2142, 1968; Jackson, J.B. and Crofts, A.R., Bromothymol blue and bromocresol purple as indicators of pH changes in chromatophores of *Rhodospirillum rubrum*, *Eur.J.Biochem.* 10,226-237, 1969; Gromet-Elhanan, Z. and Briller, S., On the use of bromthymol blue as an indicator of internal pH changes in chromatophores from *Rhodospirillum rubrum*, *Biochem.Biophys.Res.Commun.* 37, 261-265, 1969; Smith, L., Bromthymol blue as a pH indicator in mitochondrial suspensions, *Ann.N.Y.Acad.Sci* 147, 856, 1969; Lowry, J.B., Direct spectrophotometric assay of quaternary ammonium compounds using bromthymol blue, *J.Pharm.Sci.* 68, 110-111, 1979; Mashimo, T., Ueda, I., Shieh, D.D., *et al.*, Hydrophilic region of lecithin membranes studied by bromothymol blue and effects of an inhalation anesthetic, enflurane, *Proc.Natl.Acad.Sci.USA* 76, 5114-5118, 1979; Yamamoto, A., Utsumi, E., Sakane, T. *et al.*, Immunological control of drug absorption from the gastrointestinal tract: the mechanism whereby intestinal anaphylaxis interferes with the intestinal absorption of bromthymol blue in the rat, *J.Pharm.Pharmacol.* 38, 357-362, 1986; Dean, V.S., Dingley, J., and Vaughan, R.S., The use of bromothymol blue and sodium thiopentone to confirm tracheal intubation, *Anaesthesia* 51, 29-32, 1996; Gorbenko, G.P., Bromothymol blue as a probe for structural changes of model membranes induced by hemoglobin, *Biochim.Biophys.Acta* 1370, 107-118, 1998; Ramesh, K.C., Gowda, B.G., Melwanki, M.B., *et al.*, Extractive spectrophotometric determination of antiallergic drugs in pharmaceutical formulations using bromopyrogallol and bromothymol blue, *Anal.Sci.* 17, 1101-1103, 2001; Rahman, N., Ahmed Khan, N. and Hejaz Azmi, S.N., Extractive spectrophotometric methods for the determination of nifedipine in pharmaceutical formulations using bromocresol green, bromophenol blue, bromothymol blue and eriochrome black T, *Farmaco* 59, 47-54, 2004; Erk, N., Spectrophotometric determination of indinavir in bulk and pharmaceutical formulations using bromocresol purple and bromothymol blue, *Pharmazie* 59, 183-186, 2004;

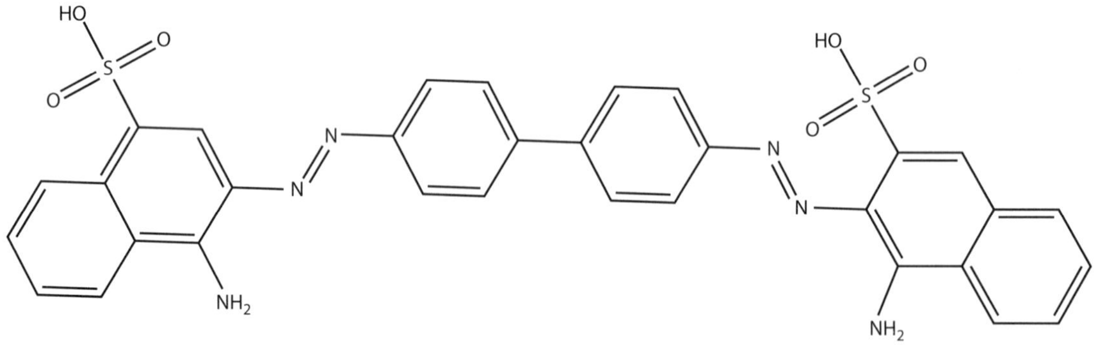

SOME BIOLOGICAL STAINS AND DYES (Continued)

Name	Description

Congo Red (C.I. direct red 28); 696.7 as the disodium salt — Developed as acid-base indicator (Congo Red paper, Riegel's paper); more recent use to detect amyloid peptide aggregates and other fibril structures (crossed β structures); also cellulose surfaces. See also thioflavin T

Glenner, G.G., The basis of the staining of amyloid fibers: their physico-chemical nature and the mechanism of their dye-substrate interaction, *Prog. Histochem.Cytochem.* 13, 1-37, 1981; Elghetany, M.T. and Saleem, A., Methods for staining amyloid in tissues: a review, *Stain Technol.* 63, 201-212, 1988; Lorenzo, A. and Yankner, B.A., Amyloid fibril toxicity in Alzheimer's disease and diabetes, *Ann.N.Y.Acad.Sci.* 777, 89-95, 1996; Sipe, J.D. and Cohen, A.S., Review: history of the amyloid fibril, *J.Struct.Biol.* 130, 88098, 2000; Piekarska, B., Konieczny, L, Rybarska, J., *et al.*, Intramolecular signaling in immunoglobulins—new evidence emerging from the use of supramolecular protein ligands, *J.Physiol.Pharmacol.* 55, 487-501, 2004; Nilsson, M.R., Techniques to study amyloid fibril formation in vitro, *Methods* 34, 151-160, 2004; Ho, M.R., Lou, Y.C., Lin, W.C., *et al.*, Human pancreatitis-associated protein forms fibrillar aggregates with a native-like conformation, *J.Biol.Chem.* 281, 33566-33576, 2006; Hatters, D.M., Zhong, N., Rutenber, E., and Weisgraber, K.H., Amino-terminal domain stability mediates apolipoprotein E aggregation into neurotoxic fibrils, *J.Mol.Biol.* 361, 932-944, 2006; McLaughlin, R.W., De Stigter, J.K., Sikkink, L.A., *et al.*, The effects of sodium sulfate, glycosaminoglycans, and Congo red on the structure, stability, and amyloid formation of an immunoglobulin light-chain protein, *Protein Sci.* 15, 1710-1722, 2006; Sladewski, T.E., Shafer, A.M., and Hoag, C.M., The effect of ionic strength on the UV-vis spectrum of congo red in aqueous solution, *Spectrochim.Acta A Mol.Biomol.Spectrosc.* 65, 985-987, 2006; Eisert, R., Felau, L., and Brown, L.R., Methods for enhancing the accuracy and reproducibility of Congo Red and thioflavin T assays, *Anal.Biochem.* 353, 144-146, 2006; Sereikaite, J. and Bumelis, V.A., Congo red interaction with α-proteins, *Acta Biochim.Pol.* 53, 87-92, 2006; Frid, P., Anisimov, S.V. and Popovic, N., Congo red and protein aggregation in neurodegenerative diseases, *Brain Res.Brain Res.Rev.* 53, 135-160, 2007; Goodrich, J.D. and Winter, W.T., Alpha-chitin nanocrystals prepared from shrimp shells and their specific surface area measurement, *Biomacromolecules* 8, 252-257, 2007; Lencki, R.W., Evidence for fibril-like structure in bovine casein micelles, *J.Dairy Sci.* 90, 75-89, 2007;

Diaminobenzidine (DAB); 3,3'-dimethyl-aminobenzidine; 3,3',4,4'-tetraaminobiphenyl; MW 214; 360 as the tetrahydrochloride — Histochemical demonstration of peroxidases, oxidases, catalases where DAB serves as electron acceptor forming a polymeric brown product; also used for Western blotting.

Seligman, A.M., Karnovsky, M.J., Wasserkrug, H.L., and Hanker, J.S., Nondroplet ultrastructural demonstration of cytochrome oxidase activity with a polymerizing osmiophilic reagent, diaminobenzidine (DAB), *J.Cell Biol.* 38, 1-14, 1968; Novikoff, A.B., and Goldfischer, S., Visualization of peroxisomes (microbodies) and mitochondria with diaminobenzidine, *J.Histochem.Cytochem.* 17, 675-680, 1969; Ekes, M., The use of diaminobenzidine (DAB) for the histochemical demonstration of cytochrome oxidase activity in unfixed plant tissues, *Histochemie* 27, 103-108, 1971; Herzog, V. and Fahimi, H.D., A new sensitive colorimetric assay for peroxidase using 3,3'-diaminobenzidine as hydrogen donor, *Anal.Biochem.* 55, 554-562, 1973; Nishimura, E.T. and Cooper, C., Peroxidatic reaction of catalase-antibody complex of leukocyte demonstrated by diaminobenzidine, *Cancer Res.* 34, 2386-2392, 1974; Pelliniemi, L.J., Dym, M., and Karnovsky, M.J., Peroxidase histochemistry using diaminobenzidine tetrahydrochloride stored as a frozen solution, *J.Histochem.Cytochem.* 28, 191-192, 1980; van Bogaert, L.J., Quinones, J.A., and van Craynest, M.P., Difficulties involved in diaminobenzidine histochemistry of endogenous peroxidase, *Acta Histochem.* 67, 180-194, 1980; Perotti, M.E., Anderson, W.A., and Swift, H., Quantitative cytochemistry of the diaminobenzidine cytochrome oxidase reaction product in mitochondria of cardiac muscle and pancreas, *J.Histochem.Cytochem.* 31, 351-365, 1983; Bosman, F.T., Some recent developments in immunocytochemistry, *Histochem.J.* 15, 189-200, 1983; Deimann, W., Endogenous peroxidase activity in mononuclear phagocytes, *Prog. Histochem.Cytochem.* 15, 1-58, 1984; Kugler, P., Enzyme histochemical applied in the brain, *Eur.J.Morphol.* 28, 109-120, 1990; Deitch, J.S., Smith, K.L., Swann, J.W., and Turner, J.N., Parameters affecting imaging of the horseradish-peroxidase-diaminobenzidine reaction product in the confocal scanning laser microscope, *J.Microsc.* 160, 265-278, 1990; Ludany, A.,Gallyas, F., Gaszner, B. *et al.*, Skimmed-milk blocking improves silver staining post-intensification of peroxidase-diaminobenzidine staining on nitrocellulose membrane in immunoblotting, *Electrophoresis* 14, 78-80, 1993; Fritz, P., Wu, X., Tuczek, H., *et al.*, Quantitation in immunohistochemistry. A research method or a diagnostic tool in surgical pathology?, *Pathologica* 87, 300-309, 1995; Werner, M., Von Wasielewski, R. and Komminoth, P., Antigen retrieval, signal amplification and intensification in immunohistochemistry, *Histochem.Cell Biol.* 105, 253-260, 1996; Horn, H., Safe diaminobenzidine (DAB) disposal, *Biotech.Histochem.* 77, 229, 2002; Kiernan, J.A., Stability and solubility of 3,3'-diaminobenzidine (DAB), *Biotech. Histochem.* 78, 135, 2003; Rimm, D.L., What brown cannot do for you, *Nat.Biotechnol.* 24, 914-916, 2006

SOME BIOLOGICAL STAINS AND DYES (Continued)

Name	Description
Dichlorofluorescin diacetate (2′,7′-dichlorofluorescin diacetate) MW 487	Histochemical demonstration of peroxidases and esterases; detection of reactive oxygen species (ROS). Not be confused with the fluorescein derivatives

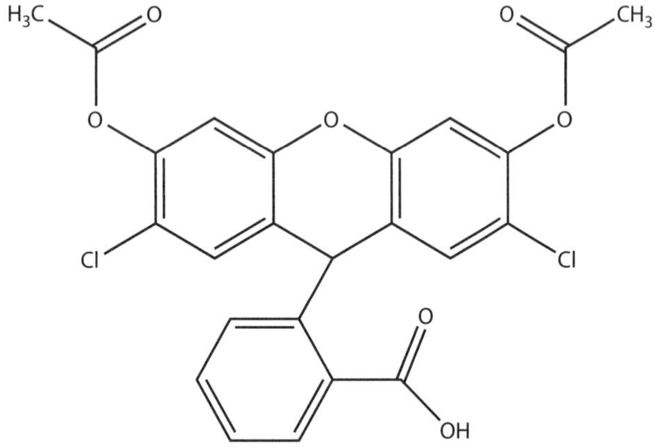

Hassan, N.F., Campbell, D.E., and Douglas, S.D., Phorbol myristate acetate induced oxidation of 2′,7′-dichlorofluorescin by neutrophils from patients with chronic granulomatous disease, *J.Leukoc.Biol.* 43, 317-322, 1988; Rosenkranz, A.R., Schmaldienst, S., Stuhlmeier, K.M., *et al.*, A microplate assay for the detection of oxidative products using 2′,7′-dichlorofluorescin diacetate, *J.Immunol.Methods* 156, 39-45, 1992; Royall, J.A. and Ischiropoulos, H., Evaluation of 2′,7′-dichlorofluorescin and dihydrorhodamine 123 as fluorescent probes for intracellular H_2O_2 in cultured endothelial cells, *Arch.Biochem.Biophys.* 302, 348-355, 1993; Kooy, N.W., Royall, J.A., and Ishiropoulos, H., Oxidation of 2′,7′-dichlorofluorescin by peroxynitrite, *Free Radic.Res.* 27, 245-254, 1997; van Reyk, D.M., King, N.J., Dinauer, M.C., and Hunt, N.M., The intracellular oxidation of 2′.7′-dichlorofluorescin in murine T lymphocytes, *Free Rad.Biol.Med.* 30, 82-88, 2001; Burkitt, M.J. and Wardman, P., Cytochrome C is a potent catalyst of dichlorofluorescin oxidation: implications for the role of reactive oxygen species in apoptosis, *Biochem.Biophys.Res.Commun.* 282, 329-333, 2001; Chignell, C.F. and Sik, R.H., A photochemical study of cells loaded with 2′,7′-dichlorofluorescin: implications for the detection of reactive oxygen species generated during UVA irradiation, *Free Rad.Biol.Med.* 34, 1029-1034, 2003; Afzal, M. Matsugo, S., Sesai, M. *et al.*, Method to overcome photoreaction , a serious drawback to the use of dichlorofluorescin in evaluation of reactive oxygen species, *Biochem. Biophys.Res. Commun.* 304, 619-624, 2003; Lawrence, A., Jones, C.M., Wardman, P., and Burkitt, M.J., Evidence for the role of a peroxidase compound I-type intermediate in the oxidation of glutathione, NADH, ascorbate, and dichlorofluorescein by cytochrome c/H_2O_2. Implications for oxidative stress during apoptosis, *J.Biol.Chem.* 278, 29410-29419, 2003; Myhre, O., Andersen, J.M., Aarnes, H., and Fonnum, F., Evaluation of the probes 2′,7′-dichlorofluorescin diacetate, luminol, and lucigenin as indicators of reactive species formation, *Biochem.Pharmacol.* 65, 1575-1582, 2003; Laggner, H., Hermann, M., Gmeiner, B.M., and Kapiotis, S., Cu^{2+} and Cu^+ bathocuproine disulfonate complexes promote he oxidation of the ROS-detecting compound dichlorofluorescin (DCFH), *Anal.Bioanal.Chem.* 385, 959-961, 2006; Matsugo, S., Sasai, M., Shinmori, H. *et al.*, Generation of a novel fluorescent product, monochlorofluorescein from dichlorofluorescin by photo-irradiation, *Free Radic.Res.* 40, 959-965, 2006

Name	Description
3,3′-Dimethyl-9-methyl-4,5,4′,5′-dibenzothiacarbocyanine; DBTC; "Stains all"; MW 560	A cationic dye with a broad specificity for interaction including glycosaminoglycans and nucleic acids. DBTC has been used to stain for calmodulin and other calcium-binding proteins.

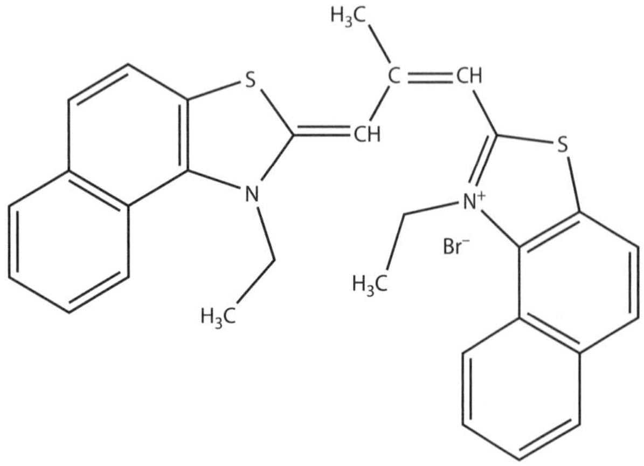

SOME BIOLOGICAL STAINS AND DYES (Continued)

Name	Description

Scheres, J.M., Production of C and T bands in human chromosomes after heat treatment at high pH and staining with "stains-all", *Humangenetik*, 23, 311-314, 1974; Green, M.R. and Pastewka, J.V., The cationic carbocyanine dyes Stains-all, DBTC, and Ethyl-stains-all, DBTC-3,3'9-triethyl, *J.Histochem.Cytochem.* 27, 797-799, 1979; Caday, C.G. and Steiner, R.F., The interaction of calmodulin with the carbocyanine dye (Stains all), *J.Biol. Chem.* 260, 5985-5990, 1985; Caday, C.G., Lambooy, P.K., and Steiner, R.F., The interaction of Ca^{2+}-binding proteins with the carbocyanine dye stains-all, *Biopolymers* 25, 1579-1595, 1986; Sharma, Y., Rao, C.M. Rao, S.C., *et al.*, Binding site conformation dictates the color of the dye stains-all. A study of the binding of this dye to the eye lens proteins crystallins, *J.Biol.Chem.* 264, 20923-20927, 1989; Lu, M., Guo, Q., Seeman, N.C. and Kallenbach, N.R., Drug binding by branched DNA: selective interaction of the dye stains-all with an immobile junction, *Biochemistry* 29, 3407-3412, 1990; Nakamura, K., Masuyama, E., Wada, S., and Okuno, M., Applications of stains' all staining to the analysis of axonemal tubulins: identification of beta-tubulin and beta-isotubulins, *J.Biochem.Biophys.Methods* 21, 237-245, 1990; Gruber, H.E. and Mekikian, P., Application of stains-all for demarcation of cement lines in methacrylate embedded bone, *Biotech.Histochem.* 66, 181-184, 1991; Sharma, Y., Gapalakrishna, A., Balasubramanian, D., *et al.*, Studies on the interaction of the dye, stains-all, with individual calcium-binding domains of calmodulin, *FEBS Lett.* 326, 59-64, 1993; Lee, H.G. and Cowman, K., An agarose gel electrophoretic method of analysis of hyaluronan molecular weight distribution, *Anal.Biochem.* 219, 278-287, 1994; Myers, J.M., Veis, A., Sabsay, B., and Wheeler, A.P., A method for enhancing the sensitivity and stability of stains-all for Phosphoproteins separated in sodium dodecyl sulfate-polyacrylamide gels, *Anal.Biochem.* 240, 300-302, 1996; Goldberg, H.A. and Warner, K.J., The staining of acidic proteins on polyacrylamide gels: enhanced sensitivity and stability of "Stains All" staining in combination with silver nitrate, *Anal.Biochem.* 251, 227-233, 1997; Volpi, N., and Maccari, F., Detection of submicrogram quantities of glycosaminoglycans on agarose gels by sequential staining with toluidine blue and Stains All, *Electrophoresis* 23, 4060-4066, 2002; Volpi, N., Macari, F., and Titze, J., Simultaneous detection of submicrogram quantities of hyaluronic acid and dermatan sulfate on agarose-gel by sequential staining with toluidine blue and Stains All, *J.Chromatog.B. Analyt. Technol. Biomed.Life Sci.* 820, 131-135, 2005

Evans Blue (C.I. Direct Blue 53); MW 961 as tetrasodium salt.

Early use as a method for determining blood volume; histochemical use as a protein stain, extensive use as a vital stain; more recent use to demonstrate vascular leakage and surgical dye.

Gibson, J.G., and Evans, W.A., Clinical studies of the blood volume. I. Clinical application of a method employing the azo dye "Evans Blue" and the spectrophotometer, *J.Clin.Invest.* 16, 301-316, 1937; Morris, C.J., The determination of plasma volume by the Evans blue method: the analysis of haemolyzed plasma, *J.Physiol.* 102, 441-445, 1944; Morris, C.J., Chromatographic determination of Evans blue in plasma and serum, *Biochem.J.* 38, 203-204, 1944; McCord, W.M. and Ezell, H.K., Cell volume determinations with Evans blue, *Proc.Soc.Exptl.Biol.Med.* 76, 727-728, 1951; Caster, W.O., Simon, A.B., and Armstrong, W.D., Evans blue space in tissues of the rat, *Am.J.Physiol.* 183, 317-321, 1955; Clausen, D.F. and Lifson, N., Determination of Evans blue dye in blood in tissues, *Proc.Soc.Exp.Biol.Med.* 91, 11-14, 1956; Larsen, O.A. and Jarnum, S., The Evans Blue test in amyloidosis, *Scand.J.Clin.Lab.Invest.* 17, 287-294, 1965; Rabinovitz, MK. and Schen, R.J., The characteristics of certain alpha globulins in immunoelectrophoresis of human serum, using Evans blue dye, *Clin.Chim.Acta* 17, 499-503, 1967; Crippen, R.W. and Perrier, J.L., The use of neutral red and Evans blue for live-dead determination of marine plankton (with comments on the use of rotenone for inhibition of grazing), *Stain Technol.* 49, 97-104, 1974; Fry, D.L., Mahley, R.W., Weisgraber, K.H. and Oh, S.Y., Simultaneous accumulation of Evans blue dye and albumin in the canine aortic wall, *Am.J.Physiol.* 233, H66-H79, 1977; Shoemaker, K., Rubin, J., Zumbro, G.L., and Tackett, R., Evans blue and gentian violet: alternatives to methylene blue as a surgical marker dye, *J.Thorac.Cardiovasc.Surg.* 112, 542-544, 1996; Skowronek, M., Roterman, Konieczny, L., *et al.*, The conformational characteristics of Congo Red, Evans blue and Trypan blue, *Comput.Chem.* 24, 429-450, 2000; Hamer, P.W., McGeachie, J.M., Davies, M.J., and Grounds, M.D., Evans Blue dye as an *in vivo* marker of myofibre damage: optimizing parameters for detecting initial myofibre membrane permeability, *J.Anat.* 200, 69-79, 2002; Kaptanoglu, E., Okutan, O., Akbiyik, F., *et al.*, Correlation of injury severity and tissue Evans glue content, lipid peroxidation and clinical evaluation in acute spinal cord injury in rats, *J.Clin.Neurosci.* 11, 879-885, 2004; Green, M., Frashid, G., Kollias, J. *et al.*, The tissue distribution of Evans blue dye in a sheep model of sentinel node biopsy, *Nucl.Med.Commun.* 27, 695-700, 2006;

SOME BIOLOGICAL STAINS AND DYES (Continued)

Name	Description
Fast Green (Fast Green FCF; C.I. food green 3) M.W. 809 as the trisodium salt	A histochemical cytoplasmic counterstain; a stain for protein; used to demonstrate histones; marker dye; a food dye FD & C fast green 3).

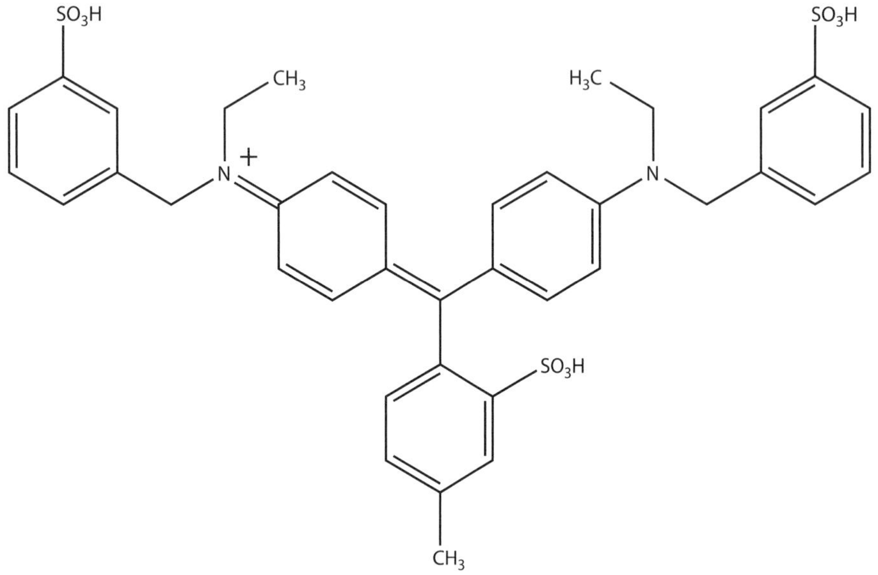

Bryan, J.H., Differential staining with a mixture of safranin and fast green FCF, *Stain Technol.* 30, 153-157, 1955; Garcia, A.M., Studies on deoxyribonucleoprotein in leukocytes and related cells of mammals. VII. The fast green histone content of rabbit leukocytes after hypertonic treatment, *Stain Technol.* 30, 153-157, 1955; Hunt, D.E. and Caldwell, R.C., Use of fast green in agar-diffusion microbiological assays, *Appl.Microbiol.* 18, 1098-1099, 1969; Gorovsky, M.A., Carlson, K., and Rosenbaum, J.L., Simple method for quantitative densitometry of polyacrylamide gels using fast green, *Anal. Biochem.* 35, 359-370, 1970; Entwhistle, K.W., Congo red-fast green fcf as a supra-vital stain for ram and bull spermatozoa, *Aust.Vet.J.* 48, 515-519, 1972; Noeske, K., Discrepancies between cytophotometric alkaline Fast Green measurements and nuclear histone protein content, *Histochem.J.* 5, 303-311, 1973; McMaster-Kaye, R. and Kaye, J.S., Staining of histones on polyacrylamide gels with amido blank and fast green, *Anal. Biochem.* 61, 120-132, 1974; Medugorac, I., Quantitative determination of cardiac myosin subunits stained with fast green in SDS-electrophoretic gels, *Basic Res. Cardiol.* 74, 406-416, 1979; Glimore, L.B. and Hook, G.E., Quantitation of specific proteins in polyacrylamide gels by the elution of Fast Green FCF, *J.Biochem.Biophys.Methods* 5, 57-66, 1981; Smit, E.F., de Vries, E.G., Meijer, C., *et al.*, Limitations of the fast green assay for chemosensitivity testing in human lung cancer, *Chest* 100, 1358-1363, 1991; Li, Y.F., Huang, C.Z., and Li, M., A resonance light-scattering determination of proteins with fast green FCF, *Anal.Sci.*18, 177-181, 2002; Tsuji, S., Yoshii, K., and Tonogai, Y. Identification of isomers and subsidiary colors in commercial Fast Green FCF (FD & C Green No. 3, Food Green No. 3) by liquid chromatography-mass spectrometry and comparison between amounts of the subsidiary colors by high-performance liquid chromatography and thin-layer chromatography-spectrometry, *J.Chromatog.A.* 1101, 214-221, 2006; Luo, S., Wehr, N.B., and Levine, R.L., Quantitation of protein on gels and blots by infrared fluorescence of Coomassie glue and Fast Green, *Anal.Biochem.* 350, 233-238, 2006; Ali, M.A. and Bashier, S.A., Effect of fast green dye on some biophysical properties of thymocytes and splenocytes of albino mice, *Food Addit.Contam.* 23, 452-561, 2006

SOME BIOLOGICAL STAINS AND DYES (Continued)

Name	Description
Fluorescein (C.I. solvent yellow 94) MW as free acid is 332 while the disodium salt is 376.	Fluorescein is an acidic hydroxyxanthene which exists in a variety of forms. Depending on pH, fluorescein is in equilibrium the free acid and a dianion form. When the dianion is present as the sodium salt, the term uranin has been used to describe the molecule. Fluorescein has a broad range of use including use in large quantities to trace environment water flow. Fluorescein has been used to vascular flow with particular interest in ophthalmology. Fluorescein has been modified to include reactive functional groups (isothiocyanate; for covalent insertion of fluorescent probes into protein and other biological macromolecules. Care must be taken to avoid confusion with fluorescin.

Fluorescein

Uranin

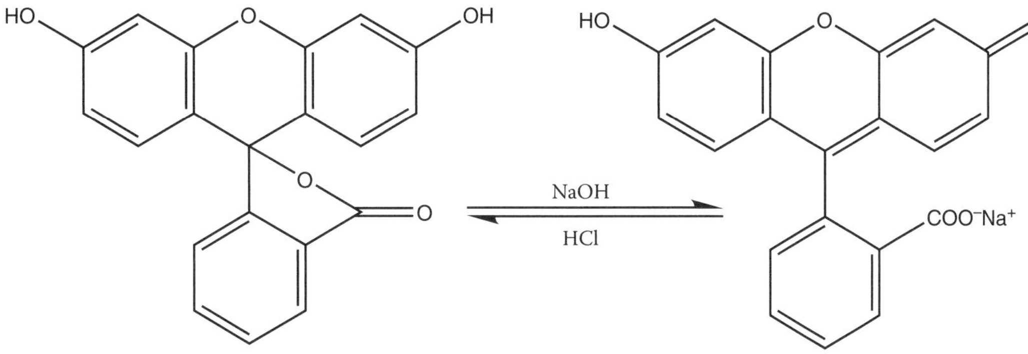

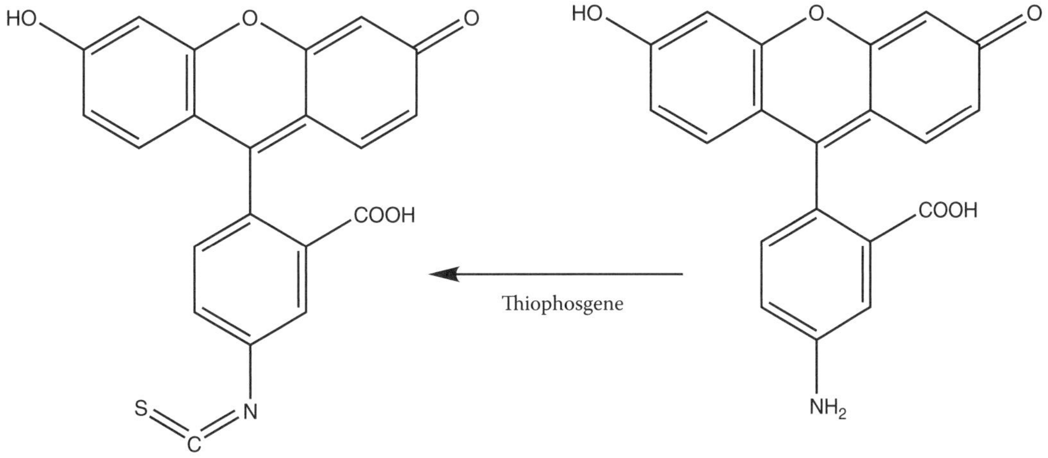

NaOH

HCl

Thiophosgene

Fluorescein Isothiocyanate

SOME BIOLOGICAL STAINS AND DYES (Continued)

Name	Description

Ray, R.R. and Binkhorst, R.D., The diagnosis of papillary block by intravenous injection of fluorescein, *Am.J.Ophthamol.* 61, 480-483, 1966; Hill, D.W., Fluorescein angiography in fundus diagnosis, *Br.Med.Bull.* 26, 161-165, 1970; Gass, J.D., Fluorescein angiography. An aid in the differential diagnosis of intraocular tumors, *Int.Ophthalomol.Clin.* 12, 85-120, 1972; Ohkuma, S., Use of fluorescein isothiocynate-dextran to measure proton pumping in lysosomes and related organelles, *Methods Enzymol.* 174, 131-154, 1989; Klose, A.D. and Gericke, K.R., Fluorescein as a circulation determinant, *Ann. Pharmacother.* 28, 891-893, 1994; Wischke, C. and Borchert, H.H., Fluorescein isothiocyanate labelled bovine serum albumin (FITC-BSA) as a model protein drug: opportunities and drawbacks, *Pharmazie* 61, 770-774, 2006; Berginc, K., Zakelj, S., Levstik, L., *et al.*, Fluorescein transport properties across artificial lipid membranes, Caco-2 cell monolayers and rat jejuna, *Eur.J.Pharm.Biopharm.*, in press, 2006.

Malachite Green (Victoria Green, C.I. basic green 4); MW leukobase, 330; carbinol free base as the hydrochloride, 383)

A cationic diaminotriphenylmethane which is used to stain a variety of cells including bacterial spores, pH indicators, stain for phospholipid; measurement of inorganic phosphate

Malachite Green
Victoria Green B

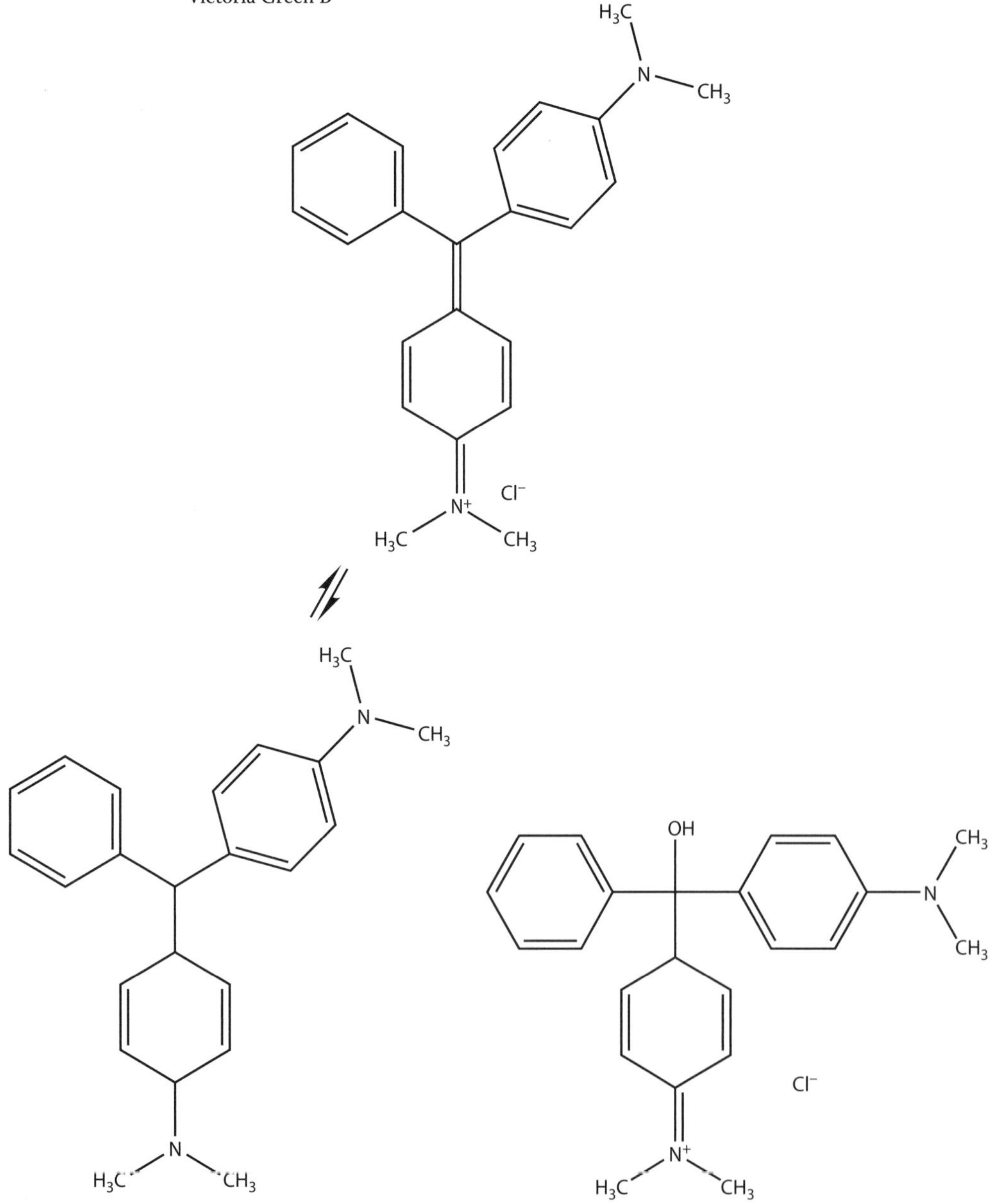

Leukomalachite green
Leuko form

Malachite Green Carbinol Chloride

SOME BIOLOGICAL STAINS AND DYES (Continued)

Name Description

Norris, D., Reconstitution of virus X-saturated potato varieties with malachite green, *Nature* 172, 816, 1953; Kanetsuna, F., A study of malachite green staining of leprosy bacilli, *Int.J.Lepr.* 32, 185-194, 1964; Solari, A.A., Herrero, M.M., and Painceira, M.T., Use of malachite green for staining flagella in bacteria, *Appl.Microbiol.* 16, 792, 1968; Teichman, R.J., Takei, G.H., and Cummins, J.M., Detection of fatty acids, fatty aldehydes, phospholipids, glycolipids and cholesterol on thin-layer chromatograms stained with malachite green, *J.Chromatog.* 88, 425-427, 1974; Singh, E.F., Moawad, A., and Zuspan, F.P., Malachite green—a new staining reagent for prostaglandins, *J.Chromatog* 105, 194-196, 1975; Nefussi, J.R., Septier, D., Sautier, J.M. *et al.*, Localization of malachite green positive lipids in the matrix of bone nodule formed in vitro, *Calif.Tissue Int.* 50, 273-282, 1992; Henderson, A.L., Schmitt, T.C., Heinze, T.M., and Cerniglia, C.E., Reduction of malachite green to leucomalachite green by intestinal bacteria, *Appl.Environ.Microbiol* 63, 4099-4101, 1997; Nguyen, D.H., DeFina, S.C., Fink, W.H., and Dieckmann, T., Binding to an RNA aptamers changes the charge distribution and conformation of malachite green, *J.Am.Chem.Soc.* 124, 15081-15084, 2002; Dutta, K., Bhattacharjee, S., Chauduri, B., and Mukhopadhyay, S., Oxidative degradation of malachite green by Fenton generated hydroxyl radicals in aqueous acidic media, *J.Environ.Sci.Health A. Tox. Hazard Subst. Environ.Eng.* 38, 1311-1326, 2003; Jadhav, J.P. and Govindwar, S.P., Biotransformation of malachite green by *Saccharomyces cerevisiae* MTCC 463, *Yeast* 23, 315-323, 2006;.

Methylene Blue, MW 374 as the chloride monohydrate

A weakly hydrophilic cationic diaminothiazine. Methylene blue is used for a variety of purposes including development of the Romanowsky stain, bacterial staining, assay of redox reactions, vital staining, photooxidation of proteins, surgical marker (this use appears to undergoing serious reconsideration), and determination of cell wall permeability. Methylene blue with UV irradiation is used for the inactivation of pathogens in blood plasma.

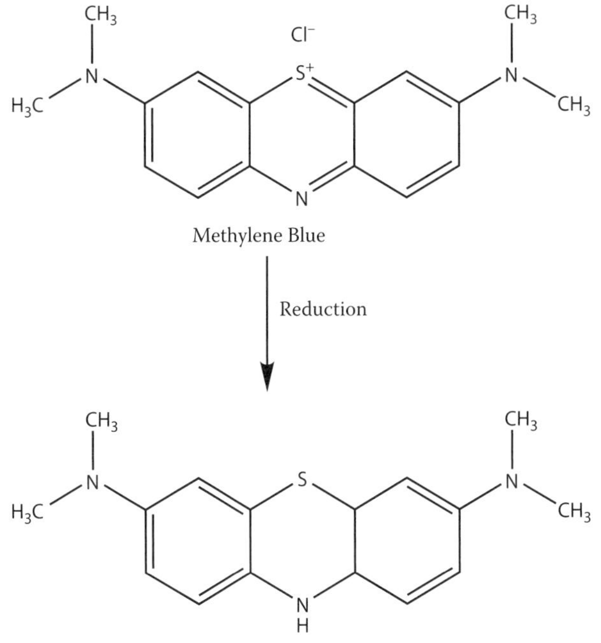

Methylene Blue

Reduction

Methylene Blue Leuko Form

SOME BIOLOGICAL STAINS AND DYES (Continued)

Name   **Description**

Wishart, G.M., On the reduction of methylene blue by tissue extracts, *Biochem.J.* 17, 103-114, 1923; Whitehead, H.R., The reduction of methylene blue in milk: The influence of light, *Biochem.J.* 24, 579-584, 1930; Worley, L.G., The relation between the Golgi apparatus and "Droplets" in the cell stainable vitally with methylene blue, *Proc.Natl.Acad.Sci.USA* 29, 228-231, 1943; Weil, L., Gordon, W.G., and Buchert, A.R., Photooxidation of amino acids in the presence of methylene blue, *Arch.Biochem.* 33, 90-109, 1951; Moore, T., Sharman, I.M., Ward, R.J., The vitamin E activity of substances related to methylene blue, *Biochem.J.* 54, xvi-xvii, 1953; Yamazaki, I., Fujinaga, K., and Takehara, I., The reduction of methylene blue catalyzed by the turnip peroxidase, *Arch.Biochem.Biophys.* 72, 42-48, 1957; Borzani, W. and Vairo, M.L., Adsorption of methylene blue as a means of determining cell concentration of dead bacteria in suspensions, *Stain Technol.* 35, 77-81, 1960; Tinne, J.E., A methylene blue medium for distinguishing virulent mycobacteria, *Scott.Med.J.* 4, 130-132, 1959; Barbosa, P. and Peters, T.M., The effects of vital dyes on living organisms with special reference to methylene blue and neutral red, *Histochem.J.* 3, 71-93, 1971; Bentley, S.A., Marshall, P.N., and Trobaugh, F.E., Jr., Standardization of the Romanowksy staining procedure: an overview, *Anal.Quant.Cytol.* 2, 15-18, 1980; Wittekind, D.H. and Gehring, T., On the nature of Romanowsky-Giemsa staining and the Romanowsky-Giemsa effect. I. Model experiments on the specificity of azure B-eosin Y stain as compared with other thiazine dye – eosin Y combinations, *Histochem.J.* 17, 263-289, 1985; Schulte, E. and Wittekind, D., The influence of Romanowsky-Giemsa type stains on nuclear and cytoplasmic features of cytological specimens, *Anal.Cell Pathol.* 1, 83-86, 1989; Tuite, E.M. and Kelly, J.M., Photochemical interactions of methylene blue and analogues with DNA and other biological substrates, *J.Photochem.Photobiol.B.* 21, 103-124, 1993; Bradbeer, J.N., Riminucci, M., and Bianco, P., Giemsa as a fluorescent stain for mineralized bone, *J.Histochem.Cytochem.* 42, 677-680, 1994; Wainwright, M. and Crossley, K.B., Methylene blue – a therapeutic dye for all seasons?, *J.Chemother.* 14, 431-443, 2002; Inamura, K., Ikeda, E., Nagayasu, T., *et al.*, Adsorption behavior of methylene blue and its congeners on a stainless steel surface, *J.Colloid Interface Sci.* 245, 50-57, 2002; Floyd, R.A., Schneider, J.E., Jr., and Dittmern, D.P., Methylene blue photoinactivation of RNA viruses, *Antiviral Res.* 61, 141-151, 2004; Rider, K.A. and Flick, L.M., Differentiation of bone and soft tissues in formalin-fixed, paraffin-embedded tissue by using methylene blue/acid fuchsin stain, *Anal.Quant.Cytol.Histol.* 26, 246-248, 2004; Rider, K.A. and Flick, L.M., Differentiating of bone and soft tissues in formalin-fixed, paraffin-embedded tissue by using methylene blue/acid fuchsin stain, *Anal. Quant.Cytol. Histol.* 26, 246-248, 2004; Papin, J.F., Floyd, R.A., and Dittmer, D.P., Methylene blue photoinactivation abolishes West Nile virus infectivity *in vivo*, *Antiviral Res.* 68, 84-87, 2005; Dilgin, Y. and Nisli, G., Fluorometric determination of ascorbic acid in vitamin C tablets using methylene blue, *Chem.Pharm.Bull.* 53, 1251-1254, 2005; Itoh, K., Decolorization and degradation of methylene blue by *Arthrobacter globiformis*, *Bull. Environ.Contam.Toxicol.* 75, 1131-1136, 2005; D'Amico, F., A polychromatic staining method for epoxy embedded tissues: a new combination of methylene blue and basic fuchsine for light microscopy, *Biotech.Histochem.* 80, 207-210, 2005; Cheng, Y., Liu, W.F., Yan, Y.B., and Zhou, H.M., A nonradiometric assay for poly(a)-specific ribonuclease activity by methylene blue colorimetry, *Protein Pept.Lett.* 13, 125-128, 2006; Jurado, E., Fernandez-Serrano, M., Nunez-Olea, J., *et al.*, Simplified spectrophotometric method using methylene blue for determining anionic surfactants: applications to the study of primary biodegradation in aerobic screening tests, *Chemosphere* 65, 278-285, 2006; Dinc, S., Ozaslan, C., Kuru, B., *et al.*, Methylene blue prevents surgery-induced peritoneal adhesions but impairs the early phase of anastomotic wound healing, *Can.J.Surg.* 49, 321-328, 2006; Appadurai, I.R. and Scott-Coombes, D., Methylene blue for parathyroid localization, *Anaesthesia* 62, 94, 2007; Mihai, R., Mitchell, E.W., and Warwick, J., Dose-response and postoperative confusion following methylene blue infusion during parathyroidectomy, *Can.J.Anaesth.* 54, 79-81, 2007; McCullagh, C. and Robertson, P., Effect of polyethyleneimine, a cell permeabiliser, on the photo-sensitised destruction of algae by methylene blue and nuclear fast red, *Photochem.Photobiol.*, in press, 2007;

Methyl Orange, Orange III, Gold   An azo dye with limited use in histology; use for the assay of redox reactions. Used for the study of protein
Orange, C.I. Orange 52; MW 327   conformation; early use for assay of albumin; binding to cationic proteins
as sodium salt

Wetlaufer, D.B. and Stahmann, M.A., The interaction of methyl orange anions with lysine polypeptides, *J.Biol.Chem.* 203, 117-126, 1953; Colvein, J.R., The adsorption of methyl orange by lysozyme, *Can.J.Biochem.Physiol.* 32, 109-118, 1954; Lundh, B., Serum albumin as determined by the methyl orange method and by electrophoresis, *Scand.J.Clin.Lab.Invest.* 17, 503-504, 1965; Barrett, J.F., Pitt, P.A., Ryan, A.J., and Wright, S.E., The demethylation of *m*-methyl orange and methyl orange *in vivo* and *in vitro*, *Biochem.Pharmacol.* 15, 675-680, 1966; Shikama, K., Denaturation and renaturation of binding sites of bovine serum albumin for methyl orange, *J.Biochem.* 64, 55-63, 1968; Lang, J., Auborn, J.J., and Eyring, E.M., Kinetic studies of the interaction of methyl orange with beta-lactoglobulin between pH 3.7 and 2.0, *J.Biol.Chem.* 246, 5380-5383, 1971; Browner, C.J. and Lindup, W.E., Decreased plasma protein binding of *o*-methyl red methyl orange and phenytoin (diphenylhydantoin) in rats with acute renal failure,

SOME BIOLOGICAL STAINS AND DYES (Continued)

Name **Description**

Br.J.Pharmacol. 63, 367P, 1978; Ford, C.L. and Winzor, D.J., A recycling gel partition technique for the study of protein-ligand interactions: the binding of methyl orange to bovine serum albumin, *Anal.Biochem.* 114, 146-152, 1981; Chung, K.T., Stevens, S.E., Jr., and Cerniglia, C.E., The reduction of azo dyes by the intestinal microflora, *Crit.Rev.Microbiol.* 18, 175-190, 1992; Yang Y. Jung, D.W., Bai, D.G., *et al.*, Counterion-dye staining method for DNA in agarose gels using crystal violet and methyl orange, *Electrophoresis* 22 855-859, 2001; Nam, W., Kim, J., and Han, G., Photocatalytic oxidation of methyl orange in a three-phase fluidized bed reactor, *Chemosphere* 47, 1019-1024, 2002; Marci, G., Augugliaro, V., Bianco Prevot, A., *et al.*, Photocatalytic oxidation of methyl-orange in aqueous suspension: comparison of the performance of different polycrystalline titanium dioxide, *Ann.Chim.* 93, 639-648, 2003; Del Nero, J., de Araujo, R.E., Gomes, A.S., and de Melo, C.P., Theoretical and experimental investigation or the second hyperpolarizabilities of methyl orange, *J.Chem. Phys.* 122, 104506, 2005; de Oliveira, H.P., Oliveira, E.G., and de Melo, C.P., *J.Colloid Interface Sci.* 303, 444-449, 2006; Bejarano-Perez, N.J. and Suarez-Herrera, M.F., Sonophotocatalytic degradation of congo red and methyl orange in the presence of TiO$_2$ as a catalyst, *Ultrason.Sonochem.* in press, 2006;

Methyl Red (2-[[4-(dimethylamino)-phenyl]azo] benzoic acid; C.I. acid red 2) MW 270; 306 as the hydrochloride

pH indicator; use as signal in redox reactions; rare use in histochemistry; some use for staining protozoa and other bacteria; standard for spectroscopy

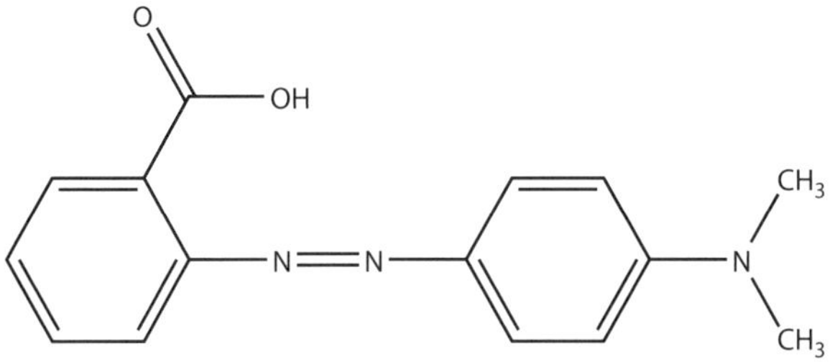

Cowan, S.T., Micromethod for the methyl red test, *J.Gen.Microbiol.* 9, 101-109, 1953; Ljutov, V., Technique of methyl red test, *Acta Pathol.Microbiol. Scand.* 51, 369-380, 1961; Barry, A.L., Berhsohn, K.L., Adams, A.P., and Thrupp, L.D., Improved 18-hour methyl red test, *Appl.Microbiol.* 20, 866-870, 1970; Korzun, W.J. and Miller, W.G., Monitoring the stability of wavelength calibration of spectrophotometers, *Clin.Chem.* 32, 162-165, 1986; Chung, K.T., Stevens, S.E., Jr., and Cerniglia, C.E., The reduction of azo dyes by the intestinal microflora, *Crit.Rev.Microbiol.* 18, 175-190, 1992; Miyajima, M., Sagami, I., Daff, S., *et al.*, Azo reduction of methyl red by neuronal nitric oxide synthase: the important role of FMN in catalysis, *Biochem.Biophys.Res. Commun.* 275, 752-758, 2000; Kashida, H., Tanaka, M., Baba, S., *et al.*, Covalent incorporation of methyl red dyes into double-stranded DNA for their ordered clustering, *Chemistry* 12, 777-784, 2006; Kalyuzhnyi, S., Yemashova, N., and Fedorovich, V., Kinetics of anaerobic biodecolourisation of azo dyes, *Water Sct.Technol.* 54, 73-79, 2006; Katsuda, T. Ooshima, H., Azuma, M., and Kato, J., New detection method for hydrogen gas for screening hydrogen-producing microorganisms using water-soluble Wilkinson's catalyst derivative, *J.Biosci.Bioeng.* 102, 220-226, 2006; Hsueh, C.C. and Chen, B.Y., Comparative study on reaction selectivity of azo dye decolorization by *Pseudomonas luteola,*, *J.Hazard.Mater.*, in press, 2006

Methyl Violet (C.I. Basic violet 1); Molecular weight depends on the degree of methylation

Methyl violet is a mixture of *N*-methylated pararoanilines which is used a pH indicator and a biological stain. The term gentian violet is used for this mixture in Europe and for crystal violet (methyl violet 10B, the hexamethyl derivative) in the United States. There is some use of methyl violet for staining DNA. The industrial use is for dyeing fabrics blue and violet. The depth of color is dependent on the degree of methylation.

Methyl Violet

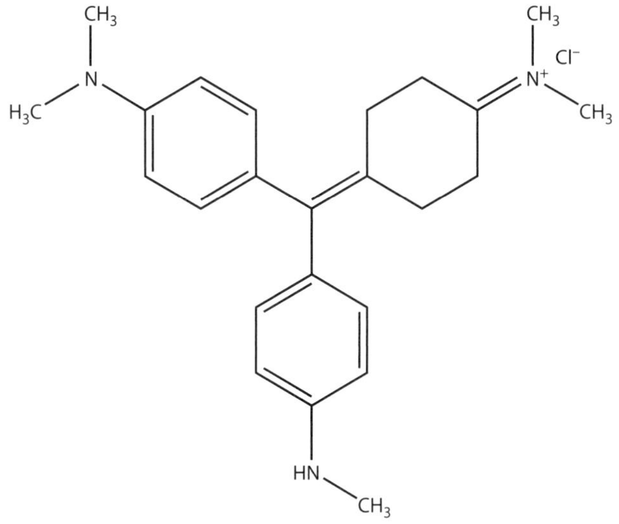

SOME BIOLOGICAL STAINS AND DYES (Continued)

Name   **Description**

Bancroft, J.D., Methyl green as a differentiator and counterstain in the methyl violet technique for demonstration of amyloid in fresh cryostat sections, *Stain Technol.* 38, 336-337, 1963; Campbell, L.M. and Roth, I.L., Methyl violet: a selective agent for differentiation of *Klebsiella pneumonia* from *Enterobacter aerogenes* and other gram-negative organisms, *Appl.Microbiol.* 30, 258-261, 1975; Dutt, M.K., Staining of depolymerized DNA in mammalian tissues with methyl violet 6B and crystal violet, *Folia Histochem.Cytochem.* 18, 79-83, 1980; Liu, Y., Ma, C.Q., Li, K.A., *et al.*, Rayleigh light scattering study on the reaction of nucleic acids and methyl violet, *Anal.Biochem.* 268, 187-192, 1999; Dogan, M. and Aikan, M., Removal of methyl violet from aqueous solution by perlite, *J.Colloid Interface Sci.* 267, 32-41, 2003; Jin, L.T. and Choi, J.K., Usefulness of visible dyes for the staining of protein of DNA in electrophoresis, *Electrophoresis* 25, 2429-2438, 2004;

Neutral Red (C.I. Basic Red 5; nuclear fast red; 3-amino-7-dimethylamino-2-methylphenazine hydrochloride; toluylene red as the unprotonated form), MW 289 as the hydrochloride

pH indicator, histological stain for Golgi, nuclear. Also a vital stain for intracellular organelles and cytotoxicity testing.

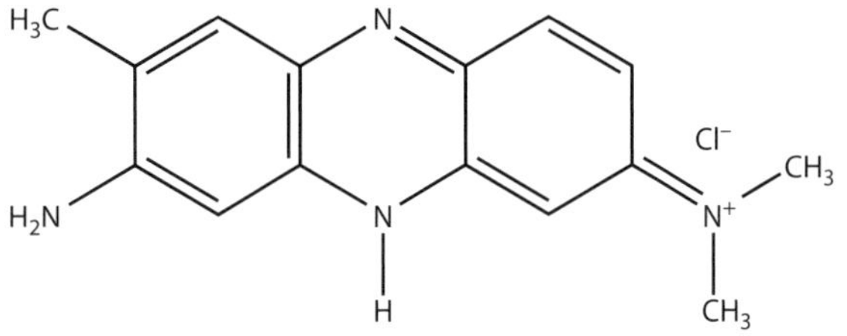

Lepper, E.H. and Martin, C.J., The protein error in estimating pH with neutral red and phenol red, *Biochem.J.* 21, 356-361, 1927; Morse, W.C., Dail, M.C., and Olitzky, I., A study of the neutral red reaction for determining virulence of *Mycobacteria*, *Am.J.Public Health* 43, 36-39, 1953; Vivian, D.L. and Belkin, M., Unexpected anomalies in the behavior of neutral red and related dyes, *Nature* 178, 154, 1956; Darnell, J.E., Jr., Lockart, R.Z., Jr., and Sawyer, T.K., The effect of neutral red on plaque formation in two virus-cell systems, *Virology* 6, 567-568, 1958; Crowther, D. and Melnick, J.L., The incorporation of neutral red and acridine orange into developing poliovirus particles making them photosensitive, *Virology* 14, 11-21, 1961; Boyer, M.G., The role of tannins in neutral red staining of pine needle vacuoles, *Stain Technol.* 38, 117-120, 1967; Sawicki, W., Kieler, J., and Briand, P., Vital staining with neutral red and trypan blue of ^{3}H-thymidine-labeled cells prior to autoradiography, *Stain Technol.* 42, 143-146, 1967; Barbosa, P. and Peters, T.M., The effects of vital dyes on living organisms with special reference to methylene blue and neutral red, *Histochem.J.* 3, 71-93, 1971; Gutter, B., Speck, W.T., and Rosenkranz, H.S., Light-induced mutagenicity of neutral red (3-amino-7-dimethylamino-2-methylphenazine hydrochloride), *Cancer Res.* 37, 1112-1114, 1977; Nemes, Z., Dietz, R., Luth, J.B., *et al.*, The pharmacological relevance of vital staining with neutral red, *Experientia* 35, 1475-1476, 1979; Gray, D.W., Millard, P.R., McShane, P., and Morris, P.J., The use of the dye neutral red as a specific, non-toxic, intra-vital stain of islets of Langerhans, *Br.J.Exp.Pathol.* 64, 553-558, 1983; LaManna, J.C., Intracellular pH determination by absorption spectrophotometry of neutral red, *Metab.Brain Dis* 2, 167-182, 1987; Elliott, W.M. and Auersperg, N., Comparison of the neutral red and methylene blue assays to study cell growth in culture, *Biotech.Histochem.* 68, 29-35, 1993; Fautz, R., Husein, B., and Hechenberger, C., Application of the neutral red assay (NR assay) to monolayer cultures of primary hepatocytes: rapid colorimetric viability determination for the unscheduled DNA synthesis test (UDS), *Mutat.Res.* 253, 173-179, 1991; Kado, R.T., Neutral red: a specific fluorescent dye in the cerebellum, *Jpn.J.Physiol.* 43(Suppl 1), S161-S169, 1993; Ishiyama, M., Tominaga, H. Shiga, M., A combined assay of cell viability and in vitro cytotoxicity with a highly water-soluble tetrazolium salt, neutral red and crystal violet, *Biol.Pharm.Bull.* 19, 1518-1520, 1996; Ciapetti, G., Granchi, D., Verri, E., *et al.*, Application of a combination of neutral red and amido black staining for rapid, reliable cytotoxicity testing of biomaterials, *Biomaterials* 17, 1259-1264, 1996; Sousa, C., Sá e Melo, T., Geze, M., *et al.*, Solvent polarity and pH effects on the spectroscopic properties of neutral red: application to lysosomal microenvironment probing in living cells, *Photochem.Photobiol.* 63, 601-607, 1996; Baker, C.S., Crystallization of neutral red vital stain from minimum essential medium due to pH instability, *In Vitro Cell.Dev.Biol.Anim.* 34, 607-608, 1998; Okada, D., Neutral red as a hydrophobic probe for monitoring neuronal activity, *J.Neurosci.Methods* 101, 85-92, 2000; Zuang, V., The neutral red release assay: a review, *Altern.Lab.Anim.* 29, 575-599, 2001; Svendsen, C., Spurgeon, D.J., Hankard, P.K., and Weeks, J.M., A review of lysosomal membrane stability measured by neutral red retention: is a workable earthworm biomarker?, *Ecotoxicol.Environ. Saf.* 57, 20-29, 2004.

SOME BIOLOGICAL STAINS AND DYES (Continued)

Name	Description

Nile Red (Nile pink); MW 318

A lipophilic benzooxazone used for lipid staining and as a hydrophobic probe for proteins; Nile red has been used as a stain for protein on acrylamide gel. The fluorescence of Nile red is very dependent on solvent. Nile red is poorly soluble in aqueous systems but the recent development of water-soluble derivatives has improved utility.

Greenspan, P., Mayer, E.P., and Fowler, S.D., Nile red; a selective fluorescent stain for intracellular lipid droplets, *J.Cell Biol.* 100, 965-973, 1985; Fowler, S.D. and Greenspan, P., Application of Nile red, a fluorescent hydrophobic probe, for the detection of neutral lipid deposits in tissue sections: comparison with oil red O, *J.Histochem.Cytochem.* 33, 833-836, 1985; Sackett, D.L. and Wolff, J., Nile red as a polarity-sensitive fluorescent probe of hydrophobic protein surfaces, *Anal.Biochem.* 167, 229-234, 1987; Brown, W.J., Warfel, J., and Greenspan, P., Use of Nile red stain in the detection of cholesteryl ester accumulation in acid lipase-deficient fibroblasts, *Arch.Pathol.Lab.Med.* 112, 295-297, 1988; Brown, W.J., Sullivan, T.R., and Greenspan, P, Nile red staining of lysosomal phospholipid inclusions, *Histochemistry* 97, 349-354, 1992; Brown, M.B., Miller, J.N., and Seare, N.J., An investigation of the use of Nile red as a long-wavelength fluorescent probe for the study of alpha 1-acid glycoprotein-drug interactions, *J.Pharm. Biomed.Anal.* 13 1011-1017, 1995; Alba, F.J., Bermudez, A., Bartolome, S. and Daban, J.R,. Detection of five nanograms of protein by two-minute nile red staining of unfixed SDS gels, *BioTechniques* 21, 625-626, 1996; Daban, J.R., Fluorescent labeling of proteins with Nile red and 2-methoxy-2,4-diphenyl-3(2H)-furanone: physicochemical basis and application to the rapid staining of sodium dodecyl sulfate polyacrylamide gels and western blots, *Electrophoresis* 22, 874-880, 2001; Hendriks, J., Gensch, T., Hviid, L., *et al.,* Transient exposure of hydrophobic surface in the photoactive yellow protein monitored with Nile red, *Biophys.J.* 82, 1632-1643, 2002; Prokhorenko, I.A., Dioubankova, N.N., and Korshun, V.A., Oligonucleotide conjugates of Nile Red, *Nucleosides Nucleotides Nucleic Acids* 23, 509-520, 2004; Yablon, D.G. and Schilowitz, A.M., Solvatochromism of Nile Red in nonpolar solvents, *Appl.Spectros.* 58, 843-847, 2004; Genicot, G., Leroy, J.L., Soom, A.V., and Donnay, I., The use of a fluorescent dye, Nile red, to evaluate the lipid content of single mammalian oocytes, *THeriogeneology* 63, 1181-1194, 2005; Sebok-Nagy, K., Miskoczy, Z. and Biczok, L., Interaction of 2-hydroxy-substituted Nile red fluorescent probe with organic nitrogen compounds, *Photochem.Photobiol.* 81, 1212-1218, 2005; Thomas, K.J., Sherman, D.B., Amiss, T.J., *et al.,* A long-wavelength fluorescent glucose biosensor based on bioconjugates on galactose/glucose binding protein and Nile red derivatives, *Diabetes Technol.Ther.* 8, 261-268, 2006; Jose, J. and Burgess, K., Syntheses and properties of water-soluble Nile red derivatives, *J.Org.Chem.* 71, 7835-7839, 2006; Mukherjee, S., Raghuraman, H., and Chattopadhyay, A., Membrane localization and dynamics of Nile red: effect of cholesterol, *Biochim.Biophys.Acta* 1768, 59-66, 2007

Oil Red O (Sudan Red 5B; C.I. Solvent Red 27)

An extremely lipophilic dye; used for the demonstration of lipid depositions.

Moran, P. and Heyden, G., Enzyme histochemical studies on the formation of hyaline bodies in the epithelium of odontogenic cysts, *J.Oral Pathol.* 4, 120-127, 1975; Merrick, J.M., Schifferle, R., Zadarlik, K., *et al.,* Isolation and partial characterization of the heterophile antigen of infectious mononucleosis from bovine erythrocytes, *J.Supramol.Struct.* 6, 275-290, 1977; Anderson, L.C. and Garrett, J.R., Lipid accumulation in the major salivary glands of streptozotocin-diabetic rats, *Arch.Oral Biol.* 31, 469-475, 1986; Chastre, J., Fagon, J.Y., Soler, P., *et al.,* Bronchoalveolar lavage for rapid diagnosis of the fat embolism syndrome in trauma patients, *Ann.Intern.Med.* 113, 583-588, 1990; Aleksic, I., Ren, M., Popov, A., *et al.,* In vivo liposome-mediated transfection of HLA-DR alpha-chain gene into pig hearts, *Eur.J.Cardiothorac.Surg.* 12, 792-797, 1997.

SOME BIOLOGICAL STAINS AND DYES (Continued)

Name   **Description**

Rhodamine B (C.I. Basic Violet 10)   An aminoxanthene which is a lipophilic cation in acid solution and a lipophilic anion in basic solution. Metal
MW 479   binding reagent, a neutral stain; a fluorescent lipid stain; stain for Phosphoproteins for gel electrophoresis.
                                                                    Isothiocyanate derivative used for labeling proteins.

Martin, G., Colorimetric determination of zinc by thiocyanate derivatives; a new method using rhodamine B, *Bull.Soc.Chim.Biol.* 34, 1174-1177, 1952; Webb,
J.M., Hansen, W.H., Desmond, A., and Fitzhugh, O.G., Biochemical and toxicological studies of rhodamine B and 3,6-diaminofluroan, *Toxicol.Appl.Pharmacol.*
3, 696-706, 1961; Miketukova, V., Detection of metals on paper chromatograms with Rhodamine B., *J.Chromatog.* 24, 302-304, 1966; Liisberg, M.F., Rhodamine
B as an extremely specific stain for cornification, *Acta Anat.* 69, 52-57, 1968; Shelley, W.B., Fluorescent staining of elastic tissue with Rhodamine B and related
xanthene dyes, *Histochemie* 20, 244-249, 1969; Oshima, G.T. and Nagasawa, K., Fluorometric method for determination of mercury(II) with Rhodamine B,
Chem.Pharm.Bull. 18, 687-692, 1970, Zahradnicek, L., Kratochvila, J., and Garcis, A., Determination of inorganic phosphorus using Rhodamine B by the
continuous-flow technique, *Clin.Chim.Acta* 80, 431-433, 1977; Wessely, Z., Shapiro, S.H., Klavins, J.V., and Tinberg, H.M., Identification of Mallory bodies with
rhodamine B fluorescence and other strains for keratin, *Stain Technol.* 56, 169-176, 1981; Debruyne, I., Inorganic phosphate determination: colorimetric assay
based on the formation of a rhodamine B-phosphomolybdate complex, *Anal.Biochem.* 130, 454-460, 1983; Debruyne, I., Staining of alkali-labile
Phosphoproteins and alkaline phosphatases on polyacrylamide gels, *Anal.Biochem.* 133, 110-115, 1983; Balcerzak, M., Sensitive spectrophotometric
determination of osmium with tin(II) chloride and rhodamine B after flotation using cyclohexane, *Analyst* 113, 129-132, 1988; Glimcher, M.J. and Lefteriou, B.,
Soluble glycosylated phosphoproteins of cementum, *Calcif.Tissue Int.* 45, 165-172, 1989; Fernandez-Busquets, X. and Burger, M.M., Use of rhodamine B
isothiocyanate to detect proteoglycan core proteins in polyacrylamide gels, *Anal.Biochem.* 227, 394-396, 1995; Jung, D.W., Yoo, G.S. and Choi, J.K., Mixed-dye
staining method for protein detection in polyacrylamide gel electrophoresis using calconcarboxylic acid and rhodamine B, *Electrophoresis* 19, 2412-2415, 1998;
Pal, J.K., Godbole, D., and Sharma, K., Staining of proteins on SDS polyacrylamide gels and on nitrocellulose membranes by Alta, a colour used as a cosmetic,
J.Biochem.Biophys.Methods 61, 339-347, 2004; dos Santos Silva, A.L., Joekes, I., Rhodamine B diffusion in hair as a probe for structural integrity, *Colloids
Surf.B.Biointerfaces* 40, 19-24, 2005; Moreno-Villoslada, I., Jofre, M., Miranda, V., *et al.*, pH dependence of the interaction between rhodamine B and the
water-soluble poly(sodium 4-styrenesulfonate), *J.Phys.Chem.B Condens. Mater. Surf.Interfaces Biophys.* 110, 11809-11812, 2006.

Rose Bengal (Rose Bengal B; rose   An acidic hydroxyxanthene soluble in water used as a cellular stain; diagnostic aid in ophthalmology; source of
bengale; C.I. acid red 94; the disodium   singlet oxygen in photooxidation; use as a probe for surface protein structure; early use of the radioactive
salt is referred to as rose Bengal   derivative as a tracer for liver function.
extra); MW 1049 as dipotassium salt
and 1018 as disodium salt

SOME BIOLOGICAL STAINS AND DYES (Continued)

Name   Description

Conn, H.J., Rose Bengal as a general bacterial stain, *J.Bacteriol.* 6, 253-254, 1921; Forster, H.W., Jr., Rose Bengal test in diagnosis of deficient tear formation, *AMA Arch.Ophthalmol.* 45, 419-424, 1951; Rippa, M. and Picco, C., Rose Bengal as a reporter of the polarity and acidity of the TPN binding site in 6-phosphoglucoonate dehydrogenase, *Ital.J.Biochem.* 19, 178-192, 1970; Lyons, A.B. Ashman, L.K., The Rose Bengal assay for monoclonal antibodies to cell surface antigens: comparisons with common hybridoma screening methods, *J.Immunoassay* 6, 325-345, 1985; Hederstedt, L. and Hatefi, Y., Modification of bovine heart succinate dehydrogenase with ethoxyformic anhydride and rose Bengal: evidence for essential histidyl residues protectable by substrates, *Arch.Biochem.Biophys.* 247, 346-354, 1986; Allen, M.T., Lynch, M., Lagos, A. *et al.*, A wavelength dependent mechanism for rose bengal-sensitized photoinhibition of red cell acetylcholinesterase, *Biochim.Biophys.Acta.* 1075, 42-49, 1991; Feenstra, R.P. and Tseng, S.C., What is actually stained by rose bengal?, *Arch.Ophthalmol.* 110, 984-993, 1992; Shan, M.A. and Ali, R., Modification of pig kidney diamine oxidase with ethoxyformic anhydride and rose bengal: evidence for essential histidine residue at the active site, *Biochem.Mol.Biol.Int.* 33, 9-19, 1994; Tseng, S.C. and Zhang, S.H., Interaction between rose bengal and different protein components, *Cornea* 14, 427-435, 1995; Singh, R.J., Hogg, N. and Kalyanaraman, B., Interaction of nitric oxide with photoexcited rose bengal: evidence for one-electron reduction of nitric oxide to nitroxyl anion, *Arch.Biochem.Biophys.* 324, 367-373, 1995; Bottiroli, G., Croce, A.C., Balzarini, P., *et al.*, Enzyme-assisted cell photosensitization: a proposal for an efficient approach to tumor therapy and diagnosis. The rose bengal fluorogenic substrate, *Photochem.Photobiol.* 66, 374-383, 1997; Perez-Ruiz, T., Martinez-Lozano, C., Tomas, V. and Fenoll, J., Determination of proteins in serum by fluorescence quenching of rose bengal using the stopped-blow mixing technique, *Analyst* 125, 507-510, 2000; Lin, W., Garnett, M.C., Davis, S.S., *et al.*, Preparation and characterization of rose Bengal-loaded surface-modified albumin nanoparticles, *J.Control. Res.* 71, 117-126, 2001; Posadez, A., Biasutti, A., Casale, C., *et al.*, Rose Bengal-sensitized photooxidation of the dipeptides L-tryptophyl-L-phenylalanine, L-tryotophyl-L-tyrosine and L-tryptophyl-L-tryptophan: kinetics, mechanism and photoproducts, *Photochem.Photobiol.* 80, 132-138, 2004; Luiz, M., Biasutti, M.A., and Garcia, N.A., Effect of reverse micelles on the Rose Bengal-sensitized photo-oxidation of 1- and 2-hydroxynaphthalene, *Redox Rep.* 9, 199-205, 2004; Khan-Lim, D. and Berry, M., Still confused about rose bengal?, *Curr.Eye Res.* 29, 311-317, 2004; Soldani, C., Bottone, M.G., Croce, A.C., *et al.*, The Golgi apparatus is a primary site of intracellular damage after photosensitization with Rose Bengal acetate, *Eur.J.Histochem.* 48, 443-449, 2004; de Lima Santos, H., Forest Rigos. C., Claudio Tedesco, A., and Ciancaglini, P., Rose Bengal located within liposome do not affect the activity of inside-out oriented Na,K-ATPase, *Biochim.Biophys.Acta* 1715 96-103, 2005; Miller, J.S., Rose bengal-stimulated photooxidation of 2-chlorophenol in water using solar simulated light, *Water Res.* 39, 412-422, 2005; Shimizu, O., Watanabe, J., Naito, S. and Shibata, Y., Quenching mechanism of Rose Bengal triplet state involved in photosensitization of oxygen in ethylene glycol, *J.Phys.Chem.A Mol.Spectrosc.Kinet.Environ.Gen.Theory* 110, 1735-1739, 2006; Seitzman, G.D., Cevallos, V. and Margolis, T.P., Rose bengal and lissamine green inhibit detection of herpes simplex virus by PCR, *Am.J.Ophthalmol.* 141,756-758, 2006; Fini, P., Loseto, R., Catucci, L., *et al.*, Study on the aggregation and electrochemical properties of Rose Bengal in aqueous solution of cyclodextrins, *Bioelectrochemistry* 70, 44-49, 2007.

SITS (stilbene isothiocyanate sulfonic acid; 4-Acetamido-4'-isothiocyanostilbene-,2'-disulfonic acid); MW 498 as disodium salt

Fluorescent stain; use on fixed tissues. Reactive groups permits use as a label for antibodies. Also used as a vital stain and as general cytoplasmic stain; used to characterize membrane anion-transport processes

Benjaminson, M.A. and Katz, I.J., Properties of SITS (4-acetamido-4'-isothiocyanostilbene-2,2'-disulfonic acid): fluorescence and biological staining, *Stain Technol.* 45, 57-62, 1970; Rothbarth, P.H., Tanke, H.J., Mul, N.A., *et al.*, Immunofluorescence studies with 4-acetamido-4'-isothiocyanato stilbene -2-2'-disulphonic acid (SITS), *J.Immunol.Methods* 19,101-109, 1978; Schmued, L.C. and Swanson, L.W., SITS: a covalently bound fluorescent retrograde tracer that does not appear to taken up by fibers-of-passage, *Brain Res.* 249, 137-141, 1982; Gilbert, P., Kettenmann, H., Orkland, R.K., and Schachner, M., Immunocytochemical cell identification in nervous system culture combined with intracellular injection of a blue fluorescing dye (SITS), *Neurosci.Lett.* 34, 123-128, 1982; Ploem, J.S., van Driel-Kulker, A.M., Goyarts-Veldstra, L., *et al.*, Image analysis combined with quantitative cytochemistry. Results and instrumental developments for cancer diagnosis, *Histochemistry* 84, 549-555, 1986; Pedini, V., Ceccarelli, P., and Gargiulo, A.M., A lectin histochemical study of the zygomatic salivary gland of adult dogs, *Vet.Res.Commun.* 19, 363-375, 1995; Papageorgiou, P.. ,Shmukler, B.E., Stuant-Tilley, A.K., *et al.*, AE anion exchangers in atrial tumor cells, *Am.J.Physiol.Heart Circ.Physiol.* 280, H937-H945, 2001; Quilty, J.A., Cordat, E., and Reithmeier, R.A., Impaired trafficking of human kidney anion exchanger (kAE1) caused by hetero-oligomer formation with a truncated mutant associated with distal renal tubular acidosis, *Biochem.J.* 368, 895-903, 2002. For SITS-sensitive anion transport see Villereal, M.L. and Levinson, C., Chloride-stimulated sulfate efflux in Ehrlich ascites tumor cells: evidence for 1:1 coupling, *J.Cell Physiol.* 90, 553-563, 1977; Kimelberg, H.K., Bowman, C.L., and Hirata, H., Anion transport in astrocytes, *Ann.N.Y.Acad.Sci.* 481, 334-353, 1986; Montrose, M., Randles, J., and Kimmich, G.A., SITS-sensitive Cl⁻ conductance pathway in chick intestinal cells, *Am.J.Physiol.* 253, C693-C699, 1987; Bassnett, S., Stewart, S., Duncan, G., and Crogha, P.C., Efflux of chloride from the rat lens: influence of membrane potential and intracellular acidification, *Q.J.Exp.Physiol.* 73, 941-949, 1988; Ishibashi, K., Rector, F.C., Jr., and Berry, C.A., Chloride transport across the basolateral membrane of rabbit proximal convoluted tubules, *Am.J.Physiol.* 258, F1569-F1578, 1990; Wilson, J.M., Laurent, P., Tufts, B.L. *et al.*, NaCl uptake by the branchial epithelium I freshwater teleost fish: an immunological approach to ion-transport protein localization, *J.Exp.Biol.* 203, 2279-2296, 2000; Small, D.L. and Tauskela, J., and Xia, Z., Role for chloride but not potassium channels in apoptosis in primary rat cortical cultures, *Neurosci.Lett.* 334, 95-98, 2002.

SOME BIOLOGICAL STAINS AND DYES (Continued)

Name	Description
Texas Red (TR, sulphorhodamine 101 acid chloride; MW 625	The presence of a reactive sulfonyl chloride function allows the labeling of protein amino groups. General probe for proteins and other molecules including DNA and carbohydrate to follow cellular transit.

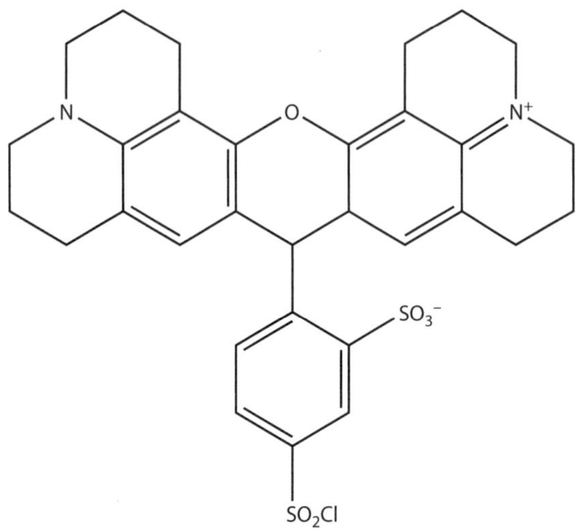

Titus, J.A., Haugland, R., Sharrow, S.O., and Segal, D.M., Texas Red, a hydrophilic, red-emitting fluorophore for use with fluorescein in dual parameter flow microfluorimetric and fluorescence microscopic studies, *J.Immunol.Methods* 50, 193-204, 1982; Schneider, H., Differential intracellular staining of identified neurones in Locusta with texas red and lucifer yellow, *J.Neurosci.Methods* 30, 107-115, 1989; Leonce, S. and Cudennec, C.A., Modification of membrane permeability measured by Texas-Red during cell cycle progression and differentiation, *Anticancer Res.* 10, 369-374, 1990; Srour, E.F., Lemmhuis, T., Brandt, J.E, *et al.,* Simultaneous use of rhodamine 123, phycoerythrin, Texas red, and allophycocyanin for the isolation of human hematopoietic progenitor cells, *Cytometry* 12, 179-183, 1991; Wessendorf, M.W. and Brelje, T.C., Which fluorophore is brightest? A comparison of the staining obtained using fluorescein, tetramethylrhodamine, lissamine rhodamine, Texas red, and cyanine 3.18, *Histochemistry* 98, 81-85, 1992; Belichenko, P.V. and Dahlstrom, A., Dual channel confocal laser scanning microscopy of Lucifer yellow-microinjected human brain cells combined with Texas red immunofluorescence, *J.Neurosci.Methods* 52, 111-118, 1994; Brismar, H., Trepte, O. and Ulfhake, B., Spectra and fluorescence lifetimes of lissamine rhodamine, tetramethylrhodamine isothiocyanate, texas red, and cyanine 3.18 fluorophores: influences of some environmental factors recorded with a confocal laser scanning microscope, *J.Histochem.Cytochem.* 43, 699-707, 1995; Anees, M., Location of tumour cells in colon tissue by Texas red labelled pentosan polysulphate, an inhibitor of a cell surface protease, *J.Enzyme Inhib.* 10, 203-214, 1996; Simon, S., Reipert, B., Eibl, M.M., Steinkasserer, A., Detection of phosphatidylinositol glycan class A gene transcripts by RT in situ PCR hybridization. A comparative study using fluorescein, Texas Red, and digoxigenin-11 dUTP for color detection, *J.Histochem.Cytochem.* 45, 1659-1664, 1997; Larramendy, M.L., El-Rifai, W., and Knutila, S., Comparison of fluorescein isothiocyanate- and Texas red-conjugated nucleotides for direct labeling in comparative genomic hybridization, *Cytometry* 31,174-179, 1998; Hembry, R.M., Detection of focal proteolysis using Texas-red-gelatin, *Methods Mol.Biol.* 151, 417-424, 2001; Kahn, E., Lizard, G., Frouin, F., *et al.,* Confocal analysis of phosphatidylserine externalization with the use of biotinylated annexin V revealed with streptavidin-FITC, -europium, -physcoerythrin or −Texas Red in oxysterol-treated apoptotic cells, *Anal.Quant.Cytol.Histol.* 23, 47-55, 2001; Alba, F.J. and Daban, J.R., Detection of Texas red-labelled double-stranded DNA by non-enzymatic peroxyoxalate chemiluminescence, *Luminescence* 16, 247-249, 2001; Watanabe, K. and Hattori, M., Real-time dual zymographic analysis of matrix metalloproteinases using fluorescein-isothiocyanate-labeled gelatin and Texas-red-labeled casein, *Anal.Biochem.* 307, 390-392, 2002; Tan, H.H., Thornhill, J.A., Al-Adhami, B.H., *et al.,,* A study of the effect of surface damage on the uptake of Texas Red-BSA by schistosomula of *Shistosoma mansoni*, *Paristology* 126, 235-240, 2003; Wippersteg, V., Ribeiro, F., Liedtke, S., *et al.,* The uptake of Texas Red-BSA in excretory system of schistosomes and its colocalisation with ER60 promoter-induced GFP in transiently transformed adult males, *Int.J.Parasitol.* 33, 1139-1143, 2003; Unruh, J.R., Gokulrangan, G., Wilson, G.S., and Johnson, C.K., Fluorescence properties of fluorescein, tetramethylrhodamine and Texas Red linked to a DNA aptamer, *Photochem.Photobiol.* 81, 682-690, 2005.

Thioflavin T (Basic Yellow 1; 318.9)	Fluorescent dye use to detect β-sheet structures such as amyloid protein; also used to detect conformational changes in therapeutic proteins on drying.

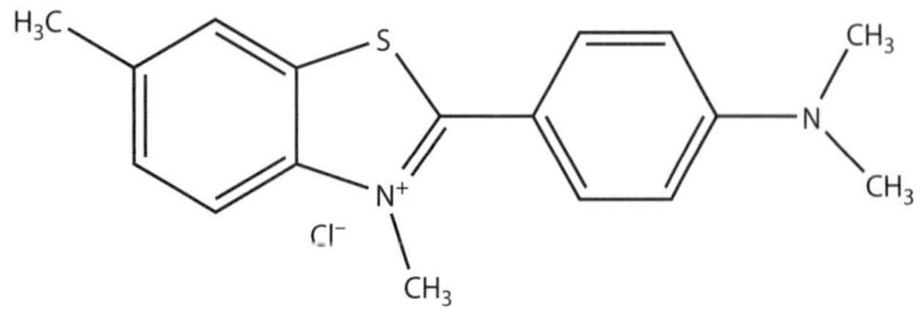

SOME BIOLOGICAL STAINS AND DYES (Continued)

Name	Description

Rogers, D.R., Screening for amyloid with the thioflavin-T fluorescent method, *Am.J.Clin.Pathol.* 44, 59-61, 1965; Saeed, S.M. and Fine, G., Thioflavin-T for amyloid detection, *Am.J.Clin.Pathol.* 47, 588-593, 1967; De Ferrari, G.V., Mallender, W.D., Inestrose, N.C. and Rosenberry, T.L., Thioflavin T is a fluorescent probe of the acetylcholinesterase peripheral site that reveals conformational interactions between the peripheral and acylation sites, *J.Biol.Chem.* 276, 23282-23287, 2001; Mathis, C.A, Bacskai, B.J., Kajdasz, S.T., *et al.*, A lipophilic thioflavin-T derivative for positron emission tomography (PET) imaging of amyloid in brain, *Bioorg. Med.Chem.Lett.* 12, 295-298, 2002; Kramenburg, O., Bouma, B., Kroon-Batenberg, M.J., *et al.*, Tissue-type plasminogen activator is a multiligand cross-β structure receptor, *Curr.Biol.*12, 1833-1839, 2002: Bouma, B., Loes, M., Kroon-Batenberg, M.J., *et al.*, Glycation induced formation of amyloid cross-β-structure, *J.Biol. Chem.* 278, 41810-41819, 2003; Krebs, M.R., Bromley, E.H., and Donald, A.M., The binding of thioflavin-T to amyloid fibrils: localization and implications, *J.Struct.Biol.* 149, 30-37, 2005; Khurana, R., Coleman, C., Ionescu-Zanetti, C., *et al.*, Mechanism of thioflavin T binding to amyloid fibrils, *J.Struct.Biol.* 151, 229-238, 2005; Inbar, P., Li, C.Q., Takayama, S.A., *et al.*, Oligo (ethylene glycol) derivatives of Thioflavin T as inhibitors of protein-amyloid interactions, *Chem.Bio. Chem* 7, 1563-1566, 2006; Okuno, A., Kato, M., and Taniguchi, Y., The secondary structure of pressure- and temperature-induced aggregation of equine serum albumin studied by FT-IR spectroscopy, *Biochim.Biophys.Acta* 1764, 1407-1412, 2006; Groenning, M., Olsen, L,. van de Weert, M., *et al.*, Study on the binding of Thioflavin T to β-sheet-rich and non-β-sheet cavities, *J.Struct.Biol.*, in press, 2006; Maskevich, A.A., Stsiapura, V.I., Kuzmitsky, V.A., *et al.*, Spectral properties of Thioflavin-T in solvents with different dielectric properties and in a fibril-incorporated form, *J.Proteome Res.*, in press, 2007; Maas, C., Hermeling, S., Bouma, B., *et al.*, A role for protein misfolding in immunogenicity of biopharmaceuticals, *J.Biol.Chem.* 282, 2229-2236, 2007

Name	Description
Toluidine Blue (C.I. Basic Blue 17; Methylene Blue T5O, TBO, toluidine blue O, tolonium chloride) MW 306	A diaminothiazine which can exist in a cation form which is hydrophilic. Used for metachromatic staining of biological molecules such as mucins and macromolecular structures. Used for demarcation of tumor tissue for surgery. Very early use for heparin neutralization in blood coagulation

Haley, T.J. and Rhodes, B., Effect of toluidine blue on the coagulation of fibrinogen by thrombin, *Science* 117, 604-606, 1953; Ball, J. and Jackson, D.S., Histological, chromatographic and spectrophotometric studies of toluidine blue, *Stain Technol.* 28, 33-40, 1953; Gustafsson, B.E. and Cronberg, S., The effect of hyaluronidase and toluidine blue on the mast cells in rats and hamsters, *Acta Rheumatol.Scand.* 5, 179-189, 1959; Schueller, E., Peutsch, M., Bohacek, L.G., and Gupta, R.K., A simplified toluidine blue stain for mast cells, *Can.J.Med.Technol.* 29, 137-138, 1967; Itzhaki, R.F., Binding of polylysine and Toluidine Blue to deoxyribonucleoprotein, *Biochem.J.* 121, 25P-26P, 1971; Sakai, W.S., Simple method for differential staining of paraffin embedded plant material using toluidine blue O, *Stain Technol.* 48, 247-249, 1973; Koski, J.P. and McGarvey, K.E., Toluidine blue as a capricious dye, *Am.J.Clin.Pathol.* 73, 457, 1980; Busing, C.M. and Pfiester, P., Permanent staining of rapid frozen section with toluidine blue, *Pathol.Res.Pract.* 172, 211-215, 1981; Drzymala, R.E., Liebman, P.A. and Romhanyi, G., Acid polysaccharide content of frog rod outer segments determined by metachromatic toluidine blue staining, *Histochemistry* 76, 363-379. 1982; O'Toole, D.K., The toluidine blue-membrane filter method: absorption spectra of toluidine blue stained bacterial cells and the relationship between absorbance and dry mass of bacteria, *Stain Technol.* 58, 357-364, 1983; Waller, J.R., Hodel, S.L., and Nuti, R.N., Improvement of two toluidine blue O-mediated techniques for DNase detection, *J.Clin.Microbiol.* 21, 195-199, 1985; Lior, H. and Patel, A., Improved toluidine blue-DNA agar for detection of DNA hydrolysis by campylobacters, *J.Clin.Microbiol.* 25, 2030-2031, 1987; Paardekooper, M., De Bruijne, A.W., Van Steveninck, J., and Van den Broek, P.J., Inhibition of transport systems in yeast by photodynamic treatment with toluidine blue, *Biochim.Biophys.Acta* 1151, 143-148, 1993; Passmore, L.J., and Killeen, A.A., Toluidine blue dye-binding method for measurement of genomic DNA extracted from peripheral blood leukocytes, *Mol.Diagn.* 1, 329-334, 1996; Korn, K., Greiner-Stoffele, T. and Hahn, U., Ribonuclease assays utilizing toluidine blue indicator plates, methylene blue, or fluorescence correlation spectroscopy, *Methods Enzymol.* 341, 142-153, 2001; Sanchez, A., Guzman, A., Ortiz, A.. *et al.*, Toluidine blue-O of prion protein deposits, *Histochem.Cell Biol.* 116, 519-524, 2001; Prat, E., Camps, J., del Rey, J., *et al.*, Combination of toluidine blue staining and in situ hybridization to evaluate paraffin tissue sections, *Cancer Genet. Cytogenet.* 155, 89-91, 2004; Kaji, Y., Hiraoka, T., and Oshika, T., Vital staining of squamous cell carcinoma of the conjunctive using toluidine blue, *Acta Ophthalmol.Scand.* 84, 825-826, 2006; Missmann, M., Jank, S., Laimer, K., and Gassner, R., A reason for the use of toluidine blue staining in the presurgical management of patients with oral squamous cell carcinomas, *Oral Surg.Oral Med. Oral Pathol. Oral Radiol.Endod.* 102, 741-743, 2006.

SOME BIOLOGICAL STAINS AND DYES (Continued)

Name	Description
Trypan Blue (diamine blue; C.I. Direct Blue 14) MW 961 as tetrasodium salt	A tetrasulfated anionic dye composed of a large planar aromatic system. Moderately soluble in water, more soluble in ethylene glycol but essentially insoluble in ethanol. Used a biological stain with particular interest as vital stain. Used in the early polychrome stains. Current use also as surgical stain for cataract surgery.

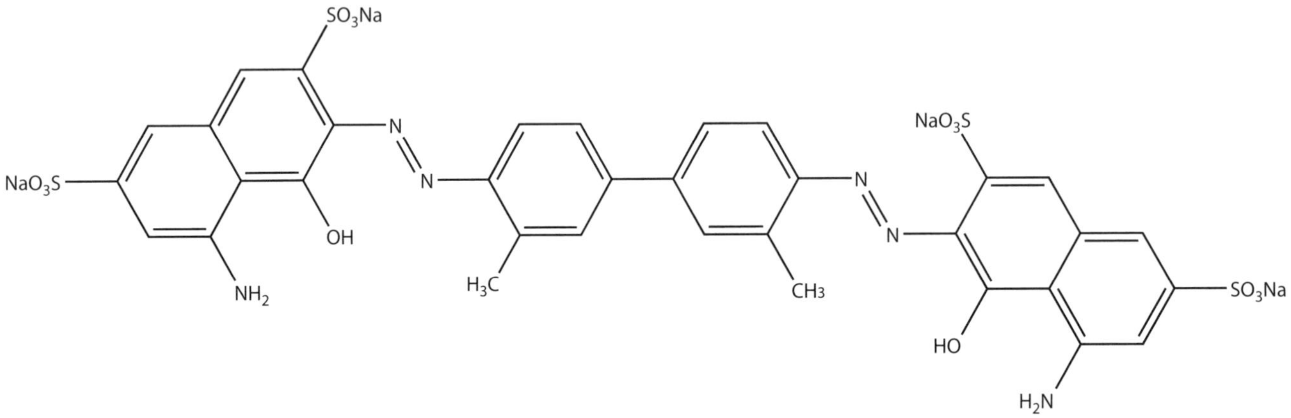

Menkin, V., Effects of ACTH on the mechanism of increased capillary permeability to trypan blue in inflammation, *Am.J.Physiol.* 166, 518-523, 1951; Wislocki, G.B. and Leduc, E.H., Vital staining of the hematoencephalic barrier by silver nitrate and trypan blue, and cytological comparisons of the neurohypophysis, pineal body, area postrema, intercolumnar tubercule and supraoptic crest, *J.Comp.Neurol.* 96, 371-413, 1952; Auskaps, A.M. and Shaw, J.H., Vital staining of calcifying bone and dentin with trypan blue, *J.Dent.Res.* 34, 452-459, 1955; Ferm, V.H., Permeability of the rabbit blastocyst to trypan blue, *Anat.Rec.* 125, 745-759, 1956; Kelly, J.W., Staining reactions of some anionic disazo dyes and histochemical properties of the red impurity in trypan blue, *Stain Technol.* 33, 89-94, 1958; Tennant, J.R., Evaluation of the trypan blue technique for determination of cell viability, *Transplantation* 2, 685-694, 1964; Holl, A., Vital staining by trypan blue: its selectivity for olfactory receptor cells of the brown bullhead, *Ictalurus natalis*, *Stain Technol.* 40, 269-273, 1965; Estupinan, J. and Hanson, R.P., Congo red and trypan blue as stains for plaque assay of Newcastle disease virus, *Avian Dis.* 13, 330-339. 1969; Lloyd, J.B. and Field, F.E., The red impurity in trypan blue, *Experientia* 26, 868-869, 1970; Dickinson, J.P. and Apricio, S.G., Trypan blue: reaction with myelin, *Biochem.J.* 122, 65P-66P, 1971; Davies, M., The effect of triton WR-1339 on the subcellular distribution of trypan blue and ^{125}I-labelled albumin in rat liver, *Biochem.J.* 136, 57-65, 1973; Davis, H.W. and Sauter, R.W., Fluorescence of Trypan Blue in frozen-dried embryos of the rat, *Histochemistry* 54, 177-189, 1977; Loike, J.D. and Silverstein, S.C., A fluorescence quenching technique using trypan blue to differentiate between attached and injested glutaraldehyde-fixed red blood cells in phagocytosing murine macrophages, *J.Immunol. Methods.* 57, 373-379, 1983; Boiadjieva, S., Hallberg, C., Hogstrom, M., and Busch, C., Methods in laboratory investigation. Exclusion of trypan blue from microcarriers by endothelial cells: an in vitro barrier function test, *Lab.Invest.* 50, 239-246, 1984; Lee, R.M., Chambers, C., O'Brodovich, H., and Forrest, J.B., Trypan blue method for the identification of damage to airway epithelium due to mechanical trauma, *Scan.Electron Microsc.* (Pt. 3), 1267-1271, 1984; Shen, W.C., Yang, D., and Ryser, H.J., Colorimetric determination of microgram quantities of polylysine by trypan blue precipitation, *Anal.Biochem.* 142, 521-524, 1984; Perry, S.W., Epstein, L.G., and Gelbard, H.A., *In situ* trypan blue staining of monolayer cell cultures for permanent fixation and mounting, *Biotechniques* 22, 1020-1024, 1997; Mascotti, K., McCullough, J., and Burger, S.R., HPC viability measurement: trypan blue versus acridine orange and propidium iodide, *Transfusion* 40, 693-696, 2000; Igarashi, H. Nagura, K., and Sugimura, H., Trypan blue as a slow migrating dye for SSCP detection in polyacrylamide gel electrophoresis, *Biotechniques* 29, 42-44, 2000.

Name	Description
Zincon; 2-[1-(2-hydroxy-5-sulfonatophenyl)-3-phenyl-5-formazano] benzoic acid, sodium salt MW 462	A metal complexing formazan dye which is used in histochemistry to demonstrate the presence of zinc; zincon is also used to stain for zinc proteins on solid support electrophoresis. Zincon is also used to demonstrate the presence of zinc in solution using spectrophotometric estimation of the zinc-dye complex

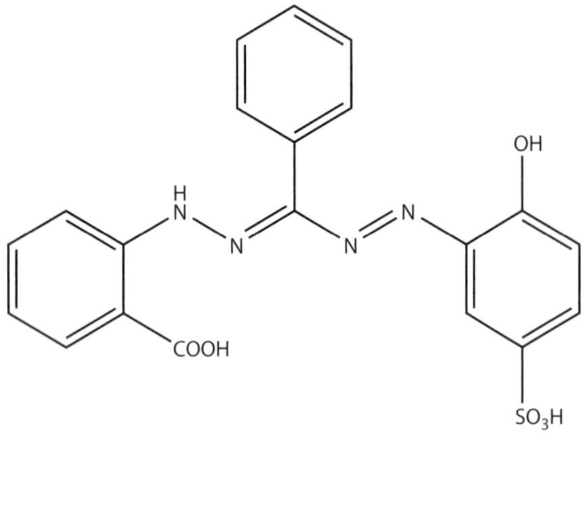

SOME BIOLOGICAL STAINS AND DYES (Continued)

Name	Description

Corns, C.M., A new colorimetric method for the measurement of serum calcium using a zinc-zincon indicator, *Ann.Clin.Biochem.* 24, 591-597, 1987; Richter, P., Toral, M.I., Tapia, A.E., and Fuenzalida, E., Flow injection photometric determination of zinc and copper with zincon based on the variation of the stability of the complexes with pH, *Analyst.* 122, 1045-1048, 1997; Choi, J.K., Tak, K.H., Jin, L.T., *et al.,* Background-free, fast protein staining in sodium dodecyl sulfate polyacrylamide gel using counterion dyes, zincon and ethyl violet, *Electrophoresis* 23, 4053-4059, 2002; Smejkal, G.B. and Hoff, H.F., Use of the formazan dye zincon for staining proteins in polyacrylamide gels, *Biotechniques* 34, 486-468, 2003; Choi, J.K., Chae, H.Z., Hwang, S.Y., *et al.,* Fast visible dye staining of proteins in one- and two-dimensional sodium dodecyl sulfate-polyacrylamide gels compatible with matrix assisted laser desorption/ionization-mass spectrometry, *Electrophoresis* 25, 1136-1141, 2004; Morais, I.P., Souto, M.P., and Rangel, A.O., A double-line sequential injection system for the spectrophotometric determination of copper, iron, manganese, and zinc in waters, *J.AOAC Int.* 88, 639-644, 2005; Ribiero, M.F., Dias, A.C., Santos, J.L., *et al.,* Fluidized beds in flow analysis: use with ion-exchange separation for spectrophotometric determination of zinc in plant digests, *Anal.Bioanal.Chem.* 384, 1019-1024, 2006.

Biological stains and dyes – general references

The Chemistry of Synthetic Dyes and Pigments, ed. H.A. Lubs, American Chemical Society, Reinhold Publishing, New York, NY, USA, 1955

Kiernan, J.A., Classification and naming of dyes, stains and fluoro-chromes, *Biotech.Histochem.* 76, 261-278, 2001

Conn's Biological Stains a Handbook of Dyes, Stains, and Fluorochromes for use in Biology and Medicine, ed. R.W. Horobin and J.A. Kiernan, Bios, Oxford University Press, Oxford, UK, 2002.

MORDANT DYES

Mordant dye is a dye that interacts with a tissue, cell, or subcellular organelle via an interaction with a substance, a mordant, which interacts with the substrate and the dye. The interaction may involve covalent interaction. A mordant can be defined as a substance which interacts with both the substrate (for example tissue, cell, subcellular organelle) and the dye[1]. Most mordants are metal salts such ferric chloride and those of chromium and vanadium. Other compounds such as tannic acid[1] and galloylglucoses[2] can serve as mordants. Hematoxylin used in Hematoxylin and eosin (H and E) staining which contains a mixture of chelates formed between hematein and aluminum ions[3,4]. Mordant blue 3 (Chromoxane cyanine R) uses iron[5]. There has been a significant increase in our understanding of mordant dyes[4,6].

References

1. Afzelius, B.A., Section staining for electron microscopy using tannic acid as a mordant: a simple method for visualization of glycogen or collagen, *Microsc.Res.Tech.* 21, 65-72, 1992.
2. Simionescu, N. and Simionescu, M., Galloylgluose of low molecular weight as mordant in electron microscopy. I. Procedure, and evidence mordanting effect, *J.Cell Biol.* 70, 608-621, 1976.
3. Bettinger, C. and Zimmermann, H.W., New investigations on hematoloxylin, hematein and hematein-aluminum complexes. 2. Hematein-aluminum complexes and hemalum staining, *Histochemistry* 96, 215-228, 1991.
4. Horobin, R.W., Biological staining: mechanisms and theory, *Biotechnic & Histochemistry* 77, 3-13, 2002.
5. Kiernan, J.A., Chromoxane cyanine R. I. Physical and chemical properties of the dye and of some of its iron complexes, *J.Microsc.* 134, 13-23, 1984.
6. Dapson, R.W., Dye-tissue interactions: mechanisms, quantification and bonding parameters for dyes used in biological staining, *Biotech. Histochem.* 80, 49-72, 2005.

General references for mordant dyes

Conn's Biological Stains, 10th edn., ed. R.W. Horobin and J.A. Kiernan, Bios Scientific Publishers, Oxford, UK, 2002

Mordant Dyes

Dye Name	Metal Ion Specificity[a]	Comments
Mordant Blue 1; chrome azurol A; chromoxane pure blue B	Al, B, Ca,	Calcium staining in histochemistry
Mordant Blue 3; eriochrome cyanine R; chromoxane cyanine R	Al, Cr, Fe	Iron staining in histochemistry; also Al
Mordant Blue 10; Gallocyanine;	Cr, Fe, Pb	Chromium complex for staining DNA and RNA
Mordant Blue 14; Celestine blue;	Fe	Complex with iron use as nuclear stain
Mordant Blue 29; chromoxane pure blue BLD	Be	Be staining in histochemistry
Mordant Blue 45; Gallamine blue	Al, Ca	Al complex for staining nuclei
Mordant Blue 80; Chromotrope 2R	Cr	Cytoplasmic counterstain
Mordant Red 3; Alizarin red S	Al, Ca	Calcified tissue stain
Mordant Violet 39; aurin tricarboxylic acid (ATA); aluminon; chrome violet TG	Al, Be, Cr, Fe	Aluminum staining in histochemistry

[a] Metal ion specificity is broad with most mordant stains so other metal ions can be complexed by the respective mordant stain.

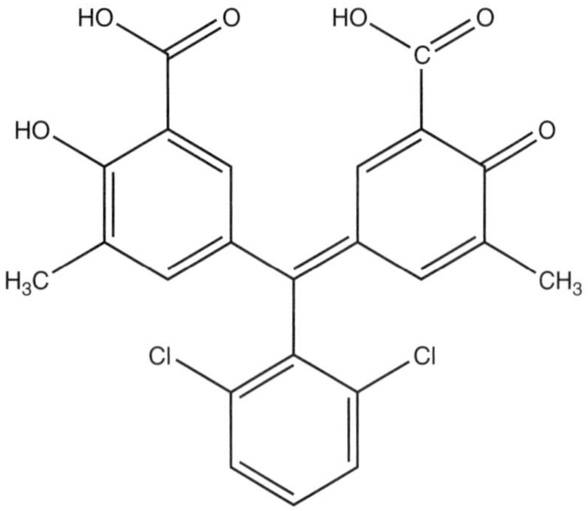

Mordant blue 1; Chrome azurol B

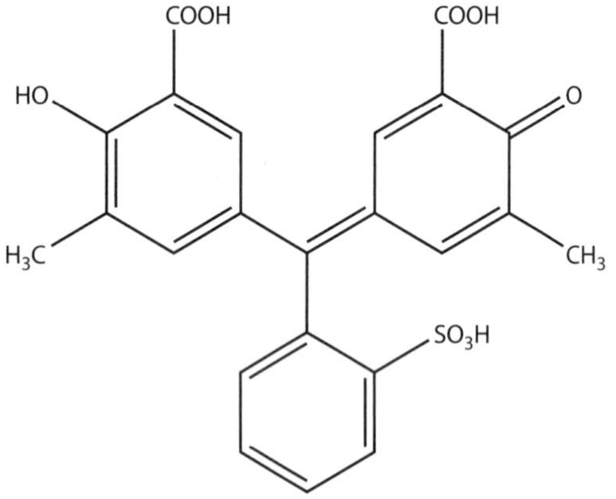

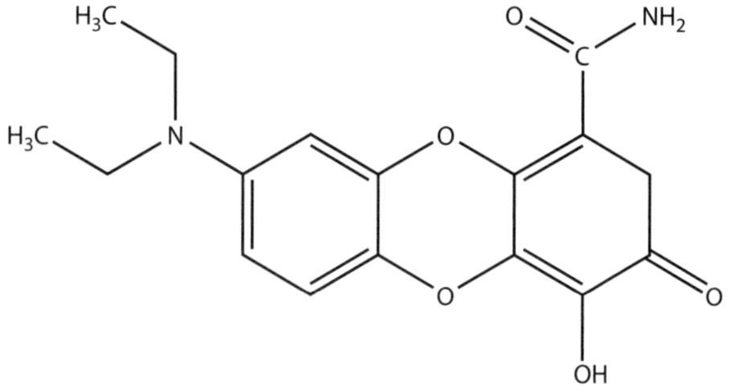

Mordant blue 14; Celestine blue

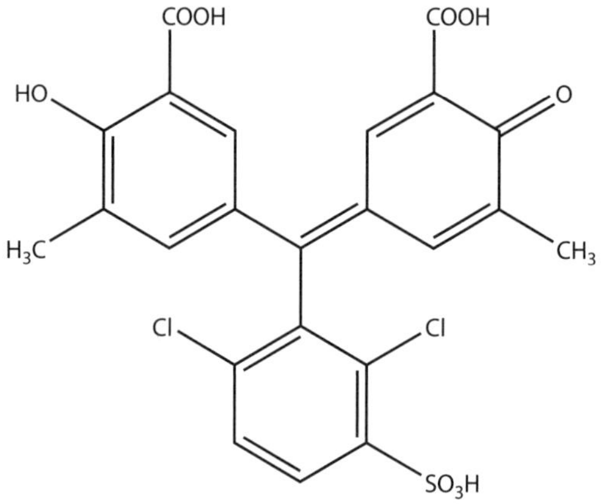

Mordant blue 3; Eriochrome cyanine R

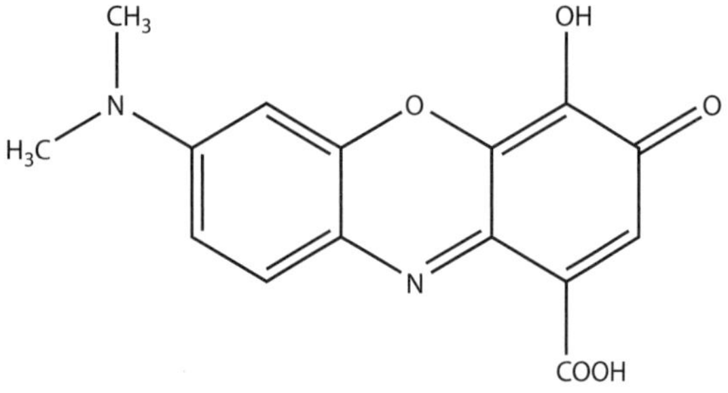

Mordant blue 29; Chromoxane pure blue

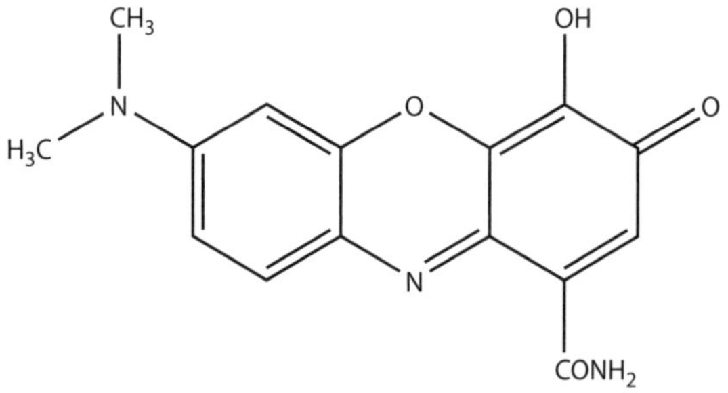

Mordant blue 10; Gallocyanine

Mordant blue 45; Gallamine blue

Mordant blue 80; Chromotrope 2R

Mordant Violet 39; Aurin Tricarboxylic Acid

Mordant red 3; Alizarin red S

METAL CHELATING AGENTS[a]

Name	M.W.	Description
BAPTA [1,2-bis(*o*-aminophenoxy) ethane, *N,N,N',N'*-tetraacetic acid	476.4	Chelating agent with higher affinity for zinc than calcium can be used as with EGTA for chelation of intracellular metal ions. BAPTA-AM, the acetoxymethyl ester, is a useful derivative.

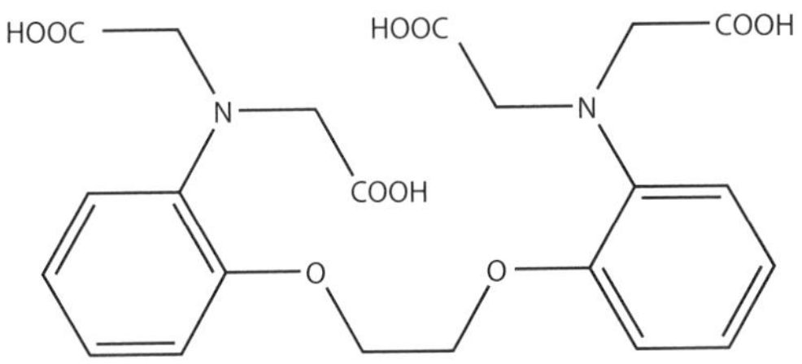

Harrison, S.M. and Bers, D.M., The effect of temperature and ionic strength on the apparent Ca-affinity of EGTA and the analogous Ca-chelators BAPTA and dibromo-BAPTA, *Biochim.Biophys.Acta.* 925, 133-143, 1987; Minta, A., Kao, J.P., and Tsien, R.Y., Fluorescent indicators for cytosolic calcium based on rhodamine and fluorescein chromophores, *J.Biol.Chem.* 264, 8171-8178, 1989; Csermely, P., Sandor, P., Radics, L., and Somogyi, J., Zinc forms complexes with higher kinetic stability than calcium, 5-F-BAPTA as a good example, *Biochem.Biophys.Res.Commun.* 165, 838-844, 1989; Marks, P.W. and Maxfield, F.R., Preparation of solutions with free calcium concentration is the nanomolar range using 1,2-*bis*(*o*-aminophenoxy)ethane-*N,N,N',N'*-tetraacetic acid, *Anal. Biochem.* 193, 61-71, 1991; Brooks, S.P. and Storey, K.B., Bound and determined: a computer program for making buffers of defined ion concentrations, *Anal. Biochem.* 201, 119-126, 1992; Dieter, P., Fitzke, E., and Duyster, J., BAPTA induces a decrease of intracellular free calcium and a translocation and inactivation of protein kinase C in macrophages, *Biol.Chem.Hoppe Seyler* 374, 171-174, 1993; Natarajan, V., Scribner, V.M., and Taher, M.M., 4-Hydroxynonenal, a metabolite of lipid peroxidation, activates phospholipase D in vascular endothelial cells, *Free Radic.Biol.Med.* 15, 365-375, 1993; Bers, D.M., Patton, C.W., and Nuccitelli, R., A practical guide to the preparation of Ca^{2+} buffers, *Methods Cell Biol.* 40, 3-29, 1994; Oiki, S., Yamamoto, T., and Okada, Y., A simultaneous evaluation method of purity and apparent stability constant of Ca-chelating agents and selectivity coefficient of Ca-selective electrodes, *Cell Calcium* 15, 199-208, 1994; Aballay, A., Sarrouf, M.N., Colombo, M.I., *et al.*, Zn^{2+} depletion blocks endosome fusion, *Biochem.J.* 312, 919-923, 1995; Britigan, B.E., Rasmussen, G.T., and Cox, C.D., Binding of iron and inhibition of iron-dependent oxidative cell injury by the "calcium chelator" 1,2-bis(2-aminophenoxy) ethane, *N,N,N',N'*-tetraacetic acid, *Biochem.Pharmacol.* 55, 287-295, 1998; Kim-Park, W.K., Moore, M.A., Hakki, Z.W., and Kowolik, M.J., Activation of the neutrophil respiratory burst requires both intracellular and extracellular calcium, *Ann.N.Y.Acad.Sci.* 832, 394-404, 1997; Barbouti, A., Doulias, P.T., Zhu, B., *et al.*, Intracellular iron, but not copper, plays a critical role in hydrogen peroxide-induced DNA damage, *Free Radic.Biol.Med.* 31, 490-498, 2001; Nielsen, A.D., Fuglsang, C.C., and Westh, P., Isothermal titration calorimetric procedure to determine protein-metal ion binding parameters in the presence of excess metal ion or chelator, *Anal.Biochem.* 314, 227-234, 2003; Gee, K.R., Rukavishnikov, A. and Rothe, A., New Ca^{2+} fluoroionophores based on the BODIPY fluorophore, *Comb.Chem.High Throughput Screen.* 6, 363-366, 2003; Swystum, V., Chen, L., Factor, P., *et al.*, Apical trypsin increases ion transport and resistance by a phospholipase C-dependent rise of Ca^{2+}, *Am.J.Physiol.Lung Cell.Mol.Physiol.* 288, L820-L830, 2005; Lazzaro, M.D., Cardenas, L., Bhatt, A.P., *et al.*, Calcium gradients in conifer pollen tube; dynamic properties differ from those seen in angiosperms, *J.Exp.Bot.* 56, 2619-2628, 2005

Name	M.W.	Description
α-**Benzoin oxime** (cupron); 2-hydroxy-1,2-diphenylethanone oxime	227.3	Complexes cupric ions, molybdenum, chromium, lead, or tungsten as well as other metal ions. Used for the analysis of these metal ions. More recently, *α*-Benzoin oxime has been immobilized and used as a chelating resin for concentration and separation of metal ions. Copper complexes have been used as biocidal agent.

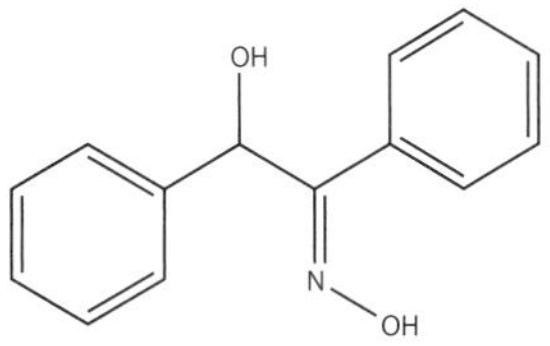

alpha-benzoin oxime

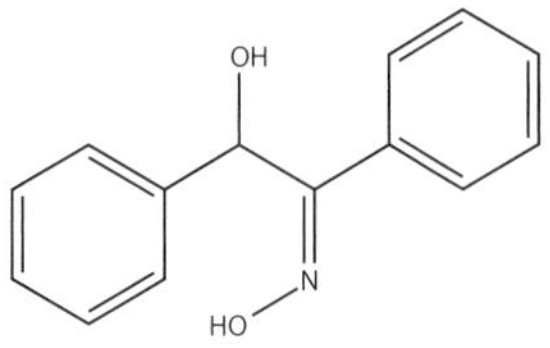

beta-benzoin oxime

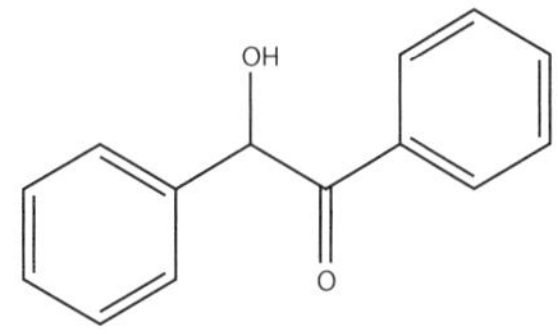

benzoin

METAL CHELATING AGENTS (Continued)

Name	M.W.	Description

Borkow, G. and Gabbay, J., Putting copper into action-impregnated products with potent biocidal activities, *FASEB J.* 18, 1728-1730, 2004; Borkow, G. and Gabbay, J., Copper as a biocidal tool, *Curr.Med.Chem.* 12, 2163-2175, 2005; Ghaedi, M., Asadpour, E., and Vafaie, A., Sensitized spectrophotometric determination of Cr(III) ion for speciation of chromium ion in surfactant media using α-benzoin oxime, *Spectrochim.Acta A Mol. Biomol.Spectrosc.* 63, 182-188, 2006; Soylak, M. and Tuzen, M., Dianion SP-850 resin as a new solid phase extractor for preconcentration-separation of trace metal ions in environmental samples, *J.Hazard.Mater.* 137, 1496-1501, 2006

Chromotropic acid; 4,5-dihydroxy-2,7-naphthalenedisulfonic acid	320.3	Chelation and fluorometric determination of aluminum and other metal ion. Also used for the analysis formaldehyde.

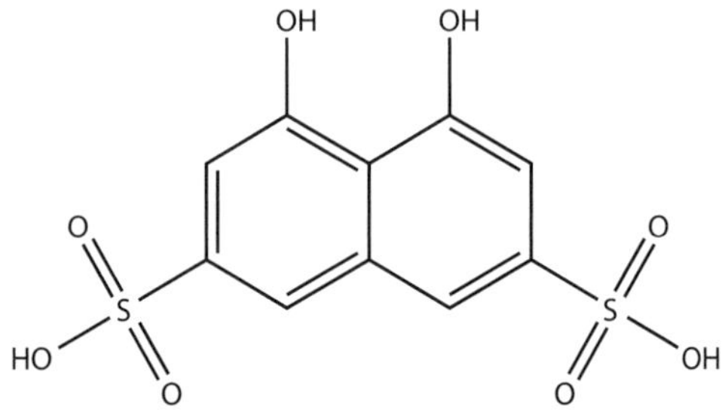

Metal Ions: Durham, A.C. and Walton, J.M., A survey of the available colorimetric indicators for Ca^{2+} and Mg^{2+} ions in biological experiments, *Cell Calcium* 4, 47-55, 1983; Prestel, H., Gahr, A., and Niessner, R., Detection of heavy metals in water by fluorescence spectroscopy: on the way to a suitable sensor system, *Fresenius J.Anal.Chem.* 368, 182-191, 2000; Destandau, E., Alain, V., and Bardez, E., Chromotropic acid, a fluorogenic chelating agent for aluminum(III), *Anal.Bioanal.Chem.* 378, 402-410, 2004; Themelis, D.G. and Kika, F.S., Flow and sequential injection methods for the spectrofluorometric determination of aluminum in pharmaceutical products using chromotropic acid as chromogenic reagent, *J.Pharm.Biomed.Anal.* 41, 1179-1185, 2006; Lemos, V.A., Santos, L.N., Alves, A.P., and David, G.T., Chromotropic acid-functionalized polyurethane foam: A new sorbent for on-line preconcentration and determination of cobalt and nickel in lettuce samples, *J.Sep.Sci.* 29, 1197-1204, 2006; Formaldehyde: Manius, G.J., Wen, L.F., and Palling, D., Three approaches to the analysis of trace formaldehyde in bulk and dosage from pharmaceuticals, *Pharm.Res.*10, 449-453, 1993; Flyvholm, M.A., Tiedmann, E., and Menne, T., Comparison of 2 tests for clinical assessment of formaldehyde exposure, *Contact Dermatitis* 34, 35-38, 1996; Pretto, A., Milani, M.R., and Cardoso, A.A., Colorimetric determination of formaldehyde in air using a hanging drop of chromotropic acid, *J.Environ.Monit.* 2, 566-570, 2000

Citric acid (2-hydroxy-1,2,3-propanetricarboxylic acid	192.1	Moderately strong chelating agent. Used frequently for calcium and iron

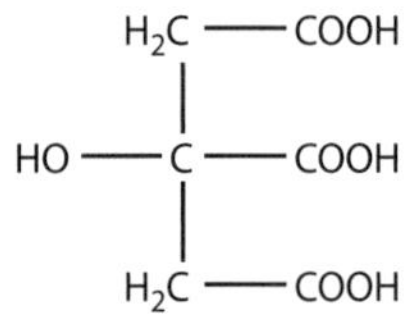

Hopkins, E.W. and Herbst, E.J., An explanation for the apparent chelation of calcium by tetrodotoxin, *Biochem.Biophys.Res.Commun.* 30, 528-533, 1968; Steinmetz, W.L., Glick, M.R., and Oei, T.O., Modified aca method for determination of iron chelated by deferoxoamine and other chelators, *Clin. Chem.* 26, 1593-1597, 1980; Ford, W.C. and Harrison, A., The role of citrate in determining the activity of calcium ions in human semen, *Int.J.Androl.* 7, 198-202, 1984; Lund, A.J. and Aust, A.E., Iron mobilization from asbestos by chelators and ascorbic acid, *Arch.Biochem.Biophys.* 278, 61-64, 1990; Francis, B., Seebart, C., and Kaiser, I.I., Citrate is an endogenous inhibitor of snake venom enzymes by metal-ion chelation, *Toxicon* 30, 1239-1246, 1992; Morgan, J.M., Navabi, H., Schmid, K.W., and Jasani, B., Possible role of tissue-bound calcium ions in citrate-mediated high temperature antigen retrieval, *J.Pathol.* 174, 301-307, 1994; Rhee, S.H. and Tanaka, J., Effect of citric acid on the nucleation of hydroxyapatite in a simulated body fluid, *Biomaterials* 20, 2155-2160, 1999; Engelmann, M.D., Bobier, R.T., Hiatt, T., and Cheng, I.F., Variability of the Fenton reaction characteristics of the EDTA, DTPA, and citrate complexes of iron, *Biometals* 16, 519-527, 2003; Fernandez, V. and Winkelmann, G., The determination of ferric iron in plants by HPLC using the microbial iron chelator desferrioxamine E, *Biometals* 18, 53-62, 2005; Matinaho, S., Karhumäki, P., and Parkkinen, J., Bicarbonate inhibits the growth of *Staphylococcus epidermidis* in platelet concentrates by lowering the level of non-transferrin-bound iron, *Transfusion* 45, 1768-1773, 2005; Reynolds, A.J., Haines, A.H., and Russell, D.A., Gold glyconanoparticles for mimics and measurement of metal ion-mediated carbohydrate-carbohydrate interactions, *Langmuir* 22, 1156-1163, 2006

METAL CHELATING AGENTS (Continued)

Name	M.W.	Description
Cupferron; *N*-hydroxy-*N*- nitrosobenzene, ammonium salt	155.2	Chelation of and precipitation of iron, copper, zinc, vanadium

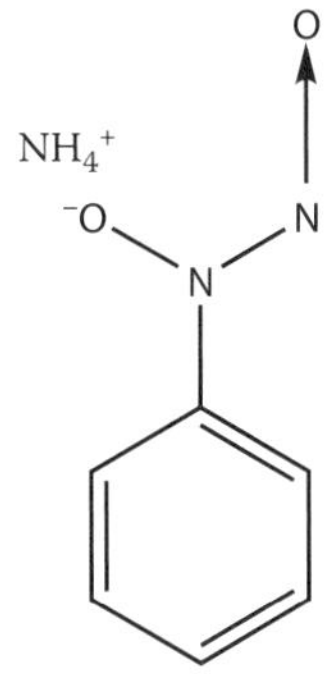

Kolthoff, I.M. and Liberti, A., Amperometric titration of copper and ferric iron with cupferron, *Analyst* 74, 635-641, 1949; Ahuja, B.S., Kiran, U., and Sudershan, *In vivo* & *in vitro* inhibition of mung bean superoxide dismutase by cupferron, *Indian J. Biochem.Biophys.* 18, 86-87, 1981; Walsh, K.A., Daniel, R.M., and Morgan, H.W., A soluble NADH dehydrogenase (NADH: ferricyanide oxidoreductase) from *Thermus aquaticus* strain T351, *Biochem.J.* 209, 427-433, 1983; Danzaki, Y., Use of cupferron as a precipitant for the determination of impurities in high-purity iron by ICP-AES, *Anal. Bioanal.Chem.* 356, 143-145, 1996; Heinemann, G. and Vogt, W., Quantification of vanadium in serum by electrothermal atomic absorption spectrometry, *Clin.Chem.* 42, 1275-1282, 1996; Oztekin, N. and Erim, F.B., Separation and direct UV detection of lanthanides complexed with cupferron by capillary electrophoresis, *J.Chromatog.A.* 895, 263-268, 2000; Hou, Y., Xie, W., Janczuk, A.J., and Wang, P.G. *O*-Alkylation of cupferron: aiming at the design and synthesis of controlled nitric oxide releasing agents, *J.Org.Chem.* 65, 4333-4337, 2000; Bourque, J.R., Burley, R.K., and Bearne, S.L., Intermediate analogue inhibitors of mandelate racemase: *N*-hydroxyformanilide and cupferron, *Bioorg.Med.Chem.Lett.* 17, 105-108, 2007

Name	M.W.	Description
Sodium diethyldithiocarbamate; Diethyldithiocarbamate, sodium salt (Dithio carb sodium); diethyl- carbamothioic acid, sodium salt; diethyldithiocarbamic acid sodium salt	171.2	Chelation of zinc, copper, mercury, and nickel

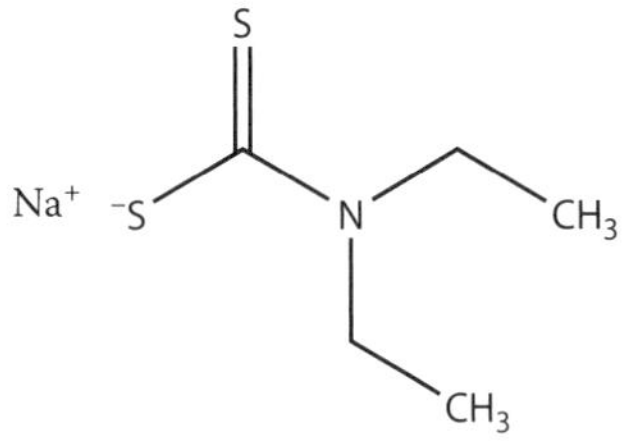

Rigas, D.A., Eginitis-Rigas, C. and Head, C., Biphasic toxicity of diethyldithiocarbamate, a metal chelation, to T lymphocytes and polymorphonuclear granulocytes: reversal by zinc and copper, *Biochem.Biophys.Res.Commun.* 88, 373-379, 1979; Khandelwal, S., Kachru, D.N., and Tandon, S.K., Chelation in metal intoxication. IX. Influence of amino and thiol chelators on excretion of manganes in poisoned rabbits, *Toxicol.Lett.* 6, 131-135, 1980; Marciani, D.J. Wilkie, S.D., and Schwartz, C.L., Colorimetric determination of agarose-immobilized proteins by formation of copper-protein complexes, *Anal.Biochem.* 128, 130-137, 1983; O'Callaghan, J.P. and Miller, D.B., Diethyldithiocarbamate increases distribution of cadmium to brain but prevents cadmium-induced neurotoxicity, *Brain Res.* 170, 354-358, 1986; Khandelwal, S., Kachru, D.N., and Tandon, S.K., Influence of metal chelators on metalloenzymes, *Toxicol.Lett.* 37, 213-219, 1987; Tandon, S.K., Hashmi, N.S., and Kashru, D.N., The lead-chelating effects of substituted dithiocarbamates, *Biomed.Environ.Sci.* 3, 299-305, 1990; Borrello, S., De Leo, M.E., Landricina, M. *et al.*, Diethyldithiocarbamate treatment up regulates manganese superoxide dismutase gene expression in rat liver, *Biochem.Biophys.Res.Commun.* 220, 546-552, 1996

Name	M.W.	Description
Dimethylglyoxime; 2,3-butanedionedioxime; diacetyldioxime	116.1	Primary for chelation and detection of nickel; also for separation of lead and chelation of copper

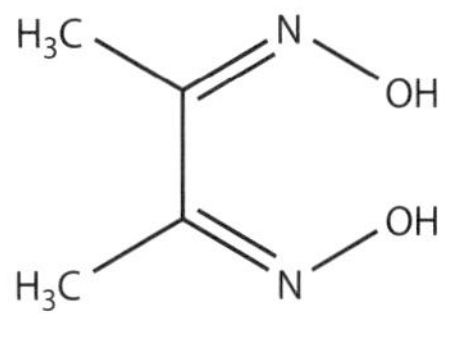

METAL CHELATING AGENTS (Continued)

Name	M.W.	Description

Lee, D.W. and Halmann, M., Selective separation of nickel (II) by dimethylglyoxime-treated polyurethane foam, *Anal.Chem.* 48, 2214-2218, 1976; Dixon, N.E., Gazzola, C., Asher, C.J. *et al.*, Jack Bean urease (EC 3.5.1.5)-II. The relationship between nickel, enzymatic activity, and the "abnormal" ultraviolet spectrum. The nickel content of jack beans, *Can.J.Biochem.* 58, 474-480, 1980; Huber, K.R., Sridhar, R., Griffith, E.H., *et al.*, Superoxide dismutase-like activities of copper(II) complexes tested in serum, *Biochim.Biophys.Acta* 915, 267-276, 1987; Heo, J., Staples, C.R., Halbleib, C.M., and Ludden, P.W., Evidence for a ligand CO that is required for catalytic activity of CO dehydrogenase from *Rhodospirilum rubrum*, *Biochemistry* 39, 7956-7963, 2000; Celo, V., Murimboh, J., Salam, M.S., and Chakrabarti, C.L., A kinetic study of nickel complexation in model systems by adsorptive catholic stripping voltammetry, *Environ.Sci.Technol.* 35, 1084-1089, 2001; Ponnuswamy, T. and Chyan, O., Detection of Ni^{2+} by a dimethylglyoxime probe using attenuated total-reflection infrared spectrometry, *Anal.Sci.* 18, 449-453, 2002

Dithizone; diphenylthiocarbazone   256.3   Chelating and measurement of mercury, zinc, cobalt, copper, and lead. Extensive use for the histochemical detection of zinc.

Landry, A.S., Optimum range for maximum accuracy in biological lead analyses by dithizone, *Ind.Health Mon.* 11, 103, 1951; Mager, M., McNary, W.F., Jr., and Lionetti, F., The histochemical detection of zinc, *J.Histochem.Cytochem.* 1, 493-504, 1953; McNary, W.F., Jr., Dithizone staining of myeloid granules, *Blood* 12, 644-648, 1957; Butler, E.J., and Newman, G.E., An absorptiometric method for the determination of traces of copper in biological materials with dithizone, *Clin.Chim.Acta* 11, 452-460, 1965; Shendriker, A.D. and West, P.W., Microdetermination of lead with dithizone and the ring-oven technique, *Anal.Chim.Acta* 61, 43-48, 1972; Nabrzyski, M., Spectrophotometric method for copper and mercury determination in the same food sample using dithizone and lead diethyldithiocarbamate, *Anal.Chem.* 47, 552-553, 1975; Song, M.K., Adham, N.F., and Rinderknecht, R., A simple, highly sensitive colorimetric method for the determination of zinc in serum, *Am.J.Clin.Pathol.* 65, 229-233, 1976; Holmquist, B., Elimination of adventitious metals, *Methods Enzymol.* 158, 6-12, 1988; Goldberg, E.D., Eschenko, V.A., and Bovt, V.D., Diabetogenic activity of chelators in some mammalian species, *Endocrinologie* 28, 51-55, 1990; Kawamura, C., Kizaki, M., Fukuchi, Y., and Ikeda, Y., A metal chelator, diphenylthiocarbazone, induces apoptosis, in acute promyelocytic leukemia (APL): cells mediated by a caspase-dependent pathway with a modulation of retinoic acid signaling pathways, *Leuk.Res.* 26, 661-668, 2002; Shaw, M.J., Jones, P., and Haddad, P.R., Dithizone derivatives as sensitive water soluble chromogenic reagents of the ion chromatographic determination of inorganic and organo-mercury in aqueous matrices, *Analyst* 128, 1209-1212, 2003; Santos, I.G., Hagenbach, A., and Abram, U., Stable gold(III) complexes with thiosemicarbazone derivatives, *Dalton Trans.* (4), 677-682, 2004; Khan, R., Ahmed, M.J., and Bhanger, M.I., A rapid spectrophotometric method for the determination of trace level lead using 1,5-diphenylthiocarbazone in aqueous micellar solutions, *Anal.Sci.* 23, 193-199, 2007

α,α-Dipyridyl; 2,2'-dipyridyl; bipyridyl;   156.2   Used for chelation of ferrous iron and other divalent metal cations
BIPY

Fredens, K. and Danscher, G., The effect of intravital chelation with dimercaprol, calcium disodium edentate, 1-10-phenanthroline and 2,2'-dipyridyl on the sulfide silver strainability of the rat brain, *Histochemie* 37, 321-331, 1973; Evans, S.A. and Shore, J.D., The role of zinc-bound water in liver alcohol dehydrogenase catalysis, *J.Biol.Chem.* 255, 1509-1514, 1980; Rao, G.H., Cox, A.C., Gerrard, J.M., and White, J.G., Effects of 2,2'-dipyridyl and related compounds on platelet prostaglandin synthesis and platelet function, *Biochim.Biophys.Acta* 628, 468-479, 1980; Ikeda, H., Wu, G.Y., and Wu, C.H., Evidence that an iron chelator regulates collagen synthesis by decreasing the stability of procollagen mRNA, *Hepatology* 15, 282-287, 1992; Hales, N.J. and Beattie, J.F., Novel inhibitors of prolyl 4-hydroxylase. 5. The intriguing structure-activity relationships seen with 2,2'-bipyridine and its 5,5'-dicarboxy acid derivatives, *J.Med.Chem.* 36, 3853-3858, 1993; Henley, R. and Worwood, M., The enhancement of iron-dependent luminal peroxidation by 2,2'-dipyridyl and nitrilotriacetate, *J.Biolumin.Chemilumin.* 9, 245-250,1994; Nocentini, G. and Barzi, A., The 2,2'-bipyridyl-6-carbothioamide copper (II) complex differs from the iron(II) complex in its biochemical effects in tumor cells, suggesting possible differences in the mechanism leading to cytotoxicity, *Biochem.Pharmacol.* 52, 65-71, 1996; Romeo, A.M., Christen, L, Niles, E.G., and Kosman, D.J., Intracellular chelation of iron by bipyridyl inhibits DNA virus replication: ribonucleotide reductase maturation as a probe of intracellular iron pools, *J.Biol.Chem.* 276, 24301-24308, 2001; Slingsby, R.W., Bordunov, A, and Grimes, M., Removal of metallic impurities from mobile phases in reversed-phase high-performance liquid chromatography by the use of an in-line chelation column, *J.Chromatog. A.* 913, 159-163, 2001; Huang, K., Dai, J., Fournier, J., *et al.*, Ferrous ion autooxidation and its chelation in iron-loaded human liver HepG2 cells, *Free Radic.Biol.Med.* 32, 84-92, 2002; Pfister, A. and Fraser, C.L., Synthesis and unexpected reactivity of iron tris(bipyridine) complexes with poly(ethylene glycol) macroligands, *Biomacromolecules* 7, 459-468, 2006

METAL CHELATING AGENTS (Continued)

Name	M.W.	Description
EDTA (ethylenediaminetetraacetic acid)	292	General chelation agent[b] Structure serves as the basis for therapeutic agents which deliver radioactive materials to tumor cells. EDTA is used to detach adherent cells from tissue culture surfaces. Frequently included in protease inhibitor "cocktails." The affinity for metal ions decrease as pH decreases and there may be a change in the relative affinity for specific metal ions[b].

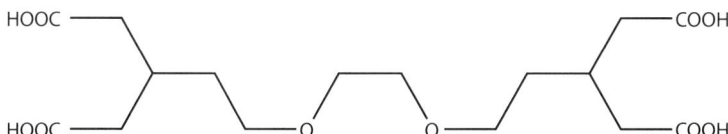

Sakabe, I., Paul, S., Mitsumoto, T., *et al.*, A factor that prevents EDTA-induced cell growth inhibition: purification of transethyretin from chick embryo brain, *Endocr.J.* 44, 375-391, 1999; Rocken, C. and Roessner, A., An evaluation of antigen retrieval procedures for immunoelectron microscopic classification of amyloid deposits, *J.Histochem.Cytochem.* 47, 1385-1394, 1999; Sciaudone, M.P., Chattopadhyay, S., and Freake, H.C., Chelation of zinc amplifies induction of growth hormone mRNA levels in cultured rat pituitary tumor cells, *J.Nutr.* 130, 158-163, 2000; Welch, K.D., Davis, T.Z., and Aust, S.D., Iron autoxidation and free radical generation: effects of buffers, ligands, and chelators, *Arch.Biochem.Biophys.* 397, 360-369, 2002; Breccia, J.D., Andersson, M.M., and Hatti-Kaul, R., The role of poly(ethyleneimine) in stabilization against metal-catalyzed oxidation of proteins: a case study with lactate dehydrogenase, *Biochim.Biophys.Acta* 1570, 165-173, 2002; Cowart, R.E, Reduction of iron by extracellular iron reductases: implications for microbial iron acquisition, *Arch.Biochem.Biophys.* 430, 273-281, 2002; Geebelen, W., Vangronaveld, J., Adriano, D.C., *et al.*, Effects of Pb-EDTA and EDTA on oxidative stress reactions and mineral uptake in *Phaseolus vulgaris*, *Physiol.Plant* 115, 377-382, 2002; Powis, D.A. and Zerbes, M., *In situ* chelation of Ca(2+) in intracellular stores induces capacitative Ca(2+) entry in bovine adrenal chromaffin cells, *Ann.N.Y.Acad.Sci.* 971, 150-152, 2002; Tarasov, K.A., O'Hare, D., and Isupov, V.P., Solid-state chelation of metal ions by ethylenediaminetetraacetate intercalated in a layered double hydroxide, *Inorg.Chem.* 42, 1919-1927, 2003; Nichols, N.M., Benner, J.S., Martin, D.D., and Evans, T.C., Jr., Zinc ion effects on individual Ssp DnaE intein-splicing steps: regulating pathway progression, *Biochemistry* 42, 5301-5311, 2003; Conzone, S.D., Hall, M.M., Day, D.E., and Brown, R.F., Biodegradable radiation delivery system utilizing glass microspheres and ethylenediaminetetraacetate chelation therapy, *J.Biomed.Mater.Res.A.* 70, 256-264, 2004; Reynolds, A.J., Haines, A.H., and Russell, D.A., Gold glyconanoparticles for mimics and measurement of metal ion-mediated carbohydrate-carbohydrate interactions, *Langmuir* 22, 1156-1163, 2006; Fernandes, C.M., Zamuner, S.R., Zuliani, J.P., *et al.*, Inflammatory effects of BaP1: a metalloproteinase isolated from *Bothrops asper* snake venom: leukocyte recruitment and release of cytokines, *Toxicon* 47, 549-559, 2006; Pajak, B. and Orzechowski, A., Ethylenediaminetetraacetic acid affects subcellular expression of clusterin protein in human colon adenocarcinoma COLO 205 cell line, *Anticancer Drugs* 18, 55-63, 2007

EGTA (ethyleneglycoltetraacetic acid)   380.4   Chelating agent with much greater affinity for calcium ions than for magnesium.

Schor, S.L., The effects of EGTA and trypsin on the serum requirements for cell attachment to collagens, *J.Cell Sci.* 40, 271-279, 1979; Bers, D.M., A simple method for the accurate determination of free [Ca] in Ca-EGTA solutions, *Am.J.Physiol.* 242, C404-408, 1982; Miller, D.J. and Smith, G.L., EGTA purity and the buffering of calcium ions in physiological solutions, *Am.J.Physiol.* 246, C160-C166, 1984; Sulakhe, P.V. and Hoehn, E.K., Interaction of EGTA with a hydrophobic region inhibits particular adenylate cyclase from rat cerebral cortex: a study of an EGTA-inhibitable enzyme by using alamethicin, *Int.J.Biochem.* 16, 1029-1035, 1984; Bryant, D.T. and Andrews, P., High-affinity binding of Ca^{2+} to bovine α-lactalbumin in the absence and presence of EGTA, *Biochem.J.* 220, 617-620, 1984; Smith, G.L. and Miller, D.J., Potentiometric measurements of stoichiometric and apparent affinity constants of EGTA for protons and divalent cations including calcium, *Biochim.Biophys.Acta* 839, 287-299, 1985; Marini, M.A., Evans, W.J., and Berger, R.L., The determination of binding constants with a differential thermal and potentiometric titration apparatus. II. EDTA, EGTA and calcium, *J.Biochem.Biophys.Methods* 12, 135-146, 1986; Harrison, S.M. and Bers, D.M., The effect of temperature and ionic strength on the apparent Ca-affinity of EGTA and the analogous Ca-chelators BAPTA and dibromo-BAPTA, *Biochim.Biophys.Acta* 925, 133-143, 1987; Guan, Y.Y., Kwan, C.Y., and Daniel, E.E., The effects of EGTA on vascular smooth muscle contractility in calcium-free medium, *Can.J.Physiol.Pharmacol.* 66, 1053-1056, 1988Youatt, J., Calcium and microorganisms, *Crit.Rev.Microbiol.* 19, 83-97, 1993; Yingst, D.R., and Barrett, V.E., Binding and elution of EGTA to anion exchange columns: implications for study of (Ca+Mg)-ATPase inhibitors, *Biochim.Biophys.Acta* 1189, 113-118, 1994; Lee, Y.C., Fluorometric determination of EDTA and EGTA using terbium-salicylate complex, *Anal.Biochem.* 293, 120-123, 2001; Rothen-Rutishauser, B., Riesen, F.K., Braun, A., *et al.*, Dynamics of tight and adherens junctions under EGTA treatment, *J.Membr.Biol.* 188, 151-162, 2002; Chen, J.L., Ahluwalia, J.P., and Stamnes, M., Selective effects of calcium chelators on anterograde and retrograde protein transport in the cell, *J.Biol.Chem.* 277, 35682-35687, 2002; Fisher, A.E., Hague, T.A., Clarke, C.L., and Naughton, D.P., Catalytic superoxide scavenging by metal complexes of the calcium chelator EGTA and contrast agent EHPG, *Biochem.Biophys.Res. Commun.* 323, 163-167, 2004; Dweck, D., Reyes-Alfonso, A., Jr., and Potter, J.D., Expanding the range of free calcium regulation in biological solutions, *Anal.Biochem.* 347, 303-315, 2005; Ellis-Davies, G.C. and Barsotti, R.J., Tuning caged calcium: photolabile analogues of EGTA with improved optical and chelation properties, *Cell Calcium* 39, 75-83, 2006; Zhou, J.L., Li, X.C., Garvin, J.L., *et al.*, Intracellular ANG II induces cytosolic Ca^{2+} mobilization by stimulating intracellular AT1 receptors in proximal tubule cells, *Am.J.Physiol.Renal Physiol.* 290, F1382-1390, 2006

METAL CHELATING AGENTS (Continued)

Name	M.W.	Description
Oxalic Acid	126.1 as hydrate	Modest chelator; history of use in blood and in dentistry.

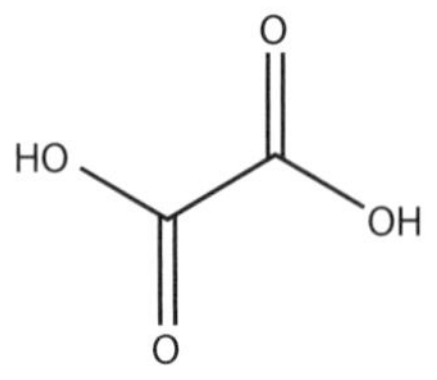

Lu, H., Mou, S., Yan, Y., *et al.*, On-line pretreatment and determination of Pb, Cu and Cd at the μ l^{-1} level in drinking water by chelation ion chromatography, *J.Chromatog.A.* 800, 247-255, 1998; Bruer, W, Ronson, A., Slotki, I.N., *et al.*, The assessment of serum nontransferrin-bound iron in chelation therapy and iron supplementation, *Blood* 95, 2975-2982, 2000; Salovaara, S., Sandberg, A.S., and Andlid, T., Combined impact of pH and organic acids on iron uptake by Caco-2 cells, *J.Agric.Food Chem.* 51, 7820-7824, 2003; Gerken, B.M., Wattenbach, C., Linke, D., Tweezing-absorptive bubble separation. Analytical method for the selective and high enrichment of metalloenzymes, *Anal.Chem.* 77, 6113-6117, 2005

***o*-Phenanthroline;** 1,10-phenanthroline	198.2 as hydrate	Moderately strong chelating agent. Historical use for zinc. More recent studies have copper and iron complexes as specific nuclease activity and there is some evidence to indicate protease activity. Possible role of metal complexes as oxidizing agents[d]

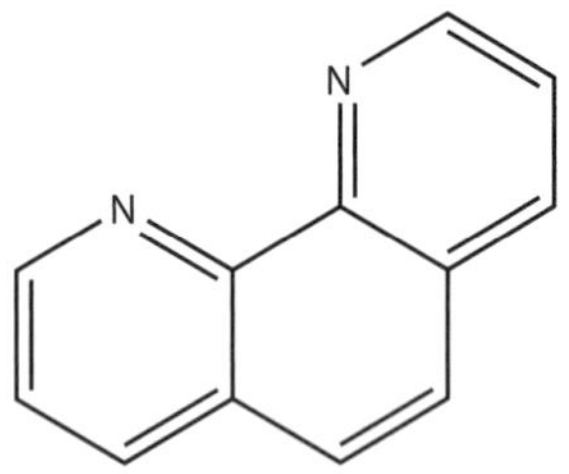

o-phenanthroline; 1,10-phenanthroline

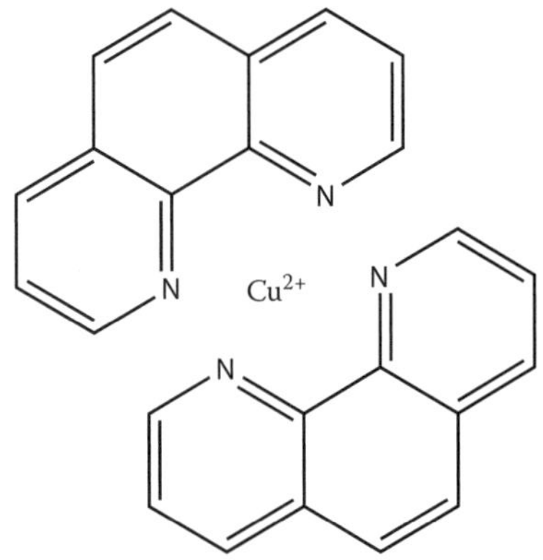

Sytkowski, A.J. and Vallee, B.L., Chemical reactivities of catalytic and noncatalytic zinc or cobalt atoms of horse liver alcohol dehydrogenase: differentiation by their thermodynamic and kinetic properties, *Proc.Nat.Acad.Sci.USA.* 73, 344-348, 1976; Kidani, Y. and Hirose, J., Coordination chemical studies on metalloenzyme. II. Kinetic behavior of various types of chelating agents towards bovine carbonic anhydrase, *J.Biochem.* 81, 1383-1391, 1977; Evans, C.W., The spectrophotometric determination of micromolar concentrations of Co^{2+} using *o*-phenanthroline, *Anal.Biochem.* 135, 335-339, 1983; Wu, H.B. and Tsou, C.L., A comparison of Zn(II)_ and Co(II) in the kinetics of inactivation of aminoacylase by 1,10-phenanthroline and reconstitution of the apoenzyme, *Biochem.J.* 296, 435-441, 1993; Auld, D.S., Removal and replacement of metal ions in metallopeptidases, *Methods Enzymol.* 248, 228-242, 1995; Leopold, I. and Fricke, B., Inhibition, reactivation and determination of metal ions in membrane metalloproteases of bacterial origin using high-performance liquid chromatography coupled on-line with inductively coupled plasma mass spectrometry, *Anal.Biochem.* 252, 277-285, 1997; Ciancaglini, P., Pizauro, J.M., and Leone, F.A., Dependence of divalent metal ions on phosphotransferase activity of osseous plate alkaline phosphatase, *J.Inorg.Biochem.* 66, 51-55, 1997

METAL CHELATING AGENTS (Continued)

Name	M.W.	Description
8-Quinolinol; 8-hydroxyquinoline	145.2	Metal chelating agent with higher affinity for Fe, Cu, and Zn. Lower affinity for Ca and Mg. Therapeutic use as antiseptic/bacteriostatic agent.

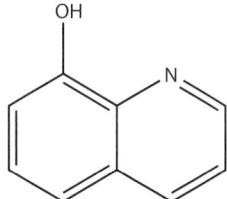

Fayez, M. and El-Tarras, M., Potentiometric titration of 8-hydroxyquinoline with Cu(II) using Cu(II)sulphide-ion selective electrode, *Pharmazie* 30, 799, 1975; Eskeland, T., The effect of various metal ions and chelating agents on the formation of noncovalently and covalently linked IgM polymers, *Scand.J.Immunol.* 6, 87-95, 1977; Albro, P.W., Corbett, J.T. and Schroeder, J.L., Generation of hydrogen peroxide by incidental metal ion-catalyzed autoxidation of glutathione, *J.Inorg.Biochem.* 27, 191-203, 1986; Yasui, T., Yuchi, A., Wade, H., and Nagagawa, S., Reversed-phase high-performance liquid chromatography of several metal-S-quinolinethiol complexes, *J.Chromatog.* 596, 73-78, 1992; Vieira, N.E., Yergey, A.L., and Abrams, S.A., Extraction of magnesium from biological fluids using 8-hydroxyquinoline and cation-exchange chromatography for isotopic enrichment using thermal ionization mass spectrometry, *Anal.Biochem.* 218, 92-97, 1994; Zachariouis, M., and Hearn, M.T., Adsorption and selectivity characteristics of several human serum proteins with immobilized hard Lewis metal ion-chelate adsorbents, *J.Chromatog.A.* 890, 95-116, 2000; Anfossi, L., Giraudi, G., Grassi, G., *et al.,* Binding properties of a polyclonal antibody directed against toward lead complexes, *Ann.Chim.* 93, 499-512, 2003; Yamada, H., Hayashi, H., and Yasui, T., Utility of 1-octanol/octane mixed solvents for the solvent extraction of aluminum(III), gallium(III), and indium(III) with 8-quinolinol, *Anal.Sci.* 22, 371-376, 2006; Song, K.C., Kim, J.S., Park, S.M., *et al.,* Fluorogenic Hg^{2+}-selective chemodosimeter derived from 8-hydroxyquinoline, *Org. Lett.* 8, 3413-3416, 2006; Mittal, S.K., Kumar, A., Gupta, N., *et al.,* 8-Hydroxyquinoline based neutral tripodal ionophore as a Cu(II) selective electrode and the effect of remote substituents on electrode properties, *Anal.Chim.Acta* 585, 161-170, 2007

---

[a] Metal chelating systems can be viewed as metal ion buffer systems
- MY = metal chelate complex; M is metal and Y is chelating agent. Metal ion are electron acceptors can be viewed as Lewis acids while chelating agents donate electrons and are Lewis bases. Also, chelating agents are polydentate
  - $MY^{n-m} = M^{+n} + M^{-m}$
  - $K_{MY} = [MY^{n+m}]/[M^{+n}][Y^{-m}]$
  - $pM = \log K_{MY} + \log [Y^{-m}]/[MY^{n-m}]$

The values for stability(association constants presented below are dependent of solvent conditions such as pH as well as relative amounts of chelating agent and metal ion.

[b] Stability of some metal ion complexes with EDTA

Metal Ion	Log K (MY)
Ca^{2+}	10.96
V^{+2}	12.70
Mn^{+2}	14.04
Fe^{2+}	14.33
Fe^{3+}	25.1
Co^{2+}	16.31
Ni^{2+}	18.62
Cu^{2+}	18.80
Zn^{2+}	16.50

Adapted from Mingos, D.M.P., *Essential Trends in Inorganic Chemistry*, Oxford University Press, Oxford, United Kingdom, 1998.

[c] Influence of pH on pM for EDTA where pK_1, 2.0; pK_2 2.67; pK_3, 6.16; pK_4 11.26.

pH	Cu(II)	Zn(II)	Mg(II)	Ca(II)	Mn(II)	Fe(III)
4	8.4	6.7	2.0	2.3	4.5	15.7
6	12.6	10.5	3.8	5.5	7.3	19.5
8	15.0	12.8	5.4	7.3	10.1	21.8
10	16.8	14.6	7.2	9.2	12.0	23.7

Adapted from Chaberek, S. and Martell, A.F., *Organic Sequestering Agents. A Discussion of the Chemical Behavior and Applications of Metal Chelate Compounds in Aqueous Systems*, John Wiley & Sons, London, UK, 1959

d Nuclease activity: Que, B.G., Downey, K.M., and So, A.G., Degradation of deoxyribonucleic acid by a 1,10-phenanthroline-copper complex: the role of hydroxyl radicals, *Biochemistry* 19, 5987-5891, 1980; Goldstein, S. and Czapski, G., The role and mechanism of metal ions and their complexes in enhancing damage in biological systems or in protecting these systems from the toxicity of O2-, *J.Free Radic.Biol.Med.* 2, 3-11, 1986; Sigman, D.S. and Chen, C.H, Chemical nucleases: new reagents in molecular biology, *Annu.Rev.Biochem.* 59, 207-236, 1990; Sigman, D.S., Chemical nucleases, *Biochemistry* 29, 9097-9105, 1990; Montenay-Garestier, T., Helene, C., and Thuong, N.T., Design of sequence-specific bifunctional nucleic acid ligands, *Ciba. Found.Sym.* 158, 147-157, 1991; Pan, C.Q., Landgraf, R., and Sigman, D.S., DNA-binding proteins as site-specific nucleases, *Mol.Microbiol.* 12, 335-342, 1994. Protease activity: Kito, M. and Urade, R., Protease activity of 1,10-phenanthroline-copper systems, *Met.Ions Biol.Syst.* 38, 187-196, 2001. Oxidation: McArdle, J.V., Gray, H.B., Creutz, C., and Sutin, N., Kinetic studies of the oxidation of ferrocytochrome c from horse heart and *Candida krusei* by tris(1,10-phenanthroline)cobalt(3), *J.Amer.Chem.Soc.* 96, 5737-5741, 1974; McArdle, J.V., Coyle, C.L., Gray, H.B., Yoneda, G.S., and Holwerda, R.A., Kinetics studies of the oxidation of blue copper proteins by tris(1-,10-phenanthroline)cobalt(III) ions, *J.Amer.Chem.Soc.* 99, 2483-2389, 1977; Lau, O.W. and Luk, S.F., Spectrophotometric determination of ascorbic acid in canned fruit juices, cordials, and soft drink with iron(III) and 1,10-phenanthroline as reagents, *J.Assoc.Off.Anal.Chem.* 70, 518-520, 1987; Mandal, S., Kazmi, N.H., and Sayre, L.M., Ligand dependence in the copper-catalyzed oxidation of hydroquinones, *Arch.Biochem.Biophys.* 435, 21-31, 2005; Hung, M. and Stanbury, D.W., Oxidation of thioglycolate by [Os(phen)3]3+: an unusual example of redox-mediated aromatic substitution, *Inorg.Chem.* 44, 9952-9960, 2005; Ishrat, Q.U. and Iftikhar, A., Kinetics of copper(II) catalyzed oxidation of iodide by iron(III)orthophenanthroline complex in aqueous solution, *Pak.J.Pharm.Sci.* 18, 20-24, 2005; Ozyurek, M., Guglu, K., Bektasoglu, B., and Apak, R., Spectrophotometric determination of ascorbic acid by the modified CUPRAC method with extractive separation of flavonoids-La(III) complexes, *Anal.Chim.Acta* 588, 88-95, 2007

Some Stability Constants for Divalent Metal Ion Chelate Complexes (log k)

Chelating Agent	Ca	Mg	Zn	Fe	Cu
BAPTA	6.8			6	
EDTA	10.6	8.7	16.4	14.2	18.8
EGTA	11.0	5.2	12.9	11.8	12.9
8-Hydroxyquinoline	3.3	4.7	8.5	8.0	8.5
1,10-Phenanthroline	0.5	1.5	6.4	5.8	6.3
Citrate	3.6	3.6	5.0	4.4	5.9
Oxalate	3.0	2.6	4.9	4.7	4.4

General references for metal chelating agents

Analytical Uses of Ethylene Diamine Tetraacetic Acid, D. Van Nostrand, Princeton, New Jersey, USA, 1950

Martell, A.E. and Calvin, M., *Chemistry of the Metal Chelate Compounds*, Prentice-Hall, Englewood Cliffs, NJ, USA, 1952

Chelating Agents and Metal Chelates, ed. F.P. Dwyer and D.P. Mellon, Academic Press, New York, NY, USA, 1964

Data for Biochemical Research, ed. R.M.C. Dawson, Clarendon Press, Oxford, UK, 1986

Handbook on Metals in Clinical and Analytical Chemistry, ed. H.G. Seiler, A. Sigel, and H. Sigel, Marcel Dekker, New York, NY, USA, 1994

Bertini, I., Gray, H.B., Strefel, E.I., and Valentine, J.S., *Biological Inorganic Chemistry. Structure and Reactivity*, University Science Books, Sausalito, CA, USA, 2007

WATER

Water Purity and Water Purity Classification

Water Type	Resistance[a] Megohm @25°C	Bioburden (cfu)[b]	Dissolved Solids (mg/L)
Deionized	0.05	—	10
Purified	0.2	100	1
Apyrogenic	0.8	0.1	1
High Purity	10	1	0.5
Ultrapure	18	1	0.005

[a] Resistance (R), determined by conductivity(A/V[c]), measured in siemans (S), R is measured in ohms (V/A[d]) (Ω)

[b] Colony-forming units

[c] SI unit for conductivity – $m^{-2} \cdot kg^{-1} \cdot s^{-3} \cdot A^2$; one S equal to the conductance of one ohm^{-1}

[d] SI unit for resistance – $m^{-2} \cdot kg^{-1} \cdot s^{-3} \cdot A^{-2}$; one R is equal to one ohm

Some Definitions of Pharmaceutical Water (FDA/ORA)

- Non-potable
- Potable
- USP purified
- USP water for injection (WFI)
- USP sterile water for injection
- USP bacteriostatic water for injection
- USP sterile water for irrigation

See 2007 USP/NF. *The Official Compendia of Standards*, US Pharmacopeia, Rockville, MD, USA, 2007, http://www.usp.org;

British Pharmacopoeia 2007, TSO Norwich, Norwich, UK, 2007; http://www.pharmacopoeia.org.uk

Process water is defined as that water which is used during the pharmaceutical manufacturing process. A higher purity water may be required for formulation of the final drug product and/or reconstitution of final product prior to use. A high purity water is required for a product which is to be injected as opposed to an oral administered product. http://www.fda.gov/ora/inspect_ref/itg/itg46.html

Meltzer, T.H., *High Purity Water Preparation*, Tall Oaks Publications, Littleton, CO, 1993

Fischbacher, C., Quality assurance in analytical chemistry, in *Encyclopedia of Analytical Chemistry*, ed. R.A. Meyers, Wiley, New York, NY, USA, pps 13563-13587, 2000

Environmental water quality issues are a separate issue with different classification issues

- Kannel, P.R., Lee, S., Lee, Y.S., *et al.*, Application of water quality indices and dissolved oxygen as indicators for river water classification and urban impact assessment, *Environ. Monit. Assess.*, in press, 2007
- Kowalkowski, T., Zhytniewski, R., Szpejna, J., and Buszewski, B., Application of chemometrics in river water classification, *Water Res.* 40, 7544-752, 2006

WATER PURIFICATION

The Following is a Representation of Estimates of Effectiveness.

Technique[a]	Effectiveness of Removal of Contaminant/Impurity			
	Inorganic or Ionized Organic	Organic	Pyrogens	Particulates
Activated Carbon	Partial	Partial	No	No
Ion Exchange	Yes	No	No	No
Distillation	Yes	Yes[b]	Yes	Yes
Reverse Osmosis	Partial	Partial	Yes	Yes
Ultrafiltration	Partial	Partial	Good	Yes

[a] In general, a combination of technologies is required–For example,
1. Input water (outside line)
2. Filtration/settling (depends on quality of input water)
3. Distillation
4. Reverse osmosis
5. Terminal filtration (e.g. 0.2 µ filter)
Other techniques such as ultraviolet irradiation may be used

[b] Effectiveness of separation depends on vapor pressure (boiling point) differences between water and specific contaminant impurity and lack of formation of a azeotropic mixture. http://www.fda.gov/ora/inspect_ref/igs/high.html

PROPERTIES OF WATER

Absorption of Light (adapted from Morton, R.A., *Biochemical Spectroscopy*, Wiley, New York, NY, USA, 1975)

Wavelength (nm)	Absorbance[a]	Wavelength(nm)	Absorbance[a]
220	1.1	500	0.02
250	0.8	600	0.1
280	0.4	700	0.8
300	0.3	800	2.4
320	0.2	1000	40.0
340	0.2	1500	194.0
400	0.1	1800	170.0

Water quality and Water analysis references

Water general references

Dorsey, N.E., *Properties of Ordinary Water-Substance*, Reinhold Publishing Company, New York, NY, USA, 1940

Eisenberg, D. and Kauzmann, W., *The Structure and Properties of Water*, Oxford University Press, New York, NY, USA, 1969

Water: A Comprehensive Treatise, ed. F. Frank, Plenum Press, London, 1972

Water and Aqueous Solutions, ed, G.W. Nelson and J.E. Enderby, Adam Hilger, Bristol, UK, 1985

Robison, G.W., Zhu, S.B., Singh, S., and Evans, M.W., *Water in Biology, Chemistry, and Physics Experimental Overviews and Computational Methodologies*, World Scientific Press, Singapore, 1996

Frank, F., *Water, 2nd edition, A Matrix of Life*, Royal Society of Chemistry, Cambridge, UK, 2000

Water purity

Swaddle, T.W., *Applied Inorganic Chemistry*, University of Calgary Press, Calgary, Alberta, Canada, Chapter 12 (Water conditioning), 1990

Afshar, A., Zhao, X., Heckert, R.A., and Trotter, H.C., Suitability of autoclaved tap water for preparation of ELISA reagents and washing buffer, *J.Virol.Methods* 46, 275-278, 1994

Stewart, K.K. and Ebel, R.E., *Chemical Measurements in Biological Systems*, John Wiley & Sons, New York, NY, USA, Chapter 2 (Water, pH, and Buffers), 2000

Mabic, S. and Kano, I., Impact of purified water quality on molecular biology experiments, *Clin.Chem.Lab.Med.* 41, 486-491, 2003.

Regnault, C., Kano, I., Darbouret, D., and Mabic, D., Ultrapure water for liquid chromatography-mass spectrometry studies, *J.Chromatog.A.* 1030, 289-295, 2004.

Bennett, A., Process Water: Analyzing the lifecycle cost of pure water, *Filtration & Separation*, March, 2006.

Water analysis

Water analysis, in *Encyclopedia of Analytical Sciences*, ed A. Townshend, Academic Press, London, UK, Volume 9, pps. 5445-5559, 1995

Fischbacher, C., Quality assurance in analytical chemistry, in *Encyclopedia of Analytical Chemistry*, ed. R.A. Meyers, Wiley, New York, NY, USA, pps 13563-13587, 2000

Reid, D., Water determination in food, in *Encyclopedia of analytical chemistry*, ed. R.H. Meyers, Wiley, New York, NY, USA, pps. 4318-4332, 2000

Spectroscopy of water

Morton, R.A., *Biochemical Spectroscopy*, Wiley, New York, NY, USA, 1975

Symans, M.C.R., Spectroscopic studies of water and aqueous solutions, in *Water and Aqueous Solutions*, ed G.W. Nelson and J.E. Enderby, Adam Hilger, Bristol, UK, pps. 41-55, 1985

Googin, P.L. and Carr, C., Far infrared spectroscopy and aqueous solutions, in *Water and Aqueous Solutions*, ed. G.W. Nelson and J.E., Enderby, Adam Hilger, Bristol, UK, pps. 149-161, 1985

Mehrotra, R., Infrared spectroscopy, gas chromatography/infrared in food analysis, in *Encyclopedia of Analytical Chemistry*, ed. R.H. Meyers, pps. 4007-4024, 2000

STABILITY OF SOLUTIONS FOR GLP AND cGMP USE

A. There are three factors which influence the stability of solutions
   1. Purity of solvent used for the preparation of the solution. This is usually water and should be of USP or higher grade. There is an assumption of sterility of water obtained in a GLP or cGMP environment. The presence of trace amounts of metal ions in solution can contribute to stability issues for certain solutes.
   2. Storage conditions – plastic container, glass container, under nitrogen, temperature of storage
   3. Lability of the solute in solutions. Solutes such as sodium chloride are indefinitely stable while solutions of carbohydrates such as glucose or glycerol can undergo oxidation and mutarotation

B. There is little information of the storage stability of solutions while there is considerable information on the storage stability of reference standards and specific laboratory reagents[1-6]. The several pharmacopeia (US, UK) discuss the stability of certain reagents and standards but do not contain information about solution stability. Compendia of chemical data[7,8] will state if a given chemical decomposes in solution.

C. It is possible to make the following recommendations with the caveat that there will be exceptions. The preferable approach to stability would involve (1) defining the critical attributes of the solvent for the specific process or analysis and (2) measurement of such property or properties over a period of one year. Storage of a laboratory solvent for more than a period of one year is not acceptable. It is assumed that the solutions are prepared with USP quality or higher water under clean conditions.

No solution should be retained for a period of time longer than 3 months unless it has been documented that effectiveness is retained beyond that time period. There may be some solutions such as dilute phosphate buffers at neutrality that are stable for much shorter periods of time.

Solutions should be stored in the cold (2-4°C) unless it has been documented that storage at 23°C (room temperature) is effective.

Any dating of an solvent used in a GLP or cGMP process shall be consistent with information in the SOP for the preparation of such solvent and information in any run sheet/analytical SOP which uses such solvent.

References

1. Urone, P.F., Stability of colorimetric reagent for chromium, S-diphenylcarbazide, in various solvents, *Analyt. Chem.* 27, 1354-1355, 1955.
2. Grant, D.R., Reagent stability in Rosen's ninhydrin method of analysis for amino acids, *Analyt.Biochem.* 6, 109, 1963.
3. Peterson, R.C., Stablity of Folin phenol reagent, *J.Pharm.Sci.* 55, 523, 1966.
4. Durham, B.W., Reagent stability, *Analyt.Chem.* 51, A922, 1979
5. Beck, J., Coleman, P., and Grzesiak, J., Cholesterol rate reagent with extended stability, *Clin.Chem.* 31, 949, 1985.
6. Georghiou, P.E., Winsor, L., Shirtliff, C.J., and Svec, J., Storage stability of formaldehyde containing paraosaniline reagent, *Analyt.Chem.* 59, 2432-2435, 1987
7. *Reagent Chemicals. Specifications and Procedures,* 10th edn., ed. P. A. Bovis, American Chem.Soc, Washington, DC, 2006
8. *CRC Handbook of Chemistry and Physics,* 86th edn., ed. D.Lide, CRC Press, Boca Raton, FL, 2006

GENERAL INFORMATION ON SPECTROSCOPY

Spectroscopy – the study of the interaction of electromagnetic radiation with matter – excluding chemical effects.

Spectrophotometer – an instrument which measures the relationship between the absorption of light by a substance and the wavelength of the incident light

Spectrometer – an instrument which measures the distribution of wavelengths in electromagnetic radiation; also an instrument which measures the energies and masses in a distribution of particles as in a mass spectrometer

Beer's Law (Beer-Lambert Law or Beer-Lambert-Bouguer Law) – states that while the relationship between transmittance and concentration is non-linear, the relationship between absorbance and concentration is linear. The practical consequence is that with high absorbance values, one is measuring small differences in large numbers with attendant inaccuracies.

I/I_o (Transmittance)	%Transmittance ($I/I_o \times 100$)	Absorbance (A)
1.0	100	0
0.1	10	1.0
0.01	1	2.0
0.001	0.1	3.0
0.0001	0.01	4.0
0.00001	0.001	5.0
0.000001	0.0001	6.0

$\log_{10} (I/I_0) = \varepsilon cl = A$; where I is the intensity of transmitted light; I_o is the intensity of incident light; ε is the molar extinction coefficient (L mol^{-1} cm^{-1}), c is concentration (mol L^{-1}), l is pathlength (cm); and A is absorbance.

Electromagnetic Radiation Ranges Frequently Used in Biochemistry and Molecular Biology

Definition	Range (Wavelength)	Range (Wavenumber)	Comments
Ultraviolet (UV)	190–360 nm		Qualitative and Quantitative
Visible (Vis)	360–780 nm		Qualitative and Quantitative
Near Infrared (NIR)	780–2500 nm	12,800 – 4000 cm^{-1}	Qualitative and Quantitative
Infrared (IR)	2500–40,000 nm[a]	4000 – 250 cm^{-1}	Mostly Qualitative
Far Infrared (FIR)	4×10^4–10^6 nm[a]	250 – 10^{-cm}	Mostly Qualitative

[a] Usually used reciprocal centimeters (cm^{-1}) for wavelength description. This quantity is the wave number (reciprocal of the wavelength in cm^{-1})

Hyperchromic – increase in absorbance

Hypochromic – decrease in absorbance

Hypsochromic – decrease in wavelength; also known as a "blue" shift

Bathochromic – increase in wavelength; also known as a "red" shift

UV-Vis Spectroscopy
Absorbance Data for Common Solvents

Solvent	UV-VIS "Cut-Off" Wavelength[a]
CHCl$_3$	240 nm
Hexanes	200 nm
MeOH/EtOH	205 nm
H$_2$O	190 nm
Dioxane	205 nm
Acetonitrile	190 nm

[a] In this context, "cut-off" is the lowest wavelength at which the solvent can be used; solvent absorbance is sufficiently high below this wavelength to marginalize results.

Water transmits light satisfactorily between 400 nm and 800 nm; from 600 nm to 900 nm, above 900 nm, transmission decreases by a factor 50 and above 1.3 μm, transmission decreases more rapidly; transmission increases from 400 to 220 while a further decrease in wavelength results in markedly decreased transmission.

- Opticalx Properties of Water (Morton, R.A., *Biochemical Spectroscopy*, John Wiley & Sons (A Halsted Press Book), New York, New York, 1975.)

Optical Properties of Plastics Used For Microplate Assays[a,b]

Material	UV "Cut-off"[c]
Quartz	180 nm
Polystyrene[d]	300 nm
Polypropylene[e]	300 nm
Polyvinyl chloride (PVC)	300 nm

[a] These are values for the "common" microplate thickness; with thin-thickness (10 μ), the "cut-off" value of polyethylene is 180 nm (Andrady, A.L., Ultraviolet radiation and polymers, in *Physical Properties of Polymers Handbook*, ed. J.E. Mark, American Institute of Physics, AIP Press, Woodberry, NY, USA, 1996).

[b] There are a number of UV-transparent microplates available on the market. These are made of proprietary plastics; most likely unique blends. The microplates will permit use at 260 nm and 280 nm making them useful for biochemical analyses. However, since the composition of these microplates is proprietary, the microplates must be evaluated for any unique binding properties.

[c] In this context, "cut-off" is the lowest wavelength at which the microplate can be used; the zero absorbance of the microplate is sufficiently high such to marginalize results.

[d] Tends to be hydrophilic so an aqueous sample tends to "film" and stick the sides of well (surface tension effect)

[e] Tends to be hydrophobic so aqueous sample tends to "bead."

Near Infrared Spectroscopy

- The spectra of water
  - Symons, M.C.R., Spectroscopy of aqueous solutions: protein and DNA interactions with water, *Cell.Mol. Life Sci.*57, 999-1007, 2000.
  - Gregory, R.B., Protein hydration and glass transition behavior, in *Protein-Solvent Interactions*, ed. R.B. Gregory, Marcel Dekker, Inc., New York, New York, USA, Chapter 3, pps 191-264, 1995.

- Application to human tissue
  - Chance, B., Nioka, S., Warren, W., and Yurtsever, G., Mitochondrial NADH as the bellwether of tissue O_2 delivery, *Adv.Exp.Med. Biol.* 566, 231-242, 2005.
  - Cerussi, A.E., Berger, A.J., Bevilacqua, F., Sources of absorption and scattering contrast for near-infrared optical mammography, *Academic Radiology* 8, 211-218, 2001.
  - Eikje, N.S., Ozaki, Y., Aizawa, K., and Arase, S., Fiber optic near-infrared Raman spectroscopy for clinical noninvasive determination of water content in diseased skin and assessment of cutaneous edema, *J.Biomed.Opt.* 10, 14013, 2005.
  - Pickup, J.C., Hussain, F., Evans, N.D., and Sachedina, N., In vivo glucose monitoring: the clinical reality and the promise, *Biosens.Bioelectron.* 20, 1897-1902, 2005.
  - Hielscher, A.H., Bluestone, A.Y., Abdoulaev, G.S., *et al.*, Near-infrared diffuse optical tomography, *Dis. Markers* 18, 313-337, 2002.
  - Christian, N.A., Milone, M.C., Ranka, S.S., *et al.*, Tat-functionalized near-infrared emissive polymerosomes for dendritic cell labeling, *Bioconjug.Chem.* 18, 31-40, 2007.
- Process Monitoring
  - Liu, J., Physical characterization of pharmaceutical formulations in frozen and freeze-dried solid states: techniques and applications in freeze-drying development, *Pharm.Dev.Technol.* 11, 3-28, 2006.
  - Scaftt, M., Arnold, S.A., Harvey, L.M., and McNeil, B., Near infrared spectroscopy for bioprocess monitoring and control: current status and future trends, *Crit.Rev. Biotechnol.* 26, 17-39, 2006.
  - Reich, G., Near-infrared spectroscopy and imaging: basic principles and pharmaceutical applications, *Adv. Drug Deliv.Rev.* 57, 1109-1143, 2005.
  - Suehara, K., and Yano, T., Bioprocess Monitoring using near-infrared spectroscopy, *Adv.Biochem.Eng. Biotechnol.* 90, 173-198, 2004.
- General
  - Ferrari, M., Mottola, L., and Quaresima, V., Principles, techniques, and limitations of near infrared spectroscopy, *Can.J.Appl.Physiol.* 29, 463-487. 2004.
  - McWhorter, S. and Soper, S.A., Near-infrared laser-induced fluorescence detection in capillary electrophoresis, *Electrophoresis* 21, 1267-1280, 2000.
  - Symons, M.C., Spectroscopy of aqueous solutions: proteins and DNA interactions with water, *Cell Mol.Life Sci.* 57, 999-1007, 2000.
  - Nir, S., Nicol, F., and Szoka, F.C., Jr., Surface aggregation and membrane penetration by peptides: relation to pore formation and fusion, *Mol.Membr.Biol.* 16, 95-101, 1999.

Some Spectrometer Window Materials for Infrared Spectroscopy

Material	Wavelength Range (cm⁻¹)	Refractive Index	Characteristics
NaCl	40,000 – 600	1.52	Soluble in H_2O, Etoh
KBr	43,500 – 400	1.54	Soluble in H_2O, EtOH
BaF_2	66,666 – 800	1.45	Insoluble in H_2O, soluble in acid
ZnSe	20,000 – 500	2.43	Insoluble in H_2O
Si	8,333 – 33	3.42	Insoluble in H_2O

General references for spectrometry

Twyman, F. and Allsop, C.B., *The Practice of Absorption Spectrophotometry*, Adam Hilger, London, 1934.

Brode, W.R., *Chemical Spectroscopy*, John Wiley & Sons, New York, New York, 1943.

Analytical Absorption Spectroscopy, ed. M.G.Mellon, John Wiley & Sons, New York, New York, 1950.

Morton, R.A. *Biochemical Spectroscopy*, John Wiley & Sons (A Halsted Press Book), New York, New York, 1975.

Campbell, L.D. and Dwek, R.A., *Biological Spectroscopy*, Benjamin/Cummings, Menlo Park, California, 1984.

Practical Absorption Spectroscopy, ed A. Knowles and C. Burgess (UV Spectrometry Group), Chapman & Hall, London, UK, 1984.

Osborne, B.G., Fearn, T., and Hindle, P.H., *Practical NIR Spectroscopy with Applications in Food and Beverage Analysis*, Longman Scientific and Technical, Harrow, Essex, UK, 1993.

UV Spectroscopy Techniques, Instrumentation, Data Handling, UV Spectrometry Group, ed. B.J. Clark, T. Frost, and M.A. Russell, Chapman & Hall, London, UK, 1993.

Stuart, B., *Biological Applications of Infrared Spectroscopy*, ACOL Series, Wiley, Chichester, UK, 1997.

Standards and Best Practices in Absorption Spectrometry, ed. C. Burgess and T. Frost (UVSG), Blackwell Science, Oxford, UK, 1999.

Stewart, K.K. and Ebel, R.E., *Chemical Measurements in Biological Systems*, John Wiley & Sons, New York, NY, USA, 2000.

Workman, J., Jr., *Handbook of Organic Compounds. NIR, IR, Raman, and UV-Vis Featuring Polymers and Surfactants*, Academic Press, San Diego, California, 2001. Volume 1 Methods and Interpretation, Volume 2 UV-Vis and NIR Spectra, Volume 3 IR and Raman Spectra.

Near-Infrared Spectroscopy, ed. H.W. Siesler, Ozaki, Y., Kawate, S., and Heise, H.M., Wiley-VCH, Weinheim, Germany, 2002.

Stuart, B.H., *Infrared Spectroscopy: Fundamentals and Applications*, John Wiley & Sons, Ltd., Chichester, UK, 2004.

Burns, D.A. and Ciurczak, E.W., *Handbook of Near-Infrared Analysis, Third Edition*, CRC Press, Boca Raton, USA, 2007.

Workman, J. and Weyer, L., *Practical Guide to Interpretive Near-Infrared Spectroscopy*, CRC Press, Boca Raton, USA, 2007.

MICROPLATES

Attributes for Microplates Used for Assay

- Absorbance of light (see Spectroscopy)
- Adsorption of materials
- Intraplate Variation (well-to-well reproducibility)
- Vendor Reproducibility

Attributes for Microplates Used for Reactions and Cell Culture

- Adsorption of materials
- Intraplate Variation (well-to-well reproducibility)
- Vendor Reproducibility

Selected references on the use of microplates

Adsorption of materials/factors influencing adsorption to microplates

- Bulter, J.E., Ni., L., Nessler, R., *et al.*, The physical and functional behavior of capture antibodies adsorbed on polystyrene, *J.Immunol.Methods* 150, 77-90, 1992
- Davies, J., Roberts, C.J., Dawkes, A.C., *et al.*, Use of scanning probe microscopy and surface-plasmon resonance as analytical tools in the study of antibody-coated microtiter wells, *Langmuir* 10, 2654-2661, 1994
- Douglas, A.S. and Monteith, C.A., Improvements to immunoassays by use of covalent binding assay plates, *Clin. Chem.* 40, 1833-1837, 1994
- Elsner, H.I. and Mouritsen, S., Use of psoralens for covalent immobilization of biomolecules in solid-phase assays, *Bioconjugate Chem.* 5, 463-467, 1994
- Stevens, P.W. and Kelso, D.M., Estimation of the protein-binding capacity of microplate wells using sequential ELISAs, *J.Immunol.Methods* 178, 59-70, 1995
- Page, J.D., Derango, R., and Huang, A.E., Chemical modification of polystyrene's surface and its effect on immobilized antibodies, *Colloids and Surfaces A – Physicochem. Eng.Aspects* 132, 193-201, 1998
- Baumann, S., Grob, P., Stuart, F., *et al.*, Indirect immobilization of recombinant proteins to a solid phase using the albumin binding domain of streptococcal protein G and immobilized albumin, *J.Immunol.Methods,* 221, 95-106, 1998
- Ricoux, R., Chazaud, B., Tresca, J.P., and Pontet, M., Quality control of coated antibodies: New, rapid determination of binding affinity, *Clin.Chem.Lab.Med.* 38, 239-243, 2000
- Qian, W.P., Yao, D.F., Yu, F., *et al.*, Immobilization of antibodies on ultraflat polystyrene surfaces, *Clin.Chem.* 46, 1456-1463, 2000

- Bulter, J.E., Solid supports in enzyme-linked immunosorbent assay and other solid-phase immunoassays, *Methods* 22, 4-23, 2000
- Sugihara, T., Seong, G.H., Kobatake, E., and Aziawa, M., Genetically synthesized antibody-binding protein self-assembled on hydrophobic matrix, *Bioconjugate Chem.* 11, 789-794, 2000
- Julián, E., Cama, M., Martinez, P., and Luquin, M., An ELISA for five glycolipids from the cell wall of *Mycobacterium tuberculosis*: Tween 20 interference in the assay, *J.Immunol.Methods* 251, 21-30, 2001
- Peluso, P., Wilson, D.S., Do, D., *et al.*, Optimizing antibody immobilization strategies for the construction of protein microarrays, *Analyt.Biochem.* 312, 113-124, 2003
- Johnson, J.C., Nettikadan, S.R., Vengasandra, S.G., and Henderson, E., Analysis of solid-phase immobilized antibodies by atomic force microscopy, *J.Biochem.Biophys. Meth.* 59, 167-180, 2004
- Clinchy, B., Youssefi, M.R., and Håjensson, L., Differences in adsorption of serum proteins and production of IL-1ra by human monocytes incubated in different tissue culture plates, *J.Immunol.Methods* 282, 53-61, 2003
- Shrivastav, T.G., Basu, A., and Karlya, K.P., Substitution of carbonate buffer by water for IgG immobilization in enzyme linked immunosorbent assay, *J.Immunoassay Immunchem.* 23, 191-203, 2003
- Clinchy, B., Gunnerås, M., Håkensson, L., and Håkensson, L., Production of IL-1Ra by human mononuclear blood cells in vitro: Influence of serum factors, *Cytokine* 34, 320-330, 2006

Issues of well-to-well variation

- Faessel, H.M., Levasseur, L.M., Slocum, H.K., and Greco, W.R., Parabolic growth patterns in 96-well cell growth experiments, *In Vitro Cell Dev.Biol.-Animal* 35, 270-278, 1999
- Pitts, B., Hamilton, M.A., Zelver, N., and Stewart, P.S., A microtiter-plate screening method for biofilm disinfection and removal, *J.Microbiol.Methods* 54, 269-276, 2003
- Patel, M.I., Tuckerman, R., and Dong, Q., A pitfall of the 3-(4,5-dimethylthiazol-2-yl)-5(3-carboxymethonyl-phenol)-2-(4-sulfophenyl)-2H-tetrazolium (MTS) assay due to evaporation in wells on the edge of a 96 well plate, *Biotechnol.Lett.* 27, 805-808, 2005
- Heaver, M., Kopun, M., Rittgen, W., and Granzow, C., Cytotoxicity determination with photochemical artifacts, *Cancer Lett.* 223, 57-66, 2005
- Straetemanns, R., O'Brien, T., Wouters, L., *et al.*, Design and analysis of drug combination experiments, *Biometrical J.* 47, 299-308, 2005

PLASTICS

Plastics are a group of materials which are used extensively in biochemistry and molecular biology. Plastics can be defined as non-metallic polymeric materials which can be molded or extruded into a shape. A plastic can be a single polymeric component such as polypropylene, a blend of several polymers or a block copolymer consisting of joined segments of two or more individual polymers such as a block copolymer of polybutadiene and poly(ethylene oxide). The physical properties of a plastic are a combination of the polymeric composition and additives such as plasticizers. Plasticizers are high molecular weight liquids or solids melting a low temperature which are blended with thermoplastic resins such as polyvinyl chloride to change physical properties. Plasticizers including phthalate derivatives such as di(2-ethyl)hexylphthalate, derivatives of organic acids such as di-2-ethylhexyladipate, and polyglycols (polyethylene glycol). Other materials included in the manufacture of plastics include antioxidants, lubricants, stabilizers, and colorants. All of these components can influence the property of the final plastic product. It must be emphasized that the biomedical market for plastics is quite small in volume compared to the overall market. Thus, unless a vendor makes their own raw material, most suppliers to biochemistry and molecular biology purchase bulk product from a large chemical company. As a result there can be batch-to-batch and vendor-to-vendor variation in product. Another issue which confounds the use of plastics is the addition of stabilizers. Stabilizers are chemicals such as hydroxybenzophenones and hydroxyphenylbensotriazoles which are added to prevent damage from ultraviolet irradiation. These compounds do absorb ultraviolet light in the 200-400 nm range and do present problems in biochemical analyses. The careful investigator will assure the source of the plastics used in products such as microplates and incubation flasks (Clinchy, B., Youssefi, M.R., and Håkansson, L., Differences in adsorption of serum proteins and production of IL-1ra by human monocytes incubated in different tissue culture microtiter plates, *J.Immunol.Methods* 282, 53-61, 2003).

General references for plastics

Dubois, J.H., and John, F.W., *Plastics*, 5th Edn, Van Nostrand Reinhold, New York, NY, USA, 1974

Billmeyer, F.W., *Textbook of Polymer Science*, Wiley, New York, NY, 1984

Griffin, G.J.L., *Chemistry and Technology of Biodegradable Polymers*, Blackie Academic & Professional, London, UK, 1994

Physical Properties of Polymers Handbook, ed. J.E. Mark, American Institute of Physics, AIP Press, Woodberry, NY, USA, 1996

Araki, T. and Qui, T-C., *Structure and Properties of Multiphase Polymeric Materials*, Marcel Dekker, New York, NY, 1998

Polymer Data Book., ed. J.E. Mark, Oxford University Press, Oxford, UK, 1999

Plastics Additives Handbook, 5th edn., ed. H. Zweifel, Hanser Publications, Munich, Germany, 2001

Bart, J.C.J., *Plastics Additives. Advanced Industrial Analysis*, IOS Press, Amsterdam, Netherlands, 2006

Chiellini, E., *Biomedical Polymers and Polymer Therapeutics*, Kluwer Academic, New York, NY, USA, 2002

Ramakrishna, S., *An Introduction to Biocomposites*, Imperial College Press, London, UK, 2004

Carraher, C.A., Jr., *Introduction to Polymer Chemistry*, CRC Press, Boca Raton, FL, USA, 2006

Polymeric Nanofibers, ed. D.H. Reneker and H. Fong, American Chemical Society, Washington, DC, USA, 2006.

American Chemical Council; http://www.plasticsresource.com.

Plasticizers

Crawford, R.R. and Esmerian, O.K., Effect of plasticizers on some physical properties of cellulose acetate phthalate films, *J.Pharm.Sci.* 60, 312-314, 1971

Ekwall, B., Nordensten, C., and Albanus, L., Toxicity of 29 plasticizers to HeLa cells in the MIT-24 system, *Toxicology* 24, 199-210, 1982

Goldstein, D.B., Feistner, G.J., Faull, K.F., and Tomer, K.B., Plasticizers as contaminants in commercial ethanol, *Alcohol Clin.Exp.Res.* 11, 521-524, 1987

Sager, G. and Little, C., The effect of tris-(2-butoxyethyl)-phosphate (TBEP) and di-(2-ethylhexyl)-phthalate (DEHP) and the β-adrenergic receptor-blockers [3H]-(-)-dihydroalprenolol ([3H]-(-)-DHA) and [3H-(-)-CGP 12177 were tested for their ability to interact with β-adrenergic binding to α 1-acid glycoprotein and mononuclear leukocytes, *Biochem.Pharmacol.* 38,2551-2557, 1989

Baker, J.K., Characterization of phthalate plasticizers by HPLC/thermospray mass spectrometry, *J.Pharm.Biomed.Anal.* 15, 145-148, 1996

Wahl, H.G., Hoffman, A., Haring, H.U., and Liebich, H.M., Identification of plasticizers in medical products by a combined direct thermodesorption—cooled injection system and gas chromatography—mass spectrometry, *J.Chromatog.A* 847, 1-7, 1999

Cano, J.M., Marin, M.L., Sanchez, A., and Hernandis, V., Determination of adipate plasticizers in poly(vinyl chloride) by microwave-assisted extraction, *J.Chromatog.A* 963, 401-409, 2002

Siepmann, F., le Brun, V., and Siepmann, J., Drugs acting as plasticizers in polymeric systems: a quantitative treatment, *J.Control.Release* 115, 298-306, 2006

Some Properties of Plastics Used in Biochemistry and Molecular Biology

Structures for these plastics may be found on p. 866–868

Plastic	Uses	References
Acrylonitrile-Butadiene elastomers (Nitrile rubber)	Soft rubber applications such as gloves and pharmaceutical stoppers	1-4
Nylon (aliphatic polyamides)	Matrix for tissue engineering, suture material, matrix for adsorptive technologies	6-10
Polyacrylamide	Primary use as a matrix for protein electrophoresis (PAGE, polyacrylamide gel electrophoresis). Early use as gel filtration matrix; more recent use as a hydrogel and implant material	11-18
Polyacrylate (PA; polyacrylic acid)[1]	Dental cement, use for manufacture of microparticles and nanoparticles	19-24
Polybutadiene	Chromatographic matrix such as polybutadiene-coated zirconia, microarray plates	25-31
Polycarbonate	Matrix for biological assays; material for tissue culture flasks; implants; filters; use as "solid" component of copolymers; tyrosine-derived polycarbonate used in tissue engineering	32-42
Polyethylene	Implants, tubing	43-45
Polyethylene oxide (PEO)	Component of hydrogels; some use in implant biology; use in copolymers	46-51
Poly(ethylene terephthalate) (PET)	Matrix for tissue engineering; cell culture matrix; used in block copolymers	52-57

Some Properties of Plastics Used in Biochemistry and Molecular Biology (Continued)

Plastic	Uses	References
Poly(methacrylate)[1]	Used in monolithic chromatographic columns	58
Polypropylene	fiber used in surgery	59-61
Polypropylene oxide	Hydrogels, block copolymers,	62-64
Polystyrene	Microplates, can be modified by irradiation for covalent binding of probes; chromatographic matrices; microbeads for assays; frequently a copolymer with divinylbenzene	65-82
Poly(vinyl chloride)	Use in biosensors; use as flexible tubing when phthalate stabilizers are used; use as copolymer for encapsulation	83-88

[1] The nomenclature for acrylic acid and derivatives can be confusing. The following definitions are used in this text.
Acrolein (2-propenal)
Acrylamide (2-propenamide)
Acrylic acid (propenoic acid)
Acrylonitrile (2-propenenitrile)
Methacrylic acid (2-methylpropenoic acid)
Methyl methacrylic acid (2-methylpropenoic acid methyl ester)

References

1. Shanker, J., Gibaldi, M., Kanig, J.L. *et al.,* Evaluation of the suitability of butadiene-acrylonitrile rubbers as closures for parenteral solutions, *J.Pharm.Sci.* 56, 100-108, 1967

2. Williams, J.R., Permeation of glove materials by physiologically harmful chemicals, *Am.Ind.Hyg.Assoc.J.* 40, 877-882, 1979

3. Parker, S. and Braden, M., Soft prosthesis materials based on powdered elastomers, *Biomaterials* 11, 482-490, 1990

4. Walsh, D.L., Schwerin, M.R., Kisielewski, R.W., *et al.,* Abrasion resistance of medical glove materials, *J.Biomed.Mater.Res.B Appl. Biomater.* 68, 81-87, 2004.

5. McConway, M.G. and Chapman, R.S., Application of solid-phase antibodies to radioimmunoassay. Evaluation of two polymeric microparticles, Dynospheres and nylon, activated by carbonyldiimidazole or tresyl chloride, *J.Immunol.Methods* 95, 259-266, 1986

6. Absolom, D.R., Zingg, W., and Neumann, A.W., Protein adsorption to polymer particle: role of surface properties, *J.Biomed.Mater.Res.* 21, 161-171, 1987

7. Alicata, R., Mantaudo, G., Puglisi, C., and Samperi, F. Influence of chain end groups on the matrix-assisted laser desorption/ionization spectra of polymer blends, *Rapid Commun.Mass.Spectrom.* 16, 248-260, 2002

8. Zhu, X., Cai, J., Yang, J., *et al.,* Films coated with molecular imprinted polymers for the selective stir bar sorption extraction of monocrotophos, *J.Chromatog.A.* 1131, 37-44, 2006.

9. Dennes, T.J., Hunt, G.C., Schwarzbauer, J.E., and Schwartz, J., High-yield activation of scaffold polymer surfaces to attach cell adhesion molecules, *J.Am.Chem.Soc.* 129, 93-97, 2007

10. Friedrich, J., Zalar, P., Mororcic, M., *et al.,* Ability of fungi to degrade synthetic polymer nylon-6, *Chemosphere,* in press, 2007

11. Hjerten, S. and Mosbach, R., "Molecular-sieve" chromatography of proteins on columns of cross-linked polyacrylamide, *Anal.Biochem.* 3, 109-118, 1962

12. Goodfriend, T., Ball, D., and Updike, S., Antibody in polyacrylamide gel, a solid phase reagent for radioimmunoassay, *Immunochemistry* 6, 481-484, 1969

13. John, M., Skrabel, H., and Dellweg, H., Use of polyacrylamide gel columns for the separation of nucleotides, *FEBS Lett.* 5, 185-186, 1969

14. Bovin, M.V., Polyacrylamide-based glycoconjugates as tools in glycobiology, *Glycoconj. J.* 15, 431-446, 1998

15. Patrick, T., Polyacrylamide gel in cosmetic procedures: experience with Aquamid®, *Semin.Cutan.Med.Surg.* 23, 233-235, 2004.

16. Plieva, F., Bober, B., Dainiak, M., *et al.,* Macroporous polyacrylamide monolithic gels with immobilized metal affinity ligands: the effect of porous structure and ligand coupling chemistry on protein binding, *J.Mol.Recognit.* 19, 305-312, 2006.

17. Sairam, M., Babu, V.R., Vijaya, B.. *et al.,* Encapsulation efficiency and controlled release characteristics of crosslinked polyacrylamide particles, *Int.J.Pharm.* 320, 131-136, 2006.

18. Sefton, M.V. and Broughton, R.L., Microencapsulation of erythrocytes, *Biochim.Biophys.Acta* 717, 473-477, 1982

19. Stevenson, W.T. and Sefton, M.V., Graft copolymer emulsions of sodium alginate with hydroxylalkyl methacrylates for microencapsulation, *Biomaterials* 8, 449-457, 1987

20. Laemmli, U.K., Characterization of DNA condensates induced by poly(ethylene oxide) and polylysine, *Proc.Nat.Acad.Sci.USA* 72, 4288-4292, 1975

21. Sefton, M.V. and Nishimura, E., Insulin permeability of hydrophilic polyacrylate membranes, *J.Pharm.Sci.* 69, 208-209, 1980

22. Svensson, A., Norrman, J., and Piculell, L., Phase behavior of polyion-surfactant ion complex salts: effects of surfactant chain length and polyion length, *J.Phys.Chem.B Condens.Matter Mater. Surf. Interface Biophys.* 110, 10332-10340, 2006

23. Turos, E., Shim, J.Y., Wang, Y., *et al.,* Antibiotic-conjugated polyacrylate nanoparticles: new opportunities for development of anti-MRSA agents, *Bioorg.Med. Chem Lett.* 17, 53-56, 2007

24. Herdt, A.R., Kim, B.S., and Taton, T.A., Encapsulated magnetic nanoparticles as supports for proteins and recyclable biocatalysts, *Bioconjug.Chem.* 18, 183-189, 2007

25. Sun, L. and Carr, P.W., Chromatography of proteins using polybutadiene-coated zirconia, *Anal.Chem.* 67, 3717-3721, 1995

26. Alvarez, C., Strumia, M. and Bertorello, H., Synthesis and characterization of a biospecific adsorbent containing bovine serum albumin as a ligand and its use for bilirubin retention, *J.Biochem.Biophys. Methods* 49, 649-656, 2001

27. Davoras, E.M. and Coutsolelos, A.G., Efficient biomimetic catalytic epoxidation of polyene polymers by manganese porphyrins, *J.Inorg. Biochem.* 94, 161-170, 2003

28. Erhardt, R., Zhang, M., Boker, A., *et al.,* Amphiphilic Janus particles with polystyrene and poly(methacrylic acid) hemispheres, *J.Am. Chem.Soc.* 125, 3260-3267, 2003

29. Xu, J. and Zubarev, E.R., Supramolecular assemblies of starlike and V-shaped PB-PEO amphiphiles, *Angew.Chem.Int.Ed.Engl.* 43, 5491-5496, 2004

30. Geng, Y., Discher, D.E., Justynska, J., and Schlaad, H., Grafting short peptides onto polybutadiene-block-poly(ethylene oxide): a platform for self-assembling hybrid amphiphiles, *Angew.Chem.Int.Ed.Engl.* 45, 7578-7581, 2006

31. Kassu, A., Taguenang, J.M., and Sharma, A., Photopatterning of butadiene substrates by interferometric ultraviolet lithography: fabrication of phospholipid microarrays, *Appl.Opt.* 46, 489-494, 2007

32. Chandy, T. and Sharma, C.P., Changes in protein adsorption on polycarbonate due to L–ascorbic acid, *Biomaterials* 6, 416-420, 1985

33. Hough, T., Singh, M.B., Smart, I.J., and Knox, R.B., Immunofluorescent screening of monoclonal antibodies to surface antigens of animal and plant cells bound to polycarbonate membranes, *J.Immunol. Methods* 92, 103-107, 1986

34. Thelu, J., and Ambroise-Thomas, P., A septate polycarbonate cell culture unit used for *Plasmodium falciparum* and hybridomas, *Trans.R.Soc.Trop.Med.Hyg.* 82, 360-362, 1988

35. Bignold, L.P., Rogers, S.D., and Harkin, D.G., Effects of plasma proteins on the adhesion, spreading, polarization in suspension, random motility and chemotaxis of neutrophil leukocytes on polycarbonate (Nucleopore) filtration membranes, *Eur.J.Cell Biol.* 53, 27-34, 1990

36. Lee, J.H., Lee, S.J., Khang, G., and Lee, H.B., Interaction of fibroblasts on polycarbonate membrane surfaces with different micropore sizes and hydrophilicity, *J.Biomater.Sci.Poly.Ed.* 10, 283-294, 1999

37. Liu, Y., Ganser, D., Schneider, A., *et al.,* Microfabricated polycarbonate CE devices for DNA analysis, *Anal.Chem.* 73, 4196-4201, 2001

38. Liu, J., Zeng, F., and Allen, C., Influence of serum protein on polycarbonate-based copolymer micelles as a delivery system for a hydrophobic anti-cancer agents, *J.Control Release* 103, 481-487, 2005

39. Meechaisue, C., Dubin, R., Supaphol, P., *et al.,* Electrospun mat of tyrosine-derived polycarbonate fibers for potential use as tissue scaffolding material, *J.Biomater.Sci. Polym.Ed.* 17, 1039-1056, 2006

40. Li, Y., Wang, Z., Ou, L.M., and Yu, H.Z., DNA detection on plastic: surface activation protocol to convert polycarbonate substrates to biochip platforms, *Anal.Chem.* 79, 426-433, 2007

41. Carion, O., Souplet, V., Olivier, C., *et al.,* Chemical micropatterning of polycarbonate for site-specific peptide immobilization and biomolecular interactions, *ChemBioChem* 8, 315-322, 2007

42. Tripathi, A., Wang, J., Luck, J.A., and Suni, L.L., Nanobiosensor design utilizing a periplasmic *E.coli* receptor protein immobilized within Au/polycarbonate nanopores, *Anal.Chem.* 79, 1266-1270, 2007

43. Raff, R.A.V. and Allison, J.B., *Polyethylene*, Interscience Publishers, New York, NY, USA, 1956

44. Bhat, S.V., *Biomaterials*, Kluwer Academic Publishers, Boston, MA, USA, 2002

45. Shanbhag, A. and Rubash, H.E., *Joint Replacement and Bone Resorption: Pathology, Biomaterials, and Clinical Practice*, Taylor & Francis, New York, NY, USA, 2006

46. Desai, N.P. and Hubbell, J.A., Biological responses to polyethylene oxide modified polyethylene terephthalate surfaces, *J.Biomed.Mater. Res.* 25, 829-843, 1991

47. Lopina, S.T., Wu, G., Merrill, E.W., and Griffith-Cima, L., Heptocyte culture on carbohydrate-modified star polyethylene oxide hydrogels, *Biomaterials* 17, 559-569, 1996

48. Vereschagin, E.I., Han, D.H., Troitsky, A.W., *et al.,* Radiation technology in the preparation of polyethylene oxide hydrophilic gels and immobilization of proteases for use in medical practice, *Arch.Pharm. Res.* 24, 229-233, 2001

49. Liu, L.S. and Berg, R.A., Adhesion barriers of carboxymethylcellulose and polyethylene oxide composite gels, *J.Biomed.Mat.Res.* 63, 326-332, 2002

50. Wu, N., Wang, L.S., Tan, D.C., *et al.,* Mathematical modeling and in vitro study of controlled drug release via a highly swellable and dissoluble polymer matrix: polyethylene oxide with high molecular weights, *J.Control.Release* 102, 569-581, 2005

51. Unsworth, L.D., Sheardown, H., and Brash, J.L., Polyethylene oxide surfaces of variable chain density by chemisorption of PEO-thiol on gold: adsorption of proteins from plasma studied by radiolabeling and immunoblotting, *Biomaterials* 26, 5927-5933, 2005

52. Nair, P.D. and Sreenivasan, K., Effect of steam sterilization on polyethylene terephthalate, *Biomaterials* 5, 305-306, 1984

53. Dadsetan, M., Mirzadeh, H., Sharifi-Sanjani, N., and Daliri, M., Cell behavior on laser surface-modified polyethylene terephthalate in vitro, *J.Biomed.Mater.Res.* 57, 183-189, 2001

54. Cenni, E., Granchi, D., Ciapetti, G., *et al.,* Interleukin-6 expression by cultured human endothelial cells in contact with carbon coated polyethylene terephthalate, *J.Mater.Sci.Mater.Med.* 12, 365-369, 2001

55. Neves, A.A., Medcalf, N., and Brindle, K.M., Influence of stirring-induced mixing on cell proliferation and extracellular matrix deposition in meniscal cartilage constructs based on polyethylene terephthalate scaffolds, *Biomaterials* 26, 4828-4836, 2005

56. Basu, S. and Yang, S.T., Astrocyte growth and glial cell line-derived neurotrophic factor secretion in three-dimensional polyethylene terephthalate fibrous matrices, *Tissue Eng.* 11, 940-952, 2005

57. Alisch-Mark, M., Herrmann, A., and Zimmermann, W., Increase of the hydrophilicity of polyethylene terephthalate fibres by hydrolases from *Thermomospora fusca* and *Fusarium solani f.sp.pisi, Biotechnol. Lett.* 28, 681-685, 2006

58. Jungbauer, A. and Hahn, R., Polymethacrylate monoliths for preparative and industrial separation of biomolecular assemblies, *J.Chromatog.A.* 1184, 62-79, 2008

59. Peter, F.H., *Polypropylene*, Gordon and Breach Science Publishers, New York, NY, USA, 1968

60. Karger-Kocsis, J., *Polypropylene: Structure, Blends and Composites*, Chapman & Hall, London, UK, 1995

61. Karger-Kocsis, J., *Polypropylene an A-Z Reference*, Dordrecht, Netherlands, 1998

62. Topchieva, I.N. and Efremova, N.V., Conjugates of proteins with block co-polymers of ethylene and polypropylene oxides, *Biotechnol. Genet.Eng.Rev.* 12, 357-382, 1994

63. Newman, M.J., Actor, J.K., Balusubramanian, M., and Jagannath, C., Use of nonionic block copolymers in vaccines and therapeutics, *Crit. Rev.Ther.Drug Carrier Syst.* 15, 89-142, 1998

64. Gutowska, A., Jeong, B., and Jasionowski, M., Injectable gels for tissue engineering, *Anat.Rec.* 263, 342-349, 2001

65. Catarero, L.A., Butler, J.E., and Osborne, J.W., The adsorptive characteristics of proteins for polystyrene and their significance in solid-phase immunoassays, *Anal.Biochem.* 105, 375-382, 1980

66. Zouali, M. and Stollar, B.D., A rapid ELISA for measurement of antibodies to nucleic acid antigens using UV-treated polystyrene microplates, *J.Immunol.Methods* 90, 105-110, 1986

67. Piskin, E., Tuncel, A., Denizli, A., and Ayhan, H., Monosize microbeads based on polystyrene and their modified forms for some selected medical and biological applications, *J.Biomater.Sci.Polym. Ed.* 5, 451-471, 1994

68. Kochanowska, I.E., Rapak, A., and Szewczuk, A., Effect of pretreatment of wells in polystyrene plates on adsorption of some human serum proteins, *Arch.Immunol.Ther.Exp.(Warsz.)* 42, 135-139, 1994

69. Staak, C., Salchow, R., Clausen, P.H., and Luge, E., Polystyrene as an affinity chromatography matrix for the purification of antibodies, *J.Immunol.Methods* 194, 141-146, 1996

70. Davankov, V., Tsyurupa, M., Ilyin, M., and Pavlova, L., Hypercross-linked polystyrene and its potential for liquid chromatography: a mini-review, *J.Chromatog.A.* 965, 65-73, 2002

71. Gessner, A., Lieske, A., Paulke, B.R., and Muller, R.H., Functional groups on polystyrene model nanoparticles: influence on protein adsorption, *J.Biomed.Mater.Res.A* 65, 319-326, 2003

72. Saitoh, T., Hattori, N., and Hiraide, M., Protein separation with surfactant-coated polystyrene involving Cibacron Blue 3GA-conjugated Triton X-100, *J.Chromatog.A.* 1028, 149-153, 2004

73. van Kooten, T.G., Spijker, H.T., and Busscher, H.J., Plasma-treated polystyrene surfaces: model surfaces for studying cell-biomaterial interactions, *Biomaterials* 25, 1735-1747, 2004

74. Recknor, J.B., Recknor, J.C., Sakaguchi, D.S., and Mallapragada, S.K., Oriented astroglial cell growth on micropatterned polystyrene substrates, *Biomaterials* 25, 2753-2767, 2004

75. Turner, S.F., Clarke, S.M., Rennie, A.R., *et al.,* Adsorption of gelatin to a polystyrene /water interface as a function of concentration, pH, and ionic strength, *Langmuir* 21, 10082-10088, 2005

76. Rosado, E., Caroll, H., Sanchez, O., and Peniche, C., Passive adsorption of human antirrabic immunoglobulin onto a polystyrene surface, *J.Biomater.Sci.Polym.Ed.* 16, 435-448, 2005

77. Carvalho, R.S., Ianzer, D.A., Malavolta, L., *et al.,* Polystyrene-type resin used for peptide synthesis: application for anion-exchange and affinity chromatography, *J.Chromatog.B.Analyt.Technol.Biomed.Life Sci.* 817, 231-238, 2005

78. Jodar-Reyes, A.B., Ortega-Vinuesa, J.L., and Martin-Rodriguez, A., Adsorption of different amphiphilic molecules onto polystyrene latices, *J.Colloid Interface Sci.* 282, 439-447, 2005

79. Mitchell, S.A., Davidson, M.R., and Bradley, R.H., Improved cellular adhesion to acetone plasma modified polystyrene surfaces, *J.Colloid Interface Sci.* 281, 122-129, 2005

80. Cao, Y.C., Hua, X.F., Zhu, X.X., *et al.,* Preparation of Au coated polystyrene beads and their application in an immunoassay, *J.Immunol. Methods* 317, 163-170, 2006

81. Tirri, M.E., Wahlroos, R., Meltola, N.J., *et al.*, Effect of polystyrene microsphere surface to fluorescence lifetime under two-photon excitation, *J.Fluoresc.* 16, 809-816, 2006

82. Baker, S.C., Atkin, N., Gunning, P.A., *et al.*, Characterization of electrospun polystyrene scaffolds for three-dimensional *in vitro* biological studies, *Biomaterials* 27, 3136-3146, 2006

83. Titow, W.V., *PVC Technology*, Elsevier Applied Science, London, UK, 1984

84. *Encyclopedia of PVC*, ed. L.I. Nass and C.A. Heiberger, Marcel Dekker, New York, NY, USA, 1986

85. Cha, G.S., Liu, D., Meyerhoff, M.E., *et al.*, Electrochemical performance, biocompatibility, and adhesion of new polymer matrices for solid-state ion sensors, *Anal.Chem.* 63, 1666-1672, 1991

86. Zielinski, B.A. and Aebischer, P., Chitosan as a matrix for mammalian cell encapsulation, *Biomaterials* 15, 1049-1056, 1994

87. *Immobilization of Enzymes and Cells,* ed. G.F. Bickerstaff, Humana Press, Totowa, NJ, USA, 1997

88. Karakus, E., Pekyardimci, S., and Esma, K., Urea biosynthesis based on PVC membrane containing palmitic acid, *Artif.Cells Blood Substit.Immobil.Biotechnol.* 33, 329-341, 2005

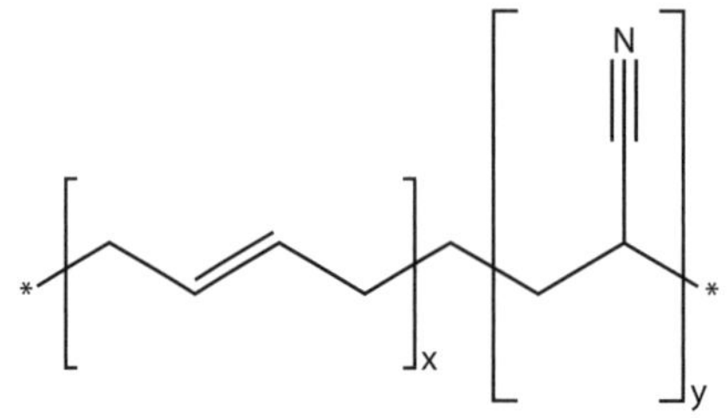

Acrylonitrile-butadiene (Nitrile)

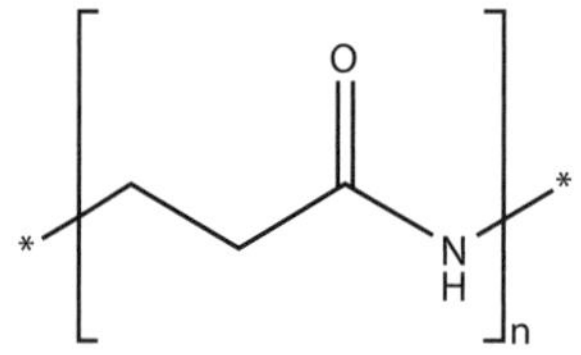

Nylon 3

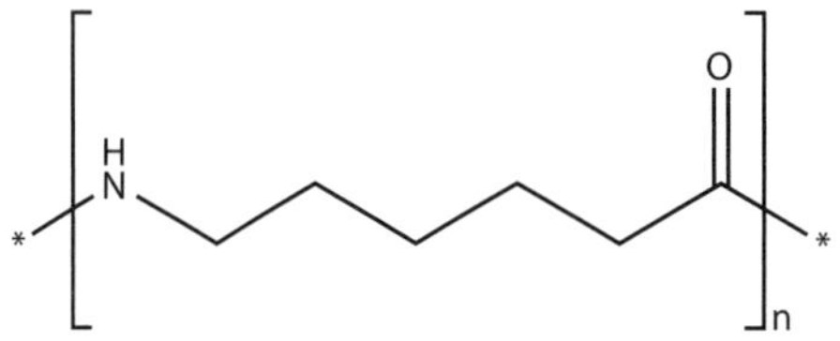

Nylon 6

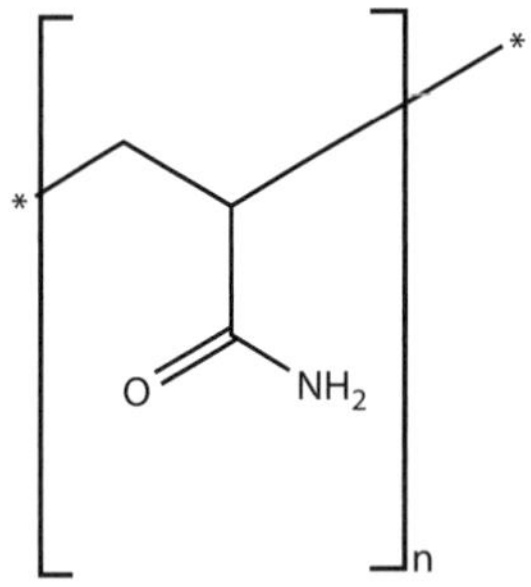

Polyacrylamide

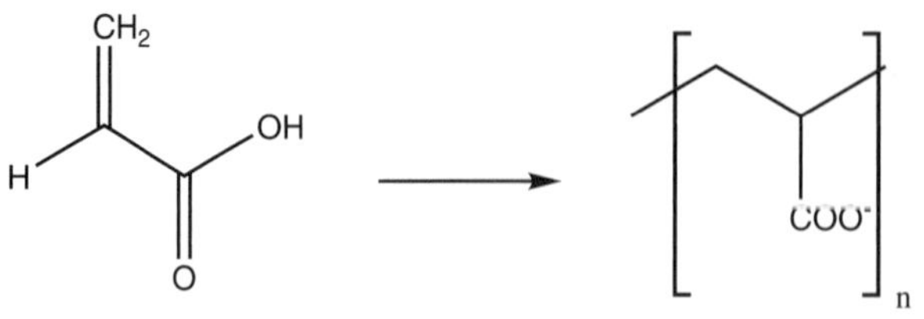

Acrylic Acid

Polyacrylic acid
(usually as sodium salt

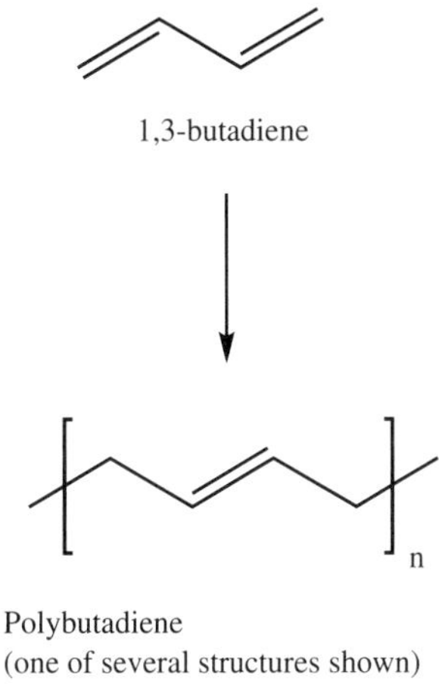

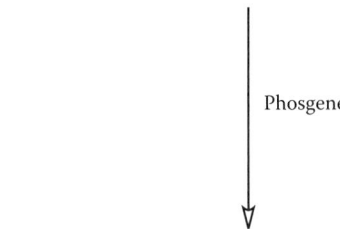

1,3-butadiene
n
Polybutadiene
(one of several structures shown)

CH_3
HO
OH
CH_3
bisphenol A
Phosgene
CH_3
O
*
O
C
O
*
CH_3
n
polycarbonate

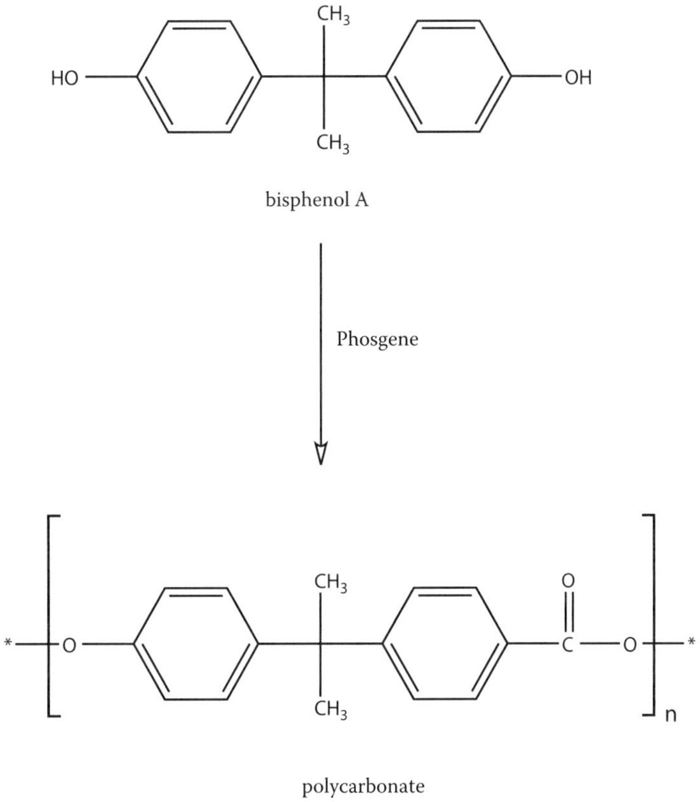

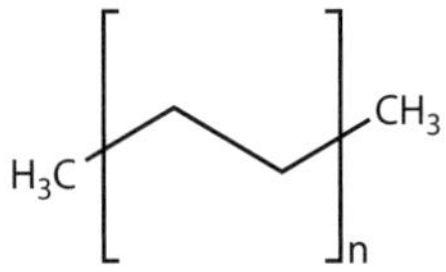

Polyethylene

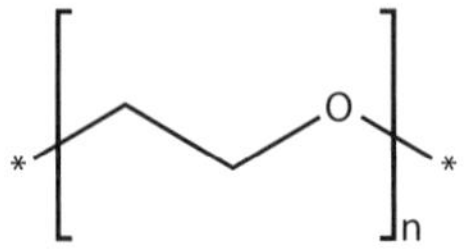

Propylene Oxide

KOH

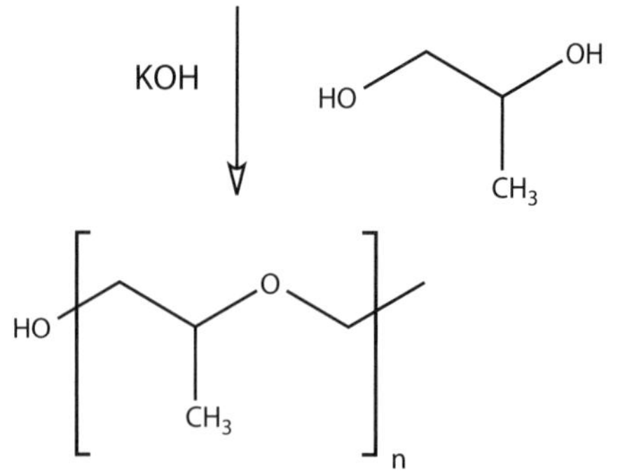

Polypropylene Oxide
(Polypropylene Glycol)

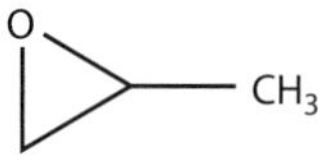

Polyethylene oxide

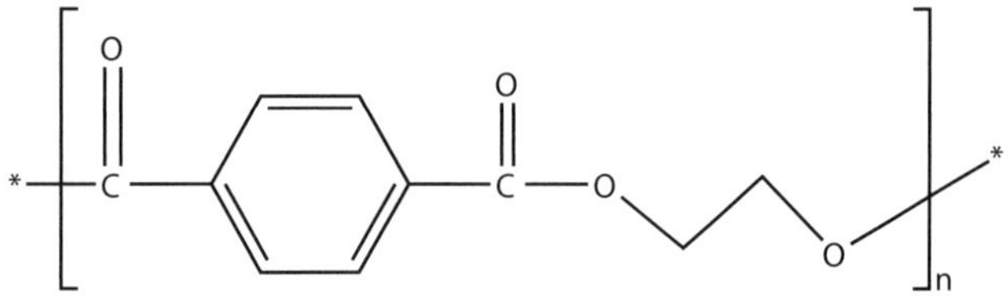

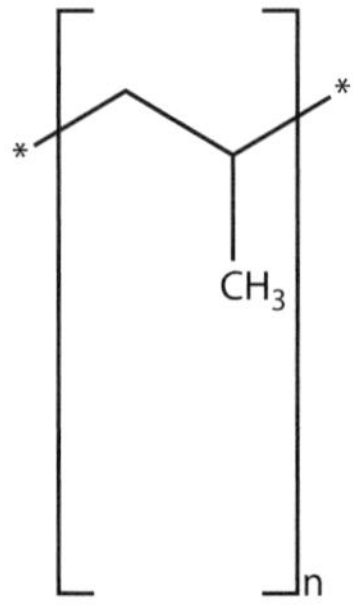

poly(ethylene)terephthalate (PET)

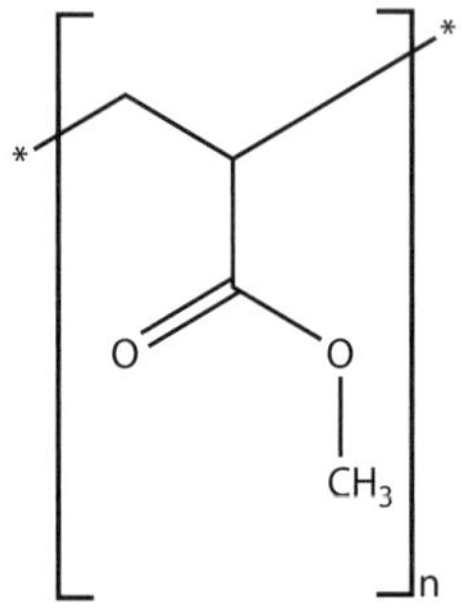

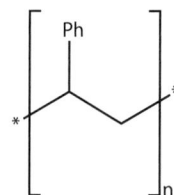

Polystyrene

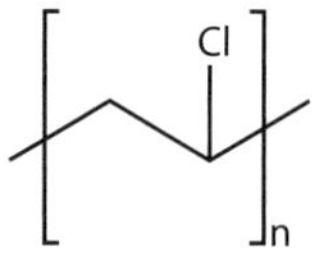

Polyvinyl chloride

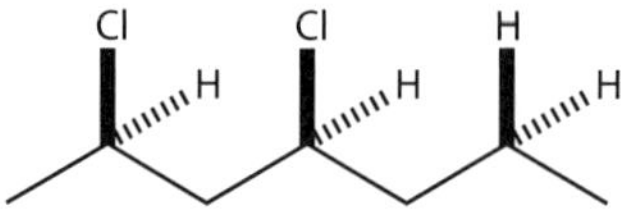

Isotactic

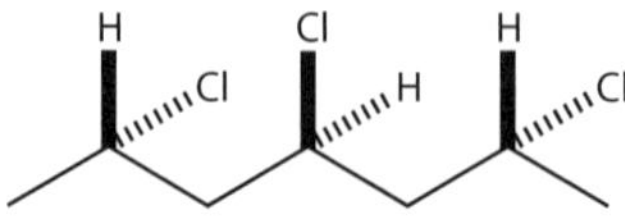

Poly(methyl acrylate)

Syndiotactic

Polypropylene

Atactic (an irregular)

CHEMICAL AND PHYSICAL PROPERTIES OF VARIOUS COMMERCIAL PLASTICS

PHYSICAL PROPERTIES

	Polyethylene Low Density	Polyethylene High Density	Polypropylene Copolymer	Polypropylene	Polymethyl-Pentene	Teflon FEP	Polycarbonate	Polyphenylene Oxide	Polystyrene General Purpose	Styrene-Acrylonitrile	Polyvinyl Chloride
Temperature Limit, °C	80	120	130	135	175	205	135	135	70	95	70
Specific Gravity	0.92	0.95	0.90	0.90	0.83	2.15	1.20	1.06	1.07	1.07	1.34
Tensile Strength, psi	2,000	4,000	2,900	5,000	4,000	3,000	8,000	9,600	6,000	11,000	6,500
Brittleness Temperature, °C	−100	−100	−40	0	—	−270	−135	—	Brittle[a]	−25	−30
Water Absorption, %	0.01	0.01	0.02	0.02	0.01	0.01	0.35	0.07	0.05	0.23	0.06
Flexibility	Excellent	Rigid	Slight	Rigid	Rigid	Excellent	Rigid	Rigid	Rigid	Rigid	Rigid
Transparency	Translucent	Opaque	Translucent	Translucent	Clear	Translucent	Clear	Opaque	Clear	Clear	Clear
Relative O$_2$ Permeability	0.40	0.08	0.20	0.11	2.0	0.59	0.15	—	0.11	0.03	0.01
Autoclavable	No	With caution	Yes	Yes	Yes	Yes	Yes	Yes	No	No	No

CHEMICAL RESISTANCE[b]

	Polyethylene Low Density	Polyethylene High Density	Polypropylene Copolymer	Polypropylene	Polymethyl-Pentene	Teflon FEP	Polycarbonate	Polyphenylene Oxide	Polystyrene General Purpose	Styrene-Acrylonitrile	Polyvinyl Chloride
Acids, inorganic	E	E	E	E	E	E	E	G	N	E	G
Acids, organic	E	E	E	E	E	E	G	E	G	E	G
Alcohols	E	E	E	E	E	E	G	E	G	G	G
Aldehydes	G	G	G	G	G	E	F	G	N	F	F
Amines	G	G	G	G	G	E	N	F	G	G	N
Bases	E	E	E	E	E	E	N	E	G	E	E
Dimethyl sulfoxide (DMSO)	E	E	E	E	E	E	N	E	N	N	N
Esters	E	E	E	E	E	E	N	F	N	N	F
Ethers	G	G	G	G	G	E	F	N	F	N	F
Foods	E	E	E	E	E	E	E	G	E	G	G
Glycols	E	E	E	E	E	E	G	E	G	G	F
Hydrocarbons, aliphatic	G	G	G	G	G	E	F	G	N	E	F
Hydrocarbons, aromatic	G	G	G	G	F	E	N	N	N	N	N
Hydrocarbons, halogenated	G	G	G	G	F	E	N	N	N	N	N
Ketones	G	G	G	G	G	E	N	N	N	N	N
Mineral oil	E	E	E	E	E	E	E	E	G	G	E
Oils, essential	G	G	G	G	G	E	G	F	N	F	N
Oils, lubricating	G	G	E	E	E	E	G	E	G	G	E
Oils, vegetable	E	E	E	E	E	E	E	E	G	E	E
Proteins, unhydrolyzed	E	E	E	E	E	E	E	G	E	E	E
Salts	E	E	E	E	E	E	E	E	E	E	G
Silicones	G	G	E	E	E	E	E	E	E	G	E
Water	E	E	E	E	E	E	E	E	E	E	E

[a] Normally somewhat brittle at room temperatures.

[b] E, Excellent. Long exposures (up to one year) at room temperatures have no effect. G, Good. Short exposures (less than 24 hours) at room temperature cause no damage. F, Fair. Short exposures at room temperature cause little or no damage under unstressed conditions. N, Not recommended. Short exposures may cause permanent damage.

By permission of Thermo Fisher Scientific.

GENERIC SOURCE-BASED NOMENCLATURE FOR POLYMERS

(IUPAC Recommendations 2001)

Prepared by a Working Group consisting of

R. E. Bareiss (Germany), R. B. Fox (Usa), K. Hatada (Japan), K. Horie (UK), A. D. Jenkins (UK), J. Kahovec (Czech Republic), R Kubisa (Poland), E. Maréchal (France), I. Meisel (Germany), W. V. Metanomski (USA), I. Mita (Japan), R. F. T. Stepto (UK), and E. S. Wilks (USA)

Prepared for publication by

E. Maréchal[1] and E. S. Wilks[2,†]

[1]*Université Pierre et Marie Curie (Paris VI), Laboratoire de Synthèse Macromoléculaire, Boîte 184, 4 Place Jussieu F-75252, Paris Cédex 05, France;*
[2]*113 Meriden Drive, Canterbury Hills, Hockessin, DE 19707 USA*

*Membership of the Commission during the preparation of this report (1993–1999) was as follows:

Titular Members: R. E. Bareiss (Germany, 1983–1993); M. Barón (Argentina, from 1996, Secretary from 1998); K. Hatada (Japan, 1989–1997); M. Hess (Germany, from 1998); K. Horie (Japan, from 1997); J. Kahovec (Czech Republic, to 1999); P. Kubisa (Poland, from 1999); E. Maréchal (France, from 1994); I. Meisel (Germany, from 2000); W. V. Metanomski (USA, 1994–1999); C. Noël (France, to 1997); V. P. Shibaev (Russia, to 1995); R. F. T. Stepto (UK, 1989–1999, Chairman to 1999); E. S. Wilks (USA, from 2000); W. J. Work (USA, 1987–1999, Secretary, 1987–1997); ***Associate Members***: M. Barón (Argentina, 1991–1995); K. Hatada (Japan, 1998–1999); J.-I. Jin (Korea, from 1993); M. Hess (Germany, 1996–1997); K. Horie (Japan, 1996–1997); O. Kramer (Denmark, from 1996); P. Kubisa (Poland, 1996–1998); E. Maréchal (France, 1991–1993); I. Meisel (Germany, 1997–1999); S. Penczek (Poland, from 1994); L. Shi (China, 1987–1995); V. P. Shibaev (Russia, 1996–1999); E. S. Wilks (USA, 1998–1999).

†Corresponding author

Abstract: The commission has already published two documents on the source-based names of linear copolymers and nonlinear polymers; however, in some cases this nomenclature leads to ambiguous names. The present document proposes a generic source-based nomenclature that solves these problems and yields clearer source-based names. A generic source-based name comprises two parts:

1) polymer class (generic) name followed by a colon
2) the actual or hypothetical monomer name(s), always parenthesized in the case of a copolymer

The formula, the structure-based name, the source-based name, and the generic source-based name of the polymer are given for each example in the document. In some cases, only generic source-based give unambiguous names, for example, when a polymer has more than one name or when it is obtained through a series of intermediate structures. The rules concern mostly polymers with one or more types of functional group or heterocyclic system in the main chain, but to some extent they are also applicable to polymers with side-groups, carbon-chain polymers such as vinyl or diene polymers, spiro and cyclic polymers, and networks.

Contents

1. Introduction

The IUPAC Commission on Macromolecular Nomenclature has published three documents [1–3] on the structure-based nomenclature for polymers that enable most polymers, except networks, to be named. The Commission has also produced two documents

Reproduced from:
Pure Appl. Chem., Vol. 73, No. 9, pp. 1511–1519, 2001.
© 2001 IUPAC

[4,5] on the source-based nomenclature of linear copolymers and nonlinear polymers. In general, source-based names are simpler and less rigorous than structure-based names. However, there are cases in which the simplicity of the source-based nomenclature leads to ambiguous names for polymers. For example, the condensation of a dianhydride (A) with a diamine (B) gives first a polyamide-acid, which can be cyclized to a polyimide; however, both products have the same name poly(A-*alt*-B) according to current source-based nomenclature. If the class name of the polymer "amide-acid" or "imide" is incorporated in the name,

differentiation is easily accomplished. Even in cases where only a single product is formed, use of the class name (generic name) may help to clarify the structure of the polymer, especially if it is very complex.

Examples of ambiguous names exist also for homopolymers. The source-based name "polybutadiene" does not indicate whether the structure is 1,2-, 1,4-*cis*-, or 1,4-*trans*-; supplementary information is needed to distinguish between the possibilities.

It is the objective of the present document to introduce a generic nomenclature system to solve these problems, and to yield better source-based names.

Most trivial names, such as polystyrene, are source-based names. Hitherto, the Commission has not systematically recommended source-based names for homopolymers because it considered that the more rigorous structure-based names were more appropriate for scientific communications. However, since the publication of "Nomenclature of Regular Single-Strand Organic Polymers" in 1976, scientists, in both industry and academia, have continued to use trivial names. Even the Commission itself adopted (1985) a source-based nomenclature for copolymers owing to its simplicity and practicality. Based on these facts, the Commission has now decided to recommend source-based nomenclature as an alternative official nomenclature for homopolymers. In this document, the rules for generating source-based names for homopolymers are described. Consequently, source-based and structure-based names are available for most polymers.

Names of the monomers in the source-based names of polymers should preferably be systematic but they may be trivial if well established by usage. Names of the organic groups, as parts of constitutional repeating units (CRU) in structure-based names, are those based on the principles of organic nomenclature and recommended by the 1993 *A Guide to IUPAC Nomenclature of Organic Compounds* [6].

2. Source-based nomenclature for homopolymers

RULE 1

The source-based name of a homopolymer is made by combining the prefix "poly" with the name of the monomer. When the latter consists of more than one word, or any ambiguity is anticipated, the name of the monomer is parenthesized.

Example 1.1

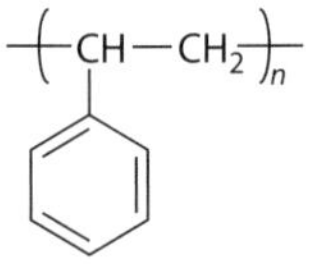

Source-based name: Polystyrene
Structure-based name: poly(1-phenylethylene)

Example 1.2

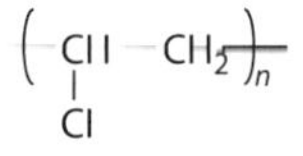

Source-based name: poly(vinyl chloride)
Structure-based name: poly(1-chloroethylene)

3. Generic nomenclature

3.1 Fundamental principles

The basic concept for generic source-based nomenclature is very simple; just add the polymer class name to the source-based name of the polymer. Addition of the polymer class name is frequently OPTIONAL; in some cases, the addition is necessary to avoid ambiguity or to clarify. However, the addition is undesirable if it fails to add clarification.

The system presented here can be applied to almost all homopolymers, copolymers, and others, such as networks. However, generic source-based nomenclature should not be considered as a third nomenclature system to be added to the other two systems of nomenclature; it must be considered as an auxiliary system and a simple extension of current source-based nomenclature. When the generic part of the name is eliminated from the name of a polymer, the well-established source-based name remains.

3.2 General rules

RULE 2

A generic source-based name of a polymer has two components in the following sequence: (1) a polymer class (generic) name (polyG) followed by a colon and (2) the actual or hypothetical monomer name(s) (A, B, etc.), always parenthesized in the case of a copolymer. In the case of a homopolymer, parentheses are introduced when it is necessary to improve clarity.

polyG: A   polyG: (B)   polyG: (A-*co*-B)   polyG: (A-*alt*-B)

Note 1 The polymer class name (generic name) describes the most appropriate type of functional group or heterocyclic ring system.

Note 2 All the rules given in the two prior documents on source-based nomenclature [4,5] can be applied to the present nomenclature system, with the addition of the generic part of the name.

Note 3 A polymer may have more than one name; this usually occurs when it can be prepared in more than one way.

Note 4 If a monomer or a pair of complementary monomers can give rise to more than one polymer, or if the polymer is obtained through a series of intermediate structures, the use of generic nomenclature is essential (see examples 2.1, 2.3, and 2.4).

Example 2.1

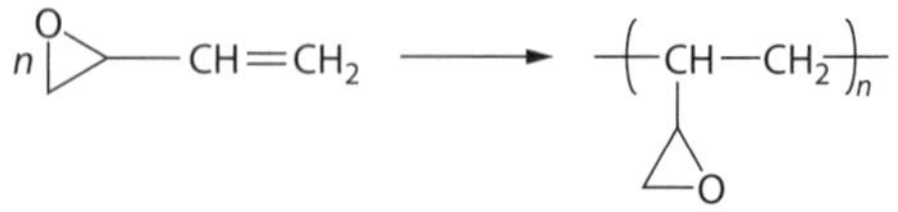

Generic source-based name:
    I. polyalkylene:vinyloxirane
    II. polyether:vinyloxirane

Source-based names:
I and II have the same source-based name:poly(vinyloxirane).

Structure-based names:
I. poly(1-oxiranylethylene)
II. poly[(oxy(1-vinylethylene)]

Example 2.2

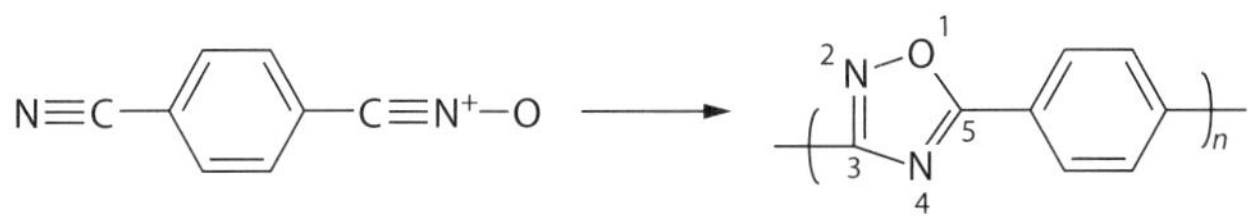

Generic source-based name:
polyoxadiazole:(4-cyanobenzonitrile *N*-oxide)

Structure-based name:
poly(1,2,4-oxadiazole-3,5-diyl-1,4-phenylene)

Example 2.3

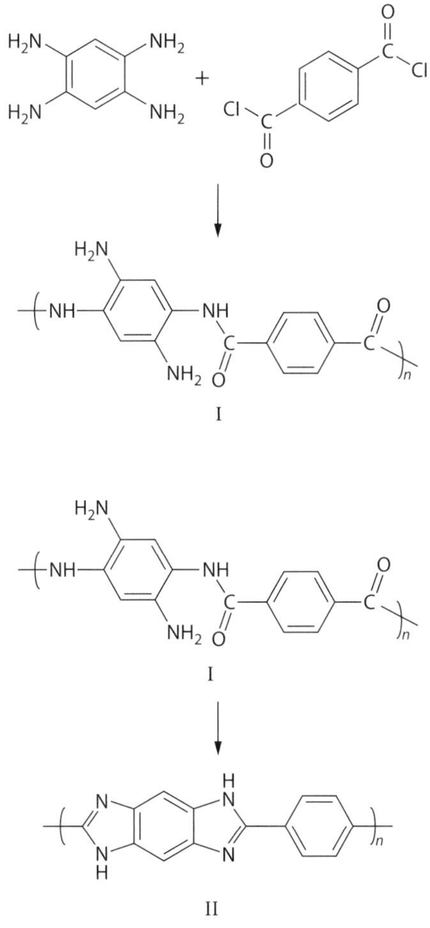

Generic source-based name:
I. polyamide:[(terephthaloyl dichloride)-*alt*-benzene-1,2,4,5-tetramine]
II. polybenzimidazole:[(terephthaloyl dichloride)-*alt*-benzene-1,2,4,5-tetramine]

Source-based name:
I and II have the same source-based name:

poly[(terephthaloyl dichloride)-*alt*-benzene-1,2,4,5-tetramine]

Structure-based names:
I. poly[imino (2,5-diamino-1,4-phenylene)iminoterephthaloyl]
II. poly[(1,5-dihydrobenzo[1,2-*d*:4,5-*d'*]diimidazole-2,6-diyl)-1,4-phenylene]

Example 2.4

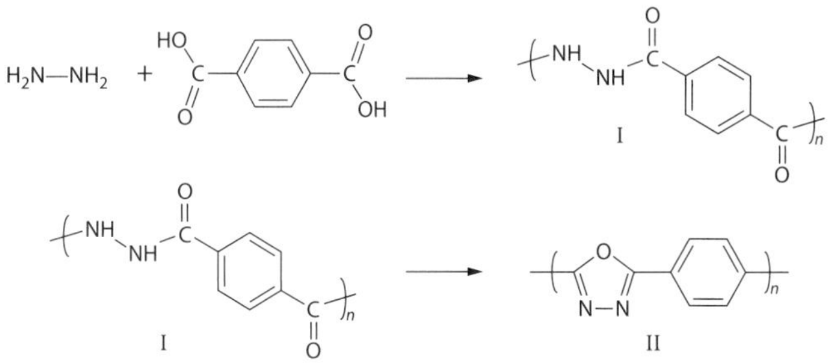

Generic source-based names:
I. polyhydrazide:[hydrazine-*alt*-(terephthalic acid)]
II. polyoxadiazole:[hydrazine-*alt*-(terephthalic acid)]

Source-based name:
I and II have the same source-based name: poly[hydrazine-*alt*-(terephthalic acid)]
Structure-based names:
I. poly(hydrazine-1,2-diylterephthaloyl)
II. poly(1,3,4-oxadiazole-2,5-diyl-1,4-phenylene)

Example 2.5

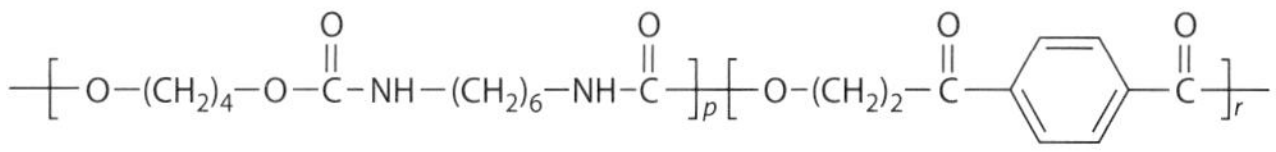

Generic source-based names:
polyurethane:[butane-1,4-diol-*alt*-(hexane-1,6-diyl diisocyanate)]-*block*-polyester:[(ethylene glycol)-*alt*-(terephthalic acid)]

Structure-based name:
poly(oxybutane-1,4-diyloxycarbonyliminohexane-1,6-diyliminocarbonyl)-*block*-poly(oxyethyleneoxyterephthaloyl)

Example 2.6

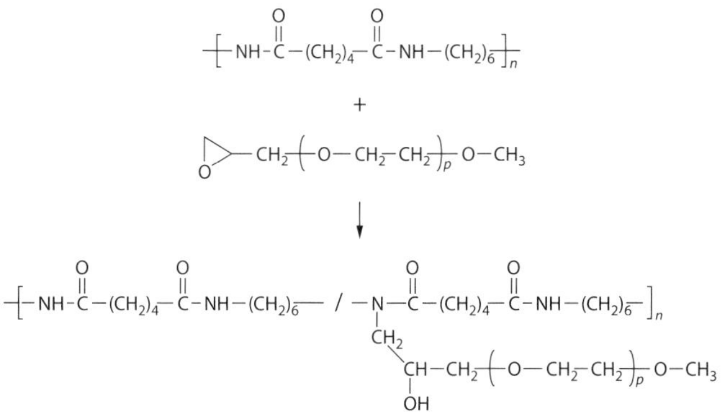

Generic source-based name:
    polyamide:[hexane-1,6-diamine-*alt*-(adipic acid)]-*graft*-
        polyether: (ethylene oxide)

Note 5   It is assumed that this reaction is limited to only one
        graft for each CRU.

RULE 3

When more than one type of functional group or heterocyclic system is present in the polymer structure, names should be
alphabetized; for example, poly(GG′):(A-*alt*-B).

Note 6   It is preferable, but not mandatory, to cite all generic
        classes.

Example 3.1

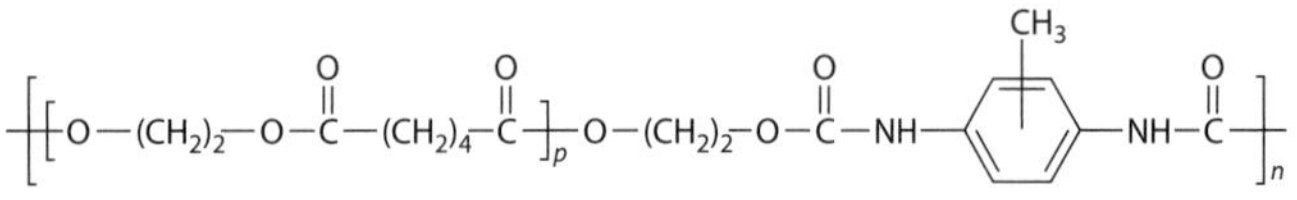

Generic source-based name:
    polyesterurethane:(*α,ω*-dihydroxyoligo[(ethylene glycol)-
        alt-(adipic acid)]-*alt*-(2,5-tolylene diisocyanate)}

Structure-based name:
    poly{[oligo(oxyethyleneoxyadipoyl)]oxyethyleneoxycar-
        bonylimino(x-methyl-1,4-phenylene)iminocarbonyl)}

Example 3.2

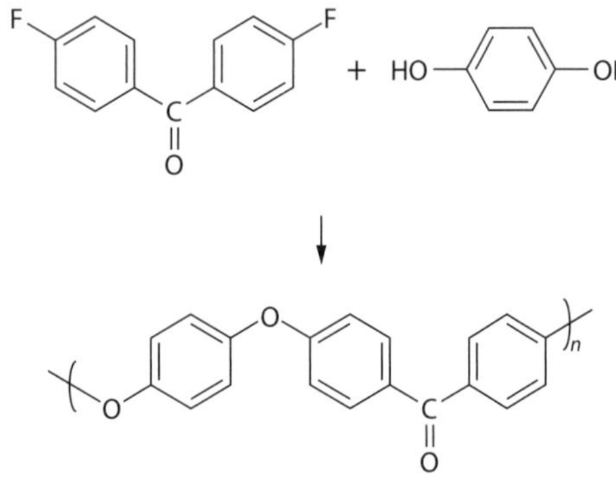

Generic source-based name:
    polyetherketone:(4,4′-difluorobenzophenone-*att*-hydro-
        quinone)

Structure-based name:
    poly(oxy-1,4-phenyleneoxy-1,4-phenylenecarbonyl-1,4-
        phenylene)

RULE 4

Polymer class names relevant only to the main chain are specified in the name; names of side-chain functional groups may also
be included after a hyphen if they are formed during the polymerization reaction.

Example 4.1

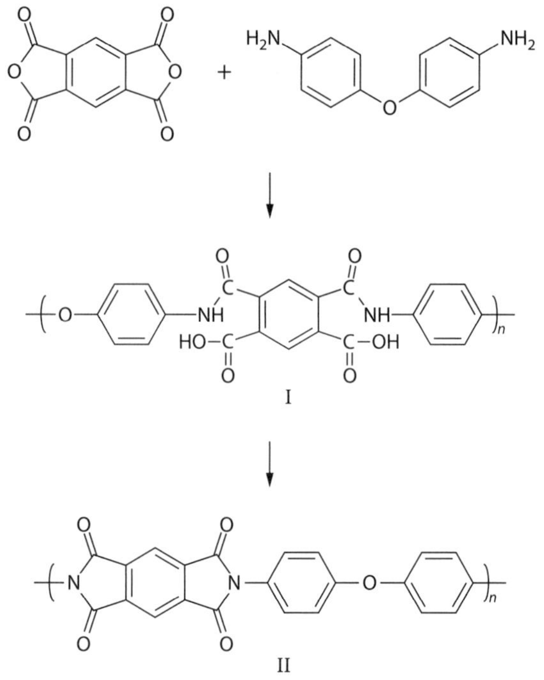

I

II

Generic source-based names:
    I. poly(amide-acid):[(pyromellitic dianhydride)-*alt*-(4,4′-
        oxydianiline)] (Both carboxy groups result from the
        polymerization reaction.)
    II. polyimide:[(pyromellitic dianhydride)-*alt*-(4,4′-oxydiani-
        line)]

Structure-based names:
    I. poly[oxy-1,4-phenyleneiminocarbonyl(4,6-dicarboxy-1,
        3-henylene)carbonylimino-1,4-phenylene]
    II. poly[[(5,7-dihydro-1,3,5,7-tetraoxobenzo[1,2-*c*:4,5-*c*′]
        dipyrrole-2,6(1*H*,3*H*)-diyl)-1,4-phenyleneoxy-1,4-phen-
        ylene]

Example 4.2

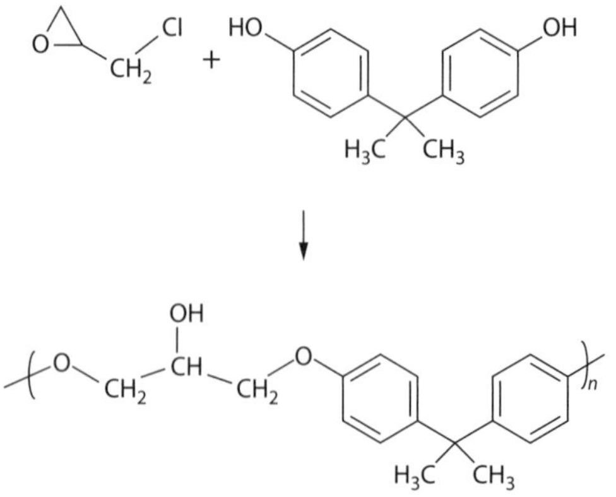

Generic source-based names:
poly(ether-alcohol):(epichlorohydrin-*alt*-bisphenol A)

Structure-based name:
poly[oxy(2-hydroxypropane-1,3-diyl)oxy-1,4-phenylene(1-methylethane-1,1-diyl)-1,4-phenylene]

RULE 5

In the case of carbon-chain polymers such as vinyl polymers or diene polymers, the generic name is to be used only when different polymer structures may arise from a given monomeric system.

Example 5.1

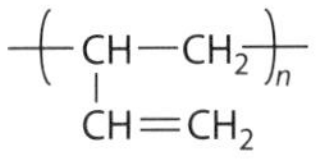

Generic source-based name:
polyalkylene:(buta-1,3-diene)

Source-based name: poly(buta-1,3-diene)

Structure-based name: poly(1-vinylethylene)

Example 5.2

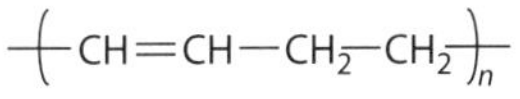

Generic source-based name:
polyalkenylene:buta-1,3-diene

Source-based name:poly(buta-1,3-diene)

Structure-based name: poly(but-1-ene-1,4-diyl)

Example 5.3

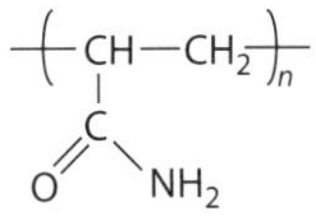

Generic source-based name:
polyalkylene: acrylamide

Structure-based name: poly[1-(aminocarbonyl)ethylene]

Example 5.4

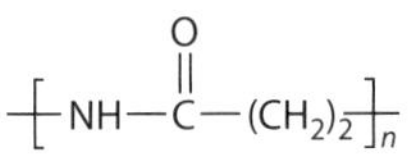

Generic source-based name: polyamide:acrylamide

Structure-based name: poly[imino(1-oxopropane-1,3-diyl)]

Note 7 The terms polyalkylene and polyalkenylene have been defined in ref. 7, p. 149.

4. Further applications of generic names

Generic source-based nomenclature can be extended to more complicated polymers such as spiro and cyclic polymers and networks.

Example 6.1

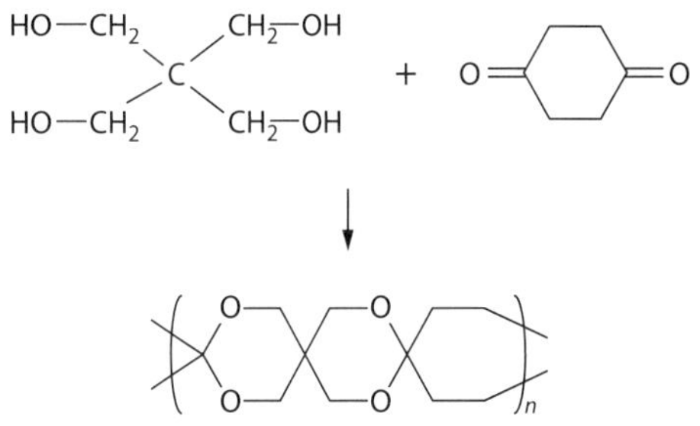

Generic source-based name:
polyspiroketal:{[2,2-bis(hydroxymethyl)-propane-1,3-diol]-*alt*-cyclohexane-1,4-dione} or polyspiroketal:(pentaerythritol-*alt*-cyclohexane-1,4-dione)

Structure-based name:
poly[2,4,8,10-tetraoxaspiro[5.5]undecane-3,3,9,9-tetrayl-9,9-bis(ethylene)]

Example 6.2

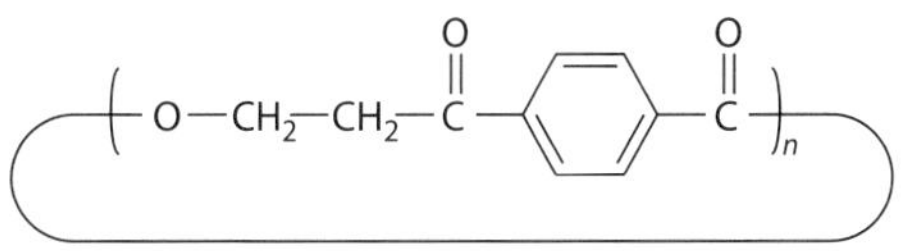

Generic source-based name:
cyclo-polyester: [(ethylene glycol)-*alt*-(terephthalic acid)]

Note 8 There is no IUPAC nomenclature for cyclic polymers.

Example 6.3

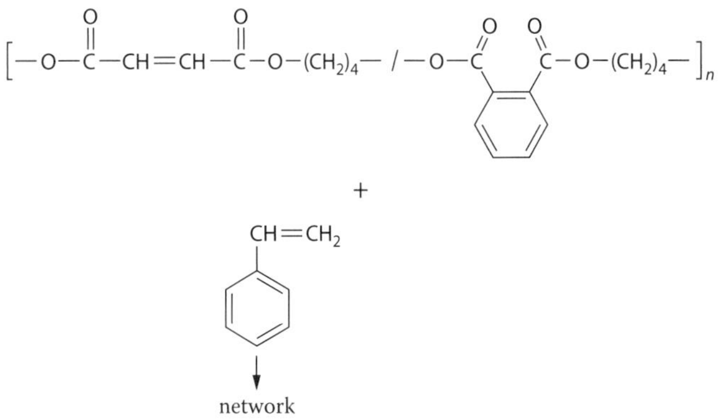

Generic source-based name:
polyester: {butane-1,4-diol-*alt*-[(maleic anhydride);(phthalic anhydride)]}-*net*-polyalkylene: (maleic anhydride)-*co*-styrene]

5. References

1. "Nomenclature of regular single-strand organic polymers, 1975", *Pure Appl. Chem.* **48**, 373–385 (1976). Reprinted as chapter 5 in Ref. 7.
2. "Nomenclature of regular double-strand (ladder and spiro) organic polymers 1993", *Pure Appl. Chem.* **65**, 1561–1580 (1993).
3. "Structure-based nomenclature for irregular single-strand organic polymers 1994", *Pure Appl. Chem.* **66**, 873–889 (1994).
4. "Source-Based Nomenclature for Copolymers 1985", *Pure Appl. Chem.* **57**, 1427–1440 (1985). Reprinted as chapter 7 in Ref. 7.
5. "Source-based nomenclature for non-linear macromolecules and macromolecular assemblies", *Pure Appl. Chem.* **69**, 2511–2521 (1997).
6. *A Guide to IUPAC Nomenclature of Organic Compounds*, R. Panico, W. H. Powell, J-C. Richer (Eds.), Blackwell Scientific Publications, Oxford (1993).
7. *Compendium of Macromolecular Nomenclature*, W. V. Metanomski (Ed.), Blackwell Scientific Publications, Oxford (1991).

DEFINITIONS OF TERMS RELATING TO REACTIONS OF POLYMERS AND TO FUNCTIONAL POLYMERIC MATERIALS

(IUPAC Recommendations 2003)

Prepared by a Working Group consisting of

K. Horie[1,‡], M. Barón[2], R. B. Fox[3], J. He[4], M. Hess[5], J. Kahovec[6], T. Kitayama[7], P. Kubisa[8], E. Maréchal[9], W. Mormann[10], R. F. T. Stepto[11], D. Tabak[12], J. Vohlídal[13], E. S. Wilks[14], and W. J. Work[15]

*Members of the Commission on Macromolecular Nomenclature (1997–2001) and the Subcommittee on Macromolecular Nomenclature (2002–2003) contributing to this report were:

G. Allegra (Italy); M. Barón (Argentina, Commission and Subcommittee Secretary); A. Fradet (France); K. Hatada (Japan); J. He (China); M. Hess (Germany, Commission and Subcommittee Chairman); K. Horie (Japan); A. D. Jenkins (UK); J.-I. Jin (Korea); R. G. Jones (UK, Subcommittee Secretary); J. Kahovec (Czech Republic); T. Kitayama (Japan); P. Kratochvíl (Czech Republic); P. Kubisa (Poland); E. Maréchal (France); I. Meisel (Germany); W. V. Metanomski (USA); G. Moad (Australia); W. Mormann (Germany); S. Penczek (Poland); L. P. Rebelo (Portugal); M. Rinaudo (France); I. Schopov (Bulgaria); M. Schubert (USA); V. P. Shibaev (Russia); S. Slomkowski (Poland); R. F. T. Stepto (UK, Commission Chairman); D. Tabak (Brazil); J. Vohlídal (Czech Republic); E. S. Wilks (USA); W. J. Work (USA, Commission Secretary); *Other Contributors:* K. Dorfner (Germany); J. M. J. Fréchet (USA); W. I. Harris (USA); P. Hodge (UK); T. Nishikubo (Japan); C. K. Ober (USA); E. Reichmanis (USA); D. C. Sherrington (UK); M. Tomoi (Japan); D. Wöhrle (Germany).

[1]Department of Organic and Polymer Materials Chemistry, Tokyo University of Agriculture and Technology, 2-24-16 Nakacho, Koganei-shi, Tokyo 184-8588, Japan; [2]Facultad de Ciências Exactas y Naturales, Universidad de Belgrano, Villanueva 1324, Buenos Aires 1426, Argentina; [3]6115 Wiscasset Road, Bethesda, MD 20816, USA; [4]State Key Laboratory of Engineering Plastics, The Chinese Academy of Sciences, Institute of Chemistry, Beijing 100080, China; [5]Fachbereich 6: Physikalische Chemie, Universitat Duisburg-Essen, D-47048 Duisburg, Germany; [6]Ústav Makromolekulární Chemie, Akademie ved Ceské Republiky, Heyrovského námestí 2, CZ-162 06 Praha 6, Czech Republic; [7]Department of Chemistry, Osaka University, Toyonaka, Osaka 560-8531, Japan; [8]Centrum Badan Molek. i Makromolek., Polska Akademia Nauk, Sienkiewicza 112, PL-90 363 Lódz, Poland; [9]Laboratoire de Synthèse Macromoléculaire, Université Pierre et Marie Curie (Paris VI), Boîte 184 - Tour 54, 4e étage, 4 place Jussieu, F-75252 Paris Cédex 05, France; [10]Makromolekulare Chemie, Universität - Gesamthochschule Siegen, Adolf Reichwein Strasse 2, D-57068 Siegen, Germany; [11]Polymer Science and Technology Group (MMSC), University of Manchester and UMIST, Grosvenor Street, Manchester M1 7HS, UK; [12]Praca Pio X, 78 Sala 1213 - Candelaria, Rio de Janerio - RJ 20091-040, Brazil; [13]Katedra Fyzikalni a Makromolekularni Chemie, Universita Karlova, Albertov 2030, CZ-128 40 Praha 2, Czech Republic; [14]113 Meriden Drive, Canterbury Hills, Hockessin, DE 19707, USA; [15]1288 Burnett Road, Huntingdon Valley, PA 19006, USA

‡Corresponding author: E-mail: horiek@cc.tuat.ac.jp

Abstract: The document defines the terms most commonly encountered in the field of polymer reactions and functional polymers. The scope has been limited to terms that are specific to polymer systems. The document is organized into three sections. The first defines the terms relating to reactions of polymers. Names of individual chemical reactions are omitted from the document, even in cases where the reactions are important in the field of polymer reactions. The second section defines the terms relating to polymer reactants and reactive polymeric materials. The third section defines the terms describing functional polymeric materials.

Contents

Introduction

Chemical reactions of polymers have received much attention during the last two decades. Many fundamentally and industrially important reactive polymers and functional polymers are prepared by the reactions of linear or cross-linked polymers and by

the introduction of reactive, catalytically active, or other groups onto polymer chains. Characteristics of polymer reactions may be appreciably different from both reactions of low-molar-mass compounds and polymerization reactions. Basic definitions of polymerization reactions have been included in the original [1] and revised [2] documents on basic terms in polymer science published by the IUPAC Commission on Macromolecular Nomenclature. Furthermore, the basic classification and definitions of polymerization reactions [3] and some polymer reactions such as degradation, aging, and related chemical transformations of polymers have been defined [4]. However, in spite of the growing importance of the field, a clear and uniform terminology covering the field of reactions and the functionalization of polymers has not been presented until now. For example, combinatorial chemistry using reactive polymer beads has become a new field in recent years. The development of a uniform terminology for such multidisciplinary areas can greatly aid communication and avoid confusion.

This document presents clear concepts and definitions of general and specific terms relating to reactions of polymers and functional polymers. The document is divided into three sections. In Section 1 terms relating to reactions of polymers are defined. Names of individual chemical reactions (e.g., chloromethylation) are omitted from this document, even in cases where the reactions are important in the field of polymer reactions, because such names are usually already in widespread use and are well defined in organic chemistry and other areas of chemistry [5]. Sections 2 and 3 deal with the terminology of reactive and functional polymers. The term "functional polymer" has two meanings: (a) a polymer bearing functional groups (such as hydroxy, carboxy, or amino groups) that make the polymer reactive and (b) a polymer performing a specific function for which it is produced and used. The function in the latter case may be either a chemical function such as a specific reactivity or a physical function like electric conductivity. Polymers bearing reactive functional groups are usually regarded as polymers capable of undergoing chemical reactions. Thus, Section 2 deals with polymers and polymeric materials that undergo various kinds of chemical reactions (i.e., show chemical functions). Section 3 deals with terms relating to polymers and polymeric materials exhibiting some specific physical functions. For definitions of some physical functions, see also *Compendium of Chemical Terminology* ("Gold Book") [6].

A functional polymer according to Definition 3.6 of the present document is a polymer that exhibits specified chemical reactivity or has specified physical, biological, pharmacological, or other uses that depend on specific chemical groups. Thus, several terms concerned with properties or the structure of polymers are included in Section 3 whenever they are closely related to specific functions.

Terms that are defined implicitly in the notes and related to the main terms are given in bold type.

1. Reactions involving polymers

1.1 Chemical amplification
Process consisting of a chemical reaction that generates a species that catalyzes another reaction and also the succeeding catalyzed reaction.

> *Note 1*: Chemical amplification can lead to a change in structure and by consequence to a change in the physical properties of a polymeric material.

> *Note 2*: The term "chemical amplification" is commonly used in photoresist lithography employing a **photo-acid generator** or **photo-base generator**.

> *Note 3*: An example of chemical amplification is the transformation of [(*tert*-butoxycarbonyl)oxy]phenyl groups in polymer chains to hydroxyphenyl groups catalyzed by a photo-generated acid.

> *Note 4*: The term "amplification reaction" as used in analytical chemistry is defined in [6], p. 21.

1.2 Chemical modification
Process by which at least one feature of the chemical constitution of a polymer is changed by chemical reaction(s).

> *Note*: A configurational change (e.g., cis–trans isomerization) is not usually referred to as a chemical modification.

1.3 Cross-linking
Reaction involving sites or groups on existing macromolecules or an interaction between existing macromolecules that results in the formation of a small region in a macromolecule from which at least four chains emanate.

> *Note 1*: See [6], p. 94 and Definition 1.59 in [2] for cross-link.

> *Note 2*: The small region may be an atom, a group of atoms, or a number of branch points connected by bonds, groups of atoms, or oligomeric chains.

> *Note 3*: A reaction of a reactive chain end of a linear macromolecule with an internal reactive site of another linear macromolecule results in the formation of a branch point, but is not regarded as a cross-linking reaction.

1.4 Curing
Chemical process of converting a prepolymer or a polymer into a polymer of higher molar mass and connectivity and finally into a network.

> *Note 1*: Curing is typically accomplished by chemical reactions induced by heating (**thermal curing**), photo-irradiation (**photo-curing**), or electron-beam irradiation (**EB curing**), or by mixing with a chemical curing agent.

> *Note 2*: Physical aging, crystallization, physical cross-linking, and postpolymerization reactions are sometimes referred to as "curing". Use of the term "curing" in these cases is discouraged.

> *Note 3*: See also Definition 1.22.

1.5 Depolymerization
Process of converting a polymer into its monomer or a mixture of monomers (see [6], p. 106 and Definition 3.25 in [2]).

1.6 Grafting
Reaction in which one or more species of block are connected to the main chain of a macromolecule as side chains having constitutional or configurational features that differ from those in the main chain.

> *Note*: See [6], p. 175 and Definition 1.28 in [2] for graft macromolecule.

1.7 Interchange reaction
Reaction that results in an exchange of atoms or groups between a polymer and low-molar-mass molecules, between polymer molecules, or between sites within the same macromolecule.

> *Note*: An interchange reaction that occurs with polyesters is called **transesterification**.

1.8 Main-chain scission
Chemical reaction that results in the breaking of main-chain bonds of a polymer molecule.

Note 1: See [6], p. 64 and Definition 3.24 in [2] for chain scission.

Note 2: Some main-chain scissions are classified according to the mechanism of the scission process: **hydrolytic, mechanochemical, thermal, photochemical**, or **oxidative scission**. Others are classified according to their location in the backbone relative to a specific structural feature, for example, α-**scission** (a scission of the C-C bond alpha to the carbon atom of a photo-excited carbonyl group) and β-**scission** (a scission of the C-C bond beta to the carbon atom bearing a radical), etc.

1.9 Mechanochemical reaction
Chemical reaction that is induced by the direct absorption of mechanical energy.

Note: Shearing, stretching, and grinding are typical methods for the mechanochemical generation of reactive sites, usually macroradicals, in polymer chains that undergo mechanochemical reactions.

1.10 Photochemical reaction
Chemical reaction that is caused by the absorption of ultraviolet, visible, or infrared radiation ([6], p. 302).

Note 1: Chemical reactions that are induced by a reactive intermediate (e.g., radical, carbene, nitrene, or ionic species) generated from a photo-excited state are sometimes dealt with as a part of photochemistry.

Note 2: An example of a photochemical reaction concerned with polymers is **photopolymerization.**

Note 3: See also Definitions 1.1, 1.18, 3.14, and 3.25.

1.11 Polymer complexation polymer complex formation
Process that results in the formation of a polymer–polymer complex or a complex composed of a polymer and a low-molar-mass substance.

1.12 Polymer cyclization
Chemical reaction that leads to the formation of ring structures in or from polymer chains.

Note 1: Examples of cyclization along polymer chains are: (a) cyclization of polyacrylonitrile, (b) acetalization of poly(vinyl alcohol) with an aldehyde, (c) cyclization of polymers of conjugated dienes such as polyisoprene or polybutadiene leading to macrocycles.

Note 2: Examples of cyclization of polymer molecules are: (a) cyclization of poly(dimethylsiloxane), (b) back-biting reaction during ionic polymerizations of heterocyclic monomers.

1.13 Polymer degradation
Chemical changes in a polymeric material that usually result in undesirable changes in the in-use properties of the material.

Note 1: In most cases (e.g., in vinyl polymers, polyamides) degradation is accompanied by a decrease in molar mass. In some cases (e.g., in polymers with aromatic rings in the main chain), degradation means changes in chemical structure. It can also be accompanied by cross-linking.

Note 2: Usually, degradation results in the loss of, or deterioration in useful properties of the material. However, in the case of **biodegradation** (degradation by biological activity), polymers may change into environmentally acceptable substances with desirable properties (see Definition 3.1).

Note 3: See Definition 16 in [4] for degradation.

1.14 Polymer functionalization
Introduction of desired chemical groups into polymer molecules to create specific chemical, physical, biological, pharmacological, or other properties.

1.15 Polymer reaction
Chemical reaction in which at least one of the reactants is a high-molar-mass substance.

1.16 Polymer-supported reaction
Chemical reaction in which at least one reactant or a catalyst is bound through chemical bonds or weaker interactions such as hydrogen bonds or donor-acceptor interactions to a polymer.

Note 1: The easy separation of low-molar-mass reactants or products from the polymer-supported species is a great advantage of polymer-supported reactions.

Note 2: Typical examples of polymer-supported reactions are: (a) reactions performed by use of polymer-supported catalysts, (b) solid-phase peptide synthesis, in which intermediate peptide molecules are chemically bonded to beads of a suitable polymer support.

1.17 Protection of a reactive group
Temporary chemical transformation of a reactive group into a group that does not react under conditions where the nonprotected group reacts.

Note: For example, **trimethylsilylation** is a typical transformation used to protect reactive functional groups such as hydroxy or amino groups from their reaction with growing anionic species in anionic polymerization.

1.18 Radiation reaction
Chemical reaction that is induced by ionizing radiation with γ-ray, X-ray, electron, or other high-energy beams.

Note 1: Radiation reactions involving polymers often lead to chain scission and cross-linking.

Note 2: A **photochemical reaction** (see Definition **1.10**) is sometimes regarded as a type of radiation reaction.

1.19 Reactive blending
Mixing process that is accompanied by the chemical reaction(s) of components of a polymer mixture.

Note 1: Examples of reactive blending are: (a) blending accompanied by the formation of a polymer-polymer complex, (b) the formation of block or graft copolymers by a combination of radicals formed by the mechanochemical scission of polymers during blending.

Note 2: Reactive blending may also be carried out as reactive extrusion or reaction injection molding (RIM).

1.20 Sol-gel process
Formation of a polymer network by the reaction of monomer(s), liquid or in solution, to form a gel, and in most cases finally to form a dry network.

Note: An inorganic polymer (e.g., silica-gel or organic-inorganic hybrid) can be prepared by the sol-gel process.

1.21 Surface grafting
Process in which a polymer surface is chemically modified by grafting or by the generation of active sites that can lead to the initiation of a graft polymerization.

Note 1: Peroxidation, ozonolysis, high-energy irradiation, and plasma etching are methods of generating active sites on a polymer surface.

Note 2: See also Definition 1.6.

1.22 Vulcanization
Chemical cross-linking of high-molar-mass linear or branched polymer or polymers to give a polymer network.

Note 1: The polymer network formed often displays rubber-like elasticity. However, a high concentration of cross-links can lead to rigid materials.

Note 2: A classic example of vulcanization is the cross-linking of *cis*-polyisoprene through sulfide bridges in the thermal treatment of natural rubber with sulfur or a sulfur-containing compound.

2. Polymer reactants and reactive polymeric materials

2.1 Chelating polymer
Polymer containing ligand groups capable of forming bonds (or other attractive interactions) between two or more separate binding sites within the same ligand group and a single atom.

Note 1: Chelating polymers mostly act as ion-exchange polymers specific to ions that form chelates with chelating ligands of the polymer.

Note 2: See [6], p. 68 for chelation.

2.2 Ion-exchange polymer
Polymer that is able to exchange ions (cations or anions) with ionic components in solution.

Note 1: See [6], p. 208 for ion exchange.

Note 2: An ion-exchange polymer in ionized form may also be referred to as a **polyanion** or a **polycation.**

Note 3: Synthetic ion-exchange organic polymers are often network polyelectrolytes.

Note 4: A membrane having ion-exchange groups is called an **ion-exchange membrane.**

Note 5: Use of the term "ion-exchange resin" for "ion-exchange polymer" is strongly discouraged.

2.3 Living polymer
Polymer with stable, polymerization-active sites formed by a chain polymerization in which irreversible chain transfer and chain termination are absent.

Note 1: See [6], p. 236 and Definition 3.21 in [2] for living polymerization.

2.4 Macromonomer
Polymer or oligomer whose molecules each have one end-group that acts as a monomer molecule, so that each polymer or oligomer molecule contributes only a single monomer unit to a chain of the product polymer.

Note 1: The homopolymerization or copolymerization of a macromonomer yields a comb or graft polymer.

Note 2: In the present definition, Definition 2.35 in [2] has been combined with Definition 1.9 in [2]. See also [6], p. 241.

Note 3: Macromonomers are also sometimes referred to as macromers. The use of the term "macromer" is strongly discouraged.

2.5 Polymer catalyst
Polymer that exhibits catalytic activity.

Note 1: Certain synthetic polymer catalysts can behave like enzymes.

Note 2: Poly(4-vinylpyridine) in its basic form and sulfonated polystyrene in its acid form are examples of polymers that can act as catalysts in some base- and acid-catalyzed reactions, respectively.

2.6 Polymer-metal complex
Complex comprising a metal and one or more polymeric ligands.

2.7 Polymer phase-transfer catalyst
Polymer that acts as a phase-transfer catalyst and thereby causes a significant enhancement of the rate of a reaction between two reactants located in neighboring phases owing to its catalysis of the extraction of one of the reactants across the interface to the other phase where the reaction takes place.

Note 1: Polymer phase-transfer catalysts in the form of beads are often referred to as **triphase catalysts** because such catalysts form the third phase of the reaction system.
Note 2: See [6], p. 299 for phase-transfer catalyst.

2.8 Polymer-supported catalyst
Catalyst system comprising a polymer support in which catalytically active species are immobilized through chemical bonds or weaker interactions such as hydrogen bonds or donor-acceptor interactions.

Note 1: Polymer-supported catalysts are often based on network polymers in the form of beads. They are easy to separate from reaction media and can be used repeatedly.

Note 2: Examples of polymer-supported catalysts are: (a) a polymer-metal complex that can coordinate reactants, (b) colloidal palladium dispersed in a swollen network polymer that can act as a hydrogenation catalyst.

Note 3: **Polymer-supported enzymes** are a type of polymer-supported catalysts.

2.9 Polymer reactant polymer reagent polymer-supported reagent

Reactant (reagent) that is or is attached to a high-molar-mass linear polymer or a polymer network.

> *Note*: The attachment may be by chemical bonds, by weaker interactions such as hydrogen bonds, or simply by inclusion.

2.10 Prepolymer

Polymer or oligomer whose molecules are capable of entering, through reactive groups, into further polymerization and thereby contributing more than one structural unit to at least one type of chain of the final polymer.

> *Note*: Definition 2.37 in [2] has been combined with Definition 1.11 in [2]. See also [6], p. 318.

2.11 Reactive polymer

Polymer having reactive functional groups that can undergo chemical transformation under the conditions required for a given reaction or application.

2.12 Redox polymer electron-exchange polymer oxidation-reduction polymer

Polymer containing groups that can be reversibly reduced or oxidized.

> *Note 1*: Reversible redox reaction can take place in a polymer main-chain, as in the case of polyaniline and quinone/hydroquinone polymers, or on side-groups, as in the case of a polymer carrying ferrocene side-groups.

> *Note 2*: See [7] p. 346.

> *Note 3*: Use of the term "redox resin" is strongly discouraged.

2.13 Resin

Soft solid or highly viscous substance, usually containing prepolymers with reactive groups.

> *Note 1*: This term was used originally because of its analogy with a natural resin (rosin) and designated, in a broad sense, any polymer that is a basic material for plastics, organic coatings, or lacquers. However, the term is now used in a more narrow sense to refer to prepolymers of thermosets (thermosetting polymers).

> *Note 2*: The term is sometimes used not only for prepolymers of thermosets, but also for cured thermosets (e.g., epoxy resins, phenolic resins). Use of the term for cured thermosets is strongly discouraged.

> *Note 3*: Use of the term "resin" to describe the polymer beads used in solid-phase synthesis and as polymer supports, catalysts, reagents, and scavengers is also discouraged.

2.14 Telechelic polymer telechelic oligomer

Prepolymer capable of entering into further polymerization or other reactions through its reactive end-groups.

> *Note 1*: Reactive end-groups in telechelic polymers come from initiator or termination or chain-transfer agents in chain polymerizations, but not from monomer(s) as in polycondensations and polyadditions.

> *Note 2*: See [6], p. 414 and the Note to Definition 1.11 in [2] for telechelic molecule.

2.15 Thermosetting polymer

Prepolymer in a soft solid or viscous state that changes irreversibly into an infusible, insoluble polymer network by curing.

> *Note 1*: Curing can be by the action of heat or suitable radiation, or both.

> *Note 2*: A cured thermosetting polymer is called a **thermoset**.

3. Functional polymeric materials

3.1 Biodegradable polymer

Polymer susceptible to degradation by biological activity, with the degradation accompanied by a lowering of its molar mass.

> *Note 1*: See also Note 2 to Definition 1.13.

> *Note 2*: See [6], p. 43 for biodegradation. In the case of a polymer, its biodegradation proceeds not only by catalytic activity of enzymes, but also by a wide variety of biological activities.

3.2 Conducting polymer

Polymeric material that exhibits bulk electric conductivity.

> *Note 1*: See [6], p. 84 for conductivity.

> *Note 2*: The electric conductivity of a conjugated polymer is markedly increased by doping it with an electron donor or acceptor, as in the case of polyacetylene doped with iodine.

> *Note 3*: A polymer showing a substantial increase in electric conductivity upon irradiation with ultraviolet or visible light is called a **photoconductive polymer**; an example is poly (*N*-vinyl-carbazole) (see [6], p. 302 for photoconductivity).

> *Note 4*: A polymer that shows electric conductivity due to the transport of ionic species is called an **ion-conducting polymer**; an example is sulfonated polyaniline. When the transported ionic species is a proton as, e.g., in the case of fuel cells, it is called a **proton-conducting polymer**.

> *Note 5*: A polymer that shows electric semiconductivity is called a **semiconducting polymer** (See [6], p. 372 for semiconductor).

> *Note 6*: Electric conductance of a nonconducting polymer can be achieved by dispersing conducting particles (e.g., metal, carbon black) in the polymer. The resulting materials are referred to as *conducting* **polymer composites** or **solid polymer-electrolyte composites**.

3.3 Electroluminescent polymer

Polymeric material that shows luminescence when an electric current passes through it such that charge carriers can combine at luminescent sites to give rise to electronically excited states of luminescent groups or molecules.

> *Note 1*: Electroluminescent polymers are often made by incorporating luminescent groups or dyes into conducting polymers.

Note 2: Electrogenerated chemiluminescence (see [6], p. 130) directly connected with electrode reactions may also be called electroluminescence.

3.4 Ferroelectric polymer
Polymer in which spontaneous polarization arises when dipoles become arranged parallel to each other by electric fields.

> *Note 1*: See [6], p. 153 for ferroelectric transition.

> *Note 2*: Poly(vinylidene fluoride) after being subjected to a corona discharge is an example of a ferroelectric polymer.

3.5 Ferromagnetic polymer
Polymer that exhibits magnetic properties because it has unpaired electron spins aligned parallel to each other or electron spins that can easily be so aligned.

3.6 Functional polymer
(a) Polymer that bears specified chemical groups

or

(b) Polymer that has specified physical, chemical, biological, pharmacological, or other uses which depend on specific chemical groups.

> *Note*: Examples of functions of functional polymers under definition (b) are catalytic activity, selective binding of particular species, capture and transport of electric charge carriers or energy, conversion of light to charge carriers and vice versa, and transport of drugs to a particular organ in which the drug is released.

3.7 Impact-modified polymer
Polymeric material whose impact resistance and toughness have been increased by the incorporation of phase microdomains of a rubbery material.

> *Note*: An example is the incorporation of soft polybutadiene domains into glassy polystyrene to produce high-impact polystyrene.

3.8 Liquid-crystalline polymer
Polymeric material that, under suitable conditions of temperature, pressure, and concentration, exists as a liquid crystalline mesophase (Definition 6.1 in [7]).

> *Note 1*: See [4], p. 235 for liquid-crystal.

> *Note 2*: A liquid-crystalline polymer can exhibit one or more liquid state(s) with one- or two-dimensional, long-range orientational order over certain ranges of temperatures either in the melt (**thermotropic liquid-crystalline polymer**) or in solution (**lyotropic liquid-crystalline polymer**).

3.9 Macroporous polymer
Glass or rubbery polymer that includes a large number of macropores (50 nm–1 μm in diameter) that persist when the polymer is immersed in solvents or in the dry state.

> *Note 1*: Macroporous polymers are often network polymers produced in bead form. However, linear polymers can also be prepared in the form of macroporous polymer beads.

> *Note 2*: Macroporous polymers swell only slightly in solvents.

> *Note 3*: Macroporous polymers are used, for example, as precursors for ion-exchange polymers, as adsorbents, as supports for catalysts or reagents, and as stationary phases in size-exclusion-chromatography columns.

> *Note 4*: Porous polymers with pore diameters from ca. 2 to 50 nm are called **mesoporous polymers.**

3.10 Nonlinear optical polymer
Polymer that exhibits an optical effect brought about by electromagnetic radiation such that the magnitude of the effect is not proportional to the irradiance.

> *Note 1*: See [6], p. 275 for nonlinear optical effect.

> *Note 2*: An example of nonlinear optical effects is the generation of higher harmonics of the incident light wave.

> *Note 3*: A polymer that exhibits a nonlinear optical effect due to anisotropic electric susceptibilities when subjected to electric field together with light irradiation is called an **electro-optical polymer.** A polymer that exhibits electro-optical behavior combined with photoconductivity is called a **photorefractive polymer.**

3.11 Optically active polymer
Polymer capable of rotating the polarization plane of a transmitted beam of linear-polarized light.

> *Note 1*: See [6], p. 282 for optical activity.

> *Note 2*: The optical activity originates from the presence of chiral elements in a polymer such as chiral centers or chiral axes due to long-range conformational order in a polymer (helicity) (see [6], p. 182 for helicity).

3.12 Photoelastic polymer
Polymer that under stress exhibits birefringence.

3.13 Photoluminescent polymer
Polymer that exhibits luminescence (i.e., fluorescence or phosphorescence arising from photo-excitation).

> *Note*: See [6], p. 304 for photoluminescence.

3.14 Photosensitive polymer
Polymer that responds to ultraviolet or visible light by exhibiting a change in its physical properties or its chemical constitution.

> *Note 1*: Examples of the changes in photosensitive polymers are a change in molecular shape (**photoresponsive polymer**), a change in its constitution (**photoreactive polymer**), and a reversible change in color (**photochromic polymer**).

> *Note 2*: Photosensitivity in photosensitive polymers means that the polymers are sensitive to the irradiated light

leading to some change in properties or structure. It is different from photo-sensitization defined in [6], p. 307.

Note 3: See [6], p. 307 for photoreaction and [6], p. 302 for photochromism.

3.15 Piezoelectric polymer

(a) Polymer that exhibits a change in dielectric properties on application of pressure

or

(b) Polymer that shows a change in its dimensions when subjected to an electric field.

3.16 Polyelectrolyte

Polymer composed of molecules in which a portion of the constitutional units has ionizable or ionic groups, or both.

Note 1: A polymer bearing both anionic and cationic groups in the same molecule is called an **amphoteric polyelectrolyte**.

Note 2: A polymer bearing acid or basic groups is called a polymer acid or a polymer base, respectively.

Note 3: A polymer acid or a polymer base can be used as a matrix for ion-conducting polymers.

Note 4: Definition 2.38 in [2] has been combined with Definition 1.65 in [2]. The present definition replaces the one in [6], p. 312.

3.17 Polymer compatibilizer

Polymeric additive that, when added to a blend of immiscible polymers, modifies their interfaces and stabilizes the blend.

Note: Typical polymer compatibilizers are block or graft copolymers.

3.18 Polymer drug

Polymer that contains either chemically bound drug molecules or pharmacologically active moieties.

Note: A polymer drug is usually used to provide drug delivery targeted to an organ and controlled release of an active drug at the target organ.

3.19 Polymer gel

Gel in which the network component is a polymer network.

Note 1: A gel is an elastic colloid or polymer network that is expanded throughout its whole volume by a fluid.

Note 2: The polymer network can be a network formed by covalent bonds or by physical aggregation with region of local order acting as network junctions.

Note 3: An example of covalent polymer gels is *net*-poly(*N*-isopropylacrylamide) swollen in water, which shows volume phase transition during heating.

Note 4: Examples of physically aggregated polymer gels are poly(vinyl alcohol) gel and agarose gel, which show reversible sol-gel transitions.

Note 5: See Definition 1.58 in [2] for network.

Note 6: The definition for gel in [6], p. 170 does not include a polymer gel.

3.20 Polymer membrane

Thin layer of polymeric material that acts as a barrier permitting mass transport of selected species.

Note: See [6], p. 251 for membrane.

3.21 Polymer solvent

Polymer that acts like a solvent for compounds of low molar mass.

Note: An example of a polymer solvent is poly(oxyethylene); it can dissolve various inorganic salts by complexation.

3.22 Polymer sorbent

Polymer that adsorbs or absorbs a certain substance or certain substances from a liquid or a gas.

Note 1: A polymer sorbent may be a **polymer adsorbent** or a **polymer absorbent.** The former acts by surface sorption and the latter by bulk sorption.

Note 2: See [6], p. 383 for sorption, [6], p. 11 for adsorption, and [6], p. 3 for absorption.

3.23 Polymer support

Polymer to or in which a reagent or catalyst is chemically bound, immobilized, dispersed, or associated.

Note 1: A polymer support is usually a network polymer.

Note 2: A polymer support is usually prepared in bead form by suspension polymerization.

Note 3: The location of active sites introduced into a polymer support depends on the type of polymer support. In a **swollen-gel-bead polymer support** the active sites are distributed uniformly throughout the beads, whereas in a **macroporous-bead polymer support** they are predominantly on the internal surfaces of the macropores.

3.24 Polymer surfactant

Polymer that lowers the surface tension of the medium in which it is dissolved, or the interfacial tension with another phase, or both.

Note: See [6], p. 409 for surfactant.

3.25 Resist polymer

Polymeric material that, when irradiated, undergoes a marked change in solubility in a given solvent or is ablated.

Note 1: A resist polymer under irradiation either forms patterns directly or undergoes chemical reactions leading to pattern formation after subsequent processing.

Note 2: A resist material that is optimized for use with ultraviolet or visible light, an electron beam, an ion beam, or X-rays is called a **photoresist** (see [6], p. 307),

electron-beam resist, ion-beam resist, or **X-ray resist,** respectively.

Note 3: In a **positive-tone resist,** also called a **positive resist,** the material in the irradiated area not covered by a mask is removed, which results in an image with a pattern identical with that on the mask. In a **negative-tone resist,** also called a **negative resist,** the non-irradiated area is subsequently removed, which results in an image with a pattern that is the complement of that on the mask.

3.26 Shape-memory polymer

Polymer that, after heating and being subjected to a plastic deformation, resumes its original shape when heated above its glass-transition or melting temperature.

Note: Crystalline *trans*-polyisoprene is an example of a shape-memory polymer.

3.27 Superabsorbent polymer

Polymer that can absorb and retain extremely large amounts of a liquid relative to its own mass.

Note 1: The liquid absorbed can be water or an organic liquid.

Note 2: The swelling ratio of a superabsorbent polymer can reach the order of 1000:1.

Note 3: Superabsorbent polymers for water are frequently polyelectrolytes.

References

1. IUPAC. *Compendium of Macromolecular Nomenclature*, (the IUPAC "Purple Book"), prepared for publication by W. V. Metanomski, Chap. 1, Blackwell, Oxford (1991); IUPAC. "Basic definitions of terms relating to polymers (1974)", *Pure Appl. Chem.* **40**, 477–491 (1974).
2. A. D. Jenkins, P. Kratochvíl, R. F. T. Stepto, U. W. Suter. "Glossary of basic terms in polymer science (IUPAC Recommendations 1996)", *Pure Appl. Chem.* **68**, 2287–2311 (1996).
3. I. Mita, R. F. T. Stepto, U. W. Suter. "Basic classification and definitions of polymerization reactions (IUPAC Recommendations 1994)", *Pure Appl. Chem.* **66**, 2483–2486 (1994).
4. K. Hatada, R. B. Fox, J. Kahovec, E. Maréchal, I. Mita, V. Shibaev. "Definitions of terms relating to degradation, aging, and related chemical transformations of polymers (IUPAC Recommendations 1996)", *Pure Appl. Chem.* **68**, 2313–2323 (1996).
5. R. A. Y. Jones and J. F. Bunnett. "Nomenclature for organic chemical transformations (IUPAC Recommendations 1989)", *Pure Appl. Chem.* **61**, 725–768 (1989).
6. IUPAC. *Compendium of Chemical Terminology: IUPAC Recommendations*, (the IUPAC "Gold Book"), 2nd ed., compiled by A. D. McNaught and A. Wilkinson, Blackwell, Oxford (1997).
7. C. Noël, V. P. Shibaev, M. Barón, M. Hess, A. D. Jenkins, Jung-Il Jin, A. Sirigu, R. F. T. Stepto, W. J. Work. "Definitions of basic terms relating to low-molar-mass and polymer liquid-crystals (IUPAC Recommendations 2001)", *Pure Appl. Chem.* **73**, 845–895 (2001).

Alphabetical Index of Terms

DEFINITIONS OF TERMS RELATED TO POLYMER BLENDS, COMPOSITES, AND MULTIPHASE POLYMERIC MATERIALS

(IUPAC Recommendations 2004)

Prepared by a Working Group consisting of
W. J. Work[1,‡], K. Horie[2], M. Hess[3], and R. F. T. Stepto[4]

[1]*11288 Burnett Road, Huntingdon Valley, PA 19006, USA;* [2]*6-11-21, Kozukayama, Tarumi-ku, Kobe 655-0002, Japan;*
[3]*Universitat Duisburg-Essen, Fachbereich 6: Physikalische Chemie, D-47048 Duisburg, Germany;*
[4]*University of Manchester and UMIST, Polymer Science and Technology Group (MMSC), Grosvenor Street, Manchester, M1 7HS, UK*

Prepared for publication by
W. J. WORK

*Membership of the Commission on Macromolecular Nomenclature (extant until 2002) during the preparation of this report (1993–2003) was as follows:

Titular Members: R. E. Bareiss (Germany, to 1993); M. Barón (Argentina, Associate Member to 1995, Titular Member from 1996, *Secretary* from 1998); K. Hatada (Japan, to 1997, Associate Member to 1999); M. Hess (Germany, Associate Member from 1996, Titular Member from 1998, *Chairman* from 2000); K. Horie (Japan, Associate Member from 1996, Titular Member from 1998); R. G. Jones (UK, Pool Titular Member to 1997, Titular Member from 1998); J Kahovec (Czech Republic, to 1999); P. Kubisa (Poland, Associate Member from 1996, Titular Member from 2000); E. Maréchal (France, Associate Member 1992–1993, 2000–2001, Titular Member 1994–1999); I Meisel (Germany, Associate Member from 1998, Titular Member from 2000); W. V. Metanomski (USA, to 1999); C. Noël (France, to 1997); V. P. Shibaev (Russia, to 1995, Associate Member to 1999); R. F. T. Stepto (UK, *Chairman* to 1999); E. S. Wilks (USA, Associate Member from 1998, Titular Member from 2000); W. J. Work (USA, *Secretary* to 1997).

Associate Members contributing to this report: J.-I. Jin (Korea, National Representative to 1993, Associate Member from 1994); T. Kitayama (Japan, from 2000); S. Penczek (Poland, from 1994); J. Vohlídal (Czech Republic, from 2000). ***National Representatives contributing to this report***: W. Mormann (Germany, from 2000).

**Membership of the Subcommittee on Macromolecular Terminology (extant from 2002) during the preparation of this report (1993–2003) was as follows:

M. Hess (Germany, *Chairman*); M. Barón (Argentina, *Secretary*); G. Allegra (Italy); A. Fradet (France); J. He (China); K. Horie (Japan); A. D. Jenkins (UK); J.-II Jin (Korea); R. G. Jones (UK); J. Kahovec (Czech Republic); T. Kitayama (Japan); P. Kratochvíl (Czech Republic); P. Kubisa (Poland); I. Meisel (Germany); W. V. Metanomski (USA); G. Moad (Australia); W. Mormann (Germany); S. Penczek (Poland); L. P. Rebelo (Portugal); M. Rinaudo (France); I. Schopov (Bulgaria); M. Schubert (USA); V. P. Shibaev (Russia); S. Slomkowski (Poland); R. F. T. Stepto (UK); D. Tabak (Brazil); J. Vohlídal (Czech Republic); E. S. Wilks (USA); W. J. Work (USA).

Other contributors to this report: S. Akiyama (Japan); P. Avakian (USA); K. Binder (Germany); C. Bucknall (UK); R. Gilbert (Australia); J. He (China); J. S. Higgins (UK); T. Inoue (Japan); B.-J. Jungnickel (Germany); R. Koningsveld (Netherlands); J. Lertola (USA); T. Nishi (Japan); T. Nose (Japan); D. Paul (USA); I. Plotzker (USA); L. A. Utracki (Canada); B. Wood (USA).

‡Corresponding author

Abstract: The document defines the terms most commonly encountered in the field of polymer blends and composites. The scope has been limited to mixtures in which the components differ in chemical composition or molar mass and in which the continuous phase is polymeric. Incidental thermodynamic descriptions are mainly limited to binary mixtures although, in principle, they could be generalized to multicomponent mixtures.

The document is organized into three sections. The first defines terms basic to the description of polymer mixtures. The second defines terms commonly encountered in descriptions of phase domain behavior of polymer mixtures. The third defines terms commonly encountered in the descriptions of the morphologies of phase-separated polymer mixtures.

Reproduced from:
Pure Appl. Chem., Vol. 76, No. 11, pp. 1985–2007, 2004.
© 2004 IUPAC

Contents

Introduction

It is the intent of this document to define the terms most commonly encountered in the field of polymer blends and composites. The scope has been limited to mixtures in which the components differ in chemical composition or molar mass or both and in which the continuous phase is polymeric. Many of the materials described by the term "multiphase" are two-phase systems that may show a multitude of finely dispersed phase domains. Hence, incidental thermodynamic descriptions are mainly limited to binary mixtures, although they can be and, in the scientific literature, have been generalized to multi-component mixtures. Crystalline polymers and liquid-crystal polymers have been considered in other documents [1,2] and are not discussed here.

This document is organized into three sections. The first defines terms basic to the description of polymer mixtures. The second defines terms commonly encountered in descriptions of phase-domain behavior of polymer mixtures. The third defines terms commonly encountered in the descriptions of the morphologies of phase-separated polymer mixtures.

General terms describing the composition of a system as defined in ref. [3] are used without further definition throughout the document. Implicit definitions are identified in boldface type throughout the document.

1. Basic terms in polymer mixtures

1.1 Polymer blend

Macroscopically homogeneous mixture of two or more different species of polymer [3,4].

Notes:

1. See the Gold Book, p. 312 [3].
2. In most cases, blends are homogeneous on scales larger than several times the wavelengths of visible light.
3. In principle, the constituents of a blend are separable by physical means.
4. No account is taken of the miscibility or immiscibility of the constituent macromolecules, i.e., no assumption is made regarding the number of phase domains present.
5. The use of the term "polymer alloy" for "polymer blend" is discouraged, as the former term includes multiphase copolymers but excludes incompatible polymer blends (see **1.3**).
6. The number of polymeric components which comprises a blend is often designated by an adjective, viz., binary, ternary, quaternary,

1.2 Miscibility

Capability of a mixture to form a single phase over certain ranges of temperature, pressure, and composition.

Notes:

1. Whether or not a single phase exists depends on the chemical structure, molar-mass distribution, and molecular architecture of the components present.
2. The single phase in a mixture may be confirmed by light scattering, X-ray scattering, and neutron scattering.
3. For a two-component mixture, a necessary and sufficient condition for stable or metastable equilibrium of a homogeneous single phase is

$$\left(\frac{\partial^2 \Delta_{\mathrm{mix}} G}{\partial \phi^2}\right)_{T,p} > 0$$

where $\Delta_{\mathrm{mix}}G$ is the Gibbs energy of mixing and ϕ the composition, where ϕ is usually taken as the volume fraction of one of the components. The system is unstable if the above second derivative is negative. The borderline (spinodal) between (meta)stable and unstable states is defined by the above second derivative equalling zero. If the compositions of two conjugate (coexisting) phases become identical upon a change of temperature or pressure, the third derivative also equals zero (defining a critical state).
4. If a mixture is thermodynamically metastable, it will demix if suitably nucleated (see **2.5**). If a mixture is thermodynamically unstable, it will demix by spinodal decomposition (see **2.8**) or by nucleation and growth if suitably nucleated, provided there is minimal kinetic hindrance.

1.3 Miscible polymer blend
homogeneous polymer blend

Polymer blend that exhibits miscibility (see **1.2**).

Notes:

1. For a polymer blend to be miscible, it must satisfy the criteria of miscibility (see **1.2**).
2. Miscibility is sometimes erroneously assigned on the basis that a blend exhibits a single T_{g} or optical clarity.
3. A miscible system can be thermodynamically stable or metastable (see note 4 in **1.2**).
4. For components of chain structures that would be expected to be miscible, miscibility may not occur if molecular architecture is changed, e.g., by crosslinking.

1.4 Homologous polymer blend

Mixture of two or more fractions of the same polymer, each of which has a different molar-mass distribution.

1.5 Isomorphic polymer blend

Polymer blend of two or more different semi-crystalline polymers that are miscible in the crystalline state as well as in the molten state.

Notes:

1. Such a blend exhibits a single, composition-dependent glass-transition temperature, T_g, and a single, composition-dependent melting point, T_m.
2. This behavior is extremely rare; very few cases are known.

1.6 Polymer-polymer complex

Complex, at least two components of which are different polymers [3].

Notes:

1. See the Gold Book, p. 313 [3].
2. A **complex** is a molecular entity formed from two or more components that can be ionic or uncharged (see the Gold Book, p. 81) [3].
3. Although the intrinsic binding energy between the individual interacting sites giving rise to the complex is weaker than a covalent bond, the total binding energy for any single molecule may exceed the energy of a single covalent bond.
4. The properties of a complex defined here differ from those given in ref. [3] because, owing to the repeating nature of a polymer molecule, many interacting sites may be present, which together will provide stronger bonding than a single covalent bond.

1.7 Metastable miscibility

Capability of a mixture to exist for an indefinite period of time as a single phase that is separated by a small or zero energy barrier from a thermodynamically more stable multiphase system.

Notes:

1. See the Gold Book, p. 255 [3].
2. Mixtures exhibiting metastable miscibility may remain unchanged or they may undergo phase separation, usually by nucleation or spinodal decomposition.

1.8 Metastable miscible polymer blend

Polymer blend that exhibits metastable miscibility.

Note: In polymers, because of the low mobility of polymer chains, particularly in a glassy state, metastable mixtures may exist for indefinite periods of time without phase separation. This has frequently led to confusion when metastable miscible polymer blends are erroneously claimed to be miscible.

1.9 Interpenetrating polymer network

Recommended acronym: IPN
Polymer comprising two or more polymer networks which are at least partially interlaced on a molecular scale, but not covalently bonded to each other and cannot be separated unless chemical bonds are broken [4].

Notes:

1. See the Gold Book, p. 205 [3].
2. A mixture of two or more preformed polymer networks is not an interpenetrating polymer network.
3. An IPN may be further described by the process by which it is synthesized. When an IPN is prepared by a process in which the second component network is polymerized following the completion of polymerization of the first component network, the IPN may be referred to as a **sequential IPN**. When an IPN is prepared by a process in which both component networks are polymerized concurrently, the IPN may be referred to as a **simultaneous IPN**.

1.10 Semi-interpenetrating polymer network

Recommended acronym: SIPN
Polymer comprising one or more polymer network(s) and one or more linear or branched polymer(s) characterized by the penetration on a molecular scale of at least one of the networks by at least some of the linear or branched chains [4].

Notes:

1. See the Gold Book, p. 372 [3].
2. Semi-interpenetrating polymer networks are different from interpenetrating polymer networks because the constituent linear-chain or branched-chain macromolecule(s) can, in principle, be separated from the constituent polymer network(s) without breaking chemical bonds, and, hence, they are polymer blends.
3. Semi-interpenetrating polymer networks may be further described by the process by which they are synthesized. When an SIPN is prepared by a process in which the second component polymer is polymerized or incorporated following the completion of polymerization of the first component polymer, the SIPN may be referred to as a **sequential SIPN**. When an SIPN is prepared by a process in which both component polymers are polymerized concurrently, the SIPN may be referred to as a **simultaneous SIPN**. (This note has been changed from that which appears in ref. [4] to allow for the possibility that a linear or branched polymer may be incorporated into a network by means other than polymerization, e.g., by swelling of the network and subsequent diffusion of the linear or branched chain into the network.).

1.11 Immiscibility

Inability of a mixture to form a single phase.

Notes:

1. Immiscibility may be limited to certain ranges of temperature, pressure, and composition.
2. Immiscibility depends on the chemical structures, molar-mass distributions, and molecular architectures of the components.

1.12 Immiscible polymer blend
heterogeneous polymer blend

Polymer blend that exhibits immiscibility.

1.13 Composite

Multicomponent material comprising multiple different (nongaseous) phase domains in which at least one type of phase domain is a continuous phase (see **3.12**).

> *Note*: Foamed substances, which are multiphased materials that consist of a gas dispersed in a liquid or solid, are not normally considered to be composites.

1.14 Polymer composite

Composite in which at least one component is a polymer.

1.15 Nanocomposite

Composite in which at least one of the phases has at least one dimension of the order of nanometers.

1.16 Laminate

Material consisting of more than one layer, the layers being distinct in composition, composition profile, or anisotropy of properties.

> *Notes*:
>
> 1. Laminates may be formed by two or more layers of different polymers.
> 2. Composite laminates generally consist of one or more layers of a substrate, often fibrous, impregnated with a curable polymer, curable polymers, or liquid reactants.
> 3. The substrate is usually a sheet-like woven or nonwoven material (e.g., glass fabric, paper, copper foil).
> 4. A single layer of a laminate is termed a **lamina**.

1.17 Lamination

Process of forming a laminate.

1.18 Delamination

Process that separates the layers of a laminate by breaking their structure in planes parallel to those layers.

1.19 Impregnation

Penetration of monomeric, oligomeric, or polymeric liquids into an assembly of fibers.

> *Notes*:
>
> 1. The term as defined here is specific to polymer science. An alternative definition of "impregnation" applies in some other fields of chemistry (see the Gold Book, p. 197) [3].
> 2. Impregnation is usually carried out on a woven fabric or a yarn.

1.20 Prepreg

Sheets of a substrate that have been impregnated with a curable polymer, curable polymers, or liquid reactants, or a thermoplastic, and are ready for fabrication of laminates.

> *Notes*:
>
> 1. See **1.16** notes 2 and 3.
> 2. During the impregnation the curable polymer, curable polymers, or liquid reactants may be allowed to react to a certain extent (sometimes termed **degree of ripening**).

1.21 Intercalation

Process by which a substance becomes transferred into pre-existing spaces of molecular dimensions in a second substance.

> *Note*: The term as defined here is specific to polymer science. An alternative definition of "intercalation" applies in some other fields of chemistry (see the Gold Book, p. 202 [3]).

1.22 Exfoliation

Process by which thin layers individually separate from a multi-layered structure.

> *Note*: In the context of a nanocomposite material, the individual layers are of the order of at most a few nanometers in thickness.

1.23 Wetting

Process by which an interface between a solid and a gas is replaced by an interface between the same solid and a liquid.

1.24 Adhesion

Holding together of two bodies by interfacial forces or mechanical interlocking on a scale of micrometers or less.

1.25 Chemical adhesion

Adhesion (see **1.25**) in which two bodies are held together at an interface by ionic or covalent bonding between molecules on either side of the interface.

1.26 Interfacial adhesion

Adhesion (see **1.25**) in which interfaces between phases or components are maintained by intermolecular forces, chain entanglements, or both, across the interfaces.

> *Notes*:
>
> 1. Interfacial adhesion is also referred to as **tack**.
> 2. **Adhesive strength** (recommended symbol: F_a, unit: N m^{-2}) is the force required to separate one condensed phase domain from another at the interface between the two phase domains divided by the area of the interface.
> 3. **Interfacial tension** (recommended symbol: γ, unit: N m^{-1}, J m^{-2}) is the change in Gibbs energy per unit change in interfacial area for substances in physical contact.
> 4. Use of the term **interfacial energy** for interfacial tension is not recommended.

1.27 Interfacial bonding

Bonding in which the surfaces of two bodies in contact with one another are held together by inter-molecular forces.

> *Note*: Examples of intermolecular forces include covalent, ionic, van der Waals, and hydrogen bonds.

1.28 Interfacial fracture

Brittle fracture that takes place at an interface.

1.29 Craze

Crack-like cavity formed when a polymer is stressed in tension that contains load-bearing fibrils spanning the gap between the surfaces of the cavity.

Note: Deformation of continua occurs with only minor changes in volume; hence, a craze consists of both fibrils and voids.

1.30 Additive
Substance added to a polymer.

Notes:

1. The term as defined here is specific to polymer science. An alternative definition of "additive" applies in some other fields of chemistry (see the Gold Book, p. 10) [3].
2. An additive is usually a minor component of the mixture formed and usually modifies the properties of the polymer.
3. Examples of additives are antioxidants, plasticizers, flame retardants, processing aids, other polymers, colorants, UV absorbers, and extenders.

1.31 Interfacial agent
Additive that reduces the interfacial energy between phase domains.

1.32 Compatibility
Capability of the individual component substances in either an immiscible polymer blend (see **1.12**) or a polymer composite (see **1.14**) to exhibit interfacial adhesion (see **1.27**).

Notes:

1. Use of the term "compatibility" to describe miscible systems is discouraged.
2. Compatibility is often established by the observation of mechanical integrity under the intended conditions of use of a composite or an immiscible polymer blend.

1.33 Compatibilization
Process of modification of the interfacial properties in an immiscible polymer blend that results in formation of the interphases (see **3.6**) and stabilization of the morphology, leading to the creation of a polymer alloy.

Note: Compatibilization may be achieved by addition of suitable copolymers or by chemical modification of interfaces through physical treatment (i.e., irradiation or thermal) or reactive processing.

1.34 Degree of compatibility
Measure of the strength of the interfacial bonding between the component substances of a composite or immiscible polymer blend (see **1.12**).

Notes:

1. Estimates of the degree of compatibility are often based upon the mechanical performance of the composite, the interphase thickness (see "Interfacial region interphase"), or the sizes of the phase domains present in the composite, relative to the corresponding properties of composites lacking compatibility.
2. The term **degree of incompatibility** is sometimes used instead of degree of compatibility. Such use is discouraged

as incompatibility is related to the weakness of interfacial bonding.

1.35 Compatible polymer blend
Immiscible polymer blend (see **1.12**) that exhibits macroscopically uniform physical properties throughout its whole volume.

Note: The macroscopically uniform physical properties are usually caused by sufficiently strong interactions between the component polymers.

1.36 Compatibilizer
Polymer or copolymer that, when added to an immiscible polymer blend (see **1.12**), modifies its inter-facial character and stabilizes its morphology.

Note: Compatibilizers usually stabilize morphologies over distances of the order of micrometers or less.

1.37 Coupling agent adhesion promoter
Interfacial agent comprised of molecules possessing two or more functional groups, each of which exhibits preferential interactions with the various types of phase domains in a composite.

Notes:

1. A coupling agent increases adhesion between phase domains.
2. An example of the use of a coupling agent is in a mineral-filled polymer material where one part of the coupling agent molecule can chemically bond to the inorganic mineral while the other part can chemically bond to the polymer.

1.38 Polymer alloy
Polymeric material, exhibiting macroscopically uniform physical properties throughout its whole volume, that comprises a compatible polymer blend (see 1.**35**), a miscible polymer blend (see **1.3**), or a multiphase copolymer (see **3.3**).

Note: See note 5 in **1.1**.

1.39 Dispersion
Material comprising more than one phase where at least one of the phases consists of finely divided phase domains (see **3.2**), often in the colloidal size range, distributed throughout a continuous phase domain.

Notes:

1. The term as defined here is specific to polymer science. An alternative definition of "dispersion" applies in some other fields of chemistry (see the Gold Book, p. 118) [3].
2. Particles in the colloidal size range have linear dimensions [3] between 1 nm and 1 μm.
3. The finely divided domains are called the dispersed or discontinuous phase domains (see **3.13**).
4. For a definition of continuous phase domain, see **3.12**.
5. A dispersion is often further characterized on the basis of the size of the phase domain as a **macrodispersion** or a **microdispersion**. To avoid ambiguity when using these terms, the size of the domain should also be defined.

1.40 Dispersing agent dispersing aid dispersant
Additive (see **1.30**), exhibiting surface activity, that is added to a suspending medium to promote uniform and maximum

separation of extremely fine solid particles, often of colloidal size (see note 2 in **1.39**).

> *Note*: Although dispersing agents achieve results similar to compatibilizers (see **1.36**), they function differently in that they reduce the attractive forces between fine particles, which allows them to be more easily separated and dispersed.

1.41 Agglomeration aggregation
Process in which dispersed molecules or particles form clusters rather than remain as isolated single molecules or particles.

> *Note*: See the Gold Book, p. 13 [3].

1.42 Agglomerate aggregate
Clusters of dispersed molecules or particles that results from agglomeration (see **1.41**).

> *Note*: The term as defined here is specific to polymer science. An alternative definition of "aggregate" is used in some other fields of chemistry (see the Gold Book, p. 13) [3].

1.43 Extender
Substance, especially a diluent or modifier, added to a polymer to increase its volume without substantially altering the desirable properties of the polymer.

> *Note*: An extender may be a liquid or a solid.

1.44 Filler
Solid extender.

> *Notes*:
>
> 1. The term as defined here is specific to polymer science. An alternative definition of "filler" applies in some other fields of chemistry (see the Gold Book, p. 154) [3].
> 2. Fillers may be added to modify mechanical, optical, electrical, thermal, flammability properties, or simply to serve as extenders.

1.45 Fill factor

Recommended symbol: ϕ_{fill}
Maximum volume fraction of a particulate filler that can be added to a polymer while maintaining the polymer as the continuous phase domain.

1.46 Thermoplastic elastomer
Melt-processable polymer blend or copolymer in which a continuous elastomeric phase domain is reinforced by dispersed hard (glassy or crystalline) phase domains that act as junction points over a limited range of temperature.

> *Notes*:
>
> 1. The behavior of the hard phase domains as junction points is thermally reversible.
> 2. The interfacial interaction between hard and soft phase domains in a thermoplastic elastomer is often the result

of covalent bonds between the phases and is sufficient to prevent the flow of the elastomeric phase domain under conditions of use.
> 3. Examples of thermoplastic elastomers include block copolymers and blends of plastics and rubbers.

2. Phase domain behavior

2.1 Miscibility window
Range of copolymer compositions in a polymer mixture, at least one component substance of which is a copolymer, that gives miscibility (see **1.2**) over a range of temperatures and pressures.

> *Notes*:
>
> 1. Outside the miscibility window immiscible mixtures are formed.
> 2. The compositions of the copolymers within the miscibility window usually exclude the homopolymer compositions of the monomers from which the copolymers are prepared.
> 3. The miscibility window is affected by the molecular weights of the component substances.
> 4. The existence of miscibility windows has been attributed to an average force between the monomer units of the copolymer that leads to those units associating preferentially with the monomer units of the other polymers.

2.2 Miscibility gap
Area within the coexistence curve of an isobaric phase diagram (temperature vs. composition) or an isothermal phase diagram (pressure vs. composition).

> *Note*: A miscibility gap is observed at temperatures below an upper critical solution temperature (UCST) (see **2.15**) or above the lower critical solution temperature (LCST) (see **2.14**). Its location depends on pressure. In the miscibility gap, there are at least two phases coexisting.

2.3 Flory–huggins theory
flory–huggins–staverman theory
Statistical thermodynamic mean-field theory of polymer solutions, formulated independently by Flory, Huggins, and Staverman, in which the thermodynamic quantities of the solution are derived from a simple concept of combinatorial entropy of mixing and a reduced Gibbs-energy parameter, the "χ interaction parameter" (see **2.4**).

> *Notes*:
>
> 1. See the Gold Book, p. 158 [3].
> 2. The Flory–Huggins theory has often been found to have utility for polymer blends; however, there are many equation-of-state theories that provide more accurate descriptions of polymer–polymer interactions.
> 3. The present definition has been modified from that which appears in ref. [5] to acknowledge the contributions of Staverman and to further clarify the statistical basis of the theory.

2.4 χ Interaction parameter
Recommended symbol: χ
Interaction parameter, employed in the Flory–Huggins theory (see **2.3**), to account for the contribution of the noncombinatorial

entropy of mixing and the enthalpy of mixing to the Gibbs energy of mixing.

Notes:

1. The definition and the name of the term have been modified from that which appears in ref. [5] to reflect its broader use in the context of polymer blends. In its simplest form, the χ parameter is defined according to the Flory–Huggins equation for binary mixtures

$$\frac{\Delta_{mix}G}{RT} = n_1 \ln \phi_1 + n_2 \ln \phi_2 + \chi x_1 n_1 \phi_2,$$

for a mixture of amounts of substance n_1 and n_2 of components denoted 1 and 2, giving volume fractions ϕ_1 and ϕ_2, with the molecules of component 1 each conceptually consisting of x_1 segments whose Gibbs energy of interaction with segments of equal volume in the molecules of component 2 is characterized by the interaction parameter χ.
2. The χ interaction parameters characterizing a given system vary with composition, molar mass, and temperature.
3. B is an alternative parameter to χ, where $B = \chi RT/V_m$, in which V_m is the molar volume of one of the components of the mixture.

2.5 Nucleation of phase separation

Initiation of phase domain formation through the presence of heterogeneities.

Notes:

1. See the Gold Book, p. 277 [3].
2. In a metastable region of a phase diagram (see **1.2**), phase separation is initiated only by nucleation.

2.6 Binodal binodal curve coexistence curve

Curve defining the region of composition and temperature in a phase diagram for a binary mixture across which a transition occurs from miscibility of the components to conditions where single-phase mixtures are metastable or unstable (see note 4 in **1.2**).

Note: Binodal compositions are defined by pairs of points on the curve of Gibbs energy of mixing vs. composition that have common tangents, corresponding to compositions of equal chemical potentials of each of the two components in two phases.

2.7 Spinodal spinodal curve

Curve defining the region of composition and temperature for a binary mixture across which a transition occurs from conditions where single-phase mixtures are metastable to conditions where single-phase mixtures are unstable and undergo phase separation by spinodal decomposition (see **2.8**).

Notes:

1. The spinodal curve for a binary mixture is defined as the geometrical locus of all states with

$$\left(\frac{\partial^2 \Delta_{mix}G}{\partial \phi^2}\right)_{T,P} = 0 \text{ (see } \textbf{1.2}, \text{ note 4)}$$

2. In the unstable region bounded by the spinodal curve, phase domain separation is spontaneous, i.e., no nucleation step is required to initiate the separation process.

2.8 Spinodal decomposition spinodal phase-demixing

Long-range, diffusion-limited, spontaneous phase domain separation initiated by delocalized concentration fluctuations occurring in an unstable region of a mixture bounded by a spinodal curve.

Note: Spinodal decomposition occurs when the magnitude of Gibbs energy fluctuations with respect to composition are zero.

2.9 Cloud point

Experimentally measured point in the phase diagram of a mixture at which a loss in transparency is observed due to light scattering caused by a transition from a single- to a two-phase state.

Notes:

1. The phenomenon is characterized by the first appearance of turbidity or cloudiness.
2. A cloud point is heating rate- or cooling rate-dependent.

2.10 Cloud-point curve

Curve of temperature vs. composition defined by the cloud points (see **2.9**) over range of compositions of two substances.

Note: Mixtures are observed to undergo a transition from a single- to a two-phase state upon heating or cooling.

2.11 Cloud-point temperature

Temperature at a cloud point (see **2.9**).

2.12 Critical point

Point in the isobaric temperature-composition plane for a binary mixture where the compositions of all coexisting phases become identical.

Notes:

1. An alternative definition of "critical solution point" refers strictly to liquid-vapor equilibria (see the Gold Book, p. 93) [3].
2. Unless specified atmospheric pressure is assumed.
3. In a phase diagram, the slope of the tangent to the spinodal is zero at this point.
4. At a critical point, binodals and spinodals coincide.
5. Although the definition holds strictly for binary mixtures, it is often erroneously applied to multicomponent mixtures.
6. See note 3 in **1.2**.

2.13 Lower critical solution temperature

Recommended acronym: LCST

Critical temperature below which a mixture is miscible.

Notes:

1. See the Gold Book, p. 93 [3].
2. Below the LCST and above the UCST (see **2.14**), if it exists, a single phase exists for all compositions.

3. The LCST depends upon pressure and the molar-mass distributions of the constituent polymer(s).
4. For a mixture containing or consisting of polymeric components, these may be different polymers or species of different molar mass of the same polymer.

2.14 Upper critical solution temperature

Recommended abbreviation: UCST

Critical temperature above which a mixture is miscible.

Notes:

1. See the Gold Book, p. 93 [3].
2. Above the UCST and below the LCST (see **2.13**), if it exists, a single phase exists for all compositions
3. The UCST depends upon the pressure and molar-mass distributions of the constituent polymer(s).
4. For a mixture containing or consisting of polymeric components, these may be different polymers or species of different molar mass of the same polymer.

2.15 Phase inversion

Process by which an initially continuous phase domain becomes the dispersed phase domain and the initially dispersed phase domains become the continuous phase domain.

Notes:

1. See the Gold Book, p. 299 [3].
2. Phase inversion may be observed during the polymerization or melt processing of polymer blend systems.
3. The phenomenon is usually observed during polymerization of a monomer containing a dissolved polymer.

2.16 Interdiffusion

Process by which homogeneity in a mixture is approached by means of spontaneous mutual molecular diffusion.

2.17 Blooming

Process in which one component of a polymer mixture, usually not a polymer, undergoes phase separation and migration to an external surface of the mixture.

2.18 Coalescence

Process in which two phase domains of essentially identical composition in contact with one another form a larger phase domain.

Notes:

1. See the Gold Book, p. 75 [3].
2. Coalescence reduces the total interfacial area.
3. The flocculation of a polymer colloid, through the formation of aggregates, may be followed by coalescence.

2.19 Morphology coarsening phase ripening

Process by which phase domains increase in size during the aging of a multiphase material.

Notes:

1. In the coarsening at the late stage of phase separation, volumes and compositions of phase domains are conserved.

2. Representative mechanisms for coarsening at the late stage of phase separation are: (1) material flow in domains driven by interfacial tension (observed in a co-continuous morphology), (2) the growth of domain size by evaporation from smaller droplets and condensation into larger droplets, and (3) coalescence (fusion) of more than two droplets. The mechanisms are usually called (1) Siggia's mechanism, (2) Ostwald ripening (or the Lifshitz-Slyozov mechanism), and (3) coalescence.
3. Morphology coarsening can be substantially stopped by, for example, vitrification, crosslinking, and **pinning**, the slowing down of molecular diffusion across domain interfaces.

3. Domains and morphologies

Many types of morphologies have been reported in the literature of multiphase polymeric materials. It is the intent of this document to define only the most commonly used terms. In addition, some morphologies have historically been described by very imprecise terms that may not have universal meanings. However, if such terms are widely used they are defined here.

3.1 Morphology

Shape, optical appearance, or form of phase domains in substances, such as high polymers, polymer blends, composites, and crystals.

Note: For a polymer blend or composite, the morphology describes the structures and shapes observed, often by microscopy or scattering techniques, of the different phase domains present within the mixture.

3.2 Phase domain

Region of a material that is uniform in chemical composition and physical state.

Notes:

1. A phase in a multiphase material can form domains differing in size.
2. The term "domain" may be qualified by the adjective microscopic or nanoscopic or the prefix micro- or nano- according to the size of the linear dimensions of the domain.
3. The prefixes micro-, and nano- are frequently incorrectly used to qualify the term "phase" instead of the term "domain"; hence, "microphase domain", and "nanophase domain" are often used. The correct terminology that should be used is **phase microdomain** and **phase nanodomain**.

3.3 Multiphase copolymer

Copolymer comprising phase-separated domains.

3.4 Domain interface domain boundary

Surface forming a boundary between two phase domains.

Note: A representation of the domain interface as a two-dimensional surface over-simplifies the actual structure. All interfaces have a third dimension, namely, the interphase or interfacial region (see **3.6**).

3.5 Domain structure

Morphology of individual phase domains in a multiphase system.

Note: Domain structures may be described for phase domains or domains that are themselves multiphased structures.

3.6 Interfacial region interphase

Region between phase domains in an immiscible polymer blend in which a gradient in composition exists.

Note: See the Gold Book, p. 205 [3].

3.7 Phase interaction

Molecular interaction between the components present in the interphases of a multiphase mixture.

Note: The **interphase elasticity** is the capability of a deformed interphase to return to its original dimensions after the force causing the deformation has been removed.

3.8 Interfacial-region thickness interphase thickness interfacial width

Linear extent of the composition gradient in an interfacial region.

Notes:

1. See the Gold Book, p. 203 [3].
2. The width at half the maximum of the composition profile across the interfacial region (see **3.6**) or the distance between locations where $d\phi/dr$ (with ϕ the composition of a component and r the distance through the interfacial region) has decreased to $1/e$ are used as measures of the interfacial-region thickness.

3.9 Hard-segment phase domain

Phase domain of microscopic or smaller size, usually in a block, graft, or segmented copolymer (see **3.11**), comprising essentially those segments of the polymer that are rigid and capable of forming strong intermolecular interactions.

Note: Hard-segment phase domains are typically of 2–15 nm linear size.

3.10 Soft-segment phase domain

Phase domain of microscopic or smaller size, usually in a block, graft, or segmented copolymer (see **3.11**), comprising essentially those segments of the polymer that have glass transition temperatures lower than the temperature of use.

Note: Soft-segment phase domains are often larger than hard-segment phase domains and are often continuous.

3.11 Segmented copolymer

Copolymer containing phase domains of microscopic or smaller size, with the domains constituted principally of single types of structural unit.

Note: The types of domain in a segmented copolymer usually comprise hard- and soft-segment phase domains.

3.12 Continuous phase domain matrix phase domain

Phase domain (see **3.2**) consisting of a single phase in a heterogeneous mixture through which a continuous path to all phase domain boundaries may be drawn without crossing a phase domain boundary.

Note: In a polymer blend, the continuous phase domain is sometimes referred to as the **host polymer, bulk substance**, or **matrix**.

3.13 Discontinuous phase domain discrete phase domain dispersed phase domain

Phase domain in a phase-separated mixture that is surrounded by a continuous phase but isolated from all other similar phase domains within the mixture.

Note: The discontinuous phase domain is sometimes referred to as the **guest polymer**.

3.14 Dual phase domain continuity co-continuous phase domains

Topological condition, in a phase-separated, two-component mixture, in which a continuous path through either phase domain may be drawn to all phase domain boundaries without crossing any phase domain boundary.

3.15 Core-shell morphology

Two-phase domain morphology, of approximately spherical shape, comprising two polymers, each in separate phase domains, in which phase domains of one polymer completely encapsulate the phase domains of the other polymer.

Note: This morphology is most commonly observed in copolymers or blends prepared in emulsion polymerization by the sequential addition and polymerization of two different monomer compositions.

3.16 Cylindrical morphology

Phase domain morphology, usually comprising two polymers, each in separate phase domains, in which the phase domains of one polymer are of cylindrical shape.

Notes:

1. Phase domains of the constituent polymers may alternate, which results in many cylindrical layers surrounding a central core domain.
2. Cylindrical morphologies can be observed, for example, in triblock copolymers.

3.17 Fibrillar morphology

Morphology in which phase domains have shapes with one dimension much larger than the other two dimensions.

Note: Fibrillar phase domains have the appearance of fibers.

3.18 Lamellar domain morphology

Morphology in which phase domains have shapes with two dimensions much larger than the third dimension.

Note: Plate-like phase domains have the appearance of extended planes that are often oriented essentially parallel to one another.

3.19 Microdomain morphology
Morphology consisting of phase microdomains.

Notes:

1. See **3.2**.
2. Microdomain morphologies are usually observed in block, graft, and segmented copolymers.
3. The type of morphology observed depends upon the relative abundance of the different types of structural units and the conditions for the generation of the morphology. The most commonly observed morphologies are spheres, cylinders, and lamellae.

3.20 Nanodomain morphology
Morphology consisting of phase nanodomains.

Note: See **3.2**.

3.21 Onion morphology
Multiphase morphology of roughly spherical shape that comprises alternating layers of different polymers arranged concentrically, all layers being of similar thickness.

3.22 Ordered co-continuous double gyroid morphology
Co-continuous morphology in which a set of two gyroid-based phase domains exhibits a highly regular, three-dimensional lattice-like morphology with Ia3d space group symmetry.

Notes:

1. The domains are composed of tripoidal units as the fundamental building structures.
2. The two domains are interlaced.

3.23 Multicoat morphology
Morphology observed in a blend of a block copolymer with the homopolymer of one of the blocks and characterized by alternating concentric shells of the copolymer and the homopolymer.

Note: The morphology is identical to onion morphology (see **3.21**) within a matrix of homopolymer [6].

3.24 Rod-like morphology
Morphology characterized by cylindrical phase domains.

3.25 Multiple inclusion morphology salami-like morphology
Multiphase morphology in which dispersed phase domains of one polymer contain and completely encapsulate many phase domains of a second polymer that may have the same composition as the continuous phase domain (see **3.12**).

References

1. IUPAC. "Definitions of terms relating to crystalline polymers (IUPAC Recommendations 1988)" *Pure Appl. Chem.* **61**, 769–785 (1989).
2. IUPAC. "Definitions of basic terms relating to low-molar-mass and polymer liquid crystals (IUPAC Recommendations 2001)", *Pure Appl. Chem.* **75**, 845–895 (2001).
3. IUPAC. *Compendium of Chemical Terminology* (the "Gold Book"), compiled by A. D. McNaught and A. Wilkinson, 2nd ed., Blackwell Science, Oxford (1997).
4. IUPAC. "Glossary of basic terms in polymer science (IUPAC Recommendations 1996)", *Pure Appl. Chem.* **68**, 2287–2311 (1996).
5. D. K. Carpenter. "Solution properties", in *Encyclopedia of Polymer Science and Engineering*, Vol. 15, 2nd ed., J. I. Kroschwitz (Ed.), pp. 419–481, Wiley Interscience, New York (1989).
6. J. M. G. Cowie. "Miscibility", in *Encyclopedia of Polymer Science and Engineering*, 2nd ed., J. I. Kroschwitz (Ed.), Supplement, pp. 455–480, Wiley Interscience, New York (1989).

Bibliography

1. IUPAC. "Definitions of terms relating to degradation, aging, and related chemical transformations of polymers (IUPAC Recommendations 1996)", *Pure Appl. Chem.* **68**, 2313–2323 (1996).
2. *ASTM Glossary of ASTM Definitions*, 2nd ed., American Society for Testing and Materials, Philadelphia, PA (1973).
3. IUPAC. *Compendium of Macromolecular Nomenclature* (the "Purple Book"), prepared for publication by W. V. Metanomski, Blackwell Scientific, Oxford (1991).
4. A. N. Gent and G. R. Hamed. "Adhesion", in *Encyclopedia of Polymer Science and Engineering*, Vol. 1, 2nd ed., J. I. Kroschwitz (Ed.), pp. 476–517, Wiley Interscience, New York (1985).
5. L. Leibler. "Phase transformations", in *Encyclopedia of Polymer Science and Engineering*, Vol. 11, 2nd ed., J. I. Kroschwitz, (Ed.), pp. 30–45, Wiley Interscience, New York (1988).
6. J. Koberstein. "Interfacial properties", in *Encyclopedia of Polymer Science and Engineering*, Vol. 8, 2nd ed., J. I. Kroschwitz (Ed.), pp. 237–279, Wiley Interscience, New York (1987).
7. D. W. Fox and R. B. Allen. "Compatibility", in *Encyclopedia of Polymer Science and Engineering*, Vol. 3, 2nd ed., J. I. Kroschwitz (Ed.), pp. 758–775, Wiley Interscience, New York (1985).
8. R. A. Orwoll. "Solubility of polymers", *Encyclopedia of Polymer Science and Engineering*, Vol. 15, 2nd ed., J. I. Kroschwitz (Ed.), pp. 380–402, Wiley Interscience, New York (1989).
9. L. H. Sperling. "Microphase structure", in *Encyclopedia of Polymer Science and Engineering*, Vol. 9, 2nd ed., J. I. Kroschwitz (Ed.), pp. 760–788, Wiley Interscience, New York (1987).
10. D. R. Paul, J. W. Barlow, and H. Keskkula. "Polymer blends", in *Encyclopedia of Polymer Science and Engineering*, Vol. 12, 2nd ed., J. I. Kroschwitz (Ed.), pp. 399–461, Wiley Interscience, New York (1988).
11. D. R. Paul and S. Newman. *Polymer Blends*, Academic Press, New York (1978).
12. D. R. Paul and C. B. Bucknall. *Polymer Blends: Formulation and Performance*, John Wiley, New York (1999).
13. L. A. Utracki. *Polymer Alloys and Blends*, Hanser Publishers, New York (1990).

Alphabetical Index of Terms

ORGANIC NAME REACTIONS USEFUL IN BIOCHEMISTRY
AND MOLECULAR BIOLOGY

Akabori Amino Acid Reaction

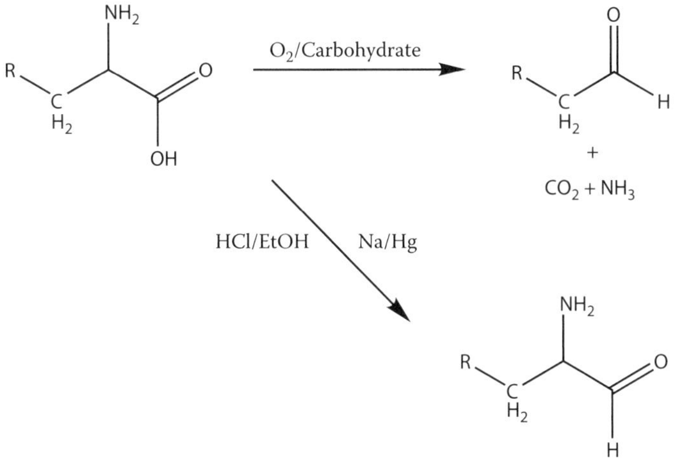

Reaction in the presence of hydrazine yields hydrazides which can be coupled to aromatic aldehydes

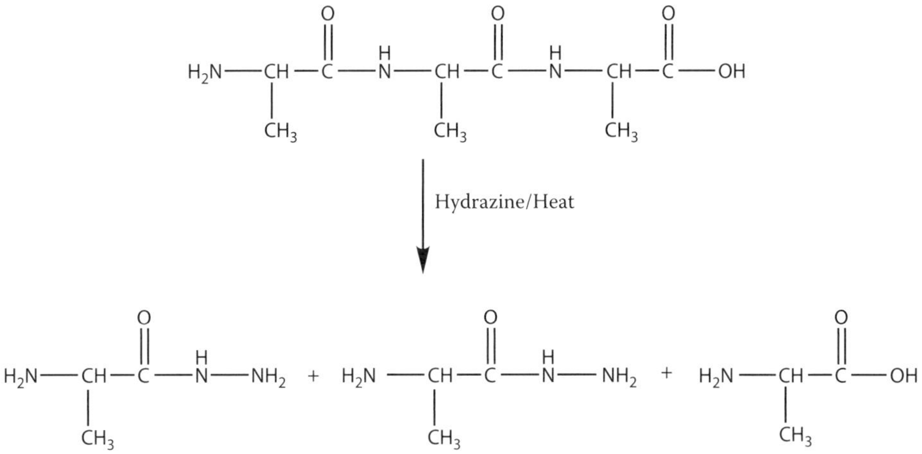

Bose, A.K., *et al.*, Microwave enhanced Akabori reaction for peptide analysis, *J.Am.Soc.Mass Spectrom.* **13**, 839–850, 2002

Originally devised as a method for the conversion of amino acids or amino acid esters to aldehydes. The Akabori reaction has been modified for use in the determination of C-terminal amino acids by performing the reaction in the presence of hydrazine and for the production of derivatives useful for mass spectrometric identification. Ambach, E. and Beck, W., Metal-complexes with biologically important ligands. 35. Nickel, cobalt, palladium, and platinum complexes with Schiff-bases of α-amino acids – A contribution to the mechanism of the Akabori reaction, *Chemische Berichte-Recueil* 118, 2722-2737, 1985; Bose, A.K., Ing, Y.H., Pramanik, B.N., *et al.*, Microwave enhanced Akabori reaction for peptide analysis, *J.Am.Soc.Mass Spectrom.* 13, 839-850, 2002; Pramanik, B.N., Ing, Y.H., Bose, A.K., *et al.*, Rapid cyclopeptide analysis by microwave enhanced Akabori reaction, *Tetrahedron Lett.* 44, 2565-2568, 2003; Puar, M.S., Chan, T.M., Delgarno, D., *et al.*, Sch 486058: A novel cyclic peptide of actinomycete origin, *J.Antibiot.* 58, 151-154, 2005.

ORGANIC NAME REACTIONS USEFUL IN BIOCHEMISTRY AND MOLECULAR BIOLOGY (Continued)

Aldol Condensation

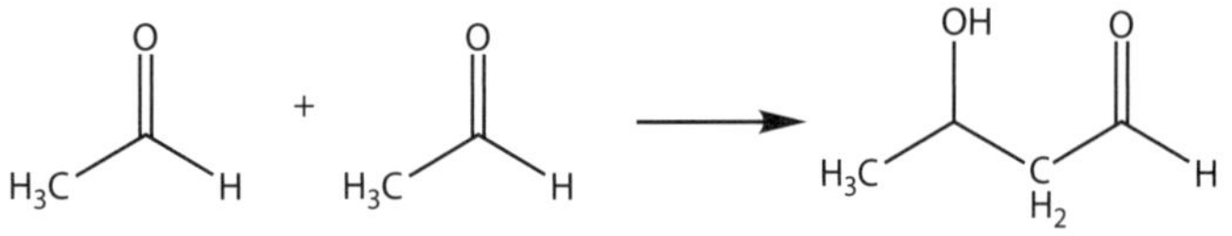

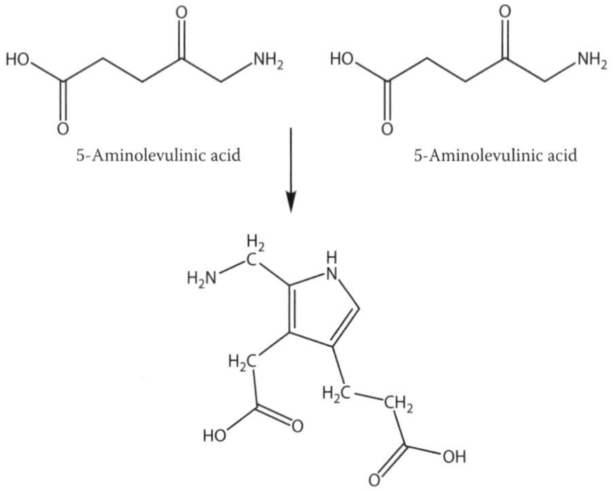

5-Aminolevulinic acid   5-Aminolevulinic acid

Porphobilinogen

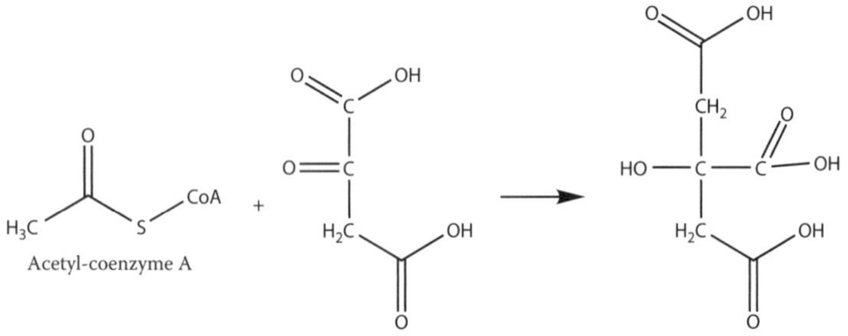

Acetyl-coenzyme A   +   Oxaloacetic acid   →   Citrate

Citrate synthase
an aldol-like condensation

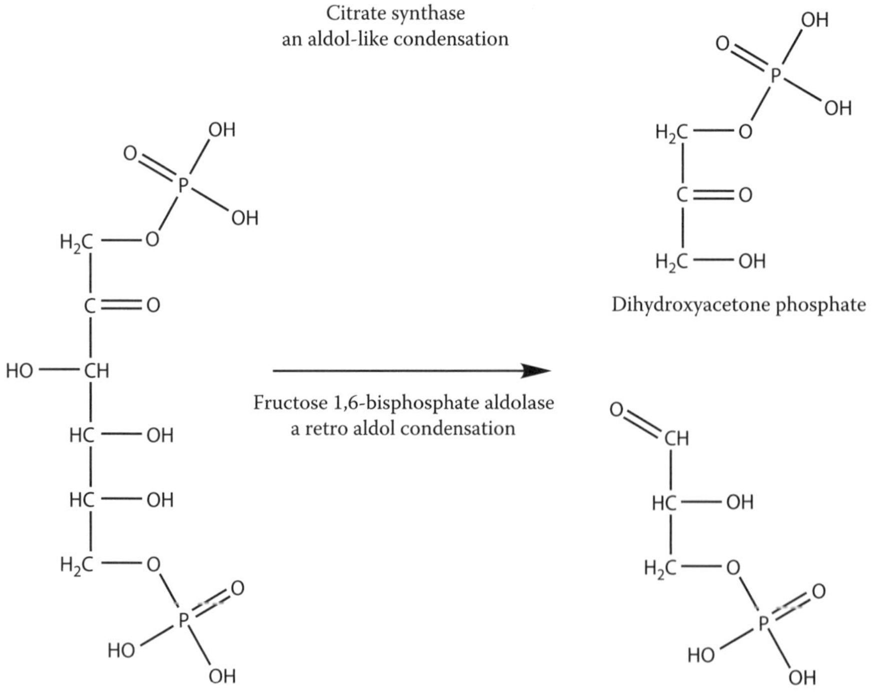

Dihydroxyacetone phosphate

Fructose 1,6-bisphosphate aldolase
a retro aldol condensation

Glyceraldehyde-3-phosphate

Condensation of one carbonyl compound with the enol/enolate form of another to form a β-hydroxyaldehyde; the base-catalyzed reaction proceeds via the enolate form while the acid-catalyzed reaction proceeds via the enol form. The basic chemistry of the aldol condensation is observed in several enzymatic reactions including citrate synthase, fructose-1,6-bisphosphate aldolase, and 2-keto-4-hydroxyglutarate aldolase. See Lane, R.S., Hansen, B.A., and Dekker, E.E., Sulfhydryl groups in relation to the structure and catalytic activity of 2-oxo-4-hydroxyglutarate aldolase from bovine liver, *Biochim.Biophys.Acta* 481, 212-221, 1977; Evans, D.A. and McGee, L.R., Aldol diastereoselection. Zirconium enolates. Product selective, enolate structure independent condensations, *Tetrahedron Lett.* 21, 3975-3978, 1980; Grady, S.R., Wang, J.K., and Dekker, E.E., Steady-state kinetics and inhibition studies of the aldol condensation reaction catalyzed by bovine liver and *Escherichia coli* 2-keto-4-hydroxyglutarate aldolase, *Biochemistry* 20, 2497-2502, 1981; Rokita, S.E., Srere, P.A., and Walsh, C.T., 3-Fluoro-3-deoxycitrate: A probe for mechanistic study of citrate-utilizing enzymes, *Biochemistry* 21, 3765-3774, 1982; Frere, R., Nentwich, M., Gacond, S., *et al.*, Probing the active site of *Pseudomonas aeruginosa* porphobilinogen synthase using newly developed inhibitors, *Biochemistry*, 45, 8243-8253, 2006; Dalsgaard, T.K., Nielsen, J.H., and Larsen, L.B., Characterization of reaction products formed in a model reaction between pentanal and lysine-containing oligopeptides, *J.Agric.Food Chem.* 54, 6367-6373, 2006. A crossed aldol refers to a condensation reaction with two different aldehydes/ketones; the second aldehyde frequently is formaldehyde as it cannot react with itself although this is not a requirement (Kiehlman, E. and Loo, P.W., Orientation in crossed aldol condensation of chloral with unsymmetrical aliphatic ketones, *Canad.J.Chem.* 49, 1588, 1971; Findlay, J.A., Desai, D.N., and McCaulay, J.B., Thermally induced crossed aldol condensations, *Canad.J.Chem.* 59, 3303-3304, 1981; Esmaelli, A.A., Tabas, M.S., Nasseri, M.A., and Kazemi, F., Solvent-free crossed aldol condensation of cyclic ketones with aromatic aldehydes assisted by microwave irradiation, *Monatshefte fur Chemie* 136, 571-576, 2005).

Amadori Rearrangement

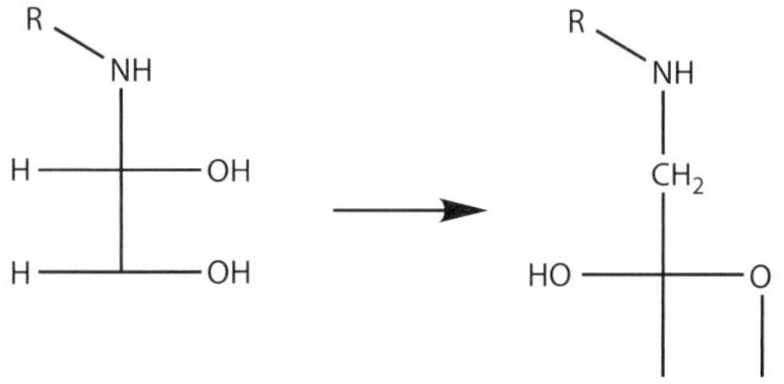

Amadori rearrangement

A reaction following the formation of the unstable reaction product between an aldehyde (reducing sugar) and an amino group (formation of a Schiff base, an aldimine) which results in a more stable ketoamine. The Amadori rearrangement is part of the Maillard reaction which is also called the Browning reaction and can result in the formation of advanced glycation end products. See Amadori, M. *Atti.Accad.Nazl.Lincei* 2, 337, 1925; Hodge, J.E., The Amadori rearrangement, *Adv.Carbohydrate Chem.* 10, 169-205, 1955; Acharya, A.S. and Manning, J.M., Amadori rearrangement of glyceraldehyde-hemoglobin Schiff based adducts. A new procedure for the determination of ketoamine adducts in proteins, *J.Biol.Chem.* 255, 7218-7224, 1980; Acharya, A.S. and Manning, J.M., Reaction of glycoaldehyde with proteins: latent crosslinking potential of α-hydroxyaldehydes, *Proc. Natl.Acad.Sci.USA* 80, 3590-3594, 1983; Roper, H., Roper, S., and Meyer, B., Amadori- and *N*-nitroso-Amadori compounds and their pyrolysis products. Chemical, analytical and biological aspects, *IARC Sci.Publ.* (57), 101-111, 1984; Baynes, J.W., Watkins, N.G., Fisher, C.I. *et al.*, The Amadori product on protein: structure and reactions, *Prog.Clin.Biol.Res.* 304, 43-67, 1989; Nacharaju, P. and Acharya, A.S., Amadori rearrangement potential of hemoglobin at its glycation sites is dependent on the three-dimensional structure of protein, *Biochemistry* 31, 12673-12679, 1992; Zyzak, D.V., Richardson, J.M., Thorpe, S.R., and Baynes, J.W., Formation of reactive intermediates from Amadori compounds under physiological conditions, *Arch. Biochem.Biophys.* 316, 547-554, 1995; Khalifah, R.G., Baynes, J.W., and Hudson, B.G., Amadorins: novel post-Amadori inhibitors of advanced glycation reactions, *Biochem.Biophys.Res.Commun.* 257, 251-158,1999; Davidek, T., Clety, N., Aubin, S., and Blank, I., Degradation of the Amadori compound *N*-(1-deoxy-D-fructos-1-yl)glycine in aqueous model system, *J.Agric.Food Chem.* 50, 5472-5479, 2002.

Baeyer-Villiger Reaction

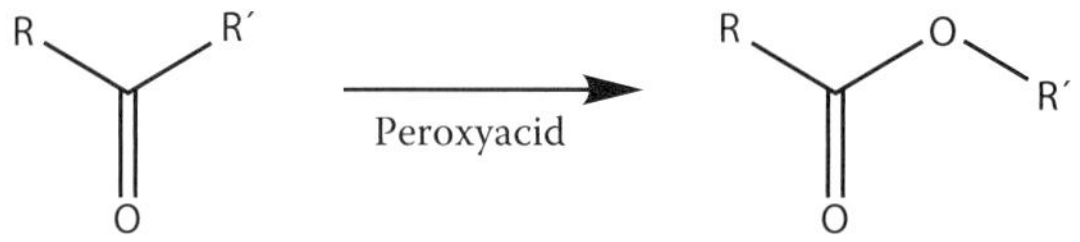

Baeyer–Villiger reaction

ORGANIC NAME REACTIONS USEFUL IN BIOCHEMISTRY AND MOLECULAR BIOLOGY (Continued)

The oxidation of a ketone by a peroxy acid to yield an ester. This reaction is catalyzed by bacterial monooxygenases and has proved useful in preparation of optically pure esters and lactones. See Ryerson, C.C., Ballou, D.P. and Walsh, C., Mechanistic studies on cyclohexanone oxygenase, *Biochemistry* 21, 2644-2655, 1982; Bolm, C., Metal-catalyzed asymmetric oxidations, *Med.Res.Rev.* 19, 348-356, 1999; Zambianchi, F., Pasta, P., Carrea, G., *et al.*, Use of isolated cyclohexanone monooxygenase from recombinant *Escherichia coli* as a biocatalyst for Baeyer-Villiger and sulfide oxidations, *Biotechnol.Bioeng.* 78, 489-496, 2002; Alphand, V., Carrea, G., Wohlgemuth, R., *et al.*, Towards large-scale synthetic application of Baeyer-Villiger monooxygenase, *Trends Biotechnol.* 21, 318-323, 2003; Walton, A.Z. and Stewart, J.D., Understanding and improving NADPH-dependent reactions by nongrowing *Escherichia coli* cells, *Biotechnol.Prog.* 20, 403-411, 2004; Malito, E., Alfieri, A., Fraaije, M.W., and Mattevi, A., Crystal structure of a Baeyer-Villiger monooxygenase, *Proc.Nat.Acad.Sci.USA* 101, 13157-13162, 2004; ten Brink, G.J., Arends, I.W., And Sheldon, R.A., The Baeyer-Villiger reaction: new developments toward greener procedures, *Chem.Rev.* 104, 4105-4124, 2004; Boronat, M., Corma. A., Renz, M., *et al.*, A multisite molecular mechanism for Baeyer-Villiger oxidations on solid catalysts using environmentally friendly H_2O_2 as oxidant, *Chemistry* 11, 6905-6915, 2005; Mihovilovic, M.D., Rudroff, E., Winninger, A., *et al.*, Microbial Baeyer-Villiger oxidation: stereopreference and substrate acceptance of cyclohexanone monooxygenase mutants prepared by directed evolution, *Org.Lett.* 8, 1221-1224, 2006; Baldwin, C.V. and Woodley, J.M., On oxygen limitation in a whole cell biocatalytic Baeyer-Villiger oxidation process, *Biotechnol.Bioeng.* 95, 362-369, 2006.

Beckmann Rearrangement

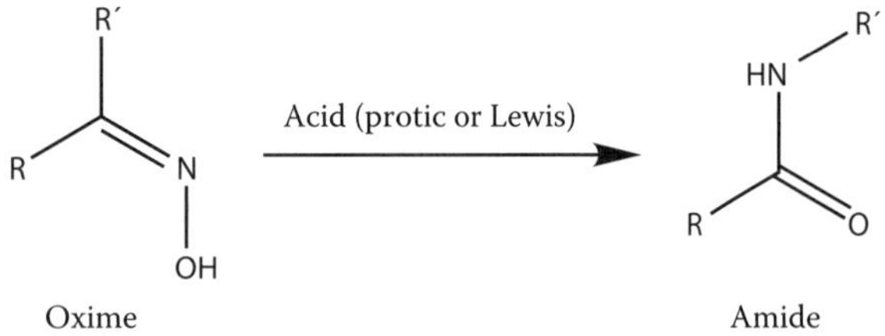

Beckmann rearrangement

An acid (protic or Lewis)-catalyzed conversion of an oxime to a substituted carboxylic amide. See Darling, C.M. and Chen, C.P., Rearrangement of *N*-benzyl-2-cyano(hydroxyimino)acetamide, *J.Pharm.Sci.* 67, 860-861, 1978; Gayen, A.K., and Knowles, C.O., Penetration and fate of methomyl and its oxime metabolite in insects and two spotted spider mites, *Arch.Environ.Contam.Toxicol.* 10, 55-67, 1981; Mangold, J.B., Mangold, B.L., and Spina, A., Rat liver aryl sulfotransferase-catalyzed sulfation and rearrangement of 9-fluorenone oxime, *Biochim.Biophys.Acta* 874, 37-43, 1986; De Luca, L., Giacomelli, G., and Procheddu, A., Beckmann rearrangement of oximes under very mild conditions, *J.Org.Chem.* 67, 6272-6274, 2002; Torisawa, Y. Nishi, T., and Minamikawa, J., A study on the conversion of indanones into carbostyrils, *Bioorg.Med.Chem.* 11, 2205-2209, 2003; Furuya, Y., Ishihara, K., and Yamamoto, H., Cyanuric chloride as a mild and active Beckmann rearrangement catalyst, *J.Am.Chem.Soc.* 127, 11240-11241, 2005; Yamabe, S., Tsuchida, N., and Yamazaki, S., Is the Beckmann rearrangement a concerted or stepwise reaction? A computational study, *J.Org.Chem.* 70,10638-10644, 2005; Ichino, T., Arimoto, H., and Uemura, D., Possibility of a non-amino acid pathway in the biosynthesis of marine-derived oxazoles, *Chem. Commun.* (16), 1742-1744, 2006.

Benzoin Condensation

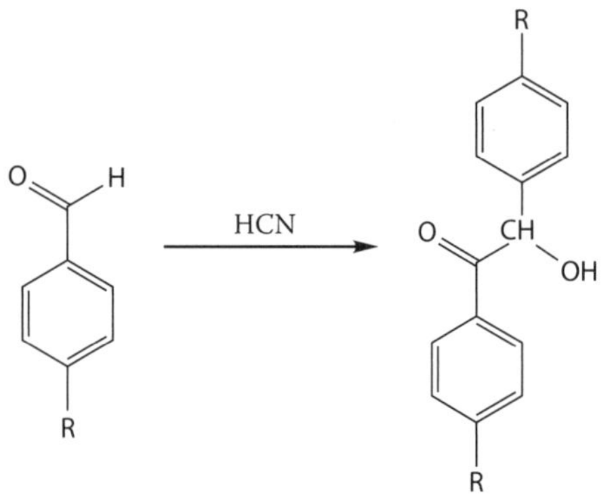

The conversion of benzaldehyde to benzoin (aromatic α-hydroxyketones) via cyanide-mediated condensation; other aromatic aldehydes can participate in this reaction. See Iding, H., Dunnwald, T., Greiner, L., *et al.*, Benzoylformate decarboxylase from *Pseudomonas putida* as stable catalyst for the synthesis of chiral 2-hydroxy ketones, *Chemistry* 6, 1483-1495, 2000; White, M.J. and Leeper, F.J., Kinetics of the thiazolium ion-catalyzed benzoin condensation, *J.Org.Chem.* 66, 5124-5131, 2001; Dunkelmann, P., Kolter-Jung, D., Nitsche, A. *et al.*, Development of a donor-acceptor concept for enzymatic cross-coupling reactions of aldehydes: the first asymmetric cross-benzoin condensation, *J.Am.Chem.Soc.* 124, 12084-12085, 2002; Pohl, M., Lingen, B., and Muller, M., Thiamin-diphosphate-dependent enzymes: new aspects of asymmetric C-C bond formation, *Chemistry* 8, 5288-5295, 2002; Wildemann, H., Dunkelmann, P., Muller, M., and Schmidt, B., A short olefin metathesis-based route to enantiomerically pure arylated dihydropyrans and α, β unsaturated δ valero lactones, *J.Org.Chem.* 68, 799-804, 2003; Murry, J.A., Synthetic methodology utilized to prepare substituted imidazole p38 MAP kinase inhibitors, *Curr.Opin.Drug Discov.Devel.* 6, 945-965, 2003; Reich, B.J., Justice, A.K., Beckstead, B.T., *et al.*, Cyanide-catalyzed cyclizations via aldimine coupling, *J.Org.Chem.* 69, 1357-1359, 2004; Sklute, G., Oizerowich, R., Shulman, H., and Keinan, E., Antibody-catalyzed benzoin oxidation as a mechanistic probe for nucleophilic catalysis by an active site lysine, *Chemistry* 10, 2159-2165, 2004; Breslow, R., Determining the geometries of transition states by use of antihydrophobic additives in water, *Acc.Chem.Res.* 37, 471-478, 2004.

ORGANIC NAME REACTIONS USEFUL IN BIOCHEMISTRY AND MOLECULAR BIOLOGY (Continued)

Cannizzaro Reaction

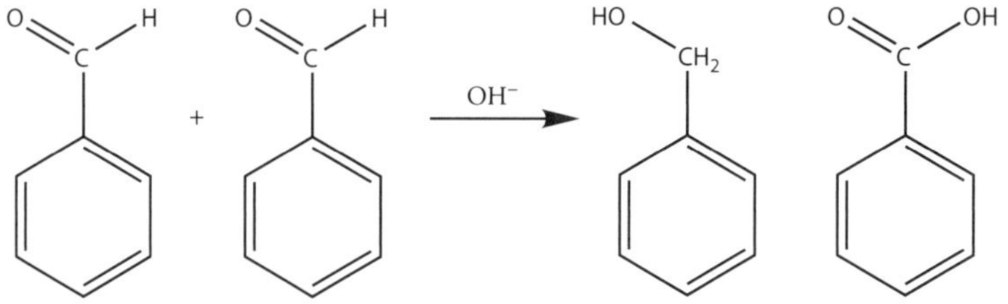

Cannizzaro reaction

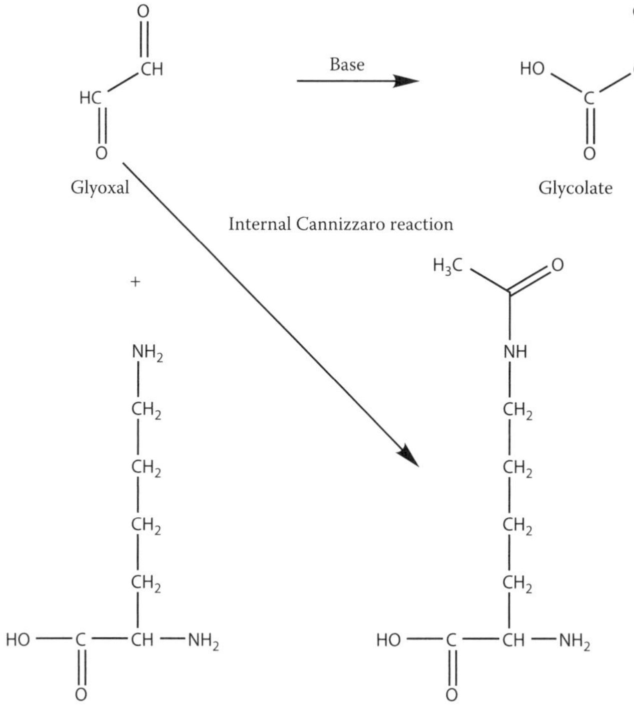

Glyoxal

Internal Cannizzaro reaction

Glycolate

Lysine

Carboxymethyllysine

Base-catalyzed disproportionation of an aldehyde to yield a carboxylic acid and the corresponding alcohol; if an α-hydrogen is present, an aldol condensation is a competing reaction. See Hazlet, S.E. and Stauffer, D.A., Crossed Cannizzaro reactions, *J.Org.Chem.* 27, 2021-2024, 1962; Entezari, M.H. and Shameli, A.A., Phase-transfer catalysis and ultrasonic waves. I. Cannizzaro reaction, *Ultrason.Sonochem.* 7, 169-172, 2000; Matin, M.M., Sharma, T., Sabharwal, S.G., and Dhavale, D.D., Synthesis and evaluation of the glycosidase inhibitory activity of 5-hydroxy substituted isofagomine analogues, *Org.Biomol.Chem.* 3, 1702-1707, 2005; Zhang, L., Wang, S., Zhou, S., *et al.*, Cannizzaro-type disproportionation of aromatic aldehydes to amides and alcohols by using either a stoichiometric amount or a catalytic amount of lanthanide compounds, *J.Org.Chem.* 71, 3149-3153, 2006. Intramolecular Cannizzaro reactions have been described (Glomb, M.A., and Monnier, V.M., Mechanism of protein modification by glyoxal and glycoaldehyde, reactive intermediates of the Maillard reaction, *J.Biol.Chem.* 270, 10017-10026, 1995; Russell, A.E., Miller, S.P., and Morken, J.P., Efficient Lewis acid catalyzed intramolecular Cannizzaro reaction, *J.Org.Chem.* 65, 8381-8383, 2000; Schramm, C. and Rinderer, B., Determination of cotton-bound glyoxal via an internal Cannizzaro reaction by means of high-performance liquid chromatography, *Anal.Chem.* 72, 5829-5833, 2000).

Claisen Condensation

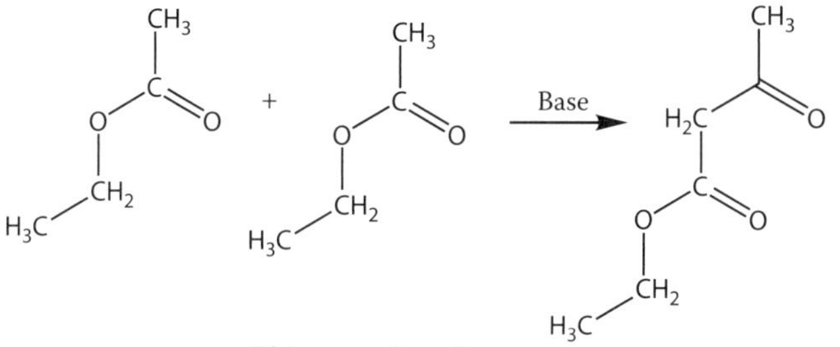

Claisen condensation

ORGANIC NAME REACTIONS USEFUL IN BIOCHEMISTRY AND MOLECULAR BIOLOGY (Continued)

The base-catalyzed condensation of two moles of an ester to give a β-keto ester. Claisen condensations are more favorable with thioesters. This reaction is of great importance in the biosynthesis of fatty acids and polyketides (Haapalainen, A.M., Meriläinen, G., and Wierenga, R.K. The thiolase superfamily: condensing enzymes with diverse reaction specificities, *Trends Biochem.Sci.* 31, 64-71, 2006). For general issues see Dewar, M.J. and Dieter, K.M., Mechanism of the chain extension step in the biosynthesis of fatty acids, *Biochemistry* 27, 3302-3308, 1988; Clark, J.D., O'Keefe, S.J., and Knowles, J.R., Malate synthase: proof of a stepwise Claisen condensation using the double-isotope fractionation test, *Biochemistry* 27, 5961-5971, 1988; Nicholson, J.M., Edafiogho, I.O., Moore, J.A., *et al.*, Cyclization reactions leading to β-hydroxyketo esters, *J.Pharm.Sci.* 83, 76-78, 1994; Lee, R.E., Armour, J.W., Takayama, K., *et al.*, Mycolic acid biosynthesis: definition and targeting of the Claisen condensation step, *Biochim.Biophys.Acta* 1346, 275-284, 1997; Shimakata, T. and Minatogawa, Y., Essential role of trehalose in the synthesis and subsequent metabolism of corynomycolic acid is *Corynebacterium matruchotil, Arch.Biochem.Biophys.* 380, 331-338, 2000; Olsen, J.G., Madziola, A., von Wettstein-Knowles, P., *et al.*, Structures of β-ketoacyl-acyl carrier protein synthase I complexed with fatty acids elucidate its catalytic machinery, *Structure* 9, 233-243, 2001; Klavins, M., Dipane, J., and Babre, K., Humic substances as catalysts in condensation reactions, *Chemosphere* 44, 737-742, 2001; Heath, R.J. and Rock, C.O., The Claisen condensation in biology, *Nat.Prod.Rep.* 19, 581-596, 2002; Takayama, K., Wang, C., and Besra, G.S., Pathway to synthesis and processing of mycolic acids in *Mycobacterium tuberculosis, Clin.Microbiol.Rev.* 18, 81-101, 2005; Ryu, Y., Kim, K.J., Roessner, C.A., and Scott, A.I., Decarboxylative Claisen condensation catalyzed by in vitro selected ribozymes, *Chem.Commun.* (13), 1439-1441, 2006; Kamijo, S. and Dudley, G.B., Claisen-type condensation of vinylogous acyl triflates, *Org.Lett.* 8, 175-177, 2006.

Claisen Rearrangement

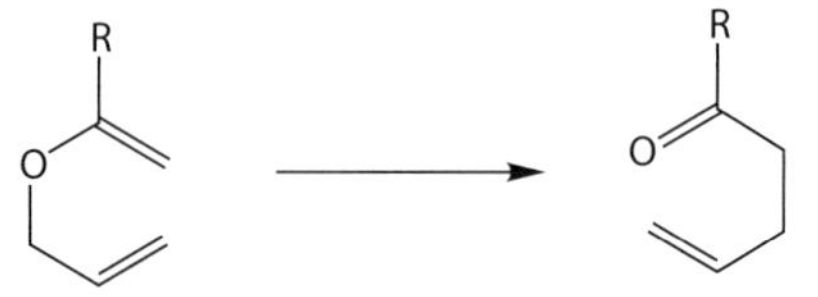

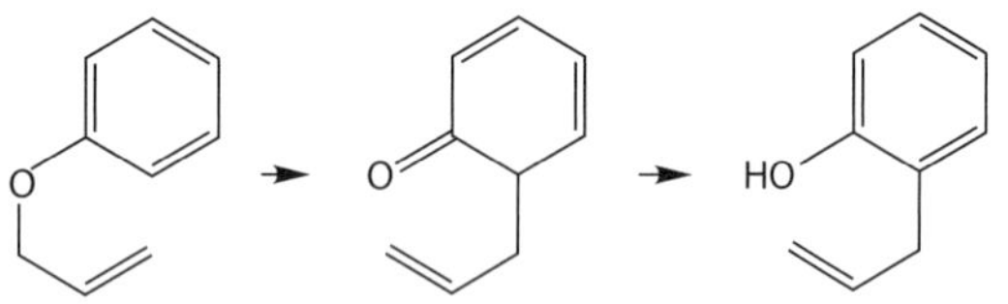

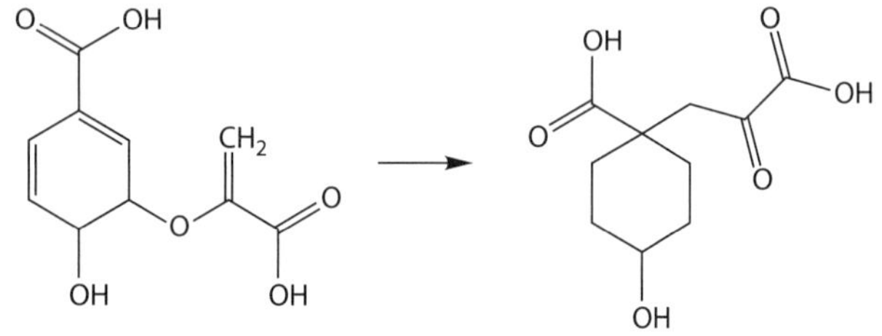

Chorismic Acid   Prephenic Acid

Zhang Z. and Bruice T.C., Temperature dependence of the structure of the substrate and active site of the *Thermus thermophilus* chorismate mutase E-S complex, *Biochemistry* **45**, 8562–8567, 2006

The rearrangement of an allyl vinyl ether or the nitrogen or sulfur analogue or allyl aryl ether to yield a γ, δ-unsaturated ketone or an *o*-allyl substituted phenol. See Hilvert, D., Carpenter, S.H., Nared, K.D., and Auditor, M.T., Catalysis of concerted reactions by antibodies: the Claisen rearrangement, *Proc.Nat.Acad.Sci.USA* 85, 4953-4955, 1988; Campbell, A.P., Tarasow, T.M., Massefski, W., *et al.*, *Proc.Nat.Acad.Sci.USA* 90, 8663-8667, 1993; Swiss, K.A. and Firestone, R.A., Catalysis of Claisen rearrangement by low molecular weight polyethylene(1), *J.Org.Chem.* 64, 2158-2159, 1999; Berkowitz, D.B., Choi, S., and Maeng, J.H., Enzyme-assisted asymmetric total synthesis of (-)-podophyllotoxin and (-)-picropodophyllin, *J.Org.Chem.* 65, 847-860, 2000; Itami, K. and Yoshida, J., The use of hydrophilic groups in aqueous organic reactions, *Chem.Rec.* 2 213-224, 2002; Martin Castro, A.M., Claisen rearrangement over the past nine decades, *Chem.Rev.* 104, 2939-3002, 2004; Sparano, B.A., Shahi, S.P., and Koide, K., Effect of binding and conformation on fluorescence quenching in new 2',7'-dichlorofluorescein derivatives, *Org.Lett.* 6, 1947-1949, 2004; Davis, C.J., Hurst, T.E., Jacob, A.M., and Moody, C.J., Microwave-mediated Claisen rearrangement followed by phenol oxidation: a simple route to naturally occurring 1,4-benzoquinones. The first synthesis of verapliquinones A and B and Panicein, A., *J.Org.Chem.* 70, 4414-4422, 2005; Wright, S.K., DeClue, M.S., Mandal, A., *et al.*, Isotope effects on the enzymatic and nonenzymatic reactions of chorismate, *J.Am.Chem.Soc.* 127, 12957-12964, 2005; Declue, M.S., Baldridge, K.K., Kast, P., and Hilvert, D., Experimental and computational investigation of the uncatalyzed rearrangement and elimination reactions of isochorismate, *J.Am.Chem.Soc.* 128, 2043-2051, 2006; Zhang, X. and Bruice, T.C., Temperature dependence of the structure of the substrate and active site of the *Thermus thermophilus* chorismate mutase E-S complex, *Biochemistry* 45, 8562-8567, 2006.

Criegee Reactions

Peroxyacid

Criegee Intermediate

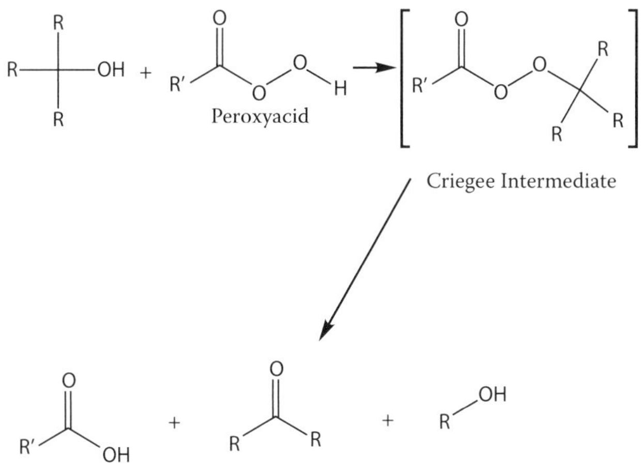

Mostly the reaction of a peroxyacid with a tertiary alcohol to form a ketone and an alcohol. The intermediate peroxyester is an intermediate (Criegee adduct or Criegee intermediate) in the Baeyer-Villiger reaction. The Criegee intermediate is important in the ozonolysis of alkenes including fatty acids. See Leffler, J.E. and Scrivener, F.E., Jr., The decomposition of cumyl peracetate in nonpolar solvents, *J.Org.Chem.* 37, 1794-1796, 1978; Srisankar, E.V. and Patterson, L.K., Reactions of ozone with fatty acid monolayer: a model system for disruption of lipid molecular assemblies by ozone, *Arch. Environ.Health* 34, 346-349, 1979; Grammer, J.C., Loo, J.A., Edmonds, C.G., *et al.*, Chemistry and mechanism of vanadate-promoted photooxidative cleavage of myosin, *Biochemistry* 35, 15582-15592, 1996; Krasutsky, P.A., Kolomitsyn, I.V., l Kiprof P., *et al.*, Observation of a stable carbocation in a consecutive Criegee rearrangement with trifluoroperacetic acid, *J.Org.Chem.* 65, 3926-3992, 1996; Carlqvist, P., Eklund, P., Hult, K., and Brinck, T., Rational design of a lipase to accommodate catalysis of Baeyer-Villiger oxidation with hydrogen peroxide, *J.Mol.Model.* 9, 164-171, 2003; Deeth, R.J. and Bugg, T.D., A density functional investigation of the extradiol cleavage mechanism in non-heme iron catechol dioxygenase, *J.Biol.Inorg.Chem.* 8, 409-418, 2003; Krasutsky, P.A., Kolomitsyn, I.V., Krasutsky, S.G., and Kiprof, P., Double- and triple-consecutive *O*-insertion into *tert*-butyl and triarylmethyl structures, *Org.Lett.* 6, 2539-2542, 2004.

Curtius Rearrangement

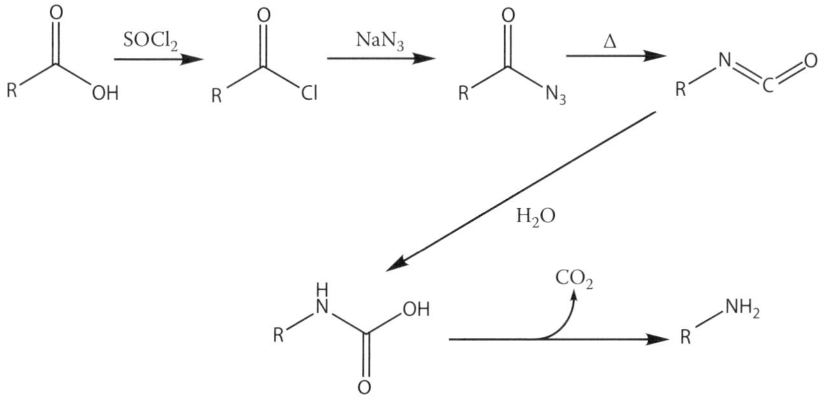

The conversion of a carboxylic acid to an amine via an acid intermediate. See Inouye, K., Watanabe, K., and Shin, M., Formation and degradation of urea derivatives in the azide method of peptide synthesis. Part 1. The Curtius rearrangement and urea formation, *J.Chem.Soc.* (17), 1905-1911, 1977; Chorev, M., and Goodman, M., Partially modified retro-inverso peptides. Comparative Curtius rearrangements to prepare 1, 1-diaminoalkane derivatives, *Int.J.Pept.Protein Res.* 21, 258-268, 1983; Sasmal, S., Geyer, A., and Maier, M.E., Synthesis of cyclic peptidomimetics from aldol building blocks, *J.Org.Chem.* 67, 6260-6263, 2002; Kedrowski, B.L., Synthesis of orthogonally protected (R)- and (S)-2-methylcysteine via an enzymatic desymmeterization and Curtius rearrangement, *J.Org.Chem.* 68, 5403-5406, 2003; Englund, E.A., Gopi, H.N., and Appella, D.H., An efficient synthesis of a probe for protein function: 2,3-diaminopropionic acid with orthogonal protecting groups, *Org.Lett.* 6, 213-215, 2004; Spino, C., Tremblay, M.C., and Gobout, C., A stereodivergent approach to amino acids, amino alcohols, or oxazolidinones of high enantiomeric purity, *Org.Lett.* 6, 2801-2804, 2004; Brase, S., Gil, C., Knepper, K., and Zimmerman, V., Organic azides: an exploding diversity of a unique class of compounds, *Angew.Chem.Int. Ed.Engl.* 44, 5188-5240, 2005; Lebel, H. and Leogane, O., Boc-protected amines via a mild and efficient one-pot Curtius rearrangement, *Org.Lett.* 7, 4107-4110, 2005.

ORGANIC NAME REACTIONS USEFUL IN BIOCHEMISTRY AND MOLECULAR BIOLOGY (Continued)

Dakin Reaction

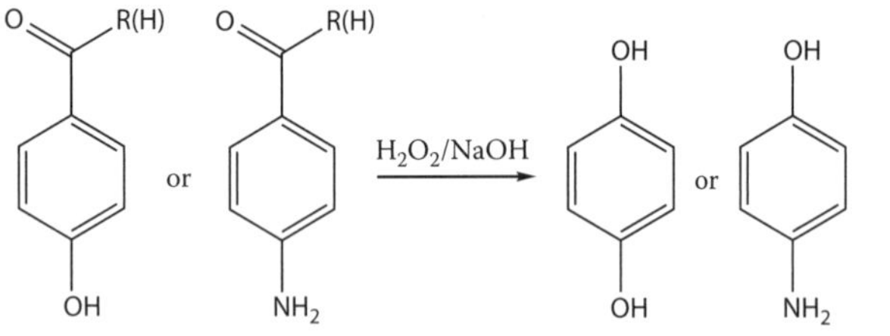

Conversion of an aromatic ketone or aldehyde to a phenolic derivative with alkaline hydrogen peroxide. The mechanism is thought to be similar to the Baeyer-Villiger reaction possibly proceeding through a peroxyacid intermediate. The presence of an amino group or a hydroxyl group in the position para to the carbonyl function is required. See Corforth, J.W. and Elliott, D.F., Mechanism of the Dakin and West reaction, *Science* 112, 534-535, 1950.

Dakin-West Reaction

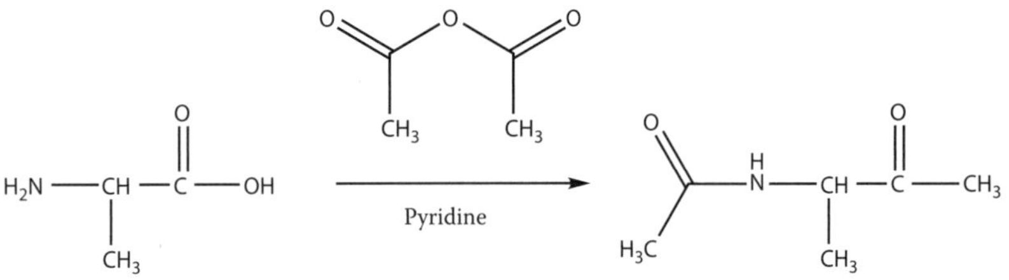

Dakin—West reaction

Conversion of amino acids to acetamidoketones via the action of acetic anhydride in base where a carboxyl group is replaced by an acyl group in a reaction proceeding through an oxazolone intermediate. This reaction has been used for the synthesis of enzyme inhibitors and receptor antagonists. See Angliker, H. Wikstrom, P., Rauber, P., *et al.*, Synthesis and properties of peptidyl derivatives of arginylfluoromethanes, *Biochem.J.* 256, 481-486, 1988; Cheng, L., Goodwin, C.A., Schully, M.F., *et al.*, Synthesis and biological activity of ketomethylene pseudopeptide analogues as thrombin inhibitors, *J.Med.Chem.* 35, 3364-3369, 1992; Godfrey, A.B., Brooks, D.A., Hay, L.A., *et al*, Application of the Dakin-West reaction for the synthesis of oxazole-containing dual PPARα/γ agonists, *J.Org.Chem.* 68, 2623-2632, 2003; Loksha, Y.M., el-Barbary, A.A., it-Barbary, M.A., *et al.*, Synthesis of 2-(aminocarbonylmethylthio)-1*H*-imidazoles as novel Capravirine analogues, *Bioorg.Med.Chem.* 13, 4209-4220, 2005

Diels-Alder Condensation

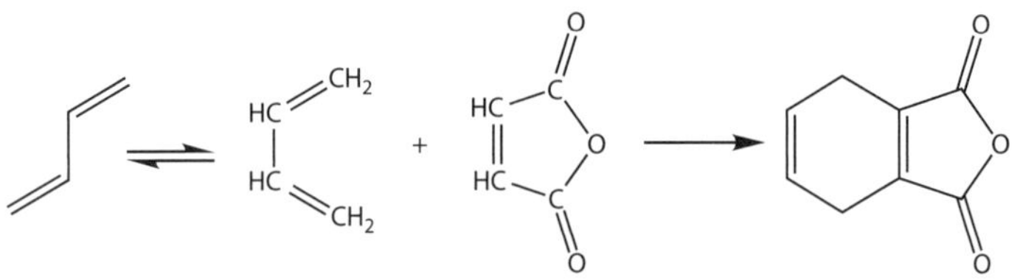

trans-butadiene   *cis*-butadiene   Maleic anhydride
                diene   dienophile

Diels—Alder condensation

A cycloaddition reaction between a conjugated diene and an alkene resulting in the formation of alkene ring; construction of a six-membered ring with multiple stereogenic centers resulting in a chiral molecule. See Wasserman, A., *Diels-Alder Reactions. Organic Background and Physico-Chemical Aspects*, Elsevier, Amsterdam, Netherlands, 1965; Fringuelli, F., and Taticchi, A., *The Diels-Alder Reaction. Selected Practical Methods*, John Wiley & Sons, Ltd., Chichester, UK, 2002; Stocking, E.M. and Williams, R.M., Chemistry and biology of biosynthetic Diels-Alder reactions, *Angew.Chem.Int.Ed.* 42, 3078-3115, 2003. See also Waller, R.L. and Recknagel, R.O., Determination of lipid conjugated dienes with tetracyanoethylene-[14]C: significance for study of the pathology of lipid peroxidation, *Lipids* 12, 914-921, 1977; Melucci, M., Barbarella, G., and Sotgiu, G., Solvent-free, microwave-assisted synthesis of thiophene oligomers via Suzuki coupling, *J.Org.Chem.* 67, 8877-8884, 2002; Breslow, R., Determining the geometries of transition states by use of antihydrophobic additives in water, *Acc.Chem.Res.* 37, 471-478, 2004; Conley, N.R., Hung, R.J., and Willison, C.G., A new synthetic route to authentic *N*-substituted aminomaleimides, *J.Org.Chem.* 70, 4553-4555, 2005; Boul, P.J., Reutenauer, P., and Lehn, J.M. Reversible Diels-Alder reactions for the generation of dynamic combinatorial libraries, *Org.Lett.* 7, 15-18, 2005. Catalytic antibodies have been used for Diels-Alder reactions (Suckling, C.J., Tedford, C.M., Proctor, G.R., *et al.*, Catalytic antibodies: a new window on protein chemistry, *Ciba Found.Symp.* 159, 201-208, 1991; Meekel, A.A., Resmini, M., and Pandit, U.K., Regioselectivity and enantioselectivity in an antibody catalyzed hetero Diels-Alder reaction, *Bioorg.Med.Chem.* 4, 1051 1057, 1996; Romesberg, F.E., Spiller, B., Schultz, P.G., and Stevens, R.C., Immunological origins of binding and catalysis in a Diels-Alderase antibody, *Science* 279, 1934-1940, 1998; Romesberg, F.E. and Schultz, P.G., A mutational study of a Diels-Alderase catalytic antibody, *Bioorg.Med. Chem.Lett.* 9, 1741-1744, 1999; Chen, J., Deng, Q., Wang, R., *et al.*, Shape complementarity binding-site dynamics, and transition state stabilization: a theoretical study of Diels-Alder catalysis by antibody IE9, *ChemBioChem* 1, 255-261, 2000; Kim, S.P., Leach, A.G., and Houk, K.N., The origins of noncovalent catalysis of intermolecular Diels-Alder reactions by cyclodextrins, self-assembling capsules, antibodies, and RNAses, *J.Org.Chem.* 67, 4250-4260, 2002; Cannizzaro, C.E., Ashley, J.A, Janda, K.D., and Houk, K.N., Experimental determination of the absolute enantioselectivity of an antibody-catalyzed Diels-Alder reaction and theoretical explorations of the origins of stereoselectivity, *J.Am.Chem.Soc.* 125, 2489-2506, 2003.

ORGANIC NAME REACTIONS USEFUL IN BIOCHEMISTRY AND MOLECULAR BIOLOGY (Continued)

Edman Degradation

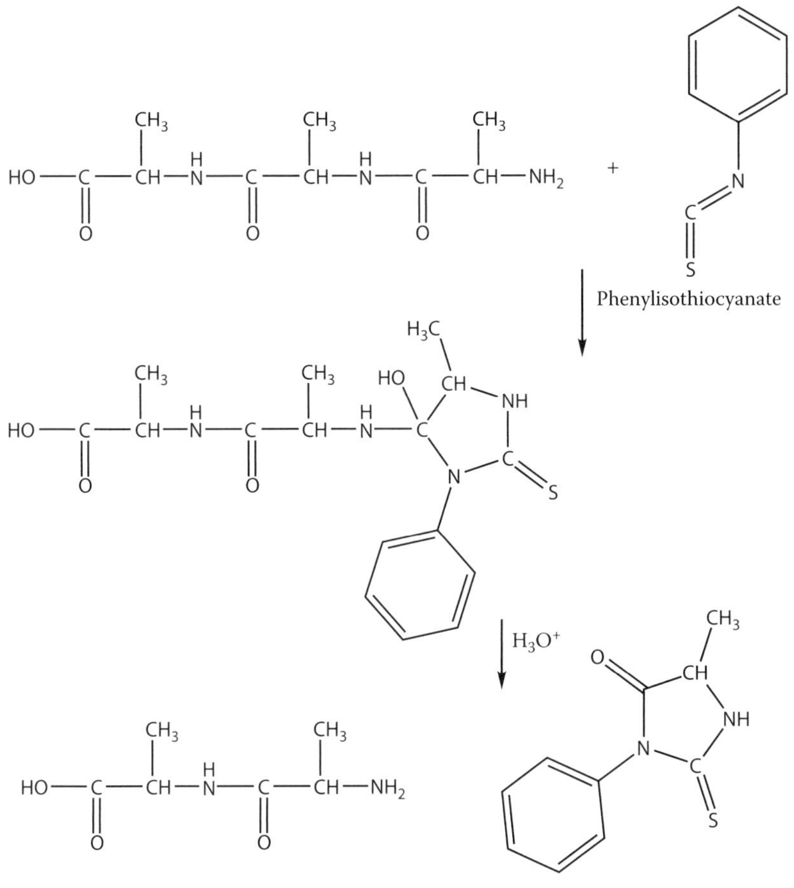

Phenylisothiocyanate

Phenylthiohydantoin

The stepwise degradation of a peptide chain from the amino-terminal via reaction with phenylisothiocyanate. This process is used for the chemical determination of the amino acid sequence of a peptide or protein. See Edman, P., Sequence determination, *Mol.Biol.Biochem.Biophys.* 9, 211-255, 1970; Heinrikson, R.L., Application of automated sequence analysis to the understanding of protein structure and function, *Ann.Clin.Lab.Sci.* 8, 295-301, 1978; Tsugita, A., Developments in protein microsequencing, *Adv.Biophys.* 23, 91-113, 1987; Han, K.K. and Martinage, A., Post-translational chemical modifications of proteins – III. Current developments in analytical procedures of identification and quantitation of post-translational chemically modified amino acid(s) and its derivatives, *Int.J.Biochem.* 25, 957-970, 1993; Masiarz, F.R. and Malcolm, B.A., Rapid determination of endoprotease specificity using peptide mixtures and Edman degradation analysis, *Methods Enzymol.* 241, 302-310, 1994; Gooley, A.A., Ou, K., Russell, J., *et al.*, A role for Edman degradation in proteome studies, *Electrophoresis* 18, 1068-1072, 1997; Wurzel, C., and Wittmann-Liebold, B., A wafer based micro reaction system for the Edman degradation of proteins and peptides, *J.Protein Chem.* 17, 561-564, 1998; Walk, T.B., Sussmuth, R., Kempter, C., *et al.*, Identification of unusual amino acids in peptides using automated sequential Edman degradation coupled to direct detection by electrospray-ionization mass spectrometry, *Biopolymers* 49, 329-340, 1999; Lauer-Fields, J.L., Nagase, H. and Fields, G.B., Use of Edman degradation sequence analysis and matrix-assisted laser desorption/ionization mass spectrometry in designing substrates for matrix metalloproteinases, *J.Chromatog.A.* 890, 117-125, 2000; Hajdu, J., Neutze, R., Sjogren, T., *et al.*, Analyzing protein functions in four dimensions, *Nat.Struct.Biol.* 7, 1006-1012, 2000; Shively, J.E., The chemistry of protein sequence analysis, *EXS* 88, 99-117, 2000; Wang, P., Arabaci, G., and Pei, D., Rapid sequencing of library-derived peptides by partial Edman degradation and mass spectrometry, *J.Comb.Chem.* 3, 251-254, 2001; Brewer, M., Oost, T., Sukonpan, C., *et al.*, Sequencing hydroxylethyleneamine-containing peptides via Edman degradation, *Org.Lett.* 4, 3469-3472, 2002; Sweeney, M.C. and Pei, D., An improved method for rapid sequencing of support-bound peptides by partial Edman degradation and mass spectrometry, *J.Comb.Chem.* 5, 218-222, 2003; Buda, F., Ensing, B., Gribnau, M.C., and Baerends, E.J., O_2 evolution in the Fenton reaction, *Chemistry* 9, 3436-3444, 2003; Liu, Q., Berchner-Pfannschmidt, U., Moller, U., *et al.*, A Fenton reaction at the endoplasmic reticulum is involved in the redox control of hypoxia-inducible gene expression, *Proc.Nat.Acad.Sci. USA* 101, 4302-4307, 2004; Maksimovic, V., Mojovic, M., Neumann, G., and Vucinic, Z., Nonenzymatic reaction of dihydroxyacetone with hydrogen peroxide enhanced via a Fenton reaction, *Ann.N.Y.Acad.Sci.* 1048, 461-465, 2005; Lu, C. and Koppenol, W.H., Inhibition of the Fenton reaction by nitrogen monoxide, *J.Biol.Inorg.Chem.* 10, 732-738, 2005; Baron, C.P., Refsgaard, H.H., Skibsted, H., and Andersen, M.L., Oxidation of bovine serum albumin initiated by the Fenton reaction—effect of EDTA, *tert*-butylhydroperoxide and tetrahydrofuran, *Free Radic.Res.* 40, 409-417, 2006; Thakkar, A., Wavreille, A.S., and Pei, D., Traceless capping agent for peptide sequencing by partial Edman degradation and mass spectrometry, *Anal.Chem.* 78, 5935-5939, 2006.

ORGANIC NAME REACTIONS USEFUL IN BIOCHEMISTRY AND MOLECULAR BIOLOGY (Continued)

Eschweiler-Clarke Reaction

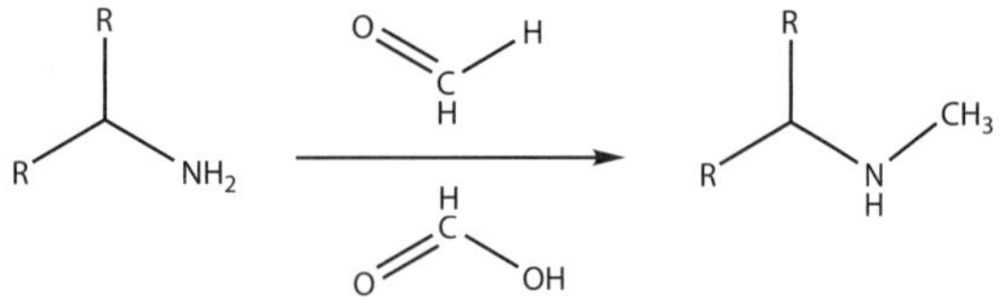

Eschweiler–Clarke reaction

The reductive methylation of amines with formaldehyde in the presence of formic acid. See Lindeke, B., Anderson, B., and Jenden, D.J., Specific deuteromethylation by the Escheweiler-Clarke reaction. Synthesis of differently labelled variants of trimethylamine and their use of the preparation of labelled choline and acetylcholine, *Biomed.Mass Spectrom.* 3, 257-259, 1976; Boldavalli, F., Bruno, O., Mariani, E., *et al.*, Esters of *N*-methyl-*N*-(2-hydroxyethyl or 3-hydroxypropyl)-1,3,3-trimethylbicyclo[2.2.1] heptan-2-endo-amine with hypotensive activity, *Farmaco* 42, 175-183, 1987; Lee, S.S., Wu, W.N., Wilton, J.H., *et al.*, Longiberine and *O*-methyllogiberine, dimeric protoberberine-benzyl tetrahydroisoqunioline alkaloids from *Thalictrum longistrylum*, *J.Nat.Prod.* 62, 1410-1414, 1999; Suma, R and Sai Prakash, P.K., Conversion of sertraline to *N*-methyl sertraline in embalming fluid: a forensic implication, *J.Anal.Toxicol.* 30, 395-399, 2006. The reaction can be accomplished with sodium borohydride or sodium cyanoborohydride and is related to the reductively methylation/alkylation of lysine residues in proteins (Lundblad, R.L., *Chemical Reagents for the Modification of Proteins*, 3rd edn., CRC Press, Boca Raton, FL, 2004).

Favorskii Rearrangement

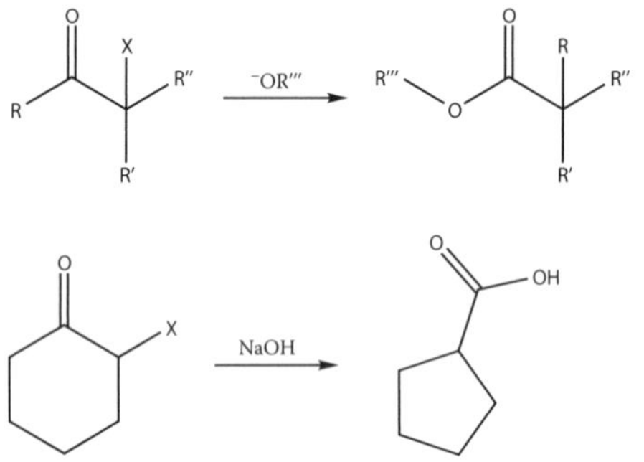

Favorskii rearrangement

The rearrangement of an α-ketone in the presence of an alkoxide to form a carboxylic ester; cyclic α-ketone undergo ring contraction. March, J. *Advanced Organic Chemistry. Reactions, Mechanisms, and Structures*, 3rd edn. John Wiley & Sons, New York, New York, 1985; Gardner, H.W., Simpson, T.D., and Hamberg, M., Mechanism of linoleic acid hydroperoxide reaction with alkali, *Lipids* 31, 1023-1028, 1996; Xiang, L., Kalaitzis, J.A., Nilsen, G. *et al.*, Mutational analysis of the enterocin favorskii biosynthetic rearrangement, *Org.Lett.* 4, 957-960, 2002; Zhang, L. and Koreeda, M., Stereocontrolled synthesis of kelsoene by the homo-favorskii rearrangement, *Org.Lett.* 4, 3755-3788, 2002; Grainger, R.S., Owoare, R.B., Tisselli, P., and Steed, J.W., A synthetic alternative to the type-II intramolecular 4 + 3 cycloaddition, *J.Org.Chem.* 68, 7899-7902, 2003.

Fenton Reagent/Reaction

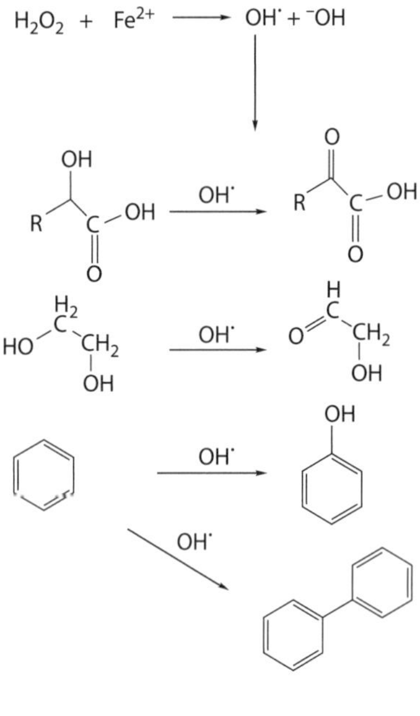

ORGANIC NAME REACTIONS USEFUL IN BIOCHEMISTRY AND MOLECULAR BIOLOGY (Continued)

The reaction of ferrous ions and hydrogen peroxide to yield a hydroxyl radical. See Aust, S.D., Morehouse, L.A., and Thomas, C.E., Role of metals in oxygen radical reactions, *J.Free Radic.Biol.Med.* 1, 3-25, 1985; Goldstein, S., Meyerstein, D. and Czapski, G., The Fenton reagents, *Free Radic.Biol.Med.* 15, 435-445, 1993; Wardman, P. and Candeias, L.P., Fenton chemistry: an introduction, *Radiat.Res.* 145, 523-531, 1996; Held, K.D., Sylvester, F.C., Hopcia, K.L., and Biaglow, J.E., Role of Fenton chemistry in the thiol-induced toxicity and apopotosis, *Radiat.Res.* 145, 542-553, 1996; Merli, C., Petrucci, E., Da Pozzo, A., and Pernetti, M., Fenton-type treatment: state of the art, *Ann.Chim.* 93, 761-770, 2003; Groves, J.T., High-valent iron in chemical and biological oxidations, *J.Inorg.Biochem.* 100, 434-447, 2006.

Fischer Carbene Complexes

Fischer carbene complex

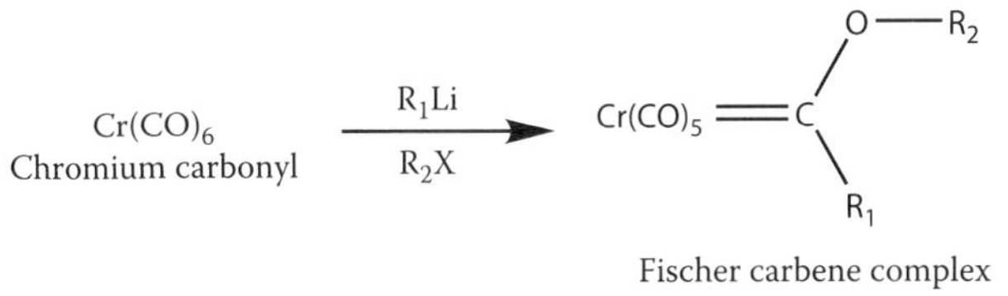

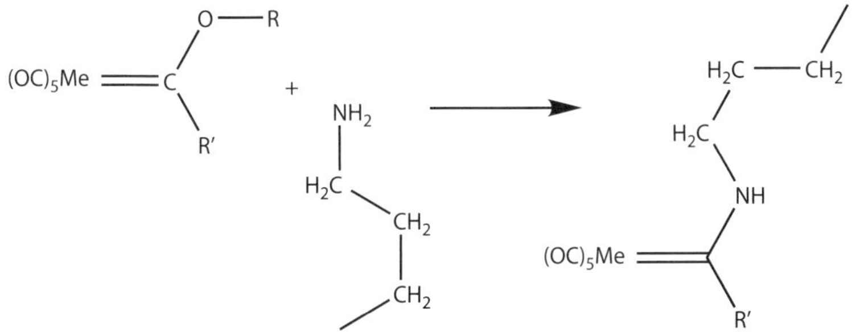

A Fischer carbene complex consists of a transition metal with a formal carbon-metal bond containing a carbene in the singlet state; stabilization of the carbene is provided by the metal interaction. The Fischer carbene complex is electrophilic as the carbene carbon as opposed to the Schrock complex which is in the triplet state and nucleophilic at the carbene carbon. The Fischer carbene complex is high reactivity and is used in many synthetic procedures. A example is provided by the α,β-unsaturated carbenepentacarbonylchromium complex (de Meijere, A., Schirmer, H., and Duetsch, M., Fischer carbene complexes as chemical multitalents: The incredible range of products from carbenepentacarbonylmetal α,β-unsaturated complexes, *Angew.Chem.Int.Ed.* 39, 3964-4002, 2000). See also Salmain, M., Blais, J.C., Tran-Huy, H., *et al*, Reaction of hen egg white lysozyme with Fischer-type metallocarbene complexes. Chacterization of the conjugates and determination of the metal complex binding sites, *Eur.J.Biochem.* 268, 5479-5487, 2001; Merlic, C.A. and Doroh, B.C., Amine-catalyzed coupling of aldehydes and ketenes derived from Fischer carbene complexes: formation of beta-lactones and enol ethers, *J.Org.Chem.* 68, 6056-6069, 2003; Barluenga, J., Santamaria, J., and Tomas, M., Synthesis of heterocycles via group VI Fischer carbene complexes, *Chem.Rev.* 104, 2259-2283, 2004; Barluenga, J., Fananas-Mastral, M., and Aznar, F., A new synthesis of allyl sulfoxides via nucleophilic addition of sulfinyl carbanions to group 6 Fischer carbene complexes, *Org.Lett.* 7, 1235-1237, 2005; Lian, Y. and Wulff, W.D., Iron in the service of chromium: the *o*-benzannulation of *trans,trans*-dienyl Fischer carbene complexes, *J.Am.Chem.Soc.* 127, 17162-17163, 2005; Barluenga, J., Mendoza, A., Dieguez, A., *et al.,* Umpolung reactivity of alkenyl Fischer carbene complexes, copper enolates, and electrophiles, *Angew.Chem.Int. Ed.Engl.* 45, 4848-4850, 2006; Samanta, D., Sawoo, S. and Sarkar, A., *In situ* generation of gold nanoparticles on a protein surface: Fischer carbene complex as reducing agent, *Chem.Commun.* (32), 3438-3440, 2006; Rawat, M., Prutyanov, V., and Wulff, W.D., Chromene chromium carbene complexes in the syntheses of naphthoyran and naphthopyrandione units present in photochromic materials and biologically active natural products, *J.Am.Chem.Soc.* 128, 11044-11053, 2006. For general information on carbenes including Fischer carbene complexes and Schrock carbene complexes, see *Carbene Chemistry. From Fleeting Intermediates to Powerful Reagents*, ed. G. Bertrand, Fontis Media/Marcel Dekker, New York, New York, 2002.

ORGANIC NAME REACTIONS USEFUL IN BIOCHEMISTRY AND MOLECULAR BIOLOGY (Continued)

Fischer Indole Synthesis

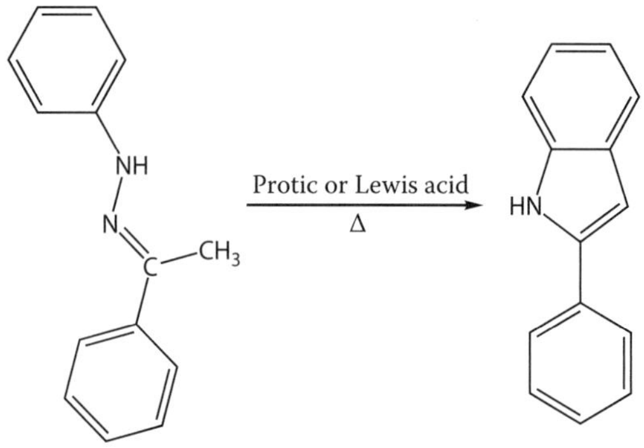

Fischer indole synthesis

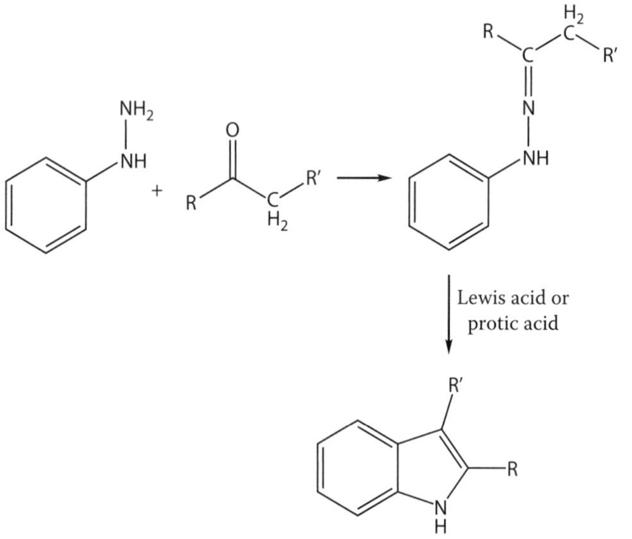

The thermal conversion of arylhydrazones in the presence of a protic acid or a Lewis acid to form an indole ring. See Owellen, R.J., Fitzgerald, J.A., Fitzgerald, B.M., *et al.*, The cyclization phase of the Fischer indole synthesis. The structure and significance of Pleininger's intermediate, *Tetrahedron Lett.* 18, 1741-1746, 1967; Kim, R.M., Manna, M., Hutchins, S.M. *et al.*, Dendrimer-supported combinatorial chemistry, *Proc.Natl.Acad.Sci.USA* 93, 10012-10017, 1996; Brase, S., Gil, C., and Knepper, K., The recent impact of solid-phase synthesis on medicinally relevant benzoannelated nitrogen heterocycles, *Bioorg.Med.Chem.* 10, 2415-2437, 2002; Rosenbaum, C., Katzka, C., Marzinzik, A., and Waldmann, H., Traceless Fischer indole synthesis on the solid phase, *Chem.Commun.* (15), 1822-1823, 2003; Mun, H.S., Ham, W.H., and Jeong, J.H., Synthesis of 2,3-disubstituted indole on solid phase by the Fischer indole synthesis, *J.Comb.Chem.* 7, 130-135, 2005; Narayana, B., Ashalatha, B.V., Vijaya Raj, K.K., *et al.*, Synthesis of some new biologically acivie 1,3,4-oxadiazolyl nitroindole and a modified Fischer indole synthesis of ethyl nitro indole-2-carboxylates, *Bioorg.Med.Chem.* 13, 4638-4644, 2005; Schmidt, A.M. and Eilbracht, P., Tandem hydroformylation-hydrazone formation-Fischer indole synthesis: a novel approach to tryptamides, *Org.Biomol.Chem.* 3, 2333-2343, 2005; Linnepe Nee Kohling, P., Schmidt, A.M., and Eilbracht, P., 2,3-Disubstituted indoles from olefins and hydrazines via tandem hydroformylation-Fischer indole synthesis and skeletal rearrangement, *Org.Biomol.Chem.* 4, 302-313, 2006; Landwehr, J., George, S., Karg, E.M., *et al.*, Design and synthesis of novel 2-amino-5-hydroxyindole derivatives that inhibit human 5-lipooxygenase, *J.Med.Chem.* 49, 4327-4332, 2006.

ORGANIC NAME REACTIONS USEFUL IN BIOCHEMISTRY AND MOLECULAR BIOLOGY (Continued)

Friedel-Crafts Reaction

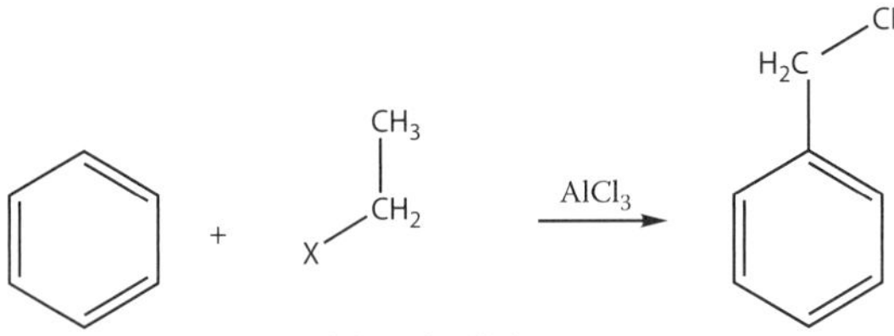

Friedel-Crafts alkylation

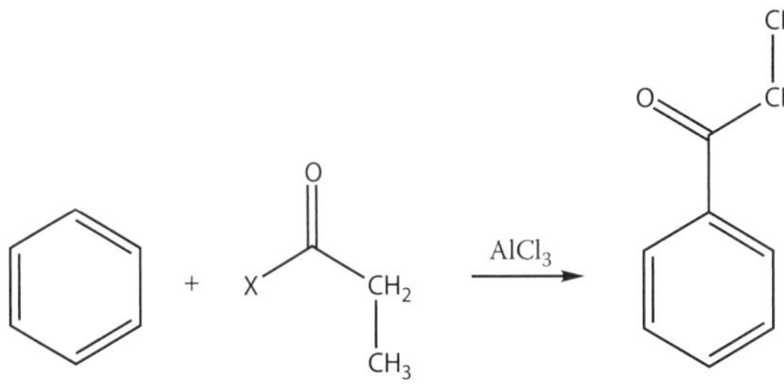

Friedel-Crafts acylation

The alkylation of an aromatic ring by an alkyl halide (order of reactivity F>Cl>Br>I) in the presence of a strong Lewis acid such as aluminum chloride; the acylation of an aromatic ring by an acyl halide (order of reactivity usually is I>Br>Cl>F) in the presence of a strong Lewis acid. Acids and acid anhydrides can replace the acyl halides. A related reaction is the Derzen-Nenitzescu ketone synthesis. See Olah, G.A., *Friedel-Crafts Chemistry*, John Wiley & Sons, New York, New York, 1973; Roberts, R.M. and Khalaf, A.A., *Friedel-Krafts Alkylation Chemistry: A Century of Discovery*, Marcel Dekker, New York, New York, 1989. See also Retey, J., Enzymatic catalysis by Friedel-Crafts-type reactions, *Naturwissenschaften* 83, 439-447, 1996; White, E.H., Darbeau, R.W., Chen, Y. *et al.*, A new look at the Friedel-Crafts alkylation reaction(1), *J.Org.Chem.* 61, 7986-7987, 1996; Studer, J., Purdie, N., and Krouse, J.A., Friedel-Crafts acylation as a quality control assay for steroids, *Appl.Spectrosc.* 57, 791-796, 2003; Retey, J., Discovery and role of methylidene imidazolone, a highly reactive electrophilic prosthetic group, *Biochim.Biophys.Acta* 1647, 179-184, 2003; Bandini, M., Melloni, A., and Umani-Ronchi, A., New catalytic approaches in the stereoselective Friedel-Crafts alkylation reaction, *Angew.Chem.Int.Ed.Engl.* 43, 550-556, 2004; Poppe, L. and Retey, J., Friedel-Crafts-type mechanism for the enzymatic elimination of ammonia from histidine and phenylalanine, *Angew.Chem.Int. Ed.Engl.* 44, 3668-3688, 2005; Keni, M., and Tepe, J.J., One-pot Friedel-Crafts/Robinson-Gabriel synthesis of oxazoles using oxazolone templates, *J.Org. Chem.* 70, 4211-4213, 2005; Movassaghi, M. and Ondrus, A.E., Enantioselective total synthesis of tricyclic myrmicarin alkaloids, *Org.Lett.* 7, 4423-4426, 2005; Paizs, C., Katona, A., and Retey, J., The interaction of heteroaryl-acrylates and alanines with phenylalanine ammonia-lyase form parsley, *Chemistry* 12, 2739-2744, 2006. Cuprous ions have been observed to promote a Friedel-Crafts acylation reaction (Kozikowski, A.P. and Ames, A., Copper(I) promoted acylation reactions. A transition metal mediated version of the Friedel-Crafts reaction, *J.Am.Chem.Soc.* 102, 860-862, 1980)

Friedländer Synthesis

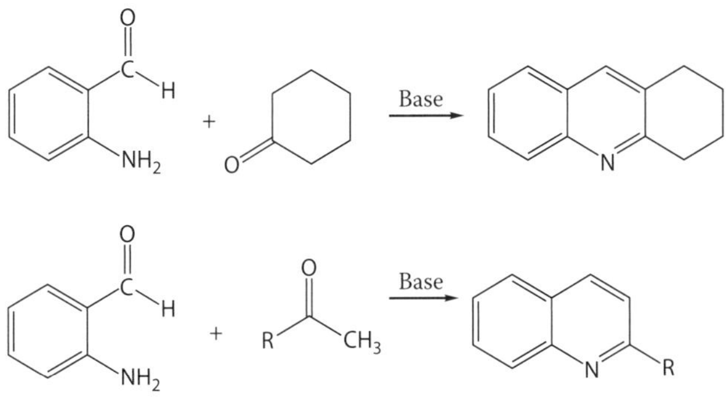

Friedlander synthesis

The base-catalyzed formation of quinoline derivatives by condensation of an *o*-aminobenzaldehyde with a ketone; also referred to as the Friedländer quinoline synthesis. The general utility of the reaction is somewhat limited by the availability of *o*-aminobenzaldehyde derivatives. See Maguire, M.P., Sheets, K.R., McVety, K., *et al.*, A new series of PDGF receptor tyrosine kinase inhibitors: 3-substituted quinoline derivatives, *J.Med.Chem.* 37, 2129-2137, 1994; Lindstrom, S., Friedlander synthesis of the food carcinogen 2-amino-1-methyl-6-phenylimidazo[4,5-*b*]pyridine, *Acta Chem.Scand.* 49, 361-363, 1995; Gladiali, S., Chelucci, G., Mudadu, M.S., *et al.*, Friedlander synthesis of chiral alkyl-substituted 1,10-phenanthrolines, *J.Org.Chem.* 66, 400-405, 2001; Patteux, C., Levacher, V., and Dupas, G., A novel traceless solid-phase Friedlander synthesis, *Org.Lett.* 5, 3061-3063, 2003; McNaughton, B.R. and Miller, B.L., A mild and efficient one-step synthesis of quinolines, *Org.Lett.* 5, 4257-4259, 2003; Yasuda, N., Hsiao, Y., Jensen, M.S., *et al.*, An efficient synthesis of an $\alpha_v\beta_3$ antagonist, *J.Org.Chem.* 69, 1959-1966, 2004.

ORGANIC NAME REACTIONS USEFUL IN BIOCHEMISTRY AND MOLECULAR BIOLOGY (Continued)

Fries Rearrangement

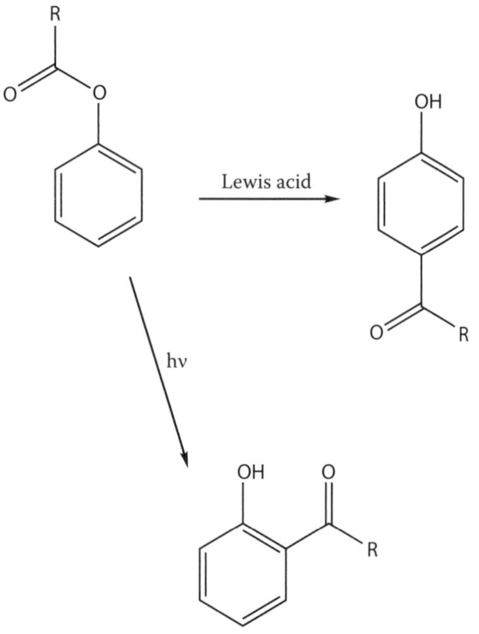

Rearrangement of a phenolic ester to yield *o*- and *p*-acylphenols. The distribution of products between the *ortho* and *para* acyl derivates depends on reaction conditions. The presence of solvent and a Lewis acid, the *para* product is preferred; with the photolytic process or at high temperature in the absence of solvent, the *ortho* derivative is preferred. See Sen, A.B. and Bhattacharji, S., Fries' rearrangement of aliphatic esters of *β*-naphthol, *Curr.Sci.* 20, 132-133, 1951; Iwasaki, S., Photochemistry of imidazolides. I. The photo-Fries-type rearrangement of *N*-substituted imidazoles, *Helv.Chim.Acta* 59, 2738-2752, 1976; Castell, J.V., Gomez, M.J., MIrabet, V., *et al.*, Photolytic degradation of benorylate: effects of the photoproducts on cultured hepatocytes, *J.Pharm.Sci.* 76, 374-378,1987; Climent, M.J. and Miranda, M.A. Gas chromatographic-mass spectrometric study of photodegradation of carbamate pesticides, *J.Chromatog.A.* 738, 225-231, 1996; Kozhevnikova, E.F., Derouane, E.G., and Kozhevnikov, I.V., Heteropoly acid as a novel efficient catalyst for Fries rearrangement, *Chem.Commun.* (11), 1178-1179, 2002; Dickerson, T.J., Tremblay, M.R., Hoffman, T.Z. *et al.*, Catalysis of the photo-Fries reaction: antibody-mediated stabilization ofhigh energy states, *J.Am.Chem.Soc.* 125, 15395-15401, 2003; Seijas, J.A., Vazquez-Tato, M.P., and Carballido-Reboredo, R., Solvent-free synthesis of functionalized flavones under microwave irradiation, *J.Org.Chem.* 70, 2855-2858, 2005;Canle Lopez, M., Fernandez, M.I., Rodriguez, S., *et al.*, Mechanisms of direct and TiO$_2$-photocatalyzed degradation of phenylurea herbicides, *Chemphyschem* 6, 2064-2074, 2005; Slana, G.B. de Azevedo, M.S., Lopes, R.S., *et al.*, Total syntheses of oxygenated brazanquinones via regioselective homologous anionic Fries rearrangement of benzylic *O*-carbamates, *Beilstein J.Org.Chem.* 2, 1, 2006.

Gabriel Synthesis

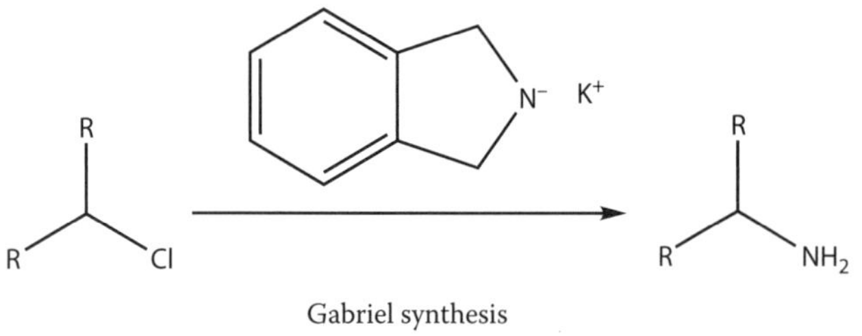

Gabriel synthesis

The conversion of an alkyl halide to alkyl amine mediated by potassium phthalimide. The intermediate product of the reaction of the alkyl halide and phthalimide is hydrolyzed to the product amine by acid or by reflux in ethanolic hydrazine. See Mikola, H. and Hanninen, E., Introduction of aliphatic amino and hydroxy groups to keto steroids using *O*-substituted hydroxylamines, *Bioconjugate Chem.* 3, 182-186, 1992; Groutas, W.C., Chong, L.S., Venkataraman, R., *et al.*, Mechanism-based inhibitors of serine proteinases based on the Gabriel-Colman rearrangement, *Biochem.Biophys.Res. Commun.* 194, 1491-1499, 1993; Konig, S., Ugi, I., and Schramm, H.J., Facile syntheses of C$_2$-symmetrical HIV-1 protease inhibitor, *Arch.Pharm.* 328, 699-704, 1995; Zhang, X.X. and Lippard, S.J., Synthesis of PDK, a novel porphyrin-linked dicarboxyate ligand, *J.Org.Chem.* 65, 5298-5305, 2000; Scozzafava, A. Saramet, I., Banciu, M.D., and Supuran, C.T., Carbonic anhydrase activity modulators: synthesis of inhibitors and activators incorporating 2-substituted-thiazol-4-yl-methyl scaffolds, *J.Enzyme Inhib.* 16, 351-358, 2001; Nicolaou, K.C., Hao, J., Reddy, M.V., *et al.*, Chemistry and biology of diazonamide A: second total synthesis and biological investigations, *J.Am.Chem.Soc.* 126, 12897-12906, 2004; Remond, C., Plantier-Royon, R., Aubry, N., and O'Donohue, M.J., An original chemoenzymatic rotue for the synthesis of *β*-D-galactofuranosides using an *α*-L-arabinofuranosidase, *Carbohydr.Res.* 340, 637-644, 2005; Pulici, M., Quartieri, F., and Felder, E.R., Trifluoroacetic acid anhydride-mediated solid-phase version of the Robison-Gabriel synthesis of oxazoles, *J.Comb.Chem.* 7, 463-473, 2005.

ORGANIC NAME REACTIONS USEFUL IN BIOCHEMISTRY AND MOLECULAR BIOLOGY (Continued)

Greiss Reaction

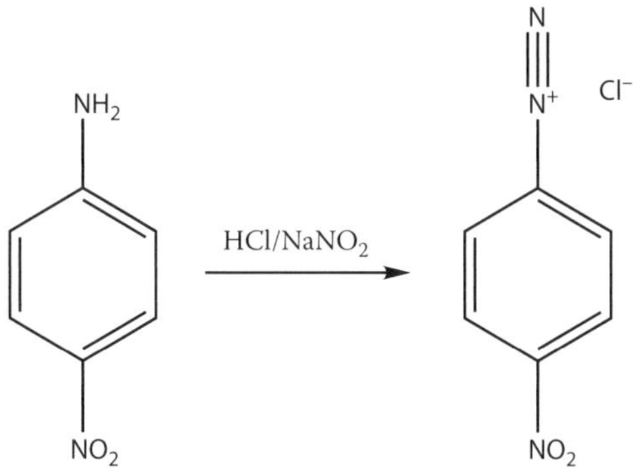

Greiss reaction

Greiss reaction as used for the measurement of nitrite

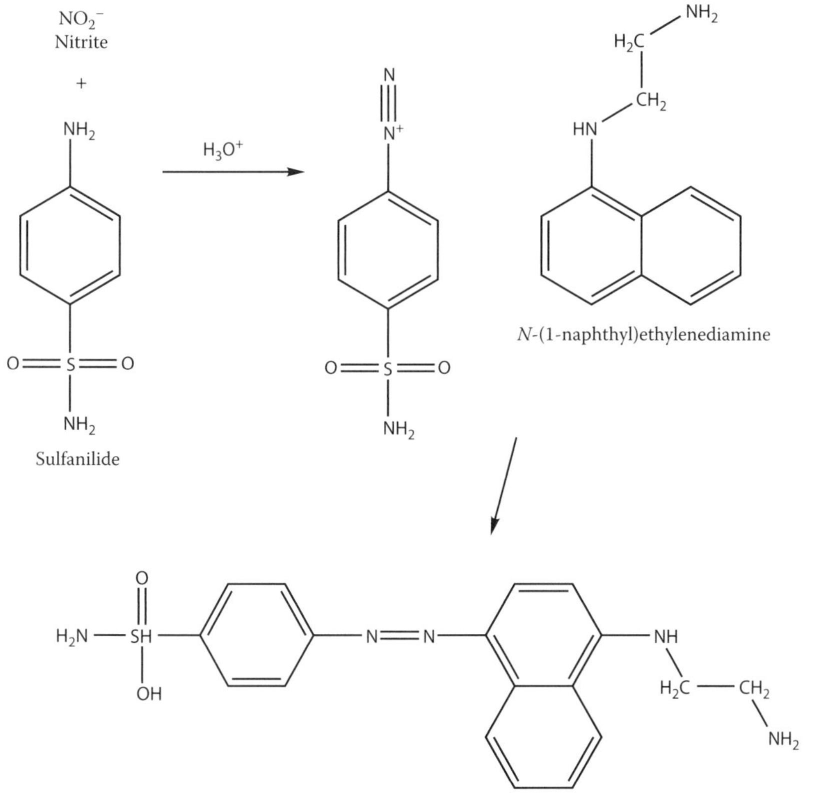

N-(1-naphthyl)ethylenediamine

Sulfanilide

Azo product measured at 520 nm

Diazotization of aromatic amines; used for the assay of nitrites in nitric oxide research. The assay for nitrates uses diazotization of sulfanilamide with subsequent coupling to an aromatic amine (*N*-1-naphthylethylenediamine) to form an chromophoric azo derivative. See Greenberg, S.S., Xie, J., Spitzer, J.J. *et al.*, Nitro containing L-arginine analogs interfere with assays for nitrate and nitrite, *Life Sci.* 57, 1949-1961, 1995; Pratt, P.F., Nithipatikom, K., and Campbell, W.B., Simultaneous determination of nitrate and nitrite in biological samples by multichannel flow injection analysis, *Anal.Biochem.* 231, 383-386, 1995; Tang, Y., Han, C. and Wang, X., Role of nitric oxide and prostaglandins in the potentiating effects of calcitonin gene-related peptide on Lipopolysaccharide-induced interleukin-6 release from mouse peritoneal macrophages, *Immunology* 96, 171-175, 1999; Baines, P.B., Stanford, S., Bishop-Bailey, D., *et al.*, Nitric oxide production in meningococcal disease is directly related to disease severity, *Crit.Care. Med.* 27, 1163-1165, 1999; Rabbani, G.H., Islam, S., Chowdhury, A.K., *et al.*, Increased nitrite and nitrate concentrations in sera and urine of patients with cholera or shigellosis, *Am.J.Gastroenterol.* 96, 467-472, 2001; Lee, R.H., Efron, D., Tantry, U. and Barbul, A., Nitric oxide in the healing wound: a time-course study, *J.Surg.Res.* 101, 104-108, 2001; Stark, J.M., Khan, A.M., Chiappetta, C.L., *et al*, Immune and functional role of nitric oxide in a mouse model of respiratory syncytial virus infection, *J.Infect.Dis.* 191, 387-395, 2005; Bellows, C.F., Alder, A., Wludyka, P., and Jaffe, B.M., Modulation of macrophage nitric oxide production by prostaglandin D2, *J.Surg.Res.* 132, 92-97, 2006. Diazotization of aromatic amines is also used for the the modification of proteins (Lundblad, R.L., *Chemical Reagents for Protein Modification*, CRC Press, Boca Raton, FL, 2004; Kennedy, J.H., Kricka, L.J., and Wilding, P., Protein-protein coupling reactions and the application of protein conjugates, *Clin.Chim.Acta* 70, 1-31, 1976; Sinnott, M.L., Affinity labeling via deamination reactions, *CRC Crit.Rev.Biochem.* 12, 327-372, 1982; Blair, A.H. and Ghose, T.I., Linkage of cytotoxic agents to

ORGANIC NAME REACTIONS USEFUL IN BIOCHEMISTRY AND MOLECULAR BIOLOGY (Continued)

immunoglobulins, *J.Immunol.Methods* 59, 129-143, 1983). While alkyl azides are unstable, carbonyl azides such as diazoacetyl derivatives have been used in the modification of proteins (Lundblad, R.L. and Stein, W.H., On the reaction of diazoacetyl compounds with pepsin, *J.Biol.Chem.* 244, 154-160, 1969; Keilova, H. and Lapresle, C., Inhibition of cathepsin E by diazoacetyl-norleucine methyl ester, *FEBS Lett.* 9, 348-350, 1970; Giraldi, T. and Nisi, C., Effects of cupric ions on the antitumour activity of diazoacetyl-glycine derivatives, *Chem.Biol.Interact.* 11,59-61, 1975; Kaehn, K., Morr, M., and Kula, M.R., Inhibition of the acid proteinase from *Neurospora crassa* by diazoaetyl-DL-norleucine methyl ester, 1,2-epoxy-3-(4-nitrophenoxy) propane and pepstatin, *Hoppe Seylers Z. Physiol.Chem.* 360, 791-794, 1979; Ouihia, A., René, L., Guilhem, J., *et al.*, A new diazoacylating reagent: Preparation, structure, and use of succinimidyl diazoacetate, *J.Org.Chem.* 58, 1641-1642, 1993.

Grignard Reagent or Grignard Reaction

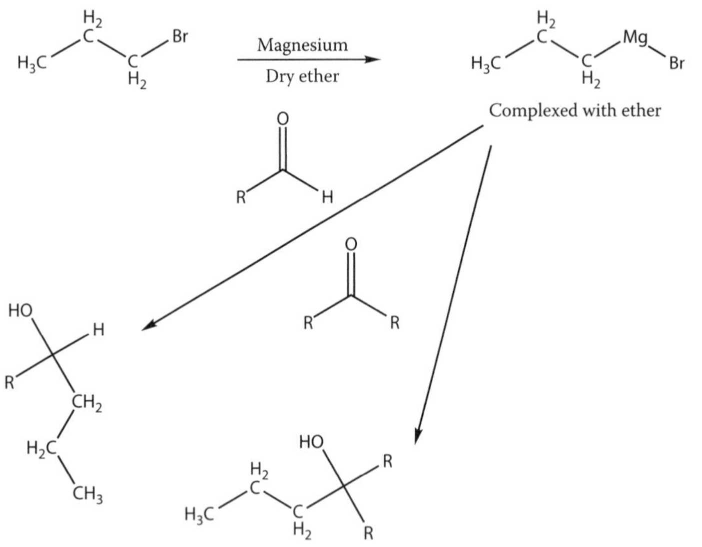

The reaction of alkyl or aryl halides with magnesium in dry ether to yield derivatives which can be used in a variety of organic synthetic reactions. See Nagano, T. and Hayashi, T., Iron-catalyzed Grignard cross-coupling with alkyl halides possessing beta-hydrogens, *Org.Lett.* 6, 1297-1299, 2004; Querner, C., Reiss, P., Bleuse, J., and Pron, A., Chelating ligands for nanocrystals' functionalization, *J.Am.Chem.Soc.* 126, 11574-11582, 2004; Agarwal, S. anad Knolker, H.J., A novel pyrrole synthesis, *Org.Biomol.Chem.* 2, 3060-3062, 2004; Hatano, M., Matsumara, T., and Ishihara, K., Highly alkyl-selective addition to ketones with magnesiumate complexes derived from Grignard reagents, *Org.Lett.* 7, 573-576, 2005; Itami, K., Higashi, S., Mineno, M., and Yoshida, J., Iron-catalyzed cross-coupling of alkenyl sulfides with Grignard reagents, *Org.Lett.* 7, 1219-1222, 2005; Wang, X.J., Zhang, L., Sun, X., *et al.*, Addition of Grignard reagents to aryl chlorides: an efficient synthesis of aryl ketones, *Org.Lett.* 7, 5593-5595, 2005; Hoffman-Emery, F., Hilpert, H., Scalone, M., and Waldmeier, F., Efficient synthesis of novel NK1 receptor antagonists: selective 1,4-additional of Grignard reagents to 6-chloronicotinic acid derivatives, *J.Org.Chem.* 71, 2000-2008, 2006; Werner, T. and Barrett, A.G., Simple method for the preparation of esters from Grigard reagents and alkyl 1-imidazolecarboxylates, *J.Org.Chem.* 71, 4302-4304, 2006; Demel, P., Keller, M., and Breit, B., *o*-DPPB-directed copper mediated and –catalyzed allylic substitution with Grignard reagents, *Chemistry* 12, 6669-6683, 2006.

Knoevenagel Reaction or Knoevenagel Condensation

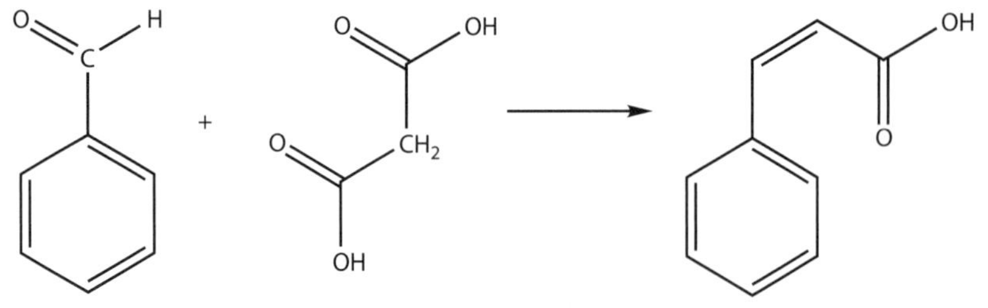

Knoevenagel condensation

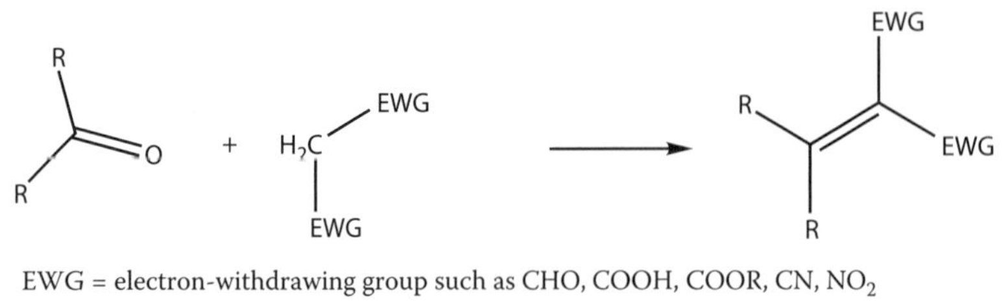

EWG = electron-withdrawing group such as CHO, COOH, COOR, CN, NO_2

ORGANIC NAME REACTIONS USEFUL IN BIOCHEMISTRY AND MOLECULAR BIOLOGY (Continued)

An amine-catalyzed reaction between active hydrogen compounds of the type Z-CH$_2$-Z where Z can be a CHO, COOH, COOR, NO$_2$,SOR, or related electron withdrawing groups and an aldehyde or ketone. For example, the reaction of malonic acid or malonic acid esters and an aldehyde or ketone to yield an α,β-unsaturated derivative. With malonic acid (Z is carboxyl group), decarboxylation occurs *in situ*. See March, J. *Advanced Organic Chemistry. Reactions, Mechanisms, and Structure,* 3rd edn., John Wiley & Sons, New York, New York, 1985; Klavins, M., Dipane, J., and Babre, K., Humic substances as catalysts in condensation reactions, *Chemosphere* 44, 737-742, 2001; Lai. S.M., Martin-Aranda, R., and Yeung, K.L., Knoevenagel condensation reaction in a membrane bioreactor, *Chem.Commun.* (2), 218-219, 2003; Pivonka, D.E. and Empfield, J.R., Real-time *in situ* Ramen analysis of microwave-assisted organic reactions, *Appl.Spectrosc.* 58, 41-46, 2004; Strohmeier, G.A., Haas, W., and Kappe, C.O., Synthesis of functionalized 1,3-thiazine libraries combining solid-phase synthesis and post-cleavage modification reactions, *Chemistry* 10, 2919-2926, 2004; Wirz, R., Ferri, D. and Baiker, A., ATR-IR spectroscopy of pendant NH$_2$ groups on silica involved in the Knoevenagel condensation, *Langmuir* 22, 3698-3706, 2006.

Leuckart Reaction

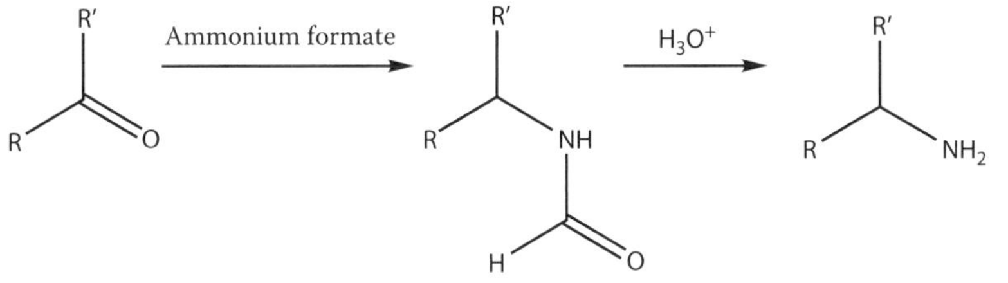

Leuckart reaction

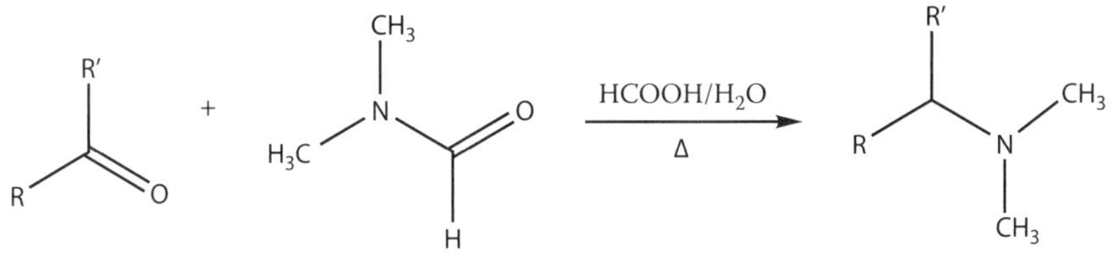

The reductive amination of carbonyl groups by ammonium formate or amine salts of formic acid; formamides may also be used in the reaction. See Matsueda, G.R. and Stewart, J.M., A *p*-methylbenzhydrylamine resin for improved solid-phase synthesis of peptide amides, *Peptides* 2, 45-50, 1981; Agwada, V.C. and Awachie, P.I., Intermediates in the Leuckart reaction of benzophenone with formamide, *Tetrahedron Lett.* 23, 779-780, 1982; Loupy, A., Monteux, D., Petit, A., *et al.,* Toward the rehabilitation of the Leuckart reductive amination reaction using microwave technology, *Tetrahedron Lett.* 37, 8177-8180, 1996; Adger, B.M., Dyer, U.C., Lennon, I.C., *et al.,* A novel synthesis of *tert*-leucine via a Leuckart type reaction, *Tetrahedron Lett.* 38, 2153-2154, 1997; Lejon, T. and Helland, I., Effect of formamide in the Leuckart reaction, *Acta Chem.Scand.* 53, 76-78, 1999; Kitamura, M.. Lee, D., Hayashi, S., *et al.,* Catalytic Leuckart-Wallach type reductive amination of ketones, *J.Org.Chem.* 67, 8685-8687, 2002; Swist, M., Wilamowski, J., and Parczewski, A., Basic and neutral route specific impurities in MDMA prepared by different synthesis methods. Comparison of impurity profiles, *Forensic Sci. Int.* 155, 100-111, 2005; Tournier, L. and Zard, S.Z., A practical variation on the Leuckart reaction, *Tetrahedron Lett.* 46, 971-973, 2005.

ORGANIC NAME REACTIONS USEFUL IN BIOCHEMISTRY AND MOLECULAR BIOLOGY (Continued)

Lossen Rearrangement

Active Site Serine

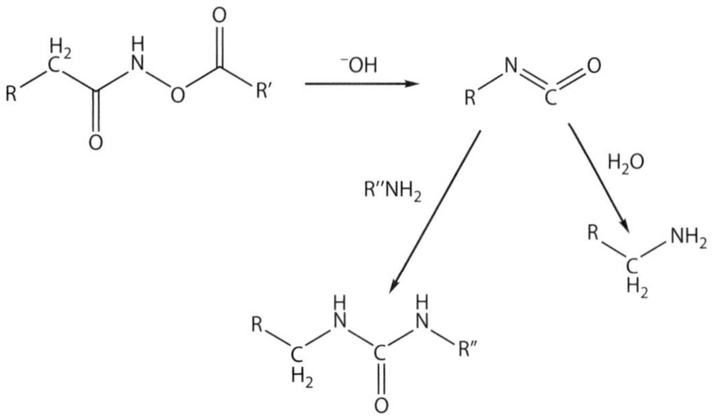

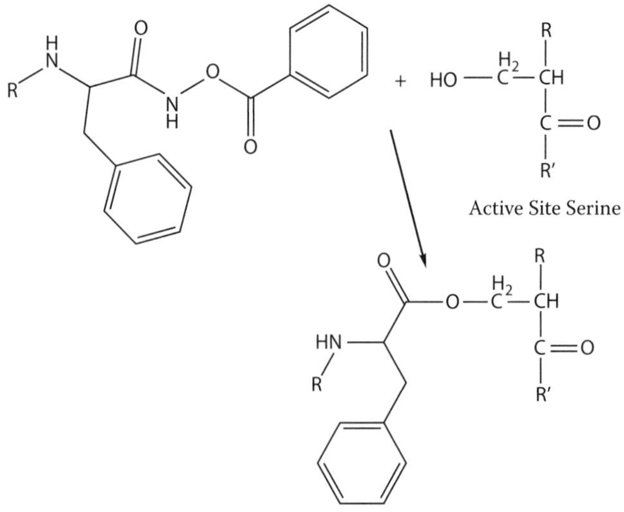

The formation of isocyanates on heating of *O*-acyl derivatives of hydroxamic acids or treatment by base. The isocyanate frequently adds water *in situ* to form an amine one carbon shorter that the parent compound; in the presence of amines, there is the formation of ureas. Andersen, W., The synthesis of phenylcarbamoyl derivatives by Lossen rearrangement of dibenzohydroxamic acid, *C.R.Trav.Lab.Carlsberg.* 30, 79-103, 1956; Gallop, P.M., Seifter, S., Lukin, M., and Meilman, E. Application of the Lossen rearrangement of dintirophenylhydroxamates to analysis of of carboxyl groups in model compounds and gelatin, *J.Biol.Chem.* 235, 2619-2627, 1960; Hoare, D.G., Olson, A., and Koshland, D.E., Jr., The reaction of hydroxamic acids with water-soluble carbodiimides. A Lossen rearrangement, *J.Am.Chem.Soc.* 90, 1638-1643, 1968; Dell, D., Boreham, D.R., and Martin, B.K., Estimation of 4-butoyphenylacetohydroxamic acid utilizing the Lossen rearrangement, *J.Pharm.Sci.* 60, 1368-1370, 1971; Harris, R.B. and Wilson, I.B., Glutamic acid is an active site residue of angiotensin I-converting enzyme. Use of the Lossen rearrangement for identification of dicarboxylic acid residues, *J.Biol. Chem.* 258, 1357-1362, 1983; Libert, R., Draye, J.P., Van Hoof, F., *et al.,* Study of reactions induced by hydroxylamine treatment of esters for organic acids and of 3-ketoacids: application to the study of urines from patients under valproate therapy, *Biol.Mass.Spectrom.* 20, 75-86, 1991; Neumann, U. and Gutschow, M., *N*-(sulfonyloxy)phthalimides and analogues are potent inactivators of serine proteases, *J.Biol.Chem.* 269, 21561-21567, 1994; Steinmetz, A.C., Demuth, H.U., and Ringe, D., Inactivation of subtilisin Carlsberg by *N*-[(*t*-butoxycarbonyl) alanylprolyl-phenylalanyl]-*O*-benzoylhydroxyl-amine: formation of a covalent enzyme-inhibitor linkage in the form of a carbamate derivative, *Biochemistry* 33, 10535-10544, 1994; Needs, P.W., Rigby, N.M., Ring, S.G., and MacDougall, A.J., Specific degradation of pectins via a carbodiimide-mediated Lossen rearrangement of methyl esterified galacturonic acid residues, *Carbohydr.Res.* 333, 47-58, 2001.

Maillard Reaction

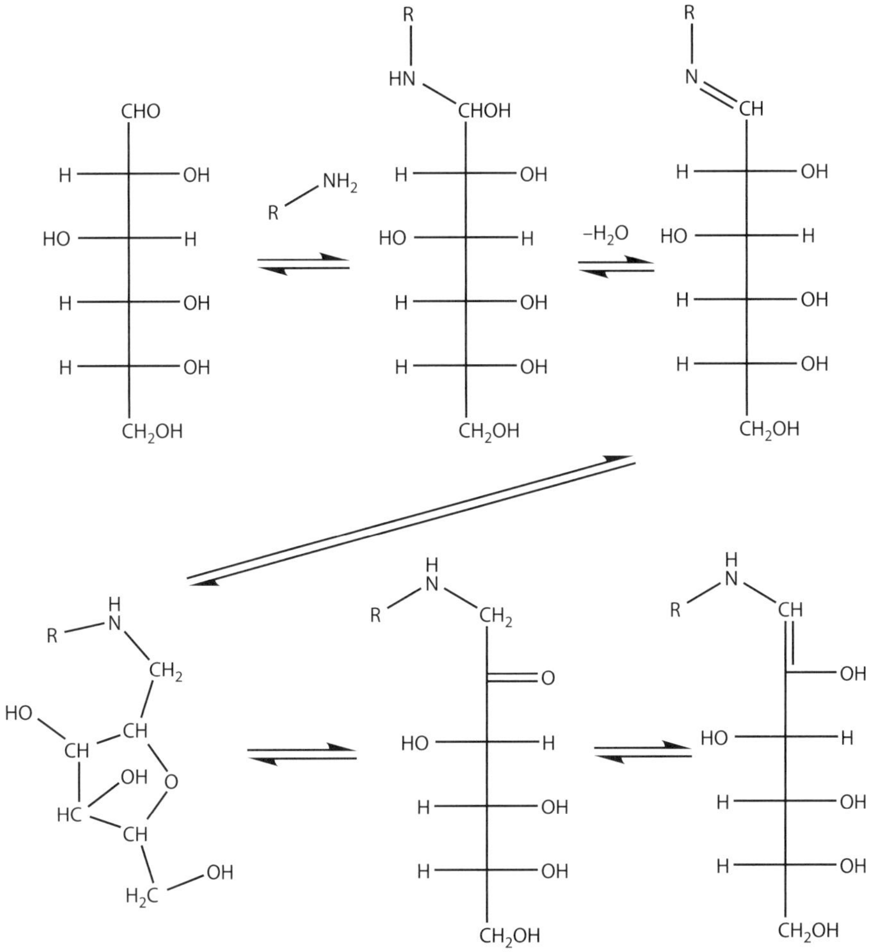

N-substituted glycosamine

Amadori product
N-substituted 1-amino-2-deoxy-2-ketose

The reaction of amino groups with carbonyl groups resulting in the formation of complex products. This process is involved in the tanning of leather and the Browning reaction which is considered unique to the reaction of carbohydrates with proteins and is a critical aspect of food preparation. The Maillard reaction involves the nonenzymatic reaction of sugars with proteins and the formation of advanced glycation end products (AGE products). The Maillard reaction results in the formation of a number of reaction products. See Dills, W.J., Jr., Protein fructosylation: fructose and the Maillard reaction, *Am.J.Clin.Nutr.* 58(Suppl 5), 779S-787S, 1993; Chuyen, N.V., Maillard reaction and food processing. Application aspects, *Adv.Exp.Med.Biol.* 434, 213-235, 1998; van Boekel, M.A., Kinetic aspects of the Maillard reaction: a critical review, *Nahrung* 45, 150-159, 2001; Horvat, S. and Jakas, A., Peptide and amino acid glycation: new insights into the Maillard reaction, *J.Pept.Sci.* 10, 119-137, 2004; Fay, L.B. and Brevard, H., Contribution of mass spectrometry to the study of the Maillard reaction in food, *Mass Spectrom.Rev.* 24, 487-507, 2005; Yaylayan, V.A., Haffenden, L., Chu, F.L., and Wnorowski, A., Oxidative pyrolysis and post pyrolytic derivatization techniques for the total analysis of Maillard model systems: investigations of control parameters of Maillard reaction pathways, *Ann.N.Y.Acad.Sci.* 1043, 41-54, 2005; Monnier, V.M., Mustata, G.T., Biemel, K.L., *et al.*, Cross-linked of the extracellular matrix by the Maillard reaction in aging and diabetes: an update on "a puzzle nearing resolution", *Ann.N.Y.Acad.Sci.* 1043, 533-544, 2005; Matiacevich, S.B., Santagapita, P.R., and Buera, M.P., Fluorescence from the Maillard reaction and its potential applications in food science, *Crit. Rev.Food Sci.Nutr.* 45, 483-495, 2005; van Boekel, M.A., Formation of flavour compounds in the Maillard reaction, *Biotechnol.Adv.* 24, 230-233, 2006.

ORGANIC NAME REACTIONS USEFUL IN BIOCHEMISTRY AND MOLECULAR BIOLOGY (Continued)

Malaprade Reaction

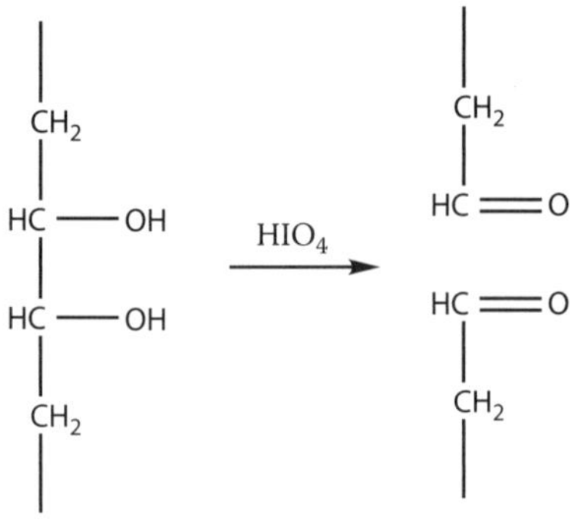

Malaprade reaction

Periodic cleavage of a diol; although this term is seldom used for this extremely common reaction, it would appear to be the correct terminology. Periodic acid is used for the diol cleavage in aqueous solvent while lead tetraacetate can be used in organic solvents. The reaction also occur an amine group vicinal to a hydroxyl function. It would appear that the term Malaprade reaction has been used more in description of analytical techniques for organic diols such as gluconic acid or in the assay of periodate. See Belcher, R., Dryhurst, G., and MacDonal, A.M., Submicro-methods for analysis of organic compounds. 22. Malaprade reaction, *Journal of the Chemical Society*,(July) 3964, 1965; Chen, K,P., Determination of calcium gluconate by selective oxidation with periodate, *J.Pharm.Sci.* 73, 681-683, 1984; Verma, K.K., Gupta, D., Sanghi, S.K., and Jain, A., Spectrophotometric determination of periodate with amodiaquine dihydrochloride and its application to the indirect determination of some organic-compounds via the Malaprade reaction, *Analyst* 112, 1519-1522, 1987; Nevado, J.J.B. and Gonzalez, P.V., Spectrophotometric determination of periodate with salicyaldehyde guanylhydrazone – indirect determination of some organic compounds using the Malaprade reaction, *Analyst* 114, 243-244, 1989; Jie, N,Q., Yang, D.L., Zhang, Q.N., *et al.*, Fluorometric determination of periodate with thiamine and its application to the determination of ethylene glycol and glycerol, *Anal.Chim.Acta* 359, 87-92, 1998; Guillan-Sans, R. and Guzman-Chozas, M., The thiobarbituric acid (TBA) reaction in foods, A review, *Crit.Rev.Food Sci.Nutrition* 38, 315-330, 1998; Pumera, M., Jelinek, I., Jindrich, J., *et al.*, Determination of cyclodextrin content using periodate oxidation by capillary electrophoresis, *J.Chromatog. A* 891, 201-206, 2000; Afkhami, A. and Mosaed, F., Kinetic determination of periodate based on its reaction with ferroin and its application to the indirect determination of ethylene glycol and glycerol, *Microchemical J.* 68, 35-40, 2001; Afkhami, A. and Mosaed, F., Sensitive kinetic-spectrophotometric determination of trace amounts of periodate ion, *J.Anal.Chem.* 58, 588-593, 2003; Mihovilovic, M.D., Spina, M., Muller, B., and Stanetty, P., Synthesis of carbo- and heterocyclic aldehydes bearing an adjacent donor group – Ozonolysis versus OsO₄/KIO₄-oxidation, *Monatshefte für Chemie* 135, 899-909, 2004.

Malonic Ester Synthesis

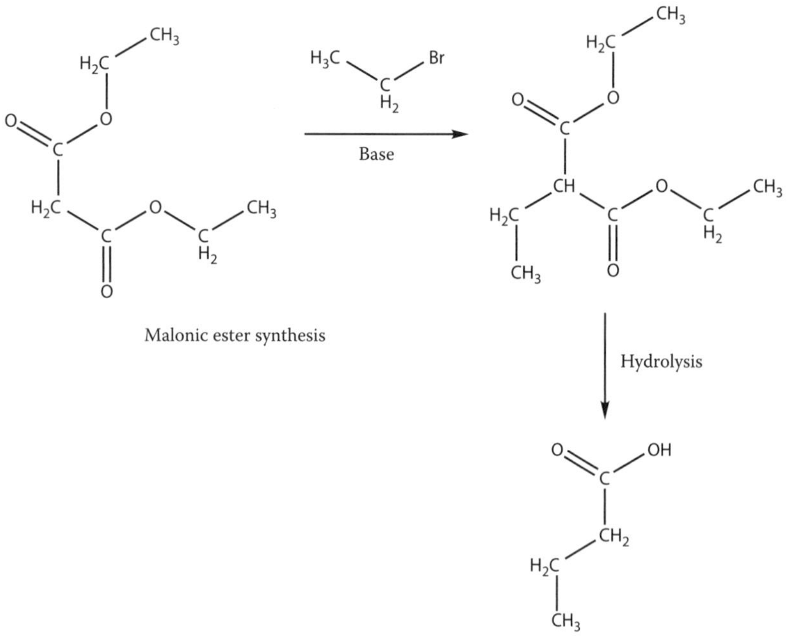

Malonic ester synthesis

The synthesis of a variety of derivatives taking advantage of the reactivity (acidity) of the methylene carbon in malonic esters. The malonic ester synthesis is related to the acetoacetic ester synthesis and the Knoevenagel synthesis. See Mizuno, Y., Adachi, K., and Ikeda, K., Studies on condensed systems of aromatic nitrogenous series. XIII. Extension of malonic ester synthesis to the heterocyclic series, *Pharm.Bull.* 2, 225-234, 1954; Beres, J.A., Varner, M.G., and Bria, C., Synthesis and cyclization of dialkylmalonuric esters, *J.Pharm.Sci.* 69, 451-454, 1980; Kinder, D.H., Frank, S.K., and Ames, M.M. Analogues of carbamyl asparate as inhibitors of dihydroorotase: preparation of boronic acid transition-state analogues and a zinc chelator

ORGANIC NAME REACTIONS USEFUL IN BIOCHEMISTRY AND MOLECULAR BIOLOGY (Continued)

carbamylhomocysteine, *J.Med.Chem.* 33, 819-823, 1990; Groth, T. and Meldal, M., Synthesis of aldehyde building blocks protected as acid labile *N*-boc-*N.O*-acetals: toward combinatorial solid phase synthesis of novel peptide isosteres, *J.Comb.Chem.* 3, 34-44, 2001; Hachiya, I., Ogura, K., and Shimizu, M., Novel 2-pyridine synthesis via nucleophilic addition of malonic esters to alkynyl imines, *Org.Lett.* 4, 2755-2757, 2002; Strohmeier, G.A., Haas, W., and Kappe, C.O., Synthesis of functionalized 1,3-thiazine libraries combining solid-phase synthesis and post-cleavage modification methods, *Chemistry* 10, 2919-2926, 2004.

Mannich Reaction

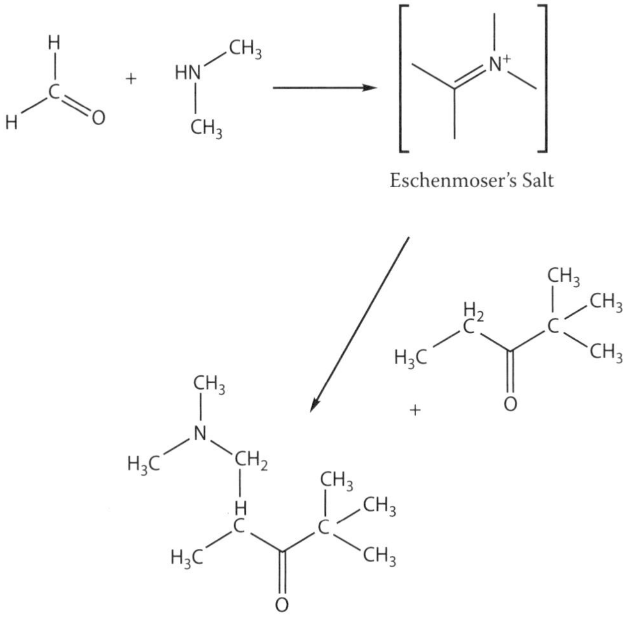

Eschenmoser's Salt

Condensation of an amine with an carbonyl compound which can exist in an enol form, and a carbonyl compound which can not exist as an enol. The reaction frequently use formaldehyde as the carbonyl compound not existing as an enol for condensing with a secondary amine in the first phase of the reaction. See Britton, S.B., Caldwell, H.C., and Nobles, W.L., The use of 2-pipecoline in the Mannich reaction, *J.Am.Pharm.Assoc.Am.Pharm.Assoc.* 43, 641-643, 1954; Nobles, W.L., and Thompson, B.B., Application of the Mannich reaction to sulfones. I. Reactive methylene moiety of sulfones, *J.Pharm. Sci.* 54, 576-580, 1965; Thompson, B.B., The Mannich reaction. Mechanistic and technological considrations, *J.Pharm.Sci.* 57, 715-733, 1968; Nobles, W.L. and Potti, N.D., Studies on the mechanism of the Mannich reaction, *J.Pharm.Sci.* 57, 1097-1103, 1968; Delia, T.J., Scovill, J.P., Munslow, W.D., and Burckhalter, J.H., Synthesis of 5-substituted aminomethyluracils via the Mannich reaction, *J.Med.Chem.* 19, 344-346, 1976; List, B., Pojarliev, P., Biller, W.T., and Martin, H.J., The proline-catalyzed direct asymmetric three-component Mannich reaction: scope, optimization, and application to the highly enantioselective synthesis of 1,2-amino alcohols, *J.Am.Chem.Soc.* 124, 827-833, 2002; Palomo, C., Oiarbide, M., Landa, A., *et al.*, Design and synthesis of a novel class of sugar-peptide hybrids: *C*-linked glyco β-amino acids through a stereoselective "acetate" Mannich reaction as the key strategic element, *J.Am.Chem.Soc.* 124, 8637-8643, 2002; Cordova, A., The direct catalytic asymmetric Mannich reaction, *Acc.Chem.Res.* 37, 102-112, 2004; Azizi, N., Torkiyan, L., and Saidi, M.R., Highly efficient one-pot three-component Mannich reaction in water catalyzed by heteropoly acids, *Org.Lett.* 8, 2079-2082, 2006; Matsuo, J., Tanaki, Y., and Ishibashi, H., Oxidative Mannich reaction of *N*-carbobenzoxy amines 1,3-dicarbonyl compounds, *Org.Lett.* 8, 4371-4374, 2006. Another important example of the Michael addition in biochemistry and molecular biology is the reaction of 4-hydroxynon-2-enal with amines and sulfydryl groups (Winter, C.K., Segall, H.J., and Haddon, W.F., Formation of cyclic adducts of deoxyguanosine with the aldehyde *trans*-4-hydroxy-2-hexenal and *trans*-4-hydroxy-2-nonenal *in vitro*, *Cancer Res.* 46, 5682-5686, 1986; Sayre, L.M., Arora, P.K., Iyer, R.S., and Salomon, R.G., Pyrrole formation from 4-hydroxyonenal and primary amines, *Chem.Res.Toxicol.* 6, 19-22, 1993; Hartley, D.P., Ruth, J.A., and Petersen, D.R., The hepatocellular metabolism of 4-hydroxynonenal by alcohol dehydrogenase, aldehyde dehydrogenase, and glutathione-*S*-transferase, *Arch.Biochem. Biophys.* 316, 197-205, 1995: Engle, M.R., Singh, S.P., Czernik, P.J., *et al.*, Physiological role of mGSTA4-4, a glutathione *S*-transferase metabolizing 4-hydroxynonenal: generation and analysis of mGst4 null mouse, *Toxicol.Appl.Pharmacol.* 194, 296-308, 2004).

Meerwein Reaction

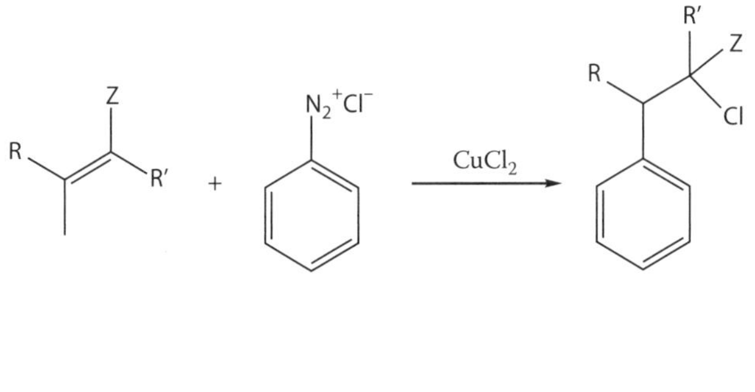

ORGANIC NAME REACTIONS USEFUL IN BIOCHEMISTRY AND MOLECULAR BIOLOGY (Continued)

The reaction of an aryl diazonium halide with an aliphatic unsaturated compound to yield an α-halo-β-phenyl alkene and alkanes. The reaction is performed in the presence of cupric ions. The presence of an electron-withdrawing group is useful in promoting the reactivity of the alkene. See Kochi, J.K., The Meerwein reaction. Catalysis by cuprous chloride, *J.Am.Chem.Soc.* 77, 5090, 1955; Moraes, L.A. and Eberlin, M.N., The gas-phase Meerwein reaction, *Chemistry* 6, 897-905, 2000; Riter, L.S., Meurer, E.C., Handberg, E.S., *et al.*, Ion/molecule reactions peformed in a miniature cylindrical ion trap mass spectrometer, *Analyst* 128, 1112-1118, 2003; Meurer, E.C., Chen, H., Riter, L.S., *et al.*, Meerwein reaction of phosphonium ions with epoxides and thioepoxides in the gas phase, *J.Am.Soc.Mass Spectrom.* 15, 398-405, 2004; Meurer, E.C. and Eberlin, M.N., The atmospheric pressure Meerwein reaction, *J.Mass Spectrom.* 41, 470-476, 2006.

Michael Addition (Michael Condensation)

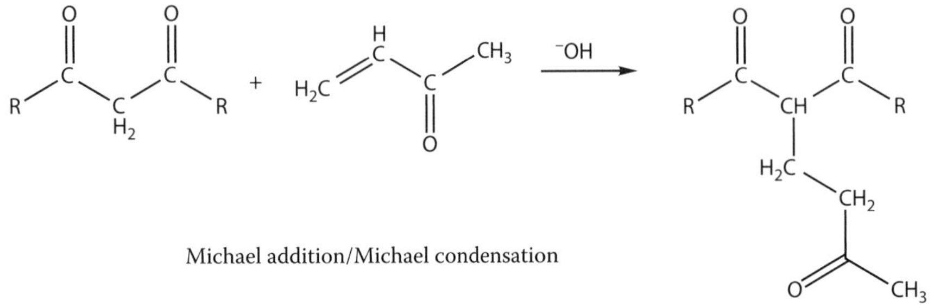

Michael addition/Michael condensation

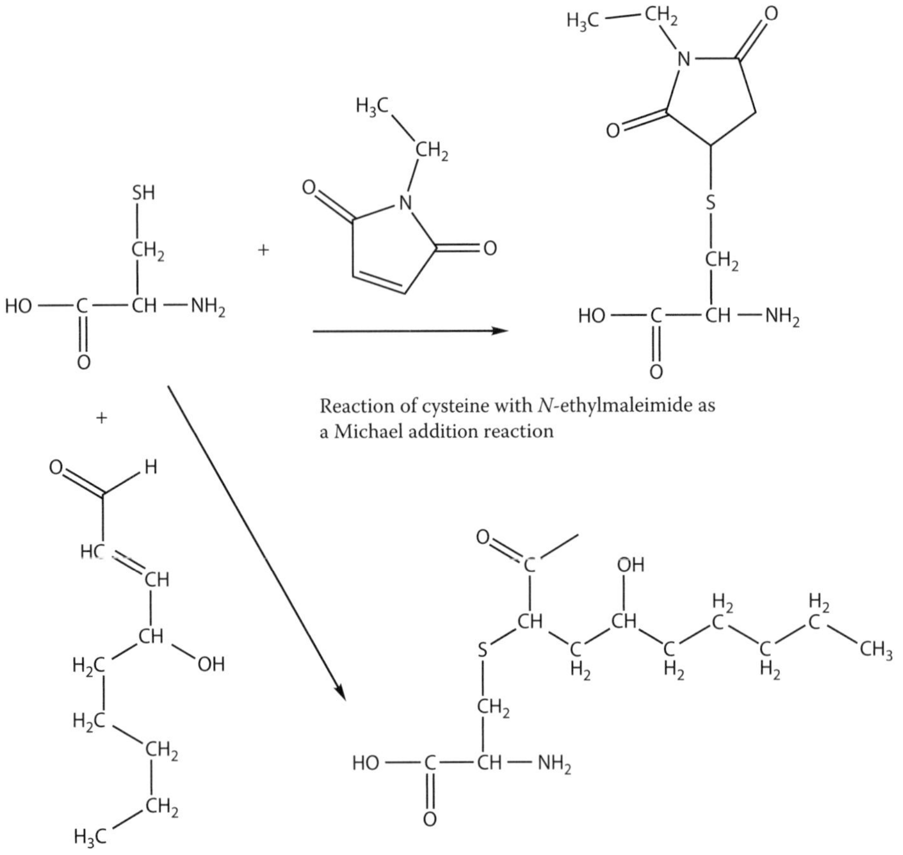

Reaction of cysteine with *N*-ethylmaleimide as
a Michael addition reaction

4-HNE

Formally a 1, 4 addition/conjugate addition of a resonance-stabilized carbanion (the reaction of an active methylene compound such as a malonate and an α,β-unsaturated carbonyl compound or the reaction of a nucleophile with an "activated unsaturated system; a carbanion defined as an anion with an even number of electrons). The addition of a nucleophile to a conjugated double bond. See Flavin, M. and Slaughter, C., Enzymatic elimination from a substituted four-carbon amino acid coupled to Michael addition of a β-carbon to an electrophilic double bond. Structure of the reaction product, *Biochemistry* 5, 1340-1350, 1966; Fitt, J.J. and Gschwend, H.W., α-Alkylation and Michael addition of amino acid—a practical method, *J.Org.Chem.* 42, 2639-2641, 1977; Powell, G.K., Winter, H.C., and Dekker, E.E., Michael addition of thiols with 4-methyleneglutamic acid: preparation of adducts, their properties and presence in peanuts, *Biochem.Biophys.Res.Commun.* 105, 1361-1367, 1982; Wang, M., Nishikawa, A. and Chung, F.L., Differential effects of thiols on DNA modifications via alkylation and Michael addition by α-acetoxy-*N*-nitrosopyrrolidine, *Chem.Res.Toxicol.* 5, 528-531, 1992; Jang, D.P., Chang, C.W., and Uang, B.J., Highly diastereoselective Michael addition of α-hydroxy acid derivatives and enantioselective synthesis of (+)-crobarbatic acid, *Org.Lett.* 3, 983-985, 2001;

ORGANIC NAME REACTIONS USEFUL IN BIOCHEMISTRY AND MOLECULAR BIOLOGY (Continued)

Naidu, B.N., Sorenson, M.E., Connolly, T.P., and Ueda, Y., Michael addition of amines and thiols to dehydroalanine amides: a remarkable rate acceleration in water, *J.Org.Chem.* 68, 10098-10102, 2003; Ooi, T., Doda, K., and Maruoka, K., Highly enantioselective Michael addition of silyl nitronates to α,β-unsaturated aldehydes catalyzed by designer chiral ammonium bifluorides: efficient access to optically active γ-nitro aldehydes and their enol silyl ethers, *J.Am.Chem.Soc.* 125, 9022-9023, 2003; Weinstein, R., Lerner, R.A., Barbas, C.F., 3rd, and Shabat, D., Antibody-catalyzed asymmetric intramolecular Michael additional of aldehydes and ketones to yield the disfavored *cis*-product, *J.Am.Chem.Soc.* 127, 13104-13105, 2005; Ding, R., Katebzadeh, K., Roman, L., *et al.,* Expanding the scope of Lewis acid catalysis in water: remarkable ligand acceleration of aqueous ytteribium triflate catalyzed Michael addition reactions, *J.Org.Chem.* 71, 352-355, 2006; Pansare, S.V. and Pandya, K., Simple diamine- and triamine-protonic acid catalysts for the enantioselective Michael addition of cyclic ketones to nitroalkenes, *J.Am.Chem.Soc.* 128, 9624-9625, 2006; Dai, H.X., Yao, S.P., and Wang, J., Michael addition of pyrimidine with disaccharide acrylates catalyzed in organic medium with lipase M from *Mucor javanicus, Biotechnol. Lett.* 28, 1503-1507, 2006. One the best examples in biochemistry is the modification of cysteine residues with *N*-alkylmaleimide derivatives (Lundblad, R.L., *Chemical Reagents for Protein Modification*, 3rd edn., CRC Press, Boca Raton, FL, 2004; Heitz, J.R., Anderson, C.D., and Anderson, B.M., Inactivation of yeast alcohol dehydrogenase by *N*-alkylmaleimides, *Arch.Biochem.Biophys.* 127, 627-636, 1968; Smyth, D.B. and Tuppy, H., Acylation reactions with cyclic imides, *Biochim.Biophys.Acta* 168, 173-180, 1968; Lusty, C.J. and Fasold, H., Characterization of sulfhydryl groups of actin, *Biochemistry* 8, 2933-2939, 1969; Bowes, T.J. and Gupta, R.S., Induction of mitochondrial fusion of cysteine-alklyators ethacrynic acid and *N*-ethylmaleimide, *J.Cell Physiol.* 202, 796-804, 2005).

Reformatsky Reaction

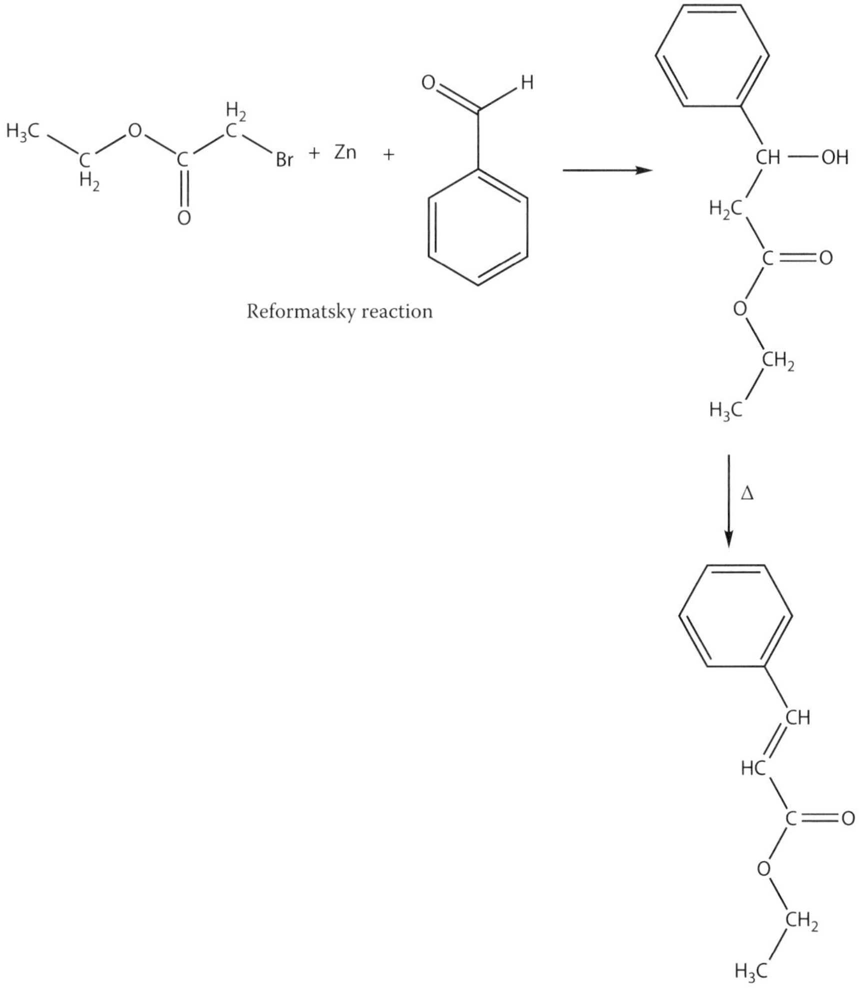

Reformatsky reaction

Formation of a complex between zinc and an α-bromoester followed by condensation with an aldehyde yielding a β-hydroxyester; an α,β-unsaturated ester via dehydration following the condensation reaction. See Tanabe, K., Studies on vitamin A and its related compounds. II. Reformatsky reaction of β-cyclocitral with methyl γ-bromosenecioate, *Pharm.Bull.* 3, 25-31, 1955; Ross, N.A. and Bartsch, R.A., High-intensity ultrasound-promoted Reformatsky reactions, *J.Org.Chem.* 68, 360-366, 2003; Jung, J.C., Lee, J.H., Oh., S., Synthesis and antitumor activity of 4-hydroxycoumarin derivatives, *Bioorg.Med.Chem.Lett.* 14, 5527-5531,2004; Kloetzing, R.J., Thaler, T., and Knochel, P., An improved asymmetric Reformatsky reaction mediated by (-)-*N,N*-dimethylaminoisoborneol, *Org.Lett.* 8, 1125-1128, 2006; Moume, R. Laavielle, S., and Karoyan, P., Efficient synthesis of β_2-amino acid by homologation of α-amino acids involving the Reformatsky reaction and Mannich-type imminium electrophile, *J.Org.Chem.* 71, 3332-3334, 2006.

ORGANIC NAME REACTIONS USEFUL IN BIOCHEMISTRY AND MOLECULAR BIOLOGY (Continued)

Rittter Reaction

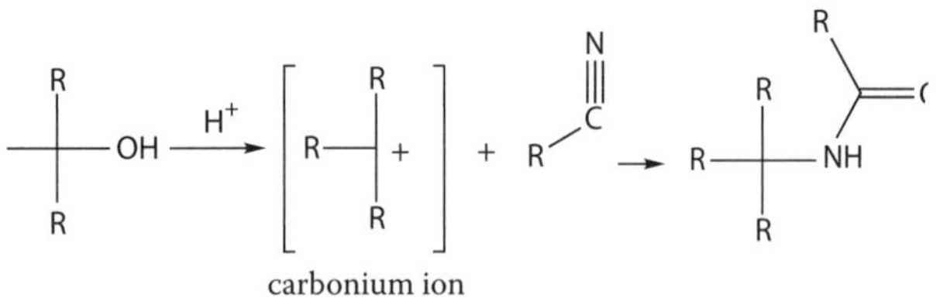

carbonium ion

Acid-catalyzed nucleophilic addition of a nitrile to a carbenium ion generated from alcohol (usually tertiary, primary alcohols other than benzyl alcohol will not react) yielding an amide. Sanguigni, J.A. and Levine, R., Amides from nitriles and alcohols by the Ritter reaction, *J.Med.Chem.* 53, 573-574, 1964; Radzicka, A. and Konieczny, M., Studies on the Ritter reaction. I. Synthesis of 3-/5-bartbituryl/-1propanesulfonic acids with anti-inflammatory activity, *Arch.Immunol.Ther.Exp.* 30, 421-432, 1982; Van Emelen, K., De Wit, T., Hoornaert, G.J., and Compernolle, F., Diastereoselective intramolecular Ritter reaction: generation of a *cis*-fused hexahydro-4a*H*-indeno[1,2-*b*] pyridine ring system with 4a,9b-diangular substituents, *Org. Lett.* 2, 3083-3086, 2000; Concellon, J.M., Reigo, E., Suarez, J.R., *et al.*, Synthesis of enantiopure imidazolines through a Ritter reaction of 2-(1-aminoalkyl)azirdines with nitriles, *Org.Lett.* 6, 4499-4501, 2004; Feske, B.D., Kaluzna, I.A., and Stewart, J.D., Enantiodivergent, biocatalytic routes to both taxol side chain antipodes, *J.Org.Chem.* 70, 9654-9657, 2005; Crich, D. and Patel, M., On the nitrile effect in L-rhamnopyranosylation, *Carbohydr.Res.* 341, 1467-1475, 2006; Fu, Q. and Li, L., Neutral loss of water from the b ions with histidine at the *C*-terminus and formation of the c ions involving lysine side chains, *J.Mass.Spectrom.* 41, 1600–1607, 2006.

Schiff Base

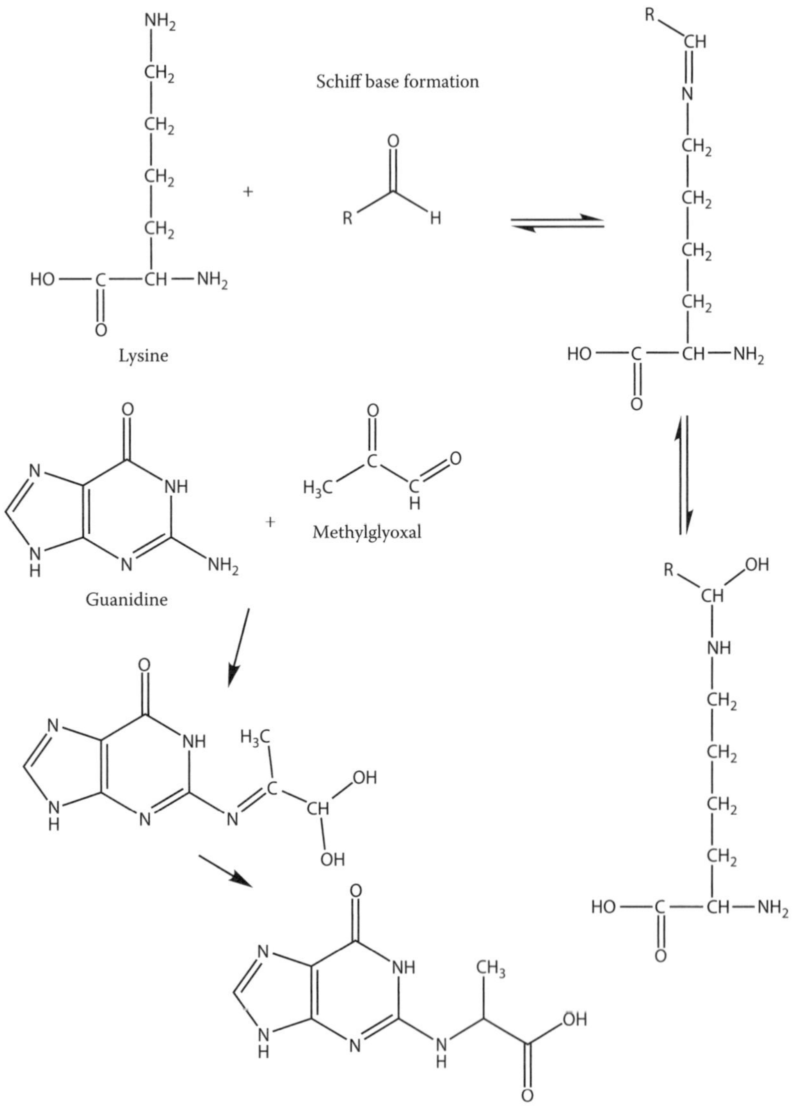

Schiff base formation

Lysine

Guanidine

Methylglyoxal

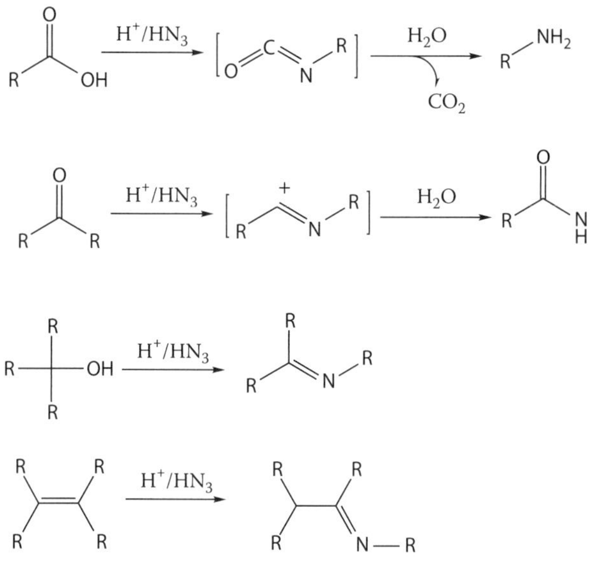

The formation of an unstable derivative generally between an carbonyl(usually an aldehyde) and an amino group. The Schiff base can be converted to a stable derivative by reduction with sodium borohydride or sodium cyanoborohydride; Schiff bases appear to be resistant to reduction with sulfhydryl-base reducing agents such as 2-mercaptoethanol or dithiothreitol and phosphines. Schiff bases are involved in a diverse group of biochemical events including the interaction of pyridoxal phosphate with proteins, the interaction of reducing carbohydrates with proteins in reaction leading to AGE products, and reductive alkylation of amino groups in proteins. See Feeney, R.E., Blankenhorn, G., and Dixon, H.B., Carbonyl-amine reactions in protein chemistry, *Adv.Protein.Chem.* 29, 135-203, 1975; Metzler, D.E. Tautomerism in pyridoxal phosphate and in enzymatic catalysis, *Adv.Enzymol. Relat.Areas Mol.Biol.* 50, 1-40, 1979; Puchtler, H. and Meloan, S.N., ON Schiff's bases and aldehyde-fuchsin: a review from H.Schiff to R.D. Lillie, *Histochemistry* 72, 321-332, 1981; O'Donnell, J.P., The reaction of amines with carbonyls: its significance in the nonezymatic metabolism of xenobiotics, *Drug.Metab.Rev.* 13, 123-159, 1982; Stadtman, E.R., Covalent modification reactions are marking steps in protein turnover, *Biochemistry* 29, 6232-6331, 1990; Tuma, D.J., Hoffman, T. and Sorrell, M.F., The chemistry of aldehyde-protein adducts, *Alcohol Alcohol Suppl.* 1, 271-276, 1991; Hargrave, P.A., Hamm, H.E., and Hofmann, K.P., Interaction of rhodopsin with the G-protein, transducin, *Bioessays* 15, 43-50, 1993; Chen, H. and Rhodes, J., Schiff base forming drugs: mechanisms of immune potentiation and therapeutic potential, *J.Mol.Med.* 74, 497-504, 1996; Yim, M.B., Yim, H.S., Lee, C., *et al.*, Protein glycation: creation of catalytic sites for free radication generation, *Ann.N.Y.Acad.Sci.* 928, 48-53, 2001; Gramatikova, S., Mouratou, B., Stetefeld, J. *et al.*, Pyridoxal-5'-phosphate-dependent catatlytic antibodies, *J.Immunol.Methods* 269, 99-110, 2002; Schaur, R.J., Basic aspects of the biochemical reactivity of 4-hydroxynonenal, *Mol.Aspects Med.* 24, 149-159, 2003; Kurtz, A.J. and Lloyd, R.S., 1, N^2-deoxyguanosine adducts of acrolein, crotonaldehyde, and *trans*-4-hydroxynonenal cross-link to peptides via Schiff base linkage, *J.Biol.Chem.* 278, 5970-5975, 2003; Kandori, H., Hydration switch model for the proton transfer in the Schiff base region of bacteriorhodopsin, *Biochim.Biophys.Acta* 1658, 72-79, 2004; Hadjoudis, E. and Mavridis, I.M., Photochomism and thermochromism of Schiff bases in the solid state: structural aspects, *Chem.Soc.Rev.* 33, 579-588, 2004; Stadler, R.H., Acrylamide formation in different foods and potential strategies for reduction, *Adv.Expt.Med.Biol.* 561, 157-169, 2005. There is some interesting chemistry on Schiff bases in inorganic chemistry (Nakoji, M., Kanayama, T., Okino, T., and Takemoto, Y., Chiral phosphine-free Pd-mediated asymmetric allylation of prochiral enolate with a chiral phase-transfer catalyst, *Org.Lett.* 2, 3329-3331, 2001; Walther, D., Fugger, C. Schreer, H., *et al.*, Reversible fixation of carbon dioxide at nickel(0) centers: a route for large organometallic rings, dimers, and tetramers, *Chemistry* 7, 5214-5221, 2001; Benny, P.D., Green, J.L., Engelbrecht, H.P., Reactivity and rhenium(V) oxo Schiff base complexes with phosphine ligands: rearrangement and reduction reactions, *Inorg.Chem.* 44, 2381-2390, 2005).

ORGANIC NAME REACTIONS USEFUL IN BIOCHEMISTRY AND MOLECULAR BIOLOGY (Continued)

Schmidt Reaction/Schmidt Rearrangement

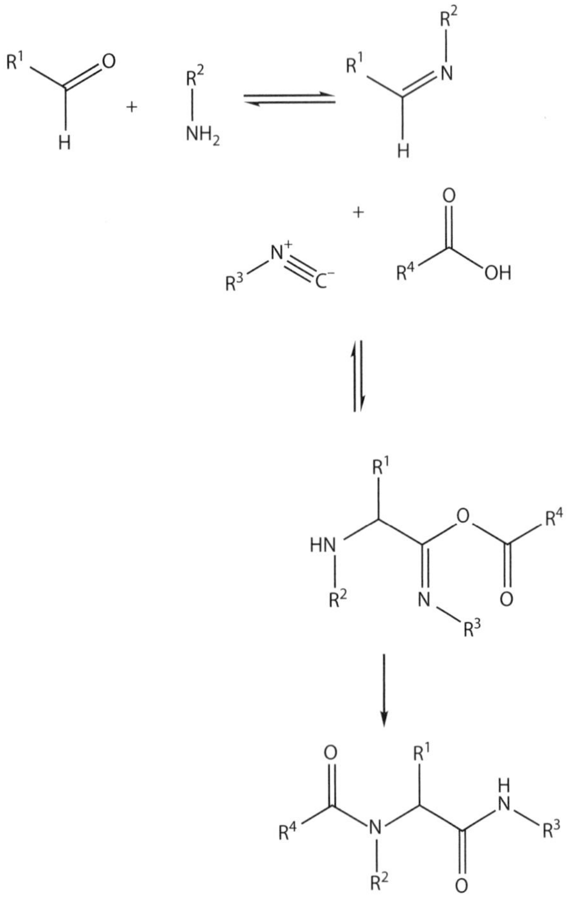

Used to describe the reaction of carboxylic acids, aldehyde and ketones(carbonyl compounds), and alcohols/alkenes with hydrazoic acid. Reaction with carboxylic acids yields amines, carbonyl compounds yield amides in a reaction involving a rearrangement, and alcohols/azides yield alkyl azides. See Rabinowitz, J.L., Chase, G.D., and Kaliner, L.F., Isotope effects of in the decarboxylation of 1-^{14}C-dicarboxylic acids studied by means of the Schmidt reaction, *Anal.Biochem.* 19, 578-583, 1967; Iyengar, R., Schildknegt, K., and Aube, J., Regiocontrol in an intramolecular Schmidt reaction: total synthesis of (+)-aspidospermidine, *Org.Lett.* 2, 1625-1627, 2000; Sahasrabudhe, K., Gracias, V., Furness, K., *et al.*, Asymmetric Schmidt reaction of hydroxyalkyl azides with ketones, *J.Am.Chem.Soc.* 125, 7914-7922, 2003; Wang, W., Mei, Y., Li, H., and Wang, J., A novel pyrrolidine imide catalyzed direct formation of α,β-unsaturated ketones from unmodified ketones and aldehydes, *Org.Lett.* 7, 601-604, 2005; Brase, S., Gil, C., Knepper, K., and Zimmerman, V., Organic azides: an exploding diversity of a unique class of compounds, *Angew.Chem.Int.Ed.Engl.* 44, 5188-5240, 2005; Lang, S. and Murphy, J.A., Azide rearrangements in electron-deficient systems, *Chem.Soc.Rev.* 35, 146-156, 2006; Zarghi, A., Zebardast, T., Hakimion, F., *et al.*, Synthesis and biological evaluation of 1,3-diphenylprop-2-en-1-ones possessing a methanesulfonamido or an azido pharmacophore as cyclooxygenase-1/-2 inhibitors, *Bioorg.Med.Chem.* 14, 7044-7050, 2006.

Ugi Condensation: A four component (aldehyde, amine, isocyanide and a carboxyl group) condensation resulting in an α-aminoacyl amide. See Liu, X.C., Clark, D.S., and Dordick, J.S., Chemoenzymatic construction of a four-component Ugi combinatorial library, *Biotechnol.Bioeng.* 69, 457-460, 2000; Bayer, T., Riemer, C., and Kessler, H., A new strategy for the synthesis of cyclopeptides containing diaminoglutaric acid, *J.Pept.Sci.* 7, 250-261, 2001; Crescenzi, V., Francescangeli, A., Renier, D., and Bellini, D., New cross-linked and sulfated derivatives of partially deacylated hyaluronan: synthesis and preliminary characterization, *Biopolymers* 64, 86-94, 2002; Liu, L., Ping Li, C., Cochran, S., and Ferro, V., Application of the four-component Ugi condensation for the preparation of glycoconjugate libraries, *Bioorg.Med.Chem.Lett.* 14, 2221-2226, 2004; Bu, H., Kjoniksen, A.L., Knudsen, K.D., and Nystrom, B., Rheological and structural properties of aqueous alginate during gelation via the Ugi multicomponent condensation reaction, *Biomacromolecules* 5, 1470-1479, 2004; Tempest, P.A., Recent advances in heterocycle generation using the efficient Ugi multiple-component condensation reaction, *Curr.Opin.Drug Discov.Devel.* 8, 776-788, 2005.

ORGANIC NAME REACTIONS USEFUL IN BIOCHEMISTRY AND MOLECULAR BIOLOGY (Continued)

Wittig Olefination

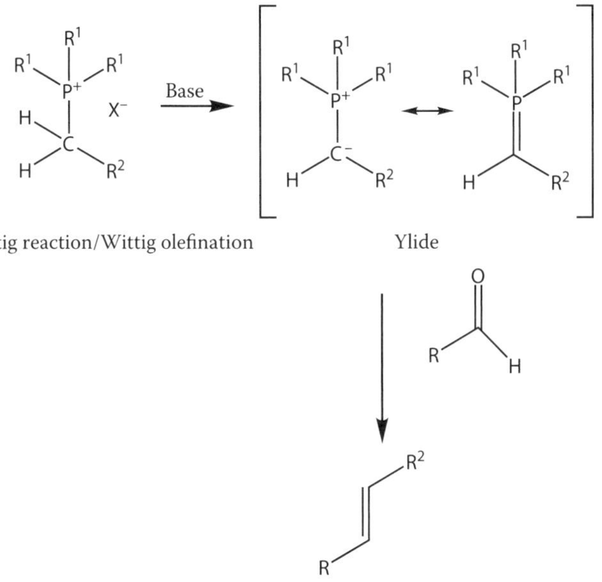

Wittig reaction/Wittig olefination   Ylide

Synthesis of an alkene from the reaction of an aldehyde or ketone with an ylide (internal salt) generated from a phosphophonium salt. See Jorgensen, M., Iversen, E.H., and Madsen, R., A convenient route to higher sugars by two-carbon chain elongation using Wittig/dihydroxylation reactions, *J.Org. Chem.* 66, 4625-4629, 2001; Magrioti, V., and Constantinou-Kokotou, V., Synthesis of (S)-α-amino oleic acid, *Lipids* 37, 223-228, 2002; van Staden, L.F., Gravestock, D., and Ager, D.J., New developments in the Peterson olefination reaction, *Chem.Soc.Rev.* 31, 195-200, 2002; Han, H., Sinha, M.K., D'Sousa, L.J., *et al.*, Total synthesis of 34-hydroxyasimicin and its photoactive derivative for affinity labeling of the mitochondrial complex I, *Chemistry* 10, 2149-2158, 2004; Rhee, J.U. and Krische, M.J., Alkynes as synthetic equivalents to stabilized Wittig reagents: intra- and intermolecular carbonyl olefinations catalyzed by Ag(1), BF$_3$, and HBF$_4$, *Org.Lett.* 7, 2493-2495, 2005; Ermolenko, L. and Sasaki, N.A., Diastereoselective synthesis of all either *l*-hexoses from L-ascorbic acid, *J.Org.Chem.* 71, 693-703, 2006; Halim, R., Brimble, M.A., and Merten, J., Synthesis of the ABC tricyclic fragment of the pectenotoxins via stereocontrolled cyclization of a γ-hydroxyepoxide appended to the AB spiroacetal unit, *Org.Biomol.Chem.* 4, 1387-1399, 2006; Phillips, D.J., Pillinger, K.S., Li, W., *et al.*, Desymmerization of diols by a tandem oxidation/Wittig olefination reaction, *Chem.Commun.* (21), 2280-2282, 2006; Modica, E., Compostella, F. Colombo, D., *et al.*, Stereoselective synthesis and immunogenic activity of the *C*-analogue of sulfatide, *Org. Lett.* 8, 3255-3258, 2006.

ENZYMES IN SYNTHETIC ORGANIC CHEMISTRY[a,b]

The specificity of enzymes has proved useful in organic synthesis where stereochemistry is critical for success.

Aldolase/Aldol Condensation   Catalysis of Aldol Condensation

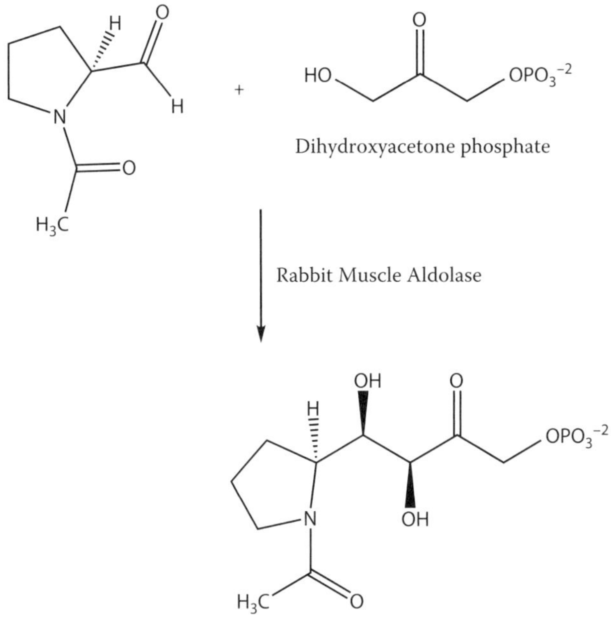

Dihydroxyacetone phosphate

Rabbit Muscle Aldolase

Austin, M.B., Izumikawa, M., Bowman, M.E., Crystal structure of a bacterial type III polyketide synthase and enzymatic control of reactive polyketide intermediates, *J.Biol.Chem.* 279, 45162-45174, 2004; Xiang, L., Kokaitzis, J.A, and Moore, B.S., EncM, a versatile enterocin biosynthetic enzyme involved in Favorskii oxidative rearrangement, aldol condensation, and heterocyclic-forming reactions, *Proc.Nat.Acad.Sci.* 101, 15609-15614, 2004; Suzuki, H., Ohnishi, Y. Fursho. Y., *et al.,* Novel benzene ring biosynthesis from C(3) and C(4) primary metabolites by two enzymes, *J.Biol.Chem.* 281, 36944-36951, 2006; Zhang, W., Watanabe, K., Wang, C.C., and Tang, Y., Heterologous biosynthesis of amidated polyketides with novel cyclization regioselectivity from oxytetracycline polyketide synthase, *J.Natl.Prod.* 69, 1633-1636, 2006; Williams, G.J., Woodhall, T., Farnsworth, L.M., *et al.,* Creation of a pair of stereochemically complementary biocatalysts, *J.Am.Chem.Soc.* 128, 16238-16247, 2006; Schetter, B. and Mahrwald, R., Model aldol methods for the total synthesis of polyketides, *Angewandte Chem. Int.Ed.* 45, 7506-7535, 2006; Suzuki, H., Ohnishi, Y., Furusho, Y., *et al.,* Novel benzene ring biosynthesis from C_3 and C_4 primary metabolites by two enzymes, *J.Biol.Chem.* 281, 36944-39511, 2007.

---

[a] *Handbook of Enzyme Biotechnology*, 2nd edn., ed. A. Wiseman, Ellis Horwood, Ltd., Chichester, UK, 1985; Laskin, A.T., *Enzymes and Immobilized Cells in Biotechnology*, Benjamin Cummings, Menlo Park, CA, USA, 1985; *Biocatalysis in Organic Media*, ed. C. Laane and J. Trayser, Elsevier, Amsterdam, Netherlands, 1987; Halgaš, J. *Biocatalysis in Organic Synthesis*, Elsevier, Amsterdam, Netherlands, 1992; Holland, H.L., *Organic Synthesis with Oxidative Enzymes*, VCH Publishers, New York, NY, USA, 1992; *Enzyme Catalysis in Organic Synthesis. A Comprehensive Handbook*, ed. K. Drauz and H. Waldman, VCH Verlagesellschaft, Weinheim, Germany, 1995; *Bioorganic Chemistry Peptides and Proteins*, ed S.M. Hecht, Oxford University Press, New York, NY, USA, 1998; Faber, K., *Biotransformations in Organic Chemistry*, 5th edn., Springer-Verlag, Berlin, Germany, 2005

[b] Ribozymes have been used for chiral synthesis – See Schlatterer, J.C., Stuhlman, F., and Jäschke, A., Stereoselective synthesis using immobilized Diels-Alderase ribozymes, *ChemBioChem.* 4, 1089-1092, 2003

Enzymes in Synthetic Organic Chemistry[a,b] (Continued)

Hydrolases (Esterases)   A group of enzymes which catalyze the cleavage of ester and amide bonds with addition of water. Esterases, of which lipases are a singularly important group, are important in synthetic organic chemistry. Of particular importance is the stereoselectivity of the reaction[c]. A racemic mixture of an ester can be resolved can be resolved into enantiomeric pairs by stereospecific hydrolysis. Butyrylcholine esterase and cocaine esterase are listed with the therapeutic enzymes.

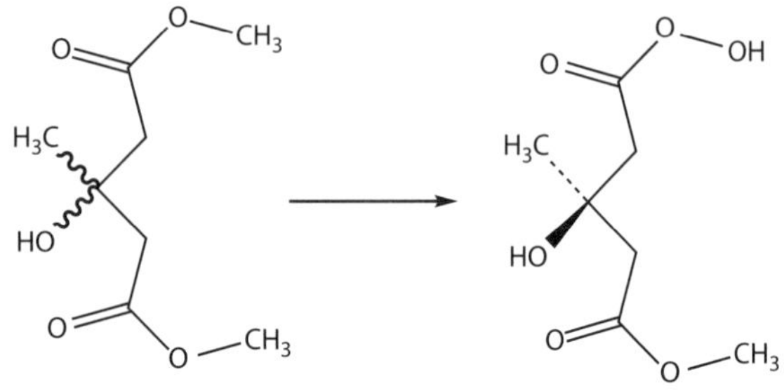

dimethyl-beta-hydroxymethylglutarate

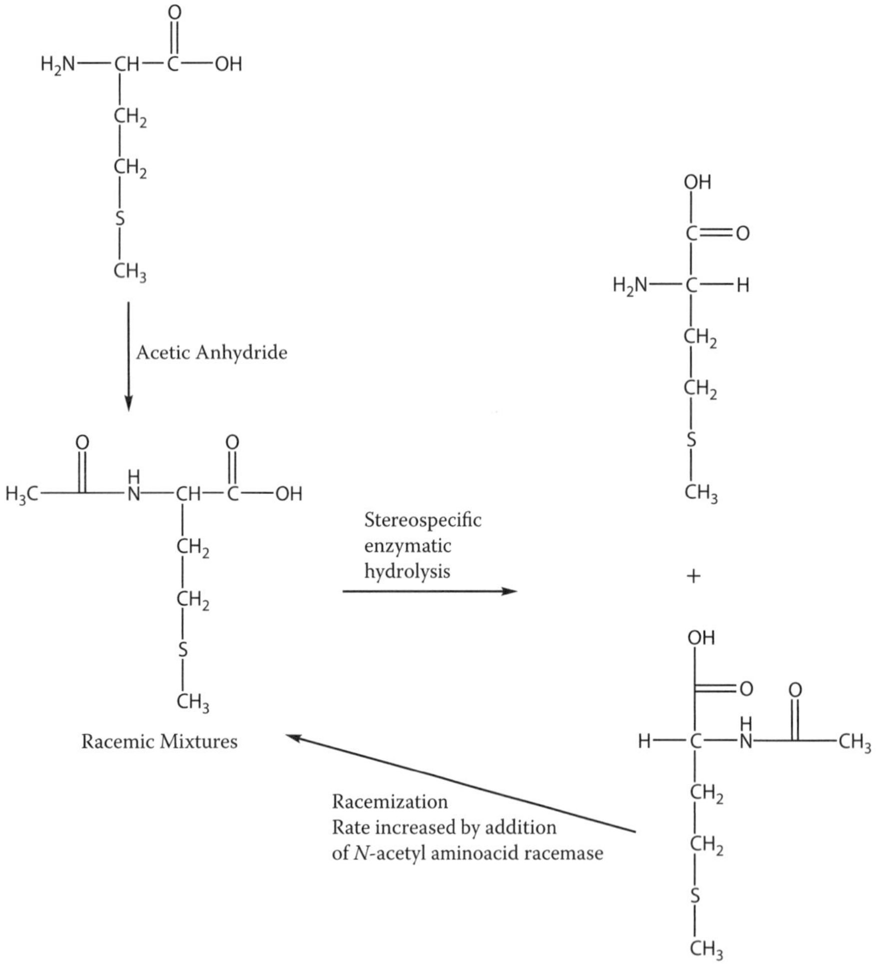

Acetic Anhydride

Stereospecific enzymatic hydrolysis

Racemic Mixtures

+

Racemization
Rate increased by addition
of N-acetyl aminoacid racemase

---

[c] A carbon center may be asymmetric in having four different substituents (other atoms such as sulfur can also be asymmetric centers). In the case of carbon, this can be result in a mixture of the two optical isomers resulting in a racemic mixture. Generally there is no driving force for the formation of a racemic mixture so there are equal forms of the D and L isomers. With an enzyme stereoselectivity can be achieved and the quality of an asymmetric mixture may be expressed as enantiomeric excess (enantiomeric excess = (moles of major enantiomer - moles of other enantiomer / Total moles of both enantiomers) × 100 and is usually expressed as a percentage.

Enzymes in Synthetic Organic Chemistry[a,b] (Continued)

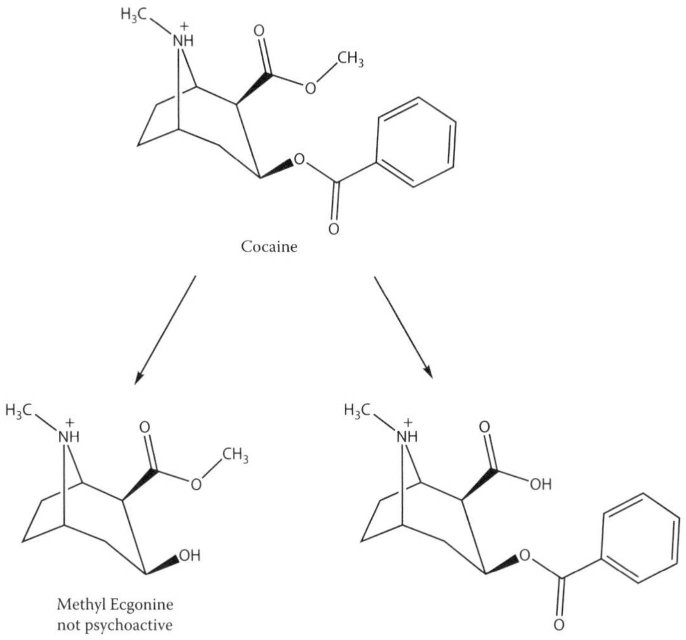

Cocaine

Methyl Ecgonine
not psychoactive

Benzoyl ecgonine
psychoactive

Venkatachalam, T.K., Samuel, P., Li, G., *et al.,* Lipase-mediated stereoselective hydrolysis of stampidine and other phosphoroamidate derivatives of stavudine, *Bioorg.Med.Chem.* 12, 3371-3381, 2004; Li, Y., Aubert, S.D., Maes, E.G., and Raushel, F.M., Enzymatic resolution of chiral phosphinate esters, *J.Am.Chem.Soc.* 126, 8888-8889, 2004; Kim, S. and Lee, S.B., Thermostable esterase from a thermoacidophilic archaeon: purification and characterization for enzymatic resolution of a chiral compound, *Biosci.Biotechnol.Biochem.* 68, 2289-2298, 2004; Molinari, F., Romano, D., Gandolfi, R., *et al.,* Newly isolated *Streptomyces* spp. As enantioselective biocatalysts: hydrolysis of 1,2-O-isopropylidene glycerol racemic esters, *J.Appl. Microbiol.* 99, 960-967, 2005; Hu, S., Martinez, C.A., Yazbeck, D.R., and Tao, J., An efficient and practical chemoenzymatic preparation of optically active secondary amines, *Org.Lett.* 7, 4329-4331, 2005; Nowlan, C., Li, Y., Hermann, J.C., *et al.,* Resolution of chiral phosphate, phosphonate, and phosphinate esters by an enantioselective enzyme library, *J.Am.Chem.Soc.* 128, 15892-15902, 2006; Gadler, P. and Faber, K., New enzymes for biotransformations: microbial alkyl sulfatases displaying stereo- and enantioselectivity, *Trends Biotechnol.* 25, 83-88, 2007.

Enzymes in Synthetic Organic Chemistry[a,b] (Continued)

Lipases   A group of hydrolytic enzymes that catalyze the release of fatty acids from triglycerides; more specifically the catalysis of the hydrolysis of ester bonds between alkanoic acids and glycerol. Phospholipases are a subclass which uses phospholipids as substrates. The ability to use alcohols and amines as acceptors of the fatty acid hydrolytic product permits the synthesis of chiral products. There has been recent interest in the use of lipases for the synthesis of combinatorial libraries[d].

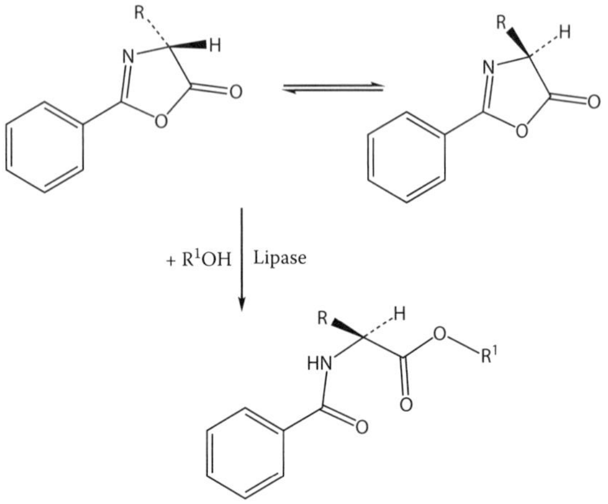

+ R^{1}OH   Lipase

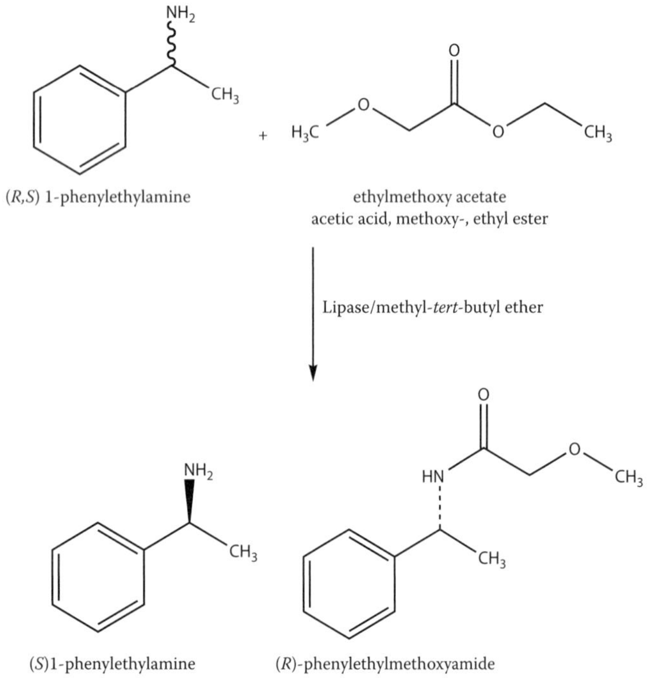

(R,S) 1-phenylethylamine   ethylmethoxy acetate
                                  acetic acid, methoxy-, ethyl ester

Lipase/methyl-*tert*-butyl ether

(S)1-phenylethylamine   (R)-phenylethylmethoxyamide

---

[d] Use of lipases in the manufacture of combinatorial libraries: Liu, K.-C., Clark, D.S., and Dordick, J.S., Chemoenzymatic construction of a four-component Ugi combinatorial library, *Biotechnol.Bioeng.* 69, 457-460, 2000; Reetz, M.T., Lipases as practical biocatalysts, *Curr.Opin.Chem.Biol.* 6, 145-150, 2002; Rich, J.G., Michels, P.C., and Khmeinitsky, Y.L., Lipases as practical biocatalysts, *Curr.Opin.Chem.Biol.* 6, 161-167, 2002; Secundo, F., Garrea, G., De Amici, M., *et al.,* A combinatorial biocatalysis approach to an array of cholic acid derivatives, *Biotechnol.Bioeng.* 81, 392-396, 2003; Kumar, R., Bruno, F., Parmar, V.S., *et al.,* "Green"-enzymatic synthesis of pegylated phenolic macromer and polymer, *Chem.Commun.* (7), 862-863, 2004; Rege, K., Hu. S., Moore, J.A., Chemoenzymatic synthesis and high-throughput screening of high-affinity displacers and DNA-binding ligands, *J.Am.Chem.Soc.* 126, 12306-12315, 2004; Vongvilal, P., Angelin, M., Larsson, R., and Ramstrom, O., Dynamic combinatorial resolution: direct asymmetric lipase-mediated screening of a dynamic nitroaldol library, *Angew.Chem.Int.Ed.Engl.* 46, 948-950, 2007.

Enzymes in Synthetic Organic Chemistry[a,b] (Continued)

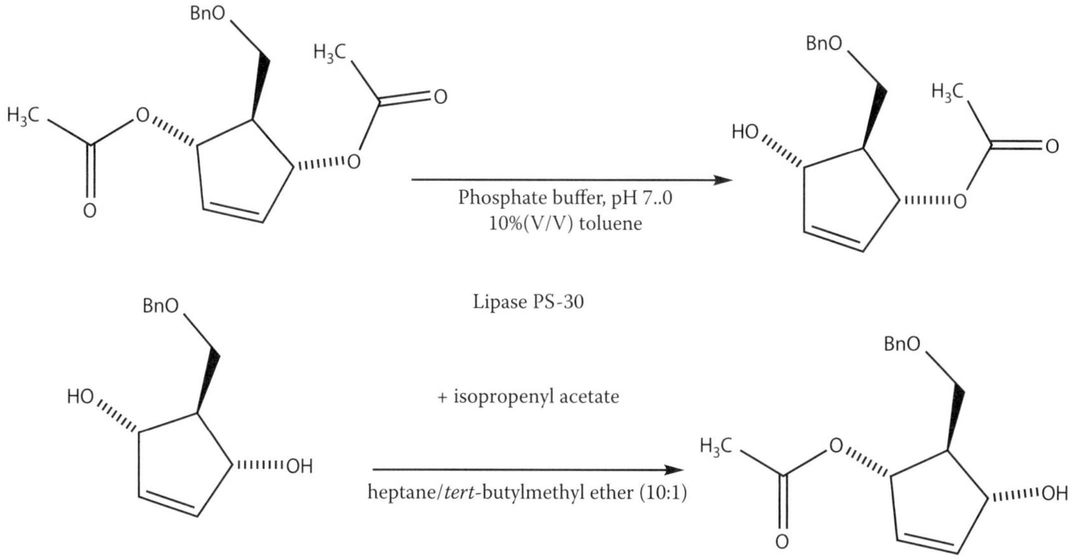

Tuomi, W.V., and Kazlauskas, R.J., Molecular basis for enantioselectivity of lipase from *Pseudomonas cepacia* towards primary alcohols. Modeling, kinetics, and chemical modification of tyr29 to increase or decrease enantioselectivity, *J.Org.Chem.* 64, 2638-2647, 1999; Ghorpade, S.R., Khani, R.K., Ioshi, R.R., *et al.*, Desymmetrization of *meso*-cyclopentene-*cis*-1,4-diol to 4-(*R*)-hydroxycyclopent-2-en-1-(*S*)-acetate by irreversible transesterification using Chirazyme®, *Tetrahedron Asymm.* 10, 891-899, 1999; Chen, J.-W. and Wu, W.-T., Regeneration of immobilized *Candida anartica* lipase for transesterification, *J.Biosci.Bioeng.* 95, 466-469, 2003; Gupta, R., Gupta, N., and Rathi, P., Bacterial lipases: an overview of production, purification and biochemical properties, *Appl.Microbiol.Biotechnol.* 64, 763-781, 2004; Kijima, T., Sato, N., Izumi, T., Lipase-catalyzed enantioselective esterification of mono-aza-benzo-15-crown-5-ether derivatives in organic media, *Biotechnol.Lett.* 26, 1505-1509, 2004; Domínguez de Maria, P., Carboni-Oerlemans, C., Tuin, B., *et al.*, Biotechnological applications of *Candida antarctica* lipase A: state-of-the-art, *J.Molec.Catal.B:Enzymatic* 37, 36-46, 2005; Sharma, J., Batovska, D., Kuwamori, Y., and Asano, Y., Enzymatic chemoselective synthesis of secondary-amide surfactant from *N*-methylethanol amine, *J.Biosci. Bioeng.* 100, 662-666, 2005; Patel, R.N., Banerjee, A., Pendri, Y.R., *et al.*, Preparation of a chiral synthon for an HBV inhibitor: enzymatic asymmetric hydrolysis of (1α,2β,3α)-2-(benzyloxymethyl)cyclopent-4-ene-1,3-diol diacetate and enzymatic asymmetric acetylation of (1α, 2β, 3α)-2-(benzyloxymethyl)cyclopent-4-ene-1,3-diol, *Tetrahedron Asymm.* 17, 175-175, 2006; Otero, C., Lopez-Herandez, A, Garcia, H.S., *et al.*, Continuous enzymatic transesterification of sesame oil and a fully hydrogenated fat: effects of reaction conditions on product characterization, *Biotechnol.Bioeng.* 94, 877-887, 2006.

Enzymes in Synthetic Organic Chemistry[a,b] (Continued)

Monooxygenase	Insertion of oxygen into organic frameworks such as the oxidation of olefins (Baeyer-Villiger Reaction); hydroxylation of alkanes and aromatics. The cytochrome P-450-dependent monooxygenase is one of the better-known examples. Monooxygenases also oxidize organic sulfur

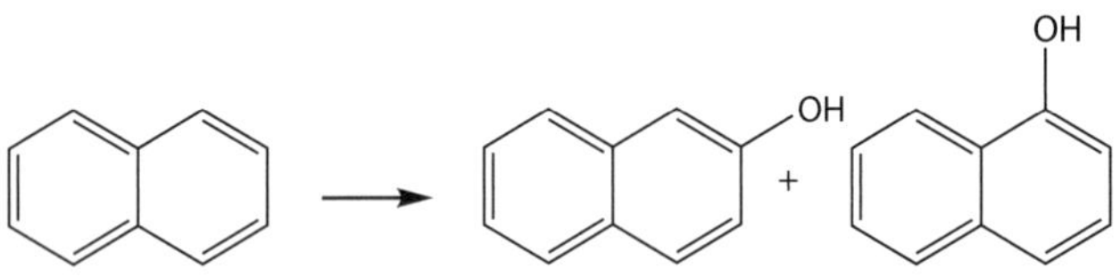

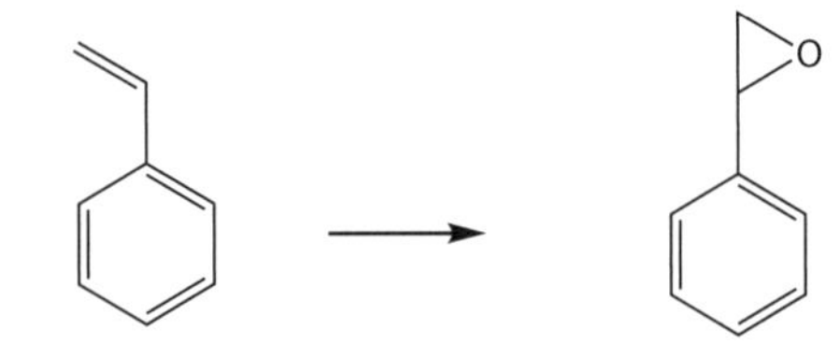

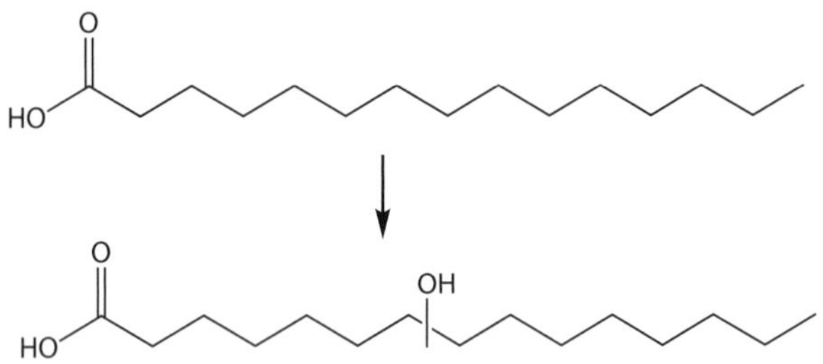

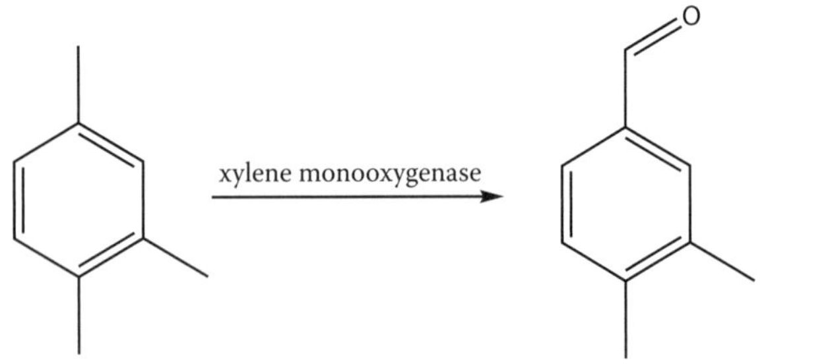

Pseudocumene   3,4-dimethylbenzaldehyde

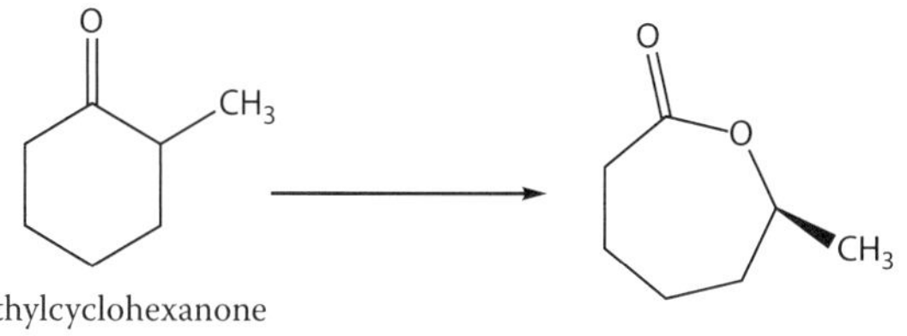

2-methylcyclohexanone

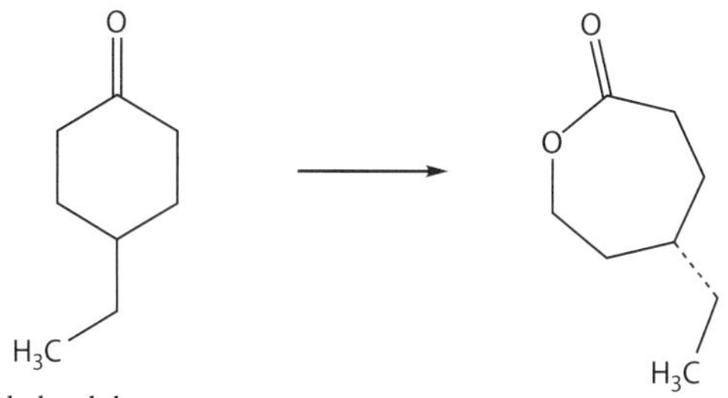

4-ethylcyclohexanone

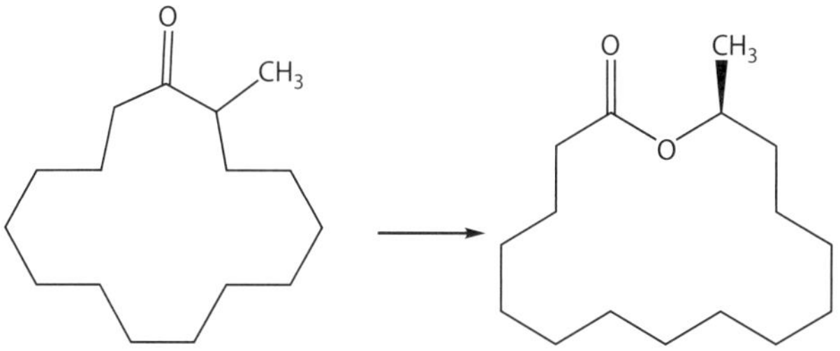

2-methylcyclopentadecanone

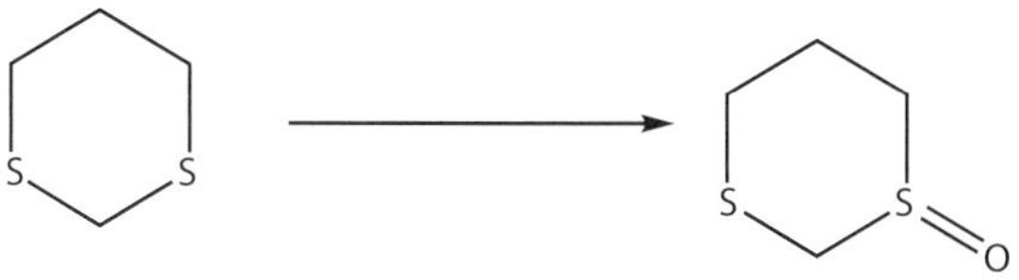

(*R*) 81%

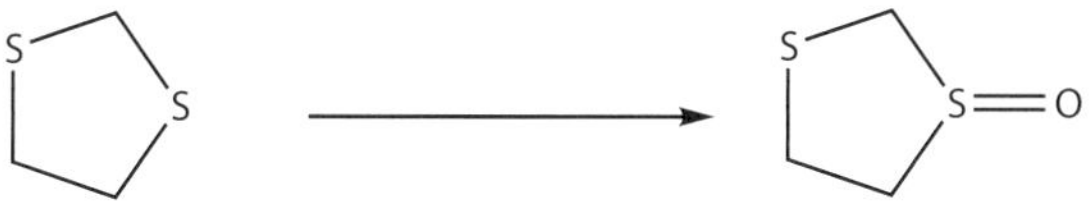

(*R*) 94%

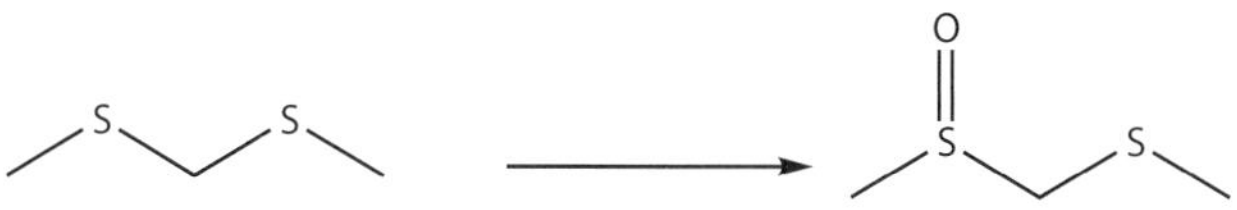

(*R*) 92%

Enzymes in Synthetic Organic Chemistry[a,b] (Continued)

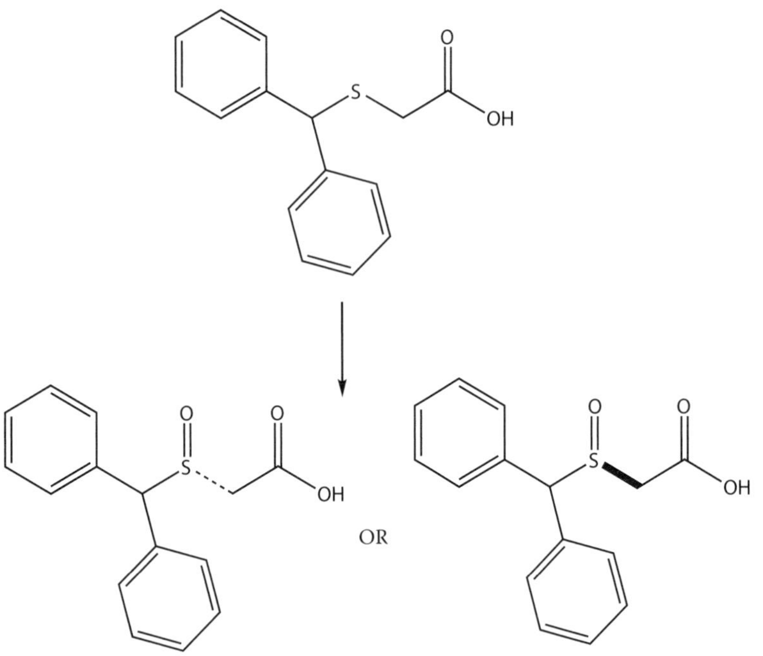

Sphingomonas sp.; 40% enantioselective   *Beauuveria bassiana*; 99% enantioselective

Ogawa, J. and Shimizu, S., Microbial enzymes: new industrial applications from traditional screening methods, *Trends Biotechnol.* 17, 13-21, 1999; Stewart, J.D., Organic transformations catalyzed by engineered yeast cells and related systems, *Curr.Opin.Biotechnol.* 11, 363-368, 2000; Mihovilovic, M.D., Muller, B., and Stanetty, P., Monooxygenase-mediated Baeyer-Villiger oxidations, *Eur.J.Org.Chem.* (22), 3711-3730, 2002; Lee, W.H., Park, Y.C., Lee, D.H., Simultaneous biocatalyst production and Baeyer-Villiger oxidation for bioconversion of cyclohexanone by recombinant *Escherichia coli* expressing cyclohexanone monooxygenase, *Appl.Biochem.Biotechnol.* 121-124, 827-836, 2005; Han, J.H., Yoo, S.K., Seo, J.S., *et al.*, Biomimetic alcohol oxidations by an iron (III) porphyrin complex: relevance to cytochrome P-450 catalytic oxidation and involvement of the two-state radical rebound mechanism, *Dalton Trans.* (2), 402-406, 2005; Kagawa, H., Tatkahashi, T., Ohta, S., and Harigaya, Y., Oxidation and rearrangements of flavanones by mammalian cytochrome P450, *Xenobiotica* 34, 797-810, 2004; Gillam, E.M., Exploring the potential of xenobiotic-metabolising enzymes as biocatalysts: evolving designer catalysts from polyfunctional cytochrome P450 enzymes, *Clin.Exp.Pharmacol.Physiol.* 32, 147-152, 2005; Bocola, M., Schultz, F., Leca, F., *et al.*, Converting phenylacetone monooxygenase into phenylcyclohexanone monooxygenase by rational design: towards practical Baeyer-Villiger monooxygenases, *Adv.Synth.Catal.* 347, 979-986, 2005; Urlacher, V.B. and Eiben, S., Cytochrome P450 monooxygenases: perspectives for synthetic application, *Trends Biotechnol.* 24, 324-330, 2006; Iwaki, H., Wang, S. ,Grosse, S., *et al.*, Pseudomonad cyclopentadecanone monooxygenase displaying an uncommon spectrum of Baeyer-Villiger oxidations of cyclic ketones, *Appl.Environ.Microbiol.* 72, 2707-2720, 2006; Mihovilovic, M.D., Enzyme mediated Baeyer-Villiger oxidations, *Curr.Org.Chem.* 10, 1265-1287, 2006.Application to the stereospecific oxidation of organic sulfur: Dodson, R.M., Newman, N., and Tsuchiya, H.M., Microbiological transformations. XI. The properties of optically active sulfoxides, *J.Am.Chem.Soc.* 27, 2707-2708, 1962; Light, D.R., Waxman, D.J., and Walsh, C., Studies on the chirality of sulfoxidation catalyzed by bacterial flavoenzyme cyclohexanone monooxygenase and hog liver flavin adenine dinucleotide containing monooxygenase, *Biochemistry* 21, 2490-2498, 1982; Waxman, D.J., Light, D.R., and Walsh, C., Chiral sulfoxidation catalyzed by rat live cytochrome P-450, *Biochemistry* 21, 2499-2507, 1982; Colonna, S., Gaggero, N., Pasta, P., and Ottolina, G., Enantioselective oxidation of sulfides to sulfoxides catalyzed by bacterial cyclohexanone monooxygenase, *Chem.Commun.* (20), 2303-2307, 1996; Mata, E.G., Recent advances in the synthesis of sulfoxides from sulfides, *Phosphorus, Sulfur and Silicon and the Related Elements* 117, 231-286, 1996; Hamman, M.A., Haehner-Daniels, B.D., Wrighton, S.A., *et al.*, Stereoselective sulfoxidation of sulindac sulfide by flavin-containing monooxygenase-Comparison of human liver and kidney microsomes and mammalian enzymes, *Biochem.Pharmacol.* 60, 7-17, 2000; Reetz, M.T., Daligault, F., Brunner, B., *et al.*, Directed evolution of cyclohexanone monooxygenases: enantioselective biocatalysts for the oxidation of prochiral thioethers, *Angewandte Chem.Int.* 43, 4078-4081, 2005; Legros, J., Dehli, J.R., and Bolm, C., Applications of catalytic asymmetric sulfide oxidations to the syntheses of biologically active sulfoxides, *Adv.Synthes.& Catal.* 19-31, 2005; Olivo, H.F., Osorio-Lozada, A., and Peeples, T.L., Microbial oxidation/amidation of benzhydrylsulfanyl acetic acid. Synthesis of (+)-modafinil, *Tetrahedron Asymmetry* 16, 3507-3511, 2005.

Enzymes in Synthetic Organic Chemistry[a,b] (Continued)

Dioxygenase An enzyme activity which inserts an oxygen molecule into an organic substrate. Where hydroxyl function(s) is the terminal reaction product, the overall reaction is the sum of two separate enzymatic reactions; an oxidation followed by a reduction.

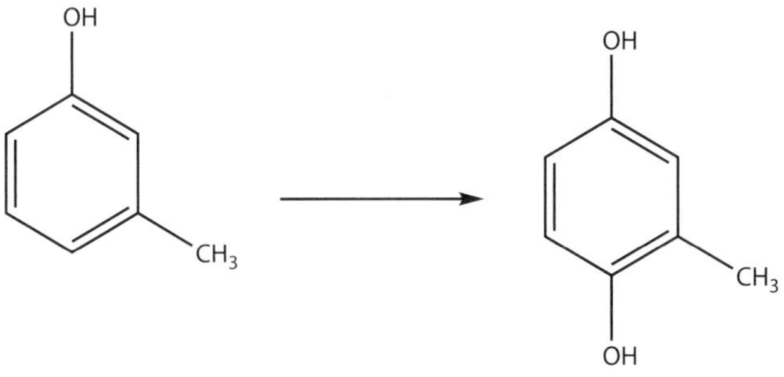
OH
CH3
OH
CH3
OH

m-cresol methyl hydroquinone

(1) 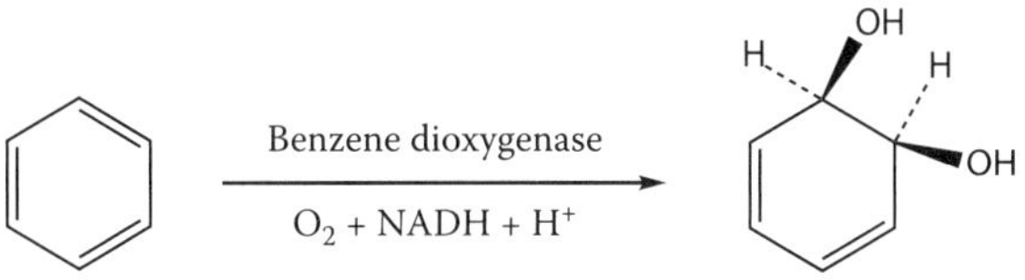
Benzene dioxygenase
O2 + NADH + H+
H
OH
H
OH

(2) 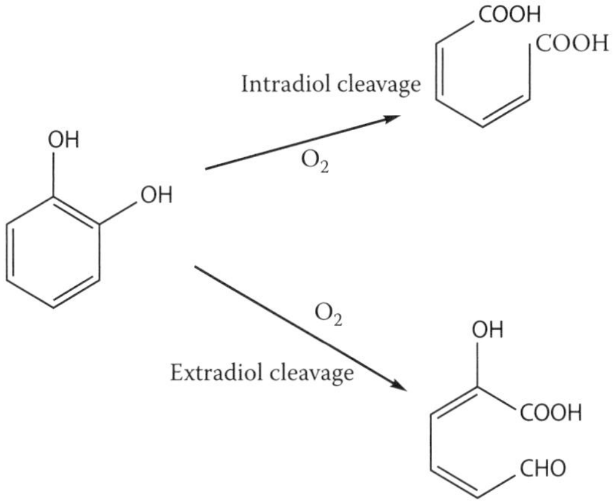
Intradiol cleavage
O2
COOH
COOH
OH
OH
O2
Extradiol cleavage
OH
COOH
CHO

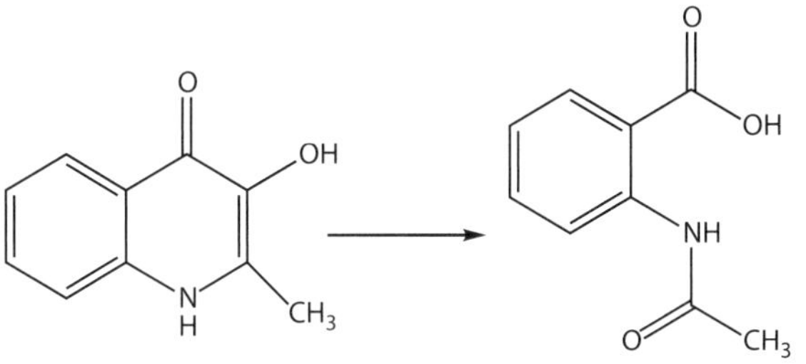
O
OH
N
H
CH3
O
OH
NH
O
CH3

1*H*-3-hydroxy-4-oxoquinaldine

N-acetylanthranilic acid

Enzymes in Synthetic Organic Chemistry[a,b] (Continued)

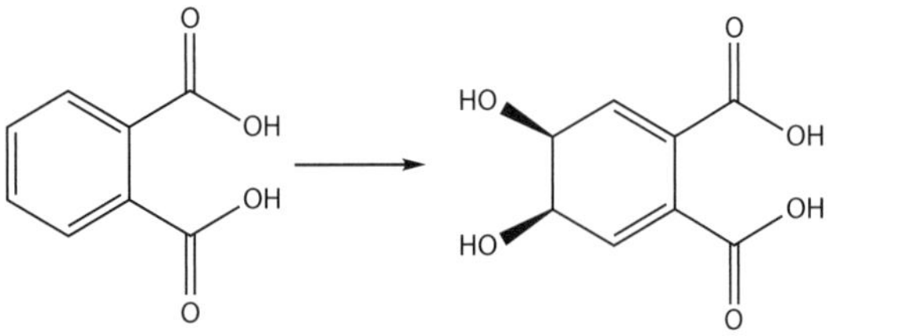

Phthalic Acid   *cis*-4,5-dihydro-4,5-dihydroxyphthalic acid

Nakata, H., Yamuchi, T., and Fujisawa, H., Studies on the reaction intermediate of protocatechuate 3,4-dioxygenase. Formation of enzyme-product complex, *Biochim.Biophys.Acta* 527, 171-181, 1978; Gassner, G.T., Ludwig, M.L., Gatti, D.L., *et al.*, Structure and mechanism of the iron-sulfur flavoprotein phthalate dioxygenase reductase, *FASEB J.* 9, 1411-1418, 1995; Miyauchi, K., Adachi, Y, Nagata, T., and Takagi, M., Cloning and sequencing of a novel *meta*-cleavage dioxygenase gene whose product is involved in degradation of γ-hexachlorocyclohexane in *Sphingomonas paucimobilis*, *J.Bacteriol.* 181, 6712-6719, 1999; Calderone, V., Trabucco, M., Menin, V. *et al.*, Cloning of human 3-hydroxyanthranilic acid dioxygenase in *Escherichia coli*: characterization of the purified enzyme in its *in vitro* inhibition by Zn²⁺, *Biochim.Biophys.Acta* 1596, 283-292, 2002; Johnson-Winters, K., Purpero, V.M., Kavana, M., *et al.*,(4-Hydroxyphenyl)pyruvate dioxygenase from *Streptomyces avermitilis*: the basis for ordered substrate addition, *Biochemistry* 42, 2072-2080, 2003; Frerichs-Deeken, U., Ranguelova, K., Kappl, R., *et al.*, Dioxygenases without requirement for cofactors and their chemical model reaction: compulsory order ternary complex mechanism of 1H-3-hydroxy-4-oxyquinaldine 2,4-dioxygenase involving general base catalysis by histidine 251 and single-electron oxidation of the substrate dianion, *Biochemistry* 43, 14485-14499, 2004; Yin, C.X. and Finke, R.G., It is true dioxygenase or classic autoxidation catalysis? Re-investigation of a claimed dioxygenase catalyst based on a Ru(2)-incorporated, polyoxometalate precatalyst, *Inorg.Chem.* 44, 4175-4188, 2005; Matsumura, E., Ooi, S., Murakami, S., *et al.*, Constitutive synthesis, purification, and characterization of catechol 1,2-dioxygenase from the aniline-assimilating bacterium *Rhodococcus* sp. An-22, *J.Biosci.Bioeng.* 98, 71-76, 2004; Lee, K., *p*-Hydroxylation reactions catalyzed by naphthalene dioxygenase, *FEMS Microbiol.Lett.* 255, 316-320, 2006; Suvorova, M.M., Solyanikova, I.P., and Gobovleva, L.A., Specificity of catechol *ortho*-cleavage during *para*-toluate degradation by *Rhodococcus opacus* 1cp, *Biochemistry*(Mosc) 71, 1316-1323, 2006.

Ketone Reductases;   Enantiomeric and diastereoisomeric reductions of ketones and β-keto esters
engineered yeast cells;
alcohol dehydrogenases

Enantioselective reactions

Diastereoselective reactions

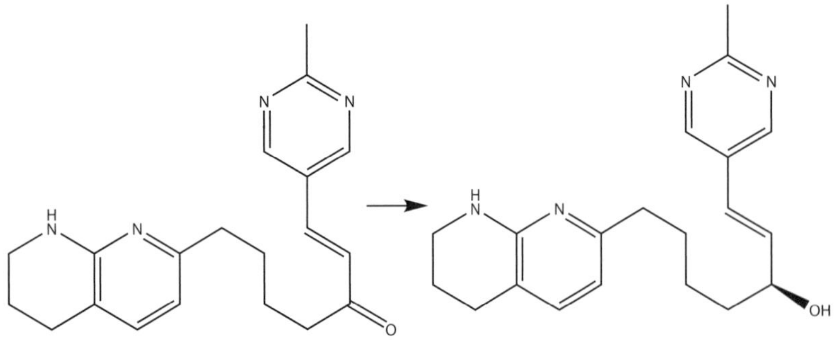

Enone   (*R*)allylic alcohol

Stewart, J.D., Organic transformations catalyzed by engineered yeast cells and related systems, *Curr.Opin.Biotechnol.* 11, 363-368, 2000; Habrych, M., Rodriguez, S., and Stewart, J.D., Purification and identification of an *Escherichia coli* β-keto ester reductase as 2,5-diketo-D-gluconate reductase YquE, *Biotechnol.Progress* 18, 257-261, 2002; Katz, M., Sarvary, I., Frejd, T., *et al.*, An improved stereoselective reduction of a bicyclic diketone by *Saccharomyces cerevisiae* combining process optimization and strain engineering, *Appl.Microbiol.Biotechnol.* 59, 641-648, 2002; Ravot, G., Wahler, D., Favre-Bulle, O., *et al.*, High throughput discovery of alcohol dehydrogenase for industrial biocatalysis, *Adv.Synthesis Catalysis* 345, 691 694, 2003; Lou, W.Y., Zong, M.H., Zhang, Y.Y. and Wu, H., Efficient synthesis of optically active organosilyl alcohol via asymmetric reduction of acyl silane with immobilized yeast, *Enzyme Microb.Technol.* 35, 190-196, 2004; Rodrigues, J.A.R., Moran, P.J.S., Conceicao, G.J.A., and Fardelone, L.C., Recent advances in the biocatalytic asymmetric reduction of acetophenones and α,β-unsaturated carbonyl compounds, *Food Technol.Biotechnol.* 42, 295-303, 2004; Pollard, D.J., Telari, K., Lane, J., *et al.*, Asymmetric reduction of α,β-unsaturated ketone to (R) allylic alcohol by *Candida chilenis*, *Biotechnol.Bioengineer.* 93, 674-686, 2006.

Enzymes in Synthetic Organic Chemistry[a,b] (Continued)

Cephalosporin Acylase　　　Conversion of cephalosporin C or adipyl-7-aminodesacetoxycephalosporonic acid to derivatives useful in the synthesis of semi-synthetic β-lactam antibiotics.

Sio, C.F., Otten,L.G., Cool, R.H., and Quax, W.M., Analysis of substrate specificity switch residue of cephalosporin acylase, *Biochem.Biophys.Res. Commun.* 312, 755-760, 2003; Sio, C.F., and Quax. W., Improved β-lactam acylases and their use as biocatalysts, *Curr.Opin.Biotechnol.* 15, 349-355, 2004; Sonawane, V.C., Enzymatic modifications of cephalosporins by cephalosporin acylase and other enzymes, *Crit.Rev.Biotechnol.*26, 95-120, 2006.

Penicillin Acylase　　　Catalyzes the conversion of benzylpenicillin (penicillin G) to 6-aminopenicillinic acid for the production of β-lactam antibiotics. Penicillin acylase also converts cephalosporin C to 6-aminopenicillinic acid. There has been considerable work on the engineering and stabilization of the enzyme. Penicillin acylase can catalyze the reverse reaction resulting in a condensation.

Penicillin G

Penicillin Acylase

Phenylacetic Acid　　　　　6-aminopenicillanic acid

Mahajan, P.B., Penicillin acylases. an update, *Appl.Biochem.Biotechnol.* 9, 538-554, 1984; Andersson, E. and Hahn-Hagerdal, B., Bioconversion in aqueous two-phase systems, *Enzyme Microb.Technol.* 12, 242-254, 1990; Valle, F., Balbas, P., Merino, E., and Bolivar, F., *Trends Biotechnol.* 16, 36-40, 1991; Zaks, A., Industrial biocatalysis, *Curr.Opin.Chem.Biol.* 5, 130-136, 2001; Arroyo, M. de la Mata, I., Acebal, C, and Castillon, M.P., Biotechnological application of penicillin acylases: state-of-the-art, *Appl.Microbiol.Biotechnol.* 60, 507-514, 2003; Albian, O., Mateo, C., Fernandez-Lorente, G., *et al.,* Improving the industrial production of 6-APA: enzymatic hydrolysis of penicillin G in the presence of organic solvents, *Biotechnol.Prog.* 19, 1639-1642, 2003; Calleri, E., Temporini, C., Massolini, G., and Caccilanza, G., Penicillin G acylase-based stationary phases: analytical applications, *J.Pharm.Biomed.Anal.* 35, 243-258, 2004; Sio, C.F. and Quzx, W.J., Improved beta-lactam acylases and their use as industrial biocatalysts, *Curr.Opin.Biotechnol.* 15, 349-355, 2004; Guranda, D.T., Volovik, T.S., and Svedas, V.K., pH stability of penicillin acylase from *Escherichia coli, Biochemistry* 69, 1386-1390, 2004; Girelli, A.M. and Maltei, E., Application of immobilized enzyme reactor in on-line high performance liquid chromatography: a review, *J.Chromatog B. Analyt.Technol.Biomed.Life Sci.* 819, 3-16, 2005; Torres, R., Pessela, B., Fuentes, M., *et al.,* Stabilization of enzymes by multipoint attachment via reversible immobilization on phenylboronic activated supports, *J.Biotechnol.* 120, 396-401, 2005; Nigam, V.K., Kundu, S., and Ghosh, P., Single-step conversion of cephalosporin-C to 7-aminocephalosporonic acid by free and immobilized cells of *Pseudomonas diminuta, Appl.Biochem.Biotechnol.* 126, 13-21, 2005; van Roon, J.L., Boom, R.M., Paasman, M.A., *et al.,* Enzyme distribution and matrix characteristics in biocatalytic particles, *J.Biotechnol.* 119, 400-415, 2005; Giorando, R.C., Ribeiro, M.P., and Giordano, R.L., Kinetics of beta-lactam antibiotic synthesis by penicillin G acylase (PGA) from the view of the industrial enzyme reactor optimization, *Biotechnol.Adv.* 24, 27-41, 2006. Narayanan, N. Xu, Y., and Chou, G.P., High-level gene expression for recombinant penicillin acylase production using the *araB* promoter system in *Escherichia coli, Biotechnol.Prog.* 22, 1518-1523, 2006; De Leon-Rodriguez, A., Rivera-Pastrana, D., Medina-Rivero, E., *et al.,* Production of penicillin Acylase by a recombinant *Escherichia coli* using cheese whey as substrate and inducer, *Biomol.Eng.* 23, 299-305, 2006; Wang, L., Wang, Z., Xu., J.H., *et al.,* An eco-friendly and sustainable process for enzymatic hydrolysis of penicillin G in cloud point system, *Bioprocess Biosyst.Eng.* 29, 157-162, 2006; Aguilar, O., Albiter, V., Serrano-Carreon, L., and Rito-Palomares, M., Direct comparison between ion-exchange chromatography and aqueous two-phase processes for the partial purification of penicillin acylase produced by E.coli, *J.Chromatog.B. Analyt.Techol. Biomed.Life Sci.* 835, 77-83, 2006; Shah, S. ,Sharma, A. and Gupta, M.N., Preparation of cross-linked enzyme aggregates by using bovine serum albumin as a proteic feeder, *Anal.Biochem.* 351, 207-213,2006.

Synthetic reaction: Nam, D.H. and Ryu, D.D., Biochemical properties of penicillin amidohydrolase from *Micrococcus luteus, Appl.Environ.Microbiol.* 38, 35-38, 1979; Youshko, M.I., van Langen, L.M., de Vroom, E., *et al.,* Highly efficient synthesis of ampicillin in an "aqueous solution-precipitate" system; repetitive addition of substrates in a semicontinuous process, *Biotechnol.Bioengineer.* 73, 4260430, 2001; Youshko, M.I., van Langen, L.M., de Vroom, E., *et al.,* Penicillin acylase-catalyzed ampicillin synthesis using a pH gradient: a new approach to optimization, *Biotechnol.Bioeng.* 78, 589-593, 2002; Goncalves, L.R., Fernandez-Lafuente, R., Guisan, J.M., *et al.,* Inhibitory effects in the side reactions occurring during the enzymic synthesis of amoxicillin: *p*-hydroxyphenylglycine methyl ester and amoxicillin hydrolysis, *Biotechnol. App..Biochem.* 38, 77-85, 2003; Alkema, W.B., de Vries, E., Floris, R., and Janssen, D.B., Kinetics of enzyme acylation and deacylation in the penicillin acylase-catalyzed synthesis of beta-lactam antibiotics, *Eur.J.Biochem.* 270, 3675-3683, 2003; Alfonso, I. and Gotor, V., Biocatalytic and biomimetic aminolysis reactions: useful tools for selective transformations or polyfunctional substrates, *Chem.Soc.Rev.* 33, 201-209,. 2004' Gabor, E.M. and Janssen, D.B., Increasing the synthetic performance of penicillin acylase PAS2 by structure-inspired semi-random mutagenesis, *Protein Eng.Des.Sci.* 17, 571579, 2004.

An unusual synthetic application was removal blocking groups from synthetic insulin; Svoboda, I., Brandenburg, D., Barth, T., *et al.,* Semisynthetic insulin analogues modified in positions B24, B25 and B29, *Biol.Chem.Hoppe Seyler* 375, 373-378, 1994.

THERAPEUTIC ENZYMES[a]

Asparaginase

Catalyzes the hydrolysis of asparagine to aspartic acid; used for the treatment of acute lymphoblastic leukemia. Normal cells can synthesize asparagine while tumor cells in acute lymphoblastic leukemia cannot synthesize asparagine

Hill, J.M., Roberts, J., Loeb, E., *et al.*, L-Asparaginase therapy for leukemia and other malignant neoplasms. Remission in human leukemia, *JAMA* 202, 882-888, 1967; Broome, J.D., Studies on the mechanism of tumor inhibition by L-asparaginase. Effects of the enzyme on asparaginel levels in the blood, normal tissue, and 6C3HED lymphomas of mice: differences in asparagine formation and utilization in asparaginase-sensitive and −resistant lymphoma cells, *J.Exp.Med.* 127, 1055-1072, 1968; Adamson, R.H. and Fabro, S., Antitumor activity and other biologic properties of L-asparaginase (NSC-109229)-a review, *Cancer Chemother.Rep.* 52, 617-626, 1968; Keating, M.J., Holmes, R., Lerner, R., and Ho, D.H., L-Asparaginase and PEG asparaginase −past, present, and future, *Leuk.Lymphoma* 10(Suppl) 153-157, 1993; Davis, F.F., PEG-adenosine deaminase and PEG-asparaginase, *Adv. Exp.Med.Biol.* 519, 51-58, 2003; Pinheiro, J.P. and Boos, J., The best way to use asparaginase in childhood acute lymphatic leukemia—still to be defined?, *Br.J.Haematol.* 125, 119-127, 2004

Blood Coagulation Factor VIIa

Treatment of Blood Coagulation factor VIII inhibitors, potentially a general intravascular hemostatic agent

Siddiqui, M.A. and Scott, L.J., Recombinant factor VIIa (Eptacog Alfa): A review of its use in congenital or acquired haemophilia and other congenital bleeding disorders, *Drugs* 65, 1161-1177, 2005; Franchini, M., Zaffanello, M., and Veneri, D., Recombinant factor VIIa. An update on its clinical use, *Thromb.Haemostas.* 93, 1027-1035, 2005; Margaritis, P. and High, K.A., Advances in gene therapy using factor VIIa in hemophilia, *Semin.Hematol.* 43(Suppl 1), S101-S104, 2006; Farrugia, A., Assessing efficacy and therapeutic claims in emerging indications for recombinant Factor VIIa: regulatory perspectives, *Semin.Hematol.* 43(1 Suppl 1), S64-S69, 2006; Bosinski, T.J. and El Solh, A.A., Recombinant factor VIIa, its clinical properties, and the tissue factor pathway of coagulation, *Mini Rev.Med.Chem.* 6, 1111-1117, 2006

Butyryl cholinesterase

Detoxification of neurotoxic agents related to DFP; bioscavenger of anticholinesterase agents; Also used for treatment of cocaine overdoses

Doctor, B.P., Raveh, L., Wolfe, A.D., *et al.*, Enzymes as pretreatment drugs for organophosphate toxicity, *Neurosci.Biobehav.Rev.* 15, 123-128, 1991; Broomfield, C.A., Maxwell, D.M., Solana, R.P., *et al.*, Protection by butyrylcholinesterase against organophosphorus poisoning in nonhuman primates, *J.Pharmacol.Exp.Ther.* 259, 633-638, 1991; Grunwald, J., Marcus, D., Papier, Y., *et al.*, Large-scale purification and long-term stability of human butyrylcholinesterase: a potential bioscavenger drug, *J.Biochem.Biophys.Methods* 34, 123-135, 1997; Lynch, T.J., Mattes, C.E., Singh, A., *et al.*, Cocaine detoxification by human plasma butyrylcholinesterase, *Toxicol.Appl.Pharmacol.* 145, 363-371, 1997; Mattes, C.E., Lynch, T.J., Singh, A., *et al.*, Therapeutic use of butyrylcholinesterase for cocaine intoxication, *Toxicol.Appl. Pharmacol.* 145, 372-380, 1997; Browne, S.P., Slaughter, E.A., Couch, R.A., *et al.*, The influence of plasma butyrylcholinesterase concentration on the in vitro hydrolysis of cocaine in human plasma, *Biopharm.Drug.Dispos.* 19, 309-314, 1998; Yuan, H.J., Yu, W.Y., Shi, C.H., and Sun, M.J., Characteristics of recombinant human butyrylcholinesterase, *Zhongguo Yao Li Xue Bao* 20, 74-80, 1999; Chambers, J. and Oppenheimer, S.F., Organophosphates, serine esterase inhibition, and modeling of organophosphate toxicity, *Toxicol.Sci.* 77, 185-187, 2004; Guven, M., Sungur, M., Eser, B. *et al.*, The effects of fresh frozen plasma on cholinesterase levels and outcomes in patients with organophosphate poisoning, *J.Toxicol. Clin.Toxicol.* 42, 617-623, 2004; Fischer, S., Arad, A., and Margalit, R., Liposome-formulated enzymes for organophosphate scavenging: butyrylcholinesterase and Demeton-S, *Arch.Biochem.Biophys.* 434, 108-115, 2005; Saez-Valero, J., de Gracia, J.A., and Lockridge, O. Intraperitoneal administration of 340 kDa human plasma butyrylcholinesterase increases the level of the enzyme in the cerebrospinal fluid of rats, *Neurosci.Lett.* 383, 93-98, 2005; Lockridge, O., Schopfer, L.M., Winger, G., and Woods, J.H., Large scale purification of butyrylcholinesterase from human plasma suitable for injection into monkeys: A potential new therapeutic for protection against cocaine and nerve agent toxicity, *J.Med.Chem.Biol.Radiol.Def.* 3:nihms5095, 2005; Gardiner, S.J. and Begg, E.J., Pharmacogenetics, drug-metabolizing enzymes, and clinical practice, *Pharmacol.Rev.* 58, 521-590, 2006; Lucic Vrdoljak, A., Calic, M., Radic, B., *et al.*, Pretreatment with pyridinium oximes improves antidotal therapy against tabun poisoning, *Toxicology* 228, 41-50, 2006

Cocaine Esterase

Cocaine Detoxification

Brzezinski, M.R., Abraham, T.L., Stone, C.L., *et al.*, Purification and characterization of a human liver cocaine carboxylesterase that catalyzes the production of benzoylecgonine and the formation of cocaethylene from alcohol and cocaine, *Biochem.Pharmacol.* 48, 1747-1755, 1994; Turner, J.M., Larsen, N.A., Basran, A., *et al.*, Biochemical characterization and structural analysis of a highly proficient cocaine esterase, *Biochemistry* 41, 12297-12307, 2002; Ascenzi, P., Clementi, E., and Polticelli, F., The *Rhodococcus* sp. cocaine esterase: a bacterial candidate for novel pharmacokinetic-based therapies for cocaine abuse, *IUBMB Life* 55, 397-402, 2003; Rogers, C.J., Mee, J.M., Kaufmann, G.F., *et al.*, Toward cocaine esterase therapeutics, *J.Am. Chem.Soc.* 127, 10016-10017, 2005; Rogers, C.J., Eubanks, L.M., Dickerson, T.J., and Janda, K.D., Unexpected acetylcholinesterase activity of cocaine esterase, *J.Am.Chem.Soc.* 128, 15364-15365, 20006; Cooper, Z.D., Narasimhan, D., Sunahara, R.K., *et al.*, Rapid and robust protection against cocaine-induced lethality in rates by bacterial cocaine esterase, *Mol.Pharmacol.* 70, 1885-1891, 2006

---

[a] Targeting of therapeutic enzymes is a challenge and is the subject of current study. See Ribeiro, C.C. Barrias, C.C., and Barbosa, M.A., Calcium phosphate-alginate microspheres as enzyme delivery matrices, *Biomaterial* 25, 4363-4373, 2004; Vogler, C., Levy, B., Grubb, J.H., *et al.*, Overcoming the blood-brain barrier with high-dose enzyme replacement therapy in murine mucopolysaccharidosis VII, *Proc.Nat.Acad.Sci.USA* 102, 14777-14782, 2005; Fukudo, T., Ahearn, M., Roberts, A., *et al.*, Autophagy and mistargeting of therapeutic enzymes in skeletal muscle in Pompe disease, *Mol.Ther.* 14, 831-839, 2006; Lee, S., Yang, S.C., Hefferman, M.J., *et al.*, Polyketal microparticles: a new delivery vehicle for superoxide dismutase, *Bioconjug.Chem.* 18, 4-7, 2007.

THERAPEUTIC ENZYMES (Continued)

DNAse

Originally used for the resolution of localized abscesses by viscosity reduction due to hydrolysis of high-molecular weight DNA arising from tissue damage. There is more recent use in the treatment of cystic fibrosis as Dornase™ alpha. A combination of streptokinase and streptodornase was developed as well and is still used as Varidase[b]

Sherry, S., Johnson, A., and Tillett, W.R., The action of streptococcal desoxyribose nuclease (Streptodornase): *In vitro and* on purulent pleural exudations of patients, *J.Clin.Invest.* 29, 1094-1104, 1949; Johnson, A.J., Cytological studies in association with local injections of streptokinase-streptodornase into patients, *J.Clin.Invest.* 29, 1376-1386, 1950; Bryson, H.M. and Borkin, E.M., Dornase alpha. A review of its pharmacological properties and therapeutic potential in cystic fibrosis, *Drugs* 48, 894-906, 1994; Thomson, A.H., Human recombinant DNAse in cystic fibrosis, *J.R.Soc. Med.* 88 (suppl 25), 24-29, 1995; Davies, J. ,Trindale, M., Wallis, C. *et al.,* The clinical use of rhDNAse, *Pediatr.Pulmonol.Suppl.* 16, 273-274, 1997; Goa, K.L. and Lamb, H., Dornase alpha. A review of pharmacoeconomic and quality-of-life aspects of its use in cystic fibrosis, *Pharmacoeconomics* 12, 409-422, 1997; Jones, A.F. and Wallis, C.E., Recombinant human deoxyribonuclease for cystic fibrosis, *Cochrane Database Syst.Rev.* (3), CD001127, 2000; Suri, R., The use of human deoxyribonuclease (rhDNAse) in the management of cystic fibrosis, *BioDrugs* 19, 135-144, 2005; Fayon, M., CF-emerging therapies: Modulation inflammation, *Pediatr.Respir.Rev.* 7 Suppl 1, S170-S174, 2006

Digestive Enzymes

Usually a crude homogenate of pancreas using to treatment pancreatic disease. Lipase is an individual enzyme which is used to treat steatorrhea[c]. An artificial saliva is available to treat salivary gland dysfunction[d].

Gullo, L., Indication for pancreatic enzyme treatment in non-pancreatic digestive diseases, *Digestion* 54(suppl 2), 43-47, 1993; Kitagawa, M., Naruse, S., Ishiguro, H., and Hayakawa, T., Pharmaceutical development for treating pancreatic disease, *Pancreas* 16, 427-431, 1998; Nakamura, T., Takeuchi, T., and Tando, Y., Pancreatic dysfunction and treatment options, *Pancreas* 16, 329-336, 1998; Divisi, D., Di Tomaso, S., Salvemini, S., *et al.,* Diet and cancer, *Acta Biomed.* 77, 118-123, 2006

Glucocerebrosidase (Acid β-glucosidase/ lysosomal β-glucosidase/lysosomal β-glucosidase) (Ceredase®/ Cerezyme®)

Replacement of a lysosomal enzyme deficiency which results in Gaucher's Disease

Wiltink, E.H. and Hollak, C.E., Alglucerase (ceredase), *Pharm.World Sci.* 18, 16-19, 1996; Bijsterbosch, M.K., Donker, W., van de Bilt, H., *et al.,* Quantitative analysis of the targeting of mannose-terminal glucocerebrosidase. Predominant uptake by liver endothelial cells, *Eur.J.Biochem.* 237, 344-349, 1996; Grabowski, G.A., Leslie, N., and Wenstrup, R., Enzyme therapy for Gaucher disease: the first 5 years, *Blood Rev.* 12,115-133, 1998; Barranger, J.A. and O'Rourke, E., Lessons learned from the development of enzyme therapy for Gaucher disease, *J.Inherit.Metab.Dis.* 24(suppl 2), 89-96, 2001; Charrow, J., Anderson, H.C., Kaplan, P., *et al., J.Pediatr.* 144, 112-120, 2004; Beutler, E., Enzyme replacement in Gaucher disease, *PLoS Med.* 1, e21, 2004; Connock, M., Burls, A., Frew, E., *et al.,* The clinical effectiveness and cost-effectiveness of enzyme replacement therapy for Gaucher's disease: a systemic review, *Health Technol.Assess.* 10, iii-iv, ix-134, 2006; vom Dahl, S., Poll, L., di Rocco, M., *et al.,* Evidence-based recommendations for monitoring bone disease and the response to enzyme replacement therapy in Gaucher's patients, *Curr.Med.Res.Opin.* 22, 1045-1064, 2006

Lactase

Used as an oral formulation for the treatment of lactose intolerance

Ramirez, F.C., Lee, K., and Graham, D.Y., All lactase preparations are not the same: results of a prospective, randomized placebo-controlled trial, *Am.J.Gastroenterol.* 89, 566-570, 1994; Gao, K.P., Mitsui, T., Fujiki, K., *et al.,* Effect of lactase preparations in asymptomatic individuals with lactase deficiency—gastric digestion of lactose and breath hydrogen analysis, *Nagoya J.Med.Sci.* 65, 21-28, 2002; Erasmus, H.D., Ludwig-Auser, H.M., Paterson, P.G., *et al.,* Enhanced weight gain in preterm infants receiving lactose-treated feeds: a randomized, double-blinded, controlled trial, *J.Pediatr.* 141, 532-537, 2002; Tan-Dy, C.R. and Ohlsson, A., Lactase treated feeds to promote growth and feeding tolerance in preterm infants, *Cochrane Database Syst.Rev.* 18, CD004591, 2005; Montalto, M., Curigliano, V., Santoro, L., *et al.,* Management and treatment of lactose malabsorption, *World J.Gastroenterol.* 12, 187-191, 2006; O'Connell, S. and Walsh, G., Physicochemical characteristics of commercial lactases relevant to their application in the alleviation of lactose intolerance, *Appl.Biochem.Biotechnol.* 134, 179-191, 2006

---

[b] Tillett, W.S. and Sherry, S., The effect in patients of streptococcal fibrinolysin (streptokinase) and streptococcal deoxyribonuclease on fibrinous, purulent, and sanguinous pleural exudations, *J.Clin.Invest.* 28, 173-190, 1949; Tillett, W.E., Sherry, S., and Read, C.T., The use of streptokinase-streptodornase in the treatment of postneumonic empyema, *J.Thorac.Surg.* 21, 275-297, 1951; Miller, J.M., Ginsberg, M., Lipin, R.J., and Long, P.H., Clinical experience with streptokinase and streptodornase, *J.Am.Med.Assoc.* 145, 620-624, 1951; Nemoto, K., Hirota, K., Ono, T., *et al.,* Effect of varidase (streptokinase) on biofilm formed by *Staphylococcus aureus, Chemotherapy* 46, 111-115, 2000; Light, R.W., Nguyen, T., Mulligan, M.E., and Sasse, S.A., The *in vitro* efficacy of varidase versus streptokinase or urokinase for liquefying thick purulent exudative material from loculated empyema, *Lung* 178, 13-18, 2000; Rutter, P.M., Carpenter, B., Hill, S.S., and Locke, I.C., Varidase: the science behind the medicament, *J.Wound Care* 9, 223-226, 2000; Zhu, E., Hawthorne, M.L., Guo, Y., *et al.,* Tissue plasminogen activator combined with human recombinant deoxyribonuclease is effective therapy for empyema in a rabbit model, *Chest* 129, 1577-1583, 2006.

[c] Greenberger, N.J., Enzymatic therapy in patients with chronic pancreatitis, *Gastroenterol.Clin.North Am.* 28, 687-693, 1999; Layer, P., Keller, J., and Lankisch, P.G., Pancreatic enzyme replacement therapy, *Curr.Gastroenterol.Rep.* 3, 101-108, 2001; DiMagno, E.P., Gastric acid suppression and treatment of severe exocrine pancreatic insufficiency, *Best Pract.Res.Clin.Gastroenterol.* 15, 477-486, 2001; Layer, P. and Keller, J., Lipase supplementation therapy: standards, alternatives, and perspectives, *Pancreas* 26, 1-7, 2003.

[d] Gaffar, A., Hunter, C.M., and Mirajkar, Y.R., Applications of polymers in dentifrices and mouthrinses, *J.Clin.Dent.* 13, 138-148, 2002; Brennan, M.T., Shariff, G., Lockart, P.B., and Fox, F.C., Treatment of xerostomia: a systematic review of the therapeutic trials, *Dent.Clin.North Am.* 46, 847-856, 2002; Guggenheimer, J. and Moore, P.A., Xerostomia: etiology, recognition and treatment, *J.Am.Dent.Assoc.* 134, 61-69, 2003; Porter, S.R., Scully, C., and Hegarty, A.M., An update of the etiology and management of xerostomia, *Oral.Surg.Oral.Med.Oral.Pathol.Radiol.Endod.* 97, 28-46, 2004; Urquhart, D. and Fowler, C.E., Review of the use of polymers in saliva substitutes for symptomatic relief of xerostomia, *J.Clin.Dent.* 17, 29-33, 2006.

THERAPEUTIC ENZYMES (Continued)

Superoxide Dismutase   Superoxide toxicity; treatment of inflammatory disorders; Also gene therapy[e] target in amyelotropic lateral sclerosis[f]

Omar, B.A., Flores, S.C., and McCord, J.M., Superoxide dismutase, in *Therapeutic Proteins. Pharmacokinetics and Pharmacodynamics*, ed. A.H.C. King, R.A. Baughman, and J.W. Larrick, W.H. Freeman and Company, New York, New York, Chapter 14, pps. 295-315, 1993; di Napoli, M. and Papa, F., M-40403 Metaphore Pharmaceuticals, *IDrugs* 8, 67-76, 2005; Leite, P.F., Liberman, M., Sandoli de Brito, F., and Laurindo, F.R., Redox processes underlying the vascular repair reaction, *World J.Surg.* 28, 331-336, 2004; Hernanez-Savedra, D., Zhou, H., and McCord, J.M., Anti-inflammatory properties of a chimeric recombinant superoxide dismutase: SOD2/3, *Biomed.Pharmacother.* 59, 204-206, 2005; St. Clair, D., Zhao, Y., Chaiswing, L., and Oberly, T., Modulation of skin tumorigenesis by SOD, *Biomed.Pharmacol.* 59, 209-214, 2005; Emerit, J., Samuel, D., and Pavio, N., Cu-Zn superoxide dismutase as a potential antifibrotic drug for hepatitis C related fibrosis, *Biomed.Pharmacother.* 60, 1-4, 2006; Yasui, K. and Baba, A., Therapeutic potential of superoxide dismutase (SOD) for resolution of inflammation, *Inflamm.Res.* 55, 359-363, 2006

Streptokinase   Plasminogen activator derived from β-hemolytic *Streptococcus* (groups A,C, and G); approximately 40-50 kD; streptokinase does not have any known catalytic activity but functions by formation of a complex with plasminogen which results in plasminogen activation. Now available as a recombinant product. Streptokinase is used in combination with DNAse(streptodornase) to treat abscesses and empyema[b].

de la Fuente Garcia, J. and Estrade, M.P., Experimental studies with recombinant streptokinase, in *Therapeutic Proteins. Pharmacokinetics and Pharmacodynamics*, ed. A.H.C. King, R.A. Baughman, and J.W. Larrick, W.H. Freeman and Company, New York, New York, Chapter 13, pps. 283-293, 1993; Konstantinides, S., Should thrombolytic therapy be used in patients with pulmonary embolism?, *Am.J.Cardiovasc.Drugs.* 4, 69-74, 2004; Capstick, T. and Henry, M.T., Efficacy of thrombolytic agents in the treatment of pulmonary embolism, *Eur.Respir.J.* 26, 664-674, 2005; Ueshima, S. and Matsuo, O., Development of new fibrinolytic agents, *Curr.Pharm.Des.* 12, 849-857, 2006; Caceres-Loriga, P.M., Perez-Lopez, H., Morlana-Herandez, K., Facundo-Sanchez, H., Thrombolysis as first choice therapy in prosthetic heart valve thrombosis. A study of 68 patients, *J.Thromb.Thrombolysis* 21, 185-190, 2006

Tissue Plasminogen Activator (tPA)   A serine protease which activate plasminogen resulting in fibrinolysis. tPA is used for the treatment of myocardial infarction and stroke. The first recombinant protein was manufactured in CHO cells; more recently an engineered form (Reteplase®) has been developed in *Escherichia coli* and is seeing clinical use.

Collen, D. and Lijnen, H.R., Tissue-type Plasminogen activator. Mechanisms of action and thrombolytic properties, *Haemostasis* 16(Suppl 3), 25-32, 1986; Anderson, J.L. Recent clinical developments in thrombolysis in acute myocardial infarction, *Drugs* 33(Suppl 3), 22-32, 1987; Hollander, J.J., Plasminogen activators and their potential in therapy, *Crit.Rev.Biotechnol.* 6, 253-271, 1987; Grossbard, E.B., Recombinant tissue plasminogen activators: a brief review, *Pharm.Res.* 4, 375-378, 1987; Montaner, J., Stroke biomarkers: Can they help us to guide stroke thrombolysis?, *Drug News Perspect.* 19, 523-532, 2006; Khaja, A.M. and Grotta, J.C., Established treatments for acute ischaemic stroke, *Lancet* 369, 319-330, 2007; Simpson, D., Siddiqui, M.A, Scott, L.J., and Hilleman, D.E., Spotlight on reteplase in thrombotic occlusive disorders, *BioDrugs* 21, 65-68, 2007

Thrombin   Therapeutic action based on the clotting of fibrinogen and aggregation of blood platelets. Thrombin is a component in fibrin sealant which is used as a tissue adhesive and is used as free-standing product as a suture support and for the treatment of vascular pseudoaneurysms.

Lundblad, R.L., Bradshaw, R.A., Gabriel, D., *et al.*, A review of the therapeutic uses of thrombin, *Thromb.Haemost.* 91, 851-860, 2004; Hagberg, R.C., Safi, H.J., Sabik, J., *et al.*, Improved intraoperative management of anastomotic bleeding during aortic reconstruction: results of a randomized controlled trial, *Am.Surg.* 70, 307-311, 2004; Aziz, O., Athanasiou, T., and Darzi, A., Haemostasis using a ready-to-use collagen sponge coated with activated thrombin and fibrinogen, *Surg.Technol.Int.* 14, 35-40, 2005; Valbonesi, M., Fibrin glues of human origin, *Best Pract.Res.Clin.Haematol.* 19, 191-203, 2006; Evans, L.A. and Morey, A.F., Hemostatic agents and tissue glues in urologic injuries and wound healing, *Urol.Clin.North Am.* 33, 1-12, 2006; Stone, P.A., AbuRhama, A.F., Flaherty, S.K., and Bates, M.C., Femoral pseudoaneurysms, *Vasc.Endovascular Surg.* 40, 109-117, 2006; Gabay, M., Absorbable hemostatic agents, *Am.J.Health Syst.Pharm.* 63, 1244-1253, 2006; Drobnic, M., Radosavljevic, D., Ravnik, D., *et al.*, Comparison of four techniques for the fixation of a collagen scaffold in the human cadaveric knee, *Osteoarthritis Cartilage* 14, 337-344, 2006

Urate oxidase (Rasburicase)   Catalyzes the oxidation of urate to 5-hydroxyisourate. Used for the treatment of hyperuricemia (excess uric acid). Specifically for tumor lysis syndrome, gout.

Bessmertny, O., Robitaille, L.M., and Cairo, M.S., Rasburicase: a new approach for preventing and/or treating tumor lysis syndrome, *Curr.Pharm.Des.* 11, 4177-4185, 2005; Oldfield, V. and Perry, C.M., Rasburicase: a review of its use in the management of anticancer therapy-induced hyperuricaemia, *Drugs* 66, 529-545, 2006; Oldfield, V. and Perry, C.M., Spotlight on rasburicase in anticancer therapy-induced hyperuricemia, *BioDrugs* 20, 197-199, 2006; Lee, S.J. and Terkeltaub, R.A., New developments in clinically relevant mechanisms and treatment of hyperuricemia, *Curr.Rheumatol.Rep.* 8, 224-230, 2006; Teng, G.G., Nair, R., and Saag, K.G., Pathophysiology, clinical presentation and treatment of gout, *Drugs* 66, 1547-1563, 2006; Higdon, M.L. and Higdon, J.A., Treatment of oncologic emergencies, *Am.Fam.Physician* 74, 1873-1880, 2006; Sood, A.R., Burry, L.D., and Cheng, D.K., Clarifying the role of rasburicase in tumor lysis syndrome, *Pharmacotherapy* 27, 111-121, 2007

---

[e] In this context, gene therapy can refer to gene augmentation therapy, gene correction therapy, and RNA silencing.

[f] Xu, Z. and Xia, X.G., RNAi therapy: dominant disease gene gets silenced, *Gene Ther.* 12, 1159-1160, 2005; Hino, T., Yokota, T., Ito, S., *et al.*, *In vivo* delivery of small interfering RNA targeting brain capillary endothelial cells, *Biochem.Biophys.Res. Commun.* 340, 263-267, 2006; Zemlyak, I., Nimon, V., Brooke, S., *et al.*, Gene therapy in the nervous system with superoxide dismutase, *Brain Res.* 1088, 12-18, 2006; Azzouz, M., Gene therapy for ALS: progress and prospects, *Biochim.Biophys.Acta* 1762, 1122-1127, 2006; Xia, X., Zhou, H., Huang, T., and Zu, Z., Allele-specific RNAi selectively silences mutant SOD1 and achieves significant therapeutic benefit *in vivo*, *Neurobiol.Dis.* 23, 578-586, 2006; Miller, T.M., Smith, R.A., and Cleveland , D.W., Amyotropic lateral sclerosis and gene therapy, *Nat.Clin.Pract.Neurol.* 2, 462-463, 2006; Davis, A.S., Zhao, H., Sun, G.H. *et al.*, Gene therapy using SOD1 protects striatal neurons from experimental stroke, *Neurosci.Lett.* 411, 32-36, 2007; Qi, X., Nauswirth, W.W., and Guy, J., Dual gene therapy with extracellular superoxide dismutase and catalase attenuates experimental optic neuritis, *Mol.Vis.* 13, 1-11, 2007; Epperly, M.W., Wegner, R., Kanai, A.J., Effects of MnSOD-plasmid liposome gene therapy on antioxidant levels in irradiated murine oral cavity orthotopic tumors, *Radiat.Res.* 167, 289-297, 2007.

THERAPEUTIC ENZYMES (Continued)

Urokinase　　　　　　　　　　　　　Originally isolated from urine, urokinase is now available a recombinant protein and is used for the treatment of thrombosis in myocardial infarction and stroke; tPA is more often used for stroke. Urokinase acts by converting plasminogen to plasmin[g].

Maksimenko, A.V. and Tischenko, E.G., New thrombolytic strategy: bolus administration of tPA and urokinase-fibrinogen conjugate, *J.Thromb. Thrombolysis* 7, 307-312, 1999; Stepanova, V.V., and Tkachuk, V.A., Urokinase as a multidomain protein and polyfunctional cell regulation, *Biochemistry* 67, 109-118, 2002; Bourekas, E.C., Slivka, A.F., and Casavant, M.J., Intra-arterial thrombolysis of a distal internal carotid artery occlusion in an adolescent, *Neurocrit.Care* 2, 179-182, 2005; Roychoudhury, P.K., Khaparde, S.S., Mattisson, R., and Kumar, A., Synthesis, regulation and production of urokinase using mammalian cell culture: a comprehensive review, *Biotechnol.Adv.* 24, 514-526, 2006; Bansal, V. and Roychoudhury, P.K., Production and purification of urokinase: a comprehensive review, *Protein Expr.Purif.* 45, 1-14, 2006; Juttler, E., Kohrmann, M., and Schellinger, P.D., Therapy for early reperfusion after stroke, *Nat.Clin.Pract.Cardiovasc.Med.* 3, 656-663, 2006; Mullen, M.T., McGarvey, M.L., and Kasner, S.E., Safety and efficacy of thrombolytic therapy in postoperative cerebral infarctions, *Neurol.Clin.* 24, 783-793, 2006.

---

[g] Plasmin is a serine protease which digests fibrin. An acyl-plasmin was developed for therapeutic use – Smith, R.A., Dupe, R.J., English, P.D., and Green, J., Fibrinolysis with acyl-enzymes: a new approach to thrombolytic therapy, *Nature* 290, 505-508, 1981; Dupe, R.J., English, P.D., Smith, R.A., and Green, D.J., Acyl-enzymes as thrombolytic agents in dog models of venous thrombosis and pulmonary embolism, *Thromb.Haemost.* 51, 249-253, 1984; Tomiya, N., Watanabe, K., Awaya, J., *et al.*, Modification of acyl-plasmin-streptokinase complex with polyethylene glycol. Reduction of sensitivity to neutralizing antibody, *FEBS Lett.* 193, 44-48, 1985; Kalindjian, S.B. and Smith, R.A., Reagents for reversible coupling of proteins to the active centres of trypsin-like serine proteinases, *Biochem.J.* 249, 409-413, 1987; Teuten, A.J., Cooper, A., Smith, R.A., and Dobson, C.M., Binding of a substrate analogue can induce co-operative structure in the plasmin-serine proteinase domain, *Biochem.J.* 293, 567-572, 1993; Lijnen, H.R., van Hoef, B., Smith, R.A., and Collen, D., Functional properties of *p*-anisolylated plasmin-staphylokinase complex, *Thromb.Haemost.* 70, 326-331, 1993.

WEIGHTS OF CELLS AND CELL CONSTITUENTS

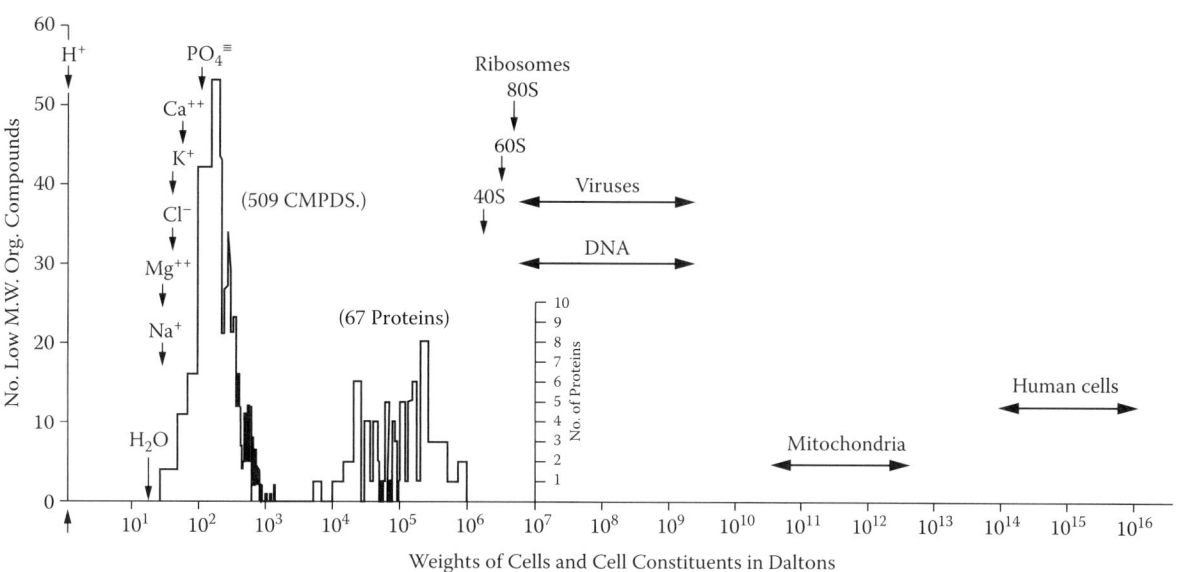

Contributed by Norman G. Anderson.

This figure originally appeared in Sober, Ed., *Handbook of Biochemistry and Selected Data for Molecular Biology,* 2nd ed., Chemical Rubber Co., Cleveland, 1970.

PARTICLE DIAMETER

Particle Diameter

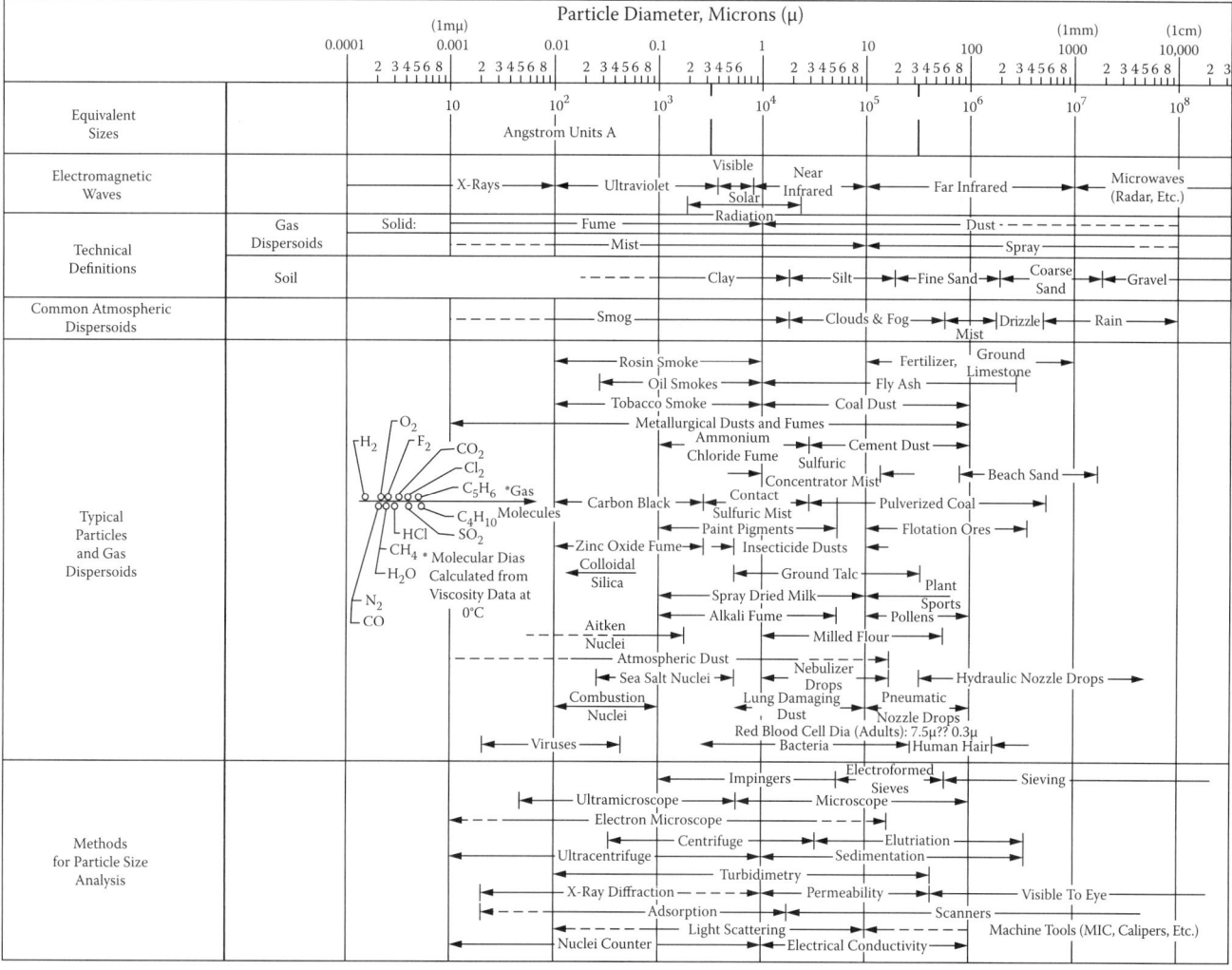

This figure originally appeared in Sober, Ed., *Handbook of Biochemistry and Selected Data for Molecular Biology*, 2nd ed., Chemical Rubber Co., Cleveland, 1970.

APPENDIX A: ABBREVIATIONS AND ACRONYMS

2D-DIGE: two-dimensional difference gel electrophoresis

2DE: two-dimensional electrophoresis

A: absorbance

A23187: a calcium ionophore, Calcimycin

AAA: abdominal aortic aneurysm; AAA+. ATPases associated with various cellular activities:

AAAA: Association Against Acronym Abuse

AAG box: an upstream *cis*-element

AAS: aminoalkylsilane; atomic absorption spectroscopy

AAT: amino acid transporter; alpha-1-antitrypsin

AAV: adenoassociated virus

ABA: Abscisic acid, a plant hormone

ABC-Transporter Proteins: ATP-Binding Cassette Transporter Proteins

ABC: ATP-Binding Cassette; antigen-binding cell

ABE: acetone butanol ethanol

Abl: retroviral oncogene derived from Abelson murine leukemia

ABRC: ABA response complex

ABRE: ABA response element

7-ACA: 7-aminocephalosporanic acid

ACES: 2-[(2-amino-2-oxyethyl)amino]-ethanesulfonic acid

ACSF: Artificial cerebrospinal fluid

ACS: active sequence collection

Ach (AcCho): acetylcholine

AChR (AcChoR): acetylcholine receptor

ACME: arginine catabolic mobile element

ACTH: adrenocorticotropin

Can: acetonitrile

Acrylodan: 6-acryloyl-2-(dimethylamino)-napthalene

ADA: adenosine deaminase; antidrug antibody

ADAM: a disintegrin and metalloproteinase

ADAMTS: a subfamily of disintegrin and metalloproteinase with thrombospondin motifs.

ADCC: antibody-dependent cell-mediated cytotoxicity as in NK cells attacking antibody-coated cells

ADH: alcohol dehydrogenase; antidiuretic hormone

ADME: adsorption, distribution, metabolism, excretion

ADME-Tox: ADME-Toxicology

AdoMet: *S*-adenosyl-L-methionine

AEC: alveolar epithelial cell

AFLP: amplified fragment length polymorphism

AFM: atomic force microscopy

AGE: advanced glycation endproducts

AGO: argonaute protein family

AGP: acid glycoprotein

AID: activation-induced cytodine deaminase

AKAP: A-kinase anchoring proteins

Akt: a protein kinase

Akt: a retroviral oncogene derived from AKT8 murine T cell lymphoma

Alk: anaplastic lymphoma kinase; receptor member of insulin superfamily

ALL: acute lymphocytic leukemia

ALP: alkaline phosphatase

ALS: anti-lymphocyte serum

ALT: alanine aminotransferase

ALV: avian leukosis virus

AML: acute myeloid leukemia

AMPK: AMP-activated protein kinase

AMS: accelerator mass spectrometry

AMT: accurate mass tag

ANDA: Abbreviated New Drug Application

ANOVA: analysis of variables (factorial analysis of variables)

ANS: 1-anilino-8-napthlenesulfonate; autonomic nervous system

ANTH: AP180 N-terminal homology as in ANTH-domain

2-AP: 2-aminopyridine

6-APA: 6-aminopenicillanic acid

APAF1: apoptotic protease activating factor 1

Apg1: a serine/threonine protein kinase required for vesicle formation which is essential for autophagy

APL: acute promyelocytic leukemia

ApoB: apolipoprotein B

AQP: adenosine tetraphosphate

ARAP3 : a dual Arf and Rho GTPase activating protein

ARD: acute respiratory disease; acireductone dioxygenase; automatic relevance determination; acid rock drainage

ARE: AU-rich elements

ARF: ADP-ribosylation factor

ARL: Arf-like

ARM: arginine-rich motif

ARS: automatic replicating sequence or autonomously replicating sequence

ART: mono-ADP-ribosyltransferase; family of proteins, large group of A-B toxins

AS: antisense

ASD: alternative splicing database; http: //www.ebi.ac.uk/asd

ASPP: ankyrin-repeat, SH3-domain and proline-rich-region-containing proteins

AST: aspartate aminotransferase

ATC: aspartate transcarbamylase domain

ATCase: aspartate transcarbamylase

ATP: adenosine-5'-triphosphate

ATPγS: adenosine-5'-3-*O*-(thiotriphosphate)

ATR-FTIR: attenuated total reflectance-Fourier transform infrared

ATR-IR: attenuated total reflection infrared

AVT: arginine vasotocin

Axl: anexceleko; used in reference to a receptor kinase related to the Tyro 3 family

BA: betaine aldehyde

BAC: bacterial artificial chromosome; also blood alcohol concentration

BAD: a member of the Bcl02 protein family considered to be a proapoptotic factor

BADH: betaine aldehyde dehydrogenase

BAEC: bovine aortic endothelial cells

BAEE: Benzoyl-arginine ethyl ester[a]

BALT: bronchial associated lymph tissue

BBB: blood brain barrier

B-CAM: basal cell adhesion molecule

BCG: bacille-Calmette-Guérin

BCR: breakpoint cluster region; B-cell receptor

BCR-ABL: BCR-ABL is the fused gene that results from the *Philadelphia chromosome*, The BCR-ABL gene produces the *BCR-ABL tyrosine kinase*,

Bcl-2: protein family regulating apoptosis

BCIP: 5-bromo, 4-chloro, 3-indoyl phosphate

BCS: biopharmaceutical classification system for describing the gastrointestinal absorption of drugs; also Budd-Chiari syndrome
BDH: D-β-butyrate dehydrogenase
BDNF: brain-derived growth factor
BEBO: an unsymmetrical cyanine dye for binding to the minor grove of DNA; 4-[(3-methyl-6-(6-methyl-benzothiazol-2-yl)-2,3,-dihydro(benzo-1,3-thiazole)-2-methylidene)]-1-methyl-pyridinium iodide.
BET: refers to an isotherm for adsorption phenomena in chromatography; acronym derived from Stepen Brunauer, Paul Emmet, and Edward Teller
B/F: bound/free
bFGF: basic fibroblast growth factor
BFP: blue fluorescent protein
BGE: background electrolyte
Bicine: *N,N*-bis(2-hydroxyethyl)glycine
BiFC: bimolecular fluorescence complementation
BIND: biomolecular interaction network database
BiP: immunoglobulin heavy chain-binding protein
Bis-**TRIS:** 2,2-bis-(hydroxymethyl)-2,2′,2″ nitriloethanol
BLA: Biologic License Application
BLAST: basic local alignment search tool
BME: 2-mercaptoethanol; β-mercaptoethanol
BMP: bone morphogenic protein
BopA: a secreted protein required for biofilm formation
BPTI: bovine pancreatic trypsin inhibitor
BCRA-1: breast cancer 1; a tumor suppressor gene associated with breast cancer
BRE-luc: a mouse embryonic stem cell line used to study bone morphogenetic protein.
BRET: bioluminescence resonance energy transfer, see FRET.
BrdU: bromodeoxyuridine
Brig: polyoxyethylene lauryl ether
BSA: bovine serum albumin
bZIP: basic leucine zipper transcription factor
C1INH: C1 inhibitor; inhibitor of activated complement component 1, missing in hereditary angioneurotic edema.
CA125: cancer antigen 125; a glycoprotein marker used for prognosis in ovarian cancer; also referred to as MUC16
CAD: multifuntional protein and initiates and regulate *de novo* pyrimidine biosynthesis; caspases-activated DNAse.
CAK: Cdk-activating kinase
CALM: clathrin assembly lymphoid myeloid leukemia as in CALM gene
CAM (CaM): calmodulin; cell adhesion molecule
CAMK: Ca²⁺/calmodulin-dependent protein kinase
CAPS: cleavable amplified polymorphic sequences; also cationic antimicrobial peptide
CArG: a promoter element [CC(A/T)₆G] gene for smooth muscle α-actin
CASP: critical assessment of structural prediction
CASPASE: cysteine-dependent asparatate-specific protease
CAT: catalase; chloramphenicol acetyl transferase
CATH: class, architecture, topology, homologous superfamily; hierarchical classification of protein domain structure.
CATP: Chloramphenicol resistance gene, caged ATP, cation transporting p-type
Cbl: a signal transducing protein downstream of a number of receptor-couple tyrosine kinases; a product of the *c-cbl* proto-oncogene
Cbs: chromosomal breakage sequence
CBz: carbobenzoxy

CCC: concordance correlation coefficient
CCD: charge couple device
CCK: choleocystokinin
CCV: clathrin-coated vesicles
CD: clusters of differentiation; circular dichroism; cyclodextrin
CDC: complement-dependent cytotoxicity; complement-mediated cell death
CDK (cDK): cyclin-dependent kinase
cDNA: complementary DNA
CDR: complementary determining region
CDTA: 1,2-cyclohexylenedinitriloacetic acid
CE: capillary electrophoresis
CEC: capillary electrochromatography
CE-SDS: capillary electrophoresis in the presence of sodium dodecyl sulfate
CELISA: cellular enzyme-linked immunosorbent assay; enzyme-linked immunosorbent assay on live cells
CERT: ceramide transport protein
CEPH: Centre d'Etude du Polymorphisme Humain
CEX: cation exchange
CFA: complete Freund's adjuvant
CFP: Cyan Fluorescent Protein
CFTR: cystic fibrosis transmembrane conductance region
Cfuc: colony forming unit
CGE: capillary gel electrophoresis
CGH: comparative genome hybridization
CGN: *cis*-Golgi network
CH: calponin homology
CHAPS: 3-[(3-cholamidopropyl)dimethylammonio]-1-propane-sulfonic acid
CHCA: α-cyano-4-hydroxycinnamic acid
CHEF: chelation-enhanced fluorescence
CHES: 2-(*N*-cyclohexylamino)ethanesulfonic acid
ChiP: chromatin immunoprecipitation
CHO: chinese hamster overy; carbohydrate
CID: collision-induced dissociation; collision-induced dimerization
CIDEP: chemically induced dynamic electron polarization
CIDNP: chemically induced dynamic nuclear polarization
CIEEL: chemically initiated electron exchange luminescence
CLIP: class-II-associated invariant chain (Ii) peptide
CLT: clotvinazole [1-(α2-chlorotrityl)imidazole]
CLUSTALW: A general purpose program for structural alignment of proteins and nucleic acids http://www.ebi.ac.uk/clustalw/
cM: centimorgan
CM: carboxymethyl
CaMK: calmodulin kinase, isoforms I, II, III
CMCA : competitive metal capture analysis
CML: chronic myelogenous leukemia
CML: carboxymethyl lysine
CAN: bacterial cell wall collagen-binding protein
Cn: calcineurin
CNC: Cap'n'Collar family of basic leucine zipper proteins
CNE: conserved non-coding elements
dCNE: duplicated CNE
COACH: comparison of alignments by constructing hidden Markow models
CoA: coenzyme A
COFFEE: consistency based objective function for alignment evaluation
COFRADIC: combined fractional diagonal chromatography
COG: conserved oligomeric Golgi; cluster of orthologous groups
COPD: chronic obstructive pulmonary disease

COX: cytochrome C oxidase
Cp: ceruloplasmin
CPA: carboxypeptidase A
CPB: carboxypeptidase B
CPD: cyclobutane pyrimidine dimer
CPDK: calcium-dependent protein kinase
CpG: cytosine-phosphate-guanine
CpG-C: cytosine-phosophate-guanine class C
Cdpk4: a Ca^{2-} protein kinase
CPP: cell penetrating peptide; combinatorial protein pattern
CPSase: carbamoyl-phosphate synthetase
CPY: carboxypeptidase Y
CRAC: calcium-release activated calcium (channels)
CRE: cyclicAMP response element
Cre1: cytokine response 1; a membrane kinase
CREA: creatinine
CREB: cAMP-response element binding protein
CRM: certified reference material
CRP: C-reactive protein; also cAMP receptor protein
CRY: chaperone
CS: chondroitin sulfate
CSF: colony stimulating factor
CSP: cold-shock protein
CSR: cluster-situated regulator; class-switch recombination
CSSL: chromosome segment substitution lines
Ct: chloroplast
CT: charge transfer
CTB: cholera toxin B subunit
CTD: C-terminal domain
CTL: cytotoxic T lymphocytes
CTLA: cytotoxic T lymphocyte-associated antigen
CTLL: cytotoxic T-cell lines
CTPSase: CTP synthetase
CtrA: a master regulator of cell cycle progression
CV: coefficient of variation
Cvt: cytosome to vacuole targeting
CW: continuous wave(non-pulsed source of electromagnetic radiation)
CYP: cytochrome P450 enzyme
Cst3: cystatin 3
CZE: capillary zone electrophoresis
D: diffusion
D_{ax}: axial dispersion coefficient
DAB(p-dab): p-dimethyl amino azo benzene
dABs: domain antibodies
DABSYL: *N,N*-dimethylaminoazobenzene-4'-sulfonyl-usually as the chloride, DABSYL chloride
DAD: diaphanous-autoregulatory domain
DAF: decay accelerating factor
DAG: diacyl glycerol
DALI: distance matrix alignment; http: //www2.ebi.ac.uk/dali/
DANSYL: 5-dimethylaminonapthalene-1-sulfonyl; usually as the chloride (DANSYL chloride)
DAP: DNAX-activation protein; also diaminopimelic acid;
DAP12: DNAX activating protein of 12kDa mass
DAS: distributed annotated system; downstream activation site
DBMB: Dulbecco's modified Eagles Medium
DBD-PyNCS: 4-(3-isocyanatopyrrolidin-1-yl)-7-(*N,N*-dimethylaminosulfonyl)-2-benzoxadiazole
DBTC: "Stains All"; 4,5,4',5'-dibenzo-3,3'-diethyl-9-methylthiacarbocyanine bromide
DC: dendritic cell
DCC: dicyclohexylcarbodiimide

DCCD: *N,N'*-dicyclohexylcarbodimide
DDBJ: DNA Databank of Japan; http: //www.ddbj.nig.ac.jp
DDRs: discoidin domain receptors (DDR1, DDR2)
DDR1: discoidin domain receptor1, CAK, CD167a, PTK3, Mck10
DDR2: discoidin domain receptor2, NTRK3, TKT, Tyro10
DEAE: diethylaminoethyl
DEG: differentially expressed gene(s)
DEX: dendritic-cell-derived exosomes
DFF: DNA fragmentation factor
DFP: diisopropylfluorophosphate; diisopropylphosphorofluoridate
DHFR: dihydrofolate reductase
DHO: dihydroorotase domain
DHOase: dihydroorotase
DHPLC(dHPLC): denaturing HPLC
DHS: DNase I hypersensitivity site
DIP: database of interacting proteins - http: //dip.doe-mbi.ucla.edu; also dictionary of interfaces in proteins – http: //drug-redesign.de/superposition.html
Dipso: 3-[*N,N.*-bis(2-hydroxyethyl)amino]-2-hydroxypropane-sulfonic acid
DLS: dynamic light scattering
DM: an accessory protein located in the lysosome associated with MHC-class-II antigen presentation. It is located in the endosomal/lysosomal system of APC
DMBA: 7,12-dimethylbenzo[α]anthracene
DMD: Duchenne muscular dystrophy; also Doctor of Dental Medicine
DMF: dimethylformamide; decayed, missing, filled(denistry)
DMS: dimethyl sulfate
DMSO: dimethyl sulfoxide
DMT1: divalent metal transporter 1
ssDNA: single-stranded DNA
DNAa: a bacterial replication initiation factor
DNAX: DNAase III, tau and gamma subunits
dNPT: deoxynucleoside triphosphate
DO: an accessory protein located in the lysosome associated with MHC-class-II antigen presentation. DO has an accessory role to DM
DOTA: tetraazacyclodedecanetetraacetic acid
DPE: downstream promoter element
DPI: dual polarization interferometry
DPM: disintegrations per minute
DPN: diphosphopyridine dinucleotide (currently NAD);
DPPC: dipalmitoylphosphatidylcholine
DPPE: 1,2-dipalmitoyl-*sn*-glycerol-3-phosphoethanolamine
DPTA: diethylenetriaminepentaacetic acid
dsDNA: double-stranded DNA
dsRNA: double-stranded RNA
dsRBD: double-stranded RNA binding domain
DRE: dehydration response element; dioxin response element
DRT: dimensionless retention time (a value for chromatography)
DSC: differential scanning calorimetry
DSP: downstream processing
DTAF: dichlorotriazinyl aminofluorescein
DTE: dithioerythritol
DTNB: 5,5'-dithio-bis(2-nitrobenzoic acid) Ellman's Reagent
DTT: dithiothreitol
DUP: a duplicated yeast gene family
DVDF: polyvinyl difluoride
E1: ubiquitin-activating enzyme
E2: ubiquitin carrier protein
E3: ubiquitin-protein isopeptide ligase

E-64: *trans*-epoxysuccinyl-L-leucylamino-(4-guanidino)-butane, proteolytic enzyme inhibitor
EAA: excitatory amino acid
EBA: expanded bead adsorption
EBV: Epstein-Barr virus
EDC(EADC): 1-ethyl-3-(3-dimethylaminopropyl) carbodiimide; *N*-ethyl-*N'*-(3-dimethyl-aminopropyl) carbodiimide
ECF: extracytoplasmic factor; extracellular fluid
ECM: extracellular matrix
EDC: 1-ethyl-(3-dimethylaminopropyl)-carbodiimide
EDI: electrodeionization
EDTA: ethylenediaminetetraacetic acid, Versene, (ethylenedinitrilo) tetraacetic acid;
EEO: electroendoosmosis
EEOF: electroendoosmotic flow
EF: electrofiltration
EGF: epidermal growth factor
EGFR: epidermal growth factor receptor; Erbβ-1; HER1
EGTA: ethyleneglycol-bis(β-aminoethylether)-*N,N,N',N'*-tetraacetic acid;
eIF: eukaryotic initiation factor
EK: electrokinetic
EKLF: erythroid Krüppel-like factor
ELISA: enzyme-linked immunosorbant assay
EMBL: European Molecular Biology Laboratory
EMCV: encephalomyocarditis virus
EMF: electromotive force
EMMA: enhanced mismatch mutation analysis
EMSA: electrophoretic mobility shift assay
ENaC: epithelial Na channel
EndoG: endonuclease G
ENTH: epsin *N*-terminal homology as ENTH-domain
ENU: *N*-ethyl-*N*-nitrosourea
EO: ethylene oxide
EOF: electroosmotic flow
Eph: a family of receptor tyrosine kinases; function as receptors/ligands for ephrins
EPL: expressed protein ligation
Epps: 4-(2-hydroxyethyl)-1-piperazinepropanesulfonic acid
EPR: electron paramagnetic resonance
ER: endoplasmic reticulum
ERAD: endoplasmic reticulum-associated protein degradation
ErbB2: epidermal Growth Factor Receptor, HER2
ErbB3: epidermal Growth Factor Receptor, HER3
ErbB4: epidermal Growth Factor Receptor, HER4
ERK: Extracellular-regulated kinase
Erk ½: P 42/44 extracellular signal-regulated kinase
Ero1p: a thiol oxidase which generates disulfide bonds inside in the endoplasmic reticulum
ERSE: endoplasmic reticulum(ER) stress response element
ES: embryonic stem as in embryonic stem cell
ESI: electrospray ionization
ESR: electron spin resonance; also erthyrocyte sedimentation rate
ESS: exonic splicing silencer
EST: expressed sequence tag
ETAAS: electrothermal atomic absorption
5,6-ETE: 5,6-epoxyeicosatrienoic acid
ETS: family of transcription factors
EUROFAN: European functional analysis network - http://mips.gsf.de/proj/eurofan/; also European Programme for the Study and Prevention of Violence in Sport
Exo1: exonuclease 1

EXP1: expansion gene
FAAH: fatty acid amide hydrolase
Fab: antigen binding fragment from immunoglobulin
FAB: fast atom bombardment
FAB-MS/MS: fast atom bombardment-mass spectrometry/mass spectrometry
FACE: fluorophore-assisted carbohydrate electrophoresis
FACS: fluorescence-activated cell sorting
FADD: Fas association death domain
FAD: flavin adeninine dinucleotide
FAK: focal adhesion kinase
FBS: fetal bovine serum
Fc: Fc region of an immunoglobulin representing the C-terminal region
FcγR: cell surface receptor for the Fc domain of IgG
FDA: fluorescein diacetate
FDC: follicular dendritic cells
FCCP: carbonyl cyanide *p*-trifluoromethoxyphenyl-hydrazine
FEAU: 2'-fluoro-2'-deoxy-β-D-arabinofuranosyl-5-ethyluracil
FERM: as in FERM-domain (four-point-one; ezrin, radixin, moesin).
FFAT: two phenylalanyl residues in an acidic tract
FIAU: 2'-fluoro-2'-deoxy-β-D-arabinofuranosyl-5-iodouracil
FecA: ferric citrate transporter
FEN: flap endonuclease
***Fes*:** retroviral oncogene derived from ST and GA feline sarcoma
FFPE: formalin-fixed, paraffin-embedded
FGF: fibroblast growth factor
FGFR: fibroblast growth factor receptor
***Fgr*:** retroviral oncogene derived from GR feline sarcoma
FIGE: field-inversion gel electrophoresis
FITC: fluoroscein isothiocyanate
FTIR: Fourier-transformed infrared reflection
FTIR-ATR: Fourier-transformed infrared reflection-attenuated total reflection
FLAG™: an epitope "tag" which can be used as a fusion partner for recombinant protein expression and purification.
FlhB: a component of the flagellum-specific export apparatus in bacteria.
FLIP: fluorescence loss in photobleaching
FLK-1: vascular endothelial growth factor receptor (VEGFR)
FLT-1: vascular endothelial growth factor receptor (VEGFR)
fMLP(FMLP): *N*-formyl methionine leucine phenylalanine
fMOC: 9-Fluorenzylmethyloxycarbonyl
***Fms*:** retroviral oncogene derived from SM feline sarcoma
Fok1: a type IIS restriction endonuclease derived from *Flavobacterium okeanokoites*
***Fos*:** retroviral oncogene derived from FBJ murine osteosarcoma
FOX: forkhead box
FpA: fibrinopeptide A
FPC: fingerprinted contigs
***Fps*:** retroviral oncogene from Fujiami avian sarcoma
FRAP: fluorescence recovery after photobleaching
FRET: fluorescence resonance energy transfer; Förster resonance energy transfer
FT: Fourier Transform
FSSP: fold classification based on structure alignment of proteins; http://www2.ebi.ac.uk/dali/fssp/fssp.html
FU: aluorescence unit
5-Fu: 5-fluorouracil
Fur: ferric uptake receptor
***Fur*:** gene for Fur

FYVE: a zinc-binding motif; acronym derived from four proteins containing this domain

G: guanine

Gα: heterotrimeric G protein, α-subunit

Gβ: heterotrimeric G protein, β-subunit

Gγ: heterotrimeric G protein, γ-subunit

G-6-PD: glucose-6-phosphate dehydrogenase

GABA: gamma (γ)-aminobutyric acid

GAG: glycosaminoglycan

GALT: gut-associated lymphoid tissues

GalNac: *N*-acetylgalactosamine

GAPDH: glyceraldehyde 3-phosphate dehydrogenase

GAPS: GTPase activating proteins

GAS6: a protein, member of the vitamin K-dependent protein family

GASP: Genome Annotation Assessment Project; http: //www.fruitfly.org/GASP1/; also growth advantage in stationary phase

GBD: GTPase binding domain

GC: gas chromatography; granular compartment

GC-MS: gas chromatography-mass spectroscopy

GC-MSD: gas chromatography-mass selective detector

GcrA: A master regulator of cell cycle progression

G-CSF: granulocyte colony stimulating factor

GCP: Good Clinical Practice

GDH: glutamate Dehydrogenase

GDNF: glial-derived neurotrophic factor

GdnHCl: guanidine hydrochloride

GEFs: guanine nucleotide exchange factors

GF-AAS: graphite furnace atomic absorption spectroscopy

GFP: green fluorescent protein

GGDEF: a protein family

GGT: gamma-glutamyl transferase

GGTC: German Gene Trap Consortium; a reference library of gene trap sequence tags (GTST) http: //www.genetrap.de/

GHG: greenhouse gas

GI: gastrointestinal; genomic islands

cGK: cyclic GMP (cGMP)-dependent protein kinase

GlcNac: *N*-acetylglucosamine

GLD: gelsolin-like domain

GLP: Good Laboratory Practice(s)

GlpD: glyceraldehyde-3-phosphate dehydrogenase

GLUT: a protein family involves in the transport of hexoses into mammalian tissues

Glut4: facilitative glucose transporter which is insulin-sensitive

Glut5: a fructose transporter, catalyzes the uptake of fructose

GM: genetically modified

GMP-PDE(cyclic GMP-PDE): cyclic GMP-phosphodiesterase

GM-CSF: granulocyte-macrophage colony stimulating factor

cGMP: current Good Manufacturing Practice

GNSO: 5-nitrosoglutathione

GPC: gel permeation chromatography

GPCR: G-protein coupled receptor

GPI: glycosyl phosphatidylinositol

GRIP: a Golgi-targeting protein domain

GRP: glucose-regulated protein

Grp78: a glucose regulated protein; identical with BiP

GSH: glutathione

GST: glutathione-*S*- transferase; gene trap sequencing tag

GTF: general Transcription factor

GTST(GST): gene trap sequence tags

GUS: beta-glucuronidase

GXP(s): A generic acronymy for good .... practices including but not limited to good clinical practice, good laboratory practice, good manufacturing processes.

HA: hemaglutin-A; hyaluronic acid; hydroxyapaptite, $Ca_{10}(PO_4)_6(OH)_2$

HABA: [2-(4'-hydroxyazobenzene)] benzoic acid;

HAS: human serum albumin; hyaluron synthase

HAT: histone acetyltransferase; hypoxanthine, aminopterin and thymidine

HBSS: Hanks Balanced Salt Solution

H/D: hydrogen/deuterium exchange

HAD: heteroduplex analysis

HDAC: histone deacetylase

HDL: high-density lipoprotein

HDLA: human leukocyte differentiation antigen

HD-ZIP: homeodomain-leucine zipper proteins

HEPT: height equivalent to plate number

HERV: human endogenous retrovirus

20-HETE: 20-hydroxyeicosatetraenoic acid

HETP: plate height (chromatography);

HexNac: *N*-acetylhexosamine

HGP: human genome project

HH: hereditary hemochromatosis

HHM: hidden Markov models

His-Tag(His$_6$; H$_6$): histidine tag – a hexahistidine sequence

HLA: human leukocyte associated antigen

HLA-DM: enzyme responsible for loading peptides onto MHC class II molecules

HLA-DO: protein factor which modulates the action of HLA-DM

HMGR: 3-hydroxy-3-methylglutamyl-coenzyme A reductase

HMP: herbal medicinal product(s)

HMT: histone

hnRNA: heterologous nuclear RNA

HOG: high-osmolarity glycerol

HOPE: Hepes-glutaminic acid-buffer mediated organic solvent protein effect

HOX(*HOX, hox*): describing a family of transcription factors

HPAED-PAD: high performance anion-exchange chromatography-pulsed amperometric detection

5-HPETE: 5-hydroperoxyeicosatetranenoic acid

HPRD: human protein reference database

HPRT: hypoxanthine phosphoribosyl transferase

HRP: horse radish peroxidase

HS: heparan sulfate

HSB: homologous synteny blocks

HSC: hematopoietic stem cell

HSCQ: heteronuclear single quantum correlation

HSE: heat shock element

Hsp: heat-shock protein

Hsp70: heat shock protein 70

5-HT: 5-hydroxytryptamine

HTF: *Hpa*II tiny fragments; distinct fragments from the *Hpa*II digestion of DNA; *Hpa*II is a restriction endonuclease

HTH: helix-turn-helix

HTS: high-throughput screening

htSNP: haplotype single nucleotide polymorphism

HUGO: human genome organization

HUVEC: human umbilical vein endothelial cells

IAA: iodoacetic acid

IAEDANS: *N*-iodoacetyl-*N'*-(5-sulfo-1-napthyl) ethylenediamine

IBD: identical-by-descent; also inflammatory bowel disease

IC: ion chromatography

ICAM: intercellular adhesion molecule

ICAT: isotope-coded affinity tag

ICH: intracerebral hemorrhage; a gene related to *Ice* involved in programmed cell death; historically, international chick unit; International Conference for Harmonisation

ICPMS: inductively coupled plasma mass spectrometry

ID: internal diameter

IDA: interaction defective allele

IDMS: isotope dilution mass spectrometry

IEC: ion-exchange chromatography

IEF: isoelectric focusing

IES: internal eliminated sequences

IFE: immunofixation electrophoresis

IFN: interferon

Ig: immunoglobulin

IGF: insulin-like growth factor

IGFR: insulin-like growth factor receptor

Ihh: indian hedgehog

IκB: NF-κB inhibitor

IκK: IκB kinase

IL: Interleukin

iLAP: integrated lysis and purification

ILGF: insulin-like growth factor

ILGFR: insulin-like growth factor receptor

ILK: integrin-linked kinase

IMAC: immobilized metal-affinity chromatography

IMINO: Na$^+$-dependent alanine-insensitive proline uptake system (SLC6A20)

IMP: integrin-mobilferrin pathway membrane protein system involves in the transport of ferric iron; also inosine-5'- monophosphate

IPG: immobilized pH gradient

IPTH: isopropylthio-β-D-galactopyranoside

IP$_3$: inositol 1,4,5-triphosphate

IPTG: Isopropylthio-β-D-galactosidase

IR: inverted repeat; insulin receptor

IRES: internal ribosome entry site

IRS: insulin receptor substrate

ISE: ion-specific electrode

ISO: International Standards Organization

ISS: immunostimulatory sequence; also intronic splicing silencer

ISS-ODN: immunostimulatory sequence-oligodeoxynucleotide

ISSR: inter-simple sequence repeats

IT: isotocin

ITAF: IRES trans-acting factor

ITAM: immunoreceptor tyrosine-based activation motif

ITC: isothermal titration calorimetry

iTRAQ: isobaric tags for relative and absolute quantitation of proteins in proteomic research

JAK: Janus Kinase

JNK: *c*-Jun *N*-terminal kinase

KARAP: killer cell activating receptor-associated protein

Kb, kb : Kilobase

KDR: kinase insert domain-containing receptor; KDR is the human homolog of the mouse FLK-1 receptor. The KDR and FLK-1 receptors are also known as VEGFR2. See VEGFR

Kit: mast/stem cell growth factor receptor, CD 117

Kit: retroviral oncogene derived from HZ4 feline sarcoma

KLF5: Kruppel-like factor 5, a transcription factor

LAK: lymphokine-activated killer cells

LATE-PCR: linear-after-the-expotential-PCR

LB: Luria-Bertani

Lck: member of the Src family of protein kinases

LC$_{50}$: median lethan concentration in air

LC-MS: liquid chromatography-mass spectrometry

LCR: low-copy repeat; locus control region; low complexity region

LCST: lower critical solution temperature

LD: as in LD motif, a leucine/aspartic acid-rich protein-binding domain; also used to refers to peptidases without stero-specificity; also longin domain; linkage disequilibrium; lactate dehydrogenase

LD$_{50}$: median lethal dose

LDL: low-density lipoprotein

LECE: ligand exchange capillary electrophoresis

LED: light emitting diode

Lek: lymphocyte-specific protein tyrosine kinase

LFA: lymphocyte function-associated antigen

LGIC: ligand-gated ion channel

LH: luteinizing hormone

LIF: laser-induced fluorescence

LIM : a domain involved in protein-protein interaction, originally described in transcription factors LIN1, ISL1, and MED3.

LINE: long interspersed nuclear element

LLE: liqud-liquid extraction

LLOD: lower limit of detection

LLOQ: lower limit of quantification

Lnr: initiator element

lnRNP: large nuclear ribonucleoprotein

LOD: limit of detection; log$_{10}$ of odds

LOLA: list of lists - annotated

LOQ: limit of quantitation

LP: lysophospholipid

LPA: lysophosphatidic acid

LPH: lipotropic hormone

LPS: lipopolysaccharide

LTB$_4$: leukotriene B$_4$

LTH: luteotropic hormone

Ltk: leukocyte tyrosine kinase

LRP: low-density lipoprotein receptor-related protein

LSPR: localized surface plasmon resonance

LTR: long terminal repeat

LUCA: last universal cellular ancestor

M13: a bacteriophage used in phage display

MΦ: macrophage

Mab, MABq: monoclonal antibody

MAC: membrane attack complex

MAD: multiwavelength anomalous diffraction

Maf: retroviral oncogene derived from AS42 avian sarcoma

MAGE: microarray and gene expression

MALDI-TOF: matrix-assisted laser desorption ionization time of flight

MAP: mitogen-activated protein; usually referring to a protein kinase such MAP-kinase

MAPK: MAP-kinase

MAPKK: MAP-kinase kinase

MAPKKK: MAP-kinase kinase kinase

MAR: matrix attachment region

Mb, mb: megabase (10^6)

MB: molecular beacon

MBL: mannose-binding lectin; maltose-binding protein

MBP: myelin basic protein

MCA: 4-methylcoumaryl-7-acetyl

MCAT: mass coded abundance tag

MCD: magnetic circular dichroism

MCM: mini-chromosome maintenance

MCS: multiple cloning site

M-CSF: M-colony stimulating factor (Macrophage-colony stimulating factor)

MDA: malondialdehyde

MDMA: 3,4-methylenedioxymethamphetamine

MDCK: Madin-Darby canine kidney

MEF: mouse embryonic fibroblasts

MEF-2: myocyte enhancer factor 2

MEGA-8: octanoyl-*N*-methylglucamide

MEGA-10: decanoyl-*N*-methylglucamide

MEK: mitogen-activated protein kinase/extracellular signal-regulated kinase kinase; also methylethyl ketone

MELC: microemulsion liquid chromatography

MELK: multi-epitope-ligand-kartographie

MEM: minimal essential medium

Mer: a receptor protein kinase; also Mertk, Mer tyrosine kinase

MES: 2-(*N*-morpholinoethanesulfonic acid)

Met: receptor for hepatocyte growth factor

MFB: membrane fusion protein

MGO: methylglyoxal

MGUS: monoclonal gammopathy of undetermined significance

MHC: major histocompatibility complex

MIAME: minimum information about a microarray experiment

Mil: retroviral oncogene derived from Mill Hill-2 chicken carcinoma

MIP: molecularly imprinted polymer; macrophage inflammatory protein; methylation induced premeiotically

MIPS: Munich Information Center for Protein Sequences

MIS: Mullerian Inhibiting Substance

MLCK: myosin light chain kinase

MLCP: myosin light chain phosphatase

MMP: matrix metalloproteinase

MMR: mismatch repair

MMTV: mouse mammary tumor virus

MOPS: 3-(*N*-morpholino)propanesulfonic acid; 4-morpholinopropanesulfonic acid

MOPSo: 3-(*N*-morpholino)-2-hydroxypropanesulfonic acid

Mos: retroviral oncogene derived from Moloney murine sarcoma

MPD: 2-methyl-2,4-pentanediol

MPSS: massively parallel signature sequencing

MR: magnetic resonance

MRI: magnetic resonance imaging

mRNA: messenger RNA

MRP: migratory inhibitory factor-related protein

MRTF: myocardin-related transcription factor

MS: mass spectrometry, also mechanosensitive (receptors), multiple sclerosis

MS/MS: mass spectrometry/mass spectrometry

MS³: tandem mass spectrometry/mass spectrometry/mass spectrometry

MSP: macrophage stimulating protein

Mt: mitochondrial

mt-DNA: mitochondrial DNA

MTBE: methyl-*t*-butyl ether

MTOC: microtubule organizing center

mTOR: a eukaryotic regulatory of cell growth and proliferation. See TOR

MTSP: membrane type serine proteases

MTT: methylthiazoletetrazolium

MTX: methotrexate

MU: Miller Units

Mu: Mutator

MUSK: muscle skeletal receptor tyrosine kinase

MuDPiT: multidimensional protein identification technology

MuLV: Muloney leukemia virus

MWCO: molecular weight cut-off

My: million years

Myb : retroviral oncogene derived from avian myeloblastosis

Myc: retroviral oncogene derived from MC29 avian myelocytomatosis

MYPT: myosin phosphatase-targeting

Mys: myristoylation site

NAA: neutron activation analysis

Nabs: neutralizing antibodies

nAChR (nAcChoR): nicotinic acetylcholine receptor

NAD: nicotinamide adenine dinucleotide (DPN)

NADP: nicotinamide adeninine dinucleotide phosphate (TPN)

NAO: non-animal origin

NAT: nucleic acid amplification testing; nucleic acid testing.

Nbs₂: Ellman's Reagent; 5,5'-dithiobis(2-nitrobenzene acid)

NBD: nucleotide-binding domain

NBD-PyNCS: 4-(3-isothiocyanato pyrrolidin-1-yl)-7-nitro-2,1,3-benzoxadiazole

NBS: *N*-bromosuccinimide

NBT: nitroblue tetrazolium

NCBI: National Center for Biotechnology Information

NCED: 9-*cis*-epoxycarotenoid dioxygenase

NDA: New Drug Application

NDB: nucleic acid databank

NDMA: *N*-methyl-D-aspartate

NDSB: 3-(1-pyridinio)-1-propanesulfonate (non-detergent sulfobetaine)

NEM: *N*-ethylmaleimide

NEO: Neopterin

NEP: nucleous-encoded polymerase (RNA polymerase)

NeuAc: *N*-acetylneuraminic acid

NeuGc: *N*-glycolylneuraminic acid

NF: National Formulary

NFAT: nuclear factor of activated T cells, a transcription factor

NF-κB: nuclear factor kappa B, a nuclear transcription factor

NGF: nerve growth factor

NGFR: nerve growth factor receptor

NHS: *N*-hydroxysuccinimide

NK: natural killer (as in cytotoxic T cell)

NKCF: natural killer cytotoxic factor

NKF: *N*-formylkynurenine

Ni-NTA: Ni^{2+}-nitriloacetate

NIR: near infrared

NIRF: near infrared fluorescence

NIST: National Institute of Standards and Technology

NMDA: *N*-methyl-D-aspartate

NMM: nicotinamide mononucleotide

NMR: nuclear magnetic resonance

NO: nitric oxide

NOE: nuclear Overhauser effect

NOESY: nuclear Overhauser effect spectroscopy

NOHA: N^w-hydroxy-L-arginine

NORs: specific chromosomal sites of nuclear reformulation

NOS: nitric oxide synthetase

iNOS: inducible oxide synthetase

NPC: nuclear pore complex

*p***NPP:** *p*-Nitrophenyl phosphate

NSAID: non-steroid anti-inflammatory drug(s)
NSF: *N*-ethylmaleimide sensitive factor; National Science Foundation; *N*-ethylmaleimide-sensitive fusion
Nt,nt: nucleotide
NTA: nitriloacetic acid
NTPDases: nucleoside triphosphate diphosphohydrolases; also known as apyrases, E-ATPases
NuSAP: nucleolar spindle-associated protein
OCED: Organization for Economic Cooperation and Development
ODMR: optically detected magnetic resonance
ODN: oligodeoxynucleotide
OFAGE: orthogonal-field-alternation gel electrophoresis
OHQ: 8-hydroxyquinoline
OMG: Object Management Group
OMIM: online Mendelian Inheritance in Man (database) OMIM220100 located at http: //www.ncbi.nlm.nih.gov
OMP: outer membrane protein; A protein family associated with membranes
OMT: outer membrane transport
OPG: Osteoprotegerin
ORC: origin recognition complex
ORD: optical rotatory dispersion
ORF: open reading frame
ORFan: orphan open-reading frame
ORFeome: the protein-coding ORFs of an organism
uORF: upstream open reading frame
OSBP: oxysterol-binding proteins
OVA: Ovalbumin
OXPHOS: oxidative phosphorylation
OYE: Old Yellow Enzyme
p53: A nuclear phosphoprotein which functions as a tumor suppressor.
PA: peptide amphiphile
PAC: P1-derived artificial chromosome
PACAP: pituitary adenyl cyclase-activating polypeptide
PAD: peptidylarginine deiminase; protein argininine deiminase (EC 3.5.5.15)
PADGEM: platelet activator-dependent granule external membrane protein; GMP-140
PAGE: polyacrylamide gel electrophoresis
PAH: polycyclic aromatic hydrocarbon
PAK: P21-activated kinase
PAO: A redundant gene family (seripaoparin)
PAR: protease-activated receptor
PAS: preautophagosomal structure
PAT1: H+-coupled amino acid transporter (slc36a1)
PAZ: a protein interaction domain; PIWI-argonaute-zwille
PBS: phosphate-buffered saline
PBST: phosphate-buffered saline with Tween-20
PBP: periplasmic binding protein
PC: polycystin; phosphatidyl choline
PCAF: p300/CBP-associated factor, a histone acetyltransferase
PCNA: proliferating Cell Nuclear Antigen; processing factor
PDB: Protein Data Bank
PDE: phosphodiesterase
PDGF: platelet-derived growth factor
PDGFR: platelet-derived growth factor receptor
pDNA : plasmid DNA
PDI: protein disulfide isomerase
PDMA: polydimethylacrylamide
PDMS: polydimethylsiloxane
PE: phycoerythrin; polyethylene
PEC: photoelectrochemistry

PECAM-1: platelet/endothelial cell adhesion molecule-1
PEI: polyethyleneimine
PEND protein: a DNA-binding protein in the inner envelope membrane of the developing chloroplast
PEP: phosphoenol pyruvate
PEPCK-C: phosphoenolpyruvate carboxykinase, cytosolic form
PERK: double-stranded RNA-activated protein kinase-like ER kinase
PES: photoelectron spectroscopy
PET: positron emission tomography
PEP: plastid-encoded polymerase (RNA polymerase)
Pfam: a protein family database; protein families database of alignments
PFGE: pulsed-field gel electrophoresis;
PFK: phosphofructokinase
PFU: plaque forming unit
PG: phosphatidyl glycerol; prostaglandin,
3-PGA: 3-phospho-D-glycerate
PGO: phenylglyoxal
PGP-Me: archaetidylglycerol methyl phosphate
PGT box: an upstream *cis*-element
PGx(PGX): pharmacogenetics (PGx) is the use of genetic information to guide drug choice; prostaglandins(PGX) include thromboxanes and prostacyclins.
PH: pleckstrin homology
PHD: plant homeodomain
pHB(*p*-HB): 4-hydoxybenzoic acid (*p*-hydroxybenzoate)
PI: propidium iodide
PIC: pre-initiation complex – complex of GTFs
PINCH: PINCH-protein; particularly interesting cis-his-rich protein
PIP$_3$: phosphatidylinositol-3,4,5-triphosphate
PIP$_n$: polyinositol polyphosphate
PIP$_n$S: polyinositol polyphosphates
Pipes: 1,4-piperzainediethanesulfonic acid
PIRLβ: paired immunoglobulin-like type 2 receptor β
PKA: protein kinase A; cAMP-dependent kinase; pKa, acid dissociation constant
PKC: protein kinase C
Pkl: paxillin kinase linker
PLL: poly-L-Lysine
PLP: pyridoxal-5-phosphate
PMA: phenyl mercuric acetate; phorbol-12-myristate-13 acetate;
PMCA: plasma membrane Ca^{2+} as PMCA-ATPase, a PMCA pump
PMSF: phenylmethylsulfonyl fluoride
PNA: peptide nucleic acid; *p*-nitroanilide
PNGase: endoglycosidase
PNP: *p*-nitrophenol (4-nitrophenol)
POD: peroxidase
POET: pooled ORF expression technology
POINT: prediction of interactome database
Pol II: RNA polymerase II
POTRA: polypeptide translocation associated
PP: polypropylene
PPAR: peroxisome proliferator activated receptor
PPase: phosphoprotein phosphatase
PQL: protein quantity loci
PS: position shift polymorphism
PSG: pregnancy-specific glycoprotein(s)
PS-1: presenilin-1
PSI: photosystem I

PSI-BLAST: position specific interative BLAST; position-shife iterated BLAST (software program)

PSII: photosystem II

PTB: polypyrimidine-tract-binding protein; a repressive regulator of protein splicing; also pulmonary tuberculosis

PTD: protein transduction domain

PTEN: phosphatase and tensin homolog deleted on chromosome 10

PTFE: polytetrafluoroethylene

PTH: phenylthiohydantoin

PTGS: post-transcriptional gene silencing

PTK: protein-tyrosine kinase

PTPase: protein tyrosine phosphatase

PVA: polyvinyl alcohol

PVDF: polyvinylidine difluoride

QA: quality assurance

QC: quality control

QSAR: Quantitive Strucure-Activity Relationship(s)

Q-TOF: quadrupole time-of-flight

QTL: quantitative trait loci

R$_f$: retardation factor

RA: rheumatoid Arthritis

RAB-GAP: Rab-GTPase-activing protein

RACE: rapid amplification of cDNA ends

Raf: retroviral oncogene derived from 3611 murine sarcoma

RAGE: receptors for Advanced Glycation Endproducts; receptors for AGE; recombinase-activated gene expression

RAMP: receptor activity modified protein

RANK: receptor activator of NF-κB

RANK-L: receptor activator of NF-κB ligand

Rap: a family of GTPase-coupled signal transduction factors which are part of the RAS superfamily

Rap1: a small GTPase involved in integrin activation and cell adhesion

RAPD : randomly amplified polymorphic DNA

RARE: RecA-assisted restriction endonuclease

RAS: GTP-binding signal transducers

H-*ras*: retroviral oncogene derived from Harvey murine sarcoma

K-*ras*: retroviral oncogene derived from Kirsten murine sarcoma

RC: recombinant cogenic

RCA: rolling circle amplification

RCCX: RP-C4-CYP21-TNX module

RCFP: reef coral fluorescent protein

RCR: rolling circle replication

RDP: receptor component protein

rDNA: ribosomal DNA

REA: restriction enzyme analysis

Rel: avian reticuloendotheliosis

REMI: restriction enzyme-mediated integration

RET: receptor for the GDNF family

RF: a transcription factor, RFX family

Rfactor: final crystallographic residual

RFID: radio frequency identification device

RFLP: restriction fragment length polymorphism

RGD: a signature peptide sequence-arginine-glycine-aspartic acid found in protein which bind integrins

RGS: regulator of G-protein signaling

RHD: Rel homology domain

Rheb: Ras homologue enriched in brain

RhoA: Ras homologous; signaling pathway

RI: random integration

RIP: repeat-induced point mutation

RIS: radioimmunoscintigraphy

RISC: RNA-induced silencing complex

RIT: radioimmunotherapy

RM: reference material

RNAi: RNA interference

dsRNA: double-stranded RNA

hpRNAi: hairpin RNA interference

ncRNA: non-coding RNA

rRNA: ribosomal RNA

shRNA: small hairpin RNA

siRNA: small interfering RNA

snRNA: small nuclear RNA

snoRNA: small nuceolar RNA

stRNA: small temporal RNA

RNAse/RNAase: ribonuclease

RNAse III: a family of ribonucleases (RNAses)

RNC: ribosome-nascent chain complex

snRNP: small nuclear ribonucleoprotein particle

RNS: reactive nitrogen species

RO: reverse osmosis

ROCK (ROK): Rho kinase

ROESY: rotating frame Overhauser effect spectroscopy

Ron: receptor for macrophage stimulating protein

ROS: reactive oxygen species

Ros: retroviral oncogenes derived from UR2 avian sarcoma

RP: reverse-phase; also a nuclear serine/threonine protein kinase

RP-CEC: reverse-phase capillary electrochromatography

RP-HPLC: reverse-phase high performance liquid chromatography

RPA: replication protein A

RPEL: a protein motif involved in the cytoskeleton

RPC: reverse-phase chromatography

RPMC: reverse phase microcapillary liquid chromatography

RPMI 1640: growth media for eukaryotic cells

RPTP: receptor protein-tyrosine kinase

RRM: RNA-recognition motif

RRS: Ras recruitment system; resonance Raleigh scattering

R,S: designating optical activity of chiral compounds where R is rectus (right) and S is sinester (left)

RSD: root square deviation

RT: reverse transcriptase; also room temperature

RTD: residence time distribution

RTK: receptor tyrosine kinase

RT-PCR: reverse transcriptase-polymerase chain reaction

RTX: repeat in toxins; pore-forming toxin of *E.Coli* type (RTX toxin); also rituximab, resiniteratoxin, renal transplantation

Rub1: a ubiquitin-like protein, Nedd8

S1P: sphingosine-1-phosphate

S100 : S100 protein family

SA: salicylic acid

SAGE: serial analysis of gene expression

SALIP: saposin-like proteins

SAM: self-assembling monolayers

SAMK: a plant MAP kinase

SAMPL: selective amplification of microsatellite polymorphic loci

Sap: Saposin

SAP: sphingolipid activator protein; also serum amyloid P, shrimp alkaline phosphatase

SAR: scaffold associated region; structure-activity relationship

SATP: heterobifunctional crosslinker; *N*-succinimidyl-*S*-acetylthiopropionate

SAXS: small angle x-ray scattering

scFv: single chain Fv fragment of an antibody

SCID: severe combined immunodeficiency

SCOP: structural classification of proteins; http://scop.mrc-lmb.cam.ac.uk/scop

SCOPE: structure-based combinatorial protein engineering

SDS: sodium dodecyl sulfate

SEC: secondary emission chamber for pulse radiolysis; size exclusion chromatography

Sec-: secretory – usually related to protein translocation

SELDI: surface-enhanced laser desorption/ionization

SELEX: systematic evolution of ligands by expotential enrichment

SERCA: sarco/endoplasmic reticulum Ca^{2+} as in SERCA-ATPase, a calcium pump

SFC: supercritical fluid

SHAP: serum-derived hyaluron-associated protein

SH2: *Src* homology domain 2

SH3: *Src* homology domain 3

Shh: sonic hedgehog

shRNA: small hairpin RNA

SHO: yeast osmosensor

SILAC: stable-isotope labeling with amino acids in cell culture

SIMK: a plant MAP kinase

SINE: short interspersed nuclear element

SINS: sequenced insertion sites

SIPK: salicylic-acid induced protein kinase

Sis: retroviral oncogene derived from simian sarcoma

SISDC: Sequence-independent site-directed chimeragenesis

Ski: retroviral oncogene derived from avian SK77

Skp: a chaperone protein

SLAC: serial Lectin Affinity Chromatography

SLE: systemic lupus erythematoses

SLN1: yeast osmosensor

S/MAR: scaffold and matrix attachment region

SMC: smooth muscle cell

SNAREs: soluble *N*-ethylmaleimide-sensitive fusion (NSF; *N*-ethylmaleimide-sensitive factor) protein attachment protein receptors; can be either R-SNAREs or Q-SNARES depending on sequence homologies

SNM: SNARE motif

SNP: Single nucleotide polymorphism

snRNA: small nuclear RNA

snoRNA: small nucleolar RNA

snRNP: small nuclear ribonucleoprotein particle

SOC: soil organic carbon; store-operated channel

SOCS: suppressors of cytokine signalling

SOD: superoxide dismutase

SOD1s: CuZn-SOD enzyme (Intracellular)

SOP: standard operating procedure

SOS: response of a cell to DNA damage; salt overly sensitive(usually plants); son of Sevenless (signaling cascade protein)

SPA: scintillation proximity assay

SPC: statistical process control

SPECT: sporozite mineneme protein essential for transversal; also single-photon emission computed tomography

SPIN: surface properties of protein-protein interfaces (database)

SPR: surface plasmon resonance

SQL: structured query language

SR: as in the SR protein family (serine- and arginine-rich proteins); also sarcoplasmic reticulum; also scavenger receptor

SRCD: synchrotron radiation circular dichroism

SRP: signal recognition particle

SRF: serum response factor, a ubiquitous transcription factor

SRPK: SR protein kinase

SRS: sequence retrieval system; SOS recruitment system

SRWC: short rotation woody crop

SSC: saline sodium citrate

ssDNA: single-stranded DNA

SSLP: simple sequence length polymorphism

SSR: simple sequence repeats

STAT: signal transducers and activators of transcription

STC: sequence-tagged connector

STM: sequence-tagged mutagenesis

STORM: systematic tailored ORF-data retrieval and management

STR: short tandem repeats

STREX: stress-axix related exon

stRNA: small temporal RNA

SUMO: small ubiquitin-like (UBL) modifier; small ubiquitin-related modifier; sentrin

SurA: a chaperone protein

SV40: simian virus 40

SVS: seminal vesicle secretion

$S_{w,20}$: sedimentation coefficient corrected to water at 20°C

SWI/SNF: Switch/sucrose-non fermenting

TAC: transcription-competent artificial chromosome

TACE: tumor necrosis factor $-\alpha$-converting enzyme; also trans-catheter arterial chemoembolization

TAFs: TBP-associated factors

TAFE: transversely alternating-field electrophoresis

TAG: triacyl glycerol

TAME: tosyl-arginine methyl ester

TAP: tandem affinity purification; also transporter associated with antigen processing

TAR: transformation-associated recombination; *trans*-activation response region

TAT: *trans*-activator of transcription

TATA: as in the TATA box which is a TATA-rich region located upstream from the initiation RNA-synthesis initiation site in eukaryotes and within the promoter region for the gene in question. Analogous to the Pribnow box in prokaryotes.

TBA-Cl: tetrabutylammonium chloride

TBP: TATA-binding protein; telomere-binding protein

TCA: trichloroacetic acid; tricarboxylic acid

TCR: T-cell receptor

TE: therapeutic equivalence; transposable elements

TEA: triethylamine

TEAA: triethylammonium acetate

TEF: toxic equivalency factor

TEM: transmission electron microscopy

TEMED (TMPD): *N,N,N',N'*-tetramethylethylenediamine

TEV: tobacco etch protease

TF: tissue factor; transcription factor

TFA: trifluoroacetic acid

TFIIIA: transcription factor IIIA

TGN: *trans*-Golgi network

TGS: transcriptional gene silencing

TH: thyroid hormone

THF: tetrahydrofuran

TIGR: The Institute for Genomic Research

TIM: translocase of inner mitochondrial membrane

TIP: tonoplast intrinsic protein(s)

TIR: toll/IL-1 receptor

TI-VAMP: tetanus neurotoxin-insensitive VAMP

TLCK: tosyl-lysyl chloromethyl ketone

TLR: toll-like receptor
T$_m$: tubular membrane
TM: transmembrane
TMAO: trimethylamine oxide
TMD: *trans*-membrane domain
TMS: trimethylsilyl; thimersol
TMV: tobacco mosaic virus
TNA: treose nucleic acid
TNB: 5-thio-2-nitrobenzoate
TNBS: trinitrobenzenesulfonic acid
TnC: troponin C
TNF: tumor necrosis factor
Tnl: troponin l
TnT: troponin T
TNF-α (TNFα): tumor necrosis factor-α
TNR: transferrin receptor
TNX: tenascin-X
TPD: temperature programmed desorption
TOC: total organic carbon
TOCSY: total correlated spectroscopy
TOF: time-of-flight
TOP: 5' tandem oligopyrimidine (terminal oligopyrimidine) tract
TOPRIN: Topoisomerase and Primase in reference to a domain
TOR: target of rapamycin; mTOR, mammalian target of rapamycin; dTOR, *Drosophila* target of rapamycin
TOX: toxicology
TPCK: tosylphenylalanylchloromethyl ketone
TPEN: *N',N'*-tetrakis-(2-pyridyl-methyl)ethylenediamine
TPN: triphosphopyridine dinucleotide (now NADP)
TRADD: a scaffold protein
TRAP: tagging and recovery of associated proteins as in RNA-TRAP; also thrombin receptor activation peptide
TRE: thyroid hormone response elements
TRH: thyrotropin-releasing hormone
TRI: as in TRI reagents such as TRIZOL™ reagents used for RNA purification from cells and tissues
Tricine: *N*-(2-hydroxy-1,1-bis(hydroxymethyl)ethyl) glycine
TRIF: TIR domain-containing adaptor-inducing interferon-β
Tris: tris-(hydroxymethyl)aminomethane methane; 2-amino-2-hydroxymethyl-1,3-propanediol
bis-Tris: 2-[bis(2-hydroxyethyl)amino]-2-(hydroxymethyl) propane-1,3-diol
Trk: neurotrophic tyrosine kinase receptor
TRL: time-resolved luminescence
TRP: transient receptor potential as in TRP-protein
TRs: thyroid receptors
TSP: thrombospondin; traveling salesman problem
TTSP: transmembrane type serine proteases
TUSC: Trait Utility System for Corn
Tween: polyoxyethylsorbitan monolaurate
TX: thromboxane, also treatment
TyroBP: tyro protein tyrosine kinase binding protein, DNAX-activation protein 12, DAP12, KARAP
UAS: upstream activation site
UBL: ubiquitin-like modifiers
UCDS: universal conditions direct sequencing
UDP: ubiquitin-domain proteins; uridine diphosphate

UDP-GlcNAc: uidine-5'-diphospho-*N*-acetylglucosamine
UNG: uracil DNA glycosylase
UPA: universal protein array; urokinse-like plasminogen activator
UPR: unfolded protein response
URL: uniform resource locator
URS: upstream repression site
USP: United States Pharmacopeia
USPS: ubiquitin-based split protein sensor
UTR: untranslated region
VAMP: vesicle-associated membrane protein
VAP: VAMP-associated protein
VDAC: voltage-dependent anion-selective channel
VCAM: vascular cellular adhesion molecule
VDJ: variable diversity joining; regions of DNA joined in recombination during lymphocyte development; see VDJ recombination.
VDR: vitamin D receptor
VEGF: vascular endothelial growth factor
VEGFR: vascular endothelial growth factor receptor
VGH: non-acronymial use; a neuronal peptide
V$_H$: variable heavy chain domain
VICKZ: a family of RNA-binding proteins recognizing specific *cis*-acting elements
VIGS: virus-induced gene silencing
VIP: vasoactive intestinal peptide
VLP: virus-like particle
VLDL: very low density lipoprotein
VNC (VNBC): viable, but not-cultivatable (bacteria)
VNTR: variable number tandem repeat; variable number of tandem repeats
VOC: volatile organic carbon
VPAC: VIP PACAP receptors
VSG: variable surface glycoproteins
VSP: vesicular sorting pathway
Vsp10: a type I transmembrane receptor responsible for delivery of protein to lysozyme/vacuole
***vsp*10:** gene for Vsp10
WGA: whole-genome amplification
WT, Wt: wild type
XBP: x-box binding protein
XO: xanthine oxidase
Y2H: yeast two-hybrid
YAC: yeast artificial chromosome
YCp: yeast centromere plasmid
YEp: yeast episomal plasmid
YFP: yellow fluorescent protein
Z: benzyloxycarbonyl
ZDF: Zucker diabetic factor
Zif: zinc finger domain peptides(i.e. Zif-1, Zif-3)
ZIP: leucine zipper
b-ZIP: basic leucine zipper
ZZ Domain: a tandem repeat dimer of the immunoglobulin-binding protein A from *Staphylococcus aureus*

---

[a] All compounds are the L isomer unless otherwise indicated

APPENDIX B: GLOSSARY OF TERMS USEFUL IN BIOCHEMISTRY

Abbreviated New Drug Application (ANDA)

This document contains data that, when submitted to FDA's Center for Drug Evaluation and Research (CDER), Office of Generic Drugs, provide for the review and ultimate approval of a generic drug product. This document does not contain preclinical or clinical data but must demonstrate that the drug is question is bioequivalent to the currently licensed drug which is also referred to as the innovator drug. See http://www.fda.gov/cder/drugsatfda/glossary.htm

ABC transporter

The ATP-binding cassette transporter family consists of a large number of membrane proteins involved in the transport of variety of substances including ions, steroids, metabolites and drugs across extracellular and intracellular membranes. A defect in an ABC transporters is important in cystic fibrosis. See Schwiebert, E.M., ABC transporter-facilitated ATP conductive transport, *Am.J.Physiol.* 276, C1-C8, 1999; Dean, M., Rzhetsky, A., and Allikmets, R., The human ATP-binding cassette (ABC) transporter superfamily, *Genome Res.* 11, 1156-1166, 2001; Dean, M., Hamon, Y., and Chimini, G., The human ATP-binding cassette (ABC) transporter superfamily, *J.Lipid Res.* 42, 1007-1017, 2001; Georujon, C., Orelle, C., Steinfels, E. *et al.*, A common mechanism for ATP hydrolysis in ABC transporter and helicase superfamilies, *Trends Biochem.Sci.* 26, 539-544, 2001; Schmitt, L., The first view of an ABC transporter: the X-ray crystal structure of MsbA from E.coli, *Chembiochem* 3, 161-165, 2002; Holland, I.B., Schmitt, L., and Young, J., Type 1 protein secretion in bacteria, the ABC-transporter dependent pathway, *Mol.Membr.Biol.* 22, 29-39, 2005; Blemans-Oldehinkel, E., Doeven, M.K., and Poolman, B., ABC transporter architechture and regulatory roles of accessory domains, *FEBS Lett.* 580, 1023-1035, 2006; Frelet, A. and Klein, M., Insight in eukaryotic ABC transporter function by mutation analysis, *FEBS Lett.* 580, 1064-1084, 2006; Crouzet, J., Trombik, T., Fraysse, A.S., and Boutry, M., Organization and function of the plant pleiotropic drug resistance ABC transporter family, *FEBS Lett.* 580, 1123-1130, 2006.

Ablation

A multifunctional word derived from the latin ablatus (to carry away). In medicine, refers to the surgical removal of tissue or the elimination of cells by irradiation or immunological approaches. The surgery approach is used extensively in cardiology (Gillinov, A.M. and Wolf, R.K., Surgical ablation of atrial fibrillation, *Prog.Cardiovasc.Dis.* 48, 169-177, 2005) while irradiation or immunological approaches are used in oncology (Appelbaum, F.R., Badger, C.C., Bernstein, I.D., *et al.*, Is there a better way to delivery total body irradiation, *Bone Marrow Transplantation* 10, Suppl. 1., 77-81, 1992; van Bekkum, D.W., Immune ablation and stem-cell therapy in autoimmune disease. Experimental basis for autologous stem-cell transplantation, *Arthritis Res.* 2, 281-284, 2000). It also refers to the reduction of particles into smaller sizes during erosion by other particles or the surrounding fluid (see Lindner, H., Koch, J., Niema, K., Production of ultrafine particles by nanosecond laser sampling

using orthogonal prepulse laser breakdown, *Anal.Chem.* 77, 7528-7533, 2005). It also has a definition in aerospace technology for the dissipation of heat generated by atmospheric friction upon reentry of a space vehicle.

Abscisic Acid

A plant hormone. See Leung, J. and Giraudet, J., Abscisic acid signal transduction, *Annu.Rev.Plant Physiol.Plant Mol.Biol.* 25, 199-221, 1998; Finkelstein, R.R., Gampala, S.S. and Rock, C.D, Abscisic acid signaling in sees and seedlings, *Plant Cell* 14 Suppl., S15-S45, 2002.

Absorption

Generally refers to the ability of a material to absorb another substance (hydration) or energy (the ability of a substance to absorb) light. See adsorption.

Absolute Oils

See essential oils

Abzymes

See catalytic antibodies

Accuracy

The difference between the measured value for an analyte and the true value. Absolute error is the difference between the measured value and the true value while the relative error is that fraction that the absolute error is of the measured amount and is usually expressed as a percentage or at ppt/ppm. See *Handbook of Analytical Chemistry*, ed. L. Meites, McGraw-Hill, New York, New York, 1963; *Analytical Chemistry Handbook*, J.A. Dean, McGraw-Hill, New York, New York, 1995; *Dean's Analytical Chemistry Handbook*, McGraw-Hill, New York, New York, 2005.

Active Ingredient

Any component of a final drug product that provides pharmacological activity or other direct effect in the diagnosis, cure, mitigation, treatment, or prevention of disease or to affect the structure on any function of the body of man or animals. Sometimes referred to as the active pharmaceutical ingredient (API). See http://www.fda.gov/cber; http://www.ich.org – See Q7, Good Manufacturing Guide for Active Pharmaceutical Ingredients.

Activity-based Proteomics

Identification of proteins in the proteome by the use of reagents which measure biological activity. Frequently the activity is measured by the incorporation of a "tag" into the active site of the enzyme. The earliest probes were derivatives of alkyl-flurophosphonates which were well-understood inhibitors of serine proteases. The technical approach is related to enzyme histochemistry/histocytochemistry. Most often used for enzymes where functional families of proteins can be identified. See See Liu, Y., Patricelli, M.P., and Cravatt, B.F., Activity-based protein profiling: the serine hydrolases, *Proc.Nat.Acad. Sci.USA* 96, 14694-14699, 1999; Adam, G.C., Sorensen, E.J., and Carvatt, B.F., Chemical strategies for functional proteomics, *Mol.Cell.Proteomics* 1, 781-790, 2002; Speers, A.E. and Cravatt, B.F., Chemical strategies for activity-based proteomics, *Chembiochem.* 5, 41-47, 2004; Kumar, S., Zhou, B., Liang, F., Activity-based probes for protein tyrosine phosphatases, *Proc. Natl.Acad.Sci.USA* 101, 7943-7948, 2004; Berger, A.B., Vitorino, P.M., and Bogyo, M., Activity-based protein profiling: applications to biomarker discovery, in vivo imaging and drug discovery, *Am.J.Pharmacogenomics* 4, 371-381, 2004; Willams, S.J., Hekmat, O., and Withers, S.G., Synthesis and testing of mechanism-based protein-profiling probes for retaining endo-glycosidases, *Chembiochem* 7, 116-124, 2006; Sieber, S.A. and Cravatt, B.F., Analytical platforms for activity-based protein profiling – exploiting the versatility of chemistry for functional proteomics, *Chem.Commun.* (22), 2311-2318, 2006; Schmidinger, H., Hermetter, A., and Birner-Gruenberger, R., Activity-based proteomics: enzymatic activity profiling in complex proteomes, *Amino Acids* 30, 333-350, 2006.

Accurate Mass Tag (AMT)

A peptide of sufficiently distinctive and accurate mass and elution time from liquid chromatography which can be used a single identifier of a protein. See Conrads, T.P., Anderson, G.A., Veenstra, T.D. *et al., Anal.Chem.* 72, 3349-3354, 2000; Smith, R.D., Anderson, G.A., Lipton, M.S., *et al.,* An accurate mass tag strategy for quantitative and high-throughput proteome measurements, *Proteomics* 2, 513-523, 2002; Strittmatter, E.F., Ferguson, P.L., Tang, K., and Smith, R.D., Proteome analyses using accurate mass and elution time peptide tags with capillary LC time-of-flight mass spectrometry, *J.Am.Soc.Mass Spectrom.* 14, 980-991, 2003; Shen, Y., Tolic, N., Masselon, C. *et al.,* Nanoscale proteomics, *Anal.Bioanal.Chem.* 378, 1037-1045, 2004; Zimmer, J.S., Monroe, M.E., Qian, W.J., and Smith, R.D., Advances in proteomics data analysis and display using an accurate mass and time tag approach, *Mass Spectrom.Rev.* 25, 450-482, 2006.

Active Sequence Collection (ACS)

A collection of active protein sequences or protein fragments or subsequences, collected in the form of function-oriented databases, http://bioinformatica.isa.cnr.it/ACS/
AIRS – Autoimmune Related Sequences
BAC – Bioactive Peptides
CHAMSE – Chameleon Sequences; sequences which can adopt both an alpha helix and beta sheet conformation
DORRS – Database of RGD Related Sequences
DVP – Delivery Vector Peptides
SSP – Structure Solved Peptides
TRANSIT – Transglutamination Sites

Acute Phase Proteins

Proteins which are either *de novo* or markedly elevated after challenge by infectious disease, inflammation or other challenge to homeostasis. Another definition is any protein whose blood concentration increases (or decreases) by 25% or more during certain inflammatory disorders. Acute phase proteins include C-reactive protein, fibrinogen, and α-1-acid glycoprotein. Acute phase proteins are part of the acute phase response. Some acute phase proteins have been used for diagnostic of specific disorders such as C-reactive protein and cardiovascular disease. See Bowman, B.H., *Hepatic Plasma Proteins: Mechanisms of Function and Regulation*, Academic Press, San Diego, CA, USA, 1993; Mackiewicz, A. and Kushner, I., *Acute Phase Proteins: Molecular Biology, Biochemistry, and Clinical Applications*, CRC Press, Boca Raton, FL, 1993; Kerr, M.A. and Thorpe, R., *Immunochemistry Labfax*, Bios Scientific Publishers, Oxford, UK, 1994; Sutton, H.E., The haptoglobins, *Prog.Med.Genet.* 7, 163-216, 1970; Gordon, A.H., Acute-phase proteins in wound healing, *Ciba Found.Symp.* 9, 73-90, 1972; Black, S., Kushner, I., and Samols, D., C-reactive protein, *J.Biol.Chem.* 279, 48487-48490, 2004; Du Clos, T.W. and Mold, C., C-reactive protein: an activator of innate immunity and a modulator of adaptive immunity, *Immunol.Res.* 30, 261-277, 2004; Garlanda, C., Bottazzi, B., Bastone, A., and Mantovani, A., Pentraxins at the crossroads between innate immunity, inflammation, matrix deposition, and female fertility, *Annu.Rev.Immunol.* 23, 337-366, 2005; Ceron, J.J., Eckersall, P.D., and Martynez-Subiela, S., Acute phase proteins in dogs and cats: current knowledge and future perspectives, *Vet.Clin.Pathol.* 34, 85-99, 2005; Sargent, P.J., Farnaud, S., and Evans, R.W., Structure/function overview of proteins involved in iron storage and transport, *Curr.Med.Chem.* 12, 2683-2693, 2005; Bottazzi, B., Garlanda, C., Salvatori, G., *et al.,* Pentraxins as a key component on innate immunity, *Curr.Opin. Immunol.* 18, 10-15, 2006; Vidt, D.G., Inflammation in renal disease, *Am.J.Cardiol.* 97, 20A-27A, 2006; Armstrong, E.J., Morrow, D.A., and Sabatine, M.S., Inflammatory biomarkers in acute coronary syndromes: part II: acute-phase reactants and biomarkers of endothelial cell activation, *Circulation* 113, e152-e155, 2006. See also heat shock proteins.

ADAMTS

A disintegrin and metalloproteinase with thrombospondin motifs. A family of multidomain metalloproteinases with a variety of biological activities. ADMETS are part of the reprolysin family. ADAMTS13 which is involved in the processing of von Willebrand Factor is the best known member of this family. See Hooper, N.M., Families of zinc metalloproteases, *FEBS Lett.* 354, 1-6, 1994; Hurskainen, T.L., Hirohata, S., Seldin, M.F., and Apte, S.S., ADAM-TS5, ADAM-TS6, and ADAM-TS7, novel members of a new family of zinc metalloproteases. General features and genomic distribution of the ADAM-TS family, *J.Biol.Chem.* 274, 2555-2563, 1999; Sandy, J.D. and Verscharen, C., Analysis of aggrecan in human knee cartilage and synovial fluid indicates that aggrecanase (ADAMTS) activity is responsible fo the catabolic turnover and loss of aggrecan whereas other protease activity is required for C-terminal processing *in vivo., Biochem.J.* 358, 615-626, 2001; Fox, J.W. and Serrano S.M., Structural considerations of the snake venom metalloproteinases, key members of the M12 reprolysin family of metalloproteinases, *Toxicon* 45, 969-985, 2005.

Adjuvant

A substance which increases an immune response. Frequently a component of the excipients in the formulation of vaccine. See Brown, L.E. and Jackson, D.C., Lipid-based self-adjuvanting vaccines, *Curr.Drug Deliv.* 2, 283-393, 2005; Gluck, R., Burri, K.G., and Metcalfe, I., Anjuvant and antigen delivery properties of virosomes, *Curr.Drug Deliv.* 2, 395-400, 2005; *Topics in Vaccine Adjuvant Research*, ed. D.R. Spriggs and W.C. Koff, CRC Press, Boca Raton, FL, 1991; *Vaccine Design: The Subunit and Adjuvant Approach*, ed. M.F. Powell, Plenum Press, New York, NY, 1995; *Therapeutic Proteins: Methods and Protocols*, ed. M.C. Smales and D.C. James, Humana Press, Totowa, NJ, 2005; *Immunopotentiation in Modern Vaccines*, ed. V.E.J.C. Schijns and D.T. O'Hagan, Elsevier, Amsterdam, Netherlands, 2006.

Adrenomedullin

Adrenomedullin is a peptide originally isolated from a phenochromocytoma (Kitamura, K., Kangawa, K., Kawamoto, M., *et al.*, Adenomedullin: a novel hypotensive peptide isolated from human phenochromocytoma, *Biochem.Biophys.Res.Commun.* 192, 553-560, 1993). Adrenomedullin elevated intracellular cAMP in platelets and caused hypotension. Since its discovery, adrenomedullin has been found in a variety of cells and tissues (Hinson, J.P., Kapas, S., and Smith, D.M., Adrenomedullin, a multifunctional regulatory peptide, *Endocrine Rev.* 21, 138-167, 2000). Adrenomedullin has been suggested to have a variety of physiological activities. See Poyner, D., Pharmacology of receptors for calcitonin gene-related peptide and amylin, *Trends Pharmacol.Sci.* 16, 424-428, 1995; Muff, R., Born, W., and Fischer, J.A., Calcitonin, calcitonin gene-related peptide, adrenomedullin and amylin: homologous peptides, separate receptors and overlapping biological actions, *Eur.J.Endocrinol.* 133, 17-20, 1995; Richards, A.M., Nichools, M.G., Lewis, L., and Lainchbury, J.G., Adrenomedullin, *Clin. Sci.* 91, 3-16, 1996; Massart, P.E., Hodeige, D., and Donckier, J., Adrenomedullin: view on a novel vasodialatory peptide with naturetic properties, *Acta Cardiol.* 51, 259-269, 1996; Hay, D.L. and Smith, D.M., Adrenomedullin receptors: molecular identity and function, *Peptides* 22, 1753-1763, 2001; Julian, M., Cacho, M., Garcia, M.A., *et al.*, Adrenomedullin: a new target for the design of small molecule modulators with promising pharmacological activities, *Eur.J.Med.Chem.* 40, 737-750, 2005; Shimosawa, T. and Fujita, T., Adrenomedullin and its related peptides, *Endocr.J.* 52, 1-10, 2005; Zudaire, E., Portal-Núñez. S., and Cuttitta, F., The central role of adrenomedullin in host defense, *J.Leuk.Biol.* 80, 237-244, 2006. Hamid, S.A. and Baxter, G.F., A critical cytoprotective role of endogenous adrenomedullin in acute myocardial infarction, *J.Mol.Cell Cardiol.* 41, 360-363, 2006.

Adsorption

The transfer of a substance from one medium to another such as the adsorption of a substance from a fluid onto a surface. The **adsorbent** is the substrate onto which material is adsorbed. The **adsorbate** is the material adsorbed onto a matrix.

Advanced Glycation Endproducts (AGE)

A heterogeneous group of products resulting from a series of chemical reactions starting with the formation of adducts between reducing sugars and protein nucleophiles such as nitrogen bases. Reaction with nucleic acid is also possible but has not been extensively described. The reactions involved are complex involving the Amadori reaction and the Maillard reaction/Some products include triosidines, N^ε-carboxymethyl-lysine, and pentosidine-adducts. These products can undergo further reactions to form cross-linked products; advanced glycation

endproducts are involved in the generation of reactive oxygen species (ROS). See Deyl, Z. and Mikšík, I., Post-translational non-enzymatic modification of proteins. I. Chromatography of marker adducts with special emphasis to glycation reactions, *J.Chromatog.* 699, 287-309, 1997; Bonnefont-Rousselot, D. D., Glucose and reactive oxygen species, *Curr.Opin.Clin.Nutr.* 5, 561-568, 2002; Tessier, F.J., Monnier, V.M., Sayre, L.M., and Kornfield, J.A., Triosidines: novel Maillard reaction products and cross-links from the reaction of triose sugars with lysine and arginine residues, *Biochem.J.*, 369, 705-710, 2003: Thornally, P.J., Battah, S., Ahmed, N., Karachalias, N., Agalou, S., Babaei-Jadidi, R. and Dawnay, A., Quantiative screening of advanced glycation endproducts in cellular and extracellular proteins by tandem mass spectrometry, *Biochem. J.* 375, 581-592, 2003; Ahmed, N., Advanced glycation endproducts—role in pathology of diabetic complications, *Diabetes Res. Clin.Pract.* 67, 3-21, 2005.

Aeration

The dispersion and/or dissolution of a gas into a liquid; generally refers to the process of dispersing air or an oxygen-gas mixture into a liquid such as culture media (Wang, D.I. and Humphrey, A.E., Developments in agitation and aeration of fermentation systems, *Prog.Ind.Microbiol.* 8, 1-34, 1968; Papoutsakis, E.T., Media additives for protecting freely suspended animal cells against agitation and aeration damage, *Trends Biotechnol.* 9, 316-324, 1991; Barberel, S.I. and Walker, J.R., The effect of aeration upon the secondary metabolism of microorganisms, *Biotechnol. Genet.Eng.Rev.* 17, 281-323, 2000). Also refers to the process of air dispersion in the pulmonary system which can include both the inspiratory process and the exchange between the pulmonary system and the vascular bed; most frequently the later (Newman, B. and Oh, K.S., Abnormal pulmonary aeration in infants and children, *Radiol.Clin.North Am.* 26, 323-339, 1988; Kothari, N.A. and Kramer, S.S., Bronchial diseases and lung aeration in children, *J.Thorac.Imaging* 16, 207-223, 2001).

Aerosol

A colloid-like dispersion of a liquid or solid material into a gas. There is considerable interest in the use of aerosols as a drug delivery vehicle. See *Aerosol Science*, ed. C.N. Davies, Academic Press, London, UK, 1996; Sanders, P.A., *Aerosol Science*, Van Nostrand Reinhold, New York, NY, 1970; Sanders, P.A., *Handbook of Aerosol Techology*, Van Nostrand Reinhold, New York, NY, 1979; Adjei, A.L. and Gupta, P.K., *Inhalation Delivery of Therapeutic Peptides and Proteins*, Marcel Dekker, New York, NY, 1997; Macalady, D.L., *Perspectives in Environmental Chemistry*, Oxford University Press, New York, NY, 1998; Hinds, W.C., *Aerosol Technology: Properties, Behavior, and Measurement of Airborne Particles*, Wiley,. New York, NY, 1999; Roche, N. and Huchon, G.J., Rationale for the choice of an aerosol delivery system, *J.Aerosol.Med.* 13, 393-404, 2000; Gautam, A., Waldrep, J.C., and Densmore, C.L., Aerosol gene therapy, *Mol. Biotechnol.* 23, 51-60, 2003; Densmore, C.L., The re-emergence of aerosol gene delivery: a viable approach to lung cancer therapy, *Curr.Cancer Drug Targets* 3, 275-286, 2003. See also colloid.

Affibody

A phage-selected protein developed using a scaffold domain from Protein A. Such a protein can be selected for specific binding characteristics. See Ronnmark, J., Hansson, M., Nguyen, T.,

Uhlen, M., Robert, A., Stahl, S., and Nygren, P.A., Construction and characterization of affibody-Fc chimeras produced in *Escherichia coli*, *J.Immunol.Meth.* 261, 199-211, 2002; Eklund, M., Axelsson, L., Uhlen, M., and Nygren, P.A., Anti-idiotypic protein domains selected from protein A-based affibody libraries, *Proteins* 48, 454-462, 2002; Renberg, B., Shiroyama, I., Engfeldt, T., Nygren, P.A., and Karlström, A.E., Affibody protein capture microarrays: synthesis and evaluation of random and directed immobilization of affibody molecules, *Anal.Biochem.* 341, 334-343, 2005; Orlova, A., Nilsson, F.Y., Wikman, M., *et al.*, Comparative in vivo evaluation of technetium and iodine labels on a n anti-HER2 affibody for single-photon imaging of HER2 expression in tumors, *J.Nucl.Med.* 47, 512-519, 2006; Wahlberg, E. and Hard, T., Conformational stabilization of an engineered binding protein, *J.Am.Chem.Soc.* 128, 7651-7660, 2006; Lendel, C., Dogan, J., and Hard, T., Structural basis of molecular recognition in an affibody: affibody complex, *J.Mol.Biol.*, 359, 1293–1304, 2006.

Affinity Proteomics

The use of affinity reagents for the study of the proteome. The concept of the design and use of affinity labels for the study of proteins is well understood (see Plapp, B.V. and Chen, W.S., Affinity labeling with omega-bromoacetamido fatty acids and analogs, *Methods Enzymol.* 72, 587-591, 1981; Plapp, B.V., Application of affinity labeling for studying structure and function of enzymes, *Methods Enzymol.* 87, 469-499, 1982; Fan, F. and Plapp,B.V., Probing the affinity and specificity of yeast alcohol dehydrogenase I for coenzymes, *Arch.Biochem.Biophys.* 367, 240-249, 1999). For application of affinity technology to proteomics, see Larsson, T., Bergstrom, J., Nilsson, C. and Karlsson, K.A., Use of an affinity proteomics approach for the identification of low-abundant bacterial adhesins as applied on the Lewis(b)-binding adhesin of *Helicobacter pylori*, *FEBS Lett* 469, 155-158, 2000; Agaton, C., Falk, R., Hoiden Guthenberg, I., Gostring, L., Uhlen, M., and Hober, S., Selective enrichment of monospecific polyclonal antibodies for antibody-based proteomics efforts, *J.Chromatog. A*, 1043, 33-40, 2004; Stults, J.T. and Arnott, D., Proteomics, *Methods Enzymol.* 402, 245-289, 2005; Monti, M., Orru, S., Pagnozzi, D., and Pucci, P., Interaction proteomics, *Biosci.Rep.* 25, 45-56, 2005; Schou, C. and Heegaard, N.H., Recent applications of affinity interactions in capillary electrophoresis, *Electrophoresis* 27, 44-59, 2006; Niwayama, S., Proteomics in medicinal chemistry, *Mini Rev.Med.Chem.* 6, 241-246, 2006; *Capillary Electrophoresis of Proteins and Peptides*, ed. M.A.Strege and Lagu, A.L., Humana Press, Totowa, NJ, 2004; *Chemical Genomics: Reviews and Protocols*, ed. E.D. Zanders, Humana Press, Totowa, NJ, 2005; *New and Emerging Proteomics Techniques*, Humana Press, Totowa, NJ, 2006. See also activity-based proteomics

Agar/Agarose

Agar is a heterogeneous natural product derived from algae/seaweed. It is used a "thickening" agent in cooking as a gelatin-like material. Agar is also used as matrix for growing microorganisms. See Johnstone, K.I., *Micromanipulation of Bacteria: The Cultivation of Single Bacteria and Their Spores by the Agar Gel Dissection Techniques*, Churchill-Livingston, Edinburgh, UK, 1973; Watanabe, T., *Pictorial Atlas of Soil and Seed Funfi: Morphologies of Cultured Fungi and Key to Species*, CRC Press, Boca Raton, FL, 1973; Wilkinson, M.H.F., *Digital Image Analysis of Microbes: Imaging, Morphometry, Fluorometry, and Motility Techniques and Applications*, Wiley, Chichester, UK, 1988; Turner, H.A., Theory of assays performed by diffusion in agar gel. I: General considerations, *J.New Drugs* 41, 221-226, 1963; Rees, D.A., Structure, conformation, and mechanism in the formation of polysaccharide gels and networks, *Adv.Carbohydr. Chem.biochem.* 24, 267-332, 1969; Metcalf., D., Clinical applications of the agar culture technique for haematopoietic cells, *Rev.Eur.Etud.Clin.Biol.* 16, 855-859, 1971; Holt, H.M., Gahrn-Hansen, B., and Bruun, B., *Shewanella algae* and *Shewanella putrefaciens*: clinical and microbiological characteristics, *Clin. Microbiol.Infect.* 11, 347-352, 2005; Discher, D.E., Janmey, P. and Wang, Y.L., Tissue cells feel and respond to the stiffness of their substrate, *Science* 310, 1139-1143, 2005. Agar is composed of two primary components, agarose which is a gelling component and agaropectin which is a sulfated, non-gelling component. Agarose is used as a matrix for the separation of large molecules such as DNA. See *Electrophoresis of Large DNA Molecules: Theory and Applications*, ed. Lai, E.H.C. and Birren,B.W., Cold Spring Harbor Laboratory Press, Cold Spring Harbor, NY, 1990; Birren, B.W. and Lai, E.H.C., *Pulsed Field Gel Electrophoresis: A Practical Guide*, Academic Press, San Diego, CA, 1993; Bickerstaff, G.F., *Immobilization of Enzymes and Cells*, Humana Press, Totowa, NJ, 1997; *Electrophoresis in Practice: A Guide to Methods and Applications of DNA and Protein Separations*, Wiley-VCH, Weinheim, Germany, 2001.

Aggregation

The process of forming a ordered or disordered group of particles, molecules, bubbles, drops, or other physical components which bind together in an undefined fashion; a common physical analogy is concrete or brick. Aggregation is used to measure macromolecular interactions and the interactions of cells such as platelets and frequently involves nephelometry. Agglutination is a term used to described the aggregation or clumping of blood cells or bacteria caused by antibody or other biological or chemical factors. Aggregation of proteins is thought to involved in the pathogenesis of diseases such as Parkinson's disease and Alzheimer's disease; these diseases are thought to be conformation diseases of proteins resulting in disorder structure and aggregation. Aggregation of blood platelets is an initial step in the hemostatic response. See Born, G.V., Inhibition of thrombogenesis by inhibition of platelet aggregation, *Thromb.Diath. Haemorrh.Suppl.* 21, 159-166, 1966; Zucker, M.B., ADP- and collagen-induced platelet aggregation in vivo and in vitro, *Thromb.Diath.Haemorrh.Suppl* 26, 175-184, 1967; Luscher, E.F., Pfueller, S.L., and Massini, P., Platelet aggregation by large molecules, *Ser.Haematol.* 6, 382-391, 1973; Harris, R.H. and Mitchell, R., The role of polymers in microbial aggregation, *Ann.Rev.Microbiol.* 27, 27-50, 1973; Harrington, R.A., Kleimna, N.S., Granger, C.B., *et al.*, Relation between inhibition of platelet aggregation and clinical outcomes, *Am.Heart J.* 136(4 Pt 2 Su), S43-S50, 1998; Holyaerts, M.F., Oury, C., Toth-Zamboki, E., and Vermylen, J., ADP receptors in platelet activation and aggregation, *Platelets* 11, 307-309, 2000; Kopito, R.R., Aggresomes, inclusion bodies and protein aggregation, *Trends Cell Biol.* 10, 524-530, 2000; Savage, B., Cattaneo, M., and Ruggeri, Z.M., Mechanisms of platelet aggregation, *Curr.Opin.Hematol.* 8, 270-276, 2001; Valente, J.J., Payne, R.W., Manning, M.C., *et al.*, Colloidal behavior of proteins: effects of the second virial coefficient on solubility, crystallization and aggregation of proteins in aqueous solution, *Curr.Pharm.Biotechnol.* 6, 427-436, 2005;

Schwarzinger, S., Horn, A.H., Ziegler, J., Sticht, H., Rare large scale subdomain motions in prion protein can initiate aggregation, *J.Biomol.Struct.Dyn.* 23, 581-590, 2006; Elllis, R.J. and Minton, A.P., Protein aggregation in crowded environments, *Bio. Chem.* 387, 485-497, 2006; Estada, L.D. and Soto, C., Inhibition of protein misfolding and aggregation by small rationally-designed peptides, *Curr.Pharm.Des.* 12, 2557-2567, 2006.

Agonist

Generally a compound or substance which binds to a receptor site which could be on a cell membrane or a protein and elicits a positive physiological response. See Gowing, L., Ali, R., and White, J., Opioid antagonists with minimal sedation for opioid withdrawal, *Cochrane Database Syst.Rev.* (2), CD002021, 2002; Bernardo, A. and Minghetti, L., PPAR-gamma agonists as regulators of microglial activation and brain inflammation, *Curr.Pharm.Des.* 12, 93-109, 2006; Bonuccelli, U. and Pavese, N., Dopamine agonists in the treatment of Parkinson's disease, *Expert Rev. Neurother.* 6, 81-89, 2006; Thobois, S., Proposed dose equivalence for rapid switch between dopamine receptor agonists in Parkinson's diseases: a review of the literature, *Clin. Ther.* 28, 1-12, 2006; Schwartz, T.W. And Holst, B., Ago-allosteric modulation and other types of allostery in dimeric 7TM receptors, *J.Recept.Signal Transduct.Res.* 26, 107-128, 2006.

Algorithm

The underlying iterative method or mathematic theory for any particular computer programming technique; a precisely described routine process that can be applied and systematically followed through to a conclusion; a step-by-step procedure for solving a problem or accomplishing some end. There is a variety of algorithms ranging from defining clinical treatment protocols to aligning and predicting sequences of biopolymers. See Rose, G.D. and Seltzer, J.P., A new algorithm for finding the peptide chain turns in a globular protein, *J.Mol. Biol.* 113, 153-164, 1977; Gotoh, O., An improved algorithm for matching biological sequences, *J.Mol.Biol.* 162, 705-708, 1982; Dandekar, T. and Argos, P., Folding the main chain of small proteins with the genetic algorithm, *J.Mol.Biol.* 236, 844-861, 1994; Rarey, M., Kramer, B., Langauer, T., and Klebe, G., A fast, flexible docking method using an incremental construction algorithm, *J.Mol.Biol.* 261, 470-489, 1996; Jones, G., Willett, P., Glen, R.C. *et al.*, Development and validation of a genetic algorithm for flexible docking, *J.Mol.Biol.* 267, 427-448, 1997; Samudrala, R. and Moult, J., A graph-theoretic algorithm for comparative modeling of protein structure, *J.Mol.Biol.* 279, 287-302, 1998; Chacon, P. Diaz, J.F., Moran, F., and Andreu, J.M., Reconstruction of protein form with X-ray solution scattering and a genetic algorithm, *J.Mol.Biol.* 299, 1289-1302, 2000; Mathews, D.H. and Turner, D.H., Dyalign: an algorithm for finding the secondary structure common to two RNA sequences, *J.Mol.Biol.* 317, 191-203, 2002; Herrmann, T., Guntert, P., and Wuthrich, K., Protein NMR structure determination with automated NOE assignment using the new software CANDID and the torsion angle dynamics algorithm DYANA, *J.Mol.Biol.* 319, 209-227, 2002; Andronescu, M., Fejes, A.P., Hutter, F., *et al.*, A new algorithm for RNA secondary structure design, *J.Mol.Biol.* 336, 607-624, 2004; Fang, Q. and Shortle, D., Protein refolding in silico with atom-based statistical potentials and conformational search using a simple genetic algorithm, *J.Mol.Biol.* 359, 1456-1467, 2006.

Albumin

A protein, most notably derived from plasma or serum and secondarily from egg (ovalbumin). It is the most abundant protein in blood/plasma constituting approximately half of the total plasma protein. It functions in establishing plasma colloid strength which preserves the fluid balance between the intravascular and extravascular space (Starling, E.H., On the absorption of fluids from the connective tissue spaces, *J.Physiol.* 19, 312-326, 1896). Albumin, particularly bovine serum albumin (BSA), is used as a model protein and as a standard for the measurement of protein concentration. See Foster, J.F., Plasma albumin, in *The Plasma Proteins*, Vol. 1 ed. F.W. Putnam, Academic Press, New York, NY, Chapter 6, pp. 179-239, 1960; Tanford, C., Protein Denaturation, *Adv.Protein Chem.* 23, 121-282, 1968; Peters, T., Jr., Serum albumin, *Adv.Clin.Chem.* 13, 37-111, 1970; Gillette, J.R., Overview of drug-protein binding, *Ann.N.Y.Acad.Sci.* 226, 6-17, 1973; Peters, T., *All about Albumin: Biochemistry, Genetics, and Medical Applications*, Academic Press, San Diego, CA, 1996; Vo-Dinh, T., Protein nanotechnology: the new frontier in biosciences, *Methods Mol.Biol.* 300, 1-13, 2005; Quinlan, G.J., Martin, G.S., and Evans, T.W., Albumin: biochemical properties and therapeutic potential, *Hepatology* 41, 1211-1219, 2005; Rasnik, I., McKenney, S.A., and Ha, T., Surfaces and orientation: much to FRET about? *Acc.Chem.Res.* 38, 542-548, 2005; *Therapeutic Proteins: Methods and Protocols*, ed. C.M. Smales and D.C., James, Humana Press, Totowa, NJ, 2005; Yamakura, F. and Ikeda, K., Modification of tryptophan and tryptophan residues in proteins by reactive nitrogen species, *Nitric Oxide* 14, 152-161, 2006; Chuang, V.T. and Otagiri, M., Stereoselective binding of human serum albumin, *Chirality* 18, 159-166, 2006; Ascenzi, P., Bocedi, A., Notari, S., *et al.*, Allosteric modulation of drug binding to human serum albumin, *Mini Rev.Med.Chem.* 6, 483-489, 2006. Albumin was the first protein biopharmaceutical (Newhauser, L.R. and Loznen, E.L., Studies on human albumin in military medicine: the standard Army-Navy package of serum albumin (concentrated), *U.S.Navy Med.Bull.* 40, 796-799, 1942; Heyl, J.T., Gibson, J.G., 2nd, and Janeway, C.W., Studies on the plasma proteins. V. The effect of concentrated solutions of human and bovine serum albumin in man, *J.Clin.Invest.* 22, 763-773, 1943) and is used for a variety of clinical indications (*Albumin and the Systemic Circulation*, ed. Blauhut, B. and Lundsgaard-Hansen, P., Karger, Berlin, 1986), including use in extracorporeal circulation as "bridge-to-transplant (Sen, S. and Williams, R., New liver support devices in acute liver failure: a critical evaluation, *Semin.Liver Dis.* 23, 283-294, 2003; Tan, H.K., Molecular absorbent recirculating system (MARS), *Ann.Acad.Med.Singapore* 33, 329-335, 2004; George, J., Artificial liver support systems, *J.Assoc.Physicians India* 52, 719-722, 2004; Barshes, N.R., Gay, A.N., Williams, B., *et al.*, Support for the acutely failing liver: a comprehensive review of historic and contemporary strategies, *J.Am.Coll.Surg.* 201, 458-476, 2005. Albumin is also noted for its ability to interact with various dyes and the binding of bromcresol green is an example of a clinical assay method for albumin (Rodkey, F.L., Direct spectrophotometric determination of albumin in human serum, *Clin.Chem.* 11, 478-487, 1965; Hill, P.G., The measurement of albumin in serum and plasma, *Ann.Clin.Biochem.* 22, 565-578, 1985; Doumas, B.T. and Peter, T., Jr., Serum and urine albumin: a progress report on their measurement and clinical significance, *Clin.Chim.Acta* 258, 3-20, 1997; Duly, E.B., Grimason, S., Grimaon, P. *et al.*, Measurement of serum albumin by capillary zone electrophoresis, bromocresol green, bromocresol purple, and immunoassay methods,

J.Clin.Pathol. 56, 780-781, 2003). Albumin is a general designation to describe a fraction of simple proteins which are soluble in water and dilute salt solutions as opposed to the globulin fraction which is insoluble in water but soluble in dilute salt solutions. This is a old classification has many exceptions(Taylor, J.F., The isolation of proteins, in *The Proteins. Chemistry, Biological Activity and Methods*, Volume I. Pt.A, ed. H.Neurath and K.Bailey, Chapter 1, pp. 1-85, 1953; Albumins also migrate faster than globulins on electrophoresis which resulted in the development of the classification of plasma proteins as albumins and globulins (Cooper, G.R., Electrophoretic and ultracentrifugal analysis of normal human serum, in *The Plasma Proteins*, ed. F.W. Putman, Academic Press, New York, NY, 1960, Chapter 3, pp. 51-103, 1960.

Alloantibody

Also an isoantibody. An antibody directed against a cell or tissue from an individual of the same species. Transplantation antibodies, transfusion antibodies and antibodies against blood coagulation factors such as factor VIII inhibitors are examples of alloantibodies. See Glotz, D., Antoine, C., and Duboust, A., Antidonor antibodies and transplantation: how to deal with them before and after transplantation, *Transplantation* 79 (Suppl 3), S30-S32, 2005; Colvin, R.B. and Smith, R.N., Antibody-mediated organ-allograft rejection, *Nat.Rev.Immunol.* 5, 807-817, 2005; Moll, S. and Pascual, M., Humoral rejection of organ allografts, *Am.J.Transplant.* 5, 2611-2618, 2005; Waanders, M.M., Roelen, D.L, Brand, A., and Class, F.H., The putative mechanism for the immunomodulating effect of HLA-DR shared allogeneic blood transfusion on the alloimmune response, *Transfus.Med.Rev.* 19, 281-287, 2005.

Alloantigen

An antigen present in some, but not all members of a species or strain. The histocompatitibility locus antigens (HLA) is an example. See Schiffman, G. and Marcus, D.M., Chemistry of the ABH blood group substances, *Prog.Hematol.* 27, 97-116, 1964; Race, R.R., Contributions of blood groups to human genetics, *Proc.R.Soc.Lond.B.Biol.Sci.* 163, 151-168, 1965; Dausset, J., Leucocye and tissue groups, *Vox Sang.* 11, 263-275, 1966; Amos, B., Immunologic factors in organ transplantation, *Am.J.Med.* 55, 767-775, 1968; March, D.M., The ABO and Lewis bloodgroup system. Immunochemistry, genetics and relation to human disease, *N.Engl.J.Med.* 280, 994-1006, 1969; Bach, F.H., Histocompatibility in man – genetic and practical considerations, *Prog.Med.Genet.* 6, 201-240, 1969; Drozina, G., Kohoutek, J., Janrane-Ferrat, N., and Peterlin, B.M., Expression of MHC II genes, *Curr.Top.Microbiol.Immunol.* 290, 147-170, 2005; Serrano, N.C., Millan, P., and Paez, M.C., Non-HLA associations with autoimmune diseases, *Autoimmun.Rev.* 5, 209-214, 2006; Koehn, B., Gangappa, S., Miller, J.D., Ahmed, R., and Larsen, C.P., Patients, pathogens, and protective immunity: the relevance of virus-induced alloreactivity in transplantation, *J.Immunol.* 176, 2691-2696, 2006. Turesson, C. and Matteson, E.L., Genetics of rheumatoid arthritis, *Mayo Clin.Proc.* 81, 94-101, 2006.

Allosteric

Originally a term which described the interaction of small molecules with an enzyme at a site physically distant from the active site where such interaction influenced enzyme activity. These small molecules were generally related to the substrate or product of the enzyme action. More recently, it has been used to describe the modulation of enzyme activity by the binding of a large or small molecule to a site distant from the active site. See Changeux, J.-P. and Kvamme, E., *Regulation of Enyzme Activity and Allosteric Interactions*, Academic Press, New York, 1968; Kurganov, B.I., *Allosteric Enzymes: Kinetic Behavior*, J.Wiley, Chichester, UK, 1982; Perutz, M.F., *Mechanisms of Cooperativity and Allosteric Regulation in Proteins*, Cambridge University Press, Cambridge, UK, 1990; Segal, L.A., *Biological Kinetics*, Cambridge University Press, Cambridge, UK, 1991; Changeux, J.P., Allosteric interactions interpreted in terms of quaternary structure, *Brookhaven Symp.Biol.* 17, 232-249, 1964; Monod, J., From enzymatic adaptation to allosteric transitions, *Science* 154, 475-483, 1966; Stadtman, E.R., Allosteric regulation of enzyme activity, *Adv.Enzymol.Relat.Areas Mol.Biol.* 28, 41-154, 1966; Frieden, C., Protein-protein interaction and enzymatic activity, *Annu.Rev.Biochem.* 40, 653-696, 1971; Matthews, B.W. and Bernhard, S.A., Structure and symmetry of oligomeric enzymes, *Annu.Rev.Biophys.Bioeng.* 2, 257-317, 1973; Hammes, G.G. and Wu, C.W., Kinetics of allosteric enzymes, *Annu.Rev.Biophys. Bioeng.* 3, 1-33, 1974; Ostermeier, M., Engineering allosteric protein switches by domain insertion, *Protein Eng.Des.Sel.* 18, 359-264, 2005; Horovitz, A. and Willison, K.R., Allosteric regulation of chaperonins, *Curr.Opin.Struct.Biol.* 15, 646-651, 2005; Ascenzi, P., Bocedi, A., Notari, S., *et al.*, Allosteric modulation of drug binding to human serum albumin, *Mini Rev.Med.Chem.* 6, 483-489, 2006.

Alternative Splicing

Alternative splicing is a process by which biological diversity can be increased without change in DNA content. Alternative splicing is a mechanism by a single pre-mRNA is processed is different ways (different splicing sites) to yield a diverse group of messenger RNA molecules. See Choi, E., Kuehl, M. and Wall, R., RNA splicing generates a variant light chain from an aberrantly rearranged kapp gene, *Nature* 286, 776-779, 1980; Mariamn, E.C., van Beek-Reinders, R.J., and van Venrooij, W.J., *J.Mol. Biol.* 163, 239-256, 1983; Lerivray, R., Mereau, A., and Osborne, H.B., Our favorite alternative splice site, *Biol.Cell.* 98, 317-321, 2006; Florea, L., Bioinformatics of alternative splicing and its regulation, *Brief Bioinform.* 7, 55-69, 2006; Xing, Y. and Lee, C., Alternative splicing and RNA selection pressure – evolutionary consequences for eukaryotic genomes, *Nat.Rev.Genet.* 7, 499-509, 2006. Alternative trans-splicing has also been demonstrated. See Maniatis, T. and Tasic, B., Alternative pre-mRNA splicing and proteome expansion in metazoans, *Nature* 418, 236-243, 2002; Garcia-Blanco, M.A., Messenger RNA reprogramming by spliceosome-mediated RNA trans-splicing, *J.Clin.Invest.* 112, 474-480, 2003; Kornblihtt, A.R., de la Mata, M., Fededa, J.P., *et al.*, Multiple links between transcription and splicing, *RNA* 10, 1489-1498, 2004; Horiuchi, T. and Aigaki, T., Alternative trans-splicing: a novel mode of pre-mRNA processing, *Biol.Chem.* 98, 135-140, 2006. The production of variants of fibronectin (is one of the better known examples of alternative splicing. See Schwarzbauer, J.E., Paul, J.T., and Hynes, R.O., On the origin of species of fibronectin, *Proc.Natl.Acad.USA* 82, 1424-1428, 1985; ffrench-Constant, C., Alternative splicing of fibronectin—many different proteins but few different functions.

Aminophospholipids

Amino-containing phospholipids such as phosphatidyl ethanol-amine and phosphatidyl serine. Phosphatidyl serine is involved in specific membrane functions and change in membrane distribution producing asymmetry are considered important for function. These are enzymes described as flippases, floppases, transporters, scramblaese, and aminophospholipid translocase which are responsible for this asymmetry which results in aminophospholipids on the cytoplasmic side of the membrane and cholines and sphingolipids on the outer surface. See Devaux, P.F., Protein involvement in transmembrane lipid asymmetry, *Annu.Rev.Biophys.Biomol.Struct.* 21, 417-439, 1992; Schelgel, R.A., Callahan, M.K., and Williamson, *Ann.N.Y.Acad.Sci.* 926, 271-225, 2000; Daleke, D.L. and Lyles, J.V., Identification and purification of aminophospholipid flippases, *Biochem.Biophys. Acta.* 1486, 108-127, 200; Balasubramanian, K., and Schroit, A.J., Aminophospholipid asymmetry: a matter of life and death, *Annu.Rev.Physiol.* 65, 701-734, 2003; Daleke, D.L.. Regulation of transbilayer plasma membrane phospholipid asymmetry, *J.Lipid.Res.* 44, 233-242, 2003.

Ambisense

A genome or genome segment which contains regions that are positive-sense for some genes and negative-sense(antisense) for other genes as in an ambisense RNA as viral ssRNA genome or genome segment. See Bishop, D.H., Ambisense RNA viruses: positive and negative polarities combined in RNA virus genomes, *Microbiol.Sci.* 3, 183-187, 1986; Ngugen, M. and Naenni, A.L., Expression strategies of ambisense viruses, *Virus Res.* 93, 141-150, 2003; van Knippenberg, I., Goldbach, R., and Kormelink, R., *Tomato spotted wilt virus* S-segment mRNAs have overlapping 3'-ends containing a predicted stem-loop structure and conserved sequence motif, *Virus Res.* 110, 125-131, 2005; Barr, J.N., Rodgers, J.W., and Wertz, G.W., The Bunyamwera virus mRNA transcription signal resides within both the 3' and the 5' terminal regions and allows ambisense transcription from a model RNA segment, *J.Virol.* 79, 12602-12607, 2005.

Amorphous Powder

A solid form of a material which does not have a definite form such as a crystal structure. Differing from a crystal form, an amorphous form is thermodynamically unstable and does not have a defined melting point. The physical characteristics of an amorphous powder make it the desired physical state for drug after lyophilization. See Jennings, T.A., *Lyophilization Introduction and Basic Principles,* Interpharm Press, Denver, Colorado, 1999; Izutsu, K., Yoshioka, S., and Kojima, S., Increased stabilizing effects of amphiphilic excipients on freeze-drying of lactate dehydrogenase (LDH) by dispersion into sugar matrices, *Pharm.Res.* 12, 838-843, 1995; Royall, P.G., Huang, C.Y., Tang, S.W., et al., The development of DMA for the detection of amorphous content in pharmaceutical powdered material, *Int.J.Pharm.* 301, 181-191, 2005; Stevenson, C.L., Bennett, D.B., and Lechuga-Ballesteros, D., Pharmaceutical liquid crystals: the relevance of partially ordered systems, *J.Pharm.Sci.* 94, 1861-1880, 2005; Skakle, J., Applications of X-ray power diffraction in materials chemistry, *Chem.Rec.* 5, 252-262, 2005; Farber, L., Tardos, G.I., and Michaels, J.N., Micro-mechanical properties of drying material bridges of pharmaceutical excipients,

Int.J.Pharm. 306, 41-55, 2005; Jovanovic, N., Bouchard, A., Hofland, G.W., et al., Distinct effects of sucrose and trehalose on protein stability during supercritical flud drying and freeze-drying, *Eur.J.Pharm.Sci.* 27, 336-345, 2006; Reverchon, E. and Atanacci, A., Cyclodextrins micrometric powders obtained by supercritical fluid processing, *Biotechnol.Bioeng.*, 94, 753–761, 2006; Jorgensen, A.C., Miroshnyk, I., Karjalainen, M., et al., Multivariate data analysis as a fast tool in evaluation of solid state phenomena, *J.Pharm.Sci.* 95, 906-916, 2006; Shah, S., Sharma, A., and Gupta, M.N., Preparation of cross-linked enzyme aggregates by using bovine serum albumin as a proteic feeder, *Anal.Biochem.* 351, 207-213, 2006.

Amplicon

(Usually) the DNA product of a PCR reaction, usually an amplified segment of a gene or DNA. An RNA amplicon would be an RNA sequence can be obtained by transcription mediated amplification. See Bustin, S.A., Benes, V., Nolan, T., and Pfaffi, M.W., Quantitative real-time RT-PCR – a perspective, *J.Mol. Endocrinol.* 34, 597-601, 2005; Sarrazin, C., Highly sensitive hepatitis C virus RNA detection methods: molecular backgrounds and clinical significance, *J.Clin.Virol.* 25, S23-S29, 2002). This also refers to herpesvirus vectors for gene therapy (Oehmig, A., Fraefel, C., and Breakfield, X.O., Update on herpesvirus amplicon vectors, *Molecular Therapy* 10, 630-643, 2004).

Amphipathic (amphiphilic)

A compound which has both hydrophilic (lyophilic) and hydrophobic (lyophobic) properties. This property is important for the interaction of protein with lipids and for the properties of cell-penetrating peptides. Detergents are amphipathic molecules. See Fasman, G.D., *Prediction of Protein Structures and the Principles of Protein Conformation,* Plenum Press, New York, NY, 1989; Epand, R.M., *The Amphipathic Helix,* CRC Press, Boca Raton, FL, 1993; Langel, U., *Cell-Penetrating Peptides: Processes and Applications,* CRC Press, Boca Raton, FL, 2002; *Peptide-Lipid Interactions,* ed. S.A. Simon and T.J. McIntosh, Academic Press, San Diego, CA, 2002; Scow, R.O., Blanchette-Mackie, E.J., and Smith, L.C., Transport of lipid across capillary endothelium, *Fed.Proc.* 39, 2610-2617, 1980; Corr, P.B., Gross, R.W., and Sobel, B.E., Amphipathic metabolites and membrane dysfunction in ischemic myocardium, *Circ.Res.* 55, 135-154, 1984; Anantharamaiah, G.M., Brouillette, C.G., Engler, J.A., et al., Role of amphipathic helixes in HDL structure/function, *Adv.Exp.Med.Biol.* 285, 131-140, 1991; Segrest, J.P., Garber, D.W., Brouillette, C.G., et al., The amphipathic alpha helix: a multifunctional structural motif in plasma apolipoproteins, *Adv.Protein Chem.* 45, 303-369, 1994; Lester, J.B. and Scott, J.D., Anchoring and scaffold proteins for kinases and phosphatases, *Recent Prog.Horm.Res.* 52, 409-429, 1997; Lesieur, C., Vecsey-Semjen, B., Abrami, L. et al., Membrane insertion: the strategies of toxins, *Mol.Membr.Biol.* 14, 45-64, 1997; Johnson, J.E. and Cornell, R.B., Amphitropic proteins: regulation by reversible membrane interactions, *Mol.Membr.Biol.* 16, 217-235, 1999; Tossi, A., Sandri, L., and Giangaspero, A., Amphipathic, alpha-helical antimicrobial peptides, *Biopolymers* 55, 4-30, 2000; Garavito, R.M. and Ferguson-Miller, S., Detergents as tools in membrane biochemistry, *J.Biol.Chem.* 276, 32403-32406, 2001; El-Andaloussi, S., Holm. T., and Langel, U., Cell-penetrating peptides: mechanisms and applications, *Curr.Pharm.Des.* 11, 3597-3611, 2005; Deshayes, S., Morris, M.C., Divita, G., and Heitz, F., Interactions of primary amphipathic cell penetrating peptides with model membranes: consequences on the mechanism of intracellular delivery of therapeutics, *Curr.Pharm.Des.* 11, 3629-3638, 2005.

Ampholyte

An amphoteric electrolyte. In proteomics, this term is used to describe small multi-charged organic buffer used to establish pH gradients in isoelectric focusing. See Righetti, P.G., Isoelectric focusing as the crow flies, *J.Biochem.Biophys.Methods* 16, 99-108, 1988; Patton, W.F., Pluskal, M.G., Skea, W.M., *et al.*, Development of a dedicated two-dimensional gel electrophoresis system that provides optimal pattern reproducibility and polypeptide resolution, *Biotechniques* 8, 518-527, 1990; Hanash, S.M., Strahler, J.R., Neel, J.V., *et al.*, Highly resolving two-dimensional gels for protein sequencing, *Proc.Natl.Acad. Sci.USA* 88, 5709-5713, 1991; Cade-Treyer, D., Cade, A.,Darjo, A., and Jouvion-Moreno, M., Isoelectric focusing and titration curves in biomedicine and in agrofood industries: a multimedia teaching program, *Electrophoresis* 17, 479-482, 1996; Stoyanov, A.V. and Pawliszyn, J., Buffer composition changes in background electrolyte during electrophoretic run in capillary zone electrophoresis, *Analyst* 129, 979-982, 2004; Gorg, A. Weiss, W. and Dunn, M.J., Current two-dimensional technology for proteomics, *Proteomics* 4, 3665-3685, 2004; Kim, S.H., Miyatake, H., Ueno, T., *et al.*, Development of a novel ampholyte buffer for isoelectric focusing: electric charge-separation of protein samples for X-ray crystallography using free-flow isoelectric focusing, *Acta Crystallogr.D Biol.Crystallogr.* 61, 799-802, 2005; Righetti, P.G., The Alpher, Bethe, Gamow of isoelectric focusing, the alpha-Centaury of electrokinetic methods, *Electrophoresis* 27, 923-938, 2006.

Amphoteric

Referring to a molecule such as protein, peptide, amino acid capable of having a positive charge, negative charge, or zero net charge. When at a zero net charge, it is also referred to as a zwitterion. See Haynes, D., The action of salts and non-electrolytcs upon buffer solutions and amphoteric electrolytes anad the relation of these effects to the permeability of the cell, *Biochem.J.* 15, 440-461, 1921; Akabori, S., Tani, H., and Noguchi, J., A synthetic amphoteric polypeptide, *Nature* 167, 1591-160, 1951; Coway-Jacobs, A. and Lewin, L.M., Isoelectric focusing in acrylamide gels: use of amphoteric dyes as internal markers for determination of isoelectric points, *Anal.Biochem.* 43, 294-400, 1971; Chiari, M. Pagani, L. and Righetti, P.G., Phsico-chemical properties of amphoteric, isoelectric, macroreticulate buffers, *J. Biochem.Biophys.Methods* 23, 115-130, 1991; Blanco, S., Clifton, M.J., Joly, J.L., and Peltre, G., Protein separation by electrophoresis in a nonsieving amphoteric medium, *Electrophoresis* 17, 1126-1133, 1996; Tulp, A., Verwoerd, D., and Hart, A.A., Density gradient isoelectric focusing of proteins in artificial pH gradients made up of binary mixtures of amphoteric buffers, *Electrophoresis* 18, 767-773, 1997; Akahoshi, A., Sato, K., Nawa, Y. *et al.*, Novel approach for large-scale, biocompatible, and low-cost fractionation of peptides in proteolytic digest of food protein based on the amphoteric nature of peptides, *J.Agric.Food Chem.* 48, 1955-1959, 2000; Matsumoto, H., Koyama, Y., and Tanioka, A., Interaction of proteins with weak amphoteric charged

membrane surfaces: effect of pH, *J.Colloid Interface Sci.* 264, 82-88, 2003; Fortis, F., Girot, P., Brieau, O., *et al.*, Amphoteric, buffering chromatographic beads for proteome prefractionation. I: theoretical model, *Proteomics* 5, 620-628, 2005; Kitano, H., Takaha, K., and Gemmei-Ide, M., Raman spectroscopic study of the structure of water in aqueous solutions of amphoteric polymers, *Phys.Chem.Chem.Phys.* 8, 1178-1185, 2006.

Amyloid

A wax-like translucent insoluble material consisting largely of protein and may or may not contain carbohydrate and is associated with tissue degeneration. Amyloid peptides/proteins are thought to be associated with Alzheimer's disease. Glenner, G.G., The pathogenetic and therapeutic implications of the discovery of the immunoglobulin origin of amyloid fibrils, *Hum. Pathol.* 3, 157-162, 1972; Franklin, E.C. and Zucker-Franklin, D., Current concepts of amyloid, *Adv.Immunol.* 15, 249-304, 1972; Glenner, G.G. and Terry, W.D., Characterization of amyloid, *Annu.Rev.Med.* 25, 131-135, 1974; Glenner, G.G. and Page, D.L., Amyloid, amyloidosis, and amyloidogenesis, *Int.Rev.Exp.Pathol.* 15, 1-92, 1976; Gorevic, P.D., Cleveland, A.B., and Franklin, E.C., The biologic significance of amyloid, *Ann.N.Y.Acad.Sci.* 389, 380-394, 1982; Reinhard, C., Herbert, S.S., and De Strooper, B., The amyloid-beta precursor protein: integrating structure with biological function, *EMBO J.* 24, 3996-4006, 2005; Meersman, F. and Dobson, C.M., Probing the pressure-temperature stability of amyloid fibrils provides new insights into their molecular properties, *Biochem. Biophys.Acta* 1764, 452-460, 2006; Tycko, R., Solid-state NMR as a probe of amyloid structure, *Protein Pept.Lett.* 13, 229-234, 2006; Torrent, J., Balny, C. and Lange, R., High pressure modulates amyloid formation, *Protein Pept.Lett.* 13, 271-277, 2006; Gorbenko, G.P. and Kinnuen, P.K., The role of lipid-protein interactions in amyloid-type protein fibril formation, *Chem. Phys.Lipids* 141, 72-82, 2006; Catalano, S.M., Dodson, E.C., Henze, D.A., *et al.*, The role of amyloid- beta derived diffusible ligands (ADDLs) in Alzheimer's disease, *Curr.Top.Med.Chem.* 6, 597-608, 2006.

Anaphylatoxin(s)

Fragment(s) of complement proteins released during complement activation. See Corbeil, L.B., Role of the complement system in immunity and immunopathology, *Vet.Clin.North Am.*8, 585-611, 1978; Hugli, T.E. and Muller-Eberhard, H.J., Anaphylatoxins: C3a and C5a, *Adv.Immunol.* 26, 1-53, 1978; Hugli, T.E., The structural basis for anaphylatoxin and chemotactic functions of C3a, C4a, and C5a, *Crit.Rev.Immunol.* 1, 321-366, 1981; Hawlisch, H., Wills-Karp, M, Karp, C.L, and Kohl, J., The anaphylatoxins bridge innate and adaptive immune responses in allergic asthma, *Mol.Immunol.* 41, 123-131, 2004; Ali, H. and Panettieri, R.A., Jr., Anaphylatoxin C3a receptors in asthma, *Respir.Res.* 6, 19, 2005; Sunyer, J.O., Boshra, H., and Li, J., Evolution of anaphylatoxins, their diversity and novel roles in innate immunity: insights from the study of fish complement, *Vet.Immunol.Immunopathol.* 108, 77-89, 2005; Schmidt, R.E. and Gessner, J.E., Fc receptors and their interactions with complement in autoimmunity, *Immunol. Lett.* 100, 56-67, 2005; Chaplin, H., Jr., Review: the burgeoning history of the complement system 1888-2005, *Immunohematol.* 21, 85-93, 2005; Lambrecht, B.N., An unexpected role for the analphylatoxin C5a receptor in allergic sensitization, *J.Clin. Invest.* 116, 626-632, 2006.

Anergy

Lack of an immune response to an allergen (antigen); can refer to an individual cell such as a B-cell or a T-cell, tissue, or intact organism; however, it is used most frequently with respect to B-cells or T-cells and immunological tolerance. See. Kantor, F.S., Infection, anergy, and cell-mediated immunity, *N.Engl.J.Med.* 292, 629-634, 1975; Bullock, W.E., Anergy and infection, *Adv. Intern.Med.* 21, 149-173, 1976; Dwyer, J.M., Anergy. The mysterious loss of immunological energy, *Prog.Allergy* 35, 15-92, 1984; Brennan, P.J., Saouaf, S.J., Greene, M.I., and Shen. Y., Anergy and suppression as coexistent mechanisms for the maintenance of peripheral T cell tolerance, *Immunol.Res.* 27, 295-302, 2003; Macian, F., Im, S.H., Garcia-Cozar, F.J., and Rao, A., T-cell anergy, *Curr.Opin.Immunol.* 16, 209-216, 2004; Mueller, D.L., E3 ubiquitin ligases as T cell anergy factors, *Nat.Immunol.* 5, 883-890, 2004; Faria, A.M. and Weiner, H.L., *Immunol.Rev.* 206, 232-259, 2005; Akdis, M., Blaser, K., and Akdis, C.A., T regulatory cells in allergy, *Chem.Immunol.Allergy* 91, 159-173, 2006; Ferry, H.. Leung, J.C., Lewis , G. *et al.*, B-cell tolerance, *Transplantation* 81, 308-315, 2006.

Angiopoietin

A protein family that binds to endothelial cells; specific for Tie2 receptor kinase. See Plank. M.J., Sleeman, B.D., and Jones, P.F., The role of the angiopoietins in tumor angiogenesis, *Growth Factors* 22, 1-11, 2004; Oike, Y., Yasunaga, K., and Suda, T., Angiopoietin-related/angiopoietin-like proteins regulate angiogenesis *Int.J.Hematol.* 80, 21-28, 2004; Giuliani, N., Colla, S., Morandi, F., and Rizzoli, V., Angiopoietin-1 and myeloma-induced angiognesis, *Leuk.Lymphoma* 46, 29033, 2005; Dhanabal, M., Jeffers, M., LaRochelle, W.J., and Lichenstein, R.S., Angioarrestin: a unique angiopoietin-related protein iwht anti-angiogenic properties, *Biochem.Biophys.Res.Commu.* 333, 308-315, 2005; Endothelial/pericyte interactions, *Circ.Res.* 97, 512-523, 2005.

Anisotropy

A difference in a physical property such as melting point when measured in different principle directions; antonym, isotropy. Anisotropy is also defined as the property of being anisotropic as in the case of light transmission where different values are obtained when along axes in different directions Time-resolved fluorescence anisotropy decay measures the time dependence of the depolarization of light emitted from a fluorophore experiencing angular motions. In botany, anisotropy is defined as assuming different positions in response to the action of external stimuli. See Kinosita, K., Jr., Kawato, S., and Ikegami, A., Dynamic structure of biological and model membranes: analysis by optical anisotropy decay measurement, *Adv.Biophys.* 17, 147-203, 1984; Kinosita, K., Jr. and Ikegami, A., Dynamic structure of membranes and subcellular components revealed by optical anisotropy decay methods, *Subcell.Biochem.* 13, 55-88, 1988; Bucci, E. and Steiner, R.F., Anisotropy decay of fluorescence as an experimental approach to proteins, *Biophys.Chem.* 30, 199-224, 1988; Matko, J., Jenei, A., Matyus, L. Ameloot, M., and Damjanovich, S., Mapping of cell surface protein-patterns by combined fluorescence anisotropy and energy transfer measurements, *J.Photochem.Photobiol.B* 19, 69-73, 1993; Rachofsky, E.L. and Laws, W.R., Kinetic methods and data analysis methods for fluorescence anisotropy decay, *Methods Enzymol.* 321, 216-238,

2000; Santos, N.C., Prieto, M., and Castanho, M.A., Quantifying molecular partition into model systems of biomembranes: an emphasis on optical spectroscopic methods, *Biochim.Biophys. Acta* 1612, 123-135, 2003; Vrielink, A. and Sampson, N., Sub-angstrom resolution x-ray structures: is seeing believing?, *Curr. Opin.Struct.Biol.* 13, 709-715, 2003; Wang, J., Cao, Z., Jiang, Y., *et al.*, Molecular signaling aptamers for real-time fluorescence analysis of proteins, *IUBMB Life* 57, 123-128, 2005; Dmitrienko, V.E., Ishida, K., Kirfel, A., and Ovchinnikova, E.N., Polarization anisotropy of X-ray atomic factors and 'forbidden' resonant reflections, *Acta Crystallogr.A* 61, 481-493, 2005; Baskin, T.I., Anisotropic expansion of the plant cell wall, *Annu.Rev.Cell Dev. Biol.* 21, 203-222, 2005; Heilker, R., Zemanova, L., Valler, M.J., and Nienhaus, G.U., Confocal fluorescence microscopy for high-throughput screening of G-protein coupled receptors, *Curr.Med. Chem.* 12, 2551-2559, 2005; Guthrie, J.W., Hamula, C.L., Zhang, H., and Le, X.C., Assays for cytokines using aptamers, *Methods* 38, 324-330, 2006.

Ankyrin-repeat domains/proteins

A domain or motif, named after ankydrin, a cytoskeletal protein, is found in a large number of proteins. This domain which was first described in a yeast cell cycle regulator (Swi6/cdc 10) and *Drosphilia* (notch protein) consists of approximately 30 amino acids and is involved in protein-protein interactions. See Liou, H.C. and Baltimore, D., Regulation of the NF-kappa b/rel transcription factor and I kappa B inhibitor system, *Curr.Opin.Cell Biol.* 5, 477-487, 1993; Dedhar, S. and Hannigan, G.E., Integrin cytoplasmic interactions and bidirectional transmembrane signalling, *Curr.Opin.Cell Biol.* 8, 657-669, 1996; Sedgwick, S.G. and Smerdon, S.J., The ankyrin repeat: a diversity of interactions on a common structural framework, *Trends Biochem.Sci.* 24, 311-316, 1999: Yoganathan, T.N., Costello, P., Chen, X., *et al.*, Integrin-linked kinase (ILK): a "hot" therapeutic target, *Biochem. Pharmacol.* 60, 1115-1119, 2000; Hryniewicz-Jankowska, A., Czogalla, A., Bok, E., and Sikorsk, A.F., Ankyrins, multifunctional proteins involved in many cellular pathways, *Folia Histochem.Cytobiol.* 40, 239-249, 2002; Lubman, O.Y., Korolev, S.V., and Kopan, R., Anchoring notch genetics and biochemistry; structural analysis of the ankyrin domain sheds light on existing data, *Mol.Cell.* 13, 619-626, 2004; Mosavi, L.K., Cammett, T.J., Desosiers, D.C., and Peng, Z.Y., The ankyrin repeat as molecular architecture for protein recognition, *Protein Sci.* 13, 1435-1448, 2004; Tanke, H.J., Dirks, R.W., and Raap, T., FISH and immunocytochemistry: toward visualizing single target molecules in living cells, *Curr.Opin.Biotechnol.* 16, 49-54, 2005; Trigiante, G. and Lu, X., ASPPs and cancer, *Nat.Rev.Cancer* 6, 217-226, 2006; Legate, K.R., Montañez, E., Kudlacek, O. and Fässler, R., ILK, PINCH and parvin: the tIPP of integrin signalling, *Nat.Rev.Mol. Cell Biol.* 7, 20-31, 2006.

Annotation

Information added to a subject after the initial overall definition. Most frequently used in molecular biology for the addition of information regarding function to the initial description of a gene/ gene sequence in a genome. See Brent, M.R., Genome annotation past, present, and future: how to define an ORF at each locus, *Genome Res.* 15, 177-1786, 2005; Boutros, P.C. and Okey, A.B., Unsupervised pattern recognition: an introduction to the whys and wherefores of clustering microarray data, *Brief Bioinform.* 6, 331-343, 2005; Boeckman, B., Blatter, M.C., Famiglietti, L., *et al.*,

Protein variety and functional diversity: Swiss-Prot annotation in its biological context, *C.R.Biol.* 328, 882-899, 2005; Koonin, E.V., Orthologs, paralogs, and evolutionary genomics, *Annu.Rev. Genet.* 39, 309-338, 2005; Cahan, P., Ahmed, A.M., Burke, H., *et al.*, List of list-annotated (LOLA): a database for annotation and comparison of published microarray gene lists, *Gene* 360, 78-82, 2005; Dong, Q., Kroiss, L., Oakley, F.D., Wang, B.B., and Brendel, V., Comparative EST analyses in plant systems, *Methods Enzymol.* 395, 400-418, 2005; Crockett, D.K., Seiler, C.E., 3rd, Elenitoba-Johnson, K.S., and Kim, M.S., *J.Biomed.Tech.* 16, 341-346, 2005; Hermida, L., Schaad, O., Demougin, P., Descombes, P., and Primig, M., MIMAS: an innovative tool for network-based high density oligonucleotide microarray data management and annotation, *BMC Bioinformatics* 7, 190, 2006; Huang, D., Wei, P., and Pan, W., Combining gene annotation and gene expression data in model-based clustering weighted method, *OMICS* 10, 28, 2006; Snyder, K.A., Feldman, H.J., Dumontier, M., Salama, J.J., and Hogue, C.W., Domain-based small molecule binding site annotation, *BMC Bioinformatics* 7, 152, 2006.

Anoikis

Apoptosis following loss of attachment to a matrix or specific anchorage site. See Grossman, J., Molecular mechanisms of "detachment-induced apoptosis—Anoikis.", *Apoptosis* 7, 247-260, 2002; Zvibel, I., Smets, F., and Soriano, H., Anoikis: roadblock to cell transplantation? *Cell Transplant.* 11, 621-630, 2002; Valentijn, A.J., Zouq, N., and Gilmore, A.P., Anoikis, *Biochem. Soc.Trans.* 32, 421-425, 2004; Zhan, M., Zhao, H., and Han, Z.C., *Histol.Histopathol.* 19, 973-983, 2004; Reddig, P.J. and Juliano, R.L., Clinging to life: cell to matrix adhesion and cell survival, *Cancer Metastasis Rev.* 24, 425-439, 2005; Rennebeck, G., Martelli, M., and Kyprianou, N., Anoikis and survival connections in the tumor microenvironment: is there a role in prostate cancer metastasis? *Cancer Res.* 65, 11230-11235, 2005.

ANTH-Domain

A protein domain similar to the ENTH-domain and contained in proteins involved in endocytotic processes. See Stahelin, R.V., Long, F., Petter, B.J., *et al.*, Contrasting membrane interaction mechanisms of AP180 N-terminal homology (ANTH) and epsin N-terminal homology (ENTH) domains, *J.Biol.Chem.* 278, 28993-28999, 2003; Sun, Y., Kaksonen, M., Madden, D.T., *et al.*, Interaction of Slap2p's ANTH domain with PtdIns(4,5)P2 is important for actin-dependent endocytotic internalization, *Mol. Biol.Cell* 16, 717-730, 2005; Yao, P.J., Bushlin, I., and Petralia, R.S., Partially overlapping distribution of epsin1 and HIP1 at the synapse: analysis by immunoelectron microscopy, *J.Comp. Neurol.* 494, 368-379, 2006.

Antibody

A protein synthesized and secreted by a plasma cell. A plasma cell or antibody-secreting cell is derived from an undifferentiated B-cell. Antibodies are designated as the humoral immune response as opposed to the cellular immune response. Antibodies are usually synthesized and secreted in response to a foreign protein or bacteria. Natural antibody preparation are polyclonal in that such preparations are derived from a population of plasma cells. A monoclonal antibody is derived from a signal plasma cell clone. Antibodies can be formed against self; such antibodies are referred to as autoantibodies. Disease resulting from the formation of antibodies are called autoimmune diseases and can result from disorders of the humoral immune system or the cellular immune system Antibodies are classified as IgG, IgM, IgE, IgA, and IgD. There are unusual naturally-occuring antibodies such as camelid antibodies and artificial derivatives such as Fab fragments and scFv fragments.

Antibody Valency

Antibody valency refers to the number of antigen binding sites there are on a single antibody molecule. An IgG molecule which consists of two heavy chain and two light chains (a dimer of heterodimers) has two antibody binding sites and hence is bivalent have a valency of two. IgM which is a pentamer of IgG has a valency of 10. An scFv fragment is monovalent. An antibody with increased valence is considered to have greater avidity. See Marrack, J.R., Hoch, H., and Johns, R.G., The valency of antibodies, *Biochem.J.* 48, xxi-xxii, 1951; Sela, M., Antibodies: shapes, homogeneity, and valency, *FEBS Lett.* 1, 83-85, 1968; van Regenmortel, M.H., Which value of antigenic valency should be used in antibody avidity calculations with multivalent antigens, *Mol.Immunol.* 25, 565-567, 1988; Gerdes, M., Meusel, M., and Spener, F., Influence of antibody valency in a displacement immunoassay for the quantitation of 2,4-dichlorophenoxyacetic acid, *J.Immunol.Methods* 223, 217-226, 1999; Hudson, P.J. and Kortt, A.A., High avidity scFv multimers: diabodies and tri-abodies, *J.Immunol.Methods* 231, 177-189, 1999; Hard, S.A. and Dimmock, N.J., Valency of antibody binding to virions and its determination by surface plasmon resonance, *Rev.Med.Virol.* 14, 123-135, 2004; Scallon, B., Cai, A., Radewonuk, J., and Naso, M., Addition of an extra immunoglobulin domain to two anti-rodent TNF monoclonal antibodies substantially increased their potency, *Mol.Immunol.* 41, 73-80, 2004; Adams, G.P., Tai, M.S., McCartney, J.E. *et al.*, Avidity-mediated enhancement of in vivo tumor targeting by single-chain Fv dimers, *Clin.Cancer Res.* 12, 1599-1605, 2006.

Antibody-dependent cellular cytotoxicity (ADCC)

The process of cell death. This is usually the process by which an organism destroys bacterial and viral pathogens but also is the mechanism by which tumor cells are lysed secondary to treatment with antibodies. The process involves the recognition of epitopes by the Fab region of the IgG on the target cell surface resulting in the binding of the antibody. The Fc domain is then recognized by a phagocytic cell such as a natural killer (NK) cell. The Fc region is critical for this process. See Dallegri, F. and Ottonello, D., Neutrophil—mediated cytotoxicity against tumor cells: state of the art, *Arch.Immunol.Ther.Exp.* 40, 39-42, 1992; Muller-Eberhard, H.J., The molecular basis of target cell killing by human lymphocytes and of killer cell self-protection, *Immunol.Rev.* 103, 87-98, 1981; Morretta, L., Moretta, A., Canonica, G.W., *et al.*, Receptors for immunoglobulins on resting and activated human T cells, *Immunol.Rev.* 56, 141-162, 1981; Santonine, A., Herberman, R.B., and Holden, H.T., Correlation between natural and antibody-dependent cell-mediated cytotoxicity against tumor targets in the mouse. II. Characterization of the effector cells, *J.Natl.Cancer Inst.* 63, 995-1003, 1979; Sissons, J.G. and Oldstone, M.B., Antibody-mediated destruction of virus-infected cells, *Adv.Immunol.* 29, 209-260, 2000; Perussia, B. and Loza, M.J., Assays for antibody-directed cell-mediated cytotoxicity (ADCC) and reverse ADCC (redirected cytotoxicity) in human natural killer cells, *Methods Mol.Biol.* 121, 179-192, 2000; Villamor, N., Montserrat, E., and Colomer, D., Mechanism of action and resistance to monoclonal antibody therapy, *Semin.Oncol.* 30, 424-433, 2003; Casadevall, A and Pirofski, L.A., Antibody-mediated regulation of cellular immunity and the inflammatory response, *Trends Immunol.* 24, 474-478, 2003; Mellstedt, H., Monoclonal antibodies in human cancer, *Drugs Today* 39 Supp C, 1-16, 2003; Gelderman, K.A., Tomlinson, S., Ross, G.D., and Gorter, A., Complement function in mAb-mediated cancer immunotherapy, *Trends Immunol.* 25, 158-164, 2004; Schmidt, R.E. and Gessner, J.E., Fc receptors and their interaction with complement in autoimmunity, *Immunol.Lett.* 100, 56-67, 2005; Iannello, A. and Ahmad, A. Role of antibody-dependent cell-mediated cytotoxicity in the efficacy of therapeutic anti-cancer monoclonal antibodies, *Cancer Metastasis Rev.* 24, 487-499, 2005.

Antibody-Proteomics/Antibody Based Proteomics

The systematic generation and use of antibodies for the analysis of the proteome. An example would be the use of an antibody-based protein microarray. See Agaton, C., Falk, R., Holden Guthenberg, I. *et al.*, Selective enrichment of monospecific polyclonal antibodies for antibody-based proteomics efforts, *J.Chromatog.A* 1043, 33-40, 2004; Nielsen, A.B. and Geierstanger, B.H., Multiplexed sandwich assays in a microwave format. *J.Immunol.Methods* 290, 107-120, 2004; Uhlen, M. and Ponten, F., Antibody-based proteomics for human tissue profiling, *Mol.Cell.Proteom.* 4, 384-393, 2005; Stenvall, M., Steen, J., Uhlen, M., *et al.*, High-throughput solubility assay for purified recombinant protein immunogens, *Biochim.Biophys.Acta.* 1752, 6-10, 2005; Uhlen, M., Bjorling, E., Agaton, C. *et al.*, A human protein atlas for normal and cancer tissues based on antibody proteomics, *Mol.Cell.Proteomics* 4, 1920-1932, 2005. See also immunoproteomics.

Antigen

A material which can be of diverse substance and origin such as protein or microorganism which elicits an immune response. An immune response can be the formation of an antibody directed against the antigen (humoral response; B-cell response) as well as a cellular response (T cell response). Antigens can be separated into immunogens (complete antigens) which can elicit an immune response and haptens or incomplete immunogens which do not by themselves elicit and immune response but can reaction with antibodies. Haptens require combination with a larger molecule such as protein to elicit antibody formation. See Nossal, G.J.V., *Antigens, Lymphoid Cells, and the Immune Response*, Academic Press, New York, USA, 1971; Langone, J.J., *Antibodies, Antigens, and Molecular Mimicry*, Academic Press, San Diego, USA, 1989; *Fundamental Immunology*, ed. W.Paul, Raven Press, New York, 1993; Cruse, J.M., Lewis, R.E., and Wang, H., *Immunology Guidebook*, Elsevier, Amsterdam, Netherlands, 2004.

Antigenic Determinant

An antigenic determinant is also an epitope; this is the region of an antigen which binds to the reactive site of an antibody which is referred to as a paratope. The antigenic determinant which elicits the antibody response. There are linear or continuous determinants which would be a continuous amino acid sequence in a protein antigen and conformation or discontinuous determinants where, for example, with a protein, the epitope is formed by protein folding. A linear determinant is recognized by T-cells as well as B-cell and antibodies while a discontinuous determinant is recognized only by B-cells and antibodies. See Eisen, H.N., The immune response to a simple antigenic determinant, *Harvey Lect.* 60, 1-34, 1966; Kabat, E.A., The nature of an antigenic determinant, *J.Immunol.* 97, 1-11, 1966; Franks, D., Antigens as markers on cultured mammalian cells, *Biol.Rev.Camb.Philos. Soc.* 43, 17-50, 1968; Stevanovic, S., Antigen processing is predictable: from genes to T cell epitopes, *Transpl.Immunol.* 14, 171-174, 2005; McRobert, E.A., Tikoo, A., Gallicchio, M.A., Cooper, M.E., and Bach, L.A., Localization of the ezrin binding epitope for glycated proteins, *Ann.N.Y.Acad.Sci.* 1043, 617-624, 2005; Lovtich, S.B., and Unanue,E.R., Conformational isomers of a peptide-class II major histocompatibility complex, *Immunol. Rev.* 207, 293-313, 2005; Phillips, W.J., Smith, D.J., Bona, C.A., Bot, A., and Zaghouani, H., Recombinant immunoglobulin-based epitope delivery: a novel class of autoimmune regulators, *Int.Rev.Immunol.* 24, 501-517, 2005; De Groot, A.S., Knopp, P.M., and Martin, W., De-immunization of therapeutic proteins by T-cell epitope modification, *Dev.Biol.(Basel)* 122, 171-194, 2005; Burlet-Schlitz, O., Claverol, S., Gairin, J.E., and Monsarrat, B., The use of mass spectrometry to identify antigens from proteosome processing, *Methods Enzymol.* 405, 264-300, 2005.

Anti-idiotypic

Usually in reference to antibodies whose specificity is directed against the idiotypic region of an antibody; most frequently with naturally occurring antibodies. Since receptors and antibodies share common binding characteristics, this term is sometimes used to described antibodies directed against receptors. See Couraud, P.O. and Strosberg, A.D., Anti-idiotypic antibodies against hormone and neurotransmitter receptors, *Biochem.Soc. Trans.* 19, 147-151, 1991; Erlanger, B.F., Antibodies to receptors by an auto-anti-idiotypic strategy, *Biochem.Soc.Trans.* 19, 138-143, 1991; Greally, J.M. Physiology of anti-idiotypic interactions: from clonal to paratopic selection, *Clin.Immunol.Immunopathol.* 60, 1-12, 1991; Friboulet, A., Izadyar, L., Avalle, B., Abzyme generation using an anti-idiotypic antibody as the "internal image" of an enzyme active site, *Appl.Biochem.Biotechnol.* 47, 229-237, 1994; Hebert, J. and Boutin, Y., Anti-idiotypic antibodies in the treatment of allergies, *Adv.Exp.Med.Biol.* 409, 431-437, 1996.

Antisense

Generally refers a nucleotide sequence that is complementary to a sequence of messenger RNA which is the product of the non-coding sequence of DNA. It also refers to the peptide products from the antisense sequence referred to as antisense peptides. Antisense peptides have been investigated for biological activity. siRNA are based on the processing of antisense RNA. See Korneev, S. and O'Shea, M., Natural antisense RNAs in the nervous system, *Rev.Neurosci.* 16, 213-222, 2005. See also MicroRNA, siRNA, antisense peptides. aptamers

Antisense Peptides

The products from the translation of antisense RNA. Some antisense peptides have been demonstrated to show affinity properties that appear to be unique to that sequence and not seen in scrambled sequences. See Schwabe, C., New thoughts on the evolution of hormone-receptor systems, *Comp.Biochem.Physiol.A* 97, 101-106, 1990; Chaiken, I., Interactions and uses of antisense peptides in affinity technology, *J.Chromatog.* 597, 29-36, 1992; Labrou, N. and Clonis, Y.D., The affinity technology in downstream processing, *J.Biotechnol.* 36, 95-119, 1994; Root-Bernstein, R.S. and Holsworth, D.D., Antisense peptides: critical minireview, *J.Theoret.Biol.* 190, 107-1199, 1998; Siemion, I.Z., Cebrat, M., and Kluczyk, A., The problem of amino acid complementarity and antisense peptides, *Curr.Protein Pept.Sci.* 5, 507-527, 2004.

Apical

In reference to a differentiated cell, that portion or apex of the cell that is pointed toward the lumen as for example in endothelial cells. The membrane protein distribution is frequently different between the apical domain and the basolateral domain. See basolateral. See Alfalah, M., Wetzel, G., Fischer, I. *et al.*, A novel type of detergent-resistant may contribute to an early protein sorting event in epithelial cells, *J.Biol.Chem.* 280, 42636-42643, 2005; Kellett, G.L., and Brot-Laroche, E., Apical GLUT2: a major pathway of intestinal sugar absorption, *Diabetes* 54, 3056-3062, 2005; Ito, K., Suszuki, H., Horie, T., and Sugiyama, Y., Apical/basolater surface expression of drug transporters and its role in vectorial drug transport, *Pharm.Res.* 22, 1559-1577, 2005; Anderson, J.M., Van Itallie, C.M., and Fanning, A.S., Setting up a selective barrier at the apical junction complex, *Curr.Opin.Cell Biol.* 16, 140-145, 2004.

Apoptosis

Programmed cell death; an organized process by which cell undergo degradation and elimination. See Tomei, L.D. and Cope, F.O., *Apoptosis: The Molecular Basis of Cell Death*, Cold Spring Harbor Laboratory Press, Plainview, NY, 1991; Studzinski, G.P., *Cell Growth and Apoptosis: A Practial Approach*, IRL Press ata Oxford University Press, Oxford, UK, 1995; Christopher, G.D., *Apoptosis and the Immune Response*, Wiley-Liss, New York, NY,1995; Kumar. S., *Apoptosis: Mechanisms and Role in Disease*, Springer, Berlin, 1998; Lockshin, R.A. and Zakeri, Z., *When Cells Die: A Comprehensive Evaluation of Apoptosis and Programmed Cell Death*, Wiley-Liss, New York, NY, 1998; Jacobson, M.D., and McCarthy, N.J., *Apoptosis*, Oxford University Press, Oxford, UK, 2002; *Apoptosis Techniques and Protocols*, 2nd edn., ed. A.C. LeBlanc, Humana Press, Totowa, NJ, 2002; *Cell Proliferation and Apoptosis*, ed. D. Hughes and Hm Mehmet, Bios, Oxford, UK, 2003; Potten, C.S. and Wilson, J.W., *Apoptosis: The Life and Death of Cells*, Cambridge Universtiy Press, Cambridge, UK, 2004.

Apoptosome

A multiprotein complex which contains capases-9 and is thought to represent a holoenzyme involved in apoptosis. See Tsujimoto, Y., Role of Bcl-2 family proteins in apoptosis: apoptosomes or mitochondria, *Genes Cells*, 3, 697-707,1998; Adrain, C. and Martin, S.J., The mitochondrial apoptosome: a killer unleashed by the cytochrome seas, *Trends Biochem.Sci.* 26, 390-397, 2001; Salvesen, G.S. and Renatus, M., Apoptosome: the seven-spoked death machine, *Dev.Cell.* 2, 256-257, 2002; Cain, K., Bratton, S.B., and Cohen, G.M., The Apaf-1 apoptosome: a large caspase-activating complex, *Biochemie* 84, 203-214, 2002; Shi, Y., Apoptosome: the cellular engine for the activation of caspase-9, *Structure* 10, 285-288, 2002; Reed, J.C., Apoptosis-based therapies, *Nat. Rev.Drug Disc.* 1, 111-121, 2002; Adams, J.M., and Cory, S.,

Apoptosomes: engines for caspase activation, *Curr.Opin.Cell Biol.* 14, 715-720, 2002; Hajra, K.M. and Liu, J.R., Apoptosome dysfunction in human cancer, *Apoptosis* 9, 691-704, 2004.

Aprotinin

A small protein (single chain protein, MW 6.5 kDa; 58 amino acids) also known as basic pancreatic trypsin inhibitor (BPTI) or the Kunitz pancreatic trypsin inhibitor. It is best known as an inhibitor of tryptic-like serine proteases such as plasma kallikrein and plasmin. Aprotinin is also used as model for protein folding. See Kellermeyer, R.W. and Graham, J.C., Jr., Kinins-possible physiologic and pathologic roles in man, *N.Eng.J.Med.* 279, 754-759, 1968; Schachter, M., Kallikreins (kininogenases) – a group of serine proteases with bioregulatory actions, *Pharmacol.Rev.* 31, 1-17, 1979; Creighton, T.E., Experimental studies of protein folding and unfolding, *Prog.Biophys.Mol.Biol.* 33, 231-297, 1978; Fritz, H. and Wunderer, G., Biochemistry and applications of Aprotinin, the kallikrein inhibitor from bovine organs, *Arzneimittelforschung* 33, 479-494, 1983; Sharpe, S., De Meester, I., Hendriks, D. *et al.*, Proteases and their inhibitors: today and tomorrows, *Biochimie* 73, 121-126, 1991; Creighton, T.E., Protein folding pathways determined using disulphide bonds, *Bioessays* 14, 195-199, 1992; Day, R. and Daggett, V., All-atom simulations of protein folding and unfolding, *Adv.Protein Chem.* 66, 373-403, 2003.

Aptamer

Aptamers are relatively short oligonucleotides (generally 100 bp or less) which have the property of acting as relatively specific ligands to a broad range of targets. Aptamers are generally selected by combinatorial chemistry techniques. See Ellington, A.D. and Szostak, J.W., In vitro selection of RNA molecules that bind specific ligands, *Nature* 346, 818-822, 1990; Burke, J.M., and Berzal-Herranz, A., In vitro selection and evolution of RNA: applications for catalytic RNA, molecular recognition, and drug discovery, *FASEB J.* 7, 106-112, 1993; Stull, R.A. and Szoka, F.C., Jr., Antigene, ribozyme and aptamer nucleic acid drugs: progress and prospects, *Pharm.Res.* 12, 465-483, 1995; Uphoff, K.W., Bell, S.D., and Ellington, A.D., In vitro slection of aptamers: the dearth of pure reason, *Curr.Opin.Struct.Biol.* 6, 281-288, 1996; Collett, J.R., Cho, E.J. and Ellington, A.D., Production and processing aptamers microarrays, *Methods* 37, 4-15, 2005; Nutiu, R. and Li, Y., Aptamers with fluorescence-signaling properties, *Methods* 37, 16-25, 2005; Proske, D.,Blank, M., Buhmann, R., and Resch, A., Aptamers—basic research, drug development, and clinical applications, *Appl.Microbiol.Biotechnol.* 69, 367-374, 2005; Pestourie, C., Tavitian, B., and Duconge, F., Aptamers against extracellular targets for in vivo applications, *Biochimie* 87, 921-930, 2005. It is noted that the term intramer is used to describe intracellular aptamers (see Famulok, M., Blind, M., and Mayer, G., Intramers as promising new tools in functional proteomics, *Chem.Biol.* 8, 931-939, 2001; Famulok, M. and Mayer, G., Intramers and aptamers: applications in protein-function analyses and potential for drug screening, *ChemBioChem* 6, 19-26, 2005).

Aquaporin

Water-specific membrane pores facilities osmosis. See van Lieburg, A.F., Knoers, N.V., and Deen, P.M., Discovery of aquaporins: a breakthrough in research on renal water transport, *Ped. Nephrol.* 9, 228-234, 1995; Sabolic, I. and Brown, D., Water transport in renal tubules is mediated by aquaporins, *Clin.Investig.*

72, 689-700, 1994; King, L.S. and Agre, P. Pathophysiology of the aquaporin water channels, *Annu.Rev.Physiol.* 58, 619-648, 1996; Chaumont, F., Moshelion, M., and Daniels, M.J., Regulation of plant aquaporin activity, *Biol.Cell.* 97, 749-764, 2005; Castle, N.A., Aquaporins as targets for drug discovery, *Drug.Discov.Today* 10, 485-493, 2005.

Arabidopsis thaliana

A small plant in the mustard family that is the model for studies of the plant genome. Meinke, D.W., Cheng, D.M., Dean, C., Rounsley, S.D., and Koorneeft, M., *Arabidopis thaliana:* A model plant for genome analysis, *Science* 282, 662-682, 1998; *Methods in Plant Molecular Biology and Biotechnology*, ed. B.R.Glick and J.E.Thompson, CRC Press, Boca Raton, FL, 1993; *Aribidopsis*, ed. E.R. Meyerowitz and C.R. Somerville, Cold Spring Harbor Laboratory Press, Cold Spring Harbor, NY, 1994; *Aribidopsis*, ed. M. Anderson and J.A. Roberts, Sheffield Academic Press, Sheffield, UK, 1998; *Aribidopsis Protocols*, ed. J. Salinas and J.J. Sánchez-Serrano, Humana Press, Totowa, NJ, 2006.

ARF Family GTPases

The ADP-ribosylation family of GTPases. The ADP-ribosylation factor small GTPases have a role in the regulation of vesicular function via the recruitment of coat proteins and regulation of phospholipid metabolism. See Goud, B., Small GTP-binding proteins as compartment markers, *Semin.Cell Biol.* 3, 301-307, 1992; Kjeldgaard, M., Nyborg, J., and Clark, B.F., The GTP binding motif: variations on a theme, *FASEB J.* 10, 1347-1368, 1996; Donaldson, J.G. and Jackson, C.L., Regulators and effectors of the ARF GTPases, *Curr.Opin.Cell Biol.* 12, 475-482, 2000; Takai, Y., Sasaki, T., and Matozaki, T., Small GTP-binding proteins, *Physiol.Rev.* 81, 153-208, 2001; *ARF Family GTPases*, ed. R.A., Kahn, Kluwer Academic Publisher, Dordrecht, Netherlands, 2003; Munro, S., The Arf-like GTPase Arl1 qnd its role in membrane traffic, *Biochem.Soc.Trans.* 33, 601-605, 2005; Kahn, R.A., Cherfils, J., Elias, M., *et al.*, Nomenclature for the human Arf family of GTP-binding proteins: ARF, ARL, and SAR proteins, *J.Cell. Biol.* 172, 645-650, 2006; Nie, Z. and Randazzo, P.A., Arf GAPs and membrane traffic, *J.Cell Sci.* 119, 1203-1211, 2006; D'Souza-Schorey, C. and Chavrier, P., ARF proteins: roles of membrane traffic and beyond, *Nat.Rev.Mol.Cell Biol.* 7, 347-358, 2006. Also includes ARL and SAR proteins.

Arrhenius Energy of Activation

An operationally defined quantity that relates rate constants to temperature by the following equation: $k = Ae^{-Ea/R}$ where k is a rate constant; A is a constant, R is the gas constant, T the absolute temperature. A plot of ln k vs 1/T (Arrhenius plot) yields the Arrhenius energy of activation. See Van Tol, A., On the occurrence of a temperature coefficient (Q10) of 18 and a discontinuous Arrhenius plot for homogeneous rabbit muscle fructosediphosphatase, *Biochem.Biophys.Res.Communn.* 62, 750-756, 1975; Ceuterick, F., Peeters, J., Heremans, K., *et al.*, Involvement of lipids in the break of the Arrhenius plot of *Azobacter nitrogenase*, *Arch.Int.Physiol. Biochim.* 84, 587-588, 1976; Ceuterick, F., Peeters, J., Heremans, K., *et al.*, Effect of high pressure, detergents and Phospholipase on the break in the Arrhenius plot of *Azobacter* nitrogenase, *Eur.J.Biochem.* 87, 401-407, 1978; Stanley, K.K. and Luzio, J.P., The Arrhenius plot behaviour of rat liver 5'-nucleotidase in different lipid environments, *Biochim.Biophys.Acta* 514, 198-205, 1978; De Smedt, H., Borghgraef, R., Ceuterick, F., and Heremans, K., The role of lipid-protein interactions in the occurrence of a non-linear Arrhenius plot for (sodium-potassium)-activated ATPase, *Arch. Int.Physiol.Biochim.* 87, 169-170, 1979; Biosca, J.A., Travers, F., and Barman, T.E., A jump in an Arrhenius plot can be the consequence of a phase transition. The binding of ATP to myosin subfragment 1, *FEBS Lett.* 153, 217-220, 1983; Haeffner, E.W. and Friedel, R., Induction of an endothermic transition in the Arrhenius plot of fatty acid uptake by lipid-depleted ascites tumor cells, *Biochim. Biophys.Acta* 1005, 27-33, 1989; Muench, J.L., Kruuv, J., and Lepock, J.R., A two-step reversible-irrreversible model can account for a negative activation energy in an Arrhenius plot, *Cryobiology* 33, 253-259, 1996; Rudzinski, W., Borowieki, T.., Panczyk, T., and Dominko, A., On the applicability of Arrhenius plot methods to determine surface energetic heterogeneity of adsorbents and catalysts surfaces from experimental TPD spectra, *Adv.Colloid Interface Sci.* 84, 1-26, 2000.

Atomic Force Microscopy

A high resolution form of microscopy which involves a probe or tip moving over a surface (alternatively the sample can move with a static tip; the detection method is the same) and as the probe changes position in response to sample topography, the movement is tracked deflection of a laser beam which is recorded by a detector (Gadegaard, N., Atomic force microscopy in biology: technology and techniques, *Biotechnic and Histochem.* 81, 87-97, 2006). See Hansma, P.K., Elings, V.B., Marti, O., and Bracker, C.E., Scanning tunneling microscopy: application to biology and technology, *Science* 242, 209-216, 1988; Yang, J., Tamm, L.K., Somlyo, A.P., and Shao, Z., Promises and problems of biological atomic force microscopy, *J.Microsc.* 171, 183-198, 1993; Hansma, H.G. and Hoh, J.H., Biomolecular imaging with the atomic force microscope, *Annu.Rev.Biophys. Biomol. Struct* 23, 115-139, 1994; Tendler, S.J., Davies, M.C., and Roberts, C.J., Molecules under the microscope, *J.Pharm.Pharmacol.* 48, 2-8, 1996; Lekka, M., Lekki, J., Shoulyarenko, A.P., *et al.*, Scanning force microscopy of biological samples, *Pol.J.Pathol.* 47, 51-55, 1996; Ivanov, Y.D., Govorum, V.M., Bykov, V.A., and Archakov, A.I., Nanotechnologies in proteomics, *Proteomics* 6, 1399-1414, 2006; Connell, S.D. and Smith, D.A., The atomic force microscope as a tool for studying phase separation in lipid membranes, *Mol.Membr.Biol.* 23, 17-28, 2006; Bai, L., Santangelo, T.J., and Wang, M.D., Single-molecule analysis of RNA polymerase transcription, *Annu.Rev.Biophys.Biomol. Struct.* 35, 343-360, 2006; Guzman, C., Jeney, S., Kreplak, L., *et al.*, Exploring the mechanical properties of single vimentin intermediate filaments by atomic force microscopy, *J.Mol.Biol.* 360, 623-630, 2006; De Jong, K.L., Incledon, B., Yip, C.M., and Defelippis, M.R., Amyloid fibrils of glucagon characterized by high-resolution atomic force microscopy, *Biophys.J.* 91, 1905-1914, 2006; Xu, H., Zhao, X., Grant, C. *et al.*, Orientation of a monoclonal antibody adsorbed at the solid/solution interface: a combined study using atomic force microscopy and neutron reflectivity, *Langmuir* 22, 6313-6320, 2006.

Atomic Radius

A measurement of an atom which is not considered precise; generally half the distance between adjacent atoms of the same type in a crystal or molecule. It may be further described as a covalent radius, an ionic radius or metallic radius. The inability of cysteine to effectively substitute for serine in serine proteases is due, in part, to the increased atomic radius of sulfur compared to oxygen. See Alterman, M.A., Chaurasia, C.S., Lu, P., and Hanzlik, R.P., Heteroatom substitution shifts regioselectivity of lauric acid metabolism from omega-hydroxylation to (omega-1)-oxidation, *Biochem.Biophys.Res.Commun.* 214, 1089-1094, 1995; Zhang, R., Villeret, V., Lipscomb, W.N., and Fromm, H.J., Kinetics and mechanisms of activation and inhibition of porcine liver fructose-1,6-bisphosphatase by monovalent cations, *Biochemistry* 35, 3038-3043, 1996; Wachter, R.M. and Brachaud, B.P., Thiols as mechanistic probes for catalysis by the free radical enzyme galactose oxidase, *Biochemistry* 35, 14425-14435, 1996; Wagner, M.A., Trickey, P., Chen, Z.W., *et al.*, Monomeric sarcosine oxidase: 1. Flavin reactivity and active site binding determinants, *Biochemistry* 39, 8813-8824, 2000; Lack, J.G., Chaudhuri, S.K., Kelly, S.D. *et al.*, Immobilization of radionucleotides and heavy metals through anaerobic bio-oxidation of Fe(II), *Appl. Environ.Microbiol.* 68, 2704-2710, 2002; Hamm, M.L., Rajguru, S., Downs, A.M., and Cholera, R., Base pair stability of 8-chloro- and 8-iodo-2'-deoxyguanosine opposite 2'-deoxycytidine: implications regarding the bioactivity of 8-oxo-2'-deoxyguanosine, *J.Am.Chem.Soc.* 127, 12220-12221, 2005.

Autoantigen

A component of self which is able to elicit an immune response, an autoimmune reaction; frequently with pathological complications such as the destruction of pancreatic beta cells (Islets of Langerhans) resulting in Type 1 diabetes. See Sigurdsson, E. and Baekkeskov, S., The 64-kDa beta cell membrane autoantigen and other target molecules of humoral autoimmunity in insulin-dependent diabetes mellitus, *Curr.Top.Microbiol.Immunol.* 164, 143-168, 1990; Werdelin, O., Autoantigen processing and the mechanisms of tolerance to self, *Immunol.Ser.* 52, 1-9, 1990; Manfredi, A.A., Protti, M.P., Bellone, M., *et al.*, Molecular anatomy of an autoantigen: T and B epitopeson the nicotinic acetylcholine receptor in myasthenia gravis, *J.Lab.Clin.Med.* 120, 13-21, 1992; Sedgwick, J.D., Immune surveillance and autoantigen recognition in the central nervous system, *Aust.N.E.J.Med.* 25, 784-792, 1995; Utz, P.J., Gensler, T.J., and Anderson, P., Death, autoantigen modifications, and tolerance, *Arthritis Res.* 2, 101-114, 2000; Narendran, P., Mannering, S.I., and Harrison, L.C., Proinsulin-a pathogenic autoantigen in type 1 diabetes, *Autoimmun.Rev.* 2, 204-210, 2003; Gentile, F., Conte, M., and Formisano, S., Thyroglobulin as an autoantigen: what can we learn about immunopathogenicity from the correlation of antigenic properties with protein structure?. *Immunology* 112, 13-25, 2004; Pendergraft, W.F., 3rd, Pressler, E.M., Jennette, J.C., *et al.*, Autoantigen complementarity: a new theory implicating complementary proteins as initiators of autoimmune disease, *J.Mol. Med.* 83, 12-25, 2005; Wu, C.T., Gershwin, M.E., and Davis, P.A., What makes an autoantigen an autoantigen? *Ann.N.Y.Acad.Sci.* 1050, 134-1045, 2005; Wong, F.S., Insulin-a primary autoantigen in type 1 diabetes?, *Trends Mol.Med.* 11, 445-448, 2005; Jasinski, J.M. and Eisenbarth, G.S., Insulin as a primary autoantigen for type 1A diabetes, *Clin.Dev.Immuol.* 12, 181-186, 2005.

Autocoid

A internal, physiologically secretion of uncertain or unknown classification. Adenosine is one of the better examples as, apart from its role as purine base in RNA and DNA, it has diverse physiologic functions. See Yan. L., Burbiel, J.C., Maass, A., and Muller, C.E., Adenosine receptor agonists: from basic medicinal chemical to clinical development, *Expert Opin.Emerg.Drugs* 8, 537-576,

2003; Tan, D.X., Manchester, L.C., Hadeland, R., Lopez-Burillo, S., Mayo, J.C., Sainz, R.M., and Reiter, R.J., Melatonin: a hormone, a tissue factor, an autocoid, a paracoid, and an antioxidant vitamin, *J.Pineal Res.* 34, 75-78, 2003; Boyan, B.D., Schwartz, Z., and Swain, L.D., Cell maturation-specific autocrine/paracrine regulation of matrix vesicles, *Bone Miner.* 17, 263-268, 1992; Polosa, R., Holgate, S.T. and Church, M,K., Adenosine as a pro-inflammatory mediator in asthma, *Pulm.Pharmacol.* 2, 21-26, 1989.

Autocrine

Usually in reference to a hormone or other biological effector such as a peptide growth factor or cytokine which has an effect on the cell or tissue responsible for the synthesis of the given compound. Differentiated from endocrine or paracrine phenomena. See Sporn, M.B. and Roberts, A.B., Autocrine, paracrine, and endocrine mechanisms of growth control, *Cancer Surv.* 4, 627-632, 1985; Heldin, C.H. and Westermark, B., PDGF-like growth factors in autocrine stimulation of growth, *J.Cell Physiol.*(Suppl 5), 31-34, 1987; Ortenzi, C., Miceli, C. Bradshaw, R.A., and Luporini, P., Identification and initial characterization of an autocrine pheromone receptor in the protozoan cilitate *Euplotes raikovi*, *J.Cell.Biol.* 111, 607-614, 1990; Vallesi, A., Giuli, G., Bradshaw, R.A., and Luporini, P., Autocrine mitogenic activity of pheromones produced by the protozoan ciliate *Euplotes raikovi*, *Nature* 376, 522-524, 1995; Bischof, P., Meissner, A., and Campana, A., Paracrine and autocrine regulators of trophoblast invasion—a review, *Placenta* 21(Suppl. A), S55-S60, 2000; Bilezikjian, L.M., Blount, A.L., Leal, A.M., *et al.*, Autocrine/paracrine regulation of pituitary function of activin, inhibin and folllistatin, *Mol. Cell.Endocrinol.* 225, 29-36, 2004; Singh, A.B. and Harris, R.C., Autocrine, paracrine and juxtacrine signaling by EGFR ligands, *Cell Signal.* 17, 1183-1193, 2005; Ventura, C. and Branzi, A., Autocrine and intracrine signaling for cardiogenesis in embryonic stem cells: a clue for the development of novel differentiating agents, *Handb.Exp.Pharmacol.* (174), 123-146, 2006.

Autophagy

A pathway for the physiological degradation of cellular macromolecules and subcellular structures mediated by intracellular organelles such as lysosomes. It can be considered to be a process by which there is a membrane reorganization to separate or sequester a portion of the cytoplasm or cytoplasmic contents for subsequent delivery to an intracellular organelle such as a lysosome for degradation. This pathway of "self-destruction" is separate from proteosome-mediated degradation of macromolecules internalized from outside the cell. See Kroemer, G. and Jaattela, M., Lysosomes and autophagy in cell death control, *Nat.Rev.Cancer* 5, 886-897, 2005; Deretic, V., Autophagy in innage and adaptive immunity, *Trends Immunol.* 26, 523-528, 2005; Baehrecke, E.H., Autophagy: dual roles in life and death? *Nat.Rev.Mol.Cell Biol.* 6, 505-510, 2005; Wang, C-W., and Klianksy, D.J., The molecular mechanism of autophagy, *Molec. Med.* 9, 65-76, 2003. Klinosky, D.J., Autophagy, *Curr.Biol.* 15, R282-F283, 2005;

Autophosphorylation

A process by which a substrate protein, usually a receptor, catalyzes self-phosphorylation usually at a tyrosine residue. The mechanism can be either intramolecular (*cis*) or intermolecular (*trans*) although at least one system has been described with both *cis* and *trans* processes. See Cobb, M.H., Sang, B.-C., Gonzalez, R., Goldsmith, E., and Ellis, L., Autophosphorylation activates the soluble cytoplasmic domain of the insulin receptor in an intermolecular reaction, *J.Biol.Chem.* 264, 18701-18706, 1989; Frattali, A.L., Treadway, J.L., and Pessin, J.E., Transmembrane signaling by the human insulin receptor kinase. Relationship between intramolecular β subunit *trans-* and *cis-* autophosphorylation and substrate kinase activation, *J.Biol.Chem.* 267, 19521-19528, 1992; Rim, J., Faurobert, E., Hurley, J.B., and Oprian, D.D., *In vitro* assay for trans-phosphorylation of rhodopsin by rhodopsin kinase, *Biochemistry* 36, 7064-7070, 1997; Cann, A.D., Bishop, S.M., Ablooglu, A.J., and Kobanski, R.A., Partial activation of the insulin receptor kinase domain by juxtamembrane autophosphorylation, *Biochemistry* 37, 11289-11300, 1998; Iwasaki, Y., Nishiyama, H., Suzuki, K., and Koizumi, S., Sequential *cis/trans* autophosphorylation in TrkB tyrosine kinase, *Biochemistry* 36, 2694-2700, 1997; Cohen, P., The regulation of protein function by multisite phosphorylation – a 25 year update, *Trends in Biochem.Sci.* 25, 596-601, 2000; Wick, M.J., Ramos, F.J., Chen, H., Quon, M.J., Dong, L.Q., and Liu, F., Mouse 3-phosphoinositide-dependent protein kinase-1 undergoes dimerization and *trans*-phosphorylation in the activation loop, *J.Biol.Chem.* 278, 42913-42919, 2003; Wu, S. and Kaufman, R.J., *trans*-Autophosphorylation by the isolated kinase domain is not sufficient for dimerization or activation of the dsRNA-activated protein kinase PKR, *Biochemistry* 43, 11027-11034, 2004.

B-Lymphocytes

Also called B-cells, named derived from original studies involving the cells from the bursa of chickens. B-cells are best known for the production of antibodies but recent studies are showing increased complexity of function. See *T and B Lymphocytes: Origins, Properties and Roles in Immune Responses,* ed. M.F. Greaves, J.J.T. Owen, and M.C. Raff, Excerpta Medica, New York, NY, 1973; *B and T Cells in Immune Recognition,* ed. Loor, F. and Roelants, G.E. Wiley, New York, NY, 1977; *Cells of Immunoglobulin Synthesis,* ed. B. Pernis and H.J. Vogel, Academic Press, New York, NY, 1979; *T and B Lymphocytes: Recognition and Function,* ed. F.H. Bach, Academic Press, New York, NY, 1979; Cambier, J.C., *B-Lymphocyte Differentiation,* CRC Press, Boca Raton, FL, 1986; *B Lymphocytes and Autoimmunity,* ed. N. Chiorazzi, New York Academy of Sciences, New York, NY, 1997; Cruse, J.M. and Lewis, R.E., *Atlas of Immunology,* CRC Press, Boca Raton, FL, 1999; *Fundamental Immunology,* ed. W.E. Paul, Lippincott, Williams, and Wilkins, Philadelphia, PA, 2003.

Backflushing or Back Flushing

A method for cleaning filters involving reverse flow through the membrane; is occasionally used for cleaning large-scale chromatographic columns. See Tamai, G., Yoshida, H., and Imai, H., High-performance liquid chromatographic drug analysis by direct injection of whole blood samples. III. Determination of hydrophobic drugs adsorbed on blood cell membranes, *J.Chromatog.* 423, 163-168, 1987; Kim, B.S. and Chang, H.N., Effects of periodic backflushing on ultrafiltration performance, *Bioseparation* 2, 23-29, 1991; Dai, X.P., Luo, R.G. and Sirkar, K.K., Pressure and flux profiles in bead-filled ultrafiltration/microfiltration hollow fiber membrane modules, *Biotechnol.Prog.* 16, 1044-1054, 2000; Seghatchian, J. and Krailadsiri, P., Validation of different enrichment strategies for analysis of leucocyte subpopulations: development and application of a new approach,

based on leucofiltration, *Transfus.Apher.Sci.* 26, 61-72, 2002; Kang, I.J., Yoon, S.H. and Lee, C.H., Comparison of the filtration characteristics or organic and inorganic membranes in a membrane-coupled anaerobic bioreactor, *Water Res.* 36, 1803-1813, 2002; Nemade, P.R. and Davis, R.H., Secondary membranes for flux optimization in membrane filtration of biologic suspensions, *Appl.Biochem.Biotechnol.* 113-116, 417-432, 2004.

Bacterial Artificial Chromosome

A bacterial artificial chromosome (BAC) is a DNA construct based on a fertility plasma and is used for transforming and cloning in bacterial. It has an average insert size of 150 kbp with a range of approximately 100 kbp to 300 kbp. Bacterial artificial chromosomes are frequently used to sequence genomes where the PCR reaction is used to prepare a region of genomic DNA and then sequenced; in other words, a BAC is a vehicle based on the bacteria *Escherichia coli* that is used to copy, or clone, fragments of DNA that are 150,000 to 180,000 base pairs (bp) long. These DNA fragments are used as starting material for DNA.sequencing. See Schalkwyk, L.C., Francis, F., and Lehrach, H., Techniques in mammalian genome mapping, *Curr.Opin.Biotechnol.* 6, 37-43, 1995; Zhang, M.B. and Wing, R.A., Physical mapping of the rice genome with BACs, *Plant Mol.Biol.* 35, 115-127, 1997; Zhu, J., Use of PCR in library screening. An overview, *Methods Mol.Biol.* 192, 353-358, 2002; Ball, K.D. and Trevors, J.T., Bacterial genomics: the use of DNA microarrays and bacterial artificial chromosomes, *J.Microbiol.Methods* 49, 275-284, 2002; Miyake, T. and Amemiya, C.T., BAC libraries and comparative genomics of aquatic chordate species, *Comp.Biochem.Physiol.C Toxicol.Pharmacol.* 138, 233-244, 2004; Ylatra, B., van den Ijssel, P., Carvalho, B., Brakenhoff, R.H., and Maijer, G.A., BAC to the future! or oligonucleotides: a perspective for micro array comparative genomic hybridization (array CGH), *Nucleic Acids Res.* 34, 445-450, 2006.

Balanced Translocation

A chromosomal relocation which does not involved the net gain or loss of DNA; also referred to as reciprocal translocation. See Fraccaro, M., Chromosome abnormalities and gamete production in man, *Differentiation* 23(Suppl), S40-S43, 1983; Davis, J.R., Rogers, B.B., Hagaman, R.M., Thies, C.A., and Veomett, I.C., Balanced reciprocal translocations: risk factors for aneuploid segregant viability, *Clin.Genet.* 27, 1-19, 1985; Greaves, M.F., Biological models for leukaemia and lymphoma, *IARC Sci. Publ.* 157, 351-372, 2004; Benet, J., Oliver-Bonet, M., Cifuentes, P., Templado, C. and Navarro, J., Segregation of chromosomes in sperm of reciprocal translocation carriers: a review, *Cytogenet. Genome Res.* 111, 281-290, 2005. Aplan, P.D., Causes of oncogenic chromosomal translocation, *Trends Genet.* 22, 46-55, 2006.

Basolateral

Literally located on the bottom opposite from the apical end of a differentiated cell. See apical. See Terada, T. and Inui, K., Peptide transporters: structure, function, regulation and application for drug delivery, *Curr.Drug.Metab.* 5, 85-94, 2004; Brone, B. and Eggermont, J., PDZ proteins retain and regulate membrane transporters in polarized epithelial cell membranes, *Am.J.Physiol.Cell Physiol.* 288, C20-C29, 2005; Rodriquez-Boulan, E and Musch, A., Protein sorting in the Golgi complex: shifting paradigms, *Biochim.Biophys.Acta* 1744, 455-464, 2005; Vinciguerra, M., Mordasini, D., Vandewalle, A., and Feraille, E.,

Hormonal and nonhormonal mechanisms of regulation of the NA,K-pump in collecting duct priniciple cells, *Semin.Nephrol.* 25, 312-321, 2005.

Bathochromic Shift

A shift in the absorption/emission of light to a longer wavelength ($\lambda > \lambda_o$); a "red" shift. See Waleh, A. and Ingraham, L.L., A molecular orbital study of the protein-controlled bathochromic shift in a model of rhodopsin, *Arch.Biochem.Biophys.* 156, 261-266, 1973; Heathcote, P., Vermeglio, A., and Clayton, R.K., The carotenoid band shift in reaction centers from the *Rhodopseudomonas sphaeroides, Biochim.Biophys.Acta* 461, 358-364, 1977; Kliger, D.S., Milder, S.J., and Dratz, E.A., Solvent effects on the spectra of retinal Schiff bases—I. models for the bathochromic shift of the chromophore spectrum in visual pigments, *Photochem. Photobiol.* 25, 277-286, 1977; Cannella, C., Berni, R., Rosato, N., and Finazzi-Agro, A., Active site modifications quench intrinsic fluorescence of rhodanese by different mechanisms, *Biochemistry* 25, 7319-7323, 1986; Hermel, H., Holtje, H.D., Bergemann, S., *et al*, Band-shifting through polypeptide beta-sheet structures in the cyanine UV-Vis spectrum, *Biochim.Biophys.Acta* 1252, 79-86, 1995; Zagalsky, P.F., β-Crustacyanin, the blue-purple carotenoprotein of lobster carapace: consideration of the bathochromic shift of the protein-bound astaxanthin, *Acta Chrystallogr. D Biol.Chrystallogr.* 59, 1529-1531, 2003.

Betaine

Glycine betaine, (carboxymethyl)trimethylamonnium inner salt, Cystadane®. Derived from choline; serves as methyl donor in the synthesis of methionine from homocysteine. Also functions as an osmoprotectant and this function is similar to trehalose in plants. See Chambers, S.T., Betaines: their significance for bacteria and the renal tract, *Clin.Sci.* 88, 25-27, 1995; Nuccio, M.L., Rhodes, D., McNeil, S.D., and Hanson, A.D., Metabolic engineering of plants for osmotic stress resistance, *Curr.Opin.Plant Biol.* 2, 129-134, 1999; Zou, C.G. and Banerjee, R., Homocysteine and redox signaling, *Antioxid.Redox.Signal.* 7, 547-559, 2005; Craig, S.A., Betaine in human nutrition, *Am.J.Clin.Nutr.* 80, 539-549, 2004; Fowler, B., Homocysteine: overview of biochemistry, molecular biology, and role in disease processes, *Semin.Vasc.Med.* 5, 77-86, 2005; Ueland, P.M., Holm, P.I. and Hustad, S., Betaine: a key modulator of one-carbon metabolism and homocysteine status, *Clin. Chem.Lab.Med.* 43, 1069-1075 2005.

Bibody

One scFv fragment coupled to the C-terminus of the C_{H1} domain of a Fab fragment. See Schoonjans, R., Willems, A., Schoonooghe, S., *et al.*, Fab chains as an efficient heterodimerization scaffold for the production of recombinant bispecific and trispecific antibody derivatives, *J.Immunol.* 165, 7050-7057, 2000. See also tribody, diabody.

Bicoid Protein

A transcription-factor protein produced in *Drosophila*. See background to bicoid, *Cell* 54, 1-2, 1988; Stephenson, E.C. and Pokrywka, N.J., Localization of bicoid message during Drosophila oogenesis, *Curr.Top.Dev.Biol.* 26, 23-34, 1992; Johnstone, O. and Lasko, P., Translational regulation and RNA localization in Drosophila oocytes and embryos, *Annu.Rev.Genet.* 35, 365-406,

2001; Lynch, J. and Desplan, C., Evolution of development: beyond bicoid, *Curr.Biol.* 12, R557-R559, 2003.

BIND

Biomolecular Interaction Data Base which is designed to store full descriptions of interactions, molecular complexes, and metabolic pathways. See Bader, G.D., Donaldson, I., Wolting, C., Ouellette, B.F., Pawson, T., and Hogue, C.W., BIND—the biomolecular interaction network database, *Nucleic Acids Res.* 29, 242-245, 2001; Alfarano, C. Andrade, C.E., Anthony, K. *et al.*, The biomolecular interaction network database and related tools 2005 update, *Nucleic Acids Res.* 33, D418-D424, 2005; Shah, S.P., Huang, Y., Xu, T., Yuen, M.M. Ling, J., and Ouellette, B.F., Atlas – a data warehouse for integrative bioinformatics, *BMC Bioinformatics* 6, 34, 2005; Aytuna, A.S., Gursoy, A., Keskin, O., Prediction of protein-protein interactions by combining structure and sequence conservation in protein interfaces, *Bioinformatics* 21,2850-2855, 2005; Gilbert, D., Biomolecular interaction network database, *Brief Bioinform.* 6, 194-198, 2005.

Bioassay

Generally used to describe an assay for a drug/biologic after administration to subject. As such, a bioassay usually involves the sampling of a biological fluid such as blood. The term bioassay is also used to describe an assay which uses a biological substrate such as a cell or organism. The term bioassay does not define a technology. See Yamamoto, S. Urano, K., and Nomura, T., Validation of transgenic mice harboring the human prototype c-Ha-ras gene as a bioassay model for rapid carcinogenicity testing, *Toxicol.Lett.* 28, 102-103, 1998; Colburn, W.A. and Lee, J.W., Biomarkers, validation and pharmacokinetic-pharmacodynamic modelling, *Clin.Pharmacokinet.* 42, 997-1022, 2003; Tuomela, M., Stanescu, I., and Krohn, K., Validation overview of bio-analytical methods, *Gene Ther.* 22 (Suppl 1), S131-138, 2005; Indelicato, S.R., Bradshaw, S.L., Chapman, J.W. and Weiner, S.H., Evaluation of standard and state of the art analytical technology-bioassays, *Dev.Biol.*(Basal) 122, 102-114, 2005.

Bioequivalence

Similarity of biological properties; used in the characterization of pharmaceuticals to demonstrate therapeutic equivalence. See Levy, R.A., Therapeutic inequivalence of pharmaceutical alternatives, *Am.Pharm.* NS23, 28-39, 1985; Durrleman, S. and Simon, R., Planning and monitoring of equivalence studies, *Biometrics* 46, 329-336, 1990; Schellekens, H., Bioequivalance and the immunogenicity of biopharmaceuticals, *Nat.Rev.Drug.Disc.* 1, 457-462, 2002; Lennernas, H. and Abrahamsson, B., The use of biopharmaceutical classification of drugs in drug discovery and development: current status and future extension, *J.Pharm.Pharmacol.* 57, 273-285, 2005; Bolton, S., Bioequivalence studies for levothyroxine, *AAPS J.* 7, E47-E53, 2005.

Bioinformatics

The use of information technology to analyze data obtained from proteomic analysis. An example is the use of data bases such as SWISSPROT to identify proteins from sequence information determined by the mass spectrometric analysis of peptides. See Baxevanis, A.D. and Ouellette, B.F.F., *Bioinformatics: A Practical Guide to the Analysis of Genes and Proteins*, Wiley, Hoboken, NJ.,

USA, 2005; Buehler, L.K. and Rashidi, H.H., *Bioinformaics basics: applications in biological sciences and medicine*, Taylor & Francis, Boca Raton, Florida, USA, 2005; Evans, W.J., *Statistical methods in bioinformatics: an introduction*, Springer, New York, New York, USA, 2005; Wang, J.T.L., *Data mining in bioinformatics*, Springer, London, United Kingdom, 2005; Lesk, A.M., *Introduction to bioinformatics*, Oxford, New York, New York, 2005; Englbrecvht, C.C. and Facius, A., Bioinformatics challenges in proteomics, *Comb.Chem.High Throughput Screen.* 8, 705-715, 2005; Chandonia, J.M. and Brenner, S.E., The impact of structural genomics: expectations and outcomes, *Science* 311, 347-351, 2006; Chalkley, R.J., Hansen, K.C., and Baldwin, M.A., Bioinformatic methods to exploit mass spectrometric data for proteomic applications, *Methods Enzymol.* 402, 289-312, 2005; Allison, D.B, Cui, X., Page, G.P., and Sabripour, M., Microarray data analysis: from disarray to consolidation and consensus, *Nat.Rev.Genet.* 7, 55-65, 2006; Brent, M.R., Genome annotation past, present, and future: how to define an ORF at each locus, *Genome Res.* 15, 1777-1786, 2005.

Biologicals

A biological product is any virus, serum, toxin, antitoxin, blood, blood component or derivative, allergenic product, or analogous product applicable to the prevention, treatment, or cure of diseases or injury. Biologic produts are a subset of "drug products" distinguished by their manufacturing processes (biological process vs. a chemical process). In general, the term "drugs" includes biological products. Within the United States, the regulation of biological is the purview of the FDA Center for Biologicals Evaluation and Research (CBER) and drugs within the FDA Center for Drug Evaluation and Research (CDER). There has been a recent shift of some drug products which were traditionally in CBER such as monoclonal antibodies and peptide growth factors to CDER. Vincent-Gattis, M., Webb, C., and Foote, M., Clinical research strategies in biotechnology, *Biotechnol.Annu.Rev.* 5, 229-267, 2000; Steinberg, F.M. and Raso, J., Biotech pharmaceuticals and biotherapy: an overview, *J.Pharm.Pharm.Sci.* 1, 48-59, 1998; Stein, K.E. and Webber, K.O., The regulation of biologic products derived from bioengineered plants, *Curr.Opin.Biotechnol.* 12, 308-311, 2001; Morrow, K.S. and Slater, J.E., Regulatory aspects of allergen vaccines in the US, *Clin.Rev.Allergy Immunol.* 21, 141-152, 2001; Hudson, P.J. and Souriau, C., Recombinant antibodies for cancer diagnosis and therapy, *Expert Opin. Biol.Ther.* 1, 845-855, 2001; Monahan, T.R., Vaccine industry perspective of current issues of good manufacturing practices regarding product inspections and stability testing, *Clin.Infect.Dis.* 33(Suppl.4), S356-S361, 2001; Hsueh, E.C. and Morton, D.L., Angiten-based immunotherapy of melanoma: Canvaxin therapeutic polyvalent cancer vaccine, *Semin.Cancer Biol.* 13, 401-407, 2003; Miller, D.L. and Ross, J.J., Vaccine INDs: review of clinical holds, *Vaccine* 23, 1099-1101, 2005; Sobell, J.M., Overview of biologic agents in medicine and dermatology, *Semin.Cutan.Med.Surg.* 23, 2-9, 2005; Morenweiser, R., Downstream processing of viral vectors and vaccines, *Gene Ther.* 12 (Suppl 1), S103-S110, 2005.

Biomarker

A change in response to an underlying pathology; current examples of molecular changes include C-reactive protein, fibrin D-dimer, and troponin; the term biomarker is also used to include higher level responses such as behavior changes or anatomic changes. See Tronick, E.Z., The neonatal behavioral assessment scale as a biomarker of the effects of environmental agents on the newborn, *Environ.Health Perspect.* 74, 185-189, 1987; Salvaggio, J.E., Use and misuse of biomarker tests in "environmental conditions", *J.Allergy Clin.Immunol.* 94, 380-384, 1994; Den Besten, P.K., Dental fluorosis: its use as a biomarker, *Adv.Dent.Res.* 8, 105-110, 1994; Lohmander, L.S. and Eyre, D.R., From biomarker to surrogate outcome to osteoarthritis-what are the challenges?, *J.Rheumatol.* 32, 1142-1143, 2005; Vineis, P. and Husgafvel-Pursiainen, K., Air pollution and cancer: biomarker studies in human populations, *Carcinogenesis* 26, 1846-1855, 2005; Seligson, D.B., The tissue micro-array as a translational research tool for biomarker profiling and validation, *Biomarkers* 10(Suppl 1), S77-S82, 2005; Danna, E.A. and Nolan, G.P., Transcending the biomarker minset: deciphering disease mechanisms at the single cell level, *Curr.Opin.Chem.Biol.* 10, 20-27, 2006; Felker, G.M., Cuculich, P.S., and Gheorghiade, M., The Valsalva maneuver: a bedside "biomarker" test for heart failure, *Am.J.Med.* 119, 117-122, 2006; Allam, A. and Kabelitz, D., TCR trans-rearrangments: biological significance in antigen recognition vs the role as lymphoma biomarker, *J.Immunol.* 176, 5707-5712, 2006.

Biopharmaceutical classification system

The biopharmaceutical classification system (BCS) provides a classification of gastrointestinal absorption. See Amidon, G., Lennernas, H., Shah, V.P., and Crison, J.A., A theoretical basis for a biopharmaceutic drug classification: the correlation of *in vitro* drug product dissolution and *in vivo* bioavailability, *Pharm.Res.* 12, 413-420, 1995; Wilding, I.R., Evolution of the biopharmaceutics classification system (BCS) to oral modified release (MR) formulations: what do we need to consider? *Eur.J.Pharm.Sci.* 8, 157-159, 1999; Dressman, J.B. and Reppas, C., In vitro-in vivo correlations for lipophilic, poorly water-soluble drugs, *Eur.J.Pharm. Sci.* 11(Suppl 2), S73-S80, 2000; Taub, M.E., Kristensen, L. and Frokjaer, S., Optimized conditions for MDCK permeability and turbidimetric solubility studies using compounds representative of BCS class I-IV, *Eur.J.Pharmaceut.Sci.* 15, 311-340, 2002; Huebert, N.D., Dasgupta, M., and Chen, Y.M., Using *in vitro* human tissues-to predict pharmacokinetic properties, *Curr.Opin. Drug.Disc.Dev.* 7, 69-74, 2004; Lennernas, H. and Abrahamsson, B., The use of biopharmaceutic classification of drugs in drug discovery and development: current status and future extension, *J. Pharm.Pharmcol.* 57, 273-285, 2005.

Bone Morphogenetic Protein(s) (BMP)

A group of peptide/proteins which are multifunctional growth factors that are members of the TGFβ superfamily. There are multiple forms of bone morphogenetic proteins which all function as differentiation factors for the maturation of mesenchymal cells into chondrocytes and osteoblasts. See Hauschka, P.V., Chen, T.L., and Mavrakos, A.E., Polypeptide growth factors in bone matrix, *Ciba Found.Symp.* 136, 207-225, 1988; Wozney, J.M., Bone morphogenetic proteins, *Prog.Growth Factor Res.* 1, 267-280, 1989; Rosen, V. and Thies, R.S., The BMP proteins in bone formation and repair, *Trends Genet.* 8, 97-102, 1992; Wang, E.A., Bone morphogenetic proteins (BMPs): therapeutic potential in healing bony defects, *Trends Biotechnol.* 11, 379-383, 1993; Kirker-Head, C.A., Recombinant bone morphogenetic proteins: novel substances for enhancing bone healing, *Vet.Surg.* 24, 408-419, 1995; Ramoshibi, L.N., Matsaba, J., Teare, L., *et al.*, Tissue engineering: TGF-β superfamily members and delivery systems in bone regeneration, *Expert Rev.Mol.Med.* 2002, 1-11, 2002; Monteiro, R.M., de Sousa Lopez, S.M., Korchynskyi, O., *et al.*, Spatio-temporal activation of Smad1 and Smad5 in vivo: monitoring transcriptional activity of Smad proteins, *J.Cell Sci.* 117, 4653-4663, 2004; Canalis, E., Deregowski, V., Pereira, R.C., and Gazzero, E., Signals that determine the fate of osteoblastic cells, *J.Endocrinol.Invest.* 28(Suppl 8), 3-7, 2005; Franceschi, R.T., Biological approaches to bone regeneration by gene therapy, *J.Dent.Res.* 84, 1093-1103, 2005; Ripamonti, U., Teare, J., and Petit, J.C., Pleiotropism of bone morphogenetic proteins: from bone induction to cementogenesis and periodontal ligament regeneration, *J.Int.Acad.Periodontol.* 8, 23-32, 2006; Logeart-Avramoglou, D., Bourguignon, M., Oudina, K., Ten Dijke, P., and Petite, H., An assay for the determination of biologically active bone morphogenetic proteins using cells transfected with an inhibitor of differentiation promoter-luciferase construct, *Anal.Biochem.* 349, 78-86, 2006.

Bottom-up Proteomics

Identification of unknown proteins by analysis of peptides obtained from unknown proteins by enzymatic (usually trypsin) hydrolysis. , See Brock, A., Horn, D.M., Peters, E.C., *et al.*, An automated matrix-assisted laser desorption/ionization quadrupole Fourier transform ion cyclotron resonance mass spectrometer for "bottom-up" proteomics, *Anal.Chem.* 75, 3419-3428, 2003; Wennder, B.R. and Lynn, B.C., Factors that affect ion trap data-dependent MS/MS in proteomics, *J.Am.Soc.Mass Spectrom.* 15, 150-157, 2004; Amoutzias, G.D., Robertson, D.L., Oliver, S.G., and Bornberg-Bauer, E., Convergent evolution of gene networks by single-gene duplications in higher eukaryotes, *EMBO Rep.* 5, 274-279, 2004; Ren, D., Julka, S., Inerowicz, H.D., and Regnier, F.E., Enrichment of cysteine-containing peptides from tryptic digests using a quaternary amine tag, *Anal.Chem.* 76, 4522-4530, 2004; Listgarten, J. and Emili, A., Statistical and computational methods for comparative proteomic profiling using liquid chromatography-tandem mass spectrometry, *Mol.Cell.Proteomics* 4, 419-434, 2005; Slysz, G.W. and Schriemer, D.C., Blending protein separation and peptide analysis through real-time proteolytic digestion, *Anal.Chem.* 77, 1572-1579, 2005; Zhong, H., Marcus, S.L., and Li, L., Microwave-assisted acid hydrolysis of proteins combined with liquid chromatography MALDI MS/MS for protein identification, *J.Am.Soc.Mass Spectrom.* 16, 471-481, 2005; Putz, S., Reinders, J., Reinders, Y., and Sickmann, A., Mass spectrometry-based peptide quantification: applications and limitations, *Expert Rev. Proteomics* 2, 381-392, 2005; Riter, L.S., Gooding, K.M., Hodge, B.D., and Julian, R.K., Jr., Comparison of the Paul ion trap to the linear ion trap for use in global proteomics, *Proteomics* 6, 1735-1740, 2006.

Brand Name Drugs

A brand name drug is a drug marketed under a proprietary, trade-marked-protected name.

BRET

Bioluminesence Resonance Energy Transfer. Similar to FRET in BRET is technique which can be used to measure physical interactions between molecules. Intrinsic bioluminescence is used in this procedure such as different fluorescent protein (e.g. green fluorescent protein and blue fluorescent protein). See De, A. and Gambhir, S.S., Noninvasive imaging of protein-protein interactions from live cells and living subjects using bioluminescence resonance energy transfer, *FASEB J.* 19, 2017-2019, 2005.

Brownian Movement

The random movement of small particles in a suspension where the force of collision between particles is not lost but retained in part by the particle. The practical effect is to set the lower limit of particle size for settling from a suspension. Brownian movements is usually restricted to particles 1 μm in diameter and is not observed with particles of 5 μm.

Bulk Solution

Bulk solution which would be defined as any macroscropic volume of a substance. In the case of an electrolyte. a bulk solution is charge neutral; intracellular and extracellular solutions possess a neutral charge even the presence of a membrane potential. The term bulk solution is also used to describe the difference between water structure in the hydration layer immediately around a macromolecule such as protein and the bulk solvent space. See Nakasako, M., Large-scale networks of hydration water molecules around proteins investigated by cryogenic X-ray crystallography, *Cell.Mol.Biol.* 47, 767-790, 2001; Lever, M., Blunt, J.W., and Maclagan, R.G., Some ways of looking at compensatory kosmotropes and different water environment, *Comp.Biochem.Physiol. A Mol. Integr.Physiol.* 130, 471-486, 2001; Halle, B., Protein hydration dynamics in solution: a critical survey, *Philos.Trans.R.Soc. Biol.Sci.* 359, 1207-1223, 2004; Levicky, R. and Horgan, A., Physicochemical perspectives on DNA microarray and biosensor technologies, *Trends Biotechnol.* 23, 143-149, 2005.

CAD

A multifunctional protein that initiates and regulates *de novo* pyrimidine biosynthesis. See Carrey, E.A., Phosphorylation, allosteric effectors and inter-domain contacts in CAD: their role in regulation of early steps of pyrimidine biosynthesis, *Biochem. Soc.Trans.* 21, 191-195, 1993; Davidson, J.N., Chen, K.C., Jamison, R.S., Musmanno, L.A., and Kern, C.B., The evolutionary history of the first three steps in pyrimidine biosynthesis, *Bioessays* 15, 157-164, 1993; Evans, D.R. and Guy, H.I., Mammalian pyrimidine biosynthesis: fresh insights into an ancient pathway, *J.Biol.Chem.* 279, 33035-33038, 2005.

Cadherins

A group of cell adhesion proteins which enable cells to interact with other cells and extracellular matrix components. See Obrink, B., Epithelial cell adhesion molecules, *Exp.Cell Res.* 163, 1-21, 1986; Takeichi, M., Cadherins: a molecular family important in selective cell-cell adhesion, *Annu.Rev.Biochem.* 59, 237-252, 1990; Geiger, B. and Ayalon, O., Cadherins, *Annu.Rev.Cell Biol.* 8, 307-332, 1992; Tanoue, T. and Takeichi, M., New insights into fat cadherins, *J. Cell Sci.* 118, 2347-2353, 2005; Lecuit, T., Cell adhesion: sorting out cell mixing with echinoid? *Curr.Biol.* 15, R505-R507, 2005; Gumbiner, B.M., Regulation of cadherin-mediated adhesion in morphogenesis, *Nat.Rev.Mol.Cell Biol.* 6, 622-634, 2005; Bamji, S.X., Cadherins: actin with the cytoskeleton to form synapses, *Neuron* 47, 175-178, 2005; Junghans, D., Hass, I.G., and Kemler, R., Mammalian cadherins and protocadherins: about cell death, synapses and processing, *Curr.Opin.Cell Biol.* 17, 446-452, 2005; Cavallaro, U., Liebner, S., and Dejana, E., Endothelial cadherins and tumor angiogenesis, *Exp.Cell Res.* 312, 659-667, 2006; Redies, C., Vanhalst, K., and Roy, F., delta-Protocadherins: unique structures and functions, *Cell Mol.Life Sci.* 62, 2840-2852,

2005; Collona, M., Cytolytic responses: cadherins put out the fire, *J.Exp.Med.* 203, 289-295, 2006; Chan, A.O., E-cadherin in gastric cancer, *World J. Gastroenterol.* 12, 199-203, 2006.

Caenorhabditis elegans

A free-living roundworm which has been used extensively for genomic studies. It is notable for the discovery of RNA silencing/RNA interference (Fire, A., Xu, S., Montgomery, M.K., *et al.,* Potent and specific genetic interference by double-stranded RNA in *Caenorhabditis elegans, Nature* 391, 806-811, 1998). For general aspects of *Caenorhabditis elegans* see Zuckerman, B.M. and Merton, B., *Nematodes as Biological Models,* Academic Press, New York, 1980; *C.elegans* see Emmons, S.W., Mechanisms of C.elegans development, *Cell* 51, 881-883, 1987; Blumenthal, T. and Thomas, J., *Cis* and *trans* splicing in *C.elegans, Trends Genet.* 4, 305-308, 1988; Wood, W.B., *The Nematode Caenorhabditis elegans,* Cold Spring Harbor Labortory Press, Cold Spring Harbor, NY, 1988; Greenwald, I., Cell-cell interactions that specify certain cell fates in *C.elegans* development, *Trends Genet.* 5, 237-241, 1989; Coulson, A., Kozono, Y., Lutterbach, B., *et al.,* YACs and the *C.elegans* genome, *Bioessays* 13, 413-417, 1991; Plasterk, R.H., Reverse genetics of *Caenorhabditis elegans, Bioessays* 14, 629-633, 1992; Burglin, T.R. and Ruvkun, G., The *Caenorhabditis elegans* homeobox gene cluster, *Curr.Opin.Genet.Dev.* 3, 615-620, 1993; Selfors, L.M. and Stern, M.J., MAP kinase function in *C.elegans, Bioessays* 16, 301-304, 1994; Stern, M.J. and DeVore, D.L., Extending and connecting signaling pathways in *C.elegans, Dev.Biol.* 166, 443-459, 1994; Kayne, P.S. and Sternberg, P.W., Ras pathways in *Caehorhabditis elegans, Curr.Opin.Genet.Dev.* 5, 38-43, 1995; Hope, I.A., *C. elegans; A Practical Approach,* Oxford University Press, Oxford, UK, 1999; Brown, A., *In the Beginning was the Worm: Finding the Secrets of Life in a Tiny Hermaphrodite,* Columbia University Press, NY, 2003; Filipowicz, W., RNAi: the nuts and bolts of the RISC machine, *Cell* 122, 17-20, 2005; Grishok, A., RNAi mechanisms in *Caenorhabditis elegans, FEBS Lett.* 579, 5932-5939, 2005. Hobert, O. and Loria, P., Uses of GFP in *Caenorhabditis elegans, Methods Biochem.Anal.* 47, 203-226, 2006; Hillier, L.W., Coulson, A., Murray, J.I., *et al.,* Genomics in *C.elegans;* so many genes, such a little worm, *Genome Res.* 15, 1651-1660, 2005; http://www.nematodes.org/Caenorhabditis; http://www.wormbook.org; http://elegans.swmed.edu.

Calcineurin

A protein phosphatase which is involved in the activation of IL-2 transcription; IL-2 stimulated the T-Cell response. Calcineurin is inhibited by immuosuppressive drugs such as cyclosporine and FK506(tacrolimus). See Pallen, C.J. and Wang, J.H., A multifunctional calmodulin-stimulated phosphatase, *Arch. Biochem.Biophys.* 237, 281-291, 1985; Klee, C.B., Draetta, G.F., and Hubbard, M.J., Calcineurin, *Adv.Enzymol.Relat.Areas Mol. Biol.* 61, 149-200, 1988; Siekierka, J.J. and Sigal, N.H., FK-506 and cyclosporine A: immunosuppressive mechanism of action and beyond, *Curr.Opin.Immunol.* 4, 484-552, 1992;Groenendyk, J., Lynch, J., and Michalak, M., Calreticulin, Ca²⁺, and calcineurin – signaling from the endoplasmic reticulum, *Mol.Cells* 30, 383-389, 2004; Michel, R.N., Dunn, S.E. and Chin, E.R., Calcineurin and skeletal muscle growth, *Proc.Nutr.Soc.* 63, 341-349, 2004; Im, S.H. and Rao, A., Activation and deactivation of gene expression by Ca²⁺/calcineurin-NFAT-mediated signaling, *Mol.Cells* 16, 1-9, 2004; Bandyopadhyay, J., Lee, J. and Bandopadhyay, A., Regulation of calcineurin, a calcium/calmodulin-dependent

protein phosphatase in *C.elegans, Mol.Cells* 18, 10-16, 2004; Taylor, A.L., Watson, C.J., and Bradley, J.A., Immunosuppresive agents in solid organ transplantation: mechanisms of action and therapeutic efficacy, *Crit.Rev.Oncol.Hematol.* 56, 23-46, 2005; Crespo-Leiro, M.G., Calcineurin inhibitors in heart transplantation, *Transplant.Proc.*37, 4018-4020, 2005.

Calcium Transients

Physiology phenomena resulting from changes in calcium concentration across membranes such as the diversity in Ca⁺⁺-stimulated transcriptional phenomona. See Spitzer, N.C., Lautermilch, N.J., Smith, R.D., and Gomez, T.M., Coding of neuronal differentiation by calcium transients, *Bioessays* 22, 811-817, 2000; Afroze, T. and Husain, M., Cell cycle dependent regulation of intracellular calcium concentration in vascular smooth muscle cells: a potential target for drug therapy, *Curr.Drug Targets Cardiovasc.Haematol. Disord.* 1, 23-40, 2001; Thorner, M.O., Holl, R.W., and Leong, D.A., The somatotrope: an endocrine cell with functional calcium transients, *J.Exp.Biol.* 139, 169-179, 1988; Morgan, K.G., Bradley, A., and DeFeo, T.T., Calcium transients in smooth muscle, *Ann.N.Y.Acad.Sci.* 522, 328-337, 1988; Fumagalli, G., Zacchetti, D., Lorenzon, P., and Grohovaz, F., Fluorimetric approaches to the study of calcium transients in living cells, *Cytotechnology* (Suppl 5) 1, 99-102, 1991; Komura, H., and Kumada, T., Ca²⁺ transients control CNS neuronal migration, *Cell Calcium* 37, 387-393, 2005; Fossier, P., Tauc, L. and Baux, G., Calcium transients and neurotransmitter release at an identified synapse, *Trends Neurosci.* 22, 161-166, 1999.

CALM-domain/CALM protein

Clathrin assembly lymphoid myeloid-domain, related to ANTH-domain proteins and involved in endocytosis, formation of clathrin-coated pits; binds to lipids. See Kim, J.A., Kim, S.R., Jung, Y.K., *et al.,* Properties of GST-CALM expressed in *E.coli. Exp. Mol.Med.* 32, 93-99, 2000; Kusner, L. and Carlin, C., Potential role for a novel AP180-related protein during endocytosis in MDCK cells, *Am.J.Physiol.Cell Physiol.* 285, C995-C1008, 2003; Archangelo, L.F>. Glasner, J., Krause, A., and Bohlander, S.K., The novel CALM interactor CATS influences the subcellular localization of the leukemogenic fusion protein CALM/AF10, *Oncogene,* 25, 4099–4109, 2006.

Calnexin

A lectin protein associated with the endoplasmic reticulum which functions as a chaperone. See Cresswell, P., Androlewicz, M.J. and Ortmann, B., Assembly and transport of class I MHC-peptide complexes, *Ciba Found.Symp.* 187, 150-162, 1994; Bergeron, J.J., Brenner, M.B., Thomas, D.Y., and Williams, D.B., Calnexin: a membrane-bound chaperone or the endoplasmic reticulum, *Trends Biochem.Sci.* 19, 124-128, 1995;, Parham, P., Functions for MHC class I carbohydrates inside and outside the cell, *Trends Biochem.Sci.* 21, 472-433, 1996; Trombetta, E.S. and Helenius, A., Lectins as chaperones in glycoprotein folding, *Curr.Opin.Struct. Biol.* 8, 587-592, 1998; Huari, H.. Appenzeller, C., Kuhn, F., and Nufer, O., Lectins and traffic in the secretory pathway, *FEBS Lett.* 476, 32-37, 2000; Ellgaard, L. and Frickel, E.M., Calnexin, calreticulin, and ERp57: teammates in glycoprotein folding, *Cell. Biochem.Biophys.* 39, 223-247, 2003; Spiro, R.G., Role of *N*-linked polymannose oligosaccharides in targeting glycoproteins for endoplasmic reticulum-associated degradation, *Cell.Mol.Life*

Sci. 61, 1025-1041, 2004; Bedard, K., Szabo, E., Michalak, M., and Opas, M., Cellular functions of endoplasmic reticulum chaperones calreticulin, calnexin, and ERp57, *Int.Rev.Cytol.* 245, 91-121, 2005; Ito, Y., Hagihara, S., Matsuo, I., and Totani, K., Structural approaches to the study of oligosaccharides in glycoprotein quality control, *Curr.Opin.Struct.Biol.*15, 481-489, 2005.

Calponin

A family of actin-binding proteins which exist in various isoforms. As with other protein isoforms or isoenzymes, the expression of the isoforms is tissue-specific. The interaction of calponin with actin inhibits the actomyosin Mg-ATPase activity. See Winder, S and Walsh, M., Inhibition of the actinomyosin MgATPase by chicken gizzard calponin, *Prog.Clin.Biol.Res.* 327, 141-148, 1990; Winder, S.J., Sutherland, C. and Walsh, M.P., Biochemical and functional characterization of smooth muscle calponin, *Adv.Exp.Med.Biol.* 304, 37-51, 1991; Winder, S.J. and Walsh, M.P., Calponin: thin filament-linked regulation of smooth muscle contraction, *Cell Signal.* 5, 677-686, 1993; el-Mezgueldi, M., Calponin, *Int.J.Biochem.Cell Biol.* 28, 1185-1189, 1996; Szymanski, P.T., Calponin (CaP) as a latch-bridge protein—a new concept in regulation of contractility in smooth muscle, *J.Muscle Res.Cell Motil.* 25, 7-19, 2004; Lehman, W. Craig, R., Kendrick-Jones, J., and Sutherland-Smith, A.J., An open or closed case for the conformation of calponin homology domains on F-actin?, *J.Muscle Res. Cell Motil.* 25, 351-358, 2004; Ferjani, I., Fattoum, A., Maciver, S.K. *et al.*, A direct interaction with calponin inhibits the actin-nucleating activity of gelsolin, *Biochem.J.* 396, 461-468, 2006.

Calreticulin

A 50-60 kDa protein found in the endoplasmic reticulum. Calreticulin binds calcium ions tightly and it thought to play a role in calcium homeostasis. Calreticulin also functions as chaperone. See Koch, G.L. and Smith, M.J., The analysis of glycoproteins in cells and tissues by two-dimensional polyacrylamide gel electrophoresis, *Electrophoresis* 11, 213-219, 1990; Krause, K.H., Ca(2+)-storage organelles, *FEBS Lett.* 285, 225-229, 1991; Herbert, D.N., Simons, J.F., Peterson, J.R., and Helenius, A., Calnexin, calreticulin, and Bip/Kar2p in protein folding, *Cold Spring Harbor Symp.Quant.Biol.* 60, 405-415, 1995; Groenendyk, J., Lynch, J. and Michalak, M., Calreticulin, Ca²⁺, and calcineurin – signaling from the endoplasmic reticulum, *Mol.Cells* 17, 383-389, 2004; Michalak, M., Guo, L., Robertson, M., Lozak, M., and Opas, M., Calreticulin in the heart, *Mol.Cell.Biochem.* 263, 137-142, 2004; Gelebart, P., Opas, M., and Michalak, M., Calreticulin, a Ca²⁺-binding chaperone of the endoplasmic reticulum,, *Int.J.Biochem.Cell Biol.* 37, 260-266, 2005; Bedard, K., Szabo, E., Michalak, M., and Opas, M., Cellular functions of endoplasmic reticulum chaperones calreticulin, calnexin, and ERp57, *Int. Rev.Cytol.* 245, 91-121, 2005; Ito, Y,, Hagihara, S., Matsuo, I., and Totani, K., Structural approaches to the study of oligosaccharides in glycoprotein quality control, *Curr.Opin.Struct.Biol.* 15, 481-489, 2005; Garbi, N., Tanaka, S., van den Broek, M. *et al.*, Accessory molecules in the assembly of major histocompatibility complex class I/peptide complexes: how essential are they for CD(+) T-cell immune responses? *Immunol.Rev.* 207, 77-88, 2005; Cribb, A.E., Peyrou, M., Muruganandan, S., and Schneider, L., The endoplasmic reticulum in xenobiotic toxicity, *Drug.Metab. Rev.* 37, 405-442, 2005; Hansson, M., Olsson, I., and Nauseef, W.M., Biosynthesis, processing and sorting of human myeloperoxidase, *Archs.Biochem.Biophys.* 445, 214-224, 2006.

Camelid Antibodies

Unique antibodies from members of the *Camelidae* family. The antibody structure consists of a heavy chain consisting of a variable region V$_H$H but no light chain. The structure also misses the the first constant domain (C$_H$1) but retains the other constants regions C-terminal from the hinge region. Thus, while a classical IgG is a dimer of heterodimers, the camelid antibody described herein is a homodimer. See Ghahroudi, M.A., Desmyter, A., Wyns, L., Hamers, R., and Muyldermans, S., Selection and identification of single domain antibody fragments from camel heavy-chain antibodies, *FEBS Lett.* 414, 521-526, 1997; Nguyen, V.K., Desmyter, A., and Muylderman, S., Functional heavy-chain antibodies in *Camelidae*, *Adv.Immunol.* 79, 261-296, 2001; Muyldermans, S., Single domain camel antibodies: current status, *J.Biotechnol.* 74, 277-302, 2001; Nguyen, V.K., Su., C., Muyldermans, S., and van der Loo, W., Heavy-chain antibodies in *Camelidae*, a case of evolutionary innovation, *Immunogenetics* 54, 39-47, 2002; Conrath, K.E., Wernery, U., Mulydermans, S., and Nguyen, V.K., Emergence and evolution of functional heavy-chain antibodies in *Camelidae*, *Dev.Comp.Immunol.* 27, 87-103, 2003; Rahbarisadeh, F., Rasaee, M.J., Forouzandeh, M., *et al.*, The production and characterization of novel heavy-chain antibodies against the tandem repeat region of MUC1 mucin, *Immunol.Invest.* 34, 431-452, 2005.

Cap

Structure at the 5'-end of eukaryotic RNA, introduced after transcription by linking the terminal phosphate of 5'-GTP to the terminal base of the mRNA. The guanine base can be nucleated. $^{7Me}G^{5'}$-ppp^{5}Np See Banerjee, A.K., 5'-terminal cap structure in eukaryotic messenger ribonucleic acids, *Microbiol.Rev.* 44, 175-205, 1980; Miura, K., The cap structure in eukaryotic RNA as a mark of a strand carrying protein information, *Adv.Biophys.* 14, 205-238, 1981; Lewin, B., *Genes IV*, Oxford University Press, Oxford, United Kingdom, 1990; Cougot, N., van Dijk, E., Babajko, S., and Seraphin, B., 'Cap-tabolism', *Trends Biochem.Sci.* 29, 436-444, 2004; Gu, M. and Lima, C.D., Processing the message: structural insights into capping and decapping mRNA, *Curr.Opin.Struct. Biol.* 15, 99-106, 2005; Bentley, D.L., Rules of engagement: co-transcriptional recruitment of pre-mRNA processing factors, *Curr.Opin.Cell Biol.* 17, 251-256, 2005; Liu, H. and Kiledjian, M., Decapping the message: a beginning or an end, *Biochem.Soc.Trans.* 34, 35-38, 2006; Simon, E., Camier, S., and Seraphin, B., *Trends Biochem.Sci.* 31, 241-243, 2006. CAP is also acronym for catabolic activator protein.

Catabolic activator protein

A transcription-regulating protein which binds to DNA in the promoter loop. See Benoff, B., Yang, H., Lawson, C.L., *et al.*, Structural basis of transcription activation: the CAP-alpha CTD-DNA complex, *Science* 297, 1562-1566, 2002; Balaeff, A., Mahadevan, L., and Schulten, K., Structural basis for cooperative DNA binding by CAP and lac repressor, *Structure* 12, 123-132, 2004; Akaboshi, E., Dynamic profiles of DNA: analysis of CAP- and LexA protein-binding regions with endonucleases, *DNA Cell Biol.* 24, 161-172, 2005.

Cathepsins

A family of intracellular thiol proteases involved in lysosomal digestion of proteins. See Janoff, A., Mediators of tissue damage in human polymorphonuclear neutrophils, *Ser.Haematol.*

3, 96-130, 1970; Harris, E.D., Jr. and Krane, S.M., Collagenases, *N.Engl.J.Med.* 291, 605-609, 1974; Larzarus, G.S., Hatcher, V.B., and Levine, N., Lysosomes and the skin, *J.Invest.Dermatol.* 65, 259-271, 1975; Ballard, F.J., Intracellular protein degradation, *Essays Biochem.* 13, 1-37, 1977; Barrett, A.J., Cathepsin D: the lysosomal aspartic proteinase, *Ciba Found.Symp.* (75), 37-50, 1979; Barrettt, A.J. and Kischeke, H., Cathepsin B, Cathepsin H, and Cathepsin L, *Methods Enzymol.* 80, 535-561, 1981; Groutas, W.C., Inhibitors of leukocyte elastase and leukocyte cathepsin G. Agents for the treatment of emphysema and related ailments, *Med.Res.Rev.* 7, 227-241, 1987; Stoka, V., Turk, B., and Turk, V., Lysosomal cathepsin proteases: structural features and their role in apoptosis, *IUBMB Life* 57, 347-353, 2005; Roberts, R., Lysosomal cysteine proteases: structure, function and inhibition of cathepsins, *Drug News Perspect.* 18, 605-614, 2005; Chwieralski, C.E., Welte, T., and Buhling, F., Cathepsin-regulated apoptosis, *Apoptosis* 11, 143-149, 2006. There is particular interest in the role of cathepsins in antigen processing (Honey, K. and Rudensky, A.Y., Lysosomal cysteine proteases regulate antigen presentation, *Nat.Rev.Immunol.* 3, 472-482, 2003; Bryant, P. and Ploegh, H., Class II MHC peptide loading by the professionals, *Curr.Opin.Immunol.* 16, 96-102, 2004; Liu, W. and Spero, D.M., Cysteine protease cathepsin S as a key step in antigen presentation, *Drug.News Perspect.* 17, 357-363, 2004; Hsing, L.C. and Rudensky, A.Y., The lysosomal cysteine proteases in MHC Class II antigen presentation, *Immunol.Rev.* 207, 229-241, 2005.

CASP

Critical Assessment of Structure Prediction describes a process for the evaluation of protein model building. See Moult, J. Predicting protein three-dimensional structure, *Curr.Opin. Biotechnol.* 10, 583-588, 1999; Moult, J., Fidelis, K., Rost, B., Hubbard, T., and Tramontano, A., Critical assessment of methods of protein structure prediction (CASP) – round 6, *Proteins* 61 Suppl. 7, 3-7, 2005; Giorgetti, A., Raimondo, D., Miele, A.E., and Tramontano, A., Evaluting the usefulness of protein structure models for molecular replacement, *Bioinformatics* 21(Suppl. 2), ii72-ii76, 2005; Espejo, F. and Patarroyo, M.E., Determining the 3D structure of human ASC2 protein involved in apoptosis and inflammation, *Biochem.Biophys.Res.Commun.* 340, 860-864, 2006; Moult, J., Rigorous performance evaluation in protein structure modeling and implications for computational biology, *Philos.Trans.R.Soc.Lond. B Biol.Sci.* 361, 453-458, 2006.

Caspases

A family of intracellular cysteine proteases that are involved in the process of apoptosis (programmed cell death). Caspases are synthesized as precursor or zymogen forms which required activation prior to function. One such activation process involves granzymes. Caspases also function in other intracellular processes. See Jacobson. M.D. and Evan, G.I., Apoptosis. Breaking the ice, *Curr.Biol.* 4, 337-340, 1994; Patel, T., Gores, G.BJ., and Kaufmann, S.H., The role of proteases during apoptosis, *FASEB J.* 10, 587-597, 1996; Alnemri, E.S., Mammalian cell death proteases: a family of highly conserved aspartate specific cysteine proteases, *J.Cell Biochem.* 64, 33-42, 1997; Zhivotovsky, B., Caspases: the enzymes of death, *Essays Biochem.* 39, 25-40, 2003; Twomey, C. and McCarthy, J.V., Pathways of apoptosis and importance in development, *J.Cell Mol.Med.* 9, 345-359, 2005; Ashton-Rickardt, P.G., The granule pathway of programmed cell death, *Crit.Rev. Immunol.* 25, 161-182, 2005; Yan, N. and Shi, Y., Mechanisms of apoptosis through structural biology, *Annu.Rev.Cell Dev.Biol.* 21, 35-56, 2005; Harwood, S.M., Yaqoob, M.M., and Allen, D.A., Caspase and calpain function in cell death: bridging the gap between apoptosis and necrosis, *Ann.Clin.Biochem.* 42, 415-431, 2005; Ho, P.K. and Hawkins, C.J., Mammalian initiator apoptotic caspases, *FEBS J.* 272, 5436-5453, 2005; Fardeel, B. and Orrenius, S., Apoptosis: a basic biological phenomenon with wide-ranging implications in human disease, *J.Intern.Med.* 258, 479-517, 2005; Cathelin, S., Rébe. C., Haddaoui, L., *et al.*, Identification of proteins cleaved downstream of caspases activation in monocytes undergoing macrophage differentiation, *J.Biol.Chem.* 281, 17779-17788, 2006.

Catalomics

The study of the enzymes in a proteome; the study of catalysis in a proteome. See Hu, Y., Uttamchandani, M., and Yao, S.Q., Microarray: a versatile platform for high-throughput functional proteomics, *Comb.Chem.High Throughput Screen.* 9, 201-212, 2006.

Catalytic Antibodies

Antibodies which demonstrate catalytic activity. The early development of these antibodies was based on the use of haptens which mirrored transition state intermediates for enzyme-catalyzed reactions. Catalytic antibodies can be referred to as abzymes. See Kraut, J., How do enzymes work? *Science* 242, 533-540, 1988; Lerner, R.A. and Tramontano, A., Catalytic antibodies, *Sci.Am.* 258, 65-70, 1988; Green, B.S., Catalytic antibodies and biomimetics, *Curr.Opin.Biotechnol.* 2, 395-400, 1991; Jacobs, J.W., New perspectives on catalytic antibodies, *Biotechnology* 9, 258-262, 1991; Blackburn, G.M., Kingsbury, G., Jayaweera, S., and Burton, D.R., Expanded transition state analogues, *Ciba Found.Symp.* 159, 211-222, 1991; O'Kennedy, R. and Roben, P., Antibody engineering: an overview, *Essays Biochem.* 26, 59-75, 1991; Stewart, J.D., Krebs, J.F., Siuzdak, G., *et al.*, Dissection of an antibody-catalyzed reaction, *Proc.Nat.Acad.Sci.USA* 91, 7404-7409, 1994; Posner, B., Smiley, J., Lee, I., and Benkovic, S., Catalytic antibodies: perusing combinatorial libraries, *Trends Biochem.Sci.* 19, 145-150, 1994; Kikuchi, K. and Hilvert, D., Antibody catalysis via strategic use of hepatenic charge, *Acta Chem.Scand.* 50, 333-336, 1996; Wentworth, P., Jr., and Janda, K.D., Catalytic antibodies: structure and function, *Cell. Biochem.Biophys.* 35, 63-87, 2001; Ostler, E.L., Resmini, M., Brocklehurst, K., and Gallacher, G., Polyclonal catalytic antibodies, *J.Immunol.Methods* 269, 111-124, 2002; Hanson, C.V., Nishiyama, Y., and Paul, S., Catalytic antibodies and their applications, *Curr.Opin.Biotechnol.* 16, 631-666, 2005. There has also been considerable interest in catalytic antibodies in pathological processes and as potential therapeutic agents. See Lacroix-Demazes, S., Kazatchkine, M.D. and Kaveri, S.V., Catalytic antibodies to factor VIII in haemophilia A., *Blood Coag.Fibrinol.* 14(Suppl 1), S31-S34, 2003; Poloukhina, D.I., Kanyshkova, T.G., Doronin, B.M., *et al.*, Hydrolysis of myelin basic protein by polyclonal catalytic IgGs from the sera of patients with multiple sclerosis, *J.Cell.Mol.Med.* 8, 359-368, 2004; Paul, S., Nishiyama, Y.,. Planque, S., *et al.*, Antibodies as defensive enzymes, *Springer Semin.Immunopathol.* 26, 485-503, 2005; Ponomarenko, N.A., Vorobiev, I.I., Alexandrova, E.S., *et al.*, Induction of a protein-targeted catalytic response in autoimmune prone mice: antibody-mediated cleavage of HIV-1 glycoprotein GP120, *Biochemistry* 45, 324-330, 2006; Lacroix-Desazes, S., Wootla, B., Delignat, S., *et al.*, Pathophysiology of catalytic antibodies, *Immunol.Lett.* 103, 3-7, 2006.

CATH

A classification process for protein domain structures based on class(C), architecture(A), topology(T), and homology superfamily (H). See Orengo, C.A., Michie, A.D., Jones, S., Jones, D.T., Swindells, M.B., and Thornton, J.M., CATH –a hierarchic classification of protein domain structures, *Structure* 5, 1093-1108, 1997; Bray, J.E., Todd, A.E., Pearl, F.M., Thornton, J.M., and Orengo, C.A., The CATH dictionary of homologous superfamilies (DHS): a consensus approach for identifying distant structural homologues, *Protein Eng.* 13, 153-165, 2000; Ranea, J.A., Buchan, D.W., Thornton, J.M., and Orengo, C.A., Evolution of protein superfamilies and bacterial genome size, *J.Mol.Biol.* 336, 871-887, 2004; Velazquez-Muriel, J.A., Sorzano, C.O., Scheres, S.H., and Carazo, J.M., SPI-EM: towards a tool for predicting CATH superfamilies in 3D-EM maps, *J.Mol.Biol.* 345, 759-771, 2005. Sillitoe, I., Dibley, M., Bray, J., Addou, S., and Orengo, C., Assessing strategies for improved superfamily recognition, *Protein Sci.* 14, 1800-1810, 2005.

CELISA

Enzyme-linked immunoassay on live cells. See Geraghyty, R.J., Jogger, C.P., and Spear, P.B, Cellular expression of alphaherpesvirus gD interferes with entry of homologous and heterologous alphaviruses by blocking access to a shared gD receptor, *Virology* 68, 147-156, 2000; Lee, R.B., Hassone, D.C., Cottle, D.L., and Picket, C., Interactions of *Campylobacter jejuni* cytolethal distending toxin subunits Cdta and Cdtc with HeLa cells, *Infect. Immun.* 71, 4883-4890, 2003.

Cell-based assays

This is a broad classification for assays where cells are used as the substrate or indictor for the action of a drug. Examples include platelet aggregation, cell-based ELISA(see below), gene expression assays, receptor-ligand interactions, etc. See Nuttall, M.E., Drug discovery and target validation, *Cells Tissues Organs* 169, 265-271, 2001; Bhadriraju, K. and Chen, C.S., Engineering cellular microenvironments to improve cell-based drug testing, *Drug Discov. Today* 7, 612-620, 2002; Indelicato, S.R., Bradshaw, S.L., Chapman, J.W., and Weiner, S.H., Evaluation of standard and state of the art analytical technology-bioassays, *Dev.Biol.* 122, 103-114, 2005; Stacey, G.N., Standardisation of cell lines, *Dev.Biol.* 111, 259-272, 2002; Qureshi, S.A., Sanders, P., Zeh, K., *et al.*, A one-arm homologous recombination approach for developing nuclear receptor assays in somatic cells, *Assay Drug Dev.Technol.* 1, 767-776, 2003; Wei, X., Swanson, S.J., and Gupta, S., Development and validation of a cell-based bioassay for th detection of neutralizing antibodies against recombinant human erythropoietin in clinical studies, *J.Immunol.Methods* 293, 115-126, 2004; Pietrak, B.L., Crouthamel, M.C., Tugusheva, K., *et al.*, Biochemical and cell-based assays for characterization of BACE-1 inhibitors, *Anal. Biochem.* 342, 144-151, 2005; Chen. T,, Hansen, G., Beske, O., *et al.*, Analysis of cellular events using Cell Card System in cell-based high-content multiplexed assays, *Expert Rev. Mol.Diagn.* 5, 817-829, 2005.

Cell-based ELISA

Cell-based ELISA are indirect or direct ELISA systems which use intact cells as the antigen sample. Cells may be dried onto the microplate surface or a microplate surface treated with polylysine,

chemical fixed with glutaraldehyde or similar reagents, or pelleted onto the surface. Hoffman, T. and Herberman, R.B., Enzyme-linked immunosorbent assay for screening monoclonal antibody production: use of intact cells as antigen, *J.Immunol.Methods* 39, 309-316, 1980; Krakauer, H., Hartman, R.J., and Johnson, A.H., Monoclonal antibodies specific for human polymorphic cell surface antigens. I. Evaluation of methodology. Report of a workshop, *Human Immunol.* 4, 167-181, 1982; Bishara, A., Brautbar, C., Marbach, A., Bonvida, B., and Nelken, D., Enzyme-linked immunosorbent assay for HLA determination on fresh and dried lymphocytes, *J.Immunol.Methods* 62, 265-271, 1983; Sharon, R., Duke-Cohan, J.S., and Galili, U., Determination of ABO blood group zygosity by an antiglobulin resetting technique and cell-based enzyme immunoassay, *Vox Sang.* 50, 245-249, 1986; Zhao, Q., LU, H., Schols, D., de Clercq, E., and Jiang, S., Development of a cell-based enzyme-linked immunosorbent assay for high-throughput screening of HIV type enzyme inhibitors targeting the coreceptor CXCR4, *Aids Res.Human Retrovirus* 19, 947-955, 2003; Yang, X.Y., Chen, E., Jiang, H. *et al.*, Development of a quantitative cell-based ELISA, for a humanized anti-IL-2/IL-15 receptor beta antibody (HuMikbeta(1)), and correlation with functional activity using an antigen-transferred murine cell line. In some cases, a cell homogenate could be used as the sample. See Franciotta, D., Martino, G., Brambilla, E., *et al.*, TE671 cell-based ELISA for anti-acetylcholine receptor antibody determination in myasthenia gravis, *Clin.Chem.* 45, 400-405, 1999. The cell-based ELISA is distinct from the ELISPOT assay where there a capture antibody on the membrane (Arvilommi, H., Elispot for detecting antibody-secreting cells in response to infections and vaccination, *APMIS* 104, 401-410, 1996).

Cell Culture

The maintenance of dispersed animal or plant cells in a specialized media (cell culture media). In biotechnology manufacturing, cell culture is used for the production of protein biopharmaceuticals using cells such as Chinese hamster ovary (CHO) cells or baby hamster kidney (BHK) cells. The use of the term cell culture differentiates such a process from fermentation. See Mantell, S.H. and Smith, H., *Plant Biotechnology*, Cambridge University Press, Cambridge, UK, 1983; *Applications of Plant Cell and Tissue Culture*, Chichester, UK, 1988; Freshney, R.I., *Animal Cell Culture: A Practical Approach*, IRL Press at Oxford University Press, Oxford, UK, 1992; Morgan, S.J. and Darling, D.C., *Animal Cell Culture*, Bios/Biochemical Society, London, UK, 1993; Davis, J.M., *Basic Cell Culture: A Practical Approach*, IRL Press at Oxford University Press, Oxford, UK, 1994; Dodds, J.H. and Roberts, L.W., *Experiments in Plant Tissue Culture*, Cambridge University Press, Cambridge, UK, 1995; Spier, R., *Encyclopedia of Cell Technology*, Wiley-Interscience, New York, NY, 2000; Hesse, F. and Wagner, R., Development and improvements in the manufacture of human therapeutics wit mammalian cell culture, *Trends Biotechnol.* 18, 173-180, 2000; James, E. and Lee, J.M., The production of foreign proteins from genetically modified plant cells, *Adv.Biochem.Eng.Biotechnol.* 72, 127-156, 2001; Kaeffer, B., Mammalian intestinal epithelial cells in primary culture: a mini-review, *In Vitro Cell Dev.Biol.-Animal* 38, 128-134, 2002; Ikonomou, L., Schneider, Y.J., and Agathos, S.N., Insect cell culture for industrial production of recombinant proteins, *Appl.Microbiol.Biotechnol.* 62, 1-20, 2003; Kallos, M.S., Sen, A., and Behie, L.A., Large-scale expansion of mammalian neural stem cells: a review, *Med.Biol.Eng.Comput.* 41, 271-282, 2003; Schiff, L.J., Review: production, characterization, and testing of banked mammalian cell substrates used to produce biological products, *In Vitro Cell Dev.Biol.-Animal* 41, 65-70, 2005; Evan, M.S., Sandusky, C.B. and Barnard, N.D., Serum-free hybridoma culture: ethical, scientific and safety considerations, *Trends Biotechnol.* 24, 105-108, 2006.

Cell Penetrating Peptide

Cell-penetrating peptides are relatively small peptides, usually less than 30 amino acids in length, which have the ability to pass through or translocate the cellular membrane in via a mechanisms which appears to be both receptor-independent as well as distinct from an endocytotic process. Such peptides have been demonstrated to "transport" diverse cargo and are being evaluated for drug delivery. See Lundberg, P. and Langel, U., A brief introduction to cell-penetrating peptides; Temsamani, J. and Vidal, P., The use of cell-penetrating peptides for drug delivery, *Drug Discovery Today* 9, 1012-1019, 2004; Gupta, B., Levchenko, T.S., and Torchilin, V.P., Intracellular delivery of large molecules and small particles by cell-penetrating proteins and peptides, *Adv.Drug.Deliv.Rev.* 57, 637-651, 2005; Deshayes, S., Morris, M.C., Divta, G., and Heitz, F., Cell-penetrating peptides: tools for intracellular delivery of therapeutics, *Cell.Mol.Life.Sci.* 62, 1839-1849, 2005. See also amphipathic.

CentiRay

A measure of the frequency of chromosome breakage between DNA markers in radiation-reduced somatic cell hybrids (radiation hybrids). One centiRay is equivalent to a 1% probability that a chromosome break (CentiRay distances are generally proportional to physical distance and are measured in centimorgans. See Hukriede, N.A., Joly, L., Tsang, M., *et al.*, Radiation hybrid mapping of the zebrafish genome, *Proc.Nat.Acad.Sci.USA* 96, 9745-9750, 1999; Hamasima, N. Suzuki, H., Mikawa, A. Morozumi, T., Plastow, G., and Mitsuhashi, T., Construction of a new porcine whole-genome framework map using a radiation hybrid panel, *Anim.Genet.* 34, 216-220, 2003; Voigt, C., Moller, S., Ibrahim, S.M., and Serrano-Fernandez, P., Non-linear conversion between genetic and physical chromosomal distances, *Bioinformatics* 20, 1966-1977, 2004.

Centimorgan

A measure of genetic distance that tells how far apart physically two genes are based on the frequency of recombination or crossover between the two gene loci. A frequency of 1% recombination in meiosis is one centimorgan and equals about 1 million base pairs. See Southern, E.M., Prospects for a complete molecular map of the human genome, *Philos.Trans.R.Soc.Lond. B Biol.Sci.* 319, 299-307, 1988; White, R., Lalauel, J.M., Leppert, M., Lathrop, M., Nakamura, Y., and O'Connell, P., Linkage maps of human chromosomes, *Genome* 31, 1066-1072, 1989; Smith, L.H., Jr., Overview of hemochromatosis, *West.J.Med.* 153, 296-308, 1990; Crabbe, J.C., Alcohol and genetics: new models, *Am.J.Med.Genet.* 114, 969-974, 2002.

Chameleon Sequences

Identical sequences in a protein which can adopt either an alpha helical conformation or a beta sheet conformation: see Minor, D.L., Jr. and Kim, P.S., Context-dependent secondary structure formation of a designed peptide sequence, *Nature* 380, 730-734,

1996; Mezei, M., Chameleon sequences in the PDB, *Protein.Eng.* 11, 411-414, 1998; Tidow, H., *et al.*, The solution structure of a chimeric LEKTI domain reveals a chameleon sequence, *Biochemistry* 43, 11238-11247, 2004.

Chaotropic

Describing a reagent which disrupts the structure of water and macromolecules such as proteins. Chaotropic is sometimes confined to uncharged molecules such as urea or thiourea but is usually extended to include reagents such as guanidine hydrochloride and sodium thiocyanate. See Dandliker, W.B., Alonso, R., de Saussure, V.A., Kierszenbaum, F., *et al.*, The effect of chaotropic ions on the dissociation of antigen-antibody complexes, *Biochemistry* 6, 1460-1467, 1967; Hanstein, W.G., Davis, K.A., and Hatefi, Y., Water structure and the chaotropic properties of haloacetates, *Arch.Biochem.Biophys.* 147, 534-544, 1971; Sawyer, W.H. and Puckridge, J., The dissociation of proteins by chaotropic salts, *J.Biol.Chem.* 248, 8429-8433, 1973; Hatefi, Y. and Hanstein, W.G., Destablization of membranes with chaotropic ions, *Methods Enzymol.* 31, 770-790, 1974; McLaughlin, S., Bruder, A., Chen, S., and Moser, C., Chaotropic anions and the surface potential of bilayer membranes, *Biochim.Biophys. Acta* 394, 304-313, 1975; Stein, M., Lazaro, J.J., and Wolsiuk, R.A., Concerted action of cosolvents, chaotropic anions and thioredoxin on chloroplast fructose-1,6-bisphosphatase. Reactivity to iodoacetate, *Eur.J.Biochem.* 185, 425-431, 1989; Lever, M., Blunt, J.W., and MacLagan, R.G., Some ways of looking at compensatory kosmotropes and different water environments, *Comp. Biochem.Physiol.A Integr.Physiol.* 130, 471-486, 2001; Pilorz, K. and Choma, I., Isocratic reversed-phase high-performance liquid chromatographic separation of tetracyclines and flumequine controlled by a chaotropic effect, *J.Chromatog.A.* 1031, 303-306, 2004; Moelbert, S., Normand, B., and De Los Rios, P., Kosmotropes and chaotropes: modelling preferential exclusion, binding and aggregate stability, *Biophys.Chem.* 112, 45-57, 2004; Salvi, G., De Los Rios, P., and Vendruscolo, M., Effective interactions between chaotropic agents and proteins, *Proteins* 61, 492-499, 2005; LoBrutto, R. and Kazakevich, Y.V., Chaotropic effects in RP-HPLC, *Adv.Chromatog.* 44, 291-315, 2006.

Chaperone

An intracellular factor, most frequently a protein, which guides the intracellular folding/assembly of another protein. Examples include heat shock proteins, chaperoinins, See Gregerson, N.. Bolund, L., and Bross, P., Protein misfolding, aggregation, and degradation in disease, *Mol.Biotechnol.* 31, 141-150, 2005; Anken, E., Braakman, I. and Craig, E., Versatility of the endoplasmic reticulum protein folding factory, *Crit.Rev.Biochem.Mol.Biol.* 40, 191-288, 2005; Macario, A.J., and Conway de Marcario, E., Sick chaperones, cellular stress, and disease, *New Eng.J.Med.* 353, 1489-1501, 2005; Weibezahn, J., Schlieker, C., Tessarz, P., Mogk, A., and Bukau, B., Novel insights into the mechanism of chaperone-assisted protein disaggregation, *Biol.Chem.* 386, 739-744, 2006.

Chemical Biology

The application of chemical techniques to problems in biology – the emphasis is directed toward study of the interaction of small molecules with proteins and other macromolecules. See Li, C.H., Current concepts on the chemical biology of pituitary hormones,

Perspect.Biol.Med. 11, 498-521, 1968; Malmstrom, B.G. and Leckner, J., The chemical biology of copper, *Curr.Opin.Chem. Biol.* 2, 286-292, 1998; Bertini, I. and Luchinat, C., New applications of paramagnetic NMR in chemical biology, *Curr.Opin. Chem.Biol.* 3, 145-151, 1999; Volkert, M., Wagner, M., Peters, C. and Waldmann, H., The chemical biology of Ras lipidation, *Biol.Chem.* 382, 1133-1145, 2001; Hahn, M.E. and Muir, T.W., Manipulating proteins with chemistry: a cross-section of chemical biology, *Trends Biochem.Sci.* 30, 26-34, 2005; Cambell-Valois, F.X., and Michnick, S., Chemical biology on PINs and NeeDLes, *Curr.Opin.Chem.Biol.* 9, 31-37, 2005; Doudna, J.A., Chemical biology at the crossroads of molecular structure and mechanism, *Nat.Chem.Biol.* 1, 300-303, 2005.

Chemical Proteomics

Use of chemical modification to identify enzymes in the proteome and to identify signalling pathways. See Jeffery, D.A. and Bogyo, M., Chemical proteomics and its application to drug discovery, *Curr.Opin.Biotechnol.* 14, 87-95, 2003; Daub, H., Godl, K., Brehmer, D., *et al.,* Evaluation of kinase inhibitor selectivity by chemical proteomics, *Assay Drug Dev.Technol.* 2, 215-224, 2004; Piggott, A.M. and Karuso, P., Quality, not quantity: the role of natural products and chemical proteomics in modern drug discovery, *Comb.Chem.High Throughput Screen.* 7, 607-630, 2004; Beillard, E and Witte, O.N., Unraveling kinase signaling pathways with chemical genetic and chemical proteomic approaches, *Cell Cycle* 4, 434-437, 2005; Sem, D.S., Chemical proteomics from a nuclear magnetic resonance spectroscopy perspective, *Expert Rev. Proteomics* 1, 165-178, 2004; Daub, H., Characterization of kinase-selective inhibitors by chemical proteomics, *Biochim. Biophys.Acta* 1754, 183-190, 2005; Verdoes, M., Berkers, C.R., Florea, B.I., *et al.,* Chemical proteomics profiling of proteosome activity, *Methods Mol.Biol.* 328, 51-69, 2006.

Chemokines

A large family of cytokines having a wide variety of biological actions but are generally associated with inducing mobilization and activation of immune cells; a contraction of chemotactic cytokines. See Horuk, R., *Chemokine Receptors*, Academic Press, San Diego, CA, USA, 1997; Vaddi, K., Keller, M., and Newton, R.C., *The Chemokine Factbook*, Academic Press, San Diego, CA, USA, 1997; Hebert, C., *Chemokines in Disease: Biology and Clinical Research*, Humana Press, Totowa, NJ, USA, 1999; Proudfoot, A.E.I. and Well, T.N.C., *Chemokine Protocols*, Humana Press, Totowa, NJ, USA, 2000; Schwiebert, L.M., *Chemokines, Chemokine Receptors, and Disease*, Elsevier, Amsterdam, NL, 2005; Atkins, P.C. and Wasserman, S.I., Chemotactic mediators, *Clin.Rev.Allergy* 1, 385-395, 1983; Hayashi, H., Honda, M., Shimokawa, Y., and Hirashima, M., Chemotactic factors associated with leukocyte emigration in immune tissue injury: their separation, characterization, and functional specificity, *Int.Rev.Cytol.* 89, 179-250, 1984; Bignold, L.P., Measurement of chemotaxis of polymorphonuclear leukocytes in vitro. The problems of the control of gradients of chemotactic factors, of the control of the cells and of the separation of chemotaxis from chemokinesis, *J.Immunol.Methods* 108, 1-18, 1988; Schwarz, M.K., and Wells, T.N.C., New therapeutics that modulate chemokine networks, *Nat.Rev.Drug.Disc.* 1, 342-358, 2002; White, F.A., Bhangoo, S.K., and Miller, R.J., Chemokines: integrators or pain and inflammation, *Nat.Rev.Drug Discov.* 4, 834-844, 2005; Steinke, J.W. and Borish, L., Cytokines and chemokines, *J.Allergy Clin.Immunol.* 117 (Suppl 2), S441-S445, 2006;

Charo, I.F., and Ranosohoff, R.M., The many roles of chemokines and chemokine receptors in inflammation, *N.Engl.J.Med.* 354, 610-621, 2006; Laudanna, C. and Alon, R., Right on the spot. Chemokine triggering of integrin-mediated arrest of rolling leukocytes, *Thromb.Haemostas.* 95, 5-11, 2006.

Chemoproteomics

The use of small molecules as affinity materials for the discovery of specific binding proteins in the proteome; the application of chemogenomics for proteomic research. See Beroza, P., Villar, H.O., Wick, M.M., and Martin, G.R., Chemoproteomics as a basis for post-genomic drug discovery, *Drug Discov.Today* 7, 807-814, 2002; Gagna, C.E., Winokur, D., Lambert, W.C., Cell biology, chemogenomics and chemoproteomics, *Cell Biol.Int.* 28, 755-764, 2004; Shin, D., Heo, Y.S., Lee, K.J., *et al,* Structural chemoproteomics and drug discovery, *Biopolymers* 80, 258-263, 2005; Hall, S.E., Chemoproteomics-driven drug discovery: addressing high attrition rates, *Drug Discov.Today* 11, 495-502, 2006.

Chondrocyte

A cartilage cell. See von der Mark, K. and Conrad, G., Cartilage cell differentiation: review, *Clin.Orthop.Relat.Res.* (139), 195-205, 1979; Serni, U. and Mannoni, A., Chondrocyte physiopathology and drug efficacy, *Drug Exp.Clin.Res.* 17, 75-79, 1991; Urban, J.P., The chondrocytes: a cell under pressure, *Br.J.Rheumatol.* 33, 901-908, 1994; Yates, K.E., Shortkroff, S., and Reish, R.G., Wnt influence on chondrocyte differentiation and cartilage function, *DNA Cell Biol.* 24, 446-457, 2005; Wendt, D., Jakob, M., and Martin, I., Bioreactor-based engineering of osteochondral grafts: from model systems to tissue manufacturing, *J.Biosci.Bioeng.* 100, 489-494, 2005; Goldring, M.B., Tsduchmochi, K., and Ijiri, K., The control of chondrogenesis, *J. Cell Biochem.* 97, 33-44, 2006; Ruano-Ravina, A. and Diaz, M.J., Autologous chondrocytes implantation: a systematic review, *Osteoarthritis Cartilage* 14, 47-51, 2006; Toh, W.S., Yang, Z., Heng, B.C., and Cao, T., New perspectives in chondrogenic differentiation of stem cells for cartilage repair, *ScientificWorldJournal* 6, 361-364, 2006.

Chromatin

Chromatin consists of a repeating fundamental nucleoprotein complex, the nucleosome; DNA wrapped around histones where the histones mediate the folding of DNA into chromatin. see Wolfe, A., *Chromatin. Structure and Function*, 3rd Ed., Academic Press, San Diego, CA, 1998; Woodcock, C.L., Chromatin architecture, *Curr.Opin.Struct.Biol.* 16, 213-220, 2006; Aligianni, S. and Varga-Weisz, P., Chromatin-remodelling factors and the maintenance of transcriptional states through DNA replication, *Biochem.Soc.Symp.* (73), 97-108, 2006; de la Serna, I.L., Ohkawa, Y., and Imbalzano, A.N., Chromatin remodelling in mammalian differentiation: lessons from ATP-dependent remodellers, *Nat. Rev.Genet.* 7, 461-473, 2006; Mersfelder, E.L., and Parthun, M.R., The tail beyond the tail: histone core domain modifications and the regulation of chromatin structure, *Nucleic Acids Res.* 34, 2653-2662, 2006.

Chromatin Remodeling

The dynamic structural change in chromatin by nucleosome sliding or post-translational modifications (acetylation, methylation) of the histones. See Becker, P.B., The chromatin accessibility complex: chromatin dynamics through nucleosome sliding, *Cold Spring Harb.Symp.Quant.Biol.* 69, 281-287, 2004; Henikoff, S. and Ahmed, K., Assembly of variant histones into chromatin, *Annu.Rev.Cell Dev.Biol.* 21, 133-153, 2005; Dhananjayan, S.C., Ismail, A., and Nawaz, Z., Ubiquitin and control of transcription, *Essays Biochem.* 41, 69-80, 2005; Lucchesi, J.C., Kelly, W.G., and Panning, B., Chromatin remodeling in dosage compensation, *Annu.Rev.Genet.* 39, 615-651, 2005; Saha, A., Wittmeyer, J.. and Cairns, B.R., Chromatin remodelling: the industrial revolution of DNA around histones, *Nat.Rev.Mol.Cell Biol.* 7, 437-447, 2006.

Chromatography

The physical separation of two or more components of a solution mixture based on the distribution of said individual components between a stationary phase and a mobile phase. Chromatography can occur within a enclosed column or tube (column chromatography, gas chromatography being a variant of column chromatography with a gaseous mobile phase) or a planar surface as in paper chromatography or thin-layer chromatography. A **chromatogram** is (usually) a graphical representation of specific solute concentration at a given moment either in time or elution volume. In the case of planar chromatography, the term chromatography can refer to the actual paper or layer on which separation has occurred. The stationary phase may be a solid, gel, or liquid adsorbed onto a solid matrix. The mobile phase may be liquid or gaseous in nature. See Ettre, L.S., Nomenclature for chromatography, *Pure Appl.Chem.* 65, 819-872, 1993; Lederer, E. and Lederer, M., *Chromatography: A Review of Principles and Applications*, Elsevier, Amsterdam, 1957; Bobbit, J.M., *Thin-Layer Chromatography*, Reinhold, New York, NY, USA, 1963; Zweig, G. and Sherma, J., *CRC Handbook of Chromatography*, CRC Press, Cleveland, OH, USA, 1972; Snyder, L.R., Kirkland, J.J., and Glajch, J.L., *Practical HPLC- Method Development*, 2nd edn., John Wiley & Sons, Inc., New York, NY, USA, 1997; Miller, J.M., *Chromatography: Concepts and Contrasts*, Wiley, New York, NY, USA, 2005; Wall, P.E., *Thin-Layer Chromatography: A Modern Practical Approach*, Royal Society of Chemistry, Cambridge, UK, 2005; Cazes, J., *Encyclopedia of Chromatography*, Taylor & Francis, Boca Raton, FL, USA, 2005; Perssen, P., Gustavsson, P.-E., Zacchi, G., and Nilsson, B., Aspects of estimating parameter dependencies in a detailed chromatography model based on frontal experiments, *Process Biochem.* 41, 1812-1821, 2006; Alpert, A.J., Chromatography of difficult and water-soluble proteins with organic solvents, *Adv.Chromatog.* 44, 317-329, 2006; Lundanes, E. and Greibrokk, T., Temperature effects in liquid chromatography, *Adv.Chromatog.* 44, 45-77, 2006.

Circadian

Used to describe an approximate 24 hour period; a phenomena has demonstrates a circadian variation if it occurs with a certain frequency within an approximate 24 hour period. See Mills, J.N., Human circadian rhythms, *Physiol.Rev.* 46, 128-171, 1966; Brady, J., How are insect circadian rhythms controlled?, Nature 223, 781-784, 1969; Menaker, M., Takahashi, J.S., and Eskin, A., The physiology of circadian pacemakers, *Annu.Rev.Physiol.* 40, 501-526, 1978; Soriano, V., The circadian rhythm embraces the variability that occurs within 24 hours, *Int.J.Neurol.* 15, 7-16, 1981; Gardner, M.J., Hubbard, K.E., Hatta, C.T., *et al.*, How plants tell the time, *Biochem.J.* 397, 15-24, 2006; McClung, C.R., Plant circadian rhythms, *Plant Cell* 18, 792-803, 2006; Brunner, M. and Schafmeier, T., Transcriptional and post-transcriptional

regulation of the circadian clock of cyanobacteria and Neurospora, *Genes Dev.* 20, 1061-1074, 2006; Hardin, P.E. and Yu, W., Circadian transcription: passing the HAT to CLOCK, *Cell* 125, 424-426, 2006; Lewy, A.J., Emens, J., Jackman, A., and Yuhas, K., Circadian uses of melatonin in humans, *Chronobiol. Int.* 23, 403-412, 2006; Rosato, E., Tauber, E., and Kyriacou, C.P., Molecular genetics of the fruit-fly circadian clock, *Eur.J.Hum. Genet.* 14, 729-738, 2006.

Cis-element; *cis*-locus; *cis*-factors

A region or regions on a DNA molecule which affects activity of DNA sequences on its own DNA molecule; an intramolecular effect; usually but not always does not code for the expression of protein; A *cis*-element or regulatory region can be complex and may contain several regulatory sequences. See Gluzman, Y., *Eukaryotic Transcription: The Role of Cis- and Trans-Acting Elements in Initiation*, Cold Spring Harbor Laboratory Press, Cold Spring Harbor, New York, USA, 1985; Hames, B.D. and Higgins, S.J., *Gene Transcription: A Practical Approach*, IRL Press at Oxford, Oxford, UK, 1993; Tanaka, N. and Taniguchi, T., Cytokine gene regulation: regulatory *cis*-elements and DNA binding factors involved in the interferon system, *Adv.Immunol.* 52, 263-281, 1992; Galson, D.L., Blanchard, K.L., Fandrey, J., Goldberg, M.A., and Bunn, H.F., *Cis* elements that regulate the erythropoietin gene, *Ann.N.Y.Acad.Sci.* 718, 21-30, 1994; Hapgood, J.P., Riedemann, J., and Scherer, S.D., Regulation of gene expression by GC-rich DNA *cis*-elements, *Cell Biol. Int.* 25, 71-31, 2001; Tumpel, S., Maconochie, M., Wiedmann, L.M., and Krumlauf, R., Conservation and diversity in the *cis*-regulatory networks that integrate information controlling expression of Hoxa2 in hindbrain and cranial neural crest cells in vertebrates, *Dev.Biol.* 246, 45-56, 2002; Moolla, N., Kew, M., and Arbuthnot, P., Regulatory elements of hepatitis B virus transcription, *J.Viral Hepat.* 9, 323-331, 2002; Manna, P.R., Wang, X.J., and Stocco, D.M., Involvement of multiple transcription factors in the regulation of steroidogenic acute regulatory protein gene expression, *Steroids* 68, 1125-1134, 2003; Gambari, R., New trends in the development of transcription factor decoy (TFD) pharmacotherapy, *Curr.Drug Targets* 5, 419-430, 2004; McBride, D.J. and Kleinjan, D.A.. Rounding up active *cis*-elements in the triple C corral: combining conservation, cleavage and conformation capture for the analysis of regulatory gene domains, *Brief Funct. Genomic Proteomic* 3, 267-279, 2004.

Circular Dichroism

The differential absorption of plane polarized light passing through a solution and is expression as molar ellipticity $[\theta]_m$. See Greenfield, N.J., Analysis of circular dichroism data. *Meth. Enzymol*, 383, 282-317, 2004; Bayer, T.M., Booth, L.N., Knudsen, S.M., and Ellington, A.D., Arginine-rich motifs present multiple interfaces for specific binding by RNA, *RNA* 11, 1848-1857, 2005; Miles, A.J., and Wallace, B.A., Synchrotron radiation circular dichroism spectroscopy of proteins and applications in structural and functional genomics, *Chem.Soc.Rev.* 35, 39-51, 2006; Paramonov, S.E., Jun, H.W., and Hartgerink, J.D., Modulation of peptide-amphiphile nanofibers via phospholipid inclusions, *Biomacromolecules* 7, 24-26, 2006; Harrington, A., Darboe, N., Kenjale, R., *et al.*, Characterization of the interaction of single tryptophan containing mutants of IpaC from *Shingella flexneri* with phospholipid membranes, *Biochemistry* 45, 626-636, 2006.

Claisen Condensation

Base-catalyzed reaction of an ester with an α-carbon hydrogen with another ester (same or different) to yield a β-keto ester. A model for thiolase reactions. See Claisen, L and Lowman, O., *Berichte* 20, 651, 1887; Clark, J.D., O'Keefe, S.J., and Knowles, J.R., Malate synthase: proof of a stepwise Claisen condensation using the double-isotope fractionation test, *Biochemistry* 27, 5961-5971, 1988; Modia, Y. and Wierenga, R.K., A biosynthetic thiolase in complex with a reaction intermediate: the crystal structure provides new insight into the catalytic mechanism, *Structure* 7, 1279-1290, 1999; Watanabe, A. and Ebizuka, Y., Unprecedented mechanism of chain length determination in fungal aromatic polyketide synthases, *Chem.Biol.* 11, 1101-1106, 2004; Veyron-Churlet, R., Bigot, S., Guerrini, O. *et al.*, The biosynthesis of mycolic acids in *Mycobacterium tuberculosis* relies on multiple specialized elongation complexes interconnected by specific protein-protein interactions, *J.Mol.Biol.* 353, 847-858, 2005; von Wettstein-Knowles, P., Olsen, J.G., McGuire, K.A, and Henriksen, A., Fatty acid synthesis. Role of active site histidines and lysine in Cys-His-His-type beta-ketoacyl-acyl carrier protein synthases, *FEBS J.* 273, 695-710, 2006; Ryu, Y., Kim, K.J., Rosennser, C.A., and Scott, A., Decarboxylative Claisen condensation catalyzed by in vitro selected ribozymes, *Chem.Commun.* 7, 1439-1441, 2006.

Class Switch Recombination

A process by which one constant region gene segment is switched with another gene segment during B-cell development when immunoglobulin production changes from IgM to IgA, IgE, or IgG. See Davis, M.M., Kim, S.K., and Hood, L.E., DNA sequences mediating class switching in alpha-immunoglobulin, *Science* 209, 1360-1365, 1980; Geha, R.S., Jabara, H.H., and Brodeur, S.R., The regulation of immunoglobulin E class-switch recombination, *Nat.Rev.Immunol.* 3, 721-732, 2003; Yu, K. and Lieber, M.R., Nucleic acid structures and enzymes in the immunoglobulin class switch recombination mechanism, *DNA Repair* 2, 1163-1174, 2003; Diamant, E. and Melamed, D., Class switch recombination in B lymphopoiesis: a potential pathway for B cell autoimmunity, *Autoimmun.Rev.* 3, 464-469, 2004; Min, I.M. and Selsing, E., Antibody class switch recombination: roles for switch sequence and mismatch repair proteins, *Adv.Immunol.* 87, 297-328, 2005.

Classical Proteomics

Proteomic analysis based on the direct analysis of the expressed proteome such an extract obtained from lysis of a cell; also referred to as forward proteomics as compared to reverse proteomics. More generally, classically proteomics is taken to mean protein separation followed by characterization. See Klade, C.S., Proteomics approaches toward antigen discovery and vaccine development, *Curr.Opin.Mol.Ther.* 4, 216-223, 2002; Vondriska, T.M. and Ping, P., Functional proteomics to study protection of the ischaemic myocardium, *Expert Opin.Ther.Targets* 6, 563-570, 2002; Thiede, B. and Rudel, T., Proteome analysis of apoptotic cells, *Mass Spectrom.Rev.* 23, 333-349, 2004; Gottlieb, D.M., Schultz, J., Bruun, S.W., *et al.*, Multivariate approaches in plant science, *Phytochemistry* 65, 1531-1548, 2004.

Clinomics

Application of oncogenomics to cancer care. See Workman, P. and Clarke, P.A., Innovative cancer drug targets: genomics, transcriptomics, and clinomics, *Expert Opin.Pharmacother.* 2, 911-915, 2001.

Clone

A cell or organism descended from and genetically identical to a single common ancestor. Clone is also used to refer to a DNA sequence encoding a product or an entire gene sequence from an organism which is replicated by genetic engineering. Such material can be transferred to another organism for the expression of such cDNA or gene. See Cunningham, A.J., Antibody formation studied at the single-cell level, *Prog.Allergy* 17, 5-50, 1973; Hamer, D.H. and Thomas, C.A., Jr., Molecular cloning, *Adv. Pathobiol.* (6), 306-319, 1977; von Boehmer, H., Haas, W., Pohlit, H., Hengartner, H., and Nabholz, M., T cell clones: their use for the study of specificity, induction, and effector-function of T cells, *Springer Semin.Immunopathol.* 3, 23-37, 1980; Fung, J.J., Gleason, K., Ward. R., and Kohler, H., Maturation of B-cell clones, *Prog. Clin.Biol.Res.* 42, 203-214, 1980; Veitia, R.A., Stochasticity or the fatal 'imperfection' of cloning, *J.Biosci.* 30, 21-30, 2005; Kettman, J.R., from clones of cells to cloned genes and their proteinpaedia, *Scand.J.Immunol.* 62, Suppl 1, 119-122, 2005; Vats, A., Bielby, R.C., Tolley, N.S., Nerem, R., and Polak, J.M., Stem cells, *Lancet* 366, 592-602, 2005; Wells, D.N., Animal cloning: problems and prospects, *Rev.Sci.Tech.* 24, 251-264, 2005; Diep, B.A., Gill, S.R., Chang, R.F., *et al.*, Complete genome sequence of USA300, an epidemic clone of community-acquired methicillin-resistant *Staphylococcus aureus*, *Lancet* 367, 731-739, 2006.

Clonal Selection

Literally, the selection of a clone. Most often used to describe the process by which a B-cell is challenged by a specific antigen to produce a committed plasma cell or the differentiation of T-cells. More generally, the selection a stem cell to become committed to a specific antigen. See Williamson, A.R., The biological origin of antibody diversity, *Annu.Rev.Biochem.* 45, 467-500, 1976; D'Eustachio, P., Rutishauser, U.S., and Edelman, G.M., Clonal selection and the ontogeny of the immune response, *Int.Rev.Cytol. Suppl.* (5), 1-60, 1977; Coutinho, A., Beyond clonal selection and network, *Immunol.Rev.* 110, 63-87, 1989; Cohen, I.R., Antigenic mimicry, clonal selection and autoimmunity, *J.Autoimmun.* 16, 337-340, 2001; Defrance, T., Casamayor-Palleja, M., and Krammer, P.H., The life and death of a B cell, *Adv.Cancer Res.* 86, 195-225, 2002; van Boehmer, H., Aifantis, I., Gounari, F., *et al.*, Thymic selection revisited: how essential is it?, *Immunol. Rev.* 191, 62-78, 2003, McHeyzer-Williams, L.J. and McHeyzer-Williams, M.G., Antigen-specific memory B cell development, *Annu.Rev.Immunol.* 23, 487-513, 2005; Bock, K.W. and Kohle, C., Ah receptor- and TCDD-mediated liver tumor promotion: clonal selection and expansion of cells evading growth arrest and apoptosis, *Biochem.Pharmacol.* 69, 1403-1408, 2005; Burnet, F.M., *The Clonal Selection Theory of Acquired Immunity*, Vanderbilt University Press, Nashville, TN, USA, 1959; Mazumdar, P.M.H., *Immunology 1930-1980: Essays on the History of Immunology*, Wall & Thompson, Toronto, Canada, 1989; Podolsky, S.H. and Tauber, A.I., *The Generation of Diversity: Clonal Selection Theory and ther Rise of Molecular Immunology*, Harvard University Press, Cambridge, MA, USA, 1997.

Coefficient of Linear Thermal Expansion (CLTE)

Ration of the change in length per degree C to length at O°C. The coefficient of linear thermal expansion (CTLE) is used to described the changes in the structure of proteins and other polymers as a function of temperature; the CTLE has also been used to describe thermal changes in micelles. See Frauenfelder, H., Hartmann, H., Karplus, M., *et al.*, Thermal expansion of a protein, *Biochemistry* 26, 254-261, 1987; Schulenberg, P.J., Rohr, M., Gartner, W., and Braslavsky, S.E., Photoinduced volume changes associated with the early transformations of bacteriorhodopsin: a laser-induced optoacoustic spectroscopy study, *Biophys.J.* 66, 838-843, 1994; Marsh, D., Intrinsic curvature in normal and inverted lipid structures and I membranes, *Biophys.J.* 70, 2248-2255, 1996; Daniels, B.V., Schoenborn, B.P., and Korszun, Z.R., A low-resolution low-temperature neutron diffraction study of myoglobin, *Acta Crystallogr.D.Biol. Crystallogr.* 53, 544-550, 1997; Cordier, F. and Grzesiek, S., Temperature-dependence of protein hydrogen bond properties as studied by high-resolution NMR, *J.Mol.Biol.* 317, 739-752, 2002; Pereira, F.R., Machado, J.C., and Foster, F.S., Ultrasound characterization of coronary artery wall in vitro using temperature-dependent wave speed, *IEEE Trans Ultrason. Ferroelectr.Freq.Control* 50, 1474-1485, 2003; Bhardwaj, R., Mohanty, A.K., Drzal, L.T., *et al.*, Renewable resource-based composites from recycled cellulose fiber and poly(3-hydroxybutyrate-co-3-hydroxyvalerate) bioplastic, *Biomacromolecules* 7, 2044-2051, 2006.

Cold-chain product

A product or reagent which must be kept cold during transit and storage; most often between 4° and 8°C. See Elliott, M.A. and Halbert, G.W., Maintaining the cold chain shipping environment for phase I clinical trial distribution, *Int.J.Pharm.* 299, 49-54, 2005; Streatfield, S.J., Mucosal immunization using recombinant plant-based oral vaccines, *Methods* 38, 150-157, 2005.

Cold Shock Protein

A group of proteins which are synthesized by plant cells, prokaryotic and eukaryotic cells in response to cold stress. It has been suggested that cold shock proteins (CSPs) function as "chaperones" for mRNA. Graumann, P.L. and Marshiel, M.A., A superfamily of proteins that contain the cold-shock domain, *Trends.Biochem.Sci.* 23, 286-290, 1998; Phadtare, S., Alsina, J., and Inouye, M., Cold-shock response and cold-shock proteins, *Curr.Opin.Microbiol.* 2, 175-180, 1999; Sommerville, J., Activities of cold-shock domain proteins in translational control, *Bioessays* 21, 319-325, 1999; Graumann, P.L. and Marahiel, M.A., Cold shock response in *Bacillus subtilis, J.Mol.Microbiol. Biotechnol.* 1, 203-209, 1999; Loa, D.A. and Murata, N., Responses to cold shock in cyanobacteria, *J.Mol.Microbiol. Biotechnol.* 1, 221-230, 1999; Ermolenko, D.N. and Makhatadze, G.I., Bacterial cold-shock proteins, *Cell.Mol.Life.Sci.* 59, 1902-1913, 2002; Alfageeh, M.B., Marchant, R.J., Carden, M.J., and Smales, C.M., The cold-shock response in cultured mammalian cells: harnessing the response for the improvement of recombinant protein production, *Biotechnol.Bioeng.* 93, 829-835, 2006; Al-Fageeh, M.B. and Smales, C.M., Control and regulation of the cellular response to cold shock: the responses in yeast and mammalian systems, *Biochem.J.* 397, 247-259, 2006; Fraser, K.r., Tuite, N.L., Bhagwat, A., and O'byrne, C.P., Global effects of homocysteine on transcription in *Escherichia coli*; induction of the gene for the major cold-shock protein, CspA, *Microbiology* 152, 221-2231, 2006; Magg, C., Kubelka, J., Holtermann, G., *et al.*, Specificity of the initial collapse in the folding of the cold shock protein, *J.Mol. Biol.* 360, 1067-1080, 2006; Sauvageot, N., Beaufils, S., Maze, A., Cloning and characterization of a gene encoding a cold-shock

protein in *Lactobacillus casei, FEMS Microbiol.Lett.* 254, 55-62, 2006; Narberhaus, F., Waldminghous, T., and Chowdhury, S., RNA thermometers, *FEMS Microbiol.Lett.* 30, 3-16, 2006.

Colloid

The term colloid refers to a particle with dimensions between 1 nm and 1 μm although it is not necessary for all three dimensions to be in this size range. For example, a thin fiber might only have two dimensions in this size range. A colloidal dispersion is a system where colloid particles are dispersed in a continuous phase of a different composition such as a suspension (particles in a liquid), a emulsion (colloids of one liquid are suspended in another liquid where the two liquids are immiscible such as oil and water), a foam (gas dispersed in a liquid or gel), or an aerosol (a colloid in a gas such as air; a fog is a liquid colloid dispersed in a gas). See Tolson, N.D., Boothroyd, B., and Hopkins, C.R., Cell surface labelling with gold colloid particulates: the use of aviden and staphylococcal protein A-coated gold in conjunction with biotin and fc-bearing ligands, *J.Microsc.* 123, 215-226, 1981; Rowe, A.J., Probing hydration and the stability of protein solution—a colloid science approach, *Biophys.Chem.* 93, 93-101, 2001; Bolhuis, P.G., Meijer, E.J., and Louis, A.A., Colloid-polymer mixtures in the protein limit, *Phys.Rev.Lett.* 90:068304, 2003; Zhang, Z and van Duijneveldt, J.S., Experimental phase diagram of a model colloid-polymer mixture in the protein limit, *Langmuir* 22, 63-66, 2006; Xu, L.C. and Logan, B.E., Adhesion forces between functionalized latex microspheres and protein-coated surfaces evaluated using colloid probe atomic force microscopy, *Colloids Surf. B. Biointerfaces* 48, 84-94, 2006.

Colloid Osmotic Strength/Colloid Osmotic Pressure Combination Electrode

An ion-selective electrode and an external reference electrode combined into a single functional unit. A separate reference electrode is not required.

Combination Product

A regulatory term used to describe a final drug product composed of, for example, two separate drugs, a drug and a biologic or a drug and a device. See Leyden, J.J., Hickman, J.G., Jarratt, M.T., *et al.*, The efficacy and safety of a combination benzoyl peroxide/clindemycin topical gel compared with benzoyl peroxide alone and a benzoyl peroxide/erythromycin combination product, *J.Cutan.Med.Surg.* 5, 37-42, 2001; Bays, H.E., Extended-release niacin/lovastatin: the first combination product for dyslipidemia, *Expert Rev.Cardiovasc.Ther.* 2, 485-501, 2004; anon, Definition of the primary mode of action of a combination product. Final rule, *Fed.Regist.* 70, 49848-49862, 2005.

Complement

A combination or system of plasma/serum proteins which interact to form a membrane attack complex which results in the lysis of bacterial pathogens and other cell targets such as tumor cells. There are three pathways of complement activation; the classical pathway, the alternative pathway, and the MBLectin (mannose-binding lectin; a plasma protein) pathway. The classical pathway is activated by an antigen-antibody complex (free antibody does not activate complement) via the Fc domain of the antibody; there are other mechanisms for classical pathway activation which make minor contributions. The alternative pathway is activated by direct recognition of foreign materials in an antibody-independent manner and is driven by the autocatalytic action of C3b. The alternative pathway is thought the oldest of the three pathways is phyllogenetic development. The MBlectin pathway is initiated by the interaction of the MBlectin with a bacterial cell surface polysaccharide. The activation of complement component C3 is common to all three pathways. It is noted that there similarities to the blood coagulation cascade. See Prodinger, W.M., Würznen, R., Erdei, A., and Dierich, M.P., Complement, in *Fundamental Immunology*, ed. W.E.Paul, Lippincott-Raven, Philadelphia, USA, Chapter 29, pps. 967-995, 1999; *Activators and Inhibitors of Complement*, ed. R.B. Sim, Kluwer Academic, Dordrecht, Netherlands, 1993; *Complement in Health and Disease*, 2nd edn., ed. K Whaley, M. Loos, and J. Weiler, Kluwer Academic, Dordrecht, Netherlands, 1993; *The Complement System*, 2nd edn., ed., K. Rother, G.O. Till, and G.M Hansch, Springer, Berlin, 1998; *The Human Complement System in Health and Disease*, ed. J.E. Volanakis and MM. Frank, Marcel Dekker, New York, 1998; *Therapeutic Interventions in the Complement System*, ed. J.D. Lambis and K.M., Holer, Humana, Totowa, New Jersey, 2000; Szebeni, J., *The Complement System Novel Roles in Health and Disease*, Kluwer Academic, Boston, 2004.

Complement Fixation

Complement fixation refers to the binding of the first component of the complement pathway, C1, to an IgG- or IgM-antigen complex. The antigen is usually a cell surface protein. Free antibody does not fix complement. Productive binding of the antigen-antibody complex (binding involves the Fc portion of the antibody and a minimum of two Fc domains is required; thus two intact antibody molecules) results in complement activation. An antibody that activates complements is describes as having fixed complement. Complement fixation has formed the basis for many serological tests but most have been replaced by ELISA assays for the diagnosis of infectious disease. See Juji, T., Saji, H., Sataki, M., and Tukinaga, K., Typing for human platelet alloantigens, *Rev.Immunogenet.* 1, 239-254, 1999; Pappagianus, D., Serological studies in coccidiomycosis, *Semin.Respir.Infect.* 16, 242-250, 2001; Nielsen, K., Diagnosis of brucellosis by serology, *Vet.Microbiol.* 90, 447-459, 20002; Al-Dahouk, S., Tomaso, H., Nackler, E. *et al.*, Laboratory-based diagnosis of brucellosis - review of the literature. Part I. Techniques for direct detection and identification of *Brucella sp.*, *Clin.Labl.* 49, 387-404, 2003; Taggart, E.W., Hill, H.R., Martins, T.B., and Litwin, C.M., Comparison of complement fixation with two enzyme-linked immosorbent assays for the detection of antibodies to respiratory viral antigens, *Amer.J.Clin.Path.* 125, 460-466, 2006. Complement fixation is usually measured by the lysis of sensitized cells (e.g hemolysis of sensitized sheep red blood cells; CH_{50} assay), See Morgen, P.B., Complement, in *Immunochemistry*, ed. C.J. van Oss and M.C.H. van Regenmortel, Marcel Dekker, New York, 1994, Chapter 34, pp. 903-923, 1994. The concept of complement fixation is still discussed with respected to *in vivo* antigen-antibody reactions such as those seen with transplantation antigens and alloantibodies. See Feucht, H.E., Felber, E., Gokel, M.J., *et al.*, Vascular deposition of complement-split products in kidney allografts with cell-mediated rejection, *Clin.Exp.Immunol.* 86, 464-470, 1991; Feucht, H.E., Complement C4d in graft capillaries—the missing link in the recognition of humoral alloreactivity, *Am.J.Transplant.* 3, 646-652, 2003; Colvin, R.B. and Smith, R.N., Antibody-mediated organ-allograft rejection, *Nat.Rev.Immunol.* 5, 807-817, 2005;

Rickert, R.C., Regulation of B lymphocyte activation by complement C_3 and the B cell coreceptor complex, *Curr.Opin.Immunol.* 17, 237-243, 2005.

Confocal Microscopy

A fluorescent microscopy technique which uses a highly focused beam of light with suppression of fluorescence above and below the point of optimum focus. An image is obtained by oving the excitation beam and measurement aperature over the sample with point-by-point measurement. See Cherry, R.J., *New Techniques of Optical Microscropy and Microspectroscopy*, CRC Press, Boca Raton, FL, 1991; Stelzer, E.H., Wacker, I., and De Mey, J.R., Confocal fluorescence microscopy in modern cell biology, *Sermin.Cell Biol.* 2, 145-152, 1991; Stevens, J.K. and Mills, L.R., *Three-Dimensional Confocal Microscopy: Volume Investigation of Biological Specimens*, Academic Press, San Diego, CA, 1994; Smith, R.F., *Microscopy and Photomicrography: A Working Manual*, CRC Press, Boca Raton, FL, 1994; Pawley, J.B., *Handbook of Biological Confocal Microscropy*, Plenum Press, New York, NY, 1995; Fay, F.S., Optical methods in cell physiology, in *Handbook of Physiology, Section 14, Cell Physiology*, ed. J.F. Hoffman and J.D. Jamieson, Oxford University Press (for the American Physiological Society), New York, NY, 1997; Paddock, S.W., *Confocal Microscopy Methods and Protocols*, Humana Press, Totowa, NJ, 1999; Brelje, T.C. Wessendorf, M.W., and Sorenson, R.L., Multicolor laser scanning confocal immunofluorescence microscopy: practical application and limitations, *Methods Cell Biol.* 70, 165-244, 2002; Bacia, K. and Schwille, P., A dynamic view of cellular processes by in vivo fluorescence auto- and cross-correlation spectroscopy, *Methods* 29, 74-85, 2003; Miyashita, T., Confocal microscopy for intracellular co-localization of proteins, *Methods Mol.Biol.* 261, 399-410, 2004; Heilker, R., Zemanova, L., Valler, M.J., and Nienhaus, G.U., Confocal fluorescence microscopy for high-throughput screening of G-protein coupled receptors, *Curr.Med.Chem.* 12, 2551-2559, 2005; Becker, B.E. and Gard, D.L., Visualization of the cytoskeleton in Xenopus oocytes and eggs by confocal immunofluorescence microscopy, *Methods Mol.Biol.* 322, 69-86, 2006.

Conjugate Vaccine

Coupling of a weak immunogen such as a polysaccharide to a protein to improve/enhance immunogenicity. See Cryz, S.J., Jr., Furer, E., Sadoff, J.C., *et al.*, Use of *Pseudomonas aeruginosa* toxin A in the construction of conjugate vaccines and immunotoxins, *Rev.Infect.Dis.* 9(Suppl. 5), S644-S649, 1987; Garner, C.V. and Pier, G.B. Immunologic considerations for the development of conjugate vaccines, *Contrib.Microbiol.Immunol.* 10, 11-17, 1989; Dintzis, R.Z., Rational design of conjugate vaccines, *Pediatr.Res.* 32, 376-385, 1992; Ellis, R.W. and Douglas, R.G., Jr., New vaccine technologies, *JAMA* 272, 929-931, 1994; Lindberg, A.A. and Pillai, S., Recent trends in the developments of bacterial vaccines, *Dev.Biol.Stand.* 87, 59-71, 1996; Zimmer, S.M. and Stephens, D.S., Meningococcal conjugate vaccines, *Expert Opin. Pharmacother.* 5, 855-863, 2004; Finn, A., Bacterial polysaccharide-protein conjugate vaccines, *Br.Med.Bull.* 70, 1-14, 2004; Shape, M.D. and Pollard, A.J., Meningococcal polysaccharide-protein conjugate vaccines, *Lancet Infect.Dis.* 5, 21-30, 2005; Finn, A. and Heath, P., Conjugate vaccines, *Arch.Dis.Child.* 90, 667-669, 2005; Jones, C., NMR assays for carbohydrate-based vaccines, *J.Pharm.Biomed.Anal.* 38, 840-850, 2005; Whitney, C.G, Impact of conjugate pneumococcal vaccines, *Pediatr.Infect. Dis.* 24, 729-730, 2005; Lee, C.J., Lee, L.H., and Gu, X.X., Mucosal immunity induced by pneumococcal glycoconjugate, *Crit.Rev. Microbiol.* 31, 137-144, 2005.

Connexins

A protein subunit of connexon which form gap junctions critical for intercellular communication. Mutations in the connexins are responsible for a diversity of diseases, including deafness, skin disorders and idiopathic atrial fibrillation. Connexins have been designated by their molecular mass while another system separate connexins on the basis of sequence homology. See Beyer, E.C., Paul, D.L., and Goodenough, D.A., Connexin family of gap junction proteins, *J.Membr.Biol.* 116, 187-194, 1990; Revel, J.P., Nicholson, B.J., and Yancey, S.B., Chemistry of gap junctions, *Annu.Rev.Physiol.* 47, 263-279, 1985; Revel, J.P., Yancey, S.B., Nicholson, B., and Hoh, J., Sequence diversity of gap junction proteins, *Ciba Found.Symp.* 125, 108-127, 1987; Gollob, M.H., Cardiac connexins as candidate genes for idiopathic atrial fibrillation, *Curr.Opin.Cardiol.* 21, 155-158, 2006; Stains, J.P. and Civitelli, R., Gap junctions in skeletal development and function, *Biochim.Biophys.Acta* 1719, 69-81, 2005; Anand, R.J., and Hackam, D.J., The role of gap junctions in health and disease, *Crit.Care Med.* 33 (Suppl 12), S535-S535, 2005; Michon, L., Nlend Nlend, R., Bavamian, S., *et al.*, Involvement of gap junctional communication in secretion, *Biochim.Biophys.Acta* 1719, 82-101, 2005; Vinken, M., Vanhaecke, T., Papeleu, P., *et al.*, Connexins and their channels in cell growth and cell death, *Cell Signal.* 18, 592-600, 2006; Petit, C., From deafness genes to hearing mechanisms: harmony and counterpoint, *Trends Mol.Med.* 12, 57-64, 2006; Evans, W.H., De Vuyst, E., and Leybaert, L., The gap junction cellular internet: connexin hemichannels enter the signalling limelight, *Biochem.J.* 397, 1-14, 2006.

Contig

The term contig was originally defined as a set of overlapping DNA sequences and has been expanded to include a set of overlapping DNA clones. Specifically, it refers to a set of gel bands which can be related to each other by overlaps sequences – see http://staden.sourceforge.net/contig.html See Staden, R., A new computer method for the storage any manipulation of DNA gel reading data, *Nucleic Acids Res.* 8, 3673-3694, 1980; Presting, G.G., Budiman, M.A., Wood, T., *et al.*, A framework for sequencing the rice genome, *Novartis Found.Symp.* 236, 13-24, 2001; Dodgson, J.B., Chicken genome sequence: a centennial gift to poultry genetics, *Cytogenet.Genome Res.* 102, 291-296, 2003; Schalkwyk, L.C., Francis, F., and Lehrach, H., Techniques in mammalian genome mapping, *Curr.Opin.Biotechnol.* 6, 37-43, 1995; Carrano, A.V., de Jong, P.J., Branscomb, E., *et al.*, Constructing chromosome- and region-specific cosmid maps fo the human genome, *Genome* 31, 1059-1065, 1989.

Contour Length

End-to-end length of a stretched DNA molecule (see Wellauer, P., Weber, R., and Wyler, T., Electron microscopic study of the influence of the preparative conditions on contour length and structure of mitrochondrial DNA of mouse liver, *J.Ultrastruct.Res.* 42, 377-393, 1973; Geller, K. and Reinert, K.E., Evidence for an increase of DNA contour length at low ionic strength, *Nucleic Acids Res.* 8, 2807-2822, 1980; Motejlek, K., Schindler, D., Assum, G., and Krone, W.. Increased amount and contour length distribution of small polydisperse circular DNA (spcDNA) in Fanconi anemia, *Mutat.Res.* 293, 205-214, 1993; Gast, F.U. and Sanger, H.L., Gel dependence of electrophoretic mobilites of double-stranded and viroid RNA and estimation of the contour length of a viroid by gel electrophoresis, *Electrophoresis* 15, 1493-1498, 1994; Sanchez-Sevilla, A., Thimonier, J., Marilley, M., *et al.*, Accuracy of AFM measurements of the contour length of DNA fragments adsorbed on mica in air and in aqueous buffer, *Ultramicroscopy* 92, 151-158, 2002) although the term has been used to describe very long proteins such as titan (Helmes, M., Trombitas, K., Centner, T., *et al.*, Mechanically driven contour-length adjustment in rat cardiac titin's unique N2B sequence: titin is an adjustable spring, *Circ. Res.* 84, 1339-1352, 1999).

Core Promoter

A region immediately (+/−30 bp) around the transcription start site that contain consensus sequence elements (TATA boxes, lnr, DPEs); *in vitro*, the core promoter is the minimal required sequence that is recognized by general transcription factors that activate correct transcription by RNA polymerase II. See Gill, G., Transcriptional initiation, *Curr.Biol.* 4, 374-376, 1994; Gill, G., Regulation of the initiation of eukaryotic transcription, *Essays Biochem.* 37, 33-43,2001; Butler, J.E. and Kadonaga, J.T., The RNA polymerase II core promoter: a key component in the regulation of gene expression, *Genes Dev.* 16, 2583-2592, 2002; Kadnoaga, J.T., The DPE, a core promoter element for transcription by RNA polymerase II, *Exp.Mol.Med.* 34, 259-264, 2002; Smale, C.T. and Kadonaga, J.T., The RNA polymerase II core promoter, *Annu.Rev. Biochem.* 72, 449–479, 2003; Lewis, B.A. and Reinberg, D., The mediator coactivator complex: functional and physical roles in transcriptional regulation, *J.Cell Sci.* 116, 3667-3675, 2003; Mulle, F. and Tora, L., The multicolored world of promoter recognition complexes, *EMBO J.* 23, 2-8, 2004; Chen, K., Organization of MAO A and MAO B promoters and regulation of gene expression, *Neurotoxicity* 25, 31-36, 2004; Hasselbach, L., Haase, S., Fischer, D., Kolberg, H.C., and Sturzbecher, H.W., Characterization of the promoter region of the human DNA-repair gene Rad51, *Eur.J.Gynecol. Oncol.* 26, 589-598, 2005.

Cosolvent

A miscible solvent added a primary solvent to enhance salvation or stability of a specific solute. Such solvents have been used extensively in studies on enzymes where cosolvents were required to dissolve the substrate. Cosolvents are also used in the formulation of pharmaceuticals and in liquid chromatography. See Tan, K.H. and Lovrien, R., Enzymology in aqueous-organic cosolvent binary mixtures, *J.Biol.Chem.* 247, 3278-3285, 1972; Richardson, N.E. and Meaekin, B.J., The influence of cosolvents and substrate substituents on the sorption of benzoic acid derivatives by polyamides, *J.Pharm.Pharmcol.* 27, 145-151, 1975; Pescheck, P.S. and Lovrien, R.E., Cosolvent control of substrate inhibition I cosolvent stimulation of beta-glucuronidase activity, *Biochem.Biophys.Res.Commun.* 79, 417-421, 1977; Bulone, D., Cupane, A. and Cordone, L, Conformational and functional properties of hemoglobin in water-organic cosolvent mixtures: effect of ethylene glycol and glycerol on oxygen affinity, *Biopolymers* 22, 119-123, 1983; Rubino, J.T. and Berryhill, W.S., Effects of solvent polarity on the acid dissociation constants of benzoic acids, *J.Pharm.Sci.* 75, 182-186, 1986; Buck, M., Trifluoroethanol and colleagues: cosolvents come of age. Recent studies with peptides and proteins, *Q.Rev.Biophys.* 31, 297-355,

1998; Jouyban-Gharamaleki, A., Valaee, L., Barzegar-Jalali, M., et al., Comparison of various cosolvency models for calculating solute solubility in water-cosolvent mixtures, Int.J.Pharm. 177, 93-101, 1999; Lee, J.C., Biopharmaceutical formulation, Curr.Opin.Biotechnol. 11, 81-84, 2000; Moelbert, S., Normand, B, and de los Rios, P., Kosmotropes and chaotropes: modeling preferential exclusion, binding and aggregate stability, Biophys. Chem. 112, 45-57, 2004; Scharnagl, C., Reif, M., and Friedrich, J., Stability of proteins: temperature, pressure and the role of solvent, Biochim.Biophys.Acta. 1749, 17-213, 2005.

Coupled Enzyme Systems

Most metabolic systems are composed of enzymes in a pathway where there is the sequential transformation of a substrate into a product through a series of separate enzyme-catalyzed reactions. One of the more simple coupled systems is the detoxification of ethyl alcohol (Plapp, B.V., Rate-limiting steps in ethanol metabolism and approaches to changing these rates biochemically, Adv.Expt.Biol.Med. 56, 77-109, 1975) or more complex (Brooks, S.P.J., Enzymes in the cell. What's really going on? in Function and Metabolism, ed. K.B. Storey, Wiley-Liss, Hoboken, NJ, USA, Chapter 3, pp. 55-86, 2004). Coupled enzyme systems are also used extensively in clinical chemistry where they are also referred to an indicator enzyme systems (Russell, C.D. and Cotlove, E., Serum glutamic-oxaloacetic transaminase: evaluation of a coupled-reaction enzyme assay by means of kinetic theory, Clin.Chem. 17, 1114-1122, 1971; Bais, R. and Pateghini, M., Principles of clinical enzymology, in Tietz Textbook of Clinical Chemistry and Molecular Diagnostics, ed. C.A. Burtis, E.R. Ashwood, and D.E. Bruns, Elsevier/Saunders, St. Louis, MO, USA, Chapter 9, pp. 191-218, 2006). The assay for creatine kinase is a coupled enzyme system as are some of the assays for glucose oxidase. An enzyme assay system is coupled to an immunological reaction in many solid-phase immunoassays such as ELISA assays (Kircks, LJ., Selected strategies for improving sensitivity and reliability of immunoassays, Clin.Chem. 40, 347-357, 1994). See Wimmer, M.C., Artiss, J.D., and Zak, B., Peroxidase-coupled method for kinetic colorimetry of total creatine kinase activity in serum, Clin.Chem. 31, 1616-1620, 1965; Shin, T., Murao, S., and Matsumura, E., A chromogenic oxidative coupling reaction of laccase: applications for laccase and angiotensin I converting enzyme assay, Anal.Biochem. 166, 380-388, 1987.

Creatine

A nitrogenous compound which is synthesized from arginine, glycine and S-adenosylmethionine (Van Pilsum J.F., Stephens, G.C., and Taylor, D., Distribution of creatine, guanidinoacetate and the enzymes for their biosynthesis in the animal kingdom, Biochem.J. 126, 325-345, 1972; Walker, J.B. and Hannan, J.K., Creatine biosynthesis during embryonic development. False feedback suppression of liver amidinotransferase by N-acetimdoylsarcosine and 1-carboxymethy-2-iminoimdazolidine (Cyclocreatine), Biochemistry 15, 2519-2522, 1976; Walker, J.B., Creatine: biosynthesis, regulation, and function, The Enzymes 50 (ed. A.Meister, Academic Press, New York, NY), 177-242, 1979; Wyss, M. and Wallimann, T., Creatine metabolism and the consequences of creatine depletion in muscle, Mol.Cell.Biochem. 133-134, 51-66, 1994; Wu, G. and Morris, S.M., Jr., Arginine metabolism: nitric oxide and beyond, Biochem.J. 336, 1-17, 1998; Brosnan, M.E. and Brosnan, J.T., Renal arginine metabolism, J.Nutr. 134(Suppl 10), 2791S-2795S, 1994; Morris, S.M., Jr., Enzymes of arginine metabolism, J.Nutr.(Suppl. 10), 2743S-2747S, 1994). Creatine is used as a biomarker for erthyrocytes (Beyer, C. and Alting, I.H., Enzymatic measurement of creatine in erythrocytes, Clin. Chem. 42, 313-318, 1996; Jiao, Y., Okumiya, T., Saibara, T., et al., An enzymatic assay for erythrocyte creatine as an index of the erythrocyte life time, Clin.Biochem. 31, 59-65, 1998; Takemoto, Y., Okumiya, T., Tsuchida, K., et al., Erythrocyte creatine as an index of the erythrocyte life span and erythropoiesis, Nephron 86, 513-514, 2000; Okumiya, T., Ishikawa-Nishi, M., Doi, T., et al., Evaluation of intravascular hemolysis with erythrocyte creatine in patients with cardiac valve prostheses, Chest 125, 2115-2120, 2004). There is increased use of creatine as a nutritional supplement (Korzun, W.J., Oral creatine supplements lower plasma homocysteine concentrations in humans, Clin.Lab.Sci. 17, 102-106, 2004; Pearlman, J.P. and Fielding, R.A., Creatine monohydrate as a therapeutic aid in muscular dysthrophy, Nutr. Rev. 64, 80-88, 2006; Hespel, P., Maughan, R.J., and Greenhaff, P.L., Dietary supplements for football, J.Sports Sci. 24, 749-761, 2006; Shao, A. and Hathcock, J.N., Risk assessment for creatine monohydrate, Regul.Toxicol. Pharmacol., 45, 242–251, 2006).

Creatine Kinase

Adenosine triphosphate: creatine N-phosphotransferase (EC 2.7.3.2) also creatine phosphokinase. Creatine kinase is found in muscle and is responsible for the formation of creatine phosphate from creatine and adenosine triphosphate; creatine phosphate is a higher energy source for muscle contraction. Creatine kinase is elevated in all forms of muscular dystrophy. Creatine kinase is dimer and is present as isozymes (CK-1, BB; CK-2, MB; CK-3, MM) and Ck-mt(mitochondrial). Creatine kinase is also used for measure cardiac muscle damage in myocardial infarction. See Bais, R. and Edwards, J.B., Creatine kinase, CRC Crit. Rev.Clin. Lab.Sci. 16, 291-355, 1982; McLeish, M.J. and Kenyon, G.L., Relating structure to mechanism in creatine kinase, Crit. Rev.Biochem.Mol.Biol. 40, 1-20, 2005.

Creatinine

A catabolic product of creatine which should be in blood as a constant quantity. An increase in urinary creatinine is associated with a loss of kidney function. See Hodgkinison, A. and Edwards, N.A., Laboratory determinations of renal function, Biochem.Clin. 2, 77-86, 1963; Blainey, J.D., The renal excretion of higher molecular weight substances, Curr.Probl.Clin.Biochem. 2, 85-100, 1968; Cook, J.G., Factors influencing the assay of creatinine, Ann.Clin. Biochem. 12, 219-232, 1975; Greenberg, N., Smith, T.A., and VanBrunt, N., Interference in the Vitros CREA method when measuring urine creatinine on samples acidified with acetic acid, Clin.Chem. 50, 1273-1275, 2004.

Price, C.P., Newall, R.G., and Boyd, J.C., Prediction of significant proteinuria: a systematic review, Clin.Chem. 51, 1577-1586, 2005; Verhoeven, N.M., Salmons, G.S., and Jakobs, C., Laboratory diagnosis of defects of creatine biosynthesis and transport, Clin.Chim.Acta 361, 1-9, 2005; Wishart, D.S., Metabolomics: the principles and potential applications to transplantation, Am.J.Transplant. 5, 2814-2820, 2005; Seron, D., Fulladosa, X., and Moreso, F., Risk factors associated with the deterioration of renal function after kidney transplantation, Kidney Int.Suppl. 99, S113-S117, 2005; Schrier, R.W., Role of diminished renal function in cardiovascular mortality: marker or pathogenic factor? J.Am. Coll.Cardiol. 47, 1-8, 2006.

Critical Pressure

The minimum pressure required to condense gas to liquid at the critical temperature.

Critical Temperature

The critical point (end of a vapor pressure curve in a phase diagram); above this temperature, a gas cannot be liquefied.

Crowding

A term used to described general effect of polymers including proteins and polysacchardies on the solution properties of proteins. See Zimmerman, S.B., Macromolecular crowding effects on macromolecular interactions: some implications for genome structure and function, *Biochim.Biophys.Acta* 1216, 175-185, 1993; Minton, A.P., Molecular crowding: analysis of effects of high concentrations of inert cosolutes on biochemical equilibria and rates in terms of volume exclusion, *Methods Enzymol.* 295, 127-149, 1998; Johansson, H.O., Brooks, D.E., and Haynes, C.A., Macromolecular crowding and its consequences, *Int.Rev.Cytol.* 192, 155-170, 2000; Ellis, R.J., Macromolecular crowding: obvious but underappreciated, *Trends Biochem.Sci.* 26, 597-604, 2001; Bernardo, P., Garcia de la Torre, J., and Pons, M., Macromolecular crowding in biological systems: hydrodynamic and NMR methods, *J.Mol.Recognit.* 17, 397-407, 2004; Martin, J., Chaperon function – effects of crowding and confinement, *J.Mol.Recognit.* 17, 465-472, 2004; Minton, A.P., Influence of macromolecular crowding upon the stability and state of association of proteins: predictions and observations, *J.Pharm.Sci.* 94, 1668-1675, 2005; del Alamo, M., Rivas, G., and Mateu, M.G., Effect of macromolecular crowding agents on human immunodeficiency virus type 1 capsid protein assembly in vitro, *J.Virol.* 79, 14271-14281, 2005; Despa, F., Orgill, D.P., and Lee, R.C., Molecular crowding effects on protein stability, *Ann.N.Y.Acad.Sci.* 1066, 54-66, 2006; Szymanski, J., Patkowski, A., Gapinski, J. *et al.*, Movement of proteins in an environment crowded by surfactant micelles: anomalous versus normal diffusion, *J.Phys.Chem.B.Condens.Matter Mater.Surf.Interfaces Biophys.* 110, 7367-7373, 2006; Derham, B.K. and Harding, J.J., The effect of the presence of globular proteins and elongated polymers on enzyme activity, *Biochim. Biophys.Acta* 1764, 1000-1006, 2006; Grailhe, R., Merola, F., Ridard, J. *et al.*, Monitoring protein interactions in the living cell through the fluorescence decays of the cyan fluorescent protein, *Chemphyschem.* 7, 1442-1454, 2006. McPhie, P., Ni, Y.S., and Minton, A.P., Macromolecular crowding stabilizes the molten globule form of apomyoglobin with respect to both cold and heat unfolding, *J.Mol.Biol.* 361, 7-10, 2006.

Crown Gall Disease/Crown Gall Tumors

Crown Gall is caused by a bacteria (*Agrobacterium tumefaciens*). These galls begin with tumor-like cell growth at or just below the soil's surface, near the base of the plant and commonly on bud unions. Galls usually begin as green, pliable tissue; then develop into dark, crusty growths. Crown Gall Disease has been used to study transformation with relevance to tumor formation. See Knoft, U.C., Crown-gall and *Agrobacterium tumefaciens*: survey of a plant-cell-transformation system of interest to medicine and agriculture, *Subcell.Biochem.* 6, 143-173, 1978; Zhu, J., Oger, P.M., Schrammeijer, B. *et al.*, The bases of crown gall tumorigenesis, *J.Bacteriol.* 182, 3885-3895, 2000; Escobar, M.A., and

Dadekar, A.M., *Agrobacterium tumefaciens* as an agent of disease, *Trends Plant.Sci.* 8, 380-386, 2003; Brencic, A. and Winans, S.C., Detection of and response to signals involved in host-microbe interactions by plant-associated bacteria. *Micobiol.Mol.Biol.Rev.* 69, 155-194, 2005;

Cryosection

A tissue section cut from a frozen specimen; in this situation, ice i)s the supporting matrix. See Yamada, E. and Watanabe, H., High voltage electron microscopy of critical-point dried cryosection, *J.Electron Microsc.* 26(Suppl), 339-342, 1977; Maddox, P.H., Tay, S.K., and Jenkins, D., A new fixed cryosection technique for the simultaneous immunocytochemical demonstration of T6 and S100 antigens, *Histochem.J.* 19, 35-38, 1987; Sod, E.W., Crooker, A.R. and Morrison, G.H., Biological cryosection preparation and practical ion yield evaluation for ion microscopic analysis, *J.Microsc.* 160, 55-65, 1990; Lewis Carl, S.A., Gillete-Ferguson, I., and Ferguson, D.G., An indirect immunofluorescence procedure for staining the same cryosection with two mouse monoclonal primary antibodies, *J.Histochem.Cytochem.* 41, 1273-1278, 1993; Jensen, H.L. and Norrild, B., Easy and reliable double-immunogold labelling of herpes simplex virus type-1 infected cells using primary antibodies and studied by cryosection electron microscopy, *Histochem.J* 31, 525-533, 1999; Gou, D. and Catchpoole, D.R., Isolation of intact RNA following cryosection of archived frozen tissue, *Biotechniques* 34, 48-50, 2003; Rieppo, J., Hyttinen, M.M. Jurvelin, J.S., and Helminen, H.J., Reference sample method reduces the error caused by variable cryosection thickness in Fourier transform infrared imaging, *Appl.Spectrosc.* 58, 137-140, 2004; Takizawa, T. and Robinson, J.M., Thin is better! Ultrathin cryosection immunocytochemistry, *J.Nippon Med.Sch.* 71, 306-307, 2004.

Cyanine Dyes (CyDyes)

A family of fluorescent polymethine dyes contain containing a -CH= group linking two nitrogen-containing heterocyclic rings; developed as sensitizer for photographic emulsions. Used in biochemistry and molecular biology on nucleic acid probes for DNA microarrays and for labeling proteins for electrophoretic analysis. See Ernst, L.A., Gupta, R.K., Mujumdar, R.B., and Waggoner, A.S., Cyanine dye labeling reagents for sulfydryl groups, *Cytometry* 10, 3-10, 1989; Mujumdar, P.S., Ernst, L.A., Mujumdar, S.R., and Waggoner, A.S., Cyanine dye labeling reagents containing isothiocyanate groups, *Cytometry* 10, 11-19, 1989; Southwick, P.L., Ernst, L.A., Tauriello, E.W., *et al.*, Cyanine dye labeling reagents—carboxymethylindocyanine succinimidyl esters, *Cytometry* 11, 418-430, 1990; Mujamdar, R.B., Ernst, L.A., Mujumdar, S.R., *et al.*, Cyanine dye labeling reagents: sulfoindocyanine succinimidyl esters, *Bioconjug.Chem.* 4, 105-111, 1993; Benchaib, M., Delorme, R., Pluvinage, M., *et al.*, Evaluation of five green fluorescence-emitting streptavidin-conjugated fluorochromes for use in immunofluorescence microscopy, *Histochem.Cell Biol.* 106, 253-256, 1996; Mujumdar, S.R., Mujumdar, R.B., Grant, C.M., and Waggoner, A.S., Cyanine-labeling reagents: sulfobenzindocyanine succinimidyl esters, *Bioconjug.Chem.* 7, 356-362, 1996; Karp, N.A. and Lilley, K.S., Maximizing sensitivity for detecting changes in protein expression: experimental design using minimal CyDyes, *Proteomics* 5, 3105-3115, 2005; Heilmann, M., Margeat, E., Kasper, R., *et al.*, Carbocyanine dyes as efficient reversible single-molecule optical switch, *J.Am.Chem.Soc.* 127, 3801-3806, 2005; Wu, T.L., Two-dimensional difference gel electrophoresis, *Methods Mol.Biol.*

328, 71-95, 2006; Boisseau, S., Mabrouk, K., Ram, N., *et al.*, Cell penetration properties of maurocalcine, a natural venom peptide active on the intracellular ryanodine receptor, *Biochim.Biophys. Acta.* 1758, 308-319, 2006. There is also use of these dyes for the measurement of membrane potentials. See Miller, J.B. and Koshland, D.E., Effects of cyanine dye membrane probes on cellular properties, *Nature* 272, 83-84, 1978; Klausner, R.D. and Wolf, D.E., Selectivity of fluorescent lipid analogues for lipid domains, *Biochemistry* 19, 6199-6203, 1980; Kragh-Hansen, U., Jorgensen, K.E., and Sheikh, M.I., The use of potential-sensitive cyanine dye for studying ion-dependent electrogenic renal transport of organic solutes. Spectrophotometric measurements, *Biochem.J.* 208, 359-368, 1982; Johnstone, R.M., Laris, P.C. and Eddy, A.A., The use of fluorescent dyes to measure membrane potentials: a critique, *J.Cell Physiol.* 112, 298-300, 1982; Toyomizu, M., Okamoto, K., Akiba, Y. *et al,*, Anacardic acid-mediated changes in membrane potential and pH gradient across liposomal membranes, *Biochim.Biophys. Acta* 1558, 54-62, 2002.

Cyclitols

Term used to describe derivatives of hexahydroxyhexane (1,2,3,4,5,6-hexahydroxyhexane). An analogue to saccharides and serves a matrix for the development of inhibitors and activators based on saccharide structure. See Tentative rules for cyclitol nomenclature, *Biochim.Biophys.Acta* 165, 1-21, 1968; Orthen, B. and Popp, M., Cyclitols as cryoprotectants for spinach and chickpea thylakoids, *Environ.Exp.Bot.* 44, 125-132, 2000; Pelyvas, I.F., Toth, Z.G., Vereb, G., *et al.*, Synthesis of new cyclitol compounds that influence the activity of phosphatidylinositol 4-kinase isoforms, PI4K230, *J.Med.Chem.* 44, 627-632, 2001; Sureshan, K.M., Shashidhar, M.S. and Varma, A.J., Cyclitol-based metal-complexing agents. Effect of the relative orientation of oxygen atoms in the ionophoric ring on the cation-binding ability of myo-inositol-based crown ethers, *J.Org.Chem.* 67, 6884-6888, 2002; Freeman, C., Liu, L., Banwell, M.G., *et al.*, Use of sulfated linked cyclitols as heparin sulfate mimetics to probe the heparin/heparin sulfate binding specificity of proteins, *J.Biol.Chem.* 280, 8842-8849, 2005; Cochran, S., Li, C.P., and Bytheway, I., An experimental and molecular-modeling study of the binding of linked sulfated tetracyclitols to FGF-1 and FGF-2, *ChemBioChem* 6, 1882-1890, 2005.

Cytochrome P-450 Enzymes (CPY)

A family of enzymes which have monooxygenase activity and are involved in the metabolism/catabolism of drugs. Cytochrome P450 proteins are found in high concentration in the liver. See Jung, C., Schunemann, V., and Lendzian, F. Freeze-quenched iron-oxo intermediate in cytochrome P450, *Biochem.Biophys. Res.Commun.* 338, 355-364, 2005; Johnson, E.F. and Stout, C.D., Structural diversity of human xenobiotic-metabolizing cytochrome P450, *Biochem.Biophys.Res.Commun.* 338, 331-336, 2005; Tang, W., Wang, R.W., and Lu, A.Y., Utility of recombinant cytochrome p450 enzymes: a drug metabolism perspective, *Curr. Drug Metab.* 6, 503-517, 2005; Krishna, D.R. and Shekar, M.S., Cytochrome P450 3A: genetic polymorphisms and inter-ethnic differences, *Methods Find.Exp.Clin.Pharmacol.* 27, 559-567, 2005; Sarlis, N.J. and Gourgiotis, L, Hormonal effects on drug metabolism through the CYP system: perspectives on their potential significance in the era of pharmacogenomics, *Curr.Drug Targets Immune Endocr.Metabol.Disord.* 5, 439-448, 2005; Grengerich, E.P., Cytochrome P405 enzymes in the generation of commercial products, *Nat.Rev.Drug.Disc.* 1, 359-366, 2002.

Cytokeratin

Cytokeratin describes intermediate filament keratins found in epithelial tissue. There are two types of cytokeratins the acidic type I cytokeratins and the basic or neutral type II cytokeratins. Cytokeratins are thought to play role in the activation of plasma prekallikrein and plasminogen. See Crewther, W.G., Fraser, R.D., Lennox, F.G., and Lindley, H., The chemistry of keratins, *Adv. Protein Chem.* 20, 191-346, 1965; Masri, M.S. and Friedman, M., Interactions of keratins with metal ions: uptake profiles, mode of binding, and effects on the properties of wool, *Adv.Exp.Med. Biol.* 48, 551-587, 1974; Fuchs, E. and Green, H., Multiple keratins of cultured human epidermal cells are translated from different mRNA molecules, *Cell* 17, 573-582, 1979; Fraser, R.D. and Macrae, T.P., Molecular structure and mechanical properties of keratins, *Symp.Soc.Exp.Biol.* 34, 211-246, 1980; Moll, R., Franke, W.W., Schiller, D.L., *et al.*, The catalog of human cytokeratins: patterns for expression in normal epithelia, tumors and cultured cells, *Cell* 31, 11-24, 1982; Lazarides, E., Intermediate filaments: a chemically heterogeneous, developmentally regulated class of proteins, *Annu.Rev.Biochem.* 51, 219-250, 1982; Gonias, S.L., Hembrough, T.A., and Sankovic, M., Cytokeratin 8 functions as a major plasminogen receptor in select epithelial and carcinoma cells, *Front. Biosci.* 6, D1403-D1411, 2001; Kaplan, A.P., Joseph, K., and Silverberg, M., Pathways for bradykinin formation and inflammatory diseases, *J.Allergy Clin.Immunol.* 109, 195-209, 2002; Shariat-Madar, Z., Mahdi, F. and Schmaier, A.H., Assembly and activation of the plasma kallikrein/kinin system: a new interpretation, *Int.Immunopharmacol.* 2, 1841-1849, 2002; Langbein, L. and Schweizer, J., Keratins of the human hair follicle, *Int.Rev.Cytol.* 243, 1-78, 2005; Gusterson, B.A., Ross, D.T., Heath, V.J., and Stein, T., Basal cytokeratins and their relationship to the cellular origin and functional classification of breast cancer, *Breast Cancer Res.* 7, 143-148, 2005; Skakle, J., Applications of X-ray powder diffraction in materials chemistry, *Chem.Rec.* 5, 252-262, 2005. See also keratin.

Cytokines

Non-antibody proteins secreted by immune system cells. This is a large category and include the various interferon and interleukins as a well as other protein substances. See DeMaeyer, E.M. and Demaeyer-Guignard, J., *Interferons and other Regulatory Cytokines*, Wiley, New York, NY, USA, 1988; Plotnkoff, N.P., *Cytokines: Stress and Immunity*, CRC Press, Boca Raton, FL, USA, 1999; Cruse, J.M. and Lewis, R.E., *Atlas of Immunology*, CRC Press, Boca Raton, FL, 1999; Rott, I.M. and Brostoff, J., *Immunology*, Mosby, Edinburgh, UK, 2001; Keisari, Y. and Ofek, I., *The Biology and Pathology of Innate Immunity Mechanisms*, Kluwer Academic, New York, NY, 2002; Henle, W., Interference and interferon in persistent viral infections of cell cultures, *J.Immunol.* 91, 145-150, 1963; Isaacs, A., Interferon, *Adv.Virus Res.* 10, 1-38, 1963; Baron, S. and Levy, H.B., Interferon, *Annu.Rev.Microbiol.* 20, 291-318, 1966; Silverstein, S., Macrophages and viral immunity, *Semin. Hematol.* 7, 185-214, 1970; Bloom, B.R., In vitro approaches to the mechanism of cell-mediated immune reactions, *Adv.Immunol.* 13, 101-208, 1971; Granger, G.A., Lymphokines-the mediators of cellular immunity, *Ser.Hematol.* 5, 8-40, 1972; Valentine, F.T., Soluble factors produced by lymphocytes, *Ann.N.Y.Acad.Sci.* 221, 317-323, 1974; Ward, P.A., Leukotaxis and leukotactic disorders. A review, *Am.J.Pathol.* 77, 520-538, 1974; Salazar-Mather, T.P. and Hokeness, K.L., Cytokine and chemokine networks: pathways to viral defense, *Curr.Top.Microbiol.Immunol.* 303, 29-46, 2006; Akira, S., Uematsu, S., and Takeuchi, O., Pathogen recognition and innate immunity, *Cell* 124, 783-801, 2006; Tedgui, A. and Mallat, Z., Cytokines in atherosclerosis: pathogenic and regulatory pathways, *Physiol.Rev.* 86, 515-581, 2006.

Cytokinesis

Cell division; the division of the cytoplasm of a cell following the division of the nucleus. See Robinson, D.N. and Spudich, J.A., Mechanics and regulation of cytokinesis, *Curr.Opin.Cell Biol.* 16, 181-188, 2004; Mayer, U. and Jurgens, G., Cytokinesis: lines of division taking shape, *Curr.Opin.Plant Biol.* 7, 599-604, 2004; Albertson, R., Riggs, B., and Sullivan, W., Membrane traffic: a driving force in cytokinesis, *Trends Cell Biol.* 15, 92-101, 2005; Glotzer, M., The molecular requirements for cytokinesis, *Science* 307, 1735-1739, 2005; Burgess, D.R., Cytokinesis: new roles for myosin, *Curr.Biol.* 15, R310-R311, 2005; Darenfeld, H. and Mandato, C.A., Wound-induced contractile ring: a model for cytokinesis, *Biochem.Cell Biol.* 83, 711-720, 2005; Konopka, C.A., Scheede, J.B., Skop, A.R., and Bednarek, S.Y., Dynamin and cytokinesis, *Traffic* 7, 239-247, 2006.

Cytomics

The molecular analysis of heterogeneous cellular systems; Davies, E., Stankovic, B., Azama, K., *et al.*, Novel components of the plant cytoskeleton: a beginning to plant 'cytomics', *Plant Sci.* 160, 185-196, 2001; Bernas, T., Gregori, G., Asem, E.K., and Robinson, J.P., Integrating cytomics and proteomics, *Mol.Cell.Proteomics* 5, 2-13, 2006; Van Osta, P., Ver Donck, K., Bols, L., and Geysen, J. Cytomics and drug discovery, *Cytometry A* 69, 117-118, 2006; Tarnok, A., Slide-based cytometry for cytomics – a minireview, *Cytometry A* 69, 555-562, 2006; Herrera, G., Diaz, L., Martinez-Romero, A., *et al.*, Cytomics: a multiparametric, dynamic approach to cell research, *Toxicol. In Vitro*, 21, 176–182, 2006; Valet, G., Cytomics as a new potential for drug discovery, *Drug Discov.Today* 11, 785-791, 2006.

Cytoskeleton

The internal framework of the cell; the cytoskeleton is composed largely of actin filaments and microtubules. See Wasteneys, G.O. and Yang, Z., New views on the plant cytoskeleton, *Plant Physiol.* 136, 3884-3891, 2004; Moller-Jensen, J. and Lowe, J., Increasing complexity of the bacterial cytoskeleton, *Curr.Opin.Cell Biol.* 17, 75-81, 2005; Smith, L.G. and Oppenheimer, D.G., Spatial control of cell expansion by the plant cytoskeleton, *Annu.Rev.Cell Dev. Biol.* 21, 271-295, 2005; Munro, E.M., PAR proteins and the cytoskeleton: a marriage of equals, *Curr.Opin.Cell Biol.* 18, 86-94, 2006; Boldogh, I.R. and Pon, L.A., Interactions of mitochondria with the actin cytoskeleton, *Biochim.Biophys.Acta* 1763, 405-462, 2006; Larsson, C., Protein kinase C and the regulation of the actin cytoskeleton, *Cell Signal.* 18, 276-284, 2006; Logan, M.R and Mandato, C.A., Regulation of the actin cytoskeleton by PIP2 in cytokinesis, *Biol.Cell.* 98, 377-388, 2006; Sheetz, M.P., Sable, J.E., and Dobereiner, H.G., Continuous membrane-cytoskeleton adhesion requires continuous accommodation to lipid and cytoskeleton dynamics, *Annu.Rev.Biophys.Biomol.Struct.* 35, 417-434, 2006; Becker, B.E. and Gard, D.L.,Visualization of the cytoskeleton in *Xenopus* oocytes and eggs by confocal immunofluorescence microscropy, *Methods Mol.Biol.* 322, 69-86, 2006; Popowicz, G.M., Scheicher, M., Noegel, A.A., and Holak, A.A., Filamins: promiscuous organizers of the cytoskeleton, *Trends Biochem.Sci.* 31, 411-419, 2006.

Cytotoxic T-cells; cytotoxic T-lymphocytes

Also known as killer cells, killer T-cells, null cells, A differentiated T cell (CD8 positive) that attacks and lyses target cells bearing specific antigens. Used in patient-specific immunotherapy with cells grown in culture. See Gillis, S., Baker, P.E., Ruscetti, F.W., and Smith, K.A., Long-term culture of human antigen-specific cytotoxic T-cell lines, *J.Exptl.Med.* 148, 1093-1098, 1978.

Database of Interacting Proteins (DIP)

The database of interacting proteins integrates the experimental evidence available on protein interactions into a single on-line resource: http://dip.doe-mbi.ucla.edu See Xenarious, I., Fernandez, E., Salwinski, L., Duan, X.J. *et al.*, DIP: the database of interacting proteins: 2001 update, *Nucleic Acids Res.* 29, 239-241, 2001; Deane, C.M., Salwinski, L, Xenarios, I., and Eisenberg, D., Protein interactions: two methods for assessment of the reliability of high throughput observations, *Mol.Cell Proteomics* 1, 349-356, 2002; Salwinski, L., Miller, C.S., Smith, A.J., *et al.*, *Nucleic Acids Res.* 32, D449-451, 2004; Han, D., Kim, H.S., Seo, J., and Jang, W., A domain combination based probalistic framework for protein-protein interaction prediction, *Genome Inform.Ser.Workshop Genome Inform.* 14, 250-259, 2003; Espadaler, J., Romero-Isart, O., Jackson, R.M., and Oliva, B., Prediction of protein-protein interactions using distant conservation of sequence patterns and structure relationships, *Bioinformatics* 21, 3360-3368, 2005.

Deconvolution

An algorithm used in electrospray mass spectrometry to translate the spectra of multiply charged ions into a spectrum of molecular species.

Dendrimers

A novel polymeric material cotaining a highly branched and well-defined structure. Dendrimers have been used for drug delivery, a biological matrix and for model drug distribution studies. Dedrimers are similar to dendrites which are branched crystals in which branches of crystallization proceeds at different rates. See Meldal, M. and Hilaire, P.M., Synthetic methods of glycopeptides assembly, and biological analysis of glycopeptides products, *Curr. Opin.Chem.Biol.* 1, 552-563, 1997; Sadler, K. and Tam, J.P., Peptide dendrimers: applications and synthesis, *J.Biotechnol.* 90, 195-229, 2002; Turnbull, W.B., and Stoddart, J.F., Design and synthesis of glycodendrimers, *J.Biotechnol.* 90, 231-255, 2002; Kobayashi, H. and Brechbiel, M.W., Dendrimer-based macromolecular MRI contrast agents: characteristics and application, *Mol.Imaging* 2, 1-10, 2003; Lee, C.C., MacKay, J.A., Frechet, J.M., and Szoka, F.C., Designing dendrimers for biological applications, *Nat.Biotechnol.* 23, 1517-1526, 2005; Qiu, L.Y. and Bae, Y.H., Polymer architechture and drug delivery, *Pharm.Res.*23, 1-30, 2006. Gupta, V., Agashe, H.B., Asthana, A., and Jain, N.K., Dendrimers: novel polymeric nanoarchitecture for solubility enhancement, *Biomacromolecules* 7, 649-658, 2006; Söntjens, S.H.M., Nettles, D.L., Carnahan, M.A., *et al.*, Biodendrimer-based hydrogel scaffolds for cartilage tissue repair, *Biomacromolecules* 7, 310-316, 2006.

Desorption

Process by which molecules in solid or liquid form are transformed into a gas phase.

Detergent Perturbation

Treatment of total human plasma proteins with sodium cholate and subsequent removal; resulting in "remodeling" of the lipoproteins; see Pownall, H.J., Remodeling of human plasma lipoproteins by detergent perturbation, *Biochemistry* 44, 9714-9722, 2005.

Deterministic Series

A series or model which contains no random or probabilistic elements. *The Cambridge Dictionary of Statistics*, ed. B.S. Everitt, Cambridge University Press, Cambridge, United Kingdom, 1998.

Diabodies

An engineered non-covalent dimer of an scFv fragment which has two antigen binding sites which may be homologous or heterologous. The normal linker engineered between the V_H and V_L domains is 15 residues (usually glycine and serine to promote maximum flexibility) which yields as monomer; if the linker is reduced to 10 residues, a dimer (diabody) is formed while with no linker there is a trimer or higher order polymer. See Atwell, J.L., Breheney, K.A., Lawrence, L.J. *et al.*, scFv multimers of th anti-neuraminidase antibody NC10: length of the linker between V_H and V_L domains dictates precisely the transition between diabodies and triabodies, *Protein Eng.* 12, 597-604, 1999. Todorovska, A., Roovers, R.C., Dolezal, O., *et al.*, Design and application of diabodies, triabodies and tetrabodiese for cancer targeting, *J.Immunol.Methods* 248, 47-66, 2001; Holliger, P. and Hudson, P.J., Engineered antibody fragments and the rise of single domains, *Nature Biotechnol.* 23, 1126-1136, 2005. While diabodies are non-covalent complexes of engineered scFv constructs based on the association of the V_H domain with the most available V_L domain. A covalent diabody was observed with an engineered anti-carcinoembryonic antigen (CEA) diabody with cysteine residues inserted for coupling with a radiolabel. The formation of a disulfide-linked diabody was an unexpected consequence. See Olafsen, T., Cheung, C.-w., Yazaki, P.J. *et al.*, Covalent disulfide-linked anti-CEA diabody allows site-specific conjugation and radiolabeling for tumor targeting applications, *Prot.Eng.Des.Sel.* 17, 21-27, 2004. Note: It has been observed that if the order of the variable regions are switch in scFv construct (V_L-V_H instead of V_H-V_L), the engineered scFv with a zero-length linker formed a dimer (diabody) instead of the expected trimer – See Arndt, M.A.E., Krauss, J., and Rybak, S.M., Antigen binding and stability properties of non-covalently linked anti-CD22 single-chain Fv dimers, *FEBS Lett.* 578, 257-261, 2004. See single chain Fv fragment, bibody, triabody.

Diapedesis

The migration of a leukocyte through the interendothelial junction space and the extracellular matrix/basement membrane to the site of tissue inflammation; a process driven by chemotaxis.

Dicer

Dicer is an RNAse III nuclease (class III) which is specific for double-stranded RNA and yields siRNAs. Structurally it consists of an amino terminal helicase domain, a PAZ domain, two RNAse III motifs, and a dsRNA binding motif. See Carmell, M.A., and Hannan, G.J., RNAse III enzymes and their initiation of gene silencing, *Nat.Struct.Mol.Biol.* 11, 214-218, 2004;

Myers, J.W. and Ferrell, J.E., Jr., Silencing gene expression with Dicer-generated siRNA pools, in *RNA Silencing. Methods and Protocols*, ed. G.G. Carmichael, Humana Press, Totowa, New Jersey, 2005; Hammond, S.M., Dicing and slicing. The core machinery of the RNA interference pathway, *FEBS Lett.* 579, 5822-5829, 2005.

Dictionary of Interfaces in Proteins (DIP)

The dictionary of interfaces in proteins is a database collecting the 3D structures of protein domains involved in interactions (patches). See Preissner, R., Goode, A., and Frommel, C., Dictionary of interfaces in proteins (DIP). Data bank of complementary molecules, *J.Mol.Biol.* 280, 535-550, 1998.

Frommel, C., Gille, C., Goede, A., *et al.*, Accelerating screening of 3D protein data with a graph theoretical approach, *Bioinformatics* 19, 2442-2447, 2003.

Differential Scanning Calorimetry (DSC)

A physical technique for the study of conformation based on measuring changes in heat capacity of a molecule under various conditions. See Zecchinon, L., Oriol, A., Netzel, U., *et al.*, Stability domains, substrate-induced conformational changes, and hinge-bending motions in a psychrophilic phosphoglycerate kinase. A microcalorimetric study, *J.Biol.Chem.* 280, 41307-41314, 2005;

Dipolar Couplings

Also residual dipolar couplings, measures the interaction between nuclei in an applied magnetic field; used for the determination of the solution structure of peptides, proteins, nucleic acids, and carbohydrates; also ligand binding, see Post, C.B., Exchange-transferred NOE spectroscopy and bound ligand structure determination, *Curr.Opin.Struct.Biol.* 13, 581-588, 2003; Bush, C.A., Martin-Pastor, M., and Imberty, A., Structure and conformation of complex carbohydrates of glycoproteins, glycolipids, and bacterial polysaccharides, *Ann.Rev.Biophys.Biomol.Struct.* 28, 269-293, 1999; MacDonald, D., and Lu, P., Residual dipolar couplings in nucleic acid structure determination, *Curr.Opin.Struct.Biol.* 12, 337-343, 2002.

Directed Library

Also focused library. A screening library of chemical compounds which may be prepared by parallel synthesis, combinatorial chemistry, phage displayed or similar multiplexed technologies. See Miller, J.L., Recent developments in focused library design: targeting gene-families, *Curr.Top.Med.Chem.* 6, 19-29. 2006; Xu, Y., Shi, J., Yamamoto, N., *et al.*, A credit-card library approach for disrupting protein-protein interactions, *Bioorg.Med.Chem.*, 14, 2660–2673, 2005; Subramanian, T., Wang, Z., Troutman, J.M., *et al.*, Directed library of anilinogeranyl analogues of farnesyl diphosphate via mixed solid- and solution-phase synthesis, *Org.Lett.* 7, 2109-2112, 2005; McGregor, M.J., and Muskal, S.M., Pharmacophore fingerprinting. 1. Application to QSAR and focused library design, *J.Chem.Inf.Comput.Sci.* 39, 569-574, 1999.

Distributed Annotation System (DAS)

The distributed annotation system is a communication protocol for the exchange of biological annotations (In genetics, the process of identifying the locations and coding regions of genes in a genome and determining what those genes do. An annotation is note added with comment on the function of the gene and/or coding region). See Hubbard, T., Biological information: making it accessible and integrated (and trying to make sense of it), *Bioinformatics* 18 (Suppl 2):S140, 2002; Olason, P.I., Integrating protein annotation resouces through the Distributed Annotation System, *Nucleic Acids Res.* 33, W468-470, 2005; Prlic, A., Down, T.A., and Hubbard, J.T., Adding some SPICE to DAS, *Bioinformatics* 21 (Suppl 2), ii40-ii41, 2005; Stamm, S., Riethovan, J.J., Le Texier, V., Gopalakrishnan, C., Kumanduri, V., Tang, Y., Barbosa-Morais, N.L., and Thanaraj, T.A., ASD: a bioinformatics resource on alternative splicing, *Nucleic Acids Res.* 32, D46-D55, 2006. See also http://www.cbs.dtu.dk/; http://www.cbs.dtu.dk/cgi-bin/das

DNA Fingerprinting

This procedure is also referred to as chromosomal fingerprinting, restriction enzyme analysis (REA). This is process where DNA is cleaved by a restriction endonuclease (restriction enzyme). The resulting DNA fragments are separated by gel electrophoresis ad detected by specific and non-specific probes. DNA fingerprinting is extensively used for forensic purposes. See Gazit, E. and Gazit, E., DNA fingerprinting, *Isr.J.Med.Sci.* 26, 158-162, 1990; Owen, R.J., Chromosomal DNA fingerprinting –a new method of species and strain identification applicable to microbial pathogens, *J.Med.Microbiol.* 30, 89-99, 1989; Cawood, A.H., DNA fingerprinting, *Clin.Chem.* 35, 1832-1837, 1989; de Gouyon, B., Julier, C., Avner, P., Georges, M, and Lathrop, M., Human variable number of tandem repeat probes as a source of polymorphic markers in experimental animals, *EXS* 58, 85-94, 1991; Webb, M.B. and Debenham, P.G., Cell line characterization by DNA fingerprinting: a review, *Dev.Biol.Stand.* 76, 39-42, 1992; Debenham, P.G., Probing identity: the changing face of DNA fingerprinting, *Trends Biotechnol.* 10, 96-102, 1992; McClelland, M. and Welsh, J., DNA fingerprinting by arbitrarily primed PCR, *PCR Methods Appl.* 4, S59-S65, 1994; Kuff, E.L. and Mietz, J.A., Analysis of DNA restriction enzyme digests by two-dimensional electrophoresis in agaraose gels, *Methods Mol.Biol.* 31, 177-186, 1994; Caetano-Anolles, G., Scanning of nucleic acids by in vitro amplification: new developments and applications, *Nat.Biotechnol.* 14, 1668-1674, 1996.

DNA Footprinting

DNA is incubated with a putative binding protein and then modified with dimethyl sulfate. Methylation of DNA bases occurs at regions not protected by the protein binding. The DNA can be cleaved at guanine residues which are then cleaved by piperidine. Footprinting can also be achieved by the use of DNAse I hydrolysis, reaction with hydroxyl radicals or with metal ion-chelate complexes. With either enzymatic or chemical fragmentation, the DNA is end-labeled with ^{32}P-phosphate to permit identification by autoradiography. This has been used to identify the sites of transcription factor binding to *cis*-regions on DNA. See Guille, M.J. and Kneale, G.G., Methods of the analysis of DNA-protein interactions, *Mol.Biotechnol.* 8, 35-52, 1997; Gao, B. and Kunos, G., DNase I Footprinting analysis of transcription factors recognizing adrenergic receptor gene promoter sequences, *Methods Mol.Biol.* 126, 419-429, 2000; Cappabianca, L, Thomassin, H., Pictet, R. and Grange, T., Genomic footprinting using nucleases, *Methods Mol. Biol.* 199, 427-442, 1999; Angelov, D., Khochbin, S., Dimitrov, S., US laser Footprinting and protein-DNA crosslinking. Application

to chromatin, *Methods Mol.Biol.* 119, 481-495, 1999; Brenowitz, M., Chance, M.R., Dhavan, G., and Takamoto, K., Probing the structural dynamics of nucleic acids by quantitative time-resolved and equilibrium hydroxyl radical "Footprinting". *Curr.Opin. Struct.Biol.* 12, 648-653, 2002; Knight, J.C., Functional implications of genetic variation in non-coding DNA for disease susceptibility and gene regulation, *Clin.Sci.* 104, 493-501, 2003.

DNA Methylation

Modification (methylation) of DNA catalyzed by DNA methyltransferase enzymes. Modification occurs at cytosine and adenosine. In multicellular organisms, methylation appears to be confined to cytosine residues. See van Steensel, B. and Henikoff, S., Epigenomic profiling using microarrays, *Biotechniques* 35, 346-350, 2003; El-Maarri, O., Methods: DNA methylation, *Adv. Exp.Med.Biol.* 544, 197-204, 2003; Gut, I.G., DNA analysis by MALDI-TOF mass spectrometry, *Hum.Mutat.* 23, 437-441, 2004; Kapoor, A., Agius, F., and Zhu, J.K., Preventing transcriptional gene silencing by active DNA demethylation, *FEBS Lett.* 579, 5889-5898, 2005; Klose, R.J. and Bird, A.P., Genomic DNA methylation: the mark and its mediators, *Trends in Biochem.Sci.* 31, 81-97, 2006.

DNAse I Hypersenstivity site

Preferred site(s) of DNA I cleavage; typically at regions where clusters of transcriptional activators bind to DNA and usually reflect a change in chromatin structure. See McGinnis, W., Shermoen, A.W., Heemskerk, J., and Beckendorf, S.K., DNA sequence changes in an upstream DNAse I-hypersensitive region are correlated with reduced gene expression, *Proc.Nat.Acad. Sci.USA* 80, 1063-1067, 1983; Cereghini, S., Saragosti, S., Yaniv, M. and Hamer, D.H., SV40-alpha-globulin hybrid minichromosomes. Differences in DNase I hypersensitivity of promoter and enhancer sequences, *Eur.J.Biochem.* 144, 545-553, 1984; Rothenberg, E.V. and Ward, S.B., A dynamic assembly of diverse transcription factors integrates activation and cell-type information for interleukin 2 gene regulation, *Proc.Nat.Acad.Sci.USA* 93, 9358-9365, 1996; Ishii, H., Sen, R., and Pazin, M.J., Combinatorial control of DNase I-hypersensitive site formation and erasure by immunoglobulin heavy chain enhancer-binding proteins, *J.Biol. Chem.* 279, 7331-7338, 2004; Hermann, B.P. and Heckert, L.L., Silencing of Fshr occurs through a conserved, hypersensitive site in the first intron, *Mol.Endocrinol.* 19, 2112-2131, 2005; Sun, D., Guo, K., Rusche, J.J., and Hurley, L.H., Facilitation of a structural transition in the polypurine/polypyrimidine tract within the proximal promoter region in the human VEGF gene by the presence of potassium and G-quadruplex-interactive agents, *Nucleic Acids Res.* 33, 6070-6080, 2005.

DNAzyme

A DNA molecule that contains a catalytic motif that cleaves bound RNA in a hydrolytic reaction; also known as deoxyribozymes or DNA enzymes. See Joyce, G.F., Directed evolution of nucleic acid enzymes, *Annu.Rev.Biochem.* 73, 791-836, 2004; Achenbach, J.C., Chiuman, W., Cruz, R.P., and Li., Y., DNAzymes: from creation *in vitro* to application *in vivo. Curr.Pharm.Biotechnol.* 5, 321-336, 2004; Sioud, M. and Iversen, P.O., Ribozymes, DNAzymes and small interfering RNAs as therapeutics, *Curr.Drug.Targets* 6, 647-653, 2005; Fiammengo, R. and Jaschke, A., Nucleic acid enzymes, *Curr.Opin.Biotechnol.* 16, 614-621, 2005.

Domain

A contiguous (usually) series of monomer units (amino acids in proteins; nucleic acid bases in nucleic acids; monosaccharide in oligosaccharides/polysaccharides. A domain can be continuous or discontinuous and is identified by a unique function such as catalysis or binding; domains are frequently identified by homology and used to group proteins into families.

Domain Antibodies

Antibodies containing a single antigen-binding domain, most ofter the V_H region or the highly variable regions from the V_H and V_L regions. These antibodies are naturally occurring in camelids (members of the order *Camelidae* which includes llamas and camels). See Dick, H.M., Single domain antibodies, *BMJ* 300, 959, 1990; Riechman, L. and Muyldermans, S., Single domain antibodies: comparison of camel VH and camelized human VH domains, *J.Immunol.Methods* 231, 25-38, 1999; Stockwin, L.H. and Holmes, S., Antibodies as therapeutic agents: vive la renaissance! *Expert Opin.Biol.Ther.* 3, 1133-1152, 2003; Holt, L.J., Herring, C., Jespers, L.S., Woolven, B.P., and Tomlinson, I.M., Domain antibodies: proteins for therapy, *Trends Biotechnol.* 21, 484-490, 2003.

Drosha

A member of the RNAse III family of double-stranded specific endonucleases. Drosha is a member of Class II in which each member contains tandem RNAse III catalytic motifs and one C-terminal dsRNA-binding domain. Class I proteins contain only one RNAse III catalytic domain and a dsRNA binding domain. Class III (see Dicer) contain a PAZ domain, a DUF283 domain, the tandem nuclease domains and a dsRNA-binding domain. See Carmell, M.A., and Hannan, G.J., RNAse III enzymes and their initiation of gene silencing, *Nat.Struct.Mol.Biol.* 11, 214-218, 2004.

Drug

A drug is defined as (1) a substance recognized by an official pharmacopoeia or formulary; (2) a substance intended for use in the diagnosis, cure, mitigation, treatment, or prevention of disease; (3) a substance (other than food) intended to affect the structure or any function of the body; (4) a substance intended for use as a component of a medicine but not a device or a component, part or accessory of a device; (5) biological products are included within this definition and are generally covered by the same laws and regulations, but differences exist regarding their manufacturing processes (chemical processes vs. biological processes).

Drug Master File

Drug Master Files (DMF) contain information on the processes and facilities used in drug or drug component manufacture and storage and are submitted to the FDA for examination and approval.

Drug Product

The final dosage form which contains a drug substance or drug substances as well as inactive materials which are also considered as excipients. The drug product is differentiated from the drug substance but may or may not be the same as the drug substance. See http://www.fda.gov/cder/drugsatfda/glossary.htm; http://www.fda.gov/cder/ondc/Presentations/2002/01-10-19_DIA_JS.pp.

Drug Targeting

The ability to target a compound to a specific organ or cell type within an organism. The compound can be a drug/pharmaceutical or it can be a compound, such as a radioisotope, which can be used as a diagnostic. See Muzykantov, V.R., Biomedical aspects of targeted delivery of drugs to pulmonary endothelium, *Expert Opin.Drug Deliv.* 2, 909-926, 2005; Weissig, V., Targeted drug delivery to mammalian mitochondria in living cells, *Expert Opin. Drug Deliv.* 2, 89-102, 2005; Hilgenbrink, A.R. and Low, P.S., Folate-receptor-mediated drug targeting: from therapeutics to diagnostics, *J.Pharm.Sci.* 94, 2135-2146, 2005.

Dye(s)

A chemical compound with a structure which yields a color (a chromophore) which can be coupled either covalently or noncovalently to a substrate matrix. The ability of the compound to yield color is based on its ability to absorb light in the visible spectrum (400 – 700 nm). Dyes can be classified by various characteristics including mechanism/chemistry (e.g. basic dyes, acid dyes; acid/base indicators/redox dyes), structure (nitroso, acridine dyes, thiazole dyes) and process use (e.g. vat dyes). A dyes is a colorant (a substance which yields color) as are pigments. A dye is chemically different from a pigment which is a particle suspended in a medium as particles in paint. More recently, the term dye has expanded to include fluorescent compounds. See Venkataraman, K., *The Chemistry of Synthetic Dyes*, Academic Press, New York, NY, 1952; Conn, H.J., *Biological Stains: A Handbook on the Nature and Uses of the Dyes Employed in the Biological; Handbook*, Williams & Wilkins, Baltimore, MD, 1961; Gurr, E., *Synthetic Dyes in Biology, Medicine, and Chemistry*, Academic Press, London, UK, 1971; Venkataraman, K., *The Chemistry of Synthetic Dyes*, Academic Press, New York, NY, 1978; Egan, H. and Fishbein, L., *Some Aromatic Amines and Azo Dyes in the General and Industrial Environment*, International Agency for Research on Cancer, Lyon, France, 1981; Clark, G. and Koastan, F.H., *History of Staining*, 3rd edn., Williams & Wilkins, Baltimore, MD, 1983; Zollinger, H. *Color Chemistry. Syntheses, Properties, and Applications of Organic Dyes and Pigments*, 2nd edn., VCH, Weiheim, Germany, 1991; *Physico-Chemical Principles of Color Chemistry*, ed. A.T. Peters and H.W. Freeman, Blackie Academic and Professional, London, UK, 1996; Mason, W.T., *Fluorescent and Luminescent Probes for Biological Activity: A Practical Guide to Technology for Quantitative Real-Time Analysis*, Academic Press, San Diego, CA, 1999; *Conn's Biological Stains. A Handbook of Dyes, Stains, and Fluorochromes for Use in Biology and Medicine*, 10th edn., ed. R.W. Horobin and J.A. Kiernan, Bios, Oxford, UK, 2002; Kasten, F.H., Cytochemical studies with acridine orange and the influence of dye contaminants in the staining of nucleic acids, *Int.Rev.Cytol.* 21, 141-202, 1967; Meyer, M.C. and Guttman, D.E., The binding of drugs by plasma proteins, *J.Pharm.Sci.* 57, 895-918, 1968; Adams, C.W. Lipid histochemistry, *Adv.Lipid Res.* 7, 1-62, 1969; Horobin, R.W., The impurities of biological dyes: their detection, removal, occurrence and histochemical significance—a review, *Histochem.J.* 1, 231-265, 1969; Biswas, B.B., Basu, P.S., and Pai, M.K., Gram staining and its molecular mechanism, *Int.Rev.Cytol.* 29, 1-27, 1970; Anumula, K.R., Advances in fluorescence derivatization methods for high-performance liquid chromatographic analysis of glycoprotein carbohydrates, *Anal.Biochem.* 350, 1-23, 2006; Waggoner, A., Fluorescent labels for proteomics and genomics, *Curr.Opin. Chem.Biol.* 10, 62-66, 2006; Chen, H., Recent advances in azo dye degrading enzyme research, *Curr.Protein Pept.Sci.* 7, 101-111, 2006;

Mondal, K. and Gupta, M.N., The affinity concept in bioseparation: evolving paradigms and expanding range of applications, *Biomol. Eng.* 23, 59-76, 2006.

Ectodomain

The extracellular domain of a transmembrane protein. The proteolysis of the ectodomain regions of specific proteins is described as ectodomain shedding and is catalyzed by ADAM proteases. See Rapraeger, A. and Bernfield, M., Cell surface proteoglycan of mammary epithelial cells. Protease releases a heparan sulfate-rich ectodomain from a putative membrane-anchored domain, *J.Biol. Chem.* 260. 4103-4109, 1985; Johnson, J.D., Wong, M.L., and Rutter, W.J., Properties of the insulin receptor ectodomain, *Proc.Natl. Acad.Sci.USA* 85, 7516-7520, 1988; Schaefer, E.M., Erickson, H.P., Federwisch, M., *et al.*, Structural organization of the human insulin receptor ectodomain, *J.Biol.Chem.* 267. 23393-23402, 1992; Attia, J., Hicks, L., Oikawa, K.. *et al.*, Structural properties of the myelin-associated glycoprotein ectodomain, *J.Neurochem.* 61, 718-726, 1993; Couet, J., Sar. S., Jolviet, A., Shedding of human thyrotropin receptor ectodomain. Involvement of a matrix metalloproteinase, *J.Biol.Chem.* 271, 4545-4552, 1996; Petty, H.R., Kindzelskii, A.L., Adachi, Y.. *et al.*, Ectodomain interactions of leukocyte integrin and pro-inflammatory GPI-linked membrane proteins, *J.Pharm. Biomed.Anal.*15, 1405-1416, 1997; Schlondorff, J. and Blobel, C.P., Metalloprotease-disintegrins: modular proteins capable of promoting cell-cell interactions and triggering signals bv protein-ectodomain shedding, *J.Cell Sci.* 112, 3603-3617,1999; Dello Sbarba, P.. and Rovida, E., Transmodulation of cell surface regulatory molecules via ectodomain shedding, *Biol.Chem*, 383, 69-83, 2002; Arribas, J. and Borroto, A., Protein ectodomain shedding, *Chem.Rev.* 102, 4627-4638, 2002; L-Selectin: mechanisms and physiological significance of ectodomain cleavage,. *J.Cell.Mol.Med.* 9, 255-266, 2005; Higashiyama, S. and Nanba, D., ADAM-mediated ectodomain shedding of HB-EGF in receptor cross-talk, *Biochim.Biophys. Acta* 1751, 110-117, 2005; Garton, K.J., Gough, P.J., and Raines, E.W., Emerging roles for ectodomain shedding in the regulation of inflammatory responses, *J.Leuk.Biol.* 79, 1105-1116, 2006.

Electrode Potential ($E°$)

The potential measured with a electrode in contact with a solution of its ions. Electrode potential values will predict whether a substance will be reduced or oxidized. Values are usually expressed as a reduction potential ($M^{n+} \rightarrow M$). A positive electrode potential would indicate that reduction is spontaneous. A negative potential for this reaction would suggest that the oxidation reaction ($M \rightarrow M^{n+}$) would be spontaneous.

Electronegativity

The tendency of an atom to pull electron towards its in a chemical bond; the difference in electronegativity between atoms in a molecule indicates polarity such that in bromoacetic acetamide permitting attack on a nucleophile such as cysteine in the protein.

Electrophoresis/MS

Proteins are separated by one-dimension or more often two-dimensional gel electrophoresis. The separated proteins are subjected to *in situ* tryptic digestion and the peptides separated by liquid chromatography and identified by mass spectrometry; Nishihara, J.C. and Champion, K.M., Quantitative evaluation of proteins in one- and two-dimensional polyacrylamide gels using a fluorescent stains, *Electrophoresis* 23, 2203-2215, 2002.

ELISA

Enzyme-Linked Immunosorbent Assay – an assay based on the reaction of antibody and antigens. There are direct, indirect, direct sandwich, and indirect sandwich assay. See Maggio, E.T., *Enzyme-Immunoassay*, CRC Press, Boca Raton, Florida, USA, 1980; Kemeny, D.M. and Challacombe, S.J., *ELISA and Other Solid Phase Immunoassays: Theoretical and Practical Aspects*, Wiley, Chichester, UK, 1988; Kemeny, D.M., *A Practical Guide to ELISA*, Pergammon Press, Oxford, UK, 1991; Kerr, M.A. and Thorpe, R., *Immunochemistry LabFax*, Bios Scientific Publishers, Oxford, UK, 1994; Law, B., *Immunoassay: A Practical Guide*, Taylor & Francis, London, UK, 1996; Crowther, J.R., *The ELISA Guidebook*, Humana Press, Totowa, New Jersey, USA, 2001; Burns, R., *Immunochemical Protocols*, Humana Press, Totowa, NJ, USA, 2005.

Elispot

The use of membranes to measure cells secreting a specific product such as an antibody or a cytokine. A membrane (nitrocellulose or PDVF) containing an antibody or other specific binding protein is placed in a microtiter plate. Cells secreting a product, such as a cytokine, are grown in this plate and the secretion of the specific product evaluated in response to stimuli. As product is secreted from an individual cell, it is captured immediately by the antibody or other specific binding protein on the membrane and subsequently detected with a probe. An individual spot then corresponds to the secretion from a single cell. There are a number of instruments designed to measure such spots. See Kalyuzhny, A., *Handbook of ELISPOT: Methods and Protocols*, Totowa, NJ, USA, 2005; Stot, D.I., Immunoblotting, dot-blotting, and ELISPOT assay: methods and applications, in *Immunochemistry*, ed. C.J. van Oss and M.H.V. van Regenmortel, Marcel Dekker, New York, New York, Chapter 35, pp. 925-948, 1994; Arvilommi, H., ELISPOT for detecting antibody-secreting cells in response to infections and vaccination, *APMIS* 104, 401-410, 1996; Stott, D.I., Immunoblotting, dot-blotting, and ELISPOT assays: methods and applications, *J.Immunoassay* 21, 273-296, 2000; Anthony, D.D. and Lehmann, P.V., T-cell epitope mapping using the ELISPOT approach, *Methods* 29, 260-269, 2003; Ghanekar, S.A. and Maecker, H.T., Cytokine flow cytometry: multiparametric approach to immune function analysis, *Cytotherapy* 5, 1-6, 2003; Letsch, A. and Scheibenbogen, C., Quantification and characterization of specific T-cells by antigen-specific cytokine production using ELISPOT assay or intracellular cytokine staining, *Methods* 31, 143-149, 2003; Herandez-Fuentes, M.P., Warrens, A.N., and Lechler, R.I., Immunologic monitoring, *Immunol.Rev.* 196, 247-264, 2003; Kalyuzhny, A.E., Chemistry and biology of the ELISPOT system, *Methods Mol.Biol.* 302, 15-31, 2005; Periwal, S.B., Spagna, K., Shahabi, K.. *et al.*, Statistical evaluation for detection of peptide specific interferon-gamma secreting T-cells induced by HIV vaccine determined by ELISPOT assay, *J.Immunol.Methods* 305, 128-134, 2005.

Embedding

Infiltration of a specimen with a liquid medium (paraffin) that can be solidified/polymerized to form a matrix to support the tissue for subsequent manipulation

Endocrine

Usually in reference to a hormone or other biological effector such as peptide growth factor or cytokine which has a systemic effect.

Endoplasmic reticulum-associated protein degradation (ERAD)

A highly specific pathway for the degradation of misfolded proteins in the endoplasmic reticulum which serves as a control mechanism for protein synthesis: see Yamaski, S., Yagishita, N., Tsuchimochi, K., Nishioka, K., and Nakajima, T., Rheumatoid arthritis as a hyper-endoplasmic reticulum-associated degradation disease, *Arthritis Res.Ther.* 7, 181-186, 2005; Meusser, B., Hirsch, C., Jarosch, E., and Sommer, T., ERAD: the long road to destruction, *Nature Cell Biol.* 7, 766-772, 2005; Werner, E.D., Brodsky, J.L., and McCracken, A.A., Proteasome-dependent endoplasmic reticulum-associated protein degradation: an unconventional route to a familiar fate, *Proc.Natl.Acad.Sci.USA* 93, 13797-13801, 1996.

Endosome

A physically distinct compartment resulting from the process of endocytosis and isolated from the rest of the cell with a permeable membrane. The endosome provides a pathway for transport of ingested materials to the lysosome. There is particular interest in this pathway for the process of antigen presentation. See Stahl, P. and Schwartz, A.L., Receptor-mediated endocytosis, *J.Clin.Invest.* 77, 657-662, 1986; Wagner, H., Heit, A., Schmitz, F., and Bauer, S., Targeting split vaccines to the endosome improves vaccination, *Curr.Opin.Biotechnol.* 15, 538-542, 2004; Boes, M., Cuvillier, A., and Ploegh, H., Membrane specializations and endosome maturation in dendritic cells and B cells, *Trends Cell Biol.* 14, 175-183, 2004; Karlsson, L., DM and DO shape the repertoire of peptide-like-MHC-class-II complexes, *Curr.Opin.Immunol.* 17, 65-70, 2005; Li, P., Gregg, J.L., Wang, N., Zhou, D., O'Donnell, P., Blum, J.S., and Crotzer, V.L., Compartmentalization of class II antigen presentation: contribution of cytoplasmic and endosomal processing, *Immunol.Rev.* 207, 206-217, 2005.

Enhancer Elements

DNA sequences which increase transcription from a linked promoter region independent of operation and position (in contrast to proximal promoter elements). Enhancer elements are located at varying distances upstream and downstream of the linked gene. See Hankinson, O., Role of coactivators in transcriptional activation by the aryl hydrocarbon receptor, *Archs. Biochem.Biophys.* 433, 379-386, 2005; West. A.G. and Fraser, P., Remote control of gene transcription, *Hum.Mol.Genet.* 14, Spec. No. 1, R101-R111, 2005; Sipos, L. and Gyurkovics, H., Long-distance interactions between enhancers and promoters, *FEBS J.* 272, 3253-3259, 2005; Zhao, H. and Dean, A., Organizing the genome: enhancers and insulators, *Biochem.Cell.Biol.* 83, 516-524, 2005.

Enlargosome

An organelle resulting from a calcium-dependent, cholesterol-independent non-secretory event; Perret, E., Lakkaraju, A., Deborde, S., Schreiner, R., and Rodriguez-Boulan, E., *Curr.Opin. Cell Biol.* 17, 423-434, 2005.

Ensembl

A data base (http://www.ensembl.org) maintained by the European Bioinformatics Institute (EMBL). This data base organizes large amounts of biological information around the sequences of large genomes. See Birney, E., Andrews, T.D., Bevan, P., *et al.*, An overview of Ensembl, *Genome Res.* 14, 925-928, 2004; Baxevanis, A.D., Using genomic databases for sequence-based biological discovery, *Mol.Med.* 9, 185-192, 2003; Stabenau, A., McVicker, G., Melsopp, C., Procter, G., Clamp, M., and Birney, E., The Ensembl core software libraries, *Genome Res.* 14, 929-933, 2004; Yanai, I., Korbel, J.O., Boue, S., McWeeney, S.K., Bork, P., and Lercher, M.J., Similar gene expression profiles do not imply similar tissue functions, *Trends Genet.* 22, 132-138, 2006.

Ensemble Theory

A proposition that several discrete compounds (proteins, nucleic, acids, carbohydrate) form a structural whole or functional whole. The term ensemble is frequently used to describe the population of discrete intermediates during the process of protein folding. See Thirumalai, D. and Hyeon, C., RNA and protein folding: common themes and variations, *Biochemistry* 44, 4957-4970, 2005; Dietrich, A., Buschmann, V., Muller, C., and Sauer, M., Fluorescence resonance energy transfer (FRET) and competing processes in donor-acceptor substituted DNA strands: a comparative study of ensemble and single-molecule data, *J.Biotechnol.* 82, 211-231, 2002; Sridevi, K., Lakshmikanth, G.S., Krishnamoorthy, G., and Udgaonkar, J.B., Increasing stability reduces conformational heterogeneity in a protein folding ensemble, *J.Mol.Biol.* 337, 699-711, 2004.

Enthalpy (ΔH°)

This is the energy change or heat of reaction for either synthetic or degradative reaction in the standard state. See standard free energy.

Entropy (s)

A thermodynamic quantity that is a measure of the "disorder" or randomness in a system. For example, a crystal structure going to a liquid is associated with an increase in entropy as, for example, the melting of ice crystals to form water under standard conditions. Entropy increases for a spontaneous process. S° refers to entropy values in standard states of substances.

Eosinophil

"Acid" staining leukocyte; associated with allergic inflammation. See Lee, J.J. and Lee, N.A, Eosinophil degranulation: an evolutionary vestige or a universally destructive effector function, *Clin.Exp.Allergy* 35, 986-994, 2005.

Eph Receptors/Ephrin

Eph receptors are the largest family of receptor tyrosine kinases. The structure of Eph receptors is comprised of an extracellular domain, an intracellular domain which are linked by a transmembrane segment. Ephrin ligands bind to Eph receptors which are classified on the quality of the ephrin ligand; ephrin-A ligands bind to EphA receptors while ephrin-B ligands bind to EphB receptors. Eph receptors and ephrin ligand are integral components of cell surfaces and their interactions mediate growth and development. See Foo, S.S., Turner, C.J., Adams, S., *et al.*, Ephrin-B2 controls cell motility and adhesion during blood-vessel-wall assembly, *Cell* 124, 161-173, 2006; Zhang, J., and Hughes, S., Role of the ephrin and Eph receptor tyrosine kinase families in angiogenesis and development of the cardiovascular system, *J.Pathol.* 208, 453-461, 2006; Haramis, A.P. and Perrakis, A., Selectivity and promiscuity in eph receptors, *Structure* 14, 169-171, 2006; Chrencik, J.E., Brooun, A., Recht, M.I., *et al.*, Structure and thermodynamic characterization of the EphB4/Ephrin-B2 antagonist peptide complex reveals the determinants for receptor stability, *Structure* 14, 321-330, 2006.

Epistasis

Masking of a phenotype caused by mutation of one gene by a mutation in another gene; Epistasis analysis can be used to define order of gene expression in a genetic pathway.

Epitome

All epitopes present in the antigenic universe; Also defined as example, paradigm; a brief presentation or statement in most dictionaries.

Erk ½

p42/44 extracellular signal-regulated kinase, phosphorylated as a result of GPCR activation. A number of GPCR appear to converge at Erk 1/2. See Dhillon, A.S. and Kolch, W., Untying the regulation of the Raf-1 kinase, *Arch.Biochem.Biophys.* 404, 3-9, 2002; Chu, C.T., Levinthal, D.J., Kulich, S.M., *et al.*, Oxidative neuronal injury. The dark side of ERK1/2, *Eur.J.Biochem.* 271, 2060-2066, 2004; Clark, M.J. and Traynor, J.R., Assays for G-protein-coupled receptor signaling using RGS-insensitive Galpha subunits, *Methods Enzymol.* 389, 155-169, 2004; Clark, A. and Sugden, P.M., Signaling through the extracellular signal-regulated kinase 1/2 cascade in cardiac myocytes, *Biochem.Cell Biol.* 82, 603-609, 2004.

Essential Oils

A heterogeneous mixture of lipophilic substances obtained from a plant. Also referred to as absolute oils. Originally referred to as the steam distillate of the rinds of certain citrus fruits but extends for more recently used materials such as tea tree oil which is suggested to have some pharmacological use. These products are also used in aromatherapy. See Halcon, L. and Milkus, K., *Staphyloccus aureus* and wounds: a review of tea tree oil as a promising antimicrobial, *Amer.J.Infect.Control* 32, 402-408, 2004; Kalemba, D. and Kunicka, A., Antibacterial and antifungal properties of essential oils, *Curr.Med.Chem.* 10, 813-829, 2003; Ranganna, S., Govindarajan, V.S., and Ramana, K.V., Citrus fruits—varieties, chemistry, technology, and quality evaluation. A Chemistry, *Crit.Rev.Food Sci.Nutri.* 18, 313-386, 1983.

EUROFAN

EUROFAN (European Functional Analysis Netwood) was established to elucidate the physiological and biochemical functions of open reading frames in yeast; http://mips.gsf.de/proj/euro-fan/ See Sanchez, J.C., Golaz, O., Frutiger, S., *et al.*, The yeast SWISS-2DPAGE database, *Electrophoresis* 17, 556-565, 1996; Dujon, B., European Functional Analysis Netword (EUROFAN) and the functional analysis of the *Saccharomyces cerevisiae*

genome, *Electrophoresis* 19, 617-624, 1998; Bianchi, M.M., Ngo, S., Vandenbol, M., *et al.*, Large-scale phenotypic analysis reveals identical contributions to cell functions of known and unknown yeast genes, *Yeast* 18, 1397-1412, 2001; Avaro, S., Belgareh, N., Sibella-Arguelles, C., *et al.*, Mutants defective in secretory/vacuolar pathways in the EUROFAN collection of yeast disruptants, *Yeast* 19, 351-371, 2002; Castrillo, J.I., Hayes, A., Mohammed, S., Gaskell, S.J., and Oliver, S.G., An optimized protocol for metabolome analysis in yeast using direct infusion electrospray mass spectrometry, *Phytochemistry* 62, 929-937. 2003; Davydenko, S.G., Juselius, J.K., Munder, T., Bogengruber, E., Jantti, J., and Keranen, S., Screening for novel essential genes of *Saccharomyces cerevisiae* involved in protein secretion, *Yeast* 21, 463-471, 2004.

Eutectic

A mixture of components in such proportions that said mixture melts and solidifies as a single temperature lower than the melting points of the constituents or any other mixture thereof; a minimum transformation temperature between a solid solution and a mechanical mixture. This is an issue with cryobiology and therapeutic protein processing processes such as lyophilization. See Gutierrez-Merino, C., Quantitation of the Forster energy transfer for two-dimensional systems. II. Protein distribution and aggregation state in biological membranes, *Biophys. Chem.* 14, 259-266, 1981; Gatlin, L.A. and Nail, S.L., Protein purification process engineering. Freeze drying: a practical overview, *Bioprocess Technol.* 18, 317-367, 1994; Nail, S.L., Jiang, S., Chongprasert, S., and Knopp, S.A., Fundamentals of freeze-drying, *Pharm. Biotechnol.* 14, 281-360, 2002; Han, B., and Bischof, J.C., Thermodynamic nonequilibrium phase change behavior and thermal properties of biological solutions for cryobiology applications, *J.Biomech.Eng.* 126, 196-203, 2004.

Exosome

A precise definition is a work in program but an exosome can be considered to be an intracellular membrane vesicle derived from fusion of endosomes with the plasma membrane. It is suggested that exosomes are involved in the intracellular transfer of molecules. See Févier, B. and Raposo, G., Exosomes: endosomal-derived vesicles shipping extracellular messages, *Curr.Opin.Cell Biol.* 16, 415-421, 2004; de Gassart, A., Géminard, C., Hoekstra, D., and Vidal, M., Exosome secretion: the art of reutilizing nonrecycled proteins? *Traffic* 5, 896-903, 2004; Chaput, N., Taïeb, Schartz, N., Flament, C., Novault, S., André, F., and Zitvogel, L. The potential of exosomes in immunotherapy of cancer, *Blood Cells, Molecules, and Diseases* 35, 111-115, 2005; Seaman, M.N.J., Recycle your receptors with retromer, *Trends Cell Biol.* 15, 68-75. 2005; Lencer, W.I. and Blumberg, R.S., A passionate kiss, then run: exocytosis and recycling of IgG by FcRn, *Trends Cell Biol.* 15, 5-9, 2005.

Exotoxicogenomics

Study of the expression of genes important in the adaptive responses important in adaptive responses to toxic exposures.

Expansins

Family of plant proteins essential for acid-induced cell wall loosening. See Cosgrove, D.J., Relaxation in a high-stress environment: The molecular basis of extensible cell walls and cell enlargement, *Plant Cell* 9, 1031-1041, 1997.

Expressed Sequence Tag

Usually an incomplete DAN sequence which can be "read" from either end of a gene fragment which is used as a "marker" or a "window" of gene presence in a genome; a short strand of DNA (approximately 200 base pairs long) which is usually unique to a specific cDNA and therefore can be used to identify genes and map their position in a genome. See Hartl, D.L., EST!EST!!EST!!!, *Bioessays* 18, 1021-1023, 1996; Gerhold, D. and Caskey, C.T., It's the genes! EST access to human genome content, *Bioessays* 18, 973-981, 1996; Wilcox, A.S., Khan, A., Hopkins, J.A., and Sikela J.M., Use of 3' untranslated sequences of human cDNA for rapid chromosome assignment and conversion to STS's: implications for an expression map of the genome, *Nucl.Acid Res.* 19, 1837-1842, 1991; Hoffman, M., Gene expression patterns in human and mouse B cell development, Curr.Top.Microbiol.Immunol. 294, 19-29, 2005; Robson, P., The maturing of the human embryonic stem cell transcriptome profile, *Trends Biotech.* 22, 609-612, 2004.

Expression Profiling

The measurement or determination of DNA expression by the measurement RNA (transcriptomics); also used to refer to protein expression as determined by proteomic technology.

Expressional Leakage

A concept where the functionally important expression of one gene can result in the ectopic expression of a neighboring gene resulting in apparent expression similarity between tissues. See de Marco, A. and de Marco, V., Bacteria co-transformed with recombinant proteins and chaperones cloned in independent plasmids are suitable for expression tuning, *J.Biotechnol.* 109, 45-52, 2004; Yanai, I., Korbel, J.O., Boue, S., McWeeney, S.K., Bork, P., and Lercher, M.J., Similar gene expression profiles do not imply similar tissue functions, *Trends Genet.* 22, 132-138, 2006.

Families of Structurally Similar Proteins (FSSP)

A database based on three-dimensional comparisons of protein structures - http://ekhidna.biocenter.helsinki.fi/dali/start See Holm, L., Ouzounis, C., Sander, C., Tuparev, G., and Vriend, G., A database of protein structure families with common folding motifs, *Protein Sci.* 1, 1691-1698, 1992; Holm, L. and Sander, C., The FSSP database: fold classification based on structure alignment of proteins, *Nucleic Acids Res.* 24, 206-209, 1996; Notredame, C., Holm, L., and Higgins, D.G., COFFEE: an objective function for multiple sequence alignments, *Bioinformatics* 14, 407-422, 1998. Hadley, C. and Jones, D.T., A systematic comparison of protein structure classifications: SCOP, CATH and FSSP, *Structure* 7, 1099-1112, 1999; Getz, G., Vendruscolo, M., Sachs, D. and Domany, E., Automated assignment of SCOP and CATH protein structure classifications from FSSP scores, *Proteins* 46, 405-415, 2002; Edgar, R.C. and Sjolander, K., A comparison of scoring functions for protein sequence profile alignment, *Bioinformatics* 20, 1301-1308, 2004; Edgar, R.C. and Sjolander, K., COACH: profile-profile alignment of protein families using hidden Marikov models, *Bioinformatics* 20, 1309-1318, 2004.

Fenton Reaction

Ferrous ion-dependent formation of hydroxyl radical from hydrogen peroxide; can be coupled with the oxidation of hydroxyl function to ketone/aldehydes; see Fenton, H.J.H., Oxidation of certain organic acids in the presence of ferrous salts, *Proc.Chem. Soc.* 15, 224-228, 1899; Goldstein, S., Meyerstein, D., and Czapski, G., The Fenton reagents, *Free Rad.Biol.Med.* 15, 435-445, 1993; Odyuo, M.M. and Sharan, R.N., Differential DNA strand breaking abilities of ·OH and ROS generating radiomimetic chemicals and γ-rays: study of plasmid dNA, pMTa4, *in vitro, Free Rad.Res.* 39, 499-505, 2005. Stadtman, E.R., Role of oxidized amino acids in protein breakdown and stability, *Meth.Enzymol.* 258, 379-393, 1995.

Fermentation

The controlled aerobic or anaerobic process where a product is produced by yeast, molds, or bacteria from a substrate. Historically, fermentation was used to describe the action of a leavan (yeast) on a carbohydrate (saccharine) as in the production of beers and wines or a dough such as in making bread. In biotechnology manufacturing, the term fermentation is used to described the product of a biopharmaceutical by yeast or bacteria while the term cell culture is used to describe the use of animal cells or plants cells. See Wiseman, A., *Principles of Biotechnology*, Chapman and Hall, New York, NY, 1983; Sinclair, C.G., Kristiansen, B., and Bu'Lock, L.D., *Fermentation Kinetics and Modelling*, Open University Press, New York, NY, 1987; *The Encyclopedia of Bioprocess Technology*, ed. M.C.Flickinger and S.W. Drew, Wiley, New York, NY, 1999; *Molecular Biology and Biotechnology*, ed. J.M. Walker and R. Rapley, Royal Society of Chemistry, Cambridge, UK, 2000; *Fermentation Biotechnology*, ed. S.C. Badal, American Chemical Society, Washington, DC, 2003.

Ferredoxin

A small protein which functions in the transport of electrons (reducing potential) in a variety of organisms. There are several classes of ferredoxins based on the nature of the chemistry of iron binding; Fe_2S_2; Fe_3S_4, Fe_4S_4. The iron is bound to cysteine residues in a cluster which also contains inorganic sulfur. See Mortenson, L.E., Nitrogen fixation: role of ferredoxin n anaerobic metabolism, *Annu.Rev.Microbiol.* 17, 115-138, 1963; Knaff, D.B., and Hirasawa, M., Ferredoxin-dependent chloroplast enzymes, *Biochim.Biophys.Acta* 1056, 93-125; Dai, S., Schwendtmayer, C., Johansson, K. *et al.*, How does light regulate chloroplast enzymes? Structure-function studies of the ferredoxin/thioredoxin system, *Q.Rev.Biophys.* 33, 67-108, 2000; Schurmann, P., Redox signaling in the chloroplast: the ferredoxin/thioredoxin system, *Antioxid. Redox.Signal.* 5, 69-78, 2003; Carrillo, N., and Ceccarelli, E.A., Open questions in ferredoxin-NADP+ reductase catalytic mechanism, *Eur.J.Biochem.* 270, 1900-1915, 2003; Karplus, P.A. and Faber, H.R., Structural aspects of plant ferredoxin: NADP+ oxidoreductases, *Photosynth.Res.* 81, 303-315, 2004; Glastas, P., Pinotsis, N., Efthymiou, G., *et al.*, The structure of the 2[4Fe-4S] ferredoxin from *Pseudomonas aeruginosa* at 1.32-A resolution: comparison with other high-resolution structures of ferredoxins and contributing structural features to reduction potential values, *J.Biol.Inorg.Chem.* 11, 445-458, 2006; Eckardt, N.A., Ferredoxin-thioredoxin system plays a key role in plant response to oxidative stress, *Plant Cell* 18, 1782, 2006.

Ferret Diameter

The longest chord of the project of a regular or irregular object as specific angles maximum, minimum, and average Ferret diameters can be determined by successive measurements. A value used in particle characterization; see M. Levin, Particle characterization – Tools and Methods, *Laboratory Equipment*, November, 2005.

Fibrillation

The process of forming fibers from small, soluble polymeric materials. It is observed with amyloid fibrils in Alzheimer's disease and with proteins during pharmaceutical processing. The term fibrillation was used in the 19th century to describe the physical changes in blood before the elucidation of fibrinogen clotting. The term fibrillation is also used to describe physical changes in structural materials with ligaments and tendons. See Arvinte, T., Cudd, A., and Drake, A.F., The structure and mechanism of formation of calcitonin fibrils, *J.Biol.Chem.* 268, 6415-6422, 1993; Ghosh, P. and Smith, M., The role of cartilage-derived antigens, pro-coagulant activity and fibrinolysis in the pathogenesis of osteoarthritis, *Med.Hypotheses* 41, 190-194, 1993; Bronfman, F.C., Garrido, J., Alvarez, A., Morgan, C. and Inestrosa, N.C., Laminin inhibits amyloid-beta-peptide fibrillation, *Neurosci.Lett.* 218, 201-203, 1996; Martin, J.A. and Buckwalter, J.A., Roles of articular cartilage aging and chondrocytes senesence in the pathogenesis of osteoarthritis, *Iowa Orthop.J.* 21, 1-7, 2001; Seyferth, S. and Lee, G., Structural studies of EDTA-induced fibrillation of salmon calcitonin, *Pharm.Res.* 20, 73-80, 2003; Librizzi, F. and Rischel, C., The kinetic behavior of insulin fibrillation is determined by heterogeneous nucleation pathways, *Protein Sci.* 14, 3129-3134, 2005; Westermark, P., Aspects on human amyloid forms and their fibril polypeptides, *FEBS J.* 272, 5942-5949, 2005; Pedersen, J.S., Dikov, D., Flink, J.L., *et al.*, The changing face of glucagon fibrillation: structural polymorphism and conformational imprinting, *J.Mol. Biol.* 355, 501-523, 2006.

Fibroblast Growth Factor

A group of peptide growth factors which regulate cell growth and proliferation.

Fibroblast Growth Factor Receptor(s)

Receptor kinases which are activated by dimerization after ligand binding; Include FGFR1, FGFR2, FGFR3, FGFR4, FGFR5.

FixJ-FixL

A two-component transcription regulatory system which is a global regulator of nitrogen fixation in *Rhizobium meliloti, Mol. Microbiol.* 5, 987-997, 1991. Sousa, F.H.S., Gonzalez, G. and Gilles-Gonazalez, M.-A., Oxygen blocks the reaction of FixL-FixJ complex with ATP but does not influence binding of FixJ or ATP to FixL, *Biochemistry* 44, 15359-15365, 2005.

FLAG™

FLAG™ has the sequence of AspTyrLysAspAspAsp-AspLys which includes an enterokinase cleavage site. This epitope tag can be used as a fusion partner for the expression and purification of recombinant proteins. See Einhauer, A. and Jungbauer, A.,

The FLAG™ peptide, a versatile fusion tag for the purification of recombinant proteins, *J.Biochem.Biophys.Methods* 49, 455-465, 2001; Terpe, K., Overview of tag protein fusions: from molecular and biochemical fundamentals to commercial systems, *Appl. Microbiol.Biotechnol.* 60, 523-533, 2003; Lichty, J.J., Malecki, J.L., Agnew, H.D., Michelson-Horowitz, D.J., and Tan, S., Comparison of affinity tags for protein purification, *Protein Exp.Purif.* 41, 98-105, 2005.

FLAP-Endonuclease (FEN; FEN-1)

An enzyme with endonuclease and exonuclease activity encoded by the *FEN1* gene. Cleaves branched DNA strucures including the 5'-end of Okazaki fragments. See Kunkel, T.A., Resnick, M.A., and Gordenin, D.A., Mutator specificity and disease: looking over the FENce, *Cell* 88, 155-158, 1997; Shen, B., Qiu, J., Hosfield, D., and Tainer, J.A., Flap endonuclease homologs in archaebacteria exist as independent proteins, *Trends Biochem.Sci.* 23, 171-173, 1998; Henneke, G., Freidrich-Heineken, E., and Hubscher, U., Flap endonuclease 1: a novel tumour suppressor protein, *Trends Biochem.Sci.* 28, 384-390, 2003; Kao, H.I. and Bambara, R.A., The protein components and mechanism of eukaryotic Okazaki fragment maturation, *Crit.Rev.Biochem.Mol. Biol.* 38, 433-452, 2003; Garg, P. and Burgers, P.M., DNA polymerases that propagate the eukaryotic DNA replication fork, *Crit.Rev.Biochem.Mol.Biol.* 40, 115-128, 2005; Olivier, M., The invader assay for SNF genotyping, *Mutat.Res.* 573, 103-110, 2005; Shen, B., Singh, P., Liu, R., *et al.*, Multiple but dissectible functions of FEN-1 nucleases in nucleic acid processing, genome stability and diseases, *Bioessays* 27, 717-729, 2005.

Flux

Flux is the continuous flow of a substance. Flux can occur with electrons (Gutman, M., Electron flux through the mitochondrial ubiquinone, *Biochim.Biophys.Acta* 594, 53-84, 1980) and protons (Wang, J.H., Coupling of proton flux to the hydrolysis and synthesis of ATP, *Annu.Rev.Biophys.Bioeng.* 12, 21-34, 1983) as well and ions and other substances. Schwartz, A., Cell membrane Na^+, K^+-ATPase and sarcoplasmic reticulum: possible regulators of intracellular ion activity, *Fed.Proc.* 35, 1279-1282, 1976; Meissner, G., Monovalent ion and calcium ion fluxes in sarcoplasmic reticulum, *Mol.Cell.Biochem.* 55, 65-82, 1983; Jones, D.P., Intracellular diffusion gradients of O_2 and ATP, *Am.J.Physiol.* 250, C663-675, 1986; Hunter, M., Kawahara, K., and Giebisch, G., Calcium-activated epithelial potassium channels, *Miner.Electrolyte Metab.* 14, 48-57, 1988; , *Cation Flux across Biomembranes*, ed. Y. Mukohata, and L. Packer, Academic Press, New York, NY, 1979; Weir, E.K. and Hume, J.R., *Ion Flux in Pulmonary Vascular Control*, Plenum Press, New York, NY, 1993. Flux is defined in several ways: unidirectional influx is defined as the molar quantity of a solute passing across 1 cm^2 membrane in a unit period of time; unidirectional efflux is defined as the molar quantity of a solute crossing1 cm^2 membrane outward from a cell in a unit period of time. Net flux is the difference between unidirectional influx and unidirectional efflux in a unit period of time. Understanding net flux is of importance in the design and interpretation of microdialysis studies (Schuck, V.J., Rinas, I., and Derendorf, H., In vitro microdialysis sample of docetaxel, *J.Pharm.Biomed.Anal.* 36, 807-813, 2004; Cano-Cebrian, M.J., Zornoza, T., Polache, A., and Granero, L., Quantitative in vivo microdialysis in pharmacokinetic studies: some reminders,

Curr.Drug.Metab. 6, 83-90, 2005; Abrahamsson, P. and Winso, O., An assessment of calibration and performance of the microdialysis system, *J.Pharm.Biomed.Anal.* 39, 730-734, 2005.

Focal Adhesion

A membrane area for cellular adhesion via actin filaments to the extracellular matrix/fibronectin resulting from the clustering of integrins. The interaction with fibronectin results in the formation of fibrillar adhesions considered to be more mature structures. Other intracellular proteins such as vincullin and focal adhesion kinase (FAK) are recruited to the actin cytoskeleton structure. See Otey, C.A. and Burridge, K., Patterning of the membrane cytoskeleton by the extracellular matrix, *Semin. Cell Biol.* 1, 391-399, 1990; Arikama, S.K., Integrins in cell adhesion and signaling, *Hum.Cell* 9, 181-186, 1996; Bershadsky, A.D. Balaban, N.Q., and Geiger, B., Adhesion-dependent cell mechanosensitivity, *Annu.Rev.Cell Dev.Biol.* 19, 677-695, 2003; Wozniak, M.A., Modzelekska, K., Kwong, L., and Keeley, P.J., Focal adhesion regulation of cell behavior, *Biochim.Biophys.Acta* 1692, 103-119, 2004; Small, J.V. and Resch, G.P., The comings and goings of actin: coupling protrusion and retraction in cell motility, *Curr. Opin.Cell Biol.* 17, 517-523, 2005; Wu, M.H., Endothelial focal adhesions and barrier function, *J.Physiol.* 569, 359-366, 2005; Cohen, L.A. and Guan, J.L., Mechanisms of focal adhesion kinase regulation, *Curr. Cancer Drug Targets* 5, 629-643, 2005; Romer, L.H., Birukov, K.G., and Garcia, J.G., Focal adhesions: paradigm for a signaling nexus, *Circ.Res.* 98, 606-616, 2006; Legate, K.R., Montañez, O. and Fässler, R., ILK, PINCH and parvin: the tIPP of integrin signalling, *Nat.Rev.Mol.Cell.Biol.* 7, 20-31, 2006.

Fok1 Restriction endonuclease

A type II restriction endonuclease isolated from *Flavobacterium okeanokoites* which has been used to identify DNA polymorphisms. There has been extensive use in the study of the vitamin D receptor gene (VDR gene). See Sugisaki, H. and Kanazawa, S., New restriction endonucleases from Flavobacterium okeanokoites (FokI) and Micrococcus luteus (MluI), *Gene* 16, 73-78, 1981; Kato, A., Yakura, K., and Tanifuji, S., Sequence analysis of Vicia faba repeated DNA, the FokI repet element, *Nucleic Acids Res.* 24, 6415-6426, 1984; Kita, K., Kotani, H., Sugisaki, H., and Tanami, M., The foci restriction-modification system. I. Organization and nucleotide sequences of the restriction and modification genes, *J.Biol.Chem.* 264, 5751-5756, 1989; Posfai, G. and Szybalski, W., A simple method for locating methylated based in DNA using class-IIS restriction enzymes, *Gene* 74, 179-181, 1988; Aggarwal, A.K. and Wah, D.A., Novel site-specific DNA endonucleases, *Curr.Opin.Struct.Biol.* 8, 19-25, 1998; Kovall, R.A. and Matthews, B.W., Type II restriction endonucleases: structural, functional and evolutionary relationships, *Curr.Opin.Chem.Biol.* 3, 578-583, 1999; Akar, A., Orkunoglu, F.E., Ozata, M., Sengul, A., and Gur, A.R., Lack of association between vitamin D receptor FokI polymorphism and alopecia areata, *Eur.J.Dermatol.* 14, 156-158, 2004; Guy, M., Lowe, L.C., Bretherton-Watt, D., *et al.*, Vitamin D receptor gene polymorphisms and breast cancer risk, *Clin.Cancer Res.* 10, 5472-5481, 2004; Claassen, M., Nouwen, J., Fang, Y., *et al.*, Staphylococcus aureus nasal carriage is not associated with known polymorphism in the vitamin D receptor gene, *FEMS Immunol.Med.Microbiol.* 43, 173-176, 2005; Bolu, S.E., Orkunoglu Suer, F.E., Deniz, F., *et al.*, The vitamin D receptor foci start codon polymorphism and bone mineral density in male hypogonadotrophic hypogonadism, *J.Endocrinol.Invest.* 28, 810-814, 2005.

Foldamers

Single chain polymers which can adopt secondary structure in solution and thus mimic proteins, nucleic acids, and polysaccharides; polymeric backbones which have well-defined and predictable folding properties in the solvent of choice. See Appella, D.H., Christianson, L.A., Klein, D.A., *et al.*, Residue-based control of helix shape in beta-peptide oligomers, *Nature* 387, 381-384, 1997; Tanatani, A., Mio, M.J., and Moore, J.S., Chain length-dependent affinity of helical foldamers for a rod-like guest, *J.Amer.Chem.Soc.* 123, 1792-1793, 2001; Cubberley, M.S. and Iverson, B.L., Models of higher-order structure: foldamers and beyond, *Curr.Opin.Chem.Biol.* 5, 650-653, 2001; Hill, D.J., Mio, M.J., Prince, R.B., Hughes, T.S. and Moore, J.S., A field guide to foldamers, *Chem.Rev.* 101, 393-4012, 2001; Martinek, T.A. and Fulop, F., Side-chain control of beta-peptide secondary structures, *Eur.J.Biochem.* 270, 3657-3666, 2003; Sanford, A.R., Yamato, K., Yang, X., Yuan, L., Han. Y., and Gong, B., Well-defined secondary structures, *Eur.J.Biochem.* 271, 1416-1425, 2004; Cheng, R.P., Beyond de novo protein design—de novo design of non-natural folded oligomers, *Curr.Opin.Struct. Biol.* 14, 512-520, 2004; Stone, M.T., Heemstra, J.M., and Moore, J.S., The chain-length dependence test, *Acc.Chem.Res.* 39, 11-20, 2006; Schmitt, M.A., Choi, S.H., Guzei, I.A., and Gellman, S.H., New helical foldamers: heterogeneous backbones with 1:2 and 2:1 alpha:beta-amino acid residue patterns, *J.Am.Chem.Soc.* 128, 4538-4539, 2006.

Fragnomics

The use of smaller molecules in the drug discovery process; Zartler, E.R. and Shapiro, M.J., Fragnomics: fragment-based drug discovery, *Curr.Opin.Chem.Biol.* 9, 366-370, 2005.

Frass

Frass is debris or excrement produced by insects. This material is thought be involved with role of cockroaches in the development of asthma (See Page, K., Hughes, V.S., Bennett, G.W., and Wong, H.R., German cockroach proteases regulate matrix metalloproteinase-9 in human bronchial epithelial cells, *Allergy* 61, 988-995, 2006.

Free Radical/radical

A molecule containing an unpaired electron can be electrically neutral. Free radicals may be created by the hemolytic cleavage of a precursor molecule free radicals can be formed by thermolytic cleavage, photolysis (ultraviolet light photolysis of hydrogen peroxide to form hydroxyl radical), radiolysis (ionizing radiation of water to form hydroxyl radical) or by homolytic cleavage with the participation of another molecule (i.e. Fenton Reaction). Perkins, J., *Radical Chemistry: The Fundamentals*, Oxford University Press, Oxford, United Kingdom, 2000.

FRET (Fluorescence Resonance Energy Transfer)

A technique for assaying the proximity of region by observed energy transfer between fluorophores. A concept similar to fluorescence quenching. With two-photon excitation, studies can be extended to the study of *in vivo* interactions with microscopy. See also BRET. See Milligan, G. and Bouvier, M., Methods to monitor the quaternary structure of G protein-coupled receptors,

FASEB J. 272, 2914-2925, 2005; Rasnik, I., McKinney, S.A., and Ha, T., Surfaces and orientations: much to FRET about? *Acc. Chem.Res.* 38, 542-548, 2005; Gertler, A., Biener, E., Ramamujan, K.V., Dijiane, J., and Herman, B., Fluorescence resonance energy transfer (FRET) microscopy in living cells as a novel tool for the study of cytokine action, *J.Dairy Res.* 72, Spec. No., 14-19, 2005; Cudakov, D.M., Lukyanov, S. and Lukyanov, K.A., Fluorescent proteins as a toolkit for in vivo imaging, *Trends Biotechnol.* 23, 605-613, 2005; Zal, T. and Gascoigne, N.R., Using live FRET imaging to reveal early protein-protein interactions during T cell activation, *Curr.Opin.Immunol.* 16, 674-683, 2004.

Freund's Adjuvant

A mixture of killed/lyophilized *Mycobacterium bovis* or *Mycobacterium tuberculosis* cells and oil resulting in an emulsion (referred to as Complete Freund's adjuvant) used with an antigen to improve the immune response (antibody formation secondary to B-cell activation). Incomplete Freund's adjuvant does not contain the bacterial cells and is used to avoid an inflammatory response. See White, R.G., Factor affecting the antibody response, *Br.Med.Bull.* 19, 207-213, 1963; White, R.G., Antigen adjuvants, *Mod.Trends Immunol.* 2, 28-52, 1967; Myrvik, Q.N., Adjuvants, *Ann.N.Y.Acad.Sci.* 221, 324-330, 1974; Osebold, J.W., Mechanisms for action by immunologic adjuvants, *J.Am.Vet. Med.Assoc.* 181, 983-987, 1982; Warren, H.S., Vogel, F.R., and Chedid, L.A., Current status of immunological adjuvants, *Annu. Rev.Immunol.* 4, 369-388, 1986; Claassen, E., de Leeuw, W., de Greeve, P., Hendriksen, C., and Boersma, W., Freund's complete adjuvant: an effective but disagreeable formula, *Res.Immunol.* 143, 478-483, 1992; Billiau, A. and Matthys, P., Modes of action of Freund's adjuvants in experimental models of autoimmune diseases, *J.Leukoc.Biol.* 70, 849-860, 2001; Cachia, P.J., Kao, D.J., and Hodges, R.S., Synthetic peptide vaccine development: measurement of polyclonal antibody affinity and cross-reactivity using a new peptide capture and release system for surface plasmon resonance spectroscopy, *J.Mol.Recog.* 17, 540-557, 2004; Stills, H.F., Jr., Adjuvants and antibody production: dispelling the myths associated with Freund's complete and other adjuvants, *ILAR J.* 46, 280-293, 2005; Miller, L.H., Saul, A., and Mahanty, S., Revisiting Freund's incomplete adjuvant for vaccines in the developing world, *Trends Paristol.* 21, 412-414, 2005.

Functional Genomics

Functional genomics refers to establishing a verifiable link between gene expression and cell/organ/tissue function/dysfunction. Functional genomics refers to establishing a verifiable link between gene expression and cell/organ/tissue function/dysfunction. See Evans, M.J., Carlton, M.B., and Russ, A.P., Gene trapping and functional genomics, *Trends Genet.* 13, 370-374, 1997; Schena, M., Heller, R.A., Theriault, T.P., *et al.*, Microarrays: biotechnology's discovery platform for functional genomics, *Trends Biotechnol.* 16, 301-306, 1998; Holtorf, H., Guitton, M.C., and Reski, R., Plant functional genomics, *Naturewissenschaften* 89, 235-249, 2002; Bader, G.D., Heilbut, A., Andrews, B., *et al.*, Functional genomics and proteomics: charting a multidimensional map of the yeast cell, *Trends Cell Sci.Biol.* 13, 344-356, 2003; Kemmeren, P. and Holstege, F.C., Integrating functional genomics data, *Biochem.Soc.Trans.* 31, 1484-1487, 2003; Werner, T., Proteomics and regulomics: the yin and yang of functional genomics, *Mass Spectrom.Rev.* 23, 25-33, 2004; Brunner, A.M., Busov, V.B., and Strauss, S.H., Poplar genome sequence: functional

genomics in an ecologically dominant plant species, *Trends Plant Sci.* 9, 49-56, 2004; Hughes, T.R., Robinson, M.D., Mitsakakis, N., and Johnston, M., The promise of functional genomics: completing the encyclopedia of a cell, *Curr.Opin.Microbiol.* 7, 546-554, 2004; Kramer, R. and Cohen, D., Functional genomics to new drug targets, *Nat.Rev.Drug Discov.* 3, 965-972, 2004; Vanhecke, D. and Janitz, M., Functional genomics using high-throughput RNA interference, *Drug Discov.Today* 10, 205-212, 2005; Sauer, S., Lange, B.M., Gobom, J., *et al.*, Miniaturization in functional genomics and proteomics, *Nat.Rev.Genet.* 6, 465-476, 2005; Stoeckert, C.J., Jr., Functional genomic databases on the web, *Cell Microbiol.* 7, 1053-1059, 2005; Foti, M., Grannuci, F., Pelizzola, M., *et al.*, Dendritic cells in pathogen recognition and induction of immune response: a functional genomics approach, *J.Leukoc. Biol.* 79, 913-916, 2006; Hunt, S.P. and Livesey, R., *Functional Genomics: A Practical Approach*, Oxford University Press, Oxford, UK, 2000; *Functional Genomics Methods and Protocols*, ed. M.J.Brownstein and A.B. Khodursky, Humana Press, Totowa, NJ, 2003; Grotewold, E., *Plant Functional Genomics*, Humana Press, Totowa, NJ, 2003; Zhou, J., *Microbial Functional Genomics*, Wiley-Liss, Hoboken, NJ, 2004.

Functional Proteomics

A broad area of enquiry encompassing the study of the function of proteins in the proteome, study of changes in protein expression within the proteome, and the use of reactive chemical probes to identify enzymes in the proteome. This short list is not meant to be wholly inclusive. See Lawrence, D.S., Functional proteomics: large-scale analysis of protein kinase activity, *Genome Biol.* 2, REVIEWS1007, 2001; Famulok, M., Blind, M., and Mayer, G., Intramers as promising new tools in functional proteomics, *Chem.Biol.* 8, 931-939, 2001; Guengerich, F.P., Functional genomics and proteomics applied to the study of nutritional metabolism, *Nutr.Rev.* 59, 259-263, 2001; Strosberg, A.D., Functional proteomics to exploit genome sequences, *Cell.Mol.Biol.* 47, 1295-1299, 2001; Yanagida, M., Functional proteomics: current achievements, *J.Chromatog.B.Anal.Technol.Biomed.Life Sci.* 771, 89-106, 2002; Hunter, T.C., Andon, N.L., Koller, A., Yates, J.R., and Haynes, P.A., *J.Chromatogr.B.Analyt.Technol.Biomed.Life Sci.* 782, 165-181, 2002; Graves, P.R. and Haystead, T.A., A functional proteomics approach to signal transduction, *Recent Prog. Horm.Res.* 58, 1-14, 2003; Ilag, L.L., Functional proteomic screens in therapeutic protein drug discovery, *Curr.Opin.Mol.Ther.* 7, 538-542, 2005; Wagner, V., Gessner, G., and Mittag, M., Functional proteomics: a promising approach to find novel components of the circadian system, *Chronobiol.Int.* 22, 403-415, 2005; Monti, M., Orru, S., Pagnozzi, D., and Pucci, P., Functional proteomics, *Clin.Chim.Acta* 357, 140-150, 2005.

Furin

Furin is a subtilisin-like regulatory protease (subtilisn-like pro-protein convertases located in the trans-Golgi network which functions in processing precursor proteins in the secretory pathway. See Molloy, S.S., Bresnahan, P.A., Leppla, S.H. *et al.*, Human furin is a calcium-dependent serine endoprotease that recognizes the sequence Arg-X-X-Arg and efficiently cleaves anthrax protective antigen, *J.Biol.Chem.* 267, 16396-16402, 1992; Yanagita, M., Hoshimo, H., Nakayama, K., and Takeuchi, T., Processing of mutated proteins with tetrabasic cleavage sites to mature insulin reflects the expression of furin in nonendocrine cell lines, *Endocrinology* 133, 639-644, 1993; Brennan, S.O. and

Nakayama, K., Furin has the proalbumin substrate specificity and serpin inhibitor properties of an in situ convertase, *FEBS Lett.* 338, 147-151, 1994; Roebroek, A.J., Creemers, J.W., Ayoubi, T.A., and Van de Ven, W.J., Furin-mediated proprotein processing activity: involvement of negatively charged amino acid residues in the substrate binding site, *Biochemie* 76, 210-216, 1994; Denault, J.B. and Leduc, R., Furin/PACE/SPC1: a convertase involved in exocytic and endocytic processing of precursor proteins, *FEBS Lett.* 379, 113-116, 1996; Nakayama, K., Furin: a mammalian subtilisin/Kex2p-like endoprotease involved in processing of a wide variety of precursor proteins, *Biochem.J.* 327, 625-635, 1997; Rockwell, N.C., Krysan, D.J., Komiyama, T., and Fuller, B.S., Precursor processing by kex2/furin proteases, *Chem. Rev.* 102, 4525-4548, 2002; Fugere, M., Limperis, P.C., Beaulieu-Audy, V., Gagnon, F., *et al.*, Inhibitory potency and specificity of subtilisin-like pro-protein convertase (SPC) prodomains, *J.Biol.Chem.* 277, 7648-7656, 2002; Rockwell, N.C. and Thorner, J.W., The kindest cuts of all: crystal structures of kex2 and furin reveal secrets of precursor processing, *Trends Biochem.Sci.* 29, 80-87, 2004; The first proprotein processing proteins was described as Kex2 protease(kexin) in *Saccharomyces cerevesiae* (Leibowtiz, M.J. and Wickner, R.B., A chromosomal gene required for killer plasmid expression, mating, and spore maturation in *Saccharomyces cerevisiae*, *Proc.Natl.Acad.Sci.USA* 73, 2061-2065, 1976; Rogers, D.T., Saville, D., and Bussey, H., Saccharomyces cerevisiae expression mutant kex2 has altered secretory proteins and glycoproteins, *Biochem.Biophys.Res. Commun.* 90, 187-193, 1979; Julius, D. Brake, A., Blair, L., *et al.*, Isolation of the putative structural gene for the lysine-arginine-cleaving endopeptidase required for processing of yeast preproalpha-factor, *Cell* 37, 1075-1089, 1984). Furin is important for the secretion of recombinant proteins in mammalian cell lines (Mark, M.R., Lokker, N.A., Zioncheck, T.F., *et al.*, Expression and characterization of heptocyte growth factor receptor-IgG fusion proteins: effects of mutations in the potential proteolytic cleavage site on processing and ligand binding, *J.Biol.Chem.* 267, 26166-26171, 1992; Bristol, J.A., Freedman, S.J., Furie, B.C., and Furie, B., Profactor IX: the propeptide inhibits binding to membrane surfaces and activation by factor IXa., *Biochemistry* 33, 14136-14143, 1994; Groskreutz, D.J., Sliwkowski, M.X., and Gorman, C.M., Genetically engineering proinsulin constitutively processed and secreted as mature, active insulin, *J.Biol. Chem.* 269, 6241-6245, 1994; Lind, P., Larsson, K., Spira,J., *et al.*, Novel forms of B-domain deleted recombinant factor VIII molecules. Construction and biochemical characterization, *Eur.J.Biochem.* 232, 19-27, 1995; Ayoubi, T.A., Meulemans, S.M. Roebroek, A.J., and Van de Ven, W.J., Production of recombinant proteins in Chinese hamster ovary cells overexpressing the subtilisin-like proprotein converting enzyme furin, *Mol.Biol. Rep.* 23, 87-95, 1996; Chiron, M.F., Fryling, C.M., and Fitzgerald, D., Furin-mediated cleavage of Pseudomonas exotoxin-derived chimeric toxins, *J.Biol.Chem.* 272, 31707-31711, 1997. Furins are functionally related to secretases in being protein precursor processing enzymes (Anders, L., Mertins, P., Lammich, S., *et al.*, Furin-, ADAM 10-, and γ-secretase-mediated cleavage of a receptor tyrosine phosphatase and regulation of β-cateinin's transcriptional activity, *Mol.Cell.Biol.* 26, 3917-3924, 2006). There has been some work on the possible role of furin in the processing of β-secretase (Bennett, B.D., Denis, P., Haniu, M., *et al.*, A furin-like convertase mediates propeptide cleavage of BACE, the Alzheimer's β-secretase, *J.Biol.Chem.* 275, 37712-37717, 2000; Creemers, J.W.M., Dominguez, D.I., Plets, E., *et al.*, Processing of β-secretase by furin and others members of the proprotein

convertase family, *J.Biol.Chem.* 276, 4211-4217, 2001; Pinnix, I., Council, J.E., Roseberry, B., *et al.*, Convertases other than fuin cleave β-secretase to its mature form, *FASEB J.* 15, 1810-1812, 2001.

Gamma(γ)-secretase

A membrane-associated regulatory protease responsible for the cleavage of amyloid precursor protein and notch protein. Gamma(γ)-secretase is composed of four subunits, presenilin, nicastrin, Aph-1 and Pen-2. Presenilin is responsible for the catalytic gamma(γ)-secretase activity and nicastrin and Aph-2 have a function in substrate recognition and complex stabilization while Pen-2 assists in catalytic function. Gamma(γ)-secretase is a therapeutic treatment for Alzheimer's disease. See Mundy, D.L., Identification of the multicatalytic enzyme as a possible γ-secretase for the amyloid precursor protein, *Biochem.Biophys.Res. Commun.* 204, 333-341. 1994; Wolfe, M.S. and Haass, C., The role of presenilins in gamma-secretase activity, *J.Biol.Chem.* 276, 5413-5416, 2001; Sisodia, S.S. and St. George-Hyslop, P.H., γ-Secretase, Notch, Aβ and Alzheimer's disease: Where do the presenilins fit in?, *Nat.Rev.Neurosci.* 3, 281-290, 2002; Kimberly, W.T. and Wolfe, M.S., Identify and function of gamma-secretases, *J.Neurosci.Res.* 74, 353-360, 2003; Iwatsubo, T., The gamma secretase complex: machinery for intramembrane proteolysis, *Curr.Opin.Neurobiol.* 14, 379-383, 2004; Raemakers, T., Esselens, C., and Annaert, W., Presenilin 1: more than just gamma-secretase, *Biochem.Soc.Trans.* 33, 559-562, 2005; De Strooper, B., Nicastrin: gatekeeper of the gamma-secretase complex, *Cell* 122, 318-320, 2005; Churcher, I., and Beher, D., Gamma-secretase as a therapeutic target for the treatment of Alzheimer's disease, *Curr.Pharm.Des.* 11, 3363-3382, 2005; Barten, D.M., Meredith, J.E.,Jr., Zaczek, R., *et al.*, Gamma-secretase inhibitors for Alzheimer's disease: balancing efficacy and toxicity, *Drugs R & D* 7, 87-97, 2006. Wolfe, M.S., The γ-secretase complex: membrane-embedded proteolytic ensemble, *Biochemistry* 45, 7931-7939, 2006.

G protein

A heteromeric protein which functions in signal transduction via modulation by G protein-coupled receptors (GPCRs). See Spiegel, A.M. and Downs, R.W., Jr., Guanine nucleotides: key regulators of hormone receptor-adenylate cyclase, *Endocr.Rev.* 2, 275-305, 1981; Cooper, D.M. and Londos, C., GTP-stimulation and inhibition of adenylate cyclase, *Horiz.Biochem.Biophys.* 6, 309-333, 1982; Poste, G., New insights into receptor regulation, *J.Appl. Physiol.* 57, 1297-1305, 1984; Cuatrecasas, P., Hormone receptors, membrane phospholipids, and protein kinases, *Harvey Lect.* 80, 89-128, 1984-1985; Neer, E.J., Guanine nucleotide-binding proteins involved in transmembrane signaling, *Symp.Fundam. Cancer Res.* 39, 123-136, 1986; Spiegel, A.M., Signal transduction by guanine nucleotide binding proteins, *Mol.Cell.Endocrinol.* 49, 1-16, 1987; Bockaert, J., Homburger, V., and Rouot, B., GTP binding proteins: a key role in cellular communication, *Biochimie* 69, 329-338, 1987; Zhang, Z., Melia, T.J., He, F., *et al.*, How a G protein binds a membrane, *J.Biol.Chem.* 279, 33937-33945, 2004; Gavi, S., Shumay, E., Wang, H.Y., and Malbon, C.C., G-protein-coupled receptors and tyrosine kinases: crossroads in cell signaling and regulation, *Trends Endocrinol.Metab.* 17, 48-54, 2006; Sato, M., Blumer, J.B., Simon, V., and Lanier, S.M., Accessory proteins for G proteins: partners in signaling, *Annu.Rev.Pharmcol. Toxicol.* 46, 151-187, 2006; Houslay, M.D. and Milligan, G., *G-Proteins as Mediators of Cellular Signaling Processes*, Wiley, Chichester, UK, 1990; Naccache, P.H., *G Proteins and Calcium*

Signaling, CRC Press, Boca Raton, FL, 1990; Johnson, R.A. and Corbin, J.D., *Adenyl Cyclase, G Proteins, and Guanylyl Cyclase*, Academic Press, San Diego, CA, 1991; Ravi, I., *Heterotrimeric G Proteins*, Academic Press, San Diego, CA, 1994; Watson, S.P., and Arkinstall, S., *The G-Protein Linked Receptor Factbooks*, Academic Press, London, UK, 1994; Siderovski, D.P., *G Proteins and Calcium Signaling*, Elsevier Academic Press, Amsterdam, Netherlands, 2004.

G-Protein Coupled Receptor (GPCR)

A membrane receptor which is functional linked to the activation of a trimeric G protein complex characterized by the presence of seven transmembrane segments.

Gα Protein

The alpha-subunit of the heterotrimeric G protein which separates into a Gα-protein-GTP complex when GTP replaces GDP. See Albert, P.R. and Robillard, L., G protein specificity: traffic direction required, *Cell Signalling* 14, 407-418, 2002; Kurose, H., Gα_{12} and Gα_{13} as key regulatory mediator in signal transduction, *Life Sci.* 74, 155-161, 2003; Kostenis, E., Waelbroeck, M., and Milligan, G., Techniques: promiscuous Gα proteins in basic research and drug discovery, *Trends Pharmacol.Sci.* 26, 595-602, 2005; Herrman, R., Heck, M., Henklein, P., *et al.*, Signal transfere from GPCRs to G proteins: role of the Gα N-terminal region in rhodopsin-transducin coupling, *J.Biol.Chem*, 281, 30234–30241, 2006.

Gelsolin; Gelsolin-like domains

Gelsolin is signature protein for a family of protein which interact with actin and influence the structure of the cytoskeleton. Gelsolin is a calcium-dependent actin-binding protein that modulates actin filament length. See Yin, H.L., Hartwig, J.H., Maruyama, K., and Stossel, T.P., Ca^{2+} control of actin filament length. Effects of macrophage gelsolin on actin polymerization, *J.Biol.Chem.* 256, 9693-9697, 1981; Matasudaira, P., Jakes, R., and Walker, J.E., A gelsolin-like Ca^{2+}-dependent actin-binding domain in villin, *Nature* 315, 248-250, 1985; Dixon, R.A.F., Kobilka, B.K., Strader, D.J., Cloning of the gene and cDNA for mammalian β-adrenergic receptor and homology with rhodopsin, *Nature* 321, 75-79, 1986; Libert, F., Parmentier, M., Lefort, A., *et al.*, Selective amplification and cloning of four new members of the G protein-coupled receptor family, *Science* 244, 569-572, 1989; Yu, F.X., Zhou, D.M., and Yin, H.L., Chimeric and truncated gCap39 elucidate the requirements for actin filament severing and end capping by the gelsolin family of proteins, *J.Biol.Chem.* 266, 19269-19275, 1991; Wen, D., Corina, K., Chow, E.P., *et al.*, The plasma and cytoplasmic forms of human gelsolin differ in disulfide structure, *Biochemistry* 35, 9700-9709, 1996; Isaacson, R.L., Weeds, A.G., and Fersht, A.R., Equilibria and kinetics of folding of gelsolin domain 2 and mutants involved in familial amyloidosis-Finnish type, *Proc.Natl.Acad.Sci.USA* 96, 11247-11252, 1996; Liu, Y.T. and Yin, H.L., Identification of the binding partners for flightless I, A novel protein bridging the leucine-rich repeat and the gelsolin superfamilies, *J.Biol. Chem.* 273, 7920-7927, 1998; Benyamini, H., Gunasekaran, K., Wolfson, H., and Nussinov, R., Conservation and amyloid formation: a study of the gelsolin-like family, *Proteins* 51, 266-282, 2003; Uruno, T., Remmert, K., and Hammer, J.A., 3rd., CARMIL is a potent capping protein antagonist: identification of a conserved CARMIL domain that inhibits the activity of capping protein and uncaps capped actin filaments, *J.Biol.Chem.* 281, 10635-10650, 2006.

General Transcription Factors

These are protein cofactors for RNA polymerase II which are required for the function of the basal transcription apparatus. The basal apparatus can be described as the functional unit required for the accurate transcription of DNA and is directed to the 5'-end of a transcriptional unit by the core promoter. See Zheng, X.M., Moncollin, V., Egly, J.M., and Chambon, P., A general transcription factor forms a stable complex with RNA polymerase B(II), *Cell* 50, 361-368, 1987; DeJong, J., Bernstein, R. and Roeder, R.G., Human general transcription factor TFIIA: characterization of a cDNA encoding the small subunit and requirement for basal and activated transcription, *Proc.Natl.Acad.Sci. USA* 92, 3313-3317, 1995; Zaid, A., Li, R., Luciakova, K., *et al.*, On the role of the general transcription factor Sp1 in the activation and repression of diverse mammalian oxidative phosphorylation genes, *J.Bioenerg.Biomembr.* 31, 129-135, 1999: Smale, J.T., Core promoter architecture for eukaryotic protein-coding genes, in *Transcription: Mechanism and Regulation*, ed. R.C. Conoway and J.W. Conoway, Raven Press, New York, NY, Chapter 5, pps. 63-81, 1999; Serizawa, H., Conoway, J.W. and Conaway, R.C., Transcriptional initiation by mammalian RNA polymerase II, in *Transcription: Mechanism and Regulation*, ed. R.C. Conoway and J.W.Conoway, Raven Press, New York, NY, Chapter 3, pps. 27-43, 1999; Pugh, B.R., RNA polymerase II transcription machinery, in *Transcription Factors*, ed. J. Locker, Bios/Academic Press, Oxford, UK, Chapter 1, pp. 1-16, 2001; Reid, J., Murray, I., Watt, K., *et al.*, The androgen receptor interacts with multiple regions of the large subunit of general transcription factor TFIIF, *J.Biol. Chem.* 277, 41247-41253, 2002; Vullhorst, D. and Buonanno, A., Characterization of general transcription factor 3, a transcription factor involved in slow muscle-specific gene expression, *J.Biol.Chem.* 278, 8370-8379, 2003; Takagi, Y. and Kornberg, R.D., Mediator as a general transcription factor, *J.Biol.Chem.* 281, 80-89, 2006; Choudhry, M.A., Ball, A., and McEwan, I.J., The role of the general transcription factor TFIID in androgen receptor-dependent transcription, *Mol.Endocrinol.*, 20, 2052–2061, 2006.

Generic Drug

A generic drug is the same as a brand name drug is dosage, safety, and efficacy. Prior to approval, a generic drug must demonstrate bioequivalence or therapeutic equivalence.

Gene Expression Domain

A genomic region that contains a gene and all of the *cis*-acting elements that are required to obtain the homeostatic level and timing of gene expression *in vivo*. Gene expression domains are generally defined by their ability to function independently of the site of integration into a transgene.

General Transcription Factors

A group of *trans*-acting factors which have a central role in the initiation of transcription by RNA polymerase II (pol II). The components are likely similar to the earlier described basal transcription factors. Greenblatt, J., RNA polymerase-associated transcription factors, *Trends Biochem.Sci.* 16, 408-411, 1991; Corden, J.L, RNA polymerase II transcription cycles, *Curr.Opin.*

Genet.Dev. 3, 213-218, 1993; Travers, A., Transcription: building an initiation machine, *Curr.Biol.* 6, 401-403, 1996; Reese, J.C., Basal transcription factors, *Curr.Opin.Genet.Dev.* 13, 114-118, 2003; Asturias, F.J., RNA polymerase II structure, and organization of the preinitiation complex, *Curr.Opin.Struct.Biol.* 14, 121-129, 2004; Boeger, H., Bushnell, D.A., Davis, R., *et al.*, Structural basis of eukaryotic gene transcription, *FEBS Lett.* 579, 899-903, 2005; Szutarisz, H., Dillon, N., and Tora, L., The role of enhancers as centres for general transcription factor recruitment, *Trends Biochem.Sci.* 30, 593-599, 2005; Gross, P. and Oelgeschlager, T., Core promoter-selective RNA polymerase II transcription, *Biochem. Soc.Symp.* (73), 225-236, 2006.

Generic Drug

A generic drug is the same as a brand name drug is dosage, safety, strength, administration, quality, and intended use. The suitability of a generic drug is based on "therapeutic equivilance." By law, a generic product must contain the identical amounts of the same active ingredient(s) as the brand name product. See Verbeeck, R.K., Kanfer, I., and Walker, R.B., Generic substitution: the use of medicinal products containings different salts and implications for safety and efficacy, *Eur.J.Pharm.Sci.* 28, 1-6, 2006; Devine, J.W., Cline, R.R., and Farley, J.F., Follow-on biologics: competition in the biopharmaceutical marketplace, *J.Am.Pharm.Assoc.* 46, 193-201, 2006.

Genome

The complete gene complement of any organism, contained in a set of chromosomes in eukaryotes, a single chromosome in bacteria, or a DNA or RNA molecule in viruses; the complete set of genes inside the cell or virus. Singer, M. and Berg, P., *Genes & Genomes: A Changing Perspective*, University Science Book, Mill Valley, CA, USA, 1991; Murray, T.H. and Rothstein, R.A., *The Human Genome Project and the Future of Health Care*, Indiana University Press, Bloomington, Indiana, USA, 1996; Brown, T.A., *Genome*, Bios Scientific Publishers/Wiley-Liss, New York, NY, 1999; Ridley, M., *Genome:The Autobiography of a Species of 23 Chapters*, HarperCollins, New York, NY, 1999.

Genome-Based Proteomics

Gene-based analysis of the proteome; analytical strategies based on the knowledge of the genome. See Rosamond, J. and Allsop, A., Harnessing the power of the genome in the search for new antibiotics, *Science* 287, 1973-1976, 2000; Agaton, C., Uhlen, M., and Hober, S., Genome-based proteomics, *Electrophoresis* 25, 1280-1288, 2004; Wisz, M.S., Suarez, M.K., Holmes, M.R., and Giddings, M.C., GFSWeb: a web tool for genome-based identification of proteins from mass spectrometric samples, *J.Proteome Res.* 3, 1292-1295, 2004; Romero, P., Wagg, J., Green, M.L., *et al.*, Computational prediction of human metabolic pathways from the complete human genome, *Genome Biol.* 6, R2, 2005; Ek. S., Adreasson, U., Hober, S., *et al.*, From gene expression analysis to tissue microarrays: a rational approach to identify therapeutic and diagnostic targets in lymphoid malignancies, *Mol.Cell. Proteomics* 5, 1072-1081, 2006.

Genomic Databases

See Baxevaris, A.D., Using genomic databases for sequence-based biological discovery, *Molec.Med.* 9, 185-192, 2003.

Genomics

The study of the structure and function of the genome, including information about the sequence, mapping, and expression, and how genes and their products work in the organism; the study of the genetic composition of organisms.

Genotype

The internally coded, inheritable information carried by all living organisms; the genetic constitution of an organism.

Glass/Glasses

A large inhomogenous class of materials with highly variable mechanical and optical properties that solidify from the molten state without crystallization. The cooling of the melt must occur without crystallization. Glasses are most frequently derived from silicates by fusing with boric oxide, aluminum oxide, or phosphorus pentoxide. Glasses are generally hard, brittle, and transparent or translucent, and are considered to be supercooled liquids rather than true solids. See Santoro, M., Gorelli, F.A., Bini, F., *et al.*, Amorphous silica-like carbon dioxide, *Nature* 441, 857-860, 2006; Huang, W., Day, D.E., Kittiratanapiboon, K., and Rahaman, M.N., Kinetics and mechanisms of the conversion of silicate (45S5), borate, and borosilicate glasses to hydroxyapatite in dilute phosphate solutions, *J.Mater.Sci.Mater.Med.* 17, 583-596, 2006; Abraham, S., Mallia, V.A., Ratheesh, K.V., *et al.*, Reversible thermal and photochemical switching of liquid crystalline phases and luminescence in diphenylbutadiene-based mesogenic dimers, *J.Am.Chem.Soc.* 128, 7692-7698, 2006; Lehner, A., Corbineau, F., and Bailly, C., Changes in lipid status and glass properties in cotyledons of developing sunflower seeds, *Plant Cell Physiol.*, 47, 818–828, 2006; Chang, R. and Yethiraj, A., Dynamics of chain molecules in disordered materials, *Phys. Rev.Lett.* 96, 107802, 2006; Katritzky, A.R., Singh, S., Kirichenko, K. *et al.*, In search of ionic liquids incorporating azolate anions, *Chemistry* 12, 4630-4641, 2006.

Glass Transition/Glass Transition Temperature

The glass transition generally refers to change of a polymer from an amorphous material to an brittle material. The glass transition of a non-crystalline material is the critical temperature at which the material changes its behavior from being a glass or brittle material to an amorphous rubber-like material. For lyophilization, it is a critical temperature during the drying cycle which is important to the final product cake. See MacKenzie, A.P., Non-equilibrium freezing behavior of aqueous systems, *Philos. Trans.R.Soc.Lond.B.Biol.Sci.* 278, 167-189, 1977; Schenz, T.W., Israel, B., and Rosolen, M.A., Thermal analysis of water-containing systems, *Adv.Exp.Med.Biol.* 302, 199-214, 1991; Craig, D.Q., Royall, P.G., Kett, V.L., and Hopton, M.L., The relevance of the amorphous state to pharmaceutical dosage forms: Glassy drugs and freeze dried systems, *Int.J.Pharm.* 179, 179-207, 1999; Oliver, A.E., Hincha, D.K., and Crowe, J.H., Looking beyond sugars: the role of amphiphilic solutes in preventing adventitious reactions in anhydrobiotes at low water contents, *Comp.Biochem. Physiol.A Mol.Integr.Physiol.* 131, 515-525, 2002; Nail, S.L., Jiang, S., Chongprasert, S., and Knopp, S.A., Fundamentals of freeze-drying, *Pharm.Biotechnol.* 14, 281-360, 2002; Franks, F., Scientific and technological aspects of aqueous glasses, *Biophys.Chem.* 105, 251-261, 2003; Vranic, E., Amorphous pharmaceutical solids,

Bosn.J.Basic Med. Sci. 4, 35-39, 2004; Hilden, L.R. and Morris, K.R., Physics of amorphous solids, *J.Pharm.Sci.* 93, 3-12, 2004.

Global Proteomics

Analysis of all proteins in a cell or a tissue or an organism. See Hancock, W.S., Wu. S.L., Stanley, R.R., and Gombocz, E.W., Publishing large proteome datasets: scientific policy meets emerging technologies, *Trends Biotechnol.* 20(Suppl 12), S39-S44, 2002; Godovac-Zimmermann, J. and Brown, L.R., Proteomics approaches to elucidation of signal transduction pathways, *Curr. Opin.Mol.Ther.* 5, 241-249, 2003; Kumar, G.K. and Klein, J.B., Analysis of expression and posttranslational modification of proteins during hypoxia, *J.Appl.Physiol.* 96, 1178-1186, 2004; Hoskisson, P.A. and Hobbs, G., Continuous culture – making a comeback? *Microbiology* 151, 3153-3159, 2005.

Globulin

A classic definition of proteins which are insoluble in water and soluble in dilute salt solutions and migrate more slowly than albumin in an electrophoretic system. See Cooper, G.R., Electrophoretic and ultracentrifugal analysis of normal human serum, in *The Plasma Proteins*, ed. F.W. Putnam, Academic Press, New York, NY, Chapter 3, pp. 51-103, 1960; . The globulins were separated into several fractions including the the γ-globulins which contain the various immunoglobulin fractions and were defined as the most slowly moving protein fraction on electrophoresis at pH 8.6 (Porter, H.R., γ-Globulins and antibodies, in *The Plasma Proteins*, ed. F.W. Putnam, Academic Press, New York, NY, Chapter 7, pp. 241-277, 1960). See Gehrke, C.W., Oh, Y.H., and Freeark, C.W., Chemical fractionation and starch gel-urea electrophoretic characterization of albumins, globulins, gliadins, and glutenins in soft wheat, *Anal.Biochem.* 7, 439-460, 1964; Nilsson, U.R. and Mueller-Eberhard, H.J., Isolation of beta IF-globulin form human serum and its characterization as the fifth component of complement, *J.Exp.Med.* 122, 277-298, 1965; Sun, S.M. and Hall, T.C., Solubility characteristics of globulins from *Phaseolus* sees in regard to their isolation and characterization, *J.Agric. Food Chem.* 23, 184-189, 1975; Hauptman, S.P., Macromolecular insoluble cold globulin (MICG): a novel protein form mouse lymphocytes—I. Isolation and characterization, *Immunochemistry* 15, 415-422, 1978.

Glucose Oxidase

An flavoprotein(FAD) enzyme(EC 1.1.3.4; β-D-glucose:oxygen 1-oxidoreductase) which catalyzes the oxidation of β-D-glucose to glucolactone/gluconic acid and hydrogen peroxide. The enzyme is highly specific for this form of glucose (Keilin, D. and Hartree, E.F., The use of glucose oxidase (Notatin) for the determination of glucose in biological material and for the study of glucose-producing systems by mannometric methods, *Biochem.J.* 42, 230-238, 1942; Sols, A. and de la Fuente, G., On the substrate specificity of glucose oxidase, *Biochim.Biophys.Acta* 24, 206-207, 1957; Wurster, B. and Hess, B., Anomeric specificity of enzymes for D-glucose metabolism, *FEBS Lett.* 40(Suppl), S112-S118, 1974) and is the basis of most of the assays for glucose in blood and bioreactors. The vast majority of assays measure the hydrogen peroxide released in the reaction (Kiang, S.W., Kuan, J.W., Kuan, S.S., and Guilbault, G.G., Measurment of glucose in plasma, with use of immobililized glucose oxidase and peroxidase, *Clin.Chem.* 22, 1378-1382, 1976; Chua, K.S. and Tan, I.K., Plasma glucose measurement with the

Yellow Springs glucose analyzer, *Clin.Chem.* 24, 150-152, 1978; Artiss, J.D., Strandbergh, D.R., and Zak, B., On the use of a sensitive indicator reaction for the automated glucose oxidase-peroxidase coupled reaction, *Clin.Biochem.* 1, 334-337, 1983; *Tietz Textbook of Clinical Chemistry and Molecular Diagnostics*, 4th Edn., ed. C.A. Burtis, E.R. Ashwood, and D.F. Bruns, Elsevier-Saunders, St. Louis, MO, USA, 2006). Glucose oxidase was discovered the early 1900's and originally described an antibacterial factor derived from moulds such as *Pencilliium notatum and Aspergillus* niger (Coulthard, C.E., Michaealis, R., Short, W.F. *et al.*, Notatin: an antibacterial glucose aerodehydrogenase from *Penicillium notatum* and *Penicillium resitculosum* sp. nov, *Biochem.J.* 39, 24-36, 1945). Glucose oxidase has subsequently been identified as the antibacterial/antibiotic activity in honey (White, J.W., Jr., Subers, M.H., and Schepartz, A.I., The identification of inhibine, the antibacterial factor in honey, as hydrogen peroxide and its origin in a honey glucose-oxidase system, *Biochim.Biophys.Acta* 73, 57-70, 1963; Schepartz, A.T. and Subers, M.H., The glucose oxidase of honey. I. Purification and some general properties of the enzyme, *Biochim.Biophys.Acta* 85, 228-237, 1964; Bang, L.M., Bunting, C., and Molan, P., The effect of dilution on the rate of hydrogen peroxide production in honey and its implications for wound healing, *J.Alternative Complementary Med.* 9, 267-273, 2003; Badawy, O.F., Shafil, S.S., Tharwat, E.E., and Kamal, A.M., *Rev.Sci.Tech.* 23, 1011-1022, 2004) and a critical component of the honey bee invertebrate immune system (Xang, X. and Cox-Foster, D.L., Impact of an ectoparasite on the immunity and pathology of an invertebrate: evidence for host immunosuppression and viral amplification, *Proc. Natl.Acad.Sci.USA* 102, 7470-7475, 2005. Glucose oxidase is also involved in herbivore offense in plants (Musser, R.O., Cipollini, D.F., Hum-Musser, S.M. *et al.*, Evidence that the caterpillar salivary enzyme glucose oxidase provides herbivore offense in solanaceous plants, *Archs.Insect Biochem.Physiol.* 58, 128-137, 2005.

Glucose regulated protein, 78kD

Grp78; glucose regulated protein, identical with BiP, a chaperone-like protein which was also described as the immunoglobulin heavy-chain-binding protein. See Munro, S. and Pelham, H.R., An Hsp70-like protein in the ER: identity with the 78 kd glucose-regulated protein and immunoglobulin heavy chain binding protein, *Cell* 46, 291-300, 1986; Hendershot, L.M., Ting, J., and Lee, A.S., Identity of the immunoglobulin heavy-chain-binding protein with the 78,000 dalton glucose-regulated protein and the role of posttranslational modifications in its binding function, *Mol.Cell Biol.* 8, 4250-4256, 1988; Haas, I.G., BiP (Grp78), an essential hsp70 resident protein in the endoplasmic reticulum, *Experientia* 50, 1012-1020, 1994; Kleizen, B. and Braakman, I., Protein folding and quality control in the endoplasmic reticulum, *Curr.Opin.Cell Biol.* 16, 343-349, 2004; Okudo, H., Kato, H., Arakaki, Y. and Urade, R., Cooperation of ER-60 and BiP in the oxidative refolding of denatured proteins in vitro, *J.Biochem.* 138, 773-780, 2005; Sorgjerd, K., Ghafouri, B., Jonsson, B.H. *et al.*, Retention of misfolded mutant transthyretin by the chaperone BiP/GRP78 mitigates amyloidogenesis, *J.Mol.Biol.* 356, 469-482, 2006; Panayi, G.S., and Corrigall, V.M., BiP regulates autoimmune inflammation and tissue damage, *Autoimmune Rev.* 5, 140-142, 2006; Li, J. and Lee, A.S., Stress induction of GRP78/BiP and its roles in cancer, *Curr.Mol. Med.* 6, 45-54, 2006; Tajima, H., and Koizumi, N., Induction of BiP by sugar independent of a cis-element for the unfolded protein response in *Arabidopsis thaliana, Biochem.Biophys.Res. Commun.* 346, 926-930, 2006.

Glucosyltransferase

A glycosyltransferase specific for the transfer of glucosides. See Doyle, R.J. and Ciardi, J.E., *Glucosyltransferases, Glucans, Sucrose and Dental Caries*, IRL Press, Washington, DC, USA, 1983; Bleicher, R.J. and Cabot, M.C., Glucosylceramide synthesis and apoptosis, *Biochim.Biophys.Acta* 1585, 172-178, 2002; Yang, J., Hoffmeister, D., Liu, L., *et al.*, Natural product glycorandomization, *Bioorg.Med.Chem.* 12, 1577-1584, 2004; Lorenc-Kukula, K., Korobczak, A., Aksamit-Stachurska, A., *et al.*, Glucosyltransferase: the gene arrangement and enzyme function, *Cell Mol.Biol.Lett.* 9, 935-946, 2004; Trombetta, E.S. and Parodi, A.J., Glycoprotein reglucosylation, *Methods* 35, 328-337, 2005.

GLUT

A family of membrane transporters which mediate the uptake of hexoses in mammalian cells. See Gould, G.W. and Holman G.D., The glucose transporter family—structure, function and tissue-specific expression, *Biochem. J.* 295, 329-341, 1993; Yang, J., Dowden, J., Tatibouet, A., Hatanaka, Y., Holman, G.D., Development of high-affinity ligands and photoaffinity labels for the D-fructose transporter GLUT5, *Biochem.J.* 367, 533-539, 2002;

Glycome

The total carbohydrates within an organism. See Feizi, T., Progress in deciphering the information content of the 'glycome'—a crescendo in the closing years of the millennium, *Glycoconj.J.* 17, 553-565, 2001; Hirabayashi, J., Arata, Y., and Kasai, K., Glycome project: concept, strategy, and preliminary application to *Caenorhabditis elegans*, *Proteomics* 1, 295-303, 2001; Loel, A., Glycome: a medical paradigm, *Adv.Exp.Biol.Med.* 546, 445-451, 2004; Hsu, K.L., Pilobello, K.T., and Mahal, L.K., Analyzing the dynamic bacterial glycome with a lectin microarray approach, *Nat.Chem.Biol.* 2, 125-126, 2006; Freeze, H.H., Genetic defects in the human glycome, *Nat.Rev.Genet.* 7, 537-551, 2006.

Glycomics

The study of the structure, function, and interactions of carbohydrates within the gycome. See Drickhamer, K. and Taylor, M.E., Glycan arrays for functional glycomics, *Genome Biol.* 3, REVIEWS1034, 2002; Love, K.R. and Seeberger, P.H., Carbohydrate arrays as tools for glycomics, *Angew.Chem.Int.Ed.Engl.* 41, 3583-3586, 2002; Hirabayashi, J., Oligosaccharide microarrays for glycomics, *Trends Biotechnol.* 21, 141-143, 2003; Feizi, T., Fazio, F., Chai, W. and Wong, C.H., Carbohydrate microarrays - a new set of technologies at the frontiers of glycomics, *Curr.Opin.Struct.Biol.* 13, 637-645, 2003; Morelle, W. and Michalski, J.C., Glycomics and mass spectrometry, *Curr.Pharm.Des.* 11, 2615-2645, 2005; Raman, R., Raguram, S., Venkataraman, G., *et al.*, Glycomics: an integrated systems approach to structure-function relationships of glycans, *Nat.Methods* 2, 817-824, 2005.

Glycosidase

An enzyme that hydrolyzes glycosidic bonds; most often in oligosaccharides and polysaccharides. See Allen, H.J. and Kisailus, E.C., *Glycoconjugates: Composition, Structure, and Function*, Dekker, New York, New York, 1992; *Guide to Techniques in Glycobiology*, ed. W.J. Lennarz and G.W. Hart, Academic Press, San Diego, CA,

1994; Bucke, C., *Carbohydrate Biotechnology Protocols*, Humana Press, Totowa, NJ, 1999; Himmel, M.E. and Baker, J.O., *Glycosyl Hydrolases for Biomass Conversion*, American Chemical Society, Washington, DC, 2001.

Glycosyltransferase

An enzyme which synthesizes compounds with glycosidic bonds by catalyzing the transfer of glycosyl groups. See Carib, E., Carbohydrate metabolism, *Annu.Rev.Biochem.* 32, 321-354, 1963; Heath, E.C., Complex polysaccharides, *Annu.Rev. Biochem.* 40, 29-56, 1971; Honjo, T. and Hayashi, O., Enzymatic ADP-ribosylation of proteins and regulation of cellular activity, *Curr.Top.Cell Regul.* 7, 87-127, 1973; *Glycoimmunology*, ed. Alavi, A. and Axford, J.S., Plenum Press, New York, New York, 1995; *Molecular Glycobiology*, ed. M. Fukuda and O. Hindsgaul, Oxford University Press, Oxford, UK, 1994; Endo, T., Aberrant glycosylation of alpha-dystroglycan and congenital muscular dystrophies, *Acta Myol.* 24, 64-69, 2005; Serafini-Cessi, F., Monti, A., and Cavallone, D., *N*-Glycans carried by Tamm-Horsfall glycoprotein have a crucial role in the defense against urinary tract diseases, *Glycoconj.J.* 22, 383-394, 2005; Milewski, S., Gabriel, I., and Olchowy, J., Enzymes of UDP-GlcNAc in yeast, *Yeast* 23, 1-14, 2006; Millar, C.M. and Brown, S.A., Oligosaccharide structures of von Willebrand factor and their potential role in von Willebrand disease, *Blood Rev.* 20, 83-92, 2006; Koch-Nolte, F., Adriouch, S., Bannas, P., *et al.*, ADP-ribosylation of membrane proteins: unveiling the secrets of a crucial regulatory mechanism in mammalian cells, *Ann. Med.* 38, 189-199, 2006.

Goblet Cell

A type of cell found in the epithelium with high occurrence in respiratory/digestive tracts which secrete mucus. See Rogers, D.F., Motor control of airway goblet cells and glands, *Respir.Physiol.* 125, 129-144, 2001; Jeffery, P. and Zhu, J., Mucin-producing elements and inflammatory cells, *Novartis Found.Symp.* 248, 51-68, 2002; Rogers, D.F., The airway goblet cell, *Int.J.Biochem.Cell. Biol.* 35, 1-6, 2003; Kim, S. and Nadel, J.A., Role of neutrophils in mucus hypersecretion in COPD and implications for therapy, *Treat.Respir.Med.* 3, 147-159, 2004; Bai, T.R. and Knight, D.A., Structural changes in the airways in asthma: observations and consequences, *Clin.Sci.* 108, 463-477, 2005; Rose, M.C., and Voynow, J.A., Respiratory tract mucin genes and mucin glycoproteins in health and disease, *Physiol.Rev.* 86, 245-278, 2006: Lievin-Le Moal, V. and Servin, A.L., The front line of enteric host defense against unwelcome intrusion of harmful microorganisms: mucins, antimicrobial peptides, and microbiota, *Clin.Microbiol. Rev.* 19, 315-337, 2006.

Golgi Apparatus

A subcellular organelle consisting of a series of membrane structures; the Golgi apparatus can be considered as a single membrane structure containing a number of membrane-bound vesicles. The Golgi apparatus functions in the protein secretory pathway by transporting and packing of proteins for distribution elsewhere in the cell. The Golgi has a *cis*-side facing the endoplasmic reticulum and a *trans*-side which interfaces with the plasma membrane and components of the endocytotic pathway. See Whaley, W.B., *The Golgi Apparatus*, Springer-Verlag, New York, NY, 1975; Pavelka, M., *Functional Morphology of the*

Golgi Apparatus, Springer-Verlag, Berlin, Germany, 1987; Loh, Y.P., *Mechanisms of Intracellular Trafficking and Processing of Preproteins*, CRC Press, Boca Raton, FL, 1993; *Guidebook to the Secretory Pathway*, ed J. Rothblatt and Novak, P., Oxford University Press, Oxford, UK, 1997; *The Golgi Apparatus*, ed. Berger, E.G. and Roth, J., Birkhäuser Verlag, Basel, 1997; Robinson, D.G., *The Golgi Apparatus and the Plant Secretory Pathway*,, CRC Press, Boca Raton, FL, 2003; Northcote, D.H., The Golgi apparatus, *Endeavor* 30, 26-33, 1971; Shnitka, T.K. and Seligman, A.M., Ultrastructural localization of enzymes, *Annu.Rev.Biochem.* 40, 375-396, 1971; Schachter, H., The subcellular sites of glycosylation, *Biochem.Soc.Trans.* 40, 47-71, 1974; Novikoff, A.B., The endoplasmic reticulum: a cytochemist's view, *Proc.Nat.Acad.Sci.USA* 73, 2781-2787, 1976; Hawes, C. and Satiat-Jeunemailtre, B., The plant Golgi apparatus – going with the flow, *Biochim.Biophys.Acta* 1744, 466-480, 2005; Meyer, H.H., Golgi reassembly after mitosis: the AAA family meets the ubiquitin family, *Biochim.Biophys.Acta* 1744, 481-492, 2005; Toivola, D.M., Tao, G.Z., Hbtezion, A., Liao, J., and Omary, M.B., Cellular integrity plus: organelle—related and protein-targeting functions of intermediate filaments, *Trends Cell Biol.* 15, 608-617, 2005; Jolliffe, N.A., Craddock, C.P., and Frigerio, L., Pathways for protein transport to see storage granules, *Biochem. Soc.Trans.* 33, 1016-1018, 2005; Ungar, D., Oka, T., Kreiger, M., and Hughson, F.M., Retrograde transport on the COG railway, *Trends Cell Biol.* 16, 113-120, 2006; Quatela, S.E. and Phillips, M.R., Ras signaling on the Golgi, *Curr.Opin.Cell Biol.* 18, 162-167, 2006; D'Souza-Schorey, C. and Chavrier, P., ARF proteins: roles in membrane traffic and beyond, *Nat.Rev.Mol.Cell Biol.* 7, 347-358, 2006.

Golgins

A family of proteins found in the Golgi apparatus. The members of this protein family are characterized by the presence of a long region of coiled-coil segments thus having a tendency to form long rod-like structures. See Fritzler, M.J., Hamel, J.C., Ocha, R.L., and Chan, E.K., Molecular characterization of two human autoantigens: unique cDNAs encoding 95- and 160-kD proteins of a putative family in the Golgi complex, *J.Exp.Med.* 178, 49-62, 1993; Kjer-Nielsen, L., Teasdale, R.D., van Vliet, C., and Gleeson, P.A., A novel Golgi-localization domain shared by a class of coiled-coil peripheral membrane proteins, *Curr.Biol.* 9, 385-388, 1999; Munro, S. and Nichols, B.J., The GRIP domain – a novel Golgi-targeting domain found in several coiled-coil proteins, *Curr.Biol.* 9, 377-380, 1999; Pfeffer, S.R., Constructing a Golgi complex, *J.Cell Biol.* 155, 873-883, 2001; Barr, F.A. and Short, B., Golgins in the structure and dynamics of the Golgi apparatus, *Curr.Opin.Cell Biol.* 15, 405-413, 2003; Darby, M.C., vanVliet, C., Brown, D. *et al.*, Mammalian GRIP domain proteins differ in their membrane binding properties and are recruited to distinct domains of the TGN, *J.Cell Biol.* 177, 5865-5874, 2004; Fridmann0Sirkis, Y., Siniossoglou, S., and Pelham, H.R., TMF is a golgin that binds Rab6 and influences Golgi morphology, *BMC Cell Biol.* 5, 18, 2004; Malsam, J., Satch, A., Pelletier, L., and Warren, G., Golgin tethers define subpopulations of COPI vesicles, *Science* 307, 1095-1098, 2005; Short, B., Haas, A., and Barr, F.A., Golgins and GTPases, giving identity and structure to the Golgi apparatus, *Biochim.Biophys.Acta* 1744, 383-395, 2005; Satoh, A., Beard, M., and Warren, G., Preparation and characterization of recombinant golgin tethers, *Methods Enzymol.* 404, 279-296, 2005.

Granzyme

Granzymes are exogenous serine proteases that are contained in cytoplasmic granules in cytotoxic T cells and natural killer cells. Granzyme enter the target cell through pores created by perforin and induce apoptosis through a variety of mechanisms including caspace-dependent and caspace-independent pathways. See Jenne, D.E. and Tchopp, J., Granzymes, a family of serine proteases released from granules of cytolytic T lymphocytes upon T cell receptor stimulation, *Immunol.Rev.* 103, 53-71, 1988; Smyth, M.J. and Trapani, J.A., Granzymes: exogenous proteinases that induce target cell apoptosis, *Immunol.Today* 16, 202-206, 1995; Lieberman, J. and Fan, Z., Nuclear war: the granzyme A-bomb, *Curr.Opin.Immunol.* 15, 553-559, 2003; Andrade, F., Casciola-Rosen, L.A., and Rosen, A., Granzyme B-induced cell death, *Acta Haematol.* 111, 28-41, 2004; Waterhouse, N.J., Clarke, C.J., Sedelies, K.A., Teng, M.W., and Trapani, J.A., Cytotoxic lymphocytes; instigators of dramatic target cell death, *Biochem. Pharmacol.* 68, 1033-1040, 2004; Ashton-Rickardt, P.G., The granule pathway of programmed cell death, *Crit.Rev.Immunol.* 25, 161-182, 2005; Bleackely, R.C., A molecular view of cytotoxic T lymphocyte induced killing, *Biochem.Cell Biol.* 83, 747-751, 2005.

Growth

Can be defined as weight or mass increases with age in a multiplicative way from Medawar, P., Size, shape and Age; *Essays in Growth and Form presented to D'Arcy Wentworth Thompson*, Clarendon Press, Oxford, United Kingdom, p. 708, 1945 as cited by Smith, R.W. and Ottema, C., Growth, oxygen consumption and protein and RNA synthesis rates in the yolk sac larvae of the African catfish (*Clarias gariepinos*), *Comp.Biochem.Physiol. Part A* 143, 315-325, 2006.

GTP-Binding Protein

Intracellular proteins which bind GTP and have a wide variety of functions including signal transduction and in turn protein synthesis and cell proliferation. These proteins are "active" when GTP is bound; on hydrolysis of the GTP to GDP, "activity" is lost. See Rouot, B., Brabet, P., Homberger, V., *et al.*, Go, a major brain GTP binding protein in search of a function: purification, immunological and biochemical characterization, *Biochimie* 69, 339-349, 1987; Obar, P.A., Shpetner, H.S., and Vallee, R.B., Dynamin: a microtubule-associated GTP-binding protein, *J.Cell Sci.*(Suppl) 14, 143-145, 1991; Lillie, T.H. and Gomperts, B.D., A cell-physiological description of GE, a GTP-binding protein that mediates exocytosis, *Ciba Found.Symp.* 176, 164-179, 1993; Kjeldgaard, M., Nyborg, J. and Clark, B.G., The GTP binding motif: variations on a theme, *FASEB J.* 10, 1347-1386, 1996; Im, M.J., Russell, M.A., and Feng, J.F., Transglutaminase II: a new class of GTP-binding protein with new biological functions, *Cell Signal.* 9, 477-482, 1997; Ridley, A.J., The GTP-binding protein Rho, *Int.J.Biochem. Cell Biol.* 29, 1225-1229, 1997; Sugden, P.H. and Clerk, A., Activation of the small GTP-binding protein Ras in the heart by hypertrophic agonists, *Trends Cardiovasc.Med.* 10, 1-8, 2000; Caron, E., Cellular functions of the Rap1 GTP-binding protein: a pattern emerges, *J. Cell Sci.* 116, 435-440, 2003; Gasper, R., Scrima, A., and Wittinghofer, A., Structural insights into HypB, a GTP-binding protein that regulates metal binding, *J.Biol.Chem.*, 281, 27492–27502, 2006.

Haber-Weiss Reaction

A cycle consisting of the reaction of hydroxyl radicals with hydrogen peroxide generating the superoxide with the subsequent reaction of superoxide with peroxide generating hydroxyl anion and hydroxyl radical; it is possible that this second reaction is catalyzed by ferric ion: see Kehrer, J.P., The Haber-Weiss reaction and mechanisms of toxicity, *Toxicology* 149, 43-50, 2000; Koppenol, W.H., The Haber-Weiss cycle – 70 years later, *Redox Rep.* 6, 229-234, 2001.

Heat Capacity (C$_\rho$)

The quantity of thermal energy needed to raise the temperature of an object by 1°C; C$_\rho$ = mass x specific heat; see specific heat. Heat capacity in proteins is measured with techniques such as differential scanning calorimetry and isothermal titration calorimetry. An understanding of heat capacity is important in understanding the glass transition in the lyophilization of proteins. See Cooper, A., Heat capacity effects in protein folding anad ligand binding: a re-evaluation of the role of water in biomolecular thermodynamics, *Biophys.Chem.* 115, 89-97, 2005; Prabhu, N.V. and Sharp, K.A., Heat capacity I proteins, *Annu.Rev.Phys.Chem.* 56, 521-548, 2005; Lemaster, D.M., Heat capacity-independent determination of differential free energy of stability between structurally homologous proteins, *Biophys. Chem.* 119, 94-100, 2006; van Teeffelen, A.M., Melinders, M.B., and de Jongh, H.H., Identification of pitfalls in the analysis of heat capacity changes of beta-lactoglobulin A, *Int.J.Biol.Macromol.* 30, 28-34, 2005; Kozlov, A.G. and Lohman, T.M., Effects of monovalent anions on a temperature-dependent heat capacity change for *Escherichia coli* SSB tetramer binding to single-stranded DNA, *Biochemistry* 45, 5190-5205, 2006; Gribenko, A.V., Keiffer, T.R., and Makhatadze, G.I., Amino acid substitutions affecting protein dynamics in eglin C do not affect heat capacity change upon unfolding, *Proteins* 64, 295-300, 2006.

Heat Shock Proteins

Heat shock proteins (HSP) are a family of proteins with chaperone activity. Heat shock proteins are involved in protein synthesis and folding, vesicular trafficking, and antigen presentation. Glucose-regulated protein 78 kDA (GRP78) which is also known as immunoglobulin heavy chain binding protein (BiP) is one of the better known members of this family and is constitutively expressed in the endoplasmic reticulum (ER) in a wide variety of cell types. Heat shock proteins were first described as part of the response of the cell to heat shock and other stress situations such as hypoxia. See Tissières, A., Mitchell, H.K., and Tracy, U.M., Protein synthesis in salivary glands of *Drosophila melanogaster*: relation to chromosome puffs, *J.Mol.Biol.* 84, 389-398, 1974; Schedl, P., Artavanis-Tsakonas, S., Steward, R., *et al*,, Two hybrid plasmids with *D. melanogaster* DNA sequences complementary to mRNA coding for the major heat shock protein, *Cell* 14, 921-929, 1978; Artavanis-Tsakonas, S., Schedl, P., Mirault, M.E., *et al.*, Genes for the 70,000 dalton heat shock protein in two cloned D. melanogaster DNA segments, *Cell* 17, 9-18, 1979; McAlister, L. and Finklestein, D.B., Heat shock proteins and thermal resistance in yeast, *Biochem. Biophys.Res.Commun.* 93, 819-824, 1980; Wang, C., Gomer, R.H., and Lazarides, E., Heat shock proteins are methylated in avian and mammalian cells, *Proc.Natl.Acad. Sci. USA* 78, 3531-3535, 1981; Roccheri, M.C., Di Bernardo, M.G., and Giudice, G., Synthesis of heat-shock proteins in developing sea urchins, *Dev.Biol.* 83, 173-177, 1981; Lindquist, S., Regulation of protein synthesis during heat shock, *Nature* 283, 311-314, 1981; Loomis, W.F., Wheeler, S., and Schmidt, J.A., Phosphorylation of the major heat shock protein of *Dictyostelium discoideum, Mol.Cell Biol.* 2, 484-489, 1982; Neidhardt, F.C., VanBogelen, R.A., and Vaughn, V., The genetics and regulation of heat-shock proteins, *Annu.Rev.Genet.* 18, 295-329, 1984; Schlesinger, M.J., Heat shock proteins: the search for functions, *J.Cell Biol.* 103, 321-325, 1986; Lanks, K.W., Modulators of the eukaryotic heat shock response, *Exp.Cell Res.* 165, 1-10, 1986; Lindquist, S. and Craig, E.A., The heat-shock proteins, *Annu. Rev. Genet.* 22, 631-677, 1988; Tanguay, R.M., Transcriptional activation of heat-shock genes in eukaryotes, *Biochem.Cell Biol.* 66, 584-593, 1988; Pelham, H.R., Control of protein exit from the endoplasmic reticulum, *Annu.Rev.Cell Biol.*5, 1-23, 1989; Bukau, B., Weissman, J., and Horwich, A., Molecular chaperones and protein quality control, *Cell* 125, 443-451, 2006; Panyai, G.S. and Corrigal, V.W., BiP regulates autoimmune inflammation and tissue damage, *Autoimmun.Rev.* 5, 140-142, 2006. More recently, there has been interest in heat shock proteins as therapeutic targets in oncology. See Dai, C. and Whitesell, L., HSP90: a rising star on the horizon of anticancer targets, *Future Oncol.* 1, 529-540, 2005; Li, J. and Lee, A.S., Stress induction of GRP78/BiP and its role in cancer, *Curr.Mol.Med.* 6, 45-54, 2006; Kim, Y., Lillo, A.M., Steiniger, S.C.J., *et al.*, Targeting heat shock proteins on cancer cells: selection, characterization, and cell-penetrating properties of a peptidic GRP78 ligand, *Biochemistry* 45, 9434-9444, 2006.

Hedgehog

A family of proteins important in tissue formation during embryonic development; generally expressed on exterior of cell and bind to receptor on adjacent cells. Sonic hedgehog is a glycoprotein important as signal molecule during differentiation. See Lum, L. and Beachy, P.A., The Hedgehog response network: sensors, switches, and routers, *Science* 304, 1755-1759, 2004; Ishibashi, M., Saitsu, H., Komada, M., and Shiota, K., Signaling cascade coordinating growth of dorsal and ventral tissues of the vertebrate brain, with special reference to the involvement of Sonic Hedgehog signaling, *Anat.Sci.Int.* 80, 30-36, 2005; Hooper, J.A. and Scott, M.P., Communicating with Hedgehogs, *Nat.Rev. Mol.Cell Biol.* 6, 206-317, 2005; Kalderon, D., The mechanism of hedgehog signal transduction, *Biochem.Soc.Trans.* 33, 1509-512, 2005; Echelard, Y., Epstein, D.J., St-Jacques, B., *et al.*, Sonic hedgehog, a member of a family of putative signaling molecules, is implicated in the regulation of CNS polarity, *Cell* 75, 1417-1430, 1993; Bumcrot, D.A. and McMahon, A.P., Somite differentiation. Sonic signals somites, *Curr.Biol.* 5, 612-614, 1995; Johnson, R.L., Riddle, R.D., Laufer, E., and Tabin, C., Sonic hedgehog: a key mediator of anterior-posterior patterning of the limb and dorsoventral patterning of axial embryonic structures, *Biochem.Soc. Trans.* 22, 59-574, 1994. Davy-Grosjean, L. and Couve-Privat, S., Sonic hedgehog signaling in basel cell carcinomas, *Cancer Lett.* 225, 181-192, 2005.

Heterochromatin

"Condensed" or modified chromatin not conducive to gene transcription. See Hyde, B.B., Ultrastructure in chromatin, *Prog.Biophys.Mol.Biol.* 15, 129-148, 1965; Brown, S.W., Heterochromatin, *Science* 151, 417-425, 1966; Back, F., The variable condition of h euchromatin and heterochromatin, *Int.Rev. Cytol.* 45, 25-64, 1976; Lewis, J. and Bird, A., DNA methylation and chromatin structure, *FEBS Lett.* 285, 155-159, 1991; Wu, C.T., Transvection, nuclear structure, and chromatin proteins,

J.Cell.Biol. 120, 587-590, 1993; Karpen, G.H., Position-effect variegation and the new biology of heterochromatin, *Curr.Opin.Genet. Dev.* 4, 281-291, 1994; Kornberg, R.D. and Lorch, Y., Interplay between chromatin structure and transcription, *Curr.Opin.Cell Biol.* 7, 371-375, 1995; Zhimulev, I.F., Polytene chromosomes, heterochromatin, and position effect variegation. *Adv.Genet.* 37, 1-566, 1998; Martin, C. and Zhang, Y., The diverse functions of histone, lysine methylation, *Nat.Rev.Mol.Cell Biol.* 6, 838-849, 2005 Wallace, J.A., and Orr-Weaver, T.L., Replication of heterochromatin: insights into mechanisms of epigenetic inheritance, *Chromosoma* 114, 389-402, 2005; Hiragami, K. and Festenstein, R., Heterochromatin protein 1: a pervasive controlling influence, *Cell.Mol.Life Sci.* 62, 2711-2726, 2005.

Heterolytic Cleavage, Heterolysis

An uneven division of a molecule such as $HCl \rightarrow H^+ + Cl^-$ which usually generates ions. The hydrogenase reaction and the oxygen radical oxidation of fatty acids are example of heterolytic cleavages. See Gardner, H.W., Oxygen radical chemistry of polyunsaturated fatty acids, *Free Radic.Biol.Med.* 7, 65-86, 1989; Fontecilla-Camps, J.C., Frey, M., Garcin, E., *et al.*, Hydrogenase: a hydrogen-metabolizing enzyme. What do the crystal structures tell us about its mode of action? *Biochimie* 79, 661-666, 1997; Richard, J.P. and Amyes, T.L., Proton transfer at carbon, *Curr. Opin.Chem.Biol.* 5, 626-633, 2001; Solomon, E.I., Decker, A., and Lehnert, N., Non-heme iron enzymes: contrasts to heme catalysis, *Proc.Natl.Acad.Sci.USA* 100, 3589-3594, 2003; Zampella,G., Bruschi, M., Fantucci, P., and De Gioia, L., Investigation of H_2 activation by [M(NHPnPr3)('S3')] (M = Ni, Pd). Insight into key factors relevant to the design of hydrogenase functional models, *J.Amer.Chem.Soc.* 127, 13180-13189, 2005.

His-Tag

Generally a hexahistidine sequence which can be attached to the carboxyl-terminal or amino-terminal end of an expressed protein. This tag can be used for the affinity purification or separation of a protein by binding to an IMAC (immobilized metal affinity chromatography) column. The tag can also be used to provide an affinity site for interaction with another molecule in solution. See Sigal, G.B., Bamdad, C., Barberis, A., Strominger, J., and Whitesides, G.M., A self-assembled monolayer for the binding and study histidine-tagged proteins by surface plasmon resonance, *Anal.Chem.* 68, 490-497, 1996; Hengen, P., Purification of His-Tag fusion proteins from *Escherichia coli, Trends Biochem. Sci.* 20, 285-286, 1995; Müller. K.M., Arndt, K.M., Bauer, K., and Plückthun, A., Tandem immobilized metal-ion affinity chromatography/immunoaffinity purification of His-tagged proteins-evaluation of two anti-his-tag monoclonal antibodies, *Anal. Biochem.* 259, 54-61, 1998; Altendorf, K., Stalz, W., Greie, J., Deckers-Hebestreit, G., Structure and function of the F(o) complex of the ATP synthase from *Escherichia coli, J.Exptl.Biol.* 203, 19-28, 2000; Terpe, K., Overview of tag protein fusions: from molecular and biochemical fundamentals to commercial systems, *Appl.Microbiol.Biotechnol.* 60, 523-533, 2003; Jenny, R.J., Mann, K.G., and Lundblad, R.L., A critical review of the methods for cleavage of fusion proteins with thrombin and factor Xa, *Protein Expr.Purif.* 31, 1-11, 2003; Meredith, G.D., Wu, H.Y., and Albritton, N.L. Targeted protein functionalization using his tags, *Bioconjugate Chem.* 15, 969-982, 2004; Zhao, Y., Benita, Y., Lok, M., *et al.*, Multi-antigen immunization using IgG binding domain ZZ as carrier, *Vaccine* 23, 5082-5090, 2005.

Hofmeister Series

Also known as the lyotropic; the order of certain ions to "salt out" or precipitate certain hydrophilic materials from aqueous solution. Polyvalent anions such as citrate and sulfate tend to precipitate while monovalent anions such as chloride and thiocyanate tend to solubilize. A similar series exists for cations. It is thought that this phenomena is related to the ability of the various ions to bind water – hence the term "salting out." See Cacace, M.G., Landau, E.M., and Ramsden, J.J., The Hofmeister series: salt and solvent effects on interfacial phenomena, *Quart.J.Biophys.* 30, 241-277, 1997; Boström, M., Tavares, F.W., Finet, S., Skouri-Panet, F., Tardieu, A., and Ninham, B.W., Why forces between proteins follow different Hofmeister series for pH above and below PI, *Biophys.Chem.* 117, 217-224. 2005.

Holliday Junction

A transient structure formed between two adjoining DNA molecules during homologous recombination which provides for the transfer of DNA sequence between the adjacent strands. See Symington, L.S. and Kolodner, R., Partial purification of an enzyme from *Saccharomyces cerevisiae* that cleaves Holliday junctions, *Proc.Natl.Acad.Sci.USA* 82, 7247-7251, 1985; Churchill, M.E., Tullius, T.D., Kallenbach, N.R., and Seeman, N.C., Holliday recombination intermediate is twofold symmetric, *Proc.Natl. Acad.Sci.USA* 85, 4653-4656, 1988; Dukett, D.R., Murchie, A.I., Diekmann, S., *et al.*, The structure of the Holliday junction, and its resolution, *Cell* 55, 79-89, 1988; Jeyaseelan, R. and Shanmugam, G., Human placental endonuclease cleaves Holliday junctions, *Biochem.Biophys.Res. Commun.* 156, 1054-1060, 1988; Sharples, G.J., Ingleston, S.M., and Lloyd, R.G., Holliday junction processing in bacteria: insights from the evolutionary conservation of RuvABC, RecG, and RusA, *J.Bacteriol.* 181, 5543-5550, 1999; Sharples, G.J., The X philes: structure-specific endonuclease that resolve Holliday junctions, *Mol.Microbiol.* 39, 823-834, 2001; Ho, P.S. and Eichman, B.F., The crystal structures of DNA Holliday junctions, *Curr.Opin.Struct.Biol.* 11, 302-308, 2001; Heyer, W.D., Ehmsen, K.T., and Solinger, J.A., Holliday junctions in the eukaryotic nucleus: resolution in sight? *Trends Biochem. Sci.* 28, 548-557, 2003; Heyer, W.D., Recombination: Holliday junction resolution and crossover formation, *Curr.Biol.* 14, R56-R58, 2004; Khuu, P.A., Voth, A.R., Hays, F.A., and Ho, P.S., The stacked-X DNA Holliday junction and protein recognition, *J.Mol. Recognit.* 19, 234-242, 2006.

Holoenzyme

The intact function enzyme unit which could consist of a protein, metal ions, coenzymes and other protein components. This term was originally used to describe the combination of a coenzyme or other low-molecular weight cofactor such as metal ion with a protein component designated as the apoenzyme to form the holoenzyme. More recently, the term holoenzyme has been used to describe DNA and RNA polymerases. See Hokin, L.E., Purification and molecular properties of the (sodium + potassium)-adenosinetriphosphatase and reconstitution of coupled sodium and potassium transport in phospholipid vesicles containing purified enzyme, *J.Exp.Zool.* 194, 197-205, 1975; Dalziel, K., McFerran, N.V., and Wonacott, A.J., Glyceraldehyde-3-phosphate dehydrogenase, *Philos.Trans.R.Soc.Lond.B Biol.Sci.* 293, 105-118, 1981; McHenry, C.S., DNA polymerase III holoenzyme. Components, structure, and mechanism of a true replicative complex,

J.Biol.Chem. 266, 19127-19130, 1991; Ishihama, A., A multi-functional enzyme with RNA polymerase and RNase activities: molecular anatomy of influenza virus RNA polymerase, *Biochimie* 78, 1097-1102, 1996; Greenblatt, J., RNA polymerase II holoenzyme and transcriptional regulation, *Curr.Opin.Cell Biol.* 9, 310-319, 1997; Amieux, P.S. and McKnight, G.S., The essential role of RI alpha in the maintenance of regulated PKA activity, *Ann.N.Y.Acad.Sci.* 968, 75-95, 2002; Taggart, A.K. and Zakian, V.A., *Curr.Opin.Cell Biol.* 15, 275-280, 2003; Borukhov, S. and Nudler, E., RNA polymerase holoenzyme: structure, function and biological significance, *Curr. Opin.Microbiol.* 6, 93-100, 2003; McHenry, C.S., Chromosomal replicases as asymmetric dimers: studies of subunit arrangement and functional consequences, *Mol.Microbiol.* 49, 1157-1165, 2003.

Holotype

The single specimen or illustration designated as the type for naming a species or subspecies when no type was specified. See Crickmore, N., Zeigler, D.R., Feitelson, J. *et al.*, Revision of the nomenclature for the *Bacillus thuringiensis* pesticidal crystal proteins, *Microbiol.Mol.Biol.Rev.* 62, 807-813, 1998; Pecher, W.T., Robledo, J.A., and Vasta, G.R., Identification of a second rRNA gene unit in the *Parkinsus andrewsi* genome, *J.Eukaryot. Microbiol.* 51, 234-245, 2004.

Homeobox

A brief sequence of nucleotides whose base sequence is virtually identical in all the genes that contain said sequence. Originally described in *Drosphilia*, it has now been found in many organisms including *Homo sapiens*. In the fruit fly, a homeobox appears to determine when particular groups of genes are expressed during development. Homobox regions encode proteins containing homeodomain regions. See Gehring, W.J. and Hiromi, Y., Homeotic genes and the homeobox, *Annu.Rev. Genet.* 20, 147-173, 1986; Stern, C.D. and Keynes, R.J., Spatial patterns of homeobox gene expression in the developing mammalian CNS, *Trends Neurosci.* 11, 190-192, 1988; Kappen, C., Schughart, K., and Ruddle, F.H., Organization and expression of homeobox genes in mouse and man, *Ann.N.Y.Acad.Sci.* 567, 243-252, 1989; Wray, G.A., Transcriptional regulation and the evolution of development, *Int.J.Dev.Biol.* 47, 675-684, 2003; Del Bene, F. and Wittbrodt, J., Cell cycle control by homeobox genes in development and disease, *Semin.Cell Dev. Biol.* 16, 449-460, 2005; Samuel, S. and Naora, H., Homeobox gene expression in cancer: insights from developmental regulation and deregulation, *Eur.J.Cancer* 41, 2428-2437, 2005.

Homeodomain

A domain in a protein that is encoded for by a homeobox; these proteins are transcription factors; the homeodomains are approximately 60 amino acids in length and are composed of three α-helices and bind DNA. See Scott, M.P., Tamkun, J.W., and Hartzell, G.W., 3rd, The structure and function of the homeodomain, *Biochim.Biophys.Acta.* 989, 25-48, 1989; Affolter, M., Schier, A., and Gehring, W.J., Homeodomain proteins and the regulation of gene expression, *Curr.Opin.Cell Biol.* 2, 485-495, 1990; Izpisua-Belmonte, J.C. and Deboule, D., Homeobox genes and pattern formation in the vertebrate limb, *Dev.Biol.* 152, 26-36, 1992; Yates, A. and Chambers, I., The homeodomain protein Nanog and pluripotency in mouse embryonic stem cells, *Biochem.Soc.Trans.* 33, 1518-1521, 2005; Towle, H.C., Glucose as

regulator of eukaryotic gene transcription, *Trends Endocrinol. Metab.* 16, 489-494, 2005.

Homeotic

A shift in structural development as in a major shift in the developmental fate of an organ or body. See Dessain, S. and McGinnis, W., Regulating the expression and function of homeotic genes, *Curr.Opin.Genet.Dev.* 1, 275-282, 1991; Morata, G., Homeotic genes of *Drosophila*, *Curr.Opin.Genet.Dev.* 3, 606-614, 1993; Doboule, D. and Morata, G., Colinearity and functional hierarchy among genes of the homeotic complexes, *Trends Genet.* 10, 358-364, 1994; Mann, R.S., The specificity of homeotic gene function, *Bioessays* 17, 855-863, 1995; Duncan, I., How do single homeotic genes control multiple segment identities? Bioessays 18, 91-94, 1996; Graba, Y., Aragnol, D., and Pradel, J., Drosophila Hox complex downstream targets and the function of homeotic genes, *Bioessays* 19, 379-388, 1997; Reichert, H. and Simone, A., Conserved usage of gap and homeotic genes in patterning the CNS, *Curr.Opin.Neurobiol.* 9, 589-595, 1999; Irish, V.F., The evolution of floral homeotic gene function, *Bioessays* 25, 637-646, 2003; Zubko, M.K., Mitochondrial tuning fork in nuclear homeotic functions, *Trends Plant Sci.* 9, 61-64, 2004. See also *HOX* genes.

Homolytic Cleavage, Homolysis

An even division of a molecule such as $HCl \rightarrow H \cdot + Cl \cdot$ which generates free radicals. The decomposition of a precursor molecules can proceed via either a homolytic pathway, a heterolytic pathway, or both. See White, R.E., Sligar, S.G., and Coon, M.J., Evidence for a homolytic mechanism of peroxide oxygen—oxygen bond cleavage during substrate hydroxylation by cytochrome P-450, *J.Biol. Chem.* 255, 11108-11011, 1980; Yang, G., Candy, T.E., Boaro, M. *et al.*, Free radical yields from the homolysis of peroxynitrous acid, *Free RadicBiol.Med.* 12, 327-330, 1992; Correia, M.A., Yao, K., Allentoff, A.J. *et al.*, Interactions of peroxy quinols with cytochromes P450 2B1, 3A1, and 3A5: influence of the apoprotein on heterocyclic versus hemolytic O-O bond cleavage, *Arch.Biochem. Biophys.* 317, 471-478, 1995; Barr, D.P., Martin, M.V., Guengerich, F.P., and Mason, R.P., Reaction of cytochrome P450 with cumene hydroperoxide: ESR spin-trapping evidence for the homolytic scission of the peroxide O-O bond by ferric cytochrome P450 1A2, *Chem.Res.Toxicol.* 9, 318-325, 1996; Marsh, E.N. and Ballou, D.P., Coupling of cobalt-carbon bond homolysis and hydrogen atom abstraction in adenosylcobalamin-dependent glutamate mutase, *Biochemistry* 37, 11864-11872, 1998; Licht, S.S., Booker, S., and Stubbe, J., Studies on the catalysis of carbon-cobalt bond homolysis by ribonucleoside triphosphate reductase: evidence for concerted carbon-cobalt bond homolysis and thiyl radical formation, *Biochemistry* 38, 1221-1233, Vlasie, M.D. and Banerjee, R., Tyrosine 89 accelerates Co-carbon bond homolysis in methylmalonyl-CoA mutase, *J.Am.Chem.Soc.* 125, 5431-5435, 2003; Lymar, S.V., Khairutdinov, R.F., and Hurst, J.K., Hydroxyl radical formation by O-O bond homolysis in peroxynitrous acid, *Inorg.Chem.* 42, 5259-5266, 2003; Rees, M.D. and Davies, M.J., Heparan sulfate degradation via reductive homolysis of its *N*-chloro derivatives, *J.Am.Chem.Soc.* 128, 3085-3097, 2006.

Homotype

A structure having the same general/function as another which may or may not be opposing. For example, the left arm is a homotype of the right arm. A selectin can be a homotype of another selectin. Homotypic is a descriptor referring to homotype. See Rouhandeh, H., Yau, T., and Lang, P.A., Homotypic and heterotypic interference among picornovirus ribonucleic acids, *Arch. Gesamte Virusforsch.* 27, 236-243, 1969; Bendini, C., Lanfranchi, A., Nobili, R., and Miyake, A., Ultrastructure of meiosis-inducing (hetero-typic) and non-inducing (homotypic) cell unions in conjugation of *Blepharisma*, *J.Cell Sci.* 32, 31-43, 1978; Daunter, B., Immune response: tissue specific T-lymphocytes, *Med.Hypotheses* 37, 76-84, 1992; Wagner, M.C., Molnar, E.E., Molitoris, B.A., and Goebl, M.G., Loss of the homotypic fusion and vacuole protein sorting or golgi-associated retrograde protein vesicle tethering complexes results in gentamicin sensitivity in the yeast *Saccharomyces cerevisiae*, *Antimicrob.Agents Chemother.* 50, 587-595, 2006; Karaulanov, E.E., Bottcher, R.T., and Niehrs, C., A role for fibronectin-leucine-rich transmembrane cell-surface proteins in homotypic cell adhesion, *EMBO Rep.* 7, 283-290, 2006; Brandhorst, D., Zwilling, D., Rizzoli, S.O., et al., Homotypic fusion of early endosomes: SNAREs do not determine fusion specificity, *Proc.Natl.Acad.Sci.USA* 103, 2701-2706, 2006; Decker, B.L. and Wickner, W.T., Enolase activates homotypic vacuole fusion and protein transport to the vacuole in yeast, *J.Biol.Chem.*, 281, 14523–14528, 2006; Stroupe, C., Collins, K.M., Fratti, R.A., and Wickner, W., Purification of active HOPS complex reveals its affinities for phosphoinositides and the SNARE Vam7p, *EMBO J.* 25, 1579-1589, 2006; Brereton, H.C., Carvell, M.J., Asare-Anane, H., et al,, Homotypic cell contact enhances insulin but not glucagon secretion, *Biochem.Biophys.Res.Commun.* 344, 995-1000, 2006.

Hoogsteen Bond

The hydrogen bonds formed in the hybridization of DNA chains to form a triple helix. See Searle, M.S. and Wickham, G., Hoogsteen versus Watson-Crick A-T basepairing in DNA complexes of a new group of 'quinomycin-like' antibiotics, *FEBS Lett.* 272, 171-174, 1990; Raghunathan, G., Miles, H.T., and Sasisekharan, V., Symmetry and structure of RNA and DNA triple helices, *Biopolymers* 36, 333-343, 1995; Soliva, R., Luque, F.J., and Orozco, M., Can G-C Hoogsteen-wobble pairs contribute to the stability of d(G, C-C) triplexes, *Nucleic Acids Res.* 27, 2248-2255, 1999; Li, J.S., Shikiya, R., Marky, L.A., and Gold, B., Triple helix forming TRIPside molecules that target mixed purine/pyrimidine DNA sequences, *Biochemistry* 43, 1440-1448, 2004.

Hormonology

The term hormonology is used to describe the study of hormones and has been proposed as a substitute for endocrinology. See Ross, J.W., Hormonology in obstetrics, *J.Natl.Med.Assoc.* 46, 19-21, 1954; Swain, C.T., Hormonology, *N.Engl.J.Med.* 280, 388-389, 1969; Kulinskii, V.I. and Kolesnichenko, L.S., Current aspects of hormonology, *Biochemistry(Mosc.)* 62, 1171-1173, 1997; Holland, M.A., Occam's razor applied to hormonology (Are cytokines produced by plants?), *Plant Physiol.* 115, 865-868, 1997; Hadden, D.R., 100 years of hormonology: a view from No. 1 Wimpole Street, *J.R.Soc.Med.* 98, 325-326, 2005; Hsueh, A.J.W., Bouchard, P. and Ben-Shlomo, I., Hormonology: a genomic perspective on hormonal research, *J.Endocrinol.* 187, 333-338, 2005.

HOX genes

Encodes a family of transcription factors, Hox proteins. See Morgan, S., *HOX* genes: a continuation of embryonic patterning? *Trends in Genetics* 22, 67-69, 2006; Hoegg, S. and Meyer, A., HOX

clusters as models for vertebrate genome evolution, *Trends in Genetics* 21, 421-424, 2005; Sekimoto, T., Yoshinobu, K., Yoshida, M., et al., Region-specific expression of murine *Hox* genes implies the *HOX* code-mediated patterning of the digestive tract, *Genes to Cells* 3, 51-64, 1998.

Hydrogels

An easily deformed pseudo-solid mass formed from largely hydrophilic colloids dispersed in an aqueous medium (dispersion medium or continuous phase). There is considerable interest in the use of hydrogels for drug delivery. See Jhon, M.S. and Andrade, J.D., Water and hydrogels, *J.Biomed.Mat.Res.* 7, 509-522, 1973; Roorda, W., Do hydrogels contain different classes of water, *J.Biomater.Sci. Polym.Ed.* 5, 383-395, 1994; Omidian, H., Rocca, J.G., and Park, K., Advances in superporous hydrogels, *J.Control.Release* 102, 3-12, 2005; Frokjaer, S. and Otzen, D.E., Protein drug stability: a formulation challenge, *Nat.Rev.Drug Discov.* 4, 298-306, 2005; Kashyap, N., Kumar, N., and Kumar, K.N., Hydrogels for pharmaceutical and biomedical applications, *Crit.Rev.Ther.Drug Carrier Syst.* 22, 107-149, 2005; Fairman, R. and Akerfeldt, K.S., Peptides as novel smart materials, *Curr.Opin.Struct.Biol.* 15, 453-463, 2005; Young, S., Wong, M., Tabata, Y., and Mikos, A.G., Gelatin as a delivery vehicle for the controlled release of bioactive molecules, *J.Control.Release* 109, 256-274, 2005; Dusek, K., *Reponsive Gels: Volume Transitions*, Springer-Verlag, Berlin, 1993; Dumitriu, S., *Polymeric Biomaterials*, Marcel Dekker, New York, 1994; Zrinyl, N., *Gels*, Springer, Darmstadt, 1996; McCormick, C.L., *Simuli-Responsive Water Soluble and Amphiphilic Polymers*, American Chemical Society, Washington, DC, 2001; Dumitriu, S., *Polymeric Biomaterials*, Marcel Dekker, New York, 2002.

Hydrophobic, Hydrophobic Effect, Hydrophobic Forces

Literally, the tendency of a molecular structure to avoid water which results in an association or clustering of hydrophobic groups. The term nonpolar is frequently used to describe such groups or molecules. Polar and nonpolar groups or functions can exist in the same molecule; for example, the ε-amino group of lysine is polar but the methylene carbon chain between the ε-amino group and the α-carbon is nonpolar. See Chander, D., Interfaces and the driving force of hydrophobic assembly, *Nature* 437, 640-647, 2005; Kauzmann, W., Some forces in the interpretation of protein denaturation, *Adv.Prot.Chem.* 14, 1-63, 1959; Tanford, C., The hydrophobic effect and the organization of living matter, *Science* 200, 1012-1018, 1978; Kumar, S. and Nussinov, R., Close-range electrostatic interactions in proteins, *ChemBioChem* 3, 604-617, 2002; Kyte, J., The basis of the hydrophobic effect, *Biophys.Chem.* 100, 193-203, 2003; Lesk, A.M., Hydrophobicity-getting into hot water, *Biophys.Chem.* 105, 179-182, 2003; Seelig, J., Thermodynamics of lipid-peptide interactions, *Biochim. Biophys.Acta* 1666, 40-50, 2004; Hofinger, S. and Zerbetto, F., Simple models for hydrophobic hydration, *Chem.Soc.Rev.* 34, 1012-1020, 2005.

Hydrophobins

Hydrophobins are secreted proteins functioning in fungal growth and development. Hydrophobins self-assemble at hydrophilic/ hydrophobic interfaces forming amphipathic membranes. See Wessels, J., De Vries, O., Asgeirsdottir, S.A., and Schuren, F., Hydrophobin genes involved in formation of aerial hyphae and fruit bodies in *Schizophyllum*, *Plant.Cell*, 3, 793-799, 1991; Wessels, J.G., Hydrophobins: proteins that change the nature of the fungal surface, *Adv.Microb.Physiol.* 38, 1-45, 1997; Ebbole, D.J., Hydrophobins and fungal infection of plants and animals, *Trends Microbiol.* 5, 405-408, 1997; Wosten, H.A., Hydrophobins: multipurpose proteins, *Annu.Rev.Microbiol.* 55, 625-646, 2001; Linder, M.B., Szilvay, G.R., Nakari-Setala, T., and Penttila, M.E., Hydrphobins: the protein-amphiphiles of filamentous fungi, *FEMS Microbiol.Rev.* 29, 877-896, 2005.

Hypsochromic

A shift of light absorption or emission to a shorter wavelength ($\lambda < \lambda_o$); a "blue" shift. See Crescitelli, F. and Karvaly, B., The gecko visual pigment: the anion hypsochromic shift, *Vision Res.* 31, 945-950, 1991; Zalis, S., Sieger, M., Greulich, S., *et al.*, Replacement of the 2,2'-bipyridine by 1,4-diazabutadiene acceptor ligands: why the bathochromic shift for [N empty set N)IrCl(C5Me5)] + complexes bu the hypsochromic shift for (N empty set N)Ir(C5Me5)? *Inorg.Chem.* 42, 5185-5191, 2003; Meier, H., Gerold, J., Kolshrn, H., and Muhling, B., Extension of conjugation leading to bathochromic or hypsochromic effects in OPV series, *Chemistry* 23, 360-370, 2004; de Garcia Ventrini, C., Andreaus, J., Machado, V.G., and Machado, C., Solvent effects in the interaction of methyl-β-cyclodextrin with solvatochromic merocyanine dyes, *Org.Biomol. Chem.* 3, 1751-1756, 2005; Kidman, G. and Northrop, D.B., Effect of pressure on nucleotide binding to yeast alcohol dehydrogenase, *Protein Pept.Lett.* 12, 495-497, 2005; Li, Y., He, W., Dong, Y. *et al.*, Human serum albumin interaction with formononetin studied using fluorescence anisotropy, FT-IR spectroscopy, and molecular modeling methods, *Bioorg.Med.Chem.* 14, 1431-1436, 2006; Schonefeld, K., Ludwig, R., and Feller, K.H., Fluorescence studies of host-guest interaction of a dansyl amide labelled calyx[6] arene, *J.Fluoresc.* 16, 449-454, 2006; Correa, N.M. and Levinger, N.E., What can you learn from a molecular probe? New insights on the behavior of C343 in homogeneous solutions and AOT reverse micelles, *J.Phys.Chem. B Condens. Matter Surf. Interfaces Biophys.* 110, 13050-13061, 2006.

Idiotypic

Idiotypic as in referring to idiotype where idiotype is that portion of the variable region of an antibody which confers specificity. See Bigazzi, P.E., Regulation of autoimmunity and the idiotypic network, *Immunol.Ser.* 54, 39-64, 1991; Schoenfeld, Y., Idiotypic induction of autoimmumity: do we need an autoantigen? *Clin. Exptl.Rheumatol.* 12(Suppl 11), S37-S40, 1994; Schoenfeld, Y. and George, J., Induction of autoimmunity. A role for the idiotypic network, *Ann.N.Y.Acad.Sci.* 815, 342-349, 1997; Bianchi, A. and Massaia, M., Idiotypic vaccination in B-cell malignancies, *Mol.Med.Today* 3, 435-441, 1997; Lacroix-Desmazes, S., Bayry, J., Misra, N., *et al.*, The concept of idiotypic vaccination against factor VIII inhibitors in haemophilia A, *Haemophilia* 8 (Suppl 2), 55-59, 2002; Coutinho, A., Will the idiotypic network help to solve natural tolerance? *Trends Immunol.* 24, 53-54, 2003.

IMAC (Immobilized metal affinity chromatography)

A chromatographic fractionation procedures which uses a matrix consisting of a metal ion tightly bound to a matrix. Nickel is the most common metal ion used but there is use of copper and other transition metals. See Porath, J. and Olin, B., Immobilized metal ion affinity adsorption and immobilized metal ion affinity chromatography of biomaterials. Serum protein affinities for gel-immobilized iron and nickel ions, *Biochemistry* 23, 1621-1630, 1982; Porath, J., Immobilized metal ion affinity chromatography, *Protein Expr.Purif.* 3, 263-281, 1992; Skerra, A., Engineered protein scaffolds for molecular recognition, *J.Mol.Recongnit.* 13, 167-187, 2000; Gaberc-Proekar, V. and Menart, V., Perspectives of immobilized-metal affinity chromatography, *J.Biochem.Biophys. Methods* 49, 335-360, 2001; Ueda, E.K., Gout, P.W., and Morganti, L., Current and prospective applications of metal ion-protein binding, *J.Chromatog.A.* 988, 1-23, 2003.

Inhibin

Inhibin is a dimeric glycoprotein secreted by the follicular cells of the ovary and the Sertoli cells of the testis that regulates secretion of follicle stimulating hormone from the anterior pituitary. Inhibin has received recent attention as a biomarker for ovarian cancer. See Chari, S., Chemistry and physiology of inhibin—a review, *Endokrinologie* 70, 99-107, 1977; Grady, R.R., Charlesworth, M.C., and Schwartz, N.B., Characterization of the FSH-suppressing activity in follicular fluids, *Recent Prog.Horm. Res.* 38, 409-456, 1982; Schwartz, N.B., Role of ovarian inhibin (folliculostatin) in regulating FSH secretion in the female rat, *Adv.Exp.Med.Biol.* 147, 15-36, 1982; Burger, H.G. and Igarashi, M., Inhibin: definition and nomenclature, including related substances, *Mol.Endocrinol.* 2, 391-392, 1988; Robertson, D.M., Stephenson, T., Cahir, N., *et al.*, Development of an inhibin α subunit ELISA with broad specificity, *Mol.Cell.Endocrinol.* 180, 79-86, 2001; Robertson, D.M., Stephenson, T., Pruysers, E., *et al.*, Inhibins/activins as diagnostic markers for ovarian cancers, *Mol. Cell.Endocrinol.* 191, 97-103, 2002; Khosravi, J., Krishna, R.G., Khaja, N., Bodani, U., and Diamandi, A., Enzyme-linked immunosorbent assay of total inhibin: direct determination based on inhibin α subunit-specific monoclonal antibodies, *Clin.Biochem.* 37, 370-376, 2004; Cook, R.W., Thompson, T.B., Jardtzky, T.S., and Woodruff, T.K., Molecular biology of inhibin action, *Semin. Reprod.Med.* 22, 269-276, 2004.

Imino Sugars

A class of carbohydrate mimetics which contain nitrogen in the place of oxygen in the ring. These sugars inhibit glycosylation reactions by acting as transition state analogues. See Paulsen, H. and Brockhausen, I., From imino sugars to cancer glycoproteins, *Glycoconjugate J.* 18, 867-870, 2001; Dwek, R.A., Butters, T.D., Platt, F.M., and Zitzmann, N., Targeting glycosylation as a therapeutic approach, *Nat.Rev.Drug Disc.* 1, 65-75, 2002; El-Ashry, el-Sh, and El Nemr, A., Synthesis of mono- and di-hydroxylated prolines and 2-hydroxymethylpyrrolidines from non-carbohydrate precursors, *Carbohyd.Res.* 338, 2265-2290, 2003; Butters, T.D., Dwek, R.A., and Platt, F.M., New therapeutics for the treatment of glycosphingolipid lysosomal storage diseases, *Adv.Exp. Med.Biol.* 535, 219-226, 2003; Butters, T.D., Dwek, R.A., and Platt, F.M., Imino sugar inhibitors for treating the lysosomal glycosphingolipidoses, *Glycobiology* 14, 43R-52R(epub), 2005.

Immunoblotting

A technique for the identification of immunoreactive substances such as proteins. Most frequently, detection by immunoblotting first involves a gel-based electrophoretic separation step followed by electrophoretic transfer to another matrix such as nitrocellulose or PVDF in a manner such that the original separation pattern is retained. The separated protein or proteins are measured by reaction with a primary probe such as antibody labeled with an enzyme or other signal; it is also possible to use a secondary probe which would react with the primary probe. A secondary probe could be an antibody with a signal such as an enzyme. directed against the primary probe. This latter situation is similar to an indirect ELISA. See *CRC Handbook of Immunoblotting of Proteins*, ed. O.J. Bjerrum and N.H.H. Heegaard, CRC Press, Boca Raton, USA., 1988; *Antibodies: A Laboratory Manual*, ed E.Harlow and D. Lane, Cold Spring Harbor, New York, USA, 1988; Manchenko, G.P., *Handbook of Detection of Enzymes on Electrophoresis*, CRC Press, Boca Raton, USA, 1994; Stot, D.I., Immunoblotting, dot-blotting, and ELISPOT assay: methods and applications, in *Immunochemistry*, ed. C.J. Van Oss and M.H.V. van Regenmortel, Marcel Dekker, New York, USA, Chapter 35, pp. 925-948, 1994; Burns, R., *Immunochemical Protocols*, Humana Press, Totowa, New Jersey, USA, 2005. Western blotting is a form of immunoblotting. See Western Blotting, ELISA, Elispot.

Immunoglobulin

A group of plasma proteins (Ig) which are synthesize by plasma cells which are formed from B-cells. There are five general class of immunoglobulins: IgA, IgE., IgD, IgG, and IgM. With the exception of some unique immunoglobulins such as camelids, immunoglobulin are based on a structure of dimers or heterodimers where the heterodimers are composed of a light chain and a heavy chain. IgM is a pentamer of this basic building block while IgA can be a monomer, dimer or trimer of the basic building block. The basic building block is bivalent in that each heterodimer can bind an antigen; IgA may be bivalent, tetravalent or hexavalent while IgM is decavalent. See B. Pernis and H.J. Vogel, *Cells of Immunoglobulin Synthesis*, Academic Press, New York, New York, 1979; Calabi, F. and Neuberger, M.S., *Molecular Genetics of Immunoglobulin*, Elsevier, Amsterdam, Netherlands, 1987; Langone, J.J., *Antibodies, Antigens, and Molecular Mimicry*, Academic Press, San Diego, CA, USA, 1989; Kuby, J., *Immunology*, W.H. Freeman, New York, New York, 1992; Cruse, J.M. and Lewis, R.E., *Atlas of Immunology*, CRC Press, Boca Raton, Florida, USA, 1999.

Immunoglobulin Superfamily

A family of cell surface glycoproteins which contain an extracellular domain homologous to immunoglobulin(Ig), a transmembrane component, and a cytoplasmic extension which interact with other cell adhesion molecules such as integrins in homotypic interactions. See Anderson, P., Morimoto, C., Breitmeyer, J.B., and Schlossman, S.F., Regulatory interactions between members of the immunoglobulin superfamily, *Immunol.Today* 9, 199-203,1988; Hunkapiller, T. and Hood, L., Diversity of the immunoglobulin gene superfamily, *Adv.Immunol.* 44, 1-63, 1989; Barclay, A.N., Membrane proteins with immunoglobulin-like domains—a master superfamily of interaction molecules, *Semin. Immunol.* 15, 215-223, 2003; Naka, Y., Bucciarelli, L.G., Wendt, T., *et al.*, RAGE axis: animal models and novel insights into the vascular complications of diabetes, *Arterioscler.Thromb.Vasc. Biol.* 24, 1342-1349, 2004; Mittler, R.S., Foell, J., McCausland, M., *et al.*, Anti-CD137 antibodies in the treatment of autoimmune disease and cancer, *Immunol.Res.* 29, 197-208, 2004; Du Pasquier, L., Zucchetti, I., and De Santis, R., Immunoglobulin superfamily receptors in protochordates: before RAG time, *Immunol.Rev.* 198, 233-248, 2004; Peggs, K.S. and Allison, J.P., Co-stimulatory

pathways in lymphocyte regulation: the immunoglobulin superfamily, *Br.J.Haematol.* 130,809-824, 2005.

Immunomics

Study of the molecular functions associated with all immune-related coding and non-coding mRNA transcripts. See Maecker, B., von Bergwelt-Baildon, Anderson, K.S., Vonderheide, R.H., and Schultze, J.L., Linking genomics to immunotherapy by reverse immunology—'immunomics' in the new millennium, *Curr.Mol. Med.* 1, 609-619, 2001. Schonbach, C., From immunogenetics to immunomics: functional prospecting of genes and transcripts, *Novartis Found.Symp.* 254, 177-188, 2003.

Immunoproteomics

Definition is a work in progress varying from the screening of two-dimensional gels for reactive antibodies to the use of mass spectrometry to study targets of the immune system; in general, the use of proteomics to study the cellular and humoral immune systems. See Haas, G., Karaali, G., Ebermayer, K., *et al.*, Immunoproteomics of *Helicobacter pylori* infection and relation to gastric disease, *Proteomics* 2, 313-324, 2002; Purcell, A.W. and Gorman, J.J., Immunproteomics: mass spectrometry-based methods to study the targets of the immune response, *Mol.Cell. Proteomics* 3, 193-208, 2004; Hess, J.L., Blazer, L., Romer, T. *et al.*, Immunoproteomics, *J.Chromatog.B.Anal.Technol Biomed.Life Sci.* 815, 65-75, 2005; Chen, Z., Peng, B., Wang, S., and Peng, X., Rapid screening of highly efficient vaccine candidates by immunoproteomics, *Proteomics* 4, 3203-3213, 2004; Paul-Satyaseela, M., Karched, M., Bian, Z., *et al*, Immunoproteomics of *Actinobacillus actinomycetemcomitans* outer-membrane proteins reveal a highly immunoreactive peptidoglycan-associated lipoprotein, *J.Med. Microbiol.* 55, 931-942, 2006; Falisse-Poirrier, N., Ruelle, V. *et al.*, Advances in immunoproteomics for serological characterization of microbial antigens, *J.Microbiol.Methods*, 67, 593–596, 2006.

Immuno Proteasome

A type of proteasome which is involved in the processing proteins for MHC class I antigen presentation. See Aki, M., Shimbara, N., Takashina, M., *et al.*, Interferon-gamma induces different subunit organizations and functional diversity of proteasomes, *J.Biochem.* 115, 257-269, 1994; Dahlmann, B., Ruppert, T., Kuehn, L., *et al.*, Different proteasome subtypes in a single tissue exhibit different enzymatic properties, *J.Mol.Biol.* 303, 643-653, 2000; Tenzer, S., Stoltze, L., Schonfisch, B., *et al*,, Quantitative analysis of prion-protein degradation by constitutive and immuno-20S proteasomes indicates differences correlated with disease susceptibility, *J.Immunol.* 172, 1083-1091, 2004; Dahlmann, B., Proteasomes, *Essays Biochem.* 41, 31-48, 2005; Ishii, K., Hisaeda, H., Duan, X., *et al.*, The involvement of immunoproteomics in induction of MHC class I-restricted immunity targeting Toxoplasma SAG1, *Microbes Infect.* 8, 1045-1053, 2006.

Immunostimulatory Sequence (ISS); Immunostimulatory Sequence Oligodeoxynucleotide (ISS-ODN)

A specific sequence non-methylated DNA containing cytosine and guanine which has immunostimulatory properties and can serve as an adjuvant. See Horner, A.A., Ronaghy, A., Cheng, P.M.

et al., Immunostimulatory DNA is a potent mucosal adjuvant, *Cell.Immunol.* 190, 77-82, 1998; Miyazaki, D., Liu, G., Clark, L., and Ono, S.J., Prevention of acute allergic conjunctivitis and late-phase inflammation with immunostimulatory DNA sequences, *Invest.Ophthalmol.* 41, 3850-3855, 2000; Horner, A.A. and Raz, E., Immunostimulatory sequence oligodeoxynucleotide: a novel mucosal adjuvant, *Clin.Immunol.* 95, S19-S29, 2000; Marshall, J.D., Abtahi, S., Eiden, J.J. *et al.,* Immunostimulatory sequence DNA linked to the Amb a 1 allergen promotes T(H) 1 cytokine expression while downregulating T(H)2 cytokine expression in PBMCs from human patients with ragweed allergy, *J.Allergy Clin.Immunol.* 108, 191-197, 2001; Horner, A.A. and Raz, E., Immunostimulatory sequence oligodeoxynucleotide-based vaccination and immunomodulation: two unique but complementary strategies for the treatment of allergic diseases, *J.Allergy Clin. Immunol.* 110, 706-712, 2002; Teleshova, N., Kenney, J., Williams, V., *et al.,* CpG ISS-ODN activation of blood-derived B cells from healthy and chronic immunodeficiency virus-infected macaques, *J.Leukoc.Biol.* 79, 257-267, 2006.

Industrial Plantation

A large-scale, usually single crop forestry or agricultural enterprise.

Infrared Spectroscopy

The common range for infrared spectroscopy is 10 – 12,800 cm^{-1} (780 – 10^6 nm). Absorption spectra are described as function of the wavenumber of the incident; the wavenumber is the reciprocal of the wavelength and has the advantage of being linear with energy. The infrared region can be divided into the near-infrared, the mid-infrared, and far-infrared regions.

Insulin Receptor

A heterotetramer consisting of two extracellular alpha subunits which bind insulin and two transmembrane beta subunits which have tyrosine kinase activity. See Chang, L., Chiang, S.-H., and Saltiel, A.R., Insulin signaling and the regulation of glucose transport, *Molec.Med.* 10, 65-71, 2004; Kanzaki, M., Insulin receptor signals regulating GLUT4 translocation and actin dynamics, *Endocr.J.* 53, 267-293, 2006; Martinez, S.C., Cras-Meneur, C., Bernal-Mizrachi, E., and Permutt, M.A., Glucose regulates Fox01 through insulin receptor signaling in the pancreatic islet β-cells, *Diabetes* 55, 1581-1591, 2006; Marine, S., Zamiara, E., Todd Smith, S., *et al.,* A miniaturized cell-based fluorescence resonance energy transfer assay for insulin-receptor activation, *Anal.Biochem.* 355, 267-277, 2006; Hao, C., Whittaker, L., and Whittaker, J., Characterization of a second ligand binding site of the insulin receptor, *Biochem.Biophys.Res. Commun.* 347, 334-339, 2006; Sisci, D., Morelli, C., Garofalo, C., Expression of nuclear insulin receptor substrate 1 (IRS-1) in breast cancer, *J.Clin.Pathol.,* 60, 633–641, 2007.

Integrins

Cell membrane glycoproteins which function as receptor for extracellular matrix components. Integrins are heterodimers containing an α-subunit and a β-subunit. The β-subunit contains RGD sequences which "recognize" ligands such as fibronectin, platatelet glycoprotein II b/IIIa, and extracellular matrix components or structural analogues or homologues. See Hynes, R.O., Integrins: a family of cell surface receptors, *Cell* 48, 549-554,

1987; Malech, H.L. and Gallin, J.I., Current concepts: immunology. Neutrophils in human diseases, *N.Engl.J.Med.* 317, 687-694, 1987; Akiyama, S.K., Yamada, K.M., and Hayashi, J., The structure of fibronectin and its role in cellular adhesion, *J.Supramol. Struct.Cell.Biochem.* 16, 345-348, 1981; Mosher, D.F., Physiology of fibronectin, *Annu.Rev.Med.* 35, 561-575, 1984; Bennett, J.S., Structure and function of the platelet integrin αIIβ3, *J.Clin.Invest.* 115, 3363-3369, 2005; Serini, G., Valdembri, D., and Bussolino, F., Integrins and angiogenesis: a sticky business, *Exp.Cell Res.* 312, 651-658, 2005; Caswell, P.T. and Norman, J.C., Integrin trafficking and the control of cell migration, *Traffic* 7, 14-21, 2006; Legate, K.R., Montanez, E., Kudlacek, O., and Fassler, R., ILK, PINCH and parvin: the tIPP of integrin signalling, *Nat.Rev.Mol.Cell.Biol.* 7, 20-31, 2006.

Integrin-mobilferrin pathway (IMP)

Pathway involved in the membrane transport of ferric iron. See Conrad, M.E,, Umbreit, J.N., Peterson, R.D., *et al.,,* Function of integrin in duodenal mucosal uptake of iron, *Blood* 81, 517-521, 1993; Wolf, G. and Wessling-Resnick, M., An integrin-mobilferrin iron transport pathway in intestine and hematopoietic cells, *Nutr.Rev.* 52, 387-389, 1994; Conrad, M.E. and Umbreit, J.N., Iron absorption and transport – an update, *Amer.J.Hematol.* 64, 287-298, 2000; Umbreit, J.N., Conrad, M.E., Hainsworth, L.N., and Simovich, M., The ferrireductase paraferritin contains divalent metal transporter as well as mobilferrin, *Am.J.Physiol.Gastrointest.Liver Physiol.* 282, G534-G539, 2002.

Intein

Intervening protein sequences which are removed by post-translational self-splicing; analogous to exon splicing. Intein regions are surrounded by an *N*-terminal extein and a *C*-terminal extein. Intein splicing has proved useful for the preparation of *N*-terminal cysteine residues which can be coupled to a matrix. See Eckenroth, B., Harris, K., Turanov, A.A., *et al.,* Semisynthesis and characterization of mammalian thioredoxin reductase, *Biochemistry* 45, 5158-5170, 2006; Hackenberger, C.P., Chen, M.M., and Imperiali, B., Expression of *N*-terminal cys-protein fragments using an intein refolding strategy, *Bioorg. Med.Chem.,* 14, 5043–5048, 2006; Sharma, S.S., Chong, S., and Harcum, S.W., Intein-mediated protein purification of fusion proteins expressed under high-cell density conditions in *E.coli,* *J.Biotechnol.,* 125, 48–56, 2006; Kwon, Y., Coleman, M.A., and Camarero, J.A., Selective immobilization of proteins onto solid supports through split-intein-mediated protein trans-splicing, *Angew.Chem.Int.Ed. Engl.* 45, 1726-1729, 2006; Colston, M.J., and Davies, E.O., The ins and outs of protein splicing elements, *Mol.Microbiol.* 12, 359-363, 1994; Cooper, A.A. and Stevens, T.H., Protein splicing: self-splicing of genetically mobile elements at the protein level, *Trends Biochem.Sci.* 20, 351-356, 1995; Paulus, H., Protein splicing and related forms of protein autoprocessing, *Annu.Rev.Biochem.* 69, 447-496, 2000; Xu, M.Q. and Evans, T.C., Jr., Intein-mediated ligation and cyclization of expressed proteins, *Methods* 24, 257-277, 2001; Durek, T. and Becker, C.F., Protein semi-synthesis: new problems for functional and structural studies, *Biomol.Eng.* 22, 153-172, 2005; Tan, L.P. and Yao, S.Q., Intein-mediated, in vitro and in vivo protein modifications with small molecules, *Protein Pep.Lett.* 12, 769-775, 2005; Anderson, L.L., Marshall, G.R., and Baranski, T.J., Expressed protein ligation to study protein interactions:

semi-synthesis of the G-protein alpha subunit, *Protein Pep.Lett.* 12, 783-787, 2005.

Intercalation

The insertion of a molecule, usually planer, between adjacent base pairs of DNA.

Interactome

The protein-protein interactions within a proteome. See Ito, T., Chiba, T., and Yoshida, M., Exploring the protein interactome using comprehensive two-hybrid projects, *Trends Biotechnol.* 19 (Suppl 10), S23-S27, 2001; Ito, T., Ota, K., Kubota, H., *et al.*, Roles for the two-hybrid system in exploration of the yeast protein interactome, *Mol.Cell.Proteomics* 1, 561-566, 2002; Vidal, M., Interactome modeling, *FEBS Lett.* 579, 1834-1838, 2005; Cusick, M.E., Klitgord, N., Vidal, M., and Hill, D.E., Interactome: gateway into systems biology, *Hum.Mol.Genetics* 15, 14 Spec.No.2, R171-R181, 2005; Ghavidel, A., Cagney, G., and Emili, A., A skeleton on the human protein interactome, *Cell* 122, 830-832, 2005.

Interleukin

A functionally defined group of small proteins which "communicate" between various immune cell types (inter + leukocytes = interleukin) (Aardem. L.A., Brunner, T.K., Creottini, J.C. *et al.*, Revised nomenclature for antigen-nonspecific T-cell proliferation and helper factors, *J. Immunol.* 123, 2928-2929, 1979; Paul, W.E., Kishimoto, T., Melchers, F., *et al.*, Nomenclature for secreted regulatory proteins of the immune system (interleukins), *Clin.Immunol.Immunopathol.* 64, 3-4, 1992; IUIS/WHO Standing Committee on Interleukin Designation, Nomenclature for secreted regulatory proteins of the immune system (interleukins): update, *Bull.World Health Org.* 75, 175, 1997). This term was developed to rationalize the nomenclature for these materials as the different terms/names where selected on the basis of activity in a particular assay system rather than an intrinsic physical or biological property; this situation is not unlike that which occurred in blood coagulation somewhat earlier. Thus, lymphocyte activating factor (LAF; mitogenic protein, B-cell differentiation factor) is IL-1 while thymocyte stimulating factor (TSF, T-cell growth factor, killer cell helper factor) is IL-2. See Watson, J. and Mochizuki, D., Interleukin 2: a class of T cell growth factors, *Immunol.Rev.* 51, 287-278, 1980; Mizel, S.B., Interleukin 1 and T cell activation, *Immunol.Rev.* 63, 51-72, 1982; Wagner, H., Hardt, C., Heeg, K. *et al.*, The in vivo effects of interleukin 2 (TCGF), *Immunobiology* 161, 139-156, 1982; Farrar, J.J., Benjamin, W.R., Hilfiker, M.L., *et al.*, The biochemistry, biology, and role of interleukin 2 in the induction of cytotoxic T cell and antibody-forming B cell responses, *Immunol.Rev.* 63, 129-166, 1982; Gillis, S., Interleukin 2: Biology and Biochemistry, *J.Clin.Immunol.* 3, 1-13, 1983; Durum, S.K., Schmidt, J.A., and Oppenheim, J.J., Interleukin 1: an immunological perspective, *Annu.Rev.Immunol.* 3, 263-287, 1985. Work on the interleukins is usually considered within the greater area of cytokines. See Porter, J.R. and Jezová, D., *Circulating Regulatory Factors and Neuroendocrine Function*, Plenum Press, New York, NY, 1990; Kimball, E.S., *Cytokines and Inflammation*, CRC Press, Boca Raton, FL, 1991; Kishimoto, T., *Interleukins: Molecular Biology and Immunology*, Karger, Basal,

1992; Thrompson, A.W., *The Cytokine Handbook*, Academic Press, London, 1994; Austen, K.F., *Therapeutic Immunology*, Blackwell Science, Malden, MA, 2001; Janeway, C.A, Travers, P., Walport, M., and Shlomchik, M., *Immunobiology 5: The Immune System in Health and Disease*, Garland Publishing/Taylor & Francis, New York, NY, 2001; Cruse, J.M., Lewis, R.F., and Wang, H., *Immunology Guidebook*, Elsevier, Amsterdam, Netherland, 2004.

Internal Standard

A compound or material which is not a analyte but is included in an unknown or standard to correct for issues in the processing or analysis of a analyte or analytes; an internal standard is not a calibration standard. See Coleman, D. and Vanatta, L., Statistics in Analytical Chemistry, Part 19-Internal Standards, *American Laboratory*, December, 2005; Julka, S. and Regnier, F., Quantification in proteomics through stable isotope coding: a review, *J.Proteome Res.* 3, 350-363, 2004; Bronstrup, M., Absolute quantification strategies in proteomics based on mass spectrometry, *Expert Rev.Proteomics* 1, 503-512, 2004.

Intrabodies

Intrabodies are intracellular antibodies or functional antibody fragments. Intrabodies can be expressed as intracellular antibodies using recombinant DNA technology and used for the study of intracellular pathways and protein-protein interactions using two-hybrid technology. See Lobato, M.N. and Rabbitts, T.H., Intracellular antibodies and challenges facing their use as therapeutic agents, *Trends Mol.Med.* 9, 390-396, 2003; Stocks, M.R., Intrabodies: production and promise, *Drug Discov.Today* 9, 960-966, 2004; Visintin, M., Meli, G.A., Cannistraci, I., and Cattaneo, A., Intracellular antibodies for proteomics, *J.Immunol.Methods* 290, 135-153, 2004; Miller, T.W. and Messer, A., Intrabody applications in neurological disorders: progress and future prospects, *Mol.Ther.* 12, 394-401, 2005; Stocks, M., Intrabodies as drug discovery tools and therapeutics, *Curr.Opin.Chem.Biol.* 9, 359-365, 2005. Antibodies or antibodies fragments can also be introduced into the cell through the use of cell penetrating peptides. Zhao, Y., Lou, D., Burkett, J., and Kohler, H., Chemical engineering of cell penetrating antibodies, *J.Immunol.Methods* 254, 137-145, 2001; Gupta, B., Levchenko, T.S. and Torchilin, V.P., Intracellular delivery of large molecules and small particles by cell-penetrating proteins and peptides, *Adv.Drgu Deliv.Rev.* 57, 637-651, 2005; De Coupade, C., Fittipaldi, A., Chagnas, V., *et al.*, Novel human-derived cell-penetrating peptides for specific subcellular delivery of therapeutic biomolecules, *Biochem.J.* 390, 407-418, 2005; Gupta, B. and Torchilin, V.P., Transactivating transcriptional activator-mediated drug delivery, *Expert Opin. Drug Deliv.* 3, 177-190, 2006.

Intron

A segment of DNA which is not transcribed into messenger RNA and is designated as non-coding DNA.

Intron Density

Average number of introns per gene over an entire genome. Grover, D., Mukerji, M., Bhatnagar, P., Kannan, K., and Brahmachari, S.K., Alu repeat analysis in the complete human genome: trends and variations with respect to genomic composition, *Bioinformatics* 20, 813-827, 2004; Niu, D.K., Hou, W.R., and Li, S.W., mRNA-mediated intron losses: evidence from extraordinary large exons, *Mol.Biol. Evol.* 22, 1475-1481, 2005; Sironi, M., Menozzi, G., Comi, G.P., *et al.*, Analysis of intronic conserved elements indicates that functional complexity might represent a major source of negative selection on non-coding sequences, *Hum.Mol.Genet.* 14, 2533-2546, 2005; Toyoda, T. and Shinozaki, K., and Tiling array-driven elucidation of transcriptional structures based on maximum-likelihood and Markov models, *Plant J.* 43, 611-621, 2005; Keeling, P.J. and Slamovits, C.H. Causes and effects of nuclear genome reduction, *Curr.Opin.Genet.Dev.* 15, 601-608, 2005; Jeffares, D.C., Mourier, T., and Penny, D., The biology of intron gain and loss, *Trends Genet.* 22, 16-22, 2006; de Cambiare, J.C., Otis, C., Lemieux, C., and Turmel, M., The complete chloroplast genome sequence of the chlorophycean green alga *Scenedesmus obliquus* reveals a compact gene organization and a biased distribution of genes on the two DNA strands, *BMC Evol.Biol.* 6, 37, 2006.

Ion Channel

An integral membrane protein(s) providing for the regulated transport of ions across a membrane via the formation of a pore-like structure. Schonherr, R., Clinical relevance of ion channels for diagnosis and therapy of cancer, *J.Membr.Biol.* 205, 175-184, 2005; Yu, F.H., Yarov-Yarovoy, V., Gutman, G.A., and Catterall, W.A., Overview of molecular relationships in the voltage-gated ion channel superfamily, *Pharmacol.Rev.* 57, 387-395, 2005; Clapham, D.E., Julius, D., Montell, C., and Schultz, G., International Union of Pharmacology XLIX. Nomemclature and structure-function relationships of transient receptor potential channels, *Pharmacol. Rev.* 57, 427-450, 2005.

Ionization Potential

Energy required to remove a given electron from its atomic orbital; value in electron volts (eV).

Ionophore

A chemical compound which binds ions and provides transport across a biological membrane. More recent work has led to the development of ion-specific electrodes and other sensors. One of the most common examples is A23187 which functions as a calcium ionophore (Painter, G.R. and Pressman, B.C., Dynamic aspects of ionophore mediated membrane transport, *Top.Curr.Chem.* 101, 83-110, 1982; Haynes, D.H., Detection of ionophore-cation complexes on phospholipid membranes, *Biochim.Biophys.Acta* 255, 406-410, 1972; Scarpa, A. and Inesi, G., Ionophore mediated equilibration of calcium ion gradients in fragmented-sarcoplasmic reticulum, *FEBS Lett.* 22, 273-276, 1972; Reed, P.W. and Lardy, H.A., A23187: a divalent cation ionophore, *J.Biol.Chem.* 247, 6970-6977, 1972; Haynes, D.H., Detection of ionophore-cation complexes on phospholipid membranes, *Biochim.Biophys.Acta* 255, 406-410, 1972; Chaney, M.O., Demarco, P.V., Jones, N.D., Occolowitz, J.L., The structure of A23187, a divalent cation ionophore, *J.Am.Chem. Soc.* 96, 1932-1933, 1974; Ferreira, H.G. and Lew, V.L., Use of ionophore A23187 to measure cytoplasmic Ca buffering and activation of the Ca pump by internal Ca, *Nature* 259, 47-49, 1976; Estensen, R.D., Reusch, M.E., Epstein, M.L., and Hill, H.R. Role of Ca^{2+} and Mg^{2+} in some human neutrophil functions as indicated by ionophore A23187, *Infect.Immun.* 13, 146-151, 1976). See Shampsipur, M., Avenes, A., Javanbakht, M., *et al.*, A 9,10-anthraquinone

derivative having two propenyl arms as a neutral ionophore for highly selective and sensitive membrane sensors for Copper(II) ion, *Anal.Sci.* 18, 875-879, 2002; Benco, J.S., Nienaber, H.A., and McGimpsey, W.G., Synthesis of an ammonium ionophore and its application in a planar ion-selective electrode, *Anal.Chem.* 75, 152-156, 2003; Kim, Y.K., Lee, Y.H., Lee, H.Y., *et al.*, Molecular recognition of anions through hydrogen bonding stabilization of anion-ionophore adducts: a novel trifluoroacetatophenone-based binding motif, *Org.Lett.* 5, 4003-4006, 2003; Grote, Z., Lehaire, M.L., Scopelliti, R., and Severin, K., Selective complexation of Li^+ in water at neutral pH using a self-assembled ionophores, *J.Am. Chem.Soc.* 125, 13638-13639, 2003; Dhungana, S., White, P.S., and Crumbliss, A.L., Crystal and molecular structures of ionophore-siderophore host-guest supramolecular assemblies relevant to molecular recognition, *J.Am.Chem.Soc.* 125, 14760-14767, 2003; Mahajan, R.K., Kaur, I., Kaur, R., *et al.*, Lipophilic lanthanide tris (beta-diketonate) complexes as an ionophore for Cl^- anion-selective electrodes, *Anal.Chem.* 76, 7354-7359, 2004; Fisher, A.E., Lau, G., and Naughton, D.P., Lipophilic ionophore complexes as superoxide dismutase mimetics, *Biochem.Biophys.Res.Commun.* 329, 930-933, 2005; Zhang, Y.L., Dunlop, J, Phung, T., *et al.*, Supported bilayer lipid membranes modified with a phosphate ionophore, *Biosens. Bioelectron.* 21, 2311-2314, 2006; Shirai, O., Yoshida, Y., and Kihara, S., Voltammetric study on ion transport across a bilayer lipid membrane in the presence of a hydrophobic ion or ionophore, *Anal. Bioanal.Chem.*, 386, 494–504, 2006; Rose, L and Jenkins, A.T., The effect of the ionophore valinomycin on biomimetic solid supported lipid DPPTE/EPC membranes, *Bioelectrochemistry*, 70, 387–393, 2006.

IQ Motif

A linear sequence of amino acids that bind calmodulin and calmodulin-like proteins where IQ are the first conserved amino acids. See Bähler, M. and Rhoads, A., Calmodulin signalling via the IQ motif, *FEBS Lett.* 513, 107-113, 2002; Greeves, M.A. and Holmes, K.C., Structural basis of muscle contraction, *Ann.Rev. Biochem.* 68, 687-728, 1999.

Isobaric

Having the same molecular mass but different chemical properties and structure; such compounds are called isobars (note: the term isobar also had a meaning in atmospheric science). Also, a process or reaction can be considered isobaric if performed under constant pressure within either space or time. See Uline, M.J. and Corti, D.S., Molecular dynamics in the isothermal-isobaric ensemble: the requirement of a "shell" molecule. II. Simulation results, *J.Chem.Phys.* 123, 164102, 2005; Rosgen, J. and Hinz, H.J., Pressure-modulated differential scanning calorimetry: theoretical background. *Anal.Chem.* 78, 991-996, 2006; Wu, W.W., Wang, G., Baek, S.J., and Shen, R.F., Comparative study of the three proteomic quantitative methods, DIGE, cICAT, and iTRAQ, using 2D gel or LC-MALDI TOF/TOF, *J.Proteome Res.* 5, 651-658, 2006; Sachon, E., Mohammed, S., Bache, N., and Jensen, O.N., Phosphopeptide quantitation using amine-reactive isobaric tagging reagents and tandem mass spectrometry: appoication to protein isolated by gel electrophoresis, *Rapid Commun.Mass Spectrom.* 20, 1127-1134, 2006; Langrock, T., Czihal, P. and Hoffman, R., Amino acid analysis by hydrophilic interaction chromatography coupled on-line to electrospray ionization mass spectrometry, *Amino Acids*, 30, 291–297, 2006.

Isocratic

A term used in chromatography to describe a stepwise elution process as opposed to a gradient elution. The term isocratic also refers to a governing system with equality. See Wang, N.W., Ion exchange in purification, *Bioprocess Technol.* 9, 359-400, 1990; Frey, D.D., Feedback regulation in preparative elution chromatography, *Biotechnol.Prog.* 7, 213-224, 1991; Coffman, J.L., Roper, D.K., and Lightfoot, E.N., High-resolution chromatography of proteins in short columns and adsorptive membranes, *Bioseparation* 4, 183-200, 1994; Hajos, P. and Nagy, L., Retention behaviours and separation of carboxylic acids by ion-exchange chromatography, *J.Chromatog.B.Biomed.Sic.Appl.* 717, 72-38, 1998; Marsh, A., Clark, B.J., and Altria, K.D., A review of the background, operating parameters and applications of microemulsion liquid chromatography, *J.Sep.Sci.* 28, 2023-2032, 2005.

Isoelectric Focusing (IEF)

An electrophoretic method for separating amphoteric molecules in pH gradients. Isoelectric focusing is an integral part of the two-dimensional analysis of proteins/peptides in proteomics using immobilized pH gradients (IPG). See Righetti, P.G. and Drysdale, J.W., Isoelectric focusing in polyacrylamide gels, *Biochim. Biophys.Acta* 236, 17-28, 1971; Haglund, H., Isoelectric focusing in pH gradients—a technique for fractionation and characterization of ampholytes, *Methods Biochem.Anal.* 19, 1-104, 1971; Righetti, P.G., and Drysdale, J.W., Small-scale fractionation of proteins and nucleic acids by isoelectric focusing in polyacrylamide gels, *Ann.N.Y.Acad.Sci.* 209, 163-186, 1973; Righetti, P.G., Molarity and ionic strength of focused carrier ampholytes in isoelectric focusing, *J.Chromatog.* 190, 275-282, 1980; Righetti, P.G., Tudor, G., and Gianazza, E., Effect of 2-mercaptoethanol on pH gradients in isoelectric focusing, *J.Biochem.Biophys.Methods* 6, 219-227, 1982; Righetti, P.G., Isoelectric focusing as the crows flies, *J.Biochem.Biophys. Methods* 16, 99-108, 1988; Strege, M.A., and Lagu, A.L., Capillary electrophoresis of biotechnology-derived proteins, *Electrophoresis* 18, 2343-2352, 1997; Korlach, J., Hagedorn, R., and Fuhr, G., pH-Regulated electroretention chromatography: towards a new method for the separation of proteins according to their isoelectric points, *Electrophoresis* 19, 1135-1139, 1998; Molloy, M.P., Two-dimensional electrophoresis of membrane proteins using immobilized pH gradients, *Anal. Biochem* 280, 1-10, 2000; Kilar, F., Recent applications of capillary isoelectric focusing, *Electrophoresis* 23, 3908-3916, 2003; Righetti, P.G., Determination of the isoelectric point of proteins by capillary isoelectric focusing, *J.Chromatog.A* 1037, 491-499, 2004; Stastna, M., Travnicek, M., and Slais, K., New azo dyes as colored point markers for isoelectric focusing in the acidic pH region, *Electrophoresis* 26, 53-59, 2005; Kelly, R.T. and Woolley, A.T., Electric field gradient focusing, *J.Sep.Sci.* 28, 1985-1993, 2005; Righetti, P.G., The Alpher, Bethe, Gamow of isoelectric focusing, the alpha-Centaury of electrokinetic methodologies. Part I, *Electrophoresis* 27, 923-938, 2006.

Isoelectric Point (I$_p$)

The pH at which an amphoteric molecule such as a protein has a net charge of zero. It is, however, possible for a protein at the isoelectric point to have localized areas or patches of positivity or negativity. See Ingram, V.M., Isoelectric point of chymotrypsinogen by a Donnan equilibrium method, *Nature* 170, 250-251, 1952; Harden, V.P. and Harris, J.O., The isoelectric point of bacterial cells, *J.Bacteriol.* 65, 198-202, 1953; Sophianopoulos, A.J. and Sasse, E.A., Isoelectric point of proteins by differential conductimetry, *J.Biol.Chem.* 240, PC1864-PC1866, 1965; Bishop, W.H. and Richards, F.M., Isoelectric point of a protein in the crosslinked crystallized state, *J.Mol.Biol.* 33, 415-421, 1968; McDonagh, P.F. and Williams, S.K., The preparation and use of fluorescent-protein conjugates for microvascular research, *Microvasc.Res.* 27, 14-27, 1984; Palant, C.E., Bonitati, J., Bartholomew, W.R., *et al.*, Nodular glomerulosclerosis associated with multiple myeloma. Role of light chain isoelectric point, *Am.J.Med.* 80, 98-102, 1986; Karpinska, B., Karlsson, M., Schinkel, H., *et al.*, A novel superoxide dismutase with a high isoelectric point in higher plants. Expression, regulation, and protein localization, *Plant Physiol.* 126, 1668-1677, 2001; Lim, T.K., Imai, S., and Matsunaga, T., Miniaturized amperometric flow immunoassay using a glass fiber membrane modified with anion, *Biotechnol.Bioeng.* 77, 758-763, 2002; Cargile, B.J. and Stephenson, J.L., Jr., An alternative to tandem mass spectrometry: isoelectric point and accurate mass for the identification of peptides, *Anal.Chem.* 76, 267-275, 2004; Shi, Q., Zhou, Y., and Sun, Y., Influence of pH and ionic strength on the steric mass-action model parameters around the isoelectric point of protein, *Biotechnol.Prog.* 21, 516-523, 2005; Sillero, A. and Maldonado, A., Isoelectric point determination of proteins and other macromoecules: oscillating method, *Comput. Biol.Med.* 36, 157-166, 2006; Proteins are usually least soluble at their isoelectric point and this has been suggested as a useful tool in crystallization. See Kantaardjieff, K.A. and Rupp, B., Protein isoelectric point as a predictor for increased crystallization screening efficiency, *Bioinformatics* 20, 2162-2168, 2004; Canaves, J.M., Page, R., Wilson, I.A., and Stevens, R.C., Protein biophysical properties that correlate with crystallization success in *Thermotoga maritime*: maximum clustering strategy for structural genomics, *J.Mol.Biol.* 344, 977-991, 2004.

Isopeptide Bond

An amide bond between a carboxyl group of one amino acid and an amino group of another amino acid where either the carboxyl or amino groups or both are not α in position; for example, the peptide bond formed between glutamine and lysine in transamidation reaction or the peptide bond formed with the β-carboxyl group of aspartic acid and the proximate amino group under acid conditions in peptides and proteins; also the bond between ubiquitin and ubiquitin-like modifiers and substrate proteins; more recently, isopeptide bonds have been described from the reaction of homocysteine lactone with ε-amino groups I proteins. See Di Donato, Λ., Ciardiello, M.A., de Nigris, M., Piccoli, R., Mazzarella, L., and D'Alessio, G., Selective deamidation of ribonuclease A. Isolation and characterization of the resulting isoaspartyl and aspartyl derivatives, *J.Biol.Chem.* 268, 4745-4751, 1993; Chen, J.S. and Mehta, K., Tissue transglutaminase: an enzyme with a split personality, *Int.J.Biochem.Cell Biol.* 31, 817-836, 1999; Pickart, C.M., Mechanisms underlying ubiquitination, *Annu.Rev.Biochem.* 70, 502-533, 2001; Perna, A.F., Capasso, R., Lombardi, C., Acanfora, F., Satat, E., and Ingrosso, D., Hyperhomocysteinemia and macromolecule modifications in uremic patients, *Clin.Chem.Lab.Med.* 43, 1032-1038, 2005.

Isosteres

Chemical compounds with the same number of valence electrons but different numbers and types of atoms See Rye, C.S. and Baell, J.B., Phosphate isosteres in medicinal chemistry, *Curr.Med.Chem.*

12, 3127-3141, 2005; Showell, G.A. and Mills, J.S., Chemistry challenges in lead optimization: silicon isosteres in drug discovery, *Drug Discov.Today* 8, 551-556, 2003; Venkatesan, N. and Kim, B.H., Synthesis and enzyme inhibitory activities of novel peptide isosteres, *Curr.Med.Chem.* 9, 2243-2270, 2002; Roy, R. and Baek, M.G., Glycodendrimers: novel glycotope isosteres unmasking sugar coding. Case study with T-antigen markers from breast cancer MUC1 glycoprotein, *J.Biotechnol.* 90, 291-309, 2002.

Isotherm

For chromatography, an arithmetic function which describes the partitioning of chromatographic solute (adsorbate) between solvent and the matrix/adsorbent. The Langmuir Isotherm is an empirical isotherm based on a postulated kinetic mechanism describing an equilibrium process for the process of adsorption based on several assumptions; assumptions include absolute uniformity of the absorbent service, all adsorption occurs by the same mechanism, and adsorbate molecules adsorb in a uniform monolayer on the absorbent. See Jacobson, J., Frenz, J., and Horvath, C., Measurement of adsorption isotherms by liquid chromatography, *J.Chromatog.* 316, 53-68, 1984; Chang, C. and Lenhoff, A.M., Comparison of protein adsorption isotherms and uptake rates in preparative cation-exchange materials, *J.Chromatog. A*, 827, 281-293, 1998; Di Giovanni, O., Mazzotti, M., Morbidell, M. *et al.*, Supercritical fluid simulated moving bed chromatography II. Langmuir isotherm, *J.Chromatog.A.* 919, 1-12, 2001; Grajek, H., Comparison of the differential isosteric adsorption enthalpies and entropies calculated from chromatographic data, *J.Chromatog.* 986, 89-99, 2003; Xia, F., Nagrath, D., and Cramer, S.M., Modeling of adsorption in hydrophobic interaction chromatography systems using a preferential interaction quadratic isotherm, *J.Chromatog.A*, 989, 47-54, 2003; Piatkowski, W., Antos, D., Gritti, F., and Guiochon, G., Study of the competitive isotherm model and the mass transfer kinetics for a BET binary system, *J.Chromatog.A*, 1003, 73-89, 2003; Lapizco-Encinas, B.H. and Pinto, N.G., Determination of adsorption isotherms of proteins by H-root method: comparison between open micro-channels and convential packed column, *J.Chromatog.A* 1070, 201-205, 2005; Cecchi, T., Use of lipophilic ion adsorption isotherms to determine the surface area and the monolayer capacity of a chromatographic packing, as well as the thermodynamic equilibrium constant for its adsorption, *J.Chromatog. A* 1072, 201-206, 2005; Cano, T., Offringa, N.D., and Willson, R.C., Competitive ion-exchange adsorption of proteins: competitive isotherms with controlled competitor concentration, *J.Chromatog.A* 1079, 116-126, 2005; Zhang, W., Shan, Y., and Seidel-Morgenstern, A., Breakthrough curves and elution profiles of single solutes in case of adsorption isotherms with two inflection points, *J.Chromatog.A.*, 1107, 215-225, 2006. This concept is also represented by the Distribution Coefficient (see *British Pharmacopoeia*, 2004). The Distribution Coefficient or Partitioning Coefficient is also used in counter-current distribution.

Isothermal titration calorimetry (ITC)

A physical method which directly measure the heat of interaction of two or more substances. Changes in temperature are measured as one substance is added another and molar heat (kcal/mol) is determined as function of the amount of material added. This information is used to calculate changes in enthalpy (ΔH). This can be applied to large molecule (ligand-receptor) and small molecule interactions. See Ciulli, A. and Abell, C., Biophysical tools to monitor enzyme-ligand interactions of enzymes involved in vitamin biosynthesis, *Biochem.Soc.Trans.* 33(4), 767-771, 2005; Holdgate, G.A. and Ward, W.H.J., Measurements of binding thermodynamics in drug discovery, *Drug Discovery Today* 10, 1543-1550, 2005; Velazquez-Campoy, A., Leavitt, S.A., and Freire, E., Characterization of protein-protein interactions by isothermal titration calorimetry, *Methods Mol.Biol.* 261, 35-54, 2004; Rudolph, M.G., Luz, J.G., and Wilson, I.A., Structural and thermodynamic correlates of T cell signaling, *Ann.Rev.Biophys. Biomol.Struct.* 31, 121-149, 2002.

Isotropy

A physical measurement such as melting point is identical when measured in different principal directions; antonym, anisotropy.

Isotype

Iso (Gr. equal); isotype usually refers to the immunoglobulin subclasses as defined by the chemical and antigenic characteristics of their constant regions. In biological terms, isotype is refers to in terms of holotype (*holo*, GR complete) where an isotype is a biological specimen that is a duplicate of a holotype.

Isotype Switching

The process where antibody class expression changes as in the rearrangment of genes in B cells resulting from the exposure of the B cell to its antigen. Naive B cells express IgA (secretory immunoglobulin) which stimulated or exposed B cells may express other immunoglobulin isotypes including IgG and IgE. Isotype switching is also referred to as antibody class switching. This is a process separate from that of somatic hypermutation which involves the variable regions of the immunoglobulins and is responsible for antibody functional diversity. See Rothman, P. Li, S.C., and Alt, F.W., The molecular events in heavy chain class-switching, *Semin.Immunol.* 1, 65-77, 1989; Whitmore, A.C., Haughton, G. and Arnold, L.W., Isotype switching in CD5 B cells, *Ann.N.Y.Acad.Sci.* 651, 143-151, 1992; Vercelli, D. and Geha, R.S., Regulation of isotype switching, *Curr.Opin.Immunol.* 4, 794-797, 1992; Rothman, P., Interleukin 4 targeting of immunoglobulin heavy chain class-switch recombination, *Res.Immunol.* 144, 579-583, 1993; Snapper, C. and Mond, J.J., Towards a comprehensive view of immunoglobulin class switching, *Immunol.Today* 14, 15-17, 1993; Diamant, E. and Melamed, D., Class switch recombination in B lymphopoiesis: a potential pathway for B cell autoimmunity, *Autoimmun.Rev.* 3, 464-469, 2004; Frasca, D. Riley, R.L., and Blomberg, B.B., *Crit.Rev.Immunol.* 24, 297-320, 2004; Fiset, P.O., Cameron, L., and Hamid, Q., Local isotype switching to IgE in airway mucosa, *J.Allergy Clin.Immunol.* 116, 233-236, 2005; Min, I.M. and Selsing, E., Antibody class switch recombination: roles for switch seqences and mismatch repair proteins, *Adv. Immunol.* 87, 297-328, 2005; Aplan, P.D., Causes of oncogenic chromosomal translocation, *Trends Genet.* 22, 46-55, 2006.

JAK (Janus-associated kinase)

A family of tyrosine kinases involved in signal transduction through cytokine receptors. There are four JAK family members, JAK1, JAK2, JAK3 and TYK2. JAK1 and JAK2 are involved in type II interferon (interferon-gamma) signaling, whereas JAK1 and TYK2 are involved type I interferon signaling. The term *Janus* is derived from the Roman god for doors and pathways frequently depicted with two faces looking in opposite directions. See Karnitz, L.M. and Abraham, R.T., Cytokine receptor signaling mechanisms, *Curr.Opin.Immunol.* 7, 320-326, 1995; Ihle, J.N., The Janus protein tyrosine kinase family and its role in cytokine signaling, *Adv.Immunol.* 60, 1-35, 1995; Gilmour, K.C. and Reich, N.C., Signal transduction and activation of gene transcription by interferon, *Gene Expr.* 5, 1-18, 1995; Gao, B., Cytokines, STATs and liver disease, *Cell.Mol.Immunol.* 2, 92-100, 2005; Yamaoka, K., Saharinen., P., Pesu, M., Holt, V.E., 3rd., Silvennoinen, O. and O'Shea, J.J., The Janus kinases (Jaks), *Genome Biol.* 5, 253 (epub), 2004.

JASPAR

An data base for transcription factor binding: see Sandelin, A., *et al.*, JASPAR: an open-access database for eukaryotic transcription factor binding profiles, *Nuc.Acid Res.* 32, D91-D94, 2004.

Karyology

Study of the nucleus of the cell, specifically the chromosomes. Used in the characterization of master cell banks and working cell banks for recombinant DNA products. See Chiarelli, A.B. and Koen, A.L., *Comparative Karyology of Primates*, Moulton, The Hague, Netherlands, 1979; Macgregor, H.C., *An Introduction to Animal Cytogenetics*, Chapman & Hall, London, UK, 1993; Petricciani, J.C. and Horaud, F.N., Karyology and tumorigenicity testing requirements: past, present, and future, *Dev.Biol.Stand.* 93, 5-13, 1998.

Katal

A International standard (SI; Systems International d'Unites) unit for enzyme activity. A katal (kat) is defined as 1 mol/s. A unit for enzyme activity is defined by the IUBMB as 1 μmol/min; then one unit of enzyme activity is equal to 16.67×10^{-9} kat or 16.67 nkat. This term is used in clinical chemistry more than basic biomedical investigation. See Dybkær, R., Problems of quantities and units in enzymology, *Enzyme* 20, 46-64, 1975; Lehmann, H.P., Metrication of clinical laboratory data in SI units, *Am.J.Clin. Pathol.* 65, 2-18, 1976; Lehman, H.P., SI units, *CRC Crit.Rev.Clin. Lab.Sci.* 10, 147-170, 1979; Bowers, G.N., Jr. and McComb, R.B., A unifying reference system for clinical enzymology: aspartate aminotransferase and the International Clinical Enzyme Scale, *Clin.Chem.* 39, 1128-1136, 1984; Powsner, E.R. SI quantities and units for American Medicine, *JAMA* 252, 1737-1741, 1984; van Assendelft, O.W., The international system of units (SI) in historical perspective, *Am.J.Public Health* 77, 1400-1403, 1987; Dybkær, R. and Storring, P.L., Application of IUPAC-IFCC recommendations on quantities and units to WHO biological reference materials for diagnostic use. International Union of Pure and Applied Chemistry (IUPAC) and International Federation of Clinical Chemistry (IFCC), *Eur.J.Clin.Chem.Clin.Biochem.* 33, 623-625, 1995; Dybkær, R., The tortuous road to the adoption of katal for the expression of catalytic activity by the general conference on weights and measures, *Clin.Chem.* 48, 586-590, 2002.

A fibrous protein found in skin, hair and surface hard tissue such as nails. Keratins are characterized by a relatively high content of sulfur-containing amino acids. See Crewther, W.G., Fraser, R.D., Lennox, F.G., and Lindley, H., The chemistry of keratins, *Adv.Protein Chem.* 20, 191-346, 1965; Roe, D.A., Sulphur metabolism in relation to cutaneous disease, *Br.J.Dermatol.* 81(Suppl 2), 49-60, 1969; Bradbury, J.H., Keratin and its formation, *Curr.Probl.*

Dermatol. 6, 34-86, 1976; Fuchs, E. and Green, H., Multiple keratins of cultured human epidermal cells are translated from different mRNA molecules, *Cell* 17, 573-582, 1979; Sun, T.T., Eichner, R., Nelson, W.C. *et al.*, Keratin classes: molecular markers for different types of epithelial differentiation, *J.Invest.Dermatol.* 81 (Suppl 1), 109S-115S, 1983; Steinert, P.M., Jones, J.C., and Goldman, R.D., Intermediate filaments, *J.Cell.Biol.* 99, 22s-27s, 1984; Virtanen, I., Miettinen, M., Lehto, V.P., *et al.*, Diagnostic application of monoclonal antibodies to intermediate filaments, *Ann. N.Y. Acad. Sci.* 455, 635-648, 1985; Dale, B.A., Resing, K.A., and Lonsdale-Eccles, J.D., Filaggrein: a keratin filament associated protein, *Ann.N.Y.Acad.Sci.* 455, 330-342, 1985.

Keratin

Fuchs, E., Keratin genes, epidermal differentiation and animal models for the study of human skin diseases, *Biochem.Soc.Trans.* 19, 1112-1115, 1991; Oshima, R.G., Intermediate filament molecular biology, *Curr.Opin.Cell Biol.* 4, 110-116, 1992; Coulombe, P.A., The cellular and molecular biology of keratins: beginning a new era, *Curr.Opin.Cell Biol.* 5, 17-29, 1993; Liao, J., Ku, N.O., and Omary, M.B., Keratins and the keratinocyte activation cycle, *J.Invest.Dermatol.* 116, 633-640, 2001; Kierszenbaum, A.L., Keratins: unraveling the coordinated construction of scaffolds in spermatogenesic cells, *Mol.Reprod.Dev.* 61, 1-2, 2002; Lane, E.B. and McLean, W.H., Keratins and skin disorders, *J.Pathol.* 204, 355-366, 2004; Zatloukal, K., Stumpter, C., Fuchsbichler, A. *et al.*, The keratin cytoskeleton in liver disease, *J.Pathol.* 204, 367-376, 2004; Gupta, R. and Ramnani, P., Microbial keratinases and their prospective applications: an overview, *Appl.Microbiol. Biotechnol.* 70, 21-33, 2006.

Kinome

The protein kinases in a proteome of an organism. See Manning, G., Whyle, D.B., Martinez, R., Hunter, T., and Sudarsanam, S., The protein kinase complement of the human genome, *Science* 298, 596-601, 2000; ter Haar, E., Walters, W.P., Pazhanisamy, S., Taslimi, P., Pierce, A.C., Bemis, G.W., Salituro, F.G., and Harbeson, S.L., Kinase chemogenomics: targeting the human kinome for target validation and drug discovery, *Mini Rev.Med.Chem.* 4, 235-253, 2004.

Kinomics

Analysis of all kinases in the proteome of a given organism. See Vieth, M., Sutherland, J.J., Robertson, D.H., and Campbell, R.M., Kinomics: characterizing the therapeutically validated kinase space, *Drug Discov.Today* 10, 839-846, 2005; Johnson, S.A. and Hunter, T., Kinomics: methods for deciphering the kinome, *Nat. Methods* 2, 17-25, 2005.

Knockdown

Suppression or inhibition of transcription by the process of RNA interference/RNA silencing.

krüppel-like Factor

A family of zinc finger transcription factors; the name is derived from the *Drosophila* Krüppel embryonic pattern regulator. See Sugawara, M., Scholl, T., Ponath, P.D., and Strominger, J.L., A factor that regulates the class II major histocompatibility complex gene DPA is a member of a subfamily of zinc finger proteins that includes a *Drosophila* developmental control protein, *Mol.Cell. Biol.* 14, 8438-8450, 1994; Kaczynski, J., Cook, T., and Urrutia, R., SpI- and Krüppel-like transcription factors, *Genome Biology* 4, article 206, 2003.

Labile Zinc

Zinc is an essential mineral for most organisms. Zn is either labile or fixed. Fixed Zn is that Zn tightly bound to metalloproteins while labile zinc is bound loosely to proteins or low molecular thiols such as glutathione. Total cellular Zn is measured by atomic absorption analysis while labile Zn can be measured, for example, with fluorophoric reagents. See Pattison, S.E. and Cousins, R.J., Zinc uptake and metabolism by hepatocytes, *Fed. Proc.* 45, 2805-2509, 1986; Truong-Tran. A.Q., Ho, L.H., Chai, F., and Zalewski, P.D., Cellular zinc fluxes and the regulation of apoptosis/gene-directed cell death, *J.Nutr.* 130(5S Suppl), 1459S-1466S, 2000; Paski, S.C., and Xu, Z., Growth factor stimulated cell proliferation is accompanied by an elevated labile intracellular pool of zinc in 3T3 cells, *Can.J.Physiol.Pharmacol.* 80, 790-795, 2002; Eide, D.J., Multiple regulatory mechanisms maintain zinc homeostasis in *Saccharomyces cerevisiae*, *J.Nutr.* 133 (5 Suppl 1), 1532S-1535S, 2003; Sauer, G.R., Smith, D.M., Cahalane, M., Wu, L.N., and Wuthier, R.E., Intracellular zinc fluxes associated with apoptosis in growth plate chondrocytes, *J.Cell.Biochem.* 88, 954-969, 2003; Roschitzki, B. and Vasak, M., Redox labile site in a Zn4 cluster of Cu4, Zn4-metallothionein-3, *Biochemistry* 42, 9822-9828, 2003; Ho, L.H., Ruffin, R.E., Murgia, C. *et al.*, Labile zinc and zinc transporter ZnT4 in mast cell granules: role in regulation of capases activation and NF-κB translocation, *J.Immunol.* 172, 7750-7760, 2004; Atsriku, C., Scott, G.K., Benz, C.C., and Baldwin, M.A., Reactivity of zinc finger cysteines: chemical modification within labile zinc fingers in estrogen receptors, *J.Am.Soc.Mass Spectrom.* 16, 2017-2026, 2005; Lee, J.Y., Hwang, J.J., Park, M.H., and Koh, J.Y., Cytosolic labile zinc: a marker for apoptosis in the developing rat brain, *Eur.J.Neurosci.* 23, 435-442, 2006; Zalewski, P.. Truong-Tran, A., Lincoln, S., *et al.*, Use of a zinc fluorophore to measure labile pools of zinc in body fluids and cell-conditioned media, *BioTechniques* 40, 509-520, 2006; Haase, H., Hebel, S., Engelhardt, G., and Rink, L., Flow cytometric measurement of labile zinc in peripheral blood mononuclear cells, *Anal.Biochem.* 352, 222-230, 2006.

Lactoferrin

Lactoferrin is an iron binding protein of very high affinity originally described in milk and other secreted biological fluids such as saliva (see Weinberg, E.D., The therapeutic potential of lactoferrin, *Expert Opin.Investig.Drugs* 12, 841-851, 2003; Van Nieuw Amerongen, A., Bolscher, J.G., and Veerman, E.C., Salivary proteins: protective and diagnostic value in cariology? *Caries Res.* 38, 247-253, 2004). Lactoferrin is also found specific granules of neutrophils. Lactoferrin is considered to play an important role in the non-specific defense process by sequestering iron required bacterial growth. See Goldman, A.S. and Smith, C.W., Host resistance factors in human milk, *J.Pediatr.* 82, 1082-1090, 1973, Bullen, J.J., Rogers, H.J., and Griffiths, E., Role of iron in bacterial infection, *Curr.Top.Microbiol. Immunol.* 80, 1-35, 1978; Reiter, B., The biological significance of lactoferin, *Int.J.Tissue React.* 5, 87-96, 1983; Birgens, H.S., The biological significance of lactoferrin in haematology,

Scand.J.Haematol. 33, 225-230, 1984; De Sousa, M., Breedvelt, F., Dynesius-Trentham, R., and Lum, J., Iron, iron-binding proteins and immune system cells, *Ann.N.Y.Acad.Sci.* 526, 310-322, 1988; Levay, P.F. and Viljoen, M., Lactoferrin: a general review, *Haematologica* 80, 252-267, 1995; Legrand, D., Elass, E., Pierce, A., and Mazurier, J., Lactoferrin and host defense: an overview of its immuno-modulating and anti-inflammatory properties, *Biometals* 17, 225-229, 2004; Yalcin, A.S., Emerging therapeutic potential of whey proteins and peptides, *Curr.Pharm Des.* 12, 1637-1643, 2006.

Latarcins

A newly defined group of antimicrobial and cytolytic peptides from spider venom. See Kozlov, S.A., Vassilevski, A.A., Feofanov, A.V., *et al.*, Latarcins, antimicrobial and cytolytic peptides from the venom of the spider *Lachesana tarabaevi* (Zodariidae) that exemplify biomolecular diversity, *J.Biol.Chem.* 281, 20983-20992, 2006.

Lectin

A protein which selectively binding carbohydrates. Lectin affinity columns can be used for the purification of carbohydrate chains and glycoproteins. Lectins are also used in histochemistry and cytochemistry. See Cohen, E., *Recognition Proteins, Receptors, and Probes: Invertebrates: Proceedings of a Symposium Entitled Recognition and Receptor Display, Lectin Cell Surface Receptors and Probes*, A.R. Liss, New York, New York, 1984; Gabius, H.J. and Gabius, S., *Lectins and Glycobiology*, Springer-Verlag, Berlin, Germany, 1993; Fukuda, M. and Kobata, A., *Glycobiology: A Practical Approach*, IRL Press at Oxford University Press, Oxford, UK, 1993; Doyle, R.J. and Shifkin, M., *Lectin-Microorganism Interactions*, Marcel Dekker, New York, New York, 1994; Brooks, S.A., Leathern, A.J.C., and Schumacher, L., *Lectin Histochemistry: A Concise Practical Handbook*, BIOS Scientific, Oxford, UK, 1997; Caron, M. and Sève, A.-P., Lectins and Pathology Harwood Academic, Amsterdam, Netherlands, 2003.

Linkage Group

A group of genes inherited as a unit such they are described a linked such that disparate phenotypic expressions are also described as linked. See Lamm, LU. and Petersen, G.B., The HLA genetic linkage group, *Transplant Proc.* 11, 1692-1696, 1979; Campbell, R.D., Dunham, I., and Sargent, C.R., Molecular mapping of the HLA-linked complement genes and the RCA linkage group, *Exp.Clin.Immunogenet.* 5, 81-98, 1988; Haig, D., A brief history of human autosomes, *Philos.Trans.R.Soc.Lond. B Biol.Sci.* 354, 1447-1470, 1999.

Lipofection

Originally described as cellular membrane translocation of DNA for gene therapy via the use of cationic lipids as a micelle. More generally the membrane translocation of RNA or DNA encapsulated in a lipid micelle and is used now for RNAi studies. The liposome and its cargo are referred to as a lipoplex. See Hart, S.L., Lipid carriers for gene therapy, *Curr.Drug.Deliv.* 2, 423-438, 2005; Zuhorn, I.S., Kalicharan, R., and Hoekstra, D., Lipoplex-mediated transfection of mammalian cells occurs through the cholesterol-dependent clathrin-mediated pathway of endocytosis, *J.Biol. Chem.* 277, 18021-18028, 2002.

Lipophilic

Affinity for hydrophobic materials such as lipids; compounds which will dissolve in organic/non-polar solvents such as benzene or cyclohexane but not in water; also hydrophobicity. This quality is frequently measured by distribution or partitioning in an octanol-water system and can be assigned a value such as Log P. Lipophilicity can also be measured by retention on an HPLC column with a suitable matrix or on thin-layer chromatography. See Markuszewski, M.J., Wiczling, P., and Kaliszan, R., High-throughput evaluation of lipophilicity and acidity by new gradient HPLC methods, *Comb.Chem.High Throughput Screen.* 7, 281-289, 2004; Klopman, G. and Zhu, H., Recent methodologies for the estimation of *n*-octanol/water partition coefficients and their use in the prediction of membrane transport properties of drugs, *Mini Rev.Med.Chem.* 5, 127-133, 2005. Mannhold, R., The impact of lipophilicity in drug research: a case report on beta-blockers, *Mini Rev.Med.Chem.* 5, 197-205, 2005; Gocan, S., Cimpan, G., and Comer, J., Lipophilicity measurements by liquid chromatography, *Adv.Chromatog.* 44, 79-176, 2006.

Liposomes

A relatively large (nano to micro) micelle composed of polar lipids. There is considerable interest in liposomes as models for biological membranes and for drug delivery. See Bangham, A.D., Lipid bilayers and biomembranes, *Annu.Rev.Biochem.* 41, 753-776, 1972; Gulik-Krzywicki, T., Structural studies the association between biological membranes components, *Biochim.Biophys.Acta* 415, 1-28, 1975; Pressman, B.C., Biological applications of ionophores, *Annu.Rev.Biochem.* 45, 501-530, 1976; Schreier, S., Polnaszek, C.F., and Smith, I.C., Spin labels in membranes. Problems in practice, *Biochim.Biophys.Acta* 515, 395-436, 1978; Hart, S.L., Lipid carriers for gene therapy, *Curr.Drug Deliv.* 2, 423-428, 2005; Zamboni, W.C., Liposomal, nanoparticle, and conjugated formulations of anticancer agents, *Clin.Chem.Res.* 11, 8230-8234, 2005; Taylor, T.M., Davidson, P.M., Bruce, B.D., and Weiss, J., Liposomal nanocapsules in food science and agriculture, *Crit.Rev.Food Sci.Nutr.* 45, 587-605, 2005; Kshirsager, N.A., Pandya, S.K., Kirodian, G.B ., and Sanath, S., Liposomal drug delivery system from laboratory to clinic, *J.Postgrad.Med.* 51(Suppl 1), S5-S15, 2005; Paleos, C.M. and Tsiourvas, D., Interaction between complementary liposomes: a process leading to multicompartment systems formation, *J.Mol. Recognit.* 19, 60-67, 2006.

Localized Surface Plasmon Resonance

The use of noble metal (Ag, Au) nanoparticles as sensors for macromolecular interactions. This phenomena is related to surface plasmon resonance and is based on the spectral properties of nanoparticles as compared to the bulk metal. See Haes, A.J. and Van Duyne, R.P., A unified view of propagating and localized surface plasmon resonance biosensors, *Anal.Bioanal. Chem.* 370, 920-930,2004; Haes, A.J., Stuart, D.A., Nie, S., and Van Duyne, R.P., Using solution-phase nanoparticles, surface-confined nanoparticle arrays and single nanoparticles as biological sensing platforms, *J.Fluoresc.* 14, 355-367, 2004; Dahlin, A., Zach, M., Rindzevicius, T., *et al.*, Localized surface plasmon resonance sensing of lipid-membrane-mediated biorecognition events, *J.Am.Chem.Soc.* 127, 5043-5048, 2005; Endo, T., Kerman, K., Nagatani, N., Takamura, Y., and Tamiya, E., Label-free detection of peptide nucleic acid-DNA hybridization using localized surface plasmon resonance based optical biosensor, *Anal.Chem.* 77, 6976-6984, 2005; Wang, Y., Chen, H., Dong, S., and Wang, E., Surface enhanced Raman scattering of *p*-aminothiophenol self-assembled monolayers in sandwich structure fabricated on glass, *J.Chem.Phys.* 124, 74709, 2006.

Locus Control Region

A *cis*-acting region on a gene locus which controls the expression of a transgene *in vivo*; a region in a gene sequence which regulatory the independent expression of a gene in transgenic mice; a regulatory element critical for globin gene expression. See Festenstein, R. and Kioussis, D., Locus control regions and epigenetic chromatin modifiers, *Curr.Opin.Genet.Dev.* 10, 199-203, 2000; Levings, P.P. and Bungert, J., The human β-globin locus region. A center of attraction. *Eur.J.Biochem.* 269, 1589-1599, 2002; Dekker, J., A closer look at long-range chromosomal interactions, *Trends Biochem.Sci.* 28, 277-280, 2003; Harrow, F. and Ortiz, B.D., The TCRα locus control region specifies thymic, but not peripheral patterns of TCRα gene expression, *J.Immunol.* 175, 6659-6667, 2005; Dean, A., On a chromosome far, far away: LCRs and gene expression, *Trends Genet.* 22, 38-45, 2006.

Longin Domain

An *N*-terminal domain in non-syntaxin soluble *N*-ethylmaleimide-sensitive factor(NSF) attachment protein receptors (SNARE) which have been suggested to regulate membrane trafficking processes. Such proteins are described as longins (longin domains are also found in YKT-like proteins). The term longin was suggested to differentiate such domains from "brevins" which are shorter *N*-terminal regions found in vesicle-associated membrane proteins See Filippini, F., Rossi, V., Galli, T., *et al.*, Longins: a new evolutionary conserved VAMP family sharing a novel SNARE domain, *Trends Biochem.Sci.* 26, 407-409, 2001; Dietrich, L.E.P., Boeddinghaus, C., LaGrassa, T.J., and Ungermann, C., Control of eukaryotic membrane fusion by N-terminal domain of SNARE proteins, *Biochim.Biophys.Acta* 1641, 111-119, 2003; Martinez-Arca, S., Rudge, R., Vacca, M. *et al.*, A dual mechanism controlling the localization and function of exocytic v-SNAREs, *Proc.Natl. Acad.Sci.USA* 100, 9011-9016, 2003; Rossi, V., Picco, R., Vacca, M. *et al.*, VAMP subfamilies identified by specific R-SNARE motifs, *Biology of the Cell* 96, 251-256, 2004; Uemura, T., Sato, M.H., and Takeyasu, K., The longin domain regulates subcellular targeting of VAMP7 in *Arabidopsis thaliana*, *FEBS Lett.* 579, 2842-2846, 2005; Schlenker, O., Hendricks, A., Sinning, I., and Wild, K., The structure of the mammalian signal recognition particle (SRP) receptor as prototype for the interaction of small GTPases with longin domains, *J.Biol.Chem.* 281, 8898-8906, 2006.

Luminescence

The emission of electromagnetic radiation from an excited molecule when an electron returns to ground state from an excited state (other than from thermal energy changes). Fluorescence, phosphorescence and chemiluminescence are forms of luminescence. The decay time for fluorescence is shorter than phosphorescence and both phenomena result from irradiation of compounds with electromagnetic irradiation while chemiluminescence is the result of a chemical reaction, usually oxidation. Phosphorescence also shows a larger shift in the wavelength of emitted light than fluorescence. Electrochemiluminescence differs from all of the above in that in that chemiluminescence is generated at the surface of an electrode. Hemmila, I. and Laitala, V., Progress in lanthanides as luminescent probes, *J.Fluoresc.* 15, 529-542, 2005; Tsuji, F.I., Role of molecular oxygen in the bioluminescence of the firefly squid, *Watasenia scintillans*, *Biochem.Biophys.Res.Commun.* 338, 250-253, 2005; Abrams, B.L. and Hollaway, P.H., Role of the surface in luminescent processes, *Chem.Rev.* 104, 5783-5801, 2004; Aslan, K., Lakowicz, J.R., and Geddes, C.D., Plasmon light scattering in biology and medicine: new sensing approaches, visions, and perspectives, *Curr.Opin.Chem.Biol.* 9, 538-544, 2005; Richter, M.M., Electrochemiluminescence, *Chem.Rev.* 104, 3003-3036, 2004. Medlycott, E.A. and Hanan, G.S., Designing tridentate ligands for ruthenium(II) complexes with prolonged room temperature luminescence lifetimes, *Chem.Soc.Rev.* 34, 133-142, 2005.

Lutheran Glycoprotein

The Lutheran glycoprotein is one of the two components of the Lutheran blood group system, a family of red blood cell antigens. The Lutheran glycoprotein together with basal cell adhesion molecule which are together recognized as CD239 are members of the immunoglobulin superfamily. See Telen, M.J., Lutheran antigens, CD44-related antigens, and Lutheran regulatory genes, *Transfus.Clin.Biol.* 2, 291-301, 1995; Daniels, G. and Crew, V., The molecular basis of the Lutheran blood group antigens, *Vox.Sang.* 83(Suppl 1), 189-192, 2002; Kikkawa, Y. and Miner, J.H., Review: Lutheran/B-CAM: a laminin receptor on red blood cells and in various tissues, *Connect.Tissue Res.* 46, 193-199, 2005; Eyler, C.E. and Telen, M.J., The Lutheran glycoprotein; a multifunction adhesion receptor, *Transfusion* 46, 668-667, 2006.

Lysosomes

An intracellular organelle in a eukaryotic cell responsible for **controlled** intracellular digestion. Lysosomal protein degradation is also involved in MHC class II antigen preparation. Lysosomes contain a variety of hydrolytic enzymes having an acid pH optimum. See Wolf. D.H., From lysosome to proteosome: the power of yeast in the dissection of proteinase function in cellular regulation and waste disposal, *Mol.Life Sci.* 61,1601-1614, 2004; Luzio, J.P., Pryor, P.R., Gray, S.R., Gratian, M.J., Piper, R.C., and Bright, N.A., *Biochem.Soc.Symp.* 72, 77-86, 2005; Bagshaw, R.D., Mahuran, D.J., and Callahan, J.W., Lysosomal membrane proteomics and biogenesis of lysosomes, *Mol.Neurobiol.* 32, 27-41, 2005; Hsing, L.C., and Rudensky, A.Y., The lysosomal cysteine proteases in MHC class II antigen presentation, *Immunol.Rev.* 207, 229-241, 2005; Hideshima, T., Bradner, J.E., Chauhan, D., and Anderson, K.C., Intracellular protein degradation and its therapeutic implications, *Clin.Cancer Res.* 11, 8530-8533, 2005. Mellman, T. Antigen processing and presentation by dendritic cells: cell biological mechanisms, *Adv.Exp.Med.Biol.* 560, 63-67, 2005; Lip, P., Gregg, J.L., Wang, N., Zhou, D., O'Donnell, P., Blum, J.S, and Crotzer, V.L., Compartmentalization of class II antigen presentation: contribution of cytoplasmic and endosomal processing, *Immunol.Rev.* 207, 206-217, 2005; Gatti, E. and Pierre, P., Understanding the cell biology of antigen presentation: the dendritic cell contribution, *Curr.Opin.Cell Biol.* 15, 468-473, 2003; Honey, K. and Rudensky, A.Y., Lysomal cysteine proteases regulate antigen presentation, *Nat.Rev.Immunol.* 3, 472-482, 2003.

Macrolide

A large ring structure with many functional groups. An example is provided by the kabiramides. See Petchprayoon, C., Swanbonriux, K., Tanaka, J., Yan, Y., Sakata, T., and Marriott, G.,

Fluorescent kabiramides: new probes to quantify actin in vitro and in vivo, *Bioconjugate Chem.* 16, 1382-1389, 2005.

Macrophage

An immune system cell which is derived from a monocyte after passage through the endothelium and is a phagocytic cell and an antigen presenting cell. See *The Macrophage*, ed. N.N. Pearsall and R.S. Weiser, Lea & Febiger, Philadelphia, PA, 1970; Carr, I., *The Macrophage: A Review of Ultrastructure and Function*, Academic Press, London, 1973; *Immunobiology of the Macrophage*, ed. D.S. Nelson and P. Alexander, Academic Press, New York, NY, 1976; Horst, M., *The Human Macrophage System: Activity and Functional Morphology*, Karger, Basel, 1988; *Macrophage Biology and Activation*, ed. S.W. Russell and S. Gordon, Springer-Verlag, Berlin, 1992; *Macrophage-Pathogen Interactions*, ed. B.S.Zwilling and T.K. Eisenstein, M. Dekker, New York, NY., 1992; *The Macrophage*, ed. B. Bernard and C.E. Lewis, Oxford University Press, Oxford, UK, 2002. Foam cells which are derived from macrophages play an important role in the pathogenesis of atherosclerosis (Schwartz, C.J., Valente, A.J., Sprague, E.A., *et al.*, The pathogenesis of atherosclerosis: an overview, *Clin.Cardiol.* 14(2 Suppl 1), I1-I16, 1991; Osterud, B. and Bjorklid, E., Role of monocytes in atherogenesis, *Physiol.Rev.* 83, 1069-1112, 2003; Linton, M.F. and Fazio, S., Macrophages, inflammation, and atherosclerosis, *Int.J.Obes.Relat.Metab.Disord.* 27(supp 3), S35-S40, 2003; Shashkin, P., Dragulev, B., and Ley, K., Macrophage differentiation to foam cells, *Curr.Pharm.Des.* 11, 3061-3072, 2005; Bobryshev, Y.V., Monocyte recruitment and foam cell formation in atherosclerosis, *Micron* 37, 208-222, 2006.

Macropinocytosis

Actin-mediated endocytotic process possessing unique ultrastructural features such as the formation of the macropinosome. See Pratten, M.K. and Lloyd, J.B., Effects of temperature, metabolic inhibitors and some other factors on fluid-phase and adsorptive pinocytosis by rat peritoneal macrophages, *Biochem.J.* 180, 567-571, 1979; Sallusto, F., Cella, M., Danieli, C. and Lanzavecchia, A., Dendritic cells use macro-pinocytosis and the mannose receptor to concentrate macromolecules in the major histocompatibilty complex class II compartment: downregulation by cytokines and bacterial products, *J.Exp.Med.* 182, 389-400, 1995; Kaplan, I.M., Wadia, J.S., and Dowdy, S.F., Cationic TAT peptide transduction domain enters cells by macropinocytosis, *J.Control.Release* 192, 247-253, 2005; Chia, C.P., Gomathinayagam, S., Schmaltz, R.J., and Smoyer, L.K., Glycoprotein gp130 of dictyostelium discodeum influences macropinocytosis and adhesion, *Mol. Biol.Cell* 16, 2681-2693, 2005; Melikov, K. and Chernomordik, L.V., Arginine-rich cell penetrating peptides: from endosomal uptake to nuclear delivery, *Cell.Mol.Life Sci.* 62, 2739-2749, 2005; Kirkham, M. and Parton, R.G., Clathrin-independent endocytosis: new insights into caveolae and non-caveolar lipid raft carriers, *Biochim.Biophys.Acta* 1746, 349-363, 2005; Zaro, J.L., Rajapaksa, T.E., Okamoto, C.T., and Shen, W.C., Membrane transduction of oligoarginine in HeLa cells is not mediated by macropinocytosis, *Mol.Pharm.* 3, 181-186, 2006; Mettlen, M., Platek, A., Van Der Smissen, P., *et al.*, Src triggers circular ruffling and macropinocytosis at the apical surface of polarized MDCK cells, *Traffic* 7, 589-603, 2006; von Delwig, A., Hilkens, C.M., Altmann, D.M. *et al.*, Inhibition of macropinocytosis blocks antigen presentation of type II collagen in vitro and in vivo in HLA-DR1 transgenic mice, *Arthritis Res.Ther.* 8, R93, 2006.

Madin-Darby canine Kidney (MDCK)

Used in reference to a mammalian cell line (MDCK cells) used to represent polarized epithelial cells. See Hidalgo, I.J., Assessing the absorption of new pharmaceuticals, *Curr.Top.Med.Chem.* 1, 385-401, 2001; Cohen, D. and Musch, A., Apical surface formation in MDCK cells: regulation by the serine/threonine kinase EMK1, *Methods* 30, 69-276, 2003; Sidorenko, Y. and Reichl, U., Structured model of influenza virus replication in MDCK cells, *Biotechnol.Bioeng.* 88, 1-14, 2004; Urquhart, P., Pang, S., and Hooper, N.M., *N*-Glycans as apical targeting signals in polarized epithelial cells, *Biochem.Soc.Symp.* 72, 39-45, 2005.

Maillard Reaction

A reaction (named after Louis-Camille Maillard) between a protein amino group, usually the epsilon-amino group of lysine and a reducing sugar/aldose. There is an initial condensation reaction to form a Schiff base which undergoes rearrangement to form an Amadori product. This process can initiate a chain reaction resulting in protein cross-linking and the formation of complex chemicals referred to as advanced glycation end products (AGE). Reaction with chemicals such as glucose and methylglyoxal results Maillard Reactions. The browning of foods and the tanning of animal skin are examples of the Maillard reaction. See Waller, G.R. and Feather, M.S., *The Maillard Reaction in Foods and Nutrition*, American Chemical Society, Washington, DC, 1983; Labuza, T.P., *Maillard Reactions in Chemistry, Food and Nutrition*, Royal Society of Chemistry, Cambridge, UK, 1994; Ikan, R., *The Maillard Reaction: Consequences for the Chemical and Life Sciences*, Wiley, Chichester, West Sussex, UK, 1996; Fayle, S.E. and Gerrard, J.A., *The Maillard Reaction*, Royal Society of Chemistry, Cambridge, UK, 2002; Nursten, H., *The Maillard Reaction. Chemistry, Biochemistry, and Implications*, Royal Society of Chemistry, Cambridge, UK, 2005; Hodge, J.E., Chemistry of the browning reactions in model systems, *J.Agric.Food Chem.* 1, 928-943, 1953; Njoroge, F.G. and Monnier, V.M., The chemistry of the Maillard reaction under physiological conditions: a review, *Prog.Clin.Biol.Res.* 304, 85-107, 1989; Kaanane, A. and Labuza, T.P., The Maillard reaction in foods, *Prog.Clin.Biol. Res.* 304, 301-327, 1989; Marko, D., Habermeyer, M., Kemény, M., *et al.*, Maillard reaction products modulating the growth of human tumor cells in vitro, *Chem.Res.Toxicol.* 16, 48-55, 2003; Takeguchi, M., Yamagishi, S., Iwaki, M., *et al.*, Advanced glycation end product (age) inhibitors and their therapeutic implications in diseases, *Int.J.Clin.Pharmacol.Res.* 24, 95-1010, 2004; Jing, H. and Nakamura, S., Production and use of Maillard products as oxidative stress modulators, *J.Med.Food* 8, 291-298, 2005; Lederer, M.O., Gerum, F., and Severin, T., Cross-linking of proteins by Maillard processes-model reactions of D-glucose or methylglyoxal with butylamine and guanidine derivatives, *Bioorg.Med.Chem.* 6, 993-1002, 1998; Oya, T., Hattori, N., Mizuno, Y. *et al.*, Methylglyoxal modification of protein. Chemical and immunochemical characterization of methylglyoxal-arginine adducts, *J.Biol.Chem.* 274, 18492-18502, 1999; Sell, D.R., Biemel, K.M., Reihl, O., Glucosepane is a major protein cross-link of the senescent human extracellular matrix. Relationship with diabetes, *J.Biol.Chem.* 280, 12310-12315, 2005.

Major Groove And Minor Groove – DNA structure

The channels or grooves formed when two complementary strands of DNA hybridize with each other to form a double helix. The major groove is ~22 Å wide and the minor groove is ~11 Å wide. See Jovin, T.M., McIntosh, L.P., Anrdt-Jovin, D.J.. *et al.*, Left-handing DNA: from synthetic polymers to chromosomes, *J.Biomol.Struct.Dyn.* 1, 21-57, 1983; Uberbacher, E.C., and Bunick. G.J., Structure of the nucleosome core particle at 8 Å resolution, *J.Biomol.Struct.Dyn.* 7, 1-18, 1989; Reddy, B.S., Sondi, S.M., and Iown, J.W., Synthetic DNA minor groove-binding drugs; Haq, I., Thermodynamics of drug-DNA interactions, *Arch.Biochem.Biophys.* 403, 1-15, 2002; Sundaralingam, M. and Pan, B., Hydrogen and hydration of DNA and RNA oligonucleotides, *Biophys.Chem.* 95, 273-282, 2002; Susbielle, G., Blatters, R., Brevet, V., Monod, C., and Kas, E., Target practice: aiming at satellite repeats with DNA minor groove binders, *Curr.Med. Chem.Anticancer Agents* 5, 409-420, 2005; Dragan, A.I.. Li, Z., Makeyeva, E.I., *et al.*, Forces driving the binding of homeodomains to DNA, *Biochemistry* 45, 141-151, 2006; Horton, J.R., Zhang, X., Maunus, R., *et al.*, DNA nicking by HikP1I endonuclease: bending, based flipping and minor groove expansion, *Nucleic Acids Res.* 34, 939-948, 2006; Lamoureux, J.S. and Glover, J.N., Principles of protein-DNA recognition revealed in the structural analysis of Ndt80-MSE DNA complexes, *Structure* 14, 555-565, 2006.

Major Histocompatibility Complex (MHC)

The major histocompatibility complex (MHC) or locus is a cluster of genes on chromosome 6 (chromosome 17 in the mouse) which encodes a family of membrane glycoproteins referred to as MHC protein or molecules. There are some non-membrane proteins such as HLA-DM and HLA-DO which are encoded by the MHC and function in the processing of peptides for delivery to the MHC membrane glycoproteins. MHC membrane glycoproteins are found in antibody presenting cells (APCs; professional antigen presenting cells, dendritic cells, macrophages, and B-cells) and "present" antigens to effector CD-4 and CD-8 T-cells. MHC membrane glycoproteins are divided into two groups; MHC class I and MHC class II. MHC class I proteins present peptides generated in the cytoplasm to CD 8 T-cells stimulating the formation of cytotoxic T-cells. Peptides for MHC class I receptors are processed by proteosomes and can be derived from virus-infected cells. MHC class II receptors present peptides to CD-4 T-cells which activate B-cells to form plasma cells which synthesize and secrete antibody. Peptides for MHC class II receptors are derived from the action of lysosomal proteases on endosomes. See Janeway, C.A., Jr., Travers, P., Walport, M., and Capra, J.D., *Immunobiology. The Immune System in Health and Disease*, 4th edn., Garland Publishing, New York, New York, 1999; Lyczak, J.B., The major histocompatibility complex, in *Immunology, Infection, and Immunity*, ed. G.B. Pier, J.B. Lyczak,and L.M. Wetzler, ASM Press, Washington, DC, 2004, Chapter 11, pp. 261-282; Krawczyk, M. and Reith, W., Regulation of MHC class II expression, a unique regulatory system identified by the study of a primary immunodeficiency disease, *Tissue Antigens* 67, 183-197, 2006; Serrano, N.C., Millan, P., and Paez, M.C., Non-HLC associations with autoimmune diseases, *Autoimmune Rev.* 5, 209-214, 2006; Drozina, G., Kohoutek, J., Jabrane-Ferrat, N., and Peterlin, B.M., Expression of MHC II genes, *Curr.Top.Microbiol. Immunol.* 290, 147-170, 2005; Hoglund, P., Induced peripheral regulatory T cells: the family grows larger, *Eur.J.Immunol.* 36, 264-266, 2006; Wucherpfennig, K.W., The structural interactions between T cell receptors and MHC-peptide complexes place physical limits on self-nonself discrimination, *Curr.Top. Microbiol.Immunol.* 296, 19-37, 2005.

MAPPER

Search engine for the identification of putative transcription factor binding sites (TFBSs): see Marinescu, V.D., Kohane, I.S., and Riva, A., MAPPER: a search engine for the computational identification of putative transcription factor binding sites in multiple genomes, *BMC Informatics* 6, 79, 2005.

Mass Spectrometer

A device that assigns mass-to-charge ratios to ions based on their momentum, cyclotron frequency, time-of-flight, or other parameters. See Kiser, R.W., *Introduction to Mass Spectrometry and Its Applications*, Prentice-Hall, Englewood Cliffs, NJ, 1965; Blauth, E.W., *Dynamic Mass Spectrometers*, Elsevier, Amsterdam, Netherlands, 1966; Majer, J.R., *The Mass Spectrometer*, Wykeham Publications, London, UK, 1977; Herbert, C.G. and Johnstone, R.A.W., *Mass Spectrometry Basics*, CRC Press, Boca Raton, FL, 2003.

Mast Cell

A type of leukocyte found in tissues with large basophilic secretory granules containing histamine and other physiological materials such as heparin during inflammatory responses. See Puxeddu, H., Ribetti, D., Crivellato, E., and Levi-Schaffer, F., Mast cells and eosinophils: a novel link between inflammation and angiogenesis in allergic diseases, *J.Allergy Clin.Immunol.* 116, 531-536, 2005. Galli, S.J., Kalesnikoff, J., Grimbaldeston, M.A., Piliponsky, A.M., Williams, C.M., and Tsai, M., Mast cells as "tunable" effector and Immunoregulatory cells: recent advances, *Annu.Rev.Immunol.* 23, 749-786, 2005; Saito, H., Role of mast cell proteases in tissue remodeling, *Chem.Immunol.Allergy* 87, 80-84, 2005; Vliagoftis, H. and Befus, A.D., Mast cells at mucosal frontiers, *Curr.Mol. Med.* 5, 573-589, 2005; Krishnaswamy, G., Ajitawi, O., and Chi, D.S., The human mast cell: an overview, *Methods Mol.Biol.* 315, 13-34, 2006.

Metabolipidomics

The study of the metabolism of lipids in the proteome. See Bleijerveld, O.B., Howeling, M., Thomas, M.J., and Cui, Z., Metabolipidomics: profiling metabolism of glycerophospholipid species by stable isotopic precursors and tandem mass spectrometry, *Anal.Biochem.* 352, 1-14, 2006.

Metabolite/Metabolic Profiling

Identification/quantification of a select group of metabolites, generally part of a specific pathway such as glycolysis or fatty acid synthesis. See Dunn, W.B. and Ellis, D.I., Metabolomics: current analytical platforms and methodologies, *Trends Anal.Chem.* 24, 285-294, 2005.

Metabolome

The total metabolites produced by the products of the genome. See Tweeddale, H., Notley-McRobb, L. and Perenci, T., Effect of slow growth on metabolism of *Escherichia coli*, as revealed by global metabolite pool ("metabolome") analysis, *J.Bacteriol.* 180, 5109-5116, 1998; Tweeddale, H., Notley-McRobb, L., and Ferenci, T., Assessing the effect of reactive oxygen species on *Escherichia coli* using a metabolome approach, *Redox.Rep.* 4, 237-241, 1999; ter

Kuile, B.H., and Westerhoff, H.V., Transcriptome meets metabolome: hierarchical and metabolic regulation of the glycolytic pathway, *FEBS Lett.* 500, 169-171, 2001; Mendes, P., Emerging bioinformatics for the metabolome, *Brief Bioinform.* 3, 134-145, 2002; Mazurek, S., Grimm, H., Boschek, C.B., Vaupel, P., and Eingebrodt, E., Pyruvate kinase type? *Phytochemistry* 62, 837-849, 2003; Nobeli, I., Ponstingl, H., Krissinel, E.B., and Thornton, J.M., A structure-based anatomy of the E.coli metabolome, *J.Mol.Biol.* 334, 697-719, 2003; Parsons, L. and Orban, J., Structural genomics and the metabolome: combining computational and NMR methods to identify target ligands, *Curr.Opin.Drug Discov.Devel.* 7, 62-68, 2004; Soloviev, M. and Finch, P., Peptidomics: bridging the gap between proteome and metabolome, *Proteomics* 6, 744-747, 2006.

Metabolomics

The study of the metabolome; ideally, the non-biased identification/quantification of all metabolites in a biological system such as a cell or organism. See Reo, N.V., NMR-based metabolomics, *Drug Chem.Toxicol.* 25, 375-382, 2002; German, J.B., Roberts, M.A., Fay, L., and Watkins, S.M., Metabolomics and individual metabolic assessment: the next great challenge for nutrition, *J.Nutr.* 132, 2486-2487, 2002; Watkins, S.M., and German, S.B., Metabolomics and biochemical profiling in drug discovery and development, *Curr.Opin.Mol.Ther.* 4, 224-228, 2002; Phelps, T.J., Palumbo, A.V., and Beliaev, A.S., Metabolomics and microarrays for improved understanding of phenotypic characteristics controlled by both genomics and environmental constraints, *Curr. Opin.Biotechnol.* 13, 20-24, 2002; Grivet, J.P., Delort, A.M., and Portais, J.C., NMR and microbiology: from physiology to metabolomics, *Biochemie* 85, 823-840, 2003; Brown, S.C., Kruppa, G., and Dasseux, J.L., Metabolomics applications of FT-ICR mass spectrometry, *Mass Spectrom.Rev.* 24, 223-231, 2005; Bhalla, R., Narasimhan, K., and Swarup, S., Metabolomics and its role in understanding cellular responses in plants, *Plant Cell Rep.* 24, 562-571, 2005; Fridman, E., and Pichersky, E., Metabolomics, genomics, proteomics, and the identification of enzymes and their substrates and products, *Curr.Opin.Plant Biol.* 8, 242-248, 2005; Rochfort, S., Metabolomics reviewed: a new "omics" platform technology for systems biology and implications for natural product research, *J.Nat.Prod.* 68, 1813-1820, 2005; Griffin, J.L., The Cinderella story of metabolic profiling: does metabolomics get to go to the functional genomics ball? *Philos.Trans.R.Soc. Lond.B Biol.Sci.* 361, 147-161, 2006.

Metagenomics

The genomic analysis of a mixed microbial population. See Schloss, P.D. and Handelsman, J., Biotechnological prospects from metagenomics, *Curr.Opin.Biotechnol.* 14, 303-310, 2003; Riesenfled, C.S., Schloss, P.D. and Handelsman, J., Metagenomics: genomic analysis of microbial communities, *Annu.Rev.Genet.* 38, 525-552, 2004; Steele, H.L. and Streit, W.R., Metagenomics: advances in ecology and biotechnology, *FEMS Microbiol.Lett.* 247, 105-111, 2005.

Metaproteomics

The proteomic analysis of mixed microbial communities. See Wilmes, P. and Bond, P.L., The application of two-dimensional polyacrylamide gel electrophoresis and downstream analyses to a mixed community of prokaryotic microorganisms, *Environ.*

Microbiol. 6, 911-920, 2004; Kan, J., Hanson, T.E., Ginter, J.M., Wnag, K., and Chen, F., Metaproteomic analysis of Chesapeake Bay microbial communities, *Saline Systems* 1, 7, 2005; Wilmes, P. and Bond, P.L., Metaproteomics: studying functional gene expression in microbial ecosystems, *Trends Microbiol.* 14, 92-97, 2006; Valenzuela, L., Chi, A., Beard, S. *et al.*, Genomics, metagenomics and proteomics in biomining microorganisms, *Biotechnol.Adv.* 24, 195-209, 2006; Ward, N., New directions and interactions in metagenomics research, *FEMS Microbiol. Ecol.* 55, 331-338, 2006.

Metazoan

Animals with differentiated cells and tissues and usually a discrete digestive tract with specialized cells.

Micelles

A small (nanoscale) particle composed of individual molecules which can be considered an aggregate. In biochemistry and molecular biology, micelle usually reference to a structure of polar lipids in aqueous solution which can form an amphipathic layer with the polar groups directed toward solvent and non-polar groups clustered toward the interior. However, a micelle can be composed of proteins or other organic materials. Casein polymers have been extensively studies because of their presence in milk. The term critical micellar concentration (CMC) defines the concentration when, for example, a lipid would move from solution phase to micelle. Micelle are used for drug delivery. The term micelle is also used to refer to small particles of materials used in humus soil formulations. See also liposomes, nanotechnology. See Hartley, G.S., *Aqueous Solutions of Paraffin-Chain Salts: A Study in Micelle Formation*, Hermann and cie, Paris, 1936; Mukerjee, P., *Critical Micelle Concentrations of Aqueous Surfactant Systems*, US National Bureau of Standards, US Government Printing Office, Washington, DC, 1971; Rosen, M.J., *Surfactants and Interfacial Phenomena* , Wiley-Interscience, Hoboken, NJ, 2004; Gentle, I and Barnes, G., *Interfacial Science: An Introduction*, Oxford University Press, Oxford, UK, 2005; Bloomfield, V.A. and Mead, R.J., Jr., Structure and stability of casein micelles, *J.Dairy Sci.* 58, 592-601, 1975; Kreuter, J., Nanoparticles and nanocapsules – new dosage forms in the nanometer size range, *Pharm.Acta Helv.* 53, 33-39, 1978; Furth, A.J., Removing unbound detergent from hydrophobic proteins, *Anal.Biochem.* 109, 207-215, 1980; Bagchi, B., Water dynamics in the hydration layer around proteins and micelles, *Chem.Rev.* 105, 3197-3129, 2005; Chandler, D., Interfaces and the driving force of hydrophobic assembly, *Nature* 437, 640-647, 2005; Aliabadi, H.M. and Lavasanifar, A., Polymeric micelles for drug delivery, *Expert Opin.Drug Deliv.* 3, 139-162, 2006.

Microarray

Generally referring to an array of probes displayed on a matrix similar to a microscope slide which are used to analyzed complex mixtures for specific analytes. The sample is usually labeled with a signal such as a fluorescent dye. Both sample preparation and analysis are complex. See DNA microarray and protein microarray. See Schena, M., Heller, R.A., Theriault, T.P., *et al,*, Microarrays: biotechnology's discovery platform for functional genomics, *Trends Biotechnol.* 16, 301-306, 1998; Gerhold, D., Rushmore, T., and Caskey, C.T., DNA chips: promising toys have

become powerful tools, *Trends Biochem Sci* 24, 168-173, 1999; Ness, S.A., Basic microarray analysis: strategies for successful experiments, *Methods Mol.Biol.* 316, 13-33, 2006; Wang, S. and Cheng, Q., Microarray analysis in drug discovery and clinical applications, *Methods Mol.Biol.* 316, 49-65, 2006; Kozarova, A., Petrinac, S., Ali, A., and Hudson, J.W., Array of informatics: applications in modern research, *J.Proteome Res.* 5, 10551-1059, 2006; Sievertzon, M., Nilsson, P. and Lundeberg, J., Improving reliability and performance of DNA microarrays, *Expert Rev. Mol.Diagn.* 6, 481-492, 2006; Sobek, J., Bartscherer, K., Jacob, A., Hoheisel, J.D., and Angenendt, P., Microarray technology as a universal tool for high-throughput analysis of biological systems. *Comb.Chem.High Throughput Screen.* 9, 365-380, 2006; Pedroso, S. and Guillen, I.A., Microarray and nanotechnology applications of functional nanoparticles, *Comb.Chem.High Througput Screen.* 9, 389-397, 2006. Microarray analysis can also be applied to the study of multiple tissue specimens with tissue microarrays (Rimm, D.L., Camp, R.L, Charette, L.A., *et al.*, Tissue microarray: a new technology for amplification of tissue resources, *Cancer J.* 7, 24-31, 2001; Bubendorf, L., Nocito, A., Moch, K., and Sauter, G., Tissue microarray (TMA) technology: Miniaturized pathology archives for high-throughput in situ studies, *J.Pathol.* 195, 72-79, 2001; Fedor, H.L. and De Marzo, A.M., Practical methods for tissue microarray construction, *Methods Mol.Biol.* 103, 89-101, 2005).

Microdialysis

A process for sampling low-molecular weight metabolites in the extracellular space of tissues. This technique is used for the study of tissue metabolism and pharmacokinetic studies. Microdialysis is accomplished through the use of a probe constructed as a concentric tube is implanted into a tissue and a perfusion fluid (a physiological solution such as Hank's Balanced Salt Solution) enters through an inner tube flowing toward the distal end and, entering the space between the inner tube and the outer dialysis membrane flow back toward the proximal end of the probe. Dialysis takes place during the passage of fluid toward the proximal end and the exiting fluid is sampled for the analyte in question. It is viewed a noninvasive method of evaluated tissue metabolism. See Lonnroth, P. and Smith, U., Microdialysis—a novel technique for clinical investigations, *J.Intern.Med.* 227, 295-300, 1990; Ungerstedt, U., Microdialysis—principles and applications for studies in animals and man, *J.Intern.Med.* 230, 365-373, 1991; Parsons, L.H. and Justice, J.B., Jr., Quantitative approaches to in vivo brain microdialysis, *Crit.Rev.Neurobiol.* 9, 189-220, 1994; Schuck, V.J., Rinas, I., and Derendorf, H., In vitro microdialysis sampling of docetaxel, *J.Pharm.Biomed. Anal.* 36, 607-613, 2004; Hocht, C., Opezzo, J.A., and Taira, C.A., Microdialysis in drug discovery, *Curr.Drug.Discov.Technol.* 1, 269-285, 2004; Rooyackeres, O., Thorell, A., Nygren, J., and Ljungqvist, O., Microdialysis method for measuring human metabolism, *Curr.Opin.Clin.Nutr. Metab.Care* 7, 515-52, 2004; Cano-Cebriaqn, M.J., Zornoza, T., Polache, A., and Granero, L., Quantitative in vivo microdialysis in pharmacokinetic studies: some reminders, *Curr.Drug.Metab.* 6, 83-90, 2005; Abrahamsson, P. and Winso, O., An assessment of calibration and performance of the microdialysis system, *J.Pharm.Biomed.Anal.* 39, 730-734, 2005; Ao, X. and Stenken, J.A., Microdialysis sample of cytokines, *Methods* 38, 331-341, 2006. Successful interpretation of microdialysis experiments will require a thorough understanding of the factor influxing membrane flux of the specific analyte or analytes.

Microsatellite

Tandem repeats of one to six nucleotides; sometimes referred to as simple sequence repeats or simple sequence lengths. These sites are use for genetic mapping and show variability in oncological disorders. Changes are measured by simple sequence length polymorphisms. See Tautz, D. and Schlotterer, C., Simple sequences, *Curr.Opin.Genet.Dev.* 4, 832-837, 1994; Albrecht, A. and Mundlos, S., The other trinucleotide repeat: polyalanine expansion disorders, *Curr.Opin.Genet.Dev.* 15, 285-293, 2005; Di Prospero, N.A. and Fischbeck, K.H., Therapeutics development for triplet repeat expansion diseases, *Nat.Rev.Genet.* 6, 756-765, 2005; Zheng, H.T., Peng, Z.H., Li, S. and He, L., Loss of heterozygosity analyzed by single nucleotide polymorphism array in cancer, *World J.Gastroenterol.* 11, 6740-6744, 2005.

Minicollagen

A small collagen usually expressed in See Mazzorana, M., Snellman, A., Kivirikkov, K.I., van der Rest, M., and Pihlajaniemi, T., Involvement of prolyl-4-hydroxylase in the assembly of trimeric minicollagen XII. Study in a baculovirus expression system, *J.Biol.Chem.* 271, 29003-29008, 1996; Milbradt, A.G., Boulegue, C., Moroder, L., and Renner, C., The two cysteine-rich head domains of minicollagen from *Hydra* nematocysts differ in their cystine framework and overall fold despite an identical cysteine sequence pattern, *J.Mol.Biol.* 354, 591-600, 2005. Minicollagen is a miniprotein derivative of collagen.

Miniprotein

The smallest portion of a protein which can fold into a unique three-dimensional structure thus differing from a peptide. See Degrado, W.F. and Sosnick, T.R., Protein minimization: Downsizing through mutation, *Proc.Natl.Acad.Sci.USA* 93, 5680-5681, 1996; Vita, C., Vizzavona, J., Drakopoulou, E., Zinn-Justin, S., Gilquin, B., and Ménez, A., Novel miniproteins engineered by the transfer of active sites to small nature scaffolds, *Biopolymers* 47, 93-100, 1998; Neuweiler, H., Doose, S., and Sauer, M., A microscopic view of miniprotein folding: enhanced folding efficiency through formation of an intermediate, *Proc.Natl.Acad.Sci.USA* 102, 16650-16655, 2005.

Modulus of Elasticity

The stress required to produce unit strain causing a change in length (Young's modulus) or a twist or shear (shear modulus) or a change volume (bulk modulus); units are dynes/cm^2. See Nash, G.B. and Gratzer, W.B., Structural determinants of the rigidity of the red cell membrane, *Biorheology* 30, 397-407, 1993; Urry, D.W. and Pattanaik, A., Elastic protein-based materials in tissue reconstruction, *Ann.N.Y.Acad.Sci.* 831, 32-46, 1997; Ambrosio, L., De Santis, R., and Nicolais, L., Composite hydrogels for implants, *Proc.Inst.Mech.Eng.* 212, 93-99, 1998; Roberts, R.J. and Rowe, R.C., Relationships between the modulus or elasticity and tensile strength for pharmaceutical drugs and excipients, *J.Pharm.Pharmcol.* 51, 975-977, 1999; Hegner, M. and Grange, W., Mechanics and imaging of single DNA molecules, *J.Muscle Res.Cell.Motil.* 23, 367-375, 2002; Carr, M.E., Jr., Development of platelet contractile force as a research and clinical measure of platelet function, *Cell.Biochem.Biophys.* 38, 55-78, 2003; Balshakova, A.V., Kiselyova, O.I., and Yaminsky, I.V, Microbial surfaces investigated using atomic force microscopy, *Biotechnol. Prog.* 20, 1615-1622, 2004; Zhang, G., Evaluating the viscoelastic

properties of biological tissues in a new way, *J.Musculoskelet. Neuronal Interact.* 5, 85-90, 2005; Seal, B.L. and Panitch A., Physical matrices stabilized by enzymatically sensitive covalent crosslinks, *Acta Biomater.* 2, 241-251, 2006.

Molecular Beacons

A DNA probe which is a single chain loop and stem structure. A fluorophore at the 5'-terminus is proximity to a quencher at te 3'-terminus is released on the binding of the probes to a specific DNA sequence. See Piatek, A.S., Tyagi, S., Pol, A.C., *et al.*, Molecular beacon sequence analysis for detecting drug resistance in *Mycobacterium tuberculosis*, *Nature Biotechnol.* 16, 35-363, 1998; Tan, W., Fang, X., Li, J., and Liu, X., Molecular beacons: a novel DNA probe for nucleic acid and protein studies, *Chem.Eur.J.* 6, 1107-1111, 2000; Fang, X., Mi, Y., Li, J.J., Beck, T., Schuster, S., and Tan, W., Molecular beacons: fluorogenic probes for living cell study, *Cell.Biochem.Biophys.* 37, 71-81, 2002; Stöhr, K., Häfner, B., Nolte, O., Wolfrum, J., Sauer, M., and Herten, D.-P., Species-specific identification of myobacterial 16S rRNA PCR amplicons using smart probes, *Anal.Chem.* 77, 7195-7203, 2005.

Molecular Clock

Places timescales on evolutionary events. See Zuckerkandl, E., On the molecular evolutionary clock, *J.Mol.Evol.* 26, 34-46, 1987; Easteal, S., A mammalian molecular clock, *Bioessays* 14, 415-419, 1992; Seoighe, C., Turning the clock back on ancient genome duplication, *Curr.Opin.Genet.Dev.* 13, 636-643, 2003; Freitas, C., Rodrigues, S., Saude, L., and Palmeirim, I., Running after the clock, *Int.J.Dev.Biol.* 49, 317-324, 2005; Renner, S.S., Relaxed molecular clocks for dating historical plant dispersal events, *Trends Plant Sci.* 10, 550-558, 2005; Ho, S.Y.W. and Larson, G., Molecular clocks: when times are a-changing, *Trends Genet.* 22, 79-83, 2006.

Monocot (monocotyledon)

Flowering plants that have embryos with only one cotyledon. See See Iyer, L.M., Kumpatla, S.P., Chadrasekharan, M.B., and Hall, T.C., Transgene silencing in monocots, *Plant Mol.Biol.* 43, 323-346, 2000; Agrawal, G.K., Iwahashi, H., and Rakwal, R., Rice MAPKs, *Biochem.Biophys.Res.Commun.* 302, 171-180, 2003; Schoof, H. and Karlowski, W.M., Comparison of rice and Arabidopsis annotation, *Curr. Opin. Plant Biol.* 6, 106–112, 2003.

Monocyte

Circulating white blood cell; an immune system cell which is a precursor to macrophage. See Nelson, D.S. and Alexander, P., *Immunobiology of the Macrophage*, Academic Press, New York, NY, 1976; Schmalzl, F. and Huhn, D., *Disorders of the Monocyte Macrophage System: Pathophysiological and Clinical Aspects*, Springer-Verlag, Berlin, 1981; Russell, S.W. and Siamon, G., *Macrophage Biology and Activation*, Springer-Verlag, Berlin, 1992; Zwilling, B.S. and Eisenstein, T.K., *Macrophage-Pathogen Interactions*, Marcel Dekker, New York, 1994; Mire-Sluis, A.R. and Thorpe, R., *Cytokines*, Academic Press, San Diego, CA, 1998; Bellosta, S. and Bernini, F., Modulation of macrophage function and metabolism, *Handb.Exp.Pharmacol.* (170), 665-695, 2005; Condeelis, J. and Pollard, J.W., Macrophages: obligate partners for tumor cell migration, invasion, and metastasis, *Cell* 124, 263-266, 2006; Noda, M., Current topics in pharmacological research on bone metabolism: regulation of bone mass by the function of endogenous modulators of bone morphogenetic protein in adult stage, *J.Pharmacol.Sci.* 100, 211-214, 2006; Lews, C.E. and Pollard, J.W., Distinct role of macrophages in different tumor microenvironments, *Cancer Res.* 66, 605-612, 2006; Bobryshev, Y.V., Monocyte recruitment and foam cell formation in atherosclerosis, *Micron* 37, 208-222, 2006; Cathelin, S., Rébe, C., Haddaoui, L., *et al.*, Identification of proteins cleaved downstream of caspases activation in monocytes undergoing macrophage differentiation, *J.Biol.Chem.* 281, 17779-17788, 2006.

Morphoproteomics

Morphoproteomics combines the technical approaches of histopathology, molecular biology, and proteomics for the study of cell biology and systems biology. See Brown, R.E., Morphoproteomic portrait of the mTOR pathway in mesenchymal chondrosarcoma, *Ann.Clin.Lab.Sci.* 34, 397-399, 2004; Brown, R.E., Morphoproteomics: exposing protein circuitries in tumors to identify potential therapeutic targets in cancer patients, *Expert Rev. Proteomics* 2, 337-348, 2005.

Mucus/Mucins

A viscous biological fluid containing mucins. Mucins are very large highly glycosylated proteins with considerable asymmetry. The carbohydrate moiety is sulfated resulting in a high negative charge which contributes to the extended conformation. It is the physical properties of mucin that are responsible for the viscosity of mucus. See Rose, M.C. and Voynow, J.A., Respiratory tract mucin genes and mucin glycoproteins in health and disease, *Physiol.Rev.* 86, 245-278, 2006; Gerken, T.A., Biophysical approaches to salivary mucin structure, conformation, and dynamics, *Crit.Rev.Oral Biol.Med.* 4, 261-270, 1993; Dodd, S., Place, G.A., Hall, R.L., and Harding, S.E., Hydrodynamic properties of mucins secreted by primary cultures of guinea-pig tracheal epithelial cells: determination of diffusion coefficients by analytical ultracentrifugation and kinetic analysis of mucus gel hydration and dissolution, *Eur.Biophys.J.* 28, 38-47, 1999; Brockhausen, I., Sulphotransferases acting on mucin-type oligosaccharides, *Biochem.Soc.Trans.* 31, 318-325, 2003; Lafitte, G., Thuresson, K., and Soderman, O., Mixtures of mucin and oppositely charged surfactant aggregates with varying charge density. Phase behavior, association, and dynamics, *Langmuir* 21, 7097-7104, 2005.

Mullerian Inhibiting Substance (Mullerian Inhibiting Hormone)

A peptide growth factor which is a member of the TGFβ family which was originally described as a large glycoprotein secreted by neonatal/fetal testis responsible for regression of the Mullerian Duct during the process of sexual differentiation. See Budzik, G.P., Powell, S.M., Kamagata, S. and Donahoe, P.K., Mullerian inhibiting substance fractionation of dye affinity chromatography, *Cell* 34, 307-314, 1983; Visser, J.A., Themmen, A.P.N., Anti-Müllerian hormone and folliculogenesis, *Molec.Cell.Endocrinol.* 234, 81-86, 2005.

Mutation Rate

The instantaneous rate at which nucleotide changes occur in the genome. Lethal or near-lethal mutations are often ignored in the calculation of maturation rate. See Ellegren, H., Smith, N.G.,

and Webster, M.T., Mutation rate variation in the mammalian genome, *Curr.Opin.Genet.Dev.* 13, 562-568, 2003; Sniegowski, P., Evolution: bacterial mutation in stationary phase, *Curr.Biol.* 14, R245-R246, 2004; Wang, C.L. and Wabl, M., Precise dosage of an endogenous mutagen in the immune system, *Cell Cycle* 3, 983-958, 2004; Pakendorf, B. and Stoneking, M., Mitochondrial DNA and human evolution, *Annu.Rev.Genomics Hum.Genet.* 6, 165-183, 2005.

Myeloid Progenitor

Precursor of granulocyte, macrophages, and mast cells; involved in chronic myeloid leukemia (CML). Development regulated by colony-simulating factors (CSFs). See Burgess, A.W. and Metcalf, D., The nature and action of granulocyte-macrophage colony stimulating factors, *Blood* 56, 947-958, 1980; Islam, A., Haemopoietic stem cells: a new concept, *Leuk.Res.* 9, 1415-1432, 1985; Cannistra, S.A. and Griffin, J.D., Regulation of the production and function of granulocytes and monocytes, *Semin. Hematol.* 25, 173-188, 1988; Grimwade, D. and Enver, T., Acute promyelocytic leukemia: where does it stem from? *Leukemia* 19, 375-384, 2004; Coulombel, L., Identification of hematopoietic stem/progenitor cells: strength and drawbacks of functional assays, *Oncogene* 23, 7210-7223, 2004; Rosmarain, A.G., Yang, Z., and Resendes, K.K., Transcriptional regulation in myelopoiesis: hematopoietic fate choice, myeloid differentiation, and luekemogenesis, *Exp.Hematol.* 33, 131-143, 2005; Marley, S.B., and Gordon, M.Y., Chronic myeloid leukemia: stem cell derived but progenitor cell driven, *Clin.Sci.* 109, 13-25, 2005.

Nanofiltration

Filtration of small (nano) particles from solvent using a filter with extremely small pores (0.001 – 0.010 micron); finer than ultrafiltration, not as fine as reverse osmosis. Used for the removal of viruses from plasma protein products. See Yaroshchuk, A.E., Dielectric exclusion of ions from membranes, *Adv.Colloid Interface Sci.* 85, 193-230, 2000; Rossano, R., D'Elia, A., and Riccio, P., One-step separation from lactose: recovery and purification of major cheese-whey proteins by hydroxyapatite—a flexible procedure suitable for small- and medium-scale preparations, *Protein Expr.Purif.* 21, 165-169, 2001; Burnouf, T. and Radosevich, M., Nanofiltration of plasma-derived biopharmaceutical products, *Haemophilia* 9, 24-37, 2003; Bhanushali, D. and Bhattacharyya, D., Advances in solvent-resistant nanofiltration membranes: experimental observations and application, *Ann.N.Y.Acad.Sci.* 984, 159-177, 2003; Weber, R., Chmiel, H., and Mavrov, V., Characteristics and application of ceramic nanofiltration membranes, *Ann.N.Y.Acad.Sci.* 984, 178-193, 2003; Tieke, B., Toutianoush, A., and Jin, W., Selective transport of ions and molecules across layer-by-layer assembled membranes of polyelectrolytes, *p*-sulfonato-calixin[n]arenes and Prussian Blue-type complex salts, *Adv.Colloid Interface Sci.* 116, 121-131, 2005; Berot, S., Compoint, J.P., Larre, C., Malabat, C., and Gueguen, J., Large-scale purification of rapeseed proteins (*Brassica napus L.*), *J.Chromatog.B Anal.Technol.Biomed.Life Sci.* 818, 35-42, 2005; Zhao, K. and Li, Y., Dielectric characterization of a nanofiltration membrane in electrolyte solutions: its double-layer structure and ion permeation, *J.Phys.Chem.B Condens. Matter Mater.Surf. Interfaces Biophys.* 110, 2755-2763, 2006; Bulut, M., Gevers, L.E., Paul, J.S., Vankelecom, I.F., and Jacobs, P.A., Directed development of high-performance membranes via high-throughput and combinatorial strategies, *J.Comb.Chem.* 8, 168-173, 2006.

Nanog

Nanog is a homeodomain transcription factor which is found in undifferentiated embryonic stem cells and is considered important for the maintenance of pluripotency. See Mitsui, K., Tokuzawa, Y., Itoh, H., *et al.*, The homeoprotein Nanog is required for maintenance of pluripotency in mouse epiblast and ES cells, *Cell* 113, 631-642, 2003; Chambers, I., Colby, D., Robertson, M., *et al.*, *Cell* 113, 643-655, 2003; Oh, J.H., Do, H.J., Yang, H.M., *et al.*, Identification of a putative transactivation domain in human Nanog, *Exp.Mol.Med.* 37, 250-254, 2005; Xu, Y., A new role for p53 in maintaining genetic stability in embryonic stem cells, *Cell Cycle* 4, 363-364, 2005; Ralston, A. and Rossant, J., Genetic regulation of stem cells origins in the mouse embryo, *Clin.Genet.* 68, 106-112, 2005; Yates, A. and Chambers, I., The homeodomain protein Nanog and pluripotency in mouse embryonic stem cells, *Biochem.Soc.Trans.* 33, 1518-1521, 2005; Silva, J., Chambers, I., Pollard, S., and Smith, A., Nanog promotes transfer of pluripotency afer cell fusion, *Nature* 441, 997-1001, 2006.

Nanotechnology

The study of particles, devices, substances having physical dimensions of 1- 100 nanometers (0.001 to 0.100 micrometers (microns). This includes the study of liposomes for drug delivery, nanofiltration, micelles, dendrimers, and quantum dots. See Wilson, M., *Nanotechnology: Basic Science and Emerging Technologies*, Chapman & Hall, Boca Raton, FL, 2002; Poole, C.P. and Owens, F.J., *Introduction to Nanotechnology*, Wiley, Hoboken, NJ, 2003; Di Ventra, M. and Evoy, S., *Introduction to Nanoscale Science and Technology*, Kluwer Academic, Boston, MA, 2004; Bhushan, B., *Springer Handbook of Nanotechnology*, Springer, Berlin, 2004; Ciofalo, M., Collins, M.W., and Hennessy, T.R., Modelling nanoscale fluid dynamics and transport in physiological flows, *Med.Eng.Phys.* 18, 437-451,1996; Seeman, N.C., DNA nanotechnology: novel DNA constructions, *Annu.Rev.Biophys. Biomol.Struct.* 27, 225-248, 1998; Melo, E.P., Aires-Barros, M.R., and Cabral, J.M., Reverse micelles and protein biotechnology, *Biotechnol.Annu.Rev.* 7, 87-129, 2001; Williams, D.J. and Sebastine, I.M., Tissue engineering and regenerative medicine: manufacturing challenges, *IEE Proc.Nanobiotechnol.* 152, 207-210, 2005; Bhattacharya, D. and Gupta, N.K., Nanotechnology and potential of microorganisms, *Crit.Rev.Biotechnol.* 25, 199-204, 2005; Eijkel, J.C. and van den Berg, A., The promise of nanotechnology for separation devices – from a top-down approach to nature-inspired separation devices, *Electrophoresis* 27, 677-685, 2006; Lange, C.F. and Finlay, W.H., Liquid atomizing: nebulizing and other methods of producing aerosols, *J.Aerosol.Med.* 18, 28-35, 2006; Langford, R.M., Focused ion beam nanofabrication: a comparison with conventional processing techniques, *J.Nanosci. Nanotechnol.* 6, 661-668, Wang, J. and Ren, J., Luminescent quantum dots: a very attractive and promising tool in biomedicine, *Curr.Med.Chem.* 13, 897-909, 2006.

Nascent peptide exit tunnel

A "tunnel"/pore starting at the peptidyl transferase center on the ribosome and ending on the solvent side on the large ribosomal subunit. See Gabashvili, I.S., Gregory, S.T., Valle, M., *et al.*, The polypeptide tunnel system in the ribosome and its gating in erythromycin resistance mutants of L4 and L22, *Mol.Cell.* 8, 181-188, 2001; Tenson, T. and Ehrenberg, M., Regulatory nascent peptides in the ribosomal tunnel, *Cell* 108, 591-594, 2002; Jenni, S. and Ban, N., The chemistry of protein synthesis and voyage through the ribosomal tunnel, *Curr.Opin.Struct.Biol.* 13, 212-219, 2003; Vimberg, V., Ziong, L., Bailey, M., Tenson, T., and Mankin, A., Peptide-mediated macrolide resistance reveals possible specific interactions in the nascent peptide exit tunnel, *Mol.Microbiol.* 54, 376-385, 2004; Baram, D. and Yonath, A., From peptide-bond formation to cotranslational folding: dynamic, regulatory and evolutionary aspects, *FEBS Lett.* 579, 948-94, 2005; Egea, P.F., Stroud, R.M., and Walter, P., Targeting proteins to membranes: structure of the signal recognition particle, *Curr.Opin.Struct. Biol.* 15, 213-220, 2005; Markin, A.S., Nascent peptide in the 'birth canal' of the ribosome, *Trends Biochem.Sci.* 31,11-16, 2006.

NAR gene, nar operon, nar promoter

Promoter region of the *nar* operon which encodes nitrate reductase in *Escherichia coli*. The promoter is generally only maximally induced under anaerobic conditions. It has been shown that the *nar* promoter in some strains of *Escherichia coli* can be induced under condition of very low oxygen tension in the presence of nitrate. This observation has been used to develop some useful processes for recombinant protein expression in *Escherichia coli*. See Li, S.F. and DeMoss, J.A., Promoter region of the *nar* operon of *Escherichia coli*: nucleotide sequence and transcription initiation signals, *J.Bacteriol.* 169, 4614-4620, 1987; Han, S.J., Chang, H.N., and Lee, J., Characterization of an oxygen-dependent inducible promoter, the *nar* promoter of *Escherichia coli*, to utilize in metabolic engineering, *Biotechnol.Bioeng.* 72, 573-577, 2001; Lee, K.H., Cho, M.H., Chung, T., *et al.*, Characterization of an oxygen-dependent inducible promoter: the *Escherichia coli nar* promoter, in gram-negative host strains, *Biotechnol.Bioeng.* 82, 271-277, 2003.

Nephelometry

Detection of electromagnetic wave scattering in a direction different from the direct path of the transmitted light; for example, electromagnetic energy scattered at a 90° angle from the incident radiation. Nephelometry is widely used for the measurement of immune complexes. See Deverilli, I. and Reeves, W.G., Light scattering and absorption—developments in immunology, *J.Immunol. Methods* 38, 191-204, 1980; Blackstock, R., In vitro methods for detection of circulating immune complexes and other solution protein-protein interactions, *Ann.Clin.Lab.Sci.* 11, 262-268, 1981; Steinberg, K.K., Cooper, G.R., Graiser, S.R., and Rosseneu, M., Some considerations of methodology and standardization of apolipoprotein A-I immunoassays, *Clin.Chem.* 29, 415-426, 1983; Price, C.P., Spencer, K., and Whicher, J., Light-scattering immunoassay of specific proteins: a review, *Ann.Clin.Biochem.* 20, 1-14, 1983; Brinkman, J.W., Bakker, S.J., Gansevoort, R.T., *et al,*, Which method for quantifying urinary albumin excretion gives what outcome? A comparison of immunonephelometry with HPLC, *Kidney Int.* Suppl. (92), S69-S75, 2004; Yeh, A.T. and Hirshburg, J., Molecular interactions of exogenous chemical agents with collagen—implications for tissue optical clearing, *J.Biomed.Opt.* 11, 014003, 2006.

NF-κB

Nuclear factor kappa B - A transcription factor which is thought to have a major role in the growth of malignant cells. See Kim, H.J., Hawke, N., and Baldwin, A.S., NF-kappaB and IKK as therapeutic targets in cancer, *Cell Death Differ.* 2006. Zingarelli, B.,

Nuclear factor kappaB, *Crit.Care.Med.* 33(Suppl 12), S414-416, 2005; Jimi, E. and Ghosh, S., Role of the nuclear factor-kappaB in the immune system and bone, *Immunol.Rev.* 208, 80-87, 2005; Bubici, C., Papa, S., Pham, C.G., Zazzeroni, F., and Froanzoso, G., The NF-kappaB-mediated control of the ROS and JNK signaling, *Histol.Histopathol.* 21, 69-80, 2006.

Nonidet P-40™

Nonidet is popular nonionic (polyoxythelene glycol derivative) detergent that has been used for membrane protein solubilization. See Dunkley, P.R., Holmes, R., and Rodnight, R., Phosphorylation of synaptic-membrane proteins from ox cerebral cortex in vitro. Preparation of fractions enriched in phosphorylated proteins by using extraction with detergents urea, and gel filtration, *Biochem.J.* 163, 369-378, 1977; Sharma, C.B., Lehle, L., and Tanner, W., *N*-Glycosylation of yeast proteins. Characterization of the solubilized oligosaccharyl transferase, *Eur.J.Biochem.* 116, 101-108, 1981; Perez-Machin, R., Henriquez-Hernandez, L., Perez-Luzardo, O., *et al.*, Solubilization and photoaffinity labeling identification of glucocorticoid binding peptides in endoplasmic reticulum from rat liver, *J.Steroid Biochem.Mol.Biol.* 84, 245-253, 2003; Shiozaki, A., Tsuji, T., Kohno, R., Proteome analysis of brain proteins in Alzheimer's disease: subproteomics following sequentially extracted protein preparation, *J.Alzheimers Dis.* 6, 257-268, 2004; Zintl, A., Pennington, S.R., and Mulcahy, G., Comparison of different methods for the solubilization of *Neospora caninum* (Phylum Apicomplexa) antigen, *Vet.Paristol.* 135, 205-213, 2006; Kalabis, J., Rosenberg, I. and Podolskyi, D.K., Vangil protein acts as a downstream effector of intestinal trefoil factor (ITF/TFF3) signaling and regulates wound healing of intestinal epithelium, *J.Biol.Chem.* 281, 6434-6441, 2006. Use in renaturing allergens after blotting (Muro, M.D., Fernandez, C. and Moneo, I., Renaturation of blotting allergens increases the sensitivity of specific IgE detection, *J.Investig.Alleregol.Clin. Immunol.* 6, 166-171, 1996) has been described as has used in a aqueous two-phase separation system (Sanchez-Ferrer, A., Bru, R., and Garcia-Carmona, F., Phase separation of biomolecules in polyoxyethylene glycol nonionic detergents, *Crit.Rev.Biochem. Mol.Biol.* 29, 275-313, 1994.

Northern Blot

A technique similar to the Southern Blot. RNA separated by electrophoresis is transferred to a PDVF membrane. Specific RNA sequences are detected with a labeled cDNA probe. See Hayes, P.C., Wolf, C.R., and Hayes, J.D., Blotting techniques for the study of DNA, RNA, and proteins, *BMJ* 299, 965-968, 1989; Dallman, M.J., Montgomery, R.A., Larsen, C.P., Wanders, A., and Wells, A.F., Cytokine gene expression: analysis using northern blotting, polymerase chain reaction and in situ hybridization, *Immunol. Rev.* 119, 163-179, 1991; Mengod, G., Goudsmit, E., Probst, A., and Palacios, J.M., In situ hybridization histochemistry in the human hypothalamus, *Prog. Brain Res.* 93, 45-55, 1992; Pajor, A.M., Hirayama, B.A., and Wright, E.M., Molecular biology approaches to comparative study of Na(+)-glucose cotransport, *Am.J.Physiol.* 263, R489-R495, 1992; Kroczek, R.A., Southern and northern analysis, *J.Chromatog.* 618, 133-145, 1993; Raval, P., Qualitative and quantitative determination of mRNA, *J.Pharmacol.Toxicol. Methods* 32, 125-127, 1994; Durrant, I., Enhanced chemiluminescent detection of horseradish peroxidase labeled probes, *Methods Mol.Biol.* 31, 147-161, 1994; Aravin, A. and Tuschi, T., Identification and characterization of small RNAs involved

in RNA silencing, *FEBS Lett.* 579, 5830-5840, 2005; Farrell, R.E., *RNA Methodologies: A Laboratory Guide for Isolation and Characterization*, Academic Press, San Diego, USA, 1993; Darling, D.C. and Brickell, P.M., *Nucleic Acid Blotting: The Basics*, Oxford University Press, Oxford, UK, 1994.

Northwestern Blot

A protein blotting technique related to the various other blotting techniques such as the Western Blot and the Northern Blot. In the Northwestern Blot, the protein mixture is separated by gel electrophoresis, transferred by electrophoresis to a PVDF or nitrocellulose membrane. Specific proteins are identified through the binding of radiolabeled or fluorophore-labeled RNA oligomers; (double-stranded RNA) dsRNA is used to identify dsRNA binding proteins. See Schiff, L.A., Nibert, M.L., Co, M.S., Brown, E.G., and Fields, B.N., Distinct binding sites for zinc and double-stranded RNA in the reovirus outer capsid protein sigma 3, *Mol. Cell.Biol.* 8, 273-283, 1988; Chen, X., Sadlock, J., and Schon, E.A., RNA-binding patterns in total human tissue proteins: analysis by northwestern blotting, *Biochem.Biophys.Res.Commun.* 191, 18-25, 1993; Kumar, A., Kim, H.R., Sobol, R.W., *et al.*, Mapping of nucleic acid binding in proteolytic domains of HIV-1 reverse transcriptase, *Biochemistry* 32, 7466-7474, 1993; Lin, G.Y., Paterson, R.G., and Lamb, R.A., The RNA binding region of the paramyxovirus SV5 V and P proteins, *Virology* 238, 460-469, 1997; Zhao, S.L., Liang, C.Y., Zhang, W.J., Tang, X.C., and Peng, H.Y., Characterization of the RNA-binding domain in the *Decrolimus punctatus* cytoplasmic polyhedrosis virus nonstructural protein p44, *Virus Res.* 114, 80-88, 2005; Sekiya, S., Noda, K., Nishikawa, F., *et al.*, Characterization and application of a novel RNA aptamers against the mouse prion protein, *J.Biochem.* 139, 383-390, 2006.

Notch

A receptor class which regulates cell differentiation and development. Notch was first identified in *Drosophila* where it is thought to be involved in long-term memory and neuronal plasticity. See Wilkin, M.B., and Baron, M., Endocytic regulation of North activation and down-regulation, *Membrane Molec.Biol.* 22, 279-289, 2005; Lai, E.C., Notch signaling: control of cell communication and cell fate, *Development* 131, 965-973, 2004; Artavanis-Tsakonas, S., The molecular biology of the Notch locus and the fine tuning of differentiation in Drosophila, *Trends Genet.* 4, 95-100, 1988; Jones, P.A., Epithelial stem cells, *Bioessays* 19, 683-690, 1997.

Nuclear Magnetic Resonance

A technique which detects nuclear-spin orientation in an applied magnetic field; detection of a nuclear magnetic moment; usually measured as the chemical shift from a standard such as such as tetramethyl silane for hydrogen and trichlorofluoromethane for fluorine. Coupling constants (spin-spin coupling, J) are also measured in two-dimensional analyses. See Roberts, J.D., *Nuclear Magnetic Resonance: Applications to Organic Chemistry*, McGraw-Hill, New York, NY, 1959; Pople, J.A., *High-Resolution Nuclear Magnetic Resonance*, McGraw-Hill, New York, NY, 1959; Dyer, J.R., *Applications of Absorption Spectroscopy of Organic Compounds*, Prentice-Hall, Englewood Cliffs, NJ., 1965; Knowles, P.R., March, D., and Rattle, H.W.E., *Magnetic Resonance of Biomolecules: An Introduction to the Theory and*

Practice of NMR and ESR in Biological Systems, Wiley, New York, NY, 1976; Leyden, D.E. and Cox, R.H., *Analytical Applications of NMR*, Wiley, New York, NY, 1977; Jardetzky, O. and Roberts, G.C.K., *NMR in Molecular Biology*, Academic Press, New York, NY, 1981; Wüthrich, K., *NMR of Proteins and Nucleic Acids*, Wiley, New York, NY, 1986; Paudler, W.W., *Nuclear Magnetic Resonance: General Concepts and Applications*, Wiley, New York, NY, 1987; Schrami, J. and Bellama, J.M., *Two-Dimensional NMR Spectroscopy*, Wiley, New York, NY, 1988; Sanders, J.K.M., and Hunter, B.K., *Modern NMR Spectroscopy: A Guide for Chemists*, Oxford University Press, Oxford, UK, 1993; Hore, P.J. and Jones, J.A., *NMR, The Tookit*, Oxford University Press, Oxford, UK, 2000; James, T.L. and Schmitz, U., *Nuclear Magnetic Resonance of Biological Molecules*, Academic Press, San Diego, CA, 2001; Lambert, J.B and Mazzola, E.P., *Nuclear Magnetic Resonance Spectroscopy: An Introduction to Principles, Applications, and Experimental Methods*, Pearson/Prentice Hall, Upper Saddle River, NJ, 2004; Mitchell, T.N. and Costisella, B., *NMR—From Spectra to Structures: An Experimental Approach*, Springer, Berlin, 2004; Friebolin, H., *Basic One- and Two-Dimensional NMR Spectroscopy*, Wiley-VCH, Weinheim, Germany, 2005.

Nuclear Pore Complex

A large transporter which spans the nuclear envelope (nuclear membrane). This structure forms a channel between the inner and outer nuclear membrane providing for the transport of materials to and from the nucleus and cytoplasm; all transport mechanisms in and out of the nucleus, active and passive, occur through a tubular element in this pore structure. The Karyopherin β family of proteins are involved in these transport processes. See Faberge, A.C., The nuclear pore complex: its free existence and an hypothesis as to its origin, *Cell Tissue Res.* 151, 403-415, 1974; Maul, G.G., Nuclear pore complexes. Elimination and reconstruction during mitosis, *J.Cell Biol.* 74, 492-500, 1977; Wozniak, R. and Clarke, P.R., Nuclear pores: sowing the seeds of assembly on the chromatin landscape, *Curr.Biol.* 13, R970-R972, 2003; Rabut, G., Lenart, P., and Ellenberg, J., Dynamics of nuclear pore complex organization through the cell cycle, *Curr.Opin.Cell Biol.* 16, 314-321, 2004; Sazer, S., Nuclear envelope: nuclear pore complexity, *Curr.Biol.* 15, R23-R26, 2005; Peters, R., Translocation through the nuclear pore complex: selectivity and speed by reduction-of-dimensionality, *Traffic* 6, 421-427, 2005; Devos, D., Dokudovskaya, S., Williams, R., *et al.*, Simple fold composition and modular architecture of the nuclear pore complex, *Proc. Natl.Acad.Sci.USA* 103, 2172-2177, 2006; van der Aa, M.A.E.M., Mastrobattista, E., Oosting, R.S., Hennink, W.E., Koning, G.A., and Commelin, D.JA., The nuclear pore complex: the gateway to successful nonviral gene delivery, *Pharmaceut.Res.* 23, 447-459, 2006.

Nucleic Acid Testing

The use of PCR technology to test for the presence of nucleic acid sequences in biological materials. This approach is receiving attention in theranostics and the screening of blood for viral pathogens. See Tabor, E. and Epstein, J.S., NAT screening of blood and plasma donations: evolution of technology and regulatory policy, *Transfusion* 42, 1230-1237, 2002; Dimech, W., Bowden, D.S., Brestovac, B., *et al.*, Validation of assembled nucleic acid-based tests in diagnostic microbiology laboratories, *Pathology* 36, 45-50, 2004; Valentine-Thon, E., Quality control in nucleic acid testing—where do we stand? *J.Clin.Virol.* 25, S13-S21, 2002.

Nucleosome

An octomer of histone proteins associated with an approximate 140 bp DNA; the octomer is composed of two each of H2A, H2B, H3, and H4. See Kornberg, R.D. and Lorch, Y., Irresistible force meets immovable object: transcription and the nucleosome, *Cell* 67, 833-836, 1991; Turner, B.M., Decoding the nucleosome, *Cell* 75, 5-8, 1993; Sivolob, A. and Prunell, A., Nucleosome conformational flexibility and implications for chromatin dynamics, *Philos. Transact.A Math.Phys.Eng.Sci.* 362, 1519-1547, 2004; Lieb, J.D. and Clarke, N.D., Control of transcription through intragenic patterns of nucleosome composition, *Cell* 123, 1187-1190, 2005; Decker, P., Nucleosome autoantibodies, *Clin.Chim.Acta* 366, 48-60, 2006; Stockdale, C., Bruno, M., Ferreira, H., *et al.*, Nucleosome dynamics, *Biochem.Soc.Symp.* (73), 109-119, 2006. Reinberg, D., and Sims, R.J., 3rd, de FACTo nucleosome dynamics, *J.Biol.Chem.* 281, 23297-23301, 2006; Segal, E., Fodufe-Mittendorf, Y., Chen, L., *et al.*, A genomic code for nucleosome positioning, *Nature* 442, 772-778, 2006; Bash, R., Wang, H., Anderson, C., *et al.*, AFM imaging of protein movements: Histone H2A-H2B release during nucleosome remodeling, *FEBS Lett.* 580, 4757-5761, 2006; Pisano, S., Pascucci, E., Cacchione, S., *et al.*, AFM imaging and theoretical modeling studies of sequence-dependent nucleosome positioning, *Biophys.Chem.*, 124, 81–89, 2006.

Nutrigenomics

Genomics of nutrition. The science of nutrigenomics seeks to provide a molecular understanding for how common dietary chemicals (i.e., nutrition) affect health by altering the expression and/or structure of an individual's genetic makeup. See van Ommen, B. and Stierum, R., Nutrigenomics: exploiting system biology in the nutrition and health arena, *Curr.Opin.Biotechnol.* 13, 517-721, 2002; Muller, M. and Kersten, S., Nutrigenomics: goals and strategies, *Nat.Rev.Genet.* 4, 315-322, 2003; Bauer, M., Hamm, A., and Pankratz, M.J., Linking nutrition to genomics, *Biol.Chem.* 385, 593-596, 2004; Davis, C.D. and Milner, J., Frontiers in nutrigenomics, proteomics, metabolomics and cancer prevention, *Mutat.Res.* 551, 51-64, 2004; van Ommen, B., Nutrigenomics: exploiting systems biology in the nutrition and health arenas, *Nutrition* 20, 4-8, 2004; Mutch, D.M., Wahli, W., and Williamson, G., Nutrigenomics and nutrigenetics: the emerging faces of nutrition, *FASEB J.* 19, 1602-1616m 2005; Corthesy-Theulaz, I., den Dunnen, J.T., Ferre, P., *et al.*, Nutrigenomics: the impact of biomics technology on nutrition research, *Ann.Nutr.Metab.* 49, 355-365, 2005; Trujillo, E., Davis, C., and Milner, J., Nutrigenomics, proteomics, metabolomics, and the practice of dietetics, *J.Am.Diet.Assos.* 106, 403-413, 2006; Afman, L. and Muller, M., Nutrigenomics: from molecular nutrition to prevention of disease, *J.Am.Diet.Assoc.* 106, 569-576, 2006; http://nutrigenomics.ucdavis.edu.

Ogston Effect

A model for the electrophoretic migration of a polymer within a fiber network which treats the fiber network or soluble polymer network as a molecular sieve and the migrating solute as an undeformable particle. The repation or biased repation model treats the migrating solute as a flexible material which can "snake" through the network. There has been interest in the application of this model to the electrophoresis of large DNA molecules. See Ogston, A.G., The spaces in a uniform random suspension of fibers, *Trans.Faraday Soc.* 54, 1754-1757, 1958; Grossman, P.D.

and Soane, D.S., Experimental and theoretical studies of DNA separations by capillary electrophoresis in entangled polymer solutions, *Biopolymers* 31, 1221-1228, 1991; Kotaka, T., Adachi, S., and Shikata, T., Biased sinusoidal field gel electrophoresis for the separation of large DNA, *Electrophoresis* 14, 313-321, 1993; Guttma, A., Lengyel, T., Szoke, M., and Sasvari-Szekely, M., Ultra-thin-layer agarose gel electrophoresis II. Separation of DNA fragments on composite agarose-linear polymer matrices, *J.Chromatog.A.* 871, 289-298, 2000; Labrie, J., Merdcier, J.F., and Slater, G.W., An exactly solvable Ogston model of gel electrophoresis. V. Attractive gel-analyte interactions and their effects on the Ferguson plot, *Electrophoresis* 21, 823-833, 2000; Slater, G.W., A theoretical study of an empirical function for the mobility of DNA fragments in sieving matrices, *Electrophoresis* 23, 1410-1416, 2002; Mercier, J.-F. and Slater, G.W., Universal interpolating function for the dispersion coefficient of DNA fragments in sieving matrices, *Electrophoresis* 27, 1453-1461, 2006.

Okazaki Fragment

Smaller fragments of DNA which are synthesized and then incorporated into larger DNA molecules showing that replication can be a discontinuous process. See Okazaki, R., Okazaki, T., Sakabe, K., Sugimoto, K., and Sugino, A, Mechanism of DNA chain growth. I. Possible discontinuity and unusual secondary structure of newly synthesized chains, *Proc.Natl.Acad.Sci.USA* 59, 598-605, 1968. Hyodo, M. and Suzuki, K.,. Chain elongation of DNA and joining of DNA intermediates in intact and permeabilized mouse cells, *J.Biochem.* 88, 17-25, 1980; Alberts, B.M., Prokaryotic DNA replication mechanisms, *Philos.Trans.R.Soc.Lond. B Biol. Sci.* 317, 395-420, 1987; Nethanel, T., Reisfeld, S., Dinter-Gottlieb, G., and Kaufmann, G., An Okazaki piece of simian virus 40 may be synthesized by ligation of shorter precursor chains, *J.Virol.* 62, 2867-2873, 1988; Egli, M., Usman, N., Zhang, S.G., and Rich, A., Crystal structure of an Okazaki fragment at 2-Å resolution, *Proc.Natl.Acad.Sci.USA* 89, 534-538, 1992; Kim, J.H., Kang, Y.H., Kang, H.J., *et al.*, In vivo and in vitro studies of Mgsl suggest a link between genome instability and Okazaki fragment processing, *Nucleic Acids Res.* 33, 6137-6150, 2005; Sporbert, A., Domaing, P., Leonhardt, H., and Cardoso, M.C., PCNA acts as a stationary loading platform for transiently interacting Okazaki fragment maturation proteins, *Nucleic Acids Res.* 33, 3521-3526, 2005.

OMP85

A protein found in gram-negative baceria which integrates proteins into bacterial outer membranes. See Gentle, I.E., Burri, L., and Littigow, T., Molecular architecture and function of the Omp85 family of proteins, *Molecular Microbiol.* 58, 1216-1225, 2005.

Oncogene

A gene that causes the transformation of normal cells into cancerous tumor cells, especially a viral gene that transforms a host cell into a tumor cell; a gene that encodes a protein product which will stimulate uncontrolled cellular proliferation. Oncogenes are derivatives of normal cellular genes. See also proto-oncogenes. See Wiman, K.G. and Hayward, W.S., Rearrangement and activation of the *c-myc* gene in avain and human B-cell lymphomas, *Tumour Biol.* 5, 211-219, 1984; Balmain, A., Transforming *ras* oncogenes and multistage carcinogenesis, *Br.J.Cancer* 51, 1-7, 1985; Newbold, R.F., Malignant transformation of mammalian cells in culture: delineation of stages and role of cellular oncogenes activation, *IARC Sci.Publ.* 67, 31-53, 1985; Ratner, L.,Josephs, S.F., and Wong-Staal, F., Oncogenes: their role in neoplastic transformation, *Annu.Rev.Microbiol.* 39, 419-449, 1985; Giehl, K., Oncogenic *Ras* in tumour progression and metastasis, *Biol.Chem.* 386, 193-205, 2005; Sanchez, P. Clement, V., Ruis, I., and Altaba, A., Therapeutic targeting of the Hedgehog-GLI pathway in prostate cancer, *Cancer Res.* 65, 2990-2992, 2005; Bellacosa, A., Kumar, C.C. Di Cristafano, A., and Testa, J.R., Activation of AKT kinases in cancer: implications for therapeutic targeting, *Adv.Cancer Res.* 94, 29-86, 2005; Kranenburg, O., The *KRAS oncogenes*: past, present, and future, *Biochim.Biophys.Acta* 1756, 81-82, 2005.

Oncogenomics

The use of molecular medicine tools such as DNA microarray and proteomics to study the oncology process, cancer genomics; study of oncogenes. See Rosell, R., Monzo, M.. O'Brate, A., and Taron, M., Translational oncogenomics: toward rational therapeutic decision-making, *Curr.Opin.Oncol.* 14, 171-179, 2002; Strausberg, R.L., Simpson, A.J., Old, L.J., and Riggins, G.J., Oncogenomics and the development of new cancer therapies, *Nature* 429, 469-474, 2004; Jain, K.K., Role of oncoproteomics in the personalized management of cancer, *Expert Rev. Proteomics* 1, 49-55, 2004; Sakamoto, K.M., Oncogenomics: dissecting cancer through genome research, *IDrugs* 4, 392-393, 2001; Lam, S.H. and Gong, Z., Modeling liver cancer using zebrafish: a comparative Oncogenomics approach, *Cell Cycle* 5, 573-577, 2006.

Onconase

Onconase is a ribonuclease isolated from amphibia. Onconase is homologous to pancreatic ribonuclease and is clinical trials as a biopharmaceutical. See Ardelt, W., Mikulski, S.M., and Shogen, K. Amino acid sequence of an anti-tumor protein from *Rana Pipiens* oocytes and early embryos. Homology to pancreatic ribonuclease, *J.Biol.Chem.* 266, 245-251, 1991; Wu, Y., Mikulski, S.M., Ardelt, W., *et al.*, A cytotoxic ribonuclease. Study of the mechanism of onconase cytotoxicity, *J.Biol.Chem.* 268, 10686-10693, 1993; Leland, P.A., Schultz, L.W., Kim, B.W., and Raines, R.T., Ribonuclease A variants with potent cytotoxic activity, *Proc.Natl. Acad.Sci.USA* 95, 10407-10412, 1998; Notomista, E., Catanzano, F., Graziano, G., *et al.*, Onconase: an usually stable protein, *Biochemistry* 39, 8711-8718, 2000; Bosch, M., Benito, A., Ribo, M., *et al.*, A nuclear localization sequence endows human pancreatic ribonuclease with cytotoxic activity, *Biochemistry* 43, 2167-2177, 2004; Kim, B.-M., Kim, H., Raines, R.T., *et al.*, Glycosylation of onconase increases its conformational stability and toxicity for cancer cells, *Biochem.Biophys.Res. Commun.* 315, 976-983, 2004; Tafech, A., Bassett, T., Sparanese, D., and Lee, C.H., Destroying RNA as a therapeutic approach, *Curr.Med.Chem.* 13, 863-681, 2006; Suhasini, A.N. and Sirdeshmukh, R., Transfer RNA cleavages by onconase reveal unusual cleavage sites, *J.Biol.Chem.* 281, 12201-12209, 2006.

Opsonization

The process by which an antigen, usually a bacterial cell, is coated with an antibody (an opsonin) and then destroyed by the subsequent process of phagocytosis. The process of opsonization use Fab' portion of the antibody recognized antigen and the Fc domain for complement activation and interaction phagocytic cells such as neutrophils. See Peterson, P.K., Kim,Y., Schemling, D., *et al.*,Complement-mediated phagocytosis of *Pseudomonas aeruginosa*, *J.Lab.Clin.Med.* 92, 883-894, 1978; Cunnion, K.M., Hair, P.S., and Buescher, E.S., Cleavage of complement C3b to iC3b on the surface of *Staphyloccus aureus* is mediated by serum complement factor I., *Infect.Immun.* 72, 2858-2863, 2004; Mueller-Ortiz, S.L., Drouin,S.M., and Wetsel, R.A., The alternative activation pathway and complement component C3 are critical for a protective immune response against *Pseudomonas aeruginosa* in a murine model of pneumonia, *Infect.Immun.* 72, 2899-2906, 2004; Coban,E., Ozdogan, M., Tuncer, M., Bozcuk, H., and Ersoy, F., The treatment of low-dose intraperitoneal immunoglobulin administration in the treatment of peritoneal dialysis-related peritonitis, *J.Nephrol.* 17, 427-430, 2004; Blasi, E., Mucci, A., Neglia, R., *et al.*, Biological importance of the two Toll-like receptors, TLR2 and TLR4, in macrophage response to infection with *Candida albicans*, *FEMS Immunol.Med.Microbiol.* 46, 69-79, 2005; Tosi, M.F., Innate immune responses to infection, *J.Allergy Clin.Immunol.* 116, 241-249,2005; Rus, H., Cudrici, C., and Niculescu, F., The role of the complement system innate immunity, *Immunol.Res.*33, 103-112,2005; Foster, T.J., Immune invasion by staphylococci, *Nat.Rev.Microbiol.* 3, 948-958, 2005; Arbo, A., Pavia-Ruz, N., and Santos, J.I., Opsonic requirements for the respiratory burst of neutrophils against *Giardia lamblia* trophozoites, *Archs.Med.Res.* 37, 465-473, 2006.

Optical Activity

The ability of (chiral) chemical compounds to change the plane of polarization of polarized light; compounds may be dextrorotatory (*d*) or levorotatory (*l*). The optical activity of a chemical compound is a chemical property and an index of stereochemical purity. Optical Rotatory Dispersion and Circular Dichroism are measurements of optical activity.

Optical Rotatory Dispersion

The measurement of the differential change in the velocity of light- and right-circularly polarized light. This technique has bee used to study the conformation of molecules. McKenzie, H.A. and Frier, R.D., The behavior of R-ovalbumin and its individual components A1, A2, and A3 in urea solution: kinetics and equilibria, *J.Prot.Chem.* 22, 207-214, 2003; Chen, E., Kumita, J.R., Woolley, G.A., and Kliger, D.S., The kinetics of helix unfolding of an azobenzene cross-linked peptide probed by nanosecond time-resolved optical rotatory dispersion, *J.Amer.Chem.Soc.* 125, 12443-12449, 2003; Chen, E., Goldbeck, R.A., and Kliger, D.S., The earliest events in protein folding: a structural requirement for ultrafast folding in cytochrome C, *J.Amer.Chem.Soc.* 126, 11175-11181, 2004; Giorgio, E., Viglione, R.G., Zanasi, R., and Rosini, C., Ab initio calculation of optical rotatory dispersion (ORD) curves: a simple and reliable approach to the assignment of the molecular absolute configuration, *J.Amer.Chem.Soc.* 126, 12968-12976, 2004.

Optical Switches

In telecommunication, an optical switch is a switch that enables signals in optical fibers or integrated optical circuits (IOCs) to be selectively switched from one circuit to another. In biology, there are several definitions – one whether an optical switch is chemical probes that undergoes a spectral transition in response to

light where such probe competes for a specific binding partner or ligand when in one but not the other spectral state – See Sakata, T., Yan, Y., and Marriott, G., Family of site-selective molecular optical switches, *J.Org.Chem.* 70, 2009-2013, 2005. Another definition is a spectral probe which is sensitive to a specific intracellular biological event – see Graves, E.E., Weissleder, R., and Ntziachristos, V., Fluorescence molecular imaging of small animal tumors, *Curr.Mol.Med.* 4, 419-430, 2004. The term optical switch is also used in conjunction with optical scissors – Feringa, B.L., In control of motion: from molecular switches to molecular motors, *Acc.Chem.Res.* 34, 504-513, 2001; Capitano, M, Vanzi, F., Broggio, C., *et al.*, Exploring molecular motors and molecular switches at the single-molecule level, *Microsc.Res.Tech.* 65, 194-204, 2004

ORFeome

The total number of protein-conding open reading frames in an organism. See Rual, J.F., Hill, D.E., and Vidal, M., ORFeome projects: gateway between genomics and omics, *Curr.Opin.Chem. Biol.* 8, 20-25. 2004; Brasch, M.A., Hartlety, J.L., and Vidal, M., ORFeome cloning and systems biology: standardized mass production of the parts from a parts list, *Genome Res.* 14, 2001-2009, 2004; Uetz, P., Rajagopala, S.V., Dong, Y.A., and Haas, J., From ORFeomes to protein interaction maps in viruses, *Genome Res.* 14, 2029-2033, 2004; Johnson, N.M., Behm, C.A., and Trowell, S.C., Heritable and inducible gene knockdown in *C.elegans* using Wormgate and the ORFeome, *Gene* 359, 26-34, 2005; Schroeder, B.K, House, B.L., Mortimer, M.W., *et al.*, Development of a functional genomics platform for *Sinorhizobium meliloti* construction of an ORFeome, *Appl.Environ.Microbiol.* 71, 5858-5864, 2005.

Organelle proteomics

Analysis of subcellular organelles such as mitochondria, nucleus, the endocytotic apparatus by proteomic techniques. See Jan van Wijk, K., Proteomics or the chloroplast: experimentation and prediction, *Trends Plant Sci.* 5, 420-425, 2000; Taylor, S.W., Fahy, E., and Ghosh, S.S., Global organellar proteomics, *Trends Biotechnol.* 21, 82-88, 2003; Huber, L.A., Pfaller, K., ad Vistor, I., Organelle proteomics: implications for subcellular fractionation in proteomics, *Circ.Res.* 92, 962-968, 2003; Dreger, M., Subcellular proteomics, *Mass Spectrom.Rev.* 22, 27-56, 2003; Brunet, S.,Thibault, P.. Gagnon, E., *et al.*, *Trends Cell Biol.* 12, 629-638, 2003; Jarvis, P., Organeller proteomics: chloroplasts in the spotlight, *Curr. Biol.* 14, R317-319, 2004; Warnock, D.E., Fahy, E., and Taylor, S.W., Identification of protein associations in organelles, using mass spectrometry-based proteomics, *Mass Spectrom.Rev.* 23, 259-280, 2004; van Wijk, K.J., Plastid proteomics, *Plant Physiol. Biochem.* 42, 963-977, 2004; Yates, J.R., 3rd, Gilchrist, A., Howell, K.E., and Bergeron, J.J., Proteomics of organelles and large cellular structures, *Nat.Rev.Mol.Cell Biol.* 6, 702-714, 2005.

Orthogonal

Two lines intersecting at right angles (mathematics). The terms is derived from the Greek *orthos* meaning straight, upright, vertical. The term orthogonal has been used to describe a variety of activities in biochemistry and molecular biology including protein purification and analysis. More recently the term orthogonal has been used to described tRNA/tRNA synthase pairs that will react with each other and not with other pairs in *Escherichia coli* permitting the incorporation of unnatural amino acids into proteins. See Guilhaus, M., Selby, D. and Miynski, V., Othogonal acceleration time-of-flight mass spectrometry, *Mass Spectrom.Rev.* 19, 65-107, 2000; Evans, C.R. and Jorgenson, J.W., Multidimension LC-LC and LC-CE for high resolution separations of biologicals, *Anal.Bioanal.Chem.* 378, 1952-1961, 2004; Righetti, P.G., Bioanalysis: its past, present, and future, *Electrophoresis* 25, 2111-2127, 2004; Speers, A.E. and Cravatt, B.F., A tandem orthogonal proteolysis strategy for high-content chemical proteomics, *J.Amer.Chem.Soc.* 127, 10018-10019, 2005; Liu, D.R., Magliery,T.J., Pastrnak, M., and Schultz, P.G., Engineering a tRNA and aminoacyl-tRNA synthetase for the site-specific incorporation of unnatural amino acids into protein *in vivo, Proc.Natl.Acad.Sci.USA* 94, 10092-10097, 1997; Chin, J.V., Martin, A.B., King, D.S., Wang, L., and Schultz, P.G., Addition of a photocrosslinking amino acid to the genetic code of *Escherichia coli, Proc.Natl.Acad.Sci.USA* 99. 11020-11024, 2002; Köhrer, C., Sullivan, E.L., and RajBhandary, U.L., Complete set of orthogonal 21st aminoacyl-tRNA synthetase-amber, ochre and opal suppressor tRNA pairs: concomitant suppression of three different termination codons in an mRNA in mammalian cells, *Nucl.Acids Res.* 21, 6200-6211, 2004.

Orthologues

Genes in different organisms which have similar function. See Lovejoy, D.A., Peptide hormone evolution: functional heterogeneity within GnRN and CRF families, *Biochem.Cell Biol.* 74, 1-7, 1996; Cole, C.N., mRNA export: the long and winding road, *Nat.Cell Biol.* 2m E55-E58, 2000; Trowsdale, J., Barten, R., Haude, A., *et al.,* The genomic context of natural killer receptor extended gene families, *Immunol.Rev.* 181, 20-38, 2001; Lieschke, G.J., Zebrafish—an emerging genetic model for the study of cytokines and hematopoiesis in the era of functional gneomics, *Int.J.Hematol.* 73, 25-31, 2001; Lamotagne, B., Larose, S., Boulanger, J., and Elela, S.A., The RNase III family: a conserved structure and expanding functions in eukaryotic dsRNA metabolism, *Curr.Issues Mol.Biol.* 3, 71-78, 2001; Stothard, P. and Pilgrim, D., Sex-determination gene and pathway evoluation in nematodes, *Bioessays* 25, 221-231, 2003; Chen, T.Y., Structure and function of clc channels, *Annu.Rev.Physiol.* 67, 809-839, 2005; Nair, V. and Zavolan, M., Virus-encoded microRNAs: novel regulators of gene expression, *Trends Microbiol.* 14, 169-175, 2006.

Osmosensor

A molecular system sensing osmotic stress and regulating pathways involved in preservation of osmotic equilibrium. See Wurgler-Murphey, S.M. and Saito, H., Two-component signal transducers and MAPK cascades, *Trends Biochem.Sci.* 22, 172-176, 1997; Urao, T., Yamaguchi-Shinozaki, K. and Shinozaki, K., Plant histidine kinases: an emerging picture of two-component signal transduction in hormone and environmental responses, *Sci.STKE* 2001(109), RE18, 2001; Reiser, V., Raitt, D.C., and Saito, H., Yeast osmosensor Sln1 and plant cytokine receptor Cre1 respond to changes in turgor pressure, *J.Cell Biol.* 161, 1035-1040, 2003; Poolman, B., Spitzer, J.J., and Wood, J.M., Bacterial osmosensing: roles of membrane structure and electrostatics in lipid-protein and protein-protein interactions, *Biochim.Biophys. Acta* 1666, 88-104, 2004; Liedtke, W., TRPV4 as osmosensor: a transgenic approach, *Pflugers Arch.* 451, 176-180, 2005; Schiller, D., Ott, V., Kramer, R., and Morbach, S., Influence of membrane composition on osmosensity by the betaine carrier BetP from *Corynebacterium glutamicum, J.Biol.Chem.* 281, 7737-7746, 2006.

Osteoclast

Bone-degrading cells, derived from macrophages; opposite function to osteoblasts. See Rifkin, B.R. and Gay, C.V., *Biology and Physiology of the Osteoclast*, CRC Press, Boca Raton, FL, 1992; Abou-Samra, A.-B. and Mundy, G.R., *Physiology and Pharmacology of Bone*, Springer-Verlag, Berlin, 1993; Massaro, E.J. and Rogers , J.M., *The Skeleton: Biochemical, Genetic, and Molecular Interactions in Development and Homeostasis*, Humana Press, Totowa, NJ, 2004; Bronner, F. and Carson, M.C., *Bone Resorption*, Spinger, London, 2005; Li, Z., Kong, K., and Qi, W., Osteoclast and its roles in calcium metabolism and bone development and remodeling, *Biochem.Biophys.Res.Commun.* 343, 345-350, 2006; Fukumoto, S., Iwamoto, T., Sakai, E., *et al.,* Current topics in pharmacological research of bone metabolism: osteoclasts differentiation regulated by glycospingolipids, *J.Pharmacol. Sci.* 100, 195-200, 2006; Wada, T., Nakashima, T., Hiroshi, N., and Penninger, J.M., RANKL-RNAK signaling in osteoclastogenesis and bone disease, *Trends Mol.Med.* 12, 17-25, 2006.

Osteoprotegerin (OPG)

A protein which suppresses the production of osteoclasts; functions in combination with RANKL regulated bone resorption. See Hofbauer, L.C. and Heufelder, A.E., Osteoprotegerin: a novel local player in bone metabolism, *Eur.J.Endocrinol.* 137, 345-346, 1997; Kong, Y.Y., Boyle, W.J., and Penninger, J.M. Osteoprotegerin ligand: a regulator of immune responses and bone physiology, *Immunol. Today* 21, 495-502, 2000; Hofbauer, L.C., Khosla, S., Dunstan, C.R., *et al.,* The roles of osteoprotegerin and osteoprotegerin ligand in the paracrine regulation of bone resorption, *J.Bone Miner.Res.* 15, 2-12, 2000; Theoleyre, S., Wittrant, Y., Tat, S.K., *et al.,* The molecular triad OPG/RANK/RANKL: involvement in the orchestration of pathophysiological bone remodeling, *Cytokine Growth Factor Rev.* 15, 457-475, 2004; Bezerra, M.C., Carvalho, J.F., Prokopowitsch, A.S., and Pereira, R.M., RANK, RANKL, and osteoprotegerin in arthritic bone loss, *Braz.J.Med.Biol.Res.* 38, 161-170, 2005; Feng, X., RANKing intracellular signaling in osteoclasts, *IUBMB Life* 57, 389-395, 2005; Kostenuik, P.J., Osteoprotegerin and RANKL regulate bone resorption, density, geometry and strength, *Curr.Opin. Pharmacol.* 5, 618-625, 2005; Neumann, E., Gay, S., Muller-Ladner, U., The RANK/RANKL/osteoprotegerin system in rheumatoid arthritis: new insights from animals models, *Arthritis Rheum.* 52, 2960-2967, 2005; Wada, T., Nakashima, T., Hiroshi, N., and Penninger, J.M., RANKL-RANK signaling in osteoclastogenesis and bone disease, *Trends Mol.Med.* 12, 17-25, 2006.

Oxyanion Hole

A feature of the active sites of hydrolytic enzymes such as lipases or chymotrypsin where the acyl carbonyl of the acyl-enzyme intermediate is stabilized by hydrogen bonding to peptide amide nitrogens on the enzyme. See Menard, R. and Storer, A.C., Oxyanion hole interactions in serine cysteine proteases, *Biol.Chem.Hoppe Seyler* 373, 393-400, 1992; Whiting, A.K. and Peticolas, W.L., Details of the acyl-enzyme intermediate and the oxyanion hole in serine protease catalysis, *Biochemistry* 33, 552-561, 1994; Johal, S.S., White, A.J., and Wharton, C.W., Effect of specificity on ligand conformation in acyl-chymotrypsins, *Biochem.J.* 297, 281-287, 1994; Cui, J., Marankan, F., Fu, W., *et al.,* An oxyanion-hole selective serine protease inhibitor in complex with trypsin, *Bioorg.Med. Chem.* 10, 41-46, 2002; Lee, L.C., Lee, Y.L., Leu, R.J.. and Shaw, J.F., Functional role of catalytic triad and oxyanion hole-forming residues on enzyme activity of *Escherichia coli* thioesterase I/protease I/Phospholipase L1, *Biochem.J.* 397, 69-76, 2006.

Palindrome

A sequence which reads the same forwards and backwards; usually refers to nucleic acid sequence where opposing strands read the same; that is the 3'→5' sequence in one strand is the same as the 5'→3' sequence in the opposing strand. Palindromic sequences are frequently present at the sites of restriction class II enzyme cleavages. See Leach, D.R., Long DNA palindromes, cruciform structures, genetic instability and secondary structure repair, *Bioessays* 16, 893-900, 1994; Beato, M., Chavez, S., Truss, M., Transcriptional regulation by steroid hormones, *Steroids* 61, 240-251, 1996; Cho-Chung, Y.S., CRE-palindrome oligonucleotide as a transcription factor decoy and an inhibitor of tumor growth, *Antisense Nucleic Acid Drug Dev.* 8, 167-170, 1998.

Paracrine

Usually in reference to a hormone or other biological effector such as a peptide growth factor or cytokine which has an effect on the cell or tissue immediately surrounding the cell or tissue responsible for the synthesis of the given compound. See Franchimont, P., *Paracrine Control*, Saunders, Philadelphia, PA, 1986; Piva, F., *Cell to Cell Communication in Endocrinology*, Raven Press, New York, NY, 1988; *Autocrine and Paracrine Mechanisms in Reproductive Endocrinology*, ed. L.C. Krey and B.J. Gulyas, Plenum Press, New York, NY, 1989; Habenicht, A., *Growth Factors, Differentiation Factors and Cytokines*, Springer-Verlag, Berlin, Germany, 1990; Hardie, D.G., *Biochemical Messengers: Hormones, Neurotransmitters, and Growth Factors*, Chapman & Hall, London, UK, 1991; Vallesi, A., Giuli, G., Bradshaw, R.A., and Luporini, P., Autocrine mitogenic activity of pheromones produced by the protozoan ciliate *Euplotes raikovi*, *Nature* 376, 522-524, 1995.

Paralogues

Genes within the same genome that have evolved by gene duplication. See Ohno, S., The one-to-four rule and paralogues of sex-determining genes, *Cell.Mol.Life Sci.* 55, 824-830, 1999; Forterre, P., Displacement of cellular proteins by functional analogues from plasmids or viruses could explain puzzling phylogenies of many DNA informational proteins, *Mol.Microbiol.* 33, 457-465, 1999; Gilbert, J.M., The evolution of engrailed genes after duplication and speciation events, *Dev.Gene.Evol.* 212, 307-318, 2002; Ferrier, D.E., Hox genes: did the vertebrate ancestor have a Hox14? *Curr. Biol.* 14, R210-R211, 2004; Tsuru, T., Kawai, M., Mizutani-Ui, Y., Uchiyama, I., and Kobayashi, I., Evolution of paralogous genes: reconstruction of genome rearrangements through comparison of multiple genomes with *Staphylococcus aureus*, *Mol.Biol.Evol.* 23, 1269–1285, 2006.

PDZ-Domains

A protein domain involved in protein-protein interactions with a preference for C-terminal regions. PDZ proteins are recognized as component of biological scaffolds. The acronym is derived from the homology of a motif in PSD-95, the *Drosophila* Discs-Large septate junction protein and the epithelial junction protein ZO-1. See Ranganathan, R. and Ross, E.M., PDZ domain proteins; Scaffolds for signaling complexes, *Curr.Biol.* 7, R770-R773, 1997; Harris, B.Z. and Lim, W.A., Mechanism and role of PDZ domains in signaling complex assembly, *J.Cell Sci.* 114, 3219-3231, 2001; Dev, K.K., Making protein interactions druggable: targeting PDZ domains, *Nat.Rev.Drug Disc.* 3, 1047-1056, 2004; Schlieker, C., Mogk, A. and Bukau, B., A PDZ switch for a cellular stress response, *Cell* 117, 417-419, 2004; Brone, B. and Eggermont, J., PDZ proteins retain and regulate membrane transporters in polarized epithelial cell membranes, *Am.J.Physiol.Cell Physiol.* 3, 1047-1056, 2005.

Pedigree Rate

Estimate of the mutation rate determined by calculating the number of nucleotide changes observed over of a known number of reproductive events. See Howell, N., Smejkal, C.B., Mackey, D.A., Chinnery, P.F., Turnbull, D.M. and Herrnstadt, C., The pedigree rate of sequence divergence in the human mitochondrial genome: there is a difference between phylogenetic and pedigree rates, *Am.J.Hum.Genet.* 72, 659-670, 2003.

Pepducins

Cell-penetrating lipidated (palmitic acid) membrane tethered peptides which act as G protein-coupled receptors: see Covic, L., *et al.*, Activation and inhibition of G protein-coupled receptors by cell-penetrating membrane-tethered peptides, *Proc.Natl.Acad. Sci.* 99, 643-648, 2002; Lomas-Neira, J. and Ayala, A., Pepducins: an effective means to inhibit GPCR signaling by neutrophils, *Trends Immunol.* 26, 619-621, 2005. See also cell penetrating peptides.

Peptidome

The peptide complement of a genome. Peptidomics is the study of the peptidome. See Schrader, M., and Schulz-Knappe, P., Peptidomics technologies for human body fluids, *Trends Biotechnol.* 19(Suppl 10), S55-S60, 2001; Jurgens, M. and Schrader, M., Peptidomic approaches in proteomic research, *Curr.Opin.Mol.Ther.* 4, 236-241, 2002; Schulze-Knappe, M., Schrader, M., and Zucht, H.D., The peptidomics concept, *Comb.Chem.High Throughput.Screen.* 8, 697-704, 2005; Zheng, X., Baker, H., and Hancock, W.S., Analysis of the low molecular weight serum peptidome using ultrafiltration and a hybrid ion trap-Fourier transform mass spectrometry, *J.Chromatog.A.*, 1120, 173–184, 2006; Hortin, G.L., The MALDI TOF mass spectrometric view of the plasma proteome and peptidome, *Clin.Chem.* 52, 1223–1237, 2006.

Peptergents

Peptide detergents; small, self-assembling peptides with detergent properties. These peptide detergents appear to quite useful for the study of membrane proteins. See Yeh, J.I., Du. S., Tortajada, A., Paulo, J., and Zhang, S., Peptergents: peptide detergents that improve stability and functionality of a membrane protein, Glycerol-3-phosphate dehydrogenase, *Biochemistry* 44, 16912-16919, 2005. Kiley, P., Zhao, X., Vaughn, M., *et al.*, Self-assembling peptide detergents stabilize isolated photosystem I on a dry surface for an extended time, *PLoS Biol.* 3, e230, 2005.

Perforin

Perforin is a protein located in the granules of CD8 T-cells (cytotoxic T cells) and natural killer cells. Upon degranulation of these cells, perforin inserts itself into the target cell's plasma membrane, forming a pore resulting in lysis of the target cell. Catalfamo, M., and Henkart, P.A., Perforin and the granule exocytosis cytotoxicity pathway, *Curr.Opin.Immunol.* 15, 522-527, 2003; Smith, M.J., Cretney, E., Kelly, J.M. *et al*, Activation of NK cell cytotoxicity, *Mol.Immunol.* 42, 501-510, 2005; Ashton-Rickardt, P.G., The granule pathway of programmed cell death, *Crit.Rev.Immunol.* 25, 161-182, 2005; Yoon, J.W. and Jun, H.S., Autoimmune destruction of pancreatic beta cells, *Am.J.Ther.* 12, 580-591, 2005; Andersen, M.H., Schrama, D., Thor Straten, P., and Becker, J.C., Cytotoxic T cells, *J.Invest.Dermatol.* 126, 32-41, 2006.

Peroxiredoxins

Group of antioxidant thioredoxin-dependent enzymes with a catalytic function in the detoxification of cellular-toxic peroxides. See Claiborne, A., Ross, R.P., and Parsonage, D., Flavin-linked peroxide reductases: protein-sulfenic acids and the oxidative stress response, *Trends Biochem.Sci.* 17, 183-186, 1992; Dietz, K-J., Horhing, F., König, J., and Baien, M., The function of chloroplast 2-cysteine peroxiredoxin I peroxide detoxification and its regulation, *J.Expt.Bot.* 53, 1321-1329, 2002; Immenschuh, S. and Baumgart-Vogt, E., Peroxiredoxins, oxidative stress, and cell proliferation, *Antioxid.Redox.Signal.* 7, 768-777, 2005; Rouhier, N. and Jacquot, J.P., The plant multigenic family of thiol peroxidases, *Free Rad.Biol.Med.* 38, 1413-1421, 2005; Rhee, S.G., Chae, H.Z., and Kim, K., Peroxiredoxins: a historical overview and speculative preview of novel mechanisms and emerging concepts in cell signaling, *Free Radic.Biol.Med.* 38, 1543-1552, 2005; Immenschuh, S. and Baumgart-Vogt, E., Peroxiredoxins, oxidative stress, and cell proliferation, *Antioxidants & Redox Sign.* 7, 768-777.

Peroxynitrite

A oxidizing/nitrating agent derived from the reaction of nitric oxide and superoxide which reacts with proteins, lipids and nucleic acids. The reactions are complex and in addition to oxidation reactions such as carbonyl formation and disulfide formation, there are reactions such as nitrosylation of cysteine and the nitration of tyrosine. See Beckman, J.S. and Crow, J.P., Pathological implications of nitric oxide, superoxide and peroxynitrite formation, *Biochem.Soc.Trans.* 21, 330-334, 1993; Pryor, W.A. and Squadrito, G.L., The chemistry of peroxynitrite: a product from the reaction of nitric oxide with superoxide, *Am.J.Physiol.* 268, L699-L722, 1995; Uppu, R.M., Squadrito, G.L., Cueto, R., and Pryor, W.A., Synthesis of peroxynitrite by azide-ozone reaction, *Methods Enzymol.* 269, 311-321, 1996; Beckman, J.S. and Koppenol, W.H., Nitric oxide, superoxide, and peroxynitrite: the good, the bad, and ugly, *Am.J.Physiol.* 271, C1424-C1437, 1996; Girotti, A.W., Lipid hydroperoxide generation, turnover, and effector action in biological systems, *J.Lipid Res.* 39, 1529-1542, 1998; Radi, R., Denicola, A., and Freeman, B.A., Peroxynitrite reactions with carbon dioxide-bicarbonate, *Methods Enzymol.* 301, 353-357, 1999; Groves, J.T., Peroxynitrite: reactive, invasive and enigmatic, *Curr.Opin.Chem.Biol.* 3, 226-235, 1999; Halliwell, B., Zhao, K., and Whiteman, M., Nitric oxide and peroxynitrite. The ugly, the uglier and the not so good: a personal view of the recent controversies, *Free Radic.Res.* 31, 651-669, 1999; Estevez, A.G. and Jordan, J., Nitric oxide and superoxide, a deadly cocktail, *Ann.N.Y.Acad.Sci.* 962, 207-211, 2002; Ohmori, H., and Kanayama, N. Immunogenicity of an inflammation-associated product, tyrosine nitrated self-proteins, *Autoimmun.Rev.* 4, 224-229, 2005; Hurd, T.R., Filipovska, A., Costa, N.J., *et al.*, Disulphide

formation on mitochondrial protein thiols, *Biochem.Soc.Trans.* 44, 1390-1393, 2005; Sawa, T. and Ohshima, H., Nitrative DNA damage in inflammation and its possible role in carcinogenesis, *Nitric Oxide* 14, 91-100, 2006; Niles, J.C., Wishnok, J.S., and Tannenbaum, S.R., Peroxynitrite-induced oxidation and nitration products of guanine and 8-oxoguanine: structures and mechanisms of product formation, *Nitric Oxide* 14, 109-121, 2006; Uppu, R.M., Synthesis of peroxynitrite using isoamyl nitrite and hydrogen peroxide in a homogeneous solvent system, *Anal.Biochem.* 354, 165-168, 2006. The reaction of tyrosine with peroxynitrite is sensitive to solvent environment with nitration favored in a hydrophobic environment as opposed to oxidation (Zhang, H., Joseph, J., Feix, J., *et al.*, Nitration and oxidation of a hydrophobic tyrosine probe by peroxynitrite in membranes: Comparison with nitration and oxidation of tyrosine by peroxynitrite in aqueous solution, *Biochemistry* 40, 7675-7686, 2001.

Pescadillo

A nuclear protein originally demonstrated in zebrafish. Pescadillo is thought to be important in a variety of nuclear activities including DNA replication, ribosome formation, and cell cycle control. See Allende, M.L., Amsterdam, A., Becker, T., *et al.*, Insertional mutagenesis in zebrafish identifies two novel genes, pescadillo and dead eye, essential for embryonic development, *Genes Dev.* 10, 3141-3155, 1996; Haque, J., Boger, S., Li, J., and Duncan, D.A., The murine Pes1 gene encodes a nuclear protein containing a BRCT domain, *Genomics* 70, 201-210, 2000; Kinoshita, Y., Jarell, A.D., Flaman, J.M., *et al.*, Pescadillo, a novel cell cycle regulatory protein abnormally expressed in malignant cells, *J.Biol.Chem.* 276, 6656-6665, 2001; Maiorana, A., Tu. X., Cheng, G., and Baserga, R., Role of pescadillo in the transformation and immortalization of mammalian cells, *Oncogene* 23, 7116-7124, 2004; Killian, A., Le Meur, N., Sesboue, R., *et al.*, Inactivation of the RRB1-Pescadillo pathway involved in ribosome biogenesis induces chromosomal instability, *Oncogene* 23, 8597-8602, 2004; Sikorski, E.M., Uo, T., Morrison, R.S., and Agarwal, A., Pescadillo interacts with the cadmium response element of the human heme oxygenase-1 promoter in renal epithelial cells, *J.Biol.Chem.* 281, 24423-24430, 2006.

Pfam

Protein family; used to describe a protein family data base. See Persson, B., Bioinformatics in protein analysis, *EXS* 88, 215-231, 2000; Bateman, A., Birney, E., Cerruti, L., *et al.*, The Pfam protein families data base, *Nucl.Acids Res.* 30, 276-280, 2002; Lubec, G., Afjehi-Sadat, L., Yang, J.W., and John, J.P., Searching for hypothetical proteins: theory and practice based upon original data and literature, *Prog.Neurobiol.* 77, 90-127, 2005; Anderston, J.N., Del Vecchio, R.L., Kannan, N., *et al.*,, Computational analysis of protein tyrosine phosphatases: practical guide to bioinformatics and data resources, *Methods* 35, 90-114, 2005.

Pharmaceutical Equivalence

Drug products can be considered to be pharmaceutical equivalents if such products (1) contain the same active ingredients, (2) are of the same dosage form and route of administration, and (3) identical in strength and concentration. The term therapeutic equivalence is also used to describe pharmaceutical equivalence. Pharmaceutically equivalent drug products may differ in attributes such as shape, color, excipients and release mechanism.

Pharmaceutical equivalence is of importance in the development of generic drugs. See http://www.fda.gov.

Pharmacogenomics

The use of genomics to study the development and utilization of drugs. Pharmacogenomics will be essential to the development of predictive medicine/personalized medicine. Wilke, R.A., Reif, D.M., and Moore, J.H., Combinatorial Pharmacogenetics, *Nat. Rev.Drug Discov.* 4, 911-918, 2005; Ginsburg, G.S., Konstance, R.P., Allsbrook, J.S., and Schulman, K.A., Implications of pharmacogenomics for drug development and clinical practice, *Arch.Intern.Med.* 165, 2331-2336,2005; Siest, G., Marteau, J.B., Maumus, S., *et al.*, Pharmacogenomics and cardiovascular drugs: need for integrated biological system with phenotypes and proteomic markers, *Eur.J.Pharmacol.* 527, 1-22, 2005; Robson, B. and Mushlin, R., Genomic messaging system and DNA mark-up language for information-based personalized medicine with clinical and proteome research applications, *J.Proteome Res.* 3, 930-948, 2004.

Pharmacophore

The collection of structural features of a compound which provide for the pharmacological properties. Uses information derived from QSAR studies. See Glennon, R.A., Pharmacophore identification for sigma-1 (sigma1) receptor binding: application of the "deconstruction-reconstruction-elaboration" approach, *Mini Rev.Med.Chem.* 5, 927-940, 2005; Guner, O.F, The impact of pharmacophore modeling in drug design, *IDrugs* 9, 567-572, 2005; Guner, O., Clent, O., Kurogi, Y., Pharmacophore modeling and three dimensional database searching for drug design using catalyst: recent advances, *Curr.Med.Chem.* 11, 2991-3005, 2004.

Pharmacoproteomics

The use of proteomics to predict individual reaction to a drug or drugs; also the use of proteomics for drug discovery and development; related to personalized medicine, theranostics, pharmacogenomics. See Di Paolo, A., Danesi, R., and Del Tacca, M., Pharmacogenetics of neoplastic diseases: new trends, *Pharmacol. Res.* 49, 331-342, 2004; Jain, K.K., Role of pharmacoproteomics in the development of personalized medicine, *Pharmacogenomics J.* 5, 331-336, 2004; Witzmann, F.A. and Grant, R.A., Pharmacoproteomics in drug development, *Pharmacogenomics J.* 3, 69-76, 2003.

Phase Diagram(s)

A graph showing the relationship between phases (i.e. solid/gas/liquid) over a range of physical conditions (usually temperature and pressure). A phase is generally defined as a homogenous part of a heterogeneous system that is clearly separated from other phases by a physical boundary; the separation between ice and water is an example of a boundary between phases. See Ohgushi, M. and Wada, A., Liquid-like state of side chains at the intermediate stage of protein denaturation, *Adv.Biophys.* 18, 75-90, 1984; Dorfler, H.D., Mixing behavior of binary insoluble phospholipid monolayers. Analysis of the mixing properties of binary lecithin and cephalin systems by application of several surface and spreading techniques, *Adv.Colloid Interface Sci.* 31, 1-110, 1990; Diamond, A.D. and Hsu, J.T., Aqueous two-phase systems for

biomolecule separation, *Adv.Biochem.Eng.Biotechnol.* 47, 89-135, 1992; Mason, J.T., Investigation of phase transitions in bilayer monolayers, *Methods Enzymol.* 295, 468-494, 1998; Crowe, J.H., Tablin, F., Tsvetkova, N. *et al.*, Are lipid phase transitions responsible for chilling damage in human platelets? *Cryobiology* 38, 180-191, 1999; Smeller, L, Pressure-temperature phase diagrams of biomolecules, *Biochim.Biophys.Acta* 1595, 11-29, 2002; Dill, K.A., Truskett, T.M., Vlachy, V., and Hribar-Lee, B., Modeling water, the hydrophobic effect, and ion salvation, *Annu.Rev.Biochem. Biomol.Struct.* 34, 173-199, 2005; Scharnagl, C., Reif, M., and Friedrich, J., Stability of proteins: temperature, pressure and the role of the solvent, *Biochim.Biophys.Acta* 1749, 187-213, 2005.

Phenotype

The physical manifestation of the genes of an organism; the collection of structure and function expressed by the genotype of an organism; the visible properties of an organism that are produced by the interaction of a genotype and the environment. See Padykula, H.A., *Control Mechanisms in the Expression of Cellular Phenotypes*, Academic Press, New York, NY, 1970; Levine, A.J., *The Transformed Phenotype*, Cold Spring Harbor Laboratory Press, Cold Spring Harbor, NY, 1984; Dewitt, T.J. and Scheiner, S.M., *Phenotypic Plasticity Function and Conceptual Approaches*, Oxford University Press, Oxford, UK, 2004; Pigliucci, M. and Preston, K., *Phenotypic Integration: Studying the Ecology and Evolution of Complex Phenotypes*, Oxford University Press, Oxford, UK, 2004.

Phospholipase C

A family of intracellular enzymes central to many signal transduction pathways via effects on Ca^{2+} and protein kinase C. Phospholipase C catalyzes the hydrolysis of phosphoinositol 4,5-bisphosphate to yield 1,4,5-inositol triphosphate and diacylglycerol. Irvine, R.F., The enzymology of stimulated inositol lipid turnover, *Cell Calcium* 3, 295-309, 1982; Farese, R.V., Phospholipids as intermediates in hormone action, *Mol.Cell Endocrinol.* 35, 1-14, 1984; Majerus, P.W., The production of phosphoinositide-derived messenger molecules, *Harvey Lect.* 82, 145-155, 1986-87; Litosch, I. and Fain, J.N., Regulation of phosphoinositide breakdown by guanine nucleotides, *Life Sci.* 39, 187-194, 1986; Putney, J.W., Jr., Formation and actions of calcium-mobilizing messenger, inositol 1,4,5-trisphosphate, *Am.J.Physiol.* 252, G149-G157, 1987; Lemmon, M.A., Pleckstrin homology domains: two halves make a hole? *Cell* 120, 574-576, 2005; Malbon, C.C., G proteins in development, *Nat.Rev.Mol.Cell Biol.* 6, 689-701, 2005; Gilfillan, A.M. and Tkaczyk, C., Integrated signaling pathways for mast-cell activation, *Nat.Rev.Immunol.* 6, 218-230, 2006; Drin, G., Dougnet, D., and Scarlata, S., The pleckstrin homology domain of Phospholipase $C\beta$ transmits enzymatic activation through modulation of the membrane-domain orientation, *Biochem.* 45, 5712-5724, 2006.

Cis-Phosphorylation or Cis-autophosphorylation

An autophosphorylation event where the kinase catalyzes the phosphorylation of itself as opposed to another molecule of the same kinase. See Frattali, A.L., Treadway, J.L., and Pessin, J.E., Transmembrane signaling by the human insulin receptor kinase. Relationship between intramolecular beta subunit *trans*- and cis-autophosphorylation and substrate kinase activation, *J.Biol.Chem.* 267, 195210-19528, 1992; Cann, A.D. and Kohanski, R.A., Cis-autophosphorylation of juxtamembrane tyrosines in the insulin receptor kinase domain, *Biochemistry* 36, 7681-7689, 1997; Cann, A.D., Bishop, S.M., Ablooglu, A.J., and Kohanski, R.A., Partial activation of the insulin receptor kinase domain by juxtamembrane autophosphorylation, *Biochemistry* 37, 11289-11300, 1998; Leu, T.H. and Maa, M.C., Tyr-863 phosphorylation enhances focal adhesion kinase autophosphorylation at Tyr-397, *Oncogene* 21, 6992-7000, 2002; Iyer, G.H., Moore, M.J., and Taylor, S.S., Consequences of lysine 72 mutation on the phosphorylation and activation state of cAMP-dependent kinase, *J.Biol.Chem.* 280, 8800-8807, 2005; Yang, K., Kim, J.H., Kim, H.J., *et al.*, Tyrosine 740 phosphorylation of discoid in domain receptor 2 by Src stimulates intramolecular autophosphorylation and Shc signaling complex formation, *J.Biol.Chem.* 280, 39058-39066, 2005.

Trans-Phosphorylation or *Trans*-autophosphorylation

An autophosphorylation event where the kinase catalyzes the phosphorylation of another molecule of the same kinase as opposed to cis-phosphorylation where the kinase phosphorylates itself. A more generic definition is the transfer of a phosphoryl function from one site to another site (transfer of a phosphate function). See Wei, L., Hubbard, S.R., Hendrickson, W.A., and Ellis, L., Expression, characterization, and crystallization of the catalytic core of the human insulin receptor protein-tyrosine kinase domain, *J.Biol.Chem.* 270, 8122-8130, 1995; McKeehan, W.L., Wang, F., and Kan, M., The heparin sulfate-fibroblast growth factor family: diversity of structure and function, *Prog. Nucleic Acid Res.Mol.Biol.* 59, 135-176, 1998; Klint, P. and Claesson-Welsh, L., Signal transduction by fibroblast growth factor receptors, *Front.Biosci.* 4, D165-D177, 1999; DiMaio, D., Lai, C.C., and Mattoon, D., The platelet-derived growth factor beta receptor as a target of the bovine papillomavirus E5 protein, *Cytokine Growth Factor Rev.* 11, 283-293, 2000; DiMaio, D. and Matoon, D., Mechanisms of cell transformation by papillomavirus E5 proteins, *Oncogene* 20, 7866-7873, 2001; Schwarz, J.K., Lovly, C.M., and Piwnica-Worms, H., Regulation of the Chk2 protein kinase by oligomerization-mediated cis- and trans-phosphorylation, *Mol.Cancer Res.* 1, 598-609, 2003; Wu, S. and Kaufman, R.J., trans-Autophosphorylation by the isolated kinase domain is not sufficient for dimerization of activation of the dsRNA-activated protein kinase PKR, *Biochemistry* 43, 11027-11034, 2004; Shi, G.W., Chen, J., Concepcion, F. *et al.*, Light causes phosphorylation of nonactivated visual pigments in intact mouse rod photoreceptor cells, *J.Biol.Chem.* 280, 41184-41191, 2005; Gao, X. and Harris, T.K., Steady-state kinetic mechanism of PDK1, *J.Biol. Chem.* 281, 21670–21681, 2006.

Phylogenetic Rate

Estimate of the substitution rate calculated by comparing the molecular sequence data obtained from different species. See Heyer, E., Zietkiewicz, E., Rochowski, A., Yotova, V., Puymirat, J., and Labuda, D., Phylogenetic and familial estimates of mitochondrial substitution rates: study of control region mutations in deep-rooting pedigrees, *Am.J.Human Genet.* 69, 1113-1126, 2001; Ritchie, P.A., Miller, C.D., Gibb, G.C., Baroni, C., and Lambert, D.M., Ancient DNA enables timing of the pleistocene origin and holocene expansion of two adelie penguin lineages in Antarctica, *Mol.Biol.Evol.* 21, 240-248, 2004.

Phytoremediation

Use of plants for remediation of toxic chemicals. This can be separated into phytoremediation/rhizofiltration which describes the removal of toxic materials from water and phytoextraction which describes the removal of toxic materials from soil by plants. See Arthur, E.L., Rice, P.J., Rice, P.J., *et al.*, Phytoremediation – An overview, *Crit.Rev. Plant Sci.* 24, 109-122, 2005.

Pinocytosis

Transcytosis across an endothelial cell through endosomic vesicles and/or a tubovesicular pathway. See Chapman-Andresen, C., *Studies on Pinocytosis in Amoebae*, Danish Science Press, Copenhagen, Denmark, 1962; LaBella, F.S., *Pinocytosis*, MSS Information Group, New York, NY, 1973; Josefsson, J.-O., *Pincytosis*, University of Lund, Lund, Sweden, 1973; Stossel, T.P., Contractile proteins in cell structure and function, *Annu.Rev. Med.* 29, 427-457, 1978; Lloyd, J.B., Insights into mechanisms of interacellular protein turnover from studies on pinocytosis, *Ciba Found.Symp.* (75), 151-165, 1979; Barondes, S.H., Lectins: their multiple endogenous cellular function, *Ann.Rev.Biochem.* 50, 207-231, 1981; Besterman, J.M., and Low, R.B., Endocytosis: a review of mechanisms and plasma membrane dynamics, *Biochem.J.* 210, 1-13, 1983; Mansilla, A.O., Arguero, R.S., Rico, F.G., and Alba, C.C., Cellular receptors, acceptors and clinical implications, *Arch.Med.Res.* 24, 127-137, 1993; Meier, O. and Greber, U.F., Adenovirus endocytosis, *J.Gene.Med.* 5, 451-462, 2003; Batahori, G., Cervenak, L. and Karadi, I., Caveolae—an alternative endocytotic pathway for targeted drug delivery, *Crit.Rev.Ther.Drug Carrier Syst.* 21, 67-95, 2004.

Piranha Solution

A mixture of concentrated sulfuric acid and hydrogen peroxide [as an example, a 7:3(v/v) ratio of 98% H_2SO_4(concentrated sulfuric acid) and 30%(w/v) H_2O_2] which is used for the cleaning of glass and other surfaces. See Seeboth, A. and Hettrich, W., Spatial orientation of highly ordered self-assembled silane monolayers or glass surfaces, *J.Adhesion Sci.Technol.* pp. 495-505, 1997; Gray, D.E., Case-Green, S.C., Fell, T.S., Dobson, P.J., and Southern, E.M., Ellipsometric and interferometric characterization of DNA probes immobilized on a combinatorial array, *Langmuir* 13, 2833-2842, 1997; Steiner, G., Möller, H., Savchuk, O. *et al.*, Characterization of ultra-thin polymer films by polarization modulation FITR spectroscopy, *J.Mol.Struct.* 563-564, 273-277, 2001; Guo, W. and Ruckenstein, E., Crosslinked glass fiber affinity membrane chromatography and its application to fibronectin separation, *J.Chromatog.B Technol.Biomed.Life Sci.* 795, 61-72, 2003; Ziegler, K.J., Gu, Z., Peng, H. *et al.*, Controlled oxidative cutting of single-walled carbon nanotubes, *J.Amer.Chem.Soc.* 127, 1541-1547, 2005; Wang, M., Liechti, K.M., Wang, Q., and White, J.M., Self-assembled monolayers: fabrication with nanoscale uniformity, *Langmuir* 21,1848-1857, 2005; Szuneritz, S. and Boukherroub, R., Preparation and characterization of thin films of SiO(x) on gold substrates for surface plamon resonance studies, *Langmuir* 22, 1660-1663, 2006; Petrovykh, D.Y., Kimura-Suda, H., Opdahl, A., *et al.*, Alkanethiols on platinum: multicomponent self-assembled monolayers, *Langmuir* 14, 2578-2587, 2006.

Plant Cell

Resulting from the capture of a cyanobacterium by a eukaryotic, mitochondria-possessing cell; the endosymbiont (cyanobacter) lost its identity and become a chloroplast. See Martin, W., Rujan, T., Richly, E., *et al.*, Evolutionary analysis of Aribidopsis, cyanobacterial and chloroplast genomes reveals plastid phylogeny and thousands of cyanobacterial genes in the nucleus, *Proc.Nat. Acad.Sci.USA* 99, 12246-12251, 2002; Grevich, J.J. and Daniell, H., Chloroplast genetic engineering: recent advances and future perspectives, *Crit.Rev.Plant Sci.* 23, 84-107, 2005.

Plasma/Serum Proteome

The identification and characterization of the proteins in the blood plasma/serum. See Lathrop, J.T., Anderson, N.L., Anderson, N.G., and Hammond, D.J., Therapeutic potential of the plasma proteome, *Curr.Opin.Mol.Ther.* 5, 250-257, 2003; Veenstra, T.D., Conrads, T.P., Hood, B.L., Avellino, A.M., Ellenbogen, R.G., and Morrison, R.S., Biomarkers: mining the biofluid proteome, *Mol. Cell.Proteomics* 4, 409-418, 2005.

Plastid

Usually refers to a chloroplast before the development of chlorophyll but can refer to any small intracellular pigmented valcuole. Plastids have been described in some bacteria. See Granick, S. and Gibor, A., The DNA of chloroplasts, mitochondria and centrioles, *Prog.Nucleic Acid Res.Mol.Biol.* 6, 143-186, 1967; Gibor, A. and Ganick, S., Plastids and mitochondria: inheritable systems, *science* 145, 890-897, 1964; Lopez-Juez, E. and Pyke, K.A., Plastids unleashed: their development and their integration in plant development, *Int.J.Dev.Biol.* 49, 557-577, 2005; Mackenzie, S.A., Plant organellar protein targeting: a traffic plan still under construction, *Trends Cell. Biol.* 15, 548-554, 2005; Miyagishima, S.Y., Origin and evolution of the chloroplast division machinery, *J.Plant.Res.* 118, 295-306, 2005; Beck, C.F., Signaling pathways for the chloroplast to the nucleus, *Planta* 222, 743-756, 2005; Toyoshima, Onda, Y., Shiina, T. and Nakahira, Y., Plastid transcription in higher plants, *Crit.Rev. Plant Sci.* 24, 59-81, 2005; Pilon, M., Abdel-Ghany, S.E., van Hoewyk, D., Ye, H., and Pilon-Smits, E.A., Biogenesis of iron-sulfur cluster proteins in plastids, *Genet.Eng.* 27, 101-117, 2006.

Plate Number

In chromatography, a plate is a separation instance or moment that a solute encounters during passage through a chromatographic column. The higher the number of plates, the more possibility for high resolution but such resolution depends on the individual behavior of solutes (see resolution). Plates may be theoretical or effective plates. The efficiency of a column is measured in the number of plates referred to as plate number (N). One equation for plate number (N). $N = 5.54(t_r/W_{1/2})^2$ where t_r is band retention time and $W_{1/2}$ is peak width at peak half-height. See Boyes, B.E. and Kirkland, J.J., Rapid, high-resolution HPLC separation of peptides using small particles at elevated temperatures, *Pept.Res.* 6, 249-258, 1993; Anspach, B., Gierlich, H.U., and Unger, K.K., comparative study of Zorbax Bio series GF 250 and GF 450 and Tsk-Gel 3000 SW and SWXL columns in size-exclusion chromatography of proteins, *J.Chromatog.* 443, 45-54, 1988; Palsson, E., Axelsson, A., and Larsson, P.O., Theories of chromatographic efficiency applied to expanded

base, *J.Chromatog.A* 912, 235-248, 2001; Mahesan, B. and Lai, W., Optimization of selected chromatographic responses using a designed experiment at the fine-tuning stage in reversed-phase high-performance liquid chromatographic method development, *Drug Dev.Ind.Pharm.* 27, 585-590, 2001; Ishizuka, N., Kobayashi, H., Minakuchi, H., *et al.*, Monolithic silica columns for high-efficiency separations by high-performance liquid chromatographic, *J.Chromatog.A.* 960, 85-96, 2002; Jandera, P., Halama, M., and Novotna, K., Stationary-phase effects in gradient high-performance liquid chromatography, *J.Chromatog.A.* 1030, 33-41, 2004; Lim, L.W., Hirose, K., Tatsumi, S.,*et al.*, Sample enrichment by using monolithic precolumns in microcolumn liquid chromatography, *J.Chromatog.A.* 1033, 205-212, 1033; Okanda, F.M. and Rassi, Z., Capillary electrochromatography with monolithic stationary phases. 4. Preparation of neutral stearyl-acrylate monoliths and their evaluation in capillary electrochromatography of neutral and charged small species as well as peptides and proteins, *Electrophoresis* 26, 1988-1995, 2005; Berezkin, V.G. and Lapin, A.B., Ultra-short open capillary columns in gas-liquid chromatography, *J.Chromatog.A.* 1075, 197-203, 2005; Donohoe, E., Denaturing high-performance liquid chromatography using the WAVE DNA fragment analysis system, *Methods Mol.Med.* 108, 173-187, 2005; Lohrmann, M., Schulte, M., and Strube, J., Generic method for systematic phase selection and method development of biochromatographic processes. Part I. Selection of a suitable cation-exchanger for the purification of a pharmaceutical protein, *J.Chromatog.A* 1092, 89-100, 2005; Chester, T.L. and Teremmi, S.O., A virtual-modeling and multivariate-optimization examination of HPLC parameter interactions and opportunites for saving analysis time, *J.Chromatog.A.* 1096, 16-27, 2005.

Pleckstrin Homology Domain

A protein domain consisting of approximately 100 amino acids which binds phosphoinositide and other activators such as heterotrimeric G proteins and participates in the process of signal transduction. The name was derived from the platelet protein pleckstrin (platelet and leukocyte C kinase substrate) identified as a substrate for protein kinase C. See Tyers, M., Rachubinski, R.A., Stewart, M.I., *et al.*, Molecular cloning and expression of the major protein kinase C substrate of platelets, *Nature* 333, 470-473, 1988. Mayer, B.J., Ren, R., Clark, K.L., A putative modular domain present in diverse signaling proteins, *Cell* 73, 629-630, 1993; Musacchio, A., Gibson, T., Rice, P., Thompson, J., and Saraste, M., The pH domain: a common piece in the structural patchwork of signalling proteins, *Trends Biochem.Sci.* 18, 343-348, 1993; Ingley, E. and Hemmings, B.A., Pleckstrin homology (PH) domains in signal transduction, *J.Cell.Biochem.* 56, 436-443, 1994; Lemmon, M.A., Ferguson, K.M., and Abrams, C.S., Pleckstrin homology domains and the cytoskeleton, *FEBS Lett.* 513, 71-76, 2002; Philip, F., Guo, Y., and Scarlata, S., Multiple roles of pleckstrin homology domains in the phospholipase Cbeta function, *FEBS Lett.* 531,29-32, 2002; Lemmon, M.A., Phosphoinositide recognition domains, *Traffic* 4, 201-213, 2003; Cozier, G.E., Carlton, J., Bouyoucef, D., and Cullen, P.J., Membrane targeting by pleckstrin homology domains, *Curr. Top.Microbiol.Immunol.* 282, 49-88, 2004; Balla, T., Inositol-lipid binding motifs: signal integrators through protein-lipid and protein-protein interaction, *J.Cell.Sci.* 119, 2093-2104, 2005; Perry, R.J. and Ridgway, N.D., Molecular mechanisms and regulation of ceramide transport, *Biochim.Biophys.Acta.* 1734, 220-234, 2005.

Pleiotropic

Having more than one phenotypic expression of a gene; an effector molecule associated with more than a single event depending on the stimulation or, in the case of regulatory proteins and peptides, more than one target receptor. See Takeda, Y., Pleiotropic actions of aldosterone and the effects of eplerenone, a selective mineralocorticoid receptor antagonist, *Hypertens.Res.* 27, 781-789, 2004; Wilkie, A.O., Bad bones, absent smell, selfish testes: the pleiotropic consequences of human FGF receptor mutations, *Cytokine Growth Factor Rev.* 16, 187-203, 2005; Staels, B. and Fruchart, J.C., Therapeutic roles of peroxisome proliferator-activated receptor agonists, *Diabetes* 54, 2460-2470, 2005; Russo, V.C., Gluckman, P.D., Feldman, E.L., and Werther, G.A., The insulin-like growth factor system and its pleiotropic functions in brain, *Endocrine Rev.* 26, 916-943, 2005; Carrillo-Vico, A., Guerrero, J.M., Lardone, P.J., and Reiter, R.J., A review of the multiple actions of melatonin on the immune system, *Endocrine* 27, 189-200, 2005.

Podosome

A specialized cell-matrix contact point which is structural distinct from focal adhesion complexes. See Linder, S., and Kopp, P., Podosomes at a glance, *J.Cell Sci.* 118, 2079-2082, 2005; McNiver, M.A., Baldassarre, M., and Buccione, R., The role of dynamin in the assembly and function of podosomes and invadopodia, *Front.Biosci.* 9, 1944-1953, 2004; Linder, S. and Aepfelbacher, M., Podosomes: adhesion hot-spots of invasive cells, *Trends Cell Biol.* 13, 376-385, 2003.

Poisson Distribution

A probability density function that is an approximation to the biomodal distribution and is characterized by its mean being equal to its variance. See Mezei, L.M., *Practical Spreadsheet Statistics and Curve Fitting for Scientists and Engineers*, Prentice Hall, Englewood Cliffs, NJ, 1990; Dowdy, S.M. and Wearden, S., *Statistics for Research*, Wiley, New York, NY, 1991; Balakrishnan, N. and Nevzorov, V.B., *A Primer on Statistical Distributions*, Wiley, Hoboken, NJ, 2003.

Polyadenylation

The attachment of 200 adenyl residues to the 3'-end of messenger of RNA protecting the mRNA from degradation by nucleases and aids in transfer of mRNA from nucleus to cytoplasm. The polyadenylation follows a specific cleavage at the termination of transcription. See Wilt, F.H., Polyadenylation of material RNA of sea urchin eggs after fertilization, *Proc.Natl.Acad.Sci.USA* 70, 2345-2349, 1973; Cooper, D.L. and Marzluff, W.F., Polyadenylation of RNA in a cell-free system from mouse myeloma cells, *J.Biol.Chem.* 253, 8375-8380, 1978; Bernstein, P. and Ross, J., Poly(A), poly(A) binding protein and the regulation of mRNA stability, *Trends in Biochem. Sci.* 14, 373-377, 1989; Manley, J.L., Polyadenylation of mRNA precursors, *Biochim.Biophys.Acta* 950, 1-12, 1988; Buratowski, S., Connections between mRNA 3' end processing and transcription termination, *Curr.Opin.Cell.Biol.* 17, 257-261, 2005.

Polymerase Chain Reaction

A method for synthesizing and amplifying a specific DNA sequence based on the use of specific oligonucleotide primers and unique DNA polymerases such as the thermostable DNA polymerase from *Thermus aquaticus* (Taq polymerase). PCR amplicons from the amplified sequence are analyzed by size or sequence. See Kleppe, K., Ohtsuka, E., Kleppe, R., *et al.*, Studies on polynucleotides XCVI. Repair replication of short synthetic DNA's as catalyzed by DNA polymerases, *J.Mol.Biol.* 56, 341-346, 1971; Saiki, R.K., Scharf, S., Faloona, F., *et al.*, Enzymatic amplification of beta-globin genomic sequences and restriction site analysis for diagnosis of sickle cell anemia, *Science* 230, 1350-1354, Saiki, R.K., Gelfand, D.H., Stoffel, S., *et al.*, Primer-directed enzymatic amplification of DNA with a thermostable DNA polymerase, *Science* 239, 487-491, 1988; Vosberg, H.P., The polymerase chain reaction: an improved method for the analysis of nucleic acids, *Hum.Genet.* 83, 1-15, 1989; Mullis, K.B., The unusual origin of the polymerase chain reaction, *Sci.Amer.* 262(4), 56-61, 1990; Taylor, G.R. and Robinson, P., The polymerase chain reaction: from functional genomics to high-school practical classes, *Curr.Opin.Biotechnol.* 9, 35-42, 1998; *PCR Protocols: A Guide to Methods and Applications*, ed. Innes, M.A., Academic Press, San Diego, 1990; *PCR Protocols: Current Methods and Applications*, ed. White, B.A., Humana Press, Totowa, NJ, 1993; Dieffenbach, C.W. and Dveksler, G.S., *PCR Primer: A Laboratory Manual*, Cold Spring Harbor Laboratory Press, Cold Spring Harbor, NY, 1995; *PCR Applications: Protocols for Functional Genomics*, ed. Sninsky, J.J. and Innes, M.A., Academic Press, San Diego, CA, 1999. See RT-PCR (reverse transcriptase-polymerase chain reaction), real-time RT-PCR, and gene expression.

Polyvinylpyrrolidone

Polyvinylpyrrolidone (PVP) is a polymer similar to poly(ethylene) glycol (PEG) is that is readily soluble in water and is used for the stabilization of proteins. Unlike PEG, PVP is useful in the lyophilization of proteins. PVP is synthesized by the free-radical polymerization of *N*-vinylpyrrolidinone (1-vinyl-2-pyrrolidinone). The final size of the polymer is controlled by choice of experimental conditions. It has a wide application in biotechnology. See Antonsen, K.P., Gombotz, W.R., and Hoffman, A.S., Attempts to stabilize a monoclonal antibody with water soluble synthetic polymers of varying hydrophobicity, *J.Biomater.Sci. Polym.Ed.* 6, 55-65, 1994; Gombotz, W.R., Pankey, S.C., Phan, D., *et al.*, The stabilization of a human IgM monoclonal antibody with poly(vinylpyrrolidone), *Pharm.Res.* 11, 624-632, 1994; Gibson, T.D., Protein stabilization using additives based on multiple electrostatic interactions, *Dev.Biol.Stand.* 87, 207-217, 1996; Anchordoquy, T.J. and Carpenter, J.F., Polymers protect lactate dehydrogenase during freeze-drying by inhibiting dissociation in the frozen state, *Arch.Biochem.Biophys.* 332, 231-238, 1996; Yoshioka, S., Aso, Y., and Kojima, S., The effect of excipients on the molecular mobility of lyophilized formulations, as measured by glass transition temperature and NMR relaxation-based critical mobility temperature, *Pharm.Res.* 16, 135-140, 1999; Sharp, J.M. and Doran, P.M., Strategies for enhancing monoclonal antibody accumulation in plant cell and organ cultures, *Biotechnol.Prog.* 17, 979-992, 2001. PVP has some direct therapeutic use (Kaneda, Y., Tsutsumi, Y., Yoshioka, Y. *et al.* The use of PVP as a polymeric carrier to improve the plasma half-life of drugs, *Biomaterials* 25, 3259-3266, 2004) and as a carrier for iodine as a disinfectant (Art, G., Combination povidone-iodine and alcohol formulations more effective, more convenient versus formulations containing either iodine or alcohol alone: a review of the literature, *J.Infus.Nurs.* 28, 314-320, 2005. An HPLC method for the analysis of PVP in pharmaceutical products has been developed (Jones, S.A., Martin, G.P., and Brown, M.B., Determination of polyvinylpyrrolidone using high-performance liquid chromatography, *J.Pharm.Biomed.Anal.* 35, 621-624, 2004.

Post-Translational Modification

A covalent modification of a protein following translation of the RNA to form the polypeptide chain. Such modification may or may not be enzyme catalyzed (γ-carboxylation versus nitration) and may or may not be reversible (phosphorylation versus γ-carboxylation).

Pre-initiation complex

A complex of general transcription factors (GTFs) that are formed at each core promoter prior to transcriptional activation and required for the action of RNA polymerase II. Recent work suggests that core promoter elements may not be an absolute requirement. See George, A.A., Sharma, M., Singh, B.N., Sahoo, N.C., and Rao, K.V., Transcription regulation from a TATA and INR-less promoter: spatial segregation of promoter function, *EMBO J.* 2006; Maag, D., Algire, M.A., and Lorsch, J.R., Communication between eukaryotic translation initiation factors 5 and 1A within the ribosomal pre-initiation complex plays a role in start site selection, *J.Mol.Biol.* 356, 724-737, 2006; Govind, C.K., Yoon, S., Qiu, H., Govind, S., and Hinnebusch, A.G., Simultaneous recruitment of coactivators by Gcn4p stimulates multiple steps of transcription *in vivo.*, *Mol.Cell.Biol.* 25, 5626-5638, 2005; Svejstrup, J.Q., The RNA polymerase II transcription cycle: cycling though chromatin, *Biochim.Biophys.Acta* 1677, 64-73, 2004.

Primase

Primase is an enzyme which catalyzes polymerization of ribonucleoside 5'-triphosphates to form RNA primers in as a sequence which is directed by DNA template. See Foiani, M, Lucchini, G., and Plevani, P., The DNA polymerase alpha-primase complex couples DNA replication, cell-cycle progression and DNA-damage response, *Trends Biochem.Sci.* 22, 424-427, 1997; Arezi, B. and Kuchta, R.D., Eurkaryotic DNA primase, *Trends Biochem. Sci.* 25, 572-576, 2000; Frick, D.N. and Richardson, C.C., DNA primases, *Annu.Rev.Biochem.* 70, 39-80, 2001; Benkovic, S.J., Valentine, A.M., and Salinas, F., Replisome-mediated DNA replication, *Annu.Rev.Biochem.* 70, 181-208, 2001; MacNeil, S.A., DNA replication: partners in the Okazaki two-step, *Curr.Biol.* 11, F842-F844, 2001; Kleymann, G., Helicase primase: targeting the Achilles heel of herpes simplex viruses, *Antivir.Chem.Chemother.* 15, 135-140, 2004; Lao-Sirieix, S.H., Pellegrini, L., and Bell, S.D., The promiscuous primase, *Trends Genet.* 21, 568-572, 2005; Lao-Sirieix, S.H., Nookala, R.K., Roversi, P., *et al.*, Structure of the heterodimeric core primase, *Nat.Struct.Mol.Biol.* 12, 1137-1144, 2005; Shutt, T.E. and Gray, M.W., Twinkle, the mitochondrial replicative DNA helicase, is widespread in the eukarotic radiation and may also be the mitochondrial DNA primase in most eukaryotes, *J.Mol.Evol.* 62, 588-599, 2006; Rodina, A. and Godson, G.N., Role of conserved amino acids in the catalytic activity of *Escherichia coli* primase, *J.Bacteriol.* 188, 3614-3621, 2006.

Promoter Elements

A regions of a segment of DNA(usually *cis*) which regulated the transcription (mRNA synthesis) from information encoded on that segment of DNA. These are elements regulating the nuclear transcription process and bind transcription factors and other regulatory factors. Kingston, R.E., Baldwin, A.S., and Sharp, P.A., Transcription control by oncogenes, *Cell* 41, 3-5, 1985; Wasylk, B., Transcription elements and factors of RNA polymerase B promoters of higher eukaryotes, *CRC Crit.Rev.Biochem.* 23, 77-120, 1988; Khokha, M.K. and Loots, G.G., Strategies for characterizing *cis*-regulatory elements in *Xenopus.*, *Brief Funct. Genomic Proteomic.* 4, 58-68, 2005; Sipos, L. and Gyurkovics, H., Long-distance interactions between enhancers and promoters, *FEBS J.* 272, 3253-3259, 2005; Fukuchi, M., Tabuchi, A., and Tsuda, M., Transcriptional regulation of neuronal genes and its effect on neural functions: cumulative mRNA expression of PACAP and BDNP genes controlled by calcium and cAMP signals in neurons, *J.Pharmacol.Sci* 98, 212-218, 2005; Anderson, S.K., Transcriptional regulation of NK cell receptors, *Curr.Top. Microbiol.Immunol.* 298, 59-75, 2006.

Protamines

A family of basic proteins associated with the chromatin in the nucleus of the cell. Protamines are characterized by a high content of arginine and replace histones in the process of spermiogenesis. See Erin-López, J.M., Frehlich L.J., and Ausió, J., Protamines, in the footsteps of linker histone evolution, *J.Biol. Chem.* 281, 1-4, 2006; Meistrich, M.L., Mohapatra, B., Shirley, C.R., and Zhao, M., Roles of transition nuclear proteins in spermiogenesis, *Chromasoma* 111, 483-488, 2003; Aoki, V.W. and Carrell, D.T., Human protamines and the developing spermatid: their structure, function, expression and relationship with male infertility, *Asian J. Androl.* 5, 315-324, 2003; Lewis, S.D. and Ausió, J., Protamine-like proteins: evidence for a novel chromatin structure, *Biochem.Cell Biol.* 80, 353-361, 2002.

Protease

A protease/proteolytic enzyme catalyzes the hydrolysis of a peptide bond in a protein. A simple classification of proteases divides these enzymes into two functional categories and four chemical categories. The functional categories are regulatory and digestive. Examples of regulatory proteolysis is proprotein processing by furin and blood coagulation while digestive enzyme include enzyme like pepsin, trypsin, and chymotrypsin found in mammalian digestive systems. Chemical categories describe the functional groups at enzyme active site and include serine proteases such as trypsin or chymotrypsin, cysteine proteases such as papain and the caspaces, aspartic acid proteases such as pepsin, and metalloproteinases such as ADAM proteases and matrix metalloproteinase (MMP). See *Regulatory Proteolytic Enzymes and Their Inhibitors*, ed. S. Magnusson, Pergammon, Press, Oxford, UK, 1978; *Mammalian Proteases: A Glossary and Bibliography*, ed. A.J. Barrett and J.K. McDonald, Academic Press, New York, NY, 1980; Polgár, L., *Mechanism of protease action*, CRC Press, Boca Raton, FL, USA, 1989; *Proteases of Infectious Agents*, ed. B.M. Dunn, Academic Press, San Diego, CA, USA, 1999; *The Proteosome-Ubiquitin Protein Degradation Pathway*, ed. P. Zwickl and W. Baumeister, Springer, Berlin, 2002; *Proteases and the Regulation of Biological Processes*, ed. J. Saklatvala, and Nagase, H., Portland Press, London, UK, 2003.

Protease-Activated Receptor (PAR)

Protease-activated receptors (PAR receptors) are a family of G-protein coupled receptors in which the ligand (tethered ligand) is intrinsic to the receptor protein and exposed by proteolysis in the N-terminal external region. These receptors may also be activated by peptides where the sequence is identical to or related to the ligand sequence in the receptor. The protease-activated receptor was first described in platelets by Coughlin and colleagues (Vu, T.K., Hung , D.T., Wheaton, V.I., and Coughlin, S.R., Molecular cloning of a functional thrombin receptor reveals a novel proteolytic mechanism of receptor activation, *Cell* 64, 1057-1068, 1991; Coughlin, S.R., Vu, T.K., Hung, D.T., and Wheaton, V.I., Expression cloning and characterization of a functional thrombin receptor reveals a novel proteolytic mechanism of receptor activation, *Semin.Thromb.Hemost.* 18, 161-166, 1992) and by another group in France (Rasmussen, U.B., Vouret-Cravieri, V., Jallet, S., *et al*, cDNA cloning and expression of a hamster alpha-thrombin receptor coupled to Ca^{2+} mobilization, *FEBS Lett.* 288, 123-128, 1991) using fibroblasts. Since the original work, four PAR receptors have been described on a wide variety of cell types. See Chen, J., Bernstein, H.S., Chen, M., *et al.*, Tethered ligand library for discovery of peptide agonists, *J.Biol.Chem.* 270, 23398-23401, 1995; Santulli, R.J., Derian, C.K., Darrow, A.L., *et al.*, Evidence for the presence of a proteinase-activated receptor distinct from the thrombin receptor in human keratinocytes, *Proc.Natl.Acad.Sci. USA* 92, 9151-9155, 1995; Ishihara, H., Connolly, A.J., Zeng, D., *et al.*, Protease-activated receptor 3 is a second thrombin receptor in humans, *Nature* 386, 502-506, 1997; Niclou, S.P., Suidan, H.S., Pavlik, A., Vejsada, R., and Monard, D., Changes in the expression of protease-activated receptor 1 and protease nexin-1 mRNA during rat nervous system development and after nerve lesion, *Eur.J.Neurosci.* 10, 1590-1607, 1998; Hou, L., Howells, G.L., Kapas, S., and Macey, M.G., The protease-activated receptors and their cellular expression and function in blood-related cells, *Br.J.Haematol.* 101, 1-9, 1998; Brass, L.F. and Molino, M., Protease-activated G protein coupled receptors on human platelets and endothelial cells, *Thromb.Haemostas.* 78, 234-241, 1997; Brass, L.F., Thrombin receptor antagonists: a work in progress, *Coron.Artery Dis.* 8, 49-58, 1997; Coughlin, S.R., Protease-activated receptors and platelet function, *Thromb.Haemost.* 82, 353-356, 1999; Cooks, T.M. and Moffatt, J.D., Protease-activated receptors: sentries for inflammation, *Trends Pharmacol.Sci.* 21, 103-108, 2000; Macfarlane, S.R., Seatter, M.J., Kanke, T., Hunter, G.D., and Plevin, R., Proteinase-activated receptors, *Pharmacol. Rev.* 53, 245-282, 2001; Bucci, M. Roviezzo, F., and Cirino, G., Protease-activated receptor-2(PAR2) in cardiovascular system, *Vascul.Pharmacol.* 43 247-253, 2005; Wang, P. and Defea, K.A., Protease-activated receptor-2 simultaneously directs beta-arrestin-1-dependent inhibition and Gαq-dependent activation of phosphatidylinositol 3-kinase, *Biochemistry* 45, 9374-9385, 2006; Oikonomopoulou, K., Hansen, K.K., Saifeddine, M., *et al.*, Proteinase-activated receptors (PARs): Targets for kallikrein signalling, *J.Biol.Chem.*, 281, 32095–35112, 2006; Wang, L., Luo, J., Fu, Y. and He, S., Induction of interleukin-8 secretion and activation of ERK1/2, p38 MAPK signaling pathways by thrombin in dermal fibroblasts, *Int.J.Biochem.Cell Biol.* 38, 1571-1583, 2006; Page, K., Hughes, V.S., Bennett, G.W., and Wong, H.R., German cockroach proteases regulate matrix metalloproteinase-9 in human bronchial epithelial cells, *Allergy* 61, 988-995, 2006.

Protease Inhibitor Cocktail

A mixture of protease inhibitors which is used to preserve protein integrity during the processing of samples for subsequent analysis. The term cocktail refers to a mixture of components. A protease inhibitor cocktail is composed of a broad spectrum of protease inhibitors intends to inhibit the diverse proteolytic enzymes found in tissue extracts and biological fluids. See Pringle, J.R., Methods for avoiding proteolytic artifacts in studies with enzymes and other proteins from yeasts, *Methods in Cell Biology* 12, 149-184, 1975; Drubin, D.G., Miller, K.G., and Botstein, D., Yeast actin-binding proteins: evidence for a role in morphogenesis, *J.Cell Biol.* 107, 2551-2561, 1988; Nanoff, C., Jacobson, C.A., and Stiles, G.L., The A2 adenosine receptor: guanine nucleotide modulation of agonist binding is enhanced by proteolysis, *Mol. Pharmacol.* 39, 130-135, 1991; Palmer, T.M., Jacobson, K.A. and Stiles, G.L., Immunological identification of A2 adenosine receptors by two antipeptide antibody preparations, *Mol.Pharmacol.* 42, 391-397, 1992; Pyle, L.E., Barton, P., Fujiwara, Y., Mitchell, A., and Fidge, N., Secretion of biologically active human proapolipoprotein A-1 in a baculovirus-insect cell system: protection from degradation by protease inhibitors, *J.Lipid Res.* 36, 2355-2361, 1995; Weidner, M.-F., Grenier, D., and Mayrand, D., Proteolytic artifacts in SDS-PAGE analysis of selected periodontal pathogens, *Oral Microbiol.Immunol.* 11, 103-108, 1996; Hassel, M., Klenk, G., and Frohme, M., Prevention of unwanted proteolysis during extraction of proteins from protease-rich tissue, *Anal.Biochem.* 242, 274-275, 1996; Salvesen, G. and Nagase, H., Inhibition of proteolytic enzymes, in *Proteolytic Enyzmes. Practical Approaches,* 2nd edn., ed. R. Benyon and J.S. Bond, Oxford University Press, Oxford, United Kingdom, Chapter 5, pps. 105-130, 2001; North, M.J. and Benyon, R.J., Prevention of unwanted proteolysis, in *Proteolytic Enyzmes. Practical Approaches,* 2nd edn., ed. R. Benyon and J.S. Bond, Oxford University Press, Oxford, United Kingdom, Chapter 9, pp. 211-232, 2001. Castellanos-Serra, L. and Paz-Lago, D., Inhibition of unwanted proteolysis during sample preparation: evaluation of its efficiency in challenge experiments, *Electrophoresis* 23, 1745-1753, 2002. Kikuchi, S., Hirohashi, T., and Nakai, M.,. Characterization of the preprotein translocon at the outer envelope membrane of chloroplasts by blue native PAGE, *Plant Cell Physiol.* 47, 363-371, 2006. The term protease cocktail is also used to refer to the combination of therapeutic protease inhibitors used in AIDS therapy. See Tamamura, H. and Fujii, N., Two orthogonal approaches to overcome multidrug resistant HIV-1s: development of protease inhibitors and entry inhibitors based on CXCR4 antagonists, *Curr.Drug Targets Infect.Disord.* 3, 103-110, 2004; Wicovsky, A., Siegmund, D., and Wajant, H., Interferons induce proteolytic degradation of TRAILR4, *Biochem.Biophys.Res.Commun.* 337, 184-190, 2005. The term cocktail is also used to describe the combination of chemicals and solvent used for liquid scintillation counting of radioisotopes. See Kobayashi, Y. and Maudsely, D.V., Practical aspects of liquid scintillation counting, *Methods Biochem.Anal.* 17, 55-133, 1969; Wood, K.J., McElroy, R.G., Surette, R.A., and Brown, R.M., Tritium sampling and measurement, *Health Phys.* 65, 610-627, 1993; Jaubert, F., Tartes, I., and Cassette, P., Quality control of liquid scintillation counting, *Appl.Radiat.Isot.*, 64, 1163–1170, 2006.

Proteasome

The proteasome is a multisubunit complex that functions in the degradation of intracellular proteins in eukaryotic cell. It is composed of catalytic subunits with different specificity and regulatory subunits. In eukaryotic cells, proteins are "marked" for proteasomal degradation by ubiquitinylation. There is a specialized proteasome which function in MHC I antigen presentation. See Arrigo, A.P., Tanaka, K., Goldberg, A.L., and Welch, W.J., Identity of the 19S 'prosome' particle with the large multifunctional protease complex of mammalian cells (the proteasome), *Nature* 331, 192-194, 1988; Falkenberg, P.E., and Kloetael, P.M.,

Identification and characterization of three different subpopulations of the Drosophila multicatalytic proteinase (proteasome), *J.Biol.Chem.* 264, 6660-6666, 1989; Dahlmann, B., Kopp, F., Kuehn, L., *et al.*, The multicatalytic proteinase (prosome, proteasome): comparison of the eukaryotic and archaebacterial enzyme, *Biomed. Biochim.Acta* 50, 465-469, 1991; Demartino, G.W., Orth, K., McCullough, M.L., *et al.*, The primary structure of four subunits of the human, high-molecular weight proteinase, macropain (proteasome), are distinct but homologous, *Biochim. Biophys.Acta* 1079, 29-38, 1991; Wlodawer, A., Proteasome: a complex protease with a new fold and a distinct mechanism, *Structure* 3, 417-420, 1995; Baumeister, W., Cejka, Z., Kania, M., and Seemuller, E., The proteasome: a macromolecular assembly designed to confine proteolysis to a nanocompartment, *Biol. Chem.* 378, 121-130, 1997; Grune, T., Merker, K., Sandig, G., and Davies, K.J., Selective degradation of oxidatively modified protein substrates by the proteasome, *Biochem.Biophys.Res.Commun.* 305, 709-718, 2003; Hartmann-Petersen, R. and Gordon, C., Proteins interacting with 26S proteasome, *Cell.Mol.Life Sci.* 61, 1589-1595, 2004; Smalle, J. and Veirstra, R.D., The ubiquitin 26S proteasome proteolytic pathway, *Ann.Rev.Plant.Biol.* 55, 555-590, 2004; Dalton, W.S., The proteasome, *Semin.Oncol.* 31(6 Suppl 16), 3-9, 2004; Qureshi, N., Vogel, S.N., Van Way, C., 3rd, *et al.*, The proteasome: a central regulator of inflammation and macrophage function, *Immunol.Res.* 31, 243-260, 2005; Glickman, M.H. and Raveh, D., Proteasome plasticity, *FEBS Lett.* 579, 3214-3223, 2005; Ye, Y., The role of the ubiquitin-proteasome system in ER quality control, *Essays Biochem.* 41, 99-112, 2005; Gao, G. and Luo, H., The ubiquitin-proteasome pathway in viral infections, *Can.J.Physiol. Pharmacol.* 84, 5-14, 2006. There is interest in the proteasome as a drug target in oncology. See Montagut, C., Rovira, A., and Albanell, J., The proteasome: a novel target for anticancer therapy, *Clin.Transl.Oncol.* 8, 313-317, 2006. There is a variation in the proteasomes involved in the presentation of MHC I antigens; see immuno proteasome.

Protein Classification

There are a number of approaches to protein classification. One simple approach is based on environmental conditions and divides proteins into three different groups (see Finkelstein, A.V. and Ptitsyn, O.B., *Protein Physics. A Course of Lectures*, Academic Press, London, UK, 2002). Fibrous proteins are usually in non-aqueous environments and usually form high regular hydrogen-bonded structures such as those see in cartilage; membrane proteins are also found in non-aqueous environments; water-soluble proteins found in the cytoplasm and extracellular fluids. Water-soluble proteins can be divided into albumins and globulins on the basis of solubility properties. Water-soluble proteins can form three-dimensional structures maintained by a variety of forces including hydrogen bonds and van der Waals forces. Water-soluble proteins can have effects on enzyme activity separate from their intrinsic activity (see Derham, B.K. and Harding, J.J., The effect of the presence of globular proteins and elongated polymers on enzyme activity, *Biochim.Biophys.Acta* 1764, 1000-1006, 2006).

Protein Disulfide Isomerase/Ero1p

Protein disulfide isomerase and Ero1p are enzymes which participate in the formation of disulfide bonds within the endoplasmic reticulum. These factors are critical for the normal formation of disulfide bonds during protein folding. See Lodi, T.,

Neglia, B. and Donnini, C., Secretion of human serum albumin by *Kluyveromyces lactis* overexpressing *KlPDIUl* and *KlERO1*, *Appl.Environ.Microbiol.* 71, 4359-4364, 2005; Kulp, M.S., Frickel, E.-M., Ellgaard, L., and Weissman, J.S., Domain architecture of protein-disulfide isomerase facilitates its dual role as an oxidase and an isomerase in Ero1p-mediated disulfide formation, *J.Biol. Chem.* 281, 876-884, 2006; Gross, E., Sevier, C.S., Heldman, *et al.*, Generating disulfides enzymatically : reaction products and electron acceptors of the endoplasmic reticulum thiol oxidase Ero1p, *J.Biol.Chem.* 281, 299-304, 2006.

Protein Profiling

The use of algorithms to determine the relationship of multiple proteins as determined by proteomic analysis such as protein microarray technology, shotgun proteomics, or SELDI-TOF-MS. See Tomlinson, I.M. and Holt, L.J., Protein profiling comes of age, *Genome Biol.* 2, REVIEWS1004, 2001; Kingamore, S.F. and Patel, D.D., Multiplexed protein profiling on antibody-based microarrays by rolling circle amplification, *Curr.Opin.Biotechnol.* 14, 74-81, 2003; Jessani, N. and Cravatt, B.F., The development and application of methods for activity-based protein profiling, *Curr.Opin.Chem.Biol.* 8, 54-59, 2004; Berger, A.B., Vitorino, P.M., and Bogyo, M., Activity-based protein profiling: applications to biomarker discovery, in vitro imaging and drug discovery, *Am.J.Pharmacogenomics* 4, 371-381, 2004; Steel, L.F., Haab, B.B., and Hanash, S.M., Methods of comparative proteomic profiling for disease diagnostics, *J.Chromatog.B.Anal. Technol. Biomed. Life Sci.* 815, 275-284, 2005; Kislinger, T. and Emili, A., Multidimensional protein identification technology: current status and future prospects, *Expert Rev. Proteomics* 2, 27-39, 2005; Katz, J.E., Mallick, P., and Agus, D.B., A perspective on protein profiling of blood, *BJU Int.* 96, 477-482, 2005; Bons, J.A., Wodzig, W.K., and van Dieijen-Visser, M.P., Protein profiling as a diagnostic tool in clinical chemistry: a review, *Clin.Chem.Lab.Med.* 43, 1281-1290, 2005.

Protein Tyrosine Phosphatases

A family of hydrolytic enzymes which catalyze the dephosphorylation of protein-bound *O*-tyrosine phosphate. Dephosphorylation of tyrosine residues can modulate biological activity and may be a specific or non-specific process. See Fischer, E.H., Tonks, N.K., Charbonneau, H., *et al.*, Protein tyrosine phosphatases: a novel family of enzymes involved in transmembrane signaling, *Adv.Second Messenger Phosphoprotein Res.* 24, 272-279, 1990; Calya, X., Goris, J., Hermann, J., *et al.*, Phosphotyrosyl phosphatase activity of the polycation-stimulated protein phosphatases and involvement of dephosphorylation in cell cycle regulation, *Adv.Enzyme Reg.* 39, 265-285, 1990; Saito, H. and Streuli, M., Molecular characterization of protein tyrosine phosphatases, *Cell Growth Differ.* 2, 59-65, 1991; Tonks, N.K., Yang, Q., and Guida, P., Jr., Structure, regulation, and function of protein tyrosine phosphatases, *Cold Spring Harbor Symp.Quant.Biol.* 56, 265-273, 1991; Lawrence, D.S., Signaling protein inhibitors via the combinatorial modification of peptide scaffolds, *Biochim. Biophys.Acta* 1754, 50-57, 2005; Boutros, R., Dozier, C., and Ducommun, B., The when and wheres of CDC25 phosphatases, *Curr.Opin.Cell Biol.* 18, 185-191, 2006; Ostman, A., Hellberg, C., and Bohmer, F.D., Protein-tyrosine phosphatases and cancer, *Nat.Rev.Cancer* 6, 307-320, 2006; Burridge, K., Sastry, S.K., and Salfee, J.L., Regulation of cell adhesion by protein-tyrosine phosphatases. I. Cell-matrix adhesion, *J.Biol.Chem.* 281, 15593-15596,

2006; Sallee, J.L., Wittchen, E.S., and Burridge, K., Regulation of cell adhesion by protein-tyrosine phosphatases. II. Cell-cell adhesion, *J.Biol.Chem.* 281, 16189-16192, 2006.

Proteome

The total expressed protein content of a genome. See Wasinger, V.C., Cordwell, S.J., Cerpa-Poljak, A., *et al.*, Progress with gene-product mapping of the Mollicutes: *Mycoplasma genitalium*, *Electrophoresis* 16, 1090-1094, 1995; Kahn, P., From genome to proteome: looking at a cell's proteins, *Science* 270, 369-370, 1995; Wilkens, M.R., Sanchez, J.C., Gooley, A.A., *et al.*, Progress with proteome projects: why all proteins expressed by a genome should be identified and how to do it, *Biotechnol.Genet.Eng.Rev.* 13, 19-50, 1996; Figeys, D., Gygi, S.P., Zhang, Y., *et al.*, Electrophoresis combined with novel mass spectrometry techniques: powerful tools for the analysis of proteins and proteomics, *Electrophoresis* 19, 1811-1818, 1998; Blackstock, W.P. and Weir, M.P., Proteomics: quantitative and physical mapping of cellular proteins, *Trends Biotechnol.* 17, 121-127, 1999; Bradshaw, R.A., Proteomics – boom or bust? *Mol.Cell.Proteomics* 1, 177-178, 2002; Bradshaw, R.A. and Burlingame, A.L., From proteins to proteomics, *IUBMB Life* 57, 267-272, 2005; Domon, B. and Aebersold, R., Mass spectrometry and protein analysis, *Science* 312, 212-217, 2006.

Proteometabolism

Metabolism of the proteome.

Proteomics

The study of the proteome; not technology limited; the qualitative and quantitative study of the proteome under various conditions including protein expression, modification, localization, and function, and protein-protein interactions, as a means of understanding biological processes.

Proto-oncogene

A normal cellular gene whose activation or modification to an oncogene is linked to malignant transformation; a progenitor of an oncogenes; a proto-oncogene can become an oncogenes either by transduction into a virus or by "disturbance" such as chromosomal translocation, amplification or point mutation at their location in a chromosome. *c-Myc* is one of the most studied of the proto-oncogenes. See Bishop, J.M., Oncogenes and proto-oncogenes, *J.Cell.Physiol.Suppl.* 4, 1-5, 1986; Cory, S., Activation of cellular oncogenes in hemapoietic cells by chromosome translocation, *Adv.Cancer Res.* 47, 189-243, 1986; Bishop, J.M., and Hannfusa, W., Proto-oncogenes in normal and neoplastic cells, in *Scientific American Molecular Oncology*, ed. J.M. Bishop and R.A. Weinberg, Scientific American, New York, New York, USA, Chapter 4, pp. 61-83, 1996. Shachaf, C.M. and Felsher, D.W., Rehabilitation of cancer through oncogenes inactivation, *Trends Mol.Med.* 11, 316-321, 2005; Barry, E.L., Baron, J.A., Grau, M.V., Wallace, K., and Haile, R.W., K-ras mutations in incident sporadic colorectal adenomas, *Cancer* 106, 1036-1040, 2006.

Protpparam

A program which allows the calculation of a number of physical and chemical properties for a protein from the known amino acid sequence. see http://www.expasy.ch/tools/protparam.html.

Proximal Promoter Element

A region located 30-200 bp upstream from the transcription start site and usually contains multiple transcription factor binding sites. See van de Klundert, F.A., Jansen, H.J., and Bloemendal, H., A proximal promoter element in the hamster desmin upstream regulatory region is responsible for activation by myogenic determination factors, *J.Biol.Chem.* 269, 220-225, 1994; Petrovic, N., Black, T.A., Fabian, J.R., *et al.*, Role of proximal promoter elements in regulation of rennin gene transcription, *J.Biol.Chem.* 271, 22499-22505, 1996; Mori, A., Kaminuma, Ogama, K., Okudaira, H. and Akiyama, K., Transcriptional regulation of IL-5 gene by nontransformed human T cells through the proximal promoter element, *Intern.Med.* 39, 618-625, 2000; Ghosh-Choudhury, N., Choudhury, G.G., Harris, M.A., *et al.*, Autoregulation of mouse BMP-2 gene transcription is directed by the proximal promoter element, *Biochem.Biophys.Res. Commun.* 286, 101-108, 2001; Rentsendorj, O., Nagy, A., Sinko, I., *et al.*, Highly conserved proximal promoter element harbouring paired Sox9-binding sites contributes to the tissue- and developmental stage-specific activity of the matrilin-1 gene, *Biochem.J.* 389, 705-716, 2005.

Pseudogenes (retropseudogenes)

Copies of cellular RNA that have been reverse transcribed and inserted into the genome. See Vanin, E.F., Processed pseudogenes: characteristics and evolution, *Annu.Rev.Genet.* 19, 253-272, 1985; Weiner, A.M., Deininger, P.L. and Efstartiadis, A., Nonviral retroposons, genes, pseudogenes, and transposable elements generated by the reverse flow of genetic information, *Annu.Rev.Biochem.* 55, 631-661, 1986; Pascual, V. and Capra, J.D., Human immunoglobulin heavy-chain variable region genes: organization, polymorphism, and expression, *Adv. Immunol.* 49, 1-74, 1991; King, C.C., Modular transposition and the dynamical structure of eurkaryote regulatory evolution, *Genetica* 86, 127-142, 1992; D'Errico, I., Gadaleta, G., and Saccone, C., Pseudogenes in metazoan: origin and features, *Brief Funct.Genomic Proteomic.* 3, 157-167, 2004; Rodin, S.N., Parkhomchuk, D.V., Rodin, A.S., Holmquist, G.P., and Riggs, A.D., Retroposition-dependent fate of duplicate genes, *DNA Cell Biol.* 24, 529-542, 2005; Pavlicek, A., Gentles, A.J., Pačes, J. *et al.*, Retroposition of processed pseudogenes: the impact of RNA stability and translational control, *Trends Genet.* 22, 69-73, 2006.

Psychogenomics

The process of applying the tools of genomics, transcriptomics, and proteomics to understand the molecular basis of behavioral abnormalities.

Psychrophilic

Functioning more efficiently at cold temperatures. See Feller, G. and Gerday, C., Psychrophilic enzymes: hot topics in cold adaptation, *Nat.Rev.Microbiol.* 1, 200-208, 2003; Bolter, M., Ecophysiology of psychrophilic and psychrotolerant microorganisms, *Cell.Mol.Biol.* 50, 563-573. 2004; Zecchinon, L.,. Oriol, A., Netzel, U., *et al.*, Stability domains, substrate-induced comformational changes and hinge-bending motions in a psychrophilic phosphoglycerate kinase. A microcalorimetric study, *J.Biol. Chem.* 280, 41307-41314, 2005.

Pullulanase

Enzyme degrading pullulan, a branched starch; pullulanase catalyzes the hydrolysis of the α-1,6-glucosidic linkage in α-glucans. Pullulanase preferentially hydrolyzes pullulan while isoamylase has a preference for glycogen and amylopectin. See Wallenfels, K., Bender, H., and Rached, J.R., Pullulanase from *Aerobacter aerogenes*; production in a cell-bound state. Purification and properties of the enzymes, *Biochem.Biophys.Res.Commun.* 22, 254-261, 1966; Hardie, D.G. and Manners, D.J., A viscometric assay for pullulanase-type, debranching enzymes, *Carbohdr.Res.* 36, 207-210, 1974.

Harada, T., Special bacterial polysaccharides and polysaccharidases, *Biochem.Soc.Symp.* 48, 97-116, 1983; Vihinen, M. and Mantsala, P., Microbial amylolytic enzymes, *Crit.Rev.Biochem. Mol.Biol.* 24, 329-418, 1989; Doman-Pytka, M. and Bardowski, J., Pullalan degrading enzymes of bacterial origin, *Crit.Rev. Microbiol.* 30, 107-121, 2004; Lammerts van Bueren, A., Finn, R., Ausio, J., and Boraston, A.B., Alpha-glucan recognition by a new family of carbohydrate-binding modules found primarily in bacterial pathogens, *Biochemistry* 43, 15633-15642, 2004; Mikami, B., Iwamoto, H., Malle, D., *et al.*, Crystal structure of pullulanase: evidence for parallel binding of oligosaccharides in the active site, *J.Mol.Biol.* 359. 690-707, 2006; Hytonen, J., Haataja, S., and Finne, J., Use of flow cytometry for the adhesion analysis of *Streptococcus pyogenes* mutant strains to epithelial cells: investigation of the possible role of surface pullulanase and cysteine protease, and the transcriptional regulation Rgg, *BMC Microbiol.* 6, 18, 2006

"Pull-Down"

The process of the capture of a protein, a protein complex, or other biological by binding to an immobilized capture reagent such as an antibody. See Cavailles, V., Dauvois, S., Danielian, P.S., and Parker, M.G., Interaction of proteins with transcriptionally active estrogen receptors, *Proc.Natl.Acad.Sci.USA* 91, 10009-10013, 1994; Magnaghi-Jaulin, L., Masutani, H., Robin, P., Lipinski, M., and Harel-Bellan, A., SRE elements are binding sites for the fusion protein EWS-FLI-1, *Nucleic Acids Res.* 24, 1052-1058, 1996; Dombrosky-Ferlan, P.M. and Corey, S.J., Yeast two-hybrid in vivo association of the Src kinase Lyn with the proto-oncogene product Cbl but not with the p85 subunit of PI 3-kinase, *Oncogene* 14, 2019-2024, 1997; Graves, P.R. and Haystead, T.A., A functional proteomics approach to signal transduction, *Recent Prog.Horm. Res.* 58, 1-24, 2003.

Pulsed-Field Gel Electrophoresis

An gel electrophoretic technique for the analysis of very large DNA molecules. It usually uses an agarose gel matrix with alternating current in that the direction of the electric field is changed (or pulsed) periodically for separation. See Cantor, C.R., Smith, C.L., and Mathew, M.K., Pulsed-field gel electrophoresis of very large DNA molecules, *Ann.Rev.Biophys.Biophys.Chem.* 17, 287-304, 1988; Lat, E., Birren, B.W., Clark, S.M., Simon, M.I., and Hood, L., Pulsed field gel electrophoresis, *Biotechniques* 7, 34-42, 1989; Olson, M.V., Separation of large DNA molecules by pulsed-field gel electrophoresis. A review of the basic phenomenology, *J.Chromatog.* 470, 377-383, 1989; Dukhin, A.S. and Dukhin, S.S., Aperiodic capillary electrophoresis method using an alternating current electric field for separation of macromolecules, *Electrophoresis* 26, 2149-213, 2005; Aires de Sousa, M. and de Lencastre, H., Bridges from hospitals to the laboratory: genetic portraits of methicillin-resistant *Staphyllococcus aureus* clones, *FEMS Immunol.Med. Microbiol.* 40, 101-111, 2004.

Pulse Radiolysis

A technique related to flash photolysis; pulse radiolysis uses very short (nanosecond) intense pulses of ionizing radiation to generate transient high concentrations of reactive species. See Salmon, G.A. and Sykes, A.G., Pulse radiolysis, *Methods Enzymol.* 227, 522-534, 1993; Bataille, C., Baldacchino, G., Cosson, R.P., *et al.*, Effect of pressure on pulse radiolysis reduction of proteins, *Biochim.Biophys. Acta* 1724, 432-439, 2005; Nakuna, B.N., Sun, G., and Anderson, V.E., Hydroxyl radical oxidation of cytochrome c by aerobic radiolysis, *Free Radic.Biol.Med.* 37, 1203-1213, 2004; Maleknia, S.D., Kieselar, J.G., and Downard, K.M., Hydroxyl radical probe of the surface of lysozyme by synchrotron radiolysis and mass spectrometry, *Rapid Commun.Mass Spectrom.* 16, 53-61, 2002.

Quadrupole Mass Spectrometry

Mass spectrometry where only electric fields are used to separate ions on the basis of mass as they pass along the central axis of four parallel rods having an applied DC charge and alternative voltage applied (Herbert, C.G. and Johnstone, R.A.W., *Mass Spectrometry Basics*, CRC Press, Boca Raton, FL, Chapter 25, 2003). These instruments are generally referred to as quadrupole/time-of-flight mass spectrometers. See Horning, E.C., Carroll, D.I., Dzidic, I., *et al.*, Development and use of analytical systems based on mass spectrometry, *Clin.Chem.* 23, 13-21, 1977; Yost, R.A. and Boyd, R.K., Tandem mass spectrometry: quadrupole and hybrid instruments, *Methods Enzymol.* 193, 154-200, 1990; Jonscher, K.R. and Yates, J.R., 3rd., The quadrupole ion trap mass spectrometry—a small solution to a big challenge, *Anal.Biochem.* 244, 1-15, 1997; Chernushevich, I.V., Loboda, A.V., and Thomson, B.A., An introduction to quadrupole-time-of-flight mass spectrometry, *J.Mass Spectrom.* 36, 849-865, 2001; Ens, W. and Standing, K.G., Hybrid quadrupole/time-of-flight mass spectrometers for analysis of biomolecules, *Methods Enzymol.* 402, 49-78, 2005; Payne, A.H. and Glish, G.L., Tandem mass spectrometry in quadrupole ion trap and ion cyclotron resonance mass spectrometers, *Methods Enzymol.* 402, 109-148, 2005.

Quantum Dots

Fluorescent semiconducting (usually CdSe surrounded by a passivation shell) nanocrystals used in the imaging of cells and subcellular particles. It is considered to have considerable advantage over other fluorescent imaging approaches. See Penner, R.M., Hybrid electrochemical/chemical synthesis of Q dots, *Acc.Chem. Res.* 33, 78-86, 2000; Lidke, D.S. and Arndt-Jovin, D.J., Imaging takes a quantum leap, *Physiology* 19, 322-325, 2004; Arya, H., Kaul, Z., Wadhwa, R., Taira, K., Hirano, T., and Kaul, S.C., Quantum dots in bio-imaging: revolution by the small, *Biochem.Biophys. Res.Commun.* 378, 1173-1177, 2005; Bentzen, E.L., Tomlinson, I.D., Mason, J., *et al.*, Surface modification to reduce nonspecific binding of quantum dots in live cell assays, *Bioconjugate Chem.* 16, 1488-1494, 2005.

Quantum Yield

Efficiency of fluorescence; percentage of incident energy emitted after absorption. The higher the quantum yield, the greater the intensity of the fluorescence, luminescence, or phosphorescence.

See Papp, S. and Vanderkooi, J.M., Tryptophan phosphorescence at room temperature as a tool to study protein structure and dynamics, *Photochem.Photobiol.* 49, 775-784, 1989; Plasek, J. and Sigler, K., Slow fluorescent indicators of membrane potential: a survey of different approaches to probe response analysis, *J.Photochem.Photobiol.* 33, 101-124, 1996; Vladimirov, Y.A., Free radicals in primary photobiological processes, *Membr.Cell Biol.* 12, 645-663, 1998; Maeda, M., New label enzymes for bioluminescent enzyme immunoassay, *J.Pharm.Biomed.Anal.* 30, 1725-1734, 2003; Imahori, H., Porphyrin-fullerene linked systems as artificial photosynthetic mimics, *Org.Biomol.Chem.* 2, 1425-1433, 2004; Katerinopoulos, H.E., The coumarin moiety as chromophore of fluorescent ion indicators in biological systems, *Curr. Pharm.Des.* 10, 3835-3852, 2004.

Quelling

A term used to describe the forceful suppression of a political uprising; to reduce to submission. In biology, quelling is suggested to uniquely describe post-translational gene silencing in *Neurospora* and, by extension, to other fungi. Quelling has some characteristic similar to RNA interference (RNAi) and cosuppression(post-translational gene silencing) in plants. Quelling involves the silencing of gene expression by segments of DNA in express of the normal number. See Morel, J.B. and Vaucheret, H., Post-transcriptional gene silencing mutants, *Plant Mol.Biol.* 43, 275-284, 2000; Fagard, M., Boutet, S., Morel, J.B., et al., AGO1, QDE-2, and RDE-1 are related proteins required for post-transcriptional gene silencing in plants, quelling in fungi, and RNA interference in animals, *Proc.Natl.Acad.Sci. USA* 97, 11650-11654, 2000; Shiu, P.K., Raju, N.B., Zickler, D., and Metzenberg, R.L., Meiotic silencing by unpaired DNA, *Cell* 107, 905-916, 2001; Pickford, A.S., Catalanotto, C., Cogoni, C., and Macino, G., Quelling in *Neurosporo crassa*, *Adv.Genet.* 46, 277-303, 2002; Goldoni, M., Azzalin, G., Macino, G., and Cogoni, C., Efficient gene silencing by expression of double stranded RNA in *Neurospora crassa*, *Fungal Genet.Biol.* 4, 1016-1024, 2004; Nakayashi, H., RNA silencing in fungi: mechanisms and applications, *FEBS Lett.* 579, 5950-5957, 2005.

Raman Spectroscopy (also called Raman Scattering)

A form of spectroscopy which uses inelastic light scattering which provides information on molecular vibrations. It is similar to infrared spectroscopy but can be used for aqueous solutions. See Warshel, A., Interpretation of resonance Raman spectra of biological molecules, *Annu.Rev.Biophys.Bioeng.* 6, 273-300, 1977; Mathlouthi, M. and Koenig, J.L., Vibrational spectra of carbohydrates, *Adv.Carbohydr.Chem.Biochem.* 44, 7-89, 1986; Ghomi, M., Letellier, R., Liquier, J., and Taillandier, E., Interpretation of DNA vibrational spectra by normal coordinate analysis, *Int.J.Biochem.* 22, 691-699, 1990; Kitagawa, T., Investigation of higher order structures of proteins by ultraviolet resonance Raman spectroscopy, *Prog.Biophy.Mol.Biol.* 58, 1-18, 1992; Loehr, T.M. and Sanders-Loehr, J., Techniques for obtaining resonance Raman spectra of metalloproteins, *Methods Enzymol.* 226, 431-470, 1993; Barron, L.D., Hecht, L., Blanch, E.W., and Bell, A.F., Solution structure and dynamics of biomolecules from Raman optical activity, *Prog.Biophys.Mol.Biol.* 73, 1-49, 2000; Blanch, E.W., Hecht, L., and Barron, L.D., Vibrational Raman optical activity of proteins, nucleic acids, and viruses, *Methods* 29, 196-209, 2003; Spiro, T.G. and Wasbotten, I.H., CD as a vibrational probe

of heme protein active sites, *J.Inorgan.Biochem.* 99, 34-44, 2005; Scheidt, W.R., Durbin, S.M., and Sage, J.T., Nuclear resonance vibrational spectroscopy—NRVS, Nuclear resonance vibrational spectroscopy—NRVS, *J.Inorgan.Biochem.* 99, 60-71, 2005; Aroca, R.F., Alvarez-Puebla, R.A., Pieczonka, N., Sanchez-Cortez, S., and Garcia-Ramos, J.V., Surface-enhanced Raman scattering on colloidal nanostructures, *Adv.Colloid Interface Sci.* 116, 45-61, 2005; Hammond, B.R. and Wooten, B.R., Resonance Raman spectroscopy measurement of carotenoids in the skin and retina, *J.Biomed.Opt.* 10:054002, 2005; Owen, C.A., Selvakumaran, J., Notingher, I., Jell, G., Hench, L.L., and Stevens, M.M., In vitro toxicology evaluation of pharmaceuticals using Raman microspectroscopy, *J.Cell.Biochem.*, 99, 178–186, 2006; Vandenabeele, P. and Moens, L., Introducing students to Raman spectroscopy, *Anal.Bioanal.Chem.* 385, 209-211, 2006.

Randomization

An unbiased process by which individual sample units (e.g., wells in microplate, experimental subjects) are assigned to experimental classes. An example is the assignment of subjects to two or more treatment groups). See Lachin, J.M., Statistical properties of randomization in clinical trials, *Control.Clin.Trials* 9, 289-311, 1988; Greenland, S., Randomization, statistics, and causal interference, *Epidemiology* 1, 421-429, 1990; Kernan, W.N., Viscoli, C.M., Makuch, R.W., et al., Stratified randomization for clinical trials, *J.Clin.Epidemiol.* 52, 19-26, 1999; Abel, U. and Koch, A., The role of randomization in clinical trials: myths and beliefs, *J.Clin.Epidemiol.* 52, 487-497, 1999. Mendelian randomization refers to the randomization of genes which are transferred from parent to offspring at the time of gamete formation (Nitsch, D., Molokhia, M., Smeeth, L., et al., Limits to causal inference based on Mendelian randomization: a comparison with randomized controlled trials, *Am.J.Epidemiol.* 163, 397-403, 2006; Zoccali, C., Testa, A., Spoto, B. et al., Mendelian randomization: a new approach to studying epidemiology in ESRD, *Am.J.Kidney Dis.* 47, 332-341, 2006.

Real-Time PCR; Real-time RT-PCR

Real time PCR permits the assay of the rate of amplicon formation during replication in the PCR reaction. Conventional PCR amplicons are measured either by size analysis or sequence analysis and while there is a relation of amplicon number to target number in the early phases, such a quantitative relationship is lost at high levels of amplification. The use of a fluorescent compounds such as SYBR Green I to bind to double-stranded DNA with an increase in fluorescence. Another approach uses FRET with a donor/acceptor pair. The use of fluorescence to measure the synthesis of amplicons permits the measurement of amplification in real time with the use of appropriate instrumentations. Real-time RT-PCR is an approach to quantitative use of the reverse transcriptase-polymerase chain reaction (RT-PCR) to measure messenger RNA and viral pathogen RNA. This is an adaptation of techniques which were based on the use of fluorescent tags to measure PCR amplicons in real time and has proved useful for the study of gene expression where real-time RT-PCR is used to "validate" other approaches to gene expression analysis such as the use of DNA microarrays. See *Real Time PCR. An Essential Guide*, ed. K. Edwards, J. Logan, and N. Saunders, *Horizon Biosciences*, Wymandham, Norfolk, UK, 2004; Lie, Y.S. and Petropoulos, C.J., Advances in quantitative PCR technology: 5' nuclease assays, *Curr.Opin.Biotechnol.* 9, 43-48, 1998; Kubista, M., Andrade, J.M., Bengtsson, M., et al.,

The real-time polymerase chain reaction, *Mol.Aspects Med.* 27, 95-125, 2006; Kuypers, J., Wright, N., Ferrenberg, J., *et al.*, Comparison of real-time PCR assays with fluorescent-antibody assays for diagnosis of respiratory virus infections in children, *J.Clin.Microbiol.* 44, 2382-2388, 2006; Diederen, B.M., de Jong, C.M., Kluytmans, J.A., *et al.*, Detection and quantification of Legionella pneumonia DNA in serum: case reports and review of the literature, *J.Med.Microbiol.* 55, 639-642, 2006; Transcriptome amplification methods in gene expression profiling, *Expert Rev. Mol.Diag.* 6, 465-480, 2006; Klein, D., Quantification using real-time PCR technology: applications and limitations, *Trends Mol. Med.* 8, 257-260, 2002; Mackay, I.M., Arden, K.E., and Nitsche, A., Real-time PCR in virology, *Nucleic Acids Res.* 30, 1292-1305, 2002; Edwards, K.J. and Saunders, K.A., Real-time PCR used to measure stress-induced changes in the expression of the genes of the aliginate pathway of *Pseudomonas aeruginoses*, J. Appl.Microbiol. 91, 29-37, 2001; Brechtbuehl, K., Whalley, S.H., Dusheiko, G.M., and Saunders, N.A., A rapid real-time quantitative polymerase chain reaction for hepatitis B virus, *J.Virol.Methods* 93, 105-113, 2001; Bustin, S.A., Benes, V., Nolan, T., and Pfaffi, M.W., Quantitative real-time RT-PCR – a perspective, *J.Mol.Endocrinol.* 34, 597-601, 2005; Bustin, S.A. and Mueller, R., Real-time reverse transcription PRC (qRT-PCR) and its potential use in clinical diagnosis, *Clin.Sci.* 109, 365-379, 2005; Delenda, C. and Gaillard, C., Real-time quantitative PCR for the design of lentiviral vector analytical assays, *Gene Ther.* 12(Suppl 1), S36-S50, 2005; Giulietti, A., Overbergh, L., Valckx, D., *et al.*, An overview of real-time quantitative PCR: Applications to quantify cytokine gene expression, *Methods* 25, 386-401, 2001; Leong, W.F. and Chow, W.T.K., Transcriptomic and proteomic analyses of rhabdomyosarcoma cells reveal differential cellular gene expression in response to enterovirus 71 infection, *Cell.Microbiol.* 8, 565-580, 2006.

Receptor Activity Modifying Proteins (RAMP)

Receptor activity modifying proteins (RAMPs) were identified as part of an effort to clone calcitonin gene-related peptide. There are three members of the family and have been demonstrated to modulate the activity of G-protein coupled receptors. See McLatchie, L.M., Fraser, M.J., Main, M.J., *et al.*, RAMPs regulate the transport and ligand specificity of the calcitonin-receptor-like receptors, *Nature* 393, 333-339, 1998; Foord, S.M. and Marshall, F.H., RAMPs: accessory proteins for seven transmembrane domain receptors, *Trends Pharmacol.Sci.* 20, 184-187, 1999; Sexton, P.M., Abiston, A., Morfis, M., *et al.*, Receptor activity modifying proteins, *Cell Signal.* 13, 73-82,. 2001; Fischer, J.A., Muff, R., and Born, W., Functional relevance of G-protein-coupled-receptor-associated proteins, exemplified by receptor-activity-modifying proteins (RAMPs), *Biochem.Soc.Trans.* 30, 455-460, 2002; Morfis, M., Christopolous, A., and Sexton, P.M., RAMPs: 5 years on. Where to now? *Trends Pharmacol.Sci.* 34, 596-601, 2003; Hay, D.L., Conner, A.C., Howitt, S.G., *et al.*, The pharmacology of CGRP-responsive receptors in cultured and transfected cells, *Peptides* 25, 2019-2026, 2004; Udawela, M., Hay, D.L., and Sexton, P.M., The receptor activity modifying protein family of G protein coupled receptor accessory proteins, *Sem.Cell Dev.Biol.* 15, 299-308, 2004; Young, A., Receptor pharmacology, *Adv.Pharmacol.* 52, 47-65, 2005.

Receptor Tyrosine Kinase

A relatively simple transmembrane protein with an extracellular ligand-binding domain and an intracellular protein kinase domain. Receptor tyrosine kinases are coupled with receptors

such as epidermal growth factor receptor, insulin receptor, etc. See Gourley, D.R., Isolation and characterization of membrane drug receptors, *Prog.Drug Res.* 20, 323-346, 1976; Adamson, E.D. and Rees, A.R., Epidermal growth factor receptors, *Mol.Cell. Biochem.* 34, 129-152, 1981; Carpenter, G., The biochemistry and physiology of the receptor-tyrosine kinase for epidermal growth factor, *Mol.Cell.Endocrinol.* 31, 1-19, 1983; Alaoui-Jamali, M.A., Paterson, J., Al Moustafa, A.E. and Yen, L., The role of Erb-2 tyrosine kinase receptor in cellular intrinsic chemoresistance: mechanisms and implications, *Biochem.Cell.Biol.* 75, 315-325, 1997; Smit, L. and Borst, J., The Cb1 family of signal transduction molecules, *Crit.Rev.Oncog.* 8, 359-379, 1997; Elchebly, M., Cheng, A., and Tremblay, M.L., Modulation of insulin signaling by protein tyrosine phosphatases, *J.Mol.Med.* 78, 473-482, 2000; Carraway, K.L, Ramsauer, V.P., Haq, B., Carrothers Carraway, C.A., Cell signaling through membrane mucins, *Bioessays* 25, 66-71, 2003; Murai, K.K. and Pasquale, E.B., Eph receptors, ephrins, and synaptic function, *Neuroscientist* 10, 304-314, 2004; Monteiro, H.P., Rocha Oliveira, C.J., Curcio, M.F., Moraes, M.S., and Arai, R.J., Tyrosine phosphorylation in nitric oxide-mediated signaling events, *Methods Enzymol.* 396, 350-358, 2005; Heroult, M., Schaffner, F., and Augustin, H.G., Eph receptor and ephrin ligand-mediated interactions during angiogenesis and tumor progression, *Exp.Cell Res.* 312, 642-650, 2006; Perona, R., Cell signaling: growth factors and tyrosine kinase receptors, *Clin.Transl. Oncol.* 8, 77-82, 2006; Li, E. and Hristova, K., Role of receptor tyrosine kinase transmembrane domains in cell signaling and human pathologies, *Biochemistry* 45, 6242-6251, 2006.

Receptor-activator of NF-κB (RANK)

A member of the neuroregulin/tumor necrosis factor superfamily. See Roundy, K., Smith, R., Weis, J.J., and Weis, J.H., Overexpression of RANKL (receptor-activator of NF-κB ligand) implicates IFN-beta-mediated elimination of B-cell precursors in the osteopetrotic bone of microphthalmic mice, *J.Bone Miner.Res.* 18, 278-288, 2003; Huang, W., Drissi, M.H., O'Keefe, R.J. and Schwarz, E.M., A rapid multiparameter approach to study factors that regulate osteoclastogenesis: demonstration of the combinantorial dominant effects of TNF-alpha and TGF-ss in RANKL-mediated osteoclastogenesis, *Calcif.Tissue Int.* 73, 584-593, 2003; Neumann, E., Gay, S., and Muller-Ladner, U., The RANK/RANKL/osteoprotegerin system in rheumatoid arthritis: new insights from animal models, *Arthritis Rheum.* 2, 3257-3268, 2005; Hamdy, N.A., Osteoprotegerin as a potential therapy for osteoporosis, *Curr.Osteoporos.Rep.* 3, 121-125, 2005; Wada, T., Nakashima, T., Hiroshi, N., Penniger, J.M., RANKL-RANK signaling in osteoclastogenesis and bone disease, *Trends Mol.Med.* 12, 17-25, 2006.

Receptors for AGE (RAGE)

Cell-surface receptors for advanced glycation endproducts (AGE). These receptors are members of the immunoglobulin superfamily and are involved in the processes of inflammation and are suggested to be involved in the pathogenesis of diseases such as diabetes and neurogenerative diseases such as Alzheimer's disease. It is also noted that RAGE are also receptors for S100/calgranulin. See Bucciarelli, L.G., Wendt, T., Rong, L., *et al.*, RAGE is a multiligand receptor of the immunoglobulin superfamily: implications for homeostasis and chronic disease, *Cell.Mol.Life. Sci.* 59, 1117-1128, 2002; Yan, S.F., Ramasamy, R., Naka, Y., and Schmidt, A.M., Glycation, inflammation, and RAGE. A scaffold for the macrovascular complications of diabetes and beyond, *Circ.Res.* 93, 1159-1169, 2003. Ramasamy, R., Vannucci, S.J., Yan, S.S., Herold, K., San, S.F. and Schmidt, A.M., Advanced glycation end products and RAGE: a common thread in aging, diabetes, neurodegeneration, and inflammation, *Glycobiology* 15, 16R-28R, 2005; Jensen, L.J., Ostergaard, J., and Flyvbjerg, A., AGE-RAGE and AGE cross-link interaction: important players in the pathogenesis of diabetic kidney disease, *Horm.Metab.Res.* 37, (Suppl. 1) 26-34, 2005; Bierhaus, A., Humpert, P.M., Stern, D.M., Arnold, B., and Nawroth, P.P., Advanced glycation end product receptor-mediated cellular dysfunction, *Ann.N.Y.Acad.Sci.* 1043, 676-680, 2005; Ding, Q. and Keller, J.N., Evaluation of rage isoforms, ligands, and signaling in the brain, *Biochim.Biophys.Acta* 1746, 18-27, 2005.

Receptorome

That portion of the proteome that function via ligand recognition. This category is subject to subdivision by receptor type as the GPCR receptorome. See Setola, V., Hufeisne, S.J., Grande-Allen, K.J., *et al.*, 3,4-methylenedioxymethamphetamine (MDMA, "Ecstacy") induces fenfluoramine-like proliferative actions on human cardiac valvular interstitial cells in vitro, *Mol.Pharmacol.* 62, 1223-1229, 2003; Armbruster, B.N. and Roth, B.L., Mining the receptorome, *J.Biol.Chem.* 280, 5129-5132, 2005; Roth, B.L., Receptor systems: will mining the receptorome yield novel targets for pharmacotherapy, *Pharmacol.Ther.* 108, 59-64, 2005.

Receptosome

An intracellular organelle resulting from receptor-mediated endocytosis of a ligand; Willingham, M.C. and Pastan, I., The receptosome: an intermediate organelle of receptor-mediated endocytosis in cultured fibroblasts, *Cell* 21, 67-77, 1980; Pastan, I.L. and Willingham, M.C., Journal of the center of the cell: role of the receptosome, *Science* 214, 504-509, 1981; Chitambar, C.R. and Zivkovic-Gilgenbach, Z., Role of the acidic receptosome in the uptake and retention of 67Ga by human leukemic HL60 cells, *Cancer Res.* 50, 1484-1487, 1990. While this specific term has not seen extensive use, there is interest in receptor-mediated endocytosis coupled with "specific" vesicular transport (Sano, H., Higashi, T., Matsumoto, K., *et al.*, Insulin enhances macrophage scavenger receptor-mediated endocytotic uptake of advanced glycation end products, *J.Biol.Chem.* 273, 8630-8637, 1998). There is some interest in receptor-mediated endocytosis for drug delivery (Selbo, P.K., Hogset, A., Prasmickaite, L., and Berg, K., Photochemical internalization: a novel drug delivery system, *Tumour Biol.* 23, 102-112, 2002).

Refractive Index (Index of Refraction)

Ration of wavelength or phase velocity of an electromagnetic wave in a vacuum to that in a substance. Changes in the refractive index of solutions have been used to measure solute concentration in techniques such as analytical ultracentrifugation and chromatography. Techniques based on refractive index have been used to study cells More recently, refractive index has provided the based for measurement of macromolecules on surfaces. See Hawkes, J.B. and Astheimer, R.W., Thermal coefficient of the refractive index of water, *Science* 110, 717, 1949; Barer, R. and Tkaczyk, S., Refractive index of concentrated protein solutions, *Nature* 173, 821-822, 1954; Barer, R. and Dick, D.A., Interferometry and refractometry of cells in tissue culture, *Exp.Cell Res.* 13(Suppl 4),

103-135, 1957; Fishman, H.A., Greenwald, D.R., and Zare, Z.N., Biosensors in chemical separations, *Annu.Rev. Biophys. Biomol. Struct.* 27, 165-198, 1998; Van Regenmortel, M.H., Altschuh, D., Chatellier, J., *et al.*, Measurement of antigen-antibody interactions with biosensors, *J.Mol.Recognit.* 11, 163-167, 1998; Eremeeva, T., Size-exclusion chromatography of enzymatically treated cellulose and related polysaccharides, *J.Biochem.Biophys.Methods* 56, 253-264, 2003; Mogridge, J., Using light-scattering to determine the stoichiometry of protein complexes, *Methods Mol.Biol.* 261, 113-118, 2004; Hut, T.S., Biophysical methods for monitoring cell-substrate interactions in drug discovery, *Assay Drug Dev.Technol.* 1, 479-488, 2003; Haes, A.J. and Van Duyne, R.P., A unified view of propagating and localized surface plasmon resonance, *Anal.Bioanal.Chem.* 379, 920-930, 2004; Stuart, D.A., Haes, A.J., Yonzon, C.R., Hicks, E.M., and Van Duyne, R.P., Biological applications of localized surface plasmonic phenomenae, *IEE Proc. Nanobiotechnol.* 152, 13-32, 2005; Yuk, J.S., Hong, D.G., Jung, J.W., *et al.*, Sensitivity enhancement of spectral surface plasmon resonance biosensors for the analysis of protein arrays, *Eur.Biophys.J.* 35, 469–476, 2006; Ogusu, K., Suzuki, K., and Nishio, H., Simple and accurate measurement of the absorption coefficient of an absorbing plate by use of the Brewster angle, *Opt.Lett.* 31, 909-911, 2006; Cardenas-Valencia, A.M, Dlutowski, J., Fries, D., and Langdebrake, L., Spectrometric determination of the refractive index of optical wave guiding materials used in lab-on-a-chip applications, *Appl.Spectrosc.* 60, 322-329, 2006; Coelho Neto, J., Agero, U., Gazzinelli, R.T., and Mesquita, O.N, Measuring optical and mechanical properties of a living cell with defocusing microscopy, *Biophys.J.*, 91, 1108–1115, 2006. Friebel, M. and Meinke, M., Model function to calculate the refractive index of native hemoglobin in the wavelength range of 250-1100 nm, *Appl.Opt.* 45, 2838-2842, 2006.

Regulators of G Protein Sgnaling (RGS)'

Regulators of G protein signaling are a family of protein which bind to the activated α-subunit of the heterotrimer G protein complex where GDP has been replaced by GTP and block signal transmission. There is increasing evidence for a broader role for RGS proteins in cell function. See Dohlman, H.G. and Thorner, J., RGS proteins and signaling by heterotrimeric G proteins, *J.Biol.Chem.* 272, 3871-3874, 1997; Berman, D.M. and Gilman, A.G., Mammalial RGS proteins: barbarians at the gate, *J.Biol. Chem.* 273, 1269-1272, 1998; Hepler, J.R., Emerging roles for RGS proteins in cell signaling, *Trends Pharmacol.Sci.* 20, 376-382, 1999; Hepler, J.R., Emerging roles for RGS proteins in cell signaling, *Trends Pharmacol.Sci.* 20, 376-382, 1999; Burchett, S.A., Regulators of G protein signaling: a bestiary of modular protein binding domains, *J.Neurochem.* 75, 1335-1351, 2000; Hollinger, S. and Hepler, J.R., Cellular regulation of RGS proteins: modulators and intergrators of G protein signaling, *Pharmacol.Rev.* 54, 527-559, 2002; Kehrl, J.H., G-protein-coupled receptor signaling, RGS proteins, and lymphocyte function, *Crit.Rev.Immunol.* 24, 409-423, 2004; Wilkie, T.M. and Kinch, L., New roles for Gα and RGS proteins: communication continues despite pulling sisters apart, *Curr.Biol.* 15, R843-R854, 2005.

Regulatory Transcription Factors

A *trans*-acting component, usually a protein or protein complex, which interacts with a *cis*-regulatory region on a gene distant from the transcription initiation site to enhance or suppress the rate of transcription. A regulatory transcription factor is not considered a part of the basal transcription apparatus. See Fujita, T., Kimura, Y.,

Miyamoto, M., *et al.*, Induction of endogenous IFN-alpha and IFN-beta genes by a regulatory transcription factor, IRF-1, *Nature* 337, 270-272, 1989.

Wingender, E., *Gene Regulation in Eukaryotes*, VCH, Weinheim, Germany, 1993; Gopakrishnan, R.V., Dolle, P., Mattei, M.G., *et al.*, Genomic structure and developmental expression of the mouse cell cycle regulatory transcription factor DP1, *Oncogene* 13, 2671-2680, 1996; Bachmaier, K., Neu, N., Pummerer, C., *et al.*, iNOS expression and nitrotyrosine formation in the myocardium in response to inflammation is controlled by the interferon regulatory transcription factor 1, *Circulation* 96, 585-591, 1997; Larochelle, O., Stewart, G., Moffatt, P., *et al.*, Characterization of the mouse metal-regulatory-element-binding proteins, metal element protein-1 and metal regulatory transcription factor-1, *Biochem.J.* 353, 591-601, 2001; Courey, A.J., Regulatory transcription factors and regulatory regions, in *Transcription Factors*, ed. J.Locker, Chapter 2, pp. 17-34, 2001; Willmore, W.G., Control of transcription in eukaryotic cells, in *Functional Metabolism: Regulation and Adaptation*, ed. K.B. Storey, Wiley-Liss, Hoboken, NJ, USA, Chapter 6, pp. 153-187, 2004.

Resolution (Chromatographic)

The chromatographic separation between components in a mixture of components. A number of equations can be developed for the expression of resolution such as $R = 2(t_2-t_1)/W_1+W_2$ where t_1 is the elution time of component 1 having peak width of W_1 and t_2 is the elution time of component 2 having peak width of W_2. See Hearn, M.T., General strategies in the separation of proteins by high-performance liquid chromatographic methods, *J.Chromatog.* 418, 3-26, 1987; Feibush, B. and Santasania, C.T., Hydrophilic shielding of hydrophobic, cation- and anion-exchange phases for separation of small analytes: direct injection of biological fluids onto high-performance liquid chromatographic columns, *J.Chromatog.* 544, 41-49, 1991; Hagan, R.L., High-performance liquid chromatography for small-scale studies of drug stability, *Am.J.Hosp.Pharm.* 51, 2162-2175, 1994; Coffman, J.L., Roper, D.K., and Lightfoot, E.N., High-resolution chromatography of proteins in short columns and adsorptive membranes, *Bioseparation* 4, 183-200, 1994; Myher, J.J., and Kuksis, A., General strategies in chromatographic analysis of lipids, *J.Chromatog.B.* 671, 3-33, 1995; Chen, H. and Horvath, C., High-speed high-performance liquid chromatography of peptides and proteins, *J.Chromatog.A.* 705, 3-20, 1995; Bojarski, J. and Aboul-Enein, H.Y., Recent applications of chromatographic resolution of enantiomers in pharmaceutical analysis, *Biomed.Chromatog.* 10, 297-302, 1996; Dolan, J.W., Temperature selectivity in reversed-phase high performance liquid chromatography, *J.Chromatog.A* 965, 195-205, 2002; Jupille, T.H., Dolan, J.W., Snyder, L.R., and Molnar, I., Two-dimensional optimization using different pairs of variables for the reverved-phase high-performance liquid chromatographic separation of a mixture of acidic compounds, *J.Chromatog.A* 948, 35-41, 2002; Pellett, J., Lukulay, P., Mao, Y. *et al.*, "Orthogonal" separations for reversed-phase liquid chromatography, *J.Chromatog.A.* 1101, 122-135, 2006; Nageswara Rao, R., Narasa Raju, A., and Nagaraju, D., Development and validation of a liquid chromatographic method for determination of enantiomeric purity of citalopram in bulk drugs and pharmaceuticals, *J.Pharm.Biomed.Anal.* 41, 280-285, 2006.

Resurrection Plants

Plants usually found in arid regions which adopt a compact shape during water deprivation and changes shape upon rehydration. See Kranner, I., Beckett, R.P., Wornik, S., Zorn, M., Preifhofer,

H.W., Revival of a resurrection plant correlates with its antioxidant status, *Plant J.* 31, 13-24, 2002; Schluepmann, H., Pellny, T., van Dijken, A., Smeeken, S., and Paul, M., Trehalose 6-phosphate is indispensable for carbohydrate utilization and growth in *Arabidopsis thaliana*, *Proc.Natl.Acad.Sci.USA* 100, 6849-6854, 2003; Jones, L. and McQueen-Mason, S., *FEBS Lett.* 559, 61-65, 2004; Helseth, L.E. and Fischer, T.M., Physical mechanisms of rehydration in *Polypodium polypodioides*, a resurrection plant, *Phys.Rev.E. Stat. Nonlin. Soft Matter. Phys.* 71(6 Pt 1):061903, epub, 2005.

Retention Time

For chromatography, the retention time (t_r) is the time from injection of solute to the apex (zenith) of the peak of the respective solute. For planer chromatography such as thin-layer chromatography or paper chromatography, the Retardation Factor (R_f) is ratio of the distance traveled by solvent (solvent front) and the distance traveled by the solute. See Palmblad, M., Ramstrom, M., Bailey, C.G., *et al.*, Protein identification by liquid chromatography-mass spectrometry using retention time prediction, *J.Chromatog.B Anal.Technol.Biomed.Life Sci.* 803, 131-135, 2004; Joutovsky, A., Hadzi-Nesic, J., and Nardi, M.A., HPLC retention time as a diagnostic tool for hemoglobin variants and hemoglobinopathies: a study of 60000 samples in a clinical diagnostic laboratory, *Clin.Chem.* 50, 1736-1747, 2004; Pierce, K.M., Wood, L.F., Wright, B.W., and Synovec, R.E., A comprehensive two-dimensional retention time alignment algorithm to enhance chemometric analysis of comprehensive two-dimensional separation data, *Anal.Chem.* 77, 7735-7743, 2005. The term retention time is also used to describe the period of time that a material resides in the digestive tract, Bernard, L. and Doreau, M., Use of rare earth elements as external markers for mean retention time measurements in ruminants, *Reprod.Nutr.Dev.* 40, 89-101, 2000; Pearson, R.A., Archibald, R.F., and Muirhead, R.H., A comparison of the effect of forage type and level of feeding on the digestibility and gastrointestinal mean retention time of dry forages given to cattle, sheep, ponies, and donkeys, *Br.J.Nutr.* 95, 88-98. 2006); retention in filtration system (Lee, Y.W., Chung, J., Jeong, Y.D. al., Backwash based methodology for the estimation of solids retention time in biological aerated filter, *Environ.Technol.* 27, 777-787, 2006); and retention time of solid waste in a bioreactor (Maase, A., Sperandio, M., and Cabassud, C., Comparison of sludge characteristics and performance of a submerged membrane bioreactor and an activated sludge process at high solids retention time, *Water Res.* 40, 2405-2415, 2006).

Retention Volume

For chromatography, the retention volume is a function of the flow rate of the mobile phase and the retention time (Frigon, R.P., Leypoldt, J.K., Uyeji, S., and Hnderson, L.W., Disparity between stokes radii of dextrans and proteins as determined by retention volume in gel permeation chromatography, *Anal.Chem.* 55, 1349-1354, 1983; Dyr, J.E. and Suttnar, J., On the increased retention volume of human hemoglobin in high-performance gel filtration, *J.Chromatog.* 408, 303-307, 1987; Griotti, F. and Guiochon, G., Influence of the pressure on the properties of chromatographic columns. III. Retention time of thiourea, hold-up volume, and compressibility of the C_{18}-bonded layer, *J.Chromatog.A.* 1075, 117-126, 2005). The term is also used to refer to urine retention (Dutkiewicz, S., Witeska, A., and Stepien, K., Relation between prostate-specific antigen, prostate volume, retention volume and age in benign prostatic hypertrophy (BPH), *Int.Urol.Nephrol.* 27, 762-768, 1995; Demaria, F., Amar, N., Blau, D., *et al.*, Prospective 3D ultrasonographic evaluation of immediate postpartum urine retention volume in 100 women who delivered vaginally, *Int. Urogynecol.J.Pelvic Floor Dysfunct.* 15, 281-285, 2004).

Retromer

A multi-protein complex thought to function in endosome-Golgi retrieval. See Pfeffer, S.R., Membrane transport: retromer to the rescue, *Curr.Biol.* 11, R109-R111, 2001; Seaman, M.N.J., Recycle your receptors with retromer, *Trends Cell Biol.* 15, 68-75, 2005; Griffin, C.T., Trejo, J., and Magnuson, T., Genetic evidence for a mammalian retromer complex containing sorting nexins 1 and 2, *Proc.Natl.Acad.Sci. USA* 102, 15173-15177, 2005; Gullapalli, A., Wolfe, B.L., Griffin, C.T., Magnuson, T., and Trejo, J., An essential role of SNX1 in lysosomal sorting of protease-activated receptor-1: evidence for retromer-, Hrs-, and Tsg101-independent functions of sorting nexins, *Mol.Biol.Cell,* 2006.

Retro-translocation (retrotranslocation)

A process by which misfolded proteins or other incorrect translation products are transported from the lumen of the endoplasmic reticulum to the cytoplasm for subsequent degradation by the proteosome. See Johnson, A.E. and Haigh, N.G., The ER translocon and retrotranslocation: is the shift into reverse manual or automatic? *Cell* 102, 709-712, 2000; Svedine, S., Wang, T., Halaban, R., and Herbert, D.N., Carbohydrates act as sorting determinants in ER-associated degradation of tyrosinase, *J.Cell. Sci.* 117, 2937-2949, 2004; Schulze, A., Sandera, S., Buerger, E., *et al.*, The ubiquitin-domain protein HERP forms a complex with components of the endoplasmic reticulum associated degradation pathway, *J.Mol.Biol.* 354, 1021-1027, 2005.

Reverse Immunology

Prediction of antigen structure based on peptide reactivity with cytotoxic T cell MHC proteins; most frequently used in the study of tumor antigens: see Boon, T. and van der Bruggen, P., Human tumor antigens recognized by T lymphocytes, *J.Exptl.Med.* 183, 725-729, 1996; Maecker, B., *et al.*, Linking genomics to immunotherapy by reverse immunology—'immunomics' in the new millennium, *Cur.Mol.Med.* 1, 609-619, 2001; Anderson, K.S. and LaBaer, J., The sentinel within: exploiting the immune system for cancer biomarkers, *J.Proteome Res.* 4, 1123-1133, 2005. see SEREX.

Reverse Micelle

A reverse micelle or inverted micelle is a stable assembly of a surfactant around an aqueous core where the lipophilic part of the surfactant is directed toward the exterior which is a non-polar solvent and the charged portion is directed toward the aqueous core. Reverse micelles have been used for the stabilization of proteins in organic solvents, for protein purification and for drug delivery. See Bernert, J.T., Jr. and Sprecher, H., Solubilization and partial purification of an enzyme involved in rat liver microsomal fatty acid chain elongation: beta-hydroxy-acyl-CoA dehydrase, *J.Biol.Chem.* 254, 11584-11590, 1979; Grandi, C., Smith, R.E. and Luisi, P.L., Micellar solubilization of biopolymers in organic solvents. Activity and conformation of lysozyme in isooctane reverse micelles, *J.Biol.Chem.* 256, 837-843, 1981; Leser, M.E., Wei, G., Luisi, P., and Maestro, M.,

Application of reverse micelles for the extraction of proteins, *Biochem.Biophys.Res.Commun.* 135, 629-635, 1986; Luisi, P.L., and Magid, L.J., Solubilization of enzymes and nucleic acids in hydrocarbon micellar solutions, *CRC Crit.Rev.Biochem.* 20, 409-474, 1986; Huruguen, J.P. and Pileni, M.P., Drastic change of reverse micellar structure by protein or enzyme addition, *Eur.Biophys.J.* 19, 103-107, 1991; Bru, R., Sanchez-Ferrer, A., and Garcia-Caroma, F., Kinetic models in reverse micelles, *Biochem.J.* 310, 721-739, 1995; Nicot, C. and Waks, M., Proteins as invited guests of reverse micelles: conformational effects, significance, applications, *Biotechnol.Genet.Eng.Rev.* 13, 267-314, 1996; Tuena de Gomez-Puyou, M. and Gomez-Puyou, A., Enzymes in low-water systems, *CRC Rev.Biochem.Mol.Biol.* 33, 53-89, 1998; Orlich, B. and Schomacker, R., *Adv.Biochem.Eng. Biotechnol.* 75, 185-208, 2002; Krishna, S.H., Srinivas, N.D., Ragnavarao, K.S., and Karanth, N.G., Reverse micellar extraction for downstream processing of proteins/enzymes, *Adv. Biochem.Eng. Biotechnol.* 75, 1190183, 2002; Marhuenda-Egea, F.C. and Bonete, M.J., Extreme halophilic enzymes in organic solvents, *Curr.Opin.Biotechnol.* 13, 385-389, 2002; Muller-Goymann, C.C., Physicochemical characterization of colloidal drug delivery systems such as reverse micelles, vesicles, liquid crystals, and nanoparticles for topical administration, *Eur.J. Pharm.Biopharm.* 58, 343-356, 2004.

Reverse Proteomics

Proteomic analysis where genomic sequence information is used to predict the resulting proteome providing the basis for experiment design. See Lamesch, P., Milstein, S., Hao, T., *et al*, *C.elegans* ORFeome version 3.1: increasing the coverage of ORFeome resources with improved gene production, *Genome Res.* 14, 2064-2069, 2004; Rual, J.F., Hirozane-Kishikawa, T., Hao, T., *et al.*, Human ORFeom version 1.1: a platform for reverse proteomics, *Genome Res.* 14, 2128-135, 2004; Gillette, W.K., Esposito, D., Frank, P.H., *et al.*, Pooled ORF expression technology (POET), *Mol.Cell.Proteom.* 4, 1647-1652, 2005.

Reverse Transcriptase

An enzymes which catalyzes the formation of DNA from an RNA template. This is an enzyme critical for the replication of RNA viruses such as HIV and is major drug targets for AIDS and other RNA viral diseases. See O'Conner, T.E., Reverse transcriptase-progress, problems and prospects, *Bibl.Haematol.* 39, 1265-1181, 1973; Wu, A.M. and Gallo, R.C., Reverse transcriptase, *CRC Crit. Rev.Biochem.* 3, 289-347, 1975; Verma, I.M., The reverse transcriptase, *Biochim.Biophys.Acta* 473, 1-38, 1977; Chandra, P., Immunological characterization of reverse transcriptase from human tumor tissues, *Surv.Immunol.Res.* 2, 170-177, 1983; Lim, D. and Maas, W.K., Reverse transcriptase in bacteria, *Mol. Microbiol.* 3, 1141-1144, 1989; Barber, A.M., Hizi, A., Maizel, J.V., Jr., and Hughes, S.H., HIV-1 reverse transcriptase: structure predictions for the polymerase domain, *AIDS Res.Hum.Retroviruses* 6, 1061-1072, 1990; Durantel, D., Brunelle, M.N., Gros, E., *et al.*, Resistance of human hepatitis B virus to reverse transcriptase inhibitors: from genotypic to phenotypic testing, *J.Clin.Vitrol.* 34(Suppl 1), S34-S43, 2005; Menendez-Arias, L., Matamoros, T., and Cases-Gonzalez, C.E., Insertions and deletions in HIV-1 reverse transcriptase: consequences for drug resistance and viral fitness, *Curr.Pharm.Des.* 12, 1811-1825, 2006; Srivastava, S., Sluis-Cremer, N., and Tachedjian, G., Dimerization of human immunodeficiency virus type 1 reverse transcriptase as an antiviral target, *Curr.Pharm.Des.* 12, 1879-1894, 2006.

Reverse Transcriptase – Polymerase chain reaction (RT-PCR)

A variation of the PCR technique in which cDNA is made from RNA via reverse transcription. The cDNA is then amplified using standard PCR protocols. See Mocharla, H., Mocharla, R., and Hodes, M.E., Coupled reverse transcription-polymerase chain reaction (RT-PCR) as a sensitive and rapid method for isozyme genotyping, *Gene* 93, 271-275, 1990; Weis, J.H., Tan, S.S., Martin, B.K., and Willwer, C.T., Detection of rare mRNAs via quantitative RT-PCR, *Trends Genet.* 8, 263-264, 1992; Akoury, D.A., Seo, J.J., James, C.D., and Zaki,S.R., RT-PCR detection of mRNA recovered from archival glass slide smears, *Mol.Pathol.* 6, 195-200, 1993; Silver, J., Maudru, T., Fujita, K. and Repaske, R., An RT-PCR assay for the enzyme activity of reverse transcriptase capable of detecting single virions, *Nucleic Acids Res.* 21, 3593-3594, 1993; Taniguchi, A., Kohsaka, H., and Carson, D.A., Competitive RT-PCR ELISA: a rapid, sensitive and non-radioactive method to quantitate cytokine mRNA, *J.Immunol.Methods* 169, 101-109, 1994; Prediger, E.A., Quantitating mRNAs with relative and competitive RT-PCR, *Methods Mol.Biol.* 160, 49-63, 2001; Lion, T., Current recommendations for positive controls in RT-PCR assays, *Leukemia* 15, 1033-1037, 2001; Joyce, C., Quantitative RT-PCR. A review of current methodologies, *Methods Mol.Biol.* 193, 83-92, 2002; Ransick, A., Detection of mRNA by in situ hybridization and RT-PCR, *Methods Mol.Biol.* 74, 601-620, 2004; Tallini, G. and Brandao, G., Assessment of RET/PTC oncogene activation in thyroid nodules utilizing laser microdissection followed by nested RT-PCR, *Methods Mol.Biol.* 293, 103-111, 2005; Ooi, C.P., Rohani, A., Zamree, I., and Lee, H.L., Temperature related storage evaluation of an RT-PCR test kit for the detection of dengue infection in mosquitoes, *Trop.Biomed.* 22, 73-76, 2005.

Rho factor

A ring-shaped homohexameric bacterial protein encoded by the *rho* gene which regulates RNA polymerase. See Lathe, R., RNA polymerase of *Escherichia coli*, *Curr.Top.Microbiol.Immunol.* 83, 37-91, 1978; Adhya, S. and Gottesman, M., Control of transcription termination, *Annu.Rev.Biochem.* 47,967-996, 1978; Aktories, K., Schmidt, G., and Just, I., Rho GTPases as targets of bacterial protein toxins, *Biol.Chem.* 381, 421-426, 2000; Anston, A.A., Single-stranded-RNA binding proteins, *Curr.Opin.Struct.Biol.* 10, 87-94, 2000; Richardson, J.P., Rho-dependent termination and ATPases in transcript termination, *Biochim.Biophys.Acta* 1577, 251-260, 2002; Banerjee, S., Chalissery, J., Bandey, I., and Sen, R., Rho-dependent transcription termination: more questions than answers, *J.Microbiol.* 44, 11-22, 2006.

Rhomboid

A family of transmembrane proteins with proteolytic activity and considered to be a regulatory of EGF signaling. Rhomboid was originally described in *Drosophila* as protease cleaving Spitz, a membrane-bound EGF. See Noll, R., Sturtevant, M.A., Gollapudi, R.R., and Bier, E., New functions of the *Drosophila* rhomboid gene during embryonic and adult development are revealed by a novel genetic method, enhancer piracy, *Development* 120, 2329-2338, 1994; Lage, P., Yan, Y.N., and Jarman, A.P., Requirement for EGF receptor signaling in neural recruitment during formation of *Drosophila* chordotonal sense organ clusters, *Curr.Biol.* 7, 166-175, 1997: Sturtevant, M.A., Roark, M., and Bier, E., The *Drosophila* rhomboid gene mediates the localized formation of wing veins

and interacts genetically with components of the EGF-R signaling pathway, *Genes Dev.* 7, 961-973, 1993; Klambt, C., EGF receptor signalling: the importance of presentation, *Curr.Biol.* 10, R399-R391, 2000; Guichard, A., Roark, M., Ronshaugen, M., and Bier, E., Brother of rhomboid, a rhomboid-related gene expressed during early *Drosophila* oogenesis, promotes EGF-R/MAPK signaling, *Dev.Biol.* 226, 255-266, 2000; Urban, S., Lee, J.R., and Freeman, M., *Drosophila* rhomboid-1 defines a family of putative intramembrane serine proteases, *Cell* 107, 173-182, 2001; Urban, S., Lee, J.R., and Freeman, M., A family of Rhomboid intramembrane proteases activates all *Drosophila* membrane-tethered EGF ligands, *EMBO J.* 21, 4277-4286, 2002; Jaszai, J. and Brand, M., Cloning and expression of Ventrhoid, a novel vertebrate homologue of the *Drosophila* EGF pathway gene rhomboid, *Mech.Dev.* 113, 73-77, 2002: Zhou, X.W., Blackman, M.J., Howell, S.A., and Carruthers, V.B., *Mol. Cell.Proteomics* 3, 565-576, 2004; Sik, A., Passer, B.J., Koonin, E.V., and Pellegrini, L., Self-regulated cleavage of the mitochondrial intramembrane-cleaving protease PARL yields Pbeta, a nuclear-targeted peptide, *J.Biol.Chem.* 279, 15323-15329, 2004; Kanaoko, M.M., Urban, S., Freeman, M., and Okada, K., An *Arabidopsis* Rhomboid homolog is an intermediate proteases in plants, *FEBS Lett.* 579, 5723-5728, 2005; Howell, S.A., Hackett, F., Johgco, A.M. *et al.*, Distinct mechanisms govern proteolytic shedding of a key invasion protein in apicomplexan pathogens, *Mol.Microbiol.* 57, 1342-1356, 2005; Uban, S. and Wolfe, M.S., Reconstitution of intramembrane proteolysis *in vitro* reveals that pure rhomboid is sufficient for catalysis and specificity, *Proc.Natl.Acad.Sci.USA* 102, 1883-1888, 2005; Nakagawa, T., Guichard, A., Castro, C.P., *et al.*, Characterization of a human rhomboid homolog, p100hRho/RHBDF1, which interacts with TGF-alpha family ligands, *Dev.Dyn.* 233, 1315-1331, 2005. Rhomboid also described a geometric shape such as a parallelogram or rhombus and, as such, has been used to describe intracellular crystal formation (Machhi, J., Kouzova, M., Komorowski, D.J. *et al.*, Crystals of alveolar soft part sarcoma in a fine needle aspiration biopsy cytology smear. A case report, *Acta Cytol.* 46, 904-908, 2002; Duan, X., Bruneval, P., Hammadeh, R., *et al.*, Metastatic juxtaglomerular cell tumor in a 52-year old man, *Am.J.Surg.Pathol.* 28, 1098-1102, 2004; Stewart, C.J. and Spagnolo, D.V. Crystalline plasma cell inclusions in Helicobacter-associated gastritis, *J.Clin.Pathol.*, 59, 851–854, 2006. The term rhomboid is an anatomical term (Dong, H.W. and Swanson, L.W., Organization of axonal projections from the anterolateral area of the bed nuclei of the stria terminalis, *J.Comp.Neurol.* 468, 277-298, 2004).

Riboswitch

A discrete RNA sequence in the leader sequences (UTR regions) of certain mRNAs which encode enzymes involved in metabolism. Earlier described as the *RFN* element (Gefland, M.A., Mironov, A.A., Jomantas, J., *et al.*, A conserved RNA structure element involved in the regulation of bacterial riboflavin synthesis genes, *Trends Genet.* 15, 439-442, 1999). Riboswitches are conceptually similar to aptamers where there is specific binding of a ligand. Binding of a ligand to a specific riboswitch influences the expression of the cognate gene at both the transcriptional and translational level. See Winkler, W., Nahvi, A., and Breaker, R.R., Thiamine derivatives bind messenger RNAs directly to regulate bacterial gene expression, *Nature* 419, 952-956, 2002; Winkler, W.C., Cohen-Chalamish, S., and Breaker, R.R., An mRNA structure that controls gene expression by binding FMN, *Proc.Natl. Acad.Sci.USA* 99, 15908-15913, 2002; Sudarsan, N., Wickiser, J.K., Nakamura, S., *et al.*, An mRNA structure in bacteria that controls gene expression by binding lysine, *Gene Dev.* 17, 2688-2697, 2003; Nudler, E. and Mironov, A.S., The riboswitch control of bacterial metabolism, *Trends Biochem.Sci.* 29, 11-17, 2004; Batey, R.T., Gilbert, S.D., and Montange, R.K., Structure of a natural guanine-responsive riboswitch complexed with the metabolite hypoxanthine, *Nature* 432, 411-415, 2004; Winkler, W.C., Riboswitches and the role of noncoding RNAs in bacterial metabolic control, *Curr.Opin.Chem.Biol.* 9, 594-602, 2005.

Montange, R.K. and Batey, R.T., Structure of the *S*-adenosylmethionine riboswitch regulatory mRNA element, *Nature* 441, 1172-1175, 2006.

Ring-Finger Proteins/Ring-Finger Domains

Ring-finger, which is related to zinc finger, describes a family of proteins defined a zinc-binding ring finger motif. This motif was first described in RING1 protein but occurs in a wide variety of proteins including proteins involved in Ubiquitinylation and c-Cbl oncoprotein. See Lovering, R., Hanson, I.M., Borden, K.L., *et al..,* Identification and preliminary characterization of a protein motif related to the zinc finger, *Proc.Natl.Acad.Sci.USA* 90, 2112-2116, 1993; Fremont, P.S., The RING finger. A novel protein sequence motif related to the zinc finger, *Ann.N.Y.Acad.Sci.* 684, 174-192, 1993; Hu, H.M., O'Rourke, K., Boguski, M.S., and Dixit, V.M., A novel RING finger protein interacts with the cytoplasmic domain of CD40, *J.Biol.Chem.* 269, 30069-30072, 1994; Borden, K.L. and Freemont, P.S., The RING finger domain: a recent example of a sequence-structure family, *Curr.Opin.Struct.Biol.* 6, 395-401, 1996; Smit, L. and Borst, J., The Cbl family of signal transduction molecules, *Crit.Rev.Oncol.* 8, 359-379, 1997; Jackson, P.K., Eldridge, A.G., Freed, E., *et al.,* The lore of the RINGs: substrate recognition and catalysis by ubiquitin ligases, *Trends Cell Biol.* 10, 429-439, 2000; Gregorio, C.C., Perry, C.N., and McElhinny, A.S., Functional properties of the titin/connectin-associated proteins, the muscle-specific RING finger proteins (MURFs), in striated muscle, *J.Muscle Res. Cell Motil.* 14, 1-12, 2006.

RNA-induced silencing complex (RISC)

The functional complex formed from the interaction of interfering RNA (RNA interference, RNAi) from small interfering RNAs (siRNA) or microRNAs (miRNAs) with mRNA-protein (Argonaut protein) to form a complex which results in post-transcriptional gene silencing as a result of mRNA cleavage from the siRNA/miRNA. See Sontheimer, E.J., Assembly and function of RNA silencing complexes, *Nat.Rev.Mol.Cell Biol.* 6, 127-138, 2005; Tang, G., siRNA and miRNA: an insight into RISCs, *Trends Biochem.Sci.* 30, 106-114, 2005; Filipowicz, W., RNAi: the nuts and bolts of the RISC machine, *Cell* 122, 17-20, 2005; Hutvagner, G., Small RNA asymmetry in RNAi: function in RISC assmble and gene regulation, *FEBS Lett.* 579, 5850-5857, 2005; Hammond, S.M., Dicing and slicing: the core machinery of the RNA interference pathway, *FEBS Lett.* 579, 5822-5829, 2005; Gilmore, I.R., Fox, S.P., Hollins, A.J., and Akhtar, S., Delivery strategies for siRNA-mediated gene silencing, *Curr.Drug Deliv.* 3, 147-155, 2006.

RNA Interference (RNAi)

The inhibition of gene transcription mediated through the production of small interfering RNA fragments (siRNA) and binding of these fragments and protein to messenger RNA. RNA interference is also referred to as RNA silencing. See Fire, A., Xu, S.,

Montgomery, M.K., *et al.,* Potent and specific genetic interference by double-straned RNA in *Caenorhabditis elegans, Nature* 391, 806-811, 1998; Bosher, J.M., Dufourcq, P, Sookhareea, S., and Labousse, M., RNA interference can target pre-mRNA: consequences for gene expression in a *Caenorhabditis elegans* operon, *Genetics* 153, 1245-1256, 1999; Tabara, H., Sarkissian, M., Kelly, W.G., *et al.,* The rde-1 gene, RNA interference, and transposon silencing in *C. elegans, Cell* 99, 123-132, 1999; Plasterk, R.H. and Ketting, R.F., The silence of the genes, *Curr.Opin.Genet. Dev.* 10, 562-567, 2000; Barstead, R., Genome-wide RNAi, *Curr. Opin.Chem.Biol.* 5, 63-33, 2001; Tuschi, T., RNA interference and small interfering RNAs, *Chembiochem* 2, 239-245, 2001; Baulcombe, D., RNA silencing, *Trends Biochem.Sci.* 30, 290-293, 2005; Filipowicz, W., Jaskiewicz, L., Kolb, F.A., and Pillai, R.S., Post-transcriptional gene silencing by siRNAs and miRNAs, *Curr.Opin.Struct.Biol.* 15,331-341, 2005; Shearwin, K.E., Callen, B.P., and Egan, J.B., Transcriptional interference-a crash course, *Trends Genet.* 21, 339-345, 2005; Sarov, M., and Stewart, A.F., The best control for the specificity of RNAi, *Trends Biotechnol.* 23, 446-448, 2005; Collins, R.E. and Cheng, X., Structural domains in RNAi, *FEBS Lett.* 579, 541-5849, 2005; Zamore, P.D. and Haley, B., Ribo-genome: the big world of small RNAs, *Science* 309, 1519-1524, 2005; Yeung, M.L., Bennasser, Y., Le, S.Y., and Jeang, K.T., siRNA, miRNA, and HIV: promises and challenges, *Cell.Res.* 15, 935-946, 2005; Fanning, G.C. and Symonds, G., Gene-expressed RNA as a therapeutic: issues to consider, using ribozymes and small hairpin RNA as specific examples, *Handb.Exp.Pharmacol.* (173), 289-303, 2006; *RNA Silencing. Methods and Protocols,* ed. G.C. Carmichael, Humana Press, Totowa, NJ, USA, 2005; Galun, E., *RNA Silencing,* World Scientific Publishing, Pte, Ltd, Singapore, 2005.

RNA Isolation ("tri-reagents")

Study of gene expression can involve the isolation of mRNA from cells and tissues for analysis by microarray technology, RT-PCR, and northern blot analysis. A variety of approaches are involved including the treatment of water with diethylpyrocarbonate (ethoxyformic anhydride), the use of RNAse inhibitors and various extraction technologies having the prefix "tri" such as Tri Reagent® and TRIzol®. These reagents use a solution of guanidine isothiocyanate and phenol (see Chomczynski, P. and Sacchi, N., Single-step method of RNA isolation by acid guanidinium thiocyanate-phenol-chloroform extraction, *Anal. Biochem.* 162, 156-159, 1987) for extraction of cells and tissue followed by a phenol extraction which yields an aqueous phase with RNA. A variety of other technologies are available and there are several excellent comparative studies. See Verhofstede, C., Fransen, K., Marissens, D., *et al.,* Isolation of HIV-1 RNA from plasma: evaluation of eight different extraction methods, *J.Virol.Methods* 60, 155-159, 1996; Chadderton, T., Wilson, C., Bewick, M., and Gluck, S., Evaluation of three rapid RNA extraction reagents: relevance for use in RT-PCR's and measurement of low level gene expression in clinical samples, *Cell.Mol.Biol.* 43, 1227-1234, 1997; Weber, K., Bolander, M.E., and Sarkar, G., PIG-B: a homemade monophasic cocktail for the extraction of RNA, *Mol.Biotechnol.* 73-77, 1998; Mannhalter, C., Koizar, D., and Mitterbauer, G., Evaluation of RNA isolation methods and reference genes for RT-PCR analyses of rare target RNA, *Clin. Chem.Lab.Med.* 38, 171-177, 2000; Deng, M.Y., Wang, H., Ward, G.B., Beckham, T.R., and McKenna, T.S., Comparison of six RNA extraction methods for the detection of classical swine fever virus by real-time and conventional reverse transcription-PCR, *J.Vet.Diagn.Invest.* 17, 574-578, 2005; Culley, D.E., Kovacik, W.P., Jr., Brockman, F.J., and Zhang, W., Optimization of RNA isolation from the archaebacterium *Methanosarcina barkeri* and validation for oligonucleotide microarray analysis, *J.Microbiol. Methods*, 67, 36–43, 2006. Prezeau, N., Silvy, M., Gabert, J., and Picard, C., Assessment of a new RNA stabilizing reagent (Tempus GLood RNA) for minimal residual disease in onco-hematology usng the EAC-protocol, *Leuk.Res.* 30, 569-574, 2006.

RNA Polymerase

The enzymes responsible for the biosynthesis of DNA-directed RNA synthesis. RNA polymerase is a nucleotide transferase that synthesizes RNA from ribonucleotides. In bacteria there is only one RNA polymerase (see Lathe, R., RNA polymerase of Escherichia coli, *Curr.Top.Microbiol.Immunol.* 83, 37-91, 1978); arachaea also has a single RNA polymerase (Geiduschek, E.P. and Ouhammouch, M., Archaeal transcription and its regulators, *Mol. Microl.* 56, 1397-1407, 2005). Eukaryotic cells have three RNA polymerases: RNA polymerase I (polI) catalyzes the synthesis of ribosomal RNA species in the form of a precursor pre-rRNA (45S) which is processed into other species such as 28 S and 18S RNAs; RNA polymerase III (polIII) synthesizes tRNA (transfer RNAs) and other smaller RNA species. RNA polymerase II (polI) (Hahn, S. Structure and mechanism of the RNA polymerase II transcription machinery, *Nature Struct.Mol.Biol.* 11, 394-403, 2004) is responsible for the synthesis of the various pre-mRNAs (messenger RNAs) which mature into the mRNA species responsible for the direction of protein biosynthesis. Viral RNA polymerases appear to be different from other RNA polymerases and appear to be derived from DNA polymerases. See *RNA Polymerase*, ed. R. Losick and M. Chamberlin, Cold Spring Harbor Laboratory Press, Cold Spring Harbor, NY, USA, 1976; *Viral Polymerases and Related Proteins*, ed. L.C. Kuo and D.B. Olsen, Academic Press, San Diego, CA, 1996; Adhya, S.L., *RNA Polymerase and Associated Factors*, Academic Press, San Diego, CA, 1996; *Eukaryotic Gene Transcription*, ed. S. Goodbourn, IRL Press at Oxford University Press, Oxford, UK, 1996; Richter, J.D., *mRNA Formation and Function*, Academic Press, San Diego, CA, 1997; Borukhov, S. and Nudler, E., RNA polymerase holoenzyme: structure, function and biological implications, *Curr.Opin.Microbiol.* 6, 93-100, 2003; Murakami, K.S. and Darst, S.A., Bacterial RNA polymerases: the whole story, *Curr.Opin. Struct.Biol.* 13, 31-39, 2003; Studitsky, V.M., Walter, W., Kireev, M., *et al.*, Chromatin remodeling by RNA polymerases, *Trends Biochem.Sci.* 29, 127-136, 2004; Bartlett, M.S., Determinants of transcription initiation by archaeal RNA polymerase, *Curr.Opin. Microbiol.* 8, 677-684, 2005; Boeger, H., Bushnell, D.A., Davis, R., *et al.*, Structural basis of eukaryotic gene transcription, *FEBS Lett.* 579, 899-903, 2005; Gralla, J.D., *Escherichia coli* ribosomal RNA transcription: regulatory roles for ppGpp, NTPs, architectural proteins and a polymerase-binding protein, *Molec.Microbiol.* 55, 973-977, 2005; Banerjee, S., Chalissery, J., Bandey, I., and Sen, R., Rho-dependent transcription: more questions than answers, *J.Microbiol.* 44, 11-22, 2006.

RNA Splicing

The removal of introns from the sequence of an mRNA following transcription to form an uninterrupted coding sequence. During this process, introns or intervening regions are removed and the remaining regions, exons, are joining together in a splicing process to form the mature RNA transcript. See Sharp, P.A.,

The discovery of split genes and RNA splicing, *Trends Biochem. Sci.* 30, 279-281, 2005; Matlin, A., Clark, F., and Smith, C.W., Understanding alternative splicing: toward a cellular code, *Nat. Rev.Mol.Cell.Biol.* 6, 386-398, 2005; Stetefeld, J. and Ruegg, M.A., Structural and functional diversity generated by alternative mRNA splicing, *Trends Biochem.Sci.* 30, 510-521, 2005.

(RNAse III) Ribonuclease III

A family of ribonucleases which are involved in RNA silencing or RNA interference. See Conrad, C. and Rauhut, R., Ribonuclease III: new sense from nuisance, *Int.J.Biochem.* 34, 116-129, 2002.

Rolling

The initial interaction between a leukocyte and the endothelium; also knows as margination.

RTX Toxins

RTX family of bacterial toxins are a group of cytolysins and cyto-toxins. Hemolysin (HlyA) is often quoted as the model for RTX toxins. See Coote, J.G., Structural and functional relationship among the RTX toxin determinants of Gram-negative bacteria, *FEMS Microbiol.Rev.* 8, 137-161, 1992.

S100 Proteins

A multifunctional family of intracellular proteins distinguished by solution in saturated ammonium sulfate (100% saturation) and their interactions with calcium ions. The seminal member of the family, S100 protein, was described as protein unique to the nervous system. See Moore, B.W., A soluble protein characteristic of the nervous system, *Biochem.Biophys.Res.Commun.* 19, 739-744, 1965; Donato, R., Perspectives in S-100 protein biology, *Cell Calcium* 12, 713-726, 1991; Passey, R.J., Xu, K., Hume, D.A., and Geczy, C.L., S100A8: emerging functions and regulations, *J.Leukocyte Biol.* 66, 549-556, 1999; Heizmann, C.W., The multifunctional S100 protein family, *Methods Mol.Biol.* 172, 69-80, 2002; Emberley, E.D., Murphy, L.C. and Watson, P.H., S100 proteins and their influence on pro-survival pathways in cancer, *Biochem.Cell Biol.* 82, 508-515, 2004.

Saposins

Sphingolipid activator proteins; small heat-stable proteins which appear to be cofactors in the hydrolysis of sphingolipids. There are four saposins, A,B,C, and D, which are generated from a common precursor, prosaposin - Vaccaro, A.M., Salivioli, R., Tatti, and Ciaffoni, F., Saposins and their interactions with lipids, *Neurochemical Res.* 24, 307-314, 1999.

Scaffold

In combinatorial chemistry or parallel synthetic strategy, the common platform which serves the core for synthesis of individual scaffold family members; also the matrix for tissue development such a bone. See Hosse, R.J., Rothe, A., and Power, B.E., A new generation of protein display scaffolds for molecular recognition, *Protein Science* 15, 14-27, 2006; Hammond, J.S., Beckingham, I.J., and Shakesheff, K.M., Scaffolds for liver tissue engineering, *Expert Rev.Med.Devices* 3, 21-27, 2006; Hollister, S.J., Porous scaffold design for tissue engineering, *Nat.Mater.* 4,

518-524, 2005; Li, C., Vepari, C., Jin, H.J., Kim, H.J., and Kaplan, D.L., Electrospun silk-BMP-2 scaffolds for bone tissue engineering, *Biomaterials*, 2006; van Lieshout, M.I., Vaz, C.M., Rutten, M.C., Peters, G.W., and Baaijens, F.P., Electrospinning versus knitting: two scaffolds for tissue engineering of the aortic valve, *J.Biomater.Sci.Polym.Ed.* 17, 77-89, 2006.

Scattering

As in light scattering. Scattering may be elastic where energy is conserved and the scattered electromagnetic waves are of the same frequency as the incident electromagnetic radiation; the frequency of the scattered radiation is different from the incident radiation, the scattering is inelastic. Reflection and refraction are types of light scattering. Turbidimetry and nephelometry are applications of light scattering. Raman spectroscopy is an example of inelastic light scattering.

Sec-dependent

Secretory protein translocation. See Sec-dependent protein translocation across biological membranes. Evolutionary conservation of an essential protein transport pathway, *Molec.Membrane Biol.* 22, 17-28, 2005.

Secretase

The term secretase is used to describe those proteolytic activities involved in the processing of amyloid precursor protein (APP) to yield the soluble circulating amyloid protein. See Stephens, B.J. and Austen, B.M., Characterization of beta-secretase, *Biochem. Soc.Trans* 26, 500-504, 1998; Wolfe, M.S., Secretase targets for Alzheimer's disease: identification and therapeutic potential, *J.Med.Chem.* 44, 2039-2060, 2001; Vassar, R., The beta-secretase, BACE: a prime target for Alzheimer's disease, *J.Mol.Neurosci.* 17, 157-170, 2001; Hooper,N.M. and Turner, A.J., The search for alpha-secretase and its potential as a therapeutic approach to Alzheimer's disease, *Curr.Med.Chem.* 9, 1107-1119, 2002; Pollack, S.J. and Lewis, H., Secretase inhibitors for Alzheimer's disease: challenges of a promiscuous protease, *Curr.Opin.Invest. Drugs* 6, 35-47, 2005. There is a more general definition as a secretase as a "sheddase" responsible for the proteolysis of type I and type II membrane proteins (Hoooper, N.M., Karran, E.H., and Turner, A.J., Membrane protein secretases, *Biochem. J.* 321, 265-279, 1997; Wolfe, M.S., and Kopan, R., Intramembrane proteolysis, *Science* 305, 1119-1123, 2004; The term secretase has been used to describe the activity responsible for the release of TNF from membranes; (Mezyk, R., Browska, M., and Bereta, J., Structure and functions of tumor necrosis factor-alpha converting enzyme, *Acta Biochim.Pol.* 50, 625-645, 2003). There is a relationship between ADAM proteases and secretases (Fahrenholz, F., Gilbert, S., Kojro, E., *et al.*, Alpha-secretase activity of the disintegrin metalloprotease ADAM 10. Influence of domain structure, *Ann.N.Y.Acad.Sci.* 920, 215-222, 2000; Higashiyama, S. and Nanba, D., ADAM-mediated ectodomain shedding of HB-EGF in receptor cross-talk, *Biochim.Biophys.Acta* 1751, 111-117, 2005) See also Gamma(γ)-secretase.

Selectivity and Selectivity Factor

The discrimination shown by a compound in reacting with two or more position on the same compound or several compounds It is quantitatively expressed by ratios of rate constants of the competing reactions, or by the decadic logarithms of such ratios. It also refers to the differential affinity of compounds to a chromatographic matrix. Chromatographic selectivity is a determining factor in resolution; it is noted that the use of selectivity is discouraged in favor of the use of the term separation factors.

SELDI

Surface-enhanced laser/desorption ionization mass spectrometry; ProteinChip®: Tang, N., Tornatore, P., and Weinberger, S.R., Current developments in SELDI affinity technology, *Mass Spectrom.Rev.* 23a, 34-44, 2004.

SELEX

Systematic evolution of nucleic acid ligands (aptamers) by exponential enrichment. This represents an approach to the development of nucleic acid ligands for affinity chromatography and therapeutic aptamers by selection from combinatorial oligonucleotide libraries directed against a putative target. See Klug, S.J. and Famulok, M., All you wanted to know about SELEX, *Mol.Biol.Rep.* 20, 97-107, 1994; Joyce, G.J.. In vitro evolution of nucleic acids, *Curr.Opin. Struct.Biol.* 4, 331-336, 1994; Gold, L., Brown, D., He, Y., *et al.*, From oligonucleotide shapes to genomic SELEX: novel biological regulatory loops, *Proc.Natl.Acad.Sci.USA* 94, 59-64, 1997; Jayasena, S.D., Aptamers: an emerging class of molecules that rival antibodies in diagnostics, *Clin.Chem.* 45, 1628-1650, 1999; Clark, S.L. and Remcho, V.T., Aptamers as analytical reagents, *Electrophoresis* 23, 1335-1340, 2002; Tuerk, C. and Gold, L., Systematic evolution of ligands by expotential enrichment: RNA ligands to bacteriophage T4 DNA polymerase, *Science* 249, 505-510, 1990; Liu, J. and Stormo, G.D., Combining SELEX with quantitative assays to rapidly obtain accurate models of protein-DNA interactions, *Nuc.Acid Res.* 33, e141, 2005; Guthrie, J.W., Hamula, C.L., Zhang, H., *et al.*, Assays for cytokines using aptamers, *Methods* 39, 324-330, 2006; Ulrich, H., RNA aptamers: from basic science towards therapy, *Handb.Exp. Pharmacol.* (173), 305-326, 2006.

Separase

A regulatory protease which initiates the metaphase-anaphase transition by cleavage of the Sec1 subunit of cohesion, a chromosomal protein complex. This is process which is regulated by shugoshin ("guardian spirit"). See Yanagida, M., Cell cycle mechanisms of sister chromatid separation: roles of Cut1/separin and Cut2/securin. *Genes Cells* 5, 1-8, 2000; Amon, A., Together until separin do us part, *Nat.Cell Biol.* 3, E12-E14, 2001; Uhlmann, F., Secured cutting: controlling separase at the metaphase to anaphase transition, *EMBO Rep.* 2, 487-492, 2001; Hearing, C.H. and Nasmyth, K., Building and breaking bridges between sister chromatids, *Bioessays* 25, 1178-1191, 2003; Uhlmann, F., The mechanism of sister chromatid cohesion, *Exp.Cell Res.* 296, 80-85, 2004; Watanabe, Y. and Kitajima, T.S., Shugoshin protects cohesion complexes at centromeres, *Philos Tran. R Soc. Lond.B.Biol.Sci.* 360, 515-521, 2005; Watanabe, Y., Shugoshin: guardian spirit at the centromere, *Curr.Opin.Cell Biol.* 17, 590-595, 2005.

Separation Factor

Designated by the term α and refers to the relative affinity of two components for a chromatographic matrix and related to the resolution. By definition the separation factor is larger than 1 and could be described by the following expression; $\alpha = t_2/t_1$ where t_2 is the elution time for the apex of the more slowly moving solute and t_1 is the elution time for the apex of the more rapidly moving solute. See Chen, Y., Kele, M., Quinones, I., Sellergren, B., and Guiochon, G., Influence of the pH on the behavior of an imprinted polymeric stationary phase – supporting evidence for a binding site model, *J.Chromatog.A* 927, 1-17, 2001; Avramescu, M.E., Borneman, Z., and Wessling, M., Mixed-matrix membrane adsorbers for protein separation, *J.Chromatog.A.* 1006, 2003; Ziomek, G., Kaspereit, M., Jezowski, J., Seidel-Morgenstern, A., and Antos, D., Effect of mobile phase composition on the SMB processes efficiency. Stochastic optimization of isocratic and gradient operation, *J.Chromatog.A* 1070, 111-124, 2005; Lesellier, E. and Tchapla, A., A simple subcritical chromatographic test for an extended ODS high performance liquid chromatography column classification, *J.Chromatog.A.* 1100, 45-59, 2005; Lapointe, J.F., Gauthier, S.F., Pouliot, Y. and Bouchard, C., Selective separation of cationic peptides from a tryptic hydrolyzate of beta-lactoglobulin by electrofiltration, *Biotechnol.Bioeng.* 94, 223-233, 2006.

SERCA

Sarcoplasmic reticulum Ca^{2+} ATPase, responsible for calcium ion transport. See Martonosi, A.N., and Pikula, S., The structure of the Ca^{2+}-ATPase of sarcoplasmic reticulum, *Acta Biochim. Pol.* 50, 337-365, 2003; Strehler, E.E. and Treiman, M., Calcium pumps of plasma membrane and cell interior, *Curr.Mol.Med.* 4, 323-335, 2004.

SEREX

Serological identification of antigens by recombinant expression cloning; see Sahin, U., *et al.*, Human neoplasms elicit multiple specific immune responses in the autologous host, *Proc.Natl.Acad. Sci.USA* 92, 11810-11813, 1995; Chen, Y.-T., *et al.*, A testicular antigen aberrantly expressed in human cancers detected by autologous antibody screening, *Proc.Natl.Acad.Sci. USA*, 94, 1914-1918, 1997; Fernandez, M.F., *et al.*, Improved approach to identify cancer-associated autoantigens, *Autoimmun. Rev.* 4, 230-235, 2005. www.licr.org/SEREX.html; www2.licr.org/CancerImmunomeDB/.

Serial Lectin Affinity Chromatography

The use of a series of two or more lectin affinity chromatography columns of known specificity for the fractionation of oligosaccharides, glycoproteins, or glycopeptides into structurally distinct groups. See Cummings, R.D. and Kornfeld, S., Fractionation of asparagine-linked oligosaccharides by serial lectin-agarose affinity chromatography. A rapid, sensitive, and specific technique, *J.Biol.Chem.* 257,. 11235-11240, 1982; Qiu, R. and Regnier, F.E., Comparative glycoproteomics of N-linked complex-type glycoforms containing sialic acid in human serum, *Anal.Chem.* 77, 7725-7231, 2005.

Serpin

A term now in own right. It was developed as an acronym for serine protease inhibitor (Carroll, R.W. and Travis, J., α-1-antitrypsin and the serpins: variation and countervariation, *Trends Biochem. Sci.* 10, 20-24, 1985). It is considered to be a structurally homologous superfamily (Hunt, L.T. and Dayhoff, M.O., A surprising new protein superfamily containing ovalbumin, antithrombin-III,

and α1-proteinase inhibitor, *Biochem.Biophys.Res.Commun.* 95, 864-871, 1980) of proteins having masses in the range of 40 kDa to 100 kDa. See Gettins, P., Patson, P.A., and Schapira, M., The role of conformational change in serpin structure and function, *Bioessays* 15, 461-467. 1993; Schulze, A.J., Huber, R., Bode,W., and Engh, R.A., Structural aspects of serpin inhibition, *FEBS Lett.* 344, 117-124, 1994; Potempa, J., Korzus, E. and Travis, J., The serpin superfamily of proteinase inhibitors: structure, function, and regulation, *J.Biol.Chem.* 269, 15957-15960, 1994; Lawrence, D.A., The role of reactive-center loop mobility in the serpin inhibitory mechanism, *Adv.Exp.Med.Biol.* 425, 99-108, 1997; Whisstock, J., Skinner, R. and Lesk, A.M., An atlas of serpin conformations, *Trends Biochem.Sci.* 23, 63-67, 1998; Gettins, P.G., Serpin structure, mechanism, and function, *Chem.Rev.* 102, 4751-4804, 2002; Huntington, J.A., Shape-shifting serpins – advantages of a mobile mechanism, *Trends Biochem.Sci.* 31, 427-435, 2006.

Shot Gun Proteomics

Identification of peptides (usually by mass spectrometry) obtained by the enzymatic or chemical digestion of the entire proteome. A naturally occurring protein mixture such as cell extract, blood plasma or other biological fluid is reduced, alkylated, and subjected to tryptic hydrolysis. The tryptic hydrolysis is fractionated by liquid chromatography and analyzed by mass spectrophotometry: Wolters, D.A., Washburn, M.P., and Yates, J.R., III, An automated multidimensional protein identification technology for shotgun proteomics, *Anal.Chem.* 73, 5683-5690, 2001; Liu, H., Sadygov, R.G., and Yates, J.R., III, A model for random sampling and estimation of relative protein abundance in shotgun proteomics, *Anal.Chem.* 76, 4193-4201, 2004.

Shugoshin

A protein family having a role in the centromeric protection of cohesion; protects the centromeric cohesion at meiosis I by inhibiting the action of separase on cohesion. See Kitajima, T.S., Kawashima, S.A., and Watanabe, Y., The conserved kinetochore protein shugoshin protects centromeric cohesion during meiosis, *Nature* 427, 510-517, 2005; Salic, A., Waters, J.C., and Mitchison, T.J., Vertebrate shugoshin links sister centromere cohesion and kinetochore microtubule stability in mitosis, *Cell* 118, 567-578, 2004; Goulding, S.E. and Earnshaw, W.C., Shugoshin: a centromeric guardian senses tension, *Bioessays* 27, 588-591, 2005; Watanabe, Y., Shugoshin: guardian spirit at the centromere, *Curr. Opin.Cell.Biol.* 17, 590-595, 2005; Stemmann, O., Boos, D., and Gorr, I.H., Rephrasing anaphase: separase FEARs shugoshin, *Chromosoma* 113, 409-417, 2005; Mcgee, P., Molecular biology: chromosome guardian on duty, *Nature* 441, 35-37, 2006.

Sigma Factor

A factor which binds to RNA polymerase and provides specificity for the transcriptional process. It also provides for DNA strand separation during the transcriptional process. Sigma factors could be considered to be subunits of the RNA polymerase enzyme. See Kazmierczak, M.J., Wiedmann, M., and Boor, K.J., Alternative sigma factors and their roles in bacterial virulence, *Microbiol. Mol.Biol.Rev.* 69, 527-543, 2005; Mooney, R.A., Darst, S.A., and Landick, R., Sigma and RNA polymerase: an on-again, off-again relationship? *Mol.Cell.* 20, 335-345, 2005; Kill, K., Binnewies, T.T., Sicheritz-Ponten, T., *et al.*, Genome update: sigma factors in 240 bacterial genomes, *Microbiology* 151, 3147-3150, 2005;

Wigneshweraraj, S.R., Burrows, P.C., Bordes, P., *et al.*, The second paradigm for activation of transcription, *Prog.Nucl.Acid Res.Mol. Biol.* 79, 339-369, 2005.

Signalosome

An endosome with an active signaling component which is transported to a juxtanuclear position; Perret, E., Lakkaraju, A., Deborde, S., Schreiner, R., and Rodriguez-Boulan, E., *Curr.Opin. Cell Biol.* 17, 423-434, 2005.

Signal Recognition Particle

A targeting chaperone involved in the transmembrane transport of proteins; involves the recognition of the signal peptide. See Pool, M.R, Signal recognition particles in chloroplasts, bacteria, yeast, and mammals, *Molec.Membrane Biol.* 22, 3-15, 2004.

Signature Domain

An amino acid sequence which is closely conserved within a group of proteins and is considered unique to that group of proteins which is also called a protein family. The sequences may or may not have homologous function (see Khuri, S., Bakker, F.T., and Dunwell, J.M., Phylogeny, function, and evolution of the cupins, a structurally conserved, functionally diverse super-family of proteins, *Mol.Biol.Evolution* 18, 593-605, 2001). In this sense, the use of the term signature is related to historical use of this term to describe a physical property or feature of a plant or other natural object as an indication of pharmacological impact because of relation of such feature to the body part (see *Oxford English Dictionary*, Oxford University Press, Oxford, United Kingdom, 1989; Webster's Third International Dictionary, Unabridged, 1996). One of the most studied example is the C1q domain (See Bérubé, N.G., Swanson, X.H., Bertram, M.J., *et al.*, Cloning and characterization of CRF, a novel C1q-related factor, expressed in areas of the brain involved in motor function, *Mol.Brain Res.* 63, 233-240, 1999; Kishore, U., Gaboriaud, C., Waters, P., *et al.*, C1q and tumor necrosis factor superfamily: modularity and versatility, *Trends Immunol.* 25, 551-561, 2004. For general considerations, see Tousidou, E., Nanopoulos, A., and Manolopoulos, Y., Improved methods for signature-tree construction, *The Computer Journal* 43, 301-314, 2000; Ye, Y. and Godzik, A., Comparative analysis of protein domain organization, *Genome Res.* 14, 343-353, 2004).

Single-chain Fv fragment (scFv)

A synthetic (usually recombinant) peptide/protein composed of the V_L and V_H domains of an antibody linked by a peptide. It is relatively small (30 kDa) and as a single peptide chain is easily expressed in bacterial systems. It is possible to express the scFv inside the cell (intracellular expression) as intrabodies for analytical and therapeutic purposes. It also possible to increase the avidity of these engineered fragments by dimerization to form diabodies and higher order polymers. Also on occasion referred to as minibodies. See Pluckthun, A. and Pack, P., New protein engineering approaches to multivalent and bispecific antibody fragments, *Immuntechnology* 3, 93-105, 1997; Hudson, P.J. and Kortt, A.A., High avidity scFv multimers; diabodies and triabodies, *J.Immunol.Methods* 231, 177-189, 1999; de Graaf, M., van der Meulen-Mulleman, I.H., Pinedo, H.M., and Haisma, H.J., Expression of scFvs and scFv fusion proteins in eukaryotic cells,

Methods Mol.Biol. 178, 379-387, 2002; Lennard, S., Standard protocols for the construction of scFv libraries, *Methods Mol. Biol.* 178, 59-71, 2002; Lunde, E., Lauvrak, V., Rasmussen, I.B., *et al.*, Troybodies and pepbodies, *Biochem.Soc.Trans.* 30, 500-506,2002; Leath, C.A., 3rd., Douglas, J.T., Curiel, D.T., and Alvarez, R.D., Single-chain antibodies: A therapeutic modality for cancer gene therapy, *Int.J.Oncol.* 24, 765-771, 2004; Visintin, M., Meli, G.A., Cannistraci, I., and Cattnaeo, A., Intracellular antibodies for proteomics, *J.Immunol.Methods* 290, 135-153, 2004; Lobato, M.N. and Rabbitts, T.H., Intracellular antibodies as specific reagents for function ablation: future therapeutic molecules, *Curr.Mol.Med.* 4, 519-528, 2004; Holliger, P. and Hudson, P.J. Engineered antibody fragments and the rise of single domains, *Nat.Biotechnol.* 23, 1126-1136, 2005; Röthlisberger, D. Honengger, A., and Plückthun, A., Domain interactions in the Fab fragment: a comparative evaluation of the single-chain Fv and Fab format engineered with variable domains of different stability, *J.Mol.Biol.* 347, 773-789, 2005; Chadd, H.E. and Chamow, S.M., Therapeutic antibody expression technology, *Curr.Opin.Biotechnol.* 12, 188-194, 2001; Fong, R.B., Ding, Z., Hoffman, A.S., and Stayton, P.S., Affinity separation using as Fv antibody fragment – "smart" polymer conjugates, *Biotechnol. Bioengin.* 79, 271-276, 2002; Kim, S.-E., Expression and purification of recombinant immunotoxin – a fusion protein stabilizes a single-chain Fv (scFv) in denaturing condition, *Prot. Express.Purif.* 27, 85-89, 2003; Kerschbaumer, R.J., Hirschl, S., Kaufmann, A., *et al.*, Single-chain Fv fusion proteins suitable as coating and detecting reagents in a double antibody sandwich enzyme-linked immunosorbent assay, *Anal.Biochem.* 249, 219-227, 1997; Sinacola, J.R. and Robinson, A.S., Rapid refolding and polishing of single-chain antibodies from *Escherichia coli* inclusion bodies, *Protein Express.Purif.* 26, 301-308, 2002; Krebs, B., Rauchenberger, R., Reiffert, S., *et al.*, High-throughput generation and engineering of recombinant human antibodies, *J.Immunol. Methods* 254, 67-84, 2001.

Small interfering RNA (siRNA)

A short-length double-stranded RNA (21-27 nucleotides in length) derived from intracellular double-stranded RNA by the action of specific endonucleases such as RNAse III (see Dicer, Drosha). The siRNA stimulates the cellular machinery to cut up messenger RNA thus inhibiting the process of transcription; this is process called knockdown. See Myers, J.W. and Ferrell, J.E., Jr., Silencing gene expression with Dicer-generated siRNA pools, in RNA Silencing. Methods and Protocols, ed. G.G. Carmichael, Humana Press, Totowa, New Jersey, Chapter 8, pp. 93-196, 2005; Aravin, A. and Tuschi, T., Identification and characterization of small RNAs involved in RNA silencing, *FEBS Lett.* 579, 5830-5840, 2005; Kim, V.N., Small RNAs: Classification, biogenesis, and function, *Molecules and Cells* 19, 1-15, 2005; Bass, B.L., Double-stranded RNA as a template for gene silencing, *Cell* 101, 235-238, 2000.

Small temporal RNA

Messenger RNAs which are expressed only a specific stage in development and encode proteins involved in specific developmental timing events (see Moss, E.G., RNA interference: It's a small RNA world, *Current Biology* 11, R772-R775, 2001; Pasquinelli, A.E., Reinhart, B.J., Slack, R., *et al.*, Conservation of the sequence and temporal expression of *let-7* heterochronic regulatory RNA, *Nature* 408, 86-89, 2000).

Small nuclear ribonucleoprotein (RNA plus protein) particle

Component of the spliceosome, the intron-removing apparatus in eukaryotic nuclei. See: Graveley, B.R., Sorting out the complexity of SR protein functions, *RNA* 6, 1197-1211, 2000; Will, C.L. and Luhrmann, R., Spliceosomal UsnRNP biogenesis, structure and function, *Curr.Opin.Cell.Biol.* 13, 290-301, 2001; Turner, I.A., Norman, C.R., Churcher, M.J., and Newman, A.J., Roles of the U5 snRNP in spliceosome dyanamics and catalysis, *Biochem.Soc. Trans.* 32, 928-931, 2004.

Smart Probes

Usually a nucleic acid probe which emits a signal only when bound to a specific target. An example is a molecular beacon. See Stöhr, K., Häfner, B., Nolte. O., Wolfrum, J., Sauer, M., and Herten, D.-P., Species-specific identification of Mycobacterial 16S rRNA PCR amplicons using smart probes, *Anal.Chem.* 77, 7195-7203, 2005. There are other example of smart probes including proteins (Wunder, A., Tung, C.-H., Müller-Ladner, U., Weissleder, R., and Mahmood, U., In vivo imaging of protease activity in arthritis, A novel approach for monitoring treatment response, *Arthritis & Rheumatism* 50, 2459-2465, 2004) and chiral compounds (Tsukube, H. and Shinoda, S., Lanthanide complexes as smart CD probes for chirality sensing of biological substrates, *Enantiomer* 5, 13-22, 2000). "Smart" contrast reagents have also been developed for magnetic resonance studies (Lowe, M.P., Activated MR contrast reagents, *Curr.Pharm.Biotechnol.* 5, 519-528, 2004).

SNARE Proteins

SNARE (soluble NSF attachment protein receptors) participate in eukaryotic membrane fusion. It is suggested that vesicle SNARE proteins fuse with target SNARE proteins during processes such as exocytosis. Most SNARE proteins have a C-terminal transmembrane domain, a substantial cytosolic domain, and a variable *N*-terminal domain (brevin domain, longin domain, YKT-domain, which regulate membrane fusion reactions. SNARE protein can be classified as Q-SNAREs or R-SNARES depending on amino acid sequence homology. There are other classification systems as well. See Ferro-Novick, S. and Jahn, R., Vesicle fusion from yeast to man, *Nature* 370, 191-193, 1994; Rothman, J.E. and Warren, G., Implications of the SNARE hypothesis for intracellular membrane topology and dynamics, *Curr.Biol.* 4, 220-233, 1994; Morgan, A., Exocytosis, *Essays Biochem.* 30, 77-95, 1995; Burgoyne, R.D., Morgan, A., Barnard, A.J., Chamberlain, L.H., Glenn, D.E., and Kibble, A.V., SNAPs and SNAREs in exocytosis in chromaffin cells, *Biochem.Soc. Trans.* 24, 653-657, 1996; Wilson, M.C., Mehta, P.P., and Hess, E.J., SNAP-25, ensnared in neurotransmission and regulation of behavior, *Biochem.Soc. Trans.* 24, 670-676, 1996; Hya, J.C. and Scheller, R.H., SNAREs and NSF in targeted membrane fusion, *Curr.Opin.Cell Biol.* 9, 505-512, 1997; Pelham, H.R., SNAREs and the secretory pathway-lessons from yeast, *Exp.Cell Res.* 247, 1-8, 1997; Whiteheart, S.W., Schraw, T. and Matleeva, E.A., *N*-ethylmaleimide sensitive factor (NSF) structure and function, *Int.Rev.Cytol.* 207, 71-112, 2001; Hay, J.C., SNARE complex structure and function, *Exp.Cell Res.* 271, 10-21, 2001; Dietrich, L.E.P., Boedinghaus, C., LaGrassa, J.T., and Ungermann, C., Control of eukaryotic membrane fusion by *N*-terminal domiains of SNARE proteins, *Biochim.Biophys.Acta* 1641, 111-119. 2003; Hong, W., SNAREs and traffic, *Biochim. Biophys.Acta* 1744, 493-517, 2005; Montecucco, C., Schiavo, G. and Pantano, S., SNARE complexes and neuroexocytosis: how many, how close? *Trends Biochem.Sci.* 30, 367-372, 2005.

Soft-Ionization

Ionization techniques such as fast atom bombardment (FAD), electrospray ionization (ESI), or matrix-assisted laser desorption/ionization (MALDI) that initiate the desorption and ionization of non-volatile thermally labile compounds such as proteins or peptides. See Fenn, J.B., Mann, M., Meng, C.K., *et al.*, Electrospray ionization for mass spectrometry of large biomolecules, *Science* 246, 64-71, 1989; Reinhold, V.N., Reinhold, B.B., and Costello, C.B., Carbohydrate molecular weight profiling, sequence, linkage, anad branching data: ES-MS and CID, *Anal.Chem.* 67, 1772-1784, 1995; Griffiths, W.J., Jonsson, A.P., Liu, S., Rai, D.K. and Wang, Y., Electrospray and tandem mass spectrometry in biochemistry, *Biochem.J.* 355, 545-561, 2001; Schalley, C.A., Molecular recognition and supramolecular chemistry in the gas phase, *Mass Spectrom.Rev.* 20, 253-309, 2001; Kislinger, T., Humeny, A., and Pischetsrider, M., Analysis of protein glycation products by matrix-assisted laser desorption ionization time-of-flight mass spectrometry, *Curr.Med.Chem.* 11, 2185-2193, 2004; Laskin, J. and Futrell, J.H., Activation of large ions in FT-ICR mass spectrometry, *Mass Spectrom.Rev.* 24, 135-167, 2005; Bolbach, G., Matrix-assisted laser desorption/ionization analysis of non-covalent complexes: fundamentals and applications, *Curr.Pharm.Des.* 11, 2535-2357, 2005; Baldwin, M.A., Mass spectrometers for the analysis of biomolecules, *Methods Enzymol.* 402, 3-48, 2005.

Somatic Hypermutation

The increased mutation in the variable region of immunoglobulin genes which allows for diversity of immune recognition. See Steele, E.J., Rothenfluh, H.S. and Both, G.W., Defining the nucleic acid substrate for somatic Hypermutation, *Immuno.Cell Biol.* 70, 129-144, 1992; Jacob, J., Miller, C. and Kelsoe, G., In situ studies of the antigen-driven somatic Hypermutation of immunoglobulin genes, *Immunol.Cell Biol.* 70, 145-152, 1992; George, J. and Clafin, L., Selection of B cell clones and memory B cells, *Semin.Immunol.* 4, 11-17, 1992; Neuberger, M.S. and Milstein, C.S., Somatic Hypermutation, *Curr.Opin.Immunol.* 7, 24-254, 1995; Hengstschlager, M., Maizels, N. and Leung, H., Targeting and regulation of immunoglobulin gene somatic Hypermutation and isotype switch recombination, *Prog.Nucleic Acid Res.Mol. Biol.* 50, 67-99, 1995; Steele, E.J., Rothenflug, H.S., and Blanden, R.V., Mechanism of antigen-driven somatic Hypermutation of rearranged immunoglobulin V(D)J genes in the mouse, *Immuno. Cell Biol.* 75, 82-95, 1997; Rajewsky, K., Clonal selection and learning in the antibody system, *Nature* 381, 751-758, 1996; Storb, U., Peters, A., Klotz, E., *et al.*, *Cis*-acting sequences that affect somatic hypermutation of Ig genes, *Immunol.Rev.* 162, 153-160, 1998; Neuberger, M.S., Ehrenstein, M.R., Klix, N. *et al.*, Monitoring and interpreting the intrinsic features of somatic Hypermutation, *Immunol.Rev.* 162, 107-116, 1998; Kuppers, R. Goossens, T., and Klein, U., The role of somatic Hypermutation in the generation of deletions and duplications in human Ig V region genes and chromosomal translocations, *Curr.Top.Microbiol. Immunol.* 246, 193-198, 1999; Harris, R.S., Kong, Q., and Maizels, N., Somatic Hypermutation and the three R's: repair, replication and recombination, *Mutat.Res.* 436, 157-178, 1999; Jacobs, H. and Bross, L., Towards an understanding of somatic hypermutation, *Curr.Opin.Immunol.* 13, 208-218, 2001; Seki, M., Gearhart, P.J., and Wood, R.D., DNA polymerases and somatic hypermutation of immunoglobulin genes, *EMBO Rep.* 6, 1143-1148, 2005; Neuberger, M.S., Di Noia, J.M., Beale, R.C., *et al.*, Somatic hypermutation at A.T pairs: polymerase error versus dUTP incorporation, *Nat.Rev.Immunol.* 5, 171-178, 2005.

Southern Blotting

The used of a complement oligonucleotide/polynucleotide to identify denatured DNA transferred by absorption from an agarose gel to another matrix, such a nitrocellulose membrane. See Southern, E.M., Detection of specific sequences among DNA fragments separated by gel electrophoresis, *J.Mol.Biol.* 98, 503-517, 1975; Southern, E.M., Detection of specific sequences among DNA fragments separated by gel electrophoresis, 1975, *Biotechnology* 24, 122-139, 1992; Southern, E.M., Blotting at 25, *Trends Biochem.Sci.* 25, 585-588, 2000; Darbre, P.D., *Introduction to Practical Molecular Biology*, Wiley, Chichester, UK, 1988; Issac, P.G., *Protocols for Nucleic Acid Analysis by Nonradioactive Probes*, Humana, Totowa, New Jersey, USA, 1994; Darling, D.C. and Brickell, P.M., *Nucleic Acid Blotting; The Basics*, Oxford, Oxford, UK, 1994; Kelly, K.F., Southern Blotting, *Proc.Nutr.Soc.* 55, 591-597, 1996; Keichle, F.L., DNA technology in the clinical laboratory, *Arch.Pathol.Lab.Med.* 123, 1151-1153, 1999; Porchet, N. and Aubert, J.P., Southern blot analysis of large DNA fragments, *Methods Mol.Biol.* 125, 313-321, 2000; Voswinkel, J. and Gause, A., From immunoglobulin gene fingerprinting to motif-specific hybridization: advances in the analysis of B lymphoid clonality in rheumatic diseases, *Arthritis Res.* 4, 1-4, 2002; Wong, L.J. and Boles, R.G., Mitochondrial DNA analysis in clinical laboratory diagnostics, *Clin.Chim.Acta* 354, 1-20, 2005; Rose, M.G., Degar, B.A., and Berliner, N., Molecular diagnostics of malignant disorders, *Clin.Adv.Hematol.Oncol.* 2, 650-660, 2004.

Southwestern Blotting

An analytical procedure used to identify the specific binding of a nucleic acid sequence to a protein which uses a technical approach similar to Southern Blot and Western Blot. A protein mixture is separated by electrophoresis and the resulting electrophoretograms is transferred to a PVDF membrane electrophoresis. The proteins are renatured on the membrane and a ^{32}P-labeled oligonucleotide probe of defined sequence is used to identify specific binding protein(s). Other labels such as cyanine dyes or fluorescein can be used for the oligonucleotide probe. See Zhu, Q., Andrisani, O.M., Pot, D.A., and Dixon, J.E., Purification and characterization of a 43-kDa transcription factor required for rat somatostatin gene expression, *J.Biol.Chem.* 264, 6550-6556, 1989; Ogura, M., Takatori, T., and Tsuro, T., Purification and characterization of NF-R1 that regulates the expression of the human multidrug resistance (MDR1) gene, *Nucleic Acids Res.* 20, 5811-5817, 1992; Kwast-Welfeld, J., de Belle, I., Walker, P.R., Whitfield, J.F., and Sikorska, M., Identification of a new cAMP response element-binding factor by southwestern blotting, *J.Biol.Chem.* 268, 19851-19585, 1993; Liu, Z. and Jacob, S.T., Characterization of a protein that interacts with the rat ribosomal gene promoter and modulates RNA polymerase I transcription, *J.Biol.Chem.* 269, 16618-16625, 1994; Handen, J.S. and Rosenberg, H.F., An improved method for Southwestern blotting, *Front.Biosci.* 2, c9-c11, 1997; Coffman, J.A. and Yuh, C.H., Identification of sequence-specific DNA binding proteins, *Methods Cell Biol.* 74, 653-675, 2004; Fedorov, A.V., Lukyanov, D.V., and Podgornaya, O.T., Identification of the proteins specifically binding to the rat LINE1 promoter, *Biochem.Biophys.Res.Commun.* 340, 553-559, 2006. There is southwestern approach used for histochemistry (Hishikawa, Y., Damavandi, E., Izumi, S., and Koji, T., Molecular histochemical analysis of estrogen receptor alpha and beta expression in the mouse ovary: in situ hybridization and Southwestern histochemistry, *Med.Electron Microsc.* 36, 67-73, 2003) and for ELISA (Fukuda, I., Nishiumi, S., Yabushita, Y., *et al.*, A new southwestern chemistry-based ELISA for detection of aryl hydrocarbon receptor transformation: application to the screening of its receptor agonists and antagonists, *J.Immunol. Methods* 287, 187-201, 2004).

Specific Heat

The amount of heat required to raise the temperature of one gram of a substance by 1°C; specific heat of water is one calorie (4.184 joule); **Heat of fusion** is the amount of thermal energy to melt one mole of a substance at the melting point; also referred to as latent heat of fusion, kcal/mole or kJ/mole. **Heat of vaporization** is the amount of energy required to convert one mole of a substance to vapor at the boiling point; also referred to as the latent heat of vaporization, kcal/mole or kJ/mole.

Specificity

In assay validation, the ability of an assay to recognize a single analyte in a sample which might contain closely related species; for example, in DNA microarray assays, specificity would be the ability of a probe to bind to a unique target sequence and produce a signal proportional of the amount of that specific target sequence only. Also referred to as selectivity. In statistics, specificity is the proportion of negative tests to the total number of negative tests.

Spectrometry/Spectroscopy

The interaction of electromagnetic radiation with materials including scattering, absorption, and emission. It does not include chemical effects such as bond formation or free radical formation. It does include some aspects of photochemistry which is a specialized form of energy transduction. See Campbell, I.D. and Dwek, R.A., *Biological Spectroscopy*, Benjamin Cummings, Menlo Park, California, USA, 1984; Stuart, B., *Infrared Spectroscopy*, John Wiley & Sons, Ltd., Chichester, United Kingdom, 2004.

Spectrum

A pattern of emissions from a particle following the application of energy. The emissions may be in form of electromagnetic waves such as observed in spectroscopy or in the form of mass such as that observed in mass spectrometry.

Sp1-like transcription factors

A family of zinc finger transcription factors in mammalian cells which binds to GC-rich promoter element. Originally described for SV40 virus. See Dynan, W.S. and Tjian, R., Isolation of transcription factors that discriminate between different promoters recognized by RNA polymerase II, *Cell* 32, 669-680, 1983; Lomberk, G. and Urrutia, R., The family feud: turning off Sp1 by Sp1-like KLF proteins, *Biochem.J.* 392, 1-11, 2005.

Spliced-Leader Trans-Splicing

Spliced-leader trans-splicing is process mediated by a spiceosome where a short RNA sequence derived from the 5'-end of a non-mRNA to an acceptor site (3'-spice acceptor site) on a pre-RNA molecule. As a result, a diverse group of mRNA molecules in an organism acquire a common 5'-sequence. This process does appear to occur in the same nuclear location as cis-splicing. See Murphey, W.J., Watkins, K.P., and Agabian, N., Identification of a novel Y branch structure as an intermediate in trypanosome mRNA processing: evidence for trans splicing, *Cell* 47, 517-525, 1986; Bruzik, J.P., Van Doren, K., Hirsh, D., and Steitz, J.A., Trans splicing involves a novel form of small nuclear ribonucleoprotein particles, *Nature* 335, 559-562, 1988; Layden, R.E. and Eisen, H., Alternate trans-splicing in *Trypanosoma equiperdum*: implication for splice site selection, *Mol.Cell.Biol.* 8, 1352-1360, 1988; Hastings, K.E.M., SL trans-splicing: easy come or easy go? *Trends Genet.* 21,240-247, 2005.

Spliceosome

A complex of RNA and protein components which function to assist the process of RNA splicing in ribosomes. Prokaryote RNA mRNA is less complex than eukaryotic mRNA and are not subject to RNA splicing. Eukaryotic RNA species that participate in spliceosome function include U1, U2, U4, U5 and U6. These RNA species are rich in uridine which recognize species sequences at the 5' and 3' sites on the pre-mRNA. The regions between these specific sites is excised that the two remaining exons are joined together in the splicing process. See: Robash, M. and Seraphin, B., Who's on first? The U1 snRNP-5' splice site interaction and splicing, *Trends in Biochem.Sci.* 16, 187-190, 1991; Garcia-Blanco, M.A., Messenger RNA reprogramming by spliceosome-mediated RNA trans-splicing, *J.Clin.Invest.* 112, 474-480, 2003; Kramer, A., Frefoglia, F., Huang, C.J., Malhaupt, F., Nesic, D., and Tanackovic, G., Structure-function analysis of the U2 snRNP-associated splicing factor SF3a, *Biochem.Soc. Trans.* 33, 439-442, 2005.

Splicing Silencer(s)

Weakly interacting *cis-* and *trans-* factors which repress constitutive and alternative splicing during mRNA processing. There are exonic spicing silencers (ESS) and intronic splicing silencers (ISS). Distinct from transcriptional silencing also known as transcriptional repression. See Staffa, A. and Cochrane, A., Identification of positive and negative splicing regulatory elements within the terminal tat-rev exon of human immunodeficiency virus type 1, *Mol.Cell.Biol.* 15, 4597-4605, 1995; Amendt, B.A., Si, Z.H., and Stoltzfus, C.M., Presence of exon splicing silencers with human immunodeficiency virus type 1 tat exon 2 and tat-rev exon 3: evidence for inhibition mediated by cellular factors, *Mol.Cell.Biol.* 15, 4606-4615, 1995: Chew, S.L., Baginsky, L., and Eperon, I.C., An exonic splicing silencer in the testes-specific DNA ligase III beta exon, *Nucleic Acids Res.* 28, 402-410, 2000; Puzzoli, U. and Sironi, M., Silencers regulate both constitutive and alternative splicing events in mammals, *Cell.Mol. Life Sci.* 62, 1579-1604, 2005; Paca-Uccaralertkun, S., Damgaard, C.K., Auewarakul, P., *et al.*, The effect of a single nucleotide substitution in the splicing silencer in the tat/rev intron on HIV type 1 envelope expression, *AIDS Res.Human Retroviruses* 22, 76-82, 2006.

SR Family of Proteins

A family of phylogenetically conserved proteins which are essential cofactors in the splicing which occurs during the maturation of messenger RNA. SR proteins are essential for both constitutive and alternative splicing events. SR proteins are characterized by the presence of an *N*-terminal RNA recognition motif or motifs and a C-terminal region characterized by repeated arginine/serine residues. See Birney, E., Kumar, S., and Krainer, A.R., Analysis of the RNA-recognition motif and RS and RBB domains: conservation in metazoan pre-mRNA splicing factors, *Nuc.Acids Res.* 25, 503-5816, 1993; Ramchatesingh, J., Zahler, A.M., Neugebauer, K.M., Roth, M.B., and Cooper, T.A., A subset of SR proteins activates splicing of the cardiac troponin T alternative exon by direct interactions with an exonic enhancer, *Mol. Cell.Biol.* 15, 4898-4907, 1995; McNally, L.M. and McNally, M.T., SR protein splicing factors interact with the Rous sarcoma virus negative regulator of splicing elements, *J. Virol.* 70, 1163-1172, 1996; Katsarou, M.E., Papakyriakou, A., Katsaros, N., and Scorilas, A., Expression of the C-terminal domain of novel human SR-A1 protein: interaction with the CTD domain of RNA polymerase II, *Biochem.Biophys.Res. Commun.* 334, 61-68, 2005; Zahler, A.M., Purification of SR protein splicing factors, *Methods.Mol.Biol.* 118, 419-432, 1999; Sanford, J.R., Ellis, J., and Cáceres, J.F., Multiple roles of arginine/serine-rich splicing factors in RNA processing, *Biochem.Soc.Trans.* 33, 443-446, 2005; Rasheva, V.I., Knight, D., Borko, P., Marsh, K., and Frolov, M.V., Specific role of the SR protein splicing factors B52 in cell cycle control in *Drosophila*, *Mol.Cell.Biol.* 26, 3468-3477, 2006.

Staining

A process by which contrast is introduced into a sample such as a tissue section or electrophoretograms. In positive staining, the item of interest is "staining"(absorbs the stain); in negatives staining, the item of interest is unreactive and the background absorbs the stain.

Standard Conditions (Standard State)

1 atm, 25°C (298°K)

Standard Electrode Potential

The value ($E°$) for the standard electromotive force of a cell in which hydrogen under standard conditions is oxidized to hydronium ions (solvated protons) at the left hand electrode. This values is used a standard to measure electrode potentials.

Standard Free Energy

A thermodynamic function designated G (After Walter Gibbs, frequently referred to as the Gibbs free energy). The change in G (ΔG) for a given reaction provides the information on the amount of energy derived from the reaction and is a product of the changes in enthalpy and entropy: $\Delta G = \Delta H - T\Delta S$. The standard free energy designated $\Delta G°$ indicates the values are those obtained for standard conditions. ΔG is negative for a thermodynamically favorable reaction. See enthalpy and entropy.

Stark Effect

The effect of an electrical field on the absorption/emission of spectra of a probe such as fluorescein or a coumarin derivative, It is derived from the interaction of the induced dipole(s) in the probe interacting with the charged group. See Sitkoff, D., Lockhart, D.J., Sharp, K.A., and Honig, B., Calculation of electrostatic effects at the amino terminal of an α helix, *Biophys.J.* 67, 2251-2260, 1994; Pierce, D.W. and Boxer, S.A., Stark effect spectroscopy of tryptophan, *Biophysical J.* 68, 1583-1591, 1995; Klymchenko, A.S., Avilov, S.V., and Demchenko, A.P., Resolution of Cys and Lys labeling of α-crystallin with site-sensitive fluorescent 3-hydroxyflavone dye, *Anal.Biochem.* 329, 43-57, 2004.

Statistical Power

The probability of detecting a true effect of a particular size; equal to 1-false negative rate. Power increases as the random error of a procedure decreases.

Steroid Hormone Receptor (SHR)

Receptors for steroid hormones located in the nucleus. These are ligand-activated transcription factors. See Lavery, D.N. and McEwan, I.J., Structure and functions of steroid receptor AF1 transactivation domains: induction of active conformations, *Biochem. J.* 391, 449-464, 2005.

Stochastic

Involving or containing random errors.

Stochastic Process

A process consisting of a series of random variables (x_t) where t assumes values in a certain range of T. *The Cambridge Dictionary of Statistics*, ed. B.S. Everitt, Cambridge University Press, Cambridge, United Kingdom, 1998.

Structural Biology

Study of the secondary, tertiary, and higher structures of proteins in the proteome including but limited to the use of crystallography, nuclear magnetic resonance and electron microscopy. See Smith, C.U.M., *Molecular Biology: A Structural Approach*, MIT Press, Cambridge, MA, 1968; Devons, S., *Biology and the Physical Sciences*, Columbia University Press, New York, NY, 1969; Rhodes, D. and Schwabe, J.W., Structural biology. Complex behavior, *Nature* 352, 478-479, 1991; Riddihough, G., Structural biology. Picture an enzyme at work, *Nature* 362, 793, 1993; Diamond, R., *Molecular Structures in Biology*, Oxford University Press, Oxford, UK, 1993; *The Encyclopedia of Molecular Biology*. ed. J.C. Kendrew and E. Lawrence, Blackwell Science, Cambridge, MA. 1994; Waksman, G. and Caparon, M., *Structural Biology of Bacterial Pathogenesis*, ASM Press, Washington, DC, 2005; Weiner, S., Sagi, I., and Addadi, L., Structural biology. Choosing the crystallization path less traveled, *Science* 309, 1027-1028, 2005; Sundstrom, S. and Martin, N., *Structural Genomcis and High Throughput Structural Biology*, Taylor & Francis, Boca Raton, FL, 2006; Chiu, W., Baker, M.L., and Almo, S.C., Structural biology of cellular machines, *Trends Cell Biol.* 16, 144-150, 2006; Aravind, L., Iyer, L.M., and Koonin, E.V., Comparative genomics and structural biology of the molecular innovations of eukaryotes, *Curr.Opin.Struct.Biol.* 16, 409-419, 2006.

Structural Genomics

Focuses on the physical aspects of the genome through the construction and comparison of gene maps and sequences, as well as gene discovery, localization, and characterization; determination of the three-dimensional structures of gene products using x-ray crystallography and NMR; known in a previous life as crystallography. See Bourne, P.E. and Weissig, H., *Structural Bioinformatics*, Wiley-Liss, Hoboken, NJ, 2003; Bernardi, G., *Structural and Evolutionary Genomcs: Natural Selection in Genome Evolution*, Elsevier, Amsterdam, Netherland, 2004; Sundstrom, M. and Norin, M., *Structural Genomics and High Throughput Structural Biology*, Taylor & Francis, Boca Raton, FL, 2006. See also Kim, S.H., Shining a light on structural genomics, *Nat.Struct.Biol.* 5(Suppl), 643-645, 1998; Skolnick, J., Fetrow, J.S., and Kolinski, A., Structural genomics and its importance for gene function analysis, *Nat.Biotechnol* 18, 283-287, 2000; Burley, S.K. and Bonnano, J.B., Structural genomics of proteins from conserved biochemical pathways and processes, *Curr.Opin.Struct.Biol.* 12, 383-391, 2002; Staunton, D., Owen, J., and Cambell, I.D., NMR and structural genomics, *Acc.Chem.Res.* 36, 207-214, 2003; Burley, S.K. and Bonanno, J.B., Structural genomics, *Methods Biochem.Anal.* 44, 591-612, 2003; Goldsmith-Fishman, S. and Honig, B., Structural genomics: computational methods for structure analysis, *Protein Sci.* 12, 1813-1821, 2003; Schmid, M.B., Seeing is believing: the impact of structural genomics on antimicrobial drug discovery, *Nat.Rev.Microbiol.* 2, 739-746, 2004; Lundstrom, K., Structural genomics of GPCRs, *Trends Biotechnol.* 12, 103-108, 2005; Chandonia, J.M. and Brenner, S.A., The impact of structural genomics: expectations and outcomes, *Science* 311, 347-351, 2006.

Structural Proteomics

Study of the primary, secondary, and tertiary structure of the proteins in a proteome; functional predictions from primary structure. See Norin, M. and Sundstrom, M., Structural proteomics: developments in structure-to-function, predictions, *Trends Biotechnol.* 20, 79-84, 2002; Mylvagenam, S.E., Prahbakaran, M., Tudor, S.S., *et al.*, Structural proteomics: methods in deriving protein structural information and issues in data management, *Biotechniques* March Suppl., 42-46, 2002; Sali, A., Glaseser, R., Earnest, T., and Baumeister, W., From words to literature in structural proteomics, *Nature* 422, 216-225, 2003; Lefkovits, I., Functional anad structural proteomics: a critical appraisal, *J.Chromatog.B.Anal.Technol.Biomed.Life.Sci.* 787, 1-10, 2003; Smith, R.D., and Veenstra, T.D., *Proteome Characterization and Proteomics*, Academic Press, San Diego, CA, 2003; Kamp, R.M., and Calvete, J.J., *Methods in Proteome and Protein Analysis*, Springer, Berlin, 2004; Jung, J.W. and Lee, W., Structure-based functional discovery of proteins: structural proteomics, *J.Biochem.Mol.Biol.* 37, 28-34, 2004; Yakunin, A.F., Yee, A.A., Savchenko, A., *et al.*, Structural proteomics: a tool of genome annotation, *Curr.Opin.Chem.Biol.* 9, 42-48, 2004; Chan, K. and Fernandez, D., Patent prosecution in structural proteomics, *Assay Drug Dev.Technol.* 2, 313-319, 2004; Liu, H.L. and Hsu, J.P., Recent developments in structural proteomics for protein structure determination, *Proteomics* 5, 2056-2068, 2005; Vinarov, D.A. and Markley, J.L, High-throughput automated platform for nuclear magnetic resonance-based structural proteomics, *Expert Rev. Proteomics* 2, 49-55, 2005.

SUMOylation

The modification of proteins with small ubiquitin-like modificr (SUMO). SUMO are ubiquitin-like proteins such as Rub1, Apg9, and Apg12 and are separate from ubiquitin domain proteins such as RAD23 and DSK2. Unlike modification with ubiquitin, sumoylation does signal protein degradation but rather appears to enhance stability and/or specific transport. See Müller, S., Hoege, C., Pyrowolakis, G., and Jenisch, S., SUMO, ubiquitin's mysterious cousin, *Nat.Rev.Molec.Cell Biol.* 2, 202-210, 2001; Watts, F.Z., SUMO modification of proteins other than transcription factors, *Semin.Cell Dev.Biol.* 15, 211-220, 2004; Gill, G., SUMO and ubiquitin in the nucleus: different functions, similar mechanisms?, *Genes Dev.* 18, 2046-2059, 2004; Navotchova, M., Budhiraja, R., Coupland, G., Eisenhaber, F. and Bachmair, A., SUMO conjugation in plants, *Planta* 220, 1-8, 2004; Bossis, G. and Melchior, F., Regulation of SUMOylation by reversible oxidation and SUMO conjugating enzymes, *Mol.Cell.* 21, 349-357, 2006.

Surface Plasmon Resonance

A technique which uses affinity binding to measure analytes in solution. Conceptually surface plasmon resonance is related to other binding assays such as ELISA assays. In surface plasmon resonance, binding is measured by the increase in mass on a target probe which is bound to a surface. Frequently gold is the surface. Incident light is refracted from the surface and measured as reflectance (surface plasmon resonance). See also localized surface plasmon resonance. See Englebienne, P., Van Hoonacker, AS., and Verhas, M., Surface plasmon resonance: principles, methods and applications in biomedical sciences, *Spectroscopy* 17, 255-273, 2003; Smith, E.A. and Corn, R.M., Surface plasmon resonance imaging as a tool to monitor biomolecular interactions in an array based format, *Appl.Spectros.* 57, 320A-332A, 2003; Lee, J.H., Yan, Y., Marriott, G., and Corn, R.M., Quantitative functional analysis of protein complexes on surfaces, *J.Physiol.* 563, 61-71, 2005; Piehler, J., New methodologies for measuring protein interactions in vivo and in vitro, *Curr.Opin.Struct.Biol.* 15, 4-14, 2005; Buijs, J. and Franklin, G.C., SPR-MS in functional proteomics, *Brief Funct.Genomic Proteomic* 4, 39-47, 2005; Pattnaik, P., Surface plasmon resonance: applications in understanding receptor-ligand interaction, *Appl.Biochem. Biotechnol.* 126, 76-92, 2005; Homola, J., Vaisocherova, H., Dostalek, J., and Piliarik, M., Multi-analyte surface plasmon resonance biosensing, *Methods* 37, 26-36, 2005.

Surface Tension

A phenomenon which occurs when tow fluids are in contact with each other due to molecular attraction between molecules of the two liquids at the surface of separation.

Surfactant

The term surfactant dates to the 1950's when it was developed as a shortened version of surface-active agent. Surfactants are amphipathic/amphiphilic molecules which tend to migrate to surfaces or interfaces in solutions (at equilibrium, the concentration of a surfactant is higher at the interface than the concentration in bulk solution). The term detergent is sometimes used interchangeably with surfactant; the purist might consider the term detergency to reflect on cleansing which is one of the several properties of surfactants. Surfactants can also be described as dispersing agents, emulsifiers, foaming agents, stabilizers, solubilizers or wetting agents depending on their performance activity and effect on final product. Surfactants can be divided into four broad chemical categories; anionic compounds such as soaps which are sodium salts of long-chain alkyl carboxylic acids (alkanoic acids); cationic compounds such as alkyl amine derivatives such as Triton™ RW;

amphoteric derivatives; and nonionic surfactants such as alkylphenol ethoxylates (Igepal™) and anhydrosorbitol esters (Tween derivatives). See *Kirk-Othmer Encyclopedia of Chemical Technology*, 3rd edn., Vol. 22, Wiley-Interscience, New York, NY, 1983. See also *Nonionic Surfactants*, ed. M.J. Schick, Marcel-Dekker, New York, New York, 1966; Attwood, D. and Florence, A.T., *Surfactant Systems: Their Chemistry, Pharmacy, and Biology*, Chapman and Hall, London, UK, 1983; Cross, J., *Anionic Surfactants Analytical Chemistry*, Marcel Dekker, New York, NY, 1998; van Oss, N.M., *Nonionic Surfactants Organic Chemistry*, Marcel Dekker, New York, NY, 1998; Holmberg, K., *Novel Surfactants Preparation, Applications, and Biodegradabilty*, Marcel Dekker, New York, NY, 1998; Kwak, J.C.T., *Polymer-Surfactant Systems*, Marcel Dekker, New York, NY, 1998; Hus, J.-P., Interfacial Forces and Fields Theory and Applications, Marcel Dekker, New York, NY, 1999; Pefferkorn, E., *Interfacial Phenomena in Chromatography*, Marcel Dekker, New York, NY, 1999; Myers, D., *Surfaces, Interfaces, and Colloid : Principles and Applications*, Wiley-VCH, New York, NY, 1999; Broze, G., *Handbook of Detergents*, Marcel Dekker, New York, NY, 1999; *Physical Properties of Lipids*, ed. A.G. Marangani and S.S. Narine, Marcel-Dekker, New York, NY, 2002. Surfactants are used extensively in the solubilization of membranes and phospholipid (Lichtenberg, D., Robson, R. J., and Dennis, E.A., Solubilization of phospholipid by detergents. Structural and kinetic aspects, *Biochim.Biophys.Acta* 737, 285-304, 1983; Dennis, E.A., Micellization and solubilization of phospholipid by surfactants, *Adv.Colloid Interface Sci.* 26, 155-175, 1986; Silvius, J.R., Solublization and functional reconstitution of biomembrane components, *Annu.Rev.Biophys. Biomol.Struct.* 21, 323-348, 1992; Henry, G.D. and Sykes, B.D., Methods to study membrane protein structure in solution, *Methods Enyzmol.* 239, 515-535, 1994; Bowie, J.H., Stabilizing membrane proteins, *Curr.Opin.Struct. Biol.* 11, 397-402, 2001; Seddon, A.M., Curow, P., and Booth, B.J., Membrane proteins, lipids and detergents: not just a soap opera, *Biochim.Biophys. Acta* 1666, 105-117, 2004. Nonionic surfactants have an effect (drag reduction) on fluid flow at low concentrations (Jacobs, E.W., Anderson, G.W., Smith, C.A., *et al.*, Drag reduction using high molecular weight fractions of poly-ethylene oxide, in *Drag Reduction in Fluid Flows. Techniques for Friction Control*, ed. R.H.J. Sellin and R.T. Moses, Ellis-Horwood, Chichester, NY, 1989; Drappier, J., Divoux, T., Amarouchene, Y., *et al.*, Turbulent drag reduction by surfactants, *Europhysics Lett.* 74, 362-368, 2006).

Surrogate Marker

A biomarker which can be used in place of a clinical indication for diagnosis or a clinical endpoint for prognosis; also referred to as surrogate endpoint. See Hilsenbeck, S.G. and Clark, G.M., Surrogate endpoints in chemoprevention of breast cancer: guidelines for evaluation of new biomarkers, *J.Cell.Biochem. Suppl.* 17G, 205-211, 1993; Morrish, P.K., How valid is dopamine transporter imaging as a surrogate marker in research trials on Parkinson's disease? *Mov.Disord.* 18, Suppl. 7, S63-70, 2003; Kluft, C., Priniciples of use of surrogate markers and endpoints, *Maturitas* 47, 293-298, 2004; Bowdish, M.E., Arcasoy, S.M., Wilt, J.S., *et al.*, Surrogate markers and risk factors for chronic lung allograft dysfunction, *Am.J.Transplant.* 4, 1171-1178, 2004; Lieberman, R., Evidence-based medical perspectives: the evolving role of PSA for early detection, monitoring of treatment response, and as a surrogate end point of efficacy for intervention in men with different clinical risk states for the prevention and progression of prostate cancer, *Am.J.Ther.* 11, 501-506, 2004; Li, Z., Chines, A.A., and Meredith, M.P., Statistical validation of surrogate endpoints:

is bone density a valid surrogate for fracture? *J.Musculoskelet. Neuronal Interact.* 4, 64-74, 2004; Kantarci, K., and Jack, C.R., Jr., Quantitative magnetic resonance techniques as surrogate markers of Alzheimer's disease, *NeuroRx* 1, 196-205, 2004; Wier, C.J. and Walley, R.J., Statistical evaluation of biomarkers as surrogate endpoints: a literature review, *Stat.Med.* 25, 183-203, 2006.

Synovial Fluid Synovial Proteome

The total protein content of synovial fluid; results obtained from the proteomic analysis of synovial fluid. See Dasuri, K., Antonovici, M., Chen, K., *et al.*, The synovial proteome: analysis of fibroblast-like synoviocytes, *Arthritis Res.Ther.* 6, R161-R168, 2004; Romeo, M.J., Espina, V., Lowenthal, M. *et al.*, CSF proteome: a protein repository for potential biomarker idenfication, *Expert. Rev.Proteomics* 2, 57-70, 2005; Hueber, W., Kidd, B.A., Tomooka, B.H., *et al.*, Antigen microarray profiling of autoantibodies in rheumatoid arthritis, *Arthritis Rheum.* 52, 2645-2655, 2005.

Systems Biology

The integration of data at the genomic, transcriptomic, proteomic, and metabolomic levels including functional and structural data to understand biological function which can be described by a mathematical function. See Ideker, T., Galitski, T., and Hood, L., A new approach to decoding life: systems biology, *Annu.Rev.Genomics Hum.Genet.* 2, 343-372, 2001; Levesque, M.P. and Benfey, P.N., Systems biology, *Curr.Biol.* 14, R179-R189, 2004; Weston, A.D. and Hood, L.J., Systems biology, proteomics, and the future of health care: toward predictive, preventative, and personalized medicine, *J.Proteome Res.* 3, 179-196, 2004; Benner, S.A. and Ricardo, A., Planetary systems biology, *Mol.Cell.* 17, 471-472, 2005; Friboulet, A. and Thomas, D., Systems biology-an interdisciplinary approach, *Biosens.Bioelectron.* 20, 2404-2407, 2005; Aderem, A., Systems biology: its practices and challenges, *Cell* 121, 511-513, 2005; Theilgaard-Monch, K., Porse, B.T., and Borregaard, N., Systems biology of neutrophil differentiation and immune response, *Curr. Opin.Immunol.* 18, 54-60, 2006; Mustacchi, R., Hohmann, S., and Nielsen, J., Yeast systems biology to unravel the network of life, *Yeast* 23, 227-238, 2006; Baker, M.D., Wolanin, P.M., and Stock, J.B., Systems biology of bacterial chemotaxis, *Curr.Opin.Microbiol.* 9, 187-192, 2006; Philippi, S. and Kohler, J., Addressing the problems with life-science databases for traditional uses and systems biology, *Nat.Rev.Genet.* 7, 482-488, 2006; Palsson, B., *Systems Biology: Properties of Reconstructed Networks*, Cambridge University Press, Cambridge, UK, 2006; *Systems Biology: Definitions and Perspectives*, ed. L.Alberghina and H.V. Westerhoff, Springer, Berlin, 2005.

Target of Rapomycin (TOR, mTOR)

A highly conserved serine/threonine kinase which is regulatory of cell growth which is target for immunosuppression in kidney transplantation with therapeutics such as sirolimus and everolimus. See Bestard, O., Cruzado, J.M., and Grinyo, J.M., Calcineurin-inhibitor-sparing immunosuppressive protocols, *Transplant Proc.* 37, 3729-3732, 2005; Mabasa, V.H. and Ensom, M.H., The role of therapeutic monitoring of everolimus in solid organ transplantation, *Ther.Drug.Monit.* 27, 666-676, 2005; Dennis, P.B., Fumagalli, S., and Thomas, G., Target of rapamycin (TOR): balancing the opposing forces of protein synthesis and degradation, *Curr.Opin.Genet.Dev.* 9, 49-54, 1999; Hall, M.N., The TOR signaling pathway and growth control in yeast,

Biochem.Soc.Trans. 24, 234-239, 1996; Avruch, J., Lin, Y., Long, X., Murthy, S., and Ortiz-Vega, S., Recent advances in the regulation of the TOR pathway by insulin and nutrients, *Curr.Opin. Nutr.Metab.Care*, 8, 67-72, 2005; Abraham, R.T., TOR signaling: an odyssey from cellular stress to the cell growth machinery, *Curr.Biol.* 15, R139-R141, 2005; Cesareni, G., Ceol, A., Gavrila, C., *et al.*, Comparative interactomics, *FEBS Lett.* 579, 1828-1833, 2005; Martin, D.E. and Hall, M.N., The expanding TOR signaling network, *Curr.Opin.Cell Biol.* 17, 158-166, 2005; Neuberger, J. and Jothimani, D., Long-term immunosuppression for prevention of nonviral disease recurrence, *Transplant.Proc.* 37, 1671-1674, 2005; dos Sarbassov, D., Ali, S.M., and Sabatini, D.M., Growing roles for the mTOR pathway, *Curr.Opin.Cell Biol.* 17, 596-603, 2005; Wullschleger, S., Loewith, R., and Hall, M.N., TOR signaling in growth and metabolism, *Cell* 124, 471-484, 2006.

Targeted Proteomics

Analysis of a defined portion of a proteome such as a glycoproteome, phosphoproteome, ribosomal proteins. See Knepper, M.A. and Masilamani, S., Targeted proteomics in the kidney using ensembles of antibodies, *Acta Physiol.Scand.* 173, 11-21, 2001; Warcheid, B. and Fenselau, C., A targeted proteomics approach to the rapid identification of bacterial cell mixtures by matrix-assisted laser desorption/ionization mass spectrometry, *Proteomics* 4, 2877-2892, 2004; Ecelbarger, C.A., Targeted proteomics using Immunoblotting technique for studying dysregulation of ion transporters in renal disorders, *Expert Rev. Proteomics* 1, 219-227, 2004; Immler, D., Greven, S., and Reinemer, P., Targeted proteomics in biomarker validation: detection and quantification of proteins using a multi-dimensional peptide separation strategy, *Proteomics* 6, 2947-2958, 2006.

Telomerase

A ribonucleoprotein complex which catalyzes the synthesis of DNA at the ends of chromosomes and confers replicative immortality to cells. It is considered to be important in the growth of cancer cells and is a therapeutic target. See Blackburn, E.H., Greider, C.W., Henderson, E., *et al.*, Recognition and elongation of telomeres by telomerase, *Genome* 31, 553-560, 1989; Lamond, A.I., Tetrahymena telomerase contains an internal RNA template, *Trends Biochem.Sci.* 14, 202-204, 1989; Greider, C.W., Telomeres, telomerase and senescence, *Bioessays* 12, 363-369, 1990; Shippen-Lentz, D. and Blackburn, E.H., Functional evidence for an RNA template in telomerase, *Science* 247, 546-552, 1990; Romero, D.P. and Blackburn, E.H., A conserved secondary structure for telomerase RNA, *Cell* 67, 343-353, 1991; Greider, C.W., Telomerase and telomere-length regulation: lessons from small eukaryotes to mammals, *Cold Spring Harb.Symp.Quant.Biol.* 58, 719-723, 1993; Harley, C.B., Kim, N.W., Prowse, K.R., *et al.*, Telomerase, cell immortality, and cancer, *Cold Spring Harb.Symp.Quant. Biol.* 59, 307-315, 1994; Rhyu, M.S., Telomeres, telomerase, and immortality, *J.Natl.Cancer Inst.* 87, 884-894, 1995; Buchkovich, K.J., Telomeres, telomerase, and the cell cycle, *Prog.Cell Cycle Res.* 2, 187-195, 1996; Shay, J.W. and Wright, W.E., Use of telomerase to create bioengineered tissues, *Ann.N.Y.Acad.Sci.* 1057, 479-491, 2005; Flores, I., Benetti, R., and Blasco, M.A., Telomerase regulation and stem cell behavior, *Curr.Opin.Cell Biol.* 18, 254-260, 2006; Wirth, T., Kuhnel, F., and Kubicka, S., Telomerase-dependent gene therapy, *Curr.Mol.Med.* 5, 243-251, 2005; Hahn, W.C., Telomere and telomerase dynamics in human cells, *Curr. Mol.Med.* 5, 227-231, 2005; Blackburn, E.H., Telomeres and telomerase: their mechanisms of action and the effects of altering their functions, *FEBS Lett.* 579, 859-862, 2005; Shin, J.S., Hong, A., Solomon, M.J., and Lee, C.S., The role of telomeres and telomerase in the pathology of human cancer and aging, *Pathology* 38, 103-113, 2006.

Tetramer

Using a protein molecule containing four identical or non-identical subunits. In the case of non-identical subunits, the tetramer is frequently a dimer of dimers such as hemoglobin (Verzili, D., Rosato, N., Ascoli, F., and Chiancone, E., Aromatic amino-acids and subunit assembly in the hemoglobins from *Scapharca inaequivalvis*; a fluorescence and Cd study of the apoproteins, *Biochim.Biophys.Acta.* 954, 108-113, 1988; Ackers, G.K. and Holt, J.M., Asymmetric cooperativity in a symmetric tetramer: human hemoglobin, *J.Biol.Chem.* 281,11441-11443, 2006) and MHC II receptors (Schafer, P.H., Pierce, S.K., and Jardetzky, T.S., The structure of MHC class II: a role for dimer of dimers, *Semin. Immunol.* 7, 389-398, 1995; Yadati, S., Nydam, T., Demian, D. *et al.*, Salt bridge residues between I-Ak dimer of dimers alpha-chains modulate antigen presentation, *Immunol.Lett.* 67, 47-55, 1999; Lindstedt, R., Monk, N., Lombardi, G., and Lechler, R., Amino acid substitutions in the putative MHC class II "dimer of dimers" interface inhibit CD4+ T cell activation, *J.Immunol.* 166, 800-808, 2001) although there examples of tetrameric proteins such as lactose dehydrogenase where there can be varying subunit composition (Pesce, A., McKay, R.H., Stolzenbach, F., *et al*,, The comparative enzymology of lactic dehydrogenases. I Properties of the crystalline beef and chicken enzymes, *J.Biol. Chem.* 239, 1753-1761, 1964; Maekawa, M., Lactate dehydrogenase isoenzymes, *J.Chromatog.* 429, 373-398, 1988). See Kosaka, M., Iishi, Y., Okagawa, K., *et al.*, Tetramer Bence Jones protein in the immunoproliferative diseases. Angioimmunoblastic lymphadenopathy, primary amyloidosis, and multiple myeloma, *Am.J.Clin.Pathol.* 91, 639-646, 1989; Meissner, G., Ligand binding and cooperative interactions among the subunits of the tetrameric Ca^{2+} release channel complex of sarcoplasmic reticulum, *Adv.Exp.Med.Biol.* 311, 277-287, 1992; Furey, W., Arjunan, P., Chen, L., *et al.*, Structure-function relationships and flexible tetramer assembly in pyruvate decarboxylase revealed by analysis of crystal structures, *Biochim.Biophys.Acta* 1385, 253-270, 1998; Grotzinger, J., Kerneback, T., Kallen, K.J., and Rose-John, S., IL-6 type cytokine receptor complexes: hexamer, tetramer or both? *Biol.Chem.* 380, 803-813, 1999; Kang, H.M., Choi, K.S., Kassam, G., *et al.*, Role of annexin II tetramer in plasminogen activation, *Trends Cardiovasc.Med.* 9, 92-102, 1999; Hamiche, A. and Richard-Foy, H., Characterization of specific nucleosomal states by use of selective substitution reagents in model octamer and tetramer structures, *Methods* 19, 457-464, 1999; Xu, X.N. and Screaton, G.R., MHC/Peptide tetramer studies of T cell function, *J.Immunol.Methods* 268, 21-28, 2002; Zimmerman, A.L., Two B or not two B? Questioning the rotational symmetry of tetrameric ion channels, *Neuron* 36, 997-999, 2002; Kita, H., He, X.S., and Gershwin, M.E., Application of tetramer technology in studies on autoimmune diseases, *Autoimmun.Rev.* 2, 43-49, 2003.

Tetrose

A four-carbon monosaccharide including *meso*-erythritol (erythritol), D-erythrose, L-erythrulose, and D-threose. See Horecker, B.L., Smyrniotis, P.Z., Hiatt, H.H, and Marks, P.A., Tetrose phosphate and the formation of aldoheptulose diphosphate,

J.Biol.Chem. 212, 827-836, 1955; Batt, R.D., Dickens, F., and Williamson, D.H., Tetrose metabolism. 1. The preparation and degradation of specifically labelled [^{14}C] tetroses and [^{14}C] tetritols, *Biochem.J.* 77, 272-281, 1960; Taylor, G.A. and Ballou, C.E., D-Glycero-tetrulose 1,4-diphosphate (D-erythulose 1,4-diphosphate), *Biochemistry* 2, 553-555, 1963; Giudici, T.A. and Fluharty, A.L., A specific method for the determination of the aldotetroses, *Anal.Biochem.* 13, 448-457, 1965; Lai, C.Y., Martinez-de Dretz, G., Bacila, M., *et al.*, Labeling of the active site of aldolase with glyceraldehyde 3-phosphate and erythrose 4-phosphate, *Biochem.Biophys.Res.Commun.* 27, 665-672, 1968; Benov, L. and Fridoich, I., Superoxide dependence of the toxicity of short chain sugars, *J.Biol.Chem.* 273, 25741-25744, 1998; Lehman, T.D. and Ortwerth, B.J., Inhibition of advanced glycation end product-associated protein cross-linking, *Biochim.Biophys.Acta* 1535, 110-119, 2001; Chaput, J.C. and Szostak, J.W., TNA synthesis by DNA polymerases, *J.Am.Chem.Soc.* 125, 9274-9275, 2003; Kempeneers, V., Vastmans, K., Rozenski, J., and Herdewijn, P., Recognition of threosyl nucleotides by DNA and RNA polymerases, *Nucleic Acids Res.* 31, 6221-6226, 2003; Ichida, J.K., Zou, K., Horhota, A., *et al.*, An in vitro selection system for TNA, *J.Am.Chem.Soc.* 127, 2802-2803, 2005; Horhota, A., Zou, K., Ichida, J.K., *et al.*, Kinetic analysis of an efficient DNA-dependent TNA polymerase, *J.Am. Chem.Soc.* 127, 7427-7434, 2005; Wamelink, M.M, Struys, E.A., Huck, J.H., *et al.*, Quantification of sugar phosphate intermediates of the pentose phosphate pathway by LC-MS/MS: application to two new inherited defects of metabolism, *J.Chromatogr.B Analyt.Technol.Biomed.Life Sci.* 823, 18-25, 2005; Weber, A.L. and Pizzarello, S., The peptide-catalyzed stereospecific synthesis of tetroses: A possible model for prebiotic molecular evolution, *Proc.Natl.Acad.Sci.USA*, 103, 12713–12717, 2006.

Theragnostic

As in theragnostic imaging described as the combined use of molecular and functional imaging to prescribe the distribution of radiations in three-dimensional space as a function of time in the process of radiation therapy. See Bentzen, S.M., Theragnostic imaging for radiation oncology: dose-painting by numbers, *Lancer Oncol.* 6, 112-117, 2005.

Theranostics

The use of diagnostic laboratory tests to guide therapeutic outcomes. Current use has emphasized "real-time" PCR assays for the identification of pathogens. See Picard, F.J. and Bergeron, M.G., Rapid molecular theranostics in infectious disease, *Drug Discov.Today.* 7, 1092-1101, 2002.

Therapeutic Equivalence (TE)

Drug products including biologics can be considered to be therapeutically equivalent if such products can be substituted for brand product/prescribed product/originator product with the full expectation that such substituted product will produce the same clinical effect and safety as the brand product/prescribed product/originator product. See Panhard, X. and Mentre, F., Evaluation by simulation of tests based on non-linear mixed-effects models in pharmacokinetic interaction and bioequivalence cross-over clinical trials, *Stat.Med.* 24, 1509-1524, 2005; Bolton, S., Bioequivalence studies for levothyroxine, *AAPS J.* 7, E47-E53, 2005; Meyer, M.C., United States Food and Drug Administration requirements for approval of generic drug products, *J.Clin.*

Psychiatry 62 Suppl 5, 4-9, 2001; Temple, R., Policy developments in regulatory approval, *Stat.Med.* 21, 2939-2048, 2002; Gould, A.L, Substantial evidence of effect, *J.Biopharm.Stat.* 12, 53-77, 2002; Chen, M.L., Patnaik, R., Hauck, W.W., *et al.,* An individual bioequivalence criterion: regulatory considerations, *Stat.Med.* 19, 2821-2842, 2000.

Thermal Conductivity

Rate of heat transfer by conduction through unit thickness, across unit area for unit difference of temperature; measured as calories/second/cm^2/cm (thickness) and a temperature difference of 1°C; unit is cal/cm sec °K or W/cm °K. See Harting, R. and Pfeiffenberger, U., Thermal conductivity of bovine and pig retina: an experimental study, *Grafes Arch.Clin.Exp.Ophthalmol.* 219, 290-291, 1982; Miller, J.H., Wilson, W.E., Swenberg, C.E. *et al,* Stochastic model of free radical yields in oriented DNA exposed to densely ionizing radiation at 77K, *Int.J.Radiat.Biol.Relat.Stud. Phys. Chem.Med.* 53, 901-907. 1988; Arkin, H., Holmes, K.R., and Chen, M.M., A technique for measuring the thermal conductivity and evaluating the "apparent conductivity" concept in biomaterials, *J.Biomech.Eng.* 111, 276-282, 1989; Cheng, J., Shoffner, M.A., Mitchelson, K.R., *et al.,* Analysis of ligase chain reaction products amplified in a silicon-glass chip using capillary electrophoresis, *J.Chromatog.A* 732, 151-158, 1996; Bhattacharya, A. and Mahajan, R.L., Temperature dependence of thermal conductivity of biological tissues, *Physiol.Meas.* 24, 769-783, 2003; Rodriguez, I., Lesaicherre, M., Tie, Y. *et al,* Practical integration of polymerase chain reaction amplification and electrophoretic analysis in microfluidic devices for genetic analysis, *Electrophoresis* 24, 172-178, 2003.

Thermophilic

Term used to describe an organism that grows at elevated temperatures which is described as thermophile. The most famous this bacteria (*Thermus aquaticus*) responsible for TAQ polymerase which is used in the PCR. See Friedman, S.M., Protein-synthesizing machinery of thermophilic bacteria, *Bacteriol.Rev.* 32, 27-38, 1968; Singleton, R., Jr. and Amelunxen, R.E., Protein from thermophilic microorganisms, *Bacteriol.Rev.* 37, 320-342, 1973; Lasa, I. and Berenguer, J., Thermophilic enzymes and their biotechnology potential, *Microbiologica* 9, 77-89, 1993; Kelly, R.M., Peeples, T.L., Halio, S.B., *et al.,* Extremely thermophilic microorganisms. Metabolic strategies, genetic characteristics, and biotechnological potential, *Ann.N.Y.Acad.Sci.* 745, 409-425, 1994; Russell, R.J. and Taylor, G.L., Engineering thermostability: lessons from the thermophilic proteins, *Curr.Opin. Biotechnol.* 6, 370-374, 1995; Kumar, S. and Nussinov, R., How do thermophilic proteins deal with heat? *Cell.Mol.Life.Sci.* 58, 1216-1233, 2001; Sambongi, Y., Uchiyama, S., Kobayashi, Y., *et al,,* Cytochrome c from a thermophilic bacterium has provided insights into the mechanisms of protein maturation, folding, and stability, *Eur.J.Biochem.* 269, 3355-3361, 2002; Radianingtyas, H. and Wright, P.C., Alcohol dehydrogenases from thermophilic and hyperthermophilic archae and bacteria, *FEMS Microbiol. Rev.* 27, 593-616, 2003; McMullen, G., Christie, J.M., Rahman, T.J., *et al.,* Habitat, applications and genomics of the aerobic, thermophilic genus *Geobacillus, Biochem.Soc.Trans.* 32, 214-217, 2004; Averhoff, B., DNA transport and natural transformation in mesophilic and thermophilic bacteria, *J.Bioenerg.Biomembr.* 36, 25-33, 2004; Pereira, M.M., Bandeiras, T.M., Fernandes, A.S., *et al.,* Respiratory chains from aerobic thermophilic prokaryotes,

J.Bioenerg.Biomembr. 36, 93-105; 2004; Nishida, C.R. and Ortiz de Montellano, P.R., Thermophilic cytochrome P450 enzymes, *Biochem.Biophys.Res. Commun.* 338, 437-445, 2005; Egorova, K. and Antranikian, G., Industrial relevance of thermophilic Archaea, *Curr.Opin.Microbiol.* 8, 649-655, 2005.

Thiolase Superfamily

A family of condensing enzymes with diverse functions such as the formation of carbon-carbon bonds in the synthesis of fatty acids and polyketides. The expression of peroxisome thiolases is regulated in the PPAR receptors. See Swartzman, E.E., Viswanathan, M.N., and Thorner, J., The PAL1 gene product is a peroxisomal ATP-binding cassette transporter in the yeast *Saccharomyces cerevisiae, J.Cell.Biol.* 132, 549-563, 1996; Aoyama, T., Peters, J.M., Iritani, N., *et al.,* Altered constitutive expression of fatty acid-metabolizing enzymes in mice lacking the peroxisome proliferator-activated receptor alpha (PPARalpha), *J.Biol. Chem.* 273, 5678-5684, 1998; Latruffe, N., Chekauoui, M.M., Nicholas-Frances, V., *et al.,* *Biochem.Pharmacol.* 60, 1027-1032, 2000; Rabus, R., Kube, M., Beck, A., Widdel, F., and Reinhardt, R., Genes involved in the anaerobic degradation of ethyl benzene in a denitrifying bacterium, strain EbN1, *Arch.Microbiol.* 178, 506-516, 2002; Kursula, P., Sikkila, H., Fukao, T., Kondo, N., and Wierenga, R.K., High resolution crystal structures of human cytosolic thiolase (CT): a comparison of the active sites of human CT, bacterial thiolase, and bacterial KAS I., *J.Mol.Biol.* 347, 189-201, 2005; Haapalinen, A.M., Merilainen, G., and Wierenga, R.K., The thiolase superfamily: condensing enzymes with diverse reaction specifications, *Trends Biochem.Sci.* 31, 64-71, 2006.

Thioredoxin(s)

A small protein function as a hydrogen donor/reducing agent in biological systems; considered to be a major regulator of redox reactions in the cell. See Holmgren, A., Thioredoxin, *Annu. Rev.Biochem.* 54, 237-271, 1985; Brot, N. and Weissbach, H., Biochemistry of methionine sulfoxide residues in proteins, *Biofactors* 3, 91-96, 1991; Martin, J.L., Thioredoxin-a fold for all reasons, *Structure* 3, 245-250, 1995; Aslund, F. and Beckwith, J., The thioredoxin superfamily: redundancy, specificity, and gray-area genomics, *J.Bacteriol.* 181, 1375-1379, 1999; Arrigo, A.P., Gene expression and the thiol redox state, *Free Radic.Biol.Med.* 27, 936-944, 1999; Holmgren, A., Antioxidant function of thioredoxin and gluaredoxin systems, *Antioxid. Redox Signal.* 2, 811-820, 2000; Arner, E.S.J. and Holmgren, A., Physiological functions of thioredoxins and thioredoxin reductase, *Eur.J.Biochem.* 267, 6102-6109, 2000; Pearson, W.R., Phylogenies of glutathione transferase families, *Methods Enzymol.* 401, 186-204, 2005; Burke-Gaffney, A., Callister, M.E., and Nakamura, H., Thioredoxin: friend or foe in human disease? *Trends Pharmacol.Sci.* 26, 398-404, 2005; Stefankova, P., Kollarova, M., and Barak, I., Thioredoxin – structural and functional complexity, *Gen.Physiol.Biophys.* 24, 3-11, 2005; Koc, A., Mathews, C.K., Wheeler, L.J., Gross, M.K., and Merrill, G.F., Thioredoxin is required for deoxyribonucleotide pool maintenance during S phase, *J.Biol.Chem.* 281, 15058-15063.

Thymosins

A group of peptide hormones (alpha, beta, and gamma) originally described in the thymus which have a broad range of functions including the modulation of actin polymerization and control of lymphoid tissue function in the development of T-cells and the maturation of B-cell to form plasma cells. See Goldstein, A.L. and White, A., Role of thymosin and other thymic factors in the development, maturation, and functions of lymphoid tissue, *Curr.Top.Exp.Endocrinol.* 1, 121-149, 1971; Bach, J.F. and Carnaud, C., Thymic factors, *Prog.Allergy* 21, 342-408, 1976; Low, T.L., Thurman, G.B., Chincarini, C., et al., Current status of thymosin research: evidence for the existence of a family of thymic factors that control T-cell maturation, *Ann.N.Y.Acad.Sci.* 332, 33-48, 1979; Ampe, C. and Vandekerchove, J., Actin-actin binding protein interfaces, *Semin.Cell Biol.* 5, 175-182, 1994; Chen, C., LI, M., Yang, H., et al., Roles of thymosins in cancers and other organ systems, *World J.Surg.* 29, 264-270, 2005; Goldstein, A.L., Hannappel, E., and Kleinman, H.K., Thymosin beta4: actin-sequestering protein moonlights to repair injured tissues, *Trends Mol.Med.* 11, 421-429, 2005.

Time-of-Flight

The term time-of-flight designates techniques and apparatus that depend on the time taken by particles to traverse a set distance, as, for example, the separation of ions according to their mass in mass spectrometry. Time-of-flight mass spectrometry (TOF-MS) measures flight time of ions; lighter ions travel a greater distance that heavier ions; mass proportional to time squared; converted to m/z by calibration with standards (Cotter, R.J., Time-of-flight mass spectrometry: an increasing role in the life sciences, *Biomed.Environ. Mass Spectrom.* 18, 513-532, 1989; Guilhaus, M., Selby, D., and Mlynski, V., Orthogonal acceleration time-of-flight mass spectrometry, *Mass Spectrom.Rev.* 19, 65-107, 2000; Belu, A.M., Graham, D.J., and Castner, D.G., Time-of-flight secondary ion mass spectrometry: techniques and applications for the characterization of biomaterial surfaces, *Biomaterials* 24, 3635-3653, 2003; Seibert, V., Wiesner, A., Buschmann, T., and Meuer, J., Surface-enhanced laser desorption ionization time-of-flight mass spectrometry (SELDI TOF-MS) and ProteinChip technology in proteomics research, *Pathol.Res.Pract.* 200, 83-94, 2004; Vestal, M.L., and Campbell, J.M., Tandem time-of-flight mass spectrometry, *Methods Enzymol.* 402, 79-108, 2005). However, the technique is used in other applications such as measurement of blood flow (Kochar, R., Khandelwal, N., Singh, P., and Suri, S., Arterial contamination: a useful indirect sign of cerebral sinovenous thrombosis, *Acta Neurol.Scand.* 114, 139-142, 2006; Han, S., Granwehr, J., Garcia, S., et al., Auxillary probe design adaptable to existing probes for remote detection NMR, MRI, and time-of-flight tracing, *J.Magn.Reson.*, 182, 260–272, 2006).

Tissue Microarray

A microarray consisting of cores (0.6 mm in diameter for example) of tissue embedded in a paraffin block. Samples are taken from existing paraffin block sections This technology allows multiple samples to be processed at the same time under the same conditions. See Moch, H., Schrami, P., Bubendorf, L, et al., High-throughput tissue microarray analysis to evaluated genes uncovered by cDNA microarray screening in renal cell carcinoma, *Am.J.Pathol.* 154, 981-986, 1999; Kallioniemi, O.P., Wagner, U., Kononen, J., and Sauter, G., Tissue microarray technology for high-throughput molecular profiling of cancer, *Hum.Mol.Genet.* 10, 657-662, 2001; Rao, J., Seligson, D., and Hemstreet, G.P., Protein expression analysis using quantitative fluorescence image analysis on tissue microarray slides, *BioTechniques* 32, 928-930, 2002; Rubin, M.A., Dunn, R., Strawderman, M., and Pienta, K.J., Tissue microarray sampling strategy for prostate cancer

biomarker analysis, *Am.J.Surg.Pathol.* 26, 312-219, 2002; Hedvat, C.V., Hedge, A., Chaganti, R.S., et al, Application of tissue microarray technology to the study of non-Hodgkin's and Hodgkin's lymphoma, *Hum.Pathol.* 33, 968-974, 2002; Kim, W.H., Rubin, M.A., and Dunn, R.L., High-density tissue microarray, *Am.J.Surg. Pathol.* 26, 1236-1238, 2002; Parker, R.L., Huntsman, D.G., Lesack, D.W., et al., Assessment of interlaboratory variation in the immunohistochemical determination of estrogen receptor status using a breast cancer tissue microarray, *Am.J.Clin.Pathol.* 117, 723-728, 2002; Giltnane, J.M. and Rimm, D.L., Technology insight: Identification of biomarkers with tissue microarray technology, *Nat.Clin.Pract.Oncol.* 1, 104-111, 2004; Zimpfer, A., Schonberg, S., Lugli, A., et al., Construction and validation of a bone marrow tissue microarray, *J.Clin.Pathol.*, 60, 57–61, 2006.

Titin

A very large protein (mass of approximately 3 million daltons) which is the third-most abundant protein in vertebrate striated muscle and has a role in providing sarcomeric alignment and recoil. Titin is a single chain protein which extends from the M line to the Z line forming a thick filament. See Maruyama, K., Connectin, an elastic filamentous protein of striated muscle, *Int.Rev.Cytol.* 104, 81-114, 1986; Trinick, J., Elastic filaments and giant proteins in muscle, *Curr.Opin.Cell.Biol.* 3, 112-119, 1991; Fulton, A.B., and Isaacs, W.B., Titin, a huge, elastic sarcomeric protein with a probable role in morphogenesis, *Bioessays* 13, 157-161, 1991; Trinick, J., Understanding the functions of titin and nebulin, *FEBS Lett.* 307, 44-48, 1992; Kellermeyer, M.S. and Grama, L., Stretching and visualizing titin molecules: combining structure, dynamics and mechanics, *J.Muscle Res.Cell Motil.* 23, 499-511, 2002; Tskhovrebova, L. and Trinick, J., Titin: properties and family relationships, *Nat.Rev.Mol.Cell Biol.* 4, 679-689, 2003; Granzier, H.L. and Labeit, S., The giant protein titin: a major player in myocardial mechanics, signaling, and disease, *Circ.Res.* 94, 284-295, 2004; Tskhovrebova, L. and Trinick, J., Properties of titin immunoglobulin and fibronectin-3 domains, *J.Biol.Chem.* 279, 46351-46354, 2005; Samori, B., Zuccheri, G., and Baschieri, R., Protein unfolding and refolding under force: methodologies for nanomechanics, *Chemphyschem.* 6, 29-34, 2005; Lange, S., Ehler, E., and Gautel, M., From A to Z and back? Multicompartment proteins in the sarcomere, *Trends Cell Biol.* 16, 11-18. 2006; Ferrari, M.B., Podugu, S., and Eskew, J.D., Assembling the myofibril: coordinating contractile cable construction with calcium, *Cell.Biochem.Biophys.* 45, 317-337, 2006.

Toll-like receptor

The term toll-like receptors is derived from the relationship of these proteins to the Toll receptor in *Drosophila* (Kuno, K. and Matsushima, K., The IL-1 receptor signaling pathway, *J.Leukoc. Biol.* 56, 542-547, 1994; Meister, M., Lemaitre, B., and Hoffman, J.A., Antimicrobial peptide defense in *Drosophila*, *Bioessays* 19, 1019-1026, 1997; Dushay, M.S. and Eldon, E.D., *Drosophila* immune responses as models for human immunity, *Am.J.Hum. Genet.* 62, 10-14; O'Neill, L.A. and Greene, C., Signal transduction pathways activated by the IL-1 receptor family: ancient signaling machinery in mammals, insects, and plants, *J.Leukoc.Biol.* 63, 650-657, 1998). Toll-like receptors in mammals are immune cell receptors which recognizing infectious agents and stimulating activation of the adaptive immune system. See Aderem, A. and Ulevitch, R.J., Toll-like receptors in the induction of the innate immune response, *Nature* 406, 782-77, 2000; Aderem, A.,

Role of the Toll-like receptors in inflammatory response in macrophages, *Crit.Care Med.* 29(7 Suppl), S16-S18, 2001; Beutler, B., Sepsis begins at the interface of pathogen and host, *Biochem.Soc. Trans.* 29, 853-859, 2001; Beutler, B. and Rietschel, E.T., Innate immune sensing and its roots: the story of endotoxin, *Nat.Rev. Immunol.* 3, 169-176, 2003; Philpott, D.J. and Girardin, S.E., The role of Toll-like receptors and Nod proteins in bacterial infection, *Mol.Immunol.* 41, 1099-1108, 2004; Jenner, R.G. and Young, R.A., Insights into host responses against pathogens from transcriptional profiling, *Nat.Rev.Microbiol.* 3, 281-294, 2005; Pasare, C. and Medzhitov, R., Toll-like receptors: linking innate and adaptive immunity, *Adv.Exp.Med.Biol.* 560, 11-18, 2005; O'Neill, L.A., How Toll-like receptors signal: what we know and what we don't know, *Curr.Opin.Immunol.* 18, 3-9, 2006; Kreig, A.M., Therapeutic potential of Toll-like receptor 9 activation, *Nat. Rev.Drug Discov.* 5, 471-484, 2006; Turvey, S.E., and Hawn, T.R., Towards subtlety: Understanding the role of Toll-like receptor signaling in susceptibility to human infections, *Clin.Immunol.* 120, 1-9, 2006.

TonB

A membrane-spanning protein which functions in receptors. See Ferguson, A.D. and Deisenhofer, J., TonB-dependent receptors – structural perspectives, *Biochim.Biophys.Acta* 1565, 318-332, 2002; Koebnik, R., TonB-dependent trans-envelope signalling: the exception or the rule? *Trends in Microbiol.* 13, 343-347, 2005.

Tonoplast

A membrane surrounding an intracellular structure or vacuole. See Barbeir-Brygoo, H. Renaudin, J.P., and Guern, J., The vacuolar membrane of plant cells: a newcomer in the field of biological membranes, *Biochimie* 68, 417-425, 1986; Bertl, A. and Slayman, C.L., Complex modulation of cation channels in the tonoplast and plasma membrane of *Saccharomyces cerevisieae*: single-channel studies, *J.Exp.Biol.* 172, 271-287, 1992; Neuhaus, J.M. and Rogers, J.C., Sorting of proteins to vacuoles in plant cells, *Plant Mol.Biol.* 38, 127-144, 1998; Luttge, U., The tonoplast functioning as the master switch for circadian regulation of crassulacean acid metabolism, *Planta* 211, 761-769, 2000.

Top-down proteomics

Mass spectrometric analysis of intact proteins as opposed to bottom-up-proteomics where mass spectrometry is used to analysis peptides derived from the proteolytic enzyme digests of proteins. Successful top-down proteomics generally requires high sophisticated instrumental approaches such as Fourier Transform Ion Cyclotron Resonance (FITCR) mass spectrometry. See Ge, Y., Lawhorn, G., ElNagger, E., *et al.*, Top down characterization of larger proteins (45 kDa) by electron capture dissociation mass spectrometry, *J.Am.Chem.Soc.* 124, 672-678, 2002; Nemeth-Cawley, J.F., Tangarone, B.S., and Rouse, J.C., "Top Down" characterization is a complementary technique to peptide sequencing for identifying protein species in complex mixtures, *J.Proteome Res.* 2, 495-505, 2003; Hirano, H., Islam, N., and Kawaski, H., Technical aspects of functional proteomics in plants, *Phytochemistry* 65, 1487-1498, 2004; Vaidyanathan, S., Kell, D.B., and Goodacre, R., Selective detection of proteins in mixtures using electrospray ionization mass spectrometry: influence of instrumental settings and implications for proteomics, *Anal.Chem.* 76, 5024-5032, 2004; Zhang, S. and van Pelt, C.K.,

Chip-based nanoelectrospray mass spectrometry for protein characterization, *Expert.Rev.Proteomics* 1, 449-468. 2004; Copper, H.J., Hakansson, K., Marshall, A.G., The role of electron capture dissociation in biomolecular analysis, *Mass Spectrom.Rev.* 24, 201-222, 2005; Godovac-Zimmerman, J., Kleiner, O., Brown, L.R., and Drukier, A.K., Perspectives in splicing up proteomics with splicing, *Proteomics* 5, 699-709, 2005; Kaiser, N.K., Anderson, G.A., and Bruce, J.E., Improved mass accuracy for tandem mass spectrometry, *J.Am.Soc.Mass Spectrom.* 16, 463-470, 2005; Williams, T.L., Monday, S.R., Edelson-Mammel, S., Buchanan, R., and Musser, S.M., A top-down proteomics approach for differentiating thermal resistant strains of *Enterobacter sakazakii*, *Proteomics* 5, 4161-4169, 2005; Demirev, P.A., Feldman, A.B., Kowalski, P., and Lin, J.S., Top-down proteomics for rapid identification of intact microorganisms, *Anal.Chem.* 77, 7455-7461, 2005; Du, Y., Parks, B.A., Sohn, S., Kwast, K.E., and Kelleher, N.L., Top-down approaches for measuring expression ratios of intact yeast proteins using fourier transform mass spectrometry, *Anal. Chem.* 78, 686-694, 2006.

Topoisomerase

A family of enzymes which the topology of DNA by catalyzing relaxation and unknotting of the double-stranded DNA complex. This is accomplished through the alteration of the supercoiling of the DNA helix. Topoisomerase I cleaves one strand of DNA while Topoisomerase II cleaves both strands of the DNA helix and there are isoforms of human topoisomerase II (Kondapi, A.K., Satyanarayana, N., and Saikrishna, A.D., A study of the topoisomerase II activity in HIV-1 replication using the ferrocene derivatives as probes, *Archs.Biochem.Biophys.* 450, 123-132, 2006). This process is critical for the transcription of the DNA to yield messenger RNA and for the replication process. There are two other classes of topoisomerases; topoisomerase III which appears to be involved in recombination while topoisomerase IV appears to involved in the separation of chromosomes. Topoisomerases are targets for cancer chemotherapy. See Glisson, B.S. and Ross, W.E., DNA topoisomerase II: a primer on the enzyme and its unique role as a multidrug target in cancer chemotherapy, *Pharmacol.Ther.* 32, 89-106, 1987; Osheroff, N., Biochemical basis for the interactions of type I and type II topoisomerases with DNA, *Pharmacol.Ther.* 41, 223-241, 1989; Gmeiner, W.H., Yu, S., Pon, R.T., Pourquier, P. and Pommier, Y., Structural basis for topoisomerase I inhibition by nucleoside analogs, *Nucleosides, Nucleotides, Nucleic Acids* 22, 653-658, 2003; Porter, A.C. and Farr, C.J., Topoisomerase II: untangling its contribution at the centromere, *Chromosome Res.* 12, 569-583, 2004; Pindur, U., Jansen, M., and Lemster, T., Advances in DNA-ligand with groove binding, intercalating and/or alkylating activity: chemistry, DNA-binding and biology, *Curr.Med.Chem.* 12, 2805-2847, 2005; Yangida, M., Basic mechanism of eukaryotic chromosome segregation, *Philos.Trans.Roy.Soc.London B Biol.Sci.* 360, 609-621, 2005; Martincic, D. and Hande, K.R., Topoisomerase II inhibitors, *Cancer Chemother.Biol.Response Modif.* 22, 101-121, 2005.

Topological Proteomics

A technology which analyzes proteins and protein interactions on a single cell level (MELK technology); study of the toponome; analysis of protein networks; Owens, J., Topological proteomics: a new approach to drug discovery, *Drug Discov.Today* 6, 1081-1082, 2001; Shubert, W. Topological proteomics, toponomics, MELK-technology, *Adv.Biochem.Eng.Biotechnol.* 82, 189-209, 2003; Han,

J.D., Dupuy, D., Bertin, N., Cusick, M.E., and Vidal, M., Effect of sampling on topology predictions of protein-protein interaction networks, *Nat.Biotechnol.* 23, 839-44, 2005; Stelzl, U., Worm, U., Lalowski, M. *et al.*, A human protein-protein interaction network: a resource for annotating the proteome, *Cell* 122, 957-968, 2005.

trans-activation

Enhancement of transcription (transcriptional activation) by a transcription factor binding to to a *cis*-factor in DNA and influencing the activity of RNA polymerase. See Roizman, B., Kristie, T., McKnight, J.L. *et al.* The trans-activation of herpes simplex virus gene expression: comparison of two factors and their cis sites, *Biochimie* 70, 1031-1043, 1988; Nevins, J.R., Mechanisms of viral-mediated trans-activation of transcription, *Adv.Virus Res.* 37, 35-83, 1989; Green, N.M, Cellular and viral transcriptional activators, *Harvey Lect.* 88, 67-96, 1992-1993; de Folter, S. and Angenent, G.C., trans meets cis in MADS science, *Trends Plant Sci.* 11, 223-231, 2006; Gomez-Roman, N., Felteon-Edkins, Z.A., Kenneth, N.S., *et al.*, Activation by c-Myc or transcription by RNA polymerases I, II, and III, *Biochem.Soc.Symp.* (73), 141-154, 2006; Campbell, K.J. and Perkins, N.D., Regulation of NF-κB function, *Biochem.Soc.Symp.* (73), 165-180, 2006; Belakvadi, M. and Fondell, J.D., Role of the mediator complex in nuclear hormone receptor signaling, *Rev.Physiol.Biochem.Pharmacol.* 156, 23-43, 2006.

trans-inactivation (transinactivation)

Gene or transgene silencing mediated by heterochromatin; gene inactivation by *trans*-inactivation is considered to be an epigenetic event; See Sabl, J.F. and Laird, C.D., Epigene conversion: a proposal with implications for gene mapping in humans, *Am.J.Hum. Genet.* 50, 1171-1177, 1992; Opsahl, M.L., Springbett, A., Lathe, R., Colman, A., McClenaghan, M., and Whitelaw, C.B., Mono-allelic expression of variegating transgene locus in the mouse, *Transgenic Res.* 12, 661-669, 2003; Sage, B.T., Jones, J.L., Holmes, A.L., Wu, M.D., and Csink, A.K., Sequence elements in cis influence heterochromatic silencing in trans, *Mol.Cell Biol.* 25, 377-388, 2005; also the suppression of the trans-phosphorylation of receptors and resulting signaling pathways. See Graness, A., Hanke, S., Boehmer, F.D., Presek, P., and Liebmann, C., Protein-tyrosine-phosphatase-mediated epidermal growth factor (EGF) receptor transinactivation and EGF receptor-independent stimulation of mitogen-activated protein kinase by bradykinin in A431 cells, *Biochem.J.* 347, 441-447, 2000; Elbaz, N., Bedecs, K., Masson, M., Sutren, M., Strosberg, A.D., and Nahmias, C., Functional trans-inactivation of insulin receptor kinase by growth-inhibitory angiotensis II AT2 receptor, *Mol.Endocrinol.* 14, 795-804, 2000; Nouet, S., Amzallag, N., Li, J.M., *et al.*, Trans-inactivation of receptor tyrosine kinases by novel angiotensin II AT2 receptor-interacting protein, ATIP, *J.Biol.Chem.* 279, 28989-28997, 2004.

Transcription

The process by which genetic information is transferred from DNA to RNA. See Hames, B.D. and Glover, D.M., *Transcription and Splicing*, IRL Press, Oxford, UK, 1988; Neidle, S., *DNA Structure and Recognition*, IRL Press, Oxford, UK, 1994; Singer, M. and Berg, P., *Exploring Genetic Mechanisms*, University Science Books, Sausalito, CA, 1997; Lewin, B., *Genes VII*, Oxford University Press, Oxford, UK, 2000; Lodish, H.F., *Molecular Cell Biology*, W.H.Freeman, New York, NY, 2000; Brown, W.M.,

and Brown, P.M., *Transcription*, Taylor & Francis, London, UK, 2002; Olson, M.O.J., *The Nucleolus*, Landes Bioscience, Georgetown, TX, 2004; Baumann, P., Qureshi, S.A., and Jackson, S.P., Transcription: new insights from studies on Archaea, *Trends Genet.* 11, 279-283, 1995; Alton, G., Schwanborn, K., Satoh, Y., and Westwick, J.K., Therapeutic modulation of inflammatory gene transcription by kinase inhibitors, *Expert.Opin.Biol.Ther.* 2, 621-632, 2002; Lee, D.K., Seol, W., and Kim, J.S., Custom DNA-binding proteins and artificial transcription factors, *Curr. Top.Med.Chem.* 3, 645-657, 2003; Mondal, N. and Parvin, J.D., Transcription from the perspective of the DNA: twists and bumps in the road, *Crit.Rev.Eukaryot.Gene Expr.* 13, 1-8, 2003; Sausville, EA. and Holbeck, S.L., Transcription profiling of gene expression in drug discovery and development: the NCI experience, *Eur.J.Cancer* 40, 2544-2549, 2004; Uesugi, M., Synthetic molecules that modulate transcription and differentiation: hints for future drug discovery, *Comb.Chem.High Throughput Screen.* 7, 653-659, 2004.

Transcription Factors

These are *trans* factors which are protein or protein complexes which bind to *cis* factors or regions which are intrinsic to the DNA sequence of the regulated gene and control the process of transcription. Transcription can be general transcription factors which are required for the basal transcription apparatus or regulatory transcription factors which may bind upstream or downstream from the transcription initiation site and either enhance or suppress the rate of transcription. See *Transcriptional Regulation*, ed. S.L. McKnight and K.R. Yamamoto, Cold Spring Harbor Press, Cold Spring Harbor, NY, 1992; Goodbourn, S., *Eukaryotic Gene Transcription*, IRL Press at Oxford, Oxford, UK, 1996; *Transcription Factor Protocols*, ed. M.J. Tymms, Humana, Totowa, NJ, 2000; *Transcription Factors*, ed. J.Locker, Academic Press(Bios), Oxford, United Kingdom, 2001; Michalik, L. and Wahli, W., Involvement of PPAR nuclear receptors in tissue injury and wound repair, *J.Clin.Invest.* 116, 598-606, 2006; Kikuchi, A., Kishida, S., and Yamamoto, H., Regulation of Wnt signaling by protein-protein interaction and post-translational modification, *Exp.Mol.Med.* 28, 1-10, 2006; Sharrocks, A.D., PIAS proteins and transcriptional regulation—more than just SUMO E3 ligases? *Genes Dev.* 20, 754-758, 2006; Campbell, K.J. and Perkins, N.D., Regulation of NF-kappaB function, *Biochem.Soc.Symp.*(73), 165-180, 2006; Russell, J. and Zomerdijk, J.C., The RNA polymerase I transcription machinery, *Biochem.Soc.Sym.*(73), 203-216, 2006; Gross, P. and Oelgeschlarger, T., Core promoter-selective RNA polymerase II transcription, *Biochem.Soc.Symp.*(73), 225-236, 2006. See NF-κB; See promoter elements, RNA polymerase, general transcription factors, regulatory transcription factors.

Transcriptional Silencing

Also known as transcription repression and results from the interaction of *cis*- and *trans*-components/sequences to inhibit the process of transcription of mRNA. Distinct from splicing silencing. See Nasmyth, K. and Shore, D., Transcriptional regulation in the yeast life cycle, *Science* 237, 1162-1170, 1987; Pannell, D. and Ellis, J., Silencing of gene expression: implications for design of retrovirus vectors, *Rev.Med.Virol.* 11, 205-217, 2001; Wanzel, M., Herold, S. and Eilers, M., Transcriptional repression by Myc, *Trends Cell Biol.* 13, 146-250, 2003; Ellis, J. and Yao, S., Retrovirus silencing and vector design: relevance to normal and cancer stem cells, *Curr.Gene.Ther.* 5, 367-373, 2005; Baniahmad, A., Nuclear hormone receptor co-repressors, *J.Steroid Biochem.Mol.Biol.* 93, 89-97, 2005; Spellman, R. and Smith, C.W.J., Novel modes of splicing repression by PTB, *Trends Biochem.Sci.* 31, 73-76, 2006.

Transcriptomics

The total RNA transcripts produced by a genome; the complete of RNA messages coded from the DNA within a cell. See Betts, J.C., Transcriptomics and proteomics: tools for the identification of novel drug targets and vaccine candidates for tuberculosis, *IUBMB Life* 53, 239-242, 2002; Hegde, P.S., White, I.R., and Delbouck, C., Interplay of transcriptomics and proteomics, *Curr. Opin.Biotechnol.* 14, 647-651, 2003; Jansen, B.J. and Schalkwijk, J., Transcriptomics and proteomics of human skin, *Brief Funct. Genomic Proteomic* 1, 326-341, 2003; Hu, Y.F., Kaplow, J., and He, Y., From traditional biomarkers to transcriptome analysis in drug development, *Curr.Mol.Med.* 5, 29-38, 2005; Viguerie, N., Poitou, C., Cancello, R., *et al.*, Transcriptomics applied to obesity and caloric restriction, *Biochemie* 87, 117-123, 2005; Seda, O., Tremblay, J., Sedova, L., and Hamet, P., Integrating genomics and transcriptomics with geo-ethnicity and the environment for the resolution of complex cardiovascular disease, *Curr.Opin.Mol. Ther.* 7, 583-587, 2005.

Transcytosis

Movement through a cell (usually an endothelial cell and vascular wall transport) as opposed to junctional transport (paracellular pathway). Involves a combination of endocytotic and exocytotic pathways. See Patel, H.M., Transcytosis of drug carriers carrying peptides across epithelial barriers, *Biochem.Soc. Trans.* 17, 940-942, 1989; Mostov, K., The polymeric immunoglobulin receptor, *Semin.Cell Biol.* 2, 411-418, 1991; Michel, C.C., Transport of macromolecules through microvascular walls, *Cardiovasc.Res.* 32, 644-653, 1996; Caplan, M.J. and Rodriguez-Boulan, E., Epithelial cell polarity: challenges and methodologies, in *Handbook of Physiology. Section 14, Cell Physiology*, ed . J.F. Hoffman and J.D. Jamieson, Oxford University Press (for the American Physiological Society), New York, NY, Chapter 17, 1997; Florence, A.T. and Hussain, N., Transcytosis of nanoparticles and dendrimer delivery systems: evoling vistas, *Adv.Drug Deliv.Rev.* 50(Suppl 1), S69-S89, 2001; Vogel, S.M. and Malik, A.B., Albumin transcytosis in mesothelium: further evidence of a transcellular pathway in polarized cells, *Am.J.Physiol.Lung Cell Mol.Physiol.* 282, L1-L2, 2002; Ghetie, V. and Ward, E.S., Transcytosis and catabolism of antibody, *Immunol.Res.* 25, 97-113, 2002; Kreuter, J., Influence of the surface properties on nanoparticle-mediated transport of drugs to the brain, *J.Nanosci.Nanotechnol.* 4, 484-488, 2004; Rot, A., Contribution of Duffy antigen to chemokine function, *Cytokine Growth Factor Rev.* 16, 687-694, 2005.

Transformation

Cell changes manifested by escape from control mechanisms generally resulting in increased growth potential, alterations in the cell surface and karyotypic abnormalities. Cell transformation generally occurs as a result of the acquisition of genetic information as by a virus entering the cell. See Enders, J.F., Cell transformation by viruses as illustrated by the response of human and hamster renal cells to Simian virus 40, *Harvey Lect.* 59, 113-153, 1965; Dulbecco, R., Transformation of cells in vitro by DNA-containing viruses, *JAMA* 190, 721-726, 1964; Black, P.M., The oncogenic DNA viruses: a review of in vitro transformation

studies, *Annu.Rev.Microbiol.* 22, 391-426, 1968; Hanafusa, H., Replication of oncogenic viruses in virus-induced tumor cells-their persistence and interaction with other viruses, *Adv. Cancer Res.*12, 137-165, 1969; Berk, A.J., Recent lessons in gene expression, cell cycle control, and cell biology from adenovirus, *Oncogene* 24, 7673-7685, 2005; Gius, D., Bradbury, C.M., Sun, L., Awwad, R.T., Huang, L., Smart, D.D., Bisht, K.S., Ho, A.S., and Nguyen, P., The epigenome as a molecular marker targer, *Cancer* 104, 1789-1793, 2005; Adhikary, S. and Eilers, M., Transcriptional regulation and transformation by Myc proteins, *Nat.Rev.Mol.Cell Biol.* 6, 635-645, 2005.

Transgene

A piece or segment of DNA, usually coding DNA, which is introduced into a cell or organism to modify the genome. Derivative animals are referred to as transgenic. See Grosveld, F.G. and Kollias, G.V., *Transgenic Animals*, Academic Press, San Diego CA, 1992; Hiatt, A., *Transgenic Plants: Fundamentals and Applications*, Marcel Dekker, New York, 1993; Gluethmann, H. and Ohashi, P.S., *Transgenesis and Targeted Mutagenesis in Immunology*, Academic Press, San Diego, 1994; Peña, L., *Transgenic Plants: Methods and Protocols*, Humana Press, Totowa, NJ, USA, 2005. Babinet, C., Morello, D., and Renard, J.P., Transgenic mice, *Genome* 31, 938-949, 1989; Dichek, D.A., Retroviral vector-mediated gene transfer into endothelial cells, *Mol.Biol.Med.* 8, 257-266, 1991; Janne, J., Hyttinen, J.M., Peura, T., *et al.*, Transgenic animals as bioproducers of therapeutic proteins, *Ann.Med.* 24, 273-280, 1992; Wright, D.C. and Wagner, T.E., Transgenic mice: a decade of progress in technology and research, *Mutat.Res.* 307, 429-440, 1994; Mittelstein, Scheid O., Transgene inactivation in *Aribidopsis thaliana*, *Curr.Top.Microbiol.Immunol.* 197, 29-42, 1995; Barry, M.A. and Johnston, S.A., Biological features of genetic immunization, *Vaccine* 15, 788-791, 1997; Patil, S.D., Rhodes, D.G., and Burgess, D.J., DNA-based therapeutics and DNA delivery systems: a comprehensive review, *AAPS J.* 7, E61-E77, 2005; Amsterdam, A. and Becker, T.S., Transgenes as screening tools to probe and manipulate the zebrafish genome, *Dev.Dyn.* 234, 255-268, 2005. Harrow, F. and Ortiz, B.D., The TCRα locus control region specifies thymic, but not peripheral patterns of TCRα gene expression, *J.Immunol.* 175, 6659-6667, 2005.

Translation

The process by which information is transferred from RNA to protein structure. See Phelps, C.S. and Arnstein, H.R.V., *Messenger RNA and Ribosomes in Protein Synthesis*, Biochemical Society, London, UK, 1982; Arnstein, H.R.V. and Cox, R.A., *Protein Biosynthesis*, IRL Press, Oxford, UK, 1992; Belasco, J.G., and Brawerman, G., *Control of Messenger RNA Stability*, Academic Press, San Diego, CA, 1993; Ilan, J., *Translational Regulation of Gene Expression 2*, Plenum Press, New York, NY, 1993; Tymms, M.J., *In Vitro Transcription and Translation Protocols*, Humana Press, Totowa, NJ, 1995; Weissman, S.M., *cDNA Preparation and Characterization*, Academic Press, San Diego, CA, 1999; Lapointe, J. and Brakier-Gingras, L., *Translation Mechanisms*, Landes Bioscience, Georgetown, TX, 2003; Schoenberg, D.R., *mRNA Processing and Metabolism: Methods and Protocols*, Humana Press, Totowa, NJ, 2004; Ochoa, S., Translation of the genetic message, *Bull.Soc.Chim.Biol.* 27, 721-737, 1967; Lewin, B., Units of transcription and translation: the relationship between heterogeneous nuclear RNA and messenger RNA, *Cell* 4, 11-20, 1975; Buetow, D.E. and Wood, W.M., The mitochondrial translation system, *Subcell.Biochem.* 5, 1-85, 1978; Kaufman, R.J., Control of gene expression at the level of translation initiation, *Curr.Opin.Biotechnol.* 5, 550-557, 1994; Yarus, M., On translation by RNAs alone, *Cold Spring Harb.Symp.Quant.Biol.* 66, 207-215, 2001; Frank, J., Towards an understanding of the structural basis of translation, *Genome Biol.* 4, 237, 2003; Huang, Y.S. and Richter, J.D., Regulation of local mRNA translation, *Curr.Opin.Cell Biol.* 16, 308-313, 2004; Kapp, L.D. and Lorsch, J.R., The molecular mechanics of eukaryotic translation, *Annu.Rev.Biochem.* 73, 657-704, 2004; Piper, M. and Holt, C., RNA translation in axons, *Annu.Rev. Cell Dev.Biol.* 20, 505-523, 2004; Noller, H.F., The driving of force for molecular evolution of translation, *RNA* 10, 1833-1837, 2004; Katz, L. and Ashley, G.W., Translation and protein synthesis: macrolides, *Chem.Rev.* 105, 499-528, 2005; Jackson, R.J., Alternative mechanisms of initiating translation of mammalian mRNAs, *Biochem.Soc.Trans.* 33, 1231-1241, 2005; Deana, A. and Belasco, J.G., Lost in translation: the influence of ribosomes on bacterial mRNA decay, *Genes Dev.* 19, 2526-2533, 2005; Pique, M., Lopez, J.M. and Mendez, R. Cytoplasmic mRNA polyadenylation and translation assays, *Methods Mol.Biol.* 322, 183-198, 2006; Schuman, E.M., Dynes, J.L., and Steward, O., Synaptic regulation of translation of dendritic mRNAs, *J.Neurosci.* 26, 7143-7146, 2006.

Translocation

The movement of a ribosome along mRNA during protein synthesis: this process involved the participation of elongation factor (EF-G) and is accompanied by GTP hydrolysis. Translocation also refers to the process of protein transport across membranes which may be assisted by chaperone. Protein secretion from the cell also is described as translocation; type II secretion (the general secretory pathway) involves a multiprotein complex referred to as the translocon. See Egae, P.F., Stroud, P.W., and Walter, P., Targeting proteins to membranes: structure of the signal recognition particle, *Curr.Opin.Struct.Biol.* 15, 213-220, 2005; Collinson, I., The structure of the bacterial translocation complex, SecYEG, *Biochem. Soc.Trans.* 33, 1225-1230, 2005; Chavan, M. and Lennarz, W., The molecular basis of coupling of translocation and *N*-glycosylation, *Trends Biochem.Sci.* 31, 17-20, 2006.

Translocation also refers to the movement of water and solutes in a plant, in particular from the roots to the shoots. See Kutchan, T.M., A role for intra- and intercellular translocation in natural products, *Curr.Opin.Plant Biol.* 8, 292-300, 2005; Yang, X., Feng, Y, He, Z., and Stoffells, P.J., Molecular mechanisms of heavy metal hyperaccumulation and phytoremediation, *J.Trace Elem.Med.Biol.* 18, 339-353, 2005; Mackenzie, S.A., Plant organellar protein targeting: a traffic plan still under construction, *Trends Cell.Biol.* 15, 548-554, 2005; Thompson, M.V., Phloem: the long and the short of it, *Trends Plant Sci.* 11, 26-32, 2006; Takahashi, H., Yoshimoto, N., and Saito, K., Anionic nutrient transport in plants: the molecular basis of the sulfate transporter gene family, *Genet.Eng.* (NY) 27, 67-80, 2006.

Translocon

A multiprotein complex (composed of several ER proteins) which mediates protein transport (cotranslational protein translocation) across membranes; interacts with single recognition particle (SRP). See Johnson, A.E. and van Waes, M.A., The translocon: a dynamic gateway at the ER membrane, *Annu.Rev.Cell Dev.Biol.* 15, 799-842, 1999; May, T. and Soll, J., Chloroplast precursor protein translocon, *FEBS Lett.* 452, 52-56, 1999; Johnson, A.E. and Haigh, N.G., The ER translocon and retrotranslocation: is the

shift into reverse manual or automatic? *Cell* 102, 709-712, 2000; White, S.H., Translocons, thermodynamics, and the folding of membrane proteins, *FEBS Lett.* 555, 116-221, 2003; Coombes, B.K. and Finlay, B.B., Insertion of the bacterial type III translocon: not your average needle stick, *Trends Microbiol.* 13, 92-95, 2006; Chavan, M. and Lennarz, W., The molecular basis of coupling of translocation and *N*-glycosylation, *Trends Biochem.Sci.* 31, 17-20, 2006.

TRANSIL

Porous silica beads which can be coated with a single phospholipid bilayers and used to study protein-lipid interactions. See Schmitz, A.A., Schleiff, E., Rohrig, C. *et al.*, Interactions of myristoylated alanine-rich kinase substrates (MARCKS)-related protein with a novel solid-supported lipid membrane system (TRANSIL), *Anal.Biochem.* 268, 343-353, 1999; Loidl-Stahlhofen, A., Hartmann, T., Schottner, M., *et al.*, Multilamellar liposomes and solid-supported lipid membranes (TRANSIL): screening of lipid-water partitioning toward a high-throughput scale, *Pharm.Res.* 18, 1782-1788, 2001; Schuhmacher, J., Kohlsdorfer, C., Buhner, K., *et al.*, High-throughput determination of the free fraction of drugs strongly bound to plasma proteins, *J.Pharm.Sci.* 93, 816-830, 2004.

Transportan

A 27 amino acid chimeric peptide with cell penetrating properties. See cell penetrating peptides. See Padiri, K. Säälik, P., Hansen, M., Koppel, K., Raid, R., Langel, U., and Pooga, M., Cell transduction pathways of transportans, *Bioconjugate Chem.* 16, 1399-1410, 2005; Pooga, M., Hällbrink, M., Zorko, M., and Langel, Ü., Cell penetration by transportan, *FASEB J.* 12, 67-77, 1998.

Trans-splicing

A process that occurs with both nucleic acids and proteins. With nucleic acids, *trans*-splicing (transsplicing) occurs as part of pre-mRNA processing increasing messenger diversity. The trans-splicing of pre-mRNA is not related the removal of introns via cis-splicing. Trans-splicing transfers RNA segments from one RNA molecule to another while cis-splicing removes introns from the same RNA molecule. See Bonen, L., Trans-splicing of pre-mRNA in plants, animals, and protists, *FASEB J.* 7, 40-46, 1993; Nilsen, T.W., Trans-splicing: an update, *Mol.Biochem.Parasitol.* 73, 1-6, 1995; Adams, M.D., Rudner, D.Z., and Rio, D.C., Biochemistry and regulation of pre-mRNA splicing, *Curr.Opin.Cell Biol.* 8, 331-339. 1996; Frantz, C., Ebel, C., Paulus, F. and Imbault, P., Characterization of *trans*-splicing in Euglenoids, *Curr.Genet.* 37, 349-355, 2000; Garcia-Blanco, M.A., Messenger RNA reprogramming by spliceosome-mediated RNA trans-splicing, *J.Clin.Invest.* 112, 474-480, 2003; Kornblihtt, A.R., de la Maya, M., Fededa, J.P. *et al.*, Multiple links between transcription and splicing, *RNA* 10, 1489-1498, 2004; Mitchell, L.G. and McGarrity, G.J., Gene therapy progress and prospects: reprogramming gene expression by trans-splicing, *Gene Ther.* 12, 1477-1485, 2005; Yang, Y. and Walsh, C.E., Spliceosome-mediated RNA trans-splicing, *Mol.Ther.* 12, 1006-1012, 2005. Cheng, G., Cohen, L., Ndegwa, D., and Davis, R.E., The flatworm spliced leader 3'-terminal AUG as a translation initiator methionine, *J.Biol.Chem.* 281, 733-743, 2006.

SL (spliced leader) trans-splicing and alternative trans-splicing are special cases of trans-splicing for nucleic acids. Trans-splicing also occurs with proteins but is most often a technique to use intein chemistry for ligation. See Shi, J. and Muir, T.W., Development of a tandem protein trans-splicing system based on native and engineered split inteins, *J.Am.Chem.Soc.* 127, 6198-6206, 2005; Khan, M.S., Khalid, A.M., and Malik, K.A., Intein-mediated protein trans-splicing and transgene containment in plastids, *Trends Biotechnol.* 23, 217-220, 2005; Kwon, Y., Coleman, M.A., and Camarero, J.A., Selective immobilization of proteiosn onto solid supports through split-intein-mediated protein trans-splicing, *Angew.Chem.Int.Ed. Engl.* 45, 1726-1729, 2006; Iwai, H., Zuger, S., Jin, J., and Tam, P.H., Highly efficient protein trans-splicing by a naturally split DnaE intein from *Nostoc punctiforms*, *FEBS Lett.* 580, 1853-1858, 2006; Muralidharan, V. and Muir, T.W., Protein ligation: an enabling technology for the biophysical analysis of proteins, *Nat.Methods* 3, 429-438, 2006. See inteins, alternative trans-splicing, spliced leader trans-splicing.

Transvection

Literally to carry over or to carry across. In mathematics, a linear function. In biology, where gene expression is influence by *trans*-interactions between alleles depending on somatic pairing between homologous chromosome regions; it can result in partial complementation between mutant alleles. See Judd, B.H., Transvection: allelic cross talk, *Cell* 53, 841-843, 1988; Rassoulzadegan, M., Magliano, M., and Cuzin, F., Transvection effects involving DNA methylation during meiosis in the mouse, *EMBO J.* 21, 440-450, 2002; Duncan, I.W., Transvection effects in *Drosophila* 36, 521-556, 2002; Coulhard, A.B., Nolan, N., Bell, J.B., and Hilliker, A.J., Transvection at the vestigial locus of *Drosophila melanogaster* 170, 1711-1721, 2005.

Triabody

An noncovalent trimer formed with scFv fragments engineered with no linker between the V_H and V_L domains. The normal linker engineered between the V_H and V_L domains is 15 residues (usually glycine and serine to promote maximum flexibility) which yields as monomer; if the linker is reduced to 10 residues, a dimer (diabody) is formed while with no linker there is a trimer or higher order polymer. See Le Gall, E., Kipriyanov, S.M., Moldenhauer, G., and Little, M., Di-, tri- and tetrameric single chain Fv antibody fragments against human CD19: effect of valency on cell binding, *FEBS Lett.* 453, 164-168, 1999; Atwell, J.L., Breheney, K.A., Lawrence, L.J. *et al.*, scFv multimers of th anti-neuraminidase antibody NC10: length of the linker between V_H and V_L domains dictates precisely the transition between diabodies and triabodies, *Protein Eng.* 12, 597-604, 1999. Todorovska, A., Roovers, R.C., Dolezal, O., *et al.*, Design and application of diabodies, triabodies and tetrabodies for cancer targeting, *J.Immunol.Methods* 248, 47-66, 2001. See also tribody, diabody, single chain Fv. Note: It has been observed that if the order of the variable regions are switch in scFv construct (V_L-V_H instead of V_H-V_L), the engineered scFv with a zero-length linker formed a dimer (diabody) instead of the expected trimer – See Arndt, M.A.E., Krauss, J., and Rybak, S.M., Antigen binding and stability properties of non-covalently linked anti-CD22 single-chain Fv dimers, *FEBS Lett.* 578, 257-261, 2004.

Tribody

A trivalent antibody construct with two scFv fragments attached to the C-terminal ends of an Fab fragment. See Schoonjans, R., Willems, A., Schoonooghe, S., Fiers, W., *et al.*, Fab chains as an efficient heterodimerization scaffold for the

production of recombinant bispecific and trispecific antibody derivatives, *J.Immunol.* 165, 7050-7057, 2000; Willems, A., Leonen, J., Schoonooghe, S., *et al.*, Optimizing expression and purification from cell culture of trispecific recombinant antibody derivatives, *J.Chromatog.B.Anal.Technol.Biomed.Life Sci.* 786, 161-176, 2003. See triabody and bibody.

"tri-reagents"

See RNA Isolation

Tris-lipidation

A process of linking a hydrophobic component to a peptide or protein to enhance membrane binding. The hydroxyl groups of Tris are esterified with fatty acids and subsequently coupled to a peptide or protein via the amino group. See Whittaker, R.G., Hayes, P.J., and Bender, V., A gentle method of linking Tris to amino acids and peptides, *Pept.Res.* 6, 125-128, 1993; Ali, M., Amon, M., Bender, V., and Manolis, N., Hydrophobic transmembrane-peptide lipid conjugation enhances membrane binding and functional activity in T-cells, *Bioconjugate Chem.* 16, 1556-1563, 2005.

Troybody

Antibody with specificity for APC which has an antigenic sequence inserted into a constant domain region. See Lunde, E., Lauvrak, V., Rasmussen, I.B., *et al.*, Troybodies and pepbodies, *Biochem.Soc.Trans.* 30, 500-506, 2002; Lunde, E., Western, K.H., Rasmussen, I.B., Sandlie, I., and Bogen, B., Efficient delivery of T cell epitopes to APC by use of MHC class II-specific troybodies, *J.Immunol.* 168, 2154-2162, 2002; Tunheim, G., Schjetne, K.W., Fredrikson, A.B., Sandlie, I., and Bogen, G., Human CD14 is an efficient target for recombinant immunoglobulin vaccine constructs that delivers T cell epitopes, *J.Leuk.Biol.* 77, 303-310, 2005.

Tubulin

A protein which polymerizes to form microtubules. Tubulin is target for anticancer therapy. See Feit, H., Slusarek, L., and Shelanski, M.L., Heterogeniety of tubulin subunits, *Proc.Natl.Acad.Sci.USA* 68, 2028-2031, 1971; Fine, R.E., Heterogeneity of tubulin, *Nat.New Biol.* 233, 283-284, 1971; Rappaport, L., Leterrier, J.F. and Nunez, J., Non phosphorylation in vitro of 6 S tubulin from brain and thyroid tissue, *FEBS Lett.* 26, 239-352, 1972; Berry, R.W. and Shelanski, M.L., Interactions of tubulin with vinblastine and guanosine triphosphate, *J.Mol.Biol.* 71, 71-80, 1972; Hemminki, K., Relative turnover of tubulin subunits in rat brain, *Biochim.Biophys.Acta* 310, 285-288, 1973; Timasheff, S.N., Frigon, R.P., and Lee, J.C., A solution physical-chemical examination of the self-association of tubulin, *Fed.Proc.* 35, 1886-1891, 1976; Mohri, H., The function of tubulin in motile systems, *Biochim.Biophys.Acta* 456, 85-127, 1976; Caplow, M. and Zeeberg, B., Stoichiometry for guanine nucleotide binding to tubulin under polymerizing and nonpolymerizing conditions, *Arch.Biochem.Biophys.* 203, 404-411, 1980; Zeeberg, B., Cheek, J., and Caplow, M., Exchange of tubulin dimer into rings in microtubule assembly-disassembly, *Biochemistry* 19, 5078-5086, 1980; Cleveland, D.W., Treadmilling of tubulin and actin, *Cell* 28, 689-691, 1982; Sternlicht, H., Yaffe, M.B., and Farr, G.W., A model of the nucleotide-binding site in tubulin, *FEBS Lett.* 214, 226-235, 1987; Oakley, B.R., γ-Tubulin, *Curr.Top.Dev.Biol.* 49, 27-54, 2000; Dutcher, S.K., Motile organelles: the importance of specific tubulin isoforms, *Curr.Biol.* 11, R419-R422, 2001; McKean, P.G., Vaughan, S., and Gull, K., The extended tubulin superfamily, *J.Cell Sci.* 114, 2723-2733, 2001; Cowan, N.J. and Lewis, S.A., Type II chaperonins, prefoldin, and the tubulin-specific chaperones, *Adv.Protein Chem.* 59, 73-104, 2001; Addinall, S.G. and Holland, B., The tubulin ancestor, FtsZ, draughtsman, designer, and driving force for bacterial cytokinesis, *J.Mol.Biol.* 318, 219-236, 2002; Szymanski, D., Tubulin folding cofactors: half a dozen for a dimer, *Curr.Biol.* 12, R767-R769, 2002; Dutcher, S.K., Long-lost relatives reappear: identification of new members of the tubulin superfamily, *Curr.Opin.Microbiol.* 6, 634-640, 2003; Caplow, M. and Fee, L., Concerning the chemical nature of tubulin subunits that cap and stabilize microtubules, *Biochemistry* 42, 2122-2126, 2003; Nogales, E., Wang, H.W., and Niederstrasser, H., Tubulin rings: which way do they curve?, *Curr.Opin.Struct.Biol.* 13, 256-261, 2003; Pellegrini, F. and Budman, D.R., Tubulin function, action of antitubulin drugs, and new drug development, *Cancer Invest.* 23, 264-273, 2005; Nogeles, E. and Wang, H.W., Structural mechanisms underlying nucleotide-dependent self-assembly of tubulin and its relatives, *Curr.Opin.Struct.Biol.* 16m, 221-229, 2006.

Tumor Suppressor Gene

A gene responsible for the encoding of products which suppress the malignant phenotype. These gene were first identified in hybrid cells resulting from cell fusion. Loss of tumor suppressor genes result in cell cycle deregulation. p53 is one of the better known tumor suppressor genes. See Wynford-Thomas, D., Oncogenes and anti-oncogenes: the molecular basis of tumor behavior, *J.Pathol.* 165, 187-201, 1991; Carbone, D.P., Oncogenes and tumor suppressor genes, *Hosp.Pract.* 28, 145-148, 1993; Skapek, S.X. and Chui, C.H., Cytogenetics and the biologic basis of sarcomas, *Curr.Opin.Oncol.* 12, 315-322, 2000; Lee, M.P., Genome-wide analysis of epigenetics in cancer, *Ann.N.Y.Acad.Sci.* 983, 101-109, 2003; Bocchetta, M. and Carbone, M., Epidemiology and molecular pathology at crossroads to establish causation: molecular mechanisms of malignant transformation, *Oncogene* 23, 6484-6491, 2004; Seth, A. and Watson, D.K., ETS transcription factors and their emerging roles in human cancer, *Eur.J.Cancer* 41, 2462-2478, 2005.

Turbidimetry

Turbidity is a measured of the light scattered from the direct path of the electromagnetic radiation; practically, it is the transmitted light. It represents electromagnetic radiation that is not absorbed as in spectroscopy but rather scattered. It is sometimes necessary to correct spectral measurements for turbidimetry or more commonly just as light scattering. The extent to which electromagnetic radiation is scattered and measured either by turbidimetry or nephelometry depends on the size of the particle and the wavelength of the incident radiation. Turbidimetry is used in clinical chemistry (Blirup-Jensen, S., Protein standardization III: Method optimization basic principles for quantitative determination of human serum proteins on automated instruments based on turbidimetry or nephelometry, *Clin.Chem.Lab.Med.* 39, 1098-1109, 2001; platelet aggregation (Cruz, W.O., Platelet determination by turbidimetry, *Blood* 9, 920-926, 1954; Jarvis, G.E., Platelet aggregation: turbidimetric measurements, *Methods Mol.Biol.* 272, 65-76, 2004) and for the assay of some enzymes (Rapport, M.M., Meyer, K., and Linker, A., Correlation of reductimetric and turbidimetric methods for hyaluronidase, *J.Biol.Chem.* 186, 615-623, 1950; Houck, J.C., The turbidimetric determination of deoxyribonuclease activity, *Arch. Biochem.Biophys.* 82, 135-144, 1959; Morsky, P., Turbidimetric determination of lysozyme with *Micrococcus lysodeikticus* cells: reexamination of reaction conditions, *Anal.Biochem.* 128, 77-85, 1983; Jenzano, J.W. and Lundblad, R.L., Effects of amines and polyamines on turbidmetric and lysoplate assays for lysozyme, *J.Clin. Microbiol.* 26, 34-37, 1988; Walker, M.B., Retzinger, A.C., and Retzinger, G.S., A turbidmetric method for measuring the activity of trypsin and its inhibition, *Anal.Biochem.* 351, 114-121, 2006). See Zattoni, A., Loli Piccolomini, E., Torsi, G., and Rschiglian, P., Turbidimetric detection method in flow-assisted separation of dispersed samples, *Anal.Chem.* 75, 6469-6477, 2003; Mori, Y., Kitao, M., Tomita, N., and Natomi, T., Real-time turbidimetry of LAMP reaction for quantifying template DNA, *J.Biochem.Biophys. Methods* 31, 145-157, 2004; Hianik, T., Rybar, P., Andreev, S.Y. *et al.*, Detection of DNA hybridization on a liposome surface using ultrasound velocimetry and turbidimetry methods, *Bioorg.Med.Chem. Lett.* 14, 3897-3900, 2004; Gonzalez, V.D., Gugliotta, L.M., Vega, J.R., and Meira, G.R., Contamination by larger particles of two almost-uniform lattices: analysis by combined dynamic light scattering and turbidimetry, *J.Colloid Interface Sci.* 285, 581-589, 2005; Stano, P., Bufali, S., Damozou, A.S., and Luisi, P.L., Effect of tryptophan oligopeptides on the size distribution of POPC liposomes: a dynamic light scattering and turbidimetric study, *J.Liposome Res.* 15, 29-47, 2005; Mao, J., Kondu, S., Ji, H.F., and McShane, M.J., Study of the near-neutral pH-sensitivity of chitosan/gelatin hydrogels by turbidimetry and microcantilever deflection, *Biotechnol. Bioeng.*, 95, 333–341, 2006.

Tyrosine Kinase

A large group of enzymes involved in intracellular signal transduction which catalyze the phosphorylation of tyrosine residues in target proteins. See Hardle, D.G., *Protein Phosphorylation: A Practical Approach*, Oxford University Press, Oxford, United Kingdom, 1993; Woodgett, J.R., *Protein Kinases*, IRL Press at Oxford University Press, Oxford, United Kingdom, 1994; *The Protein Kinase Factsbooks*, ed. D.G. Hardle and S. Hanks, Academic Press, San Diego, CA, USA, 1995; *Protein Kinase Protocols*, ed. G. Krauss, Wiley-VCH, Weinheim, Germany, 2003.

Tyrphostins

Inhibitors of protein tyrosine kinases. See Levitzki, A., Tyrphostins—potential antiproliferative agents and novel molecular tools, *Biochem.Pharmacol.* 40, 913-918, 1990; Lamb, D.J. and Shubhaba, S., Tyrphostins inhibit Sertoli cell-secreted growth factor stimulation of A431 cell growth, *Recent Prog.Homr.Res.* 48, 511-516, 1993; Wolbring, G., Hollenberg, M.D., and Schnetkamp, P.P., Inhibition of GTP-utilizing enzymes by tyrphostins, *J.Biol. Chem.* 269, 22470-22472, 1994; Holen, I., Stromhaug, P.E., Gordon, P.B., *et al.*, Inhibition of autophagy and multiple steps in asialoglycoprotein endocytosis by inhibitors of tyrosine protein kinases, (tyrphostins), *J.Biol.Chem.* 270, 12823-12831, 1995; Jaleel, M., Shenoy, A.R., and Visweswariah, S.S., Tyrphostins are inhibitors of guanylyl and adenylyl cyclases, *Biochemistry* 43, 8247-8255, 2004; Levitzki, A. and Mishani, E., Tyrphostins and other tyrosine kinase inhibitors, *Annu.Rev.Biochem.* 75, 93-109, 2006.

Ubiquitin

Ubiquitin is a small intracellular protein that serves as marker for protein degradation by the proteosome. This is a process of controlled proteolysis which is an integral part of normal cell function. Some functions of the ubiquitin-protesome system included the degradation of misfolded proteins and the production of peptides during MHC class I antigen presentation (Michalek, M.T., Grant, E.P., Gramm, C., *et al.*, A role for the ubiquitin-dependent proteolytic pathway in MHC class I-restricted antigen presentation, *Nature* 363, 552-554, 1993). Ubiquitin is linked to a protein via an isopeptide bond in a process referred to as ubiquitinylation which is catalyzed by a ubiquitin ligase (Robinson, P.A. and Ardley, H.C., Ubiquitin-protein ligases, *J.Cell.Sci.* 5191-5194, 2004; Pavletich, N.P., Structural biology of ubiquitin-protein ligases, *Harvey Lect.* 98, 65-102, 2002-2003). Ubiquitin is initially "activated" by the ubiquity ligase to form a high-energy thioester bond between the enzyme and the C-terminal glycine residue of ubiquitin; the ubiquitin is then transferred to a lysine residue on the target protein forming the isopeptide peptide. While the discovery of ubiquitin was based on its ability to target proteins to degradation, it is clear that there are other functions (Welchman, R.L., Gordon, C., and Mayer, R.J., Ubiquitin and ubiquitin-like proteins as multifunctional signals, *Nat.Rev.Mol.Cell Biol.* 6, 599-609, 2005; Hicke, L., Schubert, H.L., and Hill, C.P., Ubiquitin-binding domain, *Nature Rev.Mol.Cell Biol.* 6, 610-621. 2005; Chen, Z.J., Ubiquitin signaling in the NF-κB pathway, *Nat. Cell Biol.* 7, 758-765, 2005;). There is a ubiquitin family of proteins (Catic, A. and Ploegh, H.L., Ubiquitin — conserved protein or selfish gene? *Trends Biochem.Sci.* 30, 600-604, 2005; consisting of type I ubiquitin-like proteins and type II ubiquitin-like proteins (Pickart, C.M. and Eddins, M.J., Ubiquitin: structures, functions, mechanisms, *Biochim.Biophys.Acta* 1695, 755-72, 2004; Walters, K.J., Goh, A.M., Wang, Q., *et al.*, Ubiquitin family proteins and their relationship to the proteosome: a structural perspective, *Biochim.Biophys.Acta* 1695, 73-87, 2004). While there are few type 1 family members, they are well known with Nedd8 and SUMO (Kroetz, M.B., SUMO: a ubiquitin-like protein modifier, *Yale J.Biol.Med.* 78, 197-201, 2005. See Rechsteiner, M., Ubiquitin-mediated pathways for intracellular proteolysis, *Annu. Rev.Cell Biol.* 3, 1-30, 1987; Ciechanover, A., Gonen, H., Elias, S., and Mayer, A., Degradation of proteins by the ubiquitin-mediated proteolytic pathway, *New Biol.* 2, 227-234, 1990; Smalle, J. and Vierstra, R.D., The ubiquitin 26S proteosome proteolytic pathway, *Annu.Rev.Plant Biol.* 55, 555-590, 2004; Denison, C., Kirkpatrick, D.S., and Gygi, S.P., Proteomic insights into ubiquitin-like proteins, *Curr.Opin.Chem.Biol.* 9, 69-75, 2005; Miller, J. and Gordon, C., The regulation of proteosome degradation by multi-ubiquitin chain binding proteins, *FEBS Lett.* 579, 3224-3230, 2005; Ye, Y., The role of the ubiquitin-protesome system in ER quality control, *Essays Biochem.* 41, 99-112, 2005; Salomens, F.A., Verhoef, L.G. and Dantuma, N.P., Fluorescent reporters of the ubiquitin-proteosome system, *Essays Biochem.* 41, 113-128, 2005; Nakayama, K.I. and Nakayama, K., Ubiquitin ligases: cell-cycle control and cancer, *Nat.Rev.Cancer* 6, 369-381, 2006.

Ultraconserved Elements

A class of conserved elements in genomes between orthologous domains that share 100% identity over at least 200 bp in mammalian genomes. See Berjano, G., Pheasant, M., Makunin, I., *et al.*, Ultraconserved elements in the human genome, *Science* 304, 1321-1325, 2004.

UHF Dielectrometry

Physical technique to study the state of protein-bound water see. Hackl, E.V., Gatash, S.V. and Nikalov, O.T., Using UHF-dielectrometry to study protein structural transitions, *J.Biochem. Biophys.Methods* 64, 127-148.

Validity

External validity refers to the extent to which a specific finding from an investigation or analytical process can be generalized beyond the context of the specific investigation or analytical process. For regulatory purposes such as the manufacture of drugs and therapeutic biologicals, validity can be considered to be demonstration of the ability to reproducible repeat the process and/or assay. The validation process is the process by which an organization can demonstrate that the process is reproducible and, therefore, their process valid.

Variegation

Literally, the state of discrete diversified coloration. In biology, this can refer to the discrete coloration patterns in leaves, or the occurrence within a tissue of sectors or clones of different phenotypes. In genetics, it is taken to mean a chromosome position effect when particular loci are contiguous with heterochromatin. See Baker, W.K., Position-effect variegation, *Adv.Genet.* 14, 133-169, 1968; Henikoff, S., Position-effect variegation after 60 years, *Trends Genet.* 6, 422-426, 1990; Cook, K.R. and Karpen, G.H., A rosy future for heterochromatin, *Proc.Natl.Acad.Sci.USA* 91, 5219-5921, 1994; Martin, D.I. and Whitelaw, E., The vagaries or variegating transgenes, *Bioessays* 18, 919-923, 1996; Klein, C.G. and Costa, M., DNA methylation, heterochromatin and epigenetic carcinogenesis, *Mutat. Res.* 386, 163-180, 1997; Zhimulev, I.F., Polytene chromosomes, heterochromatin, and position effect variegation, *Adv.Genet.* 37, 1-566, 1998; Hennig, W., Heterochromatin, *Chromosoma* 108, 1-9, 1999; Shotta, G., Ebert, A., Dorn, R., and Reuter, G., Position-effect variegation and the genetic dissection of chromatin regulation in *Drosophila*, *Semin.Cell Dev.Biol.* 14, 67-75, 2003.

V(D)J Recombination

The process by which discontinuous regions of DNA become joined in lymphocytes resulting in rearrangement of the DNA germline; the process by which diversity is built into immunoglobulins. See Alt, F.W., Oltz, E.M., Young, F., Gorman, J., Taccioli, G., and Chen, J., VDJ recombination, *Immunol.Today* 13, 306-314, 1992; Jung, D. and Alt, F.W., Unraveling V(D)J recombination: insights into gene regulation, *Cell* 116, 299-311, 2004; Schatz, D.G., V(D)J recombination, *Immunol.Rev.* 200, 5-11, 2004; Jones, J.M. and Gellert, M., The taming of a transposon: V(D)J recombination and the immune system, *Immunol.Rev.* 200, 233-248, 2004; Dudley, D.D., Chaudhuri, J., Bassing, C.H., and Alt, F.W., Mechanism and control of V(D)J recombination versus class switch recombination: similarities and differences, *Adv. Immunol.* 86, 43-112, 2005; Aplon, P.D., Causes of oncogenic chromosomal translocation, *Trends Gene.* 22, 46-55, 2006.

VGF

A neuronal protein involved in cell differentiation. See Salton, S.R.J., Fischber, D.J., and Don, K.-W., Structure of the gene encoding VGF, a nervous system-specific mRNA that is rapidly and selectively induced by nerve growth factor in PC12 cells, *Mol.Cell Biol.* 11, 2335-2349, 1991.

VICKZ Proteins

A family of RNA-binding proteins recognizing specific *cis*-acting elements acting on a variety of transcriptional processes involved in cell polarity and migration. See Yisraeli, J.K., VICKZ proteins:

A multi-talented family of regulatory RNA-binding proteins, *Biol.Chem.* 97, 87-96, 2005.

Virulence Factor(s)

A factor or factors elaborated by an organism, usually as bacteria, that are response for the pathogenicity of the organism. An example would be a bacterial exotoxin. See Evans, D.J., Jr. and Evans, D.G., Classification of pathogenic *Escherichia coli* according to serotype and the production of virulence factors, with special reference to colonization-factor antigens, *Rev.Infect.Dis.* 5(Suppl 4), S692-S701, 1983 Lubran, M.M., Bacterial toxins, *Ann.Clin.Lab.Sci.* 18, 58-71, 1988; Moxon, E.R. and Kroll, J.S., Type b capsular polysaccharide as a virulence factor of *Haemophilis influenzae, Vaccine* 6, 113-115, 1988; Pragman, A.A. and Schievert, P.M. Virulence regulation in *Staphylococcus aureus*: the need for in vivo analysis of virulence factor regulation, *FEMS Immuno.Med.Microbiol.* 42, 147-154, 2004; Walker, M.J., McArthur, J.D., McKay, F., and Ranson, M., Is plasminogen deployed as a *Streptococcus pyogenes* virulence factor? *Trends Microbiol.* 13, 308-313, 2005; Lu, H., Yamaoka, Y., and Graham, D.Y., *Heliobacter pylori* virulence factors: fact and fantasies, *Curr.Opin.Gastroenterol.* 21, 653-659, 2005; Zaas, D.W., Duncan, M., Rae Wright, J., and Abraham, S.N., The role of lipid rafts in the pathogenesis of bacterial infections, *Biochim.Biophys. Acta* 1746, 305-313, 2005; Kazmierczak, M.J., Wiedmann, M., and Boor, K.J., Alternative sigma factors and their roles in bacterial virulence, *Microbiol.Mol.Biol.Rev.* 69, 527-543, 2005; Yates, S.P., Jørgensen, R., Andersen, G.R., *et al.*, Stealth and mimicry by deadly bacterial toxins, *Trends Biochem.Sci.* 31, 123-133, 2006.

Viscosity

The property of a fluid indicating resistance to change in form or resistance to flow. There is considerable interest in the viscosity of blood as its related to cardiovascular disease (Somer, T. and Meiselman, H.J., Disorders of blood viscosity, *Ann.Med.* 25, 31-39, 1993). See Kupke, D.W. and Crouch, T.H., Magnetic suspension: density-volume, viscosity, and osmotic pressure, *Methods Enzymol.* 48, 29-68, 1978; Ahmad, F. and McPhie, P., The intrinsic viscosity of glycoproteins, *Int.J.Biochem.* 11, 91-96, 1980; Harding, S.E., The intrinsic viscosity of biological macromolecules. Progress in measurement, interpretation and application to structure in dilute solution, *Prog.Biophys.Mol.Biol.* 68, 207-262, 1997; Ulker, P., Alexy, T., Meiselman, H.J., and Baskurt, O.K., Estimation of infused dextran plasma concentration via measurement of plasma viscosity, *Biorheology* 43, 161-166, 2006; Chopra, S., Lynch, R., Kim, S.H., *et al.*, Effects of temperature and viscosity on R67 dihydrofolate reductase catalysis, *Biochemistry* 45, 6596-6505, 2006; Laghaei, R., Nasrabad, A.E., and Eu, B.C., Generic van der Waals equation of state, modified free volume theory of diffusion, and viscosity of simple liquids, *J.Phys.Chem.B Condens.Matter Mater.Surf. Interfaces Biophys.* 109, 5873-5883, 2005; Brookes, R., Davies, A., Ketwaroo, G., and Madden, P.A., Diffusion coefficients in ionic liquids: relationship to the viscosity, *J.Phys.Chem.B Condens.Matter Mater.Surf. Interfaces Biophys.* 109, 6485-6490, 2005; Stillinger, F.H. and Debenedetti, P.G., Alternative view of self-diffusion and shear viscosity, *J.Phys. Chem. B Condens.Matter Mater.Surf. Interfaces Biophys.* 109, 6605-6609, 2005; Chopra, S., Lynch, R., Kim, S.H., *et al.*, Effects of temperature and viscosity on R67 dihydrofolate reductase catalysis, *Biochemistry* 45, 6596-6505, 2006; Donoso, M. and Ghaly, E.S., Use of near-infrared for quantitative measurement of viscosity and concentration of active ingredient in pharmaceutical gel,

Pharm. Dev Technol. 11, 389-397, 2006; Haidekker, M.A., Akers, W.J., Fischer, D., and Theodorakis, E.A., Optical fiber-based fluorescent viscosity sensor, *Opt.Lett.* 31, 2529-2531, 2006.

Vitamers

Different chemical structural forms of a vitamin which have the same biological activity. See Bender, D.A., *Nutritional Biochemistry of the Vitamins*, 2nd Edn., Cambridge University Press, Cambridge, United Kingdom, 2003; Voziyan, P.A. and Hudson, B.G., Pyridoxamine. The many virtues of a Maillard reaction inhibitor, *Ann.N.Y.Acad.Sci.* 1043, 807-816, 2005.

Ealker A Motif

A motif described in SKN-1, a transcription factor in *Caenorhabditis elegans*. See Walker, A.K., See, R., Batchelder, C., Kophengnavong, T., Gronniger, J.T., Shi, Y., and Blackwell, T.K., A conserved transcription motif suggesting functional parallels between *Caenorhabitis elegans* SKN-1 and Cap'n'Collar-related basic leucine zipper proteins, *J.Biol.Chem.* 275, 22166-22171, 2000.

Western Blotting

A method for identifying proteins after electrophoretic separation involving a specific probe, usually an antibody. It was derived from the earlier development of Southern Blotting and Northern Blotting. See also Southwestern Blotting and Northwestern Blotting. See Radka, S.F., Monoclonal antibodies to human major histocompatibility complex class II antigens, *Crit.Rev.Immunol.* 8, 23-48, 1987; Heerman, K.H., Gultekin, H., and Gerlich, W.H., Protein blotting: techniques and application in virus hepatitis research, *Ric.Clin.Lab.* 18, 193-221, 1988; Hayes, P.C., Wolf, C.R., and Hayes, J.D., Blotting techniques for the study of DNA, RNA, and proteins, *BMJ* 299, 965-968, 1989; Harper, D.R., Kit, M.L., and Kangrok H.O., Protein blotting: ten years on, *J.Virol. Methods* 30, 25-39, 1990; Dunn, M.J., Detection of proteins on blots using the avidin-biotin system, *Methods Mol.Biol.* 32, 227-232, 1994; Westermeier, R. and Marouga, R., Protein detection methods in proteomics research, *Biosci.Rep.* 25, 19-32, 2005; Dunbar, B.S., *Protein Blotting: A Practical Approach*, IRL Press at Oxford, Oxford, UK, 1994; Baldo, B.A. and Tovey, E.R., *Protein Blotting: Methodology, Research, and Diagnostic Applications*, Karger, Basel, 1989.

Wormbase

A public database for the genomics biology of *Caenorhabditis elegans* (a soil-dwelling nematode used extensively in biological research). See Chen, N., Harris, T.W., Antoschechkin, I, *et al.*, WormBase: a comprehensive data resource for *Caenorhabditis* biology and genomics, *Nucleic Acids Res.* 33, D383-D389, 2005; O'Connell, K., There's no place like WormBase: an indispensable resource for *Caenorhabditis elegans* researchers, *Biol.Cell.* 97, 867-872, 2005; Schwarz, E.M., Antoschechkin, I., Bastiani, C. *et al.*, WormBase: better software, richer content, *Nucleic Acids Res.* 34, D475-D478, 2006.

Wormgate

A cloning system for the expression RNAai (hairpin RNA constructs) from the *C. elegans* ORFeome library. See Johnson, N.M., Behm, C.A., and Trowell, S.C., Heritable and inducible gene knockdown in *C. elegans* using Wormgate and the ORFeome, *Gene* 359, 26-34, 2005. See also Lamesch, P., Milstein, S., Hao, T., *et al., C. elegans* ORFeome version 3.1: increasing the coverage of ORFeome resource with improved gene prediction, *Genome Res.* 14, 2064-2069, 2004(Wormbase).

Xenobiotic

A chemical found in the body of an organism which is not the biosynthetic product of said organism and is therefore from an exogenous source. Benzene is an example of a xenobiotic compound. Organisms have utilized unique metabolic pathways for the metabolism/detoxification of xenobiotic compounds. See Paulson, G.D., Lamoureux, G.L., and Feil, V.J., Advances in methods and techniques for the identification of xenobiotic conjugates, *J.Toxicol.Clin.Toxicol.* 19, 571-608, 1982; Garattini, S., Notes on xenobiotic metabolism, *Ann.N.Y.Acad.Sci.* 407, 1-25, 1983; Glatt, H., Gemperlein, I., Turchi, G., *et al.,* Search for cell culture systems with diverse xenobiotic-metabolizing activities and their use in toxicological studies, *Mol.Toxicol.* 1, 313-334, 1987-1988; Copley, S.D., Microbial dehalogenases: enzymes recruited to convert xenobiotic substrates, *Curr.Opin.Chem.Biol.* 2, 613-617, 1998; Gil, F. and Pla, A., Biomarkers as biological indicators of xenobiotic exposure, *J.Appl.Toxic.* 21,245-255, 2001; Snyder, R., Xenobiotic metabolism and the mechanism(s) of benzene toxicity, *Drug. Metab.Rev.* 36, 531-547, 2004; Pritchard, J.B. and Miller, D.S., *Toxicol.Appl.Pharmacol.* 204, 256-262, 2005; Dai, G. and Wan, Y.J., Animal models of xenobiotic receptors, *Curr.Drug Metab.* 6, 341-355, 2005; Cribb, A.E., Peyrou, M., Muruganandan, S., and Schneider, L., The endoplasmic reticulum in xenobiotic toxicity, *Drug Metab.Rev.* 37, 405-442, 2005; Janssen, D.B., Dinkla, I.J., Peolarends, G.J., and Terpstra, P., Bacterial degradation of xenobiotic compounds: evolution and distribution of novel enzyme activities, *Environ.Microbiol.* 7, 1868-1882, 2005; Gong, H., Sinz, M.W., Feng, Y., *et al.,* Animal models of xenobiotic receptors in drug metabolism and diseases, *Methods Enzymol.* 400, 598-618, 2005; Matsunaga, T., Shitani, S., and Hara, A., Multiplicity of mammalian reductases for xenobiotic carbonyl compounds, *Drug Metab.Pharmacokinet.* 21, 1-18, 2006.

Xerogel

Not a gel but rather a term used in reference to a dried, possibly open gel; a gel in which the dispersing agent has been removed as opposed to a lyogel where the dispersing agent is still present as, for example, with a hydrogel which contains a substantial amount of water. There has been interest in xerogels a drug delivery vehicles. See Kortesuo, P., Ahola, M., Karlsson, S., *et al.,* Sol-gel-processed sintered silica xerogel as a carrier in controlled drug delivery, *J.Biomed.Mat.Res.* 44, 162-167, 1999; Kortesuo, P., Ahola, M., Karlsson, S., *et al.,* Silica xerogel as an implantable carrier for controlled drug delivery—evaluation of drug distribution and tissue effects after implantation, *Biomaterials* 21, 193-198, 2000; Shamansky, L.M., Luong, K.M., Han, D., and Chronister, E.L., Photoinduced kinetics of bacteriorhodopsin in a dried xerogel glass, *Biosens.Bioelectron.* 17, 227-231, 2002; Weng, K.C., Stalgren, J.J., Duval, D.J., *et al.,* Fluid biomembranes supported on nanoporous aerogel/xerogel substrates, *Langmuir* 20, 7232-7239, 2004; Clifford, J.S. and Legge, R.L., Use of water to evaluate hydrophobicity of organically-modified xerogel enzyme supports, *Biotechnol.Bioeng.* 92, 231-237, 2005; Oh, B.K., Robbins, M.E., Nablo, B.J., and Schoenfisch, M.H., Minaturized glucose biosensor modified with a nitric oxide-releasing xerogel microarray,

Biosens.Bioelectron. 21, 749-757, 2005; Copello, G.J., Teves, S., Degrossi, J., *et al.,* Antimicrobial activity on glass materials subject to disinfectant xerogel coating, *J.Ind.Microbiol.* 33, 343-348, 2006; Xue, J.M., Tan, C.H., and Lukito, D., Biodegradable polymer-silica xerogel composite microspheres for controlled release of gentamicin, *J.Biomed.Mater.Res.B Appl.Biomater.* 78, 417-422, 2006.

Yeast Artificial Chromosomes

Yeast artificial chromosomes (YAC) are yeast DNA sequences which contain large segments of foreign recombinant DNA introduced by transformation. Yeast artificial chromosomes permit the cloning of large DNA fragments such a genes with flanking regulatory regions. See Schlessinger, D. Yeast artificial chromosomes: tools for mapping and analysis of complex genomes, *Trends Genet.* 6, 255-258, 1990; Huxley, C. and Gnirke, A., Transfer of yeast artificial chromosomes from yeast to mammalian cells, *Bioessays* 13, 545-550, 1991; Anand, R., Yeast artificial chromosomes (YACs) and the analysis of complex genomes, *Trends Biotechnol.* 10, 35-40, 1992; Huxley, C., Transfer of YACs to mammalian cells and transgenic mice, *Genet. Eng.* 16, 65-91, 1994; Schalkwyk, L.C., Francis, F., and Lehrach, H., Techniques in mammalian genome mapping, *Curr.Opin.Biotechnol.* 6, 37-43, 1995; Kouprina, N. and Larionov, V., Exploiting the yeast *Saccharomyces cerevisiae* for the study of the organization and evolution of complex genomes, *FEMS Microbiol. Rev.* 27, 629-649, 2003; Sasaki, T., Matsumoto, T., Antonio, B.A., and Nagamura, Y., From mapping to sequencing, post-sequencing and beyond, *Plant Cell Physiol.* 46, 3-13, 2005.

Zebrafish

Zebrafish (*Danio rerio*) is a freshwater fish used for research in developmental biology. see http://zfin.org; http://www.neuro. uoregon.edu/k12/FAQs.html; http://www.ncbi.nlm.nih.gov/ genome/guide/zebrafish/ See also Stemple, D.L. and Driever, W., Zebrafish: tools for investigating cellular differentiation, *Curr. Opin.Cell.Biol.* 8, 858-864, 1996; Driever, W., Stemple, D., Schier, A., and Solnica-Krezel, L., Zebrafish: genetic tools for studying vertebrate development, *Trends Genet.* 10, 152-159, 1994; Kimmel, C.B., Genetics and early development of zebrafish, *Trends Genet.* 5, 283-288, 1989; Fulwiler, C. and Gilbert, W., Zebrafish embryology and neural development, *Curr.Opin.Cell Biol.* 3, 989-991, 1991; Ingham, P.W. and Kim, H.R., Hedgehog signaling and the specification of muscle cell identity in the zebrafish embryo, *Exp. Cell Res.* 306, 336-342, 2005; Teh, C., Parinov, S. and Korzh, V., New ways to admire zebrafish: progress in functional genomics research methodology, *Biotechniques* 38, 897-906, 2005; Amsterdam, A. and Becker, T.S., Transgenes as screening tools to probe and manipulate the zebrafish genome, *Dev.Dyn.* 234, 255-268, 2005; Hsia, N. and Zon, L.I., Transcriptional regulation of hematopoietic stem cell development in zebrafish, *Exp.Hematol.* 33, 1007-1014, 2005; de Jong, J.L., and Zon, L.I., Use of the zebrafish to study primitive and definitive hematopoiesis, *Annu.Rev. Genet.* 39, 481-501, 2005; Alestrom, P., Holter, J.L., Nourizadeh-Lillabadi, R., Zebrafish in functional genomics and aquatic biomedicine, *Trends Biotechnol.* 24, 15-21, 2006.

Zeolites

An aluminum silicate cage-like compound with a negative charge which "captures" cations in the cavity. Zeolites are used as molecular sieves for drying solvents and gases (Mumpton, F.A., La roca magica: uses of natural zeolites in agriculture and industry, *Proc.Natl.Acad.Sci.USA.* 96, 3463-3470, 1999; Kaiser, L.G., Meersmann, T., Logan, J.W., and Pines, A., Visualization of gas flow and diffusion in porous media, *Proc.Nat.Acad.Sci.USA* 97, 2414-2418, 2000; Kuznicki, S.M., Bell, V.A., Nair, S., *et al.*, A titanosilicate molecular sieve with adjustable pores for size-selective adsorption of molecules, *Nature* 412, 720-724, 2001; Yan, A.X., Li, X.W., and Ye, Y.H., Recent progress on immobilization of enzymes on molecular sieves for reactions in organic solvents, *Appl.Biochem.Biotechnol.* 101, 113-129, 2002) . There has been some interest in the specific adsorption on biopolymers such as proteins on zeolites (Matsui, M., Kiyozumi, Y., Yamamoto, T., Selective adsorption of biopolymers on zeolites, *Chemistry* 7, 1555-1560, 2001; Chiku, H., Matsui, M., Murakami, S., *et al.*, Zeolites as new chromatographic carriers for proteins—easy recovery of proteins adsorbed on zeolites by polyethylene glycol, *Anal.Biochem.* 318, 80-85, 2003; Sakaguchi, K., Matsui, M., and Mizukami, F., Applications of zeolite inorganic composites in biotechnology: current status and perspectives, *Appl.Microbiol. Biotechnol.* 67, 306-311, 2005). There are suggestions for the use of zeolites in health (Pavelic, K., Hadzija, M., Bedrica, L., Natural zeolite clinoptilolite: new adjuvant in anticancer therapy, *J.Mol. Med.* 78, 708-720, 2001; Zarkovic, N., Zarkovic, K., Kralj, M., *et al.*, Anticancer and antioxidative effects of micronized zeolite clinoptilolite, *Anticancer Res.* 23, 159-1595, 2003).

Zinc Finger Motifs

Motifs in DNA- and RNA-binding proteins whose amino acids are folded into a single structural unit around a zinc atom. In the classic zinc finger, one zinc atom is bound to two cysteines and two histidines. In between the cysteines and histidines are 12 residues which form a DNA binding fingertip. By variations in the composition of the sequences in the fingertip and the number and spacing of tandem repeats of the motif, zinc fingers can form a large number of different sequence specific binding sites. Specificity of binding to the nucleic acid is achieved by recognition of an 18 bp sequence. See Gommans, W.M., Haisma, H.J., and Rots, M.G., Engineering zinc finger protein transcription factors: the therapeutic relevance of switching endogenous gene expression on or off at command, *J.Mol.Biol.* 354, 507-519, 2005; Struhl, K., Helix-turn-helix, zinc-finger, and leucine-zipper motifs for eukaryotic transcriptional regulatory proteins, *Trends Biochem. Sci.* 14, 137-140, 1989; Schleif, R., DNA binding by proteins, *Science* 241, 1182-1187, 1988; Durai, S., Mani, M., Kandavelou, K., Wu, J., Porteus, M.H., and Chandrasegaran, S., Zinc finger nucleases: custom-designed molecular scissors for genome engineering of plant and mammalian cells, *Nucl.Acid Res.* 26, 5978-5990, 2005; Chen, Y. and Varani, G., Protein families and RNA recognition, *FEBS J.* 272, 2088-2097, 2005.

Zinc Finger Nuclease

Zinc finger nucleases are engineered nucleases which consist of a zinc finger domain or domains fused to the nuclease domain from Fok1 restriction endonuclease. This nuclease domain is non-specific such that the sequence specificity cleavage of the zinc finger nucleases is provided from the zinc finger domain(s). See Urnov, F.D., Miller, J.C., Lee, Y.L., *et al.*, Highly efficient endogenous 45-52, 2006; human gene correction using zinc-finger nucleases, *Nature* 435, 646-651, 2005; Mani, M., Smith, J., Kandavelou, K., Berg, J.M., and Chandrasegaran, S., Binding of two zinc finger nuclease monomers to two specific sites is required for effective double-strand DNA cleavage, *Biochem.Biophys.Res.Commun.* 334, 1191-1197, 2005; Mani, M., Kandavelou, K., Dy, F.J., Durai, S., and Chardrasegaran, S., Design, engineering, and characterization of zinc finger nucleases, *Biochem.Biophys.Res.Commun.* 335, 447-457, 2005; Porteus, M.H., Mammalian gene targeting with designed zinc finger nucleases, *Mol.Ther.* 13, 438-446, 2006; Dhanasekaran, M., Negi, S., and Sugiura, Y., Designer zinc finger proteins: tools for creating artificial DNA-binding proteins, *Acc.Chem.Res.* 39, 45-2, 2006.

Zymography

A method for detecting enzyme activity on a matrix, usually a polyacrylamide gel or agarose gel after electrophoretic separation; See Lombard, C., Saulnier, J., and Wallach, J., Assays of matrix metalloproteinases (MMPs) activities: a review, *Biochemie* 87, 265-272, 2005; Frederiks, W.M., and Mook, O.R., Metabolic mapping of proteinase activity with emphasis on in situ zymography of gelatinases: review and protocols, *J.Histochem.Cytochem.* 52, 711-722, 2004.

Zymosan

An insoluble polysaccharide derived from the cell wall of fungi. More specifically, the term zymosan is used to refer to a specific preparation from yeast which is used in models of inflammatory disease and multi-organ dysfunction. There is evidence for specific interaction with Toll receptors on macrophages. See Fitzpatrick, F.W. and DiCarlo, F.J., Zymosan, *Ann.N.Y.Acad.Sci.* 118, 233-262, 1964; Czop, J.K. Phagocytosis of particular activators of the alternative complement pathway: effects of fibronectin, *Adv.Immunol.* 38, 361-398, 1986; Stewart, J. and Weir, D.M., Carbohydrates as recognition molecules in macrophage activities, *J.Clin.Lab. Immunol.* 28, 103-108, 1989; Takeuchi, O. and Akira, S., Toll-like receptors: their physiological role and signal transduction system, *Int.Immunopharmacol.* 1, 625-635, 2001; Levitz, S.M., Interactions of toll-like receptors with fungi, *Microbes Infect.* 6, 1351-1355, 2004; Volman, T.J., Hendriks, T., and Goris, R.J., Zymosan-induced generalized inflammation: experimental studies into mechanisms leading to multiple organ dysfunction syndrome, *Shock* 23, 291-297, 2005; Ikeda, Y., Adachi, Y., Ishibashi, K., Miura, N., and Ohno, N., Activation of toll-like receptor-mediated NF-kappa beta by zymosan-derived water-soluble fraction: possible contribution of endotoxin-like substances, *Immunopharmacol.Immunotoxicol.* 27, 285-298, 2005.

---

[a] the distinction between peptide and protein can be arbitrary.

INDEX